AF527769

VEGETABLES
Genetic Wealth and Crop Improvement

The Author

Dr. D. R. Bhardwaj was born on 15th August, 1963 at village-Narasara, P.O.-Jakkhini, District Varanasi (U.P.) and obtained his graduate (1984); post-graduate (1987) and Ph. D. degree (Horticulture with specialization in Vegetable Science) (1991) from Institute of Agricultural Sciences, B.H.U. Varanasi (U.P.). Dr. Bhardwaj has an experience of 27 years in vegetable crops improvement (research and development) under Indian Council of Agricultural Research, New Delhi. Presently, he is working as a principal scientist after joining of scientist through ARS in 1991. He focused on augmentation of genetic resources and genetic improvement of vegetable crops since 1992 and continuing till date. He has identified 7 novel/unique genetic stocks which registered/patented with ICAR-NBPGR, New Delhi. He is the first person who developed and reported gynoecism (gygene) in bitter gourd, seedless pointed gourd ("*ptgene*" parthenocarpic line), andromonoecious in sponge gourd, cluster bearing sponge gourd, and cms system in tropical carrots; which are being utilized in development of varieties/hybrids in the country. He has developed 14 varieties/hybrids in different vegetable crops which are known for high yield and quality attributes and all are very popular among the vegetable growers. He has published more than 95 research papers in reputed national and international journals, 31 papers in proceedings, 32 popular articles, 125 abstracts and 36 book chapters on vegetable crops. He is in credit of 6 books (Published by SSPH, New Delhi and Daya Publishing House; A division of Astral International Pvt. Ltd., New Delhi and ICAR-IIVR, Varanasi), 3 technical bulletins, 1 information bulletin (on Biodiversity of vegetable crops), 5 training manuals and 6 extension leaflets. He has guided 22 M.Sc. (Veg. Sci.) students and 4 Ph. D. in vegetable science. He is recipient of Dr. Rajendra Prasad Award in 2013 and 2014 conferred by ICAR, New Delhi. He is also awardee of Dr. B. R. Ambedkar Young Fellow Award (2009); ISVS Fellow award (2013); ISPGR Award (2013); Fellow Award of Horticultural Sciences (2014); Young Scientist Associate Award (2014) and Fellow of National Academy of Agricultural Sciences (2020).

VEGETABLES
Genetic Wealth and Crop Improvement

D.R. Bhardwaj
Principal Scientist (Vegetable Science)
Division of Crop Improvement
ICAR-IIVR, Varanasi - 221 305
(Uttar Pradesh)

2020
Daya Publishing House®
A Division of
Astral International Pvt. Ltd.
New Delhi – 110 002

ISBN: 978-93-89719-46-8

Published by : **Daya Publishing House®**
A Division of
Astral International Pvt. Ltd.
– ISO 9001:2008 Certified Company –
4760-61/23, Ansari Road, Darya Ganj
New Delhi-110 002
Ph. 011-43549197, 23278134
E-mail: info@astralint.com
Website: www.astralint.com

Digitiality Printed at : **Replika Press Pvt. Ltd.**

Preface

Plant genetic resources for food and agriculture from the biological basis of world food security, nutritional security and directly or indirectly support the livelihoods of every person on earth. Ever since hunter-gatherers realized some 12,000 years ago that they could save and plant seeds from season to season and about 10,000 years ago agriculture began in earnest way. Simply testing the best, people started cultivating very limited numbers of crops as staple diets and to meet the associated needs. It is estimated that only 30 crops species now provide 95 per cent of the world's human food energy and just of them-rice, wheat, maize, millet and sorghum-provide about 60 per cent; whereas over 7,000 plant species have been known to be used for food and agricultural purposes are either partly or fully domesticated in the past by human civilizations. Plants are an integral part of the world's biological diversity and in this regard global plant diversity are of great value in providing food, fuel, clothing, medicine, maintaining basic ecosystem and shelter for the whole human kind under global climate change. During this trial and error period, a few useful plants were probably selected and protected. Among different agricultural commodities vegetable is playing an important role in nutritional, economical and sociological upgrading. In vegetable crops it is estimated around 400 species, of which 80 originated in India and India is called as 'Vegetable Rich Gene Centre'. Vegetables vast diversity in brinjal, pointed gourd, ridge gourd, sponge gourd, cucumber, elephant foot yam, etc. can be observed in tropical, subtropical, temperate, desert and sea sore regions. Though, the diverse agro-climatic conditions of our country permit to grow more than 60 cultivated and 30 lesser known vegetables. Diversity is quite appreciable in introduced crops such as tomato, French bean, cowpea, leafy brassicae, amaranth, yam, bottle gourd, bitter gourd, ash gourd, snake gourd, onion and garlic. Diverse germplas are expected to give excellent hybrid vigour which provides an opportunity for selecting the diverse parents for hybridization. Such parents are expected to produce superior segregants in combination with others and thus are most valuable for breeders. Remarkable production is contributed by only a few major vegetable crops. Apart from these, a large group of vegetables having immense potential in terms of nutritional and medicinal qualities as well as climatic adaptability has not yet got attention by the growers and researchers. Aquatic vegetables are also an integral part of wetland ecosystem, which play multifaceted benefits in terms of maintaining the urban water bodies in flood control, amenity uses, wild life and numerous environmental benefits in very holistic manner. In the last few decades, India has made quantum jump in vegetable production and globally it ranks 2nd in vegetable production (1.68 million tonnes from 9.54 million hectare) after China and contributes 15.61 per cent to global area under vegetable cultivation and 14 per cent to production. There is danger that valuable plant genetic resources may not be available to future crop breeding programmes as plants are endangered by a combination of factors like over collecting, unsustainable agriculture and forestry practices, urbanization, pollution, land use changes, spread of intensive alien species etc. However, vegetable production in the national level is also threatened by land fragmentation, biotic and abiotic stresses, climate change, dwindling natural resources and uneven growth across the country. Intensive promotion of few cultivars results in biodiversity losses in farm ecosystem. These lead to displacement of traditional varieties and species and simplification of human diets. Also due to increasing pressure on population and growing human taste, the food basket has to be widened. In history it is well mentioned that 'Plant Domestication" began independently in India, Africa, China, Middle East, South and Central America and South-east Asia. The vast flora and fauna distributed on the mother earth planet from hot springs at the ocean bottom to the alpine meadows at the top of highest mountains have co-evolved with marvelous interdependence and co-existence. The development of varieties with broad genetic base will be highly useful to mitigate the various biotic and abiotic stresses.

D.R. Bhardwaj

Contents

1

Vegetable Genetic Resources: Importance, Utility and Ethnobotany

Introduction

In historical perspective, human have depended on hunting, fishing and gathering for their sustenance. Ever since the man started agriculture, he has been searching for superior and new plant types to meet the needs from available genetic resources of crops around him. Finding food was very dangerous because of the world's 2,50,000 plant species were poisonous. Over 10,000 years of trials and error, however our ancestors identified 20,000 species of plants with leaves, seeds, stems, fruits, pods, shoots, flowers, tubers or other parts that were safe to eat. They also found 2,000 edible mushrooms, and 3,500 mammals, 9,000 birds, some reptiles, amphibians (especially frog) and about 1,000 insects and used them as food materials. Most of the present day crops have been originated from wild and primitive types. Indian subcontinent is one of the twelve mega diversity centres for cultivated plants and their wild relatives. More than 15000 species of flowering plants are indigenous to this region which includes 160 species of economic importance, 320 species of wild ancestral forms and approximately 800 species of ethno-botanical origin. Global diversity in vegetable crops is estimated around 400, of which 80 originated in India. During this trial and error period, a few useful plants were probably selected and protected. Then about 10,000 years ago, agriculture began in earnest way and people started cultivating few promising plants. A very limited number of crops are cultivated as staple diets and to meet the associated needs. About 30 crops species provide 95 per cent of the world food energy whereas over 7,000 species have been known to be used for food and are either partly or fully domesticated in the past by human civilizations. Among different agricultural commodities vegetable is playing an important role in nutritional, economical and sociological upgrading. In vegetables, vast diversity can be observed in tropical, subtropical, temperate, desert and sea sore regions. Apart from these, aquatic vegetables are also an integral part of wetland ecosystem, which play multifaceted benefits in terms of maintaining the urban water bodies in flood control, amenity uses, wild life and numerous environmental benefits in very holistic manner. In the last few decades, India has made quantum jump in vegetable production and globally it ranks 2nd in vegetable production after China and contributes 15.61 per cent to global area under vegetable cultivation and 14 per cent to production. However, vegetable production in the national perspective is threatened by land fragmentation, biotic and abiotic stresses, climate change, dwindling natural resources and uneven growth across the country. Intensive promotion of few cultivars results in biodiversity losses in farm ecosystem. These lead to displacement of traditional varieties and species and simplification of human diets. Also due to increasing pressure on population and growing human taste, the food basket has to be widened. In history it is well mentioned that 'Plant Domestication" began independently in India, Africa, China, Middle East, South and Central America and South-east Asia. The vast flora and fauna distributed on the mother earth planet from hot springs at the ocean bottom to the alpine meadows at the top of highest mountains have co-evolved with marvelous interdependence and co-existence. Several families are known for potential vegetable crops. As per the vegetable genetic resources, crops can be grouped botanically as shown in Table 1.1.

Three major physical resources in the world comprise land, water and the biological diversity. Plant biodiversity in general and agro-biodiversity in particular not only maintains the diverse ecosystem on the planet but also acts as vital links of all the food chains. Biological resources of the earth are vital to humanity too because they depends upon them for clean environment, food security, health care, social needs, sources of livelihood, trade, industrial growth and economic development. Due to the ever-increasing human and domestic animal populations and demand for food, fodder, feed and fiber, conservation and sustainable use of the existing biodiversity is one of the most serious challenges of the mankind. Nearly 42.0 per cent of the world population depends on agriculture directly and indirectly, which is more so in the developing countries. In India, 17 per cent human and 11 per cent animal

Table 1.1: Botanical Classification of Vegetable Crops

	Family	*Common Name*	*Botanical Name*
I. Monocotyledoneae	**Amaryllidaceae**, Subfamily Allioideae (formerly Alliaceae)–Amaryllis Family	Elephant garlic	*Allium ampeloprasum* var. *ampeloprasum*
		Kurrat	*Allium ampeloprasum* var. *kurrat*
		Leek	*Allium ampeloprasum* var. *porrum*
		Wild leek, ramp	*Allium tricoccum*
	Aggregatum group	Shallot	*Allium cepa*
	Cepa group	Common onion	*Allium cepa*
	Proliferum group	Egyptian onion	*Allium cepa*
		Chinese onion, Chinese scallion, Oriental onion	*Allium chinense*
		Welsh onion, Japanese bunching onion	*Allium fistulosum*
		Garlic	*Allium sativum*
		Chive	*Allium schoenoprasum*
		Chinese chives	*Allium tuberosum*
Dioscoreaceae	Yam Family	Water yam, winged yam, purple yam	*Dioscorea alata*
		Air potato, varahi, kaachil	*Dioscorea bulbifera*
		Yellow yam	*Dioscorea cayenensis*
		Lesser yam	*Dioscorea esculenta*
		Chinese yam	Dioscorea *opposita*
		White yam	*Dioscorea rotundata*
		Cush-cush yam, Indian yam, napi	*Dioscorea trifida*
	Poaceae (old name Gramineae) – Grass Family	Dent corn, field corn	*Zea mays* var. *indentata*
		Flint corn, ornamental corn	*Zea mays* var. *indurata*
		Popcorn	*Zea mays* var. *everta*
		Sweet corn	*Zea mays* var. *saccharata*
	Asparagaceae (formerly Liliaceae)	Asparagus	*Asparagus officinalis*
II. Dicotyledoneae	**Amaranthaceae** (subfamily Chenopodiaceae, formerly family) – Goosefoot Family	Orach	*Atriplex hortensis*
		Chard	*Beta vulgaris* var. *cicla*
		Beet	*Beta vulgaris* var. *crassa*
		Spinach	*Spinacia oleracea*
	Asteraceae (Compositae outdated name) – Sunflower Family	Tarragon	*Artemisia dracunculus*
		Endive, escarole	*Cichorium endivia*
		Chicory, radicchio	*Cichorium intybus*
		Cardoon	*Cynara cardunculus*

	Family	Common Name	Botanical Name
		Globe artichoke	*Cynara scolymus*
		Jerusalem artichoke	*Helianthus tuberosus*
		Lettuce	*Lactuca sativa*
		Dandelion	*Taraxacum officinalis*
		Salsify, oyster plant	*Tragopogon porrifolius*
	Convolvulaceae – Morning-glory Family	Water spinach	*Ipomoea aquatica*
		Sweet potato	*Ipomoea batatas*
	Brassicaceae (Cruciferae outdated name)– Mustard Family	Horse radish	*Armoracia rusticana*
		White mustard	*Sinapis alba*
		Leaf mustard	*Brassica juncea*
	Napobrassica group	Rutabaga	*Brassica napus*
	Pabularia group	Siberian kale	*Brassica napus*
		Black mustard	*Brassica nigra*
	Acephala group	Kale, collard	*Brassica oleracea*
	Alboglabra group	Chinese kale	*Brassica oleracea*
	Botrytis group	Cauliflower	*Brassica oleracea*
	Capitata group	Cabbage	*Brassica oleracea*
	Gemmifera group	Brussels sprouts	*Brassica oleracea*
	Gongylodes group	Kohlrabi	*Brassica oleracea*
	Italica group	Broccoli	*Brassica oleracea*
	Costata group	Tronchuda cabbage	*Brassica oleracea*
	Chinensis group	Chinese cabbage (nonheading), pak-choi	*Brassica rapa*
	Pekinensis group	Chinese cabbage (heading), pe-tsai	*Brassica rapa*
	Perviridis group	Spinach mustard	*Brassica rapa*
	Rapifera group	Turnip	*Brassica rapa*
	Ruvo group	Broccoli raab, rapini	*Brassica rapa*
		Garden cress	*Lepidium sativum*
		Sea kale	*Crambe maritime*
		Watercress	*Nasturtium officinale*
		Radish	*Raphanus sativus*
	Cucurbitaceae–Gourd Family	Watermelon	*Citrullus lanatus*
		Round melon	*Praitrullus fistulosus*
	Inodorus group	Honeydew melon, casaba, crenshaw, persian	*Cucumis melo*
	Cantaloupensis group	Muskmelon, cantaloupe	*Cucumis melo*
		Long melon	*Cucumis melo* var. *utilissimus*
		Cucumber	*Cucumis sativus*
		Bar cucumber	*Cucumis anguria*
		Winter squash, pumpkin, banana squash, buttercup squash, hubbard squash	*Cucurbita maxima*
		Green cushaw, Japanese pie pumpkin, silver-seed gourd	*Cucurbita argyrosperma*
		Butternut squash, calabaza, cheese pumpkin, golden cushaw	*Cucurbita moschata*
		Pumpkin, acorn squash, summer squash, marrow	*Cucurbica pepo*
		Bottle gourd, calabash gourd	*Lagenaria siceraria*
		Smooth or sponge gourd	*Luffa aegyptiaca*

	Family	Common Name	Botanical Name
		Ridged or ribbed gourd	*Luffa acutangula*
		Satputia	*Luffa hermaphrodita*
		Chayote	*Sechium edule*
		Pointed gourd	*Trichosanthes dioica* Roxb.
		Snake gourd	*Trichosanthes cucumerina*
		Ivy gourd	*Coccinia cordifolia*
		Bitter gourd/Bitter melon	*Momordica charantia* L.
		Spine gourd	*Momordica dioica* Roxb.
		Sweet gourd	*Momordica cochinchinensis*
		Kakrol	*Momordica charantia* var. *reningera*
		Ash gourd	*Benincasa hispida*
	Euphorbiaceae–Spurge Family	Cassava, yuca	*Manihot esculenta*
	Fabaceae (Leguminosae old name –Pea or Bean Family	Chickpea, garbanzo b	*Cicer arietinum* ean
		Soybean	*Glycine max*
		Lentil	*Lens culinaris*
		Scarlet runner bean	*Phaseolus coccineus*
		Lima bean, sieva bean (butter bean)	*Phaseolus lunatus*
		Common bean (green, dry), snap bean, kidney bean	*Phaseolus vulgaris*
		Garden pea, field pea, edible-pod pea	*Pisum sativum*
		Winged bean	*Psophocarpus tetragonolobus*
		Fava bean (broad bean)	*Vicia faba*
		Urad, urd, black gram	*Vigna mungo*
		Mung bean	*Vigna radiata*
		Black-eyed pea, cowpea, asparagus bean, yard-long bean	*Vigna unguiculata*
	Malvaceae – Mallow or Cotton Family	Okra, gumbo	*Abelmoschus esculentus*
	Polygonaceae – Buckwheat Family	Rhubarb	*Rheum rhabarbarum*
		Sorrel	*Rumox acetosa*
		Dock, patience or monk's rhubarb	*Rumex patientia*
	Solanaceae – Potato or Nightshade Family	Pepper (bell, cayenne chili)	*Capsicum annuum*
		Tabasco pepper	*Capsicum frutescens*
		Tomato, cherry tomato	*Solanum lycopersicon*
		Currant tomato	*Solanum pimpinellifolium*
		Strawberry ground cherry	*Physalis pruinosa*
		Tomatillo	*Physalis philadelphica*
		Cape gooseberry	*Physalis peruviana*
		Eggplant	*Solanum melongena*
		Irish potato	*Solanum tuberosum*
	Tetragoniaceae–Carpetweed Family	New Zealand spinach	*Tetragonia tetragoniodes*
	Apiaceae (formerly Umbelliferae) – Parsley Family	Chervil	*Anthriscus cerefolium*
	Dulce group	Celery	*Apium graveolens*
	Rapaceum group	Celeriac	*Apium graveolens*
		Carrot	*Daucus carota*
		Fennel	*Foeniculum vulgare*
		Parsnip	*Pastinaca sativa*
		Parsley, turnip-rooted parsley	*Petroselinum crispum*
	Valerianaceae – Valerian Family	Corn salad, mâche	*Valerianella locusta*

Table 1.2: Crops and their Usage

Common Name	*Botanical Name*	*Usages*	*Edible Parts as Vegetable*
French bean	*Phaseolus vulgaris* L.	Food	Tender, pods and seed
Pea	*Pisum sativum*	Food	Mature and immature seeds, tender pods
Indian bean	*Dolichos lablab*	Food	Tender pods and seed
Cowpea	*Vigna unguiculata*	Food	Tender pods, and mature seed
Sward bean	*Canavalia gladiata* (Jacq.) DC.	Food	Tender pods
Broad bean	*Vicia faba*	Food	Tender pods and seed
Cluster bean	*Cyamopsis tetragonoloba*	Food	Tender pods
Jack bean	*Canavalia ensiformis*	Food	Tender pods
Winged bean	*Psophocarpus tetragonolobus*	Food	Tender pods
Yam bean	*Pachyrhizus erosus*	Food	Tuber
Velvet bean	*Mucuna puriens*	Food	Tender pods
Lima bean	*Phaseolus lunatus*	Food	Tender pods
Rice bean	*Vigna umbellata*	Food	Tender pods and seed
Sophlang	*Flemingia vestita*	Vegetable	Tuber
Tree bean	*Parkia roxburghii*	Multipurpose	Tender pods
Agathi	*Sesbania grandiflora (L.) Poiret*	Multipurpose	Tender buds
Antidote vine	*Fevillea cordifolia*	Medicinal	All parts
Jiao-gu-lan	*Gynostemma pentaphyllum*	Medicinal	All parts
Luo-ago-di	*Hemsleya amabilis Diels*	Medicinal	All parts
He-zi-cao	*Actinostemma tencrum Griff.*	Medicinal	All parts
Psendo-tritillary	*Balbostemma paniculatum*	Medicinal	All parts
White seed melon	*Cucumeropsis mannii* Naud	Food	Seed kernel
Bur gherkin	*Cucumis anguria* L.	Food	Tender fruit
Teasel gourd	*Cucumis dipsaceus* Ehrenb. ex.	Ornamental	Plant
African horned cucunber	*Cucumis metuliferus* E. Mey ex. Naud	Food	Tender fruit
Sponge plant	*Momordica angustisepala* Harms	Utilitarian	Mature fruit
Cochin chin gourd	*Momordica cochinchinensis* (Lour.) Spreng	Medicinal	Tender fruit
Balsam apple	*Momordica balsamina* L.	Medicinal	Tender fruit
Kaksa	*Momordica cymbalaria* Frenzl. ex. Hork	Food	Tender fruit
Luo-han- guo	*Siraitia grosvenovrii* (Swingle) Lu. and Zhang	Food	Tender fruit
Red hail stone	*Thladiantha dubia* Bunge	Medicinal	Root
Fluted pumpkin	*Telfairia occidentalis* Hook F.	Food	Fruit and mature seed
Oyster nut	*Telfairia pedata* (Sims)Hook	Food	Seed kernel
Hard plant	*Hodgsonia macrocarpa* (Bl.)Cogn.	Food	Tender fruit
Gymnopetlum	*Gymnopetlum cochinchinensi* (Lour.) Kurz.	Food	Tender fruit
Snake gourd	*Trichosanthes cucumerina* L.	Medicinal	Tender fruit
Chinese Snake gourd	*Trichosanthes Kirilowii* Maxim.	Medicinal	All parts
Indreni	*Trichosanthes lepiniana* (Naud.) Cogn.	Medicinal	All parts
Japanese Snake gourd	*Trichosanthes ovigera* Blume	Food	Tender fruit
Mi-Mao-qua-lou	*Trichosanthes villosa* Blume	Food	Tender fruit
Inara	*Acanthosicyos horridus* Wel. ex. Hook. f.	Food	Tender fruit
Wax gourd	*Benincasa hispida* (Thunb.) Cogn.	Food	Immature fruit
Bryoni	*Bryonia cretica* L.	Medicinal	All parts
Bryoni	*Bryonia dioica* L.	Medicinal	All parts
Colocynth	*Citrullus colocynth* (L.) Schrad	Medicinal	All parts
Ivy gourd	*Coccinia abyssinica* (L.) Cogn.	Food	Immature fruit
Little gourd	*Coccinia grandis* (L.) Voigt.	Food	Immature fruit
Lollipop climber	*Diplocyclos palmatus* (L.) C. Jeffery	Ornamental	Plant
Squirting cucumber	*Ecballium elaterium*(L.) A. Rich.	Ornamental	Plant

Common Name	Botanical Name	Usages	Edible Parts as Vegetable
Round melon	*Praecitrullus fistulosus* (Stocks) Pang	Food	Immature fruit
Cayaponica	*Cayaponica kathematophora*	Ornamental	Plant
Cayaponica	*Cayaponica ophthalmica* R.E. Schult.	Medicinal	All parts
Malabar gourd	*Cucurbita ficifolia* Bouche	Food	Immature fruit (vegetable) and seed
Casabanana	*Sicana odorifera* (Vell.) Naud.	Food	Seed kernel
Stuffing cucumber	*Cyclanthera brachybotys* (Poepp. and Endl.)	Food	Immature fruit
Stuffing cucumber	*Cyclanthera pedata*	Ornamental	Plant
Stuffing cucumber	*Cyclanthera explodens* naud.	Food	Immature fruit
Wild cucumber	*Echinocystis lobata* (Michx. Torr. and Gray)	Ornamental	Plant
Chayote	*Sechium edule* (Jacq.) Swartz	Food	Immature fruit
Chayote	*Sechium tacaco* (Pitt.) C. Jeffrey	Food	Immature fruit

populations of world are depending on agriculture. To meet the demand of increasing population, it is essential to produce more food. The food security of people does not depend merely on what is cultivated but it needs realization that what is available in a particular situation. Today, the sustainable growth in agricultural productivity is becoming a matter of prime importance in most of the countries.

Plant genetic resources (PGR) are the basic and essential materials for development of improved crop varieties designed to combine high yield potential with superior quality, resistance to diseases and pests, and also better adaptation to abiotic stress environments. Vegetable cultivation in the new world dates back to 7,000 to 5,000 BC. Root crops may have been among the earliest to be domesticated in humid tropics preceding the seed crops. Only two vegetable crops (bottle gourd and sweet potato) are known to grown in both the Old and New Worlds; before 1492. Their presence in both Western and Eastern hemispheres suggest possible pre-Columbian agricultural contacts. After the discovery of new world by Columbus, there was rapid exchange of crops among the continents. Almost one third of the world's species originated in South Eastern Asia and most of the fruit and vegetable crops did come from the East and West Asiatic Centre and Mediterranean Centre of origin, while roots and tubers and tropical fruit trees are concentrated in the Central American and Andean Centre. In India the horticulture sector presently contributes around 31 per cent of the GDP and 38 per cent of the total exports of agricultural commodities from about 14 per cent of the area. With utilization of genetic resources, production and protection technologies, investments and other need base efforts, horticultural produce during 2012-13 (268.84 MT) has surpassed the food grain production (255.36 MT) for the first time in the history of Indian Agriculture. Therefore, we can say that India has achieved self-sufficiency and a good degree of stability of food production. This has created an urgent need for providing health security to our population by supplying nutrition through balanced diet. During last few decades dedicated search is going on to find out natural source of bioactive nutritional compounds. Some of the natural sources for bioactive compounds are fruits, vegetables, pulse, cereals *etc.* Among them, vegetables are unique source of various bioactive compounds like vitamins, trace elements, antioxidants, dietary fibers, Poly-Unsaturated Fatty Acid (PUFA), which not only improve the state of health and wellbeing, they also reduce the risk of various degenerative diseases such as cancer, cardio vascular disease, macular degenerative and ageing. Now a variety of vegetables are available all the year round. In chilli, onion and pea, a number of such collections need for future exploitation.

Germplasm or Gene Pool

Germplasm or gene pool is the basic material with which a plant breeder has to initiate his breeding programme. It can be defined as:

"The sum total of genes in a crop species is referred to as genetic resources".

Or

"Gene pool refers to a whole library of different alleles of a species".

Or

"The sum of total of hereditary material i.e., all alleles various genes present in a crop species and its wild relatives is known as gene pool or genetic stock or germplasm or genetic resources".

Importance of Genetic Resources

Plant genetic resources comprises of total genetic diversity of the cultivated species and their wild relatives (Ford-Lloyd, 2001), support a lot to human and animals with different ways. Natural biodiversity is the main source of foods during famines. Food security can be assured by lessor known/lessor importance crops (nearly 80 crop species which make our basket rich) in the area of adaptation because these resources (species, strains, cultivars *etc.*) are generally resistant to pests and pathogens. Wild forms of vegetable crops with high content of mineral substances, proteins, vitamins and resistance genes for biotic and abiotic stresses assume significance for their conservation and sustainable utilization from their centres of origin and diversity (Giuseppina *et al.*, 1998). The important vegetable crops/crop groups for which good native diversity occurs in the Indian Gene Centre include beans, brinjal, cucurbits (cucumber and other gourds), leafy vegetables, okra onion and tuber crops (Taros and Yams) (Rathore *et al.*, 2005). The vegetable germplasm have significance in agriculture and food as well as nutritional security of the country. Though a number of wild and other types of germplasm in vegetable crops have been utilized for the betterment, still there are a lot of plants which need to be exploited. In this context it is also important to mention that the rich plant diversity of the Indian Himalaya is utilized by the native communities in various forms as medicine, edible/food, fodder, fuel, timber, agricultural tools, *etc.* Many wild and weedy forms have been utilized as vegetables. Not only this, several strains are being evaluated for useful chemical. Wild edible plants form an important source as a supplement/substitute food in times of scarcity for native communities. Because land holdings are small and subsistence agriculture prevails, the natives gather many wild plants for food. Over 675 wild plant species representing 384 genera and 149 families, arte used as food/edible and their various parts are either consumed raw, roasted, boiled, fried, cooked or in the form of oil, spice, seasoning materials, jams, pickles *etc.* West Himalaya shows the highest diversity (50.96 per cent) of edible plants and East Himalaya the maximum number of endemics (18 taxa) and wild relatives of economic plants (39 taxa) as reported by Samant and Dhar (1997).

During last seven-eight decades, germplasm of promising and profitable varieties have been built up and developed by different agencies, from other parts of the world. The introduction and acclimatization of vegetable germplasm from within the country and all over the world is important for exploitation and incorporation of useful characters such as early maturity, high yielding capacity, resistance to disease and pests and tolerance to different climatic hazards, *viz.* drought, heat, cold and flood situations.

1. Protection of Human Health

Natural biodiversity makes significant contributions directly to household food security, particularly in areas where such biodiversity exists and people have access to it. Plant genetic resources (PGR) meet out the basic human needs like nutrition, health care, clothing and fuel. More than 850 million peoples still continue to be underfed and undernourished. After a quantum jump in yield of principal cereals, wheat and rice, during green revolution in the 1970s near plateau has reached in their production and productivity. Substantial cultivated areas are being degraded, salinized and diverted to alternative uses. Recently phytochemical in vegetables have attracted a great deal of attention mainly due to diseases caused as a result of oxidative stress and now it is used as protective foods.

(i) Nutritional Importance

About 3 billion people in the world are malnourished due to imbalanced diets. When we analyze the nutritional situations it reflects that more than 70 per cent malnourished children are in Asia and 53 per cent children are still malnourished in India. The World Bank estimates that India is one of the highest ranking countries in the world are underweight which is double that of Sub-Saharan Africa with dire consequence for mobility, morality, productivity and economic growth. The 2011 Global Hunger Index (GHI) report ranked India 15th amongst leading countries with hunger situation. It also places India amongst the three countries where the GHI between 1996 and 2011 went up from 22.9 to 23.7 while 78 out of 81 developing countries studied, including Pakistan, Nepal, Bangladesh, Vietnam, Kenya, Nigeria, Myanmar, Uganda, Zimbabwe and Malawi, succeeded in improving hunger condition. In fact, considerable imbalances persist in respect of availability of the calories, proteins, minerals and vitamins. These deficiencies could be minimized to a considerable extent by broadening our food basket with various under-utilized/under exploited vegetables, fruits *etc.* apart from the major vegetables. Indigenous lesser known vegetables pose a vast potential to cope-up the problem of malnutrition. Progressive improvement of several unknown economic species over time by successive generations of farmers has led to evolution of today's crop plants giving rise to landraces and traditional cultivars. These precious materials along with their close wild relatives serve as building blocks for development of improved crop varieties that are used by plant breeders. They also constitute a priceless reservoir of genes conferring better adaptation and resistance to stress environments caused by biotic and abiotic factors. Biodiversity is of tremendous importance in the daily diet of human being, who depends on the bark, young shoots, buds, flowers, fruits, leaves, roots, tubers and mushrooms, especially in the season of hunger, when other cultivated crops are unavailable. The tribal peoples certainly live entirely on these foods for four to five months every year, when agriculture is impractical, or in periods before harvesting paddy. Over 90 per cent of plant species for food and agriculture are located in the economically developing parts of the world, namely, the African, Asian and Latin American continents and the Far East Islands (Madeley, 1996). It is obvious that to ensure food and nutritional security to the ever-increasing population at the present level of consumption, cereal production has to be increased by 30-50 per cent by 2025 from less area and cultivated germplasm of crop plants with narrow genetic base and higher vulnerability to biotic and abiotic stresses. Many indigenous vegetables are natural and rich source of vitamins, minerals and antioxidants. Under-utilized legumes may be a cheap, alternate source of protein and can alleviate protein malnutrition in rural areas. Nutritious pods of tree bean (*Parkia roxburghii*) are

Table 1.3: Nutritional Value of Vegetable Crops

Vegetable Crops	*Edible Parts (per 100 g)*	*Water (per cent)*	*Protein (g)*	*Fat (g)*	*Minerals (mg)*	*Fiber (mg)*	*Carbo-hydrate (per cent)*	*Calories (k-cal)*	*Vit. 'A' (IU)*	*Vit. 'C' (mg)*
Chenopods	-	89.0	3.7	0.4	2.6	0.8	2.9	30	2900	35.0
Kale	100	83.0	6.0	8.0	-	-	9.0	53.0	-	186.0
Poi	100	90.8	-	0.4	-	-	4.2	-	-	87.0
Brussels sprout	100	85.5	4.7	0.5	1.0	1.2	7.1	52.0	210	72.0
Cabbage	88	91.9	1.8	0.1	0.6	1.0	4.6	27	2000	124.0
Amaranths	39	85.7	4.0	0.5	2.7	1.0	6.3	46	10120	4.0
Colocasia leaves	89	87.9	4.0	1.2	1.7	1.0	4.2	44	17130	12.0
Coriander	70	86.3	3.3	0.6	2.3	1.2	6.3	44	11530	135.0
Fenugreek	59	86.1	4.4	0.9	1.5	1.1	6.0	49	3900	52.0
Bengal gram tips	-	68.0	8.2	0.5	3.5	-	27.2	146	6700	-
Salad	6	93.4	2.1	0.3	1.2	0.5	2.5	21	1650	10.0
Mint	45	84.9	4.8	0.6	1.9	2.0	5.8	48	2700	27.0
Mustard greens	-	89.8	4.0	0.6	1.6	0.8	3.2	34	4370	33.0
Green onions	-	90.3	1.2	0.8	1.0	1.4	5.3	33	-	-
Radish leaves	-	90.3	2.7	0.6	2.1	0.9	3.4	30	18660	103.0
Spinach	87	92.1	2.0	0.7	1.7	0.6	2.9	26	9300	28.0
Dil	-	85.2	3.4	0.7	2.5	1.5	6.7	46	-	-
Turnip leaves	51	81.9	4.8	1.5	2.2	1.0	9.4	67	15060	180.0
Carrot	95	86.0	0.9	0.2	1.1	1.2	10.6	48	3150	3.0
Colocasia	-	73.1	3.0	0.1	1.7	1.0	21.1	97	40	3.0
Onion	-	84.3	1.8	0.1	0.5	0.6	12.6	59	0	0.0
Potato	100	74.7	1.6	0.1	0.6	0.4	22.6	97	40	11.0
Radish	99	94.40	0.7	0.1	0.6	0.8	3.4	17	5	17.0
Sweet potato	-	68.5	1.2	0.3	1.0	0.8	28.2	120	10	15.0
Cassava	-	59.4	0.7	0.2	1.0	0.6	38.1	157	-	24.0
Turnip	65	91.6	0.5	0.2	0.6	0.9	6.2	29	-	25.0
Elephant Foot Yam	-	78.7	1.2	0.1	0.8	0.8	18.4	79	434	43.0
Ash gourd (Sweet making)	67	96.5	0.4	0.1	0.3	0.8	1.9	10	-	-
Ash gourd (Vegetable)	97	92.6	1.4	0.1	0.6	0.7	4.6	25	84	2.0
Bitter gourd	97	92-4	1-6	0-2	0-8	0-8	4-2	25	210	1-0
Brinjal	91	92.7	1.4	0.3	0.3	1.3	4.0	24	124	88.0
Cauliflower	70	90-8	2.6	0.4	1.0	1.2	4.0	30	51	12.0
Cluster bean	-	81.0	3.2	0.4	1.4	3.2	10.8	60	30	56.0
Arvi	86	94.0	0-3	0-3	1-2	0.6	3.6	18	174	49.0
Cowpea	-	85.9	2.6	0.2	0.8	2.0	8.5	46	941	3.0
Cucumber	83	96.3	0.4	0.1	0.3	0.4	2.5	13	-	7.0
Indian bean	93	86.1	3.8	0.7	0-9	1.8	6.7	48	312	9.0
French bean	94	91.4	1.7	0.1	0.5	1.8	4.5	26	221	14.0
Chilli	97	92.4	1.3	0.3	0.7	1.0	4.3	25	712	137.0
Knol-khol	74	92.7	1.1	0.2	0.7	1.5	3.8	21	36	85.0
Okra	84	89.6	1.9	0.2	0.7	1.2	6.4	35	88	13.0
Leek	100	78.9	1.8	0.1	0.7	1.3	17.2	77	30	11.0
Pointed gourd	95	92.0	2.0	0.3	0.5	3.0	2.2	20	225	29.0
Ridge gourd	82	95.2	0.5	0.1	0.3	0.5	3.4	17	56	5-0
Snake gourd	98	94.6	0.5	0.3	0.5	0.8	3.3	18	160	-
Round melon	9	93.5	1.4	0.2	0.5	1.0	3.4	21	23	18.0
Tomato	98	93.1	1.9	0.1	0.6	0.7	3.6	23	320	31-0

Vegetable Crops	Edible Parts (per 100 g)	Water (per cent)	Protein (g)	Fat (g)	Minerals (mg)	Fiber (mg)	Carbo-hydrate (per cent)	Calories (k-cal)	Vit. 'A' (IU)	Vit. 'C' (mg)
Ginger	-	80.9	2.3	0.9	1.2	2.4	12.3	67	67	6.0
Turmeric	-	13.1	6.3	5.1	3.5	2.6	69.4	349	50	-
Watermelon	98	95.8	0.2	0.2	0.3	0.2	3.3	16	-	1.0
Snapmelon	-	95.7	0.3	0.1	0.4	0.5	3.0	14	265	10.0
Muskmelon	80	92.8	0.6	0.1	0.6	0.5	5.4	25	450	32S.0

consumed as staple legume vegetable in the NEH Region of India. Velvet bean (*Mucuna pruriens*) is considered one of the most preferred legume vegetables in tribal belts of Uttar Pradesh and Bihar.

According to available data the malnutrition burden tends to be high in tribal areas. This is an enigma since such areas are rich in agrobiodiversity. An example is Attappadi, Kerala, where tribals are suffering from several nutritional deficiencies. An opportunity is now available for converting this malnutrition hot-spot into a leader of the nutria-farm movement because Kerala government has provided 1800 acres of fertile land to 1000 landless tribal families of Attapaddi. This land grant movement confers multiple benefits such as food, work and income. Now families are introducing specific agricultural remedies to overcome the nutritional maladies like drumstick, sweet potato *etc.* In culinary diversity and agrobiodiversity the methods of protecting community knowledge and technologies under World Trade Organizations (WTO), Geographic Indication (GI) is an important one.

Phytochemicals/Nutraceuticals/Edible Colour Pigments

Edibles colours are natural pigments found in tissue of plants. These colours are the chemical compounds produced by several biochemical pathways which give colours to the food. The colours change according to the growth stage of the plant parts or vegetable product. These include anthocyanins, betalains, carotenoids, chlorophylls *etc.* These pigments play important ecological and metabolic functions in the plants (Grotewold, 2006) and are more frequently exploited as the source of nutraceuticals to address a number of human ailments. These pigments have been implicated in regimes to maintain human health to protect against chronic disease(s) or to restore wellness by repairing tissues after disease has been established. A wide range of bioassays and tests have been forwarded to establish the biological efficacy of natural pigments in human health intervention, including *in vitro* (bioassays), in vivo, epidemiological and clinical trials (Lila, 2004). Phytochemicals are very important for human health. For example anti-gastric ulcer compounds are available in *C. frutescens* (Bhut Jolokia/Ghost chilli). Nutritionally rich coloured vegetables are becoming more and more demanding in the national and international markets. Consumers are also giving more emphasis on these types of vegetables due to health benefit.

Table 1.4: Nutrient and Health Benefit of Pigments

Sl.No.	Major Colour Pigments	Impact on Human Health
1.	Anthocyanins	Anthocyanins are natural pigments belonging to the flavonoid family. They are responsible for the blue, purple, red and orange colour of many vegetables. Anthocyanins are capable of acting on different cells involving in the development of atherosclerosis, one of the leading causes to cardiovascular dysfunction. Anthocyanins and the aglyconecyanidin were found to inhibit cyclooxygenase enzymes, which can be one marker for the initial stage of carcinogenesis. Both the Anthocyanins and cyanidiaglycone from tart cherries have been reported to reduce cell growth of human colon cancer cell lines (Kang *et al.*, 2003). On one hand they can interfere with glucose absorption and on the other hand they may have aprotective effect on pancreatic cells. The most extensively documented phytomedicinal role of anthocyanin pigments is in improving eyesight, including night vision.
2.	Betalains	Betalains have been widely used as natural colourants for many centuries, but their attractiveness for use as colourants of foods (or drugs and cosmetics) has increased recently due to their reportedly high anti-oxidative, free radical scavenging activities and concerns about the use of various synthetic alternatives.
3.	Carotenoids	Carotenoids are the second most abundant pigments in nature, and consist of more than 700 members (*Britton et., al,* 1998). Carotenoids play an important role in plant reproduction, through their role in attracting pollinators and in seed dispersal, and are essential components of humans' diets. Carotenoids provide protection to vision and eye function, and against macular degeneration and contracts. Carotenoids are credited with biological promotion of immune system response. Carotenoids are associated with inhibition of several types of cancers including cervical, esophageal, pancreatic, lung, prostate, colorectal and stomach.
4.	Chlorophylls	Chlorophyll is the most important plant pigment and a 'real life force' that nature uses to explode plants into greenery. Chlorophylls, in contrast, are typically consumed in much higher doses in a diet that incorporates green and leafy vegetables (Delgado-Vargas and Paredes-Lopez, 2003). The antimutagenic properties of chlorophylls have been demonstrated in various assays, and clearly, intake of chlorophyll has potential to act as a chemo-preventive compound in humans.

Some colour rich vegetables source(s) for different useful phytochemicals are given below:

Table 1.5: Colour Rich Vegetables

Sl. No.	*Colour*	*Pigments*	*Vegetables*
1.	Red	Lycopene	Tomatoes and watermelon
		Betacyanins	Beet root
2.	Orange	β-carotene	Carrot, cantaloupe, pumpkin, and sweet potatoes
3.	Blue/Purple	Anthocyanins	Eggplant
4.	Yellow	Lutein	Yellow corn
5.	Green	Chlorophylls	Broccoli, kale, spinach, cabbage and asparagus
6.	Black	Anthocyanins	Black carrot

Source: Kalia and Saha, 2010.

Similarly, some other useful compounds are also found abundantly in certain crops are as under:

Table 1.6: Useful Phytochemicals and their Source

Sl. No.	*Useful Phytochemicals*	*Source(s)*
1.	Lutein	Drumstick leaf, mustard green, spinach
2.	Carotenoids (β-carotene)	Carrot, pumpkin, sweet potato cv. Gouri, Red chilli
3.	Lycopene	Tomato, carrot
4.	Curcumin	Turmeric cv. Lakadong
5.	Betalains	Beet roots (Red Beets)
6.	Anthocyanins	Red cabbage, Red Radish, Beet root, sweet potato
7.	Oleoresin	Ginger, chilli, turmeric coriander, green onion
8.	Gingerol	Ginger
9.	Lutein, zeaxanthin	Capsicum Yellow
10.	Astaxanthin, lycopene	Capsicum Red
11.	Alpha, beta and gamma-carotene	Orange colour Capsicum

Source: Asati and Yadav, 2004.

If it is generalized, several nutraceuticals are important part of our diet.

Table 1.7: Vegetable Rich in Nutraceuticals

Sl. No.	*Vegetable*	*Nutraceuticals*
1.	Allium vegetables (onion, garlic, chives, leeks)	Allylsulfides
2.	Cruciferous vegetables (cauliflower, cabbage, broccoli, Brussels sprouts, kale, turnip, kholrabi)	Indol/glucosinolates, Sulfaforaphane Isothiocyanates/ thiocyanates
3.	Solanacceous vegetables (tomatoes, peppers)	Lycopene
4.	Umbelliferous vegetables (carrots, celery, parsley, parsnips)	Carotinoids, Phthalides, Polyacetylenes
5.	Compositae plants (artichoke)	Silymarin
6.	Carrot, squash, broccoli, sweet potatoes, tomatoes, kale, collards, cantaloupe and pumpkin	Vitamin A (retinol)
7.	Green peppers, broccoli, green leafy vegetables, cabbage and tomatoes	Vitamin C (ascorbic acid)
8.	Green leafy vegetables	Vitanin E

Some of the specific constituents in vegetable crops are given below:

Table 1.8: Some Specific Constituents in Vegetable Crops

Sl. No.	*Vegetables*	*Traits*	*Constituent*
1.	Tomato	Red colour	Lycopene
2.	Onion	Red coulour	Anthocynin
4.	Onion	Yellow coulour	Quercetin
5.	Chilli	Red colour	Capsanthin
6.	Turmeric	Yellow coulour	Curcumin
7.	Carrot	Red coulour	Anthocynin
8.	Carrot	Purple colour	Carotene
9.	Radish	Pungency	Isocynate
10.	Chilli	Pungency	Capsacin
11.	Garlic	Pungency	Elycin
12.	Bitter gourd	Bitterness	Momordicoside
13.	Onion	Pungency	Allel Propyll Disulphide
14.	Bottle gourd	Bitterness	Cucurbitacin (tetra-cyclic triterpinpoid cucurbitacins compounds are responsible for its bitterness and are highly toxic)
15.	Sponge gourd	Bitterness	Cucurbitacin (tetra-cyclic triterpinpoid cucurbitacins compounds are responsible for its bitterness and are highly toxic)
16.	Ridge gourd	Bitterness	Cucurbitacin (tetra-cyclic triterpinpoid cucurbitacins compounds are responsible for its bitterness and are highly toxic)
17.	Coriander leaves	Aroma	Due to 40 different organic compounds, with 82 per cent of these being aldehydes, and 17 per cent alcohols. The aldehydes are mainly those with 9-10 carbon atoms
18.	Coriander seeds	Aroma	Due to Linalool
19.	Ginger	Aroma	Due to Nerolidol
20.	Mentha	Aroma	Due to Menthol
21.	Dil	Aroma	Due to Carvone

Sl. No.	Vegetables	Traits	Constituent
22.	Spearmint	Aroma	Due to Carvone
23.	*Luffa*	Aroma	Due to Luffein
24.	Cauliflower	Bitterness	Due to 3-butenylisothio-cyanate
25.	Broccoli	Bitterness	Due to 3-butenylisothio-cyanate
26.	Leek	Bitterness	Due to 3-butenylisothio-cyanate
27.	Fenugreek	Aroma	Due to 3-Hydroxy-4,5-dimethyl-Z(5H)-furanone (sotolone) is an important volatile constituent of fenugreek

Not only this, natural pigments rich vegetable varieties have been developed and available in the market.

Table 1.9: Edible Colour Rich Varieties of Vegetable Crops

Sl. No.	Vegetables	Varieties	Pigments
1.	Carrot	Pusa Asita	Anthocyanin
		Pusa Rudhira	Lycopene
		Pusa Vrishti	Lycopene
		Pusa Vasudha	Lycopene
		Pusa Nayanjyoti	β-carotene
2.	Paprika	KTPL-19	Capsanthin
3.	Amaranth	Pusa Lal chauli	Anthocyanin
4.	Beet root	Detroit Dark Red	Betalain
5.	Red cabbage	-	Anthocyanin
6.	Purple headed broccoli	Palam Vichitra	Anthocyanin
7.	Tomato	Pusa Rohini	Lycopene
8.	Radish	Pusa Jamuni	Anthocyanin
		Pusa Gulabi	Lycopene

Edible Buds and Flowers

There are many plants which buds and flowers are eaten raw or valued in cooking. Not only this, edible flowers may be preserved for future use by adopting preservation techniques such as drying, freezing or steeping in oil. They can be used in drinks, jellies, salads, soups, syrups and are can be used in mixed form, with other main dishes. Flower flavoured oils and vinegars are made by steeping edible flower petals in these liquids. As a history, the culinary use of flowers dates back thousands of years with the 1st recorded mention being in 140 B.C (Preethi *et al.*, 2011). In the ancient Roman civilization flowers like mallow, rose, and violets were significantly used to lend aroma and flavor to the different kinds of eatables. The Chinese, Greeks, mid-eastern and Indian cultures have been consuming edible flowers from a very early period. In England particularly during the reign of Queen Victoria, edible flowers like rose were used in a wide array of dishes. Presently a number of plants are available in nature which buds and flowers are delicacies in human diets. For example buds of sunhemps, *Luffa* species, Lasora *etc.* and flowers of pumpkin, drumsticks *etc.* are very commonly available in the vegetable market. Some of the edible flowers are given in Table 1.10.

Table 1.10: Edible Flowers of Important Crop

Common Name	Botanical Name	Description
Fuchsia	*Fuchsia X hybrida*	Blooms have a slightly acidic flower. Explosive colour and graceful shape make it ideal as garnish. The berries are also edible.
Garden Sorrel	*Rumex exacetosa*	Sorrel flowers are tart, lemon tasting. So, it can be used like a lemon: on pizza, a salad topping, in sauces, over cucumber salads.
Nasturtiums	*Tropaeolum majus*	Nasturtiums rank among most similar to watercress. Whole flowers are stuffed. Leaves add peppery tong to salads. Pickled seed/pods are less expensive substitute for capers. Entire flowers can be used to garnish platters, salads, cheese tartars, open-faced sandwiches and savory appetizers.
Pansies	*Viola X Wittrockiana*	Pansy is not only a decorative plant but it is also used in phytotheropy. Because of its purifying properties infusion of wild pansy is considered particularly useful for pediatric skin problems such as acne or eczemas, and as a supportive treatment of chronic lesions of urinary system. Flowers and petals are edible. They are also used as garnishes in green salad.
Peony	*Paeonia lactiflora*	In China the fallen petals are parboiled and sweetened as a tea-time delicacy. Peony water was used for drinking in the middle ages. Peony petals can be added to salads or in punches and lemonades.
Perennial Phlox	*Phlox paniculata*	The perennial phlox is edible. It is the high-growing (taller) and low-growing (creeping) phlox that grows from 90.0-120.0 cm tall. The flowers very from reddish purple to pink, some white. Flowers are slightly spicy in taste and good in fruit salads.
Tree olive	*Osmanthus fragrans*	Tree olive is also known as fragrant olive, sweet olive but this plant is called Xî or guîhuâ in China. It flowers called guîhuâ in China which literally mean "cinnamon flower" or "cassia flower". The flowers are infused with green or black tea leaves, to create a scented tea called guîhuâchâ. In Chinese cuisine, the flowers are also used to produce osmanthus-scented jam, sweet cakes, dumplings, soups and even liquor.

Common Name	*Botanical Name*	*Description*
Tuberous Begonia	*Begonia X tuberosa*	Only petals of the tuberous begonia hybrids are edible. The bright colours and sour, fruity taste bring flavor and beauty to any summer salad. Begonia blossoms have a delicious citrus sour taste and a juicy crunch. The petals are used as a garnish and in salads. Stems also can bé used in place of rhubarb. The flowers and stems contain oxalic acid and should not be consumed by individuals suffering from gout, kidney stones or rheumatism.
Yucca Petals	*Yucca* species	The white Yucca flower is crunchy with a mildly sweet taste and can be used in salads and as a garnish.
Elderberry blossoms	*Sambucus* species	The blossoms are a creamy and have a sweet scent and sweet taste. After harvesting elder berry flowers, they should not be washed as that removes much of the fragrance and also flavor. Instead they should be checked carefully for insects. The fruit is used to made wine. The flowers, leaves, berries, bark, and roots have all been used in traditional folk medicine for centuries. All other parts of this plant except the berries are mildly toxic. They contain a bitter alkaloid and glycoside that may change into cyanide. The cooked ripe berries of the edible elders are harmless. Eating uncooked berries may cause nausea, vomiting, and diarrhea.
Angelica	*Angelica archangelica*	Depending of the variety, flower range from pale lavender to blue deep rose. It has a natural affinity with fish. The leaves have a stronger, clean taste and make an interesting addition to salads. In its native Northern Europe, even the mature leaves are used, particularly by the Lab landers, as a natural fish preservative. Many people in the cold Northern region such as Green land, Siberia and Fin land consider Angelica a vegetable and eat the stems raw, sometimes spread with butter. Young leaves can be made into a tea.
Anise Hyssop	*Agastache foeniculum*	Both flowers and leaves have a delicate anise or liquo rices flavour. Some people say the flower reminds them of root beer. The blossoms make attractive plate garnishes and or often used in Chinese-style dishes. It is excellent for use in salads.
Bee Balm	*Monarda didyma*	Bee Balm also called wild Bergamot, wild Oswego Tea, Horse mint and Moriarda. The wild Bee Balm tastes like oregano and mint. The taste of Bee Balm is reminiscent of citrus with soft mingling of lemon and orange, the red flowers have a minty flavor. The Bee Balm blossoms can be used in place of oregano. The leaves and petals can be used in salads.
Bathua	*Chenopodium album*	Inflorescences are cooked as a tasty broccoli substitute.
Daisies	*Bellisperennis*	It has astringent properties and has been used in folk medicine as an antiseptic.
Dandelions	*Taraxacum officinale*	Dandelions leaves and buds have been a part of traditional Mediterranean and Asian, most notably Chinese and Korean cuisine. Dandelions leaves can be picked in the early spring before they become tough. They are best before the flowers bloom. Later in the season the plants can be blanched, *i.e.* covered to exclude light, to improve the flavour. The flowers can be sautéed in butter or oil as a vegetable dish or dipped in tempura batter and fried. The flower petals, along with other ingredients, are used to make dandelions wine. The medicinal properties of this flower are numerous and are extensively used in treating various liver, gall bladder and urinary disorders, against gall stones. Jaundice and cirrhosis and also in treating dyspepsia. Therapeutically, dandelion is a diuretic and also reduces the cholesterol level in the body.
Fig Leaf Squash or Malabar Gourd	*Cucurbita ficifolia*	The male flowers and buds are cooked as vegetable, used in soups, stews and salads in some region of Mexico.
Pumpkin	*Cucurbita moschata*	The male flowers and buds are cooked as vegetable
Sponge gourd	*Luffa cylindrica*	Buds are fried and eaten
Ridge gourd	*Luffa acutangula*	Buds are fried and eaten
Tree bean	*Parkia roxburghii*	The fleshy part of the flower stalk may be eaten as vegetable and mature flowers are used to make curries.
Dolichos bean	*Lablab purpureus*	Flowers may be cooked and eaten.
Drumstick	*Moringa oleifera*	Flowers may be cooked and eaten
Sunhemp	*Crotolaria juncea*	Buds are cooked and eaten

(ii) Medicinal Importance and Curative Properties

The annual demand of botanical raw drugs in the country is estimated to be 4,19,5000 MT for the year 2012-2013 valued Rs. 1963 crores. Nearly 6000 plants are currently being used by the herbal industry in some form of the other. About 960 medicinal plants constitute the sources of 1289 botanical raw drugs with 178 species being consumed in volumes exceeding 100 MT per year. Most of this demand is met from the wild populations without any assessment of their actual availability on a sustainable basis.

This dependency on wild populations of plants for life saving drugs will not reduce for a long time to come for at least two reasons:

(a) Non-availability of technology for domestication of the large set of species and

(b) Growing competition for land between cultivation and developmental activities.

The exploitation of wild resources would continue unabated despite new regulations and controls. A number of underutilized leafy vegetables of Indian origins are rich source of folic acid, commonly referred as folate, which is helpful to reduce risk of cardiovascular diseases. It is abundantly available in poi/spinach (123 µg/100 g), chenopods and leaves of drumsticks. Folic acid is crucial in foetal development in the early stage of pregnancy. Accordingly nutritionist recommends that childbearing women should consume the recommended amount (400 µg/100 g) of folic acid by eating a healthy and balanced diet. Carotenoid (precursor of Vitamin A.) is the requirement of the day for everyone. Now it is realized that not only β-carotene, but also other carotenoids including lycopene may have important role in protection against major disorders including cancer and heart diseases. It is found in most of the green leafy vegetables, *Momordica cochinchinensis*, carrots, squash, melons, spinach and tomatoes which contain a highly conjugated carbon backbone and act as antioxidants. Ascorbic acid (Vitamin C) is the main water-soluble antioxidant in plasma and is readily found in many vegetables. Brinjal species *Solanum khasianum* Clarke and *Solanum mammosum* L. are wild in nature but cultivated for solasodine alkaloid. Similarly, *Solanum xanthocarpum*, *Solanum spirale* and *Solanum ferox* L. are generally used for vegetable and medicinal purpose (Table 1.11).

Similarly, medicinal applications of cucurbits are too plentiful to enumerate. Since ancient times, indigenous cultures have worldwide employed a variety of cucurbits to cure a plethora of ailments. A Chinese medicinal text written during the first century AD mentions the therapeutic benefits of wax gourd (*Benincasa hispida*), pickling melon, bottle gourd and Japanese snake gourd. In modern Chinese treatments, *Trichosanthes* species are the most important medicinal cucurbits (Yang and Walters, 1992). Some of the reported therapeutic properties, including the use of cucurbits are as purgatives, emetics, anthelmintics and wormifuge. The bitter roots of many cucurbits for example *Bryony*

Table 1.11: Nutritional and Medicinal Importance of Vegetable Crops

Sl.No.	*Species*	*Nutritional and medicinal importance*
1.	*Alocasia macrorrhiza* (L. Donn.)	Stem juice is believed to relieve scorpion and nettle sting, leaves are used as rubefacient for joint pains.
2.	*Alocasia indica* (Roxb.) Schott.	Has easily digestible starch, leaf juice is an astringent, rhizome acts as mild laxative and diuretic.
3.	*Alpinia galangal* Sw.	Rhizome used for curing catarrhal affections, fever and rheumatism
4.	*Amorphophallus dubius* Blume	Tuber cooked in butter milk and fried in ghee and used for treating dysentery. It is also used as a carminative, expectorant, and in the treatment of piles, rheumatism and hemorrhoids. Fermented juice of petiole is used for diarrhea.
5.	*Amorphophallus sylvaticus*	Fruits and seeds are crushed into a paste and used in the treatment against toothache, bruises and glandular enlargements.
6.	*Amorphophallus paeconi-folius*	It is being used as a tonic and has restorative and carminative properties. Also used in the treatment of piles, dysentery and rheumatism.
7.	*Anaphyllum wightti*	Small tubers of 1.0 g size. Maximum tuber yield 5.0-10.0 g. Used for the treatment of piles.
8.	*Arisneria pulchrum*	Used for treatment of piles.
9.	*Asparagus felicious*	Root possesses astringent properties.
10.	*Asparagus racemosus* Willd.	Roots are used as a refrigerant, demulcent, diuretic, aphrodisiac, galactogogue, anti-dysenteric and for dyspepsia. Medicated oil is recommended for rheumatic complaints.
11.	*Asparagus sarmentosus* Linn.	Roots are aphrodisiac and hence taken as a nourishing drink.
12.	*Asparagus officinalis* Linn.	Roots have demulcent, diuretic, laxative, cardiac sedative, tonic and aphrodisiac qualities. Also used in the treatment of jaundice. It contains the essential oil-asparagine and tyrosine which are used in commercial purpose.
13.	*Asparagus adscendens* Roxb.	Tubers are used as vegetable as it possessed colling and demulcent properties.
14.	*Asparagus gonocladus* Baker	Roots are aphrodisiac. It is boiled with oil and applied in cutaneous disease and treatment against gonorrhea.
15.	*Canna orientalis*	The tuber has diaphoretic, demulcent and diuretic properties. It is also a stimulant and hence used during fever and dropsy
16.	*Canna edulis* Ker-gawler.	Easily digestible starch similar to arrowroot starch
17.	*Ceropegia tuberosa*	Big tubers toxic to rats
18.	*Chlorophytum borivilianum*	Root is a vitalizer. Hence used as a health tonic, curative for pre and post natal problems, alternative to Viagra and for immunity improvement.
19.	*Coleus forskohlii* (Willd.) Briq.	Root contains Forskolin which is a bronchodilator and used as a cardiotonic in the treatment of congestive heart failure, glaucoma therapy, anti-hypertensive, remedy for metastalic condition and thrombosis
20.	*Coleus amboinicus* Lour.	Juices of the leaves which contain an aromatic carminative are used in treatment of urinary disease, dyspepsia, chronic coughs, asthma, cuts and wounds.

Sl.No.	Species	Nutritional and medicinal importance
21.	*Colocasia esculenta* (L.) Schott	Juice of the petiole is used as a stimulant and rubefacient. The corm juice is used against scorpion stings. *Colocasia* starch is used to prepare weaning baby food and cosmetics.
22.	*Corollo carpusepigens*	Used as medicine to cure skin disease.
24.	*Costus speciosus* (Koenig) Sm.	Roots used as antidote for snake bite and wounds, the rhizomes contains depurative and anthelmintic and purgative properties and used as tonic. Root powder is used in the treatment of urinary disease, body pain, fever and snake bites.
25.	*Cucrcuma aromatica*	Rhizome is used as a tonic, carminative and also for treatment of bruises and sprains
26.	*Cucrcuma caesia*	Rhizome is used as a stimulant and carminative and also in the treatment of asthma, bruises and sprains.
27.	*Cucrcuma zedoaria* Rosc.	Rhizome is a stimulant, carminative demulcent and expectorant. Also applied on bruises sprains, pains, dyspepsia, cold cough and bronchitis. Oil is used as an anti-irritant.
28.	*Cucrcuma amala*	Rhizome is a carminative, and also used for stomach.
29.	*Cucrcuma malabarica* Vel *C. zedoaria*	Collected from forests by tribals (Kadar-Sholayar forest) sold in market for starch as well as for making cheap paints. It contains easily digestible starch which has antifungal and antibacterial properties.
30.	*Cyanotis tuberosa*	Nearly 99.0 per cent watery. Used to prevent thirst.
31.	*Cyperus rotundus*	Staple food for wild folk. Broad spectrum of medicinal use.
32.	*Dioscorea deltoidea*	For production of diosgenin.
33.	*Dioscorea floribunda*	For production of diosgenin.
34.	*Dioscorea kumaonensis*	Tubers used in the treatment of arthritis and rheumatism.
35.	*Dioscorea pentaphylla* Linn.	Tubers are applied on swelling of joints and also used as a tonic to improve body immunity.
36.	*Dioscorea esculenta* (Lour.) Burk.	Tubers used for treatment of chest pain, nervous disorders and swellings.
37.	*Dioscorea hamiltonii* Hook.f.	Treatment of stomach ache.
38.	*Dioscorea hispida* Dennst.	Tuber is used as an antidote in arrow poison.
39.	*Dioscorea opposittifolia* Thunb.	Kavalakizhangu is believed to be an excellent food tonic by tribals. Tubers are used in the treatment of swelling, scorpion sting and snake bites.
40.	*Dioscorea tomentosa* Koenig.	Edible tubers eaten after boiling.
41.	*Dioscorea wallichii*	Cooked tubers could be chewed and juice consumed for treatment of jaundice.
42.	*Drosera peltata*	Highly medicinal having red coloured tubers of 1-2 g. Seen around Kalakkad Mundanthurai Tiger Reserve (KMTR) area during February. Used for the preparation of Thangabhasmam, a life drug.
43.	*Gloriosa superba* L.	Bears tubers. Red flowers. Seeds used for treatment of heart ailments. Also, called as Kalappakizhangu, Kanmanippoovu. It is also used to kill intestinal parasites.
44.	*Hedychium spicatum* Buch. Ham.	Rhizome is stomachic, carminative, stimulative and used as tonic and for the treatment of dyspepsia
45.	*Hedychium coronarium* J. Koenig	Stem juice used against swelling. Rhizome is an antirheumatics and is used as tonic and excitant.
46.	*Helianthus tuberosus* L.	Rhizome is a commercial source of levulose-sweetening agent for diabetic patients.
47.	*Ipomoea aquatica*	Root is purgative
48.	*Ipomoea batatas*	Tubers are laxative
49.	*Ipomoea carnea*	Root decoction reduces blood pressure.
50.	*Ipomoea disitata*	Root applied on swelling of joints
51.	*Ipomoea hispida*	Root used in the treatment of rheumatism, headache, epilepsy, leprosy and ulcers
52.	*Ipomoea alba*	Root is purgative
53.	*Ipomoea paniculata*	Aromatic. Tubers used for enhancing milking.
54.	*Martanta aurandinacea* L.	Rhizomes contain easily digestible starch, which are used in making face powder and special glues. It is a rubefacient and also applied on wounds and used for the treatment of ulcers.
55.	*Nelumbo nucifera* Gaertn.	Rhizome used as tonic and in the treatment of diarrhea.
56.	*Pueraria tuberosa* (Roxb.) D.C.	Roots are demulcent, used as refrigerant in fevers, reduce swellings of joints and a lactagogue
57.	*Sagittaria sagittifolia*	Roots used in treatment of skin disease, leaves used as an antidote to snake bite and insect bite and relieve itching.
58.	*Scripus grossus* L.	Tubers are astringent, laxative, cooling and have diuretic properties. Used for arresting vomiting and diarrhea. It is also used as tonic.
59.	*Scripus kysoor* L.	Tubers are used against diarrhea and vomiting.
60.	*Scripus lacustris*	Rootstock is an astringent, and used as diuretic.

Sl.No.	Species	Nutritional and medicinal importance
61.	*Scripus maritimus*	Roots are astringent and diuretic.
62.	*Scripus tuberosa*	Roots are used as laxative.
63.	*Stephania wightii*	Very big tubers, 6-7 kg, used in treating leprosy.
64.	*Tecca integrifolia*	Tubers used in making tonic and in the treatment, diarrhea and dysentery.
65.	*Tecca leontopetaloides* (L.) Kuntze.	Tubers used in the treatment of piles, rubefacient, diarrhea and dysentery.
66.	*Theriophonum divaricatum*	Tubers are used against diarrhea.
67.	*Theriophonum minutum* (Willd.) Baill.	Dried and used after steaming
68.	*Typhonium roxburghii*	Tubers used in curing skin eruptions
69.	*Typhonium trilobatum*	Tubers contains a stimulant and is used in the treatment of piles, stomach complaints, haemorrhoids *etc.*

species have cathartic effects. Leaves are employed for medicine as often dried and pounded into a powder to cure several disorders. In China, Japan and India, *Momordica* species are being tested for their per-ported analgesic, aborifacient, immuno suppressive and anti-tumor properties. Similarly, laboratory research is ongoing for *Trichosanthes* species. In 1989, compound Q; which is an active principle extracted from Chinese snake gourd, received a lot of attention when medical professionals began to study it as a treatment against the human immuno deficiency virus (HIV). Cancer is an uncontrolled proliferation of cells with poor epithelial layer differentiation. It is the presence of more than one factor, which is responsible for the cancer preventive properties of vegetables. Its consumption reduces cardiac disorders with the following reason:

a) Low Fat Content

High intake of fatty diet increased the blood cholesterol level, which in turn is linked with cardiac disorders. There is no second opinion that the animal foods are rich in proteins that is of high quality. However, it is associated with high fat content *vis-a-vis* cholesterol contents. The intake of animal base food invariably leads to higher consumption of cholesterol. Contrary to this the vegetables have very low fat which makes them an ideal source of proteins. Though the protein quality is not as good as that of animal source, nevertheless, it ranks at number one among pulses, cereals *etc.*

b) High k Content

Another factor, which has been implicated in the heart strokes, is high intake of common salt (NaCl) while the potassium (K) is known to prevent high blood pressure. Rather it is the ratio of K to Na plus Cl, which plays a crucial role. This ratio is much in favor of cucurbitaceous vegetables because of high K content. Among the cucurbits pumpkin, squashes, round melon *etc.* are rich source of K.

c) Blood Glucose Level

The hydrolysis of these polysaccharides to glucose is pre-requisite as only the latter can act as ready source of energy through metabolism. The complex carbohydrates release glucose slowly and, thereby maintain its level in the blood for a longer period. One of the most widely employed of the medicinal cucurbits is bitter gourd. Medical research has shown that this plant serves to control, though not cure, mild to moderately chronic cases of diabetes mellitus by increasing carbohydrate utilization.

d) Curative Properties

1. Cucurbitaceous vegetables being low in fat and high in proteins, minerals, vitamins and water help in checking obesity. The mobilization of reserved fats in body for production of energy requires these elements.
2. Juice of bitter *Luffa* is known to be effective medicines against jaundice.
3. Due to presence of aromatic compound these vegetables has therapeutic action in intestine.

Table 1.12: Some Specific Medicinal and Curative Properties of Cucurbits

Vegetables	Medicinal and Curative Properties
Momordica spp.	Hemorrhoids, rheumatism, parasites, anti-tumour compounds, skin disorders, diabetes
Citrullus spp.	Urinary tract infections, poor blood circulation
Cucumis spp.	Sun stroke, stomach cancer
Luffa spp.	Purgative
Trichosanthes spp.	Dysentery, reduce fever, pain, purgative, glucoside, asthma and carache, emetic febrifuge.
Cucurbita spp.	Anticancer compounds, reduce urinary problems
Benincasa hispida	Laxative, diuretic, aphrodisiac, cures urinary infections, billiousness, and blood diseases. Tonic for heart, cooling anthelmintic, dry cough, fever
Coccinia	Fleshy roots in the treatment of diabetes, sores and skin diseases, bronchitis
Lagenaria spp.	Leaves are diuretic, laxative and antibiotious. Seeds are used in dropsy; fruit has cooling, diuretic and cardiotonic effect.
Sechum edule	Diuretic, helpful in intestinal troubles, hypertension cardiovascular modifying properties.

Table 1.13: Human Health Promoting Nutritive Constituents of Vegetables

Sl.No.	Constituent	Vegetables as Source	Established or Proposed Effects on Human Wellness
1.	Vitamin C (ascorbic acid)	Broccoli, cabbage, cantaloupe, leafy greens, pepper, potato, tomato, watermelon	Prevents scurvy, aids wound healing, healthy immune-system, cardiovascular-disease
2.	Vitamin A (carotenoids)		Night blindness prevention, chronic fatigue, psoriasis, heart disease, stroke, cataracts.
3.	Vitamin K	Green onions, crucifers (cabbage, broccoli, Brussels sprouts), leafy greens	Synthesis of pro-coagulant factors, osteoporosis
4.	Vitamin E (tocopherols)	Dark green leafy vegetables	Heart-disease, LDL-oxidation, immune-system, diabetes, cancer
5.	Dietary fiber	Most fresh vegetables, cooked dry beans and peas	Diabetes, heart-disease
6.	Folate (folicin or folic acid)	Dark green vegetables (such as spinach, mustard greens, butter head lettuce, broccoli, Brussels sprouts and okra), asparagus	Birth defects, cancer, heart-disease, nervous system
7.	Calcium	Cooked vegetables (such as beans, greens, okra and tomatoes), peas, snap beans, pumpkin, cauliflower, rutabaga	Osteoporosis, muscular/skeletal, teeth, blood pressure
8.	Magnesium	Spinach, okra, potato	Osteoporosis, nervous system, teeth, immune-system
9.	Potassium	Baked potato or sweet potato, cooked greens, winter (orange) squash, cantaloupe	Hypertension (blood pressure), stroke, arterio-sclerosis

4. Extracts of bottle gourd seeds stimulate a lot
5. Bitter gourd contains bitter substances helps a lot in lowering blood sugar.
6. Tumba or Mateera (*Cirullus colocynth*) is used as a purgative during constipation. It possesses anti-tumer properties against cancer and act as a remedy for fever, urogenital disorder and other problems.
7. Seeds of squash, wax gourd, bottle gourd and other gourds are proved as anthelminic.
8. Oil extracted from several cucurbits seeds act as an antidote to counteract the poison of a snake bite. Oils also used to ward off dandruff.
9. Bitter gourd is used to combat HIV infection.

Similarly, several vegetables play a positive impact on human health (Table 1.13).

(iii) Important for Farmers

Progressive improvement of these economic species overtime by successive generations of farmers has led to evolution of today's crop plants giving rise to landraces and traditional cultivars. Tuber crops are efficient converters of solar energy, cassava producing 250 x 10^3 kcal/ha and sweet potato producing 240 x 10^3 kcal/ha as compared to 176 10^3 kcal/ha for rice, 110 x10^3 kcal/ha for wheat and 200 x 10^3 kcal/ha for maize. Tuber crops have the ability to transform a high fraction of biological yield into economic yield.

(iv) Important for Breeders

A large number of wild species have not only contributed for supply of vegetable but also as rich gene pools for important traits. These prized genetic resources along with their close relatives serve as genetic stock for plant breeders in the development of improved varieties against several biotic and abiotic stresses. The wild gene pools possess important traits and may be exploited both by direct selection or improvement through other breeding methods. Therefore, availability of PGR to plant breeders is necessary for sustaining advances in crop productivity and for stabilizing production in the country. Introduced or domesticated strain sometimes never found good enough to be utilized as a new crop or new variety but it possesses desirable genes for several economic traits like- quality.

Some promising lines introduced in vegetable crops for different traits are given below:

Table 1.14: Trait Specific Introductions of Genotypes in Vegetable Crops

Crop	Accession Number	Salient Features	Country
High yielding and desirable type			
Tomato	EC391024	Long shelf life	Australia
	EC310029	Slow ripening	U.S.A.
	EC321425-26	Paste type	Taiwan
Cucumber	EC398030	Early and determinate type	China
Chillies	EC391073	Pungent type	

A number of abiotic and biotic stresses particularly Gemini, tospo and Iris Yellow Mosaic viruses; wilt, mildews, blight and anthracnose diseases are the major limiting factors in harnessing the actual potential of improved varieties of different vegetable crops. For instance, the occurrence of tomato leaf curl virus (ToLCV), Tospo wilt virus (TsWV) in tomato, YVMV and ELCV in okra, PepLCV in chilli, downy mildew in cucumber, and bacterial wilt in brinjal cause considerable reduction in yield ranges between 50-80 per cent depending upon their severity. To meet out the above targets, it is essential to search the new genes as per desired goals. So, survey and collection of land

Table 1.15: Priorities for Developing Multiple Resistant Cultivars with Specific Premium

Sl.No.	Crop	Targeted Biotic Stresses	Prioritized Trait(s) to be Incorporated
1.	Tomato	ToLCV + TsWV + Earli blight + Bacterial wilt + RKN	Tolerant to high temperature, high TSS and lycopene
2.	Brinjal	Phomopsis + Bacterial wilt + Fruit and shoot borer	Early maturing
3.	Chilli	PepLCV + Thrips + Mites + Anthracnose	High oleoresin and less capsaicin
4.	Capsicum	Phytophthora + Thrips + Mites	Adapted to tropical regions of India
5.	Okra	ELCV + YVMV + Fruit borer + Jassid + Red Spider mite	Dark green pods with 5 ridges
6.	Onion	*Stemphyllium* + Purple blotch + Thrips	Brown bulb, white bulb with high TSS
7.	Cucumber	DM + Mosaic	Pickling type
8.	Muskmelon	PM + DM + Anthracnose + *Fusarium*	High TSS
9.	Watermelon	PM + DM + Anthracnose	High TSS and seedless fruits
10.	Cabbage	Black rot + Diamond back moth	Tolerant to high temperature
11.	Cauliflower	Black rot + Diamond back moth	Tolerant to high temperature
12.	Cowpea	Golden mosaic virus	Fleshy green pods
13.	Bottle gourd	Powdery mildew	Small seed
14.	Bitter gourd	Downy mildew	Thin fruited, less seeded
15.	Pumpkin	CMV	Early maturing, high carotene

ToLCV: Tomato leaf curl virus; TsWV: Tospo wilt virus; RKN: Root Knot Nematode; YVMV: Yellow Vein Mosaic Virus; DM: Downy mildew; PM: Powdery mildew; ELCV: Enation Leaf Curl Virus; CMV: Cucumber Mosaic Virus.

races, ecotypes and traditional grown cultivars for their quality, yield traits and resistant to biotic and abiotic stresses is must. Development of plant genetic resource management database of major vegetable crops for their effective utilization in genetic improvement programme of the crops. Hence, there is an urgent need to prioritize breeding goals for developing multiple resistant cultivars with specific premium attributes, which are given in the Table 1.15.

2. Protection of Environmental Health

Plants use solar energy for synthesis of organic products, using carbon dioxide (CO_2) from air and water as raw materials and releasing oxygen into atmosphere. Thus conservation of plant diversity is of utmost importance in ensuring protection of healthy environment over the globe.

3. Establishment of New Crops

The potential use of PGR to mankind constitutes a broad spectrum of diverse gene pools representing assemblage of landraces, primitive cultivars, varieties of traditional agriculture as well as wild and weedy relatives of crop plants. Drastic increase in the world population with increasing rate and fast depletion of natural resources has led to exploration of newer sources of vegetables and diversification of vegetable cultivation to meet out the increasing demand. Vegetables have tremendous potential for commercial exploitation but remained ignored. Apart from the cultivated vegetables, enormous diversity occurs in semi domesticated and wild types in local pockets and native people as part of their routine vegetable requirements have made selection according to their preference/choice. These crops do not require high input technology and can be raised with comparatively lower management cost. They are tolerant/resistant to several abiotic and biotic

Table 1.16: Vegetable Breeders always Think of Available Systems for Hybrid Development

Sl.No.	Crops	Hybrid Systems
1.	Cabbage	Self-incompatibility (SI)
2.	Cauliflower	Cytoplasmic/Genetic male sterility (CMS/ ms ms)
3.	Kohlrabi	Cytoplasmic/Genetic male sterility (CMS/ ms ms)
4.	Radish	Cytoplasmic/Genetic male sterility (CMS/ ms ms)
5.	Tomato	Manual or mechanical Emasculation
6.	Brinjal	Functional male sterility
7.	Chillies	Cytoplasmic/Genetic male sterility
		Positional sterility (Protruding Stigma)
8.		Male sterility (Sms ms)
9.	Okra	Manual or mechanical emasculation
10	Cucurbits	Manual pollination in female line
		Male flower pinching
		Suppression of male flowers by growth regulators (2-CPA)
		Use of gametocytes
		Use of gynoecious lines

conditions and come up well on lands, which do not support the cultivation of other vegetables. Several of them used as vegetables which are very nutritious, remunerative and have often come to the rescue of people in times of crisis. These species or its variants can be established as new crop from wild through domestication and introduction. For example in India muskmelon has been introduced as a new crop.

4. Establishment of New Varieties

Introduction of new varieties is a common phenomenon to revolutionize the vegetable crop. Several promising strains in cucumber, muskmelon water melon, bitter gourd *etc.* have been introduced in to India, not only to enrich variability but also to increase yield and quality as immediate gain.

Aesthetic Value

The vegetable grown on small or commercial scale provide income besides satisfaction to the grower and his family members. Cultivation of vegetables, right from seed sowing, sprouting, flowering, fruit formation, picking, packaging and marketing provides for joy and aesthetic value to all besides supply of fresh vegetable. Cucurbits fruits sometimes become objects of decoration and art. In the USA, ornamental gourds and squashes provide autumn table decorations and pumpkin (*Cucurbita pepo*) are carved and illuminated with candles to celebrate the holiday of Halloween. The ultimate

Table 1.17: Vegetable Varieties Established in India through Introduction

Crop	*Botanical Name*	*Introduced Varieties*	*Country*
Tomato	*Solanum lycopersicum*	Roma	South America
		Marvel	
		Sioux	U.S.
		Ponderosa	
		Marglobe	U.S.
		Best of All	
		Fireball	Canada
		La Bonita	U.S.
		Dwarf Money Maker	Isreal
		Devlin's Choice	
		Red cloud	
		Balkan	Bulgaria
Brinjal	*Solanum melongena*	Black beauty	
		Wynad Giant	
Chilli (Hot pepper)	*Capsicum annuum*	Hot Portugal	
Sweet pepper/Bell pepper	*Capsicum annuum* var. *frutesense*	Chinese Giant	
		Hungarian	
		Bull Nose	
		California Wonder	U.S.A.
		Yolo Wonder	California
		World Beater	
		R-449	Russia
Okra	*Abelmoschus esculentus* L. Monch.	Perkins Long Green	
		Ghana Red	Ghana (Africa)
Peas	*Pisum sativum*	Arkel	England
		Early Badger	U.S.
		Meteor	England
		Little Marvel	England
		Alderman	
		New Line Perfection	
		Bonneville	U.S.
		Multi-freezer	
		Thomas Laxton	
		Sylvia	Sweden
		Dwelich Commando	Wisconsin
French bean	*Phaseolus vulgaris*	Giant Stringless	Sweden
		Jampa	Mexico
		Contender	U.S.
		Kentucky Wonder	U.S.A.
		Bountyful	U.S.A.
		Top Crop	U.S.A.
		Wates	U.S.A.
		Premier	Sweden
		BKW-74	Sweden

Crop	Botanical Name	Introduced Varieties	Country
Cowpea	*Vigna unguiculata*	Philippines Early	Philippines
Cauliflower	*B. oleracea* var. *botrytis* var. *cauliflora*	D-96	
		Improved Japanese	Isreal
		Snowball-16	Holland
Cabbage	*B. oleracea* var. *capitata*	Copenhagen Market	United Kingdom
		Golden Acre	
		Jersey Wake Gield	
		Chieftain	
		September	German Democratic Republic
		Pride of India	United Kingdom
Radish	*Raphanus sativus*	Japanese White	
		White Icicle	
		Chinese Pink	
		Rapid Red White Tipped	
		Scarlet Globe	
		Pusa Chetki	Denmark
Knol-Khol		White Vienna,	
		Purple Vienna	
		King of North	
Carrot		Nantes	
		Nantes Half Long	
		Chantay	
		Danverse	
		Zeno	German Democratic Republic
Turnip		Purple Top White Globe	
		Golden Ball	
		Snow Ball	
		Early Milan	
		Red Top	
		Early Grano	
		Brown Spanish	
Watermelon	*Citrullus lanatus*	New Hampshire Midget	U.S.A.
		Charleston Grey	
		Improved Shipper	
		Asahi Yamato	Japan
		Sugar baby	U.S.A.
		Shipper	U.S.A.
Cucumber	*Cucumis sativus*	China	
		Straight Eight	
		Poinsett	U.S.A.
Potato		Magnum Bonnum	
		Craigs Defiance	
		Up-To-Date	
Summer Squash	*Cucurbita pepo*	Australian Green	Australia
		Pattypan	U.S.A.
Onion	*Allium cepa*	Early Grano	U.S.A.
		Bermuda Yellow	Philippines
		Red Granno	America
Garden beet		Detroit Dark Red	
		Crimson Globe	
		Crosby Egyptian	
		Early Wonder	
Sweet potato	*Ipomoea batatas*	FA 17 White	China

degree of artistic expression has been reached with bottle gourds. Many different cultures independently developed the custom of carving, painting and otherwise decorating the gourds that were so useful to them. The art of gourd decoration is highly developed in Peru, were even today, detailed pyroengraving techniques produce gourd craft suitable for export. Finally, some cucurbits are grown as ornamental plants. In addition to its lacy leaves and colourful fruits, the fragrant blossoms of bitter gourd make this species a very appealing trellis ornamental; it has been grown in glass house in the U.K., since Victorian times. The beautiful white, fringed petals of snake gourd and its very long serpent like fruits, dangling beneath a garden arbour, make a very picturesque sight. The spiny fruits of teasel gourd, African horned cucumber, explosive fruits of squirting cucumber and *Cyclanthera explodens* have brought these cucurbits into commercial trade as well. The most recent horticultural trend affecting cucurbits is the current passion for succulents. Collectors are beginning to learn and trade among themselves the caudiciform cucurbits of Africa (*e.g. Gerrardanthus* and *Trichomeria*), Madagascar (*Seyrigia* and *Xerosicyos*) and North America (*Marah* and *Ibervillea*).

Vegetable Oil

The seeds of many members of the group have been noted for their oil-bearing properties. However, little information has been found that describes existing methods of oil extraction. As in other oil seed crops cucurbit seed proteins, consist of mostly globulins and some albumins, have abundant nitrogen rich amino acids and are deficient in lysine and sulphur containing amino acids. Oil extracted from cucurbit seeds are primarlly composed of unsaturated fatty acids and are usually edible and suitable for various industrial purposes. In general, semi-drying oils of squash, melon, watermelon and *Luffa* seeds contain 40-60 per cent linoleic acid, 20-40 per cent oleic acid, 10-20 per cent palmitic acid and up to 15 per cent stearic acid. Seeds of fluted pumpkin (*Telfairia occidentalis*) which taste like almonds are very rich in oil composed of 37 per cent oleic, 21 per cent palmitic, 21 per cent stearic and 15 per cent linoleic acid. Buffalo gourd (*Cucurbita ficifolia*) produces seeds with linleic acid content ranging up to 82 per cent.

5. Enhancement of Genetic Stock

The wild genetic stocks serve as trait reservoirs and are used in hybridization program to interrogate gene of interest to cultivated or popular varieties.

6. Determination of Evolutionary Trends

Diversity available in a species, genus or family at place is indication for the evolutionary trends. Variability among biological entities available at a location reveals the stress on genome of species and discloses the course and rate of evolution of various loci. Vavilov (1926) proposed law of homologous series by analyzing trends of genetic variation among different species.

7. Ritual Importance

There are several vegetable crops which ritual association can be seen in the societies. For example tuber crops in Kerala:

(i) **Ettangadi:** A traditional dish associated with Thiruvathira festival. On the previous night of Thiruvathira, the women make "ettangadi" as a part of religious rituals of Hindus. Tubers of *Colocasia*, Elephant foot yam, Cassava, Chinese potato, lesser yam var. Cherukizhangu, Lesser yam var. Nanakizhangu and greater yam are baked and cut into pieces after removing skin. Into this, coconut, powdered pulses and jaggery are mixed to make ettangadi and offered to Gods Ganapathy, Siva and Goddesses Parvathi and Durga.

(ii) **Thiruvathirapuzhukku:** Elephant foot yam, *Xanthosoma*, *D. alata*, banana, Chinese potato and sweet potato are cooked with powdered green gram, coconut, red chillies, curry leaf, salt and turmeric to make this dish on the day of Thiruvathira.

(iii) **Hindu:** Hindu consumes cooked tubers of *D. alata* on the day of festival "Karthika".

(iv) **Thalukary:** Thalukary is a preparation made for giving to pilgrims at the Dhanwanthiri temple near Thakazhy in Kuttanadu, Kerala which is believed to have medicinal effect.

8. Biotic and Abiotic Adjustment through Selection of Rootstocks and Scion (Grafting)

Salinity is also a major problem in crop cultivation.

Table 1.18: Differences in the Level of Salinity Resistance of different Vegetable Crops

Crops	*Threshold Salinity Level (SY_{100})**	*SY_{50}**
Beans	$1dSm^{-1}$	$4dSm^{-1}$
Onions	$1dSm^{-1}$	$4dSm^{-1}$
Sugarbeet	$8dSm^{-1}$	$15dSm^{-1}$

Presently, it has been observed that several crops are being affected with change of climatic and edaphic conditions which can be overcome through graftings. Identification of rootstock suitable for biotic/abiotic stresses and development of grafting technique particularly in cucurbitaceous and Solanaceous vegetables is need of the day. As per the historical back ground, the production of grafted plants began in Japan and Korea in the late 1920s, when watermelon (*Citrullus lanatus*) was grafted on bottle gourd rootstock to overcome yield decline which was due to soil borne pathogens (Lee, 1994). In 1959, the scions of eggplant were grafted in large scale on rootstocks of scarlet eggplant (*Solanum integrifolium*) to avoid injury caused by soil-borne diseases such as *Verticillium* wilt, *Fusarium* wilt, bacterial wilt and nematodes. Grafting in cucumber and tomato was also started in 1959 in Japan to reduce the damage caused by soil borne diseases and to promote scion vigour. By 1990, the production of grafts in fruit bearing vegetables (eggplant, cucumber, tomato and melons) reached around 60 per cent of the area in Japan and 80 per cent in Korea (Lee, 1994; Lee *et al.*, 1998). Grafting is now commercially practiced especially in green house grown crops *viz.* watermelon, cucumber, muskmelon, bitter gourd, tomato and eggplant in Japan, Korea and other European countries. In Japan around 95 per cent of the watermelons (*Citrullus lanatus*), melons (*Cucumis melo* var. *makuwa*), green house cucumbers (*Cucumis sativus*) and Solanaceous crops are before being transplanted to the field or green house. In Japan, most of the muskmelons are grafted on squash rootstocks (*Cucurbita* species) or mixed hybrids (*C. maxima* x *C. moschata*). Inter-generic grafting is used in the production of many fruit-bearing vegetables; *i.e.*, cucumber (*C. sativus* L.) grafted on pumpkin (*Cucurbita* species), watermelon (*Citrullus lanatus*) on bottle gourd (*Lagenaria siceraria* Standl.) and melon (*Cucumis melo*) on ash gourd (*Benincasa hispida* Cogn.). Inter-specific grafting is possible in eggplant (*Solanum melongena* L.). Scarlet eggplant (*Solanum integrifolium* Poir.) and *Solanum torvum* Swartz are popular rootstocks for eggplant production.

Table 1.19: Suitable Rootstocks for Grafting of Vegetable

Sl.No.	*Scion*	*Rootstock*
1.	Cucumber	*Cucurbita moschata* (pumpkin), *C. ficifolia* (fig-leaf gourd), *C maxima*, *Sicyo sangulatus* (bur cucumber)
2.	Muskmelon (for open field)	*Cucurbita* spp., *C. moschata* x *C. maxima*, *C. melo*
3.	Muskmelon (for greenhouse)	*C. melo*, *Benincasa hispida*, *Cucurbita* spp., *C. moschata* x *C. maxima*
4.	Watermelon	*Lagenaria siceraria*, *Cucurbita* spp., *Benincasa hispida*
5.	Bitter gourd	*Cucurbita moschata*, *Lagenaria siceraria*, *Luffa cylindrica*
6.	Tomato	*Solanum pimpinellifolium*, *S. lycopersicum*, *S. nigrum*
7.	Brinjal	*Solanum torvum*, *S. integrifolium*, *S. melongena*, *S. nigrum*

Source: Oda, 1993.

Fig-leaf gourds possess good resistance to *Fusarium* wilt and low temperature conditions. It has good ability to absorb water and nutrients from the soil even at low temperature conditions. Interspecific squash rootstocks such as 'Shintozwa' are more tolerant to high temperature conditions and are commonly used for cucumber grown during summer conditions. Rootstocks for cucurbitaceous crops and related characteristics are given in Table 1.20.

Raw Materials for Industry

Today vegetables are becoming a major source and providing different kinds of raw materials to the industries. Cassava tubers are used for starch extraction in Tamil Nadu and Andhra Pradesh. The starch extracted from cassava is used for preparing sago, wafers, porridge and other items. Cassava starch is also used in textile and adhesive industries. Cassava is also called petro-crop. Good quality alcohol is extracted from cassava tubers. Of late it is called plastic crop as because biodegradable plastics are manufactured utilizing cassava starch. Tubers of sweet potato are used for starch, liquid glucose and pigments extraction. Similarly, potato, tomato, chillies, cucumber bitter gourd, bottle gourd *etc.* are has opened the room for industrial development.

9. Ethnic Uses and Value

Many cultivated and wild vegetable species date back to pre-historic times and are associated to culture of a region/country. It is presumed that vegetable were among the first plant used by man. Lesser-known vegetables have their importance mainly in the areas of their occurrence due to climatic change, physiographic diversity and adaptation. In the Indian sub-continent there is tremendous scope for their acclimatization and they not only have potential to contribute towards food and nutritional security but also are a source of rich wild gene pool for several important traits. Wild plants of economic value have played an important role in the life of people. Several local cucurbitaceous vegetable and underground roots are regular part of their diet. They are collected directly from the wild or cultivated/protected as backyard cultigens by the native people which include species *viz.*, *Citrullus*, *Momordica*, *Trichosanthes*, *Fevillea*, *Genostemma*, *Hemsleya*, *Bulbostemma*, *Thaldiantha*, *Bryonia*, *Cayaponia*, *etc.* which commonly occur in areas surrounding local habitation. Human beings have utilized plants as the main source of medicine and have learned much about the avoidance and curing illness through the consumption and application of plants and their products. Plants contain thousands of compounds, a few are beneficial and most function that are still unknown. For example buffalo-gourd has been utilized by man as food and medicine since 1000 years. Tribals use these species in herbal therapy also since the beginning of human history.

Ethno-Agriculture System

Jharkhand is the great home of genetic wisdom, indigenous plant lore with rich biodiversity occurs in ambient vegetation and natural forests. Although, there are now changes according to modern life style, concept of modern cities and urbanization of the state can be observed however still in the rural areas and hilly tracts the native people as well as forest dwellers having affinity with plant based indigenous knowledge related to meeting there day to day needs of edible and healthcare purpose. Tribal people cooked some

Table 1.20: Rootstock for Cucurbitaceous Crops and Related Characteristics

Sl.No.	Crop	Scion/Rootstocks	Cultivar	Major Characteristics	Possible Disadvantages
1.	Watermelon	Bottle gourd (*Lagenaria siceraria*)	FR Dantos, Partner, Renshi, FR Combi, Tan Tan	Vigorous root system, resistant to *Fusarium* wilt and low temperature	New *Fusarium* race susceptible to anthracnose
2.		Squash (*Cucurbita moschata*)	Chinkyo.No.8, Keumkang	Vigorous root system, resistant to *Fusarium* wilt and low temperature	Poor fruit shape and quality
3.		Interspecific hybrid squash (*Cucurbita maxima* x *C. moschata*)	Shintozwa, Shintozwa-1, Shintozwa-2, Chulgap	Vigorous root system, resistant to *Fusarium* wilt and low temperature, excellent vigour and high temperature tolerance	Reduced fertilizers requirement; reduced quality
4.		Pumpkin (*Cucurbita pepo*)	Keumsakwa, Unyong, Super Unyong	Vigorous root system, resistant to *Fusarium* wilt and low temperature	Poor fruit shape and quality
5.		Ash gourd (*Benincasa hispida*)	Lion, Best, Donga	Good for disease resistance	incompatibility
6.		Watermelon (*Citrullus lanatus*)	Kanggang, Res. 1, Tuffness, Kyohgoh, NHRI-1	*Fusarium* tolerance	Not enough vigour, disease resistance
7.		African horned cucumber (*Cucumis metuliferus*)	-	Excellent for *Fusarium* resistance and good nematode tolerance	Medium to poor compatibility
8.	Cucumber	Fig-leaf gourd (*Cucurbita ficifolia*)	Heukjong, blaci Seeded, Fig-leaf gourd	Good for low temperature tolerance and disease resistance	Fruit quality may be reduced
9.		Squash (*Cucurbita moschata*)	Butternut, Unyong-1, Super Unyong	Good for *Fusarium* tolerance and bloomless fruit skin	Affected by *Phytophthora*
10.		Interspecific hybrid squash (*Cucurbita maxima* x *C. moschata*)	Shintozwa, Keumtozwa Ferro RZ, 64-05RZ, Shinwha, Chulgap	Good for *Fusarium* and low temperature tolerance	Slightly quality reduction expected
11.		Bur cucumber (*Sicyo sangulatus*)	Andong	Good for *Fusarium* tolerance, low and high soil moisture tolerance and nematode tolerance	Reduced yield
12.		African horned cucumber (*Cucumis metuliferus*)	NHRI-1	Excellent for *Fusarium* resistance and good nematode tolerance	Weak temperature tolerance
13.	Melons-oriental melons	Squash (*Cucurbita moschata*)	Baekkukzwa, no.8, Keumkang, Hongtozwa,	Good for *Fusarium* and low temperature tolerance	*Phytophthora* infection
14.		Interspecific hybrid squash (*Cucurbita maxima* x *C. moschata*)	Shintozwa, Shintozwa-1, Shintozwa-2,	Good for *Fusarium* resistance, low and high soil temperature tolerance, and high soil moisture tolerance	*Phytophthora* infection, poor fruit quality
15.		Pumpkin (*Cucurbita pepo*)	Keumsakwa, Unyong, Super Unyong	Good for *Fusarium* resistance, low and high soil temperature tolerance, and high soil moisture tolerance	*Phytophthora* infection
16.		Melon (*Cucumis melo*)	Rootstock-1, Kangyong, Keonkak, Keumgang	Good for *Fusarium* tolerance and good fruit quality	*Phytophthora* infection
17.		African horned cucumber (*Cucumis metuliferus*)	NHRI-1	Good for *Fusarium* tolerance, low and high soil moisture tolerance and good nematode tolerance	Weak temperature tolerance

Source: Asao *et al.*, 1999; Igaraashi *et al.*, 1987; Kang *et al.*, 1992; Kim and Lee, 1989; Ko, 1999; Matsuo *et al.*, 1985; Lee, 1989, 1994, 2000; Leee *t al.*, 1998; Oda, 1999.

potherb as vegetables and growing by the farmers in own methods referred as Ethno-Agriculture system. It has been observed that some of the herbaceous annual and seasonal plant species usually cultivated in their farming and utilizing as traditional vegetable. Some of the potential species are *Boerhaavia diffusa* ('Khapra-sag'), *Centella asitica* ('Beng-sag'), *Limnophila indica* ('Hemcha-sag'), *Portulaca* spp. ('Nuni-sag', 'Jitia-sag') *etc.* have been observed utilized by almost tribal, rural and forest dwellers (Kumar and Abbas, 2016).

Immature fruits of wild types drumstick are generally bitter in taste, hence not generally used for vegetable purpose. Young fruits (10 days after anthesis; 10-15 cm long) are occasionally employed for making pickle. Flowers are less commonly used in *raita* (a dish prepared from boiled flowers and curd) preparation in Gandawal in Una district of Himachal Pradesh. At Burj village (Panchkula district), contractors are routinely collecting the flowers from wild trees (by destructive harvesting) in jute bag (moist) for sale in big markets

(Rs. 50/kg) (Pradheep *et al.*, 2011). Arora and Pandey (1996) also reported that young flowers (both pink and white forms) in packets are commonly sold during February-March in city markets of northern India. Also people seldom cooked flowers as informed in Auhar and Dagli villages in Bilaspur district of Chhattisgarh. Roots of three months old seedlings (45-60 cm long) were said to be used for pickle making and sparingly sold in market of Punjab and parts of Kangra district of Himachal Pradesh. For this, after one and a half month in nursery, seedlings were headed back to 30 cm high so as to enable root growth for another one and a half month.

Ethnobotany in Relation to Vegetable Genetic Resources

In food chain, each and every individual play their own role in the nature and by virtue of this it is truly said that "Nature keeps its natural balance itself". References to herbal medicine in ancient scriptures suggest very long history of ethno botany. This statement also indicates the ethno botanical relationship of plant and human. Yet, the world 'Ethno botany' was applied to such knowledge by Harshberger (1985) less than a century ago. In India, ethnobotanical studies has

Table 1.21: Ethnic Uses of Cucurbitaceous Vegetable Species

Common Name	Botanical Name	Frequency	Place of Cultivation	Usages
Antidote vine	*Fevillea* cordifolia	Rare	Neotropics	Medicinal
Jiao-gu-lan	*Gynostemma* pentaphyllum	Localized	Asia	Medicinal
Luo-ago-di	*Hemsleya amabilis* Diels	Localized	Asia	Medicinal
He-zi-cao	*Actinostemma tencrum Griff.*	Localized	Asia	Medicinal
Psendo-tritillary	*Balbostemma paniculatum*	Localized	Asia	Medicinal
White seed melon	*Cucumeropsis mannii* Naud	Localized	Africa	Food
Bur gherkin	*Cucumis anguria* L.	Localized	Wide spread	Food
Teasel gourd	*Cucumis dipsaceus* Ehrenb. ex.	Sporadic	Wide spread	Ornamental
African horned cucunber	*Cucumis metuliferus* E. Mey ex. Naud	Sporadic	Wide spread	Food
Sponge plant	*Momordica angustisepala* Harms	Rare	Neotropics	Utilitarian
Cochin chin gourd	*Momordica cochinchinensis* (Lour.) Spreng	Sporadic	Wide spread	Medicinal
Balsam apple	*Momordica balsamina* L.	Frequent	Wide spread	Medicinal
Kaksa	*Momordica cymbalaria* Frenzl. ex. Hork	Rare	Old world	Food
Luo-han- guo	*Siraitia grosvenovrii* (Swingle) Lu. and Zhang	Localized	Asia	Food
Red hail stone	*Thladiantha dubia* Bunge	Localized	Wide spread	Medicinal
Fluted pumpkin	*Telfairia occidentalis* Hook F.	Localized	Africa	Food
Oyster nut	*Telfairia pedata* (Sims)Hook	Localized	Africa	Food
Hard plant	*Hodgsonia macrocarpa* (Bl.)Cogn.	Infrequent	Asia	Food
Gymnopetlum	*Gymnopetlum cochinchinensi* (Lour.)Kurz.	Rare	Asia	Food
Snake gourd	*Trichosanthes cucumerina* L.	Frequent	Wide spread	Medicinal
Chinese Snake gourd	*Trichosanthes Kirilowii* Maxim.	Localized	Asia	Medicinal
Indreni	*Trichosanthes lepiniana* (Naud.) Cogn.	Localized	Asia	Medicinal
Japanese Snake gourd	*Trichosanthes ovigera* Blume	Frequent	Asia	Food
Mi-Mao-qua-lou	*Trichosanthes villosa* Blume	Localized	Asia	Food
Inara	*Acanthosicyos horridus* Wel. ex. Hook. f.	Localized	Asia	Food
Wax gourd	*Benincasa hispida* (Thunb.) Cogn.	Frequent	Wide spread	Medicinal
Bryoni	*Bryonia cretica* L.	Localized	Old world	Medicinal
Bryoni	*Bryonia dioica* L.	Sporadic	Wide spread	Medicinal
Colocynth	*Citrullus colocynth* (L.) Schrad	Sporadic	Wide spread	Medicinal
Ivy gourd	*Coccinia abyssinica* (L.) Cogn.	Localized	Africa	Food
Little gourd	*Coccinia grandis* (L.) Voigt.	Sporadic	Wide spread	Food
Lollipop climber	*Diplocyclos palmatus* (L.) C. Jeffery	Localized	Old world	Food, Ornamental
Squirting cucumber	*Ecballium elaterium* (L.) A. Rich.	Sporadic	Old world	Food, Ornamental
Round melon	*Praecitrullus fistulosus* (Stocks) Pang	Localized	Asia	Food
Cayaponica	*Cayaponica kathematophora*	Rare	Neotropics	Food, Ornamental
Cayaponica	*Cayaponica ophthalmica* R.E. Schult.	Rare	Neotropics	Medicinal
Malabar gourd	*Cucurbita ficifolia* Bouche	Localized	Wide spread	Food

Common Name	Botanical Name	Frequency	Place of Cultivation	Usages
Casabanana	*Sicana odorifera* (Vell.) Naud.	Sporadic	Neotropics	Food
Stuffing cucumber	*Cyclanthera brachybotys* (Poepp. and Endl.)	Sporadic	Neotropics	Food
Stuffing cucumber	*Cyclanthera pedata*	Sporadic	Wide spread	Food, Ornamental
Stuffing cucumber	*Cyclanthera explodens* naud.	Localized	Neotropics	Food
Wild cucumber	*Echinocystis lobata* (Michx. Torr. and Gray)	Sporadic	Wide spread	Food, Ornamental
Chayote	*Sechium edule* (Jacq.) Swartz	Common	Wide spread	Food
Chayote	*Sechium tacaco* (Pitt.) C. Jeffrey	Localized	Neotropics	Food

Source: Ram *et al.*, 1999.

undertaken in Patalkot, Abujhmarh (Bastar), Nimar/ Malwa (Madhya Pradesh, Uttar Pradesh and Himachal Pradesh hills).

Ethnobotany can be defined as:

"The total natural and traditional relationship and the interactions between man and his surrounding plant wealth".

Similarly, Faulks (1958) in his book "An Introduction to Ethnobotany" defined ethnobotany as:

"The entire realm of economic botany, i.e., the relationship of plant kingdom with the man of past, present and future".

Importance of Ethnobotanical Studies

There are several primitive populations in the world dwelling in far remote areas, still following the practices regarding the utilization of plants, which are verbally inherited through one generation to other. Harshberger (1986) suggested the following importance of ethno botanical studies:

1. Ethno botanical studies help in elucidating the cultural position of the ethnic races, who used the plants for food, shelter, and fibers;
2. An ethno botanical study throws light upon the past distribution of plants;
3. Ethno botany is useful as suggesting new lines of manufacture at present day, and
4. An ethno botanical study helps us to decide as to the ancient trade routes.

Classification of Ethnobotany

Yet, ethno botany is generally taken as the science of relationships of only the primitive or aboriginal people with plants.

The man plant relationship can be broadly classified into two groups:

(a) Abstract and

(b) Concrete

(a) Abstract

The abstract relationship of man with plants includes faith in the good or bad powers of plants, taboos, avoidances, sacred plants, worship and folklore.

(b) Concrete

The concrete relationship includes mainly the material use such as in food, medicine, house-building, agricultural operations, other domestic uses, trade or barter, plants in fine arts and culture like paintings, carvings and house decoration and the act of domestication, conservation, improvements or destruction of plants.

The tribal: As per sociological view point's tribal may be defined as:

"Tribal are a distinct ethnic group who are usually confined to definite geographical areas, form a sizeable percentage of total population, speak a common dialect are culturally homogenous and a unifying social organisaions".

The tribal people of India mostly live in forests, hills, plateaus and naturally isolated regions and differently termed as *Adivasi* (original settlers), *Adimniwasi* (oldest ethnological sector of the population), *Adimjati* (primitive caste), *Aboriginal* (indigenous), *Girijan* (hills men), *Vanyajati* (forest dwellers), Janjati (folk communities), *AnusuchitJanjati*(Schedulled tribes), and several such other names signifying either their ecological or economic, or historical, or cultural characteristics. Among these terms, the most popular is *Adivasi*, while the constitutional name for them is *AnusuchitJanjati* (Scheduled tribes). The Indian sub-continent is inhabited by over 53 million tribal people belonging to over 550 tribal communities of 227 ethnic groups (Bhatt *et al.*, 1998). They constitute about 7.7 per cent of India's population. It is estimated that the predominant tribal areas compromise about 15 per cent of the total geographical area of the country.

The tribal community can be broadly divided into three distinct zones:

(i) The tribes living in the northern and north-eastern zone in the mountain valleys and eastern frontiers of India such as Naga, Mizo, Khasi, Garo *etc.*;

(ii) The central zone of the older hills and plateaus along the dividing line between the peninsular India and the Indo-Gangetic plain spanning the states of West Bengal, Orissa, Bihar, Madhya Pradesh, Rajasthan, Maharashtra and Gujarat in which major tribal group like the *Santhal, Munda, Bhil, Gond, Baiga etc.* live and

(iii) The tribal areas of Andhra Pradesh, Karnataka, Tamil Nadu and Kerala from part of the southern zone where *Gadaba, Irula, Toda, Kota, Yenadi etc.* inhabit.

Sub-disciplinary and Interdisciplinary Approaches for Ethnobotany

Ethno botanical studies on various subgroups of the plant kingdom, like on fungi, bryophytes, pteridophytes, lichens *etc.*, are sub disciplines and have been named as ethno algology, ethno mycology, ethno bryology, ethno pteridology, ethno lichenology *etc.* but, when the inquiry in ethno botany extends beyond ordinary ream of botany and has significant input of another branch of science, like archaeology, or medicine, the work becomes interdisciplinary. More and more new lines of interdisciplinary researches are emerging under each subject.

Research Strategies based on Genetic Resources

Research on vegetable crops is taking pace with consumer preference and market demand. Now trait specific genotypes are required to be introduced.

Food Plants for Emergency Situations

The Hindustani (Indian) gene centre, with 160 cultivated species of economic plants, serves as reservoir of a large number of useful plant species. The diversity of less known cultivated food plants is confined to 56 species. Further, about 320 species represent the truly wild and weedy counterparts as components of different vegetation types (Arora and Nayar, 1984). Information on plant genetic resources of these less known cultivated food plants and their wild relatives is well documented (Arora and Nayar, 1984; Arora, 1985). There are also several introduced species which have a great potential to be exploited for various purposes in the country. An account of important underutilized plant species which await exploitation for food, fodder, energy and industrial purposes have been given earlier (Paroda and Bhag Mal, 1989; Bhag Mal, 1990). In India, there are large areas which suffer with edaphoclimatic extremes of different kinds and the people face problems for their food and other requirements. However, there exist several indigenous plant species which can support life during difficult environments and emergency situations. Information on such plant species has been well documented (Gupta and Kanodia, 1968; Bhandari, 1974; Saxena, 1979; Shankarnarayan and Saxena, 1989; Shankar, 1989 and Singh and Gupta, 1989). The important plant species which can be successfully grown for food under harsh environmental conditions are listed below:

Table 1.22: Food Plants for Emergency Situations

Situation	*Usage*	*Species*
Hot desert	Grain	*Setaria verticiliata, Haloxylon salicornicum, Citrullus colosynthis, C. lanatus, Echinochloa colonum, E. crusgalli, Abutilion indicum, Eleusine compressa*
	Leaves and tender shoots	*Amaranthus gracillis, A. hybridus, A. spinosus, Chenopodium album, Boerhavin diffusa, Euphorbia cacodifolia, Moringa oleifera, Portulaca oleracea, Aloe barbadense*
	Fruits	*Grewia tenax, Zizyphus nummularia, Cordia gharaf, Coccinia grandis, Cucumis callosus, Momordica dioica, saladora oleoides, S. persica*
	Tubers	*Cyprus rotundus, C. bulbosus, Asparagus racemosa, Ceropegia bulbosa, Butea monosperma, Portulaca tuberosa*
Cold desert	Grain	*Fagopyrum esculentum, Fagopyrum tataricum*
	Leaves	*Allium leptophyllum, Aconitum heterophyllum, Cicer songarium, Polygonum viviparum, Sedum tibeticum, Scorzonera mollis*
Semi-arid region	Leaves	*Boerhaavia diffusa, Bryonica lacimosa, Triumfetta rhombifolia, Cocculus hirsutus, Amaranthus viridis, Leptadenia reticulata*
Saline areas	Fruit	*Salvadora oleoids, S. persica*
	Foliage	*Salicornia brachiata*
	Leaves	*Chenopodium album, Portulaca oleracea, Trianthema portulacastrum*
Marshy lands/flood areas	Vegetables	*Alternanthera sessilis, Hygroryza aristata, Monochoria hastate.*

The World's Largest Vegetables

So if you exclude massive squash and pumpkins from the vegetable class, what is the world's largest vegetable? This title should be limited to vegetables in a typical human diet—because to Australian koalas, the largest vegetable would most certainly be an enormous eucalyptus tree. The 1985 "Guinness Book Of World Records" (UK Edition) lists some of the record-breaking vegetables, including a 35 pound (16 kg) turnip, a 45 pound (20 kg) red cabbage, a 28 pound (13 kg) broccoli, a 52 pound (24 kg) cauliflower, a 25 pound (11 kg) lettuce, and a remarkable 124 pound (56 kg) cabbage six feet (1.8 m) in diameter. Although this giant cabbage cited in the Guinness Book seems unbeatable for the title of "World's Largest Vegetable," there are tropical yams belonging to the genus *Dioscorea* that may be 6 to 9 feet long (2-3 m) and weigh 150 pounds (68 kg) or more, although they are usually harvested at about 2-6 pounds. These yams are not to be confused with fleshy storage roots of red sweet potatoes (*Ipomoea batatas*) of the Morning Glory Family (Convolvulaceae) which are also called yams. True yams belong to an entirely different and unrelated plant family, the Dioscoreaceae.

Table 1.23: Important Terminology

Family	:	An assemblage of genera that closely and uniformly resemble one another in general appearance and technical characters.
Genus	:	A more or less closely related and definable collection of plants that may include one or more species. The species in the genus are usually structurally or phylogenetically related, but do not routinely intercross.
Species	:	A group of similar organisms capable of interbreeding and are more or less distinctly different in morphological or other characteristics, usually reproductive parts, from other species in the genus.
Botanical Variety	:	A subdivision of a species consisting of a population with morphological characteristics distinct from other species' forms and is given a Latin name according to the rules of the International Code of Botanical Nomenclature. Variety was and continues to be used erroneously when the correct term is intended to be cultivar
Group	:	A category of cultivated plants at the subspecies level that have the same Latin binomial, but have one or more characteristics sufficiently unique to merit a name that distinguishes them from another category. The term is used for horticultural convenience and has no botanical significance. Thus, botanical variety, subspecies, and group are similar and therefore often used interchangeably.
Cultivar	:	Contraction for "cultivated variety"; a plant that is clearly distinguished by identical physical characteristics and maintains these characteristics through proper propagation means.
Cultigen	:	A plant or group of plants known only in cultivation without a determined nativity, presumably having originated in its presently known form under domestication (Rubatzky and Yamaguchi, 1007). A plant whose origin or selection is due primarily to intentional human activity is called a cultigen. (A cultivated crop species that has evolved from wild populations due to selective pressures from traditional farmers is called a landrace.)
Clone	:	A population derived from a single individual and maintained by vegetative propagation. Individual members of a cloned population are genetically identical and can be maintained essentially uniformly with relatively little selection.
Landraces	:	Plants are ideally suited to a particular region or environment as the result of natural forces.
Line	:	A uniform sexually reproduced population, usually self-pollinated, that is seed propagated and maintained to the desired standard of uniformity by selection.
Strain	:	An improved selection of a cultivar that possesses similar characteristics, but differs in some minor attribute or quality.
Lot	:	A particular batch of seeds that were produced and processed together.
Type	:	A series of cultivars of a crop that have similar characteristics without specific reference to genetic or morphological characteristics.

REFERENCES

1. Arora, R.K. 1985. Genetic resources of less known cultivated food plants. *NBPGR Sci. Monogr. No. 7*, ICAR-NBPGR, New Delhi, p. 126.
2. Arora, R.K. and Nayar, E.R. 1984. Wild relatives of crop plants in India. *NBPGR Sci. Monogr.*, No. 7, ICAR-NBPGR, New Delhi, p. 90.
3. Arora, R.K. and Pandey, A. 1996. Wild Edible Plants of India: *Diversity, Conservation and Use*. National Bureau of Plant genetic Resources, New Delhi, India, p. 226.
4. Asao, T., Pramanik, M.H.P., Tomita, K., Ohba, Y., Ohta, K., Hosoki, T. and Matsui, Y. 1999. Influence of phenolics isolated from the nutrient solution nourishing growing cucumber (*Cucumis sativus* L.) plants on fruit yield. *J. Jpn. Soc. Hort. Sci.*, **68:** 847-853.
5. Asati, B.S. and Yadav, D.S. 2004. Diversity of Horticultural crops in North Eastern Region. ENVIS Bulletin. *Himalayan Ecology*, **12(1):** 1-11.
6. Bhag Mal. 1990. Underutilized Plants: A treasure house unexplored. *Indian Fmg.*, **40(7):** 19-24.
7. Bhandari, M.M. 1974. Famine foods in Rajasthan desert. *Eco. Bot.*, **28(1):** 73-81.
8. Bhatt, T.K., Singh, B. and Sharma, P.O. 1998. Microbial degradation of tannins-a current perspective. *Bio-degradation*, **9**: 343-357.
9. Britton, G; Liaaen - Jemsen, S. and Pfander, H. 1998. Carotenoids, Biosynthesis and Metabolism, Birkauser verlag. Basel, Switzerland.
10. Faulks, J.P. 1958. *An Introduction to Ethnobotany*: London. Moredale Publications Ltd.
11. Giuseppina, M., Fabrizio, G. and Carlo, F. 1998. Horticultural Genetic Diversity: Conservation and sustainable utilization and related international agreements. World Conference on Horticultural Research, ISHS-17-20, June, Rome, Italy.
12. Gupta, R.K. and Kanodia, K.C. 1968. Plants used during scarcity and famine period in dry regions of India. *Journ. D. Agric. Trop. Et. de Bot. Applique*, **40(7 and 8):** 265-285.
13. Harshberger, J.W. (1986)-Purposes of ethnobotany. Botanical Gazette, 146-154.
14. Igaraashi, L. Kanno, T. and Karwaide, T. 1987. Disease and pest resistance of wild *Cucumis* species and their compatibility as rootstocks for muskmelon, cucumber and watermelon. *Bul. Natl. Res. Inst. Veg. Ornam. Plants and Tea*, **A1:** 173-185.
15. Kang, K.S., Choe, S.S. and Lee, S.S. 1992. Studies on the rootstocks for stable production of cucumbers. *J. Kor. Soc. Hort. Sci. Technol.*, **19:** 122-123.

16. Kim, S.E. and Lee, J.M. 1989. Effect of rootstocks and fertilizers on the growth and mineral contents in cucumber (*Cucumis sativus* L.). Kyung Hee Univ., Korea. *Res. Collection, Inst. Food Develop.*, **10:** 75-82.

17. Kumar, K. and Abbas, S.G. 2016. Utilization pattern of some wild edible plant biodiversity under Ethno-Agricultural system by the tribal, forest dwellers and rural people of Jharkhand. 1st International Agrobiodiversity Congres, held at New Delhi, 6-9 November, 2016; p. 31.

18. Lee, J.M. 1989. Studies on the cultivation of grafted plants of cucurbitaceous vegetables. *J. Kor. Soc. Hort. Sci.*, **30:** 169-170.

19. Lee, J.M. 1994. Cultivation of grafted vegetables 1. Current status, grafting methods and benefits. *HortSci.* **29:** 235-239.

20. Lee, J.M., Bang, H.J. and Ham, H.S. 1998. Grafting of vegetables. *J. Japan Soc. Hort. Sci.*, **67:** 1098-1114.

21. Madeley, J. 1996. Yours for food: Plant genetic resources and food security.

22. Oda, M. 1993. Present state of vegetable production using grafted plants in Japan. *Agri. Hort.*, **68:** 442-446.

23. Oda, M. 1999. Grafting of vegetables to improve the production under protective structure. *International Seminar on Protective Structure for Improved Crop Production.* April 12-17, Suwon, Korea.

24. Paroda, R.S. and Bhag Mal. 1989. New plant sources for food and industry in India. In: New crops for food and industry. Wickens, G.C., Heq, N. and Day, P. (eds.), p. 135-149.

25. Pradheep, K., Singh, P.K., Pandey, A. and Bhandari, D.C. 2011. Collecting genetic resources of wild *Moringa oleifera* Lam. from Western Himalaya. *Indian J. Plant Genet. Resour.*, **24(3):** 292-298.

26. Preethi, T.L., Thamaraiselvi, S.P., Padmadevi, K., Sankaranarayanan and Ponnuswamy, V. 2011. Underutilized edible flowers of hills. In: Developing the potential of underutilized horticultural crops of hill regions, pp. (Eds.)Prakash, N., Roy, S.S., Sharma, P.K. and Ngachan, S.V., Publisher: Today and Tomorrow Printers, N. Delhi, 110002.

27. Ram, D. and Srivastava, U. 1999. Some lesser known minor cucurbitaceous vegetables: Their distribution, diversity and uses. *Indian J. Pl. Genet. Resour.*, **12(3):** 307-316.

28. Rathore, D.S., Srivastava, U. and Dhillon, B.S. 2005. Management of genetic resources of horticultural crops: Issues and strategies. In: Dhillon, B.S., Tyagi, R.K., Saxena, S. and Randhawa, G.J. (Ed.). Plant Genetic Resources: Horticultural Crops. Narosa Publishing House, New Delhi, p. 1-18.

29. Samant, S.S. and Dhar, U. 1997. Diversity, endemism and economic potential of wild edible plants of Indian Himalaya. *Int. J. Sustain. Dev. World Ecol.*, **4:** 179-191.

30. Saxena, S.K. 1979. Plant foods of western Rajasthan. *Man and Environment*, **3:** 35-43.

31. Shankar, V. 1989. Life support species for emergencies in the Thar Desert of India. In: Life support plant species-diversity arid zone, plants in famine period. In: Support Plant Species-Diversity and Conservation. R.S. Paroda, Kapoor, P, Arora, R.K. and Bhag, M. (eds.), p. 37-41.

32. Shankarnarayan, K.A. and Saxena, S.K. 1989. Life supporting arid zone plants in famine period. In: Life support plant species-diversity arid zone, plants in famine period. In: Support Plant Species-Diversity and Conservation. R.S. Paroda, Kapoor, P, Arora, R.K. and Bhag, M. (eds.), p. 55-59.

33. Singh, Punjab and Gupta, J.N. 1989. Unique life support species and their populations used by local people in India under extreme environmental conditions. In: Life support plant species-diversity arid zone, plants in famine period. In: Support Plant Species-Diversity and Conservation. R.S. Parods, Kapoor, P, Arora, R.K. and Bhag Mal (eds.), p. 165-172.

34. Vavilov, N.I. 1926. Studies on the origin of cultivated plants. *Bull. Appl. Bot., Genet. and Plant Breed.*, **26(2):** 1-248, in Russian.

35. Yang, Si-Lin and Walter, Terrence, W.1992. Ethnobotany and the economic role of the Cucurbitaceae of China. Economic Botany, 46(4): 349-367.

2

Genetic Resources of Solanaceous Vegetables and Crop Improvement

Introduction

In general, Solanaceous vegetables include 75 genera worldwide but in India, the genus *Solanum* is represented by 45 species including 22 indigenous (Deb, 1979). The genus *Solanum* comprises approximately 2,000 species of which about 235 are tuber bearing (Hawkes, 1990) and belongs to section *Tuberarium*. The majority of species (about 1,800) are non-tuber bearing. These are confined mainly to the Andean region of south America.Tomato, brinjal, chillies (non-tuber bearing) and potato (tuber bearing) are most important and widely grown all over the country throughout the year for their edible fruits and tubers. Tomato, brinjal, chillies and potato are highly nutritious, acceptable to consumers and simultaniously highly remunerative providing a high return to the vegetable growers. The availability of these crops in the market throughout the year as a fresh vegetable indicates its production potentiality, popularity and palatability. A list of genus *Solanum* with important taxon, name of authority and preferred taxon name are given in Table 2.1.

Table 2.1: Taxon, Name of Authority and Preferred Taxon Name of *Solanum*

Name of Taxon	*Name of Authority*	*Preferred Taxon Name*
Solanum		*Solanum*
Solanum		*Solanum*
Solanum abancayense	Ochoa	*Solanum bukasovii*
Solanum abbottianum	Juz.	*Solanum bukasovii*
Solanum abbottii	Leonard	*Solanum lanceifolium*
Solanum abutiloides	(Griseb.) Bitter and Lillo	*Solanum abutiloides*
Solanum acanthocalyx	Klotzsch	*Solanum richardii*
Solanum acaule	Bitter	*Solanum acaule*
Solanum acaule f. *acaule*		*Solanum acaule* f. *acaule*
Solanum acaule f. *incuyo*	Ochoa	*Solanum acaule* f. *incuyo*
Solanum acaule subsp. *acaule*		*Solanum acaule* subsp. *acaule*
Solanum acaule subsp. *aemulans*	(Bitter and Wittm.) Hawkes and Hjert.	*Solanum acaule* subsp. *aemulans*
Solanum acaule subsp. *albicans*	(Ochoa) Hawkes	*Solanum albicans*
Solanum acaule subsp. *palmirense*	Kardolus	*Solanum acaule* subsp. *palmirense*

Name of Taxon	Name of Authority	Preferred Taxon Name
Solanum acaule subsp. *punae*	(Juz.) Hawkes and Hjert.	*Solanumacaule* subsp. *acaule*
Solanum acaule var. *aemulans*	(Bitter and Wittm.) Correll	*Solanum acaule* subsp. *aemulans*
Solanum acaule var. *albicans*	Ochoa	*Solanum albicans*
Solanum acaule var. *punae*	(Juz.) Hawkes	*Solanum acaule* subsp. *acaule*
Solanum acaule var. *subexinterruptum*	Bitter	*Solanum acaule* subsp. *acaule*
Solanum acerifolium	Dunal	*Solanum acerifolium*
Solanum achacachense	Cardenas	*Solanum achacachense*
Solanum acroglossum	Juz.	*Solanum acroglossum*
Solanum acroscopicum	Ochoa	*Solanum acroscopicum*
Solanum aculeastrum	Dunal	*Solanum aculeastrum*
Solanum aculeatissimum	Jacq.	*Solanum aculeatissimum*
Solanum adhaerens	Roem. and Schult.	*Solanum adhaerens*
Solanum adscendens	Sendtn.	*Solanum adscendens*
Solanum aemulans	Bitter and Wittm.	*Solanum acaule* subsp. *aemulans*
Solanum aethiopicum	L.	*Solanum aethiopicum*
Solanum agrarium	Sendtn.	*Solanum agrarium*
Solanum agrimonifolium	Rydb.	*Solanum agrimonifolium*
Solanum ajuscoense	Bukasov ex Rybin	*Solanum stoloniferum sp. stoloniferum*
Solanum alandiae	Cardenas	*Solanum alandiae*
Solanum alatum	Moench	*Solanum villosum*
Solanum alatum	Seem. and J. C. Schmidt, nom. illeg.	*Solanum robustum*
Solanum albicans	(Ochoa) Ochoa	*Solanum albicans*
Solanum albidum	Dunal	*Solanum albidum*
Solanum albornozii	Correll	*Solanum albornozii*
Solanum alternatopinnatum	Steud.	*Solanum alternatopinnatum*
Solanum alticola	Bitter	*Solanum megistacrolobum subsp. megistacrolobum*
Solanum amabile	Vargas	*Solanum bukasovii*
Solanum amayanum	Ochoa	*Solanum amayanum*
Solanum amazonicum	Ker Gawl.	*Solanum tridynamum*
Solanum ambosinum	Ochoa	*Solanum ambosinum*
Solanum americanum	Auct.	*Solanum ptychanthum*
Solanum americanum	Mill.	*Solanum americanum*
Solanum anamatophilum	Ochoa	*Solanum anamatophilum*
Solanum ancophilum	(Correll) Ochoa	*Solanum ancophilum*
Solanum ancoripae	Ochoa	*Solanum ancoripae*
Solanum andigenum	Juz. and Bukasov	*Solanum tuberosum* subsp. *andigenum*
Solanum andigenum f. *guatemalense*	Bukasov	*Solanum tuberosum* subsp. *andigenum*
Solanum andreanum	Baker	*Solanum andreanum*
Solanum anguivi	Lam.	*Solanum anguivi*
Solanum angustifolium	Mill.	*Solanum angustifolium*
Solanum anomalocalyx	Hawkes	*Solanum sparsipilum*
Solanum anomalum	Thonn.	*Solanum anomalum*
Solanum antipoviczii	Bukasov ex Rybin	*Solanum stoloniferum* subsp. *stoloniferum*
Solanum apaense	Chodat	*Solanum apaense*
Solanum appendiculatum	Humb. and Bonpl. ex Dunal	*Solanum appendiculatum*
Solanum apurimacense	Vargas	*Solanum tuberosum* subsp. *andigenum*
Solanum aracc-papa	Juz. Ex Rybin, nom. dub.	*Solanum sparsipilum*
Solanum arboreum	Humb. and Bonpl. ex Dunal	*Solanum arboreum*
Solanum argenteum	Dunal	*Solanum argenteum*
Solanum ariduphilum	Ochoa	*Solanum ariduphilum*
Solanum arnezii	Cardenas	*Solanum arnezii*

Name of Taxon	Name of Authority	Preferred Taxon Name
Solanum astleyi	Hawkes and Hjert.	*Solanum boliviense* subsp. *astleyi*
Solanum atropurpureum	Schrank	*Solanum atropurpureum*
Solanum augustii	Ochoa	*Solanum augustii*
Solanum auriculatum	Aiton	*Solanum mauritianum*
Solanum aviculare	G. Forst.	*Solanum aviculare*
Solanum avilesii	Hawkes and Hjert.	*Solanum avilesii*
Solanum ayacuchense	Ochoa	*Solanum ayacuchense*
Solanum aymaraesense	Ochoa	*Solanum aymaraesense*
Solanum bahamense	L.	*Solanum bahamense*
Solanum ballsii	Hawkes	*Solanum vernei* subsp. *ballsii*
Solanum basendopogon	Bitter	*Solanum basendopogon*
Solanum berthaultii	Hawkes	*Solanum berthaultii*
Solanum berthaultii f. *zudaniense*	(Cardenas) Correll	*Solanum tarijense*
Solanum betaceum	Cav.	*Solanum betaceum*
Solanum bill-hookeri	Ochoa	*Solanum bill-hookeri*
Solanum bitteri	Hassl.	*Solanum chacoense*
Solanum boergeri	Bukasov	*Solanum chacoense*
Solanum boerhaaviifolium	Sendtn.	*Solanum laxum*
Solanum bolivianum	Rusby	*Solanum bolivianum*
Solanum boliviense	Dunal	*Solanum boliviense*
Solanum boliviense subsp. *astleyi*	(Hawkes and Hjert.) D. M. Spooner *et al.*	*Solanum boliviense* subsp. *astleyi*
Solanum boliviense subsp. *boliviense*		*Solanum boliviense* subsp. *boliviense*
Solanum bombycinum	Ochoa	*Solanum bombycinum*
Solanum bonariense	L.	*Solanum bonariense*
Solanum bonariense var. *villaricense*	(Morong) Chodat	*Solanum guaraniticum*
Solanum boreale	(A. Gray) Bitter	*Solanum fendleri*
Solanum boyacense	Juz. and Bukasov	*Solanum phureja*
Solanum brachistotrichum	(Bitter) Rydb.	*Solanum brachistotrichum*
Solanum brachyantherum	Phil.	*Solanum brachyantherum*
Solanum brachycarpum	(Correll) Correll	*Solanum brachycarpum*
Solanum brevicaule	Bitter	*Solanum brevicaule*
Solanum brevidens	Phil.	*Solanum palustre*
Solanum brevidens var. *brevidens*		*Solanum palustre*
Solanum brevidens var. *glabrescens*	(Poepp. ex Schltdl.) Hawkes ex Correll	Solanum palustre
Solanum brevifolium	Humb. and Bonpl. ex Dunal	*Solanum brevifolium*
Solanum brevimucronatum	Hawkes	*Solanum sparsipilum*
Solanum bridgesii	A. DC., nom. illeg.	*Solanum palustre*
Solanum buesii	Vargas	*Solanum buesii*
Solanum bukasovii	Juz. ex Rybin	*Solanum bukasovii*
Solanum bukasovii f. *multidissectum*	(Hawkes) Ochoa	*Solanum bukasovii* f. *multidissectum*
Solanum bulbocastanum	Dunal	*Solanum bulbocastanum*
Solanum bulbocastanum subsp. *bulbocastanum*		*Solanum bulbocastanum* subsp. *bulbocastanum*
Solanum bulbocastanum subsp. *dolichophyllum*	(Bitter) Hawkes	*Solanum bulbocastanum* subsp. *dolichophyllum*
Solanum bulbocastanum subsp. *partitum*	(Correll) Hawkes	*Solanum bulbocastanum* subsp. *partitum*
Solanum bulbocastanum var. dolichophyllum	Bitter	*Solanum bulbocastanum* subsp. *dolichophyllum*
Solanum bulbocastanum var. glabrum	Correll	*Solanum bulbocastanum*
Solanum bulbocastanum var. partitum	Correll	*Solanum bulbocastanum* subsp. *partitum*
Solanum bullatum	Vell.	*Solanum bullatum*
Solanum burkartii	Ochoa	*Solanum burkartii*
Solanum burtonii	Ochoa	*Solanum burtonii*
Solanum bustillosii	Phil.	*Solanum etuberosum*

Name of Taxon	Name of Authority	Preferred Taxon Name
Solanum cacetanum	Ochoa	*Solanum colombianum*
Solanum caesium	Griseb.	*Solanum caesium*
Solanum cajamarquense	Ochoa	*Solanum cajamarquense*
Solanum calacalinum	Ochoa	*Solanum calacalinum*
Solanum calcense	Hawkes	*Solanum sparsipilum*
Solanum caldasii	Dunal	*Solanum ochranthum*
Solanum calidum	Bohs	*Solanum calidum*
Solanum calophyllum	Phil.	*Solanum triflorum*
Solanum calvescens	Bitter	*Solanum calvescens*
Solanum campylacanthum	Hochst.	*Solanum campylacanthum*
Solanum canasense	Hawkes	*Solanum bukasovii*
Solanum canasense var. *canasense*		*Solanum bukasovii*
Solanum canasense var. *lechnoviczii*	(Hawkes) Ochoa	*Solanum bukasovii*
Solanum canasense var. *neohawkesii*	(Ochoa) Correll	*Solanum bukasovii* f. *multidissectum*
Solanum canasense var. *xerophyllum*	(Vargas) Correll	*Solanum marinasense*
Solanum candelarianum	Bukasov	*Solanum stoloniferum* sp. *stoloniferum*
Solanum candelarianum	Cardenas, nom. illeg., pro parte majore	*Solanum avilesii*
Solanum candelarianum	Cardenas, nom. illeg., pro parte minore	*Solanum berthaultii*
Solanum candidum	Lindl.	*Solanum candidum*
Solanum candolleanum	P. Berthault	*Solanum candolleanum*
Solanum cantense	Ochoa	*Solanum cantense*
Solanum capsicastrum	Link ex Schauer	*Solanum pseudocapsicum*
Solanum capsicibaccatum	Cardenas	*Solanum circaeifolium* var. *capsicibaccatum*
Solanum capsicibaccatum var. *latifoliolatum*	Ochoa	*Solanum circaeifolium* var. *capsicibaccatum*
Solanum capsicoides	All.	*Solanum capsicoides*
Solanum cardenasii	Hawkes	*Solanum phureja*
Solanum cardiophyllum	Lindl.	*Solanum cardiophyllum*
Solanum cardiophyllum subsp. *cardiophyllum*		*Solanum cardiophyllum* subsp. *cardiophyllum*
Solanum cardiophyllum subsp. *ehrenbergii*	Bitter	*Solanum cardiophyllum* sp. *ehrenbergii*
Solanum cardiophyllum subsp. *lanceolatum*	(P. Berthault) Bitter	*Solanum cardiophyllum* subsp. *lanceolatum*
Solanum cardiophyllum var. *endoiodandrum*	Bitter	*Solanum cardiophyllum* subsp. *lanceolatum*
Solanum caribaeum	Dunal	*Solanum americanum*
Solanum caripense	Dunal	*Solanum caripense*
Solanum carolinense	L.	*Solanum carolinense*
Solanum catarthrum	Juz.	*Solanum sparsipilum*
Solanum cernuum	Vell.	*Solanum cernuum*
Solanum chacoense	Bitter	*Solanum chacoense*
Solanum chacoense f. *gibberulosum*	(Juz. and Bukasov) Correll	*Solanum chacoense*
Solanum chacoense f. *pilosulum*	Hassl.	*Solanum chacoense*
Solanum chacoense subsp. *chacoense*		*Solanum chacoense* subsp. *chacoense*
Solanum chacoense subsp. *muelleri*	(Bitter) Hawkes and Hjert.	*Solanum chacoense* subsp. *muelleri*
Solanum chancayense	Ochoa	*Solanum chancayense*
Solanum chavinense	Correll	*Solanum dolichocremastrum*
Solanum chenopodioides	Lam.	*Solanum chenopodioides*
Solanum chilliasense	Ochoa	*Solanum chilliasense*
Solanum chillonanum	Ochoa	*Solanum chillonanum*
Solanum chiquidenum	Ochoa	*Solanum chiquidenum*
Solanum chiquidenum f. *amazonense*	Ochoa	*Solanum chiquidenum* f. *amazonense*
Solanum chiquidenum var. *cachicadense*	Ochoa	*Solanum chiquidenum*
Solanum chiquidenum var. *chiquidenum*		*Solanum chiquidenum* var. *chiquidenum*
Solanum chiquidenum var. *gracile*	Ochoa	*Solanum chiquidenum* var. *gracile*

Name of Taxon	Name of Authority	Preferred Taxon Name
Solanum chiquidenum var. *porconense*	Ochoa	*Solanum chiquidenum*
Solanum chiquidenum var. *robustum*	Ochoa	*Solanum chiquidenum* var. *robustum*
Solanum chocclo	Bukasov and Lechn.	*Solanum tuberosum*
Solanum chomatophilum	Bitter	*Solanum chomatophilum*
Solanum chomatophilum f. *angustifolium*	Correll	*Solanum chomatophilum* f. *angustifolium*
Solanum chomatophilum f. *chomatophilum*		*Solanum chomatophilum* f. *chomatophilum*
Solanum chomatophilum f. *sausianense*	Ochoa	*Solanum chomatophilum* f. *sausianense*
Solanum chomatophilum var. *chomatophilum*		*Solanum chomatophilum* var. *chomatophilum*
Solanum chomatophilum var. *subnivale*	Ochoa	*Solanum chomatophilum* var. *subnivale*
Solanum chrysotrichum	Schltdl.	*Solanum chrysotrichum*
Solanum ciliatum	Lam.	*Solanum capsicoides*
Solanum circaeifolium	Bitter	*Solanum circaeifolium*
Solanum circaeifolium subsp. *quimense*	Hawkes and Hjert.	*Solanum circaeifolium* var. *capsicibaccatum*
Solanum circaeifolium var. *capsicibaccatum*	(Cardenas) Ochoa	*Solanum circaeifolium* var. *capsicibaccatum*
Solanum circaeifolium var. *circaefolium*		*Solanum circaeifolium* var. *circaefolium*
Solanum circinatum	Bohs	*Solanum circinatum*
Solanum citrullifolium	A. Braun	*Solanum citrullifolium*
Solanum clarum	Correll	*Solanum clarum*
Solanum coagulans	Forssk.	*Solanum incanum*
Solanum cobanense	J. L. Gentry	*Solanum cobanense*
Solanum cochabambense	Bitter	*Solanum cochabambense*
Solanum coelestipetalum	Vargas	*Solanum coelestipetalum*
Solanum coeruleiflorum	Hawkes	*Solanum* x *chaucha*
Solanum colombianum	Dunal	*Solanum colombianum*
Solanum colombianum var. *trianae*	Bitter	*Solanum colombianum*
Solanum commersonii	Dunal	*Solanum commersonii*
Solanum commersonii f. *malmeanum*	(Bitter) Correll	*Solanum commersonii* subsp. *malmeanum*
Solanum commersonii f. *mechonguense*	(Bukasov) Correll	*Solanum commersonii* sp. *commersonii*
Solanum commersonii subsp. *commersonii*		*Solanum commersonii* sp. *commersonii*
Solanum commersonii subsp. *malmeanum*	(Bitter) Hawkes and Hjert.	*Solanum commersonii* sp. *malmeanum*
Solanum confertiseriatum	Bitter	*Solanum confertiseriatum*
Solanum contumazaense	Ochoa	*Solanum contumazaense*
Solanum cornutum	Auct.	*Solanum rostratum*
Solanum cornutum	Lam.	*Solanum angustifolium*
Solanum corymbosum	Jacq.	*Solanum corymbosum*
Solanum coyoacanum	Bukasov ex Rybin	*Solanum cardiophyllum* subsp. *cardiophyllum*
Solanum crassifolium	Ortega, nom. illeg.	*Solanum betaceum*
Solanum crinitum	Lam.	*Solanum crinitum*
Solanum cuatrecasasii	Ochoa	*Solanum otites*
Solanum cyananthum	Dunal	*Solanum crinitum*
Solanum dasyphyllum	Schumach.	*Solanum dasyphyllum*
Solanum deflexum	Greenm.	*Solanum adscendens*
Solanum demissum	Lindl.	*Solanum demissum*
Solanum demissum f. *atrocyaneum*	Lechn. and Bukasov	*Solanum demissum*
Solanum demissum f. *tlaxpehualcoense*	Bukasov ex Rybin	*Solanum demissum*
Solanum depexum	Juz.	*Solanum acaule* subsp. *acaule*
Solanum depilatum	Kitag., nom. illeg.	*Solanum kitagawae*
Solanum diflorum	Vell.	*Solanum pseudocapsicum*
Solanum dimidiatum	Raf.	*Solanum dimidiatum*
Solanum diphyllum	L.	*Solanum diphyllum*
Solanum diploconos	(Mart.) Bohs	*Solanum diploconos*

Name of Taxon	Name of Authority	Preferred Taxon Name
Solanum dolichocremastrum	Bitter	*Solanum dolichocremastrum*
Solanum donachui	(Ochoa) Ochoa	*Solanum donachui*
Solanum donnell-smithii	J. M. Coult.	*Solanum adhaerens*
Solanum douglasii	Dunal	*Solanum douglasii*
Solanum drymophilum	O. E. Schulz	*Solanum bahamense*
Solanum dulcamara	L.	*Solanum dulcamara*
Solanum dulcamaroides	Dunal	*Solanum dulcamaroides*
Solanum duplosinuatum	Klotzsch	*Solanum dasyphyllum*
Solanum ehrenbergii	(Bitter) Rydb.	*Solanum cardio*phyllum subsp. *ehrenbergii*
Solanum elaeagnifolium	Cav.	*Solanum elaeagnifolium*
Solanum ellipticum	R. Br.	*Solanum ellipticum*
Solanum emmeae	Juz. and Bukasov	*Solanum chacoense*
Solanum endlicheri	Dunal	*Solanum suaveolens*
Solanum erianthum	D. Don	*Solanum erianthum*
Solanum estradae	L. E. Lopez	*Solanum tuberosum* subsp. *andigenum*
Solanum etuberosum	Lindl.	*Solanum etuberosum*
Solanum etuberosum var. *chillanense*	Bitter	*Solanum etuberosum*
Solanum exiguum	Bohs	*Solanum exiguum*
Solanum famatinae	Bitter and Wittm.	*Solanum spegazzinii*
Solanum fastigiatum	Willd.	*Solanum bonariense*
Solanum fastigiatum	Auct.	*Solanum guaraniticum*
Solanum fendleri	A. Gray	*Solanum fendleri*
Solanum fendleri subsp. *arizonicum*	Hawkes	*Solanum fendleri* subsp. *arizonicum*
Solanum fendleri subsp. *fendleri*	-	*Solanum fendleri* subsp. *fendleri*
Solanum fendleri var. *physaloides*	Correll	*Solanum hjertingii* var. *physaloides*
Solanum fernandezianum	Phil.	*Solanum fernandezianum*
Solanum ferox	Auct.	*Solanum lasiocarpum*
Solanum flahaultii	Bitter	*Solanum flahaultii*
Solanum flavescens	Dunal	*Solanum hirtum*
Solanum flavoviridens	Ochoa	*Solanum flavoviridens*
Solanum fragrans	Hook.	*Solanum diploconos*
Solanum fraxinifolium	Dunal	*Solanum fraxinifolium*
Solanum gandarillasii	Cardenas	*Solanum gandarillasii*
Solanum garcia-barrigae	Ochoa	*Solanum garcia-barrigae*
Solanum garcia-barrigae var. *donachui*	Ochoa	*Solanum donachui*
Solanum garciae	Juz. and Bukasov	*Solanum chacoense*
Solanum gibberulosum	Juz. and Bukasov	*Solanum chacoense*
Solanum giganteum	Jacq.	*Solanum giganteum*
Solanum gigantophyllum	Bitter	*Solanum microdontum*
Solanum gilo	Raddi	*Solanum aethiopicum*
Solanum gilo var. *pierreanum*	(Pailleux and Bois) Bitter	*Solanum gilo* var. *pierreanum*
Solanum glanduliferum	Hawkes	*Solanum infundibuliforme*
Solanum glaucophyllum	Desf.	*Solanum glaucophyllum*
Solanum goniocalyx	Juz. and Bukasov	*Solanum stenotomum* sp. *goniocalyx*
Solanum gourlayi	Hawkes	*Solanum gourlayi*
Solanum gourlayi subsp. *gourlayi*		*Solanum gourlayi* subsp. *gourlayi*
Solanum gourlayi subsp. *pachytrichum*	(Hawkes) Hawkes and Hjert.	*Solanum leptophyes*
Solanum gourlayi subsp. *saltense*	A. M. Clausen and K. A. Okada	*Solanum gourlayi* subsp. *saltense*
Solanum gourlayi subsp. *vidaurrei*	(Cardenas) Hawkes and Hjert.	*Solanum vidaurrei*
Solanum gracile	Auct. N. Amer.	*Solanum pseudogracile*
Solanum gracile	Dunal, nom. illeg.	*Solanum chenopodioides*

Name of Taxon	Name of Authority	Preferred Taxon Name
Solanum gracilifrons	Bitter	*Solanum gracilifrons*
Solanum gracilius	Herter	*Solanum chenopodioides*
Solanum grandiflorum	Ruiz and Pav.	*Solanum grandiflorum*
Solanum guamense	Merr.	*Solanum guamense*
Solanum guaraniticum	A. St.-Hil.	*Solanum guaraniticum*
Solanum guaraniticum	Hassl., nom. illeg.	*Solanum chacoense*
Solanum guerreroense	Correll	*Solanum guerreroense*
Solanum guineense	L.	*Solanum guineense*
Solanum guzmanguense	Whalen and Sagast.	*Solanum guzmanguense*
Solanum haematocladum	Dunal	*Solanum haematocladum*
Solanum hastiforme	Correll	*Solanum hastiforme*
Solanum henryi	Bukasov and Lechn.	*Solanum commersonii*
Solanum hermannii	Dunal, nom. illeg.	*Solanum anguivi*
Solanum hermannii	Auct.	*Solanum linnaeanum*
Solanum herrerae	Juz.	*Solanum tuberosum* subsp. *andigenum*
Solanum heterodoxum	Dunal	*Solanum heterodoxum*
Solanum hibernum	Bohs	*Solanum hibernum*
Solanum hintonii	Correll	*Solanum hintonii*
Solanum hirsutissimum	Standl.	*Solanum pectinatum*
Solanum hirtum	Vahl	*Solanum hirtum*
Solanum hispidum	Pers.	*Solanum hispidum*
Solanum hjertingii	Hawkes	*Solanum hjertingii*
Solanum hjertingii var. *hjertingii*		*Solanum hjertingii* var. *hjertingii*
Solanum hjertingii var. *physaloides*	(Correll) Hawkes	*Solanum hjertingii* var. *physaloides*
Solanum hondelmannii	Hawkes and Hjert.	*Solanum oplocense*
Solanum hoopesii	Hawkes and K. A. Okada	*Solanum hoopesii*
Solanum horovitzii	Bukasov	*Solanum chacoense*
Solanum hougasii	Correll	*Solanum hougasii*
Solanum huancabambense	Ochoa	*Solanum huancabambense*
Solanum huancavelicae	Ochoa	*Solanum huancavelicae*
Solanum huarochiriense	Ochoa	*Solanum huarochiriense*
Solanum humectophilum	Ochoa	*Solanum humectophilum*
Solanum hybrid		*Solanum hybrid*
Solanum hygrothermicum	Ochoa	*Solanum phureja* subsp. *hygrothermicum*
Solanum hypacrarthrum	Bitter	*Solanum hypacrarthrum*
Solanum hyporhodium	A. Braun and C. D. Bouche	*Solanum hyporhodium*
Solanum immite	Dunal	*Solanum immite*
Solanum immite var. *immite*		*Solanum immite* var. *immite*
Solanum immite var. *vernale*	Correll	*Solanum immite* var. *vernale*
Solanum incahuasinum	Ochoa	*Solanum incahuasinum*
Solanum incamayoense	K. A. Okada and A. M. Clausen	*Solanum incamayoense*
Solanum incanum	L.	*Solanum incanum*
Solanum incasicum	Ochoa	*Solanum incasicum*
Solanum incompletum	Dunal	*Solanum incompletum*
Solanum indicum	Auct.	*Solanum anguivi*
Solanum indicum	L.	*Solanum lasiocarpum*
Solanum indicum subsp. *distichum*	(Thonn.) Bitter	*Solanum anguivi*
Solanum infundibuliforme	Phil.	*Solanum infundibuliforme*
Solanum ingifolium	Ochoa	*Solanum ingifolium*
Solanum inopinum	Ewan	*Solanum inopinum*
Solanum inscendens	Rydb.	*Solanum appendiculatum*

Name of Taxon	Name of Authority	Preferred Taxon Name
Solanum insulae-paschalis	Bitter	*Solanum insulae-paschalis*
Solanum integrifolium	Poir.	*Solanum integrifolium*
Solanum interius	Rydb.	*Solanum interius*
Solanum intrusum	Soria	*Solanum scabrum*
Solanum iopetalum	(Bitter) Hawkes	*Solanum iopetalum*
Solanum irosinum	Ochoa	*Solanum irosinum*
Solanum irosinum f. irosinum		*Solanum irosinum* f. *irosinum*
Solanum irosinum f. *tarrosum*	Ochoa	*Solanum irosinum* f. *tarrosum*
Solanum jaenense	Ochoa	*Solanum jaenense*
Solanum jalcae	Ochoa	*Solanum jalcae*
Solanum jalcae var. *jalcae*		*Solanum jalcae* var. *jalcae*
Solanum jalcae var. *pubescens*	Correll	*Solanum jalcae* var. *pubescens*
Solanum jaliscanum	Greenm.	*Solanum pseudocapsicum*
Solanum jamaicense	Mill.	*Solanum jamaicense*
Solanum jamesii	Torr.	*Solanum jamesii*
Solanum jamesii subsp. *nayaritense*	Bitter	*Solanum nayaritense*
Solanum jamesii var. *brachistotrichum*	Bitter	*Solanum brachistotrichum*
Solanum jamesii var. *michoacanum*	Bitter	*Solanum* x *michoacanum*
Solanum jasminoides	Paxton	*Solanum laxum*
Solanum jubatum	Dunal, nom. illeg.	*Solanum cernuum*
Solanum juglandifolium	Dunal	*Solanum juglandifolium*
Solanum jujuyense	Hawkes	*Solanum chacoense*
Solanum kesselbrenneri	Juz. and Bukasov	*Solanum phureja*
Solanum khasianum	C. B. Clarke	*Solanum aculeatissimum*
Solanum khasianum var. *chatterjeeanum*	Sengupta	*Solanum viarum*
Solanum kitagawae	Schonb.-Tem.	*Solanum kitagawae*
Solanum knappei	Juz. and Bukasov	*Solanum chacoense*
Solanum kurtzianum	Bitter and Wittm.	*Solanum kurtzianum*
Solanum kwebense	N. E. Br.	*Solanum kwebense*
Solanum laciniatum	Aiton	*Solanum laciniatum*
Solanum lanceifolium	Jacq.	*Solanum lanceifolium*
Solanum lanceolatum	Cav.	*Solanum lanceolatum*
Solanum lanciforme	Rydb.	*Solanum cardiophyllum subsp. lanceolatum*
Solanum lapazense	Hawkes	*Solanum sparsipilum*
Solanum laplaticum	Bukasov	*Solanum chacoense*
Solanum lasiocarpum	Dunal	*Solanum lasiocarpum*
Solanum laxissimum	Bitter	*Solanum laxissimum*
Solanum laxissimum f. *rockefelleri*	(Vargas) Correll	*Solanum laxissimum*
Solanum laxum	Spreng.	*Solanum laxum*
Solanum lechnoviczii	Hawkes	*Solanum bukasovii*
Solanum lechnoviczii var. *zerophylla*	Vargas	*Solanum marinasense*
Solanum leptophyes	Bitter	*Solanum leptophyes*
Solanum leptophyes f. *gourlayi*	(Hawkes) Correll	*Solanum gourlayi* subsp. *gourlayi*
Solanum leptosepalum	Correll	*Solanum leptosepalum*
Solanum leptostigma	Juz. ex Bukasov	*Solanum tuberosum*
Solanum lesteri	Hawkes and Hjert.	*Solanum lesteri*
Solanum leucocarpon	Dunal	*Solanum leucocarpon*
Solanum lignicaule	Vargas	*Solanum lignicaule*
Solanum ligustrinum	G. Lodd.	*Solanum ligustrinum*
Solanum limbaniense	Ochoa	*Solanum limbaniense*

Name of Taxon	Name of Authority	Preferred Taxon Name
Solanum linnaeanum	Hepper and P.-M. L. Jaeger	*Solanum linnaeanum*
Solanum liriunianum	Cardenas and Hawkes	*Solanum brevicaule*
Solanum lobbianum	Bitter	*Solanum lobbianum*
Solanum lobbianum f. *multidissectum*	(Hawkes) Correll	*Solanum bukasovii* f. *multidissectum*
Solanum longibracteata	Phil.	*Solanum longibracteata*
Solanum longiconicum	Bitter	*Solanum longiconicum*
Solanum longipedicellatum	Bitter	*Solanum stoloniferum* sp. *stoloniferum*
Solanum longiusculus	Ochoa	*Solanum longiusculus*
Solanum looseri	Juz. ex Bukasov	*Solanum etuberosum*
Solanum lopez-camarenae	Ochoa	*Solanum lopez-camarenae*
Solanum luteum	Mill.	*Solanum villosum*
Solanum luteum subsp. *alatum*	(Moench) Dostal	*Solanum villosum*
Solanum lycopersicoides	Dunal	*Solanum lycopersicoides*
Solanum lycopersicum var. *cerasiforme*	(Dunal) D. M. Spooner *et al.*	*Lycopersicon esculentum* var. *cerasiforme*
Solanum macolae	Bukasov	*Solanum kurtzianum*
Solanum macranthum	Dunal	*Solanum crinitum*
Solanum macrocarpon	L.	*Solanum macrocarpon*
Solanum macrophyllum	Dunal	*Solanum lanceifolium*
Solanum macropilosum	Correll	*Solanum macropilosum*
Solanum maglia	Schltdl.	*Solanum maglia*
Solanum malacoxylon	Sendtn.	*Solanum glaucophyllum*
Solanum malinchense	Hawkes	*Solanum stoloniferum* subsp. *stoloniferum*
Solanum mamilliferum	Juz. and Bukasov	*Solanum* x *chaucha*
Solanum mammosum	L.	*Solanum mammosum*
Solanum mapiriense	Bitter	*Solanum mapiriense*
Solanum marginatum	L. f.	*Solanum marginatum*
Solanum marinasense	Vargas	*Solanum marinasense*
Solanum marinasense var. *dentifolium*	Vargas	*Solanum marinasense*
Solanum matehualae	Hjert. and T. R. Tarn	*Solanum matehualae*
Solanum maternum	Bohs	*Solanum maternum*
Solanum mauritianum	Scop.	*Solanum mauritianum*
Solanum mechonguense	Bukasov	*Solanum commersonii* subsp. *commersonii*
Solanum medians	Bitter	*Solanum medians*
Solanum medians f. *neoweberbaueri*	(Wittm.) Correll	*Solanum* x *neoweberbaueri*
Solanum medians var. *autumnale*	Correll	*Solanum medians* var. *autumnale*
Solanum medians var. medians		*Solanum medians* var. *medians*
Solanum megistacrolobum	Bitter	*Solanummegistacrolobum*
Solanum megistacrolobum f. *megistacrolobum*		*Solanum megistacrolobum* f. *megistacrolobum*
Solanum megistacrolobum f. *purpureum*	Ochoa	*Solanum megistacrolobum* f. *purpureum*
Solanum megistacrolobum subsp. *megistacrolobum*		*Solanum megistacrolobum* subsp. *megistacrolobum*
Solanum megistacrolobum subsp. *toralapanum*	(Cardenas and Hawkes) Giannattasio and D. M. Spooner	*Solanum megistacrolobum* subsp. *toralapanum*
Solanum megistacrolobum var. *toralapanum*	(Cardenas and Hawkes) Ochoa	*Solanum megistacrolobum* subsp. *toralapanum*
Solanum melanocerasum	All.	*Solanum scabrum*
Solanum melongena	L.	*Solanum melongena*
Solanum microdontum	Bitter	*Solanum microdontum*
Solanum microdontum subsp. *gigantophyllum*	(Bitter) Hawkes and Hjert.	*Solanum microdontum*
Solanum microdontum subsp. *microdontum*		*Solanum microdontum*
Solanum microdontum var. *microdontum*		*Solanum microdontum* var. *microdontum*

Name of Taxon	*Name of Authority*	*Preferred Taxon Name*
Solanum microdontum var. *montepuncoense*	Ochoa	*Solanum microdontum* var. *montepuncoense*
Solanum millanii	Bukasov and Lechn.	*Solanum commersonii* sp. *malmeanum*
Solanum miniatum	Bernh. ex Willd.	*Solanum villosum*
Solanum minutifoliolum	Correll	*Solanum minutifoliolum*
Solanum mochiquense	Ochoa	*Solanum mochiquense*
Solanum molinae	Juz.	*Solanum tuberosum* subsp. *tuberosum*
Solanum montanum	L.	*Solanum montanum*
Solanum morelliforme	Bitter and Munch	*Solanum morelliforme*
Solanum moscopanum	Hawkes	*Solanum moscopanum*
Solanum muelleri	Bitter	*Solanum chacoense* subsp. *muelleri*
Solanum multidissectum	Hawkes	*Solanum bukasovii* f. *multidissectum*
Solanum multidissectum subsp. *neohawkesii*	(Ochoa) Hawkes	*Solanum bukasovii* f. *multidissectum*
Solanum multifidum	Lam.	*Solanum multifidum*
Solanum multiflorum	Vargas	*Solanum urubambae f. urubambae*
Solanum multiinterruptum	Bitter	*Solanum multiinterruptum*
Solanum multiinterruptum f. *albiflorum*	Ochoa	*Solanum multiinterruptum* f. *albiflorum*
Solanum multiinterruptum f. *multiinterruptum*		*Solanum multiinterruptum* f. *multiinterruptum*
Solanum multiinterruptum var. *machaytambinum*	Ochoa	*Solanum multiinterruptum* var. *machaytambinum*
Solanum multiinterruptum var. *multiinterruptum*		*Solanum multiinterruptum* var. *multiinterruptum*
Solanum muricatum	Aiton	*Solanum muricatum*
Solanum myriacanthum	Dunal	*Solanum myriacanthum*
Solanum myrianthum	Rusby	*Solanum myrianthum*
Solanum naumannii	Engl.	*Solanum aethiopicum*
Solanum nayaritense	(Bitter) Rydb.	*Solanum nayaritense*
Solanum nelsonii	Dunal	*Solanum nelsonii*
Solanum nemorosum	Ochoa	*Solanum nemorosum*
Solanum neoantipoviczii	Bukasov	*Solanum stoloniferum* sp. *stoloniferum*
Solanum neocardenasii	Hawkes and Hjert.	*Solanum neocardenasii*
Solanum neohawkesii	Ochoa	*Solanum bukasovii* f. *multidissectum*
Solanum neorossii	Hawkes and Hjert.	*Solanum neorossii*
Solanum neovalenzuelae	L. E. Lopez	*Solanum neovalenzuelae*
Solanum neovargasii	Ochoa	*Solanum neovargasii*
Solanum neovavilovii	Ochoa	*Solanum neovavilovii*
Solanum nigrescens	M. Martens and Galeotti	*Solanum nigrescens*
Solanum nigricans	M. Martens and Galeotti	*Solanum nigricans*
Solanum nigrum	L.	*Solanum nigrum*
Solanum nigrum var. villosum	L.	*Solanum villosum*
Solanum nitidibaccatum	Bitter	*Solanum physalifolium var. nitidibaccatum*
Solanum nodiflorum	Jacq.	*Solanum americanum*
Solanum nubicola	Ochoa	*Solanum nubicola*
Solanum nutans	Ruiz and Pav.	*Solanum nutans*
Solanum obliquum	Ruiz and Pav.	*Solanum obliquum*
Solanum ochoae	Vargas	*Solanum bukasovii*
Solanum ochranthum	Dunal	*Solanum ochranthum*
Solanum ohrondii	Carriere	*Solanum commersonii*
Solanum okadae	Hawkes and Hjert.	*Solanum okadae*
Solanum olmosense	Ochoa	*Solanum olmosense*
Solanum opacum	A. Braun and C. D. Bouche	*Solanum opacum*
Solanum oplocense	Hawkes	*Solanum oplocense*

Name of Taxon	Name of Authority	Preferred Taxon Name
Solanum orocense	Ochoa	*Solanum orocense*
Solanum orophilum	Correll	*Solanum orophilum*
Solanum ortegae	Ochoa	*Solanum ortegae*
Solanum otites	Dunal	*Solanum otites*
Solanum oxycarpum	Schiede	*Solanum oxycarpum*
Solanum oxycarpum var. *brachycarpum*	Correll	*Solanum brachycarpum*
Solanum pachytrichum	Hawkes	*Solanum leptophyes*
Solanum palinacanthum	Dunal	*Solanum palinacanthum*
Solanum palitans	C. V. Morton	*Solanum radicans*
Solanum pallidum	Rusby	*Solanum pallidum*
Solanum palustre	Poepp. ex Schltdl.	*Solanum palustre*
Solanum pampasense	Hawkes	*Solanum pampasense*
Solanum pampasense f. glabrescens	Correll	*Solanum marinasense*
Solanum pamplonense	L. E. Lopez	*Solanum pamplonense*
Solanum panduriforme	E. Mey.	*Solanum panduriforme*
Solanum paniculatum	L.	*Solanum paniculatum*
Solanum papita	Rydb.	*Solanum papita*
Solanum paramoense	Bitter	*Solanum tuberosum subsp. andigenum*
Solanum parodii	Juz. and Bukasov	*Solanum chacoense*
Solanum pascoense	Ochoa	*Solanum chomatophilum*
Solanum paucijugum	Bitter	*Solanum paucijugum*
Solanum paucissectum	Ochoa	*Solanum paucissectum*
Solanum pearcei	Phil., nom. illeg.	*Solanum palustre*
Solanum pectinatum	Dunal	*Solanum pectinatum*
Solanum peloquinianum	Ochoa	*Solanum peloquinianum*
Solanum pendulum	Ruiz and Pav.	*Solanum pendulum*
Solanum petrophilum	F. Muell.	*Solanum petrophilum*
Solanum phureja	Juz. and Bukasov	*Solanum phureja*
Solanum phureja subsp. estradae	(L. E. Lopez) Hawkes	*Solanum tuberosum* subsp. *andigenum*
Solanum phureja subsp. hygrothermicum	(Ochoa) Hawkes	*Solanum phureja* subsp. *hygrothermicum*
Solanum phureja subsp. phureja		*Solanum phureja* subsp. *phureja*
Solanum physalifolium	Rusby	*Solanum physalifolium*
Solanum physalifolium var. nitidibaccatum	(Bitter) Edmonds	*Solanum physalifolium* var. *nitidibaccatum*
Solanum physalifolium var. physalifolium		*Solanum physalifolium* var. *physalifolium*
Solanum pierreanum	Pailleux and Bois	*Solanum aethiopicum*
Solanum pilcomayense	Morong	*Solanum pilcomayense*
Solanum pillahuatense	Vargas	*Solanum pillahuatense*
Solanum pinnatifidum	Ruiz and Pav., nom. illeg.	*Solanum multifidum*
Solanum pinnatisectum	Dunal	*Solanum pinnatisectum*
Solanum pinnatum	Cav.	*Solanum pinnatum*
Solanum piurae	Bitter	*Solanum piurae*
Solanum polyadenium	Greenm.	*Solanum polyadenium*
Solanum polytrichon	Rydb.	*Solanum polytrichon*
Solanum proteanthum	Bohs	*Solanum proteanthum*
Solanum pseudocapsicum	L.	*Solanum pseudocapsicum*
Solanum pseudogracile	Heiser	*Solanum pseudogracile*
Solanum pseudoquina	A. St.-Hil.	*Solanum pseudoquina*
Solanum ptychanthum	Dunal	*Solanum ptychanthum*
Solanum puchupuchense	Ochoa	*Solanum puchupuchense*
Solanum pulchrilobum	Bitter	*Solanum pilcomayense*

Name of Taxon	Name of Authority	Preferred Taxon Name
Solanum pumilum	Hawkes	*Solanum bukasovii*
Solanum punae	Juz.	*Solanum acaule* subsp. *acaule*
Solanum pyracanthos	Lam.	*Solanum pyracanthos*
Solanum quadriloculatum	F. Muell.	*Solanum quadriloculatum*
Solanum quinquangulare	Willd. ex Roem. and Schult.	*Solanum acerifolium*
Solanum quitoense	Lam.	*Solanum quitoense*
Solanum radicans	L. f.	*Solanum radicans*
Solanum raphanifolium	Cardenas and Hawkes	*Solanum raphanifolium*
Solanum raquialatum	Ochoa	*Solanum raquialatum*
Solanum regularifolium	Correll	*Solanum regularifolium*
Solanum retroflexum	Dunal	*Solanum retroflexum*
Solanum rhomboideilanceolatum	Ochoa	*Solanum rhomboideilanceolatum*
Solanum rhomboideilanceolatum var. *ancophilum*	Correll	*Solanum ancophilum*
Solanum richardii	Dunal	*Solanum richardii*
Solanum rickii	Correll	*Solanum sitiens*
Solanum riparium	Pers.	*Solanum riparium*
Solanum robustum	H. L. Wendl.	*Solanum robustum*
Solanum rockefelleri	Vargas	*Solanum laxissimum*
Solanum roseum	Bohs	*Solanum roseum*
Solanum rostratum	Dunal	*Solanum rostratum*
Solanum rovirosanum	Donn. Sm.	*Solanum rovirosanum*
Solanum rybinii	Juz. and Bukasov	*Solanum phureja*
Solanum salasianum	Ochoa	*Solanum salasianum*
Solanum saltense	Hawkes, nom. illeg.	*Solanum chacoense*
Solanum sanctae-rosae	Hawkes	*Solanum sanctae-rosae*
Solanum sandemanii	Hawkes	*Solanum sandemanii*
Solanum sandwicense	Hook. and Arn.	*Solanum sandwicense*
Solanum sanitwongsei	Craib	*Solanum sanitwongsei*
Solanum santolallae	Vargas	*Solanum santolallae*
Solanum sarasarae	Ochoa	*Solanum sarasarae*
Solanum sarrachoides	Sendtn.	*Solanum sarrachoides*
Solanum sarrachoides	Auct.	*Solanum physalifolium var. nitidibaccatum*
Solanum sawyeri	Ochoa	*Solanum sawyeri*
Solanum saxatilis	Ochoa	*Solanum saxatilis*
Solanum scabrifolium	Ochoa	*Solanum scabrifolium*
Solanum scabrum	Mill.	*Solanum scabrum*
Solanum scalare	C. H. Wright	*Solanum anguivi*
Solanum schenckii	Bitter	*Solanum schenckii*
Solanum schickii	Juz. and Bukasov	*Solanum chacoense*
Solanum sciadostylis	(Sendtn.) Bohs	*Solanum sciadostylis*
Solanum seaforthianum	Andrews	*Solanum seaforthianum*
Solanum sellovianum	Sendtn.	*Solanum sellovianum*
Solanum sessiliflorum	Dunal	*Solanum sessiliflorum*
Solanum sibundoyensis	(Bohs) Bohs	*Solanum sibundoyensis*
Solanum simplicifolium	Bitter	*Solanum microdontum*
Solanum simplicifolium var. *trimerophyllum*	Bitter	*Solanum microdontum*
Solanum simplicissimum	Ochoa	*Solanum simplicissimum*
Solanum sisymbriifolium	Lam.	*Solanum sisymbriifolium*
Solanum sitiens	I. M. Johnst.	*Solanum sitiens*

Name of Taxon	Name of Authority	Preferred Taxon Name
Solanum sodomeum	L.	*Solanum anguivi*
Solanum sodomeum	Auct.	*Solanum linnaeanum*
Solanum sodomeum var. hermannii	Dunal	*Solanum linnaeanum*
Solanum soestii	Hawkes and Hjert.	*Solanum soestii*
Solanum sogarandinum	Ochoa	*Solanum sogarandinum*
Solanum solisii	Hawkes	*Solanum solisii*
Solanum soukupii	Hawkes	*Solanum bukasovii*
Solanum sp.		*Solanum* sp.
Solanum sparsipilum	(Bitter) Juz. and Bukasov	*Solanum sparsipilum*
Solanum sparsipilum subsp. *calcense*	(Hawkes) Hawkes	*Solanum sparsipilum*
Solanum sparsipilum subsp. *sparsipilum*		*Solanum sparsipilum*
Solanum sparsipilum var. *llallaguanianum*	(Cardenas and Hawkes) Correll	*Solanum sparsipilum*
Solanum spectabile	(Correll) Hawkes	*Solanum hougasii*
Solanum spegazzinii	Bitter	*Solanum spegazzinii*
Solanum spinosissimum	Lodd. ex G. Don, nom. nud.	*Solanum spinosissimum*
Solanum stenophyllidium	Bitter	*Solanum stenophyllidium*
Solanum stenotomum	Juz. and Bukasov	*Solanum stenotomum*
Solanum stenotomum subsp. *goniocalyx*	(Juz. and Bukasov) Hawkes	*Solanum stenotomum* subsp. *goniocalyx*
Solanum stenotomum subsp. *stenotomum*		*Solanum stenotomum* subsp. *stenotomum*
Solanum stoloniferum	Schltdl. and Bouche	*Solanum stoloniferum*
Solanum stoloniferum subsp. *moreliae*	Hawkes	*Solanum stoloniferum* subsp. *moreliae*
Solanum stoloniferum subsp. *stoloniferum*		*Solanum stoloniferum* subsp. *stoloniferum*
Solanum stramoniifolium	Jacq.	*Solanum stramoniifolium*
Solanum stuckertii	Bitter	*Solanum stuckertii*
Solanum sturtianum	F. Muell.	*Solanum sturtianum*
Solanum suaveolens	Kunth and C. D. Bouche	*Solanum suaveolens*
Solanum subandigena	Hawkes	*Solanum tuberosum subsp. andigenum*
Solanum sublobatum	Willd. ex Roem. and Schult.	*Solanum chenopodioides*
Solanum subpanduratum	Ochoa	*Solanum subpanduratum*
Solanum subtilius	Bitter	*Solanum chacoense*
Solanum sucubunense	Ochoa	*Solanum sucubunense*
Solanum sudanense	Hammerstein	*Solanum sudanense*
Solanum surattense	Burm. f.	*Solanum virginianum*
Solanum surinamense	Steud.	*Solanum leucocarpon*
Solanum tabanoense	Correll	*Solanum tabanoense*
Solanum tacnaense	Ochoa	*Solanum tacnaense*
Solanum tacnaense f. *decurrentialatum*	(Ochoa) Correll	*Solanum tacnaense* f. *decurrentialatum*
Solanum tacnaense f. *tacnaense*		*Solanum tacnaense* f. *tacnaense*
Solanum tacnaense var. sandemanii	(Hawkes) Correll	*Solanum sandemanii*
Solanum tampicense	Dunal	*Solanum tampicense*
Solanum tapojense	Ochoa	*Solanum tapojense*
Solanum tarapatanum	Ochoa	*Solanum tarapatanum*
Solanum tarijense	Hawkes	*Solanum tarijense*
Solanum tarnii	Hawkes and Hjert.	*Solanum tarnii*
Solanum taulisense	Ochoa	*Solanum taulisense*
Solanum tejucense	Dunal	*Solanum vaillantii*
Solanum tenuifilamentum	Juz. and Bukasov	*Solanum x chaucha*
Solanum tenuisetosum	(Bitter) Bohs	*Solanum tenuisetosum*
Solanum tequilense	A. Gray ex S. Watson	*Solanum candidum*
Solanum tlaxcalense	Hawkes	*Solanum stoloniferum* subsp. *stoloniferum*

Name of Taxon	Name of Authority	Preferred Taxon Name
Solanum tomatillo	(J. Remy) F. Phil.	*Solanum ligustrinum*
Solanum topiro	Dunal	*Solanum sessiliflorum*
Solanum toralapanum	Cardenas and Hawkes	*Solanum megistacrolobum* subsp. *toralapanum*
Solanum toralapanum var. *subintegrifolium*	Cardenas and Hawkes	*Solanum megistacrolobum* subsp. *toralapanum*
Solanum torreyi	A. Gray	*Solanum dimidiatum*
Solanum torvum	Sw.	*Solanum torvum*
Solanum trachycyphum	Bitter	*Solanum trachycyphum*
Solanum trichoneuron	Lillo	*Solanum trichoneuron*
Solanum tridynamum	Dunal	*Solanum tridynamum*
Solanum trifidum	Correll	*Solanum trifidum*
Solanum triflorum	Nutt.	*Solanum triflorum*
Solanum triflorum var. *calophyllum*	(Phil.) Bitter	*Solanum triflorum*
Solanum trinitense	Ochoa	*Solanum trinitense*
Solanum tripartitum	Dunal	*Solanum tripartitum*
Solanum tuberiferum	Dunal	*Solanum montanum*
Solanum tuberosum	L.	*Solanum tuberosum*
Solanum tuberosum subsp. *andigenum*	(Juz. and Bukasov) Hawkes	*Solanum tuberosum* subsp. *andigenum*
Solanum tuberosum subsp. *sparsipilum*	Bitter	*Solanum sparsipilum*
Solanum tuberosum subsp. *tuberosum*		*Solanum tuberosum* subsp. *tuberosum*
Solanum tuberosum var. *boreale*	A. Gray	*Solanum fendleri*
Solanum tuberosum var. *guaytecarum*	(Bitter) Hawkes	*Solanum tuberosum* subsp. *tuberosum*
Solanum tucumanense	Griseb.	*Solanum tucumanense*
Solanum tunariense	Kuntze	*Solanum tunariense*
Solanum tundalomense	Ochoa	*Solanum tundalomense*
Solanum tuquerrense	Hawkes	*Solanum tuquerrense*
Solanum ugentii	Hawkes and K. A. Okada	*Solanum ugentii*
Solanum umbellatum	Mill.	*Solanum umbellatum*
Solanum unguiculatum	A. Rich.	*Solanum unguiculatum*
Solanum unilobum	(Rusby) Bohs	*Solanum unilobum*
Solanum ureyi	Cardenas	*Solanum megistacrolobum* subsp. *toralapanum*
Solanum urticans	Dunal	*Solanum urticans*
Solanum urubambae	Juz.	*Solanum urubambae*
Solanum urubambae f. *chakchabambense*	Ochoa	*Solanum urubambae* f. *chakchabambense*
Solanum urubambae f. *urubambae*		*Solanum urubambae* f. *urubambae*
Solanum urubambae f. *velutinum*	(Correll) Ochoa	*Solanum urubambae* f. *velutinum*
Solanum vaillantii	Dunal	*Solanum vaillantii*
Solanum vavilovii	Juz. and Bukasov	*Solanum wittmackii*
Solanum velardei	Ochoa	*Solanum velardei*
Solanum velascanum	Bitter and Wittm.	*Solanum kurtzianum*
Solanum venturii	Hawkes and Hjert.	*Solanum venturii*
Solanum verbascifolium	L.	*Solanum erianthum*
Solanum vernei	Bitter and Wittm.	*Solanum vernei*
Solanum vernei subsp. *ballsii*	(Hawkes) Hawkes and Hjert.	*Solanum vernei* subsp. *ballsii*
Solanum vernei subsp. *vernei*		*Solanum vernei* subsp. *vernei*
Solanum verrucosum	Schltdl.	*Solanum verrucosum*
Solanum verrucosum var. *iopetalum*	Bitter	*Solanum iopetalum*
Solanum verrucosum var. *spectabile*	Correll	*Solanum hougasii*
Solanum viarum	Dunal	*Solanum viarum*
Solanum vidaurrei	Cardenas	*Solanum vidaurrei*
Solanum villaricense	Morong	*Solanum guaraniticum*
Solanum villosum	Mill.	*Solanum villosum*

Name of Taxon	Name of Authority	Preferred Taxon Name
Solanum villuspetalum	Vargas	*Solanum urubambae* f. *urubambae*
Solanum violaceimarmoratum	Bitter	*Solanum violaceimarmoratum*
Solanum violifolium	Schott	*Solanum violifolium*
Solanum virginianum	L.	*Solanum virginianum*
Solanum virgultorum	(Bitter) Cardenas and Hawkes	*Solanum virgultorum*
Solanum viridipes	Dunal	*Solanum alternatopinnatum*
Solanum warsczewiczii	Weick ex Lambertye	*Solanum hispidum*
Solanum weberbaueri	Bitter	*Solanum medians* var. *autumnale*
Solanum weberbaueri var. *decurrentialatum*	Ochoa	*Solanum tacnaense* f. *decurrentialatum*
Solanum wendlandii	Hook. f.	*Solanum wendlandii*
Solanum wittmackii	Bitter	*Solanum wittmackii*
Solanum woodsonii	Correll	*Solanum woodsonii*
Solanum wrightii	Benth.	*Solanum wrightii*
Solanum x *ajanhuiri*	Juz. and Bukasov	*Solanum* x *ajanhuiri*
Solanum x arahuayum	Ochoa	*Solanum* x *arahuayum*
Solanum x *blanco-galdosii*	Ochoa	*Solanum* x *blanco-galdosii*
Solanum x *brucheri*	Correll	*Solanum* x *brucheri*
Solanum x *burbankii*	Bitter	*Solanum retroflexum*
Solanum x *chaucha*	Juz. and Bukasov	*Solanum* x *chaucha*
Solanum x *curtilobum*	Juz. and Bukasov	*Solanum* x *curtilobum*
Solanum x *doddsii*	Correll	*Solanum* x *doddsii*
Solanum x *edinense*	P. Berthault	*Solanum* x *edinense*
Solanum x *edinensenotho* subsp. *edinense*		*Solanum* x *edinense nothosubsp. edinense*
Solanum x *edinense* notho subsp. *salamanii*	(Hawkes) Hawkes	*Solanum* x *edinense nothosubsp. salamanii*
Solanum x hendersonii	hort. ex W. Wight	*Solanum x hendersonii*
Solanum x *indunii*	K. A. Okada and A. M. Clausen	*Solanum* x *indunii*
Solanum x *juzepczukii*	Bukasov	*Solanum* x *juzepczukii*
Solanum x *litusinum*	Ochoa	*Solanum* x *litusinum*
Solanum x *michoacanum*	(Bitter) Rydb.	*Solanum x michoacanum*
Solanum x *neoweberbaueri*	Wittm.	*Solanum* x *neoweberbaueri*
Solanum x *proteus*	H. K. A. Winkl.	*Solanum* x *proteus*
Solanum x *rechei*	Hawkes and Hjert.	*Solanum* x *rechei*
Solanum x *ruiz-lealia*	Brucher	*Solanum* x *ruiz-lealia*
Solanum x *sambucinum*	Rydb.	*Solanum* x *sambucinum*
Solanum x *semidemissum*	Juz.	*Solanum* x *semidemissum*
Solanum x *setulosistylum*	Bitter	*Solanum* x *setulosistylum*
Solanum x *sucrense*	Hawkes	*Solanum* x *sucrense*
Solanum x vallis-mexici	Juz.	*Solanum x vallis-mexici*
Solanum x *viirsooi*	K. A. Okada and A. M. Clausen	*Solanum x viirsooi*
Solanum x *zudaniense*	Cardenas	*Solanum tarijense*
Solanum x*anthocarpum*	Schrad.	*Solanum virginianum*
Solanum x*anti*	A. Gray	*Solanum xanti*
Solanum x*erophyllum*	Hawkes ex Cardenas and Hawkes	*Solanum infundibuliforme*
Solanum yabari	Hawkes	*Solanum stenotomum*
Solanum yamobambense	Ochoa	*Solanum yamobambense*
Solanum yungasense	Hawkes	*Solanum yungasense*
Solanum zuccagnianum	Dunal	*Solanum aethiopicum*

References

1. Deb, D.B. 1980. Enumeration, synonymy and distribution of the Solanaceae in India. *J. Econ. Tax. Bot.*, **1:** 33-54.

Tomato (*Solanum lycopersicum* L.)

Introduction

Tomato (*Solanum lycopersicum* (L.) Karsten ex Farw., 2n=24) is one of the most widely grown vegetables in the world either in open field or green house condition. Green house cultivation is more common in western countries. Tomatoes were initially grown only as ornamental plants: the fruits were considered to be poisonous, because of the closely related deadly nightshade (*Solanum dulcamara*). Today it is very important dietary source of vitamins, minerals and fiber, which are important for human nutrition and health. Tomatoes are consumed in its various forms like cooked, soup, salad, pickles, ketchup, puree, sauces chutneys and preserves. It also contributes substantially to dietary intake of vitamin A, B, C and essential minerals (Tigchelaar, 1986). The leaves, stems and fruits of tomato contain a steroidal glcoalkaloid "tomatine" which is responsible for the bitter taste in it (Roddick and Butcher, 1972). The red colour of tomato is due to the presence of pigment *'lycopene'* which present in raw tomato varies from 8.8-42.0 mg/100 g (Rao and Rao, 2007), serves as anti-cancerous especially against prostrate cancer. The yellow and orange colour of tomato fruit is due to the presence of *'carotene'* and 'prolycopene' pigments, respectively. Tomato has rich vitamin contents like ascorbic acid or vitamin C, vitamin A, thiamine or vitamin B_1 and riboflavin or vitamin B_2, in order (Singh *et al.*, 2001). Tomato is associated with several medicinal properties as it has antioxidants such as lycopene, β-carotene, promotes gastric secretion, acts as intestinal antiseptic and alao as a blood purifier (Hedrick and Booth, 1908). It is the world's largest vegetable crop after potato and sweet potato and tops the list of processed vegetables.

Origin, History and Distribution

The name tomato derived from the Aztec word *'Tomatl'*. Tomato is native of tropical America (Peru, Ecuador, Bolivia, Colombia and Chile) and Mexico (Sims, 1980); however, domestication took place in Mexico (Basewell, 1949). It is true that cultivated tomato originated in the New World while the Andean Zone is likely to be centre of origin of wild tomatoes. Earlier Peru was considered as centre of domestication of tomato. However, the historical linguistic and ethno-botanical evidences favour Mexico particularly Vera Cruz-Puebla area as the source of the cultivated tomatoes that were first transported to the Old World. The Aztecs of Central America called tomato as *"xitomatl"*, and wild Central American tribes called it "xiamet" (Gould, 1983). The first tomato introduced into Europe by the Spanish in 1550 shortly after 1521, when Spanish explorer Cortez conqured the Aztec city of Tenochtitlan, later to be renamed Mexico City, in 1521. The first tomatoes introduced into the Europe by Spanish (Harvey *et al.*, 2002) were yellow coloured and the red types were introduced many years later. Earlier European literature is found in an herbal written by Matthiolus in 1544 (Gould, 1983). He described tomatoes, or as they were called in Italy, pomi d'oro (golden apple), and mentioned that tomtoes were "eaten in Italy with oil, salt and pepper". In Europe it was known as 'amoris-Amorous or love apple'. It was also known as 'Poma Peruviana-apple of Peru'.Through European migrants this crop was introduced in USA and Canada. From Spain, the tomato spread to Italy, France and other European countries. The Italians grew it around 1550. It was introduced into England around 1575. From tropical America it spread to other parts of the world in the 16th century. Since the mid of 16th century tomatoes have been cultivated and consumed in southern Europe, though they only became widespread in north-western Europe by the end of the 18th century (Harvey *et al.*, 2002). In the 17th century, Europeans took the tomato to China, Soputh and South-east Asia and in the 18th century to Japan and the USA (Siemonsma and Piluek, 1993). By 1623, four types of tomatoes were known: red, yellow, orange and golden, with distinction between yellow and golden perhaps only in the mind of different authors. In the USA, Thomas Jefferson first grew it in Virginia in 1781 and brought to his table, along with French fries (a visionary). The colonists first brought tomato to North America from Britain in late 18th century but some researchers expressed his views that crop spread to North America primarily by migrating birds. Only as late as 1880, did the British finally concede that tomato is edible. It was Robert Gibbon Johanson, an ordinary farmer in the USA who first ate tomato on a hot day of August 1820 to demonstrate its edibility. From then onwards, the tomato spread throughout the world. Some experts reveal that the Portuguese introduced tomato in India, though there is no definite record of when and how it came to India. The English traders of the East India Co. introduced tomato into India in 1822. Warnock (1990) has described evolutionary pathway in Latin America prior to and after domestication. The largest concentration of wild tomatoes is in Mexico. The Portuguese introduced tomato in India, though there is no definite record. The enormous variability at species level occurs in Mexico, Central America and Coastal Peru.Tomato is cultivated throughout the world. The United States, China, Turkey, Italy and India are the top five tomato producing countries in the world. China is the largest producer of tomato in the world followed by India and several other countries like Egypt, Russia, Turkey, USA *etc.* Bihar, Karnataka, Uttar Pradesh, Odisha, Andhra Pradesh, Maharashtra, Madhya Pradesh, Punjab, Haryana and Assam are important tomato growing states in India.

Species Availability, Evolution and Diversity

Progenitors of tomatoes were herbaceous green plants with small green fruits and a center of diversity

Table 2.2: Name of *Solanum* Species (Tomato), Authority and Preferred Taxon

Name of Taxon	*Name of Authority*	*Preferred Taxon Name*
Solanum		*Solanum*
Solanum		*Solanum*
Solanum agrimoniifolium	Dunal	*Solanum hirsutum*
Solanum cheesmanii	L. Riley	*Solanum cheesmanii*
Solanum cheesmanii f. *minor*	(Hook. f.) C. H. Mull.	*Solanum cheesmanii* f. *minor*
Solanum cheesmanii var. *minor*	(Hook. f.) D. M. Porter	*Solanum cheesmanii* f. *minor*
Solanum chilense	Dunal	*Solanum Solanum chilense*
Solanum chmielewskii	C. M. Rick *et al.*	*Solanum chmielewskii*
Solanum dentatum	Dunal	*Solanum peruvianum* var. *dentatum*
Solanum esculentum	Mill.	*Solanum esculentum*
Solanum esculentum f. *pyriforme*	(Dunal) C. H. Mull.	*Solanum esculentum* var. *esculentum*
Solanum esculentum var. *cerasiforme*	(Dunal) A. Gray	*Solanum esculentum* var. *cerasiforme*
Solanum esculentum var. *commune*	L. H. Bailey	*Solanum esculentum* var. *esculentum*
Solanum esculentum var. *esculentum*		*Solanum esculentum* var. *esculentum*
Solanum esculentum var. *grandifolium*	L. H. Bailey	*Solanum esculentum* var. *esculentum*
Solanum esculentum var. *pyriforme*	(Dunal) L. H. Bailey	*Solanum Solanum esculentum* var. *esculentum*
Solanum esculentum var. *validum*	L. H. Bailey	*Solanum esculentum* var. *esculentum*
Solanum glandulosum	C. H. Mull.	*Solanum glandulosum*
Solanum habrochaites	S. Knapp and D. M. Spooner	*Solanum hirsutum*
Solanum hirsutum	Dunal	*Solanum hirsutum*
Solanum hirsutum f. *glabratum*	C. H. Mull.	*Solanum hirsutum* f. *glabratum*
Solanum hirsutum var. *typicum*	Ined.Nom. inval.	*Solanum Solanum hirsutum*
Solanum hybrid		*Solanum hybrid*
Solanum lycopersicum	(L.) H. Karst.	*Solanum esculentum* var. *esculentum*
Lycopersicon lycopersicum var. *cerasiforme*	Auct.	*Solanum esculentum* var. *cerasiforme*
Lycopersicon lycopersicum var. *pyriforme*	Auct.	*Solanum esculentum* var. *esculentum*
Solanum neorickii	D. M. Spooner *et al.*	*Solanum parviflorum*
Solanum parviflorum	C. M. Rick *et al.*	*Solanum parviflorum*
Solanum pennellii	(Correll) D'Arcy	*Solanum pennellii*
Solanum pennellii var. *pennellii*		*Solanum pennellii* var. *pennellii*
Solanum pennellii var. *puberulum*	(Correll) D'Arcy	*Solanum pennellii* var. *puberulum*
Solanum peruvianum	(L.) Mill.	*Solanum peruvianum*
Solanum peruvianum var. *dentatum*	(Dunal) Dunal	*Solanum peruvianum* var. *dentatum*
Solanum peruvianum var. *glandulosum*	Ined.	*Solanum glandulosum*
Solanum peruvianum var. *humifusum*	C. H. Mull.	*Solanum peruvianum* var. *humifusum*
Solanum peruvianum var. *peruvianum*		*Solanum peruvianum* var. *peruvianum*
Solanum pimpinellifolium	(L.) Mill.	*Solanum pimpinellifolium*
Solanum pyriforme	Dunal	*Solanum esculentum* var. *esculentum*
Solanum racemigerum	Lange	*Solanum pimpinellifolium*
Solanum sp.		*Solanum* sp.
Solanum agrimoniifolium	(Dunal) J. F. Macbr., nom. Illeg.	*Solanum hirsutum*
Solanum cheesmanii	(L. Riley) Fosberg	*Solanum cheesmanii*
Solanum chilense	(Dunal) Reiche	*Solanum chilense*
Solanum chmielewskii	(C. M. Rick *et al.*) D. M. Spooner *et al.*	*Solanum chmielewskii*
Solanum lycopersicum	L.	*Solanum esculentum* var. *esculentum*
Solanum pennellii	Correll	*Solanum pennellii* var. *pennellii*
Solanum peruvianum	L.	*Solanum peruvianum*
Solanum pimpinellifolium	L.	*Solanum pimpinellifolium*

in the highlands of Peru (Harvey *et al.*, 2002). *Solanum esculentum* var. *cerasiforme* (cherry tomato) which is indigenous throughout tropical and sub-tropical America (Siemonsma and Piluek, 1993), is an immediate ancestor of cultivated tomato *S. esculentum*. Including *Solanum lycopersicum*, there are currently 13 species recognized in *Solanum* section *Lycopersicon*, all native to South America. Three of these species *i.e. S. cheesmanii, S. galapagense*, and *S. pimpinellifolium* are fully cross compatible with domestic tomato. Four more species *i.e. S. chmielewskii, S. habrochaites, S. neorickii*, and *S. pennelli* can be readily crossed with domestic tomato though with some limitations, and five species *i.e. S. iamete, S. chilense, S. corneliomulleri, S. huaylasense*, and *S. peruvianum* can be crossed with domestic tomato with significant difficulty usually requiring embryo rescue to produce viable plants. The *Lycopersicon* section has not been fully sampled within the South American range in which wild species are found; so new species may be added in the future. *Solanum* section *Lycopersicoides* and section *Juglandifolium* are represented by two species each that are considered bridge species genetically intermediate between tomato and non-tuber bearing potato species. *S. lycopersicoides* can be crossed with domestic tomato and introgression lines have been developed. This species was significant in moving the domestic tomato from separate genus status into the *Solanum* group because it directly links the tomato into the potato family. The greatest variability occurs in Mexico, Central America and Coastal Peru.

The tomato plant has a possible centre of variability along the Pacific coast between the Equator and the north of Chile, including the Galapagos Islands, to an elevation of approximately 2,000 m in the Andes mountain range. However, the domestication and cultivation were carried out by Indian tribes that inhabited Mexico (Giordano and Silva, 2000), where the name tomato originated. In India of wild species of tomato *L. pimpinellifolium* has been found in North East region.

Table 2.3: New Nomenclature of Tomatoes and Species Distribution

Sl.No.	New Name	Lycopersicon Equivalent	Distribution
1	*Lycopersicon juglandifolium* Dunal	*Lycopersicon juglandifolium* (Dunal) J.M.H. Shaw	In montane forests from central Colombia (cordillera Central and Occidental) to south peru (Dept. Apurimac); 1900-4100 m
2	*Solanum ochranthum* Dunal	*Lycopersicon ochranthum* (Dunal) J.M.H. Shaw	Usually a plant of open areas and roadsides or the edges of forest clearings; north-east Colombia to south Ecquador in all three Cordilleras; 1200-31000 m, sometimes occurring in Paramo in south Ecqudor.
3	*Lycopersicon sitiens* I.M. Johnst	*Lycopersicon sitiens* (I.M. Johnst) J.M.H. Shaw	On the western Andean slopes in north Chile from 2350-3500 m, on rocky hillsides and dry quebradas
4	*Solanum lycopersicoides* Dunal	*Lycopersicon lycopersicoides* (Dunal in DC.) A. Child ex J.M.H. Shaw	South Peru to north Chile on the western slopes of the Andes on dry rocky hillsides, 2900-3600 m elevation
5	*Lycopersicon pennellii* Correll	*Lycopersicon pennellii* (Correll) D'Arcy	North peru (piura) to north chile (Tarapaca) in dry rocky hillsides and sandy areas from sea level to 3000 m.
6	*Lycopersicon habrochaites* S. Knapp and D.m. Spooner	*Lycopersicon hirsutum* Dunal	In a variety of forest types, from premontane forests to dry forests on the western slopes of the Andes from Central Ecquador to Central Peru, ca. 500-2500 m elevation.
7	*Lycopersicon peruvianum* described by Peralta (4 geographic races: *humifusum*, lomas, Marathon, Chotano-Yamaluc)	Part of *Lycopersicon peruvianum* (L.) Miller (Incl. var. *humifusum* and Marathon races)	Coastal and in inland Andean valleys in north Peru from ca. 100 to 2500 m. Occurs in lomas, dry quebradas and dry rocky slopes.
8	*Lycopersicon neorickii* D.M. Spooner, G.J. Anderson and R.K. Jansen	*Lycopersicon parviflorum* C.M. Rick, Kesicki, Fobes and M. Holle	South Peru to South Ecquador in dry inter-Andean valleys from 1950-2600 m. Often found trailing over rocky banks and roadsides.
9	*Lycopersicon chmielewskii* (C.M. Rick, Kesicki, Fobes and M. Holle) D.M. Spooner, G.J. Anderson and R.K. Jansen	*Lycopersicon chmielewskii* C.M. Rick, Kesicki, Fobes and M. Holle	In high dry Andean valleys from the Department of Apurimac in south Peru to Sorata in north Bolivia, from 2300-2880 m elevation.
10	*Lycopersicon comeliomuelleri* J.F. Macbr. (1 geographic race: Misti nr. Arequipa)	Part of *Lycopersicon peruvianum* (L.) Miller; also known as *Lycopersicon glandulosum* C.F. Mull	Middle to higher elevation on the western slope of the Andes from central (near Lima) to south Peru, occasionally occurs on lower slopes on the edges of landslides (huaycos) towards the southern part of the species range; (400) 1000-3000 m.
11	*Solanum peruvianum* L.	*Lycopersicon peruvianum* (L.) Miller	In lomas formations and occasionally in coastal deserts from central Peru to north Chile, sea level to 600 m. Occasionally occurs as a weak at field edges in coastal river valleys.
12	*Lycopersicon chilense* (Dunal) Reiche	*Lycopersicon chilense* Dunal	On the western slope of the Andes from the Department of Tacna in south Peru to north Chile, in hyper-arid rocky plains and coastal deserts from sea level to 2000 m.

Sl.No.	New Name	Lycopersicon Equivalent	Distribution
13	*Solanum cheesmanii* (L. Riley) Fosberg	Part of *Lycopersicon cheesmaniae* L. Riley (previously known as forma or var. *minor*)	Endemic to the Galapagos Islands, Ecuador from sea level to 500 m.
14	*Solanum galapagense* S. Darwin and Peralta		Endemic to the Galapagos Islands, particularly the western and southern islands, mostly occurring on coastal lava to within 1.0 m of high tide mark within range of sea spray (stronglt salt tolerant but also occasionally inland, for example on volcano slopes on Isabela and Femandina.)
15	*Solanum esculentum* L.	*Lycopersicon esculentum* Miller.	Known only from cultivation or escape; world wide in a variety habitats
16	*Solanum pimpinellifolium* L.	*Lycopersicon pimpinellifolium* (L.) Miller.	Apparently native to coastal areas from central Ecuador to central Chile, 0-500 m. some stray pockets in N.E. India and Kashmir where *Solanum pimpinellifolium* has been found in semi-wild state, indicating its preference to cooler conditions.

Source: www.sgn.cornel.edu (Sol Genome).

Genes and Genetics of Characters

In tomato several genes are responsible for control of characters. Time to time 'Tomato Genetics Co-operative Newsletter' has published to update list of responsible gene. In a study Peter *et al.* (1981) reported that free neck (a weak peduncle-pedicel joint) showed a monogenic inheritance and was governed by a single recessive gene '*fn*' and its dominant form fn^{+}. The locule number showed a monogenic inheritance. Low locule number was governed by a single recessive gene *lc* and its dominant form lc^{+}. Ovate fruit shape showed a monogenic inheritance and was controlled by *a* a single recessive gene 'o' and its dominant form o^{+}. The genes *fn* and 'o' were so close that the recombination percentage was 'o' between them. There is a tight linkage between genes *fn* and 'o'. List of controlling genes for striking characters are as under:

Table 2.4: List of Controlling Genes for Striking Characters

Controlling Genes	*Striking Characters*
a (*a1*)*	Anthocyanin less; stems and leaves always lack purple colour
ag	Anthocyanin gainer; purple pigment appears on cotyledon and lower sides of leaves when growth is slow.
An	Anantha; flowers greatly modifies inflorescence compound; clearly resembles ca
ap	Apetalous flowers; corolla reduced or absent; pollen scarce and non functional
as 1	Asynaptic meiosis; high pollen and ovule sterility
at	Apricot or yellow pink colour of fruit flesh
aw	Without anthocyanin; same as a
B	High β-carotene
bk	Breaked; fruits with sharp points at stylar end
by	Brachytic; internodes shorter than normal
bu	Bushy; internodes and inflorescences shortened; long petioles
c	Potato leaf; reduced number of leaf segments
ca	Cauliflower; Extremely branched inflorescence, aborted flowers
cb	Cabbage; large dark green leaves; flower leaves present
cl 1	Cleistgamous; flowers fail to open
cl 2	Cleistgamous; flowers open slightly
clm	Closed anther (governed by monogenic recessive gene)
d(*d1*)	dwarf; all parts shortened leaves dark and rugose
dl	Dialytic; stamens not united; hairs particularly suppressed
ell	Elongated fruits as in variety 'Oxheart'
ex	Exserted stigma or style twisted with anther tube.
F	Fasciated fruits; many loculed fruit as in 'Ponderosa'
fl	Fleshy calyx; sepals often curled
g	Grooved fruits; may be associated with fasciation
gs	Green stripes in epidermis of unripe fruit; golden in ripe fruit
H	Hairs absent except on hypocotyls and at growing point
j	Joint less pedicels
l	lutescent; yellowish unripe fruits; premature yellowing of leaves
Lc	Locule number reduced; associated with C
lf	Leafy; indeterminate inflorescence
lg	Light green foliage
mc	Macrocalyx; sepals leaf like
mt	Midget; all parts of plant reduced; high sterility
n(*nt*)	Nipple tips at stylar end of fruits
nc	Narrow cotyledons; slow growth
o	Ovate or pear shaped fruits
p	Peach; dullness and increased hairiness of fruit epidermis
pe	Sticky fruit epidermis
pi	Pistillate; flowers usually lack stamens and are otherwise modified
ps(*va*)	Positional sterile; corolla does not unfurt; pollen is not shed

Controlling Genes	*Striking Characters*
r	Yellow colour of fruit flesh
rc	Rolled cotyledons
ri	Ridged or rough leaves
ro	Rosette; extremely short internodes; no flowers
s	Compound inflorescence; greatly increased number of flowers
sp	Self-pruning determinate plant habit
t	Tangerine; orange colour of fruit flesh and stamens
tf(et)	Trifoliate; leaf with usually only three segments and long petiole
u	Uniform; without usual darker shoulder
v	Virescent; white seedlings
vi	Villous; stem very hairy
w	Wiry; slender thread like leaflets
wt	Wilty; leafy margins curl adaxially
* Small letter represents recessive genes, Small letter with * at top represents dominant mutant form Capital letter indicates dominance	

Genes that affect the colour and biosynthesis of carotenoids are *at*, b and its allele *ogc*, *Del*, *dps*, *gf*, *gh*, *gr*, *hp-1*, *hp-2*, *Ip* and its allele *dg*, Mo_B, *ry*, *t*, *r*, *sucr*, *etc.* (Wann *et al.*, 1985; Chalukova and Manueyan, 1991; Chetelat *et al.*, 1993; Chetelat *et al.*, 1995). Exserted stigma (positional sterility) controlled by recessive factor can also be used for the production of hybrid. If there is adequate exsertion in this case, there is no self pollination and thus such lines can be used for the production of hybrids. The exserted lines can be developed utilizing wild species especially *S. hirsutum* and *S. hirsutum* f. *glabratum*. In other collection several *ps* lines have been identified which could be economically used for the production of economic hybrid.

Many of these genes have been used in improving fruit quality, sometimes by employing marker-assisted selection. Information on gene action governing specific quantitative characters as under shown in Table 2.5.

Among the cultivated varieties a large scale of variability has been found for different traits in tomato (Table 2.6).

Table 2.5: Genetics of Traits

Character	*Type of Gene Action*	*References*
Plant height	Additive	Singh and Singh (1980), Sidhu *et al.* (1981), Alice Kurian (1990)
	Non-additive	Govindarasu *et al.* (1981), Sidhu *et al.* (1981)
	Additive and Non-additive	Peter and Rai (1980)
Days to harvest	Additive	Singh and Singh (1980); Banerjee and Kalloo (1989)
Fruits/plant	Additive	Swamy and Mathai (1982), Kalloo *et al.* (1974)
	Non-additive	Peter and Rai (1980)
	Additive and Non-additive	Raijadhav and Kale (1985)
Yield/plant	Additive	Dixit *et al.* (1980); Singh *et al.* (1988)
	Non-additive	Swamy and Mathai (1982), Chaudhary (1970)
	Additive and Non-additive	Raijadhav and Kale (1985)
	Additive and dominant	Raijadhav and Kale (1987)
Average fruit weight	Additive	Singh and Nandpuri (1974), Mittal *et al.* (1974)
	Non-additive	Mital *et al.* (1974)
	Additive and Non-additive	Raijadhav and Kale (1985)
	Additive and additive epistatic effects	Virdelwala *et al.* (1981)
Fruit shape index	Additive	Raijadhav and Kale (1987)
	Additive and Non-additive	Mital and Singh (1977)
	Additive and additive x additive	Mital and Singh (1977)
Locules/fruit	Additive	Govindarasu *et al.* (1981)
	Non-additive	Anbu *et al.* (1976)
	Dominance	Nandpuri and Tyagi (1976)
	Partial dominance	Bhutani (1981)
Pericarp thickness	Additive	Sidhu *et al.* (1981)
	Non-additive	Bhutani (1981)
Total soluble solids	Non-additive	Raijadhav and Kale (1987)
	Additive and Non-additive	Rattan *et al.* (1978)
Total solids acidity	Additive	Yada *et al.* (1988), Singh and Nandpuri (1975)
	Non-additive	Kalloo *et al.* (1974), Chaudhary (1970)
Acidity		Chaudhary (1970)
Ascorbic acid		Chaudhary (1970)

Character	Type of Gene Action	References
	Partial dominance	Bhutani (1981)
pH	Non-additive	Raijadhav and Kale (1987)
Lycopene	Non-additive	Raijadhav and Kale (1987)
	Dominance	Bhutani and Kalloo(1983)
Consistancy of juice	Additive	Kurian (1990)
Ascorbic acid	Additive	Kurian (1990)

Table 2.6: Variability for different Qualitative Traits in Tomato Genotypes

Character	Range		references
	From	to	
Fruit shape index	0.40 ('Pusa Ruby')	1.43 ('Italian Red Pear')	Roy and Choudhary (1972)
	0.70 ('Pusa Ruby')	1.42 ('Chiku Grande')	Roy and Choudhary (1981)
Pericarp thickness (cm)	0.22 ('Sabour Prabha')	0.59 ('Traveller')	Dudi *et al.* (1983)
	0.45('Pusa Ruby')	0.70 ('IIHR 674 SB')	Tikoo (1987)
Juice yield (per cent)	70.55 ('HS 101')	85.71 (Pb. Kesari)	Bhatnagar *et al.* (1987)
	80.39 ('Sel 4')	91.00 ('Prista')	Setty *et al.* (1987)
Total soluble solids (per cent)	4.20 ('Preslav')	6.00 ('Pusa Ruby')	Setty *et al.* (1987)
Total solids (per cent)	4.10 ('Mangala')	6.16 ('Indo Process II')	Sethi and Anand (1986)
Reducing sugar (per cent)	0.17 ('Camp Bell')	3.92 ('F455-D1')	Bajaj and Mahajan (1982)
Acidity (per cent)	0.92 ('Pusa Lal Meeruti')	4.05 ('B17')	Bhutani *et al.* (1983)
Acidity (per cent)	0.37 ('Pb. Kesari')	0.62 ('KS 1')	Bajaj *et al.* (1988)
Sugar/TSS acidity ratio (per cent)	6.40 ('KS 1")	12.50 ('Roma')	Bajaj *et al.* (1988)
Pulp content (per cent)	21.0 (Perslav and Hebros)	30.40 ('Pusa Ruby')	Setty *et al.* (1987)
Juice viscosity	119 ('Hebros')	250 CPS (Prista)	Setty *et al.* (1987)
Lycopene (mg/100 g)	1.57('EC154896/1230F')	5.12('EC1296604/75/122)	Madhaiah *et al.* (1986)
	1.95 ('Sel 152')	6.77 ('Pusa Ruby')	Bajaj *et al.* (1988)
Beta carotene (mg/100 g)	12.20 ('Patriot')	46.80 ('Brench')	Bajaj and Mahajan (1982)
	20.00 ('Sheetal')	60.00 ('Indo Process II')	Sethi and Anand (1986)
Ascorbic acid (mg/100 g)	10.44 ('Karnataka Hybrid')	22.10 ('AC 238')	Bajaj *et al.* (1988)
	15.00 ('Karnataka Hybrid')	30.00 ('AC 238')	Kalloo (1989)

Species Relationship

The genus differs from *Solanum* by the absence of spines in the pinnatifid leaves and anther tips are narrow and exhibit longitudinal dehiscence. In species relatyionship establishment and determination, molecular markers support a lot and considering the facts three to four major groups can be made within the genus. It reflects mating system and fruit colour, with autogamy and redorangefruit being true synapomorphs. Findings are consistent with morphological and crossability data (Rick, 1979). The traditional distinction between the *esculentum* and *peruvianum* complexes is somewhat obscured by molecular phylogenic evidence. This is because of relatively large genetic distances between *S. hirsutum* and *S. pennellii* and the remaining *Lycopersicon* species, although both species can hybridize with the highly derived *S. esculentum*, and so are considered to be part of the *esculentum* complex (Robertson and Labate, 2007). The reproductive biology of tomatoes is as under:

Table 2.7: Reproductive Biology of Tomato Species

Taxon	Self-compatible	Mating system	Cross-compatibility with cultigens
S. cheesmanii	Yes	Autogamous	Reciprocal
S. chilense	No	Allogamous	Can act as male with embryo rescue
S. chmielewskii	Yes	Facultative	Reciprocal
S. esculentum var. *esculentum*	Yes	Autogamous	Reciprocal
S. esculentum var. *cerasiforme*	Yes	Autogamous	Reciprocal
S. glandulosum	No	Allogamous	No

Taxon	*Self-compatible*	*Mating system*	*Cross-compatibility with cultigens*
S. hirsutum f. *typicum*	No. except for some biotypes	Allogamous or facultative	Can act as male
S. hirsutum f. *glabratum*	Yes	Facultative	Reciprocal
S. parviflorum	Yes	Autogamous	Reciprocal
S. pennellii	No. except for some biotypes	Allogamous or facultative	Reciprocal
S. peruvianum	No. except for some biotypes	Allogamous or facultative	No, can occasionally be overcome with technical difficulties
S. pimpinellifolium	Yes	Autogamous	Reciprocal

Source: Modified from Robertson and Labate, 2007.

Botanical Description

Kingdom : Plantae

Division : Magnoliophyta

Class : Magnoliopsida

Sub-class : Rosids

Order : Solanales

Family : Solanaceae

Genus : *Solanum* (old genus *Lycopersicon*)

Species : *lycopersicum*

Tomato belongs to the order Scrophulariales, suborder Solanineae,family Solanaceae, tribe Solaneae, genus *Lycopersicon*, subgenus *Eulycopersicon*, and species *Lycopersicon esculentum*. As it belongs to to the night shade (*Solanum nigrum* L.) family, Linnaeus called it *lycopersicum* (lykos-wolf, persicum peach, meaning thereby wolf peach) for the beauty of the fruit and its deceitful value (*lycopersicon*=wolf peach, *esculentum*=edible). In 1753, Linnaeus placed the tomato in the genus *Solanum* (alongside the potato as *Solanum lycopersicum* but in the year 1768 Philip Miller moved it to its own genus, naming it *Lycopersicon esculentum*. In the genus *Lycopersicon* the stamens forms a solid tube cone shaped with long connective tissue. Anther dehiscence is introsely longitudinal which differentiate this genus from the genus *Solanum* in which the dehiscence of the anthers is poral (Diez and Nuez, 2008). This name came into wide use, but was in breach of the plant naming rules. *S. lycopersicum* (L.) H. Karst., would be more correct, but this name (published in 1881) has hardly ever been used (except in seed catalogs, which frequently used it and still do). Genetic evidence has now shown that Linnaeus was correct to put the tomato in the genus *Solanum*, making *Solanum lycopersicum* the correct name. Both names, however, will probably be found in the literature for some time. Two of the major reasons some still consider the genera separate are the leaf structure (tomato leaves are markedly different from any other *Solanum*), and the biochemistry (many of the alkaloids common to other *Solanum* species are conspicuously absent in the tomato). Peralta and Spooner (2000), a minority of authors (Spooner *et al.*, 1993; Bohs and Olmstead, 1999; Knapp and Spooner, 1999; Olmstead *et al.*, 1999) still name tomatoes as *Solanum*. Tomato is highly self–pollinated due to homomorphism and chasmogamy. If the pollen is shed as the style grows up through the anther tube, self-fertilization is the rule. Self pollination also takes place when the style is short and the stigma is not extroverted beyond the connivant anthers. A certain degree (percentage) of cross pollination also occurs when the stigma produces outside the level of anther (extrovert). Plant habit is annual, biennial, or sometimes perennial.

Table 2.8: Botanical Description

Attributes	*Description*
Growth Habit	**Determinate type**- Inflorescence occurs almost at every internode, main axis terminates in a flower bud and is appropriately called 'self topping' or 'self pruning' type, suitable for growing in the North Indian plains where growing season is short, suitable for mechanized harvesting. Many varieties of determinate type tomato plants do not have adequate foliage to protect their fruit. Some of them fruit very early. The determinate varieties can be harvested in 2-3 harvests. **Indeterminate type**-Inflorescence occurs at every third internode, main axis continues growing indefinitely and terminates in a vegetative bud, and fruiting period of indeterminate type is prolonged suitable for growing in the hills and South India where favourable climate is available for longer period, suitable for growing in the polyhouse as well as in kitchen garden.
Root	The tomato plant can have a fibrous root system or a taproot system depending on how the plant was grown. If the plant is grown from a seed, the plant will exhibit taproot organization. When the plant is grown from cuttings, a fibrous root system will form. A taproot system is made up of a central, large root that is called the taproot. The taproot is larger in diameter than the lateral roots. Lateral roots, which originate from the pericycle, branch off from the taproot, and subsequent lateral roots can branch off other lateral roots. Taproots generally grow more deeply into the soil than do fibrous roots (Esau 1976). Unlike the taproot system, the fibrous root system is made up of thin, stringy roots that all have about the same diameter. These roots branch several times and form a complex mat under the plant that binds to the upper soil layers. Lateral roots arise from the pericycle. The adeventitious roots develop at the rate of about 33.0 mm per day penetrating 1.0 m depth and attaining a length of 1.5 m (Ward, 1964)

Attributes	*Description*
Stem	The stem is soft, brittle and hairy and woody, and capiously branched when mature. Stems 10.0-14.0 mm in diameter at base, green, pubescent and usually villous towards the apex, with a mixture of numerous simple unicellular trichomes 0.5 mm long and sparse simple uniseriate mulrticellular trichomes 3.0 mm long and composed of up to 10 cells, some gland-tipped, particularly the longer ones, giving the plant a strong scent.
Leaf	Leaves interrupted imparipinnate, sparsly pubescent with a mixture of simple unicellular and simple uniseriate multicellular trichomes like those of the stems on both surfaces; basiscopically decurrent, truncate to cordate, the margins dentate or crenate mainly near the base, rarely entire or deeply dentate or lobulated, the apex acute to attenuate; terminal primary leaflet usually larger than the laterals, 3.0-5.0 cm long, 1.5-3.0 cm wide, the petiole 0.5-1.5 cm long, the apex usually attenuate; lateral leaflets 2.0-4.5 cm long, 0.8-2.5 cm wide, the petiole 0.3-2.0 cm long; secondary leaflets present mainly acroscopically on basal leaflets, 0.2-0.8 cm long, 0.1-0.5 cm wide, sessile or with a short petiolule to 0.4 cm long, tertiary leaflets absent; interjected leaflets usually present, 6.0-10.0, 0.1-0.8 cm long, 0.1-0.6 cm eide, the petiolule 0.1-0.3 cm long; petiole 1.2-6.0 (+) cm long; pseudostipules absent.
Floral biology	Tomato is a hermaphrodite containing both male and female sex organs in the same flower. Inside anther and ovary pollen grains and eggs are formed respectively. The innermost whorl of flower contains female sex organs, *viz.*, ovary, style and stigma. Flowers with the calyx tube minute, the lobes to 0.5 cm long, linear, pubescent with long and short simple uniseriate trichomes the apex acute; corolla 1.0-2.0 cm in diameter, pentagonal, bright yellow, often fascinated and with more than 5 lobes in some cultivars, the tube 0.2-0.4 cm long, the lobes 0.5-2.0 cm long, 0.3-0.5 cm wide, narrowly lanceolate, sparsely pubescent with tangled uniseriate trichomes ca 0.5 mm long the tips and margins, the lobes spreading at anthesis; stamina column 0.6-0.8 cm long, 0.2-0.3 (-0.5) cm wide, narrowly cone shaped straight, the filaments minute to 0.5 mm long, the anthers 0.4-0.5 cm long, the sterile apicel appendage 0.2-0.3 mm long, always less than half the total anther length, ovary conical minutely glandular villous; style 0.6-1.0 cm long, <0.5 mm in diameter, usually included in the stamina column, but exerted in facultative allogamous populations; stigma capitate and green. The stigma and style in a sequentrial manner (Singh and Gill, 1987). Sepals are persistent, *i.e.*, they remain attached with the developed fruits, while corolla drops few days after fertilization.
Anthesis and pollination	In tomato anthers opening from base to top, the apex extended into an empty extremity. Anthesis takes place from 6 to 12 noon, dehiscence 8 to 11 am (dehiscence of anther occurs 1-2 days after the opening of corolla.) and shows maximum stigma receptivity on the day of anthesis between 7 am to 1 pm. Anthesis is influenced by temperature and soil moisture (Hawthon and Pollard, 1954). Anthers dehiscing introsly by longitudinal slits the tips empty. The pollens remain viable for 7-10 days and the stigma remains receptive for 6-8 days at 20 °C temperature and 70 per cent relative humidity. The fertilization and fruit formation has been found maximum, when pollination takes place at the time of anthesis. If the pollen is shed as the style grows up through the anther tube, self fertilization is rule. Self pollination also takes place when the style is short and the stigma is not extroverted beyond the connivant anthers. A certain degree of cross pollination happens when the stigma protrudes outside the level of anther (extrivert). The fruit set is clearly visible after about 96 hours of pollination (Brar, 1975 and Tiwari and Choudhary, 1993.) The optimum range of night temperature for fruit set is 15-20°C.
Fruit	Most cultivars produce red fruit, but a number of cultivars with yellow, orange, pink, purple, green, black, or white fruit are also available. Multicolored and striped fruit can also be quite striking.Cultivated tomatoes vary in size, from tomberries (berry), about 5.0 mm in diameter, through cherry tomato, about the same 1.0–2.0 cm size as the wild tomato, up to beef steak tomatoes 10.0 cm or more in diameter. The most widely grown commercial tomatoes tend to be in the 5.0–6.0 cm diameter range. Fruiting pedicels 1.0-3.0 cm long, straight or angled towards the inflorescence axis at the articulation, thickening in large fruited varieties; calyx lobes in fruit accrescent, ca. 0.8-1.0 cm long, 0.2-0.25 cm wide, somewhat to strongly reflexed.Tomatoes grown for canning and sauces are often elongated, 7.0–9.0 cm long and 4.0–5.0 cm in diameter; they are known as plum tomatoes and have a lower water content. Roma type tomatoes are important cultivars in the Sacramento Valley. Fruits are usually globose and 2-locular but often of varying shape and/or multilocular, and glabrescent.
Seed	Seeds 2.5-3.3 mm long, 1.5-2.3 mm wide, 0.5-0.8 mm thick, obovate, pale brown, pubescent with hair-like outgrowth of the lateral testa cell walls, these adpressed and giving a silky appearance to the surface or more often shaggy, narrowly winged (0.3-0.4 mm) at the apex.

Horticultural Classification

Tomato can be classified on the basis following attributes:

(a) Fruit Colour and Compatibility

According to Muller (1940a) and Luckwill (1943), the genus *Lycopersicon* (new genus *Solanum*) is divided into 2 subgenera:

(1) Eulycopersicon

Characterized by red fruited edible varieties, with carotenoid pigmentation and annual which are self-compatible.

Eulycopersicon consists of:

(a) *S.* ***esculentum***- fruits are larger, generally solitary, plant is hairy, strong odour, and

(b) *L.* ***pimpinellifolium*** **Mill. (syn.** *S.* ***pimpinellifolium, S. racemiforme, S. racemigeren*****)**–Fruits are small (diameter from 2.0-5.0 cm), in bunches, deep red colour of fruits, plant is smooth, very slender plant, a little or indifferent odour.

S. esculentum var. *pyriforme* has a pear shaped fruit, with an average length of 4.0 cm. *S. esculentum* var. *grandifolium* is known as "potato leaves"; leaces are large with few follicles. *S. esculentum* var. *validium* plants are erect and compact with very short internodes.

(2) Eriopersicon

Green-fruited varieties with anthocyanin pigmentation and self incompatible. *Eriopersicon* consists of:

(a) *S. hirsutum* Humb. and Bonpl.- This is the closet species to *S. esculentum*

(b) *S. hirsutum* var. *glabratum* Muller, 1940a)

(c) *S. peruvianum* (Muller, 1940a)

(d) *S. pissisi* (Luckwill, 1943)

(e) *S. glandulosum* C.H. (Muller, 1940a)

(f) *S. cheesmanii Riley* (Muller, 1940a)

(g) *S. cheesmanii* var. *minor*

(h) *S.cheesmanii* var. *typicus*

(i) *S. chilense* Dunal (syn. *S. peruvianum* var. *dentatum*)

(j) *S. minutum*

(k) *S. peruvianum* var. *humifusum* C.H. Muller and

(l) *S. hirsutum* var. *typicum*

Crossability with Commercial Varieties

Rick (1976) divided the genus into those species which can be easily crossed with the commercial tomato (*S. esculentum*-complex) and those which can not (*S. peruvianum* complex).

(i) Species Forming the Esculentum Complex are

(a) *S. esculentum* Mill. Mexico- is the centre of origin, the cherry tomato (*S. esculentum* var. *cerasiforme*) is the ancestor

(b) *S. pimpinellifolium*- Originated in low elevations in Peru, source of resistance to *Fusarium* wilts, bacterial speck (*Pseudomonas tomato*)

(c) *S. cheesmanii*- Originated in Galapagos Islands, self compatible

(d) *S. parviflorum* (syn. *L. minutum*)

(e) *S. chmieleswskii*

(f) *S. hirsutum* f. *typicum*- cold tolerance, *Septoria* resistance, resistance to TMV and

(g) *S. hirsutum* f. *glabratum*- less hairy, resistance to bacterial speck and root-knot nematode.

(ii) Species Forming the Peruvianum Complex are

(a) *S. chilense*- source of resistance to TMV and

(b) *S. peruvianum*

(B) Fruits Shape and Plant Spread

(a) *S. esculentum* var. *commune*-common round fruited tomato

(b) *S. esculentum* var. *grandifolium*- large leaved, potato leaved tomato

(c) *S. esculentum* var. *cerasiform* –small fruited cherry tomato-reported as the ancestral form of tomato

(d) *S. esculentum* var. *pyriforme*- pear shaped tomato and

(e) *S. esculentum* var. validum-erect tomato

(C) Growth Habit

Depending upon the growth habit, the tomato plants have been categorized into two types of growth habit *viz.* (a) Determinate and (b) Indeterminate

(a) Determinate-Inflorescence occurs almost at every internode, main axis terminates in a flower bud and is appropriately called 'self topping' or 'self pruning' type, suitable for growing in the North Indian plains where growing season is short, suitable for mechanized harvesting. Many varieties of determinate type tomato plants do not have adequate foliage to protect their fruit. Some of them fruit very early. The determinate varieties can be harvested in 2-3 harvests. The example of this group varieties are AC142, HS 101

(b) Indeterminate- Inflorescence occurs at every third internode, main axis continues growing indefinitely and terminates in a vegetative bud (leafy bud) and the fruiting period of indeterminate type is prolonged suitable for growing in the hills and South India where favourable climate is available for longer period, suitable for growing in the polyhouse as well as in kitchen garden. The example of this group varieties are Pusa Ruby, Sweet 72.

Another type of plant growth is also found in tomato known as:

Semi-determinate- In between determinate and indeterminate, number of nodes between two consecutive inflorescences approach one. The example of this group varieties are S-12, Pusa Early Dwarf *etc.*

(D) On the Basis of Fruit Set and Temperature Requirement

(a) Normal temperature set- Required 15-20°C night temperature. The example of this group varieties are Pusa Rubi, and Sweet 72.

(b) Hot temperature set- Required 20°C night temperature. The example of this group varieties are World Vegetable Centre (AVRDC) lines.

(c) Cold temperature set- Required below temperature of 15°C night temperature. The example of this group varieties are Balkan, Cold Set, Oston Ainski and Pusa Sheetal.

Augmentation of Genetic Resources and Crop Improvement

Breeding Priorities

The goals of public and private tomato breeding programmes vary widely depending on location, need and resources. In general, breeding goals in tomato have gone through four phases:

i. Breeding for yield in the 1970s

ii. For the shelf-life in the 1980s

iii. For taste in the 1990s and

iv. Nutritional quality currently.

Table 2.9: Differences between Determinate and Indeterminate Types of Tomato

Sl.No.	*Determinate Types*	*Indeterminate Types*
1.	Main stem of tomato plant terminate in flower bud	Main stem of tomato plant does not terminate in flower bud instead it possesses vegetative bud
2.	Main stem does not grow tall after flowering	Main stem remains continue in growth throughout growing season
3.	Formation of more laterals	Formation of less number of laterals
4.	There is self-topping of plant	There is no self-topping of plant
5.	Flower cluster or truss forms in many cases at every internode	Flower cluster or truss forms in many cases at every 3rd internode, separated by three leaves
6.	Comparatively having early maturity	Comparatively requires long period for maturity
7.	Incidence of sunscalding is less because fruits remain hiding under leaves	Incidence of sunscalding is more
8.	Plants do not require earthing	Plants require earthing and they are benefitted with it
9.	Plants do not staking	Plants staking
10.	Plants habit is bush type	Plant growth straight *i.e.* 'non-bush' type
11.	Not much suitable for high planting density	Suitable for high density planting
12.	**Example:** ARTH-3, Pusa Hybrid-2, NA-501, DTH-4	**Example:** Pusa Ruby, Sel.120, T-1, Sioux, Pant Bahar (AC-234), Arka Vikas, Arka Saurabh, Pant-3 (AC-238), BT-10, BT-12, ARTH-4, MTH-6, Arka Vardhan (FMH-2), KT-4, Nath-601

In broad sense following breeding priorities should be fixed:

- ☆ High yield
- ☆ High quality fruit
- ☆ Adaptability to salinity, water stress and temperature
- ☆ Resistance for insects
- ☆ Resistance for diseases
- ☆ Breeding for quality traits.

Introduction and Selection

Popular cultivars like 'Sioux' and 'La Bonita' were introduced in India and directly used as a variety for commercial cultivation. A large number of germplasm lines (cultivated and wild) of tomato available at various institutes including the ICAR-National Bureau of Plant Genetic Resources (NBPGR), New Delhi and ICAR-Indian Institute of Vegetable Research (IIVR), Varanasi (U.P.) ICAR-Indian Agricultural Research Institute New Delhi and ICAR-Indian Institute & Horticultural Research Bengaluru Karnataka. These lines also include several potential wild species like *S. pimpinellifolium*, *S. esculentum* var. *cerasiforme*, *S. peruvianum*, *S. hirsutum* f. *glabratum*, *S. pimppinellifolium etc.* ICAR-NBPGR, New Delhi has successfully augmented more than 3500 germplasm lines including wild species from various countries and from within India and in collaboration with several institute and state agricultural university has completed several exploration trips in different parts of country. Some of the promising germplasm, identified for different traits on the basis of their morphological variance are given in Table 2.10.

Nath *et al.* (2008) stated that initially most of the important varieties were introduced from the western countries and some of them became popular among the growers and consumers in different agroclimatic regions of the country depending upon their adaptability. In the beginning, ICAR-Indin Agricultural Research Institute, New Delhi had introduced several important varieties from all over the world which were evaluated during early sixties of last century for various purposes (Mital *et al.*, 1965). Varieties like 'Ponderosa' and 'Oxheart' were established earlier lines for cultivation. Some high yielding varieties like 'Pusa Ruby', 'HS 101', 'Sel. 2', Sel.

Table 2.10: Promising Germplasm of Tomato

Attributes	*Promising Accessions*
High yield	EC106285, EC118292, EC4551, EC163683, EC173856, EC170662, EC110578, EC1127, EC129081, EC129353, EC168084, EC50357, EC122063, EC1301, EC251636, EC6504, EC315481, EC315486, EC339059, EC337827, EC339074, EC357829, EC315483, TIO, P-388, P-1, P-501, NIC1122231, EC269288, IHR-2034, IHR-2030, IHR-2028, KM-1, KM-2, IC-6, Sel-148-6, EC378645, EC368838, EC378639, EC377983, EC378833, EC368830, EC368855, ECIHR-1885, IHR-1990, EC257749, EC1127, EC130053, EC163598, EC329560, RT-6, CLN-2112-86-2-10-1-7-0, CLN-1558B, LO-2455, CLN-1463-A, EC326142, BL-985, EC339061-1-1, CH-193, CLN-2413R, DR-2, HAT-142, EC357841, EC238508, EC78836, EC398702, EC398715, EC488370, EC490119, EC318193, EC251613, EC251566, EC141887, EC538408, EC538411, EC538416, Ovegen-II, Santian, Slietz, Oregan Pride, CLN-1466B, CLN-1460A, S-1-96-66, Equinox, EC400504, Arka Meghali, CH-24, EC385654, EC388943, EC174045, NO-187, T-777, AP-3, UHF-659, EC12217, Line 553, EC15990, HAT-92-2, EC416543, EC416582, EC416557, IC90882, IC249297, EC015998, LE-231, LE-1150, EC519806, EC519697, EC519796, EC519812, LO-1003, LO-2410, EC29414, EC2791, EC124371, EC114375, UHF-55, LE-231, LE-1150

120', 'T 1', 'Sweet 72', 'Pusa Early Dwarf', 'Sioux', 'Pusa Gaurav', 'Roma' (Singh *et al.*, 1977), 'L. Bonita' (Thomas *et al.*, 1976); 'Dwarf Money Maker' (Thomas *et al.*, 1979) 'Punjab Chuhara', 'KS 2', 'Pant Bahar' *etc.* are selected and recommended for commercial cultivation in different regions of the country. In China, hybrid tomato using World Vegetable Centre (AVRDC) Taiwan materials as one of the parents (CL-1131-0-0-13-0-6) locally known as Seedathip-1, Seedathip-11, Seedathip-111, Seedathip-91, have produced many promising lines of both table and processing tomato. Similarly, CL-1131 as SVRDC-4 and PT-4225; as PT4225 has been used largly in development of cultivars. Horticulture Research Centre, Bangladesh Agricultural Research Institute (BARI), has developed tomato cultivars Manik and Ratan in the year 1985 is still under cultivation. Similarly, Bangladesh Institute of Nuclear Agriculture (BINA) has developed a very good line and released as 'Bahar' in 1992. The new tomato varieties 'Arka Saurbh' and 'Arka Vikas' from ICAR-IIHR, Bengaluru (Karnataka) (Tikoo and Anand, 1983); 'CO-3' from TNAU, Coimbatore (Tamil Nadu); 'Sel-120' and 'Sel-152' from ICAR-IARI, New Delhi; Punjab Chhuhara' and 'Punjab Kesari' from Punjab Agricultural University (Punjab) (Nandpuri *et al.*, 1977); 'HS-101 and HS-102 from CCS HAU, Hisar (Haryana); 'AC-238' and 'Pant T-3' from Govind Ballabh Pant University of Agriculture and Technology, Pantnagar (Uttarakhand) and 'Sweet-72' from Gwalior (M.P.) were promising. Rai *et al.* (1987) evaluated 20 varieties of tomato for NEH-Tripura conditions and variety S-12 gave the best performance under their agroclimatic conditions. Supe *et al.* (1989) reported the performance of tomato hybrids in Haryana and concluded that the 'Suttons Bragger' gave maximum fruit weight and plant height while hybrid 'MTHH' was the best performing hybrid since it had recorded highest yield. In Nepal following varieties of tomato are commonly being grown:

Khandare *et al.* (2016) estimated the genetic diversity using 41 genotypes consisting of Pusa Sadabahar, Pusa Sheetal, Pusa Rohini, Pusa-120, Pusa Ruby, Arka meghali, arka Alok, Arka Vikas, Arka Abha, Arka Saurabh, PKM-1, Dhanashree, Phule Raja, Dantiwada Local, PBNT-1, PBNT-2, PBNT-3, PBNT-4, PBNT-5, PBNT-6, PBNT-7, PBNT-8, PBNT-9, PBNT-10, PBNT-11, PBNT-12, PBNT-13, PBNT-14, PBNT-15, PBNT-16, PBNT-17 PBNT-18, PBNT-19, PBNT-20, PBNT-21, PBNT-22, PBNT-23, PBNT-24, PBNT-25, PBNT-26 and PBNT-27. The D^2 analysis was carried out for 22 characters which partioned the forty one genotypes into 15 clusters. The maximum genetic divergence was observed between clusters IX and XIV. The maximum plant height (90.95 cm), number of primary branches (11.90), number of flowers clusters per plant (34.40), diameter of fruit (4.21 cm) and yeld/hectare (645.72q) were recorded in cluster XIV. The maximum intra cluster distance was shown by cluster VI (PBNT-12). The cluster XIV (PBNT-5) was showed maximum ascorbic acid content

Table 2.11: Some Important Cultivars and Salient Features

Varieties	*Salient Feature*	*Source Country*
NTS-64		Royal Sluis, Holland
S-901, S-808, S-807, Dynamo BFNT-R.		Sakata Seed Co. Japan
Pratibha		Pure Seed, India
Sadabahar, Gulmohar		Mahyco Seed Co., India
Chetan, BSS-40,		Beejo Sheetal Seed Co., India
Rasami, Rupali, Baishali, Naveen		Indo-American Seed Co., India
Rajani		Dhanlaxmi, India
Ajit		Pahuja Seed Co., India
EC301021	Long shelf life	Australia
EC310299	Slow ripening	USA
EC321425-26	Paste type	Taiwan
EC347359-68	F_1 hybrids	Taiwan
EC343391-95	F_1 hybrids (Cherry type)	USA
EC346011	Stamenless line (F_2)	USA
EC346013-14	Green pistillate type	USA
EC399828-38	Heat tolerance	USA
EC347359-68	Heat tolerance	Taiwan
EC321425-26	Heat tolerance	Taiwan
EC378682-83	Low irrigation requirement	Australia
EC399828-38	Multiple disease resistance	USA
EC34769-90	TMV and BW resistance	Taiwan
EC321890-982	Viral disease resistance	Taiwan

(26.65mg/100g) and tritrable acidity (0.80 per cent). Neha *et al.* (2016) evaluated 15 commercial varieties, 9 exotic genotypes and three wild species and observed high gcv and pcv for number of fruits/plant, high broad sense heritability and high genetic advance as per cent of mean was registered for number of fruits/plant. Highest fruit yield was found in EC620421 (796.09 q/ha) followed by IIHR-2620 (699.82 q/ha).

In Pakistan, cultivars like T-10, T-43, Roma, Money Maker (R), Long Tipped (I), Eva, Lima, Feston and Savio (P) are being grown commercially. University of the South Pacific/School of Agriculture (USP/SOA) has selected three cultivars:

i. **Alafua Early-**Previously tested as CL2729-0-2-1-12-0(33). This open pollinated cultivar was bred and distributed for testing by the World Vegetable Centre (AVRDC), Taiwan.

ii. **Alafua Winner-** Previously tested as CL5915-96D4-2-2-0 (40). This open pollinated cultivar was bred and distributed for testing by the World Vegetable Centre (AVRDC), Taiwan.

iii. **Alafua Large-** Previously tested as CL143-0-10-30-0-1-10 and was originally bred in Taiwan by World Vegetable Centre (AVRDC), Taiwan and later named as 'Nuu' in Western Samoa and 'Alton' in Fiji.

Hybridization and Selection

Heterosis breeding is common in tomato develop F_1 hybrids and absolute heterosis to the extent of 300 percent has been reported. In addition to productivity, hybrids bring other significant benefits such as uniformity of trait expression and scope to combine dominant alleles from two parents. In due course of time 'Sl-120 selected from 'Anahu' lines through crossing programme. A significant achievement was made at the ICAR-IARI, New Delhi in development of variety 'Pusa Ruby' which is a selection from a cross between 'Improved Meeruti' (indigenous hardy variety) and 'Sioux' (a good quality American variety). Hendrick and Booth (1907) were the 1st to report on the presence of heterosis in tomato. Presently use of high stigma type (ex) as female parents in the production of F_1 hybrids is gaining ground. 'Manaleuci' x 'Burinsma' is a hybrid developed using high stigma character of 'Bruinsma'.

The promising cross combinations were Castle Rock x *nor*-RM-1, IPA-3 x *nor*-RM-1, Nemadoro x *nor*-RM-1, and UC-82-B x *nor*-RM-1 In mid season and LT-44 x alc-IIHR-2050, LT-42 x rin-RM-2, Punjab Upma x *nor*-RM-1, IPA-3 x *nor*-RM-1 and LT-44 x rin-RM-2 in late season manifested considerable heterosis for yield/plant, marketable yield/plant, average fruit weight and shelf-life index (Garg *et al.*, 2013). Choudhury and Malhotra (2001) develop several hybrids through Line x Tester mating design and found VTL-33 x Roma as best with respect to yield and quality.

Table 2.12: Genetic Material Used as Parents in Development of Hybrids

F_1 Hybrid	*Salient Features*
'Red Current' x 'Greater Baltimore'	Early flower than mid parents
'Sanmarzano' x 'Costolute'	More than 'Sanmarzano', the better parent (17 per cent)
'Improved Meeruti' x 'Pusa Ruby'	More than the high yielding parent (heterobeltiosis 56.73 per cent)
'Marglobe' x 'Pusa Early Dwarf'	Yielded more than the high yielding parent (heterobeltiosis 33.52 per cent)
'AC142' x 'Sioux'	Early and high yield
'AC142' x 'Marglobe'	Early and high yield
'AC142' x ' Sweet-72'	Early and high yield
'Pusa Red Plum' x 'Herkness'	Higher fruit size, number and yield
'Marglobe' x 'Crack Proof'	Higher fruit size, number and yield
'Keckruth Ageti' x 'Pusa Ruby'	Higher yield

Horticulture Research Centre, Bangladesh Agricultural Research Institute (BARI) has observed that hybrids in tomato 'Chetan', Hope No.1, and Tropic Boy are dominating in several area of Bangladesh.

Table 2.13: Promising Hybrids and its Attributes

Attributes	*Hybrids*	*References*
Yield	Best of All x Pusa Early Dwarf; Pusa Ruby x Rutgers	Choudhury *et al.*, 1965
Yield	MS x 6L; MS x RD; MS x ZR	Singh and Nandpuri, 1970
Yield	Pusa Red Plum x Harkheff	Khanna *et al.*, 1972
Yield	Marglobe x CR	Chaudhury and Khanna, 1972
Yield	Punjab Tropic x Sl-120; Nematex x Angurlata	Bhutani *et al.*, 1973
Yield	Atkinson x HM82782; Ostenkinskiz x HM-82782; Atkinson x Nova	Nandpuri *et al.*, 1974
Total and early yield	Kalyanpur Kuber x HBS	Mittal *et al.*, 1974
No. of fruits and size	Sel.3-7 x HM-8278	Nandpuri *et al.*, 1975
Yiels and quality	Sel.3-13 x Sel. 1-2	Gill *et al.*, 1976
Yiels and quality	Pusa Ruby x HS-101; Pusa Ruby x Sioux; HS-101 x Pusa Early Dwarf	Singh *et al.*, 1976
Pericarp thickness	EC55055 x Punjab Tropic	Nandpuri and Tyagi, 1976
Early yield	Peru-2 x Roma; Masasi x B 2547	Swami and Mathai, 1978
Ascorbic acid	Pusa Early Dwarf x HS-101; Keckruth Ageti x Pusa Early Dwarf and HS-101 x Sioux	Singh *et al.*, 1979
Ascorbic acid	Pusa Red Plum x Punjab Chhuhara; Best of All x HS-101	Bhutani, 1980

Attributes	Hybrids	References
Yield	Gamed x Chiku grande; Roma x Punjab Chhuhara; Punjab Chhuhara x Kalyanpur Angurlata	Sidhu *et al.*, 1981
Yield	Sel. 152 x Roma; EC6050 x Lalmani; EC12156 x Lalmani; EC121193 x Gola	Jaiswal *et al.*, 1984

Kumar *et al.* (2016) developed 28 hybrids using 8 parental lines (DVRT-1, DVRT-2, CHFT-77, CHFT-60, CHFT-79, H-86, CHFT-71 and CHFT-50) and based on gca effects, identified CGFT-71 and DVRT-2 as the most promising parents for fruit yield/plant along with earliness and quality parameters. Heterosis for yield/plant ranged from -67.81 to 156.70 per cent over better parent and over standard variety (H-86) varied from -69.81 to 131.76 per cent, respectively. On the basis of heterosis (over better and standard parents) and sca effects, crosses DVRT-2 x CHFT-77, CHFT-60 x CHFT-71 and CHFT-79 x CHFT-50 were most promising for improvement of majority of traits including fruit yield and quality. Higher value of H_1 and H_2 compared to D for all the traits showed that non-additive gene effects have a greater role than additive gene effects. The positive estimate of dominance components (H_1 and H_2) also suggest that the dominance genes were in the favourable direction of dominance for all the traits. The ratio of KD/KR showed the preponderance of dominance genes in expression of yield and all other traits.

Interspecific Hybridization

On the basis of crossability tomato has been divided into two major groups, namely *esculentum* and peruvianum. The *esculentum* complex comprises two subgroups. First consisting of red-fruited tomato *Solanum esculentum*, *S. pimpinellifolium* amd *S. cheesmanii* and second subgroup contains a single green-fruited species *S. hirsutum*. These species are interfertile but *S. hirsutum* crosses only when *S. esculentum* used as female. The *peruvianum* is consisted of *S. peruvianum* and *S. chilense* species. They are not crossable with cultivated tomato unless the cross is aided by embryo rescue technique. Despite the same chromosome number (2n=24) interspecific gene flow is restricted by gene-cytoplasmic interaction, genetic incompatibility, sterility and undesirable linkage. Interspecific hybridization is done in tomato mainly to transfer disease resistance and quality characteristics from wild species to cultivated ones.

Table 2.14: Characters Introgressed into Tomato from its Wild Relatives

Characters	Species
Fungus resistance	*S. hirsutum, S. pimpinellifolium, S. peruvianum*
Virus resistance	*S. chilense, S. peruvianum*
Nematode resistance	*S. hirsutum*
Quality attributes	*S. chmielewskii*
Adaptation to adverse environments	*S. cheesmanii*
Fruit colour	*S. hirsutum*

The species *S. lycopersicum* is cross compatible with *S. hirsutum*. Two species *S. chilense* and *S. peruvianum* are difficult to cross with *S. lycopersicum* while *S. peruvianum* can be crossed with *S. esculentum* the embryos often abort. Hybrids between *S. esculentum* and *S. pimpinellifolium* show a high degree of fertility and some taxonomist regard the parents as one species. The green fruited and self incompatible *Eriopersicon* subgenus includes much more diverse group of 4 or 5 species. The diploid chromosome number of all species is 24 and interspecific hybrids normally show 12 pairs of chromosome at meiosis. When *S. lycopersicum* is crossed with species in the subgenus *Eriopersicon* the crosses are successful only if *S. lycopersicum* is used as female parent. Similarly, the F_1 interspecific hybrids can be used only as male parents in back crosses to as female parent in back crosses to the wild population of the species in the *Eriopersicon* subgenus *S.hirsutum* probably hybridizes most readly with *S. lycopersicum*. The sterility barrier that existed between *S. esculentum* and *S. peruvianum* prevented hybridization between them. Interspecific hybridization between *S. lycopersicum* and *S. peruvianum* have not been possible through traditional methods of plant breeding, but now, it is possible to transfer resistance to insect pest from *peruvianum* to *S. lycopersicum* through the embryo culture technique. The other important resistant variety developed through interspecific hybridization is 'Pusa Red Plum' by crossing *S. esculentum* x *S. pimpinellifolium*. Introgression of tolerance traits from distant wild relatives to cultivated tomato has proven difficult due to crossability barriers and linkage drag. *S. pimpinellifolium* (only red-fruited wild species of tomato) is the closest wild relative which natural introgression into the cultivated tomato and easily crossable. It is found self-compatible and bi-directional cross-compatible with the cultivated tomato (Sharma *et al.*, 2008). It is also a source of many useful genes for other traits such as yield and disease resistance. Furthermore, *S. pimpinellifolium* harbors numerous desirable genes for disease resistance, abiotic stress tolerance and good fruit quality. *S. pimpinellifolium* and *S. peruvianum* contain genes which confer resistance to Fusarium wilt and early blight. At BCKV, Kalyani (West Bengal) highly ToLCV resistant line of tomato (cross 8 F_2P_1-1) has been developed through introgression of resistant gene from *S. hirsutum*. This line is being utilized for the development of ToLCV resistant hybrids (Chattopadhyay *et al.*, 2011). The tomato wild genotype 'EC786252' (*Solanum neorickii*) shows strong resistance to early blight caused by *Alternaria solani* was crossed with susceptible cultivated tomato genotype 'Hawaii 3998' (*Solanum lycopersicum*) for genetic study and to transfer the blight resistance. The cross combinations yielded no viable seeds. In the next season, pollen from EC786252 was used to pollinate the susceptible cultivar and proembryos were collected from developing fruits of 39-49 days after pollination. Rescue of immature embryos was done on MS-medium. Pro-embryos from fruits of 42-44 days after pollination gave good response and 10 plants were regenerated. Hybridity of the regenerated plants was confirm with codominant cleaved amplified polymorphic sequences (CAPS) marker (Reddy *et al.*, 2016).

Successful distant hybridization in tomato is as under:

- *S. esculentum* x *S. pimpinellifolium*
- *S. esculentum* x *S. minutum*
- *S. esculentum* x *S. hirsutum*
- *S. esculentum* x *S. peruvianum*
- *S. pimpinellifolium* x *S. minutum*
- *S. pimpinellifolium* x *S. hirsutum*
- *S. minutum* x *S. esculentum*
- *S. peruvianum* x *S. esculentum*

S. esculentum (2n=24) x *S. peruvianum* (2n=24) and *S. esculentum* (2n=24) x *S. pennellii* (2n=24) showed unilateral incompatibility (reciprocal crosses are not possible).

Table 2.15: Somatic Hybrids Produced in Tomato

Species Combination	*Character Transferred*
Solarrum lycopersium x *S. peruvianum*	TMV, Spotted wilt virus, and cold tolerance,
Solanum lycopersicoides x *S. peruvianum*	Tomato diseases and Peass
Solanum lycopersicoides x *S. esculentum*	(Cybrid) CMS
Lycopersicon esculentum x *S. accule*	-

Source for Biotic Resistance and its Utilization

Tomato varieties susceptible to several diseases and pests in various regions of the country; which have become limiting factors in cultivation.

(i) Disease Resistance Source and its Utilization

The wild species of *Solanum* have shown to be a valuable source of resistance to various diseases and pest to quality traits in addition, which is vital requirement in any breeding programme.

The tomato wild genotype EC786252 (*Solanum neorickii*) shows strong resistance to early blight caused

Table 2.16: Genetic Nature of Resistance of Diseases in Tomato

Nature of Resistance	*Disease*
Mnogenic dominant	*Fusarium* wilt, leaf mould, *Rhizoctonia* root rot, collar rot, spotted wilt virus, tomato mosaic virus and root knot nematode
Single dominant gene with few minor genes	*Bacterial speck*
Oligogenic factors	Bacterial canker, early blight, *Rhizoctonia* root rot and spotted wilt virus
Polygenic factors	Bacterial canker, bacterial wilt, soft rot, leaf curl virus and late blight
Additive effects	Early blight of tomato (Fageria *et al.*, 1998), fruit rot of tomato (Kohli *et al.*, 1995)
Polygenically partially dominant	Bacterial wilt
Two or more recessive genes	Early blight
Br gene	Buckeye rot

by *Alternaria solani*. The resistance to buck-eye fruit rot in *S. pimpinellifolium* var. EC54725 was controlled by a dominant gene *Br* (Rattan and Saini, 1979). Resistance source against Fusarium wilt (*Fusarium oxysporum* spp. *lycopersici*) race-1 have been located in the wild species *S. hirsutum* f. *glabratum* Wir 4172, *S. hirsutum* PI13448 and *S. peruvianum* EC148898 which are immune and *S. esculentum* cvs. 'Columbia' and 'Roma' were highly resistant and 'HS-110' expressed resistant reaction (Jalali *et al.*, 1989). Two species *S. chilense* and *S. peruvianum* appear to be the most promising source of disease resistance (particularly tobacco mosaic virus) in tomatoes. Unfortunately they are also difficult to cross with *S. lycopersicum* (cultivated varieties) while *S. peruvianum* can be crossed with *S. lycopersicum*; the embryos often abort. However, this was made possible with special techniques by Majid *et al.* (1968), where few interspecific hybrids were obtained with high resistance to various diseases. *S. pimpinellifolium* is resistance to *Fusarium* wilt, early blight and leaf curl disease; *S. chilense* is tolerance to leaf curl disease; *S. hirsutum* is an excellent source of resistance to several pest and to leaf curl disease; *S. peruvianum* is resistant to root-knot nematode, tobacco mosaic virus, leaf curl and has high ascorbic acid content and *S. chilense* is good source of drought resistance. Lines 'Louisiana Pink' and 'T414' introduced from Puerto Rico showed good resistance to bacterial wilt. *S. hirsutum* (PI134437) has been noted resistant to early blight/leaf spot (Datar and Mayee, 1980). *S. pimpinellifolium*, *S. hirsutum* f. *glabratum*, *S. peruvianum* and Sel. 18-1were found resistant and associated genes were transferred into cultivated varieties and completely resistant lines P-1, H-86-7 and H-86-11 were developed (Kalloo and Banerjee, 1991d). Resistance source for late blight is noted in *S. pimpinellifolium*. PI272636 is resistance to anthracnose.

Table 2.17: Species having Resistance to Biotic Stress

Species/Lines	*Resistance Source to Diseases*
S. pimpinellifolium	*Fusarium* wilt, early blight, late blight, leaf curl disease and bacterial wilt
S. pimpinellifolium (PI 127805)	Bacterial wilt
S. peruvianum	Root knot nematode, tobacco mosaic virus, leaf curl virus and *Verticillium* wilt
S. peruvianum var. *humifusum* C.H.	Curly top virus
*S. esculentum*var. *cerasiforme*	Late blight
S. hirsutum f. *glabratum*	Early blight, *Fusarium* wilt, leaf curl disease
S. lycopersicum	Late blight

Kumar and Irulappan (1990) studied the seasonal influence of spotted wilt virus resistance in tomato varieties. Prasanna *et al.* (2002) reported developing of bacterial wilt resistant tomato hybrid with extended shelf life. High yielding bacterial wilt resistant F_1 hybrids were IIHR-2188 X IIHR-2052 (2.45 kg/plant), BWR-IF x nor-1(2.57 kg/plant) and IIHR-2199 x IIHR-1136 (2.17 kg/plant). Similarly, lines Acc.-99, Sweet-72, Acc.-151, Hyb.-54 (Tewari, 1986); plants from the segregating population of Saturn x LE-79 (Singh *et al.,* 1988) and BT-1 and BT-10 were resistant to bacterial wilt and line 'RT-1' has been released at national level for cultivation. Genotype 'Neptune' expressed multiple resistance against Bacterioal wilt, Fusarium wilt, Verticillium wilt and Grey leaf spot.

Table 2.18: Wild Species/Genotypes/Cultivars Resistant to Biotic Stress

Patho-gens	*Genotypes/Cultivars*	*Resistant Varieties*	*Wild Resistance Source*	*Resistance Gene*	*Chromosomal Location*	*Reference*
Fungal	Late blight (*Phytophthora infestans* (Mont) de Bary)	'Red Cherry', 'Sel. 1', 'WV 36', 'WV 63', 'WV 700', 'Kt 15', 'Kt 10', Ottawa-30, Ottawa-31, Fla P-14, West Virginia-36, West Virginia-63, TRB-1, EC326139, EC315481, EC162508, EC2512780, EC165700	*S. pimpinellifolium*, *S. lycopersicum* var. *cerasiforme*	-	-	-
	Late blight (*Phytophthora infestans*)	TRB-1, TRB-2	*S. pimpinellifolium*	Ph-1	7	Merk *et al.* (2012)
	Late blight (*Phytophthora infestans*)	-	*S. habrochaites*	Ph-2	10	Foolad *et al.* (2008)
	Late blight (*Phytophthora infestans*)		*S. habrochaites*	Several	All12	Brower *et al.* (2004)
	Late blight (*Phytophthora infestans*)	-	*S. pennellii*	Ph-3	9	
	Late blight (*Phytophthora infestans*)	-	*S. pennellii*	QTLs	-	Li *et al.* (2011)
	Late blight (*Phytophthora infestans*)	-	*S. pimpinellifolium*	2 QTLs	1, 10	Merk *et al.* (2012)
	Buck eye rot (*Phytophthora parasitica*)	Chamba Exh-64, Early Market Selection, Flat Large Red, Kopiah PO x 45727, Molokai, Red Cherry, San Marzano (Sharma *et al.*, 1974)	-	-	-	-
	Phytophthora root and crown rot (*Phytophthora parasitica; P. capsici*)	-	*S. lycopersicum* var. *cerasiforme*	QTLs	-	Blaker and Hewitt (1987)
	Early blight (*Alternaria solani* (Ellis and Mart.) Jones and Grout)	'College Regal', 'Step 22', 68B134 and Southland (Collar rot)	*S. hirsutum* f. *glabratum*	-	-	-
	Early blight (*Alternaria solani*)	-	*S. habrochaites*	-	-	-
	Early blight (*Alternaria solani*)	-	*S. pimpinellifolium*	-	-	-
	Alternaria stem canker	*Alternaria* spp.	*S. pennellii*	Asc	3	Scott and Gardener (2007)

Pathogens	Genotypes/Cultivars	Resistant Varieties	Wild Resistance Source	Resistance Gene	Chromosomal Location	Reference
	Fusarium wilt (*Fusarium oxysporum* race 2 Schl. Ex. Fries)	'Step 530', 'Step 534', Rustrel, No. 126915-1-8 (race1 and 2), Pan American, PI13448, EC148898, Roma, Columbia, HS-110, EC66007,	*S. pimpinellifolium, S. hirsutum, S. peruvianum*	-	-	Scott and Gardener (2007), Grattiige and O Brien (1982), Scott and Jones (1990), Huang and Lindhout (1977)
	Fusarium wilt (*Fusarium oxysporum* f. sp. *lycopersii*)	EC510445-450 (AVRDC, Taiwan) received by ICAR-NBPGR-RS, Hyderabad (Andhra Pradesh)	*S. pimpinellifolium*	1	7, 11	-
	Fusarium wilt (*Fusarium oxysporum* f. sp. *lycopersii*)	-	*S. pennellii*	1-2	-	-
	Fusarium wilt (*Fusarium oxysporum* f. sp. *lycopersii*)	-	*S. pennellii*	1-3	-	-
	Fusarium wilt (*Fusarium oxysporum* race J-3 Schl. Ex. Fries)	'IRB 301-30', 'IRB 301-31'	-	-	-	-
	Fusarium wilt and fruit rot resistant	EC160195, EC122527, EC21626, EC103598, EC116874, EC117008, EC148898, EC115947, EC117671, EC117163, EC128769, EC174023, EC174041, SGP8M2, Columbia Pant Bahar, Marglobe, Pan America, Mangla, HS-101, PI13448, and Roma.	*S. pimpinellifolium, S. hirsutum, S. peruvianum*			
	Fusarium crown and root rot (*Fusarium oxysporum* and *radices-lycopersici*	-	*S. peruvianum*	Fri	9	-
	Fusarium crown and root rot (*Fusarium oxysporum* and *radices-lycopersici*	-	*S. pennellii*	-	-	-
	Fusarium crown and root rot (*Fusarium oxysporum* and *radices-lycopersici*		*S. lycopersicum* var. *cerasiforme*			
	Anthracnose (*Colletotrichum coccodes*)	'PI 272636'	-	-	-	-
	Grey leaf spot (*Stemphyllium solani*)	EC510445-450 (AVRDC, Taiwan) received by ICAR-NBPGR-RS, Hyderabad (Andhra Pradesh)	-	-	-	-
	Grey leaf spot (*Stemphyllium* spp.)	-	*S. pimpinellifolium*	Sm	11	Scott and Gardner (2007); Parlevliet (2002)
	Leaf mold (*Cladosporium fulvum* Cooke)	'Waltham Mold', 'Proof Forcing', 'Money Maker'	*S. pimpinellifolium*	cf (26 genes)	-	Scott and Gardner (2007)
	Leaf mold (*Cladosporium fulvum* Cooke)	-	(22 genes)	1	-	Foolad *et al.* (2000)
	Leaf mold (*Cladosporium fulvum* Cooke)	-	*S. habrochaites*	1	--	

Pathogens	*Genotypes/Cultivars*	*Resistant Varieties*	*Wild Resistance Source*	*Resistance Gene*	*Chromosomal Location*	*Reference*
	Leaf mold (*Cladosporium fulvum* Cooke)	-	*S. peruvianum*	-	-	-
	Cercospora leaf spot (*Cercospora fuligera* Rolden)	'Dwarf Stone', 'Marglobe', Floradel'	-	-	-	-
	Verticillium wilt (*Verticillium alboatrum* Reinke and Benth)	'Loran Blood, 'NFVR', VR Moscow, Pant Bahar	-	-	-	-
	Verticillium wilt (*Verticillium dahliae*)	Peru wild "VEDA" (Eriopersicon)	-	Ve	9	Diwan *et al.* (1999), Stemova (2005)
	Broom rape (*Orobanche aegytiaca*)	'U 21', 'P 2VII'	-	-	-	-
	Septoria (*Septoria lycopersici*)	KT-4, EC2750, EC4555, EC6993, EC7785 (Sohi and Sokhi, 1969)	-	-	-	-
	Grey mould (*Botrytis cinerea*)	-	*S. peruvianum, S. habrochaites, S. pimpinellifolium*	R b e q 1, 2, 4	1, 2, 4	Egashira *et al.* (2000); Nicot *et al.* (2002), Finkers *et al.* (2007)
	Powdery mildew (*Odium* spp.)	-	*S. habrochaites*	01-1-01-3	6	Huang *et al.* (2000)
	Powdery mildew (*Odium* spp.)	-	*S. habrochaites*	01-05	6	Bai (2004)
	Powdery mildew (*Odium* spp.)	-	*S. esculentum* var. *cerasiforme*	01-2; 3 QTLs	4	Huang *et al.* (2000)
	Powdery mildew (*Odium* spp.)	-	*S. parviflorum*	-	6,12	Bai *et al.* (2003)
	Powdery mildew (*Odium* spp.)	-	*S. peruvianum*	01-04	6	Bai (2004)
	Southern blight (*Sclerotium rolfsi*)	-	*S. pimpinellifolium*	QTL	-	Scott (1993)
Bacterial	Bacterial Wilt (*Peudomonas solancearum* E.F. Smith)	'PI 128308', 'VC 4', 'Louisiana Pink', 'LE 79', PI127805A, PI126408, Venus, Saturn, BT-1, LE-79 (Shakti), VC-48-1, BWR-1, BWR-5, Arka Abha, Arka Alok, Pant Bahar, BT-10, BT-1, Sonali, Utkal Pallavi and Rodade.	*S. pimpinellifolium*			
	Bacterial speck (*Pseudomonas syringae* pv. *tomato*)	-	*S. pimpinellifolium*	PTO	-	Pitblado and Kerr (1980)
	Bacterial wilt (*Ralstonia solanacearum*)	EC468337-41; EC490117-43 (AVRDC, Taiwan), received by ICAR-IIHR, Bengaluru	*S. pimpinellifolium*	BWr-6	-	Hanson *et al.* (1998); Wong *et al.* (2013 a)
	Bacterial wilt (*Ralstonia solanacearum*)	-	*S. lycopersicum* var. *cerasiforme*	BWr-12	6, 12	-
	Bacterial Canker [*Corynebacterium michiganense* (Smith) Jensen]	'Bulgarian-12, 'Honma' and Utah-737	-	-	-	-
	Bacterial Canker (*Clavibacter michiganense* sub. sp. *michigonesis*)	-	*S. peruvianum*	3 QTLs	5, 7, 5	Sandlerink *et al.* (1995); Van Heuseden *et al.* (1999)

Patho-gens	*Genotypes/Cultivars*	*Resistant Varieties*	*Wild Resistance Source*	*Resistance Gene*	*Chromosomal Location*	*Reference*
	Bacterial spot (*Xanthomonas vesicatoria* (Doidge) Dye)	'108059/1XYF 145', '108059/3XYF 145'	-	-	-	-
	Bacterial spot (*Xanthomonas* species)	-	*S. pimpinellifolium*; *S. pennellii*	Rx1-3	-	Yu *et al.* (1995); Yang and Fransis (2007)
Viral	Mosaic (tobacco-mosaic virus)	'Estrelia', 'Sonato'	-	-	-	-
	Leaf curl (tobacco virus 10)	'B 2247', 'A Silvestra', Hissar Anmol, Hissar Gaurav, H-86	*S. hirsutum* f. *glabratum* 'B6013', *S. pimpinellifolium* A 1921	-	-	-
	Spotted wilt virus (Tospo virus species)	'Key to the earliness', 'Pearl Harbour	*S. cheesmanii*	Sw-5	-	Hassan *et al.* (1984)
	Spotted wilt virus (Tospo virus species)	-	*S. peruvianum*	Sw-6	-	Stevens *et al.* (1995); Rosello *et al.* (1998); Gordille *et al.* (2008); Stevens *et al.* (2007)
	Spotted wilt virus (Tospo virus species)	-	*S. chilense, S. peruvianum, S. pimpinellifolium*	Sw-7	-	-
	Spotted wilt virus (Tospo virus species)	-	*S. habrochaites, S. corneliomulleri*	-	-	Dianesc *et al.* (2011)
	(Tospo virus species)	-	*S. arcanum*	-	-	-
	Bemisia argentibolii	-	*S. pennellii*	4 QTLs	2,3,6,8,9,10,11	Heinz and Zalon (1995)
			S. peruvianum	-	-	Snyder *et al.* (1998); Nombela *et al.* (2000); Kennedy (2007); Momotaz *et al.* (2010)
	Trialeurodes vaporarioum	-	*S. habrochaites*	2 QTLs (reduced oviposition)	1, 12	Maliepaard *et al.* (1995)
			S. pennellii	3 QTLs	1, 5, 9	Berlinger *et al.* (1991), Kennedy (2007)
	Mosaic (Potato virus X)	'Ukrainskil Teplichnyi 285'				
	Curly top (Curly top virus)	'Ojo de Venado', ' Red Peach'				
	Mosaic (Potato virus Y)	'Angela'				
	Tolerant to tomato leaf curl virus and tomato mosaic virus	EC468337-41 (AVRDC, Taiwan) received by IIHR				
	Resistant to tomato mosaic virus and tomato leaf curl virus	EC497213-21(AVRDC, Taiwan) received by Hort. College and Research Institute, Coimbatore, T. Nadu				
	Resistant to tomato yellow leaf curl virus, tomato mosaic virus, race 1-2 (Various Begomo virus species)	EC510445-450 (AVRDC, Taiwan) received by ICAR-NBPGR-RS, Hyderabad (Telangana)	*S. pimpinellifolium*	QTL	-	Perez de Castro *et al.* (2007); Scott (2007)
			S. peruvianum	5 recessive genes	-	Pilowsky and Cohen (1990)

Patho-gens	*Genotypes/Cultivars*	*Resistant Varieties*	*Wild Resistance Source*	*Resistance Gene*	*Chromosomal Location*	*Reference*
			S. chilense	*Ty-1*	6, 3	Zamir *et. al.* (1994), Ji *et al.* (2008)
				Ty-3		
				Ty-4		
			S. habrochaites	*Ty-2*	11	Vidavsky and Czosnek (1998), Hanson *et al.* (2006)
			S. peruvianum	*Ty-5*	4	Anbinder *et al.* (2009)
			S. chilense	*Ty*LCV1	-	
				*Ty*LCV2	-	
				*Ty*LCV	-	
				Ty-6-1	-	
			S. cheesmanii	*Ty10-1*	-	Hassan *et al.* (1984)

Thirthamallappa *et al.* (1997) reported that tomato genotypes IHR 1939 (*S. pimpinellifolium* L4394), IHR 1816 (*S. esculentum* NCEBR-1) and IHR 1954 (*S. esculentum* var. *cerasiform* L128) showed resistance at both seedling and adult plant stages. Of these five resistant lines, IHR 1939 derived from *S. pimpinellifolium* L4394 displayed higher level of early blight resistance at both seedling (42.91 per cent CDI) and adult (37.86 per cent CDI) plant stages. Cultivar Taichung Asveg-4(F), BT-10(B), LE 79-5(B), BWR-5(B), Sunrise, Cascade, Mountain Spring, Sunbeam, (V.F) hase been reported resistance against wilt diseases (Lin and Hong, 1989; Kalloo *et al.*, 2002; Bushway, 2008).

Tomato leaf curl virus (ToLCV) is the most destructive disease of tomato causing fruit yield losses up to 90-100 per cent. Most of the cultivated tomato varieties/cultivars are susceptible to ToLCV. There are varieties which are resistant, like 'Pant Bahar' for *Verticillium* wilt and *Fusarium* wilt; 'BT-1', 'BWR-5', 'LE-79-5' and 'Arka Abhijit', 'Arka Shreshta', 'Arka Alok', and 'BT-10' and 'H-24' and 'H-36' for ToLCV are notable. The ToLCVresistance genes have been introgressed into cultivated tomato varieties mostly from its two wild relatives *Solanum chilense* and *S. habrochaites*. There are six resistance genes *Ty-1, Ty-2, Ty-3, Ty-4, Ty-5* and *Ty*-6 which are being practically utilized in breeding programmes. Seven *S. chilense* accessions (LA1938, LA2773, LA2748, LA1930, LA2879, LA2748, LA2778, LA1932) were screened under field conditions to find a new source of resiatance. All the accessions were found resistant to ToLCNDV. *Ty-1, Ty-1/Ty-3* linked five each of sequence based markers (WU_M17, WUR_M25, WU_M27, WU_M29, and WU_M31) and five SCAR markers (P6-25-F2, FLUW-25F, UF_TY3-P19, Ualty3a, Ualty3b) were validated with all the accessions for identification of novel allels for ToLCV resistance breeding programme. The markers WU_M29, P6-25-F2, FLUW-25F, UF_TY3-P19, Ualty3a,

Table 2.19: Reaction of different Species/Genotypes/Lines to Bacterial Wilt

Resistance to Diseases	*Name of Genotypes*	*References*
Resistance to Bacterial wilt (*Peudomonas solancearum*)	*L. pimpinellifolium* (PI 127805), 'La Bonita', 'CL 32d-0-1-19GS (but Celine, 1981 proved only tolerant), Oxheart' and 'CRA 66 Selection A'	Chetia and Kakati (1973), Rao *et al.* (1975), Badekar (1977), Ramachandran *et al.* (1980)
Resistance to Bacterial wilt (*Peudomonas solancearum*) race K 60, 126408-1 and Tifton 80 of race 1.	CL 32d-0-1-19GS	-
Resistance to Bacterial wilt (*Peudomonas solancearum*) isolates 126408-1.	'Venus'	-
Moderate susceptibility to *P. solancearum*	'Best of All' and 'Marglobe Supreme'	-
Withstand only pathogenic isolates and their mixture of *P. solancearum*	'Saturn' and 'PI 303811'	-
Resistance to isolates and their mixture of *P. solancearum*	'VC 9 1 UG' and 'VC 11 1 UG	-
Resistanceto Bacterial wilt (*Peudomonas solancearum*) carrying multiple recessive genes.	'CRA 66 Selection A'	-
Resistance to Bacterial wilt	'LE 79' ('Sakthi') is derivative of 'CL 32d-0-1-19GS	'Rajan and Peter, 1986

and Ualty3b were polymorphic among the accessions. In accessions LA2879 and LA1969, marker P6-25-F_2 showed that two alleles of *Ty-1* and *Ty-3* were present as compared to presence of source of single gene *Ty-1/ Ty-3* in other accessions. The marker and field screening results showed that accessions LA1938, LA2778, and LA2879 carries novel alleles resistant to ToLCNDV, potential source for introgression of ToLCNDV which can serve as resistant gene into commercial tomato varieties (Muthukumar *et al.*, 2016). *Pyl* gene conferring resistance to the fungus *Pyrenochaeta lycopersici* (Laterrot, 1978) but the *Tm-2* (Laterrot and Pecant, 1969) and *Tm-2* gene (Hall, 1980) confering resistance to tomato mosaic virus (ToMoV) used long back. *Ty-5* gene confers resistance to the tomato yellow leaf curl virus (Anbindes *et al.*, 2009). Rahman *et al.* (2002) evaluated tomato varieties for leaf curl disease of tomato and found that the resistance in varieties 'Bilah-2' and 'H-86' should be exploited in tomato improvement programme. Kulkarni and Madalangeri (2002) evaluated and reported on correlation in tomato for leaf curl virus. They found that the parents (H-36, H-86 and NDT-VR-60) and three hybrids (H-36 x NDT-VR-73, H-36 x H-86 and H-36 x L-15) were found resistant as compared with susceptible check Punjab Chhuhara.

Table 2.20a: Sources of Resistance in Wild Relatives of Tomato

Wild Relatives	*Resistance to*
Solanum pimpinellifolium (Pearl Harbor)	Spotted Wilt
S. peruvianum	Tomato Leaf Curl Virus
S. pimpinellifolium	Tomato Leaf Curl Virus
S. hirsutum f. glabratum B6013	Tomato Leaf Curl Virus
S. pimpinellifolium (LA 121)	Yellow Leaf Curl Virus
S. cheesmanii	Yellow Leaf Curl Virus
S. hirsutum f. glabratum B6013	Yellow Leaf Curl Virus
S. peruvianum	Yellow Leaf Curl Virus
S. peruvianum PI128655	Tomato Yellow Top Virus
S. peruvianum PI128655	Big Bud Disease

Sadashiva (2002) reported the breeding work of tomato for combined resistance to bacterial wilt and tomato leaf curl virus. He reported that most of the hybrids were with combined resistance to TLCV and BW were TLBR-3 x IIHR-2200 (3.88 kg/plant), TLBR-4 x IIHR-2200 (3.58 kg/plant), where as TLCV resistant hybrid Avinash-2 reported 3.3 kg/plant in bacterial wilt free soil but succumbed to complete wilt with 1.11 kg/ plant in yield during first and third seasons, respectively. This work has resulted in identification of two F_1 hybrids 'Arka Shreshtha' and 'Arka Abhijit' during 1994-95 and 'Arka Ananya' another F_1 hybrid with combined resistance to LCV and BW during 2005. Similarly in case of viruses some specific observations were made time to time are as under:

Table 2.20b: Resistance Gene and their Genetics

Virus	*Host*	*Resistance Genes*	*Genetics of Resistance*	*Reference*
Tomato mosaic	Tomato	Tm-2	Single dominant gene	Pelham, 1972
Tomato spotted wilt	Tomato	Sw-2, Sw-3, Sw-4	Recessive genes	Finlay, 1953

S. peruvianum and *S pimpinellifolium* were resistant to spotted wilt virus at MPKV, Rahuri (Joi and Summanwar, 1986). The accession *S. peruvianum* (PI126944) in particular has been described as resistant to TwSV (Paterrson *et al.*, 1989), tobacco mosaic virus (TMV) (Yamakawa and Nagata, 1975), tomato leaf curl virus (ToLCV) (Muniappa *et al.*, 1991) and to *Fusarium oxysporum* f. sp. *radices lycopersici* (Rowe and Farley, 1981). In order to better exploit the potential breeding for resistance of PI126944, Julian *et al.* (2013) were able to develop a set of introgression lines with combined resistance to TLCV and Tswv in the genetic back ground of tomato. Genotype Ohi M-29 is resistant to tomato mosaic virus (Gemini virus). Among 16 distinct virus species in the Tospovirus genus, four species including Peanut bud necrosis virus (PBNV) causing necrosis disease in tomato have been reported in India. Identification of stable sources and further utilization of wild relatives as gene sources to increase levels and diversity the bases of resistance may offer good management for the disease. A total of 13 wild species of tomato (*S. peruvianum*), two *S. pimpinellifolium*, one *S. chilense*, one *S. pennellii* and three check cultivars (*S. lycopersicum*), along with two cultivars (*S. lycopersicum*) having the Sw-5 and Sw-7 genes were evaluated under field conditions during three consecutive seasons (June to October 2008, July to December 2009, August 2010 to February 2011). Among all; a high degree of field resistance (>80 per cent) was detected in seven lines of *S. peruvianum* (L00735, L00671, L00887, L06138), *S. chilense* (TL02226) and *S. pimpinellifolium* (L03708, TL02213) lines. The field data was also supported by negative reaction against a polyclonal antiserum of the nucleocapsid protein (N) of Peanut bud necrosis (PBNV) in direct antigen coating-enzyme linked immunosorbent assay (DAC-ELISA). The cultivars with Sw-5 and Sw-7 genes were highly susceptible to PBNV (Sain and Chadha, 2016) and detail is given below (Sain and Chadha, 2016).

At CCS HAU, Hisar in screening experiment reported completely resistant among wild *S. hirsutum* f. *glabratum* whereas *S. peruvianum*, *S. pimpinellifolium* and *S. hirsutum* and among cultivated HS-101 and Punjab Chhuhara were resistant to leaf culr virus (Banerjee and KAlloo, 1991a). In another study Saikia *et al.* (1986) *S. hirsutum* (LA386, LA-1777, PI390513, *S. glandulosum* (EC68003) and *S. peruvianum* PI127830 and PI127831 were resistant. Mayee *et al.* (1974) reported that HS-110, HS-102, Nematex, T1 and Nova as tolerant to leaf curl

Table 2.21: PBNV Incidence in Tomato Entries Evaluated during 2008, 2009 and 2010-11

Entry	Species	Percent disease incidence[a]									
		Year 2008			Year 2009			Year 2010-11			Pooled Mean
		48 DAT[b]	89 DAT	AUDPC[c]	50 DAT[b]	75 DAT	AUDPC[c]	90 DAT[b]	120 DAT	AUDPC[c]	
L00671	*S. peruvianum*	04.2 (06.90)	17.5 (24.61)	16.26	13.9 (21.9)	17.8 (24.9)	18.39	2.38 (8.9)	14.29 (22.2)	16.67	16.53 (23.99)
L00673	*S. peruvianum*	16.7 (19.93)	26.7 (30.29)	19.62	19.0 (25.9)	30.2 (33.3)	22.96	19.05 (25.9)	52.38 (46.4)	10.72	36.43 (37.13)
L00678	*S. peruvianum*	06.7 (08.85)	30.0 (32.21)	21.04	-		-	-	-	-	30.0 (33.21)
L00687	*S. peruvianum*	36.7 (36.14)	43.3 (41.15)	16.02	-		-	-	-	-	43.3 (41.15)
L00688	*S. peruvianum*	21.4 (22.41)	40.5 (30.42)	25.62	-		-	-	-	-	40.5 (39.53)
L00689	*S. peruvianum*	24.1 (24.36)	31.5 (33.74)	29.61	-		-	-	-	-	31.5 (34.14)
L00737	*S. peruvianum*	20.7 (21.49)	47.8 (43.76)	33.12	-		-	-	-	-	47.8 (43.74)
L00735	*S. peruvianum* var. *humifusum*	-	-	-	09.5 (18.0)	09.5 (18.0)	14.25	2.38 (8.9)	9.52 (18.0)	19.05	9.51 (17.96)
L00738	*S. peruvianum*	20.2 (21.93)	33.3 (34.93)	25.51	-	-	-	-	-	-	33.3 (35.25)
L00882	*S. peruvianum*	18.3 (21.07)	36.9 (37.15)	28.93	30.0 (33.2)	37.8 (37.9)	32.11	4.76 (12.6)	16.7 (24.1)	14.29	30.47 (30.50)
L00887	*S. peruvianum*	04.2 (06.90)	35.6 (31.34)	31.41	05.6 (13.6)	05.6 (13.6)	10.09	4.76 (12.6)	16.7 (24.1)	14.29	16.53 (23.99)
L00890	*S. peruvianum*	20.0 (21.93)	23.3 (28.07)	26.32	-	-	-	-	-	-	23.3 (28.86)
L06138	*S. peruvianum*	10.7 (15.52)	10.7 (15.52)	39.15	16.7 (24.1)	22.2 (28.1)	20.88	14.29 (22.2)	28.57 (32.3)	3.57	20.49 (26.92)
BL1022 (Sw-5)	*S. lycopersicum*	76.7 (66.15)	100 (90.00)	63.65	84.1 (66.5)	94.4 (76.4)	75.83	59.52 (50.5)	85.71 (67.8)	47.62	93.37 (75.08)
CK-12 (Sw-7)	*S. lycopersicum*	56.7 (48.93)	92.9 (77.37)	78.58	-	-	-	-	-	-	92.9 (74.55)
TLB182 (SC)	*S. lycopersicum*	92.2 (77.37)	100 (90.00)	84.19	80.2 (63.6)	90.5 (72.0)	72.36	54.76 (47.7)	64.29 (53.3)	34.52	84.93 (67.16)
L05789	*S. pennellii*	-	-	-	-	-	-	47.62 (43.6)	47.62 (43.6)	22.62	47.62 (43.6)
L03708	*S. pimpinellifolium*	-	-	-	-	-	-	4.76 (12.6)	11.90 (20.2)	16.67	11.90 (20.2)
TL02213	*S. pimpinellifolium*	-	-	-	-	-	-	7.14 (15.5)	16.67 (24.1)	13.09	16.67 (24.1)
TL02226	*S. chilense*	-	-	-	-	-	-	21.43 (27.6)	21.43 (27.6)	3.57	21.43 (27.6)
TLCV15 (SC)	*S. lycopersicum*	-	-	-	-	-	-	33.33 (35.3)	88.09 (69.8)	35.71	88.09 (69.8)
K555 (SC)	*S. lycopersicum*	-	-	-	-	-	-	50.0 (45.0)	71.43 (57.7)	35.71	71.43 (57.7)
CD at (p at 0.05)		**26.93**	**21.77**	-	**26.62**	**27.54**	-	**17.09**	**18.64**	-	**20.18**

a: Value in parentheses are Arc sine transformed values; b: DAT= days after transplanting; c: AUPC= area under diseases progress curve.

but these varieties were highly susceptible under hisar conditions (Kalloo and Banerjee, unpublished data). Out of 166 lines only three lines *i.e.* PI3 (2247), XXXII-354-A Silvestra and 65537 exhibited fair degree of tolerance and *S. peruvianum* and *S. chilense* exhibited resistance. Gene identified in *S. pimpinellifolium* is controlled by single incomplete dominance (Banerjee and Kalloo, 1987a) where as in *S. hirsutum* f. *glabratum* it was controlled by two complementary factors (Banerjee and Kalloo, 1987b). Resistance in *S. pimpinellifolium* and *S. hirsutum* f. *glabratum* is linked with small fruit size and late maturity (Banerjee and Kalloo, 1989c).

Sadashiva (2014) able to introgress *Ty-2* gene from *S. habrochaites* into bacterial wilt and early blight

Table 2.22: Sources of TYLCV Resistance (Ty genes) in Tomato

Sources	*Species*	*Gene/alleles*	*Chromosome*	*Reference*
LA1069	*Solanum chilense*	Ty-1	6	Zimir *et al.*, 1994
B6013	*S. habrochiates*	Ty-2	11	Hanson *et al.*, 2013
LA1932	*S. chilense*	Ty-3, Ty-3a, Ty-3b	6	Ji *et al.*, 2007a
LA2779				
LA1932	*S. chilense*	Ty-4	3	Ji *et al.*, 2009
PI126926				
PI126930	*S. peruvianum*	Ty-5	4	Anbinder *et al.*, 2009
PI390681				
Ty King	*S. lycopersicum*	Ty-5	4	Hutton et al., 2012

Hanson *et al.*, 2013.

resistance lines through marker assisted selection and was successful in developing two tomato hybrids 'Arka Rakshak' and 'Arka Samrat' with triple disease resistance to TyLCKav, bacterial wilt and early blight. Muthukumar *et al.* (2016) screened *Solnum habrochaites* accessions for tomato late blight (*Phtophthora infestans*) and found lines LA1223, LA1353, LA1718, LA1777, LA2156, LA2167 and LA2556 are potential source of quantitative resistance (QTLs or QRLs). Three accessions *i.e.* LA1777, LA2167 and LA2556 showed resistance against late blight and out of three LA1777 showed a high level resistance with a PDI of 22.29, which could be potential donor for the late blight. Among the wild relatives of tomato, *S. peruvianum* is considered as the most variable. Many genes have been introgressed into tomato. Such as *Sw-5* gene that confers resistance to tomato spotted wilt virus (*TSWV*) (Stevens *et al.*, 1992).

By utilizing leaf curl tolerant lines from World Vegetable Centre (AVRDC), Taiwan, two most promising

Table 2.23: Trait Specific Introduction of Tomato Genotypes

Specific Traits	*Accessions*	*Country*	*Plant Habit*
TMV resistance	EC671595-96	Taiwan	Determinate
TMV resistance	EC675836	Taiwan	Determinate
TMV resistance (strain 1,2,3 and 4)	*S. chilense*, Ohio MR-9		
TMV, FW-2 resistance	EC671597-98	Taiwan	Determinate
BW, TMV, resistance	EC675830	Taiwan	Determinate
BW, TMV, FW-1 resistance	EC675831, EC675833	Taiwan	Determinate
GLS, BW, TMV, FW-1 resistance	EC675832	Taiwan	Determinate
TMV, LB, FW-1, FW-2, ST resistance	EC675834	Taiwan	Determinate
Resistant to GLS, BW, FW-1,TMV, Gemini virus (Ty2)	EC675835	Taiwan	Determinate
Resistant to BW, virus and Blight	EC678814-15	Taiwan	Determinate
Resistant to virus, leaf mould and Blight	EC678818	Taiwan	Determinate
Resistant to TYLCV	EC687094	Taiwan	Determinate
Resistant to TYLCV	EC687095, EC687096, EC687098	Taiwan	Determinate
Resistant to TYLCV, TMV, FW-1	EC687099	Taiwan	Semi-determinate
Resistant to Gemini virus, (Ty2), TMV, FW-1	EC687100, EC687101, EC687102, EC687103	Taiwan	Semi-determinate
Resistant to Gemini virus, (Ty2), TMV, FW-1	EC687104, EC687105,	Taiwan	Indterminate
Resistant to TMV, BW and FW-1	EC687106	Taiwan	Semi-determinate
Resistant to Gemini virus, (Ty2), TMV, FW-1	EC687107	Taiwan	Indterminate
Resistant to TMV, BW, FW-1	EC687108	Taiwan	Indterminate
Resistant to Gemini virus, (Ty1, Ty2, Ty3), TMV, FW-2	EC690981	Taiwan	Determinate
Homozygous, resistance for Gemini virus, (Ty1, Ty2, Ty3), TMV, FW-2	EC690982	Taiwan	-
Homozygous, resistance for Gemini virus, (Ty1, Ty2, Ty3), TMV, homozygous allele sysceptible to FW-2	EC690983	Taiwan	Semi-determinate

Specific Traits	Accessions	Country	Plant Habit
Homozygous allele resistance for Gemini virus, (Ty1, Ty2, Ty3), TMV, FW-2	EC690984	Taiwan	Determinate
Homozygous allele resistance for Gemini virus, (Ty1, Ty2, Ty3), TMV, FW-2	EC690985	Taiwan	Semi-determinate
Homozygous allele resistance for Gemini virus, (Ty1, Ty2, Ty3), TMV, FW-2	EC690986	Taiwan	Determinate
Homozygous allele resistance for Gemini virus, (Ty2, Ty3), TMV, homozygous allele sysceptible for Ty-1 and FW-2	EC690987	Taiwan	Determinate
Homozygous allele resistance for Ty2, Ty3 and TMV, homozygous allele sysceptible for Ty-1 and FW-2	EC690988, EC690989	Taiwan	Determinate
Homozygous allele resistance for Gemini virus, (Ty2, Ty3) and TMV	EC690990	Taiwan	Determinate
Resistant to gray leaf spot, BW, FW-1, TMV, and *begomo* virus	EC692274	Taiwan	-
Resistant to BW, FW-1 and TMV	EC692275	Taiwan	
Resistant to FW-2, gray leaf spot and early blight	EC692277	Taiwan	

hybrids *viz.* CLN2777G x BCT-59 and CLN2777A x BCT-82P have been identified for higher yield, better quality trits and possessed low PDI values for toLCV diseses at BCKV, Kalyani (West Bengal). Similarly other resistant sources were identified for further use in breeding programme.

Table 2.24: Source for Virus Resistance in Tomato

Leaf curl virus resistant	*S. pimpinellifolium*, A-1921, and *S. hirsutum* f. *glabratum*, B6013, Hisar Anmol, H-36, H-86, H-88, CLN-2123A, CLN-2116-B, CLN-21E, CLN-5915-206, AC-400, AC-441
Yellow leaf curl virus	*S. pimpinellifolium, S. hirsutum* f. *glabratum, S. peruvianum, S. cheesmanii and S. cheesmanii* sub sp. *minor* LA-1401

ToLCV resistance genotypes had more amount of total phenol than susceptible genotypes (Som and Choudhury, 1976; Banerjee and Kalloo, 1989a). Similarly susceptible genotypes had lower content of non-reducing sugar than resistant cultivars. ToLCV resistance genes have been transferred from *S. hirsutum* f. *glabratum* and *S. pimpinellifolium* to *S. esculentum* through back crossing method. From *S. hirsutum* f. *glabratum* background desirable gene has been transferred in the cultivated varieties and resistant lines like H-2, H-11, H-17, H-23 and H-24 have been developed (Kalloo and Banerjeee, 1990c, d). Tolerance to ToLCV was transferred from *S. pimpinellifolium* to cultivated varieties and tolerant lines like LCP-22, LCP-2, LCP-3 and LCP-9 have been developed (Kalloo and Banerjee, 1990b). In Cuba, Olimpia Gomez Consuegra *et al.* (2015) were succeeful in pyramiding *Ty-1* gene from *S. chilense* (LA1969) and *Ty-3* gene from *S. chilense* (A2779) and spotted wilt virus resistance gene *Sw-5* from *S. peruvianum* with the help of early and simultaneous molecular screening for resistance and productivity evaluation under field conditions.

(ii) Insect Pests Resistance Source and its Utilization

Several wild tomato species has been found potential resistance source against insect but *S. habrochaites* and *S. pennellii* are being utilized greatly. In *S. habrochaites* is conferred by trichomes IV and VI, while in *S. pennellii* the resistance is conferred by trichome of type IV. Trichomes of type IV have also been identified in *S. pimpinellifolium* (Fernandez Munmoz *et al.*, 2003). Resistance up to 16 pest species has been reported in accessions of *S. habrochaites*. One accession, PI134417 has been intensively studied, being resistant to some 12 arthropod pests of cultivated tomato. Similarly resistance against 9 arthropods has been reported in *S. pennellii*. In this species one accession LA-716, being resistant to eight of these pests. For above resistance factor in tomato or any crops is the 2-tridecanone which is found in the tips of Type VI glandular trichomes (of *S. habrochaites* var. *glabratum* (Williums *et al.*, 1980). This compound is highly toxic to many pests such as *Tetranychus urticae* (Snyder *et al.*, 2005), *Bemisia tabaci* (Channarayappa *et al.*, 1992) and *Trialeurodes vaporariorum* (Malieppaard *et al.*, 1995) and act as a repellent to others. Acc. No.-128, Acc. No.-133, Acc.-145 and Heinz-137 and 'Sabour Prabha' were less affected by fruit borer (*Heliothis armigera*). In wild resistance species *S. hirsutum* f. *glabratum* was found highly resistant to fruit borer (Kashyap *et al.*, 1982; Kashyap and Varma 1986). Leaves of *S. hirsutum* and *S. hirsutum* var. *glabratum* contained a high concentration of an ethanol extractable, apparently recessive inherited antibiotic factor which resulted in 100 per cent mortality when added to the artificial diet of *Heliothis armigera* larvae. This factor may be a useful source of resistance to instar feeding in *S. esculentum* since the species is cross compatible with *S. hirsutum*. 'Pusa Uphar' is resistant to fruit borer. *S. hirsutum*, and *S. hirsutum* var. *glabratum* both are resistance to leaf miner and fruit borer.

Table 2.25: Resistance Source against Insects in tomato

Insect Pests	Resistant Source
White fly resistant	IVT74453 (*S. hirsutum*)
Fruit borer resistant	*S. hirsutum*, *S. hirsutum* var. *glabratum*, EC110264, EC101552, EC124406, EC129599, EC129597, EC129698, EC129791, EC121451, EC160191, EC161254, EC119121, PI126449 (*L. hirsutum*), L-633, EC129354 and EC124406
Leaf miner resistant	*S. hirsutum*, *S. hirsutum* var. *glabratum*

(iii) Root-knot Nematode Resistance

In general *Meloidogyne incognita*, *M. javanica* and *M. arenaria* attack to tomato crop. *S. peruvianum* was reported as source of resistance to nematode. The *Mi* gene is associated to resistance to the main species of nematodes of the genus *Meloidogyne* (*Gilbert, 1958*). Now resistant gene *Mi* of *S. peruvianum* has been transferred into several cultivars. Some resistance genotypes like 66N, 569N-10, Monalucie, Atkinson, LP-1 and LP-2 strains of *S. peruvianum* has been isolated (Singh and Choudhury, 1972). Similarly, Acc. -7540, Heelani and Sl.-12, Anahu, NIR-1 (Dhillon and Nandpuri, 1975), Resistant Bengaluru and Nematex (Jain *et al.*, 1983) are also resistant. The resistance in Nematex and Resistant Bengaluru (Karnataka) was controlled by single dominant gene (Kalloo *et al.*, 1978). Utilizing Nematex and Resistant Bengaluru resistance source, few resistance varieties like NRT-3, NRT-8 and NRT-12 have been developed. Out of these 'Hisar Lalit' (NRT-8) has been identified/released for cultivation in root knot nematode infested areas of Haryana (Jain *et al.*, 1990). Using parents like NRT-3, NRT-8 and Sel.-12 were crossed with root knot nematode susceptible varieties promising hybrids like HS-110 x N-98, Druzuba x NRT-8 and Columbia x NRT-3 exhibiting high heterosis for yield and resistance (Jain *et al.*, 1991). In tomato, genotype 'Anahu' was released from the Hawaiian Agricultural Experiment Station which proved resistance to root knot nematode in 1955. Rao and Tikoo (1975) assessed root-rot indices after 10 weeks of inoculation and reported 'Pelican', 'Hawaii 7746' and 'Hawaii 7747'to be resistant and 'Pusa Ruby' and *S. pimpinellifolium*, highly susceptible. Singh and Choudhury (1974) screened 162 cultivars against *M. incognita*, *M. arenaria* and *M. javanica* of which 'Nematex', 'VFN 8', '65N215-1', '65N255-1' and 'Sl. 120' showed a high degree of resistance. 'Nematex' was not infected by any of the 3 species. 'VFN 8', '65N215-1', and '65N255-1' were not infected by *M. Incognita* or *M. arenaria* but *M. javanica* produced a few galls. 'Sel. 120' was infected by all three *i.e. M. incognita*, *M. arenaria* and *M. javanica* but the number of galls produced were very less. According to Mahajan *et al.* (1975), F_1 hybrid having 'Nematex' as a parent showed high degree of resistance to *M. incognita*. Sidhu and Webster (1973) reported that 'Nematex' and 'Small Fry' possess a single dominant gene for resistance while 'Cold Set 1' possesses a single recessive gene. The symble Lmi R-1, Lmi R-2, and Lmi R-3 were proposed for the resistant gene in 'Nematex', 'Small Fry' and 'Cold Set 1', respectively. Singh *et al.* (1974) reported that resistance to *M. incognita* and *M. javanica* is controlled in each cross by a dominant gene. Genotype EC6788818 introduced from Taiwan is found resistant against root knot nematode. As a result of screening 181 genotypes for resistance to root knot nematodes, the genotypes ' Pelican', 'NMR-1', 'VFN Bush', 'New Yorker', 'Healani', 'Kalohi', 'Small Fry', 'West Virgina', 'LA-121', EC118270', 'EC118276', 'EC118277' and 'Hybrid Karnataka' were found to be resistant (Mahajan *et al.*, 1979). 'Pusa-120, 'Pusa Hybrid-2' and 'Pusa Hybrid-4' are resistant to root knot nematode. Smililarly, potential geotypes like Ronita, Rossol, Sanatacruz, Venus, Satumo Atkinson, Manalucie, *S. peruvianum*, Pusa-120, PNR-7, NRT-3, NRT-8, NRT-12, TH-2312 and TH-802 are resistant to devastating nematodes *M. incognita*, *M. javanica* and *M. arneria*.

Sources for Abiotic Resistance and its Utilization

Tolerance at one stage of plant development is not necessarily correlated with tolerance to at other stage (Foolad and Lin, 1977; Foolad, 2007). Screening for stress tolerance should be done at specific ontogenetic stages such as seed germination seedling survival, initial plant growth, vegetative growth and reporoduction. Genetic variability for heat, drought and salt tolerance in cultivated tomato is limited as the genomes of cultivated tomato contain only about 5 per cent of the genetic variation of their wild relatives. In tomato several genotypes/varieties identified and selected against as a source for physiological stress. Drought resistance tomato species are *Solanum cheesmanii*, *S. pimpinellifolium*, *S. chilense* and *S. penelli* are quite usefull. For low temperature and superior seedling growth at 0°C was introgressed from high-altitude strains of *S. hirsutum*.

Table 2.26: Genotypes/Cultivars Resistant to Abiotic Stress

Abiotic Stress	Resistant Genotypes/Cultivars
Fruit cracking	'Pelican', Florida 1011'
Catface	'K 7' and 'K 10'
Puffiness	'K 7' and 'K 10'
Internal browing and greywall (High or low light intensity)	'W 1' and'W 2'
Fruit Pox	'Walter PF'
Gold fleck	'Walter PF'
Drought injury	'Earliana'
Green back	'Gamet' and 'Grenadier' (F1)
Salt injury	S. cheesmani
Sunscaled	'VF 145 F5' and 'Vivid'
Ozone injury (Higher ozone content in the atmosphere)	'Heinz 1639', New Yorker'
Resistance to low and high temperature	EC678817 (Taiwan)

Other species for this trait are *S. peruvianum*, *S. chilense* and *S. hirsutum* f. *glabratum*.

However, stress tolerance genes have been identified in several related species such as *Solanum pimpinellifolium*, *S. peruvianum*, *S. cheesmanii*, *S. habrochaites*, *S. chilense*, *S. chemielewskii* and *S. pennellii*.

Table 2.27: Genetic Control of Mineral Deficiency Resistance in Tomato

Characteristics	*Genetic Control*	*Remarks*
B-inefficiency	btl	Causes brittle stem
Fe-efficiency	Fer	Minor genes involved

Temperature Tolerance

The inheritance of seed germination at low temperature in *S. hirsutum* f. *glabratum* 'B-6013' was controlled by polygenes with dominance for inability to germinate at a mean temperature of 10°C (Kalloo and Banerjee, 12991). Low temperatures adversely affect the growth and development of plants; however, *S. hirsutum* and *S. hirsutum* f. *glabratum* are adapted to such conditions and grow well when there is low temperature of December to February of Northern Plain of India. 'Cold Set', Ostyenkinskij', 'Montfavet 63-4', 'Montfavet 63-5', 'Precoce' and 'Severianin' set fruits in winter at low temperatures whereas in other varieties there was no or very poor fruit set (Gautam *et al.*, 1981; Kalloo and Dudi, 1982a). Nandpuri *et al.* (1975) have reported 'Cold Set', 'Avalanche' and 'Illalihin' to set fruit in the winter of Punjab. However these genotypes showed poor fruit set in January. Several plants in F_2 and *S. pimpinellifolium* cross combination set fruits at 0-2 °C temperature in the month of January under Hisar condition which may be due to interaction of genes from *S. pimpinellifolium* and cultivar HS-101. The cold set gene of *S. pimpinellifolium* might have expressed in the background of cultivated type (Kalloo and Banerjeee, 1990a). Cultivar 'Pusa Rubi' and 'Sweet-72' set fruit at normal temperature (15-20 °C night temperature). There are several AVRDC lines which set fruit at hot temperature (20 °C night temperature). Cold temperature set (15 °C night temperatures) genotypes are Balkan, Cold Set, Oston Ainski and Pusa Sheetal. The reason for poor fruit set is poor pollen dehiscence at low temperature. It has been observed that few cultivars like 'Montfavet 63-4' and 'Montfavet 63-5' produces and dehisces maximum pollens; therefore fruit setting is more (Kalloo and Dudi, 1982a).

The most serious problem of high temperature (about 30°C) is the reduced size of fruits. Generally there is fruit set in heat tolerant lines but development of such fruit is very slow and poor, with the result that fruits remain smaller. Factors associated with fruit set at high temperature are anther tip burning, poor pollen dehiscence, low pollen viability, slow pollen tube growth and poor fertilization. There is a wide variation among genotypes for fruit set under high temperature conditions under studies of Delhi condition including 43 accessions (38-40°C). *S. cheesmanii* has the ability to exhibit adequate flowering at high temperatures besides setting fruits at high temperatures (Kalloo and Banerjee, 1990). *S. chilense* offered good prospects for raising level of heat tolerance in tomato (De lapena *et al.*, 2011). High temperature fruit set in cultivar PS-1 is partially dominant over non-set (Kalloo and Banerjee, 1991a). Varieties HS-101 and HS-102 had the ability to set fruit in April to May when temperature was 35-39 °C. Tewari *et al.* (2002) described a new variety of tomato named 'Pusa Sadabahar' which set fruits at 30 °C of night temperature and suitable for north India. Singh *et al.* (2002) studied the heat stress tolerance in tomato and found 'Lt-2' and 'Lt-26' had grown best under stress and non-stress conditions where as 'Lt-10' and 'Lt-16' were suitable for stress conditions and 'Lt-34' for non stress conditions. Donar germplasm lines have been identified with definite source of resistance, i.e, ability to tolerate temperature stress and drought conditions. These can be used to develop lines possessing resistance one or more stresses. Wild species show remarkable variation in their inherent adaptation to drought stress. *S. pennellii* and *S. esculentum* var. *cerasiforme* are drought tolerant genotypes. *S. hirsutum* is tolerant to cold and drought. EC130042, Sel.-28, EC65992 and *S. pimpinellifolium* (PI205009) require more days to express wilting, thus showing tolerance to drought conditions. Sel.-28, *S. cheesmanii*, K-14, EC104395, *S. pimpinellifolium* (EC65992) and *S. pimpinellifolium* (Pan American) are the potential source for drought tolerance. Similarly, EC130042, *L. cheesmanii* and EC162935 set fruits at high temperatures due to the stigma exertion of less than 1 mm, whereas other sensitive genotypes produced more than 1 mm of stigma exertion. Under Delhi conditions, heat tolerant accessions EC168070, EC130053-1 and EC165393 were recorded to have large fruit (Chandra and Gupta, 1994). Tomato genotypes 'Philippine' (Chaudhary, 1973); Punjab Tropic, Marz, P-4, (Nandpuri *et al.*, 1975); PS, 84#58, *S. cheesmanii*, EC130042, EC162935 (Rana and Kalloo, 1989a) and Land Point (84#42), PS-1, Mini Rose (84#58), PI20509 and *S. pimpinellifolium* set fruits at higher temperature (Kalloo and Banerjee, 1990e). Genotype PS-1 set fruit even at day temperature of 40 to 42 °C. However, fruits remained seedless. At BCKV, Kalyani (West Bengal) from elaborative studies on breeding heat tolerance in tomato, two lines from the genotypes of World Vegetable Centre (AVRDC), Taiwan *viz.* CLN R and CLN B were selecte which showed high promise for tolerance to high temperature stress (Chattopadhyay, *et al.*, 2011). Tolerance against chilling injury has been reported in *S. habrochaites*, *S. chilense* and *S. lycopericoides* (Robertson and Labate, 2007).

Drought Tolerance

Drought tolerance is another factor which affects tomato production in the country. Potential sources of drought tolerance during vegetative growth and reproduction has been identified in *S. chilenese* and *S. pennelli* (Rick, 1982). *S. chilenese* is able to grow in the desert because of its long primary roots and extensive secondary root system. Research programme conducted at ICAR-IIHR, Benegaluru revealed that lines 14-1, 146-2, 383, 553 and 555 were tolerant to drought and gave more yield than standard check. Similarly, at CCSHAU, Hisar genotypes EC130042, Sel.-28, *S. pimpinellifolium* (PI205009) and 84#88 remained green for longer period under drought conditions than those of other lines. Sel-28, rin, 84#76 and JC-14 yielded more than other lines (Rana and Kalloo, 1990a). Wild species *S. pimpinellifolium* (EC65992) and *S. pimpinellifolium* (Pan American) exhibited high drought index of 43 and 39 per cent respectively whereas in control (drought susceptible) drought index was only 4.4 per cent (Kalloo and Banerjee, 1991b). In general drought resistance lines possess more number of branches/plant, higher leaf area/plant and longer root length than the drought susceptible lines. Other associated morphological attributes are leaf rolling and wax on the leaf (Rana and Kalloo, 1989b). Drought resistant genotypes have higher rate of relative water content (RWC), lower rate of water saturation deficit, lower plant canopy temperatures and higher osmotic potential (less negative) than the drought susceptible lines (Rana and Kalloo, 1990c). In drought resistant cultivar Sel.-28 and wild species *S. pimpinellifolium* root length was 60 and 78.6 cm, respectively than the susceptible line Sel.-21 was only 22.3 cm (Rana and Kalloo, 1989b). Drought tolerance has been reported in *S. cheesmanii*, *S. chilense*, *S. pennellii*, *S. pimpinellifolium* and *S. esculentum* var. *cerasiforme* (Foolad, 2007). Drought tolerance at the seed germination stage is a quantitative trait and four QTLs have been identified on chromosome 1, 8, 9 and 12.

Table 2.28: Resistance Source

Drought Resistance	*S. chilense*
Drought tolerance (high yield under weekly irrigation during summer	'J 10-32', 'KT 2' and 'Roma', *L. pennelli*, IIHR-14-1, IIHR-146-2, IIHR-383, IIHR-553, IIHR-555, EC130042, EC65992, EC 520061, Sel-28, Kashi Aman, CLN1621, CO-3, WIR5032, WIR13706, WIR13708
Genetic, water stress, imbalnnce in water uptake	Heinz 1639', New Yorker'

Heat Tolerance

Genetic control of characteristics involved in heat tolerance is copmplex.

Table 2.29: Genetic Control of Characteristics Involved in Heat Tolerance

Characteristics	*Genetic Control*	*Remarks*
Per cent fruit set *†	Polygenic; additive gene	Moderate heritability
Flowers/plant *†	Polygenic	High flower number due to recessive gene; high heritability
Seed set*	Polygenic; dominance effects more important	Epistasis involved
Stigma exsertion*	Polygenic; partially dominant genes	High heritability

*: Under high temperature; †: The two traits are independent of each other.

Twenty one tomato genotypes were evaluated under high temperature stress and resulted limited fruit set. A significant increase in flower drop (19.30 per cent to 74.36 per cent), styler exertion (20.37 per cent to 83.72 per cent), decrease in pollen viability (14.78-74.55 per cent) and fruit set truss-1 (14.45 to 75.37) were noted among genotypes. On the basis of initial screening with respect to fruit set at high temperature stress, three lines *i.e.* CLN-2026, EC538439 and EC538148 emerged as high temperature tolerant genotypes, which can be used as source of high temperature tolerance genes for further breeding programs. Most of the genotypes showed stigma exertion or tip burn under high temperature condition. Fruit number and fruit weight were the important yield components, which were severely affected under high temperature stress and as a result, the yield was markedly reduced. Yield potential of the genotypes, which was dependent on fruit set and fruit weight was severely reduced under high temperature stress and yield varied from 386 g/plant (EC538380) to 1832 g/plant (CLN-2026). Results indicates that the traits like fruit setting, pollen and yield attributes traits could be used in seclection of high temperature tolerant genotypes for better fruit setting (Tiwari *et al.*, 2016). Similarly Yadav *et al.* (2016) evaluated 21 genotypes under high temperature stress and resulted significant mean sum of square due to treatments for all the traits suggested therby the substantial amount of genetic variability. Relative water content (RWC) and membrane stability index (MSI) under heat stress condition was recorded maximum in 'Pusa Sadabahar' (83 per cent and 86 per cent, respectively). Highest proline content was recorded in wild genotypes, like 'Spm' followed by 'Spr-1'. Tolerant genotypes like 'Spr-1' and Spr-2 showed high value of 'chlorophyll b' under heat stress condition as compared to sensitive genotypes. 'Chlorophyll b' may be of prime importance, as it gives a better clue for its specific role in increasing tolerance to high temperature. So relative reduction in 'chlorophyll b' may be utilized as an indicator of down regulated of photosynthetic system in general. There are several lines which are found heat tolerance in nature.

Table 2.30: Heat Tolerant Lines of Tomato

Heat tolerance	EC692274, EC692275, EC692276 (AVRDC, Taiwan), EC41824, EC7764, EC164855, EC169308, EC164653, EC122063, EC37311, EC31515, EC126955, EC130042, EC35220, EC163709, EC164322, EC154660, EC164677, Sparton Red-8, OK-7-2, OK-7-3, Saladette, TAMU Saladette, PI26934, BL-6807, S-6916, C1A-5161, VF-36, Punjab Tropic, Marzano, P-4, Avalanche, Chiko-III, Ment, C-28, Red Rock, Narcaring, Breeding Line-165, Red Cherry, KS-1, KS-2, Walter, *S. cheesmanii*, EC162935, AC-325, AC-326, PS-1, EC1127, EC130042, EC168064
Heat tolerance; received by AAU, Assam; ICAR-IIVR, Varanasi (U.P.); ICAR-NBPGR, N. Delhi	EC501573-85 (AVRDC, Taiwan)

Cold Tolerance

Genetic control of electrolyte leakage reflected in growth under low temperature and in this regard gca is more important than sca. For germination it controlled by additive gene and maternal effctes. For this trait heritability was high (69 per cent at 10°C) in growth chamber. For plant growth, it is controlled by at least three genes. For this trait heritability was high (69 per cent at 10°C) in growth chamber. Several lines of tomato which are cold tolerance in nature are given below:

Table 2.31: Cold Tolerant Lines of Tomato

Cold tolerant	Poter Narcariang, Early north, Puck, Fire Ball, Red Cloud, Cornell Breeding Lines 54-149, 54-179, PI204050, PI280597, Cold Set, Avalanche, Illalihin, Severianin, *L. pimpinellifolium*, Outdoor Gril, Immuna, Pusa Sheetal, Tempo, Oregon, AC-142 and AC-248

Superior seedling growth at 0°C was introgressed into tomato from high altitude strains of *Solanum hirsutum*. Finally, transgenes specific for cold tolerance may be expressed using genetic engineering.

Flood Tolerance

Flood tolerance has been reported in *S. esculentum* var. *cerasiforme*, *S. juglandifolium* (Rick, 1982) and *S. ochronthum* (Robertson and Labate, 2007).

Salinity Tolerance

Germination stage is the most sensitive to salinity. Two mechanisms, *i.e.* osmotic and ionic are associated with the salt resistance. In case of osmotic there is less absorption of salt as a result water stress is caused in the plant. In case ionic, salt is absorbed in large quantity but in this case many time mineral toxicity has been noticed in the plant. Till date no commercial salt resistance cultivar is available in the country.Variety 'Sabour Suphala' and wild species *S. cheesmanii*, and *S. hirsutum* f. *glabratum* are salt tolerant at seed germination stage as they exhibited highest germination at 8 mmhos/cm (Lal *et al.*, 1981). Source of salinity (salt tolerance) has been reported in *S. cheesmanii*, Sabour Suphala, EC251706 and EC151628. Other wild species like *S. pimpinellifolium*, *S. peruvianum*, *S. chilense*, *S. cheesmanii*, *S. habrochites*, *S. chmielewskii*, *S. esculentum* var. *cerasiforme* and *S. pennellii* are also major source for salt tolerance (Tal *et al.*, 1979; Asins *et al.*, 1993; Foolad and Lin 1997; Foolad, 2004; Robertson and Labate, 2007; Rao *et al.*, 2013). *S. pimpinellifolium* having good capacity to form lateral roots, readly cross compatible with S. lycopersicum and with beneficial salt tolerance traits is being utilized for salt tolerance in cultivated tomato (Foolad, 2004). Genetic of minerals like Fe, Mn and Zn uptake at different stages of growth indicated that both additive and dominance variances were responsible for uptake of these elements (Pal and Chattopadhyay, 1986).

Breeding for Processing and Quality Traits

In tomato quality parameters includes size, shape, colour, smoothness and uniformity of fruits, freedom from external defects, flavor, vitamin A and vitamin C contents, processing quality (colour, pH, tritrable acidity, total soluble solids, viscosity) *etc.* Fruit colour and carotene content (latent traits) can be introgressed by *S. hirsutum* and vit. C by *S. pimpinellifolium* and *S. peruvianum*. Randhawa *et al.* (1988) evaluated the tomato varieties for processing and found that 'Punjab Chhuhara' tops the yield as well as in product quality for processing. Tiwari *et al.* (2002) studied the suitability of different tomato varieties for processing and found that PR-3 and PR-H and Arka Ashish were found superior. Sharma *et al.* (2002) also reported the suitability of tomato hybrids for he preparation of paste and found JKTH-895 to be superior to others. Some quality attributes for fresh and processing industry are as shown in Table 2.32.

The morphological quality traits associations in tomato are as shown in Table 2.33.

Other quality traits should be considered during breeding programmes are shown in Table 2.34.

In tomato quality reflects both factors:

(i) External Quality

It includes size; colour and shape which should be homogenous. Apart from these, uniform, intense red colour, lack of cracking, small pedicel scar and flexible skin which permits easy peeling.

Indian bred cultivars like Arka Saurbh, Arka Vikash, Pusa Uphar, and Punjab Chuhara and introduced genotypes like La-Bonita, EC391024 and EC310299 are very much suitable for long distance transportation. These lines are also resorted for processing attributs. Reports for fruit size is given below:

Table 2.32: Quality Attributes for Fresh and Processing Industry

Trait	*Associated with Quality Traits*	*Remarks*
A: For fresh use		
Appearance:		
i. Size	Large	Polygenic
ii. Shape	Round	Oligogenic (allele *O*)
iii. Colour	Red	Allele og^c increases lycopene (red colour); *hp* increases carotenoids and gives excellent colour
Texture and firmness	Firm (for transportation); soft and juicy for fresh consumption	-
Flavour	Preferred	Mainly due to acids and sugars, volatile compounds, *etc.*
Nutritional value	Vitamin A	Affected by og^c and *hp*
	Vitam,in C	Some varieties have up to 50mg/100g fresh fruit
B: Processing quality		
Colour	Red	Determined by colorimeter
Fruit pH	<4.5	Important for effective sterilization
Titratable acidity	High	Expressed as per cent citric acid; polygenic
Total soluble solids	High	Mainly due to sugars; estimated as refractive index
Viscosity	High	Estimated by passing juice through viscometer or as insoluble solids

Table 2.33: Morphological Quality Traits Association in Tomato

Trait	*Associated with Quality Traits*	*Remarks*
Thick fruit skin	Better keeping quality	Not preferred by food processing industry
Indeterminate habit	More total soluble solids (TSS)	Agronomically undesirable
Deep red fruit colour	Lower pro-vit. A (β-caroten)	Preferred by consumers and food processing industry
Larger fruit	Lower total soluble solids (TSS)	Preferred by consumers

Table 2.34: Other Quality Traits in Tomato

Traits	*Preferred Range/Magnitude/Value/Type*
Tomato fruit colour	Red
pH of fruit	4.5 or below
Tritable acidity (total amount of organic acids)	High (0.65 or above)
Peeling	Easy
Fruit firmness	High
Pericarp	Thick
Total soluble solids	High
Consistency (viscosity)	High
Pedicel	Jointless
Flavour	More
Number of locules/fruit	More

Table 2.35: Some External Quality of Tomato

Accession	*Country*	*Traits of Interest*	*Plant Habit*
EC687094	Taiwan	Shape oblong, size medium to medium large, green colour stem	Determinate
EC687095, EC687096	Taiwan	Square round, medium large, green stem	Semi determinate
EC687097	Taiwan	-	Determinate
EC687098	Taiwan	Shape round, size medium, purple stem,	Determinate
EC687099, EC687100, EC687101, EC687102, EC687103	Taiwan	Red colour fruit, oblong shape	Semi-determinate

Accession	Country	Traits of Interest	Plant Habit
EC687104, EC687105	Taiwan	Red colour fruit, oblong shape	Indeterminate
EC687106	Taiwan	Red colour fruit, oblong shape	Semi-determinate
EC687107, EC687108	Taiwan	Orange colour fruit, oblong shape	Indeterminate
EC690981	Taiwan	Uniform shoulder, square round shape	Semi-determinate
EC690982	Taiwan	Uniform shoulder, square round plum shape,	Determinate
EC690983	Taiwan	Green shoulder, uniform	-
EC690984	Taiwan	Green shoulder, oblong shape	Semi-determinate
EC690985	Taiwan	Green shoulder, oblong shape	Determinate
EC690986	Taiwan	Uniform shoulder, square shape	Semi-determinate
EC690987	Taiwan	Uniform shoulder, square shape	Determinate
EC690988	Taiwan	Uniform shoulder, plum shape	Determinate
EC690989	Taiwan	Uniform shoulder, oblong shape	Determinate
EC690990	Taiwan	Uniform shoulder, plum square oblong shape	Determinate
EC692276	Taiwan	Fruit large size and globe shaped	

Table 2.36: Promising Accesions

Attributes	Promising Accesions
Large size fruit	EC163708, EC41340, EC267726, EC267728, EC231751, EC168070, EC267725, EC367857, EC361674, EC362945 and EC367856.

(ii) Internal Quality

Internal quality can be divided into:

(a) **Organoleptic quality** is related to flavor (sweetness, acidity, *etc.* which is determined by organic acids and reducing sugars), olfactory (aroma and scent) and tactile aspects (firmness, texture, *etc.*).

(b) **Nutritional quality** (TSS-4-6 °brix, ascorbic acid, pH below 4.4 *etc.*)

In tomato 400 volatile compounds have been identified and among them, approximately 30 are responsible for the aroma of fresh tomato. There are several genotypes which are known for quality traits. Varieties 'Chiku Grande', 'Roma' and 'Italian Red Pear' (Roy *et al.*, 1970), 'Ruth', 'Early Chatham', 'San Marzano' and 'Red Top' (Beer and Rane, 1976) and 'HS 110' (Pathak and Mahajan, 1978) were suitable for processing into juice or for canning of whole peel tomatoes. Based on juice yield, paste yield, colour, appearance, acidity level and lycopene content, 'EC129604 75/122, 'EC 129606/ARND and 'Selection 4' were the best for paste making (Madaiah *et al.*, 1986). For puree preparation 'Indo Process II' was the top ranking followed by 'Indo Process III' and 'Rupali' (Sethi and Anand, 1986). On the basis of organoleptic ratings 'HS 101' and 'HS 101' were very much suitable for paste and ketchup. Cultivars 'Roma', '2466-27', 'Punjab Chhuhara' and 'Sel. 152' are considered for processing types. In processing types IIHR, Bangalore has developed 'IIHR 674 SB' (UC82B) and 'Sel. 11', ISI (1967a) prescribes a minimum TSS of 25 per cent and acidity of 1.20 per cent as acetic acid in tomato ketchup to satisfy its standards. Genotype 'HW208F' with the highest ketchup yield, low energy cost, and high consistency and were more suited for paste making due to their bright red colour, since colour decides the paste quality. F_1 of cross combonation 'Sakthi' x 'HW 208F' given higest yield and consistency which hold promise as the best material for ketchup and paste. Similarly, cross combinations 'LE206' x 'Ohio 8129' and 'LE206' x 'St. 64' were found superior for quality due to enhanced colour. High ascorbic acid cultivars are Double Rich, Hisar Arun, Kashi Vishesh, Cheti Tomato, CLN-1621 and Punjab Berkha-2.

Table 2.37: Quality Traits and its Sources

Source for Quality Traits	Species and Genotypes
Rich source of ascorbic acid and carotene content	*S. pimpinellifolium*
Source of high ascorbic acid	*S. peruvianum*, Double Rich, Hisar Arun
Source of high total soluble solids (TSS)	*S. cheesmanii*, A-276, A-408 and, WIR-4360, Cheti Tomato, Red Cherry, Vaibhav
Source of high total soluble solids (TSS)	*S. chmielewskii*
High Alcohol insoluble solids (AIS)	VF-109 and 9039M.
Red colour (Lycopene and β-carotene)	High Pigment and Crimson, Cheti Tomato, Vaibhav, Punjab Berkha-2, Uttkal Pallavi, Punjab Chhuhara, Kashi Aman
Pro-vitami (β-carotene)	Caro Red
High volatile compounds (flavor characteristics)	Campbell-146, Rick High Sugar, Cherry and LA-1563

Genotype AC238 and CO-3 were found rich source for ascorbic acid. Lycopene levels in F_1s of *'rin'* (ripening inhibitor) were significantly higher than those of *'nor'* (non-ripening) F_{1s}. Heterobeltiosis for extended shelf-life was 46 per cent in 'Sel 1' x *nor* and 48 per cent in 'IIHR

Table 2.38: Resistance Source for Quality Traits

Accession	Country	Traits of Interest	Recipients
EC468370-72	USA	High β-carotene content, suitable for processing	ICAR-NBPGR, N. Delhi
EC498863-865	USA	Suitable for processing, compact determinate growth habit and crack resistant	ICAR-IIVR, Varanasi (U.P.); ICAR-NBPGR, N. Delhi

709′ x *nor*. Neha *et al.* (2016) evaluate 15 commercial varieties, 9 exotic genotypes and three wild species and observed high TSS in *Solanum chilense* (7.1 per cent) followed by Azad T-5 (7.0) and acidity in Azad T-5 (0.92 per cent) followed by IIHR-2620 (0.90 per cent). The genotypes may be used for improvement of functional fruit.

Maturity Quality

Table 2.39: Promising Accessions for different Maturity Group

Attributes	Promising Accesions
Early maturing	EC357833, EC357827, EC362956, EC362940, EC362944, EC362947, EC362948, EC362950, EC99935, EC129602, EC8591, EC135580, EC528374, EC538152, EC8822, LE-477, LE598, LE-22
Late maturing	EC357829, EC357830, EC362941, EC362942, EC362951, EC241157, EC163594, EC212469, EC241134

Transportation

In tomato crop improvement, now emphasis is being given for transportation attributes and accordingly cultivars are being developed.

Table 2.40: Promising Accessions for Transportation

Attributes	Promising Accesions
Transportation	EC161252, EC7352, EC100845, EC110268

Parthenocarpiness

Parthenocarpic varieties which do not require pollination set abundant fruits at low temperatures. 'Severianin' a parthenocarpic variety which showed abundant fruit set in December and January.

Shelf Life

Anand and Tikoo (1983) studied effect of mutants *rin* (ripening inhibitor), *nor* (non-ripening) on storage life, pigmentation and yield in heterozygous and homozygous forms along with normal varities. The crosses with *nor* had longer shelf-life than those with *rin*. F_1s of *nor* with 'Sel 1' (26 days) and 'IIHR 709' (28 days) were outstanding for shelf life. Prasanna *et al.* (2002) obserbed and reported that maximum shelf life cross combinations in 15SBSB x IIHR 2050 (32 days), 15SBSB x nor-1 (27 days) and BWR IF x nor-1 (12 days).

Male Sterility

Male sterility mechanismwas first reported by Crane (1915). In tomato different type of male sterility like genetic male sterility (due to tedious maintenance utility is limited), cytoplasmic male sterility (limited utility), functional male sterility (maximum utility), structural male sterility or stamenless type sterility are reported. Male sterile plants can be identified in the population by observing small flower size, poorly developed anthers, poor or no anther dehiscence and sterile pollen. The genic male sterile and functional male sterile lines are freely available. In genic male sterility more than 50 loci have been reported to control male sterility of which 'ms 10^{35} line' is very useful. It has been developed in Bulgaria and is of great value for production of hybrid seeds. The line is maintained under heterozygous condition (fertile) by back cross method with male sterile parent which produced 50 per cent fertile and 50 per cent sterile plants. For seed production of back cross progenies (*ms ms* x *Ms ms*) seeds are closely sown/planted which segregate 50 per cent male strile (*ms ms*) and 50 per cent male frtile (*Ms ms*). The sterile plants are used for the production of hybrids and fertile plants are removed/uprooted at the initial flowering and only male sterile plants are used as seed parent. The flowers of sterile plants are pollinated by the pollen of the male parents to produce hybrid seeds. In this case emasculation is not required and if emasculation is done it is much easier than fertile lines. This male sterility is linked with character anthocyaninless (*aa*) marker which helps in identification of male sterile plants and purity of hybrid seeds. In Europe 19 male sterile based heterotic hybrids have been released for commercial cultivation. The *ms-*10^{35} gene is linked with a recessive marker gene (*a*) responsible for absence of anthocyanin. Hence, *ms-*10^{35} sterile plant can be identified at seedling stage and fertile plant can be rouged out in the nursery itself. With respect to nuclear male sterility, 55 recessive genes have been reported *sl-2*, *ms-13* and *ms-15* which are temperature sensitive. YOC clone containing *ms-14* gene has been cloned.

- ☆ **Functional male sterility** is characterized by viable pollen which include positional sterility (*ps*), non-opening of anther (*ps2*) and short style (sh st). The *ps2* (non-opening of anther) has good potential for hybrid seed production and being exploited commercially. Functional male sterile lines are also called 'John Bay' type of male sterility. In Europe 19 male sterile based hybrid, 17 of them are based on functional sterility controlled by *ps2* gene.

☆ **Exserted stigma (positional sterility)** controlled by recessive factor can also be used for the production of hybrid. If there is adequate exertion in this case, there is no self pollination and thus such lines can be used for the production of hybrids. The exserted lines can be developed utilizing wild species especially *S. hirsutum* and *S. hirsutum* var. *glabratum*.

The sex type controlled by recessive factor has a great advantage because F_1 hybrids will produce normal styled flower without any problem of fruit set. The line '510' is a functional male sterile line. Closed anther types are being used in the production of hybrids 'Katrain Hybrid No. 1' and 'Hybrid No. 2'. Since it is easy to maintain functional male sterile lines, breeders and seed producesrs may use such types for commercial hybrid seed production which will be much advantageous to the farmers as the cost of seeds will be low. For maintaining the sterile line, fertile plants (*Ms ms*) are crossed with the sterile (*ms ms*) and resultant seeds are used for the next year hybrid seed production. Singh and Nandpuri (1970) developed hybrids MS x 6L; MS x RD; MS x 2R which were promising for yield. There are other types of male sterility in which pollen grains are fertile but there is no anther dehiscence under normal conditions thus plant remains sterile (Mittal and Thomas, 1961; Singh *et al.*, 1966). This type of sterility is easily maintained by artificial selfing and can be used more economically for the production of hybrid, provided selfing is absolutely controlled. In this case recessive marker in the seed parents like potato leaf type, anthocyaninless (*aa*) and brown seeded can be used to detect the purity of the hybrid seeds. At ICAR-IIVR, Varanasi (U.P.) nuclear male sterile gene (*ms*[15]) is being utilized. Cytoplasmic nuclear male sterility has been derived from the distinct species through protoplast but restorer gene (*Rf*) is not available.

Assemblage of Mutants and Breeding Lines

The use of *ps2* mutant has become applicable at commercial scale in few countries like Czech Republic, Moldova and Bulgaria. At Central Research Institute, Szeged, Hungary is utilizing monogenic-recessive auxotrophy functional mutants (for nutritional requirement) for the production of pure tomato hybrid seeds. In this crop utilizing thiamin dependent nutritional mutant (auxotroph; recessive homozygote; No-50-30W) as female parent and a prototroph (K509) as male parent (female auxotrophic scheme) it was demonstrate that pure F_1 seeds can be obtained even if selfing occurs in female parent *e.g.* if the male sterility is imperfect. The thiamin dependent auxotroph mutant of tomato was successfully maintained through 1-2 treatment of thiamin @5 ml/l. the utilization of auxotrophy mechanism dooes not need other marker because it is per se a genetic marker observable at seedling stage. Unfortunately, research on utilization of tomato auxotrophic mutants for hybrid seed production was discontinued. The use of chemical hybridizing agents (CHA) to produce experimental crosses of tomato has been demonstrated. A good number of genotypes having mutant genes in them have been assembled which is being used in different breeding programmes.

Table 2.41: Mutant Genotypes of Tomato

Gene Symbol	*Description of Plant Characters*
aa	Anthocyanin absent, (absence of anthocyanin in all plant part).
Aft	Anthocyanin fruit.
c	Potato leaf, reduced leaf segments.
dg	Dark green fruit and exaggerated photoresponsiveness.
ful	Fulgens, chlorophyll B mutant
hp-1	High pigment, high chlorophyll, carotenoids and ascorbic acid contents in fruit.
j	Jointless pedicel.
ms-2, ms-45	Male sterility, pollen aborted or no pollen.
og^c	Old gold crimson, high lycopene and β carotene contents in fruit.
pat 2	Parthenocarpic fruit.
ps	Positional sterile mutant, connate petals, normal viable pollens but non-dehiscent anther due to persistent stomium.
ps-2	Positional sterile-2 mutant, normal flower with viable pollens but non-dehiscent anther due to structural alterations in the zone of anther dehiscence.
rin	Ripening inhibitor, delayed ripening and bright yellow fruit at maturity.
s	Compound inflorescence.
u	Uniform ripening of fruit.
y	Unpigmented fruit epidermis

Source: Chattopadhyay *et al.*, 2011.

In tomato 4 mutants, never ripening (*Nr*), ripening inhibitor (*rin*), non-ripening (*nor*) and slow ripening alcobaca (*alc*) (usually long self life, good colour development) are available which regulate, initiate and synchronize ripening. Rick (1956) reported *Nr* to be a dominant allele which affects intensity of fruit pigmentation and softening rate. The ripening inhibitor (*rin*) alters several aspects of fruit ripening, most notably carotenoid synthesis, shelf-life and fruit softening (Robinson and Tomes, 1968). It is located on chromosome 5, very closely linked in coupling with the gene for macrocalyx (*mc*). The non-ripening (*nor*) gene, extracted from 'Italian Winter' affects carotene synthesis and fruit softening (Tigchelaar *et al.*, 1973). It is closely linked in repulsion phase with the uniform ripening gene (*u*) on chromosome 10. The *nor*^{2}, *nor*^{3} mutants were subsequently identified (Soressi, 1975). Mutant alcobaca (*alc*), a single recvessive gene controls exceptionally long shelf life. The *alc* is linked to nor (17 map units) and to *u* gene (20 map units) on chromosome 1. The crosses with *nor* had longer shelf life than those with *rin* (Anand and

Table 2.42: Mutant Varieties of Tomato Released Worldwide for Cultivation

Crop/Variety	Year of Release	Country	Mutagenic Dose/Parent Used	Character Improved
Luch 1	1965	USSR	Gamma rays [Pushkinsky]	Earliness and yield
S-12	1969	India	Gamma rays	Reduced plant height
Pusa Lal	1972	India	Gamma rays	Fruit ripening
Kyoryokureikou	1974	Japan	Gamma rays, Pollen of L. peruvianum	Resistance to TMV and wilt
PKM-1	1980	India	Gamma rays, 25 kR [Annanj]	Yield and plant type
CO-3	1981	India	EMS-0.1 per cent, seeds of Co-1	Plant type and vitamin C content
Rannii Nush	1983	-	-	-
Lal Meeruti	-	India	-	-

Tikoo, 1983). F_{1s} of *nor* with 'Sel. 1' (26 days) and 'IIHR 709' (28 days) were outstanding for shelf life. Lycopene levels in F_{1s} of *rin* were significantly higher than those of *nor* F_{1s}. heterobeltiosis for extended shelf-life was 46 per cent in 'Sel 1' x *nor* and 48 per cent in IIHR 709' x *nor*. Hybrids (F_1) developed using closed anther mutants are 'Meeruti' x 'Sioux' and 'Meeruti' x 'Red Jacket'. The other evolved mutants are' 12' from 'Sioux' and 'Pusa Meeruti' from 'Improved Meeruti'.

High Stigma Type

'Manaleuci' x 'Bruinsma' is a hybrid developed using high stigma character of 'Bruinsma'.

Genetically Modified Genotypes

There are certain genotypes *i.e.* 'Monsanto 1987' and 'Dupont 1988' which are genetically manipulated. 'Monsanto 1987' has resistance to mosaic virus, horn and fruit borer and resistance against herbicide (phosphosat). Dupont 1988'has resistance to sulphonyl urea.

Grafting

Grafting of improved cultivars of tomato over eggplant rootstocks has ability to tolerate waterlogging condition for 4 days without reduction in fruit yield (Bahadur *et al.*, 2016). In Japan several varieties are being utilized as for grafts preparation.

Table 2.43: Suitable Rootstocks for Grafting

Scion	Rootstocks	Major Variety
Tomato	*S. lycopersicum*	Mate, Hawai-7998, Joint, BF Okitsu No. 101, Helper M, PFNT No.-2, Sukuramu No.-2, Ancher T

Similarly, several rootstocks of *Solanum* were tried keeping the various targets in mind.

Table 2.44: Suitable Rootstocks for Biotic and Abiotic Stress

Scion	Rootstocks	Major Variety
Tomato	*S. lycopersicum*, *S. pimpinellifolium*, *S. habrochaites*, *S. melongena*, *S. torvum*, interspecific hybrid (*S. lycopersicum* x *S. habrochaites*)	Tolerance to bacterial wilt, fusarium wilt, root knot nematode, verticillium wilt, corky root, flooding, drought, extended fruiting period

Some potential rootstocks for biotic and abiotic stresses are given in Table 2.45.

Herbicide Tolerance

Great genotypic variation towards the herbicide sensitivity has been observed in tomato. An experiment was conducted at CCS-HAU, Hisar (Haryana) including several herbicides like pendimethalin, metribuzine, butachlor, oxadiazon and alachlor in tomato crop under field condition and maximum survival was recorded in variety 'Hisar Lalit' against al the herbicide used followed by 'Pusa Ruby', Sel.-22' and 'Sel-2'. In these groups cultivar 'Columbia' was highly sensitive to the herbicide (Banerjee *et al.*, 1990).

Genetic Resource Conservation

In tomato a large number of genetic resources were collecte and conserved by various research organization and university. Tomato Genetic Co-operative located

Table 2.45: Biotic and Abiotic Stress Resistance/Tolerance Rootstock for Tomato Grafting

Rootstock	Resistance/Tolerance	Grafting	References
S. pimpinellifolium	Nematode, salt	Tomato	Lee, 1994
S. habrochaites	Nematode, salt, cold, brown root rot	Tomato	Venema *et al.*, 2000; Venema *et al.*,2008; Venema *et al.*,1999a; Venema *et al.*,1999b; Yamakawa, 1982
S. cheesmanii	Salt	Tomato	Munns and tester, 2008
S. chilens	*Fusarium oxysporum* f.sp. *lycopersici* race-1, 2, 3, and drought	Tomato	Rai *et al.*, 2011, Reis *et al.*, 2004; O'connell *et al.*, 2007
S. lycopersicum x *S. habrochaites*	Low temperature (10-13°C), bacterial wilt	Tomato	Okimura *et al.*, 1986; King *et al.*, 2010
S. lycopersicum var. *ceresiforme*	*B. cockerelli* Aphids, White fly	Tomato	Hipolit cortez-Madrigal, 2012

in the University of California, USA, publishes genetic information on the crop and maintains a large number of collections which have already been thoroughly catalogued. The ICAR-NBPGR, New Delhi has systematically assembled over 2900 germplasm lines including wild species from diverse agro-climatic zones of world. The major donor countries include USA, Philippines, U.K., Canada, Australia, Portugal, Italy, Argentina, Cyprus, Mauritius, Israel, Netherlands, Nepal, Tanzania, Turkey, Japan, Iraq, Chile, Yugoslavia, Denmark, Switzerland, China, Trinidad, South Africa, Czechoslovakia, Ghana, Germany, Hungary, Taiwan, Poland, Bulgaria, South America, Columbia, Hong Kong, Spain, United Arab Republic, Burma, Nigeria, New Zealand and Brazil.

Table 2.46: Genetic Resource Conservations of Tomato (*Solanum species*)

Species	*No.*	*Source*
Solanum esculentum	1174	AVRDC, Taiwan, Colorado, USA
Solanum cheesmannii	04	
Solanum hirsutum	06	
Solanum pennellii	05	
Solanum pimpinellifolium	13	
Solanum peruvianum	07	
Solanum ceresiforme	08	
Solanum chielewskii	05	
Solanum parvifloram	04	
Solanum chilense	04	

REFERENCES

1. Anand, N. and Tikoo, S.K. 1983. Breeding for extended shelf life of tomato. *Proceedings of the National Seminar on the Production Technology of Tomato and Chillies*, Tamil Nadu Agricultural University, Coimbatore (Tamil Nadu), p. 41-45.
2. Anbindes, L., Reuveni, M., Azari, R., Paran, L., Nahon, S., Schlomo, H., Chen, L., Labidot, M. and Levin, I. 2009.Molecular dissection of tomato leaf curl virus resistance in tomato line TY172 derived from *Solanum peruvianum. Theor. Appl. Genet.*, **119(3):** 519-530.
3. Anbu, S., Muthukrishnan, C.R. and Irulappan, I. 1976. Line x Tester analysis in tomato. I. Combining ability. *South Indian Horticulture*, **28(4):** 128.
4. Asins, M.J., Breto, M.P., Cambro, M. and Carbonell, E.A. 1993. Salt tolerance in *Lycopersicon* species. I. Character definition and changes in gene expression. *Theor. Appl. Genet.*, **86:** 737-743.
5. Bahadur, A., Singh, M. and Singh, B. 2016. Grafting to alleviate water logging stress in tomato. *Indian Hort.*, **61(1):** 51-53.
6. Bai, Y. 2004. The genetics and mechanisms of resistance to tomato powdery mildew (*Oidium neolycopersici* in *Lycopersicon* species). Thesis, Wageningen University, The Netherlands.
7. Bai, Y., Anang, C.C., Van der Hulst, R., Meijer-Dekens, F., Bonnema, G. and Lindhout, P. 2003. QTLs for tomato powdery mildew resistance (*Oidium lycopersici*) in *Lycopersicon parviflorum* G1.1601 co-localize with two qualitative powdery mildew resistance genes. *Molecular Plant Microbe Interactions*, **16:** 169-176.
8. Bajaj, K.L. and Mahajan, R. 1982. Chemical constituents of fruits of some root knot nematode resistant varieties of tomato (*L. esculentum*). *Journal of Research, PAU.*, **19(4):** 339-342.
9. Banerjee, M.K. and Kalloo, G. 1987a.Sources and inheritance of resistance to leaf curl virus in *Lycopersicon. Theor. Appl. Genetic.*, **73:** 707-710.
10. Banerjee, M.K. and Kalloo, G. 1987b. Inheritance of potato leaf curls virus resistance in *Lycopersicon hirsutum* f. *glabratum. Euphytica*, **36:** 581-584.
11. Banerjee, M.K. and Kalloo, G. 1989c. The inheritance of earliness and fruit weight in crosses between cultivated tomatoes and two wild species of *Lycopersicon. Plant Breed.*, **102:** 148-152.
12. Basewell, V.R 1949. Our vegetable travelers. *Nat. Geogr. Mag.*, **6:**145.
13. Beerh, O.P and Rane, V.R. 1976. Physico-chemical characteristics and canning trials of some varieties of tomatoes grown in central India. *Journal of food Science and Technology*, **13(3):** 151.
14. Berlinger, M.M., Mordechi, S. and Pilowsky, M. 1991. Tomato resistance to tobacco whitefly. *Gan Ssades VaMeshek*, **9:** 56-59.
15. Bhatnagar, K., Kainsa, R.L. and Gupta, O.P. 1987. Evaluation of tomato varieties for paste and ketchup making. *Haryana Agril. Univ. Journal of Research*, **17(2):** 125-129.
16. Bhutani, R.D. 1981. Screening and diallel study for yield and quality characters in tomato (*L. esculentum*). *Abs.* **7(2):** 170-171.
17. Bhutani, R.D. and Kalloo, G. 1983. Genetics of carotenoids and lycopene in tomato (*L. esculentum*). *Genetica Agraria*, **37(1/2):** 1-6.
18. Bhutani, R.D. Kalloo, G. and Pandita, M.L. 1983. Genetic variability and scope of improvement of quality traits in tomato (*L. esculentum*). *Haryana Agril. Univ. Journal of Research*, **8(3):** 408-412.
19. Blaker, N.S. and Hewitt, J.D. 1987. A comparison of resistance to *Phytophthora partasitica* in tomato. *Phytopathology*, **77:** 1113-1116.
20. Brar, J.S. 1975. Summer Institute on Vegetable Production Techniques and Methods of Vegetable Breeding at PAU Ludhiana (Punjab).

21. Brower, D.J., Jones, E.S. and St. Clair, D.A. 2004. QTL analysis of quantitative resistance to *Phytophthora infestans* (late blight) in tomato and comparison with potato. *Genome*, **47(3):** 475-492.

22. Chalukova, M. and Manueyan, H. 1991. Breeding for carotenoids pigments in tomato. In. G. Kalloo (ed.). Genetic Improvement of tomato (Monograph on *Theor. Appl. Genet.*). Springer-Verlag, Berlin.

23. Channarayappa, G., Shivashankar, V., Muniyappa, V. and Frist, R.H. 1992. Resistance of *Lycopersicon* species to *Bemisia tabaci*: a tomato leaf curl virus vector. *Can. J. Bot.*, **70:** 2184-2192.

24. Chattopadhyay, A., Dutta, S., Hazra, P. and Sarkar, G. 2011. Conservation and utilization of vegetable germplasm at Bidhan Chandra Krishi Viswavidyalaya, Kalyani (West Bengal). Souvenir: National symposium on "Vegetable Biodiversity" held at JNKVV, Jabalpur (Madhya Pradesh); April 4-5, 2011; p. 135.

25. Chetelat, R.T., DeVerma, J.W., and Bennett, A.B. 1995. Effect of *Lycopersicon chmielewskii* sucrose accumulator gene (*sucr*) on fruit yield and quality parameters following introgression into tomato. *Theor. Appl. Genet.*, **91:** 334-339.

26. Chetelat, R.T., DeVerma, J.W., Khan, E. and Bennett, A.B. 1993. Sucrose accumulator (*sucr*), agene controlling sugar composition in fruit of *Lycopersicon chmielewskii* and *L. hirsutum*. *TGR Report*, **43:** 14-15.

27. Choudhury, D.R. and Malhotra, S.K. 2001. Studies on hybrid vigour in tomato (*L. esculentum* Mill.). *Indian J. Agric. Res.*, **35(3):** 176-180.

28. Crane, M.B. 1915. Heredity of types of inflorescence and fruits in tomato. *J. Genet.*, **5:** 1-11.

29. Datar, V.V. and Mayee, C.D. 1980. Breeding for early blight resistance in tomato. *Indian Phytopathol.*, **33:** 151.

30. De lapena, R.C., Ebert, A.W., Gniffke, P., Hanson, P. and Symoneb, R.C. 2011. Genetic adjustment to changing climate: Vegetables. In: Yadav, S.S., Redden, R.J., Hatfield, J.L., Lotze-campen, H. and Hall, A.E. (eds.). Crop adoption to climate change, first edn. John Wiley and Sons Ltd., Chichester, West Sussex, U.K., p. 396-410.

31. Deb, D.B. 1979. Solanaceae in India; The Biology and Taxonomy of Solanaceae (eds.) J.G. Hawkes, R.N. Lester and A.D. Skelding (London: Academic Press), p. 529-548.

32. Dhillon, G.S. and Nandpuri, K.S. 1975. Root-knot nematode resistance in tomato (*L. esculentum* Mill.). *Punjab Agric. Univ. J. Res.*, **12:** 373-377.

33. Dianesc, E.C., Foennseca, M.E.N., Inoue-Nagota, A.K., Resendo, R.O. and Boiteux, L.S. 2011. Search in Solanum (Section *Lycopersicon*) germplasm for sources of broad spectrum resistance to four Tospo virus species. *Euphytica*, **180:** 307-319.

34. Diwan, N., Fluhr, R., Eshed, Y. and Zamir, D. 1999. Mapping of *Ve* in tomato: a gene cofering resistance to the broad spectrum pathogen *Verticillium dahlia* race 1. *Theor. Appl. Genet.* **98:** 315-319.

35. Dixit, J., Kalloo, G., Bhutani, R.D. and Sidhu, A.S. 1980. Line x Tester analysis for the study of heterosis and combining ability in tomato (*L. esculentum*). *Haryana J. Hort. Sci.*, **9(1/2):** 56-61.

36. Dudi, B.S., Dixit, J. and Pratap, P.S. 1983. Components of variability, heritability and genetic advance studies in tomato (*L. esculentum* Mill.). *Haryana Agril. Univ. Journal of Research*, **13(1):** 135-139.

37. Egashira, H., Kawashima, A. and Imanishi, S. 2000. Screening of wild accessions resistant to grey mould (*Botrytis cinera* Pers.) in *Lycopersicon*. *Report of the Tomato Genetics Cooperative*, **50:** 14-15.

38. Fageria, M.S., Dhaka, R.S. and Jat, R.G. 1998. Field evaluation of tomato genotypes for resistance to buckeye rot and *Alternaria* blight. *J. Mycol. Pl. Pathol.* **28(1):** 52-53.

39. Fernandez Munmoz, R., Salinas, M., Alvarez, M. and Cuartero, J. 2003. Inheritance of resistance to two-spotted spider mite and glandular leaf trichomes in wild tomato (*Lycopersicon pimpinellifolium* (Juisl.) Mill.). *J. Am. Soc. Hort. Sci.*, **128:** 188-195.

40. Finkers, R., Van der berg, P., Van Berloo, R. 2007. Three QTLs for *Botrytis Cinerea* resistance in tomato. *Theoretical and Applied Genetics*, **114:** 858-593.

41. Finlay, E.W. 1953. Inheritance of spotted wilt resistance in the tomato. II. Five gene controlling spotted wilt resistance in four tomato types. *Aust. J. Biol. Sci.* **6:** 15-63.

42. Foolad, M.R. 2004. Recent advances in genetics opf salt tolerance in tomato. *Plant Cell Tissue Organ Culture*, **76:** 101-109.

43. Foolad, M.R. 2007. Tolerance to abiotic stresses. In: Razdon, M.K. and Mattoo, A.K. (eds.). Genetic improvement of solanacveous crops: tomato, Vol.2. Science Publishers, Enfield, New Hampshire, USA, p. 521-590.

44. Foolad, M.R. and Lin, G.Y. 1997. Absence of a genetic relationship between salt tolerance during seed germination and vegetative growth in tomato. *Plant Breed.*, **116(4):** 363-367.

45. Foolad, M.R., Merk, H.L. and Ashrafi, H. 2008. Genetics, genomics and breeding of late andearly blight resistance in tomato. *Crit. Rev. Plant Sci.*, **27:** 75-107.

46. Foolad, M.R., Ntahimpera, N., Christ, B.J. and Lin, G.Y. 2000. Comparison between field, green house and detached leaflet evaluation of tomato

germplasm for early blight resistance. *Plant Dis.*, **84:** 917-972.

47. Garge, N., Cheema, D.S. and Chawla, N. 2013. Manifestation of heterosis for fruit yield, quality and shelf-life in tomato (*S. esculentum* L.) hybrids incorporating *rin*, *nor* or *alc* alleles in main and late seasons of north Indian plains.

48. Gilbert, J.C. 1958. Some linkage studies with the *Mi* gene for resistance to root knot nematodes. *Tomato Genet. Coop. Rep.*, **8:** 15-17.

49. Gordille, L.F., Stevenes, M.R., Millard, M.A. and Geary, B. 2008. Screening two *Lycopersicon peruvianum* collections for resistance to tomato spotted wilt virus. *Plant Dis.*, **92:** 694-704.

50. Gould, W.A. 1983. Tomato production, Processing and Quality Evaluations. 2nd ed. AVI Publishing Company, Inc. Westport, CT.p. 3.50.

51. Govindarasu, P., Muthukrishnan, C.R. and Irulappan, I. 1981. Combining ability for yield and its components in tomato (*L. esculentum*). *Scientia Horticulturae,* **14(2):** 125-130.

52. Grattiige, R. and O' Brien, R.G. 1982. Occurrence of a third race of Fusarium wilts of tomato in Queensland. *Plant Dis.*, **66:** 165-166.

53. Hall, T.J. 1980. Resistance at the *Tm-2* locus in the tomato to tomato mosaic virus. *Euphytica*, **29:** 189-197.

54. Hanson, P., Green, S. and Kuo, G. 2006. *Ty-2*. A gene on chromosome 11 conditioning Gemini virus resistance in tomato. *Rep. Tomato Genet. Coop.*, **56:** 17-18.

55. Hanson, P.M., Licardo, O., Hanudin, Wang, J.F. and Chen, J.T. 1998. Diallel analysis of bacterial wilt resistance in tomato derived from different sources. *Plant Dis.*, **82:** 74-78.

56. Harvey, M., Quilley, S. and Beynon, H. (2002). Exploring the tomato. Transformations of Nature, Society and Economy. Edgar Publishing, Cheltenham, U.K., pp. 304.

57. Hassan, A. A., Mazyad, H., Moustate, S., Nassar, S., Makha, M. and Sims, W. 1984. Inheritance of resistance to tomato yellow leaf curl virus derived from *Lycopersicon cheesmannii* and *Lycopersicon hirsutum*. *Hort. Sci.*, **19:**574-575.

58. Hawkes, J.G. 1990. The Potato : Evolution, Biodiversity and Genetic Resources Belhaven Press, London.

59. Hedrick, U.P. and Booth, N.O. 1980. Mendelian characters in tomato. *Proc. Am. Soc. Hort. Sci.*, 5: 19.

60. Heinz, K.M. and Zalon, F.G. 1995. Variation in trichome based resistance to *Bemisia argentifolii* (*Homoptera aleyrodidae*) oviposition on tomato. *J. Econ. Entomol.*, **88:** 1494-1502.

61. Hipolit cortez-Madrigal. 2012. Grafts of crops on wild relatives as base of an integrated pest management. The tomato (*Solanum lycopersicum* L.) as example, integrated pest management and pest control-current and future Tactics, Dr. Sonia Soloneski (ed.), In: Tech., DOI: 105772/30181.

62. Huang, C.C. and Lindhout, P. 1977. Screening for resistance in wild *Lycopersicon* species to *Fusarium oxysporum* f. sp. *lycopersici* race 1 and race 2. *Euphytica*, **93:** 145-193.

63. Huang, C.C., Van de Putta, P.M. 2000. Haanstra-van der Meer, J.G., Meijer-Dekens, F. and Lindhout, P. (2000). Characterization and mapping of resistyance to *Oidium lycopersicum* in two *L. hirsutum* accessions. Evidence for close linkage of two *Ol*-genes on chromosome 6. *Heredity*, **85:** 511-520.

64. Huttan, S.F. and Scot, J.W. 2015. Recent progress in TYLCV resistance breeding and implications for tomato varieties of the future. Presentation www.1mok.ufl.edu/media./doesl./tomato./t12015/ti2015-hutton pdf.

65. ISI, New Delhi 1967a. Indian standard specification for tomato ketchup. Indian Standard Institution, New Delhi.

66. Jacobsen, E.; Daniel, M. K.; Bergervoet-van Deelen, J. E. M.; Huigen, D. J. and Ramanna, M. S. 1994. "The first and second backcross progeny of the intergeneric fusion hybrids of potato and tomato after crossing with potato". *Theoretical and Applied Genetics*, **88 (2):** 181–186. Doi: 10.1007/BF00225895.

67. Jalali, I., Banerjee, M.K. and Kalloo, G. 1989. Screening for resistance to fusarium wilt in *Lycopersicon*. Tests of Agrochemicals and cultivars 10: *Ann. Biol.*, **114(**Supplymentry**):** 160-161.

68. Ji, Y., Schuster, D.J. and Scot, J.W. 2007a. Ty-3 a begomovirus resistance locus near the tomato yellow leaf curl virus resistance locus *Ty-1* on chromosome 6 of tomato. *Mol. Breed.*, **20:** 271-284.

69. Ji, Y., Scot, J.W., Maxwell, D.P. and Schuster, D.J. 2008. *Ty-4*, a tomato yellow leaf curl virus resistance gene on chromosome 3 of tomato. *Rep. Tomato Genet. Coop.*, **58:** 29-31.

70. Joi, M.B. and Summanwar, A.S. 1986. Reaction of tomato cultivar and hybrids to tomato spotted wilt virus. *Veg. Sci.*, **13:** 183-188.

71. Julian, O., Hessair, J., Corella, S. Lolli, I.D., Soler, S., Diez, M.S. and Perez-de-castro. A. 2013. Initial development of a set of introgression lines from *Solanum peruvianum* PI126944 into tomato exploitation of resistance to viruses. *Euphytica*, **193:** 183-196.

72. Kalloo, G. 1989.Tomato (*L. esculentum* Mill.). *Indian Horticulture,* **33-34(4, 1):** 12-15.

73. Kalloo, G. and Banerjeee, M.K. 1990a. Low temperature fruit set attribute in cultivated variety of tomato derived from *Solanum pimpinellifolium*: Tomato. Eucarpia Tomato 90: XI Meeting of Eucarpia Working Group, 6-8 March, Malaga, Spain.

74. Kalloo, G. and Banerjeee, M.K. 1990b. Transfer of tolerance to tomato leaf curl virus from *S. pimpinellifolium* to *S. esculentum*. *Gartenbauwissenschaft*, **55:** 92-94.

75. Kalloo, G. and Banerjeee, M.K. 1990c. Transfer of tomato leaf curl virus resistance from *S. hirsutum* f. *glabratum* to *S. esculentum*. *Plant Breed.*, **105:**56-159.

76. Kalloo, G. and Banerjeee, M.K. 1990d. 'H-24'. A leaf curls virus resistant variety of tomato. Proposal of leaf curl virus, submitted to Director Research, Haryana Agricultural University, Hisar (Haryana).

77. Kalloo, G. Banerjeee, M.K., Singh, M. and Kumar, S. 2002. Vegetable Varieties, IIVR Technical Bulletin-3, pp. 59.

78. Kalloo, G., Jain, R.K. and Bhati, D.S. 1978. Inheritance stuidies on resistance to root knot nematode in tomato (*L. esculentum* Mill.). *Z. Pffanzenziicht*, **81:** 281-284.

79. Kalloo, G., Singh, R.K. and Bhutani, R.D. 1974. Combining ability studies in tomato (*L. esculentum* Mill.). *Theoretical and Applied Genetics*, **44(8):** 358-363.

80. Kalloo, G.and Dudi, B.S. 1982a. Development of variety for fruit set at low temperature in tomato. Annu. Rep. Dept. Vegetable Crops, CCS-HAU, Hisar, p. 16.

81. Kashyap, R.K and Varma, A.N. 1986. Screening of tomato germplasm for susceptibility to fruit borer (*Heliothis armigera* Hubner.). *Indian J. Entomol.*, **48:** 46-53.

82. Kashyap, R.K., Bhutani, R.D. and Kalloo, G. 1982. Relative susceptibility and variability studies against fruit borer in tomato. *Punjab Hort.*, **22:** 190-193.

83. Kennedy, G.C. 2007. Resistance in tomato and other *Lycopersicon* species to insect and mite pests. In: In: Razdan and M.K., Mattoo, A.K. (eds.) Genetic improvement of Solanaceous Vegetables Vol. 2. Tomato, Science Publishers, Jersey, USA. p. 488-519.

84. Khandare, V.S., Waskar, D.P. and Poobalan, V. 2016. Genetic diversity in tomato (*Solanum lycopersicon* Mill Wettsd.). 7th Indian Horticulture Congress (Abs: 75), held at ICAR-IARI, New Delhi, 15-18 November, p. 59.

85. King, S.R., Davis, A.R., Zhang, X. and Croshy, K. 2010. Genetics, breeding and selection of rootstocks for Solanaceae and Cucurbitaceae. *SciHort.*, 127: 106-111.

86. Kohli, U.K., Dev, H. and Fageria, M.S. 1995. Genetic analysis of resistance to buckeye rot (*Phytophthora nicotianae* var. *parasitica*) in tomato. *Veg. Sci.*, **22(2):** 112-115.

87. Kulkarni, G.P. and Madalageri, M.B. 2002. Evaluation and correlation studies in tomato for tomato leaf curl virus. *Intl. Conf. Veg.*, Bengaluru, 11-14th Nov. Abstract, p. 80.

88. Kumar, N. and Irulappan, I. 1990. Seasonal influence on the incidence of spotted wilt virus on tomato varieties. *South Indian Hort.*, **38(5):** 250-252.

89. Kumar, S., Singh, V. and Maurya, P.K. 2016. Studies on combining ability, heterosis and gene action in tomato (*Solanum lycopersicon* Child) for yield and its attributing traits. 7th Indian Horticulture Congress (Abs: 92), held at ICAR-IARI, New Delhi, 15-18 November, p. 71.

90. Kurian, A. 1990. Evaluation of processing characteristics in tomato and their expression in a bacterial wilt resistant genetic background. Ph.D. Thesis, Kerala Agricultural University, Vellanikkara (Kerala).

91. Laterrot, H. 1978. Resistance gux maladies. *B. pyrenochaeta lycopersici*. Report d' Active INRA station d' Ameloration desplantes Maraicheres, 1977-1978: 102-103.

92. Laterrot, H. and Pecant, P. 1969. Gene *Tm-2* new source. *Tomato Genet. Coop. Rep.*, **19:** 13-14.

93. Lee, J.M.1994. Cultivation of grafted vegetables. I. Current status, grafting methods and benefits. *HortSci.*, **29:** 235-239.

94. Li, J., Liu, L. and Bai, Y. 2011. Identification of mapping of quantitative resistance to late blight (*Phytophthora infestans*) in *Solanum habrochaites* LA1777. *Euphytica*, **179(3):** 427-438.

95. Lin, T.C. and Hong, W.T. 1989. Development of the new summer tomato variety 'Taichung-Asveg-4'. Bulletin of Taichung District Agricultural Improvement Station. **25:** 5-64.

96. Madaiah, N., Setty, R.G., Prakash, K.M.S., Swamy, N.A.M. and Patwardhan, M.V. 1986. Studies on physio-chemical characteristics of some new tomato varieties suitable for preparing tomato paste. *Indian Food Packer.* **40(3):** 6-12.

97. Mahajan, R., Ahuja, S., Kanwar, J.S. and Nandpuri, K.S. 1979. Resistance in tomato against *Meloidogyne incognita* (K and W) Chitwood, *Veg. Sci.*, **6(1):** 57-60.

98. Mahajan, R., Kanwar, J.S. and Nandpuri, K.S. 1975. Field screening of some tomato varieties and hybrids against the root-knot nematode (*M. incognita*). *Journal of Research, Punjab Agricultural University*, **12(1):** 40-42.

99. Majid, R. Swaminathan, M.S. and Iyer, R.D. 1968. Production and cytogenetic analysis of interspecific hybrids in *Lycopersicon*. *Indian J. Genet.*, **28(3):** 275-286.

100. Maliepaard, C., Bas, N., Van Heusden, S., Kos, J., Pet, G., Verker, K.R., Vrielink, R., Zabel, P. and Lindhout, P. 1995. Mapping of QTLs for grandular trichome densities and *Trialeurodes vaporariorum* (green house white fly) resistance in an F_2 from *Lycopersicon esculentum* x *L. hirsutum* f. *glabratum*. *Heredity*, **75:** 425-433.

101. Merk, H.L., Ashrat, H. and Foolad, M.R. 2012. Selective genotype to identify late blight resistance genes in an accession of tomato wild species *Solanum pimpinellifolium*. *Euphytica*, **187:** 63-75.

102. Mital, R.K., Singh, H.N., Singh, R.R. and Singh, J.B. 1974. Combining ability in tomato. *Haryana J. Horti. Sci.*, **3(3-4):** 168-176.

103. Mital, S.P., Thomas, T.A. and Srivastava, G. 1965. Assessment of genetic stocks of tomato (*Lycopersicon esculentum* Mill.) and wild relatives. *Tech. Bull. (Agric.)1*. Indian Council of Agricultural Research, New Delhi.

104. Momotaz, A., Scott, J.W. and Schuster, D.J. 2010. Identification of QTL conferring resistance to *Bemisia tabaci* in an F_2 population of *Solanum Lycopersicon* x *S. habnrochaites* accessions LA1777. *J. Amer. Soc. Hort. Sci.*, **135:** 134-142.

105. Muniyappa, J.V., Saika, S.H., Chennarayappa, A.K., Shivashankar, G., Ishwara, B.A. and Ramappa, H.K. 1991. Reaction of *Lycopersicum* cultivars and wild accessions of tomato leaf curl virus. *Euphytica*, **56:** 37-41.

106. Munns, R. and Tester, M. 2008. Mechanism of salinity tolerance. *Annual Review of Plant Biology*, **59:** 651-681.

107. Muthukumar, P., Kalia, P., Jat, G.S., Chitra, D. and Kumari, S. 2016. Screening and validation of Ty-1, Ty-3 linked markers in *Solanum chilense* accessions for resistance to tomato leaf curl virus. 1st International Agrobiodiversity Congress (Abs. Session: Trait Discovery and Enhanced Use of PGR), 6-9 November, held at New Delhi, p. 208.

108. Muthukumar, P., Kalia, P., Jat, G.S., Chitra, D. and Kumari, S. 2016. Screening of *Solanum habrochaits* accessions for tomato late blight resistance (*Phytophthora infestans*). 7th Indian Horticulture Congress (Abs: 103, held at ICAR-IARI, New Delhi, 15-18 November, p. 79.

109. Nandpuri, K.S. and Tyagi, D.D. 1976. Inheritance of some characters in tomato. *Journal of Research, Punjab Agricultural University*, **13(1):** 80-84.

110. Nandpuri, K.S., Singh, S. and Kanwar, J.S. 1977. Tomato 'Punjab Chhuhara'. *Indian Horticulture*, **22(1):** 7 and 27.

111. Nath, P., Srivastava, V.K., Dutta, O.P. and Swamy, K.R.M. 2008. Vegetable Crops- Improvement and Production. Dr. Prem Nath Agricultural Science Foundation, Bengaluru (Karnataka), 34-40.

112. Neha, P., Solanky, S.S., Akhtar, S. and Kumari, M. 2016. Assessment of genetic variation in cultivated and wild tomato genotypes for yield and quality contributing traits. 7th Indian Horticulture Congress (Abs: 08; Session: Breeding for Horticultural Crops), 15-18 November, held at ICAR-IARI, New Delhi, pp. 92.

113. Nicot, P.C., Moretti, A., Romiti, C., Bardin, M., Caranta, C. and Ferriere, H. 2002. Differences in susceptibility of pruning wounds and leaves to infection by *Botrytis cinera* among wild tomato accession. *Tomato Genet. Coop. Rep.* **52:**24-60.

114. Nombela, G., Williamson, V.M. and Muniz, M. 2000. The root knot nematode resistance gene *Mi-1.2* of tomato is responsible for resistance against the white fly *Bemisia tabaci*. *Molecular Plant Microbe Interactions*, **16:** 645-649.

115. O'connell, M.A., Medina, A.L. and Sanchez, P.N. and Trevino, M.B. 2007. Molecular genetics of drought resistance in tomato and related species. In: Razdan, M.K. and Mattoo, A.K. (eds.). Genetic improvement of Solanaeous crops, Vol.2, Tomato: Science Publishers, Enfield, USA, p. 261-283.

116. Okimura, M., Matso, S., Arai, K. and okitsu, S. 1986. Influence of soil temperature on the growth of fruit vegetable grafted on different rootstocks. *Bull. Veg. Ornam. Crops Res. Stn. Japan.*, **9:** 43-58.

117. Olimpia Gomez Consuegra, Mayte Pinon Gomez, Yamila Martinez Zubiaur. 2015. Pyramiding TYLCV and TSWV resistance genes in tomato genotypes. *Rev. Protection Veg.* 30(2), La Habana Mayo-Ago.

118. Parlevliet, J.E. 2002. Durability of resistance against fungal, bacterial and viral pathogen; present situation. *Euphytica*, **124:** 147-156.

119. Paterson, R.G., Scott, S.G., and Gergerich, R.C. 1989. Resistance in two *Lycopersicon* species to an Arkaneses isolate of tomato spotted wilt virus. *Euphytica*, **43:** 173-178.

120. Pathak, J.S. and Mahajan, P.R. 1978. Evaluation of tomato cultivars for processing. *Indian Food Packer.*, **32 (2):** 25-31.

121. Pelham, J. 1972. Strain-genotype interaction of tobacco mosaic virus in tomato. *Ann. Appl. Boil.*, **71:** 219-228.

122. Perez de Castro, A. Julian, O. and Diez, M. 2007. Genetic control and mapping of *Solanum chilense* LA1960 and LA1971-derived resistance to tomato yellow leaf curl disease. *Euphytica*, 1-12.

123. Peter, K.V. and Rai, B. 1980. Combining ability in tomato. *Indian J. Genet. Pl. Breed.*, **40(1):** 1-7.

124. Peter, K.V., Sarnaik, D.A. and Ramachandran, C. 1981. Note on the inheritance of free neck, locul

number and ovate fruit shape and their linkage estimates in tomato. *Indian J. Agri. Sci.*, **51(3):** 191-193.

125. Pilowsky, M. and Cohen, S. 1990. Tolerance to tomato yellow leaf curl virus derived from *Lycopersicon peruvianum. Plant Dis.*, **74:** 248-250.

126. Pitblado, R.E. and Kerr, E.A. 1980. Resistance to bacterial speck (*Pseudomonas syringae* pv. *tomato*) in tomato. *Acta Horticulturae*, **100:** 379-382.

127. Prasanna, K.M., Sadashiva, A.T., Madhavi Reddy, K, Krishna Reddy, M., Rawal, R.D., Balaram, M.V. and Narasimha Prasad, B.C. 2002. Development of Bacterial wilt (BW) resistant tomato (*Lycopersicon esculentum* Mill.) F_1 hybrids with extended shelf life. *Intl. Conf. Veg., Bengaluru*, 11-14th Nov. Abstract, p. 59.

128. Rahman, M.A., Vijaya, M., Neeraja, G., Chiranjeevi and Reddy, I.P. 2002. Evaluations of tomato cultivars for resistance to tomato leaf curl disease. *Intl. Conf.* Veg., Bengaluru, 11-14th Nov. Abstract, p. 271.

129. Rai, M., Dhandhar, D.G. and Varma, S.P. 1987. A comparative performance of tomato (*Lycopersicon esculentum* Mill.) varieties under Tripura condition. *Haryana J. Hort. Sci.*, **16(3-4):** 253-256.

130. Rai, N., Tiwari, S.K., Kumar, R., Singh, M. and Bhardwaj, D.R. 2011. Genetic resources of Solanaceous Vegetables in India. Natl. Symp. 'Vegetable Biodiversity' held at JNKVV, Jabalpur (M.P.), April 4-5, 2011, pp. 91-103.

131. Raijadhav, S.B. and Kale, P.N. 1985. Heterosis in root-knot nematode resistant tomato lines. *Journal of Maharashtra Agricultural University*, **10(3):** 265-268.

132. Raijadhav, S.B. and Kale, P.N. 1987.Combining ability for processing characters in tomato. *Journal of Maharashtra Agricultural University*, **12(3):** 379-380.

133. Randhawa, K.S., Saimbhi, M.S. and Gill, B.S.1988. Commercial evaluation of tomato varieties for processing. *Veg. Sci.*, **15(2):** 181-184.

134. Rao, A.V. and Rao, L.G.2007. "Carotenoids and human health". *Pharmacol. Res.*, **55(3):** 2007-216.

135. Rao, E.S., Kadirvel, P., Symonds, R.C. and Ebert, A.W. 2013. Relationship between survival and yield related traits in *Solanum pimpinellifolium* under salt stress. *Euphytica*, **191(2):** 215-228.

136. Rao, M.V.B., Sohi, H.S. and Tikoo, S.K. 1975. Reaction of wilt resistant tomato varieties and lines to *Pseudomonas solanacearum* in India. *Plant Disease Reporter*, **59:** 734-736.

137. Rattan, R.S. and Saini. S.S. 1979. Inheritance of resistance to fruit rot (*Phytophthora parasitica* Dast.) in tomato (*Lycopersocon esculentum* Mill.). *Euphytica*, **28:** 315-317.

138. Rattan, R.S., Saini. S.S. and Singh, M. 1978. Breeding tomatoes for soluble solid contents. *Veg. Sci.*, **5(2):** 78-82.

139. Reddy, S., Prasanna, H.C., Krishna, R., Kashyap, S.P., Rai, N.P. and Singh, B. 2016. Embryo rescue of *Solanum lycopersicum* x *S. neorickii* tomato cross for early blight resistance. 1st International Agrobiodiversity Congress (Session: Trait Disdcovery and Enhanced use of PGR), 6-9 November 2016; held at New Delhi; p.162

140. Reis, A., Girdano, L.b., Lopez, C. and Boiteux, L. 2004. Novel sources of multiple resistance to three race of *Fusarium oxysporum* f. sp. *lycopersici* in *Lycopersicon* germplas.*Crop Breed. Appl. Biotech.*, **4:** 495-502.

141. Rick, C.M. 1982. The potential of exotic germplasm for tomato improvement. In: Vasil, I.K., Scowcroft, W.R. and Frey, K.J. (eds.), Plant Improvement and somatic cell genetics, Academic Press, New York, USA, p. 1-28.

142. Robertson, L.D. and Labate, J.A. 2007. Genetic resources of tomato (*Lycopersicon esculentum* Mill.) and wild relatives. In: In: Razdan, M.K. and Mattoo, A.K. (eds.) *Lycopersicon chmielewskii*. Genetic improvement of Solanaceous Vegetables Vol. 2. Tomato, Science Publishers, Jersey, USA. p. 25-75.

143. Robertson, L.D. and Labate, J.A. 2007. Genetic resources of tomato (*Lycopersocon esculentum* Mill.) and wild relatives. In: Razdan, M.K. and Mattoo, A.K. (eds.) *Lycopersicon chmielewskii*. Genetic improvement of Solanaceous Vegetables Vol. II. Tomato, Science Publishers, Jersey, USA. p. 646.

144. Roddick, J.G. and Butcher, D.N. 1972. Tomatine production in cultivated excised tomato roots. *Phytochem.*, **11:** 2991-2997.

145. Rosello, S., Diez, M.J. and Nuez, F. 1998. Genetics of tomato spotted wilt virus resistance coming from Lycopersicon peruvianum. *European J. Plant Pathol.*, **104:** 499-509.

146. Rowe, R.C. and Farley, J.D. 1981. Strategies for controlling Fusarium crown and root rot in green house tomatoes. *Plant Dis.*, **65:** 107-112.

147. Roy, S.K., Krishna, K., Tewari, R.N., Choudhary, R. M. and Choudhary, B. 1970. Preliminary studies on evaluation of some tomato varieties for canning. *Indian Food Packer*, **24 (6):**11.

148. Roy, S.K.and Choudhary, B. 1972. Studies on physic chemical characteristics of a few varieties of tomato (*L. esculentum* Mill.) in relation to processing. *Journal of Food Science and Technology*, **9(3):** 151-154.

149. Sadashiva, A.T. 2014. Breeding tomato for resistance to early blight in combinations with bacterial wilt and tomato leaf curl virus presentation: ASRT-2014-Hotel Lalit Ashok, Bengaluru, India.

150. Sadashiva, A.T., Madhavi, Reddy, K., Krishna Reddy, M., Singh, T.H., Balaram, M.K., Narasimha Prasad, B.C., Prasanna, K.M., Satya Narayana, H.V. and Navven, L.R. 2002. Breeding tomato (*Lycopersocon esculentum* Mill.) for combined resistance to bacterial wilt and tomato leaf curl virus. *Intl. Conf.* Veg., Bengaluru, 11-14th Nov. Abstract, pp. 47.

151. Sain, S.K and Chadha, M.L. 2016. Response of some wild species of tomato against peanut bud necrosis virus under open-field conditions. *Indian J. Hort.*, **73(1):** 60-65.

152. Sandlerink, J.M., Van Ooijen, J.W. and Purimahua, C.C. 1995. Localization of genes for bacterial canker for resistance to Lycopersicon peruvianum using RFLPs. *Theor. Appl. Genet.*, **90:** 444-450.

153. Scot, J.W. 1993. Breeding tomatoes for resistance to high temperature, biotic and abiotic diseases. *Horticulture Brazileria*, **11:** 167-170.

154. Scott, J.W. and Gardener, R.G. 2007. Breding for resistance to fungal pathogen. In: Razdan, M.K. and Mattoo, A.K. (eds.). Genetic improvement of solanacieous crops. Vol.2: Tomato. Science Publisher, Enfield, NH, USA. p. 421-425.

155. Scott, J.W. and Jones, J.P. 1990. Soil borne fungal resistance to *Lycopersicon pennellii* accessions. *HortSci.*, **25:** 1068.

156. Sethi, V. and Anand, J.C. 1986. Quality characteristics of hybrid tomatoes for puree preparation. *Indian Food Packer*, **40 (3):** 13-18.

157. Setti, R.G., Chikkapaji, K.C., Madaiah, N., Swamy, A.N.M., Raghuvir, K.G., Dhanaraj, S. and Patwardhan, M.V. 1987. Screening of the suitability of Bulgarian varieties of tomatoes for processing into juice, ketchup and tomato paste. *Indian Food Packer*, **41 (1):** 7-15.

158. Sharma, A., Zhang, L., Nino-Liu, D., Ashrafi, H., Foolad, M.R. 2008. A *Solanum lycopersicum* x *Solanum pimpinellifolium* linkage map of tomato displaying genomic locations of R-genes, RGAs, and candidate resistance/defense response ESTs. *Int. Plant Genomics*, 2008: 926090.

159. Sharma, M., Arora, S.K.K., Singh, J. and Dhawan, S.S. 2002. Assessment of tomato hybrids for the preparation of paste. *Intl. Conf.* Veg., 11-14 November, 2002. Abstract, p. 301, Bengaluru (Karnataka).

160. Sharma, S.L., Sohi, H.S., Chowfla, S.C. and Pashin, S.N. 1974. Note on the reaction of tomato varieties in field causal organism of the buck-eye fruit rot diseases. *Indian J. Agric. Sci.*, **44:** 323-324.

161. Sidhu, A.S., Dixit, J., Kalloo, G. and Bhutani, R.D. 1981. Heterosis and combining ability in pear shaped tomato. *Haryana Agril. Univ. Journal of Research*, **11(1):** 1-17.

162. Sidhu, G. and Webster, J.M. 1973. Identification of resistance gene in tomato against root-knot nematode (*M. incognita*). *Canadian Journal of Genetics and Cytology*, **15:** 663.

163. Siemonsma, J.S. and Piluek, K. 1993. *Prosea*: Plant Resources of South-East Asia, No.8: Vegetables. Pudoc Scientific Publishers, Wageningen, The Netherlands, p. 1199-1205.

164. Sims, W.L. 1980. History of tomato production for industry around the world. *Acta Hort.* 100-125.

165. Singh, B. and Choudhury, B. 1972. Resistance in tomato to root-knot nematode. *Haryana J. Hort. Sci.*, **1:** 63-68.

166. Singh, B. and Choudhury, B. 1974. Screening tomato cultivars for resistance to *Meloidogyne* spp. *Pest Articles and News Summaries*, **20:** 319-322.

167. Singh, B., Banerjee, M.K. and Singh, K. 1974. Inheritance to root-knot nematode in tomato. *Society for Advancement in Breeding Researches in Asia and Oceana Journal*, **6:** 75-78.

168. Singh, G. and Nandpuri, K.S. 1974. Combining ability studies in tomato (*L. esculentum* Mill.) cultivar with functional male sterile lines. *Journal of Research, Punjab Agricultural University.* **11(1):** 26.

169. Singh, H. and Gill, K.S. 1987. Tomato. Procedure for breeder and foundation seed production in vegetable crops. PAU, Press, Ludhiana. p. 1-7.

170. Singh, J., Kalloo, G. and Singh, K.P. 2001. *Vegetable Crops: Nutritional Security*. Tech. Bull. No. 6, Published by ICAR-Indian Institute of Vegetable Research, Varanasi (U.P.) p. 30-37.

171. Singh, P.K., Singh, R.K., Saha, B.C. and Kumar, R. 1988. Genetic variability in tomato (*L. esculentum* Mill.). *Indain J.Agric. Sci.*, **58:** 718-720.

172. Singh, R., Sandhu, K.S., Singh, D. and Sandhu, M.S. 2002. Quality response of tomato to population density and training methods. *Haryana J. Hort. Sci.*, **29(3 & 4):** 235-236.

173. Singh, R.R. and Singh, H.N. 1980. Note on variance components in tomato. *Indian J. Agril. Sci.* **50(4):** 36-43.

174. Snyder, J.C., Simmons, A.M. and Thacker, R.R. 1998. Attractancy and ovipositional response of adult *Bemisia argentifolii* (Homoptera Aleyrodidae) to type IV trichome density on leaves of *Lycopersicon hirsutum* grown in three day length regimes. *J. Entomol. Sci.*, **33:** 270-281.

175. Sohi, H.S. and Sokhi, S.S. 1969. Relative resistance and susceptibility of different varieties of tomato to Septoria lycopersici Spek. *Indian J. Hort.*, **26:** 83-86.

176. Som, M.G. and Choudhury, B. 1976. Changes in phenol content of resistant and susceptible lines of tomato in response to leaf curl virus infection. *Soc. Experl. Agritis.*, **1(2):** 53-54.

177. Stemova, L. 2005. Resistance to *Verticillium dahlia* race-2 and its introgression into processing tomato cultivar. *Acta Horticulturae*, **695:** 257-262.

178. Stevens, M.R., Lamb, E.M. and Rhoads, D.D. 1995. Mapping the Sw-5 locus for tomato spotted wilt virus resistance in tomato using RAPD and RFLP analysis. *Theor. Appl. Genet.*, **90:** 451-456.

179. Stevens, M.R., Price, D.L., Memmott, F.D., Scott, J.W. and Obson, S.M. 2007. Identification of markers linked to Sw-7, a new tomato spotted wilt virus resistance gene, derived from *S. chilense*. Tomato Breeders Round Table, The Pennsylvania State University, Pennsylvania.

180. Stevens, M.R., Scott, S.J., and Gergerich, R.C. 1992. Inheritance of a gene for resistance to Tomato Spotted Wilt Virus (TSWV) from *Lycopersicon oeruvianum* Mill. *Euphytica*, **59:** 9-17.

181. Supe, V.S., Kale, P.N., Raijadhav, S.B., Yadav, E.D. and Patil, R.S. 1989. Performance of tomato hybrids. *J. Maharashtra Agric. Univ.*, **14(3):** 371-372.

182. Swamy, K.R.M. and Mathai, P.J. 1982. Note on combining ability in tomato. *Indian J. Agril. Sci.* **52(3):** 193-196.

183. Tal, M., Katz, A., Heikin, H. and Dehan, K. 1979. Salt tolerance in the wild relatves of the cultivated tomatoes: Proline accumulation in *Lycopersicon esculentum* Mill., *Lycopersicon peruvianum* Mill. and *Solanum pennellii* Cor. treated with NaCl and polyethylene glycol. *New Phytologist*, **82:** 349-355.

184. Tegchelaar, E.C. and Barman, R.J. 1978. Linkage of the non-ripening (*nor*) and uniform ripening (*u*) genes. *Report of Tomato Genetics Cooperative,* **28:**20.

185. Tewari, A. 1986. A note on the evaluation of tomato genotypes for resistance to bacterial wilt (*Pseudomonas solanacearum*). *Veg. Sci.*, **13:** 196-198.

186. Tewari, R.N. and Choudhary, B. 1993. Solanaceous crops. In: Vegetable Crops (Ed. T.K. Bose, M.G. Som and J. Kabir). Naya Prakash, Calcutta.

187. Thirthamallappa, H.C., Lohithaswa, A., Manjunath, A., Kulkarni, R.S. and Adiga, J.D. 1997. Resistance of *Lycopersicon* germplasm to early blight disease. *Indian J. Plant Genet. Resour.*, **10(2):** 197-204.

188. Tigchelaar, E.C. 1986. Tomato breeding. In: Bassett, M.J. (ed.), Breeding Vegetable Crops. 135-172.

189. Tikoo, S.K. and Anand, N. 1983. Annual Report. Indian Institute of Horticultural Research, Bengaluru (Karnaaka.).

190. Tiwari, R.S., Sadashiva, A.T. and Doreyappa Gowda, J.N. 2002. Stability of tomato hybrids varieties/ hybrids for processing into puree. Intl. Conf. Veg., 11-14th Nov. Abstract, p. 305, Bengaluru (Karnataka).

191. Tiwari, S. K., Singh, A.K., Singh, U., Kumar, R. and Singh, M. 2016. Crucial reporoductive traits as screening of tomato genotypes under high temperature stress. 1st International Agrobiodiversity Congress, (Sessio: Common Session), held at New Delhi, 6-9 November 2016; p. 122.

192. Van Heuseden, A.W., Koornneef, M. and Voorrips, R.E.1999. Three QTLs from *Lycopersicon peruvianum* confer a high level of resistance to *Clavibacter michiganensis* spp. *Michiganensis. Theor. Appl. Genet.*, **99:** 1068-1074.

193. Venema, J.H., Boukelien, E.D., Bax, J.E.M., Hasselt, P.R. and Elzennga, J.T.M. 2008. Grafting tomato (*Solanum lycopersicum*) onto the rootstock of a high altitude accession of *Solanum habrochaites* improves subtropical temperature tolerance. *Environmental and Experimental Botany*, **63:** 359-367.

194. Venema, J.H., Posthumus, F., and Van Hasselt, P.R. 1999a. Impact of suboptimal temperature on growth, photosynthesis, leaf pigments and carbohydrates of domestic and high altitude wild *Lycopersicon* species. *J. Plant Physiol.*, **155:** 711-718.

195. Venema, J.H., Posthumus, F., de Vries, M. and Van Hasselt, P.R. 1999b. Differential response of domestic and wild *Lycopersicon* species to chilling under low light: growth of carbohydrate content and xanthophyll cycle. *Physiol. Plant.*, **105:** 81-88.

196. Verma, T., Joshi, S., Sharma, R.K. and Pachauri, D.C. 2001. *Indian Horticulture*, **6:** 1-3.

197. Verma, T., Joshi, S., Sharma, R.K. and Pachauri, D.C. 2001. Pusa Divya: A new tomato hybrid. *Indian Horticulture*, **46 (3):** 10 and 24.

198. Vidavsky, F. and Czosnek, H. 1998. Tomato breeding lines resistant and tolerant to tomato yellow leaf curl virus issued from *Lycopersicon hirsutum*. *Phytopathology*, **88:** 910-514.

199. Virdelwala, H.A., Nanadpuri, K.S. and Singh, S. 1981. Heterosis and combining ability in tomato. *Veg. sci.*, **8(2):** 120-129.

200. Wann, E.V., Jourdain, E. L. Pressey, R. and Lyon, B.G. 1985.Effect of mutant genotypes *"hp og"* and *"dg og"* on tomato fruit quality. *J. Am. Soc. Hort. Sci.*, **110:** 212-215.

201. Warnock, S. J. 1990. Tomato evaluation and its application. *HortSci.*, **25:**139.

202. Williums, W.G., Kennedy, G.G., Yamamoto, R.T., Thacker, J.D. and Bordner, J. 1980. 2-tridecanone, naturally occurring insecticide from the wild tomato Lycopersicon hirsutum f. glabratum. *Science*, 207: 888-889.

203. Wong, J.F., Ho, F.I. and Troung, H.T.H. 2013a. Identification of major QTLsassociated with stable resistance to tomato cultivar "Hawaii 7996" to *Ralstonia solanacearum*. *Euphytica*, **190(2):** 241-252.

204. Yadav, E.D., Kale, P.N. and Wavhal, K.N. 1988. Genetic analysis of fruit dry matter in tomato. *Veg. Sci.*, **15(1):** 49-54.

205. Yadav, R.K., Kumar, M., Hussain, Z. and Yadav, P.K. 2016. Effect of high temperature on physiological and biochemical behavior of thermo sensitive genotypes of tomato (*Solanum lycopersicum* L.). 1st International Agrobiodiversity Congress, (Session: Common Session), held at New Delhi, 6-9 November 2016; p. 127.

206. Yamakawa, K. 1982. Use of rootstocks in solanceous fruit-vegetable production in Japan. *Japan Agric. Res. Qtly.*, **15:** 175-179.

207. Yamakawa, K. and Nagata, N. 1975. Three tomato lines obtained by use of chronic gamma radiation with combined resistance to TMV and Fusarium race 1-3. *Tech. News Inst. Radiat. Breed.*, **16:** 2.

208. Yang, W. and Fransis, D.M. 2007. Genetics and breeding for resistance to bacterial wilt diseases in tomato: prospects for marker assisted selection. In: Razdan, M.K. and Mattor, A.K. (eds.), Genetic Improvement of Solanaceous Crops Vol.2. Tomato. Science Publishers, Enfield, pp. 379-419.

209. Yu, Z.H., Wang, J.F., Stall, R.E. and Vallejor, C.E. 1995. Genomic localization of tomato genes that control a hypersensitive reaction to *Xanthomonas campestris* pv. *vesicatoria* (Doidge). *Dye. Genetics*, **141:** 675-682.

210. Zamir, D., Ekstein Michelson, I., Zakay, Y., Navot, N., Zeidan, M., Sarfetti, M., Eshed, Y., Harel, E., Pellban, T., Anoss, H.V., Kedar, N., Ralbin Owiteh, H.D. and Czonsnek, H. 1994. Mapping and introgression of a tomato yellow leaf curl virus tolerance gene, *Ty-1*. *Theor. Appl. Genet.*, **88:** 141-146.

Brinjal or Eggplant (*Solanum melongena* L.)

Introduction

Brinjal (Syn. eggplant, garden egg, guinea squash, cubergine (*Solanum melongena* L. 2n= 24) is an important warm-season fruit vegetable and dominate all over the world especially in Asia and Africa. Brinjal in India has many historical references in various languages (Thopmson and Kelly, 1957) with respect to cultivation. This is a typical Indian vegetable grown in many of the seasons and in more area compared to tomato and chilli in India. It is grown for its tender fruits that are usually baked, boiled, fried or prepared curries. It is also used in China, Japan and Southern European recipies. White fruit of eggplant is known for medicinal and curative properties. There are several species of *Solanum* (*S. melongena* var. *insanum*, *S. indicum* L., *S. trilobatum* L. *S. viarum* Dunal, *S. nigrum* L. and *S. xanthocarpum* Schrad and Wendl.) are used in native medicine throughout the tropics. *S. aethiopicum* and *S. macrocarpon* are cultivated for their large glabrous leaves which are consumed as vegetables. *S. incanum* is also a rich source of phenolic acid (chlorogenic acid) a potent antioxidant (Prohen *et al.*, 2013).

Origin, History and Distribution

In 'Origin of Cultivated Plants' published in 1886, De Candolle stated that *S. melongena* has been known in India since ancient times (4000 years ago) where it continues to grow wild. On the basis of variability and distribution, Vavilov (1951) and Choudhary, 1976 regarded India (Asom region) Myanmar (Burma) as the original home of brinjal due to wide diversity. Brinjal or eggplant being the crop of Indian origin is cultivated in India has many historical references in various languages (Thompson and Kelly, 1957). However, there is some uncertainty to whether the centre of orign of brinjal is India or China. In either case, it is certain that it has spread from these countries to other countries in South-East Asia, Africa, Europe and America (Chen, 1997) and this crop has developed some secondary variability in China (Vavilov, 1951). Vavilov (1928) was of the opinion that Indo-Burma is the primary centre of origin. The seeds were carried to China more than 1500 years ago where small fruited types were later developed. China is thus referred to as secondary centre of origin (Zevin and Zhukovsky, 1975). The crop is believed to have been domesticated in north-eastern India where wild forms still grow. This indicates domestication of several cultivated and wild species (Oyelana and Ogunwenmo, 2005). The crop is restricted mainly to South and South-East Asia, Southern Europe, China and Japan. The species S. *anomalum*, *S. macrocarpum* L., *S. aethiopicum* L., *S. incanum* L., *S. duplosinuatum* and several other minor species are cultivated in Africa. The species *S. quitoense* Lam., *S. muricatum* Ait. and *S. topiro* Humb and Bonpl., are cultivated in Central and South America. The species *S. aethiopicum* L. and *S. macrocarpum* L. are cultivated not only for the fruits but also for their edible leaves. In south East Asia, small fruits of *S. ferox* L. and *S. torvum* Sw. are eaten. The widely distributed species in the region (Bangladesh, Nepal and Sri Lanka) includes *Solanum torvum*, *S. indicum*, *S. insanum*, *S. surattense*, *S. pubescens* and *S. khasianum*. In India it was introduced by early traders of Arabia and Persia to the countries of the eastern and southern shores of the Mediterranean early in the middle ages. Portuguese colonoies took it to Brazil. It is now widely cultivated for its fruits in the tropical, subtropical and warm temperate zones, especially in southern Europe and the southern United States. In 1806, it was introduced to American gardens primarily as an ornamental curiosity and was probably introduced into Europe during the Moorish invasion of Spain. It gained popularity in 1890s, as minor vegetable.

The review of literature on the genus *Solanum* in India indicates that species/taxa in the southern region can be broadly classified on the basis of their origin in to indigenous and exotic viz 17 and 20 taxa, respectively (Clarke, 1883; Rama Rao, 1914; Gamble, 1921; Saldanha and Nicolson, 1976; Sharma *et al.*, 1977., Deb, 1980; Yoganarasimhan *et al.*, 1981; Mathew, 1983; Sharma *et al.*, 1984; Henry *et al.*, 1987; Subramanin *et al.*, 1987; Manilal, 1988; Ellis, 1990 and Mohanan and Henry, 1994). Out of the 37 taxa reported to occur in Southern Region of India, 27 taxa including 2 cultivated species *viz. S. melongena* and *S. macrocarpon* were collected (Velayudhan *et al.*, 1996). In India, important wild species of *Solanum* are available in the Western and Eastern Ghats, North-Eastern Hills/plains and in the *tarai* range. *S. incanum* (*S. coagulans*) and *S. melongena* var. *insanum*, are closely related to cultivated *S. melongena*. More variability in *S. incanum* occurs in *tarai* range and in *S. insanum* types in eastern peninsular tract, wherein also occurs the primitive cultivated form, the *potangi* type of *S. melongena*.

Species Availability, Evolution and Diversity

Twenty seven of the 2,000 species of genus *Solanum* are indigenous (Indian) species. There are about 28 non tuberiferous *Solanum* species growing wild in India and *S. erianthum*, *S. indicum*, *S. nigrum*, *S. stramonofolium*, *S. torvum* and *S. trilobatum* have edible fruits. Wild brinjal types are available in Southern Coastal area of Africa. A wild type of brinjal with deeply variegated leaves *S. sisimbrifolium* Lam. is considered the ancestral species of *S. melongena* L. *S. incanum* is considered as progenitor of cultivated species. It is closely related to semi-wild weedy form *S. melongena* var. *insanum*. The elctrophoretic studies also have suggested that the three species (*S. melongena*, *S. incanum* and *S. melongena* var. *insanum*) are conspecific. The ancestral form was very locally a spiny plant with small and bitter fruit occurring in India but selection for improved palatability and for relative spinelessness resulted in gradual emergence of an acceptable type (Bhaduri, 1951). India possesses rich diversity and immense variability in cultivated and closely related wild species of brinjal. Maximum diversity and distribution of *Solanum* species was noted in southern India, foothills of Himalaya and northeastern hill region (NEH). The genetic diversity of brinjal is manifested in its morphology (growth habit, plant vigour, hairiness, prickliness, fruit colour, size and shape), physiology (time of flowering, water requirement and uptake) and biochemical features (bitterness of fruit). In addition to a large number of high yielding cultivars, there are many landraces and wild species of brinjal including *S. incanum*, *S. insanum*, *S. indicum*, *S. sisymbrifolium*, *S. macrocarpon*, *S. mammosu*, *S. khasianum*, *S. aethiopicum*, *S. anomalum*, *S. integrifolium*, *S. torvum*, *S. xanthocarpum*, *S. nigrum*, *S. gilo* and *S. viarum* that have been reported. A wild type S. *melongena* var. *insanum*, with many small fruits, is found in the Bengal plains of India (Martin and Rhodes, 1979). Several wild forms with green and purple fruit colour are exhibit in Indian territory. *Solanum* species found in Asia are 38 in number and among them 22 Indian species (*S. aethiopicum* L., *S. americanum* Mill., *S. anguivi* Lam., *S. blumei* Nees., *S. dulcamara* L., *S. ferox* L., *S. incanum* L., *S. indicum* L., *S. lasiocarpum* Dunal., *S. macrocarpon* L., *S. mammosum* L., *S. marginatum* L., *S. melongena* L., *S. nigrum* L., *S. sisymbrifolium* Lam., *S. torvum* Swartz., *S. trilobatum* L., *S. undatum* Jacq.non Lam., *S. viarum* L., *S. violaceum* Ortega and *S. xanthocarpum* Schrad. Wendl.) (Choudhary, 1976).

Table 2.47: Section and Species of Subgenus Leptostemonum Present in India

Section Acanthophora Dunal	*Section Anisantherum Bitter*	*Section Cryptocarpum Dunal*	*Section Lasiocarpa (Dunal)D'Arcy*	*Section Melongena (Miller) Dunal*	*Section Monodolichopus Bitter*	*Section Oliganthes Dunal Bitter*	*Section Torva Nees*
S. aculeatissimum Jacq.	*S. pubescens* Willd.	*S. sisymbrifolium* Lam	*S. lasiocarpum* Dunal	*S. cumingii* Dunal	*S. coagulans* Forsskal	*S. aethiopicum* L.	*S. torvum* Sw.
S. viarum Dunal				*S. heteracanthum* Dunal		*S. cordatum* Forsskal	*S. barbisetum* nees
				S. incanum L.		*S. forskalii* Dunal	*S. sarmentosum* Nees
				S. incanum L.		*S. hovei* Dunal	*S. verbascifolium* L.
				S. macrocarpon L.		*S. kurzii* Prain	
				S. marginatum L.		*S. trilobatum* L.	
				S. ovigerum Dunal		*S. violaceum* Ortega	
				S. virginianum L.			

Source: Choudhary, 1976.

There is a group of 5 related ones, all prickly and diploids with 2n=2x=24, namely *S. melongena, S. coagulans, S. xanthocarpum, S. indicum and S. maccanii. The* other *Solanum* related species are *S. indicum* L., *S. trilobatum* L., *S. viarum* Dunal, *S. nigrum* L. and *S. xanthocarpum* Schrad and Wendl. These constitute a part of the genetic diversity of brinjal in India (Agarwal and Munshi, 2008,. Three species such as *S. melongena* and *S. torvum* are widely cultivated in NEH regions of India. Two African eggplants, the scarlet eggplant (*S. aethiopicum* L.) and the gboma eggplant (*S. macrocarpon* L.) are extensively cultivated there and at very limited scale in some other parts of the world.

Table 2.48 & 49: Occurance and Utility of *Solanum* Species/Wild Taxas

Sl.No.	Character	Description	Distribution
1.	*Solanum macrocarpon* Linn.		Introduced in NE region
2.	*Solanum surattense* Burm. Syn. *S. xanthocarpum* Schrad.	Used as vegetables and medicinal purpose	Common throughout India upto 2000 m in the warmer parts
3.	*Solanum indicum* Linn. Syn. *S. anguivi* Lam.	Domesticated, used as vegetable and medicine	Throughout tropical India; common in southern and eastern parts upto 1200 m.
4.	*Solanum mammosum* Linn.	Possibly introduced, ornamental with high solasodine percentage	
5.	*Solanum khasianum*	Wild and cultivated for solasodine alkaloid, sold in the market in Mizoram	NEH regions
6.	*Solanum berbisetum*	Ripe fruits are eaten	NEH regions
7.	*Solanum ferox*	Wild, leaves are used medicinally	NEH regions
8.	*Solanum spirale*	Wild but domesticated for medicinal use in Arunahcal Pradesh	NEH regions
9.	*Solanum sisymbrifolium*	Native of Africa, widely grown in Meghalaya	
10.	*Solanum kurzii*	Endemic in Garo hills, Meghalaya	
11.	*Solanum gillo* Raddi	Introduced in NE region as vegetable	
12.	*Solanum torvum* Swartz.	Wild, sold in the market of Mizorum	Throughout India; tropics except western desert area; common in the eastern plains
13.	*Solanum incanum* Linn. Syn. *S. coagulans* Forsk.	Wild	Throughout India; in the west of Shiwalik ranges, lower hills of the Western Himalaya, plains of the east; lower elevaton areas in the south.
14.	*Solanum melongena* Linn. var. *insanum*	Wild	Spread all over particularly the eastern and southern parts with its local pockets.
15.	*Solanum viarum* Dunal	rich source of solasodine	Western Ghat
16.	*Solanum surattense* Burm. F./ *Solanum virginianum* L.	Maedicinal value	Desert of Rajasthan
17.	*Solanum incanum* L., *S. indicum* L., *S. torvum* Sw. and *S. giganteum* Jacq.	Antiseptic and antidysenteric properties	Tamil Nadu, Gujarat, Madhya Pradesh, Kerala, Western Ghats, Andhra Pradesh, Odisha,
18.	*Solanum nigrum*L.	Antiseptic and antidysenteric properties	Rajasthan

Source: Asati and Yadav, 2004.

Eggplant wild relatives have much higher genetic diversity than the cultivated species (Mutegi *et al*., 2015). Maximum diversity and distribution of *Solanum* species was observed in Southern India, Foot-hills of Himalayas and North Eastern Region. The widely distributed species in the region includes *S. torvum, S. indicum, S. insanum, S. surattense, S. pubescens* and *S. khasianum*. The colour and shape of the fruit has got regional preferences and consumed in various forms. Both *S. incanum* and *S. melongena* var. *insanum* are crossable and genetic diversity in the cultivated species has been brought about by domestication, mutation, natural intercrossing and selection and human selection and hybridization. There are many other *Solanum* species *i.e. S. nigrum, S. nodiflorum etc.* are sometime cultivated or half-cultivated (*i.e.* spontaneous plants are looked after and harvested) and fruits are consumed in Africa and Asia. *S. melongena, S. incanum* and *S. insanum* (*S. melongena* complex) are very closely related with each other, but the taxonomic status of these species is disputed and it is not clear, whether *S. incanum* and *S. insanum* are the progenitors or derivatives of *S. melongena*. Meiotic behaviour of some hybrids between these species has been reported to be regular, suggesting close phylogenetic relationship among above three species. Brinjal predominantly found in eastern, northern plains, southern and north-eastern India. Primitive types are found in the Eastern Ghats. Many landraces are found in the tribal belts of Maharashtra, Madya Pradesh, and other states including some drought-tolerant landraces in West Bengal (Rana *et al*., 1994). Sheela (1982) observed

Table 2.50: Wild Gene Pool Classification in *Solanum*

Gene Pool	*Species*	*Remarks*
Primary gene pool	*Solanum incanum; S. insanum*	In primary gene pool both species are closely related (phylogenetically) to the cultivated eggplant and are part of the "eggplant Clade". Artificial cross combinations through artificial pollination between the species produce fertile F_1 hybrids successfully with complete chromosome pairing. Primary gene pool is constituted by only two species *i.e. Solanum incanum* (drought tolerant species grows in desertic environments in Middle East and North Africa); *S. insanum* (ancestor of eggplant) which provide fertile F_1 hybrids with *S. melongena.*
Secondary gene pool	*S. anguivi, S. dasyphyllum; S. lichtensteinii; S. linnaeanum; S. pyracanthum; S. aethiopicum; S. tomentosum; S.violaceum; S. campylacanthum; S.vespertlio*	In secondary gene pool members (around 20 species) are probably normally classified as different species than the crop species under consideration (the primary gene pool). Cross combination produces F_1 hybrids which have partial F_1 sterility due to incomplete chromosome pairing. Development of cross combinations through conventional method is difficult between members of the primary and secondary gene pools due to the existence of reproductive barriers and may require embryo culture. Breeding point of view this group is very important due to genetic diverseness and within it wide diversity in tolerance to abiotic stresses, resistance to pests and diseases, fruit size, fruit shape and biochemical composition.
Tertiary gene pool	*S. sisymbrifolium; S. torvum; S. macrocarpon; S. elaeagnifolium*	In tertiary gene pool admixture of species from subgenus *Leptostemonum*, including species from the Old World as well as from the New World, with which occasionally it may be possible to obtain highly sterile or low fertility hybrids after embryo rescue or colchicine treatment and somatic hybridization (Daunay and Hazra, 2012; Rotino *et al.*, 2014). *S. torvum* and *S. sisymbrifolium* are of great interest in breeding owing to its resistance to multiple diseases (Blestsos *et al.*, 2003; Gousset *et al.*, 2005). *S. macrocarpon* is resistance to shoot and fruit borer (Kumar and Sadashiva, 1996) alongwith mites, nematodes, wilts and drought (Gowda *et al.*, 1990). Another species *S. elaeagnifolium* is an invasive weed having high tolerance to drought (Christodoulakis *et al.*, 2009) may be useful genetic resource for eggplant breeding.

resistance to bacterial wilt in SM-6 and it was classified into 11 phenotypic groups based on fruit colour, shape and presence/absence of prickles on stem leaves. Some of the important landraces of north Central India include: 'Bundelkhand Desi' (drought tolerant), 'Dudhiya Baigan' (cluster bearing with milky white round to oval fruits), 'Ramnagar Baigan' [suitable for cultivation under river belt of Ganga (particularly in Varanasi and nearby disricts of Uttar Pradesh), bear large fruits], 'Jethuwa Baigan' (suitable for off season cultivation, as summer crop), 'Kuchbachia Baigan' (small size cluster bearing, suitable for long distance transportation), 'Jafrabadi Baigan' (deep purple, round oval and long fruited, weighing 250-400 g with less seeds), 'Balfahwa Jathuwa Bhanta' (tall growing 1.4-1.8 m with profuse branching, suitable for summer crop, bear fruits in cluster, better self life and "Cutttack Pinde" for better taste in Odisha).

Combination breeding indicates that *S. sisymbrifolium* and *S. elaeagnifolium*, confirming that both species are very distant from eggplant.

Gene and Genetics of Characters

In a 5x5 diallel study it was noticed that days to flower and the number of primary branches are goverened by overdominant gene action, the weight of fruits/plant by dominant gene action, plant height by additive gene action with some overdominance; flowers/inflorescence; number of long + medium styled flowers, length of fruit and equatorial perimeter of fruit by additive gene action (Peter and Singh, 1973). In brinjal several genes are reported which are associated with different traits (Table 2.51).

Phenolic acid content in brinjal is controlled by numerous genes. Breeder can select plants that produce marketable fruits with enhanced antioxidant content. Gene effects for various attributes are as shown in Table 2.52.

Table 2.51: Gene and Gene Action in Brinjal

Character	*Number of Gene*	*Type of Gene Action*
Fruit shape	3 genes	Round shape partially dominant to long
Prickly nature	Monogenic	Prickly nature dominant to non-prickly
Fruit clusterness	Monogenic	Clusterness partially dominant to solitary nature
Fruit size	Monogenic	Large fruit size partially dominant to small fruit size
Fruit colour	Monogenic	Purple dominant to green

Chadha *et al.* (1988) reported partial dominance for total chlorophyll, chlorophyll 'a', chlorophyll 'b' and anthocyanin. The components of variation showed that additive component was important for total chlorophyll, chlorophyll 'a', and chlorophyll 'b' whereas both additive and non-additive genetic components were involved in the inheritance of anthocyanin. The narrow sense heritability estimates were medium for these characters. The inheritance of various traits has been worked out which are as shown in Table 2.53.

Botanical Description

Brinjal belongs to non-tuber bearing plant. The somatic chromosome number 2n= 24. Polyploids with 2n=36, 48 have been produced with non-significant. The detail botanical descriptions are given in Table 2.54.

Table 2.52: Gene Effects on Various Attributes of Brinjal

Characteristics	*Gene Effects*	*References*
Number of days to flowering	Additive	Mital *et al.*, 1976; Bajpai, 1977; Gill *et al.*, 1976.
	Non-additive	Singh *et al.*, 1979; Singh and Mital, 1988.
	Over-dominance	Peter and Singh, 1973; Peter and Singh, 1976; Hani *et al.*, 1977.
	Dominant	Sidhu *et al.*, 1980.
Fruit weight	Additive	Singh *et al.*, 1982; Dixit *et al.*, 1984; Mital *et al.*, 1976; Peter and Singh, 1973; Peter and Singh, 1976; Salehuzzaman and Alam, 1983; Randhawa, 1987.
	Both Additive and Dominance	Dharmagowda *et al.*, 1979.
Number of fruits/plant	Additive	Gill *et al.*, 1976; Singh *et al.*, 1979; Salehuzzaman and Alam, 1983; Randhawa, 1987; Singh and Mital, 1988; Chadha and Sharma, 1989.
	Both Additive and non-additive	Dixit *et al.*, 1984.
Early yield	Additive	Kumar and Ram, 1987.
Yield	Additive	Gill *et al.*, 1976; Sharma, 1975; Madalageri, 1986; Naulsri *et al.*, 1986a; Randhawa, 1987; Kumar and Ram, 1987; Chadha and Sharma, 1989.
	Both Additive and non-additive	Borikar *et al.*, 1981; Dixit *et al.*, 1984; Dharmagowda *et al.*, 1979; Hani *et al.*, 1977; Singh *et al.*, 1979; Singh and Mital, 1988.
	Dominance	Mital *et al.*, 1976; Sidhu *et al.*, 1980; Salehuzzaman and Alam, 1983.
	Over Dominance	Singh *et al.*, 1982.
Fruit length	Additive	Peter and Singh, 1973; Peter and Singh, 1976; Singh *et al.*, 1979; Sidhu *et al.*, 1980; Dixit *et al.*, 1984; Singh and Mital, 1988.
Fruit width	Both Additive and non-additive	Singh and Mital, 1988.
Fruit size index	Additive	Gill *et al.*, 1976.
Fruit circumference	Additive	Dixit *et al.*, 1984.
Plant height	Additive	Gill *et al.*, 1976; Bajpai, 1977; Singh *et al.*, 1979; Borikar *et al.*, 1981.
	Both Additive and non-additive	Dharmagowda *et al.*, 1979; Dixit *et al.*, 1984; Singh and Mital, 1988.
Number of flowers	Dominance	Sidhu *et al.*, 1980.
	Additive	Peter and Singh, 1973; Peter and Singh, 1976.
Seed/fruit	Over Dominance	Dharmagowda *et al.*, 1979.
Number of branches	Additive	Gill *et al.*, 1976; Borikar *et al.*, 1981.
	Dominance	Bajpai, 1977; Singh and Mital, 1988.
Plant spread	Dominance	Bajpai, 1977.

Table 2.53: Inheritance of Traits in Brinjal

Attributes	*Gene Dominance*	*References*
Fruit colour	Purple>Green	Tatebe, 1944; Khan and Ramzan, 1954
	Green>White	Swamy Rao, 1970; Choudhuri, 1972; Choudhuri, 1977; More and Patil, 1982; Gopinath *et al.*, 1986; Joshi 1989
	Monogenic	Bolla, 1932; Janaki and Ammal, 1933; Swamy Rao, 1970; Choudhuri, 1972; Choudhuri, 1977; More and Patil, 1982; Patil and More, 1986
	Two genes in complementation	Thakur *et al.*, 1969
	Complex	Fukusawa, 1964; Gopinath *et al.*, 1986
	Interaction of 3 non-allelic genes	Khapre *et al.*, 1988
Plant colour	Purple>Green	Thakur *et al.*, 1969; Sharif and Habib, 1977; Nimbalkar and More, 1980; Patil and More, 1983
Stem colour	Purple>Green	Wanjari and Khapre, 1977; Patil and More, 1983
	Duplicate gene	More and Patil, 1982; Nimbalkar and More, 1980
Hypocotyl/seedling colour	Purple>Green	Wanjari and Khapre, 1977
	Monogenic	Khapre *et al.*, 1986
	Duplicate gene	Rangaswamy and Kadambavana Sundram, 1973

Attributes	*Gene Dominance*	*References*
Corolla colour	Purple>White or Light purple	Wanjari and Khapre, 1977
	Monogenic	Rangaswamy and Kadambavana Sundram, 1973
	Interaction of 3 or more non-allelic genes	Khapre *et al.*, 1986
Leaf colour	Monogenic-2	Sharif and Habib, 1977
	Complementary genes	Patil and More, 1983
Leaf Vein colour	Purple>Green	Khapre *et al.*, 1986
	3 or more genes	Patil and More, 1983
Fruit flesh colour	Green flesh>White	Wanjari and Khapre, 1977
	Monogenic	Khapre *et al.*, 1986
Fruit shape	Elongated>Round	Swamy Rao, 1970
	Round>Oval	Choudhuri, 1977
	3 genes control	Patil and More, 1983
	4 genes control	Nimbalkar and More, 1980
Bearing habit	Cluster>Single	Swamy Rao, 1970
	Single gene	Rangaswamy and Kadambavana Sundram, 1973
	Trigenic	Fukusawa, 1964
Plant Spread	Erect>Spreading polygenic	Vijay Gopal and Sathumadhavan, 1973
	2 complementary genes	Wanjari and Khapre, 1981; Khapre *et al.*, 1985
Plant height	Tall>Dwarf Monogenic	Choudhuri, 1972; Choudhuri, 1977
Prickle presence on petiole/ leaf/stem	Prickle presence>Absence Monogenic	Rangaswamy and Kadambavana Sundram, 1973
Male sterility	2 recessive nuclear genes, ms_1 and ms_2	Chauhan, 1984
Style erectness	Incurved style>Erect Monogenic	Rangaswamy and Kadambavana Sundram, 1973

Table 2.54: Botanical Description

Attributes	*Description*
Growth Habit	It is mesophytic in nature. Plants are very polymorphous, bushy, erect (60-120 cm tall), aculeate or unarmed annual herb, woody at the base and 0.5-1.5 m in height. Growth is indeterminate; however some determinate forms are also recorded.
Root	Although seedlings have tap roots, and they change to fibrous root system on transplantation.
Stem	The plant stands firm (aerial errect) herbaceous, cylindrical green and compact above the ground and usually sets out a large number of branches on the main stem. Branches are light green or tinged with violet, rather thick, densely clothed with grey, stellate tomentum.
Leaves	Leaves are alternate, in pairs, entire, lobed or pinnnatifid, solitary or shorter or longer stalks, simple, ovate or oblong-ovate-eliptic. Base is often unequalsided and oblique, rounded, truncate or cordate, apex, triangular, acuminate, obtuse or acute; irregular undulate, angular or more or less deeply lobed and above dark green. On both the sides dull, densely clothed with soft grey, stellate tomentum, flaccid, 7.0- 25.0 cm long, 5.0-15.0 cm wide, reticulate-veined and veins transparent. Petiole is robust, subterete, yellowish green, densely white stellate tomentose and 1.5-10.0 cm long.
Floral biology	Inflorescence is monochasial, scorpioid and cyme. Flowers generally emerge 40-45 days after transplanting. The flower types in brinjal (Krishnanmurthi and Subramaniam, 1954) are: (a) long styled (b) medium styled (c) true short styled (d) pseudo-short styled. Only long styled and medium styled flowers set fruit but pseudo-short styled and true short styled flowers did not set fruit. The percentage of long and medium styled flower is a varietal character. Flower is pentapetalous having white, green or purple colour. Inflorescence is solitary but sometime it constitutes a cluster of 2-5 flowers, depending on variety to variety. Flowers are rather large, stalked, lateral or leaf-opposed, the female ones solitary, the males in many flowered inforences. Pedicels are light green or strongly tinged with violet, terete, aculeate or not, densely stellate tomentose, patent or curved, 1.5-2.0 cm long. Flower is complete, actinomorphic and hermaphrodite. Calyx is 5 lobed far less than half way down, tubular campanulate, gamosepalous and persistent. It forms a cup like structure at the base. On the out side greyish green oftenly strongly tinged with purple, on both sides densely stellate tomentose and 1.5-2.0 cm long. Tube is ovoid-oblong, obtusely pentagonous, ±1.0 cm long, 0.5-0.7 cm diameter; segments narrowly oblong, acuminate, obtuse, obliquely erect, 0.5-1.5 cm long. Corolla is gamoptalous, deeply 5 lobed, stellately spreading, on the outside light violet, densely white stellite tomentose, within dark violet, glabrous, rugose and 2.5-4.0 cm in diameter. Lobes are short, ovate-oblong and rather acute. There are 5 erect stamens, ovoid-oblong, vitelline, opening at the top by 2 pores and 0.5 cm long. Stamens are free and inserted at the throat of corolla. Anthers are cone shaped (connivent in a cone) and free. Anthers dehiscing by apical pores or short slits. Ovary is hypogynous, bicarpellary, syncarpous and with basal placentation (Prasad and Prakash, 1968).

Attributes	Description
Anthesis and pollination	It is a self pollinated crop but the cross pollination (due to heterostyly) to the extent of 30 to 40 per cent. In brinjal anthesis occurs mainly in the morning *i.e.* 5.0-6.0A.M. (Aug-Sept.) and 9.30-11.15 A.M. (winter). A few flowers open at 16 hr. Vijay *et al.* (1977) reported anthesis time of brinjal as 6.0-11.30 A.M. Peak time for anthesis is 8.30-10.30 A.M.Pollen fertility on the day of anthesis to 7-10 days after anthesis and stigma receptivity one day before and 6-8 days after anthesis. Anther dehiscence occurs 15-20 minutes after flowers have opened (pollen dehiscence starts dfrom 9.30-10.0 A.M.). It is observed that the day light, temperature and humidity mainly influence anthesis and dehiscence. The receptivity of stigma could be observed from the plump and sticky appearance which gradually turns brown with loss of receptivity. Pollen is most fertile immediately after the anther dehiscence. Opening of anthers is mostly by pore or slit at or near the apex. Deshpande *et al.* (1978) reported that maximum anthesis occurred at 6.0 A.M. in Arka Shirish and at 8.0 A.M. in Arka Sheel and Arka Kusumakar.
Fruit and fruit set	The fruit is indehiscent pendent, fleshy berry, borne singly or in clusters. The optimum temperature for growth and fruit set is 15.5-21.1°C. Many of the round varieties set fruit at slightly lower temperature but are highly susceptible to frost. The long fruited varieties set fruit higher temperature and show tolerance to frost.Fruit setting in the brinjal varieties with long-styled flowers varied from 70-86 per cent while short styled flowers did not set fruit. The fruit can be oval, round or oblong, 5.0-20.0 cm in diameter. The epidermis or skin of the fruit is thin and smooth. The fruit is mostly fleshy placenta, in which the seeds are embedded in a mass of spongy tissue. Most cultivars today have purple fruits, although colours may range from green, light green, yellowish white, red and blackish purple, purple and their various combinations.Fruiting calyx is campanulate, much enlarged, bursting on one or 2 sides, pale green or tinged with purple, thinly stellatetomentose, unarmed or beset with scattered prickles, 30.-6.0 cm long and 3.0-5.0 cm in diameter. Fruiting pedicel is much elongated, much thickened towards the apex, unarmed or with scattered small prickles, pale green, nodding and 3.0-7.0 cm long.
Seed	Seeds are numerous, soft and borne on a fleshy placenta which completely fills the locular cavity. The seeds are bitter because of nicotinic alkaloids (Bose, 2002.)

Horticultural Classification

Classification Based on the Fruit Shape and Plant Spread

There are 3 main botanical varieties (based on fruit shape and plant spread) under the species *melongena* are as:

Table 2.55: Classification Based on the Fruit Shape and Plant Spread

Description	Solanum Species
Fruit is long, round or egg shaped and oval.	*S. melongena* var. *esculentum* L.
Snake brinjal' fruit is extra ordinarily long or slender types, leaf highly prickly.	*S. melongena* var. *serpentinum* L.
Plant is extremely short and dwarf.	*S. melongena* var. *depressum* L.

Source: Peter (1998).

(i) Fruit Shape

- ☆ **Long brinjal-**'Pusa Purple Long' 'Pusa Cluster'
- ☆ **Round brinjal-**'Pusa Purple Round' 'T3'
- ☆ **Oval brinjal-**'Pusa Kranti'

Table 2.56: Grouping of Eggplant on the Basis of Colour

Sl.No.	Group	Genotypes/Varieties/Strain
1.	Long purple	Pusa Purple Long, Pusa Purple Cluster, Pusa Kranti, KT-4, DBSR-31, Arka Sheel, Azad Kranti, ARU-1, S-16, Punjab Barsati, Punjab Sadabahar, Punjab-129-5, Pant Samrat, H-4, ARU-2C, H-7, NDB-25, BB-26, Sel-4, BB-13, KS-331, JB-15, NDB-28-2, Pusa Hybrid-5, NDBH-6, ARBH-201, PBH-6, Pant Hybrid-1, MHB-1, Long-13, Junagadh Long, Pusa Bhairav, IC90975-11025, IC111083
2.	Round purple	PB-91-2, T-3, K-202-9, H-8, KS-224, DBR-8, AB-1, CHBR-1, D-2-88-6, Pusa Hybrid-6, Vijay Hybrid, NDBH-1, Pusa Hybrid-9, JVH-1, MHB-2, MHB-10, Round-14, Azad Hybrid, Pusa Purple Round, Black Beauty, Jamuni Gola.
3.	Round green	Krishnanagar Purple Round, Pant Rituraj, Surti Gota, T-3, Banaras Giant, Ramnagar Giant
4.	Small round	Aruna, DBSR-44, PLR-1, DBSR-91, JB-64-1-2, ABH-1, ABH-2, Phule Hybrid-2.
5.	Long green	Arka Kusumakar, Arka Sirish, Krishnanagar Green Long.
6.	Oval or oblong purple	Arka Navneet, Junagadh Long, H-4, Pusa Anmol, Suphal, IC112341, IC112356, IC127018
7.	Round with strips (purple white)	Manjari, Manjari Gota
8.	Round white	Swarna Shree
9.	White oblong with purple stripes	IC112960, IC099747
11.	Deep purple oval	EC305049
12.	Milky white oblong	IC090963
13.	Green oval	IC126996, IC90950, EC316268
14.	Green round with stripes	EC304979
15.	Green oblong with stripes	IC99699, IC99666, IC90931

(ii) Fruit Colour

- Purple brinjal (anthocyanin present)-'Pusa Purple Long'
- Green brinjal (chlorophyll present)-'Arka Kusumakar'
- White brinjal
- Variegated brinjal

Augmentation of Genetic Resources and Crop Improvement

Breeding Priorities

- High yield
- Good quality
- Resistance to shoot and fruit borer (*Leucinodes orbonalis* Guen.)
- Resistance to *Mycoplasma* disease

Methodologies

In brinjal following breeding methodologies are being utilized for crop improvement:

- Introduction
- Mass selection
- Pedigree method

Over the years, a large number of traditionally grown cultivars have evolved in different agro-climatic zones of India. These cultivars possess valuable genes for resistance to biotic and abiotic stresses and adaption to different environments. A large number of these cultivars have been utilized directly as varieties and also as donor parents after proper characterization and evaluation (ICAR-NBPGR, N. Delhi, 1993). A literature survey has revealed that continuous selections and hybrid vigor have played an important role in the evolution and development of improved cultivars bearing various desirable characteristics. Such cultivars bear large edible fruits and are adapted to a wide range of climatic conditions. Local preferences and rapid adaptability of newly developed cultivars have paved the way for development of different genotypes (ICAR-NBPGR, N. Delhi, 2007-08; Rai *et al.*, 1995). At present, a selected number of early, high qualities, high-yielding and disease resistant varieties and hybrids are commercially cultivated and these account for almost the entire production of brinjal in India. On the basis of morphological variations, following promising accessions have been identified for different traits, are given in Table 2.57.

Horticulture Research Centre, Bangladesh Agricultural Research Institute (BARI) has developed a brinjal cultvar 'Shufala' (11-1-324) but cultivars Uttara, F_1 Tarapuri and F_1 Suktara developed in the years 1985, 1992 and 1992, respectively are very popular among the growers. In Nepal brinjal varieties like Long-3, Round-14 and Suphal developed by Indo-American Seed Co., (India) are commonly being grown. In Pakistan, cultivars like Sarhindi long, Multan Selection and Pusa Purple Long (R) are very common and being grown commercially.

Table 2.57: Important Traits and Promising Accessions/Lines/Varieties

Traits	*Promising Accessions/Lines/Varieties*
Good cluster bearing	IC201131, IC201132, IC201136, IC201139 and EC381023
Cluster bearing	IC112960, IC99747, IC111428, EC304951, NIC-3654, NIC-5874, NIC-5877. IC383190, IC5904
Very long fruits (> 30 cm)	IC201132, IC201231, NIC-10459 and NIC-6067
Long fruits (> 20-30 cm)	IC90087, IC90103, IC90125, IC90148, IC99747, IC90982, VKBR-13, BWBH-3 and 99702
Oval purple fruits	IC112343, IC127018, IC99675, IC99715, NIC-4283, NIC-6878, NIC-6885, NIC-6886, NIC-6960, N IC-9430, EC109083, CGB-75 and EC305044
Purple long fruits	IC74234, IC89905, IC90108, IC90508, IC90862, IC90764, IC112312, IC90975, IC126869, IC144038, IC144084, IC144166, IC111025, IC111083, EC304975, EC329327, EC396466, EC396505, IC89847 and IC400211
Purple round fruits	IC89832, IC89841, IC126955, IC137674, IC143995, IC144045, CGB-10, CGB-29 and NIC-5767
Deep purple oval fruits	IC90956, NIC-3603, NIC-4064, CGB-37 and EC305049
High yield	BBSR-34, BBSR-51, BBSR-104, BBSR-109, BBSR-118, CH-575, CH-378, CH-450, CH-665, Acc. No.-373, EC379246, IC90806, IC90031, HAB-884, HAB-894, 677-10-8, BB-46, 586-4, 480-1, 624-5, BBSR-135, BBSR-127, EC384565, EC379246, IC90806, 480-1, 498-8, 509-4, 540-8, 550-4-2, CGB-10, CGB-18, CGB-24, CGB-28, IC332506, IC383190, VKG-2/228, IC144060, BBSR-36, BBSR-54, BBSR-8, CH-573, CH-573, CH-665, IC127241, IC126727, IC99674, IC89974, IC249326, EC438678, EC305039, EC136240, IC111087, IC112342, IC136306, G-12, G-13, P-4, G-8, G-10, G-11, G-13, AB-2, IC74207, IC74205, IC137769, EC282490, IC113002, PB-17, PB-26, IHR-115, IHR-149, Rajendra Baigan-II and B-40, Rajendra Annapurna, EC305043, N IC-5765, B-1, LC-1, LC-2

Introduction and Selection

A large number of land races/traditionally grown cultivars have been developed in different agro-ecological zones of India and these landraces possess valuable genes for resistance to biotic and abiotic stresses and adaptation to various environments. In long fruited types 'Pusa Purple Long', 'H-H', 'Pusa Purple Cluster', 'Pusa Kranti', Azad Kranti', 'PBR 129-5' ('Pant Samrat') and in round fruited types 'Pant Rituraj', 'Jamuni Gole Baingan' 'T 3' and 'S 16' are highly productive. Similarly few varieties like 'Junagadh Long', Junagadh Oblong', 'Gujarat Brinjal Long-1', 'Gujarat oblong Brinjal -1' and Junagadh Brinjal Green Round-1 has been developed at Anand Agricultural University, Anand and Junagadh Agricultural University, Junagadh (Gujarat), are

Table 2.58: The Promising Native Eggplants Grown in different Regions of the Country

States		*Native Eggplants Genotypes*
Uttar Pradesh and North Central Region	:	Muktakeshi, Ramanagar Giant, Batia, Bholanath, Sheonath, Bundelkhand Desi, Dindhya Baigan, Ramanagar Baigan, Kucjilachi Baigan, Jafrabadi, Balfahwas Jetliwa Bhanta
Bihar	:	Hyderpur, Improved Muktakeshi
West Bengal	:	Mukta, Lokeshi, Krishnanagar Purple Round, Krishnanagar Green Long, Muktakeshi, Baromashi, Kalyani Green, Nurki
Maharashtra	:	Manjari Gota, Bourad Local, Borgona, Local Khandesh, Khed Shivapur, Muleti, Malkapuri, Shirur Khandala, Khanpur, Rajangona, Songona Kagal, Tambalwadi, Local, Chikhalgaon Local, Dorli
Gujarat	:	Junagadh Bhattu, Surti Gota, Gulabi Dorla
Karnataka	:	Coorg Local, Irangeri, West Coast Green Round
Madhya Pradesh	:	Nurki Round, Nurki Long, Banjari
Tamil Nadu	:	Wynad Giant, Pattabiram, Annamalai
Punjab	:	Ludhiana Local, Batia
Other cultivars (mainly from Bihar and Odisha)	:	Changulia, Okhla, Very Long, Lamba, Round Bangalore, Satputia Hara, Satputia Kathvatia, Grasevindy, Local Kalimpong, Gola hara, Long Saharanpur, Local Oval, Kusumsita, Behda Tupri, Bada Tupri, Raitan Mani, Burupada, Borigumma, Kutiguda, Dhenkanal, Batparsi, Balugaon, Market, Banamalipur, Narayanpur, Nisan Local, Jaganathpur (Adivasi), Malipondi, Anchla-B, Bheda Ramchandrapur, Chekkguda, Narikiri Tholi, Randha Pathuria, Chitrakonda, Rayagada Baigan, Kharagpuri, Bogma, Bhola, Bara, Mahanian, Surajmukhi, Bhanta, Bor Baingna, Pol Baigan, Cuttack Pindi

Source: Swarup, 1998.

Table 2.59: Promising Lines Identified for Useful Traits

Character		*Accessions*
Long fruits (>20 cm)	:	IC90087-1, IC90093, IC90103, IC90125, IC90148, IC90178, IC90982, IC99702, IC104086, IC11040, IC111083, IC111087, IC112342, IC112804
Fruit breadth (>20 cm)	:	IC99649, IC99666, IC99695, IC99748, IC104083, IC90958, IC90783
Early type	:	EC304980, IC90111-2, IC90135
Big round type	:	EC305042
High yield potential	:	IC127023, IC90762, IC90964, IC09701
Cluster bearing	:	IC112960, IC99747
Broad leaves	:	IC90957

Source: N. Sivaraj *et al.*, 1998, NBPGR, New Delhi.

selected and recommended for commercial cultivation in different regions of the country.

Thomas *et al.* (1990) evaluated a total of 319 germplasms of brinjal at ICAR-NBPGR-Issapur Farm, N. Delhi against 43 agro-botanical characters. Wide variability was observed in plant height (10-80 cm), number of branches/plant (3-9), spininess (spiny-non spiny), foliage colour (green purple), days to 50 per cent flowering (21-50), first fruit set (32-91), fruit maturity (56-100), number of fruits/plant (9-29), number of clusters/plant (10-29), average fruit length (4.2-25 cm), circumferences (2.7-31.2 cm) and yield in 5 plants (0.270-12.575 kg).

In Goa state two local land races of brinjal *viz.*, *Taleigao* and *Agassim* which are popular among the growers.

- ✰ The 'Taleiga type' produces violet, round fruits, the average fruit weight being 350 g and per plant yield about 2.5-35 kg. This type has hard fruit structure, which accounts to longer shelf life.
- ✰ The 'Agassium type' produces violet-purple, elongated fruit of 300 g and gives about 2.0-2.5 kg/plant. This type has soft and puffy flesh which has less self life.

Hybridization and Selection

In brinjal first time artificial hybridization was reported by Bailey and Munson (1891) and first time useful heterosis was reported by Halsted (1901). Heterotic effects were confirmed by Odland and Noll (1948).Varietal hybridization was first reported by Rao (1934). Some of them were utilized successfully in hybridization programme for getting desirable traits shown in Table 2.60.

In small group several hybrids like 'MHB 10', 'MHB 39', 'ABH 1', 'ABH 2', 'Hybrid 2', 'Sunmex 9',

Table 2.60: Genotypes Used in Hybridization and Resulting Salient Features

Female Parent	*Male Parent*	*Salient Features*
'Black Beauty'	'PI 169651'	More yield, more fruits/plant
'Pusa Purple Long'	'Hyderpur' ('Pusa Anmol')	Early, more yield
'Muktakeshi'	'Clustered White'	More yield
'Muktakeshi'	'Hyderpur'	More yield, larger fruit, superior quality
'Green Long'	'Pusa Purple Long'	More yield
'Pusa Purple Long'	'Clustered White'	More yield
'Muktakeshi'	'Banaras Giant'	More yield
'Kalyanpur Type 2'	'Pusa Purple Round'	More yield
'Black Beauty'	'Balady White Long'	Early, more yield

'Sunmex 19' and 'Aruna' were popularized. Similarly, in round group 'NDB Hybrid 1', 'Nembakar' and 'Pusa Hybrid 6' and under long fruited 'Pusa Hybrid 5', 'ARBH 2', 'Pant Hybrid 1', 'HOE 1404' and 'ARU 1' were laergely being grown by farmers. Two hybrids *i.e.* 'Anand Brinjal Hybrid-1' has been developed Anand Agricultural University, Anand and 'Gujarat Brinjal hybrid-2' has been developed by GAU (Gujarat). At BCKV, Kalyani (West Bengal) several hybrids were evaluated but outstanding hybrids based on heterosis and *per se* performance were BCC-54 x BCC-24 and BCC-1 x BCC-24 (Chattopdhayay, 2011). Gharge *et al.* (2016) found maximum heterosis for fruit yield/plant in the cross RHRB-74 x DBSR-195 (56.96 per cent) followed by JBR-2 x RHRB-77 (50.89 per cent), RHRB-77 x Kudachi (38.23 per cent) and RHRB-77 x Kudachi (37.97 per cent). The hybrid RHRB-77 x DBSR-195 also exhibited highly significant heterosis for all the characters except

Table 2.61: Heterosis for Various Attributes

Cross	*Attribute*	*References*
Black Long x Pusa Purple Cluster	Yield	Mital *et al.*, 1976.
PH-4 x S-16; PH-4 x BR-112	Yield	Dixit *et al.*, 1982.
DMP x IMK; Pusa Purple Cluster x PK; DMP x S-5	Yield	Chadha and Sidhu, 1982.
Pusa Purple Cluster x Punjab Bahar; PPI x S-373; Annamalai x S-96	Yield	Kandasamy *et al.*, 1983.
Ludhiana Local Long x Pusa Purple Cluster	Yield	Dahiya *et al.*, 1984.
Pusa Purple Cluster x S-317	Yield	Singh, 1984
White Bunches x Pusa Purple Cluster	Fruit weight	Mital *et al.*, 1976
Ludhiana Local Long x PH-4	Fruit weight	Dahiya *et al.*, 1984.
5148 x 6703	Plant height	Ram *et al.*, 1983
5148 x 6703	Fruit number	Ram *et al.*, 1983
5623 x S-4	Fruit number	Dahiya *et al.*, 1984.
Ludhiana Local Long x Pusa Purple long	Fruit number	Dahiya *et al.*, 1984.
PK x Punjab Chamkila	Total chlorophyll, chlorophyll a	Chadha *et al.*, 1988

Table 2.62: Estimates of Heterosis Based on Mid Parent (Average Heterosis) and Better Parent (Heterobeltiosis)

Crosses	*DF*	*AFW*	*PH*	*NB*	*FL*	*FD*	*FS*	*NFPP*	*YPP*
	Better	Better	Better	Better	Better	Better	Better	Better	Better
PR x PR-5	3.72	35.25*	-4.33	-17.03	-1.09	41.05*	7.33	19.23	45.38**
PR x CHBR-2	-1.87	-36.97**	1.46	-.6.08	-18.73	21.34	-30.48	53.57	29.01*
PR x BR-14	5.80	-86.20**	-2.16	-17.13	-12.28	-4.67	-29.19	37.14	-21.59
PR x PB	-11.00*	-22.95	31.46*	32.60	27.54	-26.88	-3.92	27.03	-18.39
PR x IVBL-9	0.88	-39.07*	-1.19	-6.08	23.16	-6.45	-22.11	15.15	-21.69
PR x RCMBL-1	5.54	-20.77	-20.43*	16.02	2.99	-54.84	-21.65	80.56*	-12.07
PR-5 x CHBR-2	-6.52	5.34	-3.19	19.21	2.83	26.84	1.31	14.29	29.08*
PR-5 x BR-14	-2.81	36.09	-0.86	5.59	14.04	-0.93	19.37	-2.86	-33.09*
PR-5 x PB	-8.59*	23.19	-9.58	19.21	47.21*	43.68*	29.44	51.35*	14.52
PR-5 x IVBL-9	-5.63	132.56*	20.47	9.00	-37.85*	13.16	-31.75	-21.21	38.89*

Crosses	DF	AFW	PH	NB	FL	FD	FS	NFPP	YPP
	Better	Better	Better	Better	Better	Better	Better	Better	Better
PR-5 x RCMBL-1	-6.70	186.34**	-5.46	11.80	-6.79	0.53	-5.59	-5.56	44.89**
CHBR-2 x BR-14	-10.67**	-41.35**	6.94	5.59	-24.91	-5.61	-38.86*	22.86	16.31
CHBR-2 x PB	-6.33	-77.03**	9.13	-13.91	-19.67	19.51	-27.73	0.00	-23.35
CHBR-2 x IVBL-9	3.35	-30.98*	-4.57	-19.43	-7.63	21.95	22.07	-12.12	-14.50
CHBR-2 x RCMBL-1	-6.72	-59.40**	29.50*	-6.83	5.98	-23.78	-27.30	-22.22	-50.77**
BR-14 x PB	9.55*	24.26	32.44*	18.01	40.66*	-10.28	-27.43	-18.92	-27.86
BR-14 x IVBL-9	-2.82	61.74*	37.22*	-0.47	-22.88	3.27	-16.11	-14.29	-5.91
BR-14 x RCMBL-1	-3.29	34.12	40.40*	30.43	-32.34*	-7.94	-31.85	-25.00	6.12
PB x IVBL-9	-9.18*	158.14**	3.78	-24.17**	64.69**	18.33	43.75*	37.84	11.81
PB x RCMBL-1	-7.27*	11.89	8.99	-6.83	54.35**	-7.50	16.84	27.03	6.18
IVBL-9 x RCMBL-1	-8.82*	-0.88	-20.51*	-24.17	49.73**	-16.67	8.48	-13.89	0.85
S.E.±	**1.37**	**111.45**	**7.91**	**0.82**	**1.05**	**0.62**	**8.33**	**2.00**	**0.39**

*,** Significant at 5 per cent and 1 per cent levels, respectively

DF=Days to 50 per cent flowering; AFW=Average fruit weight (g); PH=Plasnt height (cm); NB=Number of branches; FL= Fruit length (cm); FD= Fruit diameter (cm), FS= Fruit size (cm^2); NFPP=Number of fruit/plant; YPP= Yield/plant (kg)

Source: Patel *et al.*, 2013.

number of branches/plant, plant spread and number of fruits/plant.

In brinjal 10-25 per cent heterosis in yield is desirable for its commercial exploitation. In an experiment heteosis for productivity was worked out by Patel *et al.* (2013) of 21 brinjal hybrids for ten characters and the results are depicted in Table 2.62.

Pandey *et al.* (2016) evaluated 8 parents and 28 derived hybrids and found significant differences for the parents; F_1's and parents vs. F_1's for all the characters indicating wide variation except number of primary branches/plant and TSS due parents vs. F_1's. Significant heterosis over better and standard varieties were observed for all the traits. The crosses Mukta Kesi x Swarna Mani, Punjab Sadabahar x NDB-2 and Mukta Kesi x KS-224 were found significantly heterotic and common over better as well as standard varieties. Out of these common and desirable crosses Punjab Sadabahar x NDB-2 was found heterotic for long and Mukta Kesi x Swarna Mani and Mukta Kesi x KS-224 for oval fruit shape. Combining ability revealed that mean squares due to gca and sca were significant for all the characters. On the basis of gca effect and *per se* performance, parent Punjab Sadabahar followed by Swarna Mani and Mukta Kesi were found as the best general combiners for fruit yield/plant along with some other traits. The cross Punjab Sadabahar x NDB-2 was the best specific combiner for fruit yield, average fruit weight and TSS. The cross Punjab Sadabahar x KS-224 was the best specific combiner for number of fruits/plant. Gharge *et al.* (2016) workedout GCA and SCA using 8x8 diallele, excluding reciprocals and on the basis of GCA reported thatJBR-2, RHRB-74, RHRB-77 and Kudachi to be good general combiners for yield and its contributing characters. These parents had also recorded high *per se* performance for that trait. Majority of their hybrids had also recorded significant and desirable SCA effects, in addition to high *per se* performance, for the trait, indicating the suitability in breeding programmes for development of high yielding hybrids. All the hybrids exhibited significant and desirable SCA effects for yield. Of these, nine crosses had recorded desirable SCA effects in addition to high *per se* performance, for the trait. The hybrid RHRB-74 x DBSR-195, involving good and poor combiner parents for yield, had recorded maximum fruit yield, in addition to desirable SCA effects for fruit yield and several other component characters.

Interspecific Hybridization

Wild relatives of brinjal have variation for desirable traits such as disease resistance, colour, quality, fruiting habit *etc.* which could be transferred into the eggplant genome by introgression (Singh *et al.*, 2006; Daunay and Hazra, 2012). Nishio *et al.* (1984) classified *Solanum* species into 3 groups on the basis of compatability:

Table 2.63: Grouping on the Basis of Compatability

Group	Solanum Species
Group A	*Solanum melongena*
	S. incanum
	S. macrocarpon
Group B	*S. integrifolium*
	S. gilo
	S. nodiflorum
Group C	*S. indicum*
	S. mammosum
	S. torvum
	S. sisymbrifolium

Crosses were compatible within and between group A and B but were otherwise incompatible.*S indicum* when used as a male or female parent provided mostly unfilled seeds in eggplant however, it provide 50 per cent filled seeds when crossed with cultivar 'Uttara'. This will help in transferring Fusarium wilt resistance of *S. indicum* in to *S. melongena*. *S. incanum* produces fertile hybrids in crosses with *S. melongena*. The interspecific hybrid between *S. melongena* and *S. sisymbrifolium* failed to produce viable seed. The crossability of 5 species of *Solanum* was studied by Viswanathan (1975) and out of 14 cross combinations only *S. incanum* x *S. melongena* and *S. incanum* x *S. indicum* crossed successfully and produced viable seedlings. Of the 5 species studied, *S. incanum*, *S. indicum* L., *S. xanthocarpum*, *S. trilobatum* L. and *S. torvum* Swartz., are easly crossable with *S. melongena*. The non tuberiferous *Solanum* species and the cultivars have been widely used in breeding programme as sources of resistance to several biotic and abiotic stress factors (Swarup, 2006). Rajesekaran and Sivasubramanium (1971) obtained sterile F_1 hybrid between *S. zuccagninum* and *S. melongena*. At metaphase-1 the most important configuration was 10 II + IV but subsequent stages of meiosis were normal. It was suggested that the sterility may be due to segemental interchange. Rajesekaran (1971) obtained F_1 interspecific hybrid (2n=24) between *S. xanthocarpum* and *S. melongena* which was completely sterile despite normal meiosis with almost complete pairing of 12 pairs of homologous chromosomes. Only 1-2 per cent of pollens were stainable. Rangasamy and Kadambavana-Sundaram (1973) attempted hybridization between *S. indicum*, a wild relative of *S. melongena* varieties. The pollen fertility in F_1 was only 41.5 per cent in the only successful cross of *S. indicum* and SM-34, and in amphidiploid it was 35 per cent. The occurrence of petaloid anthers in some of the male sterile segregates in the F_1s and *S. melongena* back cross progenies revealed that hybrid sterility was due to diplotenic sterility the pollen fertility levels in the F_2 and backcross progenies ranged from 0-89 per cent with continuous variation. Siddiqui and Khan (1979) attempted interspecific hybridization between *S. incanum* x *S. melongena* L. The F_1 hybrids exhibited heterosis over their parents for several quantitative characters. The F_1s showed normal meiosis but a few pollen mother cells had univalent at metaphase-I. Khan *et al.* (1978) studied the morphological and cytological characters of *S. melongena* 'Pusa Purple Long', *S. integrifolium* and their F_1 hybrids. Rao and Baksh (1979) reported that F_1 hybrids between *S. melongena* varieties 'Giant of Banaras', 'Hungary 290469', 'Nurki Baigan' and 'Pusa Purple Long' and *S. integrifolium* were highly sterile despite mostly normal chromosome pairing.

Baksh and Iqbal (1979) attempted the crossesses of *S. incanum*, *S. khasianum*, *S. macranthum* and *S. mammosum* with *S. melongena*. Cross using *S. incanum* as female parent gave 201 seeds with 40 per cent germination; the F_1 hybrid showed appreciable resistance to frost and formed healthy seed. The reciprocal cross and all other crosses produced seedless fruit or no fruit. Rao (1979) crossed 'Pusa Purple Long', *S. melongena* var. *insanum* and 8 other species in all combinations. Four combinations produced parthenocarpic fruit only; and of the 39 which set seeds, 26 produced hybrid seedlings. Rao and Kumar (1980) observed that hybrids between *S. melongena* 'Pusa Purple Long' and *S. indica* were fertile and most showed normal meiosis, suggesting that there are small structural differences between the chromosomes of the parent. The resulting plants were vigorous, profusely branched, having more fruit and early maturity than 'Pusa Purple Long'. Rao (1980) studied the cytogenetic relationship and barriers to gene exchange between *S. melongena* L. and *S. hispidum*. When 100 flowers from each of the 3 heterostylic types of *S. melongena* 'Baromashi' were pollinated with pollen from *S. hispidum*, 50 per cent long styled flowers, 4 per cent of the medium styled and none of the short styled flowers set fruit. Pollen fertility in hybrids ranged from 0.71 to 1.53 per cent. Dever and Boriah (1980) crossed *S. melongena* and *S. melongena* var. *insanum* (wild). In this study some pair flowers having a common pedicel and calyx but distinct corollas were observed. Hybridization between *S. melongena* 'Pusa Purple Long' and *S. torvum* was successful only when the farmer was used as female parent and had parental habit, clustered flowers and the resistance to *Pseudomonas slanacearum*. Rao (1981) reported that the F_1 hybrids of *S. incanum* with vars. 'Pusa Purple Long' and 'Giant of Banaras' were fertile. The fertility of F_1 hybrids between *S. melongena* 'Nurki Baigan' and *S. incanum* was as high as 48.6 per cent. According to Kirti and Rao (1982), *S. integrifolium* (2n=24) was easily crossed as female parent with *S. melongena* (2n=24) and *S. melongena* var. *insanum* (2n=24) to give vigorus hybrids but the reciprocals crossed did not succeed. Kirti and Rao (1982a) studied chromosome relationship of spinous solanums in a reciprocal hybrid of *S. integrifolium* and *S. indicum*, and in F_1 of *S. integrifolium* x *S. surattense*. All the hybrids were over 95 per cent male sterile. Sharma *et al.* (1984) studied the crossability and pollination in some non-tuberous *Solanum* species. *S. melongena*, *S. khasianum*, and *S. sisymbrifolium* (all 2n=24) were intercrossed, with reciprocals and also self pollinated pollen fertility was 82 per cent in *S. melongena*, 93 per cent in *S. khasianum* and 95 per cent in *S. sisymbrifolium*. Pollen germination after 24 hours was 95 per cent in all self pollinations but ranged from 27.3 to 73 per cent in interspecific crosses. Embryos were formed in crosses *S. melongena* x *S. khasianum* and *S. melongena* x *S. sisymbrifolium* but not their reciprocals. Singh and Rao (1986) produced hybrids from the cross *S. melongena* cv. Dorli x *S. surattense* but only when the former was female parent while the cross *S. melongena* cv. Round Black x *S. surattense* was not successful in their direction. Guri and Sink (1988) attempted interspecific somatic hybrid between eggplant (*S. melongena*) and *S. torvum*. Successful hybrids were developed in *S. melongena* and

Table 2.64: Crossability between *Solanum* Wild Species (Direct and reciprocal crosses)

Male/Female	*S. melongena*	*S. melongena var. insanum*	*S. incanum*	*S. integrifolium*	*S. gilo*	*S. indicum*	*S. zuccagnianum*	*S. xanthocarpum*	*S. khasianum*	*S. sisymbrifolium*
S. melongena	N	G	G	G	G	U	U	U	R	U
S. melongena var. insanum	G	N	G	G	G	U	U	U	U	U
S. incanum	G	G	N	G	G	G	U	U	U	U
S. integrifolium	G	G	G	N	S	S	U	U	U	U
S. gilo	S	S	S	S	N	U	G	U	S	S
S. indicum	G	G	D	G	G	N	D	S	U	U
S. zuccagnianum	U	G	U	U	U	U	N	S	U	U
S. xanthocarpum	S	G	S	U	U	G	U	N	U	U
S. khasianum	U	U	O	U	U	U	U	U	N	U
S. sisymbrifolium	U	U	O	U	O	U	U	U	O	N

N=Normal seeds set on selfing; G= Normal seeds which grew well; D= Normal seeds but seedling dead; S= Normal seeds but did not grow; R=Fruits set but seeds are shrunken; O= Fruits set but parthenocarpic; U= Cross-unsuccessful.

Source: Rao, 1979.

S. torvum through embryo rescue technique (Kumchai *et al.*, 2013). *S. integrifolium* and *S. gilo* showed resistance due to hypersensitive reaction to little leaf pathogen (Chakrabarty and Chaudhari, 1975).

Solanum viarum (wild relative of brinjal and freeness from shoot and fruit borer attack) used by Praneetha *et al.* (2016) as donar parents (to introgress one or more resistance genes) to background of elite variety EP (Palur-1) against brinjal shoot and fruit borer along with quality and high yield. The recurrent parent EP-65 (Palur-1) was crossed with the donor parent *Solanum viarum* to obtain F_1 populations. Further selfing was done in F_1 progenies to raise F_2. From the segregating F_2 population selfing was done in the selected plants to raise next generation and continued till F_9 generation. In the F_9 generation two best progenies were selected (Hybrid Derivative 1 as (HD-1) and (Hybrid Derivative 2 as HD-2) based on high marketable yield along with low shoot and fruit borer infestation. In another set of crosses F_1 plants of EP-65 x *Solanum viarum* were back crossed with the recurrent parent EP-65 to eliminate the wild characters of *Solanum viarum* and to get BC_1F_1 plants. The BC_1 plants were back crossed and successive backcrossing has been successively done to generate advance breeding lines with high level of resistance and excellent fruit quality in BC_2F_1 and BC_3F_1 population. The plants of BC_3F_1 were selfed to raise BC_3F_2 and selfing continued upto BC_3F_6 plants till attaining homozygosity. Evaluation of plants showed that all the plant characters recorded very low GCV and PCV which indicated that both the population BC_3F_5 and F_8 attained homozygosity and could be to carried forward to the next generation for confirmation. In the BC_3F_6 population a best performing progeny (Hybrid Derivative 3 as HD-3) was selected on its very low infestation of shoot and fruit borer along with high marketable yield, desirable fruit qualities and devoid of any bitter taste. Molecular study with RAPD primers also revealed the introgression of the genes from donor parent *Solanum viarum* and in all the hybrid derivatives but was found to be absent in the recurrent parent EP-65. It confirmed the introgression of resistant genes of *Solanum viarum* in the hybrid derivatives. Cross compatibility studies were undertaken by Devi *et al.* (2016) using 20 cultivated genotypes of brinjal and 4 wild *Solanum* species (*Solanum incanum, S. aethiopicum, S. integrifolium* and *S. indicum*). The crossability relationship of brinjal with the wild species was determined by fruit set (per cent), number of seeds/fruit and germination (per cent) of F_1 seed. The maximum fruit set (80 per cent) was recorded in cross DBSR-91 x *S. aethiopicum*. The highest germination of F_1 seed (71 per cent) was recorded in Pusa uttam x *S. aethiopicum*, Sel-91-2 x *S. aethiopicum* and DBR-G-190 x *S. aethiopicum*. The highest number of seeds per fruit (754) was recorded in DBR-G-190 x *S. incanum*. Among the interspecific hybrids, highest number of fruits (322.0 per plant) was recorded in DBSR-52 x *S. indicum* and maximum mean yield (1.94 kg/plant) was recorded in Pusa Shyamla x *S. integrifolium*. Among the biochemical parameter, maximum ascorbic acid and phenol content (414.91 mg GAE acid/100g fresh weight) was recorded in DBSR-52 x *S. incanum*. The maximum anthocyanin content (62.32 mg/100g) was recorded in Pusa Shyamla x *S. integrifolium* and Br-112 x *S. incanum* had the highest (63.27 μmol trolox/g) antioxidant activity.

Some successful distant hybridization in brinjal

- *Solanum incanum* x *Solanum melongena*
- *S. melongena* x *S. aethiopicum*
- *S. melongena* x *S.cumingii*
- *S. giloa* x *S.indicum*
- *S. melongena* x *S. grandiflorum*
- *S. indicum* x *S. melongena*

- ☆ *S. indicum* x *S. integrifolium*
- ☆ *S. xanthocarpum* x *S. melongena*
- ☆ *S. integrifolium* x *S. incanum*
- ☆ *S. integrifolium* x *S. melongena*

S. indicum (2n=24) x *S. melongena* (2n=24), *S. demissum* (2n=72) x *S. tuberosum* (2n=24) and *S. verrucosum* (2n=24) x *S. bulbocastanum* (2n=48) showed unilateral incompatibility (reciprocal crosses are not possible).

Table 2.65: Somatic Hybrids Produced in Brinjal

Species Combination	Character Transferred
Solanum melongena x S. sanitwongsei	Nematode resitance
Solanum melongena x S. sisymbrifolium	-

Intergeneric Hybridization

Reports on intergeneric hybridization in *Solanum* through conventional hybridization are not available due to the existence of very strong crossability barriers. F_1 hybrids of this type of crosses are always sterile. However, intergeneric hybridization has been successful only in the development of the hybrids involving *Solanum melongena* and *Nicotiana tabaccum* and *Solanum melongena* crossed with an F_1 hybrid involving tomato species *S. esculentum* and *S. pennelli* by somatic hybridization (Liu *et al.*, 2015).

Ploidy Breeding

Success in hybridization depends on the ploidy level between the diploid *S. melongena* and some of the wild species.

Table 2.66: Ploidy Level of *Solanum* Species

Species	Chromosome Number	Reference
S. indicum	2n=24	Omidiji, 1975
S. aethiopicum	2n=24	Okoii, 1988
S. macrocarpon	2n=24	Rao *et al.*, 1976
S. elaegnifolium	2n=24	Heap and Carter, 199
S. sisymbrifolium	2n=24	Ugur and Kazim, 2003
S. torvum	2n=24	Vasudevan, 1975
S. erianthum	2n=24, 66	Vasudevan, 1975
S. gilo	2n=24	Rao *et al.*, 1976
S. indicum	2n=24	Krishnappa and Chenna-vaeraih, 1975
S.nigrum	2n=48, 54	Labadie, 1976

Resistance to Biotic Stresses

Therte are several species which possess resistance to biotic stresses in this crop. For example *S. incanum* is more closely related to *S. melongena* and can be used as a potential source of resistance to fruit and shoot borer, fusarium wilt, bacterial wilt and phomopsis fruit rot. Unfortunately these two species do not give fertile progenies when crossed with *S. melongena*. However, the biotechnological tools could be useful in overcoming the crossability barrier between the *S. melongena* and *S. tuberarium*.

(i) Disease Resistance Sources and its Utilization

There are several species which possess disease resistance behavior in brinjal but it is essential to understand the genetics of diseses resistance. Among the related wild species *S. sisymbrifolium* and *S. torvum* are particularly interesting on account of their resistance to two of the most serious eggplant wilt diseases (bacterial wilt and verticillium wilt). Cultivar 'SM-6-7(B)', 'PPC (B)', 'BWR-12(B)', 'BB-77', 'BB-44 (B)' and 'ARU-2C (B)' has been reported resistance against wilt diseases (Kalloo *et al.*, 2002).

Table 2.67: Genetics of Disease Resistance in Brinjal

Disease	Mode of Genetic Control	Refernce
Bacterial wilt	Single dominant gene	Swaminathan and Srinivasan, 1972
	Monogenic	Vijaygopal and Sethumadhavan, 1973
	Polygenic	Yamakawa, 1976
Fusarium wilt	Single dominant gene	-
Phomopsis blight	Recessive and polygenic	Kalda *et al.*, 1976
Anthracnose	Major gene resistance	Khan, 1974

Sheela (1982) observed resistance to bacterial wilt in SM-6. Gopalakrishnan (2004) reported release of SM-6-7, SM-6-6 and SM-141 as bacterial wilt resistant varieties and SM-6-7 x SM-116 as wilt resistant hybrid.

Kalada *et al.* (1976) studied the resistance to *Phomopsis vexans* in 399 entries, including wild *Solanum* species, eggplant cultivars and interspecific F_1 hybrids. *S. xanthocarpum, S. indicum, S. gilo, S. khasianum, S. nigrum* and *S. sisymbriifolium* were highly resistant and 2 lines of *S. melongena* were found resistant. The reactions of F_1 hybrids were variable. It was further observed that resistance to Phomopsis blight was recessive and goverened by polygenes. Dominance gene effects were more pronounced than the additive in most of the crosses (Kalada *et al.*, 1977). Decker (1951) released two cultivars by hybridization namely 'Florida Market' and 'Florida Beauty' resistant to *Phomopsis* blight. The resistance to bacterial wilt was controlled by single dominant gene (Swaminathan and Srinivasan, 1972; Vijayagopal and Sethumadhwan, 1973). 'West Coast Green Round-112-8 (WCGR-128-8)', isolated and purified from a local variety of Karnataka State was resistant to bacterial wilt (Gopinath and Madalageri, 1986). The wild eggplant *S. melongena* var. *insanum* is reported to be resistant to bacterial wilt and is highly compatible with certain cultivars of eggplant (Gopimany and George, 1979).

Table 2.68: Sources of Resistant Germplasm in Eggplant Cultivars against Diseases

Diseases	*Priority Species as Source of Resistance*	*Cultivars/Genotypes as Source of Resistance*	*Refernces*
Bacterial wilt (*Pseudomonos solanacearum* E.F. Smith.)	*S.melongena* var. *insanum* L., *S. sisymbrifolium*, *S. torvum*, *S. xanthocarpum*, *S. toxicarium*, *S. nigrum*, *S. melongena*, *S. incanum*	Sm-6-6, Sm-6-7 (Surya), BWR-7, ARU-2C, Comuy, Long Green, Sm-6, Sm-81 and Sm-164, Aroman, Sinampire, Taiwan Naga, L-7, L-19, Muktakeshi, Purple Long, Udippi, Pusa Purple Cluster, Pusa Purple Round, Pusa Anupam, Improved Muktakeshi, Annamalai, Arka Nidhi, Arka Keshav, Arka Neelkanth, Utkal Tarini, JC-1, VKBR-6, VKBR-7, VKBR-8, VKBR-10, VKBR-11, VKBR-13, VKBR-14, VKBR-15, BBSR-93, BBSR-197, BBSR-198, BBSR-200, IC090141, IC090146, BB-7 IC090982, IC099736, IC249349, Swetha, VKBR2, Haritha, JC-1, JC-2 and Pant Rituraj	
Bacterial wilt (*Pseudomonos solanacearum* E.F. Smith.)		West Coast Green Round-12-18	Gopinath and Madalageri, 1986
Bacterial wilt (*Pseudomonos solanacearum* E.F. Smith.)		'SM 6', 'SM 6-6', 'SM 141', 'Cipaye', 'Aroman', 'Gowok', 'Surya',	-
Bacterial wilt (*Pseudomonos solanacearum* E.F. Smith.)		SM48, SM56, SM71	Sheela *et al.*1984
Bacterial wilt (*Pseudomonos solanacearum* E.F. Smith.)		SM-6-7 (Surya), SM6-6, SM6-2	Sankar *et al.*1987
Dingra's Multiple Purple, Sinampiro to *Pseudomonos solanacearum*		'Pusa Purple Cluster'	Rao *et al.*, 1976; Mukherjee and Mukhopadhyay, 1982
Bacterial wilt (*Pseudomonos solanacearum* E.F. Smith.)		BB-11, BB-7 (Resistant)	Sharma and Kumar, 1995
Bacterial wilt resistant to K60 (race 1) and Wr 82 (race 3) isolates of *Pseudomonos solanacearum* E.F. Smith.)		SM-6-6	Goth *et al.*, 1983
Bacterial wilt (*Pseudomonos solanacearum* E.F. Smith.)		BWR-34; 'Pusa Purple Cluster' (Resistant)	Bora *et al.*, 1993
Bacterial wilt (*Pseudomonos solanacearum* E.F. Smith.)		BWR-54; 'Pusa Purple Cluster' (Resistant)	Singh, 1991
Phomopsis blight (*Phomopsis vexans*)		'Muktakeshi', 'Bargan', 'White Gih', 11a, 264 and 238 (Highly resistant)	Kalda *et al.*, 1976
Phomopsis blight (*Ralstonia solanacearum*)	*S. hispidum*, *S, integrifolium*, *S. melongena var. insanum*, *S. nigrum*, *S. sisymbrifolium*, *S. texanum*, *S. xanthocarpum* Schrad., *S. torvum*, *S. gilo*, *S. khasianum*, *S. testiculatum*	Surya, Swetha, Haritha, BB-7, BB-44, Arka Nidhi, Arka Keshav, WCGR, Arka Neelkanth, Annamalai, BB-49, Muktakeshi, Black Beauty, Pusa Purple Cluster, Pusa Bhairav, Pant Samarat, Florida Market, Florida Beauty, Line 119-12-2-264-1-2, PB-30, BC-1, JC-2, KS-352, IC90922, IC90047, IC90970, EC384565.	-
Fruit rot (*Phomopsis vexans* Sacc and Syd. Harter)	*S. xanthocarpum* Schrad. and Wesdl., *S. nigrum* L., *S. sisymbriifolium* Lam., *S. gilo*, *S. khasianum*, *S. melongena*, *S. incanum*		Kalda *et al.*, 1977
Phomopsis fruit rot (*Phomopsis vexans* Sacc and Syd. Harter)	*S. integrifolium*, *S. gilo*, *S. xanthocarpum* Schrad., *S. nigrum*, *S. khasianum*, *S. testiculatum*	Pusa Bhairav, Pusa Purple Cluster, Florida Beauty, Florida Market	-
Little leaf (*Mycoplasma*)	*S. indicum*, *S. sisymbriifolium* Lam.		Anjaneyulu and Ramakrishnan, 1968.
Little leaf (*Mycoplasma*)		'Pusa Purple Cluster'	Kale *et al.*, 1986; Chakraborty and Chaudhury, 1974
Little leaf (*Mycoplasma*)		Nurki, Bowad Local No. 4, Chikalgaon Local No. 1, 'Pusa Purple Cluster'	Chakraborty and Chaudhury, 1974; Mayee and Munshi, 1973
Little leaf (*Mycoplasma*)		Hisar Shyamal and H-10 (Tolerant)	Kalloo, 1990

Diseases	Priority Species as Source of Resistance	Cultivars/Genotypes as Source of Resistance	Refernces
Little leaf (*Mycoplasma*)		Sel.212-1, Sel 252-2-1 (Resistant)	Chakrabarty and Chaudhury, 1975
Little leaf (*Mycoplasma*)		SM-78 x SM-79, SM-78 x S-745, SM-77 x SM-60, SM-75 x SM-74	Ponnuswami and Irulappan, 1993
Little leaf (*Mycoplasma*)		SM-75 x SM-60 (F_2 generation of these crosses found resistant)	Ponnuswami and Irulappan, 1993
Little leaf (*Mycoplasma*)	*S. integrifolium*	'Pusa Purple Round', 'Surati', 'S 212-1' H-6, H-7, H-10, Pusa Purple Cluster, H-8, S-212-1, S-252-1	-
Verticillium Wilt (*Verticillium alboatrum* Reink and Benth)	*S. indicum* L.,	-	-
Verticillium Wilt (*Verticillium alboatrum* Reink and Benth)	*S. incanum*	'PI 164941', 'PI 174362'	
Verticillium Wilt (*Verticillium dahliae*)	*S. melongena var. insanum, S. persicum, S. integrifolium, S. sisymbriifolium* Lam., *S. torvum, S. scabrum, S. caripense*	'Florida Market', 'Harris-468 Special'	
Fusarium wilt (*Fusarium oxysporum*)	*S. incanum, S. indicum, S. integrifolium*	'K-61', 'K-72', 'Ghanal Local', 'Kopek', 'Black Beauty'	
Anthracnose (*Colletotrichum gloeosporioides* Penzig)		'Ping Tung', 'Long Red'	
Early blight [*Alternaria solani* (Ellis and Mart.) Jones and Grout]		'Pusa Purple Cluster'	
Cercospora spot (*Cercospora solani*)	*S. macrocarpon*	Improved Muktakeshi, Composite-1, Udupi Gulla	
Little leaf (MLO)	*S. viarum* (immune), *S. sisymbriifolium* Lam., *S. incanum, S. integrifolium, S. gilo*	Pusa Purple Cluster, Hisar Shyamal, Pant Rituraj, BB-2	

Table 2.69: Source of Resistant from Wild Eggplant Relatives

Pests	Source of Resistant	Reaction	Reference
Bacterial wilt	*S. melongena* var. *insanum*	Resistant	Gopimany and George, 1979
Phomopsis blight	*S. xanthocarpum, S. indicum, S. gilo, S. khasianum, S. nigrum, S. sisymbriifolium* and *S. torvum*	Highly Resistant	Kalada *et al.*, 1976
	S. melongena lines 11a and 264	Resistant	
Little leaf	*S. viarum*	Immune	Chakraborty and Chaudhury, 1974
	S. indicum and *S. sisymbriifolium*	Resistant	Chakraborty and Chaudhury, 1974
	S. integrifolium, S. gilo	Resistant	Chakraborty and Chaudhury, 1975
	S. viarum	Immune	Kale *et al.*, 1986

Little leaf disease of brinjal caused by the mycoplasma like organisms is the most devastating disease, which damages 40-80 per cent crop (Verma *et al.*, 1969; Anjaneyulu, 1969; Mote *et al.*, 1976 and Taley *et al.*, 1977). It spread in nature through insect vector *i.e.* leaf hopper (*Hishimonas phycittis* Dist.). Sulochana and Soluman (1973) observed that the internal tissues of the affected stem have crystalline induction in the cytoplasm of sieve tube and occationaly in other cells of the phloem, the conducting tissue of the plant, which in turn might have resulted in affecting the overall growth of the plant. Affected plants also show metabolic changes in their chemical constituents as compared to healthy one (Chakrabarti and Choudhury, 1974 and Mitra and Majumdar, 1977). The responsible insect vector in spread of disease has different preference with reference to chemical composition of the plant (Srinivasan and Chelliah, 1978). Nitrogen, protein and ascorbic acid contents in infected leaves of wild species as well as cultivated types were decreased as compared to healthy leaf samples (Kale *et al.*, 1986). The wild species *S. viarum* was found immune wheras *S. incanum* and *S. sisymbrifoilum* (both were resistant with 8.33 per cent little leaf reaction) as reported by Anjaneyulu and Ramakrishnan (1968) and Chakrabarti and Choudhury (1974). In case of cultivated types of *S. melongena*, 'Pusa Purple Cluster' was the only variety which was observed to be resistant (5.55 per cent) to disease while 'Nurki' (13.88 per cent), 'Bourad Local No.-4' and 'Chikhalgaon Local No.-1' (both 19.44 per cent) were moderately resistant to disease. Similar varietal response to disease reaction was observed by Mayee and Munshi (1973), Chakrabarti and Choudhury (1975), Mote *et al.* (1976) and Gill *et al.* (1978). Mayee and Munshi (1973) and Gill *et al.* (1978) have also observed that a variety, 'Pusa Purple Cluster' was resistant to little leaf disease. Genotypes 'Bourad Local No.-4', 'Nurki' and Chikhalgaon Local No.-1' and wild species *S. viarum* was immune while *S. incanum, S. sisymbrifoilum* cv. 'Pusa Purple Cluster' wre resistant (Kale *et al.*, 1986). At CCS HAU Hisar (Haryana) cultivars like 'Hisar Shyamal' and 'H-10' were found tolerant under field conditions (Kalloo, 1990). Genotype 'Pant Samrat' has expressed resistance against bacterial wilt and Phomopsis blight.

(ii) Sources for Insect Resistance and its Utilization

There are several *Solanum* species and genotypes which possess insect resistance behavior in brinjal. Identification of resistance is possible through quantifying the biochemical components like glycoalkaloid (solasodine), phenols, phenolic oxidase, enzymes namely poly phenol oxidase and peroxidase present in the brinjal genotype and these biochemical constituents possess insect resistance properties (Kalloo, 1988). As the cultivated brinjal is found to be highly susceptible to shoot and fruit borer, some other wild species possessing resistance were crossed with the brinjal to incorporate the resistance to shoot and fruit borer in the hybrids. In brinjal, interspecific crosses were done with wild forms of *Solanum* species namely *S. incanum* L. and forwarded the generation till F_4 (Rao, 1981), *S. macrocarpon* L. (Gowda *et al.*, 1990) and *S. indicum* L. (Preneetha, 2002) to obtain high yield with shoot and fruit borer resistance. Among these wild relatives, the *S. viarum* Dunal is a potential species, which yields solasodine and it is cultivated in India. The wild species *S. sisymbrifoilum* (Lal and Ahmad, 1965; Dhankar *et al.*, 1977), *S. integrifolium, S. xanthocarpum, S. nigrum* and *S. khasianum* have earlier been found free from the damage of fruit borer (*Leucinodes orbonalis* Guen) while *S. incanum* showed infestation of 8.6 per cent (Lal *et al.*, 1976). Sebastian (2000) observed resistance to fruit and shoot borer in six landraces of *S. melongena* and *S. macranthum, S. xanthocarpum* and *S. melongena* var. *insanum*. Several accessions of *S. melongena* have been screened against fruit and shoot borer and among them 'Pusa Purple Long' (Singh and Sikka, 1955; Gill and Chadha, 1979); H-128 and H-129 (Srinivasan and Basheer, 1961); Aushey (Lal and Ahmad, 1965; Dhankar *et al.*, 1977); Thorn Pendy, Black Pendy, H-165 and H-407 (Panda *et al.*, 1971); Ex Beckwai and IHR-191 (Musk Brinjal) (Krishnaiah and Vijay, 1975); SM-202; S-519, S-520, S-521 and Solan-11 (Lal *et al.*, 1976); Pusa Purple Cluster-2 (Dhankar *et al.*, 1977); S-4, H-4, Punjab Chamkila and Pusa Purple Cluster (Gill and Chadha, 1979) were tolerant. Baksh and Iqbal (1979) made cross of *S. incanum, S. khasianum,* and *S. macranthum* with *S. melongena* and resulting hybrids with *S. incanum* the F_1 hybrid showed appreciable resistance to fruit and shoot borer. Another hybrid (selected in F_4 family) between 'Pusa Purple Long' and *S. indica* showed field resistance to fruit and shoot borer (Rao and Kumar, 1980). Pradhan (1966) reported that long narrow fruited brinjal varieties were less infested than spherical fruited. Preference for egg laying was lower in resistant varieties than susceptible ones which may be one of the mechanisms for resistance. Larvae bore more successfully in round fruits than long fruited ones. Panda and Das (1975) and Mote (1979) reported that survival of borer larvae and weights gained by them on fruits were significantly less on resistant varieties.

Some resistance sources are being listed in Table 2.70.

At BCKV, Kalyani (West Bengal) more than sixty genotypes of brinjal have been evaluated elaborately over the years under natural condition and no genotype could be regarded as resistant to fruit and shoot borer. One long and slender fruited genotype, 'HE-12' emerged as the only least susceptible line to fruit and shoot borer exhibiting 25.75 fruit infestation (Chattopadhyay *et al.*, 2011). Sharma *et al.* (1980) successfully crossed *S. khasianum* and *S. melongena* by embryo culture, but further studies were impossible due to failure in germination of the F_2 seed. Field resistance to fruit and shoot borer in the progenies derived from interspecific hybridization such as *S. incanum* x *S. melongena* (Baksh and Iqbal, 1979; Rao, 1981), *S. melongena* x *S. indica* (Rao and Kumar, 1980) and *S. melongena* x *S. integrifolium* (Rao and Bakshi, 1981) has been reported. It remains to be seen whether the resistance of wild relatives can be combined with acceptable fruit quality. Prasad *et al.* (2016) screened 181 accessions of eggplant core set with 5 checks against eggplant fruit and shoot borer (EFSB) and isolated 43 promising accessions based on the percent infestation on number and weight basis. These promising accessions were screened during rainy season of 2104 in randomized block design with 3 replications. Out of 43 accessions, 22 accessions were found resistant to EFSB. These accessions were again evaluated during rainy season of 2015 with three replications at two locations *viz.*, Experimental Farm, Pusa Campus and Issapur Experimental Farm New Delhi. Results indicated the infestation level ranged from 9.58 to 55 per cent on number basis and 5.87 to 68.44 per cent on weight basis. Out of 22 accessions, 4 accessions such as EC305048, EC384970, IC111415 and IC111439, were found resistant as the infestation lwevel was less than 10 per cent both on number and weight basis copmpared to check Pusa Bindu (55 per cent).

Multiple Resistance

Multiple resistance has been located by several researchers in wild forms but the same has not been transfereed adequetly to *S. melongena* cultivars due to the incompatibility arising from the genetic make up of

Table 2.70: Priority Species of Eggplant for Resistance to Insect

Insect Pests	*Resistance Species*	*Priority Genotypes as Source of Resistance*	*References*
Shoot and fruit borer (*Leucinodes orbonalis* Guen.)	*S. khasianum, S. sisymbriifolium, S. microcarpon, S. melongena, S. incanum, S. gilo, S. xanthocarpum, S. integrifolium* and *S. indicum*	'White Long', 'Thorn Pendy', 'Black Pendy', 'W 165', 'H 407' Pusa Purple Cluster, Nimbakar Green, H-165, H-407, IIHR-191, SM-62, SM-213	Khan *et al.*, 1978; Sharma *et al.*, 1980; Chelliah and Srinivasan, 1985; Schaff *et al.*, 1982; Gowda *et al.*, 1990; Kumar and Sadashiva, 1996; Blestsos *et al.*, 2003; Padmini *et al.*, 2015; Behera *et al.*, 2002; Kalda *et al.*, 1976
Shoot and fruit borer		Banaras Giant, S-34, Arka Kusmakar, SM-215, S-258, SM-62, P-5-8, SM-2, S-2070	Pawar *et al.*, 1986
Shoot and fruit borer	*S. incanum, S. gilo, S. indicum, S. khasianum, S. sisymbriifolium, S. anomalum,*	Malkapuri, Shirur Khandala-1, Khanpur, Rajangona, Songona Kagal (moderately resistant), Aushey, Black Beauty, H-165, H-407, H-408, Annamalai, Pant Samrat, Bhagyamati, Pusa Purple Cluster, AM-62	-
Shoot and fruit borer		Tambalwadi Local, Gopuri, Pusa Purple Round, Jalagaon Local, Hingana, Dorla, Baramati, Punjab Bahar, Junagadh Bhattu (highly resistant to shoot borer infestation), Nurki, Pusa Purple Cluster, Gulabi Dorla (highly resistant)	-
Shoot and fruit borer		Araka Kusumakar, Nischintapur, Brinjal Long Green, Attapatti, Arka Shirish, Manipur, Makra and Chikon Long (relatively resistant)	Gangopadhyay *et al.*, 1996
Shoot and fruit borer		Pusa Purple Long, H-128, H-129	Gill and Chadha, 1979
Shoot and fruit borer		Aushey (tolerant)	Dhankar *et al.*, 1977
Shoot and fruit borer		Thorn Pendy, Black Pendy, H-165, H-407 (tolerant)	Panda *et al.*, 1971
Shoot and fruit borer		Ex Beckwai, IHR-191 (tolerant)	Krishnaiah and Vijay, 1975
Shoot and fruit borer		SM-202, S-519, S-520, S-521, Solan-11, Pusa Purple Cluster	Lal *et al.*, 1976
Shoot and fruit borer		S-4, H-4, Punjab Chamkila, Pusa Purple Cluster (tolerant)	Gill and Chadha, 1979
Shoot and fruit borer		SM-17-4, PBR-129-5, Punjab Barsati (tolerant)	Singh and Chadha, 1990
Shoot and fruit borer		Nischindipur Local, Muktajhuri, Shymala Dhepa, Navkiran, Banaras Long Purple, BB-1, Murshidabad local (tolerant)	Mukhopadhyay and Mandal, 1994
Shoot and fruit borer		Long Purple (tolerant)	Mehto and Lal, 1981
Shoot and fruit borer		Nimbkar Green, Arka Kusumakar, SM-213 (tolerant)	Mote, 1981
Shoot and fruit borer		Ringen Giant, Pusa Purple Cluster (tolerant)	Nathani, 1983
Shoot and fruit borer		Kuchia	Ishaq and Chaudhuri, 1984
Shoot and fruit borer		Singnath	Kabir *et al.*, 1984
Shoot and fruit borer		SM-17-4	Tewary and Moorthy, 1985
Shoot and fruit borer		IC089864A and IC144524	Sarath Babu *et al.*, 1997
Shoot and fruit borer	*S. incanum, S. xanthocarpum, S. khasianum, S. sisymbriifolium* (Immune)	Tambalwadi Local, Gopuri, Pusa Purple Round, Jalgaon Local, Hingana Dorla, Baramasi, Punjab Bahar and Junagadh Bhattu (Highly resistant to shoot borer)	Kale *et al.*, 1986
Shoot and fruit borer	*S. sisymbriifolium, S. khasianum, S. nigrum*	Resistant	Lal *et al.*, 1976
Shoot and fruit borer	*S. incanum, S. khasianum, S. macranthum, S. mammosum*	Resistant	Bakshi and Iqbal, 1979
Shoot and fruit borer	*S. macrocarpum*	Resistant	Gangopadhyay *et al.*, 1996
Shoot and fruit borer	*S. surattense, S. sisymbrifolium, S. indicum, S. viarum, S. aculeatissimum*	Resistant	Sarath Babu *et al.*, 1997

Insect Pests	*Resistance Species*	*Priority Genotypes as Source of Resistance*	*References*
Shoot and fruit borer		Arka Shirish, Hissar Sel.1-4, Shankar Vijay (resistant)	Raj and Kumarswami, 1979
Shoot and fruit borer		Shyamala Bhangar	Mukhopadhyay and Mandal, 1994
Leaf hoppers/Jassids (*Amarasca biguttula biguttula* Ishida)	*S. mammosum, S. sisymbriifolium*	H-4, Round Green, Dorly, Aushey, Jumblimulayam (resistant)	Mote, 1978
Jassids resistant		Manjari Gota	-
Leaf hoppers/Jassids		Junagadh Sel.1, Aushey, R-34, H-4, T-3 (resistant)	Bindra and Mahal, 1981
Leaf hoppers/Jassids		S-188-2, S-258, S-34, Pusa Purple Cluster, Manjri Gota, Dorli, Mukti	Pawar *et al.*, 1986
Leaf hoppers/Jassids		Borgona-1, Local Khandesh, Khedshivpur (moderately resistant)	-
Leaf hoppers/Jassids		Pusa Kranti	Mote, 1982
Aphids (*Aphis gossypii* Glover)	*S. sisymbriifolium, S. mammosum*	Sufal, Navkiran, ARM-3, Krishna, Brinjal Long Green, Suttons Long (tolerant), KB-9, Pusa Purple Round, Banaras Giant, L-13, KB-5, BB-1, KB-1, Improved Muktakeshi (tolerant)	Sambandam and Chelliah, 1983
Aphids	-	S-5	Subbaratnam and Bhutani, 1981
Spider mite (*Tetranychus urticae*)	*S. macrocarpum, S. integrifolium, S. mammosum, S. pseudocapsicum, S. sisymbriifolium*	-	Schalk *et al.*, 1975; Schaff *et al.*, 1982

different species. Wild species of *Solanum* have shown high degree of resistance besides immune reactions to several diseases and insects (Kale *et al.*, 1986). But these genes appear to have limited use in resistance breeding programme because of the difficulties in the interspecific hybridization. Mukhopadhyay and Mandal (1994) described cultivars Shyama Dhepa, Kalo Dhepa, Improved Muktakeshi, Banaras Long Purple and BB-1, as having multiple resistance against all the insect pests. Multiple resistance to the biotic stresses was seen in *S. melongena* var. *insanum* and *S. xanthocarpum* and in the land races S-1, S-13 and S-36. Literature search revealed some more cultivars of *S. melongena* having resistance to more than one pest are given in Table 2.71.

Table 2.71: Available Cultivars with Multiple Resistance

Resistance Available Against	*Cultivars*
Bacterial wilt, Little leaf and Shoot and fruit borer	Pusa Purple Cluster
Leaf hopper and Bacterial wilt	Pusa Kranti
Shoot and fruit borer and little leaf	SM-17-4
Leaf hopper and shoot and fruit borer	Pusa Purple Long, S-34, S-258, Aushey and H-4
Shoot and fruit borer Leaf hopper and Aphids	Shyama Dhepa, Kalo Dhepa, Improved Muktakeshi, banaras Long Purple and BB-1
Shoot and fruit borer and all species of *Meloidgyne*	Banaras Giant
All species of *Meloidgyne*	Vijay and Black Beauty

Table 2.72: Resistance Source to Nematodes

Nematode	*Priority Species/Genotypes as Source of Resistance*	*Reference*
Root-knot [*M. javanica* (Trueb) Chitwood]	'Kalianpur T2', 'Long Bangalore'	-
M. incognita race 1 and 2	Gulla	Ravichandra *et al.*, 1990
M. incognita	Local Round (Resistant)	Das *et al.*, 1989
M. incognita	Vijaya and Black Beauty (Resistant)	Yadav *et al.*, 1975
Less susceptible to nematode	'P 8' (Selection from Chandigarh)	-
Resistant to root knot nematode	*S. sisymbriifolium*	-
All root-knots	Sel. 419, Sel. 96-2, Olbaigan (Resistant), *S. sisymbriifolium,*	Verma and Chaudhury, 1974
All root-knots nematode resistant	Black Beauty, T-2 (UP), Vijaya, Banaras Giant	Nandwana *et al.*, 1980
Root knot nematode	*S. sisymbriifolium* Lam.	-
Less susceptible to nematode (*M. incognita*)	'P8', Black Beauty, Vijaya	-
Root knot nematode (*M. javanica*)	'Kalyanpur T-2', 'Long Bangalore', Bhanta, Black Beauty	-

Table 2.73: Available Cultivars with Multiple Resistance to Nematodes

Resistance Available Against	*Cultivars*	*References*
Shoot and fruit borer and all species of *Meloidgyne*	Banaras Giant	
All species of *Meloidgyne*	Vijay and Black Beauty	Yadav *et al.*, 1975

Table 2.74: Source of Resistance against Nematodes

Meloidogyne Javanica	*S. torvum, S. macrocarpum, S. sisymbriifolium, S warscewiczii*	Resistant	Vito *et al.*, 1992
Meloidogyne incognita	*S. khasianum, S. torvum, S. toxicarium*	Immune	Ali *et al.*, 1992; Fassuliotis and Dukes, 1972
	S. torvum, S. seaforthianum	Resistant	Shetty and Reddy, 1985
	S. torvum, S. macrocarpum, S. sisymbriifolium, S warscewiczii	Resistant	Vito *et al.*, 1992; He'bert, 1985; Ahuja *et al.*, 1987
Meloidogyne arenaria	*S. sisymbriifolium, S. torvum*	Resistant	Vito *et al.*, 1992
All root-knots (*Meloidogyne* species)	*S. torvum, S. aethiopicum, S. sisymbriifolium, Vijay, Black Beauty, Gulla*	Resistant	Chadha, 1988
	S. sisymbriifolium, S. elaegnifolium	Resistant	Verma and Chaudhury, 1974
	S. aethiopicum		Daunay and Dalmasso, 1985

Table 2.75: Trait Specific Genotypes of Brinjal

Traits	*Priority Species*	*Cultivars/Genotypes*
Frost tolerance	*S. incanum*	-
Drought	*S. macrosperma, S. gilo, S. macrocarpon*	'Atabekil', SM-19, 'SM-30', 'SM-1', 'Violette Round'
Salt tolerance	-	'JC-1', 'Pusa Bindu', 'Pragati', 'CO-1'
High temperature	-	'R-34', 'PH-4'

(iii) Resistance Sources to Nematode

Verma and Chaudhury (1974) reported that 'Sel. 96-2', 'Polbaigan' and the wild species *S. sisymbrifolium* and *S. elaegnifolium* are highly tolerant to *Meloidgyne* spp. under field and lab conditions. 'Florida Market' is a special cultivar which possess multiple resitant (root knot nematode + phomopsis fruit rot + Verticillium wilt). Among the related wild species *S. sisymbrifolium* and *S. torvum* are particularly interesting on account of their resistance to the nematode.

Multiple resistance cultivars to nematodes are the need of the day.

Shetty and Reddy (1985) studied 4 species of *Solanum, S. torvum, S. incanum, S. seaforthianum* and *S. mammosum* and 10 strains of *S. khasianum* against *M. incognita* under green house conditions. *S. torvum* and *S. seaforthianum* gave resistant reaction. A very small number of larvae invaded roots of *S. torvum* (2.0 larvae) and *S. seaforthianum* (3.3 larvae). There are several species which are considered as potential source of resistance against nematodes.

Resistance Sources to Abiotic Stresses

Thermal sensitivity was recorded for reproductive feature and high genotypic variation was evident for fruit set and fruit yield in high temperature stress condition. Pollen viability did not vary significantly over two seasons: autumn-winters and summer-rainy, although slight decrease in pollen viability was evident in summer-rainy season. Low fruit set in the summer-rainy season might have been due to high flower drop owing to non-fertilization which might have happened due to stigma exertion beyond the length of the anthers. BCB-23 emerged as promising heat tolerant line in brinajal (Chattopadhyay *et al.*, 2011). Under Hisar conditions genotypes 'R-34' and 'PH-4' were found the best to set fruit during summer (Kalloo *et al.*, 1990).

Baksh and Iqbal (1979) made cross of *S. incanum, S. khasianum*, and *S. macranthum* with *S. melongena* and resulting hybrids with *S. incanum* the F_1 hybrid showed appreciable resistance to frost. *S. macrocarpum* and *S. gilo* both are drought resistance species of brinjal.

Male Sterility

Male sterility in eggplant has been described in several reports. Functional male sterility in a population of eggplant variety 'Blackie' is reported male sterile which was developed from the cross Black Beauty by Black Nagasaki. It was due to failure of the anthers to dehisce normally. Functional male sterility controlled by single recessive gene has been reported in brinjal. In this case emasculation part is obviated and only hand pollination is required. In this case only 2-4 per cent chance selfs are expected which can be avoided by using marker especially purple plant vigour in the pollen parent. A few reports have describes cms and gms in eggplant. Two wild relative of eggplant *i.e. S. gilo* Raddi and *S. violaceum* has been used to develop cytoplasmic male sterile lines of eggplant. The cytoplasm of *S. gilo* Raddi, *S. violaceum* R. Br. and *S. virginianum* can function as a male-sterile cytoplasm for eggplant. 'Taibyo VF' a commercial Japanese rootstock cultivar released by Takii and Co., Ltd. (Kyoto, Japan) is an interspecific hybrid between *S. grandifolium* and *S. melongena* that is resistant to verticillium wilt and is generally sterile. 'Taibyo VF' has difficulty in setting seed but F_1 plants of 'Taibyo VF' x 'LS1934' (*S. melongena*) produces some seeds. F_2 plants were segregated into male sterile and fertile plants. All progeny of male sterile plants backcrossed with *S. melongena* were male sterile, indicating cms caused by the cytoplasm of *S. grandifolium*. Cytoplasmic male sterility was observed due to no dehiscence of the anthers and low pollen fertility, in the alloplasmic backcross progenies obtained by repeated backcrossing of the interspecific hybrid between *S. violaceum* R. Br. (female) and *S. melongena* (male) to *S.melongena*. This male sterility is suggested to be the result of a disharmony between the cytoplasm of *S. violaceum* R. Br. and the nucleus of *S. melongena*. Two other cases of cytoplasmic male sterility were reported in backcross progenies derived from the interspecific cross between *S. gilo* (*i.e. S. aethiopicum*, Gilo Group) and *S.melongena* (Khan and Isshiki, 2008; Saito *et al.*, 2009). At ICAR-IIVR, Varanasi (U.P.) nuclear male sterile gene (*fms*) is being utilized. Nuclear male sterility is monogenic recessive and monogenic recessive functional male (ms) sterility is available. Brinjal cmsline 'cms-14' has been developed by Ankur Seed Co. Pvt. Ltd. Nagpur (M. S.) in the year 2008 and developing restorer line using local genotye collected from Maharashtra.

Breeding for Quality Traits

In brinjal quality parameters includes fruit size, shape and colour, soft flesh and low alanine content. In internal quality is determined due to certain biochemical constituents like anthocyanins, total phenols, polyphenol oxidase activity and glycoalkloid contents. Two more species, *S. khasianum* and *S. aviculare*, having the same chromosome number have gained importance as sources of solasodine (Glycoalkaloid) which is being used for synthesis of steroid hormone (Choudhary, 1976). Twenty one genotypes were analyzed for various chemical constitutents and results indicate that long type varieties were rich in dry matter, crude protein, anthocyanin, phenolics and glycoalkloid contents while the oblong types were rich in total water soluble sugars. Among varieties the total water-soluble sugars varied from 1.00 g/100 g (PBr-129-5) to 3.63 g/100 g (Pusa Purple Round), free reducing sugars ranged from 0.38 g/100 g (PBr-129-5) to 1.73 g/100 g (Shakar Vijay Hybrid), anthocyanin from 0.16 (T_3) to 1.04 (Arka Shirish) expressed as C.D. at 550 nm in their 10 per cent (W.v) aqueous extracts. The total phenols varied from 0.44 g/100 g (Arka Navneet) to 0.138 g/100 g (Arka Shirish) and the glycoalkloid content in tern of mg/100 g varied from 6.25 (Pusa Kranti) to 20.50 (Arka Shirish). The variety Arka Shirish had the maximum dry matter, crude protein, anthocyanin, phenolic and glycoalkloid content. The polyphenol oxidase activity in arbitrary units/mg, protein/mg, orthodihydroxy phenols varied from 2.12 (ARU-1) to 45.50 (PBr-91-2) (Bajaj *et al.*, 1979, 1981). In an analysis 'Punjab Chamkila' had the maximum anthocyanin and total phenol contents at all stages of maturity (Kaur *et al.*, 1985). However, orthodihydroxy phenols are maxium in variety 'P-8' in its first picking and in 'Pusa Purple Long' in its 2nd and 3rd picking. For processing purpose, the variety 'H-4' followed by 'P-8' in its 2nd picking is very suitable keeping in view the low level of peroxidase and polyphenol oxidase activities and that of orthodihydroxy phenolic compounds which cause discoulration in cut fruits. 'Punjab Chamkila' had its maximum anthocyanin and chlorophyll contents, during November and January, respectively while 'Pusa Purple Cluster' had its maximum anthocyanin content in the beginning of January which exceeded in that of 'Punjab Chamkila'. Initially the chlorophyll content decreased in all the varieties except 'Pusa Purple Cluster' and 'S-4' then it increased in the beginning of January under Punjab condition (Bajaj *et al.*, 1988).

Berry *et al.* (1988) evaluated five cultivar namely 'BR-11', 'P-8', 'Jamuni Gola', 'Round Greeb Selection' and 'T-3' for acceptability of '*Baigan Bharta*' (local food item) quality preparation through canning in 310 x 409 lacquered cans and storing for 6 months. Results indicates that '*Baigan Bharta*' prepared from all the varieties was acceptable but it was more acceptable which was prepared from variety 'Round Green Selection' due to retaining colour, aroma, taste, mouth feel and other quality attributes. Bajaj *et al.* (1989) reported that a high yielding variety 'K-202-9', possessed the highest anthocyanin content (752.10 mg/100 g), lowest crude fibre content in another high yielding variety 'Azad Hybrid' (9.9 per cent) whereas higher (32.6 per cent) in the variety 'K-29210'. The minimum content of total phenol (0.52 mg/100 g) was found in the variety 'Azad Kranti' and of orthodihydroxy phenols (75 mg/100 g) in the variety 'ARU-1-C'. The glycoalkaloid content was least in the variety 'ARU-2-C' (0.48 mg/100 g). The mineral composition (mg/100 g) of the different varieties ranged for Zn from 1.645 to 4.000; Cu 0.65 to 1.60; Fe 5.00 to 12.5; and Mn 1.375 to 3.25. Zn and Cu were more in the round white variety of eggplant.

Exerted Lines

Several exerted lines have been reported which facilitate the production of economic hybrid seeds. These lines are PBR-91-2, PBR-202-9 and PBr-1206.

Abiotic Stress Resistance

Dhaka *et al.* (2016) studied genotype x environment interactions under two locations of Rajasthan conditions and reported significant differences and high and non-linear interactions. This means that under changeable environments the different cultivars react differently and can therefore be grouped according to the growth and earliness stability. This is very clear from the environmental mean scores, environments E_1 was more stable with lowest mean value for earliness traits and highest mean value had the highest genotype response for growth and yield traits. Seven genotypes were found to be stable across the environments for days to anthesis of first flower, 8 genotypes were found stable for days to 50 per cent flowering and 10 genotypes were also found stable for days to first fruit picking. Among the stable genotypes for earliness the Pusa Upkar and hybrid Punjab Sadabahar x Pusa Upkar were found to be stable for all the earliness traits. Thus, these stable genotypes can be recommended for commercial cultivation over wide range of environments or can be used in further breeding programme. The 42 genotypes were found stable over environments for fruit yield of which the 18 genotypes showing high yield potential and was detected for general adaptability and proposed for commercial cultivation and only two hybrids, Pusa Purple Long x Pusa Upkar and Punjab Sadabahar x Pusa Upkar exhibiting below average stability and suitable for favourable environmental conditions.

Grafting

Genera *Solanum*, *Capsicum* and *Lycopersicon* are inter graftable. Eggplant (*S. melongena*) was grafted onto scarlet eggplant (*S. integrifolium* Poir) in 1950s. Side grafting and patch budding are observed successful. These grafts can be used in growing crops in problematic areas where one genus thrives well.

In brinjal cultivation, plants raised through grafting playing an important role not only to control insect pests and diseases but also getting bumper yield per unit area in open field condtion and control conditions for longer period.

Table 2.76: Trait Specific Rootstocks as Tomato Scion

Scion	*Rootstocks*	*Purpose*
Tomato	*S. torvum, S. integrifolium, S. melongena*	Plant vigour, tolerance to bacterial wilt, verticillium wilt, fusarium wilt, low temperature and nematode.

In Japan some potential varieties has been found for grafting in brinjal. *Solanum torvum* was used as rootstock for tongue grafting a number of chosen cultivars of brinjal (*S. melongena* L.). The grafted hybrids behave as perennial and yield 3 to 4 times more. 'Taibyo VF' a commercial Japanese rootstock cultivar released by Takii and Co., Ltd. (Kyoto, Japan) is an interspecific hybrid between *S. grandifolium* and *S. melongena* that is resistant to verticillium wilt and is generally sterile.

Table 2.77: Trait Specific Rootstocks as Eggplant Scion

Scion	*Rootstock Species*	*Major Variety*
Eggplant	*Solanum torvum*	Torvum Vigor
	Solanum integrifolium	Hiranasu
	Solanum melongena	Taibyo VF, Meet, Karechen, Assist

Several useful species has been reported for different traits in brinjal as grafts (Table 2.78).

The uses of wild species in brinjal crop improvement are given in Table 2.79.

Genetic Resource Conservation

Realizing the importance, explorations were undertaken in collaboration with IPGRI covering almost all brinjal growing area of India and neighboring countries *viz.* Bangladesh, Nepal and Sri Lanka for collection of variability. Under ICAR-NBPGR, N. Delhi/ IPGRI, Roam collaborative project, 2531 collections of eggplant and 337 collections of related wild *Solanum* species from South Asian countries have been made.

Table 2.78: Useful Grafts in Brinjal

Rootstock	*Resistance/tolerance*	*Grafting*	*References*
Solanum torvum	Bacterial wilt, Fusarium wilt, Verticilium wilt, Root knot nematode, Cold stress	Eggplant	Lee, 1994; Yamakawa, 1982; King *et al.*, 2010; Okimura *et al.*, 1986
S. macnocarpon	Bacterial wilt, induces high phenolic content of grafted scion	Eggplant	Carmino *et al.*, 2011, Cappelli *et al.*, 1995
S. gilo	Fusarium wilt	Eggplant	Yamakawa, 1982
S. sisymbrifolium	Bacterial wilt, Fusarium wilt, Verticilium wilt, Root knot nematode	Eggplant	Yoshida *et al.*, 2004
S. incanum x S. melongena	Root knot nematode	Eggplant	Lester and Hasan, 1991; Gilbert *et al.*, 2011
S. melongena x S. aethiopicum	Root knot nematode	Eggplant	Behera and Singh, 2002
Solanum integrifolium	Bacterial wilt, Fusarium wilt, heat and humidity	Eggplant	Tachibana, 1994; Yamakawa, 1982
Solanum integrifolium x S. melongena	Low temperature, Bacterial wilt, Fusarium wilt, Verticilium wilt	Eggplant	Dounay *et al.*, 2001; Lee, 1994; Yamakawa, 1982

Table 2.79: *Solanum* Species Used in Eggplant Breeding for Introgression/Rootstock Purpose

Species	*Introgression for Desirable Traits/Rootstock Purpose*	*Reference*
S. anguivi Lam.	Backcross to obtain cytoplasm susstitution male-sterile lines	Khan and Isshiki, 2011
S. incanum L.	Interspecific hybrids as rootstock	Gisbert *et al.*, 2011
	Backcross introgression of high content in bioactive phenolic acids	Prohens *et al.*, 2013
	Resistance to Fusarium wilt	Rao and Kumar, 1980
	Resistance to shoot and fruit borer	Behera and Singh, 2002
S. linnaeanum Hepper and P.M.L. Jaeger	Backcross introgression of resistance to *Verticillium dahlia*	Sunseri *et al.*, 2003;, Liu *et al.*, 2015
S. sisymbrifolium	Sexual and somatic hybridization for resistance to *Ralstonia soanacearum* and *Verticillium dahlia*	Blestsos *et al.*, 2003; Collonnier *et al.*, 2003
	No backcrosses or selfing of hybrids obtained resistance to *Ralstonia soanacearum* and *Verticillium dahlia* wilt.	Collonnier *et al.*, 2003
	Resistance to fruit and shoot borers (*Leucinodes orbonalis*)	Chelliah and Srinivasan, 1985
	Resistance to root knot nematode (*Meloidogyne* species)	Ahuja *et al.*, 1987
	Resistance to spider mite	Schalk *et al.*, 1975
S. violaceum Ortega	Selection of selfing of interspecific hybrids resistant to *Fusarium* wilt	Rao and Kumar, 1980
	Backcross to obtain cytoplasm susstitution male-sterile lines	Khan and Isshiki, 2009
S. torvum Sw.	Sexual and somatic hybridization for resistance to *Ralstonia soanacearum*, *Verticillium dahlia* and root knot nematodes	Blestsos *et al.*, 2003; Jari *et al.*, 1999
	No backcrosses or selfing of hybrids obtained	Kumchai *et al.*, 2013
	Resistance to soil pathogens *Ralstonia soanacearum* and *Verticillium dahlia* and as graft stock	Jari *et al.*, 1999; Collonnier *et al.*, 2003; Gisbert *et al.*, 2011
	Resistance to root knot nematode (*Meloidogyne* species)	Ahuja *et al.*, 1987
S. macrocarpon	Successful record of fertile F2 and fertile BC1 progenies in reciprocal cross (*S. macrocarpon* x *S. melongena*) which is highly sterile hybrid for use in breeding for shoot and fruit borer resistance	Padmini *et al.*, 2015
S. viarum	Successful progenies obtained in direct cross using *S. melongena* as female parent	Pugalendhi *et al.*, 2010
Hybrids between *S. incanum* x *S. melongena*	highly vigorous rootstock	Gisbert *et al.*, 2011

Source: Padmini *et al.*, 2017.

Germplasm sampled from neighboring countries includes 175 accessions of eggplant and 27 other *Solanum* species from Bangladesh have been augmented. Some distinct species includes *S. verbascifolium, S. surattense, S.indicum, S. nigrum* and *S. viarum* have been collected from Nepal. Thirty six accessions of cultivated eggplant and 91 other *Solanum* species from Sri Lanka have also been acquired and conserved in National Gene Bank. A collaborative exploration programme was completed in 1990 by India and Bangladesh which was funded by IBPGR and 977 germplasm of *Solanum melongena* and 129 germplasm of its wild relatives (*S. melongena* var. *insanum, S. torvum, S. nigrum, S. incanum, S. indicum, S. khasianum, S. surratance, S. trilobatum, S. pubescense, S. khurzii, S. aculeatissimum, S. verbascifolium, S. frubescious, S. memosum, S. sisymbrifolium* and *S. gigantium*). Some distinct local types in eggplant and *Solanum* species like *S. verbascifolium, S. surattense, S. indicum, S. nigrum* and *S. viarum* have been collected from Nepal and 36 accessions of eggplant and 91 lines of *Solanum* species from Sri Lanka. In Sri Lanka, mostly long fruited types eggplant is grown. Rich variability in fruit morphological and plant characteristics in *S. melongena* var. *insanum* have been collected from Sri Lanka.

ICAR-NBPGR, New Delhi holds 3868 accessions belonging to 24 species of brinjal. The accessions are preserved at -18 °C and include 3,531 accessions of cultivated varieties and 337 of related wild species. Some of these varieties have exhibited natural resistance to bacterial wilt and also have certain unique features such as plant spread, number of branches, fruit weight and fruit yield per plant (Gangopadhyay, 2008; Rajam *et al.*, 2008; Kumar, 2008). Velayudhan *et al.* (1996) conducted 5 explorations in parts of Southern peninsular region cpmprising Kerala, Karnataka, Antdhra Pradesh and Tamil Nadu during 1989-1991 and some pockets of Western Ghats for collection of eggplant and its wild relatives. From the five exploration trips, a total of 351 samples belonging to 27 species were collected and team observed distribution of eggplant from sea level to elevations as high as 1800 m in Western Ghats. In India this diversity has been collected from Indo-Gangetic Plains, Western Region, North Western Deacon Plateau, northern hills and plains, North Eastern Region, Chhatishgarh region, Vidharbha and tribal belt of Central India. Eastern Ghat area was observed to be an important area for eggplant diversity. Good variability also existed in north eastern states, some highly spiny

Table 2.80: State-wise Collection of *Solanum* Species

Sl.No.	*Species name*	*Kerala*	*Karnataka*	*Tamil Nadu*	*Andhra Pradesh*	*Total*
1.	*Solanum melongena* var. *melongena*	40	45	81	50	216
2.	*S. melongena* var. *insanum*	11	3	1	1	16
3.	*S. melongena* var. incanum	6	-	-	-	6
4.	*S. pubescens*	-	-	3	6	6
5.	*S. surattense*	-	1	3	3	7
6.	*S. trilobatum*	-	2	3	1	6
7.	*S. torvum*	10	2	2	2	16
8.	*S. nigrum*	1	2	1	2	6
9.	*S. anguivi* var. *anguivi*	13	4	6	-	23
10.	*S. anguivi* var. *multiflora*	5	-	3	-	8
11.	*S. myriacanthum*	4	5	3	1	13
12.	*S. seaforthianum*	1	1	2	-	4
13.	*S. aculeatissimum*	2	1	-	-	3
14.	*S.erianthum*	1	-	4	-	5
15.	*S. mammosum*	1	-	-	-	1
16.	*S. sisymbriifolium*	-	-	1	-	1
17.	*S. caspsicoides*	-	-	1	-	1
18.	*S. elaeagnifolium*	-	-	1	-	1
19.	*S. mauritianum*	-	-	3	-	3
20	*S. giganteum*	-	-	1	-	1
.21.	*S. jasminoides*	-	-	1	-	1
22.	*S. stramoniifolium*	1	-	-	-	1
23.	*S. convolvulus*	1	-	1	-	2
24.	*S. barbisetum*	-	-	1	-	1
25.	*S. hispidum*	-	-	1	-	1
26.	*S. macrocarpon*	1	-	-	-	1
27.	*S. vagum*	-	-	1	-	1
	Total	98	66	124	63	**351**

Source: Velayudhan *et al.*, 1996.

Table 2.81: Status of Germplasm Collection

Districts	*Number of Collection*			
	S. indicum	*S. nigrum*	*S. seaforthianum*	*S. surattense*
Ajmer				01
Dungarpur	-	01	-	-
Jaipur	-	01	-	04
Rajsamand	-	-	-	02
Sirohi	01	01	01	01
Udaipur	-	01	-	01

Source: Dwivedi *et al.*, 1998.

and drought tolerant land races werec ollected from Asom.

In another collection programme Dwivedi *et al.*, 1998 collected some wild species of *Solanum* from Rajasthan which are precioius accessions for breeding programme.

A growing germplasm collection is also being maintained at several institutes (under AICRP-VC) including ICAR-Indian Institute of Vegetable Research (IIVR), Varanasi (U.P.), ICAR-Indian Institute of Horticultural Research (IIHR), Bengaluru, CSAUA and T, Klyanpur, Kanpur (U.P.), Govind Ballabh Pant University of Agriculture and Technology (GBPUA and T), Pantnagar (Uttarakhand); Punjab Agricultural University (PAU), Ludhiana Punjab; Tamil Nadu Agricultural University (TNAU), Coimbatore (Tamil Nadu), Odisha University of Agriculture and Technology, (OUA&T) Bhubaneswar (Odisha); Horticulture and

Agro-forestry Research Programme Plandu (HARP) Ranchi (Jharkhand), BCKV, Kalyani (W.B.), B.A.U, Sabour (Bihar) and other SAUs. These centres maintain landraces, advanced lines and released varieties. Gill *et al.* (1990) reported that at Odisha University of Agriculture and Technology, (OU&ST) Bhubaneswar (Odisha) in 1987-88, collected 160 different genotypes of improved, local and exotic types and evaluated for ancillary traits including incidence of diseases and insects. A catalogued for further utilization in crop improvement was developed. On the basis of difference in the phenotypic characters, 214 types were identified and their plant and fruit characteristics were recorded. Similarly, at BAU, Sabour (Bihar) in the year 1988-89 all the germplasm of brinjal was grouped considering the attributes shown in Table 2.82.

Table 2.82: Morphological Characterization of Brinjal

Sl.No.	Attributes	Description	No. of Collections
i.	Leaf colour	Green	75
		Light green	15
		Deep purple	5
ii.	Flower colour	Violet	88
		White	7
iii.	Fruit colour	Purple	70
		Green	13
		White	5
		Striped	7
iv.	Fruiting habit	Solitary	83
		Cluster	12
v.	Presence of spines	Present	7
		Absent	86
vi.	Shape of fruit	Long	58
		Round	20
		Oblong	15
		Oval	2
vii.	Weight of fruit	Below 100 g	45
		100-150 g	35
		Above 150 g	15
viii.	Wild species	*Solanum gilo*	
		S. torvum	
		S. integrifolium	
		S. xanthocarpum	

Eleven promising lines 71-9, 71-10, 71-11, 71-16, 71-19, 72-8, 73-8/81, 84-4, M-6, Green Long and White Long have been found useful for various traits. At BCKV, Kalyani (W.B.) 56 collections were grown and evaluated during the year 1988-89. Department of Vegetable Crops, CCSHAU, Hisar, (Haryana) is maintaining 150 diverse lines of eggplant. The inventory of germplasm stored in genebanks allows scientists to identify and closely analyze genetic diversity based on morphological, physiological, cytological and biochemical characteristics of cultivars with the help of DNA based marker techniques and other advanced biotechnological tools. Therefore, conservation and documentation of genetic identity of germplasm collection fulfills various curatorial and utilization needs, such as safeguard original types in germplasm repositories, determining varietal distinctiveness, pedigree and development of high yielding varieties and hybrids for registration and commercial cultivation (Tiwari, 2007). A number of morphological characters have been identified that have a direct or indirect bearing on pest resistance in eggplant.

Table 2.83: Potential Eggplant Germplasm on the Basis of Morphological Characters at ICAR-NBPGR, N. Delhi

Morphological Characters	Number of Corresponding Germplasm
Stem spininess	14
Petiole spininess	5
Leaf spininess	80
Long fruited (more than 35 cm)	14
Fruit calyx prickles (many)	2
Fruit pedicle prickles	3
More seeds/fruit (tightly arranged)	8
Number of fruits/cluster (more than 4)	14
Number of cluster/plant	18

A large number of lines with tremendous variability have been collected from Bihar. The landraces differed in plant height, maturity period, primary branches, secondary branches, plant spread, fruit weight, fruit length, width *etc.* The diversity for growth habit, pubescence, spininess, pigmentation in the different plant parts,calyx colour, fruit size and shape, fruit colour, clustering habit was also collected. In some pockets of Bihar, white coloured genotypes have high yield potential. In Bihar, farmers grow three types of fruits, *i.e.* dark green, purple and small round purple (locally known as Golagachhia). The consumer of this region has preferences for long and green-fruited types. A great deal of variability was observed in Sirsa Village also known as 'Began Gaon' (eggplant village) of Araria district in Bihar. In Kishanganj district, green fruit with white patches is most extensively grown. Harda, Ranipatra, Billouri, Subado and Chand Diwri of Purnea district (W.B.) are the hot spots for growing eggplant.

REFERENCES

1. Agarwal, R.K. and Munshi, A.D. 2008. Division of Vegetable Science, ICAR-IARI, N. Delhi.
2. Ahuja, S., Mukhopadhya, M.C., Singh, A. and Ahuja, S.P. 1987. Effects of infestation of eggplant (*Solanum melongena* L.) with root knot nematode (*Meloidogyne incognita*) of the oxidative enzymes and cell wall constituents in their roots. *Cap. Newslett.*, **6:** 98-99.

3. Ali, M., Matsuzae, N, Okub, H. and Fujida, K. 1992. Resistance of non-tuberous *Solanum* to root knot nematode. *J. Jpn. Soc. Hort. Sci.*, **60(4):** 921-926.

4. Anjaneyulu, A. 1969. Studies on little leaf disease of brinjal (*Solanum melongena* L.). Ph. D. Thesis, University of Mysore.

5. Anjaneyulu, A. and Ramakrishnan, 1968. Reaction of *Solanum* species to little leaf of brinjal. *Madras Agril. J.*, **55:**142-143.

6. Asati, B.S. and Yadav, D.S. 2004. Diversity of Horticultural Crops in North Eastern Region. ENVIS Bulletin. *Himalayan Ecology*, **12(1):** 1-11.

7. Bailey, L.H. and Munson, W.M. 1891. Experience with eggplants. New York (Cornell) *Station Bullettin*, **26:** 20.

8. Bajaj, K.L., Kaur, G. and Chadha, M.L. 1979. Glycoalkaloid content and other chemical constituents of the fruits of eggplant. *J. Plant Foods*, **3:** 199-203.

9. Bajaj, K.L., Kaur, G., Chadha, M.L. and Singh, B. 1981. Polyphenol-oxidase and other chemical constituents in fruits of some eggplant varieties. *Veg. Sci.*, **8:** 37-44.

10. Bajaj, K.L., Kaur, G. and Chadha, M.L. 1988. Seasonal variationin pigments of brinjal varieties. *Punjab Agric. J. Res.*, **25(2):** 201-205.

11. Bajaj, K.L., Singh, D. and Kaur, G. 1989. Biochemical basis of relative field resistance of eggplant (*Solanum melongena* L.) to the shoot and fruit borer (*Leucinodes orbonalis* Guen.). *Veg. Sci.*, **16:** 145-149.

12. Baksh, S. and Iqbal, H. 1979. Crossability relationship in some non-tuberous species of *Solanum*. *J. Hort. Sci.*, **54(2):** 163.

13. Behera, T.K. and Singh, N. 2002. Interspecific cross between eggplants (*S. melongena* L.) with *Solanum* species. *Scientia Horticulturae*, **95:** 165-172.

14. Bhaduri, P.N. 1951. Inter-relationship of non-tuberiferous speciesof *Solanum* with some consideration on the origin of brinjal. *Indian J. Genet. Plant Breed.* **2:** 75-86.

15. Blestsos, F., Roupakias, D., Tsaktsira, M.L., Scaltsoyjannes, A.B. and Thanassoulopoulos, C.C. 2003. Interspecific hybrids between three eggplant (*Solanum melongena* L.) cultivars and two wild species (*Solanum torvum* Sw. and *Solanum sisymbrifolium* Lam.). *Plant Breed.*, **117:** 159-164.

16. Bora, L.C., Bhattachrya, S.K., Bora, G.C. and Saikia, B.K. 1993. Screening for resistance against bacterial wilt of brinjal. *Assam. J. Agric. Sci. Soc. North East India*, **6:** 83-84.

17. Borikar, S.T., Makne, U.G. and Kulkarni, U.G. 1981. Note on diallel analysis in brinjal. *Indian J. Agric. Sci.*, **51:** 51-52.

18. Bose, T.K. 2002. Vegetable crops, 1: 265-344, Naya Prokash, Kolkata (W.B).

19. Cappelli, C., Stravato, V.M., Rotino, G.L., and Bounauria, R. 1995. Sources of resistance among *Solanum* species to an Italian isolate of *Fusarium oxysporum* f. sp. *melongenae*. EUCARPIA Meeting on Genetics and Breeding of *Capsicum*. In: *Eggplant*, **9:** 221-224.

20. Chadha, M.L. 1988. Graft brinjal on to the rootstock of *Solanum torvum* to save it from root-knot nematodes. *Indian Hort.*, **32(2):** 31-32.

21. Chadha, M.L., Hedge, R.K. and Bajaj, K.L. 1988. Heterosis and combining ability studies of pigmentation in brinjal (*Solanum melongena* L.). *Veg. Sci.*, **15:** 64-71.

22. Chadha, M.L. and Sharma, C.M. 1989. Inheritance of yield in brinjal. *Indian J. Hort.*, **46(4):** 485-489.

23. Chakrabarty, A.K. and Chaudhari, B. 1975. Breeding brinjal resistant to little leaf disease. *Proc. Ind. Nat. Sci. Acad. B.*, **41:** 379-385.

24. Chakrabarty, A.K. and Chaudhury, B. 1974. Effect of little leaf disease on metabolic changes in susceptible cultivar and resistant allied species of brinjal. *Veg. Sci.*, **1:** 12-17.

25. Chakrabarty, A.K. and Chaudhury, B. 1975. Breeding brinjal resistant to little leaf disease. *Proc. Indian Nat. Sci. Acad.* (*B*), **41(4):** 379-385.

26. Chattopadhyay, A., Dutta, S., Hazara, P. and Sarkar, G. 2011. Conservation and utilization of vegetable germplasm at Bidhan Chandra Krishi Viswavidyalaya, Kalyani (West Bengal). Natl. Symp. On 'Vegetable Biodiversity' Held at JNKVV, Jabalpur (M.P.), p. 131-140.

27. Chauhan, S.V.S. 1984. Studies in genic male sterile *Solanum melongena* L. *Indian J. Genet. Plant Breed.*, 44: 367-371.

28. Chelliah, S. and Srinivasan, K. 1985. In: Proc. Natl. Semi. Breeding Crop Plants for Resistance to Pests and Diseases (May 25-27, 1983), School of Genetics, TNAU, Coimbatore, Tamil nadu, India, pp. 43-44.

29. Chen, N.C. 1997. Cultivation and seed production of eggplant. In: Training Workshop on Vegetable Cultivation and Seed Production Technology, Shanhua, Asian Vegetable Research and Development Centre (AVRDC), Taiwan.

30. Choudhury, B. 1976. Eggplant. In: *Evolution of Crop Plants* (edn. I) p. 278-279. Simmonds N.W. (Ed) Longman, London and New York.

31. Choudhury, H.C. 1977. Genetical studies in West African eggplant. *Indian J. Genet. Pl. Breed.*, **37:** 26-33.

32. Choudhury, H.C. 1972. Genetical studies in some West African *Solanum melongena* L. *Cand. J. Genetic. Cytol.*, **14:** 446-4490.

33. Christodoulakis, N.S., Lampri, P.N. and Fasseas, C. 2009. Structural and cytochemical investigation of the leaf of silverleaf nightshade (*Solanum elaeagnifolium*), a drought resistant alien weed of the Greek flora. *Australian Journal of Botany*, **57:** 432-438.

34. Clarke, C.B. 1883. Solanaceae. In: The Flora of British India. Ed. Hooker, J.D.l. Reeve and Co., London. **4:** 228-246.

35. Collonnier, C., Mulya, K., Fock, I., Mariska, I., Servaes, A., Vedel, F., Siljak-Yakovlev, Souvannavong, V., Ducreux, G. and Sihachakr, D. 2003. Sources of resistance against *Ralstonia solanaceraum* in fertile somatic hybrids of eggplant (*Solanum melongena* L.) with *Solanum aethiopicum* L. *Plant Sci.*, **160(2):** 301-313.

36. Daunay, M.C. and Dalmasso, A. 1985. Multiplication of *Meloidogyne javanica, M. incognita* and *M. arenaria* on several *Solanum* species. *Rev. Nematol.*, **8(1):** 31-34.

37. Dhaka, S.K., Kaushik, R.A. and Sarolia, D.K. 2016. Analysis of genotype x environment interaction for growth, earliness and yield traits of in Rajasthan. 7th Indian Horticulture Congress (Abs: 06; Session: II; Breeding for Horticultural Crops), 15-18 November, held at ICAR-IARI, New Delhi, p. 90-91

38. Dhankar, B.S., Gupta, V.P. and Singh, K. 1977. Screening and variability studies for relative susceptibility to shoot and fruit borer (*L. orbonalis* Guen.) in normal and ratoon crop of brinjal (*Solanum melongena* L.). *Haryana J. Hort. Sci.*, **6:** 50-58.

39. Dhankar, B.S., Malhotra, N., Choudhary, B.D. and Pandita, M.L. 1979. Studies on genetic diversity for shoot and fruit borer resistance in nromal and ratoon crop of brinjal. *Haryana J. Hort. Sci.*, **8 (3-4):** 107-111.

40. Das, P., Pukhan, P.N. and Sarmah, D.K. 1989. Screening of some vegetable varieties against root-knot nematode (*Meloidogyne incognita*). *J. Res.*, **9(1-2):** 56-57.

41. Daunay, M.C. and Hazra, P. 2012. "Eggplant", In: *Handbook of Vegetables* (eds.) Peter, K.V. and Hazra, P. (Houston, TX: Studium Press): 257-322.

42. Daunay, M.C., Lestar, R.N., Gebhardt, C.H., Hemart, J.W., John, M., Fray, A. and Doganlar, S. 2001. Genetic reources of eggplant (*Solanum melongena* L) and allied species: A new challenge for molecular genetics and eggplant breeders (*Salanaceae*-V-(eds.). Van Den Berg, R.G., Baredse, G.W., and Mariani, Nijmegen University Press, Nijmegen, The Netherlands, p. 251-274.

43. Deb, D.B. 1980. Enumeration, synonym and distribution of the Solanaceae in India. *J. Econ. Tax. Bot.*, **1:** 33-54.

44. Decker. 1951. *Phomopsis* blight resistant eggplants. *Phytopathology*, **41:** 9. Cross compatibility study and characterization of interspecific hybrids in *Solanum* species. (Abs: 88), 7th Indian Horticulture Congress, held at ICAR-IARI, New Delhi, 15-18 November, p. 68.

45. Devi, C.P., Munshi, A.D., Behera, T.K., Choudhary, H., Tomar, B.S. and Saha, P. 2016.

46. Dharmagowda, M.V., Hiremath, K.G. and Goud, J.V. 1979. Genetic analysis of yield and its components in brinjal (*Solanum melongena* L.). *Mysore J. Agric. Sci.*, **13:** 151-155.

47. Dwivedi, N.K., Bhandari, D.C. and Bhatnagar, N. 1998.Collecting wild species of crop plants from Aravalli hills, Rajasthan (India)

48. Ellis, J.L. 1990. Flora of Nallamalais. Vol.-II. Botanical Survey of India, Calcutta.

49. Fukusawa, C.A. 1964. Genetics of clustered and solitary fruit segregants from the interspecific cross between *S. melongena* L. x *S. cumingii. Dunal. Areneta J. Agric.*, **11:** 55-75.

50. Fassuliotis, G. and Dukes, P.D. 1972. Disease reaction of *Solanum melongena* and *S. Sisymbrifolium* to *Meloidogyne incognita* and *Verticillium alboratum. J. Nematol.*, **4(4):** 222-223.

51. Gamble, J.S. 1921. Flora of the Presidency of Madras. Vol-II. Aldard and Son Ltd., London.

52. Gangopadhyay, K.K. 2008. Germplasm Evaluation Division, ICAR-NBPGR, New Delhi.

53. Gangopadhyay, C., Maity, T.K. and Mandal, S.K. 1996. Screening of brinjal germplasm against fruit and shoot borer *Leucinodes orbonalis* Guen. *Environment Ecology*, **14(4):** 834-836.

54. Gharge, C.P., Ranpise, S.A., Bawkar, S.O. and Shinde, K.G. 2016. Heterosis studies for yield and its components in brinjal (*Solanum melongena* L.). 7th Indian Horticulture Congress (Abs: 11; Session: II; Breeding for Horticultural Crops), 15-18 November, held at ICAR-IARI, New Delhi, p. 94.

55. Gharge, C.P., Ranpise, S.A., Shinde, K.G., Bhalekar, M.N. and Nimbalkar, C.A. 2016. Combining ability analysis studies and yield contributing characters in rinjal (*Solanum melongena* L.). 7th Indian Horticulture Congress (Abs: 21; Session: II; Breeding for Horticultural Crops), 15-18 November, held at ICAR-IARI, New Delhi, p. 101-102.

56. Gill, C.K. and Chadha, M.L. 1979. Resistance in brinjal to shoot and fruit borer *Leucinodes orbonalis* Guen. (Pyralidae: Lepidoptera). *Indian J. Hort.*, **36(1):** 67-71.

57. Gill, H.S., Arora, R.S. and Pachauri, D.C. 1976. Inheritance of quantitative characters in eggplant. *Indin J. Agric. Sci.*, **46:** 484-490.

58. Gill, H.S., Balakrishnan, K.A., Kataria, A.S., Singh, D., Memane, S.A., Pal, K.M., Tomar, B.A. and Verma R. 1990. Project Directorate of Vegetable, Annual Report, Part-I (1988-89 and 1989-90). Presented at XIth Workshop held at Dr. Y.S. Parmar University of Horticulture and Forestry, Solan, 4th to 7th June, 1990.

59. Gisbert, C., Prohens, J., Raigon, M.D., Stommel, J.R. and Nuez, F. 2011. Eggplant relatives as sources of variation for developing new rootstocks: Effects of grafting on eggplant yield and fruit apparent quality and composition. *Scientia Horticulturae*, **128:** 14-22.

60. Gopalakrishnan, T.R. 2004. Three decades of vegetable research in Kerala Agricultural University, Vellanikkara (Kerala), p. 153.

61. Gopimany, R. and George, M.K. 1979. Screening of brinjal varieties for wilt resistance. *Agric. Res. J. Kerala*, **17:** 7-10.

62. Gopinath, G. and Madalgeri, B.B. 1986. Genetics of yield and its components in brinjal. *Haryana J. Hort. Sci.*, **15:** 103-109.

63. Gopinath, G. and Madalgeri, B.B. 1986. Bacterial wilt (*Pseudomonos solanacearum* E.F. Smith.) resistance in eggplant. *Veg. Sci.*, **13:** 189-195.

64. Gopinath, G., Madalgeri, B.B. and Samaskar, C. 1986. A note on heredity of fruit colour in WGGR 112-B brinjal. *Curr. Research Univ. of Agric. Sci. Bengaluru*, **15:** 17-18.

65. Goth, R.W., Peter, K.V. and Webb, R.E. 1983. Bacterial wilt (*Pseudomonos solanacearum* E.F. Smith.) resistance in pepper and eggplant lines. *Phytopathology*, **73:** 808.

66. Gousset, C., Collonnier, C., Mulya, K., Mariska, I., Rotino, G.L., Besse, P., Servaes, A. and Sihachakr, D. 2005. *Solanum torvum*, as a useful source of resistance against the bacterial and fungal diseases for improvement of eggplant (*Solanum melongena* L.). *Plant Sci.*, **168:** 319-327.

67. Gowda, P.H.R., Shivashankar, T. and Joshi, S. 1990. Interspecific hybridization between *Solanum melongena* L. and *Solanum macrocarpon*: study of the F_1 hybrid plants, *Euphytica*, **48(1):** 59-61.

68. Halsted, B.D. 1901. Experiments in crossing eggplants. New Jersey Agricultural Experiment Station Annual Report, **22:** 398-400.

69. Hani, M., Badr, A.M., Khalf-Allah, M.A., EL-Shal, M.M., Kadar, Abdel and Doos, M.K. 1977. Estimation of heterosis in eggplant (*Solanum melongena* L.). *Alexandra J. Agric. Res.*, **25:** 465-471.

70. Heap, J.W. and Carter, R.J. 1999. The biology of Australian weeds. 35. *Solanum elaeagnifolium* Cav. *Plant Protection Quarterly*, **14:** 2-12

71. He'bert, Y. 1985. Comparative resistance of nine species of the genus *Solanum* to bacterial wilt (*Pseudomonas solanacearum*) and the nematode *Meloidogyne incognita*. Implications for the breeding of aubergine (*Solanum melongena* L) in the humid tropical zone. *Agronomie.*, 5: 27.

72. Henry, A.N., Kumari, G.R. and Chira, V. 1987. Flora of Tamil Nadu, India. Series I: Analysis. Vol. II. Botanical Survey of India, Coimbatore (Tamil Nadu).

73. Ishaq, N.M.S.M. and Chaudhuri, R.P. 1984. Comparative susceptibility of some varieties of eggplant to shoot and fruit borer in Assam. *Indian J. Agric. Sci.*, **54(9):** 751-756.

74. Jannaki and Ammal, E.K. 1933. The genetic analysis of some races of eggplant (*Solanum melongena* L.). 20th Ann. Meet. *Indian Sci. Cong. Patna, Sect. Bot.:* Abst: **93,** 37(PBA4: 65).

75. Jari, C.I., Rietveld, E.M. and de Hass, J.M. 1999. Transfer of fungal tolerance through interspecific somatic hybrization between *Solanum melongena* L. and *S. torvum*. *Plant Cell Rep.*, **18:** 791-796.

76. Joshi, A.K. 1989. Nature of gene effect in relation to exploitation of heterosis in brinjal (*Solanum melongena* L.). Ph. D. Thesis, PAU, Ludhiana, (Punjab).

77. Kabir, M.M., Mia, M.D., Azim, I.I., Begum, R.A. and Ahmed, A. 1984. Field screening of twelve brinjal varieties against shoot and fruit boerer, *Leucinodes orbonalis* Guen. (Lepidoptera: Pyralidae) *Bangladesh J. Zoology*, **12(1):** 47-48.

78. Kalda, T.S., Swarup, V. and Choudhary, B. 1976. Studies on resistance to phomopsis blight in eggplant (*Solanum melongena* L.). *Veg. Sci.*, **3:** 65-70.

79. Kalda, T.S., Swarup, V. and Choudhary, B. 1977. Resistance to *Phomopsis* blight in eggplant. *Veg. Sci.*, **4:** 90-101.

80. Kale, P.B., Mankar, S.W., Wangikar, P.D. and Gonge, V.S. 1986. Chemical composition of little leaf disease affected leaves of different genotypes of brinjal (*Slanum melongena* L.). *Veg. Sci.*, **13:** 199-204.

81. Kale, P.B., Mankar, S.W., Wangikar, P.D. and Gonge, V.S. 1986. Integrated effect of plant protection measures and resistance breeding for control of little leaf of brinjal. *Veg. Sci.*, **13**: 205-214.

82. Kalloo, G. 1998. Biochemical basis of insect resistance in vegetables. Vegetable Breeding Vol. II, CRC Press Inc. Boca Raton, Florida.

83. Kalloo, G. 1994. Vegetable Breeding. Combine edn. Vol.1, 2 and 3. Panima Educational Book Agency, Delhi, p. 174.

84. Kaur, G., Bajaj, K.L. and Chadha, M.L. 1985. Variation in important quality characters in

eggplant (*Solanum melongena* L.) as affected by maturity. *Veg. Sci.*, **12:** 68-73.

85. Kalloo, G. Banerjeee, M.K., Singh, M. and Kumar, S. 2002. Vegetable Varieties, ICAR-IIVR Technical Bulletin-3, p. 59.

86. Khan, A.R. and Ramzan, M. 1954. Inheritance of some important characters in brinjal (*Solanum melongena* L.). Proceedings of the 6th Pakistan Science Conference, Karachi, Part III. p. 12-14 (PBA26: 173).

87. Khan, M.M.R. and Isshiki, S. 2008. Development of a male sterile eggplant by utilizing the cytoplasm of *Solanum virginianum* and a biparental transmission of chloroplast DNA in backcrossing. *Scientia Horticulturae*, **117:** 316-320.

88. Khan, M.M.R. and Isshiki, S. 2011. Functional male sterility expressed in eggplant (*Solanum melongena* L.) containing the cytoplasm of *S. kurzii* Brace and Prain. *J. Hortic. Sci. and Biotech.*, **84:** 92-96.

89. Khan, N.A.A. 1974. Studies on *Pseudomonas solanacearum* E.F. Smith, causing wilt of brinjal, potato and tomato in Mysore state. *Mysore J. Agric. Sci.*, **8:** 472.

90. Khan, R., Rao, G.R. and Baksh, S. 1978. Cytogenetics of *S. integrifolium* and its possible use in eggplant breeding. *Indian J. Genet. Plant Breed.*, **38:** 343-347.

91. Khapre, P.R., Wanjari, K.B. and Deokar, A.B. 1988. Inheritance of fruit colour in *S. melongena* L. x *S. indicum* L. *J. Mahar. Agric. Univ.*, **13:** 97-98.

92. Khapre, P.R., Wanjari, K.B. and Deokar, A.B. 1986. Inheritance of some colour characters in the interspecific cross *S. melongena* L. x *S. indicum* L. *J. Mahar. Agric. Univ.*, **11:** 296-298.

93. King, S.R., Davis, A.R., Zhang, X. and Croshy, K. 2010. Genetics, breeding and selection of rootstocks for Solanaceae and Cucurbitaceae. *SciHort.*, 127: 106-111.

94. Krishnappa, D.G. and Chennavaeraih, M.S. 1975. Cytotaxonomy of *Solanum indicum* complex. *Cytologia.*, **40:** 322-331.

95. Krishamurthi, S. and Subramaniam, D. 1954. Some investigation on the types of flowers in brinjal based on style length and their fruit set under natural conditions and in response to 2, 4-D as plant growth regulator. *Indian J. Hort.*, **11**(2): 63-66.

96. Krishnaiah, K. and Vijay. 1975. Evaluation of brinjal varieties for resistance to shoot and fruit borer, *Leucinodes orbonalis* Guen. *Indian J. Hort.*, **32:** 84-86.

97. Kumchai, J., Wei, Y.G., Lee, C.Y., Chen, F.C. and Chin, S.W. 2013. Production of intraspecific hybrids between commercial cultivar of the eggplant (*Solanum melongena* L.) and its wild relatives *S. torvum*. *Genet. Mol. Res.*, **12(1):** 755-764.

98. Kumar, N.K.K. and Sadashiva, A.T. 1996. *Solanum macrocarpon*: a wild species of brinjal resistant to brinjal shoot and fruit borer, *Leucinodes orbonalis* (Guen.). *Insect Environment*, **2(2):** 41-42.

99. Kumar, P.A. 2008. National Research Centre on Plant Biotechnology (NRC-PB), New Delhi, Personal communication.

100. Labadie, J. 1976. In: 10PB Chromosome number reports LIV. *Taxon.*, **25:** 631-649.

101. Lal, O.P., Sharma, R.K., Verma, T.S., Bhagchandani, P.M. and Chandra, J. 1976. Resistance in brinjal to shoot and fruit borer, *Leucinodes orbonalis* Guen. (Puralidae: Lepidoptera). *Veg. Sci.*, **3:** 111-116.

102. Lal, B.S. and Ahmad, S.Q. 1965. Biology and control of brinjal fruit and shoot borer (*L. orbonalis*). *J Eco. Entomol.*, **58:** 448-451.

103. Lee, J.M.1994. Cultivation of grafted vegetables. I. Current status, grafting methods and benefits. *HortSci.*, **29:** 235-239.

104. Lester, R.N. and Hasan. 1991. Origin and domestication of the brinjal/eggplant (*Solanum melongena* from *S. incanum* in Africa and Asia: In: Hawkes, J.G., Lester, R.N., Nees, M., Estrade, N. (eds.) Solanaceae.III.): Taxonomy, chemistry, evolution. Royal Botanical Garden, Kew, p. 369-387.

105. Liu, J., Zheng, Z., Zhhou, X., Feng, C. and Zhuang, Y. 2015. Improving the resistance of eggplant (*Solanum melongena* L.) to *Verticillium* wilt using wild species *Solanum linnaeanum*. *Euphytica*, **201:** 463-469.

106. Manilal, K.S. 1988. Flora of silent valley. The Mathrubhumi (MM) Press, Calicut.

107. Martin, F.W. and Rhodes, A.M. 1979. *Euphytica*, **28:** 367-83.

108. Mathew, K.M. 1983. The Flora of the Tamil Nadu Carnatic. Vl-III, Part-2. The Rapinat Herbarium, Tiruchirapalli.

109. Mayee, C.D. and Munshi, C.D. 1973. Mycoplasma a new threat to vegetables. *Punjab Veg. Grower*, **8(2):** 12-14.

110. Mehto, D.N. and Lal, B.S. Chemical control of brinjal shoot and fruit borer. *Indian J. Entomol*, **43:** 105-106.

111. Mital, R.K., Singh, S.N., and Singh, H.N. 1976. Genetics of some characters in brinjal (*Solanum melongena* L.). *Veg. Sci.*, **3:** 79-86.

112. Mitra, D.K. and Majumdar, M. 1977. Changes in total phenolic content and poly-phenoxydase activity of brinjal as a result of little leaf infection. *Proc. Indian National Sci. Acad.*, **43:** 228-231.

113. Mohanan, M. and Henry, A.N. 1994. Flora of Thiruvananthapuram. Botanical Survey of India, Calcutta.

114. More, D.C. and Patil, S.B. 1982. Inheritance of some characters in brinjal cross SM-2 x Nimbakar Green Round. *J. Maha. Agric. Univ.*, **7:** 243-244.

115. Morra, L. 2004. Grafting in vegetable crops. In: Tognoni, F., Pardassi, A., Mensuali-sodi, A., Dimauro, B. (eds.) the production in the greenhouse after the era of the methyl bromide, Comiso, Italt, p. 147-154.

116. Mote, U.N. 1978. Studies on the varietal resistance of brinjal (*Solanum melongena* L.) to jassid under field conditions. *Veg. Sci.*, **5:** 107-110.

117. Mote, U.N. 1982. Varietal susceptibility to brinjal (*Solanum melongena* L.) to jassid (*Amaraska biguttula biguttula* Ishoda.). *J. Mahar. Agric. Univ.*, **7(1):** 59-60.

118. Mote, U.N., Joi, M.B. and Sonone, H.N. 1976. Screening of brinjal varieties to little leaf diseses under natural field condition. *J. Mahar. Agric. Univ.*, 305-306.

119. Mukherjee, R. and Mukhopadhyay, S. 1982. Effect of root exudates of brinjal on *Peseudomonas solanacearum*. *Veg. Sci.*, **9:** 122-125.

120. Mukhopadhyay, A. and Mandal, A. 1994. Screening of brinjal (*Solanum melongena* L.) for resistance to major insect pests. *Indian J. Agric. Sci.*, **64(11):** 798-803.

121. Mutegi, E., Snow, A.A., Rajkumar, M., Pasquet, R., Ponniah, H., Daunay, M.C. and Davidar, P. 2015. Genetic diversity and population structure of wild/ weedy eggplant (*Solanum insanum*; *Solanaceae*) in southern India: Implications for conservation. *Amer. J. Bot.*, **102:** 140-148.

122. Nandwana, R.P., Yadav, B.S., Verma, M.K. and Lal, A. 1980. Susceptibility of certain brinjal varieties to the root knot nematode (*Meloidogyne incognita*). *Indian J. Mycol. Pl. Path.*, **10:** 181-182.

123. Nathani, R.K. 1983. Studies on varietal susceptibility and chemical control of brinjal to shoot and fruit borer, *Leucinodes orbonalis* Guen. Thesis Abstracts, **9:** 51.

124. Nimbalkar, V.S. and More, D.C. 1980. Genetic studies in brinjal cross Muktakeshi x White Green. *J. Mahar. Agric. Univ.*, **5:** 208-210.

125. Nolla, J.A.B. 1932. Inheritance of colour in the eggplant (*Solanum melongena* L.). *J. Dept. of Agric. Puerto Rico*, **16:** 19-30 (PBA3: 127)

126. Odland, M.L. and Noll, V.J. 1948. Hybrid vigour and combining ability in eggplants. *Proc. Amer. Soc. Hort. Sci.*, **51:** 417-422.

127. Omidiji, M.O. 1975. Cytogenetic studies on the F_1 hybrid between the African eggplant, *Solanum gilo* Raddi and *Solanum melongena* L. *Z. Hortic. Res.*, **21:** 75-82.

128. Okimura, M., Matso, S., Arai, K. and Okitsu, S. 1986. Influence of soil temperature on the growth of fruit vegetable grafted on different rootstocks. *Bull. Veg. Ornam. Crops Res. Stn. Japan.*, **9:** 43-58.

129. Okoli, B.E. 1988. Cytotaxonomic studies of five West African species of *Solanum* L. (Solanaceae). *Feddes Repertorium*, **99(5-6):** 183-187.

130. Oyelana, O.A. and Ogunwenmo, K.O. 2005. Comparative assessment of induced mutants from *Solanum macrocarpon* L. (Solanaceae). *Acta Satech.*, **2:** 50-56.

131. Padmini, K., Rajasekharan, P.E., Ganga-Vishalakshi, P.N., Singh, T.H. and Upreiti, K.K. 2015. Unlocking the potential of *Slanum macrocarpon* L. for genetic enhancement in interspecific hybridization of brinjal (*Slanum melongena* L.) Challenges and approaches. In: Souveiner of lead papers and abstracts: national meet on distant hybridization in horticultural crop improvement at ICAR-IIHR, Bengaluru on 22nd and 23rd January, 2015, p. 251.

132. Panda, N.A., Mahapatra, A. and Sahoo, M. 1971. Field evaluation of some brinjal vaieties for resistance to shoot and fruit borer (*Leucinodes orbonalis* Guen.). *Indian Agric. Sci.* **41:** 597-601.

133. Pandey, P., Yadav, G.C. and Maurya, P.K. 2016. Manifestation of heterosis and combining ability in brinjal (*Solanum melongena* L.). 7th Indian Horticulture Congress (Abs: 01; Session II-Breeding of Horticultural Crops), held at ICAR-IARI, New Delhi, 15-18 November, p. 87.

134. Patel, J.P., Singh, U., Kashyap, S.P., Tiwari, S.K., Singh, D.K. and Singh, M. 2013. Heterosis for productivity traits in brinjal (*Solanum melongena* L.). *Veg. Sci.*, **40(1):** 80-82.

135. Patil, S.K. and More, D.C. 1983. Inheritance studies in brinjal. *J. Mahar. Agric. Univ.*, **8:** 43-45.

136. Pawar, D.B., Mote, U.N. and Kale, P.N. 1986. Relative susceptibility of varieties and local collections of brinjal to jassids and fruit borer. *Veg. Sci.*, **13:** 397-402.

137. Peter, K.V. and Singh, R.D. 1973. Diallel analysis of economic traits in brinjal. *Indian J. Agric. Sci.*, **43:** 452-455.

138. Peter, K.V. and Singh, R.D. 1976. Combining ability, heterosis and analysis of phenotypic variation in brinjal. *Indian J. Agric Sci.*, **44(6):** 393-339.

139. Ponnuswami, V. and Irulappan, I. 1993. Genetc assessment of resistace in eggplant inters varietal crosses to little leaf diseases. *South Indian Hort.*, **41(3):** 179-181.

140. Prasad, D.N. and Prakash, R. 1968. Floral biology of brinjal. *India J. Agric. Sci.* **38**:1053-1061.

141. Prasad, T.V., Gangopadhyay, K.K., Bhardwaj, R., Kalia, V. and Meena, B.L. 2016. Identification of sources of resistance to eggplant fruit and shoot borer. 1st International Agrobiodiversity Congress (Session: Trait Disdcovery and Enhanced Use of PGR), 6-9 November 2016; held at New Delhi; p.163

142. Preneetha, S. 2002. Breeding for shoot and fruit borer (*Leucinodes orbonalis* G.) resistance in brinjal (*Solanum melongena* L.). Ph. D. Thesis, Tamil Nadu Agricultural University, Coimbatore (Tamil Nadu).

143. Preneetha, S., Rajasree, V., Arumugum, T. and Veeraragavathatham, D. 2016. Bio-safety approach to derive resistance for brinjal (*Solanum melongena* L.) shoot and fruit borer (*Leucinodes orbonalis* Guen.) through introgression of *Solanum viarum* Gene. Abs: 63, 7th Indian Horticulture Congress, held at ICAR-IARI, New Delhi, p. 49-50.

144. Prohen, J., Whitaker, B.D., Plazzas, M., Vilonova, S., Hurta do, M., Blasco, M., Gramazio, P. and Stommel, J.R. 2013. Genetic diversity in morphological characters and phenolic acid content resulting from an interspecific cross between eggplant, *Solanum melongena* and its wild ancestors (*S. incanum*). *Ann. Appl. Biol.*, **162:** 242-257.

145. Pugalendhi, L., Veeraragavathatham, V., Natarajan, S. and Praneetha, S. 2010. Utilizing wild relative (*Solanum viarum*) as resistance source to shoot and fruit borer in brinjal (*Solanum melongena* L.). *Electronic Journal of Plant Breeding*, **1(4):** 643-648.

146. Rajam, M.V. *et al.*, 2008. Compendium of Transgenic Crop Plants: Transgenic Vegetable Crops; Blackwell Publishing, Oxford, U.K., p. 47 72.

147. Rajesekaran, S. 1971. Cytological studies on the F_1 hybrid (*S. xanthocarpum* x *S. melongena* L.) and its amphidiploids. *Caryologia*, **24:** 261-267.

148. Rajesekaran, S. and Sivasubramanium, V. 1971. Cytology of the F_1 hybrid of *S. zuccagnianum*. *Theor. Appl. Genet.*, **41:** 85-86.

149. Rama, R.M. 1914. Flowering plants of Travancore. Government Press, Trivendrum (Kerala).

150. Rana, R.S., *et al.*, 1994. Germplasm collection of okra and eggplant and their wild relatives from South Asia, ICAR-NBPGR, New Delhi.

151. Rangasamy, P. and Kadambavana-sundaram, M. 1973. A study on the inheritance of certain qualitative characters in the cross between *Solanum indicum* L. x *S. melongena* L. *South Indian Hort.*, **21:** 1-6.

152. Rangaswamy, P and Kadambavanasundarum, M. 1973. A study on the inheritance of certain qualitative characters in the cross between *Solanum indicum* L. x *Solanum melongena* L. *South Indian Hort.*, **21:** 1-6.

153. Randhawa, J.S. 1987. Genetic assessment of floral biology and productivity in brinjal (*Solanum melongena* L.). Ph. D. Thesis, PAU, Ludhiana (Punjab).

154. Rao, G.R. 1981. Results of interspecific cross pollinations between *S. melongena* L and *S. incanum* L. in eggplant breeding. *Proc. Indian natl. Sci. Acad.*, B. **47:** 893-898.

155. Rao, G.R. and Baksh, S. 1979. Cytomorphological study of the amphidiploids derived from the hybrids of the crosses between *S. melongena* and *S. integrifolium*. *Curr. Sci.*, **48:** 316-317.

156. Rao, G.R. and Baksh, S. 1981. Relationship between *S. melongena* and *S. integrifolium* Poir. *Indian J. Genet. Pl. Breed.*, **41:** 46-53.

157. Rao, G.R., Khan, A.H. and Khan, R. 1976. Interrelationship between tetraploid *Solanum nigrum* L. and *S. villosum* Mills. *Science and Culture*, **42:** 559-561.

158. Rao, M.V.B., Sohi, H.S. and Vijay, O.P. 1976. Reaction of some varieties of brinjal (*S. melongena* L.) to *Pseudomonas solanacearum*. *Veg. Sci.*, **3:** 61-64.

159. Rao, N.N. 1979. The barriers to hybridization between *Solanum melongena* and some other species of *Solanum*. In: Hawkes, J.G., Lester, R.N. and Skelding, A.D. (eds.). The Biology and Taxonomy of the *Solanaceae*. Acad. Press, London, p. 605-641.

160. Rao, G.R. and Kumar, A. 1980. Some observations on interspecific hybrids of *Solanum melongena* L. Proc. *Indian Acad. Sci.*, **89:** 47-51.

161. Rao, T.K.B. 1934. Partial sterility in the forst generation plants of crosses between wide varieties of common egg plants. *Current Science*, **2(8):** 258-286.

162. Ravichandra, N.G., Krishnappa, K., Setty, K.G.H. 1990. Interaction of *Meloidogyne javanica* and race-1, race-2 and race-3 of *M. incognita* with *Pseudomonas solanacearum* on a few brinjal cultivars/lines. *Indian J. Nematol.*, **20(2):** 138-147.

163. Rotina, G.L., Sala, T. and Toppino, L. 2014. Eggplant, in Alien Gene Transfer in Crop Plants, Vol. 2(eds.) Pratap, A., Kumar, J., New York, *Springer*, 381-409.

164. Saito, T., Matsunaga, H., Saito, A., hamato, N., Koga, T., Suzuki, T. and Yoshida, T. 2009. A novel source of cytoplasmic male sterility and a fertility restoration gene in eggplant (*Solanum melongena* L.) lines. *J. Jpn. Soc. Hort. Sci.*, **78(4):** 425-430.

165. Saldanha, C.J. and Nicolson, D.H. 1976. Flora of Hassan District of Karnataka, India. Amarind Publishing Co. Pvt. Ltd., New Delhi.

166. Salehuzzaman, M. and Alam, M.S. 1983. Genetic analysis of yield and its components in eggplant. *SABRAO J.*, **15:** 11-15.

167. Sambandam, C.N. and Chelliah, C. 1983. Breeding brinjal for resistance to *Aphis gossypii* G. In: proc. Natl. Semi. On Breeding Crop Plants for Resistance to Pests and Diseses. TNAU. Coimbatore (Tamil Nadu), p. 15.

168. Sankar, A., Jessy, M., Kuty, P.C. and Peter, K.V. 1987. Efficiency of four selection methods to improve level of wilt resistance in eggplant. *India J. Agric. Sci.*, **57(3):** 138-141.

169. Sarath Babu, B., Pandravada, S.R., Kamala, V. and Sivaraj, N. 1997. Reaction of some South Indian *Solanum* germplasm collections to shoot and fruit borer *Leucinodes orbonalis* Guen. *Indian J. Entomol. Res.*, **21(1):** 35-38.

170. Schaff, D.A., Jelenkovic, G., Boyer, C.D. and Pollack, B.L. 1982. Hybridization and fertility of hybrid derivatives of *Solanum melongena* L. and *Solanum macrocarpon* L. *Theor. Appl. Genet.*, **62:** 149-153.

171. Schalk, J.M., Stoner, A.K., Webb, R.E. and Winters H.F. 1975. Resistance in eggplant (*S. melongena* L.) and nontuber-bearing *Solanum* species to carmine spider mite. *J. Am. Soc. Hortic. Sci.*, **100:** 479-481.

172. Sebastian, S. 2000. Collection and characterization of land races of brinjal. M.Sc. thesis, Kerala Agricultural University, Vellanikkara (Kerala).

173. Sharif, R. and Habib, A.F. 1977. Inheritance studies in interspecific cross of *S. xanthocarpum* x *S. melongena* L. *Veg. Sci.*, **4:** 22-24.

174. Sharma, B.D., Singh, N.P., Raghavan, R.S. and Deshpande, U.R. 1984. Flora of India-Series: 2. Flora of Karnataka-Analysis. Botanical Survey of India, Howrah.

175. Sharma, B.D., Shetty, B.V., Vajravelu, E., Kumari, G.R., Vivekanathan, K., Chandrabose, M., Swaminathan, M.S., Chandrasekharan, R., Subba Rao, G.V., Ellis, J.L., Rathakrishnan, N.C., Karthikeyan, S., Chandrasekharan, V. and Sreenivasan, S.R. 1977. Studies on the Flora of Nilgiris, Tamil Nadu. *Biol. Mem.*, 21-186.

176. Sharma, D.R., Chaudhury, J.B., Ahuja, U. and Dhankar, B.S. 1980. Inter-specific hybridization in genus *Solanum*. A cross between *S. melongena* and *S. khasianum* through embryo culture. *Zeitschrift fur Pflanzenenziichtung*, **85:** 248-253.

177. Sharma, J.P. and Kumar, S. 1995. Evaluiation of pest resistance in brinjal to *Pseudomonas* wilt in Chotanagpur. *Indian Phytopathology*, **48(1):** 81-83.

178. Sheela, K.B. 1982. Cataloguing of brinjal germplasm to isolate lines resistant to bacterial wilt. M.Sc. thesis, Kerala Agricultural University, Vellanikkara (Kerala).

179. Sheela, K.B., Gopalakrishnan, P.K. and Peter, K.V. 1984. Resistance to bacterial wilt in a set of eggplant breeding lines. *India J. Agric. Sci.*, **54**(6): 457-60.

180. Shetty, K.D. and Reddy, D.D.R. 1985. Resistance in *Solanum* species to root knot nematode (*Meloidogyne incognita*). *India J. Nematol.*, **15:** 230.

181. Siddiqui, B.A. and Khan, I.A. 1979. Interspecific hybridization between *S. incanum* x *S. melongena* L. *Acta Bot. Indica*, **6:** 48-51.

182. Singh, A.K., Singh, M., Singh, A.K. and Singh, R. 2006. Genetic diversity within the genus *Solanum* (Solanaceae) as revealed by RAPD markers. *Curr. Sci.*, **90:** 711-716.

183. Sidhu, A.S., Bhutani, R.D., Kalloo and Singh, G.P. 1980. Genetics of yield components in brinjal (*Solanum melongena* L.). *J. Hort. Sci.*, **9:** 258-268.

184. Singh, B. 1991. Field reaction of eggplant germplasm to bacterial wilt in Nagaland. *Indian J. Agric. Sci.*, **61(9):** 694-695.

185. Singh, N.D. and Mital, R.K. 1988. Genetics of yield and its components in eggplant. *Indian J. Agric. Sci.*, **58:** 402-403.

186. Singh, S.N., Singh, H.N. and Hazarika, H.H. 1979. Fractional diallel analysis of some quantitative characters in brinjal. *Acta Horticulture*, **93:** 307-316.

187. Sivaraj, N., Rama Srinivas, T., Pandravada, S.R. and Kamla, V. 1998. Field evaluation of eggplant germplasm. *Indian J. Pl. Genet. Resour.*, **11(1):** 103-105.

188. Srinivasan, K. and Chelliah, S. 1978. Influence of time of planting and weather factor on brinjal little leaf incidence and its vector *Hithimonas phycitis* Dist. *South Indian Hort.*, **26:** 79-80.

189. Subbaratnam, G.V. and Bhutani, D.K. 1981. Screening of eggplant varieties for resistance to insect pest complex. *Veg. Sci.*, **8:** 149-153.

190. Subramanin, K.N., Venkatasubramanian, N.V. and Nallaswamy, V.K. 1987. Flora of Palghat. Bishen Singh, Mahendra Pal Singh, Dehradun (Uttarakhand).

191. Sulochana, C.B. and Soluman, J.J. 1973. The viral nature of brinjal little leaf disease. *Phyto Pathologische Zeitschirift*, **77:** 339-347.

192. Sunseri, F., Sciancalepore, A., Martelli, G., Accarri, N., Rotino, G.l., Valentina, D. and Tamietti, G. 2003. Development of RAPD-AFLP map of eggplant and improvement of tolerance to *Verticillium* wilt. *Acta Horticulturae*, **625:** 197-198.

193. Swaminathan, M. and Srinivasan, K. 1972. Studies on brinjal hybridization II. Transference of bacterial wilt resistance from a wild brinjal variety to cultivated brinjal varieties. *Agric. Res. J. Kerala*, **9:** 11-13.

194. Swamy Rao, T. 1970. A preliminary note on the inheritance of qualitative characters in brinjal. *Madras Agric., J.*, **57:** 508-509.

195. Swarup, V. 1998. Genetic Resources of Vegetable Crops in India. *Natl. Sym. on Emerging Scenario in Vegetable Research and Development*. Organized by Indian Society of Vegetable Science held at Project Directorate of Vegetable Research, Varanasi (U.P.) 12-14th December.

196. Tachibana, S., 1987. Comparision of effects of root temperature on the growth and mineral nutrient absorption in cucumber cultivars and fig leaf gourd. *J. Japan Soc. Hortic. Sci.*, **51:** 299-308.

197. Taley, Y.M., Swaarbandhe, P.A., Amin, P.W. and Sharnagat, B.K. 1977. Effect of different tetracycline and antibiotics on incidence of little leaf of brinjal (*Solanum melongena* L.). *Pestology*, **12:** 22-24.

198. Tewari, G.C. and Moorthy, P.N.K. 1985. Field respomnse of eggplant varieties to infestation by shoot and fruit borer. *Indian J. Agric. Sci.*, **55(2):** 82-84.

199. Tetbe, T. 1944. Studies on the inheritance of colour in the eggplant. *Jpn. J. Genet.*, **20:** 1-7 (PBA21: 3118).

200. Thakur, M.R., Singh, K. and Singh, J. 1969. Inheritance of some qualitative characters in brinjal (*Solanum melongena* L.). *Punjab Agric. Univ. J. Res.*, **6:** 769.

201. Thomas, T.A., Rai, M., Chandra, U., Pratibha, B., K.V. and Rana, R.S. 1990. Building up of genetic resources of vegetable crops, their characterization, evaluation, seed increase, maintenance and utilization at NBPGR (1988-90). Paper presented in Vegetable Improvement Project Workshop held at Dr. Y.S. Parmar University of Horticulture and Forestry, Nauni, Solan, from June 4-7, 1990.

202. Thompson, C. H. and Kelly, C. W., 1957, Vegetable Crops, McGraw Hill Book, Co. Inc., USA.

203. Tiwari, S.K. 2007. Development and utilization of microsatellite markers in germplasm characterization of brinjal (*Solanum melongena* L.), Division of Plant Genetic Resources, ICAR-IARI, New Delhi.

204. Ugur, B. and Kazim, A. 2003. Attempts of haploidy induction in tomato (*Lycopersicon esculentum* Mill.) via gynogenesis. I. Pollination with *Solanum sisymbrifolium* Lam. Pollen. *Pakistan J. Biol. Sci.*, **6:** 745-749.

205. Vasudevan, K.N. 1975. Contribution of the cytotaxonomy and cytogeography of the flora of the Western Himalayas (with an attempt to compare it with the flora of the Aips) Part II. *Ber. Sxchweiz Bot. Ges.*, **85:** 210-2552.

206. Vavilov, N.I. 1951. Phyto-geographical Basis of Plant Breeding. The Origin, Variation, Immunity and Breeding of Cultivated Plants.

207. Vavilov, N.I. 1951. The origin, variation, immunity and breeding of cultivated plants. *Chronica Bot.*, **13:** 2-94 (not seen in original).

208. Velayudhan, K.C., Abraham, Z., Amalraj, V.A. and Muralidharan, V.K. 1996. Collecting genetic diversity of eggplant and its wild relatives from South India. *Indian J. Plant Genet. Resour.*, **9(1):** 89-95.

209. Verma, A., Raychoudhuri, S.P., Chenulne, V.V., Prakash, N. and Rao, B.S. 1969. Mycoplasma like bodies in the tissue infected with sandle spike and brinjal little leaf. *Indian Phyto. Path.*, **22:** 289-290.

210. Verma, T.S. and Chaudhury, B. 1974. Screening of brinjal (*Solanum melongena* L.) against root-knot nematode (*Meloidogyne* spp.). *Veg. Sci.*, **1:** 55-61.

211. Vijay, O.P., Nath, P. and Jalikop, S.H. 1977. Studies on floral biology in long and round type brinjal. *India J. Agric. Sci.*, **47**: 288-291.

212. Vijaygopal, P.D. and Sethumadhava, P. 1973. Studies on the intervarietal hybrids of *S. melongena*. *Agril. Res. J. Kerala*, **11:** 43-46.

213. Vito, M.D.I., Zacheo, G. and Catalano, F. 1992. Sources of resistance to root knot nematodes (*Meloidogyne* spp.) in eggplant. *Capsicum Newsletter* (Special issue). p. 303.

214. Vishwanathan, T.V. 1975. On the occurrence of natural hybridization between *Solanum incanum* L. and *Solanum melongena* L. *Current Science*, **44:** 134.

215. Wanjari, K.B. and Khapre, P.R. 1977. Inheritance of pigmentation in *S. melongena* L. x *S. indicum* L. (In 4th International SABRAO Congress 4-8 May, 1981 at Univ. Kebangsann, Malaysia): 40.

216. Yadav, B.S., Nandwana, R.P., Lal, A. and Verma, M.K. 1975. Evaluation of certain brinjal varieties to root knot nematode (*Meloidogyne incognita*). *Indian J. Mycol. Pl. Path.*, **5:** 17.

217. Yamakawa, K. 1976. Breeding for disease resistance of vegetable crops in Japan. *Japanese Agric. Res. Qtly.*, **10:** 187-192.

218. Yamakawa, K. 1982. Use of rootstocks in solanceous fruit-vegetable production in Japan. *Japan Agric. Res. Qtly.*, **15:** 175-179.

219. Yoganarasimhan, S.N., Subramanyan, S.N. and Razi, B.A. 1981. Flora of Chikmagular District, Karnataka, India. International Book Distributors, Dehradun (Uttarakhand).

220. Yoshida, T., Monma, S., Matsunaga, H., Sakata, Y., Sato, T., Saito, T., Saito, A. and Yamada, T. 2004. New rootstocks 'Eggplant Ano2' with high resistance to bacterial wilt and fusarium wilt. Proc. 12th EUCARPIA Meeting on Genetics and Breeding of *Capsicum* and Eggplant, May 17-19, 2014, Noordwigkerhout, The Netherlands, p. 88 (Abstrct).

221. Zevin, A.C. and Zhukovsky, P.M. 1975. Dictionary of cultivated plants and their centres of diversity, PUDOC, Wageningen, p. 219.

Bitter Brinjal (*Solanum gilo*)

Introduction

Bitter brinjal (*Solanum gilo* Raddi) is also known as 'Jilo'. It is having tomato like fruit (is a kind of brinjal) which has enlarged calyx almost covering the long fruit and has the appearance between a brinjal and a tomato. It is important among the tribes of North Eastern Region. Generally this species is grown in the kitchen garden. The bitter brinjal is a prized vegetable particularly of Khasi and Mizo tribes of Meghalaya. The tribes of these states consume the immature green fruits as vegetable, which is bitter in taste (Arora and Hardas, 1977). The green fruits of bitter brinjal are a good source of protein. On the dry weight basis green fruits contain 28 per cent moisture, 5 per cent protein, 13 per cent total sugar, 4 per cent starch, 0.92 per cent free amino acids and 2.8 per cent phenolics. In African countries, these bitter fruits are liked for flavor and believed to confer various medicinal properties such as prevention of malaria or increasing lactation (Lester *et al.*, 1990). The high phenolics content is beneficial in the diet as they are known to minimize the toxicity and also have antiseptic and bacteriastatic action and regulate the pH in the intestine (West *et al.*, 1966).

Origin, History and Distribution

Bitter brinjal is probably native of Central Africa and was brought to Brazil with the slave trade. It is still grown in Africa where it is known as garden eggs. Jolo is considered as an important crop in Nigeria. In India, it is being grown in hilly areas of Meghalaya, Mizoram, Arunachal Pradesh, Assam, Tripura and Nagaland (Sharma *et al.*, 1994).

Botanical Description

Table 2.84: Botanical Description

Growth Habit	A small thornless shrub.
Stem	It grows up to 150 cm stem height with 4-5 branches.
Leaf	Petioles are adpressedly pubescent, petiole 2.5 to 3.0 cm long, leaf oblique, lobed (3 on left and 2 on right half of lamina), mid rib and secondary nerves raised on lower surface; apex acute, upper surface green and lower whitish, glabrescent, of size 15.0-20.0 cm long x 10.0-12.0 cm wide.
Floral biology	Inflorescence axillary, 3-flower at each node; flower white, gamopetelous at base, calyx-5, whitish and firmly attached.
Fruit and fruit set	Fruit peduncled, peduncle 2.0 to 2.5 cm, globose, smooth or grooved, greenish-yellow shining or pale green 2.5 to 3.0 cm in diameter and on ripening turns golden yellow or orange red; 4-locular, center pithy, flesh less juicy and bitter.
Seeds	Seeds kidney shaped, smooth, surface white, creamy or brown.

Augmentation of Genetic Resources and Crop Improvements

Introduction and Selection

In India, not much improvement work has been carried out but outside India, there are so many varieties which are popular in larger area. Few promising selections like 'Turkish Orange' 'Turkish Italian', 'Sweet Red', 'Small Ruffled Red' 'Red Ruffles' 'Hmong Red' *etc.*, are becoming popular.

Interspecific Hybridization

When *S. gilo* was used as a pollen parent crossed with different cultivars of *S. melongena* and *S. anomalum* (female parent) produced viable seeds but the hybrid plants were observed to bear small parthenocarpic fruits in cluster.

Resistance Breeding

Bitter brinjal (*Solanum gilo* Raddi) is highly resistant to shoot and fruit borer, the most destructive pest of eggplant in India and other tropical and sub-tropical parts of the world (Behra and Singh, 2002).

REFERENCES

1. Arora, R.K. and Hardas, M.W. 1977. Occurrence of *Solanum integrifolium* and *S. gilo* in North-eastern Hills. *J. Bombay Nat. Hist. Soc.*, **73:** 423-424.
2. Behra, T.K. and Singh, N. 2002. Inter-specific crosses between eggplants (*Solanum melongena* L.) with related *Solanum* species. *Scientia Horticulturae*, **95(2):** 165-172.
3. Lester, R.N., Jaeger, P.m.L., Spieirings, B., Bleijendaal, H.P.O. and Holloway, H.L.O. 1990. African eggplants-a review of collecting in West Africa. *Plant Genetic Resources Newsletter*, **18 (12):** 17-26.
4. Sharma, B.D., Samy, S.P. and Hore, D.K. 1994. *Solanum gilo* Raddi: A less known vegetable of north-Eastern India. *Indian J. Plant Genet. Resour.*, **7(1):** 109-111.
5. West, E.I., Todd, W.R., Mason, H.S. and Van Bouggen, J.T. 1966. Textbook of Biochemistry. p.538, 1218.

Solanum viarum

Steroids bearing *Solanum* (*Solanum viarum* Dunal syn. *S. khasianum* CB) Clark) is an important medicinal plant belonging to the family Solanaceae. It is much branched prickly under-shrub or shrub distributed in the Asom Khasi hills. Mature fruits are rich source of diosgenin/solasodine, an analogue of 16-DPA, which is commercially exploited as raw material for the manufactureing of sex hormones and contraceptive.

Seeds show poor germination (upto 30 per cent) at optimum temperature of 25 °C in dark which may either due to impermeable seed coat which inhibits water imbibition and thus arresting the germination or due to dormant embryo which needs triggering of hydrolytic enzymes responsible for the germination (Bewley and Black, 1978). In an experiment Gupta and Singh (1996) reported that over-night soaking in GA_3 (500 ppm) and acid scarification with sulphuric acid (25 per cent) for one hour. These treatments not only increased the germination percentage but also reduced the mean germination time significantly under laboratory conditions.

Augmentaion of Germplasm

Under *Solanum viarum* a unique line has been identified and being maintained with the name of var. Arka Sanjeevani.

REFERENCES

1. Bewley, J.D. and Black, M. 1978. Physilogy and biochemistry of seeds in relation to germination. Vol. 1. Springer-Verleg Berlin Heidelberg, New York.
2. Gupta, V. and Singh, B.B. (1996). Pre-treatments to improve germination in *Solanum viarum* Dunal. *Indian J. Plant Genet. Resour.*, **9(1):** 123-128.

Bhurat or Pea Eggplant or Turkey Berry (*Solanum torvum*)

Introduction

Other Scientific Name is *Solanum largiflorum* CT White (1917). Bhurat or Turkey berry (*Solanum torvum*) is known by several names-brihati Marathi marang in Sanskrit, bhurat, bhankatiya in Hindi, sundakkai and chundakai in Tamil. Dried berry, fried in oil or ghee, is an intrinsic part of meals in south India. Slightly bitter, the berries have a peculiar taste, which one gradually gets accustomed. A distinctive, bitter flavour, it tends to be more appreciated by older people. In Tamil Nadu and Kerala, the berries are commonly eaten. A common weed in the tropics, the plant is often gathered from the wild for local use as a food and medicine. Its popularlity increasing as a food and is occasionally cultivated as a home garden crop, being said to be especially popular in Thailand. The young shoots can be eaten raw or cooked. The fruit is eaten raw in Asia or cooked and served as a side dish with rice, or is added to stews, soups, curries *etc.* Fruit has juicy pulp, and contains many small seeds. In the West Indies, the half-grown, firm berries are boiled and eaten with foods such as yams or akees, or are added to soups and stews. An extract of the plant is sometimes used as a curdling agent. A phytopathological review by Ashok D. Agrawal and colleagues, published in 'Der Pharmacia Letter' in 2010, reveals that the fruits of *S. torvum* are used commonly in traditional medicine for its antioxidant, cardiovascular, anti-hypertensive and anti-platelet aggregation properties. *S. torvum* is also added as an ingredient in various indigenous herbal medicines. Moreover, anti-cancer compounds have also been found in the fruit and leaves of this plant. The berries have several curative powers. In Siddha medicine-one of the five alternative systems of medicine promoted by the Government of India. It is used to make what is called 'Sundavattral Choornam' which is excellent for digestion. This probably explains the 'Shiva Shiva' in the stomach, an indirect way to make people consume it as it keeps stomach ailments away.

The berries also possess anti-microbial, sedative and diuretic properties. It also contains poisonous alkaloids. Unless there are specific entries with information on edible uses, it would be unwise to ingest any part of this plant. *Solanum torvum* is considered to be a serious threat to the productivity and sustainability of pasture.

Origin, History and Distribution

Although *S. torvum* originated in the West Indies but it has been spread as a useful plant and consequently as a weed throughout the tropics and sub-tropics. It is well distributed to South America - Brazil, Colombia, and north to the Caribbean and through Central America to Mexico. A plant of the moist tropics, where it is found at elevations up to 1,600 metres, thrives well. Woodland clearings, thickets and waste places at elevations from 15-1,260 metres in Jamaica. In Nepal Naturalised on open, moist ground at elevations up to 1,000 metres. It grows best in areas where annual daytime temperatures are within the range 17-29 °C, but can tolerate 12-35 °c. It prefers a mean annual rainfall in the range 1,000 - 2,000 mm, but tolerates 700-4,200 mm. Heavy rainfall discourages fruit set. This plant succeeds in full sun and in light shade. Grows well in a range of fertile, moist well-drained soils but prefers a pH in the range 5-6 (tolerating 4.3-6.8). Plants can tolerate some drought.

Table 2.85: Distribution of *S. torvum* in different Country

Country	*Distribution*	*References*
	ASIA	
Bangladesh	Restricted	Holm *et al.*, 1991; EPPO, 2014
Brunei Darussalam	Present	Peregrine and Ahmad, 1982
Cambodia	Present	Holm *et al.*, 1991
China	Restricted distribution	Holm *et al.*, 1991; EPPO, 2014
-Guangxi	Present	Wang, 1980
-Yunnan	Present	Wang, 1980
India	Present	-
-Assam		Isahaque and Chaudhuri, 1985
-Kerala	Present	Singh and Gopalakrishnan, 1997
Indonesia	Present	Holm *et al.*, 1991; EPPO, 2014
-Java	Restricted distribution	Backer, 1973
Japan	Present	Matsuzoe *et al.*, 1996
Malaysia	Present	Narikawa *et al.*, 1988; EPPO, 2014
Philippines	Restricted distribution	Holm *et al.*, 1991; EPPO, 2014
Sri Lanka	Restricted distribution	Holm *et al.*, 1991; EPPO, 2014
Thailand	Restricted distribution	Holm *et al.*, 1991; EPPO, 2014
	AFRICA	
Cameroon	Restricted distribution	Heine, 1963; EPPO, 2014
Congo Democratic Republic	Restricted distribution	Holm *et al.*, 1991; EPPO, 2014
Côte d'Ivoire	Restricted distribution	*Holm et al.*, 1991; Goly and Tehe, 1997; EPPO, 2014
Equatorial Guinea	Present	Heine, 1963
Ghana	Restricted distribution	Holm *et al.*, 1991; Heine, 1963; EPPO, 2014
Guinea	Restricted distribution	Holm *et al.*, 1991; EPPO, 2014
Liberia	Restricted distribution	Holm *et al.*, 1991; Heine, 1963; EPPO, 2014
Malawi	Present	Holm *et al.*, 1991
Mauritius	Restricted distribution	Holm *et al.*, 1991; Anon, 1998b; EPPO, 2014
Nigeria	Restricted distribution	Holm *et al.*, 1991; Heine, 1963; EPPO, 2014
Senegal	Present	Holm *et al.*, 1991; Heine, 1963
Sierra Leone	Present	Heine, 1963
	NORTH AMERICA	
Mexico	Restricted distribution	Holm *et al.*, 1991; Breedlove, 1998; EPPO, 2014
USA	Widespread	Holm *et al.*, 1991; EPPO, 2014
-Florida	Present	Westbrooks and Eplee, 1988
-Hawaii	Present	Liquido *et al.*, 1994
-North Carolina	Present	North Carolina Dept of Agriculture and Consumer
	CENTRAL AMERICA AND CARIBBEAN	
Costa Rica	Restricted distribution	Holm *et al.*, 1991; EPPO, 2014
Cuba	Present	Almaguel *et al.*, 1984
Dominica	Present	Fournet and Hammerton, 1991
Grenada	Present	Fournet and Hammerton, 1991
Guadeloupe	Present	Fournet, 1973
Honduras	Restricted distribution	Holm *et al.*, 1991; EPPO, 2014
Jamaica	Restricted distribution	Holm *et al.*, 1991; EPPO, 2014
Martinique	Present	Fournet, 1973
Montserrat	Present	Fournet and Hammerton, 1991
Panama	Restricted distribution	Holm *et al.*, 1991; EPPO, 2014
Puerto Rico	Restricted distribution	Holm *et al.*, 1991; EPPO, 2014
Saint Kitts and Nevis	Present	Fournet and Hammerton, 1991

Country	*Distribution*	*References*
Saint Lucia	Present	Fournet and Hammerton, 1991
Saint Vincent and the Grenadines	Present	Fournet and Hammerton, 1991
EUROPE		
Italy	Present	Anon, 1998a
OCEANIA		
American Samoa	Present	Waterhouse, 1997
Australia	Restricted distribution	Holm *et al.*, 1991; EPPO, 2014
-Australian Northern Territory	Present	Hnatiuk, 1990
-New South Wales	Present	Hnatiuk, 1990
-Queensland	Present	Hnatiuk, 1990
Fiji	Restricted distribution	Holm *et al.*, 1991; Waterhouse, 1997; Parham, 1958; EPPO, 2014
French Polynesia	Present	Waterhouse, 1997
Guam	Present	Waterhouse, 1997
New Caledonia	Present	Waterhouse, 1997; MacKee, 1985
Niue	Present	Waterhouse, 1997
Papua New Guinea	Restricted distribution	Henty and Pritchard, 1975; Holm *et al.*, 1991; EPPO, 2014
Samoa	Present- Invasive	Whistler, 1983; Waterhouse, 1997; Space and Flynn, 2002
Solomon Islands	Widespread	Waterhouse, 1997; Hancock and Henderson, 1988
Tonga	Present	Whistler, 1983; Waterhouse, 1997
Vanuatu	Widespread	Waterhouse, 1997

Source: CABI, Invasive Species Compendium, Retrieve on 27-09-2016.

Species Availability, Evolution and Diversity

The species name *torvum* means harsh, fierce or sharp. There are named forms in Africa and in Thailand. At least one form is free of prickles.

Botanical Description

The plant belongs to family Solanaceae. The chromosome number is variously reported as n = 12 or 2n=24 (Symon, 1981), and the plant may be polyploid.

Domain : Eukaryota
Kingdom : Plantae
Phylum : Spermatophyta
Subphylum : Angiospermae
Class : Dicotyledonae
Order : Solanales
Family : Solanaceae
Genus : *Solanum*
Species : *Solanum torvum*

Once established, it can sprout from its roots, creating large thickets that could displace other vegetation. The plant is spread by birds that eat the fruits and spread the seeds. The fresh seed shows strong dormancy. Semi-hardwood cuttings 12-15 cm long, collected from fresh shoots and with their leaves removed, will produce roots and new shoots in 3 - 4 weeks.

Table 2.86: Botanical Description

Growth Habit	Evergreen armed shrub or small tree sparingly-armed, spreading or scrambling, which is growing up to 4 m.
Root	The root system consists of a deep and strong, woody taproot with numerous woody laterals.
Stem	The one to several soft-wooded stems are armed with stout/branched above, densely covered with fine stellate hairs and are scattered with broad-based, straight or lightly curved reddish or pale yellow prickles 2.5–10 mm long and 2-10 mm wide hooked prickles. The stems are initially green becoming brown as they mature.
Leaves	The alternate leaves occur singly along the stems and are broadly ovate and 5-20 cm long, usually with seven broad, blunt, unlobed to strongly lobed. Both surfaces are covered with very fine stellate hairs and have scattered prickles along the main veins. The upper surface is darker than the lower. The leaves have finally hairy petioles, 1 to 5 cm long, and vary considerably in shape and size depending on genetic origin and plant vigour.
Floral biology	The small, white flowers occur in large clusters. Extra-axillary corymbose cyme. Ebracteates, bisexual, ebracteolate, obliquely zygomorphic, hypogynous, cylic. **Calyx-**Cupular, lobes, pubescent; lobes, enlarged in fruit **Corolla-**Petals stellate-pubescent, acute, white, sub-rotate, plicate.
	Androecium-stamens 5, connivent around style anthers exserted, dehiscing by apical pores. Gynoecium-Ovary bilocular, superior, oblique, septum, ovules many, placenta axile (peltate).

Anthesis and pollination	The plant commences flowering when 3 - 4 months old and is said to have an economic life of 3-5 years. The plant can flower and produce fruit all year round. The inflorescence is a dense, compact, branched head consisting of 50-100 flowers at the ends of branches. These occur laterally (between two leaves) as the stems elongate. Each flower has five slender finely hairy sepals 2 to 3 mm long, five white to cream star-shaped petals each about 1 cm long, five elongate yellow stamens and a central stigma.
Fruit and fruit set	The fruits are globular berries 1-1.5 cm across and globose berry can be 10-15 mm in diameter, at first immature green and scurfy but ripening to dull yellow then orange as it ripens and containing few to many flat, woody, often reddish seeds.
Seeds	Single fruits contain 210-220 seeds. Seeds ovate to broadly ovate, broadly elliptic, or nearly circular, infrequently C-shaped in outline.

Augmentation of Genetic Resources and Crop Improvement

The plant is sometimes used as a rootstock for tomatoes (*Solanum lycopersicum*) and aubergines (*Solanum melongena*), where it conveys resistance to bacterial wilt and nematodes. *Solanum aethiopicum* cv. 'Iizuka' gives better results with tomatoes.

Genetic Resource Conservation

S. torvum is relatively resistant to *Meloidogyne* spp., and is used as a rootstock for grafting tomatoes in susceptible areas (Shetty and Reddy, 1985). It is also resistant to *Pseudomonas solanacearum* (Hebert, 1985) and phomopsis fruit rot of aubergine (Datar and Ashtaputre, 1988).

REFERENCES

1. Almaguel, L., Machado, L.R. and Caceres, I. 1984. New food-plants of the mite *Polyphagotarsonemus latus*. *Ciencia e Tecnica en la Agricultura, Proteccion de Plantas*, **7(1):** 99-108.
2. Backer, C.A. 1973. Atlas of 220 weeds of sugarcane fields in Java. In: van Steenis CGGJ, ed. Handbook for the cultivation of sugarcane and manufacturing of cane sugar in Java. Vol. 7: Atlas (final instalment). Pasuruan, Indonesia: Indonesian Sugar Experiment Station.
3. Breedlove, D.E. 1998. Floristic list for Mexico. IV. Flora of Chiapas. World Wide Web page at http// www.ibiologia.unam.mx/publicaciones/lfl4.
4. Datar, V.V. and Ashtaputre, J.U. 1988. Studies on resistance to Phomopsis fruit rot in eggplant. *Indian Phytopathology*, **41(4):** 637-638.
5. EPPO, 2014. PQR database. Paris, France: European and Mediterranean Plant Protection Organization. http://www.eppo.int/DATABASES/pqr/pqr.htm.
6. Fournet, J. 1973. Anthracnose of eggplant in the Antilles II. Methods of control. *Annales de Phytopathologie*, **5(1):**15-25.
7. Fournet, J. and Hammerton, J.L. 1991. Weeds of the Lesser Antilles. Paris, France: Department d'Economie et Sociologie Rurales, Institut National de la Recherche Agronomique.
8. Goly, P.G. and TThT, H. 1997. Effects of pineapple weeds on *Pratylenchus brachyurus* in C(te d'Ivoire. *Cahiers Agricultures*, **6(3):** 199-202.
9. Hancock, I.R., Henderson, C.P., 1988. Flora of the Solomon Islands. Research Bulletin No. 7. Honiara, Solomon Islands: Dodo Creek Research Station.
10. Hebert, Y. 1985. Comparative resistance of 9 *Solanum* species to bacterial wilt (*Pseudomonas solanacearum*) and to the nematode *Meloidogyne incognita*. Importance for breeding aubergine (*Solanum melongena* L.) in a humid tropical zone. *Agronomie*, **5(1):** 27-32.
11. Heine, H. 1963. 151. Solanaceae. In: Hutchison J, Dalziel JM, Hepper FN, eds. Flora of West Tropical Africa, Volume 2. London, UK: *Crown Agents*, 325-335.
12. Henty, E.E. and Pritchard, G.H. 1975. Weeds of New Guinea and their Control. Lp, Papua New Guinea: Department of Forests, Division of Botany, Botany Bulletin No.7.
13. Hnatiuk, R.J. 1990. Census of Australian Vascular Plants. Australian Flora and Fauna Series Number 11. Canberra, Australia: Australian Government Publishing Service.
14. Holm, L.G., Pancho, J.V., Herberger, J.P. and Plucknett, D.L. 1991. A Geographic Atlas of World Weeds. Malabar, Florida, USA: Krieger Publishing Company.
15. Isahaque N.M.M, Chaudhuri RP, 1985. A new alternate host plant of brinjal shoot and fruit borer *Leucinodes orbonalis* Guen. in Assam. *Journal of Research, Assam Agricultural University*, **4(1):** 83-85
16. Liquido, N.J., Harris, E.J. and Dekker, L.A. 1994. Ecology of *Bactrocera latifrons* (Diptera: Tephritidae) populations: host plants, natural enemies, distribution, and abundance. *Annals of the Entomological Society of America*, **87(1):** 71-84.
17. MacKee, H.S. 1985. Les Plantes Introduites et Cultivees en Nouvelle-Caledonie. Volume hors series, Flore de la Nouvelle-Caledonie et Dependances. Paris, France: Museum Nationelle d'Histoire Naturelle.
18. Matsuzoe, N., Aida, H., Hanada, K., Ali, M., Okubo, H. and Fujieda, K. 1996. Fruit quality of tomato plants grafter onto *Solanum* rootstocks. *Journal of the Japanese Society for Horticultural Science*, **65:** 73-80.

19. Narikawa, T., Sakata, Y., Komochi, S., Melor, R., Heng, C.K. and Jumali, S. 1988. Collection of solanaceous plants in Malaysia and screening for disease resistance. *Japan Agril. Res. Qtly.*, **22(2):** 101-106.

20. Parham, J.W. 1958. The Weeds of Fiji. Bulletin Fiji Department of Agriculture, 35. Suava, Fiji: Government Press.

21. Peregrine, W.T.H. and Ahmad, K. 1982. Grafting : a simple technique for overcoming bacterial wilt in tomato. *Tropical Pest Management*, **28(1):** 71-76.

22. Shetty, K.D. and Reddy D.D.R. 1985. Resistance in *Solanum* species to root-knot nematode *Meloidogyne incognita*. *Indian Journal of Nematology*, **15:** 230.

23. Space, J.C. and Flynn, T. 2002. Report to the Government of the Cook Islands on invasive plant species of environmental concern. Honolulu, USA: USAL USDA Forest Service, 146 p.

24. Wang, Z.R. 1990. Farmland Weeds in China. Beijing, China: Agricultural Publishing House.

25. Waterhouse, D.F. and Norris, K.R., 1987. Biological control: Pacific prospects. p. 454.

26. Westbrooks, R.G. and Eplee, R.E. 1988. Federal noxious weeds in Florida. Proceedings of the 42nd Annual Meeting of the *Southern Weed Science Society*, 316-321.

27. Whistler, W.A. 1983. Weed handbook of Western Polynesia. Schriftenreihe der Deutschen Gesellschaft fnr Technische Zusammenarbeit, pp. 157.

Genus *Capsicum*

Introduction

Capsicum, native to New World Tropics, now widely cultivated for use as spices or vegetables in the temperate zone. Chillies powder, red and cyenne peppers, Tabasco, paprika, sweet or bell peppers and pimentos are derived from the pod like berries of various species of *Capsicum*. The pungency of pepper is due to capsaicin or capsicutin ($C_{18}H_27NO_3$) a condensed product of 3-hydroxy-4 methoxy benzylamine and the vanilly amide of isodecylenic acid, contained in the placenta. Heiser and Pickersgill (1969) and Heiser Jr. (1976) have described different species *Capsicum*. *Capsicum* species (Chillies or pungent pepper and Bell pepper or Simla mirch) comes under genus *Capsicum*. There are about 100 species under the genus *Capsicum* but some authority reports that *Capsicum* has total-61 species and 4 species are commonly cultivated (*C. fruitescens, C. chinense, C. baccatum and C. pubescens*). Even Chillies (*C. annuum* L.) usually grown in warm to hot and humid climate throughout the country.

Table 2.87: *Capsicum* Species and Preferred Taxon Name

Name of Taxon	Name of Authority	Family	Preferred Taxon Name
Capsicum			*Capsicum*
Capsicum			*Capsicum*
Capsicum annuum	L.	Solanaceae	*Capsicum annuum*
Capsicum annuum var. *acuminatum*	Fingerh.	Solanaceae	*Capsicum annuum* var. *annuum*
Capsicum annuum var. *annuum*		Solanaceae	*Capsicum annuum* var. *annuum*
Capsicum annuum var. *aviculare*	Auct.	Solanaceae	*Capsicum annuum* var. *glabriusculum*
Capsicum annuum var. *cerasiforme*	(Mill.) Irish	Solanaceae	*Capsicum annuum* var. *annuum*
Capsicum annuum var. *conoides*	(Mill.) Irish	Solanaceae	*Capsicum annuum* var. *annuum*
Capsicum annuum var. *fasciculatum*	(Sturtev.) Irish	Solanaceae	*Capsicum annuum* var. *annuum*
Capsicum annuum var. *glabriusculum*	(Dunal) Heiser and Pickersgill	Solanaceae	*Capsicum annuum* var. *glabriusculum*
Capsicum annuum var. *grossum*	(L.) Sendtn.	Solanaceae	*Capsicum annuum* var. *annuum*
Capsicum annuum var. *longum*	Sendtn.	Solanaceae	*Capsicum annuum* var. *annuum*
Capsicum annuum var. *minimum*	(Mill.) Heiser	Solanaceae	*Capsicum annuum* var. *glabriusculum*
Capsicum anomalum	Franch. and Sav.	Solanaceae	*Tubocapsicum anomalum*
Capsicum baccatum	L.	Solanaceae	*Capsicum baccatum*
Capsicum baccatum var. *baccatum*		Solanaceae	*Capsicum baccatum* var. *baccatum*
Capsicum baccatum var. *pendulum*	(Willd.) Eshbaugh	Solanaceae	*Capsicum baccatum* var. *pendulum*
Capsicum baccatum var. *praetermissum*	(Heiser and P. G. Sm.) Hunz.	Solanaceae	*Capsicum baccatum* var. *praetermissum*
Capsicum buforum	Hunz.	Solanaceae	*Capsicum buforum*
Capsicum campylopodium	Sendtn.	Solanaceae	*Capsicum campylopodium*

Name of Taxon	Name of Authority	Family	Preferred Taxon Name
Capsicum cardenasii	Heiser and P. G. Sm.	Solanaceae	*Capsicum cardenasii*
Capsicum cerasiforme	Mill.	Solanaceae	*Capsicum annuum* var. *annuum*
Capsicum chacoense	Hunz.	Solanaceae	*Capsicum chacoense*
Capsicum chinense	Jacq.	Solanaceae	*Capsicum chinense*
Capsicum ciliatum	(Kunth) Kuntze	Solanaceae	*Capsicum ciliatum*
Capsicum coccineum	(Rusby) Hunz.	Solanaceae	*Capsicum coccineum*
Capsicum conoides	Mill.	Solanaceae	*Capsicum annuum* var. *annuum*
Capsicum cordiforme	Mill.	Solanaceae	*Capsicum annuum*
Capsicum cornutum	(Hiern) Hunz.	Solanaceae	*Capsicum cornutum*
Capsicum dimorphum	(Miers) Kuntze	Solanaceae	*Capsicum dimorphum*
Capsicum dusenii	Bitter	Solanaceae	*Capsicum dusenii*
Capsicum eggersii	Bitter	Solanaceae	*Capsicum hookerianum*
Capsicum eximium	Hunz.	Solanaceae	*Capsicum eximium*
Capsicum eximium var. *eximium*		Solanaceae	*Capsicum eximium* var. *eximium*
Capsicum eximium var. *tomentosum*	Eshbaugh and P. G. Sm.	Solanaceae	*Capsicum eximium* var. *tomentosum*
Capsicum flexuosum	Sendtn.	Solanaceae	*Capsicum flexuosum*
Capsicum frutescens	L.	Solanaceae	*Capsicum frutescens*
Capsicum frutescens var. *baccatum*	(L.) Irish	Solanaceae	*Capsicum baccatum* var. *baccatum*
Capsicum frutescens var. *cerasiforme*	(Mill.) L. H. Bailey	Solanaceae	*Capsicum annuum* var. *annuum*
Capsicum frutescens var. *conoides*	(Mill.) L. H. Bailey	Solanaceae	*Capsicum annuum* var. *annuum*
Capsicum frutescens var. *fasciculatum*	(Sturtev.) L. H. Bailey	Solanaceae	*Capsicum annuum* var. *annuum*
Capsicum frutescens var. *grossum*	(L.) L. H. Bailey	Solanaceae	*Capsicum annuum* var. *annuum*
Capsicum frutescens var. *longum*	(Sendtn.) L. H. Bailey	Solanaceae	*Capsicum annuum* var. *annuum*
Capsicum galapagoense	Hunz.	Solanaceae	*Capsicum galapagoense*
Capsicum geminifolium	(Dammer) Hunz.	Solanaceae	*Capsicum geminifolium*
Capsicum grandiflorum	Kuntze	Solanaceae	*Vassobia fasciculata*
Capsicum grossum	L.	Solanaceae	*Capsicum annuum* var. *annuum*
Capsicum hispidum var. *glabriusculum*	Dunal	Solanaceae	*Capsicum annuum* var. *glabriusculum*
Capsicum hookerianum	(Miers) Kuntze	Solanaceae	*Capsicum hookerianum*
Capsicum lanceolatum	(Greenm.) C. V. Morton and Standl.	Solanaceae	*Capsicum lanceolatum*
Capsicum leptopodum	(Dunal) Kuntze	Solanaceae	*Capsicum leptopodum*
Capsicum lycianthoides	Bitter	Solanaceae	*Capsicum lycianthoides*
Capsicum microcarpum	Cav.	Solanaceae	*Capsicum baccatum* var. *baccatum*
Capsicum microcarpum var. *tomentosum*	Chodat and Hassl.	Solanaceae	*Capsicum baccatum* var. *baccatum*
Capsicum minimum	Blanco, nom. illeg.	Solanaceae	*Capsicum frutescens*
Capsicum minutiflorum	(Rusby) Hunz.	Solanaceae	*Capsicum minutiflorum*
Capsicum mirabile	Mart. ex Sendtn.	Solanaceae	*Capsicum mirabile*
Capsicum mositicum	Toledo	Solanaceae	*Capsicum mositicum*
Capsicum parvifolium	Sendtn.	Solanaceae	*Capsicum parvifolium*
Capsicum pendulum	Willd.	Solanaceae	*Capsicum baccatum* var. *pendulum*
Capsicum petenense	Standl.	Solanaceae	*Capsicum annuum* var. *annuum*
Capsicum praetermissum	Heiser and P. G. Sm.	Solanaceae	*Capsicum baccatum* var. *praetermissum*
Capsicum pubescens	Ruiz and Pav.	Solanaceae	*Capsicum pubescens*
Capsicum rhomboideum	(Dunal) Kuntze	Solanaceae	*Capsicum rhomboideum*
Capsicum schottianum	Sendtn.	Solanaceae	*Capsicum schottianum*
Capsicum schottianum	auct.	Solanaceae	*Capsicum flexuosum*
Capsicum scolnikianum	Hunz.	Solanaceae	*Capsicum scolnikianum*
Capsicum sp.		Solanaceae	*Capsicum* sp.
Capsicum stramoniifolium	(Kunth) Kuntze	Solanaceae	*Witheringia stramoniifolia*

Name of Taxon	Name of Authority	Family	Preferred Taxon Name
Capsicum tovarii	Eshbaugh *et al.*	Solanaceae	*Capsicum tovarii*
Capsicum villosum	Sendtn.	Solanaceae	*Capsicum villosum*
Capsicum villosum var. *muticum*	Sendtn.	Solanaceae	*Capsicum villosum var. muticum*
Capsicum villosum var. *villosum*		Solanaceae	*Capsicum villosum var. villosum*
Capsicum violaceum	Kunth	Solanaceae	*Capsicum pubescens*
Tubocapsicum anomalum	(Franch. and Sav.) Makino	Solanaceae	Tubocapsicum anomalum

The large fruited bell pepper of today originated from tiny pungent and pointed fruited wild species *C. annuum*. It is considered to be the cool season crop and grown in sub-tropical and temperate region. There are four other species (*viz. C. frutenscens, C. chinense, C. baccatum* and *C. pubescens*), which are commonly grown in southern and Central America except *C. frutenscens*, which is also grown in some parts in USA. Due to continuous selection, long history of cultivation and popularity of the crop, sufficient genetic variability has been generated and now India is treated as secondary center of diversity for chillies. In hot chillies (*C. annuum* L.), rich variability in plant and fruit morphological traits occurs throughout the country particularly in south peninsular region, northeastern region, in foothills of Himalaya and Gangetic plains. The variability includes in plant type, fruit size/shape (long, short, pointed, smooth, wrinkled), bearing habit (fruits facing upwards, horizontal, downwards) and pungency in addition to annual and perennial types. However, in bell pepper less variability has been observed. Mostly introduced lines are grown particularly in mid hill of Himalaya and parts of hilly region of Central India including Karnataka and Maharashtra.

Table 2.88: Availability of *Capsicum* Genetic Resources

Species	Source
Capsicum annuum	Mayers Michale Syngenta Naples FL, USA, AVRDC, Taiwan
Capsicum fruitescens	NARI, East Coast, Guyana South America
Capsicum grossum	NBPGR, New Delhi

Even though 22 wild *Capsicum* species have been reported, only 5 species namely *C. annuum, C. baccatum, C. chinense, C. frutescens* and *C. pubscens* are cultivated. The commercial cultivars belong to *C. annuum*. Among the 31 *Capsicum* species, five are domesticated species which includes:

Table 2.89a: Names of the Cultivated *Capsicum* Species

Botanical Name	Common Name
Capsicum annuum L. var. *annuum*	Chilli pepper
Capsicum annuum L. var. *annuum*	Bell pepper
Capsicum annuum L. var. *annuum*	Paprika
Capsicum baccatum L. var. *pendulum*	Peruvian pepper
Capsicum chinense Jacq.	Bonnet pepper
Capsicum chinense Jacq.	Hobanero pepper
Capsicum frutescens L.	Tabasco pepper
Capsicum frutescens L.	Bird pepper
Capsicum pubescens Ruiz and Pavon	Rocoto
Capsicum pubescens Ruiz and Pavon	Apple chilli

Source: Heiser and Pickessgill (1969).

Other species with economic importance are:

Table 2.89b: Botanical Name of some Common Species

Botanical Name	Common Name
Capsicum annuum L. var. *globriusculum* (Dunal)	American Bird Pepper
Capsicum baccatum L. var.*baccatum*	Locoto
Capsicum cardenesii Heiser and P.G. Sm.	Utupica
Capsicum eximium (Hunz.)	Ulapuca

Capsicum chacoense (Hunz.) and *C. tovarii* (Estibaugh *et al.*, 2000) are considered semi-domesticated. Ribeiro *et al.* (2008) grouped 20 remaining species under wild taxa.

In *Capsicum* there are three complexes which include:

Table 2.90: Grouping of some Species

Sl.No.	Capsicum Species	Capsicum Grouping
1.	*C. annuum*	*C. chacoense, C. chinense, C. frutesense* and *C. galapagoense*
2.	*C. baccatum*	*C. practermissuman* and *C. tovarii*
3.	*C. pubescens*	*C. eximium and C. cardanasii*

Source: Djian-Caporalino *et al.*, 2007.

This is highly variable species in terms of morphological characters. *C. frutescens* and *C. minimum* are grown in few pockets in India (Swarup, 2006). The keys for which have been presented by Heiser and Smith (1953) and Heiser Jr. (1976) as given below:

***C. annuum* (Sweet pepper and Chilli):** Blue anthers, milky white corolla, inconspicuous calyx lobing and solitary peduncle. *Capsicum annuum* var. *annuum* is cultivatd, while species belongs to var. *minimum*.

***C. baccatum*:** Distinghuish from *Capsicum annum* by the presence of yellow, tan, or brown spots on corolla and prominent calyx teeth. Cultivation is restricted to South America. The domesticated forms are assigned

Table 2.91: Key for Identification of *Capsicum* Species

Species	Seed Color	Peduncle at Nodes	Calyx	Corolla	Anther
C. annum	Straw	One at a node	Inconspicuous lobes	Milky white	Blue
C. baccatum	Straw	One or two at a node	Prominent calyx teeth	Yellow, tan or brown spots	Yellow
C. frutescens	Straw	Two or more at some nodes	No calyx teeth	Milky greenish or yellowish white	Blue
C. chinense	Straw	Two or more at some nodes	A constriction on calyx	Purple	Blue
C. pubescens	Dark, rugose	One at a node	Calyx teeth	Purple	Purple

to *C. baccatum* var. *pendulum* and the wild types as var. *baccatum*.

***C. frutescens* (Tobacco pepper):** Blue anthers, milky greenish or yellowish white corolla and have two or more peduncles in a node.

***C. chinense*:** Cultivated in Amazon and Africa. Some cultivars are very pungent among peppers. It is supposed to be a progenitor of *C. frutescens* and a constriction below calyx is the only differentiating character between *C. chinense* and *C. frutenscens*.

C. pubescens: The Andean species is presumed to be a progenitor of one of the three American wild species, *viz. C. eximium*, *C. cardenasii* and *C. tovari* since *C. pubescens* shows affinities with them. It is distinct from other cultivated peppers with thick flesh, dark and rugose seeds.

Seven botanical varieties are also considered and it is quite doubtful whether it is justified as they all intercross and intermediates do occur (Redgrov, 1933 and Chittenden, 1956).

- ☆ **var. *abbreviatum* Fingerh:**Wrinkled peppers; fruits generally ovate, wrinkled and 5.0 cm long or less.
- ☆ **var. *acuminatum* Fingerh:** Chillies; fruits linear-oblong, over 9.0 cm long, usually pointed, pungent and widely grown in India.
- ☆ **var. *cerasiforme* (Miller) Irish:** Cherru peppers; fruits globose with firm flesh 1.2-2.5 cm in diameter, red, yellow or purple and pungent.
- ☆ **var. *conoides* (Miller) Irish:** Cone peppers; tabasco type, fruit erect, conocal, about 3.0 cm long and very pungent.
- ☆ **var. *fasciculatum* (Sturt) Irish:** Cluster pepper; fruit clustered, erect slender, about 7.5 cm long, very pungent. The fruits are not borne singly; it is probable that these are forms of *C. frutescens*.
- ☆ **var. *grossum* (L.) Sendt.:** Sweet peppers, paprika; fruits large with basal depression, inflated red or yellow, flesh thick and mild.
- ☆ **var. *longum* (DC) Sendt.:** Long peppers; fruits mostly drooping, tapering at apex, 20-30 cm long, red or yellow or ivory, often mild and calyx not embracing the fruit.

Presently *C. pendulum* is included or designated as *C. baccatum* var. *pendulum*. A number of other species are also recognized, especially with different characters with reference to corolla colour, shape, anther colour, presence or absence of calyx.

REFERENCES

1. Djian-Caporalino, C., Lefebvre, V., Sage-Daube'ze, A.M. and Palloix, A.2007. *Capsicum*. In: Singh RJ (ed). Genetic resources, chromosome engineering and crop improvement series, Vol.3, Vegetable crops. CRC Press, Florida (USA), pp 185–243.
2. Heiser, C.B. and Pickessgill, B. 1969. Names of the cultivated *Capsicum* species (Solanaceae). *Taxon*. **18:** 852-859.
3. Ribeiro, C.S.C., Lopes, C.A., Carvalho, S.I.C., Henz, G.P., Reifschneider, F.J.B. (eds.) Pimentas *Capsicum*. 2008. 200P, Embrapa Hortalices.
4. Swarup, V. 2006. Vegetable Science and Technology in India. Kalyani Publishers, New Delhi, p. 656

Chilli (*Capsicum annuum* L.)

Introduction

Chilli (*Capsicum annuum* L., 2n=2x=24) also known as hot pepper is one of the most important cash crop in India where it occupies about 0.816-0.982 million hectares with 1.36 mt production. From India an annual production of dry chilli is around 1.1 million tones. Chilli is used fresh in salads or processed as powder, paste, pickle, sauce, soup *etc.* No culinary preparation can be completed without addition of chilli. Being an indispensable item in the kitchen, it has become the most important spice crop of the world. Chilli is an important source of vitamins A, C and E. Green chilli fruits contain slightly lesser vitamin C than the red ones. It is a strong stimulant yet has no narcotic effects. Composition of per 100 g green chilli fruit is 85.7 per cent moisture, 2.9 g proteins, 0.6 g fibre, 3.0 g carbohydrates, 30 mg calcium, 80 mg phosphorus, 217 mg potassium, 1.0 mg iron, 0.19 mg thiamine, 0.39 mg riboflavin, 67 mg oxalic acid, 0.9 mg nicotinic acid, 292 IU vitamin A and 111 mg vitamin C. Recently, a Russian scientist have identified Vit. 'P' in green chillies which is considered to be important as it protects from secondary irraditations injury. Dried chillies contain about 10 per cent moisture. *C. annuum* contains 50-230 mg/100 g ascorbic acid and *C. frutescens* 2-50 mg/100 g of ascorbic acid. The pungent principle of chilli is due to an alkaloid capsaicin, $C_{18}H_{27}NO_3$, the decylenic acid derivative of vanillylamine, which is present in the placentas; commercial samples contain about 0.1 percent. Capsanthin, $C_{40}H_{48}O_3$ is the most important pigment of paprika. Chilli (*Capsicum annuum* L.) is valued as major vegetable crop around the world for its colour. Oleoresin, comprising both red colouring pigment and pungency compound, is a strong antioxidant and are used by the pharmaceutical industry. The extractable capsaicin is more in yellow coloured types than the red ones. Consumption of chilli also clears the lungs, improves blood circulation and relieves the rheumatoid arthritis patients from pain. Red colour of chilli is due to capsanthin. It is also used to make pickles.

Origin, History and Distribution

Chilli is indigenous to South America/Tropical American especially Brazil where it is still found growing in a wild state and is known from prehistoric remains in Peru (Thompson and Kelly, 1957). It is commonly believed that *Capsicum annuum* is having its origin place Mexico and Guatemala and *Capsicum frutescens* in South America (Brazil). It was unknown in Europe till the discovery of America. Columbus carried seed to Spain in 1493 (Boswell, 1951). Spaniards brought it to Europe around 1493. From Europe, chilli was introduced into United States. All the species except *C. anomalum* of Japan are native to South America. It was first cultivated around 7,000-6,000 BC. By the turn of 15th century, when Spanish and Portuguese discovered South America, it was widely cultivated. The Portuguese brought chilli (*Capsicum*) from Brazil to India during 1584 but some authors believe that it was introduced in India during 17th century (Chauhan, 1981; Singh, 1989). This Spinards and the Portuguese in their search for pepper were quick to seize this new and much more pungent pepper from the New World and it was widely distributed for cultivation. Due to availability of distinct type/variability in the North-Eastern states, it is assumed that the Christian missionaries directly introduced capsicum into NE India from South America. Now India ranks first among the chilli growing countries. Other major chilli producing countries include China, Korea, Hungary, Spain, Pakistan, Morocco, Mexico, Nigeria, Thailand, Kenya, Sudan, Uganda, Japan, Ethiopia, Indonesia and Turkey. The major chilli growing states in India are Andhra Pradesh, Karnataka, Maharashtra, Odisha, Tamil Nadu, Madhya Pradesh, West Bengal and Rajasthan. In NEH region, Manipur, Mizoram, Meghalaya, Nagaland, Tripura and Arunachal Pradesh are also known for chilli production which possesses worm to hot and humid climate. These states account for more than 80 per cent of the total area and production. Andhra Pradesh alone accounts for 25 per cent of total area and 40-50 per cent of total production.

Species Availability, Evolution and Diversity

Early taxonomic treatment of the genus resulted in more than 100 species and botanical varieties (Irish, 1989). *Capsicum annuum* includes:

- *Capsicum abyssinicum* A.Rich.
- *Capsicum angulosum* Mill.
- *Capsicum axi* Vell.
- *Capsicum bauhinii* Dunal
- *Capsicum caerulescens* Besser
- *Capsicum cerasiforme* Mill.
- *Capsicum ceratocarpum* Fingerh.
- *Capsicum cereolum* Bertol.
- *Capsicum comarim* Vell.
- *Capsicum conicum* Lam.
- *Capsicum conoide* Mill.
- *Capsicum conoides* Roem. and Schult.
- *Capsicum conoideum* Mill.
- *Capsicum cordiforme* Mill.
- *Capsicum crispum* Dunal
- *Capsicum cydoniforme* Roem. and Schult.
- *Capsicum dulce* Dunal.
- *Capsicum fasciculatum* Sturtev.
- *Capsicum fastigiatum* Blume
- *Capsicum frutescens* L.
- *Capsicum globiferum* G.Mey.
- *Capsicum globosum* Besser
- *Capsicum grossum* L.
- *Capsicum indicum* auct.
- *Capsicum longum* DC.
- *Capsicum milleri* Roem. and Schult.
- *Capsicum minimum* Mill.
- *Capsicum odoratum* Steud.
- *Capsicum odoriferum* Vell.
- *Capsicum oliviforme* Mill.
- *Capsicum ovatum* DC.
- *Capsicum petenense* Standl.
- *Capsicum pomiferum* Mart. ex Steud.
- *Capsicum purpureum* Vahl ex Hornem.
- *Capsicum pyramidale* Mill.
- *Capsicum quitense* Willd. ex Roem. and Schult.
- *Capsicum silvestre* Vell.
- *Capsicum sphaerium* Willd.
- *Capsicum tetragonum* Mill.
- *Capsicum tomatiforme* Fingerh. ex Steud.
- *Capsicum torulosum* Hornem.
- *Capsicum tournefortii* Besser
- *Capsicum ustulatum* Paxt

Linnaeus recorded 2 species of *Capsicum* in the first edition of his book 'Species Plantarum' in 1753 and 3 additional species were added in 1797. The 5 species are *C. anomalum*, *C. pubescens* Ruiz. and Pavan, *C. baccatum* (var. *pendulum*) Willd, *C. frutescens* L. and *C. annuum* L. which includes both chilli (syn. Hot pepper) and Shimla Mirch (syn. Sweet pepper). Already by 1542 three races were recognized in India. Present representation is a diverse plant group having 25 wild.

The centre of diversity of the common cultivated pepper (*C. annuum*) is probably Mexico with secondary centre at Guatemala (De Candolle, 1986). Chilli (*C. annuum*) can be spotted throughout the world particularly in India, Mexico, Japan, Ethiopia, Uganda, Nigeria, Thailand, Turkey, Indonesia, China and Pakistan. This is highly variable species in terms of morphological characters and plants are herb or sub-herbs, erect, 0.5-1.5 m tall, highly branched and bear fruits which are highly variable in size, shape, colour and degree of pungency.

The *annuum* species was domesticated in highland of Mexico and includes most of the Mexican chile (syn. chilli) peppers, most of the hot peppers of Asia and Africa and the various cultivars of sweet peppers of temperatec countries. However, due to non-adaptability of *C. annuum* in lowland tropics of Latin America,

Table 2.92: Important Species of Chilli and their Characteristics

Species	*Common Name*	*Rematks*
C. annum	Sweet pepper, chillies	Principal source of dry chillies, low pungency, fruit colour
C. annum var. *avicular*	Bird Pepper	Wild type, said to be progenitor of bell pepper
C. annum var. *grossum* Sendt.	Sweet pepper	Fruit contain less capsaicin
C. annum var. *longum* Sendt.	Pepper	Used to produce condiment
C. chinense Jacq.	Pepper or King Chilli	Closely related to *C. frutescens*. Fruit rot resistance, high oleoresin and more colour.
C. frutescens L.	Tobacco pepper, Bird chilli	Cultivated in dry region of India. Highly pungent. Fruits used for sauce preparations. High capsaicin, more pungency, resistance to damping off, Cercospora leaf spot, Sclerotinia rot, leaf curl, mosaic, Pseudomonas wilt resistance.
C. eximium Huriziken	Pepper	Wild type, pseudo self incompatible related to *C. pubescens*
C. minimum Roxb. syn. *C. fastigiatum* Bhumme	Bird-eye- chilli	Cultivated all across NE Region but at very limited scale, closely resembles to *C. annuum*
C. pubescens Ruiz. And Paron	Pepper	Introduced in Indiafor breeding purpose.
C. fasiculatum	-	*Verticillium* wilt resistance
C. angulosum	-	*Verticillium* wilt resistance
C. chacoense	-	Bacterial wilt resistance

Table 2.93: Important Species of Chillies and their Characteristics

Common Name	*Botanical Name*	*Comments*
Sweet pepper, Chillies, Hot pepper	*C. annuum* L.	Extensively cultivated in South Asia, constitutes principal source of commercial dry chillies.
Bird pepper	*C. annuum* L. var. *avicular*	Wild type, said to be progenitor of bell pepper
Sweet pepper	*C. annuum* L. var. *grossum* Sendt.	Fruit contains less capsaicin, green colour used in salad.
Pepper	*C. annuum* L. var. *longum* Sendt.	Used to produce condiment
Pepper	*C. baccatum* H.B.K. syn. *C. angulosum*	Commonly grown in Central America
-	*C. baccatum* L. var. *baccatum* Syn. *Microcarpum* Cav.	Found in Peru, Bolivia, Paraguay, North Argentina and South Brazil
	C. cardenasii Heisir and Smith	Wild type, self incompatible, related to *C. pubescens*
Pepper	*C. chinense* Jacq.	Closely related to *C. frutescens*
Pepper	*C. eximium* Hunziker	Wild type, pseudo self incompatible, related to *C. pubescens*
Tobasco pepper, Bird chilli	*C. frutescens*	Widely cultivated in dry regions of India, highly pungent, fruits used for sauce preparation
Bird-eye-chilli	*C. minimum* Roxb. Syn. *C. fastigatum* Bhumme	Cultivated throughout india but at very limited scale, closely resembles *C. annuum*
Pepper	*C. pubescens* Ruiz. and Paron	Introduced in India for breeding purpose.

C. frutescens, and *C. chinense* replaced its cultivation. *C. frutescens* is also cultivated in Africa and India for either intact fruits or for oleoresin extraction. During the exploration trip, it has been observed that it is grown in the state of Bihar. *C. frutescens* is also being grown, which bears comparatively small and conical fruits. The cultivation of *C. baccatum* and *C. pubescens* are mostly restricted to Latin America.

Verma *et al.* (2016) characterized 106 germplasm of chilli including cultivars (hot and sweet pepper) and popular land races like King-chilli, Cherry chilli, Dalle chilli and Bird Eye chilli for yield contributing traits. The results of cluster analysis revealed that the King chilli is closer to Bird eye chilli than the Dalle chilli. The popular land races were found more diverse over commercial cultivars of both chilli and capsicum. High level of genetic diversity assessed in this study at both morphological and molecular level emphasized the importance of this region for conserving *Capsicum* spp. germplasm. The pattern of genetic diversity identified in this region may be used to develop *in situ* conservation strategies for *Capsicum* spp.

Gene and Genetics of Characters

Determinate character is recessively inherited. In an experiment Pious and Peter (1986) found that cluster bearing habit was governed by 2 genes with a specific dominant and recessive epistasis.

Colour of plants, fruit shape, position, fruit apex, fruit length and thickness, pedicel length *etc.* were controlled monogenically but the ratio was not very clear.

Table 2.94: Gene and Genetics of Characters

Character	*Number of Gene*	*Type of Gene Action*
Fruit orientation	Monogenic (up+up+)	Pendulous condition dominant to erect
Fruit colour	Monogenic	Red dominant to orange
Pedicel length	Monogenic	Short dominant to long
Fruit pungency	Monogenic (CC)	Pungency dominant to non-pungency
Fruit shape	Monogenic (OO)	Oblate or round fruit shape is dominant to that of angled shape
Fruit apex	Monogenic (PP)	Pointed fruit apex incompletely dominant to blunt
Flesh texture	Monogenic (SS)	Soft flesh, fruit easily separated from calyx is dominant to fruit attached tightly with calyx
Plant habit	Monogenic	Tall dominant to dwarf
Plant spread	Monogenic	Spreading dominant to erect
Fruit lobbing	Monogenic	Multilobed dominant to bi-lobed fruit
Seedlessness	Monogenic	Seeded dominant to seedlessness
Complete sterelity	Monogenic(S-cytoplasmic) (Smsms-genetic x cytoplasmic)	Complete fertility dominant to sterelity
Female sterility	Monogenic	Female fertility dominant to sterelity

Similarly for diseases mode of genetic control has been depicted as:

Table 2.95: Diseases and Mode of Genetic Control

Diseases	*Mode of Genetic Control*
Mosaic (TEV)	Monogenic and dominant
Mosaic (CMV)	Monogenic and dominant
Mosaic (TMV)	Monogenic and dominant
Soft rot (*Erwinia carotovora*)	Oligogenic
Damping off (*Phytophthora capsici*)	Digenic
Bacterial spot (*Xanthomonas vesicatoria*)	Monogenic and dominant
Root-knot nematode (Meloidogyne incognita)	Monogenic and dominant

Botanical Description

Chilli belongs to family Solanaceae, genus *Capsicum*, *C. annuum* L., 2n=24 but polyploids with 2n= 36, 48 and aneuploids with 2n=25 have been reported. Keys to the identification of the species are:

Table 2.96: Botanical Description

Structure	*Keys for identification*
A.	Seeds black, stem and leaves rather densely pubescent: *C. pubescens*
A.	Seeds light yellow, stem and leaves rather smooth-B
B.	Corolla white with yellow or tan markings on throat: *C. pendulum*
B.	Corolla greenish-white with green markings on throat: *C. frutescens*
B.	Corolla clean white or dingy white with no coloured markings on throat: *C. annuum*

The distinguishing characteristics between *C. annuum* and *C. frutescens* are given below:

Table 2.97: Difference between *C. annuum* and *C. frutescens*

C. annuum	*C. frutescens*
Annual and herbaceous	Biennial, shrubby and tall growing
Leaves small and narrow	Leaves large and broad
Flowers small, corolla white or purple	Flowers larger, crolla greenish-white with green throat
Flower solitary pendulous	Flower in cluster, erect
Pedicel: fruit Ratio less than one	Pedicel: fruit Ratio more then one
Less pungent (0.1 per cent of capsaicin)	Highly pungent (0.22 per cent capsaicin)
Diploids and polyploids are observed	Only diploids are observed

Baily (1923) suggested keys to the classification of *C. annuum* and *C. frutescens* are given below:

Table 2.98: Classification of *C. annuum* and *C. frutescens*

Structure	*Keys for Identification*
A.	Plant annual or biennial: *C. annuum*
B	Fruits oblong, linear.C
C.	Calyx usually embracing base of fruit-D, DD, DDD.
D.	Fruits long, slender more than 10.0 cm: *C. annuum* var. *accuminatum* (syn. *C. chilense*, Nepal pepper)
DD.	Fruits long, in clusters: *C. annuum* var.*fasciculatum* (cluster chilli)
DDD.	Fruits short, erect orientation: *C. annuum* var.*connoids* (syn. *C. conoides*, coralgem, this species is called *tabasco*)
CC.	Calyx not usually embracing base of fruit-E, EE, EEE, EEEE.
E.	Base of fruit bulged, long fruits: *C. annuum* var. *longum* (syn. *C. longum*, long chilli, cayenne pepper, red pepper)
EE.	Large fruit size, less pungent, less seeded: *C. annuum* var.*grossum* (syn. *C. grossum* bell pepper, big chilli, sweet pepper)
EEE.	Coloured fruits, small fruits, cherry shaped fruits: *C. annuum* var. *cerassiforme* (cherry pepper, ornamental pepper)
EEEE.	Coloured fruits, bushy plant type, very small fruits: *C. annuum* var. *abbreviatum* (syn. *C. umbilicatum*, ornamental pepper)
AA.	Plant perennial-F
F.	Fruits very pungent *C.frutescens*var. *baccatum* (syn. *C.frutescens* var. *minima*, *C. baccatum*, bird pepper, golkonda mirapa, seema mirapa)

Modern taxonomist recognising the extent of variability have consolidated the cultivated *Capsicum* into following 5 species (Pickersgill, 1980)

1. *C. annuum* L.; syn. *C. purpureum*, *C. grossum*, *C. cerasiformoe*
2. *C. frutescens* L.; syn. *C.minimum*
3. *C.chinense* Jacq.; syn. *C. luteum*, *C. umbilicatum*, *C. sinense*
4. *C. baccatum* L.; syn. *C.pendulum*, *C. microcarpum*, *C. angulosum*
5. *C. pubescens* R. and P.

Table 2.99: Botanical Description

Growth Habit	*C. annuum* is a variable annual herb or subshrub (short lived perennial) sometimes woody at base. Plant grows straight as single stem. It is much branched, 0.5-1.5 m high
Root	Strong tap root usually broken or arrested in growth on transplanting and numerous profusely branched lateral roots develop, and extending to one meter.
Stem	After reaching a specific height determined by genetic constitution in conjunction with its specific environment, the stem normally bifurcate and then keeps bifurcating. Main shoot is radial but lateral branches are cincinnate, one of the branches at each node remaining underdeveloped and subtending bract or bacts are adnate and are carried up a lateral shoot to node above.
Leaves	Leaves are variable in size, simple and tip pointed variable in size up to 12 cm long and 7.5 cm wide. Petiole is 0.5-2.5 cm long, lamina broadly lanceolate to ovate, entire, thin, subglabrous, 1.5-122.0 x 0.5-7.5 cm, tip acuminate, and base cuneate or acute.
Floral biology	Flowers are bisexual (protogynous) usually borne singly and are terminal, but due to the form of branching, shortly 5 dentate, 10 ribbed about 2.0 mm long enlarging and enclosing base of fruits. Corolla is rotate-campanulate, deeply 5-6 partite, and 8.0-15.0 mm in diameter with white or greenish colour. Stamens are 5-6, inserted near base of corolla. Anthers are bluid, dehiscing longitudinally. Ovary is 2 celled but often multiplying under domestication. Style is simple, white or purple and stigma is capitate.
Anthesis and pollination	Chilli is a self pollinated but cross pollination also takes place to some extent. Anthesis occurs between 6.00 and 9.00 hours but in some cases it is 5 to 6 AM and completed by 8.00 AM. Flower remains open for 2 to 3 days. Anthers not connivent in a cone and not longer than the filaments. Anther dehisces through lateral slits after 30 minutes of anthesis. In general, only half to 5-6 after stigma becomes receptive. Anther dehiscence takes place from 9 to 11 AM. The stigma became receptive from the day of anthesis and remained receptive upto 2 days after antjhesis (Vijay *et al.*, 1979). Receptivity of stigma was the highest at the day of flower anthesis *i.e.* 24 hours after anthesis (Saxena *et al.*, 2003). Due to protogynous nature, chilli is a cross pollinated crop. Maximum fruit setting takes place when pollination is done at the time of opening of flower. Bees (*Exomopsis pulchella*), ants and thrips are the pollinating agents (Peter, 1998).
Fruit and fruit set	Solitary fruits are borne at nodes in the leaf axils. Fruit is indehiscent, many seeded berry, hollow and pendulous or erect (Pueseglove, 1977). Fruits are variable in size, shape, colour and degree of pungency, linear, conical or globose, 1.0-30.0 cm long, unripe fruit green or purplish, ripening to red or orange, yellow, brown, cream or purplish.
Seeds	Seeds are 3.0-5.0 mm long and pale yellow.

Horticultural Classification

Classification Based on Cultivation of Species

Table 2.100: Different Cultivars/Lines under Species

Name of Cultivated Species	*Common name*
C. annuum	Nepal pepper, Cayenne pepper, Hot pepper, Bell pepper, Sweet pepper
C. frutescens (syn. *C. baccatum*)	Tabasco, Bird pepper

Chemotaxonomy of *Capsicum*

After conducting flavanoid analysis on 3 *Capsicum* species, Lopes *et al.* (1978) suggested that there is greater affinity between *C. annuum* and *C. frutescens* than between *C. pendulum* and *C. frutescens*. Based on starch gel electrophoresis Mcleod (1978) classified 14 taxa into 5 biological species, with *C. cardenasii*, *C. eximium* and *C. pubescens* representing a single species, *C. baccatum* var. *baccatum* and *C. baccatum* var. *pendulum* and *C. practermissum* representing another and the wild and cultivated *C. annuum*, *C. frutescens*, and *C. chinense* yet another. *C. chacoense* and an unnamed species are considered valid wild species. Slightly deviating from the above classification, based on the electrophoresis of peroxidase isozyme extracted from functional leaf, Wang and MA (1987) assigned 8 *Capsicum* species to 4 groups:

Table 2.101: Grouping of Species

Groups	*Name of Species*
Group I	*C. annuum*, *C. frutescens*, and *C. chinense*
Group II	*C. chacoense* and *C. pubescens*
GroupIII	C. practermissum
GroupIV	*C. baccatum* and *C.eximium*

Smith *et al.* (1987) classified chilli cultivars on the basis if fruit shape, colour and usage:

Table 2.102: Grouping of Cultivars

Group	*Characters*	*Description*	*Cultivar's Segement with Specific Traits*
Group I	Fruit large smooth, thick-fleshed	**A-Bell group:** Fruit large, 7.5-12.5 cm long, blocky, blunt, 3-4 lobed, square to rectangular, or tapering in longitudinal section. Colour usually green when immature, red at maturity. Mostly non-pungent, although a few pungent forms are known.	Non-pungent: **(a) Green turning red (or rarely orange-yellow) when ripe-** ('California Wonder', 'Yolo Wonder', 'Key Stone Giant', 'Early Calowonder', ' Tambel 2', 'Golden California Wonder') **(b) Yellow, turning red when ripe-**('Golden Bell', 'Roumanian')
		B-Pimento group- Fruit heart shaped but pointed, 3.75-12.5 cm long, smooth, thick walled	Non-pungent- ('Pimento', 'Pimento Perfection, 'Pimento L').
Group II	Fruit broad, smooth, thin walled	A- Ancho group: Fruit 10.0-15.0 cm long, heart shaped but pointed, somewhat flattened, sweet to mildly pungent	1. Dark green turning red at maturity ('MexicanChili', 'Ancho', 'Poblano') 2. Turning brown at maturity ('Mulato')
Group III	Pods long, slender	A- Anaheim chilli group: (Long green/long red chilli). Fruit medium to dark green, smooth, 12.5-20.0 cm x 3.2-5.0 cm tapering to point, flesh medium thick, moderately pungent to sweet	1. Moderately pungent ('Sandia', 'New Mexico 9') 2. Mildly pungent ('Mild California') 3. Slightly pungent ('Mild California', 'New Mexico 6', 'Tam Mild Chilli 2') 4. Non pungent ('Paprika')
		B- Cayenne group: Fruit slender 12.5-25.0 x 1.9-2.5 cm, medium green; characteristically wrinkled and irregular in shape, thin walled and highly pungent.	1. Mature fruit red ('Cuban', 'Cubanelle', 'Aconcagua', 'Italian E-1', 'Pepperoncini)
Group IV	Fruit elongated to 7.5 cm long, green when mature	A- Jalapeno group: Fruit 3.75-5.0 cm wide, 5.0-7.5 cm long, round cylindrical shape, thick walled, dark green, smooth ('Jalapeno', 'Mild Jalapeno')	
		B- Serrano group: Fruit slender, cylindrical, often slightly constricted near middle, tapering to abrupt point, highly pungent, 1.25 x 5.0-6.25 cm ('Serrano')	
		C- Small hot group: Fruit slender, medium to thin walled <7.50 cm long, highly pungent ('Red Chilli', 'Chile de Arbol', 'Japanese Chilli', 'Santaka', 'Hontaka')	
Group V	Fruit small to 5.0 cm, globular to oblate, thick flesh	A-Cherry group:	1. Non-pungent- ('Sweet Cherry') 2. Pungent-('Large Red Cherry', 'Small Red Cherry')

Group	Characters	Description	Cultivar's Segement with Specific Traits
Group VI	Fruit yellow when mature	A- Small wax group: Fruit 7.50 cm or less in length	1. Pungent- ('Flora Gem', 'Cascabella', 'Calora') 1. Non-pungent- ('Petite Yellow Sweet Tam', ' Rio Grande Gold')
		B- Long wax group: Fruit 8.80 cm or more in length pointed or blunt.	1. Pungent- ('Hungarian Yellow Wax') 1. Non-pungent- ('Sweet Banana', 'Hungarian Sweet Wax', 'Long Yellow Sweet')
Group VII	Fruit slender, yellow turning red at maturity, 2.5-3.75 cm long, highly pungent, of the species *C. frutescens*	A-Tabasco group ('Tabasco', 'Green Leaf Tabasco')	

Common chillies groups include the following:

(i) **Paprika-** Theses chillies are used for dehydration and processing and act as a source of red pigment and flavouring. They are dark red at maturity and have no pungency, are 15.0-20.5 cm long and are conical in shape.

(ii) **Jalapeno-** These chillies are often harvested as green fruit that is used for fresh market use. A sizeable amount of the production is also sold to the processors.

(iii) **Anahein and Mexica-** These are long cylindrical chillies which are 18.0-23.0 cm long. They are harvested green and used as fresh as well as canned. They are also harvested red, in which case they are dehydrated and ground into powder for use in sauces. They range from varieties that have little pungency to varieties that have appreciable pungency.

(iv) **Wax-** These chillies are referred to as wax pod types because of their shiny pod appearance. Pods vary in size and shape and are usually yellow in colour when immature. They can be sweet or pungent and are used fresh or for pickling. Varieties include 'Calora', 'Floral Gem', 'Matador', 'Sweet Banana' and 'Santa Fe Grande'.

Classification on the Basis of Popular Name

(i) King-chilli (*Capsicum chinense*)

(ii) Cherry chilli (*Capsicum annuum* var. *annuum*)

(iii) Dalle chilli (*Capsicum chinense*)

(iv) Bird Eye chilli (*Capsicum frutescens*)

Augmentation of Genetic Resources and Crop Umprovement

Attempts have been made to breed high yielding varieties coupled with resistance to diseases, and high vitamin C content (Pal *et al.*, 1941; Cheema *et al.*, 1994; Deshpande, 1944, Parthasarthy and Kedarnath, 1950; Jeswani *et al.*, 1956; Jagadish *et al.*, 1964; Singh and Singh, 1970; Singh *et al.*, 1977). Raju (1979) and Cheema (1982) also suggested different breeding methods for the improvement of chilli. Overall, the breeding methods of improvement include, pur line selection, pedigree method of selection followed by hybridization and back crossing.

Introduction and Selection

The solitary bearing habit of chilli limits mechanical harvesting and makes harvesting process labour intensive. Nearly 20 per cent of cost of chilli cultivation is on harvesting of fruits alone. Gopalakrishnan (1985) studied 2 clustered chilli lines 'CA 33' and 'CA23'. All the clustered plants were determinate. This showed that determinate habit is pleiotropic to clustered fruting habit. The F_1s 'Jwala' x 'CA 33' and 'Pant C1' x 'CA 33' were solitary fruited with an indeterminate growth habit indicating dominance of solitary fruiting habit over clustered fruiting habit. Varieties 'Andhra Jyoti (G4)', 'Bhagyalakshmi (G 5)' and 'K 2' were identified as high yielding. Following valuable germplasm are available with ICAR-NBPGR, New Delhi.

The local varieties of 'Byadgi' of Karnataka and 'Sankeshwar' of Maharashtra are grown during rainy season. Local cultivar "Mathania Mirch" is very popular in Jodhpur (Rajasthan) for "Pakaura" preparation and GI has been granted. It has thick placenta. Raju (1979) and Usha Rani (1981) studied variability, heritability and gene action in chillies for various traits incliding ascorbic acid, pigment and pungency. Use of synthetic or composite varieties and need for reciprocal recurrent selection for simultaneous improvements for various characters have been advocated by Usha Rani (1981) and Cheema (1982). Verma and Pant (1999) evaluated the chilli germplasm for utilization in Uttar Pradesh hills. Evaluation revealed that P-32, P-1399, P-2072, EC362903-A, EC362916-B, EC362930 and KN-613 have elite types amounting to 119 accessions. Most notable genotypes include P-2072 for maximum fruit length and uniform fruiting and appealing colour and having disease tolerance. Chandra and Pandey (2001) reported a new variety of chilli 'Jawahar Mirch-283' which is suitable for areas around Jabalpur of Madhya Pradesh. This variety has been developed by JNKVV, Jabalpur (Madhya Pradesh). Plants are are dwarf (40-50 cm) with bushy habit. It produces 7-9 primary branches with dense foliage cover. The fruits are straight, firm, medium in length, thin (2.5-3.5 cm girth in the middle) and dark

Table 2.103: Promising Germplasm Introduced from different Sources

Sl.No.	Chilli Accessions	Country	Traits
1.	EC668798. EC668801, EC668803, EC668805, EC668806, EC668808, EC668812, EC668814, EC692278, EC692280	Taiwan	A line
2.	EC668799, EC668800, EC668802, EC668804, EC668807, EC668813, EC668815, EC692279, EC692281	Taiwan	B line
3.	EC668809	Taiwan	R line
4.	EC668816	Taiwan	Restorer line
5.	EC678809	China	Fresh use, green, pungent, resistance to heat and humidity
6.	EC673074	Taiwan	Cayenne type
7.	EC678805	China	Dark green, strong pungent, heat and humidity resistance
8.	EC678806	China	Fresh use, plant height 80-90 cm, light green, turn red and smooth at maturity
9.	EC678807	China	Fresh and dry use, green, high pungency
10.	EC678808	China	Fresh use, green smooth, high pungency
11.	EC678809	China	Fresh use, pungency, heat and humidity resistance
12.	EC497628-640; received by ICAR-IIVR, Varanasi (U.P.) and HPKV, Palampur (H.P.)	AVRDC	High yielding lines
13.	EC175959 to 66	Hungary	Resistant to red spider
14.	EC203581 to 601	Hungary	Tolerance to CMV
15.	EC264584 (Nigold)	Hungary	Yellow fruited, more branching, resistant to wind damage
16.	EC304963 (Marold Giant)	USA	Fruits with thick wall, spreading habit, strong upright, tolerant to TMV
17.	EC371245 and EC371246	USA	High level of resistant to root knot namatode
18.	EC260643 (*C. chinense*)	Netherlands	Resistant gene (*L-3*) for some TMV races
19.	EC260642 (C. frutescens)	Netherlands	Resistant gene (*L-2*) for all Dutch races TMV
20.	EC260644	Netherlands	Resistant gene (*L-4*) for all Dutch races TMV
21.	EC381007 to 01	Taiwan	Suitable for table purpose; fruits are cherry shaped and bear in clusters
22.	EC332333	Taiwan	Early and resistant to TMV
23.	EC354883	Taiwan	Heat tolerant and resistant to CMV
24.	EC347246 to EC347248 and EC347259	Taiwan	White fruited, tolerant to CMV
25.	EC347284	Taiwan	High level of resistant to *Phytophthora*
26.	EC347263	Taiwan	High pungent type
27.	EC391073	Taiwan	Pungent type
28.	EC391089-92	Taiwan	Early maturity and bunch type
29.	EC323333	USA	Multiple disease resistance

green in colour. Since its fruits are very attractive with high consumer preference, they fetch a premium price in the market. It is very much suitable for round the year cultivation in Madhya Pradesh. Many improved varieties are developed by pure line selection in land races or heterogenous local cultivars such as Aparna, CO-1, CO-2, Musalawadi, Arka Lohit and Kalyanpur Red.

Table 2.104: Improved Varieties of Chilli Developed in India

Cultivars	Source
G-3, Bhagyalakshmi, Andhra Jyothi, Sindhur, LCA-206B, Aparna, X-235	APAU, Lam Guntur (A.P.)
CO-1, CO-2, CO-3, K-1, K-2, PKM-1, MDU-1	TNAU, Coimbatore (Tamil Nadu)
Pusa Jwala, Pusa Sadabahar, NP-46-A, Pant C-1, Pant C-2	GBPUA&T, Pantnagar (Uttarakhand)
Punjab Lal, Punjab Surkh, CH-1, Punjab Guchhedar	PAU, Ludhiana (Punjab)
Musalwadi, Sankeshwar-32, Phule Sai, Agnirekha, Phule Jyothi, Phule Suryamukhi	MPKV, Rahuri (Maharashtra)
Arka Lohit	ICAR-IIHR, Bengaluru (Karnataka)
Ujjwala, Manjari, Anugraha, Vellayani	Kerala Agricultural Universirty Thrissur (Kerala)

Table 2.105: Trait Specific Germplasm

Traits	Promising Accessions
High yielding	EC362918, EC362922, EC362930, EC362998, EC339047, NIC-22076, NIC-22058, NIC-19150, PIC147681, PIC119193, PIC119257, PIC119676, PIC19639, N-34, NP-46A, Pusa Red, Kalyanpur Red, Kalyanpur Selection-1, Kalyanpur Yellow, Pant C-1, Pant C-2, G-2, G-3, G-4, G-5, Pusa Jwala, Arka Lohit, LCA-235, Pant C-1, LCA-218, JCA-340, CA-138, No.-31, No.-27, No.-32, Suryamukhi, Beldanga, Local, Dhnilanka, PDC-40, IC119481, IC119500, EC345655, K-119457, K-119497, K-119412, BD-213, EC345625, JCA-17, EC339565, IC204000, IC204212, NIC-9577, KCS-2013, BC-25, LCA-353, Ajeet-6, KA-2, SH-KC-12, SH-KC-9, LK-100, PBC-3, C-5, C-6, C-7, CA-138, MC-111, INGR-04052, INGR-04053, INGR-04954, BSS-128, MSL/I/14, Baljuri, Dinhata Coll-1, Dinhata Coll-2, CCA-4261 (cms line), CCA-4727 (cms line), M-12 (gms line), PCB-77, LCA-15, MM-7, EC126493, CA-83, CA-80, Jati Long, Round Chilli, SH-SP-1007, SH-SP-1004, PKM-1, CA-25, CA-66, CA-75, CA-60, CA-52, CA-48, SH-KC-23, SH-KC-12, SH-KC-16, SH-KC-6, SH-KC-18, BCC-24, BCC-1, BCC-18, BCC-11, BCC-20, Line No. 10-3, Line No. 9, IC413702, EC399550, BD-26, BD-29, BK-7, ALS—98-8, CA-141, SH-KC-46, SH-KC-55, IIVR Local Super, LCA-3531, MKR-7, MKR-1, MKR-8, MKR-19, EC587052, EC587013, EC587007, ICODCS-06-01, EC28DPS-06-01, SH-KC-1003, SH-KC-1111, LC-1, LC-2, LC-1, UHF-14,
Early flowering	IC119490, IC119497, IC19267, IC147637, IC119264, EC399566, EC399568, EC399578, Line No. 19, EC405281, IC526852, EC631776, EC631789 and EC631796
Early maturing	EC399568, KA-2, LCA-235 and Pusa Jwala
Yellow chilli	NIC-23829
Round cherry type fruits	NIC-23901
Paprika type	NIC-19953, SHP-1007, SHP-1005, SHP-1004, BD-26, BD-27, BK-7,
High pungenet type	EC347263, Bhut Jolokia (*C. chinense*)

Table 2.106: Promising Accessions of Chilli Identified at different Regional Centres of ICAR-NBPGR

Centres	Year	Identified Promising Accessions
NBPGR, N. Delhi	1989-90	89-5, CA-213, CA-282, CA-408, CA-416, CA-418, CA-420, CA-431, CA-445, CA-448, CA-953
	1996-97	EC391074, EC391079-92, EC399532-581
NBPGR, Amravati	1990-91	BDJ-224, EC103820, U-20 to U-34
	1994-95	IC119603, IC119617, IC119546, PBC-122, AD-60
	1995-96	IC119581, EC339047, NIC-19991, JBT-221C-133, IC119576, NIC-23911, NIC-23309, NIC20418
	1996-97	JBT-12/5, JBT-12/17, NIC-20924, EC382187
NBPGR, Bhowali	1994-95	P-2072, NIC-92250, NIC-72156, NIC-99916, NIC-99912
	1996-97	P-1893, P-2072, NIC-19943, EC362918, IC92123, IC119241, IC119787
NBPGR, Shilong	1995-96	NIC-22146, NIC-22058, NIC-22214, IC89727
	1996-97	NIC-22076, HM-2804, RS-8, RS-27
NBPGR, Thrissur	1990-91	IC88498, IC88503
NBPGR, Hyderabad	1996-97	IC119490, NIC23829, JBT-12/8, JBT-12/9, JBT-12/15

A list of promising accessions of chilli for different attributes is given in Table 2.105.

ICAR-National Bureau of Plant Genetic Resources (NBPGR), New Delhi and its Satellite Centre 'Amravati' are the main esponsible centre for collection, evaluation and characterization of chilli germplasm. Locally collected lines are also being maintained and evaluated at Regional Station, Meghalaya. A list of some promising accessions of chilli identified at different regional centres of ICAR-NBPGR for different attributes are given in Table 2.106.

Similarly, cultivar like TM-888 and Sky Hot by Hungnong Seed Co., Korea and Canope by Sakata, Japan has been introduced in Nepal. In Pakistan, cultivars like Gola Peshawari, Tatapuri (R), Narwala, Pusa Jwala (I), NARC-4 and Qaiser (P) are very common and being grown commercially. Verma *et al.* (1998) evaluated 119 accessions of chillies at ICAR-NBPGR, Regional Station, Bhowali, Nainital (Uttar Pradesh) and noted wide range of variability in plant height (50.0-86.0 cm), growth habit (prostrate, compact, erect), leaf and stem pubescence (glabrous, sparse, intermediate, abundant), seedling, leaf and stem colour (green, purple), density of branches, plant canopy (104.0-6348.0 sq cm), leaf length (4.2-15.9 cm), leaf width (1.2-6.8 cm), flower colour, days to 50 per cent flowering (142-172 days), pedicel position, number of pedicel per axil, days to 50 per cent fruiting (161-220 days), fruit set, calyx margin and shape, annular construction at junction of calyx and peduncle, fruit position, fruit colour at immature stage (green, yellow, orange, red, purple, brown/black), fruit shape (elongate, oblate, round, conical, companulate, bell/ blocky), fruit shape at peduncle attachment (truncate, cordate, lobate, acute, obtuse), fruit shape at blossom end (pointed, blunt, sunken), presence and absence of neck at base of fruit, fruit cross section (smooth, intermediate,

Table 2.107: Promising Accessions Identified in Chillies

Sl.No.	*Characters*	*Promising Accessions*
1.	Days to 50 per cent germination >80 days	EC362911, EC362931,
2.	Leaf length >12 cm	EC362918, EC362919, EC362922, P-1939, P-2072, N-1149, N-1382
3.	Leaf width >5 cm	EC362919, EC362925, EC362930, EC362938, P-1639, P-1713, P-1718
4.	Plant height >60 cm	P-2072, N-1160, N-1382
5.	Days to 50 per cent flowering <150 days	P-32, P-64, P-173, P-290, P-241, P-272, P-249, EC362905
6.	Days to 50 per cent fruiting <165 days	P-32, P-64, P-173, P-297, P-641, EC362905
7.	Number of fruits/plant >60	P-1399, EC362903, EC362916, EC362930
8.	Fruit length >9 cm	P-1311, P-2072
9.	Fruit width >2 cm	P-840, P-190, P-1258, P-2072, N-349, KN-613
10.	Fruit (10) green weight >60 g	P-64, P-100, P-1258, P-2072, N-349, KN-613
11.	Fruit (10) dry weight >30 g	P-64, P-190, P-1258, P-1639, N-349, N-368, N-867, N-1149
12.	Disease free accessions from leaf spot powdery mildew and fruit rot	IC99910, IC99912, IC92150, P-1718, P-1939, N-1015, EC362901, EC362910, EC362913, EC362925

Source: Verma *et al.*, 1998.

corrugated), fruit pungency, number of fruits per plant (2.1-120.0), fruit length (1.6-12.4 cm), fruit width (0.2-2.6 cm), fruit green weight/10 fruits (7.0-55.0 g), fruit dry weight/10 fruit (1.0-7.5 g), seed colour and pest and disease of varying intensity).

In Goa state a land race of chilli called 'Kholla chilli' named after its place of cultivation is a pungent chilli, exclusively used for papad making and fetches premium price in the market. Verma *et al.* (2016) characterized 106 germplasm of chilli including cultivars (hot and sweet pepper) and popular land races like King-chilli, Cherry chilli, Dalle chilli and Bird Eye chilli for yield contributing traits and also differentiate on the basis of microsatellite markers. Wide range of variation observed for all yield contributing traits.

Hybridization and Selection

In chilli several genotypes were utilized for earliness, fruit yield and their component characters. Nagaraju and Sreelathakumary (2016) recorded significant *gca* and *sca* values for most of the characters, revealing that both additive and non-additive gene effects involved in genetic control. The lower $\beta^2 gca/\beta^2 sca$ ratio indicates that the predominance of non-additive (dominance) gene action is important for all the traits except fruit girth and fruit weight. CA-32 was good general combiners for both red as well as green fruit yield per plant. CA-8 was good general combiner for red fruit yield, while CA-5 and CA-6 were the best general combiners for quality character. The highest sca effect for red fruit yield and green fruit yield per plant in F_1 generation was noted in the cross of CA-6 x CA-23 and CA-23 x CA-32. The hybrids CA-23 x CA-32, CA-8 x CA-23, CA-8 x CA-32 and CA-6 x CA-23 were best specific crosses for several characters on the basis of specific combining ability effect. Although none of the parents was a good general combiner for all the traits, some parents showed high *gca* effects for some of the economic traits, suggesting that these parental lines could be considered simultaneously while formulating a breeding programme for improving fruit yield and yield contributing characters. Deshpande (1933) demonstrated hybrid vigour in chillies with regard to total number of fruits and total weight of dry produce. Sontakke (1981) observed heterosis and heterobeltiosis for yield, branches, earliness, plant height, carotenoids, ascorbic acid, fruit size *etc.* Similarly, Pal (1945) observed increased yield and early maturity in hybrid chillies. Singh and Singh (1978) said it may be worthwhile to subject the parents to diallele

Table 2.108: Genotypes Used in Hybridization

Female Parent	*Male Parent*	*Salient Features*
'Bell Pepper'	'Round Early'	Early, higher yield, uniform fruits
'D 103'	Sivriya 600'	Early ripening, higher yield
'Jubilee 307'	'Pazardshik Pepper'	Early, higher yield
'Jubilee 307'	'Z-4/57'	Early, higher yield
'Pazardshik Pepper'	'Jubilee 307'	Early, higher yield
'Early Canning 79-68'	'Pazardshik Pepper'	Early, higher yield
'Jubilee 307'	'Krymskwhite 29-293'	Higher ascorbic acid content
'Early Canning 79-68','Maiko'	'Pazardshik Pepper'	Higher ascorbic acid content
'Jwala'	'Pant C1'	Early, higher yield (Gopalakrishnan *et al.*, 1987)
'Jwala'	'CA 23'	More pungency

selective mating for fostering greater recombination if simultaneous improvement of developmental characters related to productivity is sought in chilli.

In crosses between 'KAU Cluster' and bell pepper varieties ('Hungarian Wax', 'Early Cal Wonder', 'Hybrid Pepper Bell Boy'). Pious (1985) observed heterosis for earliness, plant height, fruit length, fruit perimeter, average fruit weight and green fruit yield. The best yielding cross was 'Bell Boy' x 'KAU Cluster' (46406 g/ plant), '672 Hungarian Wax' x 'KAU Cluster' (401.5 g/ plant) with a standard heterosis of 108.3 per cent over 'Pant C-1' for yield was the most promising (Pious and Peter, 1988). Dambe and Kawarkhe (1998) reported a new variety of chilli 'Jayanti' (AKC 86-39) which was developed by pure line selection and resulted in higher yield potential, bigger size and red colour fruits. Several varieties like Andhra Jyoti, Pusa Jwala, Pusa Sadabahar, Pant C-1, Pant C-2, Pinjab Lal *etc.* were developed by hybridization techniques. Chilli local strains were identified and evaluated under Kashmir valley conditions. The highest yield of dry chilli was recorded in CC-84 (262 g/plant) followed by CC-167(227 g/plant) and CC-39(202 g/plant). The longest fruit was recorded in CC-151 (15.3 cm) followed by CC-132 and CC-126 (15.1 cm) (Anon., 2006). Nagaraju and Sreelathakumary (2016) in evaluation of 15 hybrids (F_1s) observed that the hybrids CA-23 x CA-32 (57.90 per cent), CA-6 x CA-23 (49.67 per cent) and CA-8 x CA-32 (38.21 per cent) found to be most promising for green fruit yield, whereas CA-6 x CA-23 (22.52 per cent) and CA-8 x CA-32 (22.36 per cent) were the most promising for dry fruit yield. Among the hybrids CA-3 x CA-5 (26.74 per cent) exhibited the highest standard heterosis for capsaicin and CA-6 x CA-8 (82.14 per cent) for oleoresin. Based on the *per se* performance and standard heterosis, the hybrids CA-23 x CA-32, CA-8 x CA-32, CA-8 x CA-23 and CA-6 x CA-32 were found superior in respect of six characters *viz.*, days to first harvest, fruit length, fruit girth, fruit weight, green fruit yield per plant and dry fruit yield per plant.

Interspecific Hybridization

There are about 100 species under the genus *Capsicum*. The species cross compatibility adds considerably to genetic variability. In the interspecific hybridization *C. annuum* and *C. frutescens*, Smith and Heiser (1957) obtained 2 per cent viable seeds when *C. frutescens* was used as female parent. The F_1 plants ranged from partially fertile to completely sterile. Goud *et al.* (1970) observed varietal variation in the success of interspecific hybridization between *C. annuum* and *C. frutescens*. Kumari and Peter (1986) also reported varietal influences in the interspecific hybridization between *C. annuum* and *C. frutescens*. Cross was attemted using *C. annuum* cultivars 'Jwala' and 'K' with *C. frutescens* cultivars 'White Kanthari', 'Ornamental Type' and 'Green Chuna' as female parent and obtained perfect set between 2 species except in combinations where 'White

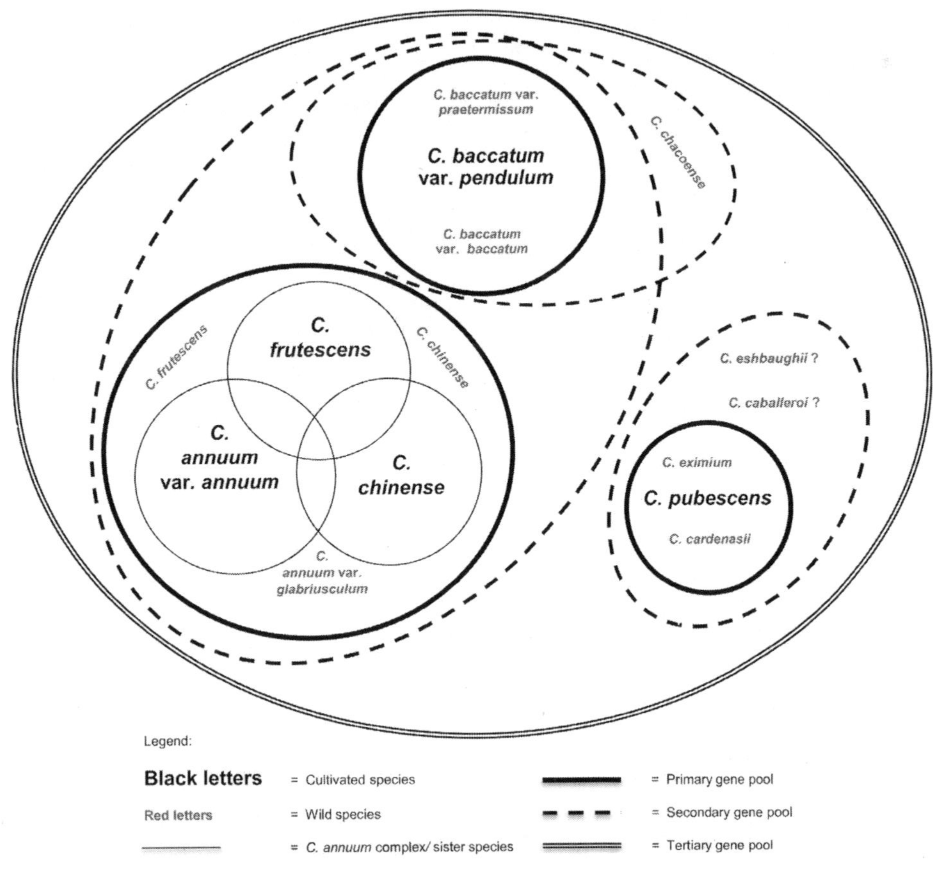

Cultivated *Capsicum* Gene Pools

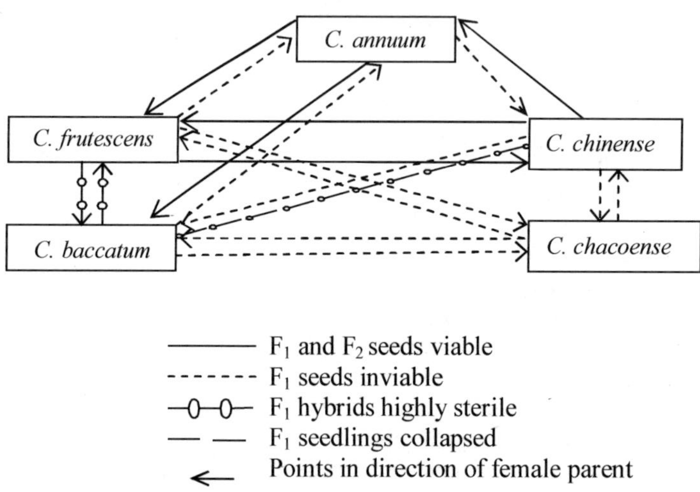

Crossability Polygon among 5 Species of *Capsicum*

Kanthari' was used as female parent. Radhakrishanan *et al.* (1977) developed a technique in which the upper parts of the gynoecium (style and stigma) were excised and a drop of 5 per cent sucrose solution was applied to cut surface prior to pollination. By this way, they got fruit set and seed set in crosses between *C. annuum* and *C. frutescens*. When *C. annuum* was used as female parent, the percentage of fruit set was considerably lower than the reciprocal. A similar result was reported by Nair *et al.* (1979) in a cross of *C. annuum* with *C. frutescens* as male parent. In a cross between *C. annuum* and *C. frutescens*, Sundaresan and Chandrasekaran (1979) reported one way incompatibility. The failure of pollen grain of *C. frutescens* in the stigma of *C. annuum* revealed the existence of pre-fertilization barrier. When *C. frutescens* was used as female parent, fruit set was 17 per cent. Peter and Mc Collum (1984) observed pollen-stigma compatibility in the cross *C. frutescens* x *C. annuum*, but plant failed to set fruits after artificial pollination in the green house, through pollen tubes reached ovule within 24 hours after pollination. Kumar (1990) worked out a crossability polygon among 5 species of *Capsicum*. Most interspecific crosses with in the same complex produce partially fertile hybrids without embryo rescue. Crossability between two species is not always reciprocal successful (Dlian-Caporalino *et al.*, 2007).

The present studies established close relationship between *C. chinense* and *C. frutescens*. There is possibility for transferring desirable genes from *C. chacoense* to the cultivars of *C. annuum*. Hybrids of *C. annuum* x *C. chinense* had high values of capsaicin (0.92 per cent), oleoresin (34.40 per cent) and colour (58.62 ASTA unit). The other important resistant variety developed through interspecific hybridization is 'Pant C-1' by crossing *C. annuum* x *C. frutescens*. Krishnakumary (1984) estimated crossability between *C. annuum* and *C. frutescens* and Mini (1989) studied that between *C. annuum*, *C. chinense* and *C. chacoense* crossing is possible. By using *C. annuum* as female parent to hybridize with *C. baccatum*, embryo rescue technique was adopted to overcome embryo abortion (post fertilization barrier). However, in cases where post fertilization embryo abortion occurred before globular stage, tri species bridge crosses were successful. *C. chinense* was used in the bridge parent to accomplish the hybridization between *C. annuum* and *C. baccatum* (Yoon and Park, 2005).

Some successful distant hybridization in chilli is as under:

- *Capsicum annuum* x *C. sinense*
- *C. frutescens* x *C. sinense*
- *C. sinense* x *C. pendulum*
- *C. pubescens* x *C. sinense* (embryo culture)

Mutants

Several variants and mutants have been observed which are mainly useful for genetic studies. However, use of close flower mutant (*cf*) and male sterile (*ms*) mutants can be used for hybrid seed production (Deshpande *et al.*, 1983a, b; Pathak *et al.*, 1983 a, b). In chillies variants have been developed by using gamma rays (12 kr) and mutant 'Pazard-shik Pepper 794' gave 100 per cent male sterile progeny. 'Albena' is another chilli variety developed through mutation which is early, high yielding and has high percentage of standard fruits. Popular and old cultivar 'MDU-1' was released as a mutant for commercial cultivation. Line *Ms12* has GMS allele *ms509* (induced through mutagenesis), and is being used for hybrid seed production in India, Korea and Hungary.

Male Sterility

In pepper more than 12 recessive genes have been reported *ms-12* (*ms-509*/*ms-10*) and *ms-3* genes are commercially utilized in India and Hungry, respectively. Punjab Agricultural University, Ludhiana has taken a lead in developing CH-1, hybrid by crossing MS-12 x LLS. MS-12 is a male sterile line which has been developed by utilizing the male sterility genes introduced from France.

CH-1 hybrid exhibits 160 per cent heterosis for yield besides having tolerance to some diseases. At ICAR-IIVR, Varanasi (U.P.) nuclear male sterile gene (*ms-3* and *ms-10*) are successfully being utilized to produce cost effective experimental crosses. The *ms-10* gene is linked with taller plant height, erect growth and dark purple anther. In pepper, cytoplasmic male sterility was first reported in Indian accession. Most of the cms lines are temperature sensitive. Occurrence of *Rf* allele is common in small fruited (usually hot pepper) and *rf* in large fruited (usually sweet pepper) lines. RAPD markers linked to *Rf* gene have been identified. In chiili hybrid seed production is carried out by both genetic male sterility (GMS) cytoplasmic genetic male sterility (CGMS).

Genetic Male Sterility (GMS)

In utilizing genetic male sterility; sterile plants are used as the female parent of a hybrid cross. The male sterile characteristic is often inherited as a single recessive gene (*ms*). Its used in hybrid seed production is limited due to the inefficiency of producing and maintaining a population of male sterile plants. In order to produce more male sterile plants one must cross a fertile plant heterozygous for the male sterile trait to the male sterile plant, and then only half the progeny from this cross will be male sterile, thus the GMS being maintained in heterozygous condition. During hybrid seed production, the heterozygous population consisting of male fertile and male sterile plants and the male parent of interest are planted in alternate rows in the plot in isolation (400-500 m from other commercial grown chilli crop). After flowering one can easily distinguish the sterility/fertility by observing for pollens through nail test (shed pollen from the flower on to the nail and if whitish powder like particles are seen–it indicates male fertility, and if no powder substance is seen, the plant is male sterile) or through pollen viability test under the microscope. The male fertile plants are removed from the plot and the remaining plants from the heterogeneous population are female fertile (male sterile). The seeds developed on the male sterile plants after open pollination from the male fertile plants (male parent of interest) are the F_1 hybrid.

Cytoplasmic Genetic Male Sterility (CGMS)

Cytoplasmic genetic male sterility (commonly call CMS system) is another mean for hybrid seed production in chilli. CMS in hot pepper plants (*C. annuum*) was first documented by Peterson (1958) in the PI164835 line introduced from India. The advantage of a CMS system is that a population of sterile plants can be generated in which all the offspring are sterile. Sterility results an interaction of nuclear and cytoplasmic genes, hence called cytoplasmic genetic male sterility. CMS is maternally inherited and is with a specific (mitochondrial) gene whose expression impairs the production of viable pollen. Since restorer of fertility (*Rf*) genes in the nucleus function to suppress the CMS phenotype, nuclear restoration allows commercial exploitation of the CMS system for the production of high yielding and heterotic hybrid seeds and avoids the need for intensive labour and extensive hand emasculation. Dominant restorer alleles have been identified in several hot pepper genotypes to help breeders to differentiate restorer lines from maintainer lines to secure a completely sterile female parent, but fully fertile hybrids. Since male part (constricted anthers with little pollen) is sterile, use of such line precludes huge cost on manual emasculation. The CMS line (also called 'A' line) is maintained by making crosses on it using pollen from maintainer line (also called 'B' line). Through such crossing, 100 percent male sterile seeds are obtained, unlike GMS line. The seedling prepared from such crosses can be transplanted as female rows (4) along with the selected restorer line (called 'R' line) as male row (1) in the hybrid seed production field. Since considerale amount of natural cross pollination occur in chilli, no hand pollination is required.

Polyploids

Natural polyploids with 2n=36, 48 are available in the bell pepper group *C. annuum* var. *grossum*. The polyploids have a low fertility rate and seed content is very low.

Resistance breeding

Table 2.109: Chilli Hybrids Developed through Male Sterility

Name of the Hybrid	*Type of Male Sterility Used*	*Developing Organization*
CH-1	Geneic male sterility (GMS)	Punjab Agricultural University, Ludhiana Punjab
CH-2	Geneic male sterility (GMS)	Punjab Agricultural University, Ludhiana Punjab
CH-3	Geneic male sterility (GMS)	Punjab Agricultural University, Ludhiana Punjab
Kashi Surkh (CCH-2)	Cytoplasmic genic male sterility (CGMS)	ICAR-Indian Institute of Vegetable Research, Varanasi (U.P.)
CCH-3	Cytoplasmic genic male sterility (CGMS)	ICAR-Indian Institute of Vegetable Research, Varanasi (U.P.)
Kashi Tej (CCH-4)	Cytoplasmic genic male sterility (CGMS)	ICAR-Indian Institute of Vegetable Research, Varanasi (U.P.)
Arka Meghna (MSH172)	Cytoplasmic genic male sterility (CGMS)	ICAR-Indian Institute of Horticultural Research, Bengaluru Karnataka
Arka Sweta (MSH-149)	Cytoplasmic genic male sterility (CGMS)	ICAR-Indian Institute of Horticultural Research, Bengaluru Karnataka
Arka Harita (MSH-172)	Cytoplasmic genic male sterility (CGMS)	ICAR-Indian Institute of Horticultural Research, Bengaluru Karnataka

Biotic Resistance

Resistance to Diseases

Backcrossing is useful in transferring a single gene controlled character into a desired variety particularly disease resistance (Swarup, 2006). Transgression of useful genes for disease resistance is possible through utilizing resistance genes from wild and weedy species of capsicum species. Information regarding genetics of resistance is vital to choose an appropriate breeding methodology.

There are several species which are known for resistance source against diseases in chillies. Source of resistance to anthracnose, cercospora leaf spot, bacterial leaf spot and powdery mildew have been identified by Cheema (1982), Sontakke (1981). The probable source of resistance to virus disease was reported by Jayarajan and Ramakrishnan (1969). Mini (1989) studied that between *C. baccatum* is resistant to anthracnose, Phytophthora rot and cucumber mosaic virus; *C.chinense* is resistant to anthracnose, leaf curl virus, TMV and Verticillium wilt and *C. frutescens* is a source of resistance to leaf curl virus.

Variety 'Pant C1' known for its resistance to leaf curl is a selection from 'NP 46A' x 'Local Kandari'. Several varieties have been identified as source of resistance to parasitic and non-parasitic diseases. 'Pant C1' was also found resistant to 4 *Pseudomonas solanacearum* isolates and moderately resistant to Phytophthora root-rot. Markose (1996) and Indira (1997) reported resistance to bacterial wilt in CA-752 and CA-517, respectively. Gopalakrishnan (2004) reported that chilli genotypes CA-33 and CA-219 are free from bacterial wilt. Hot pepper popular cultivar 'Pant C-1' has expressed resistance against damping off, root knot nematode and leaf curl virus.

Powdery mildew resistance gene available in in *C. chinense* has been transferred into a Jalapeno

Table 2.110: Genetics of Resistance to Diseases in Pepper

Diseases	*Mode of Genetic Control*
Damping off	Digenic (Smith *et al.*, 1965)
Anthracnose	Recessive, additive, dominant and epistatic effects were important (Cheema, 1985)
Cercospora leaf spot	Recessive, 3 complementay genes, additive and dominance being important (Cheema, 1985)
Bacterial wilt	Dominant gene
Bacterial leaf spot	Monogenic dominant (Cooka ans Stall, 1967)
Bacterial soft rot	Oligogenic (Bartz and Stall, 1974)
Cucumber mosaic and Potat virus Y.	Monogenic recessive (Barrios *et al.*, 1971), Monogenic recessive single pair of identical genes in *C. angulosum* and *C. microcarpum* inherited independentaly (Singh and Chenulu, 1985)
Root knot nematode	Dominant gene
Root rot	Monogenic dominant (N) (Here, 1957)

Table 2.111: Priority Species of Chilli for Economically Important Stresses

Genus	*Priority Species*	*Source of Resistance*
Capsicum	*C. baccatum*	Phytophthora rot and cucumber mosaic virus
	C. chinense	Leaf curl virus, tomato mosaic virus and verticillium wilt
	C. frutescens	*Phytophthora capsici* Leonian, *Cercospora capsici* Heald and Wolf, *Scierotinia rolfsii* Saee, *Xanthomonas vesicatoria* (Doidge) Dye, Leaf curl virus, mosaic and *Pseudomonas solanacearum* E.F. Smith.
	C. fasciculatum	*Verticillium alboatrum* Reinke C. Benth.
	C. angulosum	*Verticillium alboatrum* Reinke C. Benth.

Table 2.112: Resistance Sources to Parasitic and Non-Parasitic Diseases in Chilli

Nature	*Disease*	*Pathogen(s)*	*Resiatant Varieties/Genotypes*
Non- parasitic	Blossom-end rot		'Avelam', 'Lkeda'
Fungal	Damping off	*Phytophthora capsici* Leonian	'A 2-5', 'Kuban Early', 'Sal-gochu'
	Verticillium wilt	*Verticillium alboatrum* Reinke and Benth	'Podanok', 'Buketan'
	Fusarium wilt	Fusarium *annuum*	'College No. 9
	Leaf spot	*Cercospora capsici* Heald and Wolf	Small fruited varieties of Brazil
	Ripe rot	*Collrtotrichum capsici* (Sydow) Butter and Bisby.	'Se. 4', 'Sel. 25', 'A 126', Achar'
	Sclerotinia disease	*Sclrotium rolfsii* Sacc.	'Tabasco'
	Powdery mildew	*Oidiopsis scalia*	'Avelar'
Bacterial	Soft rot	*Erwinia carotovora* (Jones) Martinee and Kocur	'Jalapeno'
	Leaf spot	*Xanthomonas vesicatoria* (Doidge) Dye	'Santaka', 'Agronomico 8-2-17'
	Wilt	*Pseudomonas solanacearum* E.F. Smith	'Kandan', 'Pungent Pride', 'Cherry Red', 'KAU Cluster' (Manjari), 'CA 219', AAUM-1, AAUM-2, Utkal Rashmi

Nature	Disease	Pathogen(s)	Resiatant Varieties/Genotypes
	Bacterial wilt	Resistant to bacterial wilt (*Clavbacter michiganensis*)	EC497628-640 (AVRDC, Taiwan) received by ICAR-IIVR, Varanasi (U.P.); HPKV, Palampur (H. P.)
	Bacterial wilt	Resistant to bacterial wilt (*Clavbacter michiganensis*)	EC498405-417 (Korea) received by HPKV, Palampur (H. P.)
	Wilt and dieback	Resistant to wilt and dieback	Punjab Surkh, LCA-235, BC-21-2, LCA-305
Viral	Mosaic	Tobacco-mosaic virus	'Florida Vr 2', 'Rutgers'
	Mosaic	Potato virus x	'Florida Vr 2', 'P 34'
	Mosaic	Tobacco etch virus	'Florida Vr 2', 'Lr 1'
	Mosaic	Cucumber mosaic virus	'Lr 1', 'x 196'
	Leaf curl	Tobacco leaf curl virus	'Pant C 1', 'Puri Red', 'Puri Orange'
	Veinal mottling	Pepper mottle virus	'Line AV 234'
	CMV and TMV	Resistant to CMV and TMV	Punjab Lal

background with extremely low heritability suggesting multiple recessive genes or incomplete penetrance. For this disease many *C. chinense* and *C. baccatum* types seem to be either immune or poor hosts for the pathogen. *C. annuum* is also reported resistance against powdery mildew (Deshpande *et al.*, 1985). The breeding line 'White Kandari' was resistant to 3 isolates of the bacterium. These Indian chilli lines could be an additional source in multiple-disease-resistant breeding programmes. Multiple disease resistance and good horticultural characteristics make 'Pant C-1' an excellent source of these resistances. High content of ascorbic acid was observed in resistant varieties to early blight caused by *Alternaria solani*. A phytoalexin capsidiol having wide spectrum fungal toxicity has been isolated in resistant varieties to fungal diseases.

Anthracnose in *Capsicum* is caused by several *Colletotrichum* spp. like *C. acutatum*, *C. gloeosporioides*, *C. capsici* and *C. coccodes* (Park and Kin, 1992). *C. acutatum* and *C. gloeosporioides* are most destructive and wide spread (Sarath Babu *et al.*, 2011; Voorrips *et al.*, 2004) which affect leaves, stems and fruit at green and red ripe stage. *C. baccatum* and *C. chilense* (available at World

Table 2.113: Trait Specific Genotypes of Chilli

Trait Specific Diseases	Promising Resistant Accessions
Resistant to multiple diseases	EC32333, *C. baccatum*, *C. frutescens*
Fruit rot	*C. chinense*, Acc. 1555, Acc. 1554, Ac. 906, Chinese Giant and Spartan
Bacterial leaf spot	PI163192, PI260435, PI163189, PI163192, PI271322 and PI322719
Phytophthora root rot	PI201234
Root rot in C. frutescens, C.O. *Phytophthora capsici* resistant	Line No. 491, Line No.493, Chinese Giant, Yolo, Spartan Lorai, S-27, S-41-1, *C. chinenese*, Waxy Globe
Tobacco Leaf Curl Virus	Pusa Sadabahar, Pusa Jwala, Punjab Lal, *C. chinenese*, P-159236
Leaf Curl Virus	Pusa Jwala, Pant-C-1, PBC-535, BS-35
TMV	Sonnette, Keystone Resistant Giant, Yolo Wonder, Florida, VR-2, VR-2-34, XV R3-25
Fruit rot and bacterial wilt	Jati Long and Round Chilli
CMV	LIC-13, LCA-315, G-4, LCA-305, LCA-235
Bacterial wilt resistant	Kali Cluster, White Kandari and Pant C-1

Table 2.114: Genetics of Anthracnose Resistance in *Colletotrichum* spp.

Pathogens	Capsicum spp.	Resistanc	Stage of Fruit	References
Colletotrichum acutatum	*Capsicum chinense* (PBC-932)	2 complementary dominant genes	Green fruit	Lin *et al.* (2007)
		2 recessive genes	Red fruit	
	C. chinense (AR)	Single recessuive gene	Green fruit	Kim *et al.* (2008)
	C. baccatum (P159413)	Monogenic	Green fruit	Kim *et al.* (2008a)
	C. baccatum (PBC-80)	Monogenic recessive	Green fruit	Mahasuk *et al.* (2009b)
		Monogenic dominant	Red fruit	
Colletotrichum capsici	*Capsicum chinense* (PBC-232)	Single recessuive gene	Green fruit	Kim *et al.* (2008b); Mahasuk *et al.* (2009)
			Red fruit	Pakdeevaraporn *et al.* (2005)
		Single recessuive gene with epistatic effect		Cheema *et al.* (1984)
Colletotrichum capsici	*C. annuum* 83-168	Dominant		Park *et al.* (1990a and b)
	Chung Young	Recessive		Lin *et al.* (2002)

Vegetable Centre, Taiwan) are potential source and being utilized to develop resistance cultivar (Kim *et al.*, 2010; Lee *et al.*, 2010, Lin *et al.*, 2002; Pakeleevaraporn *et al.*, 2005; Voorrips *et al.*, 2004)

Against *Phytophthora* spp. two wild accessions of *C. annuum*, CM-334 and AC2258 derived from P120134 possess high level of resistance to *P. capsici* (Ares *et al.*, 2005). The resistance in CM-334 is controlled by two genes with dominant and recessive epistasis (Reifchneider *et al.*, 1992). Resistance in AC2258 is controlled by two dominant genes without additive effects (Smith *et al.*, 1967).

The peppers are being infected by 68 different types of viruses (Pernezny *et al.*, 2003) and out of these 20 viruses are causing severe loss to the crop (Moury and Verdin, 2012). Main virus groups are transmitted through aphids, white fly and thrips. In India mainly *potyvirus*, *cucumovirus*, *begomovirus* and *tospovirus* are damaging the capsicum crops (Reddy and Reddy, 2010; Lawrence *et al.*, 2014; Reddy *et al.*, 2014).

In *C. annuum*, *C. frutesens* and *C. chinense*, several monogenic and polygenic virus resistance genes have been detected but only few are commercially exploited for development of resistant cultivars.

Table 2.115: Important Viruses Infecting *Capsicum* spp. in India and its Vector

Virus Group	*Virus*	*Vector*
Potyvirus	Chilli veinal mottle virus (chivmv)	Aphids
Cucumovirus	Cucumber mosaic virus (CMV)	Aphids
Begomovirus	Chilli leaf curl virus (chiLcv)	White fly
	Pepper leaf curl Varanasi virus (Pep LCVaV)	White fly
	Tomato leaf curl Joydepur virus (ToLCJaV)	White fly
	Tomato leaf curl Karnataka cirus (ToLCKV)	White fly
	Tomato leaf curl New Delhi Virus (ToLCNDV)	White fly
Tospo virus	Tomato spotted wilt virus (TSWV)	Thrips
	Groundnut bud necrosis virus (GBNV)	Thrips
	Peanut bud necrosis virus (PBNV)	Thrips

Complex Diseases

'Myliddy-1' and 'Myliddy-2' are the potential resistant source for leaf curl complex diseases (Tobacco leaf curl virus, and Pepper mottle virus). Similarly,

Table 2.116: Virus Resistance Identified in *Capsicum* spp.

Virus Group	*Gene Q=QTL*	*Virus*	*Donor Species*	*Donor Cultivar or Accession*	*Chromosomal Location*	*Type of Gene Action*	*References*
Begomovirus	2 recessive	Pep GMV, PHYVV	*C. chinense*	BG-3821	?	Virus replication and movement	Garcia and Rivera (2011)
	1recessive	PepLCVaV	*C.frutescens x C. chinense*	Bhut Jolokia	?	-	Rai *et al.* (2014)
Cucumovirus	Cmr1	CMV-FNV	*C. annuum*	Bukang	P2	Movement epidermal cell to mesophyll	King *et al.* (2010)
	CmV.hb-4.1(2)/ CmV.hb-8.2(Q)	CMV-HB	*C. annuum*	BJ0747	P5(LG4)/ P11(LG8)	-	Yao *et al.* (2013)
	CmV11.1(Q)/ CmV13.1(Q)	CMV1 Israel	*C. annuum*	Perennial	P11/?	-	Ben *et al.* (2001)
	CmV4.1(Q)/ CmV6.1(Q)	-P4/P6	-	-	P4/P6	-	-
Tospo virus	Cac	CaCV	*C. chinense*	PI290972	?	-	Persley *et al.* (2010)
	1recessive	TNRV	*C. annuum*	PY-6423	?		Pungamalai *et al.* (2013)
				PY-6424			Pungamalai *et al.* (2013)
	Tsw	TSWW	*C. chinense*	PI152225; PI159236	P10(Pvr4, Pvr7)	HR-like localization NBS-LRR	Jahn *et al.* (2000)
Potyvirus	Both Pvr1[1] and Pvr1[2] alleles/ Pvr5	PVY	*C. annuum*	P4	-	-	Jeong *et al.* (2012)
			C. annuum				Parrella *et al.* (2002)

Virus Group	Gene Q=QTL	Virus	Donor Species	Donor Cultivar or Accession	Chromosomal Location	Type of Gene Action	References
	Pvr6 in combination with Pvr1[2]	PVMV	-	-	-	-	Caranta *et al.* (1996)
	Natural variants of Pvr6	Chi VMV, PVMV					Rubio *et al.* (2009)
	Single recessive gene Pvr3	PepMoV					Murphy *et al.* (1998)
	Dominant gene Pvr4	PVY, PepMoV					Janzac *et al.* (2009)
	Dominant gene Pvr7 tightly linked to Pvr4	PepMoV					Chaine *et al.* (1996)

Source: Lawrence *et al.*, 2014.

varieties 'A-127', 'C-1' and 'Puri Red' are the potential resistant source for mosaic complex (Potato virus Y, Tobacco mosaic virus, Tobacco etch virus, alpha alpha mosaic virus, and Potato virus V)

Insect Resistance

Table 2.117: Source for Insect Resistance

Specific Insect	Promising Resistant Accessions
Thrips	Caleapin Red, Chamatkar, NP-46A, BG-4, EC113937, LCA-348, EC15916, LEC-29, PBC-6, PBC-16 and LIC-64
Mite	LEC-1, Kalyanpur Red, Punjab Lal, LIC-13, LCA-253, LIC-36, EC12896
Aphid	Kalyanpur Red, X-1068

Nematode Resistance

'Myliddy 1' and 'Myliddy 2' are the potential resistant source for mites, thrips and aphids. Variety 'Pant C1' expressed resistant to root-knot nematode. Siomilarly, the breeding line 'White Kandari' was also found resistant to root-knot nematode.

Table 2.118: Source for Nematodes

Scientific Name	Resiatant Varieties/Genotypes
Meloidogyne arenaria	'G 2', 'Bombay 742'
Meloidogyne javanica	'Oakview Wonder', 'Santaka xs', No. 405B
Meloidogyne incognita	'Red chilli, 'Santaka xs'

Abiotic Resiatance

Natarajan *et al.* (1988) evaluated chilli genotypes under semi dry conditions at Paramakudi. Among the types, CA-42 was adjudged to be the best recording the highest yield of 82 g of dry fruits per plant with moderate fruit size. In this group CA-419 was also superior followed by CA-349 for fruit characters.

Table 2.119: Resistance Source for Abiotic Components in Chilli

1.	EC678809, EC678805	China	Resistance to heat and humidity
2.	EC678809	China	Heat and humidity resistance

Kameswari and Prasad (2005) evaluated genotypes of chilli for salt tolerance at Bapatla (Andhra Pradesh). Among chilli a variety, LCA-235 was found to be salt tolerant followed by LCA-64, and LCA-235 withstood the salinity up to 8EC. Variety LCA-334 was found to be sensitive and its plant height, leaf area and dry matter production was reduced by 46.3 per cent, 33.7 per cent and 58.7 per cent, respectively with increase in salinity from 0-8 dsm-1. Ten number of chilli cultivars were evaluated on the basis of their vegetative growth, flowering ehaviour, fruit yield and quality in the saline belt for consecutive two years during 2008-10. The experiment was designed in randomised block design with 10 treatments and 3 replications. Seedlings were spaced at 60 x 45 cm and grown with recommended cultural practices for saline belt of West Bengal. Result showed that in this agro-climatic situation, cv. Beladanga performed the best than others in both the years in relation to most of the characters like highest number of flowers and fruits per plant and highest projected yield of green and dry chilli. Regarding qualitative parameters, maximum capsaicin and ascorbic acid content in the fruit was also obtained in this cultivar. The cultivation of cv. Beladanga in this agro-climatic condition was also found most economical considering the Benefit: Cost ratio (4.03) (Karak *et al.*, 2015). Genotype LIC-7, LIC-9, EC12275 and EC12094 were found tolerant to excess moisture.

Quality Breeding

There are several genotypes which possesess the desirable quality attributes like fruit colour, pungency, taste, aroma/flavor *etc.*

In India some cultivars/genotypes has been identified with their specific purpose are as shown in Table 2.121.

Table 2.120: Quality Attributes in Chilli

Sl.No.	Accessions	Source	Salient Features
1.	EC678809	China	Fresh use, green, pungent
2.	EC673074	Taiwan	Cayenne type
3.	EC678805	China	Dark green, strong pungent
4.	EC678806	China	Fresh use, plant height 80-90 cm, light green, turn red and smooth at maturity
5.	EC678807	China	Fresh and dry use, green, high pungency
6.	EC678808	China	Fresh use, green smooth, high pungency
7.	EC678809	China	Fresh use, pungency
8.	EC468630	World vegetable centre (AVRDC), Taiwan	Cayenne type for bright red colour pepper powder, received by Nirma Agricultural Research Development Foundation, Pacchora (Maharashtra)
9.	EC501588 var. Posila type bajjo	World vegetable centre (AVRDC), Taiwan	High carotenoid content; received by ICAR-IIVR, Varanasi (U.P.); ICAR-NBPGR-RS, Hyderabad (Andhra Pradesh)

Table 2.121: Available Cultivars for Various Purposes

Sl.No.	Name of Variety	Name of Research Centre
A-Varieties suitable for export as dry chillies		
1.	G3, G4, G5, Kiran, X235	Lam (Andhra Pradesh)
2.	CO1, CO2	Coimbatore (T. Nadu)
3.	Sel.1	ICAR-IIHR, Bengaluru
4.	Sobour Anal, Angon	BAU, Sabour (Bihar)
5.	Muslabadi	MPKV, Rahuri (Maharashtra)
6.	Jawahar-218	JNKVV, Jabalpur (M.P.)
7.	Punjab Lal	PAU, Ludhiana (Punjab)
8.	Pant C-1, Kalyan Sel.-1	GBPUA&T, Pantnagar (Uttrakhand)
B- Varieties suitable for salad purpose		
1.	Andhra Jyoti (G-5)	Lam (Andhra Pradesh)
2.	RS-1	Rajasthan
3.	CO2	Coimbatore (T. Nadu)
C- Varieties suitable for capsaicin extraction		
1.	Yellow Pod Mutant (*C. frutescnes*) x 235, Aparna	Lam (Andhra Pradesh)
2.	Kalianpur Yellow	CSAUA&T, Kalianpur, Kanpur (U.P.)
3.	Punjab Lal	PAU, Ludhiana (Punjab)
D- Varieties suitable for colouring matter		
1.	LCA-206	Lam (Andhra Pradesh)
2.	Punjab Lal	PAU, Ludhiana (Punjab)
E- Varieties suitable for essence, preparation of chilli sauce and oleresin		
1.	X-235, Red Yellow (*C. frutescnes*)	Lam (Andhra Pradesh)
2.	Punjab Lal	PAU, Ludhiana (Punjab)
3.	Pant C-1	GBPUA&T, Pantnagar (Uttrakhand)
4.	Muslabadi	MPKV, Rahuri (Maharashtra)
F- Varieties suitable for green chilli production		
1.	X-235	Lam (Andhra Pradesh)
2.	Pusa Jwala, NP-46	ICAR-IARI, New Delhi
3.	Jawahar-218	JNKVV, Jabalpur (M.P.)

Similarly, some other observations recoreded are as shown in Table 2.122.

Table 2.122: Trait Specific Promising Accessions

Traits	Promising Accessions
Capsaicin (per cent)	GP276 (0.581), GP148 (0.571)
Oleoresis (per cent)	GP-89 (14.51), GP-82 (14.31)
Capsanthin (EOA)	GP-289 (43375), GP-132 (47672)

Genetic Resource Conservation

The germplasm is being maintained by ICAR-NBPGR, New Delhi, ICAR-Indian Institute of

Vegetable Research, Varanasi (Uttar Pradesh) and at several places of All India Coordinated Vegetable Improvement Project (AICRP-VC) particularly Lam (Guntur, Andhra Pradesh), Kovilpatti (Tamil Nadu) and Kalyanpur (Kapur, Uttar Pradesh). Rich diversity can be collected from different part of the country because several local genotypes are being grown by the farmers (Nath *et al.*, 2008). Most important strains have been reported from various parts of the country where it is intensively grown for dry as well as green chillies like 'Bhoot Jolokia' in Manipur and other north eastern regions. Similarly 'Byadgi' of Karnataka and 'Sankeshwar' of Maharashtra are also popular variety. In Andhra Pradesh several local cultivars like Nallapadu, Warangal, Lanka Chilli, Seema Mirapa, Golcanda Mirchi *etc.* are dominating in several pockets. Certainly different states have different cultivars that are adapted to local conditions (Swarup, 2006). The important areas of growing chillies are Andhra Pradesh, Maharashtra, Karnataka Tamil Nadu, Madhya Pradesh, Rajasthan, Bihar, Uttar Pradesh and Punjab.

REFERENCES

1. Anonymous. 2006. Annual Report 2005-2006. ICAR-Indian Institute of Horticultural Research, Bengaluru (Karnataka), p. 7-8.
2. Ares, J.l.A., Martinz, A.R. and Paz, J.F. 2005. Resistance of pepper germplasm to *Phytophthora capsici* isolates collected in northwest Spain. *Spn. J. Agric. Res.*, **3:** 429-436.
3. Barrios, E.P. Mosokar, H.I. and Black, L.L. 1971. Inheritance of resistance to tobacco etch and cucumber mosaic virus in *Capsicum frutescens. Phytopathology*, **61(10):** 1318
4. Bartz, J.A. and Stall, W.M. 1974. *Phytopathology*, **64(10):** 1290-1293.
5. Boswell, V.R. 1951. Garden peppers both a vegetable and condiment. *Natl. Geogr. Mag.* **96:** 166-167.
6. Ben, C.A., Grube, R., Lapidot, M., Jahn, M. and Paran, L. 2000. Identification of quantitative trait loci associated with resistance to cucumber mosaic virus in *Capsicum annuum* L. *Theor. Appl. Genet.*, **102:** 1213-1220.
7. Caranta, C., Palloix, A., GebreSelassie, K., Lefebvre, V., Moury, B. and Daubeze, A.M. 1996. A complementation of two genes originating from susceptible *Capsicum annuum* lines confers a new and complete resistance to pepper venal mottle virus. *Phytopathology*, **86:** 739-743.
8. Chaine, D.C., Palloix, A., Daubze, A.M., Merchoux, G., Selassie, K.G. and Pochard, E. 1996. Genetic analysis of broad spectrum resistance to Potyviruses using double haploid lines of pepper (*Capsicum annuum* L.). *Euphytica*, **88:** 231-239.
9. Chandra, A. and Pandey, R.P. 2001. Jawahar Mirch-283: A summer-set chilli. *Indian Hort.*, **46(1):** 20-21.
10. Chauhan, D.V.S. 1981. Vegetable Production in India. Ram Prasad and Sons, Agra (U.P.).
11. Cheema, D.S., Singh, D.P., Rawat, R.D. and Deshpande, A.A. 1984. Inheritance of resistance to anthracnose disease in chillies. *Capasicum Eggplant News*, **3:** 44.
12. Cheema, G.S. 1982. Studies on improvement in chillies for resistance to Anthracnose and Cercospora leaf spot diseases. *Ph. D. Thesis* submitted to UAS, Bengaluru (Karnataka).
13. Cheema, G.S. 1985. Studies on improvement in chillies for resistance to anthracnose and Cercospora leaf spot diseases. *Mysore J. Agric. Sci.*, **17(4):** 388-389.
14. Cheema, G.S., Nazareth, B. and Dhaneshwar, S.R. 1994. Improvement of chilli (*Capsicum annuum* L.) by selection in the Bombay province. *Indian J. Hort.*, **11(2):** 49-61.
15. Cooka, A.A. and Stall, R.F. 1963. Inheritance of resistance in pepper to bacterial spot. *Phytopathology*, **53:** 1060-1062.
16. De Candolle, A. 1986. Origin of Cultivated Plants. Hafner Publishing Co., New York.
17. Djian-Caporalino, C., Lefebvre, V., Sage-Daube'ze, A.M. and Palloix, A.2007. *Capsicum*. In: Singh RJ (ed). Genetic resources, chromosome engineering and crop improvement series, Vol.3, Vegetable crops. CRC Press, Florida (USA), p. 185–243.
18. Deshpande, A.A., Anand, N., Pathak, C.S. and Sridhar, T.S. 1985. New sources of powdery mildew resistance in *Capsicum* species. *Capsicum Newsl.*, **4:** 75-76.
19. Deshpande, A.A., Pathak, C.S. and Singh, D.P. 1983. Variants in chilli pepper (*Capsicum annuum* L.). *Capsicum Newsl.*, **2:** 81.
20. Deshpande, R.B. 1933. Studies in chillies III. The inheritance of some characters in *Capsicum annuum* L. *Indian J. Agric. Sci.*, **3:** 219-300.
21. Deshpande, R.B. 1944. Inheritance of bunchy habit in chilli. *Indian J. Genet.*, **4(1):** 54.
22. García-Neria M.A. and Rivera-Bustamante R. 2011. Characterization of geminivirus resistance in an accession of *Capsicum chinense Jacq. Mol. Plant. Microbe Interact.*, **24:** 172–182.
23. Gopalakrishnan, T.R. 1985. 'Inheritance of clusterness, destalkness and depend colour in chilli'. Ph.D. Thesis, Kerala Agricultural Universitry, Vellanikkara (Kerala). Goud, J.V., Nayar, K.M.D. and Rao, M.G. 1970. Interspecific hybridization in chillies (*Capsicum annuum* x *Capsicum frutescens*). *Mysore Journal of Agricultural Sciences*, **4:** 358-361.

24. Indira, P. 1997. Diversity interrelationships among *Capsicum* spp. and forms and development of paprika. Ph. D. thesis, Kerala Agricultural University, Vellanikkara, Thrissur (Kerala).

25. Irish, H.C. 1989. A revision of the genus *Capsicum* with special reference to garden varieties. *Report, of Missouri Botanical Garden,* **9:** 53-110.

26. Jagadish, C.A., Chandrasekharan, P. and Ponnaiya, B.W.X. 1964. Studies on the genus *Capsicum*. *Madras Agric. J.*, **51(8):** 36.

27. Jahn, M., Paran, L., Hoffmann, K., Radwanski, E.R., Livingstone, K.D., Grube, R.C., Afetrgoot, E., Lapidot, M. and Moyer, J.W. 2000. Genetic mapping of the Tsw locus for resistance to the tospovirus tomato spotted wilt virus in *Capsicum* spp. and its relationship to the Sw-5 gene for resistance to the same pathogen in tomato. *Mol. Plant Microbe Interactions*, **13:** 673-682.

28. Janzac, B., Faber, M.F., Palloix, A. and Moury, B. 2009. Phenotype and spectrum of action of the Pvr y resistance in pepper against *Potyvirus* and selection for virulent variants. *Plant Pathology*, **58(3):** 443-449.

29. Jeong, H.J., Kwon, J.K., Pandeya, D., Hwong, J., Hoong, N., and Bae, J.H. 2012. A survey of natural and ethyl methane sulphate induced variation of elF4E using high resolution melting analysis in *Capsicum*. *Mol. Breed.*, **29(2):** 349-360.

30. Jeswani, L.M., Deshpande, R.B. and Joshi, A.B. 1956. Inheritance of some fruit characters in chilli. *Indian J. Genet.*, **16(2):** 138-143.

31. Kameswari, P.L. and Prasad, D.M. 2005. Evaluation of chilli genotypes for salt tolerance. *Indian J. Hort.*, **62**: 46-48.

32. Karak, P.K., Pariari, A. and Karak, C. 23015. Varietal evaluation of chilli in the saline belt of West Bengal. *Journal Crop and Weed*, **11**(Special Issue): 86-89.

33. Kim, S.H., Yoon, J.B. and Park, H.G. 2008b. Inheritance of anthracnose resistance in new genetic resources, *Capsicum baccatum* PI594137. *J. Crop. Sci. Biotechnol.*, 11: 13-16.

34. Kim, S.H., Yoon, J.B., Do, J.W. and Park, H.G. 2008a. A major recessive gene associated anthracnose resistance to *Colletotrichum capsici* in chilli pepper (*Capsicum annuum*). *Breed. Sci.*, **58:** 137-141.

35. Kim, S., Kim, K.T., Kim, D.H., Yang, E.Y., Chao, M.C., Jamal, A., Cahe, Y., Pae, D.H., Oh, D.G. and Hwang, J.K. 2010. Identification of QTL associated with anthracnose resistance in chilli pepper (*Capsicum* spp.). *Korean J. Hort. Sci. Technol.*, **28:** 1014-1024.

36. King, S.R., Davis, A.R., Zhang, X. and Croshy, K. 2010. Genetics, breeding and selection of rootstocks for Solanaceae and Cucurbitaceae. *Sci. Hort.*, **127:** 106-111.

37. Krishanakumary, K. 1984. Compatibilty among varieties in *C. annuum* and *C. frutescens*. M.Sc. thesis, Kerala Agricultural University, Vellanikkara, Thrissur (Kerala).

38. Kumari, K. and Peter, K.V. 1986. Compatibility among varieties of *Capsicum annuum* L. and *Capsicum frutescens*. *Indian Journal of Genetics and Plant Breeding,* **46(3):** 531-535.

39. Kumar, P.T. 1990. Interspecific hybridization in *Capsicum*. M.Sc. Thesis Kerala Agricultural Universitry, Vellanikkara (Kerala).

40. Lawrence, D.P., Gannibal, P.B., Dugan, F.M., Pryor, B.M. 2014. Characterization of *Alternaria* isolates from the infectoria species-group and a new taxon from *Arrhenatherum, Pseudoalternaria arrhenatheria* sp. *Mycol. Prog.*, **13:** 257–276.

41. Lee, J., Hong, H., Do, J., Yoon, J., Lee, J.D., Hong, J.H., Do, J.W. and Yoon, J.B. 2010. Identification of QTLs for resistance to anthracnose to two *Colletotrichum* species in pepper. *J. Crop Sci. Biotechnol.*, **13:** 227-233.

42. Lin, S.W., Gniffke, P.A. and Wang, T.C. 2007. Inheritance of resistance for pepper anthracnose caused by *Colletotrichum acutatum*. *Acta Hortic.*, **760:** 329-334.

43. Lin, Q., Kanchana-udomkam, C., Jaunet, T., Mongkolporn, O., Lin, Q., Chutchamas, K.U. and Orarat, M. 2002. Genetic analysis of resistance caused by *Colletotrichum capsici*. *Thai. J. Agric. Sci.*, **35:** 259-264.

44. Mahasuk, P., Khumpeng, N., Waree, S., Taylor, P.W.J. and Mongkolporn, O. (2009a). Inheritance of resistance to anthracnose at seedling and green and red chilli fruit stages. *Plant Breed.*, **128:** 701-706.

45. Mahasuk, P., Taylor, P.W.J. and Mongkolporn, O. (2009b). Identification of two new genes conferring resistance to *Colletotrichum acutatum* in *Capsicum baccatum*. *Phytopathology*, **99:** 100-1104.

46. Markose, B.L. 1996. Genetic and biochemical basis of resistance to bacterial wilt in chilli. Ph. D. thesis, Kerala Agricultural University, Vellanikkara, Thrissur (Kerala).

47. Mc Leod, M.J. 1978. A systematic and evolutionary study of genus *Capsicum*. *Dissertation Abstract International B.*, **38(9):** 82-84.

48. Mini, C. 1989. Evaluation of ornamental chillies and selection for dwarf clustered plant types. Ph. D. thesis, Kerala Agricultural University, Vellanikkara, Thrissur (Kerala).

49. Mouri, B. and Verdin, E. 2012. Viruses of pepper in the Mediterranean basins: A remarkable stasis. *Adv. Virus Res.*, **89:** 127-162.

50. Murphy, J.F., Blauth, J.R., Livingstone, K.D., Lackney, V.K. and John, M.K.1994. Genetic mapping of pvr1 locus in *Capsicum* spp. and evidence that distinct Potyvirus resistance loci control response that differs at the whole plant cellular levels. *Mol. Plant Microbe Interactions*, **11:** 943-951.

51. Nagaraju, M.M. and Sreelathakumary, I. 2016. Combining ability analysis in chilli (*Capsicum annuum* L.) for yield and yield attributing traits (Abs: 59). 7th Indian Horticulture Congress, held at ICAR-IARI, New Delhi, p. 46-47.

52. Nagaraju, M.M. and Sreelathakumary, I. 2016. Heterosis studies for yield and quality characters in chilli (*Capsicum annuum* L.) (Abs: 60). 7th Indian Horticulture Congress, held at ICAR-IARI, New Delhi, p. 47-48.

53. Nair, K.M.D., Rao, M.G. and Reddy, B.G.S. 1979. New interspecific chilli strains of Karnataka. *Current Research*, **8(9):** 158-160.

54. Natarajan, S., Papaiah, C.M., Rangaswamy, P. and David P.V.N., 1988, Evaluation of chilli genotypes under semi-dry condition. *South Indian Hort.*, **36 (1):** 8-12.

55. Pakeleevaraporn, P., Wasee, S., Taylor, P.W.J. and Mongkolporn, D. 2005. Inheritance of resistance to anthracnose caused by *Colletotrichum capsici* in *Capsicum. Plant Breed.*, **124:** 206-208.

56. Pal, B.P. 1945. Studies in hybrid vigour II. Notes on the manifestation of hybrid vigour in gram, sesamum and chillies. *India J. Genet.*, **5:** 106-121.

57. Pal, B.P.; Ramanujam, S. and Raju, A.B. 1941. Colchicine induced polyploidy in crop plants-chilli. *Indian J. Genet.*, **1:** 128-139.

58. Park, H.K., Kim, B.S. and Lee, W.S. 1990a. Inheritance of resistance to anthracnose (*Colletotrichum* spp.) in pepper (*Capsicum annuum* L.).I. Genetic analysis of anthracnose resistance by diallel crosses. *J. Kor. Soc. Hortic. Sci.*, **31:** 91-105.

59. Park, H.K., Kim, B.S. and Lee, W.S. 1990b. Inheritance of resistance to anthracnose (*Colletotrichum* spp.) in pepper (*Capsicum annuum* L.).I. Genetic analysis of resistance to *Colletotrichum dematium. J. Kor. Soc. Hortic. Sci.*, **31:** 207-201.

60. Park, K.S. and Kin, C.H. 1992. Identification, distribution and etiological characteristics of anthracnose fungi of red pepper in Korea. *Korean J. Plant Pathol.*, **8:** 61-69.

61. Parrella, G., Ruffel, S., Moretti, A, Morel. C., Palloix, A. and Caranta, C. 2002. Recessive resistance genes against Potyviruses are localized in collinear genomic regions of the tomato (*Lycopersicon* spp.) and pepper (*Capsicum* spp.) genomes. *Theor. Appl. Genet.*, **105:** 855-856.

62. Parthasarthy, N and Kedarnath, S. 1950. Varietal differences in the breeding behavior of colchine induced auto-tetraplois of chilli. *Indian J. Genet.*, **10 (1):** 14-20.

63. Pathak, C.S., Singh, D.P. and Deshpande, A.A. 1983a. Male and female sterility in chilli pepper (*Capsicum annuum* L.). *Capsicum Newsl.*, **2:** 102.

64. Pathak, C.S., Singh, D.P. and Deshpande, A.A. 1983b. Close flower mutant in *Capsicum annuum* L. *Capsicum Newsl.*, **2:** 106.

65. Pernezny, K., Robert, P.D., Murphy, J.F., Goldberg, N.P. 2003. Compendium of pepper diseases. *The American Phytopathology Society Minnesto*, 24-25.

66. Persley, D.M., McGroth, D., Sharman, M. and Walker, I.O. 2010. Breeding for tospo virus resistance to capsicum and tomato in Australia. IXth International symposium on Thysanoptera and Tospo viruses, 31st August-4th Sept. 2009., *Journal of Insect Science*, **10:** 166.

67. Peter, K.V. 1998. Genetics and Breeding of Vegetables. Indian Council of Agricultural Research, New Delhi.

68. Pickersgill, B. 1980. Some aspects of interspecific hybridization in *Capsicum. Fourth Meeting of the EUCARPIA Capsicum* Working Group, held during 14-16 Oct., 1980 at Wageningen, the Netherlands.

69. Pious, T. 1985. 'Transfer of clusterness to bell peppers (*Capsicum annuum* L. var. *grossum* Sendt.). M.Sc. Thesis Kerala Agricultural Universitry, Vellanikkara (Kerala).

70. Pious, T. and Peter, K.V. 1986. Inheritance of clustering in chilli (*Capsicum annuum* L.) *Indian J. Genet. and Pl. Breed.*, **46(2):** 311-314.

71. Pious, T. and Peter, K.V. 1988. Heterosis in intervarietal crosses of bell pepper (*Capsicum annuum* L. var. *grossum*) and hot chilli (*C. annuum* var. *fasciculatum*). *Indian J. Agri. Sci.*, **58(10):** 747-750.

72. Pueseglove, J.W. 1977. Tropical Crops: Dicotledones. Longman, London.

73. Pungamalai, P., Potapohn, N., Akarapisarn, A. and Pascho, H.J. 2013. Inheritance of tomato necrotic ring virus resistance in *Capsicum annuum* L. *J. Agril. Sci.*, **5(2):** 129-133.

74. Pungamalai, P., Potapohn, N., Akarapisarn, A., Cheewa C.S. and Insuan, N. 2013. Screening capsicum accession for tomato necrotic ringspot virus resistance Cmv. *J. Nat. Sci.*, **12(1):** 35.

75. Radhakrishnan, K.P., Mercy, S.T. and George, M.K. 1977. Crossability studies and analysis of incompatibility in three species of *Capsicum*. *Agricultural Research Journal, Kerala*. **15:** 124-127.

76. Rai, V.P., Kumar, R., Singh, S.P., Kumar, S., Singh, M. and Rai, M. Monogenic recessive resistance to pepper leaf curl virus in an interspecifc cross of capsicum. *Scientia Horticultuae*, **172:** 34-38.

77. Raju Narsimaha, D.N. 1979. Correlation and path coefficient analysis in capsicum (*C. annuum* L. var. *grossum*). *M.Sc Thesis*, submitted to UAS, Bengaluru (Karnataka).

78. Reddy, K.M. and Reddy, M.K. 2010. Breeding for virus resistance. In: Kumar, R., Rai, A.B., Rai, M. and Singh, H.P. (eds.). Advances in chilli research, Studium Press, New Delhi, India, p. 119-132.

79. Reddy, M.K., Srivastava, A., Kumar, S., Kumar, R., Chawda, N., Ebert, A.W. and Vishwakarma, M.N 2014. Chilli (*Capsicum annuum* L.) breeding in India: An overview. *SABRAO Journal of Breeding and Genetics*, **46(2):** 160-173.

80. Rubio, M., Nicolai, M, Caranta, C. and Palloix, A. 2009. Allele mining in the pepper gene pool provided new complementation effects between pvr2-elF4E and pvr6-elF (ISO) 4E alleles for resistance to pepper venal mottle virus. *Journal of General Virology*, **90:** 2808-2814.

81. Sarath Babu, B., Pandravada, S., Prasad, R.R., Antha K., Chakrabarthy, S. and Varaprasad, K. 2011. Global sources of pepper genetic resources against arthropods, nematodes and pathogens. *Crop Port.*, **30:** 389-400.

82. Saxena, A., Hundal, J.S. and Dhall, R.K. 2003. Floral biology studies in chilli (*C. annuum* L.). *Veg. Sci.*, **30(2):** 173-175.

83. Singh, A. and Singh, H.N. 1978. Line x tester analysis of yield in chilli. *Indian J. Genet.*, **38(1):** 52-56.

84. Singh, A., Singh, H.N. and Bajpayee, N.K. 1977. A new strain of chilli with loose calyx. *Indian Hort.*, **22(4):** 13.

85. Singh, N.B. and Singh, B. 1970. Interrelationship, heritability estimate and genetic advance in yield and other characters in chillies. *Madras Agric. J.*, **57(7):** 369-373.

86. Singh, S.P. 1989. Production Technology of Vegetable Crops. Agricultural Research Communication Centre, Karnal, Haryana (India.).

87. Smith, P.G. and Lippert, L.F. 1960. Plant breeding of commercial peppers for disease resistance. *Calif. Agric.*, **14(11):** 13.

88. Smith, P.G. and Heisser, C.B. 1957. Breeding behavior of cultivated peppers. *Proceedings of the American Society of Horticultural Science*, **70:** 286-290.

89. Smith, P.G., Kimble, K.A., Grogan, R.G. and Millet, A.H. 1967. Inheritance of resistance in peppers to *Phytophthora* root rot. *Phytopathology*, **57:** 377-379.

90. Smith, P.G., Villalon, B. and Villa, P.L. 1987. Horticultural classification of peppers grown in the United States. *Hort. Sci.*, **22:** 11-13.

91. Sontakke, M.B. 1981. Genetical studies in chillies. *Ph.D. Thesis* submitted to UAS, Bengaluru (Karnataka).

92. Sundaresan, N. and Chandrasekaran, P. 1979. The cytomorphology of an interspecific hybrid in *Capsicum*. *Madras Agricultural Journal*, **66(8):** 495-502.

93. Swarup, V. 2006. Vegetable Science and Techonology in India. Kalyani Publishers, New Delhi, p. 656.

94. Thompson, H.C. and Kelly, W.C. 1957. Vegetable Crops. Tata McGraw-Hill Publishing Co., Ltd., New Delhi.

95. Usha Rani, P. 1981.Varietal difference in chillies for pungency, pigmentation and ascorbic acid contents. M.Sc. Thesis submitted to UAS, Bengaluru (Karnataka).

96. Verma, S.K. and Pant, K.C. 1999. Evaluation and utilization of chilli (*Capsicum* sp.) germplasm in Uttar Pradesh hills. *Haryana J. Hort. Sci.*, **28(1 and 2):** 80-81.

97. Verma, S.K., Pant, K.C. and Muneem, K.C. 1998. Evaluation of chillies germplasm. *Indian J. Plant Genet. Resour.*, **11(2):** 237-239.

98. Verma, V.K., Pandey, A., Ryntathiang, I. and Jha, A.K. 2016. Genetic diversity in popular land races over cultivars of *Capasicum* spp. grown in North Eastern Region of India based on morphological and molecular analysis. 1st International Agrobiodiversity congress (Abs. Session Trait Discovery and Enhanced Use of PGR), 6-9 November, held at New Delhi, p. 176.

99. Voorrips, R.E., Finkers, R., Sanjaya, L., Groenwald, R. 2004. QTL mapping of anthracnose (*Colletotrichum* spp.) resistance in a cross between *Capsicum annuum* and *C. chinense*. *Theor. Appl. Genet.*, **109:** 1275-1282.

100. Wang, M. and Ma, D.H. 1987. Study of analytical electrophoresis on *Capsicum* germplasm. *Capsicum Newsl.*, **6:** 25.

101. Yao, M., Li, N., Wang, F. and Ye, Z. 2013. Genetic analysis and identification of QTLs for resistance to cucumber mosaic virus in chilli pepper (*Capsicum annuum* L.). *Euphytica*, **193:** 135-145.

102. Yoon, J.B. and Park, H.G. 2005. Tri-species bridge crosses (*Capsicum annuum* x *C. chinense*) x *C. baccatum* as an alternative for introgression of anthracnose resistance from *C. baccatum* into *C. annum*. *J. Kore. Sco. Hortic. Sci.*, **46:** 5-9.

Sweet Pepper/Bell Pepper (*Capsicum annuun* L. var. *grossum* Sendt.)

Introduction

Sweet paper (*Capsicum annuun* var. *grossum*, 2n=24) is prized and palatable vegetable in the family Solonaceae which is most widely cultivated species of peppers in the world for its green and coloured fruits. The word '*Capsicum*' originated from the Greek word 'kapso' means to bite. In other version, it originated from Latin '*Capsa*' meaning box, which refers to the shape of fruit. *Capsicum* terminology is confusing still today. *Capsicum* is known by various terms such as pepper, chili, chile, chilli, aji, paprika, Capsicum, Sweet pepper, Bell pepper, Simla Mirch or Sagiya Mirch are used interchangeably for plants in the genus *Capsicum*. It is normally considerd as a luxury vegetable, consumption being greater in and around the cities (Singh *et al.*, 1993). Sweet pepper fruits are generally blocky, square, rectangular, thick fleshed, 3-4 lobed and are non-pungent. Fruits are cooked as vegetable, pickled in brine, processed (baking and stuffing) or eaten raw as salad or at ripened stages because of its special flavor, aroma, colour and important sources of Vitamin A and Vitamin C. Diced green or red capsicum are some times mixed with sweet corn or other vegetables. High quality coloured bell peppers can be produced year round and especially during the high-priced off season, in protected environments such as high passively ventilated greenhouses. Fruits can be harvested with red, orange, or yellow colour depending on the pepper cultivar used.

Origin, History and Distribution

Capsicum is originated in the new world tropics and subtropics (Greanleaf, 1985). *Capsicum* species are thought to be native to Central and South America. The most cultivated species among the *Capsicum* species was domesticated in Mexico. It is believed that Columbus first introduced pungent *Capsicum* to Europe after he returned from the new World in 16th century and the sprtead elsewhere took place from Portugal (Seshadri and Chaterjee, 1996). During the Spanish conquest of Mexico, jalapenos were used as barter and to pay taxes. In Europe, peppers acquired immediate popularity as a spice, both in fresh and dehydrated forms and were often used as a substitute for black pepper. The Portuguese brought *Capsicum* from Brazil to India during 16th or 17th century (Somas, 1984) and establishes by British in Himachal Pradesh and Tamil Nadu (Neelgiris). As per historical views capsicum/bell pepper/sweet pepper is a recent introduction into India from USA, Italy, Brazil, Hungary, Bulgaria, Czechoslovakia and USSR. The sweet pepper was popular in Spain and was referred to as "pimientos" to descvribe the red colouration and sweet taste. Being a cool season crop, in India it is grown commercially in the hills of Uttarakhand, some parts of Uttar Pradesh, Himachal Pradesh, J & K and Nilgiri (Tamil Nadu) during summer months and occupied more area in Maharashtra, Karnataka during autumn winter season. It is also popular in some part of Tamil Nadu, Andhra Pradesh, West Bengal and Madhya Pradesh during autumn winter season.

Species Availability, Evolution and Diversity

There are 5 domestcated species of Capsicum, *viz. Capsicum annuum, Capsicum frutescens, Capsicum baccatum, Capsicum chinense* and *Capsicum pubescens,* which are clearly distinct by taxonomic criterion with a species complex. The bellpepper is highly polymorphic genus with much morphological genetic diversity both intra-specifically and inter-specifically. Intra-species and inter-species improvement is possible to evolve better types. Since, *Capsicum annuum* lacks the variability; hence studies on variability have been very few.

Botanical Description

Both hot and sweet peppers have been originated from *C. annuum*. The *C. annuum* is distinguished by

Table 2.123: Morphological Characteristics of Domesticated *Capsicum* Species and their Wild Relatives

Species	*Wild/Cultivated*	*Calyx Teeth*	*Corolla*	*Flowers/Node*	*Seeds*
C. annuum annuum	Cultivated	Present	White large	1	Yellow smooth
C. annuum var. *aviculare*	Wild progenitor	Absent	White small	1(3-A)	Yellow smooth
C. baccatum var. *pendulum*	Cultivated	Present	White with yellow/ green spot	1	Yellow smooth
C. baccatum var. *baccatum*	Wild progenitor	Present	White with yellow/ green spot	2-3	Yellow smooth
C. chinense	Cultivated	Absent	Dull-white	2-3	Yellow smooth
C. frutescens	Cultivated	Present	Greenish-white	2-3	Yellow smooth
C. pubescens	Cultivated	Absent	Purple	1(2-3)	Rough
C. cardenasii	Wild progenitor	Absent	Purple campanulate	2-3	Yellow smooth
C. eximium	Wild progenitor	Absent	Magenta to white with yellow/green spots	2-3	Yellow smooth

Source: Deshpande (2001).

white flowers (rarely purple), anthers with anthocyanin, calyx with distinct points or teeth and longitudinal ridges. Flowers one/node (occasionally 2 at the first flowering node), erect in the wild form and pendent in many types. Seeds are tan, waxy, pale greenish yellow flowers, anthers with anthocyanin, calyx smooth without longitudinal ridges, characterize the *C. frutescens*. Flowers 2-5/node (usually 2), held stiffly erect with unopened flowers at an angle to peduncle. Cultivated types have less variability than *C. annuum*, but may have slender fruit up to 15 cm in length. Non-pungent forms are not known. *C. frutescens* requires a tropical climate, hence restricted in habitat. In the tropics, peppers are perennials, but they are grown in sub-tropical and temperate zones as annuals. Peppers are non-climacteric which means they do not produce ethylene. They need to stay on the vine to continue the ripening process.

Table 2.124: Botanical Description

Parts	*Description*
Growth Habit	Peppers grown in temperate regions are herbaceous annual but short lived herbaceous perennials where temperatures do not drop below freezing. Pepper plants growth habit may be prostrate, compact, or erect and attain the height of 1.5m.
Root	It is deep tap root system having lot of thin and fibrous lateral roots. The spindle shaped roots will develop into fibrous secondary root systems spreading laterally and downward 91.44-121.92 cm.
Stem	The main axis develops dichotomous branching which is racemose at the bottom and cymose on the top. Side shoots are either absent or poorly develop. The length of the internodes reduces as it develops towards the growing tip and on the topmost region growth is restricted when several fruits are present. The stem is soft at early stages; later on it becomes hard and green in colour.
Leaves	The leaves of the pepper plant arise singly and are simple, entire and asymmetrical. Typical of all Solanaceous plants the leaves are arranged alternately on the stem. They are simple and alternate, shiny and glabrous and vary in shape from broadly ovate to ovate lanceolate, cauline and ramal. Generally exstipullate, petiolate, alternate, entire, reticulate, and unicostate venation. Leaves of the wild types are generally smaller and often quite pubescent.
Floral biology	The flowers are born singly (solitary) or in doubles or triplets continuously as the upper structure of the plant profliferates. The flowers are bisexual attached to the shoot by the peduncles. They arise from the leaf axils and branches, generally in the upper canopy as the plant grows. The corolla is white and 5 lobed; the anthers are 5-6, short, elliptic in shape, and yellow or bluish in colour. The plant flowers have an open anther formation and will indefinitely self pollinate. The pistil consists of ovary, style and stigma. The ovary is round or slightly elliptical and funnel shaped.
Anthesis and pollination	In sweet peppers, anthesis starts at 7.15 a.m. and continues upto 11.15 a.m., with the peack at 7.15 a.m. (Vijay *et al.*, 1979). Anthers dehisce after 30 minutes of antheis. The stigma is receptive up to 2 days after anthesis.
Fruit set and Fruit	Fruit set was maximum on the day of anthesis with around 80 per cent set and pollen fertility around 96 per cent. Fruit set in flower buds pollinated a day prior to anthesis has been observed to be good. Peppers are non-climacteric which means they do not produce ethylene. They need to stay on the vine to continue the ripening process. The fruit (large 3.5" x 2.4") of a pepper plant is classified as an inflated berry with colours ranging from green, yellow, red, purple, black, brown, white and orange. Green is an immature fruit, yet commonly eaten this way, and as the fruit matures it changes colour. Colour changes in most commercial cultivars are green to red, green to yellw or green to orange. Usually, fruits of the purple and white varieties have these colours as they develop, and therefore do not have a green stage. Finally, the fruit turns red and begins to soften. For fruit to set, the ovaries need to be fertilized. Fruits may be borne uproight, drooping or semi-erect. Fruit apex may be lobed, pointed, obtuse, or blunt depending on the cultivar. Auxin is then produced by the seeds, which determine fruit cell elongation. The number of seeds fertilized will determine the size and shape of the fruit.
Seeds	The seed develop on the interior and attach to the veins. Fully, developed seed is kidney shaped, 3-4 mm in diameter and 05-1.0 mm thick, with a rim-like thickening on the edge (Somos, 1984). Seeds are located inside the fruit on the central placenta. The weight of 1000 seeds ranges from 5-10 g. The seed is of straw colour. Temperature requirements for proper germination are higher for hot peppers, as compared to sweet peppers.

Botanical Classification

Peppers are classified into various horticultural groups or types

Classification Based on their Fruit Size and Shape

The major groups are:

(i) **Bell group-** Fruits are large, blocky and blunt. Fruit colour is green when immature and red when mature. Both green and red fruits are sold fresh.

(ii) **Anaheim Chile group-**Ancheim Chile groupbears smooth fruits, tapering to a point with medium thick flesh.

(iii) **Jalapeno group-** Fruits are small with a round, cylindrical shape, smooth and with or without a corky network on the skin. Fruit colour is green when immature and red when mature.

(iv) **Cherry group**-Fruits are small and round in shape, yellow in colour when immature. Pungent and non-pungent are sold fresh or pickled.

(v) **Tabasco group**-Fruits are quite small and slender, green when immature and turn red at maturity. Fruits are most pungent, immature fruits are pickled, while mature red fruits are used in hot pepper sauces.

Classification Based on their Growth Habit

On the basis of plant growth habit peppers can be classified in two groups:

(i) Indeterminate growth habit

(ii) Determinate growth habit

(i) Indeterminate Growth Habit

Green house production of peppers is based on indeterminate cultivars in which the plants continually develop and grow from new meristems that produce new stems, leaves, flowers and fruit. Indeterminate cultivars require constant pruning to manage their growth. In order to optimize yield, a balance between vegetative (leaves and stems) and generative (flowers and fruit) growth must be established and maintained.

(ii) Determinate Growth Habit

Field pepper cultivars are determinate; the plant grows to a certain size, produces fruit and stops growing and eventually dies.

Classification Based on the Basis of Fruit Colour

1. Red
2. Orange
3. Yellow colour
4. Dark blue/blackish

Genetic Resource Augmentation and Crop Improvement

The modern pepper breeders owe a lot to the natives who created wide variability within this horticultural group. Definetly, the developmental programme is thought to have occurred during the last 4000 years (Pickersgill, 1969) for genetic resource augmentation and crop improvement. The following are some of the more commonly used methods in bell pepper breeding programme:

- ☆ Plant Introduction
- ☆ Pure line breeding
- ☆ Pedigree selection
- ☆ Backcross breeding
- ☆ Heterosis breeding
- ☆ Use of male sterility

Introduction and Selection

Capsicum introduced into India during British days and their cultivation developed with preference towards green blocky types than to elephant trunk types. Cooler regions like Bengaluru, Hyderabad, Poona, Ranchi, Solan, Kashmir Valley *etc.*, are now growing sweet pepper. Most of the cultivars adopted by the growers are introduction from foreign countries. Some important introduced varieties and hybrids are 'California Wonder', 'Yolo Wonder', 'King of North', 'Early Giant', 'World Beater', 'Neapolian', 'Early Calwonder', 'Windsor A', 'Harris Early Giant', 'Vinedale', 'Bullnose', 'Russian-449', 'Chinese Giant', 'Ruby King', Tara, California, Oshkosh, Hungarian Wax, Elephant's Trunk, R-449 (Russian introduction), Golden Bell, Domino *etc.* (Thomas *et al.*, 1983). The promising lines of bell pepper are EC393214-2, EC203583, SH-SP-1007, SH-SP-1004, Capsicum-2, Midway and EC143570. Some useful selections like Arka Basant, Arka Mohini, Arka Gaurav, Bharat *etc.* are being cultivate largely in Indian

Table 2.125: Morphological and Quality Attributes in Capsicum

Genotypes	*Fruit Length (cm)*	*Fruit Width (cm)*	*Fruit Weight (g)*	*Fruits/Plant*	*Yield/Plant (kg)*	*Ascorbic Acid (mg/100 g)*	*Total Chlorophyll (kg/100 g)*
Feroz	7.6	17.1	53.0	20.0	0.950	56.60	74.82
SP-24	8.2	23.0	86.7	24.	1.580	444.94	54.28
DARL-70	9.6	20.6	71.7	18.0	1.470	99.87	23.73
DARL-71	9.6	25.2	103.0	14.0	1.220	83.63	40.03
EC579997	8.8	13.7	35.0	39.0	1.170	128.91	44.65
SP-701	9.1	16.6	56.6	33.0	1.380	77.38	36.95
SP-19	9.6	17.0	68.3	33.0	1.520	113.87	32.57
SP-6	10.1	21.1	88.0	20.0	1.150	104.71	52.14
SP-7	8.3	15.8	50.0	38.0	1.640	103.89	66.49
Dark Green	7.2	16.3	43.3	42.0	1.570	81.16	54.14
Capsicum Violet	5.0	15.3	30.0	28.0	0.790	58.81	50.06
California Wonder	8.6	17.1	60.0	32.0	1.62.	126.60	129.23
SEM	**0.105**	**0.649**	**4.947**	**1.076**	**0.098**	**7.644**	**0.585**
CD at 5 per cent	**0.309**	**1.905**	**14.509**	**3.156**	**0.287**	**22.418**	**0.172**
CD at 1 per cent	**0.421**	**2.589**	**19.720**	**4.290**	**0.3904**	**30.409**	**0.233**

Source: Pandey *et al.*, 2013.

conditions. Total 12 genotypes were evaluated for different horticultural traits at DIBER, Field Station, Pithoragarh (Uttarakhand) and significant difference observed (Pandey *et al.*, 2013).

Narayan *et al.* (2016) estimated genetic divergence on 18 morphological, growth and chemotypic charactristics using 20 germplasm collected from diverse locations of India. The genotypes were grouped into 7 clusters. Cluster II had maximum number of genotypes (06) followed by Cluster I and III (4 each), IV (3), whereas Cluster V, VI and VII had the minimum number (1 each) of genotypes. The highest (35.91) inter cluster distance was observed between Cluster VI and VII and the lowest (12.66) between Cluster I and II. Cluster IV (D^2 24.35) exhibited highest intra cluster distance whereas Cluster I recorded lowest intra cluster distance (D^2 6.28). On the basis of cluster means, genotypes *viz.*, CITH-M-SP-12, UHF-1 and SP-461 contributed maximum towards genetic divergence for high ascorbic acid content. Likewise, CITH-M-SP-17 contributed maximum for TSS and total sugar content, hence they may prove as better parents in hybrid development especially of enriched TSS and ascorbic acid content. Considering diversity pattern and other horticultural performance, CITH-M-SP-2 from Cluster VI may be considered as one of the better parent in hybridization programme.

In Nepal capsicum variety 'Sweet Banana' by Indo-American Seed Co. and 'Hawai' by Bejo-Sheetal Seed Co. has been introduced and largely being grown by farmers. In Pakistan, cultivars like California Wonder, Mexibell (R), Long Green and NARC-5 (P) are very common and being grown commercially.

Hybridization and Selection

Hybrid development is an attractive approach for enhancing productivity of sweet pepper. Sweet pepper is considered that it is intercrossable with chilli belonging to the same species *C. annuum*. It is believed that the large fruited bell peppers of today is product from tiny, pungent and pointed fruited wild species *C. annuum* var. *aviculare*. Besides, *C. annuum* which encompasses both, bell and chilli peppers, the other four cultivated species are *C. frutescens* L., *C. chinense* Jacquin; *C. pendulum* var. *baccatum* and *C. pubescens* Ruiz and Pavon are of having their crossing affinity. Sharma and Saini (1977) observed heterosis over better parent for yield and plant height but were of low order for number of branches and leaf area. Yolo Wonder x HC-201 gave the highest yield per plant and showed maximum heterosis over the best variety 'Golden Queen' (Joshi and Singh, 1980). Bullnose x HC-201, HC-201 x Ruby King and Yolo Wonder x Bullnos expressed 46.79, 22.9 and 23.2 percent superiority in total yield over the best parent HC-201 (Joshi, 1986). Poor x poor yielders showed greater degree of heterosis relative to the better parents but were low in absolute yield compared with high x high and high x poor combinations. Out of 24 cross combinations Sweet Banana x California Wonder, Osh Region x California Wonder and HC-201 x California Wonder expressed 34, 33 and 25 percent heterosis respectively over the better parent for fruit yield (Kaul and Sharma, 1988). HC-201, Sweet banana, Sel.-4 and Osh Region were good combiners for fruit length whereas, for fruit diameter only Early Prolific was a good combiner. Significant gca effects were noted for number of fruits per plant and fruit yield per plant in California Wonder. Predominance of non-additive gene effect and absence of non-interacting crosses for yield indicated utilization of heterosis to obtain higher production in bell peppers (Thakur, 1987). Joshi and Singh (1987) suggested that gca estimates and per se performance of lines should be considered for assessing their breeding value. For *Capsicum* improvement, parent HC-201, Golden Queen, Vinedale, Ruby King and Yolo Wonder could be used in a hybridization programme followed by selection (Joshi and Singh, 1987). They also reported tat there was general agreement for major contribution by the dominance type of gene action and suggested heterosis breeding. Pandey *et al.* (2012) estimated gca and sca value of yield and yield contributing traits in capsicum through diallel analysis and reported that analysis of variance for combining ability exhibited the significance for gca and sca effects of all the characters. The remarkable heterosis for yield and its yield attributing traits was also observed. The high performing crosses for all the characters studied showed that in general these crosses involved high x high, low x low, high x low and low x high general combiners. The low x low crosses giving high sca value may be due to genetic diversity of the parents and non allelic interactions. On the basis of heterosis and sca effects, Feroj x SP-6, DARL-70 x DARL-71, Capsicum Violet x California Wonder, SP-6 x Dark Green and Feroj x DARL-71 were identified as promisimng hybrids.

Interspecific Hybridization

Capsicum frutescens (2n=24) x *Capsicum annuum* (2n=24) showed unilateral incompatibility (reciprocal crosses are not possible).

Male Sterility

For cost effective hybrid development through CMS approach, a stable CMS line, its perfect maintainer and restorer lines with high pollen fertility are needed. The experiment was conducted by Sharma *et al.* (2016) to identify suitable maintainer and restorer lines among 25 high yielding sweet pepper genotypes based on pollen fertility in the hybrid combinations using 5 CMS lines at ICAR-IARI, Regional Station Katrain (Himachal Pradesh). The pollen fertility per cent of test hybrids for perfect pollen fertility restoration was varied from 81.00 (C x P6) to 91.30 per cent (C5 x T10). Testers P7, P9, P8, P10 and P19 behaved as an effective restorer for all the CMS lines under study except P19 which fails to restore compete fertility in C5, P7 in C1 an C5 and P8 in C4 and C5. The pollen fertility per cent of test hybrids for perfect maintainers was ranged from 0.00 (C1 x P1, C2 x P2, C3 x P3) to 0.72 (C1 x P11). The tester P19 behaved as effective maintainer with all the CMS lines except C5 and PRSM-1 with C2, C3 and C4. Out of 125 test hybrids, eight hybrid combinations C1 x P1, C2 x P2, C2 x PRSM-1, C3 x P3, C3 x PRSM-1, C4 x P4, C4 x PRSM-1 and C5 x P5 indicated 100 per cent pollen sterility and the male parents involved in these combination may be used for diversification programme for maintainer lines for the respective CS lines. Similarly, 28 cropss combinations indicated >80 per cent pollen fertility and the male parents involved in these crosses may be used for restorer line development programme for getting high yielding hybrid combinations.

In the absence of genic-cytoplasmic male sterile lines for hybrid seed production, androcides (gametocide) were tried. Kohli *et al.* (1981) used an androcides (gametocide) GA_3 at 10000 mg/l. It was sprayed thrice at the on set of flowering and another after 10 days causing complete male sterility, which lasted throughout the season.

Mutation Breeding

Mutation breeding could be profitably employed to get novel variants or break undesiarable linkages. Irulappan *et al.* (1981), in preliminary experiments on the sensitivity of sweet pepper, found that LD50 for germination percent ranged between 10 and 20 Kr. At 80 Kr the germination was only 4.15 percent. Subhash and Nizam (1973) induced by x-ray irradiation of seeds, mature fruits from flowers with multilocular ovaries which were larger than mature fruits from untreated plants. Use of directed mutagenesis may be helpful in this crop to create more variability.

Resistance Breeding

i. Disease Resistance

Genetics of disease resistance is very important to understand the nature of resistance source and its transfer to cultivated one.

Table 2.126: Genetic Control for Diseases

Disease	*Genetic Control*
Phytophthora rot	Single recessive gene
Leaf curl virus	Single recessive gene
Tobacco mosaic virus	Single dominant gene
Cucumber mosaic virus	Single recessive gene

Dishri *et al.* (2016) conducted an experiment to isolate and screen the anthracnose infected samples of capsicum hybrids collected from the field. Pathogen was cultured on Potato Dextrose Agar (PDA) media and mature fruits of capsicum were inoculated with mycelia of *Colletrotrichum capsici* cultures in plant. Disease score was measured in terms of lesion diameter that developed in and around the puncture wound after 15 days of inoculation. The hybrids were rated as resistant and susceptible based on the range of lesion area (mm) measured. Hybrids KTCH-5, KTCH-17 and KTCH-20 were found to be resistant which manifested less lesion diameter in comparision hybrids such as KTCH-7, KTCH-4, KTCH-11, KTCH-12, KTCH-16, KTCH-48 and

Table 2.127: Diseases and their Source of Resistance

Common Name	*Causal Organisms*	*Resistance Source*	*Reference(s)*
Fruit rot	*Colletotrichum capsici* Syot, Butt. Bisby	*Capsicum baccatum* var. *baccatum*	Ullasa *et al.*, 1981
	Colletotrichum gleosporodes Penz	Hungarian Wax and PI288982 (out of 298 genotypes: 16 resistant; 6 moderately resistant Elephant's Trunk—Resistant. Vinedale, Cubanelle, Kolocsari Javittot cecei, Bharat, F_1-Moderately resistant)	Singh and Thind, 1985
	Phytophthora capsici Leon	Lorai, Perennial, S-27, S-41-1, 77-16-1-2-1	Kaur and Singh, 1985
	Phytophthora nicotianae var. *nicotianae*	-	
	Alternaria alternata	-	
	Alternaria solani	-	
Leaf spots	*Cercospora capsici* Heald and Wolf	Hungarian Wax-Moderately resistant, 8 others resistant	Ullasa *et al.*, 1981
	Ph. Nicotianae var. *nicotianae*	-	
	C. gleosporoides Penz	-	
	Xanthomonas vesicatoria (Doidge) Dowson	-	
Wilt	*Pseudomonas solanacearum*	Arka Gaurav (tolerant)	Anon., 1984
Powdery mildew	*Leveillula taurica*	Out of 298 genotypes *C. baccatum* var. *baccatum* PI288982-moderately resistant 6-resistant 213 moderately resistant IHR-703 *C. frutescens*-Resistant *C. baccatum* and *C. frutescens*	Ullasa *et al.*, 1981; Anand *et al.*, 1987; Deshpandee *et al.*, 1985
Viruses	PVY	Resistant; Perennial and S-41-1. Resistance governed by a single recessive gene.	Sharma *et al.*, 1989

KTCH-13. Srivastava *et al.* (2016) repeated the findings and reported that KTCH-4, KTCH-5, KTCH-6, KTCH-10, KTCH-14, KTCH-22and KTCH-24 were found to be resistant with less area of lesion wheras hybrids such as KTCH-7, KTCH-8, KTCH-9, KTCH-11 and KTCH-13 showed larger lesion. There are several diseases which attack sweet pappers on various stages of crop.

Bell peppers cross easily with several chillies species. Therefore, in the absence of availability of highly resistant sources within large fruited types, resistant material could be taken from chillies background. Resistance to PVY was identified in IHR-996, a large fruited collection from Brazil. However, none of the bell pepper lines tested was resistant to CMV. Only Puri Red, Perennial, IHR-328-9 among the chilli lines were found to be resistance (George, 1989). Resistance to powdery mildew in IHR-517A and 517B (*C. baccatum* var. *pendulum*) reported by Deshpande *et al.* (1984).

ii. Abiotic Resistance

A very important capsicum line EC334206 was introduced in India from USA for water submergence.

Genetic Resource Conservation

A large number of germplasm are being maintained by ICAR-IIVR, Varanasi, ICAR-NBPGR, New Delhi, ICAR-IIHR, Bengaluru; TNAU, Coimbatore, HPKV, Palampur (H.P.), and CITH, Srinagar.

REFERENCES

1. Anand, N., Deshpande, A.A. and Ramchander, P.R. 1987. Intra group geometary in *Capsicum annuum* L. *Genetica Agraria*, **41(4):** 453-460.
2. Anonymous, 1984. *ICAR-IIHR News*, **5(2):** 2.
3. Deshpande, A.A. 2001. Chilli and Capsicum. In: Vegetables, Tuber crops and Spices. (Eds. S. Thamburaj and Narendra Singh). DIPA-ICAR, New Delhi.
4. Deshpande, A.A., Anand, N., Pathak, C.S. and Sridhar, T.S. 1985. New sources of powdery mildew resistance in *Capsicum* species. *Capsicum Newsl.*, **4:** 75-76.
5. Deshpande, A.A., Rawal, R.D., Singh, D.P. and Pathak, C.S. 1984. Chilli pepper genotypes resistant to *Cercospora* leaf spot and powdery mildew. *Tropical Pest Management*, **30:** 470-471.
6. Dishri, M., Kumar, A.B. and Srivastava, A. 2016. Screenimng of sweet pepper (*Capsicum annuum* L.) hybrids for resistance to anthracnose. 7th Indian Horticulture Congress (Abs: 74), 15-18 November, held at ICAR-IARI, New Delhi, p. 58-59.
7. Irulappan, I., Ponnuswamy, V. and Vadivel, E. 1981. Research note on the sensitivity of sweet pepper (*Capsicum annuum* L.). *S. Indian Hort.*, **29:** 223.
8. Joshi, S. 1986. Results of heterosis breeding on sweet pepper (*Capsicum annuum* L.). *Capsicum Newsl.*, **5:** 33-34.
9. Joshi, S. and Singh, B. 1980. A note on hybrid vigour in sweet pepper (*Capsicum annuum* L.). *Haryana J. Hort. Sci.*, **9:** 90-92.
10. Joshi, S. and Singh, B. 1987. Results of combining ability studies in sweet pepper (*Capsicum annuum* L.). *Capsicum Newsl.*, **6:** 49-50.
11. Kaul, B.L. and Sharma, P.P. 1988. Heterosis and combining ability studies for some characters in bell pepper (*Capsicum annuum* L.). *Veg Sci.*, **15:** 171-180.
12. Kaur, S. and Singh, J. 1985. Resistance to fruit rots diseases under field condition in pepper (*Capsicum annuum* L.). *Capsicum Newsl.*, **4:** 67.
13. Narayan, R., Kumar, A., Devnath, S., Singh, D.B., Mer, M.S., Kishor, A. and Jeena, A.S. 2016. Genetic divergence in capscicum (*Capscicum annuum* L. var. *grossum* Sendt.) based on morpho-chemical traits under polyhouse. 7th Indian Horticulture Congress (Abs: 89), 15-18 November, held at ICAR-IARI, New Delhi, p. 68-69.
14. Pandey, V., Chura, A., Arya, M.C. and Ahmed, Z. 2012. Diallel analysis for yield and yield attributing traits in capscicum (*Capscicum annuum* L. var. *grossum* Sendt.). *Veg. Sci.*, **39(2):** 37-39.
15. Pandey, V., Chura, A., Pandey, H.K., Meena, H.S., Arya, M.C. and Ahmed, Z. 2013. Variability parameters for quantitative and qualitative traits in sweet pepper in mid hills of western Himalaya. *Veg. Sci.*, **40(1):** 136-139.
16. Pickersgill, B. 1969. The archeological record of chilli peppers (*Capsicum* spp.) and the sequence of plant domestication in Peru. *Amer. Antiq.*, **34:** 53-61.
17. Seshadri, V.S. and Chaterjee, S.S. 1996. The history and adaptation of some introduced vegetable crops in India. *Veg. Sci.* **23(2):** 114-129.
18. Sharma, O.P., Sharma, P.P. and Chowfla, P.C. 1989. Inheritance of resistance to potato virus Y in garden pepper (*Capsicum annuum* L.). *Euphytica*, **42:** 31-33.
19. Sharma, P.P. and Saini, S.S. 1977. Heterosis and combining ability study for yield and agronomic characters in pepper *C. annuum* L. *Veg. Sci.*, **4:** 43-48.
20. Sharma, V.K., Chandel, C., Kumar, R., Dey, S.S., Prakash, C., Bhatia, R., Dabral, M., Mishra, V. and Mawliya, M. 2016. Identification of potential maintainers and restorers using CMS lines in sweet pepper. 7th Indian Horticulture Congress (Abs: 18; Session II: Breeding for Horticultural Crops), 15-18 November, held at ICAR-IARI, New Delhi, p. 99-100.

21. Singh, J. and Thind, T.S. 1985. Search for anthracnose and dieback resistant varieties of chillies in Punjab, *Punjab Vegetable Grower*, **15:** 37-38.

22. Singh, O.P., Anand, N. and Deshpandey, A.A. 1993. Impovement of bell pepper. In: K.L. Chadha and G. KAlloo (eds.) Advances in Horticulture (5), Malhotra Publishing House, New Delhi, p. 87-104.

23. Somas, A. 1984. The paprika, Akademiiai kiado-budapest.

24. Srivastava, A., Dishri, M. and Kumar, A.B. 2016. Screenimng of sweet pepper hybrids against anthracnose (*Colletotrichum capsici*). 7th Indian Horticulture Congress (Abs: 03; Session: Breeding for Horticultural Crops), 15-18 November, held at ICAR-IARI, New Delhi, p. 88.

25. Subhash, K. and Nizam, J. 1973. Preliminary study on the effect of X-ray irradiation on *Capsicum annuum*. *Science and Culture*, **39(8):** 350.

26. Thakur, P.C. 1987. Gene action. An index for heterosis breeding in sweet pepper. *Capsicum Newsl.*, **6:** 41-42.

27. Thomas, T.A., Singh, R. and Prasad, R. 1983. Genetic resources for improvement of vegetable crops. *Indian Hort.*, **31:** 59-73.

28. Ullasa, B.A., Rawal, R.D., Sohi, H.S., Singh, D.P. and Joshi, M.C. 1981. Reaction of sweet papper genotypes to anthracnose, Cercospora leaf spot and powdery mildew. *Plant Dis.*, **65:** 600-601.

Bird Eye Chilli (*Capsicum frutescence* var. *minimum*)

Introduction

Bird's Eye Chilli (*Capsicum frutescens* Syn. *C. fastigiatum* or *C. minimum* Roxb. Syn. *fastigiatum*) is also called as Kanthari Mulagu (Malayalam), Sohmynken Khnai (Khasi), Khut Jolokia (Assamese) *etc.* It is often used to denote small sized, pointed chilli of high pungency, whose small fruits separate easily from the calyx and are dispersed by birds. It is often added to tonics and is said to be unequalled for warding off diseases. If the throat is very sensitive it can be given in pill form- generally made with 1-10 grains powder. The capsaicin content ranges from 0.26 per cent to 1.20 per cent w/w or 1,00,000 to 1,50,000 Scoville units (indicates amount of water required to neutralize the burn) and the ASTA (American Spice Trade Association) colour value is 41.7. It can be used as chutney. Bird Eye Chilli is rich in β-carotene, ascorbic acid and contains tocopherol as vital antioxidant is well known (Mathew, 2002). Most Bird Eye Chillies are processed to extract the oleoresins (colouring matter and capsaicin) for sale to the food and pharmaceutical industries due to its high pungency. It eliminates microbial contaminations; provide uniformity of colour, flavor and economical transport and storage over bulky whole or powdered spices (Singh *et al.*, 2001). Capsaicin is used as a chemical in many pain balms, carminative tonics *etc.* and as colour additives in food industry (Senthilkumaran and Vadivel, 2002). A powerful local stimulant, with has no narcotic effect largely used in hot climates as a condiment and most useful in atony of the intestines and stomach. For persons addicted to drink it seems to be useful possibly by reducing the dilated blood vessels and thus reliving chronic congestion. It is also used to cure Ranikhet disease of poultry at initial stage (Bose *et al.*, 1993). It is used extremely for the treatment of tonsillitis, diphtheria and snake bites. It is also used in lozenge form which helps in curing sore throat and hoarseness. In Asom, Bird Eye Chilli is used traditionally used to treat mouth blisters by rubbing it on the blisters. Indonesians use it to treat acute stomach ache, loss of appetite and low blood pressure. A powerful local stimulant with no narcotic effect, bird-eye-chilli is largely used in hot climates as a condiment and in agony of the intestines and stomach. It is also used in the treatment of cancer patients, under going chemotherapy to reduce nausea. It relieve (temperorarily) of blocked sinuses, a congested throat or lungs. For persons addicted to drink, it seems to be useful possibly by reducing the dilated blood-vessels and thus relieving chronic congestion. Studies have also shown that capsaicin works with protein molecules to decrease cholesterol absorption in the body and to speed up the metabolism. But, probably the best thing about eating Bird Eye Chilli is that it enhance the release of endoephins, a substance in the brain that cause one to feel pleasure. A non conventional use of it is in the self defence sprays gaining popularity among working women who have to travel along and work late and senior citizens of USA. The spray consists of oleoresins of hot chilli at ultra high emission rate which temporarily immobilizes the attacker (Thamburaj and Singh, 2001). It is also becoming an indispensable war-weapon. That is why Defence Research Laboratories are assigned to find out its improvement in all aspects (Baruah and Baruah, 2004).

Origin, History and Distribution

Bird Eye Chilli is originated from Central and South America and closely resembles with *C. annum* (Bukusov, 1930); Shoemaker, 1953). The Bird Eye Chilli is grown in pockets of Africa and Asia, and is widely distributed throughout the tropical and subtropical America. It has now become naturalized in many parts of the tropics. It is believed that Bird Eye Chilli has been introduced into India by Portuguese in the 17th century (Premnath and Singh, 1976). In wild, the seeds are dispersed by birds are found abundantly in a half wild about villages and especially in limestone rock districts. In India, this grows in coastal Kerala, Karnataka, Vijayanagar districts of Andhra Pradesh and North Eastern states. In North Eastern Region though it is cultivated in all states at very limited scale but it is widely grown in Mizoram (approximately 140 ha with annual production of 560 tonnes), Manipur (approximately 122 ha with annual production of 488 tonnes) and Asom (approximately 280 ha with annual production of 1120 tonnes).

Species Availability, Evolution and Diversity

Botanical Description

The common features of all Fruitescens cultivars are their greenish flowers and their upward directed pedicels. The flowering begins 1-2 months after planting and it takes again one month from fruiting.

Table 2.128: Botanical Description

Growth Habit	A branching/shrubby perennial plant of 60.0-180.0 cm height and having life span of 2-3 years.
Root	Bird chilli has vigorous tap root.
Stem	Stem woody at base. Branches are angular, usually enlarged and slightly purple at the nodes.
Leaf	Leaves are unequal in shape, oval or oblong acute apex 1.5 to 10 cm in length and 0.5 to 5.0 cm wide. Petioles are medium.
Floral biology	Greenish flower in a 2 or 3 number are borne where branches form and are erect having blue anthers and pentamerous (Chadha and Kalloo, 1993). var. *baccatum* has globose flowers. Calyx is small cup-shaped, 5 toothed yellow green, petal yellow or green white, rotate, 0.5 to 1.0 cm in diameter, stamens-5; ovary-2 celled; style-slender, and stigma capitate.
Anthesis and pollination	Both self and cross pollination occur, the latter being about 16 percent (Odland and Porter, 1941; Murthy and Murthy, 1962; Free, 1975). The flowers open in the morning between 6.0-1.0 A.M. Anthesis takes place sometimes after the flowers have opened. Flowers remain open for 2-3 days. Anthers dehisce an hour after the flower opening, however, flower opening and anther dehiscence to a large extent depened upon the weather conditions. During cold as well as cloudy days, the opening of the flower is delayed. The percentage of fruit setting is 40-50 (Pursglove, 1977). Bees, ants and thrips are the possible agents of pollination. The extent of natural crossing in *Capsicum frutescens* is reported to about 16.5 per cent (Muthukrishnan *et al.*, 1993). Crosses can be made any time during the day but morning hours are preferred.
Fruit and fruit set	Fruit usually bears 2 or more erect berries at fruiting nodes, small and narrow, 0.7-2.5 x 0.3-1.0 cm. (sometimes 7.5 cm long and 1.0 cm in diameter) red or orange yellow (when ripe), and extremely pungent. Upward directed pedicels. Peduncles are slender, often in pairs, and longer than the fruit. Fruits keep their upright position until maturity appearing above the foliage. Fruit colour ranges from creamish yellow to orange scarlet to blood red.
Seeds	Seeds are small, flat, and 10-30 in number per fruit. The cuticle of the pericarp is uniformly striated and in this particular is distinct from other species. Seeds are 250 per gram.

REFERENCES

1. Baruah, S.J.N. and Baruah, Meghali. 2004. Bird's Eye Chilli-A forex earner for North East. *Spice India*, **17(4):** 40-43.
2. Bose, T.K., Som, M.G. and Kabir, J. 1993. Vegetable crops. Naya Prakash, Calcutta. 334-358.
3. Bukusov, S.M. 1930. Bull. Appl. Bot. Genenet. *Plant Breed.* (Suppl.) **47:** 261-273.
4. Chadha, K.L. and Kalloo. G. 1993. Advances in Horticlture: Vol.5 Vegetable Crops Part I. Malhotrapubl. House, New Delhi, pp. 69-95.
5. Mathew, A.G. 2002. Why chilli is charmingly red. *Indian Spice*, **39(2):** 12-13.
6. Premnath, V.S. and Singh, D.P. 1976. Vegetable for tropical region, ICAR, New Delhi.
7. Shoemaker, J.S. 1953. Vegetable growing. Chapman and Hall Ltd. London.
8. Singh, R. Gill, B.S. and Hundal, J.S. 2001. Studies on extraction of chilli oleoresin. *Capsicum News Letter*, **38(1):** 2-3.
9. Thamburaj, S. and Singh, N. 2001. Text Book of Vegetables, Tuber crops and spices, ICAR, New Delhi, p. 49-75.

King Chilli (*Capsicum chinense*)

Introduction

King chilli is also known as Bhut Jolokia, Bih Jolokia (Asomose), Naga Jolokia, Naga Mircha (Nagamese), Raja Mircha *etc.* It is very–very hot. In Asom, it is called as Bhut Jolokia (Bhut means Ghost, Jolokia means chilli in Asomose which is probably due to its ghostly bite), Bih Jolokia (poision chilli), and Naga Jolokia (due to acute hotness representing the aggressive temperament of the warriors of naga community). In Nagaland and Manipur it is known as Raja Mircha/Raja chilli (Raja means King *i.e.* King Chilli). The certificate of *"Guineess World Records"* for the hottest of all spices (February, 2007) brought still more laurels for India as a predominant chilli growing and exporting country.

King chilli is located with vitamin, a potent antioxidant and boost to the immune system. As the pods mature and darken high quantities of vitamin C are gradually replaced with β-carotene and the capsaicin levels are at their highest. Due to these capsaicin levels, some believe that eating chillis may have an extra thermic affect. Since it temporarily speed up the metabolic rate, hence burning off calories at a faster rate. The alkaloids from the capsaicin stimulate the action of stomach and intestine improving the whole digestion process. King chilli is used as a spice in food or eaten alone. One seed from a bhut Jolokia can sustain intense pain sensation in the mouth for up to 30 minutes before subsiding. Extreme care should be taken when ingesting the pepper and its seeds, so as to not get it in the eyes. It is used as a cure for stomach ailments. It is also used as a remedy to summer heat, presumably by inducing perspiration (Anonymous, 2007). In North Eastern India the peppers are smeared on fences or used in smoke bombs as a safety precaution to keep wild elephants at a distance (Wasbir, 2007). The hot flavor of chillies is due to the presence of a group of seven closely related compounds called capsainoids, but capsaicin (8-methyl-N-vanilly-6-nonenamide) and dihydrocapsaicin are responsible for approximately of the pungency. Due to its extra oprdinary pungency level and irritating properties it has also been used as lachrymatory agent (a chemical compound that irritates the eyes to cause tears, pain and even temporary blindness). Nagas are known to have used this chilli as biological weapon in ancient warfare to get rid of enemies and also used to smoke out fox and rodents in their fields. The oleoresins capsaicin (°C) extract from the chilli is popularly in use for riots control, crowed control and personal self-defence. The general concepts is that irritating compounds from the natural product will be most acceptable from the human point of view and environment-friendly than synthetically produced compounds CN (1-chloroacetaphenon), CS (O-chlorobazyladene malononitrile) and CR (Dibenz (b,f)-1,4-oxozepine).

Origin, History and Distribution

This landrace of chilli originates from the northeast of India, particularly Assam, Nagaland and Manipur. It is cultivated in North Eastern Regions (Asom, Nagaland and Manipur) of India, Sri Lanka and Bangladesh. King chilli or Bhut Jolokia has created history when it captured the attention of the world as the world's hottest chilli variety. Disagreement has arisen on whether it is a *Capsicum frutescens* or a *Capsicum*. The Indians claim it is a *C. frutescens* (Mathur *et al.*, 2000) but the derived cultivar Dorset Naga was assessed as a *C. chinense*. Recent DNA tests have found both *C. chinense* and *C. frutescens* genes (Zoschke *et al.*, 2007).

In August, 2000 Indian scientists from Defence Research Laboratory published a scientific correspondence in the Journal *"Current Science"* in which it was reported that the 'Naga Jolokia' grown in North Eastern hills is the hottest chilli in the world with 8,55,000 Scovil Heat Unit (SHU)(Mathur *et al.*, 2000; Anonymous, 2006a) against 5,77,000 SHU of world champion 'Red Savina', which would mean it is almost twice as hot as the 'Red Savina' pepper and roughly equal to the similar-looking 'Dorest Naga'. This statement from Indian Researchers altered 'Frank Garcia' of GNS species, the California developer and grower of the 'Red Savina'. Garcia also questioned the reliability of the test results, possibilirty of adultering the samples with oleoresin, the extremely hot capsaicin extract *etc.* Therefore, he opted for testing the samples of Naga Jolokia in some American Laboratories to compare the hotness with 'Red Savina'. Dr. Paul Bosland, the noted chilli breeder at New Mexico State University also questioned about the preparation of chilli samples like whether Indian scientist weighed the sample before extracting or were the seed, pericarp and placenta ground together or only the hot parts used for testing. Dr. Paul Bosland and his colleagues collected some samples of 'Naga Jolokia'/'Bhut Jolokia' for conducting a large scale experiments. An extensive study was undertaken to compare the heat level of 'Red Savina', 'Bhut Jolokia' and regular 'Habanero'; to find out whether 'Bhut Jolokia' truly had a higher level of heat than 'Red Savina' and to determine the species designation of the 'Bhut Jolokia' during 2005. After observing standard field and laboratory procedures, Dr. Paul Bosland found some interesting and surprising results. The HPLC analysis revealed the 'Habanero' had a mean heat level of 3,57,729 SHU, for 'Bhut Jolokia', the analysis revealed that it possessed an extremely high heat level of 10,01,304 SHU (Shaline, 2007) and surprisingly the test result for scored a rather low heat level of just 2,48,556 SHU. This result proved that the SHU value for 'Bhut Jolokia' was four times higher than 'Red Savina' which was enough for establishing 'Bhut Jolokia' as a world champion. Based on Dr. Paul Bosland result, it was officially declared on 9th September, 2006 by *"Guineess World Records"* that 'Bhut Jolokia' is the hottest of all chillies species.

Botanical Description

Table 2.129: Botanical Description

Growth Habit	The plant is sub shrub with multiple stems and may grow up to 45-120 cm.
Root	Tap root system
Stem	Green stem
Leaf	Green leaf colour. The length of the leaf may vary from 10.65-14.25 cm with width 5.4-7.5 cm.
Floral biology	The flowers develop 120-130 days after transplanting. The colour of the flower is yellow green with pale blue anther.
Fruit and fruit set	Fruiting take place 135-150 days after transplanting. The fruit is berry, consisting of a pericarp, placenta, seeds and dry cavity. The immature fruit is green in colour which turns in to red or deep orange colour. The shape of the fruit is sub-conical to conical with length of 5.95-8.54 cm and width of 2.5-2.95 cm. the fruit surface is rough and uneven. A matured fruit may weight up to 6.95-8.97 g.
Seeds	A fruit contain 19-34 numbers of light brown seeds.

Augmentation and Utilization of Genetic Resources (Abiotic and Biotic Resistance, Quality, Graftings)

Dorset Naga cultivar was developed as a hybrid of Naga Morich (another name of Naga Jolokia) by a couple in Dorset, England. The cultivar Dorset Naga pepper (cultivar status in process) is grown in west Bexington, Dorset, England (Anonymous, 2006b, Anonymous, 2006c)

REFERENCES

1. Anonymous, 2006a.Indian Defence Research Laboratory (html).
2. Anonymous, 2006b. Dorset claims world's hottest chilli. Daily Telegraph, 2006-04-01.
3. Anonymous, 20062c. The chilli so hot you need gloves, The Times, 2006-04-01.
4. Anonymous, 2007. Ghost Chilli burns away stomach ills-Diet and nutrition-MSNBC.com (html). Associated Press.
5. Mathur, R., Dangi, R.S., Dass. S.C. and Malhotra, R.C. 2000. The hottest chilli variety in India. *Current science*, **79(3):** 287-288.
6. Shaline, l.l. 2007. NMSU is home to the world hottest chile pepper (html).
7. Wasbir, H. 2007. World hottest chilli used as elephant repellent. National Geographic.
8. Zoschke, H. Dave, DeWitt, and Bosland, P. 2007. Naga Jolokia-searching for the new world hottest chile. (html).

3

Genetic Resources of Cucurbitaceous Vegetables and Crop Improvement

Introduction

Cucurbits belong to the family Cucurbitaceae. Cucurbitaceae is divided into two subfamilies and eight tribes representing varying degree of circumscriptive cohesiveness with about 118 genera and 825 species, according to the last taxonomic treatment of Jeffrey (1990) but Yamaguchi (1983) and Chakravarty (1982) earlier reported 100 genera and over 750 species and 110 genera and 640 species in the world, respectively. Out of them 36 genera and 100 species are found in India which includes 38 endemic species and 38 non-endemic species. About half of the cucurbits cultivated in the country are indigenous to India or Indian sub-continrnt itself. Archaeological evidence has indicated that cucurbits were present in ancient and prehistoric cultures. Genus *Lagenaria* was associated with man as early as 12,000 BC in Peru (Esquinas-Alcazar and Gulick, 1983). Archaeological expeditions in the Oaxaca region of Mexico have reported *Cucurbita pepo* to be associated with man as early as 8500 BC and cultivated by 4050 BC (Esquinas-Alcazar and Gulick, 1983). Written Chinese records describing the use of cultivated cucurbits have been found from as early as 685 BC (Herklots, 1972). American Indians cultivated squash in pre-Columbian times (Whitaker and Davis, 1962), and chayote was a common vegetable among the Aztecs prior to the Spanish conquest (Herklots, 1972). In addition, some cucurbits have been used for ornamental purpose (*e.g.* gourds), for utensils (*e.g.* bowls, ladles, sponges, boxes, bird houses, musical instruments), and for fuel and pharmacological uses in certain areas of the world. Cucurbits are among the most important plant families supplying edible products and useful fibers to humans being. Among the cucurbits bottle gourd, bitter gourd, ash gourd, muskmelon, watermelon, pumpkin and cucumbers are the most popular. The genetic diversity in cucurbits extends to both vegetative and reproductive characteristics.

Table 3.1: Occurrence of Genera and Species in Cucurbits

Cucurbit Genera	World	India
Citrullus	4	2
Coccinia	35	1
Cucumis	50	5
Luffa	9	6
Momordica	60	7
Neo. luffa	1	1
Trichosanthes	44	22
Lagenaria.	6	1
Cucurbita	27	5
Sechium	8	1
Benincasa	1	1
Solena amlexicaulis	1	1

Squash, derived from Algonquin Indian "askoot asquash" which means "eaten green" is generic term to describe cultivars of four *Cucurbita* species: *C. argyrosperma* (=*C. mixta*), *C. maxima*, *C. moschata*, and *C. pepo*. These species are new World in origin, with all but *C. maxima* originating in central to southern

Mexico (*C. maxima* originated in South America and was the only species not cultivated in the United States until post-Columbian times). Traditional forms of the *Cucurbita* species include summer squash, winter squash, pumpkins, and gourds.

The important Cucurbit species for which the Old World especially India is the centre of origin include cucumber (*Cucumis sativus* L.), muskmelon (*Cucumis melo* L.), gherkin (*Cucumis anguria* L.), bitter gourd (*Momordica charantia* L.), ivy gourd (*Coccinia grandis* (L.) Voigy) and ridge gourd [*Luffa acutangula* (L.) Roxb.]. For all the above species, the subcontinent is the primary centre of diversity, with occurrence of tremendous variability in both cultivated and wild forms with their specific features. The most important genera of family Cucurbitaceae in India are *Citrullus, Cucumis* and *Cucurbita*. Some minor genera like *Momordica, Trichosanthes, Benincasa, Coccinia, Praecitrullus, Cyclanthera* and *Sechium* have their importance mainly in the areas of their adaptation. Cucurbits are well known for their nutritional, medicinal, curative uses and as a potential source of crop diversity. Depending upon the species, virtually all parts of the plant can be used for food, including leaves, shoots, roots, flowers, seeds, and immature and mature fruits. Starch can be extracted from roots, and the seeds are a rich source of oils and proteins (Jacks *et al.*, 1972).

Table 3.2: Origin and Geographical Distribution of differtent Cucurbits

Species	*Common Name*	*Origin*	*Geographical Distribution*	*Plant Parts Used*	*Growth Habit*
Old World					
Benincasa hispida	Wax gourd, Ash gourd, White gourd, Petha	South Eastern Asia	Tropical Asia	Young leaves, flower buds, seeds, immature fruit	Annual
Citrullus colocynthis L. Schrad	Colocynth	Africa	Tropical Africa, North Western India	Seeds, leaves, shoots	Annual
Coccinia grandis Wight and Arn.	Little gourd, Ivy gourd, Kundru	Tropical Africa	Tropical Asia and Africa	Immature fruits	Semi-perennial
Cucumeropsis mannii Naud.	Egusi melon	Tropical America	Tropical Africa	Mature fruits	Annual
Cucumis anguria L.	West Indian Gherkin Maroon cucumber, West Indian gourd, cackery	Africa	Brazil, West Indies	Immature fruits	Annual
Cucurbita ficifolia Bouche.	Fig leaf, Malabar gourd	Central Mexico	Central Mexico to Chile	Fruits, seeds	Annual
Cucurbita foetidissima Kunth.	Buffalo gourd	Mexico South Western USA	Arid South Western USA and Northern to Central Mexico	Fruits, seeds	Annual
Cyclanthera pedata (L.) Schrad.	Stuffing cucumber	Tropical America	Andean countries	Immature fruits	Annual
Hodgsonia macvrocarpia (Bl.) Cogn.	Chinese lord plant	South Asia	South and South Eastern Asia		Annual
Lagenaria siceraria (Mol.) Standl.	Bottle gourd, Lauki	Africa	All tropical and subtropical areas especially in Africa	Young fruits, shoots, leaves, seeds	Annual
Luffa acutangula (L.)Roxb.	Ridge gourd	India	India, East Asia	Immature fruits, leaves	Annual
Luffa cylindrica (L.) Roem, M. J.	Sponge gourd, Chikani torai	India	Throughout tropics especially in India	Immature fruits	
Momordica charantia L.	Bitter gourd, Karela	Tropical Asia	Throughout tropics especially in South and South East Asia	Immature fruits, young shoots, young leaves	Annual
Momordica cochinchinensis (Lour.) Spreng.	Cochinchin gourd, Sweet gourd	India	South Asia	Immature fruits, fleshy roots	Annual
Momordica dioica Roxb. Ex. Willd.	Khaksa, Spine gourd, Kakrol	Southern Tropical Asia	Tropical Asia and Africa	Immature fruits, fleshy roots	Annual
Praecitrullus fistulosus (Stocks) Pang.	Round melon, India squash, Tinda	India	North West India	Mature fruit, seeds	Annual
Sechium edule (Jacq.) Swartz	Chayote, Chow-Chow, Mirliton	Mexico, Guatemala, Brazil, Peru	Throughout tropics especially in Latin America	Immature fruits	Annual
Sicana odorifera (Vell.) Naud	Casabanana	Brazil, Peru	Central America, Northern South America	Seeds used as food	Annual

Species	Common Name	Origin	Geographical Distribution	Plant Parts Used	Growth Habit
Telfairia occidentalis Hook.f.	Fluted pumpkin, fluted pumpkin	West Africa	Tropical Asia	Female shoots, seeds	Perennial
Telfairia pedata (Sims.) Hook	Oyster nut, fluted pumpkin, Zanzibar vine	Africa	Tropical Africa	Seeds	Perennial
Trichosanthes cucumerina L.	Snake gourd, Chichinda	India	South and South East Asia	Immature fruits, young shoots, young leaves	Annual
Trichosanthes dioica Roxb.	Pointed gourd, Parwal	India	India	Immature fruits	Perennial
Trichosanthes kirilowii Maxim.	Chinese snake gourd		Asia	Medicinal	
Trichosanthes lepiniana (Naud.) Cogn.	Indreni		Asia	Medicinal	
Trichosanthes ovigera Blume	Japanese snake gourd		Asia	Food	
Trichosanthes villosa Bl.	Mi-mao-gua-lou		Asia	Food	
Acanthosicyas horridus Welw. ex. Hook. f.	Inara		Africa	Food	
New World					
Cucurbita ficifolia Bouche	Fig-leaved gourd, Malabar gourd	Central Mexico	-	Mature fruits, immature fruits	Annual
Cucurbita foetidissima H.B.K.	Buffalo gourd, mock orange, stinking wild gourd, chilicote	Mexico and Southern US	-	Mature fruits, roots	Perennial
Cyclanthera pedata Schrad.	Korila, caihua, chayote, cho-cho, christophine	South America	-	Immature fruits, shoots	Annual
Sechium edule So. (Jacq.) Sw.	Choke, choko, sou-sou, chaka plant, chayotl vegetable pear mirliton	Mexico and Central America and	-	Fruits, leaves, tubers, young seeds, tendrils	Perennial

Source: Modified from Chakravarty, 1990; Herklots, 1972; Tindal 1983; Whitaker, 1990; Whitaker and Davis, 1962 and Yamaguchi, 1983, Sirohi *et al.*, 2005.

World Scenario

Cucurbits are an economically important vegetables and fruits world wide. India is among the largest producers of cucurbits next only to China. Fruits are eaten when immature (summer squash) or mature (watermelon and muskmelon). Fruits can be baked (squash), pickled (cucumber), candied (watermelon), or consumed fresh in salads (cucumber) or dessert (melon). Seeds, flowers (squash and pumpkins) and roots (chayote, spine gourd and sweet gourd) are also being consumed by humans in several forms. Seeds or fruit parts of some cucurbits are reported to possess purgatives, emetics and antihelmintics properties due to the secondary metabolite cucurbitacin content (Robinson and Decker-Walters, 1997). Cucurbits are also produced for other uses than food. Fruits (bottlegourd) are used for storage, drinking containers, bottles, utensils, smoking pipes, musical instruments, gourdcraft decoration, masks, floats for fish net, and other items. The fiber of a mature loofah fruit can be used as a sponge for personal hygiene, household cleaning and various other purposes, including filtration. India is the primary centre of origin for cucurbits like cucumber, ridge gourd, sponge gourd, ash gourd, pointed gourd and secondary centre of origin for watermelon, bottle gourd and ivy gourd. There is considerable range in the monoploid (x) chromosome number (Jeffrey, 1990), including 7 (*Cucumis sativus*), 11 (*Citrullus* spp., *Momordica* spp., *Lagenaria* spp., *Sechium* spp., and *Trichosanthes* spp.), 12 (*Benincasa hispida*, *Coccinia cordifolia*, *Cucumis* spp. other than *C. sativus*, and *Praecitrullus fistulosus*), 13 (*Luffa* spp.) and 20 (*Cucurbita* spp.).

The north eastern egion and soputhern parts of the country are considered as the major hot spot of plant biodiversity. Both the zones are richest reservoir of genetic variability and diversity of various kinds of cucurbitaceous vegetables. Among the wild species, *Cucurbita lundelliana* Bailey, *C. martinezii* and *C. okeechobeensis* Bailey are resistant to powdery mildew. *C. martinezii* also possess resistance to squash mosaic and watermelon mosaic virus whereas *C. foetidissima* Kunth is resistant to drought. *C. pedatifolia* Bailey is a species intermediate between xerophytic and mesophytic species. Wild species of *Momordica*, *Trichosanthes* and *Cucumis* (*Cucumis hardwickii*) occurs predominantly in the North-Eastern and Peninsular tracts extending to the hills. Considerable diversiy exists in *Cucumis*,

Table 3.3: Cucurbits and their Chromosome Number (2n=)

English Name	*Hindi Name*	*Botanical Names*	*Chromosome No. (2n=)*
Bitter gourd	Karela	*Momordica charantia* L.	22
Sweet gourd	Kakrol	*Momordicacochinchinensis* spreng.	28
Spine gourd	Kartoli	Momordica dioica	28
Bottle gourd	Lauki/Ghiya	*Lagenaria siceraria* (Mol.) Standl.	22
Pointed gourd	Parwal	*Trichosanthes dioica Roxb.*	22
Snake gourd	Chichinda	*Trichosanthes cucumerina* L. (Syn. *T. anguina* L.	22
Sponge gourd	Ghiya Tori (Nenua)	*Luffa aegyptiaca* (Roxb.) L.	26
Ribbed or Ridge gourd	Kali Tori/Ara Torai	*Luffa acutangula*	26
Satputia	Sathputia	*Luffa hermaphrotida*	26
Roundmelon or Indian squash	Tinda	*Praecitrullus fistulosus* (Stocks.) *Pang.* (Syn. *C. vulgaris* var. *fistulosus, C. fistulosus* Duthie and Fullerus	24
Watermelon	Tarbuz	*Citrullus lanatus* (*Thunb.*) Matsumura and Nakai (Syn. *C. vulgaris* Schrad.	22
Mateera	Tumba	*Citrullus colocynthis* (L.) O. Kutz.)	
Muskmelon	Kharbuza	*Cucumis melo* L.	24
Snapmelon	Phoot	*Cucumis melo* L. var. *momordica*	24
Long or serpent melon	Kakari	*Cucumis melo* L.var. *utilissimus* (Syn. *Cucumis melo* L. var. *flexuosus* Naud.)	24
Midget Karela, Mokha	Balsam apple	*Momordica balsamina* Linn	22
Cucumber	Kheera	*Cucumis sativus*	14
Kachri	Kachra or Machri	*Cucumis callosus*	24
West African Gherkin	Gherkin	*Cucumis anguria*	24
Ivy or little or scarlet gourd	Kundru/Tondli	*Coccinia cordifolia Cogn.* (*Syn. C.* Wight. And Arn.) *C. grandis* (L.) Voigt.	24
Bankundri	Melothria	*Solena amplexicaulis* (Lamk) Gandhi	
Snap melon	Phoot	*Cucumis melo var. momordica*	24
Send, Astu, Pehato	Mangomelon	*Cucumis melo var. cultus*	24
Chow-chow	Chayote	*Sechium edule* (Jacq.) Swartz.	24
Pumpkin	Sitaphal or Kashiphal	*Cucurbita moschata* Duch.ex Poir	40
Summer Squash	Chappan Kaddu	*Cucurbita pepo* L.	40
Winter Squash	Squash	*Cucurbita maxima* Duch.ex Poir	40
Fig leaf gourd	Fig leaf gourd	*Cucurbita fecifolia,* Baiche	40
Buffalo gourd	Buffalo gourd	*Cucurbita foetidissima* H.B. and K	40
Sweet bitter gourd	Stuffing cucumber	*Cyclanthera pedata* (L.) Schrad.	32

Source: Jeffrey, 1990.

Momordica, Trichosanthes, Benincasa, and *Cucurbita,* with reference to plant type, morphological/physiological characteristics, reaction to diseases and pest, adaptability and distribution (Table 3.4).

Telfairia occidentalis

A dioecious perennial grown at elevation up to 2,000 m in West Africa, the fluted gourd is drought tolerant and is usually trellised. Shoots from the female plants can be cooked and eaten (Lucas 1988). The fruits are large (up to 13 kg) and inedible, but the seeds contain up to 30 per cent protein and can be boiled and eaten, or ground into powder for soupe. Seeds can also be fermented for several days and eaten as slurry (Badifu and Ogunsua, 1991).

Telfairia pedata

The oystor nut is a perennial cucurbit grown in Central and east Africa. It is drought tolerant, can grow at elevatuions up to 2,000 m. It requires 18 months to flowering, and is usually trellised. It produce very large, long, flat seeds which taste similar to almonds when roasted.

Sex Expression and Sex Modification

Cucurbits *viz.* gourds, melons, pumpkins and squashes form the largest group of summer and rainy season vegetable crops grown throughout India. There is a wide variation in sex forms of cucurbits. Each species is characterized by a specific sex expression but variations do exhibit.

Table 3.4: Diversities of Cucurbits in North East India

Cultivated Species	*Area of Concentration for Diversities*	*Range of Diversities*
Cucurbita maxima	Throughout the country	Extensive
Cucurbita moschata	Hilly areas	Moderate
Cucurbita ficifolia	Megahalaya	Introduced and naturalized
Cucurbita pepo	Megahalaya and Mizoram	Limited
Coccinia grandis	Asom, West Bengal	Limited
Cucumis sativus	Throughout the country	Wide
Cucumis callosus	Foot hills of Asom	Confined to limited pockets
Luffa acutangula	Tropical areas of Asom	Wide
Luffa cylindrica	Tropical and sub-tropical areas of Asom, Meghalaya, Manipur, West Bengal	Moderate
Momordica charantia	Throughout the country	Moderate
Momordica cochinchinensis	Asom, Meghalaya, Manipur, Tripura and West Bengal	Limited
Momordica dioica	Garo hills	Rare
Trichosanthes anguina	Meghalaya, Tripura, Asom, and West Bengal	limited
Trichosanthes dioica	Tropical areas of Asom, and Tripura	Limited
Cyclanthera pedata	Hiils of Meghalaya, Manipur, Nagaland and Arunachal	Moderate
Benincasa hispida	Asom, Nagaland and Meghalaya	Wide
Lagenaria siceraria	Throughout the country	Wide
Sechium edule	High hills of Meghalaya, Manipur, Mizoram, Nagaland, Sikkim and Darjeeling of West Bengal	Moderate

Following sex forms have been observed in cucurbits:

1. **Monoecious.** Staminate and pistillate flowers are found on the same plant separately. This is most common sex forms in majority of cucurbits *e.g.* ash gourd, bottle gourd, sponge gourd, ridge gourd, bitter gourd, cucumber, pumpkin, summer squash, snapmelon.
2. **Andromonoecious.** Staminate and hermaphrodite flowers are found on the same plant separately. This is predominant sex form of muskmelon and very rare in some watermelon, ridge gourd, smooth gourd and cucumber in varying proportion
3. **Dioecious.** Staminate and pistillate flowers are produced on separate plants. Pointed gourd, ivy gourd, sweet gourd and spine gourd are among the cultivated cucurbits that exhibit this sex form.
4. **Gynomonoecious.** Pistillate and perfect flowers are produced separately on the same plant. This sex form is reported in certain stock of cucumber and also in segregates from crosses between the two sexes.
5. **Gynoecious.** Plants bear only pistillate flowers. The gynoecious sex form is prevalent in cucumber and also reported in bitter gourd (IC296539) from ICAR-IIVR, Varanasi (Ram *et al.*, 20020).
6. **Androecious.** Only staminate flowers are produced. It is rarely recorded and usually occurs as a segregate from andromonoecious form. It has been recorded in segregating population of ridge gourd, cucumber and muskmelon.
7. **Trimonoecious or Gynoandromonoecious.** Staminate, pistillate and hermaphrodite flowers are produced on the same plant in varying proportion. This is highly unstable sex form. Such form is met within some breeding line of watermelon, muskmelon, cucumber and ridge gourd and is considered to be the from which monoecious plants have been evolved.
8. **Hermaphrodite.** Staminate and pistillate organs are found on the same flower. It is found in satputia a variety of ridge gourd. This sex form has also been recorded in cucumber and muskmelon.

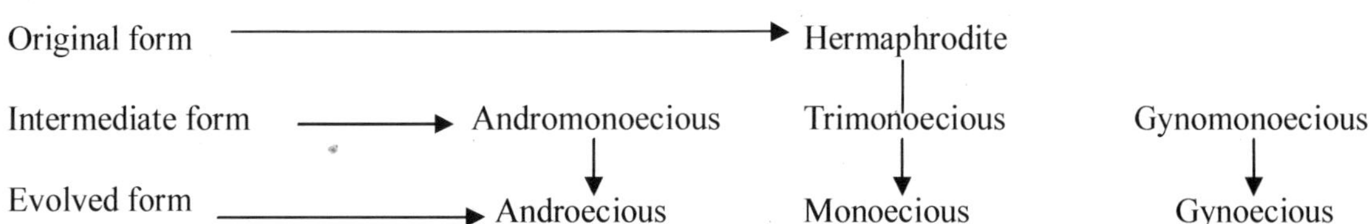

Evolution of Sex Expression in Cucurbits.

It is also a fact that most of the cucurbits originated in tropical and subtropical parts of the world. Therefore, the responsibility of exploration, conservation, characterization, and exploitation through crop improvement of cucurbits lies more on the countries confined to these regions.

REFERENCES

1. Badifu, G.I.O. and Ogunsua, A.O. 1991. Chemical composition of kernels from some species of Cucurbitaceae grown in Nigeria. *Plant Foods Human Nutr.*, **41:** 35-44.
2. Chakravarty, H.L. 1982. Fascicles of Flora of India, Fascicle II, Cucurbitaceae, Botanical Survey of India, Howrah, India.
3. Chakravarty, H.L. 1990. Cucurbits of India and their role in the development of vegetable crops: (In: Biology and Utilization of Cucurbitaceae. Bates, D.M., Robinson, R.W. and Jeffrey, C. (Eds.), Cornell University Press. Ithaca, New York (USA), p. 325-334.
4. Esquinas-Alcazar, J.T. and Gulick, P.J. 1983. Genetic resources of Cucurbitaceae. Int. Board for Plant Genet. Resources, Rome.
5. Herklots, G.A.C. 1972. Vegetable in south East Asia. George Allen and Unwin, London.
6. Jacks, T.H., Hensarling, T.P and Yatsu, L.Y. 1972. Cucurbits seeds: I. Characterization and uses of oils and proteins, A. review: *Econ. Bot.*, **26:** 135-141.
7. Jeffrey, C. 1990. Systematics of the Cucurbitaceae: an overview. (In) Biology and Utilization of Cucurbitaceae. Bates, D.M., Robinson, R.W. and Jeffrey, C. (Eds.), Cornell University Press. Ithaca, New York, USA, p.3-9.
8. Lucas, E.O. 1988. The potential of leaf vegetables in Nigeria. *Outlook Agr.*, **17:** 163-168.
9. Sirohi, P.S., Munshi, A.D., Kumar, G. and Behera, T.K. 2005. In: Dhillon, B.S., Tyagi, R.K., Saxena, S. and Randhawa, G.J. (eds.) Plant Genetic Resources: Horticultural Crops. Narosa Publishing House, New Delhi.
10. Tindal, H.D. 1983. Vegetable in the tropics. AVI, Westport, CT.
11. Whitaker, T.W.and Davis, G.N. 1962. Cucurbits. Inter Science Publishers Inc. New York (USA).
12. Whitaker, T.W. 1990.Cucurbits of potential economic importance: (In) Biology and Utilization of Cucurbitaceae. Bates, D.M., Robinson, R.W. and Jeffrey, C. (Eds.), Cornell University Press. Ithaca, New York (USA), pp. 318-324.
13. Yamaguchi. M. 1983. World Vegetables. AVI, Westport.

Genus *Cucurbita*

Introduction

The genus *Cucurbita* comprises of about 27 species or species group, both wild and cultivated (having same chromosome number of 2n=2x=40); mostly concentrated in the Bew World specially tropical and sub-tropical regions of Central and South America. The origin of the diversification of *Cucurbita* is southern Mexico to Guatemala, according to Whitaker and Bemis (1975). *Cucurbita* or yellow flowered cucurbit is considered to be one of the most morphologicaly origins and evolution of cultivated cucurbits. Variable genera can be observed in the entire plant kingdom (Robinson *et al.*, 1976; Whitaker and Davis, 1962). The 22 wild and five cultivated species are extremely diverse in fruit color, size, and shape. The cultivated species (*C. maxima*, *C. pepo*, *C. ficifolia* and the current addition *C. argyrosperma* syn. *C. mixta*) are reproductively isolated from each other by genetic barriers as well as a convincing number of morphological features of the trichomes, leaves, calyxes, corollas, stamens, pedicels and seeds (Whitaker and Bemis, 1964; Whitaker and Bemis, 1975; Nee, 1990). The constant and relatively high chromosome number (2*n* = 40) as well as the complex isozyme pattern suggest an allopolyploid origin for the genus (Singh, 1979; Kirkpatrick *et al.*, 1985). Archaeological records of the New World suggest that *Cucurbita* was one of the first plant to be domesticated (Nee, 1990). *Cucurbita*-corn-bean complex formed the nutritional basis for pre-Columbian civilizations in the Western Hemisphere (Whitaker and Bemis, 1975). One of the first species to be domesticated in the New World was *C. pepo*. Cultivation by the inhabitants of Guila Naquitz cave dated between, 10,000 to 8,000 before present (BP), predating corn and beans by more than 4,000 years (Smith, 1997). The origin and early spread of all *Cucurbita* species was in the Americas. *Cucurbita ficifolia* was the most wide spread cultivated species with a native range in the mountains from Mexico to northern Chile and Argentina (Whitaker and Bemis, 1975; Wilson *et al.*, 1992). *Cucurbita maxima*, was the only cultivated species with a native range restricted to South America, in the warm temperate areas of Uruguay and Argentina. *Cucurbita moschata* was native to the low lands of tropical and sub-tropical America (Mexico and South America), *C. argyrosperma* to the pacific coast ranging from Mexico to Nicaragua, and *C. pepo* to the high elevations of Mexico and northern Central America (Nee, 1990; Wilson *et al.*, 1992). Also, *C. moshata* was unique in being spread in two distinct native areas, a major one in Mexico and a minor one in the northern South America (Whitaker and Bemis, 1975).

The cultivated species of *Cucurbita* can be divided into:

(i) Mesophytic annuals (*C. maxima*, *C. argyrosperma*, *C. moschata*, and *C. pepo*) or

(ii) Mesophytic perennial (*C. ficifolia*) (Whitaker and Bemis, 1964).

The wild forms of *Cucurbita* are classified into

(i) Xerophytic forms- *C. digitata*, *C. palmata*, *C. cylindrica*, *C. foetidissima* and

(ii) Mesophytic forms- *C. ficifolia*, *C. sororia*, *C. lundelliana*.

Cucurbita moschata appears to be more closely related to xerophytic forms and is considered as the axis through which the species under xerophytic forms are related to each other. The mesophytic forms are mostly cross compatible and *Cucurbita lundelliana* is seen as the key species in this group.*Cucurbita lundelliana* and *Cucurbita mortinezii* are to some extent cross compatible with the cultivars. The plants are vigorous, bearing small fruits with tough rind, bitter, coarse and stringy flesh and abundant seeds. The evidence that the complex of 5 cultivated species derived form the small bitter gourds is suggested but by no means decisive. Three species have defined ancestors. *Cucurbita andreana* Naud., a weedy species, is the ancestor of *C. maxima*; *C. sororia* Bailey is the ancestor of *C. argyrosperma* (Nee, 1990); and *C. fraternal* Bailey and/or *C. texana* (Scheele) Gray are the possible ancestors of *C. pepo* (Decker, 1988; Nee, 1990). Allozyme analysis showed an independent domestication of *C. pepo* in the eastern United States and in Mexico, from divergent populations of the original and respective wild progenitor's *C. fraterna* and *C. texana* (Decker-Walters, 1990). The high level of gene flow between *C. texana* and *C. pepo* in field experiments suggested a long-term of interspecific hybridizations and confirmed *C. texana* as ancestor of *C. pepo* (Kirkpatrick and Wilson, 1988). Among cultivated species, *C. moshata* was the most variable and closely related species and nearest the common ancestor of the genus, because of the high interspecific compatibility (Whitaker and Bemis, 1975). Isozyme study showed high allelic diversity in *C. pepo* and *C. moschata*. *C. pepo* shares a common ancestor with *C. moschata* and *C. argyrosperma*, but not with *C. maxima* (Decker walters *et al.*, 1990). There are only 5 domesticated species of *Cucurbita* (*C. maxima*, *C. pepo*, *C. moschata*, *C. ficifolia* and the current addition *C. argyrosperma* (syn. *C. mixta*), Nee, 1990) which have been domesticated in the New World and for thousands of years they have been cultivated or at least handled by American societies and they are separated by sterility barriers as well as a convincing number of morphological features of the trichomes, leaves, calyxes, corollas, stamens, pedicels and seeds.

Status of Taxon

In this genus, few species are of commercial importance *viz.* field pumpkin (*Cucurbita moschata* Duch. Ex. Poir.), winter squash pumpkins (*C. maxima* Duch.), summer squash or vegetable marrow (*C. pepo* L.), winter squash pumpkin (*C. mixta* Pang. Syn. *C. argyrosperma*) and Malabar gourd or fig leaf gourd (*C. ficifolia* Bouche). The vegetable marrow (*C. pepo* L.), is an introduced crop which withstands in cooler climate. *C. texa* Gray grows wild in Texas.

Cucurbita species has been classified in several groups and subgroups which are given as shown in Table 3.7.

Table 3.5: Taxon Name and Preferred Taxon Name of *Cucurbita*

Name of Taxon	Name of Authority	Family	Preferred Taxon Name
Cucurbita			Cucurbita
Cucurbita andreana	Naudin	Cucurbitaceae	*Cucurbita maxima* subsp. *andreana*
Cucurbita argyrosperma	C. Huber	Cucurbitaceae	*Cucurbita argyrosperma*
Cucurbita argyrosperma subsp. *argyrosperma*		Cucurbitaceae	*Cucurbita argyrosperma* subsp. *argyrosperma*
Cucurbita argyrosperma subsp. *sororia*	(L. H. Bailey) L. Merrick and D. M. Bates	Cucurbitaceae	*Cucurbita argyrosperma* subsp. *sororia*
Cucurbita argyrosperma var. *argyrosperma*		Cucurbitaceae	*Cucurbita argyrosperma* var. *argyrosperma*
Cucurbita argyrosperma var. *callicarpa*	L. Merrick and D. M. Bates	Cucurbitaceae	*Cucurbita argyrosperma* var. *callicarpa*
Cucurbita argyrosperma var. *palmeri*	(L. H. Bailey) L. Merrick and D. M. Bates	Cucurbitaceae	*Cucurbita argyrosperma* var. *palmeri*
Cucurbita argyrosperma var. *stenosperma*	(Pangalo) L. Merrick and D. M. Bates	Cucurbitaceae	*Cucurbita argyrosperma* var. *stenosperma*
Cucurbita citrullus	L.	Cucurbitaceae	*Citrullus lanatus* var. *lanatus*
Cucurbita cylindrata	L. H. Bailey	Cucurbitaceae	*Cucurbita cylindrata*
Cucurbita digitata	A. Gray	Cucurbitaceae	*Cucurbita digitata*
Cucurbita ecuadorensis	H. C. Cutler and Whitaker	Cucurbitaceae	*Cucurbita ecuadorensis*
Cucurbita evodicarpa	Hassk.	Cucurbitaceae	*Sicana odorifera*
Cucurbita ficifolia	(Bouche) Britton	Cucurbitaceae	*Cucurbita ficifolia*
Cucurbita ficifolia	Bouche	Cucurbitaceae	*Cucurbita ficifolia*

Name of Taxon	*Name of Authority*	*Family*	*Preferred Taxon Name*
Cucurbita foetidissima	Kunth	Cucurbitaceae	*Cucurbita foetidissima*
Cucurbita fraterna	L. H. Bailey	Cucurbitaceae	*Cucurbita pepo* var. *fraterna*
Cucurbita galeottii	Cogn.	Cucurbitaceae	*Cucurbita pepo*
Cucurbita hispida	Thunb.	Cucurbitaceae	*Benincasa hispida*
Cucurbita hybrid		Cucurbitaceae	*Cucurbita hybrid*
Cucurbita kellyana	L. H. Bailey	Cucurbitaceae	*Cucurbita argyrosperma* subsp. *sororia*
Cucurbita lagenaria	L.	Cucurbitaceae	*Lagenaria siceraria*
Cucurbita leucantha	Duchesne	Cucurbitaceae	*Lagenaria siceraria*
Cucurbita longa	hort.	Cucurbitaceae	*Lagenaria siceraria*
Cucurbita luffa	hort.	Cucurbitaceae	*Luffa aegyptiaca*
Cucurbita lundelliana	L. H. Bailey	Cucurbitaceae	*Cucurbita lundelliana*
Cucurbita mammeata	Molina	Cucurbitaceae	*Cucurbita pepo*
Cucurbita martinezii	L. H. Bailey	Cucurbitaceae	*Cucurbita okeechobeensis* subsp. *martinezii*
Cucurbita maxima	Duchesne	Cucurbitaceae	*Cucurbita maxima*
Cucurbita maxima subsp. *andreana*	(Naudin) Filov	Cucurbitaceae	*Cucurbita maxima* subsp. *andreana*
Cucurbita maxima subsp. *maxima*		Cucurbitaceae	*Cucurbita maxima* subsp. *maxima*
Cucurbita maxima var. *turbaniformis*	(M. Roem.) L. H. Bailey	Cucurbitaceae	*Cucurbita maxima* subsp. *maxima*
Cucurbita melanosperma	Gasp.	Cucurbitaceae	*Cucurbita ficifolia*
Cucurbita melopepo	L.	Cucurbitaceae	*Cucurbita pepo* var. *ovifera*
Cucurbita mixta	Pangalo	Cucurbitaceae	*Cucurbita argyrosperma* subsp. *argyrosperma*
Cucurbita mixta var. *stenosperma*	Pangalo	Cucurbitaceae	*Cucurbita argyrosperma* var. *stenosperma*
Cucurbita moschata	Duchesne	Cucurbitaceae	*Cucurbita moschata*
Cucurbita odorifera	Vell.	Cucurbitaceae	*Sicana odorifera*
Cucurbita okeechobeensis	(Small) L. H. Bailey	Cucurbitaceae	*Cucurbita okeechobeensis*
Cucurbita okeechobeensis subsp. *martinezii*	(L. H. Bailey) T. W. Walters and D. S. Decker	Cucurbitaceae	*Cucurbita okeechobeensis* subsp. *martinezii*
Cucurbita okeechobeensis subsp. *okeechobeensis*		Cucurbitaceae	*Cucurbita okeechobeensis* subsp. *okeechobeensis*
Cucurbita ovifera	L.	Cucurbitaceae	*Cucurbita pepo* var. *ovifera*
Cucurbita palmata	S. Watson	Cucurbitaceae	*Cucurbita palmata*
Cucurbita palmeri	L. H. Bailey	Cucurbitaceae	*Cucurbita argyrosperma* var. *palmeri*
Cucurbita pedatifolia	L. H. Bailey	Cucurbitaceae	*Cucurbita pedatifolia*
Cucurbita pepo	L.	Cucurbitaceae	*Cucurbita pepo*
Cucurbita pepo var. *citrulina*	Filov, nom. nud.	Cucurbitaceae	*Cucurbita pepo* var. *pepo*
Cucurbita pepo var. *fraterna*	(L. H. Bailey) Filov	Cucurbitaceae	*Cucurbita pepo* var. *fraterna*
Cucurbita pepo var. *giraumontia*	Filov, nom. nud.	Cucurbitaceae	*Cucurbita pepo* var. *pepo*
Cucurbita pepo var. *melopepo*	(L.) Harz	Cucurbitaceae	*Cucurbita pepo* var. *ovifera*
Cucurbita pepo var. *moschata*	Duchesne	Cucurbitaceae	*Cucurbita moschata*
Cucurbita pepo var. *ovifera*	(L.) Harz	Cucurbitaceae	*Cucurbita pepo* var. *ovifera*
Cucurbita pepo var. *ozarkana*	D. S. Decker	Cucurbitaceae	*Cucurbita pepo* var. *ozarkana*
Cucurbita pepo var. *patisson*	Filov, nom. nud.	Cucurbitaceae	*Cucurbita pepo* var. *ovifera*
Cucurbita pepo var. *pepo*		Cucurbitaceae	*Cucurbita pepo* var. *pepo*
Cucurbita pepo var. *styriaca*	Greb.	Cucurbitaceae	*Cucurbita pepo* var. *pepo*
Cucurbita pepo var. *texana*	(Scheele) Filov	Cucurbitaceae	*Cucurbita pepo* var. *texana*
Cucurbita radicans	Naudin	Cucurbitaceae	*Cucurbita radicans*
Cucurbita scabridifolia	L. H. Bailey	Cucurbitaceae	*Cucurbita scabridifolia*
Cucurbita siceraria	Molina	Cucurbitaceae	*Lagenaria siceraria*
Cucurbita sororia	L. H. Bailey	Cucurbitaceae	*Cucurbita argyrosperma* subsp. *sororia*
Cucurbita species.		Cucurbitaceae	*Cucurbita species*
Cucurbita texana	(Scheele) A. Gray	Cucurbitaceae	*Cucurbita pepo* var. *texana*
Cucurbita turbaniformis	M. Roem.	Cucurbitaceae	*Cucurbita maxima* subsp. *maxima*
Unident-Cucurbitaceae species.		Cucurbitaceae	Unident-Cucurbitaceae species.

Table 3.6: Common Name and Centre of Origin

Common Names	*Botanical N*	*Chromosome No. (2n=)*	*Centre of Origin*
Fig-leaf gourd	*Cucurbita ficifolia* Bouché.	40	Mexico, Central and South Americas
Winter squash	*Cucurbita argyrosperma* Huber. syn. *C. mixta* Pang.	40	Mexico, Central America
Winter squash, pumpkin	*Cucurbita maxima* Duch.	40	South America
Winter squash Duch. and Poir	*Cucurbita moschata* (Duch. ex. Lam.)	40	Mexico, South America
Summer squash, pumpkin, marrow	*Cucurbita pepo* L.	40	North and Central Americas

Reaction of *Cucurbita* Species against Viruses

C. eucuadronsis is resistant to the most important and common diseses of *C.pepo* in France: Powdery mildew, cucumber mosaic virus, watermelon mosaic virus using *C.pepo* as pistillate parent (*C. eucuadronsis* produced no pistillate flowers).

REFERENCES

1. Whitaker, T.W. and Bemis, W.P. 1975. Bull. *Torrey Bot. Club.*, **102:** 363-367.

Table 3.7: *Cucurbita* Species of Economic Importance

Crops	*Species*
Cucurbita spp.	*C. moschata, C. maxima, C. pepo, C. ficifolia, C. mixta or C. argyrosperma*
Group I	*C. digitata, C. palmata, C. californica, C. cylindrica, C. cordata*
Group II	*C. lundelliana, C. okeechobensis, C. martinezii, C. mixta, C. palineri, C. sororia, C. grassicolor*
Cone specific group	*C. pepo, C. texana*
Group –5	*C. maxima, C. andrina, C. pedatifolia, C. moschata, C. foetidissima, C. ficifolia, C. ecuadorensis*

Table 3.8: Distinguishing Morphological Characters among Species of *Cucurbita*

Cultivated Species	*Common Name*	*Peduncle*	*Stem*	*Skin Texture*	*Fruit Flesh Colour and Texture*	*Leaf*	*Stage of Edible Fruit*	*Seed*
C. moschata	Pumpkin	Hard, smoothly angled, broadly flared	Hard, smoothly grooved	Hard rind, smooth surface	Fine grained, gelatinous substance sticking to seed, yellow flesh	Nearly round to moderately lobed, soft	Ripe, strong flavor, long keeping quality	Dull white to brown, surface smooth to somewhat rough, margin prominent, typically ragged and often dark, seed scar slightly oblique.
C. pepo	Summer squash	Hard, angular, sometimes slightly flared right next to fruit	Hard, angular	Hard rind, angular, grooved, prickly	Coarse grained, deep yellow flesh	Palmately lobed, often deeply prickly	Immature, poor keeping quality	Dull white to tan, surface smooth, margin prominent but usually smooth, seed scar square or rounded.
C. mixta	Winter squash	-	-	Hard rind	Coarse grained, deep yellow flesh	-	Immature	-
C. maxima	Winter squash	Generally sopft, round, often corky, not flared	Soft round	Hard rind	Fine grained without gelatinous substance sticking to seed	Usually unlobed, nearly round, soft.	Immature	White to brown, often plump, surface sometimes split or wrinkled margin very narrow seed scar oblique.
C. ficifolia	-	Hard, smoothly angled with grooved	Hard, smoothly grooved	-	-	Lobed, nearly round, slightly prickly	-	Usually black, sometimes tam, surface often minutely pitted, margin smooth and narrow
C. argyrosperma	-	Hard angular but becoming round at maturity, corky only slightly flared at fruit maturity.	Hard, angular grooved	-	-	Moderately lobed, softly pubescent	-	White to brown, often plump, surface sometimes split or wrinkled, margin very narrow, seed scar oblique.

Table 3.9: Resistant Behavior for different Viruses

Species	CMV	TRSV	BYMV	TmRSV	WMV-1	WMV-2
C. andreana	S	R	O	S	S	S
C. cordata	R	R	O	S	S	S
C. cylindrata	R	R	S	R	S	S
C. digitata	R	S	S	R	S	S
C. eucuadronsis	R	R	S	S	O	O
C. foetidissima	R	R	O	S	O	O
C. gracilior	R	R	O	R	S	S
C. lundelliana	R	S	S	S	S	S
C. martinezii	R	R	O	S	S	S
C. okeechobeensis	R	R	O	S	S	S
C. palmata	R	R	R	R	S	S
C. palmeri	R	R	R	R	S	S
C. sororia	S	S	R	R	S	S
C. texana	S	R	S	S	S	S
C. maxima	S	R	R	S	S	S
C. moschata	S	R	O	S	S	S
C. pepo	S	R	S	S	S	S

Pumpkin (*Cucurbita moschata* Duchesne ex Poiret.)

Introduction

Among the cucurbits, pumpkin (*Cucurbita moschata* Duchesne ex Poiret, 2n=40) is one of the most popular warm season vegetables grown commercially all over India. Pumpkin is also known as *kashiphal* or *'sitaphal'*or *'lal kaddu'* (Hindi); pumpkin, winter squash, musky squash, cushaw (English); Nahuatltamalayota (Mexico, Colombia [Guerrerol]); Spanish calabaza (Mexico), ayote (Guatemala to Costa Rica), auyama (Panama to Venezuela), zapallo (Ecuador, Peru), and joko (Bolivia). The name pumpkin originated from the Greek word *'pepon'* used for large melon like fruit. The French name 'potiron' was modified to 'pampion' turned the name *'pepon'* into 'pompon' and the English changed that into *'pumpion'* means eaten cooked by the sun or when ripe, and the present name *'pumpkin'* was chosen by the Native American and colonists as it had been one of the major ingredients of their staple diet. Apart from these uses pumpkins are grown today in the USA more for decoration than for food. In the greater part of *C. moschata* native area, its flowers, young stems and young-ripe fruits eaten as a vegetable. In India, it is extensively grown in different parts of the country for its immature and mature fruits which is used as cooked vegetable, stewed, baked or boiled and also processed to make jam, sauce, candies, sweets and puddings like *'halwa'*. It is of great significance due to its high carotene content good keeping quality and low energy content. Per 100 g fresh fruit contains 92.6 g moisture, 4.6 g carbohydrates, 1.4 g protein, 0.1 g fat, 0.7 g fibre, 0.6 g minerals (0.7 mg iron, 10.0 mg calcium and 30.0 mg phosphorus), 50 µg carotene, 0.06 mg thiamine, 0.04 mg riboflavin, 0.5 mg niacin and 2.0 mg vitamin C. The seeds are eaten whole, roasted or toasted and are ground into different stews. They have high oil and protein contents.

Squash derived from Algonquin Indian 'askoot asquash' which means eaten green is a generic term to describe cultivars of four *Cucurbita* species:

(i) *C. mixta* (*C. argyrosperma*)

(ii) *C. maxima*

(iii) *C. moschata* and

(iv) *C. pepo*

Squash and pumpkin have no precise botanical identity and the terms are used interchangeably and with some confusion. Another speciality crop is calabaza (*C. moschata*), a hard-shelled squash with bright orange, fine-grained flesh and execellent nutritive properties (Wessel-Beaver and Varela, 1991).

Origin, History and Distribution

Cucurbita moschata, like *C. ficifolia,* was of Asiatic origin. The variation in this species does not suggest any region in particular as the centre of origin, since this species is extremely variable in the morphology of its fruit and seeds. Few researchers are in opinion that species are New World in origin *viz.*, tropical and sub-tropical Central or southern America. However, it is now evident that it was domesticated in Latin America, although it is still unclear what the precise area of domestication of either species. Archaeological evidences suggest that *Cucurbita moschata* was widely distributed in both North and South America (Mexico and Peru), being the primary centers of origin. It is generally thought to have been domesticated first in Central America or Mexico. In this region its occurrence dates from 4900 BC to 3500 BC; in Guatemala from 2000 BC to AD 850 and in Peru 3000 BC. The cultivated species were staple food in diet of the American Indians long before 1942. The geographical distributions of the known archeological remains of *C. moschata* indicate that it has been cultivated for more than 5000-6000 years. Its spread to other countries, both within Latin America and outside the continent, was certainly very early. This is shown by the existence of the variety called 'Serminole Pumpkin', grown since pre-columbian times by indigenous groups of Florida in the United States, and also by its appearance in 16th century botanical illustrations. Though, pumpkins appeared in Europe in mid 16th century. Such an early spread must have been very continuous and intense since, in the last decade of the 19th century, the species was cultivated in India, Java, Angola and Japan. It was introduced to India from South America by the foreign navigators and emissaries. It is widely cultivated in South America (Mexico and Peru), North America (the USA), Asia (India) and Africa. In India, it is cultivated in all parts of the country. Area and production wise, Odisha ranked number one followed by Asom, Rajasthan and Punjab. The genomic structure of pumpkin (*C. moschata*) is AABB indicates that it is an amphidiploid.

Species Availability, Evolution and Diversity

The genus *Cucurbita* comprised closed group of about 27 species, both wild and cultivated mostly concentrated in the tropical regions of Central and South America. Commonly 'pumpkins' are defined as the edible fruits of all five *Cucurbita* species that are harvested at mature stage and usually not used as baked vegetable. The difference between 'pumpkins' and 'squash' is that the pumpkins have larger and rounded fruits than squashes. As *C. lundelliana* has wide spectrum of compatibility with the cultivated species particularly *C. moschata*. This suggests that *C. lundelliana* is involved in the ancestory of the cultivated types. It has been well established that the advanced civilization of the Americas, prior to the European influx were founded upon a food complex of maize, beans and squashes. The fruits of many wild species of genus *Cucurbita* are of same size and colour to attract the attention of primitive man. The fruits of primitive and wild types had hard and tough rinds with excessively bitter flesh. The seeds, however, are non-bitter, tasty and nutritious. It is assumed that early man, in sampling fruits of wild *Cucurbita* species for seeds found mutants lacking in bitter principle and thus began the long process of selection that presumably resulted in the modern domesticated species. In India, field pumpkins (*C. moschata*) are grown widely than other cultivated species of *Cucurbita*. *C. moschata* can tolerate hotter field conditions than other species. The pumpkin fruits are usually harvested at maturity, and can be stored for a few months. The fruits weight range from less than 0.5 kg to more than 50 kg. It has ability to grow to such a large size makes it the largest fruit in the plant kingdom. Summer squashes or vegetable marrows have smaller fruits than pumpkins and generally sausage shaped. They are usually harvested when they are still very small and immature. It has very short storage life.Widespread cultivation in India has caused the development of wide genetic diversity in different area.

The wide range of altitudes at which *C. moschata* is cultivated within the American continent, the considerable morphological diversity of its seeds and fruit (colour, shape, thickness and durability of the fruit's skin), the existence of varieties with life cycles of different duration as well as the existence of numerous cultivars developed in other parts of the world and of local varieties with excellent agronomic characteristics, clearly indicate that the genetic variation of this species is very extensive. Some interesting regional varieties for Latin America are those existing on the Yucatan peninsula (and possibly in other regions of Latin America), with two life cycles of different duration, and those cultivated in Guanajuato and Chiapas, in which resistance to some viral disease was recently found. Among the former, the short-cycle variety commonly grown in Mayan vegetable gardens is of great interest, since it was certainly from this that the most commercially important variety in the region was derived, naturally. Guanajuato and Chiapas both are currently being used in genetic improvement of pumpkin.

Botanical Description

Mature ripe fruits have pale reddish brown skin with ribbed or smooth surface. Its immature fruits of greenish or greenish yellow skinned ones are also harvested and marketed.

Botanical description of pumpkin (*C. moschata*) is as under:

Table 3.10: Botanical Description

Growth Habit	Plants are annual monoecious, robust, tendril bearing climber, usually creeping over the ground and fetid when bruised.
Root	Deep tap root system, and laterals just below surface.
Stem	Stem is robust, thick, moderately hard, obtusely quiquangular or longitudinally furrowed, 5-angled, densely clothed with very stiff and rough hairs, densely gland-dotted, light green and 5.0-25.0 m long.
Tendril	Tendrils are inserted beside the leaves, on robust stalks, 3-4 fid clothed with patent unequal, white hairs, light or yellowish green, 10.0-40.0 cm long, inclusive and stalks are 1.5-7.0 cm long.

Leaf	Leaves are distichous (shallowly lobed not pinnatifid), long stalked, petioles of 30 cm or more, simple, in outline orbicular-reniform or broadly ovate base deeply and widely cordate (measure 20-25 x 25 to 30 cm), apex shortly acuminate, obtuse or rounded; 5-7 lobed, densely toothed; teeth obtuse, glandular; at the base coarsely serrate or lobed, hispid along the margins, light or dark green and feebly shining above, light green and dull beneath, on both surfaces but specially above densely clothed with patent soft or rough hairs, beneath with scattered, pale yellow or greenish, knod shaped glands, 3- nerved, the 2 lateral main-nerves repeatedly branched on one side or pedately 5-9 nerved, with white transparent veins, 7.0-35.0 cm long and 10.0-35.0 cm wide. petiole is very robust, fistular, with numerous longitudinal faint grooves, densely clothed with patent, very rough hairs, light or dark green, 4.0-30.0 cm long and 0.5-1.0 cm thick. Stiple is absent or replaced by a gland.
Floral biology	Flowers are peduncled, large, unisexual-monoecious, pentamerous, solitary or fasciculate in the leaf-axils, for the greater part male, very hairy, dotted with many glands and 6.0-15.0 cm in diameter. Peduncle is robust, light green, densely clothed with patent, unequal white, rough hairs and with gland-tipped hairs. Stalk of male flowers (pedicels) are longer *i.e.* 16 to 18 cm, and thinner than that of the female ones, often subtertete, 2.0-25.0 cm long and 0.5-0.7 cm thick. Stalk of female flowers is distinctly 5-angular and longitudinally furrowed, 2.0-7.0 cm long, 0.75-1.0 cm thick, much lengthened after the anthesis. Short calyx (lemon yellow to deep orange) segments are 5, obliquely patent, narrowly linear-lanceolate or spathulate, their tips foliaceous, acute or obtusely acuminate, irregular lobed-dentate, light green, on both sides clothed with patent, rough, white hairs and with gland-tipped hairs, somewhat fleshy; in the male flowers 2.0-5.0 cm long, and 0.25-,030 cm wide, at the top sometimes 0.5-1.0 cm wide; in the female flowers 3.0-6.0 cm long, at the top 0.5-3.0 cm wide. Corolla is companulate, gamopetalous, 5-lobed less than half way down or about to the middle, on the outside light yellow with very prominent, yellowish green longitudinal ribs or vines, clothed with long, patent, white rough hairs, within light yellow in the tube, higher up orange-coloured or salmon pink, beset with scattered (or at the base crowded) patent, light yellow or orange-coloured long papillae or hairs. The female crolla is 7.0-15.0 cm long, 10.0-15.0 cm diameter. Lobes are broadly ovate or broadly rhomboid, acutely acuminate with light green tops, widely patent, 2.5-5.0 cm or in the female flower 4.0-7.5 cm long and wide. Male flowes are with entirely connate, stamens of 2.0-3.0 cm long; stamina column robust, with a broad base, sulphureous, 0.5-1.25 cm long; anthers oblong, dorsally connate, ochreous, 1.5-2.5 cm long. Ovary in the female flower is large, globose ovoid or oblong, light green or greyish green, densely irregular gibbose or warty, with white spots; style columnar, angular, at the apex 2-3 partite, yellowish white, glabrous, 1.5-2.0 cm long; stigma thick. Deeply two lobed, ovoid or oblong, dark orange-coloured, densely papillose and 1.25-2.0 cm long.
	Receptacle and staminodes are curving upwards in the shape of a wide cup, faintly 3-lobed, yellowish white; tops of the lobes emarginated.
Anthesis and pollination	Pumpkin is highly cross pollinated crop and entomophilous. Anthesis of flowers takes place between 4.30 to 6.00 a.m., while pollen dehiscence takes place before anthesis. The stigma is receptive on the day of anthesis. Therefore, pollination must be done on the day of anthesis to ensure maximum fruit set. Honey bees are the pollinating agent.
Fruit and fruit set	The fruits are long robust stalks, pendulous, hard and tough rinds very variables in shape, colour and dimensions, rotate, ovoid or oblong, round, lanceolate-ellipsoid, more or less deply longitudinally furrowed or lobed, blotched ormarbled, and sordily crimson orange to pale yellow coloured when quite ripe. The fruits are fleshy, flesh yellow to orange and flesh thickness from 3.5 to 5.5 cm. Fruit weight varies from 1.0-20.0 kg. The fruits are many seeded pepo.
Seeds	Seeds are numerous and compressed. Seeds are white or brown, apex oblique, with broad rim, containing 40-50 percent oil and 30 percent protein.

Horticultural Classification

1. **Classification on the Basis of Adaptability**: On the basis of adaptability, the wild forms of *Cucurbita* are classified into:
   - **Xerophytic Forms**: It includes *C. digitata, C. palmata, C. cylindiata* and *C. foetodissima*. Pumpkin (*C. moschata*) appears to be more closely related to xerophytic forms and is considered as the axis through which the species under xerophytic forms are related to each other.
   - **Mesophytic Forms**: It includes *C. ficifolia, C. sororia* and *C. lundelliana*. Themesophytic forms are mostly cross compatible and *C. lundelliana* is seen as the key species in this group.
2. Classification on the basis of fruit flesh colour, texture and stage of edible fruit
3. Classification on the basis of certain other morphological characters

Gene and Genetics of Characters

There are certain characters which can be observed on the plant and others organs, helps in identification of species. The assigned gene symble for specific traits are as under:

Table 3.11: Gene and Genetics of Characters

Gene Symbol	*Characters*	*Species*
a or aa	Androecious	*C. pepo*
AA	Monoecious	*C. pepo*
Bi	Bitter fruit, high cucurbitacin in fruit	*C. pepo*
Bu	Bush habit, short internodes	*C. pepo, C. maxima*
Cr	Green corolla	*C. moschata*
D	Dark green stem, dominant to light green stem	*C. pepo*

Gene Symbol	Characters	Species
Di	Disc fruit shape, dominant to spherical	*C. pepo*
Fr	Resistant to fruitfly	*C. maxima*
G	Gynoecious sex expression	*C. foetidissima*
Gr	Green rind, dominant to buff skin of mature fruit	*C. moschata*
lo	Lobbed leaves	*C. maxima*
lt	Leafy tendril, tendrils with laminae	*C. pepo*
M	Mottled leaves, silver grey areas in axils of leaf veins	*C. maxima, C. moschata*
n	Naked seeds, lacking a lignified seed coat	*C. pepo*
r	Recessive white, white fruit colour	*C. pepo*
s	Sterile, male flowers small, without pollen, female flower sterile	*C. maxima*
Wt	Warty fruit, dominant to smooth	*C. pepo*

Augmentation of Genetic Resources and Crop Improvement

Breeding Priorites

- ✰ High female to male ratio
- ✰ High yield
- ✰ Earliness
- ✰ Narrow seed cavity
- ✰ High β-carotene and thick flesh
- ✰ Uniform fruit shape and size
- ✰ Transport quality
- ✰ Resistant to diseases (powdery mildew, downy mildew and viral diseses) and insect pests (red pumpkin beetle and fruit fly)

Introduction and Selection

Another part of the genetic stock of *C. moschata* is represented by the numerous commercial cultivars that have developed, mainly in the United States and to a lesser extent in Brazil. Prominent among these are 'Butternut Squash', 'Golden Cushaw', 'Large Cheese', 'Tennessee Sweet Potato', 'Kentucky Field', 'Menina Brasileira' and others. Varieties like 'Emerald', 'Butternut' and 'Japanese Pie' in squashes are introductions to India. Variety 'CO-1' has been evolved through selection. 'Ambali (CM 14)' and 'Arka Chandan' were identified promising for horticultural traits.'Arka Suryamukhi (squash variety)' was found high yielder. 'CO-1', 'CO-2'and 'Pusa Viswas' were also reported for yield and fruit quality. Nagar *et al.* (2016) evaluated 76 genotypes during spring-summer season and found significant differences among the genotypes for all the traits. On the basis of mean performance for different mineral and yield traits, the genotype DPU-6, DPU-26 and DPU-51-3 were found most promising. Further, high heritability estimates coupled with high genetic gain were observed for all the traits under study, which indicated that these traits are under additive gene effects and more reliable for effective selection. The correlation coefficients (phenotypic and genotypic) among different minerals along with fruit yield per plant depicts that fruit yield per plant has expressed significantly positive association with calcium (0.187 and 0.216) and magnesium (0.149-0.155). Further, path coefficient analysis at genotypic level depicted that iron has maximum positive direct effect (0.374) on marketable fruit yield per plant, followed by calcium (0.198) and copper (0.037). Besides this, maximum positive indirect effect of manganese (0.331) and copper (0.265) *via* iron was observed on the fruit yield per plant. Popular pumpkin cultivars like 'CO-1', 'CO-2' and 'CM-14' wre developed from the use of Indigenous materials through selection with progeny testing.

Hybridization and Selection

In few F_1 hybrids *i.e.* 'Volge Grey 92' x 'Kit 599' and 'High Vitamin' x 'High Carotene 102' yield increase upto 80 per cent have been observed. Doijode *et al.* (1982) found significant heterobeltiosis for fruit weight in cross combinations 'CM 12' x 'Arka Chandan' and 'CM 12' x 'IIHR 8'. Similarly, Sirohi *et al.* (1986) found heterotic effect for yield/plant, fruits/plant and fruit weight in cross combination 'S 93' x 'CM 12'. Fifteen F_1 hybrids were developed by using 'Australian Green', 'EC 27050', 'Vegetable Marrow', 'Sel 1-P1-8', 'Sel 9-P1-2' and 'Early Yellow Prolific'genotypes in all possible combinations and F_{1s} displayed considerable heterosis for yield. Hybrid Kashi Shishin (VRPKH-1) has been developed from ICAR-IIVR, Varanasi (U.P.).

Interspecific Hybridization

Whitaker and Robinson (1986) have elaborated interspecific hybridization information on pumpkin. Four *Cucurbita* species, *viz.*, *C. moschata*, *C.pepo*, *C. maxima*, *C. argyrosperma* (formerly called *C. mixta*) and *C. ficifolia* are treated as a unit and express limited inter-crossing (do not hybridize readly with each other) does not affect the quality of seeds. Based upon species crossability, Whitaker and Davis (1962) concluded that *C. moschata* occupies a central position among annual species and can be crossed with difficulty with *C.pepo*, *C. maxima* and *C. mixta* producing partially fertile hybrids. The interspecific crossing is highly cultivar dependent, usually produces few viable seeds (Robinson and Decker-Walter, 1999). They indicated that likelihood of success varied with cultivars of the 'Butternut' group of *C. moschata* with *C.pepo* 'Delicata'. They were also in opinion that 'Butternut' and 'Delicata' may be less promising as parents to achieve this interspecific cross than some cultivars previously used such as 'Large Cheese' and 'Connecticut Field'. Munger (1990)

Table 3.12: Some Promising and Useful Introduction of Pumpkin

Crop Species	Accessions	Country	Promising and Useful Attributes	Indentor
Cucurbita pepo	EC-380995	USA	Unique, small, oblong early maturing, the flesh comes out in strings, can be bakes or used in breads	IIVR, Varanasi
	EC- 380996	USA	Hull-less seeded pumpkin; seeds can be popped and eaten like snack, high in nutrition and good source of Zinc.	IIVR, Varanasi
	EC-516625, EC-516626, EC-516627, EC-516628, EC-516629, EC-516630, EC-516631, EC-516632, EC-516633, EC-516634, EC-516635, EC-516636, EC-516637, EC-516638, EC-516639, EC-516640, EC-516641, EC-516642, EC-516643, EC-516644, EC-516645, EC-516646, EC-516647, EC-516648, EC-516649, EC-516650, EC-516651, EC-516652, EC-516653, EC-516654, EC-516655, EC-516656, EC-516657, EC-516658, EC-516659, EC-516660, EC-516661, EC-516662, EC-516663, EC-516664, EC-516665, EC-473294, EC-473295, EC-473296	-	-	-

maintained interspecific population obtained through tissue culture from 'Yantee Hybrid', a yellow straight neck summer squash (*C.pepo*) pollinated by 'Butternut' (*C. moschata*). The wild ancestor *C. lundelliana* can be used as a bridging species to effective interspecific gene transfer among cultivated species of *Cucurbita*. The cross compatibility among cultivated *Cucurbita* and the wild ancestral species *C. lundelliana* are given in table 3.13.

The possiobilities of hybridization that *C. moschata* has shown with other cultivated species (for example, *C. maxima*) enable us to affirm that there are good prospects for the improvement of the crop. To obtain fruits and fertile seeds from F_1 plants of interspecific crosses, additional techniques like repeated pollination, bud pollination, mixed pollen pollination, embryo culture, amphidiploid and adjustment of environment of flowering are adopted (Bemis, 1973; Cheng *et al.*, 2002; Hiroshi, 1963; Shirfriss, 1987 and Rakho *et al.*, 2012). However, these techniques could not releave the male sterility/incompatibility problems in subsequent generations (Stebbins, 1956). *C. lundeliana,* when used as a female parent is cross compatible with nearly all cultivated species and used as a bridge to facilitate gene flow. C. argyrosperma also has been used as genetic bridge to transfer gene in less compatible Cucurbita species (McCondless, 1998; Wessel-Beaver *et al.*, 2004). In a trait specific cross *C. pepo* (*C. lundeliana, C. moschata*) the bush habit of *C. pepo* can be transferred to *C. moschata* (Rhodes, 1959). The perennial species *C. okeechobeensis* and *C. martinezii* are cross compatible with *C. moschata* (Whitaker and Bemis, 1964, Hurd *et al.*, 1971). Successful attempts have been made by Quizhang *et al.* (2012), in the development of interspecific bridge lines among *C. pepo*, *C. moschata* and *C. maxima* with normal sexual compatability by varietal recombination and successive selection towards increasing compatability and interspecific crosses and the successively derived generations. Japanese Company- Sakata Seed Company has commercially exploited interspecific hybrids of *C. moschata* x *C. maxima.*

Distant Hybridization

Some successful hybridization in pumpkin/squashes is as under:

- ✰ *Cucurbita moschata* x *C. maxima*
- ✰ *C. moschata* x *C. pepo* (tissue culture)
- ✰ *C. lundelliana* x *C. moschata*
- ✰ *C. lundelliana* x *C. pepo*
- ✰ *C. lundelliana* x *C. mixta*
- ✰ *C. lundelliana* x *C. maxima*
- ✰ *C. lundelliana* x *C. ficifolia*

C. lundelliana has wide cross compatibility.

Table 3.13: Cross Compatibility among Cultivated Species of *Cucurbita*

Species	C. moschata	C. pepo	C. mixta	C. maxima	C. ficifolia	C. lundelliana
C. moschata	SC	****	-	*	***	CC (*)
C. pepo	****	SC	****	*		CC +
C. mixta		****	SC	*	***	CC +
C. maxima	*	*	+	SC	**	CC (x)
C. ficifolia		**	***	**	SC	CC (x)
C. lundelliana	CC (x)	CC +	CC +	CC (x)	CC (x)	SC

SC-Self-compatible; CC-Cross-compatible;*-F_1viable but self sterile; **-a few viable F_1seeds, weak F_1 plants; ***-F1 sparigly viable, self sterile (embryo culture can be adopted); ****- Fruit formation occurs but no seed is formed; +-small F_1seeds, seedling very weak and fail to come up; (x)-F_1self fertile.

Polyploidy Breeding

Tetraploids (2n=80) of 4 Japanese varieties of *C. moschata* were produced through chemical mutation (colchicines treatment, one drop of 0.2-0.4 per cent solution was applied for 5-7 days) expressed larger epidermal and stomatal cells, greater thickness, fruits for times heavier, larger seed, increased sugar content, delayed ripening and increased resistance to cold and drought. At the same time a reduction of 35 per cent fertility was also observed in the population.

Breeding for Quality Attributes

Cucurbita moschata is reported to possess keeping quality, hard rind and strong flavor. Cultivars like Arka Chandan and Jammu Special were reported for high carotene and keeping quality. 'Arka Suryamukhi and Chanani Green were found very much suitable for transport quality. *C. lundelliana* has perennial habit.

Disese Resistance Breeding

With regard to the sources of variation of *C. moschata* cultivars developed outside its area of origin, the best example is that of a cultivar, native to Nigeria, which represents the only source of resistance to certain viral diseases.

Fungal Disease Resistance

Sources of resistance in *Cucurbita* was other allied *C. lundelliana*, *C. ecuadorensis*, and *C. martinezu* for powdery mildew and cucumber mosaic virus (CMV). *C. pepo* also possesses powdery mildew resistance. Genotype 'Butternut Squash', 'Golden Cushaw', 'Large Cheese', 'Tennessee Sweet Potato', 'Kentucky Field', 'Menina Brasileira' and other commercial cultivars have different levels of resistance and/or susceptible to certain diseases, which is indicative of the wide genetic variation of this species.

Viral Resistance

The high level of resistance is not available against squash mosaic virus. All species become infected, however *C. ecuadorensis*, *C. okeechobeensis* and *C. martinezii* could recovered from initial systemic symptoms (Provvidenti, 1990). *Cucurbita ecudorensis* was immune or resistant to 4 viruses infecting cucurbits and it was compatible with *C. maxima* which would be a good source of resistance in breeding programme (Providenti *et al.*, 1978). They also reported that *C. foetidissima* is a good source of resistance to 3 viruses infecting cucurbits and that *C. martinezzi* was resistant only to 2 viruses but could be used to transfer resistance to CMV to *C. moschata*. Herrington *et al.* (1991) were able to transfer resistance of WMV-2, PRV-W, ZyMV and CMV of *C. ecudorensis* to *C. maxima* to produce cultivar 'Redlands Traiblazer' *C. foetidissima* is equally good in virus resistance but it is difficult to hybridize it with other member of genus *Cucurbita*. *C. martinezii* on the otherhand is not resistant to many viruses but can be used to transfer resistance to CMV and powdery mildew to *C. moschata* and the F_1 can be used as a bridge to transfer these resistant to *C. pepo* (Munger, 1976). Pumpkin varieties 'CM 14', 'King of the Mammottis', and 'Large Red' to be resistant to yellow-vein mosaic (Jayasree, 1984; Suresh Babu, 1989). Pumpkin mosaic virus is not transmitted through seeds or by the whitefly. The virus is transmitted through mechanical means, grafting and by aphids (*Aphis gossypii* and *A. craccivora*). In varietal screening of *C. moschata* genotypes to pumpkin mosaic virus, no susceptibility observed (Umamaheswaran, 1985). Similarly, in inheritance studies of resistance to Zucchini-yellow mosaic virus and reported that a single gene when homologous in *C. moschata* confers a high level of resistance to yellow-mosaic virus (Munger and Provvidenti, 1987; Balakrishnan, 1988; Paris *et al.*, 1998). In varietal evaluation 'CM 67' and 'Arka Suryamukhi' were the least susceptible to pumpkin mosaic virus. Provvidenti (1990) observed that *C. ecuadorensis*, *C. foetidissima* and *C. ficifolia* were the best sources for multi-resistance being immune to WMV1 and WMV2 and resistance to CMV and TRSV. *Cucurbita moschata*, *C. ecudorensis*, *C. foetidissima* and *C. martinezzi* are reported to be resistance to yellow mosaic virus.

C. pepo variety 'Freedom II', carrying resistance to WMV and ZYMV. In pumpkin following sources can be utilized as resistance source in breeding programme

Table 3.14: Multiple Viral Resistance in *Cucurbita* Species

Cucurbita Species	Multiple viral resistance
Cucurbita ecudorensis	CMV, TRSV, WMV-1 or PRSV, WMV-2
C. foetidissima	CMV, TRSV, WMV-1, WMV-2
C. lundelliana	CM, PM and viruses (Bohn and Whitaker, 1964)
C. martinezii	CMV, TRSV, PM and viruses
C. maxima	TRSV, BYMV, PRSV-W
C. moschata	TRSV, PRSV-W, ZYMV
C. pepo	TRSV
C. ficifolia	WMV-1, WMV-2, CMV, TRSV, BYMV-S, TmRSV

Table 3.15: Resistance Source against some Diseases

Crop	Disease	Resistance Source
Pumpkin	Powdery Mildew (PM) and Viruses	Cucurbita lundelliana
	Zucchini Yellow Mosaic Virus (ZYMV)	C. ecuadorensis
	Watermelon Mosaic Virus (WMV)	C. foetidissima

Insect Resistance

In *Cucurbita* spp. insect and their sources of resistance against pumpkin beetle are BG-5, BG-98, and BG-102. *C. pepo* has expressed against red pumpkin beetle.

Breeding for Quality Traits

C. pepo has tendency to bear more female flowers and possess early maturity quality.

Grafting

In pumpkin CM 214 (Nigerian Local, Immune) did not develop any disease symptom even after grafting on the YVMV infected symptomatic rootstock. The environmental fluctuations also did not influence the resistance expression of Nigerian Local against PMV and YVMV (Kuriakose, 1998). In field tests on ground infected with *Fusarium oxysporum* of 4 self rooted varieties of melon and 2 of those grafted to ash gourd Rugeri (1968) observed that 90 per cent of plants belonging to each of the ungrafted varieties became infected while those grafted on ash gourd escaped infection.

Table 3.16: Diseases and their Sources of Resistance

Crop	Disease	**Resistance source**
Cucurbita spp.	Cucumber Mosaic Virus	*Cucurbita pepo cv cinderella, Gold Rush*
	Watermelon Mosaic Virus	*Cucurbita ecuadorensis, Cucurbita joetidissima*
	Bacterial wilt	*Cucurbita ficifolia, Cucurbita lundelliana*
	Powdery Mildew	*Cucurbita lundelliana, Cucurbita okeechobeensid*
	Powdery mildew, squash and watermelon virus	*Cucurbita martinezii*
	Pumpkin mosaic virus	CM 214 (Nigerian Local, Immune)

Source: Bhardwaj, 2010.

Genetic Resource Conservation

C. moschata is the best represented of Cucurbita in the gene banks of America, in which more than 2000 accessions have been deposited. These come chiefly from Mexico and Central America and to a lesser degree from South America and other regions of the world. The most important accessions are those from the United States and Costa Rica. Collectively, the accessions are made up of American material, mainly from Central America. For its part, the CIEAP collection in Mexico is possibly the most representative of *C. moschata* variation in that country. There has also been recent interest in edible pumpkin seed, particularly in genotypes with the hull-less trait (Loy, 1990).

Table 3.17: Rootstock Tolerance to Biotic and Abiotic Stresses

Rootstock	*Resiatance/tolerance*	*Grafting*	*References*
Cucurbita maxima x *C. moschata*	*Fusarium oxysporum* f. sp. *niveum, F. oxysporum* f.sp. *melonis*, cold, heat and drought stress	Watermelon, muskmelon, cucumber	Davis *et al.*, 2008; King *et al.*, 2010, Lee, 1994; Rouphael *et al.*, 2008
C. moschata	Bloomless cucumber	Cucumber	Sakata *et al.*, 2007
C. ficifolia	Fusarium wilt, cold stress	Cucumber	Lee, 1994; Tachibana, 1982; Marukawa and Takatsu, 1969

REFERENCES

1. Balakrishnan, S. 1988. Mosaic disease of pumpkin in Kerala. *Research Highlights on Vegetables*, Kerala Agricultural University, Thrissur (Kerala), p. 44-46.
2. Bhardwaj, D.R. 2010. Breeding for pest and disease resistance in cucurbits. In: Compendium of Winter School "Emerging trends and strategies for the management of pests and diseases in vegetable crops. Published by ICAR-IIVR, Varanasi, p. 102-109.
3. Bemis, W.P. 1973. Interspecific aneuploidy in *Cucurbita. Genet. Res.*, **21:** 221-228.
4. Cheng, Y.G., Zhang, B.K., Zhang, E.H. and Zhao, Z.L. 2002. Germplasm innovation by interspecific crosses in pumpkin. *Cucurbit Genet. Coop. Rep.*, **25:** 56-57.
5. Davis, A.R., Perkins-Veazie, P., Dakata, Y., Lopez-Galarza, S., Cohen, R. and Lee, J.M. 2008. Cucurbit grafting. *Crit. Rev. Plant Sci.*, **27:** 50-74.
6. Doijode, S.D., Nath, P. and Sulladmath, U.V. 1982. Hybrid vigour in pumpkin (*C. moschata*). *Genetica Agrarian*, **36:** 87-94.
7. Hiroshi, H. 1963. Cucurbit crosses: Flower ages and reciprocal cross compatability in *C. pepo, C. moschata* and *C. maxima. Japanese Jouranal of Breeding*, **13:** 159-167.
8. Hurd, Jr P.D., Linsley, *E.G.* and Whitaker, T.W. 1971. Squash and gourd bees (*Peponapis, Xenoglossa*) and the origin of the cultivated *Cucurbita. Evolution*, **25:** 218-234.
9. Jayasree, P.K. 1984. Yellow vein mosaic disease of pumpkin in Kerala. M.Sc. thesis, Kerala Agricultural University, Vellanikkara Thrissur (Kerala).
10. King, W.H., Hoong, N.H., Yang, H.B., Kwon, J.K., Jo, S.H., Seo, J.K., Choi, D. and Kang, B.C. 2010. Molecular mapping and characterization of a single dominant gene controlling CMV resistance in pepper (*Capsicum annuum* L.). *Theor. Appl. Genet.*, **120:** 1587-1596.
11. Lee, J.M. 1994. Cultivation of grafted vegetables. I. Current status, grafting methods and benefits. *HortSci.*, **29:** 235-239.
12. Loy, J.B. 1990. Hull-less seeded pumpkin: a new edible snackseed crop. Janick, J. and Simon, J.E.

(eds.). Advances in new crops. Timber Press, Portland, OR, pp. 403-407.

13. Marukawa, S. and Takatsu, L. 1969. Studies on the selection of *Cucurbita* species as cucumber rootstock. *Bul. Ibaraki Hort. Expt. Sta.*, **3:** 11-18.
14. McCondless, L. 1998. Geneva release 'Whitekar' summer squash at the NYS vegetable Conference, 2011.
15. Munger, H.M. 1976. *Cucurbita martinezii* as a source of disease resistance. *Veg. Imp. Newsletter*, **18:** 4.
16. Munger, H.M. and Provvidenti, R. 1987. Inheritance of resistance to Zuchhini Yellow Mosaic virus in *C. moschata*. *Cucurbit Genet. Coop. Rep.*, **10:** 80-81.
17. Nagar, A., Sureja, A.K., Munshi, A.D., Bhardwaj, R., Kar, A., Gopalakrishnan, S. and Kumar, S. 2016. Genetic variability and character association for yield and quality traits in pumpkin. (Abs: 61). 7th Indian Horticulture Congress, held at ICAR-IARI, New Delhi, p. 48
18. Nee, M. 1990. The domestication of *Cucurbita* (Cucurbitaceae). *Economic Botany*, **44 (3):** 56-58.
19. Paris, H.S., Cochen, S., Burger, R.Y. and Yoseph, R. 1998. Single gene resistance to Zucchini yellow-mosaic virus in *C. moschata*. *Euphytica*, **37:** 27-29.
20. Provvidenti, R. 1990. Viral disease and genetic sources of resistance in *Cucurbita* species. Biology and Utilization of the Cucurbitaceae (D.M. Bates, R.W. Robinson and Ch. Jeffery, eds.) Comstock Publ. Assoc. Cornell University Press, Ithaca and Londan.
21. Provvidenti, R., Rbinson, R.W. and Munger, H.M. 1978. Multiple virus resistance in *Cucurbita* species. *Cucurbit Genet. Coop. Rep.*, **1:** 26-27.
22. Rhodes, A.M. 1959. Species hybridization and interspecific gene transfer in the genus *Cucurbita*. *Proc. Amer. Soc. Hort. Sci.*, **74:** 546-552.
23. Robinson, R.W. and Decker-Walter, D.S. 1999. Cucurbits. CAB International, New York.
24. Rouphael, Y.M., Carderelli, M. and Colla, G. 2008. Yield, mineral composition water relations and water use efficiency of grafted mini watermelon under deficit irrigation. *HortSci.*, 43: 730-736.
25. Sakata, Y., Ohara, T. and Sugiyama, M. 2007. The history and present status of grafting of cucurbitaceous vegetables in Japan. In: McConchie, G.R. (ed.), Proc. of the 3rd International Symposium on cucurbits. *Acta Hortic.*, 159-170.
26. Shirfriss, O. 1987. Interspecific hybridization: Notes on squash breeding. *Cucurbit Genet. Coop. Rep.*, **10:** 93.
27. Sirohi, P.S., Kumar, T.S. and Choudhary, B. 1986. Studies on combining ability in pumpkin (*C. moschata* Duch. ex Poir). *Indian Journal of Horticulture*, **43:** 98-104.
28. Stebbins, G.L. 1956. Artificial polyploidy as a tool in plant breeding. Book: Genetics in Plant Breeding. Brook-*haven Symposium in Biology*. p. 37-52.
29. Suresh Babu, V. 1989. 'Divergence studies in pumpkin (*C. moschata* Duch. ex Poir) M.Sc. thesis (Agrculture), Kerala Agricultural University, Vellanikkara Thrissur (Kerala).
30. Tachibana, S. 1982. Comparision of effects of root temperature on the growth and mineral nutrient absorption in cucumber cultivars and fig leaf gourd. *J. Japan Soc. Hortic. Soc.*, **51:** 299-308.
31. Umamaheswaran, K. 1985. 'Transmission, physical properties and host range of pumpkin mosaic virus. M.Sc. (Agrculture) thesis, Kerala Agricultural University, Vellanikkara Thrissur (Kerala).
32. Wessel-Beaver, L. and Varela, F. 1991. Performance of parents and progenies in Caribbean x temperate crosses of *C. moschata*. *HortSci.*, **26:** 740 (Abstr.).
33. Wessel-Beaver, L., Luevas, H.E., Andres, T.C. and Piperno, D.R. 2004. Genetic compatibility between *Cucurbita moschata* and *C. argyrosperma*. Progress in cucurbit genetics and breeding research, Proc. of Cucurbitaceae, 2000. The 8th *EUCARPIA Meeting on Cucurbit Genetics and Breeding*, Palceky University, Olomone Czech Republic, p. 393-400.
34. Whitaker, T.W. and Bemis, W.P. 1964. Evolution in the genus *Cucurbita*. *Evolution*, **18:** 553-559.
35. Whitaker, T.W. and Davis, G.N. 1962. *Cucurbits: Botany, Cultivation and Utilization*, p. 250. Inter Science Publisher, New York.
36. Whitaker, T.W. and Robinson, R.W. 1986. Squash breeding. In: Bassett, M.J. (ed.). Breeding Vegetable Crops, p. 209-242. AVI, Publishing Co.

Buffalo Gourd (*Cucurbita ficifolia* Bouché or *F. foetidissima*)

Introduction

Fig Leaf Squash or Malabar gourd *Cucurbita melanosperma* (A. Braun (1824), *Pepo ficifolia* or *C. ficifolia* Bouche, 2n=24) is also known as cidra, sidra (English), Lacoyote (Spanish: Peru, Bolivia and Argentina), Chiverri (Honduras, Costa Rica), and Victoria (Colombia). All parts of the buffalo gourd plant are extremely bitter because of water soluble glycosides known as cucurbitacins. This vegetable is very popular for its edible seeds, fruits and greens. Fig leaf gourd is mainly grown for its large unripe fruits (boiled as a vegetable). The tender immature fruits are used like summer squash or cucumber. As indicated by its white colour, the flesh of the fruit is deficient in β-carotene, and has a moderate quantity of carbohydrates and a low vitamin and mineral content. The flesh of the mature fruit is often impregnated with sugar and the resulting product used as candy. It may also be fermented to make an alcoholic beverage (Bates *et al.*, 1990). The young leaves and vines tips may be prepared as a green vegetable in some regions of Mexico (and perhaps other countries of the continent), and the young stems (or "runner tips"). The male flowers and buds are used in soups, stews and salads. The Indians used the buffalo gourd for food, utilizing the seeds which were usually ground up and eaten in the form of a cooked mush. The composition of *C. ficifolia* is comparable to mature fruits of other *Cucurbita* species, and lies in the following range (per 100 g edible portions): Water 85-91 g, protein 0.8 – 2.0 g, fat 0.1-0.5 g, carbohydrates 3.3–11.0 g, Vitamin A 340-7800 IU, Vitamin B1 0.07 – 0.14 mg, Vit. B_2 0.01 – 0.04 mg, niacin 0.5 – 1.2 mg, Vitamin C_6=21 mg, Ca 14-48 mg, Fe 7.0 mg, Mg 16-34 mg, and P 21-39 mg. The energy value is 85-170 kj/100 g. The fruit pulp contains a proteolytic enzyme that has potential value in the food industry. Recent research in Chile has shown that some proteolytic enzymes from the flesh of *C. ficifolia* fruit can be used to treat waste water from the industrial processing of foods derived from fish. Mature fruits can be preserved. This discovery is of great intrest because of the reducing in costs that these industries could achieve by using enzymes which would replace those imported at present. In Latin America, the flesh is impregnated with sugar to make a candy or it can be fermented to make beer (Whitaker, 1990). The ripe fruit is used as fodder for domestic animals. Fig-leaf gourd is also cultivated for its ornamental watermelon like fruits and abundant foliage. The seeds are the most nutritious part of the fruit which is edible in nature and are also, are rich in protein and oil. In some areas, the raw or roasted seeds are eaten as a snack ford. The seeds from the fruit do not contain this bitter chemical, but unless they are washed well they will be coated with it because it is present throughout the pulp of the fruit. The seeds are also greatly valued and in Mexico as they are used with honey to prepare desserts known as *'palanquetas'*. Besides eating them cooked, the Indians extracted oil from the ground up seeds by rinsing them in hot water. The turgid pepos, and to lesser extent the root were used domestically as hand wash and laundry soap, shampoo, and stain remover, a turgid pepo cut in half and added to laundry results in a detergent action. As medicinally, the roots were used by Indians to cure saddle sores on their horses, as a laxative, and as a treatment for hemorrhoids.

Origin, History and Distribution

This feral species of *Cucurbita*, the genus that includes the cultivated squash and pumpkins, is a New World species indiogenous to the arid and semi-arid regions of western North America (Bemis *et al.*, 1978a). The genus *'Cucurbita'* evolved into many species-tropical, mesophytic, and a group of highly specialized species adapted to xerophytic; the buffalo gourd was one of these. The association between the aboriginal Americans and the buffalo gourd has excisted for as long as 9000 years (Cutler and Whitaker, 1961). Although it was never a cultivated plant, it was an excellent "camp follower", being well adapted to perennial growth on trash heaps around centers of habitation. It can be spotted today growing on dump heaps around the mesa top pueblos of the Hopi Indians in some of the oldest continuously inhabited settlements in America (Nabhan, 1980). During this period of its evolution, the buffalo gourd may have become dependent upon man for its survival (Bohrer, 1975), as it is not found mainly on disturbed soil areas such as abandoned cultivated fields, pastures, fence rows and rail road and highway right–of-ways.

At the end of the last and the beginning of 20th century, some authors were suggesting an Asiatic origin for *C. ficifolia*. Since the middle of that century, the consensus has been that it is of American origin (Paris, 1989). Some researchers felt that it may be a cultigens which probably originated in central Mexico and mainly spread to South America with an ecological preference for high land areas (high plateaus of Central America and along the Andes to Central Chile) as suggested by Andres (1990). However, its centre of origin and domestication are still unknown. Biosystematics studies have been unable to support the Mexican origin suggested by the distribution of common names derived from Natural throughout America. The first fruits to reach Europe apparently took a circuitous route from South America to the Malabar Coast of India along the lunch traveled trade routes in the 16th and 17th centuries hence the vernacular names in English and French. In south East Asia it is only grown in the high lands of Luzon, the Philippines, at an elevation of 1700 m above sea level. Its spread to Europe (France and Portugal for example) and Asia (India) appartently began in the sixteenth and seventeenth centuries when its fruit reached the Old world from South America and India. The main production area of *C. ficifolia* is central and south America, northern Mexico to Argentina and Chile (Merrick, 1990). It is only of local importance in South East Asia, *i.e.*, in the Philippines. Since then, its cultivation has spread to many other parts of the world (Germany, France, Japan and the Philippines) as mentioned by Baily (1978).

C. ficifolia is grown over a wide distribution area from 1000 to almost 3000 m in practically all the mountain ranges of Latin America (earlier cultivated plants). Archaeologicalevidences it was cultivated in Peru around 3000 BC (Herklots, 1972). The restriction of cultivation to areas of considerable altitude is a distinctive characteristic of *C. ficifolia* with other cultivated species of the genus, which can generally grow in a wide range of ecological conditions (in the case of *C. pepo* and *C. moschata*, from 0-2300 m).

Species Availability, Evolution and Diversity

The discovery indicates that *Peponapis atrata* does not appear to be a specific pollinator of *Cucurbita ficifolia* has lead to the suggestion that the wild ancestor of this species might have been a still undiscovered species whose habitat could be the eastern region of the Andes. In the taxonomic literature, 3 characteristics are often mentioned to distinguish *C. ficifolia* from other cultivated *cucurbita* species; perennial growth habit, leaf shaped like a leaf of the fig (*Ficus carica* L.) and black seeds. Those characteristic can be misleading because *C. ficifolia* does not differs in longevity from the other squash species, the fig leaf form also occurs in other species and the seeds are not always black. Most diagnostic are the following characteristics: the presence of trichomes on the filaments in the male flowers, and the shape of the seeds (length to width ratio of 3:2 which is broader than in other cultivated cucurbits).

C. foetidissima identified as an underexploited tropical crop by the National Academy of Science (1975), the buffalo gourd has multiple and non-food uses (Bemis *et al.*, 1975, Gathman and Bemis, 1990). It is a perennial and is found growing wild in marginal lands in the southwestern United States. Some plants have been reported to be over 40 years old. It has a very large, fleshy storage root which can grow to depths up to 5 m and weigh as much 30 kg after two growing seasons. Roots of older plants can weigh over 100 kg. Buffalo gourd primarily reproduces by asexual reproduction, but also produces small yellow, hard shelled fruits which are considered inedible. American and Indians have used the ripe fruit as a soap substitute and as ceremonial rattles. The seeds, which contain an abundant quantity of polyunsaturated fats and protein, are edible. The large storage roots contain large amounts of starch (56 per cent of the dry weight), and can also be used as fuel. Air-dried roots burn with the heat equivalent of wood and are being tested in Afghanistan as an alternative fuel to decrease deforestation (Winrock International, 1991). Means, ranges and coefficients of variations of seven fruit and seed characters of 85 buffalo gourd accessions.

Table 3.18: Variations of Fruit and Seed Characters of Buffalo Gourd

Characteristics	*Mean*	*Range*	*CV*
Fruit diameter (cm)	6.5	5.2-7.7	9.1
Seed weight/100 seeds (g)	3.8	1.1-5.5	20.0
Seed weight/fruit (g)	8.4	2.2-18.8	36.9
Seed number/fruit	225	87-386	38.3
Embryo in seed (per cent)	67.4	42.1-76.3	9.7
Crude fat in seed (per cent)	32.9	21.1-43.1	14.3
Crude protein in seed (per cent)	30.7	19.5-35.4	8.7

Among the cultivated *Cucurbita* species, *C. ficifolia* has the most restricted habitat. In the high altitude tropics where it is grown, it will often maintain vigour through cool humid periods while the other species perish because of limited cold tolerance. In South and Central America it is usually grown between 1000-2800 m above sea level. It requires a short photoperiod for flower initiation, but day neutral cultivars have also been reported. *C. ficifolia* is moderately tolerant to acid conditions and grown successfully in fertile, well drained soils with a pH range of 6.5-7.5. Chen *et al.* (2000) reported that *C. ficifolia* it used a grafted seedlings. The higher photosynthetic rate can be obtained as compared with own root seedlings under low temperature stress due to higher chlorophyll content, stomatal conductance and initial activity of RuBP case.

Gene and Genetics of Characters

A dominant mutant gene causing abortion of the male flower buds resulting gynoecious (all female) plants was found to be widespread in native populations.

Botanical Description

The vines of Buffalo Gourd or Fig Leaf Squash or Malabar Gourd (*C. ficifolia* Bouché or *F. foetidissima*) is frost-sensitive and killed by temperature below 0°C, but the roots may survive winter air temperature as low as -25°C, particularly when the soil has insulation of snow cover. The primary mode of reproduction is asexual by the development of adventitious roots produced at the nodes of the vines. Large homogenus colonies of plants are produced in this manner, and annual vine growth is extensive.

Table 3.19: Botanical Description

Growth Habit	It is a short live (annual although persistent for a certain period, giving the impression of being a short-lived perennial by virtue of its exceedingly large fleshy storage roots) creeping or climbing plant, monoecious, herbaceous but becoming somewhat woody. It is villose to softly pubscent, with some short, sharp spines dispersed over the veghetative parts.
Root	Taproot upto 2.0 m long, lateral roots forming a network slightly below the soil surface.Without swollen reserve roots (Whitaker and Davis, 1962). They have denticulate margins and 3-4 ramified tendrils. A single root may reach fresh weights of over 40.0 kg in three or four seasons of growth. The root is so extremely bitter that it was not used as a food.
Stem	Stem with numerous runners up to 10.0 m long, prickly or spiny dispersed over the vegetative parts, hard smoothly 5 angled (slightly angular) to round often rooting at the nodes; tendrils long, branched. The vine growth is extremely prolific. A single root in its second season of growth will produce from 6-20 vine initials thart are capable of producing vines about 6.0 m in length. These initial vines also branch at many of the nodes resulting in a dense mat of vine growth.

Leaf	Leaves simple, alternate, leaf blade circular ovate to nearly reniform in outline, 18-25 cm in diameter, 5-25 cm petioles (ovate-cordate to suborbicular-cordate), sinuate to lobed and with obtuse sinuses, margins apicular serrate to entire. The base width is 10.0-13.0 cm and a midrib length of 20.0-25.0 cm. Leaves are with or without white spots on the surface, and have 3-5 rounded or obtuse, apiculate lobules, the central one bigger than the lateral ones. They have denticulate margins and three to four ramified tendrils.
Floral biology	The large unisexual flowers (pistillate or staminate) are borne singly at most nodes of the vine. Flowers solitary, pentamerous, solitary, and axillary, yellow to light orange upto 7.5 cm in diameter, calyx (5.0-10.0 mm long and almost as wide, 5.0 to 15 x 1.0-2.0 mm) linear sepals and corolla campanulate with short tube (it is rather broader towards the base, 6.0-12.0 cm long and yellow to pale orange), staminate flowers (long and pedicellate) with short thick and columnar androecious, filaments with trichomes more than 1.0 mm long; pistillate flowers on short, ridged pedicels with small, smooth pale yellow gynoecium. They have three stamens. The female flowers have sturdy peduncle, 3.0-5.0 cm long, ovoid to elliptical, multilocular ovary. They have thickened style and have 3 lobate stigmas.
Fruit and fruit set	Fruit a pepo, globular to cylindrical, 15.0-20.0 cm long, 5.0-7.0 cm in diameters, white to green with white stripes and blotches, rind smooth, hard, flesh white, coarse, tough, fibrous and rather dry fruit stalk hard, round to 5 angled without cork development, not or only slightly enlarged at point of fruit attachment. A single plant is capable of producing as many as 200 fruits in a single growing season.
Seeds	The number of seeds per fruit ranges from 200-300. Seed flattened oblong ellipsoidal, 1.5 to 2.5 cm long, length to width ratio of 3:2 hard without a spongy epidermis, black or sometimes light but creamy white coloured. The weight of 1000 seeds is 180-250 g. The seed contains 30-40 per cent edible oil and 30-35 per cent protein.

Botanical Classification

Fruits have three colour patterns:

(i) Light or dark green, with or without longitudinal white lines or stripes towards the apex;

(ii) Mainly spotted white and green;

(iii) White, cream or flesh white.

Augmentation of Genetic Resources and Crop Improvement

In view of the reproductive incompatibility of *C. ficifolia* with the other species of the genus, it may be said that its genetic stock is limited to itself. Howeve, in other respects this species is much less different than are other cultivated species of the genus and there are no commercial cultivars. *C. ficifolia* is a resistance source against cucumber green mottle mosaic virus. There has been no breeding programmes for the improvement of *C. ficifolia*. Although, the cultivated *Cucurbita* species have developed a sequence of sterility barriers that prevents gene flow, *C. ficifolia* is potentiality valuable for squash breeding. It is resistant to several viruses, shows some tolerance to powdery mildew has the ability to grow under cool moist conditions and the mature fruits have long storage life without refrigeration. Fig Leaf Squash or Malabar Gourd shows considerable biochemical differences from the other members of its genus, and does not hybridize readly with them. It became popular in western countries (Japan and Germany) as a rootstock for winter production of green house cucumber due to its cold tolerance and good resistance to soil borne pathogens. It is resistant to low temperatures but not to severe frosts.

Interspecific Hybridization

C. ficifolia can be hybridized with the other squash species by the use of embryo culture. Attempts at obtaining hybrids beyond the first generation with the other four cultivated species have failed and the few results obtained have required the use of special techniques such as embryo cultivation. These results have been corroborated by other studies which reveal that *C. ficifolia* shows considerable isoenzymatic and chromosomic differences compared with all the taxa of the genus. Possibility of using wild (or cultivated) species in future programmes for the genetic improvement of this crop and their use in the improvement of other cultivated species of the genus is still remote. Considering the importance of wild species and following suitable breeding methodology in improvement programmes lies in the fact that collections have been identified which are resistant or completely immune to the attack of different viruses that severely affect other cultivated species.

Genetic Resource Conservation

C. ficifolia is the leaf variable of the cultivated *cucurbita* species. It is represented by varying numbers of accessions in the *Cucurbita* collections of many institutions in particular in central and South America (CATIE), Turrialba, Costa Rica, San Carlos University, Guatemala, INIFAP, Celaya, Mexico in the United States (Southern Regional Plant Introduction Station Georgia) and in Russia (Vavilov Institute of Plant Industry, Petersburg). The one sizeable collection in South East Asia is at the Institute of Plant Breeding, the Philippines. Accession numbers and collection locations of lines selected for breeding and seed production are given in Table 3.20.

Table 3.20: Accession Numbers and Locations of Collected Lines

Accession Number	*Collection Location*
140	Forest Dale, Arizona
142	Jerome, Arizona
156	Greaterville, Arizona
158	Patagnia, Arizona
162	Lochiel, Arizona
185	Pearce, Arizona
250	Hatch, New Mexico
300	Saltillo, Mexico

The buffalo gourd is currently being grown in Australia where several hundred acres have been seeded under dry-land agriculture.

REFERENCES

1. Andrés, T.C. 1990. Biosystematics, theories on the origin and breeding potential of *Cucurbita ficifolia.* In: Bates, D.M., Robinson, R.W. and Jeffrey, C., eds. *Biology and Utilization of the Cucurbitaceae*, p. 102-199. Ithaca, N.Y., USA, Cornell University Press.
2. Bates, D.M., Robinson, R.W. and Jeffrey, C., eds. *Biology and Utilization of the Cucurbitaceae*, p. 102-199. Ithaca, N.Y., USA, Cornell University Press.
3. Baily, L.H. 1978. The domesticated cucurbits. First Paper. *Genet. Herb.*, **2:** 23-34.
4. Bemis, W.P., Curtis, L.C., Weber, C.W., Berry, J.W. and Nelson, J.M. 1975. The buffalo gourd (*Cucurbita foetidissima* HBK.): A potential crop for the production of protein, oil, and starch on arid lands. Office Agr., Tech. Assistance Bureau, USAID Washington, DC.
5. Cutler, H.C. and Whitaker, T.W. 1961. History and distribution of the cultivated cucurbits in the America. *Am. Antiq.*, **26:** 469-485.
6. Gathman, A.C. and Bemis, W.P. 1990. Domestication of buffalo gourd (*Cucurbita foetidissima* HBK.). In: Bates, D.M., Robinson, R.W. and Jeffrey, C., eds. *Biology and uutilization of the Cucurbitaceae*, p. 318-324, Ithaca, N.Y., USA, Cornell University Press.
7. Herklots, G.A.C. 1972. Vegetable in South East Asia. George Allen and Unwin, London.
8. Merrick, L. C. 1990. Systematics and evolution of a domesticated squash, *Cucurbita argyrosperma* and its wild and weedy relatives. In: Bates, D.M., Robinson, R.W. and Jeffrey, C., eds. *Biology and uutilization of the Cucurbitaceae*, p. 102-199. Ithaca, N.Y., USA, Cornell University Press.
9. Nabhan, G. 1980. *Ammobroma sonoraean* endangered Parasitic Plant in extremely Arid in North America. Tuscan: The University of Arizona. *Desert Plants*, **2(3):** 188-196.
10. National Academy of Sciences.1975. Underexploited tropical plants with promising economic value. Natl. Acad. Sci., Washington, DC.
11. Paris, H. 1989. Historical records, origin and development of the edible cultivar groups of *Cucurbita pepo* (Cucurbitaceae). *Econ. Bot.*, **43:** 423-443.
12. Whitaker, T.W. 1990. Cucurbits of potential economic importance. In: Bates, D.M., Robinson, R.W. and Jeffrey, C., eds. *Biology and uutilization of the Cucurbitaceae*, pp. 318-324. Ithaca, N.Y., USA, Cornell University Press.
13. Whitaker, T.W. and Davis, G.N. 1962. Cucurbits botany, cultivation and utilizations. London, Leonard Hill.
14. Winrock International. 1991. Development: the changing landscape-Winrock International Annual Report 1990. Winrock International, Morrilton, AR.

Summer Squash or Zucchini (*Cucurbita pepo* L. var. *cylindrica*)

Introduction

This prized vegetable is known as pumpkin, vegetable marrow, summer pumpkin, autumn pumpkin (English), Spanish: calabaza (Mexico) and Huicoy (Guatemala). Like the other species of the genus, the mature or young fruit and the seeds of *C. pepo*, as well as to a lesser extent the flowers and young tips of the stems, are eaten in many parts of its native distribution area and in other regions of the world. At present, the fruit of some cultivars (for example, Zucchini and Cocozzelle) has a nutritional and commercial role in several regions of the world. The use of seeds as dried fruit is common in some areas of Mesoamerica and almost unknown in others. The seeds are a good source of protein and oil, and their industrial preparation and marketing should be investigated. The word squash may have been derived from an American Indian Word for the *Cucurbita* fruit.

Origin, History and Distribution

According to archaeological recordings, *C. pepo* appears to be one of the first domesticated species. The oldest remains have been found in Mexico, in the Oaxaca valley (8750 BC to AD 700) and in the caves of Ocampo, Tamaulipas (7000 to 500 BC). Its presence in the United States also dates back a long time, as the recordings in Missouri (4000 BC) and Mississippi (1400 BC) indicates. *Cucurbita pepo* is said to be originated from America probably in Northeastern Mexico, where archeological evidence has been found which suggests that its fruits were eaten as early 9000 year ago. It has been widely distributed and some of its cultivars withstand cooler climate than the other species of *Cucurbita*.

This species may have been domesticated at least on two occasions and in two different regions:

(i) In Mexico

(ii) Eastern United States

In each case having *C. fraterna* and *C. texana*, respectively, as possible progenitors. The question of the origin of the cultivars native to the Mayan area in the middle and low areas of Chiapas and the Yucatan peninsula still has to be resolved. These cultivars, whose fruits are either without or with unpronounced ribs and have rather rounded and oval seeds, are grown from a little above sea level to nearly 1800 m. The distribution of outside America is possibly the best documented of this genus. It is known that some cultivars reached Europe approximately half a century after 1492 and it is even said that others originated on that continent. In contrast to *C. pepo*'s long-established presence in the Old World, it seems that its arrival in South America was very recent. Traditionally, *C. pepo* is cultivated from North America to Central America and in some parts of South America, although it is generally said to be a crop of high areas. Like *C. moschata*, this species covers a fairly wide range of altitudes. In Mexico, there are native varieties which grow from very close to sea level and in semi-dry climates, such as the variety called tsol in Yucatan, to others which are cultivated at altitudes greater than 2000 m, such as those called guiches in Oaxaca. In Guatemala, the native cultivars, commonly called guicoy, are grown above 1000 m, while tsols are sown in the low, hot humid parts of the Peten below 500 m.

Species Availability, Evolution and Diversity

In cultivated group, this crop possesses pumpkin (*C. pepo* L. var. *pepo* L. Bailey), scallop (*C. pepo* L. var. *clypeata* Alefield), acorn (*C. pepo* L. var. *turbinate* Paris), crookneck (*C. pepo* L. var. *torticollia* Alefield), straightneck (*C. pepo* L. var. *recticollis* Pans), vegetable marrow (*C. pepo* L. var. *fastigata* Paris), cocozzelle (*C. pepo* L. var. *longa* Paris) and Zuchini (*C. pepo* L. var. *cylindrica*). The primary genetic stock of *C. pepo* is formed by the groups of edible cultivars (ssp. *pepo*) and ornamental cultivars (ssp. *ovifera*) as well as wild taxa (*C. fraterna* and *C. texcana*).

Botanical Description

Table 3.21: Botanical Description

Growth Habit	Creeping plants which are compact or semi-shrubby, annual, monoecious and pubescent-scabrous. Tendrils have two to six branches, or are simple and little developed tendrils in the semi-shruby types.
Leaf	Broadly ovate-cordate to triangular-cordate leaves, 20 to 30 x 20 to 35 cm, with or without white spots, often with three to five deep lobules, and denticulate to serrate-denticulate margins.
Floral biology	Pentamerous, solitary, axillary flowers, the males of which have pedicels 7b to 20 cm in length, a campanulate calyx of 9 to 12 mm, linear sepals of 12 to 25 x 1 to 2 mm, a tubular/campanulate corolla, 5 to 10 cm long, which is divided into five for up to one-third or more of its length; and three stamens. The female flowers have sturdy, subcated pedicels of 2 to 5 cm; the ovary is globose, oblate, ovoid, cylindrical, rarely piriform, smooth, ribbed or verrucose and multilocular; and the calyx is very small.
Fruit and fruit set	The fruit is very variable in size and shape: smooth to heavily obbed, oftein verrucose and rarely smooth, with a rigid skin varying in colour from light to dark green, plain to minutely speckled with cream or green contrasting with yellow, orange or two-coloured. The flesh is cream to yellowish or pale orange; it ranges from soft and not bitter to fibrous and bitter, has numerous seeds which are narrowly or broadly elliptical or rarely orbicular, slightly flattened and 3 to 20 x 4 to 12 mm.

Horticultural Classification

With regard to traditional cultivars, it is common to see a fair representation of cultivars with characteristics similar to those of each of the commercial groups in one single field cultivated by Mesoamerican peasants. Eight groups of edible cultivars of *C. pepo* are known:

(i) Pumpkin (*C. pepo* L. var. *pepo* L. Bailey) includes cultivars of creeping plants which produce spherical, oval or oblate fruit that is rounded or flat at the ends. The fruit of this group is grown to be eaten when ripe and sometimes is used as fodder.

(ii) Scallop (*C. pepo* L. var. *clypeata* Alefield) has a semi-shrubby habit, the fruit ranges from flat to almost discoidal, with undulations or equatorial margins, and it is eaten before maturity.

(iii) Acorn (*C. pepo* L. var. *turbinata* Paris) is both a shrubby and creeping plant with fruit which is obovoid or conical, pointed at the apex and longitudinal costate-grooved. The rind is soft; hence the fruit can be eaten in the ripe state.

(iv) Crookneck (*C. pepo* L. var. *torticollia* Alefield) is a shrubby type, with yellow, golden or white fruit which is claviform and curved at the distal or apical end and generally has a verrucose rind. It is eaten unripe since the rind and the flesh harden when ripe.

(v) Straightneck (*C. pepo* L. var. *recticollis* Pans) is a shrubby plant with yellow or golden fruit and a verrcose rind similar to that of var. *torticollia*.

(vi) Vegetable marrow (*C. pepo* L. var. *fastigata* Paris) has creeper characteristics as a semi-shrub and has short cylindrical fruit that is slightly broader at the apex, with a smooth rind which hardens and thickens on ripening and which varies in colour from cream to dark green.

(vii) Cocozzelle (*C. pepo* L. var. *longa* Paris) has cylindrical, long fruit that is slender and slightly bulbous at the apex; it is eaten in the unripe state and one of the most common names is Cocozzelle.

(viii) Zuchini (*C. pepo* L. var. *cylindrica*) is the most common group of cultivars at present. Like the previous group, the zucchini group has a strong affinity with the vegetable marrow and its origin is also recent (nineteenth century). Its plants are generally semi-shrubby and its cylindrical fruit does not broaden or else broadens only slightly. It is eaten as a vegetable in the unripe state.

Breeding Priorities

- ✰ High yield
- ✰ Earliness
- ✰ Narrow seed cavity
- ✰ High β-carotene and thick flesh

- ☆ Uniform fruit shape and size
- ☆ Transport quality
- ☆ Resistant to diseases (Powdery mildew, downy mildew and viral diseses) and insect oests (red pumpkin beetle and fruit fly)

Introduction and Selection

In summer squash cultivars like 'Patty Pan' and 'Australian Green' were introduced and used for commercial cultivation directly. In India some cultivars like Punjab Chapan Kaddoo No.-1 (PAU, Ludhiana), Pusa Alankar (ICAR-IARI, New Delhi) and Australian Green (Introduction from Australia) are commercial being grown. Popular cultivars like Punjab Chappan Kaddu-1 and Early Yellow prolific were developed by use of Indigenous material through selection with progeny testing. New selection has been made in summer squash (pepo) Kashi Subhangi (VRSS-06-12) from ICAR-IIVR, Varansi (U.P.).

In Nepal following varieties of Zucchini are commonly being grown:

Table 3.22: Isolation of Zucchini Cultivars by Hognong Seed Co., Korea

Varieties	*Source*
Bulam House, Hungnong Zucchini, Rondo, Golden	Hognong Seed Co., Korea

Resistance source: In *C. pepo* line PI438699 is resistant to cucumber onosaic virus (CMV) as reported by Lebeda and Kristkova (2007). Similarly against *Fusarium oxysporum* and powdery mildew. CV. Jhinghu No.1 is resistant (Littaizhen *et al*; 2005).

Genetic Resource Conservation

There are a great number of commercial cultivars with particular characteristics which, together with the local varieties referred to, which are grown mainly in Mexico, constitute an extraordinary genetic stock. However, in contrast to other species, this diversity does not represent an important source of genes resistant to pests and disease. Since *C. pepo* (including *C. texana*) is probably the species with the greatest susceptibility to the most important viral diseases that attack cultivated species of *Cucurbita*. Species that might represent a secondary genetic stock are scarce, as most of the attempts at hybridizing *C. pepo* with other wild or cultivated species have required special techniques such as embryo culture; however, good results with hybridization have been achieved in Mexico and the United States. Data obtained from the USDA gene banks show that *C. pepo* is the species of the genus with the second highest number of accessions (1135). However, this refers only to cultivated and edible forms, since those corresponding to the two closest wild relatives are very scarce and, in fact, those of *C. fraterna* were completed only very recently. The gene banks with the greatest representation of *C. pepo* cultivars are in the United States, Mexico and Costa Rica. In Yucatan, the tsol or 'mensejo' (short cycled) variety is grown, generally in vegetable gardens. Production has recently increased with specialty forms such as spaghetti squash, whose internal flesh texture resembles strands of spaghetti following cooking.

There is still much to be done in terms of the collection, conservation, evaluation and use of regional or local varieties. These tasks are feasible. As the range of these crops are still to be found in the rural communities of the New World.

Cushaw (*Cucurbita argyrosperma* Huber)/(=*C. mixta*)

Introduction

Cushaw (*Cucurbita argyrosperma* Huber); =*C. mixta*) is known as cushaw (United States); Spanish: calabaza, calabaza pinta, calabaza pipiana (Mexico), pipian (Mexico, El Salvador, Nicaragua, Costa Rica), and saquil, pipitoria (Guatemala). The flowers, young stems, young fruit and ripe fruit of *C. argyrosperma* are eaten as vegetables. The ripe fruit is occasionally used to prepare sweets although it is used most frequently as feed for live stock and poultry. The seeds are eaten whole, roasted, toasted or ground and are the main ingredient of sauces used to prepare various stews (for example- *pipian*-meat cooked in thick chilli sauce, *mole verde*-a stew prepared with chilli and green tomatoes). The seeds are the most important product, chiefly because of their oil (39 per cent) and protein (44 per cent), and their consumption in urban areas of Mexico and other countries of Central America is fairly common. In some regions of Mexico, the seeds and also the unripe fruit of wild taxa are used as food. The latter are eaten after being washed and boiled several times to rid them of the bitter taste deriving from the cucurbiacins present in the flesh and placenta, while the seeds are simply washed, seasoned with salt and roasted or toasted. On the Yucatan peninsula, peasants use the flesh of the cultivated varieties fruit to treat burns, sores and skin eruptions, while the seeds are prepared with water and used as an anesthetic and to stimulate women's milk production for breastfeeding.

Origin, History and Distribution

According to the age of the archacological remains discovered thus far, it has been suggested that domestication of *Cucurbita argyrosperma* must have occurred in southern Mexico more than 7000 years ago. As a history, the different degrees of variation in the nutritionally important parts of the three cultivated varieties of the complex *argyrosperma* suggest a strong association with human intrests. The relatively large seed size of the var. *argyrosperma* indicates that it was mainly selected to obtain seeds, while the great diversity of shapes, colours and size of the fruits and seeds of var. *stenosperma* and *callicarpa* indicate that selection had a double aim:

(i) to obtain flesh
(ii) to obtain seeds

Unlike with the rest of the cultivated species of *Cucurbita*, data on the distribution of cultivated *Cucurbita argyrosperma* varieties outside America are very scarce and there is no certainty that this species was cultivated at any time in the Old World or even outside its general area of domestication. In South America, it is grown in Peru and Argentina, although it appears to involve very recent introductions of certain cultivars which can be classified within the var. *callicarpa*. In the United States, some cultivars of var. *callicarpa* are cultivated on a very low scale for nutritional purposes, and one cultivar of var. *argyrosperma*, Silver Seed Gourd, is occasionally grown as a horticultural curiosity. The reasons for the sparse world distribution of this species are not known; although the situation is not surprising, given the low quality of the fruit's flesh compared with that of *C. moschata* or *C. pepo* and the size of the seeds of all the cultivated varieties that may have been attractive to the first Europeans who learned of them. The three cultivated varieties of *C. argyrosperma* are found in a relatively wide range of altitudes (0 to 1800 m), generally in areas with a hot, fairly dry climate or a well-defined rainy season. The species does not tolerate very low temperatures, which limits its cultivation to the altitudes mentioned. Each cultivated variety has a fairly defined distribution model, although there are some areas where two varieties can be found cultivated simultaneously.

Species Availability, Evolution and Diversity

Cucurbita argyrosperma is one of the cultivated species of the genus which has undergone the most profound study in recent years. There are two subspecies:

(i) *argyrosperma*, comprising four varieties *argyrosperma, callicarpa, stenosperma* and *palmieri*- three of which include all the cultivated types, while the fourth corresponds to spontaneous populations of northeastern Mexico that are generally known as *Cucurbita palmieri* L. Bailey.

(ii) *sororia*, which includes wild populations with a wide distribution from Mexico to Nicaragua, originally described under the name *C. sororia* L. Bailey. This subspecies has been designated as the wild ancestor of the group.

The characteristics that were most transformed in the process of domestication of the ssp. *argyrosperma* were, as in the crops, mainly those connected with handling and preferred uses. For example, relatively uniform germination; a reduction in size and abundance of trichomes; an increase in the size of parts and organs used, such as fruits, and seeds; and a reduction in the bitter taste of the flesh. It is considered that var. *argyrosperma* is the least specialized or most primitive of the group and that var. *callicarpa*, on the other hand, is the most recent or specialized. In Mexico, the var. *argyrosperma* is grown on the slope of the gulf (Tamaulipas, San Luis Potosi, Puebla, Veracruz, Tabasco, Chiapas and Yucatan). In Central America it has been recorded in Belize, Guatemala, Honduras, El Salvador, Nicaragua, Costa Rica and Panama. The var. *callicarpa* is found mainly on the Pacific slope, from the southern United States to central Mexico (Sonora, Sinaloa, Chihuahua, Zacatecas, Guanajuato, Nayarit and Jalisco). The var. *stenosperma* is endemic to Mexico and is grown in the central and southeastern states (Guerrero, Morelos, Michoacan and Oaxaca) as well as some areas of the gulf slope (Veracruz and Yucatan).

Botanical Description

Table 3.23 : Botanical Description

Growth Habit	Very vigorus plant. Villous to pubescent and which may be hirsute, with short, rigid and rather enlarged and sharp trichomes.
Root	It has fibrous roots and slightly angular stems.
Leaf	Its ovate-cordate leaves have petioles of up to 30 cm and measure 10 to 30 x 15 to 40 cm. They have white spots, number three to five and are lobulated with triangular or elliptical lobules. The margins are denticulate to serrate-denticulate. There are 2 to 4 ramified tendrils and pentamerous, solitary, axillary flowers.
Floral biology	The male flowers are on pedicels of 10 to 20 cm and have a campanulate calyx of 5 to 20 x 8 to 25 mm. There sepals are linear-lanceolate or (rarely) foliaceous and are 10 to 35 mm long. They have a tubular-campanulate corolla that is yellow to orange, 6 to 12 cm long with 5 lobules for up to one-third of its total length and they have three stamens. The female flowers grow on sturdy peduncles of 2 to 3.5 cm; have a globose, ovoid-elliptical, botuliform or piriform, multilocular ovary, a small calyx and a corolla that is somewhat bigger than that of the male flowers. They have 3 stigmas.
Fruit and fruit set	The fruit is short or long and piriform, straight or curved in the thinnest part and 11 to 50 cm long. It has a hard rind which is smooth to slightly ribbed and is white with longitudinal green reticulate stripes or completely white. The flesh is white, yellow or orange, the seeds elliptical and slightly inflated, measuring 15 to 30 x 8 to 16 mm, with a white, smooth and even testa.

Cucurbita argyrosperma is a monoecious plant.

Augmentation of Genetic Resources and Crop Improvement

Genetic relatios of *C. argyrosperma* Huber and the consequent inclusion and definition of wild and cultivated taxa within its taxonomic limits have considerably widened the species genetic stock. This includes: the local races of the cultivated vrieties in the southe eastern United States. Mexico and Central America; the two wild taxa of the complex (var. *palmieri* and ssp. *sororia*): and, in the United States, Green Striped Cushaw, White Cushaw, Magdalena Striped, and Papago, Silver Seed gourd, Japanese Pie, Hopi, Taos, Parral Cuishaw and Veracruz Pepita.

Interspecific Hybridization

(i) First hybridization experiments of the taxa belonging to *Cucurbita argyrosperma* with other wild and cultivated taxa of the genus and some field observations have revealed that, with the cultivated species, *C. moschata* has the highest degree of compatibility.

(ii) A second level of compatibility consists of the wild and cultivated taxa of *C. pepo*, some cultivars of *C. maxima* and the wild perennial species of *C. foetidissima* H.B.K.

(iii) A third group is formed by *C. lundelliana* L. Bailey, and *C. martinezii* L. Bailey, with which crossing only produced fruit without viable seeds.

(iv) The fourth and last group includes the perennial species *C. pedatifolia* L. Bailey, *C. digitata* A. Gray, in the broad sense and *C. radicans* Naudin with which only a few fruits were able to be obtained but generally without viable seeds.

However, because of two factors relating to interspecific crossings, some obstacles will have to be overcome before a successful plant improvement programme including all the taxa is carried out:

- ✰ Hybrids with viable seeds were not always produced.
- ✰ When some success was achieved, the plant receiving the pollen was always from the complex *C. argyrosperma* and, in the case of the species of the last two groups; it always involved var. *palmieri* or ssp. *sororia*.

Biotic Resistance

The wild species that have shown some degree of compatibility with the taxa of the complex *C. argyrosperma* possess genes resistant to some viral diseases that have a high incidence in the cultivated species.

Genetic Resource Conservation

Germplasm collections are not scarce for the cultivated taxa but they are in very short supply in the case of the wild ones. Most of the accessions come from Mexico and are deposited in the CIFAP's gene bank in Mexico and in the USDA's in the United States.

Genus *Cucumis*

Introduction

Presently *Cucumis* includes over 40 species of which 8 are Indian (Chakravarthy, 1959) and are available either in wild or in cultivated forms in Africa and Asia (Swarup, 2006). All these species are monoecious. It is an annuals and herbaceous perennials of the Old World tropics and warm temperate regions. Earlier the reports were available about only 30 (Swarup, 2006). The major crops of economic importance are cucumber (*Cucumis sativus*), muskmelon (*C. melo*), snapmelon (*C. melo* var. *momordica*), long melon (*C. melo* var. *utilissimum*). The Indian sub-continent is said to be the center of origin for *C. sativus* and center of diversity for *C. melo*. The genus *Cucumis* comprises of about several species.

Jeffrey (1980) classified the entire genus into two subgenera:

1. Subgenus *Cucumis*, x=7, 3 or 4 Sino-Himalayan species including *C. sativus* L., *C. hystrix*, Chakra. (syn. *C. mauriculatus* Chakra.).
2. Subgenus *Melo*, x=12, 25 species mostly in tropiocal and South Africa.
   - *(i)* *anguria* group-dioecious, monoecioous or andromonoecious perennials or annuals with yellowish or brown striped browny fruit (20 sp.), closely related and partially fertile interspecific hybrids, *C. anguria*, *C. dipsaceus*, *C. prophetarum*, *C. myriocarpus* and *C. sacleuxii*.
   - *(ii)* *metuliferus* group- monoecious, annuals with red spiny fruits, *C. metuliferus* Naud.
   - *(iii)* *melo* group-monoecious, andromonoecious perennials or annuals with smooth fruits, *C. melo*, *C. sagittatus* Peyr. (syn. *C. anglolensis*), *C. dinteri* Cogn; *humifructus* Stent.
   - *(iv)* *hirsutus* group-dioecious perennials with smooth orange fruit. *C. hirsutus* Sond.

Table 3.24: Name of Taxon and Preferred Taxon Name of *Cucumis* species

Name of Taxon	Name of Authority	Family	Preferred Taxon Name
Cucumis		Cucurbitaceae	*Cucumis*
Cucumis		Cucurbitaceae	*Cucumis*
Cucumis aculeatus	Cogn.	Cucurbitaceae	*Cucumis aculeatus*
Cucumis acutangulus	L.	Cucurbitaceae	*Luffa acutangula*
Cucumis africanus	L. f.	Cucurbitaceae	*Cucumis africanus*
Cucumis angolensis	Hook. f. ex Cogn.	Cucurbitaceae	*Cucumis sagittatus*
Cucumis anguria	L.	Cucurbitaceae	*Cucumis anguria*
Cucumis anguria var. *anguria*		Cucurbitaceae	*Cucumis anguria* var. *anguria*
Cucumis anguria var. *longaculeatus*	J. H. Kirkbr.	Cucurbitaceae	*Cucumis anguria* var. *longaculeatus*
Cucumis anguria var. *longipes*	(Hook. f.) A. Meeuse	Cucurbitaceae	*Cucumis anguria* var. *anguria*
Cucumis anguria var. *longipes*	auct.	Cucurbitaceae	*Cucumis anguria* var. *longaculeatus*
Cucumis asper	Cogn.	Cucurbitaceae	*Cucumella asper*
Cucumis baladensis	Thulin	Cucurbitaceae	*Cucumis baladensis*
Cucumis carolinus	J. H. Kirkbr.	Cucurbitaceae	*Cucumis carolinus*
Cucumis chito	C. Morren	Cucurbitaceae	*Cucumis melo* var. *chito*
Cucumis collosus	(Rottler) Cogn.	Cucurbitaceae	*Cucumis melo* subsp. *melo*
Cucumis colocynthis	L.	Cucurbitaceae	*Citrullus colocynthis*
Cucumis dinteri	Cogn.	Cucurbitaceae	*Cucumis sagittatus*
Cucumis dipsaceus	Ehrenb. ex Spach	Cucurbitaceae	*Cucumis dipsaceus*
Cucumis dudaim	L.	Cucurbitaceae	*Cucumis melo* var. *chito*
Cucumis dudaim var. *aegyptiacus*	Sickenb.	Cucurbitaceae	*Cucumis melo*
Cucumis ficifolius	A. Rich.	Cucurbitaceae	*Cucumis ficifolius*
Cucumis figarei	Delile ex Naudin, nom. illeg.	Cucurbitaceae	*Cucumis ficifolius*
Cucumis figarei	auct.	Cucurbitaceae	*Cucumis pustulatus*
Cucumis figarei var. *dissectus*	Naudin	Cucurbitaceae	*Cucumis prophetarum* subsp. *dissectus*
Cucumis figarei var. *echinophorus*	Naudin	Cucurbitaceae	*Cucumis prophetarum* subsp. *dissectus*
Cucumis flexuosus	L.	Cucurbitaceae	*Cucumis melo* var. *flexuosus*
Cucumis globosus	C. Jeffrey	Cucurbitaceae	*Cucumis globosus*
Cucumis hardwickii	Royle	Cucurbitaceae	*Cucumis sativus* var. *hardwickii*
Cucumis hastatus	Thulin	Cucurbitaceae	*Cucumis hastatus*
Cucumis heptadactylus	Naudin	Cucurbitaceae	*Cucumis heptadactylus*
Cucumis hirsutus	Sond.	Cucurbitaceae	*Cucumis hirsutus*
Cucumis hookeri	Naudin	Cucurbitaceae	*Cucumis africanus*
Cucumis humifructus	Stent	Cucurbitaceae	*Cucumis humifructus*
Cucumis hystrix	Chakrav.	Cucurbitaceae	*Cucumis hystrix*
Cucumis insignis	C. Jeffrey	Cucurbitaceae	*Cucumis insignis*
Cucumis jeffreyanus	Thulin	Cucurbitaceae	*Cucumis jeffreyanus*
Cucumis kalahariensis	A. Meeuse	Cucurbitaceae	*Cucumis kalahariensis*
Cucumis leptodermis	Schweick.	Cucurbitaceae	*Cucumis myriocarpus* subsp. *leptodermis*
Cucumis longipes	Hook. f.	Cucurbitaceae	*Cucumis anguria* var. *anguria*
Cucumis maderaspatanus	L.	Cucurbitaceae	*Mukia maderaspatana*
Cucumis meeusei	C. Jeffrey	Cucurbitaceae	*Cucumis meeusei*
Cucumis melo	L.	Cucurbitaceae	*Cucumis melo*
Cucumis melo subsp. *agrestis*	(Naudin) Pangalo	Cucurbitaceae	*Cucumis melo* subsp. *agrestis*
Cucumis melo subsp. *conomon*	(Thunb.) Greb., nom. inval.	Cucurbitaceae	*Cucumis melo* var. *conomon*
Cucumis melo subsp. *melo*		Cucurbitaceae	*Cucumis melo* subsp. *melo*
Cucumis melo var. *acidulus*	Naudin	Cucurbitaceae	*Cucumis melo*
Cucumis melo var. *aegyptiacus*	(Sickenb.) Hassib	Cucurbitaceae	*Cucumis melo*
Cucumis melo var. *agrestis*	Naudin	Cucurbitaceae	*Cucumis melo* subsp. *agrestis*
Cucumis melo var. *ameri*	Gabaev	Cucurbitaceae	*Cucumis melo*

Name of Taxon	Name of Authority	Family	Preferred Taxon Name
Cucumis melo var. *cantalupensis*	Naudin	Cucurbitaceae	*Cucumis melo* var. *cantalupensis*
Cucumis melo var. *chito*	(C. Morren) Naudin	Cucurbitaceae	*Cucumis melo* var. *chito*
Cucumis melo var. *conomon*	(Thunb.) Makino	Cucurbitaceae	*Cucumis melo* var. *conomon*
Cucumis melo var. *dudaim*	(L.) Naudin	Cucurbitaceae	*Cucumis melo* var. *chito*
Cucumis melo var. *duripulposus*	Filov, nom. inval.	Cucurbitaceae	*Cucumis melo*
Cucumis melo var. *flexuosus*	(L.) Naudin	Cucurbitaceae	*Cucumis melo* var. *flexuosus*
Cucumis melo var. *hibernus*	Filov, nom. inval.	Cucurbitaceae	*Cucumis melo*
Cucumis melo var. *inodorus*	H. Jacq.	Cucurbitaceae	*Cucumis melo* var. *inodorus*
Cucumis melo var. *makuwa*	Makino	Cucurbitaceae	*Cucumis melo*
Cucumis melo var. *microspermus*	Nakai ex Kitam., nom. illeg.	Cucurbitaceae	*Cucumis melo*
Cucumis melo var. *momordica*	(Roxb.) Duthie and J. B. Fuller	Cucurbitaceae	*Cucumis melo* var. *momordica*
Cucumis melo var. *reticulatus*	Naudin, nom. illeg.	Cucurbitaceae	*Cucumis melo* var. *cantalupensis*
Cucumis melo var. *texanus*	Naudin	Cucurbitaceae	*Cucumis melo* var. *texanus*
Cucumis melo var. *utilissimus*	(Roxb.) Duthie and J. B. Fuller	Cucurbitaceae	*Cucumis melo* var. *conomon*
Cucumis membranifolius	Hook. f.	Cucurbitaceae	*Oreosyce africana*
Cucumis metuliferus	E. Mey. ex Naudin	Cucurbitaceae	*Cucumis metuliferus*
Cucumis microspermus	Nakai	Cucurbitaceae	*Cucumis melo*
Cucumis momordica	Roxb.	Cucurbitaceae	*Cucumis melo* var. *momordica*
Cucumis myriocarpus	Naudin	Cucurbitaceae	*Cucumis myriocarpus*
Cucumis myriocarpus subsp. *leptodermis*	(Schweick.) C. Jeffrey and P. Halliday	*Cucurbitaceae*	*Cucumis myriocarpus subsp. leptodermis*
Cucumismyriocarpus subsp. *myriocarpus*		Cucurbitaceae	*Cucumis myriocarpus* subsp. *myriocarpus*
Cucumis naudinianus	Sond.	Cucurbitaceae	*Acanthosicyos naudinianus*
Cucumis pepo	(L.) Dumort.	Cucurbitaceae	*Cucurbita pepo*
Cucumis prolatior	J. H. Kirkbr.	Cucurbitaceae	*Cucumis prolatior*
Cucumis prophetarum	L.	Cucurbitaceae	*Cucumis prophetarum*
Cucumis prophetarum subsp. *dissectus*	(Naudin) C. Jeffrey	Cucurbitaceae	*Cucumis prophetarum* subsp. *dissectus*
Cucumis prophetarum subsp. *prophetarum*		Cucurbitaceae	*Cucumis prophetarum* subsp. *prophetarum*
Cucumis pubituberculatus	Thulin	Cucurbitaceae	*Cucumis pubituberculatus*
Cucumis pustulatus	Hook. f.	Cucurbitaceae	*Cucumis pustulatus*
Cucumis quintanilhae	R. Fern. and A. Fern.	Cucurbitaceae	*Cucumis quintanilhae*
Cucumis rigidus	E. Mey. ex Sond.	Cucurbitaceae	*Cucumis rigidus*
Cucumis rostratus	J. H. Kirkbr.	Cucurbitaceae	*Cucumis rostratus*
Cucumis sacleuxii	Pailleux and Bois	Cucurbitaceae	*Cucumis sacleuxii*
Cucumis sagittatus	Peyr.	Cucurbitaceae	*Cucumis sagittatus*
Cucumis sativus	L.	Cucurbitaceae	*Cucumis sativus*
Cucumis sativus var. *anatolicus*	Gabaev	Cucurbitaceae	*Cucumis sativus* var. *sativus*
Cucumis sativus var. *anglicus*	L. H. Bailey	Cucurbitaceae	*Cucumis sativus* var. *sativus*
Cucumis sativus var. *cilicicus*	Gabaev	Cucurbitaceae	*Cucumis sativus* var. *sativus*
Cucumis sativus var. *europaeus*	Gabaev	Cucurbitaceae	*Cucumis sativus* var. *sativus*
Cucumis sativus var. *falcatus*	Gabaev	Cucurbitaceae	*Cucumis sativus* var. *sativus*
Cucumissativus var. *hardwickii*	(Royle) Gabaev	Cucurbitaceae	*Cucumis sativus* var. *hardwickii*
Cucumis sativus var. indo-*europaeus*	Gabaev	Cucurbitaceae	*Cucumis sativus* var. *sativus*
Cucumis sativus var. *irano-turanicus*	Gabaev	Cucurbitaceae	*Cucumis sativus* var. *sativus*
Cucumis sativus var. *izmir*	Gabaev	Cucurbitaceae	*Cucumis sativus* var. *sativus*
Cucumis sativus var. *sativus*		Cucurbitaceae	*Cucumis sativus* var. *sativus*
Cucumis sativus var. sikkimensis	Hook. f.	Cucurbitaceae	*Cucumis sativus* var. sikkimensis
Cucumis sativus var. *squamosus*	Gabaev	Cucurbitaceae	*Cucumis sativus* var. *sativus*
Cucumis sativus var. *testudaceus*	Gabaev	Cucurbitaceae	*Cucumis sativus* var. *sativus*
Cucumis sativus var. *tuberculatus*	Gabaev	Cucurbitaceae	*Cucumis sativus* var. *sativus*

Name of Taxon	Name of Authority	Family	Preferred Taxon Name
Cucumis sativus var. *vulgatus*	Gabaev	Cucurbitaceae	*Cucumis sativus* var. *sativus*
Cucumis sativus var. *xishuangbannanesis*	Ined.	Cucurbitaceae	*Cucumis sativus* var. *xishuangbannanesis*
Cucumis sp.		Cucurbitaceae	*Cucumis* sp.
Cucumis thulinianus	J. H. Kirkbr.	Cucurbitaceae	*Cucumis thulinianus*
Cucumis trigonus	Roxb.	Cucurbitaceae	*Cucumis melo* subsp. *melo*
Cucumis trilobatus	L.	Cucurbitaceae	*Psiguria trilobata*
Cucumis utilissimus	Roxb.	Cucurbitaceae	*Cucumis melo* var. *conomon*
Cucumis x hytivus	J. F. Chen and J. H. Kirkbr.	Cucurbitaceae	*Cucumis x hytivus*
Cucumis zeyheri	Sond.	Cucurbitaceae	*Cucumis zeyheri*

Table 3.25: Distribution of *Cucumis* Species

Species	Distribution
C. hardwickii	Much variability occurs in the Western Himalaya, for the former in the Western Himalaya and for the latter. It is cold adaptable in nature.
C. trigonus	The entire Himalayan belt and in the peninsular hills.
C. hystrix	Largely distributed in the dry north-western plains, Rajasthan and adjoining belt.
C. setosus	Eastern India, upper gangetic plains, West Bengal and peninsular region
C. trigonus Roxb. Syn. *C. callosus* (Rottl.) Cogn.	Widely distributed in Eastern and Western Ghats, north-eastern region, extending to the Himalaya; *turbinatus* types possess edible fruits.

In *Cucumis* for and, while *C. seotsus* is restricted to eastern plains, *C. hystrix* extends its range from eastern plains to north-eastern hills in Asom, Tura range in Meghalaya and Mishmi hills. By and large, the wild types related to cultivated cucumber (*Cucumis sativus*), *C. trigonus* and *C. hardwickii*, occur mostly in the Himalayan belt. Localized variability at species level occurs in the Western Ghats for *C. hystrix* and *C. setosus* in comparison to drier north-western plains, where only *C. prophetarum* occurs.

Cucumber (*Cucumis sativus* L.)

Introduction

Among them cucumber (*Cucumis sativus* L., 2n=14) is on of the most important commercial crop grown extensively throughout world in open field condition or in green house. It is well adapted in temperate and humid climate. The fruits of cucumbers are usually eaten fresh or in salads. Besides, fruits are pickled, cooked or are used to make preserves. Young leaves and stems are cooked in Southern Asia. Decorticated seeds and extracted oils are used by confectioners in a variety of ways. Nutritionally, the cucumber is low in calories (113-148K Cal.) and low in mineral and vitamin food value. The low calorie content of cucumbers makes them indispensable part of the diet of those suffering from obesity. Its excellent flavor and texture make cucumbers among the most prized vegetables. Cucumber contains a total of 4-6 per cent of dry substances, approximately 2 per cent of sugar, 1 per cent of albuminous substances, 0.7 per cent of cellular tissue and 0.1 per cent fat. There are some other species like *Cucumis sativus* var. *xishungbanesis* from China has orange flesh and more β-carotene than other cucumbers. Medicinal applications of cucumbers are too plentiful to enumerate. Its fresh consumption is known far diuretic and antipyretic properties and sharply reduces gastric acidity. It plays an important part in regulating the work of intestine by boosting its function. It also allows for the supernumerary amount of cholesterol to be cleared from the body. Mashed cucumber (100 g on empty stomach) is recommended for those suffering from spastic colitis. The crystalline compound of the fruit serves sanitary purpose. The cucumber is used in folk remedies in the treatment of gout, tumors as well as lungs and kidney diseases. It is a good source of iodine. A ferment of the approximate composition of insulin has been developed from the cucumber fruits, which makes a dietary product of particular value. The fresh taste and aroma of the cucumber are determined by free organic acid and essential oils present in it. The juice of cucumbers is used as a cosmetic preparation like health and beauty products including perfumes, lotions, soaps and shampoos. Indigenous practitioners create medical concoctions from the roots, leaves, stems and seeds.Takei and Ono (1939) attributed that flavor of cucumber is said to be derive mainly from the 2, 6 nonadienal, with assistance from 2-hexenal, the more astringent taste being contributed by 2-noenal. Kemp *et al.* (1974) identified some saturated straight chain aldehyde volatile compounds *i.e.* inonanol-trans-2-nonen-1-01, cis-3-nonen-1-01, cis-6-nonen-1-01, trans, cis- 2-6-nonadien-1-01, cis-3, 6-nonadien-1-01, cis-6-nonenal and C_{10}-C_{15} in cucumber which were not reported earlier. Cucurbitacins are free glycosides or in bound form has tetracyclic triterpenes compounds which are responsible for bitterness in cucumber, greatly found in fruits and roots with leave being normally. The pollen also contains bitter principles and hence when bitter pollen fertilizes non-bitter ovules, the resulting fruit becomes bitter. This phenomenon is called metaxenia (Seshadri, 1986).

Origin, History and Distribution

Cucumber has been originated in southern Asia and Africa in broder prospective but the cucumber believed to have been indigenous to India (De Candolle, 1967), as mentioned *"urvaruka"* in the *"Rigveda"* or *"ervaruka"* in the "Charaka Samhita" and its cultivation dates back to 3000 years in the South and East of the Himalayas, forming the Asiatic group (Kroon *et al.*, 1979; Ramachandran and Narayan, 1985). Burma could be regarded as a secondary centre of origin of this crop but few researchers explained as China is the secondary centre of origin. Cucumber is known from descriptions in Iraq around 600 BC and in Mediterranean region around 200 BC. The plant was carried west ward to Asia Minor, North America and Africa long before written history (Whitaker and Davis, 1962). Cucumber was introduced to Northern Europe by the Greco-Roman culture. Its cultivation also dates back for over 3000 years. Sir Flinders Petrie an Egyptologist found the remains of the cucumber in tombs of Egypt about 2300 B.C. and established the species as one of the most ancient cultivated species by man. In the Caucasus region there are names for the cucumber that predate Sanskrit. The chief evidence offered for this viewpoint is the finding of *Cucumis hardwickii* Royle, at the foot of the Himalayas, which is supposed to be progenitor of *Cucumis sativus.* In due course of time from India, it spread to China, and even earlier and more rapidly to the West where it was much appreciated by the Italins, Greeks and the Romans. Latter on it was introduced throughout the Europe by 'Greek' and 'Roman' empires. Cucumbers have been cultivated in France since 9th century. It was first introduced to England (14th century) during the region of King Edward III. By the end of 17th century, the cucumber became fairly common in England. Great settlers Columbus grew the crop in 1494 on the island of Haiti and Spanish introduced cucumber in North America (16th century) where now it is adopted as a popular commercial crop(Swiader *et al.*, 1992).

Cucumber is extensively grown throughout the world especially the slicing types. China is the largest producer of cucumbers and produces 51.0 per cent of the total world production. The other important cucumber producing countries include Iran, Turkey, the USA, India and Russia. In north Europe and Japan, most of the cucumber cultivation is in green houses. In USA, cucumbers have been traditionally grown for either pickling or slicing purpose. Newer forms which are increasing in importance include hot house cucumbers, which are elongated, seedless, and "burpless" (putatively reducing eructation). Pickling types are not popularly grown in India but are grown in Europe, North America and Japan. Recently, pickling types cucumbers are being grown in South India for export to Europe.

Species Availabillty, Evolution and Diversity

According to a recent comprehensive biosystematic monograph of Kirkbride (1993), the genus *Cucumis* includes 32 annual and perennial species divided in to two very distinct groups defined by geographic origin and chromosome number (African $2n$ = 24 and Asiatic group $2n$ = 14 chromosomes). The African group includes melon (*C. melo*) and the Asiatic group includes cucumber (*C. sativus*) and its probable ancestor is *C. sativus* var. *hardwickii* (Royle) or simply *C. hardwickii* (Perltreves and Galun, 1985). *C. hystrix* is the only *Cucumis* species with 2n=24 and is native to Asia. It bears a morphological resemblance and biochemical affinity to *C. sativus* while its chromosome number is the same as that of *C. melo* (Chen *et al.*, 1997a). Studies based on isozymes, chloroplast DNA and restriction fragment length polymorphism supported the distinction between melon and cucumber (Perltreves and Galun, 1985; Perltreves *et al.*, 1985). There are two theories that attempt to explain the existence of a haploid chromosome number n = 7 and a close relative with a haploid chromosome number n = 12. The haploid chromosome number may have been increased to (n = 12) by fragmentation or reduced (n = 7) by fusion. Fusion is supported by the increased amount of heterochromatin content in *C. sativus* (Ramachandran and Narayan, 1985) and by the presence of chromosomal alterations (karyotypic) in *Cucumis* speciation (Singh and Roy, 1974). Comparative genomics between *C. melo* and *C. sativus* may clarify the phylogeny of these species (Danin Poleg, *et al.*, 2000). In genus *Cucumis*, 8 species are of Indian origin (Chakravarthy, 1959). The wild species (*C. hardwickii*) is an endemic species found growing in natural habitats in the foothills of Himalayas (particularly Western Himalayas). It is a cold adaptable, small bitter cucumber with sparse and stiff spines. Free hybridization with cultivated *C. sativus* with no reduction of fertility in F_2 generation might be a feral form. Cytological studies were carried out for *C. callosus* (2n=14), *C. hardwickii* (2n=14), *C. melo* (2n=20, 22, 24), *C. prophaterum* (2n=14) and *C. dispaceus* (2n=24) as reported by Singh and Roy (1974). The studies revealed that the genus is dibasic with both 7 and 12 as the base chromosome number. Bhaduri and Bose (1947) considered that the evolution of the base number 12 in the genus is from the base number 7 by fragmentation. Origin of base number 7 from 12 by fusion has also been suggested (Trivedi and Roy, 1970). Its closet species *C. trigonus* Roxb. (syn. *C. callosus*), also has 2n=14.*C. anguria* L.(2n=12), *e.g.* West Indian gherkin; *C. metuliferus* E. Mey. ex Naud. (2n=24), African horned cucumber; *C. dispaceus* ehrenb. ex spach. (2n=24), *e.g.* hedge hog or teasel gourd; and *C. hirsutus* (perennial and dioecious). India being the primary center of origin, bestowed with a wide range of variability.

C. hardwickii (progenitor of cucumber) is a short day plant which flowers only when the photoperiod is less than 12 hours at 30 °C/20 °C night temperature. The short day nature of *C. hardwickii* has restricted its use in genetic studies as well as in population developmemt by random mating in the field. It is annual monoecious species resembling *C. sativus* hybridizes with *C. sativus* producing a fertile F_1 with no observed reduction of fertility in the F_2. It sets larger number of fruit per plant than *C. sativus* and does so in a manner not typical of *C. sativus* fruit with developing seed do not inhibit growth and development of latter fruit as occure in *C. sativus*. Plant of *C. hardwickii* line can be as indeterminate because seem to lack apical dominamnce and have more and larger laterals than lines of *C. sativus*. *C. hardwickii* plants are typically slow growing in early stage of growth and remain vegetative for extended periods of time. After the slow growth period, *C. hardwickii* enters a rapid phase of growth surpassing total vine length of *C. sativus* cultivars. Plants grew well vegetatively but did not flower until early September regardless of whether planted in May or July. *C. hardwickii* germplasm might result in genotype with higher yields per plant than present *C. sativus* cultivars.

The expression of sex forms in cucumber is controlled by a series of multiple alleles at the *'acr'* locus interacting with genes controlling photoperiodic reaction. Male flowers are produced first and then the female flowers in monoecious cucumber. Abundance of light tends in increased number of male flowers while reduction of light increases pistillate flowers.

Boss (1955) classified *C. sativus* into sub species based on origin and fruit shape:

(i) *C. sativus* ssp. *rigidus*-The English long fruited cucumber of East Asiatic type.

(ii) *C. sativus* ssp. *agrestis*-The Soviet Far East type.

(iii) *C. sativus* ssp. *gracillis*

(iv) *C. sativus* ssp. *sphaerocarpus* (syn. *C. sphaerocarpus*)

Gene and Genetics of Characters

Table 3.26: Gene List for Cucumber (as proposed by Pierce and Wehner, 1989)

Gene Symbol	Character
a or aa	Androecious (single recessive gene 'a' influence by 'acr' locus)
mmff	Andromonoecious/Hermaphrodite
Aa	Intensifies the maleness with ;stst'
Ar	Resistance to anthracnose
bi	Bitter free
Bt	Bitter fruit
Bw	Resistance to bacterial wilt
de	Determinate
df	Delayed flowering
dw	Dwarf
fba	Flower bud abortion
gc	Golden cotyledon
Gene Symbol	Character
gy	Gynoecious
ll	Little leaf
mp	multi-pistillate
ns	Numerous spines
pc(p)	Parthenocarpy
pr	Protruding ovary
rc	Revolute cotyledon
ss	Small spines
Tr	Trimonoecious
w	White immature fruit colour
Yp	Yellow plant
Gene symbol	Character
ap	Apetalous
Gynoecism	Interaction of M and F of allels at two loci -MMFF produce all female flowers -M locus controlling pistillate flowers; strong linkage
F; Acr, st, acr(F), gy and m	Complete femaleness ('*Acr*' gene accelerate the rate of sex conversion from male to female)
B	Black brown spine
bla	Blunt leaf
bu	Bush
cl	Closed flower
cr	Crinkled leaf
dl	Delayed growth
fa	Fasciated
g	Golden leaves
gl	Glabrous
l	Locule number
m	Andromonoecious
n	Negative geotropic peduncle response
opp	Opposite leaf arrangement
pm-1	Resistance to powdery mildew-1
R	Red fruit
ro	Rosette
sp	Short petiole
ul	Umbrella leaf
wf	White flesh
yg(gr)	Yellow-green immature fruit colour

Botanical Description

Table 3.27: Botanical Description

Growth Habit	Plant is normally an annual, tendril bearing herb, creeping or trailing or climbing in habits producing both male and female flowers separately on the same plant (monoecious). Plant is scabrous.
Root	Extensive, mainly shallow
Stem	Stem is robust, vines 60.0-300.0 cm long covered with stiff bristly hairs. The roughly triangular/ quinqangular (4-angled), densely hispidulous, yellowish green and 50-250 cm long.

Tendril	Tendrils inserted beside the leaves, robust, patent, peduncled, simple, unbranched, hispid, yellowish green and 10.0-30.0 cm long.
Leaf	Leaves (7.0-20.0 cm in length) are distichous, alternate, born on 3-10 inches long petioles which permit the leaves to over shadow the prostrate branches, flowers and fruit. The cucumber *C. sativus var. sativus* has 3-5 deeply or shallow lobed leaves not pinnatifid (simple in outline broadly ovate and angular); however, *Cucumis sativus* var. *sikkimensis* has 7-9 lobed leaves with deeply cordate base, apex and acuminate acute. Leaves have unequal glandular teeth, hispid along the margins, above light or dark green, beneath pale green or yellowish green, dull on both surface, densely patently hispid, very scabrid, 5-7 nerved, the lowest main-nerves branched on the outer side, lateral nerves white, transparent: 7-18 cm long and 7-15 cm wide. Lobes are triangular, acuminate, acute dentate. Petiole is robust, farctate, furrowed on the upper side, clothed with numerous, patent, very stiff hairs, yellowish green, 3-20 cm long and stiple wanting.
Floral biology	Cucumber flowers are axillary, shortly stalked, unisexual-monoecious, rather large, densely hispid, 3-4 cm diameter, for the greater part male, the female ones solitary, the male ones in fascicles of 3-7. Stalk of the male flowers is thin, lengthening after the anthesis, clothed with patent hairs with much swollen and ovoid base, 0.3-0.5 cm long and thick. The staminate ones are borne in clusters, each flower on a slender peduncle or stem. The pistillate ones are usually borne solitary on a stout peduncle. Corolla rotate, deeply 5 parted and small.
	Ovary in the cucumber is sparsely covered with spiny wortlike growths. The yellow (in some species deep orange), wrinkled petals are small. Receptacle is campanulate or tubular, 0.8-1.0 cm high and densely tomentose. Calyx is segmented on the margin of the receptacle, 5-partite more than half-way down, with widely patent segments, vitalline or sulphureous, pale on the outside, with numerous longitudinal ribs or nerves which are clothed with scattered long, stiff hairs, between the nerves densely and very shortly hispidulous, wrinkled on the inside, denely clothed with minute gland-tipped hairs and 2-3 cm diameter. Segments are ovate-oblong, lanceolate or obovate, acute, veined, 1.5-2.0 cm long, and 1.0-1.5 cm wide. Male flowers are with a broadly ovoid, yellowish white, 0.5-0.7 cm long column of anthers, which are supported by 2-3 very short, free, white filaments. Female flowers are with an ellipsoid or fusiform, pruinose ovary, clothed with scattered hairs with swollen, ovoid or wart-like bases and style simple. The pistillate flower has 3-5 thick stigma lobes atop a short broad style (Heimlich, 1927). The staminate flowers usually appear about 10 days before the first pistillate flowers appear. They normally out-number the pistillate flowers about 10+01 (Alex, 1957), but this ratio has been known to reach 100+01, and there are seasonal variation in the ratio (Currence, 1932, Edmond, 1931). This ratio can be altered also by the application of certain phenomone chemicals (Me Murray and Miller, 1968, Robinson *et al.,* 1968, Sims and Gledhill, 1969).In the recently developed "gynoecious" plants the flowers are predominantly pistillate (Peterson, 1960, Peterson and Anhder, 1960, Peterson and de Zeeux, 1963, Peterson and Weigle, 1958). Different floral abnormalities like mixed inflorescence, hermaphroditism, fusion, dimorphic female flowers, reduction and increase in the floral parts were observed, particularly in exotic collections.
	*-The staminate flowers can be recognized by their pedicels. Since hermaphroditic flowers mostly afternoon for anthesis as well as pistil, their column at the top of the ovary is enlarged in comparision with the pistillate flowers and can therefore easily be distinguished from them.
Anthesis and pollination	Opening and closing of the male flowers are mainly influenced by the sun-rise and sun-set, that is by light and the time of the day. Anther in all varieties dehisced between temperature range of 20.5-21.5°C. Rise in temperature causes early drying of stigmatic secretion. Pollen fertility was considerable up to noon and by afternoon (2.00 p.m.), fertility was greatly reduced, and it was negligible by the evening. Connor (1969) found that the best time of day for effective cucumber pollination was from 10 am to 3 pm. Stigmatic receptivity was of very short duration and pollination should be carried out within 2 hours after anthesis. He also found that pollination was about equally effective whether the pollen was placed on one lobe of the stigma or on all the lobes. Seaten *et al.* (1936) also stated that the stigma is receptive throughout the day but most receptive in the early morning and that several hundred pollen grains should reach the stigma for most effective pollination. About 30-35 per cent natural self-pollination has been reported under field condition in cucumber. Bees are the main pollinating agents.
Fruit and fruit set	Ovary usually 3 placentiferous with fruits oblong, obscurely trigonus or cylindrical however *Cucumis sativus* var. *sikkimensis* has ovary often 5 placentiferous. Fruits are pendulous, on robust stalks (2.0-5.0 cm long), very variables as to shape and size, usually oblong-lanceolate or linear-cylindrical, small or large, straight or curved, at first green or whitish green, pruinose, with 5-10 more or less conspicuous longitudinal, white or yellowish white streaks, especially towards the apex and often with scattered spinous tubercles and warts, sordidly yellow or sordidly orange-coloured and glabrous when ripe, and many seeded pepo. Fruits are 10-20 cm long or longer, 3-7 cm in diameter and flesh pale green. In monoecious types, fruits are produced progressively at nodes. Fruits of hermophroditic flowers *i.e.* of perigynous flowers are more round than the fruits of epigynous flowers. A developing fruit set at a lower node may inhibit or delay fruits from setting at subsequent nodes. The size and shape at the fruit is usually related to the number of seeds produced by pollination, with each seed requiring one or more pollen grains. Poor pollination is one of the main causes of fruit abortion, mis-shapen fruit or poor fruit set in cucumber flower should be visited 8-10 times by bees for satisfactory fruit set, but the number of seeds and weight of fruit increases up to 40 to 50 visits. Connor (1969) also found that as many as eight visits per flower were necessary for maximum set and seed production was significantly greater with 20 or more visits than with 10 visits.Some cucumber cultivars are able to set fruit without pollination as parthenocarpic fruits and are used primarily in greenhouse production. Parthenocarpic fruits do not have seeds, even though; occasionally some seeds may be present. In order to obtain normal fruit growth and development, parthenocarpic cultivars must be sprayed with a fruit setting hormone, such as chlorflurenol. Some newer cultivars contain a gene for parthenocarpy and do not require hormone spray for fruit set.
Seeds	Seeds are flat, white, ovate-oblong, rather acute at both ends, sharply margined, 0.8-1.0 cm long and 0.3-0.5 cm broad. Approximately 1000 seed weighs about 20.0 g.

Horticultural Classification

Classification Based on Utility of Edible Part

Based on utility of edible part, cucumber can be divided into (Whitaker and Davis, 1962):

(i) Slicing or table types-Slicing or tables types are larger sized and have thin skin. Most of the cultivars have white spined. Used for salad purpose.

(ii) Pickling or processing types-Pickling types are smaller sized and often have prominent thick warty skin than slicers. The length to width ratio, usually about three to one, is important for pickling cucumbers. Pickling types may have either white or black spines. Cultivars of this group with white spines are becoming more popular bercause they retain green colour longer. Used after cooking.

Classification Based on Season of Cultivation

Based on season, cucumber can be divided into (Purseglove, 1974):

(i) Hot weather cultivars

(ii) Rainy season cultivars

Classification Based on Sex Form

On the basis of sex form cucumber can be classified as:

(i) Monoecious- As general rule

(ii) Andromonoecious

(iii) Gynomonoecious

(iv) Hermaphrodite

(v) Androecious

(vi) Gynoecious-Two gynoecious line PPC-2 and GPC-1 (available at ICAR-IARI, N. Delhi) which was used in crossing programme with monoecious line PPC x Pusa Uday and GPC-1 x Pusa Uday expressed 64.51 and 55.61 percent heterosis. (Gograj *et al.*, 2015).

Classification Based on Regions

(i) **European-American-** Some of the English cultivated varieties belong to this group.

(ii) **West Asiatic-** It is a very xerophytic species.

(iii) **Chinese-** Fruits are very long and large with semi-glossy rind.

(iv) **Indo-Japan-** Himalayan and hermaphrodite types

Augmentation of Genetic Resources and Crop Improvement

Breeding Priorites

- ☆ High yield
- ☆ Earliness
- ☆ Crispness and free from bitterness
- ☆ Uniform fruit shape, size, and tenderness
- ☆ Light green cylindrical fruits without spine and crook neck
- ☆ Transport quality
- ☆ Resistant to diseses and pests

Introduction and Selection

It is true that India is the primary centre of origin for cucumber but several genotypes/varieties/hybrids like 'Poinsette', 'Japanese Long Green', 'Straight Eight' and 'Crina', have been introduced which are high yielding and very popular for quality fruits. Similarly introduced genotypes are known for resistance to diseases. Popular cultivar 'Sheetal' was developed by use of Indigenous source through selection with progeny testing.

Table 3.28: Introduction of Cucumber Germplasm in India

Name of Germplasm	*Salient Feature*	*Source Country*
EC398030	Early determinate type	China, USA
EC382737-39	Gynoecious line	USA
EC329300	Gynoecious line (G4)	USA
EC320526	Multiple disease resistance	

The availability of parthenocarpic cultivars are generally produced by European seed companies, however a few of companies have sales and technical support representative in other parts of world and Asia. Some of the popular parthenocarpic varıties are: Kalunga, Bellissma, Millagon, Discover, Marianna, Fitness, aramon, Fidelio, 90-0048, e1828, B1157, Sweet Success (AAs), County fair, Bush Crop, Space Master, Patio Pick, Bush Whopper, Bush Champion, Bush Pickler, Euro-American, Adrian, AI Rashid F_1, Mustang, Bronco, Sandra, Boneva, Daleva, Padex, Fertila, factum, Fanspot, Fem France, toska 70, Farbio, Corona, Sweet Slice, Radja, Bella, and La Reine. In Nepal cucumber variety 'Priya' developed by Indo-American Seed Co., India is commonly being grown. In Pakistan some cultivars like Sialkot Selection No. 1 (R) and PARC-1 are very common and being grown commercially.

Gynoecism and Maintenance

In cucumber gynoecism is very common and being utilized greately for hybrid seed production. Gynoecy has been reported in both the types of cucumber and parthenocarpic genes are also linked with the Gynoecy. Thus hybrids developed by utilizing gynoecious lines are totally female with parthenocarpic genes and does not require pollination for fruit set. Generally in India, gynoecious lines introduced in this country have been developed in temperate regions, thus requires the incorporation of tropical and sub-tropical attributes for their better adaptability. In India, cucumber hybrid 'Pusa Sanyog' (Japanese Gynoecious lines x Green long Naples) has been developed by ICAR-IARI, New Delhi; identified by All India Coordinated Research Project (Vegetable Crops) and recommended for commercial cultivation.

Several plant growth regulators (PGR) are being used to maintain the gynoecism in cucurbits. Gibberellic acid, promote staminate flower formation in gynoecious cucumber lines discovered by Peterson and Anhder (1960) and Galun (1973). They advocated that mixture of gibberellins A4 and A7 were more effective than GA_3 alone for maintaining gynoecious lines. Different gynoecious inbreds may vary in response to GA application (Shifriss and George, 1964). Silver ions inhibit ethylene action and therefore promote male sex expression in gynoecious cucumber (Beyer, 1976). Monoecious cucumbers treated with ethephon at the seedling stage produced only pisttilate flowers on the first 18 noces (Robinson *et al.*, 1969). Silver nitrate and gibberellins A4/A7 (GA4/7) were applied on a gynoecious cucumber in the form of foliar spray. Silver nitrate at 100, 200 and 400 ppm induced significantly more staminate flowers per plant than did GA4/7. The widely used rate 50 ppm makes gynoecious x gynoecious hybrid seed production feasible. The strong induction of all plants treated with silver nitrate should minimize the genetic shift toward maleness observed from one generation to the next when GA is used to increase seed of gynoecious parental line (Tolla and Peterson, 1979). Institute of Vegetable and Flowers, China, is pioneer in working with this issue and now there are 5 F_1 hybrids which are being produced from gynoecious lines. For propagation of gynoecious line the androecious flowers were induced by spraying $AgNO_3$ solution 300-500 ppm at the stage of 2-4 true leaf. Isolation distance should be more than 1000 m. female to male should be 4:1.

Hybridization and Selection

In cucumber several genotyps were used for development of superior varieties/hybrids having desirable traits like high yield, fruit quality, crispness, *etc.* Dogra and Kanwar (2011) developed cross combinations among eight parents (including two gynoecious lines) and evaluated with raised 28 F_{1s}, 8 parents and one check in two locations *i.e.* L1-Nauni and L2-Chambaghat. Sufficient genetic diversity among parents and F_{1s} for all the traits infers the scope of improvement through selection in the parental material. Heterobeltiosis and standard heterosis for earliness was maximum in EC173934 x LC-40 and G_2 x Gyn-1, respectively. The cross K-90 x G_2 exhibited maximum heterobeltiosis and standard heterosis for yield and number of fruits. Cucumber hybrid Kashi Nutan (VRCUH-01) from ICAR-IIVR and Pusa seedless cucumber-6 (DPaC) ICAR-IARI has been developed.

Table 3.29: Important Cross Combination and its Salient Features

Cross Combinations	*Salient Features*
'VIR 501' x 'VIR 505'	Intense photosynthetic activity, high-yielding
'Conqueror' x 'Vijazhiki 37'	Early, high-yielding
'Don 175' x 'Boston'	High germination, growth rate, better pollen viability
'MSU 7135' x 'Spartan 27'	Early, high-yielding
'Koga Aomoga Fushinavi' (Japanese Gynoecious line) x 'Green Long of Naples'), 'Pusa Sanyog'	Early, high-yielding
'Udipi' x 'Kasargod'	High-yielding
'Nileswar' x 'Udipi'	High-yielding

The first F_1 processing cucumber developed from a gynoecious line ('MSU 71305') is 'Spartan Dawn'.

Interspecific Hybridization

Whitaker (1930) reported the occurrence of interspecific cross compatabilities in *Cucumis*. However, successful cross between the cucumber and melon has been reported (De Ruiter, 1973). In general among wild species are frequently possible but all attempts to cross any of these wild species with the two cultivated species: *C. sativus* and *C. melo* failed (Deakin *et al.*, 1971). Interspecific hybridization has been done among various *Cucumis* species-*sativus* and *melo*; to transfer high level of resistance from wild *Cucumis* species to cultivated ones (Leppick, 1996). *Cucumis sativus* (2n=14) is crossable with other species of genus *Cucumis* having same chromosome number (2n=14). Conspecific (belong to the same species) which are cross compatible in nature and produce F_1 hybrids are given below:

Table 3.30: Cross Compatible Conspecific Species

Conspecific Species	*References*
Cucumis anguria and *Cucumis longipes*	Peral-Treves *et al.*, 1985
Cucumis denteri and *Cucumis sagittatus*	Dane *et al.*, 1980; Deakin *et al.*, 1971; Kroon *et al.*, 1979; Kho *et al.*, 1980
Cucumis sativus and *Cucumis hardwickii*	Peral-Treves *et al.*, 1985; Dane *et al.*, 1980; Deakin *et al.*, 1971
Cucumis myriocarpus and *Cucumis leptodermis*	Deakin *et al.*, 1971; Peral-Treves *et al.*, 1985;

The free hybridization with cultivated *C. sativus*, with no reduction of fertility in F_2 generation suggested that *Cucumis hardwickii* (2n=14) is likely cultivated cucumber. The West Indian Gherkin (*Cucumis anguria* L.) and the African horned cucumber or Jelly melon (*Cucumis metuliferus*) are far more limited in their use. Diallel reciprocal cross pollinations were made among cucumber, muskmelon and 19 wild species of *Cucumis*.

The only species that cross with the cucumber was *C. hardwickii*, *C. trigonus* and *C. callosus* were the only species to cross with the muskmelon. In each of these crosses there was no sterility barrier; the cross was easily made and the F_1 was fully fertile. In morphology as well as in compatability *Cucumis hardwickii* appeared to be *C. sativus* and *C. trigonus* and *C. callosus* appeared not to be valid species but rather sub species of *C. melo*. Fruit set was obtained on cucumber but not on muskmelon, when pollinated with other *Cucumis* species. No seed was produced however and embryo culture attempts were unsuccessful. Viable seed was produced by *C. metuliferus*, but the progeny of interspecific matings were identical to the maternal parent probably due to apomixes. Gene inter change is possible among *C. anguria*, *C. longipes*, *C. hookeri*, *C. leptodermis* and *C. zeyheri*, since crosses were made among all these species and the hybrids were fertile. Crosses were also made between *C. africanus* and *C. membranifolius*; *C. prophetarum* and *C. pustulatus* and between *C. dipsaceus* and *C. prophetarum*. Chen *et al.* (1995) reported that based on the isozyme analysis of *C. hysteria*, *C. melo* and *C. sativus* led to the hypothesis that a triangular poly genetic relationship occurs among the species. Successful crosses has been made but no significant hybrids have been reported from these combinations. Some successful attempts of distant hybridization in cumber are *C. sativus* x *C. hardwickii* and *C. sativus* x *C. trigonus* (*C. callosus*). *C. metuliferus* (2n=24) x *C. melo* (2n=24) showed unilateral incompatibility (reciprocal crosses are not possible).

Table 3.31: Wide-cross Attempted between Cultivated and Wild *Cucumis* Species

Cross	Result	Reference
C. sagitatus x *C. melo*	Embryos only	Deakin *et al.*, 1971
C. metuliferus x *C. melo*	Embryos only	Fassuliotis, 1977
C. sativus x *C. melo*	Globular stage	Niemirowicz-Szczytt and Kubici 1979
	Embryos only	-
C. metuliferus x *C. melo*	Fertile F_1	Norton and Granberry, 1980
C. propheterum x *C. melo*	Fruit with non-viable seeds	Singh and Yadav, 1984b
C. zeheri x *C. sativus*	Fruit with non-viable seeds	Custers and Nijs, 1990
C. sativus x *C. metuliferus*	Embryos only	Franken *et al.*, 1988
C. melo x *C. metuliferus*	Embryos only	Soria *et al.*, 1990
C. sativus x *C. hystrix*	Sterile plants (2n+4n)	Chen *et al.*, 199

The incompatibility between cultivated *Cucumis* species and their wild relatives is mainly characterized by:

i. Delayed growth of pollen tube
ii. Arrested pollen tube growth on the stigma
iii. Inability of the pollen tube to reach the ovule (Kishi and Fukishte, 1969)
iv. Lack of cell division of the zygote and abortion of the endosperm (Kishi and Fukishte, 1969)

Approaches to overcome compatibility barriers:

i. Use of growth regulator (Beharav and Cohn, 1994)
ii. Use of menter pollen (Kho *et al.*, 1980)
iii. Bud pollination (Chatterjiee and More, 1991)
iv. Embryo rescue (Chen *et al.*, 1997a)

Intergeneric Hybridization

Plants were obtained from open pollination of *C. sativus* and *Luffa* had glabrous stem, leaf, petiole, flower, tendril and fruit; and grew well in hot climate. The intergeneric crosses would have only a limited use.

Polyploidy Breeding

Two species *C. heptadactylus* and *C. ficifolia* Rich., are natural hexaploids (2n=42). Tetraploid cucumbers have been developed through colchicines treatment but they are low yielder, and at the same time fruits obtained are thicker. Triploids have also been produced by crossing tetraploids with diploids.

Parthenocarpy

Parthenocarpic cucumber is very popular in many areas of world. The fruits of parthenocarpic cucumber are mild in flavor, seedless and have thin edible skin that requires no peeling. The parthenocarpic varieties require no pollination for fruit setting. If pollination occurs, the fruit will form seed, shape of fruit wiil be destroyed and bitter tasting fruit will develop. It is therefore essential to prevent bees and other pollinators from entering in the green houses and carry pollen from field planting cucumber. Almost all slicing cucumber varieties cultivated in greenhouse in Western Europe are parthenocarpic, and parthenocarpic pickling cucumber varieties are also available in the market. In Germany the use of parthenocarpic varieties in production for the processing industry is a major part of the total cultivation area of pickling cucumber. Parthenocarpic fruit setting results in an earlier and much more regular production rhythm (Zwinkels, 1987). Parthenocarpy circumvents the inhibitory effect of seed formation on subsequent fruit development. Parthenocarpy must be combined with constant femaleness, because the fruits formed after fertilization in genetic parthenocarpic varieties become misshappen, are of no economical value and lead to loss of production. It is possible to induce artificially parthenocarpic fruit set by treatment with monphactin (Chlorfluorenol) as reported by Robinson and Cantliffe (1971).

Parthenocarpic varieties do not require pollination to set the fruit and fruits are seedless. So the maintenance is major problem in these lines. Tha parthenocarpic lines are usually maintained by use of plant growth regulators and chemicals *e.g.* silver nitrate and GA_3. Kasrawi (1988) studied the effect of silver nitrate on sex expression and pollen viability in parthenocarpic cucumber. He investigated the effect of $AgNO_3$ concentration, number of sprays and time of application on sex expression and pollen viability of parthenocarpic cucumber. He found that greatest number of staminate flowers are produced on plants when $AgNO_3$ sprayed twice with 300 ppm concentration at one week interval with initial spray at first true leaf stage. He also found that 2-3 sprays of 200-500 ppm $AgNO_3$ starting at first leaf stage produce the maximum number of staminate nodes. When the plants were sprayed at first true leaf stage production of staminate flowers began with the 1st node and at the time of onset of flowering. Hallidri (through internet searh) reported the effect of $AgNO_3$ induction of staminate flowers of gynoecious parthenocarpic flowers and found that $AgNO_3$ concentration and number of sprays affect the induction of staminate flowers. He also found that all the treatments using one spray were ineffective for producing staminate flowers. The treatment with 100 ppm $AgNO_3$ failed to induce staminate flowers. The greatest numbers of staminate nodes were produced on plant by 2-3 sprays of 400-500 ppm $AgNO_3$. The the initial spray was applied at first true leaf stage and subsequent treatment was applied at one week interval. He also reported that plants showed injury a few days after spray of 400-500 ppm $AgNO_3$ and recovered within 7-10 days. Some plant growth regulators may be used for induction of male flower in gynoecious parthenocarpic line *e.g.* GA_3. At GBPU and T, Pantnagar, staminate flowers were induced in parthenocarpic line through use of plant growth regulators like GA_3 and $AgNO_3$ as four sprays at 4 days interval (Singh and Ram, 2004).

Mutation Breeding

In cucumber seed treatment of 'Cum-15' with 0.05 per cent ethylamine for 24 hours has given 7.6 per cent plant population resistant to *Meloidogyne* spp.

Breeding for Processing and Quality Traits

Following quality traits should be considered during breeding programmes:

Table 3.32: Quality Traits

Traits	*Preferred Range/Magnitude/Value/Type*
Rind colour	Yellowish green with brown streaks
Rind thickness	More for pickling type
Placenta hollowness	Less for salad type
Pulp core	Absent
Core	Small
TSS	More

Resistance Source for Biotic Factors and Utilization

(i) Diseses Resistance Source and its Utilization

Nature and genetics of disease resistance is important components for breeding resistant variety.

Table 3.33: Genetics of Disease Resistance in Cucumber

Disease	*Genetic Control*
Anthracnose	Single recessive gene
Powdery mildew	Two gene interaction
Abgular leaf spot	Polygenic
Melon mosaic	Single dominant gene
Cucumber green mottle mosaic virus	Single dominant gene (c gene)

There are several genotypes which are resistant to many diseases. Scab resistant lines have been developed through back cross breeding. Cultivars like Pusa Sanyog, Pant Sankar Kheera-1, Priya, Solan Hybrid, Poinsette (introduction) are found resistant to downy mildew, powdery mildew and anthracnose (Table 3.34).

Table 3.34: Diseases and their Sources of Resistance

Crop	*Disease*	*Resistance Source*
Cucumber	Anthracnose	PI 175111, PI 175120, PI 179676, PI 182445, Wise-2757 (USA), EC 398968-70, Poinsette (Donor), Hybrid 517,PI, 200815, PI, 20818, *C. hardwickii*
	Downy mildew	B-184, B-159, Wise-2757 (USA), EC 398974-90, *Cucumis hardwickii*, Palmetto, Ashley, Chinese Long, Stono PI,179676, Clinton, PI, 235628, Painsett-76
	Powdery mildew	PI 200815, PI 200818, *Cucumis hardwickii,* Wise-2757 (USA) Polaris, Ambra, Yamaki
	CMV, Scab and Powdery mildew	EC 320556
	Multiple diseases with gynoecism	EC 329300
	Angular leaf spot	EC 398966-67
	Fruit rot	EC 398971-73
	Leaf spot	EC 398991-399007, Hivergreen and Biveryel
	Cucumber mosaic virus(CMV)	Poinsette, Wisc SMR-12, SMR-15, SMR-18, Wise 2757
	Scab	Wisconsin SR 12, Beleanto and Ashe Fletcher
	Fusarium wilt (*Fusarium oxysporum*)	Hybrids

Crop	*Disease*	*Resistance Source*
Cucumis sp.	CMV and Powdery mildew	EC 399914-37
	Downy mildew and Powdery mildew	PI 197087 (Donor)
	Cucumber scab and Angular leaf spot	Poinsette, Yomaki, PI 179376
	Angular leaf spot	Poinsette, Dixie Gemin
	Bacterial wilt	Wisconsin SMR-9, Poinsette, MSU 9402, PI 169400, PI200815, PI200816, PI 200817 PI200818, PI 196477 (Wilsom *et al.* 1996).
	Cucumber Green Mottle Mosaic	*Cucumis anguria, Cucumis africanus, Cucumis ficifolius, Cucumis asper, Cucumis dinteri* and *Cucumis sagittatus*
	Gummy Stem Blight	M-17 (Wehner *et al.*, 1996)

Source: Bhardwaj, 2010.

Table 3.35: Sources of Resistance to *Cucumis* Species

Cucumis Species	*Resistance to*	*References*
Cucumis metuliferus	Nematode, CGMMV, WMV, SqMV	Den Nijs and Custers, 1990; Wehner *et al.*, 1990
C. anguria	Nematode, spider mites, CGMMV	Laboda, 1996; Laboda 1999; Laboda 2007; MeCarthy *et al.*, 2001
C. africanus	Nematode, spider mites, CGMMVt	
C. trigonus	Nematode, CGMMV	
C. myriocarpus	Mites	
C. zeyheri	Mites	
C. hardwickii	Nematode, CGMMV	
C. callosus	Fruit fly	
C. melo (Mp-1)	Downy mildew	

Table 3.36: Resistance Sources against Various Diseases

	Genotypes/Varieties	*Source*	*Resistant Source and Special Feature*
Fungal	'Wantee'	USA	Resistant to downy mildew (*Pseudoperonospora cubensis*)
	Wisconsin', 'SMR 12', 'Stono'	USA	High-yielding, resistant to downy mildew
	'Polaris', 'Ambra', yamaki'	-	Powdery mildew (*Erysiphe cichoracearum*)
	Wisconsin', 'SMR 12', 'SMR 9','Stono', 'Beleanto', 'Ashe Fletcher'	USA	Resistant toscab (*Cladosporium cucumerinum*)
	'Hybrid 517', PI197087	-	Resistant to anthracnose (*Colletotrichum lagenarium*)
	Hybrids	-	Fusarium wilt (*Fusarium oxysporum*)
	'Hivergreen', 'Biveryel'	-	Leaf spot (*Cercospora melonis*)
Bacterial	'PI 200816', 'PI 200817', 'PI 196477'	-	Bacterial wilt (Erwinia tracheiphila)
	'Poinsett', 'Dixie Gemin'		Angular leaf-spot (*Pseudomonas lachrymana*)
Viral	'Chinese Long'	China	Resistant to tobacco-mosaic virus
	'Tokyo Long Green', Ohayo Lines'	Japan	Resistant to mosaic virus
	'Wisconsin', 'SMR 9'	USA	Resistant to cucumber- mosaic virus
	'Kyoto-3-Feet', 'Table Green 65', 'Market More'	-	Resistant to mosaic

Table 3.37: Promising and Useful Attributes and Accessions Introduced

Species	*Accessions*	*Country*	*Promising and Useful Attributes*	*Indentor*
C. sativus	EC-399914 -37	UK	Resistant to CMV, downy mildew	PAU, Ludhiana
	EC-398030	China	Early maturing, determinate type, bear fruits in cluster	ICAR-IIVR, Varanasi
	EC-398966-67	USA	Resistant to angular leaf spot	Sandoz India
	EC- 398968-70	USA	Resistant to Anthracnose	Sandoz India
	EC- 398971 73	USA	Resistant to fruit rot	Sandoz India
	EC- 398974-90	USA	Resistant to downy mildew	Sandoz India
	EC- 398991-9007	USA	Resistant to leaf spot	Sandoz India
	EC-329300	USA	Gynoecious line (G4)	-

Species	Accessions	Country	Promising and Useful Attributes	Indentor
	EC-320526, WI2757	-	Multiple disease resistance	-
	EC-382737-39	USA	Gynoecious lines	ICAR-IARI, New Delhi, PAU, Ludhiana
	GPC-1 and PPC-2	-	Gynoecious lines	ICAR-IARI, New Delhi
	EC-497645, EC-497646, EC-497647			

Cucumber cultivar 'Jinyou No. 15 (F_1) is reported resistance against wilt disease (Li *et al.*, 2000) (Table 3.35).

An introduction 'PI 197087' is known resistant source for anthracnose disease (Table 3.36).

In cucumber genotype 'WI2757' carrying resistance to as many as nine diseases *i.e.* powdery mildew, downy mildew, scab, anthracnose, angular leaf spot, bacterial wilt, target leaf spot, cucumber mosaic virus (cmv) and *Fusarium* wilt. Some new introductions are as shown in Table 3.37. Cucumber genotype 'SC-2B4-1 was found resistant against powdery mildew, downy mildew, anthracnose and cucumber mosaic. Recently evaluated cucumber lines for resistance are given in Table 3.38.

Table 3.38: Resistance Source against some Diseases

Disease	Resistance Source
Anthracnose	PI 175111, PI 175120, PI 179676, PI 182445
Powdery mildew	PI 200815, PI 200818
CMV	Wisc SMR-12, SMR-15, SMR-18

(ii) Insect Resistance Source and its Utilization

Species *C. trigonus* Roxb. (syn. *C. callosus*) and *C. hardwickii* are potential source of resistance to fruitfly (*Dacus cucurbitae*). Similarly, reaction against whitefly resistance has been worked out in various *Cucumis* species are as under: Resistance source for to spotted spider miles are PI 220860, Hybrid Long Green Pickle, PI178885, Ohio MR-200 and Taipei No. (Robert, 1994; Labada, 1996)

The three species with the greatest density of trichoms *C. dinteri, C. angolensis* and *C. asper* were the species most resistant to white flies. Cultivars like 'All season, 'Gy-604', 'improved green long', 'Korea Dutch *etc.* are habe been developed which are less susceptible to red pumpkin beetle. Resistance source for two spotted spider mites are PI 220860, Hybrid Long Green Pickle, PI 178885, Ohio MR-200 and Taipei No.1 (Robert, 1994, Labada, 1996).

Table 3.39: Reaction of Whitefly in different *Cucumis* Species under different Stage

Sl.No.	Species	Trichoms	Eggs	Larvae	Pupae	Adult
1.	*C. prophetarum*	34	108	19	12	9.0
2.	*C. trigonus*	24	175	236	148	9.0
3.	*C. sativus*	33	151	177	168	8.8
4.	*C. heptadactylus*	14	257	197	18	8.5
5.	*C. ficifolius*	18	64	119	10	8.5
6.	*C. leptodermis*	34	173	144	98	8.5
7.	*C. longopes*	38	160	166	133	8.2
8.	*C. metuliferus*	26	82	119	63	8.0
9.	*C. hookeri*	40	102	144	72	8.0
10.	*C. dipsaceus*	62	123	99	28	7.5
11.	*C. anguria*	45	234	226	74	7.5
12.	*C. zeyheri*	43	106	121	40	7.1
13.	*C. pustulatus*	30	46	70	19	7.0
14.	*C. myriocarpus*	32	52	36	35	6.5
15.	*C. africanus*	52	107	103	38	5.8
16.	*C. melo*	56	88	77	17	5.0
17.	*C. asper*	110	44	44	2	2.0
18.	*C. angolensis*	117	44	86	13	2.0
19.	*C. dinteri*	200	12	4	0	1.5

(iii) Nematode Resistance Source and its Utilization

West Indian gherkin, *C. metuliferus* E. Mey. Ex Naud. (2n=24), *C. anguria*, *C. africanus*, *C. trigonus*, C. zeyheri and *C. hardwickii* have high level of nematodal resistance (ded Nijs and Custers, 1990). Similarly, Lucia Monteo and Shelby are resistance source for nematode (Walters and Wehner, 1997).

Abiotic Resistance

C. hardwickii has potential genes for wider adaptability and drought tolerance. In cucumber genetic control of salinity expressed as leaf necrosis, is governed by one dominant major gene and several minor genes which heritability is 41-86 per cent. Germination under chilled condition it is controlled by additive gene and variance predominant. For this trait heritability was high at 17°C in growth chamber. There are certain genotypes which useful in breeding programme related to abiotic stresses are mentioned in Table 3.40. How the adverse abiotic conditions can modulate with simultaneous drought and pathogen (Ramegowda and Senthil-Kumar, 2015; Choudhury *et al.*, 2017). The salt stress-induced stomata closure signaled by root-produced ABA was intensified after intection, likely via SA/H_2O_2 related mechanism which prolonged inhibition of PSII under combined stress (Chojak *et al.*, 2017).

Table 3.40: Abiotic Condition and Resistant Genotypes

Abiotic Condition	*Resistant Genotypes*
Low temperature injury	'Azerbarjan'
Gummosis	'Super OE 48'
Drought	'Hanski 264'

Grafting

Now modern grafting techniques have been reported to successful method for plant development which imparts resistance against *Fusarium* wilt in several cucurbits. In case of cucumber some specific targets has been fixed to solve the problems through graftings.

Cucumbers are grafted on pumpkin (*Cucurbita moschata*) to give resistance against *Fusarium* spp. *Sicos angulatus* (Bur cucumber) is is excellent rootstock as developing coldstress resistance/tolerance for cucumber grafting (Venema *et al.*, 2008). *C. metuliferous* (African horned cucumber) when used as rootstock for cucumber grafting results resistance/tolerance to root knot nematode (*Meloidogyne incognita*) (Marukawa and Takatsu, 1969).

Conservation of Genetic Resources

Extensive germplasm collections are being maintained by the USDA Plant Germplasm Sytem at

Table 3.41: Useful Rootstocks for Grafting

Scion	*Rootstocks*	*Purpose*
Cucumber	*Cucurbita moschata*, *C. ficifolia* (fig-leaf gourd), *C. maxima*, *Sicyos angulatus* (bur cucumber)	Vigorous root system, tolerance to fusarium wilt, low temperature, Phytophthora, high soil moisture, nematode

Table 3.42: Important Diseases and Pests of Vegetables Controlled by Grafting

Scion	*Rootstocks*	*Disease/Pests*	*References*	*Major Variety*
Cucumber	Squash cv. Kurodane and Shin-Tosa	Phytophthora wilt (*Phytophthora melonis*)	Miyata *et al.* (1976)	
	C. moschata Shin-Tosa F_1	Fusarium wilt	Yoneyama (1976)	
	C. ficifolia	*Fusarium oxysporum* f. *cucumerinum*	Klose *et al.* (1980)	Kurodane
	KJ-100 (*Sicycos angulatus*)	Meloidogyne incognita	Uffelen (1983)	
	Pumpkins (cv. Yunnan Black Seeded)	Nematode	Liu *et al.* (1998)	
Cucumber	*Cucurbita* species	-	-	Hikari-Power*, Super-Unryu*, Kitora*, Strong-Ikki*

* Rootstock for bloomless fruit.

Table 3.43: Valuable Resources available in Various Countries

Sl.No.	*Conservation Centre*	*Country*	*Germplasm Conserved*	*Evaluation for Specific Traits*
1.	Maritsa Institute for Vegetable Crops Brezovsko Shessa, Plovdiv	Bulgaria	Commercial cultivars and inbred lines	Earliness, resistance to *Pseudoperonospora cubensis* (CMV-1)
2.	Institute for Horticultural Plant Breeding 91VT, Wageningn	Netherland	Land races old and commercial cultivars (mainly Dutch worldwide representation)	Disease resistane, stress tolerance and quality
3.	National Seed Storage Laboratory (NSSL) USDA, ARS Colorado State University	USA	Land races and commercial cultivars	Disease resistane, stress tolerance and quality

the Plant Introduction Station in Iowa (Clark *et al.*, 1991) and by the Vavilov Institute in Leningrad, USSR (Robinson, 1989). Another major germplasm repository is maintained by the Peoples' Republic of China and smaller genebanks are located in Mexico, India, Spain, Nigeria, Costa Rica and Phillippines (Esquinas-Alcazar and Gulick, 1983). These germplasm collections represent a valuable resource for breeding adapted cultivars of these exotic cucurbits for domestic production.

REFERENCES

1. Alex, A.H. 1957. Honeybees aid pollination of cucumbers and cantaloupes. *Tex. Agr. Expt. Sta. Prog. Rpt.*, 1936, 4 p.
2. Beharav, A. and Cohn, Y. 1994. Callus formation from cotyledon and hypocotyls of *Cucumis melo* L. and *Cucumis metuliferus*. *Cucurbit Genet. Coop. Rep.*, **17:** 97-100.
3. Bhaduri, P.N. and Bose, P.C. 1947. Cytogenetical investing actions in some common cucurbit with special reference to fragmentation of chromosome as physical basis of speciation. *Journal of Genetics*, **48:** 237-256.
4. Bhardwaj, D.R. 2010. Breeding for pest and disease resistance in cucurbits. In: Compedium of Winter School "Emerging trends and strategies for the management of pests and diseases in vegetable crops. Published by ICAR-IIVR, Varanasi (U.P.), p. 102-109
5. Chakravarthy, H. 1959. Monograph on Indian Cucurbitaceae. *Records of Botanical Survey of India*, **14(1):** 1-29.
6. Chatterjiee, M. and More, T.A. 1991. Interspecific hybridization in *Cucumis* spp. *Cucurbit Genet. Coop. Rep.*, **14:** 69.
7. Chen, J.F., Isshiki, S., Tashiro, Y. and Miyazaki, S. 1995. Studies on a wild cucumber from China (*Cucumis hystrix* Chakr.).I. Genetic distances between *C. hystrix* and two cultivated *Cucumis* species (*C. sativus* L. and *C. melo* L.) based on isozyme analysis, *J. Jpn. Soc. Hort. Sci.*, **64(2):** 264-265.
8. Chen, J.F., Isshiki, S., Tashiro, Y. and Miyazaki, S. 1997a. Successful interspecific hybridization between *Cucumis sativus* L. and *Cucumis hystrix*. *Euphytica*, **96:** 413-419.
9. Clarke, R.L., Widrlechner, M.P., Reitsma, K.R. and Block, C.C. 1991. Cucurbit germplasm at the North Central Regional Plant Introduction Station, Ames, Iowa. *HortSci.*, **26 (326):** 450-451.
10. Connor, L.J. and Martin, E.C. 1969. Honeybee pollination of cucumbers. *Pickle Pak.*, **29:** 3.
11. Currence, T.M. 1932. Nodal sequence of flower type in the cucumber. *Amer. Soc. Hort. Sci. Proc.*, **29:** 477-479.
12. Custers, J.B.M. and den Nijs, A.P.M. 1990. Effect of amino ethoxy vinyl glycine (AVG), environment and genotype in overcoming hybridization barriers between *Cucumis* species. *Euphytica*, **35:** 639-647.
13. Dane, F., Denna, D.W. and Tsuchiya, T. 1980. Evolutionary studies of wild species in the genus *Cucumis*. *Zptanzenzucht*, **85:** 89-109.
14. Danin-Poleg, Y; Reis, N. and Baudracco-Arnas 2000. Simple sequence repeats in *Cucumis* mapping and map merging. Genome, 43: 963-974.
15. De Candolle, A. 1967. Origin of Cultivated Plants (edn.1), p. 264. Hafnar Publishing Co., New York.
16. Deakin, J.R., Bohn, G.W., Witaker, T.W. 1971. Interspecific hybridization in *Cucumis*. *Econ. Bot.*, **25:** 195-211.
17. den Nijs, A.P.M. and Custers, J.B.M. 1990. Introducing resistance into cucumber by interspecific hybridization. In: Bates, D. *et al.* (eds.): Biology and Utilization of Cucurbitaceae. Cornell University Press, Ithaca, NewYork, 382-395.
18. Dogra, B.S. and Kanwar, M.S. 2011. Exploiting heterosis for yield and horticultural traits in cucumber (*Cucumis sativus* L.). *Indian J. Plant Genet. Resour.*, **24(3):** 332-339.
19. Edmond J.B. 1931. Seasonal variation in sex expression of certain cucumber varieties. J. *Amer. Soc. Hort. Sci. Proc;* **27**:329-332.
20. Esquinas-Alcazar, J.T. and Gulick. 1983. Genetic resources of Cucurbitaceae. Int. Board for Plant Genet. Resources, Rome.
21. Fassuliotis, S.G. 1977. Self fertilization of *Cucumis metuliferus* (Neud.) and its cross compatibility with *C. melo*. *J. Amer. Soc. Hort. Sci.*, **102:** 336-339.
22. Franken, J., Custors, J.B.M. and Bino, R.J. 1988. Effects of temperature on pollen tube growth and fruit set in reciprocal cross between *Cucumis sativus* and *C. metuliferus*. *Plant Breed.*, **100:** 150-153.
23. Heimlich, L.F. 1927. The development and anatomy of the staminate flower of the cucumber. *Amer. Jour. Bot.*, 14:227-237.
24. Kirkbride, J.H. Jr. 1993. Biosystematic monograph of the genus *Cucumis* (Cucurbitaceae) North Carolina: Parkway, Boone, P. 159.
25. Kasrawi, M.A. 1988. Effect of silver nitrate ($AgNO_3$) on sex expression and pollen viability in parthenocarpic cucumber (*Cucumis sativus* L.). *Dirast (Jorolan)*, **15(2):** 69-78.
26. Kemp, T.R., Knavel, D.E. and Stolz, L.P. 1974. 3,6-Nonadien-1-ol from *Citrullus vulgaris* and *Cucumis melo*. Phytochemistry 13 1167-1170.
27. Kemp, T.R., Knavel, D.E. and Stolz. L.P. 1974. Identification of some volatile compounds from cucumber. Journal of Agricultural Food and Chemistry, 22: 717-718.

28. Kho, Y.D., den Nijs, A.P.M. and Franken, J. 1980. Interspecific hybridization in *Cucumis* L. 2. The crossability of species on investigation of *in vitro* pollen tube growth and seed set. *Euphytica*, **29:** 661-672.

29. Kishi, Y. and Fukishte, N. 1969. Studies on interspecific hybridization in the genus *Cucumis*. I. Pollen germination and pollen tube growth in selfing and incompatibility crossing. *J. Jpn. Soc. Hort. Sci.*, **28:** 329-334.

30. Kishi, Y. and Fukishte, N. 1970. Studies on interspecific hybridization in the genus *Cucumis*.II. Pollen tube growth, fertilization and embryogenesis. *J. Jpn. Soc. Hort. Sci.*, **39:** 329-334.

31. Kroon, G.H., Custers, J.B.M., Kho, Y.O., den Nijs, A.P.M. and Varekamp, H.Q. 1979. Interspecific hybridization in *Cucumis sativus* L. Need for genetic ariation, biosystematics relations and possibilities to overcome crossability barriers. *Euphytica*, **28:** 723-728.

32. Labada, A. 1996. Resistance in wild *Cucumis* species to two spotted spider mites (*Tetranychus urticae*). *Acta Phytopathol. Entom. Hung.*, **31:** 247-252.

33. Labada, A., Kubalakova, M., Kriskova, E., Navratilova, B., Dolazel, K., Dolazel, J. and Lysak, M. 1999. Morphological and physiological characteristics of plant issued from interspecific hybridization of *Cucumis sativus* x *C. melo*. *Acta Hort.*, **492:** 149-155.

34. Labada, A., Windrlechner, M.P., Staub, J., Ezura, H., Zalapa, J. and Kriskova, E. 2007. Cucurbits (*Cucurbitaceae, Cucumis* species, *Cucurbita* species). Chapter 8, 271-376. In: Singh, R. (ed.) Genetic resources, chromosome engineering and crop improvement. Series, Vol.3-Vegetable Crops. CRC Press, Boca Raton, FL, USA.

35. Leppick, E.E. 1996. Searching gene centres of the genus *Cucumis*. *Eyphytica*, **15:** 51-57.

36. Li, J.W., Zhang, W.J., Sun, Z.K. and Wang, L.J. 2000. A new cucumber variety for solar green house production 'Jinyou No. 5 (F)'. *Acta Horticulturae Sinica*, 27: 232.

37. Marukawa, S. and Takatsu, L. 1969. Studies on the selection of *Cucurbita* species as cucumber rootstock. *Bul. Ibaraki Hort. Expt. Sta.*, **3:** 11-18.

38. Mc Carthy, W.H., Wehner, T.C., Xie, J. and Daub, M.E. 2001. Isolation and callus production from cotyledon protoplast of *Cucumis metuliferus*. *Cucurbit Genet. Coop. Rep.*, **24:** 102-106.

39. Me Murray, A.L. and Miller, C.H. 1968. Cucumber sex expression modified by 2-chloroethanephosphonic acid. *Science*, 1397-1398.

40. Niemirowicz-Szczytt, K. and Kubici, B. 1979. Cross fertilization between cultivated species of genera *Cucumis* (L.) and *Cucurbita* (L.). *Genetica Polonica*, **20:** 117-125.

41. Norton, J.D. and Granberry, D.M. 1980. Characteristics of progeny from an interspecific cross of *Cucumis melo* with *Cucumis metuliferus*. *J. Amer. Soc. Hort. Sci.*, **105:** 174-180.

42. Peral-Treves, R. and Galun, E. 1985. The *Cucumis* plastome: physical map, intrageneric variation and phylogenetic relationships. *Theor. Appl. Genet.*, **71:** 417-429.

43. Peral-Treves, R., Zamir, D., Navot, N. and Galun, E. 1985. Phylogeny of *Cucumis* based on isozyme variability and its comparison with plastome phylogeny. *Theor. Appl. Genet.*, **71:** 430-436.

44. Peterson, C.E. 1960. A gynoecious inbred line of cucumber. Mich. *Agr. Expt. Sta. Quart. Bul.*, **43:**40-42.

45. Peterson, C.E. and Andher, L.D. 1960. Induction of staminate flowers on gynoecious cucumbers with Gibberellin A3. *Science*, **131:**1673-1674.

46. Peterson, C.E. and de Zeeux, D.J. 1963. The hybrid pickling cucumber, Spartan Dawn, Mich. *Agr. Expt. Sta. Quart. Bul.*, **46:**267-273.

47. Peterson, C.E. and Weigle, J.L. 1958. A new method for producing hybrid cucumber seed. Mich. *Agr. Expt. Sta. Quart. Bul.*, **40:** 960-965.

48. Pierce. L.K. and Wehner, T.C, 1989. Gene list of cucumber. *Cucurbit Genet. Coop. Rep.* **12:** 91-97.

49. Ramachandran, C. and Narayan, R.K.J. 1985. Chromosomal DNA variation in *Cucumis*. *Theork* Appl Genet; 69:497-502.

50. Robinson, R.W. 1989. Genetic resources of the Cucurbitaceae, P85a-85J. In: C.E. Thomas (Ed.) Proc. Cucurbitaceae, 89: Evaluation and enhancement of cucurbit germplasm. USDA Vegetable Lab., Charleston, SC.

51. Robinson, R.W. and Cantliffe, D.J. 1971. Morphactin-iduced parthenocarpy in cucumber. *Science*, **171:** 1251-1255.

52. Robinson, R.W. Shannon, S. and Guardia, M. De. La. 1968. Regulation of sex expression in the cucumber. *Bioscience*, **19:**141-142.

53. Sheshadri, V.S. 1986. Cucurbits. Vegetable crops in India. (eds. Bose, T.K. and SOO, M.G.) Haya Prokash, Calcutta, P. 93-96.

54. Seaton, H.L. Hutson, R. and Muncie, J.H. 1936. The production of cucumbers for pickling process Mich. *Agr. Expt. Sta. Spec. Bul.*, 273,131 pp.

55. Singh, G., Munshi, A.D., Behera, T.K., Choudhary, H. and Dev, D. 2015. Exploitation of heterosis in cucumber for earliness, yield and yield components utilizing gynoecious lines. *Indian J. Hort.*, **72(4):** 494-499.

56. Sims, W.L. and Gledhill, B.L. 1969. Ethrel effects in sex expression and growth development in pickling cucumber. *Calif. Agri.*, **23(2)**:4-6.

57. Singh, D.K. and Ram, H.H. 2004. Isolation of parthenocarpic cucumber line. Presented in First Indian Horticulture Congress on "Improving Productivity, Quality, Post Harvest Management and Trade in Horticultural Crops" Organized by Indian Society of Horticulture, held at ICAR-IARI, New Delhi, 6th-9th November, 2004. Book of Abstracts p. 87.

58. Singh, A.K. and Roy, R.P. 1974. Karyological studies in *Cucumis* L. *Caryologia*, **27:** 183-160.

59. Singh, A.K. and Yadav, K.S. 1984b. Cytogenetics of *Cucumis*: IV. Comparative study of natural and induced polyploids. *Cytologia*, **49:** 183-192.

60. Soria, C., Gomez-Guillamen, M.L., Esteva, J. and Nuez, F. 1990. Ten interspecific crosses in the genus *Cucumis*: A preparatory study to seek crosses reistant to melon yellow disease. *Cucurbit Genet. Coop. Rep.*, 13: 31-33.

61. Swarup, V. 2006. Vegetable Science and Technology in India, Kalyani Publishers, Ludhiana. p. 378-385.

62. Takei, S. and Ono, M. 1939. Leaf alcohol III. Fragrance of cucumber. *Journal of Agriculture Chemsitry Society of Japan*, 15:193.

63. Swaiden, J.M. Ware, G.W. and Mc Collum, J.P. Producing vegetable cros Dan Ville, Illinois : Interstate P. 626.

64. Tolla, G.E. and Peterson, C.E. 1979. Comparison of gibberellin A4/A7 and silver nitrate for induction of staminate flowers in gynoeious cucumber (*Cucumis sativus*) line. HortSci. 14(4): 542-544.

65. Trivedi, R.N. and Ray, R.P. 1970. Cytological studies in *Cucumis* and *Citrullus*, *Caryologia*, **35:** 561-569.

66. Venema, J.h., Boukelien, E.D., Bax, J.E.M., Hasselt, P.R.V. and Elzenga, J.T.M. 2008. Grafting tomato (*Solanum lycopersicum*) onto the rootstock of a high altitude accession of *Solanum habrochaites* improves subtropical temperature tolerance. *Environmental and Experimental Botany*, **63:** 359-367.

67. Wehner, T.C., Cade, R.M. and Locy, R.D. 1990. Cell, tissue and organ culture techniques for genetic improvement of cucurbits. In: Bates, D. *et. al.* (eds.): Bilogy and Utilization of the *Cucurbitaceae*. Cornell University Press, Ithaca, New York, p. 367-381.

68. Whitaker, T.W., 1930. Chromosome number in cultivated cucurbits. *Am. J. Bot.*, **17:** 1033–1040.

69. Whitaker, T.W. and Davis, G.N. 1962. Cucurbits, edn 1, p. 1-136. Inter Science Publishers Inc., New York.

70. Zwinkles, L.A.M. 1987. The present situation and fruit development in breeding of parthenocarpic pickling cucumber. *Acta Horticulture*, **220:** 129-130.

Gherkin (*Cucumis sativus* var. *anguria*)

Introduction

Gherkin (*Cucumis sativus* var. *anguria*) (2n=14) called as 'pickling cucumber' is an important crop of family Cucurbitaceae. The less important cultivated species (*Cucumis anguria* L. 2n=24) commonly called as West Indian gherkin (but sometimes erroneously referred to as pickling cucumber) has been known in semi-wild state in West Indies and is now considered to conespecific with *Cucumis longipes* Hook of Africa and probably became establishes after its introduction by slave trade from Africa. The wild form was retained as var. *longipes* and cultivated variety as *anguria*. It is used in a similar fashion as the cucumber.

Gherkin has gained importance in India due to its export potential (to Europe and USA in pickling industry) and as its consumption within the country is almost negligible. The immature fruits are used for the preparation of pickles, eaten as a cooked vegetable and also used in curries and salad (Purseglove, 1969). It is good source of energy, carbohydrate, sodium, calcium and beta carotene.

Origin, History and Distribution

Gherkin was introduced into the United States in the early 1800s, but remains cultivated to only a limited extent. It is commonly grown in USA, Australia and Sri Lanka. The crop was introduced in India in only 1990 and it can be cultivated round the year unlike other major producing countries like Mexico, Hungary and Madagascar, where it can be cultivated only for 3 months in a year (Anonymous, 2005). In India, it is cultivated in Karnataka, Tamil Nadu and Andhra Pradesh.

Botanical Description

Table 3.44: Botanical Description

Growth Habit	An annual climber, trailing or climbing vine of 90 days duration.
Root	Tap root system with profuse adventitious roots growing in all directions.
Stem	Vine branching freely with slender, rough hairy, 4 angled hairy stems with simple tendrils.
Leaf	Leaf is triangular, 3-5 lobed, hairy, long petiole (5-15 cm).
Floral biology	Flowers are monoecious and yellow in colour. Calyx and corolla are 5-10 lobed. Stamens are 5, ovary inferior and spiny and 3 loculed.

Anthesis and pollination	The opening and closing of flower, pollen dehiscence is goverened by photoperiod and temperature. The stigma becomes receptive 12 hours before and remains so for 24 hours after opening of flowers at moderate temperature. About 30-35 per cent natural self pollination has been reported under field conditions (Jenkins, 1992).
Fruit and fruit set	Fruits are in many shape and size (oval in shape with a round cross-section) and are smaller than those of cucumber. The perfect gherkin fruit is said to have a length to diameter ratio of 3:1 with perfect length being between 5-9 cm, which facilitating its storage in typical canning jars. It has a highly warted skin, long spines and a large cavity with many seeds. While serving gherkin in North America it should preferably exhibit seven warts per square inch (Europeans prefer them wartless). Fruits, which are spiny, green, oval and about 5 cm long, are pickled, eaten fresh as salad or cooked.Fruit is pepo.
Seeds	Seeds are flat and 1g contains 50 seeds.

REFERENCES

1. Anonymous, 2005. Gherkin Agriculture Export Zone-Karnataka. Report prepared by National Bank for Agriculture and Rural Development, Karnataka Regional Office, Bangalore. *Evaluation Study Series* No., 13.
2. Jenkins, J.M. 1992. Natural self-pollination in cucumbers. *Proc. Am. Soc. Hort. Sci.*, **40:** 411.
3. Purseglove, J.W. 1969. Tropical rops Dicotyledons-I. Longmans Green and Co. Ltd., p. 109-110.

Kachari (*Cucumis callosus*)

Introduction

Kachari is also known as Kachariya, Pathya, Sengha, Goradi *etc.*in different parts of India. In Sanskrit it is known as chitra, chitraphala, mrigakshi *etc.*whereas in English it is regarded as "Cucumber pubescent" (Vijayvergia, 1999). Under hot arid agro-climate where soil is low in organic matter and available plant nutrients are great importance in increaseing yield by the balanced plant nutrients NPK (40kg N+20kg of P_2O_2 + 20kg K_2O along with 50 tornnes/ha FYM in Kachari for getting 11.308 yield (Jatav *et al.* 2016.)

Origin, History and Distribution

Kachari is a feral species of India and found in Rajasthan, Punjab, Haryana, parts of Maharashtra, the North Western Province and the Sind area (now in Pakistan). Kachari thrive well both in arid and semi-arid conditions. Mostly it is found growing during rainy season in its wild form in almost all parts of Rajasthan but its occurrence is more particularly in western district like Bikaner, Nagaur, Churu, Jaisalmer, Jodhpur, Barmer *etc.* and some adjoining district of Haryana. Now at some places of Bhilwara district its cultivation has also started.

Botanical Description

Table 3.45: Botanical Description

Floral biology	It is monoecious in nature. It is highly cross pollinated crop. Due to this mechanism great variation exists in this plant with respect to fruit, leaf, quality *etc.* male flower appears early which begin from 25-30 days after sowing. Female flowers come 30-35 days after sowing. Flowers are yellow in colour.
Anthesis and pollination	Anthesis starts between 5.0 AM to 9.0AM.
Fruit and fruit set	Fruit is generally small, egg shaped, sub globose having 3.0-125.0 g weight (Pareek and Samadia, 1998). The skin colour varies from completely white to dark green and its various colour varies from completely white to dark green and its various shades. At immature stage, it is generally green and bitter in taste but turns to pale yellow and sweeting in ripening.

Augmentation of Genetic Resources and Crop Improvement

Kachar possess genes for resistance to fruit fly and leaf-eating caterpillers. AHK-119 and AHK-200 has been developed by Samadia and Pareek (2000).

Table 3.46(a): Collection of *C. callosus* and *C. hystrix* from Parts of Aravalli Hills in Rajasthan

Districts	*Number of Collection*	
	C. callosus	*C. hystrix*
Dungarpur	03	-
Rajsamand	02	-
Sikar	01	-
Sirohi	04	02
Udaipur	18	04

Source: Dwivedi *et al.*, 1998.

Table 3.46(b): ICAR-NBPGR, Collection of Kachari

IC No.	*Date of Collection*	*Species*	*Cultivar Name*	*Location of Collection*
IC 329897	09 August 2001	*Cucumis callosus*	Kakri	Rajsamand
IC 329896	09 August 2001	*C. callosus*	Kachari	Rajsamand
IC 376067	25 November 2002	*C. callosus*	Kachari	Ahmedabad
IC 373582	28 October 2002	*C. callosus*	Kachari	Jodhpur
IC 373541	24 October 2002	*C. callosus*	Kachri	Churu
IC 373531	24 October 2002	*C. callosus*	Kachari	Ganganagar
IC 373516	23 October 2002	*C. callosus*	Kacharari	H. Garh
IC 373443	19 October 2002	*C. callosus*	Kachari	Nagour
IC 370479	04 October 2002	*C. callosus*	Chimdiya	Barmer
IC 369872	08 October 2002	*C. callosus*	Kacjari	Pale
IC 369849	07 October 2002	*C. callosus*	Kachari	Bharatpur
IC 369833	05 October 2002	*C. callosus*	Desi Kachari	Tonk
IC 369832	05 October 2002	*C. callosus*	Kachari	Sawai Madhopur
IC 369827	04 October 2002	*C. callosus*	Kachari	Kota
IC 369801	02 October 2002	*C. callosus*	Kachari	Ahmedabad
IC 258143	21 October 2002	*C. callosus*	Kachari	Dungerpun
IC 258141	21 October 2002	*C. callosus*	Kachari	Ahmedabad
IC 258134	19 October 1999	*C. callosus*	Kachari	Chittourgarh
IC 258130	19 October 1999	*C. callosus*	Kachari	Udaipur
IC 258128	18 October 1999	*C. callosus*	Kachari	Udaipur
IC 258126	18 October 1999	*C. callosus*	Kachari	Udaipur
IC 258125	18 October 1999	*C. callosus*	Kachari	Udaipur
IC 258115	15 October 1999	*C. callosus*	Chibadi	Sirohi
IC 258113	14 October 1999	*C. callosus*	Kachari	Jalor
IC 373575	28 October 2002	*C. callosus*	Kachari	Tonk
IC 272826	25 October 1999	*C. callosus*	Kachari	Udaipur
IC 272825	25 October 1999	*C. callosus*	Kachari	Udaipur
IC 272824	25 October 1999	*C. callosus*	Kachari	Udaipur
IC 385209	04 October 2002	*C. callosus*	Kachari	Sikar
IC 385200	02 October 2002	*C. callosus*	Kachari	Hanumangarh
IC 325842	13 October 2001	*C. callosus*	Kachari	Pali
IC 325807	12 October 2001	*C. callosus*	Kachari	Pali
IC 325751	11 October 2001	*C. callosus*	Kachari	Pali
IC 325741	11 October 2001	*C. callosus*	Kachari	Pali
IC 550169	27 October 2006	*C. callosus*	Chibariya	Pali
IC 550170	27 October 2006	*C. callosus*	Chibari	Sirohi
IC 550171	27 October 2006	*C. callosus*	Chibari	Sirohi
IC 550172	28 October 2006	*C. callosus*	Chibari	Sirohi
IC 550173	28 October 2006	*C. callosus*	Chibari	Sirohi
IC 550173, 75,76,77	28 October 2006	*C. callosus*	Chibari	Sirohi
IC 550181	29 October 2006	*C. callosus*	Chibari	Sirohi
IC 550183,84,85	29 October 2006	*C. callosus*	Chibariyan	Sirohi
IC 550186,87	29 October 2006	*C. callosus*	Kachadi	Sirohi
IC 550190,91,92	30 October 2006	*C. callosus*	Kachadi	Sirohi
IC 550195	31 October 2006	*C. callosus*	Kachadi	Sirohi
IC 550200	01 November 2006	*C. callosus*	Kachari	Durgapur
IC 550201	01 November 2006	*C. callosus*	Kachari	Durgapur
IC 550206	03 November 2006	*C. callosus*	Kachari	Udaipur

Conservation of Genetic Resources

A large number of genetic resources of this crop is being maitatined at CIAH, Bikaner (Rajasthan).

REFERENCES

1. Dwivedi, N.K., Bhandari, D.C. and Bhatnagar, N. 1998. Collecting wild species of crop plants from Aravalli hills, Rajasthan (India).
2. Jatav, M.,Sharma, B.D., Samadia, D.K. and Meena, S.R-2016. Effect of different sources of nitrogen on kachari performance in the arid region. Annals of Arid Zone 55 (1&2): 35-39.
3. Samadia, D.K. and Pareek, O.P. 2000. Two *Cucumis* sp. (Kachari), varieties of arid region. *ICAR News,* **6(2):** 15-16.
4. Vijayvergia, V.A. 1999. Prospectus for fighting poverty, hunger and malnutrition. International conference on Indigenous Vegetables and Legumes.

Muskmelon (*Cucumis melo* L.)

Introduction

Among the melons, muskmelon (*Cucumis melo* L., 2n=24) is considered as major crop in the family cucurbitaceae. The words muskmelon and cantaloupe are used rather loosely to mean the same. As a matter of fact, all canatloups are muskmelon but all muskmelons are not cantaloupe. The same cultigens represent as cantaloupe in the USA is known as muskmelon in India. In fact netted cultivars developed in the USA to meet shipping requirements are not the true cantaloupes but belong to the botanical variety *reticulates*. Immature fruits are often used as a cooked vegetable (non-desert types) but the delicious fruits after ripening are used as desert fruit and are an unavoidable item of western diet. The desert types quench thirst and add to the nutrient content of main diet. Edible portion of melons contain water (90 per cent), protein (0.6 per cent), fat (0.2 per cent), and carbohydrate (3.50 per cent). Apart from these elements sufficient amount of calcium (32 mg), phosphorus (14 mg), iron (1.4 mg) carotene (16 mg) and vitamin C (26 mg) are found in 100 g edible portion (Akyroyd, 1963). The TSS content varies from 8-17 per cent. The carbohydrate is essentially all sugar. Certain melons are good sources of vitamin A. The yellow and orange fleshed melons contain more than 350 mg of β-carotene, a precursor of vitamin A. Cantaloupe contains 45 mg and honey dew 32 mg of vitamin C per 100 g of edible portion. In muskmelon, the aromatic compound is predominantly ester of acetic acid. Nerson *et al.* (1990) reported on the development of melofon, a genotype of *C. melo* which is suitable for pickle production.

Origin, History and Distribution

The origin of muskmelon is in dispute. A few researchers consider tropical Africa (more specifically in the region, south of Sahara region) as the centre of origin for many *Cucumis* species, since 40 or more wild species of *Cucumis* are found to occure in the tropics and subtropics of Africa (Whitaker and Davis, 1962). China is the primary centre of origin of muskmelon. Whitaker and Bemis (1976) thought that Central Asia comprising some parts of Southern Russia, Iran, Afghanistan, Pakistan and North West India (Indo-Burma) may be regarded as a secondary centre of muskmelon. Archeological remains indicated that melon was cultivated in Iran 5,000 BP. Ancient melon that was distributed throughout the Middle East and Asia (India, Iran, Afghanistan and China) originated the genetic diversity that exists in the area (Robinson and Deckerwalters, 1997). Some researchers suggested that India is the centre of domestication because it has cultivated there for centuries and inedible form grows wild in India. Some believe that muskmelon domestication began from Iran. Melon can be considered as the most highly developed types of ancient cultivated species and, through many changes, melon could get into those elite forms that exist today (Mallick and Masui, 1986). Melon was introduced in Central America in 1516, in Virginia 1609 and New York in 1629 (Ware and Mc Collum, 1980). It has been introduced to Europe and from Europe to the USA by the early travelers. Muskmelon had been cultivated in Egypt since 2400 BC. Cultivation of muskmelon in Asia is of very ancient date. It was introduced into India, probably during Mughal invasion around 14th century, from the Central Asian region comprising Uzbekistan, Iran, Afghanistan and parts of Pakistan. Muskmelon entered into China from India around beginning of the Christian era. Its cultivation started in Italy around 14th century after collapse of Roman Empire. Melons were introduced to England around the turn of 16th century. Its extensive cultivation in France started around 1629. The cantaloupes derived their name from a place near Rome called Cantaluppi, where it had been cultivated for long. Columbus carried the seeds during his second voyage to the New World. Around the same time muskmelon was introduced into central and northern Europe from Italy. Around 1600, muskmelon was grown in North America from Florida to New England. However, it was only after 1865 that muskmelon became a commercial crop in the United States. Muskmelon is cultivated throughout the world, comprising tropical and sub-tropical regions. China, Iran, Turkey and the USA together produces 57 per cent of the total world production. It is an important truck and garden crop of India and is popularly cultivated throughout the country. It has become an important cash crop of north India, especially Punjab, Uttar Pradesh and Rajasthan. River beds of Jamuna, Ganges and Narmada in the North and Kaveri, Krishna and Godavari in the south are the major centres of muskmelon cultivation. They are popular desert and preserving melons cultivated especially in the South India.

Species Availability, Evolution and Diversity

The huge diversity of this polymorphic species *Cucumis melo* has been recognized as a complex of cultigenic species with free possible inter-crossing. Infact Russian botanist has raised *Cucumis melo* to genetic status of *melo* species retaining cucumber under *Cucumis*. According to them, no African wild species could be suspected to be an ancestor of the cultivated or wild melons. But studies conducted on alloenzyme variation on wild and cultivated species of *Cucumis* totaling 30 species distributed over two distinct geographical areas:

(i) South East of Himalayas is an important region of Asiatic group with basic Chromosome of x= 7 to which cucumber belong and

(ii) African group, comprising part of Africa, Miiddle East, Central Asia extending to Pakistan and south Arabia. Most species with basic chromosome number x=12 are found in the region with few tetraploid and hexaploids.

Mostly melons occure in this region. It is suggested that India rather than Africa, is the centre of origoin of cultivate melons. The African group (melon group) has 30 species divided into six subgroups (Kirkbride, 1993). Melon and other $2n = 24$ species were originally distributed across a large part of Africa and Middle East up to Pakistan and South Arabia. However, some species also occurred in the Asiatic group range (Kroon *et al.*, 1979; Ramachandran andNaryan, 1985). This is the case of *C. hystrix* Chakr., which is the only $2n = 24$ native to Asia. This species is of particular interest because of morphological and biochemical characteristics similar to *C. sativus*and chromosome number equal to *C. melo*, indicating a possible bridge between the two species (Chen andAdelberg, 2000). In India, particularly the oriental pickling melon (*C. melo* var. *conomon*) and snap melon (*C. melo* var. *momordica*) arec unique and have considerable variability in the Western Ghats. On the basis of extensive cytogenetical studies of 43 wild species from South, East and West Africa, consisting of annual, and perennials, diploid, tetraploid and hexaploid, however South Africa is considered to be the primary gene centre of the genus.

Table 3.47: Botanical Varieties of *C. melo*

Common Name	*Botanical Groups*
C. melo var. *acidulus*	Sour melon
C. melo var. *aestivales*	Summer tough melon
C. melo var. *agrestis*	Weed melon
C. melo var. *albida*	New melon
C. melo var. *autumnales*	Fall melon in central Asia
C. melo var. *bucharici*	Summer soft melon
C. melo var. *cantalupa*	Cantaloupe melon
C. melo var. *cantalupensis*	Rock melon
C. melo var. *chadalak*	Chandalak melon
C. melo var. *chate*	Ajjuol melon
C. melo var. *chito*	Lemon melon
C. melo var. *conomon*	Pickling melon
C. melo var. *dudaum*	Pocket melon
C. melo ar. *Duripulposus*	Europe summer melon
C. melo var. *erythraeus*	-
C. melo var. *flava*	Ogon melon
C. melo var. *flexuosus*	Snake melon
C. melo var. *ginmakuwa*	Yellow sweet melon
C. melo var. *Grubek*	Girubek melon
C. melo var. *hassanbey*	Winter casaba melon
C. melo var. *hibernus*	Winter melon in central Asia
C. melo var. *hime*	Orange melon
C. melo var. *hiemalis*	Gimofuka melon
C. melo var. *inodorus*	Winter melon
C. melo var. *kikumelon*	Yuki melon
C. melo var. *Makuwa*	Oriental sweet melon
C. melo var. *melitensis*	Glatte melon
C. melo var.. *momordica*	Snap melon
C. melo var. *monoclinus*	-
C. melo var. *neomakuwa*	-
C. melo var. *microspermus*	Seikan melon
C. melo var. *praecox*	Russian early melon
C. melo var. *reticulatus*	Netted melon
C. melo var. *rokkiford*	Rokkiford melon
C. melo var. *saccharinus*	Pineapple melon
C. melo var. *tomago*	Egg melon
C. melo var. *tarra*	Tarra melon
C. melo var. *utilissimus*	Kakri
C. melo var. *zhukovskii*	Summer casaba melon

Source: Mallick and Masui, 1986.

Gene and Genetics of Characters

Crop improvement can be affected by applying appropriate breeding methods based on the information of genetic control of various qualitative and quantitative traits, which is given as under:

Table 3.48: Gene List for Melons

Gene Symbol	*Characters*
a or aG	Andromonoecious
Interaction of 'a' and 'g' genes	Trimonoecious -Gene 'A' convert into pistillate - Gene 'G' convert into staminate
Ag	Tolerant to *Aphis gossypii*
dc	Resistant to *Dacus cucurbitae*
gf	Green flesh colour
gp	Green petals
jf	Juishy flesh
msl	Male sterile 1
o	Oval fruit
r	Red stem
s	Surures

Gene Symbol	Characters
sp	Spherical fruit
wf	White flesh
yg(y)	Yellow-green
Gene symbol	Characters
Af	Resistant toAulacophora specis
AG	Monoecious
g or Ag	Gynomonoecious
ag	Hermaphrodite
Bi	Bitter
gl	Glabrous
gyc	Greenish-yellow corolla
lmi	Long main stem internode
n	Nectorless
p	Pentamerous
ri	Ridge
So	Sour
st	Striped pericarp
wi	White fruits (immature)
y	Yellow pericarp

Yellow rind colour is monogenic and partially dominant over cream, while green rind is dominant over yellow. Striped and mottled pericarp is dominant over white. Flesh colour has complex inheritance with indications of some material effects. Orange flesh colour is monogenecally dominant over light green, while white colour is monogeneically dominant over green. Thin flesh is partially dominant over thick flesh and monogenically controlled. Less juiciness is monogeneically dominant over juiciness. Fruit slipping is monogencnically dominant over non-slipping. Rind cracking which is an undesirable trait is controlled by two genes with complementary and supplementary gene action. Netting on the rind which imparts firmness to the fruit, is governed digenically with netting being dominant over smoothness. Gene "S" suppress netting when present in homozygous condition and gene "N" is epistatic over 'S" and so was "n" over "s". Netting is also reported to be monogenically controlled, which is partially dominant over smoothness

Botanical Description

Muskmelon belongs to family Cucurbitaceae, genus *Cucumis* and botanical name is *C. melo* L.

Table 3.49: Botanical Description

Growth Habit	Muskmelon is annual with climbing, creeping or trailing vines of length up to 3.0 m.
Root	Highly branched tap root system, medium deep rooted.
Stem	Trailing stem grows up to 300.0 cm, ridged with simple tendril.
Leaf	The leaves of muskmelons are deep or 3-7 shallowly lobed, oval or kidney shaped and more or less rounded (circular), 8.0-15.0 cm in diameter, not pinnatifid, dentate, cordate at base, hairy on under surface and petioles 4.0-10.0 cm in length
Floral biology	The flowering starts after 45 days of sowing. Flowers are large, solitary; flowers lemon yellow to deep orange, fragrant, corolla rotate deeply parted and small flowers. Flowers are monoecious or may be andromonoecious (hermaphrodite and staminate flowers occurring on same plant), gynoecious (only pistillate flowers), or monoecious (pistillate and staminate flowers), monoecious and andromonoecious are most common (Rosa, 1928). Staminate flowers appear first as a cluster on main and secondary branches, but hermaphroditic flowers appear alone on secondary branches; pistillate flowers solitary, 1.2-3.0 cm in diameter, yellow, pedicels short; calyx 5-lobed, 60.0-8.0 mm in length; corolla 5-partite, petals round, 2.0 cm in length, fused at base; stamens 3, free; pistil with 3-5 placentas and stigmas; ovary inferior.
Anthesis and pollination	Melons are highly cross pollinated (natural cross pollination is 3.10-95.0 per cent). Flowers anthesis takes place at around 5:30-6.30A.M. and dehiscence of anther takes place between 5:00 to 6:00 A.M. Stigma is receptive 2 hours before to the day of anthesis. Duration of pollen fertility is 5:0 A.M.-2:00 P.M.For maximum fruit setting early morning pollination is more useful. For natural crossing pollinating agent are honeybees (Nandpuri and Brar, 1966).
Fruit and fruit set	Fruit is many seeded pepo. Fruits are extremely variable in size and shape *i.e.* fleshy, ellipsoid to globose, smooth with or without longitudinal grooves and smooth to rough and reticulate (netting). Skin may be soft or hard yellow, green, cream or orange in colour and plain, netted or echinate (prickly) in texture. Flesh inside varies from white to cream-yellow, orange or green. At present most of the commercial types have thin skin and thick orange pulp.
Seeds	Flattened, white, cream or buff coloured and smooth, 5.0-15.0 mm in length, the edible seed kernel contains approximately 46 percent oil.

Horticultural Classification

Out of the seven melons, *C. melo* var. *agretis*, *C. melo* var. *cantaloupensis* Naud. (muskmelon), *C. melo* var. *utilissumum*, *C. melo* var. *momordica*, and *C. melo* var. *flexuosus* Naud are available in Indian Sub- continent. In muskmelon the netted varieties developed in the USA to meet shipping requirements are not the true cantaloupes but belong to the botanical variety reticulates. The cantaloupes *i.e. C. melo* var. *cantaloupensis* are grown in the Europe. Forty botanical varieties of *C. melo* were listed by Mallick and Masui (1986).

Classification on the Basis of Sex Forms

Many sex forms-are present in melons leading to complete allogamy.

(i) Monoecious,

(ii) Gynomonoecious,

(iii) Andromonoecious and

(iv) Perfect

Generally two sex forms *viz.* adromonoecious and monoecious have been observed but muskmelon having predominantly adromonoecious plants (Rosa, 1928). It is well kown that all forms hybridize readly and there are many intermediate forms available in nature (Mathew 1984).

Classification on the Basis of Distribution of Diversity

Muskmelon (*C.melo*) based on the distribution of diversity can be grouped into seven sub sets (Munger and Robinson, 1991):

Table 3.50: Species-wise Description of *C. melo*

Species	Common Name	Salient features
C. melo var. *agrestis*	Kachri	It is Wild type with slender vines and small inedible fruits, probably synonym of *C. melo* var. *callosus* (*C. callosus*) and *C. melo* var. *trigonus* (*C. trigonus*).
C. melo var. cantaloupensis Naud.	Musk melon or cantaloupe	Medium sized fruits with netted, warty or scaly surface, flesh usually orange but sometimes green, aromatic or musky flavour. Fruits dehiscent at maturity, usually andromonoecious.
C. melo var. flexuosus Naud. (*C. melo var. utilissumum*)	Snake melon, snake cucumber, Tar-kakari	Fruits long and slender consumed at immature stage, monoecious.
C. melo var. momordica (*C. melo var. momordica*)	Snap-melon or phut	Monoecious crop grown in India and other Asian countries. It has white to pale orange, less sweet pulp. The smooth surface of the fruit starts cracking at the time of maturity.
C. melo var. *conomon*	Sweet or pickling melon	Generally andromonoecious in nature and bears small fruits with skin, white flesh.
C. melo var. inodorous Naud.	Winter melon	The fruit is smooth or wrinkled surface with white or green flesh, and lacking musky odor. It is also andromonoecious in nature and usually required more time for maturity.
C. melo var. dudaim Naud.	Mango melon, Pomegranate melon	Fruits small, globular, smooth mottled but not netted, flesh with acid flavor, picking type.

Classification on the Basis of Fruit Shape

Melons can be classified on the basis of fruit shape:

Table 3.51: Species and Salient Features of Fruit Shape

Species	Salient Features of Fruit Shape
C. melo var. *agrestis* Naud.	Fruit is small, round and inedible, grown at higher altitudes up to 1,100 m.
C. melo var. *cultas*	Fruit is large and edible
C. melo var. *reticulatus*	Netted or nutmeg muskmelon, cantaloupe (the USA), Persian melon-Green background, white netted, skin thick.
C. melo var. *canataloupensis*	Cantaloupe-tetraploid (2n=48) Cantaloupe (Europe).
C. melo var. *inodorus*	Winter melon 'Honey Dew', white skinned summer; casaba melon, fruit is smooth, with musky odour.
C. melo var. *conomon*	Fruit is very large, thick flesh, no odour, oriental pickling melon
C. melo var. *flexuosus*	Serpent melon, snake melon, fruit shape is long and peculiar
C. melo var. chicko	Mango melon, garden melon- fruit has odour like mango fruits, fruits are like cucumber
C. melo var. *dudain*	Pomegranate melon, Queen Anne's Pocket melon-Fruit has deep orange coloured flesh
C. melo var. *momordica* Duthie and Fuller	Snapmelon, skin cracks on maturity
C. melo var. *utilissimus* Duthie and Fuller	Long melon, fruits are long and slender, fruits are ribbed.

Classification on the Basis of Common Cultivation

Table 3.52: Cultivated Group and Salient Features

Cultivated Group	Salient Features
Cantaloupe melon of Europe	Thick, scaly, rough rind, often deeply grooved fruits
Muskmelon grown in the USA	This has smaller fruits and rinds which are finely netted to nearly smooth and very shallow ribs.
Casaba or Winter melon	This produces large fruits which mature late and have good storage quality. The rind is usually smooth, often striped or splashed and yellow in colour. Flesh is firm with little musky odour or flavor. The cultivar 'Honey Dew' of USA is one of these.
Oriental pickling melons	A number of types often with elongate fruits resembling cucumbers are grown in India and the Far East and are used as vegetables.

Classification on the Basis of Growing Region at Global Level

Table 3.53: Species and Salient Features of Fruit Shape

Species	Growing Region at Global Level
C. melo ssp. *rigidus*	Melons growing in hot dry regions (Central Asia and Iran)
C. melo ssp. *orientale*	Melons growing in hot and humid regions (Asia minor)
C. melo ssp. *eurpees*	Melons growing in cool, dry regions (Asia minor)
C. melo ssp. *flexuosus*	Semi feral melons- Snake melons
C. melo ssp. *chinensis*	Chinense melons (East Asia)
C. melo ssp. *spontaneum*	Wild melons

Classification on the Basis of Season

(i) Summer melon

(ii) Spring melon

Classification on the Basis of Actual Fruit Taste

Table 3.54: Species and Taste of Fruit

Species	*Actual Fruit Taste*
C. melo var. *callsus*	Bitter
C. melo var. *acidiulus*	Sour
C. melo var. *cantalupensis*	Bland
C. melo var. *reticulatus*	Sweet

Mallick and Mausi (1986) classified *Cucumis melo* in 40 botanical varieties, which are as under:

Table 3.55: Species and Common Name

Species	*Common Name*
C. melo var. *acidulus*	Sour melon
C. melo var. *aestivales*	Summer tough melon
C. melo var. *agrestis*	Weed melon
C. melo var. *albida*	New melon
C. melo var. *autumnales*	Central Asia, Fall melon
C. melo var. *bucharici*	Summer soft melon
C. melo var. *cantalupa*	Cantalupe melon
C. melo var. *cantaloupensis*	Rock melon
C. melo var. *chadalak*	Chandalok melon
C. melo var. *chate*	Ajjuol melon
C. melo var. *chito*	Lemon melon
C. melo var. *conomon*	Pickling melon
C. melo var. *dudaum*	Pocket melon
C. melo var. *duripulposus*	Europe summer melon
C. melo var. *erythraeus*	-
C. melo var. *flava*	Ogon melon
C. melo var. *flexuosus*	Snake melon
C. melo var. *ginmakuwa*	Yellow sweet melon
C. melo var. *grubek*	Girubek melon
C. melo var. *hassanbey*	Winter casaba melon
C. melo var.*hibernus*	Central Asia, Winter melon
C. melo var. *hime*	Orange melon
C. melo var. *hiemalis*	Gimobuka melon
C. melo var. *inodorus*	Winter melon
C. melo var. *kikumelon*	Yuki melon
C. melo var. *makuwa*	Oriental sweet melon
C. melo var. *melitensis*	Glatte melon
C. melo var. *momordica*	Snapmelon
C. melo var. *monoclinus*	-
C. melo var. *neomakuwa*	-
C. melo var. *microspermus*	Seikan melon
C. melo var. *praecox*	Russian early melon
C. melo var. *reticulatus*	Netted melon
C. melo var. *rokkiford*	Rokkiford melon
C. melo var. *saccharinus*	Pineapple melon
C. melo var. *tamago*	Egg melon
C. melo var. *tarra*	Tarra melon
C. melo var. *utilissimus*	Kakari
C. melo var. *zhukovskii*	Summer casaba melon

Augmentation of Genetic Resources and Crop Improvement

Breeding Priorities

Table 3.56: Traits and Descriptions

Traits	*Descriptions*
Early maturity	The number of node at which first pistillate (or hermaphrodite) flower appears, gives a fair indication of earliness.
Sex ratio	A high female to male ratio is desirable
Fruit shape	The fruit shape varies in different cultivars of the crop, like flat or oblong
Fruit surface	Netting and thick rind are useful in transportable quality/long distance haulage.
Flesh colour and thickness	Flesh colour in muskmelon may be white or greenish or salmon orange. The flesh should be thick with small seed cavity.
Sweetness	Sweetness in melons is judged as total soluble solids content read by squeezing the juice in a hand refractometer, subject to varying conditions of environment. Generally 10 percent TSS may be taken as a minimum standard for selection of fruits to meet consumer preference. In international market more than 13 percent TSS is acceptable.
Flavour	High aroma in muskmelon is preferred by consumers.
Yield	Yioeld in terms of weight and number of fruits of marketable quality will be the deciding criterian in selection. Early yield, especially of higher percentage of total yield is a desirable character.
Shelf-life	Longer shelf-life is desirable. Muskmelon varieties with netted thick rind and medium in juiciness have better shelf-life.

Earliness is predominantly governed by dominance variance which is more than additive variance and indicaions of the presence of over-dominance are also available. Presence of both dominance and additive gene effects in the control of earliness in some crosses while presence of dominance as well as epistasis in other crosses was also recorded. Rind thickness is governed by additive, dominance and epistasis gene action. Flesh thickness is governed by partial dominance with adequate additive genetic variance. Some reports indicated the role of additive, dominance and epistasis in the expression of flesh thickness. TSS content which is a premium trait in muskmelon is governed by partial dominance coupled with adequate additive genetic variance. More than seven groups of genes with over-dominance govern the inheritance of TSS content. Presence of additive, dominance and epistasis has also been recorded in respect of TSS content. Fruit yield is predominantly governed by dominance variance or even over-dominance. However, role of additive gene effects for yield has also been indicated. Number of fruits/vine is governed by partial dominance alongwith adequate additive genetic variance. Presence of additive, dominance and epistasis gene effects has also been observed. Fruit weight which is an important component of yield is reported to be governed by thirty nine genes with geometrically cumulative effects, the small fruitedness being partially dominant over large fruitedness. Vine length is controlled by additive and dominance genetic variance, dominance variance being higher than additive variance.

Introduction and Selection

Inheritance of various qualitative traits *viz.*, rind colour, stripes and mottling on pericarp, flesh colour and thickness, juiciness of flesh, fruit slipping, rind cracking and netting have been included in selection of cultivars. In muskmelon several genotypes were introduced from various countries for various traits.

Table 3.57: Introduction of Muskmelon Germplasm in India

Name of Germplasm	Salient Feature	Source Country
EC348140-43	F_1 hybrid (early and sweet)	USA
EC382726-36	Male sterile gene (MS2 to MS5)	France
EC350603	Sulpher tolerance and WMB tolerance	USA
EC342751-54	Heat tolerance	Japan
EC350602	Aphid resistance	USA
EC330956	Multiple disease resistace	USA

The muskmelon varieties 'Arka Jeet', 'Arka Rajhans' and 'MH 1' were developed by single plant selection and latter maintained by mass selection. In Nepal, muskmelon cultivar 'Swarna' is very popular among the growers. In Pakistan cultivars like T-96 (R) and Ravi (I) are very common and being grown commercially. Patel *et al.* (2106) evaluated 12 promising genotypes and found significant differences for all the traits under study. The components out of 11 showed maximum amount of variation (89.87 per cent). Principal component (PC) 1, accounted variation for seed cavity followed by TSS, pH, β carotene and fruit yield which was contributed by genotypes MC-2012-12, MM-202 and MS-1. Similarly, PC 2 was governed by flesh thickness, rind thickness, days to first female flower opening and days to 1st fruit ripening due to entries Farmer's Glory, MM-314 amd MH-27. Biplot analysis showed fruit yield, seed cavity, titrable acidity and average fruit weight were correlated in comparison to other traits, while among genotypes, MC-2012-12, Punjab Sunehri and MM-314 were found diverse with respect to traits in the study. There was least effect of environment in all studied traits. Fruit yield was positively correlated with average fruit weight, seed cavity and titrable acidity but negative correlated with pH and fruit firmness at both genotypic and phenotypic level. TSS was negatively associated with fruit yield, average fruit weight but positively associated with β carotene, pH and fruit firmness. To enhance yield and quality of muskmelon, association between different traits must be kept in view during selection and genotypes showing maximum diversity could be used in hybrid development with desirable heterosis for particular traits.

Table 3.58: Muskmelon Varities Developed by Selection with Progeny Testing

Variety	Source
RM-43	Indigenous
Durgapura Madhu	Indigenous
Arka Rajhans	Indigenous
Arka Jeet	Indigenous
Pusa Madhuras	Indigenous

New cultivars Pusa Modhuriama (DM-159-1) and Pusa Sarda (DHM-163) has been developed by ICAR-IARI, New Delhi.

Use of Gynoecism and its Maintenance

Availability of gynoecism in muskmelon can facilitate easier, cheaper and efficient method of hybrid seed production. The stable gynoecious line G Y-4 was derived from a segregating population Wiscosin 998. (Peterson *et al.* 1983). Similarly 'Gylan' was another useful gynoecious line (cohenn *et al.* (1993). Gynoecious line namely WI-998 and GM-6E-7 can be exploited for commercial hybrid seed production. However, stability of gynoecious sex expression must be ensured in order to make the hybrid seed production dependable. Further the lines must be improved upon for desirable fruit quality so as to obtain commercially acceptable F_1 hybrids. The gynoecious line can be maintained by inducing anthers in the pistillate flowers by spraying growth regulators namely silver nitrate @ 100-200 ppm, ethrel and Alar or silver thiosulphate @ 500 ppm. Effect of Silver thiosulphate (STS) (100, 200, 400 ppm) was applied at three stages of plant growth (two true leaf stage, four true leaf stage, 5-10 pistillate flowers anthesized) on a pure gynoecious muskmelon, WI-998. STS 400 ppm was found to be the most effective which produced maximum number of hermaphrodite flowers (45, 72 averaged over stages per plant). Concentrations of STS had no influence on the node number of first hermaphrodite or pistillate flower. Average effectiveness of STS estimated to be 15-31 days. It would be possible to induce hermaphrodite flowers on true breeding gynoecious muskmelon with the help of STS (100, 400 ppm) application at any stage of plant growth (More and Seshadri, 1987).

Hybridization and Selection

Dixit and Kalloo (1983) reported heterosis for total fruit yield per plant to the extent of 46.7 per cent and 38.1 per cent in the cross combinations Pusa Sharbati x Sarda Melon and Pusa Sharbati x Punjab Sunehri, respectively. Cross combinations Pusa Madhuras x Ravi, Pusa Sharbati x Pusa Madhuras, and Pusa Madhuras x Hara Madhu were observed to be the most promising for yield perplant. The best performing F_1 hybrid Pusa Madhuras x Ravi, which recored 20.15 per cent higher yield than the best commercial check Pusa Madhuras, was suitable for commercial cultivation. The muskmelon varieties 'Pusa Sharbati' ('Kuttana' x 'American Cantaloupe') and 'Punjab Sunehari' ('Hara Madhu' x 'Edisto') were developed by pedigree method. Long back many hubrids were produced using monoecious lines as materanal parents and among them 'M_4' x 'Pusa Madhuras' and 'M_3' x 'Durgapura Madhu' were promising. Heterosis for yield was found 46.7 per cent in 'Pusa Sharbati' x 'Sarda Melon', 54.3 per cent for number of fruits in 'Punjab Sunehari' x 'Sel 1', 39.7 per cent for fruit weight in 'Arka Jeet' x 'Pusa Sharbati' and 26.1 per cent for TSS in 'Arka Jeet' x Sarda Melon'.

Table 3.59: Cross Combinations and Salient Features

Cross Combinations	*Salient Features*
'Persian 5' x 'Plovdissko Banan'	Early, high yielding
'Mono 2' x 'Sharbati'	Sweeter taste
'Mono 2' x 'Arka Jeet'	Sweeter taste
'PMR 5' x 'Kelly Sweet'	Early maturity
'Smith Perfect' x 'PR 45'	Early maturity

Male Sterility

In muskmelon male sterile 'ms-1/ms-1' was first reported by Bohn and Whitaker but due to certain reasons it could not be utilized in hybrid development. A second occurrence of male sterility was observed as a variant in a powdery mildew resistant cantaloup breeding line (La Jolla 40460) as reported by Bohn and Principe, 1964). A male sterility gene ms1 was discovered in a cross 'Georgia 47' x 'Smith Perfection' but it was difficult to identify the plants. Male sterility now beeing utilized in the exploitation of heterosis. Two American male sterile lines 'MS1' and 'MS2' are used to develop F_1 hybrids using 'Earls Favourite' and 'Pearl' as pollen parents. The F_1 hybrid between a male sterile line 'MS1' and Hara Madhu was highly heterotic and was released for commercial cultivation in the Punjab state under the name "Punjab Hybrid" is based on GMS. Apart from yield, it was early maturing and superb in fruit quality in terms of high and stable TSS, juiciness and flavor. Mishra and Seshadri (1985) working with male sterility recorded 313.8 per cent heterosis over better parent for total yield, 135.6 per cent for average fruit weight, 130.6 per cent for number of fruits/vine and 44.1 per cent for TSS content. They identified F_1 MS-2 x Pusa Madhuras for TSS, MS-1 x Harela for yield and MS-1 x Persiiki-5 for earliness.

Interspecific Hybridization

The botanical varieties of *C. melo* are inter-crossable. East Africa, adjoining Arabian Peninsula and eastern part of Mediterranean have greatest concentration of several wild species (2x=24, 48, 72) that showed phytogenetically old characters. There are several primitive forms with xerophytic and perennial habit and there are some tetraploids. Many of these species are resistant to pests and also those of South Africa. Most of the species are tolerant to aphids and powdery mildew and *C. anguris*, *C. ficifolius* and *C. metuliferus* are resistant to nematode and there are several species which carry resistant to bean spider mite and whitefly. Some studies showed that all the wild African species did not cross with *C. melo* either at diploid or tetraploid level. The *Cucumis* spp. grouped under 5 groups, where *C. sativus* and other 14 chromosome spcies stand apart in one group and cultivated melo in another. The *anguria* group consisted of *C. longipes, C. africanus, C. dipraceus, C. leptodermis, C. myriocarpus, C. prophetarum, C. zeyheri* and tetraploid (2n=48) *C. ficifolius* and *C. heptadactylus*. The fourth group consisted of *C. dinteri* and its conspecific *C. sagittatas, C. humifructus* and the fifth group was typified by *C. metuliferus*. The latter investigations that included cytological studies confirmed the conspecific between *C. dinteri* and *C. sagittatas* and between *C. myriocarpus* and *C. leptodermis*. The cross compatible group listed include annual species *C. africanus, C. anguria, C. dipsaceus, C. leptodermis* and *C. myriocarpus*, perennial ones *C. ficifolius* and *C. zeyheri*; perennial tetraploid *C. aculeatus*,*C. heptadactylus* and*C. zeyheri* and perennial hexaploid *C. figarea* Mathew (1984) found successful compatibility (crossability among each other) among oriental preserving melon (*C.melo* var. *conomon* Mak), muskmelon (*C.melo* var. *inodorus* Naud), long melon (*C. melo* var. *utilissimus* Duth. and Full.), snake melon (*C. melo* var. *flexuosus* Naud) and snapmelon (*C. melo* var. *momordica* Duth. and Full.). In this experiment muskmelon and snake melon had the maximum genetic distance of 14.49 but the long melon and snapmelon were the closest (D^2=0.28). Attempts have been made to cross muskmelon with cucumber but no success achieved. Chatterjee and More (1991) attempted interspecific hybridization in *Cucumis* spp. to transfer high level resistance to cucumber green mottle mosaic virus (CGMMV) from 5 wild *Cucumis* species, *viz. C. figarei, C. zeyheri, C. meeusii, C. myriocarpus* and *C. africanus*. Besides in the first 4 species, a complete resistance to Fusarium wilt caused by *Fusarium oxysporum* sp. *melonis* was confirmed. Sixteen interspecific crosses were made by crossing 4 wild species with 2 cultivated forms of melon ('Pusa Madhuras' and 'Monoecious 4') and their 8 reciprocal crosses were attempted under field conditions by conventional hybridization procedure. Fruit set was not observed in any of the crosses. Inability of pollen of *C. melo* to germinate on the stigma of *C. figarei, C. meeusii* and *C. zeyheri* and the pollen of these species on *C. melo* stigmas, even up to 72 hours after pollination, indicated that a pre fertilization barrier is involved in the failure of interspecific by hybridization in *Cucumis*. Chatterjee and More (1991) suggested that barriers for interspecific hybridization can be removed through bud pollination, use of mentor pollen, brush pollination, growth regulator (benzyl adenine) application, bio-technological interventions and treatment of stigma with organic solvent (n-hexane). In the cross *C. figarei* x *C. melo*

(M_4) 2 per cent fruit set was obtained when pollination was done after washing the stigma with organic solvent (n-hexane). "Prince Melon" a hybrid variety between a cantaloupe (*C. melo* L. *cantalupensis* group) and Oriental melon (*C. melo* L. *makuwa* group), was released in 1962. Because of its sweetness, it has completely replaced Oriental melon. However, "Prince Melon" was not equal in quality to Western melon (*C. melo* L. reticulatus group) in flavor. After the success of "Prince Melon", new breeding trials with cross-combination of Oriental and Western melon have been carried out. The interspecific crosses utilizing six genetically diverse inbred melons *viz.*, IC0599709 (monoecious lines of muskmelon), AHS-82 (var. *momordica*), AHLM-2 (var. *utilissimus*), Arya (var. *chate*), and AHK-119 (*C. callosus*) were made by conventional crossing during summer season of 2014 at ICAR-CIAH, Bikaner Rajasthan and evaluated during summer seasonof 2015 in RBD replicated thrice. Among the crosses, AHS-82 x AHK-119 took minimum days to produce 50 per cent female flowering (42 days) followed by AHLM-2 x IC0599709 (47.67 days). The maximum fruit weight (1142 g) was recorded in AHLM-2 x IC0599709 followed by AHS-82 x AHLM-2 (979g) whereas, it was minimum in AHLM-2 x AHS-82 (76.4g) followed by AHLM-2 x Arya (77g). AHLM-2 x AHS-82 resulted in maximum fruit length (40.2 cm). The maximum TSS was observed in IC0599709 (11.26 per cent). The cross, AHLM-2 x AHLM-2 produced 10 fruits/plant followed by AHLM-2 x IC0599709 (8.4). All evaluated F_1's showed monoecious sex form and the fruits of the crosses AHLM-2 x IC0599709, AHS-82 x AHLM-2, AHS-82 x AHK-119, AHK-119 x IC0599709 were found bitter at immature stage and become edible at full ripening stage. Splitting of fruits at ripening stage was observed in AHLM-2 x AHS-82, AHLM-2 x IC0599709 and AHS-82 x AHLM-2 (Choudhary and Singh, 2016).

Some successful distant hybridization is as under:

- ✰ *Cucumis metuliferus* x *C. melo*
- ✰ *C. africans* x *C. dipsaceus*
- ✰ *C. africans* x *C. anguria*
- ✰ *C. anguria* x *C. prophetarum*
- ✰ *C. anguria* x *C. zeyheri*
- ✰ *C. dipsaceus* x *C. figarei*

Intergeneric Hybridization

Attempts have been made to cross muskmelon (*Cucumis melo*) with watermelon (*Citrullus lanatus*) and round melon (*Praecitrullus fistulosus*), without substantial success.

Breeding for Quality Traits

Muskmelon quality parameters includes attractive (round/spherical fruit shape), thick flesh with attractive colour, small seed cavity and tough netted skin. It is highly relished as desert because of its unique aromatic musky flavor, sweet taste, juiciness, high total soluble solids (TSS not less than 10 per cent) and being a rich source of vitamins and minerals. Untill recently, melon production has been limited in most parts of the United States to the reticulated (netted), orange-fleshed muskmelon. However, the smooth-skinned, green fleshed honeydews have increased in popularity over the past decades, and varied displays of casaba and canary (smooth-skinned with yellow or mottled rinds, white-fleshed), Persian (lightly-netted, pink-fleshed), and Crenshaw (smooth-skinned, pale orange-fleshed) melons are becoming an increansingly common sight in major markets. The genetic diversity within the species for fruit characteristics has resulted in recent cultivar develoments such as orange-fleshed honeydews and green-fleshed netted melons. Additionaly, producers are showing interest in forms cultivated in other countries, including the smooth-skinned, delicately-fleshed "Charentais" types in Europe and the dark green smooth-skined "Hami Qua" melons of Asia. Genetic resources of *Cucumis melo* from 3 horticultural groups (*inodorus*, *cantaloupensis* and *momordica*) were augmented and evaluated at IARI, New Delhi for yield, fruit quality (TSS, acidity) and phytonutrients (P, K, Ca, Mg, Mn, Zn, Na, Cu, Fe, Vit. C, total carotenoids) by Choudhary *et al.* (2016). High genetic variation for Fe, total carotenoids, Zn, Cu, Mg, Mn, K, Ca, Na, P and low variation for TSS, Vit. C and acidity were recorded among 60 genotypes. The first variety Pusa Sarda of *C. melo* var. *inodorous* from public sector in India was developed for protected cultivation. *C. melo* var. *cantaloupensis* possess desirable traits like thick rind and good keeping quality. *C. melo* (more than 40 botanical varieties) is known source for diverse genes of economic value such as fruit shape, size and colour *etc.* Pusa Madhurima for open field has unique lobed fruit shape. Inter varietal hybrid DMH-5 can be grown in open field condition which yield is significantly higher (25.2 t/ha), with TSS 12.8 °brix and improved post harvest life over Pusa Madhuras. Indian bred varieties like Durgapur Madhu, Pusa Madhuras and Hara Madhu were reported for more sweetness and juiciness. Cultivar 'Arka Ajit' was reported for high vit. C content.

Breeding for Shelf-Life

Melon fruit varies in shelf-life depending upon the varieties. Some melon varieties having poor shelf-life

Table 3.60: Rootstocks and their Usable Purpose

Scion	Rootstocks	Purpose
Muskmelon	*Cucurbita moschata*, *C. pepo*, Intetspecific hybrid squash (*C. maxima* x *C. moschata*), *Cucumis melo*, *Cucumis metuliferus* (AH Cucumber)	Vigrous root system, tolerance to fusarium wilt, low and high temperature, Phytophthora diseases, high soil moisture nematodes, fruit quality.

run a huge market loss within a short time between harvest and consumption. Poor keeping quality limits their wide commercial acceptance and thus long shelf-life has become an indispensable character in modern cultivars of melon. In fact, the variation of shelf-life existed among varieties of *C. melo*. In early studies on shelf-life, researchers generally focused on a few varieties such as *inodorus*, *reticulatus* and *cantalupensis*. Liu *et al.* (2004) evaluated shelf-life of some other varieties *viz.* var. *saccharinus*, *asidulus* and *makuwa* and identified characters which are closely related with shel-life. Varieties *asidulus* and *makuwa* both of which belong to the Oriental melon are closely related because of their short growth duration, small seed, thin pericarp and poor shelf-life. American cantaloupe (var. *reticulatus*) and European cantaloupe (var. *cantalupensis*) are rather closely related due to their climacteric fruit with orange flesh, slipped peduncle and rapid yellowing of epidermis at maturity, which are associated with their shel-life. Variety *saccharinus* is closer to var. *inodorus* than to the other varieties, due to their requirement of long period for development, large size of seed and fruit and half or non-slipped peduncle. Shelf-life of melon fruit is significantly correlated with quality of flesh, size of seed and fruit, abscission of peduncle, development of periods of plant and fruit, yellowing of epidermis at maturity, Brix value and colour of flesh and epidermis. Accessions with good shelf-life are mostly found in var. *saccharinus* and *inodorus*. For improving upon the shelf-life of muskmelon fruit, the ripening process has been strongly retarded in the transgenic melon (*C. melo* L.) cv. "Amarillooro" which is an important cv. muskmelon widely cultivated in Spain. It is *Agrobacterium tumefaciens*, mediated transferred cultivar. Cultivars like Arka Rajhans and Punjab Hybrid-1 were found for good keeping quality.

Grafting

There are severe problems of soil borne diseases and pests and it is strenuous to incorporate resistance to such stresses in the varieties. Some specific problems can be overcome through in muskmelon.

Muskmelon is grafted over the same species or bottle gourd. In muskmelon grafting is successful and this technique can be used not only to increase, yield, quality, earliness *etc.* but also can be effectively utilized in management of several diseases. Grafting technique may represent a quick though expensive, method of controlling *Fusarium* race 1 and 2. The rootstock can greatly influence plant growth, yield and are often detrimental. However, these vary greatly with the scion cultivar. Rootstocks *Benincasa hispida*, *Cucumis metuliferus*, *Cucumis ficifolius*, *Cucurbita maxima*, *Cucurbita moschata* and *Lagenaria siceraria* are resistant to race 1, and 2 of *Fusarium*. Grafts on *B. hispida* negatively influence both yield and fruit quality while *C. metuliferus* and *C. zeyheri* exert negative impacts on productivity and fruit quality, respectively.

Table 3.61: Rootstocks and Major Variety

Scion	*Rootstocks*	*Major Variety*
Melons for green house	*Cucumis melo*	Oi, Enken No.2, Bernet Hill Favourite
	Benincasa hispida	Lion
	Cucurbita species	Unryu
	C. moschata x C. maxima	Shintosa
Melons for open field	*C. moschata x C. maxima*	Shintosa
	Cucumis melo	Enken No.2, Kenkyaku

Breeding for Resistance

Muskmelon is attacked by a number of diseases and insect-pests. Among the diseases, downy mildew, powdery mildew, fusarium wilt, virus and nematodoes are very important while among the insect-pests, red pumpkin beetle, fruit fly, mite, aphids and leaf miner are the most serious ones.

Disease Resistance

It has been observed that before grouping of genotypes in muskmelon it is utmost important that to understand the genetics of disese resistance.

Table 3.62: Genetics of Disease Resistance in Muskmelon

Disease	*Genetic Control*
Fusarium wilt	Single dominant gene
Cucumber mosaic virus	Complex (many dominant genes involved)
Powdery mildew	Single dominant gene

Fusarium wilt resistance is a major problem in muskmelon growing and to cope up the problem Indian snap melon germplasm DSM-11-6 has been identified by Choudhary *et al.* (2016). Genetics of resistance has been studied in cross with susceptible parents Kashi Madhu and Hara Madhu and it was found to be a single dominant gene.

A notable milestone in plant breeding history occurred in 1973 when Jaggar and Scott (1937) reported research that led to the release of powdery mildew resistant Cantaloupe-45. They also observed resistance in PMR-5, PMR-6, PMR-7, Arka Rajhans, Jacumba, Edisto, Gulf Stream and Florifold. In 1925, Professor J.T. Rosa of the University of California, Davis, received a shipment of muskmelon seed from India through one of his farmer students. Dr. Rosa in collaboration with I.C. Jagger of the USDA found a few plants in this collection that showed field resistance to powdery mildew. These Indian muskmelons had all sorts of objectionable characters. This was used as a source of resistance in several back crosses to the commercial type. In 1936 a superior cultivar 'powdery mildew resistant Cantaloupe-45' was released to seedsmen. The release of PMR-45 caused a major revolution in the cantaloupe industry. Even today

PMR-45 is one of the most widely grown cultivars of cantaloupe for shipping to distant markets. 'Campo' a cantaloupe variety from USA, is resiatant to powdery mildew race prevalent in Jamuna riverbed. There are other varieties *i.e.* Campo, Jacumba, Perlita and PMR-6 to be highly resistant to powdery mildew (Shiakami and Choudhury, 1974). In their findings they isolated 3 lines *i.e.* PMRS-4, PMRS-2 and PMRS-3 using 'Campo' as resistant parent. 'Gulf Stream' and 'FM-5'(Cornell Breeding Line 83-275-6L) were found to be resistance to powdery mildew (Kesavan and More, 1987) and 'FM-5' was found to be a better parent since it is monoecious in nature and have better horticultural traits.

Table 3.63: Gene Action in Muskmelon

Disease	*Mode of genetic control*
Downy mildew	Duplicate type of gene action, where two dominant genes were involved for resistance in Phoot (*C. melo* var. *momordica*) x Monoecious-3 and Phoot x Pusa Madhuras crosses whereas; two recessive genes governed the downy mildew resistance in Phoot x Lucknow Safeda.
Powdery mildew	Resistant dominant over susceptibility (Nath *et al.*, 1973), resistance recessive to susceptibility (Chaudhary and Sivakami, 1972)
Cucumber mosaic	Incompletely dominant or recessive (Karchi, *et al.*, 1975)

Some indigenous sources of muskmelon, especially 'Phoot' snapmelon a land race carrying high degree of resuistance to downy mildew (*Pseudoperonospora cubensis*) has been to breed a moderately downy mildew resistant muskmelon variety MR-12 named as 'Punjab Rasila' by PAU, Ludhiana. Sambandam *et al.* (1979) identified land races 'Nakkadosokaya', Budamkaya' and 'Phoot' asresistant to downy mildew besides imported cantaloupes 'Gulf Stream', 'Planters Jumbo' and 'Earli Gold'. Waraitch *et al.* (1977) also identified Arka Rajhans, Jacumba, Dulce, PMR-5 and Gulfstream to exhibit field resistance to powdery mildew. Breeding line PI124111 which is a seventh generation selection derived from PI124111 is the only breeding line possessing multiple race resistance *e.g.* resistance to *P. cubensis* pathotype-3, powdery mildew race-1 and race-2 of *Sphaerotheca fuligenea* and *Fusarium* wilt *F. oxysporum*. Some wild species also did not show any disease symptoms. The non-dessert subspecies *viz. C. melo* var. *callosus* (wild melon) and *C. melo* var. *momordica* (snapmelon) were found to be highly susceptible to powdery mildew. In respect of downy mildew, genotypes Buduma Type-1, -2, -3; Phoonte, Goomuk, Nakkadosa, EX-2, Annamalai, Edisto and Harvest Queen were found resistant. Excellent source for downy mildew, powdery mildew and Fusarium wilt resistant found in PI2411F (Cohen and Eyal, 1987).

Muskmelon variety 'Longtianxueguan (F)' is reported to be resistance to wilt diseases (Wang, 2000).

Cucumis zeyheri, *C. myriocarpus* and *C. meeusii* are found potential resistance source against *Fusarium* wilt.

Muskmelon cultivar 'Punjab Hybrid' is moderately resistant to powdery mildew. The original source of resistance to diseases like- powdery mildew, downy mildew, watermelon virus 1, and muskmelon mosaic can be obtained from the Indian origin (Jagger, 1937). Pedigrees of a few resistant lines are given in Table-3.65.

Multiple resistant lines of cantaloupe have been developed carrying resistance to powdery and downy mildew, *Alternaria* and watermelon mosaic virus in the USA.

Table 3.64: Resistance Source against Diseases and their Utilization

Diseses	*Resistance Source and Utilization*
Downy mildew	Resistant to downy mildew has been reported in some exotic accessions *viz.* PI312105, PI124111F, PI124002, PI180280, PI223637, PI164323 and PI165449. Snapmelon genotypes *viz.*, SP-1, SP-2, SP-3, KP-1, KP-7, KP-9 and PI163888 have also been reported to be resistant. Resistance in wild melon (*C. callosus*) has also been reported. Somkuwar and More (1996) reported two dominant genes governing resistance in snapmelon when crossed with monoecious-3 and Pusa Madhuras whereas two recessive genes in snapmelon crossed with Lucknow Safeda. Divergent reaction of a plant material indicates that different races of the pathogen do exist. Reddy *et al.* (2002) reported some biochemical associations of downy mildew. Peroxidase activity was significantly higher in resistance lines. Phenol, soluble protein, and lignin content were significantly higher in resistant lines. These may be used as marker for screening germplasm for identifying resistant ones. According to Balass *et al.* (1993) resistant genes PC-1 and PC-2 which conferred resistance to downy mildew in PI124111F were specifically inactivated at the lower temperature (12-15 °C.). Resistance to downy mildew is governed by two complementary dominant genes and two partially dominant genes. Resistance has been utilized to develop resistant/tolerant varieties *viz.*, Smith's Perfect, Florida Dew, Campo, Jacumba, Seminole, Georgia-47, Cinco and MR-1. In India, downy mildew resistant/ tolerant varieties like Punjab Rasila, Hisar Madhur, DMDR-1 and DMDR-2 have been developed.
Powdery mildew	The disease is caused by two fungi *i.e. Erysiphe cichoracearum* and *Sphaerotheca fuliginea*. Seven races of *S. fuliginea* and two races of *E. cichoracearum* have been identified using differential set of *C. melo* consisting of the genotypes 'Iran H', 'Vedrantais', PMR-45', 'Nantais Oblong', 'WMR-29', 'Edisto-47', 'PMR-5', 'MR-1', 'PI124112' and 'PI414723'. Accessions PI-124112 and MR-1, derived from the accession PI124111, are reported to be resistant to the powdery mildew species. Accession PI124112 carries four different resistance genes that are closely linked. PI124111 carries resistance to race 1 and resistance is governed by gene *Pm-3*, accession PI414723 carries *Pm-x* gene with resistance to *S. fulginea*. Differential responses *i.e.* incomplete resistance or susceptibility of *C. melo* genotypes to MR-1 and PI124112 to powdery mildew have been reported by Kristkova *et al.* (2004)

Diseses	*Resistance Source and Utilization*
Fusarium wilt	It is common disease affecting muskmelon. The causal organism has been identified as *Fusarium oxysporum* f. *melonis.* Use of resistant varieties is the best way to control the disease. Four races of the pathogen (0, 1, 2 and 1, 2) have been identified. Only the first three are efficiently controlled by single dominant genes *Fom-1* and *Fom-2* which are both already present in commercial F_1 hybrids. Race 1, and 2 is not yet suitably controlled by appropriate resistance.
Viral diseases	Virus diseases are the main cause of serious yield losses in muskmelon. More than 50 viruses have been reported as infecting cucurbit crops. Among them cucumovirus Cucumber mosaic virus (CMV), and the poty viruses water melon mosaic virus (WMV), Zuccchini yellow mosaic virus (ZYMV) and Papaya ringspot virus-W are some of the most widespread in melon crop. Breeding for resistance is the best option for virus control. Various level of resistance to CMV has been incorporated to some commercial melon cultivars. Resistance to PRSV has been described in melon and there are some commercial cvs. Carrying the 'prv' gene. A monogenic dominant and an oligogenic resistance to ZYMV have been reported in melons, although no resistant commercial melon variety to that virus has been produced-yet. Resistance to cucurbit aphid borne yellow virus located in the Indian line "PI124112" is governed by double recessive genes. Regarding WMV, some level of resistance has been observed in melon, but the resistance is not complete. However, Gilbert *et al.* (1994) stated that resistance to watermelon mosaic virus poty virus-2 was governed by a single dominant gene. In melon, indirect resistance to viruses can be incorporated by using the "Vat" gene which confers resistance to virus transmission by *Aphis gossypii.* The "Vat" gene was originally reported from the Korean melon line PI161375 and then from five melon cultivars from India and Japan. This gene has been introduced into some commercial melon cultivars. In preliminary A. gossyppi-mediated inoculation assays carried out with the Zimbabwean melon accession 'TGR-1551', transmission of CMV failed. As 'TGR-1551' is susceptible to CMV by mechanical inoculation, failure of CMV transmission by vector indicates the possible resistance to virus transmission by this pahid species in the accession. This resistance is conferred by a dominant gene and is fdifferent from "Vat" gene. Oligogenic resistance to CMV has been located in the Korean line "PI161375". Rajamony *et al.* (1990) observed that resistance to CGMM tobacco virus was polygenic in nature. Resistance to all strains of Melon Necrotic Spot Virus (MNSV) is governed by a single recessive gene.

Table 3.65: Source of Resistance

Variety/Genotypes	*Pedigree*
'Compo', 'Jacumba'	Indian collection x Hales Best ⟶ PMR 50 ⟶ PMR 45 ⟶ (resistant to race 1 of powdery mildew) PMR 5 and PMR 6 (resistant to race 1 and 2) ⟶ 'Campo' and 'Jacumba'
'Georgia 47'	'PI124112' ⟶ 'Georgia'
'B 665'	'PI 180280' ⟶ 'B 665'

Table 3.66: Resistant Species/Varieties/Genotypes of Muskmelon

Disease	*Resistant Varieties/Genotypes*
Powdery mildew (PM)	'Campo', 'Jacumba', PMR-45, PMR-450, PMR-5, PMR-6, PI124111, EC399934, Edisto. Georgia-47, ΓM-5
Downy mildew	'Georgia 47', 'Home Garden', 'Gulf Stream', 'Planters Jumbo', MR-1, PI414723, DMDR-2, EC399917, EC399924-31, Seminole, Edisto, Harvest Queen, *Cucumis callosus,* Phoontee, Annamalai, Budama Types-1, 2 and 3, Goomuk, Georgia-47, 'Home Garden', Gulf Stream, Planters jumbo
Watermelon virus 1	'B 665'
WMV	PI414723, EC350603
Muskmelon mosaic Virus (WMV)	'Oriental Pickling Melon'
Cucumber green mottle mosaic virus (CGMMV)	DVRM-1, DVRM- 2, *C. africanus*
Cucumber mosaioc virus (CMV)	PI161375, EC399914, EC399932 (Tolerance), Dutgapura Madhu, Kabul Melon, Ludhiana Col.2
Multiple resistance line against PM, DM, muskmelon mosaic Virus	MR-12
Watermelon mosaic Virus (WMV) and Powdery mildew (PM)	EC350604
Fusarium	EC382734, *C. melo* var. *chito, C. inodorus, C. flexuosus, C. eticulates* (Donor)
Powdery mildew (PM), Muskmelon mosaic Virus (WMV) and Blight	EC178496 (Donor)
Muskmelon mosaic Virus (WMV), Green mottle mosaic virus and Gummy Stem Blight	PI414723, *C. metuliferus* and Gulf Coast (Donor)
Muskmelon mosaic	Oriental Pickling Melon

Source: Bhardwaj, 2010.

Table 3.67: Some Promising and Useful Introduction

Crop Species	*Accessions*	*Country*	*Promising and Useful Attributes*	*Indentor*
Cucumis melo	EC-399866-212	UK	Resistant to downy mildew, powdery mildew and CMV	-
	EC 348140-43	USA	F_1 hybrid (early and sweet)	
	EC- 382726-36	France	Germplasm with male sterility genes MS1, MS2, MS3, MS4 and MS5, white fruit (gene), Fusarium wilt resistant (genes form-1, form-3)	IARI, New Delhi
	EC 350603	USA	Sulpher tolerance and WMB tolerance	
	EC 350602	USA	Aphid resistance	
	EC 330956	USA	Multiple disease resistance	
	EC342751-54	Japan	Heat tolerance	
	EC-468986, EC-477977, EC-477978, EC-477979, EC-477980, EC-477981	—	Quality and multiple disease resistance.	—

Some Indian bred cultivars are very popular and carrying high resistance. Among the indigenous collections, Pusa Sharbati was moderately resistant to powdery mildew. Genotype MR-12 was found resistant against downy mildew, powdery mildew and muskmelon mosaic virus.

Table 3.68: Varietal Source for Resistance

Disease	*Resistance Varieties Developed*
Arka Rajhans	Powdery mildew
Punjab Rasila	Downy mildew
DVRM-1 and DVRM-2	Cucumber green mottle mosaic virus
DMDR-2	Downy mildew

Insect Resistance

Work on breeding melon varieties resistant/ tolerant to insect-pests has been scanty. Major insect pests attacking muskmelon are red pumpkin beetle, (*Rhaphidopalpa foveicollis*), fruit fly (*Dacus cucurbitae*) and spider mite (*Tetranychus urticae*). Apart from direct damage, aphids and whitefly also serve as vectoe for the spread of viral diseases in muskmelon. Home garden variety resistant to aphids and downy and powdery mildew has been bred in the USA. In respect of fruit fly, almost all varieties are susceptible. However, resistance has been identified in wild melon (*C. callosus*). Resistance to fruitfly in wild melon is dominant. *C. melo* var. *callosus* is resistance to fruit and leaf caterpillar. Wild melon is also resistant to *Fusarium* wilt and fruit damaging caterpillars of *Margoronia indica*. Vashishta and Choudhury (1972a) reported that genotype 'Casaba' exhibited a high degree of tolerance to red pumpkin beetle. Asian land races have been found to be a good source of resistance to some insect like aphids. Tolerance to aphid in genotype LJ20234 has been found to be governed by a single dominant gene, while tolerance in line LJ90230 is complex in nature. Genotype LJ90234, an inbred of PI371395 from India, possesses resistance to the western biotypes of aphid. Varieties Hales Best, Jumbo, AR-5 and AR Top Mark with resistance to aphids have been developed. Resistance to red pumpkin beetle is governed by a single dominant gene. Muskmelon gonotypes AHMM/BR-1, RM-50 and AHMM/BR-8 were resistant; MHY-5 Durgapura Madhu and Pusa Sarbati moderately resistant; AHMM/BR-13, Pusa Madhuras and Arka Ject susceptible; while Arka Rajhans and GMM-3 were highly susceptible genotypes to fruit fly infestation (Shravan *et al.* 2015).

Table 3.69: Source for Insect Resistance

Insect	*Resistant Varieties/Genotypes*
Aphids	EC350602, EC178496 (Donar), LJ90234 an inbred of PI 371795
Fruit fly	*Cucumis callosus*
White fly	*C. asper, C. denery, C. dipsaceus, C. sagitatus*
Bemisia tabacci	EC399923
Mites	Durgapura Madhu, Sel.1, Punjab Sunehri, Pusa Sharbati, Kocha-4
Red Pumpkin beetle	Casaba and PI170683

Source: Bhardwaj, 2010.

Nematode Resistance

Nematode resistance has been found in African wild species; which could not be transferred because of barriers. Non-commercial species of *Cucumis* exhibit resistance to root knot nematodes (*Meloidogyne incognita*). They are *C. anguria*, *C. ficifolious*, *C. longipes*, *C. metuliferus* and *C. heptadactylus* (Singh *et al.*, 1974). *Cucumis metuliferus* is the potential source for nematode resistance. The resistance to root-knot nematode in muskmelon (*C. melo* L.) can be incorporated from its two related species, *C. metuliferus* and *C. anguria* through the embryo culture technique. Difficulties with infertile hybrids hinder the transfer of root-knot nematode resistance from *C. anguria* and *C. metuliferus* to muskmelon. Advanced selective techniques could ensure the incorporation and expression of genes conferring nematode resistance into cvs. of *C. melo*. Transgenic muskmelon plants containing foreign genes produced via Agrobacterium mediated transformation, have been reported by Fang and Grumet (1990).

Genetic Resource Conservation

There are several organizations which are maintaining valuable germplasm for present use and future posterity.

Table 3.70: Germplasm Conservation in Various Countries

Conservation Centre	Country	Germplasm Conserved
Taiwan Agricultural Research Institute, Wufeng, Taichung	China	Cultivars from China and India
North Central Regional Plant Introduction Station, Iowa; State University USDA-ARS, Ames, Iowa	USA	Land races, cultivars and populations
Southern Regional Plant Introduction, USDA, Griffin, Georgia	USA	Cultivars from Afghanistan, India and other countries.

REFERENCES

1. Balass, M., Cohen, Y. and Bar, J.M. 1993. Temperature dependent resistance to downy and powdery mildew in muskmelon: structural responses. *Physio. Molecular Pl. Path.*, **43:** 11-20.
2. Bhardwaj, D.R. 2010. Breeding for pest and disease resistance in cucurbits. In: Compedium of Witer School "Emerging trends and strategies for the management of pests and diseases in vegetable crops. Published by ICAR-IIVR, Varanasi (U.P.) p. 102-109.
3. Bohn, G.W. and Principe, J.A. 1964. A second male sterility gene in the muskmelon. Journal of Heridity 55(5):; 211-215.
4. Chattejee, M. and more, T.A. 1991. Techniques to overcome barrien of interspectic hybridization in *Cucumis* Cue. Gent. Coop. 14:66-68.
5. Chen, J.F. and Adelberg, J. 2000. Interspecific hybridization in *Cucumis:* Progress, Problems, and Perspectives. *HortSci.,* **35(1)**: 11-15.
6. Choudhary, B. and Sivakami, N. 1972. Screening muskmelon (*Cucumis melo*) for breeding resistant to powdery mildew. 3rd Intl. Symp. Subtrop. *Trop. Horticul.*, **2:** 10.
7. Choudhary, B.R. and Singh, D. 2016. Evaluation of interspecific crosses among *Cucumis* species for horticultural traits. (Abs. 56; Session: I, Breeding of Horticultural Crops). 7th Indian Horticulture Congress-2016, held at ICAR-IARI, New Delhi, p. 44.
8. Choudhary, H., Dhar, S. and Tomar, B.S. 2016. Genetic resources of *Cucumis melo* for improvement of nutritional quality and resistance to Fusarium wilt. 1st International Agribiodiversity Congress; (Session: Trait Discovery and Enhanced Use of PGR), held at New Delhi; 6-9 November 2016; p. 195.
9. Cohen, Y. and Eyal, H. 1987. Downy mildew, powdery mildew and *Fusarium* wilt resistant muskmelon breeding line PI 124111F. Photoparasitica 5:187-195.
10. Cohen, Y. Eyal, H. and Cohen, A. 1993. Gylan-A gynoecious muskmelon. Hortsci; 28(8): 855.
11. Dixit, J. and Kalloo. 1983. Heterosis in muskmelon (*Cucumis melo* L.). *Haryana Agric. Univ. J. Res.*, **13(4):** 549-553.
12. Fang, G. and Grumet, R. 1990. *Agrobacterium tumefaciens* mediated transformation and regeneration of muskmelon plants. *Plant Cell Reports*, **9:** 160-164.
13. Gilbert, R.Z., Kyle, M.M., Munger, H.M. and Gray, S.M. 1994. Inheritance of resistance to watermelon mosaic virus in *Cucumis melo* L. *HortSci.*, **29:** 107-110.
14. Jagger, I.C. and Scott, G.W. 1937. Development of powdery mildew resistant cantaloupe No.45. *U.S. Deptt. Cir. Agr.*, 141.
15. Karchi, Z., Cohen, S. and Govers, A. 1975. Inheritance of resistance to cucumber mosaic virus in melons. *Phyt.*, **65:** 497-481.
16. Kesavan, P.K. and More, T.A. 1987. Powdery mildew resistance in F_1 muskmelon. In National Symposium on Heterosis Exploitation, Accomplishment and prospects: Panbhani, Maharastra, October. 15-17, P.54.
17. Kirkbride, J.H. 1993. Biosystematic monograph of the genus *Cucumis*. Parkway publishers Boone, North Carolina; 159.
18. Kristkov, E., Lebeda, A. and Sedlakova, B. 2004. Virulence of Czech cucurbit powdery mildew isolates on *Cucumis melo* genotypes MR-1 and PI124112. *Scientia Horticulturae*, **99:** 257-265.
19. Kroon, G.H., Custers, J.B.M. and Kho, Y.O. 1979. Interspecific hybridization in muskmelon (*Cucumis melo* L.). Need for genetic variation, biosystematic relations and possibilities to overcome crossability barriers. *Euphytica*, **28:** 723-728.
20. Liu Li, Kakihara, F. and Kato, M. 2004. Characterization of six varieties of *Cucumis melo* L. based on morphological and physiological characters, including shelf-life of fruit. *Euphytica*, **135:** 305-313.
21. Mallick, M.F.R. and Masui, M. 1986. Origin, distribution and taxonomy of melons. *Scientia Horticulturae*, **28:** 252- 261.
22. Mathew, S.M. 1984. Compatibility among *Cucumis melo* varieties *inodorus, conomon, flexuosus, momordica* and *utilissmus*. M.Sc. (Horticulture) thesis, Kerala Agricultural University, Vellanikkara (Kerala).
23. Mishra, J.P. and Seshadri, V.S. 1985. Studies on male sterility in muskmelon (*Cucumis melo* L.).II. Studies on heterosis. *Genet. Agr.*, **39:** 367-376.

24. Nandpuri, K.S. and Brar, J.S. 1966. Studies on floral biology in muskmelon (*Cucumis melo*). Journl of Research Ludhiana, 3:395-399.

25. Nath, P., Dutta, O.P., Vijay, O.P. and Subramanian, S. 1973. 'Arka Chandan' the carotene rich pumpkin. *Indian Hort.*, **17:** 19-20.

26. Nerson, H., Paris, H.S. and Edelstein, M. 1990. Melofon: a new crop for concerted yield of pickles. In: Janick, J. and Simon, J.E. (eds.), Advances in new crops. Timber Press, Portland, OR., pp. 399-402.

27. Patel, A.S., Vashisht, V.K. and Dhatt, A.S. 2016. Pre-breeding in muskmelon (*Cucumis melo* L.). (Abs. 63; Session: Breeding of Horticultural Crops II). 7th Indian Horticulture Congress-2016, held at ICAR-IARI, New Delhi, p. 133-134.

28. Peterson, C.E. Owens, K.W. and Rowe, P.R. 1983. Wisconin-998 muskemlon germplasm. Hortsci. 18:116.

29. Rajamony, L., More, T.A. and Seshadri, V.S. 1990. Inheritance of resistance of cucumber green mottle mosaic virus in muskmelon (*Cucumis melo* L.). *Euphytica*, **47:** 93-97.

30. Ramachandran, C. and Naryan, R.K.J. 1985. Chromosomal DNA variation in *Cucumis. Theor Appl. Genet.*, **69**: 497-502.

31. Reddy, N.S., Dutta, O.P., Reddy, K.M., Krishnappa, K.S. and Pitchaimuthu, M. 2002. Role of peroxidase activity, soluble protein, phenol and lignin in downy mildew resistance of (*Cucumis melo* L.). Abst. Intl. Confe. On Vegetables, Nov. 11-14. Bengaluru (Karnataka), No. 148, p.108.

32. Robinson, R.W and Deckerwalters, D.S. 1997. Cucurbits. New York Cab International. 226. (Crop Production Science in Horticulture No.6).

33. Rosa, J.T. 1928. Inheritance of flower types in *Cucumis melo* and *Citrullus*. Hillgardia 3:253-250.

34. Shravan, M.H; Choudhary, B.R. and Bhargava, R. 2015. Antixenotic resistance traits of muskmelon (*Cucumis melo* L.) against fruit fly (*Bactocera cucurbitae*) in arid region of India. Indian Journal of Entomology 29(2): 81-87.

35. Singh, B., Bhatti, D.S. and Singh, K. 1974. Resistance to root knot nematodes (*Meloidogyne* spp.) in vegetable crops. PANS, 20: 58-67.

36. Somkuwar, R.G. and More, T.A. 1996. Inheritance of downy mildew resistance in muskmelon (*Cucumis melo* L.). *Indian J. Hort.*, **53:** 50-52.

37. Vashishta, R.N. and Choudhury, B. 1972a. Studies on growth and yield potentials of muskmelon cultivars resistant to red pumpkin beetle. *Haryana J. Hort. Sci.*, **1:** 55-61.

38. Wang, X.Q. 2000. A new muskmelon variety 'Longtianxueguan (F)'. *China Vegetables*, **6:** 25-26.

39. Waraitch, K.S., Munshi, G.D., Nandpuri, K.S. and Lal, T. 1977. Screening of muskmelon, wild melon and snapmelon for resistance to powdery mildew (*Sphaerotheca fuliginea*). *Phytopathol. Meditr.*, **16:** 37-39.

40. Ware, G.W and Mc Collum, J.P. 1980. Producing vegetable crops Danville, Illinois: Interstate, 607.

41. Whitaker, T.W. and Davis, G.N. 1962. Cucurbits (edn.1), p. 1-136. Inter Science Publishers Inc.; New York.

Genus *Citrullus*

Introduction

The genus, *Citrullus* has two species of economic importance *viz.* water melon (*C. lanatus* (Thunb.) and round gourd (*C. vulgaris* Schrad var. *fistulosus*). Based on the evidence and linguistic data, *C. lanatus* originated in Africa and reached India in prehistoric times. Cultivation of large watermelon *C. lanatus* var. *citrodes* is comparatively recent and Soviet varieties grown today still have shape of their African ancestors (*C. lanatus* var. *caffer*). *C. vulgaris, C. colocynthis, C. ecirrhosus* and *C. nandinianus* are related and cross compatible with each other, while *C. colocynthoides* is probably the ancestor of watermelon. *C. colocynthoides* exhibits much variation in north-western plains.

Table 3.71: Taxon and Preferred Taxon Name

Name of Taxon	Name of Authority	Family	Preferred Taxon Name
Citrullus			Citrullus
Citrullus aedulis	Pangalo, nom. illeg.	Cucurbitaceae	Citrullus lanatus var. *lanatus*
Citrullus colocynthis	(L.) Schrad.	Cucurbitaceae	*Citrullus colocynthis*
Citrullus ecirrhosus	Cogn.	Cucurbitaceae	*Citrullus ecirrhosus*
Citrullus fistulosus	Stocks	Cucurbitaceae	*Praecitrullus fistulosus*
Citrullus lanatus	(Thunb.) Matsum. and Nakai	Cucurbitaceae	*Citrullus lanatus*
Citrullus lanatus var. *caffer*	(Schrad.) Mansf.	Cucurbitaceae	*Citrullus lanatus* var. *lanatus*
Citrullus lanatus var. *citroides*	(L. H. Bailey) Mansf.	Cucurbitaceae	*Citrullus lanatus* var. *citroides*

Name of Taxon	Name of Authority	Family	Preferred Taxon Name
Citrullus lanatus var. *lanatus*		Cucurbitaceae	*Citrullus lanatus* var. *lanatus*
Citrullus naudinianus	(Sond.) Hook. f.	Cucurbitaceae	*Acanthosicyos naudinianus*
Citrullus rehmii	De Winter	Cucurbitaceae	*Citrullus rehmii*
Citrullus sp.		Cucurbitaceae	*Citrullus* sp.
Citrullus vulgaris	Schrad.	Cucurbitaceae	*Citrullus lanatus* var. *lanatus*
Citrullus vulgaris var. *citroides*	L. H. Bailey	Cucurbitaceae	*Citrullus lanatus* var. citroides
Praecitrullus fistulosus	(Stocks) Pangalo	Cucurbitaceae	Praecitrullus fistulosus

Grouping of *Citrullus* Species

Table 3.72: Grouping of *Citrullus* Species

Citrullus lanatus	*C. lanatus* var. *lanatus, C. lanatus* var. *spermus, C. lanatus* var. *vulgaris, C. lanatus* var. *fistulosus,*
Citrullus colocynthis (L.)	-
Citrullus ecirrhosus	-
Citrullus naudinianus	-
Forage watermelon	*C. lanatus* var. *citroides, C. lanatus* var. *caffer*

Watermelon (*Citrullus lanatus* (Thunb.)

Introduction

Watermelon (*Citrullus lanatus* (Thunb., 2n=22) is one of the important cucurbitaceous crops. This crop is locally known as 'Tarbuz'. It is a tender annual vine crop which is grown for its juicy and luscious fruits that are immensely liked by people allover the world. The fruits are delicious and sweet when ripe. The sweetness and colling effect of the juicy pulp of watermelon fruits is relished both by rich and poor during hot summer months of tropics. Watermelon juice added with salt and pepper is also a popular drink in hot weather. But very seldom, it is cooked as vegetable when immature. The commercial fruit cultivation has gained popularity due to its long term returns and maximum yield per unit area. Watermelon is very low in cholesterol and sodium. It is very good source of magnesium and a good source of phosphorus, manganese and zinc. Next to tomato, watermelon is a rich source of lycopene, a powerful antioxidant. One hundred grams of watermelon flesh contain 91.5 g water, 7.2 g carbohydrates, 0.6 g protein, 0.5 g fibre, 0.2 g fat, 8.0 mg calcium, 0.2 mg iron, 11.0 mg magnesium, 9.0 mg phosphorus, 116.0 mg potassium, 2.0 mg sodium, 9.6 mg vitamin C, 0.14 mg vitamin B6, 366 IU vitamin A and 0.2 mg niacin.

Watermelon is generally consumed as a desert fruit, however its flesh is also processed for delicious sweet pickles/candy making and rind for making pickle. Watermelon is also served in fruit salads. Fruits serve as an important source of water in Kalahari Desert of Africa during dry months of the year, when there is no surface water available. The leaves and young fruits are cooked as a vegetable. Imperfect fruits are used as livestock feed and immature fruits are used for preparing squash. Jam, sorbet and juice are some other nutritious and delicious products made from watermelon. The deep pink and light pink colour of watermelon flesh is due to the presence of *Anthocyanin* and *Lycopene* (powerful antioxidant for preventing cancer, Singh *et al.*, 2001) pigments respectively. Watermelon is high in lycopene, second only to tomatoes but low in calories. The rind has high silicon content and is used to treat those suffering from diabetes and hypertension. Being diuretic, watermelon seeds stimulate functioning of kidneys and bladder. Seeds are relatively rich in protein (25-32 per cent) with high edible oil content (oil is extracted from seed is around 20-40 per cent), lipids, and carbohydrates. The roasted watermelon seeds are seasoned and eaten as a snack food in China, Israel and elsewhere. In Africa, seeds are grounded into coarse flour.

Origin, History and Distribution

Watermelon originated in tropical Africa (Kalahari Desert of South Africa) and India (Mallick and Masui, 1986; Seringe, 1928). Its wild forms occur in sandy dry areas of South Africa chiefly Kalahari Desert. Watermelon [(*Citrullus lanatus* (Thunb.) Matsum and Nakai] with diploid chromosome number 2n=22, is indigenous to India (Pangalo, 1930; 1944, 1955), tropical and subtropical Africa. Linnaeus reported it to be a native of Southern Italy. It has been grown widely on both sides of equator. Watermelon is an important crop in warmer parts of Russia and other parts of Asia Minor, the Near East, China and Japan. This crop was not taken to China until 10th or 11th Century AD. Description of watermelon in Europe dates back to the 13th century. In the New World, cultivation began in Massachusetts as early as 1629 (Mohr, 1986). Watermelon was brought to America by Spanish and quickly became very popular crop (Robinson and Decker-Walters, 1997). It had been cultivated in Egypt and west Asia around 2000 BC and from there it had spread to other countries like Italy, Greece *etc.* along the Mediterranean Sea. Modern watermelons derive from a North African vine, *Citrullus vulgaris,* which has been cultivated in the Nile Valley for centuries. It still grows wild in the desert interior supplying water to natives during drought. There are reports that *C. vulgaris* has been found in wild state on both side of the equator. A Sanskrit name testifies to its early introduction to Asia. It reached India in prehistoric times *i.e.* 4th century AD. India is considerred as the second home of watermelon. The watermelon is grown in about one hundred tropical and sub-tropical countries of the world but is more popular in the United States than elsewhere. China is the largest producer of watermelons in the world. The other important watermelon growing countries include Turkey, USA, Iran, Russia and India. The top four watermelon growing countries *viz.* China, Turkey, Iran and the USA produce 50 per cent of the total world production. The crop has been under cultivation in India from ancient times and due to this reason, India is often considered as its secondary centre of origin.Watermelon is grown in almost all the states (from the lower Himalayan regions to South India) in our country except in north western region. Watermelon is commercial crop in Rajasthan, Punjab, Madhya Pradesh, and Uttar Pradesh.

Species Availability, Evolution and Diversity

The genus *Citrullus* consists of eight species and sub-species. Watermelon, the only cultivated species of the genus, is a diploid with 22 chromosomes ($2n = 22$) (Mallick and Masui, 1986). The ancestor of watermelon is the bitter-fruit form of *C. vulgaris* Schrader. (Mohr, 1986). Whitaker (1933) considered *Citrullus colocynthoides* as the probable ancestor of cultivated watermelon and the *Citrullus vulgaris* was renamed as *Citrullus lanatus*. There are two genera: *Citrullus* and under the *Cucurbitaceae* this family. The African distribution of the genus has been recognized and four species have been listed. *C. vulgaris* Schrad., which include cultivated watermelon (now renamed *C. lanatus* (Thunb.) Matsumara and Nakai), *C. colocynthis* L. *C. ecirrhosus* Cogn. and *C. naudiniaus* (Sond.) Hook. There were of the opinion that there is one group of closely related species, *viz. C. vulgaris, C. colocynthis* and *C. ecirrohsus,* in which cucurbitacin E is the main bitter substance and the other group of isolated species, *viz. C. naudiniaus* with cucurbitacin B together with E and their derivatives. From the results of morphological and cytological studies, it is observed that all four species are cross compatible with each other and several factors like geographical isolation, differences in flowering habit, genetic differences and structural changes in chromosomes would have contributed to the maintenance of their identity.

Both bitter and non-bitter wild races of *Citrullus vulgaris* are distinguished and treated as *C. edulis* and *C. colocynthoides*. Cultivated watermelon can be divided into var. *lanatus* and *citroides*, the latter comprising the citron or preserving melon, producing fruits with hard in-edible flesh and greenish or tan seeds. Morphologically the *colocynth*, which is found in India (North-Western plains, fallow lands/fields, drought hardy type) and used in Ayurvedic medicine, has similar characters with *vulgaris*, but fruits are bitter and seeds are small. But African investigators consider bitter forms, *Citrullus vulgaris* as the ancestor of cultivated watermelon. Further this view was supported on the basis of same chromosome number, free crossability and their cohabitation through out Africa and Asia. Good variability of *C. colocynthis* can be observed in Rajasthan and Gujarat. The citron or preserving melon (var. *citroides* Bailey) can be considered as intermediate stage between primitive bitter and cultivated sweet forms.

Table 3.73: Other *Citrullus* Species

Citrullus Species	*Nature*
Citrullus rehmii	Annual
Citrullus colocynthis	Perennial
Citrullus ecirrhosus	Perennial
Citrullus naudinianus	Perennial

Gene and Genetics of Characters

Information on the genetic make-up of various qualitative and quantitative traits is required for determining the most appropriate breeding procedures to be adopted for watermelon improvement. Gene list reported in watermelon are as under (Henderson, 1991) shown in Table 3.74.

Guner and Wehner (2004) have published the latest version of the gene list for watermelon. Three major genes which control seed coat colour are 'r' for red 't' for tan and 'w' for white seed coat. These genes interact to produce six phenotypes. Two genes *viz.* 's' for short and 'l' for long seed length which control seed size. Several genes govern leaf or foliage traits of watermelon. Non-lobed leaf has the entire leaf rather than the lobed leaf of the typical watermelon. Four dwarf genes have been identified that affect stem length and plant habit. Golden yellow mutant is governed by a single recessive gene (*go*). One benefit of this gene is that the fruit becomes golden yellow as it matures, possibly useful as a maturity indicator. The flesh colour changes from pink to red. A single recessive gene (*su*) eliminates bitterness in fruit of *C. lanatus* and appears to be the same as the dominant

Table 3.74: Gene List for Watermelon

Gene Symbol	*Characters*
a	Andromonoecious, recessive to monoecious
Ar-1	Anthracnose resistance to race 1 of *Glomerella ingulate* var. *orbiculare*
C	Canary yellow flesh, dominant to pink
d	Dotted seed coat, black dotted seeds when dominant for r, t and w
f	Furrowed fruit surface, recessive to smooth
Fo-1	Dominant gene for resistance to race 1 of *Fusarium oxysporum* f. sp. *niveum*
Fwr	Resistance to fruit fly in watermelon, dominant to susceptibility to *Dacus cucurbitae*
g	Light green skin, light green fruit recessive to dark green
gs	Striped green skin, recessive to dark green but dominant to light green skin
go	Golden, yellow colour of older leaves and mature fruit
l	Long seed, long recessive to medium length of seed; interacts with s
m	Mottled skin, greenish-white mottling of fruit skin
nl	Non-lobbed leaves. Leaves lack lobing, dominance incomplete
o	Elongate fruit, incompletely dominant to spherical
p	Pencilled lines on skin, inconspicuous, recessive to netted fruit
pm	Powdery mildew susceptibility, red seed coat, interacts with w and t
s	Short seeds, epistatic to 1
t	Tan seed coat, interacts with r and w
w	White seed coat, interacts with *r* and *t*
Wf	Whitish flesh, *Wf* is epistatic to a second gene (y) which conditions yellow and red flesh. *Wf-Yy* and *wf-Y*-are white fleshed; *wfwfY-* is red fleshed. *Wfwfyy* is red fleshed
yo	Orange flesh from 'Tender Sweet Orange Flesh'. Orange flesh (YoYo) epistatic to yellow flesh (*yy*) from 'Golden Honey' and hypostatic to red flesh (*y*-)
Y1	Yellow leaf (from 'Yellow Skin'), incompletely dominant to green leaf

gene (*Su*) for bitter flavor in the fruit of *C. colocynthis*. Fruit shape (elongate, oval or spherical) is governed by a single incompletely dominant gene. Thirteen genes have been identified that affect fruit colour either in the skin or flesh. Watermelon flesh colour is controlled by several genes to produce red, orange, salmon yellow, canary yellow or white.

Botanical Description

The genus belongs to the sub tribe Beneicasinae (Ser.) C. Jeffr., tribe Benincaseae Ser. and sub family cucurbitoideae. The watermelon falls under genus *Citrullus* and species *Lanatus* under the family *Cucurbitaceae*.

Table 3.75: Botanical Description

Growth Habit	The diclinous plant of watermelon is typically a trailing (vine) annual but recently dwarf forms, which may be described as 'bush' have been located and are being used to develop cultivars suitable for growth in a limited space.
Root	The deep tap root system is very extensive but shallow, consisting of taproot andmany lateral roots within the top 60.0 cm of soil. Early destruction of the taproot (as in transplanting) may be advantageous in getting superior yields. The roots are within 60.0 cm of the surface.
Stem	Stems are angular in cross section slender hairy with tendrils having good branching habit.Tendrils (2-3) are branched in watermelon. The dwarf may attain the vine length as 60.0 cm whereas the more rampant vine cultivars may be 900.0 cm wityh long white hairs. Tendrils are 2-3 and branched. Dwarfing is primarily related to shortened internodes (Mohr, 1956). In the typical vine, there is a delay in the development of lateral branches, with a dominant single 'runner' reaching several feet in length before branching is initiated and maintaining dominance for sometime thereafter. In some of the dwarf types, multiple branching occur simultaneously from the crown of the plant when it is quite young, which provides more potemntial bearing areas concentrated fruit set.
Leaf	Leaves alternate and 5.0-10.0 cm long. The leaves are cordate at the base and deeply pennated into three or four pairs of lobes (some times 5 of lobes), which are sub-divided and toothed with broad apices. The base is cordate. Petiole is 1.0-10.0 cm in length.
Floral biology	Flowers are unisexual, monoecious and hermaphrodite. The first female and hermaphrodite flowers are formed on the main vine. Flowers are mostly borne single and the female flowers may occur at every 3rd, 4th, 9th and 10th node. The male flowers open first in the axil of 4th and 12th leaf. Male flowers generally appear on the plants several days before the female and hermaphrodite flowers, those of early cultivars in the axil of 5th to 7th leaf and in late cultivars in the axil of 25th to 30th leaf. The male: female ratio ranged from 6to10:1 in ealy and 20 to 25:1 in late cultivars. More female flowers are formed in spring than in autumn. Flowers are solitary. A male flower has a slender stem, while the base of the female flower has a small swollen undeveloped fruit fruit. The female flower is easily recognized by the presence of a miniature fruit below the flower petals. Staminate flowers are borne on the nodes which bear no female flowers (Munger, 1942.).
	The male flowers has calyx 5-lobed, corolla deeply 5 partite, petals are pale-yellow in colour, 2.5-3.0 cm diameter, petals1.0-1.5 cm bearing 3 long stamens 3-5 with short free filaments, anthers almost free, female flower with 3 staminodes, ovary ovoid with wooly hairs, style 0.5 cm long with 3 lobes. Female flower has globose ovary and the fruits are globular or oblong, mostly glabrous rind which is hard but not durable.

Anthesis and pollination	Crop having sex form of monoecious (separate male and female flowers on same plant) and andromonoecious type. Watermelon is highly cross pollinated in nature. Anthesis takes place at 6.00-7.30 a.m. in autumn, and 5.30 and 7.30 am in spring and the best pollination time is between 6.30-9.00 a.m. The stigma becomes receptive 2 hours before flowering and remains so for 3 hours after flowering. At high temperature stigmatic fluid starts drying and stigma becomes non-receptive by 3 p.m. Pollen from hermaphrodite flowers germinate rather slowely than that from female flowers but was equally effective for fertilization. Bees are the pollinators. In watermelon crosses, higher seed set was observed when pollination occurred in the early morning immediately after flower opening. Seed set was considerably lower when pollination was carried out in the afternoon. Pollen kept at room temperature for 5 hours gave a lower seed set than freshly collected pollen and after 30 hours completely lost viability.
Fruit and fruit set	The fruit shape varies from globular to oblong (size up to 60.00 cm) with tough, smooth and hairless rind. Rind mostly glabrous, green or cream, striped or mottled green, hard but not durable. Rind ranges from brittle to tough and flexible, from very thin to excessively thick and form very soft to very hard. Fruit may or may not have strips. The fruit is many seeded pepo which is essentially an inferior berry. The flesh consists of highly developed placental tissue within the fruit. The flesh colour varies from white, creamy, green, honey, yellow, saffron, red to scarlet. Flesh texture varies from fine grained and 'melting' to firm coarse, and fibrous. Maximum fruit setting is recorded during the first hours after anthesis and minimum 2 hours before anthesis (Salunkhe *et al.*, 1987 and Singh, 1989). Sweetness (TSS) varies from 9-17 Brix. The greater part of the fruit is the edible flesh, which is mostly derived from the placenta.
Seeds	The seeds are borne with partial placentation. Seeds range in size from very small (size of tomato seed) to large (size of pumpkin seed). Seeds usually range in colour from white to black or reddish, green, brown mottled, flat and smooth (Sutariya, 1961; Bailey, 1964; Purseglov, 1977 and Seshadri, 1993) having size of 0.6-1.5 x0.5-0.7 cm. Approximately 1000 seed weighs arounf 60.0 g.

Horticultural Classification

Shimotsuma (1965) classified watermelon as *Citrullus lanatus* (Thunb.) Mans f., *C. ecirrohsus*, *C. colocynthis* (L.) Schrad., possible ancestor of watermelon *C. naudinaus*, *C. fistulosus* (roundmelon) and *C. vulgaris* var. *citroid* (Bailey) Mansf. (Citron melon).

Egusi *C. colocynthis* (L.) give high yield when environment is dry, sunny periods and reduced during periods of excessive rainfall and high humidity. It is suitable for production in "marginal growing areas." The fruits are extremely bitter, but the seeds can be removed and roasted as an edible commodity (Soliman *et al.*, 1985). The seeds are rich in oils, which can be extracted for cooking purpose, and the seeds can also be ground into a powder and used as a soup thickner or flavoring agent (Badifu and Ogunsua, 1991).

Classification on the Basis of Rind Colour, Fruit Shape and Seed Colour

Watermelon is classified into 4 major categories based on rind colour, fruit shape and seed colour:

(a) Rind ivory white with conspicuous green stripes:
   (i) Fruits cylindrical, seeds white, and dark tanned
   (ii) Fruits cylindrical, brownish tinge or black seeds
(b) Rind dark green or bluish fruit size variable: Seeds are brown with markings
(c) Rind is grey, seeds are white, fruit size varying
(d) Rind white or pale green with dark green stripes, seeds white

Classification on the Basis of Chromosome Number

The chromosome number is 2n= 22, x= 11. Chromosome number of other species are *C. edulis* 2n=22, 24; *C. naudinus* 2n=22, *C. colocynthes* 2n=22, and *C. fistulosus* 2n=24.

Augmentation of Genetic Resources and Crop Improvement

Breeding Priorities

High yield is a major goal for watermelon breeders. Earlier efforts in watermelon breeding were involved in development of new cultivars of different types with high fruit quality and early maturity. Now a days, high yielding cultivars have become a major goal for public and private breeders. In watermelon, TSS content, flesh colour and fruit shape have been identified to be the premium attributes. While making selection for improvement of watermelon, it is essential to score the plant characters as accurately as possible. Selection is based on the following attributes:

Table 3.76: Selection Guidelines for Attributes

Characters	*Description*
Plant type	Plant type bearing first pistillate (or hermaphrodite) flower at early node number. it gives fair indication of earliness.
Rind	Tough rind of the fruit which is better suited for long distance haulage.
Fruits	The intermediate fruit shape or smaller oval shaped fruits is preferred over typical long and round fruit shape, since consumer preference are changing mainly due smaller family size.
	And also this shape more preferred due to easy handling of fruit from market to home and more convenient storage in home refrigerator. Fruits with 10 per cent or higher TSS content so as to meet consumer preference.
Flesh colour	Red, yellow or light red as per demand of the market

Characters	Description
Seed	Fruits with fewer and smaller seeds and attractive deep red flesh.
Yield	High fruit yield in terms of weight and number of fruits of marketable quality.
Resistance	Resistance to pathogens and insect pests. Among the diseases, data on reaction to three diseases *viz.*, powdery mildew, downy mildew and Tospo virus have to be recorded for identifying best performing varieties.

Of course, successful cultivars of watermelon must have traits other than high yield. High fruit quality is of major importance and includes bright flesh colour, firm flesh texture, high sugar content and proper fruit shape. Surprisingly, among the open pollinated varieties, Sugar Baby is the most popular variety for its high fruit quality (deep red and crisp flesh of sweet and distinctive flavour) and acceptable small fruit size, although it is low yielder. Apart from high yield and better fruit quality, disease resistance should also be incorporated for developing new cultivars.

Mini Watermelon

The decrease in size of families in the latter part of the 20th century, led to a demand for small fruit known as 'ice box watermelon' in the US. For example In the USA, demand is also there for mini watermelons (upto 3.5 kg per fruit), seedless yellow-flesh type and seedless dark-solid rind types. Improvements in small watermelons were made with the introduction of 'Minilee' and 'Micltylee' in 1986 by J.M. Crall. The first true mini watermelon cultivar 'New Hampshire Midget' was developed by A.F. Yeager at the University of New Hampshire and introduced in 1951. These important new watermelons are sold under the 'Pure Heart' brand in the United States and 'Solinda' brand in Europe. An array of rind patterns is available and all are characterized by thin rinds and deep red seedless flesh. Other companies quickly seized on this concept and began marketing seeds of their own mini watermelon cultivars. These watermelons have become extremely popular based on a combination of superb flavor, crisp flesh, seedlessness and conveniet size.

Introduction and Selection

Popular cultivar 'Sugar Baby' was introduced in India and directly used as variety for commercial cultivation. Genotype x environment interaction plays a great role in the expression of characters days to first female flower, days to maturity, yield per vine, fruit weight and number of fruits per vine. Genotypes Sugar Baby, Arka Jyoti, RSIIC and Special No.1 have been identified to exhibit average stability. It has also been observed that late maturing varieties, through high yielding, had poor stability. 'New Hampshire Midget', 'Asahi Yamato', 'Sugar Baby', and 'Durgapura Meetha' are some popular introduction and selection. Popular cultivars like 'Durgapura Meetha' and 'Durgapura Keasar' were developed by use of Indigenous source through selection with progeny testing. There are several other cultivars locally grown which are named after the region in which they are grown, such as 'Furrukhabadi', 'Moradabadi', 'Faizabadi' of Uttar Pradesh and Mau apple. 'Charleston Grey' an introduction of watermelon from USDA is characterized by its oblong to cylindrical shape, is a leading cultivar suited to long distance transportation. Horticulture Research Centre, Bangladesh Agricultural Research Institute (BARI) has developed hybrid F_1Sugar. The centre has observed that hybrids in watermelon 'Top Yield' and 'Glory' are dominating in several area of Bangladesh. In Nepal, watermelon cultivars 'Madhu' and 'Milan' are very popular among the growers. In Pakistan, some cultivars like T-14 and 'Sugar Baby' (R) are being grown commercially.

Table 3.77: Introduction of Watermelon Germplasm in India

Name of Germplasm	Salient Feature	Source Country
EC380989-92	Yellow fleshed, early maturing	USA
EC393240-43	Red fleshed, round and small	USA
EC384809	Fusarium wilt resistance	USA
EC378523-24	Fusarium wilt resistance	USA
EC333659-69	Wilt resistance	USA
EC217073-74	Fusarium and Gummy blight resistance	USA

Hybridization and Selection

Hybrids are popular among private breeders for protection of intellectual property and manifestation of adequate heterosis in watermelon. Although attempts for utilizing heterosis in watermelon date back to 1949 in Japan, significant progress was recorded only during the 1960s. Hybrid varieties both diploid and triploid are now fast replacing open-pollinated cultivars. The impact has been significant in Japan, China and the USA, as hybrid varieties occupy significant acreage. Popularity of hybrid varieties is due to their higher yield, better quality, fruit uniformity, good storage and disease resistance.

Watermelon, having predominant monoecious sex expression, is highly cross-pollinated crop and hence highly suitable for hybrid seed production. Heterosis is exhibited for early seed germination; fast seedling growth; profuse root system, thick, dark green and strong leaves; strong stem growth, good branching and better adaptability. The studies measured heterosis as well as general combining ability (gca) and specific combining ability (sca) in watermelon (Brar and Sidhu, 1977; Brar and Sukhija, 1977). Combining ability studies conducted by Sidhu and Brar (1977) revealed that F_{1s} between low x high and medium x high general combiners produced productive hybrids. Brar and Sukhija (1977) pinpointed the importance of greater diversity in origin of parental lines for obtaining enhanced magnitude of heterosis. Uniformity of the hybrids depends upon the level of genetic purity of the combining parents. For the expression of potential heterosis prevailing environmental factors also play a significant role. Major problems with those studies have been that heterosis was inconsistent across experiments and the results were based on diallel or top crosses of elite inbreds, but not a random set of lines from a population. There are several promising genotypes which were used for development of superior hybrids for high yield, earliness, disease resistant, TSS and excellent fruit quality. 'Arka Jyoti' is an important F_1 hybrid developed at the ICAR-IIHR, Bengaluru (Karnataka). The other F_1 hybrids found superior are 'Nath 101', 'Nath 102', 'MHW 6' and 'MHW 11'. Work carried out by Nandpuri *et al.* (1974; Nandpuri *et al.* (1977) led to the identification of heterotic cross combination Shipper x Striped Klondik for yield and number of fruits and Sugar Baby x Verona for early bearing of female flowers and TSS content. Sidhu and Brar (1977) observed non-consistency between parental divergence and the maginitude of heterosis in the F_1 hybrids. Gill and Kumar (1988) and Gill and Kumar (1989) identified superior hybrid combination *viz.*, Sugar Baby x Durgapura Meetha for yield and number of fruits per vine and also good performance for earliness.

Table 3.78: Promising Genotypes Used for Development of Hybrids

Cross Combination	Salient Features
'Special 1' (Round) x 'Charleston Gray' (long)	Round fruited, high yielding and high TSS
'Shipper'(Round) x 'Charleston Gray' (long)	Round fruited, high yielding and high TSS
'Sweet Princes Long' x 'Charleston Gray' (long)	Round fruited, and high yielding
'Crimpson Sweet' x 'New Hampshire'	High yielding and high TSS
'Midget IHR 6' x 'Charleston Gray'	High yielding and high TSS

Hybrids exhibit better tolerance to diseases like anthracnose. Zhang and Rhodes (1992) recorded 33.3 to 86.9 per cent heterosis for number of fruits per vine and average fruit weight. Hybrids also exhibit increased sugar content and low fiber content. Nath and Dutta (1970) reported more than 50 per cent heterosis for yield in the cross combinations IHR-6 x Charleston Grey, IHR-20 x Crimson Sweet and Sugar Baby x Crimson Sweet. High heterosis to an extent of 87 per cent for yield was attributed to more number of fruits per vine and fruit size, fruit number contributed significantly. Besides, there are number of extensively cultivated hybrids being marketed by private companies' due to appealing appearance, good quality flesh and transportability and resistance/tolerance to wilt are Madhu, Milan, Mohini, MHW-4, MHW-5, MHW-6, MHW-11, MHW-12, MHW-15, NS-246, NS-295, Suruchi, Samtrupti, Amruth and Century-2.

Interspecific Hybridization

All the *Citrullus* species (*Citrullus rehmii*-annual; *Citrullus colocynthis*-perennial; *Citrullus ecirrhosus*-perennial and *Citrullus naudinianus*-perennial) are cross copmpatible with each other. Watermelon is readily crossable with *Citrullus colocynthes*, resulting in production of variants which have possible uses in crop improvement.

Poloploidy Breeding

Poloploidy breeding programme is being used largely for development of seedless varieties which is in high demand. Seedless watermelon offers the consumer preference and high quality fruits. Flesh is firm and sweet. A relatively recent development in watermelon breeding has been the use of ploidy manipulations to produce seedless triploid genotypes (Kihara, 1951). A number of seedless cultivars have been developed, but they tend to be more susceptible to physiological problems such as poor seed germination and hollow heart. Seedless watermelon is one classical example of the practical use of colchicine and application of autopolyploidy in crop improvement. Triploid (3x) hybrid seeds are produced by crossing tetraploids (4x=44) with diploids (2x=22). 'Pusa Bedana' ('Tetra II' x 'Pusa Rasal') is a seedless watermelon variety developed at the ICAR-IARI, New Delhi. 'Alena' is a tetraploid from the diploid 'Sugar Baby' (Srivastava, 1977). Watermelon in northwest China, for edible seeds are an important crop (Zhang and Jiang, 1990). These melons are small in size (2.5 to 3.5 kg) with low soluble solids content, but have a high ratio of seed to flesh in the fruits. The seeds are roasted before eating. However, in the future, it might be possible to develop transgenic diploid seedless watermelons. In that case, the question of the advantages in using heterotic hybrids vs inbred cvs. (or open pollinated varieties) will still be important. Kihara's work was unknown outside of Japan until 1951 because of World War II. The concept was commercialized in the US by Eigsti with the introduction of 'Tri- x 313' in the 1962 and the formation of American Seedless watermelon Seed Corporation. 'Tri- x 313' is still a popular cultivar in many parts of the world and has many look alike, using the same tetraploid parent, under various names. It took many years for triploids to become a major factor in the watermelon industry because of several deficiencies in early cultivars such as poor seed germination, presence of hard seeds in the fruit, erratic performance and high seed costs.

Male Sterility

In watermelon male sterility is available. Four genes for male sterility have been reported. Glabrous male sterile is unique, with sterility associated with glabrous foliage. All male sterile genes reduce female fertility as well. These mutants have been used in hybrid development programme but have not been as successful as hoped, since they often have low seed yield. A new spontaneous male sterile mutant (ms-2) with high normal seed set has been identified, and will be more useful for hybrid production. Male sterility can be induced in watermelon by applying maleic hydrazide (MH) before anthesis. However, chemicals are not suitable for inducing transient male sterility as they exert adverse effect on female fertility and vegetative growth. Watts (1962) reported the occurrence of male sterile line in watermelon variety Sugar Baby by irradiating it with 64 Krad X-rays. The male sterility so induced was having glabrous leaves. Male sterility as well as glabrous leaf trait was found to be governed by a single recessive gene Watts (1967). Glabrous seedling marker facilities early identification and rouging of male fertile sibs from the seed parent line. However, this genetic male syerility could not be commercially exploited due to poor female fertility association with the system (Xia *et al.* (1988). Spontaneously occurring male sterile plants have been found to be controlled by a single.

Quality Traits

Watermelons types have traditionally been red-fleshed and seeded. There is genetic variation for flesh colour in the species, however, and colours can range from white or yellow to orange, depending upon the genetic constitution. Yellow-fleshed cultivars are now available, and there may be a market for white-fleshed cultivars if quality could be assured, since consumers tend to associated white flesh with immaturity.

Resistance Breeding

I. Disease Resistance

Resistance to various viral and fungal disease located in different *Citrulls* species (*Citrullus rehmii*-annual; *Citrullus colocynthis*-perennial; *Citrullus ecirrhosus*-perennial and *Citrullus naudinianus*-perennial) have been summarized in Table 3.79.

Based on the gene list given by Guner and Wehner (2004) resistance to race 1 and race 3 of anthracnose (*Colletotrichum lagenarium*) is governed by a single dominant gene. Resistance to race 2 of anthracnose is also controlled by a single dominant gene. Resistance to race 1 of *Fusarium oxysporum* f. sp. *niveum* is controlled by a single dominant gene. Most watermelons are resistant to cross of powdery mildew (*Sphaerotheca fuliginea*), but single recessive gene (*pm*) for susceptibility has been found in the accession PI269677. A high level of resistance to zucchini yellow mosaic virus available

Table 3.79: Sources of Resistance to Viral and Fungal Diseases in *Citrullus* Species

Citrulls Species	*Resistance Sources*	*Pathogen*	*References*
Citrullus lanatus var. *lanatus*	PI595203	ZYMV	Guner and Wehner, 2008
		PRSV	Strange *et al.*, 2002
		ZYMV-FL	Boyhan *et al.*, 1992
		ZYMV-ch	Xu *et al.*, 2004
		WMV	Xu *et al.*, 2004
	PI482261-1, PI482261	ZYMV-FL	Provvidenti, 1991
Citrullus lanatus var. *citroides* (Egun)	-	Watermelon mosaic virus-2 (now called WMV)	Webb, 1997
	PI244017, PI244018, PI244019, PI482342, PI482318, PI482379, PI485583	PRSV-W	Strange *et al.*, 2002
Citrullus lanatus var. *lanatus*	PI296341-FR	*Fusarium oxysporum* f. sp. *niveum* Race-0' Race-1 and Race-2	Martyn and Netzer, 1991
Citrullus colocynthis	PI386016	WMV	Gillaspie and Wright, 1993
	PI386024	WMV	Gillaspie and Wright, 1993
	PI386025	ZYMV	Guner and Wehner, 2008
	WMV	WMV	Gillaspie and Wright, 1993
	PI386026	ZYMV	Guner and Wehner, 2008
		WMV	Gillaspie and Wright, 1993
	PI388770	WMV	Gillaspie and Wright, 1993
	PI494528	ZYMV-FL	Provvidenti, 1991
		WMV	Lecoq *et al.*, 1998
	PI494532	ZYMV	Lecoq *et al.*, 1998
		WMV	Provvidenti, 1989; 1991
	-	ZYMV	Boyhan *et al.*, 1992

Source: Dinesh and Sankaran, 2017.

Table 3.80: Some Promising and Useful Introduction of *Citrullus lanatus*

Accessions	*Country*	*Promising and useful attributes*	*Indentor*
EC-393240-43	USA	Small medium round fruits, red fleshed for breeding purpose	IIVR, Varanasi
EC-402549	USA	Multiple disease resistant to race-2, anthracnose, fusarium wilt and gummy stem blight with early maturity	-
EC-384809	USA	Fusarium wilt resistance	
EC-380989-92	USA	Yellow fleshed, early maturing	-
EC-378523-24	USA	Fusarium wilt resistance	
EC-333659-69		Wilt resistance	
EC-217073-74	USA	Fusarium and gummy stem blight esistance	
EC-380989-91	USA	Yellow fleshed water melon, early maturing, sweet crisp flesh	IIVR, Varanasi
EC-382753	USA	Breeding line SSDL resistant to Fusarium wilt and anthracnose	IIVR, Varanasi
EC-378523-23	USA	Fusarium wilt resistant	IIVR, Varanasi
EC-509468, EC-477972, EC-477973, EC-477974	USA		
EC-393240-43	USA		
EC-402549	USA		
EC678820 and EC678822	Taiwan	Heat tolerance, strong resistance to bacterial wilt (BW) and blight and rainfall	
EC678820	Taiwan	Heat tolerance, strong resistance to bacterial wilt (BW) and blight	

in four landraces of *C. lanatus,* is governed by a single recessive gene. Genotypes 'Caltham Gray', 'Summit', 'Shipper', 'White Hope' and 'Verons' are resistant to *Fusarium oxysporum* f. sp. *niveum*. Lines 'PI 189225' and 'PI 271778' are resistant to *Mycosphaerella citrullina*, and 'Fairfax' to *Alternaria cucumerina*. Accession PI595203 has multiple resistance to ZYMV, PRSV and WMV. *Citrullus lanatus* var. *citroides* (PI296341-FR) shows resistance to *Fusarium oxysporum* f. sp. *niveum* race-0, race-1 and race-2.

Table 3.81: Resistance Source against Diseases

Diseases	*Resistance Source*
Fusarium wilt	Summitte, Conqueror, Charleston Gray, Dixilec, Crimson Sweet, EC 333659-69, Citron, Shipper, Smokylee, Shipper, White Hope, Verons
Anthracnose	Fair Charleston Gray, Congo, PI 189225, African Citron-8, Charleston Gray, Congo, Crimson Sweet, Jubilee
Fusarium and Anthracnose	EC382753, EC 402549
Fusarium	EC 383809
Powdery mildew	Arka Manik
Water melon bud necrosis virus (WBNV)	Citrullus colocynthis
Alternaria cucumerina	Fairfax

Source: Bhardwaj, 2010.

In watermelon the main source of resistance was the 'citron types'. Indian bred varieties 'Arka Manik' is popular variety of South India and carrying high degree of multiple resistance against Anthracnose, powdery mildew, downy mildew. Screening work done at Punjab Agricultural University, Ludhiana (Punjab) led to the identification of varieties 'Asahi Yamato' and 'Sugar Baby' to be moderately resistant to powdery mildew. Watermelon cultivars 'Qingfa-8 (F_1) and Summit (F_1) were reported resistance (Zhang *et al.*, 1999; Schenck, 1961) against wilt diseases.

II. Insect Resistance

Resistance has been reported in watermelon in respect of fruit fly (*Dacus cucubitae*) which is governed by a single dominant gene and red pumpkin beetle (*Aulacophora foveicollis*) which is also controlled by a single dominant gene.

Table 3.82 Resistance source against insects

Citrulls species	*Resistance sources*	*Pathogen*	*References*
Citrullus colocynthis	PI388770	Spotted spider mites	Gillaspie and Wright, 1993
	PI525080	Spotted spider mites	Cantu, Hector Jr, 2014
	PI537300	Spotted spider mites	Cantu, Hector Jr, 2014
	PI494532	Resistant to spider mites	Lopez *et al.*, 2005

III. Nematode Resistance

Table 3.83. Resistance source against nematode

Citrulls Species as Resistance Sources	*Pathogen*	*References*
Citrullus lanatus var. *citroides* (Egun)	Resistant to root knot nematodes	Theies and Levi, 2007

Breeding for Abiotic Stress

Citrullus colocynthis a relative of watermelon is native to tropical Africa and highly drought tolerant. An

Table 3.84: Rootstocks and their Quality

Scion	Rootstocks	Purpose
Watermelon	*Lagenaria siceraria*, *Cucurbita moschata*, interspecific hybrid squash (*C. maxima* x *C. moschata*), *Benincasa hispida*	Vigorous root system, tolerance to fusarium wilt, wilting due to physiological disorders, low and high temperature, *Benincasa hispida* for multiple disease tolerance.

Table 3.85: Important Diseases and Pests of Vegetables Controlled by Grafting

Scion	Rootstocks	Disease/pests	References	Major variety
Watermelon	*Lagenaria siceraria*	Fusarium wilt (*Fusarium oxysporum*)	Balaz (1982)	-
	Lagenaria siceraria	*Fusarium oxysporum* f.sp. *niveum* wilt	Mondal *et al.* (1994)	-
	Lagenaria siceraria	-	-	Don K, Kachidoki No.2, Aioi, FR-10, FR-Choju
	Cucurbita moschata	-	-	No.-8
	Benincasa hispida			Best

Table 3.86: Conservation of Genetic Resources

Sl.No.	Conservation Centre	Country	Germplasm Conserved	Evaluation for Specific Traits
1.	Institute of Introduction and Plant Genetic Resources 'K Malkov' Plovdiv, Sadovo	Bulgaria	Land races, old cultivars, breeding material	Resistance to diseases and Pest, chemical composition
2.	Deptt. of Field and Vegetable Crops, The Faculty of Agriculture Hebrew University of Jerusalem	Israel	-	Diseases and insect resistance, shelf life, yield and stress
3.	National Plant Genetic Resources Laboratory (NPGRL), University of Philippines, Los Bonas College, Languna	Philippines	From different countries	-
4.	South Regional Plant Introduction, USDA, Griffin, Georgia	USA	From different countries	

introduction EC678820 EC678820 and EC678822 from Taiwan found heat tolerant.

Grafting

Inter-generic grafting has been applied to the production of fruit-bearing vegetables. Presently the specific problems can be encountered through the adapting of grafting techniques.

The production of grafted vegetable plants was 1st started in Japan and Korea in the late 1920s with watermelon (*Citrullus lanatus* et Nakai) grafted onto pumpkin (*Cucurbita moschata*) rootstock (Lee, 2003, Oda, 1999). Soon after, watermelon (*Citrullus lanatus* et. Nakai) were grafted onto bottle gourd (*Lagenaria siceraria* L.) rootstocks (Oda, 1999) and rarely on pumpkin. The shape and taste of the fruit of plant grafted on pumpkin degrade markedly in watermelons. Grafting has been reported to successful method for imparting resistance against *Fusarium* wilt in several cucurbits. In pumpkin CM 214 (Nigerian Local, Immune) did not develop any disease symptom even after grafting on the YVMV infected symptomatic rootstock. The environmental fluctuations also did not influence the resistance expression of Nigerian Local against PMV and YVMV (Kuriakose, 1998). In field tests on ground infected with *Fusarium oxysporum* of 4 self rooted varieties of melon and 2 of those grafted to ash gourd Rugeri (1968) observed that 90 per cent of plants belonging to each of the ungrafted varieties became infected while those grafted on ash gourd escaped infection. When *C. lanatus* var. *citroids* used rootstock for grafting watermelon proved resistance/tolerance against *Fusarium oxysporum* f. sp. *niveum*, *F. oxysporum* f. sp. *melonis*, *F. oxysporum* f. sp *cucumerinum*, root knot nematode, and gummy stem blight (Davis *et al.*, 2008; Theis and Levi, 2003; Theis *et al.*, 2015)

Conservation of Genetic Resources

Singh and Chauhan (2003) have visited south west region (Mathura, Agra, Etawah, Kanpur, Farrukhabad and Hathras district) of Uttar Pradesh and found uniform distribution of watermelon all across to Ganga river bank.

All India Coordinated Research Project (Vegetable Crops) has entrusted the responsibility for collecting, maintaining and characterizing the germplasm of watermelon to ICAR-IIHR, Bengaluru, ARS, Durgapura and PAU, Ludhian centres.

REFERENCES

1. Badifu, G.I.O. and Ogunsua, A.O. 1991. Chemical composition of kernels from some species of Cucurbitaceae grown in Nigeria. *Plant Food Human Nutr.*, **41:** 35-44.
2. Bailey, L.H. 1964. Manual of cultivated plants. The MacMillan Company, New York.

3. Boyhan, G., Norton, J.D., Jacobson, B.J. and Abrahems, B.R. 1992. Evaluation of watermelon and related germplasm for resistance to Zucchini yellow mosaic virus. *Plant Dis.*, **76:** 251-252.

4. Brar, J.S. and Sidhu, A.S. 1977. Heterosis and combining ability of earliness and quality characters in watermelon [*Citrullus vulgaris* (Thunb.) Mansf.]. *II J.Res. Punjab Agric. Univ.*, **14:** 272-278.

5. Bra, J.S and Sukhija, B.S. 1977. Hybrid vigour in intervarietal crosses in watermelon [*Citrullus vulgaris* (Thunb.) Mansf.]. *Indian J. Hort.*, **34:** 277-283.

6. Bhardwaj, D.R. 2010. Breeding for pest and disease resistance in cucurbits. In: Compedium of Wnter School "Emerging trends and strategies for the management of pests and diseases in vegetable crops. Published by ICARIIVR, Varanasi (U.P.), p. 102-109.

7. Cantu, Hector Jr, 2014. An evaluation of watermelon (*Citrullus* species) germplasm for additional sources of resistance to the two spotted spider mite (*Tetranychus urticae* Koch.). Thesis, in Agronomy and Horticulture, p. 80.

8. *Ciência Rural*. Santa Maria, 2002. **32(5):** 715-723.

9. Davis, A.R., Perkins-Veazie, P., Dakata, Y., Lopez-Galarza, S., Cohen, R., and Lee, J.M. 2008. Cucurbit grafting. *Crit. Rev. Plant Sci.*, **27:** 50-74.

10. Dinesh, M.R. and Sankaran, M. 2017. Distant hybridization in horticultural crops. Daya Publishing House A division of Astral International Pvt. Ltd. p.133-134.

11. Gill, B.S. and Kumar, J.C. 1988. Combining ability analysis in watermelon [*Citrullus lanatus* (Thunb.) Mansf.]. *Indian J. Hort.*, **45:** 104-109.

12. Gill, B.S. and Kumar, J.C. 1988. Phenotypic stability in watermelon [*Citrullus lanatus* (Thunb.) Mansf.]. *Indian J. Agric. Sci.*, **59:** 145-148.

13. Gillaspie, A.G. and Wright, J.M. 1993. Evaluation of *Citrullus* species germplasm for resistance to watermelon mosaic virus-2. *Plant Dis.*, **77:** 352-354.

14. Guner, N. and Wehner, T.C. 2004. The genes of watermelon. *HortSci.*, **39(6):** 1175-1182.

15. Guner, N. and Wehner, T.C. 2008. Overview of potyvirus resistance in watermelon. In: Proc.IXth EUCARPIA meeting on genetics and breeding of Cucurbitaceae. Pitrat, M. (ed.) INRA, Avignon (France), May 21-24th, 2008.

16. Henderson, W.R. 1991. Gene list for watermelon. *Cucurbit Gene. Coop. Rep.*, **14:** 129-134.

17. Kihara, H. 1951. Triploid watermelons. Proc. *Amer. Soc. Hort. Sci.*, **58:** 217-230.

18. Lecoq, H., Wisler, G. and Pitrat, M. 1998. Cucurbit viruses: the classics and the emerging. In: McCreight, J.D. (ed.). Cucurbitaceae. Evaluation and enhancement of cucurbits germplasm, ASHS Press, Alexandria, p. 126-142.

19. Lee, J.M. and Oda, M. 2003. Grafting of herbaceous vegetable and ornamental crops. *Horticultural Review*, **28:** 61-124.

20. Lopez, R., Amnon, L., Shepard, B.M., Simmons, A.M. and Jackson, D.M. 2005. Sources of resistance to two spotted spider mite (*Acarii tetranychidae*) in *Citrullus* spp. *HortSci.*, **40(6):** 1661-1663.

21. Mallick, M.F.R. and Masui, M. 1986, distibution and taxonomy of melons. *Scientica Horticulturae*, 28:251-216.

22. Martyn, R.D. and Netzer, D. 1991. Resistance to races 0, 1 and 2 of Fusarium wilt of watermelon in *Citrullus* sp.PI 129341-FR. *HortSci.*, **26:** 429-432.

23. Mohn. H.C. 1986. Watermelon breeding. In: Bassett, M.J, editor, ed. Breeding Vegetable Crops. Westport CT: Avi Publishing, 37-66.

24. Nandpuri, K.S., Kumar, J.C. and Dhillon, G.S. 1974. Heterosis in watermelon. *Punjab Hort. J.*, **14:** 75-83.

25. Nandpuri, K.S., Kumar, J.C. and Dhillon, G.S. 1974. Combining ability estimates in a set of top crosses in watermelons [*Citrullus lamnatus* (Sch.)]. *J. Res. Punjab Agric. Univ.*, **15:** 65-70.

26. Nath, P. and Dutta, O.P. 1970. Heterosis in watermelon. *Indian J. Hort.*, **27:** 176.

27. Oda, M. 1999. Grafting of vegetables to Improve Greehouse Production. College of Agricultural Education, Osaka Prefecture University, Japan, p. 1-11.

28. Pangalo, K.I. 1930. Watermelon of the northern hemisphere-in Russian englis summart. Bull. *Appl. Genet. and Pl. Breed.*, **23:** 41-84.

29. Pangalo, K.I. 1944. A new genus of the cucurbitaceae, *Praecitrullus* M., an ancestor of the contemporary watermelon (*Citrullus lamnatus* Forsk.) in Russian English summary. *J.Bot. De 1' USSR*, **29:** 200-204.

30. Pangalo, K.I. 1955. Origin and evolutionary trends in cucurbitaceous crops. *Problemy Botaniki, Akad, Nank. USSR*, **2:** 329-338.

31. Provvidenti, R. 1989. Sources of resistance to viruses in cucumber, melon, squash and watermelon. p. 29-36. In: Proc. Cucurbitaceae. Evaluation and Enhancement of cucurbit germplasm: Dept. Agric. Res. Serv. U.S. Vegetable Laboratory, Charleston, SC.

32. Provvidenti, R. 1991. Inheritance of resistance to the Florida strain of Zucchini yellow mosaic virus in watermelon. *HortSci.*, **26:** 407-408.

33. Purseglov, J.W. 1977. Tropical Crops, Dicotyledons. Longman Group Ltd., London.

34. Robinson, R.W. and Deckev-Walters, D.S. 1997. Cucurbits. Wallingford: CAB International. 65-97,152-163.

35. Salunkhe, D.K., Desai, B.B. and Bhat, N.R.1987. Vegetable and Flower Seed Production. Asgricole Publishing Academy, New Delhi, p. 233-264.

36. Schenck, N.C. 1961. 'Summit (F)' *Florida Agricultural Experiment Station Journal* Series. No. **1338:** 183-186.

37. Seshadri, V.S. 1993.Cucurbits. In. Vegetable Crops (Ed. Bose T.K., Som, M.G. and Kabir, J.) Naya Prakash, Calcutta, p. 40-109.

38. Sidhu, A.S. and Brar, J.S. 1977. Heterosis and combining ability of yield and its components in watermelon [*Citrullus vulgaris* (Thunb.) Mansf.]. *J. Res. Punjab Agric. Univ.*, **14:** 52-58.

39. Singh, J., Kalloo, G. and Singh, K.P. 2001. Vegetable Crops: Nutritional Security. Tech. Bull. 6. Published by ICAR-Indian Institute of Vegetable Research. p. 30-37.

40. Singh, K.P. and Chauhan, S.V.S. 2003. Diversity of watermelon (*Citrullus vulgaris* Linn.) from Agra and Kanpur region. *National J.Plant Genet. Resour.*, **16(2):** 100-104.

41. Singh, S.P. 1989. Production technology for vegetable crops. Agricultural Research Communication Centre, Kernal, Haryana (India).

42. Soliman, M.A., ElSawy, A.A., Fadel, H.M., Osmon, F. and Gad, A.M. 1985. Volatile compounds of roasted *Citrullus colocynthis* var. *colocynthoides*. *Agr. Biol. Chem. Tokyo*, **49:** 269-275.

43. Srivastava, L.S. 1977. Watermelon without seeds. *Intensive Agriculture,* **11(2):** 20-22.

44. Strange, E.B., Guner, N., Pesic-VanEsbroek, Z. and Wehner, J.C. 2002. Screening the watermelon germplasm collection for resistance to papaya ringspot virus type-w. *Crop Sci.*, **42:** 1324-1330.

45. Sutariya, R.N. 1961. A Text Book of Systematic Botany. Ramnik, P. Kothariya, Gandhi Road, Ahmedabad.

46. Thies, J.A and Levi, A. 2003. Resistance of watermelon germplasm to the peanut root knot nematode. *HortSci.*, **38:** 1417-1421.

47. Thies, J.A and Levi, A. 2007. Characterization of watermelon (*Citrullus lanatus* var. *citroides*) germplasm for resistance to root knot nematode. *HortSci.*, **42:** 1530-1533.

48. Thies, J.A., Levi, A., Jennifer, J.A. and Richard, L.H. 2015. RKVL-318, a root nematode resistant watermelon line as rootstock for grafted watermelon. *HortSci.*, **50(1):** 141-142.

49. Watts, V.M. 1962. A marked male sterility mutant in watermelon. *Proc. Am. Soc. Hort. Sci.*, **81:** 498-505.

50. Watts, V.M. 1967. Development of disease resistance and seed production in watermelon stocks carrying *msg* gene. *J.Am. Soc. Hort. Sci.*, **91:** 579-590.

51. Webb, R.E. 1997. Resistance to watermelon mosaic virus in *Citrullus lanatus* (Absts). Proc. *Am. Phytopathol. Soc.*, **4:** 220.

52. Xia, X., Lin, Y., Liu, W. and Chen, A. 1988. Selection of watermelon [*Citrullus vulgaris*) male sterile line G17 AB. *J. Shenyang Agric. Univ.*, **19:** 9-13.

53. Xu, Y., Kang, D., Shi, Z., Shen, H., and Wehner, T.C. 2004. 2004. Inheritance of resistance to Zucchini yellow mosaic virus and watermelon mosaic virus in watermelon. *J. Hered.* **96:** 498-502.

54. Zhang, X.P. and Jiang, Y. 1990. Edible seed watermelon (*Citrullus lanatus* (Thunb.) Matsum. and Nakai) in northwest China. *Cucurbit Genet. Coop. Rpt.* **13:** 40-42.

55. Zhang, X.P. and Rhodes, B.B. 1992. Watermelon variety improvement in China. *Cucurbit Genet. Coop. Rep.*, **15:** 76-79.

56. Zhang, Z.F., Xu, X.L., Shi, X.J., Li, J.W. and Chen, X. 1999. 'Qingfa No.8 (F)'-a new watermelon hybrid variety. *Acta Horticulturae Sinica*, **26:** 349.

Genus *Praecitrullus*

Round Melon or Indian Squash (*Praecitrullus fistulosus* (Stocks) Chakra)

Introduction

Round melon (*Praecitrullus fistulosus* Stocks, 2n=22) is very popular vegetable crop and locally known as Indian round gourd or Indian squash. It is a squash like fruit and entire immature fruit is used as a cooked vegetable due to high nutritive value. Round melon is very much popular in South Asia. The seeds can also be removed and eaten. It is also known with the local names like tinda, dilpasand, tindu, tendsi, tendus, *etc.* in different parts of the country, and commercially grown in India and Pakistan. It is most popular summer vegetable in North India, especially in the Punjab, Haryana, Delhi, Western Uttar Pradesh, Maharashtara, Gujrat and Rajasthan. The immature, tender, and light green fruits of Indian melon are used as a cooked, boiled and fried vegetable curry and are pickled or candied. The seeds after removing seed coat are roasted or fried and eaten as a snack. As a medicinal value the fruits are also considered good for dry cough and improving blood circulation in the body (Nadkarni, 1927). Edible tender fruits of round melon provide a fair amount of carbohydrates, protein, calcium, phosphorus, vitamin A and C. Its consumption gives cooling effect to the body.

Origin, History and Distribution

This crop is indigenous of India. This was originally thought to be a subspecies of watermelon as *Citrullus vulgaris* var. *fistulosus* (Stocks) Chakra and later raised to an independent species as *Citrullus fistulosus* Stocks. Several known botanists like Pangalo, (1930, 1944), Guljaev (1963), Trivedi and Roy (1970) and Dutta and Nath (1970) suggested that this taxon was long back considered to be *Citrullus lanatus* but it should be placed in a separate genus *Praecitrullus* much earlier so the present accepted botanical name is *Praecitrullus fistulosus* Pang. The relationship with watermelon has been conclusively disproved by cytogenetical studies of Shimotsuma (1963) and Khoshoo (1955). In some places it is mentioned that roundmelon is more related to muskmelon owing to same chromosome number (Khoshoo and Vij, 1963; Shimotsuma, 1965). They recommended that it could be taken out of the genus *Citrullus* since it is not at all found in their original home, *i.e.*, Africa. The chromosome number of roundmelon (2n= 24) is different from that of watermelon (2n= 22), even though, no marked morphological differences were found for dimensions among *Citrullus lanatus, C. colocynthis and Praecitrullus fistulosus* (Shelepina and Turkov, 1974). Electrophoretic analysis of 26 enzyme-coding genes was conducted on accessions of three *Citrullus* species and the related *Praecitrullus fistulosus* [*Citrullus vulgaris* var. *fistulosus*] and *Acanthosicyos naudinianus* and an isozyme phylogeny was constructed based on pair wise measurements of genetic distance. In this scheme, *P. fistulosus* and *A. naudinianus* form two distinct out groups to *Citrullus*, which is characterized by two main cluster. The first includes two *Citrullus colocynthis* races and the second *Citrullus lanatus* and *Citrullus lanatus* var. *citroides*, which are more closely related to each other than they are to *Citrullus ecirrhosus*. *Citrullus lanatus* var. *citroides* is regarded as the progenitor of cultivated *Citrullus lanatus* (Whitaker, and Bemis, 1976). The isozyme phylogeny was consistent with the variability in six seed protein bands and with the cross ability relations among the examined species (Navot and Zamir, 1987). *Praecitrullus fistulosus* (X= 12) roots and hypocotyls were subjected to isozyme analysis to investigate its affinities with *Citrullus* and *Cucumis* species. Seven peroxidase isozymes differed in electrophoretic mobility from six in *Citrullus lanatus* and eight in *Cucumis melo*. Three bands of glutamate aloacetate transaminase [aspartate aminotransferase] of *Praecitrullus fistulosus* were different from two in *Citrullus* and four in *Cucumis*, which support the separation of *Praecitrullus* into a discrete genus (Sujatha and Seshadri 1989). Similarly, trichomal studies were also conducted to know the relationship among *Cucumis, Citrullus* and *Praecitrullus* (Khan and Rajput, 1993). Clear-cut histological differences between *Citrullus lanatus* ssp. *mucosospermus* and the other species studied also lend support to the view that it is distinct specie (*Citrullus mucosospermus*). Forms of *Citrullus colocynthis* with fruits free from bitter principle are intermediate between *Citrullus colocynthis* and watermelon and may be hybrids. *Citrullus colocynthis* shows resemblances to *Citrullus ecirrhosus*, *Citrullus naudinianus* to the genera *Benincasa* and *Acanthosicyos*, on the other hand, certain resemblances between *Praecitrullus fistulosus* and the watermelon are regarded as fortuitous but no close resemblance was found between *Citrullus* and *Cucumis* species (Fursa, 1972, Fursa, 1974, Fursa and Gavrilyuk, 1990). This crop is commercially grown in India and Pakistan for vegetable purpose. It is most popular summer vegetable in North India, especially in the Punjab, Haryana, Delhi, Western Uttar Pradesh, Maharashtara, Gujarat and Rajasthan.it is not commercially cultivated in West Bengal.

Species Availability, Evolution and Diversity

The round melon was long considered to be *Citrullus lanatus* but was recently given its own taxonomic category due in part to its difference in monoploid chromosome number (Sujatha and Seshadri, 1989).

Name of Taxon	*Name of Authority*	*Botanical Family*	*Preferred Taxon Name*
Praecitrullus fistulosus	(Stocks) Pangalo	Cucurbitaceae	Praecitrullus fistulosus

Botanical Description

Table 3.87: Botanical Description

Growth Habit	A diffuse annual herbaceous creeping or climbing plant.
Leaf	Light green hairy leaves.
Floral biology	Roundmelon is commonly a monoecious plant that bears staminate and pistillate flowers on separate nodes, however, few andromonoecious cultivars also exist in nature, producing perfect as well as staminate flowers on the same plant but on separate positions. Flowers are small, less showy, and located solitary. The pistillate flowers normally occur in every seventh leaf axil, while staminate flowers on the plant mostly occupy the intervening axils. The sex ratio staminate to pistillate flowers varies 9:1 to 22:1. Male flower appear first then female flowers.
Anthesis and pollination	Anthesis takes place in morning from 6.00 to 8.30 a.m., and anther dehiscence starts one hour before anthesis and continues until 7.00 a.m. Pollen is visually evident in sticky masses adhering to anther. The stigma becomes receptive two hours before to three hours after anthesis. As the day advances and temperature increases, stigmatic fluid starts drying, and as a result, stigma becomes non-receptive. Maximum fruit set has been noticed during first hour after anthesis. Honeybees are the main pollinating agent (Chadha, *et al.*, 1993). Breeding roundmelon for quality and yield is an utmost need of the day for improvement of the crop (Nath *et al.*, 1977, Dahiya *et al.*, 2001).
Fruit and fruit set	It bears small pale or dark green colour fruit having hairs or sometimes smooth
Seeds	Blackish seeds

Relationship of Indian Squash or Round Melon with Watermelon

It is pertinent to have about the relationship of 'tinda' or Indian squash or round melon with watermelon. This was originally thought to be a subspecies of watermelon as *Citrullus vulgaris* var. *fistulosus* (Stocks) Chakra., and later raised to an independent species as *fistulosus* Stocks. The relationship with watermelon neither been conclusively disproved by cytological studies nor is it related to muskmelon inspite of the same chromosome number. It is recommended that it be taken out of the genus *Citrullus* since it is not at all found in Africa. The Russian botanist Pangalo has raised this taxon to a new independent genus *Praecitrullus* much earlier. The present botanical name is *Praecitrullus fistulosus* Pang. Genetic diversity of this crop is rather limited suggesting its partial or recent domestication.

Augmentation of Genetic Resources and Crop Improvement

A field experiment was conducted at Horticulture Farm, Institute of Agriculture, Sriniketan (West Bengal) during summer 2016 with seven commercial cultivars (Ludhiana Special, Tinda Dil Pasand, Tinda Ludhiana Special, Mahy-1, Chitra, Mahy Tinda and Golden Tinda) available in India to see the sutability of growing under Red and Laterite zone of West Bengal (Mohanta and Mandal, 2016). Significant variation among the cultivars was noticed for various growth, flowering, yield and yield attributes. The cultivar Golden Tinda was found early in all flowering traits. However, Tinda Dil Pasand was superior to other cultivars in various growth and yield parameters and yield. Findings indicate that Tinda can be susccessfully grown under Red and Laterite zone of West Bengal and its commercial cultivation may be tried. Tinda Dil Pasdand, the best performing cultivar, may be promoted in this region because of adaptability. Popular cultivars like 'Arka Tinda' and 'Tinda S-48' were developed by use of Indigenous source through selection with progeny testing.

REFERENCES

1. Aykrod, W.R. 1963. *Health Bulletin.* ICMR, Special Report Series. No. 42.
2. Bose, T.K. and Som, M.G., 1986. *Vegetable Crops in India.* Naya Prakash, 206 Bidhan Sarani, Calcutta-91, India.
3. Vhadha, M.L., Lal, Tarsem, Lal, T. and Chadha, K.L. 1993. Improvement of Cucurbits. *Advances in Horticulture: Vegetable Crops,* **5:** 137-179.
4. Dahiya, M.S., Baswana, K.S. and Tehlan, S.K. 2001. Genetic variability studies in roundmelon (*Praecitrullus fistulosus* Pang.).
5. Dhesi, N.S., Padda, D.S. and Malik, B.S., 1964. Effect of nitrogen, phosphorus and potash on yield of squash melon. *Punjab Hort. J.*, **4:** 45-47.
6. Dutta, O.P. and Nath, P., 1970. Crossability in melons. *Indian J. Hort.*, **27:** 60-67.
7. Nasir, E. and Ali, S.I., 1970. Cucurbitaceous Vegetables. Flora of West Pakistan.
8. Fursa, T.B., 1972. The evolution of the genus *Citrullus* Schrad. *Botanicheskii-Zhurnal.* **57 (11):** 1365-1372.
9. Fursa, T.B., 1974. The seed coat of watermelon as a taxonomic character. *Trudy-po-Prikladnoi-Botanike,-Genetiki-i-Selektsii.* **51:** 3, 39-48.
10. Fursa, T.B. and Gavrilyuk, I.P. 1990. Phylogenetic relation in the genus *Citrullus* Schrad. According to data from immunochemical analysis of the seed proteins. Sbornik-Nauchnykh-Trudov-po-Parikadnoi- Botanika, Genetika-I-Selektsii. **133:** 19-26.
11. Gopalan, C., Ramasastri, B.V. and Balasubramanian, S.C., 1982. *Nutritive Values of Indian Foods.* National Institute of Nutrition, Hyderabad (A.P.).
12. Guljaev, V.A., 1963. Comparative embryology of the cucurbitaceous and its significance for the taxonomy of the family. *Bot. J. Moskva*, **48:**80-85.
13. Hazara, P. and Som, M.G. 1999. Technology for Vegetable Production and Improvement. Naya Prakash, 206, Bidhan Sarani, Calcutta. India.
14. Khan, M.I. and and Rajput, M.T.M. 1993. Trichomal studies in cucurbitaceae: in the genera of *Cucumis, Citrullus* and *Praecitrullus.Tehqiqe*, **9(2):** 41-54.
15. Khoshoo, T.N., 1955. Cytotaxonomy of Indian species of *Citrullus. Curr. Sci.*, **24:** 377-378.
16. Khoshoo, T.N. and Vij, S.P., 1953. Biosystematics of *Citrullus vulgaris* var. *fistulosus*. *Caryologia*, **16:** 541-552.
17. Khurana, S.C., Mangal, J.L. and Singh, G.R. 1988. Chemical weed control in cucurbits: A review. *Agriculture Research,* **9(2):** 69-80.
18. Kore, S.S. and Kharwade, L.R. 1987. Fruit rot of round gourd caused by *Fusarium oxysporum. Indian Phytopath.*, **40(2):** 251.
19. Mahakal, K.G., Joshi, A.T., Deshmukh, P.P. and Pawar, P.R., 1977. Effect of N.P. and K on tinda (*Citrullus vulgaris* var. *fistulosus*). *Orisa J. Hort.*, **5(1-2):** 62-63.
20. Mangal, J.L., Batra, B.R. and Singh, G.R., 1985. Studies on nitrogen fertilization under various soil moisture regimes on growth and productivity of round melon. *Haryana J. Hort. Sci.*, **14**(3-4): 232-236.
21. Mathur, K. and. Gujur, R.B.S., 2002. Drying Tinda (*Praecitrullus fistulosus*). *The Hindu*, Thursday, 14, by A.R.S. Durgapura, Jaipur.
22. Mohanta, S. and Mandal, J. 2016. Performance of Tinda (tinda (*Citrullus vulgaris* var. *fistulosus* Stocks) cultivars under Red and Laterite zone of West Bengal. (Abs.62; Session II: breeding of

Horticultural Crops). 7th Indian Horticulture Congress, 15-18 November 2016, held at ICAR-IARI, New Delhi, p. 132.

23. Nadkarni, K.M. 1927. *Indian Materia Medica.* Nadkarni and Co. Bombay.

24. Nath, P. and Dutta, O.P., 1971a. A note on the improvement in round melon. *Indian J. Hort.*, **28:** 55-56.

25. Nath, P. and Dutta, O.P., 1971b. New high yielding varieties in cucurbits. *Indian Hort.*, **15**: 11-14.

26. Nath, P., Dutta, O.P., Velayudhan, S. and Nath, P. 1977. Breeding roundmelon for quality and yield. *Veg. Sci.*, **4(1):** 11-18.

27. Navot, N and Zamir, D., 1987. Isozyme and seed protein phylogeny of the genus *Citrullus* (Cucurbitaceae). *Plant Systematic and Evolution*, **156(1-2):** 61-67.

28. Pangalo, K.I., 1930. Watermelons of the northern hemisphere in Russian. English Summary. *Bull. Appl. Bot. Genet. Pl. Breed.*, **23**: 41-84.

29. Pangalo, K.I., 1944. A new genus of the cucurbitaceae, *Praecitrullus,* an ancestor of the contemporary water melon (*Citrullus* Forsk). *J.Bot. de'l. U.S.S.R.* **29**: 200-204.

30. Saimbhi, M.S. and Randhawa, K.S. 1977. *Haryana J. Hort. Sci.*, **6**: 91-93.

31. Seshadri, V.S. 1986. Cucurbits (in): Vegetable Crops in India, Bose, T.K. and Som, M.G. eds. Naya Prakash, 206 Bidhan Sarani, Calcutta-91, India.

32. Shakir 1994. Fungi associated with tinda (*Praecitrullus fistulosus*) seed, their effect on seed germination and their chemical control. *Pakistan J. Phytopatho.*, **6(2):** 120-124.

33. Shelepina, G.A. and Turkov, V.D., 1974. Caryotypic characteristics of some species of cucumber. *Trudy po Prikladnoi Botanike Genetiki i Selektsii*, **51**(3): 65-76.

34. Shimotsuma, M., 1965. Classification of *Citrullus* species. *Seiken Ziho.*, **15**: 24-34.

35. Sidhu, A.S.; Pandita, M.L. and Arora S.K. 1981. Effect of Maleic hydrazide on round melon. *Haryana Agric. Univ. J. Res.*, **11(4):** 485-489.

36. Singh, K.P, Pandita, M.L. and Singh, G.R., 1983. Nitrogen and Phosphorus nutrition in seed production of round melon. *H.A.U. J. Res. Hisar*, **13(3):** 413-415.

37. Singh, N. and Joshi, P. 1993. Khira Vargiya Sabjiyan, ICAR, Publication, New Delhi.

38. Singh, S.V.; Mishra, A.; Bisen, R. and Malik, Y.P. 2000. Host preference of red pumpkin beetle (*Aulacophora foveicollis*) and melon fruit fly (*Dacus cucurbitae*). *Indian J. Ento.*, **62(3)**: 242-246.

39. Sujatha, V.S. and Seshadri, V.S., 1989. Taxonomic position of round melon (*Praecitrullus fistulosus*). *Cucurbit Genet. Coop. Rep.*, **12**: 86-88.

40. Trivedi, R.N. and Roy, R.P. 1970. Cytological studies in *Cucumis* and *Citrullus*. *Cytologia*, **35:** 561-569.

41. Whitaker, T.W. and Bemis, W.P. 1976. Cucurbits (in) Evolution of Crop Plants, edn 1, p. 64-69.

42. Yadav, J.L, Banerjee, M.K. and Arora, S.K. 1990. Insect pest of Kharif vegetable crops and their control measures. *Farmer and Parliament*, **45(7):** 5 and 25.

43. Yawalkar, K. S. 1985. Vegetable Crops of India. Published by Agro- Horticultural Publishing House, Nagpur.

Genus *Luffa*

The genus *Luffa* (Miller) is one of the most important and widely adapted vegetable of the humid tropics with several health benefits. The name Luffa or Luffah is of Arabic origin because sponge characteristics have been described in Egyptian writing and Chinese name Szkuadishcloth gourd or towel gourd, its mention in early Chinese literature. Sanskrit name Koshataki indicates its early cultivation in India. Four well-differentiated species from the Old World (*L. echinata* Roxb., *L. acutangula, L. aegyptiaca,* and *L. graveolens* Roxb.) and three species from the New World (*L. quinquefida* (Hook. and Arn.) Seem., *L. operculata* (L.) Cogn., and *L. astorii* Svens.) (Heiser and Schilling, 1990). The edible species are *Luffa acutangula, Luffa cylindrica* and *Luffa hermaphrodita*. These species are cross compatible which makes them amenable to traits transferability for genetic enhancement of *Luffa* species utilizing the prevailing diversity.

All species have 26 chromosomes ($2n = 26$) (Dutt and Roy, 1990; Heiser and Schilling, 1990). Cytological and hybridization studies suggest a close relationship between the two cultivated species (*L. aegyptiaca* and *L. acutangula*) and their derivation from either *L. graveolens* or an unidentified common ancestor (Dutt and Roy, 1990). While Old World species are well differentiated from each other and from the American species, the American species are rather similar to each other. *Luffa aegyptiaca* is the most extensively cultivated species (Heiser and Schilling, 1990). The early spread of the genus *Luffa* was in the New and Old World, but both cultivated species originated in India (Heiser and Schilling, 1990). The Indian gene center has rich diversity in genetic resources of *luffa* species. The genus comprised of 9 species in the world and out of this 7 species (*L. acutangula, L. cylindrica,*

L. echinata, L. graveolens, L. hermaphrodita, L. tuberosa and *L. umbellata*) is native to India (Chakravarty, 1982). There is ambiguity with regards *L. tuberosa* and *L. umbellata* because they are considered synonym to species of *Momordica* and *Cucurbita*, respectively. Thus there is only four undisputedly established species. They are *L. acutangula, L. cylindrica, L. echinata,* and*L. graveolens.* Of these, cultivated and wild taxa have been raised to specific rank of *L. hermaphrodita* by Singh and Bhandari, 1963, Prakash *et al.* 2013. Mabberly (1987), however, records only 6 species. According to Singh (1990) *Luffa* consists of 7 species; four occur in old world from Africa to India, Australia, and and three occur in Neotropics. In *Luffa* most of the species occur in disturbed sites, forest openings *etc. Luffa hermaphrodita* (locally known as Satputia) is characterized by short viny growth, earliness and most importantly, bears small smooth fruits in clusters. These traits can be utilized to enhance the genetic architecture of *Luffa acutangula* and *Luffa cylindrica* as they are commercially cultivated.

The sponge gourd (*L. cylindrica*) and ridge gourd (*L. acutangula*) has rich diversity throughout India. Rich diversity in vine and fruit morphological characteristics occurs in North-Eastern region including Sikkim, West Bengal, Western, Central and Southern India. *L. hermaphrodita* considered to be originating from *L. graveolans* is another potential species distributed in parts of north-central India. *L. acutangula* var. *amara* grows in peninsular India, is a wild relative of sponge gourd (*L. cylindrica*) and *L. echinata* in natural habitats is in Western Himalayas, Central India and Gangetic plains. Other related species are *L. operculata* (L.) Cogn., *L. gigante, L. scabra* and *L. narylandica.*

Table 3.88: Old World and New World Species of *Luffa*

Old world species of *Luffa*	*L. acutangula* (L.) Roxb.
	L. cylindrica (L.) Roem.
	L. echinata Roxb.
	L. graveolens Roxb.
	L. hermaphrodita
	L. tuberosa Roxb.
	L. umbellate (Klein) Roem
New world species of *Luffa*	*L. astorii* Suens
	L. operculata (L.) Cogn.
	L. guinguefida (Hook and Arn.)Seem.

The *L. echinata* can be identified with reniform, suborbicular, and deeply 5-lobed having round apex

Table 3.89: Taxon and Preferred Taxon Name of *Luffa* Species

Name of Taxon	*Name of Authority*	*Family*	*Preferred Taxon Name*
Luffa			*Luffa*
Luffa acutangula	(L.) Roxb.	Cucurbitaceae	*Luffa acutangula*
Luffa aegyptiaca	Mill.	Cucurbitaceae	*Luffa aegyptiaca*
Luffa astorii	Svenson	Cucurbitaceae	*Luffa astorii*
Luffa cylindrica	M. Roem.	Cucurbitaceae	*Luffa aegyptiaca*
Luffa graveolens	Roxb.	Cucurbitaceae	*Luffa graveolens*
Luffa operculata	(L.) Cogn.	Cucurbitaceae	*Luffa operculata*
Luffa pentandra	Roxb.	Cucurbitaceae	*Luffa aegyptiaca*
Luffa quinquefida	(Hook. and Arn.) Seem.	Cucurbitaceae	*Luffa quinquefida*
Luffa sp.		Cucurbitaceae	*Luffa* sp.
Luffa sphaerica	Sond.	Cucurbitaceae	*Lagenaria sphaerica*
Luffa tuberosa	Roxb.	Cucurbitaceae	*Momordica cymbalaria*

Table 3.90: Identification Keys for *Luffa* Species

Traits	*L.cylindrica*	*L. acutangula*	*L.umbellata*	*L. echinata*	*L. tuberosa*	*L. graveolens*	*L. hermaphrodita*
Nature	Monoecious, annual herbs	Monoecious, annual herbs	Monoecious, annual herbs	Dioecious	Dioecious, perennial herbs	Monoecious, annual herbs	Annual herbs, bisexual (hermaphrodite flowers)
Leaves	Palmately 5-lobed, angular with acute apex, as broad 15.0-25.0 cm long	Palmately 5-lobed, angular with acute apex	Palmately 5-lobed, angular with acute apex, as broad 3.0-5.0 cm long	Reniform, suborbicular, and deeply 5-lobed having round apex, as broad 4.0-10.0 x 5.0-11.0 cm long	Palmately 5-lobed, angular with acute apex, as broad 2.0 x 4.5 x 12.5-5.0 cm long	Almost round, as broad 4.0-12.0 cm long	Palmately 5-lobed, angular with acute apex, as broad 6.0-12.0 cm long

Traits	L.cylindrica	L. acutangula	L.umbellata	L. echinata	L. tuberosa	L. graveolens	L. hermaphrodita
Petiole	10.0-12.0 cm long	8.0-12.0 cm long	6.0-12.0 cm long	4.0-12.0 cm long	2.0-5.0 cm long	4.0-7.0 cm long	8.0-12.0 cm long
Tendril	Generally 3-5 fids	Generally 3-5 fids	Generally 3-5 fids	Generally 3-5 fids	Simple and filiform	Generally 3-5 fids	Generally 3-5 fids
Probacts	Fleshy with glistering gland	Fleshy with glistering gland	No probracts	Fleshy with glistering gland	No probracts	Fleshy with glistering gland	Fleshy with glistering gland
Flowers	Large	Medium	Medium	Small	Medium	Small	Medium
Flower colour	Yellow	Yellow	White	white	Yellow	Yellow	Yellow
Fruit size	15.0-50.0 x 6.0-10.0 cm,	15.0-30.0 x 6.0-10.0 cm,	-	4.0-7.0 mm,	2.0-4.0 x 1.5 cm	4.0-6.0 x 2.0-3.0 cm,	5.0-8.0 x 3.5-5.0 cm,
Fruit skin	Not angular fruits	angular fruits	Short, ecostate, densely long spinous, spines densely wooly	Ovoid, base dilated, densely covered with bristles	Angular, pyriform with 8 sharp ridges, attenuated ast apex and base	Ovoid, shortly villose, sparsely shortly tuberculate or spinous	Angular, oblong-ellipsoidal, obscurely 10 ribbedor smooth
Seed size	12.0 x 8.0-9.0 mm	10.0-14.0 x 5.0-8.0 mm	-	6.0 x 4.5 mm	-	6.0-7.0 x 4.0-5.0 mm	9.0-11.0 x 5.0-6.0 mm

of leaves. This species has 4.0-12.0 cm long petiole. The species *L. graveolens* has almost round leaves and has 4.0-7.0 cm long petiole. The tendril is simple and filiform in case of *L. tuberosa*. This species has 2.0-5.0 cm long petiole. The *L. umbellata* has 6.0-12.0 cm long petiole. Flowers are small (*L. echinata* and *L. graveolens*). In *L. echinata* probracts are fleshy with glistering glands. Probracts have not been reported in *L. tuberosa* and *L. umbellata*. Fleshy probacts with secreting glands attract ants and insects and probably help in pollination.

The cultivated species *i.e.* ridge gourd (*Luffa acutangula*) and smooth gourd (*Luffa cylindrica*) is closely related but there are differences between them are as under:

Table 3.91: Difference between Ridge Rourd (*Luffa acutangula*) and Smooth Gourd (*Luffa cylindrica*)

Ridge Gourd (Luffa acutangula)	Smooth Gourd (Luffa cylindrica)
Six sex forms (androecious, gynoecious, gynomonoecious, andromonoecious, monoecious and hermaphrodite) are seen	Generally monoecious sex forms are seen
Leaves are shallowly lobbed and larger	Leaves are deeply lobbed and smaller
No white patches on leaf	White patches on leaf
Male flowers smaller	Male flowers larger
Disc shaped and not tomentose ovary	Cylindrical and tomentose ovary
Angled and 10 ribbed fruits	Smooth fruits
Seeds colour black and pitted	Seeds are white flat, black and non-pitted
Anthesis starts from afternoon	Anthesis starts from early morning

Pollination fairly rigid within the species, pollination is enforced by the pollinating agents with the result that no natural hybrids are observed where different species are growing in the same locality. In nature, therefore, the four species of *Luffa* appear to be sexually well isolated. However, when ridge gourd (*Luffa acutangula*) and smooth gourd (*Luffa cylindrica*) were crossed artificially, about 30-35 per cent of pollen fertility is observed in the F_1 hybrids but in nature these species do not cross even when grow side by side (Dutt and Roy, 1971).

Gene and Genetics of Characters

The differentiation of sex in the dioecious species *L. echinata* is genic since all chromosome pairs are homomorphic. Sex expression in *Luffa* is controlled by two independent suppressor genes, A and G with multiple alleles. Gene 'A' suppressing the male where as 'G' suppressing the female. 'A' factors determining the inheritance of both the sexes where as 'B' factors determining the inheritance of female sex only. The genotype of various sex forms have been designated as:

Table 3.92: Genetics of Sex Forms in *Luffa* Species

Sex Forms	Genetic Makeup
Monoecios	AaGg, Aa,Gg, Aagg AaGg
Andromonoecious	aaGg, aaGg, aagg, aaGg
Gynomonoecious	Aagg, Aagg
Hermaphrodite	aagg, aagg, aagg
Gynoecious	Aagg
Androecious	aaGG, aaGG, aaGg

Taxonomic and Phylogenetic Status of Genus *Luffa*

The taxonomic position of *Luffa* aroused much interest after Jeffrey (1961) created a new subtribe Luffinae under Cucurbitaceae to include the genus. Jeffrey (1962) suggests that the dry operculated fruits of *Luffa* are approached elsewhere in the family only on the tribe Cyclantherae and so the subtribe Luffinae might be regarded as a link between the cucurbitaceae and cyclantherae. Dunnill and Fowden (1965) have supported this relationship on the basis of occurrence of certain amino acids in seeds of *Luffa* species and those of *Cyclanthera, Echinocystis* and *Marah*. Singh (1964) has justified the position of *Luffa* in Jeffrey's classification on the basis of comparative study of the endosperm haustorium in the family. Jeffrey (1980) while reviewing the Cucurbitaceae exposed the view that *Luffa* is exceptional in the tribe Cucurbitaceae in its fruit structure, operculated dehiscence and seed coat anatomy (Singh, 1971). In both these respects it resembles the tribe Cyclantherae and it has been suggested that it may form a link between this tribe and the Benincaseae. This possibility is supported by the seed amino acids (Dunnill and Fowden, 1965). *Luffa* and Cyclantherae have only citrulline and m-carboxyphenylalanine while the latter is absent in Benincaseae (except *Congiauxia*) which have ß-pyrazol-1-ylanine. However, basechromosome number (n=13) differs from those of the Benincaseae (n=11, 12, 13) and Cyclantherae (n=8).

Luffa tuberosa has been transferred to *Momordica* under *M. cymbalaria* Fenzl. By Clarke in Hooker's Flora of British India (2:618, 1879). Congiaux recognized *M. tuberosa* (Roxb.) Cogn., based on Roxburgh's *Luffa tuberosa*. The fruit is a specific character in *Luffa* and there is no reason to shift this species to *Momordica* which has either muriculate or echinate fruits but never angular (Chakravarty, 1982). Roxburgh's (Fl. Ind.) note in the original description is interesting. He stated that the fruit is exactly like *Luffa amara* Roxb. (*Luffa acutangula* var. *amara*) but without the stopple.Chakravarty (1982) based on his observations of fruits of *Luffa amara* supported Roxburgh's idea. The leaves of all the species of *Momordica* contain true cystoliths on the lower surface. Cystoliths are absent in *Luffa tuberosa*. Floiaceous bracts which are common features in *Momordica* are also absent in this species (Chakravarty, 1959). The seed coat anatomy does not support the inclusion of *M. cymbalaria* unde *Luffa*.

The delimination of species in the genus is debatable. Regarding two cultivated species, L. acutangula and *L. cylindrica*, Haines (1925) concludes that "the characters given to separate the two species are not at all constant". Chakravarty (1959) differtentiates the two species on the number of stamens and the nature of fruit, but variations in these characters are known to occur. Singh and Bhandari (1963) reported bisexual flowers in *L. hermaphrodita*. According to Chakravarty (1982) record of bisexual flowers in the species is interesting and deserved further studies. The morphological and cytological data of the interspecific hybrids has helped in accessing the relationship and evolutionary trends among various species of the genus *Luffa* (Dutt and Roy, 1969, 1971; Roy *et al.*, 1970, 1972). Among the perennial species, *L. graveolens* and *L. echinata* are wild (dioecious) and *L. acutangula* is cultivated. Of the two wild species, *L. graveolens* is monoecious (prime species which gave rise to cultivated and wild species of *Luffa* as reported by Dutt and Roy, 1971) and *L. echinata* is dioecious. *L. acutangula* is monoecious but it is also has hermaphrodite cultivars (*L. hermaphrodita*). The hermaphrodite variety when crossed to the monoecious type, forms regular bivalents in F_1 and is fully fertile (Dutt and Roy, 1971). Therefore it is highly debatable how far it is justified to elevate it to the rank of the species (Pandey *et al.*, 2009).

Table 3.93: Distribution of *Luffa* Species

Species	*Distribution*
L. hermaphrodita	It is being cultivated throughout Bihar during rainy season. However, cultivars vary in their fruits morphology (shape, size, number of fruits per inflorescence). Number of female flowers in aninflorescence varies from 4-15 and all the ovaries do not develop into fruits. In Bhojpur (Bihar) it has been found that cultivars of *L. hermaphrodita* bear seven fruits in each inflorescence and this appears to be constant features of the cultivars. On the basis of development of seven fruits per inflorescence, *L. hermaphrodita* is locally known as 'Satputia' meaning seven childrens.
L. echinata	It is a wild dioecious species is confined to north-western Himalaya and upper Gangetic plains. North Bihar especially in Purnea district. Haines (1925) reported this species as monoecious. In this sopecies in norther India profuse flowering can be observed during September-October.
L. graveolens	This species can be seen in Rajmahal hills of Bihar (Haines, 1925) but it is extremely rare. Some researchers reported that this species is widely distributed, in north-eastern plains, extending south to Tamil Nad; sporadic elsewhere in Eastern Huimalaya.
L. acutangula var. *amara* (Roxb.) Clarke	Mianly distributed in peninsular tract.
L. umbellate M. (Klein) Roem.	Confined largely to the eastern coast/ Coromandal belt.
Neoluffa sikimensis Chak.	Confined largely to Sikkim Himalaya, (allied to *L. acutangula* var. *amara*).

REFERENCES

1. Chakravarty, H.L. 1959. Monograph of Indian Cucurbitaceae. *Rec. Bot. Surv. India.*, **17:** 1-234.
2. Chakravarty, H.L. 1982. Cucurbitaceae. Fascicles of Flora of India. Fascicles II. Botanical Survey of India. Howrah.
3. Dunnill, P.M. and Fowden, L. 1965. The amino acids of seed of the Cucurbitaceae. *Phytochemistry*, **4:** 933-944.
4. Dunnil, P.M. and Fowden, L. 1965. The amino acid obseeds of the cucurbitaceal. Phytochem. 4:933-944.
5. Dutt, B. and Roy, R.P. 1971. Cytogenetic investigation in Cucurbitaceae. Interspecific hybridization in *Luffa*. *Genetica*, **42:** 139-156.

6. Heisen, C.B. and Schilling, E.E. 1990. The genus Luffa. A problem in phytogeography, Bilogy and Utilization of the Cucurbitaceae.

7. Jeffrey, C. 1961. Notes on Cucurbitaceae including a proposed new classification of the family. *Kew Bull.*, **15:** 337-371.

8. Jeffrey, C. 1962. Notes on Cucurbitaceae including a proposed new classification of the family Kew Bull; 15:337-371.

9. Jeffrey, C. 1980. A review of the Cucurbitaceae. *Bot. J. Linn. Soc.*, **81:** 233-247.

10. Mabberly, D.J. 1987. The Plant Bopok. Cambridge Univ., Press, Cambridge.

11. Prakash, K., Pandey, A; Radhamni J. and Bisht, I.S. 2013. Morphological variability in cultivated and wild species of *Luffa* (Cucurbitaeae) from India. Genet. Resour Crop Evol, 60:2319-2329.

12. Pandey, A.K., Varma, S.K. and Md. Ali, A. 2009. *Luffa* genus in India: Diversity and Conservation. In: Biodiversity in India. pp. 264-270.

13. Singh, D. 1964. A further contribution to the endosperm of the Cucurbitaceae. *Proc. Indian Acad. Sci.*, **60:** 399-413.

14. Singh, D. and Bhandari, M.M. 1963. The identity of an imperfectly known hermaphrodite *Luffa*, with a note on related species. *Baileya*, 11: 132-141.

15. Singh, A.K. 1990. Cytogenetics and evolution in the cucurbitaceae In: Bates, D.M., Rohinson R.W. Jeffrey, C. Biology and Utilization of the cucurbitaceae Ithaca and London: Cornell University p. 10-28, 485 p.

16. Singh, D. and Bhanari M.M 1963. The identity of an imperfectly known hermaphrodite *Luffa*, with a note on related species. Bailey 11:132-141.

Ridge Gourd (*Luffa acutanguala* L.)

Introduction

Ridge gourd is also called as Chinese okra or angled luffa, four-angled gourd, angular loofah, fluted loofah or angled loofah (*Luffa acutangula* L. Roxb.) is the prominent and important warm-season vegetable of family cucurbitaceae. The fruits of ridge gourd are ribbed therefore it also called as 'Nasdar tori' and due to dark colour it is called as 'Kali Tori'. It is known by several vernacular names like 'Ara Torai' (Hindi), 'Jhinga' (Bengali), 'Pekanaki' (Tamil), 'Birakaya' (Telugu), 'Pechinga' (Malayalam), 'Maup khia' (Vietnam), 'Oyong gambas' (Java), 'Ketola' (Philippines), 'Buyo-buyo' (Cambodia) and 'Ronoongchrung' (Laos). Immature fruits, which are dark green with tender ridges, are used in soups and curries or as a cooked vegetable. Spongy flesh of ridge gourd contains 0.5 per cent protein, 3.04 per cent carbohydrates, 37 mg carotene and 18.0 mg vitamin C per 100 g edible portion (Aykroyd, 1963). The fibrous material inside the fruit is used in herbal preparations. As medicinal properties, it is associated with the lung, stomach and liver. Ridge gourd is known to remove chest pain and congestion, muscular pain, and mastitis, among other conditions.

Origin, History and Distribution

There have a long history of ridge gourd cultivation in the hot and humid tropical countries of Asia and Africa. It is presumed that ridge gourd originated from India, Java and Sudan. Its major clue falls in India where wild forms still occur in the North-West of India. It is now cultivated in south and South East Asia and occassionlly in other tropical and subtropical areas. It sometimes grows wild as an escape from cultivation. It is widely cultivated throughout India, Sri Lanka, Malayasia, Japan, Brazil, Indonesia, China, and South Africa at elevation up to 1000 m above mean sea level. Ridge gourd is commonly cultivated in southern regions, like Tamil Nadu, Karnataka, Gujarat, Maharashtra, West Bengal, Asom, Punjab, Haryana, Rajasthan, Madhya Pradesh and Western Uttar Pradesh.

Species Availability, Evolution and Diversity

More diversity on ridge gourd (*L. acutangula* L., 2n=26) was found in western, central and southern India and is regarded as the wild form of cultivated species. *Luffa acutangula* var. *amara* occurs in peninsular India and is the wild relative of cultivated *Luffa cylindrica. Luffa echinata* (dioecious in nature) occurs in western Himalayas, Central India and upper gangetic plains and *Luffa graveolans* consider a wild progenitor of *L. hermaphrodita* (bisexual in nature) found in India (Bihar, U.P., M.P., Chattishgarh, Jharkhand, northwards to Sikkim and south of it to Tamil Nadu), Australia and Indonesia. *L. hermaphrodita* is endemic to India. *L. tuberosa* (perennial and dioecious in nature) is found in India and tropical Africa (Chakravarty, 1982). *L. umbellata* is confined to the eastern coast.

A much higher frequency of univalent at metaphase I in the cross *L. echinata* x *L. graveolens* and higher frequency of bivalents at metaphase I in the cross *L. acutangula*x *L. graveolens* indicated that *L. graveolens*is closely more related to *L. acutangula* than to *L. echinata*. Scientific report of the IARI (1963) contained the following information on inheritance of sex in ridged gourd.

P → monoecious x Hermaphrodite

F_1 monoecious

F_2 Monoecious : androecious : gynoecious : Monoecious

9 → 3 3 1

The above findings indicated the action of 2 pairs of genes in *L. acutangula* in the expression of sex.

Luffa acutangula var. *amara* occurs in peninsular India and is the wild related typ of the cultivated smooth gourd. *Luffa acutangula* can also be spotted in wild, in forests of Santhal Pargana and north Champaran.

Botanical Description

It is only genus of the subtribe Luffinae C, Jeffr., tribe Benincaseae Ser., sub-family Cucurbitoideae. The genus includes seven species: four paleotropic and three neotropic in distribution. Two species are cultivated *i.e.* *Luffa acutangula* (L.) Roxb. (2n=2x=26 ridge gourd, ribbed gourd, angled *Luffa*) and *Luffa cylindrica* M.J. Roem. (2n=2x=26 sponge gourd, towel gourd, smooth loofah, vegetable sponge dish cloth gourd, syn. *L. aegyptiaca* Mill.). Other related species are *Luffa operculata* (L.) Cogn. (grown in tropical America), *Luffa graveolens, Luffa umbellata, Luffa pentandra, Luffa gigante, Luffa scabra* and *Luffa narylandica*. Both sponge gourd and ridge gourd contain a gelatinous compound called 'luffein'.

Table 3.94: Botanical Description

Growth Habit	An annual vine (climber) crop. The plants are fast growing or trailing vines with stems and branched tendrils. Tendrils hairy and 3 fid or more.
Root	Tap root system
Stem	Stem actually 5 angled, upto 10.0 m in length, slightly hairy, and 3 branched tendrils.
Leaf	Palmate leaves are 5-7 angled ovate or sometimes shallowly lobed, 10.0-25.0 cm x 10.0-25.0 cm, acute apex scabrous, glabrous, dentate, surface rough, apex acute and pale green. No white patches on leaf. Petiole 8.0-12.0 cm long. Petiole without glands at the apex (ridge gourd), emitting strong odour when rubbed.
Tendril	Tendril is 3-5 fids.
Floral biology	Flowers are medium. The sex form in ridge gourd are monoecious (staminate and pistillate flowers separately in the same plant), androecious, gynoecious, gynomonoecious, and andromonoecious and hermaphrodite. Inflorescence is a clustered racemose flower mostly unisexual. Usually staminate and pistillate flowers are borne in different axils but some times pistillate flowers may be borne in same leaf axils (singly in short peduncles) as male flowers the same node. Male inflorescence in racemes of several flowers. Peduncle of male flowers 15.0-35.0 cm long, bracteate, female flowers solitary, borne axillarily; flowers 4-5 cm in diameter, 5-merous, fragrant and pale yellow. Male flowers smaller, disc shaped and not tomentose ovary. Flower has campanulate showy corolla, calyx 5, partite forming a perianth tube, calyx lobes alternating with corolla lobes, petals yellow, obovate, 2.0-2.5 cm long, filament free, two stamens (sometimes 3) having two locules and one unilocular. Ovary inferior, filiform with 10 longitudinal ribs on which are swolleds, style short, stigma trifid, and flowers are open in evening. Style is persistant. Hermaphrodite form, which bears only bisexual flowers, like 'Satputia' cultivar of *Luffa acutangula*' is rare. Probracts are fleshy with glistering glands. Fleshy probacts with secreting glands attract ants insects and probably help in pollination.
Anthesis and pollination	The anthesis (afternoon and evening, 5.0-8.0 p.m.) and remain open throughout the night. Anther dehiscence is between 17-20 hr. Pollen fertility is maximum on the day of anthesis and lasts till 2-3 days after anthesis in winter and one-and-a-half days in the rainy season.
	Stigma is receptive 6 hr before to 84 hr after anthesis (Singh, 1957). Stamens appearing as 3, 2 double 2-thecous, 1 single 1-thecous. Therefore pollination takes place in early morning/forenoon. In winter and summer or rainy season, pollens grain viable for 2-3 days and 1-1.5 days respectively. Bees and ants are the main pollen vectors.
Fruit and fruit set	Fruit is essentially a (inferior) berry, even though called 'pepo'. Fruit are club shaped, angled, clavate-oblong, acutely, 10 ribed, 15-50 cm x 5-10 cm, apex obtuse or slightly acute, crowned by enlarged calyx (sepals) and style. Fruits are fibrous inside. They are trilocular usually with a stopple, apex usually operculately dehiscent, many seeded. The fruits are no doubt angular in the beginning but as the fruits attain maturity, their angular nature slowly appears and fruit becomes shorter in length. It has been observed that fruits of some cultivars become bitter in taste.
Seeds	Seed ellipsoid, 41-1.3 cm x 0.7-0.9 cm pitted black without wing like margin.

Though cultivated varieties are monoecious in nature, different sex forms *viz.*, androecious, gynoecious, gynomonoecious, andromonoecious and hermaphrodite plants are also reported (Choudhary and Thakur, 1966). Zeven and Zhukovsky (1975) mention a trimonoecious species, *L. hermaphrodita* by Singh and Bhandari (1963) as the cultivar 'Satputia' commonly cultivated in Bihar and Bengal and bordering area of Nepal.

Horticultural Classification

In ridge gourd (*Luffa acutangula* L.) 4 groups are distinguished:

- cv. Group Angled Loofah (syn: var. *acutangula*) the large cultivated forms.
- var. *amara* (Roxb.)C.B. Clarke, a wild or feral form confined to India (Peninsular region) with small, extremely bitter fruits.
- var. *forskohlii* (Horms.) Heiser and Schilling, a wild form confined to Yemen but possibly developed from escapes of the cultivated form.
- cv. group *Satputia* (var. *hermaphrodita*) Indian cultivar, borne in cluster, hermaphrodite, common in north-eastern U.P.

Augmentation of Genetic Resources and Crop Improvement

Breeding Priorities

1. Bitter free varieties
2. Early and high yielding varieties/hybrids
3. Less fibrous
4. Fleshy and less seeded type

Table 3.95. Range, Mean, S.D., GCV, Heritability, Genetic Advance and Genetic Gain

Characters	Range	Mean	S.D.	GCV	Heritability	Genetic Advance	Genetic Gain
Days to emergence of first fertile male flower	28.00-45.25	33.26	5.64	11.88	0.98	2.02	6.07
Days to emergence of first female flower	28.00-56.00	38.87	8.04	14.56	0.99	2.04	5.25
Node to first fertile male flower,	3.25-14.75	6.58	3.08	31.97	0.93	1.92	29.22
Node to first female flower	9.63-38.63	18.93	11.04	41.14	0.99	2.05	10.82
Pollen fertility (per cent)	61.00-99.75	86.43	10.15	8.28	0.99	2.05	2.37
Number of fruits per plant	17.00-50.63	33.99	12.65	24.13	1.00	2.05	5.55
Average fruit weight (g)	80.00-425.50	233.86	118.26	35.76	1.00	2.06	0.88
Average fruit length (cm)	9.75-73.50	31.45	20.28	45.58	1.00	2.06	6.54
Average fruit girth	14.13-27.00	19.28	3.59	12.84	0.95	1.96	10.16
days to first harvest	47.00-81.75	55.89	9.83	12.39	0.99	2.05	3.66
Number of harvests	3.50-6.75	5.21	0.93	6.44	0.38	0.79	15.13
Yield per plant (kg)	3.05-7.73	5.50	1.29	13.03	0.66	1.37	24.82

Source: Hegade, 2009.

Introduction and Selection

Ridge gourd has a wide range of variability for maturity, yield and fruit characters like shape and size. Significant variability among the landraces were reported with respect to days to emergence of first fertile male flower, days to emergence of first female flower, node to first fertile male flower, node to first female flower, pollen fertility, number of fruits per plant, fruit weight, fruit length, fruit girth, days to first harvest, number of harvest, yield per plant, number of seeds per fruit, and seed yield per plant (Deshpande *et al.*, 1980; Anitha, 1998; Ahmed *et al.*, 2006). Hegade (2009) studied range, mean, standard deviation, genotypic coefficient of variation, heritability, genetic advance and genetic gain values of ridge gourd (Table 3.95). The uniform character of all the cultivars is ridged fruits.

In ridge gourd variety 'Pusa Nasdar' was selected and recommended for commercial cultivation. In an

Table 3.96: Development Varieties and Hybrids of Ridge Gourd

Variety	Organization	Pedigree	Brief Description
CO-1	TNAU, Coimbatore	Selection	It is vigorous in vegetative growth, bears 10-12 fruits per vine (3-4 kg). Fruits are 60 to 70 cm long with 3.0 cm diameter. Duration of crop is 125 days. Released by Tamil Nadu Agriculture University, Coimbatore.
CO-2	TNAU, Coimbatore	Selection	It bears 8-10 fruits/vine. Fruits are green, very large (up to 1.0m). Duration of crop is 120 days with an average yield of 25.0 t/ha.
PKM-1	TNAU, Coimbatore	Selection	This variety is developed through induced mutation. Fruits are dark green, 60-70 cm long with a hallow grooves. Crop duration is 160 days with an average yield of 25.0-28.0 t/ha.
Pusa Nasdar	ICAR-IARI, New Delhi	Selection	More suited for rainy season cultivation, fruits club shaped and light green. Average yield is 16.5 t/ha. Flowering in 50 days and bears 25-20 fruits per vine.
Arka Sumeet	ICAR- IIHR, Bangalore	Pure line selection	Vines are medium vigorous, grow upto 4-5 m length. Fruits are 55 cm long with 2.5 cm girth with prominent ridges and delicate aroma. It takes 52 days for first picking. Average fruit weight 380 g, crop duration 120 days. Fruits are lush green, cylindrical with slow seed maturity.
Arka Sujat	ICAR-IIHR, Bangalore	Pure line selection	Fruits are lush green and tender, medium long (35-45 cm) with an average fruit weight of 350 g. Its yield potential is 63.0 t/ha. Good transport and keeping quality. Duration of crop is 100 days. It is moderately resistant to downy mildew.
Punjab Sadabahar	ICAR- PAU, Ludhiana	Pure line selection	Fruits long, 3-5 cm thick in diameter, straight, green and tender. Best sowing time of this variety is May to July for rainy season crop. Average yield per hectare is 10.0 t/ha.
Swarna Manjari	ICAR-HARP, Ranchi (Jharkhand)	Pure line selection	Fruits are elongated, medium sized, highly ridged, green with soft pulp; tolerant to powdery mildew. First picking starts 65-70 days after sowing. It gives 18-20 t/ha yields in 140-150 days of crop duration.
Swarna Uphar	ICAR-HARP, Ranchi (Jharkhand	Pure line selection	Fruits are elongated, medium sized (200 g), weak ridged with soft flesh and less fiber. First picking starts 65-70 days after sowing. It gives 20-30 t/ha yields in 140-150 days of crop duration.

Variety	Organization	Pedigree	Brief Description
Pant Torai-1	GBPUA&T. Pantnagar	Pure line selection	It bears 25.0 cm long and club shaped fruits. First picking starts about 65 days after sowing. It gives an average yield of 10 t/ha; specifically suitable for rainy season cultivation.
Haritham	College of Horticulture, Vellanikkara (Kerala)	Selection	Fruits are cylindrical, bright green with typical ridges and tapering towards the base. Average length, girth and weight of fruit are 46.5 cm, 20.0 cm and 650 g, respectively. Average yield is about 13 t/ha.
Konkan Harith	KKV, Maharashtra	Selection	Fruits are medium long and attractive. Average yield is about 12 t/ha.
Deepthi	KAU, Kerala	Selection	Fruits are long and light green. Average yield is about 11t/ha.

Satputia cultivar Kashi Khushi (VRS-1) has been developed from ICAR-IIVR, Varanasi

evaluation of ridge gourd under the dry condition of Karnataka region, Pusa Nasdar attended maximum vine length followed by Arka Sumeet. The number of branches/vine, node of first female flower appearance and days to first male flower appearance did not vary between cultivars. Arka Sumeet expressed the highest fruit length and fruit girth while Local had the lowest. Fruit yield/plant was greatest for Arka Sumeet followed by Local and Pusa Nasdar (1753, 1207 and 789 g/plant) respectively (Anjanappa *et al.*, 1999). Sahni *et al.* (1987) recorded high expected genetic advance for fruit number per stem, female flower per node, fruit diameter and fruit yield per plant in ridge gourd. Popular cultivars like Pusa Nasdar, CO-1 and CO-2 were developed by use of Indigenous materials through selection with progeny testing.

Ram *et al.* (2006) evaluated 24 accessions of *Luffa hermaphrodita* on 12 economic characters and observed considerable genetic diversity among the accessions for aggregate effect of all the characters. Genotype SP-59 recorded highest edible fruit yield/ha (107.9 q) while DR/Y-1 expressed earliness with respect to days to first hermaphrodite flower anthesis (40.6) and maximum mean value for number of fruits per plant (185.3). High hetritability for number of seeds per fruit, seed weight and edible fruit yield per hectare suggests that characters phenotypic value could provide fair and reliable measure of good genotype. High hetritability and high genetic advance for number of seeds per fruit and edible fruit yield/ha indicated the presence of high amount of flexible variation among genotypes for these traits, which provides ample scope for varietal improvement through hybridization and selection in this species. Khatoon *et al.* (2016) evaluated 40 ridge gourd genotypes collected from different part of India. Multivariate analysis followed Mahalanobis D^2 analysis revealed that the genotypes of ridge gourd have considerable genetic divergence between Cluster IV and Cluster V. The Cluster IV with genotype CHFRG-8, CHFRG-12, CHFRG-6, CHFRG-10 and CHFRG-22 were the more divergent for improving vine length, number of node per vine, crop duration, pedicel length, fruit length, fruit diameter, number of fruit per plant and fruit yield. Cluster VI with 9 genotypes are found to be promising lines for improving number of node to first pistillate flower appearance, days to first pistillate flower anthesis, days to first staminate flower anthesis, and days to first fruit harvest. The genotype CHFRG-13 in Cluster V was identified as genetically divergent for average fruit weight and hence, genotypes in these clusters can be utilized in crop improvement as donor parents for improving the desirable characters. Shivanand *et al.* (2015) evaluated eighteen genotypes of ridge gourd for physiological, yield, quality and fruit characters under the field condition revealed that PCV was higher than the GCV for most of traits. High heritability with moderate to high GCV and genetic gain was recorded for leaf area, leaf area index, specific leaf weight at 45 days after sowing (DAS), absolute growth rate, crop growth rate, relative growth rate of leaf, tendril and vine at 45-90 DAS, number of fruits per vine and yield per vine indicated that these characters could be improved by simple selection. Correlation coefficient analysis revealed that leaf area during 45 and 90 DAS (0.5350 and 0.4689, respectively), leaf area index during 45 DAS (0.5714), AGR of leaf at the period of 45 to 90 DAS (0.7765), CGR of leaf at the period of 45 to 90 DAS (0.6095) had positive significant correlation with the total fruit yield per vine but NAR of vine at the period of 45 to 90 DAS (-0.4796) had negative significant correlation. Narasannavar *et. al.* (2004) also established correlation with yield.

Hybridization and Selection

Fifty-one F_1 crosses of ridge gourd in a Line × Tester set involving 20 parents were evaluated to study heterosis for growth, earliness, yield and quality parameters. Maximum standard heterosis for vine length (9.20 per cent) was observed in KRG-7 x ASM, for number of leaves in KRG-11 x PN (172.12 per cent), for number of branches in KRG-11 x PN (141.38 per cent), for days to first male flower appearance in KRG-3 x PN (-19.48 per cent), for node to first female flower appearance in KRG-9 x ASJ (-56.92 per cent), for node to first female flower appearance in KRG-2 x ASM (-19.35 per cent), for sex ratio in KRG-17 x PN (-40.34 per cent), for per cent fruit set in KRG-16 x PN (36.45 per cent), for number of fruits per vine in KRG-10 x PN (13.95 per cent), for average fruit weight in KRG-6 x ASJ (36.34 per cent), for fruit length in KRG-1 x ASJ (36.91 per cent), for fruit diameter in KRG-12 x PN (55.09 per cent), for fruit yield per plot in KRG-3 x ASM (26.53 per cent), for fruit yield per hectare in KRG-3 x ASM (26.46 per cent), for rind thickness in KRG-12 x PN (8.07 per cent), for flesh thickness in KRG-12 x PN (70.27 per cent), for number of seeds per fruits in KRG-16 x ASM (238.80 per cent) and for seed yield per fruit in KRG-4xASM (201.11 per cent). The four best performing F1 hybrids *viz.*, the cross KRG-3×ASJ (23.61 per cent) followed by KRG-3×PN (13.80 per cent), KRG-10×PN (12.45 per cent) and KRG-11×PN (6.00 per cent) exhibited the highest standard heterosis for total yield per vine in order of merit. The higher yield recorded by these hybrids could be due to increased per cent fruit set and number of fruits per plant (Narasannavar *et al.*, 2004).

The economic significance of hybrid generations obtained from intercrosses of *Luffa acutangula* L. were shown by Richharia (1952). Thakur and Choudhury (1965) reported heterosis for quantitative characters like yield per plant (41.29 per cent) and girth of fruit (17.69 per cent) in the hybrids obtained from long fruited and short fruited ridge gourds. Thakur and Singh (1968) also observed heterotic effects on yield and maturity in hybrids obtained from crosses involving gynoecious plant with four monoecious plants. Kohle (1972) obtained 57 crosses from 16 parents and reported that all the combinations were superior to their mid parents and better parents. Out of these, four hybrids *viz.*, B-24 x M-33, B-24 x Co-18, B-20 x B-24 and B-24 x Co-18 manifested heterosis to the magnitude of 11 to 50 per cent for yield. Jamsandekar (1982) using 16 lines and two testers analysis in ridge gourd observed appreciable heterosis for various characters studied except number of fruits per vine. He reported desired positive heterobeltiosis for weight of fruits per plant, fruit length, fruit diameter, vine length, average fruit weight, total soluble solids and desired negative hetrobeltiosis for first female flower. Likewise, Kadam (1989) also reported significant and positive heterosis over top parent for yield per vine, vine length, fruiting nodes per vine, number of fruits per vine and significant and negative heterosis over top parent for female flower node. He also pointed out that highly heterotic crosses did not necessarily record high *per se* performance. He did not observe heterobeltiosis for number of branches per vine, fruit length and fruit diameter. Forty five F_1 hybrids in a diallel set involving 10 inbred lines were studied by Abusaleha and Dutta (1994) to assess the extent of hybrid vigour in ridge gourd. They noted that the range of heterosis over better heterotic F_1 hybrids exhibited the highest percentage of heterosis when compared to their respective better parental lines included IIHR-6 x IIHR-54 (-20.87 per cent) for days to first female flowering, IIHR-6 x IIHR-54 (-15.77 per cent) for days to first harvest, IIHR-6 x Coimbtore Long (51.77 per cent) for vine length, Pusa Nasdhar x IIHR-54 (30.16 per cent) for fruit girth as well as flesh thickness (29.76 per cent), IIHR-7 x IIHR-54 (93.90 per cent) for fruits per vine and IIHR-6 x Coimbtore Long (63.69 per cent) for average fruit weight and total yield (85.80 per cent). Kadam *et al.* (1995) in heterosis studies reported significant heterobeltiosis to the extent of 73 per cent for yield per plant (Parala Shirala x RG-109), 63 per cent for number of fruits per vine (Parala Shirala x RG-109), 52 per cent for fruit weight (Parala Shirala x Goleswar), 49 per cent for number of fruiting nodes (Shahapur x Pusa Nasdhar), -37 per cent for female node number (Shahapur x Goleswar) and -48 per cent for powdery mildew infestation (Shravani x Parala Shirala).Twenty eight F_1 crosses of ridge gourd in diallel set involving eight parents (excluding reciprocals) were evaluated by Rao *et al.* (2000) during *kharif* and *summer* seasons. They noted appreciable and significant heterosis over the better parental lines consistently over the season for all the traits like node to first male flower, number of fruits per vine, branches per vine, fruits per branch, length of the vine and yield per vine except node to first female flower and node to first fruit. The best three performing hybrids *viz.*, LA-46 x LA-99, LA-12 x LA-76 and LA-32 x LA-37 also showed 51.8 and 125.9 per cent, 81.8 and 121.1 per cent, and 99.1 and 200.1 per cent higher yield over better parent in *kharif* and *summer* season, respectively. The high yield recorded in these hybrids had been directly attributed to increase number of fruits per vine and fruits per branch. Appreciable amount of heterosis was recorded by Mole *et al.* (2001) over better parent and standard check (LA-86) in ridge gourd for various characters *viz.*, vine length, number of primary branches, days to first female flower, days to vegetable maturity, days to first harvest, number of fruits per plant, yield per plant, average fruit weight, fruit length, fruit girth, duration of crop and seeds per plant. For yield per plant up to 50.81 per cent heterosis was observed over better parent and 99.71 per cent over standard parent in the cross LA-81 x LA-44. Shala and Kale (2003) studied on heterosis in ridge gourd. They reported that the crosses (Atigre x Pusa Nasdar, Tendoli x Pusa Nasdar, Inamangari x Pusa Nasdar) were the best as they exhibited significant heterobeltiosis for fruit weight (yield) and number of fruits per vine. The crosses Kawalapur x Tendoli and CO-1 x Pusa Nasdar for fruit diameter, Atigre x Punjab Sadabahar, Punjab Sadabahar x CO-1, Atigre x Kawalapur, Panvel x Tendoli and Kawalapur x CO-1 for average fruit weight and Atigre x Inamangari for vine length recorded significant heterobeltiosis. The crosses Panvel x Pusa Nasdar, Atigre x CO-1, Jaipur x CO-1 and Panvel x Punjab Sadabahar which exhibited high heterobeltiosis for first female flower at lower node also expressed high heterobeltiosis for days to first harvest.

Interspecific Hybridization

Luffa acutangula is not crossable with other species in nature therefore interspecific hybrids to be not of much practical value. The crosses involving 'T28' (ridge gourd, 2n=26) and 'T-21' (sponge gourd, 2n=26) as male survived, forming 13 bivalents at meiosis.

The morphological features of F_1 plants (2n=26) were intermediate between parents. Richaria (1948a) analysed a cross between a monoecious selection of *Luffa acutangula* and *Luffa hermaphrodita* (hermaphrodite selection). The F_1 was monoecious and the F_2 segregated into 3 forms-female, monoecious and hermaphrodite. For creation of variability interspecific hybridizations were made between *L. acutangula* x *L. cylindrica*, *L. acutangula* x *L. graveolens*, and *L. echinata* x *L. graveolens*. *Luffa hermaphrodita* easily crosses with *L. acutangula*. Although the fruits of bisexual flowers were generally smaller than those of unisexual flowers, the overall yield per plant may actually be higher because more fruits were produced. Attempts were also made to effect a 3-way cross between (*L. cylindrica* x *L. acutangula* x *Trichosanthes cucumerina*). Chakravarty (1990) reported that crosses between monoecious forms of *L. acutangula* and *Luffa hermaphrodita* (Satputia) produced five times the yield of either cultivar.

Polyploid Breeding

Colchicine induced tetraploids of Taiwan variety 'Tokado' tended to have thicker and shorter fruits, stouter fibres and lower fertility than the corresponding diploids.

Male Sterility

Natural mutant male sterile lines are reported from India by Pradeepkumar *et al.*, 2007 which was isolated from the variety 'Haritham'. It can be utilized in commercial hybrid seed production.

Genetic Resource Coservation

ICAR-National Bureau of Plant Genetic Resoureces, New Delhi, ICAR-Indian Institute of Vegetable Research, Varanasi (U.P.), ICAR-Indian Institute of Horticultural Research, Bengaluru (Karnataka) and several state agricultural university are maintain the germplasm of ridge gourd.

REFERENCES

1. Abusaleha and Dutta, O.P. 1994. Manifestation of heterosis in ridge gourd. *Indian J. Hort.*, **51**(4): 389-392.
2. Anjanappa, M., Reddy, S.S., Murali, K., Krishnappa, K.S. and Pitchaimuthu, M. 1999. Performance of certain ridge gourd varieties under southern dry region of Karnataka. *Karnataka J. Agril Sci.*, 12: 246-247.
3. Anitha, C.A. 1998. Variability in ridge gourd (*Luffa acutangula* (L.) Roxb.). M.Sc. (Hort.) thesis, Kerala Agricultural University, Vellanikkara Thrissur (Kerala).
4. Aykroyd, W.R. 1963. Package of practices for vegetable crops. DEE, SKUAST, Jammu, p. 41-42.
5. Chakravarty, H.L. 1982. Monograph of Indian Cucurbitaceae. *Rec. Bot. Surv. India*, **17:** 1-234.
6. Chakravarty, H.H. 1990. Cucurbits in India and their role in development of vegetable crops 325-348 (in) D.M. Bates, R.W. Robinson and C. Jeffrey (Eds.), Biology and Utilization of the Cucurbitaceae. Cornel University Press, Ithaca, New York.
7. Choudhary, B. and Thakur, M.R. 1966. Inheritance of sex forms in *Luffa*. *Indian J. Genet. Plant Breed.*, **25:** 188-197.
8. Deshpande, A.A., Bankapur, V.M. and Venkatasubbaiah, K. 1980. Floral biology of ridge gourd (*Luffa acutangula* Roxb.). *Mysore J. Agric. Sci.*, **14:** 5-7.
9. Jamsandekar, S.M. 1982. Genetic studies in ridge gourd (*Luffa acutangula* (Roxb.) L.). M.Sc. (Agri.) Thesis, Mahtma Phule Krishi Vidyapeeth, Rahuri, India.
10. Kadam, P.Y., Desai, U.T. and Kale, P.N. 1995. Heterosis studies in ridge gourd. *J. Maharashtra Agric. Univ.*, **20**(1): 119-120.
11. Khatoon, U., Dubey, R.K., Singh, V. and Kushwaha, J.K. 2016. Genetic divergence in ridge gourd (*Luffa acutangula* (L.) Roxb.) accessions based on D^2 analysis. 7th Indian Horticulture Congress (Abs. 97), held at ICAR-IARI, New Delhi, 15-18 November, p. 75.
12. Kolhe, A.K. 1972. Exploitation of hybrid vigour in cucurbits. *Indian J. Hort.*, **29**(1): 78-80.
13. Mole, T.J., Devi, S.N., Rajan, S. and Sadhankumar, P.G. 2001. Heterosis and combining ability in ridge gourd (*Luffa acutangula* (Roxb.)L.). *Veg. Sci.*, **28** (2): 165-167.
14. Narasannavar, A.R., Gasti, V.D., Shantappa, T., Mulge, R., Allolli, T.B. and Thammaiah, N. 2014. Heterosis studies in ridge gourd [*Luffa acutangula* (L.) Roxb.]. *Karnataka J. Agric. Sci.*, **27 (1):** 47-51.
15. Pradeepkumar, T., Sujatha, R., Krishnaprasad, B.T. and Johnkutty, I. 2007. New source of male sterility in ridge gourd (*Luffa acutangula* (L.) Roxb.) and its maintenance through in vitro culture. *Cucurbit Genet. Coop. Rpt.*, **30:** 60-63.
16. Ram, D., Rai, M., Verma, A. and Singh, Y. 2016. Genetic variability and association analysis in *Luffa* species. *Indian J. Hort.*, **63(3):** 294-297.
17. Rao, B.N., Venkata Rao, P. and Muralimohan Reddy, B. 2000. Heterosis in ridge gourd (*Luffa acutangula* (Roxb.) L.). *Haryana J. Hort. Sci.*, **29**(1 and 2): 96-98.
18. Richharia, R.H. 1952. Inter varietal hybridization in *Luffa acutangula* (Roxb.) L. and its economics significance. *Proc. Bihar, Acd. Agric. Sci.*, **1:** 1-9.
19. Shaha, S.R. and Kale, P.N. 2003. Diallel analysis for combining ability in ridge gourd. *J. Maharashtra Agric. Univ.*, **28**(3): 252-254.

20. Shaha, S.R. and Kale, P.N. 2003. Heterosis studies in ridge gourd. *J. Maharashtra Agric. Univ.*, **28**(3): 334-336.

21. Sahni, G.P., Singh, R.K. and Saho, B.C. 1987. Genotypic and phenotypic variability in ridge gourd. *Indian J. Agril. Sci.*, **57(9):** 666-668.

22. Shivanand, B.K., Mukesh, L., Chavan, Jagadeesha R.C., Vijaykumar R., Jayappa, J. and Sumangala K. 2015. Genetic variation and correlations among the physiological growth attributing characters in ridge gourd (*Luffa acutangula* Roxb.) with reference to yield. *International Journal of Advanced Research*, **3(4):** 961-971.

23. Singh, S.N. 1957. Studies in the biology of *Luffa*. *Horticultural Advances*, **1:** 86.

24. Thakur, M.R. and Choudhury, B. 1965. Inheritance of some quantitative characters in *Luffa acutangula* (Roxb.)L. *Indian J. Hort.*, **22**(2):188-189.

25. Zeven, A.C. and Zhukovsky, P.M. 1975. Dictionary of cultivated plants and their centres of diversity. Centre from Agricultural Publishing and Documention, Wageningen, The Netherlands.

Sponge Gourd (*Luffa aegyptiaca* Mil.)

Introduction

Sponge gourd or Smooth gourd (dish rag gourd) are (*Luffa cylindrica* (L.) M.J. Roem syn. *Luffa aegyptiaca* Mil., 2n=26), is an important warm-season fruit vegetable of family Cucurbitaceae and falls under genus *Luffa*. Sponge gourd fruit are smmoth and called as *'Ghiya Tori, 'Ghewda', 'Nenua', 'Kandoli' etc.* It is also known by several local names like- Kali Torai, Ram Torai, Nemia, Ghesali, Gilki, Nunibria, Patola, Dhundhul, Trada, Bhol, Rajkosataki and Kayee. It is grown in Malaysia, Ceylon, Japan and Brazil. Sponge gourd (*L. cylindrica* (L.) M.J. Roem syn. *L. aegyptiaca* Mill.) is grown principally for sponge production earlier as the name implies but now it is one of the most important vegetable. Immature fruits are eaten fresh, cooked, or in soups. The mature fruits are the source of the spongy reticulated material known as the domestic loofah. These loofahs are used for sponges and filters, and for stuffing pillows, saddles, and slippers. They can also be used for insulation and are attractive sources for packing material because of their biodegradability. There is an increasing intrest in domestic production (Davis, 1991) since the United States is the major market and imports millions of loofahs from Asia each year. Normally, mature fruits are left on vine to dry and the dry, thin outer skin is removed. The fruit is then soaked in running water for several days, after which the softer tissue is removed. After further soaking, then drying, the seeds are shaken out and the loofah is bleached either chemically or by the sun prior to marketing.

Origin, History and Distribution

It is presumed that sponge gourd originated probably from South Asia (North West India) or Africa because the wild races are still available in the above area. It is found wild in the North-West of India. It is well distributed throughout the tropical area. In a separate evolutionary path *L. cylindrica* evolved from it, the monoecious form of *L. acutangula*. Other related species are *L. operculata* (L.) Cogn. grown in tropical America, *L. graveolens, L. umbellate, L. achinata, L. pentandra, L. gigante, L. scabra* and *L. narylandica*. The Indian gene centre has rich diversity in genetic resources of sponge gourd; more especially in eastern peninsular tracts, indo-gangetic plains, north-eastern region and in tribal dominated belt of central India. Wild forms of *L. cylindrica* (syn. *L. aegyptica*) occur from Mayanmar to the Philippines and Southwards to North-eastern Australia and Tahiti. *L. cylindrica* is more widespread than *L. acutangula* and grows wild in many parts of the tropics. It has been cultivated for centuries in the Middle East and India, China, Japan and Malaysia (Porterfield, 1955).

Species Availability, Evolution and Diversity

The Indian gene center has rich diversity in genetic resources of *luffa* species. The genus comprised of 9 species in the world and out of this 7 species *i.e. Luffa acutangula, Luffa cylindrica, Luffa echinata, Luffa graveolens, Luffa hermaphrodita, Luffa tuberosa* and *Luffa umbellata* are native to India. There is ambiguity with regards *Luffa tuberosa* and *Luffa umbellata* because they are considered synonym to species of *Momordica* and *Cucurbita* respectively (Chadha and Lal, 1993). The sponge gourd (*Luffa cylindrica*) and ridge gourd (*Luffa acutangula*) has rich diversity throughout India. Rich diversity in vine and fruit morphological characteristics occurs in North-Eastern Region including Sikkim, West Bengal, Western, Central and Southern India. *Luffa hermaphrodita* considered to be originate from *Luffa graveolans*, is another potential species distributed in parts of north-central India. *Luffa acutangula* var. *amara* grows in peninsular India, is a wild relative of sponge gourd (*Luffa cylindrica*) and *Luffa echinata* in natural habitats is in Western Himalayas, central India and Gangetic plains.

Botanical Description

It is only gemnus of the subtribe *Luffinae* C, Jeffr., tribe Benincaseae Ser., sub-family Cucurbitoideae. The genus includes seven species: four paleotropic and three neotropic in distribution. Two species are cultivated *i.e. Luffa acutangula* (L.) Roxb. (2n=2x=26 ridge gourd, ribbed gourd, angled *Luffa*) and *Luffa cylindrica* M.J. Roem.

(2n=2x=26 sponge gourd, towel gourd, smooth loofah, vegetable sponge dish cloth gourd, syn. *L. aegyptiaca* Mill.). Other related species are *Luffa operculata* (L.) Cogn. (grown in tropical America), *L. graveolens, L. umbellate, L. pentandra, L. gigante, L. scabra* and *L. narylandica*. Both sponge gourd and ridge gourd contain a gelatinous compound called 'luffein'.

Horticultural Classification

Broadly two groups are distinguished:

- cv. Group smooth Loofah (var. *cylindrica* syn var. *aegyptiaca*) the long fruited, less bitter cultivated forms, very commonly found in India and adjoining countries with different cultivars for the production of the best vegetable or the best sponges.
- var. *leicarpa* (Naudin) Heiser and Schilling, the wild forms occurring from Myanmar to the Philippines to north-eastern Australia and Tahiti.

Table 3.97: Botanical Description

Growth Habit	An annual herbaceous vine (extensive climbers) and monoecious vegetable crop.Other sex form like androecious, gynoecious gynomonoecious, and andromonoecious and hermaphrodite are also reported.
Root	Tap root system
Stem	Stem 5 angled up to 15.0 m long; slightly hairy and tendrils 2-6 fid.
Leaf	Leaves are deeply lobbed, 5-7 lobed 6.0-26.0 cm x 8.0-27.0 cm, ovate to reniform, scabrous, cordate at base, dentate apex acute, drak green; petiole 10.0-12.0 cm long, hairy and some time white patches on leaf.Leaf emits odour when bruised.
Tendril	Tendril is 3-5 fids.
Floral biology	Flowers are monoecious. Male flowers are larger, cylindrical and which is born in 12-13 cm long racemes. Female flowers born in same leaf-axils as male flowers and having tomentose ovary. Flowers 5-10 cm in diameter pentamerous and deep yellow. Probracts are fleshy with glistering glands. Fleshy probacts with secreting glands attract ants insects and probably help in pollination. Petals 5, obovate, stamens 5, free; ovary inferior, cylindrical; stigmas 3; and flowers open early in the day.
Anthesis and pollination	Spong gourd flowers in the early morning (4.0-8.0 a.m.) and during this period maximum anther dehiscence take place. Whole day pollination continues. Pollen grains are highly fertile.Fertility is high on the day of anthesis and lasts till 2-3 days after anthesis in winter and one-and-a-half days in the rainy season. The stigma has a longer duration of receptivity (6 hr before to 84 hr after anthesis). The percentage of good pollen ranged from 18-40 per cent. Bees and ants are the main pollen vectors.
Fruit and fruit set	Fruits are sub-cylindrical pepo (berry) smooth or not prominently ribbed, fusiform, obtuse, 30-60 cm long, crowned by enlarged calyx and style. They are trilocular usually with a stopple, apex usually operculately dehiscent, many seeded. Fruits are strongly fibrous inside.
Seeds	Seed broadly ellipsoid, 1-1.5 cm long-smooth, black sometimes white, with a narrow wing-like margin, flat and not pitted. The seed contains 46 percent oil and 40 percent protein. Approximately 1000 seed weighs around 90.0 g.

There is still controversy about the correct botanical nomenclature of sponge gourd/smooth loofah. In most of the South Asian countries, its name as *L. aegyptiaca* is adopted but in India *L. cylindrica* (L.) M.J. Roomer is mostly used.

Augmentation of Genetic Resources and Crop Improvement

Breeding Priorities

1. Bitter free varieties
2. Early and high yielding varieties/hybrids
3. Less fibrous
4. Fleshy and less seeded type

Introduction and Selection

In Anand Agricultural University, Anamnd (Gujarat) interspecific hybridization was attemted in *Luffa* species, using sponge gourd, ridge gourd and satputia (*L. hermaphrodita*). The crosses were attempted between ridge gourd x sponge gourd and it's reciprocal as well as ridge gourd x satputia. The F_2 and back cross materials were also advanced for further study.

Table 3.98: Developed Varieties and Hybrids in Sponge Gourd

Variety	*Organization*	*Pedigree*	*Brief Description*
Pusa chikni	IARI New Delhi	Selection	Early maturity, flowers in 60 days, suitable for summer and rainy season cultivation. Fruits are smooth, dark green in colour and cylindrical in shape. Single plant bears 15-17 fruits.
Pusa Sneha	IARI New Delhi	Selection	Fruits are dark green, non-hairy, smooth and straight with hard skin and tender flesh. Fruits become ready for first picking in 50-55 days after sowing in spring- summer and 40-45 days in rainy season. Average yield is 12 t/ha.
Rajendra Nenua-1	R.A.U., Sabour (Bihar)	Selection	Fruits are long, greenish white, smooth and thick. It is suitable for summer and rainy season cultivation. Average yield is 25t/ha.

In sponge gourd, Panwar *et al.* (1977) and Singh and Srivastava (1977) observed significant and positive correlation coefficient at genotypic level between yield and number of fruits/plant, while yield showed significant but negative correlation with days to flower and age of the edible fruit. Panwar *et al.* (1977) observed high estimates of heritability and genetic advance for length of fruit and days to first flowering in sponge gourd. Popular cultivar like Pusa Chikni was developed by use of Indigenous material through selection with progeny testing.

Hybridization and Selection

Fifteen diverse sponge gourd inbreds *viz.*, VRSG-1, VRSG-5, VRSG-6, VRSG-8, VRSG-11, VRSG-12, VRSG-41, VRSG-42, VRSG-44, VRSG-45, VRSG-46, VRSG-49, VRSG-52 and VRSG-61 were crossed with VRSG-2, VRSG-3 and VRSG-66 in Line x Tester mating design and evaluated 45 F_1s. the results indicated that early yield per plant and total marketable yield per plant were most heterotic traits. The highest positive heterosis was reported for early yield per plant in VRSG-1 x VRSG-66 (73.31 per cent) followed by total marketable yield per plant in VRSG-4 x VRSG-66 (72.84 per cent). Only cross combination VRSG-8 x VRSG-66 exhibited significant and positive heterosis (15.53 per cent) for fruit length. The characters as node on which first female flower appeared and days to anthesis of first female flower for which negative heterosis is desirable was shown by the crosses VRSG-6 x VRSG-66 (-45.09 per cent) and VRSG-5 x VRSG-2 (-22.37 per cent), respectively (Singh *et al.*, 2009).

Newly released sponge gourd cultivars-Kashi Rakshita-VRSGH-1, Kashi Shreya-VRSS-194, Kashi Saumya-VRSGH-3 and Kashi Jyoti-VRSG-1-17 from ICAR-IIVR, Varansai (U.P.). Hybrid cultivar Pusa Shreshtha (DSGH-9) has been developed from ICAR-IARI-New Delhi.

Gynoecism

The occurrence of gynoecism in sponge gourd [*Luffa cylindrica* (Roem) L.] is very rare, although population with high proportion of pistillate flowers has been developed and utilized for hybrid development. The gynoecious and male sterile populations are always been desired by the breeders across the globe. A cluster bearing sponge gourd genotype VRSG-52-1 and from this genotype of sponge gourd four plants *viz.*, Gy25S, Gy27S, Gy28S and Gy29S with absolete expression of gynoecism obtained, which are being maintained through sibmating. The gynoecious sponge gourd popluations will help in the commercial production of hybrid seed of sponge gourd (Singh *et al.*, 2010).

Resistance Breeding

In sponge gours EC-339891 has been reported as resistant to downy mildew.

Genetic Resource Conservation

ICAR-National Bureau of Plant Genetic Resoureces, New Delhi, ICAR-Indian Institute of Vegetable Research, Varanasi (U.P.), ICAR-Indian Institute of horticultural Research, Bengaluru (Karnataka) and several state agricultural university are maintain the Germplasm of ridge gourd.

REFERENCES

1. Chadha, M.L. and Lal, T. 1993. Improvements of Cucurbits. In: *Advances in Horticulture*, Vol.5 Vegetable Crops; Chadha, K.L. and Kalloo, G., Malhotra Publishing House, New Delhi.
2. Davis, J.M. 1991. Development of a production system for *Luffa* sponge gourd. *HortSci.*, **26:** 708.
3. Heisen, C.B. and Schilling E.E. 1990. The genus *Luffa*. A problem in phytogeography, Biology and Utilization of Cucurbitaceae, 120-133.
4. Singh, P.K., Choudhary, B.R., Singh, R., Bhardwaj, D.R. and Rai, M. 2010. Developing sponge gourd [*Luffa cylindrica* (Roem) L.] populations with a very high proportion of pistillate flowers. *Acta Hort.*, 938, ISHS2012; p. 491-493.
5. Singh, P.K., Choudhary, B.R., Bhardwaj, D.R. and Singh, R. 2009. Heterotic response in sponge gourd [*Luffa cylindrica* (Roem) L.]. *Indian Journal of Arid Horticulture*, **4(2):** 118-119.

Genus *Lagenaria*

The archaeological evidences suggest that *Lagenaria* is not a monotypic genus and has an ancient pan tropical distribution.

Table 3.99: Taxon and Preferred TaNon Name

Name of Taxon	*Name of Authority*	*Family*	*Preferred Taxon Name*
Lagenaria			*Lagenaria*
Lagenaria			*Lagenaria*
Lagenaria abyssinica	(Hook. f.) C. Jeffrey	Cucurbitaceae	*Lagenaria abyssinica*
Lagenaria breviflora	(Benth.) Roberty	Cucurbitaceae	*Lagenaria breviflora*
Lagenaria lagenaria	(L.) Cockerell, nom. inval.	Cucurbitaceae	*Lagenaria siceraria*
Lagenaria leucantha	Rusby, nom. illeg.	Cucurbitaceae	*Lagenaria siceraria*
Lagenaria rufa	(Gilg) C. Jeffrey	Cucurbitaceae	*Lagenaria rufa*
Lagenaria siceraria	(Molina) Standl.	Cucurbitaceae	*Lagenaria siceraria*
Lagenaria sp.		Cucurbitaceae	*Lagenaria* sp.
Lagenaria sphaerica	(Sond.) Naudin	Cucurbitaceae	*Lagenaria sphaerica*
Lagenaria vulgaris	Ser.	Cucurbitaceae	Lagenaria siceraria

Bottle Gourd (*L. siceraria* L.)

Introduction

Bottle gourd (*Lagenaria siceraria* L., 2n=22) an important vegetable crop is grown for its immature fruits. In world literature the bottle gourd is also called birdhouse gourd, trumpet gourd, calabash gourd, and white-flowered gourd. The name "bottle gourd" is especially appropriate, because this plant species is one of the few from which useful and lasting containers can be made. In the older literature it is often referred to as *L. vulgaris* (common) or *L. leucantha* (white flowered gourd) but it is now generally agreed that the correct name for bottle gourd is *L. siceraria* which is derived from word '*sicraria*' meaning 'drinking vassels'. Linnaeus called the plant of bottle gourd as *Cucurbita lagenaria*. The bottle gourd belongs to the genus *Lagenaria* that is derived from lagena, meaning, 'bottle'. Bottle gourd (syn. white flowered gourd) is an important warm-season fruit vegetable. In India bottle gourd is known as Lowki (Hindi); Lau (Bengali and Oriya); Dudhi (Gujrati); Soreki (Kannada); Zeeth (Kashmiri); Charanga (Malayalum); Pandhara-bhopla (Marathi); Ghai (Punjabi); Suraikai (Tamil); and Anapakaya (Telegu). Young fruits are used as a cooked vegetable similar to zucchini. The flesh is white, firm, and has an excellent texture and a mild taste. Young shoots and leaves can be cooked, and seeds can be used in soupe. Flesh of immature fruits can also be used in making icing for cakes, and the hard skin is sometimes sliced into thin, dry strips for cooking. Some forms of *L. siceraria* are grown for non-food uses. Mature fruits, whose inside may be poisonous, contain an extremely hard and waterproof rind when dried. They can be used as multipurpose containers (bowls, boxes, water jugs, cups, planters, *etc.*), utensils (ladles, pipes), musical instruments (*e.g.* sitars), floats for fishnets and rafts, or for ornamental purpose such as masks or native artifacts. It is a rich source of potassium, vitamin C, protein, sulphur and phosphorous. It is ideal for people suffering from biliousness and indigestion (Thumburaj and Singh, 2003).

Origin, History and Distribution

The bottle gourd is probably originated in Africa (Singh, 1990), although archaeological evidence has placed it in Peru around 12000 BC, in Thailand about 8000 BC, and in Zambia around 2000 BC (Esquinas-Alcazar and Gulick, 1983). From there it was widely distributed in pre-Columbian times, perhaps by floating on the seas (salt water) due to hard, dry skin of the mature for the better part of a year without any loss in seed viability (Herklots, 1972; Tindal, 1983). The dried gourds with viable seeds have survived in seawater for at least 200 days. It traveled to India, where it has evolved into numerous local varieties, and from India to China, Indonesia, and as far as New Zealand. *Lagenaria* also traveled to the New World. Bottle gourd has been found in wild form in India, the Moluccas and Ethiopia (South Africa). The centre of origin has been located as the coastal areas of Malabar (north Kerala) and the humid forests of Dehradun (north India). It has been used by people for thousands of years in the tropics and subtropics. It seems that bottle gourd was originated from India because its wild races are still found in high humid area of Dehradoon (Uttrakhand) and Malabar costal area. Old Indian script reveals its cultivation 2000 B.C. Bottle gourd appears to have been domesticated independently in Asia, Africa and the New World (Heiser and Smith, 1973). The fossil records indicate its culture in India even before 2000 BC. Archaeological remains show that the bottle gourd was used in Egypt about 3500 to 3300 BC. Remains found in Mexico date from 7000 to 5500 BC and in Peru from about 10,000 BC probably no region has as much diversity and so many named varieties of bottle gourd as India. It has been used in the New World for at least 12,000-15,000 years back. Plants are domesticated in America. The bottle gourd is thus an ancient crop, widespread and well used, from warm parts of the temperate zone throughout the dry and wet tropics. It is the only crop known to have been cultivated in pre-Columbian times in both the Old and New World. Recently, seeds and other remains of the gourd have turned up in several archaeological digs in Florida. In India bottle gourd is mainly cultivated in Uttar Pradesh, Punjab, Gujrat, Asom, Meghalaya and Rajasthan. The major distribution of genus including 6 species is in Africa, Madagascar, Indonesia, Malaysia and the neotropics.

Species Availability, Evolution and Diversity

Under genus *Lagenaria*, 6 species are reported out of these *L. abyssinica, L. siceraria*, and *L. leucantha* are common. Among these species only one *i.e. L. siceraria* is generally cultivated as annual in all tropical parts of the world, especially in India and few African countries, other are wild perennial and dioecious in nature. The wild species occur only in East Africa and Madagascar. *Lagenaria siceraria* (Molina) Standl. (Syn. *Lagenaria vulgaris* Ser. *Lagenaria leucantha* Duch. Rusby; 2n=2x=22) has been independently domesticated in the old and new world. The flowers of *Lagenaria abyssinica* are open during the day and have a very pleasant aroma, which is entirely lacking in the other bottle gourd species. It bears male and female flowers on separate plants (dioecious nature). Genus *L. sphaerica* produced both male and female flowers on same plant. Two wild species *i.e. L. abyssinica*, and *L. bravifolia* are perennial in nature. There are suggestions that *Lagenaria* occurs in wild form in South America and in India. Pathak and Singh (1950) mentioned about wild *Lagenaria* but they did not give their criteria for determining between wild and domesticated forms, except the bitter and unpalatable taste in wild and sweet and good taste in domesticated form.Bottle gourd previously considered to be monotypic *Lagenaria siceraria* (Molina) Standl. Related species of *Lagenaria* are as under:

- *Lagenaria siceraria* ssp. *siceraria*
- *Lagenaria sphaerica* (Perennial and monoecious)
- *Lagenaria breviflora* (Perennial and produced only female flowers)
- *Lagenaria leucantha*
- *Lagenaria guineensis*
- *Lagenaria abyssinica* (Perennial and monoecious, Only species having very pleasant aroma)
- *Lagenaria rufa*

Occurrence of considerable diversity in the morphology of fruits in the archaeological sites suggested a steady influx of genetic materials. There are suggestions that *Lagenaria* species still occurs in wild form in South America and in India (de Candolle 1882). Pathak and Singh (1950) mentioned about wild *Lagenaria* but they did not give their criteria for determining between wild and domesticated forms, except the bitter and unpalatable taste in wild and sweet and good taste in domesticated form.

Gene and Genetics of Characters

In bottle gourd, fruit colour is monogenically inherited with green dominant over white. Fruit shape is also monogenecally inherited with long dominant over round. The time of maturity and seed size are polygenecally inherited.

Botanical Description

It belongs to the subtribe Benincasinae (Ser.) C. Jeffr., tribe Benincaseae (Ser.), subfamily Cucurbitoidae.

Table 3.100: Botanical Desscription

Growth Habit	It is an annual, tendril bearing and disagreeably running vine or climber. It grows fast and may begin to flower only 2 months after seeding.
Root	Tap root system
Stem	The branched vines and tendrils spread or climb 3.0-15.0 m in length. The thick stem is robust, obtusely quinquangular, deeply longitudinally furrowed (grooved), and light green. Tendrils (2 branched-one being short and the other long) are inserted beside the leaves, on robust stalks of 3.0-6.0 cm long, cleft, densely clothed with soft, patent, white, gland-tipped hairs and 15-20 cm long.
Leaf	Large leaves and a lush appearance. The foliage is covered with soft hairs and has a foul musky odor when crushed. Leaves have a velvety texture because of the fine hairs, especially on the undersurface. Leaves are distichous, long petioled, simple in outlined broadly cordate or reniform. Base is widely cordate, apex obtuse or rounded, suddenly caudate; faintly 3-5-7 lobed or undulate, patently acutely dentate, above dark green, beneath light green, on both side dull, on both surfaces and along the edges densely clothed with soft, white, gland-tipped hairs, 5-7 nerved or pedately 9-15 nerved, with white transparent veins and 10.0-40.0 cm in diameter.
	Petiole is robust, subfarctate or narrowly fistular, subterete, longitudinally furrowed, with a narrow grooved on the upper side, densely clothed with patent, soft, white, gland-tipped hairs, light hreen at the apex with 2 dark green knob-shaped glands, 5-30 cm long and stipule absent.
Floral biology	Flowers are borne singly on the axils of the leaves, the males on long peduncles and the females on short peduncles. The flowers are large, solitary, chalky white, unisexual-monoecious, axillary, the two sexes on different branches and attractive, up to 7-10 cm in diameter, with spreading petals. Male flowers are very long stalked, and faintly fragrant. Otherwise, the male and female flowers are similar in appearance. Peduncle is terete, yellowish green, densely clothed with soft, patent, gland-tipped hairs; longitudinally furrowed, obtusely quiquangular, jointed at the top and 15-25 cm long.
	Receptacle is campanulate or copular, whitish green or yellowish green, longitudinally veined, on both sides densely clothed with white gland-tipped hairs and 1.5-3.0 cm long. Sepals are 5, widely patent or obliquely erect, narrowly triangular or lanceolate, narrowed towards the apex acute, green, 1-1.5 cm long and 0.2 0.3 cm wide. Petals are 5, patent broadly obovate, emarginated, cuspidate, undulate-crenulate, 5-nerved, densely wooly on both sides, white with 5 yellowish green, swollen nerves and 2.5-4.0 cm long. Filaments are 3, short and robuststalks. Stalks is terete, light gren, densely clothed with wooly, wooly white, gland tipped hairs, not jointed, lengthening after the anthesis and 2-3 cm long. Calyx and corolla are as in the male flowers. The ovary is inferior, ellipsoid, densely clothed, with soft wooly, gland-tipped hairs light green or yellowish green and 2.5-3.0 cm long. Style is short, stigma 3, thick, clavate or curved and citrine. The anthers are borne on short filaments grouped at the center of the flower. The stigmas are short, thickened, and branched.
Anthesis and pollination	Flowers open at night. Being a monoecious crop, bottle gourd is strictly cross pollinated. Bees are the major pollinators. The time of anthesis in bottle gourd was observed between 5-8 PM under Punjab conditions. Stigma becomes receptive 24 hr before anthesis and remains up to 24 hr after anthesis. Stigma is receptive 36 hrs before anthesis at the ICAR-IIHR, Bengaluru, indicated that Anthesis starts from 9.0 AM and gets completed by 7.0 PM on the same day or opens in the evening and remains open until the following midday. Anther dehiscence begins at 11.0 AM and is completed at 2.0 PM. Both the male and female flowers open at the same time (Joshi and Gour, 1971). Male flowers remain open only for a few hours, after word the petals withered, thus the flowers are short lived. It is a highly cross pollinated crop due to its monoecious and andromonecious nature (Swiander *et al.*, 1994) and has wide genetic diversity. The maximum pollen germination occurs in 7 per cent sucrose solution. The pollen germination was accomplished in about half to 2 hr in bottle gourd. After about 24 hr of pollination, the petals of female flower withered, the stigma lost its shine and became brownish. In pollen experimental studies on pollen germination, the maximum germination occurs in 7 per cent sucrose solution. The pollen germination was accomplished in about half to 2 hrs in bottle gourd. After about 24 hrs of pollination, the petals withered, the stigma lost its shine and become brownish and increased by about 3.0 mm in length, showing the start of fruit formation.

Fruit and fruit set	Fruits are born on long peduncle, big, fleshy-water and many seeded pepo. Fruits shape may be round, bottle-shaped, dumbbell- shaped, crook- necked, coiled or spoon shaped. The background color of the fruit is either light green or dark green. The size of the fruit varies from 20-100 cm in length and 10-15 cm in diameter. Fruits are green or whitish green, often with scattered grey or white blotches, yellowish when ripe. The fruit can have a sterile (seedless) neck that varies from a few to 15 inches in length and from 1-2 inches in width. Wider necks usually contain seeds. The inner flesh is white. Immature fruits are tender and covered with soft hairs. Fruits are harvested about 7-10 days after anthesis.
Seeds	Seed matures about 90 days after fertilization. The brownish seeds are numerous in a whitish green pulp. Each seed is somewhat rectangular oblong-obovate or oblong- triangular, truncate or emarginated at the apex, white or brown (tan) and 0.7-2.0 cm long and width 8.0 mm. Average weight of 1000 seed is 150.0 g. Oil content in seed is 45 per cent.

Horticultural Classification

Classification on the basis of fruit shape.

i. **Round bottle gourd-**The fruits are spherical or globular with smooth and soft surface.
ii. **Flat round bottle gourd-**The fruits are flatty, spherical in shape, smooth and thin with soft pericarp
iii. **Tuma bottle gourd-**The fruits are large pyriform and looks like a tuma hence called tumari. The pericarp is tough and the thick.
iv. **Long bottle gourd-**the fruits are elongated and cylindrical, smooth surface, thin and soft pericarp.

Augmentation of Genetic Resources and Crop Improvement

Breeding Priorities

1. Early and high yielding varieties/hybrids
2. Fleshy and less seeded type varieties/hybrids
3. Better fruit size

In bottle gourd, Prasad and Prasad (1978a) reported that length and diameter of fruit showed high heritability and high expected genetic advance. Prasad and Prasad (1979b) further studied heritability and genetic advance in bottle gourd and reported high estimates of heritability and genetic advance for plant length, fruit length and fruit diameter. Sharma *et al.* (1983) reported that the estimates of heritability for days to opening of fruit male and female flowers and marketable maturity were high. These estimates were low for number of fruits, fruit weight and total yield. The characters having high heritability failed to express high genetic advance which was assigned to non-additive action of gene. Prasad (1985) found heritability estimates were consistently higher for length and width of cotyledonary leaves, days to appearance of first male and first female flowers, node number of first female flower, number of male flowers on main axis, number of male and female flower on primary laterals, number of male and female flower per vine, sex ratio, number of primary laterals per vine, length of vine at start of flowering, total vine length at maturity, node number of first fruit set, number of fruits per vine, average weight of fruit and yield per plant. He also recorded genetic advance as per cent of mean was maximum for yield per plant following by number of male flowers on primary laterals, number of male flowers per vine, total length of vine at maturity and number of fruits of vine at maturity and number of fruits per vine. Genetical analysis in long fruited bottle gourd revealed that narrow-sense heritability was high or days to opening of first male flower, days to opening of fruit female flower, fruit length, girth, weight and number of fruits per plant (Sirohi *et al.*, 1986). Sirohi *et al.* (1988) reported that narrow-sense heritability was quite low for vine length, days to opening of first male and first female flower, days to first harvest, fruit diameter, number of fruits per plant and yield per plant. Maurya (1991) observed narrow-sense heritability in the range of 5.92 per cent (yield per plant) to 95.96 per cent (node number at which first female flower appear) in bottle gourd. Genetic advance ranged from 2.25 (100-seed weight) to 42.15 (se ratio). High heritability in narrow-sense coupled with high genetic advance as per cent of mean was observed only for node number at which first female flower appears and sex ration, indicating their by these traits were less influenced by environment and were mainly under the control of additive genes. In bottle gourd, Tyagi (1972) observed that yield was positively and significantly correlated with the number of seeds per fruit and negatively associated with length of main shoot. Length of main shoot was significantly associated with number of branches and number of seeds per fruit. Number of branches was significantly associated with number of seeds per fruit. Prasad and Prasad (1978b) also observed that yield per plant had positive and significant correlations with number of fruits per plant both at genotypic and phenotypic levels in bottle gourd. Prasad (1985) estimated correlation coefficients at genotypic and phenotypic levels in bottle gourd. In general, genotypic correlation coefficient was higher than phenotypic correlation coefficients. The correlation coefficient of yield per plant was positive and significant with number of fruits per vine followed by average weight of fruit, number of male flowers on primary laterals, number of female flowers on primary laterals, number of female flowers per vine and node number at which first male flower, both at genotypic and phenotypic levels. It showed significant and positive correlation with number of primary laterals per vine, length of vine at maturity and days to appearance of first female flower, only at phenotypic level. Naryan *et al.* (1996) reported that yield was positively and significantly correlated with average weight of edible fruits, number of fruits per plant and number of primary branches per plant but negative significant association was found with number of days required for anthesis of first male flower only at phenotypic level. Singh *et al.* (2012) evaluated 14 genotypes and reported highly significant variations in seven charactes. Range of variation was highest for fruit length (11.33-44.000 cm). Phenotypic and genotypic coefficient of variation was highest for fruit breadth (30.04, 28.39, respectively). Phenotypic coefficients of variation were greater than genotypic coefficient of variation indicating influence of environment. Days to 50 per cent flowering, fruit length and fruit breadth showed high heritability values. All characters recorded high genetic gain except days to 50 per cent flowering and node at which first female flower appears. Hence characters with high genotypic coefficients of variation, high heritability and high genetic gain are controlled by additive gene action and hence direct selection is effective. Genotypic correlation revealed that yield/plant showed significant positive correlation with number of fruits/plant. Therefore this single trait should be taken into consideration while making selection for improvement in fruit yield. Days to 50 per cent flowering and fruit breadth exhibited negative significant correlation with fruit yield implying selection for earliness leads to lower yield. Genotypic correlation with yield/plant showed significant positive correlation with number of fruits/plant (Ram *et al.*, 2005 and Raja *et al.*, 2006).

Table 3.101: Bottle Gourd Cultivars Developed by Selection with Progeny Testing

Variety	Source
Pusa Summer Prolific Long	Indigenous
Pusa Summer Prolific Round	Indigenous
Punjab Long	Indigenous
Punjab Round	Indigenous
Arka Bahar	Indigenous
Pusa Naveen	Indigenous
CO-1	Indigenous

Hybridization and Selection

It is true combining ability, gene action and relative amount of additive and non-additive components of genetic variance present cross combination help a lot to determine the feasibility of its utilization and identification of best combiners. Sivakami *et al.* (1987) studied combining ability for yield/plant and 8 related characters in diallel crosses of 10 bottle gourd (*Lagenaria siceraria*) lines. The g.c.a and s.c.a. effects were significant for all 9 characters. General combining of ability effects were predominants over s.c.a. effect for these characters suggesting that recurrent selection would be effective in improving those characters. Janakiram and Sirohi (1988) studied combining ability of quantitative characters in 10 x 10 diallel cross of round fruited bottle gourd and reported that the variances of general combining ability were larger than those for s.c.a. for all the characters except for day to open first male and female flowers and fruit polar diameter, S-42 and S-54 were the best combiners for a number of characters including yield/plant. Maurya *et al.* (1993) reported high s.c.a. variance than g.c.a. variance for yield per plant, vine length, number of branches/plant number of fruits/plant, average weight of edible fruits days to anthesis of first male and female flower, days to first harvest and male, female flower ratio, indicating thereby possibility for exploitation of heterosis in bottle gourd. Maurya (1994) observed that in general in the expression of yield and yield attributing trait of bottle gourd variances due to g.c.a. were of lower magnitude than the s.c.a. variances in unfavourable environment while the situation was just reverse in favourable environment. The result therefore indicated that non-additive type of gene action is more important in determining the expression of major economic traits in unfavourable. Singh *et al.* (1996) studied combining ability analysis in bottle gourd [*Lagenaria siceraria* (Mol.) Standl.] and reported that among the parents Rajendra Chamatkar, Dholi Safed and Pusa Naveen proved to be the best general combiners and were also most suitable for rainy season cropping whereas hybrids Rajendra Chamatkar x Faizabad and Dholi Safed x Faizabadi had good s.c.a. for several characters during both seasons. Kushwaha ana Ram (1997) noted additive gene action for days to anthesis, first female flower, node number to first female flower, fruit diameter, fruit length, fruit weight and primary branches per plant. For fruit diameter and fruit length, both additive and non additive gene actions were predominant. Most of GCA and SCA effects were non-significant. BOG-22, BOG-7, BOG-40 and NDBG-1 were identified as good general combiners for fruit length and BOG-68 and BOG-13 fruit diameter. Sharma *et al.* (1983) studied the genetic make up of days to opening of first male and female flower, dyas to marketable maturity, number of fruits, fruit weight, total and marketable yields and fruiting period in bottle gourd. The degree of dominance indicated the existence of over dominance for fruit number, fruit weight and total yield. The generaton mean analysis further showed that the dominance and epistatic effects were more important than additive for almost all of the traits. But some of the effects based on generation means alone may not be much reliable. However, the components variance analysis revealed that the situation was reversed and the genetic variances were more important for days to opening of first male and female flowers, marketable maturity and marketable yield. Whereas, dominance effects were predominant for fruit number, fruit weight and total yield. Thus, both additive and non-additive gene effects were considered to be equally important in the inheritance of various economic characters in bottle gourd. Janakiram and

Sirohi (1987) studied gene action in round fruited bottle gourd and reported that over dominance was present in vine length, days to opening of first male flower, days to first harvest, fruit equatorial diameter fruits per plant, fruit weight and yield per plant, dominance for days to opening of first female flower and partial dominance for fruit polar diameter. Epistatis was pronounced in all the traits. Dominant alleles were more frequent for all the characters including yield per plant and presence of dominance genes in the parents for most of the characters revealed that the above type of gene action could be exploited by developing F_1 hybrids. In a diallel, incorporating ten bottle gourd (long) genotypes and their direct crosses Sivakami *et al.* (1987) observed highly significant mean sum of squares of general combining ability as well as specific combining ability for all the nine characters, namely, vine length, days to opening of first male and female flowers, days to first harvest, fruit length, number of fruits per plant, fruit weight and total yield per plant. This indicated the importance of both additive and non-additive variances for all the expression of these traits. The ratio of components of genetic variance indicated preponderance of additive gene action over non additiveness for these traits. In diallel of nine diverse bottle gourd lines Maurya and Singh (1994) observed preponderance of non-additive gene action for all the characters including yield number of fruits average weight of edible fruit, days to anthesis of first male and female flowers, days to first harvest, number of branches and vine length. They also observed that the crosses having high s.c.a. effects and *per se* performance were not necessarily the product of parents having high g.c.a. effects and high *per se* performance. Pal *et al.* (2004) estimated combining ability effects for different characters in a line x tester method comprising 30 hybrids obtained by crossing 10 lines and 3 testers. Significant differences were observed among the parents and crosses for general combing ability (g.c.a.) and specific combing ability (s.c.a.), respectively. The higher estimate of dominance variance as compared to additive variance was recorded for all the characters, except fruit diameter and number of fruits/plant. Varances due to general and specific combing abilities were significant for most of the characters. The partents, VRBG-112, NDBG-312, NDBG-140, IC92362, IC92462, Pusa Naveen and PSPL were found good general combiners. Therefore, these parents can be exploited for hybridization for producing desirable recombinants in the segregating generations. High s.c.a. effects for yield and other traits were exhibited by the crosses NDBG-140 x Pusa Naveen, VRBG-88 x NDBG-56, VRBG-105 x Pusa Naveen, PBOG-22 x PSPL, VRBG-88 x Pusa Naveen, IC92362 x Pusa Naveen, NDBG-132 x NDBG-156, VRBG-112 x NDBG-56, PBOG-22 x NDBG-56 and NDBG-132 x PSPL. Most of the crosses, good x good, poor x poor, poor x good, average x good, good x poor, good x average and average x poor combinations indicating the role of additive, epistatic, dominance, additive, dominance, additive and dominance gene action. High s.c.a. in the crosses involving goo x good general combiners might be due to additive x additive type of interaction which is partially fixable. High s.c.a. effects in the crosses involving poor x poor combining parents were possibly due to intra and inter allelic interactions.

Rajendran (1961) recorded heterosis of 266.51 per cent in the yield of the best hybrid combination of bottle gourd over better parent. Roy (1964) reported 81.42 and 78.60 per cent heterosis in bottle gourd over their respective better parents in yield of two best F_1 hybrids. Chaudhury and Singh (1971) developed two high yielding bottle gourd hybrids which gave 75 per cent and 106 per cent higher yield over their respective better parent. Eight hybrids and their parents were studied to investigate the extent heterosis and out of these hybrids seven showed significant increase for the number of female flowers over better parent by Tyagi (1973). The maximum significant increase of 33.33 per cent for number of fruits was exhibited by the cross Type 1 x 6022 over mean of the parents. The significant increase of 41.72 per cent for length of fruit was exhibited by the cross 5902 x 5716. The maximum heterosis of 38.29 and 13.64 per cent for weight of fruits was exhibited by the cross Type 1 x 6022 over means of the parents and better parent, respectively. Significant heterosis of 17.14 per cent for number of seeds per fruit was noted by the cross 5717 x Type 1 over mean of the parents. Length, breadth and 100-seed weight showed positive heterosis in different crosses but the increase was not significant. Sharma *et al.* (1983) studied the parents F_1 and F_2 and back crosses (BC_1 and BC_2) of two crosses of bottle gourd and observed heterosis over better parent for fruit number, fruit weight and total yield in both the crosses. Pal *et al.* (1984) observed more rapid germination (2-4 vs. 5-6 days) and earlier fruit maturity (10-11 days) in F_1 hybrids over the better parent. The females were borne at lower nodes and hybrids had 17.28 per cent higher flesh thickness. It was noted that hybrid gave 20 per cent higher early yield than parents and had a longer harvesting period (65-71 days) than the latter (55-65 days). Sirchi *et al.* (1985) reported heterosis of 0.44 to 21.90 per cent for fruit length, 1.38 to 64.71 per cent for number of fruits per plant, 1.13 to 22.91 per cent for fruit weight and 0.90 to 76.46 per cent for yield per plant in bottle gourd. The best F_1 hybrid S-48 x S-47, which gave 115 per cent higher yield over the commercial cultivar PSPL and 32.70 per cent over the best parental line S-48 was suggested for commercial cultivation. Sirohi *et al.* (1987) studied 66 F_1 hybrids in diallel set involving 12 parents excluding reciprocals to investigate the manifestation of heterosis in bottle gourd for yield and its components. Significant heterosis was recorded over the respective parental line for vine length, fruit number and yield. For yield of the F_1 hybrids P4 x P10, P3 x P9 and P2 x P12 manifested 48.01, 54.61 and 29.03 per cent heterosis, respectively over the best check P11. In a diallel set involving nine parents of bottle gourd, Maurya *et al.* (1993) studied the performance of 36 F_1 hybrids, excluding reciprocals under low temperature condition during winter season. They noted that the highest yielding cross took only 33.33 days for its first picking compared to 111.33 days needed by commercial cultivar PSPL. Heterosis of the highest yielding cross over the top parent for yield per plant was 80.51 per cent. Janakiram and Sirohi (1989) studied heterosis for yield components in 45 F_1 hybrid namely S-46 x S-54, S-10 x S-52-7 and S-54 x S-52-7 showed 84.5, 80 and 80 per cent heterosis, respectively for yield over the best parental line, S-41. The higher yields in these crosses were attributed to increase number of fruits/plant, fruits weight and fruit size. The cross S-46 x S-54 gave 148.97 per cent higher yield than the commercial cultivar Pusa Summer Prolific Round. Janakiram and Sirohi (1992) reported significant heterosis over the better parent for eight characters in bottle gourd. The best performing hybrid for yield were S-36-1 x NC59812-1 and S-89-1 x S-1-3 with 76.4 and 58.1 per cent heterosis over better parent respectively. The highest yielding parents were S-9-1 and NC5981-1 (5.13 and 7.69 kg/plant, respectively). In a diallel set involving nine parent of bottle gourd. Maurya *et al.* (1993) studies the performance of 36 F_1 hybrid excluding reciprocals under low temperature condition during winder season. They noted that the highest yielding cross took only 83.33 days for its first picking compared to 111.33 days needed by commercial cultivar PSPL. Heterosis of the highest yielding crook over the top parent for yield per plant was 80.51 per cent. Sharma *et al.* (1993) derived information on heterosis from data on yield components in the parents and F_1 hybrids from 11 line x 3 tester crosses. The cross Summer Long Green Selection 2 x Faizabadi Long had the highest heterosis over the control cultivar PSPL for number of fruit (106.63 per cent) and total yield (110.33 per cent). Hissar Local x PSPL had the highest heterosis (22.93 per cent) over the control cultivar for fruit length. Pitchaimuthu and Sirohi studied heterosis in bottle gourd and respected appreciable amount of heterosis for all the characters except day to first harvest. The F_1 hybrid Pusa Naveen x S-10, Pusa Naveen x PSPL and PSPL x S-10 were observed to be the best performing for yield and they showed 64.62, 60.38 and 53.20 per cent heterosis for yield over top parent S-10. The better performing F_1 hybrid was 'Pusa Naveen x S-10' recorded 64.02 per cent heterosis over the standard commercial cultivar 'Pusa Naveen'.

Newly released bottle gourd cultivars Kashi Kundal-VRBOG-16, and Kashi Kirti-VRBOG-63-02 are winter fruited type and Kashi Kiran-VRGB-4-round type are from ICAR-IIVAR-Varanasi (D.P.). Similarly bottle gourd hybrid-Gujarat Anand Bottle gourd Hybrid-1 (GABGH-1) - ABGH-08-08 from GAU, Anand, Gujarat.

Pal *et al.* (2005) conducted experiment with 30 hybrids through line x tester (10 lines and 3 testers) analysis to estimate the extent of heterosis over mid and better parents for yield and its contributing traits in bottle gourd. Analysis of variance indicated highly significant differences for almost all characters except vine length suggesting presence of wide genetic variability. The heterosis exhibited for fruit characters and fruit yield was mostly positive. The results revealed that the crosses between IC92362 x Pusa Naveen and IC92462 x Pusa Navven exhibited high heterosis for yield and also other yield components *viz.*, fruit diameter, vine length and number of branches/plant indicating the additive or synergistic effect of the components characters on fruit yield. Both crosses can be considered as superior among all the 30 cross combinations evaluated. The cross combinations can be utilized for commercial exploitation and determining the strategies for future bottle gourd hybrid development. Singh *et al.* (2016) found highest significant and positive heterosis over average parent for fruit yield/plant was recorded by Arka Bahar x Pusa Naveen followed by Arka Bahar x Kasi Ganga and Pusa Santushti x PSPL while, higher heterobeltiosis was recorded for fruit yield/plant was by Arka Bahar x Kasi Ganga and Arka Bahar x Pusa Naveen. Arka Bahar x N. Dharidhar and Punjab Komal x PSPL were having maximum sca value for fruit yield/plant which shows the preponderance of dominance variance in the expression of this trait by the above cross combination and it can be exploited for the production of potential hybrids. Maurya *et al.* (2016) obtained significant mean sum of squares due to genotypes, parents, hybrids and parents vs. hybrids was observed for most of the traits. Heterosis for fruit yield/plant ranged from -33.75 to 44.25 per cent over better parent and -24.84 to 39.63 per cent over standard variety (Pusa Naveen). The best cross combinations over better as well as standard parent were NDBG-5009 x NDBG-512 (44.25 per cent and 39.63 per cent) andPusa Naveen x NDBG-104 (20.49 per cent and 20.49 per cent), which may be exploited as commercial hybrid after their stability test as they are not only high yielder but also possessed attractive fruit shape and colour as per present market demand. The estimates of gca and sca variances were significant indicating importance of additive and non-additive gene actions for all the traits. Narendra Rashmi was found the best gca effects for number of fruits and fruit yield/plant. The crosses NDBG-5009 x NDBG-512 (1.95), NDBG-5005 x NDBG-104 (1.31) and NDBG-104 x NDBG-11 (0.93) were found to be good specific combiners for fruit yield and its related contributing traits.

Interspecific Hybridization

Though hybridization of *Lagenaria siceraria* (Molina) Standl.with at least one perennial wild species *Lagenaria sphaerica* (Sond.) Naud., is known, the hybrid is sterile and possibility of transfer of desirable traits to cultivated species needs investigation.

Insect Resistance

In bottle gourd very less research works related to insect resistance has been carried out.

Table 3.102: Development of Cultivars

Variety	*Organization*	*Pedigree*	*Brief Description*
Kashi Ganga	IIVR, Varanasi	IC-092465 x DVBG-151	It bears light green fruits of 30.00 cm length having 7.00 cm diameter and fruit weight is about 800-900 g. An average yield is about 48-55 t/ha. This is tolerant to anthracnose and suitable for cultivation during both kharif and summer seasons.
Kashi Bahar	IIVR, Varanasi	VR-104 x VR-4	It is hybrid and bears about 12 light green straight fruits of 30-32 cm length having 780-850 g individual fruit weight. An average yield is 50-55 t/ha.
Arka Bahar	IIHR, Bangalore	Selection	Fruits straight, without crookneck, medium sized, average fruit weight is 1 kg, light green, very good cooking and keeping quality. Yield is 40-45 t/ha. Crop duration is 120 days. Tolerant to blossom end rot.
Pusa Summer Prolific Round	IARI, New Delhi.	Selection	Fruit is round, green, 15-18 cm in girth, prolific bearer. Suitable for both summer and kharif seasons. First picking in 60-65 days after sowing with average. An average yield is 25 t/ha.
Pusa Summer Prolific long	IARI, New Delhi.	Selection	Suitable for both summer and kharif season fruits are long (40-45 cm), green, prolific bearer. First picking in 60 days and yield 25.0-27.5 t/ha. IARI, New Delhi.
Punjab Round	P.A.U. Ludhiana	Selection	Plants are vigorous and prolific bearer, fruit spherical, tender and shining. Recommended for cultivation under Punjab condition. First picking starts 65 days after seed sowing. It yields about 30 t/ha.
Punjab Long	P.A.U. Ludhiana	Selection	Fruits are small, long, tender, light green, vines are vigorous and profusely branched. Recommended for cultivation in Punjab. First harvesting 70-80 days after sowing or 10-12 days after fruit setting. Tolerant to cucumber mosaic virus (CMV). Average yield is 40 t/ha.
Pusa Naveen	IARI, New Delhi.	Selection	Suitable for spring summer and Kharif season in northern plains. Fruits are cylindrical, straight; average fruit weight is 850 g. It is free from crook-neck and gives an average yield of 35 t/ha.
Punjab Komal	P.A.U. Ludhiana	LC-11 X LC-5	It is an early variety, bears oblong, medium sized, light green, pubescent, 10-12 fruits on a vine. It takes 70 days from sowing to first marketable yield. Tolerant to cucumber mosaic virus (CMV). Average yield is 50 t/ha.
Pusa Sandesh	I.A.R.I. New Delhi	Selection	Vines are medium long, short internodes, fruits medium sized, green, round, weighing 600 g each and leaves deep oblate. Suitable for both summer and kharif season. Average yield is 32 t/ha. First picking 55-60 days in kharif and 60-65 days in summer.
Pusa Hybrid-3	I.A.R.I. New Delhi	-	Fruits are green, slightly club shape without neck with good keeping quality. First picking starts on 50-55 days after seed sowing. It gives an average yield of 425.0-475.0 q/ha.
Narendra Rashmi	N.D.U.A. and T., Faizabad	Selection	Plants are moderately tolerant to red pumpkin beetle, powdery mildew and downy mildew. Fruits are bottle shaped, having shallow neck and white flesh. Average fruit weight is 1 kg with 400-500 seeds. Average yield is 300 q/ha. This variety is recommended for Punjab, U.P. and Bihar
Narendra Jyoti (NDBG-104)	N.D.U.A. T. Faizabad	Selection	Fruits are cylindrical, long and early maturing. Average fruit weight is 1.00 kg. Average yield is 300.0 q/ha.
Narendra Shishir)	N.D.U.A. T. Faizabad	Selection	It is a winter fruited variety and bears fruit for longer duration. It is tolerant to anthracnose and Downey mildew.
Narendra Dharidar	N.D.U.A. T. Faizabad	Selection	Fruits are bottle shaped, striped and green in colour. Average yield is 350.0 q/ha.
Narendra Sankar Lauki-4	N.D.U.A. T. Faizabad	-	It is prolific bearer with near cylindrical attractive fruits. It gives an average yield of about 400.0-600.0 q/ha.
CO-1	TNAU, coimbatore	Selection	Fruits are round at the base with prominent bottleneck at the top, medium sized, light green with average fruit weight of 2-2.5 kg. Average yield is 360 q/ha. In 135 days crop duration.
Azad Nutan	CSAUA&T, Kanpur	Selection	Fruits are medium long, light green, smooth and fruit weight is about 1.0-1.5 kg.
Pant Sankar Lauki-1 (PBOGH-2)	GBPUA&T, Pantnagar		Fruits are 35.0 cm long, green, medium sized and somewhat cylindrical. First picking staryts in about 60 days after seed sowing. It gives an average yield of 400.0-420.0 q/ha.
Pant Sankar Lauki-2 (PBOGH-2)	GBPUA&T, Pantnagar		Fruits are 40.0 cm long, green medium sized and cylindrical. First picking starts in about 65 days after seed sowing. It gives an average yield of 400.0-410 q/ha.

Table 3.103: Insect Pest and their Sources of Resistance

Crop	*Insects*	*Resistance Source*
Bottle gourd	Fruit fly (*Dacus cucurbitae*)	NB-29
	Red Pumpkin Beetle	NB-28

Source: Bhardwaj, 2010.

Diseases Resistance

In cucurbits, bottle gourd [(*Lagenaria siceraria* (Monila) Standl.)] is a popular vegetable widely cultivated in summer/kharif season across all over India, however, its cultivation is threatened by several disease particularly by Downy Mildew (*Pseudoperonospora cubensis* (*Berk. and Curtis., Rostovzev*) which account yield loss of 40-48 per cent. Seeing the worsen situation, it is essential to develop a consistent and reliable method to screen bottle gourd genotypes for identifying resistance source to this melody. Screening were conducted in 51 germplasm with 3 plants per germplasm as per assay and repeated two times. The plants were inoculated with a suspension of *P. cubensis* (10^4 *sporangia* ml^{-1}) at the two cotyledon stage. The adaxial and abaxial surfaces of the cotyledons were sprayed with the inoculum suspensions until incipient run-off. Later, inoculated plants were placed in the dark at 20° C and high humidity for 21 hours. Subsequently, they were transferred to the greenhouse under ambient conditions of 24-28° C and 80-90 per cent relative humidity. After 7 days, the disease reactions on the host plants were recorded. A set of the host plants without fungal inoculation were used as a control. Plants were rated on a scale of 0–5 in which 0 stands for immune and 5 stands for highly susceptible. Among all accession lines screened, No single line found immune, only one was found resistant (VRBG-12). Besides, 3 others (VRBG-26, VRBG-47, and VRBG-17) showed moderately resistant and four others were moderately susceptible to the fungus. The remaining 43 lines were range from susceptible to highly susceptible for the fungus. The most resistant and most susceptible cultigens were retested, to verify their reaction. The identification of resistant line and highly tolerant or moderately resistant bottle gourd lines in this study points at the ICAR-IIVR germplasm collections of potential materials that could be included in further disease resistant breeding programme (Bhardwaj *et al.*, 2016). Fifty-one accessions were screened against gummy stem blight and none were found immune/ resistant under field condition. Out of 51 lines least severity (28.8%) was observed in VRBG-556, 13 were moderately susceptible with PDI ranging between 28.80 to 49.44, 32 were found to be susceptible (50.4-74.4%) and 6 highly susceptible (75.2-87.2%) as reported by Bhardwaj *et al.*, 2018.)

Table 3.104: Diseases and their Sources of Resistance

Disease	*Resistance source*
CMV,SMV,WMV	PI 271353
Fusarium wilt	Taiwan variety Renshi

Source: Bhardwaj, 2010.

Bhardwaj *et al.* (2016) reveal that gummy stem blight caused by *Dydimella bryoniae* (Syn. *Mycosphaerella melonis*) is one of the most important limiting biological factors for the production of cucurbits reducing both yield (36-40 per cent) and fruit quality (80-90 per cent) worldwide. Score the disease on 5 plants on 5 leaves each using an index of necrotic tissue on the following 0-5 rating scale (0: no infection visible; 1: >0 per cent β20 per cent leaf area necrotic; 2: >20 per cent β40 per cent leaf area necrotic; 3: >40 per cent β 60 per cent leaf area necrotic; 4: >40 per cent 80 per cent leaf area necrotic and 5: >leaf area necrotic).PDI was ranging from 28-80 per cent to 90.40 per cent across the germplasm accessions. It has been found that among the 75 germplasm accessions screened, VRBG-556 recorded least PDI of 28.8 per cent followed by VRBG-52 with 32 per cent PDI categorized under moderately susceptible. Also VRBG-18 recorded the highest PDI of 90.4 per cent. None of the germplam accessions were found to be resistant or moderately resistant in field condition. On the basis of the report these genotypes can be further used for disease breeding programme.

Abiotic Stress

Tolerant to a wide range of rainfall, it may be grown either on the ground or trellised.

REFERENCES

1. Bhardwaj, D.R. 2010. Breeding for pest and disease resistance in cucurbits. In: Compendium of Wnter School "Emerging trends and strategies for the management of pests and diseases in vegetable crops. Published by ICAR-IIVR, Varanasi, p. 102-109.
2. Bhardwaj, D.R., Singh, B., Saha, S., Gautam, K.K. and Karmakar, P. 2016. Screening of Bottle gourd germplasm against Downy Mildew (*Pseudoperonospora cubensis*) through artificial inoculation. International Agribiodiversity Congress (Abs. SessionL Trait Discovery and Enhanced Use of PR), 6-9 November 2016; held at New Delhi, p. 173.
3. Bhardwaj, D.R., Nagendran, K., Gautam, K.K., Singh, B., Singh, A.K., Chaube, T. and Mishra, A. 2016. Field screening of bottle gourd Germplasm accessions against gummy stem blight (*Dydimella bryoniae*) (Abs: 73), 7th Indian Horticulture Congress, held at ICAR-IARI, New Delhi, 15-18 November, p. 58.
4. Bhardwaj, D.R., Gautam, K.K., Saha, S., Nagendran, K., Pnadey, K.K., Singh, A.K., Singh., P.M. and Singh, B. 2018. Mining the sources of resistance for downy mildew and gummy stem blight in bottle gourd (*Lagenaria siceraria*) accessions. *Indian J. Agril. Sci.*, 88(5): 746-750.
5. De candolle, A. 1882. Origin of cultivated plants. Kessinger, Publ. New York.
6. Esquinas-Alcazar, J.T. and Gulick, P.J. 1983. Genetic resources of Cucurbitaceae. Int. Board for Plant Genet. Resources, Rome.
7. Herklots, G.A.C. 1972. Vegetable in South East Asia. George Allen and Unwin, London.
8. Joshi D.P. and Gour, S.K.S. 1971. Floral biology studies of *Lagenaria siceraria* Standl. Journal of Research Punjab Agricultural University 8: 420-426.

9. Maurya, I.B. 1994. Heterosis combining ability and stability analysis in bottle gourd (*Lagenaria siceraria.* Ph.D. Thesis NDUA&T, Faizadaba (U.P.).

10. Maurya, P.K., Yadav, G.C., Kumar, S. and Pandey, P. 2016. Estimation of heterosis and combining ability in bottle gourd [*Lagenaria siceraria* (Mol.) *Standl.*]. 7th Indian Horticulture Congress (Abs: 108, held at ICAR-IARI, New Delhi, 15-18 November, p. 83.

11. Prasad, A. and Prasad, B 1978. Correlation studies, flowering characters and yield in bottle gourd [*Lagenaria sicerariac* (mol.) Standl]. Plant science 10:25-28.

12. Pathak, G.N. and Sing S.M. 1950. Genetical studies in *Lagenaria*. Indian journal of Gentics and Plant Breeding, 10:25-35.

13. Pal, S.N., Ram, D., Pal, A.K. and singh, G. 2004. Combining ability stidies for certain metric traits in bottle gourd [Lagenaria siceraria (Mol.) Standl.]. Indian J. Hort., 61(1): 46-50.

14. Pal, S.N., Ram, D., Pal, A.K. and Rai, M. 2005. Heterosis studies in bottle gourd [Lagenaria siceraria (Mol.) Standl.]. Indian J. Hort., 62(3): 253-256.

15. Prasad, A. And Prasad, R. 1978. Genetic studies in bottle gourd (Lagenaria siceraria (maol.) standl. Plant science, 10:9-16.

16. Raja, S., Bagle, B.G. and Dhandar, D.G. 2006. Identification of yield attributes in bottle gourd for raifed conditions. Veg. Sci., 33(1): 106-108.

17. Ram, D., Singh, R.S., Pandey, S. and Rai, M. 2005. Study of polygenic traits in offseason bottle gourd [Lagenaria siceraria (Mol.) Standl.]. Veg. Sci., 32(2): 189-191.

18. Singh, A.K. 1990. Cytogenetics and evolution in the Cucurbitaceae. Cornell University, London, pp 10-28.

19. Singh, R.P., Jain, P.K., Sengupta, S.K., Mehta, A.K. and Verma, A.K. 2016. Heterosis and combining ability in bottle gourd (Lagenaria siceraria Mol. Standl.) (Abs: 67). 7th Indian Horticulture Congress, held at ICAR-IARI, New Delhi, 15-18th November 2016, pp. 53.

20. Sirohi, P. S; Sivakami, N. and Choudhury B. 1988. Genetic analysis in bottle goud Ann. Agvic Res; 9(1): 1-5.

21. Sirohi, P. S; Sivakami N. and Choudhury B. 1986. Genetic analysis in long fruited bottle gourd. Indian Journal of Agricultural Science 56623-625.

22. Swiander JM, Ware GW, Maccollum J.P. 1994. Vegetable crops. Interstate Publishers, pp 323-340.

23. Thumburaj S, Singh N, 2003. Vegetables and Tuber Crops and Spices. ICAR, New Delhi, pp 271-272.

24. Tindal, H.D. 1983. Vegetable in the tropics. AVI, Westport, CT.

Genus Trichosanthes

Introduction

Genus *Trichosanthes* has 44 species of which 22 species occur in India (Chakravarty, 1982) and the major zones of species concentration are along the Malabar coast in Western Ghats and low and medium elevation zones (up to 1500 m) in Eastern Ghats and north-eastern region. Thirty nine species of the genus *Trichosanthes* are found in China, Japan, Western India and Australia. As many as 20-24 species recognized by Chakravarty (19590 which are spread to North Australia in one direction and China to Japan in another. The other related species are *T. japonica, T. multiloba, T. bracteata* (syn. *T. palmata*), *T. lobata, T. cordata, T. cucumeroides, T. shikokiana* and *T.nervifolia*. Trichosanthes is the eargest genus of the Cucurbitaceae with species (Jeffrey, 1980, 1990; Yueh and Cheng, 1980; Huan *et al*. 1997, 1998, 2007; Rugayah and de Wilde, 1999, de Boer and Thulin 2012).

Table 3.105: Taxon and Preferred Taxon Name

Name of Taxon	*Name of Authority*	*Family*	*Preferred Taxon Name*
Trichosanthes			*Trichosanthes*
Trichosanthes anguina	L.	Cucurbitaceae	*Trichosanthes cucumerina var. anguina*
Trichosanthes bracteata	(Lam.) Voigt	Cucurbitaceae	*Trichosanthes tricuspidata*
Trichosanthes cucumerina	L.	Cucurbitaceae	*Trichosanthes cucumerina*
Trichosanthes cucumerina var. *anguina*	(L.) Haines	Cucurbitaceae	*Trichosanthes cucumerina* var. *anguina*
Trichosanthes cucumerina var. *cucumerina*		Cucurbitaceae	*Trichosanthes cucumerina* var. *cucumerina*
Trichosanthes cucumeroides	(Ser.) Maxim.	Cucurbitaceae	*Trichosanthes ovigera*
Trichosanthes dioica	Roxb.	Cucurbitaceae	*Trichosanthes dioica*

Name of Taxon	Name of Authority	Family	Preferred Taxon Name
Trichosanthes globosa	Blume	Cucurbitaceae	*Trichosanthes globosa*
Trichosanthes heteroclita	Roxb.	Cucurbitaceae	*Hodgsonia macrocarpa*
Trichosanthes himalensis	C. B. Clarke	Cucurbitaceae	*Trichosanthes ovigera*
Trichosanthes hybrid		Cucurbitaceae	*Trichosanthes hybrid*
Trichosanthes japonica	Regel	Cucurbitaceae	*Trichosanthes kirilowii*
Trichosanthes kirilowii	Maxim.	Cucurbitaceae	*Trichosanthes kirilowii*
Trichosanthes macrocarpa	Blume	Cucurbitaceae	*Hodgsonia macrocarpa*
Trichosanthes ovigera	Blume	Cucurbitaceae	*Trichosanthes ovigera*
Trichosanthes palmata	Roxb., nom. illeg.	Cucurbitaceae	*Trichosanthes tricuspidata*
Trichosanthes quinquangulata	A. Gray	Cucurbitaceae	*Trichosanthes quinquangulata*
Trichosanthes rosthornii	Harms	Cucurbitaceae	*Trichosanthes rosthornii*
Trichosanthes sp.		Cucurbitaceae	*Trichosanthes* sp.
Trichosanthes tricuspidata	Lour.	Cucurbitaceae	*Trichosanthes tricuspidata*
Trichosanthes uniflora	K. S. Hao	Cucurbitaceae	*Trichosanthes rosthornii*
Trichosanthes villosa	Blume	Cucurbitaceae	*Trichosanthes villosa*

The genus *Trichosanthes* underwent great changes in its species classification. Kundu (1942) divided the genus into 2 sections:

(i) Eutrichosanthes- This section include 23 species

(ii) Pseudotrichosanthes-This section include 3 species

Trichosanthes has 22 species and is reported to be of Indian origin/parts of Tropical Asia or Indo-Malayan region. Among these, there are 2 cultivated species under this genus:

(i) Trichosanthes anguina L.or *Trichosanthes cucumerina* (snake gourd)

(ii) Trichosanthes dioica Roxb. (Pointed gourd)

The related species are *T. cucumerina, T. bracteata* (*syn. T. palmata*), *T. lobata, T. cordata, T. multiloba, T. cucumeroides, T. japonica, T. shikokiana* and *T. nervifolia* (Varghese, 1973).

Varghese (1973) in a comparison the karyotype of *T. anguina* (n=11), *T. cucumerina* (n=11) and *T. palmate* (n=11) revealed that the former 2 species show closer resemblance. A comparision of their idiograms (Dutta and Basu, 1978) shows that they differ from each other in number of types of chromosome and the number of each type represented in them and in total chromatin length. The three taxa *T. anguina, T. lobata* and *T. cucumerina* may therefore be related in one of the two ways:

1	*T. cucumerina*	®	*T. lobata*	®	*T. anguina*
2	*T. cucumerina*	®	*T. anguina*		

Table 3.106: Chromosome Number in the Genus *Trichosanthes*

Name of Species	(n=)	(2n=)	Reference
T. anguina L.	11	22	Dutta and Basu (1978)
T. cucumerina L.	11	22	Dutta and Basu (1978) (Mckay 1931) Beevy and Kuriachen
T. lobata	11	22	Varghese (1971)
T. palmata Roxb. (male)	11	22	Varghese (1971) Sinha and Roy (1970a)
T. palmata Roxb. (female)	11	22	Varghese (1979)
Male	22	44	Varghese (1971)
T. dioica Roxb.	11	22	Sen (1976)
T. bracteata (Lam.) Voigt	22	44	Thakur (1973), Thakur and Sinha 1976
T. cucumeroides	22 (XY)	-	Nakajima (1973)
T. japonica	11 (XY)	-	Nakajima (1973), Sinoto (1929)
T. multiloba C.B. Clarke	-	22 (XX, XY)	Kurita (1939)
T. shikokiana	-	22	Kurita (1939)
T. anamalaiensis	-	22	Beevy and Kuriachany (1996)
T. truncata	-	22	Huang *et al.* 1994
T. brevifolia	-	22	Beevy and Kuriachan (1996)
T. lepiniana	22	44	Huang *et al*; (1994)
T. kirilowii	44	88	Hung *et al.* (1994)

Distribution of *Trichosanthes* Species

In India total 21 different species are available in this crop. The related species of potential importance '*Trichosanthes cucumerina*' occurs in the north-eastern region, along with semi-wild *Trichosanthes dioica*. Another widely distributed species is *Trichosanthes bracteata* occurring in eastern India, extending to the south, and sporadically in the Himalaya (1500 m). *Trichosanthes cordata* (related to *Trichosanthes anguina*), occurs in the peninsular region extending to north-eastern plains and hills.

Table 3.107: Species and their Distribution

Species	*Distribution*
Trichosanthes bracteata (Lam.) Voigt. var. tomentosa (Heyne ex Clarke) Chak.	Occurs in the peninsular region and north-eastern zone; var. *bracteata* occurs all over, largely in eastern India, in the Himalaya upto an elevation of 1500 m and extending southwards to the Andaman Islands.
Trichosanthes cucumerina Linn.	Occurs throughout India and may be the wild form of *T. anguina*.
Trichosanthes cordata Roxb.	Occurs in the peninsular India and north-eastern plains and hills.
Trichosanthes dioica Roxb.	Semi-wild populations occur in Assam plains, Brahmaputra valley

REFERENCES

1. Beevy, S.S. and P. Kuriachan. 1996. Chromosome numbers of south Indian Cucurbitaceae and a note on the cytological evolution in the family. *J. Cytol. Genet*, 31: 65-71.
2. Chakrvarty, H.L. 1959. Monograph on Indian Cucurbitaceae. *Rec. Bot* Surv. India, 17: 1-234.
3. Chakrvarty, H.L. 1982. Cucurbitaceae. Fascicles of flora of India 2. Botanical survey of India, Howrah.
4. De Boer, H.J. and Thulin, M. 2012. Synopsis *Trichosanthes* (Cucurbitaceae) based on recent molecular phylogenetic data. *Phytokeys*, 12:23-33.
5. Dutta, S.K. and Basu, R.K. 1978. Cytomorphological, biochemical and palynological studies in *Trichosanthes anguina* L. and *T. cucumeriana* *Cytologia*, 43:107-117.
6. Huang L. Yue, C.X; Yang, B; Cheng, C.Y., Li, N.H. Ruan, C; Q4, L.J.and Gu, H.Y. 1998. Systematic studies on *Trichosanthes* L. (Cucurbitaceae). P. 75-83.
7. Huang, L.G. Yue, C. and Li, M. 1994. Chromosome numbers of 8 species in *Trichosathes. Acta. Bot. Yunnanica,* 16(1): 95-96.
8. Huang, La. Yang, B. Yue, C.X. and Cheng, J.R. 1997. Pollen morphology of *Trichosanthes* and its taxonomic significance. *Acta Phytotax.* Sin; 35(2) : 125-135.
9. Huang, Li, Lu, A. and Jeffrey, C. 2007. *Trichosanthes.* In. Flora of China, vol. 19. Draft a vailable online.
10. In: Chiu, S.T. and Peng, C.I. (eds.) Proc. Cross strait Symp. Flov. Div. Conserv. Taipei, 1996, Natt. Mus. Nat. Sci. Taichung (Taiwan).
11. Jeffrey, C. 1980. A review of the Cucurbitaceae. *Bot. J. Linnean soc;* 81:233-247.
12. Jeffrey, C. 1990. An outline classification of the Cucurbitaceae p.449-463.
13. Kurita M. 1939. Geschlachts chromosomen bei *Trichosanthes multiloba* Miq. and chromosomen zahlen bei einigen anderen Arten von Cucurbita ceen. *Bot. Mag. Tokyo,* 53:505-0510.
14. Nakajima. G. 1937. Cytological studies in some dioecious plants. Cytologia Fuji Jub, vol. 282.
15. Rugayah, E.A. and de Wilde, W.J.J.O. 1999. Conspectus of Trichosanthes (cucurbitaceae) in Malaysia, Reinwardita, 11:227-280.
16. Sen R. 1976. Cytogenic investigation in some Indian cucurbits. Ph.D. dessertation Univ. Kalyani West Bengal, India.
17. Sinoto, Y. 1929. Chromosome studies in some dioecious plants with special reference to the allosomes. Cytologia, 1: 109-119.
18. Thakur, G.K. 1973. A natural tetraplaid in the genus *Trichosanthes* from Bihar. Proc. 60th Indian science Congress, Chandigarh, p. 324.
19. Verghese, B.M. 1971. Cytology of *trichosanthes palmata* Roxb. Cytologia, 36: 205-209.

Pointed Gourd (*Trichosanthes dioica* Roxb.)

Introduction

Pointed gourd (*Trichosanthes dioica* Roxb., 2n=2x=22) is an important, high- priced vegetable commodity not only in India but in other country also. It is known by several other names such as parwal, palwal, patar, parora or parmal in different parts of India and Bangladesh. Pointed gourd is being exported to several countries and there it is sold with great profit. It is one of the important vegetables which hold a coveted position in the vegetable market of our country, particularly during summer and rainy season. The fruits are available from February to November with the lean period of December and January. The crop also has potential for production in home gardens where it can provide a nutritious vegetable for an extended period. The green tender fruit is cooked as a vegetable in soups, stews or curries in various ways either alone or in combination with other vegetables or meats and is prepared as a sweet dish. However, the tender shoots and leaves are also used as potherbs particularly in West Bengal. Parwal fruit is a good source of carbohydrates, vitamin A and vitamin C. It also contains major nutrients and trace elements (magnesium, potassium, copper, sulphur, and chlorine) which are needed only in small quantities, but proven to play a very important role in human physiology (Singh, 1989). It is easily digestible, diuretic and laxative. It also invigorates the heart and brain. It is purported that pointed gourd possesses the medicinal property of lowering total cholesterol and blood sugar. These claims are supported by preliminary clinical trials with rats (Chandrasekar *et al.*, 1988) and rabbits (Sharma and Pant, 1988; Sharma *et al.*, 1988). Pointed gourd is useful in the disorders of the circulatory system. The fruits show some prospects in the control of certain cancer like conditions. Due to richness in vitamins and minerals, pointed gourd is called as king of gourd. It contains Mg, 9.0 mg, Na 2.6 mg, K 83.0 mg, Cu 1.1 mg, S 17.0 mg, protein 2.0 g, fat 0.3 g, carbohydrate 2.2 g, fiber 3.0 g, phosphorus 40.0 mg, calcium 30.0 mg, chlorine 4.0mg, iron 1.7 mg, vitamin A 255 I.U., vitamin C 29mg, Riboflavin 0.06mg, thiamine 0.05mg, nicotinic acid 0.5mg and oxalic acid 7.0mg per 100 g edible part. Protein content of the fruit is 10 times that of bottle gourd and 4 times that of snake gourd, ridge gourd and wax gourd. Vitamin A content is about 5 times that of ridge gourd, 3 times of pumpkin and almost 100 times of bottle and wax gourd. Its vitamin C content is much higher than that of any gourds (Singh, 1989).

Origin, History and Distribution

The center of origin is not precisely known but most authors agree that Indian subcontinent or Indo-Malayan region as its original home (Seshadri, 1986). Its rich diversity of cultivated form in Asom-Bengal also proved its primary centre of origin in India because its wild forms are also available throughout north India (Choudhury, 1979). This crop is extensively cultivated in eastern Uttar Pradesh, Bihar, West Bengal and Asom and to a lesser extent (in small pockets of hilly tracts) in Odisha, Madhya Pradesh, Maharashtra, Andhra Pradesh, Tamil Nadu and Gujarat. Recently, it has been introduced in and around Hyderabad and Bangalore area. It is grown in the uplands as irrigated crop in Uttar Pradesh, Bihar and West Bengal. In the Diara lands (river basins prone to flood inundation) of above states the crop is rain fed.

Species Availability, Evolution and Diversity

There are 22 species among them 9 related species (*Trichosanthes cucumerina* L. (syn. *T. anguina* L.), *T. bracteata* (Lamk.) Voigt. (syn. *T. palmata* Roxb.), *T. lobata*, *T. cordata* Roxb., *T. wallichiana* Wight (syn. *T. multiloba* Clarke), *T. cucumeroides*, *T. japonica*, *T. shikokiana* and *T. nervifolia* L.) of potential importance. Among these *T. cucumerina* (snake gourd) occurs in the north-eastern region, along with semi-wild *T. dioica* Roxb (pointed gourd). Another widelydistributed species is *T. bracteata* occurring in eastern India, extending to the south, and sporadically in the Himalaya (1500 m). *T. cordata* (related to *T. anguina*), occurs in the peninsular region extending to north-eastern plains and hills. The major zone for distribution of diversity for *T. dioica* is north-central and northeastern India including West Bengal.

Botanical Description

Pointed gourd is a true diploid with 2n=2x=22, and with apparently a little genetic or no chromosomal divergence. It has got the unique heterochromosomes. *T. dioica* Roxb. is a male heterogametic XY type (20 +XY). No marker characters are available to predict sex in the crop. RAPD marker associated with female that is not present in males of *T. dioica* Roxb (Singh *et al.*, 2002) and finding will be helpful in early assessment of gender. Flowering and fruiting in pointed gourd starts from February and continued up to November. Due to dioecious nature, the crop is highly cross pollinated.

Table 3.108: Botanical Description

Growth Habit	A perennial/semi-perennial creeper remains dormant during winter. Vines are pencil thick in size with dark green heart-shaped simple leaves (cordate). Normally, a node in the male plant bears a leaf, a simple or bifid tendril, a vegetative shoot and a glandular bract.
Root	Roots are tuberous in nature with long tap root system which develops many lateral branches.
Stem	Semi perennial trailing plant
Leaf	Leaves ovate, the basal lobes being rounded and narrow. Thev tendrils are normally unbranched.
Floral biology	The sex form is dioecious *i.e.* male (staminate) and female (pistillate) plants are separate. Male flowers are not racemed and are woolly outside. One or two solitary male flowers on distinct pedicels. The node of the female plant also bear the same parts except that male flowers are replaced by one or two (sometimes even four) females flowers on very short and indistinct peduncles. The female flower is 1-5 (usually 3) carpellary, thick and short styled, which is terminated by three bilobed or divided stigmas. Flowers are tubular white. Calyx tube is 3.0 cm long, narrow and erect, anthers are free. The pollen is sticky and not suitable for wind pollination. Occasional hermaphrodite flowers were functionless.
Anthesis and pollination	Flower opens during early morning and closes before sunset. Anthesis occure16–19 days after initiation of bud for pistillate flowers and 10–14 days for staminate flowers. Low temperature can delay the full development of flowers by 8-9 days. The flowers (both male and female) open at early hours of night (7.0-9.0 p.m.). At full blooming, the petals are separated from each other, straightened and acquire somewhat reflexed condition. Pollen dehiscence takes place with the anthesis. The pollens are perfectly round in shape, 39-78 micron in diameter with 3-week spots for the emergence of pollen tube. The stigma of female flower becomes receptive 4 hours before anthesis and the receptivity remains normal up to 12 hours after anthesis (*i.e.* up to 8-8.30 a.m. in the morning) and 40–70 per cent of flowers set fruit (Singh *et al.*, 1989). The crop is strictly entomophilous. Maximum pollen germination percentage is seen at annhesis *i.e.*, at night, which gradually declines.

	So, hand pollination must be completed as early as possible in the morning preferably within 5-5.30 a.m. Pollens of one male flower is sufficient to pollinate 8-10 female flowers. Simple touching the pollen laden male flower to stigma of the female flowers is sufficient to ensure effective pollination. Because of dioecy, the crop is highly cross pollinated and strictly entomophilous.
Fruit and fruit set	Fruit size is variable, ranging from small and round to thick and long (2.0" to 6.3" long). Fruits are globose, oblong, smooth, 5-12 cm long and 2-6 cm in breadth, striped or no stripes, stripes are light green on young fruits. The fruit is essentially a "pepo" and the edible portion is mostly pericarp with little mesocarp. Parthenocarpy is also observed in this crop (Ram *et al.*, 2003).
Seeds	The seeds are globose, brown and contains edible oil and protein.
Propagation	The propagation of pointed gourd can be done by various ways but it is usually propagated through vine cuttings and root suckers. **Seed**-Multiplication through seed is possible but in general seeds are not used in planting because seeds germinate poorly, and 50 per cent of those that do germinate produce non-fruiting male plants. Plants developed by seeds look weak with small leaves and take about a year to fruit. **Root suckers**-To propagate pointed gourd from root suckers, roots of one year old plants are dug in the early spring and sub-divided. Dip them in the solution prepared with copper sulphate and plant them in the field directly or in the nursery in sand medium for root development. Regular watering is required till it produces new sprouts. **Vine cuttings**-Both pre-rooted and fresh vine cuttings are also used for propagation. Cuttings of 60-90 cm having 10-15 nodes should be taken from mature vines in October when fruiting is almost over. In the temperate area vine cuttings made in the fall of previous year and rooted during winter are planted when danger from frost is over in the spring in order to obtain a crop in the same year. Current year vine cuttings are also planted to establish the crop during the summer, but optimum plant yield is only obtained during the next year. Fresh vines used for field planting should have 8–10 nodes per cutting and should be partially or fully defoliated to check transpiration. For raising nursery, planting bag/tube of 100 gauges having 15.0 x 10.0 cm size is used. These polyethylene bags are filled with equal amount of soil, sand and well decomposed FYM or compost. There should be 4-5 holes on upper side and 2-3 holes in bottom of polyethylene bags for air circulation and drain of excess water. Composted saw dust (4 months old) can also be used for cucumber nursery production. Collect 30.00 cm long vine and treat with the rooting hormone. In each bag, insert the treated portion of the stem in polythene bags filled with a potting mixture in such a way that 2-3 nodes are outside the bag. Apply water with rose can sparingly and put them in shade or plastic house. Sprouting take place within 30 days but root development take place after 45 days after planting in bags. These can be transplanted in the main field after 3 months (Nath *et al.*, 1987). This method can ensure cent per cent rooting, producing a large number of plants of one variety in one place.
	They can easily be transported to the main field or distant places. During transplanting, polyethylene bags/tubes should be detached by cutting sharp blade or knife in vertical ways. Generally, transplanting should be done in evening, with whole soil ball. Just after transplanting, light irrigation is essential so that plants get established well in the main field. **Tissue culture-**Rapid *in vitro* multiplication of CV. Swarna Rekha and Swarna Alaukik was achieved at ICAR-IIHR, Bangalore by culturing shoot and nodal explants on Murashige and Skoog medium containing indol-3 acetic acid (1.0µM) and indol-3 (0.2µM). Results revealed that Swarna Alaukik showed highest rate (6.6 cuttings per explant) followed by Swarna Rekha (4.9). The basal node stumps and plantlets from one week of culture were acclimatized and established *ex vitro* with 75-95 per cent efficiency in different genotypes (Mythili and Thomas, 1999). Good establishment of regenerated plantlets (over 70 per cent) in soil medium could also be achieved using physiologically mature seeds as explants cultured in different hormones (BA, NAA *etc.*) media (Hoque, *et al.*, 1998). **Ratio of male and female plants and pollination-** Pointed gourd is dioecious in nature means male and female flowers appear on separate plants. Due to lack of pollination, sometimes fruits remain undeveloped. The male plants are unproductive but supply pollens for fertilization and fruit set. So it is essential for better pollination and yield. Male and female plant population or ratio should be balanced in the field. Male and female ratio of 1:10 has been found suitable for maximum fruit set (Maurya *et al.*, 1985). Male and female plant can be identified only when they are in flowering and fruiting stage. On the basis of flowers and fruits, stems and root suckers are selected separately for propagation.

Detail botany and horticultural aspects are given by Kumar and Singh (2012).

Augmentation of Genetic Resources and Crop Improvement

Breeding Priorities

The goals for cultivar development in pointed gourd include increases productivity (increased branching, earlier maturity and more fruit per plant), high fruit quality (proper size, darker colour, reduced seed content, longer self-life) and high insect and disease resistacne. Specific proritties are:

1. Early and high yielding varieties/hybrids
2. Seedless/less seeded type varieties
3. Early identification of male and female plant
4. Thick flesh

Existing diversity with respect to yield and yield attributing traits in pointed gourd is the basic requirement in order to select the desirable types. Singh (1983) working on 18 genotypes of pointed gourd reported high genotypic and phenotypic coefficients of variation for yield/plant, moderate coefficients

Table 3.109: Varietal Description of Pointed Gourd

Varieties	Organization	Pedigree	Description
Swarna Alaukik	HARP, Ranchi	Clonal selection	Fruit is oval in shape, without strips with light green colour. Fruits are fleshy and less seeded, medium in size (5-7 cm long). Average yield is about 22-28 t/ha.
Swarna Rekha	HARP, Ranchi	Clonal selection	Fruits are spindle shaped, light green colour with white stripes. Fruits are long (8-10 cm) with average fruit weight of 30-35 g. fruits are fleshy and soft seeded. Every node bears fruits. Average yield is about 20-26 t/ha.
Kashi Alankar (VRPG-1)	ICAR-IIVR, Varanasi	Clonal selection	Fruits are medium sized and have some strips at the end. Average yield is 38.0t/ha.
Kashi Suphal (VRPG-2)	ICAR-IIVR, Varanasi	Clonal selection	Fruits are medium sized and have some strips at the end. Average yield is 30.0t/ha.
Kasihi Jagriti (IIVRPG-2)	ICAR-IIVR, Varanasi	Clonal selection	It is an early maturing variety. Fruiting starts from 4th to 6th node onwards. Fruits are light green with soft flesh. Average yield is 28.0-30.0 t/ha.
Kashi Pragya	ICAR-IIVR, Varanasi	Clonal selection	It is a seedless variety. Fruits are medium size and bear more in number
Rajendra Parwal-1	RAU, Pusa (Bihar)	Clonal selection	Fruits are long, green with white strips and tapering at both ends. Average fruit weight is 40 g. Suitable for long distance transportation and tolerant to fruit fly
Rajendra Parwal-2	RAU, Pusa (Bihar)	Clonal selection	Fruits are drum shaped, whitish green with very light self strips and soft flesh. Average fruit weight is 30 g. It is tolerant to stem and fruit rot as well as fruit fly. Average yield is 16.0t/ha.
Narendra Parwal-260	NDUA and T. Faizabad	Clonal selection	Fruits are light green, striped, 13-15 cm in length with thick flesh. It produces an average fruit yield of 22.5t/ha. When trained on bamboo stakes. It is tolerant to wilt disease.
Narendra Parwal-307	NDUA and T. Faizabad	Clonal selection	Fruit size is small, round shaped and colour is dark green with strips. It has very good self life and gives an average fruit yield of 22.5 t/ha.
Narendra Parwal-604	NDUA and T. Faizabad	Clonal selection	Fruits are medium sized, light green and without strips. It gives an average fruit yield of 22.5 t/ha.

of variation for fruit length, number of seeds/fruit, diameter of fruits and vine length. However, variability in number of days taken for first fruit set, number of days taken for first flowering and number of days taken for first picking were low. In this crop Singh (1983) recorded high heritability for fruit number, days to first picking, days to first flowering and yield/plant. Higher values of genetic advance were recorded for number of fruits/plant followed by yield/plant and length of the fruit. Singh *et al.* (1993) reported that yield/plant is positively correlated with number of fruits/plant both at genotypic and phenotypic levels. Similarly it has been observed that yield/plant was significantly and positively correlated with number of fruit/plant, number of seeds/fruit and length of vine, both at genotypic and phenotypic levels but negative and significant correlation was observed only at genotypic level with number of days taken for first flowering and number of days taken for first picking in pointed gourd (Chandra *et al.*, 1995). In pointed gourd, Singh *et al.* (1993) reported that yield per plant was positively correlated with number of fruits per plant both at genotypic and phenotypic levels.

Clonal Selection

Clonal part is genetically uniform material derived from single individual and propagated by vegetatively means. The most important step in clonal selection is creation of variability by assemblage of wide range of clones from different sources. Among collected clones, superior clones are selected on the basis of phenotype and other desirable traits after multi-locational testing. The best clone is released as new variety. Some varieties of pointed gourd *i.e.* 'Kashi Alankar (VRPG-1), 'Kasi Suphal (VRPG-2)', Kashi Amul and Seed less *i.e.* VRPG-105 have been developed by clonal selection at IIVR, Varanasi (Ram *et al.*, 2003). Clustering of genotypes revealed that genetic diversity was independent of geographic origin and grouped the genotypes under cluster I (BPS-1, BPS-3, BPS-4, BPS-10), II (BPS-2, BPS-5, BPS-6, BPS-7), III (BPS-8, BPS-9) and IV (BPS-11); (Dora *et al.*, 2002). Existing diversity with respect to yield and yield attributing traits in pointed gourd is the basic requirement in order to select the desirable types. Kashi Amulya (VRPG-89) is new clonal selection from ICAR-IIVR, Varansi (U.P.).

Hybridization and Selection

In pointed gourd hybridization has been made to develop the F_1 for better yield, quality and adoption. The superior progeny can be selected after evaluation and further maintained through vegetable reproduction, to get uniform expression of characters. In pointed gourd, hybridization is mostly done to improve the quality, yield and resistance to biotic and abiotic factors. In pointed gourd, one variety named CHES Hybrid-1 has been developed at HARP, Ranchi by this method. Few hybrids evaluated by Ara *et al.* (2018) in Bangladesh and selected 5 hybrids.

Coservation of Genetic Resources

Extensive collections of pointed gourd germplasm are being maintained at multiple locations in India.

Presently pointed gourd germplasm are being maintained bu ICAR-Indian Institute of Vegetable

Research, Varanasi (U.P.), ICAR-HARP, Ranchi (Jharkhand), NDUA & T, Faizabad, BAU, Sabour, and BCKV, Kalyani (West Bengal).

REFERENCES

1. Ara N; Moniruzzaman, M; Rahman, K.S., Moniruzzaman, M. and Sultan, R. 2018. Performance of hybrid lines of pointed gourd (*Trichosanthes dioica* Roxb.) Bangladesh J. Agril Res. 43(3):383-393.
2. Chandrasekar, B; Mukherjee, B. and Mukherjee, S.K. 1988. Blood sugar lowereng effect of *Trichosanthes dioica* Roxb. In experimental rat models. *Intl. J. Crude Drug Res*; 26:102-106.
3. Choudhury, B.1979. Vegetables. National Book Trust, New Delhi, p. 160-162.
4. Hoque, M.E., Bhowmik, A. and Khalequzzaman, M. 1998. *In vitro* culture of pointed gourd (*Trichosanthes dioica* Roxb.). *Thai. J. Agric. Sci.*, **31:**369-374.
5. Kumar, S. and Singh, B.D. 2012. Pointed gourd : Botany and Horticulture. Horticultural reviews, 39(1): 203-238.
6. Kundu, B.C. 1942. *Journal of Bombay Natural History,* **43:** 362-368.
7. Kuritha, M. 1939. Geschlechts chromosome bei *Trichosanthes multiloba* Miq. and chromosome zahlen bei einegen anderen Arten von Cucubitaceen. *Botanical Magazine, Tokyo,* **53:** 505-510.
8. Maurya, K.R., S. Barooah, R.K. Bhattacharya, and R.K. Goswami. 1985. Standardization of male plant population in pointed gourd. *Ann. Agr. Sci.*, **30:** 1405–1411.
9. Nakajima, G. 1937. Cytological studies in some dioecious plants. *Cytologia*, 282-292.
10. Nath, P., Velayudhan, S. and singh, D.P. 1987. Vegetable for Tropical Regios. ICAR, New Delhi, p. 63-67.
11. Ram, D., Kalloo, G., Banerjee, M.K. and Singh, B. 2003. Seedless and long duration fruiting pointed gourd (*Trichosanthes dioica* Roxb.) IIVRPG-105 (INGR No. 03035; IC296492). *Indian J. Plant Genet. Resour.*, **16(2):** 163-164.
12. Sen, R. 1976. Cytogenetic investigation in some Indian cucurbits. Ph.D. thesis, University of Kalyani, West Bengal.
13. Seshadri, V.S., 1986. Cucurbits. In: Vegetable Crops in India (Eds. Bose, T.K. and Som, M.G.) Naya Prokash, 206, Bidhan Sarani, Calcutta, India. p. 91-164.
14. Sharma G; and Pant, M.C. and G. Sharma. 1988. Preliminary observations on serum biochemical parameters of albino rabbits fed on *Trichosanthes dioica* (Roxb.) Indian J. Med. Res., 87:398-400.
15. Sharma, G. and Pant, M.C. 1988. Effect of feeding *Trichosanthes dioica* (Parwal) on blood glucose serum triglyceride, phospholipid cholesterol and high density lipoprotein-cholesteral levels in the normal albino rabbit. *Curr. Sci.* 57: 1085-1087.
16. Singh, A.K., R.D. Singh, and J.P. Singh. 1989. Studies on floral biology of pointed gourd (*Trichosanthes dioica* Roxb.). *Veg. Sci.*, **16:**185–190.
17. Singh, K. 1989. Pointed gourd (*Trichosanthes dioica* Roxb.). *Indian Hort.*, **33:**35-38.
18. Singh, M., Kumar, S., Singh, A.K., Ram, D. and Kalloo, G. 2002. Female sex associated RAPD marker in pointed gourd (*Trichosanthes dioica* Roxb.). *Current Science,* **82(2):** 131-132.
19. Singh, V.P. 1983. Genetic variability and correlation studies in Parwal (*Trichosanthes dioiea* Roxb.) M. Sc. Thesis, NDUA&T. Faizabd.
20. Varghese, B.M. 1973. Studies on cytology and evolution of South Indian Cucurbitaceae. Ph.D. thesis, Kerala Agricultural University, Trivendrum (Kerala).

Snake Gourd (*Trichosanthes cucumerina* L.)

Introduction

Snake gourd (*Trichosanthes cucumerina* L. var. *anguina*, 2n==2x=22) is considered life support plant. This crop is also known as Chichinda, serpent gourd, and padwal. It is named 'snake gourd' as a result of its slightly twisted fruit shape when allowed to grow naturally. Chinese name of snake gourd is "Manka" means "cucumber of the southern barbarians", indicated India. Its cultivation is being done only in kitchen garden or homestead garden for its strikingly long immature fruit which is used as a vegetable, mature for soups and for medicine. Now it occupies a pride place among vegetables grown particularly in South India. In north India, it is grown only as a rainy season crop. Immature and tender fruits (0.3 to 0.4 m long) are harvested and cooked. Sometimes fruit attain the length up to 1.5 m. Its soft bland, somewhat mucilaginous flesh is similar to that of the *Luffa* and the calabash. Sometimes fruits are peeled, sliced or cut into pieces and boiled and eaten likew green beans. Fruits are also used in curries, '*sambar*' and stews. Snake gourd is an important source of protein, minerals and vitamins A, B, and C. In per 100 g fresh fruit weight, the constituents are 92.00 g moisture, 2.0 g protein, 0.3 g fat, 0.5 g minerals (30.0 mg calcium, 40.0 mg phosphorus and 1.7 mg iron), 155 μg carotene, 0.05 mg thiamine, 0.06 mg riboflavin, 0.5 mg niacin and 29.0 mg vitamin C. The bright red pulp around the mature seeds is extracted and used in cooking the same way as the tomatoes are used. Occasionally, the tender leaves and shoots are also cooked as a potherb due to better source of protein, carbohydrates, minerals and fat than the fruits. As medicinal properties, the fruits are believed to improve appetite and act as a tonic. The roots and seeds are used to cure diarrhea, bronchitis and fever (Chopra *et al.*, 1986). Ojiako and Igwe (2008); Gopalan *et al.* (1989).

Origin, History and Distribution

Snake gourd (*T. anguina*) was originally wild in India. Indian Archipelago considered to be a cultivated state of *T. cucumerina* from which it differes in the fruit. As the botanist of 17th century received the plant from China, they imagined that the plant was indigenous to China. Chinese name of snake gourd 'Manka' means 'cucumber of the southern barbarians', indicated India as the possible centre of origin. At present snakegourd is also being cultivated in Mauritius, and Central and East Java. *T. anguina* or *T. cucumerina* is distributed throughout the country. Most gourd specilists agreed that the name *T. anguina* be reserved for the wild species. Snake gourd cultivation can be spotted in India, South Asia, China, Japan, West Africa, Latin America, the Caribbean and tropical Australia. Its distribution can be observed in West Bengal, Punjab, Delhi, Uttaar Pradesh, Uttarakhand, Bihar, Gujarat and Maharashtra.

Species Availability, Evolution and Diversity

Genus *Trichosanthes* has 40 species in India and among them snake gourd (*Trichosanthes anguina*) is most popular vegetable. Apart from *T. anguina* or *T. cucumerina*, 39 other species of the genus *Trichosanthes* are available in East, South East, China, Japan, Western India, Fiji and tropical Australia. This species can also be spotted in Southeast Asia, extending through Malaya to North Australia in one direction and China to Japan in another. A new species of *Trichosanthes i.e. Trichosanthes mianyangensis* reported as a perennial herbaceous and climbing plant (Yueh and Liao, 1972). Two cultivated species under the genus are *T. anguina* and *T. dioica*. The related species of *Trichosanthes* are *T. cucumerina*. *T. bracteata* (Lam.) Voigt.(syn. *T. palmata* Roxb.), *T. lobata*.*T. cordata*, *T. wallichiana* Wight (syn. *T. multiloba* Clarke), T. *cucumeroides*, *T. japonica*, *T. shikokiana* and *T. nervifolia*. Similarly, *Trichosanthes dafangensis* was newly identified from Dafang country (China), which flowers are similar to those of *Trichosanthes hylonoma* and pubescence and bracteoles to those of *Trichosanthes anguina* (*Trichosanthes cucumerina*) and *Trichosanthes cucumerina* are degenerated pistils in male flower stamens in female flowers and cylindrical fruits more than 20.0 cm in length with 80 seeds/fruit (Ye and Li, 1989). A widely distributed species is *Trichosanthes bracteata* occurring in eastern India, extending to the south, and sporadically in the Himalaya (1500 m). Snake gourd (*Trichosanthes anguina*), occurs in the peninsular region extending to north- eastern plains and hills. In a comparison the karyotype of *Trichosanthes anguina* (2n=2x=22), *Trichosanthes cucumerina* (2n=2x=22), and *Trichosanthes palmata* (2n=2x=22) revealed that the former two species show closer resemblance. In a comparison of their ideograms it is clear that they differ from each other in number of types of chromosomes and the number of each type represented in them and total chromatid length. The three taxa *Trichosanthes anguina*, *Trichosanthes lobata* and *Trichosanthes cucumerina* may therefore be related in one of the two ways either (i) *Trichosanthes cucumerina* ® *Trichosanthes lobata* ® *Trichosanthes anguina* or (2) *Trichosanthes cucumerina* ®*Trichosanthes anguina*.

In India the major zones of this species are along the Malabar Coast in Western Ghats and low and medium elevation zones (up to 1500 m) in Eastern Ghats and north eastern region. Genetic variability status is medium in India. Its genetic erosion status still not reported. In snake gourd distribution and extent of diversity (wild and cultivated) is lying mainly tribal areas of India. More diversified nature of this crop may

Table 3.110: Genera and their Wild Relatives (Interspecific, intraspecific and infraspecific categories)

Crop	Genus	Cultivars (Species)	Wild Relatives (Species)	No. of Taxa available in the Region
Snake gourd	Trichosanthes	T. palmata	T. palmata	T. palmata
		T. multilova	T. multilova	T. wallichiana
		T. cordata	T. cordata	T. multilova
		T. truncata	T. truncata	T. cordata
		T. himalensis	T. himalensis	T. truncata
		T. dicaelosperma	T. dicaelosperma	T. dioica
		T. dioica	T. nervifolia	T. anguina
		T. nervifolia	T. cucumeriana	
		T. cucumeriana	T. lobata	
		T. anguina	T. integrifolia	
		T. lobata		
		T. integrifolia		

Details of *Trichosanthes* Species

Table 3.111: Characteristics of different Species

Trichosanthes palmata	Leaves 3-7 inches across, usually palmately 3-7 lobed; Flowers 1.5-2.5 inches across, male flowers in elongate racemes, female flowers on short peduncles; Fruit 1.5-2 inches across, ellipsoid or globose, at first red with orange streaks.
T. multiloba	Leaves 3-6 inches across, glabrous or scabrous with scattered bristly hairs on the nerves; Flowers dioecious; Fruit bright red with orange stripes.
T. cordata	Leaves 6-8 inches across, entire, angled or obscurely lobed; Flowers dioecious; Fruit globose.
T. truncata	Leaves 4-5.5 by 2.5-3 inches, ovate acuminate; male flowers white, female plant not seen.
T. himalensis	Leaves 5 inches, irregular serrate; Fruit 3-4 by $1\text{-}1^1/_4$ inches, nearly circular in section.
T. dicaelosperma	Leaves 4 by $3^1/_2$ inches; Fruits pubescent with 10 paler vertical bands
T. nervifolia	Leaves 3 by $1^1/_2$-2 inches, membranous, not lobed; Fruit 2-3 inches, circular in section, about twice as long as broad.
T. cucumerina	Plant is annual and closely related to cultivated *T. anguina* (snake gourd) but differs in fruit length, shape and absence of stripes on fruit surface. Leaves 2-4 inches diameter, usually 5-lobed about half-way down, lobes obtuse; Flower momnoecious, male peduncles in pairs, the earlier 1 flowered the later racemed.
T. lobata	Leaves pubescent beneath and with flaccid hairs; Fruit 2-4 by 1-2 inches, acute at both ends.
T. integrifolia	Leaves $2^1/_2$-6 inches; Fruit 2 inches in diameter, red in colour

be observed northern plains, north eastern regions in southern tract particularly in the wild form. Snake gourd is distributed throughout the country, however, the rich diversity in *T. anguina* has been observed in north-eastern states, West Bengal, Malabar Coast and in Eastern Ghats in low and mid hills. More diversified nature of this crop may be observed northern plains, North Eastern regions in southern tract particularly in the wild form. A good variability exists in India. Genetic variability status is medium in India.

Botanical Description

Snakegourd belongs to family Cucurbitaceae, genus *Trichosanthes*, and botanical name is *T. anguina* L.

Table 3.112: Botanical Description

Growth Habit	An annual, fast growing climbing cucurbit. A long growing season is necessary.
Root	Non-tubrous tap root system
Stem	Stems are slender, 5 angled, grooved and pubescent with tendrils that are branched having glandular hairs.
Leaf	Leaves are simple, hairy, angular or 5-7 lobed, dentate, 10.0-25.0 cm in length, 15.0 cm in diameter and emitted foetid odor when damaged. Leaf has dark green on upper surface and light green on lower surface and emit foetid odour when rubbed.
Floral biology	The flowers open late in the afternoon. Flowers are monoecious (individual flowers are either male or female, but both sexes can be found on the same plant) and white small (up to 5.0 cm in diameter), very aromatic, unisexual, axillary, white, female flowers solitary, and sessile. Ovary long, narrow, densely hairy, male inflorescence on long, slender densely pubescent stalks 10.0-20.0 cm long, 5 to many flowered, floweres in axils of short bracts, pedicels long, filiform, densely pubescent, 1.0-2.5 cm long, calyx companulate, 5 partile, densely pubescent, crolla 5 partile, ovate-oblong, long fimbriate. Flowers occur in leaf axils in 3 different ways: (i) Male (staminate) solitary (ii) Male (staminate) solitary and male (staminate) inflorescence
	(iii) Female solitary and male(staminate) inflorescence Ratio of male: female flowers range from 31.38: 1 to 225.23:.1. Days to first male flower appearance range from 36 to 45 days after sowing and female flower appearance from 45 to 51 days.
Anthesis and pollination	The crop is highly cross-pollinated due to monoecy. Flowers open in the leaf axil after (as a bud) 8-12 days in pistillate flowers and 13-16 days in staminate flower. Flowers are fragrant. Plants bloom at early hours of night and anther dehiscence occurs before flower opening or tends to go together. Deshpande *et al.* (1980) reported that anthesis begins between 05.15 PM. taking 1.0 hour for completion and continued until 09.30 PM. Pollen-grains remain viable 10 hours before dehiscence and 46-49 hours after dehiscence. Stigma remains receptive from 10 hours before opening till 51 hours after opening. Flower open late in the day (early evening) and at night, suggesting moth (insect) pollination.
Fruit and fruit set	Fruits attain vegetable maturity 7-10 days after flowering. Fruits are cylindrical with waxy surface, slender, narrow, cylindrical, tapering at both the ends, 40.0-150.0 cm in length 4.0-10.0 cm in diameter. Surface colours are greenish white and may be stripped.Orange when ripe, pulp red at maturity. Interior flesh is white. Individual fruit weight is about 200-400 g or more.
Seeds	In general, seeds are generally brown with rough surface, indented edges, tapering at one end and 10.0-15.0 mm in length. Approximately 40-70 seeds/ fruit and 4-5 seeds/g.

Horticultural Classification

In snake gourd, variation is not seen only due to length of fruit but also seen on the basis of stripness. In some genotypes stripes are found on the whole fruit in length but in some genotypes it is not prominent. On the basis of fruit colour and stripes developed on it may be divided into two groups.

1. Light green white stripes
2. Dark green yellow or creamy stripes

Augmentation of Genetic Resources and Crop Improvement

The southern peninsular tract of India was identified as the first priority region for collection of *T. cucumeriana* var. *anguina* based on germplasm distribution, variability, extent of gene ersion, and a survey of existing collections (Toll and Vausloten, 1982). Wide variability of snake gourd exists in south India, and institutions like Tamil Nadu Agricultural University, Coimbatore and Kerala Agriculural University, have made a diverse collections of germplasom. A wide range of variability exists in Bangladesh (Ahmed *et al*; 2000; Rahman, 2004; Khatun *et al.* 2010) and in India (Phnadath, 1978; Mathew 1996; Ashok, 2000). N.I. Varilov Research Institute of Plant Industry (VIR), St. Petersburg, Russian Federation collected 15 genotypes of *Trichosanthes cucumerina* Var. *anguina*. Wild snake gourd grows in scrub vegetation along forest edges and open forest, from sea level to 1,500 m altitude.

Gene and Genetics

In snake gourd a multiple allelic series controls fruit skin colour, which segregated in a monohybrid ratio in the F_2 generation. Deep green was dominant over green, yellow and white; green was dominant over yellow and white; green was dominant over yellow and white and yellow was domiant over white (Sardar and Mukherjee, 1987). Mini Raj *et al.* (1993) reported monogenic inheritance of fruit colour-white (WW) dominant over green-white stripes (ww), fruit size (partial dominance), and fruit pericarp colour (deep green dominant over green, yellow, and white; green dominant over yellow and white and yellow dominant over white). Characters such as node of first pistillate flower, fruit length, fruit weight, fruit diameter, number of pistillate flowers fruit length, fruit weight, fruit diameter, number of pistillate flowers per plant, and fruit per plant are polygenic and positively correlated with yield. Kumaresan *et al.* (2006b) reported that days to first staminate flower opening, days to first pistillate flower opening, main vine length, number of fruit per vine, fruit length, fruit diameter, fruit weight, and yield per vine were controlled by both additive and non-additive gene action with predominace of gene action. (Bharathi *et al.* 2013). Path coefficient analysis showed that individual fruit weight, fruit diameter, number of fruit per plant, and node of first pistillate flowers are the most important characteristics contributing to yield due to high direct effects (Pynadath, 1978; Kumaresan *et al.*, 2006a; Naryanan Kutty *et al.*, 2006). On the bais of correlation and path coefficient analysis, proper weight must be given to pistillate flowers per plant number of fruit per plant, and fruit weight by virtue of their high contribution to yield (Kondalrag *et al.*, 1984). Analysis of combining ability showed significance of general and specific combining ability variances for all characters excepts sex ratio, number of fruit per plant, and crop duration, indicating the role of both additive and non-additive gene action (Varghese, 1991).

Breeding Priorities

1. Straight fruit
2. Early and high yielding varieties/hybrids
3. Good storability
4. Fleshy and less seeded type

A good variability of the crop exists in India. Significant positive correlations were observed between days to first male flower and days to first female flower, days to first male flower and node at which first female flower appeared and days to first female flower and node at which first female flower appeared. The early

Table 3.113: Varietal Description of Snake Gourd

Varieties	Organization	Pedigree	Description
CO-1	TNAU, Coimbatore	Selection	Fruits are long (160-180 cm), dark green with white strips and whitish green flesh. It gives an average yield of 18 t/ha.
CO-2	TNAU, Coimbatore	Selection	Fruits are greenish-white, average fruit weight is 400-600 g. It gives an average yield of 36t/ha. In 105 days crop duration.
PKM-1	TNAU, Coimbatore	Mutation	Fruits are small and having an individual fruit weight of 45.65 g with green strips. It gives an average yield of 34 t/ha in crop duration of 150-155 days.
MDU-1	TNAU, Coimbatore	Selection	This cultivan produce 13 fruits/vine. Fruits weigh about 7.15 kg. The fruits are medium-long with white strips. It gives an average yield of 30.0-35.0 t/ha.
Kaumudi	KAU, Vellanikkara	Selection	Fruits are large (100 cm long and 6.9 cm girth), uniform in size with white acute tip. Average weight of fruit is 1.34 kg. It gives an average yield of 50 t/ha.
Baby	KAU, Vellanikkara	Selection	Fruit are 36.5 cm long and white with good keeping quality. Average weight of fruit is 475 g. It gives an average yield of 9.8 t/ha.
Manusree	KAU, Vellanikkara	Selection	Fruit are 32 cm long and white with good keeping quality.Average weight of fruit is 405 g. It gives an average yield of 10 t/ha.
H-8	CCS HAU, Hisar	Selection	-
H-371	CCS HAU, Hisar	Selection	-
H-372	CCS HAU, Hisar	Selection	-
TA-19	KAU	Selection	-
IIHR-16A	ICAR-IIHR	Selection	-

varieties carried female flowers in lower nodes. Over exploitation and use improved varieties appears to have taken a heavy toll on the genetic resources of this crop.

Ara *et al.* (2013) reported wide range of variability for days to first male flower anthesis, days to first female flower anthesis, vine length, node number at first female flower, number of shoot/plant, and days to first harvest. However, with regard to quantitative trails like number of fruit/plant, the highest number (45.6) was obtained from the genotype TAISD028 which was statistically similar to TAISD009, TAISD011, TAISD019 TAISD027, TAISDO 30 and TAISD032 also performed well under the agroclimatic conditions of Pabna (Bangladesh).

Introduction and selection

Many varieties were evolved through selection. Among them 'CO1' and 'CO2' from the Tamil Nadu Agricultural University, Coimbatore (Tamil Nadu) and 'TA 19' from the Kerala Agricultural University, Vellanikkara (Kerala), are based on the single-plant selection. At TNAU, under the vegetable improvement scheme three selections, *i.e.*, H-8, H-371 and H-375 were reported very promising. Similarly in MPKV, Rahuri, (MS) 30-2-2 has been selected an excellent yielding line. ICAR-IIHR has also selected an early and better yielding variety, *i.e.*, IIHR-16-A (Nath and Seenappa, 1973). Popular cultivars like and TA-19 and Baby developed by use of Indigenous material through selection with progeny testing.

Hybridization and Selection

Natarajan *et al.* (1985) reported a hybrid snakegourd 'MDU 1', a cross between 'Panni Pudal' and 'Sel 1'. Prakash (1953) reported that intervarietal F_1 hybrids of snake gourd are superior to their parents in vigour, number and size of fruits.

Mutation

Pillai *et al.* (1979) exposed dry seeds of 'H 375' to gamma rays at 10, 20, 40 and 50 kr. One of the mutants from 20 kr treatment was selected promising ('PKM 1') after evaluation for 5 generations. 'PKM 1' is very vigourous, the vine flowered in 55 days after sowing and 1st harvest could be done in 65 days Pillai *et al.* 1979. Seed treatment of snake gourd by 18 kr X-rays has given superior mutant for fruit size, yield and high punicic acid content in seed oil (Datta 1994; Datta 1997).

Polyploidy

Datta (1977) reported that colchicines treatment resulted in larger pollen. The tetraploids had larger pollen-grains than diploids and triploids. Sardar and Mukherjee (1988) observed flower on tendril (FOT) plants in which the female flower was attached to the tendril, on treatment with colchicines. No useful colchiploids were detected.

Coservation of Genetic Resources

The southern penisulan tract of India was indentified as the first priority region for collection of *T. cucumerina* var. *anguina* based on germplasm distribution, variability, extent of gene erosion and survey of existing collections (Toll and Vausloten, 1982).

TNAU Coimbatore and KAU Kerala have made a rich collection of germplasm and maintaining them for further use in breeding programme. KAU Kerala has collected 25 land races from different parts of the country. ICAR-NBPGR, New Delhi has given emphasis in maintaining and conserving the valuable lines of snake gourd (144) in National Gene Bank/National Active Germplasm Sites

REFERENCES

1. Ara, N; Hossain, M.F. and Akhter, B. 2013. Characterization and evaluation of snake gourd (*Trichosanthes anguina* L.) genotypes. *Wudpecker* Journal of Agricultural research, 2(12): 330-334.
2. Bharathi, L.K.; Behera, T.K., Sureja, A.K. John, K.J. And Wehner, Todd-C. 2013. Snake gourd and pointed gourd: Botany and Horticulture, 41:457-495.
3. Chopra, R.N., Nayar, S.L. and Chopra, I.C. 1986. Glossary of Indian Medicinal plants (including the supplement). Council of Scientific and Industrial Research, New Delhi.
4. Datta, S.K. 1977. Studies on an X-ray induced yellow fruit mutant of *Trichosanthes anguina* L. *Transactions of Crop Improvement,* **14:** 182-184.
5. Datta, S.K. and Basu, R.K. 1978. Cytomorphologiocal, biochemical and palynilogical studies in *Trichosanthes anguina* and *T. cucumerina. Cytologia,* **43:** 103-117.
6. Datta, S.K. 1994. Induction and analysis of short thick fruit in *Trichosanthes anguina* L. *Indian J. Genet.* Plant Breed; 54(1) : 13-17.
7. Datta, S.K. 1997. Studies on an X-ray indeced yellow fruit mutant of *Trichosanthes anguina* L. crop. Improv, 14: 182-184.
8. Gopalan, C., Rama Sastri, B.V. and Balasubramanian, S.C. 1989. Nutritive value of Indian foods. NIN, Hyderabad, p. 1-156.
9. Natarajan, S., Shanmugavelu, K.G., Nambisan, K.M.P. and Krishnan, B.M. 1985. New varieties of horticultural crops released by Tamil Nadu Agricultural University, Coimbatore (Tamil Nadu), during 1985. *South Indian Horticulture,* **33:** 454.
10. Ojiako, O.A. and Igwe, C.U. 2008. The nutritive, anti-nutritive and hepatotoxic properties of the *Trichosanthes anguina* (snake tomato) fruits from Nigeria. *Pakistan J. Nutr.* 71(1): 85-89.
11. Pillai, O.A.A; Vellu Kutty, B. and Jeyapal, R. 1979. PKM-1 a new snake gourd (*Trichosanthes anguina* L.) South Indian Hort; 27: 371-382.
12. Toll, J. and Vausloten, D.H. 1982. IBPGR directory of germplasm collections-Vegtables, IBPGR, Rome.

Genus *Momordica*

Introduction

The genus *Momordica* has 60 species reported to occur in the world and 7 species in India but only four (*Momordica balsamina, M. charantia, M.chochinchinensis, M. dioica*) are commonly available and cultivated in one or other part of north eastern region.

About six species are cultivated for the fruits, used as a vegetable, the most common being bitter gourd (*Momordica charantia*) grown all over the country in tropical and sub-tropical climate. *Momordica dioica* (Kartoli) grows all over West Bengal, Assam, parts of Bihar and adjoining area; *Momordica cochinchinensis* (kakrol or sweet gourd or gol kakora) is most popular particularly in Tripura, Asom and West Bengal. Species of *Momordica* occurring in India are as under:

Indian species can be grouped under two headings: *Momordica charantia* L. and *Momordica balsamina* L. representing the monoecious group and *Momordica dioica* Roxb., *Momordica sahyadrica* Joseph and Antony, *Momordica cochinchinensis* (Lour.) Spreng. and *Momordica subangulata* Blume (ssp. *renigera* (G. Don) W.J.J. deWilde) representing the dioecious group. Although, the genus, as circumscribed here, does not includes *M. cymbalaria* Fenz. [*Luffa cymbalaria* = *Momordica tuberosa* (Roxb.) Cogn.], some workers still treat it under *Momordica* (Joseph John, 2005). Its occurrence is restricted to the Western Ghats, Maharshtra southwards, with only sporadic availability in the eastern peninsular region. *M. dioica* Roxb. and *M. cochinchinensis* occur wild/ semi-wild in the Gangetic plains extending eastwards. *M. subangulata* and *M. macrophylla* occur largely in the north-eastern region; *M. subangulata* also exhibits sporadic distribution in Deccan plateau, extending to the Eastern Ghats. *M. denundata* is lagely confined to eastern peninsular tract.

Indian bitter gourd is classified into two botanical varieties based on fruit size, shape, color, and surface texture:

1. *M. charantia* var. *charantia* has large fusiform fruits, which do not taper at both ends, and possess numerous triangular tubercles giving the appearance of a "crocodile's back";
2. *M. charantia* var. *muricata* (Wild), which develops small and round fruits with tubercles, more or less tapering at each end (Chakravarty, 1990). Both varieties are widely cultivated throughout tropical and subtropical regions of India.

Table 3.114: Taxon and Preferred Taxon Name

Name of taxon	Name of authority	Family	Preferred taxon name
Momordica			Momordica
Momordica			Momordica
Momordica balsamina	L.	Cucurbitaceae	Momordica balsamina
Momordica charantia	L.	Cucurbitaceae	Momordica charantia
Momordica cochinchinensis	(Lour.) Spreng.	Cucurbitaceae	Momordica cochinchinensis
Momordica cylindrica	L.	Cucurbitaceae	Luffa aegyptiaca
Momordica cymbalaria	Hook. f.	Cucurbitaceae	Momordica cymbalaria
Momordica dioica	Roxb. ex Willd.	Cucurbitaceae	Momordica dioica
Momordica elaterium	L.	Cucurbitaceae	Ecballium elaterium
Momordica foetida	Schumach.	Cucurbitaceae	Momordica foetida
Momordica grosvenorii	Swingle	Cucurbitaceae	Siraitia grosvenorii
Momordica luffa	L.	Cucurbitaceae	Luffa aegyptiaca
Momordica muricata	Willd.	Cucurbitaceae	Momordica charantia
Momordica operculata	L.	Cucurbitaceae	Luffa operculata
Momordica ovata	Cogn.	Cucurbitaceae	Momordica ovata
Momordica pedata	L.	Cucurbitaceae	Cyclanthera pedata
Momordica peteri	A. Zimm.	Cucurbitaceae	Momordica peteri
Momordica racemosa	Steud.	Cucurbitaceae	Echinopepon racemosus
Momordica schimperiana	Naudin	Cucurbitaceae	Momordica schimperiana
Momordica sp.		Cucurbitaceae	Momordica sp.
Momordica trifoliolata	Hook. f.	Cucurbitaceae	Momordica trifoliolata
Momordica tuberosa	(Roxb.) Cogn., nom. illeg.	Cucurbitaceae	Momordica cymbalaria
Momordica zeylanica	Mill.	Cucurbitaceae	Momordica zeylanica

Table 3.115: Occurrence of *Momordica* Species in India

Genera and Species	*Description*
Momordica charantia L. (*Momordica muricata* DC.)	It is totally domesticated. Occurs naturally and in cultivation throughout India, Sri Lanka, Malaya, China, tropical Africa and America. Flowers are monoecious. Two varieties *Momordica charantia* (fruits large, not tapering at both ends) and *Momordica muricata* (fruits small, tapering at both ends) are vine type and has variation in fruits shape and size. Leaves 1-3 inches across, deeply 5-7 lobed; flowers monoecious and yellow in colour; fruit 1-5 inches long, tapering at both ends, is characterized by its tuberculate-muricate skin and sculptured seed surface; full mature unripe fruit at tender stage is used for vegetables, in pickles and curries and also in salted salad. The fruit becomes yellow when ripes.
Momordica dioica Roxb. (*Momordica balsamina* Wall., non L.)	Occurs naturally throughout India, Sri Lanka, Burma, China and Malaya; cultivated in the Deccan. Flowers dioecious and tubers are also edible. It is also found in cultivated/semi wild form. It is dioecious, perennial in nature having tuberous roots. The green fruit is extensively used as vegetable by cooking or frying. Leaves 1.5-4 inches long, cordate, acute more or less 3-5 lobed; Flowers large, dioecious and yellow in colour; Fruit 1-3 inches long, shortly beaked, densely covered with soft spines.
Momordica balsamina L. (non Wall.)	Occurs naturally in Malaysia, China, Polynesia, Australia, western Asia and tropical Africa; introduced into America and cultivated in Sind (Pakistan). In Idia, it occurs in semi-dry north-western plains, and only sporadically elsewhere in upper Gangetic region and in the northern parts of Western and Eastern ghats. It is cultivated in Punjab, Maharashtra, Rajasthan and Uttar Pradesh but rare in Deccan. Flowers are monoecious. Leaves $1^1/_2$-3 inches diameter, palmately 3-5 lobed to about the middle, and fruit 1-3 inches, red, usually quite smooth. Immature tender fruits are are used as vegetable or pickled.
Momordica subangulata Blume	Occurs naturally in India (Karnataka, Meghalaya and Sikkim), Malaya and Java but rare. Flowers are dioecious and perennial in nature with short annual vines.
Momordica cochinchinensis (Lour.) Spreng.	Occurs naturally in India (Asom, Nagaland, Tamilnadu, Uttar Pradesh and West Bengal), Burma, Malaya, Formosa, China, Philippines and Pacific islands; cultivated in north Kanara. Flowers are dioecious. The plant is perennial with tuberous roots. Leaves 4-5 inches across, usually 3 lobed; flowers dioecious, large, whitish; fruit 3-5 inches long, ovate or oblong, covered with conical points, and bright red when ripes. Hard spines are found on fruit surface. The mature fruit is used as vegetable.
Momordicaa denudata (Thw.) Cl.	West coast of peninsular India and Sri Lanka. Leaves 2-4 by 1-2 inches, deeply cordate, and 3-5 lobed.
Momordica macrophylla Gage	India (Asom), Burma and Malaysia.
M. cymbalaria	Leaves 1-2 inches broad, 5-lobed; Fruit, few, shortly obovoid, smooth, and shining.

Yang and Walters (1992) classified bitter gourd into three horticultural groups or types:

1. A small-fruited type where fruit are 10 to 20 cm long, 0.1 to 0.3 kg in weight, usually dark green and very bitter.
2. A long-fruited type (most commonly grown commercially in China), where fruit are 30 to 60 cm long, 0.2 to 0.6 kg in weight, light green in color with medium-size protuberances, and are only slightly bitter.
3. A triangular-fruited type where cone-shape fruit are 9.0 to 12.0 cm long, 0.3 to 0.6kg in weight, light to dark green with prominent tubercles, and moderately to strongly bitter.

More recently, Reyes *et al.* (1994) reclassified Indian and southeast Asian *M. charantia* botanical varieties based on fruit diameter (*M. charantia* var. *minima* Williams and Ng < 5 cm and *M. charantia* var. *maxima* Williams and Ng > 5.0 cm). The morphological difference among six cultivated species of *Momordica* is described in Table 3.117. Cytogenetic studies confirmed the diploid chromosome number (2n = 22) of *M. charantia*. Another species of *Momordica* (*M. dioica*; 2n = 28), a dioecious cucurbit, has a different karyotype from the other two species (*M. charantia, M. balsamina*; 2n = 22) but the meiosis in these species is regular (Bhaduri and Bose, 1947; Roy *et al.*, 1966; Trivedi and Roy 1972). Crosses of *M. charantia* and *M. balsamina* (2n = 22) with *M.* dioica (2n = 28) are sexually incompatible (Singh, 1990). Likesvise, the cross *M. charantia* x *M. dioica* (and reciprocal), failed to set fruit when normal pollen was used (Vahab and Peter, 1993).

Table 3.116: Occurrence/Domestication of *Momordica* Species

Species	*Diversity/Domestication/Distributon*
M. charantia	Throughout India
M. balsamina Linn.	Semi-dry northwestern plains, Upper Gangetic region, northern parts of Western and Eastern Ghats
M. dioica Roxb. ex Willd.	Occurs wild near habitations, open forests, cleared sites Western Ghats, U.P., Bihar, Lower foot hills of Uttaranchal, Eastern-central peninsular tract, extending north-east
M. cochinchinensis (Lour.) Spr.	Meghalaya, Asom, West Bengal, Eastern peninsular tract, humid tropical forests, in openings also north-eastern region
M. cymbalaria Fenzl. ex Hook. f.	Mianly in Western Ghats, Eastern peninsular tract, Maharashtra southwards, and in eastern peninsular tract
M. subangulata Bume	Deccan plateau, North-eastern hills, Eastern Ghats
M. macrophylla	North- eastern region
M. denundata (Thwait) C.B. Clarke.	Mainly in eastern peninsular tract
M. denticulata	Eastern peninsular tract

Table 3.117: Morphological Variation among 8 *Momordica* Species Including *M. sahyadrica*

Species	*Parts*	*Description*
M. charantia L.	Plant	Annual, slender climber, 2.0-4.0 m high, scarcely to densely pubescent (tender pans wooly), monoecious climber.
	Stem	Stems are thin, acutely quinquangular with broad longitudinal furrows, white pubescent, light green, internodes 5.0-6.0 cm. Tendrils are inserted beside the leaves, on long, robust stalks, simple or cleft, faintly 5-angular, yellowish green 15.0-30.0 cm longanddelicate.
	Root	Deep tap root with many lateral branches.
	Leaf	Deeply and palmately 5-9 lobed, reniform to orbicular or sub orbicular in outline,2.5-8.0 x 4.0-10.0 cm, alternate, long stalkedcordate at base, acute or acuminate at apex, lobes ovate or obovate, narrowed at base, margins sinuate to undulate, mucronate; petioles 1.5-5.0 cm long.
	Flower	Male flower stalks slender with bract midway or toward base; peduncle 2.0-5.0 cm long; bract reniform, 5-11 mm dim., green, pedicel 2-6 cm long; receptacle-tube cup shape, 2-4 mm long and 2-3 mm wide; sepals ovate-elliptic, 4-6 x 2-3 mm, pale green; petals obovate, 10-20 x 7-15 mm, mucronate at apex, scales 2; filaments 1.5-2 mm long, inserted in the throat of the receptacle tube: anthers coherent- Female flower peduncle 1-6 cm long;
		bract 1-9mm diameter; pedicel 1-8 cm long: sepals narrow, oblong lanceolate, 2-5 mm long; petals smaller than or equal to that in male, 7-10 mm long; ovary fusiform, narrowly rostrate, 5-11 x 2-3 mm, muricate, tuberculate or longitudinally ridged; style 2 rare long.
	Anthesis and pollination	Bitter gourd is monoecious crop, with male and female flowers being borne on the same plant. Varieties differ in days to flower, time of anthesis and pollen viability (Deshpande *et al.,* 1979). Anthesis occurred between 3.30 and 7.30 AM and anther dehiscence between midnight and complete opening of the flower (7.30 AM). Pollen viability was high in all varieties (90-99 percent), while pollen germination from 66-68 percent. The stigma remained receptive for 24 hrs before and after anthesis. Pollination is brought by bees. The appearance of first pistillate flower at a lower node number is an indication of the earliness of the variety (Trivedi, 1983).
	Fruit	Pendulous, stalk 2-8 cm long; fruit discoid, ovoid, ellipsoid to oblong or blocky, often narrowed at ends, sometimes finely rostrate, 3-20 x 2-5 cm, white or green turning orange on maturity, soft tuberculate with 8-10 broken or continuous ridges, splitting from base in to 3 irregular valvate at the top7-30 cm long and flesh blood red.
	Seed	Seeds are numerous (5-30), flattened, squarish rectangular, ends subtridentete, faces compressed, sculptured, 5-9 x 3-6 rare, margins grooved and testa brown or black.Seeds are warty, 1.0-1.5 cm long and 0.6-0.9 cm broad. Seed contains 32 percent oil. Approximately 1000 seed weighs about 60.0 g.
M. balsamina L.	Plant	Annual, slender, trailing herb, 1.5-3.0 m high, sub glabrous. monoecious.
	Stem	Round, internodes 5.5-6 cm; tendrils delicate, 11-13 cm long, basal 1-1.5 cm
	Leaf	Lobed (5-7), sub circular in outline, 4-6 cut diem., base cordate with a cuneate petiole-blade juncture, apex mucronate, lobes rhomboid, margins acutely 3-7 lobulate; petiole 1-4 cm long, slender, puberulous.
	Flower	Staminate flowers larger than pistillate; peduncles slender 3-5 Got long; bract subapical, suborbiculnr, up to 0.6 x 0.5 cm, pale green, cordate at base, margins finely dentate; pedicel 0.3-0.4 cm long, receptacle tube cup-shaped, up to 0.2 mm long; sepals ovate, up to 0.7 x 0.3 mm, obtuse, pubescent; petals obevate, 1-1.3 x 0.7-0.9 cm, pale yellow to creamish-yellow, undulate margins, scales in 2 petals only; filaments up to 0.2 nun long, inserted on the rim of the receptacle tube, anthers up to 1.2-1.8 rain long. Pistillate flowers 1.7-1.8 cm across; peduncles 0.2-0.3 cm long: pedicels 0.4-0.6 cm long; bract small; calyx minute, thread like, thin; petals 0.8 x 0.8 cm, pale yellow to creamish-yellow undulate margins; ovary ovoid to fusiform. 5-7 mm long, style short, slender, whitish yellow.
	Fruit	Ovoid to ellipsoid, bulged at middle, 2.5-3.5 (4.0) cm long, 1.8-2.0cm in circumference and stalk 1-2 cm long, shortly rostrate, ashy-olive green with 2-3 white tubercles in lines across the whole length of fruits; fruits turning orange and later scarlet red on ripening; pericarp thin.
	Seed	3-5, covered by deep, red sarcotesta, ovate oblong, compressed, 8.5-9.5 x 5.9-6.2 mm, and margins finely grooved, crenulate; tests grey or light brown.
M. dioica Willd.	Plant	Vine climbing up to 3-10 m high, tuberous roots, dioecjous
	Stem	Slender, the internodes 3-8 cm long: Tendrils axillary, 4-12 cm long, the basal 2-4 cm straight and the rest spiral.
	Leaf	Thin, light green to green, ovate-cordate, nearly triangular in outline, lobed and sub lobed to various degrees or cordate and cuneate at base, the margins entire, undulate, irregularly or coarsely; the upper surface and margins with scattered short hairs, the lower surface densely short hairy; petiole slender to medium thick, 3-7 cm long, 1-1.5 Dim in diameter, longitudinally grooved.
	Flower	Staminate flowers solitary; peduncles 3-7.5 cm long (usually 5-6 cm), light green, thin; pedicels sub sessile, 2-3 mm long, whitish yellow, subtended by and protected inside a reniform clasping swollen bract, 4-5 x 8-10 mm, light green; calvx funnel shaped, lobes 5, light green, narrow acute, up to 6 x 1mm; petals 5, free, pale yellow, glandular, oblong-lanceolate 12-22 x 5-8 mm. Stamens 5, two of them with a pair of anthers and the otherwith a single anther, filaments 2-3mm long, anthers sub triangular, 2-3 mm long, yellowish brown on Inner side. Pistillate flowers solitary in leaf axils; peduncles thin, very short 0.5-2.Ocm long; pedicels thin. 2-4 cm long, subtended by a small bract of 3-4 x 2-S mm; bracts reniform with acute tip just like in male but of small size; sepals 5, semi persistent, green, narrow, 3-6 x 0.8 mm, acute at apex; petals 5; ovary oblong-ovoid, 6-9* 2-3 mm, rounded at base; styles short, up to 4mm long, glandular hairy.

Species	*Parts*	*Description*
	Fruit	Oblong, rounded atbase, abruptly conical with rostrate lip at apex, 3-4 x 2-3cm, the entire surface covered with soft short spines (except the beak), light green or dark green, turning uniformly orange on ripening, splitting from base into three irregular pieces and rolling back exposing scarlet red arils (seeds).
	Seed	2-3mm across, black lustrous and golden-lined (when fresh), sculptured on surfaces, small round to slightly oval or shortly stellate (round-ovate and smooth in Central Indian specimens), seed coat brittle, shell hard, membrane thin, whitish, endosperm oily with characteristic aromatic odor when crushed.
M. sahyadrica	Plant	Robust climber, vines up to 5-6 m high, tuberous roots with outer skin brownish and inner flesh whitish yellow, dioecious.
	Stem	Stout, the internodes 5-10cm long, nodes quadrangular, blackish green, distinctly long hairy. Tendrils medium thick, unbranched, 8-15 cm long, 4-S cm of base uncoiled, remaining coiled.
	Leaf	3-5 lobed or entire, 10-16 x 8-18 cm, deeply cordate at base with a subangulate juncture with petiole, petiole 3-8 cm long, 1-1,5 mm. in diameter; blades medium thick, ovate, broadly triangular in outline, sometimes hastate, acute, or acuminate at apex, margins highly variable, entire, undulate or coarsely crenulate, lateral veins 5-7 pairs, the lower pair running close to the margin of the subangulate petiole juncture, hairs short, scattered without, snowy white within.
	Flower	Staminate flowers axillary, solitary; peduncles 2-5 cm. long; dark green, pedicels short, 0.5-1 cm, long, whitish green subtended, and covered by an inflated bract, up to 1 x1.5 cm, reniform; calyx base funnel shape, up to 5mm long and 1cm across; petals 5, free, fleshy, obovate, up to 4 x 1.5 cm, bright yellow with a greenish yellow narrow base; stamens 3, two of them with a pair of anthers, the other with a single anther, filaments up to 3mm. Pistillate flowers solitary in leaf axils; peduncles 0.5-2,0 cm, pedicel short, up to 0,5 cm long, subtended by a small rudimentary bract, 1-3 x 0.5-5 mm; sepals 5, green, persistent, 0.8-1.3 '< 1-3 mm, equal, lanceolate, acuminate at apex densely glandular hairy within and without; petals 5, free, fleshy, up to 4 x 2cm, narrow, greenish yellow, widening toward middle, bright yellow; ovary inferior, oblong-ovoid, 1-1.5 x 0.2-0.4 cm; style up to 6mm long, whitish yellow, stigma up to 4.0 x 9.0 mm, cushiony, trifid.
	Fruit	Broadly ellipsoid, ovoid to fusiform with round blossom end and rostrate distal end, 5-7.5 x 3-4.2 cm in size, 9-12 cm in circumference, 35-50 gm in weight, dark green turning bright orange on ripening, densely clothed with soft short spines; spines 2-4 mm long; arils sweet taste, ripe fruits aromatic and slightly bitter.
	Seed	Black, shining, losing its luster on drying, round stellate to slightly cog wheel shape, warty-dentate on margins sculptured on faces with irregular furrows and ridges, 0.2-0.3 x 0.2-0.3 cm, seed coat brittle, hard shell-like, the membrane very thin, smooth, blackish green, conspicuously veined, endosperm oily, distinctly aromatic when crushed.
M. subangulata	Plant	Vine climbing up to 6-10 m high, tuberous roots, dioecious. Stem: Stout, the internodes 7-11 cm, Tendrils simple, axillar y, 15-17 cm long, basal 5-7 cm erect, the rest when uncoiled.
	Leaf	Light green, ovate cordate, unlobed, 6-12 x 7-l1 cm., acuminate at apex, cuneate at base, the basal flaps almost touching the petiole, the margins undulate; veins 3-5, ascending and many pinnate from midrib ending up in fine network of areoles, 4-5 mm across, glabrous above, glandular hairy below; petioles 7-10 cm long, thick, channeled longitudinally, margins finely ridged.
	Flower	Staminate flowers large, solitary, axillary, showy, creamish yellow, up to 9 cm across; peduncles 4-6 cm long, pedicel 0.5-1 ran long, subtended and covered inside a reniform bract, 2 x 2.5 cm, light green, sepals 5, greenish crimson, united at the base; petals 5.5-6 x 3-4 cm, free, fleshy, 3 inner petals with blackish purple blotch of 2 x 6 m size and long glandular hairs; nectary, orange yellow, enclosed in calyx cup: stamens 3, two of them with a pair of anthers, the other with a single anther, filaments up to 4 mm long, black on sides. Pistillate flowers with peduncles short, 1-1.3 con long, pedicels 10-17 cm long; bracts minute, rudimentary, near axil, often ascar of 2 x 1 m size; sepals 5,5-9 x 1-1.5 mm persistent, acute at apex; corolla and scales as in male; ovary oblong ovoid, dark green. 1.5-2 x 0.6 cm., rounded at base, finely echinulate on surface; stylii 5 to 7 m long, pale yellow, stigma cushiony, up to 4 x 6mm, trilobod.
	Fruit	Broadly ellipsoid, with dome-shape ends, 7-8 x 13-14 cm, each weighing 50-80 g; densely softly echinate, rarely with remnant ridges at base, spines 2-3rare long, light green turning yellow and finally bright orange on ripening, exposing the seeds (35-50 per fruit) by basal splitting of the fruit and rolling hack of the split lobes: flesh thick (5-6 mm), aril deep red.
	Seed	Flat, sub orbicular to subtridentate, rectangularly –cog wheel shape, 6x 3 mm and up to 4 mm thick, sculptured on faces with grooves and dented edges, margins with a double row of wart like small protuberances.
M. cochinchinensis (Lour.) Spreng	Plant	Stout perennial climber up to 20 m high, roots tuberous, woody, all parts glabrous, dioecious.
	Stem	Round, internodes 5-6 cm; tendrils delicate, 11-13 cm long, basal 2-2.5 cm uncoiled.
	Leaf	Entire or 3-5, palmately lobed, or 3 foliolate (leaflets ± elliptic with minute petiole), broadly ovate or suborbicularin outline, up to 10 x 16 cm, base cordate (sometimes with 2-4 glandular beadlike projections toward cordate margin), acute or acuminate at apex or acuminate, margins entire, undulate or remotely dentate; petiole 5-12 cm long.

Species	Parts	Description
	Flower	Solitary, axillary, staminate sometimes in a loose fascicle of 5-7, with a separate basal one. Staminate flowers with subapical bract; peduncles B-12 cm long, bract cucullate, sub orbicular or reniform, 20-40 mar wide, rounded at base, acute at apex, margins undulate; pedicels 5-8mm long, receptacle tube saucer shape, 4-5 x 8-12 mm, blackish outside; sepals 0-12 x 4-8 mm, ovate- oblong or triangular, acute at apex, blackish; petals sub elliptic, 2.5-4 x purple bull's-eye mark at base, filaments short, fleshy, 5-6 min long, inserted at the base of the receptacle tube, anthers variable in size, '5' shape, connective swollen. Female flower with small or rudimentary bract; sepals linear oblong, 4-10 mm long; petals as in male; ovary ellipsoid oblong, 12-15 race long, densely soft muricate; style 8-9 mm long.
	Fruit	Ovoid or oblongoid, bulged at middle, 10-15 x 6-10 cm, rostrate at base and stalk 5-12 cm long; pericarp densely tuberculate with uniformly short round conical structures or interspersed with larger tubercles; single fruit weighing between 350-500 g or more, green turning orange on ripening and bursting irregularly.
	Seed	Many, variable in size, 1-1.5 x 0.8-1.2cm, broadly ovate hexa-octagonal with flat sculptured surfaces, sub tridentate at ends and margins, testa black.
M. foetida Schumach	Plant	Dioecious, perennial herb, trailing or climbing with simple or bifld tendrils.
	Stem	Grows up to 4.5 m long, with dark green flecks when young, woody when old, rooting at the nodes.
	Leaf	Alternate, simple; stipules absent; petiole 1.5-17 cm long; blade broadly ovate-cordate to triangular-cordate, 1.5-16 can 1.5-17cm, base deeply cordate.
	Flower	Unisexual, regular, 5-merous; calyx with obconic tube end lobes up to 11 mm long; petals free, obovate-lingulate, up to 3.5 cm long, white, pale yellow to orange-yellow, 3 with scales inside at base; staminate flowers 1-9 together in fascicles on peduncle 2-23 cm long, with 3 stamens, anthers coherent in center of flower; female flowers solitary in leaf axils, with inferior, ovoid ovary, and stigma 3-lobed.
	Fruit	Ellipsoid berry up to 7 x 5 cm with a long-stalked, orange when ripe, densely and softly spiny, dehiscing with 3 valves and exposing the many seeds embeddedin scarlet pulp.
	Seed	Oblong, flattened, 0.1cm long, brown, testa sculptured, margins 2-grooved.
M. rostrata Zimm.	Plant	Dioecious, perennial herb with tuberous roots, trailing or climbing with simple tendrils.
	Stem	Grows up to 7 m becoming woody with gray bark.
	Leaf	Alternate, pedately 9-foliolate; stipules absent; petiole up to 2.5 cm long; central leaflet elliptical to almost circular, 1-4.5 cm 1-3 cm, lateral leaflets smaller.
	Flower	Regular, 5-merous; male flowers in axillary, 1-14-flowered, umbel-like clusters with peduncle up to 10 cm long, sepals triangular, 2-4 mm long, petals oblong, 7-13 rare long, rounded, pale orange-yellow, stamens 3, free; female flowers solitary, sub sessile, sepals triangular-lanceolate, 1.5-2 nun long, petals c. S x 4mm, ovary inferior, narrowly ovoid, 12-14 mm long and 2.5-3 mm across.
	Fruit	Ovoid berry 3-7 cm x 1.5-3 cm, beaked, rounded or slightly 8-angled, bright red, with many seeds embedded in yellow pulp.
	Seed	Broadly ovate, ca. 14 mm long, blackish brown, testa sculptured and margins grooved.

Source: De Wilde and Duyfjes (2002), Joseph and Antony (2007).

Bitter Gourd (*Momordica charantia* L.)

Introduction

This crop is known as Karela, Kash (Bengali); Balsam pear, Bitter gourd (English); Karela, Kareli (Hindi); Halal (Kannad); Kaippa, Kaippakka, Kaippavalli, Mitipagal, Pavel, Pullayini, Rajavalli (Malayalam); Karle (Marathi); Karkotakee, Karavalli, Karavella, Karavellaka, Katika, Sushavi, Vishakantaki (Sanskrit); Pagel, Pavakkai (Tamil) and Kakara kayi (Telagu). The bitter gourd is adapted to a wide variation of climates, although production is best in hot, humid areas such as tropical Asia. The fruits are bitter (except in *M. dioica and M. cochinchinensis*) but it is used as boiled, fried or cooked vegetable. Volatile components released during cooking enhance the flavor (Binder *et al.*, 1989). It is a rich source of phosphorous and iron. It also contains minerals like calcium, phosphorous, iron, copper and potassium. Young shoots and leaves are an excellent source of Vitamin A, and a fair source of protein, thiamin, calcium and phosphorus (Vinning, 1995). It has twice the amount of beta-carotene than in broccoli and twice the calcium content of spinach. Bitter gourd leaves are packed into 5 kg mixed produce boxes and sold to Melbourne and Sydney, as a medicinal plant (Peter Mc Laughlin, 1999). The seeds of the ripe fruit are used as condiments in India. The alkaloid momordicasoides impart the bitter taste to the fruit. Bitter gourd has rich nutritional and medicinal value. The immature fruit is valued for its bitter flavor. The alkaloid momordicasoides imparts the bitter taste to the fruit. The fruit is highly nutritious due to the iron and ascorbic acid content. The bitter gourd is cooked and eaten as a vegetable in India and the Far East. The bitterness is reduced by steeping the peeled fruit in salt water before cooking It is usually eaten fresh (stuffed and/or sliced) but can also be pickled, and has been canned in brine. Immature bitter gourd fruits are the best for cooking when the skin is bright green in color, the flesh inside is white, and the seeds are small and tender. Bitter gourd contains vitamin A, B1, B2, and C. It is a rich source of phosphorous and iron. It also contains minerals like calcium, phosphorous, iron, copper and potassium. Young shoots and leaves are an excellent source of Vitamin A, and a fair source of protein, thiamin, calcium and phosphorus (Vinning, 1995). It has twice the amount of beta-carotene than in broccoli and twice the calcium content of spinach. Bitter gourd leaves are packed into 5 kg mixed produce boxes and sold to Melbourne and Sydney, as a medicinal plant (Peter Mc Laughlin, 1999). The seeds of the ripe fruit are used as a condiment in India. Species of *Momordica* have been in use in indigenous medical systems in various countries in Asia and Africa. The bitter gourd has excellent medicinal virtues as many native herbolarios practice. It helps in purifying blood, enhances digestion, and stimulates the liver, activates spleen and liver. It is a purgative, appetizer, digestive, anti-inflammatory and has healing capacity. It has antidotal, antipyretic tonic, appetizing, stomachic, antibilious and laxative properties. From the ayurvedic perspective, bitter gourd is excellent for balancing Kapha. Leaf juice is used for curing cough, as a purgative and antihelminthic to expel intestinal parasites and for healing wounds. It became popular in the news recently because of its putative medicinal value especially against HIV/AIDS (compound Q). Some forms have bright red seeds due to high lycopene content. Yen and Hwang (1985) have proposed using this pigment as an artificial food colortant.

Origin, History and Distribution

Bitter gourd or bitter cucumber or balsam pear or La-Kwa, or Chinese bitter melon, (*M. charantia* L. *syn. M. muricata* Terra.) is an important tropical monoecious crop originated in China or India (Indo-Burma center of origin). The regions of eastern India and southern China are suggested possible centers of domestication (Sands, 1928). Bitter gourd is widely grown in India, Indonesia, Sri Lanka, Malaysia, the Philippines, China, tropical Africa, North and South America and the Caribbean. The smaller (or more elongate) and greener variety is more bitter than the larger (or plumper) pale green ones that are usually found in Asian-American markets. The bitter gourd is a common vegetable cultivated extensively all over India. Area and production wise, the important bitter gourd states of India are Tamil Nadu, Uttar Pradesh, Maharashtra, Kerala, Andhra Pradesh and Odisha.

Wild *M. charantia* var. *abbreviate*, a native of Asia, may be the progenitor to the domesticate (Degener, 1974). Based on karyological and morphological studies on *M. charantia* (n=11) and *M. tuberosa* (n=11), a possible origin of the cultivated *M. charantia* from the wild *M. tuberosa* through karyotype alterations including deletion of satellites in the first pair and a peri-centric inversion in the 11th pair of chromosome is visualized (Varghese, 1973).

Species Availability, Evolution and Diversity

Momordica is a genus of herbaceous annual or perennial climbers with 45 species native to tropical Asia and Africa. About six species are cultivated for the fruits used as a vegetable, the most common being *M. charantia*. The related species of this genus are *M. dioica* Roxb. ex Wild., *M. cochinchinensis* Spreng., *M. tuberosa* (Roxb) Cogm. (syn. *M. cymbalaria*), *M. balsamina* L. and *M. cabriei*, first 2 species *i.e.* (*M. dioica* Roxb. Ex. Wild. and *M. cochinchinensis* Spreng.,) are edible in nature and cultivated. Other wild species are *Momordica subangulata* Bl., *M. denudata* and *M. macrophylla*. Tetraploid lines with 2n=44, were obtained by treatment of seedlings with 0.2 per cent colchicines for 18 hr to the shoot tips, and triploids by crossing 4n with 2n plants. A natural triploid (2n=33) of the cultivated *M. charantia* was also reported from India (Roy, 1985). Species of *Momordica* occurring in India are:

1. *M. charantia* L. (*M. muricata* DC.): occurs naturally and in cultivation throughout India, Sri Lanka, Malaya, China, tropical Africa and America. Flowers are monoecious. Two varieties, var. *charantia* (fruits large, not tapering at both ends) and var. *muricata* (Willd.) Chakravarty, (fruits small, tapering at both ends) occur.
2. *M. dioica* Roxb. (*M. balsamina* Wall., non L.): occurs naturally throughout India, Sri Lanka, Burma, China and Malaya; cultivated in the Deccan. Flowers dioecious and tubers are also edible.
3. *M. balsamina* L. (non Wall.): occurs naturally in Malaysia, China, Polynesia, Australia, western Asia and tropical Africa; introduced into America; cultivated in Sind (Pakistan), Punjab, Maharashtra, Rajasthan and Uttar Pradesh but rare in Deccan. Flowers are monoecious.
4. *M. subangulata* Blume: occurs naturally in India (Karnataka, Meghalaya and Sikkim), Malaya and Java but rare. Flowers are dioecious.
5. *M. cochinchinensis* (Lour.) Spreng.: occurs naturally in India (Asom, Nagaland, Tamilnadu, Uttar Pradesh and West Bengal), Burma, Malaya, Formosa, China, Philippines and Pacific Islands; cultivated in north Kanara. Flowers are dioecious.
6. *M. denudata* (Thw.) Cl.: West coast of peninsular India and Sri Lanka.
7. *M. macrophylla* Gage: India (Asom), Burma and Malaysia.

Species in Africa

In addition to the species mentioned, above *M. foetida* Schum. and Thonn., occurs wild and in cultivation in Nigeria and neighboring countries.

Excluded Species

M. cymbalaria Fenzl ex Naudin (*M. tuberosa* (Roxb.) Cognaux): occurs naturally in western peninsular India and tropical Africa; cultivated in Sholapur, Iagatpuri and Sattara. Flowers are monoecious. This species should be excluded from *Momordica* as it was restored to the genus *Luffa*, as *Luffa tuberosa* Roxb., by Chakravarty (1982) who accepted the treatment by Roxburgh (1832).

Table 3.118: Availability of *Momordica* Species

Species	*Availability*
Momordica balsamina	North Western Indo-Gangetic Plains
Momordica cochinchinensis	Humid Tropical Forests of Peninsular Region and North East
Momordica cymbalaria	Western Ghats and Maharashtra
Momordica denudata	Eastern Peninsular Region
Momordica dioica	Central Peninsular Region
Momordica macrophylla	Noerth East Region
Momordica subangulata	Noerth Eastern Hills, Eastern Ghats and Deccan Peninsula

Due to high cross pollination behavior diverse visual variation can be observed in fruit and vegetyative characters. Resmi and Sreelathakumary (2016) collected 33 genotypes from different agro-climatic regions of the country and catalogued wide biodiversity in morphological and qualitative characters *e.g.* vine length (103.75-620.00 cm), internodal length (1.25-5.58 cm), fruit length (4.91-38.83 cm), fruit girth (6.96-25.53 cm), fruit number (8.75-34.25), average fruit weight (4.26-578.75 g), β-carotene (52.58-138.96 µg/100 g), vitamin C (62.54-124.29 mg/100 g), iron content (2.38-6.88 mg/100 g), chlorophyll A (0.00110-0.54365 mg/g), chlorophyll B (0.00045-0.03075 mg/g), total chlorophyll (0.003-0.12 mg/g) and bitterness value (1227.50-10400 units/g). Leaf size varied from small to large with ovate pedate or reniform shape. Variability was more pronounced for flower and fruit characters. Flower and fruit size ranged from small to very large. Fruit form showed wide variation. Skin texture was either smooth, rough or with spines. Majority of genotypes had green skin coloured fruits. Seed quantity per fruit ranged from very few to many with small to very large seed size.

Gene and Genetics of Characters

Days to first harvest and fruits/plant are largely controlled by additive and additive x additive components (Sirohi and Choudhury, 1979). Studies also indicated that green fruit colour, dark brown seeds and seed size are monogenically inherited. Alcazar and Gulick (1973) reported that white epicarp is controlled by a single gene 'w', white being recessive to green. Vahab (1989) observed that inheritance of fruit colour and surface are monogenic, green and spiny fruits being dominant over white and smooth fruits respectively. Singh and Joshi (1980) recorded observations on yield, stem length, number of primary branches per plant, fruit length, number and weight of fruits per plant for five inbred of bitter gourd. They found that these characters were governed mainly by additive gene action. They also recorded high g.c.a. values for BWL1 for fruit length, number and weight of fruits per plant. Hybrids of BWL1 with BB1, BWM and BS1 had high s.c.a. values for weight of fruits per plant.

Botanical Description

Table 3.119: Botanical Description

Growth Habit	Annual, slender climber, 2.0-4.0 m high, scarcely to densely pubescent (tender pans wooly), monoecious climber.
Root	Deep tap root with many lateral branches.
Stem	Stems are thin, acutely quinquangular with broad longitudinal furrows, white pubescent, light green, internodes 5.0-6.0 cm. Tendrils are inserted beside the leaves, on long, robust stalks, simple or cleft, faintly 5-angular, yellowish green 15.0-30.0 cm long and delicate.
Leaf	Deeply and palmately 5-9 lobed, reniform to orbicular or sub orbicular in outline, 2.5-8.0 x 4.0-10.0 cm, alternate, long stalked cordate at base, acute or acuminate at apex, lobes ovate or obovate, narrowed at base, margins sinuate to undulate, mucronate; petioles 1.5-5.0 cm long.
Flower	Male flower stalks slender with bract midway or toward base; peduncle 2.0-5.0 cm long; bract reniform, 5-11 mm diameter., green, pedicel 2-6 cm long;
	receptacle-tube cup shape, 2-4 mm long and 2-3 mm wide; sepals ovate-elliptic, 4-6 x 2-3 mm, pale green; petals obovate, 10-20 x 7-15 mm, mucronate at apex, scales 2; filaments 1.5-2 mm long, inserted in the throat of the receptacle tube: anthers coherent- Female flower peduncle 1-6cm long; bract 1-9mm diameter; pedicel 1-8cm long: sepals narrow, oblong lanceolate, 2-5 mm long; petals smaller than or equal to that in male, 7-10mm long; ovary fusiform, narrowly rostrate, 5-11 x 2-3 mm, muricate, tuberculate or longitudinally ridged; style 2 rare long. Long days cause the male flowers to bloom up to 2 weeks before the female flowers, while short days have the reverse effect (Huyskens *et al.*, 1992). Spraying with 50 mg/L dikegulac at the 6-8 leaf stage increases female flower numbers and may double the number of fruit (Basu *et al.*, 1994). Spraying with Gibberellic acid at 25-100 mg/L at the 2-4 leaf stage increases female flower number and can last for up to 80 days (Wang and Zeng, 1996). Treatment with 2,4-D, maleic hydrazide and Cycocel (chlormequat) has also increased female flowers and yields, but reduced vine length and leaf area (Kabir *et al.*, 1989) Pruning lower laterals increases the total number of flowers per plant by increasing the number of flowers on higher laterals (Rasco and Castillo, 1990).

Anthesis and pollination	Bitter gourd is monoecious crop, with male and female flowers being borne on the same plant. Varieties differ in days to flower, time of anthesis and pollen viability (Deshpande *et al.,* 1979). Anthesis occurred between 3.30 and 7.30 AM and anther dehiscence between midnight and complete opening of the flower (7.30 AM). Pollen viability was high in all varieties (90-99 percent), while pollen germination from 66-68 percent. The stigma remained receptive for 24 hrs before and after anthesis. Pollination is brought by bees. The appearance of first pistillate flower at a lower node number is an indication of the earliness of the variety (Trivedi, 1983).
Fruit	Pendulous, stalk 2.0-8.0 cm long; fruit discoid, ovoid, ellipsoid to oblong or blocky, often narrowed at ends, sometimes finely rostrate, 3-20 x 2-5 cm, white or green turning orange on maturity, soft tuberculate with 8-10 broken or continuous ridges, splitting from base in to 3 irregular valvate at the top7-30 cm long and flesh blood red.
Seed	Seeds are numerous (5-30), flattened, squarish rectangular, ends subtridentete, faces compressed, sculptured, 5-9 x 3-6 rare, margins grooved and testa brown or black.Seeds are warty, 1.0-1.5 cm long and 0.6-0.9 cm broad. Seed contains 32 percent oil. Approximately 1000 seed weighs about 60.0 g.

Source: Yang and Walters, 1992.

Horticultural Classification

There are two types of varieties, cultivated by growers. The large kind is long, oblong and pale green in color. The other kind is small, little oval and dark green. Both the types are bitter in taste. Long fruited type cultivars are more vigorous in early growth resulting in a greater proportion of vegetation on the trellis and greater fruit set.

Augmentation of Genetic Resources and Crop Improvement

Breeding Priorities

1. Self-staked varieties/hybrids
2. Early and high yielding varieties/hybrids
3. Fleshy and less seeded type varieties/hybrids
4. Resistant to fruit fly

Introduction and Selection

Singh *et al.* (2014) evaluated 25 diverse bitter gourd genotypes collected from different part of India and reported significant differences among the genotypes for the traits examined. A wide range of variation was observed for all chacters which represents the variation was high and genotypes were rich in diversity. The wide variation in these yield related traits will be useful for selecting the best genotypes with high yield. The long sized fruits with wide diameter generally consisted more number of seeds as compare to normal and small fruited genotypes because it consisted more space for seed development within the flesh. Cluster analysis of 25 genotypes based on 10 quantitative traits was indicated that all the genotypes were resolved into two major clusters. The genotypes from these distant clusters may be used in crop improvement programmes to produce populations with wide varaibility with transgressive segregants possessing high yielding and nutrition rich fruits. Moharana *et al.* (2016) evaluated 20 genotypes and found significant differences for all the characters under study. High heriatability coupled with high genetic

Table 3.120: Groups of Bitter Melon Varieties

Group	*Length (cm)*	*Weight (kg)*	*Salient Features*
Small fruit	10.0– 20.0	0.1 - 0.3	Usually dark green, and fruits are very bitter
Long fruit	30.0– 60.0	0.2 - 0.6	Light green with medium ridges, and slightly bitter. It is most commonly grown and commercial group in China
Triangular fruit	9.0– 12.0	0.3 - 0.6	Light to dark green with prominent ridges, and moderately to strongly bitter. Fruits are cone shaped.

Source: Yang and Walters, 1992.

Table 3.121: Popular Varieties of Bitter Gourd in India

Variety	*Organization*	*Method Used*	*Brief Description*
Arka Harit	ICAR-IIHR, Bangalore	Selection	A pure line selection from a local collection (IIHR-4) from Rajasthan. Thin stem with light green, deeply lobed foliage. Fruits spindle shaped, glossy green skin colour without tubercles.Duration 100-110 days. Yield 12 t/ha.
Priya (VK-1)	Kerala Agricultural University, Kerala	Selection	Bears long green spiny fruits tinged white at tip. The fruit is 40 cm long and 17 cm in girth with flesh thickness of 5-9 mm and TSS 3.12 percent. Each fruit contains 24 seeds weighing 6 g (100 seeds weight is 24 g). Average yield is 30 t/ha.
Coimbatore Long	TNAU, Coimbatore	Selection	Fruit is long, tender, white, and suitable for rainy season cultivation. Average yield is 20 t/ha at proper cultural practices.
CO-1	TNAU, Coimbatore.	Selection	Fruits are dark green, 25-30 cm long with prominent protuberance. Duration of the crop is 115 days with average yield of 14 t/ha.
CO-2	TNAU, Coimbatore.	Selection	It is early in flowering. Fruits are 35-40 cm long and thick. Each fruit weighs 40 g. It gives an average yield of 30.0 t/ha.

Variety	Organization	Method Used	Brief Description
Pusa Do Mausmi	ICAR-IARI. New Delhi	Selection	Suitable for both spring-summer and rainy season sowing. Fruit long medium, dark green, thick, club shaped with 7-8 continuous ridges. Fruit length at edible stage is about 15-18 cm, average fruit weight is 110 g. The average fruit yield is 15 t/ha.
Punjab-14	PAU, Ludhiana	Selection	Suitable for both rainy and summer season cultivation. Bushy plant habit and average fruit weight is about 35 g. Fruit colour is light green. Average yield is 14 t/ha.
Pusa Vishesh	ICAR-IARI New Delhi	Selection	Fruits are glossy green, thick, medium long in size. Vines short hence more number of plant can be accommodated per unit area. Average fruit weight is 120 g. Average yield is 15.0 t/ha. This variety is released from IARI, New Delhi. Suitable for spring summer season; first picking in 55-60 days after sowing.
Kalyanpur Sona	CSAUA&T, Kalyanpur, Kanpur.	Selection	Fruits medium sized, green, and average yield is 12-13 t/ha
Kalyanpur Baramasi	CSAUA&T, Kalyanpur, Kanpur	Selection	It is vigorous creeper. Fruits are long (30-50 cm) light green, thin and tapering. Yield potential is nearly 20.0 t/ha in about 120 days. It is reported to be tolerant to fruitfly and mosaic. Suitable for Kharif season cultivation.
Hirkani	MPKV,Rahuril		Fruits are dark green and 15-20.0 cm long. Each vine produces 20-22 fruits, each weighing 100-120 g. it yields 13.8 t/ha.
Pant Karela-1	GBPUA&T, Pantnagar		The fruits are long (15.00 cm) with tapering ends. It takes 55 days to first picking and the yield potential is 15.0 t/ha. It is suitable for cultivation in hills.

advance associated with average fruit weight, number of fruits/plant and number of primary branches/plant suggest their applicability for identification of high yielding lines as these traits had shown high direct positive effect on yield. In relation to yield, the genotypes like Meghana No.2, Improved Kathai and Vikas were found to be the high yielders. On the basis of days to first fruit harvest, the genotypes like Jharli, Special Bolder Ucha and Meghana No.2 can be recommended for early capuring of market to fetch higher price. As Meghana No.2 possess both high yielding nature along with early days to first harvest so it can be the genotype of choice for higher return.

There are several varieties developed for commercial cultivation. Few cultivars are listed below:

MC-23, MC-84 and Priyanka are high yielding varieties released from Kerala Agricultural University, Thrissur (Kerala) (Gopalakrishnan, 2004).

Table 3.122: Bitter Gourd Cultivars Developed by Selection with Progeny Testing

Variety	Source
Coimbatore long	Indigenous
Pusa Do Mausmi	Indigenous
Arka harit	Indigenous
Priya (VK-1)	Indigenous
CO-1	Indigenous
MC-23	Indigenous
Pusa Vishes	Indigenous
Punjab BG-14	Indigenous
NDB-1	Indigenous
Phule BG-6	Indigenous
Kalyanpur Sona	Indigenous

There are few varities identified thorough All India Coordinated Research Project (Vegetable Crops) for different zones.

In Pakistan bitter gourd variety' Lyallpur No.-1 is very common and being grown commercially.

Hybridization and Selection

In bitter gourd, Srivastava and Srivastava (1976) observed high estimates of heritability and genetic advance for fruit number, fruit weight and yield. While studying genetic variability and correlation Singh *et al.* (1977) recorded high genotypic coefficient of variation along with high estimates of heritability and genetic advance for yield, number of fruits per plant and fruit length. Ramachandran and Gopalakrishnan (1980) found high or moderate estimates of heritability for content of vitamin C, TSS, phosphorus and iron. It was suggested that these traits were controlled by additive genes, while protein and potassium content controlled by non-additive genes. Mangal *et al.* (1981) reported high estimates of heritability along with high genetic advance for yield, number of fruits and weight in this crop. In bitter gourd Srivastava and Srivastava (1976) observed that yield per plant was highly and positively correlated with number of fruits per plant, number of lateral branches per plant and number of female flowers per plant. Number of fruits per plant was also positively associated with number of lateral branches and number of female flowers per plant. Mangal *et al.* (1981) reported that significant positive correlation coefficient between yield and number of fruits and weight of fruit but negative correlation with days to flowering in bitter gourd. Rajput *et al.* (1995) working with 21 genotype of bitter gourd, observed that fruit yield/vine was positively correlated with number of fruits/vine. Paranjape and Rajput (1995) evaluated 21 bitter gourd genotypes and revealed that yield was mainly contributed by number of fruits/vine, average fruit weight, fruit length and number of female flowers. The physiological attributes like vine length, primary branches and average leaf area were mutually associated and had effects on yield. Early genotypes produced

Table 3.123: Varieties Identified through All India Coordinated Research Project (Vegetable Crops)

Crop	Variety	Centre	Zone	Year
Bitter gourd	Priya	KAU, Kerela	II, VII, VIII	1992
	RHRBG-4-1	MPKV, Rahuri	IV, VII	1998
	KBG-16	CSAUA&T, Kanpur	IV	1998
	PBIG-3	GBPUA&T, Pantnagar	VII, IV	2000

higher yield and selection for such genotypes is possible in bitter gourd. However, vine length, primary branches, nodes on main axis, leaf area, fruit length, number of fruits/vine and seed content indirectly contributed towards yield. Significant gca and sca effects were recorded for days to flowering, fruit per plant in several parental breeding lines of bitter gourd in a set of 10 x 10 diallel (Srivastava and Nath, 1983). Ram *et al.* (1999) estimated the combining ability in 8 lines x 3 testers for 13 characters and noted significant differences for all the characters. The variances due to sca were higher than the gca for all the characters indicating the predominance of non-additive gene effects. "Narendra" and "VRBT-46" were good general combiners for marketable fruit yield. "VRBT-46" was good general combiner for leaf length, fruit length, fruit girth and number of seeds/ fruit also. The crosses Faislabad x VRBT-46, IC44410B x VRBT-46, IC50516 x VRBT-77, MC-48 x VRBT-78 and Narendra x VRBT-78 combined well for earliness. For marketable yield, the good cross combinations were Narendra x VRBT-46, Arka Harit x VRBT-78, MC-63 x VRBT-77 and IC50516 x VRBT-77. Genetic variability and interrelationship was worked out considering five F_3 progenies of cross Hirkani x Phule Ujwala and found significant differences among the progenies for all the trits. High value of genotypic and phenotypic coefficient of variability and heritability estimates were associated with greater values of genetic advance as a percent mean for vine length at last harvest, number of primary branches/vine, number of fruits/vine, average fruit yield/vine, average weight of fruit, average length of fruit and yield (t/ha). Fruit yield/ha showed highly significant positive correlation with yield and yield contributing characters such as vine length at last harvest, number of primary branches/vine, number of fruits/vine, average weight of fruit, average length of fruit and average diameter of fruit and fruit yield/vine (Alekar *et al.*, 2016).

Sirohi and Chaudhury (1978) studied 28 F_1 hybrids of bitter gourd in a diallel set involving 8 parents, excluding reciprocals to investigate the extent of heterosis for all the characters namely vine length, fruit length, fruit diameter, fruit rind thickness, fruits per plant, fruit weight, seeds per fruit, seed weight per fruit and total yield per plant except for the number of days to first harvest. The minimum and maximum heterosis observed for days to first harvest were 30.20 and 128.41 per cent, respectively. The increase in the yield over their respective better parents in heterosis hybrids ranged from 0.08 to 128.11 per cent. Ram *et al.* (1997), observe presence of non-allelic gene interaction among the majority of the crosses for most of the characters. Dominance and dominance x dominance components were predominant in most the crosses. Duplicate type of epistasis was present in most of the characters. Intermating the fruit weight had maximum direct bearing on yield of selective individuals in early generations and recurrent selection is suggested for the improvement of yield and yield related traits in bitter gourd. Mishra *et al.* (1998) observed a frequent occurrence of dominant alleles and both additive and non-additive gene action was involved in character expression. Unequal distribution of genes at loci exhibiting dominance was common in the parental publication, indicating an interaction between partial dominance and dominance in character expression. Presence of dominance and complementary gene action combined with low estimates of narrow-sense heritability for yield indicates that heterosis breeding would be advantageous for higher yields. Additive and non-additive gene effects were observed for seeds/fruit, 100 seed weight and seedling length. Over dominance was also observed for all characters except seeds/fruit. It is concluded that production and use of F_1 hybrid seeds is appropriate to improve seed yield and quality in bitter gourd (Devadas and Ramdas, 1997). Pal *et al.* (1983) conducted a line x tester analysis of combining ability using 5 lines ('Arka Harit', 'Monsoon Miracle', 'Largest', 'Uchhe' and 'China') and 2 testers ('Holly Green' and 'Indian Prime'). Many F_1 hybrids were synthesized.

Table 3.124: Performance of F_1 Hybrids

F_1 hybrid	Special Feature	References
'Panipat Local' x 'Ambala Local'	High yield	Pal and Singh (1946)
'B-15' x 'J 21'	High yield	Khola (1972)
'Green Local' x 'White Local'	High yield	Lal *et al.* (1976)
'Pusa Do Mousmi' x 'S 144', 'pusa Do Mousmi' x 'S 63', 'Coimbatore Long' x 'S 63'	High yield	Sirohi and Choudhury (1976)
'MDU 1' x 'VK1-Priya'	High yield	Gopalakrishnan (1986)
'C 96' x 'Green', 'MC 49' x 'MC 69'	High percentage of female flower and yield	Vahab (1989)

Table 3.125: There are Few Hybrids which Performed better on Grower's Field

Variety	Organization	Brief Description
Pusa Hybrid-1	ICAR-IARI New Delhi	Fruits are medium thick and long, glossy green, suitable for picking and dehydration. It gives higher yield than pusa vishesh and Pusa do Mausami; first pickling in 55-60 days, suitable for growing in spring summer season. It yield 200.0 q/ha.
Pusa Hybrid-2	ICAR-IARI New Delhi	Fruits are medium long and glossy green. It gives yield about 200.0 q/ha.
RHRBGH-1	MPKVV, Rahuri	Fruits are dark green, and about 20.0 cm long. It gives an average yield of 200.0 q/ha and tolerant to downy mildew.

Ram *et al.* (2000) carried out an experiment to detect epistasis and to estimate components of genetic variation for nine quantitative traits by crossing 8 lines with two testers. The epistasis was important for plant height, fruit length, fruit diameter, fruit weight and yield/plant analysis variance for sums and differences were significant for all the chacters and indicating the importance of additive and dominant components of variation for all the characters. The estmate of directional component (F) was significant and positive for days to male flower anthesis, days to female flower anthesis, fruit length, fruits/plant, fruit weight and yield/plant indicating the isodirectional nature of dominance. The degree of dominance was in the range of partial dominance for most of the characters.

Evaluation of parents based on *per se* and gca effects revealed that the parents Preethi, CO-1, MC-30, Uchha Bolder, Green Long, and MC-105 were identified as the best genotypes for improvement of yield combined with quality characters. The hybrids *viz.*, Preethi x MC-30, KR x USL, MC-105 x MC-10 and Priyanka x CO-1 had registered favourable values of mean, significant sca and standard heterosis for yield and quality parameters. Hence these hybrids are recommended for commercial exploitation of heterosis. A comparison of the parental gca and sca of hybrids revealed that hybridization between good x good and medium x poor or medium x medium combiners had given rise to hybrids with significant sca effects. Considering the per se performance, sca and the standard heterosis, the hybrids Preethi x MC-30 had registered favourable values for the most important characters like earliness, number of fruits, yield of fruits and quality parameters. Further the top performing F_1 hybrids can be tested in different seasons over different locations for assessing their stability for high yield and quality (Thangamani *et al.*, 2011).

Gynoecism

First time gynoecious plants of bitter gourd were identified with absolute expression of gynoecism (Ram *et al.* (2002a) in India and using gynoecisms several populations with very high proportion of pistillate flowers bearing lines were developed. The gynoecism was under the control of few major recessive genes (Ram *et al.*, 2000a). Further, utilizing segregating generations from the gynoecious plants, several high frequency pisttilate lines were developed and its use was advocated in cost effective hybrid seed production in bitter gourd (Ram *et al.*, 2002b). These high frequency pistillate lines (333/1, 333/2. 333/3. 333/4 and 323/4) had more than 90 per cent pistillate flowers and desirable fruit size and shape has provided scope to use them as female parent in developing cost effective hybrids (Ram *et al.*, 2003). Plants of all these populations are able to produce pistillate flowers at lower node and the staminate flowers appear 30 days after the appearance of pistillate flowers. Thus, in hybrid seed production field, all the fruits developed before staminate flower appearance on selected population (as female plant) will be hybrid seeds derived from the selected good combiner (as male parent). The inheritance of gynoecism in bitter gourd was worked out in a 100 per cent gynoecious line (Gy263B) by Ram *et al.* (2006) and proposed gy-1 as a gene symbol for gynoecism in bitter gourd. Jatav and Singh (2016) evaluated 24 genotypes and and reported PBIG-22 is predominantly female (sub-gynoecious with desirable agronomic traits having a 1:5.5 ratio of male and female flowers with highest fruit yield per plant (2.32 kg), yield/ha (193.52q), vine length (3.30 m), high number of fruits plant (33.00), fruit length (12.33 cm), fruit diameter (3.73 cm), fruit weight (68.89 g), number of branches per vine (9.00), lower number of node at which first female flower emerged (8.57) and days taken to first harvest (63.06).

Interspecific Hybridization

Roy (1985) reported that all attempts in interspecific crosses in *Momordica* failed completely. Fluorescence microscopy studies under-taken to investigate the causes for the failure of embryo formation in cross *M. dioica* X *M. charantia* showed heavy deposition of callose at the tops of pollen tubes which obstructed transfer of male gamete and fertilization. Vahab (1989) also reported complete incompatibility among *M. charantia*, *M. dioica* and *M. cymbalaria*. Pathak *et al.* (2016) reported that crosses between *Momordica balsamina* (Varh Karela) and *Momordica charantia* (Karela) was not successful due to poor fruit retention (<1 per cent) and also the two F_1 seeds obtained resulted into albino seedlings, which later on died. In other cases, seeds were either hollow or contained shriveled embryos with papery cotyledons and did not germinate. Therefore, 19-21 day old embryos from *M. charantia* Acc. PAUBG-56 x *M. balsamina* Acc. MB-3 produced at a frequency of 0.43 per cent. However, all the plants were observed to be albino and eventually died. Bitter gourd specific SSR markers were used to confirm the hybridity of interspecific *Momordica* species hybrids. These preliminary results indicate the presence of post-fertilization barrier(s) for the development of F_1 hybrid between these *Momordica* species.

Intergeneric Hybridization

Patrude and Krishnamurthy (1934) crossed *M. charantia* with *Trichosanthes anguina*. Over 50 per cent of the pollinated flowers developed into fruits. Of the seeds, 50 per cent germinated well, producing a normal crop. The floral and vegetative characters of *M. charantia* were dominant. The above observation needs further experiments for confermation.

Mutation Breeding

In bitter gourd, gamma radiation and EMS (ethyl methane sulphonate) treatment lead to an increase in genetic variability (Mallaiah and Nizam 1986). Narang and Prakash (1983) studied effects of gamma radiation on seed germination and seedling growth in 'Coimbatore'. Rajasekharan and Shanmugavelu (1984) irradiated local cultivar 'MC 103' with gamma rays and a promising line was isolated in M_2. It was subsequently stabilized for its characters for 6 generation and released as 'MDU 1'.

Polyploidy Breeding

Wanjari and Phadnis (1971) induced tetraploidy by colchicine treatment. Soaked seeds (24 hr) and sprouted seeds were treated separately with 0.1, 0.2 and 0.3 per cent colchicine for 6, 12 and 24 hr soaked seed treatment for 24 hr and sprouted seed treatment for 24 hr were effective. Auto tetraploid had a fewer female flower, larger flowers and petals and flowered later than diploids. Trivedi and Roy (1973) studies cytogenetics of *M. charantia* and to polyploids.

Quality Breeding

Five cultivars ('BG12', 'BG13', 'BG14', 'C96', 'and 'Sel 1') were assessed for suitability for processing (Kalra *et al.*, 1983). 'BG14 was the best for canning and dehydration. Awasthi and Jaiswal (1988) identified 'Coimbatore Long', as suitable genotype for processing. Resmi and Sreelathakumary (2016) analysed qualitative traits of 33 genotypes collected from different agroclimatic regions of the country and found variability in β-carotene (52.58-138.96 μg/100 g), vitamin C (62.54-124.29 mg/100 g), iron content (2.38-6.88 mg/100 g), chlorophyll A (0.00110-0.54365 mg/g), chlorophyll B (0.00045-0.03075 mg/g), total chlorophyll (0.003-0.12 mg/g) and bitterness value (1227.50-10400 units/g) content.

Resistance Breeding

Arunachalam (2002) reported resistance to bitter gourd leaf distortion mosaic virus in nine genotypes collected from northern and central parts of Kerala.

Bhardwaj *et al.* (2019) reported that out of 30 genotypes only 6 genotypes (VRBTG-36, VRBTG-47-2, VRBTG-5-2, DVBTG-3, DVBTG-4 and DVBTG-5) were found least infestested (less than 10%) for powdery mildew.

Tissue Culture

Halder and Gadgil (1982) attempts tissue culturing programme in bittergourd. Two-year-old calluses derived from cotyledons were subjected to NAA (0-1mg/liter) + Adenine(13.5-67.5mg/liter) and IBA + BA (Both 0.25 -10.0 mg/liter) treaments. The NAA(1.0 mg/liter) + Adenine (33.75 mg/liter) induced buds in bitter gourd. Fresh hypocotyl callus cultures did not induce into buds and plantlets, but 23-month-old cultures could be induced to buds and plantlats by the Addition of NAA (1.0 mg/liter) and Adenine (33.75mg/liter).

Table 3.126: Important Diseases and Pests of Vegetables Controlled by Grafting

Crop	*Disease/Pests*	*Rootstocks*	*References*
Bitter gourd	Fusarium wilt (*Fusarium oxysporum* f.sp. *momordicae* (FOM)	*Luffa aegyptica* (R) bitter gourd (S), Summer squash (S)	Lin *et al.* (1998)

Grafting

Grafting has been reported to successful method for imparting resistance against *Fusarium* wilt in several cucurbits. In pumpkin CM 214 (Nigerian Local, Immune) did not develop any disease symptom even after grafting on the YVMV infected symptomatic rootstock. The environmental fluctuations also did not influence the resistance expression of Nigerian Local against PMV and YVMV (Kuriakose, 1998). In field tests on ground infected with *Fusarium oxysporum* of 4 self rooted varieties of melon and 2 of those grafted to ash gourd Rugeri (1968) observed that 90 per cent of plants belonging to each of the ungrafted varieties became infected while those grafted on ash gourd escaped infection.

REFERENCES

1. Alekar, A.N., Khamkar, M.B., Shinde, K.G. and Bhalekar, M.N. 2016. Genetic variability and interrelationship among different traits in F_3 progenies of bitter gourd (*Momordica charantia* L.). 7th Indian Horticulture Congress (Abs: 89), held at ICAR-IARI, New Delhi, 15-18 November, p. 69-70.
2. Arunachalam, P. 2002. Breeding for resistance to leaf distortion mosaic virus in bitter gourd. Ph.D. thesis, Kerala Agricultural University, Vellanikkara Thrissur (Kerala).
3. Awasthi, C. P. and Jaiswal, R. C., 1986: Biochemical composition and nutritional quality of fruits of

bitter gourd grown in Uttar Pradesh. *Progressive Horticulture*, **18(3-4):** 265-269.

4. Bhardwaj, D.R., Chaube, T. Gautam, K.K. Verma, S.K. Singh, P.M. Singh A.K. Moharana, D.P. and Kumar, S. 2019. Reaction of bitter gourd (*M. charantia* L.) genotypes to the incidence of powdery mildew in Midgangetic plains of India Abs. VEGCON-2019, p. 67.
5. Binder, R.G., Flath, R.A. and Mon, T.R. 1989. Volatile components of bittermelon. *J. Agr. Food Chem.*, **37:** 418-420.
6. Deshpande, A.A; Venkatasubbaiah, K; Bankapur, V.M., and Nalawadi, U.G. 1979. Studies on floral biology of bitter gourd (*Momordica charantha* L.) Mysore Journal of Agriculture Science, 13: 156-159.
7. Gopalakrishnan, T.R. 2004. Three decades of vegetable research in Kerala Agricultural University, Vellanikkara Thrissur (Kerala).
8. Haldar, T. and Gadgil, V.N.1982. Morphogenesis in some plant species of the family cucurbitaceae. In: Rao, A.N. (ed.) 1982. Tissue culture of economically important plants, Proc. of Intl. Symp. Singapore, 1981; COSTED and ANBS, p. 98-103.
9. Jatav, V. and Singh, D.K. 2016. Identification of novel sub-gynoecious line in bitter gourd (*Momordica charantia* L.). 7th Indian Horticulture Congress (Abs: 107, held at ICAR-IARI, New Delhi, 15-18 November, p. 82-83.
10. Lin, Y.S., Hwang, C.H. and Soong, S.C. (1998). Resistance of bitter gourd loofah grafts to *Fusarium oxysporum* f. sp. *momordicae* and their yield. *Plant Prot. Bullet.*, **40 (2):** 121–132.
11. Moharana, D.P., Syamal, M.M. and Singh, A.K. 2016. Genetic evaluation for yield and yield attributes in elite genotypes of bitter gourd (*Momordica charantia* L) in Northern Indo-Gangetic plains. 7th Indian Horticulture Congress (Abs: 09; Session: Breeding for Horticulturaml Crops), 15-18 November, held at ICAR-IARI, New Delhi, p. 92-93.
12. Pathak, M., Kaur, N., Batra, N. and Pahwa, K. 2016. Development of interspecific *Momordica* species hybrid: A step towards introgression of root-knot nematode resistance. 1st International Agrobiodiversity Congress (Abs. Session: Trait Discovery and Enhanced Use of PGR), 6-9 November, held at New Delhi, pp. 197.
13. Ram, D., Kalloo, G. and Singh, M. 1999. Combining ability of quantitative characters in bitter gourd (*M. charantia*) using modified triple test cross. *Indian Journal of Agricultural Scences*, **69(2):** 122-125.
14. Ram, D., Kalloo, G. and Singh, M. 2000. Genetic analysis in bitter gourd (*M. charantia*) using modified triple test cross. *Indian Journal of Agricultural Scences*, **70(10):** 971-673.
15. Ram, D., Kumar, S., Banerjee, M.K. and Kalloo, G. 2002. Occurrence, identification and preliminary characterization of gynoecism in bitter gourd (*M. charantia*). *Indian Journal of Agricultural Scences*, **72(6):** 348-349.
16. Ram, D., Kumar, S., Singh, M., Rai, M. and Kalloo, G. 2006. Inheritance of gynoecism in bitter gourd (*M. charantia* L.). *Journal of Heridity*, **97(3):** 294-295.
17. Resmi, J and Sreelathakumary, I. 2016. Genetic cataloguing of bitter gourd (*Momordica charantia* L.). 1st International Agrobiodiversity Congress (Abs. Session: Trait Discovery and Enhanced Use of PGR), 6-9 November 2016, held at New Delhi, p. 174.
18. Reyes, M.E.C., Gildemacher, B.H. and Jansen, G.J. 1994 *Momordica* L. In: Siemonsma, J.S., Piluek. K. (Eds) Plant resources of South-east Asia: Vegetables, Pudoc Scientific Publishers, Wageningen, The Netherlands p. 206-210.
19. Singh, M.K., Bhardwaj, D.R., Solankey, S.S. and Pandey, A.K. 2014. Morphological analysis defines the genetic diversity of Indian bitter gourd (*M. charantia* L.). *VEGETOS*, **27(1):** 170-173.
20. Thangamani, C., Pugalendhi, L., Sumathi, T., Kavitha, C. and Rajashree, V. 2011. Estimation of combining ability and heterosis for yield and quality characters in bitter gourd (*Momordica charantia* L.). *Electronic Journal of Plant Breeding*, **2(1):** 62-66.
21. Trivedi, D.P. 1983. Morphological and genetical studies in bitter gourd (*Momordica charantia* L.) Ph.D. Thesis CSADA&T, Kanpur (U.P.).
22. Yang, S.L. And walters, L.W. 1992. Ethino botany and economic role of the cucurbitaceae of China. Econ. Bot; 46:349-367.
23. Yen, G.C. and Hwang, L.S. 1985. Lycopene from the sseeds of ripe bittermelon (*M. charantia*) as a potential red food colorant. II. Storage stability, preparation of powdered lycopene and food application. *J. Chin. Agr. Chem. Soc.*, **23:** 151-161.

Sweet Gourd (*Momordica cochinchinensis* Spreng.) and Spine gourd (*M. dioica* Roxb.)

Sweet Gourd (*Momordica cochinchinensis* Spreng)

Introduction

Sweet gourd (*Momordica cochinchinensis* Roxb.) is also called as Kakrol, Baby jackfruit, Spiny bitter gourd, Cochinchinesis gourd but in Asom it called as 'Bhat Karela'. Now it is designated as *Momordica subangulata* subsp. *reningera*. It is well known under-utilized/wild vegetable (tender fruits and leaves) for food, nutritional, medicinal, economic and of ecosystem value. People like it due to devoid of bitterness in fruit. Ripe fruit yields 'gac pulp' used in Vietnamese cooking as a natural food colorant, flavouring and Vitamin-A source. It contains extraordinarily high levels of β-carotene (641μg/g) and lycopene pigments (408.4μg/g) and often considered as fruit from the Heaven (Bauernfield,

1971; Voung, 2000; Aoki *et al.*, 2002). It is true that on an average, total carotenoids varied from 716 to 832 µg/g and β-carotene concentrations ranged from 133.28 to 141.17µg/g (Bharathi *et al.*, 2013). In north eastern region, during summer this vegetable fetches a high food value. In Traditional Chinese Medicine (TCM), the seeds are thought to have resolvent and cooling properties, and are used for liver and spleen disorders, wounds, hemorrhoids, bruises, swelling and pus (De Shan *et al.*, 2001), while in India the seeds are used for treating anemic and arthritis (Nayak, 1993). In Vietnam, bright red seed pulp (aril) is used in the preparation of 'xoi gac', where the freshly harvested aril and seeds are cooked with glutinous rice to produce nutritious food with orange colour and nutty flavor (Voung *et al.*, 2006). Nutritional trials in Vietnam have shown that the children fed with 'xoi gac' increased serum vitamin A levels significantly compared to those who received synthetic β-carotene powder or none. Increased in plasma retinol, alpha-carotene, zeaxanthin, and lycopene levels were also significantly higher in children given 'xoi gac' (Voung *et al.*, 2002). The fatty acid in sweet gourd arils improves the bioavailability of β-carotene than that of the synthetic form (Voung *et al.*, 2002). This crop gives a high economic return and profit to the farmers. Now sweet gourd can be seen in different cities because some traders of the region have started marketing. The nutritional properties and antioxidant capacity of peels, pulp and arils of *M. coginchinesis* Spreng fruit grown in Malaysia, was workedout by Nazri *et al.* (2016). The proximate copmposition of pulp had significantly highest the moisture content (94.9g/100g) compared to arils (90.7g/100g) and peels (88.1g/100g), on weight basis. Pulp also found to have high ash content at the range of 17.3g/100 g to 23.5 g/100 g followed by peels (14.0g to 16.0/100g) and arils (6.3g to 13.1g/100g). High available carbohydrate content was determined with 55.6 g/100 g, 30.9 g/100 g and 19.3g/100 g in arils, pulp and peels respectively. Meanwhile, the peels contained the highest protein content (6.2 g/100g) followed by arils (5.8g/100g) and pulp (4.6g/100g). Neverthless, arils found to contain significant amount of fat content at 22.3g/100g and low percentage for both peels and pulp (1.6g/100g). The peel and pulp of the fruits were categorized as low energy dense foods with the energy content 107.4kcal/100g and 147.4kcal/100g dry weight, respectively. Meanwhile, arils were categorized as high energy dense foods with the energy content 446.3 kcal/100g that mostly contributed by fat content. According to student t-test, the proximate and FRAP assay showed significantly different between the peels, pulp and arils ($p<0.05$), except for the DPPH assay. Several factors such as geographicl and climate differences and stage of maturity might influence the nutritional contents of the *M. cochinchinesis* Sprengh fruit.

The difference between Kakrol or sweet gourd (*M. cochinchinensis* Spreng.) and Kartoli or spine gourd (*M. dioica* Roxb.) are as under:

Table 3.127: Difference between Sweet Gourd and Spine Gourd

Kakrol or Sweet Gourd (Momordica cochinchinensis Spreng)	*Kartoli or Spine Gourd (Momordica dioica Roxb.)*
Roots develop bigger tubers	Roots develop small tubers
Leaves are bigger	Leaves are small
Large flowers and white to yellow in colour	Small flowers and yellows in colour
There are umblicate glands in lamina base of leaves.	-
Generally 1-5 glands in their petioles	-
There are 3 small deep black or blue circular dots at the base of petals	No circular dots at the base of petals
Anthesis occur during early morning	Anthesis occur during evening
Flowers take 72 minutes to open	Flowers take 7-72 minutes to open
Pollen remain viable for 36 hours	Pollen remain viable for 36 hours
Stigma receptivity remains up to 18 hours	Stigma receptivity remains up to 18 hours
Fruits are abnormally large and oblong	Fruits are round to oval
Individual fruit weight is around 60-80 g and can attain up to 500 g.	Individual fruit weight is around 10-15 g and can attain up to 30 g.
Fruit ripening starts from periphery to inner side	Fruit ripening starts from inner side to periphery
Fruits are light green to light yellow	Fruits are dark green in colour
Tough spine on fruits	Smooth and false spines
It takes 26 days to reach edible maturity	It takes 20 days to reach edible maturity
Flowering and fruiting occur for a short period	Flowering and fruiting continue for a longer period

Source: Ram *et al.*, 2002.

Sweet gourd is a crop in the domestication interface, still more of a wild gathered vegetable of tribal communities (Joseph and Bharathi, 2008) in India; is reported to be grown as a vegetable in China and Japan (Jeffrey, 2001).

Origin, History and Distribution

Centre of origin of sweet gourd (M. *cochinchinensis*; 2n=56) is probably Indo-Burma region particularly East and South East Asia (de Wilde and Duyfjes, 2002) which distribution is extended to Pacific Islands. It occurs as a natural component of coastal scrub jungles and partially opened forests especially stream banks in south, middle and North Andaman Islands (Joseph, 2005; Bharathi *et al.*, 2006). The crop is entirely spread in Assam, the Garo Hills of Meghalaya, Madhya Pradesh, Gujarat, Andaman Island and some of the south Indian States. It has many economically exploitable large fruited forms in Myanmar, Philippines, Vietnam and Formosa.

Botanical Description

Table 3.128: Botanical Description

Growth Habit	Sweet gourd and spine gourd both bears dioecious stout perennial climber. Plant of sweet gourd attains 30 m height whereas spine gourd attains only 2-5 m.
Root	Sweet gourd and spine gourd both bears tuberous roots. Plants of 2-3 years old develop more number of tuberous roots.
Stem	Stems are weak, green and possess simple tendril. Its main creeper attains a height of 2.0-4.0 m. In each plant numerous primary and seconadary branches develop at its peak period of growth.
Leaf	Generally one plant possesses 200-600 leaves. Leaves are cordate with acute tip. Leaves are simple, alternate, petiolate, exstipulate with palmately reticulate venation. There are some varation found among the leaves of different cultivars mostly in the margin, such as some are coarsely serrated and others are sharply serrated.
Floral biology	It is dioecious in nature bearing male and female flowers in separate plants. Dubey *et al.* (2007) studied floral biology of spine gourd. They found that the male flowers appeared in the second week of August while the female flowers appeared in the third week of August. The male male and female plant can distinguished from each other with the help of flowers only. Flowers are large, solitary and unisexual. The male flower possess 5 stamens, united in two pairs each stamen is 2-lobed and twisted. The female flower is having 1 style and 3 stigmas which are united. Vijay *et al.* (1977) studied the floral biology under hot and humid conditions of Coorg and found that the male flower buds took 22-24 days from initiation to full bloom and the female flowers 19-22 days.
Anthesis and pollination	In spine gourd maximum anthesis was noted at 8.30 pm at a temperature range of 36.8 to 27 °C. Anthesis began at 05.50 hours in the female and 05.20 hours in the male, with a peak at 05.50 hours in the female and 06.50 hours in the male. Flowering continued until 07.20 hours. In spine gourd dehiscence started with appearance of a slit on the anther lobe which resulted in the liberation of pollen grains on the anther surface. Anther dehiscence began at 22.45 hours, reached a peak at 23.00 hours and was complete by midnight. The stigma was receptive for 18 hours before and after anthesis, but most receptive for 12 hours before and after. Ovary is therefore, unilocular and having partial placenration.
Fruit and fruit set	The ripe fruits are ovoid or oblong or round in shape. In immature stage the fruits are light green to dark green in colour whereas at full maturity they turn to slightly yellowish green colour with bright red arils. Atripe stage the fruits become yellow.
Seeds	The seed colour varies from dark-grey to black and sculptured. Number of seeds found in the fruit may vary from 20-35. Its seed contain oleostearic acid, a kind of fatty acids (Ishida *et al.*, 2004). Seeds have a long period of dormancy and low germination rate (50 per cent).

Botanical Classification

Though there are two types of sweet gourd locally available namely:

On the Basis of Presence of Fruit Shape

(i) Oblong fruits type which is 6.00-10.00 cm long.

(ii) Round fruit which is 4.0-6.0 cm long.

On the Basis of Presence of Spines Fruits can be grouped into two:

(i) Spines are smooth and dense

(ii) Spines are hard and thinly arranged.

Augmentation of Genetic Resources and Crop Improvement

Control of root-knot nematodes and Cucumber Stunting Virus are some of the immediate research supports needed for popularizing sweet gourd. The fresh weight, number of seeds fresh weight and dry weight of seeds and seed: fruit weight ratio were determined for 20 fruits of *M. cochinchinensis*, and observed that apart from seed: fresh weight ratio, all the parameters were positively correlated to fresh weight. There was a positive correlation between the number of seeds/fruit and seed: fresh weight fruit ratio. Seed fresh weight and dry weight were also positively correlated with seed: fresh fruit ratio (Puzari, 1999).Twenty nine kakrol (*M. dioica* Roxb.) accessions and its one wild relative *M. cochinchinensis*, were evaluated considering 29 morphological characters. Considerable variations were observed among the genotypes with dissimilarity values ranging from 4.6 to 58.6. Cluster analysis based on this variation is effective hence proved its usefulness in classification of primarily classified *M. dioica* and *M. cochinchinensi* in two groups and found five clusters in which one cluster contained 24 *M. dioica* accessions. No relationship was found between genetic divergence and geographical distribution of the accessions indicating no regional gene pool was responsible solely for clustering. Morpho-physiological variation in *Momordica* species (Rasul *et al.*, 2008) can be seen with the add of PGR like pathenocarpiness. Variability and heritability variances were high for internodal length, petiole length, leaf area, fruit weight, total number of fruits/ plant and yield/plant in 34 spine gourd (*Momordica dioica*) genotypes (Maharana, *et al.*,1995). The fresh weight, number of seeds, fruit weight and dry weight of seeds and seed: fruit ratios were determined for 20 fruits of *M. cochinchinensis* by Puzari (1999). He obtained positive correlation between the number of seeds/fruit and seed:fruit weight ratio. Seed fruit weight and dry weight were also positively correlated with seed: fruit weight ratio.

Sweet gourd is least attended crop by scientist. Therefore, improvement related research work is almost negligible. In almost all parts of the country only local races are being grown till date. ICAR Research Complex for NEH Region, Umiam, Shillong has collected and evaluated 45 accessions. Some of the genotypes given yield of >3.0kg/plant

Table 3.129: Germplasm as Source of Insect Resistance

Crop	Disease/ Insects	Organism	Accessions	Reference
Bitter gourd	Red pumpkin beetle	*Epilachna septima*	BG98 and BG102 (R)	Khaire *et al.* (1987)

Coservation of Genetic Resources

Augmentation of germplasm through introduction of elite types from South-East Asian countries, collection of diversity from North-East India, Andaman Islands, Eastern India and Jharkhand, on farm conservation in homestead is essential lead. India holds a sizeable population of sweet gourd in the wild state particularly in north eastern region of the country. Few collections of sweet gourd and spine gourd are being maintained at RRS Kushinagar, ICAR-Indian Institute of Vegetable Research, Varanasi (U.P.). MPKV, Rahuri (Maharashtra) is also maintaning some spine gourd lines.

Spine Gourd (*Momordica dioica* Roxb.)

Introduction

Kartoli (*Momordica dioica* Roxb., 2n=28) can be spotted as wild in the field naturally but survive well under natural fertility condition of the soil with the support of organics. It is devoid of hazardous agro-chemical like fertilizer, insecticide, fungicide, weedicide, plant growth regulator *etc.*, due to no or poor domestication. Tribal and local folk harvest the tender green fruits from the naturally growing plant for their own use and also supply sufficient amount in the nearby market. Today no vegetable is costlier than kartoli in cucurbits, which is generally sold @ Rs.40.00 to 60.00 per kg during fruiting season (July to October). Kartoli support the livelihood of poor people by providing better price. This prized vegetable can be greatly exploited not only for domestic need but also for fetching high return by export potentialities like high yield, better keeping quality, suitability for processing *etc.* As a processed product, its roots/tubers are powdered and used in the preparation of high quality of "Tikhur". Tender fruits can be canned easily in the 8 per cent brine solution. A new glycoprotein has been isolated from the cotyledonary tissues of *M. dioica* seeds which agglutinate blood and plant protoplasts activities (Ghosh *et al.*, 1981). Similarly, two new aliphatic constituents, characterized as 6-methyl tritriacont-5on-28-ol and 8-methyl hentriacont-3-ene were isolated for the first time fruit rind of *M. dioica* along with the known sterol pleuchiol. Momordica urssenol an unknown pentactriterpene isolated from the seed was identified as urs-12, 18 (19)-dien-3 beta-ol on the basis of spectral data and chemical analysis (Ali and Srivastava, 1998). Similarly, seven constituents were isolated from roots of *M. dioica*. Their structures were elucidated from spectral data. Two new compounds were established to be 3 beta-O-benzoyl-11-oxo-ursolic acid and 3beta–O-benzoyl-6-oxo-ursolic acid. Five known compounds were identified as oleanolic acid, gypsogenin, hederagenin, alpha-spinasterol and stearic acid (Luo *et al.*1997). Kartoli and kakrol is a purely indigenous vegetable and its wild occurrence is quite common in several regions of the country. Biodiversity degradation of this valuable crop is exceeding greatly by human intervention as well as by wild and domestic animals. Damage causes to this crop greatly by several enemies to foliage, fruit and underground tubers. Tubers are eaten by wild animals like Porcupine, wild boar, rabbits, mice *etc.* due to its good taste; immature tender and mature fruits are eaten by monkeys, squirrel and tender leaves and shoots are eaten by goat, sheep, cow *etc.* Considering the above fact the germplasm building efforts are being made consistently through plant exploration under different programme like NATP on Plant Biodiversity and ACIRP (Vegetable Crops). Since there is lack of systematic work and scientific information's on the biodiversity of this species.

Origin, Distribution and Diversity

Spine gourd can be spotted largely in several districts of Andhra Pradesh and Telangana and othe states of (India. it is also cultivated on commercial scale in West Godavari district over pendals under irrigation and over bushes under rainfed conditions around Kathipudi, Prathipadu, villages near Annavaram in East Godavari district of Andhra Pradesh. For collecting the diversity available in this species an extensive survey and exploration is essential, specially in different areas of Rajmahal, Hazaribagh, Tikigora hills of Santhal Pargana, Lalmattia, Chandana and nearby forest of Ranchi, Jaunpur, Mirzapur, Sonbhadra, Chandauli, Varanasi, Mahoba, Jhansi, Jalaun, Chittarkoot and Allahabad of U.P.; Rewa, Sidhi, Satana, Chhatarpur of Madhya Pradesh, Chattishgarh and eastern parts of Bihar during 1996 to 2003.Unavailability of planting material (tuber) is the major bottleneck associated with the kartoli cultivation. With the above facts this crop required adequate attention for finding place in cultivation for diversification of vegetables.

The wild form *Momordica dioica* which occurs in Bihar, Maharashtra and Rajasthan, India, is diploid with a chromosome number of 2n=28; the form that grows in Assam is tetraploid with 2n=56 (*M. cochinchinensis*). Compared wuth the diploid the tetraploid has larger leaves, flowers and fruits, and is more vigorous and earlier. The flowers of the tetraploid open in the morning, those of diploid at night. The triploid, derived from crossing the diploid and tetraploid, has characters which are intermediate between those of its parents (Mishra and Sahu, 1983). Working with spine gourd (*Momordica dioica*), Ram *et al.* (2001) recoded considerable range of variation for plant height (1.50 m to 3.95 m), number of fruits/plant (32-120), individual fruit weight (8.20-19.4 g), number of seeds/fruit (10-23) and yield/plant (1.00-2.85 kg). They also found significant response of tuber size on flowering fruiting, seed maturity and plant vigour.

Rao (2016) reported that spine gourd seeds will be in dormant for about 5-6 months from harvest (November to next May), and germinate in May/June with early rains. The germination percentage was found to be 33 per cent. Out of the germinated seedlings, the female plants were found to be again 30-35 per cent others being the male plants. Among the germinated female plants, the percentage of vines with single lobed, trilobed and pentalobed laves were found to be 30, 55, 15 per cent, respectively. Based on this, for planting one hectare area with 2 x 2 m spacing, nearly 2,500 female plants, are needed for which, about 2 kg seed is required per hectare, which gives vines of 825 single lobed, 1512 trilobed one and 412 penta lobed leaves, respectively.

Horticultural Classification of Spine Gourd

Three phenotypes have been observed in the spine gourd *viz.* (Rao, 2016):

1. Vines with single lobed leaves-The fruit yield potential of the vines with single lobed leaves was 2-3 kg (per season) found high (with cylindrical fruitsof 6.5 x 2.5 cm size, with beak at tip).
2. Trilobed leaves- The fruit yield of the vines was intermediate *i.e.*1-2 kg (per season).
3. Penta lobed leaves- The fruit yield of the vines was lowest with the vines *I.e.* 0.5-1.0 kg (per season).

In all the three phenotypes male and female vines exist separately. Nitrogen @ of 375 kg N/ha with 2x 2 m spacing (2,500 vines/ha) found to increase the yield/vine from 1 to 1.5 kg. On an average a net return of 0.5 lakh (rainfed) and 1.0 lakh (irrigated) can be obtained in 6 months period (from July/August to October/November) upto 3 years @ rs. 80/-per kg out of one hectare area.

Momordica subangulata subsp. *subangulata*

Momordica subangulata subsp. *subangulata* is a rare/critically endangered Indo-chinese taxon, collected (six samples) for the first time in India from a 10 km stretch in Tuensang district of Nagaland. A ten kilometer stretch in Tuensang district of Nagaland from Kuttur to Tuensang on Tuensang-Kifri route, by the side of NH 202 was found to have more than 200 plants comprising male and female of all age cohorts. Restricted to a small stretch of secondary forest in Ngangpong, Chingmelon and Kuttur villages in a 10 km radius of Tuensang. Excavated dormant tubers broke their dormancy within two weeks when brought to Thrissur (Kerala) and have profuse vegetative growth, but shy flowering in the hot and humid summer months (John *et al.*, 2016). It is a critically endangered taxon demanding both ex situ, in situ and on-farm conservation. Tender fruits, young leaves and clippings are cooked and eaten and is a valuable genetic resources for improving bitter gourd, spine gourd and teasel gourd.

REFERENCES

1. Aoki, H., Kieu, N.T.M., Kuze, N., Tomisaka, K. and Chuyen, V.N. 2002. Carotenoid pigments in gac fruit (*Momordica cochinchinensis* Spreng.). *Biosci. Biotech. Biochem.*, **66:** 2479-2484.
2. Bauernfield, J.C. 1971. Carotenoid vitamin A precursors and analogs in foods and feeds. *J. Agric. Fd. Chem.*, **20:** 456-473.
3. Bharathi, L.K., Behera, T.K., Singh, R. and Singh, A. 2013. Carotenoid contents in sweet gourd (*Momordica cochinchinensis* Spreng.) accessions of India. *Indian J. Hort.*, **70(2):** 165-169.
4. Bharathi, L.K., Naik, G. and Vishal Nath. 2006. An exquisite vegetable variety. *ICAR News*, **12:** 14-15.
5. De Shan, M., Hu, L.H.B. and Chen, Z.L. 2001. A new multiflorane triterpenoid ester form *Momordica cochinchinensis* Spreng. *Nat. Prod. Let.*, **15:** 139-145.
6. de Wilde, W.J.J.O. and Duyfjes, B.E.E. 2002. Synopsis of *Momordica* (Cucurbitaceae) in South East Asia and Malaysia. *Bot. Zhurn.*, **57:** 132-48.
7. Dube, A.K; srivastava, J.P. and Singh, N.P. 2007. Studies on floral bilogy of spine gourd (*Momordica diolca* Roxb.) *Acta Hortic*; 752:453-458.
8. Ghosh, B.N; Dasgupta, B. and Sircar, P.K. 1981. Purification of lectin from a tropical plant, *Momordica dioica* Roxb. Indian. J. Exp. Biol. 19(3): 253-255.
9. Ishida, B.K., Turner, C., Chapman, M.H., McKeon, T.A. 2004. Fatty acids and carotenoid composition in gac (*Momordica cochinchinensis* Spreng.) fruit. *Journal of Agricultural and Food Chemistry*, **52:** 274-279.
10. Jeffrey, C. 2001. Cucurbitaceae. In: Encyclopedia of Agricultural and Horticultural Crops. Vol.3, Hanelt, P. (Ed.), Springer, Berlin, pp., 1510-1557.
11. John, J.K. and Bharathi, L.K. 2008. Sweet gourd (*Momordica cochinchinensis* (Lour) Spreng.). In: Underutilized and Underexploited Horticultural Crops. Vol.4, Peter, K.V. (Ed.), New India Pub. Agency, New Delhi, p. 185-191.

12. John, J.K., Rajasekharan, P.E., Harisdh, G.D., Pitchaimuthu, M. and Pradheep, K. 2016. 1st International Agrobiodiversity Congress (Abs. Session: III. Crop Wild Relatives: Back to the Wild to Save the Future), 6-9 November, held at New Delhi, pp. 324.

13. Joseph, J.K. 2005. Studies on ecogeography and genetic diversity of the genus *Momordica* L. in India. Dissertation, Mahatma Gandhi University, Kottayam, Kerala, India.

14. Nayak, H. 1993. Vrikshalatara Chamatkar Guna (in Oriya.). Pustak Bhandar Publishers, Cuttack, India, p. 31.

15. Nazri, A.R., Amin, I., Desa, H.M. and Tan, S.H. 2016. Nurtitional composition and antioxidant capacity of peels, pulp and arils of spiny gourd (*Momordica cochinchinensis* Spreng) fruits grown in Malaysia. 1st International Agrobiodiversity Congress (Abs. Session: Agrobiodiversity for nutrition and health), 6-9 November, held at New Delhi, p. 221.

16. Puzari, N.N. 1999. Effect of plant growth regulators on quality traits of spine gourd (*Momordica cochinchinensis* Roxb.) Indian J. Hill Farm. 12:62-64.

17. Ram, D., Kalloo G. and Banerjee, M.K. 2002. Popularizing kakrol and kartoli: the indigenous nutritious vegetables. *Indian Horticulture*, 6-9.

18. Rao, M.B.N. 2016. The growth and yield of spine gourd (*Momordica dioica* Roxb.) influenced by its phenotypes. (Abs: 82), 7th Indian Horticulture Congress, held at ICAR-IARI, New Delhi, 15-18 November, p. 64.

19. Rao, M.B.N. 2016. The phenotypic and sex ratio in spine gourd (*Momordica dioica* Roxb.) and its seeds dormancy. (Abs: 81), 7th Indian Horticulture Congress, held at ICAR-IARI, New Delhi, 15-18 November, p. 63.

20. Rasul, M.G; Mian, M.A.K. Cho, Y; Ozaki, Y. and Okubo, H. 2008. Application of plant growth on the parthenocarpic fruit development in teasel gourd (kakrol. *Momordica dioica* Roxb.) J. Fac. Agric Kyushu Univ. 53:39-42.

21. Vijay, O.P., Jalikop, S.H., Nath P. and Prem Nath. 1977. Studies on floral biology in kakrol (*Momordica cochinchinensis* Spreng.). *Indian Journal of Horticulture*, **34(3):** 284-288.

22. Voung, L.T. 2000. Under-utilized beta-carotene rich crops of Vietnam. *Fd. Nutr. Bull.*, **2:** 173-181.

23. Voung, L.T., Dueker, S.R. and Murphy, S.P. 2002. Plasma β-carotene and retinol concentrations of children increase after a 30-d-supplementation with the fruit *Momordica cochinchinensis* Spreng. (gac). *American J. Clin. Nutr.*, **75:** 872-879.

24. Voung, L.T., Franke, A.A., Custer, L.J. and Murphy, S.P. 2006. *Momordica cochinchinensis* Spreng. (gac) fruit carotenoids re-evaluated. *J. Fd. Comp. Anal.*, **19:** 664-68.

Genus *Benincasa*

Ash Gourd or Wax Gourd (*Benincasa* species)

Introduction

Ash gourd (*Benincasa hispida*) is also known as hairy melon, winter melon, fuzzy melon, ash pumpkin, green pumpkin white pumpkin, Chinese preserving melon, white gourd (Wikipedia, 2011), wax gourd and petha. It is an important vegetable crop of India, China, the Philippines and many other parts of Asian countries (Pandey *et al.*, 2015). The name wax gourd refers to thick white/chalky deposition on the fruit surface at the time of maturity which provides long shelf life. Although the fruit is referred to as a "melon", the fully grown crop is not sweet. It was one of two cucurbit species identified by the National Academy of Sciences (1975) as being an underexploited tropical crop. It is occasionally used as vegetable but mostly used by confectioners and Ayurvedic medicinal preparation (Indira and Peter, 1987). The immature fruits of ash gourd are used as vegetable which contain various minerals and vitamins, having a relatively high level of K and Na, and low calorific value and no fat. 'Petha' a preserved candy prepared by its mature fruit is famous sweet of India particularly in Agra region of U.P., where it is commercially grown. Mature fruits can be stored for as long as a year without refrigeration. In some part of our country, its young shoots, leaves and flowers are also used as a vegetable. An enzyme extracted from ash gourd juice has been successfully used instead of calf rennent in the production cheddar cheese. Its fruits also possess several medicinal properties like laxative, diuretic and cooling and is said to be helpful for patients suffering from nervous disorders. Sometimes seeds of mature fruits are extracted, fried and eaten like pumpkin seeds. Usually whole fruits are sold in domestic market, but are commonly displayed and sold by the slice in Asian markets. Somewhat bland in flavor when eaten fresh, the flesh is often used to make soup stock. Canned winter melon soup and dehydrated winter melon slices represent two of the processed produce made from this species. The genotypes under cultivation are nondescript ones. Apart from the vegetable (immature fruit), mature fruits are used in making candy, preserves, sweet and pickles. Extract of ash gourd has acid neutralizing property and recommended in Ayurveda for curing peptic ulcer and epilepsy (Grover *et al.*, 2001, Ramesh *et al.*, 1989) and lowering blood sugar (Chandrasekar *et al.*, 1989). The nutritive value of fruits are 98 per cent edible portion, 93.8 g water, 0.7 g protein, 0.1 g fat, 2.0

Table 3.130: Taxon and Preferred Taxon Name

Name of Taxon	Name of Authority	Family	Preferred Taxon Name
Benincasa	-	-	*Benincasa*
Benincasa cerifera	Savi	Cucurbitaceae	*Benincasa hispida*
Benincasa hispida	(Thunb.) Cogn.	Cucurbitaceae	*Benincasa hispida*

g carbohydrate, 12 mg Ca, 250 mg potassium, 2 mg sodium, 69 mg vitamin-C and 0.2 mg niacin in 100 g edible portion (Wills *et al.*, 1984). Besides it contains 0.38 ppm iodine and 3.5 ppm of fluorine. A small fruited medicinal ash gourd is also grown in Kerala which is good for people suffering from nervousness.

Origin, History and Distribution

Benincasa hispida Cogn. (Monotypic genus) is reported to be native of Java and Japan (Indo-Malayan area of South East Asia and Japan) but is now found throughout tropics. In India this vegetable crop can be spotted throughout the plains and on the hills up to 1200 m altitude. It is cultivated for its fruits throughout Burma and Ceylan, on hills up to 4000 feet. Since the existence of wild population of ash gourd is unconfirmed, therefore centre of origin remains uncertain. Originally cultivated in South-East Asia, the winter melon is now widely grown in East-Asia and South Asia as well. The winter melon has been reported to have been grown as a vegetable in China since 500 AD; even today, however, it is cultivated little outside of Asia. Indo-China area is the center of greatest diversity. As the diversity of cultivar expressed in India, China suggested that this crop may be indigenous to southern China (Yang and Walters, 1992). It is grown upto an altitude of 1500 m. It is supposed to be indigenous to Asian subtropics. Some researchers presumed that ash gourd is originated from Malaysia. Simonds (1976) reported that ash gourd is indigenous to Asia tropics and distributed from Japan to India. Its availability can be spotted in Burma and Ceylon to an altitude of 1500 m. It is also believed that ash gourd was introduced to India from Japan and Java by foreign navigators and missionaries. Chinese scriptures of AD V and VI refer to the ash gourd. Its cultivation in Latin America and the Caribbean become popular only due to migrants of Chinese descent. It was domesticated in India during pre-historic time. Ash gourd is widely grown through out India, particularly in tropical and sub-tropical climates. Besides our country; it is also grown extensively in China, Thailand, Vietnam Philippines and elsewhere in Asia. Exhibiting relatively rapid growth, *B. hispida* grows best in temperate climates with adequate but not excessive rainfall. In Sri Lanka, the plant produces fruit from seed in two months during the rainy season. It is also grown in Latin America and the Caribbean.

Species Availability, Evolution and Diversity

The generic name of ash gourd *Benincasa* has been named by famous famous Italian botanist, Gaetano Savi, in 1818 to honour Giuseppe *Benincasa*, an Italian patron of Botany. He called the plant as *Benincasa cerifera* "wax bearing'. The term *hispida* refers to the foliage and immature fruits covered with long rigid hairs or bristles on whole parts of plant. It is a monotypic genus and the only cultivated species is *hispida*. De Candole found its wild form on the seashore of Java and it has spread north words to Japan and also to Central America and West Indies. It possesses variability in fruit morphological characteristics. Ash gourd was earlier classified under genus *Cucurbita* but on the basis of its 3 stamens which are separate whereas in *Cucurbita* more or less united, therefore it was placed under *Benincasa*.

Botanical Description

Ash gourd (*B. hispida* (Thunb.) Cogn.(syn. *B. cerifera* Savi.) belongs to monospecific genus *Benincasa*. The genus *Benincasa* belong the tribe Benincaseae Ser., subfamily Cucurbitoideae.

Table 3.131: Botanical Description

Growth Habit	Ash gourd is a monoecious, annual herb, vigorous, hispid and extensive climbing/trailing cucurbit.
Root	Tap root system.
Stem	Stem stout, prominently furrowed/angular ribbed, light green with scattered rough hairs. It is climbing, growing to several meters in length, tendrils 2-3 branched; ovate stipules, upto 2.5 cm in length, opposite to tendrils.
Leaf	Leaves are simple, disticous, dark green; hairy petiole 10.0-20.0 cm in length, lamina 10.0-25.0 x 10.0-20.0 cm, 5-11 angled or lobed, dentate, base cordate, emitting a foetid smell when bruised and tendril bifid.
Floral biology	Flowers are monoecious and the fine parted yellow flowers are solitary in the leaf axils, measuring 46 cm in diameter and 3.0-5.0 cm in length excluding the height of ovary. It has long male flowers (5.0-15.0 cm in length) with long pedicels (5.0-15.0 cm) whereas female flowers are short stalked, almost sessile (1-3.8.0 cm length of pedicel) are 8.0-10.0 cm in diameter. Calyx is foliaceous with 5 lobes; corolla rotate, and petals 5 almost free. Ovary densely hairy, 2.0-4.0 cm in length; stigmas 3, curved and bilobed. The distribution of staminate and pistillate flowers is influenced by temperature and daylength.

Anthesis and pollination	Flower opening takes place in the early hours (4:30 to 07:30) with anther dehiscence between 03 and 05 hours. The stigma is receptive eight hours before anthesis until eighteen hours after anthesis. Pollen fertility is maximum (95 per cent) at anthesis. Fruit a large, stalked berry (pepo), covered with white silky hairs.
Fruit and fruit set	The mature fruit is the primary harvested plant part. Fruits are large, stalked, 20.0-35.0 x 15.0-20.0 cm, spherical to long, as large as 80 cm in length, oblong, rough hairy (Heiser, 1920) at first becoming glabrous with maturity (losing hairs at mautrity), dark green, covered with a waxy coating (chalky) deposit which is easily removed; flesh white and spongy in nature. Chalky wax deters microorganisms and helps impartant extraordinary longevity to the melon. Fruits may weight up to 35 kg and consist of more than 96 per cent water.
Seeds	Numerous, oval, flat, smooth with narrow base, white, yellow or pale brown, 1.0-1.5 cm x 0.5-0.7 cm.

Cytology and Cytogenetics

Chromosome number of ash gourd is 2n=24 (x=12). Verghese (1973) observed secondary association of chromosomes which suggested that *Benincasa hispida* though apparently a diploid; is in fact a stable tetraploid derived from an ancestor with x=6. Meiotic preparations from root tip cells had 24 chromosomes. Analysis of karyotype revealed that the chromosome length varies from 2.0 to 3.0 μ. The compliment consists of 8 pairs of median and 4 pairs of sub terminal chromosome. According to their comparative length and arm ratio, 12 pairs of chromosomes could be arranged into 6 groups. Total length of complement and mean length of chromosomes were estimated to be 57.36 and 2.39 μ respectively. All stages of meiosis were normal. Pollen fertility was 97.8 per cent.

Sex Expression and Modification

The response to climatic factors on sex expression in ash gourd was studied by Kamalanathan *et al.* (1972). Staminate flowers always appeared first at the lower nodes, followed by pistillate flowers at the higher nodes. Long day lengths and high temperatures delayed flowering, but relative humidity (RH) and rainfall had little influence. Short day lengths (5.7 to 6.1 h), low maximum temperatures (29.4 to 29.9 ºC), low minimum temperature (18.2 to 18.9 ºC), high RH and cloudiness all enhanced pistillate flower production. Advantage may be taken of the plant's climatic preferences by raising crops during the *Kharif* season (June to August) as flowering during the cooler months results in earliness and the production of a high percentage of pistillate flowers with a narrow sex ratio, desirable characters for a high yielding crop. Ghosh and Bose (1972) stated that foliar spray of B. nine (daminozide) at 1250 ppm sprayed to seedlings at 2-leaf stage and again at 4-leaf stage, ratio of female to male flowers increased due to suppression of staminate flowers. In general, 1 female flower to 38 male flowers has been observed.

Horticultural Classification

In our country, few major categories of ash gourd have been recognized *viz.*

1. Unridged winter melon group
2. Ridged winter melon group
3. Fuzzy gourd group
4. Wax gourd group and
5. Waxless gourd group

Four major cultivar groups are recognized:

(i) A late maturing winter melon group that have unridged seeds and fruits are cylindrical (50.0-100.0 cm long), dark green with sparse wax.

(ii) A ridge winter melon group that is similar except the seeds are ridged.

(iii) A fuzzy gourd group with ridged seeds and narrow cylindrical, long (20.0-25.0 cm), green, hairy, sparsely waxed fruit and

(iv) The wax group that also ridged seeds, and fruits are oblong 10.0+60.0 dianeter,light green colour, waxy and sometimes hairy fruit.

Further, wax gourd group is prominent with 3 cultivars

1. Fruits nearly round and essentially hairless
2. Fruits nearly round and hairy
3. Fruits oblong and hairy

Augmentation of Genetic Resources and Crop Improvement

Breeding Priorities

1. Bushy behavior
2. High protein content
3. Linear seed arrangements
4. High flesh recovery

George (1981) and Menon (1998) conducted biometrical studies to estimate genetic variability, correlations path coefficients and selection indices in ash gourd. Thirty two ash gourd genotypes studied were significantly different for 28 polygenic characters studied. Genotype 'BH-21' was found superior for yield. Average fruit weight had the highest value of genetic advance in the next cycle of selection, followed by weight of first mature fruit and yield/plant. Fruit yield/plant was positively correlated with main vine length, female flowers/plant, average fruit weight and weight of first mature fruit. Average fruit weight had maximum direct effect on fruit yield/plant, followed by main vine length. Length main vine was positively correlated with node at which first female flower appeared, node on main vine and leaves/plant. Fruit circumference had the maximum direct effect on first mature fruit weight, followed by fruit

Table 3.132: Varietal Description of Ash Gourd

Varieties	Organization	Pedigree	Description
CO-1	TNAU, Coimbatore	Selection	The vine is moderately vigorous, trailing (up to 4.0 m) with dark green leaves. Fruits are globular, green, large, oblong, 35 cm long, 22 cm girth with an average weight of about 6.8 kg. First harvesting is done at about 100 days after sowing. It gives an average yield 20-25t/ha in 140-150 days of crop duration.
CO-2	TNAU, Coimbatore	Selection	Fruits are small-long, spherical with average weight of 3.00 kg. The flesh is green and it give an average yield of 34t/ha in 120 days of crop duration.
Mudliar	TNAU, Coimbatore	Selection	Fruit are big and pale-green in colour
Indu	College of Horticulture, Vellanikkara	Selection	Fruits are round, 24.3 cm long, 23.78 cm breadth with average weight of 4.82 kg., tolerant to mosaic disease gives an average yield of 24.5t/ha.
KAU Local	College of Horticulture, Vellanikkara	Selection	Fruits are medium sized, oval to oblong, green at tender stage. The fruits are 45-55 cm long, length: breadth ratio is 2.05 with average weight of 6-8 kg. Flesh thickness is 5.1-6.2 cm, it gives an average yield of 28t/ha. The crop duration is 105-120 days for mature fruit production.
Kashi Dhawal (IVAG-502)	ICAR-IIVR, Varanasi	Selection	The vine length is 4.5-5.0 m. Fruits are oblong with average weight of 12-13 kg. Flesh is white and seed arrangement is linear; gives an average yield of 30-35t/ha.
PAG-3	PAU, Ludhiana	Selection	Fruits are globular, medium sized with average weight of 8-10 kg. Crop duration is about 145 days but picking of fruits is started 60 days after fruit setting. It gives an average yield of 70-75 t/ha.

length. The discriminate function indicated maximum selection response for yield when main vine length, female flowers/plant, average fruit weight and weight of first mature fruit were used together in selection with equal weightage. Randhawa *et al.* (1983) identified 'EC 1' with long medium sized fruits containing a few seeds, and 'S-5' and 'LCA' with round large fruits from a collection of 8 lines based on dry matter, protein, fat, minerals, crude fibre and carbohydrates. The released varieties are KAU Local, Indu, CO-1, CO-2, APAU Shakthi, Karikumbala, Boodikumbala and IVAG-502 (Kashi Dhawal) were developed by selection through progeny testing.

Resistance Breeding

Diseases Resistance

Screening of germplasm for reaction to mosaic disease has shown that AG-1, AG-22, AG-50 and AG-53 were free from diseases (Gopalakrishnan, 2004).

Table 3.133: Diseases and their Sources of Resistance

Diseases	Resistance Source
Mosaic	AG-1 (Indu), AG-22, AG-50 and AG53 (can be obtained from KAU, Kerala)
Fusarium oxysporum f. sp. Melonis race-1 and race-2	Kogannenaski Golden Pear
WMV Race 1	PI 391544 and PI 391545

Source: Bhardwaj, 2010.

Table 3.134: Germplasm as Source of Resistance

Crop	Disease/Insects	Organism	Accessions	Reference
Ash gourd	*Fusarium* wilt	*Fusarium semitectum, Fusarium oxysporum, Fusarium equiseti* (R)	Koganenaski Golden Pear	Kaur *et al.* (1985)
	Virus	WMV1And WMV2	PI391544, PI391545 (R to WMV1 and T to WMV2)	Providenti (1966)

REFERENCES

1. Bhardwaj, D.R. 2010. Breeding for pest and disease resistance in cucurbits. In: Compedium of Wnter School "Emerging trends and strategies for the management of pests and diseases in vegetable crops. Published by ICARIIVR, Varanasi, p. 102-109.
2. Chandrasekar, B; Mukerjee, B. and Mukharjee, S.S. 1989. Blood sugar lowering potentialities of selected Cucurbitaceae plants of Indian origin. Indian Journal of Medical Research, 90:300-305.
3. Gopalakrishnan, T.R. 2004. Three decades of vegetable research in Kerala Agricultural University, Vellanikkara, Thrissur (Kerala).
4. Grover, J.K; Adiga, G; Vats, V. and Rathi S.S. 2001. Extracts of *B. hispida* prevents devlopment of experimental ulcrs. J. Ethnopharmocology, 78:159-164.
5. George, T.E. 1981. Biometrical studies in ash gourd (*Benincasa hispida* (Thunb.) Cogn.). M. Sc. (Ag.) Thesis. Kerala Agricultural University, Trissur, Trivendrum.
6. Heiser, B. Charles. *The Gourd Book*, p. 61.
7. Indira, P. and Peter, K.V. 1987. Underexploited Tropical Vegetables, pp. 4-8. Kerala Agril. Univ., Vellanikkara.

8. Kamalanathan, S., Thamburaj, S., Sundararajan, S; Subbiah, R. and Seemanthani, B. 1972. Influence of time of sowing on flowering and fruiting behaviour of pumpkin (*Cucurbita moschata* L.) and ash gourd (*Benincasa hispida* L.), Madra Agri. J. 59:231-236.
9. Menon, M.P. 1998. Cataloguing and identification of promising ash gourd ecotypes in relation to season and maturity. M.Sc. thesis, Kerala Agricultural University, Vellanikkara, Thrissur (Kerala).
10. Pandey, A.K., Bhardwaj, D.R., Dubey, R.K., Singh, V. and Pandey, S. 2015. Botany, diversity, utilization and improvement of ash gourd (*Benincasa hispida* Thunb. EX Murray Cogn.) : A review. *Annals of Horticulture*, **8(1):** 1-15.
11. Ramesh, M; Gayathri, V, Rao AVNA, Prabhakar, M.C. and Rao, C.S. 1989. Pharmacological action of fruit juice of *Benincasa hispida*. Fitoterapia, 60: 241-247.
12. Simmonds, N.W. 1976. Evolution of crop plants. edn. 1. 306 p. Longman, New York.
13. Verghese, B.M. 1973. Studies on the cytology of South Indian Cucurbitaceae. Ph.D. Thesis, Kerala Agric Univ. Trivandrum.
14. Wills R.B. H; Wong, A.W.K; Scriven, F.M. and Greenfield, H. 1984. Nutrient Compositon of Chinese vegetables, Journal of Agriulture and Food Chemsitry; 32:413-416.
15. Winter Melon, edited on 23rd October, 2011, available from Wikipedia, the free encyclopedia, http:// en.wikipedia.org/wiki/Winter-melon.
16. Yang, S. L. and Walters, T.W. 1992. Ethnobotany and the economic role of the Cucurbitaceae of China, *Econ. Bot.* 46:349-367.

Genus *Coccinia*

Ivy Gourd (*Coccinia cordifolia*)

Introduction

Ivy gourd is an important vegetable crop of the family Cucurbitaceae which is known as coccinia, kundru, little gourd or tondil. It is an under exploited cucurbit. Sometimes, immature fruits of species *Coccinia adenensis* are bitter in taste due to the presence of cucurbitacin '*b*' in the form of glycoside (Bhakuni *et.al*; 1962) but lose their bitterness rapidly during ripening (Rehm *et al.*, 1957). There is an ample scope for the export of this vegetable (Grubben, 1977). The leaves, shoots, and immature fruits are cooked and eaten; mature fruits are sometimes preserved. There is needs to develop lines which are early, have bold and tender fruits with good storability and resistance to pests and diseases (Grubben, 1977).

Origin, History and Distribution

The original home of ivy gourd or little gourd or scarlet-fruited gourd (*Coccina grandis* L.) is believed to be originated in India (Nath, 1966). The nomenclature is derived from the Latin word *Coccineus*, meaning scarlet, in reference to fruit color (Wagner *et al.*, 1999). In Indian scriptures, this vegetable has also been mentioned extensively as *Bimba* fruit (*Coccinia*) of which ripe red colour fruit has been compared to woman's lips in several places, like *Mahabharata* and also referred to by *Kalidasa* in his Sanskrit plays. It has about 35 species distributed in Tropical Africa and Asia, out of which only single species, *i.e., Coccina grandis* is under cultivation. Its synonym is *Coccinia cordifolia* (L.) Cogn., *Cephalandra indica* and *Bryonia grandis* L. (Pier, 2001). The related species are *Coccinia histella* and *Coccinia sessilifolia*, which are widely cultivated in several countries, like Africa, Central America, China, Malaysia, Australia and other tropical Asian countries (PIER, 2001; Wagner *et al.*, 1999). It has become a vigorous weed in several countries, like Hawaii, Australia, Spain, Texas, and Florida. Under Indian conditions, ivy gourd has naturalized in central and peninsular area. It can be spotted easily in kitchen garden of several tribal families. Northern plains, tarai range, extending to peninsular regions are major hub for ivy gourd (cultivated and wild) distribution. This vegetable can be spotted in southern and eastern states. In India ivy gourd is cultivated commercially in Madhya Pradesh, Chhattisgarh, Gujarat (Khera and Badodara), Andhra Pradesh, Karnataka, Uttar Pradesh, Odisha, West Bengal, Tamil Nadu and Asom.

Species Availability, Evolution and Diversity

Coccinia has about 35 species distributed in tropical Africa and Asia can be said to be old world, most are in Africa. Out of these, only one species *i.e. Coccinia indica* Wight. and Arn. (syn. *Coccinia cordifolia* Cogn., *Ciphlandra indica* Naud.) is under cultivation and occuring naturally throughout India and tropical Africa. The related species are *C. histella and C. sessilifolia*, which is widely cultivated in several countries like Africa, Central America, China, Malaysia, Australia and other tropical Asian countries. The related species of *C. indica* are *C. histella* and *C. sessilifolia*. In Ghana, a monoecious species is also cultivated (Whitaker and Bemis, 1976).

In this crop considerable variability can be observed in Bihar, South Karnataka and West Bengal.

Botanical Description

The genus is included in the subtribe Bemincasinae (Ser.) C. Jeffr., tribe Benincaseae Ser., and sub-family Cucurbitoideae. It consists of distinct male and female plants (dioecious). Gynodioecious sex forms consisting of female and hermaphrodite flowers have also been observed in nature. It is highly cross-pollinated crop. one of the few dioecious cucurbits, with a heteromorphic

Table 3.135: Taxon and Preferred Taxon Name

Name of Taxon	*Name of Authority*	*Family*	*Preferred Taxon Name*
Coccinia abyssinica	(Lam.) Cogn.	Cucurbitaceae	*Coccinia abyssinica*
Coccinia cordifolia	(L.) Cogn.	Cucurbitaceae	*Mukia maderaspatana*
Coccinia cordifolia	Auct.	Cucurbitaceae	*Coccinia grandis*
Coccinia grandis	(L.) Voigt	Cucurbitaceae	*Coccinia grandis*
Coccinia indica	Wight and Arn., nom. illeg.	Cucurbitaceae	*Coccinia grandis*
Coccinia rehmannii	Cogn.	Cucurbitaceae	*Coccinia rehmannii*
Coccinia sessilifolia	(Sond.) Cogn.	Cucurbitaceae	*Coccinia sessilifolia*
Coccinia sp.		Cucurbitaceae	*Coccinia* sp.

(XY) chromosome pair determining sex. The somatic chromosome number is 2n=24. Triploids with 2n=36 have been observed. Distinct 22 autosomes and 2 sex chromosomes have been seen in *Coccinia*. The male plant is heteromorphic for sex chromosomes (22A + XY). The female plant has 22A + XX but is highly sterile. A male triploid with 33A + XXY has been discovered. The Y chromosome influences expression of maleness Lal *et. al.* (1963). It is a strictly cross-pollinated crop. Breeding and research work on ivy gourd is very meager.

Table 3.136: Botanical Description

Growth Habit	Ivy gourd is dioecious and semi-perennial/perennial glabrous herbaceous much branched creeper. It grows best under conditions of adequate rainfall and high humidity.
Root	Plant develop branched tap roots which are tuberous in nature.
Stem	Stems up to several meter long (6.0-25.0 m), thin terete, slightly scarbid, longitudinally furrowed when young (becomes round and woody when older), dark green, hispid mostly glabrous, produced annually from a tuberous rootstock.Stems are semi-perennial glabrous herbaceous much branched creeper with trailing habit.Tendrils are formed at each node, simple auxiliary and unbranched.
Leaf	Plants bear simple, alternate, broadly ovate to oblong leaves with angular basal lobes, 5-10 cm in diameter. Leaves broadly ovate in outline with a basal sinus, blades 3-10 cm long, 4-10 cm wide, 5 angled to palmately 3-7 lobed, the lobes shallow to deep and lobulate, upper surface glabrate, lower surface hispid and bearing 3-8 glands near attachment of petiole and major vein branching, margins denticulate, apex obtuse, acuminate, mucronate, and petioles 1-3 cm long. Lobes are broadly ovate to narrowly linear lanceilate, irregular patently serrate and ciliate, above dark green and feebly shining, beneath light green, when young oubescent on the nerves beneath, afterwards glabrous on both surface, thinly fleshy, on the upper surface densely studded with subnken transparent dots, beneath at the base with several dark green glands, 2.5-1.0 cm long and 2.0-9.0 cm wide. Petiole is subterete, narrowly furrowed on the upper side, thinly pubescent or subglabrous, dark green and 0.75 cm long. Tendrils are leaf opposed, simple yellowish green and 7.0-25.0 cm long.
Floral biology	Flowers are dioecious, large, white, solitary in the leaf axils, rather large and long peduncled. Peduncle is subcompressed, faintly curved, shining, glabrous yellowish green and 1.0-5.0 cm long. Calyx is companulate, greenish white, glabrous, 5.0 cm about half way down, 1.0-1.25 cm long, 1.5-2.0 cm diameter; faintly constructed some what below the limb. Segments are spreading or slightly reflexed, ovate oblong, thinly pubscent on the inner side, with yellowish brown tips, 0.5-0.75 cm long and 0.2-0.25 cm wide. Corolla is companulate, limb 4-6 partite hardly down to the middle, white, on the outside with strong, yellowish green veins, on both sides but especially with in scattered long white hairs, 3.5-5.0 cm long. 4.0-7.0 cm diameter. Lobes are white, broadly ovate, 1.5-2.0 cm long and acute. Male flowers have 3-5 stamens with conduplicate, connate anthers of 0.5-0.75 cm long and connective is yellowish white in colour. Female (pistillate) flowers solitary with a densely pilose torus; staminoide 3, ovate-oblong, pubscent, whitish, one of them half as broad as the two others. Ovary is inferior; narrowly oblong, broadened above the base, and narrowed upwards, yellowish green, shining, with several longitudinal, undulate, interrupted faintly raised ridges, glabrous, 1.5-2.0 cm long and 0.3-0.5 cm in diameter. Style is columnar, simple, terete, yellowish green and 0.3-0.5 cm long. Stigma is 3 lobed, light citrine; lobes of the stigma patent, forked halfway down, densely beset with hair like papillae.
Anthesis and pollination	Among the climatologically factors, relative humidity was found to be correlated with the opening time of the male and female flowers and anther dehiscence, while the minimum and maximum temperatures were not found influencing these biological processes. Various insects of Hymenoptera group have been observed on flowers, including bees (*Apis melifera*) and ants, which are quite helpful in pollination.
Fruit and fruit set	Fruits are developed from flowers arising from the leaf axils of new growth. The fruits are dark green or green cylindrical with white streaks in the younger stages and become red or deep orange on maturity (Jeffrey, 1990). Fruit ovoid to ellipsoid, 25-60 mm long, 15-35 mm in diameter, glabrous, on stalks 10-40 mm long, pulp red. The fruit is a botanically pepo sub globose and round at both end and green with white stripes or no stripes. The pulp is creamish white or yellowish green in colour when immature. When fruit ripe at first orange coloured, finally scarlet, many seeded 3.0-6.5 cm long and 1.5-2.5 cm in diameter. Peduncle is 1.0-5.0 cm long.
Seeds	Seeds are transverse, obovate-oblong, with a narrowed base and flat, white or tan, hairy, flattened, with broad rim. Seeds 6.0-7.0 mm long, margins thickened." (Wagner *et. al.* 1999).

Augmentation of Genetic Resources and Crop Improvement

Breeding Priorities

1. Early and high yielding varieties/hybrids
2. Bold and tender fruits
3. Good storability
4. Fleshy and less seeded type
5. Resistant to pests and diseases

Correlation coefficients were estimated in 35 genotypes of ivy gourd, at genotypic and phenotypic levels. In general, genotypic correlation coefficients were higher than phenotypic correlation coefficients. Yield/plant showed positive and significant correlation with number of fruits/plant and diameter of fruits at genotypic and phenotypic levels. The yield was also significantly and positively correlated with the length of intern ode at genotypic levels (Sarnaik *et al.*, 1999).

Sulabha (CG-23), a high yielding variety, released by Kerala Agricultural University Vellanikkara, Thrissur (Kerala) produces light green, cylindrical long fruits measuring 9.25 cm length and 6.70 cm girth. The average fruit weight is 18.4 g. Fruits have continuous striations on it. Plants come to flower in 37 days, fruits set parthenocarpically and the first harvest can be made in 45 days after planting. It gives an average yield of 18.16 kg/plant/year. From IGKV, Raipur (Chattishgarh) two high yielding genotypes were released (Sarnaik *et al.*, 2005). They are Indira Kunduru-5 which produces fruits of 4.30 cm length, 2.63 cm diameter and yield of 22.94 kg fruits/plant/year and Indira Kunduru-35 which produces fruits of 6.0 cm length, 2.43 cm diameter and yield of 21.08 kg fruits/plant/year.

Table 3.137: Attributes of Cultivar IIVRK-1

Varieties	*Organization*	*Pedigree*	*Description*
IIVRK-1	IIVR, Varanasi	Clonal selection	This is early maturing variety. Fruits are long striped and fruiting period is comparatively long. Average yield is 25-30 t/ha.

Conservation of Genetic Resources

There is considerable scope for survey, collection and maintenance of the germplasm. Ivy gourd germplasm is being maintained by ICAR-Indian Institute of Vegetable research, Varanasi (U.P.) and Indira Gandhi Krishi Vishwavidalaya, Raipur (Chattishgarh).

REFERENCES

1. Bhakuni, D.S. 1962. Gopalan, C., Ramasastri, B.V. and Balasubramanian, S.C. 1982. *Nutritive value of Indian Foods*. National Institute of Nutrition, Hyderabad.
2. Bhakuni, D.S. Srivastava D.S; Sharma, S.N; Kaul, V.N. and Kaul, K.N. 1962. Chemical examination of the fruits of *Coccinea indica*. *J. Sci. Industr. Res.*, **21**:237-238.
3. Gurben, G.J.H. 1977. Tropical vegetables and their genetic resources. International Board for Plant Genetic Resources, Rome.
4. Jeffrey,1990. Biology and utilization of the *Cucurbitaceae*. Ithaca and London: Cornell University, p.167-182.
5. Nath, P., Velayudhan, S. and Singh, D.P. 1987. *Vegetable for the Tropical Regions*. ICAR. New Delhi.
6. Pier. 2001. "Invasive plant species: *Coccinia grandis*." Pacific Island Ecosystems at Risk. United States Department of Agriculture, Natural Resources Conservation Services. The Maui Weekly Wrap. *Maui News*, pp 10-18.
7. Rhem, S., P.R. Ensliv, A.D.J. Meevse and J.H. Wessels (1957). *J. Sci. Food Agric.*, **12:** 676.
8. Sarnaik, D.A., Satish, K.V. and Sharma G.L. 2005. Evaluation of ivy gourd (*Coccinia grandis* L.) germplasm. Proceedings of National Seminar on Cucurbits Breeding and Production Technology, 22-23 September, Held at GB Pant University of Agriculture and Technology, Pantnagar (Uttarakhand).
9. Wagner, W.L., D.R. Herbst, and S.H. Sohmer. 1999. *Manual of the Flowering Plants of Hawaii*. Bishop Museum Special Publication **83** (Vol-2.), University of Hawai and Bishop Museum Press, Honolulu, HI.
10. Whitaker, T.W. and Bemis, W.P. 1976. Cucurbits. (In: Evolution of Crop Plants), p. 64-69

Genus *Sechium*

Chayote or Chow-chow (*Sechiumedule*)

Introduction

Chow-chow (*Sechium edule*; 2n=24) is important single seeded cucurbits known as chayote, cho-cho, miriliton, mango squash, Spanish citrayota, gayto, askas and vegetable pear. In hilly areas it is very much popular as a vegetable (typically cooked). The word chayote has been derived from chayotli, its name in the ancient Aztec word chayotli and it was well known to the Aztec civilization of Mexico. It is easy to produce both for home consumption and city markets. It is cultivated in empty patches, door yards/kitchen garden/back yards and market gardens. The immature green fruits are used as vegetable like other squashes. Fruit possess mild flavor. In the food processing industries its fruits are also used as filler for tomato ketchup and other baby foods. Plant of chow-chow develops large succulent edible roots which are good source of easily digestible starch. In Mexico, people have several preparations like boiled, roasted, fried or candied in sugar. Every part of chayote is useful, but it is mainly grown as a fruit vegetable. Its immature fruits, young leaves, shoot and tendrils are cooked as greens which rich in vitamins A, B and C, calcium and iron. Fruits are very rich source of calcium (70 mg/100 g) while its swollen starchy roots resemble to those of the sweet potato. The fruits vary in flavor according to cultivar, from bland to sweetish or starchy. Bland cultivars are also used industrially as a food, filler for pastes and sauces and as a substitute for apple in pies and tarts. Tuberous roots are also consumed. The starch obtained by tuberous roots is used in soups and stews and sometimes candied. Owing to carbohydrate and minerals, chow-chow fruits can also be used for preparation of pickles, candy and flour products. The fruit and particularly the seeds are rich in amino acids such as aspartic acid, glutamic acid, alanine, arginine, cysteine, phenylalanine, glycine, histidine, isoleucine, leucine, methionine (only in the fruit), proline, serine, tyrosine, threonine and valine. Fruits of Chow-chow are considered diuretic and seeds are said to be helpful in intestinal problems. Fruits, shoots and tubers are also used as fodder. Fibers of the stem are used locally to make hats and baskets. It is used to create shade by spreading on bower and check soil erosion during rain. Seed is a good source of protein and said to be helpful in intestinal problems. Starch obtained from the roots can be used in textile industry and is good source for manufacturing the biodegradable plastics.

Origin, History and Distribution

Genus *Sechium* said to be originated in mountainous region of humid tropical regions of Central America and Mexico whehere it was first domesticated by the Aztecs (Singh, 1990). In these area, it has been grown for centuries and remains popular today (Thomson and Kely, 1957; Fairchild, 1947; Hernándo and León, 1994) in theses area but some reaearchers feel that chow-chow was originally in Central America and neighboring Islands (Hoover, 1923, Bailey and Bailey, 1969 and Mac Gillivray, 1953). Lionti (1959) stressed upon its introduction to Tuscany. Whatley (1962) has mentioned about the cultivation of this crop in Ghana and Bokareva and Fogel (1965) in Sukumi.Chayote were domesticated in Mexico and Guatemala in pre-Columbian times either from wild forms of the species or from its closest relative *S. compositum* (Sm.) Jeffe. (Newstrom, 1990) where it has been grown from centuries and remains popular to day. After America colonization, chayote spread rapidly to all tropical areas of the New World and became a popular staple item in the diet for people of the Old World (Whitaker and Davis, 1962). No pollen grains or other structure of this species have been identified on archaeological sites. Today, chayote is cultivated throughout tropical and subtropical regions of the world (Newstrom, 1990). In India principally it was confined to Tropical and sub-tropical climate (Asia, Africa, Australia and South America) and particularly in mid hill conditions. Maximum variability occurs in Sikkim. In Meghalaya and Mezoram it is grown on commercial scale and is popular in Darjeeling Hills, also. Putative wild populations of *Sechium edule* and related wild species are found in Central America. Chayote was already a common cultivated vegetable among the Aztecs in pre-Colombian times. Chayote's centre for genetic diversity suggests it was first domesticated in Mexico and Central America but this variation could have been secondarily derived. However, inclusive these 3 lines of evidence are they do converge on the same conclusion supported by new botanical evidence. Chayote's closest wild relative, *Sechium compositum* from Mexico and Guatemala and two potentially ancestral forms of Chayote recently discovered outside cultivation in Mexico, demarcate Mexico and Guatemala as the most likely centre of origin for this ancient crop (Newstrom, 1991). Chow-chow has now spread throughout the tropics and sub tropics including South- East Asia, Africa, Australia and South America. In the Old World it is cultivated at least in India (Chakravarty, 1990; Kavitarani and Bhagirath, 1991), Nepal (Sharma *et al.*, 1995), China, Papua New Guinea, Southeast Asia, Taiwan (Engels and Jeffery, 1993) and European countries such as Italy. It is very much popular in all south-East Asian countries as a cheap vegetable. This crop is widely cultivated throughout America.Chow-chow was originally considered as monotypic genus without wild relative. Since the 1970, wider genus concepts have been proposed, including 3-9 species, all indigenous to Central America but recently 8 wild species have been identified in central America. *S. compositum* (Donn. smith) C. Jeffery, occurring in Southern Mexico (Chiapas) and in Guatemala, is considered the closest wild relative of *S. edule*.In the opinion of other reasearches, the closest relatives to *S. edule* are so-called wild forms of *S. edule*, the taxonomic positions of which are unresolved since they are distributed in an apparently natural way in the Mexican states of Veracruz, Puebla, hidalgo, Oaxaca and Chiapas. *S. hintonii*, a species endemic to Mexico, until recently considered to be extinct and which grows in the states of Mexico and Guerrero and possibly in Jalisco.

Though, it is a native of Mexico but considerable diversity is found in NEH region (between 800-2000 m altitudes) particularly, Meghalaya, Mizoram, and Sikkim (Rai *et al.*, 2002, Rai *et al.*, 2005, Yadav *et al.*, 2005). In India chow-chow is still a neglected crop species found in mountainous regions from Sikkim Himalayas to Central Himalayas. In India it is commonly grown in the North eastern regions and in the coastal belt and low hills (1000-2000 m) of Maharashtra, Karnataka and

Tamil Nadu. It is also cultivated in Mandi district of the Himachal Pradesh. In Maharashtra Chow-chow do well in the Koraput hills (Kolhe, 1962). Now its commercial cultivation is going on in the lower Palney hills of Madurai and Nilgiris districts in Tamil Nadu at elevations of 850-1700 m. and some parts of Karnataka like Bangalore (Seemanthini, 1964). Mizoram is the leading state with an estimated area of 845 ha and 10,985 metric tonnes production.

Species Availability, Evolution and Diversity

Sechium was previously considered monotypic, but now includes as many as eight species (Newstrom, 1990). Chayote (*Sechium edule* (Jacq.) Swartz.), is the only cultivated perennial cucurbit in the sub-family *Sicyoideae*. The cultivated chayote is diploid with 24 chromosomes (2*n* = 24). This species and all their wild relatives were native to the New World (Whitaker and Davis, 1962). The closest wild relative is *S. compositum* (Donn.Sm.) C. Jeffrey, seems to be propgenitor of chow-chow. Some important species are *Microsechium compositum, Microsechium gintonii, Sechium compositum, Sechiumedule, S. hintoni* and *S. jamaicense.*

Table 3.138: Taxon and Preferred Taxon Name

Name of Taxon	*Name of Authority*	*Preferred Taxon Name*
Microsechium compositum	Donn. Sm.	*Sechium compositum*
Microsechium gintonii	Paul G. Wilson	*Sechium hintonii*
Sechium compositum	(Donn. Sm.) C. Jeffrey	*Sechium compositum*
Sechium edule	(Jacq.) Sw.	*Sechium edule*
Sechium hintonii	(Paul G. Wilson) C. Jeffrey	*Sechium hintonii*
Sechium jamaicense	J. St.-Hil.	*Sechium jamaicense*

Exploration records concur in the finding that the widest variation of *S. edule* under cultivation is found between southern Mexico and Guatemala. A wide variation in the *S. edule* chromosome no. has been documented in the literature. Some studies agree that the haploid and diploid no. of this species are n= 12 and 2n = 24 respectively (Sugiura, 1940, Sobti and Singh, 1961) while other report accounts of 2n = 28 (Giusti *et al.,* 1938) or 2n =22 (Singh, 1990). *Sechium edule,* originally discovered by Browne (1756) and in 1763 it was classified as *Sicyos edulis* by Jacquin and as *Chocho edulis* by Adanson. Then Swartz (1800) include this species in *Sechium* and chow chow presently known as *Sechium edule* (Jacq.) Swartz. A taxonomic classification for the Chayote (*Sechium edule*) is proposed that it includes two subspecies *Sechiumedule* subsp. *edule* and *Sechium edule* subsp. *sylvestre* (Lira *et al.,* 1999). In another study two species of frantza are transferred to *Sechium* (*Sechium panamens* and *Sechium verosum.*). A new species in this region *Sechium chintalense from the low land* of Oxacais also reported (Lira and Chiang, 1991). Lira and Soto (1992) supported that *Sechium hintonii* has its back ground in Mexico more than 30 years after its original description provided morphological and polynological data. They also added *Sechium compositum, Sechium tobacco, Sechium talamancensis* and *Sechium edule.* The most significant morphological differences between cultivated and wild chow-chow are the difference in size of the vegetative and reproductive structures. Wild plants are more robust and their leaves, flowers and staminate inflorescence are bigger than those found in cultivated plants. There are different in their chromosome number. The haploid number reported 2n =12 (Palacios, 1987) and 2n =13 has been determined (Lira and Chiang, 1992) whereas *Sechium edule,* diploid are 2n=22, 2n=24 and 2n =26.

Botanical Description

Chow-Chow belongs to the subtribe Sicyinae C. Jeffr., tribe Sicyeae Schrad; sub-family *Cucurbitoideae.*

Table 3.139: Botanical Description

Growth Habit	The plant of Chow-chow is a vigorously growing perennial vine/hairy climber bearing tendril, making of growth of about 15.0-20.0 m in one season from May to December.
Root	Chow-chow develops succulent tuberous roots which helps the plant to survive under unfavorable conditions. The roots stored good amount of starch and become ready for harvest after 3-4 years of planting.
Stem	It possesses 12.0-15.0 m or more longer vines which require training. The stem is usually slightly compressed, sometimes subterete, longitudinally furrowed, light green and glaborous. Tendrils large, 2-5 branched are inserted beside the leaves, robust, 2-5 partite, light green, glabrous, 15.0-40.0 cm long and is nearly hairless.
Leaf	Leaves simple, distichous, spirally arrange, petiole 3.25 cm long; leaf- blades broadly ovate and circular in outline, 7.25 cm in diameter, base deeply cordate, 3-7 angular or lobed, apex triangular, angular or faintly lobed, patently shortly dentate, cariaceous, above dark green, clothed with very rough, short hairs, beneath light green, very shortly hairy, somewhat scabrous, 5-7 nerved, with transparent veins, 7.0-25.0 cm in diameter. Petiole is robust, succulent, slightly angular or terete, longitudinally furrowed, light green sub glabrous or patently hairy beneath, curved at the base, 2.5-15.0 cm long.
Floral biology	Chow-chow is monoecious in nature; therefore emergence of male and female flower occurs on the same plant. Flowering occurs 100-120 days after planting in the tropical and sub-tropical regions having 12 to 12.5 hour day length. Flowers are small, unisexual-monoecious, axillary, male and female inflorescence annually together.

	Five merous flowers, hypanthium saucer-shaped, with 10 pouch like nectarines on the bottom; male recemes with peduncle 6.0-30.0 cm long, 10-30 flowered; stamens, 5 filaments united; female flowers usually solitary on short pedicels, in same axil as male; corolla calyx. 2 mm in diameter; connate style and stigmas, forming a small head. Calyx is campanulate, deeply 5-partite, light green, denely shortly pubescent. Segments are thick, lanceolate, erect or widely patent. Petals are 5, adnate at the base to the calyx-tube, falling off as a whole, ovate-oblong, denely shortly pubescent, ciliate, yellowish-greenish white. Male flowers are racemed; racemes on peduncles of 1.5-5.4 cm length, many-flowered and globose. Stamens are five, thick, cirtrine, 4 of them connate in pairs; anthers conduplicate. Female flowers are axillary, solitary or fascicled, on shorter and longer stalks, which are sometimes connate in pairs, much larger than the male flowers. Peduncles are 0.5-4.0 cm long. Ovary is inferior, large, oblong-obovoid or pyriform, with 5 longitudinal furrows, densely covered with a mealy pubescence; style 1, stigmas 2-3, curved and thick.
Anthesis and pollination	Its male flowers appear in clusters in a 1.5-4.0 cm long peduncle. Female flowers are solitary. Usually male and female flowers are born in the same axil. Flowers possess pouch like nectarines, which attract pollinators. Pollination is generally completed through bees.
Fruit and fruit set	Single seeded fruits are viviparous germination (the seed germinate inside the fruit still on the vine), pendulous, large, pyriform (pear shaped), soft with several deep longitudinal furrows and a deeply emarginated apex, hence often 5-10 lobed. Fruits colour also varies from white to dark green or yellowish green, shining, thinly pubescent and more or less spinose. Fruit stalk is 2.0-3.0 cm long. The fruit is either smooth or thinly pubescent (soft spines) 10.0-20.0 cm long, 5.0-8.0 cm in diameter and weight ranges from 100 g to 500 g. flesh is very thick (2-3 cm), fleshy succulent, white or greenish white and 2.0-3.0 cm thick. Stalk is thin and 2.0-3.0 cm long.
Seeds	The flat large seed (8.0 cm x 3.0 cm size) has the unusual property of germinating while still inside the pear-shaped fruit (viviparous in nature). The seed is ovoid-oblong, elliptic, flat shining white or milky white and compressed with a soft and smooth testa. Seeds are 3.0-4.0 cm long, 2.0-2.5 cm wide and 1.0-1.5 cm thick.

Augmentation of Genetic Resources and Crop Improvement

In India therte is no improved variety yet released for commerical cultivation. The few varieties released outside India are Florida Green and Monticello White. ICAR Research omplex for NEH Region, Umiam, Meghalaya collected about 50 genotypes from different states of NEH region, evaluated them for yield and quality characters. Some of the genotypes produce up to 40.0 kg fruits/plant (Sanwal, *et al.*, 2008).

Genetic Resource Conservation

Chow-Chow genetic resources are being maintained ICAR Research Complex, Umium Barapani (Meghalaya).

REFERENCES

1. Browne, P. 1756. Civil and Natural History of Jamaica. London, England.
2. Cruz- León, A. and Querol-Lipcovich, D. 1985. Catalogo de recursos genéticos de chayote (Sechium edule Sw.) en el Centro Regional Universitario Oriente. Chapingo, Mexico, UACH.
3. Cruz-León, A. 1985-86. Chayote o cruzas intergenéricas? Hallazgo y características. *Rev. Geogr. Agric.*, **9-10:** 100-106.
4. Flores, E. 1989. El chayote, (Sechium edule Swartz) (Cucurbitaceae). *Rev. Biol. Trop.*, **37(1):** 1-54.
5. Giusti, L., M. Resnik, T.del V. Ruiz and A. Grau. 1978. Notas acerca de la biologia de *Sechium edule* (Jacq) swartz (Cucurbitaceae). *Lilloa*, **35:** 5-13
6. Hernándo Bermejo J.E. and León J. (eds.) (1994). Neglected Crops: Plant Production and Protection Series No. 26. FAO, Rome, Italy. pp. 79-84.
7. Jeffrey, C. 1978. Further notes on Cucurbitaceae. IV. Some New World taxa. *Kew Bull.*, **33:** 347-380.
8. Lira, R. 1988. Cucurbitaceae de la península de Yucatán: taxonomía y etnobotánica. Instituto Nacional de Investigaciones sobre Recursos Bióticos, Jalapa, Veracruz. (thesis)
9. Lira, R. 1990. Estudios taxonómico y ecogeográfico de las Cucurbitáceas de Latinoamérica. 1st biannual report (Jan.-Aug. 1990). IBPGR, Rome.
10. Lira, R. 1991. Estudios taxonómico y ecogeográfico de las Cucurbitáceas de Latinoamérica. 2nd biannual report (Aug. 1990-Jan. 1991). IBPGR, Rome.
11. Lira, R. and Chiang, F. 1992. Two new combinations in *Sechium* (Cucurbitaceae) from central America and a new species from Oaxaca, Mexico, *Novon*, 2:227-231.
12. Lira, R. and Soto, C. 1991. *Sechium hintonii* (P.G. Wilson) C. Jeffrey (Cucurbitaceae), Re-discovery and observations. FAO/IBPGR Plant Genet. Res. Newsletter, 87: 5-10.
13. Maffioli, A. 1981. Recursos genéticos de chayote, Sechium edule (Jacq.) Swartz (Cucurbitaceae). Turrialba, Costa Rica, CATIE-GTZ.
14. Newstrom, L. 1989. Reproductive biology and evolution of the cultivated chayote *Sechium edule*: Cucurbitaceae. In G.H. Bock and Y.B. Linhart, eds. Evolutionary ecology of plants, p.491-509. Boulder, Colo., USA, Westview.
15. Newstrom, L. 1990. Origin and evolution of chayote, (*Sechium edule*). In D.M. Bates, R.W. Robinson and C. Jeffrey, eds. Biology and utilization of the Cucurbitaceae, p. 141-149. Ithaca, N.Y., CornellUniversity Press.

16. Palacios, R. 1987. Estudio Exploratoriodel Numero cromosomico del chayote *Sechium edule* Sue. Tesis Lie. Ciencias Agricolas. Universidad Veracruzana, p. 59.

17. Rai, N., Yadav, D.S., Nath, A. and Yadav, R.K. 2002. Chow-chow: a poor man vegetable in NEH region. *Indian Farming*, **52**: 18-20.

18. Rai, N., Yadav, R.K. and Yadav, D.S. 2005. Genetic variability studies in chow-chow (*Sechium edule* SW.) *Veg. Sci.*, 32(2): 184-186.

19. Sanwal, S.K; Yadav, R.K. Singh, P.K; Rai, N. 2008. Variability and gentic diversity studies in indigenous chow-chow genotypes of northeast India. Indian J. Hortic 65(2) : 167-170.

20. Singh, A.K. 1990. Cytogenetics and evaluation in the Cucurbitaceae. p. 10-28. In Biology and utilization of the Cucurbitaceae. (D.M. Bates, R.W. Robinson and C.Jeffrey, eds.). Cornell University Press, Ithaca, NY.

21. Sobti, S.N. and S.D. Singh, 1961. A chromosome survey of Indian medicinal plants. Part I. Proc. *Indian Acad. Sci.*, **54:** 138-144

22. Sugiura, T. 1940. Studies on the chromosome numbers in higher vascular plants. *Citologia*, **10:** 363-370.

23. Whitaker, T.W. and Davis, G. 1962. Cucurbits: Botany Cultivation, and Utilization. New York. Swartz, O. 1800. Flora Indiae Occidentalis.

24. Yadav, R.K., Rai, N., Yadav, D.S. and Asati, B.S. 2005. Genetic variability and correlation studies for fruit characters in chow-chow (*Sechium edule*). *The Hort. J.*, **18(2):** 106-109.

Genus *Cyclanthera*

Stuffing Cucumber or Meetha Karela (*Cyclanthera pedata*)

Introduction

This crop is known as Meetha Karela (Kumaoni), Ram Karela, Wild cucumber, Koreilla, Korila Achoccha Caihua, Lady's slipper, Slipper gourd, sparrow gourd and stuffing cucumber. It is relatively cold tolerant and adapted to elevations up to 2,000 m. It is also easy to cultivate in the tropics and sub tropics.Tender young fruits are eaten raw or cooked as vegetable. From the hardened fruits the seeds are removed and the resulting seed cavity is quite suitable for stuffing with meat and spices. In the fresh farm, the taste is like cucumber. Young shoots and leaves are also edible. It is largely being used to make salad or soup. The stuffing cucumber is a rich source of minerals (Ca, Mg, and Fe). Its 100 g of eduible portion contain 94.0 g water, 0.6 g protein, 0.1 g fat, 4.0 g carbohydrate, 0.7 g fibre, 14.0 mg calcium, 14.0 mg phosphorus, 0.8 mg Fe, 0.04 mg thiamin, 0.04 mg riboflavin, 14.0 mg ascorbic acid and 0.03 mg niacin (Rubatzky and Yamaguchi, 1997). The young shoots and leaves may also be eaten as greens. In case of disease or/and obesity it can be eatyen as juice extract during fast. It is also effective as dehydrated powder. Ripe fruit contains peptin, galacturonic acid, dihydroxitriptamine, pierine, resins, minerals (phosphorus), vitamins (thiamine, vit.C), lipoproteins and steroidal compound (systosterol and 3 β- D glucoside) with hypoglicaemic and antilipemic (against cholesterol LDL) action, low density lipoproteins (Montoro, 2001). It also has the reputation of inflammatory (more potent than Ibuprofin), hypocholesterolaemic and hypoglycemic. A series of studies on chemical constituents of alimentary plants*Cyclanthera pedata* from South America has been done. Nine triterpenoid saponins including 6 new compounds were isolated from the Me OH extract of *Cyclanthera pedata* (De Tammosi *et al.*, 1996).

Origin, History and Distribution

Ram Karela or Kaji Karel is native to Andean South America. It is an escape from cultivation. Its cultivation is largely done from Mexico to Peru and Equador. Its cultivation can be spotted occasionly in the old World tropics (*e.g.* Malaysia, Nepal, often found in Northern Himalayas, Bhutan, Bolivia and Taiwan). It is currently cultivated in the Caribbean and in Central and South America. It is distributed in warmer parts of South Asia. Sometimes it is found as an escape in northern Himalayas, Bhutan and Nepal. In India, this vegetable grows in backyards during warm season.

Species Availability, Evolution and Diversity

There are about 30 species of *Cyclanthera* that are native to warm-temperature and tropical America. The *Cyclanthera pedata* is commonly cultivated species. The closely related species *i.e. Cyclanthera brachystachia* (Ser.) Cogn.(Syn. *C. explodens* Naudin) originating from the same area is also cultivated for its edible fruits. The fruits, seeds *etc.* can be obtained from IPIFA (Instituto Peruano Investigaciones Fitoterapica Andina and also conserved in Herbario de Museo de Historia Natural "J. Prado" Un. Lima (Peru).

Botanical Description

Table 3.140: Botanical Description

Growth Habit	It is an annual climber and vine. Climbing with the help of long forked tendrils.
Stem	Weak to vigorous stem herb vine up to 5.0 m long. The stem is branched and glabrous.

Leaf	Leaves are on glabrous petioles 1.0-8.0 cm long. Leaves ovate to sub orbicular in outline ca 17.5 x 20.0 cm or palmately divided. In general leaves are 3-5 foliolated. The lower leaves may be further divided into three four minor lobes (the three centre lobes are undivided) which gives the leaf an appearance of having 9 to 11 lobes.
Floral biology	The flowers are monoecious (individual flowers are either male or female, but both sexes can be found on the same plant). Male flowers are small, yellow, in cymes arising from leaf axils/panicles with a few flowers confined to upper half or less of peduncle, recepacle. Female flowers solitary.
Anthesis and pollination	Pollination takes place through insects.
Fruit and fruit set	Fruits are white green or pale greensoft tapering semiflattened, oblong with longitudinal lines and sparsely spined, 10.0-15.0 cm in length (Popenoe, 1992). In general, fruits are 5.0-10.0 cm in width (Huxley, 1990) and 2.0-3.0 cm in thickness. The interior of the fruit is hollow (much like a bell-pepper), fleshy rind is 3.0-4.0 mm thick, the inner tissues is a white spongy pulp containing several black seeds attached to the placenta. Fruits look like a cucumber.
Seeds	Several black seeds

REFERENCES

1. Huxley, A. 1992. The New RHS Dictionary of Gardening. MacMillan Press.
2. Montoro, P. 2001. Studies on the constituents of *Cyclanthera pedata* fruits: isolation and structure elucidation of new flavonoid glycosides and their antioxidant activity. *J. Agric. Food Chem.*, **49(11):** 5156-5160.
3. Popenoe, H. 1990. Lost Crops of the Incas. National Academy press.
4. Rubatzky, V.E. and Yamaguchi, M. 1997. World vegetables: principals, production and nutritive values. 2nd Edition. Chapman and Hall, New York, United States. p. 843.
5. De Tommasi, N; De Simone, F; Speranza, G. and Pizza, C. 1996. Studies on the constituents of *Cyclanthera pedata* (Caigua) seeds: Isolation and characterization of six new cucurbitacin glycosides. *J. Agric.* Food Chem. 44(8): 2020-2025.

Ecballium Species

Taxonomic Position of Genera *Ecballium*

Name of Taxon	*Name of Authority*	*Botanical Family*	*Preferred Taxon Name*
Ecballium elaterium	(L.) A. Rich.	Cucurbitaceae	*Ecballium elaterium*
Ecballium sp.		Cucurbitaceae	*Ecballium* sp.

4

Genetic Resources of Legume Vegetables and Crop Improvement

INTRODUCTION

Legumes, broadly defined by their unusual flower structure, podded fruit and the ability of 88 per cent of the species examined to date to form nodules with rhizobia, are second only to the gramineae in their importance to humans (Graham and Vance, 2003). The legume family is the third largest flowering plants family with approximately 670 to 750 genera and nearly 20,000 species of herbs, shrubs, trees and climbers (Polhill *et al.*, 1981; Bhardwaj and Gautam, 2016) include important grain, pasture and agroforestry species. The sub-family Papilionoideae has about 480 genera and 12,000 species distributed throughout the world. Among them only about 30 species are cultivated, but only a few (soybean, groundnut and peas) are widely grown across the continents. Diet dish of any vegetarian becomes balance diet-dish with addition of legume vegetable either in fresh or dry form. Legumes are important sources of proteins, vitamins and minerals and are popularly known as "Poor man's meat" and "rich

Table 4.1: List of Recorded indigenous Edible Legume Vegetables from Bay Islands

Sl.No.	*Common Names*	*Botanical Name*	*Economic Part*
1.	African Yam bean	*Sphenostylis stenocarpa*	Tuber
2.	Agathi	*Sesbania sesban*	Leaves and flowers
3.	Black henna	*Indigofera tinctoria*	Flowers and pods
4.	Butterfly pea creeper	*Clitoria ternatea*	Flowers
5.	Butterfly tree	*Bahunia purpurea*	Young pods and seeds are edible
6.	Climbing wattle	*Acacia pennata*	Shoots are used in soups and curries
7.	Cluster bean	*Cyamopsis tetragonolobus*	Pods
8.	Cowpea	*Vigna unguiculata*	Pods
9.	Faba bean	*Vicia faba*	Pods
10.	Fenugreek	*Trigonella foenum-graecum*	Leaves and seeds
11.	French bean	*Phaseolus vulgaris*	Pods and dry seeds
12.	Golden shower	*Cassia fistula*	Pods
13.	Kudzu tree	*Pueraria thunbergiana*	Flowers, leaves, stems and roots
14.	Nicker bean	*Entada scandens*	Seed
15.	Sangri	*Prosopis cineraria*	Pods
16.	Sea bean	*Dioclea reflexa*	Seed powder as a thickener in curries

Sl.No.	Common Names	Botanical Name	Economic Part
17.	Dolichos bean/Sem	*Lablab purpureus*	Pods
18.	Sun hemp	*Crotolaria juncea*	Flowers
19.	Sword bean	*Canavalia gladiata*	Pods and seed
20.	Tamarind	*Tamarindus indica*	Leaf and pods
21.	Tree bean	*Parkia roxburghii*	Flowers, pods and seed
22.	Vegetable pigeon pea	*Cajanus cajan*	Pods
23.	Velvet bean	*Mucuna pruriens*	Seed
24.	Winged bean	*Psophocarpus tetragonolobus*	Pods, seeds, flowers, shoots, leaves and root
25.	Yam bean	*Pachyrhizus erosus*	Tuber

Source: Divya *et al.*, 2016.

Table 4.2: List of Wild Relatives of Legume Vegetables

Sl.No.	Genus	Wild Taxon	Remarks
1.	*Atylosia*	*Atylosia cajanifolia*	Progenitor of pigeonpea
		A. grandiflora	Flowers and pods are edible
		A. nivea	Pods are edible
2.	*Canavalia*	*Canavalia stocksii*	Seeds are edible
3.	*Dolichos*	*Dolichos bracteatus*	Pods are edible
4.	*Mucuna*	*Mucuna gigantea*	Boiled seeds and raw pods are eaten by Nicobarese
5.	*Vigna*	*Vigna marina*	Possesses salt tolerance
		V. heretiella	-
		V.curcosca	-
		V. adenantha	Seeds, pods, stem and leaves are edible. Fiber used for weaving
		V. capensis (wild cowpea)	Tuber is edible

Source: Divya *et al.*, 2016.

Table 4.3: Known Genetic Resources in Legumes

Genus	No. of Species	Cultigens	Gene pools				
Cajanus (*Atylosia*)	2+35	*Cajanus cajan*	+	?	+	?	?
Cicer	40	*Cicer arietinum*	+	+	+		+
Lallab	1	*L. niger*	+	+	?+	?+	?
Macrotyloma	24	*M. geocarpum*	+	+	?	?	?
		M. uniflorum	a	+?	?+	?+	?
Lathyrus	1505	*L. sativus*	+	+	?	?	?
Lens		*L. culinaris*	+	+	+	?	?
Phaseolus	50	*P. acutifolius*	+	+		+	+
		P. coccineus	+	+	+	+	+
		P. lunatus	+	+		+	+
		P. vulgaris	+	?+	?	?	+
Pisum	2	*P. sativum*	+	+	?+	?+	+
Psophyocarpus	9	*P. tetragonolobus*	+	?+	?	?	+
Vicia	140	*V. faba*	+				
Vigna	150	*V. aconitifolia*	+	+	?	+	?
		V. angularis	+	+	+	+	?
		V. mungo	+	+	+	+	?
		V. radiata	+	+	+	+	?
		V. subterranea	+	+	?	?	?
		V. umbellata	+	+	+	+	?
		V. unguiculata	+	+			?

Source: Singh *et al.*, 1994.

Table 4.4: Primary Areas of Diversity and Ancestral Relationships in some Common Legumes

Sl.No.	Crop	Botanical Name	Tribe	Primary Areas of Diversity	Wild Ancestor/Near Relatives
1.	Chick pea	*Cicer arietinum*	Viciea	Asian near East	*C. reticulantum, C. echinospermum*
2.	Pigeon pea	*Cajanus cajan*	Phaseolae	E. Africa and Indian subcontinent	*Atylosia lineata*
3.	Mung bean	*Vigna mungo*	Phaseolae	India, South East Asia	*Vigna radiata* var. *subulata*
4.	Urid bean	*Vigna radiata*	Phaseolae	India, South East Asia	*Vigna radiata* var. *subulata*
5.	Lentil	*Lens culinaris*	Vicieae	South West Asia and Mediterranean	*L. oriantalis, L. nigricans*
6.	Field pea	*Pisum sativum*	Vicieae	South West Asia	*P. elatius, P humills*
7.	Cowpea	*Vigna unguiculata*	Phaseolae	West Africa, India	*V. pubescens*
8.	Rajmah (Common bean)	*Phaseolus vulgaris*	Phaseolae	Mexico and Guatemala	*P. aborigineus*
9.	Horsegram	*Dolichos biflorus*	Phaseolae	South East Asia	-
10.	Moth bean	*Phaseolus aconitifolius*	Phaseolae	-	-
11.	Khesari	*Lathyrus sativus*	Vicieae	South West Asia	Wild *L. sativus*
12.	Adzuki bean	*Vigna angularis*	Phaseolae	China, Japan	Not known
13.	Lima bean	*Phaseolus lunatus*	Vicieae	South and Central America	Wild *Phaseolus*
14.	Broad bean	*Vicia faba*	Vicieae	Asia and Mediterranean	Not identified

Source: Singh *et al.*, 1994.

man's vegetable". In nature, large numbers of legume crops are available which provides edible seeds, oils, vegetables (pods, leaf *etc.*) and contributed significantly in increasing human, animal and soil health. India is endowed with rich diversity of legums which are grown from temperate region to humid tropic region between the altitudes from sea level to snow line. The statuses of the crops are important next only to cereals. These are also known as "poor man's meat" because these contain more than double protein content (17-38 per cent) as compared with grains and also their amino acid composition complements to that of cereals. In India, the Andaman and Nicobar group of Islands has its own unique combination and considered as richest reservoirs of genetic variability and diversity of various kinds of legume vegetables.

There are several wild relatives of legume vegetables which are widely distributed in Andaman and Nicobar Islands.

Some legumes (pulse crops) are also the oldest to be cultivated by early man. *Pisum arvense*, lentil and *Lathyrus sativus* were domesticated in India in Neolithic and Chalcolithic periods (Vishnu Mittre, 1974). The legumes are well adapted to diversity in various ecologic, edaphic and climatic factors through the process of natural selection and domestication. Most of the legumes have their primary diversity confined to limited geographical region (West Africa to Indian subcontinent) except Lima bean, common bean (America) and Adzuki bean (Chinese centre) *etc.* (Adams and Pipoly, 1980).

The primary areas of diversity and ancestral relationships in legume crops are also essential for effecrive utilization of genetic resources in crop improvement programme.

Table 4.5: Common and Botanical Name of Underutilized Leguminous Vegetables

Common Name	Botanical Name
Winged bean	*Psophocarpus tetragonolobus*
Cluster bean	*Cyamopsis tetragonoloba*
Sword bean	*Canavalia gladiata*
Jack bean	*Canavalia ensiformis*
Moth bean	*Phaseolus acontifolius*
Scarlet runner bean	*Phaseolus coccineus*
Lima bean	*Phaseolus lunatus*
Yam bean	*Pachyrahizus erosus*
Potato bean	*Pachyrrhizus tuberosus*

Legume vegetables are major source of proteins for human being and animals. Presently much attention has been given in India for introduction and production of major (French bean and garden peas) and minor legume (cowpea, Indian bean, *etc.*) legume vegetables. The other beans (cultivated locally or regionally) need attention for improvement includes tree bean (*Parkia roxburghii*), winged bean (*Psophocarpus tetragonolobus*), sword bean (*Cannavalia gladiata*), faba bean (*Vicia faba*), jack bean (*Cannavalia ensiformis*), velvet bean or mucuna bean (*Mucuna pruriens*), cluster bean (*Cyamopsis tetragonoloba*) *etc.* The underutilized legumes are Adzuki bean (*Vigna angularis*), rice bean (*Vigna umbellata*), and lablab bean (*Lablab purpureus*) also need attention. Serious threat to the genetic diversity of these species (cultivated in the area more than 200-300 years) found while they are important to food, fodder, nutritional and economic security. Genetic variation in major and underutilized legumes vegetables (weedy and wild relatives) is of prime importance for the successful breeding of improved cultivar with added value and desirable resistance of diseases and pest.

REFERENCES

1. Adams, M.W. and Pipoly, I.J. 1980. Biological classification and distinction of economic legumes. In: Advances in Legume Science. R.J. Summerfield and R.H. Bunting (eds.). Royal Botanical Gardens, Kew, p. 1-17.

2. Bhardwaj, D.R. and Gautam, K. 2016. Sword bean (*Canavalia gladiata* L.). Natl. Symp. On Vegetable Legumes for Soil and Human Health, Org. by ICAR-IIVR, Varanasi (U.P.), 12-14th February, 2016.

3. Divya, P., Jaisankar, I., Gautam, R.K. and Roy, S.D. 2016. Status and prospects of legume vegetables in Andaman and Nicobar Islands. In: Souvenir and Abstracts, Natl. Sym. on 'Vegetable Legumes for Soil and Human health. Held at ICAR-IIVR, Varanasi (U.P.), 12-14th February, p. 38-42.

4. Graham, P.H. and Vance, C.P. 2003. Legumes: Importance and Constraints to Greater Use. *Plant Physiology*, **131(3):** 872-877.

5. Polhill, R.M., Raven, P.H. and Stirton, C.H. 1981. Evolution and systematics of the Leguminosae. In: Advances in Legume Systematics Part-I (eds.) Polhill, R.M. and Raven, P.H. (Royal Botanic gardens, Kew, UK), p. 1-26.

6. Singh, B., Kochhar, S. and Rana, R.S. 1994. Germplasm collection, evaluation, documentation, enhancement and utilization programme in pulses. In: Plant Genetic Resources: Exploration, Evaluation and Maintenance. Rana, R.S., Singh, B., Koppar, N.M., Rai, M., Kochhar, S. and Duhoon, S.S. ICAR-NBPGR, New Delhi.

7. Vishnu, Mittre. 1974. The beginning of agriculture: Palaeobotanical evidence in India. In: J.B. Hutchinson (ed.), Evolutionary studies in World Crop Diversity and Change in Indian sub-continent, Cambridge.

Genus *Pisum*

Garden Pea (*Pisum sativum* var. *sativum* L.)

INTRODUCTION

Garden pea (*Pisum sativum* var. *sativum* L., 2n=14) is a very nutritious legume vegetable grown throughout the world. Among the grain legume it ranks 3rd at global level. In India, it is grown as winter vegetable in the plains of North India and as summer vegetable in the hills ranked second largest producer of green peas after China. Green peas are eaten fresh, canned and frozen. Some varieties have thin pod membranes (snap pea) and are consumed whole, while others must be shelled. Mature peas are usually dried and split and often used in soups and purees. Mushy peas are dried peas that have been re-hydrated and mashed, and they are also a popular accompaniment to fish and chips or meat pies in England and Australia. Immature peas have higher sugar content, tastier and are more eaten as green vegetables. In many Asian countries, peas are roasted, salted and eaten as a snack food. In some parts of Asia and Africa, leaves of pea plant are cooked and eaten like spinach. Field peas are also used as animal feed, with plants commonly grazed after the pods on the first few nodes have filled. Seeds of both the green pods and dry are rich in protein (25.6 per cent) and are execellent sources of vitamins and minerals (Salunkhe and Desai, 1984). Nutrional values of green peas per 100 g contain 14.5 g carbohydrates, 5.7 g sugar, 5.1 g dietary fiber, 0.4 g fat, 5.4 g protein, 449 µg Beta carotene, 0.3 mg Thiamin (Vitamin B_1), 0.1 mg Riboflavin (Vitamin B_2), 2.1 mg Niacin (Vitamin B_3), 0.1 mg Pantothenic Acid (Vitamin B_5), 0.2 mg Vitamin B_6, 65µg Folate(Vitamin B_9), 40.0 mg Vitamin C, 25.0 mg Calcium, 1.5 mg Iron, 33.0 mg Magnesium, 108 mg Phosphorus, 244.0 mg Potassium, 1.2 mg Zinc, and 80 kcl (340 kj) energy. Consumption of peas improve immune system, maintain bone health, promote cardiovascular health, reduce the risk of age-related macular degeneration and contracts and solves the constipation problem of an individual. Recent studies reveal that a mixture of small protein-called pea protein hydrolysate-from the yellow garden pea could be effective utilized to reduce blood pressure and avoid chronic kidney diasease (American Chemical Society, 2009).

Origin, History and Distribution

According to Vavilov (1926) pea has different centres of origin. The wild pea is restricted to the Mediterranean basin and the Near East. Area comprising central Asia, the near East, Abyssinia and the Mediterranean is the centre of origin based on genetic diversity. Cultivated garden pea is not seen in the wild state and it might have been originated from wild field pea or other related species. Pea or garden pea is a very popular winter season crop and is native of Europe and western Asia (Thompson and Kelly, 1957). Pea (*Pisum sativum* L.) originated in the Near East and Mediterranean region and has been grown for food and feed since early Neolithic times (Zohary and Hopf, 1973). Janick *et al.* (1969) described the association of genus *Pisum* with man at least from Stone Age. Renfrew (1973) reported that in Austria its discovered from Bronze age, whereas in Germany they were grown in Iron Age. Evidence from carbonized remains indicates that pea has been cultivated with cereal crops such as wheat and Barley since its domestication in 6-7000 BC. Soon after domestication pea and other legumes spread to other parts of the world with the movement and activities of man (Marx,

1977). The earliest archeological finding of peas come from Neolithic Syria, Turkey and Jordan. In Egypt, early findings date from ca. 4800-4400 BC in the Nile delta area, and from ca. 3800-3600 BC in Upper Egypt. The pea was also present in Georgia in the 5th millennium BC. Farther east, the finds are younger. Peas were present in Afghanistan ca. 2000 BC, in Harappa, Pakistan, and in Northwest India in 2250-1750 BC. In the second half of the 2nd millennium BC this pulse crop appears in the Gangetic basin and southern India (Jauhari *et al.*, 2000). Archaeological records in the sub-region have shown evidence of very old tradition of cultivating peas in the Neolithic, Chalcolithic (Harappan and Post-Harappan), and Pre-Historic era. In India carbonised seeds have been discovered from Chirand, Neolithic site in Bihar and some scattered seeds were found mixed with wheat seed from Navadatoli-Maheshwar a Chacolithic site in Madhya Pradesh and Alsi in Osmanabad in Maharashtra (Vishne Mittre, 1974). There is a Sanskrit equivalent for pea, mentioned in "Susruta Samhita" (3rd to 4th century A.D.) as "satina" (Ray *et al.*, 1980).

Oldest records are available for peas in the archeological sites from Indus plains, in India and Pakistan for more than 4000 years. The geographic distribution of wild species, the center of genetic variability of the crop and archeological findings suggest that the pea is typical genus belonging to the old world. According to L.I Govorov (1973), peas originated in the mountainous regions of Southwest Asia (Afghanistan, India) Transcausia and Ethiopia. He has presumed that the secondary center of variability of the cultivated forms of these crops is located in the Mediterranean region. According to Blixt (1970), the Mediterranean region is the primary centre of diversity with secondary centers in Ethiopia and the Near East. De Candolle (1896) in his book "Origin of cultivated plants" stated that Aryans introduced it into Europe but the species existed in northern India before the arrival of Aryans in India. Pea had been domesticated in the Near East where it had been found in archeological sites dating back to 7500 BC. The characteristic features of domestication became evident by 6500 BC. Pea cultivation spread to Central Europe by 4000 BC. By 2000 BC, it was known throughout Europe. Dry peas were used for food in Europe from very early days, though green peas were not used until the 16th century. The edible podded types evolved in more recent times. Columbus brought the pea to North America where it quickly spread to other parts. It was introduced into India soon after Columbus and a winter type was introduced from Austria in 1922. Pea is predominantly grown in Europe, North America and Central Asia. Important pea growing countries include France, Russia, Ukraine, Denmark and United Kingdom in Europe; China and India in Asia; Canada and the USA in North America; Ethiopia in Africa and Australia.

India is second largest producer of green peas after China. Cultivation of pea increased from 2.9 per cent during 1991-92 to 4.6 per cent during 2013-2014. During the corresponding period, the pea production has increased from 1.7 to 2.4 per cent of the total vegetable production (total area 4.33 lakh hectare and 38.68 lakh tonnes). The productivity has also increased from 4.8t/ha during 1991-92 to 8.9t/ha during 2013-2014.

In India, Uttar Pradesh, Bihar, Haryana, Punjab, Himachal Pradesh, Odisha and Karnataka are important pea growing states. Uttar Pradesh alone produces approximately 60 per cent of the total national production. Uttarakhand is also the emerging one as new vegetable pea growing state and there the farmers are taking three crops in a year. In north eastern hills of India, pea is grown mostly as garden pea. This vegetable is commercially grown in smaller extent of Nilgiris (Ooty), Dindugl (Kodaikanal), Erode (Thalavadi) and Krishnagiri (Hosur) district of Tamil Nadu.

Species Availability, Evolution and Diversity

Genus *Pisum* belongs to the family Fabaceae. There are 8 species *Pisum formosum* (Ornamental Pea), *Pisum auch* (Aucheri Pea), *Pisum fulvum* (Red Yellow Pea), *Pisum abyssinicum* (Abyssinian Pea), *Pisum syriacum* (Syrian Pea), *Pisum elatius* (Meditarian Pea), *Pisum arvense* (Field Pea) and *Pisum sativum* (Garden Pea) under this genus. There are three kinds of vegetable peas *viz.* garden pea or green pea (*Pisum sativum* var. *sativum*), snow pea or sugar pea and snap pea (both *Pisum sativum* var. *macrocarpon*). The cultivated pea had originated from the wild species *P. elatius*, which has pods that are small and tough when immature and split open at maturity. Some commonly available species are *Pisum abyssinicum, P. arvense, P. elatius, P. formosum, P. fulvum, P. humile, P. jomardii, P. maritimum, P. ochrus, P. sativum, P. syriacum, P. transcaucasicum etc.*

Species Relationship

L.I. Govorov (1973) has proposed that out of the four wild species, two could have a direct relationship to the development of cultivated field peas: *Pisum elatius* Stev. (The Mediterranean pea) and *Pisum fulvum* Subth. (the red-yellow pea). He presumed that these two species could have existed together in adjacent areas over a widespread zone in the ancient Mediterranean. As a result of mutation and natural hybridization, populations were created in different zones which gave rise to other species: *P. abyssinicum* and *P. syriacum* (Berger) Lehm, as well as cultivated forms. Thus, it is believed that North West Asia including Transcaucasia, Asia Minor and Iran were the principal area of origin and cultivation of peas as the area of its primary introduction.

This genus contains one to five species, depending on taxonomic interpretation; the International Legume Database (ILDIS) accepts three species, one with two subspecies.

Table 4.6: Name of Taxon and Preferred Taxon of Genus *Pisum*

Name of Taxon	Name of Authority	Family	Preferred Taxon Name
Lemuropisum edule	H. Perrier	Fabaceae	*Lemuropisum edule*
Pisum	H. Perrier	Fabaceae	*Pisum*
Pisum abyssinicum	A. Braun	Fabaceae	*Pisum sativum* subsp. *abyssinicum*
Pisum abyssinicum var. *vacilaianum*	H. Perrier	Fabaceae	*Pisum sativum* subsp. *abyssinicum*
Pisum arvense	L.	Fabaceae	*Pisum sativum* var. *arvense*
Pisum arvense subsp. *humile*	Holmboe	Fabaceae	*Pisum sativum* var. *pumilio*
Pisum commune	Clavaud	Fabaceae	*Pisum sativum* subsp. *sativum*
Pisum elatius	Steven ex M. Bieb.	Fabaceae	*Pisum sativum* var. *elatius*
Pisum formosum	(Steven) Alef.	Fabaceae	*Vavilovia formosa*
Pisum fulvum	Sm.	Fabaceae	*Pisum fulvum*
Pisum fulvum var. *amphicarpum*	Warb. and Eig	Fabaceae	*Pisum fulvum*
Pisum humile	Boiss. and Noe, nom. illeg.	Fabaceae	*Pisum sativum* var. *pumilio*
Pisum jomardii	Schrank	Fabaceae	*Pisum sativum* subsp. *abyssinicum*
Pisum maritimum	L.	Fabaceae	*Lathyrus japonicus* subsp. *maritimus*
Pisum marifimum var. *glabrum*	Ser.	Fabaceae	*Lathyrus japonicus* subsp. *maritimus*
Pisum ochrus	L.	Fabaceae	*Lathyrus ochrus*
Pisum sativum	L.	Fabaceae	*Pisum sativum*
Pisum sativum subsp. *abyssinicum*	(A. Braun) Govorov	Fabaceae	*Pisum sativum* subsp. *abyssinicum*
Pisum sativum subsp. *asiaticum*	Govorov	Fabaceae	*Pisum sativum* subsp. *asiaticum*
Pisum sativum subsp. *commune*	(Clavaud) Govorov	Fabaceae	*Pisum sativum* subsp. *sativum*
Pisum sativum subsp. *elatius*	(Steven ex M. Bieb.) Asch. and Graebn.	Fabaceae	*Pisum sativum* subsp. *elatius*
Pisum sativum subsp. *hortense*	(Neilr.) Asch. and Graebn.	Fabaceae	*Pisum sativum* var. *sativum*
Pisum sativum subsp. *humile*	(Holmboe) Greuter *et al.*	Fabaceae	*Pisum sativum* var. *pumilio*
Pisum sativum subsp. *sativum*		Fabaceae	*Pisum sativum* subsp. *sativum*
Pisum sativum subsp. *syriacum*	A. Berger	Fabaceae	*Pisum sativum* var. *pumilio*
Pisum sativum subsp. *transcaucasicum*	Govorov	Fabaceae	*Pisum sativum* subsp. *transcaucasicum*
Pisum sativum var. *arvense*	(L.) Poir	Fabaceae	*Pisum sativum* var. *arvense*
Pisum sativum var. *brevipedunculatum*	P. H. Davis and Meikle	Fabaceae	*Pisum sativum* var. *brevipedunculatum*
Pisum sativum var. *elatius*	(Steven ex M. Bieb.) Meikle	Fabaceae	*Pisum sativum* var. *elatius*
Pisum sativum var. *hortense*	Neilr.	Fabaceae	*Pisum sativum* var. *sativum*
Pisum sativum var. *macrocarpon*	Ser.	Fabaceae	*Pisum sativum* var. *macrocarpon*
Pisum sativum var. *pumilio*	Meikle	Fabaceae	*Pisum sativum* var. *pumilio*
Pisum sativum var. *sativum*		Fabaceae	*Pisum sativum* var. *sativum*
Pisum sativum var. *speciosum*	(Alef.) Makasheva	Fabaceae	*Pisum sativum* var. *sativum*
Pisum sp.		Fabaceae	*Pisum* sp.
Pisum syriacum	(A. Berger) C. O. Lehm.	Fabaceae	*Pisum sativum* var. *pumilio*
Pisum transcaucasicum	(Govorov) Stankov	Fabaceae	*Pisum sativum* subsp. *transcaucasicum*

(i) *Pisum abyssinicum* (syn. *P. sativum* subsp. *abyssinicum*)

(ii) *Pisum fulvum*

(iii) *Pisum sativum*-pea

- (a) *Pisum sativum* subsp. *elatius* (*P. elatius, P. syriacum*)
- (b) *Pisum sativum* subsp. *sativum*

Besides *Pisum sativum* various species of pea are found in wild forms as stated below: *Pisum sativum* var. *macrocarpon* is commonly known as the snow pea *Pisum sativum* var. *macrocarpon* Ser. cv.is known as the sugar or snap pea.

Botanical Description

Pea is annual and diploid with 2n=14 which comes under genus *Pisum* of family Fabaceae, synonymously with family Leguminaceae. It is a cool season crop grown in many parts of the world; seeding can take place from winter to early summer depending on location. Germination is epigeal.

Kingdom : Plantae
Kingdom : Plantae
Sub Kingdom : Angiosperms
Division : Magnoliophyta
Class (unranked) : Eudicots
Sub class : Rosids
Order : Fabales
Family : Fabaceae
Sub Family : Faboideae
Tribe : Vicieae
Genus : *Pisum*
Species : *sativum*

Source: Wikipedia, 2017

Table 4.7: Botanical Descriptions

Parts	Description
Growth Habit	Plants are short lived, herbaceous annual, glabrous that are either bushy or climbing (2.0-3.0 m tall), with a life cycle of one year. Plants climbed by leaflet tendrils. The cultivar may be dwarf, semi-dwarf or tall.
Root	Taproot well developed but not strong, up to 1.2 m long, with many lateral roots and sub branches which are colonized by symbiotic nitrogen fixing Rhizobium (*Rhizobium leguminosarum*) to form nodules.
Stem	Pea has generally slender, circular and weak stems upto 150.0 cm long. Stem with no or few basal branches, internodes hollow. Dwarf cultivars with plant height only upto 60-70 cm have been developed and they do not need staking.
Leaf	Leaves ovate alternate; compound leaves have 1-3 pairs of leaflets ovate (pinnate) that is usually covered with a wax like coating. This coating helps limit water loss and protect the plant against diseases. Whitish area often appear on the leaf where the outermost layer of cells, the epidermis, separates from the inner layers, forming an air pocket. The leaves terminate in branched curling tendrils (terminal branched) that help the plant to climb. Stipules leaf like, up to 8-10 cm (large) and petiole (2.0-7.5 cm) long. Leaflets shortly staked, ovate up to 6 in numbers, elliptical, 1.5-8.0 cm x 0.5-4.0 cm entire to toothed, and sometimes converted into tendrils.
Floral biology	Inflorescence an axillary, white, pink or lavender flowers (1-3) develop on stalks in elongated clusters called racmes. The flowers are solitary or up to three flowers per raceme. Flowers bisexual, papilionaceous; bracts are very small, calyx (oblique and lobes unequal) with tube 4-8 mm long, lobes as long or longer than tube, corolla white, pink or purple, standard 1-3 cm x 2.5-4.5 cm, wings a little shorter than standard, keel much shorter, incurved and obtuse. Stamens 10 (9 united and 1 free) diadelphos, filaments broad and anthers uniform. Ovary superior, 1-celled, style falcate/curved, flattened, and bearded on inner surface longitudinally grooved. Stigma is minute and terminal. A developing floral bud could be distinguished from a developing vegetative by its highly compressed sepals forming a cone. A period of 7-9 days are generally taken from this stage to bud initiation (indistinguishable whether floral or vegetative) till the first stage, having thus 12-14 days for competition of the development.
Anthesis and pollination	Pea flowers are self-pollinated, with usually less than 1 per cent outcrossing. Anthesis starts about 5.45 AM with the unfolding of axilla and wings, exposing the keel. This process continues till 6.30 to 7.30 AM at stage III or IV of a bud. Dehiscence takes place by longitudinal splitting of anther lobes. Three longer stamens dehisce first followed by others. The stigma is receptive 48 hours before anthesis and remained 24 hours after it. The stigma retained good receptivity for only one day. Temperature is a factor, which affects the duration of dehiscence. It came down to 88 per cent at anthesis and further reduces to 50-53 per cent in 24 hours anthesis. Optimum temperature for pollen tube growth is about 25°C. Exposure to low temperature of about 0°C has no injurious effect on subsequent development of pollen tubes. Germination percentage of pollen grains is mostly 75-80 per cent at anthesis and it gradually comes down to zero after 24 hours.
Pod and Pod set	The flowers on each raceme produce pods that develop at different times, with one at the tip maturing last. Pods are swollen or compressed, straight or curved on short stalk upto 12.0-15.0 cm with as many as 2-11 seeded. Pod an oblong-ovate pod 3.5-15.0 cm x 1.0-2.5 cm, pendant.
Seeds	Seeds are angular or globose 4-10 in numbers, Smooth or sometimes wrinkled, 5-8 mm in diameter, varying in colour from uniform yellow (sugar pea), green (wrinkled garden pea) to purple or spotted or cream-white. The mainly consists of two cotyledons, which provide initial growth and support to embryo with a primordial root or radical stem tissue and embryonic growing point or plumule. The seed has a hilum with which it has attached to the maternal pod tissue during development and near this is a micropyle, where the germinating radical emerges. Sometimes mottled, seed weight upto 250 mg (Smartt, 1976).

Gene and Genetics of Characters

The highly heritable polygenic characters are plant height, earliness, number of pods per plant, pod length, seeds per pod and 100 seed weight. The number of pods per plant, number of seeds per plant, 100 seed weight, number of days to maturity and plant height had direct effect on yield. Genetics of characters are being given in Table 4.8.

The inheritance of protein content is polygenecally controlled, and mainly by recessive factors for high protein content. The varieties GC-195 and the local cultivar, Kinnauri have high soluble protein content due to the presence of a very high number of dominant genes.

Horticultural Classification

(1) Classification on the Basis of Species

(i) ***Pisum saccharatum***-No or very thin lining membrane in the pod walls.

(i) Types with thin pod walls

I. Type with smooth seeds

II. Type with wrinkled seeds

Table 4.8: Genetics of Characters in Pea

Characters	*Responsible gene*	*References*
Anthovyanin in flower colour	Two basic complementary genes A (gene for leaf axil colour) and Aw (does not influence the leaf axil colour)	Wellensiek, 1948 b
	Am and Aw were identical. Combination of Am Aw was is responsible for purple-flower colour and am tends to suppress this character and results in orange tinge and lighter pigmentation.	Lamprecht, 1954a
Grey spotting on the leaves	Gene *Fl* produced grey spotting on the leaves	
Normal and broad leaves	The L at was responsible for the normal condition of stipules, leaflets and pods while its recessive form caused broad ones. The *tl*[pat] at the *Tl* locus resulted in leaves of the *Acacia* type with long petioles. The effect of Obo gene, responsible for obovate leaves and accompanied complete sterility. The length/breadth index was due to additive gene action. Nine genes, each with pleiotropic effects on length/breadth ratio of leaflets, stipules and sepals were reported.	Lamm, 1957, 1958 a, 1960 a
Flowering time	Controlled by two major genes with the possible presence of unspecified number of modifiers	Hansel, 1954
Photoperiod response	Conferred by the joint presence of the complementary genes Sn, Dne and Ppd	-
Pod colour	Gene *Pa* control light green pods and its recessive allele (pa) was responsible for dark green pods.	Nilsson, 1954
Pod shape	A recessive gene *te* (tenuis) goverened narrow pods and its allele (*Te*) was incompletely dominant. Leaf accentuates the effect of *Te* on pod breadth. The recessive gene, te and laf individually reduce pod breadth by about 25 per cent and jointly by 35 per cent while it increases pod breadth by about 25 per cent. Reduction effect in pod breadth of lines carrying both *lt* and *Lt.* The gene co in presence of con controls the development of convex pod and bopth of these genes independently were responsible for curved pods and the effect on con was prominent than of co. The plants with conco produced convex pods and ConCo or Conco gave rise to straight pods.	Lamprecht, 1953a, Lamprecht, 1954b, Lamprecht, 1957b, Lamprecht, 1955
	Straight pod shape was monogenic dominant to curved pod shape.	Srivastava and Singh, 1973
Pod thickness	Nn determined pod thicknerss.	
Seed surface	Seed form, round vs. wrinkled goverened by a single gene	Mendel, 1865
	Gene A and a caused indented and smooth testa, respectively. R and r determined smooth and wrinkled conditions, respectively and di gave rise to dimpling in pressure of a gene. The 'a' and 'di' were epistatic to the smooth phenotype of R. A gene Ro controlled flattened seed shape.	Wellensiek, 1942
Seed colour	A new gene, lob was described which together with Dem and Cal influences the colour pattern of seed carrying the genes A, Z and mp. The gene gri resulted in coloured stripes near the caruncula and passing down the back of the seed. The recesive gene cor, possibly an allelomorph of GL caused an ochre ring round the hilum. Large number of additive genes are responsible for the colour of the testa.	Hillman, 1955

Table 4.9: Some Important Qualitative Genes of Pea

Character	*Symbol*	*Description*
Plant height	cry	Influences length of internodes and plant height along with la and le
	la	Influences length of internodes and plant height with *cry* and *le*
	le	Influences length of internodes and plant height along with *cry* and *la*
Branching	Fr	With *fru* determines number of basal branches
	Fru	With *Fr* determines number of basal branches
	ram	Increase number of branches
Leaves and stipules	*Af*	Leaflets converted into tendrils
	Lat	Double leaflet and stipule area
	Tac	Tendrils present on acacia leaves
	Tl	Leaves with extra leaflets and no tendril
Wax	*wa*	Without wax on pods, upper and lower stipules, surfaces and underside of leaflets
	wb	Pods without wax, little was on rest of plant
	Wel	Wax absent from all parts of the plant
Colour	*A*	Absence of anthocyanin; dominant allele for anthocyanin production in plant, flower and seed
	ch-l	Plant light yellowish green
	D	Green leaf axil; dependant on a for manifestation of colour
	Pa	Dark green immature seed and foliage
	Vm	Effect similar to *pa*

Character	Symbol	Description
Inflorescence, number of flowers	*Fn*	With *fna* determines number of flowers on the infloresence, greatly influenced by environment
	fna	With *fn* determines number of flowers on the infloresence, greatly influenced by environment
Flower colour	*b*	Flower pink; dependent on a for manifestation of colour
	Ce	Flower rose; dependent on a for manifestation of colour
Seeds	*com*	Sides of seeds flattened
	Di	Small dimpled depressions in seed; observable only
	R	With a r seed
	Rb	Seed cotyledons wrinkled
	Gty	Seed cotyledons wrinkled
	I	Gritty seed surface
	Pl	Green cotyledons;/produces yellow cotyledons; Hilum black with a ar b
Pods	*it*	Increase pod width 25 per cent
	Bt	Apex of pods blunt
	Con	Affects curvature of pods
	N	Pods wall thick
	Dpo	Pods tough and leathery; readly dehisce at mautrity
	p	Reduces or eliminates sclerenchymatous membrane on inner pod wall
	V	Same as *p*
	Gp	Young pod yellow
Resistance	*fnw*	Resistant to *F. oxysporum* race 2
	fw	Resistant to *F. oxysporum* race 1
	sbm	Resistant to pea seed borne mosaic virus
	cr-1	Resistant to *E. polygoni*
	wil	Wilts quickly under moisture stress

Source: Gritton, 1986.

(ii) Types with thick pod walls

I. Type with smooth seeds

II. Type with wrinkled seeds

(ii) *Pisum pachylobum*- Well developed lining membranous in the pod walls.

(i) Types with thin pod walls

I. Type with smooth seeds

II. Type with wrinkled seeds

(ii) Types with thick pod walls

I. Type with smooth seeds

II .Type with wrinkled seeds

The cultivars in pea are grouped on the basis of various characters (Chauhan, 1968; Mac Gillivray, 1961)

(2) Classification on the Basis of Pod Characters

There are three kinds of vegetable peas:

(i) Garden or Green Pea (*Pisum sativum* Linn. var. *sativum*)

Garden or green pea are internally lined with tough sheath and only grains are consumed. Green pea is an important cool weather crop grown throughout the world for green pods. In India, it is grown as a winter vegetable in plains and as a summer and autumn season vegetable at higher altitudes.

(ii) Snow or Sugar Pea (*Pisum sativum* Linn. var. *saccharatum*)

Sugar peas are flat, tender, and stringless and the whole pod is consumed as a salad or steamed. They are usually stir-fried in Chinese dishes, steamed or cooked like snap beans. If the seeds are allowed to develop, they are shelled and cooked as garden pea. However, shelling is more tedious compared with garden pea.

(iii) Snap Pea (*Pisum sativum* Linn. var. *macrocarpon*)

Snap peas are also called as edible podded pea which do not have sclerenchymatous membrane or inner pod walls. In this group development of seed is very slow; some time very small or rudimentary. The pods are round with well-developed grains and whole pod is consumed as a salad, snapped or steamed/cooked like snap beans or shelled as garden pea.

(b) Classification on the Basis of Flower Colour (Singh and Joshi, 1970; Singh *et al.*, 1974)

(i) White flowered garden pea includes *Pisum sativum* L.

(ii) Pink flowered garden pea includes *Pisum arvense* L.

(iii) Purple flowered garden pea includes *Pisum arvense* L.

(C) Classification on the Basis of Growth Habits (Height of Plant)

(i) **Dwarf type-**Determinate type, bush type, smaller plants, a maximum of 6 nodes.

(ii) **Semi pole or semi-dwarf type-** Runner type, not very prolific, 4-8 nodes

(iii) **Pole or Tall type-**Indeterminate type requiring support

(D) Classification on the Basis of Maturity Period

(i) Early-65-80 days

(ii) Medium-90-100 days

(iii) Main season-100-120 days

(E) Classification on the Basis of Use of Pods

(i) Fresh market types

(ii) Canning types

(iii) Dehydration types

Augmentation of Genetic Resources and Crop Improvement

In the early phase of vegetable pea improvement programme, a large number of germplasm lines and cultivars were collected from the major pea growing areas of India. These were evaluated for various desirable attributes like resistance to diseases and insect-pests, adaptability, yield potential and morphological traits *etc.* which were incorporated in breeding programmes. Breeding objectives of garden pea are high pod yield, high shelling percentage, suitable for freezing, resistance/tolerance to diseases pests. Breeding systems include pedigree method, back cross method, mutation breeding, bulk method, individual plant selection and single seed descent method (Nath *et al.*, 2008).

Introduction and Selection

Popular cultivars like Bonneville and Arkel were introduced in India and directly used as a variety for commercial cultyivation. In garden pea, there are several introduction and selection which became very popular among the growers. In north eastern hill regions beside local land races, popular varieties grown are Arkel, Azad Pea-1, Azad Pea-3, Azad Pea-4, Prakash and TSX-10. The green pod yield ranged 8.7-10.5 t/ha (cv. TSX-10) under the mid hills of Sikkim (Lepcha *et al.*, 2015). Devi and Kumar (2016) evaluated 43 diverse pea genotypes/accessions including 2 standard check varieties collected from different sources under mid-hill conditions of Himachal Preadesh and found significant differences among the genotypes for all studied characters. Mean performance of genotypes *viz.*, 14/PEVAR-6, 14/PEVAR-2, 14/PEVAR-1 and EC598628 were observed to be better over other genotypes including standard

Table 4.10: Different Maturity Groups, Salient Features and Promising Genotypes

Groups	*Salient Features*	*Promising Genotypes*
Early groups	The first flowering takes place in about 25-40 days after sowing. These varieties become ready for first picking in about 55-70 days after seed sowing; and two to three pickings at an interval of 10-15 days can be obtained.	Kashi Nandini (VRP-5), Kashi Uday (VRP-6), Kashi Mukti (VRP-22), Asauji, Hara Bona, Arkel, Azad Pea-3, Ageta (E-6), Early badger, Meteor, Little Marvel, PM-2, VL Ageti Matar-7, Early December (JM-4), Harbhajan (EC33866)
Mid groups	Varieties of this group flower at about 45-60 days and give first picking in about 75-80 days after sowing. Total three pickings are obtained at the interval of 10-15 days.	Kashi Shakti (VRP-7), Bonneville, Perfection New Line, P-35, Azad Pea-1, JM-1, Pant Uphar (IP-3), Punjab-89, Punjab-88, VL-3, Vivek Matar-8, NDVP-8, NDVP-10 and Ooty-1
Late groups	Varieties of this flower in 60-70 days after sowing and first picking are obtained in about 90 days.	Kashi Samrath (VRPMR-9), Kashi Samridhi (VRPMR-11), NP-29, Multi-freezer, Lincoln, JP-19, Arka Ajit, Arka Karthik, Jawahar Pea-83, Jawahar Pea-71 and Vivek Matar-11

Table 4.11: List of Pea Group and Prominent Cultivars/Lines

Group	*Cultivars/Lines*
Early wrinkled seeded	Arkel, Early Badger, Early December, Early Giant, Little Marvel, VL Ageti-Matar-7 (dimpled seeded), Hisar Harit, Jawahar Matar-4 (JM-4 or GC-195).
Early smooth seeded	Alaska, Asauji, Lucknow Bonia, Meteor
Mid to late maturing wrinkled seeded	Alderman, Bonneville, Jawahar Matar-1, Jawahar Matar-2, Khapar Kheda, Lincoln, N.P.-29, Perfection New Line, Pant Uphar (IP-3)
Mid to late maturing smooth seeded	Rachna
Green seeded varieties	Alaska, Arkel, Bonneville, Field pea-48, Jawahar Matar-1, Jawahar Matar-4, Thomas Laxton, VL Ageti-Matar-7, Hisar Harit, Pant Uphar (IP-3)
Double-podded varieties	Bonneville, Kanwari, Khapar Kheda, T-19, VL Ageti-Matar-7,
Edible podded varieties	Aparna, Matar Ageti-6, Mithi Phali, Sylvia

checks for all studied characters. High PCV, GCV estimates for days to 50 per cent flowering (number), days to marketable maturity (number), total soluble solids (per cent) and pod yield (kg/plot, q/ha) revealed that sufficient variability is present among the studied genotypes and will respond to selection. High heritability coupled with high to moderate genetic gain was observed for days to 50 per cent flowering, days to marketable maturity, weight of 100 pods and total phenols indicating prominent role of additive gene effects and less influence of environment in governing these traits.

Table 4.12: Introduction and Selection of Garden Pea

Introduction/ Variety	Description
Asauji	Selection from material collected from Amritsar. Smooth seeded
Early Supperb	It is an introduction. Smooth seeded.
Harbhajan	It is an introduction. Susceptible to powdery mildew
Little Marvel	An introduction from England (cross of Chelsea Gem x Suttons Alaska.). Very sweet and fine in quality
Meteor	An introduction from England. It is an early maturing variety
Bonneville	An introduction from USA. This is a mid-season variety. Plants are medium tall and pods are light green and straight. Average green pod yield is 8.5 t/ha.Wrinkled seeded, highly susceptible to powdery mildew.
Arkel	An introduction from UK. This is an early maturing variety. Plants are dwarf, bear double pods at flower nodes. Pods are attractive, dark green, 8.8 cm long, curved towards suture at distal end (sickle shaped) and well filled. Average green pod yield is 7.5 t/ha.Wrinkled seeded, suitable for fresh market and dehydration
Azad Pea-3	Introduction and selection
Lincon	Introduction from USA and selection made at ICAR-IARI Regional Station Katrain (H.P.)

Table 4.13: List of Indian and Exotic Germplasm of Pea

Place of Collection	Accession Numbers
Bihar, India	IC240779
Madhya Pradesh, India	IC240778, IC209136, IC209137
Uttar Pradesh, India	IC209117, IC209119, IC209120, IC209122 IC209127, IC209131, IC209132, IC219020, IC21906, IC240777, IC240780, IC240781
Rajasthan, India	IC209092, IC209093, IC209095, IC209096, IC209102, IC209103, IC209107, IC209108, IC209109, IC209110, IC209111, IC209112, IC209113, IC219005, IC240773, IC240774, IC240775, IC240776
USA	EC8257, EC8495, EC8762, EC8763, EC341765, EC341766, EC341771, EC341787, EC341793, EC341986, EC341988, EC341998, EC342007, EC381953, EC381854, EC381855, EC381856, EC381857, EC381858, EC381859, EC381860, EC381861, EC381862, EC381864, EC381865, EC381866
UK	EC6620, EC385244, EC385245, EC385246
France	EC269322, EC269538, EC269540, EC269581
The Netherlands	EC9123

Source: Sardana *et al.*, 1998.

Horticulture Research Centre, Bangladesh Agricultural Research Institute (BARI) has identified a potential cultivar of garden pea GP0001 which can be used fresh use and canning. In Pakistan, cultivars like H-57, Meteor, Samrina Zard, Climax Improved (R), FC-3954, Green Feast (I) and 5 Lines (P) are being very common and grown commercially.

Genotypic coefficient of variation revealed high to moderate variability for all the characters studied, except for days to 50 per cent flowering and days to maturity, which showed low variability. Seed yield/ plant, plant height, number of primary branches per plant and seed weight expressed high heritability (broad

Table 4.14: Mean, Range, Mean Sum of Squares (MSS) and Standard Error of Mean

Sl.No.	Characters	Mean Square		Range		Mean ± SEM
		Treatment	Error			
	(df)	72	144			
1.	Seed yield/plant	173.57**	3.82	2.92 (IC240781)	37.39 (EC6620)	7.55±1.59
2.	Number of primary branches/plant	18.87**	1.45	2.0 (IC240774)	13.70 (IC240781)	6.09±0.98
3.	Number of pods/plant	2013.03**	364.33	19.30 (EC381862)	120.00 (IC209026)	63.21±15.58
4.	Number of clusters per plant	1623.27**	192.81	17.00 (EC385242)	106.00 (EC8257)	54.18±11.33
5.	Pod length (cm)	3.66**	0.17	4.40 (IC240781)	10.60 (EC342007)	5.97±0.34
6.	Number of seeds/pod	4.37**	0.81	2.60 (IC240775)	7.80 (IC209102)	5.38±0.73
7.	100-seed weight (g)	58.91**	0.01	4.35 (IC240781)	26.50 (IC209122)	17.52±0.09
8.	Plant height (cm)	3171.58**	171.98	40.00 (EC341793)	158.00 (IC209122)	84.42±10.70
9.	Days to 50 per cent flowering	119.48**	16.37	77.00 (EC381861)	109.00 (IC240779)	98.65±3.30
10.	Days to maturity	51.40**	0.63	137.0 (EC381862)	156.00 (EC8763)	150.38±0.65

** Significant at 1 per cent level.

Table 4.15: Estimates of Genetic Parameters for some Agro-morphological Traits in Pea Germplasm

Sl.No.	Characters	Coefficient of Variation			Heritability in Broad Sense (Per cent)	Genetic Advance as Per cent of Mean
		Phenotypic	Genotypic	Environmental		
1.	Seed yield/plant	102.88	99.57	25.86	93.68	198.54
2.	Number of primary branches/plant	44.19	39.54	19.74	80.05	72.87
3.	Number of pods/plant	47.83	37.09	30.19	60.13	59.24
4.	Number of clusters/plant	47.78	40.30	25.63	71.21	70.06
5.	Pod length (cm)	20.20	18.94	7.01	87.95	36.59
6.	Number of seeds/pod	26.27	20.26	16.75	59.48	32.19
7.	100-seed weight (g)	25.29	25.28	0.65	99.93	52.17
8.	Plant height (cm)	40.55	37.46	15.53	85.32	71.27
9.	Days to 50 per cent flowering	7.22	5.94	4.10	67.73	10.07
10.	Days to maturity	2.79	2.74	0.53	96.43	5.53

Source: Sardana *et al.*, 1998.

sense) coupled with high genetic advance; number of clusters per plant, number of pods/plant, pod length and number of seeds/pod showed moderate to high values whereas days to 50 per cent flowering and days to maturity had low estimates. The wide genetic variability observed in the accessions for yield and yield related traits could be useful for breeding superior genotypes. WEC6620, EC8257, IC240775, IC240779 and IC209107 were identified as superior accessions for utilization.

Seed yield per plant had highest coefficient of variation, both at the phenotypic and genotypic levels and the magnitude was considerably higher than he corresponding values at environment level.

Occurrence of sufficient genetic variability for yield *per se* and its contributing traits in the present set of germplasm can be exploited for pea improvement and also for characterization of traits during selection or validation. Seed yield/plant, plant height, and 100-seed weight were observed to be most reliable selection parameters. EC6620, IC240779, EC8257, IC209107, IC240775 and IC209026 were recorded as superior lines, along with some others for yield *per se* and its traits, which can be used in further breeding programme (Table 4.16).

Hybridization and Selection

In garden pea, there are several cross combination which resulted better varieties. Heterosis and inbreeding studies by Narsinghani and Singh (1979) and Narsinghani and Rao (1982) confirmed the non-significant relationship between genetic divergence and hybrid performance in peas. Heterosis and combining ability which identify the hybrids with high yield are the most important criteria in breeding programmes. Kumar *et al.* (2000) observed appreciable heterosis over better parent and mid parent for most of the yield and quality characters except TSS and protein content.

Interspecific Hybridization

In *Pisum* two crossability groups exist. One containing *Pisum sativum* and two wild taxa namely, *P. elatius* and *P. humile* and the second group is monospecific having *P. fulvum*. *Pisum sativum* readly crosses with *P. elatius* and *P. humile* in both the directions. However, *P. sativum* (2n=14) is one way crossable with *P. fulvum* (2n=14) parent, showing unilateral incompatibility.

Table 4.16: Donors for Various Morpho-agronomic Characters in Pea Germplasm

Characters	Accessions					
	EC6620	EC8257	IC240779	IC209107	IC240775	IC209026
Seed yield/plant	37.39	34.49	32.08	30.00	31.82	28.96
100-seed weight (g)	22.10	18.91	22.83	23.20	24.99	13.83
Plant height (cm)	115.00	125.00	124.33	77.67	68.00	158.00
Number of primary branches/plant	9.67	9.00	7.67	5.67	5.33	11.00
Number of clusters/plant	47.67	106.30	63.67	67.67	51.00	99.33
Number of pods/plant	48.00	104.00	70.67	85.67	69.33	121.67
Pod length (cm)	5.52	4.40	6.27	6.73	5.2	5.50
Number of seeds/pod	5.67	5.00	5.00	5.33	3.00	3.33
Days to 50 per cent flowering	105.00	120.00	109.30	100.33	96.67	103.33
Days to maturity	155.00	151.00	154.00	152.00	149.00	149.00

Source: Sardana *et al.*, 1998.

Table 4.17: Garden Pea Varieties Developed through Hybridization

Azad Pea-1	Developed from the cross between 6416 x 6405. Wrinkled seeded
Hara Bona	Pure line selection from the local field pea variety
Jawahar Matar-1	Developed from the cross between T-19 x Greater Progress. Wrinkled seeded
Jawahar Matar-2	Developed from the cross between Russian-2 x Greater Progress. Better keeping quality suitable for transportation
Early December or Jawahar Matar-3	Developed through T-19 x Early Badger. This is very early maturing variety. Plants are dwarf (50 cm), produce 7.0 cm long light green, roundish-oval pods. Average green pod yield is 4.0 t/ha.
Jawahar Matar-4	Developed from the cross between T-19 x Little Marvel. Wrinkled seeded
Jawahar Peas-4	Developed from the cross between Local Yellow Batri x (6588 x 46C). Suitable to hilly areas and can be planted in August
Jawahar Peas-71	Developed from the double cross between (Arkel x JP-829) x (JP-501 x JM-1). Powdery mildew resistance.
Jawahar Peas-83	Developed from the double cross between (Arkel x JP-829) x (46C x JP-501). Mid-season powdery mildew resistant garden pea
Madhu	Developed from the cross between 6126 x Sylvia. Susceptible to powdery mildew
Punjab-87	Developed from the cross between Pusa-2 x Morrasis-55. Wrinkled seeded.
Vivek-6	Developed from the cross between Pant Upahar x VL Matar-3. Tolerant to cold and moisture stress conditions.
Vivek Matar-8	Hybridization (Azad Pea-1 x PRS-18-6-4-5-1) followed by pedigree selection
Vivek Matar-11	Hybridization (Perfection x Bonneville) followed by pedigree selection
Punjab-88	This is an early maturing variety selected from cross 'Pusa-2' x 'Morasis-55'. Plants are dwarf, vigorous, erect with dark green foliage. Pods are dark green, long and slightly curved. Average green pod yield is 15.0 t/ha. Powdery mildew resistance.
PM-2	This is an early maturing variety developed from a cross of 'Early Badger' x 'IP-3'. Pods are small. Average green pod yield is 10.0 t/ha.
VL-3	This is medium maturing variety developed from a cross of'Old Sugar' x 'Early Wrinkled Dwarf 2-2-a'. Plants are determinate with light green foliage. Pods are light green, straight and 6.8 cm long. Average green pod yield is 10.0 t/ha. Susceptible to powdery mildew.
VL Ageti Matar-7	Hybridization (Pant Uphar x Arkel) followed by pedigree selection
Azad P-2 (PRS-4)	Developed through the cross between Bonneville x 6587. Plants are tall (130-450 cm), erect and pods are medium. Average green pod yield is 12.0 t/ha. Wrinkled seeded. This is a powdery mildew resistant variety.
Kashi Nandini	Hybridization (P1542 x VT-2-1) followed by pedigree selection
Kashi Uday	Hybridization (Arkel x FC-1) followed by pedigree selection
Kasi Mukti	Hybridization (Cross No.7 x PM-5) followed by pedigree selection
Kashi Shakti	Hybridization (Hara Bona x NDVP-8) followed by pedigree selection
Kashi Samrath	Hybridization (VL-3 x Arkel) followed by pedigree selection
Kashi Samridhi	Hybridization (FC1 x PM-5) followed by pedigree selection
Arka Ajit	Hybridization (Bonneville x IIHR-209) x (Freezer-656) followed by pedigree selection
Arka Karthik	Hybridization (Arka Ajit x IIHR-554) followed by pedigree selection
Arka Poorva	Hybridization (Oregon Sugar pod x IIHR-18) followed by pedigree selection

Male Sterility

Male sterility is present in *Pisum* germplasm; however, due to limitations associated with cross pollination in the male sterile parent, development of hybrid cultivars has not been successful. Some of the genetic information with regard to economic characters and male sterility has been reported (Joshi and Singh, 1959; Brahmappa and Singh, 1977; Narsinghani and Singh, 1979; Narsinghani *et al.*, 1979; Pal *et al.*, 1982).

Mutation Breeding

Mutation breediong is the process of exposing seeds to chemical or radiation in order to generate mutants with desirable traits to be bred with other cultivars. Generally, doses and concentration reducing 30 to 50 per cent of germination are suitable for mutational studies. A dose of 10-15 krad of gamma rays is appropriate for seeds. Among chemicals, Ethylmethane sulfonate (EMS), N-nitroso-N-ethyl urea (NEU), ethylene imine (EI), N-nitroso-N-methyl urea (NMU) and sodium azide seem to be the most efficient mutagens for peas. Physical and chemical mutagens can be combined to increase the effectiveness and efficiency of mutagens. Recommended concentration of certain chemical mutagens for pea seeds are given in Table 4.18.

Table 4.18: Recommended Concentration of Certain Chemical Mutagens for Pea

Mutagens	*Concentration (Per cent)*
Ethylmethane sulfonate (EMS)	0.05-0.3
N-nitroso-N-ethyl urea (NEU)	0.20-0.40
Ethylene imine (EI)	0.05-0.15
NMH	0.01-0.03
Diethyl sulfate	0.03

Resistance Breeding

(i) Resistance to Diseases

Eleven diseases of pea are reported in India and they cause about 30-40 per cent yield loss annually. Among the eleven diseases; powdery mildew caused by *Erysiphe polygoni* De C. is more dreaded followed by rust caused by *Uromyces fabae* de Barry. It is important to group the genotypes but always consider the genetics of disease resistance in pea. *P. fulvum* is one of the important sources for powdery mildew resistance (*er*3). Pea cultivars 'Kamelot', 'CO-59196' and 'CO-33142' were reported to posses resistance against Fusarium wilt (Kreuzman, 2001). Details of single dominant genes linked polygenes with a recessive gene confer resistance to several diseases are given in Table 4.19.

Table 4.19: Genetics of Disease Resistance in Pea

Disease	*Responsible Gene*	*Mode of Genetic Control*
Resistance to Powdery mildew (*Erysiphe polygoni*)	er	Resistance controlled by single recessive gene
Resistance to Powdery mildew (*Erysiphe polygoni*)	er-2	Resistance controlled by single recessive gene (Mishra and Shukla, 1984) and Singh *et al.*, 1986b.
Resistance to Powdery mildew (*Erysiphe polygoni*)	er-3	Resistance controlled by single recessive gene
Resistance to enation mosaic virus	en	Resistance controlled by single dominant gene
Resistance to *Fusarium oxysporum f. pisi* race 2, Near Wilt	fnw	Resistance controlled by single dominant gene
Resistance to *Fusarium oxysporum f. pisi* race 1, Fusarium Wilt	*fw*	Resistance controlled by single dominant gene
Resistance to pea seed borne mosaic virus	*sbm*	Resistance controlled by single dominant gene
Resistance to brown root of peas, *Fusarium solani f. sp. pisi,*		Resistance controlled by Monogenic, dominant
Resistance to downy mildew, *Perenospora pisi,*		Resistance controlled by Monogenic, dominant
Resistance to rust, *Uromyces fabae,*		Resistance controlled by Monogenic, dominant
Resistance to Ascochyta blight (*Ascochyta pisi*)		Resistance controlled by Single or 2 or 3 dominant or recessive gene; duplicate factor; or single dominant gene
Resistance to bean yellow mosaic virus	*mo*	Resistance controlled by Single recessive gene
Resistance to top yellow virus		Resistance controlled by Monogenic recessive gene
Resistance to pea leaf roll virus		Resistance controlled by Single recessive gene
Resistance to pea streak virus		Resistance controlled by Single recessive gene
Resistance to pea mosaic virus	*pmv*	Resistance controlled by Single recessive gene
Resistance to bean virus-2		Resistance controlled by Single recessive gene

Maro, Proteus, Porta, Bib Ben, Roi Rone and VUT have been found resistant against *Aphanomycess euteiches* (root rot), and Fusarium wilt (Sementsova and Bulyntsev, 1982). Variety 'Kala Nagni' was immune to the New Delhi strain and Alaska was resuistant to Fusarium wilt (Phal and Choudhary, 1983). Three genotypes GP-6, GP-55 and GP-942 were to be highly resistant while genotypes GP-17, GP-48, GP-473, and GP-941 wre resistant to fusarium wilt (Shubha *et al.*, 2016). Resistant to powdery mildew can be determined by late planting in the field. Resistance was observed in HUVP1, HUVP2, PMR lines, and JP lines (Jalali *et al.*, 1986). A wild type accession JI85 from Afghanistan has been resistant to downy mildew (Mathews *et al.*, 1985). 'Wisconsin Perfection' has been found highly resistant to yellow bean mosaic virus (Hegedorn, 1951). A number of lines have been identified as source of resitance and tolerance to various diseases. Genotype 'FP-186' has been found to root and powdery mildew.

Table 4.20: Available Resistance Source against Various Diseases

	Types	*Resistance Source*	
		Dwarf	*Tall*
Powdery mildew	Pulse type	**Resistant**-Apama, Uttara, Jayanti, Sapna, Malviya Matar 15, Swati, Pusa Prabhat, IPFD 1-10, IPFD 99-13, Pant Pea-42, HFP9907 B	**Resistant**-Rachna, Pant P-5, Malviya Matar 2, JP 885, Alankar, Shikha, Ambika, IPF 99-25, VL Matar-42
	Vegetable type	Kashi Samridhi, Kashi Samrath, Kashi Mukti, Arka Ajeet, P-185, P-388, P-6585, P-6587, Sel.1, Morassis-55, T-10, T-56, JP-501, Azad Pea-1, Palam Priya, FP-186, PMR-21, DPP-62, DMR-7, HFP-4, HFP-12, JP-4, JP-83 and JP-885.	-
Rust	Pulse type	**Tolerant**- PJ 207508, C 12, Wisconsin, DMR3, Pant P 5, Pant P 8, Pant P 9, HFP 8711, HUDP 15 **Resistant**- Pant P 5, Swati (KFPD-24), Malviya Matar 15 (HUDP-15) and IPFD 1-10 (Chaudhary and Naimuddin, 2000)	

Several exotic pea varieties like Alaska, Early Giant, Sylvia, Kelvedon Monarch and a few others as well as some Indian varieties, like T-17, Sel.-1, GC-66, GC-468, Lokar, Pusa Vipasha, and the local varieties Kala Nagini were found to resistant to *Fusarium* wilt when tested under epiphytotic conditions. However, wilt resistance varied at different locations due to race problem. Two races of *Fusarium* wilt have been reported. A *Fusarium* wilt resistance source, JP-501A/2, was identified at JNKVV, Jabalpur (Madhya Pradesh). It was also resistant to powdery mildew. The variety JP-179 was developed which is resistant to *Fusarium* wilt along with resistance to powdery mildew and tolerant to rust. Generally, the powdery mildew resistant varieties are susceptible to rust. However, the variety Arka Ajeet developed at ICAR-IIHR, Bengaluru (Karnataka) and two lines VRP-174 and VRP-343 developed from ICAR-IIVR, Varanasi (U.P.) has combined resistance to rust and powdery mildew. There are also two pea lines, JP Batri Brown-3 and JP Batri Brown-4, resistant to rust. However, both these lines are highly susceptible to powdery mildew. These two lines are being used widely by breeders in developing improved rust resistant cultivars with other desired attributes.

Table 4.21: Introduction of Resistance Source of Pea

Accessions	Source Country	Salient Features	Distribution/ Recipient
EC499761	USA	Resistant to powdery mildew (*Erysiphe polygonii*) and Fusarium wilt race-1 (*Fusarium oxysporum*)	ICAR-IIPR, Kanpur; ICAR-NBPGR, N.Delhi
EC499762	USA	Resistant to Fusarium wilt race-1 (*Fusarium oxysporum*), pea enation mosaic virus (PSbMV)	ICAR-IIPR, Kanpur; ICAR-NBPGR, N.Delhi

Rust and powdery mildew resistant varieties have been rleased and variety 'Arka Ajit' and 'Arka Karthik', possessing high yield and rust resistance characters also been released from ICAR-IIHR, Bengaluru (Karnataka).

(ii) Resistance to Insects

Diverse accessions of pea were screened for 2 years and the minimum incidence of leaf minor was observed in LMR4, LMR 10 and LMR 20 (Pesho *et al.*, 1977) and also in JP-169, JP-179 and JP-747. Improved cultivars having resistance to leaf miner have not been developed yet. They have reported varietal differences in pea weevil incidence and a low degree of incidence was recorded in PI 196027, PI 244241, PI244254, PI 244263, PI 273036 and PI 280612. In pea variety 'Mammoth Melting' and 'Hybrid Early Onward' x 'Mammoth Melting Sugar' were resistant to cowpea seed beetle (Waly *et al.*, 1985.). Bruchids are a serious problem in pea growing regions where they occur feeding on pollen and then go on to cause damage to the developing seed where the larvae feed and continue their development thus rendering the seed unmarketable and *P. fulvum* is one of the important source for bruchid resistance. To date, this resistance has not been successfully transferred to cultivated varieties of *P. sativum*. The two pea lines, JP-9 and JP-179 selected are said to be resistant to pea weevil and Bruchus. The crosses of *P. fulvum* with *P. sativum* hold promise for resistance to the pea seed weevil (*Bruchus pisorum*). Multiple resistant lines like JP-179 highly resistant to powdery mildew, resistant to *Fusarium* wilt, tolerant to rust and resistant to leaf miner and Bruchus; JP-9 resistant to powdery mildew and Bruchus and JP Batri Brown -3 and JP Batri Brown-4 resistant to rust and bruchus, Kashi Mukti resistant to powdery mildew and tolerant to leaf minor, VRP-343 resistant to powdery mildew, rust and tolerant to leaf minor.

Breeding for Abiotic Stress

The two major environmental constraints for pea crops were drought and high temperatures during flowering. Semi-leaflets peas with reduced leaf area are considered less sensitive to drought than conventional-leafed types. The cultivar Early Badger reported tolerant to heat and drought can be well exploited for drought resistant breeding. Similarly, cultivar 'Multifreezer' can be a good source of frost tolerance. High temperatures are particularly detrimental because they affect crops directly by imparting physiological processes and indirectly by altering plant water relations. Among the cool-season annuals, pea is very sensitive to high day temperatures with death of the plant occurring when air temperatures exceed about 35°C for sufficient duration. Plants are particularly sensitive to high temperatures during flowering. The Alaska cultivar is tolerant to heat. There are few varieties like Alderman, Early Giant, Arkel, and VL Ageti Matar-7 are very much suitable for cool climate.

Table 4.22: Introduction of Exotic Germplasm

Accession Number	Source	Salient Feature	Recipient
EC499761 var. Fallon	USA	Lodging resistant	ICAR-NBPGR, N. Delhi

Breeding for Processing and Quality Traits

Following quality traits should be considered during breeding programmes.

Table 4.23: Preferred Range/Magnitude/ Value/Type for Traits

Traits	Preferred Range/ Magnitude/Value/Type
Pod colour	Dark green
Chlorophyll content	High
TSS	High
Fresh seed colour	Dark green
AIS (Alcohol insoluble solids)	11 to 16 per cent
Flavour	High
Sugar content	High
Seed size	Bold

Pea is a cool season crop and in market as fresh condition it is available only during winter season and spring season. The off season demand can only be meet out through processed peas. To meet the year round demand, large amounts of peas are canned, frozen or dehydrated for consumption in the period of scarcity.

(i) **Dehydration-** Large sized wrinkled and dark green peas like Arkel, Kashi Uday, NP-29, New Line Perfection, JM-1, JM-2 and P-87 are suitable for dehydration. The wrinkled seed peas contain 26-33 per cent protein content and in smooth seeds it is 23-31 per cent.

(ii) **Canning-** For canning, both round and wrinkled seeded varieties like T-19, PB-87, Early Badger, Lincoln, Alaska-81 and Bonneville can be used and

(iii) **Freezing-** for freezing only wrinkled seeded varieties like Thomas Laxton, Alderman, Taichung-12 *etc.* are preferred.

Devi *et al.* (2016) assayed 22 genotypes including released varieties and advance breeding lines belonging to early, mid and late maturity group for antioxidant phytochemicals. Antioxidant activity of these, Antioxidant Capacity (CUPRAC), Trolox Equivalent Antioxidant Capacity (TEAC) and the 2, 2-diphenyl-1-picrylhydrazyl (DPPH). Among these tested genotypes, EC9485 and VRP-233 bearing purple colour flower, were found to be the highest sources of total phenol (128.64 and 104.01 mg GAE/100g) and flavonoids (45.84 and 36.84 mg CE/100g). This two genotype also showed stronge antioxidant potential (26.79 and 20.96 µ Mol TE/g). Cultivar belonging to early group like VRP-100, VRP-101, VRP-25, VRP-5, VRP-6, AP-3 and Arkel, possessed comparatively lower amount of phenolics, flavonoids and antioxidant potential. The positive correlation between total phenols, flavonoids with antioxidant activities indicates phenols and flavonoids are the main component for antioxidant activity of the pea genotype. Devi and Kumar (2016) evaluated 43 diverse pea genotypes/accessions including 2 standard check varieties and reported high PCV and GCV for for total phenols indicating prominent role of additive gene effects.

Genetic Resource Conservation

Collection, maintenance and utilization of germplasm are more important for genetic improvement of pea crop. Concentration on improvement of yield and quality in peas were leads to loss of other traits like abiotic and biotic stress resistance in many varieties. Now a day's focus is moving towards incorporation of those lost in cultivated peas from wild relatives. Several useful germplasm are available with ICAR-NBPGR, New Delhi, ICAR-Indian Innstitute of Vegetable Research, Varanasi (U.P.), GBUA and T, Pantnagar (Uttarakhand), Division of Vegetable Science, ICAR-IARI, New Delhi, JNKVV, Jabalpur (M.P.), PAU, Ludhiana (Punjab), BCKV, Kalyani (W.B.) *etc.*

References

1. American Chemical Society, 2009. "Protein from garden pea may help fight high blood pressure, kidney disease".Science Daily, 28th March, 2009. www.sciencedaily.com/release/2009/03/090322154407.htm.
2. Blixt, S. 1970. *Pisum*. In: Frankel, O.H. and Bennet, E. (eds.). Genetic Resources in Plants: Their exploration and conservation. *Int. Biol. Programme*, p. 321-326.
3. Chaudhary, R.G. and Naimuddin. 2000. Pea diseases in Indian Perspective and their Economic Management, p.47-60. In Narayan, U., Kumar, K. and Srivastva, M. (ed.). Advances in Plant Diseases Managhement, Advance Publishing Concept, New Delhi.
4. De Candolle, A. 1986. Origin of cultivated plants. Hafner Publishing Co. New York.
5. Devi, J., Sanwal, S.K., Dubey, R.K., Koley, T.K., Singh, P.M. and Singh, B. 2016. Variability and correlation studies for yield and quality traits in garden pea (*Pisum sativum* L.). 1st International Congress, held at New Delhi, 6-9 November, 2016; p. 23.
6. Devi, S. and Kumar, M. 2016. Assessment of genetic variability among pea (*Pisum sativum* L.) genotypes under mid-hill conditions of Himachal Pradesh. 7th Indian Horticulture Congress (Abs. 17; Session: II-Breeding of Horticultural Crops), held at ICAR-IARI, New Delhi, 15-18 November, p. 98-99.
7. Gritton, E.T. 1986. Pea Breeding. In Basett, M.J. (ed.). Breeding Vegetable Crops, pp. 283-319. AVI Co. Westport, Connecticut.
8. Hegedorn, D.J. 1951. The reaction of perfection type peas to Wisconsin bean virus of isolates from pea. *Phytopathology*, **41:** 494.
9. Jalali, I., Kalloo, G. and Banerjee, M.K.1986. Unpublished data.
10. Janick, J., Schety, R.W., Woods, F.W. and Ruttan, V.W. 1969. *Plant Science*. Freeman, San Francisco.
11. Kreuzman, J. 2001. Pea variety 'Kamelot'. *Czech Jouranal of Genetics and Plant Breeding*, **37:** 129-130.
12. Kumar, S., Singh, K.P. and Panda, P.K. 2000. Heterosis for green pod yield and its components in garden pea. *Haryana Journal of Agricultural Sciences*, **29:** 99-101.
13. Lepcha, B., Avasthe, R., Singh, R. and Yadav, A. 2015. Impact of front line demonstration on productivity and profitability of green pea (var. TSx-10) under mid hills of Sikkim. *Indian Research Journal of Extension Education*, **15(4):** 157-160.
14. Marx, G.A. 1977. Classification, genetics and breeding. In: Suteliffe, J.F. and Pate, J.S. (eds.). Physiology of the garden pea. Academic Press, New York, p. 21-44.

15. Mathews, P., Dow. P., Graves, K., Brittain, M. and Taylor, P. 1985. Pea pathology, In: John Innes Institute, 73rd Rep. Covering the 2 years 1983 to 1984. Norwich, United Kingdom.11.

16. Mishra, S.P. and Shukla, P. 1984. Inheritance of powdery mildew in pea. *Z. Pflanzenzucht*, **93:** 251-254.

17. Narsinghani, V.G and Rao, V.S.N. 1982. Genotype environment interaction in pea. *JNKVV Research Journal*, **12:** 40-41.

18. Narsinghani, V.G and Singh, S.P. 1979. Heterosis in mutant pea types. *Indian Journal of Agricultural Sciences*, **49:** 39-42.

19. Nath, P., Srivastava, V.K., Dutta, O.P. and Swamy, K. R.M. 2008. Vegetable Crops-Improvement and Production. PNASF, Bengaluru (Karnataka). pp. 102-108.

20. Pesho, G.R., Muehlbauer, F.J. and Herberts, W.H. 1977. Resitance of pea introduction to the pea weevil. *J. Econ. Entomol.*, **70:** 30.

21. Phal, R. and Choudhary, B. 1983. Fusarium wilt of garden pea: race situation. *Indian J. Agric. Sci.*, **53:** 863.

22. Ray, P., Gupta, H.N. and Roy, M. 1980. "Susuruta Samhita" Scientific Synopsis. *Indian Natl. Sci.*; New Delhi.

23. Renfrew, J.M. 1973. Paleoethnobotany-Metheum, London.

24. Salunkhe, D.K. and Desai, B.B. 1984. Peas and Beans. In: Post-harvest Biotechnology ofVegetables (CH.6), Vol. CRC Press, Boca Raton, p. 125.

25. Sardana, S., Kumar, D., Suneja, P. and Kochhar, S. 1998. Variability in germplasm for some agromorphological traits. *Indian J. Plant Genet. Resour.*, **11(2):** 197-202.

26. Sementsova, M.P. and Bulyntsev, S.V. 1982. Evaluation of pea varieties for resistance to pathogen causing root rots. *Turkey Prikl. Bot. Genet. Sel.*,**72:** 134.

27. Shubha, K., Shri Dhar, Choudhary, H., Dubey, S.C. and Sharma, R.K. 2016. Identification of resistant sources and inheritance of Fusarium wilt resistance in garden pea (*Pisum sativum* ssp. *hortense*). *Indian J. Hort.*, **73(2):** 356-361.

28. Singh, H.B. and Joshi, B.S. 1970. Pulse crop of India. ICAR, New Delhi, pp. 183-200.

29. Singh, U.P., Shrivastava, C.P., Singh, M.N. and Singh, R.M. 1986b. Allelic test and inheritance of resistance to powdery mildew in pea. *Veg. Sci.*, **13:** 131-136.

30. Smartt, J. 1976. *Tropical Pulses*, Longman, London, p 348. Thompson, H.C. and Kelly, W.C. 1957. Vegetable crops McGraw-Hills Book Co., Inc. New York, Toronto London.

31. Thompson, H.C. and Kelly, W.C. 1957. Vegetable Crops, McGraw Hill Book Company Inc; New York.

32. Vavilov, N.I. 1926. *Turkey Prikl. Bot. Genet. Sel.*, **20:** 1.

33. Vishne Mittre 1974. Paleobotinical evidence in India-In: 'Evolutionary Studies in World Crops'. (ed.) Hutchison Cambridge University Pres.

34. Waly, E.A., Hussein, M.H. and Abd-El-Aal, S.A. 1985. Inheritance of cowpea seed beetle infestation and its relation to total carbohydrate in garden pea. *Beit. Trop. Landwirtsch, Veterinarmewd*, **23:** 51.

35. Zohary, D. and Hopf, M. 1973. Domestication of pulses in the world. *Science*, **182:** 887-894.

Edible Podded Peas (*Pisum sativum* L. var. *saccharatum*)

The edible podded peas comprise of snow peas and snap peas. Snow pea (*Pisum sativum* L. var. *saccharatum*) is distinct botanical cultivar or subspecies of *Pisum sativum* L. which belongs to family Fabaceae. Snow peas are often referred to as sugar peas, Oriental peas or sometimes as Chinese pea pods. In France, it is also known as '*mange*-tout' meaning 'eat it all'. Snow peas are often confused with its other botanical relative sugar snap pea (*Pisum sativum L. var. macrocarpon*) and garden peas (*Pisum sativum* L. var. *hortense*) but there are slight differences between the three. In case of snow pea plant bears pods that are broad, flat, soft and translucent with tiny and tender sweet peas inside. The whole pod is edible although the tough "strings' along the edges are usually removed before eating. The modern sugar snap pea is the progeny of a cross between a snow pea and an unusual pea that was tightly podded with thick walls. The cross was made in the late 1960s by Calvin Lamborn who was hired by the Gallatin valley Seed Company (USA) as a breeder and Dr. M.C. Parker who introduced him to a bizarre fluke mutation he did found a decade and a half earlier (Agfact, 2001). Both these botanical cultivars, known as edible podded peas, differ from green shell and dry edible types as they lack 'parchment" the normal fibrous layer composed of highly lignified sclerenchyma cells in the inner pod wall (Sneddon, 1970; McGee and Baggett, 1992; Singh and Jauhar, 2005, De Ron, 2005). Snow peas (with thin pod walls) and snap peas (with thick pod walls), as well as parchment green shell and dry edible types (Garden peas or English peas), are normally characterized by a prominent "string" that is present in both pod sutures as well as especially strong in the ventral suture (McGee and Baggett, 1992).

Table 4.24: Edible Part and their Use in Three Botanical Varieties

Market Class	*Edible Part*	*Used as*
Garden (English, Shell) pea	Immature green seed	Fresh, Frozen, canned
Snow (Chinese) pea	Immature pod, thin pod walls	Fresh, Frozen
Snap (Sugar) pea	Immature pod, thic pod walls	Fresh, Frozen

Gene and Genetics of Characters

Two independent recessive genes as *p* and *v* are accountable for expression of this trait and are known to facilitate elimination of schlerenchymatous membrane and reduction of pod wall thickness, respectively (Myers *et al.*, 2001; Singh and Jauhar, 2005). Either gene present in the homozygous condition greatly reduces parchment, while pod of plants homozygous for both *p* and *v* are considered parchment-free (White, 1917). Thus, modern edible podded pea cultivars have genotype 'ppvv' which is required to make the pods suitable for consumption when the pods are large and the seeds have started to form whereas shell pea cultivars usually have genotype 'PPVV' (Deppe, 2000). An additional recessive allele 'n' at the n locus, increases pod thickness and all snap pea cultivars have genotype ppvvnn (Gritton, 1986; Gritton and Myers, 1996. Deppe, 2000). McGee and Baggett (1992) found the stringless character difficult to use in breeding because the number of stringless plants recovered in segregating generations has been fewer than expected for a presumed monogenic character and because of it, difficult to recover vigorous stringless plants with wrinkled seeds.

Augmentation of Genetic Resources and Crop Improvement

Table 4.25: Edible Podded Cultivars

Group	*Cultivars/Line*
Edible podded varieties	Aparna, Matar Ageti-6, Mithi Phali, Sylvia

Resistance Breeding

A promising line, IIHR53-6, named as Arka Sampoorna with edible pod quality and combined resistance to rust and powdery mildew was selected in F_9 generation at ICAR-IIHR, Benguluru (Karnataka).

REFERENCES

1. Agfact. 2001. Snow peas and sugar snap peas. Agfact H8.1.35, second edition 2001C. Beckingham, District Horticulturist, Division of Plant Industries, Bathurst.
2. Deppe, C. 2000. Breed your own vegetable varieties: the gardener's and farmer's guide to plant breeding and seed saving. Chelsea green Publ. River Junction, VT.
3. De Ron, A.M., magallanes, J.J., Martinez, O., Rodino, P. and Santalla, M. 2005. Identyfying superior snow pea breeding lines. *HortSci.*, **40:** 1216-1220.
4. Gritton, E.T. 1986. Pea breeding. P.283-320. In: M.J. Bassett (ed.). Breeding vegetable crops. AVI Publ. West Port, Conn.
5. Gritton, E.T. and Myers, J.M. 1996. Vegetables Cultivar description for north America. Pea. North Caroline State University (online) http://cuke.hort.ncsu.edu./cucurbit/wehner/eng/cul/pecl/greenal/html.
6. McGee, R.J. and Baggett, J.R. 1992. Inheritance of stringless pod in *Pisum sativum* L. *Amer. Soc. HortSci.*, **117(4):** 628-632.
7. Myers, J.R., Baggett, J.R. and Lamborn, C. 2001. Origin, history and genetic improvement of snap pea (*Pisum sativum* L.). *Plant Breed. Rev.*, **21:** 93-198.
8. Sardana, S., Kumar, D., Suneja, P. and Kochhar, S. 1998. Variability in pea germplasm for some agromorphological traits. *Indian J. Plant Genet. Resour.*, **11(2):** 197-202.
9. Singh, R.J. and Jauhar, P.P. 2005. Genetic Resources, Chromosome Engineering and Crop Improvement, CRC Press, p. 74.
10. Sneddon, J.L. 1970. Identification of garden pea varieties (I) grouping arrangement and use of continuous characters. *J. Natl. Inst. Agri. Bot.*, **12:** 1-16
11. White, O.E. 1917. Studies of inheritance in *Pisum*. II. Present state of knowledge. Proc. *Amer. Philoso. Soc.*, **56:** 487-588.

Genus *Vigna*

Cowpea (*Vigna* species)

INTRODUCTION

Vigna uanguiculta (L.) Walp.cultivar group *sesquipedalis*, 2n-22, the yard long bean (syn. string bean, asparagus bean) is a type of cowpea used as vegetable. The dual purpose cowpea [(*V. uanguiculta* (L.) Walp.] cultivar group *cylindrica* is used both as a pulse crop as well as a vegetable. Cowpea is grown for immature green pods are variously known as asparagus bean, snake bean or yard-long bean, while grown for dry and immature seeds are known as southern pea, black eye pea, crowder pea, Kaffir pea, lubia, niebe, coupe or frijole China pea or Southern bean. It is true that cowpea is commercially grown throughout India and used as a grain crop (dry seeds as a pulse), for animal fodder (leaves and stalks), or as a vegetable (green pods). It is an important component of farming system of arid and humid tropics. It possesses excellent nutritionnal qualities and low in antinutritional factors. Green pod is a good source of riboflavin, niacin, vitamin B6, Vitamin A, Vitamin C, Thiamin, folate, magnesium, manganese, calcium, iron, phosphorus, potassium, and copper. It is a good source of protein (grain contains 25 per cent protein). It contains high amount of lysine and tryptophan than cereals. The cowpea is well adapted to wide range of ecological condition and can tolerate to varying stresses. Its root nodules help in fixing the atmospheric nitrogen and in this way it is also helpful in improving the soil health.

Origin, History and Distribution

Cowpea is in cultivation from very ancient times in the tropics of old world but its country of origin is uncertain. The history of cowpea dates to ancient West African cereal farming, 5000-6000 years ago, where it was closely associated with the cultivation of sorghum and pearl millet. Vavilov (1939) considered India as the main centre of origin but cowpea is probably a native of tropical or Central Africa (Thompson and Kelly, 1957.) because its wild races are found even now.India and China are considered as secondary centres of origin (Singh, *et al.*, 1971; Mishra *et al.*, 1985b). The exact place of cowpea domestication is uncertain. According to Ng and Marechal (1985) the probable centers of domestication are Ethiopia, Central, South and West Africa. Cowpea, moved from Southern Africa, to East and West Africa and Asia. It is widely grown in Africa, Latin America, South-East Asia and in the southern United States. In India, it has been known since the Vedic times and commonly grown all over India. The crop spread from Africa to Asia and Europe through Egypt. It was introduced by the Spaniards in to the West Indies in the 16th century and was taken to the United States around 1700. It believed that cowpea was introduced to the Indian subcontinent from Africa approximately 2000-3500 years ago. From Asia cowpea was introduced into Europe and North America around 300 BC. The slaves from West Africa introduced it into the USA in the early 18th century. Cowpea is grown in the semi-arid tropics covering Africa, Asia, Europe, United States and Central and Southern America. The largest area under cultivation is in Central and West Africa. The largest production is in Africa with Nigeria and Niger predominating, but Brazil, Haiti, India, Myanmar, Sri Lanka, Australia, U.S.; Bosnia and Herzegovina all have significant production. The other major cowpea producers in Africa are Burkina Faso, Ghana, Kenya, Uganda, Malvi, Tanzania, Togo and Senegal. The next largest cowpea producing country is Brazil where 26.4 per cent of the world's total production is produced. The other cowpea producers are Latin America, Venezuela, Peru, Panama, El- Salvador and Haiti (Tolmasquim *et al.*, 1971). Asian production includes that of yard long bean as vegetable and area under cowpea cultivation in India, Sri Lanka, Bangladesh, Myanmar, China, Korea, Indonesia, Nepal, Pakistan, Philippines, Thailand and Malaysia has been estimated to be about 1.2 million hectares. Among the Asian countries, India alone occupies 0.5 million hectares in all forms of green pods, dry seeds, fodder, green manure and cover crops (Rachie, 1985). The only developed countries producing large amount of cowpea is, the USA, where it is grown on about 0.08 million hectares (Ferry, 1985).

Species Availability, Evolution and Diversity

Out of 170 species of genus *Vigna*, 120 are in Africa, 22 in India and Southern Asia and the rest in America and Australia. Out of the 100 species of *Vigna* (Marechal, *et al.*, 1978) only nine species have been domesticated, two of which were domesticated in Africa and seven in Asia. The African *Vigna* consists of cowpea [*V. unguiculata* (L.) Walp.] and Bambara groundnut [*V. subterranea* (L.) Verdc.] (Smartt, 1990). There are two main cultivated divisions of cowpea worldwide, *i.e.* the dominant ssp. *unguiculata*, which is widely cultivated in Africa, India, Middle East and the America for grain which is a major source of dietary protein for millions of people in developing countries (Singh, 2002) and the subspecies *sesquipedalis*, which is also known as asparagus bean or 'yardlong' bean cultivated in eastern and southern Asia for immature green pods for vegetable (Ehlers and Hall, 1997). The cultivated cowpea is polymorphic and consists of *Vigna sinensis* subsp. *sinensis* (syn. *V. unguiculata* var. *radiata, Dolichos sinensis* L.), *V. catjang* (syn. *V. sinensis* var. *cylindrica*, *V. unguiculata*. var. *cylindrica*) and *V. sesquipedalis*, (syn. *V. unguiculata* var. *sesquipedalis, Dolichos sesquipedalis*). It is stated that *V. sinensis* and *V. catjang* differentiated from a common ancestoral stock and *sesquipedalis* is a later selection from *sinensis*. Genetic similarity between the two subspecies is high as evidenced by the fact that hybrids are fully fertile and molecular marker system developed in ssp. *unguiculata* are effective and readily transferred to ssp. *sesquipedalis* (Xu *et al.*, 2010). Sen and Bhowal (1960) listed 11 wild species *i.e. V. parviflora, V. wilmsii, V. oblongizolia, V. gracillis, V. heterophylla, V. membranacea, V. vexillata, V. davyi, V. schimperi, V. marima* (Burm.) Merr., and *V. luteola* of *Vigna*. Recently all the cultivated species have been placed under a single species V. *unguiculata* (L.) Walp. Verdcourt (1970) recognized 5 subspecies of *V. unguiculata*. These are *V. unguiculata* (L.) Walp. ubsp. *unguiculata* (syn. *V. unguiculata* var. radiata); *V. unguiculata* var. *cylindrica* (L.) Verd; *V. unguiculata* subsp. *sesquipedalis* (L.) Verd; *V. unguiculata* subsp. *dekindtiana* (Harms) Verd, *V. unguiculata* subsp. *mansensis*. First 3 species are cultivated and the rest are wild species. The related species are *V. luteola*, *V. nilotica* and *V. marina* (Burma) Merr. In other classification,the genus *Vigna* was subdivided into seven sub genera, namely, *Vigna, sigmoidotropis, plectotropis, macrorhyncha, ceratotropis, Haydonia* and *Lasiocarpa* and the sub genus *Vigna* was again divided into six sections, *Vigna, Comosae Macrodontae, Reticulata, Liebrechtsia* and *Catjang*. The sub-species *unguiculata*, *cylindrica* and cv. gr. *tixtilis* *Syn. V. sinensis* var. *tixtilis*) were given cultivar status (Marechal, 1978). Steel and Mehra (1980) opined that cultivars grown for fodder and seeds are intermediate between the cultigroups *unguiculata* and *biflora* and those grown for pods as vegetable are intermediate between the cultigroups *unguiculata* and *sesquipedalis*.

Table 4.26: Taxon Name, Authority Name and Preferred Taxon Name

Name of Taxon	Name of Authority	Family	Preferred Taxon Name
Dolichovigna pilosa	(Willd.) Niyomdham	Fabaceae	*Vigna pilosa*
Pseudovigna argentea	(Willd.) Verdc.	Fabaceae	*Pseudovigna argentea*
Vigna			*Vigna*
Vigna aconitifolia	(Jacq.) Marechal	Fabaceae	*Vigna aconitifolia*
Vigna adenantha	(G. Mey.) Marechal *et al.*	Fabaceae	*Vigna adenantha*
Vigna ambacensis	Welw. ex Baker	Fabaceae	*Vigna ambacensis*
Vigna ambacensis var. *pubigera*	(Baker) Marechal *et al.*	Fabaceae	*Vigna ambacensis*
Vigna angivensis	Baker	Fabaceae	*Vigna angivensis*
Vigna angularis	(Willd.) Ohwi and H. Ohashi	Fabaceae	*Vigna angularis*
Vigna angularis var. *angularis*		Fabaceae	*Vigna angularis* var. *angularis*
Vigna angularis var. *nipponensis*	(Ohwi) Ohwi and H. Ohashi	Fabaceae	*Vigna angularis* var. *nipponensis*
Vigna antillana	(Urb.) Fawc. and Rendle	Fabaceae	*Vigna antillana*
Vigna baoulensis	A. Chev.	Fabaceae	*Vigna unguiculata* subsp. *baoulensis*
Vigna buchneri	Harms	Fabaceae	*Vigna frutescens var. buchneri*
Vigna burchellii	(DC.) Harv.	Fabaceae	*Otoptera burchellii*
Vigna calcarata	(Roxb.) Kurz	Fabaceae	*Vigna umbellata*
Vigna capensis	(L.) Walp.	Fabaceae	*Vigna* sp.
Vigna caracalla	(L.) Verdc.	Fabaceae	*Vigna caracalla*
Vigna catjang	(Burm. f.) Walp.	Fabaceae	*Vigna unguiculata* subsp. *cylindrica*
Vigna coerulea	Baker	Fabaceae	*Vigna unguiculata* subsp. *tenuis*
Vigna comosa	Baker	Fabaceae	*Vigna comosa*
Vigna comosa var. *comosa*		Fabaceae	*Vigna comosa var. comosa*
Vigna comosa var. *lebrunii*	(Baker f.) Verdc.	Fabaceae	*Vigna comosa var. lebrunii*
Vigna cylindrica	(L.) Skeels	Fabaceae	*Vigna unguiculata subsp. cylindrica*
Vigna dalzelliana	(Kuntze) Verdc.	Fabaceae	*Vigna dalzelliana*
Vigna davyi	Bolus	Fabaceae	*Vigna vexillata var. davyi*
Vigna decipiens	Harv.	Fabaceae	*Vigna decipiens*
Vigna dekindtiana	Harms	Fabaceae	*Vigna unguiculata* subsp. *dekindtiana*
Vigna elegans	(Piper) Marechal *et al.*	Fabaceae	*Vigna elegans*
Vigna filicaulis	Hepper	Fabaceae	*Vigna filicaulis*
Vigna filicaulis var. *filicaulis*		Fabaceae	*Vigna filicaulis var. filicaulis*
Vigna filicaulis var. *pseudovenulosa*	Marechal *et al.*	Fabaceae	*Vigna filicaulis* var. *pseudovenulosa*
Vigna friesiorum		Fabaceae	*Vigna friesiorum*
Vigna frutescens	A. Rich.	Fabaceae	*Vigna frutescens*
Vigna frutescens var. *buchneri*	(Harms) Verdc.	Fabaceae	*Vigna frutescens var. buchneri*
Vigna frutescens var. frutescens		Fabaceae	*Vigna frutescens var. frutescens*
Vigna galpinii	Burtt Davy	Fabaceae	*Vigna nervosa*
Vigna glabrescens	Marechal *et al.*	Fabaceae	*Vigna glabrescens*
Vigna gracilis	(Guill. and Perr.) Hook. f.	Fabaceae	*Vigna gracilis*
Vigna heterophylla	A. Rich.	Fabaceae	*Vigna heterophylla*
Vigna hosei	(Craib) Backer	Fabaceae	*Vigna hosei*
Vigna hybrid		Fabaceae	*Vigna hybrid*
Vigna kirkii	(Baker) J. B. Gillett	Fabaceae	*Vigna kirkii*
Vigna lanceolata	Benth.	Fabaceae	*Vigna lanceolata*
Vigna lancifolia	A. Rich.	Fabaceae	*Vigna oblongifolia var. oblongifolia*
Vigna lasiocarpa	(Mart. ex Benth.) Verdc.	Fabaceae	*Vigna lasiocarpa*
Vigna laurentii	De Wild.	Fabaceae	*Vigna laurentii*
Vigna lebrunii	Baker f.	Fabaceae	*Vigna comosa var. lebrunii*
Vigna linearis	(Kunth) Marechal *et al.*	Fabaceae	*Vigna linearis*

Name of Taxon	Name of Authority	Family	Preferred Taxon Name
Vigna linearis var. latifolia	(Benth.) Marechal *et al.*	Fabaceae	*Vigna linearis var. latifolia*
Vigna linearis var. linearis		Fabaceae	*Vigna linearis var. linearis*
Vigna lobatifolia	Baker	Fabaceae	*Vigna lobatifolia*
Vigna longifolia	(Benth.) Verdc.	Fabaceae	*Vigna longifolia*
Vigna lutea	(Sw.) A. Gray	Fabaceae	*Vigna marina*
Vigna luteola	(Jacq.) Benth.	Fabaceae	*Vigna luteola*
Vigna macrorhyncha	(Harms) Milne-Redh.	Fabaceae	*Vigna macrorhyncha*
Vigna marina	(Burm.) Merr.	Fabaceae	*Vigna marina*
Vigna membranacea	A. Rich.	Fabaceae	*Vigna membranacea*
Vigna mensensis	Schweinf.	Fabaceae	*Vigna unguiculata subsp. mensensis*
Vigna micrantha	Harms	Fabaceae	*Vigna comosa var. comosa*
Vigna minima	(Roxb.) Ohwi and H. Ohashi	Fabaceae	*Vigna minima*
Vigna monophylla	Taub.	Fabaceae	*Vigna monophylla*
Vigna multinervis	Hutch. and Dalziel	Fabaceae	*Vigna multinervis*
Vigna mungo	(L.) Hepper	Fabaceae	*Vigna mungo*
Vigna nervosa	Markotter	Fabaceae	*Vigna nervosa*
Vigna nilotica	(Delile) Hook. f.	Fabaceae	*Vigna luteola*
Vigna o-wahuensis	Vogel	Fabaceae	*Vigna o-wahuensis*
Vigna oblongifolia	A. Rich.	Fabaceae	*Vigna oblongifolia*
Vigna oblongifolia var. oblongifolia		Fabaceae	*Vigna oblongifolia var. oblongifolia*
Vigna oblongifolia var. parviflora	(Welw. ex Baker) Verdc.	Fabaceae	*Vigna oblongifolia var. parviflora*
Vigna ornata	Welw. ex Baker	Fabaceae	*Sphenostylis stenocarpa*
Vigna parkeri	Baker	Fabaceae	*Vigna parkeri*
Vigna parkeri subsp. maranguensis	(Taub.) Verdc.	Fabaceae	*Vigna parkeri subsp. maranguensis*
Vigna parkeri subsp. parkeri		Fabaceae	*Vigna parkeri subsp. parkeri*
Vigna parviflora	Welw. ex Baker	Fabaceae	*Vigna oblongifolia var. parviflora*
Vigna peduncularis	(Kunth) Fawc. and Rendle	Fabaceae	*Vigna peduncularis*
Vigna peduncularis var. clitorioides	(Mart. ex Benth.) Marechal *et al.*	Fabaceae	*Vigna peduncularis var. clitorioides*
Vigna peduncularis var. peduncularis		Fabaceae	*Vigna peduncularis var. peduncularis*
Vigna peduncularis var. pusilla	(Hassl.) Marechal *et al.*	Fabaceae	*Vigna peduncularis var. pusilla*
Vigna pilosa	(Klein ex Willd.) Baker	Fabaceae	*Vigna pilosa*
Vigna pubescens	R. Wilczek	Fabaceae	*Vigna unguiculata subsp. pubescens*
Vigna pubigera	Baker	Fabaceae	*Vigna ambacensis*
Vigna punctata	Micheli	Fabaceae	*Adenodolichos punctatus*
Vigna racemosa	(G. Don) Hutch. and Dalziel	Fabaceae	*Vigna racemosa*
Vigna radiata	(L.) R. Wilczek	Fabaceae	*Vigna radiata*
Vigna radiata var. glabra	(Roxb.) Verdc.	Fabaceae	*Vigna glabrescens*
Vigna radiata var. radiata		Fabaceae	*Vigna radiata var. radiata*
Vigna radiata var. sublobata	(Roxb.) Verdc.	Fabaceae	*Vigna radiata var. sublobata*
Vigna reflexopilosa	Hayata	Fabaceae	*Vigna reflexopilosa*
Vigna repens	(L.) Kuntze, nom. illeg.	Fabaceae	*Vigna luteola*
Vigna reticulata	Hook. f.	Fabaceae	*Vigna reticulata*
Vigna retusa	(E. Mey.) Walp.	Fabaceae	*Vigna marina*
Vigna schimperi	Baker	Fabaceae	*Vigna schimperi*
Vigna sesquipedalis	(L.) Fruwirth	Fabaceae	*Vigna unguiculata subsp. sesquipedalis*
Vigna sinensis	(L.) Savi ex Hassk.	Fabaceae	*Vigna unguiculata subsp. unguiculata*
Vigna sinensis subsp. sesquipedalis	(L.) Van Eselt.	Fabaceae	*Vigna unguiculata subsp. sesquipedalis*
Vigna sp.		Fabaceae	*Vigna sp.*
Vigna speciosa	(Kunth) Verdc.	Fabaceae	*Vigna speciosa*
Vigna spectabilis	(Standl.) A. Delgado	Fabaceae	*Vigna spectabilis*
Vigna strobilophora	B. L. Rob.	Fabaceae	*Ramirezella strobilophora*

Name of Taxon	Name of Authority	Family	Preferred Taxon Name
Vigna strobilophora var. buseri	(Micheli) McVaugh	Fabaceae	*Ramirezella strobilophora var. buseri*
Vigna stuhlmannii	Harms	Fabaceae	*Vigna ambacensis*
Vigna sublobata		Fabaceae	*Vigna sublobata*
Vigna subterranea	(L.) Verdc.	Fabaceae	*Vigna subterranea*
Vigna triloba	Walp., nom. illeg.	Fabaceae	*Vigna unguiculata subsp. stenophylla*
Vigna triloba var. stenophylla	Harv.	Fabaceae	*Vigna unguiculata subsp. stenophylla*
Vigna trilobata	(L.) Verdc.	Fabaceae	*Vigna trilobata*
Vigna triphylla	(R. Wilczek) Verdc.	Fabaceae	*Vigna triphylla*
Vigna truxilensis		Fabaceae	*Vigna truxilensis*
Vigna umbellata	(Thunb.) Ohwi and H. Ohashi	Fabaceae	*Vigna umbellata*
Vigna unguiculata	(L.) Walp.	Fabaceae	*Vigna unguiculata*
Vigna unguiculata subsp. baoulensis	(A. Chev.) Pasquet	Fabaceae	*Vigna unguiculata subsp. baoulensis*
Vigna unguiculata subsp. burundiensis	Pasquet	Fabaceae	*Vigna unguiculata subsp. burundiensis*
Vigna unguiculata subsp. cylindrica	(L.) Verdc.	Fabaceae	*Vigna unguiculata subsp. cylindrica*
Vigna unguiculata subsp. dekindtiana	(Harms) Verdc.	Fabaceae	*Vigna unguiculata subsp. dekindtiana*
Vigna unguiculata subsp. letouzeyi	Pasquet	Fabaceae	*Vigna unguiculata subsp. letouzeyi*
Vigna unguiculata subsp. mensensis	(Schweinf.) Verdc.	Fabaceae	*Vigna unguiculata subsp. mensensis*
Vigna unguiculata subsp. pubescens	(R. Wilczek) Pasquet	Fabaceae	*Vigna unguiculata subsp. pubescens*
Vigna unguiculata subsp. sesquipedalis	(L.) Verdc.	Fabaceae	*Vigna unguiculata subsp. sesquipedalis*
Vigna unguiculata subsp. stenophylla	(Harv.) Marechal *et al.*	Fabaceae	*Vigna unguiculata subsp. stenophylla*
Vigna unguiculata subsp. tenuis	(E. Mey.) Marechal *et al.*	Fabaceae	*Vigna unguiculata subsp. tenuis*
Vigna unguiculata subsp. unguiculata		Fabaceae	*Vigna unguiculata subsp. unguiculata*
Vigna unguiculata var. dekindtiana	(Harms) Marechal	Fabaceae	*Vigna unguiculata subsp. dekindtiana*
Vigna unguiculata var. mensensis	(Schweinf.) Marechal *et al.*	Fabaceae	*Vigna unguiculata subsp. mensensis*
Vigna unguiculata var. protracta	(E. Mey.) Verdc.	Fabaceae	*Vigna unguiculata subsp. stenophylla*
Vigna unguiculata var. pubescens	(R. Wilczek) Marechal *et al.*	Fabaceae	*Vigna unguiculata subsp. pubescens*
Vigna venulosa	Baker	Fabaceae	*Vigna venulosa*
Vigna verticillata		Fabaceae	*Vigna verticillata*
Vigna vexillata	(L.) A. Rich.	Fabaceae	*Vigna vexillata*
Vigna vexillata var. angustifolia	(Schumach. and Thonn.) Baker	Fabaceae	*Vigna vexillata var. angustifolia*
Vigna vexillata var. davyi	(Bolus) B. J. Pienaar	Fabaceae	*Vigna vexillata var. davyi*
Vigna vexillata var. macrosperma	Marechal *et al.*	Fabaceae	*Vigna vexillata var. macrosperma*
Vigna vexillata var. vexillata		Fabaceae	*Vigna vexillata var. vexillata*
Vigna vexillata var. youngiana	F. M. Bailey	Fabaceae	*Vigna vexillata var. youngiana*
Vigna wilmsii	Burtt Davy	Fabaceae	*Vigna oblongifolia var. oblongifolia*
Vigna wittei	Baker f.	Fabaceae	*Vigna wittei*

According to Ng (1995), cowpea growth habit changed from perennial to annual nature during process of evolution. Cultivated cowpea (suspp. *unguiculata*) evolved through domestication and selection of the annual wild cowpea (var. *dekindtiana*).

Gene and Genetics of Characters

Sen and Bhowal (1961) worked out genetics of *Vigna sinensis* (L.) Savi (Syn. *V. unguiculta*). They results indicate existence of multiple allelic series for the different colour patterns in unripe fruits. They proposed following genes on the basis of results. The multilple allelic series, p, green fruit; p^1, purple tip fruit, p^v, green fruit with purple ventral suture, and purple pods with green suture. The others are: Ps, green fruit with purple suture; ps, green fruit; and Pu, uniformly purple fruit; pu, green fruit with faintly purple suture; and tip, a tentative one from the phenotype of the double recessive (pupupp).

Table 4.27: Gene and Genetics of Characters

Character	Number of Gene	Type of Gene Action
Pod colour	Monogenic (GpGp)	Green dominant to white
Flower colour	Monogenic (PfPf)	Purple dominant to white
Standard petal colour	Monogenic (Ystp Ystp)	Yellow dominant to white
Grain length	Monogenic (LgLg)	Long dominant to short

Character	Number of Gene	Type of Gene Action
Leaf shape	Oligogenic (Ls1, Ls2, Ls3, Ls4)	Hastate leave dominant to rhomboid
Seed coat colour	--	Purple dominant overall other colours, Black dominant over other colours except purple
Leaf foliation	Monogenic	Trifoliate dominant to unifoliate
Cotyledon colour	Monogenic	Purple dominant to green

Source: Peter, 1998.

In cowpea there are several simply inherited qualitative characters *viz.* axillary buds, peduncle, plant growth, leaf shape and size, colour of plant, petiole inflorescence and flower, pod shape and colour pod attachment and appearance and texture, seed colour, spotting, mottling and hilum. The characters conditioned by polygenes include flowering and maturity, pod number per plant, pod length, weight and girth, seed weight and shape, pod and seed protein content, nitrogen fixation and pod yield.

Table 4.28: Gene Effects of different Characters of Cowpea

Character	Gene Action	References
Vine length	Additive	Hazra (1991)
Days to flowering	Both additive and non-additive	Chattopadhyay (1995)
Pods/plant	Both additive and non-additive	Chattopadhyay (1995)
Pod length	Additive	Hazra, *et al.* (1994); Chattopadhyay (1995)
Pod weight	Additive	Hazra, *et al.* (1994); Chattopadhyay (1995)
Seeds/pod	Both additive and non-additive	Chattopadhyay (1995)
Seed shape	Additive	Hazra, *et al.* (1996)
Seed protein content	Additive	Hazra, *et al.* (1996)
Pod yield/plant	Both additive and non-additive	Chattopadhyay (1995); Hazra, *et al.* (1996)

Plant height, commencement, completion and spread of flowering, length and width of individual pods, number of seeds/pod, weight of 100 seeds, length, bridth and thickness of seeds were inherited as quantitative characters controlled by either polygenes or by a few major genes whose action is suitably modified by minor genes (Amma 1981). Jalajakumari (1981) found additive gene action for characters number of primary branches, number of pods/plant, length and weight of individual pod, 100 seed weight, length of seed, pod and seed yield/plant. Yield of seeds/plant was highly correlated with number of pods/plant, weight of pod, number of seeds/pod, breadth of seed, thickness of seed and yield of pods/plant.

Table 4.29: Some Important Genes of Cowpea

Gene Symbol	Character
bcm	Resistance to black eye cowpea mosaic virus
Bk-2	Black pod colour
fa	Fasciated plant
Dhp	Dehiscent pod
Gr	Green shoot colour
P1	Pleiotropic gene for axil, calyx, corolla, pod tip, and seed colours
Pms	Partial male sterility
Ps	Photosensitivity
Ptc	Calyx pigmentation
Pt	Calyx pigmentation
Rds1	Resistance to drought
Rds2	Resistance to drought
Rt1	Rough seed coat texture
Rt2	Rough seed coat texture
Vsm	V-Shaped mark on leaves

Source: Singh, 2000.

Additive components played greater role in the expression of green pod yield (Kumar *et al.*, 1998) whereas both additive and dominant components were observed for earliness in vegetable cowpea (Kumar *et al.*, 2000).

Botanical Description

Cowpea [(*V. unguiculata* (L.) Walp.], an annual legume and belongs to the family Fabaceae. Studies on the structure of stomata on vegetative and floral organs have confirmed the inclusion of the cowpea species in the tribe Phaseolae (Shah and Gopal, 1969). It belongs to the botanical species (*V. unguicülta* (L.) Walp. cultivar group *sesquipedalis*, the yard long bean (syn. string bean, asparagus bean) is a type of cowpea used as vegetable under the sub-family Papilionaceae or Fabiaceae of family Leguminoceae or Fabiaceae.

Kingdom	:	Plantae
Sub-kingdom	:	Angiosperms
Class	:	Eudiocots
Subclass	:	Rosids
Order	:	Fabales
Family	:	Fabacea
Subfamily	:	Faboidea
Tribe		Phaseoleae
Subtribe		Phaseolinae
Genus		*Vigna*
Species		*unguiculata*
Botanical name		*Vigna unguiculata* (L.) Walp.
Synonyms		*Vigna sinensis*

The 11 wild species *i.e. V. parviflora, V. wilmsii, V. oblongizolia, V. gracillis, V. heterophylla, V. membranacea, V. vexillata, V. davyi, V. schimperi, V. marima* (Burm.) Merr., and *V. luteola* of *Vigna* studied reveal 2 chromosome numbers in their somatic complements, 2n=22 and 2n=20. In wild taxa, the chromosome number vary from 20-24.

Table 4.30: Botanical Description

Parts	Description
Growth Habit	Pole and bush both vegetable types are available in vegetable cowpea. Growth habit ranges from indeterminate, profusely branching types to erect, determinate, non-branching to prostrate or climbing types. Indeterminate with profuse branching can be observed.
Root	Cowpea has a well developed tap root with a considerable number of lateral roots.
Stem	The stem is almost glabrous, twisting, sometimes hollow and slightly ridged/ribbed. Stems may be green or pigmented (purple).
Leaf	The leaves are long or petioled, pinnately, trifoliate with one symmetrical terminal leaflet and two asymmetrical leaflets. Petioles are glabrous and 3.5-10.0 cm long. Leaves developed in alternate with scattered short hairs. Stipules are far above the acute base, ovate, obtuse and 1.0-1.8 cm long. Leaflets are shortly stalked, ovate-rhomboid, base broadly luneate or rounded, apex gradually narrowed, acute, entire or fainty lobed, glabrous or thinly pubscent, 6.5-16.0 cm long.
Floral biology	Pedicels rather thick. The inflorescence is an unbranched auxiliary raceme bearing several flowers at the terminal end of peduncles (20-50 cm long). The flower is bisexual white, purple, or pale violet. It is a complete flower with 5 sepals and 5 petals. The keel petal is straight; the standard petal is whitish to violet or sordidly yellow in colour and 2.0-3.0 cm in diameter. The calyx is gamosepalous, more or less plicate-wrinkled, and glabrous with small triangular acute teeth and 1.0 cm long. The two are large and three are small, which makes (2+3) arrangements of sepals. The corolla is papilionaceous with an erect standard spreading at the time of flower opening. The wings are violet, truncate at the base, adherent to the boat-shaped keel, enclosing the androecium. The stamens are diadelphous (9+1). Ovary is monocarpellary, and unilocular with many ovules.
Anthesis and pollination	Cowpea flowers bloom early in the morning (between 6.30 hr. to 7.00 hour) and close before noon and falls in the same day. The process of opening of corolla takes 45-60 minutes. Anthers mature and start dehiscing about 7-8 hr before the time of opening of corolla. The extra floral nectarines at the base of the corolla attract ants, flies and bees but a heavy insect is required to depress the wing petals and expose the stamen and stigma. Cowpea is a self-pollinated crop (Davis *et al.*, 2000) but a few lines of cowpea are cross-pollinated due to extrovert nature of pistil. Optimum time for emasculation is at 17.30 hour – 18.30 hour and hand pollination at 6.00 hour – 7.00 hour the next day. Stigma becomes receptive from 12 hr before blooming to 6 hr after blooming. Pollen remains viable for 42 hr at room temperature of 28.5 °C and relative humidity 91 per cent. Stigma becomes receptive from 12 hour before blooming to 6 hour after blooming. For hybridization of cowpea; emasculation can only be effective if it is completed 20 hrs before flower opening. The safest time for emasculation may be during the morning proceeding the day of blooming. Pollens remain viable for 24 hrs at room temperature of 28.5°C and relative humidity 91 per cent. When stored in a refrigerator at 9°C and 0 per cent relative humidituy. The period of pollen viability increased from 12 hrs before blooming to 6 hrs after blooming.
Pod and pod set	The typical vegetable cowpea is characterized by cylindrical long pods (>30cm), constricted between the seeds, stringless pods, fleshy pod pericarp, pods thin linear with a thick, obtuse beak, 10-85 cm long and 0.8-1.0 cm broad and higher monosaccharide: polysaccharide ratio.,
Seeds	The seeds are numerous, long seeds, bean-shaped, seed coat is often smooth or wrinkled and many times spotted with different colours such as brown, green yellow, white, cream, buff, red, black and mottled.

Species Relationship

Cowpea belongs to the botanical species (*Vigna unguiculata* (L.)Walp. Verdcourt (1970) divided the ciowpea species *unguiculata* in the following sub-species:

(i) *Unguiculata*

(ii) *Cylindrica*

(iii) *Sesquipedalis*

(iv) *Dekindtiana*

(v) *Mensensis*

First three sub-species are considered as cultivated sub-species, but Marechal *et al.* (1978) grouped these species into one sub-species *Vigna unguiculata* subsp. *unguiculata* and differentiate them by the interspecific category "Cultigroup". All the cultivated cowpeas are grouped under *V. unguiculata* subspecies *unguiculata* which is subdivided into for cultigroups, namely *unguiculata, biflora, sersquipedalis* and *textilis* (Ng and Marechal, 1985). The nomenclature and classification of wild taxa within *unguiculata* is complicated. Neema (1986) studied genomic relationship between different subspecies. She found that subsp. *unguiculata* and *sesquipedalis* were easily crossable. An extensive research work on characterization of different wild accessions was conducted at International Institute of Tropical Agriculture, Nigeria (Padulosi, 1993). Presently, the four cultivated sub-species of cowpea are considered under species *unguiculata*:

(a) *V. unguiculata* subsp. *cylindrica*-Catjang

(b) *V. unguiculata* subsp. *dekindtiana*

(c) *V. unguiculata* subsp. *sesquipedalis* (L.) Verdc-Yardlong bean-It is an important vegetable legume throughout the tropics and sub-tropics of Asia. Researchers believed that it has been selected and developed for their long tender pods in South-East Asia, from vegetable types of *Vigna unguiculata* introduced there from India (Steele and Mehra, 1980).It is known by several names, such as string bean, long-podded cowpea, asparagus bean, snake bean, Chinese long bean, pea-bean, boro, juro-kusasagemae (Japanese), dowgauk (Chinese) or sitaw (Filipino), yardlong beans are cultivated for their strikingly long drooping pods (30-90 cm) that are relished as a vegetable. Besides the tender pods, the young leaves and seeds are also eaten. Referred to as 'poor man's meat' in the Philippines, yardlong bean commercially important in parts of Indonesia, Thailand, Philippines, Taiwan and China, while being a minor vegetable throughout its range of distribution. It is estimated that area under yardlong bean in China alone exceeds 250, 000 ha annually (Rubatzky and Yamaguchi, 1997), while in Thailand it is grown on about 18,560-20,160 ha annually (Sarutayophat *et al.*, 2007). Yardlong beans are mainly a warm-season crop and capable of surviving extreme humidity and heat. They can be planted in a wide range of climatic conditions but are very sensitive to cold temperatures. Although cowpeas originated in Africa, this vegetable form has been introduced there only in recent times. It is now grown as a minor garden crop in many sub-tropical countries of Africa (especially West Africa) and America (*e.g.* Caribbean). It is popular in these countries also due to its beautiful, large violet flowers and the long drooping pods. In India, yardlong bean is a less known under-utilized vegetable grown in the peninsular region particularly Kerala, Tamil Nadu, coastal Andhra Pradesh and Odisha. It is also cultivated to some extent in West Bengal, Assam and the North Eastern Hill (NEH) region. It is a highly self-pollinated, vigorous climbing annual; growing upto a height of 3-4 m. it produces very long, slender and succulent pods that may be white, light green, dark green, brownish red or purple. Nutritionally, the tender green pods are rich in crude protein (3.5-5 per cent), besides being a good source of vitamin (941 IU) and C (13 mg), iron (2.5 mg), calcium (80 mg), phosphorus (74 mg) and dietary fibre (2 g) making it an excellent vegetable (Singh *et al.*, 2001).

(d) *V. unguiculata* subsp. *unguiculata*- Black-eyed pea

The cultivated cowpea is polymorphic and consists of

- ✰ *V. sinensis* subsp. *sinensis* L. (syn. *V. unguiculata* var. *radiata, Dolichos sinensis* L.)
- ✰ *V. catjang* (syn. *V. sinensis* var. *cylindrica, V. unguiculata* var. *cylindrica*)
- ✰ *V. sesquipedalis* (syn. *V. unguiculata* var. *sesquipedalis*)

Table 4.31: Desirable and Undesirable Traits of *V. unguiculata* Subspecies *unguiculata* and *sesquipedalis*

Subspecies	*Desirable Traits*	*Undesirable Traits*
sesquipedalis	Better pod quality *viz.* soft, long, pulpy and parchment free pods	Indterminate growth habit needs staking, late in flowering and fruiting, generally susceptible to cowpea golden mosaic virus *Cercospora* disease.
unguiculata	Determinate growth habit, earliness in flowerinh and fruiting	Poor pod quality *viz.* hard, short and non-fleshy pods, presence of thick parchment.

Horticultural Classification

It is stated that *sinensis* and *catjang* differentiated from a common ancestoral stock and *sesquipedalis* is a later selection from *sinensis*. The *cytotaxonomy* of cowpea is relatively simple being uncomplicated by polyploids. It is a true diploid with 2n = 22. Sen and Bhowal (1960) made cyto-taxonomic studies on *V. sinensis* (L.) Savi; *V. catjang* (Bur. M) Walp. and *V. sesquipedalis* (L.) Fruw., popularly known as cowpea, catjung bean and asparagus bean. The three cultivated species could be easily crossed among themselves. The hybrids among them were fertile like inter-varietal hybrids. Chromosome numbers of each of three cultivated species were found 2n = 22. As mentioned elsewhere all the above-cultivated species are put under *V. unguiculata*. Studies on the structure of stomata on vegetative and floral organs have confirmed the inclusion of the cowpea species in the tribe *Phaseolae* (Shah and Gopal, 1969). The cultivated cowpea is polymorphic and consists of *V. sinensis* subsp. *sinensis* (syn. *V. unguiculata* var. *radiata, Dolichos sinensis* L.), *V. catjang* (syn. *V. sinensis* var. *cylindrica, V. unguiculata* var. *cylindrica*) and *sesquipedalis* (syn. *V. unguiculata* var. *sesquipedalis, Dolichos sesquipedalis*). Recently all the cultivated species have been placed under a single species *V. unguiculata* (L.) Walp. The typical vegetable cowpea is characterized by long pods (>30.0 cm), stringless pods, fleshy pod pericarp, thin and long seeds, and higher monosaccharide: polysaccharide ratio.

Classification of Cowpea

Classification on the Basis of Utility

(i) Grain type (*V. unguiculta* (L.) sub sp. *catjang* Wall.: Syn. *V. sinensis* sub sp. *catjang* Wall

(ii) Yardlong bean (*V. unguiculata* var. *sesquipedalis* (L.) Verdcourt

(iii) The dual purpose cowpea (*V. unguiculta* (L.) Walp., belongs to cultivar group *cylindrica* var. *eseltine* which is used both as a pulse crop as well as vegetable.

(iv) Exclusively pulse type grouped as *V. unguiculata* var. *radiata*. Pole and bush types are available in vegetable cowpea.

Classification on the Basis of Edible Pod Maturity

(i) Early- Pod mature between 22-37 days to flower

(ii) Medium-Pod mature between 32-44 days to flower

(iii) Late-Pod mature between 49-62days to flower

Classification on the Basis of Plant Height

(i) Bush types-Bush types of yardlong bean have been developed 'Los Banos Bush Sitato No. 1'. Now several bush genotypes and cultivars are availableat ICAR-IIVR, Varanasi and ICAR-IIHR Bangaluru in India.

(ii) Pole types-Several pole type genotypes are available. In central and indogangetic plains of India, it is very popular and called 'Kala Jhamra'.

(iii) **Semi-pole types-** Not much common but exist in nature.

Augmentation of Genetic Resources and Crop Improvement

Introduction and Selection

At international level IITA, Nigeria has taken lead ine world for collection, evaluation and utilization of *Vigna* species.

Table 4.32: Varieties Developed at IITA, Nigeria

Traits	*Name of Varieties/Genotypes*
Extra early matuing varieties	IT 87D-879-1, IT 86D-1010, IT 90K-284-2, IT 87D-829-5, IT 86D-719, IT 87D-697-2, IT 87D-611-3, IT 87D-941-1, IT 82E-32
Extra early varieties (photo insensitive)	IT 90K-284-2, IT 90K-56, IT 88D-643-1, IT 89KD-389, IT 91K-93-10, IT 90K-59-4, IT 84S-2246-4(all are suitable for intercropping with multiple resistance to insects and diseases)
Medium matuing varieties	IT 90K-372-1-2, IT 90K-277-2, IT 88DM-363, UT 89KD-374-8, IT 89KD-374-57, IT 90K-109, IT 89KD-349, IT 88D- 867-11, IT 88DM-400, IT 90K-319
	IT 90K-277-2, IT 88DM-363 (photo insensitive), IT 90K-372-1-2, IT 89KD-374-8, IT 89KD-374-57, IAR-48 (all are suitable for intercropping with multiple resistance to insects and diseases)
Late matuing varieties	IT 81D-985, IT 89KD-252, IT 89KD-260, IT 89KD-288
Dry season varieties	T 86D-715, IAR-48, IT 86D-719, IT 90K-76, IT 90K-59, IT 90K-101, IT 89K-288, IT 89KD-374
Bush type varieties	IT 81D-1228-10, IT 81D-1228-13, IT 81D-1228-14, IT 81D-1228-15, IT 86D-880
Dual purpose varieties (feed and fodder)	IT 81D-985, IT 89KD-252

Correlation and path coefficient analysis partitioned into direct and indirect effects for the 10 yield contributing characters were studied in 47 genotypes of cowpea by Patil *et al.*, 2005. The present correlation studies revealed that seed yield per plant had positive and highly significant correlation with plant height at genotypic level and number of pods/plant, harvest index at both phenotypic and genotypic levels. The path coefficient analysis in the present investigation reveals that among the yield contributing characters *viz.*, plant height, number of seeds/pod, harvest index, protein per cent had positive direct effect on seed yield per plant at both phenotypic and genotypic levels.

Based on characterization and preliminary evaluation data, the following promising accessions were identified for various attributes, *viz.* earliness, dwarf habit, pod length, pods per plant, seeds per pod *etc.* Varieties 'Pusa Phalguni' suited for spring season, 'Pusa Barsati' for rainy season and 'Pusa Dofasali' for both the seasons habe been developed. Similarly very early genotypes 'K1552' named 'Pusa Komal' evolved which complete its life cycle within 55 days. Several indigenous cultivars have performed well in different parts of the country. Some of these indigenous lines, which have performed well, are: VRCP-1,VRCP-2, VRCP-3, VRCP-4, Pusa Komal (K-1552), Arka Garima, Cowpea Sel-263, BC-244002, NDCP-8 *etc.* There are several lines like IC559404, EC528412, IC528396, ET11902 and BCKV-1 are popular for bush habit. Dual purpose varieties are IT 89KD-288, IT 89KD-391.

In Kerala's climate, yardlong beans may be cultivated throughout the year and significant research on this crop has been undertaken by Kerala Agricultural University Kerala. Several high yielding varieties such as Kanakmany, Lola, KMV-1, Mallika, Anaswara, Varun, Kairali, Sharika, Vellayani, Jyothica and Vyjayanthi have been released for the long pods ranging from 40-60 cm. Resmi (1998) found significant difference among the 30 yardlong bean genotypes. Pod yield/plant was strongly associated with pod weight, pod length, and pods per plant. Genotypes VS-6 and VS-11 recorded the maximum selection index. Ajith (2001) reported high phenotypic and genotypic coefficients of variation for length of main stem, number of primary branches and pod weight. Arka Garima recorded highest pod yield and average pod weight while Arka Suman recorded highest number of pods/plant. At BCKV, Kalyani (W.B) has collected and evaluated bush type and pole type cowpea genotypes.

Table 4.33: Promising Accessions for different Traits Identified

Promising Traits	Vareities/Accession	Source
Dwarf habit	Pusa Komal,	ICAR-IARI, New Delhi
	VRCP-1, VRCP-2, VRCP-3, VRCP-4, VRCP-5, VRCP-6	ICAR-IIVR, Varanasi (U.P.)
Earliness	Pusa Komal,	ICAR-IARI, New Delhi
	VRCP-1, VRCP-2, VRCP-3, VRCP-4, VRCP-5, VRCP-6	ICAR-IIVR, Varanasi (U.P.)
	EC 466882, EC 472282, IC 276938	ICAR-NBPGR, New Delhi
Long pods	VRCP-2, VRCP-3, VRCP-4, VRCP-5, VRCP-6	ICAR-IIVR, Varanasi (U.P.)
	EC 472286, SKJ- 1/156, IC 253271	ICAR-NBPGR, New Delhi
More pods per plant	VRCP-2, VRCP-3, VRCP-4	ICAR-IIVR, Varanasi (U.P.)
	KG/ZH/BLM-286, IC 219871, IC 259105, IC 287428	ICAR-NBPGR, New Delhi
Seeds per pod	VRCP-1, VRCP-2, VRCP-3, VRCP-4	ICAR-IIVR, Varanasi (U.P.)
	KG/ZH/BLM-98, IC 58905, EC-20584	ICAR-NBPGR, New Delhi
Good pod quality	VRCP-2, VRCP-3, VRCP-4	ICAR-IIVR, Varanasi (U.P.)
Higher pod yield	VRCP-1, VRCP-2, VRCP-3, VRCP-4	ICAR-IIVR, Varanasi (U.P.)

Table 4.34: Trait Specific Grouping of Cowpea

Crop Group	No. of Collection	Important Traits under Consideration	References
Cowpea-Bush type	75	Plant frame, flowering habits, pod colour, shape, curvature, pigmentation in plant parts, tolerance against cowpea mosaic virus, aphid, protein content of pods, seeds, *etc.*	Chattopadhyay *et al.* (1996); Chattopadhyay *et al.* (2014)
Cowpea-Pole type	35	Plant frame, flowering habits, pod colour, shape, curvature, pigmentation in plant parts, tolerance against cowpea mosaic virus, aphid, protein content of pods, seeds, *etc.*	Chattopadhyay *et al.* (1996)

At BCKV, Kalyani (West Bengal) has been observed great diversity in vegetable cowpea and released two popular cultivars *i.e.* Bidhan Barbati-1 and Bidhan Barbati-2 (Hazra *et al.*, 2001).

Hybridiazation and Selection

In India Kamalam (1975) reported that parents S-78, S-58, S-51 and Pannithandan were the best general combiners for yield. Inasi (1980) observed heterosis to the same extent in hybrids of both genetically related

Table 4.35: Promising Accessions of Cowpea

Character	Range	Promosing Accesions
Days to 50 per cent flowering	38.00-78.00	82-1, EC243954
Peduncle length (cm)	23.03-34.80	BCCP-14, 5269
Pod length (cm)	12.20-41.51	BCCP-19, EC305827
Pod width (cm)	0.81-1.10	BCCP-14, BCCP-8
Pod weight (g)	3.10-20.42	EC305827, BCCP-17
Pods/plant	19.00-43.00	EC243954, BCCP-19
Pod yield (g/plant)	170.00-517.00	EC305827, EC243954, BCCP-19
Pod protein (per cent)	2.20-4.50	BB-1, BB-2, EC243954
Cowpea mosaic incidence	1.50-62.00	BB-1, BCCP-3, EC243954

Source: Chattopadhyay *et al.* (2016).

and unrelated parents. P-118 x GP PLS-139 was the best among the 16 hybrids produced. Several elite genotypes were utilized to develop desirable varieties of cowpea. Zaveri *et al.* (1983) observed high degree of heterosis for grain yield, number of clusters per plant and number of pods per plant, but not for pod length, grains per pod and 100 seed weight. Days to 50 per cent flowering and maturity have been found positive heterosis over the mid parent and better parent. The heterosis in grain yield could be attributed to high heterosis in number of clusters and number of pods per plant. Patil and Shete (1987) recorded highest positive heterosis over mid-parent value which occurred for number of pod clusters per plant (158 per cent), followed by seed yield per plant (131 per cent), while that over the better parent it has been observed the highest for pod cluster per plant (133 per cent) followed by number of pods per plant (89 per cent). Number of pods per plant and pod cluster per plant, supplemented by 100 seed weight, appeared to contribute to seed yield per plant. Therefore, it is suggested that segregating population from crosses in which heterosis for seed yield per plant is associated with heterosis for number of pod clutsters + number of pods per plant provided the useful material from which desirable line could be selected. Hazra *et al.* (1993) derived information for heterosis from the data on pod yield and four important components like plant height/vine length, number of pods per plant, pod length and pod yield per plant. They observed that frequency and level of heterosis are related more to s.c.a. than to the genetic divergence of the parents as estimated by "Mahalanobis", D^2 statistics. Sangwan and Lodhi (1995) reported positive heterosis for yield ranging from 28-8-84.0 per cent over better parent. Heterosis upto 81.6-35.6, 20.4 and 36.3 per cent over better parent has been observed for number of pods per plant, pod length, number of seeds per p[od and seed weight, respectively. Danam and Chauhari (2000) observed maximum positive heterosis for seed yield per plant over mid parent (MP), over better parent (BP) and standard check (SC)to the extent of 64.47, 46.45 and 63.33 per cent, respectively. They also reported that heterosis in seed yield per plant is due to the heterosis found in yield components, mainly, pods per plant, seed per pod, cluster per plant, branches per plant and test weight. Pal *et al.* (2002) observed appreciable heterosis over better parent and mid parent in cowpea. They found that F_1 hybrids namely NDCP-13 x Arka Garima, Cowpea Local x Cowpea-263, and Red Seeded x Pusa Komal showed high degree of heterosis. Considering the importance of both additive and non-additive effects in the genetic control of pod number and pod yield/ plant (Hazra *et al.*, 1993; 1994a and b; Chattopadhyay, 1995), homozygous lines equal to or better than F_1 can be developed in cowpea utilizing transgressive segregants. Amma (1981) reported that crosses GP PLS-139 x P-118 and Red Seeded Selection x Kolingipayar were identified as suitable combinations for grain and dual purpose. For vegetable purpose, crosses Pannithandan Early x Kolingipayar White, Manjari Black x Kolingipayar White and IC20729 x Red Seeded Selection were the ideal ones. Rejatha (1992) reported that Selection 104 and Selection-145 were the best general combiners for most of the characters while Selection-129 was the best general combiner for earliness. The hybrid Selection-145 x Selection-129 was the best specific combiner for mean weight of pod and nymber of seeds/pod. Selection-145 x Kurutholapayar was the best specific combiner for earliness. Shobha (1994) reported Arka Garima and Selection 2-1 as the best general combiner for yield. Hybrids VS-18 x Arka Garima and Selection 2-1 x VS-389 had high s.c.a. effects for yield. VS-18 x Arka Garima and Selection 2-1 x VS-389 were identified as promising hybrids. Mishra *et al.* (1985) found that male parent affects pod set significantly varying from 0 to 30 per cent. Heterosis in vegetable cowpea ranged up to -6.44 to -12.50 per cent for days to 50 per cent flowering, 8.2-32.31 per cent for pod length, 57.0-59.0 per cent for number of pods per plant, 47.99-81.8 per cent for green pod yield per plant, 35.5-50.41 per cent for number of seeds per pod and 13.46 per cent for protein content (Lal *et al.*, 2007; Kajale *et al.*, 2013). High values for heterosis indicates good genetic diversity among cowpea varieties indicating the possibility of isolating high yielding transgressive segregates from hybrid populations. Anilkumar (1993) conducted combining ability stidies for yield and drought tolerance in cowpea and recommended DPLC-98, IC38956, C-152 and Chharodi-1 are good combiner and the hybrids DPLC9198 x C-152,

Table 4.36: Magnitude of Relative Heterosis, Heterobeltiosis and Standard Heterosis in Vegetable Cowpea

Hybrids	*Pod Length (cm)*		*Pod Weight g)*		*Number of Pods/Cluster*		*Number of Pods/Plant*	
ICP-26 x Pusa Komal	10.24	6.89	-6.36	-25.88**	21.88**	0.38	1.59	-13.62*
ICP-26 x Arka Garima	11.68	4.05	-14.41*	-31.95**	42.50**	3.54	2.51	-11.85
ICP-26 x Indira Hari	0.18	-9.41	-17.03*	-31.25**	29.41**	27.17**	21.81*	14.83*
ICP-26 x Khalleshwari	-20.87*	-38.51**	-12.50	-27.54**	46.15**	7.65*	20.18*	8.66
ICP-38 x Pusa Komal	20.78**	2.63	-3.80	-16.41**	9.09*	7.69*	35.41**	13.18
ICP-38 x Arka Garima	23.82**	1.72	5.64	-8.74*	31.18**	10.90**	33.88**	13.13
ICP-38 x Indira Hari	-0.79	-4.71	-11.29*	-26.49**	34.38**	13.44**	43.60**	32.73**
ICP-38 x Khalleshwari	-36.61**	-55.24**	-37.38**	-59.86**	20.88**	3.97	17.17*	4.01
ICP-42 x Pusa Komal	0.98	-25.31**	0.87	-21.42**	13.75**	10.99*	34.83**	9.46
ICP-42 x Arka Garima	10.29*	-20.55**	36.14**	5.61	16.67**	1.63	50.78**	23.71**
ICP-42 x Indira Hari	9.60*	-10.61**	35.66**	1.55	40.60**	15.02**	27.85**	14.27*
ICP-42 x Khalleshwari	-47.42**	-66.27**	-44.34*	-66.22**	17.02**	3.97	26.44**	8.71
ICP-45 x Pusa Komal	19.97**	10.37*	-3.05	-7.93	23.94**	12.69**	27.74**	2.08
ICP-45 x Arka Garima	14.87*	1.81	-1.11	-6.63	33.91**	5.72	20.54*	-2.71
ICP-45 x Indira Hari	2.07	-2.85	0.70	-9.35*	46.03**	32.39**	41.79**	24.41**
ICP-45 x Khalleshwari	-31.93**	-49.11**	-38.01**	-58.37**	37.18**	10.20**	19.19*	0.69
ICP-49 x Pusa Komal	1.37	-7.39	4.63	-3.57	28.99**	14.23**	17.05	-11.58
ICP-49 x Arka Garima	2.77	-9.54	6.96	-1.94	36.47**	5.45	14.53	-12.72
ICP-49 x Indira Hari	-6.55	-10.39*	-0.82	-13.16**	23.75**	15.50**	39.68**	14.88*
ICP-49 x Khalleshwari	-29.44**	-47.50**	-35.75**	-57.45**	32.53**	3.97	27.20**	1.06
ICP-54 x Pusa Komal	7.28	-18.79**	3.22	-11.64**	5.00	2.56	37.76**	15.14*
ICP-54 x Arka Garima	7.38	-20.95**	14.00**	-2.84	18.75**	3.54	39.35**	17.72*
ICP-54 x Indira Hari	1.37	-15.02**	-5.78	-23.02**	32.84**	8.79*	74.20**	60.92**
ICP-54 x Khalleshwari	-45.97**	-64.79**	-42.21**	-63.19**	22.77**	9.07**	20.62**	7.00

*and ** Significant at 5 and 1 per cent level; RH-Relative Heterosis, HB- heterobeltiosis, SH-standard heterosis.

Source: Sharma *et al.*, 2013.

DPLC-198 x Chharodi-1, IC38956 x C-152 and IC38956 x Chharodi-1 are good yielder. He suggests that parents can be used iin hybridization and hybrids can be used for generating segtegating progenies for various traits.

Sharma *et al.* (2013) performed Line x Tester (6x4) analysis of in cowpea and obsorved exhibited significant differences among parents and their crosses for all these characters. ICP-38 x Arka Garima, ICP-45 x Pusa Komal and ICP-42 x Indira Hari exhibited maximum positive magnitude for relative heterosis, heterobeltiosis and standard heterosis, respectively for pod length. Cross combinations ICP-42 x Arka Garima and ICP-54 x Indira Hari showed maximum positive magnitude for relative heterosis, heterobeltiosis and standard heterosis for pod weight and number of pods per plant, respectively. Additive genetic variance was predominant for pod length and pod weight wheras, non-additive gene action for number of pods per cluster and number of pods per plant. ICP-42, ICP-45, Pusa Komal, Arka Garima and Indira Hari were good general combiners for pod length and pod weight. ICP-42, Arka Garima and Khalleshwari were good general combiners for pods per cluster where as, ICP-38 and ICP-54 for number of pods per plant. The cross combinations ICP-42 x Indira Hari, ICP-26 x Khalleshwari, ICP-42 x Indira Hari and ICP-42 x Arka Garima were identified as best specific combiners for yield contributing characters.

Interspecific Hybridization

From the several experiments it is clear that five species of *V. unguiculata* are cross-comptible and using them is restored to transfer desirable characters to cultivated species of *V. unguiculata*. In an experiment, the 11 wild species of *Vigna* were crossed with one cultivated species *V.catjang* (syn. *V. unguiculata*) by Sen and Bhowal (1960). A few crosses with the species *V. luteola* gave shrunken seeds but all crosses with the rest of wild species were unsuccessful. The hybrid seeds of the cross *V. luteola* x *V. catjang* were shrunken. Another interspecific cross between two wild species *V. luteola* x *V. parviflora* gave some seeds which were very crumpled. Fatokun and Singh (1987) reported interspecific hybridization between *V. unguiculata* (L.) Walp. and a hairy wild relative (*V. pubescens*) was achieved by embryo rescue technique, the F_1 plants were vigorous in growth, partially sterile, slightly hairy and showed some intermediate features between the two parental types. Gomathinayagam *et al.* (1998) reported successful crosses between *V. vexillata* and *V. unguiculata* using embryo culture. They grew the F_1 hybrids and harvested F_2 seeds that were planted and then backcrossed to *V. unguiculata*. The resulting backcross seeds looked closer to *V. vexillata*. Tyagi and Chawla (1999) also reported successful crosses between *V. radiata* and *V. unguiculata* using *in vitro* culture techniques. Gibberellic acid treatment sustained the pods for 9-10 days, which were then used for embryo culture. About 10 per cent of total embryos cultured resulted in plantlet formation. Neema (1986) studied genomic relationship between different subspecies. She found that subsp. *unguiculata* and *sesquipedalis* were easily crossable. The chromosome number of subsp. *unguiculata* is 2n=22 and that of subsp. *sesquipedalis* is 2n=24. F_1 hybrid had a chromosome number 2n=23. Length of meiotic and somatic metaphase chromosomes were more in subsp. *sesquipedalis*, lowest in subsp. *unguiculata* and intermediate in the F_1 hybrid. Pollen fertility was more in subsp. *sesquipedalis*.

Resistance Breeding

Several cultivated varieties/lines are resistant/ tolerant source of biotic stresses. Most of these lines can be utilized to develop resistant varieties against specific biotic stresses. Many valuable genes like male sterility, high protein and methionine content, resistance to thrips, aphids, pod borers, root-knot nematodes, anthracnose and golden mosaic virus are available which need further utilization.

Resistance to Diseases

Incidence of diseases directly or indirectly lowering the production and productivity of crops and sometimes the entire crop is damaged epidemically by them. The extent of damage due to various diseases has not been estimated except for a few diseases. Although tremendous progress has been made in breeding of improved cowpea varieties for high yield and quality, available resistant lines have not been fully exploited.

- ☆ Very early genotypes 'K1552' named 'Pusa Komal' is resistant to bacterial blight.
- ☆ Tumwegamire *et al.* (1998) study the genetics of resistanc to scab by using a half-diallel cross set of 10 resistant lines. Broad-sense heritability for foliar resistance was 93.8 per cent and for pod resistance it was 97 per cent and 84.5 per cent, respectively.
- ☆ Rodriguez *et al.* (1997) found L-198 and CN x 377-1E to be resistant to *Macrophomina*. Zohri (1993) artificially inoculated 16 cowpea varieties with *Aspergillus flavus* to monitor aflatoxin production. He found that two cowpea varieties from IITA, IT82E-16 and IT81D-1032, did not support Aspergillus growth and therefore no aflotoxin production was observed on these varieties. This indicates the possibility of breeding for resiatance to *Aspergillus flavus*.
- ☆ **Anthracnose (*Colletotrichum lindemuthianum*)**: Anthracnose is a very serious problem of cowpea. As per the protein profile of Kanakamany (immune), Arimbra Local (resistant), Pusa Komal (susceptible) and Lola (susceptible) were catyegorized at KAU, Kerala.

Screening of cowpea genotypes for resistance to Aphid Borne Mosaic Virus Disease (CAMV) showed that out of 59, two cultures namely V-317 and V-276 were found resistant under field conditions (Sudhakumari, 1993).

Resistance to Insects

Incidence of insect pests also lowered the production and productivity of cowpea in the field and grains in the stores.

Cowpea Aphid

Cowpea aphid (*Aphis craccivora* Koch.) is a cosmopolitan species of considerable importance in cowpea which built-up is very fast due to parthenogenesis and telescoping of generations. It also affect cowpea yield indirectly by transmitting cowpea virus diseases. As a vector its association causes alfalfa mosaic, bean common mosaic, bean yellow mosaic, cowpea aphid-borne mosaic, cowpea-banding mosaic, cowpea-mild mottle, pea-leaf roll and chickpea stunt. In India it was first reported in Tamil Nadu, Punjab, Maharashtra, Bihar, West Bengal, Odisha, Andhra Pradesh, Kaenataka and Kerala (David, 1957). Pal *et al.* (1978) concluded ideal conditions for the outbreak of aphids at about 80 per cent relative humidity, 27.5-28.5°C air temperature and low sunshine hours. The aphid population reduced considerably with an increase in number of rainy days

Table 4.37: Resistance Source for various Diseases in Cowpea

Name of Diseases	*Variety/Accessionas a Source of Resistance*	*References*
Bacterial blight	Pusa Komal (vegetable type)	Peter, 1998
	IT90K-284-2, IT 90K-277-2, IT91K-93-10, IT91K-118-20, IT 86D-715, IT 86D-719, IT 81D-1228-14 (grain type), IT95K-398-14, IT95K-193-12, IT81D-1228-14, IT95K-1133-6, IT97K-556-4, IT97K-1069-8, IT90K-284-2, IT91K-93-1,	Wydra and Singh, 1998; Singh *et al.*, 1997; Singh, 1998 and Singh, 1999a
Cowpea mosaic virus	Goit, Dixies-Cream, Alabunch	Peter, 1998
Cowpea golden mosaic virus	CNC0434, TVu 612, CE-315 (TVu 2331) and BR-1-Poty	Santos *et al.*, 1986
Resistance against six viral diseases	IT 82E-16, IT 82E-18, IT 82D-807, IT 82D-828, IT 82D-889, IT 82D-950	
	VRCP-3 and VRCP-4	Anonymous, 2004
Fusarium wilt	Grant	Peter, 1998
	EC 496737	Anonymous,2002
Phytophthora stem rot	Chinese Red	Peter, 1998
Root rot	V 265	Burman and Lodha, 1996
Anthracnose	TVx3236	
Cercospora leaf spot	VRB-10	Booker and Umaharan, 2007
Cercospora leaf spot	IT89KD-288, IT97K-1021-15, IT97K-463-7, IT97K-478-10, IT97K-1069-8, IT97K-556-4	Singh *et al.*, 1997; Singh, 1998; and Singh, 1999a
Smut	IT97K-556-4, IT95K-1090-12, IT95K-1091-3, IT95K-1106-6, IAR-48, IT97K-506-6	
Rust (Uromyces)	IT97K-1042-8, IT97K-569-9, IT97K-556-4, IT97K-1069-8, IT95K-238-3, IT97K-819-118, IT90K-277-2, IT97K-1021-15, IT96D-610, IT86D-719	Singh *et al.*, 1997; Singh, 1998; and Singh, 1999a
Septoria	TVu12349, TVu11761, IT95K-398-14, IT90K-284-2, IT95K-1090-12, IT97K-1021-15, IT98K-205-8, IT98K-476-8, IT97K-819-118, IT95K-193-12, TVu 1234, IT95K-1090-12	
Scab	IT98K-476-8, IT97K-1069-8, TVx3236, IT95K-398-14, IT97K-1021-15, IT95K-1133-6	
Ascochyta	TVu11761	Singh *et al.*, 1997; Singh, 1998; and Singh, 1999a
Bacterial blight	IT95K-398-14, IT95K-193-12, IT81D-1228-14, IT95K-1133-6, IT97K-556-4, IT97K-1069-8, IT90K-284-2, IT91K-93-1, IT91K-118-20	
Cowpea yellow mosaic, Blackeye cowpea mosaic, and Cowpea aphid borne mosaic (multiple resistant)	IT96D-659, IT96D-660, IT97K-1068-7 and IT95K-52-34	Singh and Hughes (1998, 1999)
Cowpea severe mosaic virus and cowpea aphid borne mosaic (multiple resistant)	TVu379, TVu382, TVu966 and TVu3961	Lima *et al.*, 1986

Source: Singh *et al.*, 2000.

Table 4.38: Trait Specific Disease Resistance Germplasm of Cowpea

Accessions	*Source*	*Salient Features*	*Distribution/Recipient*
EC493644, EC493648-49, EC493652, EC493655, EC493658-61 (*V. unguiculata*)	UK	Resistant to black eye cowpea mosaic virus (BICMV)	Kerala Agricultural University, Thrissur, Kerala
UKEC493645-57	UK	Resistant to cowpea aphid-borne mosaic virus	Kerala Agricultural University, Thrissur, Kerala
EC496737 (var. California black)	USA	Resistant to fusarium wilt (Fusarium oxysporum) and root knot nematode (*Meloidgyne* sp.)	Project Coordinatoe (Arid, Legumes), CAZRI; ICAR-NBPGR, N.Delhi

and average relative humidity (Sulochana, 1984). The aphid population is decreased from mid May to end of June in Punjab due to high temperature (Peter, 1998). Mode of inheritance of insect resistance has been worked out (Table 4.39). The pubescence of resistant and susceptible lines of cowpea were electronically scanned and found that the resistant line was the most pubescent line with mixed type of trichomes, which are straight with sharp tips and the tender most shoot tips of the resistant lines are also densely pubescent, thereby protecting the growing tips and making them unsuitable or non-preferred for aphid feeding, locomotion and colonization (Salikutty, 1990). Non preference mechanism acts in conjunction with other biochemical or physiological factors to impart host resistance to aphids. 'Tvu1889' as source of resistance to aphids can be used broadely (Sulochana, 1984; Sulochana and Peter, 986). Joseph (1990) identified three aphid resistant genotypes *viz.*, TVU-1889, VS-350, VS-438, and VS-452. She found that growing tips of resistant lines were highly pubescent. The resistant lines had more reducing sugars and lower non-reducing and total sugars when compared to susceptible lines. Fecundity of aphids was significantly higher on susceptible lines when compared to resistant lines. Total phenol content and orthodihydroxy phenols were high in resistant lines and low in susceptible lines. Trypsin inhibitors were also observed in resistant lines. It was also found that aphid resistant was goverened by a single dominant gene. The seed mixture VS-438+Kanakmany were promising in yield with 83.13 per cent resistance under field conditions. Yunes *et al.* (1998) observed that the 7s-storage protein, "vicillin" is responsible for bruchid resistance in cowpea lines related to 'Tvu2027'.

Bruchids

Jyothi (2001) observed that genotype having thicker seed coat showed better resistance to pulse beetle attack and this character was found to be independent in gene action. Genotype 'Kanakamany' and 'EC390231'were found to be the most promising genotypes. For evolving hybrids showing resistance to pulse beetle C-152 x BC-367711 and V-240 x IC201092 were rated as the best.

Leaf Beetle

Pandey *et al.* (1995) reported 'TVU-908' to be resistant to leaf beetles.

Table 4.39: Varieties/Accessions Identified for Resistance to Insects

Name of Insect	*Resistant Variety/Accession*	*Reference*
Aphids	TVu 408 P_2, Tvu 46, Tvu 3417, Tvu 2740 and Tvu 3509	Singh, 1977 at IITA Nigeria
	PI 473, and PI 476	Dhanorkar and Daware, 1980
	Mandya Local, MS 370, Tvu 2740, P 912 and PI 475	Jayappa and Lingappa, 1988
	Tvu 1989	Sulochana and Peter, 1986
	Vs 350, Vs 438 and Vs 452 are highly resistant but 'Vs 306', 'Vs 307', 'Vs 147', 'Vs 456, 'Vs 457' and 'Vs 458' are moderately resistant	Salikutty, 1990
Heliothis armigera (*Helicoverpa armigera*)	TVX 7	Prasad, 1996
Spotted pod borer [*Maruca testulalis* (*M. vitrata*)]	GC 82-7	Prasad, 1996
Psidia tikora (*Cydia tychora*)	DLPC 16	Prasad, 1996
Bruchids	C 152, C 190 and GC 82-7; IT 97K-207-15, IT 95K- 398-14 and 98K-506-1	Prasad, 1996; Singh, 1999b
Root-knot nematode	Mississipi 57-1, Iron	Peter, 1998
	IITA 3, Habana-82, Incarita-1, IITA 7, P 902, IT 86D364, IT 87D14638, Vinales-144	Rodriguezetd, 1996
Multiple resistant for Aphid, Thrips and Bruchid	IT90K-76, IT90K-59 and IT90K-277-2	Singh *et al.,* 1996
Varieties having combined resuistance to root-knot nematode, aphid and bruchid	IT84S-2246-4, IT 89KD-288, IT 90K-59, IT 90K-76	

Table 4.40: Mode of Insect Resistance in Cowpea

Insect Pest	*Inheritance*	*Reference*
Aphid (*Aphis craccivora*)	Resistance is a dominant trait (Ac 1, Ac 2, Roc)	Ombakho *et al.* (1989)
Weevil (*Callosobruchus maculates*)	Resistance has additive, dominance and maternal components	IITA (1984)
Bruchids and thrips	Resistance was a recessive trait and digenically inherited	IITA (1984)

Legume Pod Borer (*Maruca testularis*)/Spotted Pod Borer [(*M. vitrata* (Fab.)]

The larvae of spotted pod borer [(*Maruca vitrata* (Fab.)] bore into buds, flowers and feed on internal tissues. Their concealed nature protects the larvae from natural enemies as well as from application of insecticides. Anithakumari (1992) screened several genotypes but none of the genotypes were noted resistant to pod borer but genotype V-30 expressed moderately resistant behavior. Forty eight cowpea accessions were screened for resistance to this pest. The extent of borer infestation varied from 0.0-41.04 per cent. Five accessions were identified as resistant (Nasiya, 2015). Jagginavan *et al.* (1995) observed cowpea lines P120 and C11 to be resistant to (least damaged) *Maruca*. Similarly, Veeranna and Hussain (1997) found Tvx7 to be most resistant to *Maruca* and have a high density of trichomes (21.41/mm^2). Veeranna (1998) screened 45 cowpea lines for resistance to *Maruca* pod borer and observed that the tolerant lines had higher phenol and tannin contents compared to susceptile lines. Cowpea varieties having pigmented calyx, petioles, pods and pod tips suffer less damage due to *Maruca*. Panicker (2000) reported that genotypes VS-27 as most resistant and VS-14as the highest yielder. Fifty yardlong bean genotypes were screened for pod borer resistance and three genotypes namely Kuruppumthara Local, Kanichar Local and KMV-1 were found to have less pod borer incidence (Jithesh, 2012). Screening vegetable cowpea for legume pod borer resistance resulted in the identification of VS-27 as most resistant and VS-14 as the highest yielder (Panicker, 2000). Vidya (2000) reported that VS-42 was suitable for cultivation in legume pod borer endemic areas.

Serpentine Leaf Miner (*Liriomyza trifolii*)

Serpentine leaf miner (*Liriomyza trifolii*) is becoming now a serious threat to cowpea production. For searching the source of resistance; Reji (2002) screened 20 accessions (pole and bush types) and observed that trailing types were more susceptible to the pest. VU-12 was least susceptible among the trailing types while Sharika and Vyjayanthi were highly susceptible. Among the bush types, VU-4, VU-6, VU-3 and Arka Garima were less suisceptible.

Table 4.41: Sources of Resistance to different Insect Pest in Cowpea

Insect Pest	*Sources of Resistance*
Leaf hoppers	Tvu59, Tvu123, VITA3
Aphid	Tvu36, Tvu408, Tvu410, Tvu2896, Tvu3000
Bruchids	Tvu2027, Tvu11952, Tvu11953
Thrips	Tvu1509

Source: Singh *et al.* (1983).

Resistance to Nematodes

Severral sources of esistance to netodes were identified including some of the improved breeding lines with high yield potential (Rodriguez *et al.*, 1996). Some of the varieties with high yield and nematode resistance are IT849-2049, IT89KD-288, IT86D-634, IT87D-1463, IOT95K-398-14, IT96D-772, IT96D-748, IT95K-222-5, IT96D-610, IT87K-818-18, and It97K-556-4. Among these varieties, IT89KD-288 was found to be resistance to four strains of *Meloidogyne incognita* in USA (Ehlers *et al.*, 2000a). Singh *et al.* (1996) found IT89KD-288 to be high yielding and highly resistant to nematode in the trials conducted at Kano (Nigeria), where nematode attack is very high in the dry season planting with irrigation. At BCKV, Kalyani (West Bengal) COPB VAR-1 showed root knot nematode tolerant and ACP-1, and BCCP-3 expressed tolerant to cowpea mosaic/Golden mosaic virus alongwith resistance to *Cercospora* leaf spot. Similarly, at ICAR-IIVR, Varanasi (U.P) cultivar Kashi Kanchan and Kashi Nidhi showed resistance to cowpea mosaic/Golden mosaic virus.

Table 4.42: Root Knot Resistant Accession of Cowpea

Accessions	*Source*	*Salient Features*	*Distribution/ Recipient*
EC496737 (var. California Black)	USA	Resistant to root knot nematode (*Meloidgyne* sp.)	Project Coordinatoe (Arid legumes), CAZRI; ICAR-NBPGR, N.Delhi

Weed Resistance

Table 4.43: Striga Resistant Genotypes

Weeds	*Resistance Source*
Striga resistant varieties	IT 88D-867-11, IT 90K-59, IT 90K-76, IT 90K-82-2

Abiotic Stress

Mereena (1989) reported that drought tolerant cowpea should be early flowering with deep and wide spread root system, high root/shoot ratio, low proline content, fewer number of stomata, long grain filling period and medium duration, resulting in the production of more number of pods per plant and high harvest index. Among the sixteen cowpea genotypes/ varieties evaluated UPC124, DPLC-198 and DPLC-216, were drought tolerant. Anilkumar (1993) conducted combing ability studies for yield and drought tolerance in cowpea and recommended DPLC-98, IC38956, C-152

Table 4.44: Sources of Tolerance to Abiotic Stresses in Cowpea

Abiotic Stresses	Resistant Source	Reference
Soil salinity/alkalinity	Genotypes 210856, 211557 and Asebot (from IBC and MARC, Ethiopia)	Gogile *et al.*, 2013
	Sonorense, CB-3	Murillo-Amador *et al.*, 2006
	IT86D-715	Padilla *et al.*, 2009
Drought tolerant	Kanannado, Dan Illa, IT88D-867-11 and IT90K-59-2	Singh and Matsui, 2002
	IT98K-412-13 and IT93K-452-1	Ishiyaku and Aliyu, 2013
	IT93K-503-1	Barrera-Figueroa *et al.*, 2011
	Sh-50, UC-524B, Namuesse, Inhaca-I, INIA-25, Yacine, INIA-36H and IT83D-338-1	Chiulele *et al.*, 2011
	D83970 (CPRD8), D83971 (CPRD14) and D83972 (CPRD22)	Iuchi *et al.*, 1996
	AB030293 (VuNCED1), AB030294(CPRD86) and AB030295 (VuABA1)	Iuchi *et al.*, 2000
	U61379 (VucAPX), AY466858 (VupAPX), AY484492 (VupAPX) and AY484493 (VupAPX),	Lameta *et al.*, 2006
	AF165891 (VuPAP-α) and AF171230 (VuPAP-β)	Marcel *et al.*, 2000
Heat tolerant	Tvu 4552, PI204647 and Prima	Warrag and Hall, 1983
	CB27	Ehlers *et al.*, 2000a
	Mouride, B89-600, IT89KD-252, IT89KD-245, IT89DM-400, IT89KD-107-5, TN88-63, IT89KD-58-57, IT89KD-355, Suvita-2	Ehlers and Hall, 1998
	IC39888, IC39890, IC39856	Kumar, 2005
	NGVU-05-24	Singh *et al.*, 2007
	H36, H8-9 and DLS99	Ismail and Hall, 1999
	EC496737 var. California black (Source USA)	Recipient ICAR-IIPR, Kanpur (U.P); ICAR-NBPGR, new Delhi

and Chharodi-1 and the hybrids DPLC-198 x C-152, DPLC-198 x Chharodi-1, IC38956 x C-152 and IC38956 x Chharodi-1 for further utilization.

Mereena (1989) reported that UPC-124, DPLC-198 and DPLC-216, were drought tolerant and tolerance was associated with earliness, deep and widespread root system, high root/shoot ration, low proline content, fewer number of stomata, long grain filling period, medium duration and high harvest index.

Mutation Breeding

Protruded stigma types and male sterile lines have been developed through mutation breeding.'VITA 25' is an erect breeding line carrying male sterility genes (ms_2ms_2). Adu-Dapaah *et al.* (1999) reported a fascinated mutant and Singh and Adu-Dapaah (1998) reported a partial sterile mutant, both of which originated spontaneously. The fascinated mutant does not have much breeding value but the partial sterile mutant can be used for facilitating hybridization in cowpea. John (1999) reported 50 Kr of gamma rays to be most effective for inducing mutations in cowpea and Odeigah *et al.* (1996) obtained several male sterile mutants using gamma rays, ethyl methane sulphonate (EMS), and sodioumazide. Saber and Hussein (1998) reported induced mutants using gamma rays showing resistance to rust. Gunasekaran *et al.* (1998) treated seeds of the cowpea variety CO-4 with gamma rays and ethidium bromide and analyzed M_1 and M_2 progenies for different agronomic traits. They observed a great deal of variation in M_2 population for different traits and further noticed that gamma rays were more effective in inducing mutation than ethidium bromide. Nair (1964) treated seeds of cowpea variety, 'African' with X-ray and observed no effect on germination and fruit set but number of seeds were reduced. Kumaran (1965) studied X_3 and X_4 generations and selected four pure breeding progeny lines with respect to pod and seed colour. Natrajan (1966) studied X_4 mutants and isolated two promising mutants namely T_7 and T_6. The favourable combination of all the productive characters together with a shorter flowering phase and uniform maturity of pods was observed in T_7. T_6 recorded the maximum weight of haulm due to its prolonged vegetative phase and profuse growth. Oomen (1980) induced variability by gamma rays (5-30 k rad) EMS (0.1-0.7 per cent) in the variety 'New Era'. Among the gamma ray doses employed, 10 k rad was the most efficient on the basis of lethality or sterility. 30 k rad was the most efficient dose on the basis of injury. Ethyl Methane Sulphonate (0.4 per cent) had the highest mutagenic efficiency. Mini (1989) subjected seeds of yardlong bean to four doses of gamma rays (10-40 k rad) and four doses of EMS (0.5 to 2 per cent) and studied their effects in the M_1 and M_2 generations. The mutagenic effectiveness in inducing chlorophyll mutations was high at the lowest dose of both mutagens. On the basis of lethality, 20 k rad of gamma rays was the most efficient, while on the basis of injury and sterility, 10 k rad high at the lowest dose of both mutagens. On the basis of lethality, 20 k rad of gamma rays was the most efficient.

Quality Breeding

There is a need to evolve high essential amino acid (high protein and methionine content) lines in cowpea. Dry seeds of cowpea contain 20-25 per cent protein. Due to its good protein quality and high nutritional value especially rich in vitamins and minerals, cowpea is often referred to as the poor mans meat. Dhar *et al.* (2016) estimated genetic variability, heritability, genetic advance, correlation and path coefficient for 13 yield contributing traits and 5 biochemical nutritional quality traits using 36 diverse genotypes of cowpea. The study indicated existence of considerable amount of genetic variability for all the characters studied.Chugh *et al.* (2016) estimated phenolic content and antioxidant activity Indian six cowpea cultivars and found diffrences in total phenolic content, total flavonoids and antioxidant activity among different cultivars of cowpea under the influence of different extraction solvents. CS-88 showed the highest total phenolic content in the 80 per cent acetone extract and Russian Joint with least in 100 per cent methanol extract.TFC was found maximum in Russian Joint (80 per cent acetone extract) and minimum amount was found for HC-46 (70 per cent ethanol). Overall, TPC extracted by selected solvents were found in the order from high to low: 80 per cent acetone> 70 per cent ethanol >100 per cent methanol. Pusa Komal showing the highest antioxidant activity (80 per cent acetone extract) and selection (70 per cent ethanol). The FRAP value order from high to low: 80 per cent acetone> 70 per cent ethanol >100 per cent methanol was found for all the cultivars except selection (80 per cent acetone> 100 per cent methanol >70 per cent ethanol).

Genetic Resource Conservation

International Institute of Tropical Agriculture (IITA)comes under Consultative Group on International Agriculture Research (CGIAR) established in 1968 by Ford and Rockefeller Foundation at Ibadan, Nigeria. Many valuable genes like male sterility, high protein and methionine content, resistance to thrips, aphids, pod borers, root knot nematode, Anthracnose and cowpea mosaic virus, are available which need further utilization. Bush types of yard long bean have been developed (Los Banos bush Sitao No.1).

The Indian region is the secondary centre of diversity for cowpeas. Main mission of ICAR-NBPGR, New Delhi is germplasm collection of grain legumes, particularly of cowpea. Regular collection, evaluation, characterization and maintenance of cowpea germplasm are being carried out in India by various ICAR Institutes, State Agricultural Universities and other organizations. A total number of 369 and 450 germplasm were evaluated and maintained by NBPGR, New Delhi during the year 2001 (Anonymous, 2001) and 2004 (Anonymous, 2004), respectively. Besides, thirty four *Vigna* spp. were collected from Uttarakhand, H.P., Odisha, Chhattisgarh, Madhya Pradesh, Tamil Nadu and Kerala by ICAR-NBPGR, New Delhi during 2001. Important wild relatives represented from diverse habitats were *V. unguiculata, V. trilobata, V. hainiana, V. vexillata, V. umbellata, V. radiata* var. *sublobata, V. radiata* var. *setulosa* and *V. delizilliana*were collected and conserved (Anonymous, 2001). Similarly, two hundred ten lines have been evaluated for various traits and all are being maintained at ICAR-Indian Institute of Vegetable Research, Varanasi (U.P.). At ICAR-NBPGR, New Delhi, 3,000 accessions are conserved in the National Gene Bank (NGB)

REFERENCES

1. Adu-Dapaah, H.K., Singh, B.B. and Fatokun, C. 1999. A fascinated mutant in cowpea (*Vigna unguiculata* (L.) Walp.). *Acta Agronomica Hungarica*, **47:** 371-376.
2. Ajith, P.M. 2001. Variability and path analysis in bush type vegetable cowpea (*Vigna unguiculata*). M.Sc. Thesis, Kerala Agricultural University, Vellanikkara, Thrissur (Kerala).
3. Amma S.B. 1981. Investigation on inter varietal F_2 hybrids in cowpea. M.Sc. Thesis, Kerala Agricultural University, Vellanikkara, Thrissur (Kerala).
4. Anilkumar, G. 1993. Combining ability for yield and drought tolerance in (*Vigna unguiculata* (L.) Walp.) M.Sc. Thesis, Kerala Agricultural University, Vellanikkara, Thrissur (Kerala).
5. Anithakumari, V. 1992. Host resistance in cowpea (*Vigna unguiculata* (L.) Walp.) to the pod borer (*Maruca testularis*). M.Sc. Thesis, Kerala Agricultural University, Vellanikkara, Thrissur (Kerala).
6. Barrera-Figueroa, B.E., Gao, L., Diop, N.N., Wu, Z., Ehlers, J.D., Roberts, P.A., Close T.J., Zhu, J.K. and Liu, R. 2011. Identification and comparative analysis of drought-associated micro RNAs in two cowpea genotypes. *BMC Plant Biology*, **11:** 127, DOI: 10.1186/1471-2229-11-127.
7. Booker, H.M. and Umaharan, P. 2007. Identification of resistance to *Cercospora* leaf spot of cowpea. *European J. Plant Pathology*, **118(4):** 401-410.
8. Chattopadhyay, A. 1995.Yield components, nodulation, adaptability and gene action in vegetable cowpea. Ph.D. thesis submitted to BCKV, Mohanpur, Nadia (West Bengal).
9. Chattopadhyay, A., Chakraborty, K., Dasgupta, T., Som, M.G. and Hazra, P. 1996. Evaluation of genotypes for agronomic and morphological characters in vegetable cowpea. *Indian J. Hort.*, **53(4):** 304-308.
10. Chattopadhyay, A., Dasgupta, T., Hazra, P. and Som, M.G. 1996. Genetic analysis of pod yield and its component employing a diallel cross in cowpea. *Veg. Sci.*, **28(1):** 69-74.

11. Chattopadhyay, A., Dutta, S. and Hazra, P. 2016. Current status of leguminous vegetables research in West Bengal. Natl. Sym. on "Vegetable Legumes for Soil and Human Health.Souvenir and Abstracts, p. 26-37.

12. Chattopadhyay, A., Pandiarana, N., Tania, S., Sibsankar, D. Chatterjee, S and Dutta, S. 2014. Identification of selection indices and choosing of parents for vegetable cowpea (*Vigna unguiculata* cv. gr. *sesquipedalis*) breeding programme. *Legume Res.*, **37(1):** 19-25.

13. Chiulele, R.M., Mwangi, G., Tongoona, P., Ehlers, J.D. and Ndeve, A.D. 2011. Assessment of cowpea genotypes for variability to drought tolerance. *African Crop Science Conference Proceedings*, **10:** 531-537.

14. Chugh, C., Mishra, S. and Sarika, K. 2016. Assesment of polyphenols and antioxidant activity in cowpea (*Vigna unguiculata* L. Walp.) genotypes. (Abs: 80), 7th Indian Horticulture Congress, held at ICAR-IARI, New Delhi, 15-18 November, p. 62-63.

15. Danam, S. B. and Chauhari, F.P. 2000. Heterosis studies in cowpea. *Annals of Agricultural Research*, **21:** 248-252.

16. David, K.S. 1957. Notes on South Indian aphids III. Lachininse to aphidinae. *Indian Journal of Entomology*, **19:** 171-180.

17. Davis, D.W., Oelke, E.A., Oplinger, E.S., Doll, J.D., Hanso, C.V. and Putnam, D.H. 2000. Alternative field crops manual. http//www.Hort.purdue. newcrop/afem/cowpea.html.

18. Dhanorkar, B.K. and Daware, D.G. 1980. Differences in number of aphids found in lines of cowpea in replicated trial. *Tropical Grain Legume Bulletin*, **19:** 3-4.

19. Dhar, S., Singh, S.K. and Choudhary, H. 2016. Evaluation and characterization of cowpea genotypes for fresh pod nutritional quality and yield traits. 1st International Agrobiodiversity Congress; (Session: Trait Discovery and Enhanced Use of PGR), held at New Delhi; 6-9 November 2016; p. 193.

20. Dharam, S. and Patel, P.N. 1977. Studies on resistance in crops to bacterial diseases in India. VII. Inheritance of resistance to bacterial, blight disease in cowpea. *Indian Phytopathology*, **30(1):** 99-102.

21. Ehlers, J.D. and Hall, A.E. 1997. Cowpea (*Vigna unguiculata* (L.) Walp.). *Field Crops Res.*, **53:** 187-204.

22. Ehlers, J.D. and Hall, A.E. 1998. Heat tolerance of contrasting cowpea lines in short and long days. *Field Crops Res.*, **55(1):** 11-21.

23. Ehlers, J.D., Hall, A.E., Patel, P.N., Roberts, P.A. and Matthews, W.C. 2000b. Registration of 'California blackeye 27' cowpea. *Crop Sci.*, **40:** 854-855.

24. Ehlers, J.D., Matthews, W.C., Hall, A.E., and Roberts, P.A. 2000b. Inheritance of a broad based form of nematode resistance in cowpea. *Crop Sci.*, **40:** 611-618.

25. Fatokun, C.A. and Singh, B.B. 1987. Interspecific hybridization between *Vigna pubscens* and *V. unguiculata* (L.) Walp. through embryo rescue. *Plant Cell, Tissue and Organ Culture*, **9:** 229-233.

26. Gogile, A., Andargie, M and Muthuswamy, M. 2013. Screening selected genotypes of cowpea (*Vigna unguiculata* (L.) Walp.) for salt tolerance during seedling growth stage. *Pak. J. Biol. Sci.*, **16(14):** 671-679.

27. Gomathinayagam, P, Ram, S.G., Rathnaswamy, R. and Ramaswamy, N.M. 1998. Interspecific hybridization between ***V.*** *unguiculata* (L.) Walp. and *V. vexillata* (L.). A rich through *in vitro* embryo culture. *Euphytica*, **102(2):** 203-209.

28. Gunasekaran, Selvaraj, M.U and Raveemdram, T.S. 1998.Induced polygenic mutations in cowpea (*Vigna unguiculata* (L.) Walp.). *South Indian Horticulture*, **46(1-2):** 13-17.

29. Hazra, P., Chattopadhyay, A., Dasgupta, T. and Som, M.G. 2001. 'Bidhan Barbati-1' and 'Bidhan Barbati-2': New high yielding and non-viny vegetable cowpea. *Indian Hort.*, **46(1):** 13-15.

30. Hazra, P., Das, P.K. and Som, M.G. 1994a. Inheritance of pod weight in cowpea. *Indian J. Genet.*, **54:** 175-178.

31. Hazra, P., Som, M.G. and Das, P.K. 1993. Studies on heterosis and combining ability for forage and seed yield characters in Ph. D. Thesis, Rajasthan Agriculture University, Udaipur, Rajasthan.

32. Hazra, P., Som, M.G. and Das, P.K. 1994b. Inheritance of pod length in cowpes. *Indian J. Hort.*, **50:** 358-363.

33. Hazra, P., Som, M.G. and Das, P.K. 1996. Combining ability for pod yield and seed protein in cowpea. *Indian J. Genet.*, **56:** 516-518.

34. IITA, Ibadan. 1984. Research highlights, 1983. International Institute of Tropical Agriculture, Ibadan, Nigeria, p. 79-80.

35. Ishiyaku, M.F. and Aliyu, H. 2013. Field evaluation of cowpea genotypes for drought tolerance and Striga resistance in the dry Savanna of the north-west Nigeria. *Inter. J. Pl. Breed. and Genet.*, **7:** 47-56.

36. Ismail, A.M. and Hall, A.E. 1999. Reproductive-stage heat tolerance, leaf membrane thermostability and plant morphology on cowpea. *Crop Sci.*, **39(6):** 1762-1768.

37. Iuchi, S., Kobayashi, M., Shinozaki, Y.K. and Shinozaki, K. 2000. A stress-inducible gene for 9-cis-epoxicartenoid dioxygenase involved in abscisic acid biosynthesis under water stress in drought tolerant cowpea. *Plant Physiology*, **123:** 553-562.

38. Iuchi, S., Yamaguchi-Shinozaki, K., Urao, T., Tereo, T. and Shinozaki, K. 1996. Novel drought inducible genes in the highly drought-tolerant cowpea: cloning of cDNA and analysis of their gene expression. *Plant Cell Physiology*, **37:** 1073-1082.

39. Jalajakumari, M.B. 1981. Variability studies in cowpea (*Vigna unguiculata* (L.)Walp.). M.Sc. Thesis, Kerala Agricultural University, Vellanikkara, Thrissur (Kerala).

40. Jayappa, B.G. and Lingappa, S. 1988. Screening of cowpea germplasm for resistance to Aphis craccivora Koch in India. *Tropical Pest Management*, **34(1):** 62-64.

41. Jithesh, V.G. 2009. Genetic analysis of resistance to pod borers and yield in yard long bean. Ph.D. Thesis, Kerala Agricultural University, Vellanikkara, Thrissur (Kerala).

42. John, S.A. 1999. Mutation frequency and chlorophyll mutations in parents and hybrid of cowpea following gamma irradiation. *Indian Journal of Genetics and Plant Breeding*, **59(3):** 357-361.

43. Joseph, S. 1990. Genetics of resistance to aphids and utility of line mixtures in cowpea. Ph.D. Thesis, Kerala Agricultural University, Vellanikkara, Thrissur (Kerala).

44. Jyothi, C. 2001. Genetics of bruchid (*Calloso bruchus* sp.) resistance and yield in cowpea. M.Sc. Thesis, Kerala Agricultural University, Vellanikkara, Thrissur (Kerala).

45. Kajale, D.B., Ravindrababu, Y., Prajapati, D.B. and Khule, A.A. 2013. Heterosis in cowpea (*Vigna unguiculata* (L.)Walp.). *Veg. Sci.*, **40(2):** 237-239.

46. Kamalam, N. 1975. Diallel analysis in cowpea (*Vigna unguiculata* (L.)Walp.). M.Sc. Thesis, Kerala Agricultural University, Vellanikkara, Thrissur (Kerala).

47. Kandoria, J.L., Janwal, R. and Singh, G. 1984. Seasonal activity and host range of *Aphis gossypi* Glover in the Pumjab. *Journal of Insect Science,* **2(1):** 68-70.

48. Kumar, D. 2005. Simple screening method for drought tolerance in cowpea. Indian Journal of *Pulses Research*, **18(2):** 187-191.

49. Kumar, P., Prakash, R. and Haque, M.F. 1976. Floral biology of cowpea.*Tropical Grain Legume Bulletin,* **6:** 9-11.

50. Kumar, R., Singh, S.P. and Joshi, A.K. 1998. Genetics of green pod yield in cowpea (*Vigna unguiculata* (L.) Walp.). *Veg. Sci.*, **25(1):** 36-39.

51. Kumar, R., Singh, S.P. and Joshi, A.K. 2000. Genetics of earliness in in cowpea (*Vigna unguiculata* (L.) Walp.). *Veg. Sci.*, **27(1):** 28-30.

52. Kumaran, K. 1965. Studies on the morphological and cytological behavior of F_3 and F_4 generations in cowpea (*Vigna sinensis* L.). M.Sc. Thesis, Kerala Agricultural University, Vellanikkara, Thrissur (Kerala).

53. Lal, H., Singh, A.P., Verma, A., Rai, M., Singh, S.N., Nath, V. and Ram, D. 2007. Heterosis and inbreeding depression in cowpea (*Vigna unguiculata* (L.) Walp.). *Veg. Sci.*, 34(2): 145-149.

54. Lameta, D.A., Iliou, R.F., Ansel, D.C., Phamthi, A.T. and Fodil, Z.Y. 2006. Isolation and characterization of four ascorbate peroxidase cDNA responsive to water deficit in cowpea leaves. *Ann. Bot.*, **97:** 133-140.

55. Lima, J.A.A., Santos, C.D.G. and Silveria, L.E.S. 1986. Comportamento de genotipos de caupi em relacao aos dois principais virus que ocorrem no Ceara. *Fitopatologia Brasileira*, **11:** 151-161.

56. Marcel, G.C.F., Matos, A., Lameta, D.A., Kader, J.C., Fodil, Z.Y. and Thi, A.P. 2000. Two novel plant cDNAs homologous to animal type-2 phosphatidate phosphatase are expressed in cowpea leaves and are differently regulated by water deficits. *Biochem Soc. Trans.*, **28:** 915-917.

57. Marechal, R., Mascherpa, J.M. and Stainer, F. 1978. Etude taxonomique d'un groupe complexe d'speces des genres *Phaseolus* et *Vigna* (Papilionaceae) sur la base de donnees morphologiques et polliniques, traitees par l' analyse informatique. *Biossiera*, **28:** 1-273.

58. Mereena, M. 1989. Potential for drought tolerance in cowpea. M.Sc. Thesis, Kerala Agricultural University, Vellanikkara, Thrissur (Kerala).

59. Mini, V. 1989. Induced mutations in cowpea cultivar "Kurutholapayar". M.Sc. Thesis, Kerala Agricultural University, Vellanikkara, Thrissur (Kerala).

60. Mishra, S.N., Rastogi, R. and Verma, J.S. 1985. Note on parental selectivity- a factor affecting success in cowpea crossing. *Trop. Legume Bull.*, **31:** 24-26.

61. Mishra, S.N., Verma, J.S. and Jayasekhra, J.B.A. 1985b. Breeding cowpeas to suit Asian cropping systems and consumer tastes. In: Singh, S.R. and Rachie (eds.), Cowpea: Research, Production and Utilization, John Wiley and Sons, p. 117-123.

62. Murillo-Amador, B., Troyo-Die guez, E., Garcya-hernandez, J.L., Lopez-Aguilar, R., Avila-Serrano, N.Y., Zamora-Salgado, S., Rueda-Puenta, E.O., and Kaya, C. 2006. Effect of NaCl salinity in the genotypic variation of cowpea (*Vigna unguiculata* (L.) during early vegetative growth. *Scientia Horticulturae*, **108:** 423-431.

63. Nair, R.P.N. 1964. Investigation on the effect of X-rays on cowpea (*Vigna sinensis* L.). M.Sc. Thesis,

Kerala Agricultural University, Vellanikkara, Thrissur (Kerala).

64. Nasiya, B. 2015. Morphological and biochemical bases of resistance to spotted pod borer [Maruca vitrata (Fab.)] (Lepidoptera: Crambidae) in cowpea. M.Sc. Thesis, Kerala Agricultural University, Vellanikkara, Thrissur (Kerala).

65. Natrajan, E.N. 1966. A comparative study on X4 mutants of cowpea (*Vigna sinensis* L. Savi) with reference to quantitative characters. M.Sc. Thesis, Kerala Agricultural University, Vellanikkara, Thrissur (Kerala).

66. Neema, V.P. 1986. Genomic relationship in *Vigna* sp. cowpea (*Vigna sinensis* L.). M.Sc. Thesis, Kerala Agricultural University, Vellanikkara, Thrissur (Kerala).

67. Ng, N.O. and Marechal, R. 1985.Cowpea taxonomy, origin and germplasm. In: Singh, S.R. and Rachie (eds.), Cowpea: Research, Production and Utilization, John Wiley and Sons. pp. 11-21.

68. Odeigah, P.G.C., Osanyin Peju, A.O. and Myers, G.O. 1996. Induced male sterility in cowpea. *Journal of Genetics and Plant Breeding*, **50:** 171-175.

69. Ombakho, G.A., Tyagi, A.P. and Pathak, R.S. 1989. Inheritance of resistance to the cowpea aphid in cowpea. *Theor. Appl. Genet.*, **74(6):** 817-819.

70. Padila, *e.g.*, Sanchez, R.C.L., Eichler-Loebermann, B., Fernandez-pascual, M. Barrero, K.A. and Martinez, L.A. 2009. Salt stress effects on cowpea (*Vigna unguiculata* (L.)Walp.) varieties at different growing stages. Conference on international research on Food Security, Natural resource Management and rural Development, University of Humburg, October 6-8, 2009.

71. Pal, A.K., Ram, D., Maurya, A.N. and Rajput, C.B.S. 2002. Heterosis over mid and better parent under Line x Tester design in cowpea. *Indian J. Hort.*, **59:** 279-287.

72. Pal, S.K., Rao, G.G.S.N. and Kushawaha, K.S. 1978. A note on the pest fluctuation in relation to weather elements in eight varieties of moong (*Vigna radiata*) under arid conditions. *Indian Journal of Entomology*, **39:** 375-376.

73. Pandey, K.C., Hasan, N., Bhaskar, R.B., Ahmed, S.T. and Kohli, K.S. 1995. Genetic evaluation of cowpea (*Vigna unguiculata* (L.)Walp.) lines for multiple pest resistance. *Indian Journal of Genetics and Plant Breeding*, **55(2):** 198-203.

74. Panicker, P.R. 2000. Evaluation of vegetable cowpea (*Vigna unguiculata* spp. *sesquipedalis* (L.) Verdcourt) for legume pod borer (*Maruca vitrata*) resistance and yield. M.Sc. Thesis, Kerala Agricultural University, Vellanikkara, Thrissur (Kerala).

75. Pathak, R.S. 1988. Genetics of resistance to aphid in cowpea. *Crop Sci.*, **28:** 474-476.

76. Patil, A.R., Bendale, V.W., Bhave, S.G. and Mehta, J.L. 2005. Correlation and path analysis studies of yield contributing characters in cowpea (*Vigna unguiculata* (L.) Walp.). *The Orissa Journal of Horticultuture*, **33(1):** 35-38.

77. Patil, R.B. and Shete, M.M. 1987. Heterosis in cross of seven genotypes of cowpea. *Journal of Maharashtra Agricultural University*, **12:** 51-54.

78. Peter, K.V. 1998. Leguminosae-French bean. In: Genetics and Breeeding of Vegetable Crops. ICAR, New Delhi, p.149-176.

79. Purseglove, J.W. 1969. *Tropical Crops, Dicotyledons*, Vol. 1, 2, pp. 321-328. The English Language Book Society and Longman, London.

80. Rejatha, V. (1992. Combining ability in vegetable cowpea. M.Sc. Thesis, Kerala Agricultural University, Vellanikkara, Thrissur (Kerala).

81. Reji, G.V. 2002. Management of the American serpentine leaf miner, *Liriomyza trifolii* (Burgess) Dietars on cowpea. M.Sc. Thesis, Kerala Agricultural University, Vellanikkara, Thrissur (Kerala).

82. Resmi, P.S. 1981. Genetic variability in yard long bean. M.Sc. Thesis, Kerala Agricultural University, Vellanikkara, Thrissur (Kerala).

83. Rodriguez, I., Rodriguez, M.G., Sanchez, L. and Iglesias, A. 1996. Expression of resistance to *Meloidogyne incognita* in cowpea cultivars (*Vigna unguiculata*). *Revista de Protection Vegeta.*, **111(1):** 63-65.

84. Rodriguez, V.J.L.B., Menezes, M., Coelho, R.S.B. and Miranda, P. 1997. Identification of resistance sorces on genotypes of cowpea (*Vigna unguiculata* (L.)Walp.) a *Macrophomina phaseolina* (Tass.) Goid., em condicoes de casa-de-vegetacao. *Summa Phytopathologica*, **23(2):** 170-172.

85. Rubatzky, V.E. and Yamaguchi, M. 1997. World Vegetables: Principles, Production, and Nutritive values. Chapman and Hall, New York, p.572.

86. Sangwan, R.S. and Lodhi, G.P. 1995. Heterosis for grain characters in cowpea (*Vigna unguiculata* (L.) Walp.), *Legume Research*, **18:** 75-80.

87. Santos, A.A., Dos, F.R. and Freire Filho, E. 1986. Genotipos de caupi (*Vigna unguiculata* [L.] Walp.) Com Resistencia de campo ao virus do mosaic dourado do caupi. In: Seminario de pesquisa agropecuaria do piaui, 4. Teresina, Anais. Teresina: EMBRAPA-Uepae de Teresina, p. 191-203.

88. Sarutayophat, T., Nualsri, C., Santipracha, Q. and Saereprasert. 2007. Characterization and genetic relatedness among 37 yardlong bean and cowpea accessions based on morphological characters and

RAPD analysis. *Songklanakarin J. Sci. Technol.*, **29:** 591-600.

89. Sen, N.K. and Bhowal, J.G. 1960. Cytotaxonomy studies on *Vigna*. *Cytologia*, **25:** 195-207.

90. Sen, N.K. and Bhowal, J.G. 1961. Genetics of *Vigna sinensis* (L.) Savi. *Genetics*, **32(3):** 247-266.

91. Shah, G.L. and Gopal, B.V. 1969. Stomatal ontogeny of the vegetative and floral organs of some Papilionaceae. *Aust. J. Bot.*, **17:** 81.

92. Sharma, D., Mehta, N., Singh, J. and Gupta, C.R. 2013. Genetic study for some pod characters in vegetable cowpea (*Vigna unguiculata* (L.) Walp.). *Veg. Sci.*, **40 (1):** 73-76.

93. Shobha, P.P. 1964. Variability and heterosis in bush type vegetable cowpea. M.Sc. Thesis, Kerala Agricultural University, Vellanikkara, Thrissur (Kerala).

94. Singh, B.B. 199**8**. Sources of resistance to septoria, scab, bacterial blight and Cercospora leaf spot. IITA Annual Report 1998, Project 11, p 24-27.

95. Singh, B.B. 1999a. Evaluation of new improved breeding lines without insecticides sprays. **IITA** Annual Report 1999, Project 11, p 26.

96. Singh, B.B. 1999b. Improved breeding lines with resistance to bruchid. IITA Annual Report 1999, Project 11, p 29-30.

97. Singh, B.B. and Adu-Dapaah, H.K. 1998. A partial male sterile mutant in cowpea. *Afr. Crop Sci. J.*, 6: 97-101.

98. Singh, B.B. and Hughes, J.D. 1998. Sources of multiple virus resistance. IITA Annual Report 1998, Project 11, p 24-27.

99. Singh, B.B. and Hughes, J.D. 1998. Sources of multiple virus resistance. IITA Annual Report 1999, Project 11, p. 30.

100. Singh, B.B. and Matsui, T. 2002. Cowpea varieties for drought tolerance. In; Challenges and opportunities for enhancing sustainable cowpea production, (Fatokun, C.A., Tarawali, S.A., Singh, B.B., Kormawa, P.M. and Tamo, M. (eds.), p. 287-300.

101. Singh, B.B., Asante, S.K., Florini, D., Jackai, L.E.N., Fatokun, C and Wydra, K. 1997. Breeding for multiple disease and insect resistance. IITA Annual Report 1997, Project 11, p. 22.

102. Singh, B.B., Asante, S.K., Jackai, L.E.N. and Hughes, J.D. 1996. Screening for resistance to parasitic plants, virus, aphid and bruchid. IITA Annual Report 1996, Project 11, p. 24.

103. Singh, B.B., Asante, S.K., Jackai, L.E.N., and Hughes, J.D. 1996. Screening for resistance to parasitic plants, virus, aphid and bruchid. IITA Annual Report, 1996 Project 11, p. 24.

104. Singh, B.B., Ehlers, J.D., Sharma, B., and Freire Filho, F.R. 2000. Recent progress in cowpea breeding. International Institute of Tropical Agriculture, Kano Station, PMB3112, Kano, Nigeria.

105. Singh, B.B., Olufajo, O.O., Ishiyaku, M.F., Adelleke, R.A., Ajeigbe, H.A. and Mohammed, S.G. 2007. Registration of 'NGVU-05-24' cowpea. *Journal of Plant Registration*, **1(1):** 48-49.

106. Singh, C.B., Mehra, K.L. and Kohli, K.S. 1971. Evaluation of a world collection of *V. sinensis* (L.) savi for pod characters. *SABRAO Newsletter*, **3:**11-16.

107. Singh, J., Kalloo, G. and Singh, K.P. 2001. Vegetable Crops: Nutritional Security. Indian Institute of Vegetable Research, Varanasi (U.P.), p. 56.

108. Singh, S.R. 1977. Cowpea cultivars resistant to insect pests in world germplasm collection. *Tropical Grain Legume Bulletin*, 9: 3-7.

109. Singh, S.R., Singh, B.B., Jackai, L.E.N., Natare, B.R. 1983. Cowpea research at International Institute of Tropical Agriculture, Ibadan, Nigeria. *IITA Information Series*, **14:** 1-20.

110. Smartt, J. 1990. Grain legumes: evolution and genetic resources. Cambridge: Cambridge University Press.

111. Steele, W.M. 1976. Cowpea. In: *Evolution of Crop Plants*, edn.I, pp. 183-185. Simmonds, N.W. (Ed.), Longman, London.

112. Steele, W.M. and Mehra, K.L. 1980. Structure, evolution and adaptation to farming systems and environments in *Vigna*. In: Advances in legume sciences. Edited by R.J. Summerfield and A.H. Bunting. Royal botanic Gardens, Kew, UK, p. 393-404.

113. Sudhakumari. 1993. Screening of cowpea (*V. unguiculat* (L.) Walp.) genotypes for resistance to mosaic disease. M.Sc. Thesis, Kerala Agricultural University, Vellanikkara, Thrissur (Kerala).

114. Sulochana, K.A. 1984. Screening for resistance to Aphis craccivora in vegetable cowpea (*Vigna unguiculata* var. *sesquipedalis* and var. *cylindrica*) and evaluation of early types. M.Sc. Thesis, Kerala Agricultural University, Vellanikara (Kerala).

115. Sulochana, K.A. and Peter, K.V. 1986. Evaluation for host reaction to *Aphis craccivora* Koch in cowpea. *Agricultural Research Journal, Kerala*, **24(1):** 63-65.

116. Sulochana, K.A. and Peter, K.V. 1987.Phenotypic stability in cowpea. *Indian Jouranal of Agricultural Sciences*, **57(6):** 425-426.

117. Tumwegamire, S., Rubaihayo, P.R., and Adipala, E.1998. Genetics of resistance to Sphaceloma scab of cowpea. *African Crop Science Journal*, **6(3):** 227-240.

118. Tyagi, D.K. and Chawala, H.S. 1999. Effect of season and hormones on crossability barriers and *in vitro* hybrid development between *Vigna radiata* and

Vigna unguiculata. Acta Agronomica Hungarica, **47:** 147-154.

119. Vavilov, N.I. 1939. The new systematic of cultivated plants, In: Huxley, J.Ed; The New Systematic, Oxford University Press, London, 549-566.
120. Veeranna, R. and Hussain, M.A. 1997. Trichomesas barriers for cowpea pod borer *Maruca testulalis Geyer. (Lepidoptera: Pyralidae) Insect Enviroment, 3:15.*
121. Veeranna, R. 1998. Phenol and tannin reduce the damage of copea pod borer *Manica testulalis, Insect. Environment*, 4:5-6.
122. Verdcourt, B. 1970. Studies in the Leguminosae/ Papilionaceae for the flora of tropical East Africa. *Kew Bulletin*, **24:** 507-569.
123. Vidya, C. 2000. Legume pod borer resistance and genetic divergence in yard long bean. M.Sc. Kerala Agricultural University, Vellanikara (Kerala).
124. Warrag, M.O.A. and Hall, A.E. 1983. Reproductive responses of cowpea to heat stress: genotypic differences in tolerance to heat at flowering. *Crop Sci.*, **23:** 1088.
125. Wydra, K. and Singh, B.B. 1998. Breeding for resistance to multiple strains of cowpea bacterial blight. IITA Annual Report, 1998. Project 11, p. 25-27.
126. Yunes, A.A., Sales, de M.T., Andrade, M.P., Morais, R.A., Fernandes, K.V.S., Gopmes, V.M., Xavier-Filho, J. and Andrade de, M.T. 1998. Legume seed cicilins (7s-storage proteins) interface with the development of the cowpea weevil (*Callosobruchus mauculates* FJ). *Journal of the Science of Food and Agriculture*, **76(1):** 111-116.
127. Zaveri, P.P., Patel, R.K., Yadavendra, J.P. and Shah, R.M. 1983. Heterosis and combining ability in cowpea. *Indian J. Agril. Sci.*, **53:**703.
128. Zohri, A.A. 1993. Studies on some cowpea cultivars. II. Suitability for *Aspergillus flavus growth* and aflatoxin production. *Qatar University Science Journal*, **13(1):** 57-62.

Yardlong Bean (*Vigna sesquipedalis* L. Fruw.)

Yardlong bean is known by several names, such as 'Yardlong Bean' (synonymously 'boro', 'long-podded cowpea', 'pea-bean', 'bodi', 'sitao', 'snake bean', or 'Chinese long bean') 'string bean', 'asparagus bean', 'juro-kusasaagemae' (Japanese), 'dowgauk' (Chinese) or 'sitaw' (Filipino) or 'kousenband' ('garter-slip') in Dutch actually belongs to *Vigna unguiculata* subspecies *unguiculata* cultigroup *sesquipedalis*. The term *sesquipedalis* meaning 'one-and-a half-foot, long, is used as vegetable though young leaves and green seeds are also used as vegetable. It is characterized by its striking long pods (30-90 cm) that are harvested at an immature 'snap' stage and its climbing growth habit that is rare in ssp. *unguiculata*. Yardlong bean is cultivated mainly for crisp and tender pods that are consumed both fresh and cooked. The crisp and tender pods are cut into small pieces and used for making many south Indian dishes like *thoran*, *mezhukkupuratti* and *aviyal*. In Odisha, they are called as *jhudunga*, used to make a variety of dishes. Besides immature pods, seeds, tender leaves and shoots are eaten. It is highly nutritive vegetable containing high percentage of digestible protein (2.8 per cent) vitamin A (865 IU) and C (18.8 mg), iron (0.47 mg), calcium (50 mg), phosphorus (59 mg) and dietary fibre (0.3 g) making it an excellent vegetable (USDA Nutrient Database, 2014). Besides, it also contain appreciable amount of zinc (0.37 mg) and selenium (1.5 µg). Fresh yardlong beans are one of the finest sources of folates and 100 g beans provide 62 mg or 15 per cent of daily requirement of folates (Hossain *et al.*, 2013).

Origin, History and Distribution

N.I. Vavilov identified the mountains of central and western China and adjacent areas as the place of origin of the asparagus bean, but now it is widely believed that the wild ancestors of domestic yardlong species came from central Africa. The original African species spread eastward to India and China northward to the Mediterranean and Europe, and westward to the Caribbean and North America. The yardlong bean is thought to have been developed 1000 of years ago in China's Yunnan Province from a strain of cowpea, which in turn had its origins in Africa. Diverse landrace forms of asparagus bean are found throughout Asia (Xu, *et al.*, 2010), and yet this species is never observed in Africa, suggesting that this form of cowpea arose in Asia from an unknown progenitor form of cowpea long ago (Xu, *et al.*, 2011a). The asparagus bean was first described in 1763 by Linnaeus, who identified it as a New World plant. This notion is no longer accepted though asparagus bean is mostly cultivated in southwest Asia where the pods can be grow to 3 feet in length, it is associated with Mediterranean cooking. In Asia a vegetable form of cowpea, yardlong bean, subsequently evolved from cowpea (Kongjaimun *et al.*, 2012). Ng (1995) postulated that cowpea cultigroup *unguiculata* was, in the first place domesticated in West Africa through this process of selection 2000 BC. Later, the selection for types with very long peduncles for fiber resulted in the cultigroup *textilis* (Ng and Marechal, 1985). The crop was brought to Europe probably through northeastern Africa around 300 BC and to India about 200BC (Singh *et al.*, 1997).

It is claimed to be one of the most important vegetable crop commercially grown in China, Indonesia, Philippines, Taiwan and Thailand (Rachie, 1985). Area under yardlong bean in China alone exceeds 250,000 ha annually (Rubatzky and Yamaguchi, 1997), while in Thailand it is grown on 18,560-20,160 ha annually (Sarutayophat *et al.*, 2007). The Indian region is the secondary centre of diversity for this crop.Yardlong bean (*V. sesquipedalis*) introduced from India are believed to have been selected and developed from vegetable types of *Vigna unguiculata* for their long tender pods in South-East Asia and the cultigroup *biflora* for its grain (Steel and Mehra, 1980). Although Asia is one of the major asparagus growing areas and rich in genetic diversity but researches are still very little (Tan *et al.*, 2012). It is one of the most popular and cosmopolitan vegetable crop grown in many parts of India including Odisha, Andhra Pradesh, Kerala, West Bengal, Uttar Pradesh, Bihar and North-East States. Some pockets of Uttar Pradesh, Bihar and West Bengal are known to cultivate yardlong beans on small scale for local consumption but this is not been commercially exploited. However, in Kerala, yardlong beans may be cultivated extensively throughout the year.

Species Availability, Evolution and Diversity

It is stated that *V. sinensis* and *V. catjang* differentiated from a common ancestoral stock and *sesquipedalis* is a later selection from *sinensis* during the evolution of the crop. Botanically classified as *V. unguiculata* subsp. *sesquipedalis* (L. Verdc.) are believed to have been selected and developed for their long tender pods in South-East Asia, from vegetable types of *V. unguiculata* introduced from India (Steele and Mehra, 1980). Genetic similarity between the two sub species (*unguiculata* and *sesquipedalis*) is high as evidenced by the fact that hybrids are fully fertile and molecular marker system developed in ssp. *unguiculata* are effective and readily transferred to ssp. *sesquipedalis* (Xu *et al.*, 2010). Pods are shorter in cowpeas while it is very slender long (60-90 cm) with beautifully delicate flavor in yardlong beans. Dwarf yardlong bean, a new horticultural crop, was bred from a cross between yardlong bean (*Vigna unguiculata* ssp. *sesquipedalis*) and cowpea (*V. unguiculata* ssp. *unguiculata*), and is currently gaining popularity due to several desirable traits including disease resistance, drought tolerance and long reproductive growth (Lausuwan *et al.*, 1996). The scientific identity of asparagus bean either as a species or as a sub-species is still a controversy. Cultivation of the crop from ancient times in Kerala has resulted in a rich and diverse domestic germplasm (Vidya *et al.*, 2002).

Improved cultivars of asparagus bean in China generally had a narrow genetic basis compared with landraces. This finding suggested that breeding programs of asparagus bean need to utilize landrace germplasm to enhance genetic variability, ensure long-term gains from selection, and reduce genetic vulnerability to pathogen or pest epidemics. Tantasawat *et al.* (2010) estimated genetic diversity and relatedness of 23 yardlong bean (*V. unguiculata* ssp. ssp. *sesquipedalis*) accessions and 7 accessions of a hybrid between cowpea (*V. unguiculata* ssp. *unguiculata*) and dwarf yardlong bean in Thailand by morphological characters, SSR and ISSR markers. Five morphological characters were diverse among most accessions. However, five groups of 2-3 accessions could not be distinguished from one another based on these morphological characters alone. The comparision of average marker index of the multilocus marker and mantel test indicated higher efficiency of IISR for estimating the levels of genetic diversity and relationships among yardlong bean and dwarf yardlong beans in the study. Xu *et al.* (2011) reported the first genetic map of asparagus bean based on SNP and SSR markers. The current map consists of 375 loci mapped onto 11 linkage groups, with 191 loci detected by SNP markers and 184 loci by SSR markers. The overall map length is 745 cm, with an average marker distance of 1.98 cm. Considerable diversity has been assembled from Kerala with respect to high genotypic and phenotypic coefficient of variation for pod yield/plant, clusters/plant, pods/plant, pods/cluster and pod weight, and therefore scope for improvement throught selection (Vidya and Oomen, 2002; Vidya *et al.*, 2002; Resmi and Gopalakrishnan, 2004; Lovely and Radhadevi, 2006; Kumar and Devi, 2009).

Gene and Genetics of Traits

Genetic analysis and mapping of the genes governing flower and seed coat colour was studied in asparagus bean which suggest monogenic inheritance of both traits, purple flower and brown seed coat are dominant over white flower and cream seed coat, respectively and the genes governing both these traits are tightly linked on LG8, 0.4cM apart (Xu *et al.*, 2001a). Aphid resistance in cowpea is inherited as a monogenic dominant trait (Bata *et al.*, 1987; Pathak, 1988). A high-density genetic map for asparagus bean employing combination of the SSR markers and EST-derived SNP markers has been developed (Xu *et al.*, 2011b). Three hundred seventy one markers targeting 375 discrete loci were mapped onto 11 linkage groups (LGs) corresponding to the chromosome pair number of asparagus bean. For consistency, the nomenclature of LG1 to LG11 was assigned according to SNP markers shared with the cowpea consensus map (Muchero *et al.*, 2009). For identification of QTLs for pod length, a genetic linkage map of yardlong bean using SSR markers have also been developed (Kongjaimun, *et al.*, 2012).

Botanical Description

It is botanically more closely related to cowpea, but it is much more a traling and climbing plant than the cowpea, often reaching 9-12 feet in height (USDA-ARS, 2012). Yardlong bean is a short day plant with many accessions that are photo insensitive (Ehlers and Hall, 1997).

Table 4.45: Botanical Description

Growth Habit	Yardlong bean is vigorous climbing annual vine attaining a height of 2-4 m.
Root	Cowpea has a well developed tap root with a considerable number of lateral roots.
Stem	The stem is almost glabrous, twisting, sometimes hollow and slightly ridged/ribbed. Stems may be green or pigmented (purple).
Leaf	Leaves are trifoliate, leaflets are ovate, 12-15 cm long and 4-7 cm wide, and acute.
Floral biology	Flowers are white, yellow or light purple mostly produced on axillary peduncle.
Anthesis and pollination	Cowpea flowers bloom early in the morning (between 6.30 hr. to 7.00 hour) and close before noon and falls in the same day. The process of opening of corolla takes 45-60 minutes. Anthers mature and start dehiscing about 7-8 hr before the time of opening of corolla. The extra floral nectarines at the base of the corolla attract ants, flies and bees but a heavy insect is required to depress the wing petals and expose the stamen and stigma. Cowpea is a self-pollinated crop (Davis *et al.*, 2000) but a few lines of cowpea are cross-pollinated due to extrovert nature of pistil. Optimum time for emasculation is at 17.30 hour – 18.30 hour and hand pollination at 6.00 hour – 7.00 hour the next day. Stigma becomes receptive from 12 hr before blooming to 6 hr after blooming. Pollen remains viable for 42 hr at room temperature of 28.5 °C and relative humidity 91 per cent. Stigma becomes receptive from 12 hrour before blooming to 6 hour after blooming.For hybridization of cowpea; emasculation can only be effective if it is complted 20 hrs before flower opening. The safest time for emasculation may be during the morning proceeding the day of blooming. Pollens remain viable for 24 hrs at room temperature of 28.5°C and relative humidity 91 per cent. When stored in a refrigerator at 9°C and 0 per cent relative humidituy. The period of pollen viability increased from 12 hrs before blooming to 6 hrs after blooming.
Pod and pod set	Pods are 30-100 cm long, slender, white, light green, dark green or brownish/wine-red in colour with purple or whitish purple mottle, smooth or rugose or crinkled, becoming inflated and fabby before ripening (Lim, 2011). The pods, which begin to form just 50 -60 days after sowing (Kongjaimun *et al.*, 2012), usually hang in pairs. Pods are very slender long (60-90 cm) with beautifully delicate flavor.
Seeds	In asparagus bean seeds are kidney-shaped (Xu *et al.*, 2011a) and distributed throughout the length of the pod, thus making it easier whereas in cowpea, seeds are crowded next to each other in the pod (Kamala *et al.*, 2014).

Horticultural Classification

Verdcourt (1970) subdivided cultivated forms of *V. unguiculata* into three subspecies (including both annual-wild and cultivated types):

(i) *unguiculata*
(ii) *sesquipedalis*
(iii) *cylindrica*

The taxonomy of cowpea has been redefined, with the four subspecies constituting the primary gene pool: *unguiculata*, the cultigen, being roughly divided into four cultigroups (Baudoin and Marechal, 1985).

1. *unguiculata* -The seed type cowpea which is the major group in Africa.
2. *biflora*- The fodder and seed type with small and erect pods which is mainly grown in Southeast Asia,
3. *sesquipedalis*- The yardlong bean or asparagus bean with long pods and climbing growth habit, a green pod vegetable grown in India, Southeast Asia and China
4. *textilis*- A small group found in West Africa which is grown for the textile fibres of the long peduncles.

Marechal *et al.* (1987) then divided *Vigna unguiculata* into 5 cultivar groups *viz.*

i. *unguiculata* (common cowpea)
ii. *sesquipedalis* group (yadlong bean and Dolichos bean)
iii. *biflora* group (catjang cowpea)
iv. *melanophthalmus* group (West African bean)
v. *textilis* group (Nigerian bean)

Once again, cowpea was classified as belonging to the same group as yardlong bean (Li *et al.*, 2001; Phansak *et al.*, 2005; Tantasawat *et al.*, 2010)

Cowpea in India is endowed with rich diversity in its two forms: *unguiculata* and *biflora* while *sesquipedalis* form is of sporadic occurrence (Chandel *et al.*, 1978; Pant *et al.*, 1978).

Augmentation of Genetic Resources and Crop Improvement

Introduction and Selection

Horticulture Research Centre, Bangladesh Agricultural Research Institute (BARI) has identified a potential cultivar 'Chinese Yardlong bean' for commercial cultivation. More than 3,000 accessions are conserved in the National Gene Bank (NGB) at ICAR-National Bureau of Plant Genetic Resources (NBPGR), New Delhi. However, only a few of these are specially designated as yardlong bean, collected from a few pockets in Odisha. Twenty genotypes collected from various parts of Mizoram were evaluated for qualitative traits and quantitative traits. The genotypes varied significantly for different horticultural triatss (Table 4.46).

Table 4.46: Performance of Mizoram Local Germplasm of Yardlong Bean

Germplasm	*Plant Length (cm)*	*Plant FW (g/plant*	*No. of Nodules/ Plant*	*Nodule FW (g/ plant)*	*No. of Seeds/ Pod*	*Pod Length (cm)*	*Pod Wt. (g)*	*No. of Pods/ Plant*	*Pod Wt. (g/plant)*	*Pod Yield (q/ha)*
MZCP-1	320.0	679.3	80.7	0.955	10.4	28.6	11.1	22.4	248.2	116.7
MZCP-2	365.8	333.0	323.0	7.171	9.7	24.6	8.9	19.1	169.6	79.7
MZCP-3	449.7	665.4	183.3	5.891	10.8	26.5	11.2	18.4	205.3	96.5
MZCP-4	343.8	315.0	83.0	2.282	11.8	26.0	10.4	12.8	133.1	62.6
MZCP-5	339.7	687.7	177.7	5.723	9.7	25.4	9.6	13.7	131.5	61.8
MZCP-6	417.2	561.8	137.4	4.430	11.4	26.3	11.1	20.4	226.8	106.6
MZCP-7	282.0	220.7	64.3	0.442	13.1	25.6	12.4	12.8	158.7	74.6
MZCP-9	301.0	168.5	64.7	0.531	16.2	52.7	19.1	18.7	357.5	168.0
MZCP-10	358.7	233.7	82.7	1.712	18.4	52.3	23.4	17.5	409.5	192.5
MZCP-11	379.7	443.0	95.7	2.858	11.2	26.1	12.2	27.6	337.8	158.8
MZCP-12	257.6	210.8	101.2	2.651	12.1	22.8	9.6	11.5	110.4	51.9
MZCP-15	272.0	434.0	155.4	4.875	9.9	46.2	18.0	12.4	223.2	104.9
MZCP-20	389.0	260.0	136.0	4.297	11.8	23.4	10.7	19.8	211.9	99.6
Avg.	**344.3**	**408.7**	**121.9**	**3.371**	**12.0**	**31.3**	**12.9**	**17.5**	**224.9**	**105.7**

Source: ICAR Research Complex for NEH Region, Mizoram Centre, Kolasib, Mizoram.

In Thailand, at least 30 varieties of yardlong bean and dwarf yardlong bean are used commercially throughout the country. Benchasri *et al.* (2007) evaluated 24 yardlong bean and cowpea genotypes in Thailand and reported that genotype IT82E-16 showed high level of resistance for aphid. In Pakistan, cultivars like AVRDC CV. (R) and NARC Sel. (P) are being very common and grown commercially.

Kerala Agricultural University has released many high yielding varieties of yardlong bean namely Malika, Sharika, KMV-1, Lola, Vellayani, Jyothika and Vijayanthi. Vijayanthi is trailing type variety with long and wine red coloured pods containing 20-22 brown coloured seeds. Average length of pods is 50.62 cm and average fruit weight is 16.17 g. Its productivity is 12.6 t/ha. Lola is a trailing type yardlong bean produce 53.38 cm long pod and give 20 t/ha. green pod yield. Arka Mangala is a photoinsensitive variety released from ICAR-IIHR, Bangaluru (Karnataka) producing marketable pods in 60 days after sowing. It gives average yield of 25 t/ha. A pure line selection from IC582850 (Jeypore, and Koraput of Odisha) was released for cultivation as 'Arka Mangla'. The variety recorded a pod yield of 25.7 t/ha with an increase of 24 per cent and 30 per cent pod yield over the check varieties 'Lola' and 'Vijayanthi', repectively. Variation in pod length ranged from 24-75 cm; in fresh pod weight (10 pods) from 60-140 g and seeds/pod from 6.2-17.0. Hundred seed weight varied from 7.3-23.2g. Pod length had the highest dispersion index (158 per cent), followed by 100-seed weight (127 per cent), fresh pod weight (89 per cent) and seeds/pod (83 per cent). Accessions IC582850 and IC582863 had the longest pods (75 cm) followed by IC582861 (45 cm), IC582862 (40 cm), IC582872 (39 cm) and IC582841 (39 cm). Highest fresh pod weight was observed in IC582872 and IC582850 (140 g) followed by IC582873 (135 g) and IC582882 (120 g) as reported by Kamala *et al.* (2014).

Hybridization and Selection

The long pod character of yardlong bean has the potential for developing bush-type cowpea by hybridizing with grain type cowpea. The bush type vegetable cowpea with semi-erect growth habit with long-succulent pods and long peduncles (40-50 cm long) protruding over canopy and holding pods above ground has an advantage of requiring no staking, thus reducing cultivation costs (Singh *et al.*, 2003).

Interspecific Hybridization

Wide crosses have been widely used to transfer agruiculturally important characters, such as pest resistance, from one gene pool to another within the genus *Vigna* (Ehlers and Hall, 1997; Hall *et al.*, 1997; Singh *et al.*, 1997). Attempts have also been made to transfer genes between the cv-gr. *sesquipedalis* (yardlong cowpea), cv-gr. *unguiculata* (cultivated cowpea) and wild cowpea within the subgenus *Vigna*, however low success has been reported (Kongjaimun *et al.*, 2012; Hazra *et al.*, 2007; Fatokun *et al.*, 1997). Crosses between *sesquipedalis* and *dekindtiana* genotypes recorded zero success (Ogunkanmi *et al.*, 2013). This strongly suggests the existence of sexual barriers preventing such intrageneric gene transfer (Fatokun *et al.*, 1997). Hazra *et al.* (2007) revealed that partial dominance of short and light pods of *unguiculata* and *cylindrica* over long and heavy pods of *sesquipedalis* genotypes may have also contributed to cross incompatability or hybrid failure. He suggested that advancement of genotypes of highly heterotic *unguiculata* x *sesquipedalis* or *dekindtiana* x *sesquipedalis* hybrids could have been modified by backcross-pedigree method which proved successful (Hazra *et al.*, 2007). He observed failure of germination of the hybrid seeds could have resulted from loss of seed viability which could have been caused series of factors such as poor seed preservation technique, unfavourable germination conditions, and genetic factors like the presence of lethal alleles in hybrid genotype (Saccardo *et al.*, 1992). Egawa (1988) reported that all the 56 seeds from a *V. mungo* x *V. radiata* cross and 5 seeds from *V. umbellata* x *V. radiata* cross failed to germinate. The reasons for failure of interspecific crosses in *Vigna* are poorly understood (Chen *et al.*, 1983). Several factors, such as specific cross combination, genetic divergence and environment, could have influenced the low success rate. Furthermore, pre-fertilization and post-fertilization barriers such as pollen-pistillate incompatibility (Chowdhury and Chowdhury, 1977), embryo abortion (Ahn and Hartmann, 1997); Fatokun *et al.*, 1997, and hybrid failure and breakdown (Chen *et al.*, 1983) could have also been responsible for the low success rates and failure of crosses. Chances of success should therefore be increased by using *in vitro* pollination and fertilization followed by embryo rescue (Fatokun and Singh, 1987; Chen *et al.*, 1977). Pollination methods such as cut-style and style grafting should also be employed alongside the bud pollination technique (Van Tuyl and De Jeu, 1997).

Genetic Resource Conservation

The potential ares for collection of yardlong bean are Kerala, Andhra Pradesh, Karnataka, Odisha (Koraput and Gajapati), Uttar Pradesh, Bihar, Jharkhand, Chattishgarh, and Madhya Pradesh. Sporadically good genotypes can be collected from NEH regions. More than 3,000 accessions are conserved in the National Gene Bank (NGB) at ICAR-NBPGR, N. Delhi but only a few specially designated as 'Yardlong bean'.

REFERENCES

1. Bata, H.D., Singh, B.B., Singh, S.R. and Ladeinde, T.A.O. 1987. Inheritance of resistance to aphid in cowpea. *Crop Sci.*, **27:** 892-894.
2. Baudin, J.P. and Marechal, R.1985. Genetic diversity in *Vigna*. In: Cowpea Research, Production and Utilization (Singh, S.R. and Rachie, K.O., eds.), John Wiley and Sons Ltd, New York, p.-3-9.
3. Benchasri, S., Nualsri, C., Santipracha, Q. and Ngampongsai, A.2007. Evaluation of aphid (*Aphis craccivora* Koch.) resistance in 24 accessions of yardlong bean and cowpea. The first Joint PSUUNS International Conference on Bioscience: Food, Agriculture and the Environment, J.B. Hotel, Hat Yai, Songkhla, pp. 215-222.
4. Chandel, K.P.S., Joshi, B.S. and Pant, K.C. 1978. Genetic Resources of *Vigna* species in India and their contribution in crop improvement. Proc. Intern. Conf. on *Cytogenetics and Crop Improvement*, Varanasi (U.P.).
5. Chen, N.C., Baker, L.R. and Honma, S. 1983. Interspecific crossability among four species of *Vigna* food legumes. *Euphytica*, **32(3):** 925-937.
6. Chen, N.C., Parrot, J.F., Jacobs, T., Baker, L.R. and Carlson, P.S. 1977. Interspecific hybridization of food legumes by unconventional methods of plant breeding. In: Proceedings, First International Mung bean Symposium. Asian Vegetable Reserch and Development Centre, Shanhua Taiwan, p. 247-252.
7. Chowdhury, R.K. and Chowdhury, J.B. 1977. Interspecific hybridization between *Vigna mungo* (L. Hepper and *Phaseolus calcaratus* Roxb.). *Indian Journal of Agricultural Sciences*, **47:** 117-212.
8. Egawa, Y. 1988. Phylogenetic differentiation between thee Asian *Vigna* species, *V. radiata*, *V. mungo* and *V. umbellata*. Bulletin of the National Institute of Agrobiological Resources, **4:** 189-200.
9. Ehlers, J.D. and Hall, A.E. 1997. Cowpea (*Vigna unguiculata* (L.) Walp.). *Field Crops Res.*, 53: 187-204.
10. Fatokun, C.A. and Singh, B.B. 1987. Interspecific hybridization between *Vigna pubscens* and *V. unguiculata* (L.) Walp. through embryo rescue. *Plant Cell, Tissue and Organ Culture*, **9:** 229-233.
11. Fatokun, C.A., Perrino, P. and Ng N.Q. 1997. Wide crossing in Africa *Vigna* species. In: Advances in cowpea research. Singh, B.B., Mohan Raj, D.R., Dashiell, K.E. and Jackai, L.E.N., eds., ITA, Ibadan.
12. Hall, A., Singh, B. and Ehlers, J. 1997. Cowpea breeding. *Plant Breeding Reviews*. **15:** 215-274.
13. Hazra, P., Chattopadhaya, A., Dasgupta, T., Kar, N., Das, P.K. and Som, M.G. 2007. Breeding strategy for improving plant type, pod yield and protein content in vegetable cowpea (*Vigna unguiculata* (L.) Walp.). *Acta Horticulture*, **752:** 275-280.
14. Hossain, M. K., Alam, N., Teixeria da Silva, J.A., Biswas, B.K. and Mohsin, G.M. 2013. Genetic relationship and diversity based on agro-morphogenic characters in yardlong bean (*Vigna sesquipedalis* L. Fruw.) germplasm. *International Journal of Plant Breeding*, **7(2):** 129-135.
15. Kamala, V., Aghora, T.S., Sivaraj, N., Rao, T., Pandravada, S.R., Sunil, N., varaprasad, K.S. and Chakrabarty, S.K. 2014. Germplasm collection and diversity analysis in yardlong bean *Vigna unguiculata* subsp. *sesquipedalis* from Coastal Andhra Pradesh and Odisha. *Indian J. Plant Genet. Resour.* **27(2):** 171-177. DOI 10.5958/0976-1926.2014.00013.8.
16. Kongjaimun, A., Kaga, A., Tomooka, N., Somta, P., Shimizu, T., Shu, Y., Isemura, T., Vaughan, D. A. and Srinives, P. 2012. An SSR-based linkage map of yardlong bean *Vigna unguiculata* (L.) Walp. subsp. *unguiculata sesquipedalis* Group) and QTL analysis of pod length. *Genome*. **55(2):** 81-92.

17. Kumar, K.M. and Devi, D.S.R. 2009. Variability studies in yardlong bean (*Vigna unguiculata* ssp. *sesquipedalis* (L.) Verdcourt). *J. Res. ANGRAU*, **37:** 58-61.

18. Lausuwan, P., Naengmai, M. and Saengoon, C. 1996. Comparison of dwarf yardlong bean varieties. *Suranaree J. Sci.*, **3:** 147-150.

19. Li, C.D., Fatokun, C.A., Benjamin, U., Singh, B.B. and Scoles, G.J. 2001.Determining genetic similarities and relationships among cowpea breeding lines and cultivars by microsatellite markers. *Crop Science*, **41:** 189-197.

20. Lim, T.K. 2011. *Vigna unguiculata* cv-gr. *sesquipedalis*. In: Edible Medicinal and non-Medicinal Plants. p. 971-975.

21. Lovely, B. and Radhadevi, D.S. 2006. Character association studies in yardlong bean (*Vigna unguiculata* ssp. *sesquipedalis* (L.) Verdcourt). *Indian, J. Plant Genet Resour.*, **19:** 80-82.

22. Marechal, R., Mascherpa, J.M. and Stainer, F. 1978. Etude taxonomique d'un groupe complexe d' speces des genres *Phaseolous* et *Vigna* (Papiilionaceae) sur la base de donnees morphologiques et polliniques, traitees par l' analyse informatique. *Biossiera*, **28:** 191-193.

23. Muchero, M., Diop, N.N., Bhat, P.R., Fenton, R.D., Wanamaker, S., Pottorff, M., Hearne, S., Cisse, N., Fatokun, C., Ehlers, J.D., Roberts, P.A. and Close, T.J. 2009. A consensus genetic map of cowpea [*Vigna unguiculata* (L.) Walp.] and synteny based on EST-derived SNPs. *Proc. Natl. Acad. Sci. USA.*, **106:** 18159-18164.

24. Ng, N.Q. and Marechal, R. 1985. Cowpea Taxonomy, Origin and Germplasm. In: Evolution of Crop Plants Smartt, J. and Simmonds, N.W. (eds.). In: 2nd Edition, Longman, Harlow, U.K., p. 11-21.

25. Ng, N.Q. 1995. Cowpea *Vigna unguiculata* (Leguminosae-Papilionoideae). Pages 326-332. In: 2nd Edition, Longman, Harlow, U.K.

26. Ogunkanmi, L.A., Fakoya, A.O. and Adekoya, K.O. 2013. Breeding for long pod trait in cowpea (*Vigna unguiculata* (L.) Walp.). *J. Sci. Res. Dev.*, **14:** 9-14.

27. Pant, K.C., Chandel, K.P.S. and Joshi, B.S. 1978. Analysis of diversity in Indian cowpea genetic resources. *SABRAO Journal*, **14:** 102-111.

28. Phansak, P., Taylor, P.W.J. and Mongkolporn, O. 2005. Genetic diversity in yardlong bean (*Vigna unguiculata* ssp. *sesquipedalis*) and related *Vigna* species using sequence tagged microsatellite site analysis. *Scientia Horticulturae*, **106:** 137-146.

29. Rachie, K.O. 1985. Introduction. In: Singh, S.R. and K.O. Rachie (eds). Cowpea research, production, and utilization. International Institute of Tropical Agriculture IITA, Ibadan, Nigeria.

30. Resmi, R. and Gopalakrishnan, T.R. 2004. Effect of plant growth regulators on the performance of yardlong bean (*Vigna unguiculata* ssp. *sesquipedalis* (L.) Verdcourt). *J. Trop. Agric.*, **42:** 55-57.

31. Rubatzky, V.E. and Yamaguchi, M. 1997. World Vegetable: Principles, Production, and Nutritive Values. Chapman and Hall, New York. pp. 572.

32. Saccardo, F., Del Giudice, A. and Galasso, I. 1992. Cytogenetics of cowpea. In: Biotechnology: enhancing research on tropical crops in Africa. Thottappily, G., Monti, L., Mohan Ral, D. R., and Morre, A.W. (eds.). CTA/IITA co-publication. IITA, Ibadan, Nigeriapp. 89-98.

33. Sarutayophat, T., nualsri, C., Santipracha, Q. and Saereeprasert, V. 2007. Characterization and genetic relationship among 37 yardlong bean and cowpea accessions based on morphological characters and RAPD analysis. *Songklanakarin J.Sci. Technol.*, **29:** 591-600.

34. Singh, B.B., Chambliss, O.L. and Sharma, B. 1997. Recent advances in cowpea beeding. Advances in cowpea research. p. 30-49.

35. Singh, B.B., Hartmann, P., Fatokun. M., Tamo, M., tarawali, S. and Ortiz, R. 2003. Recent progress in cowpea improvement. *Chron. Hort.*, **43:** 8-12.

36. Singh, B.B., Raj, D.R., Mohan D.K.E. and Jackai, L.E.N. 1997. Advances in cowpea research.Co-publication of International Institute of Tropical Agriculture (IITA) and Japan International Research Center for Agricultural Sciences (JIRCAS), IITA, Ibadan, Nigeria.

37. Singh, B.B. 2005. Cowpea (*Vigna unguiculata* (L.) Walp.). In: Singh, R.J., Jauhar, P.P. (eds.). Genetic resources, chromosome engineering and crop improvement. Taylor and Francis, LLC, p. 117-161.

38. Steel, W.M. and Mehra, K.L. 1980. Structure, evolution and adaptation to farming systems and environments in *Vigna*. In: Advances in legume science (ed.) by Summerfield, R.J. and Bunting, A.H. Royal Botanic Gardens, Kew, U.K. pp. 393-404.

39. Tan, H., Tie, M., Luo, Q., Zhu, Y., Lai, J. and Li, H. 2012. A review of molecular marker applied in cowpea [*Vigna unguiculata* (L.) Walp., breeding. *Journal of Life Sciences*, **6:** 1190-1199.

40. Tantasawat, P., Trongchuen, J. Prajongjai, T., Seehalak, W. and Jittayasothorn, Y. 2010. Variety identification and comparative analysis of genetic diversity in yardlong bean (*Vigna unguiculata* ssp. *sesquipedalis*) using morphological characters, SSR and ISSR analysis. *Scientia Horticulturae*, 124: 204-216.

41. USDA Nutrient Database, 2014. United States Department of Agriculture (USDA). Agricultural

Research Service. National Nutrient Database for Standard Reference. http://ndb,nal,usda.gov/ndb.

42. USDA-ARS, 2012. United States Department of Agriculture (USDA). Plant Guide: Yardlong bean *Vigna unguiculata* (L.) Walp. ssp. *sesquipedalis* (L.) Verdc. [Online] Available: http://plants,usda.gov/plantguide/pdf/pg_viuns2.pdf [Accessed on 12 May, 2012].
43. Van Tuyl, J.M. and De Jeu, M.J. 1997. Methods for overcoming interspecific crossing barriers. In: "Pollen biotechnology for crop production and improvement". p. 273-293. (eds. Shivanna, K.R. and Sawhney,V.K.), Cambridge University Press.
44. Verdcourt, B. 1970. Studies in the Leguminosae Papilionoideae for the "Flora of Tropical East Africa": IV. Kew Bulletin, 24, p. 542-544.
45. Vidya, C. and Oommen, S.K.2002. Correlation and path analysis in yardlong bean. *J. Trop. Agri.* **40:** 48-50.
46. Vidya, C., Oommen, S.K. and Kumar, V. 2002. Genetic variability and heritability of yield and related characters in yardlong bean. *J. Trop. Agri.*, **40:** 11-13.
47. Xu, P., Wu, X., Wang, B., Liu, Y., Jeffrey, E, Close, T., Roberts, J., Philip, A., Diop, N.N., Quin, D., Tingting, H., Lu, Zhongfu and Li, Guojing. 2011b. A SNP and SSR based genetic map of asparagus bean (*Vigna unguiculata* ssp. *sesquipedalis*) and comparision with the broader species. *PLoS ONE*, **6(1):** e15952. Doi: 10.1371/journal.pone.0015952.
48. Xu, P., Hu, T., Yuejian, Y., Wu, X., Wang, B., Liu, Y., Qin, D., Jeffrey, E, Close, T., Lu, Zhongfu and Li, Guojing. 2011a. Mapping genes governing flower and seedcoat colour in asparagus bean (*Vigna unguiculata* ssp. *sesquipedalis*) based on single nucleotide polymorphism and simple sequence repeat markers. *HortSci.*, **46(8):** 1102-1104.
49. Xu, P., Wu, X., Wang, B., Liu, Y., Qin, D., Ehlers, J.D., Close, T.J., Hu, T., Lu, Z. and Li, G. 2010. Development and polymorphism of *Vigna unguiculata* ssp. *unguiculata* microsatellite markers used for phylogenetic analysis in asparagus bean (*Vigna unguiculata* ssp. *sesquipedalis*) (L.) Verdc.). *Mol. Breeding*, **25:** 675-684.
50. Xu, P., Wu, X.H., Wang, B. G., Liu, Y.H., Qin, D. H. and Ehlers, J.D. 2011. A SNP and SSR based genetic map of asparagus bean (*Vigna unguiculata* ssp. *sesquipedalis*) and comparison with the broader species, *PLoS ONE*, **6(1):** 1-8

Other *Vigna* Species

Adzuki Bean [*Vigna angularis* (Wild.) Ohwi and Ohashi]

It is cultivated mainly in South East Asia, including China, Japan, Korea, Thailand and Vietnam. Being a short duration crop, it fits well in double cropping system. Seeds are cooked whole and in making soups and also used in specially cakes and confection. Adzuki bean is also used in oriental medicine. Research for crop improvement is minimal but there is potential to increase yield and make the crop amendable to mechanized cultivation. Thirty one accessions were evaluated at ICAR-NBPGR Regional Station, Shimla (H.P) and a wide range of variation was observed for plant height (25-49 cm), branches per plant (3-8), cluster per plant (6-10), pods per cluster (3-5), pods per plant (11-30), pod length (6-10 cm), seeds per pod (6-9) and seed yield per plant (9-16 g).

Bambara Groundnut [*Vigna subterranean* (L.) Verde.]

Bambara groundnut is originated from South Africa and cultivated mainly in the drier areas of Africa and also in Latin America, Australia and Asia. In Asia, it is cultivated in India, Indonesia, Malaysia, Philippines and Thailand. Bambara groundnut is a short day plant and grows well in 600-700 m regions. The immature seeds are boiled or roasted and eaten as snacks. Mature dry seeds are used in various food preparations. Current landraces are of long-duration and low yielding. Hence there is need for short-duration, high yielding varieties that can be grown in different farming systems profitably. Germplasm comprising 30 accessions were evaluated at ICAR-NBPGR Regional Station and a good degree of variation was observed in pod size, seed size and grain yield. Testing of 12 promising accessions revealed that 134511, EC37527 and EC27881 were high yielding types.

Rice Bean [*Vigna umbellata* (Thund.) Ohwi and Ohashi]

INTRODUCTION

Rice bean (*V. umbellata* Thunb.) Ohwi and Ohashi, (previously designated as *Phaseolus calcaratus*) is a low input warm-season underutilized crop/neglected multipurpose grain legume which provide edible bean, good fodder, protein rich seeds and green manure. It is of great potential to solve the food and nutritional problem. In most East Asian countries, rice bean is used as a substitute of Adzuki bean in various food preparations. Rice bean is grouped in the category of small grain and find place in inter crop or mixed crop with miaze (*Zea mays*), sorghum (*Sorghum bicolor*) or cowpea (*V. unguiculata*). Sometime it is cultivatesd as a sole crop in the uplands but on a very limited area. The crop is also used as nutritious green and dry fodder with a smaller proportion. It is also suitable for green mauring (Joshi *et al.*, 2006) which fixed sufficient quantity of nitrogen in depleted soils. However, research has been limited to develop improve varieties for various agro climates.

Origin, History and Distribution

Its distriobution ranges from humid tropical to sub-temperate environments. The centre of domestication is believed to be India-China. This crop mainly can be spotted between 700 and 1300 m above mean sea level, although in home gardens it is found from 200 up to 2000 m. Rachie amnd Roberts (1974) classed rice bean as adapted to sub-humid regions with 1000-1500 mm precipitation, although they noted that other factors were also involved in adaptation, for example rainfall pattern, moisture distribution, temperature, cloud cover and relative humidity, soil characteristics, pests and diseases. For subsistence it is grown in small areas of Nepal, northern and north-eastern India and parts of South East Asia. It is cultivated in tropical areas of Asia (China, Myanmar, India, Bangladesh, Malaysia, Thailand, Philippines, Laos and Vietnam), Africa, America and Australia (Tomooka *et al.*, 1991). It is thought to be derived from the wild form *V. umbellata* var *gracilis*, with which it is cross-fertile and well distributed from Southern China through the north of Vietnam, Laos and Thailand into Myanmar and India (Tomooka *et al.*, 1991). Studies of the genetic and eco-geographical relationships among the wild relatives of *Vigna* species were made by Saravanakumar *et al.* (2003). In India, the crop is mainly found in the Western and Eastern Ghats and the NE Himalayas, including Assam, Meghalay, Mizoram and Manipur (Jain and Mehra, 1978) but is also grown in the sub-temperate western Himalayas in the Uttarakhand and Himachal Pradesh hills as a good alternative to other pulse crops such as black gram and green gram.

Table 4.47: States Reported to be Growing Rice Bean in India

Spread of the Crop	*States with the Crop Reported*
High	Assam, Megalaya, Manipur, Nagaland, Mizoram, parts of Arunachal Pradesh, North Bengal hill regions, Sikkim and Odisha
Medium	Himachal Pradesh, Uttarakhand, Chattishgarh, Jharkhand, Madhya Pradesh
Low	Kerala, Tamil Nadu, Gujrat, Punjab, Maharshtra, Haryana, Jammu and Kashmir, Karnataka, Andhra Pradesh, Andaman and Nicobar Islands

Source: FOSRIN Report, 2008.

Species Availability, Evolution and Diversity

Rice bean is thought to derive from the wild form *V. umbellatea* var. *gracilis*, with which it is cross-fertile and pace in evoluation. In India *V. umbellata* Thunb. is found in wild form in natural habitats of the Khasi and Jaintia hills of Meghalaya. One of the interesting species of *Vigna* namely *V. vexillata* is grown by the tribals of Tripura. The wild forms are also wide spread from Kerala through the Western and Eastern Ghats to the NW Himalayas and the NE (Bisht *et al.*, 2005; Jain and Mehra, 1978) up to about 1500 m, with the main botanical source for the species being the humid tropical region of the Western Ghats (Arora and Chandel, 1972). There is great variability in wild forms in the NE region, as well as sporadically in the Western Himalayas and the Eastern and Western Ghats (Arora, 1991).

It is a legume cum tuber crop with much variation in edible tubers (Arora and Pandey, 1996). Department of Biotechnology, Guwahati University, Guwahati screened 30 rice bean collections and found that crude protein content varied from 14.66 per cent to 26.88 per cent. Arora (1985) noted rich diversity in cultigen in the NE region and Eastern Himalayas, in particular, in the Assam plains and Adjacent hill areas, and in North Bengal and Sikkim, and it is possible that introgression and folk selection have played a dominant role in the creation of rice bean's rich genetic variability and also in the evolution of present day landraces. Over 500 accessions from north East, Central and East Peninisular Regions as well as introduced from other countries were evaluated at the ICAR-NBPGR Regional Station, Shimala (H.P.). The analysis of inter-regional diversity patterns for agro-morphological traits, biochemical constituents and disease and pest screening indicated a rich diversity in Indian germplasm. Cultivated forms are erect/semi-erect. Tendrillous types also occur in its region of distribution particularly in Asia and Southeast Asia. Mainly intermediate forms are found in India. Considerable genetic polymorphism was observed for pod shape, size and colour. Several accessions with black, purple, red, violet, greenish, yellowish or mottled seeds were observed. Biochemical studies revealed its nutritional richness with high protein, calcium, iron and phosphorus along with comparatively high tryptophan and methoionine.

Botanical Description

Rice bean (*V. umbellata* Thunb.) Ohwi and Ohashi is an annual legume.

Kingdom	:	Plantae
(unranked)		Angiosperms
(unranked)		Eudicots
(unranked)		Rosids
Order	:	Fabales
Family	:	Fabaceae
Subfamily	:	Faboideae
Tribe	:	Phaseoleae
Genus	:	*Vigna*

The detail botanical description is given below:

Table 4.48: Botanical Description

Growth Habit	It is a short-lived perennial legume usually grown as an annual. It has a very variable habit: it can be erect, semi-erect or twining.
Root	It has an extensive taproot root system with a taproot that can go as deep as 100-150 cm.
Stem	It is usually 30-100 cm in height, but can grow up to 200 cm (Ecoport, 2014). The stems are branched/ grooved and finely haired.
Leaf	The leaves are alternate, trifoliate with stipules lanceolate 0.5 cm long leaf entire, 6-9 cm long leaflets, and petiole 5–10 cm long. Leaflets broadly ovate to ovate-lanceolate, 5–10 (–13) cm × 1.5–6(–7) cm, entire or 2–3-lobed, the lateral leaflets unequal-sided, membranous, almost glabrous
Floral biology	Inflorescence an erect axillary false raceme and 3–10 cm long, 5–20-flowered, with flowers usually in pairs; peduncle up to 20 cm long. The flowers are papillonaceous and bright yellow in colour. Flowers bisexual, papilionaceous; pedicel c. 5 mm long; calyx campanulate, c. 4 mm long, 5-toothed; corolla bright yellow, standard 1.5–2 cm in diameter; wings large and broad, enclosing the keel; keel with a curved beak and a conical pocket on one side; stamens 10, 9 connate into a tube, upper one free; ovary superior, 1-celled, style broadened and curved.
Pod and pod set	Fruit a linear-cylindrical pod, 6–13 cm x 0.3–0.6 cm, deflexed, glabrous, green when young, black-brown at maturity, 6–8(–16)-seeded
Seeds	Pods contain 6-10 oblong seeded, 5–10 mm x 2–5 mm x 3–4 mm seeds with a concave hilum. Rice bean seeds are very variable in colour, smooth, yellow, green, dark red, brown, black, speckled or mottled. Yellow-brownish types are reported to be the most nutritious. Hilum excentric, elongate, hidden by a cream-coloured rim. The red type gives its common name to the grain (Ecoport, 2014).

Augmentation of Genetic Resources and Crop Improvement

Despite having all the favourable traits; it is not much popular among the farmers due to the late maturity and indeterminate growth habit. A little genetic improvement with respect to maturity and growth habit can revive its cultivation.

Introduction and Selection

Rice bean is pest resistant and versatile crops which adjust well with the change of climate and soil condition. It is tolerant to some degree of waterlogging, although the young plant appears to be susceptible (De Carvallho and Veira, 1996). A variety RBL-1 was developed at PAU, Ludhiana (Punjab) through single progeny selection and releasd for Punjab state, and RBL-6 has also been released (Mal *et al.*, 1996). Other introductions which gave consistently superior performance in north Indian conditions included EC 93452, EC101887, PI247685 and PI247693. The variety Konkan Rice Bean-1 (RB-10, possibly the same as KRB-1), a fodder type cultivar, was developed by single plant selection through multi-environment testing in Maharashtra (Thaware *et al.*, 2005). The varieties 'BRS-1', 'BRS-2' (bred at the ICAR-NBPGR Research Station, Bhowali, Uttarakhand) and 'Naini' have been released for hill land. Another variety, Bidhan (K-1) has also been released nationally (Shukla *et al.*, 2003).

Hybridization and Selection

Hybridization at ICAR-NBPGR, New Delhi between an early variety from China, a bold seeded variety from Mysore, India, and a yellow seeded variety from Nepal led to the development of several promising lines combining earliness and high yield (Mal and Joshi, 1991). A promising selection C x $M_{12}P_{2-3}$ capable of producing up to 2.5 t/ha grain yield and hold promise for cultivation in the Shimla hills. Singh *et al.* (1999) identified combinations of parents for various breeding objectives: RBM-1 x RBM-10 for shorter height, early flowering and greater number of seeds per pod, RBM-10 x RBM-17 for increased number of primary branches per plant, RBM-11 x RBM-17 for higher number of pods per plant and seed yield per plant, and RBM-31 x RBM-17 for greater pod length and seed weight.

Interspecific Hybridization

Since variability for agronomic traits is very low in this crop, inter-specific hybridization of ricebean with other *Vigna* species can lead to its improvement. interspeicific hybridization was performed among ricebean, adzukibean, mash and mung to introgress desired traits into otherwise high yielding genotypes. Results revealed that there was no free crossing in these species. Successful pod set was observed in 10 crosses out of 72 interspecific crosses with mash (*V. mungo*) and mung (*V. radiata*). Highest crossability was observed in *V. mungo* and *V. umbellata* crosses while moderate crossability in *V. radiata* and *V. umbellata* crosses. Crossing of adzukibean with ricebean showed poor or no pod set percentage among all the mentioned cross combinations (Bhardwaj *et al.*, 2016).

Mutation Breeding

Mutation breeding using gamma radiation in three cultigens from Manipur (RBM-6, RBM-13 and RBM-31), produced two dwarf mutants from RBM-6 (less than 50 per cent of the height of the parent), and an early maturing mutant (120 days vs 154 days) of RBM-13 in the M_2 generation (Devi and Singh, 2006). Highly induced variability was seen for clusters, pods and grain yield. An indeterminate landrace treated with 100 Kr gamma rays resulted in the production of an early, photo insensitive and determinate mutant (Singh and Tomar, 1989).

Genetic Resource Conservation

The World Vegetable Centre (formerly the AVRDC), Taiwan has 197 accessions of rice bean, including 8 genotypes from Nepal and 24 from India. However, there is little or no passport data (World Vegetable Centre, 2007) other than for a Nepalese genotype (given the name Mogimass), collected at 2000 m in Bajura district. In Nepal, the Plant genetic Resources Unit of the NARC maintains a collection of some 300 accessions from various parts of the country. Large germplasm collections of rice bean are kept in China (Institute of Crop Germplasm Resources (CAAS), Beijing; 1363 accessions). No germplasm collections of rice bean are known in Africa.

In India, germplasm collection was initiated by NBPGR, New Delhi during the 1960s, particularly from the north-eastern region (Thomas and Mathur, 1991) and ICAR-NBPGR maintains the world base collection of rice bean germplasm. Between 1970 and 1976, Arora and Mehra (1978) sampled many crops including rice bean from fields, gardens and markets in north-east India (Assam, Meghalaya, Mizoram, Manipur and Arunachal Pradesh). Arora *et al.* (1980) evaluated over 300 collections of rice bean from the tribal dominated mountains in the east and northeast between 1970 and 1979. ICAR-National Bureau of Plant Genetic Resources (NBPGR), New Delhi is maintaing 902 accessions. The Indian genotypes IC 7588, IC 8229, EC 18771 and IC 7506 are noted as being less sensitive to photoperiod, but no other information is given. At present ICAR-National Bureau of Plant genetic Resources (NBPGR), New Delhi contains over 1700 accessions from a variety of Asian countries (ICAR-NBPGR, New Delhi, 2007). As well as there is a collection held at the ICAR-Indian Institute for Pulses Research, Kanpur (U.P.) and the ICAR-NBPGR station at Bhowali, Uttarakhand also maintain a collection over 300 genotypes (Negi *et al.*, 1996). Thirty land races of *V. umbellata*, collected from the hilly state of Nagaland, by Department of Biotechnology, Guwahati University, Guwahati (India). Local rice bean landraces still exist in villages of NEH regions but the area under the crop in India is considerably declining each year. Rice bean is not threatened with genetic erosion, but more germplasm collection and characterization works are required.

REFERENCES

1. Arora, R.K. and Chandel, K.P.S. 1972. Botanical sources areas of wild herbage legumes in India. *Tropical Grasslands*, **6:** 213-221.
2. Arora, R.K. and Mehra, K.L. 1978. Exploration in NE India (Cereals and Pulses). *Plant Genetic Resources Newsletter*, 34: 4-8.
3. Arora, R.K., Chandel, K.P.S., Joshi, B.S. and Pant, K.C. 1980. Rice bean: tribal pulse of eastern India. *Economic Botany*, **34:** 260-263.
4. Arora, R.K. 1985. Diversity and collection of wild *Vigna* species in India. *FAO/IBPGR Plant Genetic Resources Newsletter*, **63:** 26-35.
5. Arora, R.K. 1991. Plant diversity in the Indian gene centre, in: Plant Genetic Resources Conservation and Management-Concepts and Approaches (Paroda, R.S. and Arora, R.K.; eds.). IBPGR, New Delhi. Available online at http://www.bioversityinternational.org/publications/Web_version/174/ch06.htm.
6. Arora, R.K. and Pandey, A. 1996. Wild edible plants of India-diversity, conservation and use. ICAR-NBPGR, New Delhi, p. 294.
7. Bhardwaj, N., Kapoor, T. and Thakur, S. 2016. Crossability barriers in inter-specific hybridization of ricebean (*Vigna umbellata*) with other *Vigna* species. 1st International Agrobiodiversity Congress, held at New Delhi6-9 November 2016; p. 21.
8. Bisht, I.S., Bhat, K.V, Lakhanpaul, S., Latha, M., Jayan, P.K., Biswas, B.K. and Singh, A.K. 2005. Diversiy and genetic resources of wild *Vigna* species in India. *Genetic Resources and Crop Evolution*, **52:** 53-68.
9. De Carvallho, N.M. and Veira, R.D. 1996. Rice bean (*Vigna umbellata* (Thunb.) Ohwi and Ohasi. In: Legumes and Oilseeds in Nutrition (Nwokolo, E. and Smartt, J. eds.). Chapman and Hall, p. 222-228.
10. Devi, T.R. and Singh, N.B. 2006. Mutagenic induction of variability and selection in M_2 generation of selected rice bean [(*Vigna umbellata* (Thunb) Ohwi and Ohasi] cultivars of Manipur. *Legume Research*, **29:** 150-153.
11. Jain, H.K. and Mehra, K.L. 1978. Evolution, relationships and uses of the species of *Vigna* cultivated in India. In: Advances in Legume Science (Summerfield, R.J. and Bunting, A.H.; eds.). Royal Botanic Garden, Kew, p. 459-468.
12. Joshi, K.D., Bhandari, B., Gautam R., Bajracharya, J. and Hollington, P.A. 2006. Rice bean: A multipurpose underutilized legume. Paper presented at 5th International Symposium on New Crops and Uses: Their roles in a rapidly changing world. Organized by the Centre for Underutilized Crops, University of Southampton in partnership with National Non-food Crops Centre and the Tropical Agricultural Association.
13. Mal, B., Rana, R.S. and Joshi, V. 1996. Status of research on new crops in India. In: Proceedings of the 9th International Conference on jojoba and its uses and of the 3rd International Conference on new industrial crops and products, p. 198-204.
14. Negi, K.S., Pant, K.C., Muneem, K.C. and Mal, B. 1996. Evaluation of rice bean genetic resources. *Indian J. For.*, **19:** 156-163.
15. Rachie, K.O. and Roberts, L.M. 1974. Grain legumes of the low land tropics. *Adv. Agron.*, **26:** 1-132.

16. Saravankumar P, Tomooka N, Kaga, A and Vaughan DA (2003) Studies on wild relatives of grain legumes in Southern South Asia with particular reference to the genera *Cajanus* and *Vigna*. In: AHM Jayasuriya and DA Vaughan (eds) *Conservation and use of crop wild relatives*. Proceedings of the joint Department of Agriculture, Sri Lanka and National Institute of Agrobiological Science, Japan Workshop held on 3 February 2003.

17. Shukla, G.P., Melkania, N.P., Singh, K., Rajpali, S.K. and Arya, O.N. 2003. Genotype x location environment interaction in rice bean [*Vigna umbellata* (Thunb.) Ohwi and Ohashi] with special reference to acidic soil. *Forage Research*, **28:** 194-200.

18. Singh, V.P. and Tomar, Y.S. 1989. A photopinsensitive promising mutant of rice bean (*Vigna umbellata*). *Legume Research*, **12(1):** 47-48.

19. Thaware, B.L., Bendale, V.W. and Toro, V.A. 2005. Konkan Rice Bean-1 (RB-10), a new fodder rice bean variety for Konkan Region of Maharashtra. *Journal of Maharashtra Agricultural Universities*, **30:** 295-298.

20. Thomas, T.A. and Mathur, P.N. 1991. Germplasm evaluation and utilization. In: Plant Genetic Resources Conservation and Management-Concepts and Approach (Paroda, R.S. and Arora, R.K.; eds.). IBPGR, New Delhi. Available online at http://www.bioversityinternational.org/publications/Web_version/174/ch06.htm.

21. Tomooka, N, Kaga, A, Vaughan, DA, and Jayasuriya AHM (2003) Advances in understanding the genus *Vigna* subgenus *Ceratotropis* In: Jayasuriya, A.H.M. and Vaughan, D.A. (eds), *Conservation and use of crop wild relatives*. Proceedings of the joint Department of Agriculture, Sri Lanka and National Institute of Agrobiological Science, Japan Workshop held on 3 February 2003

22. World Vegetable Centre, Taiwan. 2007. Vegetable Genetic Resources Information Sytem.

Genus *Phaseolus*

There are several phaseolus species in nature which are greately associated with human and animals.

Table 4.49: Taxon and Preferred Taxon Name

Name of Taxon	*Name of Authority*	*Family*	*Preferred Taxon Name*
Phaseolus			*Phaseolus*
Phaseolus aborigineus	Burkart	Fabaceae	*Phaseolus vulgaris var. aborigineus*
Phaseolus aconitifolius	Jacq.	Fabaceae	*Vigna aconitifolia*
Phaseolus acutifolius	A. Gray	Fabaceae	*Phaseolus acutifolius*
Phaseolus acutifolius var. *acutifolius*		Fabaceae	*Phaseolus acutifolius* var. *acutifolius*
Phaseolus acutifolius var. *latifolius*	G. F. Freeman	Fabaceae	*Phaseolus acutifolius* var. *acutifolius*
Phaseolus acutifolius var. *tenuifolius*	A. Gray	Fabaceae	*Phaseolus acutifolius* var. *tenuifolius*
Phaseolus adenanthus	G. Mey.	Fabaceae	*Vigna adenantha*
Phaseolus altimontanus	Freytag and Debouck, nom. inval.	Fabaceae	*Phaseolus* sp.
Phaseolus amabilis	Standl.	Fabaceae	*Phaseolus parvulus*
Phaseolus amblyosepalus	(Piper) C. V. Morton	Fabaceae	*Phaseolus amblyosepalus*
Phaseolus angularis	(Willd.) W. Wight	Fabaceae	*Vigna angularis* var. *angularis*
Phaseolus angularis var. *nipponensis*	(Ohwi) Ohwi	Fabaceae	*Vigna angularis* var. *nipponensis*
Phaseolus angustifolius	Roxb., nom. nud.	Fabaceae	*Phaseolus* sp.
Phaseolus angustissimus	A. Gray	Fabaceae	*Phaseolus angustissimus*
Phaseolus anisotrichus	Schltdl.	Fabaceae	*Phaseolus leptostachyus*
Phaseolus anisotrichus subsp. *incisus*	Piper	Fabaceae	*Phaseolus leptostachyus*
Phaseolus antillanus	Urb.	Fabaceae	*Vigna antillana*
Phaseolus atropurpureus	DC.	Fabaceae	*Macroptilium atropurpureum*
Phaseolus augusti	Harms	Fabaceae	*Phaseolus augusti*
Phaseolus aureus	Roxb.	Fabaceae	*Vigna radiata* var. *radiata*
Phaseolus bolivianus	Piper	Fabaceae	*Phaseolus augusti*
Phaseolus bracteatus	Nees and Mart.	Fabaceae	*Macroptilium bracteatum*
Phaseolus brevicalyx	Micheli	Fabaceae	*Phaseolus micranthus*
Phaseolus buseri	Micheli	Fabaceae	*Ramirezella strobilophora* var. *buseri*
Phaseolus calcaratus	Roxb.	Fabaceae	*Vigna umbellata*
Phaseolus campestris	Mart. ex Benth.	Fabaceae	*Macroptilium gracile*

Name of Taxon	Name of Authority	Family	Preferred Taxon Name
Phaseolus caracalla	L.	Fabaceae	*Vigna caracalla*
Phaseolus carteri	Freytag and Debouck, nom. inval.	Fabaceae	*Phaseolus* sp.
Phaseolus chiapasanus	Piper	Fabaceae	*Phaseolus chiapasanus*
Phaseolus clitorioides	Mart. ex Benth.	Fabaceae	*Vigna peduncularis* var. *clitorioides*
Phaseolus coccineus	L.	Fabaceae	*Phaseolus coccineus*
Phaseolus coccineus subsp. *coccineus*		Fabaceae	*Phaseolus coccineus* subsp. *coccineus*
Phaseolus coccineus subsp. *darwinianus*	Hern.-Xol. and Miranda-Colin	Fabaceae	*Phaseolus coccineus* subsp. *darwinianus*
Phaseolus coccineus subsp. *formosus*	(Kunth) Marechal *et al.*	Fabaceae	*Phaseolus* coccineus subsp. *formosus*
Phaseolus coccineus subsp. *glabellus*	(Piper) A. Delgado, nom. inval.	Fabaceae	*Phaseolus glabellus*
Phaseolus coccineus subsp. *griseus*	(Piper) A. Delgado, nom. inval.	Fabaceae	*Phaseolus griseus*
Phaseolus coccineus subsp. *obvallatus*	(Schltdl.) Marechal *et al.*	Fabaceae	*Phaseolus coccineus* subsp. *formosus*
Phaseolus coccineus subsp. *polyanthus*	(Greenm.) Marechal *et al.*	Fabaceae	*Phaseolus coccineus* subsp. *darwinianus*
Phaseolus coccineus var. *purpurascens*	Freytag, nom. inval.	Fabaceae	*Phaseolus coccineus* subsp. *formosus*
Phaseolus coccineus var. *zongolicensis*	Freytag, nom. inval.	Fabaceae	*Phaseolus coccineus* subsp. coccineus
Phaseolus compressus	DC.	Fabaceae	*Phaseolus vulgaris* var. *vulgaris*
Phaseolus compressus var. *carneus*	G. Martens	Fabaceae	*Phaseolus vulgaris* var. *vulgaris*
Phaseolus compressus var. *cervinus*	G. Martens	Fabaceae	*Phaseolus vulgaris* var. *vulgaris*
Phaseolus compressus var. *ferrugineus*	G. Martens	Fabaceae	*Phaseolus vulgaris* var. *vulgaris*
Phaseolus costaricensis	Freytag and Debouck	Fabaceae	*Phaseolus costaricensis*
Phaseolus cylindricus	L.	Fabaceae	*Vigna unguiculata* subsp. *cylindrica*
Phaseolus dalzellianus	Kuntze	Fabaceae	*Vigna dalzelliana*
Phaseolus dumosus	Macfad.	Fabaceae	*Phaseolus coccineus* subsp. *darwinianus*
Phaseolus elegans	Piper	Fabaceae	*Vigna elegans*
Phaseolus ellipticus var. *albus*	G. Martens	Fabaceae	*Phaseolus vulgaris* var. *vulgaris*
Phaseolus ellipticus var. *aureolus*	G. Martens	Fabaceae	*Phaseolusvulgaris* var. *vulgaris*
Phaseolus ellipticus var. *helvolus*	Savi	Fabaceae	*Phaseolusvulgaris* var. *vulgaris*
Phaseolus ellipticus var. *mesomelos*	Haberle	Fabaceae	*Phaseolus vulgaris* var. *vulgaris*
Phaseolus ellipticus var. *pictus*	Caval.	Fabaceae	*Phaseolus vulgaris* var. *vulgaris*
Phaseolus ellipticus var. *spadiceus*	G. Martens	Fabaceae	*Phaseolus vulgaris* var. *vulgaris*
Phaseolus erythroloma	Mart. ex Benth.	Fabaceae	*Macroptilium erythroloma*
Phaseolus esperanzae	Seaton	Fabaceae	*Phaseolus pedicellatus* var. *pedicellatus*
Phaseolus falcatus	Benth. ex Hemsl., nom. nud.	Fabaceae	*Phaseolus* sp.
Phaseolus filiformis	Benth.	Fabaceae	*Phaseolus filiformis*
Phaseolus floribundus	Piper	Fabaceae	*Phaseolus pedicellatus* var. *pedicellatus*
Phaseolus formosus	Kunth	Fabaceae	*Phaseolus coccineus* subsp. *formosus*
Phaseolus gibbosifolius	Ortega	Fabaceae	*Macroptilium gibbosifolium*
Phaseolus glabellus	Piper	Fabaceae	*Phaseolus glabellus*
Phaseolus glaber	Roxb.	Fabaceae	*Vigna glabrescens*
Phaseolus gonospermus var. oryzoides	G. Martens	Fabaceae	*Phaseolus vulgaris* var. *vulgaris*
Phaseolus gonospermus var. variegatus	Savi	Fabaceae	*Phaseolus vulgaris* var. *vulgaris*
Phaseolus gracilis	Poepp. ex Benth.	Fabaceae	*Macroptilium gracile*
Phaseolus grandiflorus	Steud.	Fabaceae	*Vigna linearis* var. *latifolia*
Phaseolus grandis	Wall. ex Benth.	Fabaceae	*Dysolobium grande*
Phaseolus grayanus	Wooton and Standl.	Fabaceae	*Phaseolus pedicellatus* var. *grayanus*
Phaseolus griseus	Piper	Fabaceae	*Phaseolus griseus*
Phaseolus hastatus	Freytag and Debouck, nom. inval.	Fabaceae	*Phaseolus* sp.
Phaseolus helvolus	L.	Fabaceae	*Strophostyles helvola*
Phaseolus heterophyllus	Willd.	Fabaceae	*Macroptilium gibbosifolium*
Phaseolus hintonii	A. Delgado	Fabaceae	*Phaseolus hintonii*
Phaseolus hirsutus	Mart. ex Benth.	Fabaceae	*Vigna lasiocarpa*

Name of Taxon	Name of Authority	Family	Preferred Taxon Name
Phaseolus hybrid		Fabaceae	*Phaseolus hybrid*
Phaseolus inamoenus	L.	Fabaceae	*Phaseolus lunatus* var. *lunatus*
Phaseolus jaliscanus	Piper	Fabaceae	*Phaseolus jaliscanus*
Phaseolus juquilensis	A. Delgado	Fabaceae	*Phaseolus juquilensis*
Phaseolus kirkii	Baker	Fabaceae	*Vigna kirkii*
Phaseolus lasiocarpus	Mart. ex Benth.	Fabaceae	*Vigna lasiocarpa*
Phaseolus lathyroides	L.	Fabaceae	*Macroptilium lathyroides*
Phaseolus laxiflorus	Piper	Fabaceae	*Phaseolus pedicellatus* var. *pedicellatus*
Phaseolus leiosepalus	Piper	Fabaceae	*Phaseolus coccineus* subsp. *formosus*
Phaseolus leiospermus	Torr. and A. Gray	Fabaceae	*Strophostyles leiosperma*
Phaseolus leptophyllus	G. Don	Fabaceae	*Phaseolus leptophyllus*
Phaseolus leptostachyus	Benth.	Fabaceae	*Phaseolus leptostachyus*
Phaseolus leptostachyus subsp. *leptostachyus*		Fabaceae	*Phaseolus leptostachyus*
Phaseolus leptostachyus var. *incisus*	(Piper) Freytag, nom. inval.	Fabaceae	*Phaseolus leptostachyus*
Phaseolus leptostachyus var. *nanus*	Freytag, nom. inval.	Fabaceae	*Phaseolus leptostachyus*
Phaseolus leucanthus	Piper	Fabaceae	*Phaseolus coccineus* subsp. *darwinianus*
Phaseolus lignosus	Britton	Fabaceae	*Phaseolus lignosus*
Phaseolus ligulatus	Piper	Fabaceae	*Phaseolus ligulatus*
Phaseolus limensis	Macfad.	Fabaceae	*Phaseolus lunatus* var. *lunatus*
Phaseolus linearis	Kunth	Fabaceae	*Vigna linearis* var. *linearis*
Phaseolus linearis var. *latifolius*	Benth.	Fabaceae	*Vigna linearis* var. *latifolia*
Phaseolus longifolius	Benth.	Fabaceae	*Vigna longifolia*
Phaseolus longipedunculatus	Mart. ex Benth.	Fabaceae	*Macroptilium gracile*
Phaseolus lunatus	L.	Fabaceae	*Phaseolus lunatus*
Phaseolus lunatus var. *lunatus*		Fabaceae	*Phaseolus lunatus* var. *lunatus*
Phaseolus lunatus var. *macrocarpus*	Benth.	Fabaceae	*Phaseolus lunatus* var. *lunatus*
Phaseolus lunatus var. *silvester*	Baudet	Fabaceae	*Phaseolus lunatus* var. *silvester*
Phaseolus macrolepis	Piper	Fabaceae	*Phaseolus macrolepis*
Phaseolus macrorhynchus	Harms	Fabaceae	*Vigna macrorhyncha*
Phaseolus maculatus	Scheele	Fabaceae	*Phaseolus maculatus*
Phaseolus maculatus subsp. *ritensis*	(M. E. Jones) Freytag, nom. inval.	Fabaceae	*Phaseolus ritensis*
Phaseolus macvaughii	A. Delgado	Fabaceae	*Phaseolus macvaughii*
Phaseolus magnilobatus	Freytag, nom. inval.	Fabaceae	*Phaseolus* sp.
Phaseolus marechalii	A. Delgado	Fabaceae	*Phaseolus marechalii*
Phaseolus marinus	Burm.	Fabaceae	*Vigna marina*
Phaseolus martii	Benth.	Fabaceae	*Macroptilium martii*
Phaseolus max	L.	Fabaceae	*Glycine max*
Phaseolus metcalfei	Wooton and Standl.	Fabaceae	*Phaseolus maculatus*
Phaseolus micranthus	Hook. and Arn.	Fabaceae	*Phaseolus micranthus*
Phaseolus microcarpus	Mart.	Fabaceae	*Phaseolus microcarpus*
Phaseolus minimus	Roxb.	Fabaceae	*Vigna minima*
Phaseolus mollis	Hook. f.	Fabaceae	*Phaseolus mollis*
Phaseolus multiflorus	Lam.	Fabaceae	*Phaseolus coccineus* subsp. *coccineus*
Phaseolus mungo	L.	Fabaceae	*Vigna mungo*
Phaseolus neglectus	F. J. Herm.	Fabaceae	*Phaseolus neglectus*
Phaseolus nelsonii	Marechal *et al.*	Fabaceae	*Phaseolus nelsonii*
Phaseolus nipponensis	Ohwi	Fabaceae	*Vigna angularis* var. *nipponensis*
Phaseolus oaxacanus	Rose	Fabaceae	*Phaseolus pedicellatus* var. *oaxacanus*
Phaseolus oblongus var. *albus*	G. Martens	Fabaceae	*Phaseolus vulgaris* var. *vulgaris*
Phaseolus oblongus var. *spadiceus*	Savi	Fabaceae	*Phaseolus vulgaris* var. *vulgaris*

Name of Taxon	Name of Authority	Family	Preferred Taxon Name
Phaseolus oblongus var. *zebrinus*	G. Martens	Fabaceae	*Phaseolus vulgaris* var. *vulgaris*
Phaseolus obvallatus	Schltdl.	Fabaceae	*Phaseolus coccineus* subsp. *formosus*
Phaseolus oligospermus	Piper	Fabaceae	*Phaseolus oligospermus*
Phaseolus ovatifolius	Piper	Fabaceae	*Phaseolus ritensis*
Phaseolus pachyrrhizoides	Harms	Fabaceae	*Phaseolus pachyrrhizoides*
Phaseolus panduratus	Mart. ex Benth.	Fabaceae	*Macroptilium panduratum*
Phaseolus parvifolius	Freytag, nom. inval.	Fabaceae	*Phaseolus* sp.
Phaseolus parvulus	Greene	Fabaceae	*Phaseolus parvulus*
Phaseolus pauciflorus	Sesse and Moc. ex G. Don	Fabaceae	*Phaseolus pauciflorus*
Phaseolus pedicellatus	Benth.	Fabaceae	*Phaseolus pedicellatus*
Phaseolus pedicellatus var. *grayanus*	(Wooton and Standl.) A. Delgado ex Isely	Fabaceae	*Phaseolus pedicellatus* var. *grayanus*
Phaseolus pedicellatus var. *oaxacanus*	(Rose) A. Delgado, nom. inval?	Fabaceae	*Phaseolus pedicellatus* var. *oaxacanus*
Phaseolus pedicellatus var. *pedicellatus*		Fabaceae	*Phaseolus pedicellatus* var. *pedicellatus*
Phaseolus pedicellatus var. *polymorphus*	(S. Watson) A. Delgado, nom. inval?	Fabaceae	*Phaseolus pedicellatus* var. *polymorphus*
Phaseolus pedicellatus var. *purpusii*	(Brandegee) A. Delgado, nom. inval?	Fabaceae	*Phaseolus pedicellatus* var. *purpusii*
Phaseolus peduncularis	Kunth	Fabaceae	*Vigna peduncularis*
Phaseolus peduncularis var. *pusillus*	Hassl.	Fabaceae	*Vigna peduncularis* var. *pusilla*
Phaseolus perplexus	A. Delgado	Fabaceae	Phaseolus perplexus
Phaseolus pilosus	Kunth	Fabaceae	*Vigna lasiocarpa*
Phaseolus plagiocylix	Harms	Fabaceae	*Phaseolus plagiocylix*
Phaseolus pluriflorus	Marechal *et al.*	Fabaceae	*Phaseolus pluriflorus*
Phaseolus polyanthus	Greenm.	Fabaceae	*Phaseolus coccineus* subsp. *darwinianus*
Phaseolus polymorphus	S. Watson	Fabaceae	*Phaseolus pedicellatus* var. *polymorphus*
Phaseolus polystachios	(L.) Britton *et al.*	Fabaceae	*Phaseolus polystachios*
Phaseolus polystachios var. *polystachios*		Fabaceae	*Phaseolus polystachios* var. *polystachios*
Phaseolus polystachios var. *sinuatus*	(Nutt. ex Torr. and A. Gray) Marechal *et al.*	Fabaceae	*Phaseolus polystachios* var. *sinuatus*
Phaseolus prostratus	Benth.	Fabaceae	*Macroptilium prostratum*
Phaseolus pulchellus	Piper	Fabaceae	*Vigna* sp.
Phaseolus purpusii	Brandegee	Fabaceae	*Phaseolus pedicellatus* var. *purpusii*
Phaseolus radiatus	L.	Fabaceae	*Vigna radiata* var. *radiata*
Phaseolus retusus	Benth., nom. illeg.	Fabaceae	*Phaseolus maculatus*
Phaseolus rigidus	Piper	Fabaceae	*Phaseolus rigidus*
Phaseolus rioverdensis	Freytag and Debouck, nom. inval.	Fabaceae	*Phaseolus* sp.
Phaseolus ritensis	M. E. Jones	Fabaceae	*Phaseolus ritensis*
Phaseolus salicifolius	Piper	Fabaceae	*Phaseolus salicifolius*
Phaseolus scaberulus	Miq.	Fabaceae	*Vigna* sp.
Phaseolus scabrellus	Benth. ex S. Watson	Fabaceae	*Phaseolus pedicellatus* var. *pedicellatus*
Phaseolus schaffneri	Piper	Fabaceae	*Phaseolus pedicellatus* var. *polymorphus*
Phaseolus schottii	Benth.	Fabaceae	*Vigna longifolia*
Phaseolus scolecocarpus	Piper	Fabaceae	*Macroptilium* sp.
Phaseolus semierectus	L.	Fabaceae	*Macroptilium lathyroides*
Phaseolus semierectus var. angustifolius	Benth.	Fabaceae	*Macroptilium lathyroides*
Phaseolus sempervirens	Piper	Fabaceae	*Phaseolus jaliscanus*
Phaseolus sinuatus	Nutt. ex Torr. and A. Gray	Fabaceae	*Phaseolus polystachios* var. *sinuatus*
Phaseolus sonorensis	Standl.	Fabaceae	*Phaseolus sonorensis*
Phaseolus sp.		Fabaceae	*Phaseolus* sp.
Phaseolus speciosus	Kunth	Fabaceae	*Vigna speciosa*

Name of Taxon	Name of Authority	Family	Preferred Taxon Name
Phaseolus spectabilis	Standl.	Fabaceae	*Vigna spectabilis*
Phaseolus sphaericus var. atropurpureus	G. Martens	Fabaceae	*Phaseolus vulgaris* var. *vulgaris*
Phaseolus sphaericus var. *minor*	G. Martens	Fabaceae	*Phaseolus vulgaris* var. *vulgaris*
Phaseolus stenolobus	Standl.	Fabaceae	*Vigna linearis* var. *latifolia*
Phaseolus striatus	Brandegee	Fabaceae	*Phaseolus coccineus* subsp. *formosus*
Phaseolus striatus var. *purpurascens*	Freytag, nom. inval.	Fabaceae	*Phaseolus coccineus* subsp. *formosus*
Phaseolus strobilophorus	(B. L. Rob.) C. V. Morton	Fabaceae	*Ramirezella strobilophora*
Phaseolus sublobatus	Roxb.	Fabaceae	*Vigna radiata* var. *sublobata*
Phaseolus supinus	Wiggins and Rollins	Fabaceae	*Macroptilium supinum*
Phaseolus talamancensis	Debouck and Torres Gonzalez	Fabaceae	*Phaseolus talamancensis*
Phaseolus tenellus	Piper	Fabaceae	*Phaseolus tenellus*
Phaseolus tenuifolius	(A. Gray) Wooton and Standl.	Fabaceae	*Phaseolus acutifolius* var. *tenuifolius*
Phaseolus trilobatus	(L.) Schreb.	Fabaceae	*Vigna trilobata*
Phaseolus trilobus	Auct.	Fabaceae	*Vigna trilobata*
Phaseolus trinervius	Wight and Arn.	Fabaceae	*Vigna radiata* var. *sublobata*
Phaseolus tuberosus	Sesse and Moc. ex G. Don, nom. illeg.	Fabaceae	*Macroptilium gibbosifolium*
Phaseolus tuerckheimii	Donn. Sm.	Fabaceae	*Phaseolus tuerckheimii*
Phaseolus tunkinensis	Lour.	Fabaceae	*Phaseolus lunatus* var. *lunatus*
Phaseolus umbellatus	(Muhl. ex Willd.) Britton	Fabaceae	*Strophostyles umbellata*
Phaseolus unguiculatus	(L.) Piper	Fabaceae	*Vigna unguiculata* subsp. *unguiculata*
Phaseolus venosus	Piper	Fabaceae	*Phaseolus maculatus*
Phaseolus vexillatus	L.	Fabaceae	*Vigna vexillata* var. *vexillata*
Phaseolus viridis	Piper	Fabaceae	*Phaseolus viridis*
Phaseolus viridissimus	Ten. ex Miq., nom. inval.	Fabaceae	*Vigna* sp.
Phaseolus vulgaris	L.	Fabaceae	*Phaseolus vulgaris*
Phaseolus vulgaris var. *aborigineus*	(Burkart) Baudet	Fabaceae	*Phaseolus vulgaris* var. *aborigineus*
Phaseolus vulgaris var. *albus*	Haberle	Fabaceae	*Phaseolus vulgaris* var. *vulgaris*
Phaseolus vulgaris var. *mexicanus*	Freytag, nom. inval.	Fabaceae	*Phaseolus vulgaris*
Phaseolus vulgaris var. *nanus*	G. Martens	Fabaceae	*Phaseolus vulgaris* var. *vulgaris*
Phaseolus vulgaris var. *niger*	G. Martens	Fabaceae	*Phaseolus vulgaris* var. *vulgaris*
Phaseolus vulgaris var. *ochraceus*	Savi	Fabaceae	*Phaseolus vulgaris* var. *vulgaris*
Phaseolus vulgaris var. *variegatus*	DC.	Fabaceae	*Phaseolus vulgaris* var. *vulgaris*
Phaseolus vulgaris var. *vulgaris*		Fabaceae	*Phaseolus vulgaris* var. *vulgaris*
Phaseolus wrightii	A. Gray	Fabaceae	*Phaseolus filiformis*
Phaseolus xanthotrichus	Piper	Fabaceae	*Phaseolus xanthotrichus*
Phaseolus xanthotrichus var. *zimapanensis*	Ined.	Fabaceae	*Phaseolus* sp.
Phaseolus xolocotzii	A. Delgado	Fabaceae	*Phaseolus xolocotzii*
Phaseolus zebra var. *carneus*	G. Martens	Fabaceae	*Phaseolus vulgaris* var. *vulgaris*
Phaseolus zebra var. *purpurascens*	G. Martens	Fabaceae	*Phaseolus vulgaris* var. *vulgaris*
Phaseolus zimapanensis	A. Delgado	Fabaceae	*Phaseolus zimapanensis*

French Bean (*Phaseolus vulgaris* L.)

INTRODUCTION

French bean (*Phaseolus vulgaris* L., 2n=2x=22) is the most important among agriculturally successful crop. Five domesticated beans [(*P. vulgaris* (common bean), *P. coccineus* (scarlet runner bean), *P. lunatus* (lima bean) and *P. acutifolius* (tapary bean) and recently added *P. dumosus* (year bean)] which goes by the name of kidney bean or chhota sem or Farash bean (Katyal and Chadha, 1988) that accounts for more than 90 per cent of the cultivated *Phaseolus* worldwide (Singh, 2001; Acosta-Gallegos *et al.*, 2007). It is also known as snap bean, salad bean, greenbean, *haricotvert*, apply to varieties used as vegetable; haricot bean, dry bean, navy bean are applicable only to pulse types, string bean, dwarf bean and pole bean are applicable only to particular growth forms. In temperate countries they are mainly grown for fresh pod consumption and for processing as a frozen vegetable. Snap bean cultivars have slender, long and cylindrical pods with greately reduced fibre, thickened pod walls and smaller seeds. The extremely low pod fibre and pod shape could be used to morphologically distinguish between dry and snap bean in the archaeological records. It is consumed as tender pods, shelled green beans and dry beans. The dried seed beans are marketed under the name of *rajmash* in India. Snap bean seed can also be used in dry state like the dry bean types (Sharma and Joshi, 1993). Like other beans, French bean is a good source of folate, magnesium, manganese and dietary fiber. French bean is very low in saturated fatty acids and cholesterol. Per 100 g edible pods constitute 94 per cent water, 1.7 per cent protein, 0.1 per cent carbohydrates, 4.5 per cent fat, 12.66-34.17 per cent crude fibre and 0.1 per cent minerals. French bean is grown for green tender and stringless pods, which are cooked, whole or shelled as pea. The immature seeds are boiled or steamed and used as a vegetable. Young leaves are consumed raw or cooked as a potherb. The mature seeds are dried and stored for future use. They must be thoroughly cooked before consumption and are best if soaked in water for about 12 hours prior to this. They can be boiled, baked, pureed or ground into powder. The seeds can be sprouted and used in salads or cooked. French bean is in great demand in cities (Singh, 2001).

Origin, History and Distribution

The genus has contributed to crop plants in agriculture in both the old and the new world (Evans, 1975). *Phaseolus* beans are among the oldest establishment crop plants, and radio carbon dates show that *P. vulgaris* was domesticated in Central America about 7,000 years ago. Vavilov (1951) reported that *P. vulgaris* L. originated in South Mexico and Central American Centres of origin (primary centre of origin) as a large number of its species are found in Mesoamerica (Kaplan, 1956, Freytag and Debouck, 2002; Acosta-Gallegos *et al.*, 2007), while Ecuador-Peru-bolivia region is the secondary centre of origin (Vavilov, 1950; Smartt, 1976; Gepts, 1998; Belucci *et al.*, 2014). The Red Indian of America cultivated French bean along with maize, from them Whitemen started its cultivation (Sturtevant, 1919.). Spanish explores and introduced it into Europe (Purseglove, 1968). The hypotheis of Mesoamerican origin of the common bean is supported by the observations that the closest relatives of wild *P. vulgaris* are distributed throughout Mesoamerica (Schmit *et al.*, 1993; Delgado-Salinas *et al.*, 2006). The current distribution of the wild common bean encompasses a large geographical area: from northern Mexico to north-western Argentina. Prior to domestication, wild *P. vulgaris* had diverged into two major gene pools on the basis of geographical distribution:

(i) The Mesoamerican *i.e.* Middle America and

(ii) The Andean *i.e.* Andean South America which can be distguished at the morphological biochemical and molecular levels (Singh *et al.*, 1991a), and also display partial reproductive isolation caused by F_1 lethality (Gepts and Bliss, 1985).

With the exceptions, no successful recombination has occurred between the two major gene pools.

(i) A first exeception is provided by Chilean landraces which showed signs of introgression from the Mesoamerican gene pool based on phaseolin seed protein and allozymes (Paredes and Gepts, 1995).

(ii) The second exception is evolution of snap bean cultivars.

Although they originated in the Andean gene pool, many varieties are actually intermediate between the two gene pools as evidenced by RAPD markers (Skroch and Nienhuis, 1995). This intermediate position may be attributed to recent breeding efforts aimed at introducing disease resistance from the Mesoamerican gene pool into the snap bean cultivars (Gepts, 1998). While only these two major gene pools are recognized in the domesticated population, the geographical structure of the wild form of the common bean is more complex, with an additional third gene pool that is localized between Peru and Ecuador, and characterized by a specific storage seed protein, phaseolin type I (Debouck *et al.*, 1993; Kami *et al.*, 1995). Generally, the Mesoamerican gene pool possesses higher content of lectin, ca, P, S and Zn than the Andean gene pool but lower phaseolin and Fe (Islam *et al.*, 2002). Further, the two major gene pools in *P. vulgaris* have been divided into six races (Mesoamerican gene pool: Mesoamerica, Durango and Jalisco; and Andean gene pool: Nueva Granada, Peru and Chile) as the members of each race share distinct morphological, agronomic, physiological and biochemical traits; and differ from other races in allelic frequencies of genes controlling these traits (Singh *et al.*, 1991a). The earliest archeological evidence of domesticated bean seed is from 5500 BC in northern Peru and 5,000 BC in Mexico. It is believed that the 1st cultivar has been selected at Peru. From Peru the crop spread to Europe in 16th and 17th century (mid 1500s) after the Spanish conquest and then to Asia and then to all parts of world by trading. The French Huguenots introduced this crop into England by 1594 (Purseglove, 1968). Portuguese took the crop to Africa and other parts of the Old World. At present French beans are grown throughout the world in subtropical and temperate climates. Introduction of French bean into India is comparatively recent and then the Europeans brought it to India in the 17th century (Simmonds, 1976).

It is true that in the 19th century, the French men made green beans a household vegetable, with the name haricot verts branding them as "French bean" in the minds of many Europeans (Andrews, 2013). French bean is grown intensively in five major continental areas; North and Central America, Eastern Africa, Eastern Asia and Western and Southeastern Europe. Latin America produces roughly 30 per cent of the world production. Brazil is the largest producer of French bean in the world. In the USA it is commercially grown for processing. In India it is largely grown in hilly areas of Punjab, Haryana, Himachal Pradesh, Jammu and Kashmir and North East states during summer and autumn crop and as autumn crop in western Uttar Pradesh, Maharashtra, Karnataka, Tamil Nadu and Andhra Pradesh.

Species Availability, Evolution and Diversity

The domestication of *P. vulgaris* L. might have occurred in Brazil and northern Argentina from a wild form of *P. arborigineus*, which is progenitor of *P. vulgaris* L. (Yarnell, 1965).

The genus *Phaseolus* contains 50-70 species, all of which are natuive to the Americas (Miranda-Colin, 1966; Miranda-Colin, 1974; Verdcourt, 1980; Baudet, 1977; 1978; Marechal *et al.*, 1978). As reported by Evans (1976 and 1980) four species are cultivated through the world: *P. vulgaris* (common bean), *P. coccineus* (scarlet runner bean), *P. lunatus* (lima bean) and *P. acutifolius* (tapary bean) and recently added *P. dumosus* (year bean). Wild forms of *P. coccineus* and *P. lunatus* are perennial, but *P. vulgaris* and *P. acutifolius* are annuals (Freytag, 1975). *P. vulgaris P. acutifolius* and *P. lunatus*, self pollinate before the flower opens. French bean (*P. vulgaris*) is an important vegetable under this genus. A wild related species *P. aborigines* in North West Argentina is considered as the wild progenitor of French bean. Many are applicable to particular forms within the species only, and do not apply to the whole range of forms found in the species. The related species are *P. coccineus L.* (*syn. P. multiflorus* L.) and *P. aconotifolius* Jacq.

Table 4.50: *Phaseolus* Species Possessing Desirable Traits

Species	*Desirable Traits*
Phaseolus vulgaris var.sub-*compressus*	Disease resistance
P. coccineus	Root rot (*Macrophomina phaseoli*) and Halo blight (*Pseudomonas syringae* pv *phaseolicola*)
P. flavescens	Rust resistance
P. acutifolius	Bacterial blight and wilt resistance
P. ritensis	Disease resistance
P. lunatus	Disease resistance and quality characters
P. dumosus	Disease resistance

The traditional snap bean type of southern Europe 'Romano' (Flat pod beans) might be one of the predecessors to the contemporary snap bean. However, the stringless trait was discovered in 1970 by C.N. Keeney while working in L Roy, New York; called the "father of the stringless bean". He established a seed company at Le Roy as an outgrowth of his interest in green beans. Since then stringless trait has been widely incorporated to develop new cultivars of snap bean (Ram, 2005; Andrews, 2013). French bean has wider diversity and hence larger gene pool. Quantitative traits such as yileld, pod weight, number of pods per plant and number of seeds per pod contributes 62.26 per cent of the genetic diversity in the region (Verma *et al.*, 2014). Despite the wide diversity among common beans, the genetic base for the cultivars of green bean market classes, particularly snap bean, is narrow (Kelly, 2004). There are five cultivated species of genus *Phaseolus* which are called differently and also need different climates for existence.

Thirteen EC genotypes EC14907, EC22367, EC24646, EC39439, EC42960, EC43.36, EC57080, EC 94469, EC99539, EC10080, EC102711, EC108101 and Premier were for different horticultural traits. For green pod yield, EC57080 and EC108101 with higher mean values (127.28 and 112.86) with significant 'b' value of 1.22 and 0.9 and non-significant deviations from regression might be stable over the years in a favourable environment (Tyagi and Joshi, 1996).

Table 4.51: Common Name and Adaptation of *Phaseolus* Species

Phaseolus Species	*Common Name*	*Altitude*	*Temp. (°C)*	*Precipitation (mm/year)*	*Growth Cycle (Day)*
P. vulgaris L.	Common bean, dry bean, shell bean, snap bean, French bean	50-3000	14-26	400-1600	70-330
P. polyanthus Greenman	Year long bean	800-2600	14-24	1000-2600	110-365
P. coccineus L.	Runner bean, scarlet runner bean	1400-2800	12-22	400-2600	90-365
P. acutifolius A. Gray	Tepary bean	50-1900	20-32	200-400	60-110
P. lunatus L.	Lima bean (large seeded), sieve bean (small seeded), butter bean, Madagascar bean	50-2800	16-26	0-2800	90-365

Source: Debouck, 1991.

Table 4.52: Gene Pools, Protein and Distribution of Various Races of *Phaseolus vulgaris* L.

Sl.No.	*Race*	*Phaseolin*	*cpDNA Haplotype#*	*Distribution*
A. Mesoamerican race (Middle American race)				
A1.	Mesoamerica	S, Sd, B	K, L, I, J	Tropical lowlands and intermediate altitudes of mexico, Central America, Colombia, Venezuela and Brazil
A2.	Durango	S, Sd	K, L, J	Semiarid central and northern highlands of Mexico and south-western USA
A3.	Jalisco	S	L, K	Humid highlands of Central Mexico and Guatemala
A4.	Guatemala	S	I	Highlands of Guatemala and Chiapas, Mexico
B. Andean race (South American race)				
B1.	Nueva granda	T	C	Intermediate altitudes of the northern Andes (Colombia, Ecuador, and Peru), also in Argentina, Belize, Bolivia, Brazil, Chile, Panama and few Caribbean countries
B2.	Chile	T, C	C	Drier regions of lower altitudes in the southern Andes (southern Peru, Bolivia, Chile and Argentina)
B3.	Peru	T, C, H	C, K	Northern Colombian highlands to Argentina

* Phaseolin seed protein electrophoretic type (Gepts *et al.*, 1988; Gepts and Bliss, 1985; Koenig *et al.*, 1990; Singh *et al.*, 1991b, 1991c).

Frequency of haplotype in desending order from left to right; distribution of haplotypes based on the survey of 127 Mesoamerican and Andean landraces (Chacon *et al.*, 200).

Races of Common Bean

Common bean and its wild relatives showed a wide geographical distribution found in Middle America for the Mesoamerican race and South America for the Andean gene pools, but greater amounts of genetic variation were found among the Mesoamerican population as compared to Andean gene pool (Gepts *et al.*, 1988; Acosta-Gallegos *et al.*, 2007). Within the two domesticated gene pools, several eco-geographic races have been identified in each gene pool based on plant morphology, eco-geographic and ecological distribution, isozyme and molecular information (Singh *et al.*, 1991a, 1991b, 1991c; Acosta Gallegos *et al.*, 2007). Additional, Beebe *et al.* (2000) indicated that the races Durango and Mesoamerica could be further subdivided in two sub-class; and a distinct race was separated from the Jalisco race, namely 'Guatemala race' that includes the climbers found in the highlands of Guatemal and Chiapas, Mexico (Chacon *et al.*, 2005).

Difference between *Phaseolus* L. and *Vigna* Savi. Genera

The difference between genera *Phaseolus* L. and *Vigna* Savi the degree to which the beak of the keel was curved, the keel of *Phaseolus* being coiled or twisted and that of *Vigna* is being curved. The main distinguishing characters between *Phaseolus* L. and *Vigna* Savi are:

Table 4.53: Difference between *Phaseolus* L. and *Vigna* Savi

Phaseolus L.	*Vigna Savi*
Stipules not prolonged	Stipules prolonged
Keel in a spiral	Keel erect, in curved
Styles without apical appendage	Styles with distinct beak
Stipules truncate	Stipules cordate
Keel a spirally twisted	Keel usually straight
Fruit not septate	Fruit septate
First pair of leaves, petiolate	First pair of leaves, sessile
Fine sculptured pollen grains	Open reticulation of pollen
Type spp. *P. vulgaris* L.	Type spp. *V. luteola* Jacq.

Gene and Genetics of Characters

Red or purple stem is always associated with red flower colour, pink stem with pink flowers and green stem can have all the coloured flowers. White flowered plants bear white seeds while darker coloured ones bear types of coloured seeds. Seed coat colour is a complicated character, range of colour being coloured, colourless and mottled. The basic colour is buff colour.

Table 4.54: Character, Number of Gene and Type of Gene Action

Character	*Number of Gene*	*Type of Gene Action*
Growth habit	Monogenic	Indeterminate or climbing habit dominant to determinate habit
Pod stringness	Monogenic	Stringness dominant to stringlessness
Pod surface	Monogenic (acc acc)	Shiny node recessive to non-shiny
Earliness	Monogenic (fdfd)	Early flowering dominant to late
Groove of pods	Monogenic	Non-grooved dominant to grooved
Pod shape	Polygenic	Normal round has a general dominance over other shapes
Colour of foliage	Digenic	Green dominant to variegated, duplicate gene action
Hard seed	Monogenic	Soft seed dominant to hard seed.

Source: Peter, 1998.

Prakken, 1934 reported that pod suture stringlessness governed by a single dominant gene (*St*). Drifjhout *et al.* (1978) and Drifjhout (1978) confirmed the *St* gene and also described a temperature sensitive dominant gene (*Ts*) forms strings at higher temperature.

Table 4.55: Some Important Genes of *Phaseolus vulgaris*

Gene Symbol	*Phenotype*
Are	Resistance to anthracnose
Da	Straight pod
Ext	Interspecific gene for external stigma in *Phaseolus coccineus*
Fa	Basic gene for pod membrane
Fin (fin+)	Indeterminate vs 'fin' determinate plant growth
gas	Causes both male and female sterility
mo	Conditions resistance to bean virus
neu+	Short day flowering response
ri	Confers resistance to bean virus
T	Twining habits vs non twining
te	Pod short (5.0-8.0 cm) and narrow
To	Cell wall fiber

Source: Yarnell, 1965.

Table 4.56: Growth Habit in *Phaseolus vulgaris*

Growtt Habit	*Type of Growth Habit*	*Gene*	*Gene Remark*
Bush determinate	Type-I growth habit	llaatt	Determinate compact growth habit
Bush indeterminate	Type-II growth habit	L-aatt	Determinate prostrate growth habit
Pole determinate	Type-III growth habit	L-A-tt	Indeterminate prostrate growth with well-developed branching, but low or non-existent of climbing ability
Pole indeterminate	Type-IV growth habit	L-A-T	Indeterminate with long vine and high climbing ability

Botanical Description

Genus *Phaseolus* belongs to the family Fabaceae. Earlier it was under family Leguminosae, subfamily Lotoideae (Lawrence, 1973.). Hutchinson (1966) classified it is a member of Papilinoidae and the tribe Phaseoleae. It falls under genus *Phaseolus* and the key character to the genus being coiling of keel of corolla. It is highly self-pollinated naturally. The chromosome number is 2n=2x=22. Tetraploids (2n=44) have been identified. Monosomic (2n=21) have also been reported. A haploid genome size is estimated to be between 587-637 Mbp (Arumuganathan and Earle, 1991; Bennett and Leitch, 2010).

Table 4.57: Botanical Description

Growth habit	It is an annual herb. The bush forms, which grow in erect leafy clumps, reach 1 m tall, while twining forms, which need to be supported with poles or trellising, grow up to 4 m long.
Root	Tap roots (grows up to 120 cm) with poor nodule or no nodule formation (nitrogen fixing bacteria). Large number of adventitious roots arises from the base of the stem and tap root.
Stem	Broadly, stem could be bush type or pole type depending upon growth and twining habits. Stem growth habit is governed by three genes such as L/l (long stem >short stem), A/a (indeterminate growth habit >determinate growth habit) and T/t (twining tendency>non- twining tendency).
Leaf	Plants have pinnately trifoliate compound leaves with oval to rhombic leaflets (broad ovate or long and narrow), each up to 16 cm (9.25 in.) long, which are pubescent (covered with downy hair).
Floral biology	The flowers, which may be white, yellow, violet, or red, in loose, open unbranched clusters (racemes) that are shorter than the leaves, and develop into linear round to slightly flattened pods up to 15 cm (6.75 in.) long.
Anthesis and pollination	Anthesis occurs in the morning (5.0-7.0 AM) as described by Bhatnagar *et al.*, 1972. Anther dehiscence is from 5.30 to 7.30 AM. Pollen remains fertile on the day of anthesis. Stigma receptivity is at its peak on the day of anthesis between 7.0-8.0 AM.
Pod and pod set	The tender pods (yellow, green and purple in colour) are harvested when they are rapidly growing, bright in colour, fleshy with small seeds, generally about 9-12 days after flowering. The pod size of French bean varies from thin, pencil like linear pod up to 20 cm long and about 2.00 cm in diameter. The pods are round, flat or curved in shape and the colour varies from green to yellow, red and purple. The pod can be green, yellow, red or purple
Seeds	Single pod may contains up to 12 seeds. Seeds are smooth, plump, kidney-shaped, up to 2.0 cm long, range widely in color, and are often mottled in two or more colors.

Horticultural Classification

French bean can be distinguished by seed size, leave shape, texture and colour, flower colour, pod shape, size, curvature colour and external pigmentation and presence or absence of string. Names such as French bean, haricot bean, green bean, string bean, edible podded bean, wax bean or Romano (also known as Italian or Flat pod beans) describe subgroups or market classes of snap bean. The term string bean refers to the older cultivars of snap bean that had fibre in the pod suture or seam, which had to be removed manually before cooking. Blue Lake Green Beans, varieties of snap bean, are named for the area in which they were developed in the early 1900s, the Blue Lake area near Ukiah, California, USA. They were originally developed for canning. By the mid-to-late 1920s, the beans had been developed into stringless bean for use as a green bean, probably in Oregon, USA. A variety 'Tender Green' of snap bean came on the scene in 1925 (Andrews, 2013). The snap beans, especially vegetable type genotypes, are slender, long, fleshy, tender, soft, free from the fibrous layer found in the pod wall (inedible fibre/parchment) and contain fewer parchment strings (present along the both pod-sutures but especially strong on ventral side which is made up of lignified sclerenchyma cells) whose immature fleshy pods and seeds are consumed as vegetable (Singh *et al.*, 2014). Romano bean pods are green in colour with flat cross-sectional shape and relatively fibre-free. In contrast to the Romano, green beans have fleshy pods that are generally oval to round in cross-section. Wax bean pods have similar shape to green bean pods, but are pale yellow to golden, instead of green in colour, because chlorophyll is absent from the pods, petioles and young stems (Myers and Baggett, 1999). The traits of economic importance include colour, pod shape, length, cross-sectional shape, straightness, smoothness, fibre content, interlocular cavitation, rate of seed development and point of detachment (Silbernagel, 1986; Myers and Baggett, 1999).

French bean can be classified broadly on the following basis.

(i) Condition for the Use

The snap bean genotypes comprise a group of common beans that have been selected for succulent tender pods with reduced fibres and strings; while common beans usually for seed characteristics (colour, shape and size). As food, green bean (vegetable bean) shelled beans or snap beans or string beans or dry shelled beans, based on the content of fiber in the pods.

(ii) Colour of the Immature Pod

(i) Green pod

(ii) Wax pod

(iii) Purple pod-The purple podded genotypes are rich in anthocyanin content, 15 times higher than the normal green-podded genotypes (Singh *et al.*, 2011)

Above group may be with or without fibres. Wax podded varieties are not common and mostly confined to the home gardens.

(iii) Growth Habit

On the basis of growth, these have been classified into two broader group *viz.* bush types and pole types and some times a third group-semi pole type. It consists of:

(a) Pole or climbing beans-Indeterminate type requiring support; and bush beans (dwarf in nature). It is popular among the tribals since it is used for mix cropping with maize, the stem of which act as the support for the bean.

(b) Semi pole type- Not uncommon, runner type, not very prolific, and pods emerges on 4-8 nodes

(c) Dwarf type or Bush type- Determinate type, bush type, smaller plants stature, height upto 40-60 cm, erect stem, main axis consisting of 4-8 shortened internodes, terminating in an inflorescence leaves alternate trifoliate, and pods emerges on 6 nodes with long petiole. It is vey popular among the growers because cultivation is popsible without support and realization of more yield per unit area.

Ram (2005) reported that common bean cultivars have four prototypes with respect to their growth habits on the basis of gene responsible:

Table 4.58: Growth Habit in *P. vulgaris*

Growth Habit	*Type of Growth Habit*	*Gene*	*Remark*
Bush determinate	Type-I growth habit	llaatt	Determinate compact growth habit and these types are very popular.
Bush indeterminate	Type-II growth habit	L-aatt	Determinate prostrate growth habit
Pole determinate	Type-III growth habit	L-A-tt	Indeterminate prostrate growth with well-developed branching, but low or non-existent of climbing ability.
Pole indeterminate	Type-IV growth habit	L-A-T	Indeterminate with long vine and high climbing ability

Source: Ram, 2005.

(iv) Beans for Processing

Classification on the Basis of Growth Habit

Keys for the classification based on growth habit:

Table 4.59: Classification Based on Growth Habit

	Pole Type	*Bush Type*	*Semi Pole Type*
	Green podded	Wax podded	
Flat podded	Round podded	-	-
Flower colour			
Seed colour			
Tolerance to diseases *etc.*			

Evans (1975) observed 5 distinct habits of plants in *P. vulgaris.*

(D) Classification on the basis of Edible Parts

(i) Dry-shelled beans

(ii) Vegetable bean

(iii) Bean for processing

(iv) Seed beans

Table 4.60: Distinct Habits of Plants in *P. vulgaris*

Race	*Habits of Plants*	*Internode Length*	*Number of Internode*	*Leaf Size*
1	Indeterminate	Long	17-35	1 (a) Small
				1 (b) Large
2	Indeterminate semi climber	Long	14-30	Small
3	Indeterminate bush	Short	13-25	Small
4	Determinate climber	Short	12-15	Small
5	Determinate bush	Long	3-5	Large

(E) Classification of Cultivars

The French bean cultivars are classified as follows (*cf.* Bland, 1971).

Table 4.61: Classification of Snap Beans and their Cultivars

Growth Habit and Pod Colour		*Cultivars/Promising Genotypes*
Dwarf or bush type		
1	Green pods	Contender, Sparton Arrow, Premier, Tender Green, King Green, Processor, Tender Crop, Top crop, Cascade, Bountiful (Flat pod), Plentiful (Flat pod), Green Ruler(Flat pod), GV-50, Quick Freezing, Siegerin, Prinsa, Lancet, Bush Blue Lake-92, GP-72-122, Bush Blue Lake, Blue Lake-47, Pusa Parvati, VL Boni-1, Arka Komal, Arka Suvidha, Arka Sarath, Arka Anoop, Arka Bold, Pant Anupama, Pant Bean-2, Kasi Sampann, Kashi Rajhans, VRFBB-91, VRFBB-95
2	Wax pods	Cherokee Wax, Uranus, Golden Ruler (Flat pod), SUG-131, Natal Sugar
3	Purple pods	Red Swan
4	Streaked	Dragon's Tongue
Climbing or pole types		
1	Green pods	Canadian Wonder, Kentucky Wonder, Pusa Himlata, Green Lake, Onatra, Blue Lake, Romano(Flat pod), Phenomenal Long Poded, IC593590, IC593591, IC593592, IC593593, IC593594, IC593595, VRFBP-44
2	Wax pods	Kentuky Wonder Wax, Yellow Wax, Yellow Romano(Flat pod), Gold marie, Neckar Gold
3	Streaked	Rattlesnake

Augmentation of Genetic Resources and Crop Improvement

Breeding Priorities

- Earliness
- High yield
- Wide range of crop adaptation
- Fleshy and tenderness
- Resistance against biotic and abiotic stresses

The breeding objectives in French bean are for the improvement of vegetable types and dry bean types. The breeding procedures for the crop include hybridization, pedigree selection, single plant selection, bulk method of selection *etc.*

Table 4.62: Breeding Methods Applicable in French Bean Improvement

Character	*Methods of Breeding*
Yield	Pedgree method, mass selection, single plant selection, single cross, back cross, multiple cross, hybridization and mass sowing of individual selection (HIMSI)
Seed colour, seed size and cold resistance	Selection from crosses
Protein (per cent)	Modified pedigree selection using single seed descent
Mosaic resistance	Back crossing, selection from back crosses
Resistance to anthracnose	Pure line selection
Resistance to halo blight	Back crossing

Introduction and Selection

Many promising genotypes have been introduced and developed as varities like 'Contender (USA)', Bountiful (USA), 'Giant Stringless'(USA), Kentucky Wonder (USA), 'Big Ben Red', 'VL Boni I' and 'Pusa Parvati'. ICAR Research Complex for NEH Region, Umiam (Meghalaya) evaluated 94 germplasm and noted highest yield per plant from the pole genotype MZFB-47 (320 g), MZFB-45 (302 g), RCFB-88 (266 g), RCFB-34 (252 g), MZFB-36 (251 g) and RCMFB-1A (250 g) over check Naga Local (166 g). However in bush type the highest yield was recorded from the RCFB-3 (133 g) over check Arka Anoop (62 g). The seed weight ranged from 22 g (RCFB-5) to 60 g (Naga Local-1). High yielding anthocyanin rich purple genotype MZFB-45 (315.25 g) and MZFB-48 (290 g/plant) were found field tolerant to moisture stress at ICAR Research Centre, Mizoram. The pod of Nagaland collections were flat and bold seeded while collection of Mizoram were round podded with small seeds. However collections of Hailakandi were flat podded with long seeds. Five genetically diverse

Table 4.63: Trait Specific Introduction of French Bean Genotypes

Accessions	*Source*	*Salient Features*	*Distrbution/ Recipient*
EC502154-58	USA	Upright determinate and high yielding	ICAR-IIPR, Kanpur (U.P.); ICAR-NBPGR, N. Delhi
EC506078	USA	Upright, short vine growth habit, mid season maturity and most unique feature is retention of a bright red colour after cooking	ICAR-IIPR, Kanpur (U.P.); ICAR-NBPGR, N. Delhi

Table 4.64: Mean Performance of different Genotypes for Important Characters in French Bean

Characters	*Mini.*	*Max.*	*Mean*	*Promoising Lines*
Days to 50 per cent flowering	52.44	75.31	56.18	EC530983, PGRFB-28-07, EC530942, EC530989, PGRFB-23-07, EC530932
Number of pods/ cluster	1.66	4.17	2.72	Swarna Lata, PGRFB-41-07, PGRFB-45-07, PGRFB-37.07, PGRFB-61-07, Pant Bean-2
Number of pods/ plant	6.7	16.03	9.22	Swarna Lata, PGRFB-41-07, PGRFB-45-07, PGRFB-37.07, Swarna Lata, PGRFB-41-07, PGRFB-45-07, PGRFB-37.0762-07, Pant Bean-2
Pod length (cm)	7.38	12.43	9.80	Pant Anupama, PGRFB-35-07, PGRFB-61-07, PGRFB-45-07, PGRFB-62-07, FBDK-2
Plant height (cm)	19.97	102.17	45.92	PGRFB-41.07, PGRFB-48-07, PGRFB-62-07, PGRFB-45-07, PGRFB-27-07, Swarna Lata
No. of primary branches/plant	2.53	4.43	3.60	Swarna Lata, PGRFB-22-07, EC530873, Pant Anupama, EC530996 and PGRFB-27-07
Days to 50 per cent maturity	86.73	113.5	99.33	PGRFB-28-07, EC530933, EC531017, EC530932, EC530959, PGRFB-25-07
Pod yield/plant (g)	35.27	67.62	50.28	Swarna Lata, PGRFB-35-07, Pant Anupama, Sel.-3, EC530980, EC530938
Pod yield (q/ha)	70.54	135.23	100.57	Swarna Lata, PGRFB-35-07, Pant Anupama, Sel.-3, EC530980, EC530938
Seed yield/plant (g)	7.75	19.57	11.33	Swarna Lata, PGRFB-35-07, Pant Anupama, EC530936, Pant Bean-2, EC530939
100-seed weight (g)	16.67	39.52	27.78	EC530942, PGRFB-35-07, EC530975, EC530944, EC530940, PGRFB-49-07

parents (Anupam, Badshah, Falguni, Darjeeling Local-1 and Shillong Local-2) having contrasting growth habits were selected through multivariate analysis for further combination breeding approach through hybridization and selection of desirable segregates in advanced generation (Lyndoh, 2014).

Pant Bean-2 and YCDI are dual purpose types for both green pods and dry seeds (Nath *et al.*, 2008). Most of the genetic variations exist among the introduced varieties/strains in the grower's field where no effort was made to maintain in the genetic purity of a particular variety. Mehra and Singh (2012) evaluated 70 germplasm at GBPUA&T, Pantnagar (Uttarkhand) and results are as shown in Table 4.64 and 4:65.

Table 4.65: Estimates of Stability Parameters (mean, b and S^2d) for Green Pod Yield

Genotype	Source	Mean	b	S^2d
EC14907	USA	86.92	1.37	316.74**
EC22367	USA	89.40	0.82**	366.86**
EC 24646	Sweden	70.78	0.76**	559.74**
EC39439	USA	96.92	0.64	288.94**
EC42960	USA	87.76	1.76	498.16**
EC43036	USA	112.90	0.71	12.68
EC57080	Australia	127.28	1.22**	-20.91
EC94469	Bulgaria	99.58	2.48	35.09**
EC99539	USSR	89.48	1.32	78.59**
EC100680	Netherlands	96.72	0.006	68.27**
EC102711	Australia	85.84	0.33	-563.76**
EC108101	USA	112.86	0.91**	9.35
Premier	USA	79.58	0.31	153.85**
Grand mean-94.93; **S.Em.**=6.12; **C.D.**=16.96; **P=0.01				

Similarly, Tyagi and Joshi (1996) worked out the genetic paramers in vegetable type French bean and reported for green pod yield, EC57080 and EC108101 with higher mean values (122.28 and 112.86)with significant 'b' value of 1.22 and 0.9 and non-significant deviations from regression might be stable over the years in a favourable environments. Also the genotypes such as EC100680 and Premier showed relatively better mean values with low 'b' values (0.006 and 0.31) indicating their better performacnce in poor environments.

Lynkdoh *et al.* (2016) evaluated 18 genotypes (12 varieties and 6 indigenous collections from Shilong, Meghalaya, Darjeeling and Nadia district of West Bengal) employing 30 important characters and grouped in various clusters. Cluster IV and Cluster III were the largest clusters containing 4 genotypes each. The inter Cluster D^2 value was maximum between Cluster IV and VI and genotypes belonging to these clusters with maximum inter cluster distance are genetically more divergent and could be used in hybridization programme to obtain promising segregants. Cluster III emerged as unique because it had high mean values for maximum number of yield and total number of pods per plant. Considering the clustering pattern of both the analysis, intra and inter cluster divergence, cluster mean values for different characters the following Anupam and Badshah from Cluster IV, Falguni from Cluster III, Darjeeling Local-1 and Shillong Local-2 are the divergent genotypes which have been selected for further combination breeding approach through hybridization and selection of desirable segregates in advanced generations. Rajesh and Bahadur (2016) evaluated 15 genotypes and noticed wider variability, high heriatability and genetic advance for pod yield/ plant, number of leaves, pod flesh thickness (cm), pods/ plant, single pod weight (g), cluster/plant, seeds/pods, pods/cluster, primary branches/plant, pod width (cm), pod length (cm), plant height, days to germination, plant spread (cm), germination percentage, seed index, days to first pod picking and days to 50 per cent flowering. Maximum number of genotypes (6) was grouped in to Cluster I followed by (4) in Cluster III. The intercluster distance was higher than intracluster distance indicating wide genetic diversity among the genotypes. Maximum intercluster distance was calculated between Cluster II and Cluster III (4688.674) followed by Cluster II and Cluster IV (4137.214), Cluster I and Cluster IV (3029.85), Cluster I and Cluster II (2392.382), Cluster III and Cluster IV (2059.792), Cluster I and Cluster III (2057.825). the genotype Kashi Sampanna, Arka Anoop, Arka Komal, Arka Suvidha are superior varieties based on mean green pod yield/plant and can be used as parent for hybridization programme to develop varieties with higher yield potential.

Hybridization and Selection

Ram and Singh (1980) evaluated 150 lines of French bean and isolated 'UPF 191', 'UPF 203', 'UPF 204' and 'UPF 209'; which were found high yielder. Specific donors ('UPF 33', 'UPF 208', 'UPF 481', 'UPF 482') for earliness; 'UPF 21', 'UPF 479', UPF 481', 'UPF 487' for dwarfness; 'UPF 127', 'UPF 209', 'UPF 485' for pods/ plant and 'UPF 67", 'UPF 154' for seed yield were identified. Few cross combinations developed in this crop are given below with salient features.

Table 4.66: F_1 Hybrids and their Salient Features

F_1 Hybrid	*Salient Features*
'Caro' x 'Michelite'	More leaf area
'Pinto x 'Torbat'	More pods/plant, 100-seed weight, yield/plant
Soviet hybrids	Heterosis for yield components

Radkov (1979) found no reciprocal difference in the intervarietal hybrid of French bean. Singh and Saini reported that non-additive gene effects for most of the crosses and Guitierrez and Singh (1985) reported moderate to low heterosis for yield and yield traits. However, Savova (1985) observed significant importance of F_1 and F_2 hybrids of *Phaseolus vulgaris* L. (Savi) with stable lines of *P. vulgaris* x *P. coccineus*. Ram and Rajput (1998) observed that French bean varieties/ lines VC-1, VL-6 and Pusa Parvati has been found good general combiners and the crosses *viz.* EC43036 x HUR-30, EC108101 x HUR-30 and VC-1 x Pusa Parvati exhibited high heterosis for plant height, pod length, number of pods/plant and total green pod yield/ plant, respectively. In 1999, they also reported that the cross VC-1 x PDR-14 showed maximum heterosis for total green pod yield per plant, number of pods per pplant, pod length and earliness. They reported that heterosis for green pod yield is generally accompanied by heterosis for yield components. Singh and Saini (1983) also studies on heterosis and inbreeding depression for primary components and yield components in French bean (*Phaseolus vulgaris* L.). 'Lakshmi' is a cross between 'Contender' (bush) x Local (pole) varieties and Peary Marrow is a cross between 'Wells Red Kidney' and 'White Marrow'. Pant Anupama developed through selection and TKD-1 selection made from germplasm population at ICAR-IIHR, Bangalore. Pusa Himlata is a pure line selection of an exotic introduction and YCD-1 is a pure line selection from a local indigenous type collected from Shevroy hill range of Tamil Nadu (India). IIHR-909 is a pedigree selection from the cross Blue Crop x Contender. Ram and Rajput (1999) conducted an experiment through line x tester analysis with four female parents and 10 pollinator parents to identify suitable general and specific combiners for breeding programme. Additive gene action was observed for the main yield-contributing traits in French bean, *i.e.* number of pods/plant and green pod yield/plant, and non-additive gene action was predominant for pod thickness only. Protein content and node of first flower initiation were governed by both additive and non-additive components equally. Specific combining ability was highly operative for green-pod yield which could be profitably be exploited for production of new plant type in French bean. Among lines VC-1 and among testers Pusa Parvati were observed to be the best combiners for most of the traits.

Interspecific Hybridization

The possibility of crosses between different species of *Phaseolus* L. may involve not only an improvement in the production of fruits and seeds but also ensure genetic

Table 4.67: Species, Compatibility and Utility

Sl.No.	*Species Crosses*	*Compatibility*	*Utility*
1.	*P. vulgaris* L. x *P. coccineus* L.	F_1 seeds viable and F_1 plants fertile	Tolerance to frost, resistance to *Phaseolous* virus I and II, extended flowering
2.	*P. vulgaris* var. sub-*compressus* x *P. multifloruous* Willd. var. *albus*	Compatible	Disease resistance
3.	*P. vulgaris* x *P. dumosus*	Pod set	Disease resistance
4.	*P. vulgaris* x *P. flavescens*	Crosses readily	Resistance to *Uromyces colletotrich.*
5.	*P. vulgaris* x *P. acutifolius*	Total F_1 sterility, F_1 seed viable	Resistance to disease
6.	*P. vulgaris* x *P. ritensis*	Sterile F_1 triploids were obtained by chromosome doubling	Resistance to disease
7.	*P. vulgaris* x *P.lunatus* L.	Fertility of F_1 ranged from nil to normal	

Source: Peter, 1998.

diversity of all those properties that characterize this genus (Marechal *et al.*, 1978; Mok *et al.*, 1978; Leonard *et al.*, 1987; Hoc *et al.*, 2003; Kusova and Myers, 2005). While numerous attempts have been made to cross all four major species to create bi- and tri-specific hybrids (Coyne, 1964; Hucl and Scoles, 1985), such attempts are fraught with difficulties to the lack of reproductive affinity among species (Thomas, 1964; Al-Yasiri and Coyne, 1966; Smartt and Haq, 1972a, 1972b; Mok *et al.*, 1978; Smartt, 1979, 1981; Pratt, 1983; e to cross all four major species to create bi- and tri-specific hybrids (Coyne, 1985; Andrade-Aguilar and Jackson, 1988).

Cultivar 'SVM-1' has been developed through interspecific hybridization between *P. vulgaris* L. var. Contender and *P. multifloruous* var. PBL257. Some distant hybridization is as under:

- *Phaseolus vulgaris* x *P. coccineus*
- *P. vulgaris* x *P. dumosus*
- *P. vulgaris* x *P. flavescens*
- *P. vulgaris* x *P. acutifolius*
- *P. vulgaris* x *P. ritensis*
- *P. vulgaris* x *P.lunatus*

However, *P. coccineus* (2n=14) is one way crossable with *P. vulgaris* (2n=14) parent, showing unilateral incompatibility.

Intergeneric Hybridization

Table 4.68: Crossability of Genera

Sl.No.	Genera	Crossability
1.	*Pisum sativum* L. x *Phaseolus vulgaris* L.	Incompatible, it can be overcome by using the irradiated pollen of the pea
2.	*Vigna sesquipedalis* x *Phaseolus vulgaris* L.	No seed set

Source: Peter, 1998.

Polyploidy Breeding

Autotetraploids (2n=44) have been reported in French bean. Monosomic (2n=21) has also been identified. Amphidiploids have been created by colchicines treatment of cross between *P. vulgaris* x *P. coccineus*, which showed improved fertility than the F_1. The autotetraploids have no distinct advantages over diploids.

Mutation Breeding

Mutation breeding programme on French bean is going on several countries for creating variability and identifying superior genotypes for biotic- fungi (rust, anthracnose), bacteria and virus (common mosaic, golden mosaic) and abiotic resistance (Neto, 1990). Through mutation breeding (dry seeds having 10 per cent moisture were X-rayed) a prominent variety 'Pusa Parvathi' was released as a mutant of waxy type 'EC1906' (Swarup and Gill, 1968; Anon., 1971b.). The variety is early, bush type, with attractive round and light green pods.

Table 4.69: Mutant and their Salient Characteristics

Cultivar/ Genetic Resource	Kind of Mutagen	Salient Characteristics
'Universal'	300 R X-rays (Granda)	Early-maturing, higher yield, resistance to *Colletotrichum lindemuthianum*
'Sanilac'	X-rays (Michelite)	Bust type, early-maturing, resistant to alpha, beta and gamma races of *C. lindemuthianum* and bean common mosaic viruses 1, and 1,2,3.

Source: Neto, 1990.

Resistance Breeding

(i) Insect Resistance

Panwar *et al.* (1987) evaluated 11 strains of French bean for resistance to stemfly (*Ophiomya phaseoli*) and found that cultivars Sel.2, UPF-191 (Pant Anupama), SVM-1, Sel-9 and Sel-4, has significantly lower agromyzid population than VL Boni, Sel-5, Pusa Parvati and Contender. Varietal resistance to *Empoasca* spp. was studies and less reduction in yield was found in white seeded varieties as compared to other types. Coloured seeded lines Viva, PI165435 and PI165426 were less susceptible to larval attack of bean seed maggots. Bat-1235, Bat-1230, Porrillo-70, S-6038 and Porrillo-1 were tolerant to pod weevil.

Table 4.70: Resistance Sources against Insect Pests and root-Knot nematode

Insect Pest	Resistance Genotypes/Varieties
Root-knot nematode	'Albama No. 1', 'Albama No. 2'
Aphids	'Strogele', 'Burna'
Leafhopper	'London Horticultural', 'Contender', 'Top Crop'

(ii) Disease Resistance

Resistance to bean rust is controlled by series of genes that-to-date-all are single and dominant (Kelly *et al.*, 1996). Some of the resistant genes reported are *Ur-3, Ur-4, Ur-5, Ur-6, Ur-7, Ur-9, Ur-11, Ur-12 and Ur-13* further, BAC-6, *Ouro-negro, dorado-108, dorado-53* which are unnamed are also reported (Miklas *et al.*, 2002). Other resistant genes like *Ur-Us-3* and *Ur-Resisto* are tentatively named as *Ur-8* and *Ur-10* respectively. The genes *Ur-4, Ur-6, Ur-9* and *Ur-12* are from Andean gene pool while, *Ur-3, Ur-5, Ur-7, Ur-11* are from Middle American gene pool. All the rust resistant genes differ in their spectrum. Among them, *Ur-11* has broad spectrum of resistance and resists all known races except one which is from Honduras *i.e.*, race-108. True horizontal resistance to common bean rust remains to be proven and utilized

(Stavely, 2000). French bean varieties CO-59196 and CO-33142 were found resistance against wilt (*Fusarium*) (Rodolfo-Velasquez and Schwartz, 2000).

It is important to group the genotypes but always consider the genetics of disease resistance in French bean.

Table 4.71: Genetics of Disease Resistance in French Bean

Disease	*Mode of Genetic Control*
Fusarium yellows	In completely dominant gene
Rhizoctonia wilt	Dominant genes
Sclerotinia rot	Ws gene
Angular leaf spot	Single dominant genes
Anthracnose	Single dominant genes

There are limitations in the natural genetic variability for resistance to bacteria, and golden mosaic virus. Purple hypocotyle and black seed-coat adhering tightly to cotyledon are associated with *Rhizoctonia* resistance. Common blight is inherited quantitatively, with low heritability. *Fusarium oxysporum* sp. *phaseoli* causes 'Fusarium Yellows' in French bean. Gene *Fop 1* controlled resistance to a race from Brazil in cultivars 'Tenderette', 'Pintado' and 'Early Gallatin', whereas gene *Fop 2*, incompletely dominant, conferred resistance to race from the USA in variety 'Preto Uberabinha' as reported by Ribeiro *et al.* (1979).

Genetic resistance to Bean Common Mosaic Virus (BCMV) is conditioned by a series of independent multi-allelic loci (Drijfhout *et al.*, 1978). At least 19 different strains of BCMV have been identified and biologically authenticated. The dominant I gene, resistance to BCMV and related potyviruses discovered by Ali (1950). It conditions either an immune or temperature dependent hypersensitive resistance. Therefore, this gene has been widely backcrossed into many bean varieties. Resistance to BCMV is controlled by the dominant gene I originally found in the cv. Corbett Refugee, race specific additional genes *bc-1*, $bc\text{-}1^2$, *bc-2*, $bc2^2$ or *bc-3* have been identified and one of these genes should be combined with I gene to have broad resistance to both the necrotic and mosaic reactions (Sibernagel, 1986). A variety 'Cornell 49/242' has been reported to have dominant gene '*Are*' for resistance to anthracnose.This resistance has been overcome by new races and new genes *Mex-2* and *Mex-3* will have to be combined with 'Are' gene to have stable polygenic resistance. Cultivar PBL257 (*P. coccineus*) was found resistant to angular leaf spot (*Isariopsis griseola*) and resistance was conferred by a recessive gene (Singh and Saini, 1980). However, another report indicated that resistance to angular leaf spot was inheritaed by a single dominant factor (Cardona, 1962). Resistance sources against diseases and their inheritance are given in Table 4.73 and Table 4.74, respectively.

Table 4.72: Inheritance of Host Resistance to Diseases

Disease	*Host Resistance*	*Genes Involved*
Halo blight (*Pseudomonas phaseolicola*)	Red Mexican	Single dominant
	PI150414	Single recessive
	LHB-1	Tolerant to leaves
	PDHB-1	Tolerant to pods
Common blight (*Xanthomonas phaseoli*)	GN Nebraska1, Sel-27, PI1207262	Polygenic (low heritable)
White mold (*Sclerotina sclerotiorum*)	P. coccineus	Ws
Angular leaf spot (*Isariopsis griseola*)	PBL-257 (Angular leaf spot (*Isariopsis griseola*)	Single recessive gene
Bean Common Mosaic Virus (BCMV)	-	*bc-1*, $bc\text{-}1^2$, *bc-2*, $bc2^2$ or *bc-3*
Bean Yellow Mosaic Virus (BYMV)	GN1140	by3
Rust (*Uromyces phaseoli*)	GN1140	RB11

Source: Kalloo, 1998.

Table 4.73: Inheritance of Resistance to Anthracnose (*Colletotrichum lindemuthianum*) in French bean

Pathotype	*P. vulgaris Variety*		*Ratio in F_2*		
	White Marrow (AAbb)*	*Robust (aaBB)*	*White Marrow x Robust (F_1, AaBb)*	*Resistant*	*Susceptible*
Alpha [$P_{(b)}$]**	R	S	R	3	1
Beta [$P_{(a)}$]	S	R	R	3	1
Avirulent [$P_{(o)}$]	R	R	R	15	1
Alpha + Beta [$P_{(a)+}P_{(b)}$]	S	S	R	9	7

*letters within parenthesis denote the genotype of the host.

** $P_{(a)}$, $P_{(b)}$ and $P_{(o)}$ denote the pathotype with reference to the resistance genes *A* and *B*. $P_{(a)}$ is virulent towards resistance gene *A*, $P_{(b)}$ towards gene *B*, $P_{(o)}$ towards neither gene *A* or *B*.

Table 4.74: Sources of Disease Resistance

Disease	*Race*	*Resistance Genotypes/Varieties*
Halo blight (*Pseudomonas phaseolicola*)	Race 1	'Refugee', 'Extra Early', 'Valentine', 'French Horticultural', 'Kentucky Wonder', 'Canadian Wonder', RMUI-3, Red Mexican, GN Nebraska1, Sel 27, V4604
Bacterial wilt (*Corynebacterium flaccumfaciens*)		'Emerson'
Powdery mildew (*Erysiphe polygoni*)		'Macarrao', 'Favortis', 'Ag 480'
Rust (*Uromyces phaseoli*)		'Ricobaio 1014', 'West Ralius'
Anthracnose (*Colletotrichum lindemuthianum*, and races)	Race beta and gamma	'Michelite 62', 'Sea Way'
	beta gamma and delta	PI304110
	Alpha beta gamma and delta	'Cornell 49-242'
Bean Common Mosaic Virus (BCMV)	bc-1, bc-2, bc-3 (recessive genes) Drijfhout (1978); 1(single dominant gene) Ali (1950)	'Morena', 'Monviso'Niveo', 'Montebianco''Cranbury', 'Lady Washington', 'Pinto', 'Great Northern', 'Red Mexican', *P. coccineus*
Angular leaf spot (*Isariopsis griseola*)	-	'Albama No.1', 'California', 'Small White', 'Epicure'
White mold (*Sclerotina sclerotiorum*)		*P. coccineus*
Rust (*Uromyces phaseoli*)	US rust races	'US Pinto 5' and 'US Pinto14'
	Maryland races	'Bush Blue Lake 290', 'Custer', 'Mountaineer white', Half runner

Source: Kalloo, 1998.

Among the bacterial diseases, common bacterial blight (CBB) is one of the most serious seed-borne disease that plague been production worldwide, especially in the warmer areas with high humidity and plant wounds during and after flowering. Breeding for resistance is complex as revealed by identification of at least 24 QTL distribution across al 11 linkage groups or chromosomes, and the expression of these QTL are influenced by environment, disease pressure, plant maturity and plant organs affected (Miklas *et al.*, 2006; Yu *et al.*, 2012). However, a dominant gene Xap conferring resistance to CBB was found in the small white bean line PR 0313-58 (Zapata *et al.*, 2010). Anthracnose (*Colletotrichum lindemuthianum*) is a seed-borne pathogen, found on all the continents, especially in the areas with high relative humidity and mild temperatures. The successful development of anthracnose resistant cultivars depends on understanding of the levels of variability within and among populations of the pathogen. More than 100 races of *C. lindemuthianum* have been identified worldwide (Sicard *et al.*, 1997). The resistance is governed by monogenic independent genes CO-1 to CO-10, and multiple alleles exist at the CO-1, CO-3, CO-4 and CO-9 (Kelly and Vallejo, 2004); mendez-vigo *et al.*, 2005.

Another serious fungal disease white mold, caused by *Sclerotinia sclerotiorum*, is responsible for considerable yield loss. Wet, cool and cloudy weather, particularly winter crop during rains at the bloom period, favours infection. Resistance to white mold is quantitatively inherited (Genchev and Kiryakov, 200; Kolkman and Kelly, 2003), consisting of physiological resistance and avoidance. Plant avoidance of white mold can be due to plant architectural traits which allow a drier and warmer microclimate under the canopy (bush indeterminate plants *i.e.* Type II growth habit, resistant to lodging, stay-green stem character, and open porous canopy and branching pattern) or agronomic management practices (Sofkova *et al.*, 2010).

Common bean variety Coyne (PI 655574) has the *Ur-3* and *Ur-6* genes for resistance to rust and carries the single dominant hypersensitive *I* gene that provides resistance to all non-necrotic strains of Bean Common Mosaic virus (Urreaa *et al.*, 2009). SCAR markers identified in French bean for rust disease are listed in Table 4.76.

Abiotic Stress

Over-expression of antioxidant enzymes could play

Table 4.75: Introduced Genotypes and their Salient Features

Accessions	*Source*	*Salient Features*	*Distrbution/Recipient*
EC498445 (var. AC Scarlet)	Canada	Tolerant to common bacterial blight (*Xanthomonas axonopodis*) and bean common mosaic virus (BCMV) possessing the I gene	ICAR-IIPR, Kanpur (U.P.); ICAR-NBPGR, N. Delhi
EC502154-58	USA	Resistant to seed-borne poty virus, bean common mosaic virus (BCMV), stem blight and rust	ICAR-IIPR, Kanpur (U.P.); ICAR-NBPGR, N. Delhi
EC506078	USA	Multiple disease resistance and has dominant I gene resistant to BCMV, complete resistant to virus (CTV)	ICAR-IIPR, Kanpur (U.P.); ICAR-NBPGR, N. Delhi

Table 4.76: SCAR Markers Linked to Rust (*Uromyces phaseoli*) Resistance

SCAR Name	Size (bp)/Orientation	Sequences of SCARS	Tagged Locus	Reference
SK14	620 cis	CCC GCT ACA CAC CAA TAC CTG CCC GCT ACA CTT GAT AAA ATG TTA G	Ur-3	Haley *et al.*, 1994 Nemchin. and Stavely, 1998 *Miklas *et al.*, 2002
SA14	1079/800 Codominant	CTA TCT GCC ATT ATC AAC TCA AAC GTG CTG GGA AAC ATT ACC TAT T	Ur-4	Miklas *et al.*, 1993 Miene *et al.*, 2004 *Miklas *et al.*, 2002
SI19	460 cis	AAT GCG GGA GAT ATT AAA AGG AAA G AAT GCG GGA GTT CAA TAG AAA AAC C	Ur-5	Haley *et al.*, 1993; Melotto *et al.*, 1998 *Miklas *et al.*, 2000c
SBC6	308 cis	GAA GGC GAG AAG AAA AAG AAA AAT GAA GGC GAG AGC ACC TAG CTG AAG	*Ur-6*	Park *et al.*, 2003b, 2004b *Miklas *et al.*, 2002
SAD12	537 cis	AAG AGG GCG TGA GAT CGT CG AAG AGG GCG TCT TGA AGG TT	*Ur-7*	Park *et al.*, 2003a, 2004a, 2008
SAE19	890 trans	CAG TCC CTG ACA ACA TAA CAC C CAG TCC CTA AAG TAG TTT GTC CCT A	*Ur-11*	Johnson *et al.*, 1995 Queiroz *et al.*, 2004c *Miklas *et al.*, 2002
UR11-GT2	450 cis	CGC ACT TAG GAG CAC AAA TGG TGG GTC CCA TAT TTT G	*Ur-11*	Boone *et al.*, 1999, Miklas *et al.*, 2002
KB126	405/430 Codominant	GAA TTC AAC CTC GGC CAC TAC C TTA AAC CTT CCG GAG GAT TC	*Ur-13*	Mienie *et al.*, 2005
SF10	1072 cis	GGA AGC TTG GTG AGC AAG GA GGA AGC TTG GCT ATGATG GT	*Ouro Negro*	Correa *et al.*, 2000; *Miklas *et al.*,2002
SBA8	530 cis	CCA CAG CCG ACG GAG GAG GCC ATG TTT TTT GTC CCC	*Ouro Negro*	Correa *et al.*, 2000; *Miklas *et al.*, 2002

an imperative role in improving the effects of various types of abiotic and biotic stresses.

(i) Low temperature: The transfer of tolerance to low temperature tolerance from *Phaseolus retensis* to *Phaseolus vulgaris* has been achieved through embryo culture technique.

(ii) High temperature: High temperature stress (>35 °C) adversely affects plant physio-biochemical processes; limiting plant growth, flowering, pollens, and pod setting and its development. Anthesis and pod setting at high temperatures may result in reduced activity of pollen and anther, sluggish growth of pollen tubes,

Table 4.77: Lodging Resistant Genotype

Accessions	Source	Salient Features	Distrbution/Recipient
EC498445 (var. AC Scarlet)	Canada	Lodging resistant, high yielding	ICAR-IIPR, Kanpur (U.P.); ICAR-NBPGR, N. Delhi

slower embryo development and restrict pod enlargement and thereby leading to reductions in pod yield. Plants emply various adaptive mechanisms such as earliness, cooler canopies development (enables plants to maintain physiological functions in balance under elevated temperature), high transpiration rate, stay-green trait and reduced photosynthetic rates to adopt high temperature. Early maturity provides an escape mechanism under late incidence of high temperature which could be a good approach for snap bean breeding for North Indian Plains that suffers from terminal high temperatures from March and onwards. Maintaining higher content of leaf chlorophyll (stay-green) is also considered desirable traits as it indicates a low degree of photo-inhibition of the photosynthetic apparatus at high temperature. Genetic control of important trait like number of pods/plant (day time temperature 38-43°C in a glass house; combined with drought stress) is being carried out by 1 or 2 dominant genes and expressed as epistasis. Thermostability of cellular membranes in *P. vulgaris* L. is polygenic; additive effects and predominant. It is epistasis in some crosses and expressed hertyability as ~60 per cent.

(iii) Drought and high temperature: Tolerance to drought and high temperature has been reported in the few populations of *P. acutifolius* (Federici *et al.*, 1990) which could possibly be used in further breeding programme.

(iv) Lodging resistant

See Table 4.77.

Breeding for Processing and Quality Traits

Following quality traits should be considered during breeding programmes:

Table 4.78: Preferred Range/Magnitude/ Value/Type of Traits

Traits	Preferred Range/Magnitude/Value/Type
Pod shape	Long and fleshy
Pod colour	Green to dark green
Fiber	Low

The pod cross-section (PCS) index or shape (the ratio of thickness to width) varies greatly such as flat (elliptic: 0.400-0.700), oval (ovate: 0.850-.0950), round (circular: 0.950-1.100) and create-back (eight-shaped: 1.150-1.300). Most of the snap bean cultivars are oval to round in shape because these pods are fleshier and have higher transport durability. Inherently, the pods may be straight, curved, fish-hooked or S-shaped. The straightness of pod is also affected by plant growth habit *i.e.* upright bush types and pole beans tend to have straighter pods. Pod straightness is inversely associated with sieve-size. With respect to fiber content in the walls, three major genes control the switch from the highly fibrous dry bean type pod to a relatively fibre-free pod of the typical snap bean (Leakey, 1988). A few work aim at increasing protein content, and certain essential amino acids has been initiated.

Genetic Resource Conservation

An international institute, CIAT, Cali (Colombia) has the mandate for global germplasm collection and conservation of *Phaseolus* species. The Germplasm Bank of CIAT hosts the world's largest and most diverse collection, with more than 36000 *Phaseolus* materials, corresponding to 44 taxa from 110 countries. Majority of these collections belong to primary centre of origin in the Neotropics (Mexico, Peru, Colombia and Guatemala). Globally, more than 260000 accessions of different species of *Phaseolus* have been collected and being maintained in >245 gene banks of various countries (Table 4.79) including about 2900 accessions at ICAR-NBPGR, New Delhi.

ICAR Research Complex for NEH Region, Umiam Meghalaya collected 94 germplasm (73 pole type and 21 bush type) from different parts of the North Eastern India. Odisha University of Agriculture and Technology, Bhubaneswar (Odisha) has identified a runner type of Raikia bean (*P. vulgaris* L.) grown in Raikia, G.Udayagiri and Tikabali blocks in Kandhamal district of Odisha. It has long fleshy pods with an average productivity of 10.0 t/ha. Due to its fleshiness, less fiber content and appealing taste, it has wide market demand. It is cultivated both during *kharif* and *Rabi* season covering an area of about 4200 ha in Kandhamal district of Odisha. Some 600 collections have been made in this crop (Bhagat *et al.*, 1979). In addition to these, ICAR-IIVR, Varanasi (U.P.) is maintining 214 accessions of various *Phaseolus* species (202 of *P. vulgaris* both bush and pole types, 9 of *P. lunatus*, 2 of *P. coccineus* and one of *P. acutifolius*). The Millennium Seed Bank and the Global Crop Diversity Trust are engaged in a ten-year project, called 'Adapting Agriculture to Climate Change'. The project aims to protect, collect and prepare the wild relatives of 29 key food crops, including common bean, so that they are available to pre-breeders for the development of new varieties that are more resilient to the effects of climate change. The Millennium Seed Bank Partnership aims to save plants worldwide, focusing on those plants which

Table 4.79: Global Germplasm Collection of *Phaseolus* Species at Various Genebanks

Sl.No.	Name of Gene Bank	Accession		Type of Accessions (per cent)				
		Number	Per cent Share	WS	LR	BL	AC	OT
1	CIAT, Cali, Colombia	35891	13.70	6	85	2	7	-
2	USDA-ARS NPGS, Washington, USA	14674	5.60	6	67	3	21	4
3	CNPAF-EMBRAPA, Brazil	14460	5.52	-	-	-	-	100
4	INIPAF, Mexico	12752	4.87	17	-	-	-	83
5	IPK, Germany	8680	3.31	1	66	4	28	1
6	ICGR-CAAS, China	7365	2.81	-	-	-	-	100
7	VIR, Russia	6144	2.35	-	22	20	3	55
8	BCA, Malawi	6000	2.29	-	100	-	-	-
9	RCA, Hungary	4350	1.66	-	70	<1	<1	30
10	LBN, Indonesia	3846	1.47	-	-	-		100
11	KARI-NGBK, Kenya	3534	1.35	<1	34	3	35	28
12	IPGR, Bulgaria	3220	1.23	-	32	-	<1	68
13	DENAREF, Ecuador	3102	1.18	2	6	17	<1	75
14	ISAR, Rwanda	3075	1.17	-	-	-	-	100
15	INIACRF, Spain	3038	1.16	-	98	<1	<1	1
16	Other gene banks (231)	131832	50.32	-	30	5	13	52
Total		**261963**	**100.00**	**2**	**39**	**4**	**10**	**45**

Source: FAO, 2010.

are under threat and those which are of most use in the future. Once seeds have been collected they are dried, packaged and stored at -20°C in seed bank vault.

REFERENCES

1. Acosta-Gallegos, J.A., Kelly, J.D. and Gepts, P. 2007. Pre-breeding and genetic diversity in common bean (*Phaseolus vulgaris* L.). *Crop Sci.*, **47(3):** S44-S59.
2. Ali, M.A. 1950. Genetics of resistance to common bean mosaic virus (*bean virus-1*) in the bean (*Phaseolus vulgaris* L.). *Phyt.*, **40:** 69-79.
3. Andrews, S.A. 2013. Snapping green beans. http://humanelivingnet.net/201312/17/snapping-greenbeans.
4. Anonymous. 1971b. New ICAR-IARI Vegetable varieties. IARI Exten. Bull. No. 2, ICAR-IARI, New Delhi.
5. Arumuganathan, K. and Earle, E.D. 1991. Nuclear DNA content of some important plant species. *Plant Molecular Biology Reporters*, **9(3):** 208-218.
6. Baudet, J.C. 1977. Origine et classification des especes cultivees du genre *Phaseolus*. *Bull. Soc. R. Bot. Belg.*, **110:** 65-76.
7. Baudet, J.C. 1978. Prodrome d'une classification generique des Paplionaceae-Phaseoleae. *Bull. Jard. Bot. Natl. Belg.*, **48:** 183-220.
8. Bellucci, E., Bitocchi, E., Rau, D., Rodriguez, M., Biagetti, E., Giardini, A., Attene, G, Nanni, L. and Papa, R. 2014. Genomics of origin, domestication and evolution of *Phaseolus vulgaris* L. In: Genomics of Plant Genetic Resources-Vol.1, eds. (Tuberosa, R., Graner, A. and Frison, E.). Springer: Dordrecht, The Netherlands, p. 483-507.
9. Bennett, M.D. and Leitch, I.J. 2010. Plant DNA C-values Database (Release 5.0, Dec, 2010).
10. Bhagat, N.R., Kawalkar, T.G., Bhatia, H.P. and Hardas, M.W. 1979. Plant Introduction experiments with tender-podded French bean. *Veg. Sci.*, **6(2):** 92-97.
11. Bhatnagar, C.P., Saxena, D.K. and Chandola, R.P. 1972. A new crossing technique in *Phaseolus*. *Science and Culture*, **38:** 91.
12. Chacon, S.M.I, Pickersgill, B. and Debouck, D.G. 2005. Domestication patterns in common bean (*Phaseolus vulgaris* L.) and the origin of the Mesoamerican and Andean cultivated races. *Theoretical and Applied Genetics*, **110:** 432-444.
13. Debouck, D.G., Toro, O., Paredes, O.M., Johnson, W.C. and Gepts, P. 1993. Genetic diversity and ecological distribution of *Phaseolus vulgaris* in north western South America. *Economic Botany*, **47:** 408-423.
14. Delgado-Salinas, A., Bibler, R. and Lavin, M. 2006. Phylogeny of the genus *Phaseolus* (Leguminosae): A recent diversification in an ancient landscape. *Systematic Botany*, **31:** 779-791.
15. Drijfhout, E. 1978. Genetic interaction between *Phaseolus vulgaris* and bean common mosaic virus with implication for strain identification and breeding for resistance. Agric. Res. Rept. Gen. Agr. Pub. Doc., Wageningen, p. 1-90.

16. Drijfhout, E. 1978. Inheritance of temperature dependent string formation in common bean *Phaseolus vulgaris* L. *Netherlands Journal of Plant Pathology*, **26:** 99-105.

17. Drifjhout, E. Silbernagel, M.J.D. and Burke, W. 1978. Differentiation of strains of bean common mosaic virus. *Netherlands Journal of Plant Pathology*, **84:** 13-26.

18. Evans, A.M. 1975. Genetic improvement of *Phaseolus vulgaris*. (In): Nutritional Imporovement of Food Legume by Breeding, p. 107-115. Milner, Max (Ed.). John Wiley and Sons, New York.

19. Freytag, G.F. and Debouck, D.G. 2002. Taxonomy, Distribution and Ecology of the genus *Phaseolus* (Leguminosae-Papilionoideae) in North America, Mexico and Central America. Botanical Research Institute of Texas, Ft. Worth, Texas, USA.

20. Genchev, D. and Kiryakov, I. 2002. Inheritance of resistanceto white mold disease (*Sclerotinia sclerotiorum* (Lib.) de Bary) in a 195 *Phaseolus vulgaris*. *Bulgarian Journal of Agricultural Sciences*, **8:** 181-187.

21. Gepts, P. and Bliss, F.A. 1985. F_1 hybrid weakness in the common bean: differential geographic origin suggests two gene pools in cultivated bean germplasm. *Journal of Heredity*, **76:** 447-450.

22. Gepts, P., Kmiecik, K., Pereira, P. and Bliss, F.A. 1988. Dissemination pathways of common bean (*Phaseolus vulgaris*, Fabaceae) deduced from phaseolin electrophoretic variability I. *The Americas Economic Botany*, **42:** 73-85.

23. Gepts, P. 1998. Origin and evolution of common bean: Past events and recent trends. *HortSci.*, **33:** 1124-1130.

24. Guitierrez, J.A. and Singh, S.P. 1985. Heterosis and inbreeding depression in dry bush bean (*Phaseolus vulgaris* L.). *Canadian Journal of Plant Science*, **65:** 243-449.

25. Hoc, P.S., Espert, S.M., Drewes, S.I. and Burghardt, A.D. 2003. Polymorphism of *Phaseolus vulgaris* var. *aborigeus* (Fabaceae). Evidences of natural hybridization. *Revista de Biologia Tropical*, **51:** 725-736.

26. Islam, F.M.A., Basford, K.E., Jara, C, Redden, R.L. and Beebe, S. 2002. Seed compositional and disease resistance differences among gene pools in cultivated common bean. *Genetic Resources and Crop Evolution*, **49:** 285-293.

27. Kalloo, G. 1998. Vegetable breeding combined edition Vol.I, II and III, Panima Educational Book Agency, New Delhi, pp. 26, 46.

28. Kami, J., Velasquez, V.B., Debouck, D.G. and Gepts, P. 1995. Identification of presumed ancestral DNA sequence of phaseolin in *Phaseolus vulgaris*. Proceeding of the National Academy of Science of the USA, 92: 1101-1104.

29. Kaplan, L. 1956. The cultivated beans of pre-historic South West. *Ann. Missouri Bot. Garden*, **43:** 189-251.

30. Katyal, S.L. and Chadha, K.L. 1988. Vegetable Growing in India. Oxford and IBH Pub. Co. Pvt. Ltd., New Delhi (India) p. 157.

31. Kelly, J.D. 2004. Advances in common bean improvement: some case histories with broader applications. *Acta Horticulturae*, **637:** 99-122.

32. Kelly, J.D. and Vallejo, V.A. 2004. A comprehensive review of the major genes conditioning resistance to anthracnose in common bean. *HortSci.*, 39: 1196-1207.

33. Koenig, R., Singh, S.P. and Gepts, P. 1990. Novel phaseolin types in wild and cultivated common bean (*Phaseolus vulgaris* L., Fabaceae). *Economic Botany*, **44:** 50-60.

34. Kolkman, J.M. and Kelly, J.D. 2003. QTL conferring resistance and avoidance to white mold in common bean. *Crop Science*, **43:** 539-548.

35. Kusova, P.M. and Myers, J.R. 2005. Interspecific hybridization between *P. vulgaris* and *P. acutifolius* to transfer bruchid resistance. *Annual Report of the Bean Improvement Cooperative*, **48:** 28-29.

36. Lawerence, G.H.M. 1969. *Taxonomy of Vasacular Plants*. Second Indian Reprint, p. 545-548. Oxford and IBH Publishing Co. New Delhi.

37. Leakey, C.L.A. 1988. Genotypic and phenotypic markers in common bean. In: Genetic Resources of *Phaseolus* Beans (ed.). Kluwer Academic Publishers, Boston, USA, p. 245-327.

38. Leonard, M.F., Stephens, L.C. and Summers, W.L. 1987. Effect of maternal genotype on development of *Phaseolus vulgaris* L. x *P. lunatus* L. interspecific hybrid embryos. *Euphytica*, **36:** 327-332.

39. Lyngdoh, Y.A. 2014. Genetic variation and character association of French bean for pod, seed and quality characters. Ph.D. thesis submitted to BCKV, Kalyani (West Bengal).

40. Lyngdoh, Y.A., Thapa, Umesh, Shadap, A. and Kulshreshtha, S.K. 2016. Genetic divergence studies for yield and quality attributes in French bean (*Phaseolus vulgaris* L.). 7th Indian Horticulture Congress (Abs: 05; Session II: Breeding for Horticultural Crops), 15-18 November, held at ICAR-IARI, New Delhi, p. 89-90.

41. Marchal, R., Mascherpa, J.M. and Stainier, F. 1978. Etude taxonomique d'un groupe complete d'especes des genres *Phaesolus* et *Vigna* (Papilionaceae) sur la base de donnees morphologiques et polliniques traitees parl, analyse informatique. *Biossiera*, **28:** 1-273.

42. Miklas, P.N., Kelly, J.D., Beebe, S.E. and Blair, M.W. 2006. Common bean breeding for resistance against biotic and abiotic stresses from classical to MAS breeding. *Euphytica*, **147:** 106-131.

43. Miranda, C.S. 1974. Evolutionary genetics of wild and cultivated *Phaseolus vulgaris* L. and *P. coccineus*, Ph.D. Thesis, Univ. of Cambridge.

44. Mok, D.W.S., Mok, M.C. and Rabakoarihanta, A. 1978. Interspecific hybridization of *Phaseolus vulgaris* L. with *P. lunatus* and *P. acutifoilus*. *Theoretical and Applied Genetics*, **52:** 209-215.

45. Myers, J.R. and Baggett, J.R. 1999. Improvement of snap bean. In: Common Bean Improvement in the Twenty-first Century (Singh, S.P.; ed.). Kluwer Academic Press, Dordrecht, The Netherlands, p. 289-329.

46. Nath, P., Srivastava, V.K., Dutta, O.P. and Swamy, K.R.M. 2008. Vegetable Crops-Improvement and Production. PNASF, Bangalore, p. 110-113.

47. Neto, A.T. 1990. Genetic improvement of beans (*Phaseolus vulgaris* L.) through mutation breeding. (In): Genetic Improvement of Pulse Crop, p. 296-327. Nizam, J. (Ed.) Premier Publishing House, Hyderabad (A.P.).

48. Panwar, D.B., Kale, P.N. and Lawande, K.E. 1987. Reaction of some French bean varieties to stemfly infestation under field conditions. *J. Maharashtra Agric. Univ.*, **12:** 263.

49. Paredes, M. and Gepts, P. 1995. Extensive introgression of Middle American germplasm into Chilean common bean cultivars. *Genetic Resources and Crop Evolution*, **42:** 29-41.

50. Peter, K.V. 1998. Leguminosae-French bean. In: Gernetics and Breeeding of Vegetable Crops. ICAR, New Delhi, p.149-176.

51. Prakken, R. 1934. Inheritance of colour and pod characters in *Phaseolus vulgaris* L. *Genetica*, **16:** 177-294.

52. Purseglove, J.W. 1968. Beans. In: Evolution of crop plants (Ed: Simmonds, N.W.), pp. 169.

53. Radkov, P. 1979. Intervarietal hybrids of French bean. *Priroda Bulgaria*, **28:** 78-80.

54. Rajesh, B. and Bahadur, V. 2016. Study on genetic variability, heritability, genetic advance and cluster analysis in French beans (*Phaseolus vulgaris* L.). 7th Indian Horticulture Congress (Abs: 25; Session II: Breeding for Horticultural Crops), 15-18 November, held at ICAR-IARI, New Delhi, pp. 104-105.

55. Ram, D. and Rajput, C.B.S. 1999. Line x tester analysis in French bean (*Phaseolus vulgaris* L.). *Indian J. Hort.*, **56(2):** 169-172.

56. Ram, D. and Rajput, C.B.S. 1999. Heterosis studies in vegetable French bean (*Phaseolus vulgaris* L.). *Haryana Journal of Horticultural Sciences*, **28:** 86-88.

57. Ram, D. and Rajput, C.B.S. 1998. Studies on combining ability in French bean (*Phaseolus vulgaris* L.) for green pod yield and its attributes. *Haryana Journal of Horticultural Sciences*, **27:** 125-130.

58. Ram, H.H. 2005.Vegetable Breeding: Principle and Practices. 2nd Revision, Kalyani Publishers, Ludhiana, India, pp. 256-273.

59. Ram, H.H. and Singh, B.P. 1980. Evaluation of French bean germplasm. *Veg. Sci.*, **17:** 17-55.

60. Ribeiro, R., De, L.D. and Hagedorn, D.J. 1979. Inheritance and nature of resistance in beans to *Fusarium oxysporum* F. sp. *phaseoli*. *Phytopathology*, **69:** 859.

61. Rodolfo-Velasquez, V. and Schwartz, H.F. 2000. Resistance of two bean lines to wilt by *Fusarium oxysporum* f.sp. *phaseoli* under dofferent soil temperature. *Agro. Ciencia*, **16:** 81-86.

62. Savova, N. 1985. Genetical studies of F_1 and F_2 hybrids of *Phaseolus vulgaris* L. x *Phaseolus coccineus*. *GenetiKai Selektsiya*, **18:** 383-395.

63. Schmit, V., Jardin, P., Baudin, J.P. and Debouck, D.G.1993. Use of chloroplast DNA polymorphism or the phylogenetic study of seven *Phaseolus* taxa including *P. vulgaris* and *P. coccineus*. *Theoretical and Applied Genetics*, **87:** 506-516.

64. Sharma, P.P. and Joshi, A.K. 1993. Improvement of French bean. Advances in Horticulture Vol.5-Vegetable Crops. Part-1 (eds.), Chadha, K.L. and Kalloo, G.; Malhotra Publishing House, New Delhi.

65. Sicard, D., michalakis, Y., Dron, M. and Neema, C. 1997. Genetic diversity and pathogenic variation of *Colletotrichum lindemuthianum* in the three centers of diversity of its host, *Phaseolus vulgaris* L. *Phytopathology*, **87:** 807-813.

66. Silbernagel, M.J. 1986. Snap Bean Breeding. In: Bassett, M.J (ed.), Avi Publ. Co. Breeding Vegetable Crops, pp.243-282.

67. Simmonds, N.W. 1976. Evolution of Crop Plants. Longman, London and New York.

68. Singh, B.K., Deka, B.C. and Ramakrishna, Y. 2014. Genetic variability, heritability and interrelationships in pole-type French bean (*Phaseolus vulgaris* L.). Proceedings of the National Academy of Sciences, India Section B: *Biological Sciences*, **84(3):** 587-592.

69. Singh, A.K. and Saini, S.S. 1983. Studies on heterosis and inbreeding depression for primary components and yield components in French bean (*Phaseolus vulgaris* L.). *Indian Horticulture*, **34:** 187-195.

70. Singh, A.K. and Saini, S.S. 1980. Inheritance of resistance to Angular leaf spot (*Isariopsis griseola* Sacc.) in French bean (*Phaseolus vulgaris* L.). *Euphytica*, **29:** 175.

71. Singh, B.K., Pathak, K.A., Ramakrishna, Y., Verma, V.K. and Deca, B.C. 2011. Purple podded French bean with high antioxidant content. ICAR News: *A Science and Technology Newsletter*, 17(3): 9.

72. Singh, S.P. 2001. Broadening genetic base of common bean cultivars: A review. *Crop Sci.*, **41:** 1659-1675.

73. Singh, S.P. 2001. Seed Production of Commercial Vegetables. Agrotech Publishing Academy Udaipur (India).

74. Singh, S.P., Gepts, P. and Debouck, D.G. 1991a. Races of common bean (*Phaseolus vulgaris* L., Fabaceae). *Economic Botany*, **45:** 379-396.

75. Singh, S.P., Guitierrez, J.A., Molina, A., Urrea, C. and Gepts, P. 1991b. Genetic diversity in cultivated common bean.II. Marker-based analysis of morphological and agronomic traits. *Crop Sci.*, **31:** 23-29.

76. Singh, S.P., Nodari, R. and Gepts, P. 1991c. Genetic diversity in cultivated common bean.I. Allozymes. *Crop Sci.*, **31:** 19-23.

77. Skroch, P.W. and Nienhuis, J. 1995. Qualitative and quantitative characterization of RAPD variation among snap bean genotypes (*Phaseolus vulgaris* L.). *Theoretical and Applied Genetics*, **91:** 1078-1085.

78. Smartt, J. 1976. Tropical Pulses. Longman.

79. Sofkova, S., Poryazov, I. and Kiryakov, I. 2010. Breeding green beans (*Phaseolus vulgaris* L.). *Genetics and Breeding*, **38(3):** 77-88.

80. Sturtevant, E.L. 1919. Sturtevant's Notes on Edible Plants (Ed: Hedrick, U.P.) New York State Agril. Exp. Sta. *27th Annu. Rep.*, 2: 1-86.

81. Tyagi, R.K. and Joshi, B.D. 1996. Stability analysis of green pod yield in French bean. *Indian J. Plant Genet. Resour.*, **9(2):** 303-305.

82. Vavilov, N.I. 1950. The origin, variation, immunity and breeding of cultivated plants. *Chron. Bot.*, **13:** 1-365.

83. Verdcourt, B. 1980. The classification of *Dolichos, Lablab, Phaseolus, Vigna* and their allies. In: Adavances in Legume Science, Summerfield and Bunting, Royal Botanical Garden, Kew.

84. Verma, V.K., Jha, A.K., Pandey, A., Kumar, A., Choudhury, P. and Swer, T. 2014. Genetic divergence, path coefficient and cluster analysis of French bean (*Phaseolus vulgaris* L.) genotypes. *Indian Journal of Agricultural Sciences*, **84(8):** 925-930.

85. Swarup, V. and Gill, H.S. 1968. Mutations in French bean. *Indian J. Genet.*, **28(1):** 44-58.

86. Yarnell, S.H. 1965. Cytogenetic of Vegetable Crops-IV. Legumes. *Bot. Rev.*, **31:** 247-303.

87. Yu, K., Chun, S. and Zhang, B.L. 2012. Development and application of molecular markers to breed common bean (*Phaseolus vulgaris* L.) for resistance to common bacterial blight (CBB)-current status and future directions. In: Applied Photosynthesis (Najafpour, M., ed.). Shanghai, China, pp. 365-388. http://cdn.intechopen.com/pdfs-wm/30614.pdf.

88. Zapata, M., Beaver, J.S. and Porch, T.G. 2010. Dominant gene for common bean resistance to common bacterial blight caused by *Xanthomonas axonopodis* pv. *phaseoli*. *Euphytica*, **179:** 373-382.

Lima Bean/Butter Bean (*Phaseolus lunatus* L.)

INTRODUCTION

Lima bean (*Phaseolus lunatus*, 2n=22) is also known as Butter bean, double bean, Haba bean, Pallar bean, Burma bean, Madagaskar bean, Prolific bean, Sugar bean *etc.* It is cultivated as pole beans in the home gardens. Lima bean sometimes called 'butter beans' because of their starchy yet buttery texture, have a delicate flavor that complements a wide variety of dishes. The immature seed, fresh mature seed, mature dry seed, immature pod, tender shoot tip and tender foliage are the edible vegetable portion of the lima bean. However, tender pods as well as fully matured green beans which are cream in colour are mostly used for consumption after cooking. Nutritional value per 100 g contain 482 kJ (115 kcal) Energy, 20.88 Carbohydrates, 2.9 g Sugars, 7 g Dietary fiber, 0.38 g Fat, 7.8 g Protein, (14 per cent) 0.161 mg Thiamine (B1), (5 per cent) 0.055 mg Riboflavin (B2), (3 per cent) 0.421 mg N iacin (B3), (8 per cent) 0.422 mg Pantothenic acid (B5), (12 per cent) 0.161 mg Vitamin B6, (21 per cent) 83 µg Folate (B9), (1 per cent) 0.18 mg Vitamin E, (2 per cent) 2 µg Vitamin K, (2 per cent) 17 mg Calcium, (18 per cent) 2.39 mg Iron, (12 per cent) 43 mg Magnesium, (25 per cent) 0.516 mg Manganese, (16 per cent) 111 mg Phosphorus, (11 per cent) 508 mg Potassium, (0 per cent) 2 mg Sodium, (10 per cent) 0.95 mg Zinc and 2.2 µg floride. It is grown for its fresh seed, which is eaten as a vegetable, are rich in calcium, iron and vitamin C but may contain cyanogenic glucosides (HCN) which can be adequately removed by cooking. The glucoside content of present day cultivars, especially those with light coloured seeds, contains little or none. Nevertheless, consumption of raw lima beans is not recommended. Dried and canned lima beans are available throughout the year. Lima bean is a good source of minor elements. Immature seed is cooked and used like peas in soups, stews *etc.* The mature seed is dried and stored for future use (Purseglove, 1974). It must be thoroughly cooked before being eaten. It is best soaked for about 12 hours prior to cooking and is eaten in soups, stews or fermented and made into tempeh. The sprouted seeds are cooked and used in Chinese dishes. The dried seed can be ground into a powder then used as a thickener in soups or can be mixed with cereal flours when making bread (Facciola, 1990).Young seed pods is steamed and used as a side dish with rice, or added to soups, stews *etc.* Although Lima bean leaves are somewhat bitter but cooked (Facciola, 1990). As a medicinal properties Lima beans are very good source of cholesterol-lowering compound. In addition to lowering cholesterol, Lima beans content high amount of fiber which prevents blood sugar levels from rising too rapidly after a meal, making these beans an especially good hypoglycemia. When combined with whole grains such as rice, lima beans provide virtually far free high quality protein. Lima bean and butter beans contain linamarin, acyanogenic glucoside, although the beans are rendered safe when cooked, and low-linamarin varieties are typically used for culinary purposes. The seeds are astringent and generally used in the diet of people with fevers (Facciola, 1990).

Origin, History and Distribution

Lima bean has two main cenres of diversity, the Central America (Mexico and Guatemala) for the small-seeded forms and the Andean region (Peru and Ecuador) for the large-seeded forms. However, according to Grubben (1977), the primary centre is situated in Central America, from where different cultivar groups spread to North and South America in pre-Columbian times. Many primitive cultivars are grown in tropical countries mostly in West Africa, Malagasy and South East Asia. These can be considered as secondary centres of diversity and escaped types are sometimes found in the natural vegetation. Lima bean is of Andean and Mesoamerican origin (Hoshikawa, 1981). It is a popular vegetable in most of the tropical low land areas of Africa, Asia and Latin America.Two separate domestication events are believed to have occurred (Salgado *et al.*, 1995):

(i) **Big Lima type-**The 1st event taking place in the Andes around 2000 BC, produced a large flat seeded type (butter bean or double bean and Madagascar bean) variety. "Fordhook' is a well recognized name for the large seeded group.

(ii) **Sieva type-**The 2nd event taking place most likely in Mesoamerica around 800 AD produced a small-seeded variety. Common names identifying small to medium size flat seeded, often red coloured types and used as a pulse.

(iii) **Potato type-**In "Potato" varieties, the shape of seed approaches spherical/small round.

By 1301 AD, cultivation had spread to North America, and in the 16th century the plant arrived and began to be cultivated in the Eastern Hemisphere. The small-seeded wild form (Sieva type) is found distributed from Mexico Argentina generally below 1600 m above sea level, while the large-seeded wild form (Lima type) is found distributed in the north of Peru, between 320 and 2030 m above sea level. Its domestication was in Peru, about 7600-10,000 years ago. From its domestication centre, lima bean spread to USA, Central America, Brazil and the Peruvian coastal region. Later in the post-Columbian period, it was taken to other countries by the early European explores. From Spain it went to various countries of Asia. It was brought to Africa from Brazil through slave trade. Now it becomes an important staple pulse crop in some African and southern Asian regions. A white seeded type is known as butter bean in Kenya and South Africa. The largest area and production of lima bean is in the USA followed by the Malagasy Republic and Peru. During the Spanish Vice royalty of Peru, lima beans were exported to the rest of the Americas and Europe, since the boxes of such goods had their place of origin labeled "Lima-Peru", the beans got named as such. In India this crop is not grown on commercial scale and confined only in the hilly areas of the country.

Species Availability, Evolution and Diversity

Lima bean is classified as rosid eudicot. It is close to the wild species *P. ritensis* Jones; *P. maculatus* Scheele; *P. polystachyus* (L.) Britt., Strems and Pogy; *P. jaliscamus* Piper; and *P. salicifolius* Piper. Lima bean is not closely related to French bean (*P. vulgaris*) and Runner bean (*P. coccineus*). Disagreement about the taxonomic classification of lima beans exists relative to the species designation of *P. lunatus* and *P. limensis*. An earlier classification identified thick podded, large seeded limas as *P. limensis* and thin podded, small-seeded forms as *P. lunatus*. Different forms were assigned botanical variety designations which often are found to be spelled differently. Separate species status for the different types of limas is questionable and probably not justified because all types are inter-fertile. Currently, all limas, wild and cultivated are identified as *P. lunatus*. A fertile segment allotetraploid (2n=44) was produced from the cross between *P. lunatus* and *P. polystachyus*. The allotetraploid had late flowering and unstable fertility. Molecular markers have provided biochemical evidence for two different gene pools in lima bean, a Central American pattern and an Andean pattern of storage proteins. In Central America, the seed morphotypes are Sieva and Potato while the Andean has Big Lima morphotype.

The bush forms have little diversity while a large diversity occurs in the climbing or pole types.

Gene and Genetics of Characters

About a dozen plant and seed characters are reported to be monogenic and there are four independent linkage groups. The yield was found to be negatively correlated with early flowering and maturity, seed pod size, and 100 seed weight. Dry seed yield is positively correlated with number of pods per plant, shelling percentage and number of seeds per pod.

Botanical Description

The chromosome length varies from 1.4 μm to 2.47 μm with an average length of 1.92 μm.

Table 4.80: Botanical Description

Growth Habit	Lima bean considered perennial or long-lived annual (but cultivated as annuals), twining and perennial herbs and attend the height of 3.0-4.0 m. Some small bush types are available which attend the height of 50.0-90.0 cm.
Root	Enlarged rootstocks and roots are able to develop *Rhizobium* containing nodules.
Stem	Plants have a highly branched root system of moderate depth, often more than 1.0 m.
Floral biology	It is in flower from July to September. The flowers are hermaphrodite (have both male and female organs).
Anthesis and pollination	Lima bean is a self-pollinated crop, but due to 20 per cent out-crossing it is called as often-cross-pollinated bea. Herkogamy condition is mainly responsible for out crossing. Pollination takeplace by insects.
Pods and pod set	The length and width of the slightly curved oblong pods ranges from 5.0 to 15.0 cm long and 2.0 to 3.0 cm, respectively. Pods of some cultivars are thick; others are relatively thin.
Seeds	The mature seeds are large, 1.0-3.0 cm long flat and oval to kidney shaped/moon shaped; hence its species name. Other seed types are also flat but are more rounded and about 1.0 cm in length; seed of each types are smooth. In most of the varieties the seeds are quite flat, but in the "potato" varieties the shape approaches spherical. White seeds and light green are common, but black, red, orange and variously mottled seeds are also known. The immature seeds are uniform green. Most cultivars usually contain 2 to 4 seeds, although with others, pods may contain as many six seeds.

Botanical Classification

There are two types of Lima beans:

(i) Pole types: Pole type's cultivar attend the height of 2.0-4.0 m and take longer time to come in bearing but these cultivars are high yielder.

Table 4.81: Distinct Group of Lima Bean

Growing Habit	*Cultivars*
Pole type	King of the Garden, Carolina Butter, Challenger, Florida Butter
Semi-pole type	Hopi, Wilbur
Bush type	Handerson Bush, Burpee Bush, Fordhook-242, Baby Potato, Baby Fordhook

(ii) Bush types: Bush type's cultivar attend the height of 30.0-90.0 cm and mature earlier than the pole type varieties. These cultivars are more suitable for cooler regions.

(iii) Potato type: In "Potato" varieties, the shape of seed approaches spherical/small round.

Augmentation of Genetic Resources and Crop Improvement

Not much research work has been done on this crop. Mostly farmers grow the local material. In this crop KKL-1 variety has been developed from Horticultural Research Station, Kodaikanal, Tamil Nadu Agricultural University, Coimbatore (Tamil Nadu) through selection from a local type (collected from Vilpatti). KKL-1- butter bean variety possess high amount of protein coupled with fat content. It is pole type growing upto 2.42 m and bears pods in clusters. The pods are 11.6 cm long, and beans are 5 to 6 in number per pod. The pods are green when immature, turning creamy yellow with brownish purple streaks on the surface at maturity. The seeds are bold, globular, snow white in colour and with excellent cooking quality. The variety is suited for hilly regions of Tamil Nadu from an altitude of 1200m to 2200m above MSL. The crop will be ready for first harvest from 100 days after sowing. The harvest will continue upto 140 days. It yields 3.47 t/ha of ripe pods in 3 to 4 pickings.

Other genotypes as being used for cultivation are:

- Carolina'/'Sieva', 75 days (heirloom)
- 'Christmas'/'Giant Speckled'/'Speckled Calico', 78 days (heirloom, seeds white mottled with red)
- 'Big 6'/'Big Mama', 80 days[9]
- 'King of the Garden', 85 days (heirloom)

Genetic Resource Conservation

The world collection of germplasm is being maintained at the International Centre for Tropical Agriculture (CIAT), Cali, Columbia at request of IBPGR, Rome. Small germplasm collections are in other countries also like USA, Brazil, Costa Rica, Germany, Indonesia, Belgium and Russia. Primitive cultivars have proved to be useful in breeding programme, a nematode resistant bush cultivar 'Nemagreen' was selected in USA after crossing with a semi wild pole type. 'Fordhook-242, 'Hendersson', 'Ventura' and several other cultivars are popular bush type bean in USA. Climbing cultivars such as 'Carolina', 'Early Shell' and 'Florida Speckled Butter' are popular in home gardens. The American high yielding lima beans are now becoming widely cultivated in tropics. Almost all varieties of lima bean grown in India are introductions from abroad. Green-white cultivars are often grown for fresh consumption and commercial dwarf cultivars with large white-green seeds are used for processing purposes. Most of the local cultivars are short-day plants, whereas many improved cultivars are day-neutral. In India Horticultural Research Station, Kodaikanal, Tamil Nadu Agricultural

University, Coimbatore (Tamil Nadu) maintaining some germplasm of butter bean. In this context it is suggested that efforts should be made to collect local genotypes before it wiping out situations from the local niches.

REFERENCES

1. Facciola, S. 1990. Cornucopia-a source book of edible plants. Kampong Publications.
2. Grubben, G.J.H. 1977. Tropical Vegetables and Their Genetic Resources. International Board for Plant Genetic Resources, Rome, Italy, p. 121.
3. Hoshikawa, K. 1981. Lai mame (lima bean). In "Shokuyou Sakumotu" (food Crops). Yoken-do, Tokyo. (In: Japanese), p. 491-494.
4. Purseglove, J.W. 1974. *Phaseolus lunatus* L. In "Tropical Crops: dicotyledons" London: Longman. p. 296-301.
5. Salgado, A.G., Gepts, P. and Debouck, D.G. 1995. Evidence for two gene pools of the Lima bean, *Phaseolus lunatus* L., in the Americas. *Genetic Resources and Crop Evolution*, **42:** 15-28.

Scarlet Bean or Scarlet Runner Bean (*Phaseolus coccineus* L.)

INTRODUCTION

Scarlet bean or Runner bean (*Phaseolus coccineus* L., 2n=22); fay-see-OH-lus koh-SIN-ee-us, is a Greek word (fah-SO-lee-KOH-kee-no) meaning "Red Bean"; though the seeds are multicoloured. Scarlet Runner bean reflects the growth habit and scarlet/red flowers. It is also known as fire-bean, mammoth, red-giant, scarlet-emperor, runner bean, multiflora bean, haricot d' Espagne, Oregon Lima bean, Aycoctl by the Aztecs, Ayocote by the Spanish, case knife bean in England, and Feija da Espanha, feijaescarlata and feijatrepador in Portugal. It occupies second position after *P. vulgaris* because of its highly nutritive value for humans and animals. Its pods are eaten raw when they are young and not yet fibrous, but its pods can also be lightly steamed and used as a snack. It is estimated that 100 g edible portion of raw tender pods contains water 91.2 g, energy 93 kJ (22 kcal), protein 1.6 g, fat 0.4 g, carbohydrate 3.2 g, fibre 2.6 g, Ca 33 mg, Mg 19 mg, P 34 mg, Fe 1.2 mg, Zn 0.2 mg, carotene 145 µg, thiamin 0.06 mg, riboflavin 0.03 mg, niacin trace and ascorbic acid 18 mg (Holland *et al.*, 1991). According to Grela and Gunter (1995), the scarlet bean has slightly lower protein and higher ash content compared to common bean. The beans are used in many cuisines. A variety named 'Judion de la granja' producing large, white edible beans is cultivated in San Ildefonso, Spain. In Greece, cultivars of the runner bean with white blossom and white beans are known as *fasolia gigantes*. The beans have an important role in Greek and British cuisine, appearing in many dishes. In Austria, the coloured versions are cultivated and served as "Kaferbohnen', a dish made of the dry beans with pumpkin seed oil. It is considered a typical dish of regional Austrian cuisine, but dried runner beans are also consumed to a small extent in Germany. Dried seeds possess water 12.5 g, energy 1415 kJ (338 kcal), protein 20.3 g, fat 1.8 g, carbohydrate 62.0 g, fibre 4.8 g, Ca 114 mg, P 354 mg, Fe 9.0 mg, Thiamin 0.50 mg, riboflavin 0.19 mg, niacin 2.3 mg and ascorbic acid 2 mg (Leung *et al.*, 1968). Fleshy roots may serve as a mean of vegetative reproduction and also used as food after cooking.

Origin, History and Distribution

Its origin and domestication have resulted in the following names-ayocotl (Nahuatl origin, Central Mexico), patol (Mexico), botil (actually *P. coccineus* subsp. *darwinianus*, Mexico), chomborote and ploy (the high plateau of Guatemala), Cuba (Costa Rica), Ayocote (Spain), Oregon lima bean, haricot d' Espagne, multiflora bean, case knife bean (UK), and Feja da Espanha, feijaesccarlata and feijatrepador (Portugal). Scarlet bean native to central and northern South America and well domesticated (more than 2000 years) where suitable climate (cooler, high-altitude regions around 2000 m) is prevailing; although wild, small-podded plants are still found growing in the cool and partially shaded valleys of mixed pine-oak forests in Guatemala and Mexico. The wild forms of *P. coccineus*, unconfirmed as ancestral throughout, extend from Chihuahua in Mexico to Panama, generally between 1400-2800 m in the humid high forest. This species is native to America (Panama, Guatemala and Honduras), being cultivated during the period 9000-7000 BC. Archaecological evidences indicate that scarlet bean was domesticated in the humid high zones of Mexico *i.e.* Durango and Puebla around 900 AD (Debouck and Smartt, 1995; Delgado, 1988; Freytag and Debouck, 2002). In pre-Columbian Mexico, the people of the Anahuac had cultivated it extensively. Notably, its introduction into southern Columbia (Antioquia and Narino) was most likely before reaching to the Ethiopian highlands and to the Europe. It was introduced following Columbus's second voyage to the new World in 1493 and then spread into the eastern Mediterranean. By the 17th century it was widely cultivated there and was also being grown in northern Europe as an ornamental, with John Tradescant (gardener to King Cxharles-I) including a plant that has been identified as *P. coccineus* in a list of his plants in 1634. European first visited the Americas, taking the attractively coloured seeds back home to sow in their gardens (Zeven *et al.*, 1993). Spain is purported to be the country of introduction into Europe which is evidenced for the French name 'haricot d' Espagne' referred to runner bean. Its extravagant inflorescence may be the reason for its recent expansion as an ornamental plant in Europe. Now a days, scarlet bean is being cultivated in temperate countries (Spain, UK, Italy and Netherlands), and occasionally in the highland areas of Central and South America, Africa (Ethiopia, Kenya, Uganda, South Africa, Madagascar, east-southern tropical Africa) and Asia. In UK, France and USA the scarlet bean is cultivated both as an ornamental and as a vegetable. In United States, until the late 1970s plants were grew as an ornamental but in Europe they were grow for pods and beans for generations. The scarlet beans are one of the popular green bean in Central and South America, and UK. Historically, it was in English and early American gardens by the 1600s. In Uganda, where the crop is grown at high altitudes in Nakuru District, white-seeded cultivars are most common. In India, it can be grown in different parts like northern and north-Eastern states.

Species Availability, Evolution and Diversity

Dutch runner bean, *P. coccineus* var. *alba*, has white coloured flowers along with white seeds. *P. coccineus* subsp. *darwinianus* is a cultivated subspecies, commonly referred to as the botil in Mexico. The related species considered most useful for interbreeding with *P. coccineus* to increase its genetic diversity are *P. dumosus* and *P. vulgaris*.

Botanical Description

It belongs to the Fabaceae or Legume family. Scarlet bean (*P. coccineus*) differs from the common bean (*P. vulgaris*).

Table 4.82: Botanical Differences between Scarlet Bean (*P. coccineus*) and Common Bean (*P. vulgaris*)

Scarlet Bean (P. coccineus)	*Common Bean (P. vulgaris)*
The cotyledons stay in the ground during germination (hypogeous)	Epigeous germination
The plant is a perennial vine with tuberous-roots (though it is usually treated as an annual)	No tuberous-roots and annual
Clock-wise twining habit	-
Cross pollination (allogamous)	Self-pollination

The detail botanical descriptions are given below:

Table 4.83: Botanical Description

Growth Habit	Mostly growth habit is climbing/trailing (4.5 m tall) perennial but usually grown as annual half-hardy runners and dwarf varieties (bush). In bush type one or two long shoots are visible. Half-runner beans have a growth habit between that of bush and vine/pole type.
Root	It has fleshy and thickened tuberous-roots which are divided and generally fusiform that allows cotyledonary young shoots to re-sprout over several consecutive years
Stem	The vines may grow 5-6 m or more in length.
Leaf	The first pair of leaves is large, simple, alternate and opposite and looks very similar to common beans with dark green, green heart shaped trifoliate leaves having purple tinged veins on the undersides. Stipules triangular; petiole 6-8 to 10-15 cm long, rachis 1.5-2.5 to 4-5 cm long; stipels 0.4-0.6 cm long; leaflets ovate-rhombic, base cuneate or truncate, apex acute, thinly pubescent to glabrescent.
Floral biology	Inflorescence is an axillary or borne on terminal raceme, 30-50 cm long and cluster of flowered. Flowers are bisexual, papilionaceous; pedicel 0.5-1.5 cm long; calyx campanulate, glabrescent, the upper 2 lobes united, the lower 3 triangular; corolla attractive bright scarlet/red/pink/white, the uppermost petal *i.e.* standard hood-shaped/ circular/broadly-obovate, two petal on sides *i.e.* wings broadly obovate, two lowermost petals *i.e.* keel coiled; stamens 10, 9 fused and 1 free; ovary superior, finely pubescent, style coiled with collar of hairs below the stigma.
Anthesis and pollination	The flowers open at sunrise and fade at sunset; and are highly attractive to bees and humming birds. Runner bean, in contrast to common bean, is an allogamous plant *i.e.* cross-pollination assisted by its extrorsely stigma and nectaries and through the action of bees and humming birds. Thus, the presence of pollinating agents is extremely important for production of tender pods as well as dry seeds. It is a melliferous plant whose magnificent flowers (up to 25 pieces in an inflorescence), secrete abundant amounts of nectar and the flowering period (50-110 days) prolonging to the period of the post-harvest gap in the summer bee pollen and honeydew is very beneficial for bees (Wroblewska, 1991).
Pod and Pod set	Numerous slender pods of about 10-12 cm long and 2.5-3.0 cm wide appears during January to March on the plant. The knife-shaped pods are normally groon to unuεual purplo podε. Linoar lanceolate, straight or slightly curved pod, and laterally compressed beaked, glabrescent, rough textured with small oblique ridgs. Fewer pods are set in hot-weather.
Seeds	Multicoloured seeds are common but white flowers and white seeds are also available. Dried seed colour varies from white to shining black or mottled in various shades, and ellipsoid-oblong in shape is also found. Per pod 4-8 large seeds weighing 80-160 g per 100 seeds along with small, narrow and elliptical hilum.

Originally, the domesticated forms are usually short-day; most of the modern cultivars are day neutral to light response.

Table 4.84: Promising Cultivars and their Salient Features

Strains/Cultivars	*Salient Features*
Painted Lady	Red and white flowers
Sunset	Pink flowers
Scarlet Emperor	Scarlet flowers
Butler	Stringless
Kelvedon Wonder	Early and long pods

Botanical Classification

1. Green pod-Most of the varieties are green podded types
2. Purple pod-Purple podded runner bean is 'Aeron Purple Star'

Other cultivars include 'Aeron Purple Star', 'Black Runner', 'Case Knife', 'Hammond's Dwarf', 'Pickwick Dwarf', 'Polestar', 'Scarlet Runner', and 'White Dutch Runner'.

Augmentation of Genetic Resources and Crop Improvement

Interspecific Hybridization

With *P. coccineus*, two species *i.e. P. dumosus* and *P. vulgaris* are supposed to be most useful for interbreeding and to increase its genetic diversity. Scarlet bean is a rigorous species and good source of variability for several traits for the genetic improvement of common bean (Gepts, 1981; Vanderborght, 1983; Singh, 2001).

Resistance Breeding

This species contains several useful agronomic and disease resistance attributes such as cold tolerance; lodging resistance due to thick stem bases; presence of a tuberous root system allowing a perennial cycle; long epicotyls and racemes; a high number of pods/ inflorescence (Vanderborght, 1983); resistance/tolerance to anthracnose, angular leaf spot, powdery mildew, *Ascochyta* blight (Schmit and Baudoin, 1992), white mold (Gilmore *et al.*, 2002), common bacterial blight (Miklas *et al.*, 1994), Fusarium root rot (Wallace and Wilkinson, 1965), bean flies (*Ophiomyia* spp.)

Breeding for Abiotic Stress

Scarlet bean is more tolerant to cool conditions than the other *Phaseolus* species and beans, grows well at temperature between 15-22 °C, germinates well above 10 °C and damage occurs at temperature below 5 °C. Pod and seed setting and their development are inhibited above 25 °C temperature. The scarlet bean tolerates higher precipitations than other species of *Phaseolus*, provided that the soil has good drainage capacity. They also have tuberous roots allowing them to overwinter, if grown in an area where they need protection from too much cold and moisture. In the tropics, it is most successful at altitudes of 1500-2000 m; whereas in Kenya, it is grown at 1900-2600 m. in Ethiopia, it is successfully grown in the areas up to about 2000 m elevation having an average annual rainfall of 1500 mm.

Genetic Resource Conservation

Globally, 428 accessions are being maintained by EMBRAPA/CENARGEN (Brasilia, Brazil), 478 accessions at USDA-ARS Western Regional Plant Introduction Station (Pullman, Washington, USA), 311 accessions at Banco Nacional de Germoplasma Vegetal (Universidad Autonoma Chapingo, Mexico), 14 at Aegean Agricultural Research Institute (Menemen, Izmir, Turkey), 9 at World Vegetable Centre earlier AVRDC (Taiwan), and 6 at Division of Plant and Seed Control (Department of Agriculture, Pretoria, South Africa).

REFERENCES

1. Debouck, D.G. and Smartt, J. 1995. In: Evolution of Crop Plants (Smartt, J. and Simmonds, N.W.; eds.). 2nd Edition. Longman, London, United Kingdom, p. 287-294.
2. Delgado, A. 1988. Variation, taxonomy, domestication and germplasm, potentialities in 13 *Phaseolus coccineus*. p. 441–463. In: Gepts P (ed.) Genetic Resources of *Phaseolus* 14 Beans, Kluwer Academic Publishers, Dordrecht, Netherlands. Freytag, G. and Debouck, D.G. 2002. Taxonomy, distribution and ecology of the genus *Phaseolus* (Leguminosae-Papilionodeae) in North America, Mexico and Central America. Botanical Institute of Texas, Fort Worth, Texas, United States, p. 300.
3. Gepts, P. 1981. Genetic Resources of *Phaseolus* beans: Their Maintenance, Domestication, Evolution and Utilization. Kluwer Academiv Publishers, Dordrecht, Netherlands, p. 613.
4. Gilmore, B., J.R. Myers, and D. Kean. 2002. Completion of testing of *Phaseolus coccineus* plant introductions (PIs) for white mold, *Sclerotinia sclerotiorum*, resistance. *Annu. Rep. Bean Improv. Coop.*, **45:** 64–65.
5. Holland, B., Unwin, I.D. and Buss, D.H. 1991. Vegetables, Herbs and Spices. 4th Edition. Royal Society of Chemistry, Cambridge, United Kingdom, p. 163.
6. Leung, W.T.W., Busson, F. and Jardin, C. 1968. Food Composition, table for use in Africa. FAO, Rome, Italy, p. 306.
7. Miklas, P.N., Rosas, J.C., Beaver, J.S., Telek, L. and Freytag, G.P. 1994. Field performance of select tepary bean germplasm in the tropics. *Crop Sci.*, **34:**1639–1644.
8. Schmit, V. and Baudoin, J.P. 1992. Screening for resistance to Ascochyta blight in populations of *Phaseolus coccineus* L. and *P. polyanthus* Greenman. *Field Crops Res.*, **30:** 155-165.
9. Singh, S.P. 2001. Broadening genetic base of common bean cultivars: A review. *Crop Sci.*, **41:** 1659-1675.
10. Vanderborght, T. 1983. Evaluation of *P. vulgaris* wild types and weedy forms. *Plant genet. Resour. Newsl.*, **54:** 18-25.

11. Wallace, D. Hand R. E., Wilkinson. 1965. Breeding for *Fusarium* root rot resistance in beans *Phytopathology*, **55:** 1227-1231.

12. Wroblewska, A. 1991. Attractiveness of *Phaseolus* L. flowers for pollinating insects. *Acta Hort.*, **288:** 321-324.

Genus *Lablab*

Indian Bean, Dolichos Bean or Lablab Bean [*Lablab purpureus* (L.) Sweet]

INTRODUCTION

Dolichos bean (*Lablab purpureus* L., previously classified as *Dolichos lablab* L., 2n=2x=24) is commonly known as Indian bean, Indian bitter bean, field bean, hyacinth bean, bonavist bean (Western countries), lobia bean, lablab bean, country bean, Egyption bean, Australian pea, Tonga bean, sem and waby salad bean. This multiplicity of names is indicative of the range of forms available globally and the fact that it has long been cultivated for human food, as a green manure and cover crop (Cameron, 1988). As green or red pods are consumed as cooked or fried like snap beans. Dolichos bean is an important source of proteins, minerals and dietary fiber. The palatability of lablab surpasses that of velvet beans or jack beans. The mature seeds, however, contain anti-nutritional factors (cyanogenic glucosides) such as tannins, phylate and trypsin inhibitors, which are toxic. Removing seed coat, soaking and proper cooking minimize activity of these compounds. Per 100 g of edible green pod contain moisture 86.1 g, carbohydrates 6.7 g, protein 3.8 g, fibre 1.8 g, minerals 0.9 g (calcium 210.00 mg, phosphorus 68.00 mg, sodium 55.4 mg, potassium 74.00 mg, sulphur 40.00 mg, magnesium 34.00 mg, iron 1.7 mg) and fat 0.7 g. The young tender leaves are used in salad and the relatively old leaves are cooked as a potherb like spinach. Leaves contain up to 28 per cent protein (dry weight). Flowers may be cooked and eaten. The dried seeds are eaten cooked or processed to bean cakes or fermented as tempeh (Schaaffhausen, 1963). Seed sprouts are consumed as those of other beans (mung beans). Lablab plants may be grazed by cattle, sheep, pigs, and goats (Agishi, 1991). Young dried dead plant material may be used as animal feed. Lablab bean green material also can be made into silage. However, regional preferences are existed in the cultivation of Indian bean. For instance, green shelled seeds are mostly preferred in south India, whereas white pods are liked in eastern India. In north Indian Plains, people like green fleshy pods of Indian bean.

Origin, History and Distribution

Dolichos is a Greek word meaning long pod and *lablab* is Arabic or Egyptian word meaning dull rattle of the seed inside the dry pod. Dolichos bean is believed to have originated from the Asia and now it is confirmed that this is originated in India, as large number of indigenous strains (wild forms are also avalable) are grown in north India (Rao, 1977). From India, it might have been introduced into China, Sudan, Egypt and West Asia and then to Africa during 8th century (Choudhury, 1972; Pengelly and Maass, 2001). In Australia, lablab bean was 1st grown in Botanical Garden, Sydney in 1819. Purseglove (1968) believed that the crop was taken to Africa and has subsequently been distributed to many tropical countries. It is cultivated throughout the tropical, sub-tropical and temperate regions (up to 2000 m) of Asia, South and Central America, East and West Indies, India, China, Africa extending from Cameroon to Swaziland and Zimbabwe, throughout Sudan, Ethiopia, Uganda, Kenya and Tanzania. In India Dolichos bean is grown in Maharashtra, Madhya Pradesh, Uttar Pradesh, Andhra Pradesh, Tamil Nadu and Karnataka as a field crop. In Tamil Nadu *Lablab purpureus* is called as 'Avarai' and closely related botanical variety of this species is *D. lablab* var. *lignosus* is called as 'Mochai' in which pods are more fibrous even at mature stage and the development of grains will be fast. The garden varieties are raise in Norrthern India but common in Southern, Western, Central and Eastern India.

Species Availability, Evolution and Diversity

Indian bean is supposed to have originated in India as wild forms of this bean are found in this country (Choudhury, 1967). Under this genus *L. purpureus* var. *lignosus* (field bean) and *L. purpureus* var. *typicus* (vegetable type or garden type) are well koown. Two botanical varieties *i.e. L. niger* var. *lablab* (var. *typicus*) (syn. *Dolichos lablab* var. *typicus*) in which the long axix of the seed is parallel to the suture and *L. niger* var. *lignosus* (syn. *D. lablab* var. *lignosus*, *D. lignosus* (in which seeds are at right angles to the suture.) are used as dry seed. The two botanical varieties *lignosus* and *lablab* are cross compatible. The original ancestoral species of the crop is considered as *D. purpureus*. Some assign it to the African Centre of diversity. *D. falcatus* and *D. biflorus* are wild in nature. Bangladesh poses a wide diversity in hyacinth bean (*Lablab purpureus*) due to climatic and edaphic factors.

Islam *et al.* (2016) evaluated and characterized 150 accessions from Bangladesh and 16 other countries of Asia, Africa and Europe. High Shannon- Weaver Diversity Index (H'≥0.75) was observed in colour of hypocotyl, epicotyl, stem, leaf vein, flower as well as the shape and size of seed. The accessions were exhibited 4.92 to 16.74 cm edible pod length, 1.35 to 4.36 cm pod width, 2.10 to 17.07 g pod weight and 84 to 537 pods/ plant. The genotypic coefficient of variation was ranged from 3.59 to 40.08 per cent from leaf ratio (L:W) and

Table 4.85: Taxon Name and Preferred Taxon Name of *Lablab*

Name of Taxon	*Name of Authority*	*Family*	*Preferred Taxon Name*
Lablab leucocarpos	Savi	Fabaceae	*Lablab purpureus* subsp. *purpureus*
Lablab microcarpus	DC.	Fabaceae	*Canavalia cathartica*
Lablab niger	Medik.	Fabaceae	*Lablab purpureus* subsp. *purpureus*
Lablab purpureus	(L.) Sweet	Fabaceae	Lablab purpureus
Lablab purpureus subsp. *purpureus*		Fabaceae	*Lablab purpureus* subsp. *purpureus*
Lablab purpureus subsp. *uncinatus*	Verdc.	Fabaceae	*Lablab purpureus* subsp. *uncinatus*
Lablab vulgaris	Savi	Fabaceae	*Lablab purpureus* subsp. *purpureus*

rachis length, respectively among the 19 quantitative characters. Accessions were grouped into ten clusters using D^2 values. Number of accessions in each cluster was ranged from 8 to 25. Accessions collected from the same districts of Bangladesh or abroad were distributed into different clusters. No clear relationship was found among the accessions from different geographical region. Crosses between accessions belonging to maximum divergent clusters of CPI 106548 (India), ILRI 14437 (Zimbabwe) and TOT 7905 (Uzbekistan) from cluster IX with accessions of BD 122 (Bangladesh) and BD 8785 (Bangladesh) from cluster I and BD 8770 (Bangladesh) from cluster VI could be done for obtaining better variability to the subsequent generation.

Gene and Genetics of Characters

The genetics of some of the economic characters have been studied (Cruz and Ponnaiya, 1969; Joshi, 1971; Rajendran and Sathyanarayana, 1981). Characters like protein content of pod, number of pods/plant, breadth of pod, weight of pod and pod yield/plant exhibiting high GCV, PCV, heritability and genetic advance as per-cent of mean in pod type (Chattopadhyay and Dutta, 2010; Das *et al.*, 2015) and pod length in bush type (Das *et al.*, 2015) indicating that such situation may arise due to the action of addiotive genes controlling the characters.

The types of gene action involved for the conditioning of different traits of dolichos bean are as under:

Table 4.86: Genetics of Characters

Character	*Number of Gene*	*Type of Gene Action*
Pigmentation of vegetative parts and flower	Monogenic	Purple dominant to white
Pod colour	Monogenic	Green dominant to light green
Pod character	Monogenic (FpFp)	Flat dominant to swollen
Seed colour	Monogenic	Chocolate dominant to brown

Source: Peter, 1998.

Table 4.87: Gene Effects of different Characters of Dolichos Bean

Character	*Gene action*	*References*
Number of inflorescences per plant, number of nodes per inflorescence, number of flower buds per inflorescence and seed protein content of pod	Additive	Basu *et al.* (1999); Das *et al.* (2014)
Length of inflorescence, number of pods per inflorescence, pod length and number of seeds per pod	Non-additive	Das *et al.* (2014)
Days to first flowering, days to 50 per cent flowering, number of pods per plant, pod width, pod weight, shelling percentage of fresh pod and pod yield per plant	Both additive and non-additive	Das *et al.* (2014)

Botanical Description

Genus *Lablab* belongs to the family Fabaceae subfamily Papilionaceae, tribe Viciae and genus *Lablab*, L. *niger* Medic. (syn. *Lablab vulgaris* Savi, *Dolichos lablab* L.). The pods have a strong flavor. Some genotypes have purple colour of the pods, which disappears after cooking. The chromosome number is 2n=20, 22, 24.

Table 4.88: Botanical Description

Growth Habit	*Dolichos* bean is herbaceous vigorous (robust) trailing twining vine/bushy herbaceous annual (often) or short-lived perennial. It can attained the height up to 1.5-6.0 m.
Root	Tap root which can extract water from the soil down to a depth of about 2 metres. It produces nitrogen through fixation in the root nodules.
Stem	Stems are robust and 3.0-6.0 m in length. Varying colouration of stems.
Leaf	Petioles are long and slender. Leaves 7.5-15.0 cm long and 4.5-15.0 cm wide, alternate, pinnately trifoliate, stipules and stipels present.Thick pulvine at the base of the leaflets (leaflets ovate or ovate-rhomboid; base broadly rounded or very obtuse) and acuminate at the apex, acute; glabrous except on the nerves. Leaves are smooth above and shorthaired underneath. Varying colouration of leaves.
Floral biology	Dolichos bean have cleistogamous flower structure, so they are highly self-pollinaated. Varying colouration of flowers (bright purple, white). Inflorescence axillary, erect, many flowered (1-5 together from prominent tubercular thickenings on peduncle) racemeson elongated peduncles, flowers-white (in 'Rongai') or blue or purple on short pedicles; bracteoles at the base of the calyx 2, fugacious, 0.4-0.6 cm long; corolla white or purple; limb of standard 1.2-1.4 cm long, 1.6-2.0 cm wide;

	keel narrow, curved upwards as to form a right angle; papilonaceous, stamen10 (9+1); anthers uniform, ovary sessile, ovules ±6, style in curved with soft hair, stigma terminal glabrous capitates and style persistent. The protruded stigma (extrovert) in a few varieties leads to allogamy.
Anthesis and pollination	In the young immature bud, the anthers are far behind the level of the stigma. As the bud grows the stamina filament elongate more rapidly than the styleand the anther dehisce when the long stamen comes above or on a level with the stigma. Anthesis occurs 9.0 to 17.0 hours (Pokle and Deshmukh, 1971). The secretion of the exudates starts at tip of the stigma four days before anthesis. It is maximum at the time of pollination after which it decreases gradually. Anther dehiscence is between 5 and 14 hours. Pollen is fertile on the day of anthesis. Stigma is receptive on the day of anthesis. Indian bean is a highly self-pollinated crop but cross pollination through insects has also been reported. Pollination and seed setting is reduced in cold weather (Purseglove, 1968). Indian bean is self-pollinated but chances for natural crossing vary from 2 to 8 per cent (Agrawal, 1989; Kumar and Bhatnagar, 1989). Emasculation and pollination result in fruit set up to 33 per cent and is most successful when emasculation is carried out before noon and pollination is performed on the same or following days. Bagging after emasculation and pollination increase the percentage fruit set to 63 per cent due to the protection of pollen and stigmatic exudates from the drying effects of the sun (Chowdhary *et al.*, 1989).
Pod and pod set	Pods obliquely oblong flattened (5-15 x 1-5 cm) but pod shape is variable (broadly scimitar shaped). Varying colouration of pods (green, bright purple, creamy). Pods are smooth and beaked by the persistent style and tuberculate along margin.
Seeds	The fruit is legume pod. Pods contain 2-6 seeds (varies in size and colour). Seeds in 'Rongai' buff or pale brown coloured, ovoid, laterally compressed, with a linear white conspicuous hilum, 1.0 cm long x 0.7 cm broad. The seed colour may be white, cream, buff, reddish, brown or black or variously spectled. Generally hilum white, prominent, raised extending one-third distance around seeds. Weight of 100 seeds 25-50 g.

Horticultural Classification

Classification on the Basis of Growth Habits

On the basis of growth habits, cultivars can be classified in to two groups:

(i) Determinate types- It is a group of bushy types

(ii) Indeterminate types- It is a group of vining or twining or pole types

Classification on the Basis of Flower Colour (The key to genera is flowers)

(i) Flowers white or purple for *Lablab*

(ii) Flowers yellow for *Dolichos*

Classification on the Basis of Utility

(i) Vegetable type or garden type (*Dolichos purpureus var. typicus*)- Soft edible pods having less fibre in their pod walls. Long axis of the seed is parallel to the suture.

(ii) Field bean type (*Dolichos purpureus* var. *lignosus*)- It is field bean grown for dry seeds generally used as a pulse. Its pods have a characteristic aroma and the pod walls have high fibre content. Seeds are at right angle to the suture. The variety of *lignosus* group is popularly known as carpet legume or field bean is a perennial, semi erect, hardy and drought resistant legume.

Classification on the Basis of Specific Species

Two botanical varieties are recognized *i.e. Lablab niger* var. *lablab* (var. *typicus*) (syn. *D. lablab* var. *typicus*) in which the long axis of the seed is parallel to the suture and *Lablab niger* var. *lignosus* (syn. *D. lablab* var. *lignosus*, *D. lignosus*) in which seeds are at right angles to the suture. The vegetable type belongs to the former *Lablab niger* var. *lablab* and the latter is grown as a pulse crop. Both the botanical varieties i.e. *lignosus* and *lablab* are cross compatible.

Augmentation of Genetic Resources and Crop Improvement

Introduction and Selection

Annappan *et al.* (1989) collected and selected few genotypes. Considering high yield, reduced duration, edible nature of green and dry beans and drought tolerance, 'DL2539' was released as 'CO1 Mochai'. A viny variety 'Pusa Early Prolific', a selection made at the ICAR-IARI, New Delhi, is a promising vegetable type (long thin pods in bunches) in dolichos bean. ICAR-IARI, New Delhi has isolated Pusa Sem-2 and Pusa sem-3. Rewa' is another selection which is suitable for un-irrigated areas. Similarly, ICAR-IIHR Bengaluru has also selected Arka Jay and Arka Vijay, which are bush types. Other varieties like 'CO 1' ('DL250'), 'CO 3' 'DL269' (Subramanian, *et al.*, 1961), 'CO 4' ('DL453'), and 'CO 5' ('DL692') were developed from Tamil Nadu Agricultural University, Tamil Nadu. In Tamil Nadu Agricultural University Coimbatore (Tamil Nadu) bush type promising genotyps are 'CO 6', CO 7', 'CO 8', 'CO 9', 'CO 10', 'CO 11', 'CO 12', 'CO 13', and 'CO (Gb) 14' (Kunjamma and Rajashekaran, 1969). At BCKV, Kalyani (W.B.) five genetically diverse parents (BCDB-1, DOLP VAR5, and DOLB VAR 10. Arka Jay and DOLB VAR-2) having contrasting growth habits were selected through multivariate analysis (Das, 2013). This university has developed two potential varieties *i.e.* BCDB-1 and BCDB-5 for commercial cultivation. At ICAR-IIVR has developed Kashi Haritima (pole type) which is high yielder and possess field resistance to golden mosaic virus. At Kerala Agricultural University, Vellanikkara, Thirissur (Kerala), Biju (2000) studied variability in forty four accessions. Highest phenotype and genotypic coefficients of variation were observed for weight of pods followed by number of pods/plant, thickness of pod and pod yield/plant. High heritability and genetic advance were noted for weight of pods and pod length. DL-6 was the highest yielding and DL-8, the earliest flowering. At this university during 1985 to 1992, several bush and pole type dolichos bean genotypes were evaluated and high yielding lines were identified at AICRP (Vegetable Crops), Vellanikkara. IIHR Sel-1 (bush) and IIHR Sel-2 were high yielding among the bush type. In Karnataka 'Hebbal Avare' of UAS, Bengaluru is popular and grown commercially. DPLD-1, HD-60, Rajni, JDL-53 (pole type), JDL-79 (pole type), 'KDB-405' and 'Dasarawal' (pole type) were high yielding among pole types. Cultivars 'Hima' and 'Grace' were developed at Kerala Agricultural University, Vellanikkara, Thirissur (Kerala) for commercial cultivation. Twenty four dolichos bean genotypes from different parts of India were evaluated for ten quantitative traits in Andaman Islands. Moderate to low genotypic and phenotypic coefficients of variation were recorded for all the traits. High heritability (70.97 per cent) and high genetic advance (74.96 per cent) were observed for plant height, while high heritability (64.24 per cent) but low genetic advance (2.58 per cent) were recorded for pod length. Pod length (r=0.611), number of seeds per pod (r=0.764), number of pods/plant (r=1.00) and pod yield per plant (r=1.00) had positive and significant correlation with pod yield. Path coefficient analysis revealed maximum direct effect of days to 50 per cent flowering, pod width, number of pods per plant and pod yield per plant on pod yield indicating traits of importance for yield improvement (Singh *et al.*, 2015).

Horticulture Research Centre, Bangladesh Agricultural Research Institute (BARI) has identified a potential cultivars of Hyacinth bean HC0084, and HC0010 which are very much suitable for Bangladesh condition but Institute of Post Graduate Studies in Agriculture (IPSA), Bangladesh has released two cultivars *i.e.* Baromashi (purple) and Baromashi (white) in the year 1992.

Hybridization and Selection

Crossability percentage of dolichos bean having different growth habit is different therefore it is necessary to know the extent of pod setting, which is given in Table 4.89.

From a series of hybridization programme at BCKV Kalyani (West Bengal), two cross combinations *viz.*, Arka Jay x DOLB VAR-2 and DOLP VAR-10 x DOLB VAR-2 exhibited heterobeltiosis for pod yield per plant to the extent of 20.00 per cent and 9.20 per cent, respectively (Das *et al.*, 2014). These heterobeltiosis for pod yield per plant were reflected through heterosis in pod weight.

Table 4.89: Crossability Percentage of Dolichos Bean having different Growth Habits

Cross	*No. of Flower Buds Pollinated*	*No. of Pods Set*	*Crossing Success (per cent)*
Pole x Pole type			
BCDB-1 x DOLP VAR-5	50	15	30.00
BCDB-1 x DOLP VAR-10	50	13	26.00
DOLP VAR-5 x DOLP VAR-10	50	14	28.00
Mean	**50**	**14**	**28.00**
Pole x Bush type			
BCDB-1 x Arka Jay	50	11	22.00
BCDB-1 x DOLB VAR-2	50	13	26.00
DOLP VAR-5 x Arka Jay	50	14	28.00
DOLP VAR-5 x DOLB VAR-2	50	12	24.00
DOLP VAR-10 x Arka Jay	50	10	20.00
DOLP VAR-10 x DOLB VAR-2	50	9	18.00
Mean	**50**	**11.5**	**23.00**
Bush x Bush type			
Arka Jay x DOLB VAR-2	50	6	12.00
Mean	**50**	**6**	**12.00**

Source: Chattopadhyay *et al.*, 2016.

These crosses also exhibited highly significant positive SCA effects, included one parent with a high GCA effect for pod yield per plant and other desirable horticultural traits, and so would be expected to produce segregants of a fixable nature in segregating generations through the simple pedigree method. The two botanical varieties *lignosus* and *lablab* are cross-compatible. By artificial crossing Ramasami *et al.* (1990) developed 2 garden bean varieties 'CO9' (Bush type) and white 'Yanaikathu' (pandal or bower type). Similarly in F_6 generation one selection was made and named as 'Colt 22' which was released as 'CO11 Avarai'. A bushy variety 'CO 6' derived from the cross between 'DL 244' (garden bean) and 'DL 3196' (field bean) was selected for green pod yield and quality.

Table 4.90: Hybrids and their Salient Features

F_1 Hybrid	*Salient Features of Progeny*
'CO 5' x 'DL 3196'	Higher yield, seed size and better pod quality
'6023 A' x '6010'	Higher ascorbic content
'16019' x '6010	-
'6014 B' x '6015'	-

Source: Peter, 1998.

The estimates of heterosis were studied or six characters in 15 crosses made from six diverse lines (ACCW-116-accession of Wall No. 116 from Palgad; ACCW-16-113 from Giolangaon; ACCW-165 from Gimhvane,; ACCW-16 from Kelshi; Konkan Bhushan the lablab variety released by Dr. B.S.K.K.V., Dapoli (Maharashtra) and Arka Vijay and Hebbal-3 the two

Table 4.91: Mean Performance of Parents for Yield, Earliness and Plant Height

Characters/Parents	*ACCW-166*	*ACCW-116*	*ACCW-213*	*Konkan Bhushan*	*Arka Vijay*	*Hebbal-3*	*Mean*
Plant height (cm)	138.27	78.5	103.97	56.73	45.13	80.86	83.90
First female flower bud appearance (days)	49.20	59.49	57.33	37.07	40.67	42.07	47.67
First male flower appearance (days)	56.00	68.18	66.93	41.60	48.95	55.00	56.11
Pod maturity (days)	86.93	103.93	98.07	76.73	77.42	82.07	87.53
Overall maturity (days)	106.67	121.53	116.47	93.80	89.85	97.47	104.47
Grain yield/plant (kg)	21.13	16.73	17.73	10.41	9.18	16.00	15.20

Source: Bendale *et al.*, 2005.

Table 4.92: Heterotic Performance Range for Yield, Earliness and Dwarfness in Lablab Bean

Characters	*Heterosis Over Better Parents*		*Heterosis Over Mid Parents*		*No. of Crosses Showing Significant Heterosis Over*	
	Range	*S.E.*	*Range*	*S.E.*	*Better Parents*	*Mid Parents*
Plant height (cm)	-14.14 to 33.70	0.45	-10.54 to 128.41	0.52	1	4
First female flower bud appearance (days)	-21.00 to 4.95	0.32	-16.46 to 22.52	0.37	5	13
First male flower appearance (days)	-19.00 to 5.30	0.39	-15.57 to 35.78	0.45	7	13
Pod maturity (days)	-17.80 to 1.13	0.31	-9.76 to 15.77	0.35	6	14
Overall maturity (days)	-14.25 to 5.98	0.34	-9.53 to 23.90	0.38	6	10
Grain yield/plant (kg)	-32.17 to 101.19	0.97	-24.29 to 159.82	0.84	9	11

Source: Bendale *et al.*, 2005.

varieties from Karnataka state) of lablab bean (*Lablab purpureus* (L.) Sweet), in half diallel mating designs. Among the parents ACCW-166 had highest grain yield/plant. Highest value for heterosis was observed for grain yield/plant followed by plant height. The extent of heterosis was 159.82 per cent for grain yield over mid parent and 109.19 per cent over better parent. The extent of heterosis for plant height was 128.4 per cent over mid parent and 33.70 per cent over better parent (Bendale *et al.*, 2005).

The extent of heterosis was 159.82 per cent for grain yield over mid parent and 109.19 per cent over better parent. The extent of heterosis for plant height was 128.4 per cent over mid parent and 33.70 per cent over better parent.

Interspecific Hybridization

Variety 'CO 1' a selection from a cross between var. *typicus* and var. *lignosus* is an early with high protein and fat content.

Breeding for Quality Attributes

There are certain varities which were selected for their quality attributes like genotype 'Rewa' has a protein content of 25.11 per cent. Rai *et al.* (2014) observed significant genotypic variation in the level of protein

Table 4.93: Quality Attributes in Dolichos Bean

Variety	*Breeding Methods Used*	*Quality Attributes*	*Type of Genotype*	*Developing Source*
'Pusa Early Prolific'	Selection	Selection made for vegetable type	Pole type	IARI, New Delhi
'Rewa'	Selection	High protein content (25.11 per cent)	-	-
'CO 1'	Hybridization between var. *typicus* and var. *lignosus*	High protein and fat content	-	TNAU, Coimbatore
Arka Jay	Selection	Cooking quality	Bush type	ICAR-IIHR, Bengaluru
Arka Vijay	Selection	Tolerant to low moisture stress, Bushy	Bush type	ICAR-IIHR, Bengaluru
'Konkan Bhusan'	Hybridization between 'Hebbal Avare 3' x 'Wal 1'	Bushy	Bush type	KVK, Dapoli
'Hebbal Avare 4'	Hybridization between 'Hebbal Avare' x 'CO 80'	Compact plants	Bush type	UAS Bengaluru
'Pusa Sem 2'	Selection	Semi round stringless, tolerant to anthracnose and yellow mosaic virus	Pole type	ICAR-IARI, New Delhi
'Pusa Sem 3'	Selection	Meaty and stringless, tolerant to anthracnose yellow mosaic virus, Aphid, Jassid and pod borer	Pole type	ICAR-IARI, New Delhi

(102-635.6 mg/g), sugar (0.188-1.11 mg/g), chlorophyll (0.121-0.716 mg/g), phenol (1.7-9.67 mg/g), proline (0.02-7.06 µg/g) and carotenoids (0.04-0.231 mg/g) content in dolichos bean.

Mutation Breeding

Sivasubramanian *et al.* (1989) gamma irradiated seeds (24 kr) of a bushy variety 'CO 6' (Derivative of a cross between 'DL 244' (garden bean) and 'DL 3196' (field bean) ant its progenies screened for green-pod yield and quality. In the M_2 generation, a macromutant possessing economic values was isolated and designated as 'Mutant 3'. The mutant being higher yielder than 'CO 9' as well as possessing better values for sensory tests and biochemical analysis, it was released as 'CO 10 Avarai'.

Hybridization

Some promising hybrids are CO57 x DL3196 (higher yield, seed size and better pod quality), 6023A x 6010, 16019 x 6010 and 6014B x 6015 (higher ascorbic content). The hybrids are produced through hand-emasculation and pollination, making use of protruded stigma and use of selective differential gametocide like EW450 (Peter, 1998).

(A) Resistance for Biotic Factors

(a) Insect Resistant

(i) Culture resistant to aphid are PLS224, PLS126 and PLS12 (Rajsekharan and Babu, 1983)

(ii) Culture resistant to pod borer are PLS224, PLS16-1 (Chakravarthy, 1977).

(b) Disease Resistance

Some of the useful resistant/tolerant sources against different diseases of dolichos bean crop have been identified over the years and documented as:

Table 4.94: Promising Dolichos Bean Entries Tolerant to different Biotic Stresses

Disease	*Sources*	*References*
Dolichos yellow mosaic virus	BCDB-1, Gomchi Green, Pusa Early Prolific, BCDB-2	Anonymous (2008-09); Anonymous (2009-10); Chattopadhyay and Dutta (2010)
Cercospora leaf spot of dolichos bean	BCDB-1, SEMVAR-8	Anonymous (2008-09); Chattopadhyay and Dutta (2010)

At ICAR-IIHR Bengaluru (Karnataka) two lines *viz.*, 'IHR93' and 'IHR 140' were reported to be rust resistant and high yielders with good pod quality.

(B) Abiotic Stress

Photo Insensitive

Breeding photo insensitive varieties both as vegetable and seed type is gaining popularity and two varieties Arka Jay and Arka Vijay have been developed and recommended for commercial cultivation by ICAR-IIHR, Bengaluru (Nath *et al.*, 2008). Most of the pole types are photo-insensitive hence suited for growing in specific season like *kharif* and will not produce flowers when grown in the summer season.

Physiological Resistant

(i) Resistant to pod and seed damage are JDL60, JDL145 and JDL18 (Parsai, *et al.*, 1989)

(C) Culture Resistant to Root-Knot Nematode

Culture resistant to root-knot nematode are IL028, IL029, IL033 and IL1949 (Azami and Patil, 1989).

Genetic Resource Conservation

In Tamil Nadu Agricultural University bush type promising genotyps like 'CO-6', CO-7', 'CO-8', 'CO-9', 'CO-10', 'CO-11', 'CO-12', 'CO-13', and 'CO-14 (Gb)' and pole type genotypes like 'CO-1', 'CO-2', 'CO 3', 'CO 4', and 'CO 6' are being maintained. Wild germplasm are strongly perennial.

REFERENCES

1. Agishi, E.C. 1991. A bibliography on *Lablab purpureus*. Plant Science Division Working Document No. A6. International livestock Centre for Africa (ILCA), Addis Ababa, Ethiopia.
2. Agrawal, R.L. 1989. Seed Technology, Oxford and IBH Publishing Co., New Delhi
3. Annappan, R.S., Sivasubramanian, P., Rangaswamy, P., Soundrapandian, G., Ramanathan, T. and Iyemperumal, S. 1989. 'Col Mochai': A high yielding field Lablab. *Madaras Agricultural Journal,* **76:** 640-642.
4. Anonymous (2008-09). Annual Report-AICRP (VC), BCKV, Kalyani Centre, W.B.
5. Anonymous (2009-10). Annual Report-AICRP (VC), BCKV, Kalyani Centre, W.B.
6. Azami, M.I. and Patil, B.D. 1989. Effect of root knot nematode on carpet legume, *Dolichos lablab* var. *lignosus* with a note on screening of resistant lines, *Indian J. Nematol.*, **18:** 104.
7. Basu, A.K., Pal, D., Sasmal, S.C. and Samanta, S.K. 1999. Genetic analysis for embryo weight, cotyledon weight and seed protein in lablab bean. *Veg. Sci.*, **26:** 37-40.
8. Bendale, V.W., Kumbhar, S.D., Bhave, S.G., Mehta, J.L. and Pethe, U.B. 2005. Heterosis studies for developmental characters in lablab bean (*Lablab purpureus* (L.) Sweet). *The Orissa Journal of Horticulture*, **33(1):** 20-23.
9. Biju, M.G. 2000. Genetic variability in hyacinth bean (*Lablab purpureus*). M.Sc. thesis Kerala Agricultural University, Vellanikkara, Thirissur (Kerala).

10. Cameron, D. G. 1988. Tropical and subtropical pasture legumes. *Queensland Agricultural Journal*, March-April, 110-113.

11. Chakravarthy, A.K. 1977. Pod borer resistance in *Lablab niger* (L.) cultivars with special reference to the pod borer. *Adisura atkinsoni* (Moore) [(*Lepidoptera*)], M.Sc. (Ag.), Thersis submitted to University of Agricultural Sciences, Bengaluru (Karnataka).

12. Chattopadhyay, A. and Dutta, S. 2010. Characterization and identification of selection indices in pole type dolichos bean. *Vegetable Crops Research Bulletin*, Poland, **73:** 33-45.

13. Chattopadhyay, A., Dutta, S. and Hazra, P. 2016. Current status of leguminous vegetables research in West Bengal. In: Souvenir and Abstracts. Natl. Sym. Vegetable legumes for Soil and Human Health. Feb.12-14, 2016 at ICAR-IIVR, Varanasi (U.P.).

14. Choudhury, B. 1967. Vegetables. National Book Trust, India, New Delhi.

15. Chowdhary, A.R., Ali, M. and Quadir, M.A. 1989. Aspects of pollination and floral biology of lablab bean (*Lablab purpureus* L.) Sweet). *J. Japanese Soc. Hort. Sci.*, **58(3):** 665-671.

16. Cruz, R.D. and Ponnaiya, V.T.S. 1969. Inheritance of pod and seed colour and pod shape in garden bean. *Indian J. Genet.*, **29(1):** 139-141.

17. Das, I. 2013. Studies on characterization, yield components and gene action in dolichos bean. M.Sc. (Hort.), Thesis submitted to BCKV, West Bengal, India.

18. Das, I., Seth, T., Durwas, S.V., Dutta, S., Chattopadhyay, A. and Singh, B. 2013. Gene action and combining ability for pod yield and yield component traits in dolichos bean (*Dolichos lablab* L. var. *typicus*). *SABRAO J. Breed. Gen.*, **46(2):** 293-304.

19. Das, I., Shende, V.D., Seth, T., Yadav, Y. and Chattopadhyay, A. 2015. Genetic analysis and interrelationships among yield attributing traits in pole and bush type dolichos bean (*Lablab purpureus* L.). *J. Crop and Weed*, **11(2):** 74-77.

20. Islam, M.T., Haque, M.S., Islam, M.M. and Haque, M.M. 2016. Genetic diversity of Hayacinth bean collections in Bangladesh. 1st International Agribiodiversity Congress; (Session: Trait Discovery and Enhanced Use of PGR), held at New Delhi; 6-9 November 2016; p. 187.

21. Joshi, S.N 1971. Studies on genetic variability for yield and its components in Indian bean (*Dolichos lablab* var. *lignosus*). *Madras Agric. J.*, **58(3):** 367-371.

22. Kumar, M. and Bhatnagar, S.P. 1989. Stigmatic surface and exudate in *Lablab purpureas*. *Phytopathology*, **39(1):** 97-102.

23. Kunjamma, H.V.K. and Rajashekaran, V.P.R. 1969. CO-2, CO-3. CO-4 and CO-5 the four economic strains of garden lablab (*Lablab niger* var. *typicus*). *Madras Agric. J.*, **56(7):** 475-479.

24. Parsai, S.K., Rawat, R.R. and Gupta, M.P. 1989. Relative susceptibility of field bean varieties to bean beetle (*C. phaseoli* Gyllihal). *Agric. Sci. Digest*, **9(3):** 136.

25. Pengelly, B.C. and Maass, B.L. 2001. *Lablab purpureus* (L.) Sweet-diversity, potential use and determination. *Genetic Resources and Crop Evolution*, **48:** 261-272.

26. Peter, K.V. 1998. Leguminosae-French bean. In: Gernetics and Breeeding of Vegetable Crops. ICAR, New Delhi, p.149-176.

27. Pokle, Y.S. and Deshmukh, N.N. 1971. Anthesis and pollination in garden bean (*Dolichos lablab* L.) *Nagpur College of Agriculture Magazine*, **44:** 17.

28. Purseglove, J.W. 1968. Tropical Crops: Dicotledons. John Wiley, New York.

29. Rai, N., Rai, K.K., Tiwari, G. and Kumar, S. 2014. Nutritional and antinutritional properties and their inter-relationship, Indian bean (*Lablab purpureus*). *Indian Journal of Agricultural Sciences*, **84(9):** 1051-1055.

30. Rajendran, R. and Sathyanarayana, A. 1981. Annual Report. ICAR-Indian Institute of Horticultural Research, Bengaluru (Karnataka).

31. Rajsekharan, B. and Babu, P.C.S.1983. Screening of lablab varieties for resistance to aphid (*Aphis cracivora* Koch). *South Indian Hortic.*, **31:** 50.

32. Ramasami, P., Balasubramaniam, M., Gnanam, R. and Rangasami, P. 1990. CO-11 *Avarai*, (*Lablab purpureus* var. *typicus* (L.) Sweet. *Madaras Agricultural Journal*, **77:** 121-124.

33. Rao, M. 1977. Genetic studies in lablab [(*Lablab purpureus* (L.) Sweet] for green pod yield and its components. M.Sc. (Ag.) Thesis, Tamil Nadu Agricultural University, Coimbatore, India.

34. Schaaffhausen, R.V. 1963. Dolichos lablab or hyacinth bean: its uses for feed, food and soil improvement. *Economic Botany*, **17:** 146-153.

35. Singh, S., Singh, P.K., Singh, D.R., Pandey, V.B. and Srivastava, R.C. 2015. Genetic variability and character association study in dolichos bean. *Indian J. Hort.*, **72(3):** 343-346.

36. Sivasubramanian, P., Annappan, R.S., Iyemperumal, S., Sreerangaswamy. S.R. and Pandiyan, M. 1989. 'CO 10': a dual purpose lab-lab.*Madaras Agricultural Journal*, **76:** 411-412.

37. Subramanian, A., Premsekhar, S. and Anavardhan, L. 1961. 'DL 269': A high yielding good quality strain of garden bean *D. lablab* var. *typicus*. *Madaras Agricultural Journal*, **48:** 144-154.

Genus *Psophocarpus*

Winged Bean (*Psophocarpus tetragonolobus* (L.) DC.)

INTRODUCTION

A large number of leguminous species, which have not yet been exploited or are under exploited, have a great potential for contributing to nutritive food, feed and fodder (Rai *et al.*, 2005) needs in the tropical countries of the world where almost half of the world's under and malnourished population lives. Among them, winged bean (*Psophocarpus tetragonolobus* (L.) DC.), 2n=26), is also known by different vernacular names *viz.* flying bean, four angled bean or in India as asparagus pea, Manila bean, princess pea, and Goa bean which has drawn worldwide attention of the scientists mainly due to its rich food value and a unique vegetable. It is also called as *chara kono-sem* (West Bengal), *Clzougula sem* (Madhya Pradesh), *Clzaudari sem* (Maharashtra), *Rakki aravat* (Odisha), *Kattu avar* (Kerala) and *Marisu avar* (Tamil Nadu). This also named as a 'Super market on a stalk' because the plant combines the desirable characteristics of common bean, pea, spinach, mushroom, soybean and potato. Normally, winged bean considered a minor home garden crop in most of the countries except in areas of the Papua New Guinea highlands and Indonesia where it is grown as a field crop. It can be profitably utilized, several superlatives like 'wonder legume', 'miracle plant', 'the miracle bean', 'wonder bean', and 'backyard super market' (Thompson and Haryona, 1980) have often been used for winged bean. All plant parts of winged bean such as young leaves, flowers, immature pods (can be used as stewed, boiled, fried, roasted or made into milk), seeds are edible like soybean (Khan, 1982) and tuberous roots are also edible and used in a variety of preparations. It is very rich source of dietary protein (Claydon, 1975). The tender immature pods are used as a vegetable which are prepared like French bean and cowpea pods (Pospisil *et al.*, 1971; Masefield1, 973; Thompson and Haryono, 1980) in almost all parts of the world where the crop is grown. The unripe seeds are used in soup and curry preparations. The acceptance and consumption of the winged bean green pods as vegetable has been very good in the new areas where it has not been grown previously. Uncooked immature pods are frequently used as salad in some Indonesian prepasrations. In some areas, the young leaves and shoots of the vigorously growing planta are eaten both raw, cooked and tender leaves make good sauce and curry. Similarly, flowers (sweet taste due to nectar) and buds (like steamed mushrooms) are used as fried or lightly boiled vegetable in Papua New Guinea, eaten raw or cooked as stews in Indonesia, and eaten raw or cooked in Malaysia and as a live stock feed in Bangladesh. It consumed fresh as salad. The flowers are used to colour the dishes for the table. It prevents night blindness. The tendrils taste like asparagus. In high lands of Papua New Guinea and Myanmar, the root tubers are also consumed since it is a high protein supplement for the predominantly carbohydrate diet. Fully mature or ripe seeds are used after roasting like peanuts (Degener, 1945; Aykroyd and Doughty, 1964; Pospisil *et al.*, 1971) in many countries but to a lesser degree. Boiled, fried roasted or parched preparations are made for eating after soaking in water overnight. The cooking techniques have been developed to promote the use of winged bean (Ruberte and Martin, 1978). The greatest potential for the winged bean is in the kitchen garden or cultivation on a smaller scale, because the vines require staking, bear pods (pea like beans with four winged edges having chewy texture and a slightly sweet taste) over a long period of time and need regular harvesting. It is good source of protein, thiamin, riboflavin, vitamin B6, Calcium, iron, magnesium and potassium and a very good source of vitamin C, folate and manganese. The composition of green pods on per 100 g fresh weight basis comprises 71.2 g moisture, 2.4 g protein, 0.2-0.3 g fat, 3.1-3.8 g carbohydrates, 0.8-2.6 g fibre and 0.4-1.9 g minerals. Winged bean is low in cholesterol and sodium. Nigombam *et al.* (2012) reported highest K (8.9 mg/g), Ca (8.06 mg/g) and Mg (5.72 mg/g) in the mature pods. The National Academy of Science, the USA, called it as vegetable of the 20th century. The immature pods contain 34 per cent protein on dry-weight basis and 17 per cent oil. The slightly sweet tuberous roots contain 20 per cent protein (dry weight). The seeds possess a high nutritional value and contain about 40 per cent proteins (Garcia and Palmer, 1980) which may be used as flour and also to make a coffee-like drink. It is also used as a forage crops (Ekpenyyong and Barchars, 1978). Two other synonyms attributed to the crop are 'Soyas rival' and 'God sent vegetable'. A winged bean "milk" and flour are used as dietary treatments for protein deprived children. Its tuberous roots are edible but are not widely consume for food. The use of winged bean tubers as food appears to be restricted to Burma, Papua New Guinea and South Pacific Islands (Herklots, 1972) where the plant is apparently grown chiefly for the tubers. Tubers are generally eaten after boiling, frying, baking or roasting. Its chips are very tasty and have a desirable nut like flavor (Thompson and Haryona, 1980). Tubers are sometimes eaten fresh as well which taste slightly sweet like apple (Burkhill, 1935). Seeds are also rich source of oil. Its extraordinary potential to nodulate and fix atmospheric nitrogen with a wide spectrum of *Rhizobia*, because of the size and number of nodules produced by its roots even without proper artificial inoculation (Masefield, 1973; Masefield, 1961; Anonymous, 1975a; Elmes, 1976, Harding *et al.*, 1978a). Wide adaptability to diverse soil types, altitudes ranging from 0 to 2500 m above sea level (Singh, 1981a), and from hot humid tropics to moderately cold sub-tropics, are some of its important features. It is also reported to grow well even under shade and in areas with continued cloud cover. Inspite of having such exceptional combination of so many desirable characteristics, the winged bean plant remained obscure till recent past. It was only after the appearance of some of the recent publications on winged bean (Masefield, 1973; Anonymous, 1975a) that stimulated interest has generated among the scientist and development people for its improvement and rational utilization. Western Australian Department of Agriculture has published a monograph on winged bean in recent years (Khan, 1980). In Jawa, the plant is parasitized by a fungus (*Woroninella psophocarpi*) due to which young parts swell and become turgid. When steemed, these are prized as a delicacy (Degener, 1945). Medicinally, the leaves of winged bean plant have been used for the treatment of smallpox. The roots have been used in Malaya as a poultice to cure vertigo (Hymowitz and Boyd, 1977). It has exceptional nitrogen fixing properties and produces more nodular weight per plant than any other member of Fabaceae. Also the winged bean has shown promise as a green manure, cover and fallow restorative crop (Pospisil *et al.*, 1971). What has really excited the imagination and attention of people all over the world is the rema caby high protein content of most of the edible portions of this wonder plant (Cerny, *et al.*, 1971; Pospisil *et al.*, 1971; Anonymous, 1975a; Evans *et al.*, 1978b and Ravelli *et al.*, 1978). It is unfortunate that the dietary value of winged bean has not been fully recognized in spite of the fact that its high protein content has been known since 1929 (Agcaoili, 1929; (Harikumar, 1988).

Origin, History and Distribution

It is not clear exactly where winged bean originally came from. One of the earliest references about winged bean was made by Pickering in 1879 (cf. Hymowitz and Boyd, 1977). Pickering asserted that *Psophocarpus tetragonolobus* (L.) DC was a native of equatorial West Africa; however, considering the rainfall in that area which is far below that is required for the growth of the winged bean and the fact that *Psophocarpus tetragonolobus* (L.) have never been found in wild form in West Africa, this speculation does not to be true. Vavilov (1951) claimed India as the centre of origin for winged bean, though the likelihood of it is, remote. It is however, certain that it was introduced in India and is being grown here since very early time. Burkhil (1935) believed that this species was introduced to South-East Asia by Arab Traders during the 17th century. Burkhil (1935) thought that it originated in Mauritius, Medagascar or the African region where all the other 4 or 5 species of *Psophocarpus* are found wild. Winged bean grows well in humid (high rainfall), tropical and subtropical parts of the world (from the Philippines and Indonesia to India, Burma and Sri Lanka). The centre of origin of winged bean is South East Asia particularly India or South Pacific Island (Anonymous, 1969) and the centre of diversity is Papua New Guinea and Indonesia (Khan, 1976). Similarly Matejka (1987) also discussed that Papua New Guinea, South East Asia or Upper Burma as possible centres of origin of winged bean. Ryan (1972) ventures that Papua New Guinea may be the origin of the winged bean and that its domestication is of some antiquity in the region. Rajmohan *et al.* (1982) established South India as a possible secondary centre of the origin. It might have reached to Papua New Gunea almost at the same time (Powell, 1974). The geographical distribution comprising Thailand, Malaysia, Sri Lanka, the Philipines, Vietnam, Indonesia, Ghana, South Burma and southern states of India, is considered as the secondary centre of origin. Hag and Smartt (1973) opined that it was in cultivation for the past 2 centuries in Bangladesh. Valicer *et al.* (1987) reported that it is in cultivation in many Asian and African countries like in Philippines, India, Myanmar (Burma), Bangladesh, Thailand, Vietnam, Malaysia, Indonesia, Ghana, Nigeria, West Indies, USA and Sri Lanka since time immemorial. Papua New Guinea and Indonesia are possible candidates because there are so many different strains of winged beans. In Malaysia the crop is cultivated for its tuberous roots. It was introduced into India in 1799 and is cultivated on a limited scale in humid subtropical parts (about 1200 m elevation) like Eastern states/NEH of Assam, Tripura, Manipur, Mizoram, Nagaland, Meghalaya (Sarma, 2001), West Bengal, Tamil Nadu, Kerala, Karnataka, Goa, Odisha, Western Ghats, Maharashtra southwards and central/eastern peninsular region. Though, exact ecological requirements of the winged bean are not yet established, it is certain that the species is mainly confined to humid tropics. It most widely distributed in South-East Asia and has been reported to exist in India (Burkhill, 1935), Burma (Russell1952, Maheshwari and Singh, 1965) Thailand, Phillippines, Indonesia, China, Papua New Guinea and in several south pacific islands. Further, the species of *Psophocarpus tetragonolobus* (L.) has been introduced in a number of African and tropical American countries.

Species Availability, Evolution and Diversity

Psophocarpus is a genus of about six species. All the species, except perhaps *P. tetragonolobus* appear to be indigenous to Africa. This genus has two species *i.e. P. tetragonolobus* DC. and *P. pallustris* Desv., are used as vegetables but *P. scandenss* (Endl.) Verde., is used as cover crops. Another important species is *P. longepedunculatus*. The three species *i.e. P. tetragonolobus*, *P. palustris*, and *P. longepedunculatus*, have a common progenitor. The other species are *P. scandens*, *P. lacncifolius*, *P. grandiflorus* and *P. lecomati*. According to Harder *et al.* (1990), 8 of 9 the species of *Psophocarpus* are native to Tropical Africa. A related species, *P. tetragonolobus* Desv., is distributed widely in Africa both in cultivated and wild form, where the other members of the genus are also found (Masefield, 1935). In Papua New Guinea, where winged bean is most common, it is an ingredient of some of the tribal rituals, suggesting that it has been grown there for so many years (Anonymous, 1977). Papua New Guinea has been described as the centre of greatest diversity for winged bean. Erskine and Khan (1981) found wide variation within and between land races of winged bean in the high lands of Papua New Guinea. They concluded that strong local preferences and a low extent of cross pollination between adjacent land races could be the major factors for these variations. However, no other wild or cultivated member of the *Psophocarpus* genus is known to occur in Papua New Guinea. Eagleton *et al.* (1981) opined that bean is essentially Asian and there are two main loci for domestication-upper Burma and Papua New Guinea Highlands which have reservoir of more primitive germplasm. A large array of genetic diversity has been observed in the winged bean material, particularly from Papua New Guinea. Such a diversity is exhibited by large differences in growth, pattern, vigour, colour of the stem, flowers and seeds, shape and size of pod, leaf shape and size, pod length, seeds per pod, seed yield, tuber yield and shelling percentage. Green and purple are the two basic pigmentation of the stem with the former being most common in the Papua New Guinea material. Variations in the pigmentation on are from purple tinge on the green pigmentation to the deep purple pigmentation on the stem twigs, *etc.* Flower colours are blue and purple with former being a little more prevalent (Khan, 1976). Mutant with noticeably reddish purple flower and some other uncommon flower colours have been reported by Thompson and Haryono (1980).

Considerble variation has been observed in the winged bean material both within and between the unselected accessions. Differences in the number, size and shape of various plant parts are quite obvious. However, growth habit has always been reported to be indeterminate. There appears to be two clear groups in the winged bean germplasm, *i.e.* one from Papua New Guinea and the other from Indonesia, the latter being found also in other areas of South-east Asia. Similarly, the Nigerian material appears to have been originated from the Papua New Guinea material. The pod length in the former has been reported to exceed 70.0 cm while in the latter; it ranged from 5.8-26.4 cm (Thompson and Haryona, 1980). Seed coat colours of white, creamy, uniform black and mottled are found, brown being the most common. Indonesian material has a high level of resistance to the false rust (*Synchytrium psophocarpi*), which the Papua New Guinea material lacks. However, the latter has a higher level of resistance to yellow mosaic type virus (Thompson and Haryona, 1979). It may also be pointed out that Papua New Guinea material generally behavs as annual with the plants dying back and failing to regenerate after producing a seed crop. Contrarily, the Indonesian accessions by and large retain their vegetative vigour after fruiting and could be characterized as short lived perennials. This basic difference in growth habit results in large differences in seed and tuber yield of Papua New Guinea materials and also influence cultural recommendations. Gopalkrishnan *et al.* (1981) on the basis of somatic analysis of south Indian winged bean collection, concluded that the geographical area consist the states of Kerala, Karnataka and part of Tamil Nadu is possible the secondary centre of origin of winged bean. Chundawat *et al.* (1981) observed wide range of phenotypic variation for many of the plant characters (intermediate to semi spreading types with shorter internodal patterns) in a collection. Philip (1984) reported that phenotypic and genotypic coefficients of variation were high for root tuber weight, followed by pod yield and pods/plant. Crude protein was maximum (41.35 per cent) in seeds, followed by leaves (37-41 per cent), pods (31.28 per cent), flowers (27.45 per cent) and root tubers (25.89 per cent). The extent of variation among quantitative characters as revealed through coefficient of variation ranged from 30 per cent for seed weight and pod length to 83 per cent for seed yield per plant (Khan, 1976). Recently (Chomachalow and Pongpangam, 1981), morphological and quantitative variation among 528 accessions of winged bean consisting of 454 indigenous lines (*i.e.* from Thailand) and 74 exotics mostly from Indonesia, Pureto Rico and Papua New Guinea has been reported. Considerable variation was found for the morphological character which was in no way different than the one observed earlier and described by Khan (1976) and Chomchalow *et al.* (1980). Based on the evaluation of these above mentioned accessions, Chomchalow and Pongpangan (1981) indicated that although the primary centre of diversity of winged bean is still obscure, Thailand might be considered another secondary centre of diversity of winged bean. The range of variation for different quantitative chractes such as days to first flower (81-90), pod length (12.3-33.9 cm), seeds per pod (5-18), 100 seed weight (16.2-46.0 g); seed yield per plant (=1.-210.0 g), sheling percentage (30.4-+68.4 per cent) and tuber yield per plant (0.0-272.0 g) was appreciable and suggests ample opportunity of winged bean improvement through selection (Mohanty, 2013). Maximum variation occurred for tuber yield per plant, followed by seed yield per plant and days to first flower.

Gene and Genetics of Characters

Stem colour, controlled by single gene with purple dominant over green can be used as a marker character. Erskine (1980) measured the degree of cross pollination in the winged bean using the stem colour seedling marker the difference between the green and purple stem colour is monogenically inherited with the allele for purple coluor being dominant (Erskine and Khan, 1977).

Botanical Description

Winged bean, the only cultivar among the genus *Psophocarpus* belongs to the tribe Phaseolae and the sub-family Fabaceae of Leguminosae. From the meiotic observations, Ramirez (1960) reported n=13 as the chromosome number of winged bean. Dr R. D. Brock of

Table 4.95: Genetics of Characters

Character	*Number of Gene*	*Type of Gene Action*
Pod colour	Monogenic	Purplr dominant over green
Pod shape	Monogenic	Rectangular pod shape dominant over flat pods
Pod length	Polygenic	Partial dominance to complete dominance
Pod weight	Polygenic	Additive
Pods/plant	Polygenic	Over dominance
Seed size	Polygenic	Additive
Seed weight	Polygenic	Additive
Seeds/pod	Polygenic	Additive
Earliness (days from sowing to flower)	Polygenic (affected by photoperiodism)	Partial dominance
Time from flowering to maturity	Polygenic	Additive
Leaf size	Polygenic	Additive
Leaves/plant at flowering	Polygenic	Additive and non-additive
Shelling (per cent)	Polygenic	Dominance

Source: Peter, 1998.

C.S. I. R. O. (ef Khan, 1976) made extensive chromosome counts in the winged bean material both from mitotic metaphase and meiotic late prophase and found n=9 as the chromosome number in the P. N. G. (Papua New Guinea) material. This confirms earlier findings of Tixier (1965). Later, the chromosome number of winged bean (2n=18) firmly established by Dr. Brock and his group (Annon, 1975a). Pickersgill, (1980) conducted detailed cytological analysis of 20 accessions of *P. tetragonolobus* from Papua New Guinea, Thailand, the Phillippines and Indonesia and of two accessions of *P. scandens*, in both of the species, there were three pairs of short metacentric or sub- metacentric chromosomes and six pairs of long chromosomes. The long chromosomes in *P. tetragonolobus* were sub- metacentric to acrocentric, whereas they were mostly metacentric in *P. scandens*. This observation of Pickersgill (1980) supported previous evidence that *P. scandens* is not the immediate wild ancestor of domesticated *P. tetragonolobus*. Meiotic studies did not reveal any structural heterozygosity in *P. tetragonolobus* accessions, chiasmata did not appear to be localized and pollen stainability was high. It is considered that there could be few restriction to intra-specific recombination other than the predominance of self-fertilization. Karyotypic differences between *P. tetragonolobus* and *P. scandens* may hinder the introduction of genes from the later into the former (Pickersgill, 1980). Winged bean is a short day plant as far as flowering is concerned means cultivar will not flower until daylight length is less than 12 hours. Some researcher has isolated a day neutral type cultivars which can be grown in the higher latitudes. It is sensitive to frost.

Table 4.96: Botanical Description

Growth Habit	The plant is a twining herbaceous perennial herb but grown as an annual (short day plant). The plant grow vigorously. It is taller and notably larger than the common bean (5 m in height). All the cultivars are indeterminate in growth habit, show extensive vegetative growth like cowpea and lablab beans.
Root	Plants develop extensive tuberous root system which is fibrous in nature. The storage roots (tubers) occur in many cultivars. They are small, irregular, spindle shaped and each weighing around 50.0 g. The plants of winged bean appear to have more nodules on their roots compared with other edible legumes and fixed nitrogen *via* its root nodules. Tuberizaion is a varietal character.
Stem	Climbing stems attain the height of 3.0-4.0 m. and the basic pigmentation of the stem is green to purple in colour.
Leaf	Pinnately trifoliate green leaves similar to the genus *Phaseolus*. The majority of leaves are deltoid, the rest being lanceolate.
Floral biology	Usually flowering commences after 50-60 days of sowing. Inflorescence is axilary and is a 3-12 flowered on erect raceme. Pedicels have bracteoles at the base of calyx. Calyx tube is 5 lobed, long; limb of the standard petal is 2.5-3.5 cm in diameter, green on the back and flowers varied like light violet, blue, white or purple. The standard petal is large and broad; two irregular obviate wing petals, and an obtuse, incurved keel. They enclose ten stamens, all but one of which is united by their filaments at the base; the remaining one is free. The multi-ovulate ovary has a long incurved style and terminates in a globose stigma with dense hairs around and below it.
Anthesis and pollination	Flowering in winged bean occurs between 50-90 days after planting. Maximum flower bud opening has been observed between 10-12 A.M. In Papua New Guinea, the flowers open in the afternoon, which is contrastingly different from the situations in Ghana where they have been described to open 8 and 10 A.M. (Pospisil *et al.*, 1971). Working in Papua New Guinea, Erskine and Bala (1976) reported winged bean flowers to be cleistogamous and anthesis occurring during the night prior of flowering opening. The stigma remains receptive up to 34 hours after the flowers have opened. The anther dehiscence occurs even before the flower opening. Winged bean is cleistogamous and considered largely as self-pollinated crop with a level of cross pollination ranging from 0.0-7.6 percent. In Ghana, frequency of pod set was to the same extent in the caged and in the un-caged plants suggesting that insect pollination was not essential for pod set in the cultivated winged bean (Karikari, 1972). However, different kinds of beas have been seen visiting winged bean flowers in Indonesia and some cross pollination has been recognized (Aminah Lubis, 1980). Insect (bumble bees-*Xylocapa diffuse*) pollination has also been found beneficial for fruit setting. Erskine (1980) observed that rate of outcrossing was less than 1 per cent but in the wet season, it reached 7.6 per cent owing to the activity of carpenter bees (*Xylocapa aruama*).

Pods and pod set	Pod formation is visible 5 to 7 days after anthesis. Some varieties rarely produce more than one pod per plant inflorescence, whereas in others the occurrence of more than one pod per cluster is common. The pods mature in 3-5 months (Khan, 1976) and dry out. Pods are usually 06.0-38.0 cm long but can attain a length of 50.0 cm or so, have four longitudinal wings with protruding (frilly edges running length wise) from angles and hence the nane winged bean. The skin is waxy and the flesh partially translucent in the young pods. When the pod is fully ripe, it turns to an ash brown colour and splits open to release the seeds. Wide variation can be observed in pod colour, shape and the wing characteristics. The basic colours of the pods are green, pink or purple. The wings may be undulate, serrate or lobed.
Seeds	Seed smooth and nearly globular (Parthasarathy, 1986). The seeds, varying from 5-40/pod (Sahoo *et al.*, 2002), are round and look a little like those of soybeans. Shining seeds may be white, yellow, brown, black, purple or mottled and very in weight from 0.06-0.5 g each. It also shows variation in their testa hardness.

Horticultural Classification

(A) Classification on the Basis of Flower Colour

(i) Blue-common

(ii) Purple-common

(iii) Mutanta-Reddish purple flower or uncommon flower colour

(B) Classification on the Basis of Pod Colour

The pods are basically of three background colours *i.e.*

(i) Green-on the green background, quite often there are purple spots extending to varying degrees and at times with such intensity that the pods can be defined essentially as purple. The wings on the green pods commonly show light to deep purple colouration.

(ii) Pink-Pink and pale background pods show a close association with smaller pods, short statured plants, thin stem and small leaves (Khan, 1976).

(iii) Pale yellow-Pods with pale yellow background display green purple wings.

(C) Classification on the Basis of Pod Edges Colour

(i) Rectangular- Rectangular pods are most common

(ii) Semi-flat- Semi-flat on sides

(iii) Flat- Flat on suture pod shapes are also found

(D) Classification on the Basis of Pod Surface

Pod surface can be classified as:

(i) Rough

(ii) Semi-rough

(iii) Smooth

(D) Classification on the Basis of Seed Colours

(i) Brown- may vary from light to dark in intensity. The brown seeds often have a dark ring around the hilum.

(ii) Tan

Environment may also affect the seed colour to some extent (Khan, 1976)

Augmentation of Genetic Resources and Crop Improvement

Germplasm collection, evaluation and consequently the selection of most desirable types, from the naturally occurring genetic variability, is the foremost step for the development of an unexploited crop like winged bean. Though, the promise that winged bean holds had been shown since a long time, the plant remained obscure until recently. The various plant genetic resources activities, such as, survey, exploration and collection, classification and description, preservation and conservation, maintenance, multiplication and rejuvenation, exchange and quarantine, evaluation, documentation, information retrieval and exchange, and utilization are vital and interlinked (Singh, 1981b). Although, winged bean is known since ancient time as a backyard crop with impressive variability, particularly in Asia and Pacific Island countries, no planned work on genetic resource collection has been taken till recent past. Realizing this and the importance of winged bean as a valuable source of nutritive human food and its wider adaptability, International Board of Plant Genetic Resources (IBPGR), Rome augmented the efforts and support for extensive collection of winged bean germplasm in Thailand, Papua New Guinea, Indonesia and Bangladesh. Proposals from Malaysia and other countries of South-East Asia, for capturing and utilizing the genetic variability of winged bean are also being considered. Countrywise information on the winged bean germplasm collection with respect to the institute holding the collection, scientist-in-charge, size of collection, *i.e.* the number of accessions, geographic representation, utilization and material exchange, has been given by Singh (1981b). There are about 25 countries involved in the network on germplasm collection and the classification on is underway. International Board of Plant Genetic Resources (IBPGR) prepared a list of descriptors to be used in collection and evaluation of winged bean. In India, winged bean germplasm is available at the ICAR-National Bureau of Plant Genetic Resources (NBPGR), New Delhi. In a total of 106 accessions 15 are indigenous (mostly from north-eastern region). Among exotics, most accessions are from Indonesia and Philippines. Research development work on winged bean in India is in progress also at the Indian Institute of Horticultural Research, Bengluru (Karnataka), Tamil Nadu Agricultural University, Coimbatore (Tamil Nadu) and at Pune (Maharashtra Association for Evaluation of Science). Philip (1984) studied the extent of genetic variability, heritability and genetic advance in winged bean. The content of crude protein was high in seeds, followed by leaves, pods, flowers and tubers. The accessions PT-4 and PT-30 recorded maximum protein content in green pod (38.50 per cent) and 38.28 per cent, respectively). The genotype PT-6 and PT-25 were the best performing ones.

Introduction and Selection

Considerable variability of the crop exists in Papua New Guinea. A number of genotypes have been introduyced to India through the ICAR-NBPGR, New Delhi (1984). At ICAR-IIHR Bangalore, Sel. 21, Sel.60 and Sel. 71 were found promising and recommended for cultivation in India. Similarly 'WBC 2' from Meghalaya is found good for yield (Parthasarathy, 1986). Another variety 'Revathy' developed from KAU, Velannikara (Kerala).

(i) SLS-44-This variety has been released from Sri Lanka and can be cultivated during any season of the year.

(ii) UPS-122-This variety has been released from Sri Lanka and can be cultivated during any season of the year.

(iii) WBC-2- Advance line selected from local material of Meghalaya (Parthasarathy, 1986).

(iv) PT-62-Cultivated in Kerala

(v) PT-16-Cultivated in Kerala

(vi) PT-49-Cultivated in Kerala

(vii) PT-2 -Cultivated in Kerala

(viii) GRWB-11-Released from USA and well adapted for both young pod and seed production.

IIHR Sel-21, IIHR Sel-60 and IIHR Sel-71 cultivars are promising. Promising strains were tested in multilocational trials and four selections namely, EC38821, EC38821B, EC38955 and IIHR-13 a yield level of 10.5 t/ha for green pods has been released for cultivation as a vegetable type in winged bean growing areas in the country. Few introductions *viz.* Blue Course (EC27884), Blue Fine (EC28854) both from Ghana, Molk (EC38855) for Papua New Guinea and EC116886 and EC121818 appeared promising in grain yield. Promising early types included EC121396 from Nigeria and EC37755A, EC38823A, EC38825 and EC27886A from Papua New Guinea. Sahu *et al.* (2005) evaluated 10 winged bean varieties for three consecutive years during *kharif* seasons of 2002 to 2004 in rainfed upland condition in a sandy loam soil. 'Dwarf Mutant' (165 days) and NBRI Sel. (170 days) were early maturing varieties, whereas EC178313 (184 days) and EC142665 (186 days) were the mid-maturity varieties. Plant height was significantly highest in NBRI Sel. (195.8 cm). Length of pod was highest in EC142665 (178.3 cm). Number of pods per plant was maximum (30 per plant) in EC142665. The 1000 seed weight was maximum in EC142665 (34.15 g) and minimum in NBRI Sel. (17.86 g). The variety EC14665 produced significantly the maximum seed yield of 20.94 q/ha over other varieties while 'Dwarf Mutant' produced the lowest seed yield (10.82 q/ha).

Interspecific Hybridization

Three species of *Psophocarpus* [*P. tetragonolobus* (L.) D.C.; *P. palustris* Desv., and *P. longepedunculatus* Hassk.] are cross compatible and hence would be of used in breeding. Another related species, *P. scandens* did not show storage root formation. The root system of *P. scandens* was considerably denser than that of other winged bean accessions.

Mutation Breeding

Seeds treated with gamma rays (10, 20, 30 Kr)

Table 4.97: Growth and Yield of Winged Bean Genotypes (Pooled of three years)

Sl.No.	Genotypes	Days to 50 per cent Flowering	Days to Maturity	Plant Height (cm)	Number of Branches/ Plant	Pod Length (cm)	Number of Pods (Vegetable stage)	1000 Seed Weight (g)	Sedd Yield (q/ha)	Per cent Increase or Decrease Over Check
1.	EC178313	72	184	181.3	6.1	17.52	28	31.96	17.83	53.31
2.	EC178171	74	183	187.5	5.7	16.02	27	30.37	14.52	24.85
3.	EC178331	71	180	164.7	5.8	15.45	26	29.25	15.03	29.23
4.	EC142665	75	186	171.0	7.3	17.83	32	34.15	20.94	80.08
5.	EC38955	66	178	181.0	5.5	15.70	23	30.12	11.30	-2.84
6.	IC26945	63	178	192.8	6.4	17.63	31	32.15	18.09	55.55
7.	NBRI Sel.	61	171	195.8	5.6	15.93	22	27.86	11.58	-0.43
8.	Mysore Local	63	174	165.6	5.4	15.30	22	30.67	11.48	-1.29
9.	Dwarf Mutant	60	165	143.0	5.8	14.05	20	27.99	10.82	-6.96
10.	AKWBI (Check)	66	175	182.5	5.7	15.35	22	30.17	11.63	-
CD (0.05)		**2.8**	**4.1**	**9.6**	**0.24**	**0.28**	**3.1**	**0.46**	**3.12**	**-**

Source: Sahu *et al.*, 2005.

using Co^{60} as source of radiation. It was found that seed germination, plant height and nodulation were reduced with increase in irradiation doses. Another utility of irradiation was cause profuse flower drop, creating a self-pruned plant desirable for good tuber formation. Savithriamma (1987) reported 0.6 per cent ethyl methane sulphonate as the optimum dose for inducing mutations in winged bean. Viable mutations included dwarf, bush and early maturing types. Effect of gamma rays and EMS on winged bean was studied by Dharan (1993). He found greater variability for morphological characters in the M_1 generation. With increasing dose, growth of root and shoot were reduced linearly by gamma rays. Mutagenic effect of gamma rays and EMS was found to be high in lower doses (100 Gy units of gamma rays and 40 mM of EMS). The efficient doses of gamma rays identified on the basis of injury and lethality or sterility were 100 Gy units and 200 Gy units, respectively. In EMS, the efficiency on the basis of sterility, lethality and injury was maximum at 120, 160 and 80 mM, respectively (Deepa, 1995). University of Agricultural Sciences, Bengaluru (Karnataka) developed promising mutants with bushy, dwarf, determinate/semi-erect growth habit and early maturity (shivshankar, 1982; Veeresh, 1986; Savithriamma, 1987).

Disease Resistance

Attempt has been made to evolve winged bean lines resistant to false rust caused by *Synchytrium psophocarpi*. The resistance is found to be goverened by recessive genes.

Abiotic Stress

Winged bean can tolerate a short period of drought (Martin and Delpin, 1978).

Genetic Resource Conservation

The ICAR-National Bureau of Plant Genetic Resources, New Delhi, has initiated survey, exploration and collection, classification and description, preservation and conservation, maintenance, multiplication and rejuvenation, distribution, exchange and quarantine evaluation, documentation, information retrieval and exchange of the germplasm in winged bean. From Eastern states/NEH of Assam, Tripura, Manipur, Mizoram, Nagaland, Meghalaya, West Bengal, Tamil Nadu, Kerala, Karnataka, Goa, Odisha, Western Ghats, Maharashtra southwards and central/eastern peninsular region enormous variability in indigenous collections has been reported from these areas.

REFERENCES

1. Agcaoili, F. 1929. Segudillas bean *Philippine J. Sci.*, **40:** 513-514.
2. Aminah Lubis, H. 1980. Proceedings of the Workshop/Seminar on the development of the potential of the winged bean 1978, Phillippines Council for Agriculture and Resources Research, Los Benos.
3. Anonymous. 1975a. National Academy of Science, Washington, DC.
4. Anonymous. 1977. *New Scientist*, **75(1064):** 356.
5. Aykroyd, W.R. and Doughty, J. 1964. FAO Nutritional Studies No. 19 Rome.
6. Burkill, IH. 1935. A *Dictionary of Economic Products of Malaya Peninsula*, Vol. 2, Grown Agents for the Colonies, London. p. 2402.
7. Cerny, K. M.; Kordylas, M.; Pospisil, F.; Svabensky, O. and Zajic, B. 1971. Nutritive value of the winged bean (*Psophocarpus palustris*). *British J. Nutr.*, **26:** 293-299.
8. Chomchalow, N. and Pongpa-ngan, L. 1981. *IBHGR Newsletter*, 5: 10-12.
9. Chomchalow, N., Suputtitada, S., and Piyachoknagul, S. 1980. A preliminary report. Proc. 1st internal. Symp. Developing the potentials of the winged bean. Manila, Philippines p. 46-57.
10. Chundawat, B.S., Patel, G.L. and Jikka, S.D.S.1981. Variability, correlations and path coefficient analysis of yield in winged bean (*Psophocarpus tetragonolobus*). Second International Symposium on Developing the Potentials of winged bean, held during 1991 at Colombo, Sri Lanka.
11. Claydon, A. 1975. A review of the nutritional value of the winged bean, *Psophocarpus tetragonolobus* (L.) DC., with special reference to Papua New Guinea. *Science in New Guinea*, **3(2):** 102-114.
12. Degener, O. 1945. *J. New York Bot. Garden*, **46:** 110.
13. Deepa, T.O. 1995. Mutagenicity of gamma rays and EMS on winged bean, *Psophocarpus tetragonolobus* (L.) DC. M.Sc. thesis Kerala Agricultural University, Vellanikkara, Thirissur (Kerala).
14. Dharan, R.S. 1993. Morphological effect of gamma rays and EMS on winged bean. M.Sc. thesis Kerala Agricultural University, Vellanikkara, Thirissur (Kerala).
15. Eagleton, G.E.; Khan, T.N. and Chai, N.F. 1981. Abstracts of papers Second International Seminar on winged bean (cf. winged bean Flyer), 3(II), 9p.
16. Ekpenyyong, T.E. and Barchars, R.L. 1978. Workshop Report on Development Potential of winged bean, Los Banos, Philippines.
17. Elmes, R. P. T. 1976. *Papua New Guinea Agr., J.*, **23:** 53-57.
18. Erskine, W. and Bala, A.A. 1976. *Tropical Grain Legume Bulletin*, **6:** 32-35.
19. Erskine, W. and Khan, T.N. 1977. Inheritance of pigmentation and pod shape in winged bean

(*Psophocarpus tetrogonolobus* (L.) DC.) *Euphytica*, **26:** 829-831.

20. Erskine, W. and Khan, T.N. 1981. Variation within and between land races of winged bean (*Psophocarpus tetragonolobus*). *Field Crops Research*, **3:** 359-364.

21. Evans, I.M.; Boulter, D.; Engleshan, A.R.J. and Dart, P.J. 1977. Protein content and protein quality of tuberous roots of some legumes determined by chemical methods. *Qual. Plant. Foods for Human Nutrition*, 27: 275-285.

22. Garcia, V.V. and Palmer, J.K. 1980. Proximate analysis of five varieties of winged beans, *Psophocarpus tetragonolobus* (L.) DC. *Internationa Journal of Food Science and Technology*, **15(5):** 469-476.

23. Gopalkrishnan, P.K.; Rajemohan, K.; Prabhaharan, T. and Peter, K.V. 1981. Abstracts of papers Second International eminar on winged bean (cf. winged bean Flyer), 3(II), 7p.

24. Hag, N. and Smartt, J. 1973. *Tropical Grain Legume Bulletin*, *15*: 35-38.

25. Harder, D., Pene, O., MBuTu, Lolema and Tshisand, M.C. 1990. Uses, nutritional composition and ecogeography of four species of *Psophocarpus* in Zaire. *Economic Botany*, **44:** 391-409.

26. Harding, J., Lugo-Loper, M.A. and Escolar, R.P. 1978a. *Trop. Agric.* (Trinidad), **55:** 315-324.

27. Harikumar, K. 1998. Physiological and biochemical studies on winged bean (*Psophocarpus tetragonolobus* (L.) DC.). Thesis Department of Botany, University of Calicut, Kerala.

28. Hymowitz, T. and Boyd, J. 1977. Origin, ethnobotany and agricultural potential of the inged bean (*Psophocarpus tetragonolobus*) *Econ. Bot.*, **31:** 180-188.

29. Karikari, S. 1972. *Ghana J. Agric. Sci.*, **5:** 235-239.

30. Khan, T. N. 1976. Papua New Guinea: A centre of genetic diversity in winged bean (*Psophocarpus tetragonolobus*). *Euphytica*, **25:** 693-706.

31. Khan, T.N. 1980. A monograph, Western Australian Depatment of Agriculture.

32. Maheshwari, P. and Singh, U. 1965. Dictionary of Economic Plants in India, Indian Council of Agricultural Research, New Delhi.

33. Martin, F.W. and Delpin, H. 1978. Vegetable for Hot Humid Tropics: the Winged bean (*Psophocarpus tetragonolobus*), USDA Bulletin, New Orleans, pp. 22.

34. Masefield, G.B. 1973. Philippines, 1973? *Field Crop* (Abs.), **26:** 157-160.

35. Masefield, G.B. 1961. *Trop. Agric.* (Trinidad), **38:** 225-229.

36. Matejka, V. 1987. On the origin and climatic requirement of the winged bean. *Agricultura Tropica et Subtropica*, **20:** 33-45.

37. Mohanty, C.S., Verma, S., Singh, V., Khan, S., Gaur, P., Gupta, P., Nizar, M.A., Dikshit, N., Pattanayak, R., Shukla, A., Niranjan, A., Sahu, N., Behera, S.K. and Rana, T.S. 2013. Characterization of winged bean (*Psophocarpus tetragonolobus* (L.) DC.) based on molecular, chemical and physiological parameters. *American Journal of Molecular Biology*, **3:** 187-197.

38. NBPGR Science Mopnograph (No. 8).1984. Winged bean in India. ICAR-National Bureau of Plant Genetic Resources, New Delhi.

39. Ningombam, R.D., Singh, P.K. and Salam, J.S. 2012. Proximate composition and nutritional evaluation of underutilized legumes *Psophocarpus tetragonolobus* L. grown in Manipur. *American Journal of Food Technology*, **7(8):** 487-493.

40. Parthasarathy, V.A. 1986. Winged bean (In) Vegetable Crops in India, Edn:I. Bose, T.K. and Som, H.G. (Eds), Naya Prakash, Calcutta.

41. Peter, K.V. 1988. Genetics and Breeding of Vegetable Crops, ICAR, New Delhi, p. 149-176.

42. Philip, A. 1984. Genetic variability andc correlation studies in winged bean (*Psophocarpus tetragonolobus*), M.Sc. thesis Kerala Agricultural University, Vellanikkara (Kerala).

43. Pickersgill, B. 1980. Cytology of two species of winged bean (*Psophocarpus tetragonolabus* (L.) DC and *P. scandens* (Endl.) Verdc (Leguminosae). Botanical Journal of the Linnean Society, 80:279-291.

44. Pospisil, F., Karikary, S.K. and Boamah-Mensha, E. 1971. Investigations of winged bean in Ghana *World Crops*, **23:** 260-264.

45. Powell, J.M. 1974. *Science in New Guinea*, **2:** 48-63.

46. Rai, N., Asati, B.S., Patel, R.K., Patel, K.K. and Yadav, D.S. 2005. Underutilized Horticultural Crops in North Eastern Region. ENVIS Bulletin: *Himalayan Ecology*, 13(1).

47. Rajmohan, K., Gopalakrishnan, P.K. and Peter, K.V. 1982.Winged bean: A vegetable of promise. *World Crops*, **34(5):** 179-180.

48. Ravelli, G.P.; Zi G.K.N.; Diaby, L.; Dri, K.B.N.; Mayer, C.G. and Sylla, B.S. 1978. Proc. Workshop/ Seminar on the development of the potentials of the Winged Bean Los Banos, Philippines.

49. Ruberte, R.M. and Martin, F.W. 1978. Agr. (Univ. Puerto Rico), **62:** 321-329.

50. Russell, P.G.1952. Plant inventory No. 154, USDA, Washington DC.

51. Ryan, P. 1972. Encyclopaedia of Papua New Guinea Vol. A.-K. Melbourne University Press, Melbourne.

52. Sahoo, J. Panigrahi, R. and Moharana, T. 2002. Winged bean: A promising under exploited pulse crop for the farmers. *Indian Farming*, May, p. 26-28.

53. Sahu, P.K., Jena, S.N., Singh, G.S. and Tripathy, S. 2005. Evaluation of winged bean varieties in raifed upland condotions. *The Orissa Journal of Horticulture*, **33(1):** 51-53.

54. Sarma, B.K. 2001. Underutilized crops for hills and mountain ecosystems. Summer school on agriculture for hills and mountain ecosystem, p. 308-314.

55. Savithriamma, D.L. 1987. Studies on induced mutagenesis in winged bean (*Psophocarpus tetragonolobus*) through ethylmethane sulphonate. *Mysore Journal of Agricultural Sciences*. **21:**93.

56. Shivshankar, G. 1982. Bush type mutant of winged bean (*Psophocarpus tetragonolobus* L.) DC. *Winged Bean Flyer*, 4(2).

57. Singh, R.B. 1981a. *Editorial, IBPGR Newsletter*, **5(2):** 3-4.

58. Singh, R.B. 1981b. *IBPGR Newsletter*, **5:** 5-9.

59. Thompson, A.E. and Haryono, K. 1979. Sources of resistance to two important diseases of winged bean (*Psophocarpus tetragonolobus*) (L.) DC. *Hort. Sci.*, **14:** 532-533.

60. Thompson, A.E. and Haryono, K. 1980. Winged bean: Underexploited tropical food crop. *Hort. Sci.*, **15:** 233-238.

61. Tixier, P. 1965. Dannees cytologiques sur quelques legum euse cultivees ou spontanees de Vietnam et du Lao (The cytology of wild and cultivated legumes of vietnam and Laos). *Rev. Cytol. Biol. Veg.*, **28:** 133-135.

62. Valicer, P., Halva, B. and Mohan, N.V. 1987. Results of a variety trial with winged bean. *Agricultura Tropica et Subtropica*, **20:** 47-57.

63. Veeresh, L.C. 1986. Studies on mutagenic responses of winged bean (*Psophocarpus tetragonolobus* L.) DC. *Mysore Journal of Agricultural Sciences*, **20(4):** 334.

Genus *Canavalia*

Jack Bean and Sword Bean

Canavalia is a small genus consisting of 60 species (most of these of American origin) of which 4 species *viz., C. ensiformis* (L.), *C. gladiata* (Jacq), *C. maritime* and *C. virosa* (Roxb.) are reported from India. Of these four species *C. ensiformis* (Jack bean), and *C. gladiata* (Sword bean), occur in North Eastern Region of India and are being cultivated for the edible pods.

An analysis based on DNA data designated to distinguish legume species failed to reveal differences between *C. ensiformis* and *C. gladiata.*

Table 4.98: Difference between Jack Bean and Sword Bean

Jack bean	*Sword bean*
Scientific name of jack bean is *Canavalia ensiformis* (L.) A. DC.	Scientific name of sword bean is *Canavalia gladiata*
Jack bean is bushy in nature	Sword bean is climbing in nature
Length of the seed hilum (scar) only about one-third as long as the seed.	Length of the seed hilum (scar) is more than one-half the length of the seed
The seed colour is white	The seed colour is red
As a New World Tropic	As an old World Tropic

Jack Bean (*Canavalia ensiformis* (L.) A. DC.)

INTRODUCTION

Jack bean (*Canavalia ensiformis* (L.) A. DC. 2n=22) is commonly grown for the young pods and immature seeds which are used as food for human and feed for animals. The tribes of North Cachar and Mikir Hills grow these in jhoom field. The tender pod is used as vegetable. The young pods are usually picked about 10 day's interval for use as vegetable in curries and chutneys (fresh or steamed). The immature pods and seeds contain about 75.2 per cent water, 6.9 per cent protein, 0.5 per cent fat, 13.3 per cent carbohydrate, 3.3 per cent fibre and 0.8 per cent ash (Indira and Peter, 1988). Green pods are rich source of protein (28.31 per cent) and seeds contain carbohydrate (44.56 per cent) with a fairly digestible protein. It is also a good source of Ca, Zn, P, Mg, Cu and Ni (Akpapumum and Sefa-Dedeh, 1996). The essential amino acid profile of total seed proteins compared favourably with FAO/ WHO requirements. The leaves of the plant are used as green manure as well as fodder for animals. Possibilities exist for using this bean species as a source of purified protein and starch concentrates as food supplements or for use in livestock feeds. Sword bean is a source of an enzyme, urease, which has been exploited commercially. It is useful as a cover crop and as a green manure. Its use as raw livestock feed is limited by the growth inhibiting substances it contains.

Origin, History and Distribution

The Jack bean is a native of Central America and West Indies (Anonymous, 1950.). India is considered as secondary centre for this crop and is being cultivated in north east states.

Botanical Description

Cultivars of *Canavalia ensiformis* (L.) vary widely, particularly in the degree of twining, the size of the fruits and the number and colour of the seeds.

Table 4.99: Botanical Description

Growth Habit	The plant is perennial, herbaceous, bushy in nature but strong enough and can withstand stress. The cultivated plants may reach from 60.0-120.0 cm in height.
Root	The root is a well branched tap root, penetrating deep into the soil down up to 150.0 cm. but has surface feeding nature.
Stem	The stem is grassy, cylindrical green, branched (5-7 lateral sprouts).
Leaf	The leaves are cordate, green and large.
Floral biology	The flowers lie in axillary racemes (about 10 racemes on one plant).
Pod and pod set	The plant bears linear pods profusely and the growth of the pod is very rapid. Mature pods are fibrous and unsuitable for vegetable. Pod attend the length of 25.0-35.0 cm and width of 2.5-4.0 cm.
Seeds	The mature pod gives 8-20 white seeds. Seeds are large, 1.25-2.00 cm long and nearly broad. The hilum of jack bean is about one-third as long as the seed.

Augmentation of Genetic Resources and Crop Improvement

Thirty eight accessions of jack bean were augmented from Peninsular India and characterized during 2010-14 and considerable variation was found among the genotypes for morphological characters *viz.*, growth habit (pole and bush), leaf density (sparse, intermediate and dense), stem colour (light green, purple and dark purple), flower colour (purple white), pod shape (straight, intermediate and curved), pod colour (light green and green), pod beak length (short, medium and long), pod curvature (straight, curved and highly curved), pod surface (smooth and wrinkled) and seed colour (brown, reddish-purple, white, grayish-yellow and grayish-orange). While the protein content ranged between 16-21 per cent, the oil content ranged between 0.4-2.3 per cent with fatty acids oleic, linoleic and palmitic being predominant. The plant height, pod weight, number of pods per plant, and test weight were observed with high genetic variability and direct selection would be rewarding for jack bean improvement in India (Sivaraj *et al.*, 2016).

Genetic Resource Conservation

Not much research work has been done for genetic resource conservation of jack bean in India.

REFERENCES

1. Akpapumum, M.A. and Sefa-Dedeh, S. 1996. Jack bean (*Canavalia ensiformis*): Nutrition related aspects and needed nutrition research. *Plant Foods for Human Nutrition*, 50(2): 93-99.
2. Anonymous. 1950. The Wealth of India, Vol. II, pp. 56. Council of Scientific and Industrial Research, New Delhi.
3. Indira, R. and Peter, K.V. 1988. Jack bean: An under exploited tropical vegetables, p. 46-48, published by KAU, Kerala.
4. Sivaraj, N., Pandravada, S.R., Kamala, V. and Babu, B.S. 2016. Characterization and evaluation of jack bean [(*Canavalia ensiformis* L. (DC)], an underutilized wild legume collections from Peninsular India. 1st International Agrobiodiversity Congress, 6-9 Nov., 2016, held at New Delhi., p. 10.

Sword Bean (*Canavalia gladiata* Jacq.)

INTRODUCTION

Sword bean *Canavalia gladiata* Jacq.) DC., 2n=22, 44) is also named as dolic en sabre, pois sabre, pois sabre rouge, schwertbohne, feijão de porco, haba de burro, awara (Sinhala) and poroto sable. It is called 'Chamma kaya' and 'Tammi kaya' in Telgu. In Tamil, it is called 'Valavaraik-kay' and 'Valavaran-gai' which means the vegetable that looks like a sword. The other vernacular names are sword jack bean, Japanese jack bean, haricot sabre (Fr), Fava-contra (Po) *etc.* Sword bean (is a minor crop, generally wild and commonly grown for its immature tender green pods which is used as vegetable on limited scale in NEH (CSIR, 1950). The young pods and immature seeds are used as food for human and feed for animals. The tribes of North Cachar and Mikir Hills grow these in jhoom field. The tender pod is used as vegetable. Green pods are rich source of protein (28.31 per cent), starch (42 per cent) and seeds contain carbohydrate (44.56 per cent) with a fairly digestible protein. It is also a good source of Ca, Zn, P, Mg, Cu and Ni (Akpapumum and Sefa-Dedeh, 1996). The essential amino acid profile of total seed proteins compared favourably with FAO/WHO requirements (Vadivel and Janardhanam, 2004). The leaves of the plant are used as green manure as well as fodder for animals. Possibilities exist for using this bean species as a source of purified protein and starch concentrates as food supplements or for use in livestock feeds. Sword bean is a source of an enzyme, urease, which has been exploited commercially. It is useful as a cover crop and as a green manure. Its use as raw livestock feed is limited by the growth inhibiting substances it contains. The seeds are sown close to fences in homestead gardens or close to trees and shrubs which provide support to the trailing plant. The dry seeds though nutritious, are coarse skinned and not appetizing at all. The immature seeds are palatable. The seeds are eaten in Ethiopia and Tanzania, but often only as a famine food. The seeds contain traces of water soluble sapetoxin which can be removed through boiling and draining off water. It is sometimes grown as an ornamental with very attractive pink-white pea like flowers. In Tanzania, the leaves are used in a recipe for treating smallpox. Some toxic alkaloids of non-protein-nitrogen compounds like lectins, canavine, concanavine, canaline, *etc.*, which are heat labile and hydro soluble. The lectin reacts with the sugar components in intestinal cells causing a disruption in cell structure, which leads to abnormalities in nutrient absorption (Mello *et al.*, 1985). However, the canavine is a major toxic factor, which is present to an extent of 2-3.5 per cent of dry seed (Fearon and Bell, 1955), which also inhibits protein metabolism.

Origin, History and Distribution

Sword bean (*C. gladiata* Jacq.) DC. is probably native of Southern Asia (Indonesian and Indo-Chinese centre) or possibly Africa. India is considered as secondary centre for this crop and is being cultivated in north East states. Sword bean is cultivated widely in South and South East Asia, West Indies, Africa, especially in India, Sri Lanka, and Burma and has been introduced into tropical parts of Australia. It is well spread throughout the humid tropics but grown chiefly for local consumption. It has been introduced into tropical parts of Australia. There are many reports in the literature of the occurrence of the sword bean in tropical Africa, but there are few herbarium specimens from Africa. Many of the reports likely refer to the widespread indigenous *C. africana* Dunn (Syn. *C. virosa* auct. non. (Roxb.) Wight and Arn.). In the interior of tropical Africa at higher altitudes (over 300 m) *C. Africana* is the most common *Canavalia* species.

Species Availability, Evolution and Diversity

Canavalia has 60 species, most of these of American origin. Sword bean is probably derived from a wild ancestor *Canavalia virosa* (Roxb.) Wight and Ann., grows wild in tropical Asia and Africa. *C. porystechs* Schweinf of China is closely related to *C. gladiata* (*Jacq.*) DC. Another related species is the white seeded *C. ensiformis* (L.) DC. (Jack bean 2n=22). The wide dispersal in historic times is believed to be partly due to carrying the remarkable seeds as curios. It is known chiefly as a cultivated species now. In India, researchers have named plants *C. africana* as collected from the wild form of *C. gladiata*. The above mentioned taxa are considered by some as a single species as they cross freely and their uses and chemical composition are similar. The degree of twining, size of seed pods, number and colour of seed pods show considerable variation.

Botanical Description

It is drought resistant perennial climber. Genus *Canavalia* belongs to the family Fabaceae. Sword bean seed germinates readly and the plant is relatively fast growing.

Table 4.100: Botanical Description

Growth Habit	The plant is perennial (often cultivated as an annual), herbaceous, bushy in nature but strong enough and can withstand stress. Semi-erect forms are also available in the nature. The cultivated plant is twining in nature which may reach from 4.5 to 10 ft. in height.
Root	The root is a well branched tap root, penetrating deep into the soil down up to 150.0 cm. but has surface feeding nature.
Stem	The stem is grassy, cylindrical green, branched (5-7 lateral sprouts). The tips of the branches tend to twine under shade.
Leaf	The leaves are cordate, green and large. Leaves are stipulate and trifoliate with pulvinus base. Petiole 5-17 cm long (petioles are shorter than the leaflets with a groove above and stout with large pulvinous at base and at the base of each leaflet); leaflets with 4-7 mm long stalks, ovate, 7.5-20.0 cm x 5-14 cm, apex acuminate, shortly pubescent on both sides with large pubescent leaflets (8-10 x 6-14 cm).
Floral biology	Inflorescence is an axillary long stalked raceme with 10.0-15.0 flowers borne in groups of 3.0-5.0 swollen pedicellar glands. Several flowers develop in succession. Flowers are 3.0-5.0 cm long, white or pinkish. Flowers are bisexual, papilionaceous, often resupinate; calyx up to 1.5 cm long, 2-lipped with a large 2-fid upper lip and a much smaller 3-fid lower lip; corolla white, standard c. 3.5 cm long; stamens 10, all joined; ovary superior, style slender, curved, and stigma small. Flowers are inverted, the standard being at the bottom.
Anthesis and pollination	Flowering time is between April-October. Highly cross pollinated crop (20 per cent) due to extrovert flowers and is entomophilous, bees being the pollinators. Bagged flowers set pods and seeds indicating perfect self compatibility. Crossing is little difficult as the emasculated flowers tend to drop. Crossing should be confined to the lower 2-3 flowers of an inflorescence. Flowers are pollinated by insects and cross-pollination occurs to the extent of 20 per cent or more.
Pod and pod set	The plant bears linear pods profusely and the growth of the pod is very rapid. Mature pods are fibrous and unsuitable for vegetable. The pods are curved with more strongly developed ridges, size 15.0-24.0 cm x 3.0-5.0 cm. Peduncle 4-20 cm long and ribbed near suture.
Seeds	Generally 8-16 coarse skinned seeds are found in the pod. Generally seeds are elliptical, larger, 2.5-3.5 cm long, white or red in colour with dark brown hilum, 2.0-2.5 cm long extending almost the entire length of the seed. The hilum of sword bean is more than one-half the length of the seed. The seed has a tough thick coat

Horticultural Classification

According to Burns (1939), Roxburgh described three varieties in India:

(i) Plants with red flowers and red seeds
(ii) White flowers and red seeds and
(iii) White flowers and white seeds.

In Sri Lanka, the above mentioned three varieties are recognized and have specific common names for each variety (rathu awara, sudu awara and gan awara).

Augmentation of Genetic Resources and Crop Improvement

Not much research work has been done on this crop in India.

Introduction and Selection

One sword bean variety SPS-1 has been released by TNAU, Coimbatore (Tamil Nadu) in 1990.

Interspecific Hybridization

Hybridization between sword bean [(*C. gladiata* (Jacq.) DC, 2n=22, 44)] and Jack bean (*C. ensiformis*, 2n=22) is possible.

Genetic Resource Conservation

The world wide collection of *Canavalia* germplasm is urgently needed. Small germplasm collections of *Canavalia* are being maintained at the Australian Tropical Crops and Forages Genetic Resources Centre, Biloela, Queensland. A few accessions are available in Ethiopia (ILRI), Nigeria (IITA), South Africa, Brazil, China, Colombia and India.

REFERENCES

1. Anonymous. 1950. The Wealth of India, Vol. II, p. 56. Council of Scientific and Industrial Research, New Delhi.
2. CSIR, 1950. The Wealth of India, CSIR, New Delhi. Vol.II, p. 56.
3. Fearon, W.R. and Bell, E.A. 1955. Canavine: Detection and Occurrence in *Colutea arborescens*. *Biochem J.*, **59:** 221-224.
4. Grubben, G.J.H. 1977. *Tropical Vegetables and Their Genetic Resources*. Royal Tropical Institute, Amsterdam, p. 76.
5. Purseglove, J.W. 1977. *Tropical Crops Dicotyledons*, vols. I and II, Longman, London, p. 416
6. Vadivel, V. and Janardhanam, K. 2004. Seed Technology Laboratory, Department of Botany, Bharathiar University, Coimbatore (Tamil Nadu). The nutritional and antinutritional attributes of sword bean (*Canavalia gladiata* Jacq.): An under-utilized tribal pulse from south India. *International Journal of Food Sciences and Technology*, **39(9):** 917-926.

Genus *Vicia*

Broad Bean (*Vicia faba* L.)

INTRODUCTION

Broad bean (*Vicia faba* L.; 2n=12) is also known as Windsor bean, Faba bean (Chauhan and Mangal, 1993), field bean, tick bean or Horse bean. Faba bean is called *baqla* in Persian and that is the name in Hindi today. Interestingly, while Dara Shikoh (c.1650) mentions the word *baqla* in the Nuskha Dar Fanni-Falahat (Razia Akbar, 2000), the Ain-i-Akbari (c.1590) calls it *rajmaan*, probably from *rajmash*, which is kidney bean. Globally, it is third most important feed grain legume after soybean and pea (Mihailovic *et al.*, 2005, Filippetti and Ricciardi, 1993). It is very important cool season vegetable crop, but it is commercialy grown to a limited extent in the northeastern region mainly Manipur. It is grown as a dual purpose crops, tender pods as a vegetable and mature seed for dal purpose. Broad bean is used as green-shell, the seeds removed from the pod before maturity, or as dry beans. It is the principle protein source for poor people in NEH region. The protein content of bean seeds is high, amounting ton about 20-25 per cent, which depend on the variety (Chavan *et al.*, 1989). Being a high protein crop, it is used to replace meat in humam diet. Apart from this it is also rich in carbohydrates (51-68 per cent) in total, the malor proportion of which constituted by starch (41-53 per cent) (Cerning *et al.*, 1975); vitamins (B-group) and dietary minerals (phosphorus, potassium, calcium, sulphur and iron). Calcium content ranges from 120 to 260 mg/100 g dry mass (Chavan *et al.*, 1989).

It is not poisonous to humans in the conventional sense, however presence of anti-nutritional factors [tannins, lectin, protease inhibitor and glucocide vicine that induce favism (a haemolytic anemia) in susceptible individuals] or who have a genetically transmitted, male sex linked deficiency to the enzyme glucose-6phosphate dehydrogenase but in seed has limited its widespread use. However, selection can reduce the contents of these anti-nutritional factors. Broad bean is rich in L-dopa, a substance used medically in the treatment of Parkinson's disease. L-dopa is also a natriuretic agent; which might help in controlling hypertension (Vered *et al.*, 1997). Sometimes peoples used fava beans as a natural alternative to drugs like Viagra, citing a link between L-dopa production and the human libido. An illness some times fatal and known as "favim" is caused (Choudhury, 1967). The green plants are used as cattle feed and in green manuring. Broad bean is often grown as a cover crop to prevent erosion because they can over-winter and became as a legume, they fix nitrogen in the soil.

Table 4.101: Compositional Value of Fresh and Dried Seeds of Broad Bean

Constituents	*Fresh Seeds*	*Dried Seeds*
Protein (g)	7.1	26.2
Calcium (g)	38.0	104.0
Phosphorous (mg)	127.0	34.0
Iron (mg)	-	6.7
Vitamin A (IU)	-	130
Thiamine (mg)	0.10	0.38
Riboflavin (mg)	0.22	0.24
Niacine (mg)	-	2.1
Ascorbic Acid (mg)	140.0	16.0

Source: Duke, 1981.

Origin, History and Distribution

Although the broad bean is probably the bean of history it is not, properly specking, a bean but a vetch originating in the Mediterranean region of Africa, the Middle East and Caspian sea (Cubero, 1973). However, De Candolle (1967) indicated either North Africa or the South Caspian Sea to be the place of origin of broad bean, while Egypt was referred to by Abdalla (1979). It was reported by Cubero (1974) and Ladizinsky (1975) that the Mediterranean basin and the near and Middle Eastern regions are the probable site of origin of broad bean. Broad beans have long tradition of cultivation in old world agriculture, being among the most ancient plants in cultivation and also among the easiest to grow. Broad bean has been on man's food list since 'Stone Age'. It is believed that along with lentils, peas, and chickpeas, they became part of the eastern Mediterranean diet in around 6000 BC or earlier. In ancient Roman courtor in public bodies where questions were put to the vote, to the ballot, it was the seed of broad bean that was used, a white one representing innocent and black one guilty. Eropean living in India grew it as a garden crop (Watt, 1889). Broad bean is widely cultivated since pre-historic times and has been found in the remains of lake-dwellings in England and Switzerland. It was grown by the ancient Egyptians, Hebrews, Greeks and Romans. It was an early introduction into western Asia and reached China in remote times (Thomson and Kelly, 1983). Broad bean was the only edible bean known in Europe in pre-Columbian times and was introduced into new world after 1492 (Purseglove, 1965).

According to United Nation, Food and Agricultural Organizations (FAO), Faba bean is grown in 58 countries. The major countries are China, Ethiopia, Latin America, Egypt, Mexico and Brazil where broad bean is widely cultivated. China is Mihailovi currently the world leading producer with 60 per cent of the total production. Other important producers are northern Europe, The Mediterranean, Ethiopia, Central and East Asia and Latin America. In India it is cultivated in almost entire north eastern states during winter season (October-March) and other states like Uttar Pradesh, Bihar, Punjab, Haryana, Jammu and Kashmir, Rajasthan, Karnataka, Madhya Pradesh, Gujarat *etc.* (Baboo and Kumar, 1996). However it is minor legume crop of the country and also grown in localized area of Kashmir, Rajasthan, Uttar Pradesh, Punjab, Haryana, Karnataka, Madhya Pradesh and Bihar.

Species Availability, Evolution and Diversity

Although usually classified in the same genus *Vicia* as the vetches, some botanists treat it in a separate monotypic genus as *Faba sativa* Moench. It is related to *V. pliniana* Trabut, which grows wild in Algeria. Muratova (1931) classified *Vicia faba* L. in two subspecies with maximum number of leaflets per leaf:

(i) *Faba* (*eufaba* in nomenclature) with more than four leaflets per leaf

(ii) *Paucijuga* with fewer than four. It is subdivided into var. *minor* with small rounded seed (1 cm long), var. *equina* with medium sized seeds (1.5 cm) and var. *major* with large broad flat seeds (2.5 cm).

Table 4.102: Name of Taxon and Preferred Taxon of Genus *Vicia*

Name of Taxon	Name of Authority	Family	Preferred Taxon Name
Vicia acutifolia	Elliott	Fabaceae	*Vicia acutifolia*
Vicia aintabensis	Boiss. and Hausskn.	Fabaceae	*Vicia aintabensis*
Vicia albicans	Lowe	Fabaceae	*Vicia benghalensis*
Vicia americana	Muhl. ex Willd.	Fabaceae	*Vicia americana*
Vicia americana var. *oregana*	(Nutt.) A. Nelson	Fabaceae	*Vicia americana*
Vicia amoena	Fisch.	Fabaceae	*Vicia amoena*
Vicia amoena var. *oblongifolia*	Regel	Fabaceae	*Vicia amoena*
Vicia amphicarpa	L.	Fabaceae	*Vicia sativa* subsp. *amphicarpa*
Vicia amurensis	Oett.	Fabaceae	*Vicia amurensis*
Vicia anatolica	Turrill	Fabaceae	*Vicia anatolica*
Vicia andicola	Kunth	Fabaceae	*Vicia andicola*
Vicia angustifolia	L.	Fabaceae	*Vicia sativa* subsp. *nigra*
Vicia angustifolia var. *segetalis*	(Thuill.) W. D. J. Koch	Fabaceae	*Vicia sativa* subsp. *nigra*
Vicia apiculata	Phil.	Fabaceae	*Vicia nigricans* subsp. *nigricans*
Vicia articulata	Hornem.	Fabaceae	*Vicia articulata*
Vicia assyriaca	Boiss.	Fabaceae	*Vicia assyriaca*
Vicia atropurpurea	Desf.	Fabaceae	*Vicia benghalensis*
Vicia aurantia	(Steven ex M. Bieb.) Boiss.	Fabaceae	*Vicia crocea*
Vicia baicalensis	(Turcz.) B. Fedtsch.	Fabaceae	*Vicia venosa*
Vicia balansae	Boiss.	Fabaceae	*Vicia balansae*

Name of Taxon	Name of Authority	Family	Preferred Taxon Name
Vicia benghalensis	L.	Fabaceae	*Vicia benghalensis*
Vicia biennis	L.	Fabaceae	*Vicia biennis*
Vicia biflora	Desf.	Fabaceae	*Vicia monantha*
Vicia bijuga		Fabaceae	*Vicia bijuga*
Vicia bithynica	(L.) L.	Fabaceae	*Vicia bithynica*
Vicia bivonea	Raf.	Fabaceae	*Vicia villosa*
Vicia brachytropis	Kar. and Kir.	Fabaceae	*Vicia cracca* subsp. *tenuifolia*
Vicia calcarata	Desf.	Fabaceae	*Vicia monantha*
Vicia canescens	Labill.	Fabaceae	*Vicia canescens*
Vicia canescens subsp. *gregaria*	(Boiss. and Heldr.) P. H. Davis	Fabaceae	*Vicia canescens* subsp. *gregaria*
Vicia canescens subsp. *variegata*	(Willd.) P. H. Davis	Fabaceae	*Vicia canescens* subsp. *variegata*
Vicia cappadocica	Boiss. and Balansa	Fabaceae	*Vicia cappadocica*
Vicia caroliniana	Walter	Fabaceae	*Vicia caroliniana*
Vicia cassubica	L.	Fabaceae	*Vicia cassubica*
Vicia chinensis	Franch.	Fabaceae	*Vicia chinensis*
Vicia cirrhosa	C. Sm. ex Webb and Berthel.	Fabaceae	*Vicia cirrhosa*
Vicia cordata	Wulfen ex Hoppe	Fabaceae	*Vicia sativa* subsp. *cordata*
Vicia cornigera	Chaub.	Fabaceae	*Vicia* sp.
Vicia costata	Ledeb.	Fabaceae	*Vicia costata*
Vicia cracca	L.	Fabaceae	*Vicia cracca*
Vicia cracca subsp. *cracca*		Fabaceae	*Vicia cracca* subsp. *cracca*
Vicia cracca subsp. *gerardii*	Bonnier and Layens	Fabaceae	*Vicia cracca* subsp. *incana*
Vicia cracca subsp. *grossheimii*	(Ekutim.) Hashimov, nom. inval.	Fabaceae	*Vicia cracca* subsp. *cracca*
Vicia cracca subsp. *incana*	(Gouan) Rouy	Fabaceae	*Vicia cracca* subsp. *incana*
Vicia cracca subsp. *stenophylla*	(Velen.) P. H. Davis	Fabaceae	*Vicia cracca* subsp. *stenophylla*
Vicia cracca subsp. *tenuifolia*	(Roth) Bonnier and Layens	Fabaceae	*Vicia cracca* subsp. *tenuifolia*
Vicia cretica	Boiss. and Heldr.	Fabaceae	*Vicia cretica*
Vicia crocea	(Desf.) B. Fedtsch.	Fabaceae	*Vicia crocea*
Vicia cypria	Kotoohy	Fabaceae	*Vicia cypria*
Vicia dalmatica	A. Kern.	Fabaceae	*Vicia cracca* subsp. *stenophylla*
Vicia dasycarpa	Ten.	Fabaceae	*Vicia villosa* subsp. *varia*
Vicia dionysiensis	Mouterde	Fabaceae	*Vicia dionysiensis*
Vicia disperma	DC.	Fabaceae	*Vicia disperma*
Vicia dumetorum	L.	Fabaceae	*Vicia dumetorum*
Vicia elegans	Guss.	Fabaceae	*Vicia cracca* subsp. *stenophylla*
Vicia epetiolaris	Burkart	Fabaceae	*Vicia epetiolaris*
Vicia eriocarpa	(Hausskn.) Hal.	Fabaceae	*Vicia villosa* subsp. *eriocarpa*
Vicia ervilia	(L.) Willd.	Fabaceae	*Vicia ervilia*
Vicia faba	L.	Fabaceae	*Vicia faba*
Vicia faba var. *equina*	Pers.	Fabaceae	*Vicia faba* var. *equina*
Vicia faba var. *faba*		Fabaceae	*Vicia faba* var. *faba*
Vicia faba var. *major*	(Harz) Beck	Fabaceae	*Vicia faba* var. *faba*
Vicia faba var. *minor*	Peterm, nom. illeg.	Fabaceae	*Vicia faba* var. *equina*
Vicia faba var. *minor*	(Harz) Beck, nom. illeg.	Fabaceae	*Vicia faba* var. *minuta*
Vicia faba var. *minuta*	(Hort. ex Alef.) Mansf.	Fabaceae	*Vicia faba* var. *minuta*
Vicia filicaulis	Webb and Berthel.	Fabaceae	*Vicia filicaulis*
Vicia floridana	S. Watson	Fabaceae	*Vicia floridana*
Vicia fulgens	Batt.	Fabaceae	*Vicia fulgens*
Vicia galeata	Boiss.	Fabaceae	*Vicia galeata*
Vicia galegifolia	Andrews	Fabaceae	*Swainsona galegifolia*
Vicia gemella	Crantz	Fabaceae	*Vicia tetrasperma*

Name of Taxon	Name of Authority	Family	Preferred Taxon Name
Vicia gigantea	Hook.	Fabaceae	*Vicia nigricans* subsp. *gigantea*
Vicia globosa	Retz.	Fabaceae	*Vicia sativa*
Vicia gracilis	Loisel, nom. illeg.	Fabaceae	*Vicia parviflora*
Vicia graminea	Sm.	Fabaceae	*Vicia graminea*
Vicia grandiflora	Scop.	Fabaceae	*Vicia grandiflora*
Vicia grandiflora var. *kitaibeliana*	W. D. J. Koch	Fabaceae	*Vicia grandiflora*
Vicia gregaria	Boiss. and Heldr.	Fabaceae	*Vicia canescens* subsp. *gregaria*
Vicia grossheimii	Ekutim.	Fabaceae	*Vicia cracca* subsp. *cracca*
Vicia hajastana	Grossh.	Fabaceae	*Vicia anatolica*
Vicia hirsuta	(L.) Gray	Fabaceae	*Vicia hirsuta*
Vicia hirsuta var. *terronii*	(Ten.) Burnat	Fabaceae	*Vicia hirsuta*
Vicia hirta	Balb. ex DC.	Fabaceae	*Vicia lutea* subsp. *vestita*
Vicia humilis	Kunth	Fabaceae	*Vicia humilis*
Vicia hyaeniscyamus	Mouterde	Fabaceae	*Vicia hyaeniscyamus*
Vicia hybrid		Fabaceae	*Vicia hybrid*
Vicia hybrida	L.	Fabaceae	*Vicia hybrida*
Vicia hyrcanica	Fisch. and C. A. Mey.	Fabaceae	*Vicia hyrcanica*
Vicia incana	Gouan	Fabaceae	*Vicia cracca* subsp. *incana*
Vicia incisa	M. Bieb.	Fabaceae	*Vicia incisa*
Vicia incisiformis	Stef., nom. inval.	Fabaceae	*Vicia incisa*
Vicia incurva	Roth	Fabaceae	*Lathyrus incurvus*
Vicia japonica	A. Gray	Fabaceae	*Vicia japonica*
Vicia japonica var. *oblongifolia*	A. Gray	Fabaceae	*Vicia japonica*
Vicia kingii	Hook. f.	Fabaceae	*Vicia magellanica*
Vicia kokanica	Regel and Schmalh.	Fabaceae	*Vicia kokanica*
Vicia laevigata	Sm.	Fabaceae	*Vicia lutea*
Vicia lathyroides	L.	Fabaceae	*Vicia lathyroides*
Vicia laxiflora	Brot., nom. illeg.	Fabaceae	*Vicia parviflora*
Vicia leavenworthii	Torr. and A. Gray	Fabaceae	*Vicia ludoviciana* subsp. *leavenworthii*
Vicia leucosperma	Moench	Fabaceae	*Vicia sativa* subsp. *sativa*
Vicia lilacina	Ledeb.	Fabaceae	*Vicia lilacina*
Vicia linearifolia	Hook. and Arn.	Fabaceae	*Vicia linearifolia*
Vicia litvinovii	Boriss., nom. illeg.	Fabaceae	*Vicia loiseleurii*
Vicia loiseleurii	(M. Bieb.) Litv.	Fabaceae	*Vicia loiseleurii*
Vicia loweana	Steud.	Fabaceae	*Vicia benghalensis*
Vicia ludoviciana	Nutt.	Fabaceae	*Vicia ludoviciana*
Vicia ludoviciana subsp. *leavenworthii*	(Torr. and A. Gray) Lassetter and C. R. Gunn	Fabaceae	*Vicia ludoviciana* subsp. *leavenworthii*
Vicia ludoviciana subsp. *ludoviciana*		Fabaceae	*Vicia ludoviciana* subsp. *ludoviciana*
Vicia lutea	L.	Fabaceae	*Vicia lutea*
Vicia lutea subsp. *lutea*		Fabaceae	*Vicia lutea* subsp. *lutea*
Vicia lutea subsp. *vestita*	(Boiss.) Rouy	Fabaceae	*Vicia lutea* subsp. *vestita*
Vicia lutea var. *hirta*	(Balb. ex DC.) Loisel.	Fabaceae	*Vicia lutea* subsp. *vestita*
Vicia lutea var. *laevigata*	(Sm.) Boiss.	Fabaceae	*Vicia lutea*
Vicia lutea var. *violascens*	Rouy	Fabaceae	*Vicia lutea* subsp. *vestita*
Vicia macraei	Hook. and Arn.	Fabaceae	*Vicia nigricans* subsp. *nigricans*
Vicia macrocarpa	(Moris) Bertol.	Fabaceae	*Vicia sativa* subsp. *macrocarpa*
Vicia macrograminea	Burkart	Fabaceae	*Vicia macrograminea*
Vicia maculata	C. Presl	Fabaceae	*Vicia sativa*
Vicia magellanica	Hook. f.	Fabaceae	*Vicia magellanica*
Vicia megalotropis	Ledeb.	Fabaceae	*Vicia megalotropis*

Name of Taxon	Name of Authority	Family	Preferred Taxon Name
Vicia melanops	Sm.	Fabaceae	*Vicia melanops*
Vicia menziesii	Spreng.	Fabaceae	*Vicia menziesii*
Vicia meyeri	Boiss.	Fabaceae	*Vicia loiseleurii*
Vicia michauxii	Spreng.	Fabaceae	*Vicia michauxii*
Vicia micrantha	Lowe, nom. illeg.	Fabaceae	*Vicia benghalensis*
Vicia monantha	Retz.	Fabaceae	*Vicia monantha*
Vicia monantha subsp. *monantha*		Fabaceae	*Vicia monantha* subsp. *monantha*
Vicia monantha subsp. *triflora*	(Ten.) B. L. Burtt and P. Lewis	Fabaceae	*Vicia monantha* subsp. *triflora*
Vicia monanthos	(L.) Desf., nom. illeg.	Fabaceae	*Vicia articulata*
Vicia montbretii	Fisch. and C. A. Mey.	Fabaceae	*Vicia montbretii*
Vicia montevidensis	Vogel	Fabaceae	*Vicia montevidensis*
Vicia multicaulis	Ledeb.	Fabaceae	*Vicia multicaulis*
Vicia nana	Vogel	Fabaceae	*Vicia nana*
Vicia nana	Kellogg, nom. illeg.	Fabaceae	*Lathyrus nevadensis*
Vicia narbonensis	L.	Fabaceae	*Vicia narbonensis*
Vicia narbonensis f. *serratifolia*	(Jacq.) F. J. Herm.	Fabaceae	*Vicia serratifolia*
Vicia narbonensis var. *serratifolia*	(Jacq.) Ser.	Fabaceae	*Vicia serratifolia*
Vicia neglecta	Hanelt and Mett.	Fabaceae	*Vicia lilacina*
Vicia nervata	Sipliv.	Fabaceae	*Vicia nervata*
Vicia nigricans	Hook. and Arn.	Fabaceae	*Vicia nigricans*
Vicia nigricans subsp. *gigantea*	(Hook.) Lassetter and C. R. Gunn	Fabaceae	*Vicia nigricans* subsp. *gigantea*
Vicia nigricans subsp. *nigricans*		Fabaceae	*Vicia nigricans* subsp. *nigricans*
Vicia noeana	Reut. ex Boiss.	Fabaceae	*Vicia noeana*
Vicia obscura	Vogel	Fabaceae	*Vicia montevidensis*
Vicia ocalensis	R. K. Godfrey and Kral	Fabaceae	*Vicia ocalensis*
Vicia ochroleuca	Ten.	Fabaceae	*Vicia ochroleuca*
Vicia onobrychioides	L.	Fabaceae	*Vicia onobrychioides*
Vicia oregana	Nutt.	Fabaceae	*Vicia americana*
Vicia oroboides	Wulfen	Fabaceae	*Vicia oroboides*
Vicia orobus	DC.	Fabaceae	*Vicia orobus*
Vicia palaestina	Boiss.	Fabaceae	*Vicia palaestina*
Vicia pallida	Turcz., nom. illeg.	Fabaceae	*Vicia amurensis*
Vicia pallida	Hook. and Arn.	Fabaceae	*Vicia pallida*
Vicia pannonica	Crantz	Fabaceae	*Vicia pannonica*
Vicia pannonica subsp. *pannonica*		Fabaceae	*Vicia pannonica* subsp. *pannonica*
Vicia pannonica subsp. *striata*	(M. Bieb.) Nyman	Fabaceae	*Vicia pannonica* subsp. *striata*
Vicia parviflora	Loisel., nom. illeg.	Fabaceae	*Vicia disperma*
Vicia parviflora	Cav.	Fabaceae	*Vicia parviflora*
Vicia peregrina	L.	Fabaceae	*Vicia peregrina*
Vicia persica	Boiss.	Fabaceae	*Vicia canescens* subsp. *variegata*
Vicia picta	Fisch. and C. A. Mey.	Fabaceae	*Vicia biennis*
Vicia pilosa	M. Bieb.	Fabaceae	*Vicia sativa* subsp. *nigra*
Vicia pisiformis	L.	Fabaceae	*Vicia pisiformis*
Vicia pseudo-orobus	Fisch. and C. A. Mey.	Fabaceae	*Vicia pseudo-orobus*
Vicia pseudocracca	Bertol.	Fabaceae	*Vicia villosa* subsp. *pseudocracca*
Vicia pubescens	(DC.) Link	Fabaceae	*Vicia pubescens*
Vicia pulchella		Fabaceae	*Vicia pulchella*
Vicia pyrenaica	Pourr.	Fabaceae	*Vicia pyrenaica*
Vicia ramuliflora	(Maxim.) Ohwi	Fabaceae	*Vicia ramuliflora*
Vicia sativa	L.	Fabaceae	*Vicia sativa*
Vicia sativa subsp. *amphicarpa*	(L.) Batt.	Fabaceae	*Vicia sativa* subsp. *amphicarpa*

Name of Taxon	Name of Authority	Family	Preferred Taxon Name
Vicia sativa subsp. *cordata*	(Wulfen ex Hoppe) Batt.	Fabaceae	*Vicia sativa* subsp. *cordata*
Vicia sativa subsp. *incisa*	(M. Bieb.) Arcang.	Fabaceae	*Vicia incisa*
Vicia sativa subsp. *macrocarpa*	(Moris) Arcang.	Fabaceae	*Vicia sativa* subsp. *macrocarpa*
Vicia sativa subsp. *nigra*	(L.) Ehrh.	Fabaceae	*Vicia sativa* subsp. *nigra*
Vicia sativa subsp. *sativa*		Fabaceae	*Vicia sativa* subsp. *sativa*
Vicia sativa subsp. *segetalis*	(Thuill.) Celak.	Fabaceae	*Vicia sativa* subsp. *nigra*
Vicia sativa var. *amphicarpa*	(L.) Boiss.	Fabaceae	*Vicia sativa* subsp. *amphicarpa*
Vicia sativa var. *angustifolia*	L.	Fabaceae	*Vicia sativa* subsp. *nigra*
Vicia sativa var. *leucosperma*	(Moench) Ser.	Fabaceae	*Vicia sativa* subsp. *sativa*
Vicia sativa var. *macrocarpa*	Moris	Fabaceae	*Vicia sativa* subsp. *macrocarpa*
Vicia sativa var. *maculata*	(C. Presl) Burnat	Fabaceae	*Vicia sativa*
Vicia sativa var. *nigra*	L.	Fabaceae	*Vicia sativa* subsp. *nigra*
Vicia sativa var. *obovata*	Ser.	Fabaceae	*Vicia sativa* subsp. *sativa*
Vicia scandens	R. P. Murray	Fabaceae	*Vicia scandens*
Vicia segetalis	Thuill.	Fabaceae	*Vicia sativa* subsp. *nigra*
Vicia selloi	Vogel	Fabaceae	*Vicia graminea*
Vicia semiglabra	Rupr. ex Boiss.	Fabaceae	*Vicia semiglabra*
Vicia sepium	L.	Fabaceae	*Vicia sepium*
Vicia sericocarpa	Fenzl	Fabaceae	*Vicia sericocarpa*
Vicia sericocarpa var. *microphylla*	Boiss.	Fabaceae	*Vicia sericocarpa* var. *microphylla*
Vicia sericocarpa var. *sericocarpa*		Fabaceae	*Vicia sericocarpa* var. *sericocarpa*
Vicia serratifolia	Jacq.	Fabaceae	*Vicia serratifolia*
Vicia serratifolia f. *integrifolia*	Beck	Fabaceae	*Vicia narbonensis*
Vicia sessiliflora	Clos	Fabaceae	*Vicia sessiliflora*
Vicia setifolia var. *bonariensis*	Burkart	Fabaceae	*Vicia setifolia* var. *bonariensis*
Vicia sexajuga	Nakai	Fabaceae	*Vicia venosa* var. *cuspidata*
Vicia sicula	(Raf.) Guss.	Fabaceae	*Vicia sicula*
Vicia sinkiangensis	H. W. Kung	Fabaceae	*Vicia sinkiangensis*
Vicia sp.		Fabaceae	*Vicia* sp.
Vicia striata	M. Bieb.	Fabaceae	*Vicia pannonica* subsp. *striata*
Vicia sylvatica	L.	Fabaceae	*Vicia sylvatica*
Vicia tenuifolia	Roth	Fabaceae	*Vicia cracca* subsp. *tenuifolia*
Vicia tenuifolia subsp. *stenophylla*	Velen.	Fabaceae	*Vicia cracca* subsp. *stenophylla*
Vicia tenuifolia var. *dalmatica*	(A. Kern.) Asch. and Graebn.	Fabaceae	*Vicia cracca* subsp. *stenophylla*
Vicia tenuifolia var. *laxiflora*	Griseb.	Fabaceae	*Vicia cracca* subsp. *stenophylla*
Vicia tenuissima	auct.	Fabaceae	*Vicia parviflora*
Vicia tetrasperma	(L.) Schreb.	Fabaceae	*Vicia tetrasperma*
Vicia tetrasperma subsp. *gracilis*	(DC.) Hook. f.	Fabaceae	*Vicia parviflora*
Vicia triflora	Ten.	Fabaceae	*Vicia monantha* subsp. *triflora*
Vicia unijuga	A. Braun	Fabaceae	*Vicia unijuga*
Vicia varia	Host	Fabaceae	*Vicia villosa* subsp. *varia*
Vicia varia var. *eriocarpa*	Hausskn.	Fabaceae	*Vicia villosa* subsp. *eriocarpa*
Vicia variabilis	Freyn, nom. ambig.	Fabaceae	*Vicia cracca* subsp. *stenophylla*
Vicia variegata	Willd.	Fabaceae	Vicia canescens subsp. variegata
Vicia venosa	(Willd. ex Link) Maxim.	Fabaceae	*Vicia venosa*
Vicia venosa var. *cuspidata*	Maxim.	Fabaceae	*Vicia venosa* var. *cuspidata*
Vicia venosa var. *venosa*		Fabaceae	*Vicia venosa* var. *venosa*
Vicia venosissima	Nakai	Fabaceae	*Vicia venosissima*
Vicia vestita	Boiss.	Fabaceae	*Vicia lutea* subsp. *vestita*
Vicia vicina var. *pallida*	(Hook. and Arn.) Reiche	Fabaceae	*Vicia pallida*
Vicia villosa	Roth	Fabaceae	*Vicia villosa*

Name of Taxon	Name of Authority	Family	Preferred Taxon Name
Vicia villosa subsp. *dasycarpa*	(Ten.) Cavill.	Fabaceae	*Vicia villosa* subsp. *varia*
Vicia villosa subsp. *eriocarpa*	(Hausskn.) P. W. Ball	Fabaceae	*Vicia villosa* subsp. *eriocarpa*
Vicia villosa subsp. *pseudocracca*	(Bertol.) Rouy	Fabaceae	*Vicia villosa* subsp. *pseudocracca*
Vicia villosa subsp. *varia*	(Host) Corb.	Fabaceae	*Vicia villosa* subsp. *varia*
Vicia villosa subsp. *villosa*		Fabaceae	*Vicia villosa* subsp. *villosa*

Botanical Description

In broad bean 5 pairs are acerocentric chromosome and 1 pair is metacentric. Taxonomically, the crop belongs to Section Faba of the genus *Vicia*.

Table 4.103: Botanical Description

Growth Habit	It is a glabrous annual, rigid, erect, robust herb unbranched and plants are erect 0.3-2.0 m tall with heavy, stiff-stems. Growth is indeterminate though determinate mutants are available.
Root	There is robust tap root to 100 cm or more with profusely branched secondary roots. Small roots with large clusters of small lobed nodules.
Stem	One or more hollow stems from the base. Stout, coarse and upright stems with a square cross-section means stem is tetragonous and fistular. It is without tendrils or with rudimentary tendrils. Generally angles with 1-7 branches.
Leaf	The leaves are alternate distichous and abruptly pinnate with 2-7 leaflets (1-3 pairs), 10.0-25.0 cm long, and of a distinct glaucous grey-green colour; unlike most other vetches, the leaves do not have tendrils for climbing over other vegetation. Stipules are inserted above the base, dentate, near the apex with a black spot. Leaflets are ovate-oblong, mucronate, somewhat fleshy and 3.0-5.0 cm long.
Floral biology	Racemes are axillary, and very shortly peduncled with 1-6 flowers. Flowers are fragrant and 1.0-2.5 cm long, with five petals, the standard petal white, the wing petals, free large, born on short pedicels, white with a black spot (true black, not deep purple or blue) as is the case in many 'black' colouring and the keel petals white. Stamens are short diadelphous. Cluster of 195 on each axillary raceme usually between the 5th and 10th node. Style short bent, with tuft of hairs near tip; and stigma terminal.
Anthesis and pollination	Anthesis in broad bean ocuures early in the morning and self pollination is the general rule. About 30 per cent of the plants in a population are cross-fertilized and the main insect pollinators are bumble bees.
Pod and podset	The plant flowers profusely but only a small proportion of flowers produce pods. The broad turgid, leathery pod, coriaceous, velutinous, 10.0-20.0 cm long, green maturing blackish-brown, with a densely downy surface; in the wild species, the pods are 5.0-10.0 cm long and 1.0 cm diameter, but many modern cultivars developed for food use have pods 15.0-25.0 cm long and 2.0-3.0 cm thick. Generally 1-4 pods develop from each flower cluster.
Seeds	Each pod contains 3-8 seeds; round to oval and 5.0-10.0 mm diameter in the wild plant, usually flattened and up to 20.0-25.0 mm long, 15.0 mm broad and 5.0-10.0 mm thick in food cultivars. Each seed is separated by a spongy tissue. Seeds very variable in shape and size, strongly compressed to nearly globular, white green, buff, brown, purple or black. Weight of 100 seeds 40-180 g and 1.5 cm or more in length (Purseglove, 1968).

Horticultural Classification

Muratova (1931) classified *Vicia Faba* L. in two subspecies with different maximum number of leaflets per leaf:

(i) *faba* (*eufaba* in her nomenclature) with more than four leaflets per leaf

(ii) *paucijuga*, with fewer than four.

The sub species *faba*, on the basis of seed size included three botanic varieties (Cubero, 1969):

(a) *Major*

(b) *Equina*

(c) *Minor*

(d) *paucijuga*

In addition, botanic varieties were sub-divided into sub-varieties on the basis of geographical distinction. The species is taxonomically isolated from the genus, and although there is evidence of successful intercrosses in the genus, this is not the case for *Vicia Faba* L. (Yamamoto, 1980).

According to Purseglove (1968) some authorities divide it into 2 subspecies:

1. Sub sp. *paucijuga* Mur., comprising several Indian cultivars.
2. Sub sp. *eufaba*, which occurs both in Europe and Asia, is sub-divided into 3 varieties:
   (a) var. *minor* Beck, Tick peas or pigeon beans, with small rounded bolster- shaped seeds;
   (b) var. *equina* Pers., horse beans, with medium-sized, bolster-shaped seeds, with an average length of 1.5 cm are also known as field bean.
   (c) var. *major* Harz., the broad bean used for culinary purpose, with large broad flat seeds, with average length of 2.5 cm, and with longer pods.

Augmentation of Genetic Resources and Crop Improvement

Earlier, there were no improved varieties in Indian conditions therefore; several exotic varieties were introduced from abroad (Singh and Nath, 2011). 'Pusa Sumeet'has been developed through selection from ICAR-IARI, New Delhi. Attractive dark green pods in cluster, first picking in 65-7 days. Yield 18.0 t/ha. Other good lines are SRS-1, BR-1 and BR-2. Most of the North

European faba beans are synthetic varieties (Filippetti and Ricciardi, 1993). Throw's MS, Maris Beagle, Soravi, Avrissot, are synthetic ultivars which are very popular among growers.

Table 4.104: Grouping and Varietal Availability

Types	*Varieties/Strains*
Long type	Aquadule Claudin, Imperial White Long Pod, Masterpiece Green Long Pod, Imperial Green Long Pod, Red Epicure
Windsor type	Imperial White Windsor, Giant Four Seeded Green Windsor, Imperial Green Windsor
Local types	Bakla Local
Synthetic varieties	Throw's MS, Maris Beagle, Soravi, Avrissot

Information on diversity available in vegetable faba bean germplasm is lacking and consequently affecting the sustainable use. Sixty one accessions of exotic vegetable faba bean alongwith check cv. Pusa Sumeet were studied for assessing the genetic diversity pattern during winter season for four years (2011-13, 2014-16) at ICAR-NBPGR, New Delhi using 26 phenological characters (18 qualitative and 8 quantitative). The pooled analysis at revealed significant differences among the accessions for all the eight phenological quantitative characters *viz.*, plant height, number of branches, days to 50 per cent flowering, pod length and width, 5 fruit weight, number of fruits/plant and fruit weight/ plant. Wide range of variability for all the quantitative characters was observed. Higher coefficient of variation (>20 per cent) was observed in number of fruits per plant and fruit weight per plant. The Shannon diversity index showed higher diversity ($\geq$0.40) for three qualitative and all the eight quantitative characters. The Ward's Minimum Variance Dendrogram showed 4 clusters consisting of 25, 31, 5 and 1 accession. The principal component analysis explained 84 per cent cumulative variation by four informative principal components. PC-1 and PC-2 together explained ~57 per cent variation with greater weightage on pod length, pod width, 5 fruit weight, number of fruits/plant and days to 50 per cent flowering, respectively. The PCA biplot showed that majority of the accessions were tightly grouped and two out of four groups were distinct (Gangopadhyay *et al.*, 2016).

Genetic Resource Conservation

Filippetti and Ricciardi (1993) reported that the International Board for Plant Genetic Resources (IBPGR) has developed a network of plant genetic resources centres, each with the responsibility Rome to maintain the world collection of a particular crop. The International Centre for Agricultural Research in the Dry Areas (ICARDA) has a world mandate for the improvement of faba bean. Since its inception, ICARDA has one of the largest faba bean germplasm collections for the continuous improvement of this crop. Faba bean germplasm is maintained at ICARDA in two collections:

i. International Legume Faba Bean (ILB) collection (3305 accessions) containing the original germplasm accessions as received, usually heterogenous and heterozygous to a great extent.
ii. Faba Bean Pure Line (BPL) collection which has been derived from the ILB collection by a pre breeding process of cyclic single plant progeny rows.

The collection expeditions have also been taken place in Cyprus, Egypt and Afghanistan. Collections of faba bean (*Vicia faba*) have also been made in the Mediterranean region by the Germplasm Institute, CNR, BARI, Italy and in the Ethiopia by Ethiopian Genetic Resources Centre. The collection stored in BARI includes material from Algeria, Greece, Spain, Tunisia, Ethiopia, Italy, Egypt and Libya (Perrino, 1985). In Cyprus, the Agricultural Research Institute in collaboration with FAO and IBPGR collected 98 accessions of faba bean (Della, 1981). Afghanistan, Ethiopia, Germany, Turkey, Cyprus, Spain, Canada, UK, Morocco, Egypt, Iraq, Italy, Syria and the former USSR are the main countries for the major collections of broad bean germplasm.

Plant Breeding Institute of Cambridge, National Institute for Agricultural Research (INRA-France); ICAR-NBPGR, New Delhi *etc.* are maintaining the germplasm of faba bean.

REFERENCES

1. Abdalla, M.M.F. 1979. The origin and evolution of *Vicia Faba* L., In: *Proc. 1st Mediterranean Conf. Genet.* 713.
2. Baboo, R. and Kumar, S. 1996. Quality and pod yield of broad bean (*Vicia faba* L.) as influenced by nitrogen (with and without inoculation) and phosphorus applications. *Veg. Sci.*, **23(1):** 16-20
3. Candolle, A. De. 1967. Origin of cultivated plants, 2nd edn. Report, Hafner, New York, 1986.
4. Cerning, J., Saposnik, A. and Guilbot A. 1975. Carbohydrate composition of horsbean (*Vicia faba*) of different origins. In: Hebblethwaite, P.D. (ed.) p. 159-80. The Faba bean. Butterworths, London. *Cereal Chem.*, **52:** 125-138.
5. Chauhan, K.S. and Mangal, J.L. 1993. Agro-techniques for legume vegetables. In: Advances in Horticulture Vol. 5-Vegetable Crops: Part1. (Eds. Chadha, K.L. and Gautam, Kalloo). Malhotra Publishing House, New Delhi-110064, India.
6. Chavan, J.K., Kute, L.S. and Kadam, S.S. 1989. Broad bean: In: Handbook for world food legumes: nutritional, processing, technology and utilization, Vol. I. CRC Press, Boca Raton, Fl., p. 223-245.
7. Choudhury, B. 1967. Vegetables, National Book Trust, India, New Delhi.
8. Cubero, J.I. 1973. Evolutionary trends in *Vicia faba*. *Theor. Appl. Genet.*, **43:** 59-65.

9. Cubero, J.I. 1974. On the evolution of *Vicia faba* L., *Theor, Appl. Genet.*, **44:** 45-47.

10. Della, A. 1981. Report on the exploration and collection of broad beans (*Vicia faba* L.) in Cyprus, *Plant Genetic Resources Newsl.* FAO, Rome.

11. Duke, J.A. 1981. Handbook of legumes of world economic importance. New York, Plenum.

12. Filippetti, A. and Ricciardi, L. 1993. Faba bean (*Vicia faba* L.). In: Genetic Improvement of Vegetable Crops (Eds. Gautam Kalloo and Bergh, B.O.). Pergamon Press, Oxford, U.K.

13. Gangopadhyay, K.K., Singh, A., Pragya, R., Prasad, T.V., Srivastava, R., Phogat, B.S. and Dutta, M. 2016. Phonological diversity in exotic vegetable Faba bean germplasm. 1st International Agrobiodiversity Congress (Abs. Session: Trait Discovery and Enhances Use of PGR), 6-9 November, 2016, held at New Delhi, p. 164.

14. Ladizinsky, G. 1975. On the origin of the broad bean (*Vicia faba* L.). *Israel, J. Bot.*, **24:** 80.

15. Mihailovic, V., Matic, R., Mikic, A. and Cupina, B. 2005a. Possibilities for utilization of spring vetch for grain. *Contemporary Agriculture, LIV*, **3-4:** 318-322.

16. Mihailovic, V., Mikic, A., Cupina, B. and ERIÆ, P. 2005b. Field pea and vetches in Serbia and Montenegro. *Grain Legumes*, 44, 25-26.

17. Mihailovic, V., Mikic, A., ERIÆ, P., Vasiljevic, S., Cupina, B. and Katic, S. 2005c. Protein pea in animal feeding. *Biotechnology in Animal Husbandry*, 21, 5-6, 1, 281-285.

18. Perrino, P. 1985. Conservazione delle risorse genetiche vegetali in Italia. *Realizzazioni e programme, in Risorse Genetiche Vegetali: Autonomia di* Sviluppo, Centro. Internazionale CROCEVIA-Roma.

19. Purseglove, J.W. 1968. Tropical crops: Dicotyledons. Vol. 1, ELBS, Longman, London.

20. Vered, Y, Grosskopf, I., Palevitch, D., Harsat, A., Charach, G., Weintraub, M.S. and graff. E. 1997. The influence of broad bean (*Vicia faba* L.) seedlings in urinary sodium excretion. *Planta Med.*, **63:** 237-240.

21. Yamamoto, K. 1980. On the interspecific progenies between *Vicia pilosa* M.B. and *V. amphicarpa* Dorth, *Jpn. J. Breed.*, **30:** 351.

Genus *Mucuna*

Velvet Bean (*Mucuna prurita* syn. *Mucuna pruriens*)

INTRODUCTION

The plant is known as Common Cowitch, Velvet bean and Cowhage. The immature green pod of velvet bean (*Mucuna pruriens*) is full of nutrition. Worldwide it is one of the popular crop and important medicinal plants of India because all parts of plant possess valuable ingredients (Caius, 1989). It is constutuents of more than 200 indigenous drugs formulations. Therefore there is heavy demand of *Mucuna* in Indian and International drugs markets. The plant is an important cover crops (or green manure crop) in many parts of the world, especially among subsistence farmers (Buckles, 1995). This plant is widely used in Ayurveda, which is the ancient traditional medicinal system in India. The pods are covered with reddish-orange hairs that are readly dislodged and can cause intense irritation to the skin. Therfore, the species name "*pruriens*" (from the Latin, "itching sensation") refers to the results to be had from contact with the seed pod hairs. The seeds of velvet bean are high in protein, carbohydrates, lipids, fiber, and minerals which can be used as a feed for pig and poultry. *Mucuna pruriens* is yet another species under investigation for its medicinal value. They are also rich in novel alkaloids, saponins, and sterols. The seeds of all *Mucuna* species contain a high concentration of L-DOPA (a major free amino-acid which is used relatively expensive to produce synthetically) and are used in the treatment of "Parkinson's syndrome", a nervous disorder and in certain heart ailments. Its concentration in velvet bean seeds contains 7-10 per cent L-DOPA. Concentrations of serotonin also have been found in the pod, leaf and fruit. The stinging hairs of the seed pods contain the phytochemical 'mucunain' which is responsible for causing skin irritation and itch. It is considered a diuretic, nerve tonic, and aphrodisiac. Externally it is applied to ulcers. Velvet bean has a long history of use in Indian Ayurvedic medicine, where it is used for worms, dysentery, diarrhea, snake bite, sexual debility, cough, tuberculosis, impotence, rheumatic disorders, muscular pain, sterility, gout, menstrual disorders, diabetes, and cancer. In India it is considered an aphrodisiac,menstrual promoter, uterine stimulant, nerve tonic, diuretic and blood purifier. It is also effective in moderate to seere hypertension. In many parts of the world *Mucuna pruriens* is used as an important forage, fallow and green manure crop. The whole plant of *Mucuna pruriens* is fed to animals as silage, dried hay or dried seeds. Its silage contains 11-23 per cent crude protein, 35-40 per cent crude fiber, the dried beans 20-35 per cent crude protein, and less than 5 per cent crude fiber (Duke, 1981). The residual cake has a protein content of 15-20 per cent.

Origin, History and Distribution

Velvet bean is indigenous to tropical regions, especially Africa, India and the West Indies. It is wide spread over most of the subcontinent. Plant can be spotted in bushes, hedges, dry-deciduous, and low forests throughout the plains of India (Rastogi and Mehrotra, 1994). This crop is grown in north eastern states for vegetables as well as medicinal purpose and also occurs in wild form.

Species Availability, Evolution and Diversity

In India 14 species of *Mucuna* are found in the foothills of the Himalayas, the plains of West Bengal, Madhya Pradesh, Karnataka, Kerala, Andhra Pradesh, Uttar Pradesh, the Andaman and Nicobar Islands and Sri Lanka (Farooqi and Sree-Ramu, 2001). Under this genus *Mucuna prurita* syn. *Mucuna pruriens* is popular as vegetable crop and *Mucuna pachylobia* and *Mucuna utilis* are popular as fodder crops. Other related species are *Mucuna cochinchinensis* syn. *Mucuna nivea, Stizolobium niveum, Mucuna deeringiana* syn. *Stizolobiumdeeringianum, Mucuna monosperma, Mucuna atropurpurea, Mucuna bracteota, Mucuna capitata, Mucuna gigantean, Mucuna hirsute, Mucuna macrocarpa* and *Mucuna nigicans syn. Mucuna imbricate.* Wide range of variability was observed for the growth and yield attributing traits. *Mucuna pruriens* and *Mucuna utilis* wall. ex Wight are high value medicinal crops grown extensively in South India.

Table 4.105: Botanical Description

Growth Habit	Velvet bean is an annual twining/climbing herbaceous shrub with vines that grows 3.0-18.0 m in height. When the plant is young, it is almost completely covered with fuzzy hairs, but when older, it is almost completely free of hairs.
Leaf	The leaves are tripinnate, ovate, reverse ovate, rhombus shaped or widely ovate. The sides of the leaves are often heavely grooved and the tips are pointy. In young *Mucuna pruriens* plants, both sides of the leaves have hairs. The stems of the leaflets are 2-3 mm long. Additional adjacent leaves are present and are about 5.0 mm long.
Floral biology	The flower heads take the form of axially arrayed panicles. There is 15.0-32.0 cm long and have 2-3 or many flowers. Its flowers are white to dark purple and hang in long clusters. The accompanying leaves are about 12.5 mm long; the flower stand axes are from 2.5 -5.0 mm. The bell is 7.5-9.0 mm long and silky. The sepals are longer or of the same length as the shuttles. The crown is purplish or white. The flag is 1.5 mm long. The wings are 2.5-3.8 cm long.
Anthesis and pollination	It is self pollinated in nature (Duke, 1981).
Pod and pod set	The plant produces cluster of pods. The pods are covered with reddish-orange hairs that are readly dislodged. In the fruit ripening stage, a 4.0-13.0 cm long, 1.0-2.0 cm wide, unwinged leguminous fruit develops. There is a ridge along the length of the fruit. The husk is very hairy and carries up to seven seeds.
Seeds	Seeds are known as Mucuna beans. The seeds are flattened uniform ellipsoid, 1.0-1.9 cm long, 0.8-1.3 cm wide and 4.0-6.5 cm thick. The hilum, the base of the funiculus (connection between placenta and plant seeds) is surrounded by a significant arillus (fleshy seeds shell).

Botanical Description

Genus *Mucuna* belongs to the family Fabaceae. *Mucuna pruriens* has highly vigorous plants growing to a height of 10 to 12 feet compared to *M. utilis*, growing to a height of 6 to 7 feet with less branching habit.

Augmentation of Genetic Resources and Crop Improvement

Introduction and Selection

During 1989-91, a total of 81 diverse lines which were collected by ICAR-NBPGR, and evaluated for 12 agro-botanical traits and chemical constituents as per descriptors to identify L-DOPA lines. The germplasm were planted in the month of August during the said year, in 3m long row, spaced 75 cms apart. Data were recorded on flower colour, pod colour, seed colour, pod hairiness, pods/cluster, pod length (cm), pods/plant, seeds/pod, seed length (cm), seed width (cm), 100 seed weight (g) and L-DOPA (per cent). The range of variation for quantitative attributes was noted as below:

Table 4.106: Trait Specific Variability in Velvet Bean

Sl.No.	Traits	Range
1.	Pods/cluster	4.1-10.2
2.	Pod length (cm)	6.26-12.07
3.	Pods/plant	8.66-21.66
4.	Seeds/pod	3.0-4.99
5.	Seed length (cm)	1.17-2.19
6.	Seed width (cm)	0.83-1.31
7.	100 seed weight (g)	21.46-128.02
8.	L-DOPA,(per cent)	1.80-3.22

Source: Singh, 1994.

Days to flower initiation ranged from 48 days (UKDMP-7) to 123 days (NGLMP-25). Highest seed yield per plant was recorded from the collections WBNMP-6 (2.47 t/ha) followed by NGLMP-25 (2.42 t/ha), MZRMP-03 (2.3 t/ha), UPMP-06 (2.22 t/ha) and WBNMP-3 (2.0 t/ha). The highest L-DOPA content in seeds was in collection UKDMP-11 (6.54 per cent) followed by MGHMP-25 (5.87 per cent). *Mucuna pruriens* and *Mucuna utilis* Wall. ex Wight., are high value medicinal crops for amino-acid derivative L3, 4-dihydroxyphenyllanine (L-DOPA) present in seeds.

Hybridization and Selection

Heterosis and combining ability study was carried out using 6x6 half diallel to know the gene action and selection of superior parents of velvet bean at ICAR-IIHR Bengaluru by Polaiah *et al.* (2016). Combining ability analysis revealed that magnitude of *sca* variance was greater than *gca* variance suggesting predominance of non-additive gene action for most of the characters. For the traits to 50 per cent flowering, number of flowers per inflorescence, number of pods per bunch and test weight where *gca* variance was greater than *sca* variance indicating non-additive gene action. Hence both additive and non-additive gene actions are important for inheritance of the characters. Based on *gca* effects, Arka Dhanvantari, IIHR Sel.-8 and IIHR Sel.-2 were best general combiners for the traits such as number of flowers per plant, pod length, pod weight, number of bunches per plant, number of pods per bunch, dry pod yield and seed yield per plant which can be utilized in breeding programme. Specific combining ability studies indicated the cross IIHR Sel.-2 x IIHR Sel.-8 was best specific combiner for the economically important traits like pod length, pod weight, dry pod yield per plant and seed yield per plant. The IIHR Sel.-8 x IIHR Sel.-3 was recorded high heterotic effects for dry pod yield per plant, seed yield per plant and test weight of the seed. IIHR Sel.-2 x IIHR Sel.-8 was superior hybrid recorded high heterobeltiosis for days to 50 per cent flowering, pod length, pod weight, dry pod yield per plant and seed yield per plant and days to maturity over better parent.

Genetic Resource Conservation

Seventy nine collections of velvet bean were collected and evaluated by the ICAR Research Complex for NEH Region, Meghalaya. ICAR-NBPGR, New Delhi is maintaining huge germplasm of velvet bean. More attention is being given by ICAR-IIHR, Bengaluru in collection and evaluation of velvet bean.

REFERENCES

1. Buckles, D. 1995. Velvet bean (*Mucuna pruriens*): A new plant with a history. *Economic Botany*, **49(1):** 13-25.
2. Duke, J.A. 1981. Handbook of legumes of world economic importance. New York, Plenum.
3. Caius, J.F. 1989. The medicinal and poisonous legumes of India, Scientific Publishers, Jodhpur, p. 70-71.
4. Farooqi, A.A. and Sree-Ramu, B.S. 2001. Cultivation of medicinal and aromatic crops, Universities Press, New Delhi, p. 74.
5. Polaiah, A.C., Hima Bindu, K., Rao, N.H., Kumar, S.S. and Kumar, T.V. 2016. Heterosis and combining ability studies for yield and its quantitative and qualitative traits in velvet bean (*Mucuna pruriens* L. DC.). (Abs. 57; 7th Indian Horticulture Congress-2016), held at ICAR-IARI, New Delhi, p. 45.
6. Rastogi, R.P. and Mehrotra, B.N. 1994. Compendium of Indian medicinal plants, Vol. 5, CDRI, Lucknow (U.P.), p. 554.
7. Singh, B.M. 1994. Studies on lesser known medicinal plants. In: Plant Genetic Resources-Exploration, Evaluation and Maintenance. R.S. Rana, Singh, B., Koppar, M.N., Rai, M., Kochhar, S. and Duhoon, S.S. ICAR-NBPGR, New Delhi, p. 270.

Genus *Cyamopsis*

Cluster Bean (*Cyamopsis tetragonolobus*)

INTRODUCTION

Among the beans grown in India, cluster bean is commonly known as *'guar'* [*Cyamopsis tetragonolobus* (L.) Taub.] hitherto had been used as animal feed, food additive, fodder, medicinal, and gum/resin and also cultivated to a lesser extent as a pod vegetable crop. The demand for the produce and products of this crop species is increasing by leaps and bounds (Bewal *et al.*, 2009). Small, green, and tender pods are used as a vegetable. Green pods of cluster bean are used as a vegetable and the dry seeds as a pulse. It is also used as thickener and stabilizer in cheese and in cold-meat processing, and in salad dressing, ice-creams, lollipops, bakery products, meat and sausages. Cluster bean is an important source of protein and minerals. Per 100 g edible pods contain 81.0 g moisture, 10.8 g carbohydrates, 3.2 g protein, 1.4 g minerals, 0.4 g fat, 2.3 g fibre, 330 IU vitamins A and 49 mg vitamin C. It is a well adapting crop in varying agro-climatic and edaphic conditions. It is drought tolerant and is also grown in arid zones as a fodder crop. It is also popular as a green manure. But now it has assumed great importance as an industrial crop because of the gum contained in seed. Gum which is a polymer of galactose and mannose is contained in a thin layer of endosperm found between the seed coat and the cotyledons. The gum is major component of adhesive including those widely used on postage stamps; and to impart lusture to silk and other textiles, printing, mining, explosive, food processing, oil industries, toilet goods, paper and cosmetic industries. Some of the varieties are suitable for the extraction of gum (20 per cent on dry weight basis), which is becoming an important commodity in international trade. Its dried seeds are used as a chemotherapeutic agent against small pox and as a laxative.

Origin, History and Distribution

There is yet no unanimity about its origin. Chevalier (1939) postulated that *Cyamopsis senegalensis* probably extended up to Sind, after domestication of a few of its cultigen in India. Due to this reason researchers feel that cluster bean is of Indian origin. However, Vavilov (1951) has suggested India as a geographical centre of diversity for guar, though it is not found to occur in wild state in this region. Besides India, cluster bean is grown in Myanmar, Sri Lanka, Pakistan and arid regions of the

United States like Arizona and Texas. It is also adapted to locations with more tropical climates, such as Florida and Puerto Rico. In India this crop can be successfully cultivated in north-west and southern parts of India, where rainfall is scanty and irrigation is scare. Cluster bean is cultivated in Maharashtra, Andhra Pradesh, Tamil Nadu, Karnataka, Haryana, Uttar Pradesh, Rajasthan and Gujarat.

Species Availability, Evolution and Diversity

Cyamopsis comprises four species. *C. tetragonoloba*, commonly known as cluster bean or 'guar'. Hymowitz (1972), on the other hand, believes that the African wild species *Cyamopsis senegalensis* seems to be the immediate ancestor to cluster bean, which are not found anywhere in the wild state. Smith (1976) reported that the genus of cluster bean consists of three species, out of which *Cyamopsis tetragonolobus* is the only member of economic importance. West African *C. tetragonolobus* a shattering type, an intermediate of the wild species and the cultivated types is frequently observed (Dabas and Thomas, 1986) in western Rajasthan and its neighbouring area, which indicate that guar has perhaps domesticated in this region.

Botanical Description

Table 4.107: Botanical Description

Growth Habit	It is an erect annual growing to a height of 3.0 m with stiff erect branches.
Stem	Stems are angular.
Leaf	Leaves are trifoliate, ovate and serrate.
Floral biology	The white or pink coloured flowers are small and borne on axillary raceme. Androecium is monoadelphous (10 stamens).
Pod and pod set	The pods are small, compressed, linear, erect and clustered, double ridged on the dorsal side single ridge below, length 4.0-10.0 cm and beaked.
Seeds	Per pod seeds are 5-12, white to grey or black in colour, 5.0 mm long with an average weight of 0.06 g (Smartt, 1976).

Botanical Classification

(i) **Fodder type-** Fodder type varieties are quite late in pod bearing and have more branching tendency.

(ii) **Vegetable type-**Vegetable type varieties are early in pod bearing and have less branching tendency.

Augmentation of Genetic Resources and Crop Improvement

ICAR-National Bureau of Plant Genetic Resources (ICAR-NBPGR) New Delhi to collect, evaluate, document and utilize the natural available genetic variability.

Table 4.108: Number of Germplasm Collections in Cluster Bean

Years	*Number of Accessions*
Up to 1960	63
1961-1965	475
1966-1970	912
1971-1975	59
1976-1980	1446
1981-1985	1052
1986-1990	674
1991-1993	92
Total	**4773**

Source: Dabas and Phogat, 1994.

In addition, 44 accessions including the two wild species namely *C. senegalensis* and *C. serrata* were introduced from USA. Efforts have also been made at the State Agricultural Universities to strenghthen the germplasm collections. There is still need to explore some pockets particularly in Madhya Pradesh, Uttar Pradesh, Uttarakhand, Peninsular India, Pakistan and tropical Africa.

REFERENCES

1. Bewal, S., Purohit, J., Kumar, A., Rajkumar, K. and Rao, S.R. 2009. Cytogenetical investigation in colchicines-induced tetraploids of *Cyamopsis tetragonolobus* L. *Czech J. Genet. Plant Breed.*, **45:** 143-154.
2. Dabas, B.S. and Phogat, B.S. 1994. Collection, evaluation and prospects of arid legume. In: Plant Genetic Resources: Exploration, Evaluation and Maintenance. Rana, R.S., Singh, B., Koppar, M.N., Rai, M., Kochhar, S. and Duhoon, S.S. (eds.) Pub. by ICAR-NBPGR, New Delhi. p 184-190.
3. Dabas, B.S. and Thomas, T.A. 1986. Shattering guar [*Cyamopsis tetragonolobus* (L.) Taub.] and its significance. *Intern. J. Trop. Ayril.*, **2:** 185-187.
4. Chevalier, A. 1939. Research on spices of the genus *Cyamopsis*, forage plants for tropical and semi-arid countries. *Res. Bot. Appl.*, **19:** 242-249.
5. Hymowitz, T.1972. The trans-domestication concept as applied to guar. *Eco. Bot.*, **26:** 49-60.
6. Vavilov, N.I. The origin, variation, immunity and breeding of cultivated plants. *Chron. Bot.*, 13.

5

Genetic Resources of Brassica Vegetables and Crop Improvement

INTRODUCTION

Brassica crops are most popular vegetables grown during winter season in the world. All the *Brassica* crops are hardy and thrive best under conditions of cool climate. They are grown in the plains during the winter season and can be cultivated through out the year in the hilly regions of the country. Among the *Brassica* crops, cauliflower and cabbage are the most important crops, while knol-khol is limited only in hills and sub-tropical parts while broccoli is also gaining momentum in some areas of the country due to its high medicinal value. Curly kale and knoll-khol were the first known cultivated types and other crops were added to the list in the later centuries. Cole crops are rich in nutrients including several carotenoids (β-carotene, lutein, zeaxanthin); vitamins C, E and K, folate and minerals).

Table 5.1: Evolution of Cultivated *B. oleracea* Crops

Possible Sequence of Evolution	*Scientific Name (B. oleracea var.)*	*Common Name*	*Ancestor*
1	var. *sylvestris* L.	Wild cabbage	-
2	var. *ramosa* DC.	Thousand-head kale, branching bush kale	1
3	var. *gemmifera* DC	Brussels sprout	2
4	var. *dalechampii*	-	3
5	var. *costata* DC.	Portuguese tree kale; tronchuda kale	1
6	var. *medullosa* Thell.	Marrow-stem kale	1
7	Intermediate between 6 and 8	-	6
8	var. *gongylodes* L.	Kohlrabi	7
9	var. *sabauda* L.	Savoy cabbage	5
10	var. *capitata* L.	White cabbage	9
11	var. *capitata* L.	Red cabbage	10
12	var. viridis L.; *sabellica* L., var. *palmifolia* DC.	Kale and collards	1
13	var. *italica* Plencks	Broccoli, Calabrase	12
14	var. *botrytis* L.	Cauliflower (biennial)	13
15	var. *botrytis* L.	Cauliflower (annual)	14
16	var. *botrytis* L.	Cauliflower (Indian) or Tropical Cauliflower	15

Source: Prakash *et al.*, 2011.

In addition, these cruciferous vegetables contain a group of substances known as glucosinolates which are responsible for the pungent aroma and bitter flavours.

Brassica oleracea is group of a highly differentiated plants having 18 number of diploid and somatic chromosomes (2n=2x=18) and these are grown all over the world from arctic to tropical climatic conditions. *B. oleracea* is a triple tetrasomic for chromosome types B, C and E. comprising the genome formula BBCCDEEF with 6 basic genomes and showing some secondary pairing. The word 'cole' seems to have been derived from the abbreviation of the word 'caulis' meaning stem/cabbage/stalk, and it was variously spelt as kale (English), Kohl (German), Kool (Dutch), Kal (Scandinavian), Kaali (Finnish), Kaol and Kol (Breton), Chou (French), Col (Spanish), Cal (Irish), Cavolo (Italian), and Couve (Portuguese); but Usually, the word cole is more recognized in the literature worldwide. Variation within and between subspecies of *B. oleracea* and present day cultivated cole crops (kale, broccoli, cauliflower, cabbage, kohlrabi and Brussels sprout) have evolved after a long time of natural/artificial hybridization, mutation, selection and domestication. All these forms are descended from a common kale like ancestor, the wild cabbage (*B. oleracea* L. var. *sylvestris* L.) which is still prevalent in the western and southern Europe and North Africa.

Among the *Brassica* crops Red cabbage (*B. oleracea* sub group *rubra*), Savoy cabbage (*B. oleracea* sub group *sabauda*), Broccoli sprouting (*B. oleracea* group *italica*), Broccoli Chinese (*B. oleracea* group *alboglabra*), Calabrese (*B. oleracea* group *italica*), Kale (*B. oleracea* group *acephala*), Brussels sprout (*B. oleracea* group *gemmifera*), Kohlrabi (*B. oleracea* group *gongylodes*), Komatsuna (*B. oleracea* group *perviridis*), Mizuna greens (*B. rapa* group *nipposinica*), Pakchoi (*B. rapa* group *chinensis*), Chinese cabbage (*B. campestris* group *pekinensis*), Mustard greens (*B. juncea*) *etc.* are grown commercially. *Brassica* crops are grown for botanically different edible parts (floral bud, leaves, stems and immature pods) but their cultural practices are almost similar. All *Brassica* species contain glucosinolates, which in crushed leaves are broken down by the enzyme myrosinase to give bitter taste and goitrogenic substances; hence selection at early stage of domestication must have been for plants with less bitter taste than usual.

For bolting and flowering, most of the cole crops are generally performed on mature vegetative plants either *in situ* or *ex situ* at temperatures between 5-10ºC for 30-30 days *i.e.* vernalization other than tropical varieties/genotypes (Table 5.2).

Table 5.2: Low Temperature Requirements (5-10 °C for 30-60 days) for Cole Crops

Crop	Development of Edible Parts	Flowering
Broccoli, winter type	Yes	Yes
Broccoli, summer type	No	Yes
Broccoli, tropical type	No	No
Brussel sprouts	No	Yes
Cabbage	No	Yes
Cabbage, tropical type	No	No
Cauliflower, winter type	Yes	Yes
Cauliflower, summer type	No	Yes
Cauliflower, tropical type	No	No
Chinese broccoli	No	No
Collards	No	Yes
Kale	No	Yes
Kale, tropical type	No	No
Kohlrabi	No	Yes

Genetic Designations of *Brassica*

The genus *Brassica* includes about 100 species. The genetic designations of varietal or sub-specific taxa and it morphotypes which exist among several species are given in Table 5.3).

Table 5.3: Genetic Designations of Varietal or Sub-specific Taxa

Species (n)	Sub-species or Variety	Common Name
Brassica nigra (L.) (bb=16)	-	Black mustered
B. oleracea	-	cole crops
	acephala	Kales
	alboglabra	Chinese kale, Kailan
	botrytis	Cauliflower, Heading brocooli
	capitata	Cabbage
	castata	Purtuguese cabbage
	gemmifera	Brussels sprouts
	gongylodes	Kohlrabi
	italica	Broccoli, Calabresse
	medullosa	Marrow stem kale
	palmifolia	Tree cabbage
	ramosa	Savoy cabbage

Species (n)	Sub-species or Variety	Common Name
	sabellica	Collards
	selensia	Borecole
B. rapa (L.emend. Metzger) (aa=20)	(syn. *B. campestris* Hook. f. Thoms. *chinensis*)	Pakchoi
	narinosa	-
	nipposinica	-
	oleifera	Turnip rape, toria
	parachinensis	Choy sum
	pekinensis	Chinese cabbage, petsai
	perviridis	Tendergreen, komatsuna, mustard, spinach
	rapifera	Turnip
	trilocularis	Sarson
	utilis	Broccoli aabb
B. carinata A. Br. (bbcc=34) *B. juncea* (L.)	-	Ethiopian mustard
	capitata	Head mustard
	crispifolia	Cut leaf mustard
	faciliflora	Broccoli mustard
	lapitata	Large petiole mustard
	oleifera	Indian mustard, raya
	rapifera	Root mustard
	rugosa	Leaf mustard
	spicea	Mustard
	tsa-tsai	Big stem mustard
B. napus L. (aacc=38)	-	Fodder rape
	oleifera	Oil rape
	rapifera	Swede, ruta baga
Raphanus sativus L. (rr=18)	-	Radish
	radicola	Radish, dicon
	oleifera	Oil radish
	caudatus	Rat tail radish

Majority of which are native to the Mediterranean region. All these species developed from *B. oleracea* L. var. *sylvestris*, commonly known as wild cabbage, cliff cabbage or colewart, from which the name 'cole' crops has probably been derived. Sikka (1940) discussed the role played by gene mutations, structural changes of chromosome and hybridization in species formation in *Brassica*. Of these, hybridization has played the most important part, as amphiploidy has frequently occurred in the genus.

Table 5.4: Taxon of *Brassica* and Preferred Taxon Name

Name of Taxon	Name of Authority	Botanical Family	Preferred Taxon Name
Brassica alba	(L.) Rabenh.	Brassicaceae	*Sinapis alba* subsp. *alba*
Brassica alboglabra	L. H. Bailey	Brassicaceae	*Brassica oleracea* var. *alboglabra*
Brassica amplexicaulis	(Desf.) Pomel, nom. illeg.	Brassicaceae	*Brassica souliei* subsp. *amplexicaulis*
Brassica arvensis	L.	Brassicaceae	*Moricandia arvensis*
Brassica arvensis	(L.) Rabenh., nom. illeg.	Brassicaceae	*Sinapis arvensis* subsp. *arvensis*
Brassica assyriaca	Mouterde	Brassicaceae	*Brassica assyriaca*
Brassica atlantica	(Coss.) O. E. Schulz	Brassicaceae	*Brassica insularis*
Brassica aucheri	Boiss.	Brassicaceae	*Brassica aucheri*
Brassica balearica	Pers.	Brassicaceae	*Brassica balearica*
Brassica barrelieri	(L.) Janka	Brassicaceae	*Brassica barrelieri*
Brassica barrelieri subsp. *barrelieri*		Brassicaceae	*Brassica barrelieri*
Brassica barrelieri subsp. *oxyrrhina*	(Coss.) Regel	Brassicaceae	*Brassica oxyrrhina*
Brassica barrelieri subsp. *sabularia*	Maire	Brassicaceae	*Brassica barrelieri*
Brassica besseriana	Andrz.	Brassicaceae	*Brassica juncea*

Brassica blancoana	Boiss.	Brassicaceae	*Brassica repanda* subsp. *blancoana*
Brassica bourgeaui	(Webb ex H. Christ) Kuntze	Brassicaceae	*Brassica bourgeaui*
Brassica brachyloma	Boiss. and Reut.	Brassicaceae	*Brassica gravinae* var. *brachyloma*
Brassica cadmea	O. E. Schulz	Brassicaceae	*Brassica cadmea*
Brassica campestris	L.	Brassicaceae	*Brassica rapa* subsp. *campestris*
Brassica campestris var. oleifera	DC.	Brassicaceae	*Brassica rapa* subsp. *oleifera*
Brassica campestris f. annua	Schubl. and G. Martens	Brassicaceae	*Brassica napus* var. *napus*
Brassica campestris f. biennis	Schubl. and G. Martens	Brassicaceae	Brassica napus var. *napus*
Brassica campestris subsp. campestris		Brassicaceae	*Brassica rapa* subsp. *campestris*
Brassica campestris subsp. chinensis	(L.) Makino	Brassicaceae	*Brassica rapa* subsp. *chinensis*
Brassica campestris subsp. napus	(L.) Hook. f. and T. Anderson	Brassicaceae	*Brassica napus* var. *napus*
Brassica campestris subsp. rapifera	(Metzg.) Sinskaya	Brassicaceae	*Brassica rapa* subsp. *rapa*
Brassica campestris var. dichotoma	(Roxb.) G. Watt	Brassicaceae	*Brassica rapa* subsp. *dichotoma*
Brassica campestris var. pabularia	DC.	Brassicaceae	*Brassica napus* var. *pabularia*
Brassica campestris var. rapa	(L.) Hartm.	Brassicaceae	*Brassica rapa* subsp. *rapa*
Brassica campestris var. sarson	Prain	Brassicaceae	*Brassica rapa* subsp. *trilocularis*
Brassica campestris var. toria	Duthie and J. B. Fuller	Brassicaceae	*Brassica rapa* subsp. *dichotoma*
Brassica carinata	A. Braun	Brassicaceae	*Brassica carinata*
Brassica caulorapa	(DC.) Pasq.	Brassicaceae	*Brassica oleracea* var. *gongylodes*
Brassica chinensis	L.	Brassicaceae	*Brassica rapa* subsp. *chinensis*
Brassica cossoniana	Boiss. and Reut.	Brassicaceae	*Brassica fruticulosa* subsp. *cossoniana*
Brassica cretica	Lam.	Brassicaceae	*Brassica cretica*
Brassica cretica subsp. *aegaea*	(Heldr. and Hal.) Snogerup *et al.*	Brassicaceae	*Brassica cretica* subsp. *aegaea*
Brassica cretica subsp. *atlantica*	(Coss.) Onno	Brassicaceae	*Brassica insularis*
Brassica cretica subsp. *cretica*		Brassicaceae	*Brassica cretica* subsp. *cretica*
Brassica cretica subsp. *laconica*	M. A. Gust. and Snogerup	Brassicaceae	*Brassica cretica* subsp. *laconica*
Brassica cretica var. *aegaea*	Heldr. and Hal.	Brassicaceae	*Brassica cretica* subsp. *aegaea*
Brassica deflexa	Boiss.	Brassicaceae	*Brassica deflexa*
Brassica deflexa subsp. *deflexa*		Brassicaceae	*Brassica deflexa* subsp. *deflexa*
Brassica deflexa subsp. *leptocarpa*	(Boiss.) Hedge	Brassicaceae	*Brassica deflexa* subsp. *leptocarpa*
Brassica deserti	Danin and Hedge	Brassicaceae	*Brassica deserti*
Brassica desnottesii	Emb. and Maire	Brassicaceae	*Brassica desnottesii*
Brassica dimorpha	Coss. and Durieu	Brassicaceae	*Brassica dimorpha*
Brassica drepanensis	(Caruel) Ponzo	Brassicaceae	*Brassica drepanensis*
Brassica dubiosa	L. H. Bailey	Brassicaceae	*Brassica dubiosa*
Brassica elata	Ball	Brassicaceae	*Erucastrum elatum* var. *elatum*
Brassica elongata	Ehrh.	Brassicaceae	*Brassica elongata*
Brassica elongata subsp. *elongata*		Brassicaceae	*Brassica elongata* subsp. *elongata*
Brassica elongata subsp. *imdrhasiana*	Quezel	Brassicaceae	*Brassica elongata* subsp. *imdrhasiana*
Brassica elongata subsp. *integrifolia*	(Boiss.) Breistr.	Brassicaceae	*Brassica elongata* subsp. *integrifolia*
Brassica elongata subsp. *pinnatifida*	(Smal'g.) Greuter and Burdet	Brassicaceae	*Brassica elongata* subsp. *pinnatifida*
Brassica elongata subsp. *subscaposa*	(Maire and Weiller) Maire	Brassicaceae	*Brassica elongata* subsp. *subscaposa*
Brassica elongata var. *integrifolia*	Boiss.	Brassicaceae	*Brassica elongata* subsp. *integrifolia*
Brassica elongata var. *pinnatifida*	Smal'g.	Brassicaceae	*Brassica elongata* subsp. *pinnatifida*
Brassica eruca	L.	Brassicaceae	*Eruca sativa*
Brassica erucoides	Roxb., nom. illeg.	Brassicaceae	*Eruca sativa*
Brassica fruticulosa	Cirillo	Brassicaceae	*Brassica fruticulosa*
Brassica fruticulosa subsp. *cossoniana*	(Boiss. and Reut.) Maire	Brassicaceae	*Brassica fruticulosa* subsp. *cossoniana*
Brassica fruticulosa subsp. *fruticulosa*		Brassicaceae	*Brassica fruticulosa* subsp. *fruticulosa*
Brassica fruticulosa subsp. *glaberrima*	(Pomel) Batt.	Brassicaceae	*Brassica fruticulosa* subsp. *glaberrima*
Brassica fruticulosa subsp. *mauritanica*	(Coss.) Maire	Brassicaceae	*Brassica fruticulosa* subsp. *mauritanica*

Brassica fruticulosa subsp. *numidica*	(Coss.) Maire	Brassicaceae	*Brassica fruticulosa* subsp. *numidica*
Brassica fruticulosa subsp. *pomeliana*	Maire	Brassicaceae	*Brassica fruticulosa* subsp. *pomeliana*
Brassica fruticulosa subsp. *radicata*	(Desf.) Batt.	Brassicaceae	*Brassica fruticulosa* subsp. *radicata*
Brassica fruticulosa var. *mauritanica*	Coss.	Brassicaceae	*Brassica fruticulosa* subsp. *mauritanica*
Brassica fruticulosa var. *numidica*	Coss.	Brassicaceae	*Brassica fruticulosa* subsp. *numidica*
Brassica glaberrima	Pomel	Brassicaceae	*Brassica fruticulosa* subsp. *glaberrima*
Brassica glabrescens	Pold.	Brassicaceae	*Brassica glabrescens*
Brassica gravinae	Ten.	Brassicaceae	*Brassica gravinae*
Brassica gravinae f. *djurdjurae*	Batt.	Brassicaceae	*Brassica gravinae* var. *djurdjurae*
Brassica gravinae var. *brachyloma*	(Boiss. and Reut.) O. E. Schulz	Brassicaceae	*Brassica gravinae* var. *brachyloma*
Brassica gravinae var. *djurdjurae*	(Batt.) Maire	Brassicaceae	*Brassica gravinae* var. *djurdjurae*
Brassica gravinae var. *gravinae*		*Brassicaceae*	*Brassica gravinae var. gravinae*
Brassica hilarionis	Post	Brassicaceae	*Brassica hilarionis*
Brassica hirta	Moench	Brassicaceae	*Sinapis alba* subsp. *alba*
Brassica incana	Ten.	Brassicaceae	*Brassica incana*
Brassica insularis	Moris	Brassicaceae	*Brassica insularis*
Brassica integrifolia	(H. West) Rupr.	Brassicaceae	*Brassica juncea* var. *integrifolia*
Brassica japonica	Makino, nom. illeg.	Brassicaceae	*Brassica rapa* subsp. *nipposinica*
Brassica jordanoffii	O. E. Schulz	Brassicaceae	*Brassica jordanoffii*
Brassica juncea	(L.) Czern.	Brassicaceae	*Brassica juncea*
Brassica juncea subsp. *rugosa*	(Roxb.) Prain	Brassicaceae	*Brassica juncea* var. *rugosa*
Brassica juncea var. *agrestis*	Prain	Brassicaceae	*Brassica juncea* var. *juncea*
Brassica juncea var. *capitata*	N. Tsen and S. N. Lee	Brassicaceae	*Brassica juncea* var. *rugosa*
Brassica juncea var. *crispifolia*	L. H. Bailey	Brassicaceae	*Brassica juncea* var. *crispifolia*
Brassica juncea var. *cuneifolia*	(Roxb.) Kitam.	Brassicaceae	*Brassica juncea* var. cuneifolia
Brassica juncea var. *foliosa*	L. H. Bailey	Brassicaceae	*Brassica juncea* var. *foliosa*
Brassica juncea var. *integrifolia*	(H. West) Sinskaya	Brassicaceae	*Brassica juncea* var. *integrifolia*
Brassica juncea var. *japonica*	(Thunb.) L. H. Bailey	Brassicaceae	*Brassica juncea* var. *japonica*
Brassica juncea var. *juncea*		Brassicaceae	*Brassica juncea* var. *juncea*
Brassica juncea var. *longidens*	L. H. Bailey	Brassicaceae	*Brassica juncea* var. *longidens*
Brassica juncea var. *megarrhiza*	N. Tsen and S. N. Lee	Brassicaceae	*Brassica juncea* var. *napiformis*
Brassica juncea var. *multiceps*	N. Tsen and S. N. Lee	Brassicaceae	*Brassica juncea* var. *multiceps*
Brassica juncea var. *napiformis*	(Pailleux and Bois) Kitam.	Brassicaceae	*Brassica juncea* var. *napiformis*
Brassica juncea var. *oleifera*	Prain	Brassicaceae	*Brassica juncea* var. *juncea*
Brassica juncea var. *rugosa*	(Roxb.) N. Tsen and S. N. Lee	Brassicaceae	*Brassica juncea* var. *rugosa*
Brassica juncea var. *strumata*	N. Tsen and S. N. Lee	Brassicaceae	Brassica juncea var. strumata
Brassica juncea var. *tsatsai*	ined.	Brassicaceae	Brassica juncea var. tumida
Brassica juncea var. *tumida*	N. Tsen and S. N. Lee	Brassicaceae	*Brassica juncea* var. *tumida*
Brassica kaber	(DC.) L. C. Wheeler	Brassicaceae	*Sinapis arvensis* subsp. *arvensis*
Brassica kaber var. *pinnatifida*	(Stokes) L. C. Wheeler	Brassicaceae	*Sinapis arvensis* subsp. *arvensis*
Brassica latisiliqua	Boiss. and Reut.	Brassicaceae	*Brassica repanda* subsp. *latisiliqua*
Brassica leptocarpa	Boiss.	Brassicaceae	*Brassica deflexa* subsp. *leptocarpa*
Brassica loncholoma	Pomel	Brassicaceae	*Eruca loncholoma*
Brassica macrocarpa	Guss.	Brassicaceae	*Brassica macrocarpa*
Brassica maurorum	Durieu	Brassicaceae	*Brassica maurorum*
Brassica montana	Pourr.	Brassicaceae	*Brassica montana*
Brassica napiformis	(Pailleux and Bois) L. H. Bailey	Brassicaceae	*Brassica juncea* var. *napiformis*
Brassica napobrassica	(L.) Mill.	Brassicaceae	*Brassica napus* var. *napobrassica*
Brassica napus	L.	Brassicaceae	*Brassica napus*
Brassica napus f. *annua*	(Schubl. and G. Martens) Thell.	Brassicaceae	*Brassica napus* var. *napus*
Brassica napus f. *biennis*	(Schubl. and G. Martens) Thell.	Brassicaceae	*Brassica napus* var. *napus*
Brassica *napus* subsp. *napobrassica*	(L.) Jafri	Brassicaceae	*Brassica napus* var. *napobrassica*

Brassica napus subsp. *oleifera*	(Delile) Sinskaya	Brassicaceae	*Brassica napus* var. *napus*
Brassica napus subsp. *rapifera*	(Metzg.) Sinskaya	Brassicaceae	*Brassica napus* var. *napobrassica*
Brassica napus var. *annua*	W. D. J. Koch	Brassicaceae	*Brassica napus* var. *napus*
Brassica napus var. *biennis*	(Schubl. and G. Martens) Rchb.	Brassicaceae	*Brassica napus* var. *napus*
Brassica napus var. *dichotoma*	(Roxb.) Prain	Brassicaceae	*Brassica rapa* subsp. *dichotoma*
Brassica napus var. *glauca*	(Roxb.) O. E. Schulz	Brassicaceae	*Brassica rapa* subsp. *trilocularis*
Brassica napus var. *napobrassica*	(L.) Rchb.	Brassicaceae	*Brassica napus* var. *napobrassica*
Brassica napus var. *napus*		Brassicaceae	*Brassica napus* var. *napus*
Brassica napus var. *oleifera*	Delile	Brassicaceae	*Brassica napus* var. *napus*
Brassica napus var. *pabularia*	(DC.) Rchb.	Brassicaceae	*Brassica napus* var. *pabularia*
Brassica napus var. *rapifera*	Metzg.	Brassicaceae	*Brassica napus* var. *napobrassica*
Brassica narinosa	L. H. Bailey	Brassicaceae	*Brassica rapa* subsp. *narinosa*
Brassica nigra	(L.) W. D. J. Koch	Brassicaceae	*Brassica nigra*
Brassica nigra var. *abyssinica*	A. Braun	Brassicaceae	*Brassica nigra*
Brassica nipposinica	L. H. Bailey	Brassicaceae	*Brassica rapa* subsp. *nipposinica*
Brassica oleracea	L.	Brassicaceae	*Brassica oleracea*
Brassica oleracea subsp. *robertiana*	(J. Gay) Rouy and Foucaud	Brassicaceae	*Brassica montana*
Brassica oleracea var. *acephala*	DC., pro parte	Brassicaceae	*Brassica oleracea* var. *viridis*
Brassica oleracea var. *acephala*	DC., p.p.	Brassicaceae	*Brassica oleracea* var. *sabellica*
Brassica oleracea var. *alboglabra*	(L. H. Bailey) Musil	Brassicaceae	*Brassica oleracea* var. *alboglabra*
Brassica oleracea var. *botrytis*	L.	Brassicaceae	*Brassica oleracea* var. *botrytis*
Brassica oleracea var. *bullata*	DC., pro parte minore	Brassicaceae	*Brassica oleracea* var. *gemmifera*
Brassica oleracea var. *bullata*	DC., pro parte majore	Brassicaceae	*Brassica oleracea* var. *sabauda*
Brassica oleracea var. *capitata*	L.	Brassicaceae	*Brassica oleracea* var. *capitata*
Brassica oleracea var. *caulorapa*	DC.	Brassicaceae	*Brassica oleracea* var. *gongylodes*
Brassica oleracea var. *chinensis*	(L.) Prain	Brassicaceae	*Brassica rapa* subsp. *chinensis*
Brassica oleracea var. *costata*	DC.	Brassicaceae	*Brassica oleracea* var. *costata*
Brassica oleracea var. *gemmifera*	Zenker	Brassicaceae	*Brassica oleracea* var. *gemmifera*
Brassica oleracea var. *gongylodes*	L.	Brassicaceae	*Brassica oleracea* var. *gongylodes*
Brassica oleracea var. *italica*	Plenck	Brassicaceae	*Brassica oleracea* var. *italica*
Brassica oleracea var. *medullosa*	Theil.	Brassicaceae	*Brassica oleracea* var. *medullosa*
Brassica oleracea var. *napobrassica*	L.	Brassicaceae	*Brassica napus* var. *napobrassica*
Brassica oleracea var. *oleracea*		Brassicaceae	*Brassica oleracea* var. *oleracea*
Brassica oleracea var. *palmifolia*	DC.	Brassicaceae	*Brassica oleracea* var. *palmifolia*
Brassica oleracea var. *ramosa*	DC.	Brassicaceae	*Brassica oleracea* var. *ramosa*
Brassica oleracea var. *sabauda*	L.	Brassicaceae	*Brassica oleracea* var. *sabauda*
Brassica oleracea var. *sabellica*	L.	Brassicaceae	*Brassica oleracea* var. *sabellica*
Brassica oleracea var. *tronchuda*	L. H. Bailey	Brassicaceae	*Brassica oleracea* var. *costata*
Brassica oleracea var. *viridis*	L.	Brassicaceae	*Brassica oleracea* var. *viridis*
Brassica orientalis	L.	Brassicaceae	*Conringia orientalis*
Brassica oxyrrhina	Coss.	Brassicaceae	*Brassica oxyrrhina*
Brassica parachinensis	L. H. Bailey	Brassicaceae	*Brassica rapa* var. *parachinensis*
Brassica pe-tsai	L. H. Bailey	Brassicaceae	*Brassica rapa* subsp. *pekinensis*
Brassica pekinensis	(Lour.) Rupr.	Brassicaceae	*Brassica rapa* subsp. *pekinensis*
Brassica perviridis	(L. H. Bailey) L. H. Bailey	Brassicaceae	*Brassica rapa* var. *perviridis*
Brassica pinnatifida	Desf.	Brassicaceae	*Eruca pinnatifida*
Brassica procumbens	(Poir.) O. E. Schulz	Brassicaceae	*Brassica procumbens*
Brassica purpurascens	Banks and Sol.	Brassicaceae	*Chorispora purpurascens*
Brassica rapa	L.	Brassicaceae	*Brassica rapa*
Brassica rapa c. *oleifera*	DC.	Brassicaceae	*Brassica rapa* subsp. *oleifera*
Brassica rapa subsp. *campestris*	(L.) A. R. Clapham	Brassicaceae	*Brassica rapa* subsp. *campestris*

Brassica rapa subsp. *chinensis*	(L.) Hanelt	Brassicaceae	*Brassica rapa* subsp. *chinensis*
Brassica rapa subsp. *dichotoma*	(Roxb.) Hanelt	Brassicaceae	*Brassica rapa* subsp. *dichotoma*
Brassica rapa subsp. *narinosa*	(L. H. Bailey) Hanelt	Brassicaceae	*Brassica rapa* subsp. *narinosa*
Brassica rapa subsp. *nipposinica*	(L. H. Bailey) Hanelt	Brassicaceae	*Brassica rapa* subsp. *nipposinica*
Brassica rapa subsp. *oleifera*	(DC.) Metzg.	Brassicaceae	*Brassica rapa* subsp. *oleifera*
Brassica rapa subsp. *pekinensis*	(Lour.) Hanelt	Brassicaceae	*Brassica rapa* subsp. *pekinensis*
Brassica rapa subsp. *rapa*		Brassicaceae	*Brassica rapa* subsp. *rapa*
Brassica rapa subsp. *rapifera*	Metzg.	Brassicaceae	*Brassica rapa* subsp. *rapa*
Brassica rapa subsp. *sarson*	(Prain) Denford	Brassicaceae	*Brassica rapa* subsp. *trilocularis*
Brassica rapa subsp. *silvestris*	(Lam.) Janch.	Brassicaceae	*Brassica rapa* subsp. *campestris*
Brassica rapa subsp. *trilocularis*	(Roxb.) Hanelt	Brassicaceae	*Brassica rapa* subsp. *trilocularis*
Brassica rapa subvar. *pe-tsai*	(L. H. Bailey) Kitam.	Brassicaceae	*Brassica rapa* subsp. *pekinensis*
Brassica rapa var. *amplexicaulis*	Y. Tanaka and Ono	Brassicaceae	*Brassica rapa* subsp. *pekinensis*
Brassica rapa var. *chinensis*	(L.) Kitam.	Brassicaceae	*Brassica rapa* subsp. *chinensis*
Brassica rapa var. *dichotoma*	(Roxb.) Kitam.	Brassicaceae	*Brassica rapa* subsp. *dichotoma*
Brassica rapa var. *narinosa*	(L. H. Bailey) Kitam.	Brassicaceae	*Brassica rapa* subsp. *narinosa*
Brassica rapa var. *parachinensis*	(L. H. Bailey) Hanelt	Brassicaceae	*Brassica rapa* var. *parachinensis*
Brassica rapa var. *perviridis*	L. H. Bailey	Brassicaceae	*Brassica rapa* var. *perviridis*
Brassica rapa var. *purpuraria*	(L. H. Bailey) Kitam.	Brassicaceae	*Brassica rapa* var. *purpuraria*
Brassica rapa var. *septiceps*	L. H. Bailey	Brassicaceae	*Brassica rapa* subsp. *rapa*
Brassica rapa var. *silvestris*	(Lam.) Briggs	Brassicaceae	*Brassica rapa* subsp. *campestris*
Brassica rapa var. *trilocularis*	(Roxb.) Kitam.	Brassicaceae	*Brassica rapa* subsp. *trilocularis*
Brassica repanda	(Willd.) DC.	Brassicaceae	*Brassica repanda*
Brassica repanda subsp. *africana*	(Maire) Greuter and Burdet	Brassicaceae	*Brassica repanda* subsp. *africana*
Brassica repanda subsp. *almeriensis*	Gomez-Campo	Brassicaceae	*Brassica repanda* subsp. *almeriensis*
Brassica repanda subsp. *blancoana*	(Boiss.) Heywood	Brassicaceae	*Brassica repanda* subsp. *blancoana*
Brassica repanda subsp. *cadevallii*	(Font Quer) Heywood	Brassicaceae	*Brassica repanda* subsp. *cadevallii*
Brassica repanda subsp. *cantabrica*	(Font Quer) Heywood	Brassicaceae	*Brassica repanda* subsp. *cantabrica*
Brassica repanda subsp. *confusa*	(Emb. and Maire) Heywood	Brassicaceae	*Brassica repanda* subsp. *confusa*
Brassica repanda subsp. *dertosensis*	Molero and Rovira	Brassicaceae	*Brassica repanda* subsp. *dertosensis*
Brassica repanda subsp. *galissieri*	(Giraudias) Heywood	Brassicaceae	*Brassica repanda* subsp. *galissieri*
Brassica repanda subsp. *gypsicola*	Gomez-Campo	Brassicaceae	*Brassica repanda* subsp. *gypsicola*
Brassica repanda subsp. *latisiliqua*	(Boiss. and Reut.) Heywood	Brassicaceae	*Brassica repanda* subsp. *latisiliqua*
Brassica repanda subsp. *maritima*	(Willk.) Heywood	Brassicaceae	*Brassica repanda* subsp. *maritima*
Brassica repanda subsp. *nudicaulis*	(Lag.) Heywood	Brassicaceae	*Brassica repanda* subsp. *africana*
Brassica repanda subsp. *repanda*		Brassicaceae	*Brassica repanda* subsp. *repanda*
Brassica repanda subsp. *saxatilis*	(DC.) Heywood	Brassicaceae	*Brassica repanda* subsp. *saxatilis*
Brassica repanda subsp. *silenifolia*	(Emb.) Greuter and Burdet	Brassicaceae	*Brassica repanda* subsp. *silenifolia*
Brassica repanda subsp. *turbonis*	(P. Monts.) J. M. Monts. and Romo	Brassicaceae	*Brassica repanda* subsp. *turbonis*
Brassica rerayensis	Ball	Brassicaceae	*Erucastrum elatum* var. *rerayense*
Brassica rifana	Emb. and Maire	Brassicaceae	*Erucastrum rifanum*
Brassica robertiana	J. Gay	Brassicaceae	*Brassica montana*
Brassica rugosa	(Roxb.) L. H. Bailey	Brassicaceae	*Brassica juncea* var. *rugosa*
Brassica rupestris	Raf.	Brassicaceae	*Brassica rupestris*
Brassica ruvo	L. H. Bailey	Brassicaceae	*Brassica ruvo*
Brassica sabularia	Brot., nom. illeg.	Brassicaceae	*Brassica barrelieri*
Brassica saxatilis subsp. *africana*	Maire	Brassicaceae	*Brassica repanda* subsp. *africana*
Brassica saxatilis subsp. *confusa*	Emb. and Maire	Brassicaceae	*Brassica repanda* subsp. *confusa*
Brassica saxatilis subsp. *silenifolia*	Emb.	Brassicaceae	*Brassica repanda* subsp. *silenifolia*
Brassica saxatilis var. *cadevallii*	Font Quer	Brassicaceae	*Brassica repanda* subsp. *cadevallii*
Brassica saxatilis var. *cantabrica*	Font Quer	Brassicaceae	*Brassica repanda* subsp. *cantabrica*

Brassica saxatilis var. *turbonis*	P. Monts.	Brassicaceae	*Brassica repanda* subsp. *turbonis*
Brassica septiceps	(L. H. Bailey) L. H. Bailey	Brassicaceae	*Brassica rapa* subsp. *rapa*
Brassica sinapistrum	Boiss.	Brassicaceae	*Sinapis arvensis* subsp. *arvensis*
Brassica souliei	(Batt.) Batt.	Brassicaceae	*Brassica souliei*
Brassica souliei subsp. *amplexicaulis*	(Desf.) Greuter and Burdet	Brassicaceae	*Brassica souliei* subsp. *amplexicaulis*
Brassica souliei subsp. *souliei*		Brassicaceae	*Brassica souliei* subsp. *souliei*
Brassica sp.		Brassicaceae	*Brassica* sp.
Brassica spinescens	Pomel	Brassicaceae	*Brassica spinescens*
Brassica subscaposa	Maire and Weiller	Brassicaceae	*Brassica elongata* subsp. *subscaposa*
Brassica suffruticosa	Desf.	Brassicaceae	*Moricandia arvensis* subsp. *suffruticosa*
Brassica tournefortii	Gouan	Brassicaceae	*Brassica tournefortii*
Brassica trilocularis	(Roxb.) Hook. f. and Thomson	Brassicaceae	*Brassica rapa* subsp. *trilocularis*
Brassica varia	Durieu	Brassicaceae	*Erucastrum varium* subsp. *varium*
Brassica vesicaria	L.	Brassicaceae	*Eruca vesicaria*
Brassica villosa	Biv.	Brassicaceae	*Brassica villosa*

Group of *Brassica* Based on Aneuploidy

On the basis of an aneuploidy *Brassica* can be grouped into following:

Table 5.5: Group of *Brassica* Based on Aneuploidy

Group	*n=*	*Species*
I	n=8	*B. nigra*
II	n=9	*B. oleracea, B. alboglabra, B. sinapistrum, B. atlantica* (origin of Algeria, Tunis), *B. balearica* (origin of Majorca), *B. cretica* (origin of Greece, Crete, Cyprus), *B. incana* (origin of Adriatic sea coasts, Southern Italy, Sicily), *B. montana* (origin of Spain, south France, Italy), *B. rupestris* (origin of Sicily, southern Italy), *B.villosa* (origin of Sicily)
III	n=10	*B. campestris, B. rapa, B. chinensis, B. japonica, B. pekinensis, B. yriocularis, B. toumeforti*
IV	n=12	*B. wrightri, B. monensis*
V	n=17	*B. carinata*
VI	n=18	*B. juncea, B.cemua*
VII	n=19	*B. napus, B. napella, B. rugosa*
VIII	n=24	*B. cheiranthus*

Classification of Cultivated *Brassica oleracea*

See the Tables 5.6 and 5.7.

Gene List of *Brassica oleracea*

ABBCCDEEF-Data obtained from studies on secondary association in cabbage indicated that the minimum number of secondary association groups which included all chromosomes is three. This implies that the basic chromosome number x=3.

Crossability of *Brassica* Species

See the Table 5.9.

Potential Genes for Insect Resistance in Brassicas

Genetic engineering of Brassicas vegetable for pest and disease resisance offers a safe and attractive strategy to acieve a long lasting resistance. Introduction of genes coding for *Bacillus thuringiensis* (Bt) insecticidal crystal protein, cowpea and soyabean trypsin inhibitors and

Table 5.6: Baily (1930) Classified the Cultivated *Oleracea* into 7 Divisions

Class	*Key Points*	*Group*	*Details of Characters*	*Species*
I	The races grown for the open foliage	1	Branching shrubby kales	*B. oleracea* var. *fructicasa*
		2	Common biennial mostly unbranched kales	*B. oleracea* var. *acephala*
		3	Low cabbage like or Portuguese kales	*B. oleracea* var. *tronchuda*
II	The head races grown for the compact leaf-heads or buds	4	Edible product a terminal head, the cabbage	*B. oleracea* var. *capitata*
		5	Edible product latyeral small heads or buds, the Brussels sprouts	*B. oleracea* var. *gemmifera*
III	The inflorescence races grown for the modified thickened flower-shoots or malformed flowers	6	Flowers only partially if at all modified	*B. oleracea* var. *italica*
		7	Flowers thickened and malformed into a head or several heads	*B. oleracea* var. *botrytis*

Table 5.7: The Main Cultivated Forms of *Brassica oleracea*

Common Name	Botanical Name
Cauliflower	*B. oleracea* var. *botrytis* var. *cauliflora*
Sprouting broccoli	*B. oleracea* var. *botrytis* sub var. *cymosa*
White cabbage	*B. oleracea* var. *capitata* f. *alba*
Red cabbage	*B. oleracea* var. *capitata* f. *rubra*
Savoy cabbage	*B. oleracea* var. *sabauda*
Brussels sprouts	*B. oleracea* var. *gemmifera*
Kohlrabi	*B. oleracea* var. *gongylodes*
Curly kale	*B. oleracea* var. *acephala* sub var. *lacinata*
Smooth leaved kale	*B. oleracea* var. *acephala* sub var. *plana*
Thousand head kale	*B. oleracea* var. *acephala* sub var. *milicapitata*
Tree kale	*B. oleracea* var. *acephala* sub var. *palmifolia*
	B. oleracea var. *acephala* sub var. *medullosa*

cytokinin biosynthesis enzymes are known to provide substantial protection against the insect damage to the crops.

Brassica suffers from a variety of fungal doseases at all stages of growth and development. For example black rot is most destructive disease of *Brassica* vegetable worldwide, causing severe damage to the crop. Kumar *et al.* (2016) evaluated 105 lines of *Brassica* vegetables (cauliflower, cabbage and kale) for obtaining new sources

Table 5.8: Gene List of *Brassica*

Gene Symbol	Characters
A	Anthocyanin development, several alleles
ac1	Albugo candida race 1 susceptibility
c1	Anthocyanin free plant, anthocyanin suppressor
cf	Cluster flower, several flowers grow together to form a cluster at end of raceme
cp	Crinkled petal, petals in bud are folded and remain crinkled in open flower
cr	Cream petal, petals cream coloured
fc	Fused cotyledons, cotyledons fused to form a funnel shaped structure
foc.1	*Fusarium oxysporum* f. sp. *conglutinans* race 1 resistance
fn	Fern leaf, partial dominant, margin of leaf more or less deeply cut
g1	Glossy plant, waxy bloom absent
Hr 1	Hairy leaf, margins of first leaf have hairs
ms 1	Male sterility, absence of pollen, anthers shrunken and frequency brownish
pg 1	Pale green, pale green true leaves, plant development is weak.
ps	Persistent sepals
rf	Reflexed petal or turned down to be parallel with axis
s-2,-14,-25,-29,-63	Self compatibility, allelic series of varying expressivity

Table 5.9: Crossability of *Brassica* Species

Species	*Group*	*Turnip*	*Swede*	*Chinese Cabbage*	*Knol-khol*	*Cabbage*
B. rapa (turnip)		SI	CCx	CC (X)	CC	CC
B. carinata		CC	SI	CC	CC	CC
	Swede (2n=34)	X	SC	-	-	-
B. chinensis Juslen		CC	CC	SI	CC	CC
	non Duthie and Fuller/*B. pekinensis* Rupr. (2n=20)	(X)	-	SC	-	-
B. oleracea L. Var. *gongylodes* (2n=18)		CC	CC	CC	SI	CC
		-	-	-	SC	-
	B. oleracea var. *capitata* (2n=18)	CC	CC	CC	CC	SI
		-	-	-	-	SC

Table 5.10: Common Name, Scientific Name and Potential Gene

Common Name	Scientific Name	Potential Genes
Diamond backmoth	*Plutella xylostella*	*Cry 1A*
Cabbage butterfly	*Pieris brassicae*	*Cry 1A*
Stem borer	*Hellula undalis*	*Cry 1A*
Hairy catterpillar	*Spilosoma obliqua*	*Cry 1A*
Cut worm	*Agrotis ipsilon*	*Cry 1A*
Aphid	*Brevicorne brassicae*	*Snow drop lectin*
Thrips	*Caliothrips indicus*	-
	Thrips tabaci	*Cry 2A*
Leaf miner	*Phytomyza horticola*	-

Source: Sharma, 1999.

Table 5.11: Useful Genes having Fungal Resistance in Brassicas

Disease	*Causal Organism*	*Useful Genes*
Downy mildew	*Pernospora parasitica*	Chitase
Wire-stem	*Rhizoctonia solani*	Glucanase
Leaf-spot	*Alternaria brassicae*	Thionin
Cabbage yellows	*Fusarium oxysporum*	RIP
Damping off	*Pythium* sp.	Permatins
White rust	*Albugo candida*	Osmotin
Black-rot	*Xanthomonas compestris*	Chitinase
Soft-rot	*Erwinia carotivira*	Chitinase

Source: Sharma, 1999.

of black rot resitance having durable performance. Thirty days old seedlings of all the genotypes were artificially inoculated with the culture of race *Xcc4* bacterium and were kept in the moist chamber at 90-95 per cent relative humidity for nine days. Thereafter, inoculated plants were kept in glass house and observed daily for disease development, till the leaves become senescent. The severity of symptoms was assessed on a six point scale of 0 to 9 based on the relative lesion size and per cent disease intensity (PDI) worked out. In the investigations, two genotypes of cauliflower *viz.*, DC-76 (2.22 per cent) and 1385 (4.44 per cent) were found resistant, while three other genotypes *viz.*, C-8-204 (6.66 per cent) [Cabbage], Supermex Early (8.88 per cent) [Cauliflower] and S-208 (8.88 per cent) [Cabbage], were found to have mild resistance. A wide range of biotechnological approaches to control fungal diseases are available. Source of resistance in C genome of *Brassica* have been reported by many workers, but none of them durable.

REFERENCES

1. Kumar, S., Praksh, C., Dey, S.S., Singh, R., Kumar, A., Kumar, R. and Dhiman, M.R. 2016. Screening of *Brassica* vegetables for resistance against black rot (*Xanthomonas campestris* pv. *campestris Xcc4*) disease. 7th Indian Horticulture Congress (Abs: 07; Session II: Breeding of Horticultural Crops), 15-18 November, held at ICAR-IARI, New Delhi, p. 91.
2. Prakash, S., Wu, X.M. and Bhat, S.R. 2011. History, evolution and domestication of *Brassica* crops. *Plant Breeding Reviews*, **35:** 19-84.
3. Sharma, R.P.1999. Role of biotechnological options for vegetable breeding. In: Narendra Singh (ed.). Summer school on advanced technologies in improvement of vegetable crops including cole crops. IARI, New Delhi

Cauliflower (*Brassica oleracea* L. var. *botrytis*)

INTRODUCTION

The word cauliflower (*Brassica oleracea* L. var. *botrytis* L.; 2n=2x=18) comes from Latin term *caulis* (cabbage), *floris* (flower) and *botrytis* (budding) meaning stem or stalk, flower and bud, respectively. Cauliflower synonimicaly known as heading broccoli is one of the most popular *Brassica* vegetables after cabbage worldwide in different climatic conditions (temperate to tropics). It grows at varying latitudes from 45 degree New Zealand to 65 degree N in Scandinavian countries in different climatic conditions, ranging from temperate to tropics (5-8°C to 25-28°C) during most of cropping seasons and is available round the year in the market. In its vegetative growth period, it may stand temperature as low as 10°C and as high as 40°C for a few days (Swarup and Chatterjee, 1972). In India, cauliflower is grown both in the hills and in the plains and from 11°N to 35°N. India is the second major cauliflower producing country after China in the world. Its worldwide total area and production is 23, 73,818 ha and 68840531 metric tonnes, respectively during 2011 (http;//faostat.fao.org/site/5671/Desktop Default.aspx?PageID=567) while it occupied an area of 3,69,000 ha with production of 67,45,000 metric tonnes in India during the year 2011 (NHB, 2011). Its white and tender head/curd formed by the shortened undifferentiated flower structure (charactristics of both the vegetative and reproductive apices; David, 1978). Cauliflower is generally used as cooked vegetable either singly or mixed with potato, carrot and peas. In raw form, it is also mixed with green salad or its pieces are dipped into sauces. It is also used in the preparation of pickles or mixed pickle with other vegetables. Tender curds are used for curries, soups, *etc.* cauliflower is low in calories, but is a good source of ascorbic acid and contains substantial amount of protein, and nutrients like phosphorus, calcium, and iron. Curds are harvested from late-August or early September to late-February or early-March in the north Indian plains and from March to November in the hills.

Origin, History and Distribution

Brassica oleracea L. grows wild in primitive form in Atlantic coasts of Europe. Cyprus and areas around Mediterranean coast are considered the primary centres of origin of cauliflower (Swarup and Chatterjee, 1972). A polyphyllectic origin by incorporation of genes into the *Brassica oleracea* L. genome from different wild Mediterranean species was suggested (Gustafsson, 1979; Snogerup, 1980). According to Hyams (1971), cauliflower was first noticed, selected and propagated in Syria. In Mediterranean and Middle East region it is under cultivation since 600 BC. Cauliflower was first described as a *Brassica florida* var. *botrytis* in 1576 and as *Brassica cauliflora* in 1583. In some literature it is mentioned that cauliflower originated from a related species *B. cretica* (as the probable progenitor). The taxonomical studies

Table 5.12: Species Verification with Basic Chromosome Number and Genome

	Species	Cultivated Type	Basic Chromosome Number (x)	Genome
Monogenomic specie	*B. nigra* (L.) Koch.	Black mustared	8	b
	B. oleracea L.	Cole group	9	c
	B. campestris Hook. f. and Thomas	Turnip group	10	a
Digenomic specie	*B. carinata* A. Br.	Abyssinian mustard	17	bc
	B. juncea (L.) Czem. and Coss.	Brown mustard and Leaf mustard	18	ab
	B. napus L.		19	ac

suggested that the progenitor of *B. oleracea* exists in the 9 chromosomes wild *B. oleracea* kale (Snogerup, 1980). RFLP studies also revealed that a primitive cultivated *B. oleracea* might have evolved from wild *B. oleracea* (Song *et al.*, 1988a and b; Figdore *et al.*, 1988). Cultivated form of cauliflower might have originated gradually from the wild cabbage through mutation, human selection and adaptation and suppose to have been domesticated in the eastern Mediterranean region (Helm, 1963).

Bosewell (1949) revealed that cauliflower is originated in the Island of Cyprus from where it moved to other areas like Syria, Turkey, Egypt, Italy, Spain and North Western Europe. In the 12th century, three varieties were described in Spain, as introduction from Syria, where it had been cultivated from thousands of years. It was introduced into Italy around 1490 by Genoese from Levent or Cyprus and into France and other European countries around 16th century. Historical literarature also cleared that in Turkey and Egypt it was cultivated in 16th century. In England, in the year 1586 cauliflower was referred to as 'Cyprus colewart', suggesting its recent introduction from Cyprus. This crop was first grown in North America in late 17th century. At the beginning of 17th century it reaches Germany, France and England. Cauliflower was introduced to India in 1822 by a botanist Dr. Jemson, a botanist from Kew Garden, London, when he took the charge of the Company Bag (United Province, Saharanpur in the northern plains) to carry out some horticultural experiments on cauliflower during the period of 'East India Company' (Swarup and Chatterjee, 1972) and at the same time it was grown for seed along the Gulf of Naples. The Royal Agri-Horticultural Society also introduced seeds of cauliflowers. After that it has undergone accilimatization and selection as a result of which the Indian cauliflower today has attained a characteristically different as compared to other temperate types of the world. The main difference is tolerance to high temperature and rainfall and earliness. Its further development and improvement were achived in North and North-Western Europe extending its cultivation up to 60°N latitude. The development on Indian cauliflower types made it possible to extend its growing area in the tropics and sub-tropics of the world. The crop is presently cultivated from 11°N to 60°N latitude. As per historical records, seeds were also imported from South Africa and England and were distributed to farmers in North India. This prized vegetable is grown in many countries like India, China, Italy, Europe, America, *etc.* apart from India and China; the other major producers of cauliflower are France, United Kingdom, Spain, Poland, Germany and Pakistan.

Species Availability, Evolution and Diversity

Cauliflower is originated from single progenitor wild cliff cabbage known as 'colewart' (*Brassica oleracea* var. *sylvestiris*) as a result of mutations. The wild 'colewarts' are still found in the wild state in sea coasts of England and Denmark and in noth west France and various localities from Greece to Great Britain. The related species are shown in Table 5.12.

The other species related to *B. oleracea* L. are given in Table 5.13:

Table 5.13: Species Origin and Somatic Chromosome Number

Species	Origin	Somatic Chromosome Number (2n=)
B. alboglabra	-	18
B. atlantica	Algeria, Tunis	18
B. balearica	Majorca	18
B. cretica	Greece, Crete, Cyprus, Syria, Lebanon	18
B. incana	Adriatic sea coasts, Southern Italy, Sicily	18
B. montana	Spain, South France, Italy	18
B. rupestris	Sicily, Southern Italy	18
B. villosa	Sicily	18

During a period of about one hundred years (1822-1929), these introduced cauliflower varieties underwent selection by local growers when seed production was attemted by them in North Indian Plains.

Gene and Genetics of Characters

Basic chromosome number x=5, making cauliflower a modified amphidiploids from a cross between two primitive 5 chromosome species with subsequent loss of one pair of chromosomes. In another study 3 groups of 2 bivalents each and 3 groups of one bivalent each have repeatedly been found. From this secondary pairing of chromosomes, it has been concluded that they all derive from a basic species with x=6. The six chromosomes are designated by the letters A, B, C, D, E and F. Later it was indicated that the minimum number of secondary association groups which included all chromosomes was three. This implied that the basic chromosome number is three.

Table 5.14: Genetic Behaviour of Traits

Character	*Number of Genes*	*Type of Gene Cction*
Plant height	Monogenic (TT)	Tallness dominant to dwarfness
Cotyledon size	Monogenic (LcLc)	Large dominant to small
Leaves in curd	Monogenic (LL)	Absence of leaf in curd is dominant to presence of leaf
Earliness	Monogenic (EaEa)	Earliness is dominant to late
Puffiness of bud	Monogenic (pbpb)	Nonpuffy bud dominant to puffy bud
Blindness (absence of curd)	Monogenic (bcbc)	Presence of curd dominant to absence
Male sterility	Monogenic	Fertility is dominant to sterility
Curd colour	Monogenic	White dominant to coloured

The genetics of quantitative characters of cauliflower are as under (Table 5.15):

Table 5.15: The Genetics of Quantitative Characters of Cauliflower

Character	*Nature of Gene Action*	*References*
Curd weight	Dominance and epistasis	Swaroop and Pal, 1966
	Pronounced over	Singh *et al.*, 1975
	Dominance and epistasis	Singh *et al.*, 1976a
	Additive and dominance	Jyoti and Vashistha, 1986; Gangopadhayay *et al.*, 1997; Sharma *et al.*, 1988
Curd to plant ratio	Partial dominance	Kale *et al.*, 1979
Curd diameter	Predominance of dominanc	Lal *et al.*, 1979
Curd size index	Predominance over dominanc	Singh *et al.*, 1975
	Dominance and epistasis	Swaroop and Pal, 1966; Lal *et al.*, 1979
	Additive and dominance	Singh *et al.*, 1976; Sharma *et al.*, 1988
	Partial dominance	Kale *et al.*, 1979
Curd angle	Pronounced and additive	Lal *et al.*, 1979
	Additive and dominance	Dadlani, 1979; Chand, 1980
Curd compactness	Dominance and additive	Nieuwhof and Garretson, 1961; Lal *et al.*, 1979
	Additive gene action	Vashistha *et al.*, 1985
Maturity earliness	Partial dominance	Watts, 1964; Swaroop and Pal, 1966
	Dominance and epistasis	Singh *et al.*, 1975; 1976b
	Predominance of additive	Lal *et al.*, 1979; Mahajan *et al.*, 1996
	Additive gene action	Gangopadhayay *et al.*, 1997
Lateness	Additive and dominance	Kale *et al.*, 1979; Sandhu and Singh, 1977
	Recessive polygenes	Sharma *et al.*, 1988; Watts, 1963

Botanical Description

Cauliflower is an important member in the genus *Brassica* and botanically named as *B. oleracea* L. convar. *botrytis* (L.) Alef. var. *botrytis* (syn. *B. botrytis, B. florida botrytis, B. cauliflora*). The edible portion of cauliflower is the white curd (mass composed of a close aggregation of abortive flowers); has been described as a pre-floral structure (thick bunches of the inflorescence), which has the characteristics of both the vegetative and reproductive apices (Sadic, 1962; Margara and David, 1978).

Table 5.16: Botanical Description

Growth Habit	Cauliflower is herbaceous annual for vegetable production and biennial for seed production.
Root	The root system consisted of a large number of delicate white roots less than 1 millimeter in diameter. Many of these ran horizontally just beneath the soil surface, some to a distance of 15 inches. Others ran outward and downward and still others descended almost vertically. The greatest depth was 15 inches. Laterals 2 to 4 inches long had grown from the older parts of the largest roots at the rate of 8 to 16 per inch. Some of these gave rise to short sublaterals. A similar branching rate was evident throughout the plant, but tertiary branches had not yet been formed on most of the rootlets. The thorough occupancy of the surface soil was possible because of very shallow cultivation.
Stem	The curd is made up of numerous divided hypertrophic branches which terminate the main stem of the plant and is highly suppressed with no part of flower appernt there (Kato, 1964).
Leaf	The vegetative shoots follows the 5 to 8 phyllotaxy of leaves, but the leaf development reduced so that only bracts are formed. Cauliflower has long and linear leaves.
Floral biology	The whole shoot system is much shortened and thicker and can give rise to the future inflorescence. Curd is the early stage of inflorescence development without any axillary sprouts. The flower has four green 7.0-10.0 mm long sepals, while the cruciferous corolla consist of four yellow petals measuring 12.0-15.0 mm. The stamens are tetradynamous and the anthers are dorsifixed and introse. The gynoecium is bicarpellary and syncarpus.

	The superior ovary is 7.0-8.0 mm long and quadrangular measuring 1.0 mm across. There are about 24 ovules on the partial placenta in the two locules. The stylar neck is 1.0-1.5 mm long and the surmounted stigma measuring about 1.0-1.5 mm.
Anthesis and pollination	The crop is mostly self-compatible and combat self in compatibility bud pollination procedure is adopted for self-pollination. The late group consists of long day plants which are self-incompatible and hence highly cross pollinated and pollination take place through insects and honey bees. Male sterility is another reason for allogamy in cauliflower. Mostly cytoplasmic male sterility is found. The microsporangium dehisces by longitudinal slits releasing a mass of yellow elliptical pollen grains. Anthers are being emasculated by physical removing the anthers for controlled hybridization. Anthers are dehisce before the flower is fully open and emasculation can safely be left until a day before anthesis, even though the stigma may be level with the anthers within the flower bud.
Curd and curdt set	The lateral buds of the shoot meristem elongate and are much branched whose apices form the surface of the curd. Botanically, the edible part of cauliflower is hypocotyls branches or pre-floral fleshy epical meristem, called as curd and surrounded by leaves which are longer but narrower than those of cabbage.
Pods	Pods (siliqua) linear and indehiscent at maturity. Siliqua measuring about 8.0 cm contains seeds that are 1.0 mm in diameter.

Horticultural Classification

Swarup and Chaterjee (1972) classified the cauliflower varieties in various groups as shown in Table 5.17.

Crisp (1982) has also classified the groups of cauliflower according to their phylogeny in the following manner shown in Table 5.18.

Further cauliflower varieties can also be classified on the basis of several traits (Singh, 1971; Swarup and Chatterjee):

(A) Classification on the Basis of Curd formation

(i) Early maturity-August-September

(ii) Mid maturity -November-December

(iii) Late maturity –January-March

(B) Classification on the Basis of Compatibility Early Group has Two Distinct Groups

1. Self-compatible as well as cross- compatible and self-compatible as well as cross- incompatible
2. Self- incompatibility in late group of sporophytic type where pollen reaction depends on the male and female sporophytes. Male sterility is another reason for allogamy

Table 5.17: Various Groups of Cauliflower Varieties Classified by Swarup and Chaterjee

Cauliflower Types	*Country of Origin*	*Probable Period of First Cultivation*	*Salient Features*
Italians or original	Mediterranean	16th Century	Plants are short; leaves erect broad with rounded tips, bluish green; curds good not protected by leaves.
Cornish	England	Early 19th Centuary	Plants vigorous; long stalked; leaves loosely arranged, broadly wavy; curds flat; irregular, loose, not protected by leaves, yellow highly flavoured
Northerns	England	19th Century	Leaves petioles, broad, very wavy, serrated; curds good and well protected.
Roscoff	France	19th Century	Plants shorts; leaves long erect, slightly wavy with pointed tip, midrib prominent, bluish green; and curds are white or creamy. Curds are hemispherical and well protected.
Angers	France	Late 19th Century	Leaves are very wavy, serrated, greyish green; curds solid, well protected.
Erfurt and Snowball	Germany and Netherlands	18th Century	Plants dwarf; leaves short, erect, glaucous green, curds solid and well protected.
Indian cauliflower	India	Late 19th Century	Plants short, long stalked; leaves; loosely arranged, broadly wavy; curds flat, somewhat loose, yellow to creamy, not protected and highly flavoured.

Table 5.18: Cauliflower Varieties Classified by Crisp According to their Phylogeny

Group Name	*Salient Features*	*Common Types*
Italian	Very diverse, include both annual and biennials and curds with peculiar conformations and colours	Jezi, Naples (Autumn Giant), Romanesco and Flora Blanca
North West European biennials	Derived within the last 300 years from Italian material	Old English, Walcheran, Roscoff, Angers and St. Malo
North European annuals	Developed in Northern Europe for at least 400 years. Origin unknown, perhaps Italian or Eastern Mediterranean	Lecerf, Alpha. Mechelse, Erfurt and Danish
Asian	Recombinants of European annuals and biennials develop within 250 years, adapted to tropics	Four maturity groups are recognized by Swarup and Chaterjee (1972)
Australian	Recombinants of European annuals and biennials and perhaps Italian stock, developing during the last 200 years.	Not yet been recognized

Table 5.19: Cauliflower has been Grouped on the Basis of Specific Temperatrure Requirement for Curding

Group	Curding Temperature °C	Varities	Ideal Time for Sowing
Group I (a) Mid September to mid-October (extra early; early-September maturity)	25	'Pusa Early Synthetic'	15th, May
October maturity group	20-25	'Pusa Deepali'	25th -30th, May
Group I (b) Mid-October to Mid-November (Medium), November maturity group	20-25	Pant Gobhi-2, Pusa Hybrid-2, Pusa Hybrid	
Mid Early	16-20	Improved Japanese	
Group II Mid Late	16-20	'11-C', '12-C',	15th -20th, July
Group III December maturity and mid-December to early-January maturity group	12-16	'Pusa Synthetic', Pusa Subhra	15th -20th, August
Late Snowball	10-16	Pusa Snowball-I, Pusa Snowball-II	

(C) Classification on the Basis of Temperature Requirement for Curding

See the Table 5.19

(D) Classification of Early Group on the Basis Self-compatiblity

Self-compatible early group has two distinct groups:

(i) Self-compatible as well as cross-compatible and self- compatible as well as cross-incompatible.

(ii) Self- incompatibility in late group is of sporophytic type where pollen reaction depends on the male and female sporophytes.

(E) Classification on the Basis of Curd Maturity

Most of the early Indian varieties of cauliflower are called by name of the month in which the curds are mature and become ready for harvest, *viz.* 'Kunwari' (September to October), 'Katki' (October to November maturity), 'Agahani' (November maturity), 'Poosi' (December maturty), and 'Maghi' (January maturity). Most of these varieties are highly heterozygous for maturity, plant growth and curd traits. The curds are grouped as:

Table 5.20: Curd Maturity Base Grouping

Group	Curd Maturity Season	Prominent Cultivar
Group I-A	Mid September to mid-October	'Pusa Katki' and 'Early Kunwari'
Group I-B	Mid-October to Mid-November	'328-5-10-1', Pant Shubra', 'Synthetic I', and 'Pusa Deepali'
Group III.	December maturity and mid-December to early-January maturity group	'Improved Japanese', 'Hissar I', 'Pusa Synthetic' and 'Pusa Shubra
Group IV.	Late maturity group snow ball group	'Pusa Snowball K-1', 'Sel. 7', EC12012', 'Sel. 1' Snow Ball-16', 'Pusa Snowball-1'

(F) Classification on the Basis of Curd Colour

- **White-** White cauliflower is the most common color of cauliflower.
- **Orange-** Orange cauliflower (*B. oleracea* L. var. *botrytis*) contains 25 per cent more vitamin A than white varieties (Cohen *et al.,* 20090. This trait originated from a natural mutant found in a cauliflower field in Canada (Dickson *et al.,* 1988). Cultivars include 'Cheddar' and 'Orange Bouquet'.
- **Green-** Green cauliflower, of the *B. oleracea botrytis* group, is sometimes called broccoflower. It is available in the normal curd shape and with a fractal spiral curd called Romanesco broccoli. Both have been commercially available in the U.S. and Europe since the early 1990s. Green-curded varieties include 'Alverda', 'Green Goddess' and 'Vorda'. Romanesco varieties include 'Minaret' and 'Veronica'.
- **Purple-** The purple color in the cauliflower is caused by the presence of the antioxidant group anthocyanins, which can also be found in red cabbage and red wine (Chiu *et al.,* 2005). Varieties include 'Graffiti' and 'Purple Cape'.

Augmentation of Genetic Resources and Crop Improvement

Introduction and Selection

In India perhaps 'Early' and 'Main Crop Patana' and 'Early''Main Crop Banaras' were the first early genotypes in the world and have greatly contributed for developing improved varieties adapted to hot weather conditions in different countries, *viz.* 'Pua Kea' in Hawaii, 'Compinus' in Brazil. 'Improved Japanese' and '96-D' in Israel and 'Extra Early' in Taiwan. For earliness the selection of variety 'Comet' from 'Erfurt Dwarf' which takes only 60 days from transplanting to curd harvest. The other selections like 'Ilocos Early Patana', 'Early Patana', and 'Early Benaras' are popular among vegetable growers (Swarup and Chatterjee, 1972). Popular cultivars like 'Snowball-16' and 'Improved Japanese' were introduced in India and directly used as a variety for commercial cultivation.

Table 5.21: Introduction of Cauliflower Germplasm in India

Name of Genotypes/ Cultivars	*Source Country*	*Salient Feature*
Puakia	Hawaii	
Compinus	Brazil	Good quality curd
Improved Japanese and 96-G	Isrial	Big and compact curd
EC348166	USA	High yielding early type
EC164414-16	USA	Heat tolerance
EC175800-06	USA	Multiple disease resistance

In the beginning selection were made for early maturity and wider adaptability to hot and humid weather conditions. These types are commonly known as Indian or Tropical cauliflower which are good for early planting and early harvest. In later season, temperate types are gown which are known as snowball or late cauliflower or main season cauliflower. High yielding varieties like Pusa Deepali, Line 328, Line 12C *etc.* are very popular among the growers. In cauliflower popular variety 'Pusa Katki' was developed from the use of indigenous source by mass selection method.

Table 5.22: Different Maturity Group of Cauliflower

Maturity Group	*Cutlivars (Open Pollinated and Hybrids)*
Extra Early group	Early Kunwari, Pusa Early Synthetic, Early Selection, Pusa Meghna, Atoi Shighra (F_1) (NS-1351), Parijat, Himdhara, Himangi, Himlata, Himdev, Summer King, Basant (F_1) (NS-245), NS 60N (F_1),
Early group	Pusa Katiki, Pusa Deepali, Arka Kanti, Pant Gobhi-2, Pant Gobhi-3, Summer Special-10, Rimzim, Sharavani, Hybrid-212, Pawas
Medium group	Improved Japanese, Pusa Himjyoti, Pusa Sharad, Pusa Hybrid-2, Pant Shubra, Pant Gobhi-4, Swarna, Shweta, Chandramukhi (open pollinated), Basanti (open pollinated), Somersat, Amazing, Hybrid-218, NS-60 (F_1), NS-50 (F_1), Puspanjali
Mid-late group	Pusa Synthetic, Pusa Shubra, Marghashira, Hisar-1, Amarabati, Aparajita, Goodman, Fastman, Lateman, NS-66 (F_1), NS-75 (F_1), NS-76 (F_1), Romanesco, Cassius, Anarkali, Puspagiri, himsagar (F_1)
Late group	Pusa Snowball-1, Pusa Snowball-2, Snowball-16, Pusa Snowball-K-1, Pusa Snowball-K-25, Dania, Dania Kalimpong, Punjab Giant-26, Punjab Giant-35, Himyug, Garima, Sharad Safedhi-70, Mahima-80, Sharmilee-85, Shivangi (F_1), NS-84(F_1), NS-90(F_1),

Gayen *et al.* (2016) evaluated 18 genotypes of cauliflower and found significant differences among the genotypes for all the characters. The genotypes *viz.*, ICF-1, ICF-3, ICF-12, ICF-11 and ICF-7 were found promising as they had more than one desirable quantitative and qualitative trait. ICF-1 had maxium curd yield (33.05 t/ha) while ICF-10 exhibited earliest harvestable curd maturity (107.85 days). The study showed that both phenotypic co-efficient of variation (PCV) and genotypic co-efficient of variation (GCV) were higher for net curd weight (21.52 and 24.16 g), gross plant weight (20.82 and 23.43 g). High heritability alongwith high genetic advance as per cent of mean were observed for characters like stalk length, net curd weight, marketable curd weight, curd size index and 1000 seed weight. However, lowest co-efficient of variation, heritability and genetic advance were observed for days to 50 per cent curd initiation and days to 50 per cent curd maturity. Combining ability in snowball cauliflower was worked out using CMS lines by Ram *et al.* (2016) and they observed highly significant differences among the treatments for most of the traits studied except polar diameter, equatorial diameter and gross plant weight. The mean squares due to lines and testers were significant for all the traits except gross plant weight, polar diameter, equatorial diameter and harvest index whereas due to line x tester, it was significant for length, leaf width, number of leaves/ plant, plant height, days to 50 per cent curd initiation and days to 50 per cent curd maturity. Predominance of dominance component of variance was observed for leaf length, number of leaves/plant, plant height, days to 50 per cent curd initiation, marketable curd weight and marketable curd yield. Among the CMS lines, Ogu-13-01 was the best general combiner for leaf length, leaf width and plant height; Ogu-13-85 for number of leaves/ plant; Ogu-101 for days to 50 per cent curd initiation, days to 50 per cent curd maturity and net curd weight and Ogu-119 for harvest index. The tester Kt-18 was the best general combiner for leaf width, DB-1305 for number of leaves/plant, DB-187 for leaf length, plant height, gross plant weight and days to 50 per cent curd initiation and Lalchowk Maghi for days to 50 per cent curd initiation and days to 50 per cent curd maturity. Hybrids Ogu-101 x DB-1305 and Ogu-119 x Suprimax Late were best specific combinations for marketable curd weight, marketable curd yield, net curd weight and net curd yield. Sharma *et al.* (2016) evaluated 25 genotypes at Palampur (Himachal Pradesh) and estimated genetic variability, heritability and relationship between yield and related traits. High magnitude of PCV and GCV were recorded for curd solidity depicting the presence of substantial variability. High heriatability estimates was observed for ascorbic acid content, days to initiation of curd from transplanting, curd size index, harvest index, curd depth, days to marketable curd maturity from date of transplanting, stalk length, per cent marketable curds, curd diameter, plant height, total soluble solids, marketable yield/plant, gross weight per plant and curd solidity. High heritability coupled with high genetic advance was observed for curd solidity and ascorbic acid content revealing the importance of additive gene action. Marketable yield/plant showed significant and positive correlation with curd size index, gross weight/plant, curd solidity, curd compactness, curd diameter, stalk length, number of leaves and harvest index. Based on the correlation and path coefficient analysis, curd size index, gross weight/plant, curd solidity, curd compactness, curd diameter and harvest index for marketable yield/ plant could be considered as the reliable selection parameters for evolving high yielding genotypes.

Table 5.23: Cauliflower Varieties Developed through Line or Family Selection

Cultivar	*Source*
Hisar-1	Exotic
Pusa Him Jyoti	Exotic
Snowball-16	Exotic
Pusa Snowball K-1	Exotic
Punjab Giant-26	Exotic
Punjab Giant-35	Exotic

Horticulture Research Centre, Bangladesh Agricultural Research Institute (BARI) has observed that hybrids in cauliflower Snow Crown, Snow King, Spring Snow and White Baron are dominating in several area of Bangladesh. In Pakistan, cultivars like Andes, Alpha Veralta, Vernon, Champa, Snowdrift (R) Faislabad Early, Faislabad Second Early (I) and Kathmandu Special (P) are very common and being grown commercially. In Nepal several varieties of cauliflower are commonly being grown by farmers without horticultural classification.

Table 5.24: Popular Cultivars of Cauliflower in Nepal

Varieties	*Source*
Snow King, Snow Crown, Snow Queen	Takii, Japan
Rami, Master	Denmark
Swati	Indo-American

Hybridization and Selection

Cauliflower shown positive response towards heterosis especially in high yield, uniform maturity, earliness, product uniformity, tolerance to biotic and abiotic stresses and better produce quality with respect to compactness/colour/tenderness. In cauliflower single cross, double cross, and synthetic cross methods are used in hybridization by using promising genotypes. The hybrids develop between cross combination like 'S 35' x 'S 36', and 'Autumn Giant' x 'Marses' were early, high-yielding and resistant to diseases. A few hybrids have been developed by utilizing self-incompatibility and cytoplasmic male sterility. At present cauliflower hybrids cover 13 per cent of the total area (Nath *et al.*, 2008). First time synthetic varieties in Indian cauliflowers of maturity group II and III was developed at ICAR-IARI, New Delhi (Singh, *et al.*, 1981). Among the notable synthetic cultivars are 'Early Synthetic', 'Synthetic II' and 'Synthetic II'. Few hybrids of 3-5 months maturity have also been developed by Australian Company. Some of these hybrids are 'Flanca', 'Sebal', 'Gija' and 'Nedcha'.

Table 5.25: Some Promising Hybrids of Cauliflower

Name of Hybrid	*Crossing Methods*	*Genetic Emasculation Technique*
Pusa Kartik Sankar	Single cross hybrid	Self-incompatibility (SI)
Pusa Hybrid-2	Single cross hybrid	Self-incompatibility (SI)
Pusa Snow Ball Hybrid-1	Single cross hybrid	Cytoplasmic male sterility

Self-Incompatibility

Self-Incompatibility (SI) refers to the inability of a fertile hermaphrodite plant to set seed upon self pollination. There are two types of SI namely gametophytic (incompatibility reaction of pollen and stigma is determined by the genotype of the mother plant on which pollens are produced *e.g.* tomato) and sporophytic (incompatibility reaction is determined by the genotype of the mother plant on which pollens are produced which is very much common in cole crops). It indicates that members of *Brassica* family including cole crops posses homomorphic sporophytic self-incompatibility associated with trinucleate pollen and inhibition of pollen germination at stigma surface (Bateman, 1955). In monofactorial sporophytic self-incompatibility, pollen from a compatible pollination adheres to the papillae of the stigma, imbibes, and then germinates (Christ, 1959).

There are two main reasons for the expression of self-incompatibility:

(i) Due to lack of adhesion, hydration and germination of the pollen grain, and

(ii) The failure to penetrate the papillae

In case of single cross-hybrid, the parental lines should have two different *S* alleles for strong self-incompatibility. One SI inbred is used as female parent and a good pollinator (an open pollinated variety) as male to develop top cross hybrid, while four SI parental lines having different S alleles are used to produce double cross hybrids. In cole vegetables like cabbage, cauliflower, broccoli self- incompatibility (sporophytic) mechanism is being utilized for hybrid seed production at several places in India. Thompson and Taylor (1966) reported that ancestral *Brassica oleracea* was highly self-incompatible. On this basis, it is but natural that the cole crops would be self-incompatible. In cauliflower, systematic studies on self-incompatibility was initiated by Watts (1963, 1965b). He found higher level of self-incompatibility in biennial winter and autumn types and low in European summer types (snowball, alpha and erfurt). This was also confirmed by Nieuwhof (1974), Hoser-Krauze (1979) and many other workers. Chatterjee and Mukherjee (1965) found that medium duration strains of cauliflower were more compatible than long duration strains and further reported that fully self-compatible to self-incompatible forms occur. The distribution of S-haplotypes were performed by Ruffio-Chable *et al.* (2001), ten S-haplotypes were detected by immunochemical analysis. Half of the plants analyzed (126 belonged to 82 open pollinated populations, representing the variability of the group) possessed the same haplotype designated S15 in the Ockendon nomenclature. The self-compatibility of summer and autumn types were caused by the presence of the SC-haplotype which would have lost the kinase activity of SRK (Ruffio-Chable, personal communication). Annual Indian cauliflower and biennial winter cauliflower have stronger self-incompatibility mechanism. A detailed investigation in Indian cauliflower of self-incompatibility revealed that inbreds/lines of maturity group I have strongest self-incompatibility followed by maturity group II, group III showed weak self-incompatibility (Murugiah, 1978; Vidyasagar, 1981; Chatterjee and Swarup, 1984; Sharma *et al.*, 2001). Murugiah *et al.* (1983) reported that among the identified self-incompatible alleles with varying degree of dominance, high ranked *Sd Sd* and *Sv Sv* alleles retained their incompatibility throughout flowering period whereas mid-ranked *SmSm* and low ranked *SaSa* lost it gradually in Indian cauliflower having homozygous alleles. But loss in low ranked was slow than mid-ranked lines.

Table 5.26: Sporophytic Self-incompatibility System in Cauliflower; S_1S_3 x S_1S_2

Pollen Reaction	Pistil Reaction	Compatibility
Independent	Independent	Incompatible
S_1 dominant to S_2	Independent	Incompatible
S_2dominant to S_1	Independent	Compatible
Independent S_1	Dominant to S_3	Incompatible
Independent S_3	Dominant to S_1	Compatible
S_1 dominant to S_2	S_1 dominant to S_3	Incompatible
S_1 dominant to S_2	S_3 dominant to S_1	Compatible
S_2 dominant to S_1	S_1 dominant to S_3	Compatible
S_2 dominant to S_1	S_3dominant to S_1	Compatible

Intspecific Hybridization

The interspecific hybridization is resorted to generate variability in cauliflower. Compatibility among a few species is given below:

Table 5.27: Compatibility among Species

Species	B. nigra	B. oleracea	B.campestris
B. nigra (x=8) (b)	SI	CC	CC
	SC		
B. oleracea (x=9c)	CC	SI	CC
		SC	
B.campestris(x=10) c	CC	CC	SI
			SC

Somatic Hybrids

Table 5.28: Somatic Hybrids Produced in Cauliflower

Species Combination	Character Transferred
B. oleracea var. *botrytis* x *Sinpsis alba*	Alternaria and stem rot resistance
B. oleracea var. *botrytis* x *B. carinata*	Nematode resistance

Intergeneric Hybridization

The cross between cauliflower and radish (*Raphanus sativus*) was fairly successful only when the former acted as the male parent. The failure of reciprocal cross was due to the inability of the radish pollen to penetrate the styles of the cauliflower. Moreover, inadequate taxonomic knowledge of the wider variability found in this crop and efforts made to improve it by transferring desirable traits from distant botanical varieties (cabbage or kale) instead of close relative (broccoli).

Somatic Hybridization

Protoplast isolated from root tips, cotyledons, leaves and hypocotyls have been induced to form callus in *Brassica* species. The success of callus to differentiate into somatic embryoides depends upon a number of environmental factors and medium components. In cauliflower, Jourdan *et al.* (1990) developed an improved procedure for protoplast culture. Jourdan and Earle (1989) found that genotype played a critical role in determining the success rate of leaf protoplast culture in five cruciferous species including *Brassica oleracea.* Protoplast fusion is a potential tool to create cybrids through hybridizing divergent, texas, which otherwise are non-crossable. The cybrid contains organelles from both the parents to create genetic variability. This also helps in bringing together cytoplasmic traits, which is rather impossible through conventional breeding. Schenck and Robbeln (1982) synthesized *Brassica napus* through protoplast fusion of *Brassica campestris* with *B. oleracea* for the first time. However, later, it was reproduced by many workers. Jourdan *et al.* (1989) developed cybrid plants through protoplast fusion of cauliflower and a variety of *Brassica napus* which have cytoplasmic derived resistant to atrazine. These cybrids showed no segregation in population on selfing, which confirmed the cytoplasmic inheritance transfer of this character in the offspring. Protoplast fusion has also been used for the effective transfer of Ogura type of cytoplasmic male sterility traits from radish to cauliflower and other cole crops to devoid them from associated physiological abnormalities (Kagami *et al.*, 1990).

Breeding for Quality Traits

'Green Comet', Green Homets' and 'Premium Crop' are recommended for fresh market and freezing and 'Neptune' F_1 hybrid for canning and fresh use.

Cultivars like Savale', 'Fanolz', 'Bandone-Z' and 'Fano II-Z' are known potential source for quality curds and also having better storability. Similarly, Variety 'Takii 2' is suitable for cold storage.

Male Sterility

In cauliflower genic male sterility and cytoplasmic male sterility has been developed. The genic male sterilie lines are maintained by tissue culture techniques. In nuclear male sterility, both recessive and dominant genes have been reported. So far no acceptable varieties have been developed by utilizing cytoplasmic male sterility. Cytoplasmic nuclear male sterility derived from *B. nigra* and *Raphanus sativus* Ogura type (Ogura, 1968) is being exploited. In this case only problem is of seedling yellowing (at low temperature).

Mutation Breeding

GMS as a spontaneous mutation has been reported in cauliflower (Ruffio-Chable *et al.*, 1993).

Self-incompatibility

The self-incompatible alleles present in cauliflower are generally weak in Indian cauliflower; therefore, it does not have much scope for the development of hybrids. There is the possibility of production of sibs due to weak self-incompatibility.

Resistance Breeding

i. Disease Resistance

Cauliflower is susceptible to black rot and many other diseases. There is a need to develop hybrids resistant to these diseases alongwith heat tolerance. There are several varieties and hybrids which are high yielding and disease resistance but before searching and grouping of genotypes it is essential to understand the genetics of disease resistance.

Table 5.29: Genetics of Disese Resistance in Cauliflower

Diseases	*Mode of Genetic Control*
Black rot	Dominant, monogenic or polygenic
Downy mildew	One or two dominant genes
Black spot or curd blight	Dominant and polygenic
Cabbage yellows	Dominant and polygenic
White rot (*Sclerotinia sclerotiorum*)	Polygenecally controlled and recessive in nature (Sharma *et al.*, 1997)

Cauliflower variety 'Pusa Shubhra' is resistant to black rot disease and 'Pusa Snowball K-1' is resistant to curd/inflorescence blight.

Table 5.30: Resistance Source against Diseses

Varieties and Hybrids	*Resistance to*
'Native Dwarf Erfurt', AH 46'	Mosaic
'South Specific'	*Rhizoctonia solanii*

Table 5.31: Resistance Source against Insect

Varieties and Hybrids	*Resistance to*
'Snowball A'	Striped flea beetle

Abiotic Resistance

There are several varieties and hybrids which are high yielding and disease resistance but not heat tolerane. Therefore it is essential to develop heat tolerance lines.

Table 5.32: Resistance Source against Abiotic Stress

Varieties and Hybrids	*Resistance to*
'Lecerf', M16/1	Whiptail
Savale', 'Fanolz', 'Bandone-Z' and 'Fano II-z'	Frost and cold tolerance
Savale', 'Fanolz', 'Bandone-Z' and 'Fano II-z'	Frredom from bolting
EC678824-25	Heat tolerance line introduced from China

Genetic Resource Conservation

With the adoption of improved modern cultivars, the genetic variability present in the form of land races and the primitive types are disappearing. Replacement of open-pollinated cultivars with narrow genetic based F_1 hybrids has resulted in the genetic erosion of cauliflower. So, sincere efforts are needed to preserve the germplasm. Germplasm conservation is encouraged by an international organization: International Plant genetic Resources Institute (IPGRI) Rome. At the European level, European Cooperative Program for Crop Genetic and Resources Networks (ECP-GR) has grouped 35 countries. It is true that for *Brassica*, information is centralized in the Wageningen University (The Netherlands) in a Center for Genetic Resources (CGR). The European *Brassica* collection is spread in several countries. European countries collected 1000 sample of cauliflower and deposited in genebank of Horticultural Research International Wellsbourne, U.K.; The Institute voor de veredelling van Tuinbouvelgewassen, Wageningen, the Netherlands; the Instituo del Germoplasm, Bari, Italy (Van der Meer *et al.*, 1984) and in the INRA of Le Rheu, in France. A comprehensive base collection of cultivated *Brassica oleracea* L. including cauliflower has been established at Vegetable Gene Bank of Horticulture Research International (HRI), Wellsbourne, U.K. All these samples were characterized by their type and site of origin which are the most important data for genetic resources. Further large collections are being maintained by ICAR-IARI, ICAR-NBPGR, N. Delhi (India) and United States Department of Agriculture, Plant Introduction Service. Germplasm of tropical cauliflower types and that of temperate (Snowball) types are being maintained at ICAR- IARI, N. Delhi and Dr. Y.S. Parmar University of Horticulture and Forestry,

Solan (Himachal Pradesh), respectively. Tissue culture and androgenesis are certain potential techniques to be used for the maintenance of inbred lines/germplasm.

REFERENCES

1. Bateman, A.J. 1955. Self-incompatibility system in angiosperms III Cruciferae. *Heredity*, **9:** 53.
2. Bosewell, V.R. 1949. Our Vegetable Travellers. *National Geographic Magazine*, **996:** 170-177.
3. Chatterjee, S.S. and Mukherjee, S.K. 1965. Selection and maintenance of cauliflower. *Indian Journal of Horticulture*, **22:** 60-88.
4. Chatterjee, S.S. and Swarup, V. 1984. Self-incompatibility in Indian and combining ability studies in early group of Indian cauliflower involving self-incompatible lines. *Veg Sci.*, **24:** 26-28.
5. Cohen, A. and Koenig, L. 2009. "The story behind orange cauliflower". *Saveur.com.*
6. Chiu, L., Prior, R.L., Wu, X., Li, L. (2005). "Toward identification of the candidate gene controlling anthocyanin accumulation in purple cauliflower (*Brassica oleracea* L. var. *botrytis*)". *Amer. Soc. Plant Biolog.*, Annual Meeting. July 16, 2005. p. 628.
7. Christ, B. 1959. Entwicklungsgeschichtliache und physiologische Untersuchungen uber die Selbssteriliat von *Cardamine pratensis* L. *Z. Bot.*, **47:** 88-112.
8. Dadlani, N.K.1979. Studies on selection for yield and quality in cauliflower. Ph. D. Thesis, PG School, ICAR-IARI, New Delhi.
9. Dickson, M.H., Lee C.Y., Blamble A.E. 1988. "Orange-curd high carotene cauliflower inbreds, NY 156, NY 163, and NY 165". *HortSci.*, **23:** 778–779.
10. Figdore, S.S., Kennard, W.C., Song, K.M., Slocum, M.K. and Osborn, T.C. 1988. Assessment of the degree of *Theor. Appl. Genet.*, **75:** 833-840.
11. Gangopadhayay, K.K., Gill, H.S. and Sharma, S.R. 1997. Heterosis and combining ability studies in early group of Indian cauliflower involving self-incompatible lines. *Veg. Sci.*, **24:** 26-28.
12. Gayen, R., Sahu, R.K. and Singh, J. Studies on genetic variability in Indian (*Brassica oleracea* L. var. *botrytis*)". 7th Indian Horticulture Congress (Abs: 76; Session I: Breeding of Horticultural Crops), held at ICAR-IAI, New Delhi, 15-18 November, p. 59-60.
13. Hyams, E. 1971. Plants in the Service of Man, London: p., 33-81.
14. Gustafsson, M. 1979. Biosystematics of *Brassica oleracea* L. group. In: Eucarpia Cruciferae Conference. Wageningen, p. 11-21.
15. Jourdan, P.S. and Earle, E.D. 1989. Genotypic variability in frequency of plant regeneration from leaf protoplasts of four *Brassica* species and *Raphanus sativus*. *J. Am. Soc. Hort. Sci.*, **1124:** 343.
16. Jourdan, P.S., Earle, E.D. and Murschler, M.A. 1989. Synthesis of male sterile, triazine resistant *Brassica napus* by somatic hybridization between cytoplasmic male sterile *B. oleracea* and atrazine resistant *B. campestris*. *Theor. Appl. Genet.*, **78:** 445.
17. Jourdan, P.S., Earle, E.D. and Mutschler, M.A. 1990. Improved protoplast culture and stability of cytoplasmic traits in plants regenerated from leaf protoplasts of cauliflower (*Brassica oleracea* species *botrytis*). *Plant Cell Tissue Org. Cult.*, **21:** 227.
18. Kale, P.N., Chatterjee, S.S. and Swarup, V. 1979. Heterosis and combining ability studies in early group of Indian cauliflower (*Brassica oleracea* var. *botrytis* L.). *Veg Sci.*, **6:** 11-20.
19. Kato, T. 1964. On flower bud formation and development in cauliflower. I. Ecological studies on flower bud formation and development. *J. Jap. Soc. HortSci.*, **33:** 316-326.
20. Lal, G., Chatterjee, S.S. and Swarup, V. 1979. Genetics of important characters in Indian cauliflower (*Brassica oleracea* var. *botrytis* L.). *Genet. Agr.*, **33:** 181-190.
21. Mahajan, V., Gill, H.S., Sharma, S.R. and Singh, R.1996. Combining ability studies in Indian cauliflower (*Brassica oleracea* var. *botrytis* L.). *Veg Sci.*, **23:** 166-170.
22. Margara, J. and David, C. 1978. Les etapes morphologiques du developpement du meristeme de choufleur, *Brassica oleracea* L. var. *botrytis*. *Comptes Rendues de VAcademie des Sciences, Series*, D 287: 1369-72.
23. Murugiah, S. 1978. Studies on Self-incompability in Indian cauliflower. M.Sc. Thesis, PG School, IARI, New Delhi.
24. Murugiah, S., Vidyasagar, Chatterjee, S.S. and Swarup, V. 1983. Inter allelic relationship and ability of S-alleles in Indian cauliflower. Abstract No. 44, XV International Congress of Genetics, New Delhi.
25. Nath, P., Srivastava, V.K., Dutta, O.P. and Swamy, K.R.M. 208. Vegetable Crops- Improvement and Production. PNASF, Bengaluru. p. 135-138.
26. Nieuwhof, M. 1974. The occurrence of self-incompatibility in cauliflower (*B. oleracea* var. *Botrytis* L. subvar. *cauliflora* D.C.) and possibilities to produce uniform varieties. *Euphytica*, **23:** 473-478.
27. Nieuwhof, M. and Garretson, F. 1961. The solidity of the cauliflower curd. *Euphytica*, **10:** 301-306.
28. Ogura, H. 1968. Studies on the new male-sterility in Japanese radish, with special reference to the utilization of this sterility towards the practical raising of hybrid seeds. *Mem. Fac. Agric. Kagoshima Univ.*, **6:** 39–78.

29. Ram, H., Sureja, A.K., Kalia, P., Krishnan, S.G., Dey, S.S. and Kumar, A.M.B. 2016. Combining ability for important horticultural traits in snowball cauliflower (*Brassica oleracea* var. *botrytis* L.). 7th Indian Horticulture Congress (Abs: 04; Session II: Breeding for Horticultural Crops), 15-18 November, held at ICAR-IARI, New Delhi, p. 89.

30. Ruffio-Chable, V., Bellis, H. and Herve, Y. 1993. A dominant gene for male sterility in cauliflower (*Brassica oleracea* var. *botrytis*) - phenotype expression, inheritance and use in F_1 hybrid production. *Euphytica*, **67:** 9-17.

31. Ruffio-Chable, V. and Gaude, T. 2001. S-haplotype polymorphism in *Brassica oleracea*. *Acta Horticulturae*, **546:** 257-261.

32. Sadic, S. 1962. Morphology of the curd of cauliflower. *Am. J. Bot.*, **49:** 290.

33. Sandhu, J.S. and Singh, A.K. 1977. Inheritance of maturity period and curd weight in cauliflower (*Brassica oleracea* var. *botrytis* L.). *Haryana J. Hort. Sci.*, **6:** 161-165.

34. Sharma, S.R., Gill, H.S. and Kapoor, K.S. 1997. Inheritance of resistance to white rot in cauliflower. *Indian J. Hort.*, **54(1):** 86-90.

35. Sharma, S.R., Gill, H.S. and Kapoor, K.S. 1988. Heterosis and combining ability studies in late cauliflower (*Brassica oleracea* var. *botrytis* L.). *Veg. Sci.*, **15:** 55-63.

36. Sharma, S.R., Singh, R., Vinod, Singh, D.K. and Edison, S.J. 2001. Self-Incompatibility level in different maturity groups in Indian cauliflower. Diamond Jubilee Symp.–Hundred Years of Post Mendelian Genetics and Plant Breeding, Nov. 6-9, New Delhi, p. 155-156.

37. Sharma, S., Singh, Y. and Verma, A. 2016. Genetic variability and relationship between yield and some yield contributing morpho-physiological components in late maturing cauliflower (*Brassica oleracea* var. *botrytis* L.). 7th Indian Horticulture Congress (Abs: 19; Session II: Breeding for Horticultural Crops), 15-18 November, held at ICAR-IARI, New Delhi, p. 100.

38. Singh, D.P. 1971. Genetical studies in Indian cauliflower. Ph.D. Thesis submitted to ICAR-IARI, New Delhi.

39. Singh, D.P., Swarup, V. and Chatterjee, S.S. 1981. Development of synthetic varieties in Indian cauliflowers. *Indian J. Hort.*, **38(1 and 2):** 79-82.

40. Singh, D.P., Swarup, V. and Chatterjee, S.S. 1975. Genetic studies in Indian cauliflowers (*Brassica oleracea* L. var. *botrytis*).I. Heterosis and combining ability in maturity group I. *Veg. Sci.*, **2:** 1-7.

41. Singh, D.P., Swarup, V. and Chatterjee, S.S. 1975b. Genetical studies in Indian cauliflowers (*Brassica oleracea* L. var. *botrytis*).III. Heterosis and combining ability in maturity group III. *Veg. Sci.*, **3:** 47-53.

42. Singh, D.P., Swarup, V. and Chatterjee, S.S. 1976a. Genetic studies in Indian cauliflowers (*Brassica oleracea* L. var. *botrytis*). Heterosis and combining ability in maturity group II. *Veg. Sci.*, **3:** 41-46.

43. Snogerup, S. 1980. *Brassica* Crops and Wild Allies. (eds.) Tsunoda, S., Himda, K. and Gomezcampo, C. Japan Scientific Societies Press, Tokyo.

44. Song, K.M., Osborn, T.C. and William, P.H. 1988a. *Brassica:* taxonomy based on nuclear restriction fragment length polymorphism (RFLPs). Genome evolution of diploid and amphiploid species. *Theor. Appl. Genet.*, **75:** 784-794.

45. Song, K.M., Osborn, T.C. and William, P.H. 1988b. Brassica taxonomy based on nuclear restriction fragment length polymorphism (RFLPs). Premilinary analysis of subspecies with B. rapa (Sync. *Campestris*) and *B. oleracea*. *Theor. Appl. Genet.*, **76:** 593-600.

46. Swaroop, V. and Chatterjee, S.S. 1972. Origin and genetic improvement of Indian cauliflower. *Economic Botany*, **26(4):** 381-393.

47. Swaroop, V. and Chatterjee, S.S. 1976. Improvement of Indian Cauliflower. ICAR-Indian Agricultural Research Institute, New Delhi.

48. Swaroop, V. and Pal, A.B. 1966. Gene effects and heterosis in cauliflower-I. *Indian J. Genet.*, **26:** 269.

49. Thompson, K.F. and Taylor, J.P. 1966. Non-linear dominance relationships between S-alleles. *Heredity*, **21:** 345-362.

50. Vashistha, R.N., Neog, S.J. and Pandita, M.L. 1985. Gene effects studies of curd compactness in cauliflower. *Haryana Agric. Univ. J. Res.*, **15:** 406-409.

51. Vidyasagar, 1981. Self-incompatibility and S-alleles in Indian cauliflower. Ph.D. Thesis, PG School, ICAR-IARI, New Delhi.

52. Watts, L.E. 1963. Investigations Breeding.I. Studies on self-incompatibility. *Euphytica*, **12:** 330-340.

53. Watts, L.E. 1965b. Investigations into the breeding system ofcauliflower. II. Adaptation of the system to inbreeding. *Euphytica*, **14:** 67-77

Cabbage (*B. oleracea* L. convar. *capitata* (L.) Alef. var. *capitata*)

INTRODUCTION

Cabbage (*B. oleracea* L. convar. *capitata* (L.) Alef. var. *capitata*, 2n=18) is the most important cole crop next to cauliflower (in area and production) which is generally grown during cool season (grows best between temperatures of 10-25°C) for its tender 'swollen heads'. The head is basically overlapped leaves. Therefore cabbage is consumed throughout the world as a green leafy vegetable. It possesses excellent storability. Cabbage is not photo periodically sensitive but the apical meristem of plants that have begun to form a loose head must be exposed to 4 to 6 weeks of 4-7°C for vernalization to occur eliciting production of a reproductive apical meristem. Vernalization of the seed or seedling does not cause flowering. For initian of the seed stalk to penetrate the tightly oppressed leaves of the cabbage head, the head is often quartered by a knife. The seed stalk then attains a length of 1.0-2.0 metres. It covers 4.5 per cent areas under vegetable crops in our country. The increase in area has been possible due to the introduction of heat-tolerant varieties and F_1 hybrids which give more yields per unit area as compared to traditional cultivars in the region of mild winters. The hybrids occupy 22.9 per cent of the area under cabbage (Swarup, 2006).

Red cabbage (*B. oleracea* L. var. *capitata*) is also known as purple cabbage, blue cabbage and pH indicator plant. It is nutritious and delicious vegetable that has become very popular throughout the world. Red cabbage is a known source of cyanadin-based anthocyanins which provide a wide range of health benefits including a prevention of premature aging, reducing the chamnces of cancer, improve the health of the skin and eyes, helps in weight loss, boosts the immune system, detoxifies the body and improves heart health.

Origin, History and Distribution

Most wild forms of cabbage are found along the coasts of the Mediterranean Sea which is considered its place of origin. The real headed cabbage was evolved in Germany which indicates that Greek and Roman civilizations were the first to recognize the value of cabbage. By the time the Greek and Roman empires expanded into Europe. Few researchers opined that cabbage and kale were first to be domesticated and have probably originated in Europe. Cabbage is one of the earlier cultivated forms of cole crops (*Barassica oleracea* L.) and was probably in general use, around 2000-2500. Snogerup (1980) was of the view that cabbage originated in Western Europe. Cabbage was considered a staple food that was carried with them and introduced throughout the continent. European settlers brought cabbage in the 16th and 17th centuries to North America. By the 1700s, cabbage was a common crop grown by the colonists. The vegetable, which was limited in use till 1960s has now become very popular as the awareness about its high nutritive value has increased manifolds. At present cabbages are grown in Caribbean countries, Indonesia, Malaysia, Central East and West America, the Philippines and tropics in general. Traditionally, it is grown to a large extent in the southern, eastern and coastal areas of India. But after the development of hybrids and some tropical lines it is produced almost round the year throughout the country barring 1-2 months (during very high and low temperature).

Species Availability, Evolution and Diversity

The genus *Brassica* includes about 100 species and cabbage is originated from single progenitor wild cliff cabbage known as 'cole wart' (*Brassica oleracea* L. var. *sylvestiris* L.) as a result of mutations, human selection and adaptation. Cabbage evolved in Germany (Helm, 1963) and both red and white cabbage known in Germany by 1150 AD and by 14th century in England. Nieuwhof (1969) recognized the following three cultivated forms:

(i) *Brassica oleracea* L. var. *capitata* L. f. *alba* 1DC- White cabbage

(ii) *B. oleracea* L. var. *capitata* L. f. *rubra* (L.) Jhell- Red cabbage

(iii) *B. oleracea* L. var. *sabauda* L.- Savoy cabbage

Gene and Genetics of Characters

Table 5.33: Gene and Genetics of Characters

Character	*Number of Genes*	*Type of Gene Action*
Stem pigmentation	Monogenic	Purple dominant to green
Core length	Digenic	Incompletely dominant gene action. Short core was correlated with round head and long core with flat head
Leaf glossiness (wax layer)	Digenic	Normal dominant to glossy

Botanical Description

Cabbage belongs to family Cruciferae, genus *Brassica*, *B. oleracea* L. convar. *capitata* (L.) Alef. var. *capitata*. The common cabbage grown in India is white cabbage. The red cabbage has the same botanical name except form is *rubra* (L.) Thell.

Table 5.34: Botanical Description

Growth Habit	Cabbage plant is a polymorphous, annual or biennial plant; erect, 30.0-150.0 cm high and much branched in the upper part. The first year of growth requires the production of of a vegetative head followed by flowering and seed production in the second year.
Root	Cabbage has an extensive, shallow root system. The majority of the roots are found in the top 0.4 to 0.5 m of the soil with a rapid decrease in root density with depth.

Stem	Stem is obtusely quadrangular or subterete, pruinose and glabrous.
Leaf	Petiole is lyrate or obovate, subentire or undulate, more or less deeply and irregularly lobed. Terminal lobe is orbicular or ovate, apex rounded. Leaves are mostly sessile, semi-amplexicaul, ovate- oblong or linear-lanceolate, entire or subentire. Leaves are very variable, at the base of the stem and 2.0-10.0 cm long. Leaves are green, yellowish green or red fleshy, thick, glabrous, more or less distinctly dotted.
Floral biology	Inflorescence is terminal, elongate, lax, racemiform or paniculiform, 10.0-50.0 cm long and racemes 20-40 flowered. Flowers are actinomorphic and bisexual (complete). Flowers are rather large, buds raised far above the expanded flowers, pedicels 0.5-0.8 cm long. Sepals (gamosepalous with 4 sepals) are erect, obtuse, 1.0 cm long, the outer ovate oblong and narrow, saccate at the base, the inner ones more ovate or broadly ovate oblong. Petals are oblong-spathulate (gamopetalous with 4 petals and cruciform), rounded at the tips, light yellow and 1.5-2.0 cm lobg. Stamens are subequal, erect and 1.0-1-3 cm long.
Anthesis and pollination	Cabbage is highly cross pollinated due to sporophytic self-incompatibility, determined by up to 10 alleles at tha 'S' locus. Pollination is entomophilous. Anthesis occurs 8.00-10.00 hours. Pollen fertility is maximum on the day of anthesis. Stigma is receptive 2-3 days before to the day of anthesis (Kakizaki and Kasai, 1933)
Head	The thickened main bud of cabbage is called 'head' which is tightly packed overlapping leaves manifesting a large head. The shape of the head may be round, conical, oblong, flat, or savoyed depending on the variety.
Pods	Pods on stalks are 1.5-2.0 cm long, more or less patent, thin, cylindrical or linear elongate, with a long beak with many seeded, 5.0-9.5 cm long and 0.3-0.5 cm thick.
Seeds	Seeds are 1-seriate, globose and smooth.

Horticultural Classification

Cabbages can be classified on the basis of traits showing in Table 5.35a.

Three major types of cabbage classified according to market demand are showing in Table 5.35b.

Table 5.35a: (A) Botanical Colour and Form of Heads

Head Colour	*Botanical Name*
White cabbage	*B. oleracea* var. *capitata* L.f. *alba*
Red cabbage	*B. oleracea* var. *capitata* L. f. *rubra*
Savoy cabbage	*B. oleracea* var. *sabauda*

Table 5.35b: (B) Market Demand

Types	*Description*
Fresh market types	Fresh market types usually form conical to globe heads and have a range of maturities from 50 to 95 days.
Storage types	Storage types are slow in growth and have a firm head with thin, finely veined leaves.
Sauerkraut types	Sauerkraut types form large heads that mature in about 130 days.

Chaudhary (1976) classified cabbages on the basis of place of origin:

Table 5.35c: (C) Place of Origin

Origin Place	*Botanical Name*
Mediterranean cabbage	*B. oleracea* var. *capitata* ssp. *mediterranea*
Oriental cabbage	*B. oleracea* var. *capitata* ssp. *orientalis*
European cabbage	*B. oleracea* var. *capitata* ssp. *europea*
Savoy cabbage	*B. oleracea* var. *sabauda*

Chaudhary (1976) classified cabbages on the basis of head shape

Table 5.35d: (D) Head Shape

Shape	*Varieties/Genotypes*
Round shape, head or ball head type	'Golden Acre', 'Pride of India', 'Copenhagen Market and 'Mammoth Hock Hed'
Flat head or drum head type	'Pusa Drum Head'
Conical head type	'Jersey Wakefield'
Savoy type	'Chieftain'

Myer's classification based on size and shape of head are as under (Allen, 1914):

Table 5.35e: (E) Myer's Classification Based on Size and Shape of Head

Group	*Description of Head*	*Varieties/Genotypes*
Wake-field and Winnings Kadt group	Heads are small, conical and very early	'Jersey Wakefield', 'Charleston Wakefield'
Copenhagen market group	Round heads, early, large head	'Copenhagen market', 'Golden Drum Head', 'Globe', 'Bonanza'
Flat head or drum head group	Heads flattened from both ends	'Pusa Drum Head', 'Early Round Dutch', 'All Head Early', 'Succession', All Seasons', 'Sure crops'
Savoy group	Leaves are wrinkled, heads are of very high quality, limited commercial use	'Drum Head Savoy', 'Improved American Savoy', 'Chieftain'
Danish ball head group	Leaves are thin, heads compact, solid, medium-sized, good keeping quality, fine texture	'Danish Bll Head', 'Hollander', 'Wisconsin Ball Head'
Alpha group	The earliest group, heads very small, solid head, limited commercial use	'Miniature Marrow'
Volga group	Thick leaves, shining blue coloured leaves, loose bottom	'Volga'
Red cabbage group	Similar to Danish ball head except red colour of leaves	'Red Rock', 'Red Dutch', 'Large Red'

Table 5.35f: On the Basis of Head Shape, Choudhary (1976) Classified

Head Shape	*Varieties/Genotypes*
Round shape. Head/Ballhead type	'Golden Acre', 'Pride of India', 'Copenhagen Market' and 'Mammoth Rock Red'
Flat head or Drumhead type	'Pusa Drum Head'
Conical head type	'Jersy Wakefield'
Savoy type	'Chieftan'

The round head varieties mature the earliest followed by the conical varieties, whereas Drumhead varieties with flat head and Savoy varieties are usually late.

Table 5.35g: (F) Botanical Colour and Form of Heads

Colour	*Species*
White cabbage	*B. oleracea* var. *capitata* L.f. *alba*
Red cabbage	*B. oleracea* var. *capitata* L.f. *rubra*
Savoy cabbage	*B. oleracea* var. *sabauda*

(G) Classification Based on the Basis of Maturity

Table 5.35h: Available Cultivars under different Maturity Group

Maturity Group	*Maturity Days*	*Cultivars (Open Pollinated and Hybrids)*
Early group	45-90	Golden Acre, Pride of India, Sel.8, Copenhagen Market, Express, Sri Ganesh Gol, Pusa Mukta, Pusa Ageti, Pusa Synthetic, Bajrang (BSS-50), Pusa Sambandha, Green Express, Kalimpong Eclipse Ball, Green Boy, CH-21, Quisto, Kranti, Mitra, Early Drumhead, Sudha (BSS-15), BSS-44, Hara Gola, Vaibhav, Evergreen, Super Cross, Harini, Red Green, Primero, Benson, Fresco, Gonzales, Morris, Fieldman, General, Invento, Bronco, Governor, Rotonda, Pragati, Green Ball, Green gold, Green Diamond, Green Diamond Improved, Green Samrat (F_1), Gold Star (F_1), Super Ball-50, Gray Ball-60, Goody Ball, Big Ball-75, Quick Ball-45, Golden Ball, Kopen Hagen Market Imported, Pashan, Priya (F_1), Gold Star (F_1), Red Baby, Mahima (F_1), Simran (F_1), Sumitra, Bobby, NS-22 (F_1), NS-25 (F_1), NS-26 (F_1), Mansi, Bharat (45), Ria (105), Summer Queen (F_1), Vishwas (F_1), Combinor (F_1), Equatoria (F_1), Green Globe (F_1), KS-Cross (F_1), S-107, Hybrid-123, Hybrid-124, Hybrid-125, Hybrid-126, Hybrid-133, Super Hero, S-101, Improved Savitri (Open pollinated)
Medium group	76-90	Pusa Drumhead, Bahar, Unnati, Geetanjali, Konark, Raj-2, Meenakshi, Hari Rani, Varun, Hema-II, Bhima, Vijay, Dublin, Hero, Qutsts (F_1), Gideon, Ramada, Shristi, Melisa, Drumhead Savoy, Red cabbage, Field Rocket, Jersey Wakefield, Green Samrat (F_1), MCV-1NS21 (F_1),S-103, S-601, Bejo-1857
Late group	90-120	Late Large Drumhead, Kalipong Eclipse Drumhead, September, Hari Rani Gole, Ratna, Suvarna (BSS-32)

Augmentation of Genetic Resources and Crop Improvement

Introduction and Selection

Many lines/genotypes introduced to India time to time and became promising and highly adaptable. Among them 'Golden Acre', 'August' 'September' and 'Copenhagen Market' performed better in several agroclimatic zones of India. Some low chilling types were selected in exotic introduction like 'Eclipse Drumhead', 'Danish Balhead' *etc.* from Darjeeling hills. Seeds of the exotic cabbage varieties like 'Green Express' and 'Green Boy' are being marketed by National Seeds Corporation (Nath *et al.*, 2008) for commercial cultivation in India. Other varieties of white cabbage grown in India are 'Golden Ball' and 'Early Drumhead'. Sankar and Sonvane (2016) found that the prevailing climate in hilly region of Coorg is an ideal to grow red cabbage. They evaluated six red cabbage hybrids (Scarlet O Hara, Primero, Red Acre, Red Express, Red Dynasty and Red Sky) were evaluated in randomized Block Design (RBD) with four replications at CHES, Chettalli during the winter season, 2014-2015 under open conditions. Significant differences were observed in growth, yield and yield contributing characters. Among the hybrids 'Scarlet O Hara' was recorded the highest number of leaves (52.9) and plant spread (70.0 cm) in case of 'Red Express' hybrid noticed the lowest leaves (39.8) and plant spread (53.0 cm). However, 'Red Express' recorded the early maturity (69.5 days) as compared to 'Scarlet O Hara' (89 days). The highest marketable head weight (1.750 kg/plant) and marketable yield (58.3 t/ha) was recorded in 'Primero' hybrid when compared to other hybrids.

Table 5.36: Introduction of Cabbage Germplasm in India

Name of Germplasm	*Salient Feature*	*Source Country*
EC304718-25	CMS and restorer lines	USA
EC187228-30	Male sterile lines	Canada
EC350918	Heat tolerant	USA
EC168041-42	Heat tolerant, club rot and black rot	USA

The popular cabbage cultivars 'Golden Acre' and 'September' were developed by use of exotic (Copenhagen Market) and Germany materials, respectively, through mass selection method. Pride of India was developed through family selection.

Horticulture Research Centre, Bangladesh Agricultural Research Institute (BARI) has identified a potential cultivar 'Probhati' in the year 1985 which very much suitable in Bangladesh conditions with respect to yield and qulity. But under hybrid cabbage crop K-K Cross, K-Y Cross, atlas-70 and Green Express are dominating in several area of Bangladesh. In Nepal several exported varieties of cabbage are commonly being grown in large scale.

Table 5.37: Popular Cultivar of Cabbage in Nepal

Varieties	*Source*
Green Coronet, Green Cross, Green Crown, K-K Cross, K-Y Cross, Green Stone, Southern Treasure, Tokyo Frid, Ruby Ball (Red Cabbage), Empire	Takii, Japan
Green Express, Leo, Alisa, Stone Head, Pacfica, Hercules, Czar (Caesar), Charmatat, Luna, Green Disk, Saf-Gard, Venus, Royal Vantage, Pluto, White contessa, Milkyway, White Baron, Candid Charm, Tast Bright (No. 69), Regalia, Super Set, Grandslam, Blue Vantage, Succession Yellow Leaved	Sakata, Japan
Takamine, Kenzan YR, Sawani Dori	Takamin Seed Co., Japan
Green Challenger, BM-215, BM-115	Hungnog Seed Cop., Korea
Gloria, Ratan	Denmark
Kranti, Hari Rani Gol	Mahyco Seed Co., India
BSS-50	Bejo Sheetal Co., India

In Pakistan, cultivar like 'Golden Acre (R)' is being grown commercially.

Hybridization and Selection

Hybrid cabbage a great proposed in the country due to its high yield, quality and resistance. Today hybrids are known to out perform open-pollinated varieties in marketable head yield and horticultural traits. There is a need to develop tropical cabbage hybrid along with resistance to important disease and pests. In cabbage, almost all hybrids are exotic. The area under F_1 hybrids is about 85 per cent of the total area of cabbage in the country. 'Sel 8' is a selection from 'EC10109' x 'EC 24855' made at ICAR-IARI, Regional Station, Katrain (H.P.). In India several hybrids are available among them given in Table 5.38.

Table 5.38: Available Hybrids and their Salient Features

Hybrids	*Salient Features*
'Summer Monarch', 'Autumn Monarch', 'Winter Monarch', 'Hurt', 'Ideal'	Larger heads, long storability and high yielding
'Sri Ganesh Gol", 'Hari Rani', 'Meenakshi', 'Green Boy', 'Green Express'	Larger heads, and high yielding
'Nomes Purvyj'x 'Kjure-17' (Russian)	Early, and high yielding
'Glory of Enkhusizen' x 'Pudjan'	Early, high yielding and resistant to black rot
'BRH 5' ('Pusa Mukta' x 'AC 204')	High yielding

The current interest in the imported cabbage hybrids especially in peninsular, eastern and western India is for the tolerance to high temperature and the hard compact quality of heads which have good keeping quality in long distance transportation (Seshadri and Chatterjee, 1996). Swarup and Gill (1964) suggested the use of purple pigmentation in hybrid seed production of cabbage. In cabbage the single cross hybrids and double cross hybrids both are produced and top crossing (inbred x variety) is used to produce single cross hybrids.

Table 5.39: Some Promising Hybrids of Cabbage

Name of Hybrid	*Crossing Methods*	*Genetic Emasculation Technique*
Pusa Cabbage Hybrid-1	Single cross hybrid	Self-incompatibility (SI)
Pusa Cabbage Hybrid-81	Single cross hybrid	Cytoplasmic male sterility

Synthetic Varieties

At ICAR-IARI-Regional station Katrain (H.P.) a synthetic variety 'Pusa Synthetic' was made using 7 varities ('ARU Glory', 'Pride of India', 'India Market', 'Vermas Pride', 'Pride of Asia', 'Golden Acre', and 'Pusa Mukta').This synthetic cabbage has a head size index of 143.52 cm^2, a maturity period of 98 days, and yields 20.1 tonnes/ha.

Interspecific Hybridization

Somatic hybrids *B. oleracea* var. *capitata* x *B. oleracea* was developed for cold tolerance.

Self-Incompatibility and Uses

Self-incompatibility (SI) has been reported in cabbage which is most important mechanism for the production of hybrids. Self-incompatibility possesses strong alleles which are pre-requisite for production of hybrids. In 1951, Takii Seed Company was the first in the world to raise F_1 hybrid by applying self-incompatibility system. As a utility More *et al.* (1991) identified 'H-44' ('S_2S_2'x 'Cornell 83-6') and 'H-43' ('S_2S_2'x 'Pusa Mukta') to be heterotic for yield. For maintaining self-incompatibile line CO_2 treatment @ 3.5 per cent or sodium chloride @ 2-3 per cent is used.

Male Sterility and Use in Hybrid Development

Male sterility has been reported which may be used in the production of hybrid seeds. The male sterile lines were controlled by a recessive gene. Cytoplasmic male sterility (CMS) seems to be better option to develop hybrids than sporophytic self-incompatability in cabbage. The cytoplasm of Ogura male sterile radish has been successfully utilized in cabbage. The nucleus of cabbage was introduced in to the Ogura cytoplasm of radish by back crossing. The other two sources of CMS were from *Brassica nigra* (sterile) cytoplasm and sterile Anand cytoplasm from *Brassic rapa*. The hybrids developed CMS systems are now available for commercial cultivation in cabbage and cauliflowers. Hybrids were also developed by using cytoplasmic male sterile line ('EC 173419). Line 'A' (male sterile) and 'B' (*maintainer*) was imported from Cornell, the USA. On evaluation at ICAR-IARI, Reginal Station, Katrain (H.P.), and IARI, New Delhi, the hybrids 'H-11' ('Cornell 83-23' x ;Golden Acre') and 'H-46'('Cornell 83-23' x ;Golden Acre') were found promising.

Table 5.40: Carrot Hybrids Developed through Male Sterility

Name of the Hybrid	*Type of Male Sterility Used*	*Developing Organization*
KCH-5	Ogura Cytoplasmic male sterility	ICAR-IARI, Regional Station, Katrain (H.P.)

Chadha *et al.* (2016) developed 24 hybrid combinations using eight lines (three CMS and five SI) with three testers evaluated for various horticultural traits. The hybrid SI-4-6 x Glory-7 gave the highest yield (29.99 t/ha) and was statistically at par with Palam KGAT-II SCMS x Palam Glory-1 and III-1-1 x Palam KGAT-I (29.47 t/ha). The respective marketable head weight of these top three performing hybrids was 808.18g, 780.61g and 797.90g. The cross combination SI-4-6 x Glory-7 was significantly superior over all the four checks. The marketable head yield of the check H702 was 27.79 t/ha with 771.84g head weight, whereas that of H1101 was 28.46 t/ha with 671.06g head weight. Marketable head yield of KGMR was 23.73 t/ha, while it was 19.26 t/ha in Varun. The marketable head weight of KGMR and Varun was 637.06g and 550.0g, respectively. Thakur *et al.* (2016) evaluated progenies for their sterility during the flowering regimes *viz.*, 25-50 per cent, 50-75 per cent and 75-100 per cent. All four CMS progenies had the head shape index at par with their respective maintainer lines. For days to harvest, except KGAt-1 CMS during 1st year the CMS progenies were at par with their respective fertile counterparts as well. The CMS progenies were at par with their maintainer lines for heading (per cent), marketable heads per plot, non wrapper leaves and gross weight, except one or two. While, heads compactness, net head weight and marketable head yield (kg/plot), the exceptions were even 50 per cent of the CMS progenies studied. This may be attributed because cabbage is highly cross-pollinated crop and the later traits are subject to the influence of prevailing weather conditions. All CMS progenies revealed breakdown of sterility (very mild-mild pollen grain production) in variable number of plants in all the flowering regimes during experimental years. There exists a possibility of developing stable CMS progenies through rigorous selection over the years for stable F_1 hybrid development.

Resistance Source and its Utilization

(i) Disease Resistance Source and its Utilization

In cabbage several varieties have been evolved which has expressed resistance to parasitic and non-parasitic diseases. There are several varieties and hybrids which are high yielding and disease resistance but before searching and grouping of genotypes it is essential to understand the genetics of disease resistance.

Table 5.41: Genetics of Disease Resistance in Cabbage

Diseases	*Mode of Genetic Control*
Black rot (*Xanthomonas campestris*)	Dominant, monogenic or polygenic (Tewari *et al.*, 1979)
Downy mildew	One or two dominant genes
Black spot or curd blight	Dominant and polygenic
Cabbage yellows	Dominant and polygenic
White rot (*Sclerotinia sclerotiorum*)	Polygenecally controlled and recessive in nature (Sharma *et al.*, 1997)

Mode of Inheritance of Disease Resistance

Table 5.42: Mode of Inheritance of Disease Resistance

Disease	*Mode of Inheritance*
Tip burn	Dominant
Yellows	Type A resistance-Monogenic and dominant
	Type A resistance-Polygenic
Club root	Polygenic, recessive
Rhizoctonia disease	Monogenic and dominant
Powdery mildew	Monogenic and dominant
Black rot	Monogenic and recessive

Mechanisms of disease resistance have been worked out for many of the parasitic and non- parasitic diseases:

Table 5.43: Mechanisms of Disease Resistance

Disease	*Mechanisms of Disease Resistance*
Club root	High vitamin C content and low catalase and peroxidase activity
Downy mildew	Presence of highly volatile fungi toxic isotriocyanate
Yellows	Brown resin like plugs in vessels
Black rot	Large amount of mechanical tissue in the xylem

Cabbage cultivars 'Pusa Mukta' and 'Pusa Drumhead' both are resistant to black rot disease. ICAR-IARI, Reginal Station, Katrain (H.P.) has developed variety 'Selection 8' developed by cross combination 'EC10109' x 'EC 24855' has moderately resistant to black rot. Genotype 'Sanibel' has expressed resistance against *Fusarium* wilt, powdery mildew, *Rhizoctonia* rot and cabbage mosaic virus.

Table 5.44: Some Parasitic or Non-parasitic Pathogen Associated with Cabbage and Resistant Varieties

Name of Disease	Parasitic or Non-parasitic Pathogen Association	Resistant Varieties
Tip burn	*Non- parasitic*	'Wisconsin', 'Copenhagen', 'Bonaza
Downey mildew	*Peronospora parasitica*	'January King'
Yellows	*Fusarium oxysporum f. conglutinans*	'Wisconsin Ball Head', ' Improved Wisconsin Ball Head', ' Marion Market'
Club rot	*Plasmodiophora brassica*	'Miher', 'Late Moscow', ' White Russian'
Powdery mildew	*Mycosphaerella brassicola*	'Louisiana'
Black rot	*Xanthomonas campestris*	'Louisiana', 'CBR 6', 'CBR 7', 'CBR 9'
Mosaic	Several viruses	'Badger Ball Head'

Mutation

Mutant '19p-2' of variety 'Kjure 17' obtained through irradiating seeds with 60 kr gamma rays was semi-sterile. GMS as a spontaneous mutation has been reported in cabbage (Fang *et al.*, 1997).

Breeding for Quality Traits

Prakash *et al.* (2016) evaluated 60 F_1 hybrids (derived from 5 CMS lines and 12 testers) along with 3 standard checks (Vellore, Ajanta and KTCBH-81) and observed that CMS lines 6A and 9A, and testers Chhaki-2, KIRC-8 and 83-6-MR-1 were superior on the basis of mean performance and general combining ability for quality traits. The cross combinations 9A x KIRC-8, 9A x Chhaki-6 and 6A x Sel-5-83-6 for CUPRAC, 9A x 83-6-MR-1, 9A x KIRC-8 and 9A x C-121 for FRAP, 6A x Chhaki-2, 6A x 83-5-8 and 6A x CH-5 for total carotenoids, 6A x Sel-5, 6A x CH-6 and 6A x C-122 for total phenolics and 208A x Chhaki-2, 208A x Sel-5 and 208A x CH-5 for ascorbic acid were found to be the best on the basis of specific combining ability, heterotic potential and mean performance. The variance ratio (β^2g/β^2s) was found less than unity for total carotenoids and total phenolics indicating involvement of non-additive gene action which could be exploited through the development of hybrids. While preponderance of additive gene action (s^2g) were recorded for CUPRAC, FRAP and ascorbic acid. Hence, significant advancement could be achieved in the segregating generations using simple selection procedures or conventional breeding methods such as pedigree and bulk selection, which are useful for accumulation of desirable genes for these traits.

Resistance Source for Abiotic Stress

'KY' crosses and 'KK' crosses are a few of hybrids are early, high yielder and suitable for hot climates during 1950s. EC678826, a heat tolerance line of cabbage introduced from China.

REFERENCES

1. Allen, C.L. 1914. Cabbae, Cauliflower, and Allied Vegetable. Orange Judd Publishing, New York.
2. Chadha, S., Kumar, N., Sood, S., Chauhan, Y.S. and Kanwar, S. 2016. Performance of CMS and SI based cabbage hybrids in mid-Himalayas: New vistas in hybrid development. Abs. 55; 7th Indian Horticultural congress, Session1-: Breeding of Horticultural Crops, held at ICAR-IARI, New Delhi, p. 43
3. Choudhary, B. 1976. Vegetable, edn. 4th Eidition National Book Trust of India, New Delhi.
4. Fang, Z.Y., Sun, P.T., Liu, Y., Yang, Y., Wang, X., hou, A. and Bian, C. 1997. A male sterile line with dominant gene (MsMs) in cabbage (*Brassica oleracea* L.) and its utilization for hybrid seed production. *Euphytica*, **97:** 265-268.
5. Kakizaki, Y. and Kasai, T.1933. Bud pollination in cabbage and radish. *Journal of Heredity* **24:** 359.
6. More, T.A., Joshi, S. and Verma, T.S. 1991. Self incompatible cabbage hybrids for the plains of India. Golden Jubilee Symposium, Abstr.3. *Indian Society of Genetics and Plant Breeding*. New Delhi, p.607.
7. Nath, P., Srivastava, V.K., Dutta, O.P. and Swamy, K.R.M. 2008. Vegetable Crops:Improvement and Production. PNASF, Bengaluru, p. 126-130.
8. Prakash, C., Kumar, S., Thakur, N., Kumar, S., Kumar, A., Dey, S.S., Kumar, V., Dhiman, M.R. and Kumar, R. 2016. Genetics of quality traits in cabbage (*Brassica oleracea* L.) using CMS lines. (Abs. 28, Session II: Breeding of Horticultural Crops). 7th Indian Horticulture Congress, held at ICAR-IARI, New Delhi, 15-18th November 2016, p. 107.
9. Sankar, V. and Sonvane, P. 2016. Evaluation of red cabbage hybrids under hilly region of Coorg conditions. (Abs. 56, Session II: Breeding of Horticultural Crops). 7th Indian Horticulture

Congress, held at ICAR-IARI, New Delhi, 15-18th November 2016, p. 128-129.

10. Sharma, S.R., Gill, H.S. and Kapoor, K.S. 1997. Inheritance of resistance to white rot in cauliflower. *Indian J.Hort.* **54(1):** 86-90.
11. Swarup, V. 2006. Vegetable Science and Technology in India. Kalyani Publishers, New Delhi. p. 656.
12. Seshadri, V.S. and Chatterjee, S.S. 1996. The history and adaptation of some introduced vegetable crops in India. *Veg. Sci.*, **23(2):** 114-140.
13. Swarup, V. and Gill, H.S. 1964. The use of marker gene in hybrid seed production in cabbage. *Curr. Sci.*, **33(10):** 315.
14. Tewari, R.N., Chatterjee, S.S. and Swarup, V. 1979. Inheritance of resistance to black rot (*Xanthomonas campestris* Pam) Dowson, in cabbage. *Veg. Sci.*, 6(1): 27-36.
15. Thakur, P., Singh, S. and Vidyasagar. 2016. Evaluating male sterile progenies and maintainer lines of cabbage (*Brassica oleracea* var. *capitata* L.) for yield and horticultural traits. 7th Indian Horticulture Congress (Abs: 109, held at ICAR-IARI, New Delhi, 15-18 November, p. 84

Chinese Cabbage (*Brassica pekinensis* and *Brassica chinensis*)

INTRODUCTION

Chinese cabbage (*Brassica pekinensis* and *Brassica chinensis* 2n=2x=20) grow well in winter season, as a salad crop. It is an herbaceous annual and used as pot herb. It is generally consumed as fresh in salad or cooked as vegetable like other green vegetables (Singh and Singh, 2007). It can also be used as a substitute for lettuce and celery. Chinese cabbage is a good source of protein, vitamin, carbohydrate and minerals (Bose *et al.*, 1999). The 100 g of edible portion contain: water 95.0 g, protein 1.2 g, fat 0.1 g, carbohydrate 3.0 g, phosphorus 40 mg, potassium 25.3 mg, riboflavin 0.047 mg, vitamin C 24.0 mg, calcium 43 mg, niacin 0.6 mg, energy 145 K cal and vitamin A 150 I.U.

Origin and Distribution

Chinese cabbage is indigenous to China and eastern Asia where it has been in cultivation since 5th century. It is most important and popular crop cultivated in China since 5th century. It is grown in small scale in India (Singh and Singh, 2007) and USA where this crop is considered as new vegetable (Swiader *et al.*, 1993). Chinese cabbage is well distributed in Taiwan, Korea, Japan and Mongolia as well.

Species Availability and Diversity

There are two types of Chinese cabbage-one that heading type (*B. pekinensis*) and other open leaf type (*B. chinensis*). The former forms an erect, moderately compact, usually cylindrical heads, while the later develops clusters of succulent leaves with forming head. Rich diversity in plant type occurs in foot hills of Himalaya and north eastern states including Sikkim. There are two distinct species of this vegetable namely:

1. **Heading type Pet-sai (*Brassica pekinensis*) -** Resemble cos-lettuce but produces a much larger heads which elongated and compact.
2. **Leafy type Pak-choi (*Brassica chinensis*) –**This resemble Swiss chard in growth habit. The leaves are long dark green and oblong or oval and do not form a solid head. It is also called Chinese mustard. The midrib is broad and light in colour.

Botanical Description

The plant of Chinese cabbage is annul, sometimes a biennial and has few characteristics same or common as cabbage (Thamburaj and Singh, 2003).

Horticultural classification

It is classified as:

(a) Loose leaved variety (var. *dissluta* Li.)
(b) Semi-heading variety (var. *infraction* Li.)
(c) Fluffy topped heading variety (var. *laxa* Tse et Lee)
(d) Heading variety (var. *cephalata* Tse et Lee)

This variety is further classified into three type:

(i) Ovate type (*f. ovate* Li)
(ii) Flat topped type, (*f. depressa* Li) and
(iii) Cylindrical (*f. cylindrica* Li)

Chinese cabbage varieties have also been classified into three groups according to growing seasons (Singh and Nath, 2011).

(i) Spring varieties (January to April)
(ii) Tropical varieties (April to July)
(iii) Autumn or main season varieties (August to September)

Aumentation of Genetic Resources and Crop Improvement

In Taiwan Chinese cabbage is a semi-temperate crop and is a scarce commodity during summer time.

However, the release of AVRDC's Chinese cabbage variety 'ASVEG-1', locally known as Kiew Maysa or 'April Jade', made this vegetable at an affordable price during off season. This variety has been tropicalized, improved and tested for Thailand conditions by ARC and KU researchers. It can grow and form head during hot rainy season assuring farmers of the needed income. This variety gives an average yield of around 29 t/ha during the hot rainy season, around 39 t/ha on the rainy season and as high as 51 t/during the cool season.

In Nepal following varieties of Chinese cabbage are commonly being grown:

Table 5.45: Introduction of Exotic Genotypes

Varieties	*Source*
Tropic Emperor, WR-50 Days, WR-50 Days, King, Pack-Chay Green	Taki, Japan
Cannoball (No.40), Bluse, China Pride, Satoka (R-11), Spring Triumph, Michihilli Jade Pagoda	Sakata, Japan

In Korea Chinese cultivar 'Chunpakylgu' has enabled the spring production of quality produce. Similarly, 'Konaengjiyorum' has been identified for summer production.

Hybridization and Selection

Since the introduction of the first Chinese cabbage hybrids, 'Wonye-1' and 'Wonye-2' in 1960 vegetable breeding in Korea has advanced significantly.

Male Sterility in Chinese Cabbage

GMS as a spontaneous mutation has been reported in Chinese cabbage (Van der Meer, 1987). In China the lines were improved to keep 50 per cent MS plants and 50 per cent fertile plants at Institute of Vegetable and Flowers, China. This is so called "AB line". The propagation of "AB line" was conducted under isolated condiotion with bee pollination. One row of fertile plants remained as pollen donar. The other fertile plants in each four rows would be eliminated. The seeds from male sterile plants would be kept as "AB line" parents. F_1 hybrids would be collected from male syerile plants after elimination of fertile plants at female side (50 per cent). Generally speaking, 3-4 times of female parent with dense space of 20-25 cm should be between plants planted to cross with counterpart male parents with 40 cm space of plants. Recently, Institute of Vegetable and Flowers, China, developed a system to propagate 100 per cent of MS lines by tissue culture in order to improve efficiency of seed collection. Recently it is also reported that *'Sp'* gene (dominant) interacts with *ms* gene to control male sterility of Chinese cabbage. This system has been used widely in China.

Selection of *'Sp'* gene to interacts with *ms* gene in Chinese cabbage

SpSpmsms (sterile)	x	spspmsms (fertile)	
	⇩		
	⇩		
	Spspmsms (100 sterile)		
	⇩	x spspmsms (test cross)	
	Spspmsms :	spspmsms x ⟹	No segregation

Propagation of Male Sterile Line and F_1 Hybrid Collection

SpSpmsms (sterile)	x	SpSpMsms (fertile)		spspmsms (fertile)	
	⇩			⇩	
	⇩			⇩	
SpSpMsms (fertile) 50 per cent	SpSpmsms (sterile) 50 per cent		x ⇩		
			Spspmsms (sterile) 100 per cent	x	Male inbreeding lines
				⇩	
				F_1 hybrids	

REFERENCES

1. Bose, T.K., Som, M.G. and Kabir, J. 1999. Vegetable crops (Eds. Naya Prokash, Kolkata-6.)
2. Singh, D.N. and Singh, S.P. 2007. Production Management of Underutilized vegetables. Agrihortica Publication, Junagadh (Gujarat), India, pp. 71-80.
3. Singh, D.N. and Nath, V. 2011. Varieties and Hybrids of Vegetables. Satish Serial Publishing House, Azadpur, Delhi, India, p. 41-43.
4. Swiader, J.M., Ware, G.W. and Collum, M.C. 1993. Producing vegetable Crops. International Book Distributing Co., Charbagh, Lucknow, India.
5. Thamburaj, S. and Singh, N. 2003. Vegetables, Tuber Crops amnd Spices. (eds.) Indian Council of Agricultural Research, New Delhi.
6. Van der Meer, Q.P. 1987. Chromosomal monogeneic dominant male starility in Chinese cabbage (*Brassica rapa* subsp. *pekinensis* (Lour.) Hanelt). *Euphytica*, **36:** 927-931.

Broccoli (*Brassica oleracea* L. var. *italica* Plenck)

INTRODUCTION

Broccoli (*Brassica oleracea* L. var. *italica* Plenck; 2n=2x=18) is one of the important winter vegetable crop that has high nutritional and good commercial value (Yoldas *et al.*, 2008). Broccoli is an Italian word from Latin'Brachium' meaning an arm or branch (Choudhury, 1967). It can be easily distinguished from cauliflower by having a head composed of differentiated flower buds rather than a curd (Gray, 1982). Broccoli buds or mini cauliflower are cherished for its delicious taste, flavor and nutritive value. It is a rich source of dietary antioxidants and contains higher levels of nitrate (Heaton, 2001). Broccoli is much favored in big hotels and restaurants for its use in soups and salads. It is getting popularity worldwide since last few years due to its low fat content, low in calories, high vitamin C and good source of vitamin A, vitamin B2 and calcium (Decoteau, 2000). It also contains sulphoraphane, a compound associated with reducing risk of cancer (Kalia, 1995). It is reported to prevent cancer. In India also now broccoli has become a special vegetable crop in northern plains and also in hills of the country.

Origin, History and Distribution

Broccoli is an Italian word derived from the Latin word brachium, meaning an arm or branch and well written about its esblishment in 16th century herbalists by (Nieuwhof 1969; Bosewell, 1949). Broccoli is Mediterranean in origin and is spread all over the Europe and has been cultivated continually since 2500 BC and the ancestral forms of modern cultivars were probably selected about the time of Christ (Schery, 1972). Snogerup (1980) suggested that broccoli might have originated as an abnormality by crossing between early forms of branching shrub kale and other botanical varieties, followed by human selections. However, Song *et al.* (1990) supported monophyletic origin of cultivated morphotypes in *Brassica oleracea*. Broccoli made its appearance in thirties of 20th and 19th century (Verma and Sharma, 2000). United States of America is the largest producer of sprouting broccoli. Besides USA, the other major producers are Italy, Northern Europe and cooler regions of Far East (Sharma and Singh, 2001). It is a recent introduction to India therefore broccoli has not yet gained popularity among the vegetable growers till early nineties despite having nutritional importance and high demand for export market. Italy is a centre of diversification. The ancient Romans also knew this plant and called it 'Cyma'. Cultivation spread over to U.K., Japan, USA and Europe. In India its cultivation has been started recently around the big cities like Delhi, Mumbai, Kolkata, Chennai and Bengaluru. The prevailing climate of high altitude and high rainfall areas of Kodagu region of Western Ghats is ideal to grow broccoli.

Species Availability, Evolution and Diversity

Broccoli is originated from single progenitor wild cliff cabbage known as 'cole wart' (*Brassica oleracea* var. *sylvestiris*) as a result of mutations (Gates, 1953). The green forms called as 'calabrese'is mainly cultivated for frozen food.

Botanical Description

Table 5.46: Botanical Description

Growth Habit	The plant is generally regarded as an annual and a period of cold temperature, is not required to induce flower bud initiation. Although flowering is hastened by exposure to low temperatures. Plants resemble like cauliflower with grayish-green leaves and succulent edible stems, which support large, compact heads of thickly green clustered flower buds and thick fleshy flower stalk. There are different colours of head present in broccoli. Heads may be blue-green, purple-green or green (*Italica* group). Green Broccoli does not form a central head; instead it forms many slender, flowering shoots from a central stem.
Stem	Succulent edible short stems. Stalks are larger than those of cauliflower. The internode is long.
Leaf	A rosette of leaves in early developmental stage. However, sometime, there after, elongation of the internodes begins and at a certain point the terminal axis proliferates into numerous branches.

Head	The central growing point along with axillary bud cluster (terminal inflorescence) form thick fleshy flower stem which are the consuming parts. In Sprouting broccoli, length of the flower stem is less restricted and flower primordial develop into normal buds which are united in cluster, which is commonly called as head. Head is not covered with stem. Further, head colour may vary depending upon the varieties.
Floral biology	The infloescence is made of fully differentiated flower buds. Flower buds vary from white to green depending on cultivar. Broccoli flower has yellow colour with 4 petals, 4 sepals, 6 stamens, one pistil containing 2 carpels with 10-15 ovules in each carpel. Before initiation of flowering, the head becomes loose and flowering shoots starts elongation. The central axis of head forms main flowering stalk; while lateral stalk arises from bud cluster on the periphery of the main stalk. The head becomes loose quickly and flower bud starts opening earlier than cauliflower.
Anthesis and pollination	Flowering may start at low temperature, but pollination and seed setting will not be possible unless average daily temperature goes above 15°C (Verma and Sharma, 2000). Satisfactory pollination and seed setting may be achieved at a day temperature of 20°C. At high temperature, there can be pollen sterility and low seed setting. In broccoli 95 per cent cross pollination has been reported (Watts, 1968). Bumble bees (*Bumbus* spp.) and honey bees (*Apis indica*) are the main insect pollinators in the temperate region while *Apis dorsata* and *Apis florae* are the main pollinators in the plain (Shanmungavelu, 1989.)
Fruit and fruit set	After fertilization of ovules, seeds are formed inside the fruit. The fruit is known as siliqua.
Seeds	Asiatic broccoli, early variety of broccoli can produce seed even at high temperature *i.e.* in plains while in late varieties of broccoli, seed production is possible only on hills.

Botanical Classification

(I) Classification on the Basis of Colour

Table 5.47: Several Forms of Broccoli within the *italic* a Group

Crop Type	*Description*
Purple Sprouting broccoli	Over wintering, NW Eiropean biannual branched, purple spears, variable, early and late from overlapping.
Purple Cape broccoli	Over wintering, NW Eiropean biannual branched, purple head, variable.
White Sprouting broccoli	Over wintering, NW Eiropean biannual branched, white spears distinct early and late maturing form of which the late type more strictly with in the *botrytis* group (Gray, 1982)
Purple Sicillian broccoli	Single heading, pale purple heads, also known as in horticulture as 'purple cauliflower'
Couve broccoli Roxode Cobeca	Top purple hrading/sprouting type from Portugal.
Calabrase/Green types	Green sprouting broccoli from California region, now a worldwide crop and virtually single heading. These are early, medium and late type as per their maturity. The cultivars having solid firm head with small green buds and few buds and few lateral shoots are preferred. Many hybrid cultivars are available.
Black broccoli	Highly branch, with dark green spears and sickle shape leaves from Rome area.

There are two groups (Asiatic broccoli and European broccoli) of varieties in broccoli, depending on maturity time and adaptation.

(II) Classification on the Basis of Maturity

(a) **Early cultivars-**Cultivars are ready for harvest after 40-45 days of transplanting. De Cicco, Greenbud and Spartan Early are common cultivars (Honma, 1960). Early cultivars do not perform well under low temperatures. Waltham 29, Green Mountain and Coastal Atlantic are suitable for early and late planting.

(b) **Mid cultivars-**Some mid planting cultivar is 'Green Sprouting Medium'. Generally, biennial in nature and mature in 100 days.

(c) **Late cultivars-**Some late cultivar is 'Green Sprouting Late'. Generally, biennial in nature and matured in more than 100 days.

(II) Classification on the Basis of Sprouts Formation (season)

(a) Extra early-Southern Comet, Premium Crop, Clipper and Laser

(b) Mid season-Corsair, Excalibur, Cruiser, Emerald and Corona.

(c) Late season-Late Corona, Stiff, kayak, Green surf belongs to Calabrese group. These are superior in yield, uniform in size and maturity have gained popularity and commonly grown in developed countries.

Augmentation of Genetic Resources and Crop Improvement

Introduction and Selection

In Nepal following varieties of broccoli are commonly being grown:

Table 5.48: Introduction of Exotic Genotypes

Varieties	*Source*
Premium Crop, Green Comet, Characle	Taki, Japan
Packman, Pirate	USA

Hybridization and Selection

Sankar and Sonvane (2016) evaluated seven hybrids (Fiesta, Biocarve Broccoli, Dynasty, Inspiration, Olympic, Princes and bejo Broccoli) and reported that the mean plant height ranged from 53.5-73.5 cm and plant spread varied between 39.1-57.1 cm. The highest plant height (87.2 cm) was recorded in 'Inspiration' hybrid and the lowest plant height (53.5 cm) was noticed in 'Olympic' hybrid. With regards to early maturity, 'Dynasty' hybrid registered early head initiation (65.8 days) followed by early head maturity (77.5 days) compared to other hybrids evaluated. The mean flower head yield ranged from 8.4-19.9 t/ha. The highest flower head yield (19.9 t/ha) was observed in 'Fiesta' hybrid which was produced bright green with compact flower head.

REFERENCES

1. Bosewell, V.R.1949. Our vegetable travelers. Natural Geographical Magazine, 96: 134-217.
2. Choudhury, B. 1967. Vegetables. National Book Trust, India, New Delhi.
3. Decoteau, D.R. 2000. Vegetable crops. Upper Rever Company. New Jersey, U.S.A.
4. Gates, R.R. 1953. Wild cabbage and their effect of cultivation. *J. Genet.*, **51:** 363-372.
5. Gray, A.R. 1982. Taxonomy and evolution of broccoli (*Brassica oleracea* L. var. *italica* Plenk). *Econ. Bot.*, **36:** 397.
6. Honma, S. 1960. Spartan Early: A new variety of broccoli, *Quart. Bull. Mich. St. Univ. Agric. Expt. Stn.*, **42:** 470.
7. Kalia, P. 1995. Biochemical assays and yield performance of sprouting broccoli genotypes of Himachal Pradesh. *Cruciferae Newsletter*, **17:** 90-91.
8. Nieuwhof, M. 1969. Cole crops: Botany, Cultivation and Utilization, World Crops Book-Ser, Leonard Hill, Londan.
9. Sankar, V. and Sonvane, P. 2016. Evaluation of red cabbage hybrids under hilly region of Coorg conditions. (Abs. 57, Session II: Breeding of Horticultural Crops). 7th Indian Horticulture Congress, held at ICAR-IARI, New Delhi, 15-18th November 2016, p. 129-130.
10. Schery, R.W. 1972. Plants for man, p. 508-509.
11. Shanmungavelu, K.G. 1989. Seed production techniques and multiplication in vegetable crops. In Production Technology of Vegetable Crops. Oxford and IBH Publishing Co. Ltd. New Delhi, Bombay, Culcutta. p. 184-93.
12. Sharma, S.R. and Singh, R. 2001. Brussels Sprout, Sprouting Broccoli and Knol Khol. In: (Ed. Thamburaj and Singh). Text Book of Vegetables, Tuber crops and Spices. Directorate of Information and Publication of Agriculture, ICAR, New Delhi. p. 122-146.
13. Snogerup, S. 1980. *Brassica* crops and wild allies. (eds.) Tsunoda, S., Himda, K. and Gomezcampo, C., *Japan Sci. Soc.*, Press, Tokyo.
14. Song, K., Osborn, T.C. and Williams, P.H. 1990. *Brassica* taxonomy based on nuclear restriction fragment length polymorphisms (RFLPs), 3. Genome relationships in *Brassica* and related genera and the origin of *B. oleracea* and *B. rapa* (Syn. *campestris*), *Theor. Appl. Genet.*, 79: 497.
15. Verma, T.S. and Sharma, S.C. 2000. Producing seeds of biennial vegetables in temperate regions. Directorate of Information and Publication of Ariculture, ICAR, New Delhi, p. 26-50.
16. Verma, T.S. and Sharma, S.C. 2000. Production of Seeds of Biennial Vegetables. Directorate of Information and Publication of Agriculture, ICAR, New Delhi.
17. Watts, L.E. 1968. Natural cross-pollination and the identitication of hybrids between botanical varieties of *Brassica oleracea*. *Euphytica*, **17:** 74-80.
18. Yoldas, F., Ceylan, S., Yagmur, B. and mordogan, N. 2008. Effect of nitrogen fertilizer on yield, quality and nutrient content in broccoli. *J. Plant Nutr.*, **31:** 33-43.

Knol-khol or Kohlrabi (*B. oleracea* L. convar. *acephala* (DC) Alef. var. *gongylodes* L.)

INTRODUCTION

The term 'kohlrabi' is the German name for cabbage-turnip, resembles an above ground turnip (Choudhury, 1967). Knol khol or Kohlrabi or Ganth Gobhi (2n=18) is a short duration cool season vegetable and grown for its enlarged and fleshy stem. Basically tubers are swollen stems which develop above the ground may be due to secondary growth in thickness (Kalloo, 1993). The tuber is harvested and used for vegetable before it becomes tough or woody. It is liked very much by the people of Kashmir. After the partition of India, its seed production was mainly concentrated in Jammu and Kashmir. Now its seed production is successfully being conducted in Kullu Valley of Himachal Pradesh. Knol khol knobs are generally consumed as a cooked or boiled vegetable. Leaves are also used as potherb in Kashmir valley. The knobs are also sometimes used fresh in slads and the leaves in soupe.

Origin, History and Distribution

Knol-khol or Kohlrabi has originated in Mediterranean centre of crop origin, especially in the coastal countries of Western Europe, where this crop is commercially grown at large in Mediterranean region. Knol-khol in its present form appeared during 15th century in Nothern Europe and Germany in the 16th century (Boswell, 1949; Helm, 1963). It was introduced into the USA in the 18th century. In India it was introduced long back (Nath *et al.*, 1976) from England where it was being cultivated from 1850. According to Garade, the knol-khol was discussed in Century Herbals and stated that it was grown in Italy, Spain, and Germany (Helm, 1962). In general, it is grown in temperate, subtropical and tropical countries of the world where the mean temperature is around 20°C-25°C for a period of four monthes. Knol-khol is very popular crop in West Europe, North America, the Near East, Japan and India. In India it is a more popular crop in Himachal Prades, and Kashmir Valley and is also cultivated in selected pockets of West Bengal, Uttar Pradesh, Uttarakhand and Madhya Pradesh. It is also popular as a minor crop in southern India.

Species Availability, Evolution and Diversity

According to Encyclopedia Britannica, Knol-khol (cabbage-turnip) was first described in 18th century as another cabbage of Western European origin (Shanmugavelu, 1989; Rai and Yadav, 2005). Some scientists are in opinion that this crop is originated from the wild cliff cabbage known as 'cole wart' (*B. oleracea* var. *sylvestiris*) through introgression, mutation and selection, whereas, others are of the view that it came by the cross of cabbage and turnip (Singh *et al.*, 2004). Thompson (1976) opined that when the early cultivars spread to North Western Europe, the biennial types with winter hardiness and considerable cold vernalization requirement for flowering were selected besides less fibrous, thicker stems and more succulent storage organs.

Botanical Description

Knol-khol belongs to family Cruciferae and genus *Brassica*, *B. oleracea* L. convar. *acephala* (DC) Alef. var. *gongylodes* L. (syn. *B. caulorapa* Pasq.). Knol-khol is often called turnip rooted cabbage stem which is actually the swollen stem and is formed as a result of thickening of the stem tissue above the cotyledon called the knob (Bosewell, 1949 and Helm, 1963) to which are attached the leaves and are arranged in a compressed spiral on the bulbous part. Knob is green or violet and generally round to flat round in shape. The leaves often form tough inedible tissue. In young kholrabi seedling the stem above the epicotyls shows primary growth in thickness, while growth in length is inhibited. This results in the development of an aerial tuber, borne on a short stem which consists of hypocotyls and some un-thickened internodes. The secondary growth in thickness is restricted mainly to the un-thickened basal parts of the stem.

Table 5.49: Botanical Description

Growth Habit	Knol khol is a biennial, herbaceous vegetable crop grown as an annual which produces a swollen, bulb like stem. It is temperate crop which produces edible vegetable as an annual in the plains and in the hills. For the seed production it is biennial in the hills.
Stem	Knol khol had a swollen stem (globular to slightly flatten) which is edible (Nath *et al.*, 1976). It is also called as tuber or 'knob' which is arises from the thickening of stem tissue just above the ground.
Leaf	Glaucous leaves have long slender petioles and grow out of this stem arranged in a compressed spiral on the bulbous part.
Floral biology	Flower has 4 sepals, 4 petals, 6 stamens in tetradynamous condition (2 short and 4 long stamens) and a bicarpellary ovary which is superior and has a false septum. Ovules are attached on both the sides of septum. Two active nectarines are located between the bases of short stamens and ovary. Flowers are slightly protogynous.
Anthesis and pollination	The buds open under pressure of rapidly growing petals and become fully expanded in about 12 hours. Knol khol is naturally cross pollinated due to sporophytic self-incompatibility. Pollination is brought by bees and flies. Bud pollination has been successful for the production of selfed seeds.

Augmentation of Genetic Resources and Crop Improvement

Introduction and selection

As an introduction 'White Vienna' is a variety, which is early, with globular light green, smooth, tender and medium-sized knobs. 'Purple Viena' is another

Table 5.51: Key Feature

Sl.No.	*Group*	*Varieties/strains*	*Feature*
1.	European group	Wiesmoor Forcing White, Gaugels Forcing White	Quick growing (4-6 week of planting)
2.	Short duration and resistant to low temperature	Stuicka, Polni Ranar, Czekes, Blanr Speak, Weiver Blan, Prasake Bele, Rogglis, Soroksari, Knaufs Frushweis, Prasfer Weisser Treib, Trero, Primerera (Frydrych, 1971).	-
3.	Varieties suitable for summer cultivation	Blaso, Pollux, Reliktess Bleu,	Most suitable for summer cultivation in temperate regions
4.	Moderate resistant to *Fusarium* wilt	Vienna White-1350 (Kardaeva and Katarzhine, 1979)	
5.	Resistant to *Plasmodiophora brassicae*	K240 from Bulgaria	
6.	Resistant to *Xanthomonas campestris*	Szentesi Tartos Kek (Pesti and Fodor, 1982)	
7.	Cultivars resitant to cold and free from bolting	Wiesmoor White Forcing, Kuauf's Early White and Moravia (Luzuy, 1965)	
8.	Varieties with short foliage	Wiesmoor Forcing White and Purple variety Kuauf's Ideal (Hartrath, 1977)	

variety which is late (a week), the knobs are purplish and the flesh is greenish-white (Tewari, 1975).

Hybridization and Selection

The first hybrid was developed by the firm Roggli. These hybrids were Rowel, Roblau and Rosta (Anonymous, 1984). For high yield and earliness was reported in crosses 'Prag Trub' x '1176', 'Weisnoor Weisser Trub' x '1176', and 'Rogglis Trub' x '1176'. Crossing between early and late varieties yielded more productive F_1 hybrids. It has CMS with sterile Ogura cytoplasm from radish similar to cabbage/cauliflower which can be used in hybrid seed production.

Intergeneric Hybridization

Intergeneric crosees of radish with this crop shows high heterotic value but some crosses expresses complete sterility due to absent of reproductive structure or abnormal development of seedlings.

Self-Compatible and Self-incompatible

Self-compatible and Self-incompatible lines are available in kohlarabi.

Genetic Resource Conservation

The Regional Research Centre, Katrain, Kullu Valley (H.P.) of ICAR-IARI New Delhi is the only centre, where knol-khol research programme is being carriedout (Nath *et al.*, 2008).

REFERENCES

1. Anonymous, 1984. First hybrid Kohlrabi virotis from the firm Roggli, *Gemuse*, **20(4):** 152-154.
2. Bosewell, V.R.1949. Our vegetable travelers. *Natural Geographical Magazine*, **96:** 134-217.
3. Choudhury, B. 1967. Vegetables, National book Trust India, New Delhi.
4. Frydrych, J. 1971. Photosynthetic activity of diploid and tetraploid forms of *B. oleracea*, var. *gongylodes*, *Photosynthetische*, **5:** 38.
5. Hartrath, 1977. Kohlrabi under glass: varieties and spacing experiments. *Gemuse*, **13(1):** 8-10.
6. Helm, J. 1962. Die Lacimiaten Sipp-en Von *Brassica oleracea* L. Kulturplanze; **11:** 92-210.
7. Helm, J. 1963. Morphologish-taxonomische. G. tiederungder Kultursippen von *Brassica oleracea. Kulturplanzei*, **11:** 92-210.
8. Kalloo, G. 1993. Kohlrabi '*Brassica oleracea* var. *gongylodes*. In: Genetic Improvement of Vegetable Crops (eds. G. Kalloo and B.O. Bergh). Pergamon Press, New York.
9. Kardaeva, N.S. and Katarzhine, L.A. 1979. Resistance of *Brassica* varieties to *Fusarium* wilt, *Trudy Prikl. Bot. Genet.* Sel. **64:** 141.
10. Luzuy, J. 1965. Testing the collection of kohlrabi and the F_1 generation for bolting, *Rostlinna Vyroba*, **2:** 45.
11. Nath, P., Velayudhan, S. and Singh, D.P. 1976. Vegetables for the tropical region. ICAR, New Delhi.
12. Nath, P., Srivastava, V.K., Dutta, O.P. and Swamy, K.R.M. 2008.Vegetable Crops: Improvement and Production. PNASF, Bengaluru, p. 143-145.
13. Pesti, M and Fodor, L. 1982. Study of the susceptibility of *Brassica oleracea* crops to the pathogen *Xanthomonas campestris*, pv. *campestris. Novenyvedelem*, **22:** 537.
14. Peter, K.V. 1998. Knol khol. In: Genetics and Breeding of Vegetables. ICAR, New Delhi, p.57.
15. Rai, N. and Yadav, D.S. 2005. Knol khol. In: Advances in Vegetable Production. Research Co. Book Centre, New Delhi, p. 311.
16. Shanmugavelu, K.G. 1989. Production Technology of Vegetable Crops. Oxford and IBH Publishing Co. Pvt. Ltd., New Delhi.
17. Singh, N.P., Bhardwaj, A.K., Kumar, A. and Singh, K.H. 2004. Knol khol. In: Modern Technology on Vegetable Production. International Book Distributing Co. Charbagh, Lucknow, p.139.
18. Thompson, K.F.1976. Evolution of crop plants (Ed., Simmonds, and N.W.). Longman, London and New York, p. 19-52.
19. Tewari, R.S. 1975. Grow more vegetable this winter. *Indian Farmers Digest*, **8(a):** 83-57.

Brussels Sprouts (*Brassica oleracea* L. *gemmifera* DC.)

INTRODUCTION

Brussels sprout (*Brassica oleracea* L. *gemmifera* DC.) is a very delicious cold climate winter vegetable in temperate regions after cauliflower and cabbgae. According to Bosewell (1971), Brussels sprout produce tiny edible swollen cabbage heads/buds, about one inch in diameter that form in the axils of the leaves along the much-elongated main stem. It is known as sprouts or buttons or mini cabbage (Nieuwhof, 1969). In fresh form it is a very rich source of vitamin A, B and C along with P, S, Ca, Fe, I and Cu. It can be cooked as single or mixed vegetable with potato or other vegetables. Boiled sprouts can also be consumed after adding salt, butter and black pepper powder depending upon the taste (Sharma and Singh, 2001). Usually sprouts are eaten as raw, boiled or steamed, cooked and pickled. As a local preparation, sprouts are cut into two halves are fried with '*basan*' (gram flour) paste, give excellemnt '*pakors*'.

Nowdays, Brussels sprout is being canned and used as frozen product.

Origin, History and Distribution

The origin of this plant remains obscure and has been ascribed to savoy, cabbage and various forms of kale, *e.g.*, thousand-headed (Ockendon and Smith, 1993). The earliest known reference is to spruyten, mentioned in the market regulations of Belgium towns as early as 1213 (Helm, 1963). Whether these were Brussels sprout as we known them or just forms of decapitated cabbage is uncertain. In 1696 Zwinger described a many-headed cabbage which was probably Brussels sprout (Toxopaeus, 1979). Brussels is Mediterranean (coastal area) in origin and is spread all over the Europe and has been cultivated continually since 2500 BC. Brussels sprout was developed near Brussels in Belgium during 13th or 14th century (Shoemaker, 1953). According to Bosewell (1971) Brussels sprouts in its present form appeared around 1750 in Brussels (Belgium) as the name indicates. In 1800 it reached France and England. Brussels sprout was widely known in 19th century. Brussels sprout is a very popular crop in North-West Europe (late 18th and early 19th centuries) and California State of USA. It takes around 500 years to reach English gardeners for cultivation. It is widely grown in Australia, the Netherlands with small scale in Belgium and France. In India Brussels sprout has not yet gained popularity among the vegetable growers till early nineties despite having nutritional importance and high demand for export market. However, it is getting popularity worldwide since last few years and in India also it has become popular around big cities (Delhi, Mumbai, Kolkata, Chennai and Bengaluru *etc.*) to meet the demand of big hotels catering to the foreign tourists (Gill, 1993). It is delicacy with the foreigners. Now it is cultivated in Kodaikanal and Nilgiris of Tamil Nadu and Maharashtra.

Species Availability, Evolution and Diversity

Brussels sprout is originated from single wild species, *B. oleracea* var. *sylvestris*, commonly known as Cole Wart, through single gene mutation.

Botanical Description

Brussels sprout is biennial in nature *i.e.* it requires two growing seasons for flowering. In first phase, vegetative growth/juvenile phase is completed. After that it enters into the reproductive phase but this is possible only when plants are exposed to low temperature of 2.2 to 10°C for 6 to 8 weeks, called vernalization. The plants are able to receive vernalization treatment only after formation of about 30 petiolated leaves (Nieuwhof, 1969). After low temperature treatment at this stage, floral bud initials are laid down, leaf primordial becomes progressively smaller and thereafter floral prinordia arise (Verma and Sharma, 2000). The plant bolt and flower in spring when temperature starts rising. The flower stalk arises from the terminal growing points and sometimes from upper sprouts also.

Table 5.52: Botanical Description

Growth Habit	In the seedling stage and early development, it closely resemble to common cabbage and rises two to three feet tall. Plants are tall biennial erect with good leaf protection. Late the main stem generally attains normal longitudinal growth 60-90 cm.
Root	
Stem	The plants of Brussels sprout have an erect and central stem which is unbranched. The terminal portion of plants becomes globular in shape and stem become thickened.
Leaf	The buds in the axils of usually petiolated leaves develop in to miniature heads of few centimeters in diameter and it is the buds which are called sprouts resemble the cabbage head but they are much smaller size.
Floral biology	A number of little "sprouts" that resembles miniature cabbage are densely packed along it, between the petioles of the leaves. If these sprouts left on the plant they would eventually develop into flowering shoots. The first flowering stalk develops from terminal buds and then other flowering stalks develop from axillary buds *i.e.* sprouts. Hence, to produce sufficient number of flowering stalks, upper sprouts should not be harvested. Flowering in Brussels sprout takes place in April-May but when temperature goes above 25°C, seed setting is affected due to pollen sterility and mortality. Flowers are yellow in colour with 4 petals, 4 sepals, 6 stamens, one pistil containing two carpels.
Anthesis and pollination	In Brussels sprout 72 per cent cross pollination has been reported (Watts, 1968). It has saprophytic self-incompatibility system. Bumble bees (*Bumbus* spp.) and honey bees (*Apis indica*) are the main insect pollinators in the temperate region while *Apis dorsata* and *Apis florae* are the main pollinators in the plain (Shanmungavelu, 1989.)
Head	The axillary buds along the stem develop into small heads/sprouts similar to head cabbage, but only 2.5-4.0 cm or less in diameter (Purseglove, 1974).
Seeds	After fertilization of ovules, seeds are formed inside the fruit. The fruit is known as siliqua.

According to Whitehead (1966), Brussels sprouts differ from kale in that the auxillary bud remain practically stemless and form balls or buttons. Bougoetuge (1943) reported that it looks like a miniature cabbage and grows to 60.0-90.0 cm tall.

Horticultural Classification

Brussels sprouts may be classified as:

(I) Classification on the Basis of Plant Height (Sharma and Singh, 2001)

(a) Dwarf Cultivars

They have short stem (less than 50 cm in length) and sprouts are small and crowded on the stem. These are mostly early cropper and most suitable for the areas where growing season is short. Common cultivars are

'Improved Island', 'Early Morn', 'Dwarf Improved', 'Frontier Zwerg' and Kvik.

(b) Tall cultivars

Tall cultivars are grown in areas where growing season is longer especially in England, other parts of Europe and North Indian hills. Strains of cultivar 'Eveshan Rapid' (Sweden) and Wilhelmsburg (Germany) are early in maturity but do not withstand hard winters.

(II) Classification on the Basis of Winter Tolerance

(a) **Susceptible cultivars-** Eveshan Rapid' (Sweden) and Wilhelmsburg

(b) **Hardier and late cultivars-**Red Vein, Hilds Ideal, De Rosny Polarstjernen and Amager.

Augmentation of Genetic Resource and Utilization in Crop Improvement

The first reference to the breeding of Brussels Sprout was probably by Van Mons (1820).

High yielding dwarf strains are Bollo, Cambridge, Cambridge No.-1, Cambridge No.-2, Cambridge No.-3, Cambridge No.-4, Cambridge No.-5, Dalington Early, Favourite, Goliath, Fill Basket, Ovmskirth Giant, Rouslenchlate and Wroxton. In Nepal Brussels sprout cultivars like BSS-74 and BSS-75 introduced by Bejo Sheetal Seed Co., India are quite common in cultivation.

Hybridization

In India very less research works related to hybridization is being carried out and most of the hybrids are being marketed by seed companies of Japan and Europe. For example Jade Cross, Avion Cross (1st hybrids by National vegetable Research Station, Wellesbbourne). Other important hybrids are Doreman, Alkazar, Merlon, Poster, Rampart, Fortress, Predora, Rovoka, Sonara, Ladosa, Pearl Crystal, Oliver, Valiant, Rogor, Rider, Boxer, Richard Royal Marvel, Asgard, Kundry, Rasmunda and Stephen.

(I) Insect Resistance

(a) Aphid resistant- Aphid resistant varieties are Uremo Inter, Irish Elegans and Roodnef

Table 5.53: Performance of some Varieties of Brussels Sprouts grown in Himachal Pradesh

Variety	*Days from Yransplant to Harvest*	*Plant Height (cm)*	*Sprout Size (cm)*	*Sprout Type*	*Yield/ 3 m Lines (kg)*
Hilds Ideals	123	107.50	3.6 x 3.6	Firm	1.113
Express	105	99.50	4.6 x 4.0	Firm	1.225
Fill Basket	120	100.00	3.8 x 3.7	Tightly packed	1.000
Rubine	170	150.00	3.4 x 3.5	Firm	0.600
Reger	103	53.50	3.2 x 3.3	Lightly packed	0.950
Titurel	100	36.30	3.7 x 3.4	Lightly packed	0.950
Carvalient	136	62.60	4.2 x 3.6	loose	1.153
Oliver	120	80.41	4.6 x 3.7	Uniform	0.900
Robert	123	93.60	4.4 x 4.3	Firm	0.762
BBs-74	80	65.00	2.5 x 3.0	Uniform	0.500

Source: Peshin (1998).

Table 5.54: Special Characteristics of some Important Hybrids

Sl.No.		*Name of Hybrids*	*Special Character*	*References*
1.		King Arthur	Suited to mechanical harvesting, best cropper	Wood, 1970; Nuyten *et al.*, 1973
2.		Peer Gynt	Good yielder	Thompson *et al.*, 1973
3.	i.	Royal Cross	Frost resistant	Meduldcramans, 1971
	ii.	Uremo Post	-	
	iii.	Top Grade No. 6977		
4.	i.	Topscore Dutch	Best yielder	Woykee *et al.*, 1972
	ii.	Trigomox Hybrid		
	iii.	Berner Market Swiss Hybrid		
	iv.	Jade Cross Japan Hybrid		
5.		Lanceolat, Orion	Outstanding culinary qualities	*Berrera et al.*, 1975
6.	i.	Craton	Early to mid-early	Anonymous, 1976
	ii.	Merton	Tall, large, firm sprouts, mid-early, fine hard round sprouts	Anonymous, 1976

Source: Shanmugavelu, 1989.

(II) Disease Resistance

(a) Powdery Mildew Resistant

Bin Dee, Engard 2077, Frigo Star, Groeningse Late, Seven Hills, Solid Line, Triumph are resistant to powdery mildew.

REFERENCES

1. Gill, H.S. 1993. Improvement of Cole Crops. In: Advances in Horticulture Vol. 5- Vegetable crops. Part-1 (eds. Chadha, K.L. and Kalloo, G.). Malhotra Publishing House, New Delhi, India.
2. Helm, J. 1963. Morphologish-taxonomische. G. tiederungder Kultursippen von *Brassica oleracea. Kulturplanzei*, **11:** 92.
3. Nieuwhof, M. 1969. Cole crops: Botany, Cultivation and Utilization World Crops Book-Ser Leonard Hill, London.
4. Ockendon, D.J. and Smith, B.M. 1993. Brussels sprouts *Brassica oleracea* var. *gemmifera* DC. In: Genetic Improvement of Vegetable Crops. (Eds: G. Kalloo and B.O. Bergh), Pergamon Press, New York.
5. Peshin, S.N. 1998. Promotion of Rare Exotic Vegetables. In: Production and Post Harvest Handling of Vegetables. Centre of Advance Studies, Horticulture-Vegetable (ICAR), Dept. of Veg. Crop. Dr. Y.S.P.U.H.F., Nauni, Solan, H.P., p. 80-87.
6. Purseglove, J.W. 1974. Tropical crops: Dicotledons. ELBS, Longman, London.
7. Verma, T.S. and Sharma, S.C. 2000. *Production of Seeds of Biennials Vegetables*. Directorate of Information and Publication of Agriculture, ICAR, New Delhi.
8. Shanmungavelu, K.G. 1989. Seed production techniques and multiplication in vegetable crops. In Production technology of vegetable crops. Oxford and IBH Publishing Co. Ltd. New Delhi, Bombay, Culcutta. p. 184-93.
9. Shanmungavelu, K.G. 1989. Production Technology of vegetable crops. Oxford and IBH Publishing Co. Pvt. Ltd., 66, Janpath, New Delhi.
10. Sharma, S.R. and Singh, R. 2001. Brussels Sprout, Sprouting Broccoli and Knol Khol. In: (Ed. Thamburaj and Singh). Text Book of Vegetables, Tuber crops and Spices. Directorate of Information and Publication of Agriculture, ICAR, New Delhi. p. 122-146.
11. Toxopaeus, H. 1979. The domestication of brassica crops in Europe. Proc. Eucarpia Meeting-Cruciferous crops (eds. N.P.A. Van Marrewijk and H.Toxopaeus). Wagemingen, 29.
12. Van Mons, J.B. 1820. Substance of a memoir on the cultivation and variation of Brussels Sprout, *Trans. Hort. Soc.* London, **3:** 197.
13. Watts, L.E. 1968. Natural cross pollination and the identification of hybrids between botanical varieties of *Brassica oleracea* L. *Euphytica*, 17: 74-80.

Kale (*Brassica oleracea* var. *costata* or *Brassica oleracea* var. *medullosa* L.)

INTRODUCTION

Though kale or Karam-sag (*Brassica oleracea* var. *costata* or *Brassica oleracea* var. *medullosa* L.; 2n=18) is used as vegetable but certain strains, the coloured foliage of several households' produces a highly decorative effect and the strain exhibiting these features are used for ornamental garden effect. The leaves of kale are cooked similar to cabbage (single or mixed with other vegetables) greens or in stir fries (Thompson and Kelly, 1981). Kale is used as winter vegetable when other vegetables scare. There is also a wide range in the height of several forms. In Kashmir kale is known as 'Hak'. The Sanskrit nomenclature for all edible green leaves is 'Shaka'. This Sanskrit nomenclature with the passage of time has become 'Sag' in some parts of the country and 'Hak' in Kashmir. Hak (kale) is the oldest cultivated vegetable in Kashmir. Kale is a preferred leafy vegetable for which every Kasmiri has a weakness. This vegetable is available for consumption throughout the year in Kasmir valley because of successive cropping, cooked almost everyday exclusively or mixed with other vegetables or mutton. Kale is not of much commercial importance in India except in northern states particularly J&K. Most plants sold as "Ornamental cabbage" are in fact kales. "Ornamental Kale" is as edible as any other variety, provided it has not been treated with pesticides or other harmful chemicals. When uncooked, standard kale is a popular garnish. Nutritionally it is very rich source of carotene, calcium and potassium. Per 100.0 g of edible portion contains 83.0 moisture, 6.0 g protein, 0.8 g fat, 9.0 g carbohydrates, 53 K. calorie, 249.0 mg calcium, 93.0 mg phosphorus, 75.0 mg sodium, 378 mg potassium, 10,000 IU Vit. A, 0.16 mg thiamine, 0.26 mg riboflavin, 2.1 mg niacin and 186.0 mg ascorbic acid (Watt and Meril, 1963).

Origin, History and Distribution

The origin place of kale is probably southern and western Europe (Knott, 1988). Word 'Kale' is drived from Scotish, which originated from 'Coles' or 'caulis' and largely used as cabbage by Greek and Romans. Kale is a descendant of the wild cabbage which still grows along the coasts of North Africa and Europe. Kale spread in other countries through seamens (sailors). They were the first *Brassica* to be cultivated. It is thought to have originated in Asia Minor and have been brought to Europe around 600 B.C. by group of Celtic wanderers. About kale, Catto described well in 200 BC (Decoteau, 2000). Curly kale played an important role in early European food ways having been a significant crop during ancient Roman times and a popular vegetable eaten by peasants in the Middle Ages. English settlers brought kale to the United States in the 17th century. It is an important vegetable crop in home garden in many areas in the Southern half of the United States and a crop of commercial importance in Virginia (Thompson and Kelley, 1979). Although originated in Asia and the Mediterranean, it has been the most important cuisine of north Europe. 'Dinosaur Kale' was discovered in Italy in the late 19th century. 'Ornamental Kale' originally a decorative garden plant was first cultivated commercially in the 1980s in California. Curly leaved varieties already existed along with flat leafed varieties in Greece, in the fourth century B.C. These forms were referred to by the Romans as Siberian kale and are considered to be the ancestors of modern kales. In India (introduced during 19th century), kale has not been grown as a vegetable crop for commercial use (Anon., 1980). However, it is commercially grown in Jammu and Kashmir and to a limited extent in Himachal Pradesh, Assam, Punjab, Haryana and Nilgiri hills of South India (Choudhary, 1977).

Species Availability, Evolution and Diversity

Non-heading type of kale, perhaps var. *costata*, the Portugese kale or the var. *medullosa* are found in the Mediterranean centre of crop origin (Peter, 1998). Kale is originated from single progenitor wild cliff cabbage known as 'cole wart' (*B. oleracea* var. *sylvestiris*) as a result of mutations. There are different types of kales classified as sub-varieties of *B. oleracea* var. *acephala* L. most of which bear a rosette of leaves at the top of the stem. These are curly kales having finely curled foliage (sub var. *plana*) thousand head kales with sprouted auxiliary buds forming loose leafy rosette (sub var. *millicapitata*), tree kales have more plant height up to 2.0 m (sub var.*palmifolia*) and narrow stem kale which is grown for human consumption and is one of the earliest forms of cole crops. Significant genetic variability for all yield/quality contributing traits is present in Kashmir (Khan *et al.*, 2008)

Botanical Description

Kale (*B. oleracea* var. *acephala* L.) belong to family Brassicaceae with chromosome number 2n=18. Many varieties of kale are referred to as "flowering kales" and are grown mainly for their ornamental leaves, which are brilliant white, red, pink, lavender, blue or violet in the interior or the rosette. It is true that kale doesnot bear heads like cabbage and leaves are curled.

Horticultral Classification

Kale is generally grouped as:

(i) Dwarf - Generally, plants reach to the height of 40.0 cm.

(ii) Medium tall- Generally, plants reach to the height of 40.0-80.0 cm.

Table 5.55: Botanical Description

Growth Habit	In kale, the plant growth habit has been noted as dwarf, medium tall and tall. It is biennial but cultivated as an annual. Young shoots at the top of the stem which resemble cabbage prior to heading (Nick and Bradley, 1999)
Root	Rosette of leaves.
Stem	Stems erect, undivided and rough which attains the height of 30.0-100.0 cm but tree kale can reach up to 200.0 cm., which is generally grown for decorative purpose. Thousand headed kale developed numerous shoots from the ground and generally used for fodder purpose.
Leaf	Leaves of kale are green, purple or brown, curled, smooth with an open type growth habit while of collards are smooth with upright growth habit. Large number of leaves emerge from the top the plant. Leaves of Veins of Portugalis leaves are very succulent in nature and generally grown for vegetable prrpose.
Floral biology	The flowers are hermaphrodite.
Anthesis and pollination	Bud pollination has been successful for the production of selfed seeds. There are varietal differences in the degree of self-incompatibility. Yousuf (1985) reported that anthesis commenced from 6 am and continued up to 4 pm with peak between 9 am to 10 am under temperature conditions of Solan (Himachal Pradesh) in cultivar Khanyari. Dehiscence occurred about 30 minutes after anthesis and continued up to 4 pm. Fresh pollen viability was 97.72 per cent and remained as high as 66.35 per cent even after seven days under cool storage (4°C) while under room temperature got reduced to 36 per cent on the 7th day of storage. Stigma receptivity was 3 days before and remained so 4 days after, with the peak on a day before and a day after anthesis Yousuf (1985)
Fruit and fruit set	Fruit is siliqua

(iii) Tall- Generally, plants develop above the 80.0 cm. height. Cultivars of this group shed their leaves quickely which lead to poor yield.

Augmentation and Utilization of Genetic Resources (Abiotic and Biotic resistance, Quality, Graftings)

Introduction and Selection

Table 5.56: Varietal Grouping, Cultivars and Descriptions

Group	*Cultivar*	*Description*
Dwarf varieties	Dwarf Green Curled Scotch, Dwarf Moss Curled	Generally, plants reach to the height of 40.0 cm.
	Dwarf Blue Scothch	Generally, plants reach to the height of 40.0 cm.
Medium tall	Moss Curled, Fiambergen Market	Generally, plants reach to the height of 40.0-80.0 cm.
Tall varieties	Scotish and Siberian varieties	Generally, plants develop above the 80.0 cm. height. Cultivars of this group shed their leaves quickely which lead to poor yield.

Cold Adoption

In general, Scotish and Siberian types are very suitable for cold regions.

Hybridization

Since, Kale has well adaptation to cold climate (cold tolerant gene) therefore cold tolerant cultivars of cole crops by utilizing kale, are being laegely developed.

Genetic Resource Conservation

Genetic resources of kale are largely collected and being conserved in European countries (Jansen *et al.*, 1993)

REFERENCES

1. Anonymous. 1980. The Wealth of India: Raw materials and Industrial products. Vol.3. *Council of Scientific and Industrial Research*, New Delhi.
2. Choudhary, B. 1977. Leafy vegetables. In: Vegetable National Book Trust of India, New Delhi. p. 193-94.
3. Decoteau, D.R. 2000. Vegetable crops. Prentice-Hall Inc. Upper Saddle River, New Jersey.
4. Jansen, P.C.M., Siemonsma, I.S. and Narciso, J.O. 1993. *Brassica oleracea* L. In: PROSEA. Plant Resources of South EOGT Asia (Eds.).
5. Khan, S.H., Ahmed, N., Jabeen, N., Chattoo, M.A. and Narayan, S. 2008. Evaluation of various Kale (*Brassica oleracea* var. *acephala* L.) genotypes for yield and attributing traits over environments. *Environment and Ecology* **26(3):** 1141-1145.
6. Knott, J.E. 1988. Handbook of Vegetable Growing. John Wiley and Sons, New York.
7. Nick, J.M.A. and Bradley, F.M. 1999. Growing fruits and vegetable organically, Rodale Press, Emmaus, Pennsylvania.
8. Thompson, H.C. and Kelley, W.C. 1979. Potherbs or Greens. In: Vegetable crops, Tata M. C. Graw Hill Publishing Ltd. New Delhi. pp. 214-235.
9. Thompson, H.C. and Kelley, W.C. 1981. In: Vegetable crops, Tata McGraw Hill Publishing Ltd. New Delhi.
10. Yousuf, M. 1985. Seed setting studies in Kale (*Brassica oleracea* var. *acephala* DC). Ph.D Thesis.

Collard (*Brassica oleracea* var. *viridis* L. or *Brassica oleracea* var. *acephala* L.)

INTRODUCTION

Collard (*Brassica oleracea* L. var. *acephala*, 2n= 18) is not of much commercial importance in India except in northern states particularly J&K. The term *acephala* means without head. In Kashmir, Collard is known as 'Hak'. Group of tender leaves which born on stem, are used as vegetable. It is a very popular vegetable of South America. Nutritionally collard is very important vegetable. It has very less calorie but very rich in carotene and vitamins. Alongwith several mineral matters, it has highest amount of calcium. Per 100 g of edible portion of collard contains 85.0 g moisture, 4.8 g protein, 0.8 g fat, 7.5 g carbohydrates, 45 K. calorie, 250.0 mg calcium, 82.0 mg phosphorus, 1.5 mg iron, 450.0 mg potassium, 9300 IU vit. A, 0.16 mg thiamine, 0.31 mg riboflavin, 1.7 mg niacin, and 152.0 mg ascorbic acid.

Origin, History and Distribution

Probanly, collard is originated from eastern Mediterranean region. As per historical records, its cultivation was done firstly by Greek and Roaman peoples. It was brought to Britain or France by roman or Seltes (people of Europe and Asia Minor). So, its domestication to Birtish is around 400 BC while in America collard was first introduced in the year 1669 (Decoteau, 2000). Collard is the major crop of South America where it is available from beginning of December to April in Florida, South Carolina, Georzia, Verginia and Albama. Other countries have not paid much attention for its domestication. In India its cultivation is limited to Himachal Pradesh, J&K or adjacent area.

Species Availability, Evolution and Diversity

Kale and collard both are very nearer relatives to wild cabbage. Generally, it is believed that collard developed between the cross of without head forming and head forming cabbage, as it expresses intermediae form. Some researchers felt that it may be primitive form of cabbage.

Botanical Description

Collard (*B. oleracea* var. *viridis* L. or var. *acephala*) belongs to family Brassicaceae. Like kale, it also does not form head but 'Morris Heading' forms the head like cabbage. It forms group of tender leaves on stem. It is very difficult to identify cabbage and collard during early stage of growth. Earlier, some botanists put the same botanical name for kale and collard (*B. oleracea* var. *acephala*) but latter on both were separated due to certain variation in morphology. Leaves are very much similar (shape and size) to cabbage but veins on leaf are very prominent. The flowers are hermaphrodite.

Augmentation of Genetic Resources and Crop Improvement

Introduction and Selection

See Table 5.57.

Table 5.57: Cultivars and Descriptions

Genotypes	*Description*
Georgia	'Georgia' is a bunching type cultivar and commonly grown. Plants attain the heights of 75-90 cm. generally leaves are bluish green and overlap to each other. Harvesting can be done 80 days after transplanting.
Vates	'Vates' can tolerate to low temperature and bolting process is very slow. Plants remain green for longer period. Plants are erect and attain the heights of 50-60 cm.
Morris Heading	'Morris Heading' can tolerate to low temperature and bolting process is very slow. Plants attain the heights of 75-85 cm. leaves are broader and medium green. Stem is short. It forms cabbage like head.
Lousiana	It has been developed from Lousiana State University. It is very much similar to Georgia cultivar. Leaves are thick and very tender. First harvesting can be done 75 days after transplanting.

Adaptation to Varying Climates

Collard well adapt to varying climates (high and low temperature).

REFERENCE

1. Decoteau, D.R. 2000. Vegetable crops. Prentice-Hall Inc. Upper Saddle River, New Jersey.

6

Genetic Resources of Root Vegetables and Crop Improvement

INTRODUCTION

Root vegetables are underground plant parts eaten by humans as food. Root vegetables include a number of crops which are grown for their enlarged, edible fleshy storage root. In general it can be narrated as storage organs that are enlarged to store energy in the form of carbohydrates and other important nutrients. With respect to natural adaptation, they are hardy, cool season crops with a long storage life. The major root vegetables grown commercially are grouped as:

A. Brassicaceae (Crucifer) - Radish (*Raphanus sativus*), turnip (*Brassica rapa* subsp. *rapa*), rutabaga (*B. napus* subsp. *napobrassica*) and horse radish (*Armoracia rusticana*).

B. Apiaceae (Carrot family)–Carrot (*Daucus carota*), Parsnip (*Pastinaca sativa*).

C. Chenopodiaceae (Goosefoot family)–Beet root (*Beta vulgaris*).

D. Asteraceae (Sunflower family)–Jerusalem artichoke (*Helianthus tuberosus*), salsify (*Tragopogon porrifolius*).

Although they belong to several unrelated plants and very different botanically but these crops share similar cultural requirements such as weather, soil type and cultural practices.

Genus *Daucus*

Carrot (*Daucus carota*)

INTRODUCTION

Carrot (*Daucus carota* Linn., 2n=18) is a popular cool weather root crop grown all over the world for its fleshy edible tap root which is known for excellent taste, flavor and nutritive value. It is among the top ten vegetable crops and serves as major source of carotenoids. Carrot is also a rich source of antioxidants like carotenoids, lycopene, phenolics and Vitamin C. Carrot is grown in spring, summer and autumn season in temperate climate and during winter in tropical and subtropical regions. Carrot has numerous categories of roots varying in colour, shape, size, and flavor. The root colour varied from white to black colour; but orange, red, black and yellow carrots are most popular and of commercial importance worldwide across all the climates ranging from temperate to tropical conditions. Mostly orange roots are used for table and frozen purposes; red for table, juice, halwa and pickle making; and black carrots for colour extraction and preparation of beverage called 'kanji' which is supposed to be good appetizer. Carrots are cooked alone or with other vegetables in the preparation of curries, soups and stews. Fresh grated carrot roots are used in salad and tender roots are pickled. It is very popular for *Halwa*-a kind of sweet dish (using *Khowa*-milk products because carotene dissolved in fat) and cooked with mixed vegetables. It is also preserved by salting, pickling, canning and drying. Carrots are rich sources of Ü-and β-carotene (orange carrot), lycopene (red carrot), anthocyanin (black/ purple carrot) and xanthophylls (yellow carrot). They are also good sources of carbohydrates, minerals and dietary fiber and are becoming more popular because of the neutracitical values, colour characteristics, health benefits as well as antioxidant capacity. Carrot juice is a fresh source of β-carotene, a precursor of vitamin A. it is canned and may also be deliberated. Carrots are valued as feed (roots and top) for horses and dairy cattle. It is believed that early use of the plant was medicinal, prescribed for curing stomach problems and treating wounds, ulcers and liver and kidney ailments. The seeds contain an essential oil which is used for flavouring and perfumery. Scibrja and Zemoja (1966) were of the opinion that the carotene content was higher in varieties with orange colour.

Origin, History and Distribution

Carrot is derived from the wild carrot, subsp, *carota*, which occur natural in Europe, Asia and Africa. Carrots probably originated in Afghanistan and around Northwest India. First record indicates that its cultivation for medicinal purposes began 2000 to 3000 years ago. Cultivation of roots for consumption dates back to 6th century when purple rooted was grown in the area of Afghanistan. The black and yellow types were eventually evolved and produced in Afghanistan, Turkey and Middle East areas during 10th century. Bailey (1948) reported that carrot is a native of Europe and was found as a weed in America. They were introduced into England during the 15th century in the Royal Gardens, Asia and North America and possibly North and South America (Thompson and Kelly, 1957). As per Encyclopaedia Britanica, carrot is believed to be of middle Asian origin 'Afghanistan', North West India, Iran and Turkey (primary centre of origin) and they were cultivated in the middle east and the Mediterranean lands even before the Christian era. However, Punjab and Kashmir are also considered its primary centres of origin with secondary centre in Ethiopia and North America. The first root consumed, probably purple in colour which was large and rough, was grown in Afghanistan around 600 A.D. Eventually, yellow types were isolated and grown in Syria in the 9th or 10th century. The crop was introduced to China, France and Germany in the late 13th century and to Europe many years later. White and orange types were evolved in the Europe and Middle East during 17th century. Moreover, red carrots are evolved in India, Japan and China during 18th century. Carrots are believed to have been introduced into India from Persia. Himachal Hills and Kashmir valley still abounds in its wild forms where wild animal, especially brown bear feed on its roots. In some parts of Kashmir people still eat the wild carrots (Yawalkar, 1980). In north India, highly coloured types of carrot are met which are not available in Europe. Types varying from absolute colourless to light lemon, light orange, orange and deep orange, light purple, deep purple and almost black are available. This is another indication of its being a primary centre of origin. It is known to have been cultivated in Afghanistan for about 5000 years. Purple carrot found in Egypt date back to 2000 BC. Subsequently, carrots spread to the Mediterranean, Asia and Africa. Morish invaders brought carrot from North America to Southern Europe. By 13th century, carrots were grown in China (13th -14th century), Japan (17th century), Germany and France and in England by 15th century. Encyclopaedia American pronounces that carrot is a native of Europe and Asia but wild types were found in 'New Found Land' as weeds. Bailey (1928) reported that carrot is a native of Europe and was found as a weed in America in 10th century. Orange type carrots first appeared in the Netherlands in 16th century. Carrot was also a favourite vegetable in England in the 16th century and was taken to Virginia in 1609. It was introduced into the USA about 250 years ago and Canada about 150 years later. It has since spread throughout North America. This crop was introduced into India early in its history. The yellow mutants were traced in Asia Minor in the 10th and 11th centuries, in Spain in 12th century, and in continental North Western Europe in 14th century. The crop was introduced to Japan in 17th century. There is the Sanskrit equivalent called 'Girijana' for carrot which probably indicates the derivation of Hindi name 'Gajar'. The first settlers to Virginia brought seeds and soon carrots were grown by American Indians, carrot probably were first grown in Salem gardens, about 1620 A.D. (Shoemaker, 1947). Carrot is cultivated in India, Malaysia, Indonesia, Philippines; Central, East and West Africa, South America and the Caribbean. It is the second most important vegetable in England after potato. In Australia, carrot is third most important vegetable after potato and tomato. In India, the important carrot growing states are Uttar Pradesh, Assam, Karnataka, Andhra Pradesh, Punjab and Haryana. In north India highly coloured types are met which are not available in Europe. Types varying from absolute colour less to light lemon, light orange, orange and deep orange, light purple, deep purple and almost black are available. This is another indication of its being a primary centre of origin.

Species Availability, Evolution and Diversity

Genus *Daucus* contrains 161 species worldwide of which only two (*D. carota* and *D. sativus* D. C.) are being cultivated. The species *D. carota* is native of western Asia (Afghanistan). The cultivated types were evolved as selection from inter-specific cross between *D. carota* and *D. sativus*. Some prominant species are: *D. aureus*, *D. bicolor*, *D. bocconei*, *D. brachiatus*, *D. broteri*, *D. capillifolius*, *D. carota*, *D. carota* cv. *atrorubens*, *D. carota* subsp. *carota*, *D. carota* subsp. *commutatus*, *D. carota* subsp. *drepanensis*,

D. carota subsp. *gadecaei*, *D. carota* subsp. *gummifer*, *D. carota* subsp. *hispanicus*, *D. carota* subsp. *hispidus*, *D. carota* subsp. *maritimus*, *D. carota* subsp. *maximus*, *D. carota* var. *atrorubens*, *D. carota* var. *boissieri*, *D. carota* var. *commutatus*, *D. carota* var. *sativus*, *D. crinitus*, *D. durieua*, *D. gingidium*, *etc.* The largest diversity in morphological characters of this species have been found to occur in that region. The roots vary in degree of remification, fleshiness and colour which ranges from white to anthocyaninsed. Wild types are found in South West Asia, Southern Europe, North Africa and North America. Primitive black and red types are popular in India. The orange European types having higher carotenoid content, were believed to have been introduced later during the visit of European traders. Obviously, the black and red types might have come earlier from the central Asian Region of its origin-probably Afghanistan, Persia and Southern Russia. Ain-i-Akbari (Anonymous) record carrot in Mughal cuisine which explains the spread from central Asia (Seshadri and Chaterjee, 1996).

Classification on the Basis of Cultivated Cpecies

Under genus *Daucus* two cultivated species (*D. carota* and *D. sativus*) defined as types *i.e.* Asiatic and European types are more popular. Classification of *D. carota* as proposed by Thelleng (1926) is as under:

There are 2 sub-genera under *Daucus, Eucarota* and *Gummifer.*

Eucarota

The *Eucarota* consists of mainly annuals and biennials. The species under *Eucarota* are :

D. carota L.

D. maritimus

D. major

D. sativus (Hoffm) Thell.

D. maximum

Gummiferea

The *Gummiferea* consists of mainly pertennials. The species under *Gummiferea* are :

D. cummutalis

D. hispanicus

D. fontanesii

D. boeconiel

D. gummeferea

The species *D. carota* and *D. sativus* are only cultivated. There is another view that all the above are subspecies of *D. carota* as *D. carota* spp. *carota* and *D. carota* subsp. *maritimus.*

It is probably evolved from a wild plant resembling Queen Anne's lace (Banga, 1976). The western carotene carrot (European type) has been derived from the Eastern anthocyanin carrot (tropical type) and the later had probably been developed from anthocyanin containing forms of *D. carota* ssp. *carota* such as those found in Afghanistan. Afghanistan is the centre of diversity of the purple coloured carrot (anthocyanin carrot). Integration between Asiatic and european carrots has given the present day forms which are fleshier, smoother, lesser forked, and better coloured. The

Table 6.1: Taxon and Preferred Taxon Name of *Daucus*

Name of Taxon	Name of Authority	Family	Preferred Taxon Name
Astrodaucus littoralis	(M. Bieb.) Drude	Apiaceae	*Astrodaucus littoralis*
Astrodaucus orientalis	(L.) Drude	Apiaceae	Astrodaucus orientalis
Astrodaucus orientalis var. *eriocarpus*	(Boiss.) Woronow	Apiaceae	*Astrodaucus orientalis* var. *eriocarpus*
Astrodaucus orientalis var. *orientalis*		Apiaceae	*Astrodaucus orientalis* var. *orientalis*
Daucus aureus	Desf.	Apiaceae	*Daucus aureus*
Daucus bicolor	Sm.	Apiaceae	*Daucus guttatus*
Daucus bocconei	Guss.	Apiaceae	*Daucus carota* subsp. *drepanensis*
Daucus brachiatus	Sieber ex DC.	Apiaceae	*Daucus glochidiatus*
Daucus broteri	Ten.	Apiaceae	*Daucus broteri*
Daucus capillifolius	Gilli	Apiaceae	*Daucus capillifolius*
Daucus carota	L.	Apiaceae	*Daucus carota*
Daucus carota cv. *atrorubens*	Alef.	Apiaceae	*Daucus carota* var. *boissieri*
Daucus carota subsp. *carota*		Apiaceae	*Daucus carota* subsp. *carota*
Daucus carota subsp. *commutatus*	(Paol.) Thell.	Apiaceae	*Daucus carota* subsp. *commutatus*
Daucus carota subsp. *drepanensis*	(Arcang.) Heywood	Apiaceae	*Daucus carota* subsp. *drepanensis*
Daucus carota subsp. *gadecaei*	(Rouy and E. G. Camus) Heywood	Apiaceae	*Daucus carota* subsp. *gadecaei*
Daucus carota subsp. *gummifer*	Hook. f.	Apiaceae	*Daucus carota* subsp. *gummifer*
Daucus carota subsp. *hispanicus*	(Gouan) Thell.	Apiaceae	*Daucus carota* subsp. *hispanicus*

Name of Taxon	*Name of Authority*	*Family*	*Preferred Taxon Name*
Daucus carota subsp. *hispidus*	(Arcang.) Heywood	Apiaceae	*Daucus carota* subsp. *hispidus*
Daucus carota subsp. *maritimus*	(Lam.) Batt.	Apiaceae	*Daucus carota* subsp. *maritimus*
Daucus carota subsp. *maximus*	(Desf.) Ball	Apiaceae	*Daucus carota* subsp. *maximus*
Daucus carota var. *atrorubens*	(Alef.) E. Small	Apiaceae	*Daucus carota* var. *boissieri*
Daucus carota var. *boissieri*	Schweinf.	Apiaceae	*Daucus carota* var. *boissieri*
Daucus carota var. *commutatus*	Paol.	Apiaceae	*Daucus carota* subsp. *commutatus*
Daucus carota var. *sativus*	Hoffm.	Apiaceae	*Daucus carota* var. *sativus*
Daucus crinitus	Desf.	Apiaceae	*Daucus crinitus*
Daucus durieua	Lange	Apiaceae	*Daucus durieua*
Daucus gingidium	L.	Apiaceae	*Daucus carota*
Daucus glochidiatus	(Labill.) Fisch. and C. A. Mey.	Apiaceae	*Daucus glochidiatus*
Daucus gracilis	Steinh.	Apiaceae	*Daucus gracilis*
Daucus gummifer	Lam., nom. illeg.	Apiaceae	*Daucus carota* subsp. *gummifer*
Daucus gummifer subsp. *hispidus*	Arcang.	Apiaceae	*Daucus carota* subsp. *hispidus*
Daucus guttatus	Sm.	Apiaceae	*Daucus guttatus*
Daucus halophilus	Brot.	Apiaceae	*Daucus halophilus*
Daucus hispanicus	Gouan	Apiaceae	*Daucus carota* subsp. *hispanicus*
Daucus hispidifolius	Clos	Apiaceae	*Daucus pusillus*
Daucus involucratus	Sm.	Apiaceae	*Daucus involucratus*
Daucus jordanicus	Post	Apiaceae	*Daucus jordanicus*
Daucus littoralis	Sm.	Apiaceae	*Daucus littoralis*
Daucus maritimus	Lam.	Apiaceae	*Daucus carota* subsp. *maritimus*
Daucus maximus	Desf.	Apiaceae	*Daucus carota* subsp. *maximus*
Daucus montanus	Humb. and Bonpl. ex Spreng.	Apiaceae	*Daucus montanus*
Daucus montevidensis	Link ex Spreng.	Apiaceae	*Daucus pusillus*
Daucus muricatus	(L.) L.	Apiaceae	*Daucus muricatus*
Daucus polygamus	Gouan	Apiaceae	*Daucus carota* subsp. *drepanensis*
Daucus pulcherrimus	(Willd.) W. D. J. Koch ex DC.	Apiaceae	*Astrodaucus orientalis*
Daucus pulcherrimus var. *eriocarpus*	Boiss.	Apiaceae	*Astrodaucus orientalis* var. *eriocarpus*
Daucus pumilus	(L.) Ball	Apiaceae	*Pseudorlaya pumila*
Daucus pusillus	Michx.	Apiaceae	*Daucus pusillus*
Daucus setifolius	Desf.	Apiaceae	*Daucus setifolius*
Daucus setulosus	Guss. ex DC.	Apiaceae	*Daucus guttatus*
Daucus sp.		Apiaceae	*Daucus* sp.
Daucus visnaga	L.	Apiaceae	*Ammi visnaga*

yellow coloured carotene rich forms were also thought as mutants of anthocyanin carrots. The occurrence of wild carrots (*D. carota* ssp. *carota*, regarded as a weed worldwide) is dangerous for seed production in the field due to outcrossing nature.

Gene and Genetic of Characters

Root shape in carrot is governed by 3 genes D, N and P.

Bienz (1968) reported that splitting was a varietal characteristic and was governed partially by genetic factors in addition to the effect of spacing and nitrogen application. Dickson (1966) concluded that cracking susceptibility may be governed by a single dominant gene.

Table 6.2: Gene and Genetic of Characters

Character	*Number of Gene*	*Type of Gene Action*
Root colour	Diegenic (P-Y-) two modifiers	Deep purple (*iiPPYYEE*)
		Purple (*iiPPYYee*)
		Diffused purple (*iiPPyyee*)
		Yellow (*IIppYYee*)
		Red (*iiPPyyEE*)
		Light red/orange (*iippyyee*)
Root tip	Monogenic (S-)	Blunt root incompletely dominant to pointed root
Root cracking	Monogenic	Susceptible, dominant to resistance

Botanical Description

The carrot belongs to the family Umbelliferae, genus *Daucus* and botanical name is *D. carota* (2n=18). It is dicotyledonous herbaceous crop grown for its enlarged taproot. The carrot cultivars that have been originated in Asia are annuals when grown under long days and do not have a vernalization requirement whereas, the cultivars that have been developed in the temperate regions, are biennials. Carrots are available in different colours. Most of the tropical varieties are red coloured due to the presence of the pigment lycopene. Orange colour of the European type is due to β-carotene pigment. Purple and yellow colour of carrots is due to the pigments anthocyanin and xanthophylls, respectively. Absence of all these results in white carrots grown in eastern and Western Europe as cattle feed. Temperature is the single environmental factor affecting root shape and colour in this crop.

Horticultural Classification

The morphological characteristics lead to a division of the cultivated carrot (*Daucus carota* subsp. *sativus*) into two botanical varieties:

(i) var. *atrorubens* (Tropical/Oriental/Eastern-annual in nature) includes carrots originating from the East, exhibiting purple, red or yellow storage roots due to presence of anthocyanin lycopene and xanthophyllus pigments and poorly indented, grey-green, pubescent foliage

Table 6.3: Botanical Description

Growth Habit	The carrot is an annual or biennial herb.
Root	The edible portion of carrot root is actually enlarged fleshy swollen base of taproot (hypocotyls; stem consists of a small plate like crown) which may be short and stumpy or long tapering cone. It consists primarily of phloem or cortex and core or xylum. Good quality carrot has a maximum of cortex and a minimum of core and these are also called 'coreless' cultivars. The core is small and deeply pigmented so that the cortex and core are evenly coloured. The colour of cultivated carrot roots varies from white to yellow, orange, light purple, deep red, or deep violet and the shape varies from short stumps to tapering cones. The root diameter and length may vary from 2-6 cm and 6-30 cm, respectively. Root with a smooth, tender, bright red or orange flesh, with minimum of core is preferred.
Stem	Carrot stem is erect to much branched; 30-120 cm high arising from a thick fleshy tap root. Usually the stem elongates and produces round hispid branches during the second year.
Leaf	Leaves are pinnately compound in a rosette along with long petioles expended into a stalk at the base.
Floral biology	The plant is andromonoecious (due to separation of sex). Carrot requires a period of up to 6-8 weeks at temperature below 10°C for flower induction, and induction is accelerated under long days. Involucral bracts prominent. The flowers are hermaphrodite. The carrot umbel is a compound inflorescence. Umbel development begins with a broading of floral axis and internodes elongation (Borthwick *et al.*, 1931). The first umbel which is terminal to the main stalk. Branches from main stalk form secondary umbels and subsequent branches from these form tertiary umbels. A primary umbel can contain over 1000 flowers at maturity, whereas secondary, tertiary and quaternary umbels bear successively fewer flowers. Usually several floral stalks develop from a single plant. Flowers are arranged in umbel-lets and umbel-lets into umbel. The flowers that mature first are those on the outer edges of the outer umbel-lets. The primary umbel consists mainly of bisexual flowers, but male flowers can, occur frequently in subsequent umbels. The pollens from flowers at the centre of an umbel-let are larger and more frequently fertile than that from peripheral flowers. Carrots flowers are protoandrous. The petals separate and the filaments begin unrolling to release the anthers at anthesis, although the anthers in one flower may not extrude simultaneously. After the filament is straightened, the pollen is shed and the stamen is quickly abscised. The petals then open fully and the style elongates. The carrot has a split style, which separates when the flower is receptive to pollination. The petals of male fertile plants fall soon after the split of stigma is receptive. It is interesting that petals of petaloid but not brown anther, male sterile plants are persistent until the seed ripens. Flowers are epigynous with 5 small sepals, 5 petals, 5 stamens and 2 carpels. The mature flower and consequently mature fruit are approximately 2.0 mm long. Early in development, each carpel bears two ovule primordial but only the lower one continues to grow. The carrot embryo sac is monosporic (developing from the chalazal macrospore) and 8-nucleate. Nectaries on the ovary wall are important in insect attraction for pollination. The flowers are perfect with small petals, usually white or yellowish. The calyx is entire. There are 5 stamens. The ovary is inferior and consists of two locules, each with a single ovule. On the upper surface of the ovary there is swollen nectar secreting disks, which support the style and stigma. The carrot fruit is bilocular schizocarp, which dries and splits upon maturity to yield two mesocarps (achenes) with one seed each. The mericarp are covered with spines, which must be removed, for ease of handling. It produces the first order umbel, called king umbel, second order and tertiary umbels.
Anthesis and pollination	Protandry behavior is responsible for cross-pollination (95 per cent). Pollen dehiscence may last one to two days per floret and six to seven days after anthesis and may take about 4-6 weeks for complete blooming. The different orders umbels flowers at intervals 8-12 days from each other. Anthesis in a single umbel is completed in 7-9 days. The studies on carrot of variety 'Imperator' revealed that umbels of the first two order set seeds in 100 per cent umbels and the third, fourth and onwards orders set in 82.7, 6.29 and 2.78 per cent umbels, respectively. The setting of seed in late formed umbels was extremely poor. The stigma fertility starts when dehiscence is completed in whole umbel. First king umbel comes in flower it gets pollen from secondary umbel. The secondary gets pollen from tertiary umbel. The stigma becomes receptive on the 5th day after flowers open and remain active for 8 days, but the better fruit set are from pollination on 6 to 11 days after flower opening. The temperate types start bolting by third week of April. The king umbels start flowering by the end of May. Flowering in first, second and third order umbels are almost completed by end of June. The fourth and onwards umbels flower during July and August. Over 95 per cent of cross-pollinated have been observed in carrot.
Fruit and fruit set	The carrot fruit is a bilocular schizocarp, setose, which dries and splits upon maturity to yield 2 mericarps with 1 seed each. The carrot plants generally require a cold period for bolting (flower initiation). The fruit is oblong and about 3.0 mm long.
Seeds	The carrot seed is actually a fruit, an indehiscent mericarp, containing a single seed. Infact, two mericarp pair to form the schizocarp, the true carrot fruit, which develop from a two-loculed ovary. The mericarp or seed is somewhat flattened on one side and opposite side has longitudinal ribs with bristly hairs.

(ii) var. *sativus* (Temperate/European/Western-biennial in nature) refers to carrot originating from the West and exhibiting orange or sometimes white roots because of carotene and leucocyanidin pigments and highly indented, non-pubescent, yellow-green foliage.

But many intermediate variants of these two botanical varieties exist in nature.

Classification of *D. sativus* on the Basis of Centre of Origin

i. Asiatic-*D. sativus* spp. *afghanicus, D. sativus* spp. *syriacus* and *D. sativus* spp. *cilicions*).
ii. European-*D. sativus* spp. *mediterraneus* and *D. sativus* spp. *japonicas*

D. sativus is a transitional stage species.

Classification on the Basis of Root Shape

- Long root or Desi type: Roots may be 25.0 cm or more in length, generally tapering, these perform best in comparatively light soil. Root is cylindrical. It has a genotype ddnnP- ('Nantes")
- Half long root: Where root is above/cuneate. It has a genotype ddN-P- ('Chantenay'). Root length does not usually exceed 20.0 cm.
- Roots cylindrical with straight or sloppy shoulder, *e.g.* Nantes.
- Roots tapering with blunt or semi blunt tip, *e.g.* Chanteney, Imperator
- Short-stump root: It has a genotype ddN-pp, (round varieties).
- Heart shaped, *e.g.* Oxheart
- Oval, *e.g.* Early Scarlet horn
- Round, *e.g.* French Forcing

Classification on the Basis of Root Colour

Carrot roots have wide range of colours and can be classified as:

1. Purple
2. Orange
3. Red and
4. Yellow

Classification on the Basis of Centre of Adaptation

Here, it is important to mentioned that *D. sativus* is a transitional stage species. On the basis of centre of origin *D. sativus* can be classified into:

i. Asiatic types- In Asiatic type *D. sativus* spp. *afghanicus, D. sativus* spp. *syriacus* and *D. sativus* spp. *cilicions* are included.
ii. European types- In European types *D. sativus* spp. *mediterraneus* and *D. sativus* spp. *Japonicus* are included.

Classification on the Basis of Temperature Response to Flowering

- **Temperate or European or biennial types-**These cultivars are biennial in nature and require low temperature (4.8-10.0°C) treatments for flowering to occure. These do not produce seeds in the plains, *e.g.* Nantes, Chantenay Imperator, Zeno, Pusa Yamdagni, Early Nantes and Ooty 1. *etc.*
- **Tropical, Asiaic or Oriental or Annual types-** These cultivar do not require low temperature treatment for flowering and they seed freely in the plains of India. The varieties are Pusa Kesar, Pusa Meghali, Sel-233, No. 29, Desi Red, NS 820, Hisar Gairic, Hybrid No.-1, Gajar No. 29, *etc.*

Augmentation of Genetic Resources and Crop Improvement

Breeding methods include mass selection and pedigree selections within different population.

Breeding Priorities

- High root yield
- Uniform maturity
- Earliness
- Uniformity in root shape and size
- Uniform external colour
- Better root appearance (lacking lateral roots, smooth and shiny)
- Small core size
- Self-coloured core (uniform colour of phloem and xylem)
- Root integrity (free from cracking or splitting)
- Shoulder without green colour along with flat or slightly convex shape rather than concave or sunken
- Free from bolting
- Tolerance to biotic and abiotic stresses.

Introduction and Selection

Popular cultivar 'Nantes' introduced in India and directly used a variety for commercial cultivation. It is reported that selection based on phenotypic values may prove quite effective because of high heritability estimates. Chandel and Rattan (1988) observed that both additive and dominance variances were significant for others characteristics. Gill and Kataria (1974) found mark variation in dry matter, T.S.S., and carotene content among European and Asiatic varieties. European types of carrot were superior to Asiatic types in respect of all above characters. The bigger size of core and high percentage of moisture was mainly responsible for poor nutritive value of Asiatic type of carrot. The varieties successfully introduced to India are 'Imperator', 'Danvers', 'Nantes' and 'Perfection'. 'Pusa Yamdagini' developed from ICAR-IARI, Regional Research Station, Katrain (Himachal Pradesh) is known for high yield (12.0-15.0 t/ha). In Nepal, introduced carrot cultivars like 'Elake Gross' by Denmark; BSS-68 and BSS-37 by Bejo Sheetal Seed Co. and T-390 by Takii, Japan are commonly being grown. In Pakistan, cultivars like T-29 (R) and Nantes (P) are being very common and grown commercially. Popular variety 'Hisar Gairic' was developed by use of Indigenous material through mass selection method

Hybridization and Selection

The first report of heterosis in carrot upon crossing of two inbred lines was made by Poole (1937). Moreover, Frimmel and Lauche (1938) were the first to recommend the breeding of F_1 carrot varieties. Efforts were made at ICAR-IARI, New Delhi and its sub-station in Kullu Valley to cross the useful strains of the Asiatic and exotic strains, with an intention to incorporate good root qualities of European types into high yielding Asiatic types. Adopting bulk population breeding method 'Pusa Kesar' was selected from a cross between 'Nantes' x 'Local Asiatic' which has good qualities of both Asiatic and European types with an ability to produce seeds in the plains has been developed (Singh, 1963). In 1985 another Asiatic carrot 'Pusa Meghali' was evolved from a cross between 'Pusa Kesar' and 'Nantes'. Bhagchandani and Choudhury (1971) noted the various degrees of heterosis and inbreeding depression in Asiatic carrot. There are several elite genotypes which resulted significant increase in yield and resistance when used as a parents in crossing programme:

Table 6.4: Promising Combinations and their Salient Features

Cross Combination	*Salient Features*
'Chantenay' x 'Genanda'	Increased yield (25-33 per cent)
'Italian' x 'Nantes'	Increased yield (12 per cent)
'Virginia Savoy' x 'Old Domain'	Resistant to cucumber mosaic virus

Interspecific Hybridization

The interspecific hybridization is resorted to generate useful variability.

Table 6.5. Crossability among the Species

	D. carota ssp. *sativus*	*D. gingidum*	*D. capillifolium*
D. carota ssp. *sativus*	SI	CC	CC
D. gingidum	CC	SI	NK
	-	SC	-
D. capillifolium	CC	NK	SI
			SC

SI: Self-incompatible; SC: Self-compatible; CC: Cross compatible; NK: Not known.

Table 6.6: Resistance Source for Biotic Stress and Utilization

Disease/Pest	*Resistant Varieties/Genotypes*
Cercospora carotae	'Inbred line WCRI'
Motley virus	'Sweet Crop', 'Top Weight'
Cucumber Mosaic Virus (cmv)	'Virginia Savoy'
Pythium spp.	'Nantes', Imperator', Danvers'
Heterodera carotae (nematode)	'Vilmorium 66'
Tip burn (Physiological disorder)	'Scarlet', 'Nantes'

Resistance Source for Abiotic Stress and Utilization

The Asiatic carrots (syn. tropical carrot, eastern carrots, anthocyanin carrots) are heat tolerant, deep red or purple coloured, high-yielding and low in carotene content, while the European carrot (western carrot, temperate carrot, carotene carrot) is cold tolerant, orange coloured and rich in carotene. Cytoplasmic male sterility, governed by one or more recessive genes has been reported. The major problem is 'tip burn' (Physiological disorder) which can be minimized through use of Resistant varieties/genotypes like 'Scarlet' and 'Nantes'.

Male Sterility

The male sterility in carrot was detected and analyzed by Thomson (1962) and Hanschke and Gabelman (1963) which promote hybrid breeding and production of low cost hybrid seeds. In carrot hybridization and development of hybrids depends on different types of sterility mechanisms. The heterosis breeding in carrot is generally based on three systems of CMS with different genetic backgrounds and origin. In carrot, there are two CMS system operates:

1. Petaloi (anthers turn to petals)-Petaloid type is stable and commonly used system in hybrid breeding of orange carrots but negligible in tropical carrot.
2. Brown anther (shriveled brown)

Functional male sterility is useful in hybrid seed production. Closed anther mutants and types with unrolled stamen filaments have been observed. They have 'brown anthers' (ba) ms line and petaloid stamen. The brown anther type (ba) is present in all cultivated orange coloured open pollinated varieties which results from an interaction of the 'Sa'-cytoplasm with two independent nuclear genes, *i.e.* homozygous recessive aa or dominant B- (Banga *et al.*, 1964). The phenotype of 'ba' is characterized by deformed, brown-coloured anthers without functional pollen, caused by a genetic block in meiosis. The cytoplasm of petaloid type (pt) is derived from a wild form of *Daucus carota* L and inherited by an interaction of Sp-cytoplasm and two independent dominant genes, *viz.* M1 and M2 (Morelock, 1974). The pt-type is characterized by transformation of the anthers into petals or petal-like structures which do not produce functional pollen. The deformation of androecium varies in degree an has been classified as spoon-type, strap-type or carpeloid-type. Generally, the CMS system, ba-type and pt-type suffer from instability of male sterility under high temperature, dry weather and long-day conditions. A third system of CMS, gu-type, derived from the cross of orange coloured carrot and wild carrot *D. carota gummifer* Hook. (Nothnagel, 1992). The new cytoplasmic male sterile system in an alloplasmic form of orange carrot is a cross between the wild carrot *Daucus gummifer* and the cultivated carrot *D. carota*. It is governed by an interaction of the 'gummfer' cytoplasm with a recessive nuclear gene, *gugu* (Stein and Nothnagel, 1995). Genetic of restoration is complex. The gum-type can potentially be used as a new source of CMS for carrot breeding because of the simple mode of selection and phenotype stability.

The male sterile line is maintained by crossing it with pollinator strain used as a recurrent parent in backcross, since the nuclear genotype of the pollinator is identical with that of the new male sterility line. Such a male fertile line is known as maintainer line *i.e.* B-line and male sterile line is known as A-line. The CMS system used for the development of hybrids involves three line systems:

(i) A-line (male sterile)

(ii) B-line (maintainer line, male fertile)

(iii) C-line or R-line (pollinator line, male fertile)

A-line is maintained by crossing it with corresponding B-line, while B-line and C-line are maintained by bud-pollination, selfing or sib mating. The development of the A-line and B-line is a very laborious process due to the dominant state of male sterility. Crossing, backcrossing, selfing and testing of the progenies in the following two generations, inducing isolation of the positive progenies, are characteristic steps in breeding processes. The incorporation of CMS into carrot inbred is slightly different from other crops. Generally, this process begins at the F_2 or F_3 generation with the intercrossing of individual selected fertile plants to a male sterile plant. In the next generation progeny from the male sterile are examined for male fertility in the process of backcrossing for sterile line development. Presence of male fertile plant indicates the presence of restorer allele(s) from the original male fertile plant. It has been relatively easy to identify and select male sterile maintaining carrot inbreds from a wide range of Germplasm backgrounds but the incidence of restorer varies widely. Backcrosses should be performed by use of the male-fertile genotype (Sp) m1m1m2m2 as a tester in order to find dominant alleles.

Globally CMS systems have been usually used not only to produce the quality seeds but also to harness the heterotic vigour for higher root yield, earliness (maturity, root colour, root shape, root size), self-coloured core, adaptation and tolerance to biotic and abiotic stresses. In carrot 'Pusa Nayanjyoti' was developed using petaloid CGMS at ICAR-IARI-Regional Station, Katarain (H.P.).

Table 6.7: Carrot Hybrids Developed through Male Sterility

Name of the Hybrid	*Type of Male Sterility Used*	*Developing Organization*
Arka Kirtiman	Cytoplasmic genic male sterility (CGMS)	ICAR-Indian Institute of Horticultural Research, Bengaluru
Arka Lalima	Cytoplasmic genic male sterility (CGMS)	ICAR-Indian Institute of Horticultural Research, Bengaluru
Hybrid-63	Cytoplasmic genic male sterility (CGMS)	ICAR-Indian Institute of Agricultural Research, New Delhi
Hybrid-35	Cytoplasmic genic male sterility (CGMS)	ICAR-Indian Institute of Agricultural Research, New Delhi

Singh *et al.* (2016) used petaloid CMS system in introgression of tropical carrot in 12 lines at ICAR-IARI, New Delhi. As a part, 12 CMS lines using 14 morphological traits and 6 CMS specific DN markers. All the CMS lines had green umbel, 9 pink petaloids and 6 with spoon like petals. Variation was observed in XMS line for petal size (0.66-0.95 x 0.35-0.69 cm), petaloid size (0.84-2.39 cm x 0.026-0.79 cm) and style length (0.63-1.69 cm). The CMS lines had variations for root (length, diameter) and quality attributes (β-carotene), lycopene, anthocyanin). Ten DNA markers specific to *atp1*, *atp6*, *atp9* and *cob* loci from mitochondrial genome of carrot were used and two markers based on *cob* and *atp6* genes distinguished CMS lines amplifying a dominant band, whereas four primers from *atp9* genes amplified alternate alleles. These markers are useful to identify inter and intra-population variation in CMS line(s) for use in hybrid breeding.

Breeding for Processing and Quality Traits

Morphological quality traits like deep orange root colour (high β-carotene content) is generally preferred by consumers.

Table 6.8: Following Quality Traits should be Considered during Breeding Programmes

Traits	*Preferred Range/Magnitude/Value/Type*
Sugar content	High
Root colour	Red or Deep orange
Root shape	Splender and uniform or with blunt end
Root texture	Tender
Core texture	Tender
Core colour	Light red to dark red
Fiber content	Low
Forking or splitting	Absent
TSS	10 per cent or more

Indian cultivars like Pusa Kesar, Pusa Meghali and Pusa Yamdagini were reported for high β-carotene. Carrot roots have wide range of colours, particularly purple, orange, red and yellow due to anthocyanins, β-carotene, lycopene and lutein content. Due to recent awareness of nutritional health security, there is increasing consumption of carrot by consumers as well as nutraceutical based industry due to huge bioavailability of dietary vitamins, phytonutrients and bioactive compounds of carrots. For breeding multi-nutrient rich carrot varieties, there is a need to study and understand the genetic mechanism of nutritional traits such as total solids, total soluble solids, total carotenoids, β-carotene, lycopene and total anthocyanin. The six digenic parameter model study revealed that the nutritional traits were governed by additive (d), dominance (h), additive x additive (i), additive x dominance (j) and dominance x dominance (l) interactions. The positive sign of dominance (h) and dominance x dominance (l) resulted into complementary epistatic interactions controlling moisture content and total solids (White Pale x IPC-126 cross), total soluble solids (White Pale x IPC-122, PM x IPC-126, PM x IPC-122, IPC-122 x White Pale, IPC-122 x IPC-126 and IPC-122 x PM crosses), total carotenoids and β-carotene (White Pale x IPC-122, White Pale x PM, IPC-126 x IPC-122, PM x White Pale, PM x IPC-122, IPC-122 x White Pale, IPC-122 x IPC-126 and IPC-122 x PM crosses), lycopene (in all crosses except PM x IPC-122) and total anthocyanin content (White Pale x IPC-126, IPC-126 x White Pale, IPC-126 x IPC-122, IPC-126 x PM, PM x IPC-126 and IPC-122 x IPC-126 crosses). The higher frequency of allele association [d+i] among parents suggested possibility of exploitation of heterosis breeding (Selvakumar and Kalia, 2016). Kumar *et al.* (2016) reported that none of line or tester exhibited good general combining ability for all the traits studied. Out of six lines, highest value of GCA effects were shown by the line KT10A for CUPRAC, FRAP, lycopene and β-carotene, KT-28A and KT-7A for phenols and KT-80A and 7A for ascorbic acid contents. Similarly among the testers, KS20 had highest GCA effects for FRAP and anthocyanin, KS-22 for lycopene and β-carotene and KS-21 for phenols, respectively. According to these results, lin KT10A and the tester KT-20 and KT-22 showed maximum positive GCA effects for most of the traits so these parents could be successfully used in breeding new genotypes with increased antioxidant activity in future. Significant superior SCA effects for all the traits were not shown by a single cross. Cross combination KT-7A x KS-22 exhibited significant highest SCA effects for CUPRAC, FRAP, lycopene and ascorbic acid contents, KT-10A x KS-20 SCA effects for anthocyanin, KT-28A x KS-20 for phenols and KT-80A x KT-59 for β-carotene, respectively. Among the 125 hybrids, hubrid KT-7A x KS-22 exhibited significant SCA effects for all the four characters except phenols, anthocyanin and β-carotene for which KT-28A x KS-21, KT-10A x KS-20 and KT-80A x KT-59 had the highest level of antioxidant activity.

Abiotic Stress Resistance

The Asiatic carrots are heat resistance/tolerant and European carrot is cold tolerant. Self-incompatible lines are also available. There are several varieties which are reported as resistant to diseases and pests. Recently the heat tolerant carrots like 'Kuroda', and 'New Kuroda', having deep orange coloured and stumped roots, are imported from Japan and South Korea by private companies. They are becoming popular in the southern states (Nath *et al.*, 2008).

Genetic Resource Conservation

Carrot accessions (*Daucus carota* including both wild and cultivated subspecies) available in two major collections in Europe and the USA. These collections include weedy, hybrid and breeding material in addition to the wild, landrace and cultivars listed in the following Table 6.8a.

Table 6.8a: Available Germplasm at EURISCO and ARS-GRIN

Source	*EURISCO*[1]	*ARS-GRIN*[2]
Total *Daucus carota* accessions	4379	882
Wild material	379	154
Landraces	845	207
Cultivars	1552	116

Source: [1]Data downloaded from the EURISCO database at http://eurisco.ecpgr.org.

EURISCO is a web-based catalogue that provides information about *ex situ* plant collections maintained in Europe, including that of the Warwick HRI GRU [2]Data downloaded from the Genetic Resources Information Network (http://www.ars-grin.gov), a program within the United States Department of Agriculture's Agricultural Research Service. Statistics for the NCRPIS germplasm availability is at 80 per cent with 82 per cent of the collection backed up at the National Center for Genetic Resources Preservation (NCGRP) in Ft. Collins, CO. Dr. Philipp Simon and Dr. David Spooner donated 94 new *Daucus* accessions acquired during an NPGS-sponsored collection trip in Tunisia in 2009.

REFERENCES

1. Bailey, L.H. 1948. Manual of cultivated plants. The Mac Millan Co., New York.
2. Bailey, L.H. 1928. The Standard Encyclopedia of Horticulture, **1:** 674.
3. Banga, O. 1976. *Carrot*. In: *Evolution of Crop Plants*. edn. 1, p. 291-294. Simmonds, N.W. (Ed.). Longman, London.
4. Banga, O., Petiet, J. and Van Bennekom, J.L. 1964. Genetical analysis of male sterility in carrots (*Daucus carota* L.). *Euphytica*, **19:** 263-269.

Table 6.9: Collection of Carrot Genotypes at NCRPIS

Taxon	New Accessions	PI Numbers	Ames Numbers	Total Accessions	Available	Backed up at NCGRP
D. aureus	2	7	5	12	4	4
D. broteri		10	9	19	15	15
D. capillifolius		1	1	1	1	
D. carota	84	691	145	836 (865a)	706 (716)	747
D. carota subsp. carota		41	11	52	41	42
D. carota subsp. commutatus		3	1	4	4	3
D. carota subsp. *drepanensis*		2		2		1
D. carota subsp. *fontinesii*			4	4		
D. carota subsp. *gummifer*		2		2	2	2
D. carota subsp. *hispanicus*		2		2	2	2
D. carota subsp. *major*		2	3	5	5	5
D. carota subsp. *maritimus*		2	16	18	2	2
D. carota subsp. *maximus*		9	9	18	9	9
D. carota subsp. *sativus*		2		2	2	2
D. carota var. *atrorubens*b		2		2	1	2
D. carota var. *sativus*		92	7	99	99	97
D. crinitus		5	7	12	3	3
D. durieua			1	1		
D. glochidiatus		1		1		
D. guttatus		17	8	25	24	20
D. involucratus		3		3	3	2
D. littoralis		2		2	1	2
D. muricatus	2	1	14	15	3	3
D. pusillus		6	1	7	6	6
D. sahariensis	4		7	7	3	
D. syrticus			10	10	6	3
D. unidentified species	2		31	31	26	26
Total	**94**	**903**	**289**	**1221**	**977**	**999**

a. The total number of *Daucus carota* accessions is 865 when including 29 NSSL-numbered accessions sent to Ames by NCGRP in 2006 for regeneration due to low viability. Ten of the 29 are currently available and will be assigned PI numbers.

b. Nomenclature change: *D. carota* var. *atrorubens* was formerly known as *D. carota* var. *boissieri.*

5. Bienz, D.R. 1968. Evidence for carrot splitting as an inherited tendency. *Proc. Soc. Hort. Sci.*, **93:** 429-433.
6. Bhagchandani, P.M. and Choudhury, B. 1971. Heterosis and inbreeding depression in carrot (*Daucus carota* L.), *Progressive Horticulture*, **2:** 65-75.
7. Borthwick, H.A., Phillips, M. and Robbins, W.W. 1931. Floral development in *Daucus carota. American Journal of Botany*, **18:** 784-796.
8. Chandel, K.S. and Rattan, R.S. 1988. Epistasis, additive and dominance variation in carrot. *Veg. Sci.*, 15: 31-37.
9. Dickson, M.H. 1966. The inheritance of longitudinal cracking in carrots. *Euphytica*, **15:** 99-101.
10. Frimmel, F. and Lauche, K. 1938. Heterosis-Versuehe an Kaotten. *Z. Pflanzenzuchtg*, **22:** 469-481.
11. Gill, H.S. and Kataria, A.S. 1974. Some biochemical studies in European and Asiatic varieties of carrot (*Daucus carota* L.). *Curr. Sci.*, 43: 184-185.
12. Hanschke, P.E. and Gabelman, W.H. 1963. Digenic control of male sterility in carrot (*Daucus carota* L.). *Crop Sci.*, **3:** 383-386.
13. Kumar, R., Kumar, V., Prakash, C., Dey, S.S., Chandel, C., Mishra, V., Dabral, M. and Mawlia, M. 2016. Breeding for increased antioxidant activity in temperate carrot (*Daucus carota* var. *sativus* L.). (Abs. 29, Session II: Breeding of Horticultural Crops). 7th Indian Horticulture Congress, held at ICAR-IARI, New Delhi, 15-18th November 2016, p. 108.
14. Morelock, T.E. 1974. Influence of protoplasm source expression of male sterility in carrot. PhD Thesis, Wisconsin University, USA.

15. Nath, P.; Srivastava, V.K., Dutta, O.P. and Swamy, K.R.M. 2008. Vegetable crops- Improvement and Production. PNASF, Bengaluru, p. 177-181.

16. Nothnagel, T. 1992. Results in the development of alloplasmic carrots (*Daucus carota* var. *sativus* Hoffm.). *Plant Breed.*, **109:** 67-74.

17. Poole, C.F. 1937. Year book, USDA, 379.

18. Scibrja, G. and Zemoja, A. 1966. Agricultural techniques for carrot growing for carotene production. 131-139. Abstr. 21, 55, 335. *Plant Breeding* (Abs.), **39(1):** 516.

19. Selvakumar, R. and Kalia, P. 2016. Genetic analysis of nutritional traits in tropical carrot (*Daucus carota* L.). (Abs: 87 Session I: Breeding of Horticultural Crops), 7th Indian Horticulture Congress, held at ICAR-IARI, New Delhi, 15-18 November, p. 67.

20. Seshadri, V.S. and Chaterjee, S.S. 1996. The history and adaptation of some introduced vegetable crops in India. *Veg. Sci.*, **23(2):** 114-140.

21. Shoemaker, J.S. 1947. Vegetable growing. John Wiley and sons Inc. New York.

22. Singh, H.B. 1963. 'Pusa Kesar' and 'Pusa Kanchan', carrot and turnip varieties. *Indian Hort.*, **7(4):** 21.

23. Singh, S., Kalia, P., Mangal, M., Choudhary, P. and Chetna. 2016. Conversion and characterization of CMS lines in tropical carrot for hybrid breeding. 7th Indian Horticulture Congress (Abs: 78 Session I: Breeding of Horticultural Crops), held at ICAR-IAI, New Delhi, 18 November, p. 61.

24. Stein, M. and Nothnagel, T. 1995. Some remarks on carrot (*Daucus carota* var. *sativus* Hoffm.) breeding. *Plant Breed.*, **114:** 1-11.

25. Thomson, D.J. 1962. Studies on the inheritance of male sterility in the carrot (*Daucus carota* L.). *Proc. Amer Soc. Hort. Sci.*, **78:** 332-338.

26. Thompson, H.C. and Kelly, W.C. 1957. Vegetable crops. Mc Graw Hill Book Co., Inc., USA.

27. Yawalkar, K.S. 1980. Vegetable crops of India. Second Edition, Agri-Horticultural Publishing House, Nagpur (MS).

Genus *Raphanus*

Radish (*Raphanus sativus*)

INTRODUCTION

Radish (*Raphanus sativus* L., 2n=2x=18) is very popular as salad and cooked vegetable which is cultivated for enlarged thickened and fleshy edible tender roots. Roots are largely preserved by salting, pickling, canning and drying. It contains volatile compounds *i.e.* isothiocynates. Once the plants start bolting, the flavor and quality of roots deteriorates, christened as pre-mature bolting when it is grown for root yield. It occupies an important position in commercial vegetable farming because of its round the year cultivation and market demand. It is most popular vegetable in the temperate and tropical regions of the world due to wider environmental adaptability. Radish is very often used as a garnish for food dishes such as fish and poultry, sautéed or boiled and added to soups and stews. Per 100 g radish root has 94.4 per cent moisture, 3.4 g carbohydrates, 07 g protein, 0.1 g protein, 0.06 mg thiamine, 0.02 mg riboflavin, 15 mg vitamin C, 35.0 mg calcium and 0.04 mg iron. The use of coloured radishes in the salads and their anthocyanins as colorants are gaining popularity because of the colour characteristics, health benefits as well as antioxidant activities (Singh *et al.*, 2016). It is proved that leaves (tops) are more nutritious than its roots. The leaves (rich source of calcium, phosphorus, vitamin C and protein than the root) and young pods are also used as vegetable. As a medicinal benefits radish has cooling effect, prevents constipation and increases appetite. It is recommended for patients suffering from piles, liver troubles and jaundice. Juice of fresh leaves is used as diuretic and laxative. It helps in preventing stone formation in the urinary tract. Black radish juice has tonic and laxative action on intestine and indirectly stimulates the flow of bile. It can be preserved by storage, pickling, canning or drying. Roots and leaves also serve as an excellent fodder. It is a quick growing and short duration vegetable crop which suitably fit as companion crop or inter crop between the rows of other vegetables. Sometimes it is sown to separate one plot from another by planting on ridges.

Origin, History and Distribution

Radish is native to Mediterranean region where radish was originally domesticated. The recent study indicates that east of Mediterranean; most probably China was the centre of origin of this crop. Werth (1973) reported that European type likely to be original in the area of east Mediterranean Sea to Caspian Sea. Indian group of radish including rat tail type is believed to have originated in the area of their distribution though their ancestors have been lost (Bailey and Bailey, 1956; Mansfield, 1959). The small European or table type radishes originated later than the large ones (Helm, 1957). White globular forms were developed in 18th century and later on the red ones were also developed. This way, the origin of the cultivated radish is not clear. As a journey, it is considered most ancient vegetable as evidence (inscriptions on the walls of pyramids) reveal that radish was in cultivation in Egypt since 2000 BC (Banga, 1976; Sadhu, 1993) and considered as important food. Fossils records also proved that radish was a food as early as 2700 BC. It was introduced into China during 500 BC and to Middle East and Japan around 700 AD (Sirks, 1957). In China and Korea there exist the oldest records on domestication dated to be 400 and 100 BC, respectively. This crop was known in the east Mediterranean at least 2 millennia (2000 BC) before it were known in China. Radish appears to have reached England around 1550. A long white type of radish first appeared in Europe at the end of 16th century. In the 18th century, globular forms were developed, white at first and later on red also. Ultimately, there was a great variation in shape, colour and root size of radish. Radish has been cultivated in Mexico around 1500 AD. The first English colonists brought radish in the USA. The globular form of salad radish was first reported in 19th century. The larger radish has a wide range of adaptability, withstands relatively high temperature, and is grown as an important vegetable in Japan, Korea, China, India and other Eastern countries. The introduction of radish in India has not been chronicled.'Mulaka' is the Sanskrit equivalent for radish in ancient India. Radish was considered as a medicinal plant in Puranas (Sensarma, 1989). In the 'Sutra' period (800 BC -300 BC), the mention of radish roots is made as aiding digestion when munched after meals. In 'Susruta Samhita' the medicinal treatise of 3rd -4th century AD radish has been recorded having medicinal properties (Seshadri and Chatterjee, 1996). In India, West Bengal, Bihar, Uttar Pradesh, Punjab, Assam, Haryana, Gujarat, and Himachal Pradeshare major radish producing states.

Species Availability, Evolution and Diversity

Radish does not occur in wild state. It probably evolved from closely related species such as *Raphanus raphanistrum* L., *R. maritimus*, *R. landra* and *R. rostratus* which are native to Mediterranean region. Japanese type radish *Raphanus sativus* f. *raphanistroides* which is found wild in coastal area of Japan. Some researchers also had speculation that radish have been derived from the *R. sativus* var. *niger*, seems reasonable. Indian group of radishes seems to have evolved in the Near Eastern and Indo-Burma centre of origin, from ancestors in their area of distribution which now have become extinct. The globular forms of salad radish were developed from *radicula*. The large rooted radishes are cultivar of the *niger* and *radicula*. *Raphanus raphanistrum* L. is a widely distributed weed in Europe. More than one source is thought to have contributed to it, which is evident from the ecological and morphological differences existing across the globe. The Indian type of radishes got evolved in the Near Eastern and Indo-Burma centre of origin. Its presumed ancestor has been lost from the nature. Fodder radish also bears little or no fleshy root. Genus *Raphanus* includes *R. caudatus*, *R. landra*, *R. lyratus*, *R. pterocarpus*, *R. raphanistrum*, *R. raphanistrum* subsp. *Landra*, *R. sativus*, *R. sativus* var. *caudatus*, *R. sativus* var. *longipinnatus*, *R. sativus* var. *niger*, *R. sativus* var. *oleiferus*, *R. sativus* var. *oleiformis*, *R. sativus* var. *sativus*, *R. violaceus* and *X Brassicoraphanus* species. *R. sativus* var. *niger*, seems to be progenitor of common radish. Decreasing in crop variability has been observed to China and from there to Japan. It is summarized that that it would have originated by natural or artificial crossing between *R. laldra* and *R. maritimus* (Kitamura, 1958 and Tsunoda, 1991)

Table 6.10: Taxon and Preferred Taxon Name of *Raphanus*

Name of Taxon	*Name of Authority*	*Family*	*Preferred Taxon Name*
Raphanus caudatus	L.	Brassicaceae	*Raphanus sativus* var. *caudatus*
Raphanus landra	Moretti ex DC.	Brassicaceae	*Raphanus raphanistrum* subsp. *landra*
Raphanus lyratus	Forssk.	Brassicaceae	*Enarthrocarpus lyratus*
Raphanus pterocarpus	Pers.	Brassicaceae	*Enarthrocarpus pterocarpus*
Raphanus raphanistrum	L.	Brassicaceae	*Raphanus raphanistrum*
Raphanus raphanistrum subsp. *landra*	(Moretti ex DC.) Bonnier and Layens	Brassicaceae	*Raphanus raphanistrum* subsp. *landra*
Raphanus sativus	L.	Brassicaceae	*Raphanus sativus*
Raphanus sativus var. *caudatus*	(L.) L. H. Bailey	Brassicaceae	*Raphanus sativus* var. *caudatus*
Raphanus sativus var. *longipinnatus*	L. H. Bailey	Brassicaceae	*Raphanus sativus* var. *niger*
Raphanus sativus var. *niger*	J. Kern.	Brassicaceae	*Raphanus sativus* var. *niger*
Raphanus sativus var. *oleiferus*	Stokes	Brassicaceae	*Raphanus sativus* var. *oleiformis*
Raphanus sativus var. *oleiformis*	Pers.	Brassicaceae	*Raphanus sativus* var. *oleiformis*
Raphanus sativus var. *sativus*		Brassicaceae	*Raphanus sativus* var. *sativus*
Raphanus sp.		Brassicaceae	*Raphanus* sp.
Raphanus violaceus	L.	Brassicaceae	*Orychophragmus violaceus*
X Brassicoraphanus sp.		Brassicaceae	*X Brassicoraphanus* sp.

The area of maximum diversity can be spotted in the regions extending from eastern Mediterranean to Csapian Sea.

Gene and Genetics of Characters

In radish self-incompatibility is sporophytically determined by a multiple allelic series (Si-59) with dominance relations. Pseudo self-incompatibility as observed in radish represents a breakdown of the self-incompatibility system, giving limited to full seed set after an incompatible pollination.

Botanical Description

Radish belongs to family Brassicaceae and botanical name is *Raphanus satious* L.

Kingdom	Plantae
Unranked	Angiosperms
Unranked	Eudicots
Unranked	Rosids
Order	Brassicales
Family	Brassicaceae
Genus	*Raphanus*
Species	*sativus*
Botanical name	*Raphanus sativus*

Table 6.11: Botanical Description

Growth Habit	Radish is an annual or biennial herb depending on types. It is a very quick growing crop.
Root	The fleshy edible portion is swollen tap root (fusiform) formed from both the primary root and the hypocotyls (Choudhary, 1983) which can be globular, tapering, or cylindrical. It varies in shape, size, colour and other external features. Shape may be long, half-long, globular, pear shaped, flat round, tapering and stump. Colour varies from white, red, yellow, scarlet and black. Flesh colour is white/creamy white but variation for flesh colour can be observed. A German botanist reported radishes of 45 kg and roughly 90 cm in length.
Stem	The shoot system consists of stem (including branches.
Leaf	Leaves are in rosette form covered with stiff bristles, lyrate-pinnatifid size varying from 10.0-45.0 cm in same small-rooted cultivars to as such as 45.0 cm long in some large rooted cultivars. Leaves are dark or light green, margin either serrated or straight. The leaves are usually covered with stiff bristle. Various morphological variations for leaves are recorded in segregating generation of crosses between Japanese and Asiatic types.
Floral biology	Inflorescence of radish is typical terminal raceme with white small flower, liliac or rosy in colour with purple veins in bractless racemes. The flowers are homogamous and open during 11.00-14.00 hours. Sepals are erect and clawed petals.
Anthesis and pollination	Anthesis occurs during 9.00-10.00 hours. Anther dehiscence is between 9.00 and 10.00 hours. Pollen fertility is more on the day of anthesis. When stored at 3°C, the pollen is viable for 60 days. Stigma is receptive at the time of anthesis and lasts till 4 days after anthesis (Kumar, 1982). Radishes are self-incompatible (sporophytic in which the papillae on the stigma surface form the barrier to the pollen tube penetration) and entomophilous. The self-incompatible behavior is sporophically determined by a multiple allelic series (Si-59) with dominance relations. They are hermaphrodite (have both male and female organs) and are pollinated by bees and flies (PFAF 2008).
Pod and pod set	Pod is siliqua, indehiscent, 3.0-20.0 cm long and up to 1.5 cm in diameter with 6-12 seeds which are red to brown in colour and long conical beak.
Seeds	Seed set is reported 3-12 times greater after open pollinated than selfing. Seed is globose in shape with light to dark red in colour and 3.00 mm in diameter. Radish seed when mature is at first yellowish but turning reddish brown with age. There will be 120 seeds in 1.0 kg of seeds. The seed take 7-14 days to germinate under optimum environmental conditions and remain viable for 3-5 years when stored at room temperature.

Few researchers treated the *R. sativus* as a sub-species under *R. raphanistrum* L. *R. sativus* exists in several forms. *R. radicula, R. niger, R. mougri,* and *R. oleifera* intercross freely and also hybridize with wild *Raphanus* species. The larger radish has a wide range of adaptability, withstands relatively high temperature but the small and short duration type is strictly a cool season vegetable. The cultivated radish is diploid in nature. The tap root (except in var. *mougri*) is swollen and varies from almost globular, about 1.0-2.0 cm in diameter in the salad types to as much as 1.0 m long and 15.0 cm in diameter, cylindrical or conical in shape, in the oriental types and weighing up to 15.00 kg. The flesh is normally white, though in some may be pink to red. In the salad radish, the skin is usually red (occasionally white); in the Oriental radish it is normally white. In radish red colour is due to anthocyanin and pungency is accounted by volatile isothiocyanates (trans-4-methyl-thiobutenyl isothiocyanate).

Horticultural Classification

Radish has numerous categories-varying in colour, size, shape and flavor of roots; period of maturity, leaf colour and morphology and vernalization requirement.

Classification on the Basis of Botanical Varieties, Recognized within the Species

Four botanical varieties are recognized within the species, *R. sativus* L. namely:

(i) *R. radicula,* (tuberous roots)

(ii) *R. niger,* (tuberous roots)

(iii) *R. mougri* (or *caudatus*)

(iv) *R. oleifera* (grown primarily for the oil in its seeds)

The first 2 of which are grown for their tuberous roots, whereas *oleifera* is grown primarily for the oil in its seeds (Kay, 1987)

Classification on the Basis of Climatic Adaptation

Banga (1976) divided the cultivated radish world into 4 types:

(i) Small cool season

(ii) Large with wider adaptation

(iii) Rat tail or *mougri* type with slender longer pod (20-60 cm) and forming no fleshy roots.

(iv) Fodder radish with no fleshy roots but more of leaves

Classification on the Basis of Ecological Adaptation

On the basis of ecological adaptation and plant morphology, the radishes are:

i. European or Temperate Types

The European types which are frost tolerant, rich in carotene, long-day plants, and strictly annuals, are called *R. sativus*. The roots are small in size, mildly pungent roots and small tops excellent for salad making. Roots remain edible for a few days only and developed pithiness if harvesting is delayed. The European types originated directly or through hybridization from a few wild species like *R. maritimus*, *R. landra* and *R. rostratus*. They are early (20 to 30 days or so) but cannot tolerate high temperature (due to this reason European or Temperate types are not grown in tropical and subtropical climate but for seed production, they are grown in the hills).

ii. Japanese Types

The Japanese types which are intermediate to Indian and European types, growing in the coastal regions of Japan and neighboring regions, and the turnip shaped giant radishes cultivated in the Sakurjuna Island of Japan. The Japanese types are derived from *R. sativus* f. *raphanistroides* (syn. *R. raphanistroides*).

iii. Asiatic or Tropical and Sub-tropical or Indian Types

The Indian types which are heat tolerant, pungent, and mostly used for vegetable and salad purpose but sometimes also used for pickle making are called *R. indicus*. Root development is slow and top is very heavy. Roots remain edible for longer period. The roots are pungent in taste. Genotypes of this group are good for pickle and cooking purpose. These types can produce seed both in hills and in the plains. These types of radish are grown for one season for root production and for one more season for seed production that is the biennial for seed production. They can tolerate high temperatures

The morphological differences among Indian, Japanese and European types are not due to differences in chromosome number and the luxuriance observed in certain forcing types is not due to polyploidy.

Table 6.12: Wild Species and their Common Name

Species	Common Name
R. sativus var. *niger* (Mill) Pers	Diakon type-Large rooted winter radish of Asia
R. sativus var. *radicula* DC	Common radish
R. sativus var. *caudatus* L.	Rat tail radish
R. sativus var. *olefer* Netz	Fodder radish

All are intercrossable to each other.

Classification the Basis of Root Shapes

(i) Round to Oval

(ii) Icicle

(iii) Half long

(iv) Long-These types further divided as:

(v) Conical

(vi) Cylindrical

(vii) Spindle

A longer root form, including oriental radishes, daikon or mooli, and winter radishes, grows up to 60 cm long with foliage about 60 cm high with a spread of 45 cm. Daikon refers to a wide variety of winter radishes from east Asia. While the Japanese name *daikon* has been adopted in English, it is also sometimes called the Japanese radish, Chinese radish, or Oriental radish (AMHER, 2004). In areas with a large South Asian population, it is marketed as *mooli*. Daikon commonly has elongated white roots, although many varieties of daikon exist. One well known variety is April Cross, with smooth white roots (Faust 1996; Peterson 1999). Faust (1996) describes Masato Red and Masato Green varieties as extremely long, well suited for fall planting and winter storage. The Sakurajima daikon is a hot flavored variety, which is typically grown to around four to five pounds (1.8 to 2.2 kilograms) when harvested, but which is reputed to grow as heavy as 70 pounds (32 kilograms) when left in the ground (Faust 1996)

Classification the Basis of Root Colour

(i) White

(ii) Pink

(iii) Red

(iv) Purple

(v) Black

Recent classification suggests a partition of the cultivated radishes into three botanical con-varieties (Crisp, 1995):

(i) *oleifer* Alef. (Oilseed and fodder radish),

(ii) *caudatus* (L.F.) Pistr. (Pod radish) and

(iii) *sativus* (Small and large rooted).

Classification on the Basis of Edible Root Maturity (Days)

- Quick growing-There are quick growing spring type radishes (25-30 days).
- Slow growing- There are slow growing summer and winter radishes (40-55 days).

In India, summer and winter radishes are grown in one or the other parts of the country almost throughout the year because of the varied climatic conditions, cultivars, economic importance and demand.

All con-varieties intercross freely and also with related wild species. It is a dicotyledonous biennial herbaceous plant usually grown as an annual crop for its enlarged fleshy tap root. *Raphanus sativus* (2n =2x=18) is normally a self-incompatible, Insect pollinated crop (Rubatzky and Yamaguchi, 1997). However, by rupturing the stigma surface or by applying pollen to the conducting tissue of style, it is possible to obtain normal seeds by selfing. Normal seeds may also be obtained by bud pollination (George and Evans, 1981). Cultivars have also been developed and maintained as open pollinated, outcrossing populations (Crisp, 1976; Simmonds, 1979).

Augmentation of Genetic Resources and Crop Improvement

The commonly adopted procedures for crop improvement are mass selection, recurrent selection, hybridization and heterosis.

Breeding Priorities

- High yield
- Earliness
- Non-pithiness
- Uniformity in maturity
- Better quality and tenderness
- Root shape
- Tolerance to biotic and abiotic stresses

Introduction and Selection

Popular cultivar 'Japanese White' introduced in India and directly used as a variety for commercial cultivation. Fifteen germplasm lines of radish collected from different parts of Meghalaya and other parts of India were evaluated on the basis of twenty three qualitative and quantitative traits in a randomized block design with three replications at KVK, ICAR-(RC) NEH region, Sangsangiri, Tura, West Garo Hills, Meghalaya (India) by Chakraborty *et al.* (2016). Significant variations was observed among genotypes for leaf length, leaf width, number of leaves, petiole length, crown head diameter, root length, root diameter, dyas to root harvest, root weight and yield. Five genotypes namely Tura Selection-7, Tura Selection-8, Tura Selection-4, Tura Selection-1 and Shillong Selection-1 were identified as superior genotypes on the basis of yield and root weight which could be used in breeding programmes for developing new varieties. Highest value of genotypic coefficient of variation (GCV) was observed for yield (51.46 t/ha), while leaf width exhibited the lowest (9.37 cm). All the characters showed medium to high broad sense heritability estimates during investigation. Highest heritability for yield (94.14 per cent) and lowest heritability for root diameter (38.33 per cent) was observed. However, maximum genetic advance (GA) as percentage of mean for yield (5.83) and minimum for root diameter (0.51) was noticed. 'Scarlet Globe', 'Rapid Red', 'White Icicle', and 'Japan Scarlet' are recommended for spring season cultivation; 'White Chinese', 'Japanese White', 'Pusa Desi' and 'Desi' types are evolved for winter cultivation and 'Black Radish' and 'Pusa Chetki' are suited for summer cultivation' It has oval tubers and cream white colour. 'Selection No. 73' is early, having good shape and is suited for the rainy season cultivation. 'Arka Nishant' was the variety developed at ICAR-IIHR, Bengaluru (Karnataka) in 1972 which produces high yield with the better quality roots in the southern region. Horticulture Research Centre, Bangladesh Agricultural Research Institute (BARI) has identified two potential cultivars 'Tasaki San' developed in the year 1983 and 'Pinky' (1-13-228) for the Bangladesh conditions. In Pakistan, cultivars like Yokohama (R), 40 Days, Mino, Lalpari (I) and Korea No.4 (P) are being common and grown commercially. In Korea radish cultivar 'Bulamdaehyung' has enabled the spring and summer production. 'Pyongjiyorum' is the first Korean radish that can be successfully grown under lowland summer conditions. Ola *et al.* (2016) evaluated 22 varieties (namely, Arka Nishant, Baramasi, bharsar local-2, Delhi White, Dunagiri, Evergreen, Hill Queen, J.U., Japanese White, Local-1, Local-2, Local-3, Local-4, Local-5, Local-6, Local-7, Local-8, M.E.L.W., Pusa Mirdula, Snow White, Sonali White and Pusa Chetki) and found significant differences among the varieties of radish. The growth parameters showed maximum plant height (66.33 cm) and number of leaves (23.70) in Snow White, whereas the minimum plant height (31.92 cm) and number of leaves (10.29) in Pusa Mirdula. The yield parameters showed significantly maximum root diameter (5.91 cm) was observed in Pusa Mirdula, while the minimum diameter of root (2.72 cm) was recoded in Local-2. The maximum total weight of plant (435.31 g/plant), root yield per plot (30.56 kg) and root yield per/ha (509.28q) in Evergreen, whereas the minimum weight of root (107.03g), total weight of plant (150.37g), root yield/plot (15.00 kg) and per hectare (250.05 q) was recorded in Pusa Mirdula.

Table 6.13: Radish Varieties Developed through Mass Selection

Variety	*Source*
Pusa Desi	Indigenous
Punjab Safed	Indigenous
Punjab Ageti	Indigenous
Kalyanpur-1	Indigenous
Arka Nishant	Indigenous

Popular radish cultivars like Pusa Chetki and CO-1 were developed by using exotic materials received from Denamrk and Indigenous material respectively through the family selection.

Hybridization and Selection

The F_1 hybrids are produced by single crosses or double crosses. Attempts have been made to improve upon the Asiatic strains by crossing with the exotic varieties to obtain new varieties with roots qualities of temperate types and ability to produce seeds in the plains (Singh *et al.*, 1971; Dutta and Vijay, 1970). An important achievements in this direction has been made in release of radish variety 'Pusa Himani' which is a selection from cross between Black Radish' x Japanese White' (Singh *et al.*, 1971). Similarly 'Pusa Rashmi' is also selection from a cross 'Green Top' x 'Desi'. Several promising hybrids (F_1s) were developed by utilizing elite genetic stocks like-'White Tipped' x 'Icicle', 'Rapid Red' x 'Sparkler', 'Desi White' x Japanese White', 'Icicle' x 'Round Black Winter' and 'Red White Tipped' x 'Ruby'. Horticulture Research Centre, Bangladesh Agricultural Research Institute (BARI) has observed that hybrid in radish 'Everest' is dominating in several area of Bangladesh.

Intetspecific Hybridization

Transgression of useful traits is possible by wide interspecific crosses between *Raphanus* and *Brassica* species by protoplast fusion and genetic engineering (Nath *et al.*, 2008). Certain species like *R. indicus*, *R. sativus*, *Sensu stricto* and *R. raphanistroides* are completely cross compatible. These lines are being utilized for development of heat tolerant and diseases and pests resistant genotypes. *R. radicula*, (tuberous roots), *R. niger*, (tuberous roots), *R. mongri* (or *caudatus*) and *R. oleifera* (grown primarily for the oil in its seeds) varieties intercross freely and also hybridize with wild *Raphanus* spp.

Somatic Hybrids

Application of somatic hybridization (fusion between protoplast) can be seen in production of *Raphanobrassica* by Karpechenko (1928).

Table 6.14: Somatic Hybrids Produced in Cauliflower

Species Combination	*Character Transferred*
Raphanus sativus x *Brassica napus*	Club rot resistance
R. sativus x *B. oleracea* var. *botrytis*	Club rot resistance

Intergeneric Hybridization

First classical work on intergeneric hybridization was done in crucifers. *Raphanobrassica* is an intergeneric hybrid between radish (*Raphanus sativus* L.) and cabbage (*Brassica oleracea* L.) was reported by Karpechenko (1928). Subsequently, a number of intergeneric between *Brassica* species of the 'U' triangle and *R. sativus* have been produced; the 'U' triangle refers to the ancestral A, B and C genomes that have been rise to modern *Brassica* crop species. Radish (*Raphanus sativus* L., 2n=18 crosses readily with cabbage (*Brassica oleracea* L. convar. *capitata*, 2n=18) and cauliflower (*Brassica oleracea* L. convar. *botrytis*) showing unilateral incompatibility. Varying degree of sterility was observed when radish was crossed with Abyssinian mustard (*B. corinata*), Chinese cabbage (*B. chinensis*) and turnip (*B. rapa*).

Male Sterility

Cytoplasmic male sterile (Ogura CMS) line has been identified first time in cultivar of wild Japanese radish by (Ogura, 1968) and was introduced by transferring to *Brassica oleracea* L. genome through repeated backcross with broccoli (Bannerot *et al.*, 1974; McCollum, 1981) (leaf chlorosis, floral abnormalities). In radish, Ogura CMS is controlled by recessive nuclear gene (msms) in combination with sterile cytoplasm (S-cytoplasm). Later on several types of CMS system has been characterized in *Raphanus* genus by various researchers, namely 77-01A CMS, Kosena CMS, NWB CMS, DCGMS and 805A CMS. Now it is used for hybrid seed production in several countries. Muthukrishnan and Arumugam (1971) reported that the root length is the most important contributory factor for yield. Radish with cytoplasmic sterility was found from local cultivars 'Henas' and 'Shandong' in China. *Rf* is frequent in European and Chinese cultivars. Genetic male sterile, 3 recessive ms genes reported and being commercially exploited. The genotype of male sterile plants is *S msms* and maintainer line has *N msms* genetic makeup. Usually the cytoplasm of zygote comes primarily from the eggs cell and due to this progeny of such male sterile plants would always be male sterile. The plants show cytoplasmic male sterility no matter what is a nuclear gene. It has the characteristics of the cytoplasmic inheritance, maternal inheritance, and its sterility is easy to be maintained, but not easy to be recovered. CMS could be transferred easily to a given line by using that line as a pollinator (recurrent parent *i.e.* B-line) in the successive generation of backcross programme. After 5-7 backcross, the nuclear genotype of new male sterile line (A-line) would be almost identical to that of the B-line (>98 per cent).

Resistance Breeding

i. Disease Resistance

Emphasis has also been laid on developing varieties resistant to disease. Dutta and Vijay (1970) reported radish strain 'IIHR-1' to be highly resistant to white rust (*Albugo candida*)

Table 6.15: Diseases and their Resistance Source

Disease	Pathogens	Resistant Varieties/ Genotypes
Wilt	*Fusarium oxysporum* sp. *raphani*	'Fix'
Black rot	*Aphanomyces raphani*	'Summer Wonder', Regensburg Market'
Downy mildew	*Albugo candida*	'China Rose White', 'Round Black Spanish'
Leaf spot	*Alternaria* sp.	'Punjab Safed'
Mosaic	Virus	Country's Wealth, 'High Agriculture', 'Shogani', 'Early Large', 'Nerina.

ii. Insect Resistance

In resistance breeding programme variety 'Asmer Tip Top Sparkler' is resistant to pest *Erioischia brassicae*.

Breeding for Quality Traits

Quality parameter in radish includes white, long/ stump roots; thin tap root, non-branching habit, non-pithy roots and pungency of roots as per consumer preference. The information on the genetic variability heritability and inter-relationship of total phenolics, anthocyanins and antioxidant activities in coloured tropical radish is very limited, but pre-requisite to initiate breeding programme. Singh *et al.* (2016) observed significant variability for all the antioxidants; ranged from 12.4-61.7 mg/100 g for total phenolics, 4.6-166.3 μg/g for anthocyanins content, 1.14-5.66 μmol/g for FRAP activity and 2.73-11.01 μmol/g for CUPRAC activity; and the genotypes alone account for >97 per cent of total variations. Furthermore, there were meager differences between phenotypic and genotypic coefficient of variation indicating the greater role of genotypes and lesser influence of the environment on the biosynthesis and accumulation of pigments and antioxidants. Significant positive correlations along with higher magnitude for anthocyanins content, total phenolics, FRAP activity and CUPRAC activity (0.823-0.964) could be used as indirect selection criteria for improving levels of antioxidant compounds. The estimates of heritability and genetic advance indicate the role of additive and non-additive genes for biosynthesis of antioxidants and root development, respectively; therefore, recurrent selection would be the best breeding approach to improve the both traits (yield and antioxidants) simultaneously in coloured radish.

Genetic Resource Conservation

Different useful strains of Asiatic types have been reported from different regions of the country. A good number of temperate and Asiatic varieties were also introduced and some of them have been grown as commercial varieties for decades, but their collection and conservation aspects are lacking because of seed production (require specific environment which is available in Kashmir, Kulu Valley (Himachal Pradesh) and Nilgiri Hills (Tamil Nadu), only.

REFERENCES

1. Bailey, L.H. and Bailey, E.Z. 1956. Hortus Second: A Concise Dictionary of Gardening and General Horticulture. Mac Millon Co., New York.
2. Banga, O. 1976. In, Simmonds, N.W. (Ed.) Evolution of Crop Plants. Longman, London.
3. Bannerot, H.O., Boulidard, L., Cauderon, Y. and Tempe, T. 1974. Cytoplasmic male sterility transfer from *Raphanus* to *Brassica*. In: EUCARPIA Cruciferae Conference. Scottish Horticulture Research Institute, Myinefield, Dundee, Scotland, p. 52.
4. Chakraborty, S., Barman, A., Kumar, S., Momin, B.C. and Sangma, S.M. 2016. Studies on genetic diversity of radish (*Raphanus sativus* L.) under West Garo Hills conditions (Abs: 69, Session I, Breeding for Horticultural Crops). 7th Indian Horticulture Congress, held at ICAR-IARI, New Delhi, 15-18th November 2016, p. 55.
5. Chakraborty, S., Barman, A., Kumar, S., Momin, B.C. and Sangma, S.M. 2016. Studies on genetic diversity of radish (*Raphanus sativus* L.) under West Garo Hills conditions (Abs: 20, Session II, Breeding for Horticultural Crops). 7th Indian Horticulture Congress, held at ICAR-IARI, New Delhi, 15-18th November 2016, p. 101.
6. Crisp, P., 1995. Radish, *Raphanus sativus* (Cruciferae). In: Smartt, J. and N.W. Simmonds (Eds.) Evolution of crop plants. 2nd Edition, p. 86–89. Longman Scientific and Technical, UK.
7. Crisp, P. 1976. Trends in the breeding and cultivation of cruciferous crops. In: Vaughan, J.G. Mac Leod, A.J. & Jones B.M.G. (Eds.) The bilology and chemistry of the cruciferae, p. 69-117.
8. Dutta, O.P. and Vijay, O.P. 1970. Improvement of Asiatic root crops through selection and hybridization. Annual Report. ICAR-IIHR, Bangalore (Karnataka).
9. Faust, J. Lee. 1996. Hail the speedy radish, in all its forms. *New York Times*, March 3, 1996. Retrieved July 21, 2008.

10. George, R.A.T. and D.R. Evans, 1981. A classification of winter radish cultivars. *Euphytica*, **30:** 483–492.

11. Helm, J. 1957. Ueberden typus der art *Raphanus sativus* deran Gliedevberg amnd Synonymica. *Kulture pfianze*, **5:** 41-54.

12. Karpechenko, G.D. 1928. Polyploid hybrids of *Raphanus sativus* x *Brassica oleracea* L. Z. Ind. Abst. *Verebgsl*, **48**: 1-85.

13. Kay, E.D. 1987. Root Crops. Tropical Development and Research Institute, London p. 174-186.

14. Kitamura, S. 1958. Varieties and transitions of radish. In : Japanese radish, (ed.) I. Nishiyama, *Jpn. Soc. Fron. Sci.* Tokyo, **1**.

15. Kumar, V. 1982. Pollination and seed setting studies in radish (*Raphanus sativus* L.). *M.Sc. thesis*, Himachal Pradesh Krishi Vishva Vidyalaya, Palampur (Himachal Pradesh).

16. Mansfield, R. 1959. Die Kulturpfianzen Baikeft, p.1-659. Akademic Verlog. Berlin.

17. McCollum, G.D. 1981. Induction of an alloplasmic male sterile *Brassica oleracea* L. by substituting cytoplasm from 'Early Scarlet Globe' radish (*Raphanus sativus* L.). *Euphytica*, **30(3):** 855-859.

18. Muthukrishnan, C.R. and Arumugam, R. 1971. Correlation and regression studies in radish. *Indian J. Hort.* 34: 163-165.

19. Nath, P.; Srivastava, V.K., Dutta, O.P. and Swamy, K.R.M. 2008. Vegetable crops- Improvement and Production. PNASF, Bengaluru (Karnataka), pp. 182-187.

20. Ogura, H. 1968. Studies on the new male sterility in Japanese radish with special reference to the utilization of this sterility towards the practical raising of hybrid seeds. *Mem. Fac. Agric. Kagoshima Univ.*, p. 39-79.

21. Ola, A.L., Jhajhra, M.R., Rana, D.K. and Bairwa, L.N. 2016. Performanse of radish (*Raphanus sativus* L.) varieties for growth and yield potential under valley condition of Garhwal hills. 7th Indian Horticulture Congress (Abs: 91), held at ICAR-IARI, New Delhi, 15-18 November, p. 70.

22. Peterson, C. 1999. Radishes : Easy to sprout, hard to grow right; The New York Times, May-2, 1999.

23. Plants for a Future (PFAF). 2008. *Raphanus sativus. Plants For A Future.* Retrieved July 21, 2008.

24. Rubatzky, V. E. and Yamaguchi, M. 1997. World Vegetables: Principles, Production and Nutritive Values 2nd Edition, Chapman & Hall, USA. p. 843.

25. Sadhu, M.K., 1993. Root crops. In: Bose, T.K., M.G. Som and J. Kabir (Eds) Vegetable crops, p. 470–491. Naya Prokash, Calcutta, India.

26. Rubatzky, V.E. and M. Yamaguchi, 1997. World vegetables: Principles, production and nutritive values and Edition, Chapman and Hall, USA, pp. 843.

27. Seshadri, V.S. and Chatterjee, S.S. 1996. The history and adaptation of some introduced vegetable crops in India. *Veg. Sci.*, **32(2):** 122-142.

28. Simmonds, N.W., 1979. Principles of Crop Improvement. Longman, New York.

29. Singh, B.K., Koley, T.K., Singh, B. and Singh, P.M. 2016. Colourful radishes for healthier salad. ICAR News: *A Science and Technology Newsletter*, **22(2):** 12.

30. Singh, B.K., Koley, T.K., Karmakar, P., Tripathy, A. and Singh, B. 2016. Genetic variability and inter-relationships of total phenolics, Anthocyanins and antioxidant activity in coloured radish (*Raphanus sativus* L.). 1st International Agrobiodiversity Congress, held at New Delhi; 6-9 November, 2016; p. 23.

31. Singh, J.P., Gill, H.S., Swarup, V. and Arora, R.S. 1971. 'Pusa Himani' radish. *Indian Hort.*, **16(1):** 13-14.

32. Tsunoda, S. 1991. The history of cultivated plants in the tribe Brassicae. *Agric. Hort.*, **66:** 283.

Rat Tail Radish (*Raphanus caudatus* L.)

Mougri/Moongra rat tail radish (*R. caudatus* L.) forms no fleshy root (non-tuber forming) is closely related to the Indian type of radishes. The Indian types of radishes are got evolved in the Near Eastern and Indo-Burma centre of origin. It is presumed that ancestor has been lost. It is grown in Southern Asia for its leaves and young seed pods (Siliqua) which are 20-100 cm long. Similarly, fodder radish bears little or no fleshy root and is also grown for its foliage used as fodder or green manure. This species is closely related to the Indian type of radishes.

Chinese Radish (*Raphanus sativus* L. var. *niger*)

INTRODUCTION

The Chinese radish is a favourite crop of the home gardeners because of very short duration nature (3-6 weeks after seed sowing) and simple growing practices. It is fast growing erect annual herb for root production (Kaul and Kanwar, 1997). Roots of Chinese radish is mildly pungent and an excellent vegetable for cooking. During severe winters of Ladakh, when no fresh vegetables are available (Singh and Singh, 2007) Chinese radish is available in the market. On the basis of 100 g fresh edible portion, Chinese radish contains: protein 1 g, calcium 32 mg, iron 1.4 mg, vitamin C 26 mg, potassium 400 mg and energy 30 Kcal (Knott, 1988).

Origin, History and Distribution

Chinese radish is probanly originated from the old world, cultivated long before Greeks and Romans came to know of it as 'rhaphanis' and 'radix' respectively (Thomson and Kelly, 1957). It is widely grown in USA, China, Europe and in other temperate part of the world. In India it is cultivated in Kashmir (Himachal Pradesh), Leh and Kargil (Jammu and Kashmir) and scattered in other hill regions. It is less known underutilized vegetable. It is found growing in kitchen gardens (Singh and Singh, 2007).

Botanical Description

Chinese radish is biennial but cultivated as annual for root production. Seed production takes another season. The top of the plant is heavy. It is large and heavy rooted, broad shouldered, thick skinned. The roots have an attractive pink colour and weighing on an average 1-2 kg each (Anonymous, 1972). Chinese radish require temperate climate. The roots are thick skinned therefore; they can adopt wider range of temperature adoption (Ware and McCollum, 1980).

REFERENCES

1. Anonymous, 1972. The Wealth of India: Raw Material and Industrial Products. Vol. VIII. CSIR, New Delhi.
2. Kaul, B.L. and Kanwar, H.S. 1997. Chinese radish is a promising vegetable for hills. *Indian Hort.*, **42(1):** 16-17.
3. Knott, J.E. 1988. Handbook of Vegetable Growers. John Wiley and Sons, New York.
4. Singh, D.N. and Singh, S.P. 2007. Production Management of Underutilized Vegetables. Agrihortica Publications, Junagadh (Gujarat), India.
5. Thompson, H.C. and Kelly, W.C. 1957. Vegetable crops. McGraw Hill Book Co., Inc., USA.
6. Ware, G.W. and McCollum, J.P. 1980. Producing Vegetable Crops. Interstate Printers and Publishers. INC, Illionis, Danvilla.

Genus *Armoracia*

Horseradish (*Armoracia rusticana* Garten)

INTRODUCTION

Horseradish (*Armoracia rusticana* Garten), a near relative of turnip, cabbage, and mustard. It is one of the oldest condiments (appetizing condiments) in the world and its white fleshy and tasty roots are used as commercial part. The root contains a pungent, acrid and vesicating volatile oil. The pungency is due to presence of an allyl isothiocyanate and butylthiocynate similar to mustard oil, occurring in combination with the glucoside 'sinigrin' (Choudhury, 1996). Organoleptocally active substances of horseradish are mainly due to presence of 2-propenyl isothiocynate and 2- phenyl ethyl isothiocynate (Taborsky *et al.*, 1989). It is highly priced as condiment especially with oyster and cold meats. Inhaling of horseradish essential oil or taking horseradish by mouth promotes blood flow in nasal and sinus tissues and helps possibly relieving respiratory congestion.

Origin, History and Distribution

Horseradish originated from middle Europe. It is a large leaved hardy perennial herb used as culinary herb in different European countries more than 3000 years. It is a very popular condiment for foods in Germany and surrounding areas. It is grown in Eastern Europe, U.K., U.S.A., India and Sri Lanka. In India, it is grown in gardens both in north India and hill station of south India (Shastri, 1988).

Horticultural Classification

Horseradish is generally of two types:

1. Common type
2. Bohemian type

Common type produces better quality roots.

Augmentation of Genetic Resources and Crop Improvement

'Edelkafener', 'Jugosalavischer', 'Humberger' and 'Steirischer' are a few of the ecotypes of horseradish (Nebel *et al.*, 1988).

REFERENCES

1. Choudhury, B. 1996. Vegetables. National Book Trust, India (Reprinted), pp. 223.

2. Nebel, N., Weichmann, J. and Fritz, D. 1988. Yield of different horseradish ecotypes. *Gemuse*, **24(6):** 272-273.

3. Shastri, B.N. 1988. The Wealth of India (Raw Materials). CSIR Publication (Reprinted), Vol.2, pp. 260.

Turnip [*Brassica rapa* L. (syn. *B. compestris* Hooks f. and Thomas var. *rapifera* Metzger)]

INTRODUCTION

Turnip [*Brassica rapa* L. (syn. *B. compestris* Hooks f. and Thomas var. *rapifera* Metzger), 2n=20] is an important root-tuber vegetable grown as a summer crop in temperate climates and as a winter vegetable in tropical and subtropical climates. It is basically cultivated for their fleshy, succulent and tender roots. The roots are consumed fresh in salads, processed into pickles and cooked as a vegetable. Sometimes, turnip leaves are also cooked as a potherb or stir-fried. Roots are sometimes, sun dried and stored for later use. Fodder type turnip (*B. rapa* var, *rapa*) is extensively used as a fodder in Europe and New Zealand.

Origin, History and Distribution

The exact centre of origin of turnip (*Brassica rapa* L.) is not known. Eastern Afghanistan and the adjoining areas of Pakistan are considered to be the primary centre of origin of Asiatic types with Asia minor, transcaucasus and Iran, as secondary centres (Sinskaia, 1928) whereas the Mediterranean area is thought to be the primary centre of origin of European types (Bailey, 1963). Asia Minor, Transcaucasus and Iran are considered its secondary centre of origin (Sinskaia, 1928). At present its occurrence can be spotted throughout temperate regions, subtropics and tropics where the temperature for a period of 3 months is 15.6-21.1°C. Turnip is one of the oldest cultivated vegetables. The oil seed form of turnip is mentioned in 'Sanskrit' records in India as early as 2000-1500 BC. It has been under cultivation in Europe for the last 5000 years or so. Turnip was introduced in Canada in 1540 by Cartier during his voyage of exploration, and was taken to Mexico in 1586 and Virginia in 1610 (Peter, 1998). It has been a popular livestock fodder for at least 600 years. Roman Conquest brought it to England from France. It was taken to Mexico in 1586 and to Virginia in 1610. Turnip was introduced to North America in the late 1500's and has been grown in the USA since 1622. However, it was not until early 18th century that turnip was cultivated as a field crop in England. It is widely distributed in Europe, Germany, U.K., Russia and Asian countries *i.e.* Afghanistan, Pakistan, Iran, China, and India (Purewal, 1957). It is widely cultivated in Jammu and Kashmir, Delhi, Punjab, Himachal Pradesh and Western U.P. including Haryana and Punjab. It can be spotted (successful crop) at elevation of 1500 m or above (Brar and Nandpuri, 1972).

Species Availability, Evolution and Diversity

The parents of cultivated turnip are seen in Russia, Siberia and Scandinavia in wild forms, which have potential use in its improvement (Thompson and Kelly, 1957).

Botanical Description

Turnip belongs to family Brassicaceae and botanical name is *Brassica rapa* L. emend. Metzger (syn. *B. campestris* Hook. f. and Thomas var. *rapifera* Metzger), 2n=20. The crop is highly cross pollinated due to self-incompatibility and male sterility. Bees are the pollinators.

Table 6.16: Botanical Description

Growth Habit	Turnip is herbaceous annual or biennial in nature *i.e.* first season or year for enlarged fleshy tap root and another season for seed production.
Root	The fleshy thickened underground portion of turnip is actually the hypocotyl, the colour and shape of which vary depending mainly on the cultivar. Its shape may vary from flat through globular to top-shaped and long. The below ground colour may be white or yellow, white the skin colour of above ground portion may be red, purple, white yellow or green. A distinct tap root and secondary roots arise from the lower part of the swollen hypocotyl. A report from Japan shows that the elongation of turnip hypocotyl stops by the 6th leaf stage in all cultivars tested, whereas the beginning at the 4th leaf stage, while in others, it is not evident till the 6th leaf stage. Thickening begins in the central part of the hypocotyl, followed by upper and lower parts (Iwasaki and Takeda, 1977). Generally, the edible maturity of roots reaches in 40-80 days, depending on cultivars and cultural condition.
Leaf	Leaves and petioles of turnip are hairy and coarse to the touch and yellowish green in colour. Leaves are oblong to oval and may be entire, serrate or even-pinnate depending on cultivars and time of development. Leaves formed early are less likely to be pinnate. The leaves developing on the inflorescence are alternate, glaucous, oblong or lanceolate and entire or dentate.
Floral biology	The inflorescence of turnip is a terminal raceme on the main stem. The flowers in structure and arrangement are practically identical with other cruciferous crops. Flowers are hermaphrodite in nature.
Anthesis and pollination	The crop is highly cross-pollinated due to self-incompatibility and male sterility (Peter, 1998; Yamaguchi, 1983; Nath *et al.*, 1987).

Horticultural Classification

There are two distinct groups of turnip

1. Annual, tropical or Asiatic types
2. Biennial or temperate or European types- the seeds can be produced in the hills and plains (Gill, 1993)

Shoemaker (1949) classified turnip cultivars in to several groups on the basis of morphological characteristics of the root and top (A) White fleshed (B) Yellow fleshed and (C) Foliage cultivars (Table 6.17).

Turnip can also be classified considering the certain facts shown in Table 6.18.

Table 6.17: Grouping and Subgrouping of Cultivars

Group	*Sub-group*	*Cultivars with Distinction*
(A) White fleshed	(a) Purple Topped	(i) Flat type: Purple Top Milan
		(ii) Globe type: Purple Top White Globe
	(b) Green topped	(i) Globe type: Green Top White and Green Globe
		(ii) Long type: Cow Horn
	(c) White-Topped	(i) Flat type White Milan, White Flat Dutch and Norfolk
		(ii) Globe type: White Stone and Quick Silver
		(iii) Half-long type: White Egg and White Gem
		(iv) Long type: Lily White
(B) Yellow fleshed	(a) Purple topped, globe type	Aberdeen Purple Top
	(b) Bronze or Green topped globe type	Aberdeen Green Top and Amber Globe
	(c) Yellow topped, globe type	Yellow Globe, Golden Ball and All Seasons
(C) Foliage cultivars	-	Flat Japan, Shogoin and Seven Tops

Table 6.18: Trait Specific Groups

Sl.No.	*Base of Classification*	*Group and cultivars*
(A)	On the basis of skin colour	White Yellow Purple Red Occasionally green
(B)	On the basis of flesh colour	White Creamy Yellow
(C)	On the basis of root shapes	Flat Round Spindle Long
(D)	On the basis of utility	**(i) Vegetable type-**The vegetable types are classified as *B. compestris* var. *rapifera* **(ii) Fodder type-** The fodder type is a *B. compestris* var. *rapa*
(E)	On the basis of duration	**(i) Annual-** **Asiatic type-**'Ludhiana White', Punjab Red Long' **Japanese type-** Japanese varieties **(ii) Biennial-**'Snow Ball', 'Purple Top White Globe'
(F)	On the basis of climatic requirement	**(i) European (temperate) type- 'Purple Top White Globe', 'Snow Ball', 'Golden Ball', 'Early Milan Red'** **(ii) Asiatic (tropical) type- 'Pusa Sweti'**
(G)	On the basis of growth habit	(i) Erect plant types (ii) Spreading plant types (iii) Intermediate plant types

Table 6.19: Interspecific Hybridization

	Turnip	*Swede*	*Chinese Cabbage*	*Knol-khol*	*Cabbage*
B. rapa (Turnip)	SI	CCX	CC (X)	CC	CC
B. carinata (Swede, 2n= 34)	CC X	SI SC	CC	CC	CC
B. chinensis Juslen, non Duthie and Fuller	CC	CC	SI	CC	CC
B. pekinensis Rupr (2n= 20)	(X)		SC		
B. oleracea L. var. *gongylodes* (2n= 18)	CC	CC	CC	SI SC	CC
B. oleracea var. *capitata* (2n= 18)	CC	CC	CC	CC	SI SC

SI: Mainly self-incompatible; SC: Self-compatible forms are also present; CC: Cross compatible; X: F_1 sterile and (X): F_1 fertile.

Augmentation of Genetic Resources and Crop Improvement

Turnip is a cross pollinated hence inbreeding is done in this crop. The important uses of inbreeding in cross pollinated crops are to be obtain uniformity in plant characters, to improve yield by individual plant selection and to combine suitable inbred lines in the production of hybrids and synthetics. Self-fertilization is the most intense form of inbreeding. Until 3rd or 4th generations, inbreeding depression is not much. Mass selection, progeny selection and hybridization are the common breeding systems in turnip. In Nepal turnip cultivar like 'Sakurajima Mamoth' introduced by Takii Japan is very popular among the growers. In Pakistan, cultivars like Purple Top, Golden Ball, Nankana Red (R) and Milan Red Top (P) are being very common and grown commercially.

Introduction and Selection

Popular cultivar 'Purple Top White Globe' introduced in India and directly used as a variety for commercial cultivation. Variety 'Improved Golden Ball' has been selected through mass selection from a population of 'Golden Ball'. Similarly 'Pusa Sweti' was selected (family selection) which is early-maturing tropical variety with medium–sized and round tubers. Its flesh texture is white, fine grained and of soft texture.

Hybridization

Hybridization between Asiatic and temperate types has been effective in this crop (Nath *et al.*, 2008). As utilization of elite lines, F_1 hybrid 'Pusa Chandrima' (European type), (Singh *et al.*, 1971) was developed (ICAR-IARI, New Delhi) by crossing of 'Snow Ball' x 'Japanese White' which expresses superiority over 'Snow Ball'. 'Pusa Kanchan' (Singh, 1963) is another selection from the cross 'Golden Ball' (European type) x 'Ludhiana Red' (Local Red Round Asiatic type).

Interspecific Hybridization

Interspecific hybridization is resorted to generate variability for useful traits (Table 6.19).

Inter-generic Hybridization

A triploid (2n=28) was isolated from a cross between a tetraploid fodder radish and a tetraploid turnip. The triploid had 9 bivalents and 10 univalent.

Mutation Breeding

A non-waxy mutant was developed using gamma rays on 'Pusa Sweti'. The mutant showed resistance to aphids.

Polyploidy Breeding

A triploid (2n=28) was identified from a cross between a tetraploid fodder radish and a tetraploid stable turnip. The triploid had 9 bivalents and 10 univalents.

Resistance Breeding

Varieties like 'Dutch Half Long', 'Dale Hybrid' and 'Yellow May' are resistant to club root, whereas' Charleston' showed resistance to cold wind and bolting.

Abiotic Stress

There are some important genotypes like Sel.-66, Snow Ball, and Purple Top White Globe which can be suitably grown under saline and sodic soil.

Genetic Resource Conservation

A number of good genetic stocks of Asiatic and European types are available in India. Efforts were made in the past to maintain some of the Asiatic strains, but no intensive breeding work was taken up to broaden the genetic base.

REFERENCES

1. Bailey, L.H. 1963. *The Standard Encyclopaedia of Horticulture*, **3:** 3403-3404.
2. Brar, J.S. and Nandpuri, K.S. 1972. Cultivation of root crops, Punjab Agricultural University, Ludhiana (Punjab).
3. Gill, H.S. 1993. Improvement of Root Crops-Turnip. In: Advances in Horticulture, Vol.5, Vegetable Crops: Part: 1, (Eds. Chadha, K.L. and Gautam, Kalloo) Malhotra Publishing House, New Delhi.

4. Iwasaki, F. and Takeda, Y. 1977. Morphological and histochemical observation on the thickening of the hypocotyl in *Brassica rapa* L., **46(2):** 193-200.

5. Nath, P.; Srivastava, V.K., Dutta, O.P. and Swamy, K.R.M. 2008. Vegetable crops- Improvement and Production. PNASF, Bengaluru (Karnataka), p.188-189.

6. Nath, P., Velayudhan, S. and Singh, D.P. 1987. Vegetables for the tropical region. ICAR, New Delhi.

7. Peter, K.V. 1998. Genetics and Breeding of Vegetables. Indian Council of Agricultural Research, New Delhi.

8. Purewal, S.S. 1957. Vegetable cultivation in northern India. Tech. Bull. 39, ICAR, New Delhi.

9. Shoemaker, J.S. 1949. Vegetable Growing. John Wiley and sons Inc. New York

10. Sinskaia, E.N. 1928. The oleiferous plants and root crops of the family Cruciferae. *Bull. Appl. Bot. Genet. Plant Breed.*, **19(3):** 1-648.

11. Thompson, H.C. and Kelly, W.C. 1957. Vegetable Crops. Tata McGraw-Hill Publishing Company Ltd., New Delhi.

12. Yamaguchi, M. 1983. World Vegetables. AVI, Publishing Co., Inc. West Port, USA.

Genus *Beta*

Beetroot, Garden Beet or Table Beet (*Beta vulgaris* L. sub sp. *vulgaris*)

INTRODUCTION

Beetroot, garden beet in the USA (Pink, 1993) or table beet (*Beta vulgaris* L., 2n=18) is an important vegetable grown for its swollen fleshy roots and is the horticultural form of *Beta vulgaris* L. sub sp. *vulgaris* which also includes the agricultural types sugar beet, mangold and fodder beet. The beetroot does not contain significant amount of any anti-nutritional factors such as oxalates, nitrites *etc.* Hence, this is an ideal vegetable for health conscious consumers. Most of the commercial production of beets is for processing. The leaves were eaten and the roots were used for medicinal purposes only (De Boek, 1986). In India this crop is popular among the growers for food and fodder, particularly during the winter (Yamaguchi, 1983).

Origin, History and Distribution

Beetroot (*Beta vulgaris* L.) was initially grown in eastern and central Europe (less important in Western Europe), North America and Middle East for its edible leaves. Romans were the first to grow beetroot around 3rd century AD. Beetroot was grown in France, Spain and Italy in middle Ages. It was first described in Germany in 1558 and was referred to as Roman beet and in England in 1576. From Eastern Europe and the Middle East, beetroot was carried to East Asia and was known in India in classical times and in China around 850 AD. Beetroot was introduced into USA around 1800 and was referred to as garden beet. Now it is very known crop in Germany, England, America, China, Arab, Greece, Rome and India. It is grown in almost all the states of India but not as common as other root crops (Nath *et al.*, 1987).

Species Availability, Evolution and Diversity

Beta vulgaris L. sub sp. *vulgaris* is common form available. The species also includes chard (sub sp. *cicla*) and sea beet (sub sp. *maritima*), which is thought to be the ancestor of most cultivars (Campbell, 1976), which grows wild on sea shores in Britain and through Europe and Asia to the East Indies. The earlier form of domesticated beet was leaf beet.

Botanical Description

The beet root, garden beet (*Beta vulgaris* L. sub sp. *vulgaris*) is a member of the family Chenopodiaceae. The garden beet is different from the sugar beet where emphasis is laid more on the sugar content of the root, whereas the garden beet is used for salad and cooked vegetable purpose (Nath *et al.*, 1987).

Table 6.20: Botanical Description

Growth Habit	Plants are usually biennial glabrous herbs with conspicuously swollen roots at junction with stem. For root production it is herbaceous annual.
Root	The root is thick and fleshy storing large amount of reserve food. Morphologically, the upper portion of the root develops from the hypocotyl and the lower portion, from which the secondary roots arise, develops from the tap root. The tap root may penetrate the soil to a depth of 300.0 cm (Weaver and Bruner, 1927). A number of strong lateral roots grow diagonally for a distance and then downward to depths nearly as that of the taproot. The extensive root system penetrating to a great depth indicates drought resistant property of beet. The roots are usually a deep red colour and may be globular or cylindrical (Purseglove, 1997).
Leaf	Leaves are often ovate and cordate, dark green or reddish. The rosette leaves develop in a close spiral with the oldest ones the outside. According to Magruder *et al.* (1940) temperature, time of growing, spacing and soil moisture influence the size, shape and colour of foliage. Erwin and Haber (1934) found that at relatively low temperature (10.0-15.6°C), the leaves are broad, thick and heavily savoyed with short, thick, straight and erect petioles, whereas at a temperature range of 21.1-26.7°C narrow, thin, smooth surfaced leaves with long, slender and drooping petioles are produced.
Floral biology	The inflorescence, which normally develops in the second year, is botanically a large spike. The flowers are almost sessile; these arise in clusters of 3-4 in the axils of bracts of the inflorescence axis and its secondary branches.

	The flowers are small, inconspicuous without corolla, but with green calyx, which become thicker towards base as the fruit ripens. Stamens are 5, ovary sunk in disc and stigma three in number. Although the ovary is three carpellate, there is only one ovule per ovary.
Anthesis and pollination	It is a wind cross pollinated crop having protandrous flowers.
Fruit and fruit set	Fruits are mostly aggregate formed by the cohesion of two or more fruits and held together by swollen perianth (calyx) base and thus forming an irregular dry cork-like body, known as seed ball or the so called seed. However, if there is a single flower, a single germ seed will develop.
Seeds	The true seed is small, about 3.0 mm long, kidney-shaped and shiny to reddish brown in colour. A gramme (g) of beet seed ball counts about 50 seeds, which retain viability for 5-6 years under ordinary storage condition.

Horticultural Classification

The shape and colour of roots vary with the genotype and environment (G x E) conditions such as temperature, soil, nutrition *etc.* The root skin colour varies from orange-red to dark purple-red. Red-violet colour of the roots is due to the pigment β-cyanins and yellow colour due to β-xanthins. Some white coloured beetroots are also available.

(A) Classification on the Basis of Edible Parts

Beets are generally classified into two major groups on the basis of the edible parts:

1. Beets grown mainly for fleshy roots-The beets grown mainly for their fleshy roots belong to the *crassa* or garden group and include yellow beets and the common table, or red beets.
2. Beets grown mainly for leafy vegetables-This group is grown as leafy vegetables belong to the *cicla* group and include leafy beets and Swiss Chard.

(B) Classification on the Basis of Root Shape

On the basis of root shape the beets can be group in 5 segments

Table 6.21: Classification on the Basis of Root Shape

Root Shape	*Varieties/Cultivars/Strains*
Flat	Flat Egyptian
Short top	Flattened at top and bottom with rounded sides and conical or tapered base. Varieties are Crosby Egyptian, Early Wonder and Asgrow Wonder
Round or globular	Roots are Round or globular in shape. Varieties are Detroit Dark Red and Crimson Globe
Half long	Length is shorter than long types. Varieties are Half Long Blood and Winter Keeper
Long	Roots are long, may grow as much as 40 cm, quite popular in Europe. Variety is Long Dark Blood

Augmentation of Genetic Resources and Crop Improvement

Popular cultivars like 'Crimson Globe' and 'Detroit Dark Ed' were introduced in India and directly used a variety for commercial cultivation. Garden beet is a cross pollinated crop. Selfing causes inbreeding depression. The method adopted to improve beet root are mass selection, family selection and hybridization. It produces seeds in the temperate climate. Hence no attempt was made to breed varieties in the plains. All varieties that are under cultivation are from abroad. Some of the important varieties were found to be suitable for commercial production in the plains. In this crop popular variety 'Avon Early' was developed from the use of indigenous source (population derived from a large number of cultivars of Detroit type) by mass selection method. Recently efforts were made to select suitable sugar beet varieties for extracting sugar rather than beet varieties for vegetable or table purpose (Nath *et al.*, 2008). Several Varieties are resistant to physiological disorder like cracking and canker-Long Dark Blood.

Table 6.22: Identified Cultivars from different Breeding Organizations

Variety/Hybrids	*Source*
Crimson Globe	ICAR-Indian Agricultural Research Institute, New Delhi
Detroit Dark Red	ICAR-Indian Agricultural Research Institute, New Delhi
Father Ball	United States of America
Egyptian Blood	United States of America
Edmond	United States of America
Eclipse	United States of America
Crosby Egyptian	United States of America
Early Wonder	United States of America
Ooty-1	Regional Station Ooty (Tamil Nadu)
Mahyco Lal (F_1)	MAHYCO
Action (F_1)	BSS
Pablo (F_1)	BSS
Lalima (F_1)	Golden Seeds
Bolivar (F_1)	Nunhems
Ruby Queen	Peto Seeds
Ruby Queen Improved	Tokita Seeds
Bassano	-
Corumba	-
Dayana	Ralish India
Warrior	Clause International

In Nepal 'Ruby Queen' cultivar is very much popular among growers and consumers.

Male Sterility

Some of the male sterile lines and monogram on seed varieties have also been reported for commercial utilization in the foreign countries (Stewart, 1961).

REFERENCES

1. Campbell, G.K.G. 1976. Sugar beet *Beta vulgaris* (Chenopodiaceae). In : Evolution of crop plants, Simmonds, N.W. (ed.) Longman London and New York, p. 25-28.
2. De Boek, T. S.M. 1965. The genus *Beta*: domestication, taxonomy and interspecific hybridization for plant breeding. *Acta Hort.*, 182, 335-343.
3. Nath, P.; Srivastava, V.K., Dutta, O.P. and Swamy, K.R.M. 2008. Vegetable crops- Improvement and Production. PNASF, Bengaluru.
4. Nath, P., Velayudhan, S. and Singh, D.P. 1987. Vegetables for the tropical region. ICAR, New Delhi.
5. Pink, D.A.C. 1993. Beet root (*Beta vulgaris* sub sp. *vulgaris*). In: Genetic Improvement of Vegetable Crops (Eds. G. Kalloo, and B.O. Bergh), Pergamon Press, New York.
6. Purseglove, J.W. 1997. Tropical Crops-Monocotyledons, Longman, London.
7. Yamaguchi, M. 1983. World Vegetables. AVI, Publishing Co., Inc. West Port, USA.

Parsnip (*Pastinaca sativa* Linn.)

INTRODUCTION

Parsnip (*Pastinaca sativa* Linn., 2n= 22) is basically an underutilized winter vegetable crops which fleshy roots are valued for the preparation of multiple cousins. Its flavor is not fully developed until the roots have been exposed to near-freezing temperatures for 2-4 weeks in the fall and early winter (Kalloo and Bergh, 1993). Parsnip contains moisture (72.4 per cent), protein (1.3 per cent), fat (0.3 per cent), carbohydrates (23.2 per cent), fibre (1.7 per cent) and mineral matter (1.1 per cent), calcium (50 mg), phosphorus (40 mg), iron (0.4 mg), Vitamin A (30IU), thiamine (0.31 mg), niacin (0.4mg), ascorbic acid (16 mg), and riboflavin (0.06-12 mg) per 100 g edible portion (Singh and Singh, 2007). Nick and Bradley (1994) reported that parsnip can be used in following ways:

1. The roots are consumed raw and after boiling as vegetable and salad.
2. They are also used for jam, cakes, wine and in fritters or pan cakes.
3. They make an excellent soup and wines.
4. Roots contain more sugar than sugar beet which reduces to about three quarters when cooked.
5. The roots are also fed to livestock not more than 20 per cent of the total fodder.
6. The parsnip possesses diuretic and carminative properties. The fruit is based as a folk medicine for stomach, bladder and stone troubles.

Origin, History and Distribution

Parsnip (*Pastinaca sativa* Linn.) is a native to the eastern Mediterranean region. Though, the crop was known to the ancient Romans and Greeks, grown for its large, edible roots and for medicinal purpose. It was introduced to the West Indies in 1564 and brought to the U.S. by the colonists and was cultivated in the early 17th century in Virginea and Massachusetts. It is also cultivated in western Asia for its large fleshy tap root (Knott, 1988). Parsnip is widely cultivated in Europe and America. In India, it is occasionally grown as an annual in kitchen garden in temperate regions (Anonnymous, 1996).

Botanical Description

Table 6.23: Botanical Description

Growth Habit	Parsnip is a biennial in nature but it is cultivated as annual. It is similar to carrot in anatomy and development.
Root	It is hardy, white roots store well in the ground all winter and large fleshy tap root which is economic parts.
Leaf	The plant develops rosette of leaves.
Floral biology	Flowers in terminal compound umbels.
Fruit and fruit set	Fruit (commonly called head) broad, elliptic and ridged and make of two maricarps.

REFERENCES

1. Anonnymous, 1996. The Wealth of India: Raw Material and Industrial propducts.Voll. **7:** 280-281. Council of Scientific and Industrial Research, New Delhi.
2. Kalloo, G. and Bergh, B.O. 1993. Genetic Improvement of Vegetable Crops. Pergmon Press, Oxford, U.K.
3. Knott, K.E. 1988. Handbook of Vegetable Growing. John Wiley and Sons, New York.
4. Nick, J.M.A. and Bradley, F.M. 1994. Growing Fruits and Vegetables organically. Rodale Press, Emmaus, Pennsylvania.
5. Singh, D.N. and Singh, S.P. 2007. Production Management of Underutilized Vegetables. Agrihortica Publication, Junagadh (Gujarat), India.

7

Genetic Resources of Okra (Genus *Abelmoschus*) and Crop Improvement

INTRODUCTION

Okra (*Abelmoschus esculentus* (L.) Moench, 2n=2x=130,190) is a traditional and important vegetable crop, which is commercially cultivated in West Africa, India, Southeast Asia, the southern United States, Iran, Brazil, Turkey, Yugoslavia, Bangladesh, Afghanistan, Pakistan, Burma, Japan, Malaysia, Ghana, Ethiopia, Cyprus and northern Australia. As early as the 13th century okra was also known as Lady's finger in English, gumbo in French, quimgombo source in Spanish, bhindi in Hindi and pindi in Punjabi and Dogri and bamiah in Arabic. Okra production is throughout the tropical, subtropical and Mediteranean climates and contributes about 4 per cent of total vegetable consumption in most of the developing countries (Siemonsma 1982 a) as old record but today India ranks first in the world with 4.53 million tonnes (70 per cent of the total world production) of okra produced from over 0.43 million hectare (FAOSTAT, 2009). Okra has tremendous export potential as fresh and dehydrated vegetable. It is being exported as a fresh vegetable primarily to Singapore, Mauritius, Malaysia, Sri Lanka and Middle East and accounts for about 70 per cent of foreign exchange earnings (Dhankar and Mishra, 2001). Okra is a very popular, tasty, gelatinous vegetable. Its tender and nutritious green fruits are fried, boiled and cooked as delicious vegetable. Fresh tender fruits provide dietary fiber, protein and vitamin C. It is also served as salad and soup. In addition to fruits, leaves are also consumed in some African countries. In Ivory Coast, okra is also an important leafy vegetable. In Far-East countries, a form of *Abelmoschus* ssp. is used as leafy vegetable. The fruits are fairly good in nutritive value and 100 g consumable portion contains 10.4 g dry matter, 3100 calorie energy, 1.8 g protein, 90.0 mg calcium, 1.0 mg iron, 0.1 mg carotene, 0.07 mg thiamin, 0.08 mg riboflavin and niacin, and 18.0 mg vitamin C, with almost comparable constituents, barring few, in leaves (Grubben, 1977). Dehydrated and frozen fruits are also marketed for off-season consumption. According to Berry *et al.* (1988), okra dry seeds contain 18-22 per cent edible oil (may be a substitute for edible oil) and 20-24 per cent edible crude protein. The greenish yellow edible oil has a pleasant taste and odour, and is high in unsaturated fats such as oleic acid and linoleic acid. In Egypt, seed flour is consumed mixed with maize flour. Roasted and grinded seed find their use as a coffee substitute (Martin, 1982). The fibers extracted from the stems are sometimes used as on the banks of the Niger in Mali, to make string and nets. The plant's fiber can be used by paper industry (Nadkarni, 1972; Singh, 1989). Okra is said to be very useful against genitourinary disorders, spermatorrhoea and chronic dysentery (Nadkarni, 1972). Its medicinal value has also been reported in curing ulcers and relief from haemorrhoids (Adams, 1975). The root and stem are useful for cleaning cane juice from which 'gur' or brown sugar or 'jaggery' is prepared (Chauhan, 1972).

Origin, History and Distribution

Abelmoschus esculentus is reported to be originated from different parts of tropical world, namely India-Asiatic origin (Masters, 1875) but some authority believe that okra is native to Abyssinian region-Ethiopia (De, Candolle, 1883; Vavilov, 1951), West Africa (Chevalier, 1940; Murdock, 1959), Tropical Asia (Grubben, 1977, Markosa and Peter, 1990) and a bit north to the region of the Sudan. However, Zeven and Zhukovasky (1975) believed it to have originated from India, Pakistan, and Burma where its several wild forms still occur. Domestication of okra possibly found place in Egypt; when records date back to Neolithic times (Chevalier 1940, Chamier 1984). It was also said to be growing along the Nile River. Okra was known to the Spanish Moors and was used by the Egyptian in the 12th or 13th century. In the early 1600's the black slave trade brought new delicacies to North America as well as South America and the West Indies. The name never caught on, but okra brought in by slaves from Angola was called 'Ochinggombo', later shortened to ngombo. The Indians of Louisiana discovered okra's thickening abilities as they used it to thicken a stew made of vegetables and seafood which they named" gumbo." Black slaves were also brought it to Brazil and Dutch Guiana during the 17th century. As a result, okra was incorporated into their cuisines and grows well in those tropical climates. The 19th century record of its occurrence in wild state along the White Nile in Sudan (Singh *et al.*, 1975). No dates can be given for the migration of cultivated okra from Africa to the Mediterranean region and Asia. It was apparently introduced to tropical and subtropical America during the early past Columbian period (Thakur and Arora, 1986). There are two hypothesis concerns to the geographical origin of *A. esculentus*. Some authors argue that one putative ancestor (*A. tuberculatus*) is native to Uttar Pradesh (North India), suggest that the species originated from this geographical area. Others, on the basis of ancient cultivation in East Africa and the presence of the other putative ancestor (*A. ficulneus*) suggest that the area of domestication is north Egypt or Ethiopia but no definite proof is available today.

The center of diversity of *Abelmoschus* includes West Africa (Benin, Togo, and Guinea), India and Southeast Asian countries *i.e.* Burma, Indochina, Indonesia and Thailand (Chevalier 1940; Van Borssum Waalkes 1966; Siemonsma 1982a; Hamon 1988; Hamon and Hamon 1991; Bisht *et al.*, 1995). *A. esculentus* (L.) Moench) is the commercially cultivated species of okra. However *A. caillei* is also cultivated for its fresh fruit in West Africa. The geographical distribution of *A. caillei* is from northern Guinea extending to southeast Cameroon. *A. esculentus* cultivated mostly in humid parts of Guinea, mostly in an intercropped system with *A. caillei* (Hamon and Charier 1983; Hamon 1988). At present okra is an important crop of India, Afghanistan, Burma, Japan, Malaysia, Brazil, Pakistan, Iran, Western Africa, Ghana, Ethiopia, Cyprus, Middle East, Greece, Yugoslavia, Bangladesh, Turkey, the Caribbean, Southern United States and the USA. India is the largest producer of okra in the world.

In India, it is grown under diverse conditions right from coastal plains to elevations as high as 1000 m.

India ranks first in the world with 4.53 million tonnes (70 per cent of the total world production) of okra produced from over 0.43 million ha land. In India it is commercially grown in Uttar Pradesh, Odisha, Bihar, Maharashtra, Haryana, Chhattisgarh, Assam, West Bengal, and Punjab during spring season and in Gujarat, Andhra Pradesh, Karnataka and Tamil Nadu during winter season. Other okra growing states are

Table 7.1: Classification of *Abelmoschus*

Species	Type of Species	Chromosome Number (2n=)	Geographical Distribution
A.esculentus (L.) Monech	Cultivated	108-144 mostly 124 or 130; also 66-72	1
A. caillei (A. chev.) Stev.	Cultivated	185-198	2
A. moschatus Medikus Subsp. *moschatus*	Semi-wild	72	2
A. moschatus Medikus var. *moschatus*	Cultivated	-	2
A. moschatus Subsp. *moschatus*	Wild	-	2°
A. moschatus var. betulifolius	Wild	38	2°
Subsp. *biakensis* (Hochr.) Borss.			
Subsp. *tuberrosus* (Span.) Borss.			
A. manihot (L.) Medikus	Semi-wild	60-68	3
A. tetraphyllus (Roxb. ex Hornem) R. Graham var. *tetraphyllus*	Wild	138	4
var. *pungenss* (Roxb. Hochr.)	Wild	138	4
A. tuberculatus (Pal and Singh)	Wild	58	5
A. tficulneus (L.) W. and a. ex Wight	Wild	72 (African species)	
		78 (Asian species)	
A. crinitus Wall.	Wild	-	7
A. angulosus Wall. ex. W. and A.	Wild	56	8

Source: IBPGR (1991c); Chromosome numbers and cultivation from Siemonsma (1928b), Charrier (1984), Hamon and Yapo (1986) and Koechlin (1991).

Table 7.2: Species Distribution and Crop Status

Abelmoschus Species	*Chromosome Number*	*Wild/ Cultivated*	*Distribution*
A. angulosus Wall. Ex W. and A. (wild)	56	W	India, other parts of Asia
A. tuberculatus Pal and Singh (wild)	58	W	India
A. manihot (L.) Medikus (semiwild)	66	WC	India, other parts of Asia, Africa, Papua New Guinea
A. moschatus	72	WC	India, other parts of Asia, Australia, Africa
A. ficlneus (L.)	72	W	India, other parts of Asia, Australia, Africa
A. esculentus (L.) Moench	130	C	Worldwide
A. tetraphyllus (Roxb. Ex Hornem.) Hochr.	138	W	India, other parts of Asia, Papua New Guinea
A. crinitis Wall. (wild)	138	W	India, other parts of Asia
A. pungens	138	W	India, other parts of Asia
A. caillei (A. Chev.) Stev	196	WC	India, Africa

W: Wild; C: Cultivated.

Jharkhand, Chhattisgarh, and Assam. In North India it is mainly cultivated during the spring summer (March-June) and rainy (July-September) seasons while in temperate zones, it is cultivated during summer. Andhra Pradesh have highest productivity per ha followed by Jharkhand and Bihar.

Species Availability, Evolution and Diversity

The family has 82 genera and 1,400 species. Hochruitner (1924) listed 14 species under genus *Abelmoschus*. The Index Kewensis lists about 30 species under *Abelmoschus* in the Old World and 4 in the New World. Few researchers indicate the availability of 34 species of *Abelmoschus*. In India, about 111 species belonging to more than 20 genera are reported. Sub-specific entities in *A. angulosus, A. moschatus, A. tetraphyllus* and *A. tuberculatus* have been described and almost the entities have their natural distribution in India by virtue of different microclimates and adaptive niches across various eco-geographic regions. Some researchers reported that the genus *Abelmoschus* comprises 12 species, 3 of them cultivated and the rest wild. *Abelmoschus esculentus* is found all around the world from Mediterranean to equatorial areas. Cultivated and wild species are overlapping in Southeast Asia, which is considered as the center of diversity. The spread of the other species is the result of their introduction to America and Africa. *A. moschatus* and *A. manihot* are semi wild, and show a greater diversity than the cultivated forms. *A. moschatus* is morphologically and genetically the most different from other species with probably the widest geographical distribution, which extends from the South Pacific Islands through Indochina and India and over to the central and western parts of Africa. Wild species are Indigenous to Africa, a few cultivars from Sudan display primitive characters. The cultivated form (*A. moschatus* var. *moschatus*), better known as *"musk mallow," "Jew's mallow,"* or *"ambrette"* in Africa, Asia and America has fragrant seeds used for making perfume. The subspecies *biakensis* is endemic to Papua New Guinea. The cultivation of *A. manihot* is mainly in Southeast Asia and is known under the vernacular name *"qibika"*, which is also found in India, northern Australia and less frequently in the American continent and tropical West Africa. Its cultivation is exclusively for leaf consumption. The species *A. tetraphyllus, A. ficulneus, A. crinitus, A. angulosus*, and *A. tuberculatus* are truly wild. *A. ficalneus* is found from the northern parts of Australia to Southeast Asia and from India to most parts of Africa. *A. tetraphyllus, A. crinitus, A. angulosus* and *A. tuberculatus* are exclusively of Asian origin. *A. tetraphyllus* var. *tetraphyllus* and var. *pungens* are endemic to Southeast Asia with extension of the former to Papua New Guinea and New Ireland. *Abelmoschus crinitus* is found at low altitudes in India and Southeast Asia and *A. angulosus* at high altitudes in India, Sri Lanka, Indo-China, and Indonesia. *A. tuberculatus* is endemic to the northern and western parts of India. *A. caillei*, the second edible okra species which is available only in West Africa that makes it difficult to suggests an exact origin place. It is distinguished from *A. esculentus* by fruit shape, number of epicalyx segments and orientation, and shows great variability. *A. caillei* is prized for its potentially prolific yield, vigorous growth, and tolerance to some adverse environments, serving as a source of many desirable characteristics for okra found in south India as wild. Being very photosensitive it is a short day plant so it requires selection for photo-insensitive to make it more adaptable. In India, nine species of *Abelmoschus* are available. Of these only *A. esculentus* is known under cultivation. Based on different floras, floristic accounts and other published evidence expressed main areas of concentration of wild species are as follows:

- ✰ Semi-arid tracts of north and northwestern India are for *A. ficulneus* and *A. tuberculatus.*
- ✰ *Tarai* range and the foothills of the Himalayas are for *A. crinitus, A. manihot* and *A. tetraphyllus or A. tetraphyllus* var. *pungens.*
- ✰ Western and Eastern Ghats and also peninsular tracts are for *A. manihot* (including *tetraphyllus*

type), *A. angulosus* mainly confined to hilly tracts up to 2000 m in south India with only meager distribution elsewhere.

- Northeastern regions are for *A. crinitis* and *A. manihot*, mostly var. *pungens*.
- In some of the wild taxa intra-specific variation exists and has been taxonomically identified as in the *A. manihot* var. *tetraphyllus* and *pungens*, in *A. angulosus* var. *purpureus* and *grandiflorus*, in *A. moschatus* var. *biakensis* and *rugosus*, and in *A. tuberculatus* var. *deltoidefolius*.

Table 7.3: Worldwide Distribution of *Abelmoschus* Species

Name of Species	*Distribution*
Abelmoschus angulosus	India, Sri Lanka
Abelmoschus tuberculatus	India
Abelmoschus manihot	India, Sri Lanka
Abelmoschus moschatus	India, Nepal, Sri Lanka
Abelmoschus ficulneus	India, Bangladesh
Abelmoschus esculentus	Worldwide
Abelmoschus tetraphyllus or A. tetraphyllus var. pungens	India, Nepal, Sri Lanka
Abelmoschus crinitus	India, Sri Lanka
Abelmoschus caillei	India

The Indian subcontinent where remarkable variability can be observed is an important centre of diversity for okra. Much effort has been made during the last four decades to enrich its variability, which has been utilized in developing promising cultivars that are adapted to a wide range of agro-climatic conditions and cropping patterns, possess better fruit yield and quality combined with insect-pest and disease resistance.t The nearest wild related species to cultivated okra is *A. tuberculatus* which sporadically occurs in fields/field margins. *A. tuberculatus* is confined to north/north-western plains, often along field borders and except for its tuberculate hard fruits (up to 8 cm long) look unmistakably like the cultivated okra. Variability in *A. pungens* and *A. heterophyllus* occurs in the sub-Himalayan range, for the former in north-eastern belt and for the latter, in the north-western *tarai* region. *Abelmoschus* species are also available in the Western Ghats (3 of the 6 species); *A. angulosus*, a polymorphic taxon with more variation in Karnataka and Nilgiri hills; *A. manihot* (including *A. pungens*) and *A. ficulneus* occur here.

New Species of *Abelmoschus*

- Based on morphology, inter-specific crossability and hybrid fertility, a new species of *Abelmoschus*, namely *A. enbeepeegearense* was discovered.
- *A. enbeepeegearense* (J. John, Scariah, N., Bhat, K.V. and Yadav, S.K from ICAR-NBPGR, New Delhi) is a new species occurring in the low elevation Western Ghats of India comprising Kerala, Karnataka and Tamil Nadu.
- The taxon is morphologically allied to *A. moschatus* subsp. *moschatus*, *A. moschatus* subsp. *tuberosus* and *A. crinitus*, but easily distinguishable.

Table 7.4: Taxon and Preferred Taxon Name

Name of Taxon	*Name of Authority*	*Family*	*Preferred Taxon Name*
Abelmoschus		Malvaceae	*Abelmoschus*
Abelmoschus		Malvaceae	*Abelmoschus*
Abelmoschus angulosus	Wall. ex Wight and Arn.	Malvaceae	*Abelmoschus angulosus*
Abelmoschus caillei	(A. Chev.) Stevels	Malvaceae	*Abelmoschus caillei*
Abelmoschus crinitus	Wall.	Malvaceae	*Abelmoschus crinitus*
Abelmoschus esculentus	(L.) Moench	Malvaceae	*Abelmoschus esculentus*
Abelmoschus ficulneus	(L.) Wight and Arn.	Malvaceae	*Abelmoschus ficulneus*
Abelmoschus hybrid		Malvaceae	*Abelmoschus hybrid*
Abelmoschus manihot	(L.) Medik.	Malvaceae	*Abelmoschus manihot*
Abelmoschus manihot subsp. *manihot*		Malvaceae	*Abelmoschus manihot subsp. manihot*
Abelmoschus manihot var. *pungens*	(Roxb.) Hochr.	Malvaceae	*Abelmoschus manihot* var. *pungens*
Abelmoschus manihot var. *tetraphyllus*	(Roxb. ex Hornem.) Borss. Waalk.	Malvaceae	*Abelmoschus manihot* var. *tetraphyllus*
Abelmoschus mindanaensis	Warb. ex Perkins	Malvaceae	*Abelmoschus manihot* var. *tetraphyllus*
Abelmoschus moschatus	Medik.	Malvaceae	*Abelmoschus moschatus*
Abelmoschus moschatus subsp. *moschatus*		Malvaceae	*Abelmoschus moschatus* subsp. *moschatus*
Abelmoschus moschatus subsp. *tuberosus*	(Span.) Borss. Waalk.	Malvaceae	*Abelmoschus moschatus* subsp. *tuberosus*
Abelmoschus pungens	(Roxb.) Voigt	Malvaceae	*Abelmoschus manihot* var. *pungens*
Abelmoschus sagittifolius	(Kurz) Merr.	Malvaceae	*Abelmoschus moschatus* subsp. *tuberosus*
Abelmoschus sp.		Malvaceae	*Abelmoschus sp.*
Abelmoschus tuberculatus	Pal and Singh	Malvaceae	*Abelmoschus tuberculatus*

- ✰ It can be crossed with all the three taxa with varying degree of success, but with hybrid sterility.
- ✰ The materials belonging to it were earlier identified and placed under *A. moschatus* Medik, at Madras Herbarium.
- ✰ Orthotropic branching, 3-5 angles leaves, glandular hairy plant body with whitish waxy secretions, glandular non-setose epicalyx more than 8 in numbers with ovate hirsute fruits with a short mucron at apex.

Complete elucidation of the evolutionary history of okra depends on the cytogenetic relation of okra with its 3 wild species native to India *A. tuberculatus, A. moschatus* Medic and *A. ficulneus* Wight and Arn. The cytotaxonomy of *Abelmoschus* is yet so confused that an Asian origin of whole or part of the cultigens does not seem impossible. The crop could perhaps yet turn out to be polyphyletic in origin (Joshi and Hardas, 1976).

Crop Systematic

Long before okra was placed in the genus *Hibiscus* but later on based on the caducity of the calyx, and adnations of the calyx to the petals and the staminal column, it was placed in the genus *Abelmoschus*. The genus *Abelmoschus* is easily distinguished from the *Hibiscus* by the nature of its calyx that are apically 5-toothed which splits flower longitudinally along a single suture after following dehiscent, dropping together with the petals and staminal column to which it is adnate basally. The taxonomy of *Abelmoschus* is complex. During the international workshop on okra genetic resources at ICAR-IARI New Delhi (1990), the classification of Van Borssum Waalkes (1966) as a base point and on the basis of cytological evidences, new classification is adopted (Table 7.5), it was concluded that Guinean type okra and *A. manihot* var. *caillei* (A. Chev) are the same and named *A. caillei* (A. Chev) Stevels. Sudanian type of okra is called *A. esculentus*, *A. caillei* was identified wrongly earlier as *A. manihot* subsp. *manihot* by flora of Tropical West Africa, Kew Botanic Gardens, which was utilized under Yellow Vein Mosaic Virus (YVMV) resistant varietal improvement programme in India. According to new classification, the *Abelmoschus* comprises nine species showing in Table 7.5.

Species Relationship

A. moschatus and *A. manihot* are semi wild, and show a greater diversity than the cultivated forms. *A. moschatus* is morphologically and genetically the most different from the other species with probably the widest geographical distribution, which extends from the South pacific Islands through Indo-China and India and over to the central and western parts of Africa. The cultivated form (*A. moschatus* subsp. *moschatus* var. *moschatus*) which is better known as "musk mallow", 'Jew's mallow" or "ambrette" in Africa, Asia and America, has fragrant seeds used for perfume making. The subspecies *biakensis* is endemic to Papua New Guinea. The cultivation of *A. manihot* is mainly in Southeast Asia and is known under the vernacular name "aibika"; it is also found in India, northern Australia, and less frequently in the American continent and Tropical West Africa. Cultivation is exclusively for leaf consumption. The species *A. tetraphyllus, A. ficulneus, A. crinitus, A. angulosus* and *A. tuberculatus* are truly wild. *A. ficulneus* is found from the northern parts of Australia to South East Asia and from India to most parts of Africa. *A. tetraphyllus, A. crinitus, A. angulosus* and *A. tuberculatus* are exclusively of Asian origin. *A. tetraphyllus* var. *tetraphyllus* and var.

Table 7.5: Old and Present Classification of Genus *Abelmoschus*

Van Borssum Waalkes (1966)	*As Decided in 'International Workshop on Okra Genetic Resources', New Delhi (1990)*
A. maschatlls Medikus subsp. *moschatus*	*A. maschatlls* Medikus subsp. *moschatus*
var. *moschatus*	var. *moschatus*
subsp. *moschatus*	subsp. *moschatus*
var. *betulifolius* (Mast.) Hochr.	var. *betulifolius* (Mast.) Hochr.
subsp. *biakensis* (Hochr.) Borss.	subsp. *biakensis* (Hochr.) Bross.
subsp. *tuberosus* (Span.) Borss.	subsp. *tuberosus* (Span.) Borss.
A. manihot (L.) Medikus subsp. *manihot*	*A. manihot* (L.) Medikus
subsp. *tetraphyllus* (Roxb. Ex Hornem.) Borss.	*A. tetraphyllus* (Roxb. Ex Hornem.) R. Graham var. *tetraphyllus*
var. *tetraphyllus*	-
subsp. *tetraphyllus*	var. *pungens* (Roxb.) Hochr.
var. *pungens* (Roxb.) Hochr.	*A. esculentus* (L.) Moench *A. tuberculatus* Pal And Singh
A. esculentus (L.) Moench	*A. ficulneus* (L.) W. and A.
(including *A. tuberculatus* Pal And Singh)	Van Borssum Waalkes (1966)
A. ficulneus (L.) W. and	ex Wight
A. crinitus ex Wight	*A. crinitus* Wall
A. angulosus Wall ex. W. and A.	*A. caillei* (*A. Chev*) Stevels

Table 7.6: Cytological Relations in the Genus *Abelmoschus* (Proposed compartments and relations)

Level 1	*Level 2*	*Level 3*
A. moschatus; N=36		
	A. esculentus; N=62-65	"Nori-Asa' *A. caillei;* "Guinean (TFA) N=92-99
A. ficulneus (F); N=36		
A. tuberculatus; N=29		
	A. pungens; N=69	
A. angulosus; N=28		
	A. tetraphylius; N=69	
A. manihot (A) B.; N=30-34		

Cytological relations in the genus *Abelmoschus* (proposed compartments and relations).

pungens are endemic to South East Asia with extension of the former to Papua New Guinea and New Ireland. *A. crinitus* is found at low altitudes in India and South East Asia and *A. angulosus* at high altitude in India, Sri Lanka, and Indo-China. *A. tuberculatus* is endemic to the northern and western parts of India. The relationship and genetic integrity among the cultivated and wild *Abelmoschus* species have been studied. In most cases, interspecific crosses among *Abelmoschus* species, F_1s, are sterile. *Abelmoschus esculentus* crosses readily with *A. caillei* and many produce partially fertile plants. *A. esculentus* x *A. tetraphyllus* hybrids are also possible. Three ploidy levels, where *A. moschatus* (n=36), *A. ficulneus* (n=36), *A. tuberculatus* (n=39), *A. esculentus* (n=36), *A. manihot* (n=30-34) are at ploidy level 1A; *A. esculentus* (n=62-65), *A. tetraphyllus* (n=69) are at ploidy level two; and *A. caillei* (n=99) is at ploidy level three. According to Meshram and Dhapke (1981), the hybrid between *A. esculentus* and *A. tetraphyllus* was spreading and dwarf. The genus, therefore, may be polyspecies complex where continuous interchange of genes throughout its evolution has complicated the determination of relationships.

Vigorous F_1 hybrid between *A. esculentus* and *A. manihot* ssp. *tetraphyllus* var. *tetraphyllus* was obtained by Suresh Babu (1987) with 36 and 62 which suggested that one genome was common for *A. esculentus* and *A. manihot* ssp. *tetraphyllus* var. *tetraphyllus*. Madhusoodan and Nazeer (1986) reported that hybrid made between Guinean type and Sudanian type was sterile due to difference in the chromosome number of the parental types.

Botanical Description

Okra is a tropical vegetable of the Malvaceae family (Mallow family) with varying chromosome no. 2n=74, 84, 120 and 130 (Fryxell, 1957). The distinguishing feature of the genus is the caduceus calyx where calyx, corolla, and the stamen are fused together at the base and falls off as one piece after anthesis.

Table 7.7: Botanical Description

Growth Habit	It is an erect (occasionally branched), annual, herb and becoming woody at maturity. The plant often reaches 60.0-180.0 cm in height but an extremely tall type in the central parts of Nepal reaches up to 4.0 m. Plants destitute of peltate (hairy), and robust.
Root	Well-developed tap root system.
Stem	The main stem branches at the base with usually 2 to 5 but as many as 20 predominant branches/plant. The stem is usually green or with reddish tinge.
Leaf	Leaves are simple, alternate, entire, lobed or palmatifid to palmate sect-serrate (3-7 lobed), hirsute with a symmetrical lamina of chartaceous texture and present at each node in alternate position. The base is cordate, margins toothed (serrate), and apex acute (Bhat *et al.*, 1988). Stipules narrow, caduceus; petiole hisipid, often tinged red, 15.0-35.0 cm long; lamina broadly cordate, pale green beneath 10.0-25.0 cm x 10.0-35.0 cm size, and hirsute.
Floral biology	Okra has hermaphrodite golden yellow flowers, which appear at the 3rd to 19th nodes from the base (axil of each leaf) after 34 to 67 days after sowing. Flowering continues for 40-60 days. The crown of the stem at this time bears 3-4 underdeveloped flowers but later on during the period of profuse flowering. As the stem elongate, the lower most flower buds open into flowers. There may be a period of 2, 3 or more days between the time of development of each flower never does more than one flower appear on a single stem. The buds measuring 2.82 x 1.55 cm open into flowers (Purewall and Randhawa, 1947). Solitary flowers are axillary with lobed epicalyx and fruits are born upright with about 2.0 cm long peduncle. There are 8-12 epicalyx (narrow hairy bracteoles which fall before the fruit reaches maturity) segments, linear to lanceolate, surrounding the flower. Calyx generally non-lobed, and tubular. Calyx splits longitudinally as the flower opens. Petals are 5 in number having yellow colour with crimson spots on claw and 5.0-7.0 cm long. Flowers have a style in a central position surrounded by a tubular staminal column with numerous stamens, attached to the corolla base. Carpels not separating. Ovary is superior. Stigmas, small 5-9 fids, and deep/velvety red/purplish which are joined in cultivated species, are more or less separated and located much further from anthers. Anthers are arranged in concentric superimposed circles (5-6 anthers per circle). Pollen grains of okra are polysiphonous (Nair *et al.*, 1974). The entire tuber except one step elongation and that one is the functional tube. In other tubes, the protoplast activity is arrested by the formulation of coagulated protoplasmic plug (collase plugs) at the tips.

	The middle part of stamina column from the flower collected in the middle phase of flowering, possesses the most balanced biomass and therefore the most desired for breeding experiments (Srivastava, 1982). The pollen remains viable in storage at 50 per cent RH for 55 days (Dubey and Singh, 1968.). Pollen fertility is maximum during the period between 1 hr before and 1 hr after opening of the flower. It takes 2-6 hr for fertilization after pollination.
Anthesis and pollination	Flowers open at dawn and remain open until evening. Pollen grains (about 100 per anther) are sticky polysiphonous (Nair, *et al.*, 1974) and are not necessarily in contact with the stigma at anthesis. Stigma becomes receptive 2 hours before anthesis (protogyny) and receptivity persists till 3-4 hours after anthesis with varying degree of receptivity. Contact takes place mechanically during the day by elongation of the stamen column. Stigma is most effective on the day of flower opening, anther dehiscence at the time of anthesis and dusting of the stigma with pollen being almost simultaneous. Anthesis is observed 6.00 and 9.00 AM and is preceded by maximum anther dehiscence between 7.00 AM and 11.00 AM. The stigma remains receptive on the day of anthesis. Anther dehiscence commences 15-20 minutes after anthesis and is completed in 5-10 minutes. According to Srivastava (1982), second to fourth positioned flowers carry more anthers (63 anthers/flower) while only 26 in the first flower and 10 in the 10th flower. The dehiscence of anthers is transverse (Purewal and Randhawa, 1947). It is self- pollinated crop and flowers display entomophilous features such as coloured petals, nectarine, extrose dehiscent anthers, and sticky pollen grains. Reports are also available indicating cross pollination to an extent of 12 per cent in okra mainly due to prtogyny and entomophily, bees and black ants being the pollen carriers. Okra is often cross pollinated crop and cross-pollination can reach 40 per cent depending insect pollination levels and temperature. Choudhary and Anothai (1970) reported 4-6 per cent out crossing with maximum up to 42 per cent. Hence, okra is grouped as often cross pollinated crop (5-12 per cent) with no inbreeding depression.
Fruit and fruit set	Edible fruit is formed within 3-6 days of flowering opening. Flowers bear erect fruits in each leaf axil starting from the flowering node in both main stem and lateral branches. Edible fruit is formed within 3-6 days of flower opening. Commercial cultivars have ridged (4-11 per fruit). Fruit capsular, bracteoles neither spreading nor fimbriate, capsule 5-valved, epicalyx, calyx, corolla and androecium fall as one unit after fruit set. Fruit is pyramidal oblong beaked capsule, with or without sutures/ridges, light green or sometimes red in colour, pyramidal oblong, beaked, longitudinally narrowed, 10-30 cm x 2.0-3.0 cm, longitudinally furrowed, hirsute or glabrous dehiscing longitudinally when ripe. Fruits consist of 55 to 62 per cent pericarp, 30 to 40 per cent seeds and 7 to 11 per cent axil (Singh and Agarwal, 1988).
Seeds	Dry seeds can be harvested in 30-32 days from anthesis. Seeds are relatively large and heavy (50-60 g/1000 seeds) and have a light green to dark gray/ brown colour when high in quality. Shape is round to reniform and their surface is rather glabrous or sometimes hirsute or scattered, having simple short hairs (trichomes) on the surface.

Botanical Classification

The variation in chromosome numbers suggests a number of ploidy levels. Charrier (1984) suggested three ploidy levels, where *A. moschatus* (n=36), *A. ficulneus* (n=36), *A. tuberculatus* (n=29), *A. esculentus* (n=36), *A. manihot* (n=30-34), are at ploidy level 1; *A. esculentus* (n=62-65), *A. tetraphyllus* (n=69) are at ploidy level 2; and *A. caillei* (n= 99) is at ploidy level 3. Four ploidy levels seem also possible (Hamon, 1988). The genus therefore, may be a "polyspecies complex" where continuous interchange of genes throughout its evolution has complicated the determination of relationships. The chromosome number of a few related species are *A. tetraphyllus* (2n= 138), *A. manihot* (L.) Medik (*2n*=68), *A. pungens* (2n= 138), *A. angulosus* (2n= 112), and *A. crinitus* (2n= 138).

Table 7.8: Classification of *Abelmoschus* Species on the Basis of Chromosome Number

Abelmoschus Species	*Somatic Chromosome Number*
Abelmoschus angulosus Wall. Ex W. and A. (wild)	56
Abelmoschus tuberculatus and Singh Pal	58
Abelmoschus manihot (L.) Medikus (semi wild)	60-68
Abelmoschus moschatus Medikus	
subsp. *moschatus* var. *moschatus*	72
subsp. *moschatus* var. *betulifolius* (wild)	?
subsp. *biakensis* (Hochr.) Borss. (wild)	?
subsp. *tuberosus* (Span.) Borss.	38
Abelmoschus ficlneus (L.) W. and A. ex Wight (wild)	72 (African sp.) 78 (Asian sp.)
Abelmoschus esculentus (L.) Moench	108-144 mostly 124 or 130; also 66, 72
Abelmoschus tetraphyllus (Roxb. Ex Hornem.) R. Graham	
var. *tetraphyllus* (wild)	138
var. *pungens*(Roxb.) Hochr. (Wild)	138
Abelmoschus crinitis Wall. (wild)	138, ?
Abelmoschus caillei (A. Chev.) Stev	185-198

The genetic chromosome number varies from to n=72. A Japanese line with n=36 and Ceylon line with N=33 have also been reported. Okra is described as an allopolyploid with genomes from *A. tuberculatus* (T) and an unknown species (Y). When *A. esculentus* (L.) Moench. (2n=130) was crossed with *A. tuberculatus* (2n=58), the F_1 hybrids were normal and on cytological examination revealed 29 bivalents and 36 univalents indicating the perfect homology between 29 chromosomes of *A. tuberculatus* with 29 chromosomes of *A. esculentus*. The 36 univalents were denoted as the Y set. Attempts were made to identify that set with genomes from *A. ficulneus* (n=36, F) and *A. moschatus* Medic (n=36, M).

Table 7.9: Species Combinations and their Salient Features

Step	Cross Combination with Chromosome Number			Salient Feature
Step 1	*A. tuberculatus* (2n=58)	X	*A. ficulneus* (2n=72)	F_1-T (29) + F (36)-65 univalents (bivalents ranged from 0 to 7)
Step 2	Colchicine (0.2 per cent)			
Step 3	*Amphidiploid* (2n=130)	X	*A. esculentus* (2n=130)	No viable seeds were obtained either on selfing or crossing

The above observation indicated non-homology between genomes T and F. This also showed dissimilarity between Y and F.

Table 7.10: Non-homology of Species

A. esculentus (2n=130)	X	*A. ficulneus* (2n=72)	F_1-E (65) (developed through embryo culture) Less than 36 II[s] were observed
A. esculentus (2n=130)	X	*A. moschatusus* (2n=72)	F_1-E (65) + M (36) (developed through embryo culture) Less than 36 II[s] were observed

The more number of bivalents in cross between *A. esculentus* x *A. ficulneus* as compared to *A. esculentus* x *A. moschatus* indicated closer affinity between *A. esculentus* and *A. ficulneus*.

Table 7.11. Chromosome numbers (2n=) in the genus *Abelmoschus*

Species	Chromosome Numbers (2n=)	Authors/References
A. esculentus	72	Ugale *et al.* (1976)
	108	Datta and Naug (1968)
	120	Purewall and Randhawa (1947)
	130	Suresh Babu (1987)
	132	Roy and Jha (1958)
	144	Datta and Naug (1968)
A. manihot	60	Chizaki (1934)
	66	Kamalova (1977)
	68	Kuwada (1974)
A. moschatus	72	Joshi *et al.* (1974)
A. ficulneus	72	Joshi *et al.* (1974)
A. angulosus	56	Fond (1938)
A. tuberculatus	58	Gadwal *et al.* (1968)
Others: okra from Ghana	194	Singh and Bhatnagar (1975)
A. manihot sub ssp. *manihot*	194	Jambhale and Nerker (1982)
*A. pungens**	138	Joshi and Hardas (1976)
*A. tetraphyllus**	130	Ugale *et al.* (1976)
	138	Suresh Babu (1987)

Covered by Van borssum kiaalkes (1966) as forms of *A. manihot.*

Grouping of Okra on the Basis of Regional Adaptation

(i) Saudanian type: Widely adopted and cultivated species *Abelmoschus esculentus*, also called the 'Saudanian type'.

(ii) Guinean type: Guinean type is from West Africa and is strongly distinguished from other groups on the basis of red leaf veins, very late flowering, reduced number of sub calyx bracts and large number of seeds per pod. It is considered to be natural amphidiploids of *A. esculentus* and *A. manihot*.

Gene and Genetics of Characters

Genetics of quantitative characters are given in Table 7.12.

There are several specific traits for which genetics has been worked out. One to three groups of dominant

Table 7.12: Gene and Genetics of Characters

Character	*Mode of Gene Action*	*References*
Plant height	Monogenic (tall dominant over dwarf)	Jasim (1967)
	4-5 groups of dominant genes	Kulkarni *et al.* (1976)
Plant height Pods/plant	Additive gene action	Rao and Kulkarni (1976)
Earliness and more number of fruits	Dominant and over dominant (1-3 groups of dominant genes)	Kulkarni *et al.* (1976)
More number of fruits	Additive gene action with complete dominance	Kulkarni and Thimmappaiah (1977)
Days of flower, plant height and fruits/plant	Both additive and non- additive gene action	Rao (1972)
Days of flower, plant height and fruits/plant	Additive x additive with epistatic action	Kulkarni *et al.* (1978)
Yield/plant, branches/plant and plant height	Additive gene action	Reddy *et al.* (1985)
Days to flower, Yield/plant, and fruits/plant	Additive gene action	Singh and Singh (1978)
Pods/plant	Additive gene action	Rao and Kulkarni (1976)

Table 7.13: Genetics of Characters

Character	*Control of Gene*	*Behavior of Gene*	*References*
Leaf margin	Monogenic	Cut leaves dominant to lobed leaves	
Plant height	Monogenic	Tall dominant to dwarf	
Plant height	Dominant	Gene action by 4-5 groups, non-additive gene action	Kulkarni, 1976
Plant height	Polygenic	Intermediate	Jasin, 1967
Plant height		High heritability (79 per cent)	Ramu, 1976; Lal *et al.*, 1977
Internode length		High heritability (79 per cent)	Lal *et al.*, 1977
Dark red stem		Simple inheritance	Erickson and Couto, 1963
Spininess		Intermediate	Jasin, 1967
Pubescence of foliage	Incomplete dominance		Nath and Dutta, 1970
Degree of leaf lobing		Simple inheritance	Kalia and Padda, 1962; Jasin, 1967
Degree of leaf lobing	Monogenic with incomplete dominance		Nath and Dutta, 1970
Light red petiole		Simple inheritance	Erickson and Couto, 1963
Days to flower	Dominant	Gene action by 1-3 groups	
Pigmentation of calyx and corolla	Monogenic		Kolhe and D'Cruz, 1966
Pod colour	Monogenic	White dominant to green	
Pod colour	Multiple alleles	Simple inheritance	Kalia and Padda, 1962
Pod colour	Two dominant genes		Nath and Dutta, 1970
Pod colour	One dominant genes (white/green)		Jasin, 1967
Number of fruit	Dominant	1-3 groups	
Pod shape	Oleigogenic	Angular dominant over round and epistasis has been observed	
Pod shape	Two genes		Kalia and Padda, 1962
Pod shape	Digenic with epistasis (angular/ round)		Jasin, 1967
Pod shape	Several genes		Miller and Wilson, 1939
Pubescence of pods	Complex inheritance		Miller and Wilson, 1939
	Simple inheritance		Kalia and Padda, 1962
	Incomplete dominance		Nath and Dutta, 1970
Pod spininess	Monogenic	Spininess is dominant to non-spininess	
Fruit hairiness	Monogenic	Fruit hairiness is completely dominant over non- hairiness	
Fruit colour			Kolhe and D'Cruz, 1966

Character	Control of Gene	Behavior of Gene	References
Weight of fruit	Average heritability (48 per cent)		Ngah and Graham, 1973
Pods/plant	Additive gene action	Average heritability	Rao and Sathyavathi, 1977
		High heritability	Rao and Kulkarni, 1977
Seeds/pod	Very high heritability (94 per cent)		Padda *et al.*, 1970
Seed weight	High heritability (77 per cent)		Padda *et al.*, 1970
Yioeld	High heritability (63 per cent)		Padda *et al.*, 1970
Stem colour	Monogenic	Purple stem colour dominant over green stem colour	
Resistance to Jassid (*Amraska biguttula biguttula*)	Dominant, polygenic control	Non allelic interaction of duplicate and complementary types.	Sharma and Gill, 1984
Resistance to YVMV	High heritability (70 per cent)		Padda *et al.*, 1970
Resistance to YVMV	Monogenic	Resistance governed by single dominant gene.	Jambhale and Nerker, 1981; Dutta, 1988
Resistance to YVMV	Digenic	Two recessive genes, two complementary dominant genes	Singh *et al.*, 1962; Thakur, 1976; Sharma and Dhillon, 1983
	Polygenic	Dominant, additive genes.	
Earliness	Simple inheritance		Erickson and Couto, 1963
Days to flowering	High heritability (65 per cent)		Padda *et al.*, 1970
	Non-additive gene action		Rao and Sathyavathi, 1977
Calyx colour	Dominance	Simple inheritance	Kalia and Padda, 1962
Petal blotch	Dominance	Simple inheritance	Erickson and Couto, 1963
Base colour	Dominance	Simple inheritance	Kalia and Padda, 1962
Petal venation colour	Dominance	Simple inheritance	Kalia and Padda, 1962

genes controlled days to flower and number of fruits/ plant, while 4 to 5 groups of dominant genes controlled plant height. Kulkarni (1976) observed that all the three yield component characters showed over-dominance, and further analysis of gene effects showed that duplicate and complimentary types of epistasis were more pronounced (Kulkarni *et al.*, 1973, 1978).

Augmentation of Genetic Resources and Crop Improvement

Breeding Priorities

(1) To evolve high yielding varieties having wide adaptabilities

(2) To develop cultivars of different maturity group suited to specific environments

(3) To develop multiple disease and pest resistant varieties, with special emphasis on yellow vein mosaic virus, okra leaf curl virus, shoot and fruit borer, jassids and leaf hopper and root knot nematodes.

(4) To develop varieties tolerant to abiotic stresses, especially tolerance to low temperature, drought, excessive rain, saline and alkaline soils.

(5) To develop varieties having dark green, tender, smooth pods suitable for export market.

(6) To evolve varieties with better nutritive attributes and also varieties suitable for dehydration, canning and freezing.

Major Crop Improvement Research Stations

(A) In India

At national level there are so many agricultural universities and ICAR institutes in which okra research is going on successfully. Some of them like ICAR-IIVR, Varansi; ICAR-NBPGR, New Delhi; ICAR-IARI, New Delhi; MPKV, Rahuri; PAU Ludhiana; GAU, Anand; JAU, Junagarh; BCKV, Kalyani; MAU, Parbhani; ICAR-IIHR, Bengaluru; CCH HAU, Hissar; TNAU, Coimbatore; KAU, Vellanikara; CSAU & T, Kanpur; NDUA & T, Faizabad; GBPUA&T, Pantnagar; OUA & T, Bhubaneswar; UHS, Bagalkot and several private seed companies.

(B) International Level

At international level, research on okra is being done mainly in Sri Lanka, Bangladesh, The Phillipines, Sudan, Senegal, Nigeria, Cote, D'Ivoive; Brazil, China, Indonesia, Australia, Popua New Guinea, USA, World Vegetable Centre (old name AVRDC) Taiwan.

Plant Introduction and Selection

a) Augmentation of 'Sudan type'

Germplasm base in okra is relatively narrow. The maximum variation is observed in the widely adopted cultivated species *A. esculentus*, also called the 'Sudan type'. Turkey and India are maintaining greatest collections of this group. This group is also distributed in Afghanistan, Iran, Pakistan, Bangladesh and Yugoslavia to a lesser extent.

Table 7.14: International Institutes Associated with Okra Improvement

Sl.No.	Name of the Organization	Country
1.	Bangladesh Agricultural Research Institute (BARI)	Bangladesh
2.	Orstom Institute Agronomigque Mediterranean	France
3.	Idessa Bouake Cote, D'Ivoive	
4.	Plant Genetic Resources Centre, Gannoruwa	Sri Lanka
5.	Nigerian Institute of Horticulture Research, Ibadan Rese. (NIHORT)	Nigeria
6.	University of Ibadan	Nigeria
7.	National Plant Germplasm Resources Laboratory (NBGRL)	Philippines
8.	Dakar	Senegal
9.	Horticulture Res. Sec. Ward Madani Sudan	Senegal
10.	Vegetable Genetic Resource Laboratory	China
11.	Institute Chinese Academy of Agril. Sciences (CAAS)	China
12.	Gene Bank of Beijing Academy of Agriculture Sciences	China
13.	Itaguar Experiment Station, Pesagro-Rio	Brazil
14	Universidade Fedral de Viscosa UFV	Brazil
15.	Luiz'da Queiroz Agriculural School, Piraciaba, Sao Paulo	Brazil

b) Augmentation of 'Guinean Type'

Guinean types are basically from West Africa, and is strongly distinguished from other groups on the basis of red leaf veins, very late flowering, reduced number of sub calyx bracts and large number of seeds per pod. It is considered to be natural amphidiploids of *A. esculentus* and *A. manihot*.

A joint exploration and collection programme was organized by IBGR and ICAR-NBPGR in southern region of India to augment okra genotypes (wild and cultivated) during 1989-1991.

EC305616 line (resistance to YVMV) was introduced in India from Bangladesh. The varieties like 'Clemson Spineless', 'Louisiana' and 'Velvet Round' bred in the USA have been introduced to a large number of okra growing countries both for fresh market and canning purpose. Indigenous introductions have played a significant role in improving the okra wealth of many countries. In India, the first improved variety of okra, 'Pusa Makhmali', was developed through pure line selection from local material collected from West Bengal. High yielding lines recognized as varieties are 'Lam Selection', 'Sel 2', 'R 7', 'EMS 8', and 'Beltes Five' (Suresh Babu and Peter, 1984). 'Co.1' is also a single plant selection from a heterozygous population of 'Red Wonder' and 'Gujarat Bhindi-1' is a pure line selection from an unknown bulk seed. West African okra is characterized with red leaf veins, late flowering, reduced number of sub-calyx bracts, pods mounted at right angles to stem. In Gujarat, 'Gujarat Okra-1', and 'Gujarat Okra-2', both are very popular. Thirty seven genotypes were evaluated by Verma *et al.* (2016) and found maximum variability for fruit yield/plant and plant height. The PCV were higher than that of GCV values for all the characters indicating influence of environmental effects in the expression of these characters. Maximum heritability was observed for plant height (97.32 per cent), followed by fruit yield/plant (91.26 per cent), days to first picking (88.31 per cent), fruits/plant (85.18 per

Table 7.15: Augmentation of Okra Genotypes (Wild and cultivated) from Southern Region of India

Sl.No.	Species	Common Name	Status	No. of Collection	Distribution Status			
					Kerala	Karnataka	Tamil Nadu	Andhra Pradesh
1.	*A. esculentus*	Okra Sudanian	Cultivated	191	+	+	+	+
2.	*A. caillei*	Guinean okra	Cultivated	14	+	+	+	0
3.	*A. moschatus* ssp. *tuberosus*	-	Wild	1	0	0	+	0
4.	*A. moschatus* ssp. *moschatus* var. *moschatus*	-	Wild	3	0	0	0	0
5.	A. *manihot* ssp. *tetraphyllus* var. *tetraphyllus*	-	Wild	2	+	+	0	0
6.	*A. angulosus*	-	Wild	31	+	+	+	0
7.	*A. ficulneus*	-	Wild	3	0	+	+	0
8.	Abelmoschus sp. (natural hybrid)	-	Wild	1	0	+	0	0

Source: Velayudhan *et al.*, 1996.

cent), days to 50 per cent flowering (82.80 per cent) and harvest duration (82.76 per cent).

One line BCO-1 has been developed through single plant selection of okra from local germplasm, which is characterized by short plant stature, medium long (9.5 cm) green pods having yield potential of 8.0-8.5 t/ha and highly tolerant to YVMV. Choudhury and Singh (2016) evaluated 27 okra genotypes to workout genetic divergence for yield and its attributing characters. The genotypes were grouped into 6 different non-over lapping clusters on the basis of relative magnitude of D^2 values with highest of 7 genotypes in Cluster III followed by Cluster V with 6 genotypes, Cluster IV with 5 genotypes and, Cluster I with 4 genotypes. Cluster II had only three genotypes, whereas Cluster VI presented only one genotype. The intra and inter cluster values ranged from 66.144 to 85.022 and 82.904 to 693.748, respectively. Kumar (2016) evaluated 31 genotypes for 16 tarits and screen against YVMV. The general mean of morphological data expressed considerable variation among the genotypes for most of the characters under study. High GCV, heritability and genetic advance was exhibited by per cent disease incidence, showing the high heritability. Similarly, days to 1st flowering, average fruit weight and yield also showed high heritability. High heritability coupled with high heritability and genetic advance were recorded for per cent disease incidence. Inter nodal length, a desirable traits from high yield viewpoint was found minimum in P-8 and DOV-152. Fruit length was recorded maximum in Parbhani Kranti. DOV-66 recorded significantly higher yield (141.52 q/ha) than other genotypes. General mean of data showed tha genotype P-8 was earliest in days to 1st flowering. Four genotypes namely DOV-11, DOV-66, DOV-24 and DOV-12 recorded to incidence of YVMV after 30 days of sowing. However, previously bred resistant variety, Pusa Sawani recorded maximum YVMV incidence and was found highly susceptible. The popular varieties namely, Arka Abhay and Arka Anamika were also found susceptible to YVMV undr Delhi condition as they showed very high YVMV incidence after 90 days of sowing. However, newly developed genotypes DOV-12 recorded minimum disease incidence during whole growing period followed by DOV-66 and found resistant to YVMV even after 90 days of sowing, while Pusa Sawani recorded 94 per cent incidence during same period. High heritability coupled with high genetic advance was recorded in per cent YVMV incidence and yield/plant, demonstrated the preponderance of additive genes in controlling the expression of these characters and thus, provide better opportunity for effective nd reliable selection for these cjharacters. High heriatability alongwith moderate genetic advance was observed for plant height, internodal length, fruit yield/plant and fruits/plant which may be attributed to additive and non-additive gene effects. These estimates indicated the role of dominance and epistasis and hence, these traits can be improved by intermating superior genotypes of segregating population developed from combination breeding.

Table 7.16: Release and Notified Varieties of Okra

Variety	*Notification Number*	*Date/Month/Year*
Pusa Sawani	4045	24.09-69
	716	20.02.70
	361(E)	30.06.73
	440	21.08.75
	786	02.02.76
	540(E)	24.07.85
P-13	361(E)	30.06.73
S-13	440	21.08.75
Bhindi (Kind)	441(E)	21.08.75
CO-1	13	19.12.78
Punjab Padmini	449(E)	08-07-83
MDU-1	540(E)	24.07.85
Gujarat Bhindi-1	258(E)	14-05-86
Selection-2	258(E)	14-05-86
Parbhani Kranti	867(E)	26-11-86
Pubjab-7	471(E)	05.05.88
Utkal Gaurav	6369(E)	02-09-94
Varsha Uphar	115(E)	10-02-96
Hissar Unnat	98(E)	08-02-97
Azad Bhindi-1	1135(E)	15.11.01
Gujarat Okra-2	1135(E)	15.11.01
Azad Bhindi-2	597(E)	25.04-06
Susthira (AE-286-1)	597(E)	25.04-06
Kashi Pragati	597(E)	25.04-06
Kashi Vibhuti	597(E)	25.04-06
Kashi Satdhari (IIVR-10)	597(E)	25.04-06
Phule Utkarsh	597(E)	25.04-06
Arka Anamika	2035(E)	28.11.06
Arka Abhay	2035(E)	28.11.06
Hisar Naveen	2035(E)	28.11.06
Kashi Lila	2035(E)	28.11.06

Hybridization and Selection

In India, inter-varietal hybridization followed by pedigree selection, several varieties are bred among them 'Pusa Sawani' is being cultivated widely due to high yield, and yellow vein mosaic virus (YVMV) resistance but now become susceptible in plains. Important characteristics in a hybrid combination should be, medium plant height, up-right branching, high number of fruit bearing nodes, lower position of first fruiting node, deeply lobed leaves, short internodes, dark green

Table 7.17: Suitable Parents and their Salient Features

Male Parent	Female Parent	Salient Features	References
'Pusa Sawani'	'H398'	Early flowering, early maturity, higher fruit weight and yield	Joshi *et al.* (1959)
'Pusa Makhamali'	'H398'	Early flowering, early maturity, higher fruit weight and yield	Raman (1965)
'Local 5' (Malaysian)	'Emerald' (American)	Higher germination, early flowering, more plant height, and higher yield.	Jalani and Graham (1973)
'Local 7' (Malaysian)	'Gold Coast' (American)	Higher germination, early flowering, more plant height, and higher yield.	
'Local 5' (Malaysian)	'Gold Coast' (American)	Higher germination, early flowering, more plant height, and higher yield.	
Local 7' (Malaysian)	'Emerald' (American)	Higher germination, early flowering, more plant height, and higher yield.	
'Pusa Sawani'	'Smooth long Green'	More fruits/plant and higher yield	Sharma and Mahajan (1978)
'American Seven Dhari'	'Pusa Sawani'	More plant height and higher yield	Sharma and Mahajan (1978)
'New Selection'	'AE 91'	Heterosis for yield.	Poshiya and Shukla (1986)
'Balady'	'Gold Coast'	Heterosis for plant height, fruits/plant and fruit weight	El-Maksoud *et al.* (1986)
'Seven Leaves'	'Pusa Sawani'	Yield and fruits/plant	Radhika (1988)

fruit colour, earliness, tolerance to important pests and diseases. Potential for high nutrient and water up-take in order to improve production is single important factor responsible for success of a genotype/hybrid. Earlier researchers have reported yield improvement in okra from 49-60 per cent through heterosis breeding (Joshi, *et al.*, 1959; Pratap and Dhankar, 1980). However, there is still scope for yield improvement (Hoque and Hazarika, 1993). Variety 'Pusa Sawani' a selection (symptom less carrier of yellow vein mosaic virus) from cross 'Pusa Makhmali' x 'IC 1542' gave high yield in different locations of India. Not only in India, it has been introduced to the different okra-growing regions of the world and still today it is widely cultivated. Similarly, 'Selection-2' is a derivative of multiple inter-varietal crossing. There are several elite genotypes/varieties which have been used in development of superior hybrids. 'CO-1' is a single plant selection from a heterozygous population of 'Red Wonder' and 'Gujarat Bhindi-1'.

Similarly, Selection-2 is a derivative of multiple intervarietal crossing. In Gujarat, 'Gujarat Okra Hybrid-1' and 'Gujarat Anand Okra-5' are very popular.

Table 7.18: Release and Notified Hybrids of Okra

Hybrid	Notification Number	Date/Month/Year
CO-3 (Hybrid-8)	1052(E)	26.10.99
Phule Kirti (RHROH-4)	348(E)	19.04.01
Shitla Uphar (DVR-1)	1135(E)	15.11.01
Shitla Jyoti (DVR-2)	1135(E)	15.11.01
Kashi Bhairo (DVR-3)	597(E)	25.04.06
HBH-142	597(E)	25.04-06

At BCKVV, Kalyani (West Bengal) diverse genotypes of okra have been utilized to develop commercially suitable hybrids. Some promising hybrids based on heterosis and *per se* performances were BCOH-1 (BCO-1 x BCO-6) and BCOH-2 (BCO-1 x BCO-4) having good tolerance against YVMV (Chattopadhyay *et al.*, 2011). Sharma and Singh (2102) worked out combining ability (sca and gca) in okra through line x tester mating design in SKUA T, J&K. parents and F_1 crosses differed significantly for general combining ability (gca) and specific combining ability (sca) effects for all the characters, respectively. Lines *viz.*, VRO-3, VRO-4, VRO-5, VRO-6 and Parbhani Kranti and testers *viz.*, Hisar Unnat, Arka Anamika, BS-14 and BS-12 were the best general combiners for plant height, number of branches/plant, days to 50 per cent flowering, number of seeds/fruit, average fruit weight, number of fruits/plant and fruit yield whereas, the best specific crosses were VRO-3 x BS-14, VRO-4 x Arka Anamika, VRO-5 x BS-12, Varsha Uphar x Hisar Unnat, Varsha Uphar x BS-14 and Parbhani Kranti x Hisar Unnat for plant height, number of branches/plant, node at which 1st flower appear, inter nodal length, number of seeds/fruit, fruit length, number of fruits/plant and fruit yield. Lines VRO-3, VRO-4, VRO-5, VRO-6 and Parbhani Kranti and testers Hisar Unnat, Arka Anamika, BS-14 and BS-12 were the best combiners are expected to accumulate additive genes that can be exploited in development of open pollinated varieties and the cross VRO-3 x BS-14, VRO-4 x Arka Anamika, VRO-5 x BS-12, Varsha Uphar x Hisar Unnat, Varsha Uphar x BS-14 and Parbhani Kranti x Hisar Unnat showed highest specific combining ability for green fruit yield and its components and are potential source for screening of high yielding F_1 hybrid varieties. On the basis of *per se* performance, heterosis and combining ability More and Chaudhary (2016) rported that hybrid VIO47672 x GJO-3 was the best, followed by IC111493 x GJO-3, IC045796 x GJO-3, VIO47672 x GAO-5 and IC052273 x GJO-3. For fruit yield/plant; among lines, VIO47672 followed by the lines IC045796 and IC111493 while, among testers, GJO-3 followed by tester GAO-5 was the best combiners across the seasons.

Table 7.19: Crossability Relationship at a Glance among the different Taxa of *Abelmoschus*

Crossability in Abelmoschus	*A. angulosus var. angulosus*	*A. angulosus var. grandiflorus*	*A. angulosus var. purpureus*	*A. calliei*	*A. crinitus*	*A. esculentus*	*A. ficulneus*	*A. enbeepee gearensis*	*A. moschatus var. moschatus*	*A. moschatus var. tuberosus*	*A. tetraphyllus var. tetraphyllus*	*A. tetraphyllus var. pungens*	*A. tuberculatus*	*A. mizonagensis sp. nov.*	*A. moschatus subsp. rugosus*
A. angulosus var. *angulosus*	9	-	-	-	3	5	-	3	-	-	-	3	-	-	-
A. angulosus var. *grandiflorus*	-	5	-	0	-	0	0	-	0	0	4	-	0	-	-
A. angulosus var. *purpureus*	-	-	5	-	-	-	-	-	-	-	-	-	-	-	-
A. calliei	5	6/8	-	5	0	5/8	-	-	2	0	6	0	0	8	-
A. crinitus	3	-	-	-	5	-	-	3	3	0	-	-	-	-	-
A. esculentus	5	8	-	8	3	5	2/4	-	3	2	8	8	8	8	8
A. ficulneus	-	-	-	-	0	3	5	0	0	0	4	0	7	-	-
A. enbeepee gearensis	-	-	-	-	5	-	-	5	6	6	-	-	-	-	-
A. moschatus var. *moschatus*	0	1	-	4	3	4	4	2	5	7	4	0	0	-	-
A. moschatus var. *tuberosus*	-	-	-	0	3	3	-	3	7	5	0	0	4	-	-
A. tetraphyllus var. *tetraphyllus*	7	0	-	5	-	8	0	-	-	-	5	7	3	-	-
A. tetraphyllus var. *pungens*	0	-	-	0	0	0	-	0	0	0	-	5	-	-	-
A. tuberculatus	-	-	-	3	0	8	5	3	-	-	-	3	5	-	-
A. mizonagensis sp. nov.	-	-	-	-	-	-	-	-	-	-	-	-	-	-	-
A. moschatus subsp. *rugosus* (Dehra Dun)	-	-	-	-	-	-	-	-	-	-	-	-	-	-	-

0=No fruit set; 1=Fruit set; 2=Fruit set but no seed set; 3=Fruit set and seed set; 4=Fruit set, seed set but no germination; 5= Fruit set, seed set and F_1; 6= Fruit set, seed set, F_1 but no F_2/BC_1; 7= Fruit set, seed set, F_1 but no F_2/BC_1; 8= Colchichoploids-C_1 germinated and viable seeds harvested.

Source: Joseph *et al.*, 2017.

Interspecific Hybridization

The species are incompatible and gene flow is restricted by hybrid sterility except one way cross of *A. esculentus* x *A. manihot* var. *ghana* was successful. However fertile F_1 hybrids have been obtained of the cros *A. tuberculatus* (n=29) x *A. esculentus* (n=62), *A. tuberculatus* (n=20) x *A. ficulneus* (n=36) and *A. esculentus* (n=65) x *A. manihot* ssp. *manihot* (n=97) by chromosome doubling with colchicine. Attempts to develop hybrids of okra with different wild species were struck by sterility of the F_1 hybrid.

Crosses of cultivated okra genotypes *A. angulosus* var. *grandiflorus, A. caillei, A. tuberculatus, A. tetraphyllus* and *A. mizongaensis* sp. now. (Mizoram), *A. moschatus* subsp. *rugosus* and Guinean okra with *A. esculentus, A. angulosus* var. *grandiflorus* and *A. mizongaensis* were raised but could not be advanced as the interspecific hybrid failed to produce viable seeds upon selfing and back cross. However fertility was restored in the above crosses by colchicine treatment and diploidization. Application of 1 per cent colchicine for 3 days was lethal with less than 1 per cent survival, however application of 0.15 per cent colchicine for 4 days for 4 hours from 7.00 am to 11.00 am was found effective and with over 90 per cent seedling recovery. Hybridization of okra with *A. enbepeegearensis* and *A. crinitus* were also made and the seedlings were subjected to colchiploidization for seedling development. The interspecific cross between *A. esculentus* and *A. manihot* var. *ghana* yielded seed, the reciprocal cross being a failure. *A. manihot* var. *ghana* is a symptomless carrier of yellow vein mosaic virus (YVMV). The extensively studied cross is *A. esculentus* x *A. manihot* (Pal *et al.*, 1952). Partial success was achieved only when *A. esculentus*, the species with the highest chromosome number was used as the female parent. The F_1 hybrids were vigorous but mostly sterile as most of the meiotic chromosome remained as univalents. Shrivelled or ermpty seeds were obtained in a cross between *A. ficulneus* x *A. esculentus* (Pal *et al.*, 1959). Gadwal *et al.* (1968) crossed *A. esculentus* x *A. ficulneus* and ovule culture was attempted resulting in the production of hybrid offspring. *A. tuberculatus* crosses readily with *A. caillei*

Table 7.20: Success in Development of Interspecific Hybrids

Cross Combinations	Genomic Structure	Results	References
A. esculentus x *A. tuberculatus*	2n=130 x 2n=58	Positive	Joshi *et al.* (1974)
A. esculentus x *A. manihot*	2n=72 x 2n=60	Positive	Teshima (1933)
A. esculentus x *A. manihot*	2n=130 x 2n=66	Positive	Pal *et al.* (1952)
A. esculentus x *A. ficulneus*	2n=130 x 2n=72	Negative	Joshi *et al.* (1974)
A. esculentus x *A. moschatus*	2n=130 x 2n=72	Negative	Joshi *et al.* (1974)
A. esculentus x *A. tetraphyllus*	2n=130 x 2n=138	Positive	Joshi and Hardas (1976)
A. esculentus x *A. tetraphyllus*	2n=130 x 2n=130	Positive	Ugale *et al.* (1976)
A. esculentus x okra from *Ghana*	2n=130 x 2n=194	Positive	Joshi and Hardas (1976)

and may produce partially fertile plants. *A. esculentus* x *A. tetraphyllus* hybrid has been obtained. Madhusoodan and Nazeer (1986) reported that hybrid made between Guinean type and Sudanian type was sterile due to difference in the chromosome number of parental types. Successful crosses have been obtained by crossing cultivated species *A. esculentus* with *A. tuberculatus* (Pal *et al.*, 1952; Joshi and Hardas, 1956, Joshi *et al.*, 1974) because of (a) F_1s sterility or with low fertility or were obtained with embryo rescue technique (b) with *A. manihot* (Pal *et al.*, 1952; Jambhale and Nerkar, 1986) (c) with *A. moschatus* (Hamon and Yapo, 1986; Gadwall *et al.*, 1968; Joshi *et al.*, 1974) (d) with *A. caillei* (Thakur and Arora, 1976) (e) with *A. tetraphyllus* var. *tetraphyllus* (Dutta, 1991) (f) with *A. ficulneous* and *A. angulosus* (Prabhu and Warade; 2013). Some interspecific hybridization programme attempted in *Abelmoschus* are as under shown in Table 7.20.

Interspecific hybridization was also resorted to transfer resistant gene(s) from related species to *A. esculentus*, as given in the table (7.21). Presently, successful attempt has been made to transfer the YVMV resistant gene from *A. manihot* ssp. *manihot* (Thakur, 1976; Singh and Thakur, 1979; Jambhale and Nerker, 1981; Dhillon and Sharma, 1982) and *A. manihot* ssp. *tetraphyllus* (Dutta, 1984) to *A. esculentus*. *A. esculentus* (2n=130) x *A. manihot* var. *ghana* (2n=194) and *A. tuberculatus* (2n=58) x *A. manihot* (2n=68) showed unilateral incompatibility.

Table 7.21: Some Successful Interspecific Hybridization

Cross Combinations	References
A. esculentus x *A. manihot* var. *medicus*	Arumugham *et al.* (1975)
A. esculentus x *A. manihot* var. *adotachtylus*	Arumugham *et al.* (1975)
A. esculentus x *A. tetraphyllus*	Ugale *et al.* (1976)
A. esculentus x *A. manihot* var. *manihot*	Thakur (1976)
A. esculentus x *A. tetraphyllus*	Mamidwar *et al.* (1979)
A. esculentus x *A. manihot*	Jambhale and Nerker (1983)
A. esculentus x *A. manihot* ssp. *manihot*	Jambhale and Nerker (1983)
A. esculentus x *A. tetraphyllus*	Jambhale and Nerker (1983)
A. esculentus x *A. manihot* ssp. *tetraphyllus*	Dutta (1984)
A. esculentus x *A. manihot* ssp. *tetraphyllus*	Suresh Babu (1987)

Table 7.22: Cross Compatibility in Okra

Crosses	No. of Flowers Crossed	Fruit Set (per cent)	Avg. No. of Seeds in Cross Fruits	Crossability Index (per cent)	Seed Germination per cent (for two season)
Cross compatibility studies between *A. esculentus* and *A. tuberculatus*					
Phule Utkarsh x *A. tuberculatus*-1	50	30.00	14.33	21.71	37.50
A. tuberculatus-1 x Phule Utkarsh	10	30.00	3.33	18.50	30.00
Parbhani Kranti x *A. tuberculatus*-1	50	32.00	17.69	27.21	38.75
A. tuberculatus-1 x Parbhani Kranti	15	30.30	3.50	19.44	32.00
Varsha Uphar x *A. tuberculatus*-1	40	30.00	10.00	14.70	33.75
A. tuberculatus-1 x Varsha Uphar	10	20.00	2.00	11.11	25.00
Cross compatibility studies between *A. esculentus* and *A. tetraphyllus*					
Pusa Sawani x *A. tetraphyllus*-2	50	66.00	33.38	53.83	76.88
A. tetraphyllus-2 x Pusa Sawani	20	50.00	13.00	44.06	66.50
Red Bhendi x *A. tetraphyllus*-2	50	70.00	32.75	54.46	75.63
A. tetraphyllus-2 x Red Bhendi	15	46.66	12.50	42.37	55.00
Arka Anamika x *A. tetraphyllus*-1	15	53.33	22.00	36.66	58.75

Crosses	No. of Flowers Crossed	Fruit Set (per cent)	Avg. No. of Seeds in Cross Fruits	Crossability Index (per cent)	Seed Germination per cent (for two season)
A. tetraphyllus-1 x Arka Anamika	10	50.00	10.00	35.71	40.00
Arka Anamika x *A. tetraphyllus*-2	35	71.42	26.33	43.88	76.25
A. tetraphyllus-2 x Arka Anamika	15	33.33	12.00	40.67	53.20
Parbhani Kranti x *A. tetraphyllus*-1	30	56.66	29.33	45.12	68.21
A. tetraphyllus-1 x Parbhani Kranti	15	53.33	13.00	46.42	67.50
Parbhani Kranti x *A. tetraphyllus*-2	60	80.00	32.88	50.58	75.42
A. tetraphyllus-2 x Parbhani Kranti	25	56.00	14.00	47.45	46.45
Varsha Uphar x *A. tetraphyllus*-1	20	65.00	29.66	43.61	27.50
A. tetraphyllus-1 x Varsha Uphar	10	20.00	10.30	36.78	18.50
Varsha Uphar x *A. tetraphyllus*-2	50	68.00	31.48	46.29	70.63
A. tetraphyllus-2 x Varsha Uphar	20	35.00	12.00	40.67	50.50
P7 x *A. tetraphyllus*-1	20	55.00	31.00	49.20	74.55
A. tetraphyllus-1 x P7	10	40.00	12.50	44.64	66.40
P7 x *A. tetraphyllus*-2	25	64.00	27.83	44.17	70.63
A. tetraphyllus-2 x P7	10	40.00	11.00	37.28	60.75
Phule Utkarsh x *A. tetraphyllus*-1	30	60.00	35.00	53.03	81.25
A. tetraphyllus-1 x Phule Utkarsh	20	50.00	14.50	51.78	80.00
Phule Utkarsh x *A. tetraphyllus*-2	50	84.00	36.00	54.54	77.08
A. tetraphyllus-2 x Phule Utkarsh	25	75.00	15.33	51.96	58.50
Arka Abhay x x *A. tetraphyllus*-2	50	68.00	33.66	56.10	54.38
A. tetraphyllus-2 x Arka Abhay	20	60.00	15.00	50.84	46.50
Cross compatibility studies between *A. esculentus* and *A. ficulneus*					
Phule Utkarsh x *A. ficulneus*-1	15	46.00	26.00	39.39	0.00
A. ficulneus-1 x Phule Utkarsh	10	55.00	0.00	0.00	0.00
Parbhani Kranti x *A. ficulneus*-1	15	53.33	24.00	36.92	0.00
A. ficulneus-1 x Parbhani Kranti	15	60.00	0.00	0.00	0.00
Varsha Uphar x *A. ficulneus*-1	20	55.00	27.00	39.70	0.00
A. ficulneus-1 x Varsha Uphar	20	60.00	0.00	0.00	0.00

Source: Prabhu and Warade, 2013.

Interspecific hybridization has been followed in the development of several varieties like Punjab Padmini, Winter Bush, Parbhani Kranti, P-7, Arka Anamika and Arka Abhay. The other important resistant variety developed through interspecific hybridization is Pharbhani Kranti by crossing *A. esculentus* x *A. manihot*. *A. esculentus* cultivars were found compatible both directly and reciprocally with *A. tuberculatus*, *A. tetraphyllus*, *A. caillei*, *A. manihot* (L.) Medikus, *A. manihot* ssp. *manihot* and *A. manihot* ssp. *tetraphyllus* while *A. moschata* was found incompatible on both ways with *A. esculentus*. The wild species *A. ficulneus* was found compatible but seed sterile with *A. esculentus* when later used as female parent however, *A. ficulneus* was found

Table 7.23: Quantitative and Qualitative Characters Expressed by the Promising Selections in the F_9 Generation in Comparison with Parents and Arka Anamika

Characters	F_9-5-5	F_9-6-8	F_9-8-14	F_9-9-23	A.c* (P1)	A.e* (P2)	Arka Anamika
Quantitative characters							
Plant height (cm)	100.23	120	97	105	143.7	123	140
Internodal length (cm)	4	8	6	8	7.8	12.2	10
No. of primary branches	3	4	3	3	3	3	3.3
Length of epicalyx segment (cm)	1.7	2.1	2.2	2.0	1.3	2.2	2.0
Width of epicalyx segment (cm)	0.5	0.6	0.8	0.5	1.1	0.5	0.6
Petiole length (cm)	31	25	29	27	34	30.2	34
Days to flower	42	47	52	46	47	39.2	47

Characters	F_9-5-5	F_9-6-8	F_9-8-14	F_9-9-23	A.c* (P1)	A.e* (P2)	Arka Anamika
Days to first harvest	48	52	58	51	52.6	44.2	52
No. of harvests	8	9	8	6	11	14	11
First fruiting node	4	5	5	4	4.6	4	5
Length of fruit (cm)	18	22	19	22	16.8	27.9	18
Girth of fruit (cm)	6.4	5.9	6.5	5.8	8.4	7.4	7.5
Locules per pod	5	5	5	5	6	5	6
No. of fruits/plant	25	24	17	20	12	15	15
No. of ridges/pod	5	5	5	5	6	5	5
Crop duration (days)	149	154	153	155	171.6	167.3	145
Yield/plant (g)	278	285	192	210	177.3	191.4	178.01
Pollen fertility (per cent)	98.54	97.89	97.52	98.45	99.9	99.5	96.57
Mucilage content (g/100 g)	0.27	0.59	0.49	0.53	1.12	0.25	0.33
YVMV reaction	HR	HR	HR	HR	HR	HS	MS
Qualitative characters							
Pod pubescence	NP	NP	NP	NP	NP	NP	NP
Leaf margin	DF	NF	DF	NF	NF	DF	DF
Flower colour	Yellow	Yellow	Yellow	Yellow	Yellow	Yellow	Yellow
Flower size	Medium	Large	Large	Large	Yellow	Yellow	Yellow
Purple throat at base of corolla	Present	Present	Present	Present	Present	Present	Present
Colour of leaf vein	GWPT	GWPT	GWPT	GWPT	GWPT	GWPT	GWPT
Colour of leaf base	RWGT	RWGT	RWGT	RWGT	RWGT	RWGT	RWGT
Colour of fruit	DG	G	G	DG	G	LG	G

G: Green; DG: Dark Gree; LG: Light Green; HR: Highly Resistant; GWPT: Green with Purple tinge; DF: Deeply fid; NP: Not pubescent; LP: Less pubescent; NF: Narrowly fid; HS: Highly susceptible; *A.c**: *A. caillei*; *A.e**: *A. esculentus*.

Source: Mogili *et al.*, 2013.

Table 7.24: Genetic Parameters for different Quantitative Characters

Characters	PCV	GCV	Heritability	Genetic Advance
Plant height (cm)	5.01	4.97	98.2	14.89
Internodal length (cm)	1076	10.59	96.9	2.11
No. of primary branches	5.76	5.61	94.9	0.35
Length of epicalyx segment (cm)	14.59	14.18	94.5	0.58
Width of epicalyx segment (cm)	25.64	25.58	99.5	0.37
Petiole length (cm)	4.60	4.06	77.8	2.60
Days to flower	7.83	7.70	96.9	8.02
Days to first harvest	7.5	7.34	95.7	8.43
No. of harvests	11.70	10.70	92.9	2.44
First fruiting node	8.59	8.36	94.7	0.95
Length of fruit (cm)	23	22.97	99.7	8.56
Girth of fruit (cm)	3.42	3.37	97.3	0.54
Locules per pod	4.45	4.27	91.9	0.5
No. of fruits/plant	11.93	11.78	97.5	2.83
No. of ridges/pod	4.25	4.16	95.7	0.5
Crop duration (days)	1.39	1.31	88.9	4.22
Yield/plant (g)	3.26	3.21	96.7	11.58
Pollen fertility (per cent)	63.16	63.16	99.9	3.01
Mucilage content (g/100 g)	48.29	47.85	98.2	0.45
Coefficient of infection (CI)	**135.93**	**135.92**	**99.9**	**34.17**

PCV: Phenotypic coefficient of variation; GCV: Genotypic coefficient of variation.

Source: Mogili *et al.*, 2013.

incompatible with *A. esculentus* when former used as female parent while, wild species *A. angulosus* was found compatible with *A. esculentus*, only when later used as female parent. In general, highest fruit set percentage, number of seeds per crossed fruit, crossability index and germination percentage was 66.0 observed when *A. esculentus* was used as female parent while crossing with all wild species except *A. caillei* (Prabhu and Warade, 2013).

Mogili *et al.* (2013) conducted experiment to transfer YVMV resistance from 'Susthira' (*A. caillei*) to 'Salkeerthi' (*A. esculentus*) through interspecific hybridization and evaluated in F_9 generation. The resistance was judged through artificial inoculation of YVMV adopting grafting and whitefly inoculation techniques.

In the study GCV and PCV values were exhibited incidence of YVMV (Table 7.23). The low variability noticed for these characters indicate the difficulty in improving these characters by selection.

Inter-specific hybrids of okra with *A. angulosus* and *A. mizonagensis* were found to flower and fruit continuously beyond the normal season of okra (August-October) extending to peak summer months (March) offering scope for perennial okra. However, susceptibility to African mealy bug was a problem in summer crop. Hybrid unviability in a few crosses likes *A. esculentus* x *A. ficulneus* and *A. esculentus* x *A. moschatus* need to be tackled through embryo rescue technique. Similarly failure of *A, crinitus* x *A. esculentus* needs to be reinvestigated. A few viable seeds were produced in F_1 hybrid and through a series of back crosses it was possible to transfer a symptomless carrier type of tolerance to YVMV from *A. caillei* and *A. tetraphyllus* to the conventional okra varieties 'Parbhani Kranti and Arka Anamika, respectively.

Breeding for Quality Traits

In okra quality parameters includes fruits dark green, medium long, tender, thin, smooth and 4-5 ridged.

Table 7.26 quality traits may be considered during breeding programmes.

The important characteristics that determine the quality and processing ability are yield, fruit number, length, weight, colour, pod pubescence and spines, fiber, vitamin C, protein, and total sugar content. Improvement of various quality and processing attributes of okra has never been attempted through direct breeding. *A. manihot* ssp. *manihot* contributes dark green fruit colour and more number of fruits per plant in F_1 while *A. manihot* ssp. *tetraphyllus* contributes to increase in

Table 7.25: Inter and Intra-specific Hybrids Developed at ICAR-NBPGR, Rgeional Station, Thrissur (Kerala)

Female Parent	*Male Parent*	*Fruit Set (per cent)*	*Statuis of F_1 (Untreated 2n plants)*
A. esculentus	*A. angulosus* var. *grandiflorus*	100	F_1 shy seed setting
A. angulosus var. *grandiflorus*	*A. esculentus*	16.43	Seed unfilled
A. esculentus	*A. tetraphyllus* var. *tetraphyllus*	92.31	F_1 vigorous with prolific bearing. Sterile seeds with aborted embryo. Pods spiny, 12-15 cm long. Fertility restored through colchine treatment.
A. tetraphyllus var. *tetraphyllus*	*A. esculentus*	19.23	Vigorous with prolific bearing. Sterile seeds with aborted embryo, pods spiny, 12-15 cm long.
A. esculentus	*A. moschatus* subsp. *moschatus*	57.14	Few filled seeds
A. moschatus subsp. *moschatus*	*A. esculentus*	11.54	Not germinated
A. esculentus	*A. moschatus* subsp. *tuberosus*	47.06	Fruit set but no seed set
A. esculentus	*A. caillei*	38.89	Vigorous F_1, seed set but no germination
A. caillei	*A. esculentus*	25.00	Filled seeds (Vigorous F_1s but sterile)
A. esculentus	*A. ficulneus*	35.48	Unfilled seeds
A. esculentus	*A. tetraphyllus* var. *pungens*	100.00	Fruit set but no seed set
A. esculentus	*A. tuberculatus*	30.00	F_1 vigorous with prolific bearing. Sterile seeds with aborted embryo. Pods spiny, 12-15 cm long. Fertility restored through colchine treatment.
A. tuberculatus	*A. esculentus*	85.71	F_1 vigorous with prolific bearing. Sterile seeds with aborted embryo. Pods spiny, 12-15 cm long. Fertility restored through colchine treatment.
A. esculentus	*A. crinitus*	25	F_1 sterile
A. esculentus	*A. enbeepeegearensis*	30.00	F_1 fertile
A. enbeepeegearensis	*A. esculentus*	25	F_1 prolific flowering, no fruit set resembling maternal parent

Source: Joseph *et al.*, 2017.

Table 7.26: Desired Traits and Preferred Range/ Magnitude/Value/Type

Traits	Preferred Range/ Magnitude/Value/Type
Fruit colour	Dark green
Mucilage content of fruit	Low
Fiber content of fruit	Low
Dry matter content	High (Low for canning and freezing)
Chlorophyll content	High

number of branches, fruits and ridges on fruits. High vitamin C content present in the wild species can be utilized for the improvement of *A. esculentus*. Xavier *et al.* (2016) estimated cmbining ability of eight parental lines for mineral content, protein and total dietary fibre content. The general combing ability (gca) and specific combing ability (sca) were highly significant for all characters studied. The magnitude of the gca variance appared more preponderant (σ^2gca/σ^2gca ratio <1) indicating that the additive gene effect was important in the control and inheritance of most of the traits, except total dietary fibre and calcium content. Among parents, DOV-1-2 was agood combiner for phosphorus, calcium, iron, DOV-2 for total dietary fibre, phosphorus and copper, Pusa A-4 for calcium and iron, and DOV-62 for total dietary fibre and phosphorus. The significant SCA effect in positive direction was showed by the cross DOV-1-2 x IC090491 and Pusa A-4 x DOV-2 for total dietary fibre, phosphorus potassium, calcium, iron and copper; WB Selection x IC090491and for phosphorus potassium, calcium, iron and copper. Not a single parent or cross combination showed significant gca or sca for all the characters. Results indicate that there is need of utilizing population improvement methods for biofortification in okra without comprising yield and other horticultural traits. These results points to the facts that, for a more reliable prediction of hybrid with good nutrients content in okra fruits, the role of sca is over and above to that od sca.

Mutation

Mutation breeding was attempted to generate more variability and develop early-yielding varieties resistant to yellow vein mosaic virus (Nandpuri *et al.*, 1971; Nirmala Devi, 1982; Dalia Cherian, 1986). A single dominant gene controlled the mutant. A mutant 'EMS-8' (Punjab-8) carrying resistance to YVMV and tolerance to fruit borer has been evolved through the induced mutation of Pusa Sawani treated with 1 per cent ethyl-methansulfonate.

Polyploidy Breeding

Differential ploidy level and often genotype specific variation in chromosome number has been noticed in okra as well as some of its wild relatives (Keisham *et al.*, 2014). Successful production of fertile amphi-diploids of okra with various *Abelmoschus* species except *A. tuberculatus* are new reports. Kuwada (1966) has developed an amphi-diploid of okra with *A. tuberculatus* and described it as *Abelmoschus x tuberculatus*.

Table 7.28: Ploidy Level in *Abelmoschus* Species

Abelmoschus Species	Nature	Ploidy Level
A. caillei	Cultivated	Hexaploid
A. manihot	Cultivated	Diploid
A. crinitus	Wild	Diploid
A. tetraphyllus var. *tetraphyllus*	Wild	Tetraploid
A. tetraphyllus var. *pungens*	Wild	Tetraploid
A. angulosus var. *grandiflorus*	Wild	Diploid
A. moschatus	Cultivated	Diploid
A. tuberculatus	Wild	Diploid

In okra, polyploids inhibited growth and retarded flowering along with small and irregular fruits. Amphidiploids were obtained from interspecific crossing with the related wild species. Failure of F_1s reported in crosses due to hybrid sterility and prompted to produce amphidiploids (Baby Lissy and Peter, 1990). Kuwada (1964) obtained a fertile amphidiploid by duplication of the F_1 between *A. esculentus* x *A. manihot* called 'Nori-Asa'. He later synthesized amphidiploid an *A. tuberculatus* x *A. esculentus*. Jambhale and Nerker (1982) synthesized an amphidiploids from the cross *A. esculentus* (2n=130) x *A. manihot* ssp. *manihot* (2n=98). The amphidiploid was resistant to yellow vein mosaic virus. In 1983, these researchers also made a successful attempt in inducing polyploidy in three interspecific hybrids. Kuwada (1964) obtained a fertile amphidiploids by duplication of the F_1 hybrid of the cross *A. tuberculatus* (n=29) x *A. esculentus* (n=62). Joshi *et al.* (1974) reported sterile hybrid between *A. tuberculatus* (n=20) x *A. ficulneus* (n=36). Suresh Babu (1987) also induced amphidiploids from a cross between *A. esculentus* x *A. manihot* ssp. *tetraphyllus*.

Table 7.29: Some Achievements in Okra with Respect to Amphidiploids Programme

Cross	Salient Achievements of Polyploids	References
A. tuberculatus (n=29) x *A. esculentus* (n=62)	Fertile amphidiloid by duplicate of the F_1	Kuwada (1964)
A. tuberculatus (n=20) x *A. ficulneus* (n=36)	Sterile F_1	Joshi *et al.* (1974)
A. esculentus (2n=130) x *A. manihot* ssp. *manihot* (2n=194)	Fertile amphidiloid (81.8 per cent) and against F_1 fertility (7.07 per cent)	Jambhale and Nerker (1982)
A. esculentus x *A. manihot* ssp. *tetraphyllus*	Amphidiloid showed regular meiosis	Suresh Babu (1987)

Table 7.30: Characteristics of Synthetic Amphidiploids

Sl.No.	*Amphidiploids*	*Characteristics feature*
1.	*Abelmoschus* x *esoramensis* (*esculentus* x *mizonagensis* sp. nov.)	☆ Extended bearing, beyond February ☆ Dark green fruits, good taste, short fruits ☆ Robust plants and prolific bearing
2.	*Abelmoschus* x *caizoramensis* (*caillei* x *mizonagensis* sp. nov.)	☆ Attractive ovate fruits-Baby okra ☆ Good taste, high mucilage, dark green colour
3.	*Abelmoschus* x *esculophyllus* (*esculentus* x *tetraphyllus* var. *tetraphyllus*)	☆ Intermediate between parents ☆ More of *A. tetraphyllus* appearance ☆ Basal branching (6-14 branches) ☆ Reduced in fruit length ☆ Dark green colour
4.	*Abelmoschus* x *esanguflorus* (*esculentus* x *angulosus* var *grandiflorus*)	☆ Plant to plant variation in photoperiodic sensitivity ☆ Reduced fruit length, blocky fruits ☆ Intermediate between parents ☆ Two maternal genotypes used ☆ Highly branched (25-35 tertiaries)
5.	*Abelmoschus* x *escullei* (*esculentus* x *caillei*)	☆ Oval-globular fruits-Baby okra ☆ High mucilage, good raw taste ☆ Non-photoperiod sensitive ☆ More affinity to *Abelmoschus caillei* ☆ 2-4 branches/plant
6.	*Abelmoschus* x *cangulosus* (*caillei x angulosus*)	☆ Intermediate between parents ☆ Oval fruits- Baby okra ☆ Stiff hairy ☆ Petals spot only inside ☆ Non-persistent epicalyx ☆ Non photoperiod sensitivity
7.	*Abelmoschus* x *esculentus* (*esculentus x tuberculatus*)	☆ Red and green fruited plants ☆ Studded with tubercles ☆ Prolific, above 50 fruit/plant
8.	*Abelmoschus* x *tuberculatus* (*tuberculatus x esculentus*)	☆ Fruit surface studded with tubercles ☆ 10-12 cm long fruits ☆ Prolific bearing 45-60 fruit/plant

Source: Joseph *et al.*, 2017.

Natural allopolyoids like *A. esculentus* an amphidiploid of *A. tuberculatus* (2n=58) and *A.* sp (2n=72) and West African okra an amphidiloid of *A. esculentus* (2n=130-140) and *A. manihot* (2n=60-68) are widely cultivated showing considerable out crossing in nature even at species level. Among the synthetic hybrids involving wild species, *A. miizonagensis* was found to have extended bearing, continued fruit availability due to extended bearing and flowering and fruiting in off/lean season. Some of the interspecific hybrids, not involving okra genome may serve as bridge species to transfer useful traits to cultivated okra as done in root knot nematode breeding of many horticultural crops.

Male Sterility

Male sterility is also reported in okra. Heterosis has been successfully exploited in okra using the phenomenon of male sterility. It is estimated that 70 per cent saving in time and manual labour made possible with introduction and utilization of GMS based hybridization programme as compared to conventional crossing programme in okra seed production. Genetic male sterility is usually recessive and monogenic, hence can be used used in hybrid seed production. GMS traits controlled by a single recessive gene (ms1 ms1). Inheritance of genetic male sterility was studied using F_1, F_2 and test-cross generations of crosses between GMS line (NOGMS-1 and NOGMS-2) and normal elite, YVMV tolerant, fertile genotypes *i.e.* NOL-729, NOL-788, NOL-959, NOL-1015, NOL-1022, NOL-718, NOL-728, NOL-740, NOL-745, and NOL-1062. All the F_1s were found fertile. Segregation of pollen fertility in F_2 and test-cross generations was segregated in the ratio 1fertile: 1 sterile, respectively. All the lines were converted in consecutive

Table 7.31: Prominent Resistant Genotypes to Fungal Diseases

Disease	*Casual Organism*	*Resistant Genotypes*	*Wild Resistance Source*	*References*
Cercospora leaf spot	*Cercospora abelmoschi*	'EC 32598', 'EC 41292', 'IC 10238', 'IC 8248'	*A. crinitus, A. moschatus, A. angulosus, A. pungens*	Sohi and Sokhi (1974)
Leaf spot	*Cercospora hibisci*	'IC 18960', 'IC 15055', 'Round Selection', 'Walgaon', 'EC 41292' and 'IC 1542'		
Powdery mildew	*Erysiphe* sp.	Nigeria	*A. Tetraphyllus, A. angulosus, A. manihot, A. pungens*	Sohi and Sokhi (1974)
Fusarium wilt	*Fusarium oxysporum*	'Pusa Sawani', 'Pusa Makhamali'		Grover and Singh (1970)
Microphomina	*Microphomina* spp.	IC90186, U-4307, U-4365, U-4579		

four back cross generations followed by selfing. Ratio were studied and maintained during each generation. These 10 GMS lines are showing male sterility in disease resistant backgrounds. These lines are being utilized in hybrid seed production after verification of F_1 hybrids performance. Utilization of these lines are helpful to reduce seed production costs due to higher labour wages, non-availability of skilled labour, frequent monitoring *etc.* GMS based hybrids namely NOGH-3, NOGH-7 and NOGH-8 were found superior for qualitative and quantitative traits (Deshmukh *et al.*, 2016). Dubey and Singh (1968) obtained complete male sterility in okra 'Pusa Sawani' with 0.4 per cent Maleic Hydrazide (MH) or Mendok, one week prior to floral bud initiation.

Resistance Breeding

Okra is attacked by several biotic stresses which are serious threats for quantity and quality production. The various biotic stresses *viz.* diseases-YVMV, OELCV, Cercospora leaf spot and pests, white fly, shoot and fruit borers are important ones. Among these yellow vein mosaic virus and enation leaf curl virus are very serious and causes loss to the tune of 50-90 percent. The transfer of resistance from wild relatives has been hampered by sterility problems. Two *Abelmoschus* species *i.e. A. caillei* and *A. tetraphyllus* have been utilized in the improvement of okra. West African okra (*A. caillei*, also known as Guinean type) accounts for only 5 per cent of the total world production of okra (Siemonsma and Kouame, 2004), but it is a very important crop in tropical areas of Cote d'Ivoire, Benin, Cameroon, Nigeria, Ghana and Togo. This relatively newly identified amphipolyploid species (Siemonsma, 1982) is known for possessing a gene pool of variation that may be useful for okra improvement of both temperate and tropical types (Martin *et al.*, 1981). *A. caillei* is gradually replacing common okra in the tropical-humid region because of its better adaptation under humid zone and tolerance to biotic stresses (Siemonsma, 1982). Indeed under very limited and erratic rainfall in the Sudano-Sahel, earliness of *A. esculentus* (being amphidiploid) as compared to *A. caillei* (being amphipolyploid) was preferred during early domestication. In Asia, *A. caillei* has been utilized as a resistant source to breed yellow vein mosaic virus resistant common okra variety (Nerkar and Jambhale, 1985). The inter-specific cross between *A. caillei* and *A. esculentus* is successful with the possibility of gene transfer, although the partial hybrid breakdown barrier must be overcome (Fatokun, 1987). The study on geographical distribution and extent of natural outcrossing in Benin and Togo suggests that genetic integrity of these two species is not threatened (Hamon and Hamon, 1991).

(a) Fungal Disease Resistance

There are several genotypes which are found resistant to prominent diseases of okra. A partial resistance of symptomless carrier types, observed in a multi-ridged, Bengal cultivar 'IC 1542', was used in breeding for resistance and as a result 'Pusa Sawani' was bred out against yellow vein mosaic disease. Nariani and Seth (1958) reported that *A. tuberculatus* carries symptomless type of resistance while *A. pungens* carries true resistance to the disease.

Resistance to *Fusarium oxysporum* f. sp. *vasinfectum* has been reported in *A. manihot* PI379584 while resistance against powdery mildew (*Erysiphe* species) has been reported in *A. tetraphyllus, A. manihot, A. manihot* ssp. *manihot*, and *A. moschatus*. Cultivar 'Clemson Spineless' has been reported to carry resistance to *M. arenaria*.

Table 7.32: Resistance Source for Fungal Diseases

Abelmoschus Species	*Wild Resistance Source*
A. manihot	Powdery mildew, Cercospora, Alternaria blight
A. crinitus	Powdery mildew
A. tetraphyllus var. *tetraphyllus*	Powdery mildew
A. tetraphyllus var. *pungens*	Powdery mildew
A. angulosus var. *grandiflorus*	Powdery mildew, Cercospora, Alternaria blight
A. moschatus	Powdery mildew

(b) YVMV Resistance

A partial resistance of symptomless carrier types, observed in a multi-ridged, West Bengal cultivar 'IC1542' used in breeding and as a result 'Pusa Sawani' was developed which express resistance to YVMV; the most widely cultivated symptomless carrier variety. Here contribution in breeding programme is of two recessive alleles at two loci which conferred resistance (Singh *et al.,* 1962). Sandhu *et al.* (1974) reported that one accession 'EC 31830' (Asumtemkoko) from China and another accession from Ghana introduced by the Division of Plant Introduction, ICAR-IARI, New Delhi, and identified as *A. manihot* medikus subsp. ssp. *manihot* now known as *A. callei* (a semi wild species introduced from Ghana) possessed a high degree of symptomless carrier type of resistance to YVMV and contributed to the development of resistant cultivars 'Punjab Padmini', Punjab-7' and Parbhani Kranti in India. Here two complementary dominant genes are responsible for the resistance (Thakur, 1976; Sharma and Dhillon, 1983; Sharma and Sharma, 1984). Symptomless carrier nature of YVMV resistance is also observed in *Abelmoschus* spp. 'Ghana', *A. tuberculatus*, and *A.* ssp. (West African okra). This group could be utilized for breeding YVMV resistant hybrid varieties. The YVMV field tolerant variety 'Punjab Padmini' was developed by crossing 'Pusa Sawani' with this accession *A. esculentus* 'EC31830' as male and *A. caillei* followed by selection in advanced generation (Sharma, 1982). Okra variety Punjab-7 resistance to YVMV was developed by crossing *A. esculentus* (Pusa Sawani) with *A. caillei* followed by back crossing with Pusa Sawani (Thakur and Arora, 1988). YVMV resistant cultivar 'Parbhani Kranti' was developed through back cross breeding programme, using *A. manihot* as source of resistance (Jambhale and Nerker, 1986). *A. manihot* subspecies *manihot* now known as *A. caillei*, a semi-wild species introduced from Ghana has served as a source of resistance to YVMV. Subsequently Arumugham *et al.* (1975), Thakur (1976), Ugale *et al.* (1976), Jambhale and Nerker (1983), Sharma and Dhillon (1983), Sharma and Sharma (1984) and Jambhale and Nerker (1986) utilized *A. manihot* to develop series of okra lines resistant/tolerant to YVMV. The related species *A. manihot* var. *pungens*, *A. crinitus*, *A. panduraeformis* and *A. vitifolius* and varieties 'Sel 4' and 'Arka Anamika' (*A. manihot* subsp. *tetraphyllus* contributed gene foe resistance to Arka Abhay) were observed resistant to YVMV but now 'Arka Anamika' has been susceptible to this disease. True resistance to YVMV has been reported in *A. manihot* (some forms), *A. pungens*, *A. crinitus*, *A. panduraeformis*, and *A. vitifolius*. In interspecific crosses involving *A. manihot* ssp. *manihot*, the susceptibility to YVMV was controlled by two dominant genes (Thakur, 1976; Sharma and Dhillon, 1983), while Jambhale and Nerker (1981) observed single dominant gene to control YVMV resistance in *A. manihot* and *A. tetraphyllus*. A single dominant gene controlled YVMV resistance in *A. tetraphyllus* was also observed by Dutta (1984) at ICAR-IIHR, Bangalore (India). In okra YVMV two complementary dominant genes are reported (Pullaiah *et al.*, 1998).

Table 7.33: Source of Resistance in Wild Relatives of Okra

Wild Relatives	*Resistance to*
Abelmoschus manihot ssp. *manihot*	Yellow Vein Mosaic Virus
A. tetraphyllus	Yellow Vein Mosaic Virus
A. angulosus ssp. *grandiflorus*	Yellow Vein Mosaic Virus
A. ficulneus	Cercospora Leaf Spot

Cultivars like Arka Anamika (Sel-10), Sel-2, Punjab-7, Varsha Uphar, Hissar Unnat, Utkal Gaurav, 91-4, and Barkha Uphar arereported as resistant against Yellow Vein Mosaic Virus of okra.

Table 7.34: Desirable Lines/Species for YVMV

Traits	*Resistant/Tolerant Source*	*Wild Resistance Source*
YVMV	NIC-9303A, NIC-6308, NIC-3322, NIC-9408, NIC3325, EC329375, K-4409,	*A. crinitus, A. angulosus, A. manihot, A. caillei, A. tetraphyllus* var. *tetraphyllus, A. pungens. A. tetraphyllus* var. *pungens, A. angulosus* var. *grandiflorus* and
OLCV		*A. crinitus, A. angulosus, A. manihot* and *A. tetraphyllus*
ELCV		*A. crinitus, A. manihot, A. angulosus* var. *grandiflorus A. ficulneus*

Source: Singh, 2010.

(c) Insect Pest Resistance

There are several genotypes which are found resistant to prominent pests of okra. Resistance to jassids in *A. manihot* ssp. *manihot* var. *Ghana* was under polygenic control (Sharma and Gill, 1984). Both additive and dominant gene effects were significantly but additive gene effects and dominance-dominance interactions appeared relatively more important. *A. moschatus* and *A. crinitus* both are resistance to jassids. Resistance to jassids has been reported in IHR-21 (India) while against shoot and fruit borer in *A. tuberculatus*, cvs. Red-I and Red-II. *A. panduraeformis* (2n=24) could be utilized to incorporate jassid tolerance in the progeny.

(d) Nematode

'Long Green Smooth' genotype has been found resistant to root knot nematode (*Meloidogyne javanica*).

Sources for Abiotic Factors and Utilization

a) Breeding for Drought Tolerance and Low Temperature

Improvement in okra for drought tolerance is the most efficient, cheapest and long term strategy to manage drought stress (Ludlow, 1989) which includes following aspects:

Table 7.35: Prominent Jassid Resistant Genotypes

Pests	*Casual Organism*	*Resistant Genotypes*	*Wild Resistance Source*	*References*
Cotton Jassid	*Amrasca biggutula biggutula*	'IIHR 21', EC305656, EC305694, EC305695, EC305714, EC306731	*A. moschatus, A. crinitus, A. angulosus* var. *grandiflorus*	Dutta (1971)
Leaf hopper	*Amrasca devastanse*	AE-22 (controlled by single recessive gene and action expressed due to non-preference and antibiosis)		
Shoot and fruit borer	*Earias vitella*	'Red I', 'Red II', 'Red Wonder I'		Srinivasan and Narayanaswamy (1961)
Fruit borer		'AE 22', 'AE 79', 'AE 52', 'Red Wonder II" Red I', 'Red II',	*A. tuberculatus, A. moschatus*	
Mites		EC305656, EC305664, EC305696	*A. angulosus, A. crinitus, A. angulosus* var. *grandiflorus, A. moschatus*	
Aphid		-	*A. manihot*	

Table 7.36: Prominent Nematode Resistant Genotypes

Pests	*Casual Organism*	*Resistant Genotypes*	*References*
Root knot nematode	*Meloidogyne javanica*	'Long Green Smooth'	Birat (1964)

1. Drought escape, where the crop complete its life cycle, before the onset of drought. Drought escape is associated with lower biomass and yield, therefore, it is considered undesirable in the drought resistance improvement programe.
2. The drought avoidance, where the crop maximizes its water uptake and minimizes its water loss.
3. The drought tolerance, where the crop continues to grow and function at reduced water contents.

The water relation is the stage of plant growth at which stress occurs (Salter and Goode, 1967). The previous studies showed that greatest reduction in yield occurred when stress was imposed at the flowering stage. Similarly, moisture stress during the pod filling stage resulted in more than 70 per cent reduction in fruit yield of okra, while the lowest reduction in fruit yield was observed when moisture stress occurred during the vegetative stage (Mbagwa and Adesipe, 1987). Some later studies suggested that flowering stage is more sensitive to stressful conditions (Banziger *et al.*, 1999). Selection of crop genotypes for tolerance to mid-season drought stress has been found to result in improved broad adaptation and specific adaptation to drought environment (Chapman *et al.*, 1997). The basic approach for development of drought tolerant genotypes is to select locally adapted germplasm containing genetic variability for high yield potential and drought adaptive traits (Beck *et al.*, 1990; Vasal *et al.*, 1997).

Three agronomic types (A, B and C) were identified based on flowering date, plant size and fruit yield. Local germplasm belonging to types B (Sudanian or *A. esculentus*) and C (Guinean or *A. caillei*) were crossed with a view to transfer drought resistance from type C to B by back crossing. *A. panduraeformis* (2n=24) could be utilized to incorporate drought tolerance in the progeny.

Table 7.37: Sources for Abiotic Resistance

Abelmoschus Species	*Abiotic Resistance*
A. crinitus	Drought
A. tetraphyllus var. *pungens*	Low temperature
A. angulosus var. *grandiflorus*	Low temperature

Biotechnological Interventions

Efforts have been made to develop insect resistant okra varieties against shoot and fruit borer *i.e. Earias sp.* being the most destructive pest by incorporating *cry1Ac* gene in okra from a bacterium mainly *Bacillus thuringiensis*, commonly known as Bt okra. The Bt okra developed by M/s Maharashtra Hybrid Seeds Company Limited containing *cry1Ac* gene (Event OE-17A) is under safety evaluation and confined field trials. Among viral diseases, Yellow Vein Mosaic Virus (YVMV) being major disease of okra, attempts are being made by M/s Arya hybrid Seeds Ltd., for incorporation of specific genes such as CP (coat protein) gene and antisense RNA gene for elevated viral resistance. Reports on marker development in okra are very scanty and have been limited to characterization of cultivars. An agreement between clustering patterns obtained from morphological traits and molecular markers in *Abelmoschus* spp. has been demonstrated (Mortinello *et al.*, 2001). Ninety-three accessions of common (*A. esculentus*) and West African (*A. caillei*) could be distinguished using random amplified polymorphic DNA (RAPD) markers (Aladele *et al.*, 2008). Use of sequence related amplified polymorphism (SRAP) in marker aided selection (MAS) for various traits in Turkish germplasm has been suggested (Gulsen *et al.*, 2007). Recently, 20 okra accessions from Burkina Faso were analyzed using 16 primers designed to amplify SSR regions of *Medicago truncatula*. Two accessions were found distinct from the other 18, based on the presence of an unique 440 bp fragment generated primer MT-27 and also based on presence of hairs on fruits and delayed maturity of these two accessions (Sawadogo *et al.*, 2009).

Graftings

Interspecific grafts using *A. manihot* var. *ghana* as rootstock and *A. esculentus* as scion was also successfully made with the mission to transfer desirable characteristics of *A. manihot* var. *ghana* to the cultivated okra.

Genetic Resource Conservation

Turkey and India are maintaining greatest collection of okra germplasm especially *A. esculentus*. To a lesser extent okra species *A. esculentus* are being maintained in Afghanistan, Iran, Pakistan, Bangladesh and Yugoslavia. In India, the major collection is being maintained by the ICAR-NBPGR, New Delhi, and IBPGR has assigned it global responsibility for base collection. Okra is endemic to India and Bangladesh has been given priority for collection and evaluation. Large numbers of genotypes, including wild relatives, have been collected and are being made available for further testing and utilization. Germplasm collection from Asia and Africa should be augmented for broadening the genetic base of okra for direct breeding purpose as well as for conservation of the germplasm. At present 3600 different accessions are being maintained at ICAR-NBPGR, New Delhi and Regional Station, Akola (Maharashtra). Okra germplasm endemic to India and Bangladesh has been given priority for collection and evaluation. Large numbers of genotypes, including wild relatives, have been collected and are being made available for further testing and utilization. So far the primary objective has been to identify high yielding genotypes carrying resistance to YVMV and okra leaf curl virus. Besides, long fruiting habit, resistance to various other biotic stresses *viz.* shoot and fruit borer, jassid, and high vitamin C content present in the wild species can be utilized for the improvement of *A. esculentus*. A perusal of the growth performance of wild species and fruit set upon distant hybridization indicates that the agro-climatic conditions prevailing in ICAR-NBPGR, Base Centre, Thrissur, Kerala is highly conducive for wild species maintenance and performance except a few niche-specific high altitude taxa like *A. angulosus* var. *angulosus* and *A. angulosus* var. *purpureus*. There is an urgent to adequately and systematically need to collect wild gene pool of *Abelmoschus* from across the country. Voucher herbarium and photographic documentation and seed conservation of synthetic species have been done for further utilization and reference.

REFERENCES

1. Adams, C.F. 1975. Nutritive value of American foods in common units, U.S. Department of Agriculture, *Agric. Handbook*, p. 29.
2. Arumughamn, R., Chelliah, S. and Muthukrishnan, C.R. 1975. *Abelmoschus manihot*: a source of resistant to bhindi yellow vein mosaic. *Madras Agricultural Journal*, **62(5):** 310-312.
3. Beck, D.L., Vasal, S.K. and Crossa, J. 1990. Heterosis and combining ability of CIMMYT's tropical early and intermediate maturity maize germplasm. *Maydica*, **35:** 279-285.
4. Birat, R.B.S. 1964. Variability of resistance in bhindi varieties to root galls. *Science and Culture* **30:** 244.
5. Bisht, I.S., R.K. Mahajan, and R.S. Rana. 1995 Genetic diversity in South Asian okra (*Abelmoschus esculentus*) germplasm collection. *Ann. Apl. Biol.* **126:** 539-550.
6. Candolle, A.P.D.E. 1983. Origin des plantes cultivees, Paris. p. 150-151.
7. Chapman, S., Crossa, J., Basford, k.E. and Karoonenberg, P.M. 1997. Genotype by environment effects and selection for drought tolerance in tropical maize. II. Three mode pattern analysis. *Euphytica*, **95:** 11-20.
8. Charrier, A. 1984. Genetic resources of the genus *Abelmoschus* Med. (Okra). (English translation). IBPGR, Rome.
9. Chattopadhyay, A., Dutta, S., Hazara, P. and Sarkar, G. 2011. Conservation and utilization of vegetable germplasm at Bidhan Chandra Krishi Viswavidyalaya, Kalyani (West Bengal). Natl. Symp. On '*Vegetable Biodiversity*' Held at JNKVV, Jabalpur (M.P.), p. 131-140.
10. Chauhan, D.V.S. 1972. Vegetables. National Book Trust, India, New Delhi
11. Choudhary, B and Anothai, Choomsai, M.L. 1970. Natural cross pollination in some vegetable crops. *Ind. J. Agri. Society*, **40:** 805-812.
12. Chevalier, A. 1940. L'origine, la culture et les usages de cinq *Hibiscus* de la section *Abelmoschus*. *Rev. Bot. Appl.*, 20:319-328 and 402-419.
13. Choudhury, B. and Singh, A.K. 2016. Genetic divergence studies in okra (*Abelmoschus esculentus* (L.) Moench) genotypes (Abs: 65). 7th Indian Horticulture Congress (Abs. 65; Session: I-Breeding for Horticultural Crops), held at ICAR-IARI, New Delhi, 15-18 November, p. 51-52.
14. Dalia Cherian, 1986. Radiation in interspecific hybrids involving *Abelmoschus esculentus* (L.) Moench and *A. manihot* (L.) Medic. M.Sc. (Hort.) Thesis. KAU, Vellanikkara (Kerala).
15. De Candolle, A.P. 1883. *Origine des plantes cultivees*. Noble Offer Printing, New York, p. 150-151.
16. Deshmukh, N.D., Jadhav, B.P., Halakude, I.S. and Rajput, J.C. 2016. Development of elite GMS lines in okra (*Abelmoschus esculentus* (L.) Moench) for easy and economic hybrid seed production. 7th Indian Horticulture Congress (Abs: 13; Session: II; Breding

for Horticultural Crops), held at ICAR-IARI, New Delhi, 15-18 November, p. 95-96.

17. Dhankar, B.S. and Mishra, J.P. 2001. In: Thamburaj, S. and Singh, N. Ed. Text Book of Vegetable, Tuber crops and Spices. ICAR, New Delhi, p. 222-237.

18. Dinesh, M.R. and Sankaran, M. 2017. Distant hybridization in horticultural crops. Daya Publishing House, New Delhi, p. 198.

19. Dutta. O.P. 1971. Effect of gamma irradiation on seed germination, plant growth, floral biology and fruit production in *Abelmoschus esculentus* (L.). Third International Symposium on Tropical Horticulture, held at ICAR-IIHR Bengaluru (Karnataka).

20. Dutta. O.P. 1984. Breeding okra for resistance to yellow vein mosaic virus and enation leaf curl virus. Annual Report, ICAR-IIHR Bengaluru (Karnataka), p.21.

21. Dutta. O.P. 1991. Okra germplasm utilization at ICAR-IIHR Bengaluru (Karnataka). In: Report of an International workshop on okra genetic resources held at ICAR-NBPGR, New Delhi, India, 8-12 Oct. 1990. IBPGR, Rome, p. 114-116.

22. El-Maksoud, M.A., Hetal, R.M. and Mohamed, M.H. 1986. Studies an inter-varietal cross and hybrid vigour in okra. *Annals of Agricultural Sciences,* **29(1):** 431-478.

23. Fatokun, C.A. 1987. Wide hybridization in okra. *Theor. Appl. Genet.,* **74:** 483-486.

24. Fryxell, P.A. 1957. Mode of reproduction of higher plants. *Botanical Review,* 23: 135-233.

25. Gadwal, V.R., Joshi, A.B. and Iyer, R.D. 1968. Interspecific hybrids in *Abelmoschus* through ovule and embryo culture. *Indian Journal of Genetics and Plant Breeding,* **28(3):** 269-274.

26. Gulsen, O., Karagul, S. and Abak, K. 2007. Diversity and relationships among Turkish okra germplasm by SRAP and phenotypic marker polymorphism. *Biologia Bratislava,* **62:** 41-45.

27. Grover, R.K. and Singh, G. 1970. Pathology of wilt of okra caused by *Fusarium oxysporum* f. *vasinfectum* (Atk) Synder and Hassan, its host range and histopathology. *Indian Journal of Agricultural Sciences,* **40:** 989-996.

28. Grubben, G.J.H. 1977. Okra (In): Tropical vegetables and their genetic resources, IBPGR, Rome, pp. 111-114.

29. Hamon, S., A. Charrier, J. Koechlin and van Sloten, D.H. 1991. Potential improvement okra (*Abelmoschus* spp.) through the study of its genetic resources (in French). Plant Genetic resources. *Newslett.,* **86:** 9-15.

30. Hamon, S. 1988. Organisation evolutive du genre *Abelmoschus* (Gombo): co-adaptation et evolution de deux species de gombo cultivees en afrique de l'ouest *Abelmoschus esculentus* et *Abelmoschus caillei.* Edit. ORSTOM, T.D.M.46.

31. Hamon, S. and Hamon, P.1991. Future prospects of the genetic integrity of two species of okra (*Abelmoschus esculentus* and *Abelmoschus caillei*) cultivated in West Africa. *Euphytica,* 58:101-111.

32. Hamon, S. and Yapo, A. 1986. Perturbation induced within the genus *Abelmoschus* by the discovery of a second edible okra species in West Africa. *Acta Hort.,* **182:** 133-144.

33. Hochruitner, B.P.G. 1924. Genres nouveaux et. genres discutes de la famile des Malvacees. *Candollea,* **2:** 79-90.

34. Hoque, M. and Hazarika, G.N. 1993. Hybrid vigour for some economic traits in okra. In: Heterosis Breeding in Crop Plants-Theory and Applications. Verma, M.M., Irk, D.S. and Chadha, G.S. (Eds.) Symp. Proc., Ludhiana. p.22-23

35. Jalani, B.S. and Graham, K.M. 1973. A study of heterosis in crosses among local and American varieties of okra. *Maharashtra Journal Agricultural Research,* **2(1):** 7-14.

36. Jambhale, N.D. and Nerkar, Y.S. 1982. Induced amphidiploids in the cross *Abelmoschus esculentus* (L.) Moench x *A. manihot* (L.) ssp. *manihot. Genetica Agriculture,* **36:** 19-22.

37. Jambhale, N.D. and Nerkar, Y.S. 1983. Induction of polyploidy in the *Abelmoschus* spp. hybrids. *Prog. Hort.,* **15(1/2):** 157-158.

38. Jambhale, N.D. and Nerkar, Y.S. 1986. 'Parbhani Kranti': A yellow vein mosaic resistant okra. *HortSci.,* **21(6):** 1470-1471.

39. Jasin, A.Z. 1967. Inheritance of certain characters in okra (*H. esculentus* L.) Diss. Abstr.), **28(1):** 3-13.

40. Joseph, J.K., Rou, Y.G., Latha, M. and Bhat, K.V. 2017. Wide hybridization in the genus *Abelmoschus* Medikus. In: Distant hybridization in horticultural crops (eds.): Dinesh, M.R. and Sankaran, M., ICAR-IIHR, Bengaluru (Karnataka); Pub. By Astral Internat. Pvt. Ltd., New Delhi, pp. 196.

41. Joshi, A.B., Gadwal, V.R. and Hardas, M.W. 1974. Okra (In) *Evolutionary Studies in World Crops Diversity and Change in the Indian Sub-Continent,* pp. 99-105. Hutchinson, J.D. (Ed.). Cambridge University Press, London.

42. Joshi, A.B. and Hardas, M.W. 1956. Alloploidy nature of okra. *Abelmoschus esculentus* (L.) Moench. *Nature,* **178:** 1190.

43. Joshi, A.B. and Hardas, M.W. 1976. Okra. In: Evolution of Crop Plants. p. 194-195. Simmonds, N.W. (Ed.) Longman, London.

44. Joshi, A.B., Singh, H.B. and Gupta P.S. 1959. Studies in hybrid vigour. III Bhindi, *Indian Journal of Genetics and Plant Breeding,* **18:** 57-68.

45. Joshi, A.B., Gadwal, V.R. and Hardas, M.W. 1974. Okra. In: Evolutionary studies in world crops. Diversity and change in the Indian Subcontinent. J.B. Hutchinson (ed.) Cambridge University Press, Cambridge. p. 99-105.

46. Keisham, M., Josheph, J.K., Yadav, S.R., Bhat, K.V. and Rao, S.R. 2014. Comparative analysis of heterochromatin distribution in wild and cultivated *Abelmoschus* species based on fluorescent staining methods. *Protoplasma* DOI10.1007/S00709-014-0712-2.

47. Kolhe, A.K. and D'Cruz, R. 1966. Inheritance of pigmentation in okra. *Indian J. Genet.*, **26:** 112-117.

48. Kulkarni, R.S., Swami, R.T. and Virupakshappa, K. 1976. Gene action in bhindi. *Research J. Kerala Agri. Univ.*, **14:** 13-20.

49. Kulkarni, R.S., Swami, R.T. and Virupakshappa, K. 1978. Genetics of important yield components in bhindi. *Indian J. Genet. Pl. Breed.*, **38:** 160-162.

50. Kulkarni, R.S. 1976. Gene action in bhindi. *Mysore J. Agric. Sci.*, **10:** 332.

51. Kumar, R. 2016. Performance assessment of bhendi yellow vein mosaic virus resistance genotype of okra. 7th Indian Horticulture Congress (Abs: 79; Session I: Breeding of Horticultural Crops), held at ICAR-IARI, New Delhi, 15-18 November, p. 62.

52. Kuwada, H. 1966. The new amphidiploid plant named "*Abelmoschus tuberculatus*" obtained from the progeny of the reciprocal crossing between *Abelmoschus tuberculatus* and *Abelmoschus esculentus* (L.). *Jap J. Breed.*, **16(2):** 21-30.

53. Kuwada, H. 1974. F_1 hybrids of *Abelmoschus tuberculatus x A. manihot* with reference to the genomic relationship. *Japanese Journal of Breeding*, **24(5):** 207-210

54. Mamidwar, R.B., Nerkar, Y.S. and Jambhale, N.D. 1979. Cytogenetics of interspecific hybrids in the genus *Abelmoschus*. *Indian Journal of Heredity*, **11:** 35-39.

55. Martin, F.W. 1982. Natural crossing of okra in Puerto Rico. *J. Agric. Univ. of Puerto Rico*, **67:** 50-51.

56. Markosa, B.L. and Peter, K.V. 1990. Review of research on vegetable and tuber crops (okra), pp.1. Directorate of Extension, Kerala Agriculture University, (Kerala).

57. Martin, F.W., Rhodes, A.M., Ortiz, M. and Diaz, F. 1981. Variation in okra. *Euphytica*, **30:** 697-705.

58. Martinello, G.E., Leal, N.R., Amaral, Jr. A.T., Pereira, M.G. and Daher, R.F. 2001. Comparison of morphological characteristics and RAPD for estimating genetic diversity in *Abelmoschus* spp. *Acta Hort.*, **546:** 101-104.

59. Master, M.T. 1875. In: Hooker, J.D. (Ed.). Flora of British India. *Ashford Kent*, **1:** 320-348.

60. Mbagwa, J.S.C. and Adesipe, F.A. 1987. Response of three okra (*Abelmoschus esculentus* (L.) Moench.) cultivars to irrigation at specific growth stages. *Scientia Horticultrae*, **31:** 35-43.

61. More, S.J. and Chaudhary, K.N. 2016. Line x tester analysis over environments in okra (*Abelmoschus esculentus* L. Moench). 7th Indian Horticulture Congress (Abs. 15; Session-II: Breeding of Horticultural Crops), held at ICAR-IARI, New Delhi, 15-18 November, p. 97.

62. Murdock, G.P. 1959. *Africa, Its People and Their culture History*. New York. McGraw Hill. p. 556.

63. Nadkarni, K.M., 1972. *Indian Meteria Medica*, Nadkarni and Co. Bombay.

64. Nandpuri, K.S., Sandhu, K.S. and Randhawa, K.S. 1971. Effect of irradiation on variability in okra. *Journal of Research, PAU*, **6(1):** 82-89.

65. Nariani, T.K. and Seth, M.L. 1958. *Vegetable Crops in India*, p. 606-622. Bose, T.K. and Som, M.G. (Eds). Naya Prokash, Calcutta.

66. Nerkar, Y.S. and Jambhale, N.D. 1985. Transfer of resistance to yellow vein mosaic from related species into okra (*Abelmoschus esculentus* L. Moench). *Indian J. Genet.*, **45:** 261-270.

67. Nirmala Devi 1982. Induction of variability in *Abelmoschus manihot* var. *ghana* by irradiation. M.Sc. (Hort.) thesis. KAU, Vellanikkara (Kerala).

68. Pal, B.P., Singh, H.B. and Swarup, V. 1952. Taxonomic relationship and breeding possibilities of species *Abelmoschus* related to okra (*A. esculentus*). *Botanical Gazette*, **113:** 455-464.

69. Partap, P.S. and Dhankar, B.S. 1980. Heterosis studies in okra (*Abelmoschus esculentus* (L.) Moench.). *Haryana Agric. Univ. J. Res.*, **10(3):** 336-341.

70. Poshiya, V.K. and Shukla, P.T. 1986. Combining ability analysis in okra. *Gujarat Agricultural University Research Journal*, **12(1):** 25-28.

71. Prabhu, T. and Warade, S.D. 2013. Crossability studies in genus *Abelmoschus*. *Veg. Sci.*, **40(1):** 11-16.

72. Pullaiah, N., Reddy, T.B., Moses, G.J., Reddy, B.M. and Reddy, D.R.1998. Inheritance of resistance to yellow vein mosaic virus in okra (*Abelmoschus esculentus* (L. Moench). *Indian J. Genet.*, **58(3):** 349-352.

73. Purewal, S.S. and Randhawa, G.S. 1947. Studies in *Hibiscus esculentus*. Chromosome and pollen studies. *Indian Journal of Agricultural Sciences*, **42(5):** 437-438.

74. Salter, R.J. and Goode, J.B. 1967. Crop response to water at different stages of growth. Common Wealth Agric. *Farham Royals Bucks*, England, p. 246.

75. Sandhu, G.S., Sharma, B.R., Singh, B. and Bhalla, J.S. 1974. Sources of resistance to jassid and white fly in okra germplasm. *Crop Improvement*, **1:** 77-81.

76. Sawadogo, M., Ouedraogo, J.T., Balma, D., Ouedraogo, M., Gowda, B.S., Botanga, C. and Timko, M.P. 2009. The use of cross species SSR primers to study genetic diversity of okra from Burkina Faso. *Afr. J. Biotech.*, **8**: 2476-2482.

77. Sharma, B.R. 1982. Punjab Padmini-a new variety of okra. *Prog. Farm.*, **82:** 15-16.

78. Sharma, B.R. and Dhillon, T.S. 1983. Genetics of resistance to yellow vein mosaic virus in interspecific crosses of okra. *Genetica Agriculture*, **37:** 267-276.

79. Sharma, B.R. and Mahajan, Y.P. 1978. Line x Tester analysis of combining ability and heterosis for some economic characters in okra. *Scientia Horticulturae*, **9(2):** 111-118.

80. Sharma, B.R. and Sharma, O.P. 1984. Breeding for resistance to yellow vein mosaic virus in okra. *Indian Journal of Agricultural Sciences*, **54(10):** 917-920.

81. Sharma, J.P. and Singh, A.K. 2012. Line x Tester analysis for combining ability in okra (*Abelmoschus esculentus* (L.) Moench). *Veg. Sci.*, **39(2):** 132-135.

82. Siemonsma, J.S. (1982 a). La culture du gombo (*Abelmoschus* spp.) legume fruit tropical avec reference speciale a la cote d'Ivoire. Thesis Univ. Wageningen, The Netherlands.

83. Siemonsma, J.S. and Kouamé, C. 2004. *Abelmoschus esculentus* (L.) Moench. Internet Record from Protabase. Grubben GJH, Denton OA (eds.), PROTA (Plant Resources of Tropical Africa, Wageningen, Netherlands. http://database.prota.org/search.htm.

84. Singh, B. 2010. Breeding for pest and disease resistance in okra. In: Emerging trends and strategies for the Management of pests and diseases in vegetable crops. (eds.) Rai, A.B., Shivalingaswamy, S., Satpathy, S., Pandey, K.K. and Rai, Mathura, p. 125-128.

85. Singh, H.B., Joshi, B.S., Khanna, P.O. and Gupta, P.S. 1962. Breeding for field resistance to YVMV in Bhindi. *Indian J. Genet.*, **22:** 137-144.

86. Singh, H.B., Swarup, V. and Singh, B. 1975. Three Decades of Vegetable Research in India. Technical Bulletin, ICAR, New Delhi.

87. Singh, R. and Agarwal, R.A. 1988. Okra fruit components and their proportions in okra genotypes. *Res. Rep.*, **5:** 99-100.

88. Singh, S.P. 1989. Production Technology of Vegetable Crops. *Agric Res. Comm. Centre*, Karnal, p. 70.

89. Sohi, H.S and Sokhi, S.S. 1974. Behaviour of okra varieties to damping off, powdery mildew and Cercospora blight. *Indian Phytopathology*, **27(1):** 91-92.

90. Srinivasan, P.M. and Narayanaswamy, P.S. 1961. Varietal susceptibility in bhindi to the shoot and fruit borer, *Earias fabia* and *Earias insulana*. *Madras Agricultural Journal*, **48:** 345-346.

91. Suresh Babu, K.V. and Peter, K.V. 1984. Noted genotypes of bhindi. *Indian Farmers Digest*, **17(3):** 15-16.

92. Teshima, T.K. 1933. Genetical and cytogenetical studies in an interspecific hybrid of *H. esculentus* and *H. manihot*. *Journal of Faculty of Agriculture, Hokkaido, University*, **34:** 1-55.

93. Thakur, M.R. 1976. Inheritance of Yellow Vein Mosaic Virus in a cross of okra species. *Abelmoschus esculentus x A. manihot* (L.) ssp. *manihot*. *Society for Advancement in Breeding Research in Asia and Oceana Journal*, **8:** 69-73.

94. Thakur, M.R. and Arora, S.K. 1976. Okra (In) *Vegetable Crops in India*, Edn. I, p. 606-622.

95. Thakur, M.R. and Arora, S.K. 1988.Punjab-7, a virus resistant variety of okra. *Prog. Farm.*, **24:** 13.

96. Ugale, S.D., Patil, R.C. and Khupse, S.S. 1976. Cytogenetic studies in the cross between *A. esculentus* and *A. tetraphyllus*. *Journal of Maharashtra Agriculture Universities*, **1(2-6):** 106-110.

97. Van Borssum Waalkes, J. 1966. Malesian *Malvaceae* revised. *Blumea*, **1:**1-251.

98. Vasal, S.K., Cordova, H., Beck, D.L. and Edmeades, G.O. 1997. Choices among breeding procedures and strategies for developing stress tolerant maize germplasm. In: Edmeades, G.O., Banziger, M, Mickelson, H.R., Pena-Valdiva, C.B. eds. *Developing Drought and Low N tolerant maize.*

99. Vavilov, N.I. 1951. The origin, variation, immunity and breeding of cultivated plants. *Chron. Bot.*, **13:** 1949-1950.

100. Velayudhan, K.V., Amalraj, V.A. and Thomas, T.A. 1996. Collecting germplasm of okra andits wild relatives in southern India. *Indian J. Plant Genet. Resour.*, **9(2):** 261-268.

101. Verma, A., Sood, S., Singh, Y. and Sharma, S. 2016. Assessment of genetic variability, heritability and genetic advance for fruit yield and its attributes in okra (*Abelmoschus esculentus* (L.) Moench.). 7th Indian Horticulture Congress (Abs. 98; Session I: Breeding of Horticultural Crops), held at ICAR-IARI, New Delhi, 15-18 November, p. 75-76.

102. Xavier, F., Yadav, R.K., Kumar, R. and Behera, T.K. 2016. Evaluation of combining ability for

protein, total dietary fibre and minerals in okra (*Abelmoschus esculentus* (L.) Moench.). 7th Indian Horticulture Congress (Abs. 100; Session I: Breeding of Horticultural Crops), held at ICAR-IARI, New Delhi, 15-18 November, p. 77.

103. Zeven, A.C. and Zhukovsky, P.M. 1975. Dictionary of cultivated Plants and Their Centres of Diversity. Centre of Agricultural Publishing and Documentation. Wageningen, The Netherland, p. 219.

8

Genetic Resources of Leafy Vegetables and Crop Improvement

INTRODUCTION

Leafy vegetables are the most sustainable and affordable sources of micronutrients, minerals (especially iron, zink, potassium, phosphorus, magnesium and calcium) in human diets and provide other carotenoids with promise in eye health (Seddon, 2007). Generally tender leaves and shoots are eaten as vegetable. It is considered capsule of useful vitamins (vitamins A, K, C, E and B complex), essential amino acids, antioxidants, phytochemicals and dietary fiber. One of them, the xanthophyll lutein, has been identified as an important protective agent in several *in-vitro* assays, epidemiologic studies and intervention trials examining plant food consumption and prevention of age-related cataract and macular degeneration. In addition, carotenoids are attributed preventive properties against several forms of cancer. Green leafy vegetables are the major sources of lutein, β-carotene. Carotenoids play an important role in absortption of blue light and exert antioxidant activity preventing damages caused by light and free radicles. As leafy vegetables are widely available and can be cultivated at low cost, their consumption and conservation is being promoted for increased health benefits (Johns, 2007). Yadav *et al.* (2016) reported that among leafy vegetables ascorbic acid was found maximum in amaranth cv. Pusa Kiran (67 mg/100g) followed by Chenopod cv. Bathua Local (60.6 mg/100g fresh weight). Significantly high level of total carotenoids and beta carotene was recorded in amaranth cv. Pusa Lal Chaulai (69.4 and 7.0 mg fresh weight, respectively) and Pusa Kiran (59.0 and 6.2mg fresh weight, respectively). High value of CUPRAC was recorded in chenopod cv. Bathua Local (32.7μ moltrolax/g) and Pusa Bathua-1(23.0μ moltrolax/g), while FRAP was maximum in Swiss Chard Local (4.2 μ mol ascorbic acid/g fresh weight). Polyphenol content ranged from 1134 μg gallic acid equivalent (GAE) g of amaranth cv. 'Pusa Lal Chaulai' to 154 μg GAE/g in case of palak cv. 'Pusa Bharati'. Swiss chard recorded significantly higher dry matter (22.9 per cent) than others. Several genotypes

Table 8.1: Leafy Vegetable Species and Inherent Quality Attributes

Sl.No.	*Leafy Vegetables*	*Scientific Name*	*Lutein (μg/g)*	*β-carotene (μg/g)*
1.	Sessile joy weed	*Alternanthera sessilis* L.	103	96
2.	Joseph's-coat	*Amaranthus tricolor* L.	105	96
3.	Red amaranth	*Amaranthus cruentus* L.	95	78
4.	Silver cock's comb	*Celosia argentea* L.	83	72
5.	Brown Indian hemp	*Hibiscus cannabinus* L.	142	103
6.	Bladder dock	*Rumex vesicarius* L.	51	41

Source: Latha *et al.*, 2013.

Table 8.2: Major Leafy Vegetables, Family and their Chromosome Number

Leafy Vegetables	*Botanical Name*	*Family*	*Chromosome Number (2n=)*
Spinach beet	*Beta vulgaris* var. *bengalensis* Roxb.	Chenopodiaceae	18
Spinach	*Spinacia oleracea* L.	Chenopodiaceae	12
Swiss chard	*Beta vulgaris* var. *cicla*	Chenopodiaceae	18
Chenopod	*Chenopodium album* L.	Chenopodiaceae	18
Amaranth	*Amaranthus* spp.	Amranthaceae	22, 34, 64
Fenugreek	*Trigonella foenum-graecum* L.	Fabaceae	16
Agathi	*Sesbania grandiflora* (L.)Poir	Fabaceae	
Purslane	*Portulaca oleracea* L.	Portulaceae	18, 36, 45, 54
Ceylon spinach	*Talinum triangulare* Willd.	Portulaceae	48
New Zealand spinach	*Tetragonia tetragonioides* (Pillas) O. Kuntze	Aizaceae	32
Indian spinach	*Basella rubra* L.	Basellaceae	44, 48, 60
Water cress	*Nasturtium officinale* R. Br.	Brassicaceae	64
Cress	*Lepidium sativum*	Brassicaceae	
Kale	*Brassica oleracea* var. *acephala* DC	Brassicaceae	18
Collards	*Brassica oleracea* var. *acephala* DC	Brassicaceae	18
Mustard leaf	*Brassica* spp.	Brassicaceae	
Celery	*Apium graveolens* L.	Apiaceae	22
Coriander leaf	*Coriandrum staivum* L.	Apiaceae	
Parsley	*Petroselinum hortense* Hoffm.	Apiaceae	22
Lettuce	*Lactuca sativa* L.	Compositae	16, 18
Water spinach	*Ipomoea aquatic* forsk.	Convolulaceae	30
Jamaican sorrel	*Hibiscus sabdariffa* L.	Malvaceae	72
Star gooseberry	*Sauropus androgynous* (L.) Merill.	Euphorbiaceae	
Drumstick leaf	*Moringa oleifera* Lam.	Moringaceae	88
Vegetable jute	*Corchorus* spp.	Tiliaceae	14, 28

Table 8.3: Underutilized Leafy Vegetables of Economic Importance

Common Name of Leafy Vegetables	*Botanical Name*	*Family*
Guru Bhanji	*Alternanthera sessilis* (L.) A. DC	Amranthaceae
Celosia	*Celosia argentea* Linn.	Amranthaceae
Lahsua	*Digera arvensis* Forsk.	Amranthaceae
Punrwwa	*Boerhaavia diffusa* L.	Nyctaginaceae
Caralluma	*Caralluma* spp.	Ascelepiadaceae
Winter cress	*Cardamine hirsuta* L.	Brassicaceae
Cikil sena	*Cassia tora* Linn.	Caesalpiniaceae
Brahmi	*Centella asiatica* (L.) Urban	Hydrocotylaceae
Jangali Bathua	*Chenopodium murale* L.	Chenopodiaceae
Hurhur	*Cleome viscosa* L.	Cleomaceae
Kanchara	*Commelina benghalensis* Linn.	Commelinaceae
Desomodium	*Desomodium parvifolium* DC.	Papilionaceae
Hirankhuri	*Emilia sonchifolia* L.	Compositae
Buckwheat	*Fagopyrum cymosum* Meissn	Polygonaceae
Lobelia	*Lobelia trigona* Roxb.	Campanulaceae
Lysimachia	*Lysimachia candida* Lindl.	Primulaceae
Panirak	*Malva parviflora* L	Malvaceae
Mullugo	*Mullugo cerviana* (L.) Scringe	Molluginaaceae
Monochoria	*Monochoria hastata* (L.) solms.	Pontederiaceae
Lajalu	*Neptinia oleracea* lour.	Mimosaceae
Indian sorrel	*Oxaliscorniculata*L.	Oxalidaceae

were identified as rich source of mineral *viz.*, 'Pusa Lal Chaulai' for iron content (112mg/100g), 'Pusa Kiran' for calcium (2713mg/100g), spinach 'Sel-VSPS' for zinc (12.6 mg/100g), while for potassium chenopod 'Bathua Sel.-2' (4211mg/100g). On overall basis the antioxidant capacity of leafy vegetables were in order of amaranth > chenopod > spinach > palak > fenugreek > sarson sag.

In India nearly 700 botanical species which belongs to 125 families are used as leafy vegetables (Peter and Devdas, 1989). The Indian sub-continent is considered to be the place of origin for a number of leafy vegetables like *Beta vulgaris* var. *bengalensis, Basella rubra, Cichorium endivia, Enhyndra fluctuant, Lactuca indica, Amaranthus* spp. *etc.* Most of the leafy vegetables possess good amount of diversity, which is acclimatized under extreme variation in altitude with long history of its cultivation. The wild relatives of different leafy vegetables possess valuable gene particularly for resistance to biotic and abiotic stress conditions.

Among leafy types, northern plains are the major area of variability, where wild types of *Amaranthus, Chenopodium, Rumex* and *Malva* are available as common weeds.Wide spread of *Amaranthus* spp. particularly *A. spinosus, Malva rotundifolia, M. sylvestris* and *Chenopodium album* can be noticed in several agro-ecological regions of the world. There are many underutilized leafy vegetables which are becoming popular day by day.

It is well known that there are several leafy vegetables which are well considered in diet, as per their adoption in the local environment. For example mustard leaves are quite common and broadly used in Punjab, Haryana, Himachal Pradesh *etc.* Similarly in Uttar Pradesh, Bihar, Jharkhand, Chatishgarh, Madhya Pradesh *etc.* it is used as occasional of vegetable Bathua (Chenopods) are very common apart from spinach and fenugreek.During rainy season 'Karem Sag' (*Ipomoea aquatica*) is used in Odisha, Bihar and Uttar Pradesh. In south India, 'Chekumani' (*Sauropus aquaitca*) is most important leafy vegetables. Peoples of Andhra Pradesh eat 'Gogu' (*Hibiscus sabdarifa*) largely as 'Sag' alone or used in 'Sambhar'. In higher altitude of Laddakh some crop species like *Lepidium* spp., *Oxyria digyna* and *Urtica hyperborea* are used as leafy vegetables (Chaurasia *et al.*, 2000). As per liking and adaptation, some important leafy vegetables are being listed here (Table 8.4).

The topography, altitude and rainfall pattern of India particularly NEH have made floristically rich and diverse. Many wild plants that occur naturally are also used by local people as food. Tropical leafy vegetables such as *Amaranthus, Spinacia, Basella, Raphanus, Corchorus, Cucurbita, Lagenaria, Melia, Lathyrus, Chenopodium, Enhydra, Moringa, Nymphaea, Ipomoea* and ferns like *Diplazium, Pteridium* are dominantly used in planis of the region, while temperate ones like *Brassica, Colocasia, Hibiscus, Sechium, Fagopyrum, Tetragonia, Tetrastigma, Lactuca, Sauropus, Houttuynia, Corydalis, Phytolacca, Clerodendron, Eryngium* and *Sonchus* are commonly used in the region (Hore, 2004). Soyimchiten *et al.* (2016) reported that a total of 340 species belonging to 217 genera and 81 families are used in NEH region as wild leafy vegetables. Among them, major species that are being cultivated in field and kitchen garden are *Centella asiatica* (also used in gastrointestinal problems), *Chenopodium album, Clerodendrum glandulosum* (high blood pressure), *Elsholtzia blanda* (sinusitis), *Eryngium foetidum* (indigestion), *Gynura cusimbua* (vermifgue),

Table 8.4: Area Specific Popularity of Leafy Vegetables

Sl.No.	*Name of Leafy Vegetables*	*Popularity in the Area*
1	Spinach beet	Throuogout the country but mopre pouplar in Uttar Pradesh, Bihar, Bihar, Madhya Pradesh, Haryana, Punjab, Delhi and Karnataka
2	Amaranth	Throuogout the country but mopre pouplar in Uttar Pradesh, Bihar, Tamil Nadu, Kerala, Karnataka and North Eastern Region
3	Fenugreek	Haryana, Punjab, Gujarat, Rajasthan, Maharashtra, Madhya Pradesh and Uttar Pradesh
4	Chenopod	Haryana, Punjab, Delhi Uttar Pradesh, Bihar and Madhya Pradesh
5	Indian spinach	Odisha, West Bengal, Bihar, Jharkhand and Uttar Pradesh
6	Drumstick	West Bengal, Tamil Nadu, Kerala, Andhra Pradesh and Maharashtra
7	Mustard leaf	Punjab, Himachal Pradesh, Western Uttar Pradesh, North Eastern Hill Regions, Haryana
8	Ceylon spinach	Assam, Manipur, West Bengal, Kerala
9	Star goosberry	Kerala, Tamil Nadu
10	*Colocasia*	Andhra Pradesh, Maharashtra, Tamil Nadu, Uttar Pradesh, West Bengal, Bihar and Odisha
11	*Cicer soonagaricum*	Ladakh, Lahaulspiti
12	*Lobelia trigona*	Chotta nagpur
13	*Lysimachia candida*	Himalayan regions, Nilgiri, North East States particulrly in Megalalya and Manipir
14	Water cress	Sikkim, West Bengal, Odisha
15	*Oxyria digyna*	Higher altitude of Himalayan regions, Ladakh
16	Sweet belladona	Himalayan region, Assam and North Eastern Regions
17	*Polygonum* species	Himalayan regions, Nilgiri regions and North Eastern Regions
18	*Rumex* species	Himalayan regions and North Eastern Regions
19	*Sonchus oleracea*	Plains of eastern India and hilly regions, Jamju and Kashmir
20	*Suadea maritima*	Rajasthan, Gujarat and dry regions
21	*Taraxacum officinale*	Himalayan regions
22	*Dentella repens*	West Bengal particularly Poorulia
23	*Eremurus persicus*	Jamju and Kashmir

Source: Pandy, 2005.

Table 8.5: Distribution of Leafy Vegetables in NEH Regions of India

Species	*Family*	*Origin, Diversity and Distribution*	*Preference by the Community/ Tribe(s) of the Region*	*Availability in the States(s) of the Region*
Cultivated species used as leafy vegetables (Sanwal, 2008)				
Colocasia esculenta (L.) Schott.	Araceae	SE Asia, China, Japan	All communities of the region	Throughout the the region from tropical to subtropical zones
Xanthosoma violaceum Schott.	Araceae	South America	All communities of the region	Throughout the region from tropical to subtropical zones
Basella rubra L.	Basellaceae	South Asia	Assamese, Mizo	Assam, Manipur, Mizoram, Tripura
Chenopodium album L.	Chenopodiaceae	Europe	Assamese	Assam plains, Tripura
Brassica juncea Hook. f. and Th.	Brassicaceae	Africa	Assamese, Mizo	Assam, Mizoram, Tripura
Brassica juncea (L.) Cz. and Coss. var. *cuneifolia* Roxb.	Brassicaceae	India	All hill people	Throughout the hill region
Amaranthus caudatus L.	Amaranthaceae	South America, Asia	People of Assam hills and plains and Meghalaya	Tropical to sub-tropical zones of the region
Amaranthus tricolor L.	Amaranthaceae	Tropical Asia	Assam plains	Plain areas of Assam, Tripura
Moringa oleifera Lam.	Moringaceae	India	Assam plains	Sporadic occurrence in plains of Assam
Lactuca indica L.	Asteraceae	India, Japan, Philippines and China	All hill states	Grown in kitchen garden
Allium schoenoprasum L.	Liliaceae	Europe, Asia, North America	Hills of Meghalaya, Mizoram and Nagaland	Occurrence above 4,000 ft. altitude in Christian dominated states
Tetragonia expansa Murr.	Aizoaceae	Chinese-Japanese Center	Sub-tropical, Sub-humid areas of the region	Well adapted and naturalized in Meghalaya
Fagopyrum esculentum Moench.	Polygonaceae	Central Asia	Plains to high hills of the region	Arunachal Pradesh, Meghalaya and Assam
Wild species used as leafy vegetables (Hore, 2004)				
Enhydra fluctuans Lour.	Asteraceae	India, Malayam, Eurasian	-	Meghalaya
*Sonchus oleraceous*L.	Asteraceae	Eurasian	Khasi	Arunachal Pradesh
Sauropus androgynous (L.) Merr.	Euphorbiaceae	South East Asia	People of Arunachal Pradesh and Meghalaya	Assam
Paederia foetidal L.	Rubiaceae	Tropical Asia	Assamese	Arunachal Pradesh
Eryngium foetidum L.	Apiaceae	South America	All communities of hills	Assam hills
Houttuynia cordata Thunb.	Saururaceae	Indo-China, China, Vietnam	Mizos	Mizoram
Clerodendron colebrookianum Walp.	Verbenaceae	Indo-burma	Khasi	Meghalaya
Corydalis sibiricus Pers	Fumarieaceae	Temperate Himalayas	Khasi	Mikir hills of Assam
Phytolacca acinosa Roxb.	Phytolaccaceae	Tropical Asia, China, Japan	Khasi, Arunachalis	Arunachal Pradesh
Elatostema platyphylla Wedd.	Urticaceae	Old World	High hills of Arunachal Pradesh	Arunachal Pradesh
Diplazium esculentum (Retz.) Sw.	Athyriaceae	Tropical World	Hill people of Arunachal Pradesh	Arunachal Pradesh
Tectaria coadunate (Hook. And Grev.) C. Chr.	Aspidiaceae	Pan Tropical	Sikkimese	Sikkim
Pteridium aquilinum Kuhn.	Dennstaedtiaceae	Northern Hemisphere and Africa	People of Himalayan belt and North Eastern Region	Throughout the tropical to subtropical zones of the region

Houttuynia cordata (gastritis), *Phytolacca acinosa* (root juice-sinisitis), *Plukenetia corniculata*, *Zanthoxylum acanthopodium* (fruit pericarp-condiment, pounded fruits for treating dysentery and stomachache) *etc.* Exploration and collecting of germplasm of prioritized species from diverse habitats (field/homestead/wild) across altitudinal and distributional ranges for sustainable utilization and conservation, and screening them for nutritive values and other bioactive compounds, must receive paramount attention (Soyimchiten *et al.*, 2016).

Table 8.6: Variability and Distribution Status of Leafy Vegetable Crops

Crop	Genera	CS	DS	GVS	GES	GCP
Amaranth	*Amaranthus*	C	W	H	M	M
Climbing spinach	*Basella*	C	W	M	M	M
Fenugreek	*Trigonella*	W	R	M	M	M
Palak	*Beta*	C	R	M	M	H
Coriander	*Coriandrum*	C	R	H	M	H

CS: Crop Status; DS: Distribution Status; GVS: Genetic Variability Status; GES: Genetic Erosion Status; GCP: General Crop Priority.

Germplasm Status of Leafy Vegetables

Exotic accessions-Total 139 germplasm received - From Italy, the Netherlands, Poland, USA and USSR.

Table 8.7: Germplasm Status of Leafy Vegetables

Vegetables	Number of Collection
Coriander	378
Cabbage	20
Fenugreek (*Trigonella foenum-graecum*)	183
Spinach	133
Amranths (including grain type)	3859
Buckwheat	312
Climbing	120

REFERENCES

1. Chaurasia, O.P., singh, N. and singh, B. 2000. Potentail wild leafy vegetables from Ladakh. *Indian Hort.*, **45(2):** 4-5.
2. Hore, D.K. 2004. Leafy vegetables of North East India. *Indian J. Plant Genet Resour.*, **17(1):** 10-16.
3. Johns, T. 2007. Agrobiodiversity, diet and human health. In: Managing biodiversity in agricultural ecosystems. Columbia University Press, New York, 492.
4. Latha, P., Sudhakar, P., Krishna, M.B., Begam, C.R. and Krishna, T.G. 2013. Quantification of lutein and β-carotene in underutilized green leafy vegetables using high performance liquid chromatography. *Veg Sci.*, **40(1):** 109-110.
5. Pandey A. K. 2005. Leafy vegetables. Pub. Researchco Book Centre, New Delhi, p. 4-5.
6. Peter, K.V. and Devdas, Y.S. 1989. Leafy vegetables. *Indian Hort.* **33 and 34 (4 and 1):** 8-11.
7. Sanwal, S.K. 2008. Underutilized leafy vegetables: Importance and strategies for improvement and conservation. In: *Underutilized Vegetable and Spice Crops*. Published by AGROBIOS (India), p.187-196.
8. Seddon, J.M. 2007. Multivitamin-multinutritional supplements and eye disease: Age related macular degeneration and cataract. *American Journal of Clinical Nutrition*, **1:** 304-307.
9. Soyimchiten, Pradheep, K. and Pandey, A. 2016. Wild leafy vegetables of north Eastern Hil Region of India. 1st International Agrobiodiversity Congress, 6-9 Nov., 2016, held at New Delhi., pp. 12.
10. Yadav, R.K., Pachauri, N. and Dev, B. 2016. Studies on green leafy vegetables for nutritional and antioxidant activity. 7th Indian Horticulture Congress (Abs: 12; Session: Breeding for Horticultural Crops), 15-18 November, held at ICAR-IARI, New Delhi, p. 94-95.

Genus *Beta*

Spinach Beet or Beet Leaf (Palak) (*Beta vulgaris* var. *bengalensis* Hort.)

Introduction

Spinach beet or palak (*Beta vulgaris* var. *bengalensis* 2n=2x=18) is very common and an important tropical and subtropical leafy vegetable. It is also known as Indian spinach, Spinach beet, beet leaf, Garden beet, Palong, Palang sag, Teegabatchali, Vasavyeley, Dumpsbucchale and Pasalai. Green tender succulent leaves and stems of palak are used as vegetable. It is a rich source of vitamin A, also contains high quantity of ascorbic acid (70 mg/100 g), and iron (16.2 mg/100 g). it is reported that 100 g of leaves supply as much essential amino acids as 100 g of any non-vegetarian food like meat and fish. The herbaceous parts are mildly laxative besides other medicinal values.

Table 8.8: Common Difference between *Beta vulgaris* and *Spinacia oleracea* L.

Botanical name is *Beta vulgaris*	Botanical name is *Spinacia oleracea* L.
Chromosome number is 2n=18	Chromosome number is 2n=12
Leaf possess entire margin	Leaflets are separate
Generally hermaphrodite in nature	Generally monoecious in nature

Origin, History and Distribution

Beet leaf or palak is said to have its native place of Indo-Chinese region (Nath, 1976). It was known in China as early as 647 A.D. *Beta* spp. have been used as a potherb since 2000 BC. Aristotle mentioned 'Red chards' and Theophastus mentioned light green and dark green chards used in the 4th century BC. The Romans used this crop as food for humans and feed for animals. It is closely related to beet root and Swiss chard. By the 16th century, it was widely used for feeding animals during winter. Red beet featured in Roman receipes of the second and third centuries and was recorded in English recipes of 14th century. In Germany, it was first described in 1557, where referred as Roman beet. The crop was introduced into USA in 1800 and was known as garden beet (Campbell, 1976). It is a very popular crop in Bangladesh. There is a mentioned in the medical treatises "Charak Samhita" and "Susruta Samhita" about garden spinach as "Palankya" in Sanskrit but its identification as *Spinacia oleracea* L. by Ray and Gupta (1980) and Ray *et al.* (1980) respectively, may need further checking and verification. It is grown throughout the year in many parts of India, particularly in states like Uttar Pradesh, West Bengal, Punjab, Haryana, Delhi, Madhya Pradesh, Bihar, Maharashtra, Odisha, and Gujrat. It is grown in the plains and hills of India from Punjab to North Eastern Region. Palak is also becoming popular in South India. Palak is popular only in Indian subcontinent and no other country is known to cultivate this pot herb extensively (Seshadri and Chatterjee, 1996).

Species Availability, Evolution and Diversity

The spinach beet also known as Indian spinach beet (*B. vulgaris* var. *bengalensis* Hort.) is closely related to beet root and sugar beet (*Beta vulgaris* ssp. *cicla*). *Beta vulgaris* ssp. *maritime*, is the ancestor of beet root, palak and some other cultivars. Other *Beta* species includes *Beta adanensis, B. altissima, B. atriplicifolia, B. bengalensis, B. bourgaei, B. brasiliensis, B. campanulata, B. chilensis, B. cicla, B. corolliflora, B. lomatogona, B. macrocarpa, B. macrorhiza, B. maritime, B. maritima* subsp. *danica, B. maritima* var. *atriplicifolia, B. maritima* var. *erecta, B. maritima* var. *prostrata, B. nana, B. orientalis, B. palonga, B.a patellaris, B. patula, B. perennis, B. procumbens, B. trigyna, B. trojana, B. vulgaris, B. vulgaris* cv. *conditiva, B. vulgaris* cv. *saccharifera, B. vulgaris* f. *rhodopleura, B. hybrida, B. vulgaris* sub sp. *adanensis, B. vulgaris* subsp. *cicla, B. vulgaris* subsp. *flavescens, B. vulgaris* subsp. *lomatogonoides, B. vulgaris* var. *glabra, B. vulgaris* var. *cicla, B. vulgaris* var. *crassa, B. vulgaris* var. *atriplicifolia, B. vulgaris* subsp. *macrocarpa, B. vulgaris* var. *altissima, B. vulgaris* subsp. *vulgaris, B. vulgaris* subsp. *provulgaris, B. vulgaris* subsp. *patula, B. vulgaris* subsp. *maritime, B. vulgaris* subsp. *orientalis, B. vulgaris* var. *erecta, B. vulgaris* var. *flavescens, B. vulgaris* var. *foliosa, B. vulgaris* var. *grisea, B. vulgaris* var. *macrocarpa, B. vulgaris* var. *maritime, B. vulgaris* var. *orientalis, B. vulgaris* var. *perennis B. vulgaris* var. *pilosa, B. vulgaris* var. *prostrata, B. webbiana, B. intermedia* and *Beta vulgaris* var. *trojana*.

Table 8.9: Taxon and Preferred Taxon Name

Name of Taxon	Name of Authority	Family	Preferred Taxon Name
Beta adanensis	Pamukc. ex Aellen	Chenopodiaceae	*Beta vulgaris* subsp. *adanensis*
Beta altissima	Steud.	Chenopodiaceae	*Beta vulgaris* subsp. *vulgaris*
Beta atriplicifolia	Rouy	Chenopodiaceae	*Beta vulgaris* subsp. *maritima*
Beta bengalensis	Roxb.	Chenopodiaceae	*Beta vulgaris* subsp. *maritima*
Beta bourgaei	Coss.	Chenopodiaceae	*Beta macrocarpa*
Beta brasiliensis	Hort. ex Voss, nom. inval.	Chenopodiaceae	*Beta vulgaris* subsp. *vulgaris*
Beta campanulata	Coss.	Chenopodiaceae	*Beta patellaris*
Beta chilensis	hort.	Chenopodiaceae	*Beta vulgaris* subsp. *vulgaris*
Beta cicla	(L.) L.	Chenopodiaceae	*Beta vulgaris* subsp. *vulgaris*
Beta corolliflora	Zosimovic ex Buttler	Chenopodiaceae	*Beta corolliflora*
Beta hybrid		Chenopodiaceae	*Beta hybrid*
Beta lomatogona	Fisch. and C. A. Mey.	Chenopodiaceae	*Beta lomatogona*
Beta macrocarpa	Guss.	Chenopodiaceae	*Beta macrocarpa*
Beta macrorhiza	Steven	Chenopodiaceae	*Beta macrorhiza*
Beta maritima	L.	Chenopodiaceae	*Beta vulgaris* subsp. *maritima*
Beta maritima subsp. *danica*	Krassochkin	Chenopodiaceae	*Beta vulgaris* subsp. *maritima*
Beta maritima var. *atriplicifolia*	(Rouy) Krassochkin	Chenopodiaceae	*Beta vulgaris* subsp. *maritima*
Beta maritima var. *erecta*	Krassochkin	Chenopodiaceae	*Beta vulgaris* subsp. *maritima*
Beta maritima var. *prostrata*	Krassochkin	Chenopodiaceae	*Beta vulgaris* subsp. *maritima*
Beta nana	Boiss. and Heldr.	Chenopodiaceae	*Beta nana*
Beta orientalis	Roth	Chenopodiaceae	*Beta vulgaris* subsp. *maritima*
Beta palonga	R. K. Basu and K. K. Mukh.	Chenopodiaceae	*Beta vulgaris* subsp. *maritima*
Beta patellaris	Moq.	Chenopodiaceae	*Beta patellaris*
Beta patula	Aiton	Chenopodiaceae	*Beta patula*
Beta perennis	(L.) Freyn	Chenopodiaceae	*Beta vulgaris* subsp. *maritima*
Beta procumbens	C. Sm.	Chenopodiaceae	*Beta procumbens*
Beta sp.		Chenopodiaceae	*Beta* sp.

Name of Taxon	Name of Authority	Family	Preferred Taxon Name
Beta trigyna	Waldst. and Kit.	Chenopodiaceae	*Beta trigyna*
Beta trojana	Pamukc. ex Aellen	Chenopodiaceae	*Beta vulgaris* subsp. *maritima*
Beta vulgaris	L.	Chenopodiaceae	*Beta vulgaris*
Beta vulgaris cv. *conditiva*	Alef.	Chenopodiaceae	*Beta vulgaris* subsp. *vulgaris*
Beta vulgaris cv. *saccharifera*	Alef.	Chenopodiaceae	*Beta vulgaris* subsp. *vulgaris*
Beta vulgaris f. *rhodopleura*	(Alef.) Helm	Chenopodiaceae	*Beta vulgaris* subsp. *vulgaris*
Beta vulgaris subsp. *adanensis*	(Pamukc. ex Aellen) Ford-Lloyd and J. T. Williams	Chenopodiaceae	*Beta vulgaris* subsp. *adanensis*
Beta vulgaris subsp. *cicla*	(L.) W. D. J. Koch	Chenopodiaceae	*Beta vulgaris* subsp. *vulgaris*
Beta vulgaris subsp. *flavescens*	Lam.	Chenopodiaceae	*Beta vulgaris* subsp. *vulgaris*
Beta vulgaris subsp. *lomatogonoides*	Aellen	Chenopodiaceae	*Beta vulgaris* subsp. *maritima*
Beta vulgaris subsp. *macrocarpa*	(Guss.) Thell.	Chenopodiaceae	*Beta macrocarpa*
Beta vulgaris subsp. *maritima*	(L.) Arcang.	Chenopodiaceae	*Beta vulgaris* subsp. *maritima*
Beta vulgaris subsp. *orientalis*	(Roth) Aellen	Chenopodiaceae	*Beta vulgaris* subsp. *maritima*
Beta vulgaris subsp. *patula*	(Aiton) Ford-Lloyd and J. T. Williams	Chenopodiaceae	*Beta patula*
Beta vulgaris subsp. *provulgaris*	Ford-Lloyd and J. T. Williams	Chenopodiaceae	*Beta vulgaris* subsp. *maritima*
Beta vulgaris subsp. *vulgaris*		Chenopodiaceae	*Beta vulgaris* subsp. *vulgaris*
Beta vulgaris var. *altissima*	Doll	Chenopodiaceae	*Beta vulgaris* subsp. *vulgaris*
Beta vulgaris var. *atriplicifolia*	(Rouy) Krassochkin	Chenopodiaceae	*Beta vulgaris* subsp. *maritima*
Beta vulgaris var. *cicla*	L.	Chenopodiaceae	*Beta vulgaris* subsp. *vulgaris*
Beta vulgaris var. *crassa*	Alef.	Chenopodiaceae	*Beta vulgaris* subsp. *vulgaris*
Beta vulgaris var. *erecta*	(Krassochkin) Krassochkin	Chenopodiaceae	*Beta vulgaris* subsp. *maritima*
Beta vulgaris var. *flavescens*	(Lam.) DC.	Chenopodiaceae	*Beta vulgaris* subsp. *vulgaris*
Beta vulgaris var. *foliosa*	Aellen	Chenopodiaceae	*Beta vulgaris* subsp. *maritima*
Beta vulgaris var. *glabra*	(Delile) Aellen	Chenopodiaceae	*Beta vulgaris* subsp. *maritima*
Beta vulgaris var. *grisea*	Aellen	Chenopodiaceae	*Beta vulgaris* subsp. *maritima*
Beta vulgaris var. *macrocarpa*	(Guss.) Moq.	Chenopodiaceae	*Beta macrocarpa*
Beta vulgaris var. *maritima*	(L.) Moq.	Chenopodiaceae	*Beta vulgaris* subsp. *maritima*
Beta vulgaris var. *orientalis*	(Roth) Moq.	Chenopodiaceae	*Beta vulgaris* subsp. *maritima*
Beta vulgaris var. *perennis*	L.	Chenopodiaceae	*Beta vulgaris* subsp. *maritima*
Beta vulgaris var. *pilosa*	(Delile) Aellen	Chenopodiaceae	*Beta vulgaris* subsp. *maritima*
Beta vulgaris var. *prostrata*	(Krassochkin) Krassochkin	Chenopodiaceae	*Beta vulgaris* subsp. *maritima*
Beta vulgaris var. *trojana*	(Pamukc.) Ford-Lloyd and J. T. Williams	Chenopodiaceae	*Beta vulgaris* subsp. *maritima*
Beta webbiana	Moq.	Chenopodiaceae	*Beta webbiana*
Betax intermedia	Bunge	Chenopodiaceae	Beta x intermedia

Swees chard (*Beta vulgaris* var. *cicla*) perhaps artificially selected species (Campbell, 1976). Most of the *Beta* species have their centre of diversity in the Mediterranean/European regions. There is one type generally found in Europe and known as "Perpetual Spinach" and this is different from Swiss Chard (*B. vulgaris* var. *cicla*) used as a pot herb. Probably the former is more close to Indian Spinach or "Palak" of north Indian Plains. The tropicalization of *Beta* which includes temperate sugar beet and garden (red) beet may have started possibly in Bengal, which is the probable reason for the designation of "Palak" as *B. vulgaris* var. *bengalensis*. There is however, no authentic chronological record to testify this nomenclature.

Botanical Description

Beet leaf or palak (*Beta vulgaris* var. *bengalensis*) belongs to the family Chenopodiaceae. Some important *Beta* species and their chromosome number are given in the table (8.10) as suggested by Nassar *et al.*, 1973. Fedorov, 1974 described tiotal 10 species which has diploids chromosome number 2n=18.

Table 8.10: Species and their Chromosome Number

Sl.No.	Species	Chromosome Number
1	*Beta vulgaris* ssp. *bengalensis*	18
2	*B. vulgaris* ssp. *lamatogona*	18
3	*B. macrocarpa*	18
4	*B. macrorhiza*	18

Sl.No.	Species	Chromosome Number
5	*B. maritime* L. *betanana*	18
6	*B. pateliaris*	18
7	*B. patula*	18
8	*B. parrennis*	18
9	*B. procumbens*	18
10	*B. webbiana*	18
11	*B. corlliflora*	36
12	*B. trigyna*	36

The detail of botanical descriptions is as under:

Table 8.11: Botanical Description

Growth Habit	The plants are herbaceous annual as herb (leaf production) but biennial for seed production. Plants are succulent green.
Root	Plant has tap root system with number of fibrous feeder roots.
Stem	Thick stem and rosette but enlarged latter on about 75 days of sowing which results in bolting.
Leaf	The leaves are rosette, succulent and tender at early vegetative phase with long petioles and small thick stem. The leaf margin is entire.
Floral biology	Inflorescence is racemose with bracts of cymose type (Nath, 1969; Pandita and Lal, 1993). The whole phase of flower bud development takes 35 days to come in bloom. There are six development stages through which each bud passes before it iopens into a flower. Palak is having small sessile, bisexual(hermaphrodite) perfect bracteate (leafy) flowers borne in the axils of leaves in 2-3 clustersor from spurious terminal bud from the main shoots as well as from the lateral shoots. Stamens are 5, Ovary is multilocular perigynous.Flowers propduce abundant small and light pollen grains. Studies on floral biology and pollen morphology and physiology in spinach beet have been reported (Purohit and Nath, 1970; Nath and Purohit, 1969b).
Anthesis and pollination	Due to light pollens, wind pollination is effective *i.e.* Anemophilous or wind pollinated. Therefore, it is highly cross pollinated (Dark, 1971). The anthesis takes place between 7 AM to 5 PM and extreme at 11.0 AM to 1.0 PM. Anther dehiscence of anthers takes place between 8.30 am and 6.30 pm depending upon the temperature and humidity. It has been noted that intensity is maximum between 12.30 PM to 2.30 PM. The stigma become receptive 8 hours before anthesis and remains receptive for 10 hours after anthesis (Purohit and Nath, 1970).
Fruit and fruit set	The fruit is actually called as 'seed ball'
Seeds	Each 'seed ball'containing of 2-6 seeds.

Horticultural Classification

Spinach beet is divided into three subspecies:

(i) *Beta*

(ii) *Patellaris*

(iii) *Coralline*

Coons, 1954 and Crashochin 1959 classified genus *Beta* and species *vulgaris*. Under Beta vulgaris, several strains like sugar beet, beet root, mangolds and forage *Beta* subspecies *cicla* are common.

Augmentation of Genetic Resources and Crop Improvement

Breeding Priorities

- ☆ Vigorous plant growth, quick growing, strong succulent stem, and large leaves
- ☆ High nutrition
- ☆ Low oxalate and nitrate
- ☆ Fast growing multiple cutting types
- ☆ High temperature tolerance
- ☆ Development of photo-insensitive lines
- ☆ Late bolting
- ☆ Development of male sterile line for hybrid seed production

Introduction and Selection

In India, earlier improvement through natural selection by the farmers might have been in the direction of larger succulent leaves, late bolters and ability for regeneration after repeated cuttings. ICAR-IARI, New Delhi has evolved high yielding varieties like 'All Green' (Singh and Joshi, 1960) and 'Pusa Jyoti' (Anon, 1971b). Similarly; 'HS-23' has been developed by CCS HAU, Hisar (Haryana).

The popular cultivar 'HS-23' was developed by use of Indigenous material through mass selection method. Recently a new cultivar 'Jammu Spinach Bee 07 has been released from J&K. Agri. Univ. (J&K) for commercial cultivation due to better quality high yield.

Hybridization and Selection

A hybrid selection by Seedsmen Co. between an early bolter of Garden beet and Palak, known as 'Banerjee's Giant' is very popular since nearly four decades in Bengal. A biennial cultivar 'Pusa Harit' was developed by Regional Station, Katrain (H.P.) of ICAR-IARI, New Delhi. Horticultural attributes of some popular cultivars are given in Table-8.12.

Distant Hybridization

Hybridization between Palak and Sugarbeet, and Swiss chard and Beet root has been performed successfully (Nath *et al.*, 2008).

Mutation

Palak cultivar 'Jobner Green' was releases as a mutant in India for commercial cultivation.

Table 8.12: Cultivars and their Horticultural Attributes

Variety	Horticultural Attributes	Developing Source
All Green	Uniform green leaves, multicut (6-7 cuttings) at 15-20 days interval. High yielding variety (12.5t/ha).	IARI, New Delhi
Pusa Jyoti	Developed by 2 per cent colchicines treatment for 24 hours on All Green variety. Plants are vigorous, quick growing, regenerate quickly, multicut (6-8) and yield is 29.0t/ha.	IARI, New Delhi
Pusa Bharathi	It is a selection from S44-1, suitable for summer and kharif. Leaves are cordate 25.0 cm long and 14.0 cm breadth, green, smooth and tender leavess	IARI, New Delhi
Arka Anupama	Leaves are dark green and succulent. Multicut and late bolting in nature. It is resistant to Cercospora leaf spot under field condition.	IIHR, Bengaluru (Karnataka)
Jobner Green	Evolved through spontaneous mutation and selection made from a local collection (5-5). Uniform green, large thick succulent tender leaves and yield is around 29.6t/ha.	University of Udaipur, Jobner (Rajasthan)
Ooty	It is a high yielding (15.0 t/ha) cultivar and contain very high amount of carotene. Suitable for round the year cultivation. Cawithstand frost	TNAU, Coimbatore (Tamil Nadu)
HS-23	Large, dark green and thick leaves. 1st cutting can be done within 30 days.	CCS HAU, Hissar (Haryana)

Table 8.13: Cultivars Developed through Hybridization and their Horticultural Attributes

Variety	Horticultural Attributes	Developing Source
Pusa Palak	Cross between Swiss Chard and Local Palak, green leaves and late bolting	IARI, New Delhi
Pusa Harit	Cross between sugar beet and local palak. Suitable for round the year cultivation, green slightly crinkled, giant size leaves, late bolting. Tolerant to alkaline soil condition.	IARI, New Delhi
Pant Composite	This is a high yielding composite variety. Tolerant to Cercospora leaf spot.	GBPUA&T, Pantnagar
Banerjee's Giant	Very popular variety of West Bengal which is developed through hybridization (local Palak x Sugar Beet). It produces large thick leaf and succulent stem.	Seedsmen C. West Bengal

Resistance to Abiotic Stress

At the University of Udaipur, a high yielding variety 'Jobner Green' suited to alkaline soil condition was evolved (Nath, 1967, 1969a).

REFERENCES

1. Anonymous. 1971b. New IARI Vegetable Varieties, IARI, *Ext. Bull.* No.2, IARI, New Delhi.
2. Campbell, G.K.G., 1976. Sugar beet. In: Evolution of crop plants (ed. Simmnds, N.W.) Longman, London.
3. Coons, G.H. 1975. Interspecific hybrids between *Beta vulgaris* and the wild species of *Beta*. J. Amer. Soc; Sugarbeet Technol. 18 (4) : 281-306.
4. Coons, G.H. 1954. The wild species of *Beta*. Proc. Amer. Soc. *Sugarbeet Technol*. 8(2): 142-147.
5. Dark, S.O.S. 1971. Experiment on the cross pollination of sugar beet in the fiels. *J. Nat. Inst. Agric. Bot.*, **12:** 242-246.
6. Fedorov, 1974. Chromosome number of flowering plants. Atto. Koeltz. Science Publishers, 624, Koening Stein, West Germany.
7. Nassar, N.M., Sahrigy, M.A. and Rakha, F.A. 1973. Karyotype analysis of three *Chenopodium* species. *Alexandria J. Agric. Res.*, **21:** 253-260.
8. Nath, P. 1976. Vegetables for tropical regions. ICAR Book Series No. 2, New Delhi.
9. Nath, P. 1967. Major Green leafy vegetables. *Ext. Bull.*, No. 9; University of Udaipur, Rajasthan.
10. Nath, P. 1969a. A new mutant in *Beta vulgaris* var. *bengalensis*. *Indian J. Hort.*, **26 (3 and 4):** 156-159.
11. Nath, P. and Purohit, S.P. 1969b. Studies on pollen morphology and physiology in spinach beet (*Beta vulgaris* var. *bengalensis* Hort.). *J. Palynol.*, **5(2):** 111-121.
12. Nath, P., Srivastava, V.K., Dutta, O.P. and Swamy, K.R.M. 2008. Vegetable crops: Improvement and Production. Published by *Dr. Prem Nath Agril. Sci. Foundation* (PANSF), Bengaluru, Karnataka (India).
13. Pandita, M.L. and Lal, S. 1993. Leafy vegetables. In: Vegetable crops in India (eds. Bose, Som and Kobir), Naya Prokash, Kolkata (India).
14. Purohit, S.P. and Nath, P. 1970. Studies on the floral biology in spinach beet (*Beta vulgaris* var. *bengalensis*). *Madras Agric. J.*, **57:** 98-105.
15. Ray, P., Gupta, H.N. and Roy, M. 1980. "Susruta Samhita". Scientific Synopsis. Indian Natl. Sci. Acad. New Delhi.
16. Ray, P. and Gupta, H.N. 1980. "Charaka Samhita". Scientific Synopsis. Indian Natl. Sci. Acad. New Delhi.
17. Seshadri, V.S. and Chatterjee, S.S. 1996. The history and adaption of some introduced vegetable crops in India. *Veg. Sci.*, **23(2):** 114-140.
18. Singh, H.B. and Joshi, S.A. 1960. Our leafy vegetables. ICAR *Farm Bull.* No. 56, New Delhi (India).

Swiss Chard, Palang or Seed Spinach or Wild Beet (*Beta vulgaris* spp. *cicla*)

INTRODUCTION

Swiss chard (*Beta vulgaris* spp. *cicla*, 2n=18) is also known as 'Beet Palang', 'Palamg' in Bangla and 'Beej Palak', 'Sada Palak' or 'Jangali Palak' in Hindi. Swiss chard (*Beta vulgaris* ssp. *cicla* L.) is one of the neglected vegetable species. It is very much similar to common spinach (palak) but it bears long and curved leaves. Leaf petiole is long, thick and leathery. Its leaves are used for cooking like spinach. Leaves of Swiss chard are very rich source of carotene and Vit. C. Nutritive value per 100.0 g of edible portion as reported contains moisture 91.0 g, protein 2.4 g, fat 0.3 g, carbohydrates 4.6 g, energy 25.0 K calorie, calcium 88.0 mg, iron 3.2 mg, sodium 147.0 mg, potassium 550.0 mg, Vit. A 6500 IU, Thiamine 0.06 mg, riboflavin 0.17 mg, nicine 0.5 mg and Vit. C 32.0 mg. Leaves emits odour like Asparagus but after cooking, aroma disappear. Recent analysis indicated 5 different types of flavonoids in which 6 "malonyl-2" xylosyl vitesin is a natural occurring flavonoid. Green leaves contain 2.1-2.3 mg/g flavonoids (fresh weight basis) as reported by Gil *et al.*, 1998. Swiss chard is rich source of several amino acids like arginine, histidine and lysine alongwith folic acid (320-1000 micro gram per 100.0 g). Among micronutrients, quantity of cobalt is 0.06-0.13 ppm (WHO, 1999).

Origin, History and Distribution

The origin of the adjective "Swiss" is unclear, since the Mediterranean plant is neithen native to Switzerland, nor particularly commonly cultivated there. Among beet species, Swiss chard is considered oldest one. Peoples were famlier with Swiss chard at least 300 BC. Aristotle gave detail description of 'Red chard'in 350 BC. Chinese were growing Swiss chard as early as 7th century. The wild species still today can be spotted near Canary Island and Middle East. In India, cultivation of Swiss chard is limited to hilly areas only. In plains, its cultivation is done during winter season as a vegetable crop.

Species Availability, Evolution and Diversity

B. vulgaris. ssp. *vulgaris* convar. *cicla* (leaf beet or chard) - The leaf beet group has a long history dating to the second millennium BC. The first cultivated forms were believed to have been domesticated in the Mediterranean, but were introduced to the Middle East, India, and finally China by 850 AD. These were used as medicinal plants in Ancient Greece and Medieval Europe. Their popularity declined in Europe following the introduction of spinach.

- ☆ ***B. vulgaris.* ssp. *vulgaris* convar. *cicla.* var. *cicla* (spinach beet)-** This species is widely cultivated for its leaves, which are usually cooked like spinach. It can be found in many grocery stores around the world.
- ☆ ***B. vulgaris.* ssp. *vulgaris* convar. *cicla.* var. *flavescens* (Swiss chard)-** Chard leaves have thick and fleshy midribs. Both the midribs and the leaf blades are used as vegetables, often in separate dishes. Some cultivars are also grown ornamentally for their coloured midribs. The thickened midribs are thought to have arisen from the spinach beet by mutation.
- ☆ ***B. vulgaris* ssp. *vulgaris* convar. *vulgaris* (tuberous beets)-** This grouping contains all beets grown for their thickened tubers rather than their leaves.
- ☆ ***B. vulgaris* ssp. *vulgaris* convar. *vulgaris* var. *crassa* (mangelwurzel)-** This species was developed in the 18th century for its tubers for use as a fodder crop.
- ☆ ***B. vulgaris* ssp. *vulgaris* convar. *vulgaris* var. *altissima* (sugar beet)-** The sugar beet is a major commercial crop due to its high concentrations of sucrose, which is extracted to produce table sugar. It was developed in Germany in the late 18th century after the roots of beets were found to contain sugar in 1747.
- ☆ ***B. vulgaris* ssp. *vulgaris* convar. *vulgaris* var. *vulgaris* (beetroot or garden beet)-** This is the red root vegetable that is most typically associated with the word 'beet'. It is especially popular in Eastern Europe where it is the main ingredient of borscht.

Table 8.14: Latin and Common Name of *Beta*

Latin Name	*Common Name*
Beta lomatogona	Beet
Beta trigyna	Beet
Beta vulgaris altissima	Sugar Beet
Beta vulgaris craca	Beet root
Beta vulgaris flavescens	Swiss Chard
Beta vulgaris maritima	Sea Beet

Botanical Description

Swiss chard belongs to family Chenopodiaceae

Kingdom	Plantae
Subkingdom	Viridiplantae
Infrakingdom	Streptophyta – land plants
Superdivision	Embryophyta
Division	Tracheophyta – vascular plants, tracheophytes
Subdivision	Spermatophytina – spermatophytes, seed plants, phanerogames
Class	Magnoliopsida

Superorder	Caryophyllanae
Order	Caryophyllales
Family	Amaranthaceae – pigweed, amaranthes
Genus	Beta L. – beet
Species	Beta vulgaris L. – common beet
Subspecies	Beta vulgaris ssp. cicla (L.) W.D.J. Koch–chard

Table 8.15: Botanical Description

Growth Habit	Plants are annual or biennial herbs which attainthe height of 70-90 cm.
Leaf	Leaves are long petioleted. Leaf petiole and veins are white or black in colour.
Floral biology	Flowers are white and emerge on long branches. The flowers are hermaphrodite (have both male and female organs).
Anthesis and pollination	Flowers are pollinated by wind
Fruit and fruit set	Fruits are small.
Seeds	Seeds are small.

Horticultural Classification

Ford (1986) conducted the extensive research on wild beet and putforth the detail classification based on facts finding and intraspecific variations. According to them *Beta vulgaris* spp. *vulgaris* was grouped as:

(i) Beet root
(ii) Mangel
(iii) Fodder beet
(iv) Sugar beet

Similarly under subspecies *cicla* grouping was done as:

(i) Chard
(ii) Spinach beet

Among wild species was done as:

(i) *Beta maritima*
(ii) *Beta macrocarpa*

There are two rankless cultivar groups within this subspecies: the Cicla-Group for the leafy spinach beet, and the Flavescens-Group for the stalky Swiss chard.

Augmentation of Genetic Resources and Crop Improvement

Introduction and Selection

There are few introduced genotypes which are popular as cultivar:

Table 8.16: Popular Cultivar and their Description

Name of Cultivar	*Description*
Ford Hook Giant	This is very popular variety. Leaves are curved and dark green in colour.
Dark Green White Ribbed	Leaves are broad and veins are white in colour. Leaves are quite crisp.
Rhubarb Chard	Leaves are broad and veins are red in colour.
Lucullus	This cultivar is yellowish green in colour.

Similary Pokluda and Kuben 2002 evaluated and reported the finding of some collected lines. In Pakistan, cultivars like T-29 (R) and Nantes (P) are being very common and grown commercially.

Interspecic Hybridization

Cultivated species under genus *Beta i.e.* Sugar beet, Red beet, Swish Chard *etc.* are easily crossable (cross compatible) with wild races like Sea beet (*Beta maritima*) as observed by Bartsch *et al.* (1999).

Table 8.17: Mean Parameters of Selected Cultivars of Swiss Chard

Cutlivar	*Company*	*Field Germination (Per cent)*	*Plant Weight (g)*	*Plant Height (mm)*	*Stalk Width (mm)*	*Plant Leaf Number*	*Yield (t/ha)*
Bright Lights F_1	Johnny's (USA)	88	347	513	27.4	7.8	35
Cerveny	MoravoSeed (CZ)	97	161	446	15.0	10.0	16
Fordhook Giant	Johnny's (USA)	97	248	520	21.0	6.7	25
Gator	Semo (CZ)	85	519	537	23.2	15.8	52
Genfer Selma	Mauser Samen (D)	74	363	468	33.1	7.6	36
Charlotte	Mauser Samen (D)	96	443	425	23.3	10.4	44
Listovy zeleny	MoravoSeed (CZ)	44	502	538	21.7	14.8	40
Lucullus	Semo (CZ)	86	400	522	35.5	7.8	18
Lucullus	Semena Veleliby (CZ)	86	180	523	17.6	8.3	22
Rhubarb	Johnny's (USA)	83	223	457	13.6	8.6	23
Rhubarb Chard	Kings (UK)	95	233	481	15.9	10.8	42
Swiss Chard	Kings (UK)	93	422	579	28.8	7.8	46
Zürcher Gelber	Mauser Samen (D)	95	455	574	20.2	10.0	50

Source: Pokluda and Kuben 2002.

Abiotic Stress

Shannon *et al.* (2000) consider Swiss chard as a vegetable which is more tolerant to higher soil salinity. Regarding this fact, Swiss chard accumulates higher levels of sodium in its tissues compared with other leafy vegetables. Swiss chard is thus suitable for such soil conditions that limit the growing of many other sensitive vegetable species. On the other hand, extreme salinity causes a quality and yield decrease in Swiss chard, similarly like in many cultivated crops.

Genetic Resource Conservation

In Chanigis, (Switzerland), large collection of Swiss chard (*Beta vulgaris* subsp. *provulgaris* var. *flavescens*) and Spinach beet (*Beta vulgaris* var. *cicla*) are being conserved. After evaluation and charectarization, great variability observed for many traits. Developed cultivars expressed uniform growth and high yield while local strains were stable in yield, leaf colour and unaffected with the prevailing environment (Badoux, 1995).

REFERENCES

1. Bartsch, D., Lehnem, M., Clegg, J., Pohl, Orf, M., Schuphan, I. and Ellstrand, N.C. 1999. Impact of gene flow from cultivated beet on genetic diversity of wild sea beet population. *Mol. Ecol.*, **8(10):** 1733-1741.
2. Badoux, S. 1995. Study on variation within a collection of Swiss chard and Spinach beet. *Revue-Suisse-de-viticultured' Arboriculture et.d' Horti-culture*, **27(6):** 333-337,
3. Ford-Lloyd, B.V. 1986. Intraspecific variation in wild and cultivated beets and its effect upon intraspecific classification. In: B.T. Styles (Ed.) Intraspecific classification of wild and cltivated plants. The systematic assications Vol. 29 : 331-344, Oxford.
4. Gil, M.I., Fereres, F. and Thomas Barberan, F.A. 1998. Effect of modified atmosphere packaging on the flavonoids and vitamin C content of minimally processed Swiss chard (*Beta vulgaris* ssp. *cicla*). *J. Agril. Food Chem.*, **46(5):** 2007-2012.
5. Pokluda, R. and Kuben, J. 2002. Comparison of selected Swiss chard (*Beta vulgaris* ssp. *cicla* L.) varieties. *Hort. Sci.* (PRAGUE), **29 (3):** 114–118.
6. Shannon, M.C., Grieve, C.M., Lesch, S.M., Draper, J.H., 2000. Analysis of salt tolerance in nine leafy vegetables irrigated with saline drainage water. *J. Am. Soc. Hort. Sci.*, **125:** 658–664.
7. World Health Organization, 1999. Nutrition profile of South East Asian Region. World Health Organization. South East Asia Regional Office, New Delhi.

Genus *Spinacia*

Spinach (*Spinacia oleracea* L.)

Introduction

Spinach (*Spinacia oleracea*, 2n= 2x= 12, 24) is a common cool season potherb or leafy vegetable. The name 'spinach' is derived from Spanish word *ospinaca* (Ryder, 1979). The edible part of spinach is a compact rosette of leaves. It is included in the diet for century and considered as common people vegetable (Nath *et al.*, 1987). The tender leaves and shoots are rich sources of vitamin A, iron and calcium which is largely used as fresh or processed in the USA and Europe. Leaves are rich source of carotene which pertain translutein, transplus 15, 15′-cis-β-carotene, cis-neoxanthin types in the percentaghe of 39.6, 18.4, 13.7 and 11.6, respectively (Khacik *et al.*, 1986.). The calcium is made unavailable due to precipitation as calcium oxalate because the oxalic acid content is more in this species. It is good souirce of carotenoid. Basically the available carotenoid is translutein (transplus 15, 15′-cis-β-carotene), cis-neoxanthin which pertain 39.6, 18.4, 13.7 and 11.6 per cent (Khacik *et al.*, 1986). Leaves contain 299.5 microgram Vit. K or phylloquinone (Wooth *et al.*, 1994) and 123.0 microgram Folate (in free form) per 100.0 g. Matsumoto and Kaya (1982) reported that leaves are rich source of histamine. The nutritive value of spinach reported as moisture 92.1 g, protein 2.0 g, fat 0.7 g, minerals 1.0 g, fiber 0.6g, carbohydrates 2.9 g, energy 26.0 K calorie, calcium 73.0 mg, phosphorus 21.0 mg, iron 1.6 mg, carotene 5560 microgram, thiamine 0.03 mg, riboflavin 0.26 mg, niacin 0.5 mg and Vit. C 28.0 mg, per 100.0gof edible portion.Leaves are also rich source of several micronutrients and per 100.0 g edible portion contain Magnesium 64.0 mg, sodium 58.5 mg, potassium 206 mg, copper 0.10 mg, Maganese 0.56 mg, molibdenum 0.010 mg, zinc 0.30 mg, chromium 0.005 mg, sulphur 30.0 mg and chlorine 54.0 mg. The leaves have medicinal properties which contains (1-o-β-linolenoyl-2-o (7Z, 10 Z, 13 Z)-hexadecantrienoyl-3-o-β-D-galactopyronosyl)-sn-glycerol) and 1-2 di-o-β-linolenoyl-3-o-β-D-galactopyronosyl-sn-glycerol (Wang *et al.*, 2002). Spinach is also processed, primarily canned and fried (Dhaliwal and Hira, 2001). Usually leaves, petals, stalks are used for processing.

Origin, History and Distribution

Spinach is generally thought to have originated in Iran and it comes from a central Asian (South Western Asia) gene centre, most probably Persia (Iran), where it may have originated from *Spinacia tetranda*, which is still gathered as a wild edible green in Anatolia.This was taken to Spain by Moore from where it spread to other parts of the world (Choudhury, 1967). The cultivation of spinach spread to the rest of the world from Spain (Pandita and Lal, 1986). It was known to China as early as the 7th century as well as to the ancient Greeks and Romans and grown in Persia about 2000 years ago. This crop was first cultivated by the Arabs. It has been introduced early by the Arabs to North Africa from which it was spread later into Europe (Morelock and Correll 2008). It was introduced into Europe in 15th century. It was introduced into North America by Arabs (Shoemaker, 1947) and American started growing it in 19th century. It reached North Africa through Syria and Arabia. Now this crop is widely grown in U.S.A., Europe, Rome, China, Greece, Persia, and Arabian countries. The crop is more common in the hilly areas (cooler place) than other parts of India (Kohli, 1993). However, it is popular in the southern states also (Nath *et al.*, 1987).

Species Availability, Evolution and Diversity

Spinach is a member of the Chenopodiaceae or goosefoot family. Concerning the gene pool structure of spinach usually three species are distinguished, namely cultivated spinach (*Spinacia oleracea* L.), which is inter-fertile with the wild species *S. tetrandra* Steven ex M. Bieb. (Sneep, 1957) and *S. turkestanica* Iljin (Handke *et al.*, 2000). There is two old primitive distinct species of spinach *i.e. Spinacia tetrandra* and *S. turkestancia* are described by Girenko (1968). They are diploid relative of Spinach and also have 2n=18 chromosome. The both species *tetrandra* and *turkestancia* are grown as wild in North West India and Nepal. However, Mehra and Malik (1963) have reported 2n=12 chromosome in *S. suaeda* species and concluded that the basic number of the chenopodiaceae family is X=9. It is derived from a leafy, winter annual that evolved in and near the Fertile Crescent of the Middle East. Modern forms of this crop have been selected to produce a lush and robust leafy vegetable that is more adaptive across environments and seasons, but some of the traits of spinach ancestors remain. However, many spinach varieties have been developed to be spring-planted and produce a bountiful crop before the summer day length causes bolting. The greater variability can be spotted in the plains and in foothills. Variability occurs in plant morphological characteristics, seed type and foliage yield attributing component.

Gene and Genetics of Characters

Bhagchandani *et al.* (1979) reported that additive gene effect was observed for leaf size and partial dominance for yield. The inheritance of leaf size is dependent on a minor gene. Nine gene groups were involved in the inheritance of yield.

Botanical Description

It is herbaceous annual plant for leaf production and a biennial for seed production. It belongs to family Chenopodiaceae species *Spinacia* and genus *oleracea* with chromosome number 2n=12. In spring, winter annuals grow steadily until a combination of environmental factors prompts the reproductive (bolting) phase of the life-cycle. Spinach bolting is initiated primarily by day length and the ancestral forms of this crop bolt very early under less than 14 hours of light per day. This allowed the crop to mature seed before the onset of intense heat during summer time. The monoecious plants may predominantly staminate, pistillate or pure pistillate, early but with some staminate flower late or almost equally staminate or pistillate throughout the season.

Table 8.18: Botanical Description

Parts	*Description*
Growth Habit	Herbaceous annual plant for leaf production and biennial for seed production. Usually spinach plants are dioecious and attain the height of 20.0-150.0 cm.
Root	It has tap root system.
Stem	A short term stem and when the plants enter into reproductive phase, the stem elongates and forms a flowering stalk with narrow pointed leaves (Pandey and Kalloo, 1993).
Leaf	From very near to the surface of the ground rosette of fleshy leaves is produced (from one place/point of stem) during vegetative phase. Leaves are dark green; vary in shape, thickness and smoothness *i.e.* either smooth or savoyed in the vegetative phase. The leaf petiole is 6.0-12.0 long, green or light pink in colour. The shape of leaf is quite different from that of Palak.
Floral biology	Spinach is dioecious in nature and produce very small flowers. A few monoecious (very rarely) plants with varying proportion of male and female flowers are also seen. Hermaphrodite plants are also found with bisexual flowers and a few male flowers also. Male plants flower earlier than female plants. Flowers are borne in clusters (Male flowers are seen in clusters on the spike whereas female flowers are born in cluster in leaf axils) on the central stem or branches (80.0-150.0 cm) and vary from 6-12 per cluster but do not develop simultaneously. Flowers may be staminate, pistillate or hermaphrodite. Male and female flowers are borne in separate plants (dioecious). In male plants, two categories *i.e.* extreme male (earliest flowering males) and vegetative male are found. Female plants are larger in size and remain in vegetative phase for longer period. Very rarely monoecious plants are seen with different proportion of male and pistillate flowers. Besides these, hermaphrodite plants are also found with bisexual flowers and bear few male flowers. Male plants die soon after flowering (Rosa, 1925) than female plant.

Parts	Description
Anthesis and pollination	Spinach is a cross pollinated and wind pollinated. Pollens are small in size and light, produced in abundance and carried away by air currents. Pollination takes place by wind. Unpollinated pistillate flowers remain receptive for for 2 to 3 week (Mac Gillivray, 1961; Shoemaker, 1947).
Seeds	Based on the morphology of seeds, two distinct types are classified as (i) smooth roundseeded type (summer spinach) and (ii) prickly seeded type (winter spinach).

There are 5 main sex classes observed by Pandita and Lal (1986), described as below:

1. **Extreme male-** This type of plants, produce only male flowers and are practically useless because of their bolting. These plants do not produce seeds.
2. **Vegetative male-** These plants bear male flowers and bloom at the same time as the classes which follow but do not produce seeds.
3. **Monoecious plant-** The plants bear both female and male flowers on the same plant and produce seeds.
4. **Hermaphrodite plant-** Bisexual flowers having both male and female parts are produced by these plants. They may also bear male flowers and these plants produce seeds.
5. **Female plants-** The plants bear only female flowers and produce seeds.

Sex Determining Factors in Spinach

Since spinach is dioecious in nature (male and female flowers are borne on separate plants) and it has been observed that leaves of female plant contain more sugar, chrlophyll and carotene than leaves of male plant (Sivster and Sizov, 1971). According Frankel and Galun (1977) sex in spinach is controlled by a series of alleles Y, X^m and X. Female plants are characterized by XX, made by XY and X^mY, while X^mX and $X^m X^m$ result in monoecious plant. Chromosome No.1 is heteromorphic but not in all cultivars.

Horticultural Classification

(A) Classification on the Basis of Utility and Leaf Characters

(i) **Savoy type-**This type of palak produces dark green, crinkled and curd leaves. Mostly fresh market types. They are mostly hand harvested.

(ii) **Flat or Smooth leaves-** They produce broad smooth leaves and are suited for processing and mostly machine harvested.

(iii) **Semi-Savoy-** It is hybrid variety. It produces slightly crinkled leaves having texture of Savoy type. It is grown both for fresh market and processing.

*The seed type in spinach does not indicate whether it is smooth leaved or Savoy type. Asian types are mostly prickly seeded and have triangular shaped leaves.

Geise and Farley (1929), classified spinach cultivars into 2 groups:

(a) Flat-leaved type and

(b) Savoy-leaved type

Each class was further sub-divided int 3 sub-classes:

(i) Short-season group

(ii) Long standing group

(iii) Slow-growing group

(B) Classification on the Basis of Sex Expression

There is well description about four types of plants with reference to sex expression, which are:

(i) **Extreme males-**Plants are smaller in size and fast emergence of staminate flower takes place (Roja, 1925). Seed stalk is characterized by presence of small leaves and normally flower

Table 8.19: Sex Determining Components and Descriptions

Sl.No.	Factors	Description	Reference
1.	Increase of nitrogen level	As increase in nitrogen level decrease the proportion of female plants. The female plants have more leaves than male plants. However, femaleness can be enhanced by cutting shoot of young plants and allowed to form adventitious roots.	Quagliotti, 1972
2.	Application of GA_3	The maleness is enhanced, by treating the plants with GA_3 25 mg/l after cutting the leaves. Hence, leaves play an essential role in sex expression and this role is related to gibberellins synthesis in the leaves	Chailakhyan and Khryanin, 1979
	Application of CCC and GA_3	The treatment with gibberellins and CCC influenced the formation of reproductive organs through regulation of growth processes but sex was not determined directly	Cajlahjain *et al.*, 1969
3.	Photoperiod and plant growth regulators	Photoperiod and plant growth regulators showed marked effects on the sex expression. Sex ratio (male to female) is near 1 in long days. The number of female plants flowering in long days can be, increased by gibberellins, whereas kinetin reduced flowering in male plants. Abscisic acid and Amo1618 reduced flowering in female plants. In short days gibberellins stimulated flowering in female more than in male plants and at higher temperature flowering (maleness) is reduced.	Culafic and Neskovic, 1980

earlier than other 3 vegetative males. These also bear staminate flowers. These are larger in size. Extreme males are not preferred due to their small sized leaves and early bolting nature.

(ii) **Vegetative males-**Plants are comparatively taller in size.

(iii) **Monoecious plants-**In this group basically male, female or in early stage completely female and some time latter male flower or male and female in equal quantity emerged. Monoecious plants are the most desirable types both for market and seed production.

(iv) **Female-**Plants are comparatively taller and remain longer time in vegetative stage. Late emergence of flower takes place but all the flowers appeared at the same time.

The monoecious plants are (may be) predominantly staminate, pistillate or pure pistillate early but with some staminate flower late or almost equally staminate in the plains and in foothills. Variability occurs in plant morphological characteristics, seed type and foliage yield attributing component.

(C) Classification on the Basis of Seed Types

Based on the morphology of seeds, two distinct types are classified according to Mac Gillivray (1961) as:

(i) Smooth seeded type and
(ii) Prickly seeded type.

The prickly seeded spinach is an old one and was used first in Europe in 1931, where as the smooth seeded was known in 1952.

(D) Classification on the Basis of Cultivar

According to Kinney (1986) cultivars, classified into 4 groups:

(i) Norfalk or Bloosdale
(ii) Round leaved
(iii) Thick leaved
(iv) Prickly seeded

Magruder *et al.*, 1938 classified spinach as shown in Table 8.20a.

According to Drewes (1943) spinach cultivars can be classified as shown in Table 8.20b.

Table 8.20a: Grouping of Cultivar

Sl.No.	Characters			
(I)	Leaves savoyed	(A) Leaves heavily savoyed	(i) Plants late seeding	(a) Leaves blue green
				(b) Leaves dark green
			(ii) Plants not late seeding	(a) Plants very early seeding
				(b) Plants medium-early seeding
		(B) Plants medium savoyed		
(II)	Leaves not savoyed		(A) Plants early seeding	(a) Seeds prickly
				(b) Seeds not prickly
			(B) Plants not early seeding	(a) Seeds prickly
				(b) Seeds not prickly: i- Leaves blue green ii- Leaves medium green

Table 8.20b: Grouping of Cultivar

Sl.No.	Characters			
(I)	Round sedded	1. Smooth leaved (or slightly savoyed in cool weather)	i. Early bolting or not long-standing	
			ii. Slow bolting or long-standing	
		2. Savoyed leaved	i. Early bolting or not long-standing	(a) Mosaic resistant
				(b) Not mosaic resistant
			ii. Slow bolting or long-standing	
(II)	Prickly seeded	1. Smooth leaved (or slightly savoyed in cool weather)	i. Early bolting or not long-standing	(a) Mosaic resistant
				(b) Not mosaic resistant
			ii. Slow bolting or long-standing	
		2. Savoyed leaved	i. Early bolting or not long-standing	(a) Mosaic resistant
				(b) Not mosaic resistant
			ii. Slow bolting or long-standing	

Ryder (1979) classified spinach cultivars on the basis of leaf type, into 3 groups:

(I) Savoy (wrinkled leaf)
(II) Semi-savoy
(III) Smooth

Augmentation of Genetic Resources and Crop Improvement

Introduction and Selection

No significant effort has been made to improve upon the old local strains available in various parts of the country. In India some of the varieties introduced from the western countries are being grown in certain parts where it is commercially produced.

Table 8.21: Popular Cultivars and their Descriptions

Variety	*Horticultural Attributes*
Virginia Savoy	It prickly seeded cultivar having blistered dark green large leaves with a round tip. The plant is upright and vigorous.
Early smooth	It is smooth seeded exotic cultivar producing small light green leaves. Leaves have a pointed apex.
Marathon	It is a Savoy type. Plants are large and produce dark green leaves.

Target region amplification polymorphism (TRAP) markers were used to assess genetic variability among 38 germplasm accessions and 10 commercial hybrids of spinach (*Spinacia oleracea* L.). Germplasm accessions with different geographic origins and 10 commercial hybrids were examined. For assessing genetic diversity within accessions, DNA was extracted from 12 individual seedlings from six germplasm accessions and two hybrids. A relatively high level of polymorphism was found within accessions based on 59 polymorphic TRAP markers generated from one fixed primer derived from the *Arabidopsis*-like telomere repeat sequence and two arbitrary primers. For evaluating interaccession variability, DNA was extracted from a bulk of six to 13 seedlings of each accession. Of the 492 fragments amplified by 12 primer combinations, 96 (19.5 per cent) were polymorphic and discriminated the 48 accessions from each other. The average pair-wise genetic similarity coefficient (Dice) was 57.5 per cent with a range from 23.2 to 85.3 per cent. A dendrogram indicated that the genetic relationships among the accessions were not highly associated with the geographic locations in which the germplasms were collected. The seven commercial hybrids were grouped in three separate clusters, suggesting that the phenotype-based breeding activities tended to reduce the genetic variability. This preliminary study demonstrated that TRAP markers are effective for fingerprinting and evaluating genetic variability among spinach germplasms (Hu *et al.*, 2007). List of the 38 accessions used in the current studyare as under:

Table 8.22: Introductions of Promising Genotypes and their Origin

Accession Number	*Name or ID*	*Origin*
NSL 6082	Wisconsin Blooms	New York, US
NSL 6782	Hollandia	Netherlands
PI 608762	K-17068	Thailand
Ames 20168	Da Yuan Ye Bc Cai	China
PI 608712	SPI 153/89 (S. tetrandra)	Germany
NSL 68263	Indures	New York, US
PI 164966	Harlan 42	Turkey
PI 167434	Cavallius	Belgium
PI 179507	Beledi	Syria
PI 179508	Harlan 9889	Iraq
PI 179595	Victoria	Belgium
PI 193618	Neusseelander	Ethiopia
PI 193619	Victoria	Ethiopia
PI 217425	CGN 14178	South Korea
PI 222270	Esfenaj	Iran
PI 254565	Polack	Afghanistan
PI 261787	Monstrueux de Vi	France
PI 261788	Geant d'Hiver	France
PI 262911	Espinaca nobel	Spain
PI 266926	Universal	Germany
PI 285751	Koda	Warszawa, Poland
PI 296393	CGN 9662	Iran
PI 358248	Ohridski	Yugoslavia
PI 379549	Stipski	Yugoslavia
PI 418978	Chung Pu Lao	Shaanxi, China
PI 445785	Baladi	Syria
PI 449353	Universal	Turkey
PI 499372	Ispolinskij	Former Soviet Union
PI 499373	Godir	Former Soviet Union
PI 531448	Eszkimo	Hungary
PI 531449	Hegykoi	Hungary
PI 531452	Matador Darko	Denmark
PI 604785	SPI 161/86	Mongolia
PI508504	Summer Green	South Korea
PI527332	Da Jian Ye Bo Cai	China
NSL 6088	Blight Resis Savoy	New York
PI358258	Lisicki	Yugoslavia
PI374233	Baladi	Egypt

Source: Hu *et al.*, 2007.

In Nepal spinach cultivar 'BSS-69' introduced from, Hungnong Seed Co., Korea is quite polupar among the vegetable growers. In Pakistan, cultivars like Faislabad Local (R) and Patar Selection (P) are being very common and grown commercially.

Polyploidy Breeding

Polyploidy breeding has been reported in spinach.

Table 8.23: Resistance Source against Various Diseases

Name of Disease	Resistance Source	Reference
Downy mildew	Early Giants, Hybrid-320, Hybrid-308, Vikimum, Dorema, Matares, Vital, Fall Green, Noord Brabant, West Mass, Triathon, Triade, Hermes, Trisolde, Solar, Tritan, Tristar, Correnta, Mazurka, Valta, Achilles, Atorasu, Aubade Hybrid,Capella Hybrid, Califlay Febris, Gurobarus, G-one, Nores, Orai, Sevenade, Hybrid Soloman, Samson, Sanraito, EH-7, EH-10, EH-424, EH-425	Webb and Thomas, 1976
Fusarium wilt (Moderate level of resistance)	EH-7, EH-10, Hybrid-50, EH-424, EH-425, Ozarka, Green Valley, and WRG-70-5	O'Brien and Winters, 1977
White rust (*Albugo occidentalis* Wilson)	Fall Green, Green Stand, Hybrid-178, WRG-70, Green Valley-11 x PH-1452, 4C x 119, Coho	Black and Dainello, 1986
Mosaic (CMV-1)	Dixie Market, Fall Green, Roga, Lavires and Mona Lisa	
Beet yellows	Dutch varieties, Roga and Ravires	Chod, 1985
Beet mosaic	Novera, Estivato and Mona Lisa	Chod, 1985

Hybridization

Heterosis up to 30-40 per cent has been observed in F_1 hybrids and utilized for development of downy mildew and mosaic resistant hybrids.

Resistance Breeding

Disease Resistance

Disease resistance shown in Table 8.23.

Genetic Resource Conservation

The wild species *S. tetrandra* Steven ex M. Bieb. and *S. turkestanica* Iljin, both belonging to the primary spinach gene pool, are severely under represented in current germplasm collections. To improve the composition of its spinach collection, sampling of *S. turkestanica* Iljin in Tajikistan and Uzbekistan was carried out by CGN in 2008. This mission resulted in 66 sampled populations of *S. turkestanica* Iljin and two collected landraces of *S. oleracea* L. Currently, these samples are being regenerated, after which they will be included in CGN's spinach collection and made available to the user community. A collection mission to the Caucasus region to sample *S. tetrandra* Steven ex M. Bieb., is planned by CGN for 2011. Since only 18 samples of cultivated spinach from Iran and 7 from North Africa are currently included in The ISDB, more extensive sampling from these areas is considered the second priority. Also, relatively few samples of *S. oleracea* L. from America are included in the ISDB. However, spinach varieties have been introduced to the United States relatively late and nowadays hybrids are the major type of cultivars (Morelock and Correll 2008), which are difficult to maintain by genebanks.

REFERENCES

1. Bhagchandani, P.M., Thakur, P.C. and Singh, N. 1979. Diallel analysis of leaf size and yield in Spinach. *Indian J. Agric. Sci.*, **49(5):** 364-367.
2. Black, M.C. and Dainello, F.J. 1986. Comparison of percentage leaf area with white rust lesions and two other methods for evaluating partial resistance to *Albugo* spp. in spinach, *Phytopathology*, **76:** 1087.
3. Cajlahjain, M.H., Martinvic, L.N. and Koeankov, V.G. 1969. Doklady Akademii Nauk, USSR, 189, p. 662-665.
4. Chailakhyan, M. K. and Khryanin, V.N. 1979. *Planta*, **44:** 205-207.
5. Chod, J. 1985. Susceptability of some spinach cultivars and hybrids to beet mosaic virus, beet yellows virus and cucumber mosaic virus. *Zeszyty Problemowe Postepov. Nauk., Rolnicyyeh*, **291:** 89.
6. Choudhury, B. 1967. Vegetables. National Book Trust, New Delhi, India.
7. Culafic, L. and Neskovic, M. 1980. *Physiol. Plantarum*, **48:** 588-590.
8. Dhaliwal, M. and Hira, C.K. 2001. Effect of stroage on physio-chemical and nutritional characteristics of carrot, beetroot and carrot-blackcarrot juices. J. Food. Technol; 38:343-347.
9. Frankel, R. and Galun, E. 1977. Pollination Mechanisms, Reproduction, and Plant Breeding, Springer, Berlin.
10. Girenko, M.M. 1968. Classification of spinach.
11. Geise, F.W. and Farley, H.B. 1929. Variety of spinach.
12. Handke, S., Seehaus, C., and Radies, M. 2000. Detection of a linkage of the four dominant mildew resistance genes "M1M2M3M4" in spinach from the wildtype *Spinacia turkestanica*. Gartenbauwissenschaft 65:73–78.
13. Hu, J., Mou, B. and Vick, B.A. 2007. Genetic diversity of 38 spinach (*Spinacia oleracea* L.) germplasm accessions and 10 commercial hybrids assessed by TRAP markers. *Genet Resour. Crop Evol.*, **54:** 1667. doi:10.1007/s10722-006-9175-4.
14. Khacik, F.; Beecher, G.R. and Whittaker, N.F., 1986. Separation, identification and quantification of the major carotenoid and chlorophyll constituents in extracts of several green vegetables by liquid chromatography. *J. Agric. Food Chem.*, **34:** 603.
15. Kinney, L.F. 1986. R.I. Bull., p. 41.

16. Kohli, U.K. 1993. Agro-techniques for leafy vegetables. In: Advances in Horticulture Vol.5- Vegetable Crops: Part I. (Eds. Chadha, K.L. and Kalloo, G.). Malhotra Publishing House, New Delhi, India.

17. Mac Gillivray, J.H. 1961. Vegetable Production. Mc. Graw-Hill Book Co., Inc., New York.

18. Matsumoto, T. and Kaya, T. 1982. Distribution of amine in plants. *Nippon Nogei Kagakv Kaishi.* **56(3):** 209.

19. Mehra, P.N. and Malik, P.1963. Cytology of some Indian Chenopodiaceae. *Caryologia,* 16(1) : 67-84.

20. Morelock, T.E. and Correll, J.C. 2008. Spinach. In: Prohens, J. and Nuez, F. (eds.) Handbook of plant breeding, vegetables I, Asteraceae, Brassicaceae, Chenopodicaceae, and Cucurbitaceae. Springer, New York, p. 189–217.

21. Nath, P., Velayudhan, S. and Singh, D.P. 1987. Vegetables for the Tropical Region. ICAR, New Delhi.

22. O'Brien, M.J. and Winters, H.F. 1977. Evaluation of spinach accessions and cultivars for resistance to Fusarium oxysporum, Wilt. I. Green House Bench Method, *J. Am. Soc. Hortic. Sci.,* **102:** 424.

23. Pandey, S.C. and Kalloo, G. 1993. Spinach, *Spinacia oleracea* L. in: Genetic Improvement of Vegetable Crops (Eds. Kalloo, G. and Bergh, B.O.) Pergamon Press, Oxford.

24. Pandita, M.L. and Lal, S. 1986.Leafy vegetables. In: Vegetable crops in India. (eds. Bose, T.K. and Som, M.G.) Naya Prokash, Kolkata, India.

25. Quagliotti, L. 1972. The effects of nitrogen fertilization on growth, habit and sex expression in Riccio d Asti spinach. *Sementi Elette,* **18:** 5-19.

26. Roja, R.J. Jr. 1925. Sex expression in spinach. *Hilgardia,* **1:** 259-274.

27. Ryder, E.J. 1979. Leafy Salad Vegetable, A.V.I., Publishing Co., Inc., Westport Connecticut 195, USA.

28. Shoemaker, J.S. 1947. Vegetable Growing, J. Wiley & Sons. Inc. London, U.K., p. 219-234.

29. Sivster, M.V. and Sizov, S.S. 1971. Ovochrinistvoi bashtannistv. Resp. Mizhvid, Temat, Nauk 26, No. 12, p. 60-63.

30. Sneep, J. 1957. Geschiedenis van de spinazie en het ontstaan van de rassen. Instituut voor de veredeling van tuinbouwgewassen, Wageningen.

31. Wang, R., Furumoto, T., Motoyama, K., Okazaki, K., Kondo, A. and Fukui, H. 2002. Possible antitumor promoters in *Spinacia oleracea* (spinach) and comparision of their contents among cultivars. *Biosci. Biotechnol. Biochem.,* **66(2):** 248-254.

32. Webb, R.E. and Thomas, C.E. 1976. Development of F_1 spinach hybrids. *HortSci.,* **11:** 546.

Genus *Chenopodium*

Bathua (*Chenopodium* species)

Introduction

Bathua or Chenopod (*Chenopodium album* Linn., 2n=18) is also known locally as Lamb's quarters, Fat-hen, pigweed, dung weed or more ambiguously as just Goosefoot, *etc.* but the standard English name is Fat-hen, only. In India it is known with several vernacular names like Agar Lohit, Chilika (Sanskrit); Bathua Sag (Hindi); Chandan betu (Bengali); Chil Tanko (Gujarati); Huchuchakotha (Kanadda); Chakwat (Marathi); Parapu Kirai (Tamil) and Papukura (Telgu). Chenopod is an ancient food crop (grains and greens) like amaranth. In winter season leafy vegetables, chenopod has prime position due to nutritional and medicinal qualities. In early stage whole plant and later on leaves and young shoots may be used as vegetable, either steamed, or cooked like spinach. Different food items like Bathua Roti, Bathua Paratha, Bathua sag, Bathua Bhujia, Rayata Pakoda *etc.* can be prepared. The young inflorescences are also cooked as a tasty broccoli substitute. The protein and mineral matters are quite high than the spinach and cabbage. The leaves contain about 3.95 per cent protein, 0.76 per cent fat, 8.93 per cent carbohydrate, and 3.0 per cent ash. Per 100 g edible portion chenopod contain 89.6 g moisture, 3.7 g protein, 0.4 g fat, 2.6 g minerals, 0.8 g fiber, 2.9 g carbohydrates, 30 K. calorie energy, 150.0 mg calcium, 80.0 mg phosphorus, 4.2 mg iron, 3660 microgram carotene, 0.03 mg thianine, 0.06 mg riboflavin, 0.2 mg niacin and 12.0 mg vit.C.(Gopalan *et al.,* 1999). In copmarision to other leafy vegetables, chenopod has high amount lysine amino acids and less amount of methionine. The amino acids contain (mg/g nitrogen) in chenopod are 750.0 mg lysine, 20.0 mg tryptophan, 110.0 mg phenyl alenine, 50.0 mg methionine, 170.0 mg threonine, 410.0 mg leucine, 440.0 mg isoleucine and 290.0 mg valin. Apart from these, leaves also contain 250 ppm vit.K (Ganju and Puri, 1959). With the analysis it has been observed that chenopod accumulate 3.5-5.0 per cent nitrate on dry weight basis (Prakash *et al.,* 1993). Accumulation of nitrate is more in stems than leaveas which is thrown away and only leaves are cooked. Apart from nitrate it contains 4.2 ppm nitrite and 0.9-3.9 g oxalate per 100 g (fresh weight basis) and that to mainly in stems. The seed contain about 49.0 per cent carbohydrate, 16.0 per cent protein, and 7.0 per cent ash. Seed also has high amount of protein, vitamin, calcium, phosphorus and potassium. Seeds are dried and ground into a meal and eaten raw or baked into bread and proved more nutritive than buck wheat. In Kullu valley (Himachal Pradesh), seeds are used to prepare 'phambra' dish. In Shimla it is used to prepare fermented beverage call 'Soora' and 'Ghani'. The seed can also be sprouted and added to salad. The seed should be soaked in water overnight and thoroughly rinsed before being used, in order to remove saponins.

As per the scientific report, when eaten with beans, the leaves will act as a carminative to prevent wind and bloating. It helps in removing worms (round worm, hook worm *etc.*) and improves cardic muscles. The leaves are best not eaten raw. The leaves are generally very nutritious but intake of very large quantities can disturb the nervous system and cause gastric pain. The leaves are anthelmintic, antiphlogistic, antirheumatic, mildly laxative and odontalgic. An infusion is taken in the treatment of rheumatism. The leaves are applied as a wash or poultice to bug bites, sunstroke, rheumatic joints and swollen feet, while decoction is used for carious teeth. The juice of stems is applied to freckles sunburn. Incorporation of dehydrated chenopod in various conventional food items can improve the nutritional quality of products as well as add variety in the diet (Singh *et al.*, 2007). As the common name suggest, it is also used as food (both the leaves and seeds) for chickens, hens and other poultry. However, the nitrates in the plant can be converted very efficiently to nitrites in the rumen of cattle, leading to change in hemoglobin and reducing the ruminants binding capacity.

Origin, History and Distribution

Its native range is obscure due to extensive cultivation, but includes most of Europe, from where Linnaeus described the species in 1753. Plants native in eastern Asia are included under *C. album*, but often differ from European specimens. The chenopod species *Chenopodium quinoa* originated from Andean region of South America. In general it is indigenous to Andean region of South America and said to have been originatd in the vicinity of Lake Titicaca. In India, chenopod occurs as weed in plains and *C. murale* in hills up to 3000 m, more variability in upper Gangetic plains, extending to northern hills; elsewhere only sporadic distribution. It is cultivated at an elevation of 1500 m to 3000 m in Himachal Pradesh, Uttarakhand, Uttar Pradesh hills and north-Eastern Hills. As a matter of fact species *Chenopodium quinova* is not much popular for vegetable preparation due to bitter taste and accumulation of harmful compound like saponin (Morales and Curl, 1986). Other species are very popular as leafy vegetables in Andean regions, Africa and Europe. It is widely introduced elsewhere, *e.g.* Africa, Australasia, North America, and Oceania, and now occurs almost everywhere in soils rich in nitrogen, especially on wasteland. In India, mostly it is grown in North India near to Metropolitan cities. Four types of cultivars having black, brown, red and earthen seed colours are limited in distribution and are found in hill valley in Kullu district.

Species Availability, Evolution and Diversity

Worldwide total 260 species of *Chenopodium* are available (Risi and Galway, 1984) but only 8 species are found in India (Wealth of India, 1972). In India 3 species, *i.e.*, *Chenopodium album*, *C. ambrosoides* and *C. amranticolor* dominates. *Chenopodium album* Linn. has a very complex taxonomy and has been divided in numerous micro species, subspecies and varieties, but it is difficult to differentiate between them. The following texa are accepted by the *Flora Europaea*.

- ☆ *Chenopodium album* subsp. *album*
- ☆ *Chenopodium album* subsp. *striatum* (Krajan) Murr.
- ☆ *Chenopodium album* var. *reticulatum* (Allen) Uotila.

Published names and synonyms include *C. album* var. *microphyllum*, *C. album* var. *stevensii*, *C. acerifolium*, *C. centrorubrum*, *C. giganteum*, *C. jenissejense*, *C. lanceolatum*, *C. pedunculare* and *C. probstii*.

The details of some species are given below:

Table 8.24: Dominating Species and their Distribution

Sl.No.	Species	Country
1.	*Chenopodium quinoa*	South America, Guatemala, Bolivia, chille, Argentina
2.	*Chenopodium pallidicaule*	South America
3.	*Chenopodium berlandieri*	Mexico
4.	*Chenopodium album*	India
5.	*Chenopodium ambrosoides*	India
6.	*Chenopodium amranticolor*	India
7.	*Chenopodium murale*	India
8.	*Chenopodium striatum*	Checoslovakia
9.	*Chenopodium opulifolium*	Checoslovakia
10.	*Chenopodium foliosum-polyspermum*	Finland

Source: Singhal and Kulkarni, 1998.

Selected Species

- ☆ *Chenopodium acicularis*
- ☆ *Chenopodium acuminatum* Willd.
- ☆ *Chenopodium album* – White goosefoot, nickel greens, dungweed, *bathua*, *chandali*, *chandaliya*, fat hen, lamb's quarters, pigweed
- ☆ *Chenopodium album* ssp. *amaranticolor*
- ☆ *Chenopodium allanii*
- ☆ *Chenopodium atrovirens* – Dark goosefoot, pinyon goosefoot
- ☆ *Chenopodium auricomiforme*
- ☆ *Chenopodium auricomum* – Queensland bluebush
- ☆ *Chenopodium baccatum* (Syn. *Rhagodia baccata*)

- *Chenopodium berlandieri* – Pitseed goosefoot, southern huauzontle, lambsquarters
- *Chenopodium berlandieri* ssp. *nuttalliae* (Saff.) H.D.Wilson and Heiser
- *Chenopodium berlandieri* var. *bushianum*
- *Chenopodium berlandieri* var. *zschackii*
- *Chenopodium bushianum* – Village goosefoot
- *Chenopodium candolleanum* (Syn.: *Rhagodia candolleana*)
- *Chenopodium curvispicatum*
- *Chenopodium cycloides*
- *Chenopodium desertorum* – desert goosefoot
- *Chenopodium desertorum* ssp. *anidiophyllum*
- *Chenopodium desertorum* ssp. *desertorum*
- *Chenopodium desertorum* ssp. *microphyllum*
- *Chenopodium desertorum* ssp. *rectum*
- *Chenopodium desertorum* ssp. *virosum*
- *Chenopodium desiccatum* – Narrow leaf goosefoot
- *Chenopodium drummondii* (Syn.: *Rhagodia drummondii*)
- *Chenopodium eremaea* (Syn.: *Rhagodia eremaea*)
- *Chenopodium erosum* R.Br.
- *Chenopodium ficifolium* – fig-leaved goosefoot, small goosefoot
- *Chenopodium foggii* – Fogg's goosefoot
- *Chenopodium fremontii* – Fremont's goosefoot
- *Chenopodium giganteum* D.Don – tree spinach
- *Chenopodium gigantospermum*
- *Chenopodium hastatum* (Syn.: *Rhagodia hastata*)
- *Chenopodium hians*
- *Chenopodium iljinii*
- *Chenopodium incanum* – Mealy goosefoot
- *Chenopodium latifolium* (Syn.: *Rhagodia latifolia*)
- *Chenopodium leptophyllum* – Narrow leaf goosefoot
- *Chenopodium littoreum*
- *Chenopodium missouriense* – Missouri goosefoot (sometimes considered a variety of *C. album*)
- *Chenopodium neomexicanum*
- *Chenopodium nevadense*
- *Chenopodium nitrariaceum* (F.Muell.) F.Muell. ex Benth. – nitre goosefoot
- *Chenopodium nutans* (Syn.: *Rhagodia nutans*)
- *Chenopodium nuttalliae* – Huauzontle
- *Chenopodium oahuense* – *Āheahea* (Hawaii)
- *Chenopodium opulifolium* Schrad. ex W.D.J.Koch – Grey goosefoot
- *Chenopodium pallescens*
- *Chenopodium pallidicaule* – *Kañiwa*
- *Chenopodium pamiricum*
- *Chenopodium parabolicum* (Syn.: *Rhagodia parabolica*)
- *Chenopodium petiolare*
- *Chenopodium polygonoides*
- *Chenopodium pratericola* Rydb.– Pale goosefoot, desert goosefoot, narrowleaf goosefoot
- *Chenopodium preissii* (Syn. *Rhagodia preissii*)
- *Chenopodium probstii* Aellen
- *Chenopodium purpurascens* – Purple goosefoot
- *Chenopodium quinoa* – quinoa
- *Chenopodium retusum*
- *Chenopodium salinum* – Rocky Mountain goosefoot
- *Chenopodium spinescens* (Syn. *Rhagodia spinescens*)
- *Chenopodium standleyanum* – Standley's goosefoot
- *Chenopodium strictum* Roth
- *Chenopodium subglabrum* – Smooth arid goosefoot, smooth goosefoot
- *Chenopodium suecicum* – Green goosefoot
- *Chenopodium triandrum* (Syn.: *Rhagodia triandra*)
- *Chenopodium trigonon* (Syn.: *Einadia trigonos*)
- *Chenopodium truncatum*
- *Chenopodium ulicinum*
- *Chenopodium* × *variabile* (*C. album* × *C. berlandieri*)
- *Chenopodium vulvaria* – stinking goosefoot, notchweed
- *Chenopodium watsonii* – Watson's goosefoot
- *Chenopodium wilsonii* (Syn.: *Rhagodia crassifolia*)

Excluded Species

Blitum (12 species)

- *Blitum bonus-henricus* – Good King Henry, perennial goosefoot, poor-man's asparagus, Lincolnshire spinach, markery
- *Blitum californicum* – California goosefoot, Indian lettuce
- *Blitum capitatum* – Strawberry blite, blite goosefoot, strawberry goosefoot, strawberry spinach, Indian paint, Indian ink
- *Blitum virgatum* (Syn. *Chenopodium foliosum*) – Leafy goosefoot

Chenopodiastrum (5 species)

- *Chenopodiastrum murale* – Nettle-leaved goosefoot
- *Chenopodiastrum simplex* – Giantseed goosefoot
- *Dysphania* (about 43 glandular species, as *C. botrys*, *C. carinatum*, *C. cristatum*, *C. melanocarpum*, *C. multifidium*, *C. pumilio* and more)

Lipandra **(one species)**

- *Lipandra polysperma* – Many-seeded goosefoot

Oxybasis **(5 species)**

- *Oxybasis chenopodioides* – Small red goosefoot, saltmarsh goosefoot
- *Oxybasis glauca* – Oak-leaved goosefoot
- *Oxybasis rubra* – Red goosefoot, coastblite goosefoot
- *Oxybasis urbica* – Upright goosefoot

Teloxys **(one species)**

- *Teloxys aristata*
- *Suaeda australis* – Austral seablite (as *C. australe, C. insulare*)

Botanical Description

Chenopod belongs to family Chenopodiaceae. In this family mostly plants are herbs, bushy and in lesser number trees can be seen.

Table 8.25: Botanical Description

Growth Habit	Plants are monoecious (rarely dioecious). Plants are herbaceous annual. Sometime species behave as shrub or small trees. It tends to grow upright at first reaching heights of 10.0-150.0 cm (rarely to 300 cm) but typically becomes recumbent after flowering (due to the weight of the foliage and seeds) unless supported by other plants. They are non-aromatic, but sometimes foetid
Root	Roots stout and tapering at the end. Many branches may emerge from main tap root system.
Stem	The branched stems grow erect, ascending, prostrate or scrambling. Lateral branches are alternate (the lowermost ones can be nearly opposite). The young stems and leaves are often densely covered by vesicular globose hairs, thus looking farinose. Characteristically, these trichomes persist, collapsing later and becoming cup-shaped.
Leaf	The leaves are alternate, petiolated and can be varied in appearance. The first leaves, near base of the plant, are toothed and roughly diamond shaped 3.0-7.0 cm long and 3.0-6.0 cm broad. The leaves on the upper part of the flowering stems are entire and lanceolate-rhomboid, 1.0-5.0 cm long and 0.4-2.0 cm broad; they are waxy-coated, unwettable and mealy in appearance, with a whitish coat on the underside.
Floral biology	Inflorescence is standing terminal and lateral. They consist of spicately or paniculately arranged glomerules of flowers. The small flowers are radially symmetrical and grow in small cymes (green colour) on a dense branched inflorescence of 10.0-40.0 cm long. Flowers appear in cluster and that to from the apex of the stem/branches. The flowers are hermaphrodite in nature. Flowers consist of (4–) 5 perianth segments connate.
Anthesis and pollination	The flowers are pollinated by wind.
Fruit and fruit set	In fruit, perianth segments become sometimes coloured, but mostly keep unchanged, somewhat closing over or spreading from the fruit. Pericarp membranous or sometimes succulent, adherent to or loosely covering the seed.
Seeds	Each plant produces tens thousands seeds. The horizontally orientated seeds are depressed-globular to lenticular, with rounded to subacute margin. Seeds are black in colour, almost smooth to finely striate, and having disc shape structure (rugulose or pitted). Seeds are very shining.

Augmentation of Genetic Resources and Crop Improvement

Like other leafy vegetable crops, no specific exploration has been undertaken for the collection of chenopod. However, 45 collections have been made under multi-crop explorations from north-west Himalayas. A few collections have also been made from the Darjeeling, Sikkim and Khasi Hills. A total of 70 germplasm collections have been made including 25 exotic collections from Peru, Bolevia, Eucador, USA and Poland. The University of Cajamarta, Peru in their gene bank at Puno, maintaining 1500 collections of *Chenopodium quinoa* (Tapia and Mateo, 1987). Seeing the nutritional importance, efforts should be made to collect and build up germplasm of *C. quinoa*. Efforts are currently under-way to develop and promote chenopod as a major health food crop in the Andes (Tapia and Mateo, 1987). Possibilities of introducing chenopod into United Kingdom and Scandinavian countries as food, feed and a leaf protein crop are at various stages of explorations (Carlesson, 1980). Two American organizations are promoting chenopod grain as a future health food for urban societies both through imports and cultivation in the Colorado mountain area (Wood, 1985). Since exotic material of chenopods in our germplasm collection is poorly represented, there is a need to acquire materials from various countries such as Peru, Bolevia, Eucador, Guatemala, Mexico, Argentina, Ethiopia, Nigeria, Sierra Leone, Russia, China, Nepal and Bhutan. There is also a need to introduce all the known species of this crop from different sources so as to enrich Indian collections.

Introduction and Selection

Several 'Bathua' genotypes which are found suitable with respect to leaf quality were selected for human consumption. Among them 'Pusa Bathua-1', Ootty-1 and 'Magenta' were found very promising. Recently chenopodium cultivar Pusa Green from ICAR-IARI, N. Delhi and Kashi Bathua-2 and Kashi Bathua-4 from ICAR-IIVAR, Varanasi (U.P.) has been released for commerical cultivation. Eighty four germplasm accessions were evaluated for 12 characters at ICAR-NBPGR Regional Station, Shimla (H.P.). there was a wide variability in plant height, flowering and maturity period, leaf size, inflorescence size, seed size and grain yield. An accession NC58613 was identified as the most promising. Promising types were identified based on multi-locational trials included BDJ476, BDJ480, CH/LKW10 and CH/LKW49. Tall growing strains of *C. album*, namely BDJ171, BDJ285 and BD519 from hill region showed excellent growth under water stress situations in the northern plains. Evaluation of 9 entries of *C. album* during 1991 revealed a good degree of variation for days to flowering (51-65 days), days to maturity (95-125 days), plant height (71.6-146.3 cm), cluster per plant (13.5-24.2) and grain yield (0.58-1.60 t/ha). Analysis of foliage of 10 different species of Chenopodium at NBRI, Lucknow exhibited a wide variation for protein (26.0-64.0 g/kg), carotenoid (78.0-190.0 mg/kg), vitamins (0.5-2.4 g/kg), nitrate (2.6-5.0 g/kg) and oxalate (9.0-39.0 g/kg).

Interspecific Hybridization

It also hybridizes readily with several other *Chenopodium* species, including *C. berlandieri*, *C. ficifolium*, *C. opulifolium*, *C. strictum* and *C. suecicum*.

REFERENCES

1. Carlesson, R. 1980. Quantity and quality of leaf protein concentrates from *Atriplex hortensis*, *Chenopodium quinoa* and *Amaranthus caudatus* grown in Southern Sweden. *Acta Agric. Scandinavica*, **30:** 418-426.
2. Ganju, K. and Puri, B. 1959. Bioflavonoids from Indian vegetables and fruits. *Indian J. Med. Res.*, **47:** 563.
3. Gopalan, G., Rama Sastri, B.V., Balasubramanian, S.C., Rao, B.S.N., Deosthale, Y.G. and Pant, K.C. 1999. Nutritive value of Indian Foods. National Institute of Nutrition, ICMR, Hyderabad, India.
4. Morales, P.P. and Curl, C. 1986. A physiological method for total saponin determination in quinoa samples. *Re. Boliv. Quim.*, **6(1):** 13.
5. Prakash, D., Nath, P. and Pal, M. 1993. Composition variations of nutritional contents of leaves, seed protein, fat and fatty acid profile in *Chenopodium* species. *J. Sci. Food Agric.*, **62:** 203.
6. Risi, T. and Galway, N.W. 1984. In: Advances in applied biology. In: Coaker, T.H. editors. The *Chenopodium* grain of the andes: inca crops for modern agriculture. London: Academic. p. 145-216.
7. Singh, L., Yadav, N., Kumar, A.R., Gupta, A.K., Chacko, J., Pravin, K. and Tripathy, U. 2007. Preparation of value added products from dehydrated *'bathua'* leaves (*Chenopodium album* Linn.). *Natural Radiance*, **6(1):** 6-10.
8. Singhal, R.S. and Kulkarni, P.R. 1998. Leafy vegetables. In: *Handbook of Vegetable Science* (Salunkhe, D.K. and Kadam, S.S.; Eds.). Marcel Dekker. Inc. New York, Basel, Homg Kong.
9. Tapia, M. and Mateo, N. 1987. Andean phytogenetic and zoo-genetic resources. *Mountain Agric. and Crop Genetic Resources*, IDRC, IBH, New Delhi.
10. Wealth of India, 1972. Raw Materials. Vol. IX. Publication and Information Directorate, CSIR, New Delhi.

Genus *Trigonella*

Fenugreek (*Trigonella foenum-graecum* L.)

Introduction

Fenugreek (*Trigonella foenumgraecum* or *T. corniculata*, 2n=16) is popularly known as methica, chandrica, asmandagam (Sanskrit); methi, methishak, methukan, homegriev (Bengali); methi (Marathi); methi, methini bhanji (Gujarati); menticu (Telgu); vendyam (Tamil); menthia, metisoppu, menk-palle, mente (Kannad) and uluka, venthium (Malyalam). It is one of the oldest cultivated vegetable cum condiment/spice crops in the world. Among the leafy vegetables, fenugreek leaf is most important due to its nurtional and medicinal properties and consumed as green leafy vegetable during the winter months in the country. They are aromatic, carminative, tonic and galactagogue. Per 100 g of edible portion contains 86.0 g moisture, 4.4 g protein, 0.9 g fat, 1.5 g minerals, 1.1 g fiber, 6.0 g carbohydrates, 49 K calorie, 395 mg calcium, 51.0 mg phosphorus, 1.93 mg iron, 2340 microgram carotene, 0.04 mg thiamine, 0.31 mg riboflavin, 0.8 mg niacin and 52.0 mg vit. C. Apart from this 100 g of edible leaves contain 167.0 mg selenium 0.006 mg chromium, 33.0 mg magnesium31.0 mg potassium, 0.10 mg copper, 0.23 mg manganese, 0.400 mg molybdenum, 0.36 mg zinc, 167.0 mg sulphurand 165.0 mg chlorin (Gopalan *et al.*, 1999). Coefficient value of digestible protein and biological value is 77.6 per cent and 84.0 per cent, respectively. With respect to essential amino acids available in leaf are 350.0 mg arginine, 110.0 mg histidine, 300.0 mg lysine, 80.0 mg tryptophan, 300.0 mg phenyl alenine, 90.0 mg methionine, 200.0 mg threonine, 390.0 mg leucin, 330.0 mg isoleucin and 320.0 mg valine. Leaves contain bioflavonoids like kaempferol and 3′, 4′ diquercetin, which has medicinal importance. Sprouted seeds and micro greens are used as table purpose in salads. Apart from this leaves contain disogenine and fewer amounts of tigogenin and gitogenin (Rekha and Kulkarni, 1998). India is the major producer and exporter of fenugreek. The fenugreek plant has multidimensional role in food and nutritional and spice and have versatile medicinal properties. It also serves nutritious fodder for animals. The healthy crop can fix biological nitrogen in the soil up to 283 kg per hectare (Saini, 1996). Ayurvedic and Chinese system of medicines recognized its multifarious effects on human health and advocated its regular use for sound body and mind. In 10 important medicinal plants fenugreek is one (Hadda *et al.*, 2001)

Origin, History and Distribution

Fenugreek is considered to have originated South Eastern Europe and West Asia (particularly from middle Asia). Few research workers opined that fenugreek is first originated from Ethiopia. It is found growing wild in North Western India (Bailey, 1950). It is also considered that Eastern Europe and Ethiopia to be the places of origin of Fenugreek (Choudhury, 1967.). As per historical background, India and North Africa were the first to use it as spices and greens. It is very often grown as greens and fodder for animals near Russian Federation and other subtropical regions. It is cultivated in India, Argentina, Egypt, Southern France, Morocco, Spain, Turkey, China, Pakistan and Lebanon. In India it is the 3rd largest seed spice after coriander and cumin (Thangaraj and Vijayakumaran, 2004). India is one of the major producers and exporters of methi seeds to Saudi Arabia, Japan, Sri Lanka, Korea and United Kingdom. Main states producing fenugreek are Rajasthan, Madhya Pradesh (Malwa plateau region), Gujarat, Uttar Pradesh, Maharashtra and Punjab (Som and Maity, 1993). Rajasthan alone accounts for 80 per cent seed production of India. Major destinations for exports are Japan, Singapore, Saudi Arabia, Korea DP, Kuwait, Nepal and Yemen Arab republic.

Species Availability, Evolution and Diversity

The genus *Trigonella* is having 22 species, which are having different chromosomes numbers. The 15 species having 2n=16 chromosomes includes *T. angeuna, T. arabica, T. bolansae, T. calliceras, T. coerulea, T. corniculata, T. cretica, T. foenum-graecum, T. gladiota, T. glomerate, T. gordej, T. hamosa, T. lypokyi, T. monspeliach, T. radiata:* three species have 2n=18 (*T. stellata, T. striata, T. ornithopodioides*): two species have 2n=44, (*T. geminiflora, T. grandiflora*) (Fedorov, 1974) and another mixoploid species have 2n= 44 (*T. polycerata*).

Two species of *Trigonella* are of economic importance. These are:

1. *T. foenum-graecum* L., the common Methi. Plants are semi-erect, tall, moderately branched with bold, and typically yellow grains.
2. *T. corniculata* Kasuri or Champa methi. Plants are bushy and green. The flowers are bright orange yellow medium sized and pods are small and sickle shaped. It is found growing wild in North Western parts of India (Bailey, 1950).

Both the species are thought to be originated in South-Eastern-Europe and West Asia (Gangopadhyay *et al.*, 2009). They are found growing some wild species in North-Western India.

Naula *et al.* (2016) evaluated 28 genotypes (collected from different parts of India) of fenugreek along with two standard checks, namely, 'Pant Ragini' and 'Pusa Early Bunching' and estimated nutrient contents of each genotypes at GBPUA&T, Pantnagar (Uttarakhand). Highest potassium content was observed in PFG-35 (1207 mg/100g). PFG-37 and PFG-20 was richest in phosphorus, iron and manganese content among all genotypes. The lowest zinc content was found in PFG-26 (1.19 mg/100g), while the maximum zinc content was found in PFG-28 (4.43 mg/100g). Highest content of copper was found in PFG-26 (1.97 mg/100g). PFG-39 has the highest protein content (29.60 per cent). Significant differences were observed in the total saponin among the genotypes. Saponin content ranged from 0.38 g/100g to 1.31 g/100g. Therefore, the genotypes which are rich in nutraceutical properties can be used for plant biofortification, dietary supplements and herbal products.

Botanical Description

Fenugreek belongs to class Magnoliopsida, order Fabales family Fabaceae sub family Papilionaceae, genus *Trigonella* species *foenumgraecum* (common methi) or *corniculata* (kasuri methi or champa methi). They are diploid with 2n=16. These species differ in growth habit and several other morphological characters.

Table 8.26: Botanical Description

Growth Habit	It is an annual herb attaining plant height up to 70-80 cm. it produces erect shoots
Leaf	Leaves are dark or light green and are pinnately compound with or without pink margin.
Floral biology	The plants flower in about 30-37 days after sowing and the duration of flowering phase is about 7-18 days.Flowers are borne on axil of leaves which are whitish in common methi whereas bright orange to yellow or orange in kasuri or champa methi and bisexual (hermaphrodite) in nature. They are self-compatible and self fertile.
Anthesis and pollination	Strong tendency of self-pollination (clestogamous/ destogamous). Anthesis timing 9.00 am to 6.00 pm with peak between 11.30 am and 12.30 pm. Pollen fertility ranges from 95-98 per cent in the uoopened flower buds and 67-80 per cent in the open flowers. Stigma remains receptive 12 hours before and 10 hours after flower opening (Thangaraj and Vijayakumaran, 2004) and stigma emain receptive for about 10 hours after the opening of flowers.
Pod and pod set	The pods mature within 60-70 days of sowing. Pods are slender and straw coloured ripe, beak shaped and are about 8-10 cm long.
Seeds	Each pod has 8-15 yellowish brown coloured and smooth surfaced seed.

Horticultural Classification

Table 8.27: Difference between Common Methi and Kasuri Methi

Sl.No.	*Common methi (T. foenum graecum L.)*	*Kasuri methi (T. foenum corniculata)*
1.	It is quick growing	It is slow growing
2.	It produces shoots up to 50.0-70.0 cm in height	It shows rossetting habit

Sl.No.	Common methi (*T. foenum graecum L.*)	Kasuri methi (*T. foenum corniculata*)
3.	Rejuvenating power is low, so less number of cuttings in a season	Rejuvenating power is high, so varieties are multi-cut type
4.	Leaves are pear shaped and relatively light green	Leaves are lobular, round and dark green in colour
5.	Leaves are bitter and less scented	Leaves are relatively less bitter and give fragrance
6.	Generally 2-3 white flowers at the base of each leaf	Many flowers are produced on long flower stalk
7.	Flowers are white in colour	Flowers are yellow or orange in colour
8.	Pods are longer (8.0-10.0 cm) and straight	Pods are smaller and sickle shaped
9.	Seed size is bolder	Seed size is smaller
10.	Seed yield potential is 8-12 q/ha.	Seed yield potential is 5-7 q/ha.

Table 8.28: Number of Genotypes and Association of Traits

No. of Genotypes	*Salient Remarks*	*Reference(s)*
72	High variability for seed yield, number of pods per plant and protein content while low variability was recorded for days to flowering and number of grains per pod	Chandra *et al.* (2000)
29	Significant variability for days to flowering, plant height, pods per plant, seeds perpod, test weight, seed yield per plot, protein content and the level of incidence of powdery mildew disease. High GCV and PCV obtained for protein content, pods perplant, seed yield per plot and branches per plant.	Kaushik and Dashora (2001)
15	Highly significant differences among genotypes for all the character studied. High GCV and PCV obtained for straw yield, biological yield and grain yield perplant; moderatefor branches per plant, pods per plant and harvest index and low for the remaining charactes.	Saha and Kole (2001)
-	Seed yield per plant and number of pods per plant had high estimates of GCV and PCV followed by number of branches perplant and plant height.	Koli and Karan (2002)
39	The coefficient of variation was high for the economic and biological yield of the crop.	Verma and Korla (2003)
22	High GCV and PCV obtained for stem weight, moderate for plant height, branches per plant, days to flowering, duration of flowering, shelling per cent and test weight and low for pod length	Banerjee and Kole (2004)
8	All characters except primary branches per plant showed significant variation.	Datta and Chaterjee (2004)
4	Seed yield per plant and most of the components characters like pods per plant, 100 seed weight and seeds per pod were mainly under the control of non-additive gene action.	Raje (2004)
-	The magnitude of genotypic and phenotypic coefficient of variation ranged from 3.07 to 19.20 and 5.55 to 20.42 respectively.	Datta *et al.* (2005)
36	High GCV and PCV obtained for fat content, protein content and number of pods per plant.	Balai *et al.* (2006)
20	High to moderate values of genotypic and phenotypic coefficient of variation were observed in branches per plant, pods per plant, seeds per pod, 100 seed weight, straw yield, biological yield and grain yield per plant.	Kole and Mishra (2006)
30	Significant genetic differences were observed amongst genotypes for all the traits except for pod length.	Singh and Kaur (2007)
-	Significant genetic differences were observed amongst genotypes for all the traits.	Sharma and Sastry (2008)
40	Significant genetic differences were observed amongst genotypes for plant height, primary branches, days to 50 per cent flowering, pods per plant, grains per pod, days to maturity, 1000 seed weight and seed yield per plant.	Gangopadhyay *et al.* (2009)
60	Genotypic coefficients of variability were high for grain yield per plant, number of pods per plant and number of primary branches per plant.	Prajapati *et al.* (2010)
25	Estimates of GCV and PCV were high for number of branches/plant at 30 days after seeding.	Pushpa *et al.* (2012)
-	High GCV and PCV were high for number of secondary branches/plant followed by number of pods per axis, number of pods per plant, seed yield per plot, seed yield per plant, number of primary branches/plant, nodulation after 60 days of sowing, Chlorophyll content at 60 and 45 days after sowing, number of seeds per pod and vegetative yield.	Singh *et al.* (2011)
50	Observed significant genetic variaiation among the genotypes	Jain *et al.* (2012)
-	Observed significant genetic variaiation among the genotypes for all the characters studied. Estimates of GCV and PCV were high for seed yioeld/plant.	Yadav *et al.* (2013)
-	Observed significant mean squares due to genotype x season interaction for all the characters indicated differential performance of the genotypes with the change in growing conditions. Estimates of GCV were highest for seed yield/plant followed by pods per plant, branches per plant and seeds per pod.	Kole *et al.* (2014)
-	Estimates of GCV and PCV were high for seed yioeld/plant (g) followed by number of pods/plant.	Pathak *et al.* (2014)

No. of Genotypes	Salient Remarks	Reference(s)
22	Higher magnitude of coefficient of variation at genotypic level was observed for number of pods per plant, number of branches/plant, number of seeds/pod, plant height, seed yield per plant and length of pod whereas, at phenotypic level, number of pods/plant, number of brianches per plant, number of seeds per pod, plant height, seed yield per plant and length of pod revealed high magnitude of coefficient of variation.	Singh (2014)
30	Estimates of GCV and PCV were high for dry weight per plant at flower initiation (g) followed by straw yield per plant.	Gurjar *et al.* (2015)
-	Estimates of GCV and PCV were high for dry matter content and dry weight at flower initiation.	Lodhi *et al.* (2015)
50	Estimates of GCV and PCV were high for pods/plant, seed yield per plant and biological yield per plant.	Sharma *et al.* (2015)
	Estimates wide range of variability for all the traits under studies. Genotype UM-117 recorded the maximum primary branches (8) as well as secondary branches (6) per plant. Number of seeds per pod in different fenugreek Germplasm lines ranged between 7.3 and 22.2 with its average being 15.3 seeds.	Singh *et al.* (2015)

Table 8.29: Registered Germplasm of Fenugreek

Crop	Acc. No.	National Identity	Unique Traits
Fenu greek (Methi)	Hissar Methi-346	INGR-01012	Downy mildew resistant genotype with quick germination, faster initial growth, long pod, bold seed with green tan seed coat colour and light green narrow leaves
Fenu greek (Methi)	Hissar Methi-350	INGR-01012	Powdery mildew resistant and Downy mildew tolerant genotype. Plant is tall erected; multi branched with whitish green stems and dark green narrow leaves.

Augmentation of Genetic Resources and Crop Improvement

Very little effort has been made in collection, maintenance and utilization of different genotypes for the improvement of of crop. There is emence need to evaluate and improve the existing genotypes and introduced exotic cultivars for vegetable and seed purpose. Recently, few promising cultivars like AFg-5 (Ajmer Fenugreek-5) from NRC for seed spices, Ajmer (Rajasthan) and Jammu Fenugreek-07 from J&K Agril. Univ. (Jammu) has been developed.

There are several unique genotypes which are registered by ICAR-NBPGR, New Delhi and deposited in National Gene Bank for future use. Some of the lines are given in Table 8.29.

Fifty genotypes of fenugreek along with five standard checks namely Hisar Suvarna, Hisar Sonali, RMT-361, RMt-1 and AFg-3 were evaluated by Jain *et al.* (2016) and found significant variability.

Genetic Resource Conservation

Ample amount of genetic diversity exists in the fenugreek germplasm. The major focus of fenugreek genetic improvement programmes is on its productivity. Very little efforts have been made in collection, maintain or utilization of different genotypes for the improvement of this crop. National Researc centre on Seed spices has initiated trials on exploiting the potential of fenugreek as a vegetable crop.

REFERENCES

1. Bailey, L.H. 1950. The Standard Encyclopedia of Horticulture. The Mac Millian Co., New York.
2. Balai, O.P., Singh, D. and Jain, U.K. 2006. Genetic variation and character association among yield and yield related traits in fenugreek. *Indian Journal of Agricultural Research*, **40(2):** 143-146.
3. Banerjee, A. and Kole, P.C. 2004. Genetic variability, correlation and path analysis in fenugreek (*Trigonella foenum-graecum* L.). *Journal of Spices and Aromatic Crops*, **13(1):** 44-48.
4. Chandra, K., Divakara, S.E.V. and Singh, D. 2000. Genetic variation and character association of seed yield and its component characters in Fenugreek. *Agriculture Science Digest*, **20(2):** 93-95.
5. Choudhuy, B. 1967. Vegetables. National book Trust of India, New Delhi, p. 256-260.
6. Datta, S., Chatterjee, R. and Mukherjee, S. 2005. Variability, heritability and path analysis studies in fenugreek. *Indian Journal of Horticulture*, **62(1):** 96-98.
7. Datta, S. and Chaterjee, R. 2004. Performance of fenugreek genotypes under new alluvial zone of West Bengal. *Journal of Spices and Aromatic Crops*, **13(2):** 132-134.
8. Fedorov. 1974. Chromosome number of flowering plants. Atto Koeltz. Science Publishers, 624, Koening Stein, West Germany.
9. Gangopadhyay, K.K., Yadav, S.K., Kumar, G., Meena, B.L., Mahajan, R.K., Mishra, S.K. and Sharma, S.K. 2009. Correlation, path coefficient and genetic diversity pattern in fenugreek (*Trigonella foenum-graecum* L.). *Indian Journal of Agricultural Sciences*, **79(7):** 521-526.
10. Gopalan, G., Rama Sastri, B.V., Balasubramanian, S.C., Rao, B.S.N., Deosthale, Y.G. and Pant, K.C. 1999. Nutritive value of Indian Foods. National Institute of Nutrition, ICMR, Hyderabad, India.
11. Gurjar, M., Naruka, I.S. and Shaktawat, R.P.S. 2015. Variability and correlation analysis in fenugreek

(*Trigonella foenum-graecum* L.). *Legume Research*, DOI: 10.18805/lr.v0iOF.9286; 459-465.

12. Hadda, P.S., Depot, M., Settaf, A. and Cherrah, Y. 2001. Use of antidiabetic plants in Morocco and Quebec. *Diabetes Care*, **24(3):** 608-609.

13. Jain, A., Singh, B. and Kakni, R.K. 2016. Genetic variability and character association among yield and yield contributing in fenugreek (*Trigonella foenum-graecum* L.) germplasm. 7th Indian Horticulture Congress (Abs: 104, held at ICAR-IARI, New Delhi, 15-18 November, p. 80

14. Jain, A., Singh, B., Solanki, R.K., Saxena, S.N. and Kakani, R.K. 2012. Genetic variability and character association in fenugreek (*Trigonella foenum-graecum* L.). *International Journal of Seed Spices*, **3(2):** 22-28.

15. Kaushik, S.K. and Dashora, S.L. 2001. Variability assessment in M-lines of fenugreek (*Trigonella foenum-graecum* L.). *Journal of Spices and Aromatic Crops*, **10(1):** 45-47.

16. Kole, P.C., Goswami, T. and Duary, B. 2014. Performance of some genugreek genotypes in subhumid-subtropical red lateritic belt of Eastern India. *Tropical Agricultural Research and Extension*, **16(4):** 104-106.

17. Kole, P.C. and Mishra, A.K. 2006. Pattern of variability and associations among quantitative characters in fenugreek. *Indian Agriculturist*, **50(3/4):** 93-96.

18. Koli, N.R. and Karan, S. 2002. Estimation of genetic parameters in M_2 generation of fenugreek (*Trigonella foenum-graecum* L.). *Annals of Biology*, **18(2):** 211-212.

19. Lodhi, P.S., Singh, P.P., Naruka, I.S., Kushwaha, S.S. and Singh, A.K. 2015. Genetic variability, correlation and path analysis in fenugreek (*Trigonella foenum-graecum* L.). *Indian Journal of Horticulture*, **72(2):** 429-433.

20. Naula, R., Singh, A., Raghav, M., Singh, Y.V. and Lavlesh. 2016. Diversity in the nutraceutical properties of the various genotypes of fenugreek. (Abs. 58; 7th Indian Horticulture congress-2016), held at ICAR-IARI, New Delhi, p. 46

21. Pathak, A.R., Patel, A.I., Joshi, H.K. and Patel, D.A. 2014. Genetic variability, correlation and path coefficient analysis in fenugreek (*Trigonella foenum-graecum* L.). *Trends in Biosciences*, **7(4):** 234-237.

22. Prajapati, D.B., Ravindrababu, Y. and Prajapati, B.H. 2010. Genetic variability and character association in fenugreek (*Trigonella foenum-graecum* L.). *Journal of Spices and Aromatic Crops*, **19(1/2):** 61-64.

23. Pushpa, T.N., Chandregowda, M., Srikantaprasad, D. and Mallikarjuna, Gowda, A.P. 2012. Evaluation of fenugreek (*Trigonella foenum-graecum* L.) genotypes for growth and seed yield. *Crop Research*, **43(1, 2 and 3):** 238-244.

24. Raje, R.S. 2004. Gene action for seed yield and its components in fenugreek (*Trigonella foenum-graecum* L.). *Indian Journal of Genetics*, **64(4):** 335-336.

25. Rekha, S.J. and Kulkarni, P.R. 1998. Leafy vegetables. In: *Handbook of Vegetables Science and Technology*, Salunke, D.K. and Kadam, S.S. (Eds.). Marcel Dekker Inc., New York.

26. Saha, A. and Kole, P.C. 2001. Genetic variability in fenugreek (*Trigonella foenum-graecum* L.) grown in sub-humid lateritic belt of West Bengal. *Madras Agricultural Journal*, **88(4-6):** 348-350.

27. Sharma, K.C. and Sastry, E.V.D. 2008. Path analysis for seed yield and its components characters in fenugreek (*Trigonella foenum-graecum* L.). *Journal of Spices and Aromatic Crops*, **17(2):** 69-74.

28. Sharma, N., Shekhawat, A.S. and Nagariya, N.K. 2015. Genetic variability in fenugreek (*Trigonella foenum-graecum* L.). *Annals of Biology*, **31(1):** 94-96.

29. Singh, M.K. 2014. Genetic variability, heritability, genetic advance and correlation coefficient analysis in fenugreek (*Trigonella foenum-graecum* L.). *Hort. Flora Research Spectrum*, **3(2):** 178-180.

30. Singh, P.K., Singh, B., Tomar, B.S. and Naidu, A.K. 2015. Trait variation in fenugreek. *SABRAO Journal of Breeding and Genetics*, **47(4):** 413-423.

31. Singh, P. and Kaur, A. 2007. Genetic evaluation of methi (*Trigonella foenum-graecum* L.) for seed yield and quality attributes. *Crop Improvement*, **34(1):** 90-94.

32. Singh, P.P., Naruka, I.S. and Kushwaha, S.S. 2011. Genetic variability for seed yield and its component characters in fenugreek (*Trigonella foenum-graecum* L.). Abstract of National Symposium on 'Vegetable Biodiversity' held at JNKVV from April 4-5; p. 21-22.

33. Som, M. G. and Maity, S.P. 1993. Fenugreek. In: Vegetable crops (eds. Bose, T.K., som, M.G., and Kabir, J.) Naya Prokash, Kolkata (India).

34. Thangaraj, T. and Vijayakumaran, M. 2004. Fenugreek. In: Text Book of Vegetables, Tuber Crops and Spices (Thangraj, S. and Singh, N., eds.). ICAR, New Delhi, p. 364-371.

35. Verma, R. and Korla, B.N. 2003. Genetic variability in fenugreek (*Trigonella foenum-graecum* L.) grown under mid-hills of Himachal Pradesh. *Journal of Spices and Aromatic Crops*, **12(1):** 60-62.

36. Yadav, Y., Yadav, P.S., Pandey, V.P. and Kumar, A. 2013. Genetic variability, correlation and path coefficient analysis studies in fenugreek (*Trigonella foenum-graecum* L.). *The Asian Journal of Horticulture*, **8(2):** 456-459.

Genus *Amaranthus*

Amaranth (*Amaranthus* spp.)

Introduction

The word *Amaranthus* is basically derived from the Greek word 'Anthos' (flower) which means everlasting or unwilting. Amaranth is also known as Chinese spinach and Thotakura (Telugu). Amaranths (*Amaranthus spp.*) are mostly annual weeds but few are excellent nutritive leafy vegetable (cooked as a potherb) throughout the tropics and in many warm temperate regions due to fast growing habit, high yield potential, adverse temperature adaptation *etc.* Amaranths are the leading group of plants having great potentiality to alleviate malnutrition especially in the low income food deficient countries. Amaranth is easy to grow and very suitable to be fitted in crop rotations because of its very short duration nature. In India, it is an excellent hot weather substitute for spinach beet and spinach; generally grown during summer (when no other green vegetable are available in the market) and rainy seasons. Being a C_4 plant, it is highly efficient in biomass production. The harvest index is almost one in this crop. It has been increasingly recognized in recent years that grain amaranth offer perhaps the most practical means of solving the protein malnutrition problems in the Himalayas. The lysine content of this crop is twice that of wheat, and thrice that of rice and maize proteins. The per hectare production of protein is high as compared to cereals, which can be increased greatly if the present landraces are replaced by high yielding types. Its importance can be judged as: in warm humid tropic climates the yield of leaves alone amounts to 30.0 tonnes of fresh greens or 4.5 tonnes of dry matter per ha in 4 weeks from direct sowing (Peter, 1998). As a hot season vegetables, it is cultivated throughout the year in the tropics. Due to low production cost amaranth is one of the cheapest dark green leafy vegetables in tropical markets and is often described as poor man's vegetable (Grubben, 1976). Cooked amaranth is 90 per cent digestible. Amaranth has high amount of essential micronutrients (Ca, K, Fe, Cu, Mg, P and Mn); vitamins (vitamin A, vitamin B6, vitamin, vitamin, folate and riboflavin); dietary fiber (quickens digestion, lowers cholesterol levels) and a good source of protein, carotene, lysine and folic acid. Protein present in grain amaranths would be a valuable contributor to diet. Per 100 g fresh amaranth contains 85.7 per cent moisture, 4.0 g protein, 0.5 g fat, 1.0 g fat, 6.3 g carbohydrates, 2.7g minerals, 8340 µg IU vitamin A, 99.0 mg vitamin C, 0.03 mg thiamin, 0.10 mg riboflavin, 1.2 mg niacin, 99.0 mg ascorbic acid, 247 mg magnesium, 83.0 mg phosphorus, 230 mg sodium, 397 mg calcium, 341.0 mg potassium, 1.0 mg nicotinic acid, 61.0 mg sulphur, and 305.0 mg iron as per Aykroyd (1966). Amaranths leaf protein contains more lysine than the best high lysine corn and more methionine than soybean. It is also grown as a high protein grain crop. The fresh tender leaves and stem of amaranths are delicious when cooked like other fresh leafy vegetables. However there is moderately to high content of oxalic acid inhibits the absorption of calcium and zinc, and also means that they should be avoided or eaten in moderation by people with kidney disorders, gout, and rheumatoid arthritis. Reheating cooked amaranth leaves is often discouraged, particularly for consumption by small children, as the nitrates in the leaves can be converted to nitrites, similarly to spinach. The tiny seed of grain amaranths are popped or parched and milled to make flour or gruel. In taste, nutritional value and yield, the grain compares favorably with maize and other seed or oil may be of benefit for those with hypertension and cardiovascular disease; regular consumption reduce blood pressure and cholesterol levels, while improving antioxidant status and some immune parameters (Gonor *et al.*, 2006; Martirosyn *et al.*, 2007). *A. tricolor* is a major source for yellow and green dyes extraction. The red dye extracted from amaranth leaves is used to colour alcoholic beverages. The high fiber content makes it a low glycemic index food by controlling glucose levels. Amaranth has been noted to help in boosting the body's immune system. The juice of fresh amaranth leaves maintains the hair texture, adds a gloss, makes hair soft and manageable and helps in hair loss.

Origin, History and Distribution

Detemination of origin is very difficult for amaranth but it presumed that its progenitor (weedy form) may be originated from tropical regions. At least 50 tropical countries grow vegetable amaranths in quantities far from negligible. Some research workers opined that amaranth (main vegetable type-*A. tricolor* L.) might have originated in south or south-east Asia, particularly in India and have spread to neighboring countries by traders, Buddhist, monks and others (Muslim invaders). Another vegetable type amaranth species is *A. dubius* Mart. Ex. Thell., which shows diversity in Central America, Indonesia, India and in Africa. Perhaps this may be reason for popularity in Mexico, Central America, India, China, Nepal, Eastern Africa, South Pacific Islands and the Caribbean regions. *A. lividus* in south or central Europe. *A. hypochondricus* originated in North Western and central Mexico. *A. cruents* in south Mexico and Central America and *A. caudatus* in South America (Devadas and Mallika, 1991). Species like *A. retroflexus*, *A. virdis*, *A. lividus* and *A. gracecizans* are temperate types. *A. lividus* is quite popular in south and central Europe. It is believed that all the grain amaranths originated from Central and South America while among leafy types *A. tricolor* is originated south or South East Asia, particularly India. Historical backgrounds stress that Buddhist and Muslim people sperhaps, spread this crop, to other countries. *A. lividus* seems to have been a popular vegetable in Southern and Central Europe. In Africa *A. hybridus* and *A. thunberghii* are grown. *A. hybridus* is grown for grains and leaves in South Western United States, China, Africa, Nepal, South Pacific Islands, Carribean Islands, Greece, Italy and Russia.*A. hypochondriacus* and *A. cruents* are grown in Czech Republic. *A. cruents, A. hypochondriacus* and *A. caudatus* are grown in Argentina. *Amaranthus* species is a traditional vegetable of Zambia. *A. tricolor* is extensively grown in Southern Chili. *A. cruentus* is grown as a leafy vegetable in Africa. *A. caudatus, A. spinosus, A. gracilis* and *A. graecizan* are grown in Mozambique.

Amaranths occupy a unique position in the hill agriculture where the growing season is relatively short and the soil is not suitable for cultivating towards input oriented crops. In the Himalayas, little pathches of grain amaranth with colorful terminal inflorescences are a characteristic feature of landscape during autumn months. The distribution pattern of *Amaranthus* species showed a basic difference in the phsyological adaptation. *A. tricolor, A. dubius* and *A. lividus* (syn. *A. blitum*) are important leaf type grown in India. *A. caudatus* remains confined to the higher hills, whereas *A. cruentus, A. hypochondriacus* and *A. hybridus* to the mid hills and *A. spinosus* to the lower hills. *A. tricolor* is also grown in Himachal Pradesh, Uttar Pradesh, Uttarakhand, Punjab, Gujarat and Maharashtra. It is widely cultivated in Tamil Nadu, Kerala, and Karnataka. Amaranth is commercially grown during summer and rainy season in tropical Asia and Africa. In India numbers of domesticated forms are available all over the country especially in North Eastern States, West Bengal, Odisha, Tamil Nadu, Andhra Pradesh, Karnataka and Kerala. Wide ranges of variability in plant morphological characteristics are available in different parts of the country. In Odisha and North Eastern States a special type is grown were stem is consumed after full growth and different genotypes grown in different parts of the year. Amaranths are considered as the most popular vegetable crops grown in the humid tropical lowlands of Africa, Asia and Caribbean islands for their unique food value.

Species Availability, Evolution and Diversity

Though a number of species are consumed as vegetable the two species-*Amaranthus tricolor* and *A. blitum* are most important. Both are very rich in morphological variability represented by different morphotypes and varieties, conforming two species complex "tricolor complex" and "blitum complex" respectively. Based on taxonomical studies, the family is divided into two sections (Allen, 1961):

i. Amaranthus Saucer

ii. Blitopsis dumort

With nearly an equal number of species (Thellung, 1914). About 400 species of *Amaranthus* are distributed throughout the world in temperate, subtropical and tropical climate zones (Suma *et al.*, 2002; Anon., 1992). It is estimated that there are 87 species of genus *Amaranthus*: 17 in Europe, 14 in Australia and 56 in America (Mujica and Jacobsen, 1999; Jacobson *et al.*, 2000; Mujica *et al.*, 2000; Jacobson and Mujica, 2001). About 20 species are found cultivated/wild in India (Anonymous, 1992). Amaranth is cultivated for leaf (greens) and grains. However, due to the few studies on *Amranthus* systematics the number of species is still tentative. Brenner *et al.* (1999, 2000) mention 60 and USDA, ARS (1999) 86 species. Of the American species 10 are dioecious, only found in North America and 46 monoecious (Hunziker, 1992). *A. hybridus* L. whose synonym may be *A. quitensis* is a common ancestor to all three cultivated grain species (Chan and Sun, 1997).

The genus *Amaranthus* also unique in showing wide range of sexuality ranging from gynomonoecy, monoecy to dioecy and variability in mating behavior from self compatibility to obligate outcrossing. Monoecy in amaranths might have transmitted from co-sexual species of Cheno-Amaranth alliance through gynomonoecy.

Table 8.30: Geographical Distribution of Monoecious Species of Amaranths

Geographical Distribution	*Species*
America	*A. albus, A. blitoides, A. californicus, A. caracasanus, A. crassipes, A. cruentus, A. deflexus, A. hybridus, A. lividus, A. polygonoides, A. powelli, A. spinosus, A. squamulatus, A. retroflexus, A. urceolatus, A. viridis*
Only North America	*A. acutilobus, A. brandegei, A. chihuahuensis, A. fimbriatus, A. lepturus, A. mexicanus, A. obcordatus, A. pringlei, A. pumilis, A. scariousus, A. torreyi, A. vicidulus, A. wrighii*
Only Antilles, Central America and South America	*A. asplundii, A. cardenasianus, A. crispus, A. furcatus, A. kioosianus, A. lombardoi, A. looseri, A. minimus, A. muricatus, A. persimilis, A. peruvianus, A. quitensis, A. rosengurttii, A. scariosus, A. scierantoides, A. standleyanus, A. vulgatissimus*

Taxonomic delimitation in vegetable amaranth is still not clearly resolved. The presence of large number of morphotypes, varieties, landraces, overlapping morphological features, and large number of morphotypes have made taxonomic delimitation very complicated. Any idea about putative progenitor and phylogentic linkage among them is also lacking which are very important for proper utilization of genetic resources and sexual variability may be very instrumental in this aspect. In one of the population in Sarhan area, a rare morphotype was found having club-shaped inflorescence similar to that of *A. edulis* with pale white seeds. The variation in the distribution of inflorescence colour showed that the populations in the higher hills generally have uniform inflorescence colour, whereas those in lower to mid hills express more heterogeneity. This may be due to high rate of out-crossing because of high temperature in these areas. Morphological variability was evaluated morphometrically and new varieties and species were introduced (*A. tricolor* var. *acutus* S. Das; *A. bengalense* Saubhik Das and Lamonico, *A. parganensis* Saubhik Das).

The origin of various species of cultivated amaranth is complicated, because the wild ancestors are pantropical cosmopolitan weeds. The different species can be grouped on the basis of their utilization (Sauer, 1967 and 1993):

(i) Vegetable type
(ii) Grai type
(iii) Weed type
(iv) Other species

(I) Vegetable Types

There are mainly 6 leafy vegetable types *i.e.* *Amaranthus tricolor* L., *A. blitum* L., *A. spinosus* L., *A. dubius* Mart. Ex. Thell., *A. viridis* L., and *A. cruentus* L., *A. spinosus, A. hybridus* and *A. dubius* are tropical types. *A. retroflexus, A. viridis, A. lividus* and *A. graecizans* are more temperate hot season weeds. Due to easy adaptaion, amaranths are given local name irrespective to species consideration (Nath *et al.*, 1996). *A tricolor* L. has the synonyms *A. melancholicus, A. tristis, A. mangostanus* and *A. gangeticus* L. *A. blitum* is also called as *A. blitum* L. var. *oleraceus*(Peter, 1998) and *A. lividus* L. *A. cruentus* L. has the synonyms *A. paniculatus. A. viridis* is also known as *A. gracillis*.

(a) *Amaranthus tricolor* whose synonyms are: *A. gangeticus* L., *A. tristis* L., *A. mangostanus* L.
(b) *A. blitum* L. synonymous of *A. lividus* L.
(c) *A. dubius* Mart. Ex. Thell

(II) Grain Types

The grain Amaranths includes *A. hybridus, A. caudatus* (syn. *A. gracillis*), *A. retroflexus, A. albus* and *A. graecizens* (syn. *A. blitoides, A. angustifolius*).

(a) *A. caudatus* L. whose synonyms are: *A. caulis* Spegazzini, *A. Mantegacianus* Passerini
(b) *A. hypochondriacus* L. whose synonyms are: *A. leucocarpus* S. Wats; *A. flavus* L.
(c) *A. cruentus* L. whose synonym is: *A. paniculatus* L.

(III) Weed Species

The weed species are *A. retroflexus* and *A. hybridus*.

(a) *A. viridis* L. synonyms of *A. gracilis* Desf.
(b) *A. spinosus*
(c) *A. retroflexus*
(d) *A. hybridus*
(e) *A. powellii*

(IV) Other Species

Other species of *Amaranthus* are *A. leucocarpus* and *A. caudatus* (syn. *A. mantegazzianus, A. edulis*).

Table 8.31: Taxon and Preferred Taxon Name of Genus *Amaranthus*

Name of Taxon	*Name of Authority*	*Family*	*Preferred Taxon Name*
Amaranthus acanthochiton	J. D. Sauer	Amaranthaceae	*Amaranthus acanthochiton*
Amaranthus acutilobus	Uline and W. L. Bray	Amaranthaceae	*Amaranthus acutilobus*
Amaranthus aff. *blitum*		Amaranthaceae	*Amaranthus* aff. *blitum*
Amaranthus albus	L.	Amaranthaceae	*Amaranthus albus*
Amaranthus arenicola	I. M. Johnst.	Amaranthaceae	*Amaranthus arenicola*
Amaranthus ascendens	Loisel.	Amaranthaceae	*Amaranthus blitum*
Amaranthus asplundii	Thell.	Amaranthaceae	*Amaranthus asplundii*
Amaranthus asplundii var. *asplundii*		Amaranthaceae	*Amaranthus asplundii* var. *asplundii*
Amaranthus asplundii var. *australis*	Hunz.	Amaranthaceae	*Amaranthus asplundii* var. *australis*
Amaranthus australis	(A. Gray) J. D. Sauer	Amaranthaceae	*Amaranthus australis*
Amaranthus blitoides	S. Watson	Amaranthaceae	*Amaranthus blitoides*
Amaranthus blitum	L.	Amaranthaceae	*Amaranthus blitum*
Amaranthus blitum subsp. *blitum*		Amaranthaceae	*Amaranthus blitum* subsp. *blitum*
Amaranthus blitum subsp. *emarginatus*	(Moq. ex Uline and Bray) Carretero *et al.*	Amaranthaceae	*Amaranthus blitum* subsp. *emarginatus*
Amaranthus blitum subsp. *oleraceus*	(L.) Costea	Amaranthaceae	*Amaranthus blitum* subsp. *oleraceus*
Amaranthus blitum var. *emarginatus*	(Moq. ex Uline and Bray) Costea	Amaranthaceae	*Amaranthus blitum* var. *emarginatus*
Amaranthus blitum var. *oleraceus*	(L.) Hook. f.	Amaranthaceae	*Amaranthus blitum* subsp. *oleraceus*
Amaranthus blitum var. *polygonoides*	Moq.	Amaranthaceae	*Amaranthus graecizans* subsp. *thellungianus*
Amaranthus blitum var. *pseudogracilis*	(Thell.) Costea	Amaranthaceae	*Amaranthus blitum* var. *pseudogracilis*
Amaranthus bouchonii	Thell.	Amaranthaceae	*Amaranthus powellii* subsp. *bouchonii*

Name of Taxon	*Name of Authority*	*Family*	*Preferred Taxon Name*
Amaranthus brownii	Christoph. and Caum	Amaranthaceae	*Amaranthus brownii*
Amaranthus buchtienianus	Thell.	Amaranthaceae	*Amaranthus asplundii* var. *asplundii*
Amaranthus californicus	(Moq.) S. Watson	Amaranthaceae	*Amaranthus californicus*
Amaranthus cannabinus	(L.) J. D. Sauer	Amaranthaceae	*Amaranthus cannabinus*
Amaranthus cardenasianus	Hunz.	Amaranthaceae	*Amaranthus cardenasianus*
Amaranthus caudatus	L.	Amaranthaceae	*Amaranthus caudatus*
Amaranthus caudatus subsp. *mantegazzianus*	(Pass.) ined.	Amaranthaceae	*Amaranthus caudatus*
Amaranthus caudatus subsp. *saueri*	V. Jehlik	Amaranthaceae	*Amaranthus caudatus*
Amaranthus caudatus var. *albiflorus*	Moq.	Amaranthaceae	*Amaranthus caudatus*
Amaranthus caudatus var. *alopecurus*	Moq.	Amaranthaceae	*Amaranthus caudatus*
Amaranthus chlorostachys	Willd.	Amaranthaceae	*Amaranthus cruentus*
Amaranthus chlorostachys	auct.	Amaranthaceae	*Amaranthus hybridus*
Amaranthus crassipes	Schltdl.	Amaranthaceae	*Amaranthus crassipes*
Amaranthus crispus	(Lesp. and Thevenau) N. Terracc.	Amaranthaceae	*Amaranthus crispus*
Amaranthus cruentus	L.	Amaranthaceae	*Amaranthus cruentus*
Amaranthus deflexus	L.	Amaranthaceae	*Amaranthus deflexus*
Amaranthus dubius	Mart. ex Thell.	Amaranthaceae	*Amaranthus dubius*
Amaranthus dussii	Sprenger	Amaranthaceae	*Amaranthus caudatus*
Amaranthus edulis	Speg.	Amaranthaceae	*Amaranthus caudatus*
Amaranthus emarginatus	Moq. ex Uline and Bray	Amaranthaceae	*Amaranthus blitum* subsp. *emarginatus*
Amaranthusemarginatus subsp. *pseudogracilis*	(Thell.) Hugin	Amaranthaceae	*Amaranthus blitum* var. *pseudogracilis*
Amaranthus fimbriatus	(Torr.) Benth. ex S. Watson	Amaranthaceae	*Amaranthus fimbriatus*
Amaranthus floridanus	(S. Watson) J. D. Sauer	Amaranthaceae	*Amaranthus floridanus*
Amaranthus frumentaceus	Buch.-Ham. ex Roxb.	Amaranthaceae	*Amaranthus hypochondriacus*
Amaranthus gangeticus	L.	Amaranthaceae	*Amaranthus tricolor*
Amaranthus gangeticus var. *melancholicus*	(L.) Voss	Amaranthaceae	*Amaranthus tricolor*
Amaranthus gracilis	Desf., nom. nud.	Amaranthaceae	*Amaranthus viridis*
Amaranthus graecizans	L.	Amaranthaceae	*Amaranthus graecizans*
Amaranthus graecizans	auct.	Amaranthaceae	*Amaranthus blitoides*
Amaranthus graecizans subsp. *graecizans*		Amaranthaceae	*Amaranthus graecizans* subsp. *graecizans*
Amaranthusgraecizans subsp. *silvestris*	(Vill.) Brenan	Amaranthaceae	*Amaranthus graecizans* subsp. *silvestris*
Amaranthus graecizans subsp. *thellungianus*	(Nevski) Gusev	Amaranthaceae	*Amaranthus graecizans* subsp. *thellungianus*
Amaranthus graecizans var. polygonoides	(Moq.) Aellen	Amaranthaceae	*Amaranthus graecizans* subsp. *thellungianus*
Amaranthusgraecizans var. *silvestris*	(Vill.) Asch.	Amaranthaceae	*Amaranthus graecizans* subsp. *silvestris*
Amaranthus greggii	S. Watson	Amaranthaceae	*Amaranthus greggii*
Amaranthus hybrid		Amaranthaceae	*Amaranthus hybrid*
Amaranthus hybridus	L.	Amaranthaceae	*Amaranthus hybridus*
Amaranthus hybridus subsp. *cruentus*	(L.) Thell.	Amaranthaceae	*Amaranthus cruentus*
Amaranthus hybridus subsp. *quitensis*	(Kunth) Costea and Carretero	Amaranthaceae	*Amaranthus quitensis*
Amaranthus hybridus var. *erythrostachys*	Moq.	Amaranthaceae	*Amaranthus hypochondriacus*
Amaranthus hybridus var. *hypochondriacus*	(L.) B. L. Rob.	Amaranthaceae	*Amaranthus hypochondriacus*
Amaranthus hypochondriacus	L.	Amaranthaceae	*Amaranthus hypochondriacus*

Name of Taxon	Name of Authority	Family	Preferred Taxon Name
Amaranthus incurvatus	Timeroy ex Gren. and Godr.	Amaranthaceae	*Amaranthus hybridus*
Amaranthus leucocarpus	S. Watson	Amaranthaceae	*Amaranthus hypochondriacus*
Amaranthus lividus	L.	Amaranthaceae	Amaranthus blitum subsp. *oleraceus*
Amaranthus lividus f. *pseudogracilis*	Thell.	Amaranthaceae	*Amaranthus blitum* var. *pseudogracilis*
Amaranthus macrocarpus	Benth.	Amaranthaceae	*Amaranthus macrocarpus*
Amaranthus mangostanus	L.	Amaranthaceae	*Amaranthus tricolor*
Amaranthus mantegazzianus	Pass.	Amaranthaceae	*Amaranthus caudatus*
Amaranthus melancholicus	L.	Amaranthaceae	*Amaranthus tricolor*
Amaranthus muricatus	(Moq.) Hieron.	Amaranthaceae	*Amaranthus muricatus*
Amaranthus oleraceus	L.	Amaranthaceae	*Amaranthus blitum* subsp. *oleraceus*
Amaranthus palmeri	S. Watson	Amaranthaceae	*Amaranthus palmeri*
Amaranthus paniculatus	L.	Amaranthaceae	*Amaranthus cruentus*
Amaranthus patulus	Bertol.	Amaranthaceae	*Amaranthus hybridus*
Amaranthus polygamus	L.	Amaranthaceae	*Amaranthus tricolor*
Amaranthus polygonoides	L.	Amaranthaceae	*Amaranthus polygonoides*
Amaranthus powellii	S. Watson	Amaranthaceae	*Amaranthus powellii*
Amaranthus powellii subsp. *bouchonii*	(Thell.) Costea and Carretero	Amaranthaceae	*Amaranthus powellii* subsp. *bouchonii*
Amaranthus powellii subsp. *powellii*		Amaranthaceae	*Amaranthus powellii* subsp. *powellii*
Amaranthus pumilus	Raf.	Amaranthaceae	*Amaranthus pumilus*
Amaranthus quitensis	Kunth	Amaranthaceae	*Amaranthus quitensis*
Amaranthus retroflexus	L.	Amaranthaceae	*Amaranthus retroflexus*
Amaranthus rudis	J. D. Sauer	Amaranthaceae	*Amaranthus tuberculatus*
Amaranthusscariosus	Benth.	Amaranthaceae	*Amaranthus scariosus*
Amaranthus silvestris	Vill.	Amaranthaceae	*Amaranthus graecizans* subsp. *silvestris*
Amaranthus sp.		Amaranthaceae	*Amaranthus* sp.
Amaranthus spinosus	L.	Amaranthaceae	*Amaranthus spinosus*
Amaranthus standleyanus	Parodi ex Covas	Amaranthaceae	*Amaranthus standleyanus*
Amaranthus tenuifolius	Willd.	Amaranthaceae	*Amaranthus tenuifolius*
Amaranthus thellungianus	Nevski	Amaranthaceae	*Amaranthus graecizans* subsp. *thellungianus*
Amaranthus torreyi	(A. Gray) Benth. ex S. Watson	Amaranthaceae	*Amaranthus torreyi*
Amaranthus torreyi var. suffruticosus	Uline and W. L. Bray	Amaranthaceae	*Amaranthus watsonii*
Amaranthus tricolor	L.	Amaranthaceae	*Amaranthus tricolor*
Amaranthus tricolor subsp. mangostanus	(L.) Thell.	Amaranthaceae	*Amaranthus tricolor*
Amaranthus tricolor subsp. *tristis*	(L.) Thell.	Amaranthaceae	*Amaranthus tricolor*
Amaranthus tuberculatus	(Moq.) J. D. Sauer	Amaranthaceae	*Amaranthus tuberculatus*
Amaranthus viridis	L.	Amaranthaceae	*Amaranthus viridis*
Amaranthus watsonii	Standl.	Amaranthaceae	*Amaranthus watsonii*
Amaranthus wrightii	S. Watson	Amaranthaceae	*Amaranthus wrightii*

Among the leafy types, *A. tricolor* L. is the main cultivated species in India. Other cultivated species are *A. blitum*, *A. tristis* and *A. dubius*. *A. tricolor* is an ornamental and is often grown in flower gardens.

Mallika (1987) described the morphological characters of eight species of *Amaranthus* and developed a simple key to identify these species.

Mallika (1987) developed a simple key to identify *Amaranthus* species, which are given in Table 8.33.

Cytological studies revealed that cryptic structural changes and genetic drift were involved in cytogenetic differentiation of the two species *i.e. A. viridis A. spinosus*. The primitiveness of *A. spinosus* was evident from the study and this cosmopolitan weed played a major role in the evolution of other Amaranthus species (Mallika, 1987).

Table 8.32: Morphological Characters of some *Amaranthus* Species

Sl.No.	Species	Habit	Plant Height (cm)	Leaf Colour	Stem and Petiole Colour	Laeaf Shape	Inflorescence Position	Density of Florets	Seed Colour
1.	*A. tricolor*	Erect	100	Green with deep purple centre	Green	Broad ovate	Axillary and small terminal panicle	Dense	Black
2.	*A. dubius*	Erect, often branched	120	Green	Pale green	Rhombio-ovate	Huge, terminal and axillary panicle	Dense	Black
3.	*A. lividus*	Prostrate, well branched	40	Pale green	Pale green	Rhomboid	Axillary and small terminal panicle	Dense	Black
4.	*A. hypochondriacus*	Erect, unbranched	65	Purplish green	Purplish green	Elliptic	Huge terminal panicle	Dense	Cream
5.	*A. cruentus*	Erect, unbranched	150	Red/Green	Red/Green	Ovate lanceolate	Huge terminal panicle	Dense	Browinsh
6.	*A. caudatus*	Erect, unbranched	120	Pale green	Pale green	Ovate lanceolate	Huge terminal panicle	Lax	Pinkish
7.	*A. viridis*	Erect, branched	50	Green	Green	Deltoid	Slender, terminal and axillary panicle	Lax	Black
8.	*A. spinosus*	Erect, branched	80	Dull red	Dull red	Ovate lanceolate	Slender, terminal and axillary panicle	Lax	Black

Table 8.33. Key for identification of *Amaranthus* species

			Species
A	Flowers in cymes mostly clustered in leaf axils, stamens and perianth parts-3		
	B.	Utricle circmscissile, concealed by long and bristle like perianth lobes	*A. tricolor*
	BB.	Utricle biconvex, indehiscent, partially covered by short and narrow perianth lobes, leaves emarginated	*A. lividus*
AA	Flowers in cymes mostly clustered on slender terminal panicles; a few axillary clusters also present. stamens as many as perianth lobes		
	B.	Plants unarmed, leaves deltoid-ovate, utricle rugose and indehiscent; stamens 3	*A. viridis*
	BB.	Plants armed, bracts bristle tipped, distal cymes staminate,and basal cymes pistillate, stamens 5	*A. spinosus*
AAA	Flowers in cymes mostly clustered on large terminal panicles, stamens and perianth parts-5		
	B.	Each cyme with initial staminate flower and remainder pistillate	
	C.	Inflorescence smooth, lateral branches divergent, often with axillary female flowers	*A. dubius*
	CC.	Inflorescence stiff, lateral branches ascending, style branches thick at the base, leaves elliptic	*A. hypochondriacus*
	CCC.	Inflorescence moderately stiff, lateral branches divergent, utricle neck elongated, anthers often hooded, leaf base often lyrate	*A. cruentus*
	BB.	Each cyme with more staminate flowers. Inflorescence smooth, and drooping, lateral branches ascending, female flowers cup shaped, perianth lobes overlapping, style branches spreading	*A. caudatus*

Gene and Genetics of Characters

In amaranths not much information is available on genetics of characters. Red plant pigmentation (RR) is dominant to green (rr). Spineness is dominant to non-spiny on leaf. Dominant green perianth colour could be employed as marker gene to detect the extent of cross pollination in grain type.

Table 8.34: Importyant Gene Coding for different Characters

Trait	Gene	Source
Male fertile/sterile	Ms1/ms1	*A. hypochondriacus*
Leaf blade red/green	R/r	*A. tricolor*
Stem red/green	A/a	*A. tricolor*
Stem green/white	B/b	*A. tricolor*
Seed coat black/pale	P/p	*A. hypochondriacus*
Tall/dwarf	Dw/dw	*A. caudatus*

Botanical Description

Amaranth belongs to family Amaranthaceae, fall under genus *Amaranthus*. It is an annual herb rarely perennial:

(i) **Monoecious-**Most of the cultivated species are monoecious. (subg. *Amaranthus* and *Albersia*) or

(ii) **Dioecious** (subg. *Acnida*), glabrous or pubescent.

Table 8.35: Botanical Description

Growth Habit	It is an annual herb and plants are ascending, decumbent, or prostrate. Mostly unarmed and rarely armed, scarce to profuse branched.
Root	In this crop tap root system develops which goes to deeper portion of soil.
Stem	Stem erect, usually branched, occasionally simple or nearly so; without nodal spines (except in *A. spinosus*), green to purple or with mixed shades of these two colours. Stems are glabrous and succulent.
Leaf	Leaf simple, alternate or opposite, exstipulate, entire, obovate to lanceolate, with acute tip, green to purple or with different shades of these two colours, purple colouration prominent in young leaves and fades away at maturity.
Floral biology	Inflorescences is terminal and/or axillary or exclusively terminal, compound dichasia arranged in spikes, thyrses, panicles, or glomerules; components of terminal inflorescences often subtended by reduced leaves (pseudobracts), each dichasial cymes called 'glomerule', unit subtended by persistent bracts. Bracts ovate, lanceolate, linear, subulate, deltate, or broadely triangular (in *A. acanthochiton*), or proximal bracts modified into spines (in *A. spinosus*); bracts of pistillate flowers not keeled (keeled in *A. scleropoides* and *A. crassipes*); bracteoles absent or 1-2. Flowerare small, regular, mostly unisexual, monoecious (proportion of male and female flowers varies in the inflorescence), pentamerous, bracteates and bracteolate which are scaly, perianth part 4-5, imbricate, membraneous, often persistent; stamens 2-5, placed opposite to perianth parts, often some reduces to staminodes, filament either free or united blow, anthers one or two celled; carpel2-3, syncarpous, ovary superior with commonly one campylotropus ovule, and style to 2 or 3 branched. Each glomerule consists of an initial staminate flower and an indefinite number of pistillate flowers. In monoecious species, including grain amaranths two types of arrangement of the staminate and pistillate flowers in the glomerule of the inflorescence. This pattern is seen in all species except *A. spinosus*, the first flower of each glomerule is staminate and the rest are pistillate. In *A spinosus*, each glomerule bears flowers of the same sex with the pistillate flowers in the axils of the branches and at the base of the terminal inflorescence, while the staminate flowers are terminal on the main axis and lateral branches. In *A. spinosus,* all flowers of one cluster are of same sex. The percentage of male flowers per glomerule is 0.5-1.0 in grain types and 10.0-25.0 in leafy types.
Anthesis and pollination	The extent and degree of cross-pollination (due to anemophil) is governed by the proportion of the staminate and pistillate flowers vary in an inflorescence and also the position of the inflorescence in the plant. Due to protogynous nature, allogamy is reported and isolated plants expressed autogamy. It is fact that due development of more pollen grains, self pollination take place. Grain types of amaranths favour cross-pollination, while vegetable types Blitopsis are self-pollinated. In grain type cross pollination ranges from 16-35 per cent (Walton, 1968). Pal and Khoshoo (1973) are of the opinion that vegetable amaranths are predominantly self-pollinated due to the presence of a number of male flowers per glomerale, terminal inflorescence and development of axillary glomerules. They also stated that the stigma of the pistillate flowers is receptive several days prior to opening of the grain species with colourful inflorescence are occasionally visited by bees. The grain species and their close relatives are monoecious and abundant pollen production in these species leads to a high degree of self-pollination. Arrangement and sequence of anthesis of the unisexual flowers favour a combination of self and cross-pollination (Sauer, 1976, 1980). From the breeding point of view, the small closely grouped flowers in glomerules make emasculation extremely difficult. The most satisfactory method of effecting crosses is to pollinate heavely as soon as the stigma is receptive and to remove the staminate flowers by hand.
Fruit and fruit set	Fruit is a utricle, indehiscent or circumscissile enclosed by the persistent perianth parts.
Seeds	Seeds solitary. Seed varying in colour, black, brown or white compressed, smooth, shiny with floury endosperm.

Amaranth is highly cross-pollinated due to monoecy and anemophily. Protogynous nature of plants also leads to allgamy. In isolated plants, autogamy also occurs. Abundant pollen production also leads to a high degree of self-fertilization. Singh and Thomas (1978) suggested that the genus *Amaranthus* possessed 3 types of mating system:

(i) Obligate out crossing occurred in the dioecious species

(ii) Relatively greater out crossing in monoecious species of section *Amaranthus*

(iii) Relatively greater self-pollination in section *Blitopsis*

A. dubius is a tetraploid species having chromosome number 2n=4x=64. All other species are diploid. *A. tristis* and *A. blitum* have chromosome number 2n=2x=32 whereas *A. tricolor* has chromosome number 2n=2x=34. Similarly, *A. viridis*is having chromosome number 2n=2x=34. *A. cruents* is having chromosome number 2n=2x=32. Mallika (1987) studied cytogenetics of 8 species of *Amaranthus* (*A. tricolor, A. lividus, A. viridis, A. spinosus, A. dubius, A. hypochondriacus, A. cruentus and A. caudatus*) and their hybrids. *A. tricolor, A. lividus, A. viridis* belong to section *Blitopsis* and others to section *Amaranthus*. Meiosis was normal in all sections with regular formation of bivalents. Out of 8 species, 7 were diploids with n=16 or 17. *A. dubius* was a polyploidy with n=32. A preponderance of ring bivalents over rods was observed in all the species except *A. spinosus*. All the species under section *Blitopsis* had n=17. Maximum number of chiasma was observed in *A. tricolor*. *A. dubius* behaved as an allopolyploid with 32 bivalents. The lowest chiasma frequency/bivalent and per pollen mother cell (PMC) and the highest number of rod bivalents were observed in *A. spinosus*. All this indicate the primitive wild nature of *A. spinosus*. Pollen fertility was high in all the species ranging from 83.4 per cent in *A. viriodis* to 93.8 per cent in *A. lividus* (Mallika, 1987).

There are 2 chromosome groups in amaranths:

Table 8.36: Species and their Chromosome Number

Chromosome Number N=	*Chromosome Number 2N=*	*Name of Species*
16	32	*A. cruentus, A. tristis* and *A. blitum*
17	34	*A. tricolor, A. spinosus, A. viridis* and *A.blitum*
32	64	*A. dubius* (polyploid)

Horticultural Classification

Schenz (1934) grouped Amaranth in to two types:

(i) Amaranthotypus Dumort–Sauer (1967) renamed the section Amaranthotypus Dumort as section *Amaranthus* Sauer. The section *Amaranthus* Sauer has only terminal flower clusters and includes many of the domesticated ornamentals, grain types, dye amaranths and weeds.

(ii) Blitopsis Dumort-The section Blitopsis Dumort has flower clusters in axils and called as vegetable *Amaranthus* (Pal and Khoshoo, 1972; Madhusoodanan and Pal, 1981).

Feine (1980) has given few keys to describe vegetable species of genus *Amaranthus.*

Table 8.37: Chracterization of Species on the Basis of Plants Parts

Keys	*Descriptions*	*Species*
A	Flowers unisexual	
B	Tepals 3	
C	Tepals equal to or longer than utricle; utricle circumscissile	*A. tricolor*
C	Tepals shorter than utricle; utricle indehiscent	
D	Utricle smooth	*A. blitum*
D	Utricle rugose	*A. viridis*
B	Tepals 5	
E	Tepals approximately equal in length and incurved against utricle	
F	Plants armed; inflorescence with upper cymes staminate and lower cymes pistillate	*A. spinosus*
F	Plants unarmed; cymes with initial flower staminate and rest pistillate	*A. dubius*
E	Inner tepals shorter than outer, tepals straight or recurved from utricle	
G	Bracts not exceeding style branches; inflorescence fully developed, enormous seeds usually light, sometimes dark	
I	Bracts shorter than style branches; inflorescence lax	
J	Utricle narrowing into tower at apex; style branches erect; tepals with apex acute	A. cruentus

A. Tricolor, A. dubius, A. lividis and *A. tristis* are common leafy vegetables in Asian countries.

Grouping of Amaranths on the Basis of Chromosome Number

There are 2 chromosome groups (Khoshoo and Pal, 1973) in amaranths *i.e.*

(i) n=16-The species with n=16 is *A. cruentus*.

(ii) n=17- The species with n=17 are *A. tricolor, A. spinosus, A.viridis,* and *A. blitum*.

Grouping of Amaranths on the Basis of Pollen Types

There are 3 pollen types which have been identified for grouping of amaranth species *i.e.*

(i) Type A- In type 'A' micrograins are characteristics of hybrids between *A. spinosus* and *A. dubius*.

(ii) Type B-In type 'B'pollen grains with small pores are characteristics of *A. spinosus*.

(iii) Type C-In type 'C'pollen grains with large pores arecharacteristics of *A. dubius*.

Grouping of Amaranths on the Basis of Leaf Colour

There are two types of varieties based on leaf colour *i.e.*

(i) Green leaf and

(ii) Red leaf

More work is needed to determine the taxonomical status of this variable form.

Augmentation of Genetic Resources and Crop Improvement

Breding Priorities

- Vigorous plant growth, quick growing, strong succulent stem and large leaves
- Higher yield
- High nutritional profile
- Low oxalate and nitrate
- Better and quick rejuvenation ability after cuttings

Table 8.38: The D^2 Clusters and the Entries of Amaranths

Cluster Number	*Total Accessions*	*Accessions Number*
I	1	IC35553
II	124	VDV-4/45, IC35469, IC35735, IC35454, T89-13, IC35472, IC35481, IC35438, IC35604, EC1493, V-1989, IC35389, IC35465, IC35443, IC35551, DPP/3/50, IC35470, IC35518, DPP/2/29, NKG-88/42-1, DPP/3/38, IC35461, BDJ-1-847, IC35426, IC35468, IC35463, V-2008, SDI-193, IC35406, IC35781, IC35479, IC35422, BDJ-1-459, DPP/2/46, IC35712, IC35476, IC35452, IC35408A, IC35477, IC35395, IC35499, IC35556, IC35568, IC35539, IC21796, IC35565, IC35455, T-66-49, IC21930, IC35430, IC35632, IC35440, T-98-34, IC35638, SDI-93, IC35445, DO-673, IC21964, BD/NKG-108, IC35736, IC42256, BDJ-1-520, BDJ-1-635, IC35440, BDJ/NKG-62, V-217, BDJ-1/962, N-955, IC21796A, AO-941, P-57, BDJ/NKG-253, BDJ/NKG-6, P-63, IC35411, SDI-291, IC35672, IC35766, IC35755, SDI-280, IC35577, IC35741, IC35684, IC35473, IC35570, IC35574, IC35415, IC5564, IC 35421, VDV-3/89/71, T-59-14, IC 35385, IC 35193, IC 35410, V-1958, IC 35559, IC 35421, VDV-3/89/71, T-59-14, IC 35385, IC 35193, IC 35410, V-1958, IC 355535742, IC 35421, VDV-3/89/71, T-59-14, IC 35385, IC 35193, IC 35410, V-1958, IC 355535706, IC 35421, VDV-3/89/71, T-59-14, IC 35385, IC 35193, IC 35410, V-1958, IC 355535579, IC 35421, VDV-3/89/71, T-59-14, IC 35385, IC 35193, IC 35410, V-1958, IC 355535653, IC 35421, VDV-3/89/71, T-59-14, IC 35385, IC 35193, IC 35410, V-1958, IC 355535696, IC 35421, VDV-3/89/71, T-59-14, IC 35385, IC 35193, IC 35410, V-1958, IC 355535505, BDJ/NKG-388, BDJ-11-74, BDJ-364, BDJ/NKG-145, BDJ/NKG-382, BDJ/NKG-241, BDJ/NKG-341, C1643, KPS/88/176, IC 35546, C-1723, IC35659, BDJ/89/384, BDJ/89/364, NKG/88/42-5, IC35450, BDJ/NKG-322A, BDJ/NKG-149, V-2046, IC42293, IC35364
III	12	IC35397, T-89-12, IC35564, DPP/2/20, U110, VDV3/89-46, IC35399, IC35496, IC35447, IC35655, IC35436, SD2-52
IV	1	IC35552
V	1	SDI-66
VI	1	SDI-85
VII	1	IC32181,
VIII	1	IC35572
IX	1	IC35380
X	1	Annapurna

Source: Lohitaswa *et al.*, 1994.

- ✰ Good flavor and taste
- ✰ Wider adaptability to varying climate

For example amaranths offer a great deal of diversity in terms of various traits of economic importance for collection and utilization (Joshi and Mehra, 1983). In order to preserve this diversity of amaranths several region specific explorations were undertaken in the last four decades by the ICAR-NBPGR, Regional Station, Shimla (Himachal Pradesh). During these explorations, diverse ecological regions comprising Shimla, Solan, Bilaspur, Kullu, Mandi, Lahual and Spiti, Sirmour, Chamba, Kangra and Kinnaur of Himachal Pradesh; Uttar Kasi, Pauri, Almora, Pithoragarh and Nainital of Uttarakhand; J&K; Sikkim, Darjeeling and North East Hill Regions were covered. Intensive explorations resulted collection of nine species *viz. A. hypochondriacus, A. cruentus, A. cauda, A. hybridus, A. spinosus, A. dubius, A. tricolor, A.viridis* and *A. blitum*.

A great deal of variability with respect to plant height, days to flowering, days to maturity, seed shape, size and colour was collected (Joshi, 1986). The collections representing Uttarakhand and Uttar Pradesh hills are highly yielder with large inflorescences as compared to the collections from Himachal Pradesh. Non-shattering types were collected from Shyalbagh area of Kulu district, whereas large seeded types (more than 1.00 g/1000 seed weight) were also collected from Shimla and Kinnaur districts. Dwarf types collected from Kalpa and Fagu areas would be of great importance to develop non-lodging types. In Sarhan and Kotkhai areas, natural aneuploids were also found in some population creatying a great deal of taxonomic complexities in the genus. In one of the population in Sarhan area, a rare morphotype was found having club-shaped inflorescence similar to that of *A. edulis* with pale white seeds. A total of 2810 indigenous germplasm collections were augmented with 315 exotic collections which also include 25 different species.

Introduction and Selection

'Choti chauli' and 'Badi chauli', 'Pusa Kirti' and 'Pusa Kiran' from the Indian Agricultural Research Instituite, New Delhi, and 'CO-1', 'CO-2' and 'CO-3'from Tamil Nadu Agricultural University, Coimbatore (Tamil Nadu) are few promising selection. Recently a very promising variety 'Kashi Suhanvani' or VRAM-42, has been developed fromm ICAR-Indian Institute of Vegetable Research, Varanasi (U.P.) For commercial cultivation due to high yield and quality.

Based on multi-locational trials, selection IC 42285-1 was identified the best and released as 'Annapurna' a grain type cultivar for cultivastion in the hill region. This variety possesses 15 per cent protein, drought tolerance and wider adaptability and is capable of producing 2.25 t/ha grain yield. Recently, another variety 'GA1' (grain type) with long inflorescence, bold grains and high yield

Table 8.39: Important Released Varieties of India

Variety	*Species*	*Horticultural Attributes*	*Developing Source*
CO-1	*A. dubius*	Green leaves, high yield, leaf stem ratio=2, resistant to leaf spot caused by *Rhizoctonia*	TNAU, Coimbatore (Tamil Nadu)
CO-2	*A. tricolor*	Green leaves, high yield, leaf stem ratio=1.8	TNAU, Coimbatore (Tamil Nadu)
CO-3	*A. tristis*	Green leaves, high yield (30.0 t/ha), high leaf stem ratio, multicut variety.	TNAU, Coimbatore (Tamil Nadu)
CO-4	*A. hypochon-driacus*	Green leaves, high yield (80.0 t/ha), dual purpose	TNAU, Coimbatore (Tamil Nadu)
CO-5	*A. tricolor*	Plants are medium in height. Stem and petiole are pinkish red, leaves are large and obovate and yield is 40.0t/ha	TNAU, Coimbatore (Tamil Nadu)
Choti Chauli	*A. blitum*	High yield, multicut type and green leaves.	ICAR-IARI, New Delhi
Badi Chauli	*A. tricolor*	High yield, multicut and green leavers.	ICAR- IARI, New Delhi
Pusa Kirti	*A. tricolor*	Green leaves, high yield and leaf stem ratio=3.4	ICAR- IARI, New Delhi
Pusa Kiran	*A. tricolor* x *A. tristis*	Green leaves, high yield and leaf stem ratio=4.6	ICAR- IARI, New Delhi
Pusa Lal Chauli	*A. tricolor*	Red leaved, high yield (49.0t/ha)and leaf stem ratio=5	ICAR- IARI, New Delhi
Arka Sagun	*A. tricolor*	Light green, broad leaves and succulent stem, multicut, high yield (30.0t/ha), rich in calcium and iron.	ICAR- IIHR, Bengaluru
Arka Arunima	*A. tricolor*	Broad purple leaves, multicut, and yield is around 27.0 t/ha. Tolerant to white rust. Low in oxalate and nitrate.	ICAR- IIHR, Bengaluru
Arun	*A. tricolor*	Arun was developed by mass selection from 'Palappur Local'. Maroon leaves, multicut, leaf length 25 cm and leaf width 11 cm. yield is around (20 t/ha)	KAU, Vellanikkara (Kerala)
Mohini	*A. tricolor*	Green leaves, multicut (8-10 cuttings), stem is light puirple and yield is 13.2 t/ha.	KAU, Vellanikkara (Kerala)
Renu Shree	*A. tricolor*	Green leaves and red stem, multicut, leaf: stem ratio=1.5 and yield is around 15.5 t/ha.	KAU, Vellanikkara (Kerala)
Krishna Shree	*A. tricolor*	Reddish green leaves, multicut, leaf stem ratio=1.10 and yield is 14.8 t/ha.	KAU, Vellanikkara (Kerala)
Kannara Local	*A. tricolor*	Ghort day, local cultivar having red leaves. Yield is 25 t/ha.	KAU, Vellanikkara (Kerala)

potential of 2.5t/ha has been released for cultivation in Gujarat and Maharashtra states. Two other accessions IC5564 and an exotic introduction of *A. edulis* from Taiwan were observed to be promising.

Table 8.40: Amaranth Cultivars Developed by Selection with Progeny Testing

Variety	*Source*
CO-2	Indigenous
Badi Chaulai	Indigenous
Kannara Local	Indigenous
Pusa Kiran	Indigenous
Chhoti Chaulai	Indigenous
Pusa Kirti	Indigenous
CO-1	Indigenous
CO-3	Indigenous
Arka Suguna	Indigenous

Hybridization and Selection

Selection and cross-breeding is one area that could bring rapid advances. Amaranth demonstrates high levels of variability in leaf size, leaf shape, branching, bolting pattern, growth and regrowth ability and colour. Indeed, the vast wide geographical spread of the genus has produced many landraces, and in their present undeveloped state amaranths offer more genetic diversity than do many much better understood crops. The huge gene pool in widely separated areas can be tapped for the future development of the crop. The major barrier is of smaller flower size which makes emasculation difficult and labours intensive. In amaranths several varieties have been developed through hybridization. For example Pusa Kiran has been developed by crossing *A. tricolor* x *A. tristis.*

Interspecific Hybridization

Hybridization work carred out at the National Botanical Research Institute, Lucknow revealed that Intraspecific crosses between Indian and Mexican accessions of *A. cruentus* succeeded easily in both directions. However, interspecific crosses between *A. cruentus* and *A. hypochondriacus* succeeded only with *A. cruentus* as female parent. Mallika (1987) attempted interspecific hybridization between eight species and foun that most of them as unsuccessful. Out of these seven interspecific hybrids flowered which included hybrids within and between sections *Blitopsis* and *Amaranthus*. *A. lividus* x *A. tricolor* resembled the male parent in morphological characters but was highly sterile. The interspecific hybrids within the section

Table 8.41: Status of Cross Incompatibility

Species	A. cruentus	A. hybridus	A. hypochondriacus	A. powellii	A. quitensis	A. caudatus	A. edulis
A. cruentus			X				
A. hybridus							
A. hypochondriacus		F					
A. powellii		st	st				
A. quitensis							
A. caudatus		d	ab				
A. edulis	X	d	ab		F		

F: F_1 partially fertile; **ab**: Seedling abnormal; **d**: Seedling die; **st**: Normal but sterile; **x**: Failed.

Amaranthus included *A. spinosus* x *A. hypochondriacus*, *A. spinosus* x *A. caudatus*, *A. spinosus* x *A. cruentus*, *A. cruentus* x *A. caudatus*. *A. cruentus* x *A. caudatus* produced only female flowers but the hybrid was stunted and deformed. Other hybrids showed preponderance of characters of *A. spinosus* indicated by presences of spines, reduced inflorescence size and distinct placement of male and female flowers in the inflorescence. *A. spinosus* x *A. viridis*, the first successful hybrid between the sections had short and sturdy plant type with dominating characters of *A. spinosus*.

Crossing among related species having same chromosome number is possible. The species *A. tricolor*, *A. cruentus*, and *A. dubius* are cross incompatible (Madhusoodanan and Pal, 1981).

Electrophoretic studies in 20 selected *Amaranthus* species and varieties at the college of Horticulture, Vellanikkara (Kerala) indicated wide interspecific variation in protein banding pattern. 'Kannara Local' had maximum potein bands. Protein of the lowest mobility was also shown by'Kannara Local'. *A. tristis* produced similar protein patterns as that of *A. cruentus*, *A. blitum*, *A. viridis* and *A. dubius*. *A. caudatus*, *A. tricolor* and *A. dubius* also produced similar protein pattern (Gopalakrishnan, 2004).

Resistance Source and its Utilization

(i) Resistance Source for Diseases and its Utilization

A Surinam variety 'Klaroen' showed tolerance to wet-rot (*Choanephora* species). 'Arka Arunima' (*A. tricolor*) is resistant to white rust (Varalakshmi *et al.*, 2001). In amaranth genotypes A-22, A-26, A-29 and A-61 were found resistant to leaf blight/leaf spot caused by *Rhizoctonia solani*, at KAU, Vellanikkara (Priya, 1998). Krishnakumary (2000) also screened amaranth genotypes for leaf spot disease and found that accessions A-227 and A-204 belonging to *A. hypochondriacus* were immune to the disease and accessions A-3 and Co-1 belonging to *A. dubius* were resistant. Sindhu (2002) screened 32 accessions of *A. dubius* and among them A22, A26, A29 and A61 and many accessions of *A. tricolor* were leaf blight resistance (Krishnakumary, 2000). All accessions of *A. dubius* were free from leaf blight. Artificial inoculation of accessions showed that 14 were immune, 15 were highly resistant and three were moderately resistant. Celine *et al.* (2006) evaluated 89 genotypes of *Amaranthus* sp. accessions belonging to *A. dubius* were free from leaf blight caused by *Rhizoctonia solani*. Am 37 (*A. hypochondriacus*) was also resistant. All the accessions belonging to *A. tricolor* were susceptible. Gopalakrishnan (2004) found that Amt-118 and Amt-138 were tolerant to the disease. Celine *et al.* (2006) also reported resistance to the disease in *A. dubius* accessions, *A. hypochondriacus* (Am-37) and suceptability in *A. tricolor* accessions.

(ii) Resistance Source for Insect Pests and its Utilization

The species of 3 tepal group are more resistant to stem-borer than 5 tepal group.

(ii) Resistance Source for Nematode and its Utilization

Variety 'Folite' appears to be a bad host to root knot nematode (*Meloidogyne* species)

(iii) Resistance Source for Abiotic Stress

Variety 'Karuna Local' developed from the Kerala Agricultural University, Trichur; Kerala (India) is a short-day responsive and red type. Several amaranth genotypes are available which bolt only under short-day conditions. These types can be isolated to be grown as multicut types if planted very early taking into consideration the day length of the locality at a specific month. Different species differ in their day length requirements and respond differently to change in photo-and thermo-periodism. *A. caudatus*, *A. cruentus* and *A. edulis* are short- day species, while the commonly grown grain *Amaranthus* species, *viz. A. hypochondriacus* is reported to be day neutral. *A. hybridus* is fast growing and is resistant to moisture stress. *A. thunberghii* has a more prostrate growth habit and resuistant to water stress. Close genetic morphology of *A. retroflexus*, weedy species, with cultivated species opened the possibilities of combining adaptability to cold and dry conditions of *A. retroflexus* with edible seed features of *A. cruentus*.

Quality Breeding

Breeding for low anti-nutrient factors in vegetable amaranths resulted in identification of low oxalate and low nitrate lines (Devadas *et al.*, 1984). Devadas (1982) reported 'A43', comparatively lower in nitrates (0.58 per cent) and oxalates (0.94 per cent). Krishnakumary (2000) reveal that oxalates and nitrates had no influence on incidence of leaf spot disease. High contents of total phenol, OD phenol and ascorbic acid were recorded in immune accessions A-227, A-204 (*A. hypochondriacus*) and resistant accessions A-3 and Co-1 (*A. dubius*) than susceptible accessions. Sindhu (2002) reported that nutritional value, resistance to leaf blight and high organoleptic score, the accession AD-23 was the best.

Ploidy Breeding

There is one cultivar 'Co-1' is a tetraploid and very vigorous in vegetative growth.

Mutation Breeding

Krishnakumary and Panicker (2004) studied effect of gamma rays in inducing variability in M1 generation of *Amaranthus tricolor* and *A. hypochondriacus*. They found that the plant height and number of branches were reduced with increased dose of gamma rays.

Genetic Resource Conservation

Germplasm collections were augmented with 315 exotic collections which also include 25 different species are maintained as active collections for their utilization in the medium term storage of ICAR-NBPGR, Regional Station, Shimla (Himachal Pradesh).

REFERENCES

1. Allen, P. 1961. Die Amaranthaceen Mittleleuropas Carl Verlag, Munchen.
2. Anonymous, 1992. The Wealth of India. Publications and Information Directorate, CSIR, New Delhi, India.
3. Aykroyd, W.R. 1966. The nutritive value of Indian foods and the planning of satisfactory diets, ICMR, New Delhi, p. 55, 61, 91 and 97.
4. Brenner, D.M., Baltensperger, D.D., Kulakow, P.A., Lehmann, J.W., Myers, R.L., Slabbert, M.M. and Sleugh, B.B. 2000. Genetic resources and breeding of *Amaranthus*. Plant Breeding reviews, 19: 227-285.
5. Celine, V.A., Sujathasathy, S., Seema, S., Deepa, S.N., Sreelathakumary, I. and Abdulvahab, M. 2006. Characterisation and evaluation of vegetable amaranth (*Amaranthus tricolor* L.) for high yield, quality and resistance to Rhizectonia solani. Proceedings of 1st International Symposium on Indigenous Vegetables and Legumes, (December 12-15, 2006), ICRISAT, Hyderabad (A.P.), India, pp. 104.
6. Devada, V.S. and Mallika, V.K. 1991. Review of research on vegetables and tuber cops-*Amaranthus*. Directorate of Extension. Kerala Agricultural University, Thrissur, Kerala, India.
7. Devadas, V.S. 1982. 'Screening for non-bolting types) of *Amaranthus* suited for year round planting'. M. Sc. Thesis. Kerala Agricultural University, Trichur, Kerala (India).
8. Devadas, V.S., Gopalakrishnan, P.K. and Peter, K.V. 1984. Breeding for low content of anti-nutrient factors in vegetable amaranths. *Amaranthus Newsletter*, **2:** 2-3.
9. Feine, L.B. 1980. Genetic Resources of Amaranths, edn1, Appendix1, Grubben, G.J.H. and Van, Slotten (Eds.). International Board for Plant Genetic Resources, Rome.
10. Gonar, K.V., Pogozheva, A.V., Derbeneva, S.A., Mal'tsev, GIu, Trushina, E.N. and Mustafina, O.K. 2006. The influence of a diet with including amaranth oil on antioxidant and immune status in patients with ischemic heart disease and hyper lipoproteidemia. *Vopr Pitan*. **75(6):** 30-33.
11. Gopalakrishnan, T.R. 2004. Three decades of vegetable research in Kerala Agricultural University, Vellanikkara Kerala (India), p. 105-111.
12. Grubben, G.J.H. 1976. The cultivation of amaranth as a tropical leaf vegetable with speical reference to south Dahomey. Department of Agricultural Research of the Royal Tropical Institute Amsterdam, p. 1-207.
13. Joshi, B.D. 1986. Genetic variability in grain amaranth. *Indian J. Agric. Sci.*, **56(8):** 574-576.
14. Joshi, B.D. and Mehra, K.L. 1983. Genetic variability and distribution of amaranths genetic resources in the Himalayas. XVth International Congress of Genetics, New Delhi, Dec., 12-21. Abs. No., p. 983, V.L. Chopra *et al.* (eds.), Oxford-IBH. New Delhi.
15. Khoshoo, T.N. and Pal, M. 1973. Cytogenetic patterns in *Amaranthus*. *Chromosome Today*, **3:** 252-267.
16. Krishnakumary, K. 2000. Genotypic and seasonal influence of leaf spot disease in amaranth. Ph.D. (Hort.) thesis, Kerala Agricultural University, Vellanikkara. (Kerala).
17. Krishnakumary, K. and Panicker, B.V. 2004. Effect of gamma rays in inducing variability in *Amaranthus* species. Proceedings of the Indian Horticultural Congress, New Delhi (6-9, November), p. 111.
18. Lohithaswa, H.C., Kulkarni, R.S. and Ganeshaiah, K.N. 1994. Genetic divergence in Indigenous collection of grain amaranth. *Indian, J. Plant Genet. Resour.*, **7(1):** 119-122.
19. Madhusoodanan, K.J. and Pal, M. 1981. Cytology of vegetable amaranth. *Bot. J. Linn. Soc.* **82:** 61-68.
20. Mallika, V.K. 1987. Genome analysis in the genus *Amaranthus*. Ph.D. (Hort.) thesis, Kerala Agricultural University, Vellanikkara (Kerala).

21. Mallika, V.K., Ramachandran, C. and Peter, K.V. 1988. Interspecific hybrid between *Amaranthus spinosus* (Section *Amaranthus*) and *A. viridis* (Section Blitopsis). *Current Science*, **57:** 1185-1186.

22. Martirosyan, D.M., Miroshnichenko, L.A., Kulakova, S.N. Pogojeva, A.V. and Zoloedov, V.I. 2007. Amaranth oil application for coronary heart disease and hypertension. *Lipida Health Dis.* **5(6):** 1.

23. Mujica, A. and Jacobsen, S.E. 1999. Recursos geneticos, ethobotanica Y distributucion del amaranto (*Amaranthus caudatus, A. cruentus* L. Y *A. hipochondriacus* L.) In: Jacobsen S.E. valdez A. eds. Libro de Resumen es.

24. Nath, P.; Brhmachari, V.S. and Singh, D. 1996. Sabzi Vigyan. ICAR, New Delhi.

25. Pal, M. and Khoshoo, T.N. 1972. Evolution and improvement of cultivated amaranths: Inaviability, weakness and sterility in hybrids. *J. Heredity*, **63:** 78-82.

26. Pal, M. and Khoshoo, T.N. 1973. Evolution and improvement of cultivated amaranths. VI. Cytogenetic relationship in grain types. *Theoretical and Applied Genetics*, **43:** 242-251.

27. Peter, K.V. 1998. Amaranth. In: Genetics and Breeding of Vegetables. Published by ICAR, New Delhi, p. 22-30.

28. Priya, V.P. 1998. Screening amaranths genotypes for yield, quality and resistance to biotic stress. M.Sc (Hort.) thesis, Kerala Agricultural University, Vellanikkara.

29. Sauer, J.D. 1967. The grain amaranths and their wild relative. A revised taxonomic and geographic survey. *Ann. Rep., Missouri Bot. Gard.*, **54:** 103-137.

30. Sauer, J.D. 1993. Historical geography of crop peanis: a select roster, CRC, Press, Boca Raton, F.L. USA.

31. Schenz. 1934. Amaranthaceae. In: *Die Naturlichen pflanzenfamilen*. Engler, A. and Prant, I.K. (eds.)

32. Sindhu, L. 2002. Variability in vegetable amaranth (*Amaranthus dubius*) for yield, quality and resistance to leaf blight. M.Sc. thesis, Kerala Agricultural University, Vellanikkara (Kerala).

33. Suma, S., Ambika, S.R., Kazinczi, G. and Narwal, S.S. 2002. Allelopathic Plants. No.6 *Amaranthus* spp. *Allelopathy J.*, **10:** 1-11.

34. Thellung, A. 1914. *Amaranthus*. In: F. Ascherson and P. Graebner, Synopsis der Mittleurop aischen Flora, Vol. V., p. 225-356. Ascherson, F. and Graebner, P. (eds.) Borntraeger, Leipzig (Engelmann).

35. Varalakshmi, B., Pratap Reddy, V.V., Naik, G., Celia Chalam, V. and Pal, D.K. 2001. 'Arka Arunima': A new multicut amaranth. *Indian Horticulture*, **46(1):** 29-30.

36. Walton, P.D. 1968. Genetic homoeostasis in *Amaranthus caudatus*. J. Hered; 60: 365-368.

Genus *Basella*

Ceylon Spinach or Indian Spinach or Poi (Basella Species)

Introduction

Ceylon spinach (*Basella alba* 2n=48 and *Basella rubra*, 2n=44) is also known by many common names as Poi, Pasali, Indian spinach, Malabar nightshade, Malabar spinach, Gambian spinach and Pu-tin-choi, which is grown all over the country particularly in the plains and mid-hills during hot humid season. Today, it is grown widely in the tropics as a perennial in warmer temperate regions as an annual. It is very succulent and highly mucilaginous vegetable. It is a very productive leafy vegetable suited for both home gardens and market gardens (Indira and Peter, 1987). Its glutinous texture when cooked takes some getting used to, but the strong flavor is pleasant. The plants are slow growing and send out runners, which are clothed with good sized, tasty, thick, dark green leaves. The nutritive value of the young shoots and leaves is very high in terms of salt and vitamins. It contains 93 per cent moisture, 1.2 per cent protein, 8.90 g carbohydrate, 1.0 g fat, 0.15 per cent calcium, 1.4 mg/100g iron, 3250IU/100g and 87.0 mg ascorbic acid. Typical of leaf vegetables, it is high in vitamin A, vitamin C, iron and calcium. It is low in calories by volume, and high in protein per calorie. The edible leaf contains several amino acids like arginine, leucine, isoleucine, lysine, threonine and tryptophan. The succulent mucilage is a particularly rich source of soluble fiber, thought to remove mucus and toxins

Table 8.42: Name of Taxon and Preferred Taxon Name of Genus *Basella*

Name of taxon	Name of Authority	Family	Preferred Taxon Name
Basella alba	L.	Basellaceae	*Basella alba*
Basella cordifolia	Lam.	Basellaceae	*Basella alba*
Basella rubra	L.	Basellaceae	*Basella alba*
Basella tuberosa	(Caldas) Kunth	Basellaceae	*Ullucus tuberosus*
Basella vesicaria	Lam.	Basellaceae	*Anredera vesicaria*

from the body. Leaf juice is demulscent, laxative and diuretic in nature which prevent infection of gonorrhea, balantis and cattarah. The plant is also a rich source of chlorophyll. The leaves and young stems of *Basella* are an excellent hot weather spinach substitute. The young leaves may be eaten raw in salads, the leaves cooked as a potherb mixed with vegetables, the leaves cooked separately as spinach, or added to soups (Burkil, 1935). The purple juice from the fruits may be used as a food dye to colour pastries or sweets. Boiled seeds are sometimes added to dahl in Bangladesh. The plant is often grown as an ornamental.

Origin, History and Distribution

Ceylon spinach or Poi (*Basella* species) has been originated from South Asia (tropical Asia), probably originating in India or Indonesia. It is widely grown in tropical Asia, Malaysia, West Africa, Ceylon, North East and south India from olden days. As for commercial cultivation Malaysia and Philippines both are dominating. In India, it is dominating to Assam, West Bengal and Andhra Pradesh.

Species Availability, Evolution and Diversity

The genus *Basella* includes *Basella alba, B. cordifolia, B. rubra, B. tuberosa* and *B. vesicaria*. There are three main types under cultivation. The most common type has dark green, oval or nearly round leaves. This is *B. alba* a less popular type, often referred to as *B. rubra*, has round oval leaves and red stems. This is the species which is very variable. The third type less commonly grown, has cordate shaped, dark green leaves. While rich variability in plant morphological characteristics including stem fleshiness, leaf shape and size can be spotted in south India, North eastern region including Assam, West Bengal and Odisha.

According to Bailey (1973), the flowering species are considered as form of *B. rubra: B. alba*-a white flowered form, *B. coninifolia, B. cordifolia* which heart shape leaves, 9.16-12.7 cm long and 5.0-6.25 wide, *B. arassifolia, B. japonica, B. lucida*- from India-a Chinese form, *B. ramose* and *B. volubilis*.

Botanical Description

Ceylon spinach or Indian spinach belongs to the family Chenopodiaceae (C4 plants) but few researchers put the genus *Basella*, under a separate family 'Basellaceae'. Baily (1973) described *B. alba* has narrow leaves, whitish flowers on long peduncle spikes arranged in very loose clusters. Haines (1910) *B. rubra* as the one with fleshy twining much branched herb with ovate shiny and rather fleshy leaves and small sessile pinkish fleshy flowers on distant spikes. The *B. alba* has 2n= 48 chromosome number whereas *B. rubra* has 2n= 44. It is a short perennial herb with very long and slender stem. Leaves are variable, broadly ovate, thick, entire, with cordate base. Flowers are white or red, sessile few in lax pedunculate spikes; perianth fleshy 5-fid, segments elliptic and obtuse. Fruits are red, white or black and have a size of pea.

Botanical Classification

Botanically Ceylon spinach can be grouped into three:

(i) **Common cultivated type (*B. alba*)**-This type has dark green, oval or nearly round leaves.

(ii) **Less cultivated type (*B. rubra*)**-This type has round oval leaves and red stems.

(iii) **Very less in cultivation**-This type has cordate shaped, and dark green leaves.

(iv) **Green and red tinted types**-This type of genotype is also available in nature.

On the basis of thick leaves exhibiting one or two leaf forms:

(i) Closely spaced oval to rounded leaves.

(ii) More widely spaced heart-shaped leaves.

Most authors agree, however, that the two colour/ forms of *Basella* are not separate species.

Table 8.43: Botanical Distinctness of *B. alba* and *B. rubra*

	B. alba	***B. rubra***
Chromosome number	2n= 48	2n= 44
Nature	Annual or biennial	Annual or biennial
Growth Habit	Plant is viny, succulent and slightly mucilaginous	Plant is viny, succulent and slightly mucilaginous
Stem	Fleshy twining stem, green and much branched	Fleshy twining stem and purplish in colour.
Leaf	Narrow leaves, succulent. In leaf C_4 cycle photosynthesis pathway.	Ovate shiny and rather fleshy leaves. In leaf C_4 cycle photosynthesis pathway.
Floral biology	Whitish flowers	Small sessile pinkish fleshy flowers
Anthesis and pollination	Anthers versatile, bracteoles two, adnate to the face of the perianth. Pollen grains germinate after 4-6 hours of anthesis and reach to ovary through stigma after 11-16 hours. Just after it, within 16-24 hours double fertilization process completed.generally, in 2-8 dyas 2 celled pro-embryo developed. Zygotic development is Asteroid types. And zygote pertain free nuclei (Xu, 1986).	Anthers versatile bracteoles two adnate to the face of the perianth. Pollen grains germinate after 4-6 hours of anthesis and reach to ovary through stigma after 11-16 hours. Just after it, within 16-24 hours double fertilization process completed.generally, in 2-8 dyas 2 celled pro-embryo developed. Zygotic development is Asteroid types. And zygote pertain free nuclei (Xu, 1986).
Fruit and fruit set	Fruits are round and enclosed in fleshy perianth.	Fruits are round and enclosed in fleshy perianth.

Augmentation of Genetic Resources and Crop Improvement

Introduction and Selection

In *Basella* there are mainly two distinct types, one with reddish petioles and stems and the other with green leaves, petioles and stems. Both the red and green leaves cultivars are consumed as vegetables but green-leaves cultivars are commercially cultivated and belong to the var. *alba*. The red leaved and green leaved cultivars are equally common in West Bengal, Assam and South India, while the green leaved ones are more popular in Uttar Pradesh and Punjab. Ceylon spinach or Indian spinach is free from pests and diseases (Indira and Peter, 1987). Recently 3 cultivars i.e. Kashi Poi-1 (VRV-17), Kashi Poi-2 (VRV-31) and Kashi Poi-3 (VRV-45) has been released by ICAR-Indian Institute of Vegetable Research, Varanasi (U.P.) for commercial cultivation due to better yield and quality.

Polyploidy

Pushpa *et al.* (1975) treated 10 plants of *Basella alba* with the colchicine to get polyploids and got desired results. They found autotetraploid which were broader and more succulent than diploids but pollen fertility was less.

Breeding for Quality Traits

Diversity for nutritional and antinutritional quality due to the presence of total oxalate were assessed in 40 genotypes of Indian spinach by Ranjan *et al.* (2016) at ICAR-IIVR, Varanasi (U.P.). Wide range of variation was observed for total phenol (63.8-235.67 GAE/100g), total flavonoid (10.6-36.5 mg CE/100g), total carotenoids (102-641 mg β-carotene equivalent/100g fresh weight), antioxidant activity measured through CUPRAC (16.35-38.65 μmol TE/g) and FRAP (2.9-5.4g μmol TE/g) and total oxalate (4.14-6.87 mg/g fresh weight). The genotypes rich in carotenoid content are VRB-45 and VRB-23, while for anti-oxidant activity VRB-54, VRB-61, VRB-62, VRB-17 and VRB-41 were found promising.

REFERENCES

1. Bailey, L.H. 1973. Manual of Cultivated Plants. The Mac-Millan Company, New York, p. 367.
2. Burkill, I.H. 1935. A Dictionary of the economic products of the Malay Peninsula, Crown Agents, London.
3. Haines, N.H. 1910. Forest Flora of Chota Nagpur, Part I and II., M/s Bishen Singh Mahendra Pal Singh, New Connaught Place, Dehra Dun, and M/s Periodical Experts, 42-D Vivek Vihar, Delhi, p. 382.
4. Indira, P. and Peter, K.V. 1987. *Underexploited Tropical Vegetables*, Published by Kerala Agricultural University, Vellanikkara (Kerala) p. 8-11.
5. Ranjan, P., Koley, T.K., Gupta, N., Singh, P.M. and Singh, B. 2016. Genetic variability for nutritional quality in Indian Spinach (*Basella alba* L.) germplasm. (Abs: 84), 7th Indian Horticulture Congress, held at ICAR-IARI, New Delhi, 15-18 November, p. 65.
6. Xu, H.Q. 1986. Studies on fertilization and embryogenesis in Indian Spinach (*Basella rubra* L.). *Acta Botanica Sinica*, **28(4):** 361-367.

Genus *Talinum*

Ceylon Spinach or Water Leaf (*Talinum triangulare* (Jacq.)Willd.)

Introduction

Ceylon spinach or Water leaf (*Talinum triangulare* (Jacq.) Willd., 2n=24, 48, 72) is a warm season soft mucilaginous leafy vegetable (Therra, 1973). In India it is called as Silan Pasali (Tamil) and Asamia Palang or Pireli Palang (Assam). In English it is called as Waterleaf or Surinam Purslane but in other dilect it called as Grassé, pourpier tropical (Fr). Beldroega graúda, lustrosa grande (Po). Its tender leaves and stems are used as vegetable butthe high oxalate content of leaf makes its consumption limited. Cooking takes very less time. As nutritive value, per 100 g of edible portion (fresh weight basis) contains 90.71 g moisture, 2.57 g protein, 0.19 g fat, 0.95 g fiber, 4.02 g carbohydrate, 1.56 g ash, 231.51 mg calcium, 43.20 mg phosphorus and 2.79 mg iron. Similarly on dry weight basis it contains 27.65 g protein, 2.05 g fat, 0.95 g fiber, 43.27 g carbohydrate, 16.74 g ash, 2492.03 mg calcium, 465.02 mg phosphorus and 30.03 mg iron (Saikia and Shadeque, 1994). Medicinally it is a potent herb for diabetic peoples due to curing properties (Saikia and Shadeque, 1994). Ceylon spinach is a good fodder crop but it contains high amount of oxalate and hydrocynic acid, therefore its feeding for longer time is harmful (Saikia and Shadeque, 1994). The shade and moisture loving nature of the crop, makes it suitable for cultivation under other crop canopy.

Origin, History and Distribution

Waterleaf is a cosmopolitan weed common throughout the humid tropics. As per the historical record water leaf is originated from Brazil. It is claimed to have a South American origin, but an African origin may also be possible, as several *Talinum* species including the closely related *Talinum portulacifolium* (Forssk.) Schweinf. occur in Africa. It has been recorded for several countries in West and Central Africa. From Brazil it was introduced to Java in 1915. At present it is being cultivated as a minor vegetable in India, Malaysia, Indonesia, Tropical Africa (West and Central), Arabian countries, America, West Indies, Java and Brazil. Ceylon spinach or Asamia Palang is very popular leafy vegetab in Assam. In Java it used as salads. In Africa and South America, it is a common leafy vegetable and considered as an alternative of Purslane (Kulpha). It is cultivated commercially in Nigeria and Cameroon.

Species Availability, Evolution and Diversity

Talinum comprises about 40 species, most of them found in Mexico and southern United States, and 7 species in tropical Africa. *Talinum triangulare* is sometimes confused with *Talinum portulacifolium*, but the latter differs by its paniculate inflorescence with terete axis, sepals not prominently veined and smooth seeds. Genus *Talinum*, *T. triangulare* Willd. (Syn. *Talinum fruticosum* auct. non (L.) Juss.) is common in nature and used as leafy vegetable. The related species in which leaves are eaten as vegetables are *T. arnotti*, *T. caffrum*, *T. portulacifolium* (Forsk.) Aschers. Ex. Schwf., *T. patens*, *T. crassifolium*, *T. cuneifolium* Willd and *T. indicum* (Ochse, 1977; Peter, 1998).

Botanical Description

Water leaf belongs to family Portulacaceae, genus *Talinum*, *T. triangulare* Willd. Erect shrubby plants with terminal panicle having moderate-sized purple flowers, ovary free. This crop is polyploidy in nature.

Table 8.44: Botanical Description

Growth Habit	Plants are bushy erect, glabrous perennial herb, attained the height of 40.0-100.0 cm. and usually strongly branched.
Root	Roots swollen and fleshy
Stem	Stems are succulent, breakable and light green in colour on which purple spots can be seen.
Leaf	Leavers are alternate,simple, almost sessile, succulent; stipules absent; blade obovate to spatulate, 3–15 cm x 1–6 cm, base long-tapering, apex rounded to notched, mucronate, entire, venation pinnate, indistinct thick, leathery and dark green in colour. Leaves are affected with day and night cycles, so it is called as nyctinastic movement. In this condition leaves are closed in the evening and open in the morning.
Floral biology	Inflorescence a terminal cyme on a triangular stalk up to 12 cm long. Flowers are on large stalk and each stalk has 2-3 flowers. Flowers bisexual, regular; pedicel c. 1 cm long. Flowers are actinomorphic, hypogynous and having diameter of 0.5-2.5 cm. epicalyx 2, free with 3 prominent veins; petals 5, free up to 1 cm long, pink and pointed. Stamens numerous; ovary superior, 1-celled, style slender, with 3-branched stigma exceeding stamens.
Anthesis and pollination	Ceylon spinach is self pollinated vegetable crop. The onset of flowering does not appear to affect leaf or shoot production. Plants take 40–75 days to flower from stem planting, and at flowering they have 4–9 branches and 25–90 leaves. Waterleaf is recorded as being self-pollinated with a limited degree of out-crossing.
Fruit and fruit set	Fruits are conical shape, globose to ellipsoid capsule 4–7 mm long, 3-valved, elastically dehiscent, and many-seeded. Fruiting takes 75–80 days from stem planting. Under natural conditions, plants will live for 4–6 months.
Seeds	Seeds are very small, compressed globose-reniform, c. 1 mm long, tuberculate and shining black. The seed weight of 1000 seeds are 0.3 g.

Augmentation of Genetic Resources and Crop Improvement

Waterleaf is relatively tolerant to drought. Exposed to drought it adopts a crassulacean acid metabolism (CAM), resulting in an effective use of available moisture, carbon dioxide assimilation continuing during the night and increased growth. High temperatures (>35°°C) and drought negatively affect the number of leaves, leaf area, stem size, and number of branches. Accessions collected in Nigeria and Cameroon showed very limited diversity and it was not found worthwhile to select lines that could be developed into distinct cultivars. In Nigeria the green type is most common, but occasionally plants with varying degrees of purple colouration in leaves and stems have been found in the south-eastern part of the country, where waterleaf is often cultivated.

Genetic Resource Conservation

No collections of germplasm of *Talinum triangulare* are known to be maintained. There seems to be no risk of genetic erosion because it is widespread and has a weedy habit. National Plant Genetic Resource, Laboratory, Las Banas (Philippines) is maintaining some collections.

REFERENCES

1. Akachuchu, C.O. and Fawusi, M.O.A., 1995. Growth characteristics, yield and nutritive value of waterleaf, *Talinum triangulare* (Jacq.) Willd., in a semi-wild environment. *Discovery and Innovation*, **7(2):** 163–172.
2. Bailey, L.H. 1973. Manual of Cultivated Plants. The Mac-Millan Company. p. 367.
3. Burkill, H.M., 1997. The useful plants of West Tropical Africa. 2nd Edition. Volume 4, Families M–R. Royal Botanic Gardens, Kew, Richmond, United Kingdom. p. 969.
4. Haines, N.H. 1910. Forest Flora of Chota Nagpur, PartI and II, M/s Bishen Singh Mahendra Pal Singh, New Connaught Place, Dehra Dun, and M/s Periodical Experts, 42-D Vivek Vihar, p. 382.
5. Indira, P. and Peter, K.V. 1987. *Underexploited Tropical Vegetables*. Published by Kerala Agricultural University, Trichur (India), p. 8-11.
6. Ochse, J.J. 1977. *Vegetable of the Dutch East Indies*, pp. 621. Asher and Co., B.V. Amsterdam.
7. Nyananyo, B.L.O. and Olowokudejo, J.D., 1986. Taxonomic studies in the genus *Talinum* in Nigeria. *Willdenowia*, **15:** 455–463.
8. Peter, K.V. 1998. Genetics and Breeding of Vegetables. ICAR, New Delhi.
9. Rifai, M.A., 1993. *Talinum triangulare* (Jacq.) Willd. In: Siemonsma, J.S. and Kasem Piluek (Editors). Plant Resources of South-East Asia No 8. Vegetables. Pudoc Scientific Publishers, Wageningen, Netherlands. p. 268–269.
10. Saikia, A. and Shadeque, A. 1994. Ceylon spinach: A leafy vegetable growing wild in Assam. *Indian Hort.* **38(4):** 10-11.
11. Therra, G.T.A. 1973. *Tropical Vegetables*, Edn. Royal Tropical Institute, Amsterdam I., p. 77.

Genus *Sauropus*

Chekkurmanis or Star Gooseberry (*Sauropus androgynus*)

Introduction

Chekkurmanis (*Sauropus androgynus* Merr.; syn. *S. albicons* Blume, *S. gardnevi* Wight; *S. sumatrans* Mig.), is a small shrub and becoming popular as a leafy vegetable. In Tamil Nadu, this crop is popular as 'Thavarai Murungai'. In Lepcha, it is popular as 'Segtungrung' and in Khasi 'Soh Pit'. The crop holds a unique position in the list of tropical leaf vegetables owing to its high nutritive value which has not been well-realized by many people. It tender stems and fruit are also used as vegetable. It is rich in vitamin A (9510 IU/100 g of leaves) and vitamin C (247 mg/100 g Due to its high nutritive value, it is commonly known as a 'multivitamin and multimineral packed leafy vegetable' (Indira and Peter, 1987; Shanmugavelu, 1989; Oomen and Grubben, 1978).Chekkurmanis is also a rich source of proteins and vit. A, E, and K. Leaves are sour in taste and even after cooking leaves remain green colour. Per 100 g of edible portion contains 73.06 g moisture, 6.8 g protein, 3.2 g fat, 3.4 g minerals, 1.4 g fiber, 11.6 g carbohydrates, 10.3 K. calorie, 570.0 mg calcium, 200.0 mg phosphorus, 28.0 mg iron, 5706 microgram carotene, 0.48 mg thiamine, 0.32 mg riboflavin, 2.6 mg niacin and 247 mg vit. C (Gopalan *et al.*, 1999). As a medicinal properties root decoctions is used to lower down the fever and cure the urinary problems. After child birth ladies used this leafy vegetable for better lactation. In Java sour leaves are used for soupe preparations. In Indonesia Chekkurmanis leaves are used to give a light green colour (natural dye) to pastry and colouring the fermented rice in the Dutch East India. Leaves are also useful for growing as a hedge around home gardens. In Java it is often planted in live fences and in midst of garden beds to provide light shade for other vegetables in beds. Besides other uses, leaves are used as cattle and poultry feed in certain parts of the country. In Tamil Nadu (India) its leaves are considered as good fodder. In some other places, plants are planted as a soil binder to prevent soil erosion.

Origin, History and Distribution

Chekkurmanis is believed to be originated from the Indo-Burma centre.It is found in Sikkim, Himalayas, Khasi, Abor and Arka Hills at 1200 m amsl. This vegetable crop is being grown in southern parts of India (Kerala and Tamil Nadu), Indonesia, Malayasia and Singapore. It is found in the Sikkim Himalayas, Khasi, Abor and Arka Hills at 1, 200 m elevations and in Western Ghats of Kerala (south India) from Wynad northwards at altitudes of 300-1, 200 m. It is reported that the plant was 1st introduced to Kerala from Malaya in 1953 and is known vernacularly as Madhura Cheera in Malayalam. Plants of Chekkurmanis are growing wild in Western Ghats and in Travancore area of south Kerala in the evergreen forests (Wynad Northwords at altitude of 300-1200 m amsl). It was 1st located in the Malabar area of north Kerala after its introduction in 1953 and was taken to the Agricultural Research Station of Tamil Nadu during 1955-56 where it became much popular as 'Thavarai Murungai'.

Species Availability, Evolution and Diversity

The genus *Sauropus* consists of 11 species of them some species are related to *S. androgynous* Thw., *S. retroversus* Wight., *S. rigidus* and *S. quadrangularis* Muell. Chekkurmanis (*Sauropus androgynus* Merr., syn. *S. albicans* Blume, *S. gardnerianes* Wight., *S. zeylanicus* Wight., S. indicus Wight., *S. sumatrons* Mig.) is a small shrub becoming popular as a leafy vegetable. Other newly related species are *S. assimilis* Thw., *S. retroversus* Wight., *S. rigidus* and *S. quadrangularis* Muell. Attempts to establish closeness and affinity among the above 5 species including *S. androgynous* are limited.

Botanical Description

Chekkurmanis (*Sauropus androgynus* Merr., syn. *S. albicans* Blume, *S. gardnerianes* Wight., *S. zeylanicus* Wight., *S. indicus* Wight., *S. Sumatrans* Mig.) belonging to the family Euphorbiaceae is a small perennial shrub. The crop is propagated mainly through cuttings and as such transgressive and desirable plant types can be maintained and propagated without any loss in performance.

Table 8.45: Botanical Description

Growth Habit	Chekkurmanis is perennial and bushy in nature. Plants are erect, smooth and attained the height of 1.5 m.
Leaf	Leaves are 3.0-5.0 cm long and developed in alternate fashion.
Floral biology	Flowers are small, red in colour and develop in cluster.
Anthesis and pollination	Crop is highly cross-pollinated and entomophilous due to monoecious nature. The nature of plant is protogynous.
Fruit and fruit set	Fruits are pinkish in colour. Diameter of fruit is 0.2 cm.

REFERENCES

1. Gopalan, G., Rama Sastri, B.V., Balasubramanian, S.C., Rao, B.S.N., Deosthale, Y.G. and Pant, K.C. 1999. Nutritive value of Indian Foods. National Institute of Nutrition, ICMR, Hyderabad, India.
2. Indira, P. and Peter, K.V. 1987. Underexploited Tropical Vegetables, pp. 18-20. Kerala Agricultural University, Vellanikkara (Kerala).

3. Oomen, H.A.P.C. and Grubben, G.L.H. 1978. Tropical leaf vegetables in human nutrition, Communication No. 69, Department of Agriculture Research, Royal Tropical Institute, Amsterdam.

4. Shanmugavelu, K.G., 1989. Production Technology of Vegetable Crops. Oxford and IBH Pub. Co., New Delhi, p. 330.

Genus *Portulaca*

Parslane (*Portulaca oleracea* Linn.)

Introduction

Parslane (*Portulaca oleracea* L.) is known by several vernacular name like Kulpha (Hindi and Punjabi); Ghol (Marathi and Gujarati); Nunia Sag (Bengali); KulphaSag (Odishi); Pavila (Telgu); Pasali Kirai (Tamil); Pasali Kira (Malyalam) and Doda Goni Sopu (Kannad). It is tiny creeper with small, succulent fleshy leaves and tender stems, which is cooked like spinach. The leaves do not remain fresh for any length of time. So, for continuous supply of leaf seeds are sown at 15 days interval (Bhat, 2007). The plant becomes ready for first harvest in about a month after sowing. It matures quickly. Scented twigs and leaves emmited pleasant aroma, therefore it is important gradient of salads. Leaves are rich source of protein; iron and minerals. It compasses beta carotene, folic acid, vitamin C and essentially fatty acids. Scientifically it is proved that per 100 g of edible leafy portion contains 381.0 mg of phyloquinine or viatamin K_1 which does not loss even after cooking (Lagenberg, *et al.*, 1986). In general per 100 g of edible leaves contain 90.5 g moisture, 2.4 g protein, 0.6 g fat, 2.3 g minerals, 1.3 g fiber, 2.9 g carbohydrates, 27 K. calorie, 111.0 mg calcium, 45.0 mg phosphorus, 14.8 mg iron, 2292 micro gram carotene, 0.10 mg thiamine, 0.22 mg riboflavin, 0.7 mg niacine and 99.0 mg vitamin C (Gopalan *et al.*, 1999). Similarly there are several micro nutrients like magnesium (120.0 mg), sodium (67.2 mg), potassium (716.0 mg), copper (0.19 mg), sulpher (63.0 mg) and chlorine (73.0 mg) are found in the fresh leaves. Leaves also contain 89.0 micro gram β-carotene per gram but 2.3 per cent lossed at the time of cooking (Rahman *et al.*, 1990). As per the status of carotenoids, 30.0microgram β-carotene, 0.6microgram β crypto xanthin, 48.0 microgramlutein+violoxanthin, 0.7 microgram zea xanthin, and 9.0 microgram neo xanthin in total of 89.0 are found (Mercadenate and Rodriguez, 1990). On the dry base basis parslane contain 3.5 per cent lipids in which 25 per cent is free fatty acids. In toal lipids 19 per cent sterols are available which can be fractined in 72 per cent sisterol and 145 stigmasterol (Simopoulos and Salem,1989). In compariosion to spinch, pursnane contain 7 times more β tocopherol (Simopoulos *et al.*, 1992).World health Organization has designated as an important medicinal plant (Susiarti, 1993) because it contains omega-3 poly unsaturated fatty acids in sufficient quantity (Omara *et al.*, 1991; Koch, 1988). Due to presence of l-noradrenaline (0.25 fresh weights) it is considered as nerve tonic. Herb is used in refrigerant, and acts as antscorbutic, aperient, diuretic. It is very usefull in scurvy and diseases of liver and spleen. Parslane is very usful to control urine infection and gynaecological diseases. It is well recognized as cardiotonic. Regular consumption lowers down the fever and removes constipations. Application of leaf juice on skin proved beneficial to cure eczema. Individual leaves and young shoots are harvested for use in salads in summer.

Origin, History and Distribution

Parslane (*Portulaca oleracea* L.) is said to be indigenous to Himalaya and the Middle East. It was naturalized elsewhere and is considered an invassive weed in some regions. Its centre of diversity is believed in Mexico and Australia. Its availability can be spotted in several European and Asian countries. Research workers opined that cultivated purslane was developed in old world. It was cultivated in ancient Egypt and cultivated in Europe as for back as the middle ages. In India, its distribution is North Western Himalaya.

Species Availability, Evolution and Diversity

Under genus *Portulaca* 40 species are reported worldwide and among them several are dominating in tropical and subtropical regions. In south Asia its 5 species are available. In purslane polyploidy, diploid, tetraploid and hexaploid can be observed. It occurs in several forms during hot weather, in the cultivated and wild state. The cultivated forms have broad leaves whereas in wild; both broad and small leaves are seen.

Botanical Description

Botanical name of purslane is *Portulaca oleracea* L. and belongs to family Portulaceae. It is polyploidy and having chromosome number 2n=18, 36, 45 and 54. Purslane completes their life cycle in short span of time, so it is called ephemeral plant. The crop matures in 2-4 months after seeding. Fruits mature 7-12 days after fertilization.

Table 8.46: Botanical Description

Growth Habit	This is an annual succulent herb. It is tiny straight growing creeper whch attain the height of 40.0-50.0 cm.
Leaf	Leaves are green, fleshy thick and look like spetula. Leaves are oppositely on the stem or in whorl.
Floral biology	Flowers are yellow in colour. Flower has 4-5 sepals. Stamens are 7-10 in number and stigma is 5 lobed.
Fruit and fruit set	Fruits are round (4.0-5.0 mm in size.)
Seeds	Numerous black seeds are found fruit which spread easily through wind or water.

Botanical Classification

Botanically purslane can be grouped under two:

(i) *Portulaca oleracea* ssp. *oleracea*

(ii) *Portulaca oleracea* ssp. *sativa*-It is hexaploid in nature. Plants of this group are long and have more canopies.

REFERENCES

1. Bhat, K.L. 2007. Minor Vegetables: Untapped Potential, Kalyani Publisher, Ludhiana.
2. Gopalan, G., Rama Sastri, B.V., Balasubramanian, S.C., Rao, B.S.N., Deosthale, Y.G. and Pant, K.C. 1999. Nutritive value of Indian Foods. National Institute of Nutrition, ICMR, Hyderabad, India.
3. Koch, H.P. 1988. Purslane. Omega 3 fatty acids in an old medicinal plant.*Dtsch. Apoth. Ztg*., **128(47)**:93.
4. Lagenberg, J.P., Jjaden, U.R., De Vogel, E.M. and Langerak, D.I. 1986. Determination of phyloquinine (viatamin K_1) in raw and processed vegetables using reversed phase HPLC with electroflurometric detection. *Acta Aliment*., **15(3):** 187.
5. Mercadenate, A.Z. and Rodriguez Amaya, D.B., 1990. Carotenoid composition and vitamin A value of some native Brazilian green leafy vegetables. *Int. J. Food Sci. Technol*., **25:** 213.
6. Omara-Alwda, T.R., Mebrhatu, D.E., Prior and Ezekwe, M.O. 1991. Omega-3 fatty acids in purslane (*Portulaca oleracea*) tissues. *J. Am. Oil Chem. Soc*., **63(3):** 198.
7. Rahman, M.M., Wahed, M.A. and Akabar Ali, M. 1990. β-carotene losses during different methods of cooking of green leafy vegetables in Bangladesh. *J. Food. Comp. Anal*., **3(1):** 147.
8. Simopoulos, A.P. and Salem, N. Jr. 1989. N-3 fatty acids in egg from range fed Greek chickens. *New Engl. Med*., **321(20):** 1412.
9. Simopoulos, A.P., Norman, H.A., Gillaspy, J.E. and Duke, J.A. 1992. Common purslane: a source of Omega-3 fatty acids and antioxidants. *J. Am. Coll. Nutr*. **11(4):** 374.
10. Susiarti, S. 1993. *Portulaca* L. In: Prosea. Plant Resources of South East Asia 8. Vegetables (Siemonsma, J.S. and Piluek, K. Eds.). Pudoc Scientific Publishers, Wageningen.

Genus *Anethum*

Dill or Sowa (*Anethum graveolens* var. *sowa*)

Introduction

Dill or Sowa (*Anethum graveolens* var. *sowa*, 2n=32) commonly known as Sowa Palak because Sowa is grown and sold alongwith Palak in the maeket. The genus name *Anethum* is derived from Greek word aneeson or aneeton, which means strong smelling. It is very popular among winter leafy vegetables. Leaves are aromatic and emits pleasant aroma when cooked alone or with other vegetables as mixed preparations. In few countries Dill is used as salads. Nutritively it is very important and contains sufficient quantity of carotene and iron. Per 100 g of edible leaves pertains 98.0 g moisture, 3.0 g protein, 0.5 g fat, 2.2 g minerals, 1.1 g fiber, 5.2 g carbohydrates, 37 K. calorie energy, 190.0 mg calcium, 42.0 mg phosphorus, 17.4 mg iron, 7182 micro gram carotene, 0.13 mg riboflavin and 0.20 mg niacin (Gopalan *et al*., 1999). Chawla *et al*. (1988) reported that 1.0 kg sowa leaves contain 109.0 mg iron but its *in vitro* availability is only 3.7 per cent. Quantity of lysine (450.0mg) is highest among the different types of essential amino acids like tryptophan (30.0 mg), methionine (20.0), threonine (120.0 mg), leucine (290.0 mg), isoleucine (350 mg) and valine (280.0 mg) per gram nitrogen available in the leaves. Leaves, tender stems and flowers contain 0.6 per cent volatile oil. Its common use in Ayurvedic medicine is in abdominal discomfort, colic and for promoting digestion. Ayurvedic properties of shatapushpa are katu tikta rasa, usna virya, katu vipaka, laghu, tiksna and snigdha gunas. It cures 'vata', 'kapha', ulcers, abdominal pains, eye diseases and uterine pains.

Seeds of Sowa are equally important in spices and it may be substitute of 'Sah Jira" (Sharma *et al*., 2001). Seeds contain 3.0-4.0 per cent volatile oil. Due to availability of 30-63 per cent carvone in seed oil, it is largely used in medicine and in preparation of scented food items. Dill seed's most commonly used in the United States is to flavor cucumber pickles. Use 2-3 teaspoons per quart for dill pickles. These seeds are also used in Scandinavian cooking to flavor bread potatoes, vegetables, sauerkraut. For German pork roast, use 1 teaspoon per pound of meat.

Some therapeutic uses are:

- Dill is used to sooth the digestive tract and treats heartburn
- Patient suffering from jaundice should take regular to get rid off trouble.
- Powder of fresh dill seed with ghee is taken to get desirable progeny. Even sterile woman can conceive by using this preparation.
- Decoction of dill is used in indigestion, vomiting and promoting milk in delivered woman.

Table 8.47: Popular Name of Crop in different Countries

Country	*Language*	*Name*	*Dish*
Pakistan	Urdu	Soya	-
Pakistan	Punjabi	Soa	-
India	Marathi, Konkani	Shepu	Shepuchi Bhaji
India	Hindi	Savaa	Savaa Bhaji with potato. As a flavor in: Green Kheema, Kheema samosa.
India	Kannada	sabbasige soppu	
India	Telugu	Soa-Kura	
India	Manipuri	Pakhon	Chagem pomba
India	Punjabi	Soa	
India	Gujarati	Suva	Suvaa ni Bhaji(with potato)
Iran	Persian	Shevid	Aash, Baghali Polo, Shevid Polo, Mast O Khiar
Arab world	Arabic	shabat, shabath	As flavoring in various dishes
Thailand	Thai	phak chee Lao	Gaeng om
India	Malayalam	Chatapushpa	
China	Chinese	shiluo	Baozi

Origin, History and Distribution

Dill is one of the earliest cultivated herbs, showing up in Neaolithic burial sites. As per historical records sowa was cultivated by Greek and Roman (found in ruins) people in olden days. Through roman soilders in 1000 AD, it reached to Europe, where it was called as 'Dill'. It is largely cultivated in England, Finland, Germany, Rumania, *etc.* for fresh use and industry. In India its commercial cultivation is being done in Gujarat, Rajasthan, Madhya Pradesh and Andhra Pradesh.

Species Availability, Evolution and Diversity

The Plant List includes 28 scientific plant names of species rank for the genus *Anethum*. Of these 4 are accepted species names.

Table 8.48: Accepted Species by World Community

Species	*Status*	*Source*
Anethum foeniculoides Maire and Wilczek	Accepted	WCSP (in review)
Anethum graveolens L.	Accepted	WCSP (in review)
Anethum sowa Roxb. ex Fleming	Accepted	WCSP (in review)
Anethum theurkauffii Maire	Accepted	WCSP (in review)

Botanical Description

The botanical name of Sowa or Dill is *Anethum graveolens* var. *sowa* and belongs to family Apiaceae (Umbelliferae).

Table 8.49: Botanical Description

Growth Habit	It is herbaceous plants and attain the height of 50.0-150.0 cm.
Root	The root is long, and slender.
Stem	Stems are round, slender and after drying become hallow.
Leaf	Leaves are compound and leaflets are fine like needles. Alternate leaves are finally divided three or four times into pinnate sections slightly broader than similar leave of fennel. Lower leaf petioles are larger than new leaf petioles.
Floral biology	Flowers are compound white to yellow flower deve opes in to umbel (Warrier *et al.*, 1994). Stalk of umbel is quite large. Flowers of umbel are small, rogular and yellow in colour. Flower has 5 calyx, 5 joint crolla, 5 petals, and 5 stamens. Filament is yellow in colour, ovary two celled, and inferior. Sizes of primary umbels are larger than the secondary and teriary umbel. The seed recovery is poor from the big umbel therefore for more recovery of seeds; medium size umbels are quite good (Biz, 1966).
Anthesis and pollination	Crop is self pollinated and pollination take place mainly by bees. Dehiscence takes place in the morning.
Fruit and fruit set	Dill fruits are oval, compressed, winged about one-tenth inch wide, with three longitudinal ridges on the back and three dark lines or oil cells (vittae) between them and two on the flat surface. The taste of the fruits somewhat resembles caraway.
Seeds	The seeds are not true seeds. They are the halves of very small, dry fruits called schizocarps. The seeds are smaller (4–5 mm), thick (1.0 mm), straight to slightly curve with a longitudinally ridged surface flatter and lighter than caraway and have a pleasant aromatic odor. Seeds are viable for 3–10 years.

Horticultural Classification

Anethum graveolens L. is the sole species of the genus Anethum, though classified by some botanists in the related genus *Peucedanum* as *Peucedanum graveolens* (L.), (Pulliah, 2002). A variant called east Indian dill or *Sowa* (*Anethum graveoeloens* var *sowa* Roxb. ex, Flem.) occurs in India and is cultivated for its foliage as a cold weather crop throughout the Indian sub-continent, Malaysian archipelago and Japan.

Augmentation of Genetic Resources and Crop Improvement

There are certain local strains which are popular in their area of adoption as cultivars. Seed production, 'Mehsana' and 'Rubi Local' is popular in Gujarat but Rajasthan 'Pratapgarh Local' is dominating.

REFERENCES

1. Chawla, S. Saxena, A. and Seshadri, S. 1988. *In vitro* availability of iron in various green leafy vegetables. *J. Sci. Food Agric.*, **46:** 125
2. Gopalan, G., Rama Sastri, B.V., Balasubramanian, S.C., Rao, B.S.N., Deosthale, Y.G. and Pant and K.C. 1999. Nutritive value of Indian Foods. National Institute of Nutrition, ICMR, Hyderabad, India.
3. Pulliah T. 2002. Medicinal Plants in India. Vol. 1. New Delhi: Regency Publications New Delhi; p. 55–6
4. Sharma, R.K., Bhati, D.S. and Bhatt, B.N. 2001. Masalon ke Kheti. ICAR, New Delhi.
5. Warrier, P.K., Nambiar, V.P.K., Ramakutty, C. and Arya, V.S. 1994. Vol. 1. Kottakkal Madras, India: Orient Longman Limited; 1994. Indian Medicinal Plants; p. 153–154.

Genus *Tetragonia*

New Zealand Spinach (*Tetragonia tetragonioides* (Pallas) O. Kuntze)

Introduction

New Zealand Spinach (*Tetragonia tetragonioides* (Pallas) O. Kuntze, 2n=32) is not real spinach but it is popular and utilized as spinch vegetable. Succulentleaf and tender stems/twigs are used as vegetable. It is very rich in calcium, phosphorus, iron, vit.A, B and C. Per 100 g of edible portion of New Zealand spinach contains 88.6 g moisture, 0.65 g nitrogen, 0.46 g ether extract, 1.00 g crude fiber, 2.16 g fiber, 375.9 mg calcium, 68.0mg phosphorus, 2.48 mg iron, 7.64 mg carotene, 0.03 mg thiamine, 0.19 mg riboflavin, 1.49 mg nicine and 36.3 mg ascorbic acids (Wealth of India, 1969). Apart from this New Zealand spinach on dry weight basis contains 50.9-75.6 micro gram iodine and 0.009 mg cobalt per 100 g (Gopalan *et al.*, 1999). Stems conatin saponin but in seed saponin is negligible. Saponins lost after boiling of leaves but it does not affect the quality of vegetable and other constituents. Oxalate conten in this vegetable is much higher than common spinach therefore availability of calcium in this vegetable become unavailable to human body. It is true that in New Zealand spinach content of vit. C is very high, which is not only invigorate the body but also prevents the infection of lungs and intestine.

Origin, History and Distribution

Wild strains of New Zealand spinach are available abundantly in Australia, Tasmania, New Zealand, Japan, China and Taiwan. As per historical records, it was first grown in Europe for vegetable purpose. Later on this vegetable crop spreads to other temperate and sub-temperate countries. In south Asian countries like Indonesia and Philippines, grow New Zealand spinach commercially. In India, New Zealand spinach is commercialy grown in north Bengal, Meghalya and other hilly regions like Kalimpong, Darjeeling and Kranataka in south India.

Species Availability, Evolution and Diversity

Under genus *Tetragonia* several species are reported.

Botanical Description

New Zealand spinach (*Tetragonia tetragonioides* (Pallas) O. Kuntze) belongs to family Aizoaceae.

Table 8.50: Botanical Description

Growth Habit	It is annual in nature. Plants are bushy (grow up to the height of 1.0 m) and herbaceous.
Stem	On the stem, white colour papillas are found.
Leaf	Leaves are succulent, dark green in colour, triangle shaped, 12.0 cm long and 5.0 cm broad. Leaves develops on stem and branches from all the side.
Floral biology	Flowers are small, yellowish green in colour and develop from the axil of leaves. Stamens 4-10 and ovary 3-9 celled.
Fruit and fruit set	Fruits are Drup, triangle in shape and having 3-4 ridges like swellening.
Seeds	Each fruit has 4-10 seeds.

REFERENCES

1. Gopalan, G., Rama Sastri, B.V., Balasubramanian, S.C., Rao, B.S.N., Deosthale, Y.G. and Pant, K.C. 1999. Nutritive value of Indian Foods. National Institute of Nutrition, ICMR, Hyderabad (India)
2. Wealth of India, 1969. A Dictionary of Indian Raw Materials and Industrial Products. Publication and Information Directorate, CSIR, New Delhi.

Genus *Corchorus*

Vegetable Jute Leaf (*Corchorus* spp.)

Introduction

Vegetable jute (*Corchoru solitorius* and *Corchorus capsularis*, 2n=14) is an important fiber containg vegetable which tender and smooth leaves are used for culinary purpose and attaining commercial value since 19th century. Vegetble type is not different from the fiber yielding genotypes but few strains are there which realy form distinguish characters like small and bushy plants. Jute leaves are rich source of β-carotene. Not only this, leaves also contain 18 per cent protein. As per analytical observation, leaves with bitter taste are having more protein than the sweet tatse genotypes. Jute leaves are bitter due to corchorin glucoside; therefore it is also called as 'bitter leaf plant' and the genotypes which are free from bitter principle is called 'sweet leaf plant'. Edible jute leaves per 100 g contains 81.4 g moisture, 5.1 g protein, 1.1 g fat, 2.7 g mineral, 1.6 g fiber, 8.1 g carbohydrates, 63 K. calorie energy, 241.0 mg calcium and 83.0 mg phosphorus (Gopalan *et al.*, 1999). Apart from these nutrients, jute leaves contains several amino acids in protein.

Table 8.51: Amino Acids Conten in different Species of Edible Jute (mg/g on dry weight basis)

Amino Acids	*Name of Species*	
	Corchorus tridens	*Corchorus trilocularis*
Arginine	11.98	16.73
Histidine	5.70	8.10
Lysine	11.65	12.85
Metheonine	2.99	3.04
Cystine	1.56	1.62
Phenyl Alenine	11.71	1.90
Tyrosine	8.01	8.61
Leucin	23.74	22.68
Isoleucine	14.16	11.16
Valli	13.1	13.38
Threonine	8.38	9.98
Tryptophan	2.25	2.87

Source: Oliviera, and Carvalho, 1975.

Several wild species of jute are used as salads in Africa; therefore it is an urgent need of introduction of these prize genotypes in India. The regions where main diets are basically rice (carbohydrates), jute may be an important leafy vegetable.

Origin, History and Distribution

Origin of jute is not certain but in several countries like Bangladesh, India, China, Nepal *etc.* are growing *Corchorus olitorius* and *Corchorus capsularis* at large scale (Edmonds, 1990, Purseglove, 1968). *Corchorus olitorius* is commercially grown for vegetable purpose in Middle East and Africa, South East Asia and Latin America. If we see the status of jute production in India and Bangladesh, both dominate in world by producing 95 per cent of world jute but utilization as vegetable is limited.

Species Availability, Evolution and Diversity

Under family Tiliaceae total 40 genus and 400 species are reported worldwide and under genus *Corchorus* itself 50-60 species are available. Two species *i.e. Corchorus capsularis* and *Corchorus olitorius* are largely grown for jute (fiber) purpose. *Corchorus olitorius* is more popular as vegetable crop but *C. astuans* L., *C. depressus*, *C. Christensen*, *C. fascicularis*, *C. tridens* and *C. triculars* are also used for culinary purpose in different parts of the world. Most of the species are diploids but tetraploid (2n=28) are also available.

Gene and Genetics of Characters

In *Corchorus capsularis* bitterness is prominent and show monogenic inheritance Tn-tb (bitter non bitter). The bitter gene is linked, with gene responsible for branch emergence.

Botanical Description

Jute belongs to the family Tiliaceae. Basically it is a short day plant. For flowering it needs 30 days continuously photo period of 10-12 hours.

Table 8.52: Botanical Description

Growth Habit	All the reported species may be tree; bush or herbs. The vegetable type plant attains the height of 30.0-150.0 cm. At the apex of plant branches emerge.
Leaf	Leaves are alternate, smooth, light green, stipules; leaf petiole is 1.0-2.0 cm long, and lanceiolate.
Floral biology	Flowers are axillary, calyx 5, corolla 5, yellow colour, 7.0-8.0 mm long, stamens 10, filament small, pollen sac small, ovary penta celled, stigma small and spread.
Anthesis and pollination	Jute is a self pollinated crop.
Fruit and fruit set	Fruits are capsule and having 10 stripes.
Seeds	Seeds are brownish or bluish colour and 1.0-2.0 mm long. Per gram 500 sedds are counted.

Augmentation of Genetic Resources and Crop Improvement

International Plant genetic Research Institute, Rome is maintaining a lage number of vegetable type's germplasm. Apart from this organization some germplasm are being maintained by Libia, Sudan and

Zimbabwe and same materials are also being conserved by Italy, Germany and India (Toll and Van Stolen, 1982).International Jute Research Institute, Dhanka (Bangladesh) is maintaining a large collection of jute. Horticultural Research Institute, Nigeria is conserving a large number of vegetable type jute (Denton, 1993).

Genetic Resource Conservation

ICAR-Central Research Institute for Jute and Allied Fibers (CRIJAF), Barrackpore (W.B.) maintaining huge germplasm of Jute as NAGS and also provided all the germplasm to National Gene Bank for long term storage.

REFERENCES

1. Denton, R.I. 1993. Vegetable jute (*Corchorus*) In: Underutilized crops. Pulses and vegetables (ed) Williams, J.T. Chapman and Hall. London. p. 167-174.
2. Edmonds, J.M. 1990. Herbarium Survey of African *Corchorus* L. species, systematic and ecogeographical studies on crop genepools, 4. IBPGR IJO, Rome, Italy, p. 284.
3. Gopalan, G., Rama Sastri, B.V., Balasubramanian, S.C., Rao, B.S.N., Deosthale, Y.G. and Pant, K.C. 1999. Nutritive value of Indian Foods. National Institute of Nutrition, ICMR, Hyderabad (A.P.), India.
4. Oliviera, J.S. and Carvalho, M.F.1975. Nutritional value of some edible leaves used in Mozambique. *Econ. Bot.* **29:** 255-263.
5. Purseglove, J.W. 1968. Tropical crops: Dicotyledons. Longman, London.
6. Toll, J. and Van Stolen, D.H. 1982. Directory of germplasm collection 4. Vegetables, IBPGR, Rome.

Genus *Hibiscus*

Indian Sorrel or Gogu (*Hibiscus sabdariffa* Linn.)

Introduction

Indian Sorrel or Gogu (*Hibiscus sabdariffa* Linn. 2n=4x=72) is very popular leafy vegetable during rainy season. It is known as Gogu, Rosseli, Red Sorrel, Indian Sorrel or Zamaica Sorrel. In India, it is known by several vernacular name like Gogura (Telagu); Mest Pat (Bengali); Amabadi (Gujarati and Marathi); Pundi (Kannad); Nalite Sag (Odishi), Puliche Kirai (Tamil) and Chhukiar (Assamese). Its tender stems and leaves are used for culinary purpose. In south Indian dishes like 'Rasam', soup, *etc.* are prepared by using Indian sorrel leaves because leaves are sour in taste. Apart form the tender leaves and calyxes are used as salads and preparation of local dish 'Kari'. Leaves are rich source of iron, carotene and vit.C. Per 100 g of edible Indian sorrel contains 86.4 g moisture, 1.7 g protein, 1.1 g fat, 0.9 g minerals, 9.9 g carbohydrates, 56 K. calorie energy, 172.0 mg calcium, 40.0 mg phosphorus, 2.28 mg iron, 2898 micro gram carotene, 0.07 mg thiamine, 0.39 mg riboflavin, 1.1 mg niacin and 20.0 mg vit. C.Apart from this, tender edible 100 g leaves contains 66.0 mg maganesium, 0.08 mg copper, 0.03 mg maganese, 0.27 mg zinc, 0.005 mg chromium, 60.0 mg sulpher and 19.0 mg chlorin (Gopalan *et al.*, 1999). Indian sorrel is rich source of protein and in protein generally all essential amino acids are present particularly high amount of lysine which is mixed with other cereals and used. Leaves contains 310.0 mg arginine, 90.0 mg histdine, 400 mg lysine, 40.0 mg tryptophan, 200.0 mg phenyl alenine, 70.0 mg metheonine, 210.0 threonine, 440.0 mg lecine, 320.0 mg isoleucine and 240.0 mg vallin per gram nitrogen (Gopalan *et al.*, 1999). The edible calyx contains 3.74 per cent citric acid and 3.19 per cent pectin which is used in 'jelly' preparation. Apart from this calyx contains 88.26 per cent moisture, 1.46 per cent crude protein, 1.97 per cent ether extracts, 5.86 per cent carbohydrates, 1.58 per cent crude fiber, 0.87 per cent ash, 0.018 per cent calcium, 0.052 per cent phosphorus and 0.021 per cent iron (Wealth of India, 1969).Calyx contains water soluble acids like citric acid, D Malleic acid, tartaric acid and hibiscus acid. Sufficient amount of micillage and calcium citrate and ascorbic acids are also found in petals.

Origin, History and Distribution

In India, this crop is as a vegetable crop in Punjab, Andhra Pradesh, Uttar Pradesh, West Bengal, Odisha, Maharshtra and Tamil Nadu.

Botanical Description

Indian Sorrel or Gogu (*Hibiscus sabdariffa* Linn.2n=4x=72) belongs to the family Malvaceae.

Table 8.53: Botanical Description

Growth Habit	Indian Sorrel or Gogu is annual, bushy and erect growing plants
Stem	Stems are red or green colour. In general, there is no branches emerging from the stem but sometimes from the base, branch may develop. Stems are smooth and white hairs like structure are present.
Leaf	Leaves are serated. Lower leaves are undivided but upper leaves are divided in 3-5 lobed.
Floral biology	Flowers are large, and yellow to pink in colour. Epicalyx compound and calyx are joint.
Fruit and fruit set	Fruits are conical shapes which are full of black seeds.
Seeds	Seeds are black.

Horticultural Classification

On the basis of horticultural classification, Indian Sorrel or Gogu on the basis of colour can be grouped in 4 sub species:

(i) *Ruber*
(ii) *Albus*
(iii) *Intermidius*
(iv) *Bhagal-puriensis*

Among them 'Ruber' is considered as an excellent for vegetable purpose.

In Philippine 4 edible types were isolated

(i) Victor
(ii) Rico
(iii) Archer
(iv) Temprono

Augmentation of Genetic Resources and Crop Improvement

In India, Indian Sorrel or Gogu genotypes has been collected and conserved at N.G.R., Andhra Pradesh Agriculural University (Hyderabad) and Jute and Allied Fiber Crop Research Instiotuite, Barrackpore (West Bengal). Other institutions like Austrellian Forage Genetic Resource Centre, C.S.I.R.O, Quinceland, Australlia and International Jut Organization, Dhaka (Bangladesh) ale also maintaining the germplasm f this crop.

REFERENCES

1. Gopalan, G., Rama Sastri, B.V., Balasubramanian, S.C., Rao, B.S.N., Deosthale, Y.G. and Pant, K.C. 1999. Nutritive value of Indian Foods. National Institute of Nutrition, ICMR, Hyderabad (A.P.), India.
2. Wealth of India, 1969. A Dictionary of Indian Raw Materials and Industrial Products. Publication and Information Directorate, CSIR, New Delhi.

Genus *Oxalis*

Amboti (*Oxalis corniculata* L.)

Introduction

Amboti (*Oxalis corniculata* L.) is also known as 'Indian Sorrel' which is consumed during emergency by some tribal communities of central India. Its leaves serve as the major means of food for the tribals during emergency since long back and proven to save the life of tribals during severe famine and extreme scarcity. The leaves of the plant are eaten both raw as salad and cooked. Leaves are also used for making sandwiches; chutney, pickles and is considered as cooking retrigerant by the Sahariya tribe of central India. Its leaves are used to cure fever, piles, scurvy and dysentery (Kirtikar and Basu, 1975). Nutritional composition (per 100 g dry sample) of amboti contains 82.042±0.5, carbohydrates 24.67±0.4, crude protein 22.28±0.5, crude lipids 23.75±0.5, sodioum 1.12±0.02, potassioum2.17v0.31, calcium 2.5±0.08, nitrogen 3.56±0.70 and magnesium 0.25±0.03 (Jain *et al.*, 2010). Amboti can be used as supplementary food source to deal with the problems of limited food production in the tribal areas

Origin, History and Distribution

Amboti can be spotted in central India. It is abundantly available in M.P.

Species Availability, Evolution and Diversity

The genus *Oxalis* contains approximately 800 species and is found principally in South America and South Africa. A few species such as *O. ucetosella* and *O. corniculata* are cosmopolitan extending into Europe, Asia and Australia. Knuth (1930) distinguishes 37 sections, of which 5 contain the South African species. Salter (1944) gives a revised classification of the South African material and describes 208 species in 11 sections.

Botanical Description

Amboti (*Oxalis corniculata* L.) belongs to family Oxalidaceae. The most comprehensive study of chromosome numbers in the genus is that of Heitz (1927), who listed the numbers of 26 species. Later contributions by Wulff (1934), Matsuura and Sato (1935), Kostoff *et al.* (1935), Yamashita (1935), Nakajima (1936), Warburg (1938) and Rutland (1941) made available the chromosome numbers of 12 more species. There is a fairly wide range of basic numbers *viz.* 5, 6, 7, 9, 11 (Darlington and Janaki-Ammal, 1945) with 7 as the commonest (Warburg, 1938). The chromosomes are small with little variation between species although Heitz (1927) noted that the chromosomes of *O. bupleurifolia* are considerably larger than those of other species. In the presented research paper the chromosome numbers of 24 additional species are given of which 16 are South African. These new data enable the South African and South American species to be compared cytologically. There appears to be a difference between them but it requires confirmation. The lowest most reasonable factor has been taken in evaluating the basic number for any particular species. In the case of *O. caprina* and *O. eriolepis* both n = 20 and *O. ciliaris* and *O. imbricata* both 2n = 40, it is debatable whether the basic number is 5 or 10. Also the basic number for *O. corniculata* and *O. hirta* might be 6 or 12. Both alternatives for these species are considered. Species with 2n = 30 chromosomes are fertile and are considered to be hexaploid with 5 as the basic number. There is a far greater range of basic chromosome numbers in the South.

Table 8.54: Botanical Description

Growth Habit	It is creeping herb. All the South African species are, with the exception of the annual non-bulbous *O. corniculata*, small bulbous perennial herbs (Salter, 1944). Amongst the South American species there exist a range of types from dwart woody shrubs such as *O. dispar* to small herbs like *O. chrysantha*. It is noteworthy that *O. dispar* (Fig. 3, *y*) and *O. hupleurifolia* (Fig. 3, 11) and the semi-woody *O. ortgesii* have large chromosomes. The vast majority of South African species are sun-loving plants growing on open plains and hillsides. *O. incarnata* is a shade-loving species. Geographically they are found mostly in Cape Province. The South American species occur in diverse habitats throughout South America and the southern part of North America.
Root	Rooting starts from every node.
Stem	Creeping stem
Leaf	Leaves are long petioled, leaflets with white hairs on margins and lower portion of the middle vein, petioles pubescent 2-4 cm long.

REFERENCES

1. Bailey, L.H. 1950. The Standard Encyclopedia of Horticulture. The MacMillan Co., New York.
2. Jain, A.K., Tiwari, P. and Bashir, M. 2010. Nutritive aspects of *Oxalis corniculata* L. used by tribals of Central India during scarcity of food. *Botany Research International*, **3(1):** 35-37.
3. Kirtikar, K.R. and Basu, B.D. 1975. *Indian Medicinal Plants*, Vol. I-Periodical Experts, 436-440, 645, 865.
4. Dalincton, C.D. and Janaki-Ammal, E.K. 1945. *The Chromosome Atlas of Cultivated Plants*. London.
5. Dalincton, C.D. and Lacour, L.F. 1947. *The Handling of Chromosomes*. London.
6. Heitz, E. 1927. Jber multiple und aberrante Chromosonizahlcn.*4bh. Geb. Natiinr.*, **21:** 47.
7. Jain, A.K., Tiwari, P and Bashir, M. 2010. Nutritive aspects of *Oxalis corniculata* L., used by tribals of Central India during security of food. *Botany Research International*, **3(1):** 35-37.
8. Kirtikar, K.R. and Basu, B.D. 1975. Indian Medicinal Plants, Vol.I, Periodical Experts, 436-440, 645, 865.
9. Knuth, R. 193). *Oxalidaceae. Das Pflanzenreich*. Berlin.
10. Kostoff, D., Dogadkina, N. and Tichonov.A 1935. Chromosome numbers in certain angiosperm plants. C.R. *Acad. Sci. U.R.S.S.*, **3:** 401.
11. Matsuura, H. and Sato, T. 1935. Contributicins to the idiof^ram study in phanerogamous plants I. *Y. Fac. Sci. Hokkaido Univ.*, **5:** 33.
12. Nakajima, G. 1936. Chromosome numbers in some crops and wild angiosperms. *Jap.J. Genet.*, **12:** 211.
13. Rutland, J. P. 1941. The Merton Catalogue. *New Phytol.*, **40:** 210.
14. Salter, T. M. 1944. The genus *Oxalis* in South Africa. *Suppl. J.S. Afr. Bot.*, **1:** 1.
15. Warburg, E. F. 1938. Taxonomy and relationships in the Geraniales in the light of their cytology. *New Phytol.*, **37:** 189.
16. Wulff, H. D. 1937. Chromosomen studien an der schk-swigholsteinischen Angiospermen - Flora I *Ber. Dtsch. Bot. Ges.*, **55:** 262.
17. Yamashita, K. 1935. Zytologische Studien an *Oxalis I. Jap. J. Genet.*, **11:** 36.

Mustard Leaves (*Brassica juncea*)

Introduction

Indian mustard [*Brassica juncea* (L.) Czern. and Coss. 2n=4x=36] is an oil seed crops but very popular as 'Sag'. When plants are small and leave are tender and succulents, generally used as vegetable. 'Mustard Sag' is eaten in Punjab, Haryana and wester Uttar Pradesh to a greater extent. Singh and Joshi (1960) expressed that in Punjab and Haryana, 'Mustard Sag' is eaten with bread of maize, butter, and whay. In Uttarkhand people like to grow this crop in kitchen garden. Nutritionally and medicinally it is important crop. It is a rich source of vit. C, E, fiber, folate, calcium, iron and potassium. Bioavailability of calcium in the leaf is very high. Unlike other *Brassica* leafy vegwetables, leafy mustared can tolerate high temperature and high humidity. In plains, it can be sown from end of September to middle of November but in hilly area sowing is done May to August.

Table 8.55: Nutritive Value of Mstard and Lahi Patta (per 100 g of edible portion)

Nutrients	*Ntritive Value*		
	Mustard	*Lahi Patta*	
		Stem	*Leaf*
Moisture (g)	98.8	91.4	84.9
Protein (g)	4.0	3.1	5.1
Fat(g)	0.6	0.1	0.4
Minerals (g)	1.6	1.4	2.5
Fiber (g)	0.8	-	1.2
Carbohydrates (g)	3.2	4.0	5.9
Energy (K Calorie)	34	29	49
Calcium (mg)	155.0	100	370
Phosphorus (mg)	26	100	110
Iron (mg)	16.3	1.2	12.5
Carotene (micro gram)	2622	-	1380

Nutrients	Ntritive Value		
	Mustard	Lahi Patta	
		Stem	Leaf
Thiamine (mg)	0.03	-	0.01
Riboflavin (mg)	-	-	0.9
Nicine (mg)	-	-	0.9
Vit. 'C' (mg)	33.0	-	65

Source: Leung, 1980.

Origin, History and Distribution

Since, it is amphidiploid in nature, so it clears that its origin will be place where its parental species (*B. nigra* and *B. rapa*) are found. Perhaps it is only Middle Asia. Its primary centre of diversity is Middle Asia (north western India which comprises Punjab and Kashmir) and secondary centre of diversity is middle and western regions of China, Germany, France, Australia, Pakistan, Poland, Eastern India, Mayanmar and Asia miner (Hemingway, 1976). In different places peoples grow different genotypes mustard as per liking and adaptation in the agro-climatic conditions. For example in Assam, West Bengal and NEH regions, peoples grow commercially with the name 'Lahi' and sell in the market. In Hong-Kong, Muichhoi (*Brassica juncea* var. *sareptna*) is grown as a winter vegetables. Major Indian mustard producing countries include Canada, China, and India. *B. juncea* is an amphidiploid and second most important edible oil seed crop in India after groundnut and accounts for about 30 per cent of the total oil seeds produced in the country. Indian mustard is cultivated in the states of Punjab, Rajasthan, Uttar Pradesh (UP) Assam, Gujarat, Haryana, Madhya Pradesh (MP), and West Bengal (WB) as a Rabi crop (Duhoon *et al.*, 1998; Dutta *et al.*, 2008; Misra *et al.*, 2010).

Species Availability, Evolution and Diversity

Mustard is a polymorphic crop and classified different species are as under:

Table 8.56: Classification of Mustard

Sl.No.	Botanical species	Remarks
1.	*Brassica juncea* var. *crispifolia*	Leaves are attractive, serated or curved
2.	*Brassica juncea* var. *foliosa*	Leaves are very large serated margins
3.	*Brassica juncea* var. *japonica*	Leaves are attractive, serated or curved
4.	*Brassica juncea* var. *multicepts*	It is multibranching species
5.	*Brassica juncea* var. *napiformis*	Plants develop swollen roots
6.	*Brassica juncea* var. *rugosa*	Leaves are thick like cabbage
7.	*Brassica juncea* var. *strumata*	Develop long edible petiolated leaves
8.	*Brassica juncea* var. *tumida*	Main stem is edible and swollen

The mustard (*Brassica juncea*, 2n=36, amphidiploid species) developed through natural cross between balck mustard (*B. nigra*, 2n=36) and Neep (*B. rapa* 2n=20).

Botanical Description

Indian mustard [*Brassica juncea* (L.) Czern. and Coss. 2n=4x=36] is a natural amphidiploids between *Brassica campestris* (2n=20) and *Brassica nigra* (2n=16). In India, *Brassica juncea* is a predominant species which accounts for nearly 80 per cent of the oil seed Brassicas. The botanical name is *Brassica juncea* (2n=36) belongs to family Brassicaceae.

Kingdom	:	Plantae
Division	:	Angiospermae
Class	:	Dicotyledons
Order	:	Brassicales
Family	:	Brassicaceae
Genus	:	*Brassica*
Species	:	*juncea*

The detail description is given is given below:

Table 8.57: Botanical Description

Growth Habit	Mustard is annual or biennial in nasture. Plants attained the height of 50.0-150.0 cm. In general, no branching on plants.
Root	Taproot reaching 60-80 cm in depth
Stem	Stems are white and smooth.
Leaves	Leaves are crisp and develop green and smooth petiole. Lower leaves petioled, green, sometimes with a whitish bloom, ovate to obovate, variously lobed with toothed or frilled edges; upper leaves subentire, short and petioled, constricted at intervals, sessile.
Floral biology	Plants bear numerous flowers. The flowers consist of 4 (four) yellow petals arranged in a cruciform manner, 4 (four) yellowish green sepals, a short green pistil with a knobby stigma, and tetradynamous stamens with yellow anthers.
Anthesis and pollination	They are pollinated by bees
Fruit and fruit set	Fruits are siliqua.
Seeds	Each siliqua bears 10-20 seeds which are black in colour.

Botanical Classification

In Brassica two groupes are avauilable:

(i) Oil yielding types-This group is generally grown in India, Bangladesh and china.

(ii) Leafy types- Vegetable types are dominating inn several parts of Asia. It is equally important and grown as major vegetable in Indonesia, Malaysia, Philippines and Thailand.

Augmentation of Genetic Resources and Crop Improvement

Rai *et al.* (1977) collected several genotypes of mustard (*Brassica rugosa*) from hilly regions of Almora (Uttarakhand) and developed two cultivars namely Selection-1 and Selection-902 through selection breeding process.

(i) **Selection-1-** This cultivar has 12.9 per cent dry matter and 28.3 per cent crude protein.

(ii) **Selection-902-**This cultivar has 15 per cent dry matter and 70 per cent digestability value.

In Pakistan cultivar "Local Selection (P)" is common and grown extensively.

Genetic Resource Conservation

A large number of leafy genotypes are being maintained in China. In India, the rich diversity of mustard can be spotted in NEH regions, Uttarakhand, Himachal Pradesh and Jamu and Kashmir. Apart from these countries, large collections are being maintained in Australia, Canada, Taiwan, Germany, Japan and USA.

REFERENCES

1. Hemingway, J.S. 1976. Mustard: *Brassica* spp. and *Sinapsis alba* (Brassicaceae/Cruciferae). In: Simonds (ed.) Evolution of Crop plants. Longmans, London, p. 56-59.
2. Rai, B., Tyagi, D.V.S. and Verma, R.B. 1977.*Brassica* selection for leadfy vegetables. *Indian Hort.*, **22(3):** 12-13.

Genus *Cicer*

Bengal Gram Leaves (*Cicer arietinum* Linn.)

Introduction

Bengal gram (*Cicer arietinum* Linn., 2n= 14 (brown seeded) and 2n=16 (Kabuli Chans)) is also known with the various vernacular names like Bat, Chhola (Bengali), Chana, Chania (Gujarati), Kadale (Kannad), Kadala, Kadalakka (Malyalam), Harbara (Marathi), Buto (Odishi), Chanaka, Harimantha (Sanskrit), Kadali (Tamil) and Sanagelu (Telgu). Bengal gram is a leguminous crop and generally grown for 'Dal' during *Rabi* season. When plants are small, the leaves and top portion or shoots, are used as leafy vegetables. Some people eat it without cooking. Leaves and tender shoots contains flavonoids like quercetin, iso-quercetin, ononin, Kaemoferol and 3 glucoside, 3 apiosylglucoside, 3 malonylglucoside, 3 apiosyl malonylglucoside, populnin, biochanin A-7glucoside, isorhmnetin, protensein or garbanzol and enzyme carbonanhydraze and phenolic compound rutin and caffeic, chlorogenin, protocatechuic, ferulic, β. resocylic, *p.* hydroxyl benzoic, *p.* coumaric and vanillic acids. Leaves are rich source of fiber and iron (23.8 mg per 100g).Per 100 g of edible portion, Chana leaves contains 73.4 g moisture, 7.0 g protein, 1.4 g fat, 2.1 g mineral, 2.0 g fiber, 14.1 g carbohydrates, 97 K. calories energy, 340 mg calcium, 120 mg phosphorus and 23.8 mg iron.

Origin, History and Distribution

According to Vavilov, India and Middle East Asia is probably original home of Bengal gram/Chana. It is true that introduction of 'Kabuli Chana' is new. Apart from India, it is largely cultivated in Pakistan, Mayanmar, and Bangladesh. It is also in cultivation in Etheopia, Mexico, Morocco, Turky and Iron.

Species Availability, Evolution and Diversity

Under genus Cicer 4 species are reported but *Cicer arietinum*is most important.

Botanical Description

Bengal gram/Chana, is member of Papilinaceae (Family Fabaceae). Its botanical name is *Cicer arietinum* Linn.

Table 8.58: Botanical Description

Growth Habit	Plants are erect or creeping in nature. Plants attain the height of 30.0-40.0 cm. Plants are herbs.
Stem	On whole plant glandular hairs are found which contains oxalic and malic acids, therefore leaves are sour.
Leaf	Leaves are pinnately compound which has 9-17 leaflets. Leaves are arranged in alternate position.
Floral biology	Floers are white or pink in colour.
Fruit and fruit set	Single fruit contain one or two seeds. Seeds are round and beaked. Seed coat is brown to white and green in colour.

Botanical Classification

Bengal gram/Chana, is divided in two horticultural groups:

(i) Wrinkled seeded-It is called 'Deshi Chana'

(ii) Kabuli Chana- It is round seeded cultivars.

Colocasia Leaves (*Colocasia* Species)

Introduction

In English this crop is known as Colocasia but In India, the vernacular name of this crop is Arabi Ka Patta, Arabi Ka Sag(Hindi); Kochu Sag (Bengali); Samangadde Yele (Kannad); Chembu Elagal (Malyalam); Alu Pan (Marathi); Serui (Odishi); Sevam Elegal (Tamil) and Chama Akulu (Telgu). Basically, colocasia is a tuber vegetable crop but its tender leaves are also used as leafy vegetable (Sag), 'Pakaure' and other items are prepared with the addition of 'Urid'. Its leaves and tubers both create irritation (due to calcium oxalate) in the throat. In an analysis, Nakanishi, *et al.*(1982) reported that colocasia leaves contain 0.008 per cent calcium oxalate. As nutrition, leaves contain more than 9000 micro gram β-carotene. In few areas, its leaf stalks are also used as vegetables. Sulphur base amino acids are very less in total amino acids. In Ghana (African country)' Palaver Sauce' is prepared from the colocasia leaves which are a 'National Dish' (Dako, 1981). Colocasia leaves contain apegine, 3'4 di Ome luteioline and anthocyanin flavonoids (Daniel, 1989). *Colocasia* genotypes having black leaves are richer in protein, carbohydrates, carotene and vit. C than the green leaves bearing genotypes. It is true that green leaf bearing genotypes are rich in iron.

Table 8.59: Nutritive Value of Colocasia Leaves (Black and Green) and Dry Leaves

Nutritients	*Stalk*	*Black Leaves*	*Green Leaves*	*Dry Leaves*
Moisture (g)	72.8	92.7	9.3	94.0
Protein (g)	6.8	3.91	3.7	0.3
Fat (g)	2.0	1.5	5.9	0.3
Minerals (g)	2.5	2.2	12.8	1.2
Fiber (g)	1.8	2.9	16.0	0.6
Carbohydrates (g)	8.1	6.8	42.3	3.6
Energy (K. calorie)	77	56.0	27.7	18
Calcium (mg)	460.0	227.0	1546	60
Phosphorus (mg)	125.0	82.0	308.0	20
Iron (mg)	98.0	10.0	-	0.5
Carotene (mg)	12000	10278	-	104
Thiamine (mg)	6.0	0.22	-	0.07
Riboflavin (mg)	0.45	0.26	-	0.07
Niacine	1.9	1.1	-	0.1
Vit. C.	63.0	12.0	-	3.0

Per 100 g of colocasia leaves contains 32 mg magnesium, 0.18 mg copper, 0.11 mg chromium and 72 mg chlorin (Gopalan *et al.*, 1999). Per gram nitrogen, its leaves contain various amino acids like arginine (250), histidine (130), lysine (260), tryptophan (40), phenyl alenin (180), methionine (60), threonine (150), leucine (360), isoleucine (310) and vallin (280).

Origin, History and Distribution

Colocasia probably originated from South East Asia and India (Watt, 1989; Chang, 1958) and Malayasia (Keloni, 1962). As per the ethno botanical study, India is the original home of colocasia (Spier, 1951). In China, 1000 BC, colocasia was considered as staple foods (Plucknett, 1976). In South East Asian contries, colocasia is very important and stable food commodity. Probably, coclocasia mooved from East India to Formosa and Soloman Islands (Spencer, 1966) and from there it reached to China (Plucknett, 1976). In India, its distribution can be seen in Manipur, Assam, Nagaland, Meghalaya, Gujarat, Maharashtra, Keral, Tamil Nadu and Madhya Pradesh.

Botanical Description

It is monocotyledons and belongs to the family Araceae (Aroidae) and sub family Colocasioidae which falls under Nudiflorae. *C. esculenta* is polymorphic in nature which is classified in *C. esculenta* var.*esculenta* and *C. esculenta* var. *antiquorum*. The somatic chromosome number 2n=28 (diploid) and 2n= 42 (triploid) has been reported by Sharma and Das (1954); Mookerjee (1955); Ramachandran (1978) and Kuruvilla and Singh (1981). Triploids cultivars in comparision to diploids are longer, vigorous, late maturing in natureand yield more.

Table 8.60: Botanical Description

Growth Habit	Taro is an herbaceous perennial herb that grows to the height of 1.0-2.0 m. Herb develop thick shoots from a large corm; slender stolons also often produced, along with offshoot corms.
Root	Roots are adventitious.
Leaf	Develop spathacious leaf. Leaf blades to 60.0 cm long and 50.0 cm wide, arrowhead shaped, with upper surface dark green and velvety; leaves peltate (stalked from back of blade), petioles large, succulent, often purplish near top. Venation is reticulate which is rare in monocotyledons.
Floral biology	Inflorescence on a fleshy stalk shorter than leaf petioles; part of fleshy stalk enveloped by a long yellow bract (spathe). Flowers tiny densely crowded on upper part of fleshy stalk, with female flowers (fertile) below and male flowers (sterile) above. Flowers are unisexual, protogynous, sessiole and without perianth.
Anthesis and pollination	Pollination before 24 hours of anthesis results more fruits development.
Fruit and fruit set	Fruit a small berry, in clusters (more than 100 fruits/ bunch) on the fleshy stalk.
Seeds	Each fruit contain 20-40 seeds. Colocasia seeds germinate in 7-8 days and seed has no dormancy. Potential seed germination is upto 3 months only and later on germination percentage decrease drastically and after 5th months, no germination.
Edible part	Edible part corm

Botanical Classification

On the basis tuber, taro is classified into two groups (Purseglove, 1972):

(i) **Dasheen type-** In 'dasheen' types, the corms are cylindrical and large. Generally, develop one corms (main cormsd) and secondary corms (cormels) do not develop (if develop, very less in number). *C. esculenta* var. *esculenta*has been placed (Onwueme, 1978).

(ii) **Eddoe type-** In 'eddoe' types, the main corms are small, globoid and are surrounded by several cormels (stem tubers) and daughter corms. In this group, *C. esculenta* var. *antiquorum* has been placed (Onwueme, 1978).

Augmentation of Genetic Resources and Crop Improvement

In taro, limitation in flowering, lack of synchrony, pollen sterility and protogyny are the main constraints. Only 10 per cent of the germplasm flower and only limited number of accessions set seeds.

Introduction and Selection

The taro germplasm includes cultivated and stoloniferous wild taro, which is present in Western Ghats. Elite germplasm 'Muktakeshi', was selected to be resistant to the serious taro leaf blight disease caused by *Phytophthora colocasiae*. In taro, the three varieties released are direct selection from germplasm. Sree Rashmi and Sree Pallavi are selections from Kerala and Meghalaya respectively. There are many improved varieties like Sree Rashmi, Sree Pallavi, Sree Kiran, Muktakesi, Pani Saru, Vallabh Nikki, Bilaspuri Arum, ML-1, ML-1, ML-9, *etc.* are very popular among the growers.

Table 8.61: Casava Varieties Released from CTCRI, Kerala

Sl. No.	*Variety*	*Maturity Period (months)*	*Starch (per cent)*	*Yield (t/ha)*
1	'Sree Pallavi'	7	24.0-25.0	15.0-18.0
2	'Sree Subhra'	7	15.0-16.0	15.0-20.0
3	'Muktakeshi'	5-6	16.0-18.0	20.0
4	'Sree Kiran'	7	16.0-18.0	18.0
5	'Panisaru-1'	6-7	12.0-13.0	16.0
6	'Panisaru-2'	6-7	16.0-18.0	13.0

Hybridization and Selection

In this crop first hybrid was released through hybridization of fertile diploid parents from the germplasm. Flowering inadequacies, occurrence of sterile triploid genetic stocks and protogynous nature of the spadix are the main obstacles for the genetic improvement of taro through hybridization and selection. Systematic screening of hybrid progeny resulted in releasing the first hybrid variety of taro in India under the name 'Sree Kiran' during 2004 from CTCRI.

Interspecific Hybridization

Inter crosses involving wild relatives were made to transfer leaf blight resistance to improved cultivars (Santha *et al.*, 1993).

Polyploidy Breeding

The high yielding cultivars are triploids, either non-flowering or pollen sterile; and it is difficult to improve them by conventional method.

Genetic Resource Conservation

In India, colocasia germplasm are being maintained at ICAR-CTCRI, Thiruvanantpuram (Kerala) and ICAR-NBPGR, Regional Station Thrissur (Kerala). In NEH region few genotypes are being maintained at ICAR-NBPGR, Regional Station Shillong (Meghalaya).

REFERENCES

1. Chang, T.K. 1958. Dispersal of taro in Asia. *Ann. Ass. Amer. Geogr.*, **48:** 255-256.
2. Dako, D.V. 1981. Potential of dehydrated leaves and cocoyam leaf protein in the Ghanian diet. *Nutr. Rep. Int.*, **23(1):** 181.
3. Daniel, M. 1989. Polyphenols of some Indian vegetables. *Curr. Sci.*, **58(23):** 1332.
4. Gopalan, G., Rama Sastri, B.V., Balasubramanian, S.C., Rao, B.S.N., Deosthale, Y.G. and Pant, K.C. 1999. Nutritive value of Indian Foods. National Institute of Nutrition, ICMR, Hyderabad, India.
5. Kuruvilla, K.M. and Singh, A. 1981. Karyotypic and electrophoretic studies on taro and its origin. *Euphytica*, **30:** 405-413.
6. Mookerjee, A. 1955. Cytology of different species of aroids with a view to trace the basis of their evolutuion. *Cytologia*, **7:** 221.
7. Nakanishi, Y.E., Maruyama, T.K. and Hasegawa, C. 1982. Acrid components of vegetable foods, Acrid Components of taro. *Kaseigaku Kenkyu.*, **29(1):** 75.
8. Onwueme, I.C., 1978. The Tropical Tuber Crops. John Wiley and Sons, New York, p. 234.
9. Plucknett, D.L. 1976. Edible Aroids. Evolution of crop plants (Simmonds, N.W. ed.). Longman group Ltd., England.
10. Purseglove, J.W. 1972. Tropical crops. Monocotyledons. 1. Longman Group Ltd., p. 334.
11. Sharma, A.K. and Das, N.K. 1954. Study of Karyotypes and their alteration in aroids. *Agronomia Lusitana*, 16: 23-48.

12. Ramachandran, K. 1978. Cytoplogical studies on South Indian Araceae. *Cytologia*, **43:** 289-303.

13. Spencer, J.E. 1966. Shifting cultivation in South East Asia. U. Cal. Publ. Geor., Vol. 19.

14. Spier, R.F.G. 1951. Some notes on the origin of taro. *South Western J. Anthropol.*, **7:** 69-76.

15. Watt, G. 1989. Dictionary of the economic plants of India. **2:** 509-513. Supdt. Govt. Printing, Calcutta.

Genus *Fagopyrum*

Buckwheat (*Fagopyrum* spp.)

Introduction

The name of buckwheat (*Fagopyrum esculantum*, 2n =16) may be confusing to many but as a matter of fact there is absolute no relationship of buckwheat to the common wheat. Buckwheat is classified as pseudocereal. Buckwheat belongs to the genus *Fagopyrum* of the family Polygonaceae. There are two cultivated species of buckwheat, *viz. Fagopyrum esculantum* Moench (common or Japanese buckwheat) and *F. tataricum* Gaertn (Tatary buckwheat). Buckwheat is the most important crop of the mountain region above 1600 msl both for grain and green. Buckwheat is a multipurpose crop. It occupies about 90 per cent of cultivated land in the higher Himalayas with a solid stand. It is a short duration crop 2-3 months and fits well in the high Himalayas where a crop growing season is of limited period because of early winter and snow fall. In the region of Himalayas, up to 4500 m, this is the only crop grown (Joshi and Paroda, 1991). It is an important food grain for the hill peoples of the certain regions of India and Nepal. The whole plant, young shoots, leaves, flowers and grains are used in variety of ways. The tender shoots are used as leafy vegetable. The buckwheat leaf contains 57 per cent moisture, 25.0 per cent protein, 3.3 per cent fat, 21.3 per cent minerals, 7.7 per cent fiber, 36.0 per cent carbohydrates, 278 K. calorie energy, and 2.12 per cent calcium. Apart from these constituents, per 100 g edible portion contain 574 mg phosphorus and 111.0 mg iron (Chaurasia *et al.*, 2000). The grain is used in several culinary preparations during fast and religious festivals. In India, certain people do not eat cereals on some festivals and fast but they can take buckwheat grains. Its starchy flower can be used as porridge or soup. Its flour mixed with wheat, rice or maize. Flour is used to prepare pancakes, biscuits and noodles. The flour is also used as sweet puddings for making 'Chillare', an unleavened bread fried with butter or ghee, mixed with potatoes to prepare paranthas, mixed with wheat and barley flour for making chapattis. Husked kernels are cooked as rice country liqueurs called 'Pechuwi' and 'Chhang' in Uttarakhand and made from buckwheat (Joshi anf Paroda, 1991). It is used as medicinal plant due to the presence of glucoside named rutin ($C_{27}H_{30}O_{16}$, crystalline yellow product which is less soluble in water but highly soluble in alcohol, acetone and alkaline solution), which is used as a medicine in the treatment of several diseases like apoplexy or retinol haemorrhage. Acidic hydrolysis of rutin yields quercetin, rhamnose and glucose. It also reduces the effect of radiation. Buckwheat grains are also used as livestock and poultry feed. It is also good green manure crop and improves soil texture. It is highly nutritive, unlike cereals which are deficient in lysine, one of the essential amino acids for human health (Anonymus, 1979). At present the values of nutritional and medicinal-pharmacology of buckwheat have created wide alternate from many scientist (Kayashita *et al.*, 1997). The antioxidant enzyme with SOD actually was higher in Tatary buckwheat leafs (Wang *et al.*, 1993). Now the rutin extraction from Tatary buckwheat bran are used in medicine and food products as an addition in China, but the antioxidant enzyme in the bran including peroxidase, have not been reported yet. The lutin and D-chiro-inositol produced by buckwheat has been established to be a necessary link in insulin activity to regulate blood glucose level (McGregor and Mc Killan, 1952) and in the control of type II diabetes (Wang *et al.*, 1992).

Buckwheat is an excellent source of lysine (6.2 g/100g protein) and is the only food grain, not deficient in this respect (Lyman *et al.*, 1956). Buckwheat grain contains comparatively less protein than that of wheat, barley, rye or oats but slightly more than that of maize.

Table 8.62: The Composition of Common Buckwheat and its Byproducts per cent

Byproducts	*Digestible Protein*	*Total Digestible Nutrients*	*Nutrients Ratio*	*Protein*	*Fat*	*fiber*	*Mineral Nutrients*
Grain	8.9	64.4	6.2	11.9	2.4	10.5	2.9
Flour	7.9	86.1	9.9	8.6	1.7	0.7	1.2
Midding	25.8	75.7	1.9	29.7	7.3	7.4	4.9
Food	14.7	59.1	3.0	18.6	5.0	18.3	4.1
Hulls	0.1	28.0	27.0	4.3	0.9	44.4	2.5
Straw	1.2	32.3	25.9	5.6	1.3	43.0	5.5
Green fodder	2.9	21.7	6.5	4.6	0.9	8.0	3.6

Source: Morrison, 1945.

Table 8.63: Comparison of Protein Quality of Buckwheat with other Cereals

Amino Acids	Content (g/100g protein)			
	Buckwheat Flour	Brown Rice	Wheat Flour	Maize Flour
Lysine	6.2	3.4	3.2	3.4
Arginine	10.1	5.9	4.2	4.4
Tyrosine	1.5	1.4	1.0	0.7
Methionine	1.6	2.2	1.9	2.0
Valine	5.1	4.2	4.6	4.1
Isoleucine	3.7	3.5	4.2	3.9
Cystine	1.6	1.8	1.6	2.0
Leucine	6.2	7.3	7.4	12.8
Threonine	3.7	3.3	3.5	4.2
Phenglalanin	4.2	4.5	4.5	4.3
Histidine	2.2	2.1	2.1	2.5

Table 8.64: Mineral Composition of Buckwheat and its Byproducts

Mineral	Flour	Hull	Straw and Leaves
	Mg/100 g (on dry weight basis)		
Calcium	14.5 ± 0.3	97.4 ± 6.0	894.2 ± 14.5
Magnesium	248 ± 3	112 ± 1	305 ± 9
Phosphorus	379 ± 3	127 ± 3	1132 ± 10
Potassium	411 ± 3	1267 ± 38	5139 ± 159
Zinc	2.79 ± 0.02	1.24 ± 0.06	6.69 ± 0.06
Copper	0.63 ± 0.03	0.61 ± 0.01	0.27 ± 0.03
Manganese	0.89 ± 0.01	9.16 ± 0.28	38.81 ± 0.81

Source: Fagopyrum 16: 89-94 (1999).

The biological value of protein is superior to the other food plants and is nearly equal to that of eggs. According to a number of workers (Eggum, 1980) the amino acid composition of buckwheat is much better compared to cereals.

Buckwheat grains contain 11.3 g moisture, 10.3 g protein, 2.4 g fat, 2.3 g minerals, 8.6 g fiber, 65.1 g carbohydrates, 323 K. calories energy, 64 mg calcium, 355 mg phosphorus and 115.5 mg iron (Gopalan *et al.*, 1999)

Origin, History and Distribution

The center of origin of buckwheat is believed to be in the Central Asia particularly, Himalayan region either in Western China or Northern India (Elagin, 1959). However, according to Tsvetoukhine (1952) the center of origin of buckwheat is Manchuria. The biological characteristic of the plant, its present distribution and linguistic evidence lead to the conclusion that *F. esculantum* originated in Himalayan region and it was introduced in to Southern Russia in the first century A.D. (Elagin, 1959). According to Ohinishi (1998a) southern China, specifically Yunnan and Sichuan Provinces is the center of species diversity of *Fagopyrum*. The range of distribution of buckwheat extends from the mountainous region of Pakistan to India, Nepal, China, Japan, Korea, Iran Russia and Afghanistan. Distribution of wild *Fagopyrum* species in the Himalayan hills and southern China, especially in Sichuan and Yunnan provinces has been classified and the phylogenetic relationship among the wild species has been studied (Ohnishi, 1998a and 1998b). In Soviet Russia it is a major food items. In India the distribution of *Fagopyrum* spp. has been reported from the states of Sikkim, Arunachal Pradesh, Meghalaya and Assam, Manipur, Nagaland in northeast region (Joshi, 1999). According to Joshi buckwheat cultivation was confined to Arunachal Pradesh and Sikkim. The Monpas and Sherdukpen tribes of the Tawang, district of Arunachal Pradesh use the grain, as staple food science rice cannot be grown at this altitude due to physiological problems.

Species availability, evolution and diversity

Three distinct texa of *Fagopyrum* occur in the Himalayas; *Fagopyrum esculantum* and *Fagopyrum tataricum* are cultivated species, while *Fagopyrum cymosum* is grown wild (WOI, 1956). The following are the valid species of genous *Fagopyrum.*

F. crassifolium Rafin

F. cymosum Meissn. = *F. triangulare* Meissn. Ex. Wall.

F. suffructicosum F. Schmidt.

F. esculentum Moench = *F. emarginatum* Moench.

= *F. sagittatum* Gilib

= *F. valgare* T. Nees

F. tataricum Gaertn = *F. rotundatum* Bali

The wild species *Fagopyrum cymosum* is believed to be the progenitor of the two cultivated species (Krotov, 1963).Thirteen wild *Fagopyrum* species and subspecies as well as two cultivated buckwheat *F. esculentum* ssp. (common buckwheat) and *F. tataricum* ssp. (Tartary buckwheat) are distributed in Tibet region. The wild species and subspecies like *F. esculantum* ssp. *ancestralis, F. tataricum* ssp. *Potanini, F. homotropicum, F. cymosum, F. callianthum, F. urophyllum, F. lineare, F. leptopodum, F. gracilipes, F. pleioramosum, F. capillatum, F. macrocarpum* and *F. rubifolium* are also available. In Indian condition few species like *F. esculentum, F. tataricum, F. cymosum, F. emarginatum, F. himalayanum* and *F. giganticum* were evaluated and noted great diversity at ICAR-NBPGR, Regional Station, Shimla (Himachal Pradesh).

Botanical Description

Buckwheat belongs to family Polygonaceae.

Table 8.65: Botanical Description

Growth Habit	The common buckwheat *F. esculantum*is herbaceous, erect and annual plant but *F. tataricum* is herbaceous taller and coarser annual with short internodes. Similarly The wild species *F. cymosum* Meisnn is a tall vigorous plant (attain the height of 90.0-150.0 cm.).
Stem	Hollow stem with swollen nodes.
Leaf	Alternate triangular leaves are found in common buckwheat *F. esculantum* but in *F. tataricum*, narrow, shaped leaves are found. The wild species *F. cymosum* Meisnn bears large and long stalked triangular leaves. Commonly leaves are 2.54-7.5 cm long.
Floral biology	The inflorescence is axillary racemes and terminal cyme with inconspicuous light green sepals. It is densely clustered dimorphic flowers of white, pink or red colour can be observed. The species is self-fertile.
Anthesis and pollination	The species is cross-pollinated and self-sterile.
Fruit and fruit set	The fruit is a dry, one seeded, three-seeded nut called 'achene', measuring about 6 mm x 3 mm.
Seeds	The colour of the seed is brown, gray brown or black, marked with darker spot and line.

Augmentation of Genetic Resources and Crop Improvement

Indian Council of Agricultural Research started research work on this crop in 1982 under the AICRP as underutilized crop. Since, 1982, three varieties *i.e.* Himpriya (ICAR-National Bureau of Plant Genetic Resource, New Delhi), VL-7 by ICAR-Vivekanand Parvatiya Krishi Anusandhanshala, Almora (Uttarakhand) and PRB-1 (GBPU and T, Pantnagar Uttarakhand) has been released for cultivation in India. Germplasm comprising of 408 accessios evaluated and noted wide range of variation for different agro-morphological traits at ICAR-NBPGR, Regional Station, Shimla (Himachal Pradesh) and promising types possessing specific desirable traits particularly early maturity (75-85 days), short height (65-85 cm), bold grains (3.8-5.0 g/100 seeds) and high seed yield (15-20 g/plant) were identified for further use (Joshi and Paroda, 1991).

Genetic Resource Conservation

Germplasm comprising of 408 accessios belonging to six species namely, *F. esculentum, F. tataricum, F. cymosum, F. emarginatum, F. himalayanum* and *F. giganticum* were evaluated for 31 descriptors at ICAR-NBPGR, Regional Station, Shimla and the information has bee catalogued. Several collections are being maintained at VPKAS, Almora (Uttarakhand) also.

REFERENCES

1. Chaurasia, O.P., Singh, N. and Singh, B. 2000. Potential wild leafy vegetables from Ladakh. *Indian Hort.*, **45(2):** 4-5.
2. Eggum, B.O., Javornik, B. and Kreft, I. 1981. Chemical composition and protein quality of buckwheat (*Fagopyrum esculentum* Moench.). *Qual. Plant Foods-Hum. Nutr.*, **30:** 175-179.
3. Elagin, I.N. 1959. On the problem of origin of buckwheat. *Bot. Zn. SSRR.*, **44:** 1177-1181.
4. Gopalan, G., Rama Sastri, B.V., Balasubramanian, S.C., Rao, B.S.N., Deosthale, Y.G. and Pant, K.C. 1999. Nutritive value of Indian Foods. National Institute of Nutrition, ICMR, Hyderabad (A.P.), India.
5. Joshi, B.D.1999. Study of buckwheat in India. *Fagopyrum,* **16:** 7-11.
6. Joshi, B.D. and Paroda, RS. 1991. Buckwheat in India. ICAR-NBPGR Shimla, *Sci. Monogr.*, No. 2. 117P.
7. Kayashita, J, Shimaoka, I, Nakajoh, M., Yamazaki, M. and Kato, N. 1997. Consumption of buckwheat protein lowers plasma cholesterol and raises fecal neutral sterols in cholesterol-fed rate because of its low digestibility. *J. Nutr.*, **127**: 1395-1400.
8. Krotov, A.S. 1963. *BuckwheatIzdatel styo sel skohpzajstvennj Literaturey*. Moccow-Leningrade. pp. 254. (P.B.A. **35:** 34040).
9. Lyman, C.M., Kuiken, K.A. and Hall, F. 1956. Essential amino acid content of farm feeds. *J. Agric. Fd. Chem.*, **4:** 1008-1013.
10. McGregor, WG. And McKillican, ME. 1952 Rutin content of varieties of buckwheat. *Sci. Agric.* (Ottawa) **32**(1): 48-51. (Biol. Abs. 12999).
11. Morrison, F.B. 1945. Feeds and Feeding. Morrison Publ. C. New York.

12. Ohnishi, 1998a. Search for the wild ancestor of buckwheat I. Description of new *Fagopyrum* species and their distribution in China. *Fagopyrum*, **15:** 18-28.

13. Ohnishi, 1998b. Search for the wild ancestor of buckwheat III. The wild ancestor of cultivated common buckwheat and of Tatary buckwheat. *Econ. Bot.*, **25:** 123-133.

14. Tsvetoukhine, V. 1952. Buckwheat and its improvement possibilities. *Annual Inst. Natn. Rech. Agron. Paris.*, **1:** 99-115.

15. Wang, J., Liu, S., Fu, X. and Run, M. 1992. A clinical observation on the hypoglycemic effects of Xinjiang buckwheat. Proc. 5th Int. Symp. Buckwheat at Taiyuan: 465-467.

16. Wang, Z.H., Lin, RF., Zhang, Z. and Zhou, MD. 1993. Purification and characterization of superroxide dismutase from *Tartary* buckwheat leaves. *Fagopyrum*, **13**: 31-34.

Genus *Celosia*

Silver Spiked Cockscomb (*Celosia* species)

Introduction

Silver spiked cockscomb (*Celosia argentea* or *C. cristata*) is common weed and in India, it is locally called as Pituna (Sanskrit); Swet Murga (Bengali); Lambadi, Lapadi (Gujarati); Anne Sopu (Kannad); Kurada, Kurudu (Marathi); Gangachulia (Odishi); Panna Kirai (Tamil); Gurugu, Panchechettu (Telgu); bogs Kukurzova (Assam); Gogia (Garhwal) and sarpankha, Sarwali (Punjabi). Leaves are used as vegetable due to nutrition richness. When silver spiked cockscomb is used after 5-7 week old plants, more accumulation of crude protein and vit. C has been found. *Celosia argentea* var. *argentea* or Lagos spinach (a.k.a. quail grass, Soko, Celosia, feather cockscomb) is a broad leaf annual leaf vegetable. It grows widespread across Mexico, where it is known as "Velvet flower", northern South America, tropical Africa, the West Indies, South, East and Southeast Asia where it is grown as a native or naturalized wild flower, and is cultivated as a nutritious leafy green vegetable. It is traditional fare in the countries of Central and West Africa, and is one of the leading leafy green vegetables in Nigeria, where it is known as 'soko yokoto', meaning "make husbands fat and happy". In Spain it is known as "Rooster comb" because of its appearance. As a grain, Cockscomb is a pseudo-cereal, not a true cereal. These leaves, young stems and young inflorescences are used for stew, as they soften up readily in cooking. The leaves also have a soft texture and a mild spinach-like taste. They are also pepped up with such things as hot pepper, garlic, fresh lime, and red palm oil and eaten as a side dish.

Tender leaves contains fair amount of vit. B-1 and vit. C. In Nzeria some genotypes contain high amount of vit. C (553 mg/100 g). In it hydrocynic acid (2 mg/100 g, dry weight basis) and oxalic acid (11.3 per cent dry weight basis) have been found. As per NIN, Hyderabad reports, silver spiked cockscomb contains several vitamins (Table: 8.67).

Table 8.66: Nutrient Content in Leaves of Silver Spiked Cockscomb

Nutrient	*Quantity*	
	1	*2*
Moisture	85.7 g	88.0 g
Protein	3.3 g	2.2 g
Fat	0.8 g	0.7 g
Minerals	2.7 g	2.0 g
Fiber	--	1.5
Carbohydrates	7.5 g	5.8 g
Energy	50 K. calorie	38 K. calorie
Calcium	322.0 mg	323.0 mg
Iron	16.8 mg	--

Source: (1) Gopalan *et al.*, 1999; (2) Wealth of India, 1969.

Table 8.67: Vitamin Contents in Leaves (per 100.0 g)

Vitamins	*Quantity*
Vit. B-1	29.0 mg
Vit. B-2	5.4 mg
Vit. B-6	18.2 mg
Niacin	7.7 mg

Among the peoples it is popular as herbal medicine and paste of whole plant commonly used when scorpion bite. Its flowers are used during blood mixed diarrhea and to cure menorrhagia. Plant extract is very useful in removing urinary stone and also stop stone formation. Seeds are diuretic in nature and help a lot in curing blood related troubles, tumors, mouth soath, eyes related diseaeses *etc.*

Origin, History and Distribution

Silver spiked cockscomb can be spotted from plains to hills (up to 1500 m). It is very common vegetable in Nizeria.This is a widespread and cosmopolitan family from the tropics to cool temperate regions. The Amaranthaceae (sensu strictu) are predominantly tropical, whereas the former Chenopodiaceae have their centers of diversity in dry temperate and warm temperate areas (Muller and Thomas, 2005). Many of the species are halophytes, tolerating salty soils, or grow in dry steppes or semideserts.

Species Availability, Evolution and Diversity

- ☆ *Celosia argentea* L.
- ☆ *Celosia cristata* L.
- ☆ *Celosia isertii* C.C.Towns.
- ☆ *Celosia leptostachya* Benth.
- ☆ *Celosia nitida* Vahl
- ☆ *Celosia odorata* T.Cooke
- ☆ *Celosia palmeri* S.Watson
- ☆ *Celosia plumosa* syn. C. argentea
- ☆ *Celosia spicata* L.
- ☆ *Celosia trigyna* L.
- ☆ *Celosia virgata* Jacq.

Formerly Placed Here

- ☆ *Chamissoa altissima* (Jacq.) Kunth (as *C. tomentosa* Humb. and Bonpl. ex Schult.)
- ☆ *Deeringia amaranthoides* (Lam.) Merr. (as *C. baccata* Retz.)
- ☆ *Deeringia polysperma* (Roxb.) Moq. (as *C. polysperma* Roxb.)
- ☆ *Iresine diffusa* Humb. and Bonpl. ex Willd. (as *C. paniculata* L.)

Botanical Description

Silver spiked cockscomb belongs to family Amaranthaceae. Cytological studies reported that it is tetraploid and octoploid in nature. The tetraploid cultivars easily crossed with octoploid species *Celosia cristata*. Botanically octoploids are uniform in size and having resistance to diseases and and develop multi branch pattern, therefore these traits can be easily transferred into tetraploid for developing high yielding and better quality cultivars. The basic chromosome number is (rarely 6) mostly 8-9 (rarely 17).

Table 8.68: Botanical Description

Growth Habit	Plants are erect and annual or perennial herbs. Most of these are subshrubs or shrubs; very few species are vines or trees. Some species are succulent. Many species have stems with thickened nodes. The wood of the perennial stem has a typical "anomalous" secondary growth, only in subfamily Polycnemoideae is secondary growth normal (The Amaranthaceae family at APWebsite).
Stem	Stems are hard and branched.
Leaf	Leaves are thin, complete and mostly alternate but sometimes opposite. They never possess stipules. The simple leaves are flat or terete, their shape is extremely variable, with entire or toothed margins. In some species, the leaves are reduced to minute scales. In most cases, neither basal or terminal aggregations of leaves occur. Leaves 2-15 x 0.1-3.2 cm, lanceolate-oblong to narrowly linear, acute to obtuse, shortly mucronate with the excurrent midrib, glabrous; lamina of the leaves from the centre of the main stem tapering below into an indistinctly demarcated, slender petiole, to 2 cm long; upper and branch leaves smaller, markedly reducing.
Floral biology	Flowers are white to pink terminaland joint on the stem with spike wings (flowered spike, 2.5-20 x 1.5-2.2 cm). The flowers are solitary or aggregated in cymes, spikes, or panicles and typically perfect (bisexual) and actinomorphic. Some species have unisexual flowers. Bracts and bracteoles are either herbaceous or scarious. Flowers are regular with a herbaceous or scarious perianth of (1 to) mostly 5 (rarely to 8) tepals, often joined. Branches, peduncle up to c. 20 cm long; bracts and bracteoles lanceolate or the lower deltoid, 3-5 mm, hyaline, more or less aristate with the excurrent midrib, persistent. Perianth segments 6-10 mm, narrowly elliptic-oblong, acute to rather blunt, shortly mucronate, margins hyaline. Filaments very delicate, free part subequalling or exceeding the staminal sheath, sinuses rounded with very minute intermediate teeth; anthers and filaments creamy to magenta. Ovary 4-8-ovulate, style filiform, 5-7 mm long; stigmas 2-3, very short. Capsule 3-4 mm, ovoid to globose. The re are 1-5 stamens opposite to tepals or alternating, inserting from a hypogynous disc, which may have appendages (pseudostaminodes) in some species. The anthers have 2 or 4 pollen sacs (locules). In tribe *Caroxyloneae*, antheres have vesicular appendages. The pollen grains are spherical with many pores (pantoporate), with pore numbers from a few to 250 (in *Froelichia*). 1-3 (rarely -6) carpels are fused to a superior ovary with 1 (rarely 2) basal ovule (Müller and Thomas, 2005).
Anthesis and pollination	Most of the cockscomb gets pollinated with wild cockscomb pollen. Still attractive colours and good enough to attract swifts and small blue butterflies.
Fruit and fruit set	Fruits are capsule/utricles The diaspores are seeds or fruits (utricles), more often the perianth persists and is modified in fruit for means of dispersal. More rarely the fruit is a circumscissile capsule or a berry.
Seeds	Seeds are 1.25-1.5 mm, lenticular, black, shining, very finely reticulate. Sometimes even bracts and bracteoles may belong to the diaspore. The horizontal or vertical seed often has a thickened or woody seed coat. The green or white embryo is either spirally (and without perisperm) or annular (rarely straight). One thousand seeds can weigh 1.0-1.2 grams.

REFERENCES

1. *"Celosia". Integrated Taxonomic Information System. Retrieved 2011-01-30*
2. Gopalan, G., Rama Sastri, B.V., Balasubramanian, S.C., Rao, B.S.N., Deosthale, Y.G. and Pant, K.C. 1999. Nutritive value of Indian Foods. National Institute of Nutrition, ICMR, Hyderabad (A.P.), India.

3. Kai Müller, K. and Thomas B. 2005. Phylogenetics of Amaranthaceae using matK/trnK sequence data–evidence from parsimony, likelihood and Bayesian approaches. *Annals of the Missouri Botanical Garden*, **92**: 66-102.

4. Lost Crops of Africa: Volume II: Vegetables

5. The Amaranthaceae family at APWebsite.

6. Wealth of India, 1969. A Dictionary of Indian Raw materials and Industrial Products. Publication and Information Directorate, CSIR, New Delhi.

Genus *Lepidium*

Garden Cress (*Lepidium sativum* Linn.)

Introduction

Garden cress (*Lepidium sativum* Linn.) is locally known as Haleem, Herf (Hindi), Hallem, Alveri (Bengali), Ahliva (Marathi), Assaleo, Haleem (Gujarati), Adalavitulu, Adeli, Adityal (Telagu), Aliverai (Tamil) and Alibija, Kurutige (Kannad). It is a very popular leafy salads and also used for culinary purpose. Per 100.0 g of edible leaves of garden cress contains 82.3 g moisture, 5.2 g protein, 1.0 g fat, 2.2 g minerals, 8.7 g carbohydrates, 67 K. calorie energy, 360.0 mg calcium, 110.0 mg phosphorus, 28.6 mg iron, and 0.15 mg thiamine (Gopalan *et al.*, 1999). It is rich source of uronic acid and in this acid fair amount of L-arabinose, D-galactose, L-rhamnose, D-galacturonic acid and D-glucose is present. As medicinal properties, it acts as mild stimulant. It leaves consumption cure urinary troubles and removes liver pain. Roots are used in preparation of tonic which is useful in stimulating nervous systems. Leaf poultice is used inhealing of wounds. Seeds contain gums which is called 'cress seed mucilage' and used largely to cure dysentery. Gums obtain from seeds of garden cress has 18.3 per cent cellulose.Seeds contains 25.5 per cent oil which is yellow in colour and emits unpleasant asroma. Due to this reason oil is used in soap making industries.

Origin, History and Distribution

It is believed that garden cress has been originated from higher altitude of Ethiopia and Aritriya. Europe and West Asia is noted as secondary centres of origin.

Species Availability, Evolution and Diversity

Garden cress is polymorphous species.*Lepidium sativum* Linn. is commonly cultivated but its other species *Lepidium latifolium* iscultivated in Leh-Laddakh for vegetable purpose.

Botanical Description

Garden cress belongs to family Brassicaceae

Table 8.69: Botanical Description

Growth Habit	Plants are erect and attained the height of 15.0-45 cm.
Stem	Stems are hard and branched.
Leaf	Leaves are thin, complete and oval shape.
Floral biology	Flowers are raceme and bear small white colour flowers.
Fruit and fruit set	Fruits are small and curved.

REFERENCES

1. Gopalan, G., Rama Sastri, B.V., Balasubramanian, S.C., Rao, B.S.N., Deosthale, Y.G. and Pant, K.C. 1999. Nutritive value of Indian Foods. National Institute of Nutrition, ICMR, Hyderabad, India.

Genus *Rumex*

Garden Sorrel (*Rumex acetosa* Linn.)

Introduction

Vernacular name of this crop is Khatta Palak (Hindi) and Chuka Palak (Bengali). Garden sorrel (*Rumex acetosa* Linn.) is a popular herbaceous plant which tender leaves are used as salads or used for culinary purpose. Its leaves are very rich in vit. C. Per 100 g of edible portion of garden sorrel contains 90.0 g moisture, 2.6 g protein, 0.5 g fat, 2.6 g carbohydrates, 400 mg potassium, 46 mg magnesium, 44.0 mg phosphorus, 0.59 mg manganese, 0.30 mg copper, 0.22 mg zinc, 0.007 mg iodine, 124.0 mg ascorbic acid, 11.00 mg β-carotene and 0.08 mg thiamine (W.O.I., 1969; Gopalan *et al.*, 1999). Leaves of garden sorrel contains flavon glucosides, rutin (1.28 per cent), hyperin (Quercitin 3-β-D-glactocide, $C_{21}H_{20}O_{12}$, 37 mg/100 g) and Wetezin (8 C-β-D-gluco pyronosil apizenin, $C_{21}H_{20}O_{10}$) and Oxymethyl anthraquinone is also available but in very less quantity. Leaves also have medicinal properties like curing of skin diseases, lowering fever and helps in purification of bloods. It has cooling effect and diuretic in nature. It is potent homeopathic medicine.

Origin, History and Distribution

Garden sorrel can be spotted from West Himalayan regions particularly Kumayun to Kashmir up to height (elevation) of 3600 m.

Species Availability, Evolution and Diversity

Genus Rumex has large number of species but in India, 13 species are available and most of them are suitably used as leafy vegetables.

Table 8.70: Botanical Description

Growth Habit	Garden sorrel is perennial herbaceous plants which attend the height of 30.0-75.0 cm. Development of plant is generally errect.
Stem	Stems possess someto strut.
Leaf	Leaves are 5.0-15.0 cm long and margin is pointed.
Floral biology	Flowers are dioecious in nature
Fruit and fruit set	Fruits are nut and its triangle or round shape is common. Fruits are black and shining.

Botanical Description

The botanical name of garden sorrel is *Rumex acetosa* Linn. and belongs to family Polygonaceae. Under the genus annual and perennial both are found. Plants may be herbaceous or bushy of this genus.

REFERENCES

1. Gopalan, G., Rama Sastri, B.V., Balasubramanian, S.C., Rao, B.S.N., Deosthale, Y.G. and Pant, K.C. 1999. Nutritive value of Indian Foods. National Institute of Nutrition, ICMR, Hyderabad (A.P.), India.
2. Wealth of India, 1969. A Dictionary of Indian Raw Materials and Industrial Products. Publication and Information Directorate, CSIR, New Delhi.

Yellow Dock (*Rumex crispus* Linn.)

Introduction

Yellow Dock is also known as curled dock. It is a very popular leafy vegetable. Leaves are rich source of calcium, iron and vitamins C. Per 100.0 g of edible leaves contains moisture 92.6g, 1.5 g protein, 0.3 g fat, 4.1 g carbohydrates, 0.9 g fiber, 1.5 g ash, 74.0 mg calcium, 56.0 mg phosphorus, 5.6 mg iron, 1.38 mg Vit. A, 0.06 mg thiamine, 0.08 mg riboflavin, 0.4 mg nicotinic acid and 30.0 mg ascorbic acid(WOI, 1969; Gopalan, *et al.*, 1999). In homeopathic medicine system, roots of Yellow Dockb are used for preparation of tonic. Plants available in Argentina are very famous for medicine preparation because roots contains 1,8 dihydroxy-3 methyl-9 anthrone, chrysophanic acid and emodin alongwith several secondary elements (WOI, 1969).

Origin, History and Distribution

Yellow Dock is originated fro Euirope.

Botanical Description

The genus of 'Yellow Dock' is *Rumex* and species is *crispus*. It belongs to family Polygonaceae.

Table 8.71: Botanical Description

Growth Habit	Plants are perennial in nature. Plant develops up to 100.0cm.
Leaf	Leaves are smooth both the side and attain 4.0-20.0 cm length.
Floral biology	Flowers are green in colour

REFERENCES

1. Gopalan, G., Rama Sastri, B.V., Balasubramanian, S.C., Rao, B.S.N., Deosthale, Y.G. and Pant, K.C. 1999. Nutritive value of Indian Foods. National Institute of Nutrition, ICMR, Hyderabad (A.P.), India.
2. Wealth of India, 1969. A Dictionary of Indian Raw Materials and Industrial Products. Publication and Information Directorate, CSIR, New Delhi.

Jangali Palak or Wild Spinach (*Rumex dentatus* Linn.)

Introduction

'Jangali Palak' or 'Wild Spinach' (*Rumex dentatus* Linn.) is also known as 'Ambawah' and 'Amrul' in Hindi. It is very popular leafy begetable because of high content of calcium, β-carotene and Vit.C. Per 100.0 g of edible leaves contains 89.3 g moisture, 3.5 g protein, 0.7 g fiber, 4.1 g carbohydrates, 2.0 g ash, 611.5 mg calcium, 53.8 mg phosphorus, 3.4 mg iron, 115.0 mg Vit.C and 11700 IU Vit.A (WOI, 1969; Gopalan, *et al.*, 1999). Fresh leaves (225.0) and boiled leaves (209.0) contain lysine and amino acids per gram nitrogen, respectively (WOI, 1969). Roots contains sufficient amount of 'chrysophanic acid' and 'emodin' which is used in curting skin diseases. Rutin, avicularin, quercetin and quercitrin are major flavonoids.

Origin, History and Distribution

Jangali Palak or Wild Spinach is basically temperate leafy vegetable which generally found in temperate, tropical and subtropical regions of the world. In India, it can be spotted in Himalayan regions having height of 300.0 m. It can be easily spotted from plains of Assam to south India.

Species Availability, Evolution and Diversity

Tale 8.72: Species Distribution

Species	*Distribution*
Rumex dentatus Linn.	North-western Himalaya, also in northern and central plains; more variability in hills, extending to eastern region; forms from plains are akin to *R. nigricans*.

Botanical Description

The genus of 'Jangali Palak or Wild Spinach' is *Rumex* and species is *dentatus*. It belongs to family Polygonaceae.

Table 8.73: Botanical Description

Growth Habit	Erect plant and attain the height of 30.0-100.0 cm.
Root	Roots of the plant are red in colour.
Stem	Develop grooves on the stem surface.
Leaf	Leaves are cylindrical, round or heart shape and develop 7.5-10.0 cm length.
Floral biology	Flowers are bisexual and green in colour.
Fruit and fruit set	Fruits are nut and its triangle or round shape is common. Fruits are brown and shining.

REFERENCES

1. Gopalan, G., Rama Sastri, B.V., Balasubramanian, S.C., Rao, B.S.N., Deosthale, Y.G. and Pant, K.C. 1999. Nutritive value of Indian Foods. National Institute of Nutrition, ICMR, Hyderabad (A.P.), India.
2. Wealth of India, 1969. A Dictionary of Indian Raw materials and Industrial Products. Publication and Information Directorate, CSIR, New Delhi.

Bladder Dock or Chuka (*Rumex vesicarius*)

Introduction

Bladder Dock (*Rumex vesicarius*) is also known as Chuka, Ambari, Palak, Palang Sag (Hindi); Chukakura (Telagu); Sankan Kirai (Tamil) and Khatti Tan, Saluni (Kannad). It is popular vegetable crop in south wastern part of India. Tender leaves and stems are used for culinary purpose. It is equally important for curry and chatani preparation due to sourness. Per 100.0 g of edible leaves contains 95.2 g moisture, 1.6 g protein, 1.4 g carbohydrates, 0.6 g fiber, 0.9 g minerals, 63.0 mg calcium, 17.0 mg phosphorus, 8.7 mg iron, 0.03 mg thiamine, 0.06 mg riboflavin, 0.2 mg nicotinic acid, 12.0 mg vit. C and 6100 IU Vit. A (WOI, 1969; Gopalan, *et al.*, 1999).

Origin, History and Distribution

Bladder Dock or Chuka is originated from South Western Asia and Northern Africa.

In south wastern part of India, this crop can be spotted easily.

Botanical Description

The genus of 'Bladder Dock or Chuka' is *Rumex* and species is *vesicarius*. It belongs to family Polygonaceae.

Table 8.74: Botanical Description

Growth Habit	Plants are annual, herbaceous and attent the height of 15.0-30.0 cm.
Leaf	Leaves are oblong and 2.5-7.5 cm long.
Floral biology	Flowers are small and monoecious in nature.
Fruit and fruit set	Fruits are nut and triangle in shape shape. Fruits are pink and shining.

REFERENCES

1. Gopalan, G., Rama Sastri, B.V., Balasubramanian, S.C., Rao, B.S.N., Deosthale, Y.G. and Pant, K.C. 1999. Nutritive value of Indian Foods. National Institute of Nutrition, ICMR, Hyderabad, India.
2. Wealth of India, 1969. A Dictionary of Indian Raw Materials and Industrial Products. Publication and Information Directorate, CSIR, New Delhi.

Genus *Oxyriya*

Mountain Sorrel (*Oxyriya dygina*)

Introduction

In India mountain sorrel (*Oxyriya dygina*) is known as 'Amalu' or 'Chohahuk' (Punjabi) and Lamanchu (Laddakh). In India Mountain sorrel can be spotted in J&K, Himachal Pradesh and Leh Laddakh regions. In Laddakh it is used as vegetable. Due to sourness of leaves, it is used in salad and preparation of 'Chatni'. It is rich source of vit. C and full of mineral matters like iron, Zn, *etc.* Per 100.0 g of edible portion contains 79.0 g moisture, 18.3 g protein, 3.4 g fat, 29.8 g minerals, 10.8 g fiber, 29.8 g carbohydrates, 223 K calorie energy, 4.29 mg calcium, 274 mg phosphorus and 120.0 mg iron (Chaurasia *et al.*, 2000). Seeds easily germinate and plants grow in mountain's rocky and stone lands. Leaves are available from April to June in Laddakh regions.

Origin, History and Distribution

The 'Mountain Sorrel' can be spotted in alpine regions of northern hemisphere. In India plants are available in Himalayan regions *i.e.* from Kashmir to Sikkim up to height of 3000-6000m of msl. The genus Atriplex is distributed nearly worldwide from subtropical to temperate and to subarctic regions. Most species rich are Australia,North America, South America and Eurasia. Many species are halophytes and are adapted to dry environments with salty soils

Species Availability, Evolution and Diversity

In India, only one species *O. dygina* is available.

Botanical Description

The genus of 'Mountain Sorrel' is *Oxyriya* and species is *dygina*. It belongs to family Polygonaceae. Under genus *Oxyriya,* mostly plants are perennial in nature. It can be propagated through seeds and vegetative parts. In Laddakh (India) this crop is cultivated through seeds only (Chaurasia *et al.,* 2000).

References

1. Chaurasia, O.P., Singh, N. and Singh, B. 2000. Potential wild leafy vegetables from Ladakh. *Indian Hort.*, **45(2):** 4-5

Genus *Atriplex*

Atriplex (*Atriplex hortensis*)

Introduction

Atriplex (*Atriplex hortensis*, x=9, except for *Atriplex lanfrancoi*, which is x=10) popularly known as Ustak, Phaltora saltbush and orache (or orach) in Ladakh region. The generic name originated in Latin and was applied by Pliny the Elder to the edible oraches (Quattrocchi, 2010). The name saltbush derives from the fact that the plants retain salt in their leaves; they are able to grow in areas affected by soil salination. Use of Atriplex as food is known since at least the late Epipaleolithic (Mesolithic). It is used as vegetables for local dishes with different recipes by native people of Ladakh; still it is being used as wild edible vegetables in some corners of Ladakh. In this region it is a first green to appear as vegetable after prolong winter. As nutritional value its 100 g leaves contain 17 g protein, 3 g fat, 56 g total carbohydrate, 11 g fiber, 24 g ash, and perhaps 2,000 mg Ca, 150 mg P, 10 mg Fe, 2 mg Cu, 500 g Mg, 800 mg K, 10 mg Mn, 2 g carotene equivalent and it is rich source of vitamins A. Two relatively rare bioactive compounds are isolated from the leaves of Atriplex, *i.e.* sulphated flavonoids: kaempferol 3-*0*-sulphate-7-*0*-arabinopyranoside and quercetin 3-*0*-sulphate-7-*0*-arabinopyranoside (belong the group of compounds easily soluble in water). The plant is also reported to accumulate heavy metals like lead, nickel, zinc and copper that can used as phytoremeadation to clean heavy metal contaminated areas. As folk medicines the areal part of Atriplex is being used in diseases of respiratory tract, digestive and urinary system, due to their analgestic properties, and in rheumatism. Leaves are diuretic, emetic and purgative. It is also said to be a stimulant to the metabolism and its infusion is used as a spring tonic and as a remedy for tiredness and nervous exhaustion. It has been suggested as a remedy for plethora and lung ailments. Seeds mixed with wine are said to cure yellow jaundice. They also excite vomiting. Heated with vinegar, honey and salt, Atriplex is useful in gout. Fruits are purgative and emetic. Liniments and emollients prepared from the whole plant, like the juice of the plant, are said to be remedies for indurations and tumors, especially of the throat.

Origin, History and Distribution

The genus is quite variable and widely distributed. It includes many desert and seashore plants and halophytes, as well as plants of moist environments.

Species Availability, Evolution and Diversity

Atriplex is an extremely species rich genus and comprises about 250 (Gelin *et al.,* 2003)-300 (Gudrun, *et al.,* 2010) species. The genus *Atriplex* was first described in 1753 by Carl Linnaeus. The type species (lectotype) is *Atriplex hortensis* The genus evolved in Middle Miocene, the C_4-photosynthesis pathway developed about 14.1-10.9 million years ago, when the climate became increasingly dry. The genus diversified rapidly and spread over the continents. The C_4 *Atriplex* colonized North America probably from Eurasia during the Middle/Late Miocene, about 9.8-8.8 million years ago, and later spread to South America. Australia was colonized twice by two C_4 lineages, one from Eurasia or America about 9.8-7.8 million years ago, and one from Central Asia about 6.3-4.8 million years ago. The last lineage diversified rapidly, and became the ancestor of most Australian Atriplex species (Gudrun *et al.,* 2010). After phylogenetical research, Kadereit *et al.* (2010) excluded *Halimione* as a distinct sister genus. The remaining *Atripex* species were grouped into several clades that mostly are not identical with the traditional sections (Gudrun *et al.,* 2010).

Atriplex lanfrancoi/cana-Clade

- ✰ *Atriplex lanfrancoi* (Brullo and Pavone) G. Kadereit et Sukhor. (Syn.: *Cremnophyton lanfrancoi* Brullo and Pavone): endemic in Maltaand Gozo.

- ☆ *Atriplex cana* C.A. Mey.: from Eastern European Russia to Western China

Atriplex section *Atriplex*: Annual C3-plants

- ☆ *Atriplex aucheri* Moq.: in Eastern Europe and West Asia
- ☆ *Atriplex hortensis* L. – Garden Orache, Red Orach, Mountain Spinach, French Spinach: in Asia, cultivated or naturalized in Europe
- ☆ *Atriplex oblongifolia* Waldst. and Kit. – Oblong-leaved Orache: in Eurasia
- ☆ *Atriplex sagittata* Borkh. (Syn.: *Atriplex nitens* Schkuhr): in Eurasia

Atriplex section *Teutliopsis* Dumort.: Annual C3-plants

- ☆ *Atriplex australasica* Moq.
- ☆ *Atriplex calotheca* (Rafn) Fr.: in Northern Europe
- ☆ *Atriplex crassifolia* Ledeb.: in Asia
- ☆ *Atriplex davisii* Aellen: from Southern Europe to Egypt
- ☆ *Atriplex glabriuscula* Edmondston – Northeastern Saltbush, Babington's Orache, Smooth Orache, Scotland Orache, Glabrous Orache: In Central and Northern Europe
- ☆ *Atriplex gmelinii* C.A. Mey. ex Bong. – Gmelin's Saltbush: in Asia and North America
- ☆ *Atriplex intracontinentalis* Sukhor.: from Central Europe to Asia
- ☆ *Atriplex laevis* C.A. Mey.: in Asia, naturalized in Eastern Europe
- ☆ *Atriplex latifolia* Wahlenb.: in Eurasia
- ☆ *Atriplex littoralis* L. – Grass-leaved Orache: in Eurasia and North Africa
- ☆ *Atriplex longipes* Drejer – Long-stalked Orache: in Northern Europe
- ☆ *Atriplex micrantha* C.A. Mey.: in Asia, naturalized in Europe
- ☆ *Atriplex nudicaulis* Boguslaw – Baltic Saltbush: in Eurasia
- ☆ *Atriplex patula* L. – Common Orache, Spreading Orache: in Eurasia and North Africa
- ☆ *Atriplex praecox* Hülph. – Early Orache>: in Northern Europe
- ☆ *Atriplex prostrata* Moq. – Spear-leaved Orache, Thin-leaved Orache, Triangle Orache, fat hen: in Eurasia and North Africa

C_4-*Atriplex*-Clade

Containing the majority of species. The traditional classification into sections (sect. *Obione*, sect. *Pterochiton*, sect.*Psammophila*, sect. *Sclerocalymma*, sect.*Stylosa*) did not reflect the phylogenetical relationships and was rejected by Kadereit *et al.* (2010).

- ☆ *Atriplex acanthocarpa* (Torr.) S. Watson: in North America
- ☆ *Atriplex acutibractea* Anderson: in Australia
- ☆ *Atriplex altaica* Sukhor.: in Asia
- ☆ *Atriplex angulata* Benth.: in Australia
- ☆ *Atriplex billardierei* (Moq.) Hook. f.: in Australia
- ☆ *Atriplex canescens* (Pursh) Nutt. – Chamiso, Chamiza, Four-winged Saltbush, Grey Sagebrush: in North America
- ☆ *Atriplex centralasiatica* Iljin: in Asia
- ☆ *Atriplex cinerea* Poir. – Grey Saltbush, Truganini:[12] in Australia
- ☆ *Atriplex codonocarpa* P.G. Wilson: in Australia
- ☆ *Atriplex conduplicata* F. Muell.: in Australia
- ☆ *Atriplex confertifolia* (Torr. and Frém.) S. Watson – Shadscale (Saltbush): in North America
- ☆ *Atriplex cordobensis* Gand. and Stuck.: in South America
- ☆ *Atriplex deserticola* Phil.: in South America
- ☆ *Atriplex dimorphostegia* Kar. and Kir.: in North Africa
- ☆ *Atriplex eardleyae* Aellen: in Australia
- ☆ *Atriplex elachophylla* F. Muell.: in Australia
- ☆ *Atriplex fissivalvis* F. Muell.: in Australia
- ☆ *Atriplex flabellum* Bunge ex Boiss.: in Eurasia
- ☆ *Atriplex gardneri* (Moq.) D. Dietr. – Gardner's saltbush, Moundscale: in North America
- ☆ *Atriplex glauca* L.: in Portugal, Spain and in North Africa
- ☆ *Atriplex halimus* L. – Mediterranean Saltbush, Sea Orache, Shrubby Orache: in South Europe, North Africa and Southwest Asia
- ☆ *Atriplex herzogii* Standl.: in North America
- ☆ *Atriplex holocarpa* F. Muell.: in Australia
- ☆ *Atriplex hymenelytra* (Torr.) S. Watson – Desert Holly: in North America
- ☆ *Atriplex hymenotheca* Moq.: in Australia
- ☆ *Atriplex imbricata* (Moq.) D. Dietr.: in South America
- ☆ *Atriplex inamoena* Aellen: in Eurasia
- ☆ *Atriplex intermedia* Anderson: in Australia
- ☆ *Atriplex isatidea* Moq.: in Australia
- ☆ *Atriplex laciniata* L. – Frosted Orache: In Western and Northern Europe
- ☆ *Atriplex lampa* (Moq.) Gillies ex Small: in South America
- ☆ *Atriplex lehmanniana* Bunge: in Eurasia

- ☆ *Atriplex lentiformis* (Torr.) S. Watson – Quail Bush: in North America
- ☆ *Atriplex leptocarpa* F. Muell.: in Australia
- ☆ *Atriplex leucoclada* Boiss.: in Eurasia
- ☆ *Atriplex leucophylla* (Moq.) D. Dietr.: in North America
- ☆ *Atriplex lindleyi* Moq.: in Australia
- ☆ *Atriplex moneta* Bunge ex Boiss.: in Eurasia
- ☆ *Atriplex muelleri* Benth.: in Australia
- ☆ *Atriplex nessorhina* S.W.L. Jacobs: in Australia
- ☆ *Atriplex nummularia* Lindl. – Old Man Saltbush, Giant Saltbush: in Australia
- ☆ *Atriplex obovata* Moq.: in North America
- ☆ *Atriplex pamirica* Iljin: in Eurasia
- ☆ *Atriplex parishii* S. Watson: in North America
- ☆ *Atriplex parryi* S. Watson: in North America
- ☆ *Atriplex parvifolia* Kunth: in South America
- ☆ *Atriplex patagonica* (Moq.) D. Dietr.: in South America
- ☆ *Atriplex phyllostegia* (Torr. ex S. Watson) S. Watson: in North America
- ☆ *Atriplex polycarpa* (Torr.) S. Watson – Allscale (Saltbush), Desert Saltbush, Cattle Saltbush, Cattle Spinach: in North America
- ☆ *Atriplex powellii* S. Watson – Powell's Saltbush: in North America
- ☆ *Atriplex pseudocampanulata* Aellen: in Australia
- ☆ *Atriplex quinii* F. Muell.: in Australia
- ☆ *Atriplex recurva* d'Urv.: in Eurasia, endemic to areas around the Aegean
- ☆ *Atriplex rhagodioides* F. Muell.: in Australia
- ☆ *Atriplex rosea* L. – Tumbling Orache: in Eurasia and North Africa
- ☆ *Atriplex rusbyi* Britton ex Rusby: in South America
- ☆ *Atriplex schugnanica* Iljin: in Asia
- ☆ *Atriplex semibaccata* R. Br. – Australian Saltbush, Berry Saltbush, Creeping Saltbush: in Australia
- ☆ *Atriplex semilunaris* Aellen: in Australia
- ☆ *Atriplex serenana* A. Nelson ex Abrams: in North America
- ☆ *Atriplex sibirica* L.; in Asia, naturalized in Europe
- ☆ *Atriplex sphaeromorpha* Iljin: in Russia, Ukraine and Caucasus
- ☆ *Atriplex spinibractea* Anderson: in Australia
- ☆ *Atriplex spongiosa* F. Muell.: in Australia
- ☆ *Atriplex stipitata* Benth.: in Australia
- ☆ *Atriplex sturtii* S.W.L. Jacobs: in Australia
- ☆ *Atriplex suberecta* I. Verd. – sprawling saltbush, lagoon saltbush: in Australia
- ☆ *Atriplex tatarica* Aellen: in Europe, North Africa and Asia
- ☆ *Atriplex turbinata* (Anderson) Aellen: in Australia
- ☆ *Atriplex undulata* (Moq.) D. Dietr.: in South America
- ☆ *Atriplex velutinella* F. Muell.: in Australia
- ☆ *Atriplex vesicaria* Heward ex Benth. – Bladder Saltbush: in Australia

Table 8.75: Selected Species Not Yet Investigated Phylogenetically

- ☆ *Atriplex acadiensis* – Maritime Saltbush
- ☆ *Atriplex acutiloba*
- ☆ *Atriplex alaskensis* S. Watson – Alaska Orache
- ☆ *Atriplex amnicola* Paul G. Wilson – River Saltbush, Swamp Saltbush: in Australia, naturalized in California
- ☆ *Atriplex x aptera*
- ☆ *Atriplex argentea* – Silvery Saltbush
- ☆ *Atriplex asterocarpa*
- ☆ *Atriplex barclayana*
- ☆ *Atriplex bonnevillensis*
- ☆ *Atriplex bunburyana* F.Muell.
- ☆ *Atriplex cephalantha*
- ☆ *Atriplex cordifolia*
- ☆ *Atriplex cordulata*
- ☆ *Atriplex cornigera*
- ☆ *Atriplex coronata* S. Watson: in California
- ☆ *Atriplex corrugata*
- ☆ *Atriplex coulteri*
- ☆ *Atriplex crassipes*
- ☆ *Atriplex cristata*
- ☆ *Atriplex cryptocarpa*
- ☆ *Atriplex cuneata*
- ☆ *Atriplex depressa*
- ☆ *Atriplex dioica* – Thick-leaved Orache, Saline Saltbush
- ☆ *Atriplex drymarioides*
- ☆ *Atriplex eichleri*
- ☆ *Atriplex elegans*
- ☆ *Atriplex erecticaulis*
- ☆ *Atriplex exilifolia* F.Muell.
- ☆ *Atriplex falcata* – Sickle Saltbush
- ☆ *Atriplex flabelliformis*
- ☆ *Atriplex franktonii* – Frankton's Saltbush
- ☆ *Atriplex fruticulosa*
- ☆ *Atriplex garrettii*
- ☆ *Atriplex graciliflora*
- ☆ *Atriplex griffithsii* – Griffith's Saltbush (sometimes included in *A. lentiformis* or *A. torreyi*)
- ☆ *Atriplex hastata* – Halberd-leaved Orache, Spearscale(often included in *A. patula*)
- ☆ *Atriplex heterosperma* Bunge – Two-scaled Orache: in Eastern Europe

- *Atriplex humifusa*
- *Atriplex humilis* F.Muell.
- *Atriplex hypoleuca*
- *Atriplex incrassata* F.Muell.
- *Atriplex infrequens*
- *Atriplex johnsonii* (verification needed)
- *Atriplex johnstonii* (verification needed)
- *Atriplex klebergorum*
- *Atriplex kochiana* Maiden
- *Atriplex limbata*
- *Atriplex lobativalvis* F.Muell.
- *Atriplex macropterocarpa*
- *Atriplex matamorensis*
- *Atriplex maximowicziana* – Maximowicz's Saltbush
- *Atriplex minuscula*
- *Atriplex minuticarpa*
- *Atriplex morrisii*
- *Atriplex nana*
- *Atriplex navajoensis*
- *Atriplex nuttallii* S. Watson – Nuttall's Saltbush: in North America
- *Atriplex obconica*
- *Atriplex pachypoda*
- *Atriplex pacifica*
- *Atriplex paludosa* R.Br.
- *Atriplex papillata*
- *Atriplex persistens*
- *Atriplex plebeja*
- *Atriplex pumilio* R.Br.
- *Atriplex pusilla*
- *Atriplex quadrivalvata*
- *Atriplex saccaria*
- *Atriplex spinifera*
- *Atriplex spinulosa*
- *Atriplex subspicata*
- *Atriplex subtilis*
- *Atriplex tenuissima*
- *Atriplex texana*
- *Atriplex torreyi*
- *Atriplex tridentata*
- *Atriplex truncata* – Wedge-leaved Saltbush
- *Atriplex tularensis*
- *Atriplex valdesii* Flores Olv., Mexico
- *Atriplex vallicola*
- *Atriplex vestita*
- *Atriplex wardii*
- *Atriplex watsonii*
- *Atriplex welshii*
- *Atriplex wolfii*
- *Atriplex wrightii*

Excluded Species

- *Extriplex* (Elizabeth, 2010):
- *Extriplex californica* (Moq.) E.H.Zacharias (Syn. *Atriplex californica* Moq.)
- *Extriplex joaquiniana* (A.Nelson) E.H.Zacharias (Syn. *Atriplex joaquiniana* A.Nelson)
- *Halimione* Aellen (Gudrun, 2010)
- *Halimione pedunculata* (L.) Aellen (Syn.: *Atriplex pedunculata* L.)
- *Halimione portulacoides* (L.) Aellen (Syn.: *Atriplex portulacoides* L.)
- *Halimione verrucifera* (M.Bieb.) Aellen (Syn.: *Atriplex verrucifera* M.Bieb.)
- *Stutzia* E.H.Zacharias (Elizabeth, 2010)
- *Stutzia covillei* (Standl.) E. H. Zacharias (Syn. *Atriplex covillei* (Standl.) J. F. Macbr.)
- *Stutzia dioica* (Nutt.) E. H. Zacharias (Syn. *Atriplex suckleyi* (Torrey) Rydberg, *Atriplex ovata* (Rydb.) Clem. and E. G. Clem.)

Botanical Description

Kingdom	:	Plantae
Unranked	:	Angiosperms
Unranked	:	Eudicots
Unranked	:	Core eudicots
Order	:	Caryophyllales
Family	:	Amaranthaceae
Subfamily	:	Chenopodioideae
Tribe	:	Atripliceae
Genus	:	Atriplex

Atriplex (*Atriplex hortensis*) belongs to the subfamily Chenopodioideae of the family Amaranthaceae (which include the Chenopodiaceae of the Cronquist system). It is cultivated by seeds.

Table 8.76: Botanical Description

Growth Habit	Atriplex is an annual herb, prefer cool and sunny condition and mean temperature of 21°C. Atriplex thrives in well drained sandy loamy soil having pH of 6-7
Stem	Stems are smooth.
Leaf	The leaves are variously shaped, particularly round and heart shape but somewhat oblong, comparatively thin in texture, and slightly acidic to the taste. The leaves appear in red, white and green varieties.
Floral biology	The inflorescences consist of axillary or terminal spikes or spicate panicles, or axillary clusters of glomeruled flowers. Some produce flower stalks up to 1.8 m tall, and others stay less than 0.6 m tall at maturity. The flowers are unisexual, some species are monoecious (individual flowers are either male or female), but both sexes can be found on the same plant and others dioecious.

	Male flowers consist of 3-5 perianth lobes and 3-5 stamens. Female flowers are usually lacking a perianth, but are enclosed by 2 foliaceous bracteoles, and contain an ovary with a short style and 2stigmas.
Anthesis and pollination	Crop is polliated by wind.
Fruit and fruit set	In fruit, the bracteoles can enlarge, thicken or become appendaged. They enclose the fruit tightly, without becoming connate to it. The pericarp is adnate to the vertically orientated.
Seeds	Crop matures and produces seeds in September-October. The seeds dehised after maturation and become dormant during severe winter (December-February) and reappear at the end of March. Each plant produces around 24-35 g of seeds. Thus, thousand of plant can be raised from single plant. Seeds are flattened. The seed coat is thick, leathery or hardening (Stanley, 2003).

REFERENCES

1. Carl Linnaeus (1753), "*Atriplex*", Species Plantarum **2**, Lars Salvius/Biodiversity Heritage Library, p. 1052–1054, retrieved 19 May 2015.
2. Elizabeth, H., Zacharias, and Bruce, G. B. 2010. A Molecular Phylogeny of North American Atripliceae (Chenopodiaceae), with Implications for Floral and Photosynthetic Pathway Evolution. In: *Systematic Botany*, **35**(4), p.839-857. doi:10.1600/036364410X539907.
3. Gelin, Z., Sergei, L., Mosyakin and Steven, E. 2003. Clemants: *Chenopodiaceae*: *Atriplex* - online, In: Wu Zhengyi, Peter H. Raven, Deyuan Hong (ed.): *Flora of China*, Vol. 5: *Ulmaceae through Basellaceae.*, Science Press und Missouri Botanical Garden Press, Beijing und St. Louis.
4. Gudrun, K., Evgeny V.M., Elizabeth H.Z. and Alexander, P.S. 2010. Molecular phylogeny of Atripliceae (Chenopodioideae, Chenopodiaceae): Implications for systematics, biogeography, flower and fruit evolution, and the origin of C_4 Photosynthesis. - *American Journal of Botany,* **97(10):** 1664-1687.
5. Quattrocchi, U. 2000. *CRC World Dictionary of Plant Names*. I: A-C. CRC Press. p. 232. ISBN 978-0-8493-2675-2.
6. Stanley, L.W. 2003. *Atriplex* - online, In: Flora of North America Editorial Committee (ed.): *Flora of North America North of Mexico*, Volume 4: *Magnoliophyta: Caryophyllidae,* Part 1., Oxford University Press, New York.

9

Genetic Resources of Allium Vegetables and Crop Improvement

Introduction

The Alliums (monocot) are species rich genus which includes many economically important crops such as onion, bunching onion, giant onion, ball-head onion, garlic and their relatives like leeks, shallots, chives and Chinese chives that are considered prized vegetables due to palatable food, ethnomedicinal potential, spices, condiments, ornamental value or religious purposes worldwide (Negi, 2006; Davies, 1992; Fritsch and Friesen, 2002). The genus *Allium* is represented by about 750 species distributed throughout the world (out of which more than 35-40 species occur in temperate and alpine regions of Himalayas in India) including *A. cepa*, and *A. sativum* which are widely cultivated (Friesen *et al.*, 2006; Robinwitch and Brewster, 1990; Hanelt *et al.*, 1992) but more than 800 species revealed by Fritsch *et al.* (2010); Fritsch and Friesen, 2002 and Fritsch and Abbasi (2013) which are widely distributed over the warm temperate and temperate zones of the northern hemisphere. The wild species are used as vegetable, condiment, spices, ornamental or as medicine to cure various ailments. Wild species such as *A. tuberosum*, *A. hookeri* and *A. fasciculatum* are consumed raw, cooked or pickled. Cloves of *A. ampeloprasum* are used as medicine for knee pain, arthritis and ear pain. Indian gene centre is considered fairly rich in distribution and occurrence of about 34 cultivated and wild species; among which some seven species are cultivated as major crop or as minor kitchen garden, dooryards and backyards plant *e.g. A. hookeri* and *A. tuberosum*. Regions of high species diversity occur in Turkey, Iran, North Iraq, Afghanistan, Soviet, middle Asia and West Pakistan and a second less pronounced center of species diversity occurs in Western North America. About 40 species of *Allium* are reported in India (Wealth of India, 1985) of which seven are cultivated, the main being onion and garlic. In India, wild *Allium* species have long been utilized as vegetable, spices, condiments, medicinal and ornamental value in the Indian Himalayan region. In addition to this, semi-domesticated and wild economic species are grown in a few pockets of Indian Himalayan region (temperate and alpine zones) by local tribes (Negi, 2006; Pandey *et al.*, 2008; Khosa *et al.*, 2014, Kachroo, 1977). Besides cultivated species like onion (*A. cepa*), garlic (*A. sativus*), shallot (*A. ascalonicum*) and great headed garlic, some species are semi-domesticated/lesser-known types *e.g. Allium carolineanum*, *A. chinense*, *A. consdanguineum*, *A. humile*, *A. victorialis*, *A. wallichii* and *A. semenovii* and are grown under sporadic cultivation. It is true that large scale collection and utilization of some of the wild species for edible (leaves and bulbs) and flavouring purpose has resulted in loss of the natural diversity. In Hindustani centre (Indian sub-continent), only *A. ampeloprasum* has been quoted to be present by Zeven and de Wet (1982) and reported by Koul and Gojil (1970) from Kashmir. However, Hooker (1894) described four *Allium* species *i.e. A. ascalonicum*, *A. ampeloprasum*, *A. cepa* and *A. sativum* most common in use and cultivation. Chandel and Pandey (1992) listed 27 wild species of *Allium* in India distributed in different temperate regions of the country but Polunin and Stainton, 1984; Pandey *et al.*, 2008 and Karthikeyan *et al.*, 1989 reported 35-40 species occur in temperate and Alpine regions of Himalaya including major cultivated species such as onion and garlic. Alpine-subtemperate region (2500-4500 m) of the western Himalaya with about 25 taxa of *Allium* is the zone of rich species diversity (Gohil, 1992; Pandey *et al.*, 2008). In India, only Himalayan zones of J&K, Himachal Pradesh and North-East hills are well known to most of the underutilized *Alliums*. These regions are of rich in vegetational wealth and the survival of a wide variety of economically useful endemic wild plants has enriched the genetic resources base and also the economy of the natives. The natives continue to cultivate wild and economically useful *Allium* species. Three more species have been added to the previous information on *Allium* biodiversity in Indian gene centre:

(i) ***A. caesium*** chrenk- It is common to Lahaul area in Himachal Pradesh and has marked distinction by its dark blue flowers. Flowering stem is comparatively longer than the leaves which are 1-2 mm broad; and inflorescence umbel 3-4 cm.

(ii) ***A. fasiculatum*** Rendle-The scented herb, comprising white or pale-green flowers with umbel inflorescence is found in borderland area of Indian Himalaya adjoining Western Nepal and Tibet.

(iii) ***A. prattii*** Wright-The plants have pink coloured flowers, narrow elliptic paired leaves and many flowered umbel inflorescence. Flowering stems are longer than the leaves; fibrous sheath surrounds the base of the stem.

Two of the most important *Allium* species which are of great interest to taxonomists (Pran, *Allium cepa* var. *viviparum*) and breeders (*Allium roylei* Stearn.) are found in the Indian sub-continent. Pran has attracted great interest among the onion taxonomists and breeders around the world and its parentage is still a mystery (Friesen and Klaas, 1998; Puizina *et al.*, 1997). Much diversity has built up in India in these crops as they have been cultivated from ancient times and selections have been made in different agro-ecological zones. Onion and garlic crops are known only in cultivation and while no known wild ancestor is known for *A. cepa, A. longicuspis,* endemic to central Asia, has been proposed as a progenitor for *A. sativum*. Within *A. cepa* two horticultural varieties are recognized: the dry bulb onion and the multiplier onion (shallots). In *A. sativum* despite it being propagated exclusively by the division of ground cloves or aerial bulbils, a wide range of variability is available in several traits. *A. longicuspis* is belived to wild ancestor of *A. sativum*. Great headed garlic (*A. ampeloprasum*) also known as Oriental garlic or elephant garlic is not true garlic. It is more closely related to common leek. *A. ampeloprasum* is a vigorous plant, commonly forms seed stalks and has broad, flat leaves folded lengthwise. The leeks (*A. ampeloprasum*) succeed fairly well in India thriving at elevations above 300 m and often remain fit for use until the end of rainy season. The chives (*A. shoennoprasum*) are distributed in the Northern Himalayas from Kashmir to Kumaon. Cultivated and wild types interbreed and there is plenty of variation in cultivated forms. Attention is being given on these and other wild species, as they possess many disease resistance genes potentially useful in improvement programmes. *A. ampeloprasum* for white rot resistance, leek yellow stripe virus, leek rust; *A. roylei*

Table 9.1: Taxon and Preferred Taxon Name of Genus *Allium*

Name of Taxon	*Name of Authority*	*Family*	*Preferred Taxon Name*
Allium acuminatum	Hook.	Liliaceae	*Allium acuminatum*
Allium affine	Ledeb.	Liliaceae	*Allium affine*
Allium aflatunense	B. Fedtsch.	Liliaceae	*Allium aflatunense*
Allium akaka	S. G. Gmel. ex Schult. and Schult. f.	Liliaceae	*Allium akaka*
Allium albanum	Grossh.	Liliaceae	*Allium rubellum*
Allium albidum	Fisch. ex M. Bieb.	Liliaceae	*Allium albidum*
Allium albopilosum	C. H. Wright	Liliaceae	*Allium cristophii*
Allium albovianum	Vved.	Liliaceae	*Allium albovianum*
Allium altaicum	Pall.	Liliaceae	*Allium altaicum*
Allium altissimum	Regel	Liliaceae	*Allium altissimum*
Allium altyncolicum	Frizen	Liliaceae	*Allium altyncolicum*
Allium amethystinum	Tausch	Liliaceae	*Allium amethystinum*
Allium ampeloprasum	L.	Liliaceae	*Allium ampeloprasum*
Allium ampeloprasum var. *ampeloprasum*		Liliaceae	*Allium ampeloprasum* var. *ampeloprasum*
Allium ampeloprasum var. *atroviolaceum*	(Boiss.) Regel	Liliaceae	*Allium atroviolaceum*
Allium ampeloprasum var. *babingtonii*	(Borrer) Syme	Liliaceae	*Allium ampeloprasum* var. *babingtonii*
Allium ampeloprasum var. *holmense*	Asch. and Graebn.	Liliaceae	*Allium ampeloprasum*
Allium ampeloprasum var. *porrum*	(L.) J. Gay	Liliaceae	*Allium porrum*
Allium amphibolum	Ledeb.	Liliaceae	*Allium amphibolum*

Allium amplectens	Hook.	Liliaceae	*Allium amplectens*
Allium angulosum	L.	Liliaceae	*Allium angulosum*
Allium anisopodium	Ledeb.	Liliaceae	*Allium anisopodium*
Allium artemisietorum	Eig and Feinbrun	Liliaceae	*Allium artemisietorum*
Allium asarense	R. M. Fritsch and Matin	Liliaceae	*Allium asarense*
Allium ascalonicum	auct.	Liliaceae	*Allium cepa* var. *aggregatum*
Allium ascalonicum	L., nom. ambig.	Liliaceae	*Allium hierochuntinum*
Allium aschersonianum	Barbey	Liliaceae	*Allium aschersonianum*
Allium asclepiadeum	Bornm.	Liliaceae	*Allium asclepiadeum*
Allium atropurpureum	Waldst. and Kit.	Liliaceae	*Allium atropurpureum*
Allium atrosanguineum	Schrenk	Liliaceae	*Allium atrosanguineum*
Allium atroviolaceum	Boiss.	Liliaceae	*Allium atroviolaceum*
Allium attenuifolium	Kellogg	Liliaceae	*Allium amplectens*
Allium aucheri	Boiss.	Liliaceae	*Allium aucheri*
Allium austrosibiricum	Frizen	Liliaceae	*Allium austrosibiricum*
Allium azureum	Ledeb.	Liliaceae	*Allium caeruleum*
Allium azureum var. *bulbilliferum*	Schrenk	Liliaceae	*Allium caeruleum* var. *bulbilliferum*
Allium babingtonii	Borrer	Liliaceae	*Allium ampeloprasum* var. *babingtonii*
Allium baicalense	Willd.	Liliaceae	*Allium baicalense*
Allium bakeri	Regel	Liliaceae	*Allium chinense*
Allium barsczewskii	Lipsky	Liliaceae	*Allium barsczewskii*
Allium barsczewskii f. *niveum*	Krassovsk.	Liliaceae	*Allium barsczewskii* f. *niveum*
Allium baschkyzylsaicum	Krassovsk.	Liliaceae	*Allium baschkyzylsaicum*
Allium beesianum	W. W. Sm.	Liliaceae	*Allium beesianum*
Allium bidentatum	Fisch. ex Prokh.	Liliaceae	*Allium bidentatum*
Allium bidwelliae	S. Watson	Liliaceae	*Allium campanulatum*
Allium blandum	Wall.	Liliaceae	*Allium carolinianum*
Allium bodeanum	Regel	Liliaceae	*Allium bodeanum*
Allium brevidens	Vved.	Liliaceae	*Allium brevidens*
Allium bulgaricum	(Janka) Prodan	Liliaceae	*Nectaroscordum siculum* subsp. *bulgaricum*
Allium caerulescens	Boiss., nom. illeg.	Liliaceae	*Allium aucheri*
Allium caeruleum	Pall.	Liliaceae	*Allium caeruleum*
Allium caeruleum var. bulbilliferum	(Schrenk) Ledeb.	Liliaceae	*Allium caeruleum* var. *bulbilliferum*
Allium caesium	Schrenk	Liliaceae	*Allium caesium*
Allium campanulatum	S. Watson	Liliaceae	*Allium campanulatum*
Allium canadense	L.	Liliaceae	*Allium canadense*
Allium canadense var. *canadense*		Liliaceae	*Allium canadense* var. *canadense*
Allium cardiostemon	Fisch. and C. A. Mey.	Liliaceae	*Allium cardiostemon*
Allium carinatum	L.	Liliaceae	*Allium carinatum*
Allium carinatum subsp. *pulchellum*	Bonnier and Layens	Liliaceae	*Allium carinatum* subsp. *pulchellum*
Allium carolinianum	DC.	Liliaceae	*Allium carolinianum*
Allium cepa	L.	Liliaceae	*Allium cepa*
Allium cepa cv. *solaninum*	Alef.	Liliaceae	*Allium cepa* var. *aggregatum*
Allium cepa cv. *viviparum*	(Metzg.) Alef.	Liliaceae	*Allium* x *proliferum*
Allium cepa var. *aggregatum*	G. Don	Liliaceae	*Allium cepa* var. *aggregatum*
Allium cepa var. *bulbiferum*	Regel	Liliaceae	*Allium* x *proliferum*
Alliumcepa var. *cepa*		Liliaceae	*Allium cepa* var. *cepa*
Allium cepa var. *proliferum*	(Moench) Regel	Liliaceae	*Allium* x *proliferum*
Allium cepa var. *viviparum*	(Metzg.) Mansf.	Liliaceae	*Allium* x *proliferum*
Allium cernuum	Roth	Liliaceae	*Allium cernuum*
Allium chamaemoly	L.	Liliaceae	*Allium chamaemoly*

Allium charaulicum	Fomin	Liliaceae	*Allium rupestre*
Allium chinense	G. Don	Liliaceae	*Allium chinense*
Allium chloranthum	Boiss.	Liliaceae	*Allium chloranthum*
Allium christophii	Trautv., orth. var.	Liliaceae	*Allium cristophii*
Allium ciliatum	Cirillo	Liliaceae	*Allium subhirsutum*
Allium cilicium	Boiss.	Liliaceae	*Allium scorodoprasum* subsp. *rotundum*
Allium clathratum	Ledeb.	Liliaceae	*Allium clathratum*
Allium collis-magni	Kamelin	Liliaceae	*Allium collis-magni*
Allium condensatum	Turcz.	Liliaceae	*Allium condensatum*
Allium coppoleri	Tineo	Liliaceae	*Allium pallens*
Allium costatovaginatum	Kamelin and Levichev	Liliaceae	*Allium costatovaginatum*
Allium crenulatum	Wiegand	Liliaceae	*Allium crenulatum*
Allium cristophii	Trautv.	Liliaceae	*Allium cristophii*
Allium curtum	Boiss. and Gaill.	Liliaceae	*Allium curtum*
Allium cyaneum	Regel	Liliaceae	*Allium cyaneum*
Allium cyathophorum var. *farreri*	(Stearn) Stearn	Liliaceae	*Allium cyathophorum* var. *farreri*
Allium cyrilli	Ten.	Liliaceae	*Allium cyrilli*
Allium daghestanicum	Grossh.	Liliaceae	*Allium daghestanicum*
Allium dalmaticum	A. Kern. ex Janch.	Liliaceae	*Allium guttatum* subsp. *dalmaticum*
Allium darvasicum	Regel	Liliaceae	*Allium darvasicum*
Allium decipiens	Fisch. ex Schult. and Schult. f.	Liliaceae	*Allium decipiens*
Allium delicatulum	J. F. E. Siev. ex Schult. and Schult. f.	Liliaceae	*Allium delicatulum*
Allium descendens	L.	Liliaceae	*Allium sphaerocephalon*
Allium desertorum	Forssk.	Liliaceae	*Allium desertorum*
Allium dictyoprasum	C. A. Mey. ex Kunth	Liliaceae	*Allium dictyoprasum*
Allium dioscoridis	Sm.	Liliaceae	*Allium dioscoridis*
Allium douglasii	Hook.	Liliaceae	*Allium douglasii*
Allium douglasii var. *nevii*	(S. Watson) Ownbey and Mingrone	Liliaceae	*Allium nevii*
Allium drepanophyllum	Vved.	Liliaceae	*Allium drepanophyllum*
Allium drobovii	Vved.	Liliaceae	*Allium drobovii*
Allium drummondii	Regel	Liliaceae	*Allium drummondii*
Allium dumetorum	Feinbrun and Szel.	Liliaceae	*Allium dumetorum*
Allium elatum	Regel	Liliaceae	*Allium macleanii*
Allium erdelii	Zucc.	Liliaceae	*Allium erdelii*
Allium ericetorum	Thore	Liliaceae	*Allium ericetorum*
Allium ericoleum	Vved.	Liliaceae	*Allium ericoleum*
Allium fallax	Schult. and Schult. f.	Liliaceae	*Allium senescens* subsp. *montanum*
Allium farctum	Wendelbo	Liliaceae	*Allium farctum*
Allium farreri	Stearn	Liliaceae	*Allium cyathophorum* var. *farreri*
Allium fetissovii	Regel	Liliaceae	*Allium fetissovii*
Allium fibrillum	M. E. Jones	Liliaceae	*Allium fibrillum*
Allium fibrosum	Regel	Liliaceae	*Allium fibrosum*
Allium filidens	Regel	Liliaceae	*Allium filidens*
Allium fimbriatum var. *munzii*	Ownbey ex Traub	Liliaceae	*Allium munzii*
Allium fistulosum	L.	Liliaceae	*Allium fistulosum*
Allium flavum	L.	Liliaceae	*Allium flavum*
Allium flavum subsp. *flavum*		Liliaceae	*Allium flavum* subsp. *flavum*
Allium flavum subsp. *tauricum*	(Besser ex Rchb.) Stearn	Liliaceae	*Allium flavum* subsp. *tauricum*
Allium flavum var. *pulchellum*	Ledeb.	Liliaceae	*Allium carinatum* subsp. *pulchellum*
Allium flavum var. *tauricum*	Besser ex Rchb.	Liliaceae	*Allium flavum* subsp. *tauricum*
Allium fominianum	Miscz. ex Grossh. and Schischk.	Liliaceae	*Allium gramineum*
Allium fragrans	Vent.	Liliaceae	*Nothoscordum gracile*

Allium frigidum	Boiss. and Heldr.	Liliaceae	*Allium frigidum*
Allium fuscoviolaceum	Fomin	Liliaceae	*Allium fuscoviolaceum*
Allium galanthum	Kar. and Kir.	Liliaceae	*Allium galanthum*
Allium geyeri	S. Watson	Liliaceae	*Allium geyeri*
Allium geyeri var. *geyeri*		Liliaceae	*Allium geyeri* var. *geyeri*
Allium geyeri var. *tenerum*	M. E. Jones	Liliaceae	*Allium geyeri* var. *tenerum*
Allium giganteum	Regel	Liliaceae	*Allium giganteum*
Allium glaciale	Vved.	Liliaceae	*Allium glaciale*
Allium glaucum	Schrad. ex Poir.	Liliaceae	*Allium senescens* subsp. *glaucum*
Allium globosum	M. Bieb. ex Redoute	Liliaceae	*Allium saxatile*
Allium goloskokovii	Vved.	Liliaceae	*Allium goloskokovii*
Allium goodingii	Ownbey	Liliaceae	*Allium goodingii*
Allium gracile	Aiton	Liliaceae	*Nothoscordum gracile*
Allium gracile	Albov, nom. illeg.	Liliaceae	*Allium albovianum*
Allium gramineum	K. Koch	Liliaceae	*Allium gramineum*
Allium grande	Lipsky	Liliaceae	*Allium grande*
Allium grayi	Regel	Liliaceae	*Allium grayi*
Allium guttatum	Steven	Liliaceae	*Allium guttatum*
Allium guttatum subsp. *dalmaticum*	(A. Kern. ex Janch.) Stearn	Liliaceae	*Allium guttatum* subsp. *dalmaticum*
Allium guttatum subsp. *guttatum*		Liliaceae	*Allium guttatum* subsp. *guttatum*
Allium guttatum subsp. *sardoum*	(Moris) Stearn	Liliaceae	*Allium guttatum* subsp. *sardoum*
Allium heldreichii	Boiss.	Liliaceae	*Allium heldreichii*
Allium helicophyllum	Vved.	Liliaceae	*Allium helicophyllum*
Allium hierochuntinum	Boiss.	Liliaceae	*Allium hierochuntinum*
Allium hirsutum	Zucc., nom. illeg.	Liliaceae	*Allium trifoliatum* subsp. *hirsutum*
Allium hybrid		Liliaceae	*Allium hybrid*
Allium hymenorrhizum	Ledeb.	Liliaceae	*Allium hymenorrhizum*
Allium inconspicuum	Vved.	Liliaceae	*Allium inconspicuum*
Allium inodorum	auct.	Liliaceae	*Nothoscordum gracile*
Allium inodorum	Aiton	Liliaceae	*Allium neapolitanum*
Allium jacquemontii	Kunth	Liliaceae	*Allium jacquemontii*
Allium jajlae	Vved.	Liliaceae	*Allium scorodoprasum* subsp. *jajlae*
Allium jesdianum	Boiss. and Buhse	Liliaceae	*Allium jesdianum*
Allium jodanthum	Vved.	Liliaceae	*Allium jodanthum*
Allium jubatum	J. F. Macbr.	Liliaceae	*Allium jubatum*
Allium kansuense	Regel	Liliaceae	*Allium sikkimense*
Allium karataviense	Regel	Liliaceae	*Allium karataviense*
Allium karsianum	Fomin	Liliaceae	*Allium karsianum*
Allium kaschianum	Regel	Liliaceae	*Allium kaschianum*
Allium kaufmannii	Regel	Liliaceae	*Allium monadelphum*
Allium kochii	Lange	Liliaceae	*Allium vineale*
Allium kokanicum	Regel	Liliaceae	*Allium kokanicum*
Allium kujukense	Vved.	Liliaceae	*Allium kujukense*
Allium kunthianum	Vved.	Liliaceae	*Allium kunthianum*
Allium kurrat	Schweinf. ex K. Krause	Liliaceae	*Allium kurrat*
Allium ledebourianum	Schult. and Schult. f.	Liliaceae	*Allium ledebourianum*
Allium lepidum	Kunth	Liliaceae	*Allium kunthianum*
Allium leucocephalum	Turcz.	Liliaceae	*Allium leucocephalum*
Allium libani	Boiss.	Liliaceae	*Allium libani*
Allium lineare	L.	Liliaceae	*Allium lineare*
Allium longicuspis	Regel	Liliaceae	*Allium longicuspis*
Allium macleanii	Baker	Liliaceae	*Allium macleanii*

Allium macranthum	Baker	Liliaceae	*Allium macranthum*
Allium macrostemon	Bunge	Liliaceae	*Allium macrostemon*
Allium mairei	H. Lev.	Liliaceae	*Allium mairei*
Allium margaritaceum	Sm.	Liliaceae	*Allium guttatum* subsp. *sardoum*
Allium mariae	Bordz.	Liliaceae	*Allium mariae*
Allium materculae	Bordz.	Liliaceae	*Allium materculae*
Allium maximowiczii	Regel	Liliaceae	*Allium maximowiczii*
Allium melanantherum	Pancic	Liliaceae	*Allium melanantherum*
Allium modestum	Boiss.	Liliaceae	*Allium desertorum*
Allium mogoltavicum	Vved.	Liliaceae	*Allium mogoltavicum*
Allium moly	L.	Liliaceae	*Allium moly*
Allium monadelphum	Less. ex Kunth	Liliaceae	*Allium monadelphum*
Allium monadelphum	Turcz. ex Kar. and Kir.	Liliaceae	*Allium atrosanguineum*
Allium mongolicum	Regel	Liliaceae	*Allium mongolicum*
Allium montanum	F. W. Schmidt	Liliaceae	*Allium senescens* subsp. *montanum*
Allium motor	Kamelin and Levichev	Liliaceae	*Allium motor*
Allium multibulbosum	Jacq.	Liliaceae	*Allium nigrum*
Allium munzii	(Ownbey ex Traub) McNeal	Liliaceae	*Allium munzii*
Allium mutabile	Michx.	Liliaceae	*Allium canadense* var. *canadense*
Allium narcissiflorum	Vill.	Liliaceae	*Allium narcissiflorum*
Allium neapolitanum	Cirillo	Liliaceae	*Allium neapolitanum*
Allium nevii	S. Watson	Liliaceae	*Allium nevii*
Allium nigrum	L.	Liliaceae	*Allium nigrum*
Allium nipponicum	Franch. and Sav.	Liliaceae	*Allium grayi*
Allium nutans	L.	Liliaceae	*Allium nutans*
Allium nuttallii	S. Watson	Liliaceae	*Allium drummondii*
Allium obliquum	L.	Liliaceae	*Allium obliquum*
Allium ochroleucum	Waldst. and Kit.	Liliaceae	*Allium ericetorum*
Allium odorum	L.	Liliaceae	*Allium ramosum*
Allium oleraceum	L.	Liliaceae	*Allium oleraceum*
Allium oliganthum	Kar. and Kir.	Liliaceae	*Allium oliganthum*
Allium ophioscorodon	Link	Liliaceae	*Allium sativum* var. *ophioscorodon*
Allium oreophilum	C. A. Mey.	Liliaceae	*Allium oreophilum*
Allium oreoprasum	Schrenk	Liliaceae	*Allium oreoprasum*
Allium orientale	Boiss.	Liliaceae	*Allium orientale*
Allium oschaninii	O. Fedtsch.	Liliaceae	*Allium oschaninii*
Allium ostrowskianum	Regel	Liliaceae	*Allium oreophilum*
Allium oviflorum	Regel	Liliaceae	*Allium macranthum*
Allium paczoskianum	Tuzson	Liliaceae	*Allium flavum* subsp. *tauricum*
Allium pallasii	Murray	Liliaceae	*Allium pallasii*
Allium pallens	L.	Liliaceae	*Allium pallens*
Allium paniculatum	L.	Liliaceae	*Allium paniculatum*
Allium paniculatum subsp. *paniculatum*		Liliaceae	*Allium paniculatum* subsp. *paniculatum*
Allium paradoxum	(M. Bieb.) G. Don	Liliaceae	*Allium paradoxum*
Allium pedemontanum	Willd.	Liliaceae	*Allium narcissiflorum*
Allium pendulinum	Ten.	Liliaceae	*Allium pendulinum*
Allium platyspathum	Schrenk	Liliaceae	*Allium platyspathum*
Allium polyastrum	Diels	Liliaceae	*Allium polyastrum*
Allium polyphyllum	Kar. and Kir.	Liliaceae	*Allium carolinianum*
Allium polyrhizum	Turcz. ex Regel	Liliaceae	*Allium polyrhizum*
Allium ponticum	Miscz. ex Grossh.	Liliaceae	*Allium ponticum*
Allium popovii	Vved.	Liliaceae	*Allium popovii*

Allium porrum	L.	Liliaceae	*Allium porrum*
Allium praecox	Brandegee	Liliaceae	*Allium praecox*
Allium praemixtum	Vved.	Liliaceae	*Allium praemixtum*
Allium prostratum	Trevir.	Liliaceae	*Allium prostratum*
Allium pseudoampeloprasum	Miscz. ex Grossh.	Liliaceae	*Allium pseudoampeloprasum*
Allium pseudocepa	Schrenk	Liliaceae	*Allium galanthum*
Allium pseudoflavum	Vved.	Liliaceae	*Allium pseudoflavum*
Allium pseudostrictum	Albov	Liliaceae	*Allium pseudostrictum*
Allium pskemense	B. Fedtsch.	Liliaceae	*Allium pskemense*
Allium pulchellum	G. Don	Liliaceae	*Allium carinatum* subsp. *pulchellum*
Allium pyrenaicum	Costa and Vayr.	Liliaceae	*Allium pyrenaicum*
Allium ramosum	L.	Liliaceae	*Allium ramosum*
Allium regelii	Trautv.	Liliaceae	*Allium regelii*
Allium reticulatum	Nutt. ex G. Don, nom. nud.	Liliaceae	*Allium textile*
Allium rhabdotum	Stearn	Liliaceae	*Allium rhabdotum*
Allium rosenbachianum	Regel	Liliaceae	*Allium rosenbachianum*
Allium roseum	L.	Liliaceae	*Allium roseum*
Allium rothii	Zucc.	Liliaceae	*Allium rothii*
Allium rotundum	L.	Liliaceae	*Allium scorodoprasum* subsp. *rotundum*
Allium roylei	Stearn	Liliaceae	*Allium roylei*
Allium rubellum	M. Bieb.	Liliaceae	*Allium rubellum*
Allium rubens	Schrad. ex Willd.	Liliaceae	*Allium rubens*
Allium rupestre	Steven	Liliaceae	*Allium rupestre*
Allium sabulosum	Steven ex Bunge	Liliaceae	*Allium sabulosum*
Allium sacculiferum	Maxim.	Liliaceae	*Allium sacculiferum*
Allium sardoum	Moris	Liliaceae	*Allium guttatum* subsp. *sardoum*
Allium sativum	L.	Liliaceae	*Allium sativum*
Allium sativum var. *ophioscorodon*	(Link) Doll	Liliaceae	*Allium sativum* var. *ophioscorodon*
Allium sativum var. *sativum*		Liliaceae	*Allium sativum* var. *sativum*
Allium saxatile	M. Bieb.	Liliaceae	*Allium saxatile*
Allium scabriscapum	Boiss. and Kotschy	Liliaceae	*Allium scabriscapum*
Allium schoenoprasum	L.	Liliaceae	*Allium schoenoprasum*
Allium schoenoprasum var. *alpinum*	DC.	Liliaceae	*Allium schoenoprasum*
Allium schubertii	Zucc.	Liliaceae	*Allium schubertii*
Allium scorodoprasum	L.	Liliaceae	*Allium scorodoprasum*
Allium scorodoprasum subsp. *jajlae*	(Vved.) Stearn	Liliaceae	*Allium scorodoprasum* subsp. *jajlae*
Allium scorodoprasum subsp. *rotundum*	(L.) Stearn	Liliaceae	*Allium scorodoprasum* subsp. *rotundum*
Allium scorodoprasum subsp. *scorodoprasum*		Liliaceae	*Allium scorodoprasum* subsp. *scorodoprasum*
Allium scorodoprasum var. *babingtonii*	(Borrer) Regel	Liliaceae	*Allium ampeloprasum* var. *babingtonii*
Allium scorodoprasum var. *viviparum*	Regel	Liliaceae	*Allium scorodoprasum*
Allium scorzonerifolium	Desf. ex DC.	Liliaceae	*Allium scorzonerifolium*
Allium semenovii	Regel	Liliaceae	*Allium semenovii*
Allium senescens	L.	Liliaceae	*Allium senescens*
Allium senescens subsp. *glaucum*	(Schrad. ex Poir.) Dostal	Liliaceae	*Allium senescens* subsp. *glaucum*
Allium senescens subsp. *montanum*	(Fr.) Holub	Liliaceae	*Allium senescens* subsp. *montanum*
Allium senescens subsp. *senescens*		Liliaceae	*Allium senescens* subsp. *senescens*
Allium sewertzowii	Regel	Liliaceae	*Allium sewertzowii*
Allium sibiricum	L.	Liliaceae	*Allium schoenoprasum*
Allium siculum	Ucria	Liliaceae	*Nectaroscordum siculum*
Allium sikkimense	Baker	Liliaceae	*Allium sikkimense*
Allium sp.		Liliaceae	*Allium* sp.

Allium sphaerocephalon	L.	Liliaceae	*Allium sphaerocephalon*
Allium splendens	Willd. ex Schult. and Schult. f.	Liliaceae	*Allium splendens*
Allium spurium	G. Don	Liliaceae	*Allium spurium*
Allium stamineum	Boiss.	Liliaceae	*Allium stamineum*
Allium stellatum	Fraser ex Ker Gawl.	Liliaceae	*Allium stellatum*
Allium stellerianum	Willd.	Liliaceae	*Allium stellerianum*
Allium stenopetalum	Boiss. and Kotschy	Liliaceae	*Allium stenopetalum*
Allium stenophyllum	Schrenk	Liliaceae	*Allium oliganthum*
Allium stipitatum	Regel	Liliaceae	*Allium stipitatum*
Allium stojanovii	Kov.	Liliaceae	*Allium amethystinum*
Allium strictum	Schrad.	Liliaceae	*Allium strictum*
Allium suaveolens	Jacq.	Liliaceae	*Allium suaveolens*
Allium subangulatum	Regel	Liliaceae	*Allium subangulatum*
Allium subhirsutum	L.	Liliaceae	*Allium subhirsutum*
Allium suvorovii	Regel	Liliaceae	*Allium suvorovii*
Allium szovitsii	Regel	Liliaceae	*Allium szovitsii*
Allium taeniopetalum	Popov and Vved.	Liliaceae	*Allium taeniopetalum*
Allium talassicum	Regel	Liliaceae	*Allium talassicum*
Allium tanguticum	Regel	Liliaceae	*Allium tanguticum*
Allium tataricum	L. f.	Liliaceae	*Allium ramosum*
Allium tel-avivense	Eig	Liliaceae	*Allium tel-avivense*
Allium textile	A. Nelson and J. F. Macbr.	Liliaceae	*Allium textile*
Allium thomsonii	Baker	Liliaceae	*Allium carolinianum*
Allium thunbergii	G. Don	Liliaceae	*Allium thunbergii*
Allium tibeticum	Rendle	Liliaceae	*Allium sikkimense*
Allium tolmiei	C. F. Baker ex S. Watson	Liliaceae	*Allium tolmiei*
Allium tricoccum	Aiton	Liliaceae	*Allium tricoccum*
Allium trifoliatum	Cirillo	Liliaceae	*Allium trifoliatum*
Allium trifoliatum subsp. *hirsutum*	(Regel) Kollmann	Liliaceae	*Allium trifoliatum* subsp. *hirsutum*
Allium triquotrum	L.	Liliaceae	*Allium triquetrum*
Allium tubergenii	Freyn	Liliaceae	*Allium tubergenii*
Allium tuberosum	Rottler ex Spreng.	Liliaceae	*Allium tuberosum*
Allium tulipifolium	Ledeb.	Liliaceae	*Allium decipiens*
Allium turcomanicum	Regel	Liliaceae	*Allium turcomanicum*
Allium unifolium	Kellogg	Liliaceae	*Allium unifolium*
Allium urceolatum	Regel	Liliaceae	*Allium caesium*
Allium ursinum	L.	Liliaceae	*Allium ursinum*
Allium validum	S. Watson	Liliaceae	*Allium validum*
Allium vavilovii	Popov and Vved.	Liliaceae	*Allium vavilovii*
Allium victorialis	L.	Liliaceae	*Allium victorialis*
Allium vineale	L.	Liliaceae	*Allium vineale*
Allium vineale var. *purpureum*	H. P. G. Koch	Liliaceae	*Allium vineale*
Allium viviparum	Kar. and Kir.	Liliaceae	*Allium caeruleum*
Allium vodopjanovae	Frizen	Liliaceae	*Allium vodopjanovae*
Allium wallichianum	Steud., nom. nud.	Liliaceae	*Allium* sp.
Allium wallichii	Kunth	Liliaceae	*Allium wallichii*
Allium watsonii	Howell	Liliaceae	*Allium watsonii*
Allium winklerianum	Regel	Liliaceae	*Allium winklerianum*
Allium x proliferum	(Moench) Schrad. ex Willd.	Liliaceae	*Allium x proliferum*
Allium x wakegii	Araki	Liliaceae	*Allium x proliferum*
Allium yunnanense	Diels	Liliaceae	*Allium mairei*
Allium zebdanense	Boiss. and Noe	Liliaceae	*Allium zebdanense*

for downey mildew; *A. fistulosum* for pink rot, smut and onion fly, and *A. tuberosum* for *Fusarium oxysporum* and leek moth, are particularly noteworthy.

Almost all the *Allium roylei* accessions used in research studies, in both European and American gene banks, originate from a single clone (coded C502 and PI243009) and the base material was from a few plants collected from Mussorie, India (de Vries *et al.* 1992a). other wild species like *A. hookeri*, *A. chinense* and *A. carolinianum* arenative speciesfound in wild and are used by local population as vegetables, soups, pickled and flavouring purpose (Pandey *et al.*, 2008).

Affinities within some Notable *Alliums*

Allium cepa being cultivated for a long, is related to *A. pskemense* and *A. vavilovi* which are wild in central Asia especially northwest India, Afghanistan, Uzbekistan and west Tien Shan (Vavilov, 1949). Other related wild species is *A. oschaninii* (syn. *A. cepa* var. *sylvestris*) from Pamirs, Alai and Tien Shan. Wendelbo (1971) and Mc collum (1974) suggested that these species are wild relatives of *A. cepa*. Secondary centre of origin of *A. cepa* is considered to be the Mediterranean region but said to be originated in Persia. Romans introduced onion into western and northern Europe around 300 A.D. *A. sativum* has been considered to be a native of China. Some workers consider *A. longicuspis* as the wild parental stock of garlic. *A sativum* is thought to be originated in Central Asia, while the secondary centre has been recognized in the Mediterranean region. The wild and cultivated forms of *Allium ampeloprasum* (levent garlic) are highly variable and it is considered related to *A. sativum*, *A. porrum*, *A. kurrat* and *A. scorodoprasum*. Levent garlic is an ancient crop mentioned in the Bible and most probably domesticated in the Mediterranean region. *A. porrum* (leek) is a cultivated form *A. ampeloprasum*. Few nomenclaturists consider both these species as one and synonyms. Because of a close relation of *A. kurrat* to *A. ampeloprasum*, it has been mistakably included as a variety (var. *kurrat*) either of *A. ampeloprasum* or *A. porrum*. *A. kurrat* is known from the times of Pharaohs and grown mainly in Arabia and Egypt. The centre of origin of *A. ascalonicum* (Shallot) has been probably recognized in West Asia. The earlier history indicates that crop isrelated to *A. cepa*. The presence of *A. macleani* stretches in the areas from Iran to Afghanistan. According to Tutin (1976), the species *A. scorodoprasum* (giant garlic) comprised four subspecies *viz.*:

(i) *A. scorodoprasum rotundum*

(ii) *A. scorodoprasum waldsteinii*

(iii) *A. scorodoprasum jajlae* and

(iv) *A. scorodoprasum scorodoprasum*

Probably the later sub-species were derived from sub-species *rotundum*. At present *A. tuberosum* (syn. *A. odoratum*) has been reported wild from East Monglia to Japan, Philippines and through Thailand to northern India; the primary centre of origin is unknown. It is thought to have been originated in China. Its tetraploid may derive from an amphiploidization of hybrid of two diploid species or autotetraploidization of a diploid species. Egyption onion or tree onion (*A. proliferum*) was previously described as a variety of onion as *A. cepa* var. *proliferum* and *A. fistulosum* but later, on the basis of a comprehensive C-banding study, it was declared as a true hybrid of both of these species (Fiskenjo, 1975; Vosa, 1976). *A. fistulosum* (Welsh onion, stone leek, spring onion) probably originated in north west China has been recognized to show a close affinity with *A. altaicum* from North Mongolia while the former is reported to be first cultivated in north China and *A. microbulbum* is said to be a hybrid of *A. fistulosum* and *A. altaicum*. Cultivation of Welsh onion dates back to 200 BC in China and reached to Japan around 500 AD, now it is spread throughout the south-East Asia. *A. chinense* is is native to central and eastern China and mainly cultivated in China and Japan. Because of a very polymorphous nature of *A. schoenoprasum*, its intra-generic relationship with other species is not well documented (Sharma *et al.*, 1996).

Underutilized *Alliums*

All the underutilized *Alliums* found in the Himalayan zone are consumed by natives as condiments for seasoning various dishes either in fresh or dried form. In some cases the above ground parts are collected, crushed and made into small cakes, sun dried and used during latter period or lean season. Over exploitation of the underutilized *Alliums* accompanied by a lot of biotic interference has led to considerable shrinking of the population of various species at different places. Grazing in meadows of these species is quite common. Shepherd also collect green herbage and dried for later use. The women folk of near by village used to visit world heritage 'Valley of Flowers' and other national parks and sanctuaries during the month of August-September and collected large quantity of herbage of different species of *Allium* for their local consumption and for commercial purposes. Since there is no check on these activities, it is leading to erosion in the germplasm of *Allium* species in the Himalayan region and there is every possibility of some of these populations getting totally wiped out in near future only.

Several explorations to Lahaul and Spiti (HP), Sikkim, Arunachal Pradesh and parts of Assam (North East) and Ladakh-Kargil regions of J&K were undertaken to collect, identify and maintain the wild *Allium* species. Khar *et al.* (2016) collected 28 germplasm fromLadakh J&K region out of which 5 were cultivated types and consisted of 4 accessions of *A. cepa* and one of *A. ascalonicum*. Out of the rest 23 other accessions, 2 accessions of *A. oreoprasum*, 2 of *A. cyathophorum*, 12 of *A. prszewalskianum*, 1 of *A. shoenoprasum*, 2 of *A. carolinianum* and 4 identified wild species were collected. The altitude of collection ranged from 3207 amsl to 4784 amsl. All the wild species are semi domesticated

Table 9.2: Scenario of Underutilized *Alliums* in Himalayan Zone of India

Sl.No.	Botanical Name	No. of Accessions	Vernacular Names	Remarks
Semi-domesticated				
1.	*Allium ampeloprasum* L. Syn. *A. porrum* L. Syn. *A. ampeloprasum* var. *porrum* A, DC.	2	Hargandh, Vilayati lahsoon, Sidhum; Great-headed garlic, Leek, Leavent Garlic, Kurrat	Pseudostem and leaf blades are cooked as a vegetable. Cloves are utilized as a pickles and spices. Medicinally cloves dipped in mustard oil used to treat as a message for paralytic limbs,arthritis *etc.*
2.	*Allium ascalonicum* L.Syn. A. cepa L. var. *ascalonicum* Backer syn. *A. sulvia* Buch.-Ham. Syn. *A. fragrance* Ed.	3	Dhunn, Dunna, Chapi; Shallot	Leaves and leaf bases (young bulbs) are used to adulterate in many food dishes, often fresh as garnishing. Bulbs are frequently used for flavoring food, eaten either raw or cooked or fried, and more delicate than onions. Shallot bulbs are frequently taken by the natives during their fast and considered sacred.
3.	*Allium cernuum* Roth.	4	Sikwa; Nodding Onion, Ladies leek	Bulbs and cloves are used for strong flavoring food dishes, soups and also pickled.
4.	*Allium chinense* G. Don.	2	Lahsuniya, Rakyo	Bulbs used as condiments, pickled.
5.	*Allium fistulosum* L.	2	Pyazi; Japanese Bunching Onion, Welsh Onion, Cibol, Stone leek	Young leaves cooked as vegetables. Tender bulbs and young leaves taken as a raw in salad.
6.	*Allium proliferum* (Moench.) Schrad	2	Pran; Top Onion, Tree Onion, Egyptian Onion, Catanissa Onion	Young sprout leaves, the topsets (bulbils), and bulb coocked as vegetable.
7.	*Allium tuberosum* Rottl. Ex Spreng.	2	Markua, Bang-gandina; Fragrant Onion, Chinese Chives, Nira	Fresh leaves mostly green are used as condiment in various local food dishes.
(C) Wild *Allium* species maintained at Field Gene Bank of Regional Station Bhowali of ICAR-NBPGR, New Delhi				
1.	*Allium carolinianum* D.C.	6	Ladam, Pharan	Young leaves cooked as green vegetable; dried leaves are used as condiment. Bulbs eaten raw or cooked.
2.	*Allium clarkei* Hook.	1	Pharan	Young leaves cooked as green and aromatic vegetable mixed with potato. Dried leaves are used as condiments. Bulbs are eaten raw or cooked. Dried leaves and floral spike are being sold @ rs. 10-15/- per ten g in the local market.
3.	*Allium consanguuineum* Kunth	1	Pharan	Young leaves cooked as green vegetable; dried leaves are used as condiment. Bulbs eaten raw or cooked.
4.	*Allium griffithianum* Boiss var. *grandiflorum* Boiss	2	Jamboo, Pharan	Leaves dried stored and used as condiments. Bulbs are eaten raw or cooked
5.	*Allium griffithianum* var. *parviflorum*	5	Jamboo, Pharan	Leaves dried stored and used as condiments. Bulbs are eaten raw or cooked.
6.	*Allium hookeri* Thwaites	1	Pharan	Young leaves cooked as green vegetable
7.	*Allium humile* Kunth	1	Jamboo, Jimboo, Lsanya, Pngari	Young aromatic leaves used as green vegetable; dried leaves used are transported to the plains for spices and condiments.
8.	*Allium roylei* Stern.	2	Jimboo	Leaves dried stored and used as condiments. Bulbs are eaten raw or cooked.
9.	*Allium wallichhii* Kunth	2	Dunno, Gobka, Ladu, Lainka	Young leaves cooked as green vegetable. Dried leaves used as condiments. Boiled rhizomes are taken in flatulence and bile complaints.
Under utilized *Allium* species of Indian Himalayas yet to be collected and conserved at field gene bank of Regional Station Bhowali of ICAR-NBPGR, New Delhi				
1.	*Allium atropurpureum* Waldst.	-	-	-
2.	*Alliumatrosanguineum* Schrenk.		-	-
3.	*Allium cercosporum* Schrenk.	-	-	-
4.	*Allium fedschenkoanum* Regel	-	-	-
5.	*Allium loratum* Baker	-	-	-
6.	*Allium macranthum* Baker	-	-	-
7.	*Allium platyspathum* Schrenk.	-	-	-

Sl.No.	Botanical Name	No. of Accessions	Vernacular Names	Remarks
8.	*Allium pratti* Wright Syn. *A. victorialis* L. *Allium victorialis var. angustifolium hook. F.*	-		
9.	*Allium prezwalskianum* Regel. Syn. *Allium stolozkii* Regel.	-	Kin, Jimboo, Pharna	Dried leaves used as condiments. Medicinal value, used as sudorific
10.	*Allium schrenki* Regel.	-	-	
11.	*Allium semonovii* Regel.	-	-	Young leaves cooked as green vegetable.
12.	*Allium sikimense* Baker	-	-	-
13.	*Allium stracheyi* Baker	-	Dunna, Pharan, Jambu	Young leaves and tuberous roots cooked as green vegetable. Dried leaves used for flavoring. Flowers, leaves and roots are also transported to the plains for condiments.
14.	*Allium thomsoni* Schrenk.	-	Lahsooniya	-
Wild exotic *Allium* species maintained at Field Gene bank of Regional Station Bhowali of ICAR-NBPGR, New Delhi				
1.	*Allium altaicum* Pall.	-	-	Young leaves cooked as green vegetable.
2.	*Allium angulosum* L.	-	Mouse Garlic	-
3.	*Allium ledebrourianum* Roem. Et Schult	-	-	Leaves and bulbs fresh or boiled are eaten in Japan.
4.	*Allium lineare* L.	-	-	-
5.	*Allium obliquum* L.	-	-	Leaves are used as suitable for garlic.
6.	*Allium schoenoprasum*	-	Twisted Leaf Garlic	Leaves and bulbs are used for flavouring food dishes.
7.	*Allium senescens* L.	-	-	Bulbs are eaten as a vegetable among the people near Japan.

Table 9.3: Significant Features to be Noted for Infra-generic Categories

Sl.No.	Infra-generic Categories (Subgenus)	Species	Plant parts/characters
1.	Amerallium	*A. hookeri, A. macranthum, A. fasciculatum*	Root shape and fleshy character; bulb outer coat, leaf shape
2.	Anguinum	*A. victorialis, A. pratii*	Rhizome shape/size, leaf shape and width, vein characters.
3.	Butomissa	*A. tuberosum, A. oreoprasum, A. ramosum, A. gilgiticum*	Rhizome, bulb shape/size, outer and inner coat colour, texture, leaf in section; flower-perianth shape, colour of the dorsal mid-line on perianth
4.	Allium	*A. sativum, A. ampeloprasum, A. griffithianum*	Bulb shape, number, coat texture; presence of bulblets; leaf shape, density presence of aerial bulbils
5.	Cepa	*A. atrosanguineum, A. semenovii, A. cepa, A. fistulosum, A. longistylum, A. schoenoprasum, A. chinense*	Rhizome size, bulb shape, bulb coat characters, leaf fistular
6.	Polyprasun	*A. stracheyi, A. roylei, A. consanguineum, A. carolinianum*	Bulb shape, single/clustered, rhizome size, if produced into a neck; leaf shape, texture (leathery), tunic outer colour, texture-fibrous, coriaceous, brown.
7.	Cyathophora	*A. auriculatum*	Bulb shape, coat colour, texture, leaf shape, leaf in cross section, scape, size, erectness/orientation.
8.	Melanocrommyum	*A. atropurpureum, A. chitralicum, A. loratum*	Flowers if showy, nectaries located in the lower half of the ovaries, excretion through spur or rather than short tube, excretory tubes bent downwards
9.	Reticulatobulbosa	*A. humile, A. schrenkii, A. lineare, A. sikkimense, A. tenuicaule*	Bulb shape, number, tunic, colour, texture, leaf length, flat, spathe-valve number, perianth colour, filament length, scape length, ovary, apex orientation.
10.	Rhizirideum	*A. przewalskianum*	Bulb shape, tunic character, leaf length, if channeled, tepal: filament length.

Table 9.4: Taxa of *Allium* Maintained at Gene Bank of Geneva, New York

Taxa	G*	PI	Total	Number accessions backed up NCGRP
Allium cepa var. *cepa*	221	775	996	504
Allium fistulosum	33	80	113	38
Allium total	258	866	1124	561

*Geneva local number, not yet Pled.

Table 9.5: Status of the Cultivated and Wild Relatives of Cultivated Species Accessions

Taxon	*Total*	*Backed-up*	*Available*	*Per cent available*	*Common Name/Use*
Allium altaicum	24	15	20	83.3	*A. fistulosum* wild relative
Allium ampeloprasum	207	60	104	50.2	leek
Allium angulosum	6	5	5	83.3	mouse garlic
Allium galanthum	10	9	7	70.0	*A. cepa* wild relative
Allium oschaninii	6	1	3	50.0	*A. cepa* wild relative
Allium praemixtum	2	2	2	100.0	*A. cepa* wild relative
Allium pskemense	5	4	5	100.0	*A. cepa* wild relative
Allium ramosum	22	6	15	68.2	fragrant garlic
Allium roylei	1		1	100.0	*A. cepa* wild relative
Allium schoenoprasum	31	19	28	90.3	chives
Allium tuberosum	11	10	10	90.9	Chinese chives
Allium vavilovii	10	6	9	90.0	*A. cepa* wild relative

Source: Barbara Hellier, 509-335; 3763.

and are consumed either as food or used in medicinal purpose. Leaves of *Allium* species are consumed as leafy vegetables and also used for preparation of chutney and blended with local cuisines like *Tangtur, Thukpa etc.* leaves are also dried and used for flavor blending in vegetables and *Daals.* Garlic cloves are roasted and used for treatment of local cough whereas leaves of *A. carolinianum* are burned and smoke is inhaled to get relief from ear and nose ache. A range of wild species consisting of a *A. carolinianum, A. fistulosum, A. hookeri, A. tuberosum, A. macranthum, A. przewalskianum* and other wild species were collected and describe the uses of these wild species as food item or medicinal purpose by local inhabitants were documented (Khar *et al.,* 2016).

REFERENCES

1. Chandel, K.P.S. and Pandey, R. 1992. Distribution, diversity, uses and *in vitro* conservation of cultivated and wild Alliums-A brief review. *Indian J. Plant Genet. Resour.*, **5(1):** 7-36.
2. Davies, D. 1992. *Alliums.* Batsford, London.
3. De vries, J.N., Wietsma, W.A. and de Vries, T. 1992a. Introgression of leaf blight resistance from *Allium roylei* stearn into onion (*A. cepa* L.) Euphytica 62 (2) : 127-133.
4. Fiskenjo, G. 1975. Chromosomal relationship between the species of *Allium* as revealed by C-banding. *Hereditas,* **81:** 23-31.
5. Friesen, N., Fritsch, R.M. and Blattner, F.R. 2006. Phylogeny and intragenic classiofication of *Allium.* L. (Alliaceae) based on nuclear ribosomal DNA ITS sequences. *Aliso,* **22:** 372-395.
6. Friesen N. and Klaas, M. 1998. Origin of some minor vegetatively propagated *Allium* crops studied with RAPD and GISH. *Genetic Resources and Crop Evolution,* **45:** 511-523.
7. Fritsch, R.M. and Abbasi, M. 2013. A taxonomic review of *Allium* Subg. *Melanocrommyum* in Iran. Gatersleben, Germany.
8. Fritsch, R.M., Blattner, F.R. and Gurushidze, M. 2010. New classification of *Allium* L. Subg. *Melanocrommyum* (Webb and Berthel) Rouy (Alliaceae) based on molecular and morphological characters. *Phyton.,* **49:** 145-220.
9. Fritsch, R.M. and Friesen, N. 2002. Evolution, domestication and taxonomy. In: Rabinnowitch, H.D., Currah, L. (eds.) *Allium* Crop Science: Recent Advances. CABI Publishing, Wallingford, Oxfordshire, UK, p. 5-30.
10. Gohil, R.N. 1992. Himalayan representatives of *Allium.* In: Hanelt, P., Hammer, K. and Knuptter, H. (eds.). The Genus *Allium Taxonomic Problems and Genetic Resources.* Gatersleben, Germany, p. 335-340.
11. Hanelt, P., Schulze-Motel, J., Fritsch, R., Kruse, J., Maass, H., Ohle, H. and Pistrick, K.1992. Intragenetic grouping of *Allium*-The Gatersleben Approach. In: Hanelt, P.; Hammer, K., Knupffer, H. (eds.). The genus *Allium*-Taxonomic problems and genetic resources. Gatersleben, Germany, p. 107-123.
12. Hooker, J.D. 1894. The flora of British India, Vol-6 CL. VI. Liliaceae, *Allium* L., London.
13. Kachroo, P. 1977. Flora of Ladakh-An ecological and taxonomical appraisal, Dehradun (Uttarakhand).
14. Karthikeyan, S.K., Jain, S.K.; Nayar, M.P. and Sanjappa, M. 1989. Flora of India. Series-4 Florae Indicae Enumeratio: Monocotledonae, Botanical Survey of India. Calcutta, p. 177-178.
15. Khar, A., Negi, K.S. and Mehta, P.S. 2016. Wild is the furture-story of Indian Alliums. 1st International Agrobiodiversity Congress (Abs. Session: Crop Wild Relatives: Back to the Wild to Save the Future), 6-9 November 2016, held at New Delhi, p. 328.

16. Khar, A., Negi, K.S. and Mehta, P.S. 2016. Exploration and collection of *Allium* germplasm from Ladakh, J&K region of India. (Abs. 27, Session II: Breeding of Horticultural Crops). 7th Indian Horticulture Congress, held at ICAR-IARI, New Delhi, 15-18th November 2016, p. 106-107.

17. Khosa, J.S., Dhatt, A.S., Negi, K.S. and Singh, K. 2014. Characterization of *Allium* germplasm using bulb onion SSR markers. *Crop Improv.*, **40:** 69-73.

18. Koul, A.K. and Gojil, R.N. 1970. Cytology of the tetraploid *Allium ampeloprasum* with chiasma localization. *Chromosome*, **29:** 12-19.

19. Mc collum, G. 1974. Chromosome behavior and sterility of hybrids between the common onion, *Allium cepa* and the related wild *A. oschaninili. Euphytica*, **23:** 699-709.

20. Negi, K.S. 2006. *Allium* species in Himalayas and their uses with special reference to Uttarakhand. *Ethnobotany* **18:** 53-66.

21. Pandey, A., Pandey, R., Negi, K.S. and Radhamani, J. 2008. Realizing value of genetic resources of *Allium* in India. *Genet Resour. Crop Evol.*, **55:** 985-994.

22. Polunin, O. and Stainton, A. 1984. Flowers of the Himalaya.

23. Puizina, J.J., Avornik, B.B., Ohaanec, B.S., Chweizer, D.M., Aluszynska, J. and Papes, D. 1999. Random amplified polymorfic DNA analysis, genome size, and genomic *in situ* hybridization of triploid viviparous onions. *Genome*, **42:** 1208-1216.

24. Robinwitch, H.D. and Brewster, J. 1990. Onion and allied crops: Botany, Physiology and Genetics, Vol.I. CRC Press, Boca Raton, p. 1-26.

25. Sharma, D.C., Chandra, U. and Gupta, R.K. 1996. *Allium* biodiversity and its traditional prophylactic significance in Indian sub-continent. *Indian J. Plant Genet. Resour.*, **9(2):** 201-218.

26. Tutin.1976. Flora Europea. Vol.4 Cambridge Univ. Press, p. 505.

27. Vavilov, N.I. 1949. The origin, variation immunity and breeding of cultivated crops. *Chron. Bot.*, **13:** 364.

28. Vosa, C.G. 1976. Hetrochromatic patterns in *Allium. Heredity*, **36:** 383-392.

29. Wealth of India. 1985. The Wealth of India-Raw Materials (revised edition), Publications and Information Directorate, New Delhi, p. 513.

30. Wendelbo, P. 1971. Alliaceae. In: Rechinger (ed.) *Flora Iranica*, **76:** 1-100. Akad. Druck. Graz.

31. Zeven, A.C. and de Wet, J.M.J. 1982. Dictionary of cultivated plants and their regions of diversity. Centre for Agricultural Publishing and Documentation, Wageningen, p.263.

Onion (*Allium cepa* L.)

Introduction

Onion (*Allium cepa* L., 2n=2x=16) is one of the oldest and major commercial bulbous vegetable that has earned a pivotal place in global trade. It has been in use as a food for over 5000 years (Jones, 1963). In North eastern parts foliage is also used as a substitute of onion in their day to day life. It has own feature to fit in cropping system due to short duration and quick growing nature. The onion smell (strong flavor due to presence of sulphur containing compounds-olein propyldisulphides-responsible for distinctive aroma and pungency) and taste are important diagnostic features of the genus *Allium* (Robinowitch and Currah, 2002). As onion and its several value added products find place in world agriculture market. It is available in fresh, frozen, canned, pickled, powdered, chopped and dehydrated forms and involves more than 100 countries in business transaction. Apart from all it has long storage life of bulbs, and less losses during transportation. Out of 15 vegetables listed by the Food and Agricultural Organization (FAO), onion falls second only to tomato in terms of total annual world production *i.e.* 12 percent, the first being China (Currah and Proctor, 1990; Singhal, 2003). India has potential both for onion bulb (second largest grower in the world) and seed production as it is exported in the form fresh bulb, preserved onion, dehydrated onion, flakes and powdered onion to more than 34 countries (Pruthi, 2001). India has produces 19.35 million tonnes from 1.19 million hectare in the world after China. The productivity of inion in this country is 16.24 t/ha as cpmpared to many other countries. Recent record (2015-2016) is of 1.2 million tonnes export of fresh onion of Rs. 2747 crores. Major markets for Indian onions are United Arab Emirates, Saudi Arabia, Singapore, Malaysia, Japan, Sri Lanka, Bangladesh and USSR. Onion as medicinal herb is known to ancient world, as it finds in the medicinal treaties like "Charak Samhita" of third of fourth century A.D. (Ray *et al.*, 1980). Sanskrit language equivalent signifying Vedic period and Aryan usage is available for onion as "*Palandu*" (Aiyar and Vaganarayana, 1956). In mountainous regions how local peoples are using wild species of *Allium* are reported and documented by Negi and Pant (1992). Murthy *et al.* (2011) given the importance of wild species due to presence of primary and secondary metabolites like carbohydrate, proteins, alkaloids, phenolic compounds, saponins *etc.* which posses anti-bacterial activity in *E. coli*, *S. aureus* and *P. aeruginosa* isolated from infected patients. 'Wild Onion' *Urginea indica* is helpful in the treatment of wound healing. The tuber of this species has therapeutic potential which is commonly used in traditional medicine treatment. Hence, the formulation of extract of *Urginea indica* needs to be purified using biophysical techniques towards development of useful drugs/lead molecule against microbial infection, inflammation and wound healing properties, respecively.

Origin, History and Distribution

Origin place of onion is reported to be central Asia and secondary centre in the East Mediterranean Region (Mc Collum, 1976). The Russian scientist Vavilov (1951) reported that onion originated in Pakistan. He names the middle-Asiatic center, comprising northwestern India, all of Afghanistan, the Soviet Republics of Tajik and Uzbek, and western Tiensen (earlier noted by Jones and Mnn, 1963), as the primary place of origin of the commonly cultivated onion, *Allium cepa*. As secondary centers of origin of this species he lists:

1. The Near East, which includes inner Asia Minor, the whole of Transcaucasia, Persia (Iran), and the alpine Turkmen Republic, and
2. The Mediterranean region. For the Japanese onion (*A. fistulosum*) which is cultivated extensively in the Orient, Vavilov gives the primary place of origin as the Chinese center, which comprises alpine central and western China and the adjacent lowlands.

Western Asia and the area around Mediterranean sea are its secondary centre of origin. Onion comes from the Latin "Unio" meaning one or unity which is probably acquired from the city by Onia in 1703 BC near the Gulf of Swez. In ancient Egypt, the onion was a symbol of eternity because it is a circle within-a-circle as pictured monuments (Chadha, 1990). These areas could have led to evolution of ancestral form that now is extinct. The Chinese were growing onions in their gardens 5,000 years ago, while Egyptians were cultivating them as far back as 3200-3500 BC for food, medicine and religious objects (Takholm and Drar, 1954). It is well mentioned in the Bible and Koran for fedding to the workers (Chadha, 1990). In India, onion has been grown from ancient times as mentioned in the famous early medicinal treatise 'Charaka Samhita' which inferred the teaching of an earlier authority of about 6th century BC and many virtues were attributed to the onion by Charaka Samhita (Ray and Gupta, 1980). According to Food and Agricultural Organization (FAO) estimates top ten onion producers in the world are India, China, USA, Turkey, Pakistan, Russia, South Korea, Japan, Egypt and Spain with world production. Occurrence and distribution of lesser known wild species can be seen in mountainous regions (Negi and Pant, 1992). Major states growing onion in India are Andhra Pradesh, Bihar, Gujarat, Karnataka, Madhya Pradesh, Maharashtra, Odisha, Rajasthan, Tamil Nadu and Uttar Pradesh. In the country, Rajasthan ranked fourth in area (6.13 thousand hectare) and 6th in production (960.7 thousand tonnes). At present in India one organization *i.e.* ICAR-National Research Centre for Onion and Garlic has been established at Rajguru Nagar, Pune (Maharashtra) which is also running All India Co-ordinated Project on Onion and Garlic in the country. The prominent wild species distributions in mountain region of India are *A. carolinianum* D.C. (syn. *A. bladum* Wall.), *A. chinense* G. Don. (syn. *A. bakeri* Regel), *A. consanguineum* Kunth, *A. humile* Kunth (syn. *A. govanianum* Wall. ex Baker), *A. przewalskianum* Regal (syn. *A. jacquemontii* Regel, *A. stolczkii* Regal), *A. staeacheyi* Baker, *A. victorialis* L. and *A. wallichii* Kunth. Most of them are economically potential and already been introduced as semi-domesticated cash crop into the backyards, dooryards and garden of natives of the Himalayan region (Negi, 2006; Negi *et al.*, 1995).

Species Availability, Evolution and Diversity

The genus *Allium* is a largest member of the Amaryllidaceae, a large herbaceous perennial and bulbous family with more than 57 genera and 500 species (Frodin, 2004). Some researchers indicate that *Allium* represented by about 750 species widely distributed all over the warm temperate and 34 species are mainly from alpine and temperate zones of the northern hemisphere-Himalayan area (India). *Allium cepa* L. or garden onion is the most widely cultivated sopecies. Besides cultivated species (*Allium cepa, A. sativum, A. ascalonicum* and *A. ampeloprasum*) some species are semi-domesticated/lesser known types (*Allium carolineanum, A. chinense, A. consanguineum, A. humile, A. victorialis, A. wallichii* and *A. semenovii*) under sporadic cultivation. Regions of high species diversity occur in Turkey, Iran, North Iraq, Afghanistan, Soviet, middle Asia and West Pakistan and a second less pronounced centre of species diversity occurs in Western North America. About 40 species of *Allium* are reported in India of which 7 species are cultivated, the main being onion and garlic. Much diversity has built up in India in these crops as they have been cultivated from ancient times and selections have been made in different agro-ecological zones.

Onion (*Allium cepa* L.) does not exist in wild but several related species (*A. vavilovii, A. aschaninii* and *A. pskemense*) are found. Several edible species are described by Sharma (2002). The difference among the members of the complement in *A. cepa, A. cepa* types and *A. galanthum* are less pronounced than in *A. pskemense, A. roylei* and *A. drobovii*. It seems that these species may have had a common ancestor and that chromosomal differences have arisen due to inversions, translocations and pairing in unequal chromosomes. *A. longicuspis*, endemic to central Asia, has been proposed as a progenitor for *A. sativum*. Nath *et al.* (2008) reported that onion has been grown in India for a long time and some of its derivatives are also available. Various useful strains are available in different growing areas of the country.

The species is divided into different groups which are sometimes given the rank of botanical varieties:

(i) ***Allium cepa* var. *cepa*-** Common onion with large bulbs borne singly and propagated through seeds. It is mainly used as salad or in curries.

(ii) ***Allium cepa* var. *aggregatum*-** Multiplier onion or potato onion: producing small bulbs borne in clusters and generally propagated through small bulbs. Mainly used for seasoning curries.

(iii) ***Allium cepa* var. *ascalonicum*-** Produces bulbs in clusters on surface of soil. This perennial onion rarely produces seeds and is propagated through bulbs.

(iv) ***Allium cepa* var. *viviparum/proliferum*-** This viviparous plant produces a perennial underground bulb and is not widely cultivated. The bulbs produced in place of umbel are used for pickling and seasoning. Bulb-lets are also used for propagation.

(v) ***Allium schoenoprasum* (2n=16, 24, 32) Chive-** This is a hardy perennial bunching herb grown for their hollow green leaves. It is propagated by root division and is tolerant to extreme cold and drought.

(vi) ***Allium fistulosum*-** It is a perennial bunching species. It is very common in China and Japan and is grown for its edible tops and leaf base.

The Himalayan *Allium* species comprise a rather diverse assortment of floristic elements *i.e.* cultivated food plants and other taxa (secondary gene pooles) which are of potential value as plant genetic resources for future breeding programs (Xu *et al.*, 1990; Gohil, 1973; Gohil, 1992; Nayar *et al.*, 1992; Negi *et al.*, 1991; Chandel and Pandey, 1992).

Table 9.6: *Allium* Species Possessing Desirable Traits

Species	*Desirable Traits*
Allium cepa var. *aggregatum*	Production of bulbs in clusters, high temperature tolerance
A. cepa var. *viviparum*	Viviparous habit, resistance to disease and pests
A. porrum	Cold and drought resistance
A. ampeloprasum	Large bulb size
A. schoenoprasum	Tolerant to extreme cold, drought and perennial habit.
A. kurrat	Leaf-flavor
A. fistolosum	Edible green flavor top
A. tuberosum	Flavour red green flavor top
A. vavilovi	Readly crossable with *A. cepa*

Gene and Genetics of Characters

Yellow red colour of bulb was linked with resistance to *Colletotrichum arienans*.

Central Asia is the primary centre of diversity in onion and related species as *A. vavilovii, A. aschaninii* and *A. pskemense* are abundantly found in this region (Vavilov, 1951). Onion (*A. cepa* L.,) does not exists in wild but several related species are found in mountainous regions of higher Himalayas comprising Turkmenia, Ujbekistan, Western TiensenNorth Iran, Afghanistan, Pakistan and North-West India (Ladakh region). The Near East, Western Asia and the area around Mediterranean Sea are the secondary centre of origin.

Botanical Description

Onion belongs to family Alliaceae, genus *Allium, A. cepa* L.Taxonomic of cultivated onion is:

Class	:	Monocot
Super order	:	Liliflorae
Order	:	Asparagales
Family	:	Alliaceae/Amaryllidaceae/ Liliaceae
Genus	:	*Allium*
Species	.	*cepa*

Major characteristics include the presence of bulb firmed by the attachment of swollen leaf bases to the underground part of the stem. Typical flavor due to a volatile oil known as allyl propyl disulphide and red colour is because of the pigments anthocyanin and yellow colour is due to 'Quercetin'.

Horticultural Classification

Bailey divides *Allium* species into several botanical varieties as follows:

(1) The extensively cultivated bulbing type of onion propagated by seed

(2) The potato or multiplier onion, which is propagated by branching at the base; and

(3) The top onion, which forms bulbils in the inforescence, which are often used for propagation.

Table 9.7: Character, Number of Gene and Type of Gene Action

Character	*Number of Genes*	*Type of Gene Action*
Male sterility	Chromogene-one (msm)	Male sterility is recessive to fertility
	Plasmagene-one (S), Interaction between Chromogene and Plasmagene	Maternal inheritance, expressed as in female parent
Seed colour	Monogenic (bb)	Black dominant to brown
Bulb colour	Five major genes (ICGLR)	Increasing density from light to dark colour is due to accumulation of microgenes each with a small quantitative effect: II, white; ii, yellow red; Ii-between pure white and coloured

Table 9.8: Botanical Description

Growth habit	Onion is bulbous biennial or perennial Herbs (annual for the edible bulb production and biennial for seed production), which give off a distinctive and pungent odour when the tissues are crushed.
Root	Root system is fibrous, long arising in whorls from the basal part of stem. Primary roots produce by the leading, after which all other roots are adventitious, arising from the short stem. It is superficial and extending to 10-15 cm deep in the soil and within a radius of 30 mm of stem. There are few laterals and total root growth is sparse and not much aggressive.
Stem	The edible portion is a modified stem known as bulb, which develops underground. The real stem (without distinct above ground stems) is highly reduced and underground. It bears roots at the base and bulb on the upper surface or seed stalk in case plant is being grown for seed production. Underground stem is hard and solid mass of tissue. Aerial stem is seed stalk which is hollow, green, 80-100 cm tall, cylindrical somewhat swollen at the middle bearing umbel at the top. Elongation of stem forms the flower stalk (Hawthorn and Pollared, 1954). Each plant may have 2-8 seed stalks depending on variety and management conditions followed.
Leaf	The onion leaves form a shortened crown stem (stemplate). Leaves are green, hollow tapering at both the ends but swollen at the base forming the fleshy scales constituting the main edible part, the bulb. Narrow vertical leaves check excessive rise in temperature and effectively dissipate heat in air. Thus leaves impart drought and high temperature tolerance to onions. The ultimate yield of onion is determined by the number of leaves that are formed prior to bulbing.
Floral biology	It bears bisexual flowers. Inflorescence is umbel (heads) which is an aggregate of cymes of 5-10 flowers each. The umbel is enclosed in a papery spathe (sheathed by one or more spathes) consisting of 2-3 bracts. Later umbel (aggregates of 5-10 flowers) expands and single spathe which split into segments (periamth segments are 6 in 2 whorls spreading, reflexed, free and ovate)by the pressure of developing flowering buds as the young flowers emerge and a superior ovary abnove the 6 stamens in 2 whorls and petals like perianth. Flowers are borne in simple umbels at the apex of the floral stem, which is commonly hollow when mature, round in cross section and somewhat swollen at the middle or near the base. The number of flowers per plant varies from 50-2000 depending on the variety, time of planting, size of bulb and nutritional status of the mother bulb. Flowers are white, blue or bluish in colour. Flowers are protandrous means anthers mature first shedding pollens before the stigma becomes receptive.Anthers bilocular, ovary superior, 3 locules with nectary at base, and each locule has 2 ovules.
Anthesis and pollination	Flowers continue to open for 15-20 days. Each cyme of umbel opens in a definite sequence. The style, which is approximately 1.0 mm long, when the flowers first open, is not receptive, until it elongates to a length of 5.0 mm. This requires a day or two days after anthers have all dehisced. An inflorescence may continue opening for two weeks or more and a plant may be in bloom for more than 30 days. The flowers are protaandrous in nature and highly cross pollination is favored chiefly by honeybees or some times by bumble bees and other insects visiting the crop for nectures.
	Though cross-pollinated, some selfing also takes place as the insect visit many flowers in a single umbel and also umbels of the same plant before leaving for other plant. Wind also plays role in effecting the pollination. Duration of anthesis is about 4 weeks on individual umbel. In onion anthesis begins from outer flowers and proceeds inwardly to centre in succession. Anthesis timing is 6.00-7.00 am. Anther dehiscence is between 7.00 am to 5.00 pm. Pollen fertility and stigma receptivity is highest during the day/period of anthesis (Jones, 1923).
Bulb	Edible part of the onion is bulb, composed of concentric, fleshy, enlarged leaf bases or scales. The outer leaf bases lose moisture and become scaly and the inner leaves generally thicken as bulbs develop. The scales store carbohydrate and other nutrients. Colour of the bulb is due to pigment quercetin which is quantitatively inherited. The shape of the bulb may be flat, globe, and cylindrical. Outer colour is white, yellow, red, brown, pink and green.
Fruit	The fruit is a three lobed, three celled capsule, each locule containing 1 or 2 black seeds at maturity.
Seeds	Embaryo, which is mostly made of the cotyledon is spirally twisted, cylindrical and is embedded in the endosperm, which forms the major portion of seed (Hawthorn and Pollared, 1954). Seeds are black orbrown in colour, triangular, small in size and light in weight. The number of seed stems vary from 1-20 or more dependiong before it is fully expanded is enclosed in a papery spathe, consisting of 2-3 bracts. They are used for raising the nursery whereas in case of seed crop bulb is the seed source.

Within *A. cepa*, two horticultural varieties are recognized:

(i) The dry bulb onion

(ii) The multiplier onion

Classification on the Basis of Botanical Varieties/Species

The species is divided into 3 groups which are sometimes given the rank of botanical varieties as follows (Jones and Mann, 1963; Hanelt, 1990):

Table 9.9: Comon and Botanical Name of *Alliums*

Sl. No.	*General Name*	*Botanical Name*
1.	Common onion group (propagated by seed)	*Allium cepa*
2.	Aggretgatum group-multiplier onion, potato onion, shallots (propagated by branches emerge at the base)	*A. cepa* var. *aggretgatum* G. Don.
3.	Top onion type or Proliferum group (interspecific)-tree onion (forming bulbils in the inflorescence, which are often used for propagation; Desi *et al.*, 1970; Choudhury, 1983)	*A. cepa* var. *proliferum*

Classification on the Basis of Utility

Pruthi (2001) identified six different types of onion in India namely:

(i) Green onions for vegetable purpose or used as salad.

(ii) Mild pungent onions used for vegetable as well as salad.

(iii) Pungent onions used as condiment for flavoring of a number of dishes

(iv) Pearl or small onions used for pickle purpose

(v) White or yellow onions for manufacturing dehydrated products and onion powder

(vi) Red or yellow onions-It is used for culinary purposes. However, from marketing point of view, important parameters are colour, size, shape, pungency and keeping quality of the bulb. 'Red Creole', 'Granex', 'Bombay Red', 'Red Kano', 'Patna Red', 'Texas Grano' 'Yellow Bermuda', and 'Eary Cape Flat' are suitable for export.

Classification on the Basis of Photoperiodism

Onion is also classified o the basis of photoperiodism

Table 9.10: Photoperiod Base Cultivars

Response to Photoperiodism	*Cultivars*
Short day type	The Bermuda-Granex, flat to conical shaped, mild and early-maturing group, 'Texas Early', 'Grano No. 502', 'Early Yellow Cape Flat', 'Yellow Bermuda', Red Creole', 'Tropicana, 'El Toro', 'Bombay Red'
Long day type	Globe onions ('white Globe', Sweet Spinach')
Day neutral type	'Ebenezer', 'White Portugal', the creole type, with very strong flavour

Classification on the Basis of Skin Colour

(i) Red onions/light red- In red onions, 'Punjab Selection' (1975), 'Pusa Ratnar' (1975), 'Pusa Red' (1975) and 'N2-4-1' (1985) were evolved for commercial cultivation.Arka Kalyan (Selection-14), Line-102, Agrifound Dark Red, Agrifound Light Red, Punjab Red Round

(ii) White onions-In white onions, 'Pusa White Round', 'Pusa White Flat', 'S 48' and 'N 257-9-1' were found promising.

Botanical Classification

See Table 9.11.

Classification on the Basis of Size

See Table 9.12.

Augmentation of Genetic Resources and Crop Improvement

Breeding Priorities

- ☆ High yield
- ☆ Shape
- ☆ Colour
- ☆ Size
- ☆ Date of maturity
- ☆ Attractive and intact skin
- ☆ Minimal bolting and sprouting resistance
- ☆ Single centered crisp bulb

Table 9.11: Group of *Alliums*

Group	*Botanical Name*	*Edible Parts*	*Chromosome Number*
Potato onion or multiplier onion	*A. cepa* var. *aggregatum*	Bulbs	2n-16
Shallot	*A. cepa* var. *ascolonicum*	Bulbs	2n-16
Tree onion	*A. cepa* var. *viviparum*	Bulbiles	2n-16
Leek	A. porrum	Stem, leaves	2n-32, 48
Great headed garlic	*A. ampeloprasum* var. *ampeloprasum*	Leaves and cloves	2n-32, 48
Chive	*A. schoenoprasum*	Leaves	2n-16, 24
Chinese chive	*A. tuberosum*	Green leaves, flowers	2n-32
Welsh onion	*A. fistulosum*	Stem and leaves	2n-32, 48
			2n-16, 24

Table 9.12: Size of Cultivars and Quality

Size	*Cultivars Quality*	*Cultivars*
Big onions (4-6 cm size)	Light to red colour	Agrifound Dark Red (TSS 12-13 per cent); Agrifound Light Red (TSS 13-14 per cent)
Small onion or rose onion or Krishnapuram onion (2.5-3.5 cm size)	Small size onion; generally cultivated in Kolar (Karnataka) and Gudaphoh (Amndhra Pradesh)	Agrifound Rose (TSS 16-18 per cent, good storer)
Multplier onion or Podesu and shallot (Tamil Nadu)		Agrifound Red (TSS 9-10 per cent, good storer)

- Medium pungency
- High dry-matter content
- Pest (thrips, mites *etc.*) and disease (purple blotch, stemphyllium blight, garlic mosaic virus, neck rot, basal rot, black mould *etc.*) resistance (Singh and Rana, 1994)
- Stress resistance like heat, cold, frost, drought, flood *etc.*
- Good storage properties

Challenges in Onion Breeding

- Vegetative propagation
- Highly heterozygous
- Diploid crop
- Often limited in reproductive quality

Introduction and Selection

The varieties grown in major onion growing region of India are of local types with a lot of variability for different characters. Early Grano and Red Grano were introduced from America. Cultivar 'Burmuda Yellow' was introduced from Philippines. Onion variety 'Early Red', and EC473300, were introduced from Israel for high yield under Indo-Israel Project by ICAR-IARI, New Delhi. In Pakistan onion cultivars like Local Red, Local White (R), Faislabad Early, Swat I (I) and NARC-91 (P) are very common and being grown commercially.

India has developed several varieties of onion for different agroclimatic zones (Chadha, 1990). Popular cultivars like Pusa Ratnar and Hisar-2 were developed by use of Indigenous material through family selection. Onion cultivars like Arka Kalyan, Arka Niketan, Arka

Table 9.13: Introduction of Onion Germplasm in India

Name of Germplasm	*Salient Feature*	*Source Country*
EC30802-11	High bulb yield	USA
EC378476-77	High TSS and longer shelf life	UK
EC177284-85	Yellow skin bulb, longer shelf life and multiple resistance	Holland
EC187221-25	Yellow skin bulb, longer shelf life and multiple resistance	Germany

Table 9.14: Popular Cultivars of Onion

Cultivars/Varieties	*Colour of Skin*	*Developing Organizations*
'Pusa Red'	Red	ICAR-Indian Agricultural Research Institute, New Delhi
'Pusa Madhav'	Red	ICAR-Indian Agricultural Research Institute, New Delhi
'Pusa Ratnar'	Red	ICAR-Indian Agricultural Research Institute, New Delhi
'Pusa White Round'	White	ICAR-Indian Agricultural Research Institute, New Delhi
'Pusa White Flat'	White	ICAR-Indian Agricultural Research Institute, New Delhi
'Agrifound Dark Red'	Red	ICAR-National Horticultural Research and Development Foundation, Nasik (MS)
Agrifound Light Red'	Light Red'	ICAR-National Horticultural Research and Development Foundation, Nasik (MS)
'S 48'	White	Punjab Agricultural University, Ludhiana (Punjab)
'Punjab Selection'	Red	Punjab Agricultural University, Ludhiana (Punjab)
'VL 1'	Red	ICAR-Vivekanand Parvatiya Krishi Anusandhanshala, Almora (Uttarakhand)
'VL 3'	Red	ICAR-Vivekanand Parvatiya Krishi Anusandhanshala, Almora (Uttarakhand)
'N 53'	Red	Mahatama Phule Krishi Vishwavidyalaya, Rahuri (MS)
'N 2-4-1'	Red	Mahatama Phule Krishi Vishwavidyalaya, Rahuri (MS)
'Baswant 780'	Red	Mahatama Phule Krishi Vishwavidyalaya, Rahuri (MS)
'N 257-9-1	White	Mahatama Phule Krishi Vishwavidyalaya, Rahuri (MS)
'Arka Kalyan'	Red	ICAR-Indian Institute of Horticultural Research, Bengaluru (Karnataka)
'Arka Niketan'	Red	ICAR-Indian Institute of Horticultural Research, Bengaluru (Karnataka)
'Arka Bindu'	Red	ICAR-Indian Institute of Horticultural Research, Bengaluru (Karnataka)
'Arka Pragati'	White	ICAR-Indian Institute of Horticultural Research, Bengaluru (Karnataka)
'Hissar 2'	White	Ch. Charan Singh Haryana Agriculural University, Hisar (Haryana)
'Kalyanpur Red Round'	Red	Chandra Sekhar Azad University of Agriculture and Technology, Kanpur (U.P.)
'Udaipur 101'	White	Rajasthan Agricultural University, Bikaner (Rajasthan)
'Udaipur 102'	White	Rajasthan Agricultural University, Bikaner (Rajasthan)
'Udaipur 103'	White	Rajasthan Agricultural University, Bikaner (Rajasthan)
'CO 1'	Red	Tamil Nadu Agricultural University, Coimbatore (Tamil Nadu)
'CO 2'	Red	Tamil Nadu Agricultural University, Coimbatore (Tamil Nadu)

Cultivars/Varieties	Colour of Skin	Developing Organizations
'CO 3'	Red	Tamil Nadu Agricultural University, Coimbatore (Tamil Nadu)
'CO 4'	Red	Tamil Nadu Agricultural University, Coimbatore (Tamil Nadu)
Bhima Super	Red	Directorate of Onion and Garlic Research, Pune (Maharashtra)
Bhima Dark Red	Dark Red	Directorate of Onion and Garlic Research, Pune (Maharashtra)
Bhima Red	Red	Directorate of Onion and Garlic Research, Pune (Maharashtra)
Bhima Light Red	Light Red	Directorate of Onion and Garlic Research, Pune (Maharashtra)
Bhima Sweta	White	Directorate of Onion and Garlic Research, Pune (Maharashtra)
Bhima Subhra	White	Directorate of Onion and Garlic Research, Pune (Maharashtra)

Pragati, Punjab Selection, Pusa Red, Agrifound Dark Red, CO-2, Nasik Red, Pusa White Round, N-53, Kalyanpur Red Round, Agrifound Light Red, Hisar-2 and Arka Bindu were developed through mass selection breeding programme.

Choudhary *et al.* (2016) evaluated 24 onion cultivares in arid condition of Rajasthan during *Rabi* season and significant variation in terms of bulb weight and bulb yield. Significantly highest bulb weight was recorded in RO-59 (54.1 g) followed by RO-1 (52.3 g) and Hisar Onion-3 (49.2 g). Maximum bulb yield was recorded by RO-1 (27.5 t/ha) followed by RO-59 (26.6 t/ha) and Local (25.5 t/ha). Among the colour onion varieties, RO-59, RO-252 and Local; in yellow varieties Phule Suvarna and among white varieties Bhima Shweta and Alok Safed were found promising in productivity.

Welsh onion (*A. fistulosum*) and shallot (*A. ascalonicum*) are used in the green stage as leafy vegetable or other preparation.

Table 9.15: Onion Varieties Developed through Mass Selection

Variety	Source
Arka Kalyan	IIHR-145
Arka Niketan	IIHR-153
Arka Pragati	IIHR-149
Punjab Selection	Indigenous
Pusa Red	Indigenous
Agrifound Dark Red	Indigenous
CO-2	Indigenous
Nasik Red	Indigenous
Pusa White Round	Indigenous
N-53	Indigenous
Kalyanpur Red Round	Indigenous
Agrifound Light Red	Indigenous
Hisar-2	Indigenous
Arka Bindu	Indigenous

Hybridization and Selection

In onion 15-60 per cent heterosis is recorded. Very little area is under onion hybrid cultivation in India, because of non-availability of suitable hybrids seeds for different areas. The hybrid varieties have good potential and are known for uniform in size, maturity and attractive colour which increases yield, compared to local variety. It is very necessary to test the new hybrids from different sources to recommend the hybrids for cultivation in a particular region. Many promising F_1 hybrids identified are 'Abundance', 'Aristocrat', 'Banaza', 'Champion' and 'Contender'. Introduced yellow colour hybrid 'Arad' gave best performance under Indian conditions. Other cultivars like 'Tana' (H), Suprex, Granex-55, HA-60 and Granex-429 were found successful in India. Several F_1 hybrids developed in India by using elite genotypes like 'Purple' x 'Red', 'Purple' x 'White' and 'Red' x 'White' considering weight of plants and bulb diameter. Out of 60 F_1 hybrids evaluated during *rabi*, five F_1 hybrids *viz.* DOGR Hyb.-2, DOGR

Table 9.16: Performance of F_1 Hybrids Entered in AINRPOG Trials

Entry	MY (q/ha)	Superiority Over Best Check (per cent)	ABW (g)	Bolters (per cent)	DTH	Season	Storability	Bulb Shape	Bulb Colour
DOGR Hy-1	41.30	42.84	68.20	0.0	103	Rb	Good	Flat Globe	LR
DOGR Hy-2	34.96	20.91	64.50	0.0	100	Rb	Good	Globe	DR
DOGR Hy-3	34.20	16.22	68.95	0.0	95	Kh	Medium	Globe	MR
DOGR Hy-4	34.08	15.81	80.21	0.0	94	Kh	Medium	Flat Globe	DR
DOGR Hy-5	38.73	20.11	75.21	0.0	101	Kh	Medium	Globe	DR
DOGR Hy-7	40.75	34.88	68.72	0.0	105	Kh and Rb	Very Good	Flat Globe	DR
DOGR Hy-8	54.45	23.57	83.33	3.89	118	Lk	Good	Flat Globe	MR
DOGR Hy-50	37.46	24.11	63.44	0.0	106	Rb	Very Good	Globe	DR

MY: Marketable yield; ABW: Average Bulb weight; DTH: Days to harvest; LR: Light Red; DR: Dark Red; MR: Medium red; Rb: Rabi; Kh: Kharif; LK: Late Kharif.

Hyb.-3, DOGR Hyb.-4, DOGR Hyb.-5, DOGR Hyb.-7' DOGR Hyb.-8 and DOGR Hyb.-50 showed more than 30 per cent heterosis on marketable yield over standard check 'Bhima Kiran' (25.3 t/ha) whereas 10 F_1 hybrids performed better during Kharif and showed more than 20 per cent heterosis on marketable yield over standard check Arka Lalima (20.6 t/ha).

Singh (2016) evaluated the recommended cultivars of zone V (Junagarh, Nashik, Rahuri and Pune) condition along with the standard check Agrifound Light Red (*Rabi* season cultivar). The highest bulb diameter (6.64 cm) and 20 bulb weoght (3.01 kg) were recorded in hybrids Matahari and was at par with Excalibur in respect of bulb diameter. The highest gross (656.66 q/ha) and marketable yield (594.12 q/ha) were recorded in hybrid Matahari and was at par with Excalibur in respect of marketable yield. Highest total soluble solid (15.10 per cent) was recorded in hybrid Flare which was at par with varieties NHRDF-Red-2 and Agrifound Light Red. From the findings, it cleared that the onion hybrids Matahar, Excalibur, Basic F_1 Yellow, Kristal, Rio Tinto, Gobi, Denali, Lucifer, Arad-I and Colina performed better in respect of yield and yield contributing traits (bulb diameter, 20 bulbs weight, size index of bulb *etc.*) and the varieties Flare, NHRDF Red-2 and Agrifound Light Red performed better regarding total soluble solids.

Interspecific Hybridization

In onion, mass selection, pedigree selection, back cross method, hybridization breeding methods *etc.* are quite common. In monocotyledonous onion plant, transformation by *Agrobacterium rhizogenes* has been successful (Arya, 2003). Among the species which belong to *Allium, A. cepa, A. vavilovii, A. galanthum, A. roylei, A. fistulosum, A. altaicum, A. pskemense* and *A. oschaninii* have been considered as the most important gene pools of onion considering the breeding programme (introgression of foreign gene into cultivated species) to mitigate biotic and abiotic stresses (Shigyo and Kik, 2008). Cultivated and wild types interbreed results plenty of variation in cultivated forms. The objective of interspecific hybridization (introgression of agronomically desirable genes from suitable wild relatives) are to generate variability so that useful variants with high yield, dry-matter content, good quality, good storage properties, minimal bolting and resistance to diseases can be selected out (Brewster, 1994; Kik, 2002; Kielkowska and Adamus, 2006). Interspecific hybrids can be obtained by protoplast fusion. *A.cepa* and *A. cepa* type have a wider and bigger satellite than present in the other species. The successful interspecific hybrids developed are *A. cepa* var. *cepa* x *A. fistulosum* and *A. porrum* x *A. cepa* var. *cepa*. *Allium roylei* Stearn. has been used as a bridge species for interspecific hybridization between *A. cepa* and *A. fistulosum* (Khrustaleva and Kik, 2000). Van der Meer (1984) reported successful crosses between leek and kurrat, and the subsequent backcrosses to leek were made in the context of introducing yellow-stripe virus resistance into leek. *Allium roylei* has been a source of interest to onion breeders throughout the world for resistance breeding. No doubt, by crossing with *Allium roylei* Stearn several variants has been developed. It was possible to introduce resistance against downy mildew (*Peronospora destructor*) [Berk.] Casp. into onion (Van der Meer and de Vries, 1990; Vries and Vries, 1992; Scholten *et al.,* 2007). Ornamental Alliums have also been by using interspecific crosses (Nomura *et al.,* 2002). Most of the work has been done on long day type cultivars barring some reports on Indian varieties regarding interspecific hybridization (Kohli and Chauhan, 1990; Khar *et al.,* 2000; Saneet Kumar and Banerjee, 2001) and genetic diversity assessment using molecular markers (Khar *et al.,* 2010).

Table 9.17: Interspecific Hybridization in *Allium* Species

Sl. No.	Cross Combination Developed	References
1.	*A. cepa* x *A. fistulosum*	Emsweller and Jones, 1935; Van der Meer and Van Benekom, 1978; Currah and ockendon, 1988; Bark *et al.*, 1994
2.	*A. galanthum* x *A. cepa*	Mc Collum, 1980
3.	*A. fistulosum* x *A. roylei*	Mc Collum, 1982; Khrustaleva and Kik, 1998
4.	(*A. fistulosum* x *A. cepa*) x *A. cepa*	Van der Valk *et al.*, 1991a; Van der Valk *et al.*, 1991b
5.	*A. cepa* x *A. sativum*	Ohsumi *et al.*, 1993
6.	*A. chinense* x *A. fistulosum*	Nomura *et al.*, 1994; Nomura and Makara, 1993
7.	*A. fistulosum* x *A. tuberosum*	Amagi *et al.*, 1995
8.	*A. cepa* x (*A. fistulosum* x *A. roylei*)	Khrustaleva and Kik, 1998
9.	*A. cepa* x *A. ampleprasum* (leek)	Peterka *et al.*, 1997
10.	*A. cepa* x *A. sphaerocephalon*	Bino *et al.*, 1989
11.	*A. ampleprasum* x *A. sativum*	Yanagino *et al.*, 2003
12.	*A. fistulosum* x *A. schoenoprasum*	Umehara *et al.*, 2007
13.	*A. cepa* x *A. roylei*	Chuda and Adamus, 2012
14.	*A. roylei* x *A. cepa*	Van der Meer and de Vries, 1990

Introgression between *A. cepa* and *A. galanthum* has been attempted by several onion breeders (Saini and Davis, 1967; McCollum, 1971). In general hybrids were sterile but few combinations set the seeds and fertility of the progenies in later generation was restored by back cross and in breeding programme, *A. galanthum* has contributed CMS trait in shallot (Yamashita and Tashiro, 1999), in onion (Havey, 1999) and in *A. fistulosum* (Yamashita *et al.,* 1999a). In *A. fistulosum*, alloplasmic line having *A. galanthum* cytoplasm was developed by continuous backcrossing (Yamashita *et al.,* 1999a).

In this programme male sterile populations were obtained by selection from BC_1 to BC_5. But the male sterile populations segregated into fertile and sterile in equal numbers. It indicated the fertility restorer gene was also from *A. galanthum* in the nucleus (Yamashita *et al.*, 1999b) segregating in the progenies. The location of restorer loci was located on *5f* region on *A. galanthum* chromosome (Yamashita *et al.*, 2005). Studies in onion sterile could not able to locate on the same region of *A. galanthum* (Yamashita *et al.*, 2000; Havey, 1999) but on *2A* in onion (Shigyo *et al.*, 2005)

Some successful distant hybridization are as under:

- *Allium cepa* var. *cepa* x *A. fistulosum*
- *A. cepa* var. *cepa* x *A. porrum*
- *A. porrum* x *A. cepa* var. *cepa*
- *A. porrum* x *A. ampeloprasum*
- *A. cepa* var. *cepa* x *A. cepa* var. *aggregatum*
- *A. cepa* x *A. altacum*
- *A. cepa* x *A. vavilovii*
- *A. cepa* x *A. pskemense*
- *A. cepa* x *A. galanthum*
- *A. cepa* x *A. drobovii*

Table 9.18: Wild Source for Desirable Traits

Wild Species	*Desirable Traits*
A. cepa var. *aggregatum*	Production of high quality bulb colour; tolerant to high temperature
A. cepa var. *viviparum*	Quality bulbils; resistant to disease and pest resistance
A. porrum	Cold and drought tolerance
A. ampeloprasum	Large bulb size
A. scoeloprasum	Cold and drought tolerance, perennial habit
A. currot	More leaf flavonoids
A. fistulosum	More leaf flavonoids
A. vavilovii	Easily crossable

Male Sterility

Onion was the first crop where the male sterility was recognized (Jones and Emsweller, 1936). Male sterility from Indigenous source was found in cv. 'Nasik White Globe' and in this cultivar strong cytoplasmic factor was responsible for MS. Male sterility formed by interaction of nuclear gene and cytoplasmic factor (Jones and Clarke, 1943). Exotic male sterile lines (long photoperiodic lines) found unsuitable in the short photoperiodic environment of India.

Efforts have been made to study the genetics of various economic characters and to utilize the male sterile lines for commercial hybrid seed production (Singh *et al.*, 1979). It is presumed that there are two types of cytoplasm-normal (N) and sterile (S). Plants with sterile cytoplasm 'S' in the presence of homozygous recessive nuclear genes results in male sterility (Sms ms). The plants with 'N' type of cytoplasm are always male fertile. Similarly, plants having dominant nuclear gene either in homozygous or heterozygous condition are male fertile. Another form of cytoplasm factor 'T' has been reported to control male sterility in onion (Schweiguth, 1973). In the development of hybrid, there are male sterile lines, maintainer lines and pollinator lines are available. A hybrid cultivar will result from a male sterile line and a pollinator line. Male sterile (Sms ms) lines are usually designated as 'A' line. They can be maintained by crossing with male fertile 'B' line having fertile cytoplasm (N msms) in a population (Pike, 1986). 'B' line can be identified by crossing 'A' line with several other lines and if the progeny turns out to be completely male sterile in a particular case, that particular parent will be considered as 'B' line with N (msms). Fertility of 'A' line can be maintained by crossing with 'C' line having a constitution of N (Msms) as pollen parent. Progeny segregates in the ratio of 1 male sterile: 1 male

Table 9.19: Cytoplasmic and Cytoplasmic-Genetic Male Sterility in Onion

Cytoplasm	*Restorer Genes*	*Remarks*
CMS-S (alloplasmic)	-	Thermalabile; commercially used
CMS-T*	Two independent systems: (1) Gene A, (2) Gene B and C (complementary action)	May be thermalabile, *e.g.*, 93 per cent sterile plants at 14°C, and 10 per cent at 23°C; commercially used
Nasik White Globe	Known	Commercially usable

Table 9.20: Various Combinations of Genetic Factors and Cytoplasmic Conditions in Onion

Mother Plant (Male Sterile)		*Pollen Donor (Male Fertile)*	*Progenies*	*Frequency (per cent)*
S msms	x	N MsMs	S Msms-male fertile	100
S msms	x	N Msms	S Msms-male fertile	50
			S msms-male sterile	50
S msms	x	N msms	S msms-male sterile	100
S msms	x	S MsMs	S Msms-male fertile	100
S msms	x	S Msms	S Msms-male fertile	50
			S msms-male sterile	50

fertile. 'C' line is generally designated as fertility restorer line. The 1:4 ratios (male fertile to male sterile plants) are recommended for higher productivity (Larenzon *et al.*, 1993). Species like *Allium galanthum* is known for induction of male sterility.

Table 9.21: Development of New A-line

Male Sterile Line (A-line)		*Male Fertile Line (B-line)*
S msms	x	N msms
F_1 S msms (50 per cent of B-line)	x	N msms
BC_1 S msms (75 per cent of B-line)	x	N msms
BC_2 S msms (87.5 per cent of B-line)	x	N msms
BC_3 S msms (93.75 per cent of B-line)	x	N msms
BC_4 S msms (96.87 per cent of B-line)	x	N msms
BC_5 S msms (98.44 per cent of B-line)		
New A line with nuclear genes of B-line		

A schematic plan for breeding to improved onion hybrids as outlined by Pike (1986) is given below:

Table 9.22: Procedure for Onion Improvement

Year	*Procedure*
1	Grow and select 100 bulbs, store and plant also grow supply of male sterile bulbs for use in test cross
2	Self selected bulbs and at the same time, test with known male sterile line
3	Grow out bulbs from self and F_1 test crosses, select discard poor progeny rows and their F_1 pair, store
4	Plant bulbs for seed production, observe sterility characters in F_1 lines if 100 per cent sterile, self selections and make back cross to F_1, discard pairs with fertile F_1 lines
5	Grow out A and B lines as pairs, continue to select in B line side, save best bulbs from A line for next back cross
6	Self B line of selected progenies and make back cross to sterile side of pair
7	Grow bulbs and make final selection on the basis of B line side
8	Mass B line using 10-20 bulbs in cage while making the back cross to the sterile side of pair
9-12	At this point, begin seed to seed and continue through the fifth back cross using the same procedure as in the 8th year, several A line and B line should have been developed, begin making hybrid combinations for testing.

Although India is one of the leading producers but very few researchers attempted to test different hybrid combinations for heterosis and combining ability using male sterility. Male sterile lines are required for heterosis. Micropropagation of male sterile (cytoplasmic male sterile) plants of female parent can be useful in hybrid seed production. Male sterility form indigenous source was found in 'Nasik White Globe' cultivar at ICAR-IIHR Bengaluru (Karnataka). It was transferred to six diferent genotypes which are now being used for exploiting heterosis. The two promising hybrids are 'Hybrid-1' (MS-65 x SI-13-1-1-1) and 'Hybrid-5' (MS-48 x S14-1-1-1) which gave good bulb yield of 45.5 t/ha (Pathak and Gowda, 2003). Adopting cytoplasmic male sterility system maximum area (3,925 ha) under hybrids coverage can be seen in New Zealand (Ashby, 1995). In onion, CGMS based Arka Kirtiman and Arka Lalima were developed by ICAR-IIHR, Bengaluru (Karnataka) and 'Hybrid-63' and 'Hybrid-35' developed by ICAR-IARI, New Delhi are very popular among the growers. Hybrids developed under hybrid network programme at ICAR-DOGR, Pune by using two CMS line indicatd very high percentage of heterosis over standard checks which ranged from 17 to 59 per cent over N-2-4-1 and 11 to 50 per cent over Agrifound Light Red (Lawande, 2004).

Table 9.23: Hybrids Developed through CGMS System

F_1 Hybrids	*Remarks*	*References*
Early Globe	W202A x S87-707, less foliage, poor in storage	Muro *et al.*, 2006
Quer-Rich	NOR-1A x SRG-12, 20 per cent higher quercetin content	Muro *et al.*, 2010
Baifeng	Good for dehydration, higher dry matter	Zeng *et al.*, 2011
Hybrid Nun-3001	High yield and poor in storage	Gupta *et al.*, 2011b
Hybrid Orient	High yield and poor in storage	Gupta *et al.*, 2011b
Hybrid Optima	Good commercial bulb yield and average mass of commercial bulbs	Boas *et al.*, 2012
Daekwanhwang	Manchuhwang x NIHA-5001, yellowish brown bulb and good storage	Kwon *et al.*, 2013
Arka Kirthiman*	Released from ICAR-IIHR, Bengaluru for high bulb yield and quality	Veer Gowda *et al.*, 1998
Arka Lalima*	Released from ICAR-IIHR, Bengaluru for high bulb yield and quality	Veer Gowda *et al.*, 1998
Hybrid-63*	Released from ICAR- IARI, New Delhi for high bulb yield and quality	Dhall, 2010
Hybrid-35*	Released from ICAR- IARI, New Delhi for high bulb yield and quality	Dhall, 2010
DOGR Hyb.-7*	High yield, superior over N-2-4-1, red bulb, suitable for *rabi*	Gupta *et al.*, 2011a
DOGR Hyb.-1*	Light red, flat-globe, 41.30 t/ha marketable yield which is 42.84 per cent higher than check Bhima Kiran	DOGR,2013
DOGR Hyb.-2*	Dark red, globe, 34.96 t/ha marketable yield which is 20.19 per cent higher than check 'Bhima Kiran'	DOGR, 2013
DOGR Hyb.-7*	Dark red, flat-globe, 40.71 t/ha marketable yield during *rabi* which is 34.88 per cent higher than best check 'Bhima Red'	DOGR, 2014

F_1 Hybrids	Remarks	References
DOGR Hyb.-50*	Attractive dark red, globe, 37.46 t/ha marketable yield during *rabi* which is 24.11 per cent higher than best check 'Bhima Red'	DOGR, 2014
DOGR Hyb.-5*	Dark red, globe, marketable yield during *kharif* which is 20.91 per cent higher than best check 'Bhima Dark Red'	DOGR, 2015
DOGR Hyb.-3*	Red, globe, 34.20 t/ha marketable yield during *kharif* which is 16.22 per cent higher than best check 'Bhima Super'	DOGR, 2016
DOGR Hyb.-4*	Attractive dark red, flat-globe, 34.10 t/ha marketable yield during *kharif* which is 15.81 per cent higher than best check 'Bhima Super'	DOGR, 2016

* Short day F_1 hybrids.

Haploids

Haploids have been successfully produced from cross pollinated ovules and their diploids will be useful for the development of inbred lines within a shorter time. Natural tetraploid plant with 2n=32 have been observed. Production of double haploid (DH) lines is a rapid method of obtaining completely homozygous inbred lines for numerous crops; however, practical trials testing the use of DH onion lines have been limited. DH onion lines were produced from diverse and highly heterozygous material in development within the Cornell onion breeding program. DH lines were evaluated in multiple replicated trials in onion fields in New York as lines and as parents of hybrids to assess the impact and commercial potential of DH onions. Twenty DH onion lines were compared with open-pollinated cultivars developed from the same source germplasm and with a commercial hybrid over two years. The vegetative vigor of the DH lines was comparable to that of related open-pollinated cultivars, showing minimal, if any, inbreeding depression. Two sets of hybrids were produced using the DH lines as males and two different females that are the female parents of the commercial hybrid controls. Therefore, hybrids in each set are half-sibs, and any performance differences are attributable to their DH male lines. In four replicated trials of these hybrids and controls, the experimental hybrids were either not significantly different or significantly better in measures of vegetative vigor compared with conventional half-sib hybrid controls. The vegetative vigor of DH lines, and their derived hybrids, might result from selection of plantlets without deleterious sublethal genes during gynogenesis. The shortened development time and equivalent quality of DH lines compared with the traditionally bred onion inbreds warrant their use. The increased vigor in hybrid combination could be an additional benefit for onion breeding strategies (Peter *et al.*, 2012).

Resistance Sources and its Utilizaion

Several Indian varieties, indigenous germplasm, and exotic varieties were tested for resistance to pests and diseases. Genetic analysis of resistance to anthracnose in *A. roylei* revealed that it is dominantly inherited and most probably determined by more than one gene. Anitha *et al.* (2011) reported need for introducing *Allium ampeloprasum, A. bouddhae, A. obliquum, A. nutans* and *A. odorum* possessing traits of resistance to different pests.

(i) Disease Resistance Sources and its Utilizaion

Attention is being given on these and other wild species, as they possess many disease resistance genes potentially useful in improvement programme. *Allium roylei* Stearn. has been used as a donor of genes for imparting resistance against downy mildew and leaf blight (de Vries *et al.*, 1992; Scholten *et al.*, 2007) and also been utilized as a bridge species for interspecific hybridization between *A. cepa* and A. fistulosum (Khrustaleva and Kik, 2000). Similarly, *Allium roylei* harbours resistance for downy mildew (*Peronospora destructor*), leaf blight (*Botrytis squamosa*) and Anthracnose (*Colletotrichum gloeosporides* Penz.). In case of *Alliums*, major source of resistance has been reported from *Allium fistulosum*. Japanese bunching onion (*Allium fistulosum* L.) possesses traits such as resistance to pink root (*Phoma terrestris* E.M. Hans.) (Porter and Jones, 1933), smut (*Urocystis cepulae* Frost.) (Jones and Mann, 1963), leaf rot (*Botrytis squamosa* Walker), stemphylium leaf blight (SLB) (*Stemphylium vesicarium*) and onion fly (*Hylemia antiqua* Bouche) as well as high total soluble solids and cold hardiness (Van Der Meer and Van Benekom, 1978). Semenov (1977) also reported that many species were highly resistant to *Peronospora destructor*, notably *A. altaicum, A. obliquum, A. nutans* and *A. odorum*. Partial resistance to anthracnose (*Colletotrichum gloeosporides* Penz.) was observed in accessions of *A. altaicum, A. fistulosum, A. galanthum, A. psekemense* and *A. roylei* and high level of resistance were found in accessions of *A. galanthum* and *A. fistulosum* (Galvan *et al.*, 1997). *A. cepa* var. *viviparum* has resistance to several diseases.

Table 9.24: Resistance Source for Diseases

Diseases	Resistance Species
White rot	*A. ampelprasum*
Leek yellow stripes virus	*A. ampelprasum*
Leek rust	*A. ampelprasum*
Downey mildew	*A. roylei*
Pink rot	*A. fistulosum*
Smut	*A. fistulosum*
Fusarium wilt (*Fusarium oxysporum*)	*A. tuberosum*

A. fistulosum L. is known source for purple blotch, pink root and smut. Wietsma and Kik (1989) crossed welsh onion (*A. fistulosum* L.) with wild species of *A. roylei* in order to increase resistance to downy mildew.

Scarcely any crossing experiments have been carried out between leek and its relatives. Crosses were successful between *A. fistulosum* and *A. cepa* to introgress stemphyllium leaf blight into onion lines. The varieties 'Italian Red' and 'Local Brazilian' are observed resistant to purple blotch (*Alternaria porri*). 'Granex' and 'Excel 35' are reported resistant to pink root. The line IHR-56-1 was resistant to purple blotch; Red Creole, Pusa Red, and Pusa Ratnakar. Arka Kalyan, N-2-4-1 and a few are moderately resistant to purple blotch. Yellow red colour of bulb was linked with resistant to *Colletotrichum arienans*. Pusa Red Round, Rampur Local, and Patna Red are tolerant and Telagi Red, White Large, Poona Red, Bellary Red, Patna Red, VL-1, PBR-1, PBR-5, Arka Niketan and N-257-9-1 are found resistant to basal rot. Incorporation of these traits into *A. cepa* could be valuable for the genetic improvement ofg bulb onion cultivars. Recently Bejo/De Groot en Slot Jave developed two onion cultivars that are resistant to downy mildew (*Peronospora destructor*). Semi resistance bearing genotype 'Lohru' is selected from Maharashtra is being used in crossing programme for incorporating desirable traits and resistance against *Alternaria porri*, *Stemphylium vesicarium*, *Perronospora destructor* and *Botrytis* species.

(ii) Insect Resistance Sources and its Utilizaion

Theiler and Buser (1996) worked on clones and inbreds of leek and found that progenies from somewhat tolerant plants showed a significantly lower attack by thrips than those from highly susceptible plants. They concluded that there is a genetic basis for thrips resistance in leek. *A. fistulosum* is resistance to thrips.

Table 9.25: Resistance Source for Insects

Insect	*Resistance Species*
Onion fly	*A. fistulosum*
Leek moth	*A. tuberosum*

Breeding for Quality Traits

In onion quality parameters includes size, shape, colour, pungency (amounts of S-alkyl cysteine sulfoxide precursors and the enzyme allinase determine pungency of onions), firmness, dormancy (dormancy desirable for storage) and high total soluble solids (important for dehydration industry).

- ☆ Maximizing onion bulb dry matter content (TSS)
- ☆ Improving onion bulb firmness
- ☆ Managing taste and pungency of onion bulbs
- ☆ Improving onion skin appearance (thoroughly curred and dried outer scale) and condition
- ☆ Reducing onion neck thickness
- ☆ Improving storage quality
- ☆ Onion colour and pungency

Following quality traits should be considered during breeding programmes on priority basis:

Table 9.26: Quality Attributes of Onion

Traits	*Preferred Range/ Magnitude/Value/Type*
Bulb colour	Red or pink
Bulb neck	Thin and small
Pungency	High
TSS	High (15 per cent or above)
Insoluble solid content	Not below 1 per cent

Defenitly defective, diseased, damaged bulbs caused by seed stem, tops, roots, moisture, dry sunscaled, sunburn, sprouting, mechanical or other injuries and staining, free from moulds, soft rot and insect attacks are creating trouble in export of onion. Genetic studies on onion have been reported for better keeping quality and for suitability to dehydration (Singh and Joshi, 1975; 1978; Singh *et al.*, 1976). Genotypes like Pusa Red, N2-4-1, Punjab Selection, Arka Niketan, Agrifound Light Red and Baswant-780 were identified for good keeping and storage quality. Pusa Red' and 'Punjab Red Round' were best suited for storage at room temperature (up to 5 months). 'Punjab Selection-48', Pusa White Flat, Pusa White Round, Udaipur-102, N-257-9-1 and 'N-2-4-1' were found very good for dehydration. 'Wild species *i.e.*, *A. altaicum* and *A. odorum* have been known to have high ascorbic acid content (Semenov, 1977), sugar and dry matter content (Kuznetsova and Tugolukova, 1980), saponins (Inoue *et al.*, 1995) and medicinal properties (Carotenuto *et al.*, 1997; Koch and Lawson, 1996). Islam *et al.* (2016) evaluated eight onion genotypes of four different skin colour under drip irrigation during *Kharif* season under Delhi condition. Non-significant differences were observed for the morphological traits like plant height, number of leaves, neck thickness, pseudostem height and fourth leaf length. However, bulb quality traits and yield exhibited significant differences among different genotypes, there was 2.08 and 1.44 fold increase between lowest and highest yielding genotypes for yield and TSS, respectively. Dark red coloured selection, Sel.-402 had the poorest yield performance but exhibited maximum TSS content and another dark red coloured variety Agrifound Dark Red was found superior for yield and total phenol content. Antioxidant activity, as measured by CUPRAC and FRAP, had 1.73 and 3.09 fold difference between highest and lowest performers and ranged between 1.70-2.95 μmoltrolox/g fresh weight and 0.30-0.93 μmol trolox/g fresh weight, respectively. Red coloured selection, Sel.-157 had the highest CUPRAC and FRAP values and white coloured selections, Sel.-131 and Sel.-106, had the lowest value. Bulb yield had positive association with total phenol content but negative association with TSS, CUPRAC and FRAP value. No single variety/selection was found consistently superior for all the traits studied and the study suggests the potential of systematic breeding effort for development of *Kharif* onion varieties with high yield potential and functional food value. Choudhary *et al.* (2016) evaluated 24 onion cultivares in arid condition of Rajasthan during *Rabi* season and significant variation in terms dry matter content and total soluble solids (TSS). Among the cultivars tested, RO-1 registered the highest dry matter content (14.80 per cent) while Bhima Shweta being at par with Bhima Kiran and RO-252 reported highest TSS (13.3 per cent). Several research organization developed short day white onion cultivars but do not offer TSS range more than 12 per cent but processing industries needs 20-30 per cent.

USA, China, India and Egypt are leader in processing. In India nearly 60 big to medium processing plants are established in Maharashtra, Gujarat, Madhya Pradesh and Rajasthan. Onion produced in tropical countries has got high pungency and flavor which add to processed products. Presently, India is exporting nearly 60,000 tonnes processed onion products annually. Non-availability of high TSS white onion varieties; irregular year round supply; poor seed supply and high post harvest losses are some basic problems of processing industries in India.

Resistance for Abiotic Stress

Some species *e.g.*, *A. altaicum* is known for cold tolerance.

Genetic Resource Conservation

Large scale collection and utilization of some of the wild species for edible (leaves and bulbs) and flavouring purpose has resulted in loss of the natural diversity (Negi, 2011). There is an urgent need to conserve these wild species. Only a few places in the world have authentic and readly available supply of germplasm of wild *Allium*. Some of the Himalayan *Allium* species are being maintained at CGN, Wageningen, Netherlands; Warwick Crop Centre, Wellesbourne, U.K.; NPGS (USDA), Washington, USA; IPK, Gatersleben, Germany and for perennial, long day *Allium* species at Research Institute for Vegetable Growing and Breeding, Olomouc, Czechoslovakia. In India, efforts towards exploration, collection and characterization of wild *Allium* species present in India have been undertaken from time to time (Singh and Rana, 1994; Negi, 2006; Verma *et al.*, 2008). At some places efforts were made to maintain and grow the useful indigenous strains earlier by MPKV, Rahuri (Maharshtra), ICAR-Indian Agricultural Research Institute, New Delhi, ICAR-Indian Institute of Horticultural Research, Hessaraghatta Lake, Bengaluru (Karanataka) *etc.* Presently National Research Centre for Onion and Garlic (DOGR, Pune), Rajgurunagar, Pune (M.S.) and National Horticultural Research and Development Foundation, Nasik (M.S.) is involved in collection, conservation and development of suitable cultivars. DOGR, Rajgurunagar, Pune (M.S.) is maintaining 437 lines of onion. In India, most of the conservation work is being done at ICAR-NBPGR, New Delhi ant its centre Bhowali.

Table 9.27: *Ex situ* Conservation of Wild *Allium* Species at ICAR-NBPGR, New Delhi

Species Maintained	*Number of Accessions*	*Ex-situ Conservation*
A. ampeloprasum var. *ampeloprasum*	4	FGB, IV
A. auriculatum	2	SGB, FGB
A. carolinianum	5	FGB
A. clarkei	1	FGB
A. consanguineum	2	FGB
A. stracheyi	2	SGB, FGB
A. griffithianum	3	SGB, FGB, IV
A. hookeri	2	FGB, IV
A. humile	1	FGB
A. roylei	4	FGB
A. wallichii	2	FGB
A. oreoprasum	1	SGB, FGB
A. auriculatum	1	SGB, FGB
A. chinense	2	FGB, IV
A. tuberosum	5	SGB, FGB, IV

SGB: Seed Gene Bank; FGB: Field Gene Bank; IV: *in vitro* repository.

Source: Pandey *et al.*, 2008.

The ICAR-NBPGR, New Delhi has conducted extensive plant exploration in different *Allium* growing states/regions in India and collected several wild relatives like *A. ampeloprasum*, *A. auriculatum*, *A. ascalonicum*, *A. carolinianum*, *A. chinensis*, *A. wallachi*, *A. tuberosum* and *A.* rubellum. The bureau has also introduced over 1100 accessions of *Allium* germplasm which include related *Allium* species from over 40 countries (Singh and Rana, 1994) apart from the local collections of 697. Murthy *et al.* (2011) collected white bulbs of 'Wild Onion' *Urginea indica* sample from the hillocks of Vaddagere village of Tumkur district, a southern part of Karnataka. *Allium* accessions collected in Central Asia by the 1989 USSR *Allium* Expedition, their availability and distribution within the U.S. National Plant Germplasm System which is as shown in Table 9.28.

ICAR-DOGR, Rajgurunagar, Pune is maintaining 25 number of *Allium* species with 116 numbers of lines under natural conditions. Exploration trip in Arunachal Pradesh state (West Kameng, East Kameng, Kurung, Kumey, Lower Subansiri, Upper Subansiri, West Siang and Pappum Pare) including Assam area (a part of Sumitpur and Lakhimpur) was undertaken for *Allium* species (wild and cultivated) in collaboration with ICAR-NBPGR Regional Station, Bhowali (Uttarakhand) during 17th to 29th October, 2013. In most of the villages wild *Alliums i.e. Allium fasciculatum*, *A. tuberosum*, *A. hookerii*, *A. chinense* and *A. macranthum* are being semi domesticated for day to day use as spices and condiments including vegetables.

Table 9.28: *Allium* Accessions Collected in Central Asia by the USSR 1989

Species	*Accessions Collected (No.)*	*Accessions Conserved (No.)*	*Seeds and Bulb Samples Distributed (1989-2003) (No.)*
Allium aflatunense B. Fedtsch.	2	2	19
Allium altissimum Regel	1	0	0
Allium barsczewskii Lipsky	1	1	1
Allium baschykyzylsaicum Krassovsk	1	0	0
Alliumcaeruleum Pall	2	1	8
Allium caesium Schrenk	4	4	25
Allium cepa L.	4	1	0
Allium collis-magni Kamelin	1	0	0
Allium costatovaginatum Kamelin and Levichev	1	1	1
Alliumcristophii Traute.	2	0	0
Allium drepanophyllum Vved.	1	1	0
Allium ericoleum Vved.	1	1	2
Allium giganteum Regel	2	2	0
Allium jodanthum Vved.	3	3	7
Alliumkarataviense Regel	1	1	2
Alliumkokanicum Regel	1	0	0
Allium longicuspis Regel	3	3	73
Allium monadelphum Less. Ex Kunth	1	1	3
Allium motor Kamelin and Levichev	3	3	2
Allium oreophilum C.A. Mey.	1	1	3
Allium praemixtum Vved.	1	0	0
Allium pskemense B. Fedtsch.	1	0	0
Allium ramosum L.	1	0	1
Allium regelii Trautv.	1	0	1
Allium rubellum M. Bieb.	1	1	0
Allium sativum L.	9	8	273
Alliumscabriscapum Boiss. and Kotschy	2	1	6
Allium sewertzowii Regel	1	0	1
Allium sp.	23	19	27
Allium stipitatum Regel	4	4	24
Allium talassicum Regel	2	1	1
Allium vavilovii Popov and Vved.	1	0	2
Total	83	60	482

Source: Philipp W. Simon, 2005.

Table 9.29: Germplasm Collections made by ICAR-DOGR, Rajgurunagar, Pune (Maharashtra)

Sl. No.	*Common/ Local Name*	*Botanical Name*	*Cultivation Status*	*No. of Acc. Collected*	*Area of Collection*	*Usage*
1.	Garlic, Lohru, Lahsoon	*Allium sativum* L.	Cultivated	07	Assam	Vegetable spices for cooking
2.	Leek, Lam	*Allium porrum* syn. *A. ampeloprasum* var. *porrum*	Semi-domesticated	01	Arunachal Pradesh	Vegetable, pickle and condiment/ flavor and as medicinal purpose.
3.	Multiplier onion, Punyoo/Pyaz, Punroo	*Allium cepa* var. *aggregatum*	Semi-domesticated	05	Assam	Vegetable cooking
4.	Chiense chives, Lapta, Lapsar, Narang, Dite	*Allium tuberosum*	Semi-domesticated	06	Arunachal Pradesh	Vegetable for flaor/condiment
5.	Taley, Talap	*Allium hookeri*	Semi-domesticated/ wild	04	Arunachal Pradesh	Vegetable (fibrous roots), pickle (pseudostem), foliage for flavor and for making chatany

Sl. No.	Common/ Local Name	Botanical Name	Cultivation Status	No. of Acc. Collected	Area of Collection	Usage
6.	Modi Dite	*Allium* spp.	Wild	02	Arunachal Pradesh	Roots are used for medicinal purpose for blood pressure and cold diseases besides cooking as flavor
7.	Lasun, Lapsa, Modi Byke, Modi Byako	*Allium macranthum*	Semi-domesticated	14	Arunachal Pradesh	Foliage are used for flavor in vegetable and for making chatany
8.	Rakkyo, Talap, di Talap	*Allium chinense*	Semi-domesticated	02	Arunachal Pradesh	Used as vegetable, bulbs for pickle, and condiment
9.	Zap, Leppi, Lipee,	*Allium fasciculatum*	Semi-domesticated/ wild	08	Arunachal Pradesh	Vegetable and Salad
		Total		**49**		

Source: Mahajan *et al.*, 2017.

REFERENCES

1. Aiyar, A.K. and Vaganarayana. 1956. The antiquity of some field and forest flora of India. Bangalore Printing and Publishing Co. Ltd.
2. Amagi, M., ohashi, K. and Kimura, S. 1995. Breeding of interspecific hybrid between *A. fistulosum* and *A. tuberosum* by embryo culture. *Bull. Tochigi. Agr. Exp. Stn.*, **43:** 87-94.
3. Anitha, K., Venkateswaran, K., Chakrabarty, S.K., Kumar, G.S., Babu, B.S. and Varaprasad, K.S. 2011. Onion Genetic Resources and Pest Resistance: The Indian Scenario. *Indian J. Plant Protection*, **39(2):** 81-92.
4. Arya, P.S. 2003. Vegetable Breeding-Production and Seed Production: Kalyani Publishers, New Delhi, 74.
5. Ashby, J.W. 1995. Status of vegetable production in New Zealand with special reference to vegetable hybrid technology. In: Proc.'Vegetable Research with special reference to hybrid technology in the Asia-Pacific region. FAO, p. 205.
6. Bark, H.O., Havey, M.J. and Corgan, Joe, N.1994. Restriction Fragment length Polymorphism (RFLP) analysis of progeny from an *A. fistulosum* x *Allium cepa* hybrid. *J. Amer. Soc. Hort. Sci.*, **119(5):** 1046-1049.
7. Bino, R.J., Janssen, M.G., Franken, J. and de Vries, J.N.1989. Enhanced seed development in the interspecific cross *Allium cepa* x *A. sphaerocephlon* through ovary culture. *Plant Cell Tiss. and Org. Cult.*, **16:** 135-142.
8. Boas, R.C.V., Pereira, G.M., Souza, R.J., Geisenhoff, L.O. and Lima, J.A. 2012. Development and production of two onion cultivars irrigated by drip system. *Revista Brasileira de Engenharia Agricolae Ambieental*, **16(7):** 706-713.
9. Brewster, A. 1994. The Genetics and Plant Breeding of *Allium* Crops. In: Onions and other Vegetabl *Alliums*. Brewster, J.L., pp. 41-61., CAB International, ISBN 0581995101, U.K.
10. Carotenuto, A., Fattorusso, E., Lanzotti, V., Magno, S. and Feo, De. V. 1997. *Allium sativum* L. III. Its precursor and 'essential oil of garlic'. *J Am Chem Soc.*, **67**: 1032–1033.
11. Chadha, K.L. 1990. Onion and garlic research in India: Present status and future prospects. Proceedings of the National Symposium on Onion and Garlic, held at Dr. Y.S. Parmar University of Horticulture and Forestry, Nauni, Solan (Himachal Pradesh) on 2nd June, 1990.
12. (a) Chandel, K.P.S. and Pandey, R. 1992. Distribution, diversity, uses and *in vitro* conservation of cultivated and wild *Alliums*: A brief review. *Indian J. Plant Genet. Resour.*, **5(2):** 7-36.

    (b) Choudhary, S., Choudhary, B.R. and Kumhar, S.R. 2016. Productivity and quality evaluation of onion (*Allium cepa* L.) cultivars for arid zone of Rajasthan. 7th Indian Horticulture Congress (Abs: 105; Session I: Breeding of Horticultural Crops), held at ICAR-IARI, New Delhi, 15-18 November, p. 81.
13. Chuda, A. and Adamus, A. 2012. Hybridization and molecular characterization of F_1 *Allium cepa* x *Allium roylei* plants. *Acta Biologica Cracoviensia*, **54(2):** 25-31.
14. Currah, L. and Proctor, F.J. 1990. Onions in Tropical Regions. *National Resources Institute Bulletine*, No. 35, p. 323, NRI, UK.
15. Currah, L. and ockendon, D.J. 1988. A hybrid index for *Allium cepa* and *A. fistulosum* crosses. Proc. 4th EUCARPIA, *Allium* Symp. Wellesbourne, UK.
16. De Vries, J.N., Wietsma, W.A., Jongerius, M.C. 1992a. Linkage of downy mildew resistance genes*Pd1* and *Pd2* from *Allium roylei* Stearn. in progeny of its interspecific hybrid with onion (*A. cepa*L.). *Euphytica*, **64:**131–137.
17. Dhall, R.K. 2010. Status of male sterility in vegetables for hybrid development- A review: *Advances in Horticultural Sciences*, **24(4):** 263-279.

18. DOGR, 2013. DOGR Annual Report, 2012-2013. Directorate of Onion and Garlic Research, Rajgurunagar, Pune, p. 11.

19. DOGR, 2014. DOGR Annual Report, 2013-2014. Directorate of Onion and garlic Research, Rajgurunagar, Pune, p. 41-42.

20. DOGR, 2015. DOGR Annual Report, 2014-2015. Directorate of Onion and Garlic Research, Rajgurunagar, Pune, p. 48-50.

21. DOGR, 2016. DOGR Annual Report, 2015-2016. Directorate of Onion and Garlic Research, Rajgurunagar, Pune, p. 36.

22. Emsweller, S.L. and Jones, H.A. 1935. An interspeicific hybrid in *Allium*. *Hilgardia*, **9:** 265-273.

23. Frodin, D.G. (2004). History and concepts of big plant genera. *Taxon,* **53(3):** 753-776.

24. Galván, G. A., Wietsma, W.A., Putrasemedja, S., Permadi, A.H. and Kik, C. 1997. Screening for resistance to anthracnose (*Colletotrichum gloeosporioides* Penz.) in *Allium cepa* L. and its wild relatives. *Euphytica,* **95(2):** 173–178.

25. Gohil, R.N. 1973. Cytogenetic studies of some Indian *Alliums*. Ph.D. Thesis, University of, Kashmir, J&K.

26. Gohil, R.N. 1992. Himalayan representative of *Alliums*. In: Hanelt, P., Hammer, K. Knüpffer, H. (ed.). The Genus *Allium*-Taxonomic Problems and genetic Resources. Gatersleben, Germany, p. 335-340.

27. Gupta, A.J., Mahajan, V. and Lawande, K.E. 2011a. Evaluation of different onion F_1 hybrids developed through male sterile lines. In: Natl. Symp. On *Alliums*: Current Scenario and Emerging Trends, 12-14 march, 2011. ISA, DOGR, Pune, p. 151.

28. Gupta, A.J., Mahajan, V. and Lawande, K.E. 2011b. Performance of exotic red onion hybrids under short day conditions. In: In: Natl. Symp. On *Alliums*: Current Scenario and Emerging Trends, 12-14 march, 2011. ISA, DOGR, Pune, p. 151-152.

29. Havey, M.J. 1999. Seed yield, floral morphology and lack of male fertility restoration of male sterile onion (*Allium cepa* L.) population possessing the cytoplasm of *Allium galanthum*. *J. Am. Soc. Hort. Sci.,* **124:** 626-629.

30. Hawthorn, L.H. and Pollared, L.H. 1954.Vegetable and Flower Seed Production. The Blakistan Co. Inc., New York.

31. Islam, S., Chandra, K., Patel, N. and Khar, A. 2016. Morphological and antioxidant variability studies on onion genotypes during *Kharif* season-Potential for *Kharif* onion breeding. 7th Indian Horticulture Congress (Abs: 95; Session I: Breeding of Horticultural Crops), held at ICAR-IARI, New Delhi, 15-18 November, p. 73.

32. Jones, H.A. 1923. Pollination and self-fertility in the onion. *Proc. of Am. Soc.Hortic. Sci.,* **20:** 191.

33. Jones, H. and Emsweller, S. 1936. A male sterile onion. *Proc. Amer. Soc. Hort. Sci.,* **34:** 582–585.

34. Jones, H.A. and Mann, L.K. 1963. Onions and their allies. Interscience, New York.

35. Jones, H.A. and Clarke, A. 1943. Inheritance of male sterility in the onion and the production of hybrid seeds, *Proc. Am. Soc. Hortic. Sci.,* **43:** 189.

36. Jones, H.A. and Mnn, L.K.A.E. 1963. Onions and their allies. World Crop Books. Interscience Publishers. Inc. New York.

37. Khrustaleva, L.I. and Kik, C. 2000. Introgression of *Allium fistulosum* into *A. cepa* mediated by *A. roylei. Theor Appl. Genet.,* 100:17–26.

38. Khrustaleva, L.I. and Kik, C. 1998. Cytogenetical studies in the bridge cross into *Allium cepa (Allium fistulosum* and *A. roylei). Theor Appl. Genet.,* **96:** 8–14.

39. Kik, C. 2002. Exploitation of wild relatives for the breeding of cultivated *Allium* species. In: *Allium* Crop Science: Recent Advances. Robinowitch, H.D. and Currah, L., p. 81-100, CAB Publishing, ISBN0851995101, U.K.

40. Kielkowska, A. and Adamus, A. 2006. Growth of pollen tubes from foreign species in carrot (*Daucus carota* L.) pistils. In: *Haploids and Doubled Haploids in Genetics and Plant Breeding*. Adamski, T. and Surma, M., p. 193-197, Prodruk, ISBN8389887444, Poland.

41. Koch, H.P. and Lawson, L.D. (eds.) 1996. Garlic: The science and therapeutic application of *Allium sativum* L. and related species, 2nd edn. William and Wilkins, Baltimore, p. 329.

42. Kwon, Y.S., Cho, K.S., Lee, E.H., Jang, S.W., Suh, J.T., Kim, J.S., Kim, W.B., Im J.S. and Lee, J.N. 2013. High quality and long-storability of F_1 hybrid long-day onion. *Daekwanhwang Korean Journal of breeding Science,* **45(1):** 61-65.

43. Larenzon, M.C.A., Marfinho, M.R. and Freire, J.A.H. 1993. Effect of pollination by honey bees on onion hybrid seed production. *Pesquisa Agro Pecuaria Brasileira,* **28(2):** 229-235.

44. Mahajan, V., Benke, Ashwini, Anandhan, S. and Gopal, J. 2017. Onion and garlic improvement through species utilization. In: Distant Hybridization in Horticultural Crops: (eds.) Dinseh, M.R. and Sankaran, M. Astral Publ., New Delhi. pp, 210.

45. Mc Collum, G.D. 1982. Experimental hybrids between *Allium fistulosum* and *A. roylei. Bot. Gaz.,* **143:** 238-242.

46. McCollum, G.D. 1971. Sterility of some interspecific *Allium* hybrids. *J. Am. Soc. Hort. Sci.,* **96:** 359-362.

47. Mc Collum, G.D. 1976. Evolution of Crop Plants (Ed. Simmonds).

48. Mc Collum, G.D. 1980. Development of the amphidiploids of *A. galanthum* x *A. cepa*. *J. Heredity*, **71:** 445-447.

49. Muro, T., Ito, K., Sato, Y., Abe, H., Turui, H., Asari, H., Minagawa, Y. and Harada, H. 2006. Breeding of early maturing hybrid onion variety 'Early Globe' and its characteristics. Research Bull. of the Natl. Agril. Res. Centre for Hokkaido Region, 184: 45-55.

50. Muro, T., Noguchi, Y., Morishita, M., Ito, K., Sugiyama, K., Kondo, T., Kurenuma, Y. and ono, T. 2010. 'Quer Rich' a new variety of hybrid red onion with high quercetin content. *Research Bull. of the Natl. Agril. Res. Centre for Hokkaido Region*, **192:** 25-32.

51. Murthy, P.G., Mamtharani, D.R., Tejas, T.S. and Niranjan, M.S. 2011. Phytochemical analysis, *in vitro* anti-bacterial and antioxidant activities of wild onion species. *International Journal of Pharma and Biosciences*, **2(3):** 230-237.

52. Nath, P., Srivastava, V.K., Dutta, O.P. and Swamy, K.R.M. 2008. Vegetable Crops: Improvement and Production. Dr. Prem Nath Agricultural Foundation, Bengaluru (Karnataka), p. 196-208.

53. Nayar, N.m., Ahmedulla, M. and Singh, R. 1992. *Alliums* in South Asia. Importance, ethnobotanical uses, genetic resources, enumeration of species and distribution. In: Hanelt, P., Hammer, K. Knüpffer, H. (ed.). The Genus *Allium*-Taxonomic Problems and genetic Resources. Gatersleben, Germany, p. 205-213.

54. Negi, K.S. 2006. *Allium* species in Himalayas and their uses with special reference to Uttaranchal. *Ethnobot.*, **18:** 53-66.

55. Negi, K.S. 2011. Scenario of underutilized Alliums in India. Souvenir: Nat. Symp. on Alliums: Current Scenario and Emerging Trends, 12-14, March, 2011, pp. 29-42.

56. Negi, K.S. and Pant, K.C. 2006. Less known wild species of *Allium* L. (Amaryllidaceae) from mountain region of India. *Economic Botany*, **46(1):** 112-114.

57. Negi, K.S., Gaur, R.D. and Pant, K.C. 1995. Biodiversity: untapped plant wealth in Uttarakhand. Cover I, II, IV. *Indian Hort.*, **40(2):** 32-40.

58. Negi, K.S., Pant, K.C., Koppar, M.N. and Thomas, T.A. 1991. Wild relatives of genus *Allium*. L. in Himalayas. *Indian J. Plant Genet. Resour.*, **4(1):** 732-777.

59. Nomura, Y. and Makara, K. 1993. Production of interspecific hybrids between Rakkyo (*Allium chinense*) and some other *Allium* species by embryo rescue. *Jpn. J. Breed.*, 43: 13-21.

60. Nomura, Y., Maeda, M., Tsuchiya, T. and Makara, K. 1994. Efficient production of interspecific hybrids between *Allium chinense* and edible *Allium* spp. through ovary culture and pollen storage. *Breed. Sci.*, **44:** 151-155.

61. Nomura, Y., Kazuma, T., Makara, K., Nagai, T., 2002. Interspecific hybridization of autumn-flowering *Allium* species with ornamental Alliums and the characteristics of the hybrid plants. *Sci. Hort.* **95:** 223-237.

62. Ohsumi, C.A., Kojima, K., Hinata, K., Etoh, T. and Hayashi, T. 1993. Interspecific hybrid between *Allium cepa* L. and *Allium sativum*. *Theor. Appl. Genet.*, **85:** 969-975.

63. Peter, T.H., Earle, E.D. and Mutschler, M.A. 2012. Doubled Haploid Onion (*Allium cepa* L.) lines and their impact on hybrid performance. *HortSci.*, **47(12):**1690–1695.

64. Peterka, H., Budahn, H. and Chrader, O. 1997. Interspecific hybrid between onion (*Allium cepa* L.) with S-cytoplasm and leek (*Allium ampeloprasum* L.). *Theor. Appl. Genet.*, **94:** 383-389.

65. Philipp W. Simon. 2005. Realizing value from Central Asian *Allium* germplasm collection. *HortScience*, **40(2):** 309-310.

66. Pike, L.M. 1986. Onion breeding. In: Breding Vegetable Cross (Basset, M.J. (ed.) AVI Publishing Co. Inc., Westport, Connecticut, USA, p. 357-394.

67. Porter, D.R. and Jones, H.A. 1933. Resistance of some cultivated species of *Allium* to pink root (*Phoma terrestris*). *Phytopathology*, **23:**290-298.

68. Pruthi, J.S. 2001. Minor spices and condiments, crop management and post harvest technology. ICAR, New Delhi. p. 49-84.

69. Ray, P. and Gupta, H.N. 1980. "*Charaka Samhita*" Scientific Synopsis. Indian National Science Academy, New Delhi.

70. Ray, P., Gupta, H.N. and Roy, M. 1980. "*Charaka Samhita*" Scientific Synopsis. Indian National Science Academy, New Delhi.

71. Robinowitch, H.D. and Currah, L. 2002. *Allium* Crop Science: Recent Advances, CAB, International Wallingford, U.K.

72. Saini, S.S. and Davis, G.N. 1967. Compatibility in some *Alium* species. *Proc. Am. Soc. Hort. Sci.*, **91:** 401-409.

73. Schweiguth, B. 1973. Study of a new type of male sterility in onion, *Allium cepa* L. *Am. Amelior. Plant*, **23:** 221-253.

74. Scholten, O.E., van Heusden, A.W., Khrustaleva, L.I., Burger-Meijer, K., Mank, R.A., Antonise, R.G.C., Harrewijn, J.L., Van Haecke, W., Oost, E.H., Peters, R.J. and Kik, C. 2007. The long and winding road leading to the successful introgression of downy

mildew resistance into onion. *Euphytica* **156:** 345–353.

75. Semenov, V.A., 1977. Morphological and biological features of some Altai *Allium* species and their importance for introduction. Byulleten' Vsesoyuznogo Ordena Lenina i Ordena Druzhby Narodov Instituta Rastenievodstva Imeni N I Vavilova, **70:** 60-62.

76. Shigyo, M. and Kik, C. 2008. Onion. In: Prohens, J., Nuez, F. (eds.), Vegetables II. Fabaceae, Liliaceae, Slanaceae and Umbeliferae, 121-159, Springer, New York.

77. Shigyo, M., Yamane, N., Martin, W.J., Yamauchi, N. and Havey, M.J. 2005. Assignment of onion RFLP-linkage to chromosomes by using alien monosomic conditions. *Abstr. J. Jpn. Soc. Hort. Sci.*, **74(1):** 426.

78. Singh B.P. and Rana R.S. 1994. Collection and conservation of *Allium* genetic resources: An Indian perspective. *Acta Hort.* **358**: 181–185.

79. Singh, D.P. and Joshi, M.C. 1975. Preliminary investigation on breeding onions for better keeping quality. *Haryana J. Hort. Sci.*, **4 (3 and 4):** 161-166.

80. Singh, D.P. and Joshi, M.C. 1978. Correlation studies of rotting and sprouting losses of onion with other traits during storage. *Veg. Sci.*, **5(1):** 1-3.

81. Singh, D.P., Swamy, K.R.M. and Joshi, M.C. 1976. A note on breeding onions suitable for dehydration. *Indian J. Hort.*, **33(3 and 4):** 252-253.

82. Singh, D.P., Padmaja Rao, S., Ramchandar, P.R. and Biswas, S.R. 1979. Inheritance, correlation and selection Index in onion. *Indian J. Agric. Sci.*, **49(8):** 592-595.

83. Singh, R.K. 2016. Evaluation of exotic onion (*Allium cepa* L.) hybrids at Nashik for higher yield and quality. 7th Indian Horticulture Congress (Abs: 16; Session II: Bredding of Horticultural Crops), held at ICAR-IARI, New Delhi, 15-18 November, p. 97.

84. Singhal, S.C. 2003. Export of onion-Challenges and Strategies. In: Approaches for Sustainable Development of Onion and Garlic (Eds. Kirti Singh, K.E. Lawande, U.B. Pandey, Lallan Singh and S.R. Bhonde), Published by National Horticultural Research and Development Foundation (MS), India.

85. Takholm, V. and Drar, M. 1954. *Bulletin Faculty Science Egypt Unversity*, **3:** 1-644.

86. Umehara, M., Sueyoshi, T., Shimomura, K., Hirashima, K., Shimoda, M. and Nakahara, T. 2007. Production and characterization of an interspecific hybrid between *Allium fistulosum* and *Allium schoenoprasum L. Bull. Fukuoka Agric. Res. Center*, **26:** 25-30.

87. Van Der Meer, Q.P. 1984. Breeding for resistance to yellow stripe virus in leek (*Allium porrum* L.)-a progress report. In: Wietsma, W.A. and Hollander H.D.M. (Eds.). *Proced. 3rd Eucarpia, Allium* Symposium IVT Wageningen, p. 16-19.

88. Van der Meer, Q.P. and Van Benekom, J.L. 1978. Improving the onion crop (*Allium cepa* L.) by transfer of characters from *A. fistulosum. Biuletyn Warzywniczy*, **22:** 87-91.

89. Van der Meer, Q.P. and de Vries, J.N. 1990. An interspecific cross between *Allium roylei* Stearn and *Allium cepa* L. and its back cross to *A. cepa. Euphytica.* **47:** 29-31.

90. Van der Meer, Q.P. and Van Bennekom, J.L. 1978: Improving the onion crop (*Allium cepa* L.) by transfer of characters from *Allium fistulosum* L. *Biul. Warzywniczy*, **22:** 88-91.

91. Van der Valk, P., de Vries, S.E., Everink, J.T., Verstappen, F. and de Vries, J.N. 1991a. Pre-and post fertilization barriers to backcrossing the interspecific hybrid between *Allium fistulosum* and *A. cepa* L. with *A. cepa* L. *Euphytica*, **53:** 201-209.

92. Van der Valk, P., Kik, C., Verstappen, F., Everink, J.T., and de Vries, J.N. 1991b. Independent segregation of two isozyme markers and inter-plant differences in nuclear DNA content in the interspecific cross (*Allium fistulosum* x *A. cepa* L.) x *A. cepa* L. *Euphytica*, **55:** 151-156.

93. Vavilov, N.I. 1951. The origin, variation, immunity and breeding of cultivated plants. *Chron. Bot.*, **13:** 1-6.

94. Verma, V.D., Pradheep, K., Khar, A., Negi, K.S. and Rana, J.C. 2008. Collection and characterization of *Allium* species from Himachal Pradesh. *Indian J. Plant Genet. Resour.*, **21(3):** 225-228.

95. Veer Gowda, R., Pathak, C.S., Singh, D.P. and Deshpandee, A.A. 1998. Onion hybrids: 'Arka Kirthiman' and 'Arka Lalima'. *Indian Horticulture*, **43(3):** 20.

96. Vries, J.N. de and Vries, W.A.T. de. 1993. Introgression of leaf blight resistance from *Allium roylei* Stearn into onion (*A. cepa* L.), (Centrum voor Plantenveredelings- en Reproduktieonderzoek DLO, Wageningen (Netherlands).

97. Wietsma, W.A. and Kik, C. 1989. Resistance to downy mildew within reach.*Groenten en Fruit*, **45(10):** 71.

98. Xu, J.M., Hanelt, P. and Chun-Lin, L. 1990. Key to the *Allium* of China. *Herbertia*, **46(2):** 138-164.

99. Yamashita, K., Arita, H. and Tashiro, Y. 1999a. Cytoplasm of a wild species, *Allium galanthum* Kar. et Kir., is useful for developing the male sterile line of *A. fistulosum. J. Jpn. Soc. Hort.* Sci., **68:** 788-797.

100. Yamashita, K., Arita, H. and Tashiro, Y. 1999b. Isozyme and RAPD markers linked to fertility restoring gene for cytoplasmic male sterile *Allium*

fistulosum L. with cytoplasm of *A. galanthum* Kar. et Kir. *J. Jpn. Soc. Hort.* Sci., **68:** 954-959.

101. Yamashita, K., Iino, M., Shigyo, M. and Tashiro, Y. 2000. Visualization of nucleus substitution between *Allium galanthum* and shallot (*Allium cepa* L.) by genomic *in situ* hybridization. *J. Jpn. Soc. Hort. Sci.*, **69:** 189-191.

102. Yamashita, K. and Tashiro, Y. 1999. Possibility of developing male sterile lines of shallot (*Allium cepa* L. *Aggregatum* group) with cytoplasm from *A. galanthum* Kar. Et Kir. *J. Jpn. Soc. Hort. Sci.*, **68:** 256-262.

103. Yamashita, K., Takatori, Y. and Tashiro, Y. 2005. Chromosomal location of a pollen fertility restoring gene, *rf* for CMS in Japanese bunching onion (*Allium fistulosum* L.) possessing the cytoplasm of *Allium galanthum* Kar. et Kir. Revealed by genomic *in situ* hybridization. *Theor. Appl. Genet.*, **111:** 15-22.

104. Yanagino, T., Sugawara, E. and Watanabe, M. 2003. Production and characterization of an interspecific hybrid between leek and garlic. *Theor. Appl. Genet.*, **107:** 1-5.

105. Zeng, A.S., Yan, J.Y., Yun, J.S., Song, L.X. and Gao, B. 2011. A new onion F_1 hybrid for dehydration 'Baifeng'. *China Vegetables*, **12:** 103-105.

Garlic (*Allium sativum* L.)

Introduction

Garlic (*Allium sativum* L., 2n=2x=16, diploid species) is the most important and oldest known horticultural crops which was first ascribed by Carl Linnaeus. It is known by several names such as Lahasun (Lahsoon) in Hindi, rashun in Bengali, ruhan in Kshmiri, lasguna in Sanskrit and Thorn in Dogri. Garlic has been used throughout history for both culinary and medicinal purposes. According to Indian Ayurveda, garlic is one of the most effective antimicrobial herbs with anti-bacterial, anti-fungal and anti-viral properties. Garlic oil is an item of commerce. Fresh garlic bulb and whole plant on steam distillation yield 0.1-0.2 per cent and 0.05-0.09 per cent respectively of essential oils. Up to 0.5 per cent oil has been derived from the bulb. For the recovery of the oil fresh garlic is the best. The essence consists primarily of diallyl, dimethyl and allyl methyl sulfide, disulfide and trisulfide and a few minor components including those containing an n-propyl group, all formed by thermal decomposition of allicin and its homologues (Singh and Pal, 2003). In North eastern parts foliage is used as a substitute of garlic in their day to day life.

Origin, History and Distribution

Carl Linnaeus gave the name *Allium sativum* L. to garlic in 1754. According to Choudhury (1976), Central Asia is the prime origin and the Mediterranean area is a secondary centre (Thompson and Kelly, 1957). It was known in Egypt in pre-dynastic times before 3000 BC and also to ancient Greeks and Romans. In the Old World, Egyptian and Indian cultures referred to garlic 5000 years ago and there is clear historical evidence for its use by the Babylonians 4500 years ago and by the Chinese 2000 years ago. Some writings suggest that garlic was grown in China as far back as 4000 years ago and its use was also first in China as mentioned in A.D. 510. Garlic grows wild only in Central Asia (centered in Kyrgyzstan, Tajikistan, Turkmenistan and Uzbekistan) today. Earlier in history garlic grew wild over a much larger region. *A. sativum* grows in the wild in areas where it has become naturalized and infact may have occurred in an area from China to India to Egypt to the Ukraine. The region where garlic has grown in the wild is referred to as it "centre of origin" since this is the geographic region where the crop originated and the only place where it flourished in the wild. Garlic was carried out to western hemisphere by Spenish, Portuguese and French (Purseglove, 1975). It was used in England as early as first half of 16th century (Shinde and Sontakke, 1986). The ancestry of cultivated garlic, according to Zohary and Hopf (2000), is not definitely established and difficulty in the identification of its wild progenitor because of the sterility of the cultivars. Though, it is thought to be descended from the species *A. longicuspis*, which grows wild in southwestern Asia, particularly India. Garlic is still grown in Egypt, but the Syrian genotypes are the kind most esteemed. Garlic is mentioned in the Bible and the Talmud. Hippocrates, Galen, Pliny the Elder, and Dioscorides all mentioned the use of garlic for many conditions. Garlic is grown globally, but China is by far the largest producer of garlic (70 per cent) followed by India (4.1 per cent) and South Korea (2 per cent). Egypt and Russia are in 4th place and the United States where garlic is grown primarily as cash crop in 5th place. Argentina, Myanmar and Ukraine are other contries growing it at commercial level. Garlic can be spotted throughout India from temperate to tropical regions. Garlic has been used as both food and medicine in many cultures for thusands of years, dating at least as far back as the time that the Giza pyramids were built. Besides India, garlic is grown in some other countries like China, the USA, (California, Texas, Loussiana), the Phillippines, Spain, Thailand, Korea Republic, Egypt, Italy, France, Morocco *etc.* China ranks first in garlic production followed by Turkey and India. It is cultivated extensively throughout subtropical plains to intermediate higher reaches of Jammu and Kashmir. In

India, Madhya Pradesh leads in acreage and production (32 per cent area and 28 per cent of total production) followed by Gujarat (18 per cent area and 17 per cent of total production) and Odisha (14 per cent area and 11 per cent of total production). However highest productivity is reported from Punjab.

Species Availability, Evolution and Diversity

Garlic close relatives include the onion, shallot, leek and chive. The wild garlic, crow garlic and field garlic of Britain are members of the species *A. ursinum*, *A. vineale,* and *A. oleraceum*, repectively. In *A. sativum* despite it being propagated exclusively by the division of ground cloves or aerial bulbils, a wide range of variability is available in several traits. Regel (1875) followed by De Candolle (1885) and considerably later Vavilov (1926), described the *A. longicuspis* Rgl. Subspecies as the ancient parent of garlic. It is found to be endemic to Central Asia where it is distributed over the southern mountain slopes of Turkmenistan, northeast Pamir *etc.* some authorities considered that *A. longicuspis* Rgl., which is the wild ancestor and spread in ancient times to the Mediterranean Region.

Botanical Description

A. sativum L., commonly known as garlic, is a species in the onion family Alliaceae. True garlic routinely grows in the wild without the assistance of human propagation. It is a perennial that can grow 60.00 cm high or more. It is sexually sterile diploid with 2n= 2x= 16. It is generally propagated by cloves but through inflorescence bulbils in some cases.

Kingdom	:	Plantae
(Unranked)	:	Angiosperms
(Unranked)	:	Monocots
Order	:	Asparagales
Family	:	Alliaceae
Subfamily		Allioideae
Tribe		Allieae
Genus	:	*Allium*
Species		*sativum*

Table 9.30: Botanical Description

Parts	*Description*
Growth Habit	Garlic is a frost hardy bulbous annual (bulb production) and perennial (seed production) herbaceous crop.
Root	Roots are several and fibrous (superficial adventitious) providing good anchorage.
Stem	The stem is simple, and rises about 30.0-100.0 cm in height with narrow flat banes.
Leaf	The garlic plant has 6-12 long, narrow flat and grass like leaves and sheathes thelower half of the stem.
Floral biology	At the termination of the stem is a cluster of flowers and bulbs mingled together and enclosed in a pointed spathe, which opens on one side and withers. The flowers rise direct from the bulb and are white in colour (Janick, 1979). Flowering in some garlic clones is associated with presence of bulbils formed in the place of flowers over inflorescence (Kamenetsky and Rabinowitch, 2001). The scape of garlic is smooth, round and solid for its entire length unlike onion, which is hallow. Many cloves of garlic do not flower stalks. Umbels globose, nearly always with bulbils, many flowered, long peduncle. Sepals's oblong, acuminate, greenish white or more or less tinged with purple. Stamens not exerted from the perianth, teeth of the inner stamens not longer than the antheriferous central part. The inflorescence may be partially or not at all exerted, its bulbils forming a swelling somewhere within the false stigma a few centimeters above the bulb.
Bulb	The most important part of its plant for medicinal purposes is the bulb. The bulb consists of numerous bulblets, known as 'cloves' held between the papery membrances and enclosed within a whitish skin as a sac. Each bulb is made up of 6-35 cloves (smaller bulblets), surrounded by a thin white or pinkish papery sheath and each clove weighs about 1.0 g. Its bulbils forming a swelling some where within the false stems a few cm above the bulb (Purseglove, 1975). Cloves are oblong or wedge shaped and covered with dry membranes scales with a pungent odor and a disagreeable and acrid taste (Singh and Nath, 2011).

Botanical Classification

Carl Linnaeus and Kuznetsov (1948) classified *A. sativum* L. in to two sub-species:

(A) Bolting Sub-Species

It called as *A. sativum* var. *sagitatum* Kuz. Bolting type of garlic can be spotted in all parts of the world. Again this sub-species can be grouped in to:

(i) **Asian Ecotype-**In Asian ecotype it is called as 'multiflowered' type which pseudostem attains the height of 150.0 cm or more. Generally inflorescence becomes large in which numbers of sterile flowers emerge. With due course of time more than hundred small aerial bulbils developed. Bulbils are wrapped into dry outer covering.

(ii) **Caucasian Ecotype-**Caucasian Ecotype attains the height of 60.0-80.0 cm. Generally inflorescence becomes large in which numbers of sterile flowers emerge but the bulbils developed in a size of pea grain. Planting of bulbils develop small bulbs (albeit smaller) in the next season.

(iii) **Sea side Ecotype-**In Seaside Ecotype flower stem emerge quite late which can be seen on the plant after development of 3-4 leaves and at the base two or sometimes three circles of cloves formed. In this group formation of smalller size cloves are more in number than other ecotypes.

Its preference is more due to delayed formation of the flower stem; prolonging vegetative period by 12-15 days than Asian and Seaside ecotypes. Each inflresence consist 3-5 bulbils of different shape and size.

(B) Non-bolting Sub-Species

This group can be again sub-group in to:

(i) **Continental Ecotype**-The Continental Ecotype is also called as 'later' means crop for spring/ summer. This group continue its vegetative growth for longer period, very deep biological rest under normal storage conditions (bulbs can be stored until the next harvest), pseudostem is thin, compact with 40.0 cm height; leaves are in general thin, wide open, smooth and thickly waxed and bulbs are differ in size and sterile leaves (8-10) which are tightly arranged around the bulb. Crop is raised without irrigation.

(ii) **South Russian Ecotype**-The South Russian Ecotype predominates in the steppe zones of Russia, Ukraine and some neighboring countries. The distinguishable feature is early crop maturity and good bulbs storability. The nature of this group is xerophytic and thin with tall pseudostem (up to 45.0 cm).

(iii) **East Mediterranean Ecotype**-The East Mediterranean Ecotype grown in the northern shores of Mediterranean and related seas. The nature of this group is mesophytic which greately influenced by surrounding ecological factors particularly moisture. It is usually grown with top, slightly twisted along the central nerve and downwards. They have wax layer. Pseudostems (false stem) are thick and pulpy, with a softer consistency. The average weight of bulbs are 100 g and consist of numerous cloves (upto 30) located more freely towards the bottom. The outer scales are watery and fewer in number than other ecotypes.

Classification of Garlic

Garlic is of two types:

Classification on the Basis of Neck Robustness

(i) **Softnecks**-Softnecks are so called because the whole green plant dies down to pliancy, leaving nothing but the bulb and flexible stems that are easy to braid. Softnecks are the standard garlics of commerce. They are easiest to grow in regions where the weather is mild. They keep longer than hardnecks butb they are less hardy and more prone to make small, and very strong flavored cloves.

(ii) **Hardnecks**- Hardnecks have a stiff stem in the center that terminates in a beautiful flower or cluster of little bulbs then dries to a rigid stick that makes braiding impossible. Hardnecks do best where there is a real winter and are more vulnerable to splitting or do not produce when well grown in warm climates.

Classification on the Basis of Bulb Colour

Two types of varieties are available depending on their colour *i.e.*:

i. **Red**-Red coloured varieties are mostly used for medicinal purpose. They are more pungent as compared to white varieties.

ii. **White**-White coloured varieties are more popular as they are less pungent. Amongst white varieties, some produce small-sized bulbs while some of them produce big-sized bulbs. The different varieties of these groups are as below:

   1. Small sized varieties- Example- T-56-4, Creole, Italian and Tabiti
   2. Big-sized varieties- Example-Fawari, Rajalle Gaddi, Solan

Augmentation of Genetic Resources and Crop Improvement

Some ecotypes of garlic (*Allium sativum* L.) are capable of blooming and producing botanical seeds, nevertheless the clonal selection remains to be the most valuable tool to make use of the abundant genetic diversity that have this species and its botanical varieties (var. *sativum*, var. *ophioscorodon*, var. *pekinense*). These changes and the strong interaction that this crop presents with the environment, such as the one related to thermo and photoperiod, makes that the ecotypos, biotypes, varieties and cultivars are worldwide mistaken. The clonal selection can be mass or individual. The mass selection is more effective from the point of view of the quantity of acquired 'seed', but more delayed in terms of the best responses. The individual selection reaches more rapidly the proposed goals, but it is slower to reach large volumes of 'seeds'. In the due course of six or seven years, it can be achieved a new variety of garlic, capable of overcoming in comparative essays the local control plants. After formal registration in a record, this new material enters the stage of 'seed' production. In order to formalize a program of garlic 'seed' production, one must be provided with legislation and specific norms that should protect the genetic and sanitary patrimony of the country. Serial multiplications year after year generate different categories of 'fiscal seeds' (basic, registered, and certified). On the market these are known as 'improved seeds'.

Introduction and Selection

Garlic clones exhibite a wide variation in vegetative traits, flavor and pungency; bolting capacity, and fertility. Cultivar characteristics differ considerably with the location of cultivation and climate has a significant impact on garlic bulbing, florogenesis and flavor. All cultivated garlic clones are sterile, thus an increase in garlic variation is possible only via random or induced mutation, somaclonal variation and introduction of new genetic variation using modern molecular techniques. As a nature it is cross pollinated crop usually propagated vegetatively by cloves. Though its production and consumption are high, no special effort has been made to improve upon these strains. Garlic improvement is possible by clonal selection (Nath *et al.*, 2008). In Pakistan some cultivars like G.S.I. Lassan Gulabi (R) and Swat Local (P) are very common and being grown commercially (Kamenetsky *et al.*, 2007). As a general principal breeding of garlic includes Individual Bulb Selection (IBS) and adopt IBS-1, IBS-2, IBS-3 and IBS-4. There are 5 vegetative generations Vg-1 to Vg-5.

Table 9.31: Introduction of Garlic Germplasm in India

Name of Genotype	*Salient Feature*	*Source Country*
EC367655-58	High yielding and bold clove	Israel
EC378476-77	Pink root resistance	UK

Dhall and Brar (2013) worked out genetic variability, correlation and path coefficient in garlic and reported that values of genotypic correlation coefficients were higher than those of their respective phenotypic correlation coefficients in majority of cases suggesting that genotypic correlation were stronger, reliable and free from environmental factors. Bulb weight was positively and significantly correlated with bulb equatorial diameter, clove weight, clove length, bulb polar diameter, plant height at both genotypic and phenotypic level. Therefore, bulb yield can be improved by selecting genotypes having high value of equatorial diameter, clove length, bulb polar diameter and plant height.

Path coefficient studies indicated that plant height, number of leaves per plant, bulb polar diameter, bulb equatorial diameter and clove diameter had greater direct influence on bulb weight. Shree *et al.* (2016) evaluated 27 genotypes of garlic collected locally as well as from different research centres and agricultural universities. All the germplasm differed significantly with respect to the different morphological attributes, yield and other desirable traits. The height of the plant varied from 29.46-46.96 cm. length and breadth of leaves also showed significant variations among different genotypes. As for the yield characters are concerned, equilateral diameter and polar diameter of the bulb varied from 303-3.65 and 2.78-3.36 cm, respectively. Length and average diameter of single clove ranged from 2.11-2.80 cm and 0.67-1.02 cm, respectively. The average weight of single bulb varied between 10.6-28.22 g. Biochemical constituents like potassium, phosphorus and sulphur of the genotypes were between 0.630-0.999 per cent, 0.219-0.401 per cent and 0.810-2.170 per cent, respectively. TSS of the genotypes varied between 23.8-42.2 °Brix.

Gathala *et al.* (2016) evaluated 35 genotypes of garlic and observed significant differences for all the characters suggesting the presence of good deal of variability in the genotypes. The high GCV and PCV were observed for sulpher content followed by yield/plant, yield/plot, weight of 10 cloves and volume of 10 cloves. High PCV and moderate GCV were recorded for stem diameter and number of cloves/bulb. High heritability was observed for sulpher content, yield/

Table 9.32: Genotypic and Phenotypic Correlation Coefficient of 9 Quantitative Characters in Garlic

Trait	*Plant Height (cm)*	*No. of Leaves/ Plant*	*Bulb Weight (g)*	*Bulb Equatorial Dia. (mm)*	*Clove Weight (g)*	*Clove Length (mm)*	*Clove Diameter (mm)*	*No. of Cloves/ Bulb*
No. of leaves/plant	-0.1357							
	-0.1186							
Bulb weight (g)	0.2784*	-0.1218						
	0.2773*	-0.0924						
Bulb equatorial diameter (mm)	0.3122**	-0.0737	0.8417**					
	0.2844*	-0.0819	0.7782**					
Clove weight (g)	0.1018	-0.1695	0.6030**	0.3615**				
	0.1017	-0.1437	0.5925**	0.3478**				
Clove length (cm)	-0.4183**	0.0952	-0.0463	-0.1574	0.2929*			
	-0.4009**	0.0797	-0.0425	-0.1369	0.2881*			
Clove diameter (mm)	0.0171	0.0492	0.4263**	0.1742	0.9107**	0.4350**		
	0.0219	0.0718	0.3862**	0.1601	0.8210**	0.4061**		
No. of cloves/bulb	0.2885*	0.1735	0.0476	0.3640**	-0.6295**	-0.4275**	-0.7380**	
	0.2838*	0.1403	0.0457	0.3346**	-0.6229**	-0.4242**	-0.6516**	
Bulb polar diameter (mm)	0.2707*	0.0427	0.04327**	0.4327**	0.1795	0.1296	0.1148	0.3019**
	0.2634*	0.0357	0.4429**	0.3934**	0.1637	0.1240	0.0706	0.02704*

*, ** Significant at 5 and 1 per cent levels of signifiocance, respectively.

Table 9.33: Direct and Indirect Effects of Component Characters on Bulb Weight of Garlic

Trait	Plant Height (cm)	No. of Leaves/ Plant	Bulb Polar Diameter (mm)	Bulb Equatorial Dia. (mm)	No. of Cloves/ Bulb	Clove Weight (g)	Clove Length (mm)	Clove Diameter (mm)
Plant height (cm)	0.274	0.738	-0.712	0.782	0.323	-0.446	-0.348	0.737
No. of leaves/plant	0.265	0.763	-0.999	-0.109	0.441	-0.338	-0.285	0.579
Bulb polar diameter (mm)	-0.296	0.115	0.658	0.119	-0.106	0.111	0.174	-0.277
Bulb equatorial dia. (mm)	0.947	-0.367	0.347	0.226	-0.131	0.814	0.990	-0.163
No. of cloves/bulb	-0.576	-0.219	0.457	0.193	-0.153	0.100	0.131	-0.208
Clove weight (g)	0.968	0.204	-0.578	-0.145	0.122	-0.126	-0.101	-0.220
Clove length (mm)	0.886	0.204	-0.106	-0.208	0.188	-0.119	-0.107	0.220
Clove diameter (mm)	0.903	0.197	-0.817	-0.165	0.142	-0.124	-0.106	0.224

plant, yield/plot, weight of 10 cloves, volume of 10 cloves, number of cloves/bulb, clove diameter, leaf width, stem diameter and plant height at harvest. High heritability along with high genetic advance as per cent of mean was recorded for sulphur content, yield/plant and yield/plot. These characters were governed by the additive gene action. Hence, selection would be effective for their improvement. Yield/plant exhibited positive and significant correlation with plant height at 60 days after sowing, plant height at harvest, number of leaves/plant, stem diameter, equatorial bulb diameter, number of cloves/bulb, weight of 10 cloves, volume of 10 cloves and sulphur content both at genotypic and phenotypic levels. On the basis of mean performance the genotype MPG-2 followed by GG-3 and MPG-5 was the superior in terms of yield/plant.

Interspecific Hybridization

Allium cepa and *Allium sativum* have not been hybridized sexually for a long time, because of the narrow genetic pool of fertile garlic and a large genetic distance (Kik, 2002; Keller *et al.*, 1996). A cross between *A. longicuspis* with fertile *A. sativum* as a pollen parent could able to produce few sterile F_1 progenies (Etoh, 1985). The first interspecific hybrids between onion and garlic were reported by Ohsumi *et al.* (1993) and found allicin, a characteristic compound of garlic in the hybrid. Embryos were formed when onion was used as seed parent but reciprocal crosses were not successful. Hybrids were regenerated in *in vitro* by embryo rescue which possessed 16 chromosomes in somatic cells, with clear distinguishable two satellite chromosomes. Chromosome elimination was noty observed in the study and pollen viability of the hybrids were very low (2 per cent). Hybridization between diploid garlic and tetraploid leek has been attempted by Sugimoto *et al.* (1991) for the purpose of introducing fertility into garlic but characteristics of hybrid were not found. It is true that several triploid and aneuploids close to triploids hybrids were obtained. In this direction an important work was carried out by Yanagino *et al.* (2003) with aim of increasing fertility in garlic through sexual hybridization with leek. The cause of success was only due to belonginess of both into subgenus *Allium*. Leek was used as the female parent. Interspecific hybrids were obtained with the use of embryo rescue technique. Hybrids were triploid and possessed 24 chromosomes (16 from onion and 8 from garlic); however meiosis in PMC was not examined. The genuinity of hybrids was confirmed by molecular, cytological and morphological methods but all are almost completely sterile. Protoplast fusion was used to produce somatic hybrids of onion and sterile garlic (Yamashita *et al.*, 2002). Onion and garlic have sixteen chromosomes but the two lines of somatic hybrids possessed 40 and 41 chromosome with three of each chimeric chromosomes. The chimeric chromosomes in somatic hybrids could have originated from chromosomal fusion or translocation between chromosomes of onion and garlic.

Sterility

Majority of garlic cultivars are sterile, which preclude using them as a parent in sexual crossing programme. However, occurrence of fertile garlic plants can be spotted in nature (Kononkov, 1953; Katarzhin and Katarzhin, 1978; Etoh, 1983) which can be utilized in garlic breeding and crop improvement.

Breeding for Quality Traits

Gathala *et al.* (2016) evaluated 35 genotypes of garlic and observed significant differences for al the characters suggesting the presence of good deal of variability in the genotypes. The high GCV and PCV were observed for sulpher content. High heritability was observed for sulpher content. High heritability along with high genetic advance as per cent of mean was recorded for sulphur content. Yield/plant exhibited positive and significant correlation with sulphur content both at genotypic and phenotypic levels. Indain cultivars like G-282 and G-313 (Agrifound Parvati) are very much suitable for export quality. Similarly, cultivar 'Agrifound White (G-41)' is very good for storage.

Genetic Resource Conservation

A good number of local strains are available in different regions of India. They vary in the number of cloves, which ranges from 16 to 50, and in the size of the bulbs. There is very little variation in shape. Most of the strains have white colour, but some have reddish tinge also. Conservation of garlic germplasm began about 37 years ago. Since then field gene banks were established in Germany, Israel, Poland, Czech Republic Russia and the USA. An international effort aimed at immediate collection and preservation of this heritage is imperative. It should be followed by an intensive evaluation effort and accompanied by substantial preservation projects to halt the rapid and irreversible erosion of the Central Asia genepool of garlic land races and wild populations.

REFERENCES

1. Dhall, R.K. and Brar, P.S. 2013. Genetic variability, correlation and path coefficient in garlic (*Allium sativum* L.). *Veg. Sci.*, **40(1):** 102-104.
2. Etoh, T.1983. Accomplishment of microsporogenesis in a garlic clone. *Mem. Fac. Agr. Kagoshima Univ.*, **19:** 55-132.
3. Etoh, T.1985. Studies on the sterile on garlic (*Allium sativum* L). *Mem. Fac. Agr. Kagoshima Univ.*, **21:** 77-63.
4. Gathala, S., Ameta, K.D., Dubey, R.B., Kaushik, R.A. and Singh, H. 2016. Genetic heritability, correlation and path coefficient analysis in garlic (*Allium sativum* L.). 7th Indian Horticulture Congress (Abs: 10; Session: Breeding for Horticultural Crops), 15-18 November, held at ICAR-IARI, New Delhi, p. 93-94.
5. Kamenetsky, R. and Rabinowitch, H.D. 2001. Floral development in bolting garlic. *Sex Plant Reprod.*, **13:** 235-241.
6. Kamenetsky, R., Khassanov, F., Rabinowitch, H.D., Auger, J. and Kik, C. 2007. Garlic biodiversity and Genetic Resources. Medicinal and Aromatic Plant Science and Biotechnology, p. 1-5.
7. Katarzhin, M.S. and *Katarzhin, I.M. 1978. Experimental on the sexual reproduction of garlic (in Japanese). Japan, J. Breed.* **45(2):** 310.
8. Keller, R.J., Schubert, I., Fuchs, J. and Meister, A. 1996. Interspecific crosses of onion with distant *Allium* species and characterization of the presumed hybrids by means of flow cytometry, karyotype analysis and genomic *in situ* hybridization. *Theor. Appl. Genet.*, **92:** 417-424.
9. Kik, C. 2002. Exploitation of wild relatives for the breeding of cultivated *Allium* species. In: *Allium* Crop Science: Recent Advances. Rabinowitch, H.D. and Currah, L., p. 81-100, CABI Publishing, ISBN0851995101, U.K.
10. Kononkov, P.F. 1953. The question of obtaining garlic seed (in Russian). *Sad I Ogorod*, **8:** 38-40.
11. Kuznetsov, A.V. 1948. Agrobiological studies of garlic. *Phg., Krasnodar.*
12. Nath, P., Srivastava, V.K., Dutta, O.P. and Swamy, K.R.M. 2008. Vegetable Crops- Improvement and Production. Dr. Prem Nath Agricultural Foundation, Bengaluru (Karnataka), p. 196-208.
13. Ohsumi, C.A., Kojima, K., Hinata, K., Etoh, T. and Hayashi, T. 1993. Interspecific hybrids between *Allium cepa* and *Allium sativum*. *Theor. Appl. Genet.*, **85:** 969-675.
14. Purseglove, J.W. 1975. Tropical Crops: Monocotyledons. ELBS, Longman, London, p. 52-56.
15. Shinde, N.N. and Sontakke, M.B. 1986. Bulb crops. In: Vegetable crops (Eds. T.K. Bose *et al.*). Naya Prokash, Kolkata.
16. Shree, S., Singh, V.K. and Singh, P.K. 2016. Morphological and biochemical characterization of garlic (*Allium sativum* L.) for yield and storability. 7th Indian Horticulture Congress (Abs: 93), held at ICAR-IARI, New Delhi, 15-18 November, p. 71-72.
17. Singh, D.N. and Nath, V. 2011. Varieties and hybrids of vegetables. Satish Serial Publishing House, Azadpur, Delhi (India), p. 78-80.
18. Singh, N. and Pal, N. 2003. Garlic. In: Vegetables, Tuber crops and Spices, (Eds. S. Thamburaj and Narendra Singh). ICAR, New Delhi.
19. Thompson, C.H. and Kelly, C.W.1957. Vegetable Crops. Mc. Graw Hill Book Co., Inc. USA.
20. Yamashita, K., Hisatsune, Y., Sakamoto, T., Ishizuka, K. and Tashiro, Y. 2002. Chromosome and cytoplasm analyses of somatic hybrids between onion (*Allium cepa* L.) and garlic (*Allium sativum* L.). *Euphytica*, **125:** 163-167.
21. Yanagino, T., Sugawara, E. and Watanable, M. 2003. Production and characterization of an interspecific hybrid between leek and garlic. *Theor. Appl. Genet.*, **107:** 1-5.
22. Johary, D. and Hopf, M. 2000. Domestication of plants in the Old World 3rd edn p. 316. New York: Oxford University Press.

Winter Leek or Leek (*Allium porrum* L.)

Introduction

Leek (*Allium porrum* L.; syn. *A. ampeloprasum* var. *porrum*, 2n= 32), sometimes called "the gourmet's onion"are related to onion (belongs to onion family), but have flat leaves instead of tubular (as in onion round and hollow leaves) and relatively little bulb development (non-bulb forming crop) (Thompson and Kelly, 1957). They are easy to grow and delicious, with a taste all their own, very much like a mild onion. The thick leaf bases and slightly developed bulb look like a giant green onion, and are eaten as a cooked vegetable. In Europe, it is known as "poor man's asparagus". It is hardier (sometimes milder) than onion or garlic when grown during cool season. It does not have a dormant period and is not influenced by photoperiodism. Rather than forming a tight bulb like the onion, the leek produces a long cylinder of bundled leaf sheaths which are generally blanched by pushing soil around them (trenching). The edible part of the leek plant is sometimes called a stem, though technically it is bundle of leaf sheaths. The leek has leaves very similar to garlic. The onion like layers, form around a core. The tender core may be eaten, but as the leek ages the core becomes woody and better replanted than eaten. They can also be eaten raw in salads, doing especially well when they are the prime ingredient. It is also cooked with other vegetables or used as flavouring agent in soups and stews. Per 100 g of edible portion contains 79.9 g moisture, 1.8 g protein, 0.1 g fat, 5.0 g carbohydrate, 0.7 g minerals, 1.3 g fiber, 70.0 mg phosphorus, 2.3 mg iron, 30.0 IU vitamin A, 0.23 mg thiamine and 11.0 mg vitamin (Aykroyd, 1963). Leek bulb contained very low (below 20 mg/kg fresh weight) or traces (below 2.0 mg/kg fresh weight) of quercetin and Kaempferol (Noleau, *et al.*, 1991). Leek has milder and more delicate flavor than onion, though of coarser texture. When tender, it is eaten raw. It is also cooked with other vegetables and used for flavouring in soups and stews. The leeks (*A. ampeloprasum*) succeed fairly well in India thriving at elevations above 300 m and often remain fit for use until the end of rainy season.

Origin, History and Distribution

The leek is thought to be a native of Switzerland. The leek was developed from a wild type, which is native to Wester Asia and the Mediterranean counrties (Nonnecke, 1922; Harrison *et al.*, 1969) but Chauhan (1972) is of the opinion that leek is a native of Europe and West Asia. It was known to be the ancient Greeks and Romans (Thompson and Kelly, 1957). Leek is traceable to ancient civilization of the Near East. Wild leeks were used as food during the early Bronze Age, around 4,000 BC., and were probably domesticated around 2,000 B.C. As a matter of fact, the cultivated variety of leek has not been found wild. Leek was considered only as a cultivated variety of the great-headed garlic (*A. ampeloprasum* L.) so common in the East and Mediterranean region, especially Algeria. They were part of the diet of those who built the Egyptian pyramids, and Hippocrates, the father of medicine, prescribed the leek as a cure for nosebleeds. Leeks have been cultivated in Western Europe since the middle Ages, and are particularly associated with Wales-dating back to 640 A.D. when Welsh soldiers wore pieces of leek in their helmets to distinguish themselves from their Saxon foes in battle. It is commercially cultivated in Malaysia, the Philippines, France, Britain, Northern Spain, China, Belgium, Denmark, Holland, India, Syria, Iraq, Israel and many other African countries (Harrison *et al.*, 1969). In India and Sri Lanka, it thrives well at higher altitudes but moist localities are adverse to its cultivation (Chauhan (1972). Commercial cultivbation is not followed in India and wherever it grows, it is on homescale, mainly in the kitchen gardens (Choudhury, 1967). It is the national emblem of Wales.

Species Availability, Evolution and Diversity

Leek falls under genus *Allium*, and botanical name is *A. porrum* L. [syn. *A. ampeloprasum var. porrum* (L.) Regel.] is a tetra ploid with 2n=32 (Mc Collum, 1976). It is cultigen derived from the wild, *A. ampeloprasum*. It is grown for its etiolated pseudostem formed by the leaf sheath (Jones and Mann, 1963). The botanical difference between leek and great-headed garlic lies in difference in phasic development and the lesser tendency in leek to form bulb. Leek is highly cross pollinated due to heteromorphism and male sterility. It crosses readly with wild *A. ampeloprasum* of the same ploidy level (2n=32, x=8). Leek is tetraploid (2n=32) in nature. The related species are *A.cepa* L. var. *cepa* (common onion), *A. cepa* var. *aggregatum* G. Don (multiplier onion, potato onion, and shallot), *A. viviparum* (tree onion), *A sativum* L. (garlic), *A. ampeloprasum* (great-headed garlic, kurrat), *A. fistulosum* L. (Japanese bunching or Welsh onion), *A. schoenoprasum* L (chive), *A. Chinese* G. Don. (Don Rakkyo), *A. tuberosum* Rottl. ex Spreng. (Chinese chive), *A. ladebourianum* (Asatsuki), *A. macrostemon* (Chinese garlic), and *A. nipponicum* (French, Sevat). The closest species to leek (*A. porrum*) is *A. ampeloprasum*. The above 2 species are cross compatible and the F_1s are fertile, at same tetraploidy level (2n=32). The interspecific F_1 hybrids were generated to create variability. The cross between *A. cepa* and *A. porrum* was successful (Currah, 1986; Mc Collum, 1976).

Botanical Description

Leek belongs to family Alliaceae, genus *Allium*, *A. porrum* L. [syn. *A. ampeloprasum* var. *porrum* (L.) Regel.]. Leek looks like the green spring onion but is much larger. The difference between leek and great headed garlic lies in difference in phasic development and the lesser tendency in lleek to form bulbs. It does not have a dormant period and is not influenced by photoperiodism.Commercialy propagated through seeds and crop remains in the field as long as 12 months.

Table 9.34: Botanical Description

Growth Habit	Leek is a tall, hardy, and biennial in nature, even though it is generally referred to as a biennial. It multiplies by means of small lateral growths and often develops a roundish bulb at the base of the main growth. Growth is continuous and it can be harvested over a long period of time.
Root	White narrowly ovoid bulbs
Stem	Leek develops a broad, succulent stem rather than a large bulb like an onion.

Leaf	Broad leaves. The plant has a fan-like sheathb of flat, blue green or yellow-green leaves that may grow a 30.0-60.0 cm in length, on a stalk up to 30.0 cm long. Types vary in the length and thickness of the leaf shanks (pseudo-stem).
Floral biology	Fowers are numerous, pinkish with exerted anthers. Production of flowers is influenced by low temperature. Flowers are seldom produced in the tropics. The time of formation of the flower initially corresponds with the maxima in the curves showing percentage of carbohydrate and ash in the whole plant.
Anthesis and pollination	Highly cross pollinated crop due to heteromorphism and male sterility.
Bulbs	Bulbs are not formed or somew times white narrowly ovoid bulbs formed.

The edible portion, which is cooked is the elongated base of the foliage leaves or false stem, which develop below ground in order to promote blanching.

Botanical Classification

On the Basis of Leaf Colour

1. Yellow-green and green-blue summer leeks and
2. Blue and blue-grey winter leeks

A correlation exists between leaf colour and frost resistance (winter hardiness).

On the Basis of Leaf Types

They vary from long, green narrow-leaf types with long slender white stems to long wide-leaf types with thicker shorter white stems and blue-green leaves.

On the Basis of Neck

Another basis of classification is neck (portion in between disc and area of leaf separation) length. There are 3 groups:

(i) Short neck
(ii) Medium-long neck
(iii) Long neck

Augmentation of Genetic Resources and Crop Improvement

Introduction and Selection

Populaer cultivar 'Renova' derived by selection from the variety 'Regions', has darker leaves, longer stalk and less tendency to form bulb and more uniform in size and shape. In India, very little work has been done on the improvement of cultivar of leek. There are many named varieties of leeks outside India. Few promising genotypes are 'Varna', 'King Richard', 'Americal Flag Elephant', London Flag, Mussel Burg, Goliah, Winterens, Titan, Ficus and 'Splendid'.

Hybridization and Selection

Similarly 'Miner' an autumn or early winter variety derived from crosses and family selection in the variety 'St. George' has long stem and gives high yield. Inbreeding depression has been observed from F_2 generation onwards. In leek many hybrids have been evolved and significant heterosis has been recorded. 'Regius 930/57' ('St. George' x 'Special'), F_1 hybrid is a good cropper especially in late autumn and early winter.

Interspecific Hybridization

Allium porrum L. crosses readly with wild *A. ampeloprasum* of the same ploidy level (2n=32, x=8) and F_1s are fertile. The cross between *A. cepa* and *A. porrum* was successful. Leek and Kurrat (*Allium kurrat*) are interfertile and it may be supposed that they would cross readly with wild *A. ampeloprasum* of the same (4x) ploidy level (Mc Collum, 1976).

Male Sterility

Male sterility has been reported.

Abiotic Resistance Genotypes and its Utilization

The leek has greater cold resistance than the onion; bulbs are not formed. It does not have a rest period, and it is not influenced by photoperiod (Astley *et al.*, 1982). Leek cultivar 'Special 050' is suitable for forcing. Similarly, popular genotype 'Splendid' cultivated throughout the season. 'Borde Pearl' is one such high-yielding line resistant to bolting. 'Bluevetia Early' is suitable for both summer and early autumn culture. 'Copen-hagen Market' and 'Elephant' are short necked types and 'Gennivillies' and 'St. George' are long necked types with resistant to frost (Tindal, 1975).

REFERENCES

1. Astley, D., Innes, N.L. and Van der Meer, Q.P. 1982. Genetic resources of *Allium* species- global report. International Board for Plant Genetic Resources (IBPGR), Rome, p. 38.
2. Aykroyd, W.R. 1963. The nutritive value of Indian foods and the planning of satisfactory diets, ICM R, New Delhi, p. 58, 63 and 78.
3. Chauhan, D.V.S. 1972. Vegetable production in India. Ram Prasad and Sons. Agra-3, (India) pp. 240.
4. Choudhury, B. 1967. Vegetables. National Book Trust, India, New Delhi.
5. Currah, L. 1986. Leek breeding: a review. *Journal of Horticultural Sciences,* **61(4):** 407-415.
6. Jones, H.A. and Mnn, L.K.A.E. 1963. Onions and their allies. World Crop Books. Interscience Publishers. Inc. New York.

7. Mc Collum, G.D. 1976. Onion and allies. (in) *Evolution of Crop Plants*, edn. 1, p. 186-199.

8. Mc Collum, G.D. 1976. Evolution of Crop Plants, Ed. N.W. Simmonds, Longman, London and New York.

9. Noleau, I., Richard, H. and Peyrom, A.S. 1991. Volatile compounds in leek and asufoetida. *Journal of Essential Oil Research.*, **3(4):** 241-256.

10. Nonnecke, I.B. Libner. 1922. Vegetable Production. An AVI Book Published by Van Nostr and Reinhold, Springer International Edition.

11. Tindal, H.D. 1975. Commercial Vegetable Growing, Oxford, London, edn. 1, p. 176.

12. Thompson, C.H. and Kelly, C.W.1957. Vegetable Crops. Mc. Graw Hill Book Co., Inc. USA.

Hookerii (*Allium hookerii*)

Introduction

Hookerii (*Allium hookerii*) is a small spicy herb growing gregariously in the wild form at the hill slopes of Manipur (India). The people in the Manipur state has been using it an indigenous spice or condiment since old days and also growing as pot-herb or in the kitchen garden for daily use. The leaves are also used in the salads and soups. Fish curry taste great with this herb and the roots of this herb are another delight. The roots are used basically for temering and while cooking fish. It has been used traditionaly in folklore medicines as diuretic and as wormicide.

Origin, History and Distribution

Hookerii can be spotted abundantly in the Himalayan ranges including Nepal and NEH region particularly in Manipur. In these regions Hookerii is being grown by people as a decorative plant.

Botanical Description

Table 9.35: Botanical Description

Growth Habit	It is a herb.
Root	It develops thick fibrous root system.
Stem	The compact underground stem produces side shoots.
Leaf	The leaves are simple, alternate, long, smooth and thick with prominent midrib, while the leaf bears are scaly and fleshy.
Floral biology	During summer and rainy season from tuber flower shots come up with small white flower in cluster at the tip.
Fruit and fruit set	The fruits are globular capsule.

The Chives (*Allium schoenoprasum*)

Introduction

The chive (*A. shoenoprasum*) is edible in nature (LaFray, 1987). It is grown as a garden crop. It is hardy perennial growing in dense clumps with light green or purple tinted grass like leaves and cluster of narrow weekly developed bulb at the base of the leaves. The onion flavoured leaves and bulbs are used for seasoning salad.

Origin, History and Distribution

The chive (*A. shoenoprasum*) is native of Europe and adjoining parts of Asia and in India but well distributed in the northern Himalayas from Kashmir to Kumaon. Though it is widespread in nature across much of Europe, Asia, and North America (Flora of China; *Tardiff and Morisset (1990; Flora of North America; Altervista Flora Italiana and "Allium schoenoprasum - Plant Finder").*

Botanical Description

Table 9.36: Botanical Description

Growth Habit	It is is a wild perennial herb.
Stem	Scape 20-40 cm tall, base covered with leaf bases.
Leaf	Leaves 1-3, linear, 1-2 mm broad, grooved, not fistular.
Floral biology	Umbel hemispherical, 2-3 cm across. Tepals oval, 6-8 mm long, reddish, acute. Filaments longer than the tepals, inner with 2 teeth at the base of each. Style filiform, exserted.
Bulb	It has an egg-shaped bulb up to 30 mm across, and a scape up to 40 cm tall. Bulb ovoid, 2-3 cm long, outer coats membranous-scarious, reddish-brown.

References

1. Altervista Flora Italiana, Erba cipollina, wild chives, Civette, Schnittlauch, *Allium schoenoprasum* L. includes photos, drawings, European distribution map, *etc.*

2. *"Allium schoenoprasum - Plant Finder". mobot.org.*

3. Flora of North America Vol. 26 Page 240 Chive *Allium schoenoprasum.*

4. Flora of China Vol. 24 Page 195, bei cong *Allium schoenoprasum.*

5. LaFray, Joyce (1987). *Tropic Cooking: The New Cuisine from Florida and the Islands of the Caribbean.* Oakland: Ten Speed Press. p. 292. ISBN 0-89815-234-8.

6. Tardiff, B. and Morisset, P. 1990. "Clinical morphological variation of *Allium schoenoprasum* in eastern North America". Taxon. **39** (3): 417–429. JSTOR 1223088.

7. "World Checklist of Selected Plant Families: Royal Botanic Gardens, Kew". *kew.org*.

Panchali Ganja or Jungli Pyaz (*Allium roylei* Stearn)

Introduction

The genus *Allium* comprises about 750 species including *A. cepa*, *A. fistulosum* and *A. roylei*. Onion (*Allium cepa* L.) is a valuable vegetable crop preceded on the list of worldwide cultivated vegetable species only by tomato (FAOSTAT, 2011). *Allium Roylei* Stearn is a wild species from the Indian subcontinent, belonging to family Alliaceae is a lesser known plant of economic importance and commonly known as Panchali Ganja or Jungli Pyaz1. Its leaves and inflorescence are used by tribals for seasoning dishes. Bulbs are also consumed for getting relief from headache. In addition bulbs are also given to horses for relief from colic pain (Kohli and Gohil, 2009a).

Origin, History and Distribution

Allium roylei is a plant species found high in the Himalayas of Pakistan, Afghanistan and India. As per the Hooker, the species in India is distributed along the Himalayan, sub- Himalayan ranges and Garhwal Himalayas westwards between 6000-7000 ft (Kohli *et al.*, 2011). In Jammu and Kasmir, except for some plants growing along Gilgit slopes of Ladakh, there were no reports of its occurrence till 2002. Since then, explorations to several regions have unfold two more populations from the state. Surprisingly, the species finds no mention in the floras of Jammu, Srinagar and Udhampur (Sharma and Kachroo, 1981; Dhar and Kachroo, 1983; Swami and Gupta, 1998). It, however, figures in the flora of Ladakh, according to which it is distributed in Gilgit slopes (Murti, 2001). From Jammu province, Sharma and Gohil (2002) reported its occurrence for the first time from Bani (district Kathua) and Mendhar (district Poonch). Since then, some other areas of its distribution have been recorded from this province. A new population was collected from Gourwan (district Reasi) (Kohli, 2007) and another was reported from Kishtwar High Altitude National Park (Kumar and Hamal, 2009).

Species Availability, Evolution and Diversity

- ☆ *Allium roylei* Stearn
- ☆ *Allium aaseae* Ownbey
- ☆ *Allium abbasii* R.M.Fritsch
- ☆ *Allium abramsii* (Ownbey and Aase ex Traub) McNeal
- ☆ *Allium achaium* Boiss. and Orph.
- ☆ *Allium acidoides* Stearn
- ☆ *Allium aciphyllum* J.M. Xu
- ☆ *Allium acuminatum* Hook.
- ☆ *Allium acutiflorum* Loisel.
- ☆ *Allium aegilicum* Tzanoud.
- ☆ 891 more. show full tree.

Botanical Description

Genus	:	*Allium*
Subgenus	:	*Polyprason*
Section	:	*Oreiprason*
Family	:	*Amaryllidaceae*
Subfamily	:	*Allioideae*
Tribe	:	*Allieae*
Botanical name	:	*Allium Roylei* Stearn

The details are given below:

Table 9.37: Botanical Description

Growth Habit	*Allium Roylei* Stearn is a wild perennial herb.
Stem	Scape 20-40 cm tall, base covered with leaf bases.
Leaf	Leaves 1-3, linear, 1-2 mm broad, grooved, and not fistular.
Floral biology	Umbel hemispherical, 2-3 cm across. Tepals oval, 6-8 mm long, reddish, acute. Filaments longer than the tepals, inner with 2 teeth at the base of each. Style filiform, exerted.
Bulb	It has an egg-shaped bulb up to 30 mm across, and a scape up to 40 cm tall. Bulb ovoid, 2-3 cm long, outer coats membranous-scarious, reddish-brown.

Interspecific Hybridization

The first crosses between *Allium Roylei* Stearn and *A. cepa* were made in 1985 to achieve introgression of downy mildew resistance into onion (Van der Meer and de Vries, 1990). Subsequent studies on reciprocal crosses of *A. cepa* and *Allium roylei* hybridization (Kofoet *et al.*, 1990; Kofoet and Zinkernagel, 1990) led to the conclusion that resistance to downy mildew is determined by a single dominant gene, *Pd1* which was later confirmed by AFLP marker technique combined with SCAR analysis (Van Heusden *et al.*, 2000a, b). The years of investigation on interspecific hybridization and downy mildew resistance lead to F_1BC_5 generation plants expected to be resistance to *Peronospora destructor* (Scholten *et al.*, 2007). First use of embryo rescue technique in vitro produce hybrid between *A. roylei* and *A. cepa* was reported by Chuda and Adamus (2012). Ovule culture on media

MS and BDS with 20g/l sucrose resulted in 906 well acclimatized plants out of 1768 regenerated embryos. Hybridity of plants were confirmed by genomic *in situ* hybridization where hybrid had 16 chromosomes, out of which 8 were from *A. cepa* and 8 from *A. roylei*. The hybrids were also verified by three molecular markers *viz.*, SIR, ACS and A.

Disease Resistance

The species is reportedly a repository of genes imparting resistance to various harmful diseases. *Allium roylei* Stearn harbours genes imparting resistance against dreadful fungal disease (downy mildew, leaf blight and anthracnose). Being sexually compatible with *A. cepa*, it has been successfully used as a donor of these genes for improving the common onion (Scholten *et al.*, 2007). *Allium roylei* Stearn is a wild species originating from the Indian subcontinent which harbors disease resistance against downy mildew (*Peronospora destructor*), leaf blight (*Botrytis squamosa*) and anthracnose (*Colletotrichum gloesporioides*) (van der Meer and de Vries, 1990; de Vries *et al.*, 1992; Galvan *et al.*, 1997; Scholten *et al.*, 2007; Kohli and Gohil, 2009). This species has been used to expand the genetic variation of onion since 1960 (Fritsch and Friesen, 2002). The first crosses between *A. roylei* and *A. cepa* were made in 1985 to achieve introgression of downy mildew resistance into onion

Genetic Resource Conservation

Exploitation has led to its being assigned the status of a rare species in Western Himalaya (Pandey *et al.*, 2008).

REFERENCES

1. Chuda, A. and Adamus, A. 2012. Hybridization and molecular characterization of F_1 *Allium cepa* x *Allium roylei* plants. *Acta Biologica Cracoviensia*, **54(2):** 25-31.
2. De Vries J.N., Wietsma, W.A., and de Vries, T. 1992. Introgression of leaf blight resistance from *Allium roylei* Stearn into onion (*A. cepa* L.). *Euphytica*, **62:** 127–133
3. Dhar, U. and and Kachroo, P. 1983. *Alpine Flora of Kashmir Himalaya*, Published by Scientific publishers, Jodhpur, India.
4. Fritsch, R.M. and Friesen, N. 2002. Evolution, domestication and taxonomy. In: Rabinowitch HD and Currah L [eds], *Allium Crop Science: Recent Advances*, 5–38. CABI Publishing Oxon, UK.
5. Galvan, G.A., Wietsma, W.A., Putrasemedja, S., Permadi, A.H., and Kik, C. 1997. Screening for resistance to anthracnose (*Colletotrichum gloesporioides* Penz.) in *Allium cepa* and its relatives. *Euphytica*, **95(2):** 173–178.
6. Kofoet, A. and Zinkernagel, V. 1990. Resistance to downy mildew (*Pernospora destructor* (Berk.) Casp.) in *Allium* species. *Journal of Plant Diseases and Protection*, **97(1):** 13-23.
7. Kofoet, A., Kik, C., Wietsma, W.A. and de Vries. 1990. Inheritance of resistance to downy mildew (*Pernospora destructor* (Berk.) Casp.) from *Allium roylei* Stearn in the back cross *Allium cepa* L., 9 (*A. roylei 9 A. cepa*). *Plant Breed.*, **105:** 144-149.
8. Kohli, B. and Gohil, R.N. 2009a. Need to conserve *Allium roylei* Stearn: a potential gene reservoir. *Genet. Resour. Crop Evol.*, **56:** 891-893.
9. Kohli, B. 2007. Cytological studies in three collections of *Allium roylei* Stearn. M. Phil. Dissertation. Univertsity of Jammu, Jammu, India.
10. Kohli, B., Kaul, V. and Gohil, R.N. 2011. *Allium roylei* Stearn: An uncharted precious gene reservoir of North West Himalaya. *Indian J. Plant Genet. Resour.*, **24(3):** 356-358.
11. Kumar, S. and Hamal, I.A. 2009. Wild edibles of Kishtwar High Altitude National Park in North-West Himalaya, Jammu and Kashmir (India). *Ethnobotanical Leaflets*, **13:** 195-202.
12. Murti, S.K. 2001. *Flora of Cold Deserts of Western Himalaya*. Vol.I (*Monocotyledons*), Botanical Survey of India, Dehradun, India.
13. Pandey, A., Pandey, R., Negi, K.S. and Radhamanio, J. 2008. Realizing value of genetic resources of *Allium* in India.*Genet. Resou. Crop Evol.*, **55:** 985-994.
14. Scholten, O.E., va Heusden, A.W., Khrustaleva, L.I, Burger-Meijer, K., Mank, R.A., Antonise, R.G.C., Harrewijn, J.L., van Haecke, W., Oost, E.H., Peters, R.J. and Kik, C. 2007. The long and winding road leading to the successful introgression of downy mildew resistance into onion. *Euphytica*, **156:** 345-353.
15. Sharma, G.and Gohil, R.N. 2002. Additions to the germplasm of *Allium roylei* Stearn. *Indian J. Plant Genet. Resour.*, **15:** 288-289.
16. Sharma, B.M. and Kachroo, P. 1981. *Flora of Jammu and Plants of Neighbourhood*. Vol.I, published by Bishen Singh Mahendra Pal Singh, Dehradun, India.
17. Swami, A. and Gupta, B.K. 1998. *Flora of Udhampur*, published by Bishen Singh Mahendra Pal Singh, Dehradun, India.
18. Van der Meer Q.P., and de Vries, J.N. 1990. An interspecific cross between *Allium roylei* Stearn and *Allium cepa* L., and its backcross to *A. cepa*. *Euphytica*, **47:** 29–31.
19. Van Heusden, A.W., Shigyo, M., Tashiro, Y., Vrielink-Van Ginkel, R. and Kik, C. 2000b. AFLP linkage group assignment to the chromosome of *Allium cepa* L. via monosomic addition lines. *Theor. Appl. Genet.*, **100:** 480-486.
20. Van Heusden, A.W., Van Ooijen, J.W.; Vrielink-van ginkel, M.; Verbreek, W.H.J.; Wietsma, W.A. and Kik, C. 2000a. A genetic map of an cross in *Allium* based on amplified fragment length polymorphism (AFLP) markers. *Theor. Appl. Genet.*, **100:** 118-126.

Japanes Bunching Onion (*Allium fistulosum*)

Japanes bunching onion or Welsh onion (*Allium fistulosum*) is diploid species species (2n=2x=16) which is similar to onion and belong to section *cepa* of the genus *Allium* (Jones, 1990) and subgenus *Rhiziridium* (Hanelt, 1990).

Interspecific Crosses

Progenies from generative introgression of *A. cepa* with *A. fistulosum* have been studied most extensively among all interspecific crosses in the genus *Allium*. These two species can be easily hybridized and the success rate can be enhanced by the use of embryo-rescue (Dole *et al.*, 1980). As a history, the first hybrid between *A. cepa* and *A. fistulosum* were reported by Emsweller and Jones (1935) but hybrid was sterile and seed set was occasional. Critical microscopic observations results synaptonemal complex of hybrid; showed heteromorphic bivalent. Chiasmata frequency is less compared to both parents and that are mainly interstitial and distal. It is true that hybrids between *A. cepa* and *A. fistulosum* can be produced easily but low pollen fertility, often not excedding 10 per cent (Jones and Mann, 1963; Levan, 1941; Maeda, 1937; Peffley, 1986; Van der Valk *et al.*, 1991b). Attempts to backcross (BC) F_1 hybrid (*A. cepa* L. x *Allium fistulosum* L.) to *A. cepa* L. were unsuccessful (Levan, 1941). Difficulty in this programme was due to strong pre-fertilization barrier, as the growth of onion pollen tubes in the style of the hybrid was inhibited (Van der Valk *et al.*, 1991a) and the failure was due to homeologous recombination, distorted segregation and nucleo-cytoplasmic incompatibility (Ulloa *et al.*, 1995). Maed (1973) backcross F_1 with *Allium fistulosum* of *Allium cepa* and could able to obtain viable seeds successfully. Emsweller and Jones (1945) obtained several BC progenies but only when the hybrid was used as the pollen source. Khrustaleva and Kik (1998) were introgressed genes from *Allium fistulosum* to *Allium cepa* using *A. roylei* as a bridge species in crossing. Multi-colour GISH was performed to analyze genome organization of the hybrid. Results showed 8 *cepa* chromosomes, 1 metacentric chromosome from roylei and 7 recombinant chromosome, further recombination took place only in interstitial and distal portion of chromosome of interspecific hybrid between *A. fistulosum* to *A. roylei*. Pollen fertility of the second generation bridge cross hybrid was variable, which might be caused by the use of CMS onions as the female parent and or the lack of *roylei* chromosome segments containing male sexual reproduction genes (Khrustaleva and Kik, 2000; Ulloa *et al.*, 1995). Monosomic addition lines of *fistulosum-cepa* (Shigyo *et al.*, 1998) and *cepa-fistulosum* (Peffley *et al.*, 1985) were reported. These lines are valuable to study genome organization in *A. cepa* and *A. fistulosum*. Unfortunately, subsequent backcrosses of the obtained F_1 *A. cepa*, *A. fistulosum* hybrids to *A. cepa* have proved problematic. The difficulties observed in genome transmission in advance backcross generation have explained by the sterility of F_1. *A. cepa* and *A. fistulosum* hybrids probably resulting from one of the three possible mechanisms:

- Stylar incongruity (Van der Valk *et al.*, 1991a)
- Nuclear cytoplasmic incompatibility (Ulloa *et al.*, 1995)
- Central cell nuclear cytoplasmic incompatibility (Mangum and Peffley, 2005)

Due to problems in introgression of Japanese bunching onions traits into onion, so far only four new varieties (Beltsville Bunching (tetraploid), Delta Giant (triploid), Top Onion and Wakegi Onion) have been developed. 'Top Onion', a home garden crop of Europe and north-eastern Asia has been termed as *Allium proliferum* which is a cross between *A. cepa* and *A. fistulosum* (Fritsch and Friesen, 2002) and has been confirmed by genomic studies. 'Top Onion' is commonly known as tree onion, egyption onion or walking onion as it produces bulblets or topsets where normal onion posses flowers. Introgression between *Allium fistulosum* x shallot type *A. cepa* is a Wakegi onion which is diploid hybrid. Hybridity of these species have been confirmed by molecular and chromosomal studies.

Ploidy Breeding

'Beltsville Bunching' is a tetraploid hybrid where two set of chromosome comes from *A. cepa* and two from *A. fistulosum*, which is cultivated in USA. Interspecific hybrid 'Delta Giant' is a triploid in nature and grown in USA as a minor crop (Kik, 2002). 'Top Onion' is commonly known as tree onion, egyption onion or walking onion as it produces bulblets or topsets where normal onion posses flowers. Introgression between *Allium fistulosum* x shallot type *A. cepa* is a Wakegi onion which is diploid hybrid. In Japan and China 'Wakegi Onion' is used for salad purpose.

'Pran' is also interspecific hybrid species (*Allium x cornutum*) known in Kashmir region of India; however in evolution studies seed parent of this interspecific cross is still unclear while female parent is *A. cepa* (*Havey*, 1991). It produces very slender and attractive dark green colour leaves and grown as a kitchen garden crop at Jammu, (India), Europe, Canada *etc.* (Fritsch and Friesen, 2002).

REFERENCES

1. Emsweller, S.L. and Jones, H.A. 1935. An interspecific hybrid in *Allium*. Hilgardia, Vol.9, p. 265-273.

2. Fritsch, R.M. and Friesen, N. 2002. Evolution, domestication and taxonomy. In: Rabinnowitch, H.D., Currah, L. (eds.) *Allium* Crop Science: Recent Advances. CABI Publishing, Wallingford, Oxfordshire, UK, p. 5-30.

3. Hanelt, P. 1990. Taxonomy, evolution and history. In: Onions and other Vegetabl Alliums. Brewster, J.L. (eds.), Onions and Allied Crops, Vol. I. Boca Raton, CRC Press, p. 1-26

4. Jones, H.A. and Mann, L.K. 1963. Onions and Their Allies-botany, Cultivation, and utilization. Interscience Publishers inc., ISBN100249388731, New York.

5. Khrustaleva, l. and Kik, C. 2000. Introgression of *Allium fistulosum* into *Allium cepa* mediated by *A. roylei*. *Theor. Appl. Genet.*, **100:** 17-26.

6. Kik, C. 2002. Exploitation of wild relatives for the breeding of cultyivated *Allium* species. In: Kik, C. 2002. Exploitation of wild relatives for the breeding of cultivated *Allium* species. In: *Allium* Crop Science: Recent Advances. Robinowitch, H.D. and Currah, L., p. 81-100, CAB Publishing, ISBN0851995101, U.K.

7. Levan, A. 1941. The cytology of the species hybrid *Allium cepa* x *A. fistulosum* and its polyploidy derivatives. *Hereditas*, **27:** 253-272

8. Maeda, T. 1937. Chiasma studies in *Allium fistulosum*, *Allium cepa* and their F_1, F_2 and backcross hybrids. *Japan J. Genet.*, **13:** 146-159.

9. Mangum, P.D. and Peffley, E.B. 2005. Central cell nuclear-cytoplasmic incongruity: a mechanism for segregation distortion in advanced backcross and selfed generations of [(*Allium cepa* L. *A. fistulosum* L.) *A. cepa*] interspecific hybrid derivatives. *Cytogenetic and Genome Research*, **109:** 400-407.

10. Peffley, E.B. 1986. Evidence for chromosomal differentiation of *Allium fistulosum* and *A. cepa*. *J. Am. Soc. Hort. Sci.*, **111:** 126-129.

11. Peffley, E.B., Corgan, J.N., Horak, H.G. and Tanksley, S.D. 1985. Electrophoretic analysis of *Allium* alien addition lines. *Theor. Appl. Genet.*, **71:** 176-184.

12. Shigyo, M., Imamura, K., Iino, M., Yamashita, K. and Tashiro, Y. 1998. Identification of alien chromosomes in a series of *Allium fistulosum* L. - *A. cepa* L. monosomic addition lines by means of genomic *in situ* hybridization. *Genes Genet Syst.*, **73:** 311-315.

13. Ulloa, M., Corgan, J.N. and Dunford, M. 1995. Evidence for nuclear-cytoplasmic incompatibility between *Allium fistulosum* and *Allium cepa*. *Theor. Appl. Genet.*, **90:** 746-754.

14. Van der Valk, P., Evernik, J.T., Verstappen, F and de Vries, J.N. 1991a. Pre and post fertilization barriers to backcrossing the interspecific hybrid between *Allium fistulosum* L. and *A. cepa* L. with *A. cepa*. *Euphytica*, **53:** 201-209.

15. Van der Valk, P., Kik, C., verstappen, F., Evernik, J.T. and de Vries, J.N. 1991b. Independent segregation of two isozyme markers and inter-plant differed Vries, S.E., ences in Nuclear DNA content in the interspecific cross (*Allium fistulosum* L. x *A. cepa* L.). *Euphytica*, **55:** 151-156.

Great Headed Garlic (*Allium ampeloprasum* var. *ampeloprasum* L.)

The genus *Allium* represents a large diversity with respect to morphology of bulbs, nature of bulbing and appearance of foliage. Almost all the species of *Allium* are found in Mediterranean region, the adjoining areas of Western Asia and South-eastern Europe (USDA GRIN, 2011). In Indian sub-continents, two species of *Allium* *viz.*, *A. cepa* L. and *A. sativum* have been under cultivation since ancient times. A few decades back some other species *viz. A. fistulosum* L., *A. porrum* L., *A. schenoprasum* L. *etc.* have been introduced in this country. *Allium ampeloprasum* comprises 5 different *Alliums* species such as leek, kurrat, Taree, great headed garlic and pearl onion group (Brewster, 1994) which are interfertile in nature. Leek, Kurrat, and Taree are allotetraploid *Allium* species (2n=4x=32) whereas, great-headed garlic is hexaploid in nature with 48 chromosome. It is also known as elephant garlic or Levant garlic in India or abroad and in India its occurrence is confined to restricted areas of Himalayas. *Allium ampeloprasum* species is reported to be under cultivation in South-Western England and France (Bohance *et al.*, 2005). Some *Allium* species producing umbels consisting of numerous bulbils identical to small cloves, have been identified as *A. ampeloprasum* but all the varieties of *A. ampeloprasum* essentially do not have bulbils (Block, 2010). However, some taxonomists identified such species bearing aerial bulbils as *Allium* species (Bohance *et al.*, 2005; Hirscheggar *et al.*, 2006). Block (2010) identified *A. tricoccum* and characterized it as broad leaved, monocloved wild leek, the progenitor of three cultivated types *viz.*, leek or garden leek (*A. porrum*), elephant or great headed garlic (*Allium ampeloprasum* var. *ampeloprasum* L.) and kurrat or middle-eastern leek (*A. kurrat*). The controversial taxonomic position of different phenotypes of alliums many times creates confusion in identification of certain form of bulb-bulbil- and clove-bearing *Allium* species. According to Brewster (1994), *A. ampeloprasum* does not normally produce bulbs, although the variety *ampeloprasum* (elephant garlic) has been found to produce a head of cloves like garlic (*A. sativum*) but with two sizes of cloves, bigger cloves like garlic, surrounded by smaller ones, on the outside. Uttarakhand University of Horticulture and Forestry, Ranichauri Campus, Tehri-Garhwal, Uttarakhand made several collections of *A. ampeloprasum* from different villages of Tehri, during 2012 and identified one important line 'IC0598236' and named it as great headed garlic (*Allium ampeloprasum* var. *ampeloprasum* L.) using classification base of Brewster (1994, 2008) which has been under cultivation in farmers field since long as an admixture with *A. sativum* (Mishra, 2016).

Taxonomical Description

Plants are foliaceous with green, and flattened. The plants exhibiting identical growth habits bearing long, dark green, broad and leathery leaves (46.0 cm long; 2.8 cm wide and 10-12 leaves in the whole life span), pink umbles with micro-cloves (bulbils) and big size of underground bulbs and cloves. The central growing portion of shoot develops into a flowering stalk bearing terminal umbel at crop maturity during spring-summer. The umbels consist of tiny anthers, arising from the base of ovaries and developing into micro-cloves (bulbils in the case of common garlic); by storage of food material. The reproductive system in this species is non-functional. The micro-cloves have meristematic tissues differentiating inro roots and shoots. Each umbel bears about 150-175 micro-cloves and one gram of it has 28-32 micro-cloves. The viability of micro-cloves varies from 10-12 months under ideal storage conditions. The identified line 'IC0598236' (family Alliaceae) bears pink umbels at plant maturity large bulbs and cloves. The umbels consist of small cloves or bulbils which can generate new plantlets.

Horticultural Attributes

On the basis of comparative performance the prominent key attributes are as shown in Table 9.38.

The great headed garlic line 'IC0598236' is tolerant to foliar diseases like purple blotch and downy mildew. However, incidence of root rot is realized during prolonged, and high soil moisture conditions (Table 9.39).

On the basis of of morpho-agronomic performance, reactions to biotic and abiotic stresses and, bulb yield and related traits which are beneficial characteristics, great headed garlic (*Allium ampeloprasum* var. *ampeloprasum* line 'IC0598236' holds the potential as a commercial crop in Uttarakhand.

Table 9.38: Comparative Performance of Great Headed Garlic (*Allium ampeloprasum* var. *ampeloprasum* L.) and Cultivars of Common Garlic (*A. sativum*)

Genotypes	Days to Maturity		Bulb Weight (g)		10 Clove Weight (g)		Total Soluble Solids in Cloves (per cent)		Dry Matter Content in Bulb (per cent)		Marketable Bulb Yield (q/ha)	
	2012-13	2013-14	2012-13	2013-14	2012-13	2013-14	2012-13	2013-14	2012-13	2013 14	2012-13	2013-14
A. ampeloprasum (IC0598236)	180	180	65.5	71.2	61.3	63.8	41.5	41.0	46.5	48.2	248.8	230.7
A. sativum L. (Agrifound Parvati)	170	175	30.2	27.8	29.8	31.4	39.0	38.5	45.6	46.7	168.9	179.5
A. sativum L. (Ghansali Local)	185	180	35.8	33.6	35.4	36.5	43.5	44.0	56.4	56.05	108.7	114.6
SE (m)	1.8	2.2	0.93	1.2	1.85	1.7	1.2	1.3	0.9	0.76	6.4	9.2
CD (0.05)	3.9	4.7	2.0	2.6	3.96	3.7	2.6	2.8	1.9	1.6	13.7	19.73
CV (per cent)	5.0	7.3	11.6	13.4	8.6	10.8	4.5	5.6	7.2	5.8	13.5	10.6

Source: Mishra, 2016.

Table 9.39: Reaction of Diseases in different Species/Cultivars

Genotypes	Incidence of Purple Blotch (per cent)		Incidence of Downy Mildew (per cent)		Root Rot (per cent)	
	2012-13	2013-14	2012-13	2013-14	2012-13	2013-14
A. ampeloprasum (IC0598236)	2.5	2.0	3.8	2.6	15.8	21.6
A. sativum L. (Agrifound Parvati)	23.9	19.5	12.4	11.9	8.7	13.5
A. sativum L. (Ghansali Local)	25.6	26.5	9.7	10.3	5.6	8.3
SE (m)	2.7	3.2	1.7	2.1	1.2	1.6
CD (0.05)	5.86	6.94	3.66	4.52	2.58	3.44
CV (per cent)	16.8	13.5	9.2	8.6	12.3	13.2

Source: Mishra, 2016.

Interspecific Hybridization

Interspecific hybrids between *Allium cepa* and *Allium ampeloprasum* L. were generatd with the aim of intorduce S-cytoplasm from onion to leek. Few attempts of sexual hybridization of onion and leek failed to produce F_1 progeny (Dole el *et al.*, 1980; Currah, 1986). But use of *in vitro* culture to rescue the F_1 embryos at the early stages of development (7-14 days after pollination) was successful (Peterka *et al.*, 1997). Seven hybrid plants obtained were triploids and expressed 24 chromosomes. When the hybrids plants were used for transferring CMS through backcross progenies (Peterka *et al.*, 2002) chromosome composition varied and it was found that the BC_1 progeny plants had eight chromosome from onion and 30-33 from leek. In the following generations *viz.* BC_2 and BC_3, the transmission rate of univalent onion chromosome was decreased (varied from 0-8 in BC_2) which resulted into high per cent of alloplasmatic leek plants. In another study, biased transmission of organelle DNA in somatic hybrids between leek and onion was reported by Buiteveld *et al.* (1998). The somatic hybrids had rearranged mt DNA that predominantly contained mt DNA and cpDNA of leek.

REFERENCES

1. *Allium*, USDA GRIN 2011. "*Allium ampeloprasum* L." Royal Botanic Garden, Kew, Richmond, Surrey, UK, p. 350.
2. Block, E. 2010. *Garlic and other Alliums*: The lore and the science. Royal Society of Chemistry, Cambridge, U.K., p. 454.
3. Bohance, B., Jakse, M. and Sesek, P. 2005. Genetic characterization of an unknown Chinese bulbous Leek-like accession and its relationship to similar *Allium* species. *Hort. Sci.*, **40:** 1690-1694.
4. Brewster, J.L. 1994. *Crop Production Science in Horticulture*. Vol. 3. *Onions and other Vegetable Alliums*.CABI Intl., Wallingford, Oxfordshire, UK, p. 236.
5. Brewster, J.L. 2008. Crop Production Science in Horticulture. Vol. 15. Onions and other Vegetable *Alliums*.CABI Intl., 2nd edition Wallingford, Oxfordshire, UK, p. 432.
6. Buiteveld, J., Suo, Y, Van lookeren Campagne, M.M. and Creemers-Molenaar, J. 1998. Production and characterization of somatic hybrid plant between leek (*Allium ampeloprasum* L.) and onion (*Allium cepa* L.). *Theor. Appl. Genet*, **96:** 765-775.
7. Currah, L. 1986. Leek breeding: A Review. *J. Hortic. Sci.*, **61:** 407-415.
8. Dole el, J., Nowak, E.J. and Lu ny, J. 1980. Embryo development and *in vitro* culture of *Allium cepa* and its interspecific hybrids. *Z. Pflanzenzüchtg*, **85:** 177-184.
9. Hirscheggar, P., Galmarini, C. and Bohance, B. 2006. Characterization of a novel form of fertile great headed garlic (*Allium* sp.). *Plant Breed.*, **125:** 635-637.
10. Mishra, A.C. 2016. IC0598236: A potential genotype of great headed garlic (*Allium ampeloprasum* var. *ampeloprasum* L.). *Indian J. Plant Genet. Resour.*, **29(1):** 75-78.
11. Peterka, H., Budahn, H. and Schrader, O. 1997. Interspecific hybrids between onion (*Allium cepa* L.) with S-cytoplasm and leek (*Allium ampeloprasum* L.). *Theor. Appl. Genet*, **94:** 383-389.

10

Genetic Resources of Tuber Vegetables and Crop Improvement

Introduction

In the context of climate change and low input high energy tropical tuber crops considered the most important source of food crops after cereals due to highest rate of dry matter producer per day and calorie contributors. The economically and socially important tropical tuber crops are Cassava (*Manihot esculenta*), sweet potato (*Ipomoea batatas*), yams (*Dioscorea alata, D. esculenta* and *D. rotundata*), aroids which include elephant foot yam (*Amorphophallus* spp.), taro (*Colocasia* spp.) and tannia (*Xanthosoma* spp.) and other minor tuber crops namely Chinese potato, arrow root, yam bean, *Canna etc.* Generally these tuber vegetable crops are propagated vegetatively with few exceptions.

Table 10.1: Distribution of Cultivated Tuber Crops in India

Common Name	*Scientific Name*	*Chromosome Number 2n=*	*Family*	*Vernacular Name*	*Regions of Importance in India*
Cassava	*Manihot esculenta* Crantz	36	Euphorbiaceae	Maracheeni kizhangu and Kappa (M); Maravalli kizhangu and Ezhalai kizhangu (T); Maraganasu (K); Karrapendalamu (TE)	Southern Region. Occasionally in North Eastern and Western Regions
Sweet potato	*Ipomoea batatas* (L.) Lam.	90	Convolvulaceae	Mitha alu and Shakarkand (H); Chini kizhangu and Madhura kizhangu (M); Shakkareivalli kizhangu (T); Genasu (K); Chelagada (TE); Ratalu (MR); Lal alu and Ranga alu (B)	Introduced and found all over India but mostly concentrated in Eastern U.P., Bihar, West Bengal and Odisha,
Greater Yam	*Dioscorea alata* (L.)	40, 60, 80	Dioscoreaceae	Pind Aaluk (H); Kachil and Kavithu (M); Peruvalli kizhangu and Vetrilaivalli kizhangu (T)	South and North Eastern Region
White Yam (African yam)	*Dioscorea rotundata* (Poir.)	40	Dioscoreaceae	Safed Aaluk (H); African kachil (M); African valli kizhangu (T)	Introduced to India
Lesser Yam	*Dioscorea esculenta* (Lour.) Burk	30-100	Dioscoreaceae	Kayu (H); Cheruvalli kizhangu, Cherukizhangu, Nana kizhangu, Mukkizhangu (M); Siruvalli kizhangu (T)	South, N.E. and Eastern Region
Potato Yam	*Dioscorea bulbifera* var. *sativa*	40, 60	Dioscoreaceae	Mekkachil and Erachikachil (M)	Southern, North East and Eastern Region

Common Name	Scientific Name	Chromosome Number 2n=	Family	Vernacular Name	Regions of Importance in India
Taro	*Colocasia esculenta* (L.) Schott	28, 42	Araceae	Arvi, Kachalu and Ghuiya (H); Kachu (S); Chempu (M); Seppan- kizhangu (T),Kachchi (K); Shamagadde (K); Chamadumpa and Chemagadda (TE); Alu (MR); Kachu (B)	Throughout India with greater diversity in North East, Eastern region and South
Giant taro	*Alocasia indica* (roxb.) Schott		Araceae		Southern and North Eastern Region
Tannia	*Xanthosoma sagittifolium*	26	Araceae	Palchempu (M)	South and North Eastern region
Elephant Foot Yam	*Amorphophallus paeoniifolius* (Dennst.)	28	Araceae	Zamin-kand and Gimmikand (H); Arsaghna and Balukand (S); Chena (M); Karnai-kilangu (T); Suvama gadde (K); Kanda (TE); Suran (MR); Ol (B)	South, North East and Eastern Region
Chinese potato	*Solenostemon rotundifolius* (Poir.) J.K Morton *Plectranthus rotundifolius*	60	Labiatae	Koorka kizhangu (M); Sim kizhangu (T)	Soputhern parts of India
Yam Bean	*Pachyrrhizus erosus* (L.) Urban	26		Misri Kand (H); Pachi kizhangu (M)	North Eastern Region and Bihar
Moghania	*Moghania vestita* (Benthex Bak)		Fabaceae		North Eastern Region
Winged Bean	*Psophocarpus tetragonolobus* (L.) D.C.	26	Fabaceae	Chadhura payar and Goa payar	South and North East
West Indian Arrow root	*Maranta arundinaceae* L.	48	Marantaaceae	Koova (m)	Adapted to plain areas with high rain fall; shade loving
Queensland arrow-root or Purple arrow-root	*Canna edulis Ker-Gawier*		Cannaceae	-	Introduced, adapted to high rain-fall Region
H-Hindi, **S**-Sanskrit, **M**-Malayalam, **MR**-Marathi, **TE**-Telugu, **T**-Tamil, **K**-Kannada, **B**-Bengali, **NE**-North East					

Source: Edison *et al.*, 2007.

No doubt, tuber crops become an important component in the dietary habits of small and marginal farmers. Still today, tribal populations are heavily depends on tuber crops for their food security and medicinal requirements. Most of the species are ethnically important as a secondary staple and a delicacy for tribal people in the interior areas in Western Ghats. They are rich in dietary fiber and carotenoids *viz.* β carotene and anthocyanin. In industries tuber crops are used in the production of starch, preparation of stimulants, tonics, carminatives and expectorants. Many tuber crops are used in the preparation of stimulants, tonics, carminatives and expectorants. Tuber crops are associated with several rituals and beliefs. Planting of tuber crops in Kerala coincides with festival (Sivarathri) in February/March. The tuber crops are consumed as a part of religious rituals like Thiruvanthira and Karthika. Two under-utilized tuber species *viz. Tecca bipinnatifida* and *Theriophonum minutes* are very popular for their local food value and for the starch production which is used in confectionery items. Except in the industrial belt like Salem district of Tamil Nadu and East Godavari district of Andhra Pradesh, the tuber crops are cultivated and utilized by using indigenous technologies. The brief summary on Indigenous Technical Knowledge on tuber crops collected by scientists of ICAR-CTCRI Thiruvanathapuram (Kerala) is outlined in (Table 10.2).

The Indo-Burmese region is the centre of origin of aroids and Asiatic edible Yams. India possess a rich genetic diversity of tropical root and tuber crops like Cassava, Sweet potato, Aroids, Yams and several minor tuber crops. There are five major areas of distribution of root and tuber crops in India. These are:

(i) South-Western Hilly and Coastal Region
(ii) Southern Peninsular Region
(iii) Eastern Coastal Region
(iv) North-Eastern Region and
(v) North-Western Region

The two hot spots of global biodiversity *viz.* North Eastern Himalayas and Western Ghats are particularly rich in wild relatives of tuber crops. The southern part

Table 10.2: Nutritional and Medicinal Properties of Tuber Crops

Sl.No.	Species	Indigenous Technical Knowledge for using tuber bearings species
1.	*Alocasia macrorrhiza* (L. Donn.)	Stem juice is believed to relieve scorpion and nettle sting, leaves are used as rubefacient for joint pains.
2.	*Alocasia indica* (Roxb.) Schott.	Has easily digestible starch, leaf juice is an astringent, rhizome acts as mild laxative and diuretic.
3.	*Alpinia galangal* Sw.	Rhizome used for curing catarrhal affections, fever and rheumatism
4.	*Amorphophallus dubius* Blume	Tuber cooked in butter milk and fried in ghee and used for treating dysentery. It is also used as a carminative, expectorant, and in the treatment of piles, rheumatism and haemorrhoids. Fermented juice of petiole is used for diarrhea.
5.	*Amorphophallus sylvaticus*	Fruits and seeds are crushed into a paste and used in the treatment against toothache, bruises and glandular enlargements.
6.	*Amorphophallus paeconifolius*	It used as a tonic and has restorative and carminative properties. Also used in the treatment of piles, dysentery and rheumatism.
7.	*Anaphyllum wightti*	Small tubers of 1.0 g size. Rare plant. Maximum tuber yield 5.0-10.0 g. Used for the treatment of piles.
8.	*Arisneria pulchrum*	Used for treatment of piles.
9.	*Asparagus felicious*	Root possesses astringent properties.
10.	*Asparagus racemosus* Willd	Roots are used as a refrigerant, demulcent, diuretic, aphrodisiac, galactogogue, anti-dysenteric and for dyspepsia. Medicated oil is recommended for rheumatic complaints.
11.	*Asparagus sarmentosus* Linn.	Roots are aphrodisiac and hence taken as a nourishing drink.
12.	*Asparagus officinalis* Linn.	Roots have demulcent, diuretic, laxative, cardiac sedative, tonic and aphrodisiac qualities. Also used in the treatment of jaundice. It contains the essential oil-asparagine and tyrosine which are used in commercial purpose.
13.	*Asparagus adscendens* Roxb.	Tubers are used as vegetable as it possessed colling and demulcent properties.
14.	*Asparagus gonocladus* Baker	Roots are aphrodisiac. It is boiled with oil and applied in cutaneous disease and treatment against gonorrhea.
15.	*Canna orientalis*	The tuber has diaphoretic, demulcent and diuretic properties. It is also a stimulant and hence used during fever and dropsy
16.	*Canna edulis* Ker-gawler.	Easily digestible starch similar to arrowroot starch
17.	*Ceropegia tuberosa*	Big tubers toxic to rats
18.	*Chlorophytum borivilianum*	Root is a vitalizer. Hence used as a health tonic, curative for pre and post natal problems, alternative to Viagra and for immunity improvement.
19.	*Coleus forskohlii* (Willd.) Briq.	Root contains Forskolin which is a bronchodilator and used as a cardiotonic in the treatment of congestive heart failure, glaucoma therapy, anti-hypertensive, remedy for metastalic condition and thrombosis
20.	*Coleusamboinicus* Lour.	Juices of the leaves contain an aromatic carminative are used in treatment of urinary disease, dyspepsia, chronic coughs, asthma, cuts and wounds.
21.	*Colocasia esculenta* (L.) Schott	Juice of the petiole is used as a stimulant and rubefacient. The corm juice is used against scorpion stings. *Colocasia* starch is used to prepare weaning baby food and cosmetics.
22.	*Corollocarpus epigens*	Used as medicine to cure skin disease. Locally known as 'Kollan Koval'
24.	*Costus speciosus* (Koenig) Sm.	Roots used as antidote for snake bite and wounds, the rhizomes contains depurative and anthelmintic and purgative properties and used as tonic. Root powder is used in the treatment of urinary disease, body pain, fever and snake bites.
25.	*Cucrcuma aromatica*	Rhizome is used as a tonic, carminative and also for treatment of bruises and sprains
26.	*Cucrcuma caesia*	Rhizome is used as a stimulant and carminative and also in the treatment of asthma, bruises and sprains.
27.	*Cucrcuma zedoaria Rosc.*	Rhizome is a stimulant, carminative demulcent and expectorant. Also applied on bruises sprains, pains, dyspepsia, cold cough and bronchitis. Oil is used as an anti-irritant.
28.	*Cucrcuma amala*	Rhizome is a carminative, and also used for stomach.
29.	*Cucrcuma malabarica* VelC. *zedoaria*	Collected from forest by tribal (Kadar-Sholayar forest) sold in market for starch as well as for making cheap paints. It contains easily digestible starch which has antifungal and antibacterial properties.
30.	*Cyanotis tuberosa*	Nearly 99.0 per cent watery. Used to prevent thirst.
31.	*Cyperus rotundus*	Staple food for wild folk. Broad spectrum of medicinal use.
32.	*Dioscorea deltoidea*	For production of diosgenin.
33.	*Dioscorea floribunda*	For production of diosgenin.
34.	*Dioscorea kumaonensis*	Tubers used in the treatment of arthritis and rheumatism.
35.	*Dioscorea pentaphylla* Linn.	Tubers are applied on swelling of joints and also used as a tonic to improve body immunity.
36.	*Dioscorea esculenta* (Lour.) Burk.	Tubers used for treatment of chest pain, nervous disorders and swellings.
37.	*Dioscorea hamiltonii* Hook.f.	Treatment of stomach ache.

Sl.No.	Species	Indigenous Technical Knowledge for using tuber bearings species
38.	*Dioscorea hispida* Dennst.	Tuber is used as an antidote in arrow poison.
39.	*Dioscorea opposittifolia*Thunb.	Kavalakizhangu is believed to be an excellent food tonic by tribals. Tubers are used in the treatment of swelling, scorpion sting and snake bites.
40.	*Dioscorea tomentosa* Koenig.	Edible tubers eaten after boiling.
41.	*Dioscorea wallichii*	Cooked tubers could be chewed and juice consumed for treatment of jaundice.
42.	*Drosera peltata*	Highly medicinal having red coloured tubers of 1-2 g. Seen around KMTR area during February. Used for the preparation of Thangabhasmam, a life drug.
43.	*Gloriosa superba* L.	Bears tubers. Red flowers. Seeds used for treatment of heartailments. Also, called as Kalappakizhangu, Kanmanippoovu.it is also used to kill intestinal parasites.
44.	*Hedychium spicatum* Buch. Ham.	Rhizome is stomachic, carminative, stimulative and used as tonic and for the treatment of dyspepsia
45.	*Hedychium coronarium* J. Koenig	Stem juice used against swelling. Rhizome is an antirheumaticsand is used as tonic and excitant.
46.	*Helianthus tuberosus* L.	Rhizome is a commercial source of levulose-sweetening agent for diabetic patients.
47.	*Ipomoea aquatica*	Root is purgative
48.	*Ipomoea batatas*	Tubers are laxative
49.	*Ipomoea carnea*	Root decoction reduces blood pressure.
50.	*Ipomoea disitata*	Root applied on swelling of joints
51.	*Ipomoea hispida*	Root used in the treatment of rheumatism, headache, epilepsy, leprosy and ulcers
52.	*Ipomoea alba*	Root is purgative
53.	*Ipomoea paniculata*	Aromatic. Tubers used for enhancing milking.
54.	*Martanta aurandinacea* L.	Rhizomes contain easily digestible starch, which are used in making face powder and special glues. It is a rubefacient and also applied on wounds and used for the treatment of ulcers.
55.	*Nelumbo nucifera* Gaertn.	Rhizome used as tonic and in the treatment of diarrhea.
56.	*Pueraria tuberosa* (Roxb.) D.C.	Roots are demulcent, used as refrigerant in fevers, reduce swellings of joints and a lactagogue
57.	*Sagittaria sagittifolia*	Roots used in treatment of skin disease, leaves used as an antidote to snake bite and insect bite and relieve itching.
58.	*Scripus grossus* L.	Tubers are astringent, laxative, cooling and have diuretic properties. Used for arresting vomiting and diarrhea. It is also used as tonic.
59.	*Scripus kysoor* L.	Tubers are used against diarrhea and vomiting.
60.	*Scripus lacustris*	Rootstock is an astringent, and used as diuretic.
61.	*Scripus maritimus*	Roots are astringent and diuretic.
62.	*Scripus tuberosa*	Roots are used as laxative.
63.	*Stephania wightii*	Very big tubers, 6-7 kg, used in treating leprosy.
64.	*Tecca integrifolia*	Tubers used in making tonic and in the treatment, diarrhea and dysentery.
65.	*Tecca leontopetaloides* (L.) Kuntze.	Tubers used in the treatment of piles, rubefacient, diarrhea and dysentery.
66.	*Theriophonum divaricatum*	Tubers are used against diarrhea.
67.	*Theriophonum minutus* (Willd.) Baill.	Dried and used after steaming
68.	*Typhonium roxburghii*	Tubers used in curing skin eruptions
69.	*Typhonium trilobatum*	Tubers contains a stimulant and is used in the treatment of piles, stomach complaints, haemorrhoids *etc.*
70.	*Canna indica*	Seed set during February
71.	*Ceropegla tuberosa*	Big tubers toxic to rats
72.	*Cissus setosa*	Starchy, acrid, local name 'Piranda'
73.	*Curculigo orchioides*	'Nilappanakkizhangu' the tubers used for imparting strength

Source: Edison *et al.*, 2007.

of the Indian sub-continent especially the Western Ghats comprising forests, mountain ranges, the mid lands and coastal areas distributed in the states of Kerala, Tami Nadu, Karnataka and Maharashtra, is the abode of a large bounty of genetic wealth of root and tuber crops (Edison *et al.*, 2005), of which *Tacca bipinnatifida*, *Asparagus* species, *Aponogeton* species and *Ceropegia* species are also used as vegetables. The Indo-Burmese centre probably contains the ancestral species of cultivated Asiatic yams and taro. It is estimated that these biodiversity hot spots are undergoing exceptional loss of habitat resulting in the erosion or even the loss of precious endemic species. Major portion of the genetic diversity of wild species and land races existing in the tropical forests and remote villages are yet to be explored, collected and conserved. The North-eastern hill region of India, abundant with tropical forests is another site of genetic diversity of taro and yams. Introduced crops like cassava and sweet potato also have developed several morphotypes, land races and farmer's varieties due to natural hybridization and conscientious selection by the farmers who cultivate these crops traditionally (Edison *et al.*, 2007). Considerable variability is noticed in edible yams (*D. pentaphylla*, *D. oppositifolia*, *D. wightii*, *D. spicata* and *D. belophylla*) that are endemic to these areas. Henry *et al.*, 1989, recorded the *Dioscorea* sp. distributed in Tamil Nadu. In *Colocasia*, maximum diversity was noticed in north-east followed by Kerala and North-western region. In elephant foot yam, maximum diversity has been reported from Western Ghats. Seven wild species such as *Amorphophallus hohenackeri*, *A. commutatus*, *A. paeoniifolius* var. *dubius*, *A. bulbifera*, *A. smithsonianus*, *A. bonoccordensis* and *A. sylvaticus* were located in Western Ghats in Kerala and Karnataka. The taro germplasm included cultivated and stoloniferous wild tannia present in Western Ghats. In *Coleus*, four species *viz.* *C. aromaticus*, *C. forskholii*, *C. zeylanicus* and *C. spicata* (wild or semicultivated) were present in coastal Konkan region of Maharashtra. At international level, European Union aided 'International Network for Edible Aroids (INEY)' is being continued in twenty different countries including India to develop 'taro' adaptive to climatic commercial changes, food nutrition and livelihood.

Current Status of Germplasm of Tuber Crops in India

India has trusted the responsibility to the ICAR-Central Tuber Crops Research Institute (CTCRI), Thiruvananthapuram (Kerala) for collection of biodiversity of tuber crops (cultivated and wilds) from all over India, its improvement and conservations in the year 1963. At this institute the gene bank was subsequently enriched through international exchange of germplasm from CIP, CIAT, IITA, AVRDC, *etc.* In the year 1980, the total holding of tuber crops was about 2000 but with the support of Jai Vigyan Programme of the National Agricultural Technology Project on Plant Biodiversity (NATPon Plant Biodiversity) since 1998-99 and the special efforts by the Indian Council of Agricultural Research N. Delhi was made for exploration of plant biodiversity including tuber crops. Under this project several explorations were made in several unexplored regions in Chhattisgarh, Jharkhand, Madhya Pradesh, Odisha, Tripura, Mizoram and Andaman and Nicobar Islands. Collections were also made several times from the evergreen Western Ghats. Under this project more than 1000 accessions of different tropical tuber crops were collected. In addition, more than 700 accessions were acquired through International Consultative Group on Agricultural Research (CGIAR) support during 1990s. Area under cultivation under tuber crops in India is 4.0 lakh hectare of cassava and sweet potato besides approximately 2.0 lakh hectare in elephant foot yam, *Colocasia*, *Xanthosoma*, *etc.* Under the Jai Vigyan Programme of the National Agricultural Technology Project (NATP) on Plant Biodiversity, efforts were taken to create awareness among farmers, tribals *etc.*, for the collection and conservation of genetic resources of all the vegetable crops including tuber crops. In the Western Ghats biodiversity "Hot Spot' like Kalakkad Mundanthurai Tiger Reserve (KTMR) were repeatedly explored (4 explorations) for collecting tuber bearing vegetable crops. Areas like Wayanad Wildlife Sanctuary, forest ranges of Chalakudy, Sholayar, Kulathuppuzha (Kerala), Kollur and Kudremukh (Karnataka) also subjected to exploration (Unikrishnan *et al.*, 2005). Different species of *Dioscorea* were collected from all along the Western Ghats.

Table 10.3: Exploration Sites of Western Ghats (1999-2004) for Tuberous Crops

Regions Explored	*Number of Collections*
Peppara	3
Senkottai, Manimuttar, Kodayar, Gundar, Sivagiri, Puliyankudi (KTMR)	24
Karayar, Papanasam, Manjolai (KTMR)	4
Karayar, Servilar, Kudamadi, Kalakkad, Mundanthurai, (KTMR)	37
Mundanthurai, Karayar, Papanasam, Veerapandya-puram (KTMR)	13
Anamalai, Malakkapara	3
Vazhachal Division, Vettilappara, Sholayar forest	35
Kulathuppuzha range, Onthupacha, Devagiri, Mahendragiri	12
Dakshin Kannad, Uttar Kannad, Udupi district	16
Kudremukh	33
Pechippara	10
Wayanad	4
Total	**194**

At the ICAR-Central Tuber crops Research Institute, Thiruvanathapuram (Kerala), 4840 accessions of different tuber crops germplasm are being maintained as field gene bank besides 598 accessions at the Regional Centre, Bhubaneswar (Odisha).

Table 10.4: Current Status of Germplasm of Tuber Crops in India

Crop	*ICAR-CTCRI, Kerala*	*ICAR-CTCRI, Regional Centre*	*ICAR-NBPGR Thrissur*	*ICAR-NBPGR, N. Delhi*	*Centres of AICRP (TC)*	*Total*
Cassava	1693	33	184	-	569	2479
Manihot species	27	-	-	-	-	27
Sweet potato	927	243	-	260	1858	3288
Ipomoea species	6	78	-	-	-	84
Greater Yam	318	44	195	-	211	768
White Yam	258	-	-	-	-	258
Lesser Yam	125	-	68	-	-	193
Dioscorea species	82	-	184	44	-	310
Taro	1043	120	487	49	834	2533
Colocasia (wild)	7	-	-	-	-	7
Tannia	49	-	-	3	58	110
Elephant foot yam	85	32	56	-	120	293
Amorphophallus species	5	-	95	-	-	100
Chinese potato	88	1	50	1	-	140
Coleus species	8	-	-	-	-	8
Costus	2	-	-	-	-	2
Curcuma species	11	-	-	-	-	11
Canna	5	1	7	-	-	13
Arrowroot	7	1	5	-	-	13
Yam bean	63	45	-	-	-	108
Alocasia	1	-	3	1	-	5
Other tuberizing species	27	-	-	-	-	27
Typhonium species	3	-	-	-	-	3
total	**4840**	598	1334	358	3650	**10780**

Source: Edison *et al.*, 2007, 2017.

REFERENCES

1. Edison, S., Velayudhan, K.C., Amma C.S.E., Pillai, S.V., Mandal, B.B., Sheela, M.N., Vimla, B., Unnikrishnan, M. and Hussain, Z. 2005. *Tropical Root and tubers In: Plant Genetic Resources: Horticultural Crops.* (eds.) Dhillon, B.S., Tyagi, R.K., Saxena, S. and Randhawa, G.J. Narosa Publishing House, New Delhi, p. 228-250.
2. Edison, S., Unnikrishanan, M., Vimala, B., Pillai, S.V., Sheela, M.N., Sreekumari, M.T. and Abrahan, K. 2007. Biodiversity of tropical tuber crops in India. *Biodiversity in India: Issues and Concerns* (eds. S. Kannaiyan and A. Gopalam). Associate Publishing Co., N. Delhi, p. 178-206.
3. Unnikrishnan, M., Edison, S., and Sheela, M.N. 2005. Tuber Crops and their Wild Relatives. In: R. Annamalai (ed.) Tamil Nadu Biodiversity Strategy and Action Plan. Published by Tamil Nadu forest Department, Chennai, p. 159-168.

Potato (*Solanum tuberosum* L.)

Introduction

Potato (*Solanum tuberosum* L., 2n=4x=48) is one of the four major subsistence food crops of the world, the other three being rice, wheat and maize, grown in nearly 150 countries in a wide variety of soil and climate. The potato is sometimes also called white potato or Irish potato and the potato tuber as spuds. The edible part of the potato plant is the tuber, the swollen ends of its underground stems (stolons). Potato is a versatile vegetable and is consumed in a variety of ways. It is an integral part of every vegetable culinary preparation. Potato stuffed 'Paratha' is the favourite on Indian breakfast menu. The most common ways of cooking potato are boiled, fried, baked, mashed, stewed and roasted. Potato is also processed into a variety of products like chips, flakes, French fries, flour, starch preparation *etc.* A potato tuber comprises 78.0 per cent water, 18.5 per cent carbohydrates and the rest are proteins, minerals and vitamins. The fat content is only 0.1 per cent. Though, potato is not a very rich source of protein (2.1 per cent), it is an important source of essential amino acids like lysine, methionine, cysteine, phenylalanine, tryptophan *etc.* The biological value of potato protein is 71.0 and that of the whole egg protein (Ross, 1986). Potato is also an important source of minerals like calcium, potassium, phosphorus, magnesium and iron and vitamins like B_1, B_2, B_6 and C. Daily consumption of an average sized potato provides 45.0 per cent of vitamin C, 10.0 per cent of Thiamin, 8.0 per cent of Niacin, 14.0 per cent of vitamin B_6, 14.0 per cent Folacin, 6.0 per cent phosphorus, 12 per cent magnesium and 9.0 per cent of iron requirement. Further, potato produces more dry matter and more calories per unit area of land and time than any other major food crop. Potato contains 1.7 per cent fibers (Woolfe, 1987). Being a member of the Solanaceae (nightshade family), potato was earlier considered unfit for human consumption. Worldwide, about 329.58 million tones of potatoes were produced during 2009 from an area of 186.52 million hectare with average productivity of 17.67 t/ha. Which has increased 3.2 per cent more in the year 2018.

Origin, History and Distribution

Potato is one of the oldest cultivated plants. It can be spotted in both temperate and tropical regions at elevations ranging from sea level to 40000 m msl. Based on the excavation in Mexico and elsewhere, the origin of potato is dates back to an age when maize was first cultivated in Mexico. It is originated in the the Andes of Peru and Bolivian region (near Lake Titicaca basin) in South America (Simmonds, 1976; Ross, 1986). Many related wild species (wild diploid ancestors) still occur in South America, specifically in the Andes of Peru and Bolivia. Till 16th century it was unknown to the people of Europe, Asia, Africa and North America. This crop was first domesticated in the high plateau region of the Peruvian and Bolivian Andes and has been grown for several millennia.

Distributionn of existing primitive cultivated potato, their wild species and related studies indicate that potato was 1st cultivated near Lake Titicaca basin in Peru-Bolvian region. The plant was selected as an article of food by Mayas, Incas and Red Indians in South and Central America. It was introduced into Europe by Spaniards (last quarter of the 16th century), independently 1st into Spain in 1570 (Salaman, 1949) and then into England in 1590. Potato was first introduced into North America from England via Bermuda in 1621 where, it was introduced in 1631. Peter the great brought potato into Russia from Holland at the end of 17th century. The potato's global voyage began in the 17th century. Till about the mid 18th century, it was not grown in any of the western European countries except Ireland where people relied heavily on potato as a source of food. Potato remained staple food of the Irish people till 1845 when late blight appeared in an epidemic form resulting in the 'Potato Famine of Ireland'. The famine resulted in the death of one million people due to starvation and diseases. Due to worse situation, another one million people including the now famous Kennedys and Reagans have migrated to the USA and Canada. Potato spread to Sri Lanka by late 18th or early 19th century. Captain James Cook introduced potato into New Zealand on his second expedition in 1753. The British and the German colonists introduced potatoes into East Africa in 1880. In North and West Africa, the army and the colonial personnel were forced to grow their own potatoes locally during the two world wars. Although Africa is not a major producer of potato in terms of volume, none-the-less, now more number of African countries grows potato than in any other continent. Potato cultivation is more popular in Europe and Latin America than in other parts of the world. This can be gauged from the fact that per capita per year potato consumption is 80.0 kg in Europe and 63.0 kg in Latin America. The average annual consumption in Asia and Africa is just 14.0 kg and 8.0 kg per year respectively, well below the world average of 28.0 kg. Potato is brought to India either by Portuguese or later by the British missionaries in the beginning of 17th century (Pushkarnath, 1976) from Europe. The British might have carried them to hills of north India. From the hills, potatoes spread to the plains and by the beginning of 20th century, small plots of potatoes were seen scattered throughout India near the towns. Potato was 1st cultivated in India at Nilgiri hills and potato had established throughout Indian sub-continent. Potato is grown in almost all the states in India except Kerala. In India main potato growing states are Uttar Pradesh, Bihar, and West Bengal which account 75 per cent of the total area and 80 per cent of the total production. Assam and Tamil Nadu ranked 4th and 5th for area and production respectively. The 1st scheme for potato seed production in India was started in 1941 at Shimla (Himachal Pradesh) under the aegis of Imperial Agricultural Research Institute, Pusa, Bihar now Indian Agricultural Research Institute (IARI), New Delhi. In August 1949, the ICAR-Central Potato Research Institute (CPRI) came into existence with headquarters at Patana (Bihar). The Institute's headquarter was later shifted to Shimla (Himachal Pradesh) in 1956. In total production India ranks 5th after China, Russia, Poland and Ukraine. The highest yield (75.0 tonnes/ha.) has been obtained in Belgium, the Netherlands and Sweden due to favourable climatic conditions. At present potato is cultivated to large scale in Russia, Germany, France, the USA, Britain, Spain, Belgium, the Netherlands, Sweden and Italy.

1570 Chile (*S. tuberosm*) ⟶ Europe ⟶ India
↑
South America ⟶ ⟶
↓

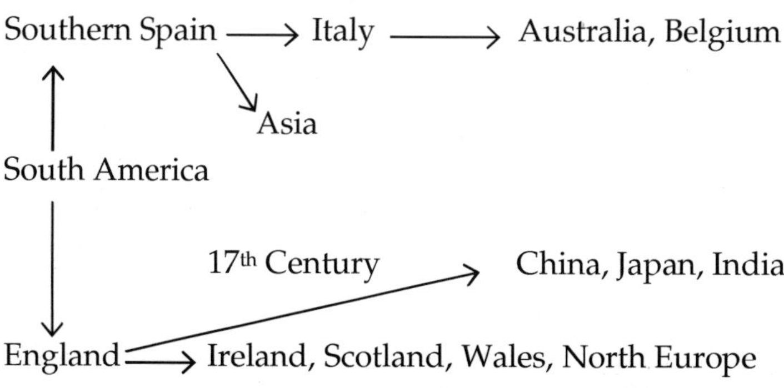

Details of the Spread of Crop are Indicated as in 1570 by Spaniards.

The wild potatoes have a wide geographical distribution and great range of ecological adaptation. Their distribution ranges from the south-western states of the United States of America through Mexico and into Guatemala, Honduras, Costa Rica and Western Panama at medium to high altitudes. They can also be spotted all along the Andes from Venezuela through Colombia, Ecuador, Peru, Bolivia and north-western Argentina, and even in the lowlands of Chile, Argentina, Uruguay, Paraguay and south-eastern Brazil. This great range of ecogeographical distribution underlines the extraordinary adaptive range of wild potatoes to diverse ecosystem from extreme cold to dry semidesert conditions and from cool temperate rain forests to coastal plains. In contrast to wild potatoes, cultivated potatoes have evolved under a very limited range of environmental conditions in cool temperate ecosystem (Pandey *et al.*, 2003). The distribution of wild tuber bearing *Solanum* species ranges from south western states of USA and into Guatemala, Hondiras, Cost Rico and western Panama. Some species are adapted to grow in the cold very high Andean regions from 3000 to 4500 m, where frosts are very common for example *S. acuale* Bitt. and *S. megistacrolobum* Bitt., whilst other occur in dry semi desert conditions for example *S. berthaultii* Hawkes, *S. tarijense* Hawkes and *S. neocardenasii* Hawkes. Due to their wide distribution and adaptations to extreme climatic conditions, wild species have become adapted to stress environments and have strong resistance to a wide range of pests and diseases. *S. ajanhuiri* (diploid), *S. phureja* (diploid), *S. stenotomum* (diploid), *S. chaucha* (triploid), *S. juzepckjukii* (triploid) and *S. curtilobum* (pentaploid) are also under cultivation but are restricted to the high Andes from Central Peru to Central Bolivia.

Species Availability, Evolution and Diversity

Potato belongs to a genetically diverse genus *Solanum*, which contains about 2,000 species of which 235 species are tuber bearing and belong to the section Tuberarium. The tuber bearing *Solanum* falls into a polyploidy series having a basic chromosome number of 12 and containing forms with diploids 2n= 2x=24 to 36, 48, 60 and 72 chromosomes (hexaploids). Chromosome counts have been made for only three fourth of the species. The commonly cultivated potato is an autotetraploid and belongs to the species *Solanum tuberosum* L. (2n=4x=48), which includes two subspecies, *viz.* ssp. *tuberosum* adapted to long days and ssp. *andigena* adapted to short days. About 73 per cent of the tuber bearing *Solanum* species is diploid, 4 per cent triploid, 15 per cent tetraploid and 6 per cent hexaploids. The cultivated potato possesses more number of related wild species than probably any other crop. The genomes

Table 10.5: Classification and Origin of the Cultivated Potatoes

Species	*Distribution*	*Origin*
Solanum tuberosum sub. sp. *tuberosum* (2n=48)	Cosmopolitan	By artificial selection in Europe, North America and Chile from introduced clones of group 2.
S. tuberosum sub. sp.*andigena* (2n=48)	Venezuela to northern Argentina; also sporadically in central America and Mexico	From groups 4 and 5 by spontaneous doubling of the chromosome number.
S. chaucha (2n=36)	Central Peru to northern Bolvia	By hybridization between groups 2 and 4 or 2 and 5.
S. phureja (2n=24)	Venezuela to Northern Bolvia	By selection for short tuber dormancy from group 5.
S. stentomum (2n=24)	Southern Peru to northern Bolvia	By natural hybridization between wild species followed by artificial selection.
S. x *Juzepczukii* (2n=36)	Central Peru to Southern Bolvia	From crosses of S. acaule with groups of 4 or 5.
S. curtilobum	Central Peru to Southern Bolvia	Crosses of *S.* x *Juzepczukii* (2n=36) with group 2

Source: Adapted from Ugent (1970).

in the *Solanum* species appear to be relatively similar and genome evolution has not progressed far. Main centres of diversity of tuber bearing *Solanum* species are Central America and Andean region of North-Western Argentina, Peru and Southern Bolivia. A total of seven cultivated species were domesticated, and they remain the most important crop in the high Andean, regions where the greatest genetic diversity of cultivated potato is concentrated (Engel, 1970). Two contrasting forms were observed, one at Chile and the other at Andean region of Peru and Bolivia. The former is botanically called *Solanum tuberosum* L. (*S. tuberosum* var. *tuberosum*) and the later *S. andigenum* (*S. tuberosum* var. *andigenum*). Once *S. tuberosum* subsp. *andigena* had been established as a tetraploid, it acted like a genetic sponge absorbing genes via multiple introgressions from wild or cultivated species, thereby constituting the tetraploid *andigena* complex. Pentaploids (2n=60) are grown in South America. Since they all have similar and common genetic background, they are all included under *S. tuberosum* (Simmonds, 1976). Potato seems to have evolved through geographical and ecological isolation. While the cultivated species were at one time confined to the Andes of South America and the lowlands of southern Chile, the related wild species were more widespread and adapted to cool temperature climates of these regions. The tetraploid *S. tuberosum* subsp. *andigena* was developd from diploid species *S. stenotomum* (2n=24), through sexual polyploidization involving an unknown diploid species, probably *S. sparsipilum* occurring at high altitude in central Peru to central Bolivia. It is believed to be the most primitive (Hawkes, 1994), having been derived from the so-called *brevicaule* complex, which contains a large number of diploid wild species including *S. brevicaule*, *S. leptophyes* or *S. canasense*, *S. sparsipilum* and *S. multidissectum* (2n=24) *etc.*, to name a few. It is also assumed that the *stenotomum* complex (containing also the diploid cultivated species of *S. phureja* occurring at lower altitudes in Andean valleys of Colombia, Ecuador, Venezuela and northern Peru) has evolved from *brevicaule* complex through selection during early domestication. Some researchers feel that at least four wild potato species (*S. acaule*, *S. sparsipilum*, *S. leptophyes* and *S. megistacrolobum*) are believed to be involved in the process of evolution. Evidence indicates that hybridization of *S. stenotomum* with the weedy species of *S. sparsipilum* either through 2n gamete or from normal haploid (n) gamete followed by subsequent chromosome doubling produced the tetraploid *S. tuberosum* subsp. *andigena* in the central Andes. This tetraploid subspecies was carried into southern Chile, where it became adapted to long photoperiod to evolve into subsp. *tuberosum*. Similar conditions produced identical results in Europe. The diploid wild species *S. magila* is considered as a putative candidate. Therefore, the evolution of cultivated potatoes occurred as a series of steps beginning at the Andean primary gene centre (Grun, 1990).This has been supported by the studies of restriction fragments of chloroplast DNA (Hosaka, 1986).

The largest genetic diversity of potato exists in Peru-Bolivian region (Ross, 1986; Hawkes, 1992). Distribution of wild species ranges from the south-western states of the USA and into Guatemala, Honduras, Costa Rico and Western Panama. Some species are adapted to grow in the cold, very high Andean regions from 3000 to 4500 m, where frosts are very common (*e.g.S. acaule* Bitt. and *S. megistacrolobum* Bitt.), whilst others occur in dry semi desert conditions (*e.g.S. berthaultii* Hawkes, *S. tarijense* Hawkes and *S. neocardenasii* Hawkes). Some are found in cool temperate rain forest (*e.g. S. violaceimarmoratum*

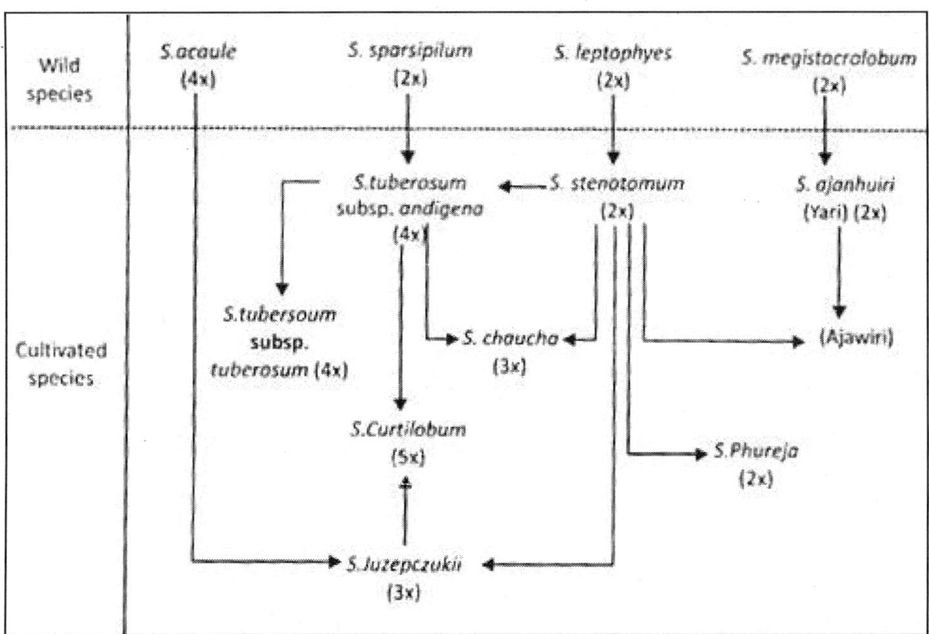

Figure 10.1: Evolutionary Relationship of Cultivated Potato and their Ploidy Levels (Hawkes, 1990).

Table 10.6: Cultivated Species of Potato

Ploidy	*Species*	*Area of Cultivation*	*Trait of Interest*
Diploids	*S. stenotomum*	South Peru and Bolivia	Frost resistant, yellow flesh, and good flavor
	S. goniocalyx	Mostly central and northern Peru	Yellow flesh, good flavor, now has been placed as a sub species of *S. stenotomum*
	S. ajanhuiri	High elevations in Peru and Bolivia	Frost resistant, and bitter
	S. phureja	South America, mostly at moderate elevations	No tuber dormancy, high dry matter, and genes for unreduced gametes
Triploids	*S. x chaucha*	Bolivia and Peru	Natural hybrids of *S. tuberosum* ssp. *andigena*
	S. x juzepczukii	High elevations of Bolivia and Peru	Natural hybrids of *S. stenotomum* and S. acaule, frost resistant, bitter
Tetraploids	*S. tuberosum* ssp. *andigena*	Andes of Venezuela, Colombia, Equador, Peru, Bolivia and North-west Argentina	Adaptation to short days conditions, variability for many traits, especially disease resistance and qualiy
	S. tuberosun ssp. *tuberosum*	Most widespread and cultivated in Europe, Asia and elsewhere	Adaptation to long days or short days conditions in subtropics, disease resistance, and good appearance
	S. hygrothermicum	Amazon basin of Peru	Adaptation to lowland tropics, said to be almost extinct
Pentaploids	*S. curtilobum*	High elevations of Bolivia and Peru	Natural hybrid of *S. tuberosum* ssp. *Andigena* and *S. juzepczukii*, frost resistant, and somewhat bitter

Source: Modified from Hoops and Plaisted, 1987.

Bitt. and *S. colombianum* Dun.), and others on the coastal plains of Argentina and surrounding countries (*S. commersonii* Dun. and *S. chacoense* Bitt.). Due to their wide distribution and adaptation to extreme climatic conditions, wild species have become adapted to stress environments and have strong resistance to a wide range of pests and diseases.

Cultivated diploid species *S. phureja* evolved from *S. stenotomum* through artificial selection by Andean farmers in lower and warmer eastern valleys and acquired shorter dormancy. Natural hybridization of *S. stenotomum* with frost-resistant species *S. megistacrolobum* gave rise to the diploid *S. ajanhuiri*. Similarly, a series of hybridizations between *S. stenotomum* and the wild tetraploid *S. acaule* gave rise to a highly sterile triploid *S. juzepczukii* and further natural cross between *S. juzepczukii* and *S. tuberosum* subsp. *andigena* produced the pentaploid *S. curtilobum*. The ploidy level of potato species varies from 2x to 6x (Hawkes, 1994) with 73 per cent of these being diploids, 4 per cent triploids, 15 per cent tetraploids, 2 per cent pentaploids and 6 per cent hexaploids. *S. hygrothermicum* is now said to be almost extinct (Ochoa, 1984). The modern cultivated potato is generally viewed as having an autopolyploid or segmental polyploidy origin.

The cultivated species either cross directly or through bridging species with several species belonging to other series. Somatic hybridization with conventionally non-crossable species has also been successfully carried out in potato. At present the germplasm repository of potato at ICAR-Central Potato Research Institute, Shimla (H.P.) has a total collection of 1400 *S. tuberosum* accessions. Creation of core set is one approach of hastening and prioritization of the germplasm characterization and ultimately its utilization for different traits specific breeding activities. Considering above, a core set of 140 accessions of *S. tuberosum* was extracted from 1400 accessions based on 19 morphological descriptors *viz.*, vegetative, tuber and sprout parameters. The validity of the core set by means and variances comparision and Shannon-Weaver diversity index and phenotypic correlations between different traits revealed that actual diversity of base collection have been conserved. The established core set forms the basis for wider exploitation of potato germplasm for effective utilization in potato improvement programme (Dalamu *et al.*, 2016).

Gene and Genetics of Characters

Gene and genetics of characters are mentioned below:

Table 10.7: Traits and Associated Gene in Potato

Character	*Number of Genes*	*Type of Gene Action*
Skin colour	Digenic (D-R-)	Complmentary D-R-red, D-rr-white, ddR-white, ddrr-white
Position of eye	Monogenic	Flat eye is incompletely dominant to deep eye
Flesh colour	Monogenic	Deep yellow colour is incompletely dominant to white, the intermediate is pale-yellow
Stolon length	Monogenic	Long stolon is dominant to short stolon
Tuber shape	Monogenic	Long tuber axis is dominant to short axis

Botanical Description

Potato belongs to family *Solanaceae*, genus *Solanum*, belongs to section *Petota* (*Tuberarium*) under subgenus *Potatoe*. Two sub-sections are recognized in section *Petota* (Hawkes, 1994). The ploidy level of potato species varies from 2x to 6x with 73 per cent of these being diploids, 4 per cent triploids, 15 per cent tetraploids, 2 per cent pentaploids and 6 per cent hexaploids (Hawkes, 1990). Besides these, six primitive species *viz. S. ajanhuiri* (diploid), *S. phureja* (diploid), *S. stenotomum* (diploid), *S. chaucha* (triploid), *S. juzepckjukii* (triploid) and *S. curtilobum* (pentaploid).

(i) The first subsection *Estolonifera* indicates the absence of stolen or tubers and contain two series namely *Etuberosa* and *Juglandifolia*. In series Etuberosa, species can be induced to form hybrids with the tuber bearing species, whereas species in *Juglandifolia* can not form hybrids and appear to have closer connections with the tomato genus *Lycopersicon* than with *Solanum*.

(ii) The second subsection Potatoe (*Hyperbasarthrum*) includes 19 series containing tuber bearing species which are of potential interest to breeders.

S. tuberosum L. which has 2n= 24, 36, 48 and a basic chromosome number of 12. *S. andigenum* Juz et. Buk. is an autotetraploid with 2n=48. The commonly cultivated potato is anautotetraploid or segmental polyploidy origin (Swaminathan and Howard, 1953) and belongs to the species *S. tuberosum* L. (2n=4x=48), which includes two subspecies *viz*. ssp. *tuberosum* adapted to long dauys and ssp. Andigena adapted to short days. Presence of male sterility with no pollen or few sterile pollen gives evidence for autoploidy nature of *S. andigenum*. Potato, a plant of temperate origin thrives well under long-day cool environment, where sufficient moisture and nourishment are available. Potato is long day plant.

Table 10.8: Botanical Description

Growth Habit	Potato is a vegetatively propagated crop but propagation through true potato seed (TPS) is common in breeding programme and commercial exploitation of TPS. In plants grown from tuber, buds on the tuber produce lateral stems giving rise to aerial stem. The aerial stem behaves as an annual but the plant is perennial because of its underground stem (dicotyledonous), which is specifically called as stolon. The young plants quickly developed stolon and produce small tubers. Under normal condition of growth of the plants are succulent and bushy but under rich fertile condition the plants produce heavy foliage mass. The plant height at flowering stage may be short, medium or tall and is quantified as <45 cm, 45-60 cm and >60 cm, respectively. The habit of the plant may be compact, medium compacty or open. **(i) Compact-**When the vegetative mass is closely packed, leaving very few gaps in between the leaves and stem. **(ii) Medium Compact-** The intermediate foliage mass may be described as medium compact. **(iii) Open-**When the leaves do not cover the entire stem and leave large gaps between themselves. The posture of the plant may be erect (stem with an angle of more than 45° with the ground), semi-erect or spreading (with an angel of 30 to 45°) or prostrate (when the stem trails on the ground). Vigour of potato plant is grades: vigorous, medium and poor. However, the plant vigour does not necessarily reflect the yielding ability of genotype. Plants with medium vegetative vigour may be high yielder than a high vigour plant.
Root	Adventitious root arising from the base of a sprout. It is basically tap root. Each sprout has its own independent root system and gives rise to independent shoot. One or several of such shoots arising out of the eyes of a seed piece constitutes a plant. Roots are rarely found on the stolons. The root zone usually spreads to 20-25 cm of the top layer of the soil. The lateral root system may be spreads up to 90 cm in rich soils. Some varieties possess deeper, bushy and fibrous root system producing long stolons and numerous small-sized tubers.
Stem	The potato stem consists of stem, leaves, stolons and tubers. Plants grown from tuber have numbers of main stems, while plants from true potato seed have one main stem. Since they develop into structures with reduced leaves occurring at the nodes, these are called stolons. The stolon develoed branches (laterally along the whole length of the stem) and branches differ in frequency, length and colour. Most of the commercial Indian potato varieties possess 6-8 thick stems andproduce sizable tubers and good yield. The variation occur in stems as to number (few, many), strength (hard, firm, week), solidity (solid, hallow), thickness (thick, medium, thin), appearance (round, angular), hairiness (slightly, moderately, very), nodes (straight or swollen), presence or absence of wing and its type (straight, wavy or crinkled) and the colouration (green or combination of green with other pigmentation-RP1, RP2, RP3, RP4, Salaman, 1926)
Leaf	The leaf of the potato is normally compound. Leaves are arranged spirally on the stem. It is pinnate with several pairs of leaflets arising in succession, along the midrib (rachis). Each rachis carries several pairs of lateral primary leaflets and a terminal leaflet. The leaves are attached to the stem through petiole. The leaf may be open, intermediate or close with stiff and erect or drooping habit. Leaflet may be attached directly (sessile) on the rachis or by means of small stalks known as petiolule. Folioles or small interjected secondary leaflets are also found in pairs on the rachis between successive pairs of lateral leaflets. Folioles are invariably present, but their number, shape and size vary with the varieties. A pair of small curved leaf-like structure, the stipules is invariably present at the junction, where base of the rachis meets the stem. The wings arise from the base of rachis and vary insize and waviness. Morphological variations are found with respect to leaflet number (few, medium, many), size (small, medium large), width (narrow, medium, broad), shape (narrow lanceolate, ovate- lanceolate, ovate or oval), colour (yellow green, grey green, dark grey green), hairiness (glabrous,glabrescent, pubescent), surface (smooth, wrinkled), glossiness (dull, glossy), texture (soft, medium, hard), fusion (distinct, fused) and the margin (entire, wavy).
Floral biology	The inflorescence of the potato may be simple or compound. (i) In simple inflorescence, the peduncle divides into secondary axis usually two in number, each of which forms a monochasial cyme. (ii) In compound inflorescence, the secondary axes divide again to form monochasial cyme. The difference between simple and compound inflorescence lies in the degree of subdivision of the main peduncle (short, medium or long) and may be green, red or purple with varying intensities.

	The poptato flowers are borne usually on long inflorescence. The inflorescence is attached to the stem through main stalk (peduncle). The flower stalk tip (pedicel) merges with into calyx. Normally pedicels bear a swollen joint where a pedicel joins the floral stalk and is known as articulation (abscission zone). A cool weather, moderate degree of humidity and long day environments are conditions favouring flowering. Potato flowers are bisexual and possess all the four essential part of a flower *i.e.* calyx, corolla, androecium (stamen) and gynoecium (pistil). Flowers are borne in compound terminal cymes. Individual flowers complete and perfect. The calyx is composed of 5 sepals united at their base to form a bell-shaped structure below the corolla. The free ends of the sepals vary in shape and size. The colour of calyx may be green, red, purple or in combination of these colours in patches or stripes. Calyx is persistent and remains attached with the berry till maturity. The crolla consists of petals joined at the base and forms a short tube with a flat five-lobed surface. Each lobe ends is a triangular point (acumen). The shape of the corolla is generally round (rotate). The colour of corolla may be white, light blue, blue, red or purple with different combinations and intensities. Androecium consists of five stamens. The stamens are composed of anther and filament joined to the corolla tube. Anthers connivent in a cone, longer than the filaments Anthers are generally fused in a conical structure or spread loosely enclosing the pistil. The colour of the anther varies from light yellow to deep orange. Pollen dehiscence is through the pores. Pollen grains are normally oval with truncated ends in cultivated varieties and somewhat elliptical in the wild species. The gynoecium of the flower is very well developed and pistil has three parts: (i) The basal swollen region (ovary) (ii) The middle thin long region (style) (iii) The flat tip (thestigma) The ovary is superior, bilocular with axile placentation. The style is an elongated portion of the pistil connecting the stigma and ovary. The style length may be longer, equal or shorter than the anther cone.
Anthesis and pollination	Potato generally flowers under long photoperiodic conditions when the crop is grown during summer in hills. Flowers open mostly in early morning. Self-pollination is the rule and cross-pollination is through wind and insect also. Anthers dehiscing by apical pores or short slits. Pollen germination is completed after 30 minutes of pollination and fertilization within 12 hours (Gopal, 1994).
Stolons	Stolons are lateral stems, which grow horizontly below ground from the buds of the underground stem. The stolons may be long, medium-long or short. Stolan has nodes and internodes and under favourable conditions develop into branches which grow more or less horizontal outwards. At the end of stolons, tuberization takes place. The tuber bearing stolons begin to develop extensive cell division takes place in the pith and much less in the outer corticle region. The vascular tissue in the tuber bulges outward due to this active division of the pith and is places very close to the peel. Growth results from the active division of the cells lying between the cortex and pith and this area becomes the chief starch containing region. The entire tuber is covered with a peel. The peel has a number of lenticels which become visible as white dots. Long stolons are not desirable and are common in wild species. The stolons are sometime branched and numerous, leading to many smaller tuber with late maturity and erect habit. Stolons are generally white in colour but occasionally bear a tinge of red or purple in patches near the heel ends of the tuber.
Tuber and tuber set	The edible part of the potato plant is the tuber, the swollen ends of its underground stems (stolons). The initian of young tubers at the tip of the stolons usually occurs 5-7 week after planting when the plants are 15-20 cm high. A tuber has two ends *i.e.* the heel or basal end attached to the stolon and the opposite end called the apical, rose or distal end. The potato tuber consists of dormant buds (eye) with detectable leaf scales (eye brows). Great variation in tuber exists withb regard to size (small, medium, large), shape (compressed, round, oval, round oval, oblong, long oblong), skin texture (smooth, russeted, scaled), skin colour (white, yellow, pale yellow, red) and in the presence and distribution of secondary skin colour. The skin (periderm) is an outer thin protective layer of the tuber.On exposure to sunlight for some days tubers normally turn green. The tuber skin also contains small lenticels to facilitate gaseous exchange during respiration.
Fruit	Fruit indehiscent, a berry develops after fertilization. The ovary developes into a berry type fruit often called the 'seed ball' or 'apple'. The shape of the berry is usually round but oval, conical or pear shaped berries are also found. The size of the berry varies from 2-4 cm in diameter depending upon age and variety. The colour of the berry is generally green but some varieties have pigmented berries. Berry has two chambers and contains 50-200 seeds depending upon the varieties.
Seeds	The seeds are embodied in the soft fleshy pulp within the berry and can be extracted easily from the mature berries after ripening. Extracted seeds from the berry are sexual seed, also known as botanical seed or true potato seed (TPS). Seed are flat-oval and small (1000 to 1500 seeds per gram). The shape of the embryo is generally curved like a 'U' and oriented towards the point of attachment to the placenta (hilum). The embryo has two poles *viz.* radicle and plumule. On germination the radicle gives rise to primordial roots and plumule to above ground plant parts.

Botanical Classification

Classification on the Basis of Species and its Utility

The commercial potatoes generally belong to two species:

(i) ***Solanum andigenum***: It is not very common. Plants have thin and long stem, small and narrow leaflets. Flowers are produced more profusely. It has long stolons and mostly coloured deep-eyed tubers.

(ii) ***Solanum tuberosum***: It is more common. Plants have shorter and thicker stem larger and wider leaflets.

In addition to two species, there are also some important species, which shows resistant to biotic and abiotic. These species are given in Table 10.9.

Table 10.9: Potato Species Resistant to Biotic and Abiotic

Species	*Utility*
Solanum demissium Lindl. (2n=72)	Resistance to late blight, virus A. virus Y.
S. stoloniferum Lindl. (2n=48)	Resistance to virus x, frost, Colorado beetle
S. vernei Bitt. et. Wittm (2n=24)	Resistance to two species of Heterodera nematode.
S. multidissectum Wawk (2n=24)	Resistance totwo species of Heterodera nematode.

Species	*Utility*
S. antipovczii Buk	Resistance to late blight
S. curtilobum Jug. Et. Buk	Resistance to frost
S. phureja Jug. Et. Buk	Non dormant type used in breeding and tolerance to high temperature.
S. chacoense Bitter	Non dormant type used in breeding.
S. anomalocalyx, S. jamessi Torr; S. saltense	Resistance to early blight

Classification on the Basis of Origin and Response to Day Length

Potato is divided on the bases of origin and response to day length.

Table 10.10: Distinctness between *S. tuberosum* and *S. andigenum*

Distinct Character	*S. tuberosum*	*S. andigenum*
Origin	Andean region	Chile region
Day-length response	Long day plants	Short day plants
Polyploidy	Many types of polyploids	Tetraploids
Distinct varieties	'Up-to-Date', 'Magnum Bonum'	All the desi varieties like 'Darjeeling Red Round', 'Phulwa' and 'Gola'

Classification on the Basis of Crop Duration

Potato is classified into early and late varieties based on crop duration. Characters of early and late varieties are given below:

Table 10.11: Classification on the Basis of Crop Duration

Character	*Early Potato*	*Late Potato*
Species	*S. tuberosum*	*S. andigenum*
Stolon length	Short	Long
Stem	Hollow	Solid
Folioles	Broader	Narrow
Stolon number	Less number	More number
Tuber colour	White	Coloured
Photoperiodism	Long days for tuber development	Short days for tuber development
Varieties	'Satha', 'Gola', 'Great Scot', 'Kufri Chandramukhi'	'Phulwa', 'Kufri Kisan', 'Kufri Safed'

Classification on the Basis of Tuber Maturity

The potato cultivars fall in 3 maturity groups *i.e.*

(i) Early maturity (75-80 days)

(ii) Medium maturity (90-100 days)

(iii) Late maturity (100-110 days)

Considering the tuber maturity following segments has been developed:

Table 10.12: Group-wise Maturity Months of Potato

Group	*Maturity Period*	*Varieties*
Early	2-2½ (months from sowing)	'Gola', 'Up-to Date', 'Great Scot', 'Kufri chandramukhi', 'Kufri Kasigaro', 'Kufri Chamatkar', 'Kufri Alankar'
Mid	2½-4½ (months from sowing)	'President', 'Kufri Sindhuri'
Late	4½-5½ (months from sowing)	'Phulwa', 'Kufri Kisan', 'Kufri Safed', Kufri Neela', 'Kufri Jeevan'

Table 10.13: Duration and Region-Wise Adaptability of the Indian Potato Varieties

Variety	*Crop Duration*			*Regions of Adaptability**				
	E	*M*	*L*	*NIH*	*NIP*	*NB and SH*	*PLT*	*SIH*
Kufri Alankar		√			√			
Kufri Anand		√			√			
Kufri Arun		√			√			
Kufri Ashoka	√				√			
Kufri Badshah		√			√		√	
Kufri Bahar		√			√			
Kufri Chamatkar			√		√			
Kufri Chandramukhi	√				√		√	
Kufri Chipsona-1		√			√			
Kufri Chipsona-2		√			√			
Kufri Chipsona-3		√			√		√	
Kufri Dewa			√		√			
Kufri Frysona		√			√			
Kufri Girdhari		√						√
Kufri Giriraj		√		√				
Kufri Himalini		√		√				
Kufri Himsona		√		√			√	
Kufri Jawahar	√				√			
Kufri Jeevan			√	√				
Kufri Jyoti		√		√	√	√	√	√
Kufri Kanchan		√				√		
Kufri Khasigaro			√					
Kufri Khyati	√				√			
Kufri Kisan			√		√			
Kufri Kuber					√		√	
Kufri Kuber	√		√					
Kufri Kumar								
Kufri Kundan		√						
Kufri Lalima		√			√			
Kufri Lauvkar	√						√	
Kufri Megha		√						
Kufri Muthu		√						√
Kufri Naveen			√					

Variety	*Crop Duration*			*Regions of Adaptability**				
	E	*M*	*L*	*NIH*	*NIP*	*NB and SH*	*PLT*	*SIH*
Kufri Neela			√					√
Kufri Neelamani			√					√
Kufri Pukhraj		√			√		√	
Kufri Pushkar		√			√		√	
Kufri Red		√			√			
Kufri Sadabahar		√			√			
Kufri Safed		√	√		√			
Kufri Shahils		√		√				
Kufri Sheetman		√			√			
Kufri Sherpa		√				√		
Kufri Sindhuri			√		√			
Kufri Surya	√				√		√	
Kufri Sutlej		√			√			
Kufri Swarna		√						√

Note: E=Early, M=Medium, L=Late, NIH=North Indian Hills; NIP= North Indian Plains; NB and SH= North Bengal Hills and Sikkim Hiolls; PLT= Plateau Region; SIH= South Indian Hills

Source: Naik et al., 2012.

Classification on the Basis of Time of Sowing

In potato, time of sowing differs in plains and hills of India. In plains potato is classified on the basis of time of sowing as:

Table 10.14: Season Base Planting Time

Season	*Months*
Monsoon crop	July-August
Early winter crop	Mid-September to mid-October
Main crop	Mid-October to mid-November
Seed crop	December-January

In hills potato is classified on the basis of time of sowing as:

Table 10.15: Avaialbility of Cultivated Varieties on the Basis Harvesting Group

Season	*Months*	*Number of Varieties Available*
Early crop	January-February	8
Main crop	March April	22
Late crop	July-August	16

Augmentation of Genetic Resources and Crop Improvement

After the establishment of ICAR-CPRI, Shimla (H.P.) in 1949, potato germplasm collection, conservation, evaluation, documentation and utilization received greater impetus. Concerted efforts to collect the genetic variability notwithstanding, only about 130 out of the 235 potato species exist in gene banks. An effort was made in 1940s to collect, characterize and catalogue the locally available potato varieties. Nearly 400 collections from different parts of the country were scrutinized on the basis of various morphological features and duplicate samples were eliminated. Potato Synonym Committee, National Institute of Agricultural botany, England, identified and characterized 16 collections as non-European variety (Desi or indigenous). Bersides, some of the known European variety adapted to long days and also survived in the hills of India.

Introduction and Clonal Selection

The early introduction of temperate long day European varieties could not adapt to the subtropical short day conditions of the Indian plains due to different agro-climatic conditions. Although a few varieties proved to be good yielders, the imported varieties could not survive the long hot summer storage periods leading to fast degeneration of seed stocks. The introductions thus made no impact on potato cultivation in subtropical Indian plains. After the establishment of the ICAR-CPRI, in 1949 at Patna and later with its headquarters at Shimla, collection, conservation, evaluation and its utilization, became a regular activity of the institute.

Table 10.16: Indigenous and Exotic Potato Varieties/Samples Assembled in India

Indigenous	*Salient Features*
Agra Red, chamba Red, Coonoor White, Coonoor Red, Darjeeling Red Round, Desi, Dhantauri, Gola Type A, Gola Type B, Gola Typo C, Phulwa, Phulwa Purple Splashed, Red LongKidney, Sathoo, Shan and Silbilati	Heat and drought tolerant, therefore cultivated predominantly in the Indian plains; tolerant to degenerative viruses; due to physiological advantage can be stored in country stores during hot Indian summers.
Exotic	*Salient Features*
Ally, Arran Counsal, Ben Cruachan, craig's Defiance, Dunbar Cavalier, Great Scot, Italian White Round, Late Carman, Magnum Bonum, Majestic, Northern Star, President, Raeburn's Gregor Cups, Red Rock, Royal Kidney and Up-to-Date	Long-day adopted; therefore suitable for the Indian hills only; multiplication was characterized with progressive accumulation of degenerative viral disease; physiological limitation on tuber storage and utilization in hot Indian summers.

Genetic resources are the collection of large number of genotypes representing genetic diversity of a crop species and its wild relatives. In potato many high yielding varieties with desirable characters have been introduced to India. Some successful introductions are: 'Magnum Bonum', 'Craigs Defiance' and 'Up-to-Date'. 'Kufri Red' is a clonal selection from 'Darjeeling Red Round'. 'K 1241' is another selection from 'Phulwa'. The exotic cultivars that have figured more frequently as parents of Indian varieties are Adina, Craigs Defiance, Ekishiraju, Katahdin, and Up-to-Date.

The state Agricultural Departments has also released some potato cultures these are: 'ON-1645', 'K-22', and 'G-4' in Uttar Pradesh; 'CO-Shimla' in Tamil

Table 10.17: Potato Germplasm Resource at ICAR-CPRI, Shimla

Description		*Number of Accessions*		
		Tuber	*in-vitro*	*Total*
Indigenous *tuberosum*				
	Cultivars bred at CPRI	44	44	44
	Advanced hybrids	86	17	86
	Indigenous varieties	51	22	51
	Indigenous samples	97	43	97
Exotic		1445	1550	1750
andigena		770	0	770
Wild and semi-cultivated species		110	972*	1082*
Total				3880

*True seeds, **118 species.

Table 10.18: Potato Varieties Developed through Clonal Selection and their Adaptation

Variety	*Parantage*	*Year*	*Tuber Characters*	*Area of Adaptation*
Kufri Red	Clonal selection from Darjeeling Red Round	1958	Medium, round, red, colour in cortex, medium deep eyes, flesh yellow	Released for cultivation in North Eastern Plains
Kufri Safed	Clonal selection from Phulwa	1958	Medium, round, white, deep and picked red-purple eyes, flesh white	Released for cultivation in North Indian Plains

Nadu and 'Gulmarg Special' and 'Gulmarg Queen' in Jammu and Kashmir. Some of these cultures *viz.* 'ON-1645' and 'K-22' are still in cultivation in small pockets of Uttar Pradesh and Bihar.

Hybridization and Selection

Hybridization relates to production of hybrid progeny following sexual mating of two parents identified based on desirable attributes. Potat is being tetraploid, heterozygous and vegetatively propagated crop, it offers a unique privilege of fixation of generated genetic variability in hybrid progeny for effecting selection. A few of the introductions have been used to develop hybrids and varieties. 'Hybrid D.N. 45' is developed through crossing of 'Katahdin' with 'President'. 'Kufri Kisan' is a multicross involving 'Ekishrazu' from Japan, 'Katahdin' from USA, 'Up-to-Date' from Scotland and 'Phulwa', a desi variety. 'Kufri Kundan' is a selection from 'Ekishrazu' x 'Katahdin'. 'Kufri Jyothi' is another selection from 'A 3069' x 'A 2814'.'Kufri Sindhuri', a selection from 'Kufri Red' x 'Kufri Kundan', has light red tubers and resistant to late blight and frost. 'Kufri Sakthi', is a sister clones from the cross cross 'Kufri Red' x 'Kufri Kundan' (Poehlman and Borthakur, 1968). 'Kufri Chandramukhi' is a selection from 'S 485' x 'Kufri Kuber' takes only 95 days from sowing to harvest. In potato heterosis has been exploited for earliness, tuber yield, tuber size and tuber weight. Selection is made in the F_1, clones Cl_1 and Cl_2 have been developed from 'Kufri Jyothi' x 'CP 1787'. There are certain old cultivars which were released earlier and proved worthiness with respect to yield and adaption after the independence.

About one third of the potato cultivars derived from *S. tuberosum* ssp. *tuberosum*, do not form berries. Since potato is propagated vegetatively, it is highly heterozygous for all the characters. Parents for hybridization are selected on the basis of characters to be incorporated and then fter hybridized seeds are collected. Desired clones are selected from thr progeny. Assessment is based on unreplicated plots of just a single plant per genotype in the seedling (F_1) generation and a few hills per genotype in first (F_1C_1) and second clonal (F_1C_2) generations.

Gopal and Kumar (2003) have suggested following procedure for an early and safe elimination of unproductive genotypes.

i. Seedlings of poor vigour may be rejected prior to transplanting in the field.
ii. Clones with undesirable tuber colour, tuber shape, eye depth and tuber cracking may be rejected from the seedlings stage onward.
iii. No rejection should be done on the basis of tuber yield, average weight or number of tubers in the seedling generation.
iv. Negative selection (rejection of the poor phenotypes) for tuber yield and tuber weight can be initiated from the first clonal generation onwards whereas, number of tubers can be considered for the rejection of undesirable types from second clonal generation onward.
v. In the subsequent generations, for simultaneous improvement in tuber yield its components, average tuber weight should be used as selection criteria after fixing a standard for the minimum number of tubers required in the selected types.

Interspecific Hybridization

In *Solanum* 150-200 species are available in the section tuberarium (tuber bearing species of *Solanum*).

Table 10.19: Potato Varieties Developed by ICAR-Central Potato Research Institute (CPRI), Shimla (H.P.)

Variety	*Parantage*	*Year*	*Tuber Characters*	*Area of Adaptation*
Kufri Kisan	Up-to-date x Sd.16	1958	Large, round, white, deep eyes with prominent eye brows flesh white. Tubers tend to crack under high dozes of nitrogen	Released for cultivation in North Indian Plains
Kufri Kuber	(*S. curtilobum* x *S. tuberosum*) x *S. andigenum*	1958	Medium, oval, tapering towards crownend, white, medium-deep eyes, flesh white. Poor keeping quality	An early variety released for cultivation in North Indian Plains and Plateau Region (Punjab, Bihar, and Maharashtra)
Kufri Kumar	Lumbri x Katahdin	1958	Medium, oval, tapering towards heel end, white, fleet eyes, flesh white	Released for cultivation in North Indian hills
Kufri Kundan	Ekishirazu x Katahdin	1958	Medium, round-oval, flattened, white, medium, deep eyes, flesh yellow	Released for cultivation in Himachal Pradesh and hills of Uttarkhand
Kufri Neela	Kàtahdin x Shamrock	1963	Medium, round, white, medium-deep eyes, flesh white	Released for cultivation in South Indian Hills
Kufri Sindhuri	Kufri Red x Kufri Kundan	1967	Medium, round, red, medium deep eyes, flesh dull white	Released for cultivation in North Indian Plains and plateau region
Kufri Alankar	Kennebec x ON 2090	1968	Large, oblong, white, fleet eyes, flesh dull white, tubers turn purple on exposure to light	Early variety released for cultivation in North Indian hills and plateau region of peninsular India
Kufri Chamatkar	Ekishirazu x Phulwa	1968	Medium, round, white, medium-deep eyes, flesh yellow	Released for cultivation in North Indian Plains
Kufri Jeevan	M 109-3 x Seedling 698-D	1968	Medium, oval, white, fleet eyes, flesh pale yellow	Released for cultivation in North Indian Hills
Kufri Chandra-mukhi	Seedling 4485 x Kufri Kuber	1968	Large, oval, slightly flattened, white, fleet eyes, flesh dull white	Released for cultivation in North Indian Plains and plateau region
Kufri Khasigaro	Taborky x Seedling 698-D	1968	Medium, round, round-oval, white, deep eyes, flesh pale yellow	Released for cultivation in North Eastern Hills
Kufri Naveen	3070-D (4) x Seddling 692-D	1968	Medium, oval, white, fleet eyes, flesh pale yellow	Released for cultivation in North Indian Hills
Kufri Neelamani	Kufri Kundan x 134-D	1968	Medium, oval, flattened, white, fleet eyes, tubers turn purple on exposure to light, flesh yellow	Released for cultivation in South Indian Hills
Kufri Sheetman	Craigs Defiance x Phulwa	1968	Medium, oval, white, fleet eyes, flesh dull white	Released for cultivation in North Western Plains
Kufri Muthu	3046 (I) x M 109-3	1971	Large, round-oval, white, medium deep eyes, flesh white	Released for cultivation in South Indian Hills
Kufri Deva	Craigs Defiance x Phulwa	1973	Medium, round, white, with purple splashes, deep and picked purple eyes, flesh white	Released for cultivation in North Indian Plains
Kufri Badashah	Kufri Jyoti x Kufri Alankar	1979	Large oval, white, fleet eyes, tubers turn purple on exposure to light, flesh dull white	Released for cultivation in North Indian Plains and plateau region
Kufri Bahar	Kufri Red x Gineke	1980	Large, round-oval, white, medium deep eyes, flesh white	Released for cultivation in North Indian plains
Kufri Lalima	Kufri Red x Ag-14 (Wis. X 37)	1982	Large to medium, round red, medium deep eyes, flesh white	Released for cultivation in North Indian plains
Kufri Sherpa	Ultimus x Adina	1983	Medium, round, flattened, white, medium deep eyes, flesh pale yellow	Released for cultivation in North Bengal Hills
Kufri Swarna	Kufri Jyoti x $(VTn)^2$ 62.33.3	1985	Large to medium, oval, white, fleet eyes, flesh white	Released for cultivation in South Indian Hills
Kufri Megha	SLB/K-31 x SLB/Z-73	1989	Medium, round, oval, white, fleet eyes, flesh white	Released for cultivation in North Eastern Hills
Kufri Jawahar	Kufri Neelamani x Kufri Jyoti	1996	Large to medium, oval, white, Medium eyes, flesh white	Released for cultivation in North Indian plains and plateau region
Kufri Ashoka	EM/C-1020 x CP1468 (Allerfuheste Gelbe)	1996	Large oval, white, fleet eyes, flesh white	Released for cultivation in North Indian Plains
Kufri Sutlej	Kufri Bahar x Kufri Alankar	1996	Medium, round-oval, creamy white, fleet eyes, flesh pale yellow	Released for cultivation in North Indian plains
Kufri Chipsona-1	Mex.750825 x MS/78-79	1998	Medium to large, oval, white, fleet eyes, flesh dull white	Released for cultivation in North Indian plains
Kufri Chipsona-2	F-6 x QB/B-92-4	1998	Medium, round, white, fleet eyes, flesh light yellow	Released for cultivation in North Indian plains

Variety	Parantage	Year	Tuber Characters	Area of Adaptation
Kufri Giriraj	SLB/J-132 x EX/A 680-16	1998	Medium to large, oval, white, fleet eyes, flesh white	Released for cultivation in North Indian Hills
Kufri Pukhraj	Craigs Defiance x JEX/B-687	1998	Large, oval, slightly tapped, yellow, fleet eyes, flesh yellow	Released for cultivation in North Indian plains and plateau region
Kufri Anand	Kufri Ashoka x PH/F-1430	1999	Large to medium, oval-long flattened white, fleet eyes, flesh white	North Indian Plains
Kufri Kanchan	SLB/Z-405 (a) x Pimpernel	1999	Medium, oval, red, eyes medium deep, flesh white	North Bengal Hills
Kufri Arun	Kufri Lalima x MS/82-797	2005	Medium to large, oval, red, shallow eyes, flesh light yellow	Released for cultivation in North Indian Plains
Kufri Pushkar	QB/A 9-120 x Spatz	2005	Medium, oval, white, shallow eyes, flesh light yellow	Released for cultivation in North Indian Plains
Kufri Shalija	Kufri jyoti x EX/A 680-16	2005	Medium, oval, white, eyes fleet, flesh dull white	Released for cultivation in North Indian Plains
Kufri Himalini	I-1062 x Bulk pollen	2006	Medium, oval, white, shallow eyes, flesh pale yellow	Released for cultivation in North Indian Hills
Kufri Chipsona-3	MP/91-86 x Kufri Chipsona-2	2006	Medium, oval, white, fleet eyes, flesh cream	Released for cultivation in North Indian Plains
Kufri Surya	Kufri Lauvkar x LT-1	2006	Medium, oval, light yellow shallow, flesh dull white	Released for cultivation in North Indian plains and plateau region
Kufri Sadabahar	MS/81-145 x PH/F-1545	2008	Medium, oval, white, cream, eyes shallow, flesh white	Released for cultivation in North Central Plains
Kufri Himsona	MP/92-35 x Kufri Chipsona-2	2008	Medium, round, white cream, shallow eyes, flesh cream	Released for cultivation in North Indian Hills
Kufri Girdhari	Kufri Megha x Bulk Pollen (10 genotypes)	2008	Medium, oval, white cream, shallow eyes, flesh white	Released for cultivation in Indian Hills
Kufri Khyati	MS/82-638 x Kufri Pukhraj	2008	Medium, oval, white cream, eyes medium, flesh cream	Released for cultivation in North Indian plains and plateau region
Kufri Frysona	MP/92-30 x MP/90-94	2009	Medium to large, long-oblong, eyes shallow, flesh white	Released for cultivation in North Indian Plains

Source: Naik *et al.*, 2012.

Crossability potential is high among species so interspecific hybridization has become a tool to develop new varieties. Genetically distant crosses with wild species are avoided, as these result in progenies with wild characters which are difficult to eliminate. The species *S. tuberosum* and *S. andigenum* are crossable. The interspecific hybrids have evolved into distinct crop varieties differing in duration from that of *S. tuberosum* and *S. andigenum*. Accordingly a mid-season group got evolved. Many varities have been evolved through interspecific hybridization in potato. 'Kufri Kuber' is a selection from *S. curtilobum* x *S. tuberosum*. Interspecific crossing between *S. demissum* and *S. tuberosum* has given a set of late blight resistant lines. An interspecific hybrid between *S. commersoniana* and *S.chacoense* has been found immune to early blight (*Alternaria solani*), and charcoal rot (*Macrophomina phaseoli*). Some successful attempts of distant hybridization in potato are as under:

- *Solanum tuberosum* x *S. andigenum*
- *S. tuberosum* x *S. stentomum*
- *S. tuberosum* x *S. sparsipilum*
- *S. tuberosum* x *S. demissum* (*via* bridge species)
- *S. curtilobum* x *S. tuberosum*
- *S. acaule* x *S.simplifolium*
- *S. demissum* x *S. rybinii*
- *S. phureja* x *S. bulbocastanum*

Interspecific sterility is common but due to vegetative mode of reproduction, crossability is enhanced by aids like tetraploidization of female parent, amphidiploidization, mixed pollination, spray of IAA on stigma, grafting of female parent on male parent and use of bridging species.

Use of Bridging Species

A number of species like *S. demissum*, *S. acaule*, *etc.* are carry desirable genes for disease and insect resistance but they will not cross readily with *S. tuberosum*. The F_1 developed through crossing between *S. demissum* and *S. tuberosum* is sterile, because of univalents during meiosis. By making use of *S. rybini* as a bridging species, meiosis could be made regular in the progenies and the commercial qualities of *S. tuberosum* maintained.

Somatic Hybridization

Somatic hybridization bypasses sexual reporoduction, and avoids crossing barriers. This technique is also useful for both compatible and

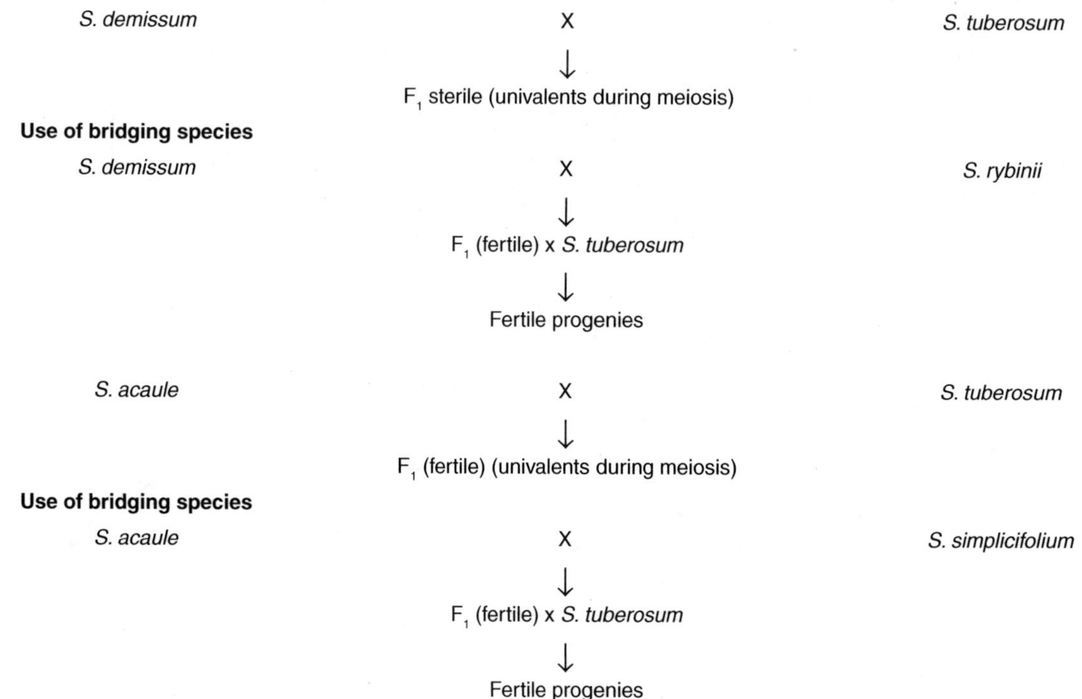

incompatible species. Both inter-generic fusion with other Solanaceae genera and intra-generic fusion with other *Solanum* species were extensively carried out in potato:

Chandel *et al.* (2016) developed interspecific potato somatic hybrids between cultivated *S. tuberosum* dihaploid 'C-13' and wild species *S. cardiophyllum* via protoplast fusion. Out of 26 regenerants, only 4 were confirmed as true somatic hybrids containing both parental genomes based on molecular markers and phenotypes. Somatic hybrids were identified by RAPD, ISSR, SSR, AFLP and cytoplasm (chloroplast and mitochondrial genomes) type molecular markers. Intermediate phenotypes of somatic hybrids were confirmed by leaf, flower and tuber traits. Late blight resistance of the hybrids was assessed by challenge inoculation of *P. infestans* under controlled conditions. Somatic hybrids were found tetraploid by flow cytometry, exhibited high pollen stainability by acetocarmine staining and formed berries and viable seeds after crossing with a common potato variety. Somatic hybrids possessed diverse cytoplasm types (W/β, W/β and T/β) as assessed by chloroplast and mitochondrial genome-specific markers. Further, cluster analysis based on the Jaccard's coefficient of molecular profiles generated by all above markers showed genetic distinctness in somatic hybrids and parents.

Table 10.20: Production of Important Somatic Hybrids

Fusion Partner A	*Fusion Partner A*	*References*
S. tuberosum	*S. chacoense*	Cheng *et al.*, 1995
S. tuberosum	*S. nigrum*	Binding *et al.*, 1982
S. tuberosum	*S. brevidens*	Fish *et al.*, 1988; Pehu *et al.*, 1989
S. tuberosum	*S. pinatisectum*, *S. bulbocastnum*	Sidorov *et al.*, 1984
S. tuberosum	*S. esculentum* (Tomato)	Melchers *et al.*, 1978; Schoenmakers *et al.*, 1994
S. tuberosum	*Nicotiana tobacum* L. (Tobacco)	Tempelaar *et al.*, 1988
S. tuberosum	*S. circalifolium*	Mattheij *et al.*, 1992
S. tuberosum	*S. pimpinellfolium*	Okamura, 1988
S. tuberosum	*S. melongena*	Li and Constabel, 1988.
S. tuberosum	*S. phureja*	Puite *et al.*, 1986
Monoploid	Monoploid *S. tuberosum* clone	Uijtewaal *et al.*, 1987
Diploid	Diploid *S. tuberosum* clone	Chaput *et al.*, 1990; Cooper-Bland *et al.*, 1994

Table 10.21: Somatic Hybrids Produced in Potato

Species Combination	*Character Transferred*
S. tuberosum x *S. chacoense*	Potato virus X
S. tuberosum x *S. brevidens*	*Phytophthora* resistance
S. tuberosum x *S. nigrum*	Potato leaf roll virus, late blight, potato virus Y, herbicide resistance
S. circalifolium x *S. tuberosum*	Frost resistance
S. circalifolium x *S. commersonii*	Nematode resistance
S. circalifolium x *S. bulbocastnum*	Pseudomonas resistance

Embryo Rescue

In potato successful rescued incompatible crosses includes (*S. microdontum* x *S. canascence*) x (*S. ocharanthan* x *S. phureja*), pH-345 x (*S. verrucosum* x *S. phureja*), V-93 x CP-1406, pH-265 x SS-1304, Kufri Chandramukhi x *S. bulbocastnum*, Kufri Jyoti *x S. microdontum* ssp. *gigantophyllum* and pH/DV-47 x *S. microdontum* (Chandra and Naik, 1993).

Resistance to Biotic Stress

(i) Resistance Source to Diseases and their Utilization

Genetics of disease resistance in potato is well documented.

Table 10.22: Genetics of Disease Resistance

Disease	*Genetic Control*
Late blight	Polygenic, many R genes
Fusarium wilt	Single dominant gene
Potato virus Y	Single dominant gene

Similarly in case of viruses some specific observations were made time to time:

Table 10.23: Virus Host Reaction and Genetics

Virus	*Host*	*Resistance Genes*	*Genetics of Resistance*	*Reference*
Potato virus X	Capsicum pendulul	-	Single dominant gene	Nagaich *et al.*, 1968
Potato virus Y	Capsicum annuum	Y^a	Incompletely dominant gene or recessive genes	Shifriss and Marco, 1980

In India, wild species *S. verrucosum* and *S. microdontum* have been used as donors of durable resistance (horizontal) to late blight. Potat cultivars like Kufri Jeevan, Kufri Jyoti, Kufri Khasigaro, Kufri Naveen, Kufri Neelamani, Kufri Muthu, Kufri Megha, Kufri Badshah and Kufri Jawahar were developed by incorporating resistance through exotic material from UK having *S. demissum* genes. All the cultivars developed for the hills and newly released medium maturing cultivars for the plains possess resistance to late blight.

Late blight resistant species are found in Mexican gene pool, where late blight fungus (*Phytophthora infestans*) has been found to reproduce sexually. Variety 'Kufri Sindhuri', a selection from 'Kufri Red' x 'Kufri Kundan' has resistant to late blight. Similarly, variety 'Kufri Badshah' ('JF 4870') has horizontal resistance to late blight. There are many races of *Phytophthora infestans* causing late blight in potato. 'Kufri Badshah'carries a set of 'R' genes imparting resistance to many races of *Phythophthora infestans*. Exploitation of resistance genes *S. tuberosum* subsp. *andigena* has resulted in development of Kufri Pukhraj, Kufri Chipsona-2 and Kufri Giriraj, which appear to have more durable resistance to late blight. Similarly, *S. chacoense* and *S. spegazinii* has been used to provide resistance to charcoal rot and *S. stoloniferum* for resistance to late blight and viruses.

Table 10.24: Indian Germplasm Resistant to Late Blight of Potato

Subsp.	*Resistant Germplasm*
Subsp. *tuberosum*	CP1425, CP1426, CP1673, CP1632, CP1690, CP1880, CP2000, CP2011, CP2013, CP2014, CP2016, CP2068, CP2073, CP2099, CP2110, CP20113, CP2137, CP2161, CP2172, CP2180, CP2184, CP2262, CP2294, CP2297, CP2333, CP2355, CP2358, CP2370, CP2373, CP2376, CP2379, CP2380, CP2381, CP2384, CP2385, CP2407, CP2415, CP2416, CP2417, CP2421, CP3072, CP3087, CP3094, CP3096, CP3147, CP3171, CP3174, CP3190, CP3198, CP3246, CP3248, CP3259, CP3269, CP3270, CP3276, CP3280, CP3287, CP3288, CP3290 and 3384.
Subsp. *andigena*	JEX/A-36, JEX/A-37, JEX/A-44, JEX/A-97, JEX/A-198, JEX/A-380, JEX/A-877, JEX/A-1244, JEX/A-1245, JEX/A-1269 and JEX/A-1270.

Species *S.andigenum*, *S. anomalocalyx*, *S. jamesii*, and *S. saltense* and a few varities of *S. tuberosum* like 'Arran Banner', 'Chanango', 'Era' and 'Placid' show a high degree of resistance to early blight (*Alternaria solani*), charcoal rot (*Macrophomina phaseoli*) is another serious disease of potato (Pushkarnath, 1976).

In potato horizontal resistance to late blight have been reported in promising clones (species) like SS1735-2, SS1736, SS1835, SS1837, SS1844-1, SS1844-2, SS1846, SS1846-2, SS1847-9, SS1848-8, SS1850, SS1850-1, SS1850-4 (*S. demissum*); SS1725-52, SS1725-54, SS1725-61, SS1725-72, SS1725-73, SS1725-84, SS1725-93, SS1725-9, SS2021-10 (*S. spegazzinii*); SS1672-6, SS1926, SS1926-9, SS1926-13 (*S. microdontum*); SS1425-5, SS1664-2, SS1724-25, SS1724-26 (*S. sparsipilum*); SS1763-9, SS1763-18, SS1763-25 (*S. albicans*); SS1764-19, SS1765-8 (*S. alandiae*); SS1784-3, SS2086-4 (*S. berthaultii*); SS551-2, SS680-6 (*S. chacoense*); SS1794-7 (*S. brevicaule*); SS508-2 (*S.aeroscopium*); SS1768-1 (*S. aamabile*); SS1878-3 (*S.guerreroense*), SS1719 (*S. hondelmannii*); SS1652-9 (*S. jamessi*) and SS1301-1 (*S.selemeri*). Some important sources of resistance to major potato diseases are listed below:

Table 10.25: Utility of Species and Chromosome Number

Species	*Chromosome No.*	*Utility of Species*
S. demissum Lindl.	2n=72	Resistance to late blight (*P. infestans* (Mont.) de Bary), nVirus A,Virus Y
S. stoloniferum Schloehtd	2n=48	Resistance to Virus x, frost, Colorado beetle
S. vernei Bitt. et. Wittam	2n=24	Resistance to 2 species of *Heterodera* nematode
S. multidissectum Wawk.	2n=24	Resistance to 2 species of *Heterodera* nematode
S. antipoviczii Buk		Resistance to late blight
S. phureja Jug. et. Buk		Non-dormant type used in breeding
S. chacoense Bitter		Non-dormant type used in breeding, tolerance to high temperature
S. anomalocalyx, S. jamessi Torr., S. saltense		Resistance to early blight

Table 10.26: Wild Species as a Source of Resistance to Various Diseases

Diseases	*Sources of Resistance*
PVX	*S.acuale, S.berthaultii, S. tuberosum* subsp. *andigena, S. sucrense, S. chacoense* and *S. curtilobum*
PVY	*S. phureja, S. demmissum, S. stoloniferum, S. chacoense, S. cardiophyllum, S.microdontum, S. polyadenium* and *S. multidissectum*
PVS	*S. tuberosum* subsp. *andigena,*
PLRV	*S.acuale,S.berthaultii, S. demmissum, S. tuberosum* subsp. *andigena, S. etuderosum,* and *S. brevidens*
Late blight (vertical)	S. demmissum, S.verrucosum, S. stoloniferum
Late blight (horizontal)	S.berthaultii, S.chacoense, S. microdontum, S. verneim
Wart	*S.acuale, S.berthaultii,*
Common scab	*S.chacoense, S. tuberosum* subsp. *andigena*
Bacterial wilt	*S.chacoense, S. tuberosum* subsp. *andigena*

Genes of *S. demissum* have been incorporated into more than 50 per cent of the world cultivars for resistance to late blight and PLRV. Cultivars Campbell-14, Mainestay, Nor King Russet and Rideau (*Verticillium*) were reported to be resistance against wilt diseases (Reeves *et al.*, 1985; Reeves *et al.*, 1997; Johansen *et al.*, 1985; Johansen *and* Rowberry, 1980). Promising cultivars like Kufri Jyoti, Kufri Kanchan, Kufri Naveen, Kufri Kuber, Kufri Megha, Kufri Jeeven and Kufri Moti are reported as resistant to late blight.

Table 10.27: Potato Varieties Released by ICAR-CPRI, Shimla (Himachal Pradesh) and Reaction to Late Blight

Cultivars/Varieties	*Year of Release*	*Reaction to Late Blight Disease*
Kufri Anand	1999	R
Kufri Ashoka	1996	R
Kufri Badshah	1979	MR
Kufri Bahar	1980	S
Kufri Chandramukhi	1968	S
Kufri hipsina-I	1998	R
Kufri hipsina-II	1998	R
Kufri Giriraj	1998	R
Kufri Jawahar	1996	MR
KufriJyoti	1968	MR
Kufri Lalima	1982	MR
Kufri Lauvkar	1972	S
Kufri Pukhraj	1998	MR
Kufri Sindhuri	1967	S
Kufri Sutlej	1996	MR
Kufri Swarna	1985	R

R= Resistance, MR= Moderately Resistance, S= Sucepatible.

The utilization of germplasm at ICAR-CPRI, Shimla (H.P.) has resulted in release of 46 improved varieties of potatoes and one true potato seed (TPS) population (92-PT-27) for commercial cultivation

Table 10.28: Attributes of Potato Cultivars

Variety	*LB***		*EB*		*X*	*Y*	*PLRV*		*W*	*CN*	*Frost*	
	*R**	*MR*	*R*	*MR*	*R*	*R*	*R*	*T*	*R*	*R*	*R*	*T*
Kufri Alankar	√											
Kufri Anand	√											
Kufri Arun		√										
Kufri Ashoka												
Kufri Badshah		√		√	√							
Kufri Bahar												
Kufri Chamatkar			√									
Kufri Chandra-mukhi												
Kufri Chipsona-1	√											
Kufri Chipsona-2	√											√
Kufri Chipsona-3		√										

Variety	LB**		EB		X	Y	PLRV		W	CN	Frost	
	R*	MR	R	MR	R	R	R	T	R	R	R	T
Kufri Dewa											√	
Kufri Frysona		√										
Kufri Girdhari	√											
Kufri Giriraj	√											
Kufri Himalini	√											
Kufri Himsona		√										
Kufri Jawahar		√										
Kufri Jeevan	√		√						√			
Kufri Jyoti		√	√						√			
Kufri Kanchan		√							√			
Kufri Khasigaro	√		√									
Kufri Khyati		√										
Kufri Kisan												
Kufri Kuber												
Kufri Kuber					√	√						
Kufri Kumar		√										
Kufri Kundan		√										
Kufri Lalima				√		√						
Kufri Lauvkar												
Kufri Megha	√											
Kufri Muthu		√										
Kufri Naveen	√								√			
Kufri Neela		√										
Kufri Neelamani		√										
Kufri Pukhraj		√	√									
Kufri Pushkar		√										
Kufri Red												
Kufri Sadabahar		√										
Kufri Safed												
Kufri Shailja	√											
Kufri Sheetman											√	
Kufri Sherpa		√		√					√			
Kufri Sindhuri				√				√				
Kufri Surya												
Kufri Sutlej		√										
Kufri Swarna	√									√		

LB=Late Blight; EB=Early blight X=Virus, Y=Virus PVRV=Virus, CN=Cyst memotades W=Wort.

(ii) Resistance Source to Insect Pest and their Utilization

(a) Colorado Potato Beetle

Through coat protein (CP) mediated strategy two potato varieties, *viz.* Atlantic "New Leaf ™", Sperior "New Leaf ™", Russet Burbank NewLeaf® -Y, and Shepody NewLeaf® -Y, have been developed by Monsanto Company, USA. These varieties posses combined resistance against Colorado potato beetle (*cry 3A* gene) and PVY (*CP-PVY*gene). In collaboration with the National Research Centre on Plant Biotechnology (NRCPB), New Delhi, synthetic *cry1Ab* gene has been transferred into Indian potato cultivars at ICAR-CPRI, Shimla. Some of these transgenic lines exhibited tolerance against *Helicoverpa armigera* (Chakrabarti *et al.*, 2000) and tuber potato moth (PTM) (Kumar *et al.*, 2010).

Table 10.29: Resistance Source to Colorado Beetle

Species	Chromosome No.	Utility of Species
S. stoloniferum Schloehtd	2n=48	Resistance to Colorado beetle

(b) Aphid

Some important source of resistance to aphid is listed below:

Table 10.30: Wild Species as a Source of Resistance to Various Diseases

Insect	Sources
Aphid	*S.berthaultii*

(iii) Resistance Source to Nematode

Resistance to cyst nematode in variety Kufri Swarna has been derived from *S. vernei*.

Table 10.31: Resistance Source to Cyst Nematode

Species	Chromosome No.	Utility of Species
S. vernei Bitt. et. Wittam	2n=24	Resistance to 2 species of *Heterodera* nematode
S. multidissectum Wawk.	2n=24	Resistance to 2 species of *Heterodera* nematode

Some important sources of resistance to nematodes are listed below:

Table 10.32: Wild Species as a Source of Resistance to Nematodes

Nematodes	*Sources*
Cyst nematode	*S. tuberosum* subsp. *andigena*, *S. berthaultii*, *S. vernei*
Root knot nematode	*S. spegazzinii*

Resistance to Abiotic Stress

Genetic control of characteristics involved in heat tolerance is complex. In potato, tuber yield under high temperature is genetically controlled and expressed as partial dominance. 'Kufri Sindhuri', a selection from 'Kufri Red' x 'Kufri Kundan' has resistant to frost. Hybrids like 'HB 827', 'HB 841' and 'HB 858' are tolerant to high temperature. Similarly, variety "Kufri Badshah ('JF 4871') has horizontal resistance to frost. *S.curtilobum* Jug. et. Buk, is resistance to frost. The variety 'Kufri Sheetman' has been specially bred for areas where frost is a problem.Varieties like 'Phulwa', 'Up-to-Date', 'Craige Defiance' and 'Hybrid 2236' respond both under short day and long day conditions and have regional adaptability.

Table 10.33: Potato Species for Trait Specific Breeding

Species	*Utility of Species*
S. phureja Jug. et. Buk	Non-dormant type used in breeding
S. chacoense Bitter	Non-dormant type used in breeding, and tolerance to high temperature

In potato some promising frost tolerance lines are SS 1636-3 (*S. goniocalyx*), SS1660-4, SS1725-22, SS1725-43, SS1725-44, SS1725-70, SS1725-74, SS1725-76, SS1725-78, SS1725-93(*S. spegazzinii*); SS1724-1 (*S. sparsipilum*); SS1732-8 (*S. demissum*); SS1770-11 (*S. arnezii*); SS1780-3 (*S. berthaultii*); SS2040, 2038-5, SS2040-13 and 2044-1 (*S. tuberosum* ssp. *andigena*). Similarly, resistance to cold induced sweetening lines are SS1763-6 (*S. albicans*); SS171780-3 (*S. berthaultii*); SS171732, SS171735-2, SS171846 (*S. demissum*); SS171652-9 (*S. jamesi*) and SS172044-5 (*S. tuberosum* ssp. *andigena*).Wild relatives as sources of freezing resistance are *S. acaule*, *S. megistacrolobum*, *S. multidisectum* and *S. vernei*.

Table 10.34: Potato Varieties Suitable for High Thermo-periods

Variety	*HT*
	T
Kufri Lauvkar	Ö
Kufri Sindhuri	Ö
Kufri Surya	Ö

HT: High thermo-periods.

Polyploidy Breeding

Potato is basically a polyploidy. Edible stem tubers and clonal propagation has offered utility of polyploidy in breeding programme and development of cultivars. Homozygous parental lines are produced through haploids (haploids are produced through crossing of tetraploid varieties with pollen from diploid varieties) and then their polyploidization is done. The homozygous lines have utility specially in disease resistance breeding where disease resistance is governed by recessive genes. For developing dihaploids *S. phureja* has been utilized as pollinator. Diploids are generally produced by their parthenogenetic development in 4x X 2x crosses between *S. tuberosum* x *S. phureja* (pollinator). Anther culture has been used to produce dihaploid in potatoes. But due to lack of response of many genotypes to anther culture, its usage is limited. Dihaploid of *tuberosum* cross readily with many diploid wild species and thus provide the opportunity to introgress useful traits from alien sources to cultivated background. The final goal of breeding at the diploid level is to develop 2x clones with specific attributes, good breeding value for agronomic traits and 2n gametes. Production of 2n (unreduced) gametes through 4x X 2x or 2x X 4x (unilaternal) or 2x X 2x (bilateral) polyploidization. The natural production of 2n gametes in many tubers bearing *Solanum* sp. is well established. This would help in development of good quality tubers.

Male Sterility

In potato breeding, male sterility is a serious constraint. Though both pollen and ovule sterility can be observed in nature. Male sterility ranging from partial to complete absence of pollen grains is common in cultivated *tuberosum* and *andigena* potato. This limits the choice of pollen parents in hybridization. In potato, male sterility is probably controlled by more than one gene, with partial dominance or by nuclear cytoplasm interactions and is also function of environmental factors.

Breeding for Quality Traits

Following attributes may be considered for processing traits:

Table 10.35: Processing Attributes for Potato

Traits	*Preferred Range/ Magnitude/Value/Type*
Glucose and reducing sugars	Low
Dry matter content	High
Phenolic content	Low
Tissue cell size	Larger
Starch and amylose content	More
Size of starch granules	Larger
Browning after cooking	Absent or low
After cooking blackening (Due to presence of chlorogenic acid and iron)	Nil to little (To avoid blacking)
N/K ratio	Low

Five varieties *viz.* Kufri Chipsona-1, Kufri Chipsona-2, Kufri Chipsona-3, Kufri Himsona and Kufri Frysona have been developed for processing. 'Kufri Sindhuri', a selection from 'Kufri Red' x 'Kufri Kundan' has light red tubers. 'Kufri Chandramukhi', a selection from 'S 485' x 'Kufri Kuber' takes only 95 days from sowing to harvest. It is white fleshed and highly suited for planting, in maize-wheat rotation. In addition to these, cultivars Kufri Jyoti, Kufri Lauvkar and Kufri Surya are also suitable for processing when grown in specific areas. A synthetic gene, producing 80 per cent essential amino acids was constructed and named *High Essential Amino Acid Encoding* (*HEAAE*) at the International Potato Center (CIP), Lima, Peru. This gene was transferred into two potato clones K-2 and K-7. Proterin analysis of transgenic plants showed that *HEAAE* protein comprise about 0.02-0.35 of the total plant protein with about 1.0 per cent increase in essential amino acids. Similarly, AmA1 gene encoding storage protein rich in essential amino acids from *Amaranthus hypochondriacus* is known to increase protein content and essential amino acids. Forty six potatoes germplasm with varying degree of flesh colour from white to yellow were evaluated for iron and zinc content in raw and peeled tubers. The iron content ranged between 14.90-67.13mg/kg dry weight and zinc between 2.78-35.40mg/kg dry weight basis in tuber flesh. A significant and positive correlation (r=0.6865) between these two nutrients suggested that breeding varieties with high content of these nutrients is feasible. Carotenoids content enhances the bioavailability of the micro nutrients but lack of correlation between tuber flesh colour index and nutrient concentration in the present study indicated that these traits are independent. The genotypes CP Nos. 3443, 1978, 2067, 3772, 1239 and 1435 are promising as parental material for developing zinc and iron rich potato varieties (Dalamu *et al.*, 2016)

Mutation Breeding

Spontaneous mutations are useful in vegetatively propagated plants. In potato several varieties have been resulted from sports (somatic cells) as proposed by Miller (1954). Some of these cultivars like Russet Burbank (mutation of smooth Burbank); Cobbler (mutation of Early Rose); Red Triumph (mutation of Triumph); Red Warba (mutation of Warba); Russet Sebago (mutation of Sebago); Red Pontiac (mutation of Pontiac) *etc.* are quite popular.

Biotechnological Approach

Development of varieties with specific characteristics such as high protein content, resistant to pests and diseases through genetic engineering is also in progress in different parts of the world. During last decade 277 patent applicatiopns on genetically modified potatoes have been filled in the world (Chakrabarti *et al.*, 2003). Field tests of transgenic potatoes have been reported in the United States of America, Canada, Europe, Argentina, Brazil, Egypt, India, Mexico, Peru, Russia, South America, and Ukraine. Transgenic potato lines resistant to tuber moth, developed at ICAR-CPRI, Shimla (H.P.) are being evaluated under limited field trials. Efforts are also on at ICAR-CPRI, Shimla (H.P.) to develop late blight and virus resistant transgenic potato lines. Nutritional quality of potato is also under improvement by transferring a protein gene from *Amaranthus* species. In potato, the technique for transfer of genes *Agrobacterium* strain C58-PB-1121 has been standardized and production of transgenic plants has been confirmed by the GVS assay. A genetic engineering target to introduce CP gene (for virus resistant) and BT gene (for insect resistant potato) has been set up by CTRI. For developing transgenic, there is a need to change in specific carbohydrates, proteins and oils in order to have improvrd seed quality. Through transgenes, in potato, there were experiments which proved useful in developing resistant transgenes for Colorado potato beetle, two viruses, *Verticillium* (fungi), bruising resistance and allied carbohydrate metabolism (Gopal and Kumar, 2003).

Export of Seed/Planting Material

Many Indian varieties have been found which favour in foreign countries as well. Indian varieties are grown in Mexico, Sri Lanka, Philippines, Medagascar, Bolivia and Vietnam. As Indian potato enjoys a high degree of consumer preference in neighbouring countries, there is an enormous scope for export of seed and table potato.

Table 10.36: Indian Potato Varieties/Hybrids being grown in other Countries

Country	Varieties
Afghanistan	Kufri Chandramukhi
Nepal	Kufri Jyoti, Kufri Sindhuri
Bhutan	Kufri Jyoti
Bangladesh	Kufri Sindhuri
Mexico	1-654 as CCM-69.1
Sri Lanka	1-822 as cv. Khrushi, 1-1085 as cv. Sita
Philippines	1-1035 as cv. Montanosa, 1-1085 as cv. BSUP-04
Madagascar	1-1035 as Mailaka
Bolivia	1-1039 as cv. India
Vietnam	1-1039 as cv. Red Skin

Genetic Resource Conservation

The wild gene pool includes more than 1500 tuber bearing *Solanum* species. The enormous diversities of cultivated potatoes and their wild relatives in potato's centre of origin were first explored by Prof. S.M. Bukasov and his group of the then N.I. Vavilov Institute of Plant Science in 1925-26. The group, collected enormous diversity and laid the foundation of the first potato gene bank in the then Leningrad (St. Petersburg). This oldest gene bank still continues to hold important potato collections at the presently renamed N.I. vavilov All-Russian Scientific Research Institute of Plant Industry (VIR). The second important potato gene bank, which came up as the Commonwealth potato collection (CPC) in Cambridge, was mainly due to the efforts of the British Empire by conducting several expeditions in South America. Today this Commonwealth Potato Collection (CPC) is stationed at the Scottish Crop Research Institute (SCRI), Scotland, with a large collection of 86 different potato species. The International Potato centre (CIP), a CGIAR organization, was established in 1971with the primary mandate of collection and conservation of potato germplasm available in the world. The US Potato Genebank (NRSP-6) at the Potato Introduction Stations, Sturgeon Bay, Wisconsin, is the repository of a large collection of potato germplasm. Other important potato genebanks are Dutch-German Potato Collection (Centre for genetic Resources, The Netherlands [CGN]), Wageningen, the Netherlands; Potato genebank of IPK-Gatersleben (GLKS), Groß Lüsewitz, Germany; The Potato Collection of INTA-Balcarce,Balcarce, Argentina;The Potato Collection of Chile, Valdivia, Chile; and The Potato Collection of the Proyecto de Investigation de la Papa (PROINPA), Cochabamba, Bolivia. Besides these major potato collection centres, different national programmes also maintain potato genetic resources.

Table 10.37: Availability of Cultivated Species of Potato at CIP

Species	*Number of Accessions*
Solanum stenotomum	264
S. phureja	131
S. goniocalyx	54
S. x *ajanhuiri*	7
Solanum 2 x *Hybrids*	62
S. x *chaucha*	101
S. x *juzepzukii*	34
S. tuberosum subsp. *andigena*	2674
S. tuberosum subsp. *tuberosum*	132
Solanum 2 x *Hybrids*	28
S. curtilobum	9

The germplasm at CIP consists of two main components:

(i) Wild species and

(ii) Primitive cultivated species.

The primitive gene pool has more than 13,000 clonal samples collected from the traditional potato growing areas in Latin America where they have been under cultivation for hundreds of years. Computerized data bank using all available data from collecting expeditions and evaluations has been organized (Huaman, 1980, 1994).

Table 10.38: The Conservation Status of Wild Potato Accessions at CIP

Country	*Number of Accessions*	*Number of Solanum taxa*
Argentina	33	8
Bolivia	420	33
Colombia	8	4
Chile	9	5
Ecuador	65	13
Mexico	78	16
Paraguay	14	1
Peru	1024	74
Uruguay	52	2
USA	5	2
Venezuela	4	2
Total	**1712**	**160**

Source: Huaman *et al.*, 2000.

In India, the Central Potato Research Institute (CPRI), Shimla (H.P.) holds the national repository of the potato germplasm and has been designated as the National Active Germplasm Site (NAGS) for potato. The early introductions were closely related to *Solanum tuberosum* ssp. *andigena* (Pushkarnath, 1969). During 1824 to 1939 systematic attempts were made to introduce new potato varieties in the country, mainly from Europe. The failure of introduced varieties in India was mainly because these varieties were primarily bred to suit temperature long days of summer in Europe, whereas, potato in India is grown during short days of sub-tropical winters. Presently, this centre is holding a modest collection of over 3,900 accessions, comprising of 2035 *tuberosum*, 803 *andigena* and 1082 clones/accessions of wild/semi-wild species. This is the largest potato collection in South-west Asia. These valuable germplasm accessions have been procured from 40 countries.

Other important collections are available with N.I. Vavilov Institute of Plant Industry, St. Petersburg, Russia; United States Potato Introduction Project (NRSP-6), Sturgeon-Bay, Wisconsin, USA; Dutch-German Potato Collection (CGN), Wageningen, The Netherlands; Institute of Plant genetics and Crop Plant Research, (GLKS), Gross-Lusewitz, Germany; commonwealth Potato Collection (CPC), Scottish Crop Research Institute, Dundee, Scotland; and Instituto Nacional de Tecnologia Agropecuaria (INTA), Balcarce, Argentina. These organizations are part of the Association of Potato Inter-genebank Collaborators (APIC), having priorities for development of global database of potato germplasm available in member gene banks.

Potato germplasm can be conserved by three methods:

Table 10.39: The Current Status of Important Potato Genebanks in the World

Country	Genebank	Total Collection	Holdings[a]
Peru	World Potato Collection, Germplasm Unit, International Potato Centre (CIP), Lima Peru	7403	World (4218), pathogen-tested (938) and research (2247)
USA	US Potato Genebank, Potato Introduction Station, Sturgeon Bay, Wisconsin	6050	Breeding stocks (201), genetic stocks (366), cultivars (314), cultivated species (1500) and wild species (3669)
Netherlands	The Dutch-German Potato Collection (BML), Centre for Genetic Resources (CGN), Wageningen UR	2704	Wild species (1947 of 117 species), primitive or traditional cultivated Andean species (744 of 5 species) and advanced material (13 of 3 lines)
United Kingdom	The Commonwealth Potato Collection (CPC), Scottish Crop Research Institute (SCRI), SARI, Dundee, Scotland	1300	86 different species
Germany	Potato Genebank of IPK Gatersleban (GLKS),Groß Lüsewitz, Germany	5800	Cultivars (2800) and wild/cultivated central and south American species (3000)
India	Central potato Research Institute (CPRI), Shimla (H.P.)	2651	Cultivars (1447), S. tuberosum subsp. andigena (834) and wild/semi-cultivated species (370)

[a]The figures within parenthesis represent the total number of accessions.

(1) *In situ* method

(2) *Ex situ* method

(3) Sexual propagation

(1) *In situ* Method

Due to practical problems of maintenance and economic viability, potato germplasm conservation through *in situ* approach has limited significance. Successful *in situ* conservation is also becoming the limiting factor due to requirement for a large population sized to avoid inbreeding and subsequent genetic drift.

(2) *Ex situ* method

Consequently, *ex situ* approach is the preferred option for the conservation of potato genetic resources. The traditional method of *ex situ* conservation entails the maintenance and multiplication of tubers in fields and glasshouse. Selected individual genotypes are conserved as vegetative clones (Roca *et al.*, 1989). However, this method, based on multiplication in every year is highly labour-intensive, require large space, exposes the germplasm to various biotic and abiotic stresses, and most importantly, results in the progressive accumulation of pathogens in tuber stocks over successive vegetative generation. It has been observed that many potato genotypes, especially the wild species, do not tuberize under certain environmental conditions. Heterozygocity in tetraploid potatoes is maintained through vegetative propagation.

(i) True Potato Seeds (TPS)

Conservation of potato genetic resources in the form of botanical (true) seed is not recommended, because they constitute a segregating population. However, the true potato seed (TPS) may be an effective way of conserving the total gene pool (Pandey and Sarkar, 2007; Thakur *et al.*, 2003).

(ii) *In vitro* Methods

There are three major strategies for *in vitro* conservation of potato germplasm:

(a) Minimal Growth or Slow Growth

Minimal growth uses differentiated tissue culture plantlets in medium-term storage. The slow growth approach using growth retardants represents a simple, efficient and relatively cheap method of germplasm conservation. It has several disadvantages, the foremost among them being the probability that physiological changes may be induced in cultures grown for protected periods in the presence of growth retardants. In some cases, it might be difficult to regenerate a physiologically normal plant at the end of the storage period. Furthermore, continuous maintenance of cultures in medium containing growth inhibitors invites the possibility of *in vitro* selection against the inhibitor. This may eventually lead to selecting lines with resistance or tolerance to the growth retardant (s) used. All these aspects may have an effect on clonal integrity which is one of the most important requirements for clonal conservation (Pandey and Sarkar, 2007).

(b) Micro-Tuber Production and Storage

Micro-tuber technique utilizes *in vitro* grown tubers for the conservation of germplasm in medium-term storage. It is a useful option, because microtubers can be induced in a large number of germplasm accessions and they can be successfully stored at reduced temperature for 12-15 months. After optimum storage period, the microtubers spontaneously sprout and regenerate into vigorous plants for further multiplication. The most important advantage underlying microtuber conservation is the convenience with which the germplasm can be exchanged and distributed in the form of microtubers as compared to the microplants.

Therefore, *in vitro* collections of microtubers have been established in many places. Despite relative advantages of microtuber conservation, this method is indirect in essence. Microtuber production is cost- and labour- intensive with additional inputs required for conservation at reduced temperature. Besides, variability in microtuber size and dormancy period may lead to difficulties in regenerating the cultures (accessions), because best regeneration results are obtained when the size range of the microtubers is 8-12 mm with a dormancy of up to 5 months. Therefore, minimal growth of differentiated microplants emplying osmotic stress (sugar and/or sugar alcohal) under reduced temperature is the most adopted method of *in vitro* conservation of potato germplasm (Naik and Sarkar, 2000). Presently, this technique is followed in all the major potato gene banks worldwide.

Both the methods enables the subculture intervals to be extended up to 12 to 36 months employing a variety of approaches, such as reduced temperature, low light intensity, use of osmotic stress, modification of the culture media and the addition of growth retardants.

(c) Cryopreservation

Cryogenic storage has been recognized as a priority option for long term conservation of potato germplasm. A high degree of gentic stability can be achieved by cryogenic storage at ultra-low temperatures in liquid nitrogen with practical storage period measurable in decades. Since liquid nitrogen is the most widely used non-hazardous cryogen (at -196°C) esposure of tissues or cells to it halts biological growth and development effectively. The main difficulties encountered in the successful cryopreservation of vegetative/*in vitro* material are cumbersome techniques and low survival rate of conserved materials. These consequently result in poor reproducibility of a protocol across a large number of germplasm accessions (Pandey and Sarkar, 2007).

Table 10.40: The National Potato Germplasm Collection at the Central Potato Research Institute, Shimla (H.P.)

Material	*Number of Accessions*			
	Tuber	*In vitro*	*True Potato Seed*	*Total*
Solanum tuberosum L. subsp. *tuberosum*				
Cultivars bred at the CPRI	41	35	-	41
Old indigenous cultivars	16	8	-	16
Parental lines	11	1	-	11
Indigenous samples	89	79	-	89
Exotic	980	1000	-	1290
S. tuberosum L. subsp. *andigena*	834	-	200	834
Wild and semi-cultivated species	124	-	246	370

Table 10.41: International Organizations Involved in Conservation of Wild and Putative Potato Species

Country	*Collection*	*Genebank Contact Details*
Argentina	The Potato Collection of INTA-Balcarce	A.M. Clausen, INTA-Balcarce, C.C. 276, 7620 Balcarce, Argentina Fax: +54-266-439101 Email: aclausen@balcarce.inta.gov.ar Info page: www.intagov.ar/balcarce
Bolivia	The Potato Collection of the Proyecto de Investigation de la Papa (PRONIPA)	Ximena Cadima, Coordinadora Recursos Geneticos, Fundacion PRONIPA, Av. Blanco Galindo km 125 Calle Prado s/n, Quillacollo, Casilla postal 4285, Cochabamba, Bolivia Fax: +591-4-4360800 Email: xcadima@pronipa.org or pronipa@pronipa.org Home page: www.pronipa.org FAO institute code: BoL038
Chile	The Potato Collection of Chile	Andres Contreras M., Instituto de Produccion y Sanidad Vegetal, Universidad Austral de Chile, Casilla 567, Valdivia, Chile Fax: +56-63-221733 Email:acontrer@uach.cl Home page:www.agrarias.uach.cl/webpage/index.html Info page:on the collection in Spanish and its english summary FAO institute code:CHL003

Country	Collection	Genebank Contact Details
Colombia	The Potato Collection of CORPOICA	G. Ligarreto, Corporation Colombiana de Investigacion Agropecuaria (CORPOICA), Apartado Aereo 151123, El Dorado, Santafe de Bogota, D.C., Colombia Fax: +57-1-2673013 Email: gligarreto@corpoica.org.co Home page:www.corpoica.org.co Info page: (in Spanish)
Germany	The Groß Lüsewitz Potato Collection (GLKS)	Klaus J. Dehmer, Institute of Plant Genetics and Crop Plant Research (IPK), Genebank Außenstelle Nord, Parkweg 1, 18190 Groß Lüsewitz, Germany Fax: +49-3820982313 Email:glksmv@t-online.de Home page:glks.ipk-gatersleben.de Search form: pgrc.ipk-gatersleben.de/potato Home page:IPK:fox-serv.ipk-gatersleben.de Search using EVA project:www.genres.de/eva/(in German) FAO institute code: DEU159
Peru	The Potato Collection of the International potato Centre (CIP)	Willy Roca, CIP, Apartado Postal 1558, Lima 12, Peru Fax: +51-1317-5326 Email:w.roca@cgiar.org Homepage:www.cipotato.org/www.singer.cgiar.org/SINGER:CGIAR System.wide Information Network for Genetic Resources FAO institute code:PER001
Peru	The Potato Collection of the University of Cusco	R. Ortega, CERRGETYR, University of Cusco, P.O. Box 295, Cusco, Peru FAO institute code:PER091
Russia	The Potato Collection of the vavilov Institute (VIR)	Stepan Kiru (Cultivated spp.) and nadezhda Zoteyeva (wild spp.), Department of Tuber Crops, N.I. vavilov All Russian Scientific Research Institute of Plant Industry (VIR), Bolshaya Morskaya Street 42, 190000 St. Petersburg, Russia Fax: +7-812-3118762 Email:s.alexanian@vir.nw.ru Home page:vir.nw.ru FAO institute code: RUS001
Netherlands	The Dutch-German Potato Collection	R. Hoekstra, Centre for Genetic Resource, The Netherlands (CGN), P.O. Box 16, 6700 AA Wageningen, The Netherlands Fax: +31-317-418094 Email:roel.hoekstra@wur.nl Home page:www.genebank.nl/potato FAO institute code:NLD037
USA	The USA Potato Genebank (NRSP-6)	J.B. Bamberg, 4312 Highway 42, Sturgeon Bay, WI54235-9620, USA Fax: +1-414-7431080 Email:nr6jb@ars-grin.gov Related pages: www.ars-grin.gov/cgi-bin/npgs/html/site.pl?NR6on-line.search Databases (download potato.zip) FAO institute code:USA004

REFERENCES

1. Binding, H., Jain, S.M., Finger, J., Mordhorst, G., Nehls, R. and Gressel, J. 1982. Somatic hybridization of an atrazine resistant biotype of *Solanum nigrum* with *Solanum tuberosum*. *Theor. Appl. Genet.*, **63:** 273-277.
2. Chakrabarti, S.K., Mandaokar, A.D., Pattanayak, P., Kumar, A., Naik, P.S. and Sharma, R.P. 2000. *Bacillus thuringiensis cry1Ab* gene confers resistance to potato against *Helicoverpa armigera* (Hubner). *Potato Research,* **43:** 143-152.
3. Chandel, P., Tiwari, J.K., Ali, N., Devi, S., Sharma, S., Sharma, S., Luthra, S.K. and Singh, B.P. 2016.

Interspecific potato somatic hybrids between *Solanum tuberosum* and *S. cardiophyllum*: A potential source of late blight resistance breeding. 7th Indian Horticulture Congress (Abs: 94), held at ICAR-IARI, New Delhi, 15-18 November, p. 72-73.

4. Chandra, R. and Naik, P.S. 1993. Potato tissue and cell culture. In: *Advances in Horticulture* Vol. 7. *Potato* (Chadha, K.L. and Grewal, J.S., eds.), Malhotra Publishing House, N.Delhi, p. 113-141.

5. Chaput, M.H., Sihachakr, D., Ducreux, G., Marie, D. and Barghi, N. 1990. Somatic hybrids plants produced by electro-fusion between dihaploid potatoes BF 15 (h11), Amirica (H6) and Cardinal (H3). *Plant Cell Rep.*, **9:** 411-414.

6. Cheng, J., Sounders, J.A. and Sinden, S.L.1995. Colorado potato beetle resistant somatic hybrid potato plants produced *via.* protoplast electro-fusion. *In Vitro,* **31:** 90-95.

7. Cooper-Blandbland, S., Maine, M.J. de, Fleming, M.L.M.H., Phillips, M.S., Powell, W. and Kumar, A. 1994. Suynthesis of interspecific somatic hybrids of *Solanum tuberosum*: Assesment of morphological biochemical and nematode (*Globodera pallida*) resistance characteristics. *J. Exp. Bot.*, **45:** 1319-1325.

8. Dalamu, Sharma, J., Sharma, V., Dua, V.K. and Singh, B. 2016. 1st International Agrobiodiversity Congress (Abs. Session: II. Agrobiodiversity for Nutrition and health), 6-9 November, 2016, held at New Delhi, p. 297.

9. Fish, N., Karp, A. and Jones, M.G.K.1988. Production of somatic hybrids by electrofusion in *Solanum*. *Theor. Appl. Genet.*, **76:** 260-266.

10. Gopal, J. 1994. Flowering behavior, male sterility and berry setting in tetraploid *Solanum tuberosum* germplasm. *Euphytica*, **72:** 133-142.

11. Gopal, J. and Raj Kumar. 2003. Strategies for successful potato breeding. In: The potato Production, Utilization in sub-tropics, p. 239. S.M. Paul Khurana, J.S. Minhas and S.K.Pandey (Eds.) Mehta Publishers, New Delhi, p. 61-68.

12. Grun, P. 1990. The evolution of cultivated potatoes. *Economic Botany*, **44:** 39-55.

13. Hawkes, J.G. 1990. The Potato, Evoloution, Biodiversity and Genetic Resources. Belhaven Press, London, p. 259.

14. Hawkes, J.G. 1992. Biosystematics of the potato. In: *The Potato Crop.* (Ed: Harris, P.M.) Chapman and Hall, New York, p. 13-64.

15. Hawkes, J.G. 1994. Origin of cultivated potatoes and species relationships. In: *Potato Genetics*. (Eds. Bradshaw, J.E. and Mackay, G.R.). CAB International, Wallingford, U.K., p. 3-42.

16. Hoops, R.W. and Plaisted, R.L. 1987. Potato. In: *Principles of Cultivar Development*. Vol. 2: *Crop Species* (Ed.Walter R. Fehr). Mac Millon Publishing Company, New York p. 385-435.

17. Hosaka, K. 1986. Who is the mother of the potato?- restriction endonucleases analysis of chloroplast DNA of cultivated potatoes. *Theor. Appl. Genet.*, 72: 606-618.

18. Huaman, Z.1994. *Ex situ* conservation of potato genetic resources at CIP. *CIP Circular*, **20(3):** 2-7.

19. Huaman, Z.1980. Utilization of CIP's primitive potato germplasm to develop improved back-up populations. Report of Planting Conference. *Utilization of the Genetic Resources of the Potato III*, 28-29.

20. Huaman, Z., Salas, A., Gomez, R., Panta, A. and Toledo, J. 2000. Conservation of potato genetic resources at CIP. In: *Potato, Global Research and Development* (Eds.) Khurana, S.M.P., Shekhwat, G.S., Singh, B.P. and Pandey, S.K., *Indian Potato Association*, Shimla, p. 102-112.

21. Johansen, R.H., Farnsworth, B., Secor, G., Nelson, D.C., Orr, P.H. and Boe, A.A. 1985. Nor-King Russet: a new potato variety. *North Dakota Farm Research*, **43:** 13-14.

22. Johansen, G.R. and Rowberry, R.G. 1980. Rideau: a new red-skinned 'Maincrop' table variety with high resistance to *Verticillium* wilt strains. *American Potato Journal*, **57:** 47-51.

23. Kumar, M., Chimote, V., Singh, R., Mishra, G.P., Naik, P.S., Pandey, S.K. and Chakrabarti, S.K. 2010. Development of *Bt.* transgenic potatoes for effective control of potato tuber moth by using *cry1Ab* gene regulated by GBSS promoter. *Crop Protection*, **29:** 121-127.

24. Li, G.G. and Constabel, F. 1988. Fusion of *Solanum tuberosum* L. cv. 'Imperial Black Beauty' (Eggplant) protoplast and subsequent callus regeneration. *Progr. Plant Protopast Res.*, 203-204.

25. Mattheij, W.M., Eijlander, R., Koning, J.R.A. de and Louwes, K.M. 1992. Interspecific hybridization between the cultivated potao *Solanum tuberosum* ssp. *tuberosum* L. and the wild species *S. circaeifolium* ssp. *circaeifolium* Bitter exhibiting resistance to *Phytophthora infestans* (Mont.) de Barry and *Globodera pallida* (Stone) Behrens. 1. Somatic hybrids. *Theor. Appl. Genet.*, **83:** 459-466.

26. Melchers, G., Sacristan, M.D. and Holder, A.A.1978. Somatic hybrid of plants of potato and tomato regenerated from fused protoplasts. *Carlsberg Res. Commun.*, **43:** 203-218.

27. Ochoa, C. 1984. *Sloanum hygrothermicum*, new potato species cultivated in the lowlands of Peru. *Econ. Botany*, **38:** 128-133.

28. Naik, P.S., Luthra, S.K. and Kumar, V. 2012. Potato (*Solanum tuberosum* L.) In: *Breeding of Field Crops* (Ed. Bharadwaj, D.N.). Published by Agrobios (India), p. 660-707.

29. Naik, P.S. and Sarkar, D. 2000. *In vitro* propagation and conservation of genetic resources in potato. *In: Biotechnology in Horticultural and Plantation Crops* (Eds.) Chadha, K.L., Ravindran, P.N. and Sahijram, L. Malhotra Publishing House, New Delhi. p. 369-406.

30. Okamura, M. 1988. Regeneration and evaluation of somatic hybrid plants between *Solanum tuberosum* and *Solanum pimpinellifolium*. *Progr. Plant Protoplast Res.*, 213-214.

31. Pandey, S.K., Kaushik, S.K. and Paul Khurana, S.M. 2003. Potato History. Papa to Alu: a fascinating journey from Andes to Himalayas. *Indian Hort.* (April-June): 46-48.

32. Pandey, S.K. and Sarkar, D. 2007. Conservation of genetic diversity in potato: Priorities and Challenges in this New Millennium. In: Agrobiodiversity Vol.1: Genetic Resouirces and Conservation. Associated Publishing Co. New Delhi. p. 162-178.

33. Pehu, E., Karp, A., Moore, k., Steele, S., Dunckley, R. and Jones, M.G.K. 1989. Molecular, cytogenetic and morphological characterization of somatic hybrids of dihaploid *Solanum tuberosum* and diploid *S. brevidens. Theor. Appl. Genet.*, **78:** 696-704.

34. Poehlman, J.M. and Borthakur, D.N. 1968. Breeding potato (in) *Breeding Asian Field Crops*, (Edn.I), p. 258-275. Oxford and IBH Publishing Company, Calcutta.

35. Puite, K.J., Roest, S. and Pijnacker, L.P. 1986. Somatic hybrid potato plants after electro-fusion of diploid *Solanum tuberosum* and *Solanum phureja*. *Plant Cell Rep.*, 5: 262-265.

36. Pushkarnath. 1969. *Potato in India: Varieties*. Indian Council of Agricultural Research, New Delhi, p. 493.

37. Pushkarnath. 1976. *Potato in Sub Tropics* (Edn.I), Orient Longman, New Delhi, p. 1-289.

38. Reeves, A.F., Cunningham, C.E., Nickson, R.L., Akeley, R.V., Murphy, H.J., Manzer, F.E. and True, R.H. 1985. Campbell-14: A *Verticillium* resistant round white potato variety. *American Potato J.*, **62:** 215-220.

39. Reeves, A.F., Porter, G.A., Work, T.M., Lambert, D.H., Davis, A.A. and Plissey, E.S. 1997. Mainesty: A high-yielding round white potato variety for fresh market. *American Potato J.*, **74:** 255-263.

40. Ross, H. 1986. Potato Breeding-Problems and Perspectives. Paul Parey, Berlin, p. 132.

41. Salaman, R.N. 1926. Potato varieties. Cambridge University Press, p. 378.

42. Salaman, R.N. 1949. The history and social influence of the potato. Cambridge University, p.685.

43. Schoenmakers, H.C.H., Wolters, A.M.A., Hann, A. de., Saiedi, A.K. and Koornneef, M.1994. Asymmetric somatic hybridization between tomato (*Solanum esculentum* Mill) and gamma-irradiated potato (*Solanum tuberosum* L.): A quantitative analysis. *Theor. Appl. Genet.*, **87:** 713-720.

44. Sidorov, V.A., Kuchko, A.A. and Gleba, U.U. 1984. Genetic modification of *Solanum tuberosum* through protoplast culture and fusion. In: (Rohlich, P. and Bascy, E., ed.), *Tissue Culture and Res.*, p. 529-534.

45. Simmonds, N.W. 1976. Potatoes. (In) Evolution of Crop Plants, Vol. I, pp. 279-284. Simmonds, N.W. (Ed.) Longman, London.

46. Subha, K., Singh, D. and Mukherjee, A. 2016. Morphological and biochemical characterization of potato (*Solanum tuberosum* L.) germplasm in *Terai* region of Uttarakhand. 1st International Agrobiodiversity Congress (Abs. Session: II. Trait Discovery and Enhances use of PGR), 6-9 November, 2016, held at New Delhi, p. 166.

47. Swaminathan, M.S. and Howard, H.W. 1953. The cytology and genetics of potato (*Solanum tuberosum* L.) and related species. *Bibliographia Genetica*, **16:** 1-192.

48. Tempelaar, M.J., Drenth-Diephuis, L.J., hesseling-Meinders, A., Hartman, R., Duyst, A. and Jacobsen, E. 1988. Inter-specific hybrids of AEC-resistant *Solanum tuberosum* and nitrate-reductase-deficient *Nicitiana plumbaginifolia*. *Prog. Plant Protoplast Res.*, 219-220.

49. Thakur, K.C., Pandey, P.C. and Pandey, S.K. 2003. Producing potato through TPS. *Indian Hort.* (April-June: 32-34.)

50. Uijtewaal, B.A., Suurs, L.C.J.M. and Jacobson, E. 1987. Protoplast fusion of monotoploid (2n=x=12) potato clones: identification of somatic hybrids using malate-dehydrogenase as biochemical marker. *Plant Sci.*, **51:** 277-284.

51. Woolfe, J.E. 1987. The potato in human Diet. Cambridge University Press, Cambridge.

Genus *Manihot*

Tapioca or Cassava (*Manihot esculenta* L.)

Introduction

In Europe and United America cassava (*Manihot esculenta* L., 2n=2x=36) is known as Manioc, Yuca and Mandioca. In India and Malayasia, it is known as tapioca. Tapioca is completely deformed word of 'tipioca' and came from 'Tupi'. Apart from its role as a subsidiary food, during the past few decades there has been growing recognition of the value of cassava roots as a low cost energy source for livestock but recently there has been a growing interest in cassava as a source of industrial starch and as a substrate for the production of fuel alcohol. This crop has the potential to produce more food per unit area, capacity to withstand adverse biotic and abiotic stresses and adaptability to the conditions of drought and marginal lands (Edison *et al.*, 2006). Among the tuber crops, cassava is the most important one in the tropics which is continuous to be a crop of food security for the millions of people especially in the developing countries of the globe (Booth-Tucker, 1908) and it ranks 4th after rice, sugarcane and maize, as a source of calories for human consumption. It is rich source of starch (carbohydrate a major source of energy), minerals and vitamins (rich in Vit. C but the thiamine, riboflavin, and niacin contents are not as high). Firstly, it is a staple food, which forms a part of diet of the people in Asia and Africa and is second most important starch producing crop of the tropics after sweet potato. Cassava provides carbohydrate as a food for about 500 million people in the world and in Africa. It is the most important source of calories in the human diet. Cassava roots are grossly deficient in proteins (1-2 per cent), fat, and some of the minerals and vitamins. The peel of cassava roots contains slightly more protein than is found in the flesh. Therefore, peeling results in loss of part of the valuable protein component of the root. The amino acid profile of the cassava root is very low in some essential amino acids, particularly lysine, methionine, and tryptophan. The cassava root contains carbohydrates, 64-72 per cent of which is made up of starch, mainly in the form of amylase (20 per cent) and amylopectin (80 per cent). In root hydrocyanic glycoside is present in varying quantity. Digestibility of starch in raw state is 48 per cent and after cooking 78 per cent. Crude fiber content varies from 1.0-2.0 g per 100 g of dry matter. The chemical composition of cassava varies in different parts of the plant, and according to variety, location, age, method of analysis and environmental conditions (Muthuswamy, 1975). About 17 per cent sucrose (in sweet varieties) and small quantities of fructose and dextrose have been reported (Anantharaman and Ramanathan, 1999). The lipid content of cassava is only 0.5 per cent. The tapioca leaves contain high amount of vegetable protein (7 per cent on fresh weight basis) than the roots and vitamins (Vit. A and B) which is consumed as a green potherbs. The tubers are easily digestible and hence can contribute to a considerable share of the energy requirements of animals and poultry.

Uses of cassava vary greatly from region to region. It is consumed as cooked/baked tubers in culinary preparations. The peeled roots are prepared by simply boiling, roasting or baking. Globally 60 per cent of the cassava production is used for human consumption, 27.5 per cent for animal feed and the rest for starch and starch based industries (Edison *et al.*, 2006). In Africa about 90 per cent of total production is used for human consumption and provides more than 50 per cent of the calorie requirement of over 200 million populations. In India about 70 per cent cassava produced is consumed in the boiled form along with fish or coconut. Baked tubers are used in Tamil Nadu and Andhra Pradesh during harvesting season. It is capable to produce more calories than rice and wheat (Ghose and Nair, 1984). The cassava starch is used for glue, laundry and pudding making. Presently it is used in making pappads. Nowadays cassava dishes are seen in big hotels and restaurants. With the prospect of production of industrial alcohol from tapioca its cultivation has been extended to other countries. Cassava fried chips is another form of utilization observed in Tamil Nadu and Kerala at cottage industries level. In the industrial level many value added products are prepared from cassava *viz.*, starch, sago, flour, chips *etc.* Cassava starch is used in textile industries as sizing agent, in pharmaceutical industries, making adhesive, dextrin manufacturing, paper industry, and many fast food preparations (Kunju, 1972). Flour is made from cassava dried chips and this finds applications in gum industry, in making 'kumkum' and in making colours applied to faces during celebrations, festivals *etc.*

The toxicity of a cultivar varies according to environmental conditions (Kay, 1987). The toxicity principle in cassava is hydrocyanic, or prussic acid, found in the roots, branches and leaves of the plant in both free and chemically bound forms. The plant contains a cyanogenetic glucoside called phaseolunatin begins to break down upon harvest into hydrocyanic acid, acetone and glucose by the action of the enzyme linase. The presence of hydrocyanic acid is easily recognized by a bitter taste. At the harvest of cassava roots, the amount of the acid in the plant varies from harmless to lethal-from a few milligrams to 250 milligrams or more per kilogram of fresh root. Investigations show that the glucoside content in the cassava plant is markedly increased by drought and by potassium deficiency (Sinha, 1969). Hydrolysis of the glucoside by the enzyme can be accelerated by soaking the roots in water, by crushing or cutting them or by heating. It was found that the hydrocyanic acid content varied little in different tubers of one plant but varied considerably in tubers obtained from different locations. The distribution of the acid in roots varied in different varieties. In sweet varieties, the major part of the acid is located in the skin and in the exterior cortical layer, while in bitter varieties the acid is uniformly distributed in all parts of the roots. In choosing a strain, the hydrocyanic acid content should be taken into account. Highly poisonous strains are preferred for plantings with the object of starch manufacture, thereby minimizing thefts by both animal and men.

Origin, History and Distribution

Cassava is native of tropical America and North-Eastern Brazil being considered as the likely centre of origin. Cassava was widely distributed in the low land tropics of South and Central America by the time of Spanish colonization. It was early domesticated and was cultivated in Peru and Mexico 4,000 and 2,000 years ago, respectively. Cassava is now cultivated throughout tropics around the world. It is believed that the slave merchants/Portuguese introduced cassava to the west coast of Africa in about 16th century. The Portuguese brought it later to their stations around the mouth of Congo River and believed that Portuguese distributed the crop from Brazil to other areas around the world. By about 1850, it was introduced from Brazil into Java, Singapore and Malaysia and later to Indonesia. The crop was introduced to Thailand from Malaysia.Portuguese also introduced the crop into India when they landed in the Malabar region, presently part of Kerals state during the 17th century from Brazil. Cassava is also an important food crop of Latin America. The Cassava is cultivated in 16.0 million hectares, spread over the continents of South America, Africa and Asia, producing 158.0 million tonnes of tubers. During the period 1919-1941, about 98 per cent of the total cassava flour was produced in Java. However, after World War II Brazil increased its production considerably. Cassava grows well in tropical and subtropical regions of the world. According to FAO estimates, 198.0 million tonnes of cassava was produced worldwide in 2012. Nigeria ranks 1st in the world with annual production of 33.0 million tonnes. Africa produces 51.0 million tonnes of cassava annually, which constitute around 40 per cent of the total world production. Brazil, Zaire, Democratic Republic of Congo, Nigeria, the USA, India, Thailand, Indonesia, Vietnam, Tanzania and Sri Lanka are other important cassava growing countries. India produced 27.42 t/h from an area of 0.24 million hectare. In India, Kerala, Andhra Pradesh, Assam, Meghalaya, Karnataka and Tamil Nadu are important cassava growing states.

Species Availability, Evolution and Diversity

Out of the 98 *Manihot* species reported, 79 species are available in Brazil but India is maintaining 8 species. Tapioca or cassava [*Manihot esculenta* (L.) Crantz., 2n=36 but polyploids with 2n=54, 72 are also available but not commonly known in the wiid state (Kay, 1987). Wild species of cassava are found in the natural habitat in Brazil but none of the existing wild species can clearly be identified as the ancestor of cassava. Some 98 species of the genus *Manihot* are found in the western hemisphere and it appears that *M. esculenta* must have arisen by mutation or hybridization. Other *Manihot* species like *M. glaziovii* Muell, *M. dichotoma* Ule., *M. saxicola etc.* are quite common. Great variation is shown in number, shape and size of tubers and the angle at which they penetrate the ground. *M. melanobasis* Ule., is a good source of protein in roots. Recently genetic variation at microsatellite loci has been used for the investigation on evolutionary and geographical origins of cassava from its wild relatives (Olsen and Schaal, 1999). The evidence from molecular analysis point towards the possible origin of cassava from Brazilian wild species *viz. Manihot tristis* and *M. esculenta* sub sp. *flabellifolia*. This finding depicts the genetic diversity in *M. esculenta* sub sp. *flabellifolia* and *M. esculenta* subsp. *peruviana* and their genetic similarity to cassava supporting the hypothesis that these two species might be the ancestors of cassava (Santha *et al.*, 1996).

Botanical Description

Tapioca or cassava belongs to family Euphorbiaceae, genus *Manihot, M. esculenta* Crantz.(syn. *M. utilissima* Pohl.). This family is recognized due to laticifers substances and produce latex.

Table 10.42: Botanical Description

Growth Habit	The plants are perennial woody slender, shrub that attain the height of about 2.0-7.0 m, not or sparingly branched or branching only near the apex shortly before the anthesis sets in.
Root	The roots are tubers radiate from the stem just below the surface of the ground. Feeder roots growing vertically from the stem and from the storage roots penetrate the soil to a depth of 50.0-100.0 cm. The number of tuberous roots and their dimensions vary greatly among the different varieties. The roots may reach a size of 30.0-120.0 cm long and 4.0-15.0 cm in diameter, and a weight of 1.0-8.0 kg or more. This capacity of the cassava plant to obtain nourishment from some distance below the surface may help to explain its growth on inferior soils.
Stem	As the plant grows, the main stem forks, usually into three branches which then divide similarly. Stems are thick with much prominent leaf scars. Young stems are yellowish green or strongly tinged with violet, pruinose and glabrescent.

Leaf	Leaves are alternate, long stalked, large, palmately 3-9 partite, and subpelate at the base. Leaves grow only toward the end of the branches. Segments are obovate-lanceeolate, narrowed at the base, acuminate, acute glaucous and shining above, beneath much lighter coloured, dul;l, 5.0-20.0 cm long and 2.0-6.0 cm wide. Slender petiole petiole is terete, glabrous, yellowish green or red and 10.0-35.0 cm long.
Floral biology	Cassava is unisexual-monoecious, male and female (base) occurs in same panicle, in laxpanicles, pedicelled and apetalous. Flowering is frequent and regular in most of the cultivars. Calyx of the male flowers is campanulate, 5 partite. Stamens are 10 in number, alternately shorter and longer. Disc is large, fleshy, stellately and double-5-lobed. Calyx of the female flowers is 5 partite. Ovary is 3 celled. Style is connate, stigma broad and many lobed.
Anthesis and pollination	Female flowers open one week earlier to male; therefore cross pollination occurs by insects particularly honeybees, which are attracted to the flowers by abundant nectar. No compatibility has been reported; self-pollination is limited but not prevented by protogynous flowering habit. The pistillate flowers opened in about mid day. Male sterility is observed at a low frequency.
Fruit and fruit set	From flowering to fruit maturity takes 3-5 months. Mature fruit is a capsule, ovoid, globose with 6 longitudinal and plicate wings having globular shape with a diameter of 1.0-1.5 cm.
Seeds	The triangular-shaped fruit contains 3 seeds having typically euphorbiaceous carunuculate seed, which are viable and can be used for the propagation of the plant. Presence of winged seed capsule is the morphological characteristics of cassava.
Root tubers	There are usually 5-10 tubers/plant, cylindrical or tapering, 3-15 cm in diameter and 15-100 cm long, occasionally longer.

Horticultural Classification

Cultivars have been distinguished on the basis of morphological characteristics like leaf shape, size, plant height and tuber shape. Earliness in maturity of the crop, root yield and cyanogenic glucoside content of roots has also been used to distinguish varieties. Colour of the young leaf bearing part of the stem is generally utilized for botanical description of cultivars. Colour and size of roots are useful for identification.

(A) Tapioca clones are often classified on the basis of taste:

(i) Sweet types-It is called as *M. esculenta.* The sweet types are commonly grown for home consumption. The glucoside is confined mainly to the peel and it is at low level. The flesh of sweet varieties is therefore relatively free of the glucoside. Sweet type has short growing season (6-9 months) and storage life is less.

(ii) Bitter types-It is called as *M. palmate* or *M. dulcis.* The bitterness is due to presence of HCN or cyanogenic glucoside, which is distributed throughout the tuber and in a high level. These are cultivated for the industrial market. Presence of winged seed capsule is the morphological characteristics of bitter types. Generally cultivars of this group require 12-18 months for maturity.

(B) Hairiness or otherwise on the upper side of the unfolding leaves and petioles, colour of the tender stem, petiole, leaf veins and mid ribs, skin of the main stem, colour of leaf scars and the proportion of width to length of the main lobe of leaf.

(C) Stamen colour (red or not), petiole length (sessile or not), colour of the tuber peel, the branches and perianth.

Augmentation of Genetic Resources and Crop Improvement

Introduction and Selection

A good number of elite genotypes have been introduced to India from Brazil, Madagascar, Nigeria, Uganda, Kenya, Ghana, Thailand, and Malaysia, Sri Lanka and Fiji. Genotype 'M4' is another important introduction from Malaysia. In cassava growing areas evolution of new varieties is a common feature. Some of the commercial varieties bear flower and set seed under natural conditions. Farmers select the outstanding seedlings and propagate them. 'Ambakadan', a high yielding cultivar in central Kerala is a typical example of such a selection. Nedunchezhiyan and Mohanty (2005) evaluated three short duration (H-65, Sree Prakash and Sree Jaya) and four long duration (M-4, H-97, H-226 and CO-2) varieties during 2001-2002 and four short duration (H-165, Sree Prakash, Sree Jaya and Sree Vijaya) and six long duration ('Sree Vishakam', 'Meg-36', 'M-4, H-97', 'H-226' and 'CO-2') during 2002-2003 at Regional Centre of Central Tuber Crops Research Institute, Dumuduma, Bhubaneswar (Odisha). The yield variation due to varieties was significant during both the years of study. Significantly higher yield was recorded with short duration varieties like 'H-165', Sree Jaya', 'Sree Vijaya' and 'Sree Prakash' than long duration varieties. However, 'Meg-36' harvested 10 months after planting recorded higher yield during 2002-03. Long post monsoon dry period during maturity phase may be affecting the yield of long duration varieties. Among the short duration varieties H-165 recorded higher yield during both the years of investigation. The tuber yield of 'Sree Prakash' was very low during the year 2002-03 compared to the year 2001-02. This showed that Sree Prakash is highly vulnerable to climatic fluctuations. Genotype 'CO-3' was developed through introduction of OP variety from Nigeria. Similarly KMC-1 was introduced fro Ramanthala. Other cultivars like 'Nidhi', 'CO-1' and 'CO-2' were developed through selection and multiplication of local germplasm.

Hybridization and Selection

In cassava, lack of flowering and pollen sterility are main constraints in germplasm utilization, as less than 50 per cent of the accessions flower, and only two or three seeds are obtained from a sigle flower. In wild species, maintenance and propagation in *ex situ* condition is difficult. However, some of the non-flowering types were found to flower in South Tamil Nadu. A hybridization garden has been established there and a few rare crosses were successfully made.

Department of Botany, University of Kerala, India had developed 1, 200 F_1 seedling plants through inter-varietal hybridization programme, and selected 10 hybrids on the basis of yield. Out of 10 hybrids 'H-105' and 'H-96' are finally released for commercial cultivation. Five promising hybrids 'H-97', 'H-165', 'H-226', 'H-1687' and 'H-2304' were isolated and released to farmers from a combination of 30, 000. 'H 1687' is named as 'Sree Visakham' and 'H-2304' as 'Sree Sahya'. In hybridization programme 'H-97' was developed between the cross of 'Manjavella' and a 'Brazilian Line'. Similarly, 'H 165' an early maturing hybrid was developed between the cross of 2 local genotypes *i.e.* 'Chadayamangalam Vella' and 'Kalikalan'. 'H-226' is another hybrid developed between the cross of 'Ethakka Karuppan' and 'M4'. 'Sree Visakham' ('H 1687') is a hybrid and developed between the cross of local variety 'Cheeni Kappa' and an exotic variety from Madagaskar. 'Sree Sahya' ('H 2304') is double hybrid. Hybrid 'H 226' is free from attack of scale insect and red mite under field conditions. Similarly, hybrid 'Sree Visakham' ('H 1687') is shows field resistance to mosaic. One high starch triploid hybrid ('Sree Harsha') and two high yielding top cross hybrids ('Sree Rekha' and 'Sree Prabha') of cassava were released for commercial cultivation.

Table 10.43: Promising Hybrids of Cassava

Cultivars	*Cross Combination Used in Development*
H-165	Chadyamangalam x Kalikalan
Sree Harsha	OP-4 x H-2304
Sree Sahya	(Ac. No. 468 x Ac. No.84) x (M-4)
H-226	Ethakkakaruppan x M-4
Sree Visakham	Ac. No. 1501 s S-2312
H-97	Manjavella x Brazilion seedling selection

Interspecific Hybridization

Interspecific crosses also were effected for imparting mosaic resistance and high protein in roots, drought tolerance *etc.*, to cultivated cassava varieties. Transfer of genes responsible for high tuber yield, resistance to virus diseases and drought are to be attained through such interspecific hybridization. Successful interspecific hybridization has been reported from many countries between tapioca and other *Manihot* species like *M. glaziovii* Muell, *M. dichotoma* Ule, and *M. saxicola* Lanj. Many of the interspecific F_1 hybrids are male sterile but with partial female fertility. The interspecific F_1 hybrids between tapioca and *M. melanobasis* Ule., are reported encouraging. The hybrid between *M. esculenta* and *M. caerulescence* is showing cassava mosaic disease (CMD) tolerance, and efforts are being made to transfer the tolerance of this hybrid to improved cultivars. Nassar (1989) hybridized two *Manihot* species, *M. neusana* and *M. anomala*, with cassava, *M. esculenta*, through controlled crosses using insects.

Table 10.44: Casava Varieties Released from CTCRI, Kerala

Variety	*Maturity Period (months)*	*Starch (Per cent)*	*Yield (t/ha)*
'H-97'	10	27.0-31.0	25.0-35.0
'H-165'	8-9	23.0-25.0	33.0-38.0
'H-226'	10	28.0-30.0	30.0-35.0
'Sree Visakham'	10	25.0-27.0	35.0-38.0
'Sree Sahya'	10-11	39.0-31.0	35.0-40.0
'Sree Prakash'	7	29.0-31.0	30.0-35.0
'Sree Harsha'	10	38.0-41.0	35.0-40.0
'Sree Jaya'	6-7	24.0-27.0	26.0-30.0
'Sree Vijaya'	6-7	27.0-30.0	25.0-28.0
'Sree Rekha'	8-10	28.0-30.0	45.0-48.0
'Sree Prabha'	8-10	25.0-29.0	40.0-45.0

Polyploidy Breeding

Polyploidy breeding is a very potent method of crop improvement in tapioca. The axillary buds are treated with 0.5-1.0 per cent colchicine in agar adding water. The tetraploids of 'H 602' and 'H 312' failed to get established in the field during the 5th clonal multiplication while the tetraploids of 'S 308' and 'M 4', though had gone through several clonal multiplications, did not lose their polyploidy vigour. Significant increase in crude protein in tubers was observed in tetraploids compared to diploids. A tetraploid of 'M 4' contained 3.97 per cent protein on dry weight basis while the diploid only 2.7 per cent.

Interspecific Grafting

Replacement of the tapioca upper canopy with graft or buds of Cera Rubber (*Manihot glaziovii* Muell) is found to improve productivity in tapioca. This system of grafting or budding of *Manihot glaziovii* Muell onto a stock of tapioca is named Mukibat system. This system is gaining popularity in East Java and other parts of Indonesia. The effect of scion on rootstock is manifesting higher productivity is an interesting matter for research work. Many Mukibat hybrids like 'H 165' and 'H 226' have been evolved.

Mutation Breeding

Mutation breeding has been of little utility in tapioca except for occurrence of spontaneous mutants in the form of chimeras. Attempts have been made to create variability in tapioca through gamma irradiation of seeds to generate useful variability for screening against tapioca mosaic.

Resistance Breeding

Disease Resistance

Cassava Mosaic Disease (CMD) is major problem associated with this crop. Some lines evaluated and screened against CMD are given in Table 10.45.

Drought Resistance

Tapioca species like *Manihot glaziovii, M. dichotoma* and *M. saxicola* are resistance to drought.

Table 10.45: Morphotype-wise Range, Mean and SD over the Years

Morphotype	No. of Accs.	Range	Mean	SD	Rating	Name of cultivar
1	2	6-8	6.8	2.891	High	Ariyan
2	18	0-5	1.9	1.474	Low	M-4
3	6	2-8	4.6	1.897	Medium	Ammayappan Kappa
4	1	-	0.0	-	Absent	Thanduchoppan
5	16	1-8	5.3	1.957	Medium	Mixture
6	2	2-2.25	2.1	0.115	Low	Block Kattan
7	6	7-9	7.5	0.359	High	Arumasan
8	1	-	8.7	-	High	Chullithandu
9	1	-	9.0	-	High	Thooavella
10	1	-	2.3	-	Low	*
11	15	6-9	7.8	1.429	High	Karuthakattan
12	1	-	8.0	-	High	Anamanthari
13	1	-	5.25	-	Medium	Vella
14	4	0-9	2.0	2.365	Low	Chullikkappa
15	5	5-9	7.4	0.784	High	Elamuriyan
16	2	2-8	5.9	1.875	Medium	Anakkomban
17	1	-	7.0		High	Padinjattivella
10	1	-	2.5	-	Low	Vellarosa
19	3	0-8	6.2	1.885	Medium	Thodan
20	1	-	8.8	-	High	Rottivarikka
21	7	0-9	6.02	1.349	Medium	Mullanthalayan
22	1	-	6.5	-	High	Karuthakattan
23	2	4-9	6.25	1.750	High	Kottaramchulli
24	8	0-9	6.5	1.587	Medium	Aambakkadan
25	4	1-7	4.9	1.021	Medium	Ethakkakamban
26	8	4-9	7.9	1.184	High	Thulakkappa
27	1	-	7.5	-	High	*
28	1	-	8.5	-	High	Ethakkakappa
29	1	-8-9	8.0	-	High	*
30	5	7-9	8.8	0.292	High	Kantharipadarpan
31	2	-0-4	8.0	0.250	High	Pathinettu
32	1	-	7.5	-	High	M-137
33	2	8-9	0.75	0.500	Low	Arumasam
34	1	-	7.0	-	High	Kadalakkamaniyan
35	4	-	8.0	0.836	High	Kalale Local
36	1	-	3.8	-	Low	Ayoorvella
37	1	-	2.8	-	Low	Ambalamkulukki
38	1	-	5.8	-	Medium	Ornamental
39	1	-	9.0	-	High	*
40	1	-	0.99	-	Low	Vakkeel Kappa

Morphotype	No. of Accs.	Range	Mean	SD	Rating	Name of cultivar
41	1	-	5.3	-	Medium	*
42	1	-	8.5	-	High	Areekkara
43	8	0-6	3.0	2.304	Low	Ramanthala
44	1	-	6.8	-	High	Karikkalapadarpan
45	1	-	7.5	-	High	*
46	1	-	8.8	-	High	Mullaanthalayan
47	2	0.0	0.0	-	Absent	Kattan
48	1	-	6.3	-	Medium	Sutharivella
49	1	-	5.3	-	Medium	Rottikkappa
50	1	-	5.3	-	Medium	*
51	1	-	8.0	-	High	Quintalkappa
52	1	-	5.8	-	Medium	Vellaelamuriyan
53	1	-	0.0	-	Absent	Chullithari
54	1	-	3.0	-	Low	Vellachulli

*Cultivar name not available.

Source: Velayudhan and Asha, 1998, ICAR-NBPGR, R.S., Vellanikkara (Kerala).

Genetic Resource Conservation

The International Institute of Tropical Agriculture, Nigeria, the ICAR-Central Tuber Crops Research Institute, Sreekaryam, Thiruvananthapuram (Kerala) India, are a few of research organizations involved actively in cassava/tapioca improvement and germplasm management. Total 1271 accessions are being maintained and conserved in the field genebank of ICAR-CTCRI, Thiruvananthapuram (Kerala), 184 landraces of cassava are conserved at ICAR-NBPGR, Thrissur (Edison, *et al.*, 2005) and 569 cassava accessions in different Agricultural Universities under the All India Co-ordinated Research Project on Root and Tuber crops (416 accessions at Tamil Nadu Agricultural University, Coimbatore). Majority of the accessions are cultivars or breeding lines. Genetic erosion of wild *Manihot* species is occurring at a fast rate in Brazil due to deforestation.

REFERENCES

1. Anantharaman, M. and Ramanathan, S. 1999. SWOT analysis of cassava in India. In: (Eds.) Balagopalan, C., Nayar, T.V.R., Sundaresan, S., Premkumar, T. and Lakshami, K.R. Tropical Tuber Crops in Food Security and Nutrition. Oxford and IBH publishing Co. Pvt. Ltd., New Delhi, p. 499-506.
2. Booth-Tucker, F. 1908. Casava as famine food. *Agricultural J. of India*, **3:** 227-230.
3. Edison, S., Anantharaman, M. and Srinivas, T. 2006. Status of Cassava in India: An overall view. *Tech. Bull.* 26. p. 172.
4. Ghose, S.P. and Nair, R.G. 1984. Tuber Crop Research and Development. *Indian Fmg.*, **5:** 8-15.
5. Kay, E. 1987. Root crops. Tropical Development and Research Institute, London. p. 30-57.
6. Kunju, U.M. 1972. Tapioca a food cum industrial crop. *Farmer and Parliament*, **7(10):** 9-10, 29.
7. Muthuswamy, P. 1975. Composition and nutritive value of certain cultivars of cassava tubers, *Manihot esculenta*, Crantz. *Madras Agricultural J.*, **62(2):** 68-70.
8. Nassar N.M.A. 1989. Broadening the genetic base of cassava, *Manihot esculenta* Crantz, by interspecific hybridization. *Can. J. Pl. Sci.*, **69:** 1071-1073.
9. Nedunchezhiyan, M.and Mohanty, A.K. 2005. Economic evaluation of cassava (*Manihot esculenta*) varieties under rainfed conditions in Orissa. *The Orissa Journal of Horticulture*, **33 (1):** 108-109.
10. Olsen, K.M. and Schaal, B.A. 1999. Evidence on the origin of cassava: Phylogeography of *Manihot esculenta*. *Proceedings of the National Academy of Science*, **96:** 5586-5591.
11. Santha, V.P., Nair, S.G., Lakshmi, K.R. and Sheela, M.N. 1996. Characterization of cassava germplasm by morphological and Isozyme markers. In: eds. C. Balagopalan, T.V.R. Nayar, S. Sundaresan, T. Premkumar and K.R. Lakshmi. Tropical Root and Tuber Crops in Food Security and Nutrition. Oxford and IBH Publishers, New Delhi. p. 112-117.
12. Sinha, S.K. 1969. Cyanogenic glucoside production and its possible control in cassava. *Indian J. of Plant Physiology*, **12(1/2):** 140-147.
13. Thampan, P.K. 1979. Cassava, (edn. I), Kerala Agricultural University, Vellanikkara (Kerala).
14. Velayudhan, K.C. and Asha, K.I. 1998. A note on incidence of cassava mosaic disease (CMD) in cassava collections. *Indian J. Pl. Genet. Resour.*, **11(1):** 97-101.

Genus *Ipomoea*

Sweet Potato [*Ipomoea batatas* (L.) Lam.]

Introduction

Sweet potato [*Ipomoea batatas* (L.) Lam.], locally known as "Shakarkand" (modified root) is one of the popular tuber crops (starchy root crops) in India and abroad owing to its yield potential and high calorific value production capacity per unit of time and area. Sweet potato has tremendous potential as a food crop (consumed after boiling or baking) and it ranks first among cultivated crop in the developing countries in terms of edible energy. It could be a better competitor as food, feed and industrial raw material (Tsou *et al.*, 1997). The orange fleshed genotypes of sweet potato are excellent source of the β carotene which is a precursor of 'vitamin-A' (Woolfe, 1992). The importance of sweet potato has increased due to the possibility of production of industrial sago, starch and ethyl alcohol from tubers. The protein present in sweet potato can reduce protein deficiency in diet. Sweet potato ranks 6th of the world crop after wheat, rice, corn, potato and barley. Presently China is the largest producer of sweet potato and accounts for 105 million tones (90 per cent) that constitutes about 83 per cent of the total world production. Vietnam, Indonesia and the Philippines are the other sweet potato growing countries. In India, it is 3rd most important tuber crop after potato. Culled roots and vines are used as animal feed.

Origin, History and Distribution

Geographically, sweet potato is being grown under latitude ranging between 40°N to 32°S and altitude up to 2200 meters above the mean sea level. The crop originated in Central America and archaeological evidence from Peru shows that domestication of sweet potato dates back to 6000 BC.Veasey *et al.* (2008) reported that sweet potato is the oldest vegetable known to mankind, having been in cultivation since about 10, 000 years ago. It was taken to Spain around 1500 and later to Belgium and England. Early Spanish explorers are believed to have taken sweet potato to the Philippines and the East Indies, from where it was introduced to India, China and Malaysia by the Portuguese travelers. From China, it was taken to Japan around 1700. The crop was introduced to Thailand from Malayasia. It is now grown almost in all the tropical and subtropical countries like China, India, Brazil, Peru, USA, Thailand and Nigeria. Sweet potatoes were reported grown in Virginia in 1748 and were in general use about the time of the American war of Independence (MacGillivary, 1961). However, due to temperate conditions, it could not be cultivated successfully in most of the European countries. Asia is the world's largest sweet potato producing region. India ranked 7th position with total area of 0.12 million hectares and production 1.44 million tones but the productivity of the crop is 9.23 metric tonnes/ha (FAOSTAT, 2017). In India the productivity of sweet potato is only 8.0-9.0 tonnes per hectare against the world average of 16.0 tonnes per hectare. Though, it is cultivated in all the states of India but the major area under this crop is confined to the states of Assam, Odisha, Bihar, Karnataka, Kerala, Madhya Pradesh, Maharashtra and Uttar Pradesh.

Species Availability, Evolution and Diversity

There are 45 genera and about 1,000 species in the family Convolvulaceae but only 142 species are identified by researchers. Among them only *Ipomoea batatas* is of major economic importance. Besides *I. batatas*, there exist 13 wild species closely related to sweet potato. The closest wild relatives are confined to American, and Mexico region is suggested as the centre of diversity for the *Batatas* section of *Ipomoea*. The evolutionary steps by which the crop originated remain uncertain. Ting and Kehr (1963) suggested hybridization between two unknown wild species followed by allohexaploidy (2n=2 x 15 + 4x 15=90). Sweet potato is highly heterozygous, and there is extensive variability within the species (Afuape *et al.*, 2011 and Afuape and Nwachukwu, 2005.). Genetic variability and relatedness among various genotypes of sweet potato using agro-morphlogical traits has been worked out by several researchers (Soegianto, *et al.*, 2011 and Tairo *et al.*, 2008). Other related species are *I. abrupta, I. alba, I. alpina, I. amnicola, I. aquatica,I. asarifolia,I. barbatisepala, I. batatas, I. batatoides, I. bona-nox, I. cairica, I. capillacea, I. carnea, I. coccinea, I. cordatotriloba, I. cordifolia, I. costata, I. costellata, I. cristulata, I. cynanchifolia, I. daturaefolia, I. demerariana, I. diversifolia, I. dumetorum, I. eggersiana, I. eggersii, I. eriocarpa, I. ghika, I. gracilis, I. graminea, I. hederacea, I. halierca, I. hederifolia, I. horrida, I. horsfalliae, I. imperati, I. incisa, I. indica, I. jalapa, I. krugii, I. lacunosa, I. leptophylla, I. leucantha, I. lindheimeri, I. littoralis, I. lobata, I. longifolia, I. macrantha, I. macrorhiza, I. marginata, I. mauritiana, I. meyeri, I. microdactyla, I. obscura, I. ochracea, I. oenotherae, I. pandurata, I. pes-caprae, I. plebeia, I. plummerae, I. polymorpha, I. pubescens, I. pulcherrima, I. purga, I. purpurea, I. quamoclit*, and *I. racemigera* having their own existence in their region of adoption.

Gene and Genetics of Characters

Table 10.46: Genetics of Characters

Character	Number of Genes	Type of Gene Action
Stem colour	Monogenic	Red stem is dominant to green stem
Tuber skin colour	Digenic	Red skin is dominant to green skin
Tuber flesh colour	Digenic	Red flesh colour is dominant to green white
Internode length	Monogenic	Long is dominant to short
Floral fertility	Monogenic (S_1S_1)	Self-fertility is dominant to sterility

Botanical Description

The cultivated sweet potato (dicot) and the closely related wild species belongs to the family Convolvulaceae, genus *Ipomoea*, subgenus Eriospermum, section Eriospermum (formerly Batatas) and series Batatas (Austin and Huaman, 1996). It is cross pollinated in nature. This crop is autohexaploid 2n=6x=90 (X=15) as reported by Afuape *et al.* (2011) while most of the wild species are 2x or 4x, although there are some species like *I. trifida* which has 2x, 3x, 4x and possibly 6x cytotypes. Austin (1988) hypothesized that *I. triloba* and *I. trifida* are the species that contribute the sweet potato genome and that *I. tiliacea* might also have been involved in the origin of sweet potato. Shiotani (1988) reported that sweet potato has the genome structure of an autohexaploid with the B genome that also exist in autotetraploids and diploids of the *I. trifida* complex. However, botanical geneology needs further research on wild *Ipomoea* species.

Sweet potato propagated vegetatively through stem cutting. The edible part in sweet potato is the modified root. The shoot and root come from each node along the main stem. Self and cross incompatibility has been observed in this crop. Sweet potato is highly cross-pollinated due to self-incompatibility and male sterility.

Table 10.47: Botanical Description

Growth Habit	It is herbaceous perennial but cultivated as annual shrub, which trails on the ground with single, entire or palmately sect leaves and adventitious root arising from each node.
Root	Adventitious and modified root. Some of the roots store food material and become tubers.Vine cuttings develop into new shoots.The edible tuberous root is long and tapered with a smooth skin.The skin colour ranges between red, purple, brown and white. The flesh colour ranges between white, yellow, orange and purple.
Stem	Sweet potato is weak stemmed and branching vine with milky juice. The vine is prostrate or ascending and twining. The vines develop roots at the joint during later part of the season. The vine is light green to purple in colour, usually angular in shape and slender to moderate in thickness.
Leaf	Entire or palmately sect leaves. Triangular leaf is borne on short to long petiole. Leaves are more or less heart-shaped but may be shouldered, toothed, entire, deeply cleft or variously lobed. The surface of leaf is glabrous or with variable pubescence. The leaves vary in colour usually green to purple.
Floral biology	The inflorescence of sweet potato is cyme, which bears single or few flowers. The corolla usually companulate (funnel shape), with 5 lobes and it is purple in colour with deeper throat and pale at margin. Sepals not aristate. There are 5 stamens which are attached near the base of corolla. The stamens are unequal in length, the longest stamens have same length or shorter than style. The pistil has a 2.0 cm long style with 2 lobed stigmas.
Anthesis and pollination	Flowering is stimulated by short days, long nights and high temperature. Sweet potato flowers readly in tropical climate. In tropics, floweing and seed weight depend on environmental conditions and season. Day length of 11 hours is ideal for flowering (Mc Clelland, 1928; Kasam, 1976). Flowers open early in the morning and wilt on the same day. Stigma becomes receptive in the bud form 6-8 PM and anthers start to dehisce from 11-12 PM (Purseglove, 1977). In a trial in India, it was observed that stamens shed their pollen around 6-7 AM with sunrise (Srinivasan, 1977). The pollens remain viable or 24 hours under normal temperature. The stigma may be receptive for few hours after the wilting of flower. Sweet potato is self-incompatible (Martin, 1965; Purseglove, 1977). Natural cross-pollintion usually occurs by bees.
Tubers	Tubers of different cultivars from different shape and size which has commercial standpoint. Too large or small size of tuber is undesirable to common consumers. The shape of tuber is fusiform and globular with smooth or ridges surface. The skin colour is commonly white, yellow, orange, yellow-reddish or purple. Root tubers develop on nodes of vine cuuting or vine which rests on the soil by the process of secondary thicking of under ground adventitious roots. Tubers consist of conducting tissues, parenchymatous storage cells, latex vessels and outer periderm.
Fruit and fruit set	Fruit dehiscent or if indehiscent thin and fragile walled.
Seeds	The seeds of sweet potato are borne on capsule. A single capsule may contain up to 4 seeds but 1 or 2 develop. The angular seeds (grains spinulose) are black in colour with hard seed coat.

Horticultural Classification

Sweet potato (cultivated and wild) has been grouped on the basis of:

(i) Chromosome number and genome analysis
(ii) Growth habit
(iii) Texture of tuber flesh
(iv) Classification on the basis of tuber maturity

(i) Chromosome Number and Genome Analysis

Nishiyama (1971) classified *Ipomoea batatas* and its wild relatives on the basis of chromosome number and genome analysis which is depicted as under:

Table 10.48: Wild Relatives and Genome Constitution

Species	*Synonyms*	*2n (x=5)*	*Genome Constitution*
I. batatas	-	90	-
I.batatas var. *batatas* f. *trifida*	*I. trifida*	90	BBBBBB
I. batatas var. *littoralis*	*I. littoralis*	60	BBBBBB
I. batatas var. *leucantha*	*I. leucantha*	30	BBBB
I. triloba	-	30	-
I. triloba var. *triloba*	*I.lacunosa*	30	AA
I. triloba var. *lacunosa*	*I. tricho-carpa*	30	AA
I. triloba var. *trichocarpa*	*I. ramoni*	30	AA
I. triloba var. *ramoni*	-	-	AA
I. gracilis R. Br.	-	60	-
I. tiliaceae (Willd,)	-	60	-

B, B genome with 15 chromosomes; A, A genome with 15 chromosomes

(ii) Classification on the Basis of Growth Habit-Vines

(a) Prostrate and spreading type- Genotype 'Porto Rico'

(b) Bunching type- Genotype

(iii) Classification on the Basis of Texture of Tuber Flesh

On the basis of texture of tuber flesh sweet potato can be grouped as;

(a) Firm, dry and mealy flesh after cooking

(b) Soft, moist and gelatinous flesh after cooking

(c) Very course flesh and tubers are suitable only for cattle feed or industrial uses.

Generally, orange, yellow-orange, white, purple and red flesh cultivars are available.

(iv) Classification on the Basis of Tuber Maturity

Collections developed as varieties having different maturity periods (days from planting to harvest) are as under:

Table 10.49: Maturity Group of available Varieties

Varieties which give Maximum Yield at			
75 Days	*90 Days*	*105 Days*	*120 Days*
'C71'	'750P9'	'CO1'	'IB700"
	'Kalmegh'	'H633'	
	'Kanghanged Local'	'H42'	
		'H620'	

(v) Sweet Potato Skin Colours

On the basis of sweet potato skin colour tubers can be grouped as creamy white, yellow-orange, tan, reddish-purple and red.

Genetic Resources Augmentation and Crop Improvement

In sweet potato, non-flowering nature of some of the accessions, compatibility barriers, frequent planting in the field genebank and insufficient *in vitro* storage facilities, are some of the constraints in utilization of germplasm.

Introduction and Selection

The germplasm of sweet potato was evaluated and categorized based on modified IPGRI descriptors at ICAR-Central Tuber Crops Research Institute, Thiruvananthapuram (Kerala). In sweet potato, in addition to 927 accessions conserved at CTCRI, Thiruvananthapuram (Kerala), AICRP centers are also involved in conservation of maintaining 1858 accessions as field gene bank. Seven hundred and sixty four sweet potato accessions were characterized, documented and the computerized data base is maintained (Rajendran *et al.*, 1993). Recently 869 accessions were evaluated for important morphological traits for assessing the spectrum of variability and association of characters for traits such as plant type, vine growth rate, petiole length, number of storage roots and tuber yield per plant (Easwari Amma, *et al.*, 1999). In sweet potato, 20 cultivars have been released at an all-India level; seven cultivars have been developed through clonal selection, and seven from open-pollinated material. Elite genotypes like 'Poro Rico', 'Triumph', 'Nancy Hall' and 'Nancy' are introductions to India. The introductions are evaluated for local adaptability, and market acceptability. High yielding clones have been developed through collection of germplasm and selection for elite plants. Variety 'C 71' is noted to yield as high as 26.0 tonnes/ha under warm humid conditions of Kerala. Varieties like 'Pusa Red' and 'Kanghangad Local' are developed through clonal selection. Other promising varieties are 'CO-2', 'OP-1' and 'CO-3'. Evaluation of germplasm resulted in the release of 'Sree Bhadra' and exotic germplasm with high yield (27.0 t ha-1). 'S-1221' an early bulking accession (75 days).

Hybridization and Selection

Six sweet potato cultivars have been developed through inter-varietal hybridization. The hybrid 'H268' is a double hybrid developed by crossing 2 exotic collection and 2 indigenous cultivars in 1966 at the ICAR-CTCRI, Trivandrum (Kerala).

Resistance Source and its Utilization

(i) Disease Resistance Source and its Utilization

Fusarium wilt (*Fusarium oxysporum*) Schlacht f. sp. *batatas* (Wr.) Snyd. and Hano, is vascular wilt of sweet potato in subtropics. Hughes *et al.* (1963) developed wilt resistant lines using 'Tinian' (PI153655) as resistant parent and 'HM15', a moderately wilt susceptible line. Struble *et al.* (1966) after evaluating '4033' sweet potato line for wilt concluded that resistance is inherited in a multi-factorial manner. 'Red Gold' and 'All Gold' with multi-factorial types of resistance to *Fusarium* wilt were released. Genotype 'Jewel' has expressed resistance against internal cork, *Fusarium* wilt, Southern root rot and Flea beetle.

Table 10.50: Sweet Potato Varieties Released from ICAR-CTCRI, Kerala

Sl. No.	Variety	Maturity Period (Days)	Starch (Per cent)	Yield (t/ha)
1	'H-41'	120	14.0-16.0	20.0-25.0
2	'H-42'	120	16.0-18.0	22.0-25.0
3	'Varsha'	120	14.0-16.0	17.0-22.0
4	'Sree Nandhini'	100-105	25.0-26.0	20.0-25.0
5	'Sree Vardhini'	100-105	13.0-15.0	20.0-25.0
6	'Sree Rekha'	90-105	22.0-23.0	20.0-26.0
7	'Sree Bhadra'	90	18.0-20.	20.0-27.0
8	'Gouri'	110-120	16.0-17.0	19.0
9	'Sankar'	120	17.0-18.0	14.0
10	'Sree Arun'	90	19.0-21.0	20.0-28.0
11	'Sree Varun'	90	21.0-22.0	20.0-28.0
12	'Sree Kanaka'	75-85	14.0-15.0	10.0-15.0
13	'Kalinga'	105-110	27.0-29.0	26.0-28.0
14	'Gautam'	105-110	24.0-26.0	18.32-32.0
15	'Sourin'	105-110	24.0-26.0	16.0-32.0
16	'Kishan'	110-120	29.0-30.0	16.0-26.0

(ii) Insect Pests Resistance Source and its Utilization

In sweet potato variety 'Kanghanged Local' and Hybrid 'H 620' were observed to be less affected by sweet potato weevil. 'Sree Bhadra' an exotic germplasm is very much suitable in nematode infested soil.

Mutation Breeding

Port Blaco' is a mutant of 'Porto Rico' and has white flesh and high starch content. Spontaneous mutants have also been isolated out of 'Contennial' and 'Early Port' varieties.

Breeding for Quality Traits

The germplasm of sweet potatoes includes about 200 carotene rich accessions. The flesh colour ranged from dark yellow to dark orange. Among the high carotene lines, the CIP accessions *viz.* 'S-1230' (CIP 440069), 'S-1355' (CIP 440056), 'S-1346' (CIP 187017) and 'S-1364' (CIP 440060) were found to be promising. Two varieties released from ICAR-CTCRI *viz.* 'Gowri' and 'Sree Kanaka' (early maturing hybrid (cross between 'S-187' and a hybrid 'H-633') with dark orange flesh colour) also had high carotene content. 'Sree Kanaka' having β-carotene to the tune of 8.8-10.0 mg/100 g fresh weight which is almost equivalent to that of carrot (11.0-12.0 mg/100 g fresh weight)

Self-Incompatibility and Male Sterility

Self-incompatibility and male sterility has been observed in sweet potatoes which are being utilized in development of hybrids. Cross-incompatibility has been reported in sweet potato.

Grafting

Many of the genotypes do not flower but by grafting sweet potato on related species of *Ipomoea*, flowering has been observed. In the finding it has been concluded that related species provide 'flower-inducing substance' once they are used as rootstock.

REFERENCES

1. Afuape, S.O. and Nwachukwu, E.C. 2005. Variability and correlation studies in some quantitative characters in selected sweet potato [*Ipomoea batatas* (L.) Lam.] genotypes. Genetics and Sustainable Agriculture. *Proc. 30th Annual Conference of Genetic Society of Nigeria* held at the University of Nigeria. Nsukka, 5th-8th, Sep., p. 124-129.
2. Afuape, S.O., Okocha, P.I. and Njoku, D. 2011. Multivariate assessment of the agro-morphological variability and yield components among sweet potato [*Ipomoea batatas* (L.) Lam.] land races. *Afr. J. P. Sci.*, **5(2):** 123-132.
3. Austin, D.F. 1988. The taxonomy, evolution and genetic diversity of sweet potatoes and related wild species. In: Exploration, maintenance and utilization of sweet potato genetic resources. Report of the first sweet potato Planning Conference 1987. International Potato Centre, Lima, Peru, p. 27-59.
4. Austin, D.F. and Huaman, Z. 1996. A synopsis of *Ipomoea* (Convolvulaceae) in the Americas. *Taxon.*, **45(1):** 3-38.
5. Easwari Amma, C.S., Sheela, M.N. and Rajendran, P.G. 1999. Evaluation for variability and Association of characters in sweet potato germplasm. In: Tropical Tuber Crops in Food Security and nutrition. Eds. Balagopalan, C., Nair, T.V.R., Sundaresan, S., Premkumar, T. and Lakshmi, K.R. p. 144-149.
6. FAOSTAT (Food and Agriculture Organization of the United Nations). 2009. Production Year Book for 2008. Rome, Italy.
7. Hughes, M.B., Steinbauer, C.E., Deonier, M.T. and Cordner, H.B. 1963. Preliminary studies on the inheritance of wilt resistance in the sweet potato. *Proc. of American Society of Horticultural Science*, **83:** 623-628.
8. Nishiyama, I. 1971. Evolution and domestication of the sweet potato. *Botanical Magazine, Tokyo*, **84:** 377-387.
9. Rajendran, P.G., Sreekumari, M.T., Nair, R.B., Pillai, K.S., Nair, N.G. and Moorthy, S.N. 1993. *Catalogue*

of Genetic Resources in Cassava. Tech. Bull. No. 7. ICAR-CTCRI, Thiruvananthapuram (Kerala)

10. Shiotani, I. 1988. Genomic structure and gene flow in sweet potato and related species. In: Exploration, maintenance and utilization of sweet potato genetic resources. Report of the first sweet potato Planning Conference 1987. International Potato Centre, Lima, Peru, p. 61-73.
11. Soegianto, A., Ardiarini, N.R. and Sugiharto, A.F. 2011. Genetic diversity of sweet potato [*Ipomoea batatas* (L.) Lam.] in East Java, Indonesia. *J. Agric. Food. Tech.*, **1(9):** 179-183.
12. Struble, F.B., Morrison, L.S. and Cordner, H.B. 1966. Inheritance of resistance to stem rot and root knot nematode in sweet potato. *Phytopathology*, **56:** 1217-1219.
13. Tairo, F., Mneney, E. and Kullaya, A. 2008. Morphological and agronomical characterization of sweet potato [*Ipomoea batatas* (L.) Lam.] germplasm collection from Tanzania. *Afr. J. P. Sci.*, **2(8):** 77-85.
14. Tng, Y.C. and Kehr, A.E. 1953. Meiotic studies in the sweet potato. *Journal of Heredity*, **44:** 207-211.
15. Tsou, C.S., Kam, K.K. and Wang, S.J. 1997. Biochemical studies on sweet potato for better utilization at AVRDC. In proceeding of International Sweet potato Workshop. VISCA Philippines. 20-25, 1987.
16. Veasey, E.A., Borges, A., Rosa, M.S., Queirozsilva, J.R., Bressan, E. de A. and Peroni, N. 2008. Genetic diversity in Brazilian sweet potato [*Ipomoea batatas* (L.) Lam.], Solanales Convolvulaceae, landraces assessed with microsatellite markers. *Genet. and Mol. Biol.*, **31(3):** 725-733.
17. Woolfe, J.A. 1992. Sweet potato: An untapped food resource. Cambridge University Press. p. 643

Genus *Dioscorea*

Yams

Introduction

In this group different types of 'Yams' (*Dioscorea* spp.), are an important under-utilized tropical monocot vegetables belonging to the family Dioscoraceae. Yam makes significant contribution in the diets of the tribal community in India. They also have immense medicinal properties due to the presence of secondary metabolites especially diosgenin (available in root and bulb of the plant) and hence is of great interest to the pharmaceuticals industry (in preparation of various steroids, such as estrogen and dehydroepiandrosterone (DHEA). It serves as a precursor for the production of corticosteroids, sexual hormones, oral contraceptives and possesses cancer chemo-therapeutic activity. In the 18th and 19th centuries, herbalists used wild yam (*Dioscorea villosa*) to treat menstrual cramps and problems related to child birth, as well as for upset stomach and coughs. The most important cultivated yam species are *Dioscorea alata* (Greater yam), *Dioscorea rotundata* (White yam) amnd *Dioscorea esculenta* (Lesser yam). Generally, yams serve as an important staple food in many parts of the world. Freshly harvested tubers of yams are consumed as boiled, baked or fried vegetables. In India yams are considered as life saving crop during famine, drought, crop failure *etc.* and commercially grown in Andhra Pradesh, Bihar, Gujarat, Kerala, Madhya Pradesh, Maharashtra, North Eastern States, Odisha, Tamil Nadu and Uttar Pradesh. Food yams are believed to have originated in the tropical areas of three separate continents:

(i) Africa
(ii) South East Asia
(iii) South America

In fact these groups are an important part of diet in Asia, Africa, South America and the Caribbean. Similarly, yams are considered as staple food in Nigeria Ghana (Africa). Wild yams and domesticated cultivars occur throughout the tropical and subtropical world. The important yams growing countries include Cameroon, Nigeria, Benin Republic, Togo, Ghana and Ivory Coast. Outside Africa, yams are grown in China and Southeast Asia. In India, yams are grown only as supplement to cereals. The edible yam species are annuals and serve as a 'life saving' plant group for the marginal farming and forest dwelling communities, during periods of food security (Roy, *et al.*, 1988; Arora and Pandey, 1996). The underground storage organs produce vines which climb on supports. It is essentially as a vegetable to supplement the main diet as well as in the preparation of different items of menu. Yams are rich sources of carbohydrates (contains 15 per cent starch) and are valuable sources of vitamins C, B, folate and minerals like potassium, iron and magnesium. They also contain 1.9-3.9 per cent protein on fresh weight basis. Yams must be cooked to remove bitter substances including dioscorine, an alkaloid and diosgenin, one of the many triterpenes. They are cooked like potato, baked, steamed, fried, roasted or boiled and mashed. Fresh yam tubers are also peeled, chipped, dried and milled into flour that is used to prepare dough called *amala* or *telibowo*. Some species of *Dioscorea* are used by the pharmaceutical industry. Diosgenin, a precursor of progesterone and cortisone is produced commercially from yams and used to treat arthritis, asthma, eczema, regulates metabolism and controls fertility. Commercially ethyl alcohol is produced from the group of these crops. Tribals eat yams for maintaining their physical health and also for medicinal purpose (Viswanathan, 2004). Consumption of wild yams rich in steroids is believed to help in the natural birth control among the tribal population.

Genus *Dioscorea* (monocot) belong to the familyDioscoreaceae and consists of about 600 species worldwide in tropical and sub-tropical regions. There are 53 species reported from China alone. Out of these only 5 are important crop plants and only 4 species *i.e. D. esculenta* (lesser yam), *D. alata* (larger yam), *D. bulbifera* (potato yam) and *D. rotundata* (white yam) are mainly used as vegetable. Prain and Burkill (1936) reported the occurrence of about 50 different *Dioscorea* species in India, largely in the west, east and northeast regions. Important species like *D. oppsitifolia, D. pubera, D. bulbifera, D. hamiltonii, D. pentaphylla, D. tomentosa, D. kalkapreshadii, D. wightii, D. wallichi, D. hispida* and *D. spicata* are available in the Western Ghats of India.

The section Enantiophyllum is characterized by the twining of vines to the right, *i.e.* in a clock-wise direction when viewed from ground upwards. The other sections are characterized twining of vines to the left, *i.e.* in an anti-clock-wise direction when viewed from ground upwards. Among the species listed above only 4 (*D. esculenta, D. alata, D. bulbifera, D. rotundata*) are mainly used as vegetable. Among other various species, *D. hamiltonii, D. belophylla, D. oppositifolia and D. pentaphylla* var. *pentaphylla* are the most acceptable ones for culinary purpose but other than *D. pentaphylla* all are very rare. Several species were found to be important source for food and medicine to the tribal people like Kadar, Paniyar, Kattunaickar Mala Pandaram and so on. In India, yams are grown in all the states but the major yam producing states are Kerala, West Bengal, Bihar, Odisha, Tamil Nadu, Assam, Rajasthan, Gujarat and Maharashtra. The estimated diversity of yams in north Eastern region 230 but diversity reported is only 200 (till the year 2000). Henry *et al.* (1989) also reported that the *Dioscorea* sp., are widely distributed in Tamil Nadu. The *Dioscorea* species *viz. D. wightii* Hook, *D. spicata* Roth and *D. belophylla* Voight., are endemic to these areas.

Considerable variability is noticed in edible wild yams (*D. pentaphylla* Linn. *D. opposittifolia* Linn.). Balakrishnan *et al.,* 2003 studied the distribution of wild species and their conservation by Kattunaikka, a predominant tribe in Western Ghats. Finding focused that about 21 different types of *Dioscorea* are present in Wayanad district of Kerala which includes a part of the mountain chain of the Nilgiris, Silent Valley and Kodagu region in the Western Ghats, a biodiversity hotspot. Among the varieties known to Kattunaikka tribes are mentioned below:

Table 10.51: Different Sections of Genus *Dioscorea*

Section	*Species*	*Twining Nature of Vines*
Section I	*Enantiophyllum* (*D. rotundata*) Poir, *D. alata* L, *D. cayenesis* Lam., *D. opposite* Thunb. and *D. japonica*	Twining of vines to the right, *i.e.* in a clock-wise direction
Section II	*Lasyophyton* (*D. dumetorum* (Kunth) Pax., *D. hispida* Dennst	Twining of vines to the left, *i.e.* in an anti-clock-wise direction
Section III	*Oxophyton* (*D. bulbifera* L.)	Twining of vines to the left, *i.e.* in an anti-clock-wise direction
Section IV	*Combilium* (*D. esculenta* (Lour.) Burk	Twining of vines to the left, *i.e.* in an anti-clock-wise direction
Section V	*Macrgynodium* (*D. trifida* L.)	Twining of vines to the left, *i.e.* in an anti-clock-wise direction

Table 10.52: Distribution of Wild Relatives of Yams

Name of Yams	*Related Species*	*Distribution*
Greater yam, Lesser yam, Potato yam	*D. hamiltonii* Hook.	All over Western Ghats at lower elevations as a trailing plant and Eastern Himalaya, Sikkim, Assam; Bihar, Chota Nagpur plateau, southern peninsular tract.
	D. pentaphylla Linn.	All over India ascending upto 4000 feet elevation in Western Ghats and 1800 m in Himalaya; Andaman Islands
	D. opposittifolia Linn.	-
	D. bulbifera Linn.	All over from coastal plains to 4000 feet in Western and Eastern himalaya
	D. tomentosa heyne	At lower elevation (1500 feet) in Western Ghats
	D. pubera Blume	-
	D. spicata Roth	In higher elevations from 3000 feet and above
	D. wightii Hook	Very rare at lower elevations in Tirunelveli hills, Tamil Nadu
	D. wallichii Hook	All over from plains to 3000 feet
	D. intermedia Thw.	At lower elevation up to 3000 feet
	D. hispida Dennst.	All over India upto an altitude of 1200 m, in Sikkim Himalaya and Khasi hills particularly at lower elevations
	D. belophylla Voight.	At 2000 to 3000 feet
	D. kalkapershadii Pr. Burk.	At the lower foot hills of Shevarayan hills, Tamil Nadu

Table 10.53: Popular Species among the Tribes

Name of Tribes	Available Species	Area
Venni kalasu	*D. hamiltonii* Hook	Predominantly seen in the interior evergreen and moist deciduous forests
Hekku kalasu	*D. belophylla* Voight.	
Kavala kalasu	*D. oppositifolia* Linn.	
Shoddi kalasu	*D. intermedia* Thw.	Predominantly seen in dry deciduous forests
Ere kalasu	*D. wightii* Hook	Predominantly seen in rocky grasslands
Noora kalasu	*D. pentaphylla* Linn.	Predominantly seen on wayside bushes
Nara kalasu	*D. wallichii* Hook	
Hendhiride kalasu	*D. bulbifera* Linn.	
Boojikkavada kalasu	*D. pubera* Blume	In marshy areas

Morphological characterization of 27 accessions belonging to 18 species of wild yams (*Dioscorea* species) was done based on 21 IPGRI descriptors by Anwar *et al.* (2016). Tuber weight/plant ranged from 40g (*D. belophylla*) to 2.860 kg (*D. hispida*) while tuber flesh colour of the accessions varied from white to red purple. The biochemical studies on starch, sugar, crude protein, crude fat and crude fiber was done keeping in mind the nutritive value of the tubers. Among the accessions, *D. belophylla* recorded the highest content of starch on dry weight basis (98 per cent) while *D. tomentosa* the lowest (25 per cent). *Dioscorea bulbifera* recorded highest protein value (1.82 fresh weights per cent) while for the fiber and fat content, *D. wallichi* and *D. floribunda* recorded the highest values of 2.2 per cent and 21.75 per cent, respectively. The species were also characterized at molecular level using 15 ISSR primers, of which, UBC 810 and $(GA)_9$ AT showed high polymorphism for all the 18 species of wild yams.

REFERENCES

1. Anwar, I., Sheela, M.N., Asha, K.I., Jyothy, A., Krishnan, B.S.P. and Abhilash, P.V. 2016. Genetic diversity analysis of wild yams in India. 1st International Agrobiodiversity Congres (Abs. Session: crop Wild Relatives: Back to the Wild to Save the Future), 6-9 November 2016, held at New Delhi, p. 327.
2. Arora, R.K. and Pandey, A. 1996. Wild Edible Plants of India, Diversity, Conservation and Use. ICAR-National Bureau of Plant Genetic Resources, New Delhi, India.
3. Balakrishnan, V., Narayanan, M.K.R. and Kumar, A.N. 2003. Ethnotaxonomy of Dioscorea among the Kattunaikka people of Wayanad District of Kerala, India. *Plant Genetic Resources Newsletter*, **135:** 24-32.
4. Burkill, I.H. 1939. Notes on the genus *Dioscorea* in the Belgian Congo. *Bull. Jard. Bot. Etat. Brux.*, **15 (4):** 345-392.
5. Henry, A.N., Kumari, G.R. and Chithra, V. 1989. Flora of Tamil Nadu, India, Volume 2, BSI, Coimbatore (Tamil Nadu).
6. Prain, D. and Burkill, I.N. 1936. An account of the genus *Dioscorea* in the East. Annals of the Royal Botanical Gardens, Culcutta; Longman, London, U.K.
7. Roy, B., Haldar, A.C. and Pal, D.C. 1988. Plants for human consumption in India. Botanical Survey India, p. 63-65.
8. Viswanathan, M.B. 2004. Ethnobotanically Important Plants. In: Tamil Nadu Biodiversity strategy and action plan (ed.) R. Annamalai, Tamil Nadu Biodiversity strategy and action plan Published by Tamil Nadu Forest Department Chennai, p. 177-231.

Larger Yam or Water Yam or Greater Yam or Winged Yam (*Dioscorea alata* L.)

Introduction

Larger yam or Water yam or Greater yam or winged yam (*Dioscorea alata* L., 2n=2x=20) is very popular vegetable tuber crop. It is the most commonly cultivated of all the edible yams. Due to square shaped with the corners compressed stems into wings hence the name is given winged yam. It is more tolerant to long dry spells than any other yam. It is considered an important climate resilient food security crops

Origin, History and Distribution

The Indo-Burma region is the centre of origin of Asiatic edible Yams (*D. alata*). The species is a native of Burma (South East Asia). From South East Asia, it was introduced into West Africa by Portuguese. Subsequently, Portuguese carried this species to the Caribbean region. It was introduced into the America by Portuguese and Spanish traders in 15000s.

Species Availability, Evolution and Diversity

The putative parents most often mentioned are *D. persimilis* and *D. hamiltonii*, which grow in the wild form in South-East Asia. Jyothy *et al.* (2016) characterized the landraces of greater yam on the basis of morphological and biochemical. The morphological variation among accession was also determined using 33 IPGRI descriptors which included 7 quantitative traits. Weight/tuber ranged from 0.120 kg (Da-53) to 6.950 kg (Da-340). The tuber flesh colour of the accessions varied from yellowish white to purple colour. The biochemical study on protein, dry matter was done to determine the nutritional quality of the selected accessions. Among the accessions Da-69 recorded highest dry matter content (48.24 per cent) and Sree Neelima recorded highest protein content (5.233 per cent).

Botanical Description

Larger yam or Water yam or Greater yam or winged yam is a dioecious species. Plants are monocot in nature.

Table 10.54: Botanical Description

Growth Habit	Annual or perennial climbing plants. It produces vigorous twining herbaceous vine. Basically it is unisexual and dioecious in nature.
Root	The root system develops from end of tubers, thick, unbranched at the begning, becomes fibrous and branched later.
Stem	Stem is 4 or more winged (square shaped with the corners compressed into wings hence the name winged yam), sinistrose and unarmed climber with tetragonousstellate in cross section, stem green, wings purple or reddish, no spines on stem. The extent and distribution of purple pigmentation (anthocyanin) on the stem and leaves are used to distinguish the cultivars.
Leaf	Leaves ovate/heart shaped, composite in arrangement, purple colouration. Lower leaves distichous, higher ones always oppsite with distinctly 5 winged. Petioles are simple ovate with a more or less deeply cordate base rounded basal lobes and a long acuminate apex, glabrous with scattered pellucid lines between the transparent reticulate veins, 7-9 nerved, 6.0-32.0 cm long, 3.0-20.0 cm wide, the nerves on the under surface namely winged and petiole is 3.0-18.0 c long.
Floral biology	Flowers are unisexual. Male flowers are usually recemiform, produced in terminal and axillary panicles. It is poniculiform, drooping, consisting of spikes. Spikes are placed 2-8 together in the axils of bracts, with a thin flexous rachis and 1.0-3.0 cm long. Perianth is light yellow or greenish yellow, glabrous and 0.15-0.20 cm long. Fertile stamens are six. Female flowers are borne on long axillary spikes (1-2 spikes arising per axil), solitary or sometimes in pairs, spiciform, lax unbranched and 12.0-50.0 cm long. Perianth is sordidly yellow, 0.2-0.25 cm long, staminodes absent and ovary glabrous.
Anthesis and pollination	Flowering is erratic. Generally there is little or marginal overlapping of the male and flowering and the female accessions flower earlier than male accessions. Male flower show consistency in flowering while no consistency is observed in female accessions. Pollination fertility ranges from 27.0 to 95.0 percent. Artificial pollination attempted in 18 female accessions showed that only 2 females were fertile (Abraham *et al.*, 1987)
Tuber	The tubers are subterraneous, usually larger, either lobed or branched or not. The tubers have variable shapes ranging from globose to conical, oblate or cylindrical. The skin colour is cork brown or purplish-black and tuber flesh white or purple, loose starch grains (watery). There is sparingly rooting. Axillary tubers are often present on the climate branchlets.
Fruit and fruit set	The fruit is a dehiscent capsule and more or less broadly pear shaped. It has 3 locules, about 2.0-3.0 cm in length. Apex is emarginated, shortly beaked, base constricted with three lobed. Lobes are broadly half or bicular, wing like and acute. Fruits mature and start drying in 3 months.
Seeds	The capsule contains up to 6 seeds, 2 in each locule. The seeds are small, flattened, round or oval in shape, 5.0-8.0 mm in diameter and about 1.0 mm thick. The seeds have well developed wings which extend all-round the seed and is suited to wind dispersal. The seeds can be sun dried for 2-3 days, dewinged of fertile seeds and stored. Establishment of sexual progeny has been a significant turning point in the breeding of larger yam. Seeds of Larger yam or Water yam or Greater yam do not have dormancy.

Genetic Resources Augmentation and Crop Improvement

In greater yam, different level of ploidy, non-flowering nature of many accessions, non-synchrony in flowering, and very low seed set are the main constraints in utilization of greater yam in breeding program.

Introduction and Selection

In larger yam 'Da 48', 'Da 80', 'Da 60', 'Da 122', and 'Da 42' were noted high yielder cultivars. 'Da 122' observed very good for medium sized round to oval tubers.

Hybridization and Selection

In greater yam, systematic efforts have helped to isolate fertile male and female germplasm collection leading to hybridization and selection. This has resulted in the development of the world's first recombinant hybrid; Sree Shilpa' of *D. alata* in 1998.

Polyploidy Breeding

Greater yam is a polyploidy, with several ploidy levels, revealing a predominance of tetraploidy. Cytological studies were conducted and variable level of polyploidy, [2n=40] (tetraploid), 60 (hexaploid) and 80 (octoploid) were observed in 198 accessions of greater yam.

Table 10.55: Greater Yam Varieties Released from ICAR-CTCRI, Kerala

Sl. No.	*Variety*	*Maturity Period (Months)*	*Starch (Per cent)*	*Yield (t/ha)*
1	'Sree Keerthi'	9-10	20.0-22.0	25.0-30.0
2	'Sree Roopa'	9-10	16.0-18.0	25.0-30.0
3	'Sree Shilpa'	8	17.0-19.0	28.0
4	'Sree Karthika'	9-10	21.0-23.0	30.0
5	'Orissa Elite'	6	20.0-21.0	22.0-25.0

REFERENCES

1. Henry, A.N., Kumar, G.R. and Chitra, V. (1989). *Flora of Tamil Nadu, India*. Vol. II, BSI, Coimbatore (Tamil Nadu).
2. Joothy, A., Sheela, M.N., Anwar, I., Radhika, N.K., Krishnan, B.S.P. and Abhilash, P.V. 2016. Genetic diversity analysis of greater yam (*Dioscorea alata* L.).

1st International Agrobiodiversity congress (Abs. Session. II. Trait Discovery and Enhanced Use of PGR), 6-9 November, held at New Delhi, p. 181.

3. Prain, D. and Burkill, I.H. (1936). An account of the Genus *Dioscorea* in the East. Annals of the Royal Botanical Gardens, Culcutta. Longman, London, U.K.

4. Roy, B., Halder, A.C., Pal, D.C. 1988. Plants for hunman consumption in India. Flora of India series 4: p. 63-65.

5. Arora, R.K. and Pandey, A. 1996. Wild edible plants of India, Diversity, Conservation and Use. ICAR-National Bureau of Plant Genetic Resources, N. Delhi.

6. Viswanathan, M.B. 2004. Ethnobotanically Important Plants. In: *Tamil Nadu Biodiversity Strategy and Action Plan* (ed.) R. Annamalai, *Tamil Nadu Biodiversity strategy and action plan* published by Tamil Nadu Forest Department Chennai, p. 177-23.1

Lesser Yam, Asiatic Yam or Chinese Yam (*Dioscorea esculenta* (Lour) Burkil

Introduction

Lesser yam or Chinese yam (*Dioscorea esculenta* (Lour) Burkil, 2n=2x=20). It is called lesser yam because it does not store well.

Origin, History and Distribution

The Chinese yam is a native of China (South Asia)/ Indo-China. The species is largely cultivated in South China for at least 1700 years (2nd century AD.). It is also cultivated largely in South East Asia, India and Western Australia. It was introduced into the USA as an ornamental or edible food crop in the 1800s. Its cultivation has now spread throughout the tropics, and even to Africa, where it was unknown until receently. Even though, it is cultivated throughout the tropics, its production is most important in South-eastern Asia, the South Pacific Islans and the West Indies. Wild type of *Dioscorea esculenta* also exists in Malaysia, Papua New Guinea, and the Philippines.

Species Availability, Evolution and Diversity

Papua New Guinea is a major center of diversity for this yam.

Botanical Description

Lesser yam or Chinese yam is a dioecious species. Plants are monocot in nature. Flowering is rare in cultivated types

Table 10.56: Botanical Description

Stem	It has a stem which twines to the left. The stem is thin, often pubescent, spiny and cylindrical.
Leaf	A few simple broadly cordate/heart shaped, light green, and arranged alternatively on the vine. Three or four prominent spines enlarge the base of petiole.
Tuber	Tuber are small, cylindrical, rounded ends/egg shape, skin is thin, smooth white or yellowish-brown, thin peel, flesh white, less fibrous texture softerand sweet in taste than that of most other yams.
	Tuber borne in clusters by each plant, unlike most other yam which usually produces only one or two large tubers per plant. Each plant of *Dioscorea esculenta* may produce 5-20 tubers. Each tuber is borne at the end of a stolon, which may be 5.0-50.0 cm long. Spines occur on the roots, but on the tuber bearing stolons. Mature tubers measure 8.0-20.0 cm in length, 2.0-5.0 cm in diameter and 100.0-200.0 g in weight. Some New Guinea varieties weight up to 3.0 kg. Starch in the tuber is more finely grained than that of other yams, cassava or sweet potato. Particles are generally less than 0.5 mm in diameter.

Horticultural Classification

The cultivars can be classified/distinguished on the basis of thorn:

(i) Very thorny
(ii) Thorny
(iii) Less thorny

Genetic Resources Augmentation and Crop Improvement

Introduction and Selection

In lesser yam, 'De 23', 'Lathani', 'De 11' ('Sree Latha'), 'Chaparia' and 'De 17' have been identified as high yielding varieties. 'De 11' ('Sree Latha') is a selection released for the first time in India.

Table 10.57: Casava Varieties Released from CTCRI, Kerala

Sl. No.	*Variety*	*Maturity Period (Months)*	*Starch (Per cent)*	*Yield (t/ha)*
1	'Sree Latha'	8	18.0-19.0	25.0-30.0
2	'Sree Kala'	7.5	23.0-25.0	20.0

Genetic Resource Conservation

In lesser yam, female accessions could not be collected so far.

Other Yams

Aerial Yam or Potato Yam (*Dioscorea bulbifera* L.)

Introduction

Aerial yam or Potato yam (*Dioscorea bulbifera* 2n=2x=20). As name indicates, this yam species produce aerial tubers, which is modification of auxillary buds into bulbils. Their aerial bulbs as well as subterranean tubers are edible.

Botanical Description

Plants are monocot in nature.

Table 10.58: Botanical Description

Growth Habit	Plants are perennial, dextrose, glabrous climber, 3.0-10.0 m long.
Stem	Stem is terete, unarmed and 0.5-0.7 cm thick. Stem has no hairs or spines
Leaf	Leaf simple, heart shaped large devoid of hairs or spines. The leaves are alternate, simple, broadly ovate or ovate oblong, the larger ones deeply cordate, and the smaller ones often only emarginated at the base. Apex is abruptly acuminate, shortly caudate acute, without or with only a few pellucid lines above dark green, shining, beneath lighter coloured, 7-13 nerved, 8.0-40.0 cm long and 6.0-30.0 cm wide. Petiole is much broadened and thickened at the base and 2.5-15.0 cm long.
Floral biology	Flowers are both sexes are sessile in drooping or reflexed spiciform inflorescence, very fragrant spikes are axillary, 2-6 together, united into a pseudo-terminal panicles. Male inflorescence are 2.0-12.0 cm long and the female ones 12.0-35.0 cm long with six fertile stamens. Female flowers with six staminodes.
Tuber	Underground tuber is small, bulbils (aerial tubers) or originated in the axil of leaves-grown for bulbils. Subterranean tubers are usually in pairs, a large and a small one, compressed globose or lobed in the shape of fan, reddish brown on the outside, entirely white with in 10.0-20.0 cm long, 20.0-30.0 cm broad and 2.58-8.0 cm thick or large.
Fruit and fruit set	Fruits reflexed, oblong with subparellel margins, rounded at the base and at the apex, yellowish brown ripe and 18.3 cm long.

Distribution

Aerial yam or Potato yam is widely distributed in tropics of Asia and Africa. It is one of the important tuber crop grown in Konkan region of Maharashtra.

Augmentation of Genetic Resources and Crop Improvement

There is good scope to exploit this crop on marginal land provided their yield potential is increase. Mhaskar *et al.* (2016) evaluated 10 promising entries received from germplasm AICRP on IFS at Centre Experiment Station, Wakawali, Dapoli (Maharashtra). Konkan Kalika (KKVDb-1) was developed by Regional Agricultural Research Station, Karjat, Balasaheb Sawant Konkan Krishi Vidyapeeth, Dapoli, Dist. Ratnagiri (Maharashtra). It yield marketable bulb about 4.85t/ha but total aerial bulb is 5.27 t/ha. It is resistant to leaf blight diseses.

REFERENCES

1. Mhaskar, N.V., Chavan, S.A., Haldankar, P.M. and mahadkar, U.V. 2016. "Konkan Kalika": Promising unique cultivar of aerial yam and wealth for farmers of Konkan Region of Maharashtra. 7th Indian Horticulture Congress (Abs. 96), held at ICAR-IARI, New Delhi, 15-18 November, p. 74.

Dioscorea villosa

In the 18th and 19th centuries, herbalists used wild yam (*Dioscorea villosa*) to treat menstrual cramps and problems related to child birth, as well as for upset stomach and coughs.

Botanical Description

D. villosa is a deciduous perennial herbaceous twiner that grows counter clockwise over small and medium-size shrubs. The upper leaves are alternate, heart-shaped and shiny with long petioles, entire margins, prominent veins and acuminated apices. The lower most leaves are usually arranged in whorls. The plants are dioecious. Small staminate (male) flowers are white and perfumed, and arranged in panicles, while carpelate (female) plants produce small solitary flowers at the leaf nodes. The fruit is a membraneous 3-valved capsule with one or two chocolate-colored winged seeds in each locule (Albrecht, 2006). The long, cylindrical seldom branched rhizomes grow to 5-10 mm in diameter, with numerous tough, slender roots attached underneath.

Distribution

D. villosa is found all over the Appalachian region. Eastern USA, most common in the central and southern regions, wet woodlands from Connecticut-Tennessee, Texas-Minnesota. It prefers moist open woods, thickets and roadsides (Gleason and Cronquist, 1991).

References

1. Albrecht, M.A. 2006. Reproductive biology of medicinal woodland herbs indigenous to the Appalachians. Ph. D. Thesis. College of Arts and Sciences of Ohio University. Accessed online at http://etd.ohiolink.edu/view.cgi/Albrecht per cent 20Matthew.pdf?ohiou1163427974. June 1st, 2011.

2. Gleason, H.A., and Cronquist, A. 1991. *Manual of Vascular Plants of Northeastern United States and Adjacent Canada*. Bronx, NY: New York Botanical Garden Press.

Cush-cush Yam (*Dioscorea trifida* L.)

Origin, History and Distribution

Cush-cush Yam (*Dioscorea trifida* L.) is originated in the northern part of South America, and is the only yam of New World origin to attain any significance as food. Cultivation is restricted to the West Indies.

Botanical Description

See Table 10.59.

Table 10.59: Botanical Description

Growth Habit	It is viny in nature which twines to the left.
Stem	The stem is rectangular in cross-section, with the angles sometimes extended to form wings. No spines on the stem.
Leaf	The leaf is simple, but is deeply divided into 3-5 lobes. The petiole is winged. No spines on the leaves.
Floral biology	Staminate flowers are borne in racemes while the pistillate flowers occur as spikes.
Tuber	Like *D. esculenta*, produces many individual small tubers per plant. The tuber flesh is white, yellow, pink or purple and is highly palatable.

Cinnamon Yam (*Dioscorea opposita* Thunb.)

Cinnamon Yam (*Dioscorea opposita* Thunb.) is native to China and tolerates a colder climate than the other economically important yams. The stem is cylindrical and twines to the right, but usually not to heights greater than 3.0 m. The leaves are simple and oppositely arranged. The tuber is thin but very long, and grows vertically downward into the soil, therby making it very difficult to harvest. Bulbils occur in the leaf axils. This species is cultivated in China, Korea and Southern Japan. It is used for medicinal purpose in China.

Dioscorea japonica Thunb.

Dioscorea japonica Thunb., is very close resembles *Dioscorea opposita* both in structure and in tolerance to colder climates. It is native to Japan and extensively cultivated in Japan and China. It is used for food and in traditional medicine.

White Yam or Guinea Yam (*Dioscorea rotundata* Poir.)

Introduction

White yam or Guinea yam (*Dioscorea rotundata* Poir., 2n=2x=20) is the most popular yam in West Africa. It is grown in larger area than any other yam species in the world.

Origin, History and Distribution

White yam is native of West Africa but some authority said that it is originated from South East Asia. It is true that it does not occur in the wild, and it probably developed in cultivation from *D. praehensilis* Benth. This is the principal yam cultivated in West Africa. This species was introduced into India in the form of seeds. Subsequently, Portuguese carried this species to the Caribbean region. Today, this species remains the principal yam cultivation in West Africa-the most important area in the world for yam production. The extent of its cultivation parallels the preference of the inhabitants of this region for this type of yam over most other kibds. Its cultivation spread to other parts of theworld-it is grown extensively in the West Indies, and to some extent in East Africa. However, there is little or no cultivation of *D. rotundata* in Asia. There is a large number of cultivars and land races of *D. rotundata* that are grown especially in West Africa where it originated, and is most widely cultivated (Waitt, 1961a). Unfortunately, the taxonomic position of most of these cultivars, and of cultivars of yam in general is somewhat confused.

Botanical Description

Plants are monocot in nature.

Table 10.60: Botanical Description

Stem	The stem is Wingless, sturdy and roughly circular (in cross section) and round, which is usually spiny at base. The stem twins to the right, and will usually grow to a length of several meters. The extent of this characteristcs varies greatly among cultivars within the species, and perhaps also within the same cultivar under different environmental conditions. No hairs present on stem.
Leaf	The leaves are small, simple, cordate/heart shaped, pointed at the tip and opposite or alternate in arrangement. No hairs present on leaf.
Floral biology	Whitish blooms are present on the stem.
Tubers	Tubers are usually spherical or cylindrical in shape with smooth; light brown skin and white creamy or yellow firm flesh. The white flesh of tubers produces a mealy starch. Starch grains in the tuber are large and ovoid.

Augmentation of Genetic Resources and Crop Improvement

Table 10.61: Casava Varieties Released from ICAR-CTCRI, Kerala

Sl. No.	*Variety*	*Maturity Period (Months)*	*Starch (per cent)*	*Yield (t/ha)*
1	'Sree Priya'	9-10	19.0-21.0	35.0-40.0
2	'Sree Subhra'	9-10	21.0-22.0	35.0-40.0
3	'Sree Dhanya'	9	22.0-24.0	20.0

Genetic Resource Conservation

The dwarf typev of white yam 'Sree Dhanya', which does not require trailing, is a novel germplasm selection from white yam.

Yellow Yam or Yellow Guinea Yam (*Dioscorea cayenesis* Lam.)

Introduction

Yellow Yam or Yellow Guinea Yam (*Dioscorea cayenesis* Lam.) is in many respects similar to *D. rotundata*. Indeed some francophone authors regard *Dioscorea* as a sub-grouping of *D. cayenesis*. However, more recent DNA studies have led to the opposite suggestion that *D. cayenesis* is a taxonomic variety of *D. rotundata* (Terauchi *et al.*, 1992). The yellow yam gets its name from the fact that its tuber flesh is usually yellow colour, although variants, which are almost white, do occur. The yellow colour is due to the presence of carotenoids.

Origin, History and Distribution

D. catyenensis is native to West Africa, where it occurs in the wild state, as well as under cultivation. It was introduced into the West Indies from West Africa during the 16th century. Today its cultivation ismainly confined to these two areas, and no significant amounts are grown in Asia.

Species Availability, Evolution and Diversity

Physiologically *D. catyenensis* differs from *D. rotundata* in having a shorter period of tuber dormancy, and a longer growing season. Thus, it can only be grown in areas (*e.g.* the forest zone) where the rainy season is slightly longer than the minimum required for *D. rotundata*.

Botanical Description

Table 10.62: Botanical Description

Stem	The stem stem twines to the right and possesses thorns, but the stem itself is usually devoid of bloom.
Leaf	The leaf is slightly bigger, and the leaf tip less pointed, than for *Dioscorea rotundata*, giving the impression of a more rounded leaf. The leaf lamina is also more deeply indented near the petiole. The vein grooves are less distinct and the lamina has more surfaces with darker and glossier than that of *Dioscorea rotundata*.
Tuber	The tuber skin is firmer and less extensively grooved than that of *Dioscorea rotundata*, and the corm at the head of the tuber is also more massive

Augmentation of Genetic Resources and Crop Improvement

Many cultivars of *Dioscorea catyenensis* exist, but very little work has yet been done to identify and described them. However, the degree of variations within this species is small in comparison with *Dioscorea rotundata* or *Dioscorea alata*, so that the number of varieties is also much smaller.

Bitter Yam or Trifoliate Yam (*Dioscorea dumetorum* (Kunth) Pax)

Introduction

Bitter Yam or Trifoliate Yam (*Dioscorea dumetorum* (Kunth) Pax) is not an important crop but grown to very lessert extent. The tubers are bitter (so, name is given bitter yam) and, in many cultivars, poisonous because of a high content of the alkaloid dioscorine or its derivatives. It is therefore necessary that the tuber be steeped in running water or salt water and boiled before it is eaten.

Origin, History and Distribution

Bitter Yam or Trifoliate Yam (*Dioscorea dumetorum*) is originated from tropical Africa, and occurs both in wild and cultivated forms. Its cultivation is still essentially restricted to West Africa. The tubers of many of the wild forms are highly poisonous, and they are therefore used in southern Africa to prepare poisons for hunting and other purpose. Variations in type occur within the species but distinct cultivars have not been identified and described.

Botanical Description

Table 10.63: Botanical Description

Stem	It has a hairy stem, which twines to the left. Spines are present on the stem.
Leaf	The leaves are trifoliate, a characteristics which distinguished this yam from the others of major economic importance. The leaf like the stem is pubescent. Each leaflet is ovate.
Tuber	The tuber is large and coarse. Each plant may produce one tuber or a cluster of them.

Genus *Pachyrrhizus*

Yam Bean or Misrikand (*Pachyrrhizus erosus* Linn.)

Introduction

Yam bean (*Pachyrrhizus erosus*Linn.) is also known as Misrikand (Hindi) or Sakaloo (Bengali). It is a high yielding root crop with good nutritional value. This species, which hails from the Americas, look like a bean plant above ground but is actually grown for the swollen roots, which developed below ground. As a legume, its root nodules harbor symbiotic colonies of nitrogen-fixing bacteria under good growing conditions. In an analysis, a season crop of yam bean can release as much as 53.0 kg/ha of usable nitrogen in the soil (Grum *et al.*, 1998). In recent decades yam bean has taken on renewed momentum and is now among the world's fastest rising new crops. In India, tender, crisp, succulent and sweet tubers are consumed as a vegetable and sometimes preferred for salad making. Tuber possesses high sugar and moderate ascorbic acid (Tadera and Taniguchi, 1984 Tader *et al.*, 1984). Per 100 g of edible portion pertains 82.38 per cent water, 1.47 per cent protein, 0.09 per cent fat, 9.72 per cent starch, 2.17 per cent reducing sugar, 3.03 per cent non-reducing sugar, 0.64 per cent fiber, and 0.50 per cent ash. It also has 0.43 mg copper, 1.13 mg Fe, 16.0 mg Ca, 0.05 mg Thiamine, 0.02 mg Riboflavin, 0.02 mg Niacin and 14.0 mg Ascorbic acid per 100 of edible portion. The tubers contain adenine, arginine, choline, and phytin. They can also be sliced and made into chips. The juicy flesh is rich in ascorbic acid. The tubers are also used as fodder and superior quality starch is obtained from mature tubers. The young pods can be used as vegetables but they are reported to become poisonous as they become mature, causing diarrhea, due to irritation of the hairs. The mature seeds have high content of alkaloids and insecticidal properties and are useful as fish poison in tropical conditions. They act as contact and as stomach poisons.

Origin, History and Distribution

P. erosus is a native of tropical America and it is widely cultivated in Mexico and South America from pre-Colombian period and has originated from hot moist region of the river Amazon. The crop is now being cultivated in India, China, Singapore, the Philippines, Hawaii, Nepal, Bhutan, Myanmar (Burma) and Indonesia. In India, it is grown in parts of West Bengal, Tripura, Bihar, Odisha and Assam. Bihar is the largest producing state of India and marketed throughout the country.

Species Availability, Evolution and Diversity

The genus includes 5 species, of which 2 are wild and 3 are cultivated types. One species is cultivated in India for its edible tuber *viz. Pachyrrhizus erosus* Linn.

Botanical Description

Yam bean belongs to the family Fabaceae (Leguminoseae).

Table 10.64: Botanical Description

Growth Habit	It is perennial, as far as underground tubers are concerned but an annual with respect to aerial portion.
Root	An edible fleshy taproot. Per plant one or more tuberous roots.
Stem	The above-ground part of the plant is herbaceous to somewhat lignified, vigorous spreading prostrate vine which reaches several feet in diameter.
Leaf	Trifoliate leaves with stipules and pinnately arranged leaflets with caduceus stipls.
Floral biology	Inflorescence is a complex to simple raceme and the flowers have a tubular calyx and a papilionaceous corolla. Flowers, either blue or white. The flowering observed in three cultivars and results revealed that flowering occur 58-68 days after sowing, and lasted 92-103 days.
Anthesis and pollination	The stigma was found to become receptive at 12 hours before and to last for 18 hours after opening of the flower. When the pollen grains were stored at room temperature their germination percentage decreased after 4 hours. On average, they remained viable for 22 hours. The ovary has a basal crenate disc-formed nectarium, the recurving style is ciliated ventrally and the vertical surface of the stigma is subglobose.
Pod and pod set	Straight pods are similar to lima beans are produced on fully developed plants.
Seeds	Seeds are square, more or less flattened, or a rounded kidney shape. Seed colour range from olive green and deep maroon to black, or from black and white to mottled cream.

Genetic Resources Augmentation and Crop Improvement

Rajendra Misrikand-1 (RM-1) which was released from Rajendra Agricultural University, now Rajendra Prasad Central Agricutural University, Pusa (Bihar), is found most suitable under Bihar and West Bengal situations. Some Mexican lines like L-19 and L-20 are also performing very well in Indian conditions.

REFERENCES

1. Grum, M., Sorensen, M., Estrella, E.J., Hamann, O.J. and Rios Ruiz, S.A. 1998. *Pachyrrhizus* DC (yam bean) symbiosis with Rhizobium bacteria: Genetic Variation and Performance. Proceedings of 2nd International Symposium on Tuberous Legumes. Guanajuato, Mexico, 5-8 August, 1996. p. 419-429.
2. Tadera, K.T., Taniguchi, T., Teramoto, M. and Arima, M. 1984. Protein and starch in tubers of winged bean, *Psophocarpus tetragonolobus* (L.) DC. and Yam Bean, *Pachyrhizus* (L.) Urban Mem Fac Agr. Kagoshima Univ. 20:73-81.

Aroids

Aroids include the tuber bearing plants belonging to the family Araceae *viz.*taro (*Colocasia*), giant taro (*Alocasia*), tannia (*Xanthosoma*), elephant foot yam (*Amorphophallus*), and swamp taro (*Cyrtosperma*).

Genus *Colocasia*

Taro (*Colocasia esculenta*)

Introduction

Taro (*Colocasia esculenta*, 2n=28 (diploid) and 2n= 42 (triploid) is also known as Cocoyams, Arvi (Hindi), Kashriew (Khasi), Kachu (Assamese) *etc.* All the plant parts *i.e.* leave, petiole, corms and cormels are eaten in some or other parts of the world. It is an important tuber food crop of West Africa and Oceania. In West Africa taro is called cocoyam. Taro is important source of food during lean period and feed for livestocks. It is primarily grown for edible corms which are easily digestible. The corms and edible cormels are consumed after being boiled, baked, roasted or fried. Taro corms are generally high in starch, minerals, and vitamins but low in fat and protein. Fresh corms of taro have about 65.0 per cent water, 13.0-29.0 per cent carbohydrates and 1.4-4.3 per cent protein. It has more protein than other tuber crops. The predominant carbohydrate is starch. Amylopectin constitutes 80.0 per cent of the starch and the rest is amylose. Its starch grains are small, therefore easy to digest and provide dietary fiber. Oxalic acid may be present in the corm especially in the leaf, and these foods should be eaten with milk or other foods rich in calcium so as to remove the risks posed by ingesting the oxalate ion, especially for people with kidney disorders, gout or rheumatoid arthritis. Calcium reacts with the oxalate to form calcium oxalate which is very insoluble. Peel contained non-starchy nutrients and can be used as fed. The leaves are sometimes cooked as a potherb due to richness in vitamins and minerals. They are good source of thiamine, riboflavin, iron, phosphorus and zinc, and a very good source of vitamin B6, and vitamin C, niacin, potassium, copper and Manganese (Aregheore and Tamni, 2002). Poi, a dish made by mashing bodied corms and by making a viscous fluid is used to settle the stomach. At present taro is most widely cultivated crop occupying 10.0 lakh hectares around the world and producing 6.61 million tons of tubers.

Origin, History and Distribution

Taro was probably first native to the low land wetlands of Malaysia (Taloes) (Plucknett, 1976).Taro is originated from South-East Asia including India (Watt, 1889). Estimates are that taro was in cultivation in wet tropical India before 5000 B.C., presumably coming from Malaysia, and from India further transported westward to ancient Egypt where it was described by Greek and Roman historians as an important crop. From it centre of origin, taro spread eastward to South-East Asia and to China, Japan, and Pacific Islands. From Asia, taro spread westward to Arabia and the Mediterranean region. The hypothesis that the North East India is the centre of origin of taro was confirmed by many Indian workers. By 100 BC it was grown in China and Egypt. It arrived in East Coast of Africa over 2000 years ago. Taro is mostly cultivated in Asia, Africa and Pacific as well as Caribbean Islands. In Pacific Islands it is an important economic crop besides being a staple food crop. The largest area under taro cultivation is in West Africa. The major taro producing countries are Ghana, Zaire, Nigeria and Cameroon in Africa; China, Japan, Philippines and Thailand in Asia and Papua New Guinea, Western Samoa, Solomon Islands, Tonga and Fiji in Oceania. Fiji, Tonga, Cook Islands, Tuvalu and Thailand are important taro exporting countries. The exports are mainly destined to the ethnic community settled in Australia and New Zealand. In India, taro is cultivated in almost all the states, right from the foot hills of the Himalayas to the coastal areas in the South.

Species Availability, Evolution and Diversity

In *Colocasia,* the richest diversity is noticed in the North-east followed by Kerala and Western region. Several feral forms are available with the tribal farmers who grow it in their 'Jhum' fields. The estimated diversity of Taros are 300 in North East region is 300 but diversity reported is 272 (till the year 2000).

Table 10.65: Distribution of Wild Relatives of Taro (*Colocasia*)

Wild Relatives	*Distribution*
C. esculenta var. *antiquorum*	All over India as wild
C. esculenta var. *esculenta*	In hills at about 3000 feet and North Eastern Region
C. esculenta var. sylvestris	Very rare, a mutant type from common wild taro
C. esculenta var.*aquatilis*	Very common in marshes, water courses and under the evergreen forests in Western Ghats
C. affinis	North east India and Kerala
C. fallax Schott.	At lower hills in Assam, West Bengal and Meghalaya

Botanical Description

It is monocotyledons and belongs to the family Araceae (Aroidae) and sub family Colocasioidae which falls under Nudiflorae of Angiopspermae.Under the genus *Colocasia* the only recognized species is *C. esculenta* which is considered as polymorphic species which is sub-classified in *C. esculenta* var. *esculenta* and *C. esculenta* var. *antiquorum*. The somatic chromosome number 2n=28 (diploid) and 2n= 42 (triploid) has been reported by Sharma and Das (1954); Mookerjee (1955); Ramachandran (1978) and Kuruvilla and Singh (1981). Triploids cultivars incomparision to diploids are longer, vigorous, late maturing in nature and yield more.

Table 10.66: Botanical Description

Growth Habit	Taro is an herbaceous perennial lactiferous, very variable herb, and grows to the height of 1.0-2.0 m. Herb develop thick shoots from a large corm; slender stolons also often produced, along with offshoot corms.
Root	Roots are adventitious.
Stem	Stem is usually thick-set, subterraneous, more or less swollen, sending forth longer or shorter subterranean and superterranean stolons which at the top develop into new plants.
Leaf	Develop spathacious large or rather large leaf. Leaf blades to 60.0 cm long and 50.0 cm wide, arrowhead shaped, with upper surface dark green and velvety; leaves peltate (stalked from back of blade), petioles larg, succulent, often purplish near top. Venation is reticulate which is rare in monocotyledons. Petiole is sheathering at the base, uniformly light or dark green with dark streaks or violet, within with many auriferous cells 40.0-150.0 cm long. Lamina is pointing downwards, ovate and peltate. The base is more or less profoundly cordate and apex is narrowe, dark green or bluishgreen and often violet at the base above, pale green or sea green beneath, pinnatinervedwith 4-7 strong, light green or violet lateral nerves. The basal nerves on the outer side with 2-5 branches, 12.0-60.0 cm long and 7.0-50.0 cm wide.
Floral biology	Inflorescences are two or three togather in the leaf axils/on a fleshy stalk shorter than leaf petioles; part of fleshy stalk enveloped by a long yellow bract (spathe).
	Spathes are oblong-lanceolate, divided by a transverse construction in two unequal parts the lower part persistent, convolute, tubular, fleshy, green, pruinose, 3.0-5.0 cm long, the upper part more tyhan three times longer than the lower part thinly, fleshy, lanceolate, acute, varying from greenish yellow to orange coloured, so convolute, deciduous or withering 10.0-32.0 cm long. Flowers tiny densely crowded on upper part of fleshy stalk, with female flowers (fertile) below and male flowers (sterile) above. Flowers are unisexual, protogynous, sessile and without perianth. Spadix is at the base over a length of 1.2-4.0 cm beset with female flowers, constricted above these and over a length of 2.0-5.0 cm be set beneath with sterile flowers,higher up with male flowers. Flowers are naked. Female flowers are consisting of a few ellipsoid or obovoid ovaries each of which with an almost sessile, capitate stigma. Male flowers are consisting of 2-6 sessile anthers, which are connate into an obconical, table topped mass (synandrium).
Anthesis and pollination	Pollination before 24 hours of anthesis results more fruits development. The stigma become receptive on the day pollen starts shedding and will remain so for two-three days thereafter. Hand pollination involves the removal of the spathe to expose the stigmas and the application of pollen with a small brush. The ovaries swell within 5 days of pollination. This is later followed by senescence and abscission of the male and sterile portions of the spadix.
Fruit and fruit set	Fruit and seed setting in taro are even rare than flowering. Fruit is a small berry (3.0-5.0 mm in diameter) and borne in clusters (more than 100 fruits/bunch) on the fleshy stalk/basal portion of the spadix. According to Strauss *et al.* (1980), the fruits will remain green for 6-8 weeks, after which period they turn orange overnight and start to emit a mild, fruity fragrance. They are harvested at this stage. After fertilization, 2-3 months are required for development of mature fruits may be collected by a cloth or paper bag. Alternatively, mature fruits may be collected just prior to shattering and dried inside well-ventilated containers.
Seeds	Many inflorescences wither without setting any seed. Berries are thin walled, green, below seeded and ± 0.4 cm long. Each fruit contain 20-40 hard seeds, contains endosperm and germinate with extreme difficulty. Colocasia seeds germinate in 7-8 days and seed has no dormancy. Potential seed germination is upto 3 months only and later on germination percentage decrease drastically and after 5th months, no germination.
Edible part	Edible part corm

Botanical Classification

There is difference opinion regarding the Asian accessions and those of Pacific Islands to which sub species it is to be ascribed. Since the classification based on floral characters is not dependable, the two major types Dasheen type and Eddoe type viz; on the basis tuber (Purseglove, 1972):

(i) Dasheen Type

In 'dasheen' types, the corm (mother tuber) none or few cylindrical and large. Generally, develop one corms (main corms) and secondary corms (cormels) donot develop (if develop, very less in number). *C. esculenta* var. *esculenta*has been placed (Onwueme, 1978).

(ii) Eddoe Type

In 'eddoe' types, the main corms are small, globoid and are surrounded by several cormels (stem tubers) and daughter corms.In this group, *C. esculenta* var. *antiquorum* has been placed (Onwueme, 1978).

Genetic Resource Augmentation and Crop Improvement

In taro, limitation in flowering, lack of synchrony, pollen sterility and protogyny are the main constraints. Only 10 per cent of the germplasm flower and only limited number of accessions set seeds.

Introduction and Selection

The taro germplasm includes cultivated and stoloniferous wild taro, which is present in Western Ghats. Elite germplasm 'Muktakeshi', was selected to be resistant to the serious taro leaf blight disease caused by *Phytophthora colocasiae*. In taro, the three varieties released are direct selection from germplasm. Sree Reshmi and Sree Pallavi are selections from Kerala and Meghalaya respectively. There are many improved varieties like Sree Rashmi, Sree Pallavi, Sree Kiran, Muktakesi, Pani Saru, Vallabh Nikki, Bilaspuri Arum, ML-1, ML-1, ML-9 *etc.* are very popular among the growers.

Table 10.67: Taro Varieties Released from ICAR-CTCRI, Kerala

Sl. No.	*Variety*	*Maturity Period (months)*	*Starch (Per cent)*	*Yield (t/ha)*
1	'Sree Pallavi'	7	24.0-25.0	15.0-18.0
2	'Sree Subhra'	7	15.0-16.0	15.0-20.0
3	'Muktakeshi'	5-6	16.0-18.0	20.0
4	'Sree Kiran'	7	16.0-18.0	18.0
5	'Panisaru-1'	6-7	12.0-13.0	16.0
6	'Panisaru-2'	6-7	16.0-18.0	13.0

Hybridization and Selection

In this crop first hybrid was released through hybridization of fertile diploid parents from the germplasm. Flowering inadequacies, occurrence of sterile triploid genetic stocks and protogynous nature of the spadix are the main obstacles for the genetic improvement of taro through hybridization and selection. Systematic screening of hybrid progeny resulted in releasing the first hybrid variety of taro in India under the name 'Sree Kiran' during 2004 from ICAR-CTCRI, Kerala.

Interspecific Hybridization

Inter crosses involving wild relatives were made to transfer leaf blight resistance to improved cultivars (Santha *et al.*, 1993).

Polyploidy Breeding

The high yielding cultivars are triploids, either non-flowering or pollen sterile; and it is difficult to improve them by conventional method.

Genetic Resource Conservation

In India, colocasia germplasms are being maintained at CTCRI, Thiruvanantpuram (Kerala) and NBPGR, Regional Station Thrissur (Kerala). In NEH region few genotypes are being maintained at ICAR-NBPGR, Regional Station Shillong (Meghalaya).

REFERENCES

1. Aregheore, E.M. and Tamni, E. 2002. The nutritional value of processed small whole taro corm (*Colocasia esculenta*) and taro peels as energy source in concentrate supplements of duckers and kids.*Thai Journal of Agricultural Sciences*, **35(3):** 309-317.
2. Kuruvilla, K.M. and Singh, A.1981. Karyotypic and electrophoretic studies on taro and its origin. *Euphytica*, **30:** 405-413.
3. Mookerjee, A. 1955. Cytology of different species of aroids with a view to trace the basis of their evolutuion. *Cytologia*, **7:** 221.
4. Onwueme, I.C., 1978. The Tropical Tuber Crops. John Wiley and Sons, New York, p. 234.
5. Plucknett, D.L.1976. Evolution of crop plants, eds. N.W. Simmonds, Longman, London and New York, p. 10-12.
6. Purseglove, J.W. 1972. Tropical crops. Monocotyledons. 1. Longman Group Ltd., p. 334.
7. Ramachandran, K. 1978. Cytoplogical studies on South Indian Araceae. *Cytologia*, **43:** 289-303.
8. Santha, Pillai, V., Thorrkappan, M. and Misra, R.S. 1993. Leaf blight resistant of taro. J. Root Crops, 19(1): 66-68.
9. Sharma, A.K. and Das, N.K. 1954. Study of Karyotypes and their alteration in aroids. *Agronomia Lusitana*, 16: 23-48.
10. Strauss, M.S; Stephens, G.C; Gonzales, C.J. and Arditti, J. 1980. Genetic variability in taro (*Colocasia esculenta* (L.) Schott (Araceae). *Ann. Bot*. 45: 429-437.
11. Watt, G. 1889. Dictionary of the economic plants of India. Supdt. Govt. Printing, Calcutta, **2:** 509-513.

Genus *Alocasia*

Giant Taro or Giant Elephant Ear (*Alocasia macrorrhiza*)

Introduction

Giant Taro or Giant Elephant Ear (*Alocasia macrorrhiza*, 2n=28) is also called elephant ear because it produces large undivided leaves. Giant Taro differs from the other large leaved arums in having an upright stem and holding its leaves upright. Some researchers referred to this plant as "Wild" taro. Ancient Polynesian (Hawaiians) called the plant as 'Ape'. It is an important rhizomes bearing plant and its modified stem (corm) is the edible economic part but shoots are equally important. The stem (a corm) is edible, but contains rapid crystals of oxalic acid that can numb and swell the tongue and pharynx (produce acridity). It has lot of. Hence it is not nuch popular. Addition of small amount of betel nut chips (*Areca catechu*) or small amount of slake lime will remove the acridity during boiling oxalate in the corm before serving or processing as a food. Leaves are also consumed as vegetable. Tiep *et al.* (2005) suggested that after processing, the *Alocasia macrorrhiza* root meal can be included at levels of up to 50 per cent in the diets of growing pigs. Petiole and corms have a lot of medicinal value. It is considered as famine food in Pacific Islands. Some hybrids, such as the Amazon Lily of the African Mask (*Alocasia* x *Amazonica*), are grown as ornamental.

Origin, History and Distribution

Giant Taro or Giant Elephant Ear is a native of Southern Asia and grows throughout the Southern Pacific. Giant Taro occurs naturally in tropical forests in Sri Lanka, India and Malaysia where it grows in the forest, understory in openings and along streams (Plucknett *et al.*, 1970). In India this crop can be spotted in Sub-Himalayan tract, north eastern region, and peninsular humid tract.

Species Availability, Evolution and Diversity

There are about 70 species of *Alocasia* occurring in Asia, Oceania and South America.

Botanical Description

Table 10.68: Botanical Description

Growth Habit	It is pertennial in nature. The whole plant can stand 37.0-4.6 m tall and spread 1.8-3.1 m across.
Leaf	Broad leaves (huge elephant ear leaves) having 1.0-1.8 m in length and 0.6-1.2 m wide born on leaf stalks (petioles) 0.6-1.2 m long. The petioles emerge from a stout upright trunk that can stand 1.8 m tall. In the typical form the leaves are glossy medium green with paler veins. They are arrow shaped at their bases and stand upright, pointing skyward, unlike other large "elephant ear" perennials. Leaves are generally green, but there is also a variegated form of giant elephant ears. Petioles are erect, but leaves are rather horizontal.
Floral biology	The greenish spathe and spadix stands 20.3-25.4 cm tall and is not particularly showy. Their beautiful araceous flowers grow at the end of short stalk, but are not conspicuous, often hidden behind the leaf petioles.

Augmentation and Utilization of Genetic Resources (Abiotic and Biotic resistance, Quality, Graftings)

'Variegata' is selection with leaves decorated with creamy white or grayish splotches. 'Violacea' has pale violet leaves.

REFERENCES

1. Plucknett, D.L., Delta Pena, R.S. and Obrero, F. 1970. *Field Crop Abstr.*, **23:** 413-426.
2. Tiep, P.S., Nguyun, V.L. and Dang, H.B. 2005. Processing and use of *Alocasia macrorrhiza* (Taro) roots for fattening pigs under mountains village conditions. Workshop-seminar, 23-35 May, Mekaran, CTU.

Genus *Cyrtosperma*

Cyrtosperma is a small genus with 12 species.

Swamp Taro (*Cyrtosperma chamissionis sy. Cyrtosperma merkusii* Schott.)

Introduction

Swamp Taro (*Cyrtosperma chamissionis* sy. *Cyrtosperma merkusii* Schott.) small species of aquatic aroids, found growing wild in fresh water swamps, rivers and lakes as weeds. During the early 20th century it was very popular crop in Southeast Asia and staple food for some people in Malay Island and Philippines but today there is hardly any cultivation of this plant except self sown in swampy areas. It is generally called 'Palauan' in Philippines. It is reach in carbohydrates but poor in other nutrients. Decoction of the spadix of this plant is used as indigenous medicine. *Cyrtosperma merkusii* or swamp taro is terrestrial usually grown in atoll pit gardens and in fresh water swamp or marsh area and is the largest of taros.

Origin and History

All species with exception of *Cyrtosperma merkusii* are confined to New Guinea and associated islands, Philippines, Borneo, Sumatra, Java, and Oceania. Originated in South East Asia, perhaps in Indinesia it spread throughout Malayan Archipelago and Pacific territories.

Corms

The corms develop by thickening at base of stem and almost cylindrical in shape and externally resemble a banana sucker.

Genus *Amorphophallus*

Elephant Foot Yam or Jimmikand [*Amorphophallus paeoniifolius* (Dennst.) Nicolson]

Introduction

In India, Elephant foot yam [*Amorphophallus paeoniifolius* (Dennst.) Nicolson Syn. *A. campanulatus* Decne] is also known as 'Suran' or 'Jimmikand'in Hindi and more prevalent in southern and eastern regions of India. It is a tropical under ground stem tuber crop and offers excellent scope for adoption in tropical countries as a cash crop due to its production potential and popularity as a vegetable in various cuisines. It has high dry matter production capability per unit area than most of the other vegetables.Tubers are cooked as other vegetables, boiled or baked. They taste acrid and cause irritation in mouth and throat due to excessive content of calcium oxalate crystals. Nutritionally it is a cheap source of carbohydrates, minerals and vitamins but poor in protein. The per 100 g edible portion contains 78.7 g moisture, 1.2 g protein, 0.1 g fat, 18.4 g carbohydrates, 0.8 g minerals, 50.0 mg calcium, 34.0 mg phosphorus, 0.6 mg iron, 4361 (I.U.) Vit. A, 0.07 mg riboflavin, 0.06 mg thiamine and 0.7m mg nicotinic acid. The leaves (petioles) and sprouts are sometimes cooked as vegetable. The tubers are also sometimes processed to make pickles. Medicinally, the root is carminative, restorative, stomachic and tonic (Chopra *et al.*, 1958). It is dried and used in the treatment of piles and dysentery (Anonymous, 1993). The tubers are believed to possess blood purifying properties. The fresh root acts as an acrid stimulant and expectorant. It is much used in India in the treatment of acute rheumatism (Chopra *et al.*, 1958). Traditionally it is also being used to cure abdominal pain, tumors, enlargement of spleen, asthma and rheumatism. The tuberous roots are also possessing tonic and appetizer properties (Kirtikar, and Basu 1994).

Origin, History and Distribution

Elephant foot yam is basically a crop from south East Asia and very important in India and Indonesia. It grows in wild form (highly acrid) in Philippines, Malaysia, Indonesia and South Eastern Asian countries but also commercially grown in Sri Lanka, South China, Malaysia, Java, Philippines, Northern Australia, and South-East Asia. In India it is cultivated on large scale in coastal areas in Andhra Pradesh, Kerala, Tamil Nadu, Odisha, Pondichery, Bihar, Gujarat, Karnataka, Maharashtra, West Bengal and North Eastern states.

Species Availability, Evolution and Diversity

The genus *Amorphophallus* consists of about 90 species indigenous to tropical Asia and Africa. There are about 14 species, which occur in India. Amongst them *Amorphophallus campanulatus, A. oncophyllus, A. variabilis* and *A. rivieri* are the edible species (Bailey, 1973). Ohtsuki (1968) reported that the wild *Amorphophallus oncophyllus, A. variabilis, A. campanulatus* and *A. bulbifer* are morphologically and chemically similar with the cultivated *A. konjac*. Tubers of the wild species are similar to those of *A. konjac* and contained mainy cells idioblasts as the storage tissue except for *A. campanulatus* and *A. bulbifer*, which is mainly stored starch. In Elephant Foot Yam, rich diversity has been reported from Western Ghats. Seven wild species such as *A.hohenackerii, A. commutatus, A. paeoniifolius* var. *dubius, A. bulbifera, A. smithsonianus, A. bonoccordensis,* and *A. sylvaticus*were located in the Western Ghats in Kerala and Karnataka.

Table 10.69: Distribution of Wild Relatives of Elephant Foot Yam (*Amorphophallus*)

Wild relatives	*Distribution*
Amorphophallus dubius Bl.	Very common in Kerala, in laterite soils in plains
A.hohenackerii Engl.	Very common in Kerala, in laterite soils in plains
A. commutatus (Schott) Engl.	In lower hills and in plains in Kerala and Coastal Karnataka
A. smithsonianus	Found in lower hills in southern districts of Kerala
A. bonoccordensis	Found in lower hills in southern districts of Kerala (rare)
A. bulbifera Bl.	Eastern peninsular tract, West Bengal, North-eastern region, Khasi Hills, Eastern Himalaya in Sikim. All over in Malabar, Waynad, South Kanara and Chikamagalur
A. sylvaticus (Roxb.) Kunth.	Karnataka
A. oncghophyllus	Andaman and Nicobar Islands
A. campanulatus (Roxb.) Bl. Ex Decne	Wild types occur in Deccan plateau

Botanical Description

The economic part is modified stem called 'corm' and occasionally it produces side cormels.

Table 10.70: Botanical Description

Growth Habit	Either a perennial or an annual, robust herb with a big subterraneous depressed globose tuber.
Stem	Well developed under ground modified stem called 'corm'.
Leaf	Generally bearing one erect, solitary broad, long petiolate leaf. Leaves are simple; petiols smooth or warty and variously spotted; blade simple, 3 parted, the divisions pinnatifid. Petiole is robust, thick massive, light green with pale spots, densely studded with small tubercles, hence very scabrous and 50.0-150.0 cm long. Leaves in plants are 1.0-2.5 cm diameter, 3-partite, each part divided into numerous pblong segments and the later large acuminate and acute.
Floral biology	The infloescence is terminal, very large, consisting of a short stalk, a spathe and a spadix. Spadix long; spathe (or flower) funnel or bell shaped at the base, springing from the great bulb like corm in advance of the leaves, the latter usually pedately compound, differs from arum and related genera by botanical characters. Flowers have disagreeable odor. The stalk is more or less scabrous, during the anthesis short, 5.0-25.0 cm long, afterwards much increasing in length and ultimately 5.0-120.0 cm long. Spathe is obliquely campanulate, before the anthesis entirely enclosing the spadix, afterwards with reflexed or flaccid and drooping, waved margins at the outside yellowish green with white spots, yellow within and studded with numerous, large, dark red, fluid secreting nipples in the upper half strongly waved and dark purple and both sides, 25.0-35.0 cm long and 15.0-30.0 cm diameter. Spadix is fetid, composed of three superposed parts, a female, a male and a neutral part. Female part of the spadix is 5.5-16.0 cm long, ovaries depressed-globose, 0.2-0.4 cm diameter each with a purple style of 0.8-1.6 cm long. Stigmas are large, yellowish white or orange-yellow, distinctly 2-3 lobed and 0.3-0.4 cm long. Male part is obconical and 5.0-9.0 cm long. Stamens are light yellow. Neutral part is shorter than the rest of the spadix, irregularly cinical, obtuse, resembling a dented cap, finely tuberculate, dark purple, shining during the anthesis, spongy within, 7.0-30.0 cm long and 6.0-33.0 cm broad. The inflorescence is covered with a tubular spathe, which open at the top. Female flowers are clustered at the bottom of the peduncle and male flowers at the top.
Anthesis and pollination	Flowering generally occurs between May and June when the corms are in 'dormant' stage having shed all vegetative part at the end of the crop season (May to February). Pollens are come out as and when corms produce spadices, which is always a single spadix per corm. The flowers are extremely protogynous, stigma receptive lasting only 24-78 hours before spathe opening. Anther dehiscence takes place only 24 hours after spathe opening thereby preventing self-pollination.
	In nature there are certain insects who love disagreeable odors of elephant foot yam flowers, which help in pollination. The opened flowers emit a nauseating small attracting for carry on flies. They lay eggs upon the bulbous crimson coloured appendix from which maggot's hatch out and feed.
Fruit and fruit set	Berries are ellipsoid, usually two seeded, ripening from the top of the female part downward, and orange red when mature.
Seeds	Ghose *et al.* (1988) reported that there is complete absence of seed set under natural conditiopns, partially due to to the very limited and erratic appearance of inflirescence under field conditions.
Tuber	The tubers are depressed-globose in shape and dark brown in colour.

Botanical Classification

Classification on the Basis of Corm Type

(i) **Smooth skinned corm type-**This type of elephant foot yam is high yielder. The corms having acridity. The planting is done by cutting the corm into pieces weighing at least 200-300 g per piece. The acridity can be reduced by boiling the corn before cooking finally.

(ii) **Rough skinned corm type-**This type of elephant foot yam is compartatively not high yielder but corms having no acridity. Therefore, its demand in market is very high. The yield is compensented by high prices. This type is propagated by whole corm.

Genetic Resources Augmentation and Crop Improvement

In elephant foot yam, one selection 'Sree Padma' was released for general cultivation. Another important selection is Gajendra, which is high yielder and free from acridily.

Table 10.71.Elephant foot yam varieties released from CTCRI, Kerala

Sl. No.	*Variety*	*Maturity Period (months)*	*Starch (Per cent)*	*Yield (t/ha)*
1	'Sree Padma'	8-9	12.0-13.0	42.0

Genetic Resource Conservation

In elephant foot yam the released promising variety from ICAR-CTCRI, Kerala is 'Sree Padma'. A large genepool is being maintained by this research station.

References

1. Bailey, L.H. 1973. Manuals of cultivated Plants, p. 367. The MacMillan Company, New York.
2. Chopra, R.N; Chopra, I.C., Handa, K.L. and Kapur, L.D. 1958. Indigenous Drugs of India (2nd Edition), Published by U.N. Dhur and Sons Private Ltd. Culcutta, P. 494-543.

3. Kirtikar, K.R. and Basu, B.D. 1994. Indian medicinal plants (Vol.4, 2nd Edn.) Dehradun Publisher Ltd. Indian, p. 2609-2610.

4. Ohtsuki, T. 1968. Studies on reserve carbohydrates of foun *Amorphophallus* species with speical reference to mannan. *Bot. Mag.* Tokyo, 81:119-126.

Genus *Xanthosoma*

Tannia (*Xanthosoma sagittifolium*)

Introduction

Tannia (*Xanthosoma sagittifolium*) closely resembles taro and is often confused with the latter.

Origin, History and Distribution

South America in particular is considered as the centre of origin of tannia.

Botanical Description

Tannia (*Xanthosoma sagittifolium*) is perennial herb. The plants are usually larger than taro and are best adapted to well drained soils.

Genus *Canna*

Canna or Queensland Arrow Root (*Canna edulis*)

Introduction

Canna or Queensland arrowroot is commercially cultivated in Australia for the production of starch whereas in India, it is grown for the edible, tuberous rhizome.

Origin, History and Distribution

Canna edulis Ker-Gawler originated in the Andean region or Peruvian coast and extended from Venezuela to northern Chile, in South America. *Canna* is widely distributed throughout the tropics and sub-tropics. In India, Canna or Queensland arrowroot is commercially grown in Odisha.

Species Availability, Evolution and Diversity

The genus *Canna* includes about 25 species of herbs.

Botanical Description

Canna belongs to the family Cannaceae. The flowers of ornamental cannas (*Canna indica*) are larger and more beautiful and variable in colour than the edible types (*Canna edulis*).

Genus *Maranta*

Arrowroot (*Maranta arundinacea*)

Introduction

Arrowroot (*Maranta arundinacea*) commonly known as 'West Indian Arrowroot' is an important herbaceous plant. It is primarily grown for its rhizomatous tubers containing starch (20 per cent) which is valued as a foodstuff, particularly for infants. The starch is often used as a thickner in all kinds of foods, dressing, soups, sauces, candies, cookies and desserts like puddings and ice-creams.

Origin, History and Distribution

Arrowroot is believed to have originated in the north western parts of South America and the Lesser Antilles. It has been widely distributed throughout the tropical countries like India, Sri Lanka, Indonesia, the Phillippines, Australia and West Indies. In India, it is grown in Uttar Pradesh, Bihar, Odisha, West Bengal, Assam, Northeastern states and Kerala as a rainfed crop.

Species Availability, Evolution and Diversity

The genus *Maranta* includes 20 species of perennial herbs. One species, *Maranta arundinacea* L. (West Indian arrowroot) is cultivated for its edible rhizomes.

Botanical Description

Arrowroot belongs to family Marantaceae. Herbaceous plant

Genus *Plectranthus* or *Solenostemon*

Chinese Potato or Coleus [*Plectranthus rotundifolius* (Poir.) Spreng.; syn. *Coleus parviflorus* Benth., *Solenostemon rotundifolius* (Poir.) J.K. Morton]

Introduction

Chinese potato [*Plectranthus rotundifolius* (Poir.) Spreng.; syn. *Coleus parviflorus* Benth., *Solenostemon rotundifolius* (Poir.) J.K. Morton] $2n=68$) or Syn. *Coleus rotundifolius* is an important underutilized tuber crop, popularly known as Chinese potato. Sometimes it is called as 'coleus' and locally it is known as 'Kurka'. It is seasonal crop cultivated for its underground edible tubers in India.Unlike other tuber crops, this is a transplanted vegetable. The tubers that resemble potato in appearance are consumed as vegetable after cooking which have an artomatic flavor on cooking and have a delicious taste. The tubers contain 76 per cent moisture, and 20-24 per cent starch. Its dry matter consisted of 91 per cent carbohydrate, 5 per cent crude protein, 4 per cent fiber, and 4 per cent ash. The small tubers of the crop are boiled dried and stored for times of food scarcity especially during planting seasons. It is also boiled and eaten as snack. Commercially, tuber size, shape and appearance are important physical quality features that determine both market value and consumer preference in Chinese potato. The most unfavourable feature of Chinese potato is the small and non-uniform size and often odd shape of the tuber due to which cleaning and de-skinning for culinary preparations becomes time consuming. It is reported to have medicinal properties due to the presence of flavonoids that help to lower the cholesterol level of blood (Horvath *et al.*, 2004; Abraham and Radhakrishnan, 2005; Sandhya and Vijayalakshmi, 2005) and enzyme inhibitors (Prathiba *et al.*, 1995). It is also used in the treatment of dysentery, blood in urine and eye disorders in Africa. Tuber contains nutritional and large amount of secondary metabolites which helps in improving body immune system and defence against disease. (Anbuselvi and Hemapriya, 2013). The crop also has a lot of socio-cultural importance, such as presentation as gifts to in-laws, served as food to mourners at funerals, and snacks at child naming ceremonies. A local alcoholic drink has also been brewed.

Origin, History and Distribution

Chinese potato is believed to be a native of Central or East Africa (Shoeninger *et al.*, 2000; Edison *et al.*, 2006) particularly from Ethiopia and spread throughout the tropics. From Ethiopia this prized vegetable spread to other areas. It is commercially cultivated in the tropical regions of India, Malaysia, Indonesia, Sri Lanka, South-east Asia and parts of Africa. In India, it is grown in most of the homestead gardens of Tamil Nadu (Tirunelveii) and Kerala in open garden or extensively in mixed cropping system. Chinese potato is mainly confined to central Kerala (large extent in Northern districts) and Malabar in laterite soils and sandy coastal soils sometimes extending to Mangalore district of Karnataka. Due to its wide acceptance as an aromatic tuber vegetable in Kerala and Tamil Nadu there is a gradual spread of the crop even to black soils in Tirunelveli district (Tamil Nadu) as an irrigated commercial crop and of late it is intensely cultivated in more than 500 hectare in SouthTamil Nadu.

Species Availability, Evolution and Diversity

The genus *Solenostemon* includes about 60 species distributed throughout the tropics. In *Coleus*, four species *viz. C. aromaticus*, *C. forskholii*, *C. zeylanicus* and *C. spicata* (wild or semi-cultivated) are present in South India.

Table 10.72: Distribution of *Coleus* spp.

Crops	*Related Species*	*Distribution*
Coleus spp.	*C. aromaticus* Benth.	All over in plains under dry situation
	C. forskholii Briq.	In hills at about 3000 feet and cultivated in Tamil Nadu. Himalayan forms are tuberising
	C. zeylanicus Benth.	In pockets of lower foot hills in Western Ghats and cultivated in pockets for its stem
	C. spicata	A wild plant in Hassan district in Karnataka

Studies on characterization and evaluation of Chinese potato in India are very limited (Muraleedharan *et al.*, 1985; Sreekumari and Abraham, 1985; Amalraj *et al.*, 1989). Mutation studies have been done in the past to introduce variability (Vasudevan and Jos, 1988, 1992; Abraham and Radhakrishnan, 2005). Though most of the morphological characters among the accessions studied showed less variation, but tuber yield was found to vary significantly among them. An extensive study on morpho-agronomic characterization of 155 accessions of Chinese potato in Ghana was done by Nanema *et al.* (2009). They also observed variability in quantitative morphological traits within the germplasm collection. The study of agro-morphological variability is the traditional method of assessing genetic diversity. Agro-morphological characterization, however, is often hindered by the limited number of polymorphic features and the need for multiple test environments (Kresovich *et al.*, 1997). Suma *et al.* (2014) conducted an experiment involving 26 accessions of Chinese potato (*Solenostemon rotundifolius* (Poir) to determine the trits and observed significant differences for the yield contributing characters between the accessions. Maximum yield was observed in IC468968 which showed highest value for both single plant yield (216.25 g) and average weight of single marketable tuber (56.63 g) and can be directly utilized in the crop improvement programme. IC412975 performed very good and had value for all characters except for length of unmarketable tuber. A positive significant correlation between all the yield and yield contributing characters except tuber length and width of both marketable and unmarketable tuber was observed. Accessions grouped in the first cluster were high yielders with more than 150 g tubers/plant, which can be further enhanced by adopting improved agronomic practices.

Botanical Description

Chinese potato (*Solenostemon rotundifolius* (Poir) belongs to family Lamiaceae (Labiatae). Chinese potato is normally propagated vegetatively bu suckers or soft wood cuttings. Hence, gene flow among the individuals in a population is limited and heterogeneity

Table 10.73: Estimates of Mean for Yield Contributing Characters of 26 Chinese Potato Accessions

Accn. No.	Single plant yield (g)	Marketable tuber weight (g)	Marketable tuber length (cm)	Marketable tuber width (cm)	Marketable Tuber Circumference (cm)	Unmarketable Tuber Weight (g)	Unmarketable Tuber Length (cm)	Unmarketable Tuber Width (cm)	Unmarketable Tuber Circumference (cm)
IC266674	179.25	49.17	4.56	4.35	13.34	3.13	1.91	1.69	5.49
IC412974	122.25	32.63	3.51	4.13	12.96	3.65	2.27	1.71	5.61
IC468942	70.75	35.31	3.63	4.11	12.88	3.25	1.62	1.75	5.39
IC468945	96.25	35.76	3.43	3.98	13.31	3.16	1.49	1.81	5.25
IC468946	108.25	40.34	4.07	4.04	12.53	3.86	2.15	1.78	5.67
IC468947	81.75	27.52	3.57	3.62	11.65	2.76	1.81	1.59	5.33
IC468948	97.25	24.01	4.44	3.28	10.14	2.80	1.88	1.55	4.98
IC468950	91.00	27.93	3.79	3.79	11.62	3.22	1.92	1.74	5.58
IC468952	137.00	49.71	4.37	4.23	13.74	3.75	2.09	1.64	5.41
IC468955	105.00	26.77	3.75	3.49	10.72	3.38	2.03	1.60	4.92
IC468956	129.25	34.11	3.82	3.71	12.33	3.62	1.99	1.61	5.16
IC468957	115.00	29.76	3.40	3.98	12.78	2.95	1.90	1.54	4.94
IC468960	137.50	45.66	4.37	4.37	12.96	2.93	1.76	1.74	5.51
IC468961	197.75	43.83	4.68	3.85	12.20	9.71	2.15	1.66	5.37
IC468963	185.75	28.53	3.77	3.63	11.32	3.61	2.02	1.70	5.66
IC468964	198.00	33.44	4.13	2.91	11.08	3.51	2.10	1.52	5.33
IC468965	156.00	44.55	4.34	4.36	13.64	4.44	2.10	1.89	6.29
IC468966	170.25	35.02	4.21	3.93	12.58	3.62	2.10	1.74	5.69
IC468967	171.50	39.21	4.87	3.74	11.73	4.22	2.52	1.66	5.39
IC468968	216.25	56.63	5.08	4.36	13.56	3.52	1.94	1.67	5.36
IC468969	206.00	40.73	5.86	3.69	10.98	4.31	2.59	1.64	5.39
IC468970	209.25	39.82	4.57	3.77	11.87	3.21	2.05	1.63	5.21
IC468972	199.00	45.28	4.53	4.20	12.98	5.82	2.79	1.95	6.13
IC469690	155.50	29.14	3.49	3.76	12.13	5.24	2.47	1.89	6.15
IC469741	187.00	33.57	4.01	3.92	12.32	4.67	2.43	1.87	6.07
Nidhi (c)	157.75	48.61	4.27	4.01	12.94	4.91	2.18	1.87	6.47

Source: Suma *et al.*, 2014.

Table 10.74: Botanical Description

Growth Habit	It is an herbaceous annual, spreading or prostratand about 30.0 cm tall.
Tubers	The tips of the adventitious roots modify into small round tubers with dark brown peel and creamish flesh produced in clusters at the base of the stem. The shallowly fibrous root systems produce tubers that vary in shape, roughly cylindrical and about 5.0 cm long. However, different shapes and sizes even within same varieties have been observed.
Stem	Succulent stems. The stem is short, stout, erect and glabrous.
Leaf	Aromatic leaves are born opposite on the stem, oval, serrate, irregular and plucked with a length of 12.7-17.8 and covered with 117 hairs.
Floral biology	The crop has a terminal spike of zygomorphic described the flowers as whitish to pink, small and borne on elongated terminal raceme, with the calyx being 21.0 mm long and campanulate. The upper tooth of the calyx is large and ovate and the lower tooth small and deltoid. The corolla is 6.35 mm long and hairy outside.
Seeds	Lack of seed setting

in population is largely due to environmental variations. Lack of seed setting in the species is also responsible for low genetic variability. Highly irregular meiosis is reported to be a reson for sterility of the species (Ramachandran, 1976; Rajmohan, 2007).

Augmentation of Genetic Resources and Crop Improvement

The research on improvement of Chinese potato is very scanty. Only few varieties have been released in this crop. Three high yielding varieties namely 'Nidhi' 'Sree Dhara' and CO-1 from Kerala Agricultural University, Thrissur, (Kerala); ICAR-Central Tuber crops Research Institute, Thiruvananthapuram, Kerala and Tamil Nadu Agricultural University, Coimbatore, (Tamil Nadu), respectively have been released as a variety are very popular among the growers.

REFERENCES

1. Abraham, M. and Radhakrishnan, C.V. 2005. Assessment and induction of variability in Coleus

(*Solenostemon rotundifolius*). *Indian J. Agric. Sci.*, **75:** 834-836.

2. Amalraj, V.A., Velayudhan, K.C. and Muraleedharan, V.K. 1989. Teratological variation in *Coleus parviflorus*. *J. Root Crops*, **15:** 61-62.
3. Anbuselvi, S. and Hemapriya, M. 2013. Nutritional and anti nutritional constituents of *Plectranthus rotundifolius*. *Int. J. Pharm. Sci. Rev. Res.*, **22:** 213-215.
4. Edison, S., Unnikrishnan, M., Vimala, B., Pillai, S.V., Sheela, M.N., Sreekumari, M.T. and Abraham, K. 2006. *Biodiversity of Tropical Tuber Crops in Indian National Biodiversity Authority*. National Biodiversity Authority, Chennai, (Tamil Nadu), India.
5. Horvath, T., Linden, A., Yoshizaki, F., Eugster, C.H. and Ruedi, P. 2004. Abietanes and a novel 20-nor-abietanoid from *Plectranthus cyaneus* (Lamiaceae). *Helv. Chim. Acta.*, 87: 2346-2353.
6. Kresovich, S., McFerson, J.R. and Westman, A.L. 1997. Using molecular markers in genebanks: identity, duplication, contamination and regeneration. In: W.G. Ayad, T. Hodgkin, A. Jaradat, V.R. Rao (eds.). Molecular genetic Techniques for Plant Genetic Resources. Report of an IPGRI Workshop, 9-11 Oct., 1995, International Plant Genetic Resources Institute, Rome, Italy, p. 23-38.
7. Muraleedharan, V.K., Velayudhan, K.C. and John, Laly. 1985. Variation in a collection of *Coleus parviflorus* Benth. In: Ramanujan, T, P.G. Rajendran, M. Thankappan, C. Balagopal and K.B., Nair (eds.). *Tropical Tuber Crops, Production and Utilization*. *Indian Society of Root Crops*, Sreekaryam. Thiruvanathapuram, Kerala, p. 83-88.
8. Nanema, R.K., Traore, E.R., Batiano/Kando, P. and Zongo, J.D. 2009. Morpho-agronomical characterization of *Solenostemon rotundifolius* (Poir.) J.K. Morton (Lamiaceae) germplasm from Burkino Faso. *Int. Biol. Chem. Sci.*, **3:** 11001113.
9. Prathiba, S., Nambisan, B. and Leelamma, S. 1995. Enzyme inhibitors in tuber crops and their thermal stability. *Plant Foods Hum. Nutr.*, **48:** 247-257.
10. Rajmohan, K. 2007. Coleus (*Solenostemon parviflorus* Benth.). In: K.V. Peter (ed.) *Underutilized and Under Explopited Horticultural Crop*. New India Publishing Agency, New Delhi, India, p. 29-36.
11. Ramachandran, K. 1976. Cytology of the genus *Coleus*. *Cytologia*, **32:** 474-480.
12. Sandhya, C. and Vijayalakshmi, N.R. 2005. Antioxidant activity of flavonoids from *Solenostemon rotundifolius* in rat fed normal and high fat diets. *Food Res. Int.*, **38:** 615-629.
13. Shoeninger, M.J., Bunn, H.T., Murray, S.S. and Marlett, J.A. 2000. Composition of tubers used by Hadza forages of Tanzania. *J. Food Compo. Anal.*, **3:** 1-12.
14. Sreekumari, M.T. and Abraham, K.1985. Variation and correlation studies in Chinese potato (*Coleus parviflorus* Benth.). *J. Root Crops*, **11:** 77-81.
15. Suma, A., Latha, M., Joseph John, K., Indira Devi, A., Nair, R.A. and Dwivedi, N.K. 2014. Evaluation of Chinese potato [(*Solenostemon rotundifolius* (Poir) Spreng.] for yield contributing traits. *Indian J. Plant Genet. Resour.*, **27(2):** 142-148.
16. Vasudevan, K. and Jos, J.S. 1988. Gamma ray induced mutants in *Colocasia*. *Mutt. Breed. Newsletter*, **32:** 5.
17. Vasudevan, K. and Jos, J.S. 1992. Variation for yield and quality in *Coleus* mutants. *Madras Agric J.*, **79:** 130-138.

Genus *Moghania*

So-h-phlang *Flemingia vestita* (synonyms *Flemingia procumbens* Roxb.; *Moghania vestita* (Benth.) ex Baker Kuntze; *Moghania procumbens* (Roxb.) Mukerjee)

Introduction

Flemingia vestita (synonyms *Flemingia procumbens* Roxb.; *Moghania vestita* (Benth.) ex Baker Kuntze; *Moghania procumbens* (Roxb.) Mukerjee) famously known as *So-h-phlang* (a lesser leguminous wild type root crop) is a nitrogen fixing herb with characteristic tuberous root, belonging to the genus *Flemingia*. The succulent root is edible and is a common vegetable in some Asian tribal communities. The tubers of sohphlag is eaten as raw either with salt and some red chilli powder and the best combination is with Nei-lieh (*Perilla frutescens*) which has been ground to a fine paste together with salt. The tubers of *Flemingia vestita* are rich in iron (2.64 mg/100 g), phosphorus (64.06 mg/100 g) and contain fair amount of protein (2.99 g/100 g), calcium (19.77g/100 g) and carbohydrates (27.02 g/100 g), (Sarma, 2001). In addition, it has been traditionally used as an anthelmintic, the basis of which is scientifically validated (NRC, 2002; *Van Valkenburg and Bunyapraphatsara, 2001*). It is rich in bioactive isoflavones such as genistein, daidzein, formononetin and pseudobaptigenin (Rao and Reddy, 1991). The tuber has been an indigenous vermifuge among the Khasis, to whom it is called *soh-phlang*. The raw tuber or the root peel is directly consumed for the treatment of soft-bodied intestinal worms (Hrckova and Velebny, 2013). Experimental investigation started in 1996 when the *in vitro* activity of tuber peel extract was tested against different helminth parasites, including the nematodes such as *Ascaris suum*, *Ascaris lumbricoides*, *Ascaridia galli*, *Heterakis gallinarum*, a cestode *Raillietina echinobothrida* and trematodes such as *Paramphistomum* sp. (Tandon *et al.*, 1997) *Artyfechinostomum sufrartyfex* and *Fasciolopsis buski* (Roy and Tondon, 1996). In 1991, an isoflavone, genistein was isolated from the tuber extract (Rao and Reddy, 1991) which was demonstrated to be the major anthelmintic principle, highly potent against trematodes and cestodes. It was further demonstrated effective against the sheep liver fluke *Fasciola hepatica* (Toner *et al.*, 2008) and human tapeworms such as *Echinococcus multilocularis* and *E. granulosus* metacestodes (Naguleswaran *et al.*, 2006).

Origin, History and Distribution

It is found as a wild herb along the mountain slopes of Himalayas. It is distributed in Sichuan and Yunnan provinces of China and Nepal. It is also sparsely found in Laos, Philippines and Vietnam. In India *Moghania vestita*is well distributed in the Khasi hills of Assam. The cultivated Himalayan types are found in the Khasi and Jaintia hills of the Meghalaya. The present day cultivation is localized to Cherrapunji near Shillong and in Jawai, south of Meghalaya in Northeast India. It is mostly grown traditionally under Jhum land as a pure as well as mixed crop during March-April to October-November. It occurs in Konkan region of Maharashtra especially in the coastal belts.

Botanical Description

Table 10.75. Botanical description

Growth Habit	*F. vestita* is a perennial herb, having a prostrate nature.
Root	The roots are tuberous (6 cm or longer).
Stem	Weak stem, measuring about ~60 cm in average. It is highly branched with hairy rhizome and hirsute stems.
Leaf	Leaves are pinnately compound with obovate-cuneate leaflets. Leaves are digitately 3-foliolate; and also pubescent like the stem. Lateral leaflets are obliquely elliptic, and slightly smaller.
Floral biology	Flowers are bright-red. It flowers during August and September Raceme is axillary or terminal, about 2-10 cm, and densely pubescent; bracts lanceolate. Calyx is 5-lobed; lobes are linear-lanceolate, lower one is longest, longer than the tube. Corolla is slightly longer than calyx and elliptical.
Fruit and fruit set	Fruits are hairy sub-cylindrical pods.
Seeds	Seed is globose, brown or black in colour.

REFERENCES

1. Chaudhri, A.B.2005. Forests Plants of Eastern India. Ashish. p. 205–206.
2. Hrckova, G. and Velebny, S. 2013. Pharmacological Potential of Selected Natural Compounds in the Control of Parasitic Diseases. Springer. p. 64.
3. Naguleswaran A, Spicher M, Vonlaufen N, Ortega-Mora LM, Torgerson P, Gottstein B, Hemphill A (2006). "*In vitro* metacestodicidal activities of genistein and other isoflavones against *Echinococcus multilocularis* and *Echinococcus granulosus*". *Antimicrob Agents Chemother,* **50(11):** 3770-3778.
4. National Research Council 2002. Tropical Legumes: Resources for the Future. Books for Business. The Minerva Group, Inc. p. 37-38.
5. Rao H.S.P., and Reddy, K.S. 1991. "Isofavones from *Flemingia vestita*". *Fitoterapia,* **62 (5):** 458.
6. Ren S., Gilbert, M.G. 2010. "FLEMINGIA Roxburgh ex W. T. Aiton, Hort. Kew., ed. 2, **4:** 349. 1812, nom. cons., not Roxburgh ex Rottler (1803)" *(PDF). Flora of China,* **10 (3):** *232-237.*
7. Roy, B. and Tandon, V. 1996. "Effect of root-tuber extract of *Flemingia vestita*, a leguminous plant, on *Artyfechinostomum sufrartyfex* and *Fasciolopsis buski*: a scanning electron microscopy study". *Parasitol Res.*, **82(3):** 248-252.
8. Sarma, B.K. 2001. Underutilized crops for hills and mountain ecosystems. Summer school on agriculture for hills and mountain ecosystem, p. 308-314.
9. Tandon, V., Pal, P., Roy, B., Rao, H.S. and Reddy, K.S. 1997. "*In vitro* anthelmintic activity of root-tuber extract of Flemingia vestita, an indigenous plant in Shillong, India". *Parasitol Res.*, **83**(5): 492-498.
10. Toner, E., Brennan, G.P., Wells, K., McGeon, J.G. and Fairweather I. 2008. "Physiological and morphological effects of genistein against the liver fluke, *Fasciola hepatica*". *Parasitology,* **135**(10): 1189-1203.
11. Van Valkenburg, J.L.C.H., Bunyapraphatsara, N. 2001. Plant Resources of South-East Asia No. 12(2): Medicinal and Poisonous Plants 2. Backhuys Publisher, Leiden, the Netherlands. p. 267-270.

Genus *Pscphocarpus*

Winged Bean (*Psophocarpus tetragonolobus*)

Introduction

The winged bean [*Psophocarpus tetragonolobus* (L.) DC.] is exceptional among all food plants in that practically all parts of the plant are edible and a usable product is provided at every stage of its life cycle, like leaves, flowers, stems, pods, roots *etc.*

Origin, History and Distribution

The origin of winged bean is uncertain, although

it is believed to be a native of either India or Mauritius or African region where all the species of this genus are found in wild. *Psophocarpus tetragonolobus* (L.) DC. is cultivated in India, Mayanmar, Tropical Africa, Asia and Madagascar for its edible pods and tuberous roots. In India it is cultivated in Maharashtra, Karnataka, Bihar, Odisha, Tripura, West Bengal, Assam, Tamil Nadu and Kerala.

Species Availability, Evolution and Diversity

The genus *Psophocarpus* includes 5 species and they are found throughout the tropics.

Botanical Description

Psophocarpus tetragonolobus (L.) DC. is a perennial climber and bear tuberous roots.

Genus *Vigna*

Under genus *Vigna*, *Vigna vexillata* or *Vigna capensis* is another unexploited edible tuber crop (Chandel *et al.*, 1978) is grown by the tribals of Tripura. It is a legume cum tuber crop with much variation in edible tubers (Arora and Pandey, 1996). Department of Biotechnology, Guwahati University, Guwahati (Assam) screened 30 rice bean collections and found that crude protein conrtent varied from 14.66 per cent to 26.88 per cent.

REFERENCES

1. Arora, R.K. and Pandey, A. 1996. Wild edible plants of India-diversity, conservation and use. ICAR-NBPGR, New Delhi, p. 294.
2. Chandel, K.P.S., Arora, R.K. and Joshi, B.S. 1978. *Vigna capensis* Walp.: An edible root legume. *Curr. Sci.*, **41:** 537.

Genus *Costus*

Introduction

It is common plant with tuberous rhizome. It is often cultivated as an ornamental plant and the rhizomes are edible.

Origin, History and Distribution

It is distributed throughout India up to an altitude of 4000 feet.

Species Availability, Evolution and Diversity

Genus *Costus* includes about 175 species.

Genus *Curcuma*

Indian Arrowroot (*Curcuma* species)

Indian Arrow root (*Curcuma* species) is grown for starch apart from West Indian arrowroot.

Genus *Typhonium*

Typhonium trilobatum (L.) Schott. belongs to Araceae family. The crop is grown in tropical and subtropical regions. It is a rich source of carbohydrates and minerals. It is used as a vegetable and for preparation of chips and pickles. The plant is hypnotic. Fresh corms are very acrid and a powerful stimulant; employed as a poultice in tumours. The corms are reported to relax the bowels and provide relief in haemorrhoids and piles. They are eaten with bananas to cure the stomach complaints.

It is a popular tuber crop in Tamil Nadu, India. With a high demand and price in the market.*Typhonium*is preferred crop in irrigated dry lands as well as command areas.

Genus *Helianthus*

Genus *Helianthus* comprises about 75 species.

Jerusalem Artichoke (*Helianthus tuberosus* L.)

Jerusalem artichoke (*Helianthus tuberosus* L.) is one of the tuber yielding crops in the genus *Helianthus*. The tuber resembles potatoes but with larger eyes. Jerusalem artichoke tubers are eaten as raw or boiled. In food value they are considered equal to potatoes. They are also pickled made into chips or ground into flour.

Genus *Tacca*

The genus Tacca includes about 30 species of perennial herbs with tuberous or creeping rhizome.

Tacca (*Tacca pinnatifida* Forstand Forst. f.)

Tacca (*Tacca pinnatifida* Forstand Forst. f.) is originated in South East Asia and is widely distributed on the moist tropics of Asia, Australia and Pacific Islands. It is a perennial herb grows to a height of 60-90 cm. the tubers are globose 15.0-20.0 cm in diameter and harvested after the lops have died down. The tubers are used for the treatment of piles. A bitter extract is prepared by washing the grated tubers in running water and that is given as rubefacient in diarrhea and dysentery.

The fresh acrid bitter tuber yields a nutritive starch having excellent culinary properties. It is known as 'Tahiti' or 'Fiji' or 'East Indian arrowroot starch'. The starch is used to prepare porridges, cakes and other sweet meals. It is also mixed with wheat flour for making bread. The starch is recommended as a food for invalids and also used as laundary starch. Tuber yield ranged from 150.0-250.0 g/plant and the shape of the tubers resembles potato. The tubers possess 22.4 per cent dry matter and 10.2 per cent starch.

Genus *Apios*

Potato Bean (*Apios americana* Medikus)

Introduction

Potat bean (*Apios americana* Medikus; 2n=2x=22) is nitrogen fixing legume that prduces edible tubers and seeds. It is also known by other names like Hopniss, Indian potato, Hodoimo, Groundnut, wild bean, America-hodoimo, American groundnut and Earthnuts. Mostly the pods and tubers are used as a vegetable. In general, the tubers of Potat bean were used as a staple food among most Native American tribes (Beardsley, 1939). Tubers can be used in soups and stews or fried like potatoes, roasted, or ground into flour for bread. Indians eat the seeds like lentils (Duke, 1983). In Ireland, it was evaluated as a possible alternative potato crop during the potato famine of 1845 (Reynolds *et al.*, 1990). Potat bean is popular for saving the pilgrims at Plymouth Rock from starvation in the winter (Gibbons, 1962; Blackmon and Reynolds, 1986; Ode, 1990). Potat bean can fix atmospheric nitrogen in its root nodules and produce an edible tuber (Putnam *et al.*, 1990)

Origin, History and Distribution

It is native to eastern North America from Canada to Southern Florida (Seabrook and Dionne, 1976; Woods, 1988) is naturally distributed from Southern Canada to central and eastern USA and southeastern Canada down through Florida and till the border of Colorado (Reynolds *et al.*, 1990). American groundnut may have reached Europe as early as 1597 (Seabrook, 1973) and was listed as a European garden crop in 1885 (Vilmorin-Andrieux, 1885). The Europeans learned to use the American groundnut from the Native Americans. As a result, the American groundnut became interwoven with the history of the American colonies and Europe. *Apios* is well-adapted to the diverse climatic conditions with a capacity to withstand prolonged flooding and moderate drought. *Apios* is now cultivated as a food crop in Japan and South Korea, because of its nutritional benefits. Before American groundnut was introduced to Japan, the people on the main island of Honshu and the Northern island of Hokkaido were already familiar with a native, wild plant there called hodoimo (*Apios fortunei*) that was occasionally eaten as an emergency food.

Species Availability, Evolution and Diversity

Two species *i.e. Apios americana* and *A. priceana* are very common

Botanical Description

Potat bean (*Apios americana* Medikus fall under Fabaceae family) is diploid, but naturally occurring triploids are also found (Seabrook and Dionne, 1976; Bruneau and Anderson, 1988) but only diploids plants are capable of producing seeds. Triploids plants produce flowers but not seeds and are morphological similar to diploids plants but are entirely dependent on tuber division for propagation; whereas, diploids can be propagated through both seeds and tubers. Other than seed production, there are no easily identifiable differences between diploid and triploids (Bruneau and Anderson, 1988).

Table 10.76: Botanical Description

Growth Habit	The Potato bean plant is a counter-clockwise-twining perennial vine.
Stem	Vine glabrous to pubescent, growing to a height of 1-5 meters.
Leaf	Leaves are odd, alternate and pinnately compound having 5-7 leaflets which are ovate, broad with sharp points and having entire margin.
Floral biology	Flowers are pink on the outside, and maroon to brown on the inside. Flowers are bilaterally produced in 7.5-13.0 cm long dence racemes arsing from leaf axils and have Papillionaceous corolla and a large concave standard with a small hood at its apex into which the narrow sickle-shaped keel is hooked containing the style and the diadelphous stamens. Plants flower during July to September.

Anthesis and pollination	Occasionally for open pollinations by insects seven or eight pods will be set sequentlly along a raceme (Bruneau and Anderson, 1988).
Pod and pod set	Fruits are botanically called as pods, generally green in colour and 5-13 cm long (Reynolds *et al.*, 1990).
Seeds	Seeds of 2-5 inch are edible when cooked.
Tubers	Plant bears edible tubers (Reynolds *et al.*, 1990). Numerous round or oval (2-8 cm diameter), brown-skinned, white-fleshed edible tubers of various sizes are arranged in chains connected by a rhizome just beneath the surface of the soil (NAS, 1979). Tuber sizes are typically smaller than 50 g, however some do average around 100 g. The tuber is also difficult to harvest because of the "beads on a string" arrangement on stolons that extend for over a meter (Cannon, 2012).

Genetic Resources Augmentation and Crop Improvement

There seeds to be a partial self-incomatibility with *Apios* breeding and manual pollinations resulting in rare seed sets.

Genetic Resource Conservation

Belamkar *et al.* (2015) phenotyped *Apios* breeding collections containing 53 genotypes in Iowa, Virginia and Pennsylvania. Six elite genotypes with an average estimated yield of 16.7t/ha-on par with other conventional tuber crops.

REFERENCES

1. Beardsley, G. 1939. The groundnut as used by the Indians of Eastern North America. Papers of the *Michigan Academy of Sciences Arts and Letters*, **25:** 507-525.
2. Blackmon, W.J. and Reynolds, B.D. 1986. The crop potential of *Apios americana*-preliminary evaluations. *HortSci.*, **21:** 1334-1336.
3. Belamkar, V; Wenger, A; kalberer, S.R., Bhattacharya, V.G., Blackmon, W.J. and Steven, B. 2015. Evaluation of phenotype variation in a collection of *Apios americana* : an edible tuberous legumes - Crop Sci; 55: 712-726.
4. Bruneau, A. and Anderson, G.J. 1988. Reproductive biology of diploid and triploid *Apios americana* (Leguminosae). *Amer. J. Bot.*, **75:** 1876-1883.
5. Cannon, S.B. 2012. "Reintroducing a novel bean species: *Apios americana* (Medikus)" (PDF). *Annual Report of the Bean Improvement Cooperative.*
6. Chandel, P., Tiwari, J.K., Ali, N., Devi, S., Sharma, S., Sharma, S., Luthra, S.K. and Singh, B.P. 2016. Interspecific potato somatic hybrids between *Solanum tuberosum* and *S. cardiophyllum*: A potential source of late blight resistance breeding. 7th Indian Horticulture Congress (Abs: 94), held at ICAR-IARI, New Delhi, 15-18 November, p. 72-73.
7. Duke, J.A. 1983. Handbook of Energy Crops.https://hort.purdue.edu/newcrop/duke_energy/*Apios americana.*
8. Gibbons, E. 1962. Stalking the Wild Asparagus. Van Rees. Press, New York.
9. National Academy of Sciences. 1979. Other root crops. In: Tropical legumes: Resources for the future. NAS, Washington, DC, pp. 32-45.
10. Ode, D. 1990. The Groundnut. *South Dakota Conservation Digest*, **57(4):** 25.
11. Putnam, D.H., Field, L. and Heichel, G.H. 1990. Inoculation, nitrogen, and cultivar effects on nodulation and tuber yield of *Apios americana.* In: In: Janick, J. and Simon, J.E. (eds.) Advances in new crops. Timber Press, Portland, Ore, p. 444.
12. Reynolds, B.D., Blackmon, W.J., Wickremesinhe, E, Wells, M.H. and Constantin, R.J. 1990. Domestication of *Apios americana*. In: Janick, J. and Simon, J.E. (eds.) Advances in new crops. Timber Press, Portland, Ore, p. 436-442.
13. Seabrook, J.A. 1973. A biosystematics study of the genus *Apios fabricius* (Leguminoseae) with special reference to *Apios americana* Medikus. MS Thesis, University of New Brunswick, Frederiction, Canada.
14. Seabrook, J.A.E. and Dionne, A.L. 1976. Studies on the genus *Apios*. I. Chromosome number and distribution of *Apios americana* and *A. priceana*. *Canadian Journal of Botany*, **54(22):** 2567-2572.
15. Vilmorin-Andrieux, M.M. 1885.Tuberous glycine. In: The Vegetable Garden. John Murry, London.

11

Genetic Resources of Aquatic Vegetables and Crop Improvement

Aquatic genetic resources (AqGR) of farmed aquatic species and their wild relatives have been used and exchanged for millennia. Today AqGR continue to provide the world with food and recreation: the world's fisheries harvest over 2000 species including fish, crustaceans, molluscs, coelenterates, echinoderms and aquatic plants (FAO, 2014). In the year 2016 FAO reported that the number of farmed aquatic species is smaller, but still extremely diverse (Table 11.1).

Table 11.1: Diversity of Aquatic Species (FAO FishStat J, 2016)

Taxon	*Wild Species*	*Number of Farmed Species (Number of families)*
Finfish	31,000	359(>90)
Molluscs	85,000	103(27)
Crustaceans	47,000	61(>13)
Other aquatic animals	**	15(>8)
Aquatic plants	13,000	~37(>22)

** These include echinoderms, coelenterates and tunicates too numerous to list, many of which have no potential as food and are all marine species, as well as few amphibian and reptiles.

The total numbers of aquatic plant species exceed 1200 and this aquatic vegetation is a valuable source of food. Aquatic vegetables predominate to water bodies like lakes, lagoons, ponds, ditches, and marshy wet places. The high rainfall regions of the world like South-East Asia is noted suitable niches for these natures' gift, which make round the year availability of green lustrous aquatic edible greens, carbohydrate rich rhizomes and nutritionally packed flowers and fruits. Micro-wetlands or satellite wetlands around a bigger wetland act as a constellation of habitat mosaic for resident and migratory waterfowl (Vijayan 1991).This is of special importance for inland wetland habitats in the flyways of migratory birds in the Indo-Gangetic plains and in Deccan peninsula. Often, the size of these micro wetlands is much smaller than 50 ha. Therefore, there is a great need to map wetlands of size smaller than 50 ha.

Table 11.2: Status of State-wise Wetland (Area in hectare)

Name of State	*:*	*Area (ha)*
Andhra Pradesh	:	366609 (scale 1:250,000)
Bihar	:	177683
Gujarat	:	209206
Jammu and Kashmir	:	406780
Karnataka	:	254015
Kerala	:	34200
Madhya Pradesh	:	294118
Maharashtra	:	284942
Odisha	:	162774
Rajasthan	:	344964
Tamil Nadu	:	161521
Uttar Pradesh	:	328690
Pondicherry	:	59
Arunachal Pradesh	:	56325 (scale 1:50,000)
Assam	:	101232

Name of State	:	Area (ha)
Delhi	:	4717
Goa	:	2145
Haryana	:	27057
Himachal Pradesh	:	54766
Manipur	:	52959
Meghalaya	:	2222
Mizoram	:	152
Nagaland	:	918
Punjab and Chandigarh	:	71879
Sikkim	:	1985
Tripura	:	9896
West Bengal	:	143859
Andaman and Nicobar Islands	:	3204
Daman Diu and Nagar Haveli	:	38
Lakshadweep	:	918
Total	:	**3558915**

Source: Pandey *et al.,* 2014.

These aquatic vegetables are embedded with vitamins minerals and phytochemicals. Consumption of aquatic vegetables is in vogue since time immemorial in several parts of the world. Earlier mentions indicate that Egyptians were fond of the consumption of rhizomes of Nymphaea lotus. Orient lotus (*Nelumbo nucifera*) has been widely cultivated and its fruits and rhizomes were used in a variety of cooked and fresh dishes for many centuries as also have the rhizomes of *Sagittaria trifolia,* the flowery seeds of gorgon fruit (*Euryale ferox*) and the starch rich fruits of *Trapa* spp. Canadian wild rice (*Zizania aquatia*) has been eaten by north American Indians for over 300 years and is now grown commercially. Water hyacinths have been used as human food in the Philippines in war time. The soft white bud of the plant was eaten either raw, or as a salad, or as an ingredient in vegetable dishes. Gorgon fruit of eastern India, lotus, water chestnut, etc. are some of the few examples which have been in use at various religious occasions since time immemorial in India.

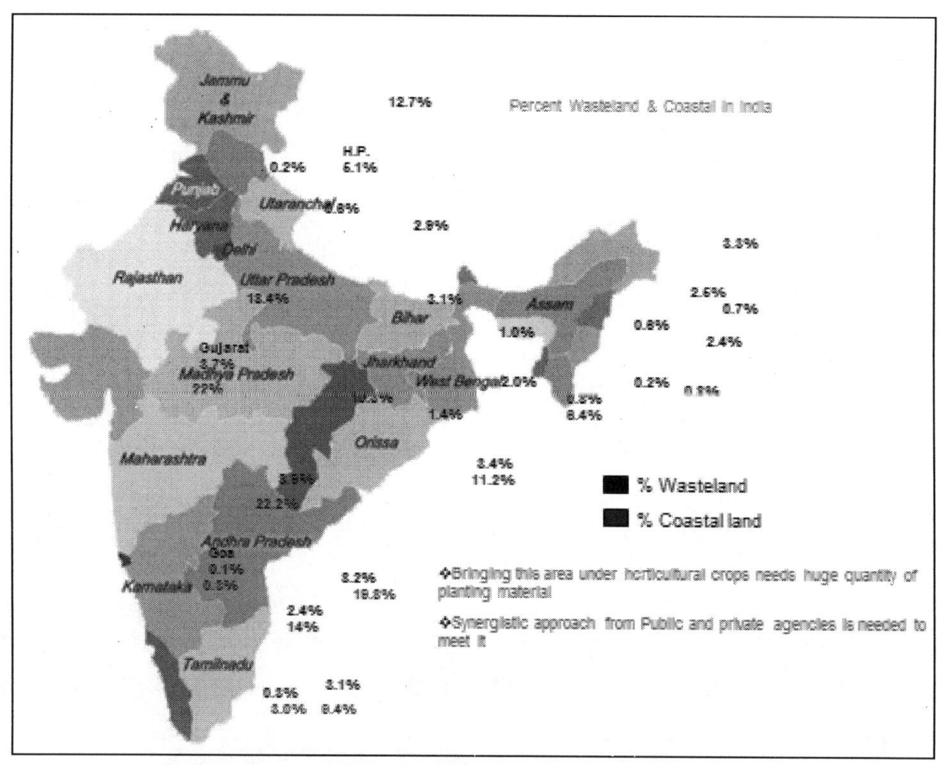

Figure 11.1: Wasteland and Coastal Land of India.

Table 11.3: Aquatic Vegetables, their Scientific Name, Family and Edible Parts

Common Name	Scientific Name	Family	Edible Parts
Velvet leaf	*Limnocharis flava* (L.)	Limnocharitaceae	Leaf vegetable
Lotus	*Nelumbo nucifera*	Nymphaeaceae	Rhizome, young leaves, flower buds
Indian blue water Lily	*Nymphaea stellate* Willd.	Nymphaeaceae	Rhizome
Pygmy water Lily	*Nymphaea tetragona* Georgi	Nymphaeaceae	Leaf buds and seeds
Indian Red water lily	*Nymphaea nouchali*	Nymphaeaceae	Rhizome
Fragrant water lily	*Nymphaea odorata*	Nymphaeaceae	Rhizome
Gorgon fruit	*Euryale ferox*	Nymphaeaceae	Popped seed

Common Name	Scientific Name	Family	Edible Parts
Water chestnut	*Trapa bispinosa*	Trapaceae	Kernel
Water lettuce	*Pistia stratiotes* L.	Araceae	Leaf vegetable
Golden club	*Orontium aquaticum* L.	Araceae	Edible root
Swamp taro	*Cyrtosperma hamissonis* (Schott) Merr.	Araceae	Rhizome
Marsilea	*Marsilea minuta*	Marsileaceae	Pot herb
Manchurian Rice	*Zizania latifolia*	Poaceae	Infected culm
Common water plantain	*Alisma plantago-aquatica* L.	Alismataceae	Roots, leaves and petioles
Water mimosa	*Neptunia oleracea* Loureiro	Fabaceae	Young shoots
Parrot's Feather	*Myriophyllum aquaticum* (J.M. da Conceicao Vellozo) Verdc.	Haloragaceae	Edible leaves
Duck lettuce	*Ottelia alismoides*	Hydrocharitaceae	Petioles and tender leaves
Aponogeton	*Aponogeton crispum*	Aponogetonaceae	Starchy tuberous rhizome
Keysoor	*Scirpus grossus* Linn. F. Tikiu	Cyperaceae	Starchy root
Achyranthes	*Achyranthes aquatica*	Amaranthaceae	Tender leaves and twigs
Hydrolea	*Hydrolea zeylanica* (L.) Vahl.	Hydroleaceae	Young shoots
Water pepper	*Polygonum hydropiper* L.	Polygonaceae	Leaf vegetable
Water fern	*Dentella repens* J.R. Forst. and G. Forst	Rubiaceae	Leaf
Giant chick weed	*Stellaria aquatic* L. Scop.	Caryophyllaceae	Young leaves and stem
Eel grass	*Vallisneria spiralis* L.	Hydrocharitaceae	Leaf vegetable
Water shield	*Brasenia schreberi* J.F. Gmel.	Cabombaceae	Leaf vegetable
Frog bit	*Hydrocharis dubia* (Blume) Backer	Hydrocharitaceae	Mucilaginous leaves and inflorescence
Paracress	*Spilanthes acmella* A.H. Moore	Asteraceae	Leaf vegetable, salad
Buffalo spinach	*Endra fluctuans* Loureiro	Asteraceae	Leaf vegetable
Water dropwort	*Oenanthe stolonifera*	Apiaceae	Pot herb
Papyrus	*Cyperus papyrus* L.	Cyperaceae	Pith of young shoots
Chinese water chestnut	*Eleocharis dulcis* (Burm. F.)	Cyperaceae	Rhizome
Chinese Arrowhead	*Sagitaria trifolia*	Alismataceae	Tuberous root
Cattail	*Typha angustifolia* L.	Typhaceae	Rhizome, petiole
Sonneratia	*Sonneratia caseolaris* (L.) Engler	Sonneratiaceae	Immature fruit
Water spinach	*Ipomoea aquatica* Forsskal	Convolvulaceae	Leaf vegetable
Monochoria	*Monochoria hastata*	Pontederiaceae	Leaf vegetable
Water cress	*Rorippa nasturtium-aquaticum*	Cruciferae	Leaf vegetable
Water sprite	*Ceratopteris thalictroides* (L.) Brongn.	Parkeriaaceae	Edible fern

Source: Pandey, 2011.

In the entire gamut of aquatic wealth, what is needed is knowledge as to how many other aquatic plants, especially those which grow in profusion and are described as weeds or underutilized edible herbage may be usefully converted to human food, because this is obviously more efficient and rewarding than having first to feed the plant material to an animal for conversion into edible products.

Some specific usage of aquatic vegetables are being given in Table 11.4.

Table 11.4: Usage of Aquatic Vegetables

Scientific Name	*Family*	*Parts Used*	*Purpose of Use*	*Status*	*Traditional Knowledge*	*Medicinal Importance*
Ipomoea aquatica	Convolvulaceae	Entire plant	Leafy vegetable	Wild and partially cultivated	Use for remedy of constipation	Considered wholesome for females suffering from nervous and general debility
Centella asiatica	Apiaceae	Leaves	Leafy vegetable and medicine	Wild	Remedy for chronic amoebiosis and blood purification	Useful tonic in diseases of skin, leprosy, nerves and blood tonic for improving memory, syphilic skin disease for internally and externally
Hygrophilla auriculata	Acanthaceae	Leaf twigs	Leafy vegetable and medicine	Wild	Use for remedy of anemia, especially female	Tonic for jaundice, dropsy, rheumatism, anasarca and disease of the urinogenital tract.

Scientific Name	Family	Parts Used	Purpose of Use	Status	Traditional Knowledge	Medicinal Importance
Bacopa monneiri	Scrophulariaceae	Leaf twig	Leafy vegetable and medicine	Wild and non-conventional management	Increase memory, especially children	Tonic useful for asthma, epilepsy, insanity, hoarseness
Mollugo cerviana	Ficoidaceae	Leaf twigs	Leafy vegetable and medicine	Wild	Remedy of skin disease and purification of blood	Febrifuge, used for promoting the flow of lochial discharges and as cure for gonorrhea
Polygonum plabejum	Polygonacae	Leaf twig	Leafy vegetable	Wild	Use for remedy of bowl complaints	Given in bowl complaints, dried and powdered taken in internally in pneumonia
Enhydra fluctuans	Astraceae	Leaf twigs	Leafy vegetable and medicine	Wild	Remedy of skin disease and purification of blood	Laxative, useful in skin and nervous affection, antibilious and dedulcent
Marsilea minuta	Marsileaceae	Foliage	Leafy vegetable and medicine	Wild	Remedy for insomnia	Tonic useful for remedy of insomnia
Alternanthera sessiles	Asteraceae	Entire plant	Leafy vegetable	Wild	Use for remedy of constipation	-
Alternanthera philoxeroides	Asteraceae	Entire plant	Leafy vegetable	Wild	Use for remedy of constipation	-

Source: Chopra *et al.*, 1986.

Aquatic vegetables have immense nutritional attributes. For example Wasabi (*Wasabia japonica*) is an important aquatic vegetable of Japan suppresses the growth of stomach cancer cell and is considered a powerful digestive stimulant, exhibiting strong antioxidant properties with fifteen powerful hydroxyl scavenging agents.

Table 11.5: Medicinal Importance of Aquatic Vegetables

Scientific Name	Family	Medicinal importance	References
Typha angustifolia	Typhaceae	The leaves are diuretic. The pollen is astringent, desiccant, diuretic, haemostatic and vulnerary and used in the treatment of nose bleeds, haematemesis, haematuria, uterine bleeding, and dysmenorrhea. Post-partum abdominal pain and gastralgia, scrofula and abscesses. The seeds are haemostatic. The rootstock is astringent and diuretic In China, 'Puhuang' is the traditional drug obtained from *Typha* plants. Plant is useful in angina cystitis, diarrhea, urethritis, menstrual cramps, burns, insect bites, rash and wounds. Pollens are dusted on bleeding wounds to prevent the flow of blood.	Duke and Ayensu, 1985; Yeung, 1985; Chopra *et al.*, 1986; Gao and Liao, 1998
Typha latipholia	Typhaceae	The leaves are diuretic. The pollen is astringent, desiccant, diuretic, haemostatic and vulnerary and used in the treatment of nose bleeds, haematemesis, haematuria, uterine bleeding, and dysmenorrhea. Post-partum abdominal pain and gastralgia, scrofula and abscesses. The seeds are haemostatic. The rootstock is astringent and diuretic. In China, 'Puhuang' is the traditional drug obtained from *Typha* plants.Plant is useful in angina cystitis, diarrhea, urethritis, menstrual cramps, burns, insect bites, rash and wounds. Pollens are dusted on bleeding wounds to prevent the flow of blood.	Duke and Ayensu, 1985; Yeung, 1985; Chopra *et al.*, 1986; Gao and Liao, 1998
Vallisneriae spiralis	Hydrocharitacea	The plant is aperitif, demulcent, refrigerant, stomachic and is also used in the treatment of women's complaints. It is also used in the treatment of leucorrhoea and is made into a tea with sesame (*Sesamum indicum*) to improve the appetite (Chopra *et al.*, 1986; Duke and Ayensu, 1985).	-
Ottelia alismoides	Hydrocharitacea	Leaves are used to cure hemorrhoids. It is claimed that this plant has rubeficient properties.	-
Alisma plantago-aquatica L.	Alismataceae	The powdered seed is an astringent, used in cases of bleeding. The seed is also said to promote sterility. The root contains an essential oil and has a wide range of medicinal uses. It is antibacterial, anticholesterolemic, diuretic and hypotensive. It lowers blood pressure, cholesterol and blood sugar levels. The root is used in the treatment of oliguria oedema, nephritis, acute diarrhea, cholesterolaemia and fatty liver (Grieve, 1985; Duke and Ayensu, 1985; Yeung, 1985; bown, 1995).	-
Sagittaria trifolia	Alismataceae	Four new biological active diterpenes, trifolianes A, B, C, and D, have been isolated from the tuber together with three new glycoside, sagitariosides a and b and arabinothalictoside. Trifoliones A, B, C, and D possess inhibitory effects on the histamine release from rat mast cells induced by compound 48/80 or calcium ionophore A-23187.	Yoshikawa *et al.*, 1993
Brasenia schreberi	Cabombaceae	Leaves contain Gallic acid and traces of vitamin B_{12}.	-

Scientific Name	Family	Medicinal importance	References
Cyperus papyrus L.	Cyperaceae	The pith is recommended for widening and drying of fistula. Its ash is useful in certain eye diseases; its ash and checks malignant ulcers from spreading in the mouth or elsewhere.	-
Scirpus grossus	Cyperaceae	Keysoor root has astringent properties, and is given to check diarrhea and vomiting. The root possesses as cooling and laxative properties and acts as a tonic to the liver. It is effective against infection and poison, diuretic and useful in burning senssations, fevers and gonorrhea.	-
Wasabia japonica	Brassicaceae	It suppresses the growth of stomach cancer cells. It is a powerful digestive stimulant and exhibits strong antioxidant properties with fifteen powerful hydroxyl radical scavenging agents. It also possesses anti-diarrhoel property. All parts of plant possess bacterial activities against Helicobacter pylori strain NCTC11637, YS27 and YS50.	(Nakayama *et al.*, 1998; Shin *et al.*, 2004)
Nymphaea odorata	Nymphaceae	The rhizomes are cooling, sweet, bitter and tonic and useful in diarrhea, dysentery, dipsia and general debility. The flowers are astringent and cardio tonic. The seeds are sweet, cooling, constipating, aphrodisiac stomachic and restorative.	-
Nymphaea stellata	Nymphaceae	The powdered rhizome is given in dyspepsia, diarrhea and piles. An infusion of rhizome and stem is considered emollient and diuretic. It is used for biennorrhagia and diseases of the urinary tract. Plant possesses hypoglycemic activity.	-
Sonneratia caseolaris	Sonneratiaceae	In Myanmar, fruits are used for poultices. Malayans used old fruit walls for worms, half-ripe fruits for coughs and pounded leaves for hematuria and small pox. Plant possesses anti-nociceptive and anti-diarrhoeal activities.	Perry, 1980
Spilanthes acmella L.	Asteraceae	The flower heads are chewed to relieve the toothache and other mouth related troubles. Leaves are used extremally in treatment of skin disease. Root decoction is used as purgative. Leaf decoction is used as diuretic and lithotriptic. Whole plant has been well documented for its uses as spices, antiseptic, anti-bacterial, anti-fungal, anti-malarial and as remedy for toothache, flu, cough, rabies disease and tuberculosis.	Burkil, 1966; Di Stasi *et al.*, 1994; Akah and Ekekwe, 1995; Singh, 1995; Storey and Salem, 1997; Ramsewak *et al.*, 1999.
Hydrolea zeylanica (L.) Vahl.	Hydroleaceae	Leaves are used as a poultice for unlcers and in traditional Cambodia medicine. The leaves are prescribed for intestinal disorders.	-
Polygonum hydropiper	Polygonaceae-Buckwheat family	The leaves are anti-inflammatory, astringent, carminative, diaphoretic, diuretic, emmenagogue, stimulant, stomachic, and styptic. They contain rutin which helps to strengthen fragile capillaries and thus helps prevent bleeding. The seed is carminative, diuretic and stimulant.	Grieve, 1984; Launert, 1981; Foster and Duke, 1990; Duke and Yuensu, 1985; Coon, 1975

Source: Pandey *et al.*, 2014.

China is home to the world's largest gene bank for aquatic vegetables. The repository, which holds more than 1,700 wild and cultivated specimens from 13 kinds of aquatic vegetables, is based in the city of Wuhan and run by the Wuhan Institute of Vegetable Science. Thirteen vegetables which include lotus, water bamboo, taro, water celery, water chestnut and watercress- are the most frequently planted aquatic vegetables in the world and out of these 11 come from China.

Table 11.6: Aquatic Genera and available Species

Genera	Species	Common Name
Achyranthes	*Achyranthes aquatica* (Roxburgh) Walich	Achyranthes
	A. abyssinica, A. acuminata, A. alba, A. albida, A. alopecuroides, A. alternifolia, A. altissima, A. amabilis, A. amaranthoides, A. ancistrophora, A. angustifolia, A. annua, A. aphylla, A. aquatica, A. arborescens, A. argentea, A. aspera, A. asperoides, A. atollensis, A. atra, A. atropurpurea, A. australis, A. avicularis, A. axillaris, A. baccata, A. bangii, A. bastosiana, A. bengalensis, A. benthamii, A. bettzickiana, A. bidentata, A. biemilleri, A. biemuelleri, A. boliviana, A. borbonica, A. brachiata, A. brasiliana, A. breviflora, A. calceolata, A. calea, A. canescens, A. capitata, A. capituliflora, A. carnosa, A. carsonii, A. chacoensis, A. chinensis, A. ciliata, A. coferta, A. cordata, A. cordobensis, A. corymbosa, A. costaricensis, A. coynei, A. crassifolia, A. crispa, A. cyclphylla, A. cylindrical, A. debilis, A. decumbens, A. diandra, A. dichotoma, A. diffusa, A. digera, A. dioica, A. echinata, A. echinocephala, A. elegantissima, A. ellipticifolia, A. elongata, A. excelsa, A. fasciculata, A. fauriei, A. ferruginea, A. ficoidea,A. filifolia, A. fischeri, A. flabellifera, A. flavicoma, A. frumentacea, A. frutescens, A. fruticosa, A. galapagensis, A. geminata, A. geniculata, A. glabella, A. glabra, A. glauca, A. glaucescens, A. globosa, A. globulifera, A. glomerata, A. gracilis, A. grandifolia, A. halimifolia, A. hamata, A. hassleriana, A. helleri, A. heudelotii, A. hirtiflora, A. hispida, A. hookeri, A. incana, A. indica, A. inermis, A. ingramicana, A. involucrata, A. jacquini, A. jacquinii, A japonica,	

Genera	Species	Common Name
	A. javanica, A. kaessneri, A. laguroides, A. lanaiensis, A. lanata, A. lanceolata, A. lanuginosa, A. lappacea, A. lehmannii, A. leiantha, A. leptostachya, A. linarifolia, A. lindaviana, A. lindenii, A. linearifolia, A. linkiana, A. longifolia, A. lorentzii, A. lupulina, A. maneleensis, A. mangarevica, A. marchionica, A. margaretarum, A. maritima, A. martii, A. martinicensis, A. mauritiana, A. megaphylla, A. minutiflora, A. mollicula, A. mollis, A. monsoniae, A. Montana, A. moorcroftiana, A. morongii, A. mucronata, A. muricata, A. mutica, A. natans, A. nelsonii, A. nervosa, A. nivea, A. nodiferta, A. nodiflora, A. nodosa, A. novae-brasiliae, A. novae-brasilianae, A. nuda, A. nudicaulis, A. oblanceolata, A. obovata, A. obovatifolia, A. obtusifolia, A. ogatai, A. okinawensis, A. olivacea, A. orbiculata, A. ovate, A. paludosa, A. panamensis, A. paniculata, A. papposa, A. paraguayensis, A. paronychioides, A. patula, A. pedicellata, A. peploides, A. persica, A. philoxeroides, A. pibosa, A. picta, A. pilosa, A. piloselloides, A. pinheirensis, A. polygonata, A. polygonoides, A. polystachya, A. porphyristachya, A. porrigens, A. portoricensis, A. praelonga, A. prizhri, A. prostrata, A. puberula, A. pubescens, A. pungens, A. purpurea, A. pycnantha, A. radicans, A. ramosissima, A. reflexa, A. reineckii, A. repens, A. rigida, A. rivinaefolia, A. robusta, A. rosea, A. rotundata, A. rotundifolia, A. rubrofusca, A. rubrolutea, A. rufa, A. rugelii, A. rugulosa, A. ryukyuensis, A. sanguinolenta, A. sarmentosa, A. scandens, A. schinzii, A. schweinfurthii, A. sequax, A. sericea, A. serpyllifolia, A. sessilis, A. setacea, A. seubertii, A. sicula, A. snodgrassi, A. snodgrassii, A. spiciflora, A. spinosa, A. splendens, A. stellata, A. stenophylla, A. strictiuscula, A. styracifolia, A. talbotii, A. tenella, A. tenuifolia, A. ternata, A.thonningii, A. thyrsiflora, A. tomentella, A. tomentosa, A. totonaca, A. triandra, A. trichotoma, A. uncinata, A. uncinulata, A. urbani, A. velutina, A. vermicularis, A. verreauxii, A. verschaffeltii, A. verticillata, A. vestita, A. villosa, A. violacea, A. virgata, A. viridis, A. watsonii, A. welwitschii, A. wightian, A. williamsii, A. winteri.	
Alisma	*Alisma plantago-aquatica* L.	Common Water Plantain
Alternanthera	*Alternanthera philoxeroides*	Alligator Weed
	A. sessilis	Ponnanganni
Aponogeton	*Aponogeton crispum* Thunb.	Aponogeton
	A. abyssinicus, A. afroviolaceus, A. ambongensis, A. angustifolium, A. angustifolius, A. appendiculatus, A. asureus, a. azureus, A. bernerianum, A. bernerianus, A. bernierianus, A. boehmii, A. bogneri, A. boivinianus, A. braunii, A. bruggenii, A. bullosus, A. capuronii, a. cordatus, A. crinifolium, A. crispum, A. crispus, A. cuneatus, A. decaryi, A. desertorum, A. dioecus, A. distachyon, a. distachyos (Cape Pond Weed), *A. distachys, A. distachyum, A. distachyus, A.eberhadtii, A. echinatum, A. echinatus, A. eggersii, A. elongates, A. elongates forma elongates forma latifolius, A. elongatus, A. elongatus forma strigosus, A. elongates* subsp. *fluitans, A. euryspermus, A. fenestrabi, A. fenestrale, A. fenestralis, A. fotianus, A. gracile, A. gracilis, A. gramineus, A. guillotii, A. henckelianus, A. hexatepalus, A.holubii, A. involucratus, A. jacobsenii, A. junceum, A. junceus, A. kimberleyensis, A. kraussianum, A. kraussianus, A. lagrangei, A. lakhonensis, A. lancesmithii, A. leptostachyos, A. leptostachyos,* var. *minor, A. leptostachys* var. *lilacina, A. leptostachys, A. longiplumulosus, a. loriae, A. madagascarensis, A. madagascarensis* var. *henkelianus, A. masoalaensis, A. monostachyon, A. monostachyos, A. monostachys, A. monostachyus, A. natalense, A. natalensis, A. natans, A. nudiflorus, A. oblongus, A. proliferus, A. quadrangulare, A. quadrangularis, A. queenslandicus, A. ranunculiflorus, A. rehmannii, A. rigidfolius, A. robinsonii, A. satarensis, A. schatzianus, A. spathaceum, A. stachyosporus, A. stachyosporus de, A. stuhlmannii, A. subconjugatus, A. tenuispicatus, A. tofus, A. troupinii, A. ulvaceous, A. ulvaceum, A. ulvaceus, A. ulvaum, A. undulatum, A. undulates, A. vallisnerioides, A. vanbruggenii, A. violaceus, A. viridis, A. womersleyi, A. zosteraefolium*	
Brasenia	*Brasenia schreberi* J.F. Gmel	Water Shield
Butomus	*Butomus umbellatus*	-
Calla	*Calla palustris*	-
Ceratopteris	*Ceratopteris thalictroides* (L.) Brongn.	Water Sprite
Cyperus	*Cyperus papyrus* L.	Papyrus
Cyrtosperma	*Cyrtosperma hamissonis* (Schott) Merr.	Swamp Taro
	C. beccarianum, C. bougainvillense, C. brassii, C. carrii, C. cuspidispathum, C. edulis, C. giganteum, C. gressittiorum, C. hambalii, C. johnstoni, C. kokodense, C. macrotum, C. matrieffianum, C. merkusii	-
Dentella	*Dentella repens* J.R. Forst. and G. Forst.; *D. serpyllifolia*	Water Fern
Eleocharis	*Eleocharis dulcis* (Burm. F.)	Chinese Water Chestnut
Enydra	*Enydra fluctuans* Loureiro	Buffalo Spinach
Euryale	*Euryale ferox* Salisb.	Gorgon Fruit
Helostoma	*Helostoma temmincki*	Kissing Gourami
Hydrocharis	*Hydrocharis dubia* (Blume) Backer	Frog Bit
Hydrolea	*Hydrolea zeylanica* (L.) Vahl	Hydrolea

Genera	Species	Common Name
	H. brevistyla, H. corymbosa, H. elatior, H. elegans, H. floribunda, H. glabra, H. macrosepala, H. megapotamica, H. multiflora, H. prostrate, H. ovate, H. paludosa, H. sansibarica, H. spinosa, H. uniflorae, H. urens, H. zeylanica.	
Ipomoea	*Ipomoea aquatic* Forsskal	-
Limnocharis	*Limnocharis flava* (L.) Buchenau	Velvet Leaf
	L. laforestii	-
Marsilea	*Marsilea minuta* L.	Marsilea
Menyanthes	*Menyanthes trifoliate* (L.) Solms	-
Monochoria	*Monochoria hastate, M. vaginalis*	Monochoria
Myriophyllum	*Myriophyllum aquaticum* (J.M. da Conceicao Vellozo) Verdc.	Parrot's Feather
Nasturtium	*Nasturtium officinale*	Watercress
Neptunia	*Neptunia oleracea* Loureiro	Water mimosa
Nelumbo	*Nelumbo nucifera*	Lotus
	N. lutea	Lotus (yellow cream flowers)
Nymphaea	*Nymphaea stellate* Willd.	Indian Blue Water Lily
	N. tetragona Georgi	Pygmy Water Lily
	N. nouchali var. *caerulea*	Indian Red Water Lily
	N. odorata	Fragrant Water Lily
	N. lotus	-
Oenanthe	*Oenanthe stolonifera*	Stolon Water Dropwort
	O. javanica (Blume) DC.	Java/Japanese Water Dropwort
	O. aquatica	Fine Leafed Water Dropwort
	O. crocata	Hemlock Water Dropwort
	O. fistulosa	Tubular Water Dropwort
	O. fluviatilis	River Water Dropwort
	O. javanica spp. *javanica*	Java Water Dropwort
	O. lachenalii	Parsley Water Dropwort
	O. laciniata	Cut Leaf Water Dropwort
	O. peucidanifolia	Cut Leaf Water Dropwort
	O. pimpinelloides	Corky Fruited Water Dropwort
	O. sarmentosa	Water Parsley
	O. silaifolia	Narroe Leafed Water Dropwort
Orontium	*Orontium aquaticum* L.	Golden Club
Osphronemus	*Osphronemus gouramy*	Giant Gourami
Ottelia	*Ottelia alismoides* (L.) Pers.	Duck Lettuce
Pistia	*Pistia stratiotes* L.	Water Lettuce
Polygonum	*Polygonum hydropipre* L.	Water Pepper
Rorippa	*Rorippa nasturtium-aquaticum, R. heterophylla*	Watercress
Sagittaria	*Sagittaria trifolia, S. latifolia*	Chinese Arrowhead
Scirpus	*Scirpus grossus* Linn.f. Tikiu, *S. yagara*	Keysoor
Sonneratia	*Sonneratia caseolaris* (L.) Engler	Sonneratia
Spilanthes	*Spilanthes acmella* A.H. Moore *S. paniculata,*	Paracress
Stellaria	*Stellaria aquatica* L. Scop.	Giantchick Weed
Talinum	*Talinum triangulare* (Jacq.) Willd.	Ceylon Spinach
	T. cuneifolium, T. arnotii, T. caffrum, T. portulacifolium (Forsk) Aschers, ex. Schwf, *T. patens, T. crassifolium, T. indicum,*	-
Trapa	*Trapa bispinosa*	Water Chestnut
	T. maximowiczii Korsh Syn. *T. quadrispinosa* auc. Non. Roxb., *T. bicornis, T. natans* Linn. Syn. *T. quadrispinosa* Roxb., *T. incisa*	-
Typha	*Typha angustifolia* L., *T. latifolia, T. davidiana*	Cattail
Vallisneria	*Vallisneria spiralis* L.	Eel Grass
Wasabia	*Wasabia japonica* (Miq.) Matsum	-
Zizania	*Zizania latifolia* (Griseb.) Turcz. Ex Stapf	Manchurian Rice

Table 11.7: Botanical Description of some Common Aquatic Vegetables

Common Name	*Name of Species*	*Botanical Description*	*Origin*
Indian Blue Water Lily	*Nymphaea stellate* Willd.	Rootstock ovoid, short, and acute. Leaves peltate 12-20 cm in diameter, orbicular or elliptic (the younger sagittate), obtuse sinuate; dentate or entir, with narrow sinus 5-7.5 cm deep, glabrous on both surface, often blotched with purple beneath; petioles long, slender, and submerged. Flowers solitary 7.5-15 cm in diameter, usually pale violet, less commonly light blue or purple, sometimes not ribbed, oblong-lanceolate, acute or sub-obtuse, and streaked with purple lines. Petals linear, oblong or lamnceolate. Filaments 10-50 dilated at the base; anthers with a lingual appendage, pollen smooth and stigmatic rays 10-30 in number without appendages. Fruits are globular and seeds are longitudinally striate.	
Pygmy Water Lily	*Nymphaea tetragona* Georgi	Rhizomes unbranched, erect, cylindricaland stolons absent. In leaves petiole glabrous. Leaf blade abaxially gren to dull purple, adaxially green, sometimes mottled with reddish brown or purple, especially on young leaves, ovate to elliptic, 3-13 (-14) x 2-11 (-13) cm, margins entire; venation radiate centrally, without weblike pattern, and principal veins 7-13 with surface glabrous. Flowers floating, 3-7.5 cm in diameter; opening and closing diurnally, only sepals and outermost (occasionally inner most) petals in distinct whorls of 4; sepals uniformity green, obscurely veined, lines of insertion on receptacle very prominent, protuberant, forming tetragons; sepals 10-17, white; stamens 30-70, yellow-orange abaxially usually suffused with purple, connective appendage projecting less than 0.2 mm beyond anther; filaments widest above middle, longer than anthers; pistil 5-10, locular appendages at margin of stigmatic disk and boat-shaped in (2-) 3-4 x 2-4 mm. Seeds ovoid, ca.2-3 x 1.5-2 mm, ca.1.3-1.5 times as long as broad and lacking papillae on surface.	
-	*Typha angustifolia* L.	Perennial plant which is 7.5-17.5 cm tall, unbranched, vegetative colonies are often formed and the root system consist of rhizomes rootstocks and fibrous roots. Plant consisting of a flowering stalk and 4 or more leaves. The flowering stalk is light green to green, glabrous, stiff, and round in cross section (terete). The leaves are 3.75-15.0 cm long and up to 1.25 across. They are green, glabrous, linear, and rather stiff. Relative to the flowering stalk and the leaves are ascending to slightly spreading. They originate from the base of the plant. The inner side of each leaf is flat to slightly concave, while the outer side is convex (planoconvex). Leaf venation is parallel. The base of each leaf is wrapped in a sheath. The stalk terminates in a spike of pistillate flowers and a spike of staminate flowers; the staminate spike is above the pistillate spike and they are separated from each other by at least 1.25 cm (usually few cm.). The staminate spike is narrowly cylindrical and densely packed with staminate flowers and abundant hairs; each staminate spike is up to 20 cm ling and 1.25 cm across. Each staminate flower bears a single grain of pollen. The pistillate spike is cylindrical and densely packed with pistillate flowers and abundant hairs; each pistillate spike is up to 30 cm long and 1.88 cm across. Each fertile pistillateflower has a stipe that is less than 1 mm long and its bears a single achene with a slender style. There are also infertile pistillate flowers that lack achenes. The staminate spike is light yellow to light brown, while the mature pistillate spike is brown or red dish brown. The staminate spike quickly withers away after shedding its pollen while the pistillate spike persists into the fall. The blooming period occurs from early to mid-summer. Pollination is by wind.	
Broadleaf Cattail	*Typha latifolia* L.	It is an aquatic or semiaquatic emergent perennial plant. Plants are normally 1-3 m tall, reed like and extensively clonal. Stems are stout, cylindrical, and unbranched. Flowering stem length is typically equal to or somewhat longer than leaf length. Leaves are thick, linear, and flat which measure 6 to 29 mm wide. Numerous tiny, dense, felty flowers occur in a terminal spike that is 1.8-5 cm thick. Male flowers make up the upper, narrower half of the spike and female flowers the lower, slightly wider half. Over 1000 flowers may be counted in the staminate spike portion of one. Male and female spikes are rarely separated. A small space of less than 1.5 cm may occur but separations of up to 4 cm are also reported (Anonymous, 1986). Spikes are typically 6 times as long as they are thick. Only female flowers are persistent. Fertilized flowers produce single-seeded, nutlike achenes up to 1.5 mm long. Long slender hairs at the base alow for wind and water transport of the 'eventually dehiscent' seeds. Broad leaf cattail has a long cylindrical spike of female flowers surmounted by a slender spike of male flowers. After flowering, the colour of the female flowers changes from yellowish-green to brown or dark brown. The substance largely responsible for the browning was identified as 5-O-trans-caffeoylshikimic acid (Ozawa and Imagawa, 1988).	
Wasabi	*Wasabia japonica* (Miq.) Matsum.	It is 20-75 cm tall herbs; glabrous or sparsely pilose on upper parts; and rhizomes fleshy which is 3 cm in diameter. Stems erect or decumbent, and simple. Basal leaves rosulate; petiole (6-) 10-12 (-26) cm, dilated at base; leaf blade cordate or reniform, (2.5-) 6-15 (-20) x (3-) 6-18 (-22) cm, base cordate, margin dentate, denticulate, shallowly crenate, repand, or subentire, with distinct apiculate callosities terminating ultimate veins, apex rounded or obtuse. Middle cauline leaves with petioles 1-5 (-8) cm; leaf blade broadly ovate to ovate-cordate, 1.5-4(-6) x 2-4 (-6) cm, palmately veined, base and margin as in basal leaves, and apex acute. Infructescence lax raceme, bracteates throughout or basally. Fruiting pedicels ascending or divaricate, slender, 1-3.5 (-5) cm. Sepals oblong, 3-4 x 2-2.5 mm, and caduceus. Petals white, oblong-spatulate, 6-8 (-9) x 2-3 mm. Filaments white white, 3.5-5 mm and anthers oblong, 0.6-0.8 mm. Ovules 6-8 per ovary.	

Common Name	Name of Species	Botanical Description	Origin
		Fruit linear, 1-2 cm x 1.5-2 mm, terete, torulose; valves with an obscure midvein; gynophore (1-) 2-5 mm; septum usually complete; style 2-3 mm. seeds are borne in siliques that are short, linear, oblong, torulose, cylindric with membranous valves and without a midrib. Each silique contains up to 8 seeds and each seed is 2-3 mm long and 1 mm wide, pointed at one end and flat on the top highly variable within varieties.	
Bufalo Spinach	*Enydra fluctuans* Loureiro	Plant is a 30-100 cm tall, aromatic and herbaceous in nature. Stem cylindrical, hollow, and hairy rooting occurs from node. Leaves are sessile, opposite, size 2-10 x 0.5-2 cm, apex narrowly obtuse, and leaf margin dentate-serrate. Small gland dots on both surface of leaf; inflorescence a globular head, diameter 1.0 cm, sessile, solitary, terminal, heterogamous, many flowered, involucres of 4 folacious longitudinal-veined bracts, longer than the flowering heads and completely covering them. Ray florets in several whorls, female, corolla with long tube and 2-5 lobed limb, white or greenish white; ovary enclosed by ciliate palea, and style exserted. Disk flower numerous in number, hermaphrodite, sometimes sterile, corolla companulate with 5 lobed rim; ovary like ray floret; style with double arms; stamens connate, and anthers exserted. Fruit an achene, oblonged app. 3 mm long, glabrous, closely clasped by two thickened palea (acting as floater) blakish, without a pappus (Ban, 1993).	
Manchurian Rice	*Zizania latifolia* (Griseb.)	Plant grows up to 4 m tall, and stolons present. Culms 100-150 cm long and spongy. Ligule an eciliate membrane and12-20 mm long. Leafblades 30-100 cm long; 20-30 mm wide and coriaceous. Leafblade surface scaberulous and rough on both sides. Leafblade apex attenuates. Inflorescence monoecious; with male and female spikelets in the same inflorescence. Inflorescence a panicle. Panicle open; lanceolate and 35-50 cm long. Sexes segregated; on unisexual branches with male below. Spikelets appressedand solitary.Fertile spikelets pedicelled. Male pikelets on the lower part of the panicle, lanceolate, 8-12 mm long, usually purplish and pedicelled (pendulous). Pedicels 3 mm long; pubescent and tip upuliform. Fertile spikelets comprising 1 fertile florets; without rhachilla extension. Spikelets oblong; laterally cpmpressed; 15-17 mm long and falling entirle. Glumes both absent and obscure. Florets fertile florets female. Fertile lemma oblong; 15-17 mm long; chartaceous; without keel; 3-veined. Lemma apex acute; awned; 1-awned. Principal lemma awn 15-50 mm long overall. Palea 1 length of lemma; 2 veined; without keels. Flower lodicules 2. Anthers 6; 5-8 mm long. Fruit caryopsis with adherent pericarp; linear and 10-20 mm long. Male spikelets distinct from female 8-12 mm long.male spikelets glumes absent. Male spikelets lemma 5-veined; awned (8-10 mm). thef	
Water Dropwort	*Oenanthe javanica* (Blume) DC.	It is perennial herb with creeping stolons and long, threadlike, white rootlets. The erect slender, hollow, green stems range from 10 cm to 150 cm high. The green coloured leaves are alternate, petiole up to 12 cm, often sheathing over most of its length with membranous margins; leaf lamina oblong to ovate in ouline, 5-30 cm x 5-15 cm, pinnate to tripinnate; ultimate segments ovate to narrowly oblong, 0.5-7.0 x 0.3-3.5 cm, with serrate or entile margins, above dark green and dull, beneath lighter coloured with transparent nerves, leaves having an odour like carrot tops. Inflorescence a compound, many flowered umbel, terminal and opposite the leaves, peduncle longitudinally furrowed, up to 25 cm long, primary rays 5-15, 0.5-3.0 cm long; involucres none or 1; bracteoles at the base of the umbellules (involucels) 2-8, linear, green; secondary rays (pedicels) 10-25, 2-5 long; flowers tiny, white, fragrant; calyx teeth 5, distinct, acute, dark gree about 0.5 mm, petals 5, obovate, up to 2 mm x 1 mm, glabrous, stamens 5, patent, filaments white, anthers small, yellowish brown; ovary glabrous; style long and persistent. Fruit a schizocarp, glabrous, mericarp 2-3 mm x 0.5-1 mm with swollen ribs, the marginal ones much more prominent than the dorsal ones (Sasmitamihardja, 1993).	
-	*Marsilea minuta* L.	Rhizome 0.4-0.8 mm thick, green to light brown, densely covered with tan hairs at apices but glabrate proximally, sometimes bearing short branches with numerous leaves at nodes, nodal and intermodal roots 0.1-0.3 mm thick, internodes 1-3.5 cm long; rhizome in water 0.5-0.8 mm thick, green to brown, glabrous and internodes 4-11 cm long. Land leaves with erect terete petioles 5-13 cm long, 0.4-0.8 mm thick, leaflets 1.2-2.0 cm long, 0.8-1.7 cm wide, cuneate to flabellate, glabrous to sparsely hairy, with concave, straight, or convex lateral margins and entire to crenulate terminal margins; leaves in water often with leaflets emergent rather than floating, with stii terete petioles 13.5-47 cm long, 0.3-0.8 mm thick, leaflets 1.7-1.8 cm long. 1.5-1.6 cm wide and hydropoten infrequent. Fertile leaves produced on land, bearing 1-4 sporocarps per leaf, on either branched unbranched peduncles attached at or slightly above the base of the petiole, unbranched peduncles and ultimate divisions of branching peduncles 2.6-6.0 mm long, the common trunk of branched peduncles 0.4 mm long. Sporocarps 2.6-4.1 mm long, 2.4-3.1 mm wide, 1.3-1.7 mm thick, nearly round to slightly oblong in lateral view, ovate or elliptical in cross section, brown to black, hairy to glabrate; raphe 1.5-2.2 mm long, inferior tooth 0.3-0.6 mm long, obtuse or truncate, or absent, superior tooth 0.3-0.6 mm long, acute; sporocarp veins ca 15, forking midway, branches often anastomosing at tips. Sori 9-12 per sprocarp, attached to a sorophore with a free acute tip, 9-14 microsporangia and 2-7 megasporangia per sorus.	

Common Name	Name of Species	Botanical Description	Origin
Monochoraia	*Monochoraia hastata*	The plant of *M. hastate* is a perennial herb with elongate creeping; spongy root leaves with very long petiole (up to 90 cm), sagittate, hastate or cordate; flower in purplish black or violet blue in colour dotted with red; in sub-umbellate racemes; capsule small, ellipsoid sub-globose or oblong.	
	M. vaginalis	*M. vaginalis* is an annual and stems are creeping. Its leaves are smaller than *M. hastata*. Plant height varies from to 2-3 m during the rains. Anthesis takes place in the morning hour and thereafter flower withers. After the completion of fertilization, young inflorescence bends downwards and the fruits develop under water. Seeds also germinate under submerged condition.	
Water Lettuce	*Pistia stratiotes* L.	It is a perennial monocotyledon with thick and soft leaves that form a rosette. The leaves can be up to 14 cm long and have no stem. Leaves are light green, with parallel veins, wavy margins and are covered in short hairs which form basket-like structures which trap air bubbles, increasing the plant growth. The flowers are dioecious and are hidden in the middle of the plant amongst the leaves. Small green berries form after successful fertilization. The plant can also undergo asexual reproduction. Primarily reproduces vegetatively. Lateral stolons (floating stems) extend out from the central axis of the plant to initiate daughter plants. plant can also reproduce via water-dispersed seeds.	
Water Sprite	*Ceraptopteris thalictroides* (L.) Brongn.	Plants usually rooted in mud, very variable in size and appearance, scales on rhizome peltate, thin, translucent, pale brown, (under a lens clear with dark cell walls) stipes 3-15 mm diameter in mature plants, spongy and air filled, sterile fronds pale green, thin, flaccid and spreading, 4-60 cm long, including a stip c. half this length, fertile fronds pale green, to brown when over mature, firm, held erect, 15-100 cm or more long, including stipe to 40 cm long, proliferous or dormant buds with overlapping dark scales sometimes present in the axils of fertile pinnae, sterile axes obviously winged, pinnae basically broad-ovate or delted with a few blunt lobes, sometimes more deeply incised, the segments 2-15 x 10.0 mm, fertile segments linear, 1-2 x 10-80 mm.	
Water Pepper	*Polygonnum hydropiper* L.	Annual asending herbs, 30-80 cm long; stems profusely branched, glabrous; red ring at base. Leaves lanceolate, 3.5-7.5 cm long, 0.8-2.0 cm wide, apex acuminate, narrowly cuneate, both surface glandular, veins strigose and glandular, upper margins cilitate. Inflorescence fascicles in spike like racemes, terminal or axillary, interrupted, to 17 cm long; 1.5-3.0 cm long; bracts funnel form, glandular, upper margin ciliate. Flowers green, pedicel exerted; perianth sefments 4 or 5, elliptic ca.2 mm long, glandular, apex styles 2-cleft, and included nectaries present. Achenes brown to dark brown, bicolour, shinning and ca.2 mm long.	
Water Fern	*Dentella repens* J.R. Forst. and G. Forst.	It is creeping, much branched herb, often forming dense mats or cushions up to 0.5 m in diameter. Stems obscurely 4-angled to terete and mostly glabrous. Leaves elliptic, oblong-lanceolate or oblong-obovate, 2-10 mm by 1-5 mm, acute at base and apex, glabrous or sparsely hairy on midrib below, lateral veins inconspicuous; petioles glabrous, 2-3 mm long; stipules truncate, and entire to c. 0.5 mm high. Flowers in leaf axils, 1 per node (flowering nodes alternating with non-flowering nodes), rarely terminal; pediocels 0.5-2.0 mm long. Calyx lobes erect, triangular-lanceolate, c. 0.5-1.5 mm long and acute. Corolla white, 5-15 mm long, glabrous on the outside; tube hairy around the throat; lobes much shorter than tube, and often with a purplish spot at the base. Stamens inserted near base of tube, and anthers subsessile, > 1 mm long. Ovary (densely) pilose, 2-3 mm high; style plus the 2 coherent stigmas c. 1.5-6.0 mm long. Fruit crowned by slightly enlarged, erect calyx lobes, globose, c. 3-4 mm in diameter, densely to sparsely cover by stiff and white hairs.	
Water Shield	*Brasenia schreberi* J.F. Gmel.	It is a slender, perennial aquatic herb, size 1-2 m tall. Rhizomes long, creeping and anchored in the muddy bottom. Stem reddish and covered in a gelatinous substance. Leaves alternate, floating, long-stalked (attached to center of leaf), green above, reddish beneath, 4-12 cm long, about half as wide, oval to elliptic. Leaf stalks and leaf undersides covered in a gelatinous substance. Flowers borne singly on a 5-20 cm long stalk, barely emergent, dull reddish purple and 2.5 cm wide, tepals six. Stamens twelve to eighteen and upright. Anthers deep purple. Fruit a capsule, 3-5 mm long, cylindrical and leathery. Seed pale brown, 2-3 mm long and egg shaped.	
Ponnanganni	*Alternanth sessilis*	It is also called alligator weed and grows commonly in submerged field of paddy, near ponds canals and other water bodies. It is well adapted to both xerophytic and hydrophytic conditions.	
Chinese Arrowhead	*Sagittaria trifolia* L.	It is aquatic, glabrous herb with thick stolons bearing tubers at tip. Leaves mostly emerged, upright, sagittate or somewhat hastate, blade ovate or linear-lanceolate (5-6.5) 8-17 cm long along midrib (1.5-2.5) 5-7 cm broad, acute, basal lobestriangular or linear-lanceolate, often longer than the blade, sharply acute; petiole 60-75 cm long and triangular. Inflorescence an unbranched scapes raceme and 30-50 cm long. Flowers in 2-6 whorls of (2-) 3 (-5) flowers each, unisexual, 1-2 cm across, white; pedicels 8-15 mm long, short in females; bracts triangular-ovate, 8-10 mm long, 3-4 mm broad and acute-acuminate. Sepals elliptic, 6-8 mm long, 5-6 mm broad, somewhat keeled margins broadly membranous and reflexed in fruit. Petals broadly elliptic-suborbicular, 12-15 mm long 12-15 mm broad, and white, without basal purple spot. Stamens c. 20, filaments lanceolate, 1.5-2.0 mm long; anthers yellow, oblong, as long as the filaments. Head of achenes subglobose, c. 1 cm across; achenes triangularly obovate, with laterally bent beak. C. 2-3 mm long and wing subcrenate to entire.	

Common Name	Name of Species	Botanical Description	Origin
		This species exhibits much variation in leaf shape and flower size *etc.* These variants do not deserve taxonomic status. It differs from the European and north Asian *Sagittaria sagittifolia* Linn. in having pure white petals with no basal purple spot, yellow rather than purple anthers, reflexed rather than spreading sepals and very acute rather than blunt tips to the basal lobes of the leaf which are often longer than the blade.	
Keysoor	*Scirpus grossus* L. f. Tikiu	A large perennial, glabrous herb, up to 3 m high, distributed through out India, up to an altitude of 700 m especially in swamps. Rootstocks stout, sometimes stoloniferous, producing in dry season, dark coloured, hard, globose tubers, clothed and matted with fibres; leaves radial, 60-90 cm x 13 mm; spikelets sub-lobosely ovoid 10 mm long in corymbiform decompound, terminal umbels, nuts obovoid, 2 mm long, and ash grey or black.	
Sonneratia	*Sonneratia caseolaris* (L.) Engler	Evergreen tree 5-15 (-20) m high without buttresses or stilt roots, with rather open spreading crown and glabrous throughout. Pneumatophores 50-90 cm high to 7 cm in diameter. Bark gray and coarsely flaky. Leaves opposite, without stipules, nearly sessile, elliptical, oblong or ovate, 5-13 cm long, 2-5 cm wide, with broad or tapering base and blunt or rounded tip, entire, with 8-12 widely spreading fine side veins on each side and leathery. Flowers 1-3 at end of drooping twigs malodorous and nocturnal. Hypanthium with 6-8 calyx lobes; petals 6-8, 2.0-3.5 cm long, 1.5-3.5 mm wide, dark or blood red, stamens numerous, with threadlike filaments 2.5-3.5 cm long, pistil with 16-21 celled ovary with many ovules; style long and stout. Fruit 3-4 x 5-7 cm, thick walled and green (Little, 1983).	
Paracress	*Spilanthes acmella* A.H. Moore	Herbs stems with adventitious roots and multicellular hairs, leaves opposite,ovate; apex acute, but rounded at tip; margins shallowly crenate and shallowly dentate with glands along margin; upper leaf surface very sparsely hairy; lower surface sparsely hairy on veins; upper leaves smaller, narrowly ovate, with margins shallowly crenate or dentate; inflorescence, heads solitary, terminal and in upper leaf axils, on erect peduncles; head with 5 involucral bracts; ray flowers 5 pistillate, each inside an involucral bracts; pappus lacking; corolla tubular basally, expanding into a yellow, 3-lobed ray; ovary inferior, flattened dorsoventrally, 3-angled, with upward-pointing bristles along ridges; style with 2 stigmas; disk flowers perfect, each inside a conduplicate, keeled receptacular bract (pale); pappus lacking; corolla tubular, expanded above base, with (3) 4-5 yellow lobes; stamens (3) 4-5, apex pointed; base of anthers not sagittate; ovary inferior, flattened with 2 ridges, with upward-pointed bristles along ridges; style with 2 yellow stigmas; receptacle cone-shaped, and achenes approximately 1.5 mm long.	
Giantchick-weed	*Stellaria aquatic* L. Scop.	Perennial herb; stems elongate, tufted, decumbent at base, 40-50 cm long, tinged with purple, usually glandular pubescent. Lower leaves long-petiolate, upper leaves sessile, ovate or broadly so, 2-5 cm long, 1.5-2.5 cm wide, acuminate or acute at the apex, cordate or nearly so at base, chartaceous margins entire and wavy, 1 nerved, glabrous on both surface or slightly hairy along veins beneath; petiole 1-2 cm long, pubescent. Flowers solitary, white axillary, 5-8 mm in diameter; pedicel 1.5-2.0 cm long, slender, glandular hairy; sepals lanceolate, 4-6 mm long; glandular hairy out side acute; petals deeply 2-lobed to base, lobes lanceolate, 5-7 mm long, obtuse; stamens 7-10, anther orange red; styles 5 and reflexed upward. Capsules broadly ovoid as long as sepals, 5-valve, bifid; seed ellipsoid, 0.7-0.8 mm long and rusty brown.	
Swamp Taro	*Crytosperma chamissonis* (Schott) Merr.	It reaches heights up to 4-6 m and produces huge green to yellowish-green stem. The swamp taro is a giant herbaceous perennial with typically 6-8 huge leaves arising from a short subterranean stem. The leaf blades are arrow shaped, 1-2 m in length, and are borne on stout petioles, 1-2 m long and tapering from about 10 cm in diameter; in some cultivars the lower parts are covered with spines. The inflorescence has a long, thick yellowish spathe and a purplish spadix, though the seeds are often not fertile. The stem thickness rapidly at the base becoming a large corm, varying in shape from cylindrical to conical or almost spherical. The size varies with cultivar and age; 15-25 kg is common, but it can weigh up to 90 kg or more in a 10 year old plant. Cormels which send up to leaves and develop into suckers are produced as side shoots on the parent corm after about three years.	South East Asia or Western Melanesia
Eel grass	*Vallisneria spiralis* L.	It is a stoloniferous herb mainly grows in ponds an ditches, leaves clustered at node, up to 35 cm long, family toothed, 5-9 nerved, sheathing at the base; male flowers many with short peduncled, lobed spathe near the base. However, at early detachment float on water surface; sepals-3, petals inconspicuous, stamens 1-3, pistil lobed absent, female flowers; staminate 3, bifid, ovary covered by spathe, style three, stigma with three lobe; fruit linear seeds fusiform enclosed with gelatinous mass (Mishra and Verma, 1989).	
Frog Bit	*Hydrocharis dubia* (Blume) Backer	Fresh water plants with 15-25 cm long stolons and fibrous roots. Leaves floating or emergent, lamina ovate-cordate to broad-ovate or suborbicular, 2-6 cm long, 2.0-7.5 cm broad, obtuse, entire, basally ± cordate or reniform, often truncate; petiole 3-15 cm long, thickened towards the base; stipule solitary, liguliform, 5-25 mm long and obtuse. Male spathe of 2 bracts, 10-30 mm long, on up to 25 mm long peduncle. Male flowers on 3-6 mm long pedicles; sepals oblong, 4-8 mm long, 2-5 mm wide, brownish mottled; petals 8-18 mm long, 6-9 mm wide, white with yellow claw; stamens often in a 6 pairs, the antipetalous 3 pairs staminodial or with minute orange-yellow anthers, filaments dilated, yellow and ribbed. Female spathe smaller than male one. Female flower: sepals and petals as in male; staminodes opposite the sepals, free or connate; ovary oblong-elliptic, 5-6 mm long; styles bibbed, and hairy. Fruit 5-10 mm long and 4-8 mm in diameter. Seed with mucilaginous hairs, broadly ellipsoidal, 1.0-1.5 mm long (Ghafoor, 2008).	

Common Name	Name of Species	Botanical Description	Origin
Papyrus	*Cyperus papyrus* L.	Plants are tall, robust, leaflets aquatic and up to 4 m in hight. Culms stout, smooth, trigonous, surrounded at base with coriaceous and large acuminate sheaths. Umbel-rays numerous, filiform, 10-45 cm long, each surrounded at base with a narrow, brown, cylindrical sheath which is up to 3 cm long. Secondary umbels 3-5 rayed, supported by narrow and elongated bracts. Spike 1-2 cm long and 6-10 mm broad. Spikelets 6-10 mm long, 1 mm broad and 6-16 flowred.	Africa, Madagascar and the Mediterranean regions
Common Water-Plantain	*Alisma plantago-aquatica* L.	It is aquatic herb. Petiole (10-) 15-30 (-40) cm long; leaf-blade ovate, (5-) 10-15 (-25) or (2.5-) 5-7 (-12) cm; base rounded to subcordate; nerves 5-9, not all radiating from the apex of the petiole. Inflorescence of about 6 whorls of branches and flowers; bracts membranous, lanceolate, those of the lowest whorl 1-2 or 0.5 cm, with several smaller bracteoles. Pedicels up to 2 cm long. Sepals being ovate and 2 mm long. Petals 4 mm long, white and delicate. Stamens 2 mm long; anthers 0.75 mm long, narrow; filaments attached above the base. Carpels ovate, compressed, less than 1 mm long; style about 1 mm long, erect or slightly curved. Achenes broadly ovate, with 2 or 3 ridges on the dorsal surface, pale brown to yellowish (Susan Carter, 1960). It is in flowerfrom June to August, and the seeds ripen from July to September. The floers are hermaphrodite and are pollinated by flies. Plant does not possess tolerance towards zinc pollution in wet land conditions (David *et al.,* 2004).	
Achyranthes	*Achyranthes aquatic* (Roxburgh) Walich	Aquatic or subaquatic pertennial 0.3-1.5 m, prostrate to straggling or erect, usually much branched, considerably rooting at the lower nodes with dense tufts of whitish rhizoids, stem near the base up to 2 cm thick, spongy and hollow. Upper stem and branches sulcate striate, glabrous for the most part but increasingly appressed pilose towards the inflorescence. Leaves lanceolate to lanceolate-oblong or oblong-ovate, cuneate or usuallty at tenuate to the base, acute to acuminate at the apex, moderately appressed-pilose on both surface (densely so when yound), lamina (2.5) 7-15 x (0.8)2-5 cm and petiole 0.4-4.0 cm long. Spike c. 4-12 cm long in flower, elongating to as much as 25 cm or occasionly even more in fruit; rhachis moderately to densely appressed pilose; peduncle short and mostly c. 1.5 cm long. Bracts deltoid lanceolate, 3-4 mm long, hyaline (drying pale brownish) with a single midrib, glabrous, finally deflexed below the hard callus left by the fallen perianth; bracteoles more or less circular hyaline, c. 1.5-2.0 mm long and glabrous. Perianth 6-8 mm long, the upper (outer) tepal, 1-3 nerved slightly longer than the remainder, with a sharper, often slightly recurved apex, and a somewhat narrower pale border; remaining tepals blunter, with up to 7 nerves. Filaments stout and 2-3 mm long. Style 1.75-2.5 mm long. Capsule c. 4 mm long and slightly broader at the base but rounded above. Seed smooth and chestnut-brown.	Africa and tropical Asia
Apon-ogeton	*Apon-ogeton crispum* Thunb.	It is perennial aquatic plants with a tuberous, elongate or cylindrical, often branched, rhizome. Leaves submerged or floating, usually with conspicuous main and cross veins. Inflorescence a long peduncled spadix, emerging above the water surface, composed of 1-11 spikes; flowers often bisexual, with 2 tepals, 6 stamens and 3 (5) ovaries.	
	Apon-ogeton lakhonensis	Rhizome elongated or obovoid up to 2 cm in diameter; leaves very variable in shape and size, submerged ones up to 25 cm x 6 cm and floating ones are smaller.	
	Apon-ogeton undulatus	Rhizome elongated, obovoid or globose, up to 2.5 cm in diameter; leaves with long petiole, up to 10-25 cm x 1-4 cm, alternately transparent or opaque in an irregular pattern, with undulating margins; the plant produces runners (resembling peduncles forming young plantlets at the top).	
Velvetleaf	*Limnocharis flava* (L.) Buchenau	It is glabrous, scapigerous and lactiferous plant. It has a short stout rhizome about 3 cm long and 3 cm in diameter and numerous fibrous roots. The leaves are erect, sheathing and curvinerved, and rise in clusters above the water. The petioles are thick, triangular, 90 cm long and contain numerous lacunae (air spaces). The leaf sheaths narrow towards the top and the leaf blade is papery, light green, variable in shape rounded, ovate or broad elliptic and 6-20 cm long and almost as equally wide. The leaf apex is generally apiculate with a small hydathode at the tip; the leaf base is cuneate to repandcordate and the leaf margin is undulated. Four to six pairs of main nerves and a marginal one sub parallel and converging towards apices, numerous secondary nerves parallel to each other and nearly perpendicular to the midrib and dense fine reticulations. One to four peduncles which are axillary, erect, triangular, flattened at the base and 120 cm long. Peduncle bears, at the apex, a cluster of flowers, a ramet or both. Inflorescences are umbellate (consisting of flowers springing from a common center and arranged in a cluster), with between two to 15 flowers supported by an involucre of bracts; sometimes one or two leaves are present between the flowers. The bracts are roundish to broad-elliptic with (usually) emarginated tops. The flowers are borne in an indeterminate umbel and each consists of three sepals, three yellow petals, and about 18 carpels surrounded by numerous stamens and staminodia. The androecium is centrifugally developed, and the last-formed members are staminodial; it is supplied by branching vascular systems. Carpels arise almost simultaneously, and prominent residual floral apex remains. The carpels are partially conduplicately closed andare also primitive in possessing laminar placentation and in lacking differentiation of a style. The gynoecious is essentially apocarpous, but there are slight fusions of adjacent carpels near their ventral margins where they are attached to the receptacle.	Tropical America

Common Name	Name of Species	Botanical Description	Origin
Goldenclub	*Orontium aquaticum* L.	Roots concentrile and rhizomes 1.5-3.0 cm in diameter. Leaves: petiole dark green to red green, 10-40 (-60) cm; blade (6-)10-30 (-45) cm, about ± 1/3 as wide (greater than 3 cm wide), mainveins from base, curving apically paralled to midvein. Inflorescence: spadix bright yellow, 2-10 cm' 5-10 mm and apex rounded. Flowers covering spadix; tepals (2-) 6, arched over ovaries, yellow; stamens (2-) 6, sometimes with 1 or 2 staminodes; ovaries 1-locular, ovules 1. Fruits 1-2 cm in diameter and pericarp thin. Seeds 1.0-1.5 cm in diameter; endosperm absent; germinating soon after falling from parent plant.	Easter North America
Duck lettuce	*Ottelia alismoides* (L.) Pers.	It is aquatic herb, rooted and completely submerged. Juvenile leaves somewhat linear, becoming oblanceolate, ovate or broadly cordate with maturity. Largest leaves from 11 to 16 cm long, their bases tapering to meet the petioles. Petioles of varying lengths (potentially up to 50 cm). Conspicuous longitudinal ribbing and cross-ribbing on the upper surface of the leaves gives a quilted effect. Lower margins of leaves and edges of petioles often shallowly serrated to sharply toothed. Flowers wrapped within spathes, cylindrical structures 2-4 cm long, composed of green bracts that are ornamented with 3 or more ryffled wings. Spathes born on long, angled stalks that become spiraled after flowering. Sepals and short lived petals of male flowers exert from the tip of the spathe just above the water surface. Spathes containing female and/or bisexual flowers are self fertile and remain submerged. Petals white, pink, blue or purple, often tinged with yellow at the base. The verrucate protrusions develop into spines during free microspore stage. A foot layer is formed by accumulation of lamellated structure. The pollen grains of *Ottelia alismoides* are inaperturate, not omniaaperturate, becvause of the well developed foot layer. The inaperturat pollen grains of this genera are characterized by the spinous protrusions and the granular foot layer (Takahashi, 1994). The resulting fleshy, encapsulated fruits contain as many as 2000 seeds (Cook *et al.,* 1984).	Tropical areas of Asia and Australia
Parrot's Feather	*Myriophyllum aquaticum* (J.M. da Conceicao Vellozo) Verdc.	Aquatic herb, either growing in water, with most of the upper leafy parts above the water level, or sometimes on mud; stems 0.45-1.2 m long, simple or sparsely branched, puberulous or glabrous, with up to ¼ emerging, mostly several from a creeping rhizome. Leaves all pinnatipartite, in whorls of 4-6, 0.6-2.5 (-3) cm long, often puberulous; lobes alternate or sub-opposite, 15-17 (-17), linear-subulate, 1.8-5.0 (-7) mm long, 0.13-0.3 mm wide, acute and entire or denticulate. Flowers unisexual, monoecious or mostly dioecious, solitary in the leaf-axils, the lower flowers and the upper. Male flowers: bracteoles 2, linear, 1.5-3.0 mm long, 0.1 mm wide, white, very finely bifid or trifid; pedicel and calyx-tube together at first 1.3-2.5 mm long, tenuous, later 3.5-5.0 mm long, 2 mm thick; lobes 4, oblong-lanceol;ate or narrowly triangular, 0.7-1.8 mm long, 0.4 mm wide, at first erect, later spreading, minutely serrate; petals 4, boat shaped, 2.5-5.0 mm long, 0.8-1.3 mm wide, clawed; stamens 4-8, filaments 0.15-5.0 mm long; anthers 1.9-3.5 mm long, 0.4 mm wide and sometimes minute rudiments of styles present. female flowers: bracteoles 2, linear, 07-1.2 mm long, 0.15 mm wide, acute, sub-membranous, toothed at the middle; pedicels 0.3-0.7 mm long, thick; calyx-tube 0.5-1.0 mm long, 0.5-0.6 mm wide, 4- angled, 4-grooved, the 4 lobes linear, (0.3-) 0.5-1.0 mm long, at first erect, later reflexed, acute, entire, minutely serrulate or pinnately 5-lobed; petals and stamens lacking; stigmas 4, short, 0.45-0.7 mm long, at first erect, later decurved, capitate or claviform, shortly papillose. Fruit ovoid,1.8 mm long, 1.2 mm in dimeter, papilose-puncatate, eventually dividing into 4 parts. (Anonymous, 1973).	Amazon River in South America
Watercress	*Rorippa nasturtium aquaticum* L. Hayek)	Watercress is an aquatic perennial herb, much branched, glabrous and tall up to 1.0 m. Stem prostrate to ascending; juicy, hollow and usually roots at the lower nodes. Leaves petiolate, autriculate, pinnate, oblong in outline, up to 10 cm long; lateral leaflets in 2-9 pairs, se-ssile, narrowly ovate, elliptical or nearly orbicular, entire to fairly dentate; terminal leaflet similar but larger. Racemes ca 10 cm long; flowers with white petals; pedicels curved and ca 1 cm long. Fruit broadly linear, 10-18 mm x 1.5-2.5 mm, more than 1:5 mm thick, often curved and torulose, 7-12 times as long as broad and seeds arranged in two distinct raws. Seed shiny, red brown and covered by a distinct reticulum.	Eastern Mediterra-nean and adjoining areas of Asia
	Rorippa heterophylla	It is an annual herb and 5-50 cm tall. Stem erect or ascending, one or more from the base, along the stem; lower ones narrowly obovate in outline, 1.5-8.0 cm long with 1-6 cm long petioles, undivided or lyrate-pinnati partite with 1-3 pairs of small lateral lobes and a much larger terminal lobe; upper leaves rather few undivided, narrowly ovate. Racemes 2-10 cm long; flowers with reduced or lacking petals, sepals green; pedicels straight and 2-8 mm long. Fruit linear, 14-25 x 0.7-1.3 mm, straight, less than 1.5 mm thick, with seed in one row only. Seed red brown and minutely foreolate.	
Chinese Water Chestnut	*Eleocharis dulcis* (Burm. F.) Trin.	Tufted perennial with slender rhizomes, sometimes bearing tubers to 10 mm diameter. Culms terete (flatten easily when dried), transversely septate, to c. 1.5 m high and 3-10 mm in diameter. Spikelet cylindrical, 25-50 mm long. Glumes very obtuse, dtriate, with a distinct midrib and hyaline margins, 6.0-6.5 mm long, straw-coloured. Bristles 6-8 rather stout, flattened and united near base, irregularly retrorsely toothed, about twice as long as nut. Stamens3 and anthers 2.5-3.0 mm long. Style trifid. Nut turgidly biconvex with angles not ribbed, obovoid, 1.5-2.0 mm long, 1.0-1.5 mm in diameter, golden brown; external cells minute, in no regular series; style base about half as long as and about two thirds as broad as nut.	Old World Tropics (Yamaguchi, 1990)

REFERENCES

1. Akah, P.A. and Ekekwe, R.K. 1995. Ethnopharmacology of some Asteraceae family used in Nigerian traditional medicine. *Fitoterapia*, **66:** 351-355.
2. Anonymous. 1973. Flora of Tropical East Africa, Vol., 1, (1973). Author: R. Boutique (Jardin botanique national de Belgique) and B. Verdcourt, Royal Botanic gardens, Kew.
3. Anonymous. 1986. Great Plains Flora Association. 1986. Flora of the Great Plains. Lawrence, K.S.: University Press of Kansas, p. 1392.
4. Ban, N.T. 1993. *Endrya fluctuans* Loureiro. In: PROSEA. Genetic Resopurces of South East Asia 8. Vegetables (Eds. J.S. Siemonsma and Kasem Piluek). Pudoc Scientific Publishers, Wageningen. p 193-194.
5. Burkil, I.H. 1966. A dictionary of the economic products of the Malay Peninsula. Vol.II. Governments of Malaysia and Singapore by the Ministry of Agriculture and Co-operatives, Kuala Lumpur.
6. Chopra, R.N., Nayar, S.L. and Chopra, I.C. 1986. Glossary of Indian Medicinal Plants, Publication and Information Directorate, CSIR, New Delhi-12.
7. Cook, C.D.K., Symoens, J. and Urmi Kong, K. 1984. A revision of the genus *Ottelia* (Hydrocharitaceae). *Aquatic Botany*, **18:** 268-274.
8. Coon, N. 1975. The Dictionary of Useful Plants. Rodale Press.
9. David, J., Mathew, B., Moran, M. and Marinus, L. O. 2004. Screening the wetland plant species *Alisma plantago-aquatica* L., *Carex rostrata* and *Phalaris arundinaceae* for innate tolerance to zinc and comparision with *Eriophorum angustifolium* and *Festuca rubra* Merlin. *Environment Pollution*, **134(2):** 343-351.
10. Di Stasi, L.C., Hiruma, C.A., Guimaraes, E.M. and Santos, C.M. 1994. Medicinal plants popularly used in Brazilian Amazon. *Fitoterapia*, **65:** 529-540.
11. Duke, J.A. and Ayensu, E.S. 1985. Medicinal plants of China: Reference Publications, Inc.
12. FAO. 2014. State of World Fisheries and Aquaculture, 2014. FAO, Rome.
13. Foster, S. and Duke, J.A. 1990. A field guide to medicinal plants. Eastern and Central N. America. Houghton Mifflin Co.
14. Gao, G. and Liao, M. 1998. Determination of flavonoids and quality evaluation of Chinese-traditional drug 'Puhuang'. *Yaoxue Xuebaio*, **33(4):** 300-303. Inst. Med. Plant, Chinese Acad. Med. Sci. and Peking Union Med. Coll., Beijing 100094, China.
15. Ghaffor, A. 2008. Flora of Pakistan, www.efloras.org.
16. Grieve. 1984. A modern Herbal. Penguin 1984 ISBN0-14-046-440-9.
17. Launert, E.1981. Edible and medicinal plants. Hamlyn.
18. Little, E.L. Jr. 1983. Common fuelwood crops: a handbook for their identification. McClain Printing Co., Parsons, WV.
19. Mishra, B.K. and Verma, B.K. 1989. Flora of Allahabad. Bishan Singh Mahendra Pal Singh, 23-A, Cannaught Place, Dehradun, India.
20. Nakayama, H. Suzuki, T. and Suzuki, Y. 1998. Effect of wasabia on ion secretion in guinea pig colon. *Nippon Nogeikagaku Kaishi*, **72(4):** 499-507.
21. Ozawa, T. and Imagawa, H. 1988. Polyphenolic compounds from female flowers of *Typha latifolia* L. *Agriculture and Biological Chemistry*, **152(2):** 595-597.
22. Pandey, A.K. 2011. Aquatic Vegetables: Introductory Consideration, Agri-Tech Publishing Academy, Udaipur, p. 3.
23. Pandey, A.K., Singh, P.M., Singh, B., Chaturvedi, A.K. and Rai, Mathura. 2014. Aquatic vegetables. In: Future Crops (ed. Peter, K.V.); Vol. 2, p. 9.
24. Ramsewak, R.S., Erickson, A.J. and Nair, M.G. 1999. Bioactive Nisobutylamides from the flower buds of *Spilanthes Acmella*. *Phytochemistry*, **51:** 729-732.
25. Sasmitamihardja, D. 1993. *Oenanthe javanica* (Blume) DC. In: PROSEA. Plant Resources of South-East Asia 8. Vegetables. Siemonsma, J.S. and Piluek. K. (Eds.) Pudoc Scientific Publishers, Wageningen.
26. Singh, V. 1995.Herbal folk remedies of morni hills (Haryana), India. *Fitoterapia*, **66:** 425-430.
27. Shin, I.S., Masuda, H. and Naohide, K. 2004. Bacterial activity of Wasabi (*Wasabi japonica*) against helicobacter pylori. *International Journal of Food Microbiology*, **94(3):** 255-261.
28. Storey, C. and Salem, J.I. 1997. Lay use of Amazonian plants for the treatment of tuberculosis. *Acta Amazonica*, **27:** 175-182.
29. Takahashi, M. 1994. Pollen development in a submerged plant, *Ottelia alismoides* (L.) Pers. (Hydrocharitaceae). *Journal of Plant Research*, **107(2):** 161-164.
30. Yamaguchi, M. 1990. World Vegetables: Principles, Production and Nutritive Values. AVI Publication. Van Nostrand Reinhold 115, fifth Avenue, New York.
31. Yeung, Him-che. 1985. Handbook of Chinese Herbs and Formulas. Institute of Chinese Medicine, Los Angeles.
32. Yoshikawa, M., Yamaguchi, S., Murakami, T., Matsuda, H., Yamahara, J. and Murakami, N. 1993. Absolute stereo structures of trifoliones A, B, C, and D new biologically active diterpenes from the tuber of *Sagitarria trifolia* L. *Chem Pharm. Bull.*, **41(9):** 1977-1679.

Genus *Ipomoea*

Water Spinach or Swamp Morning Glory (*Ipomoea aquatica* Forsk.)

Introduction

Water spinach (*Ipomoea aquatica* Forsk., 2n=30) is also known as water spinach (English); batata acuática; batata aquatic (Spanish); patate aquatique (French) and batata acuática; batata aquática (Portuguese); Kalami sag, Karmi, Patua sag (Hindi); Kalmi Shak (Bengali); Nadi Shaka, Nalini Bhanji (Marathi); Nalini Bhanji (Gujarati); Tutikura (Telgu) and VeleiKirei (Tamil). Morning glory- a name derived because of its beautiful flowers. Worldwide in 60 countries and among 20 major crops, it is known as a major weed which covers over the surface of water bodies like lakes, ponds, marshes, canal and ditches (Carlquist, 1967). It is aquatic plant and its leaves are used as vegetables in several parts of Asia (sold in tightly packed bunches). It is considered as one of the cheapest nutrient sources preferred by all groups of consumers-rich and poor. Therefore, it is cultivated as an important crop of high nutritive value essential from human health (Mandal *et al.*, 2008). In India, it can be spotted in Uttar Pradesh, Bihar, Madhya Pradesh, Jharkhand, West Bengal, Odisha, Karnataka, Assam *etc.* growing naturally. It is also a very important fodder crop to cattle and pigs and produces no ill effects. Mainly in north India during rainy season most of the leafy vegetables are not available in good condition due to heavy rain, then water spinach is only vegetable which serve the purpose. In rural area people said it as poor man's greens. Leaves are very nutritious and full of many minerals and vitamins. Carotenes found in leaves are in form of β-carotene, xanthophyll and less quantity teraxanthin. Per 100 g of fresh leaves contains 90.3 g moisture, 2.9 g protein, 0.4 g fat, 2.1 g minerals, 1.2 g fiber, 3.1 g carbohydrate, 28 K calorie, 110.0 mg calcium, 46.0 mg phosphorus, 3.9 mg iron, 1980 micro gram carotene, 0.05 mg thiamine, 0.13 mg riboflavin, 0.6 mg niacin and 37.0 mg vit. C. The nutrients availability per 100 g edible stems are 90.7 g moisture, 0.9 g protein, 0.2 g fat, 1.8 g minerals, 3.4 g fiber, 19.0 g carbohydrate, 80 K calorie, 30.0 mg calcium, and 0.8 mg phosphorus. Apart from the above nutrients, it contains about 11.8 mg per 100 g Vit.E (Wealth of India, 1972). Available lipoids contain hentriacontane, sistosterol, sistesterolglucoside and alcohol. Roots are sweet and edible in nature which generally utilized during famine. As a medicinal properties, extracts of plants lower down the toxicity appeared due to arsenic and cocaine. Leaves and stems are cold effective in nature. In Assam, peoples considered it as neuro-tonic. Water spinach is relatively rich in 5-methyl methionine (vitamin U) and is traditionally used to treat gastric and intestinal disorders. Further it has been found to have insulin like properties acting as an antihyperglycemic (Mukhopadhyay *et al.*, 1995; Villasenor *et al.*, 1998). Recently 1-hexadecanoylpyrrolidine and 1-octadeconylpyrrolidine amides have been isolated from water spinach (Tofern *et al.*, 1999).

Origin, History and Distribution

Water spinach is mainly a plant of tropical Asia and may possibly be originated from India (Westphal, 1993). Though, its plants can be spotted in south and west Asians countries, tropical Africa, south and Central America and near Oceania. Carlquist (1967) reported that water spinach spread worldwide through seeds and vegetative parts as propagation means. Cultivated and wild both strains are available in nature. In Africa, it is distributed on grass lands, lake and in river sides. In Thailand and Malaysia water spinach (white flowered cultivar 'Kong-Kong') is commercially grown crop and ranked second after 'Pakchoi'. In an estimate, in Indonesia it is cultivated in about 10, 000 hectare (Westphal, 1993). Water spinach is well adapted to growing conditions of Cuba (Pino *et al.*, 1996). Commercial seed production is being done in Hong Kong, China, Taiwan, Thailand, Philippines and Japan. Thailand alone produces nearly 700-900 tonnes seed of water spinach and sail in the world market. In India water spinach is commonly found in Uttar Pradesh, Bihar, Madhya Pradesh, Jharkhand, West Bengal, Odisha, Karnataka, and Assam.

Species Availability, Evolution and Diversity

Under genus *Ipomoea* 50 species are reported in India which is available in salty lands, water stagnated area and even hilly areas. Among them *I. aquatica* is very prominent. Poiret (1814) put forth the correct synonym as *I. reptans* which is often used.

Botanical Description

Water spinach belongs to the family Convolvulaceae (Eurypalynous family) and fall under genus *Ipomoea* and species *aquatica* Forsk.

Domain	:	Eukaryota
Kingdom	:	Plantae
Phylum	:	Spermatophyta
Subphylum	:	Angiospermae
Class	:	Dicotyledonae
Order	:	Solanales
Family	:	Convolvulaceae
Genus	:	*Ipomoea*
Species		*Ipomoea aquatica*

Leaves of water spinach accumulate more proline than *Trapa natans* and *Vallisneria spiralis*, resulting

Table 11.8: Distribution of Water Spinach in different Continent

Country	*Distribution*	*Origin*	*Invasive*	*References*
Asia				
Bangladesh	Present			Gunn and Ritchie, 1982
Brunei Darussalam	Present			Peregrine, 1974
Cambodia	Present			Holm *et al.*, 1979
China	Present			Subramanyam, 1962
-Hong Kong	Present			Subramanyam, 1962; Hodgkiss, 1978
Christmas Island (Indian Ocean)	Present			PIER, 2003
Cocos Islands	Present			Gunn and Ritchie, 1982
India	Widespread	Native		Raju and Reddy, 1993; Subramanyam, 1962; PIER, 2003
-Andaman and Nicobar Islands	Present			Cook, 1996
-Assam	Present			Cook, 1996
-Bihar	Present			Cook, 1996
-Daman	Present			Cook, 1996
-Diu	Present			Cook, 1996
-Goa	Present			Cook, 1996
-Gujarat	Present			Cook, 1996; Rajni Kant *et al.*, 1996
-Indian Punjab	Present			Cook, 1996
-Karnataka	Present			Cook, 1996
-Madhya Pradesh	Present			Cook, 1996
-Maharashtra	Present			Shingade and Chavan, 1996; Cook, 1996
-Rajasthan	Present			Usha and Pandey Alka Jain, 2002; Cook, 1996
-Tamil Nadu	Present			Cook, 1996
-Uttar Pradesh	Present			Cook, 1996
-West Bengal	Present			Cook, 1996
Indonesia	Present			Susiawaningrini and Thohari, 1981
Japan	Present			Kameoka *et al.*, 1992; Miyazaki *et al.*, 2001
-Ryukyu Archipelago	Present			Miyazaki *et al.*, 2001
Korea, DPR	Present			Gunn and Ritchie, 1982
Laos	Present			Gunn and Ritchie, 1982
Malaysia	Present			Chin and Fong, 1978; Mackeen *et al.*, 1997; Subramanyam, 1962
Pakistan	Present			Shad, 1986
Philippines	Widespread			Holm *et al.*, 1977; Tejada *et al.*, 1990; Ronato and Esquerra, 1990
Singapore	Present			Chuo *et al.*, 1981
Sri Lanka	Present			Perera *et al.*, 1999; Gunn and Ritchie, 1982; Cook, 1996
Taiwan	Present			Yen, 1982
Thailand	Present			Holm *et al.*, 1977; Stripen *et al.*, 1991
Turkey	Present			Gunn and Ritchie, 1982
Vietnam	Present			Gunn and Ritchie, 1982
Africa				
Benin	Present	Introduced	Invasive	Holm *et al.*, 1977; Holm *et al.*, 1979; Kossou *et al.*, 2001
Burkina Faso	Present	Introduced		Gunn and Ritchie, 1982
Congo Democratic Republic	Present	Introduced		Gunn and Ritchie, 1982
Côte d'Ivoire	Present	Introduced		Gunn and Ritchie, 1982
Egypt	Present	Introduced		Gunn and Ritchie, 1982
Ghana	Present	Introduced		Gunn and Ritchie, 1982

Country	Distribution	Origin	Invasive	References
Kenya	Present	Introduced		Gunn and Ritchie, 1982
Mozambique	Present	Introduced		Holm *et al.*, 1977; Holm *et al.*, 1979
Namibia	Present			USDA-ARS, 2005
Nigeria	Widespread	Introduced		Holm *et al.*, 1977; Bako *et al.*, 2002; Salawu and Afolabi, 1994
Senegal	Present	Introduced		Gunn and Ritchie, 1982
Sudan	Present	Introduced		Beshir, 1978
Tanzania	Present	Introduced		Ritoine *et al.*, 1981
Tunisia	Present	Introduced		Gunn and Ritchie, 1982
Zambia	Present	Introduced		Gunn and Ritchie, 1982
Zimbabwe	Present	Introduced		Gunn and Ritchie, 1982
NORTH AMERICA				
USA	Present	Introduced	Invasive	IFAS, 2003
Florida	Present, few occurrences	Introduced	Invasive	Westbrooks and Eplee, 1989; IFAS, 2003
Hawaii	Present	Introduced	Invasive	Haselwood and Motter, 1966; Wagner *et al.*, 1999
CENTRAL AMERICA AND CARIBBEAN				
Costa Rica	Present	Introduced		Snyder *et al.*, 1981
Cuba	Present	Introduced	Invasive	Pino *et al.*, 1996; Snyder *et al.*, 1981; Oviedo Prieto *et al.*, 2012
Jamaica	Present	Introduced		Adams, 1972
Puerto Rico	Present	Introduced		Snyder *et al.*, 1981
Trinidad and Tobago	Present	Introduced		Gunn and Ritchie, 1982
South America				
Brazil	Present	Introduced		Snyder *et al.*, 1981
Colombia	Present	Introduced		Gunn and Ritchie, 1982
Guyana	Present	Introduced		Gunn and Ritchie, 1982
Suriname	Present	Introduced		Gunn and Ritchie, 1982
Oceania				
Australia	Present	Native		Stephens and Dowling, 2002; Aston, 1973
-Australian Northern Territory	Restricted distribution	Native		Stephens and Dowling, 2002; Aston, 1973
-Queensland	Restricted distribution	Native		Stephens and Dowling, 2002; Aston, 1973
-Western Australia	Restricted distribution	Native		Stephens and Dowling, 2002; Aston, 1973
Cook Islands	Present			McCormack, 2002
Fiji	Present			Parham, 1958; Smith, 1991
French Polynesia	Present			Welsh, 1998
Guam	Present			Stone, 1970; Fosberg *et al.*, 1979; Gunn and Ritchie, 1982
Kiribati	Present			Fosberg *et al.*, 1979
Marshall Islands	Present			Gunn and Ritchie, 1982; PIER, 2003
Micronesia, Federated states of	Widespread	Introduced		Fosberg *et al.*, 1979; Gunn and Ritchie, 1982; Lorence and Flynn, 1997
Nauru	Present			Gunn and Ritchie, 1982
New Caledonia	Present			Gunn and Ritchie, 1982
Norfolk Island	Present			Gunn and Ritchie, 1982
Northern Mariana Islands	Present			Gunn and Ritchie, 1982
Palau	Present			Gunn and Ritchie, 1982; PIER, 2003
Papua New Guinea	Present			Gunn and Ritchie, 1982
Samoa	Present			Parham, 1972
Solomon Islands	Present			Gunn and Ritchie, 1982
Vanuatu	Present			Gunn and Ritchie, 1982

Source: Invasive Species Compendium; www.cabi.org retrieve on 01-07-2016.

in better survival in drought as well as in moisture. Mukherjee *et al.* (1982) concluded that plants possess an inherent capacity to accumulate proline during moisture stress and also acquire the property of drought resistance under such conditions.

Table 11.9: Botanical Description

Growth Habit	Aquatic viny herb which float or spread on muddy lands, water and grasses. Plants are annual, biennial and perennial in nature (exhibits variable habits in accordance with nature of habitats: it floats over the water surface in case of full water course area, and may creep on the ground in marshy wetland). Plants attain the height of 2.0-3.0 m. Normally its main shoot trails horizontally over the water surface or muddy soils. Often they grow vertically and profusely, covering to water surface, depending on the depth of water. This flora is basically a vine, which may form dense masses of tangled vegetation, thus developing impenetrable canopies over the water surface. Tropical and subtropical climate are ideal for its proliferation and organically enriched wetland acts as facilitator for its high proliferation (Mandal *et al.*, 2008).
Root	Rooting starts at nodes which are hollow and can float.
Stem	Shoot trails horizontally over the water surface or muddy soils. Stems are 2-3 m but sometimes a single stem can grow and extend its apical shoot up to 20 m long along with profuse branching as its apical shoot may advance 12 cm per day under suitable environment.
Leaf	Lateral shoots bears leaves. Leaves are alternate, long, elliptic or ovate oblong, cordate or hastate at the base and pointed towards the end. In general, size observed 2.5 x 8.75 cm. leaf petiole is quite long.
Floral biology	Lateral shoots bears flowers. Flowers are fundibuliform, axillary, 2.5-5.0 cm long, conical and white and pale purple with dark purple eye in colour, mostly solitary or in fine flowered cymes. Calyxes 5, joint at the base, smooth and long corolla, crolla are alternate and sepals are small, and stamens 5.
Anthesis and pollination	Anther dehiscence longitudinally. Pollens are round, prolate-spheroidal to oblate- spheroidal or spheroidal, rarely sub-prolate or prolate; tectum reticulate to punctate-scabrate or echinate. On the basis of tectal surface and opertual types, 3 distinct pollen types are recognized (Praveen *et al.*, 1998): (i) *Convolvulus arvensis* type (ii) *Cresscretica* type (iii) *Ipomoea aquatica* type Sanghi *et al.* (2001) reported size of pollen grain of *Ipomoea aquatica* ranges between 176 x 176µ, mostly prolate-spheroidal to oblate-spheroidal or spheroidal, rarely sub-prolate or prolate; tectum reticulate to punctate-scabrate or echinate and ovary bicelled.
Fruit and fruit set	Fruits are capsule and 6.0-10.0 cm long and ovoid smooth.
Seeds	Generally, 2-4 seeds per capsule densely pubescent and black in colour. Seeds may be flatty.

Botanical Classification

Botanically water spinach can be grouped as:

(i) **Light green genotypes-** Plants of this group are light green in colour. Stems are tender, smooth, round and develop leaves on long petiole. It is found near shallow water logged area. Example: Hong Kong and Chingquet-1. Both the genotypes are less affected with low temperature. In India both the species has given poor yield. 'Pak Quet' is a white flowered cultivar is very much suitable for summer cultivation because its flourish well under dry conditions. It is very good yielder and produce leaves and stems of good quality. It is a multicut cultivar and develops several side branches.

(ii) **Reddish green genotypes-** Plants of this group are red in colour. Stems are tender and smooth. Plants develop very long viny structure. Petiole and leaf length is more than other genotypes available in nature.

(iii) **Reddish genotypes-** Plants are dark red in colour. Stems are tender and smooth but quite thin in comparison to other genotypes. Leaf petiole and width of leaf is more than greenish genotypes. It is not much popular but attained limited growth.

Augmentation of Genetic Resources and Crop Improvement

Despite of rich genetic diversity available in water spinach in South East Asia, not much emphasis has been given for its improvement. Germplasm of water spinach is available at the World Vegetable Centre (formerly, Asian Vegetable Research and Development Centre -AVRDC) in Taiwan and at National Research Institute in South East Asia. A collection of at least 50 land races is available at the Kasetsar University in Bangkok (Westphal, 1993). Genetic analysis on 6 quantitative characters was carried out in 36 cultivars by Liu *et al.* (1998). They observed broad sense heritability values in order of leaf width > main stem diameter > petiole diameter > leaf length > more than 1 fruit number per inflorescence > petiole length. There are few cultivars like Hong Kong, Chingquet-1 and Pak Quet are quite popular. Efforts should be made for developing improved cultivars with acceptable quality and possessing resistance against abiotic and biotic stresses.

Horticulture Research Centre, Bangladesh Agricultural Research Institute (BARI) has identified a potential cultivar 'Gimakalmi' (*Ipomoea reptants*) in the year 1993.

Genetic Resource Conservation

Gerplasm of water spinach are being conserved at World Vegetable Centre (old name Asian Vegetable Research and Development Centre), Taiwan. Kesetsert University of Bangkok has conserved 50 strains of white flowered Kong-Kong. Under collaborative exploration and collection programme, few land races of water spinach were collected from Odisha.

Table 11.10: Characteristics of different Group of Cultivars

Cultivar Group	Habit	Leaf Shape	Lamina			Stem			Weight of Fresh 1 m Long Twig	
			Petiole Length (cm)	Length (cm)	Width (cm)	Internode Distance	Girth	Node No.	Shoot with Leaf	Leaves and Root
Light green	Spread densely in shallow water	Mostly ovate, oblong and lanceolate	3.8-9.1	5.5-10.0	2.9-3.0	4.0-7.5	0.4-0.9	18-19	13.933	3.783
Red green	Floats over water	Mostly hastate	7.0-11.2	8.5-12.5	3.8-6.5	2.5-11.5	0.4-0.6	13-24	18.267	5.797
Red stem	Creeps on moist soil	Mostly sagitate	8.3-19.5	3.5-14.0	18-6.8	5.5-12.5	0.4-0.5	12-18	8.423	3.37

Source: Tiwari and Chandra, 1985.

REFERENCES

1. Carlquist, S. 1967. The biota of long distance dispersal: V. *Dispersal to Pacific Islands Bull Torrey Bot. Club.*, **94:** 129-162.
2. Liu, Y.M., Zou, G.L. and Ke, W.D. 1998. Genetic analysis on several quantitative characters of Water spinach (*Ipomoea aquatic* Forsk.). *Adv. Horti. J.*, **2:** 622-626.
3. Mandal, R.N., Saha, G.S., Kalita, P. and Mukhopadhaya. 2008. *Ipomoea aquatica*- an aquaculture friendly macrophyte. *Aquaculture Asia*, **13(2):** 12-13.
4. Mukherjee, S.P., Kar, R.K. and Choudhuri, M.A. 1982. Proline accumulation in relation to degrees of drought resistance in plants of different ecological habitats. *Sci. and Cult.*, **48(11):** 402-404.
5. Mukhopadhyay, S.K., Buddhadeb, D., Duary, B., Dasgupta, M.K. (ed.): Ghosh, D.C. (ed.) Gupta, D.P., Majumdar, D.K., Chattopadhyay, G.n., Ganguli, P.K., Munshi, P.S. and Bhattacharya, D., 1995. Ethnobotany of some crop field weed in sub-humid agricultural tract of West Bengal, Proc. National Symposium on Sustainable Agriculture in Sub-humid zone, Sriniketan, West Bengal, March 3-5, 1995.
6. Pino, M., deLos, A., Cardo, F.E., Teesry, E. and Teram, Z. 1996. A study of water spinach (*Ipomoea aquatica* Forsk.) under different cultivation conditions. *Cultivos Tropicales*, **17(2):** 39-43.
7. Praveen, A., Quiser, M. and Praveen, A. 1998. Pollen flora of Pakistan. XII. Convolvulaceae. *Bangladesh J. Bot.*, **27(2):** 109-118.
8. Sanghi, D., Sarna, N.J. and Sanghi, D. 2001. Palynological studies of some medicinal plants. *J. Phytological Res.*, **14(1):** 83-90.
9. Tiwari, N.C. and Chandra, V. 1985. Water spinach-its varieties and cultivation. *Indian Hort.*, **30(2):** 23-24.
10. Tofern, B., Mann, P., Kalonga, M., Jenett, S.K., Witte, L. and Eiche, E. 1999. Aliphatic pyrrolidine amides from two tropical *Convolvulus* species. *Phytochemistry*, **52(8):** 1437-1441.
11. Villasenor, I.M., Cobrea, M.A., Meneses, K.B., Rivera, V.R.R. and Villasenor, R.M. 1998. Comparative antidiabetic activities of some medicinal plants. *Philippine J. Sci.*, **127(4):** 261-266.
12. Wealth of India. 1972. Raw materials. Vol. IX. Publication and Information Directorate, CSIR, New Delhi.
13. Westphal, E. 1993. *Ipomoea aquatica* Forsk. In: PROSEA. Plant Responses of South-East Asia. 8. Vegetables. Pudoc Scientific Publishers, Wageningen.

Genus *Trapa*

Water Chestnut (*Trapa bispinosa* Roxb.)

Introduction

Water chestnut (*Trapa bispinosa* Roxb.) is known by several vernacular names like Sankrit- Shringata, Sringataka, Trikonaphalam, Jalphala; Hindi, Marathi, Gujrati, Tamil and Kannad- Singhara; Bengali-Paniphal; Telgu-Kubyakama; Malayalam-Karimpolum; Punjab-Singara, and Kashmir-Gaar, Gaunri Singara. But all over India this prized crop is known as 'Singhara' or 'Paniphal'. It is grown for its starch-rich fruits. Its importance realized during fast and celebrating of other religious events. Immature fruits are used as uncooked just after peeling; while peeled immature soft kernels are cooked as vegetable and also used in other culinary preparations. Its kernels are very nutritious and rich in mineral matters (Mandal *et al.*, 2010). Per 100 g of edible portion (fresh) of water chestnut contains moisture (70 g), protein (4.7 g), crude fiber (0.6 g), fat (0.3 g), carbohydrate (23.3 g), energy (115 K cal), minerals (1.1

g), calcium (20 g), phosphorus (150 mg), iron (1.3 mg), copper (1.31mg), manganese (0.85 mg), Magnesium (72 mg), carotene (12 µg), Thiamine (0.05 mg), riboflavin (0.07 mg), nicotinic acid (0.06 mg), Vit. C (9 mg), zinc (1.56 mg) and chromium (0.011 mg). The presence of beta-amylase and a considerable amount of phosphorylase has been reported in the kernels (WOI, 1965). Processed dry nuts seem to be easily digestible and therefore are used to break fasting. Besides, its flour is used for making Abir for holi festival and for preparing bread, Halwa, biscuits and other confectionary items. As medicinal properties water chestnut contains cyto-toxic constituents, gallic acid and used in Chinese medicines for cancer (Kosuge *et al.*, 1986). It also possesses anti-tumour principles and used as a medicinal herb in several South Asian countries including China (Kosuge *et al.*, 1985; Chopra *et al.*, 1986). Certain tribes, use its flour for the treatment of diarrhea and abdominal pain.

Origin, History and Distribution

The water chestnut is actually not a nut at all, but an aquatic vegetable that grows in marshes. The name 'water chestnut' comes from the fact that it resembles a chestnut in shape and colouring. The generic name *Trapa* is derived from the Latin word for 'thistle', calcitrappa, is also another common name for the water caltrop.

Water chestnut is a probably plant of South East Asia and its several types exist in cultivation locally. Besides India, it is cultivated in China (since ancient times), Sri Lanka, countries of South East Asia, Africa and America. In India, its maximum acreage has been reported in Bihar, Uttar Pradesh, Tamil Nadu, West Bengal, Assam, Odisha, Maharashtra, Punjab, and Jammu and Kashmir, (Jaipuria and Kumar, 1994). In Bihar, it is mainly cultivated in Darbhanga, Madhubani, Muzzaffarpur, Vaishali, Patna and Bhagalpur districts. It is well distributed in Central and South-East Europe and Temperate and Tropical Asia.

Species Availability, Evolution and Diversity

Following three species are recognized under the genus:

(i) *T. bispinosa* Roxb.

(ii) *T. maximowiczii* Korsh Syn. *T. quadrispinosa* Auc. Non Roxb.

(iii) *T. natans* Linn. Syn. *T. quadrispinosa* Roxb.

However, most of the taxonomists opine that *Trapa* is a monotypic genus represented by *T. natans* L., a polymorphic species having a number of botanical varieties; of these varieties var. *bispinosa* Makino is economically the most important in India.

Botanical Description

The genus was originally placed to the family Onagraceae and later under Hydrocaryaceae. Embryological and anatomical studies, however, supports its inclusion in a separate monogenetic family, Trapaceae. Taxonomic position of the genus is given as under:

Kingdom : Plantae
Sub-kingdom : Tracheobionta
Superdivision : Spermatophyta
Division : Magnoliphyta
Class : Magnoliopsida
Sub-class : Rosidae
Order : Myrtales
Family : Trapaceae
Genus : *Trapa* L.

The red colour of stems, under side of leaves and bulbs is mainly due to anthocyanin pigments (Meksongsee and Sripen, 1987). The transverse section of water chestnut leaf shows typical features of aquatic plants like extremely reduced vascular bundles. The mesophyll is differentiated into palisade and spongy cells. The spongy parenchyma is wider than the palisade parenchyma and has extensively network of air tunnels. The stomatal distribution is mainly restricted to a baxial surface. The maximum photosynthesis had been noted with stomatal ratio factors (SMRAT) on one side of leaf surface. Full grown leaf has higher photosynthesis as compared to younger developed leaved (Chowdhury *et al.*, 2002)

Table 11.11: Botanical Description

Growth Habit	It is an annual aquatic herb.
Root	Submerged portion possessing pairs of green, pectinate, spreading organs (the pectinate organs have been differently interpreted as leaves pair of stipules, or adventitious roots),
Stem	Stems long, flexuose, and ascending in the water
Leaf	At interval below the margins of leaf scars, leaves floating, crowded at the upper parts of the stems, appearing as rosettes, rhomboidal, 5 cm x 8 cm, often somewhat 3-lobed, lower surface variegated, with long, swollen petioles
Floral biology	The inflorescence is solitary, which arises in the axil of leaf. Pedicels are about 15 cm long with woody structure. The flowers are hermaphrodite and are perigynous with a disc above the stamens. Calyx is four in number and gamosepalous. Corolla is also four in number but polypetalous with broad roundish apex. Gynoecium is inferior, bicarpellary, syncarpous, two celled and ovule is solitary in each cell. Flowers white, opening above the surface of the water in the afternoon, after pollination the petioles bend down submerging the flowers.

Anthesis and pollination	Water chestnut is self-pollinated crop which results in maintenance of genetic purity of different types in nature. A primary study about its flowering anthesis and fruit set has been made by Arima *et al.* (1999). Flowering rate changes/or correlates with the flowering conditions (microclimate/flowering time *etc.*) and species (Arima *et al.*, 1999). In Japan, flowering time and the variability of marketable fruit weight has been studied by Arima *et al.* (1992b). they have classified flowering period in 3 stages: (i) Early (from first flowering to the late August) when the ratio of marketable fruit to flowers number is high and stable (ii) Middle (last half the September) when the ratio of marketable fruits to flower (RMF) decreased and (iii) In late (after mid-September) when RMF reached zero. In the early flowering stage, the variation in the fruit size and number of fruit has small irrespective of flowering time and stem type. Also a considerable percentage of marketable fruits seemed to have started to form during the early flowering stage, since that stage has much longer than the middle flowering stage. While considering the fruit productivity of each stem in water chestnut, Arima *et al.* (1992a) observed that the lower the nodes along the mother stem from which the stem emerged, the larger the number of marketable fruit produced on the stems and the highest the productivity. Among these stems, number of fruits which had started ripening in the early flowering stage was observed to be larger. They concluded that the reduction of branching order and the increase of the number of stems emerging at an early stage are important factors for the production of a large number of marketable fruits and for increase yield.
Fruit and fruit set	Fruit is a nut, bony, turbinate, 2-4 cm long and broad, 4-angled, 2 opposite angles each with a scabrous spine, 2 other spines sometimes obsolete, indehiscent, and 1-seeded.
Seeds	Seeds white and starchy.

Botanical Classification

i. Classification on the Basis of Leaf Colour

- ✰ Finely divided feather-like submerged leaves (borne along the length of stem)
- ✰ Undivided floating leaves (borne in a rosette at the water's surface)-The floating leaves have saw-tooth edges and are ovoid or triangular in shape, 2-3 cm long, on inflated petioles 5-9 cm long, which provide added buoyancy for leaf portion.

ii. Classification on the Basis of Fruit Colour

There are two colour of fruit-coat;

1. **Red fruit-coat** -Red fruits are more acceptable in the market due to better culinary preference.
2. **Green fruit-coat** -In general, green skinned fruit possess greater weight, size and dry matter content than purple skinned fruits
3. **Purple fruit-coat**-Skinned fruit contains higher percentage of sugar and acid (Mazumdar and Jana, 1977).

Augmentation of Genetic Resources and Crop Improvement

Several types of water chestnut exist in cultivation and they are known by different names in different regions. Four distinct varieties of 'Sighara' (or *Trapa*) namely Kanpuri, Patna, Desi big and Desi Small are reported from Bihar. Similarly, four important varieties namely Green Spinous, Green Spineless, Rosy Spinous and Rosy Spineless are reported from Meerut region of Uttar Pradesh. Of these, the first two are very high yielders. Further, from Lucknow, eleven local cultivars were identified and divided into three groups according to their maturity like (i) Early-Bari Gajvee, Singree and Barkatia; (ii) Mid-Bara Chikanwa, Kanpuria Lal, Murwa and Surkha and; (iii) Late-Gulaba, Desi Lal, Saharanpuria and Kadarna. In Kashmir, Basmati, Kanger and Dogru are reported to chiefly cultivated types, the chief being Basmati. In Agra (Uttar Pradesh) a local type is quite popular. In Kota (Rajasthan) a new type with soft thin skinned nuts, sweet in taste called Kota Sudhar, has been evolved by selection from a local heterogenous material, Nagza. Kota Sudhar fruits 20 days earlier than the local types and gives a higher yield 6200 kg per hectare when compared with 4800 kg per hectare in Nagza. The nuts obtained from Kota Sudhar in the 1st week of September, can be eaten green whereas the fruits of other local types are consumed only after cooking. In group two types- Kanpuri, Patan Local Large and Local Small are quite popular. Among all four local types, fruits of all are larger except of local small type.

Genetic Resource Conservation

Despite of rich genetic wealth of this important aquatic plant, no concerted efforts have been made for its collection, characterization and maintenance of the germplasm. As most of the water bodies where its cultivation is in vogue, is under the threat due to urbanization and other eco-hazards, it is utmost essential to collect its diverse genetic wealth before they wipe out. In tissue culture study, stem segments + axiliary buds were cultured an half strength Murashige and Skoog (MS) medium containing IAA or GA at 0.5 mg + BA@ 2 mg/l and 3 per cent sucrose in light (1500-2000 lx) at 25±1.C. when 1.0 cm of a young stem segment was cultured for 2 weeks, buds formed profusely and when these were separated and sub-cultured on the original medium, new buds again formed profusely. When the buds were transformed to half strength Murashige and Skoog (MS) medium containing only IAA at 0.5 mg/litre, roots were produced after 2 weeks and the plants grew vigorously. When the plantlets were transplanted to half strength Murashige and Skoog (MS) medium without growth regulators, growth was encouraged by soaking the roots in Nanhu lake water (Zhou *et al.*, 1983).

REFERENCES

1. Arima, S., Daigoho, M. and Hoque, M.A. 1999. Flower development and anthesis behavior in water chestnut (*Trapa* asp.). *Bulletin of the Faculty of Agriculture, Saga University*, **84:**83-92.

2. Arima, S., Harada, J. and Tanaka, N. 1992a. Growth and yield performance of the water chestnut (*Trapa bispinosa* Roxb.) II. Relationship among the canopy formation of emerged leaves, flowering and the number of valuable fruits. *Japanese Journal of Crop Science*, **61(2):** 590-596.

3. Arima, S., Harada, J. and Tanaka, N. 1992b. Growth and yield performance of the water chestnut (*Trapa bispinosa* Roxb.) III. Developmental process of fruit. *Japanese Journal of Crop Science*, **61(4):** 583-589.

4. Chopra, R.N., Nayar, S.L. and Chopra, I.C. 1986. Glossary of Indian Medicinal Plants, Publication and Information Directorate, CSIR, New Delhi-12.

5. Chowdhury, S.R., Anand, P.S.B., Reddy, G.P., Sahoo, N. and Verma, H.N. 2002. Stomatal distribution, leaf anatomy and leaf age dependent behavior of water chestnut (*Trapa bispinosa*) leaves. *Orissa Journal of Horticultural*, **30(1):** 76-79.

6. Jaipuria, S. and Kumar, A. 1994. Bahu-upyogi Singhare Ki Kheti Kaise Kare. *Phal-Phul*, Januery-March, p. 29-30 and 39.

7. Kosuge, T., Yokota, M., Sugiyama, K., Okamuto, A., Saito, M. and Yamamoto, T. 1986. Studies on Chinese medicines used for cancers. III. Cytotoxic constituents against like HeLa cells in the fruit of *Trapa bispinosa* Roxb. *Yakugaku Zasshi*, **106 (2):** 183-185.

8. Kosuge, T., Yokota, M., Sugiyama, K., Yamamoto, T., Ni, M.Y. and Yan, S.C. 1985. Studies on antitumor activities and antitumor principles of Chinese herbs. *Yakugaku Zasshi*, 105 (8): 791-795.

9. Mandal, R.N., Chattopadhyay, D.N., Saha, G.S. and Mukhopadhyay, P.K. 2010. Duck Weeds-Nature's potential treasures of nourishment source for production of herbivorous fish. *Agrovet Buzz*. **3(2):** 26-31.

10. Mazumdar, B.C. and Jana, S. 1977. Physicochemical analyses of water chestnut (*Trapa bispinosa*) fruit. *Science and Culture*, 43(8): 361-362.

11. Meksongsee, L. and Sripen, K.S. 1987. Anthocyanin of *Trapa* ssp. proceedings of 11th *Asia Pacific Weed Science Society Conference*, **1:** 349-353.

12. WOI. 1965. Wealth of India, CSIR, New Delhi

13. Zhou, J.Q., Zhang, Z.P., Zhou, G.Y., Liu, S.Z. and Chen, B.Q. 1983. Culture of axiliary buds of Nanhu Singhara nut. *Plant Physiology Communication*. Zhiwu Shenglixue, Tongxun, **1:** 31.

Genus Nelumbo

Lotus (*Nelumbo nucifera* Garrtner)

Introduction

Indian lotus (*Nelumbo nucifera* Gaertn.; Synonyms: *N. speciosum, N. nelumbo, Nymphaea nelumbo, Nelumbo caspica*) is a tropical aquatic perennial species. It is invasive and harbors ponds and lakes. In different places it is called several vernacular names like in Sanskrit-Ambuja, Padma, Pankaja, Kamla; Hindi- Kawal, Kamal; Bengali- Padma, Mar-Kamal; Gujarati- Suriyakmal; Telgu-Kalung, erratamara; Tamil-Ambal, Thamarai; Kannad-Kamala, Tavaregadde; Malayalam-Thamara, Senthamara; and in Oriya-Padam. It is also called as Blue lotus, Indian lotus, sacred lotus, bean of India, East Indian lotus, sacred water lily, Oriental Lotus, Lily of Nile, Chinese arrow root, Lien Zi, Ti, Mi, On, Nadroo, Napa, Kamal, Semen, Nelumbis *etc.* The pink lotus is the national flower of India. About 70 compounds imparting scent to lotus flowers have been identified with hydrocarbons representing more than 75 per cent of the extract. 1, 4-Di-methoxy benzene, 1, 8-cineole, terpinen-4-ol and linalool are important constituents for the characterization of the scent of flowers (Omata *et al.*, 1991). It is symbolizes purity, beauty, majesty, grace, fertility, wealth, richness, knowledge and serenity. Lotus is one of the important plants of Mithila rituals. It is most commonly cultivated in water gardens. It is most commonly cultivated in water gardens. All parts of the plant consumed by human being. The rhizomes or roots are cooked alone or with other vegetables and eaten in Asian countries. Roots may be boiled and preserved in sugar or pickled in vinegar. Young leaves are boiled and cooked as vegetables. Stem is cooked to eat and taste somewhat like beet but before cooking it may be peeled. Stamens are used to flavor tea while petals can be used to garnish soups or other dishes. Seeds are eaten raw or cooked. Seeds can be popped into popcorn and ground into powder which is used to prepare bread or eaten as such. Lotus has unmatchable medicinal properties and on scale of 1-5 utility in USA, it scores 4. In Hindu mythology, lotus flower is seat of Lord Brahma, the creator of Universe. Numerous medicinal properties have been ascribed to lotus plant in the Chinese system of medicines. All parts of plant have curing properties and various uses include astringent, cardio tonic, febrifuge, hypotensive, resolvent, stomachic, styptic, tonic and vasodilator (Chopra *et al.*, 1986, Yeung, 1985, Anonymous, 1986). Now the many formulations also stand patented. Nadroo (Indian Lotus) can be called as a crop of health, wealth and happiness. Lotus is cultivated in China and Japan for its edible rhizomes and seeds.

Other Importance

- ✰ In Asia, the petals are sometimes used for decoration, while the large leaves are used as a wrap for food.
- ✰ Lotus flowers, seeds, young leaves and rhizomes are all edible.
- ✰ Various parts of the sacred lotus are also used in traditional Asian herbal; medicine.
- ✰ The tender seeds are munched happily in north east India.
- ✰ The lotus stem is eaten almost in all parts of India, and pickled too.
- ✰ Young lotus stems are used as a salad ingredient in Vietnamese cuisine.
- ✰ The distinctive dried seed heads, which resemble the spout of watering cans, are widely sold throughout the World decorative purposed and for dried flower arrangement.

Table 11.12: Usage of Lotus

Plant Parts	*Utility*
Leaf	Lotus leaves are bitter but neutral. They benefit stomach spleen and liver. They lower blood lipids. Lotus leaves are rich in nuciferine, nornuciferine, roemerine and quarcetin. The leaf juice is effective in treatment of diarrhea. They have cooling effect on body.
Flower	A decoction of lotus flowers serves a sex tonic as it prevents premature ejaculation. Flowers are recommended as a cardiac tonic. A decoction of flower receptacle is used in the treatment of abdominal cramps and bloody discharges.
Flower stalk	It is haemostatic. It is used in treatment of bleeding gastric ulcers, excessive menstruation and postpartum hemorrhage.
Fruit	A decoction of lotus fruit is used in the treatment of agitation, fever and heart complaints.
Stem	Stem contain alkaloid, resin and tannins, and has anticancer properties.
Lotus stamens	Stamens are sweet, astringent, neutral and benefit heart, kidney. They are useful in regulating urinary frequency, premature ejaculation, haemolysis, epitatis and preventing uterine bleeding.
Plumule and radical	They are useful in treating thirst in high febrile disease, hypertension, insomnia and restlessness.
Root	Lotus roots are rich in starch, tannate protein, asparaginasepyrocatchol, d-gallic-catechin, neochlorogenic acid, leucodephinidin, perioxidase, vitamin B and C. It is also rich in dietary fibre, potassium, thiamine, riboflavin, vitamin-6, phosphorus, copper and manganese while very low in saturated fat. These compounds are anticarcinogenic and antimicrobial in nature. Root is used in treatment of diarrhea, dysentery, hemorrhages, excessive menstruation and nasal bleeding (Brown, 1995).
Root nodes	These are used in treatment of nasal bleeding haemolysis, haematuria (blood in urine) and functional breeding of uterus (Yeung, 1985).
Seed	The seed is sedative and finds use as astringents in kidney and spleen tonic formulations. The sweet taste and nourishing qualities are useful to spleen. The seed has calming properties and help to overcome restlessness.

Source: Sharma, 2011.

Origin, History and Distribution

The lotus originated in India (Southern Asia and Australia) and from this country it spread to tropical and sub-tropical regions of Middle East and Asia comprising Egypt, Iran, Afghanistan, India, China, Vietnam and Australia. Some species of lotus can be spotted in Eastern United States and southwards to Colombia. The American Lotus (*Nelumbo lutea*) is native to a region stretching from south-eastern part of North America to the northern part of South America. It is smaller than the sacred lotus; bears scented, pale yellow flowers. In the west, lotus (*Nelumbo nucifera* Gaertn.) is relatively little known. However, for more than 3000 years, lotus plants have been cultivated as a crop in Far-East Asia, where they are used for food, medicine and play a significant role in religious and cultural activities (Shen, 2002). The traditional sacred lotus is only distantly related to water lily (*Nymphaea caerulea*), but possesses similar chemistry. Both contain the alkaloids nuciferine and aporphine. Mithila region of Bihar (India) is rich in its wetlands and is considered one of the natural abodes of several aquatic vegetables in including lotus (Jha *et al.*, 1994).

Species Availability, Evolution and Diversity

The genus *Nelumbo* is grouped in two extant species:

(i) The Asian lotus (*Nelumbo nucifera* Gaertn.) with pinkish to cream flower

(ii) North American *Nelumbo lutea* with the yellow cream flowers.

However, detailed study made on morphological, anatomical, playnological and phyto-geographical by Borsch and Barthlott (1994) revealed that the taxa are hardly differentiated and it is concluded the new world taxa should be considered a sub-species *Nelumbo nucifera* ssp. *lutea*. *Nelumbo lutea* is found in Eastern United States and southwards to Colombia. The American lotus (*Nelumbo lutea*) is very popular among the growers. Wei *et al.* (1994) in their study on karyotype analysis reported that *Nelumbo nucifera* has the karyotype 2n=16=9sm + 4m+3st, which belongs to the 2B type. The Barium Salt Giesma (BSG) banding method studies show that chromosome number of *Nelumbo nucifera* 2n=16 and its karyotype formula is 2n=16 =2st+4sm + 8 m+ 2m (SAT). There is a satellite on the 6th chromosome whose length ratio is 3.3; the karyotype of this chromosome belongs to the '2B' type. The composition of chromosome comparative length is 2n=16=4L + 2 M 2 + 4M1 + 6S (Yang *et al.*, 1998). Karyotype of *Nelumbo lutea* and *Nelumbo nucifera* have been found approximately the same, showing 2 chromosomes with satellites, normal meiosis and chromosome number pairing pattern 8 II. The *N. nucifera* cultivars showed some normality such as precocious segregation of bivalents at metaphase I and lagging at anaphase I. Reciprocal F_1 hybrids between the species showed regular chromosome pairing at meiosis, a pairing pattern of 8II, few chromosome abnormality and normal seed ratio, the last being positively related to that of the *Nelumbo nucifera* cultivars. Maximum nuclei numbers in inter-phase nuclei of the F_{1s}, indicates that there was no nucleolus organization region competition between the species. Thus indicates that only geographical isolation was between *N. nucifera* and *N. lutea* and it is proposed that *N. lutea* is subspecies of *N. nucifera viz. N. nucifera* ssp. *lutea* (Huang *et al.*, 1992).

Botanical Description

Earlier genus *Nelumbo* was grouped under the family Nymphaeaceae. However, the fruit wall+seed coat of *Nelumbo nucifera* constitutes a complex of polysaccharides, based on galactose and sclerotic seed coats. The distinctive nature of chemical constituents presenting propagule in walls of *Nelumbo nucifera* supports into systematic distinction of this genus in the separate family Nelumbonaceae (Van-Bergen *et al.*, 1997). As per the Dahlgren's system of classification, which appeared in 1980, family of lotus Nelumbonaceae has been put under order of Nelumbonales of super order Magnoliflorae in sub-class of Magnoliidae of class Magnoliopsida. *Nelumbo nucifera* Gaertn. is considered a primitive plant of angiosperm and has been centre of attraction of taxonomists towards its grouping under monocotyledonous and dicotyledonous.

Kingdom : Plantae
Phylum : Magnoliophyta
Class : Magnoliopsida
Order : Proteales/Nymphales
Family : Nelumbonaceae
Genus : *Nelumbo*
Species *nucifera*

It has been suggested that primitive dicotyledonous lotus may be related to more primitive monocotyledon. There has so far been no more logical evidence to support this fact. The development of chloroplast from plumules was investigated by electron microscopy mechanism of chlorophyll content, SDS-PAGE of polypeptides and measurement of florescence emission spectra. A number of differences between the development of chloroplast in *N. nucifera*, but not other angiosperm were observed. Chlorophyll content SDS-PAGE of polypeptides and study of fluorescence spectra indicates significant difference between the development of chloroplast in *N. nucifera* but not other angiosperm to synthesize chlorophyll in dark. The results indicate evolutionary characteristic intermediate between lower plants and Angiosperm (Zou *et al.*, 1992) were observed. Further, anatomical and histological studies of developing embryos of *N. nucifera* were made by Titova and Batygina (1996) and they concluded that this species has a single cotyledon primordial comprising of 2 lobes and is monocotyledons. In winter, plant remains dormant and resumes growth with onset of spring. In nature, seed germination occurs by April or May. Plants are highly resentful to root disturbances.

Table 11.13: Botanical Description

Growth Habit	*N. nucifera* Gaertn. is an aquatic herb and its large, circular or plate like leaves measuring in 60-90 cm diameter. Lotuses grow in shallow and muddy waters. The plant is an emergent aquatic that produces individual leaves and flowers directly from the root system. Root, rhizome and stem are under water while leaves either float on water surface or just few centimeters rose above water surface.
Leaves	Floating erect and stout leaves supported with long petiole place its canopy over the water surface sometimes even above the water surface. Its petiole is adapted to increase its length up to 3.0 m proportionately with increasing water level.
Rhizome	The lotus plant grows from a thick rhizome. Rhizome is edible, fleshy and 100.0-300.0 cm long, creamy to white in colour. Internally rhizomes are characterized by 8 tubular channels with solid core. Rhizome sprouts in early spring and elongated with emerging floating and upright leaves. Three to four distal rhizome internodes enlarge in late summer. Enlarged rhizomes are organs of asexual propagation. They remain dormant during unfavorable circumstances but become activated on return of congenial environment. They are source of planting material for commercial cultivation.
Root	The plant has its roots firmly in the mud and sends out long stems to which their leaves are attached. The root system has thick rhizome with fibrous roots.
Stem	Stem is thick, green up to 150 cm long in some cases up to 500.0cm. Branching pattern is multi-trunked or clumping of stem. On cooking stem tastes like beet.
Leaf	The blades of the leaves either float on the surface of the water, or they are held up to 15 cm above the water surface by their petioles. Leaves are green, spherical up to 60.0-90.0 cm in diameter. Petiole very long, smooth or with small prickles. The first leaves that appear, few in number, are flat and float on the surface. They are followed by thicker, funnel-shaped leaves with slightly wavy edges that may stand from 50-200 cm above the water. Leaf type is simple, margin undulate shape orbiculate, venation palmate, deciduous and blade length 45.00-91.00 cm. The micro-relief of plant surfaces, mainly caused by epicuticular wax crystalloids, serves different purposes and often causes effective water repellency. The particles were removed completely by water droplets that rolled off the surfaces independent of their chemical nature or size. The leaves of *N. nucifera* afford an impressive demonstration of this effect, which is, therefore, called the 'Lotus-Effect' and which may be of great biological and technological importance (Barthlott and Neinhuis, 1997).

Floral biology	Flower are solitary, large and attractive with varied colours such as snow white, yellow, creamish, light pink or rose colour *etc.* Flowers are normally appear above the water surface. They stand on strong leaflets stalks. Number of sepals in calyx and petals in corolla count approximately 20 or more. Flowers diameter vary from 15.00-25.00 cm. Stamens are numerous in number surrounding the pistils. They are hermaphrodite, and pollinated by insects. The sacred lotus attracts pollinators by producing heat through a non-phosphorylating electron transport pathway catalysed by the Alternative Oxidase (AOX) that releases energy by electron flow through an alternative respiratory pathway. Individual flowers are held up to 15.00 cm above the water surface by their peduncles (flowering stalks). Each flower is 10.00-20.00 cm across, consisting of about 15 pink or white petals, a golden yellow receptacle and a dense ring of golden yellow stamens (although the anthers are often white). The receptacle, which is located in the centre of the flower, is shaped like an upside-down cone. Along its flat upper surface, there are 15-35 short styles that look like small bumps. Each flower lasts for 2 to 5 days and darkens with age, ranging in diameter from 15-60 cm. lotus flowers are termed single when they have less than 25 petals, semi-double if they have 25-50 petals and double if they have more than 50 petals.
Anthesis and pollination	The blooming period occurs during the summer and can last 2-3 months. The short-lived flowers open up during the morning (6.00-7.00 h) and begin to lose their petals by the afternoon (12.00 h) and flowering continued for 4 days (Kasumi and Sakuma, 1992). Further, they observed that anthers dehiscence did not occur during the first flowering day. Most anther dehiscence occurred on the 2nd and 3rd day. The stigma was more receptive on the 1st day and the pollen had highest pollination ability after 1st day. They have a pleasant sweet fragrance. Each flower is replaced by a seed pod spanning 7.50-10.00 cm across and 0.60 cm deep; this seed pod becomes dark brown at maturity.
Fruit and fruit set	The nut like fruits is borne on spongy receptacle called conical pod. Fruit is torus large; fleshy but Nucifera means having hard fruit. It is round 7.5-15.0 cm. Its cover is hard, dry and brown in colour. It persists on plant.
	It bears many nut seeds through numerous holes (each of its 15-20 openings contains a fruit) in the flat surface. When the seeds are ripe, they become loose in the pod. The pod then tips down towards the water, releasing the seeds. The seed sink to the bottom and established now plants. The dried receptacles are used by florists in decoration.
Seeds	Seed is spherical, hard coated nut and edible. Due to hard coat it retains viability for many years when kept at room temperature. Nearly 1300 years old seed obtained from peat moss in Manchuria were reported capable of germination. The coat is very hard, built to prevent water and air from entering and degrading the seed. Embryo is bitter and embedded in the core of seed. The green plumule is called 'plumula Nelumbinis'.

Horticultural Classification

Generally two types of rhizomes are found:

(i) White

(ii) Red

The farinaceous rhizomes of the plant are collected and sold as vegetable (Kamal Kakadi). They measure 60-120 cm in length and 6-9 cm in diameter are white to buff orange in colour and possess a few large cavities in cross section. Shou *et al.* (1995) identified three vegetable rhizome varieties of lotus.

Augmentation of Genetic Resources and Crop Improvement

A remarkable heterosis in Chinese seed lotus has been reported by Yang *et al.* (1998). When crosses were made among geographically different varieties/ different lines from various ecological conditions, a large variation appeared in floral organs of the F_{1s}. They also have been able to develop some promising varieties. It was concluded that seed formation may be obtained by crossing between pistils of 1st day flowers with pollens from the 1st day old flowers. Further, lotus flower can successfully be stored at -20 °C ± 1°C for 14 days with silica gel as a desiccant. Seed formation is possible with stored pollen for about 1 year but the success rate greatly diminishes (Kasumi *et al.*, 2000)

Genetic Resource Conservation

Not much work has been done on development of vegetable type lotus having yield of rhizome with better quality.

REFERENCES

1. Anonymous, 1986. Oriental Herbs and Vegetables Vol. 39, No.2. Brooklyn Botanic Garden.
2. Barthlott, W. and Neinhuis, C. 1997. Purity of the sacred lotus, or escape from contamination in biological surfaces. *Planta*, **202(1):** 1-8.
3. Borsch, T. and Barthlott, W. 1994. Classification and distribution of two genus *Nelumbo adans* (Nelumbonaceae). *Beitrage zur Biologie der Pfanzen*, **62 (3):** 421-450.
4. Brown, D. 1995. Encyclopedia of herbs and their uses. Dorling Kindersley, London.
5. Chopra, R.N., Nayar, S.L. and Chopra, I.C. 1986. Glossary of Indian Medicinal Plants. CSIR, New Delhi.
6. Huang, X.Q., Chen, J.Y. and Huang, G.Z. 1992. Preliminary studies on the bio-systematics relationship between two *Nelumbo* species. *Acta Horticulturae Sinica*, **19 (2):** 164-170.
7. Jha, V., Saraswati, K.C., Mishra, S. and Basak, P.K. 1994. Plants in Mithila rituals. Ethno-biology in human welfare. Abstract of the 4th International Congress of Ethno-biology, Lucknow, U.P., India, Nov. 1994, p. 192.
8. Kasumi, M., Yashira, K. and Hayashi, M. 2000. Effectiveness of cold stored pollen on seed formation in East Indian lotus (*Nelumbo nucifera*

Gaertn.), *Journal of the Japanese Society of Horticultural Science*, **65(6):** 732-735.

9. Kasumi, M. and Sakuma, F. 1992. Flowering, pollination, fertilization and seed formation in lotus rhizome plant. *Journal of the Japanese Society of Horticultural Science*, **67(4):** 555-595.
10. Omata, A., Yomogida, K., Nakamura, S., Ohta, T., Izawa, Y. and Watanabe, S. 1991. The scent of lotus flowers. *Journal of Essential Oil Research*, **3(4):** 221-227.
11. Shen, M.J. 2002. Sacred lotus, the long living fruits of China antique. *Seed Science Research*, **12(3):** 131-143.
12. Sharma, J.P. 2011. Nadroo (Indian Lotus). (In) Quality Seed Production of Vegetable Crops: Technological Interventions Vol. II: Crop Specific Aspect, p. 516-522.
13. Shou, S.Y., cao, X.Y. and Dong, W.M. 1995. Characteristic of fine vegetable rhizome varieties of lotus in Zhejiang Province. *China Vegetable*, **4:** 21-23.
14. Titova, G.E. and Batygina, T.B. 1996. Is the embryo of Nymphaealean (Nymphaeales) dicotyledonous? *Phytomorphology*, **46(2):** 171-190.
15. Van-Bergen, P.F., Hatcher, P.G., Boon, I.J., Collinson, M.E. and DE-Leeuw, J.W. 1997. Macromolecular composition of the proagule wall of *Nelumbo nucifera*. *Phytochemistry*, **45(3):** 601-610.
16. Wei, P.H., Chen, W.P. and Chen, R.Y. 1994. Study on the karyotype analysis of Nymphaeaceae and its taxonomic position. *Acta Phytotaxonomica Sinica*, **32(4):** 293-300.
17. Yeung, Him-Che 1985. Handbook of Chinese Herbs and Formulas. Institute of Chinese Medicine, Los Angeles.
18. Yang, R.F., Guo, Q.Q., Zheng, S.X. and Chang, Y.C. 1998. Studies on the karyotype and giemsa C banding pattern of lotus (*Nelumbo nucifera*). *Journal of Human Agriculture university*, **24(3):** 191-193
19. Zou, B.Y., Li, G.Q., Tang, C.Q., Yu, Y.L., Jiang, G.Z., Kumog, T.Y. and Murata, N. 1992. A new developmental pathway of chloroplast in sacred lotus plumules related to phylogeny of plant kingdom. Research in photosynthesis Vol. 2. Proceedings of IX International Congress on Photosynthesis, Nagoya, Japan, August 30 September, 1992, 219-223.

Genus *Nymphaea*

Water Lily (*Nymphaea nouchali* Burn.f.)

Introduction

Water Lily has a worldwide distribution. It has a variety of colour and some species have fragrance. As an evidence, the goddess is said to have pointed out that the rhizomes of *Nymphaea* species were edible, and its flowers, buds and leaves are often depicted on ancient monuments, in murals, on pottery and on furniture. The blooms are in great demand for religious festivals, offerings of the flowers being made to the dead or to the gods, as well as for gifts to visiting noblemen as a gesture of friendship and goodwill. The reason for veneration lies in the belief that the beautiful blooms of the water lily, rising pure and clean from the slimy mud were comparable with the aspirations of mankind: purity and immortality. Its starchy rhizome is also eaten raw or boiled; they are sometimes baked. Flowering stalks and unripe fruits are also used as vegetable. The seeds are edible and can be eaten raw or after parching and they also be ground flour and made into a kind of bread or cooked into kanji with water.

Origin, History and Distribution

The name of the genus *Nymphaea* is a direct transliteration of Greek word which Theophrastus used to describe these plants about 300 years before the Common Era. In Egypt, *Nymphaea caerulea* (*Nymphaea nouchali* var. *caerulea*), the blue lotus and *Nymphaea lotus*, the white lotus, have been admired, painted, eaten, grown and revered for thousands of years. Monarchs and priests of ancient Egypt were laid to rest with wreaths made from the petals of the blue lotus, laid in concentric semi-circles from the chin downwards. There is also evidence, in the form of a painting in a tomb dating back to 3000-2500 BCE that *Nymphaea* species were deliberately cultivated in square, evenly spaced beds fed by canals. Both Amenhotep IV and Ramses III (1225 BCE) are known to have had them growing purely for their ornamental value in their palace gardens.

Species Availability, Evolution and Diversity

Nymphaea under the family Nymphaeaceae has four common species in India like:

i. Nymphaea alba
ii. N. rubra
iii. N. pubescens
iv. N. stellata

Botanical Description

Its taxonomic hierarchy is described as under:

Domain	:	Eukaryota
Kingdom	:	Plantae
Sub-kingdom	:	Viridaeplantae
Phylum	:	Tracheophyta

Sub-phylum	:	Euphyllphytina
Infraphylum	:	Radatopses
Class	:	Magnoliopsida
Sub-class	:	Nymphaeidae
Super order	:	Nymphaeanae
Order	:	Nymphaeales
Family	:	Nymphaeaceae
Sub-family	:	Nymphaeoideae
Tribe	:	Nymphaeeae
Genus	:	*Nymphaea*
Species	:	*nouchali*

It is a perennial plant growing up to 245.0 cm deep.

Detail description of the plant is as under (Table 11.14)

Table 11.14: Botanical Description

Growth Habit	Robust perennial aquatic herb with tuberous rhizome which grows up to 245.0 cm. deep.
Leaf	The leaves develop directly from rootstock on long petioles measured 12.5-35.5 cm across, orbicular in shape, cleft towards the middle on one side and smooth along the outer margins. Leaves 17-40 x 16-32 cm, chartaceous orbicular or subelliptic, incised-cordate, peltate, lobes slightly divergent, acute or caudate, margin sinuate-dentate, teeth acute or blunt upper surface smooth, under surface with prominent reticulate nervation, primary lateral nerves 7-9 on each side of the midrib, forming closed areas less than half way to margin, secondary nerves 6-8 pairs arising from the midrib.
Floral biology	Long flowering stalk also develops directly from rootstock and bears a single flower which opens up during morning and closes during afternoon. Flowers blue or pinkish and each flower is about 10.0-17.5 cm across (6-12 cm in diameter). Sepals 4, lanceolate, obtuse. Petals 14-20, as long as the sepals, oblong-lanceolate, obtuse or subacute. In the middle of the petals, there are many yellowish stamens (stamens c. 100) surrounding single pistil. Carpels 16-20.
Anthesis and pollination	Flowers generally open in the morning. Each flower lasts for only 4-5 days
Fruit and fruit set	Fruit are depressed-globose shape; 2x 3 cm.
Seeds	Enormous number of seeds and seeds are 1.0 mm long, subellipsoid, with longitudinal lines of hairs.

The development of reproductive cells in *Nymphaea nouchali* has been analysed, by Isabelle and Jules (1998) and their observations on embryo sac development differ from those in the literature. Megasporogenesis results in a trial of one micropylar cell and two innermost megaspores. Embryo sacs derive from a single functional megaspore and are tetranucleate. Megasporogenesis is monosporic and of the Polygonum type, but mega-gametogenesis is of Oenothera type.

REFERENCE

1. Isabelle, O. and Jules, B. 1998. Mega-gametophyte development of *Nymphaea nouchali* Burn. F. (Nymphaeaceae). *Botanical Journal of the Linean Society*, **126(4):** 339-348.

Fragrant Water Lily (*Nymphaea odorata* Ait.)

Introduction

Fragrant Water Lily (*Nymphaea odorata* Ait., 2n=28) is exceptionally beautiful aquatic plants with floating leaves and large many pelted fragrant blossoms. Its rhizomes are used as food and the young leaves and lower buds are eaten as a vegetable. Even the seeds are fried and eaten or ground into flour. Seeds are also used in sauces, condiments, spices and flavorings. The rhizomes are cooling, sweet, bitter and tonic and are useful in diarrhea, dysentery, dipsia and general debility. The flowers are astringent and cardio-tonic. The seeds are sweet, cooling, constipating, aphrodisiac, stomachic and restorative. It has found uses both as a culinary delight and starchy food staple as well as being used internally as a treatment for gastrointestinal disorder and jaundice. Leaf is used in cutaneous, subcutaneous parasitic infection, eye treatments, and pregnancy. Zhizhen Zhang *et al.* (2003) have identified two lignans, one new and (1) one known (2), together with six known flavonol glycosides (3-8). The structures of 1-8 were established by spectroscopic analysis as nymphaeoside A (1), icariside E_4 (2), kaempferol 3-O-γ-L-rhamnopyranoside (afzelin, 3), quercetin 3-O-β-L-rhamnopyranoside (4), myricetin 3-O-β-L-rhamnopyranoside (myricetin 5), quercetin 3-O- (6-O-acetyll-β-D-galac-topyranoside (6), myricetin 3-O- β-D-galac-topyranoside (7), and myricetin 3-O-(6-O-acetyl) - β-D-galac-topyranoside (8). Compounds 3, 4, and 7 showed marginal inhibitory effet against fatty acid synthase with IC_{50} values of 45, 50, and 25µg/mL, respectively. Extracts from leaf petioles, and rhizomes have allelopathic potential and may suppress the germination and growth of other aquatic species (Spence, 1998).

Origin, History and Distribution

Fragrant water lilies plants are native to the eastern United States of America including Southern Canada and it is believed that the water lily was introduced to Washington during the Alaska Pacific Yukon Exposition held in Seattle in the late 1800s. Because of their great beauty, water lilies have been intentionally planted in many Washington lakes, especially those lakes in western Washington. It has been introduced as an ornamental plant in many parts of the world and it is expected to expand.

Species Availability, Evolution and Diversity

Synonyms: *Castalia lekophylla* Small, *C. minor* (Sims) Nyar, *C. odorata* (Ait.) Wood, *C. reniformis* DC., *Nymphaea minor (Sims) DC., N. odorata* var. *gigantean* Tricker, *N. odorata* var. *godfreyi* Ward, *N. odorata* var. *minor* (Sims), *N. odorata* var.*rosea* Pursh, *N. odorata* var.*stenopetala* Fern., *N. odorata* var. *villosa* Caspary water lily.

Botanical Description

Fragrant water lily is a member of American water lily family Nymphaeaceae.

Table 11.15: Botanical Description

Growth Habit	Water lily *N. odorata* is an aquatic perennial plant. Water lilies grow in dense patches, excluding native species and even creating stagnant areas with low oxygen levels underneath the floating mats. These mats make it difficult to fish, water ski; swim, or even relatively slow-spreading, water lilies will eventually colonize shallow water depths to 2.0 m deep and can dominate the shore lines of shallow lakes. For this reason, planting water lilies in lakes is not recommended.
Root	Rhizomes are attached by adventitious roots arising in groups below the leaf bases and the rhizomes are densely covered with short black hairs.
Stem	The horizontal creeping and branching rhizomes (2-3 cm in diameter).
Leaf	Leaves are floating. The petiole leave crescent shaped scars on the rhizome when shed. Mature *N. odorata* leaves are spherical, cleft at the base, smooth to 25 cm across, and usually purple on the lower surface. Leaves are attached to under water stalks rising from thick flesh rhizomes.
Floral biology	Flowers rise on long solitary stalks and are borne at the surface of the water or elevated slightly above it. Flowers measure up to 25 cm across and have yellow centers surrounded by 25 or more petals. Flowers are fragrant and can be white or pink with yellow centers.
Anthesis and pollination	Fragrant water lily possesses an interesting pollination mechanism. Each white or pink flower has many petals surrounding both male and female reproductive parts, and is only opened during the daytime for three days. On the first morning, the flowers produce a fluid in the cup like center and are receptive to pollen from other flowers. However, they were not yet releasing pollen themselves. Pollen covered insects are attracted by the sweet smell, but the flower is designed so that when they enter the flower, they fall into the fluid. This washes the pollen of their bodies and onto the female flower parts (stigma) causing fertilization. Usually the insects manage to crawl out of the fluid and live to visit other flowers, but occasionally insect will remain trapped and die when the flower closes during the afternoon. On the second and the third days, the flowers are no longer receptive to pollen, and no fluid is produced. Instead, pollen is released from the stamens (the flexible yellow match-shaped structures in the flower center). Visiting insects pick up the pollen and transport it to flowers in the first day of the flowering cycle.
Fruit and fruit set	After the flower has finished, the stalk forms a spiral and draws the fruit below the water. The fruit is an ovoid berry-like capsule 1-2 cm in diameter containing many seeds
Seeds	After three days the flowers are brought under water by coiling their stalks. The seeds mature under water and after several weeks are released into the water. Seeds are 2-3 mm long. Water currents or ducks, which eat the seeds, distribute them to other areas.

Augmentation of Genetic Resources and Crop Improvement

The nursery industry has hybridized them and produced many colour variations. They sell tropical water lilies and hardy water lilies. It is the hardy white and (sometimes) pink lilies that have become naturalized in Washington lakes and rivers.

REFERENCES

1. Spence, S.K. 1998. Bioassay-directed isolation of the allelopathic constituents of the aquatic plant *Nymphaea odorata*. Dissertation Abstracts, International Part B: *Science and Engineering*, **58(10):** 4762.

2. Zhizhen Zhang, Elsohly Hala, N., LI Xing-Cong., Khan Shabana, I., Broedel Sheldon, E. Raulli Robert, E., Cihlar Ronald, L., Burandt Charles and Walker Larry, A.2003. Phenolic compounds from *Nymphaea odorata*. *Journal of Natural Products*, **66(4):** 548-550.

Genus *Nasturtium*

Watercress (*Nasturtium officinale*)

Introduction

Watercress carries several Latin nmes as *Rorippa nasturtium aquaticum, Nasturtium officinale* R. Br., *Sisymbrium nasturtium aquaticum* and *Radicula nasturtium aquaticum* (Thompson and Kelly, 1957). Watercress () is an important aquatic or moisture loving vegetables vegetable. It is popularly known as Piria Halim (Punjabi) and Lut Putiha (Deccan). The tender green leaves are used as salad or used for decoration of other vegetable or to make food items scented. It is used as vegetable after boiling. Leaves invigorate and improve the digestion. Water Cress is rich source of minerals and vitamins. Apart from this other nutrients like calcium, zinc, arsenic and copper are found abundantly. In comparison to cauliflower, lettuce and beans, watercress is more nutritive. Per 100 g of edible leaves contains 89.2 g moisture, 2.9 g protein, 0.2 g fat, 2.2 g minerals, 0.6 g fiber, 4.9 g carbohydrates, 33 K. calorie energy, 290.0 mg calcium, 140.0 mg phosphorus, 4.6 mg iron, 2803 microgram carotene, 0.12 mg thiamine,

0.38 mg riboflavin, 0.8 mg niacin and 13.0 mg vit. C. It is considered as best source of protein (Gopalan *et al.*, 1999). The presence of essential amino acids per gram reported as 3.0 mg leucine, 1.0 mg phenyl alanine, 1.2 mg valline, 1.5 mg lysine, 4.0 mg aspartic acid, 0.6 mg tyrosine, 1.0 mg alenine, 1.5 mg threonine, 2.7 mg glutamic acid, 0.6 mg serine, 1.0 mg sistine, 0.1 mg methionine sulpho-oxide and 0.4 mg proline (Wealth of India, 1972). Medicinally, it is very important herb to cure goiter and prevents pus formation in nose tract. Due to bactericidal effects it is useful in curing dry throat, tuberculosis and urinary infection. Watercress contains gluconasturtiin, C_{15} $H_{12}OO_9S_2NK$ which catalyzes by myrosinase enzyme and changed into Glucose phenyl isothiopcynate and potassium hydrogen sulphate. The seeds contain gluconasturtiin and 24 per cent non-drying fatty oil. In the local market, it is sold in bundles.

Origin, History and Distribution

The original home of water cress is Europe, Western Asia and probably Ethiopia. It can be spotted in higher altitude of temperate and sub-tropical climatic condition. First time water cress cultivation was started during 1750 in near Erfurt of South Germany (Manton, 1935). It is believed that for the supply of this vegetable crop to the market of London and Paris, it was started growing in England and France in the beginning of 19th century (Howard, 1976). In Germany and France, generally diploid species which are called as green cress are grown. But in in England triploid species which are called as brown cress are under cultivation because these cultivars were frost tolerant. Tetraploid species are not under cultivation due inferior quality (Pink, 1993).In India it is very common in hilly regions. In Sikkim it is grown up to the height of 1400-0600 m. *R. heterophylla* is distributed in Japan and Myanmar while *R. schlechteri* is endemic to Papua New Guinea.

Species Availability, Evolution and Diversity

In watercress *Nasturtium officinale* and *Rorippa nasturtium aquaticum* both are cultivated. In these allotetraploid species 2n=64 itself developed in nature, which is called as *Rorippa microphylla* or *nasturtium microphylla* Boenningh ex. Reichenb. They crossed to each other and believed it was evolved through triploid hybrid cultivar (2n-3x-48) R x *Sterilis*. It has another species *Rorippa schlechteri* which is called as highland cress, commonly grown in Papua New Guinean region. There are 3 cultivated species *viz.*

(i) ***R. heterophylla***- Annual herb, 5-50 cm tall. Stem erect or ascending, one or more from the base, along the stem; lower ones narrowly obovate in outline, 1.5-8.0 cm long with 1-6 cm long petioles, undivided or lyrate-pinnati partite with 1-3 pairs of small lateral lobes and a much larger terminal lobe; upper leaves rather few, undivided, narrowly ovate. Racemes 2-10 cm long; flowers with reduced or lacking petals and long. Fruit linear, 14-25 x 0.7-1.3 mm, straight, and less than 1.5 mm thick, with seed in one row only. Seed are red brown and minutely fore late (Rahmansyah, 1993).

(ii) ***R. nasturtium aquaticum***- Perennial, much branched, glabrous, aquatic herb, up to 1.0 m tall. Stems are prostrate to ascending, juicy, hollow and usually rooting at the lower nodes. Leaves petiolate, auriculate, pinnate, oblong in outline, up to 10 cm long; lateral leaflets in 2-9 pairs, sessile, narrowly ovate, elliptical or nearly orbicular, entire to fairly dentate; terminal leaflet similar but larger. Racemes ca 10 cm long; flowers with white petals; pedicels curved, ca 1 cm long. Fruit broadly linear, 10-18 mm x 1.5-2.5 mm, more than 1.5 mm thick, often curved and torulose, 7-12 times as long as broad, with seeds arranged in two distinct rows. Seed shiny, red brown and covered by a distinct reticulum (Rahmansyah, 1993).

(iii) ***R. schlechteri***

Botanical Description

Watercress is the member of family Brassicaceae. It is a short day plant and does not flower under short day photo period conditions. Palaniswamy *et al.* (1997) reported that application of enhanced photosynthetic photon flux (PPF) before harvest increased phenethyl isothiocynate in watercress under 8 h photoperiod. Watercress grows in almost all types of soil having good water holding capacity and is rich in humus.

Table 11.16: Botanical Description

Growth Habit	Watercress grows abundantly in region of high altitudes. It is perennial plant, which stems spread on the ground. Plant thrives well in running water and marshy lands.
Root	Roots developed from the lower side of stems.
Stem	Stems are branched.
Leaf	Leaves are compound and leaflets are 7-11 in number. Leaflets are oval round and pointed towards apex.
Floral biology	Flowers are small and white.
Anthesis and pollination	Watercress is self-compatible plant.
Fruit and fruit set	Fruits are siliqua and roller like.
Seeds	Seeds are small round rough and bright.

Horticultural Classification

Botanically watercress can be grouped into:

(i) Nasturtium officinale

(ii) Rorippa nasturtium aquaticum

Augmentation of Genetic Resources and Crop Improvement

Water cress tetraploid strains, *R. microphylla*, have disease resistance quality but it cross with diploid resulting triploid which is sterile.

Genetic Resource Conservation

Nearly 200 genotypes have been conserved at Horticultural Research International, Wellesbourne, U.K. (Pink, 1993). Its genetic purity can be maintained through emasculation and bagging. Johnson (1974) proposed that progeny testing method is best for this crop. In England during 1970-80 the simple mass selection was used for development desirable cultivars (Royle, 1982).

REFERENCES

1. Gopalan, G., Rama Sastri, B.V., Balasubramanian, S.C., Rao, B.S.N., Deosthale, Y.G. and Pant, K.C. 1999. Nutritive value of Indian Foods. National Institute of Nutrition, ICMR, Hyderabad, India.
2. Howard, H.W. 1976. Watercress; In: Evolution of crop plants. Ed. Simmonde, N.W., Lnodon and New York, Longman, 62.
3. Johnson, A.G. 1974. Possibilities and problems in the breeding of watercress. In papers read at a symposium on research on the water cress crop, University of Bath, 1 October, 1974.
4. Manton, I. 1935. The cytological history of watercress (*Nasturtium officinale*). *Z. Indukt. Abstamm. 4. Verb Lehre*, **69:** 132.
5. Palaniswamy, U., Macavoy, R. and Bible, B. 1997. Supplemental light before harvest increases phenethyl isothiocynate in watercress under 8 hours photoperiod. *HortSci.*, **32(2):** 222-232.
6. Pink, D.A.C. 1993. Watercress; In: Breeding of vegetable crops (Kalloo, G. and Berg, B.O.). Oxford and Cambridge Pub. p. 579-283.
7. Rahmansyah, M. 1993. *Rorippa*. In: Prosea Plant Resources of South East Asia-8, Vegetables. Pudoc Scientific Publisher, Wageningen.
8. Royle, D.C. 1982. More cross strains sorted out. *The Grower*, **97:** 5
9. Thompson, H.C. and Kelly, W.C. 1957. Vegetable Crops, Mc Graw Hill, New York.
10. Wealth of India, 1972. Raw materials. Vol. IX. Publication and Information Directorate, CSIR, New Delhi.

Genus *Euryale*

Gorgon Fruit or Makhana (*Euryale ferox* Salisbery)

Introduction

Gorgon fruit or Fox nut (*Euryale ferox* Salisb., 2n=58) is very important aquatic plant bio-resources (Mandal and Saha, 2012). The edible part is seed which is popped. The valuable popped seeds of Gorgon fruit or Fox nut are roasted and eaten and used in various kinds of delicacies with milk like *Khir*, *Payasum* and *Puddings*. As vegetable, it is important as vegetable curry and *Dal Makhani*. It is a rich source of minerals especially Fe and P. Its starch content varies between 78 and 83 per cent and has protein content of 9.95 per cent (Dutta *et al.*, 1986). It content 41.6 mg alpha tocopherol (Vit. E), in per 100 g (Su, 1993). Nath and Chakraborty (1985) reported that defatted seed powder of makhana seed contains 7.315 mg amino acid per 100 g and contains 8.51 per cent protein in addition to all 16 types of amino acids. Its fruit has widely has been used in traditional oriental medicine to treat a variety of illness. Zhao *et al.* (1989) have identified structures of three glucosylsterols in a glycolid mixture of makhana which characterized as follows: 24-methylcholest-5 ethyl-3-beta-O-pyranaglucoside, 24-ethylcholest-5-ethyl-3- beta-O-pyranaglucoside, and 24 ethylcholest-5, 22E-dienyl-3beta-O-pyranaglucoside. This plant has several medicinal uses. The fruit is highly esteemed in China as a cooling tonic food. It is also used as a source of starch. All parts of the plant are considered to be astringent, deobstruent and tonic. The seed is analgesic and aphrodisiac. It is taken internally in the treatment of chronic diarrhoea, vaginal discharge and kidney weakness (Li *et al.*, 2007; PFAF 2010; Duke 2010).

Origin, History and Distribution

Gorgon fruit plant is native to Eastern Asia and is found from India to Korea and Japan as well as parts of Eastern Russia. It can be spotted in Bangladesh; China (Anhui, Beijing, Chongqing, Fujian, Guangdong, Guangxi, Guizhou, Hainan, Hebei, Heilongjiang, Henan, Hubei, Hunan, Jiangsu, Jiangxi, Jilin, Liaoning, Nei Mongol, Ningxia, Shaanxi, Shandong, Shanghai, Shanxi, Sichuan, Tianjin, Yunnan, Zhejiang); India (Assam, Bihar, Jammu-Kashmir, Manipur, Meghalaya, Rajasthan, Tripura, Uttar Pradesh, West Bengal); Japan (Honshu, Kyushu); Korea, Republic of; Myanmar (Myanmar (mainland)); Russian Federation (Primoryi); Taiwan, and Province of China. It was cultivated for centuries as food plant in India and is widely cultivated in the ponds and lakes in the states of Bihar, West Bengal and Assam in abundance and to some extent in Tripura, Manipur, Odisha, Uttar Pradesh, Madhya Pradesh, Rajasthan (Alwar) and Jammu and Kashmir (Jha *et al.*, 1991). Bihar state accounts for approximately 90 per cent of the total area under makhana in the country. Major makhana producing district in the state of Bihar are Darbhanga, Motihari, Sitamarhi, Madhubani, Samastipur, Saharsa,

Katihar and Purnia. Approximately 70 per cent of makhana's production comes from Darbhanga and Madhubani districts alone. Presently, it is estimated that under the Gorgon fruit crop total acerage is around 13,000 hectares with the annual production of 90, 000 tonnes seed which yields around 36, 000 tonnes pop (Jha *et al.*, 2007). *Euryale ferox* Salisb. (Nymphaeaceae) is a herbaceous aquatic species distributed from India to Russian Far East, including China, Taiwan, Japan, and Korea. In South Korea, *E. ferox* is protected by law as an endangered species due to threats of destruction and reduction of habitats.

Species Availability, Evolution and Diversity

The Plant list includes 7 scientific plant names of species rank for the genus *Euryale*. Of these 1 are accepted species names.

Botanical Description

Gorgon fruit is a member of family Nymphaeaceae. Its taxonomic hierarchy is described as under (Table 11.17):

Kingdom	:	Plantae
Sub Kingdom	:	Tacheobionta
Division	:	Magnoliophyta
Class	:	Magnodiopsida
Sub-class	:	Magnoliidae
Order	:	Nymphaeales
Family	:	Nymphaeaceae
Genus	:	*Euryale*
Species	:	*Euryale ferox* Salisb.

Table 11.17: Botanical Description

Growth Habit	It is an annual aquatic herb. It grows in unused tanks, lakes, ponds with stagnant and dirty water. The depth of the habitat varies from 1-2 m.
Root	The plant is deep rooted in the soil with the help of fleshy, thick felirous roots. Roots are long, fibrous thick and fleshy
Stem	Stem rhizome short, thick and erect.
Leaf	Leaf buds are folded up in involucres which burst when leaf expands. Leaf petiolate, simple, floating and large in size, round or oval in shape, measuring 30-120 cm in diameter. The upper surface of leaf is dark green and has a few purple spines. The lower lamina is covered with strong, straight and sharp spines. Leaf float on the surface of the water like lotus.
Floral biology	Flowering bud initiate during March-April, blooms afterwards and remains open for 2-4 days. Inflorescence, cymose type, flower solitary with long pedicel. Flower large 5 cm broad, violet, blue or red in colour, showy, pedicellate, hermaphrodite, actinomorphic, complete and epigynous and prickly outside, calyx reddish, has 20-30 petals which are shorter than the calyx lotus, and stamens are numerous.
Anthesis and pollination	In makhana, flowering is indeterminate and continuous from April to June. Flower opens during day.
Fruit and fruit set	Fruits are formed within 10-15 days of anthesis. Fruit spongy round and berries covered with stout prickles, size of an orange, oval or globular or sometimes irregular in shape due to swelling of the seeds inside. In general, a plant bears 10-20 fruits. Fruit remains sub-merged in water and contains 20-50 and sometimes 40-70 seeds each. When the fruits ripen, seeds with hard seed coat develop in thin membranes, separate out and settled down at the bottom of the ponds. The edible part is starchy endosperm,
Seeds	Seeds endospermic, cerelate, embryo is well developed with fleshy cotyledons. One mature fruit contains 30-40 black seeds, spherical in shape and characterized by its hard seed coat (shell). The seeds are of the size of cherry with diameter ranging from 4.5 to 14.5 mm. average weight of 100 seed is 80-100 g and approximately 400-700 g seeds are obtained from single plant. The yield of makhana seeds varies from 2.5 to 3.0 tonnes per hectare (Jha and Prasad, 1994).

Ripening of the fruit and rotting of pedicel, petiolate leaves and spongy parts of fruits starts during August, leaving the hard seeds, enveloped in thin membrane to settle at the bottom of the pond.

Augmentation of Genetic Resources and Crop Improvement

An experiment was conducted to study genetic divergence among 36 *makhana* genotypes using D^2-statistic at ICAR-RCER, Research Centre for Makhana, Darbhanga (Bihar). The clustering pattern indicated that cluster I and cluster II were the largest ones accommodating ten genotypes each. Cluster IV and VI were observed the smallest one with single genotype. The highest inter-cluster D^2 values was observed between cluster III and cluster VII (423.20) followed by cluster cluster III and cluster IV (379.69), indicating wider genetic diversity among the groups. The highest intra-cluster D^2 values were observed for the cluster VII (75.90) and the lowest in cluster IV and cluster VII (0.00). Among 14 traits studied, 100-seed weight was observed to have highest contribution (57.30 per cent) towards genetic divergence to followed by seeds/fruit (18.25 per cent) and fruit diameter (17.46 per cent). The results suggested that genotypes BR-5 of cluster III and BR-1 of cluster VII could be the most suitable parents for development of high yielding varieties of *makhana* through hybridization. (Kumar *et al.*, 2015).

Genetic Resource Conservation

In India it has important place in the region where wet lands are available like Mithila in Bihar (Jha *et al.*, 1994). Despite of all its rich medicinal, nutritional and religious importance, no concerted efforts have been made to collect, maintain and characterize its germplasm scattered in different parts of the country. A National

Table 11.18: Clustering Pattern of 36 Makhana Genotypes on the Basis of Genetic Divergence

Cluster No.	No. of Genotypes	Name of Genotypes Included	Source
I	10	BR-8, MN-23, MN-16, MN-9, MN-20, MN-17, MN-5, MN-12, BR-10, BR-4	Bihar, Manipur
II	10	MN-7, MN-18, BR-12, MN-19, BR-2, BR-3, BR-9, MN-21, BR-7, MN-10	Bihar, Manipur
III	5	MN-8, MN-22, BR-5, MN-15, MN-2	Bihar, Manipur
IV	1	MN-24	Manipur
V	4	BR-6, MN-3, MN-6, MN-13	Bihar, Manipur
VI	5	MN-1, MN-14, MN-11, MN-13, BR-11	Bihar, Manipur
VII	1	BR-1	Bihar

Research Centre for Makhana in Bihar was sanctioned by the Department of Agriculture Research and Education, Indian Council of Agricultural Research, New Delhi during the IX plan period (1997-2002). The major objectivize of this centre is to collect, maintain, identify and document the germplasm of makhana from different growing areas. Developing high yielding genotypes with high nutritive traits and processing quality is other desired attributes. To estimate the genetic diversity and present conservation strategy for the *E. ferox* populations in South Korea, we have collected 557 individuals from 37 sites across South Korea. Sequence variation of seven plastid regions – *psbA-trnH* IGS, *psbK-psbI* IGS, *atpF-atpH*IGS, *trnL* intron, *trnL-trnF* IGS, *rpoC1* gene, and *rpoB* gene – were examined for the representatives of 20 populations, of which the *psbA-trnH* IGS and the *psbK-psbI* IGS regions only showed variation. The *psbK-psbI* IGS region showed two variations, an A9/A10 homopolymer and an A/G substitution, which comprised three haplotypes: A9/A, A9/G, and A10/G. We sequenced all the 557 samples for the *psbK-psbI* IGS regions and obtained 311 (56.8 per cent) of A9/A, 20 (3.6 per cent) of A9/G, and 217 (39.6 per cent) of A10/G. Mostly, only one haplotype was observed per site, while two haplotypes co-occurred from eight sites. The A9/G-A10/G was observed from six sites, the A9/G-A10/G and the A9/A-A9/G from only one site, respectively. The A9/A haplotype dominated for 12 sites and the A10/G for 15 sites, respectively. The A9/G haplotype is the rarest of the three and occurred only in four sites along the western coast of South Korea. These results suggest that (1) there are at least three maternal founders for the *E. ferox* populations in Korea, (2) the lack of genetic diversity indicates that self-pollination is more common than cross-pollination (which is also indicated by the lack of chasmogamous flowers in the field), (3) those sites with the rarest haplotype A9/G have priority in conservation in Korea, and (4) further research using nuclear markers is necessary to unveil genetic structure of *E. ferox*.

REFERENCES

1. Dutta, R.N., Jha, S.N. and Jha, U.N. 1986. Plant contents and quality of makhana (*Euryale ferox*). *Plant and Soil*, **96(3):** 429-432.
2. Jha, B.K., Janardan Ji and Sikka, A.K. 2007. Makhana utpadan: Vipanan Evam Niryat. Smarika. Kshetriya Krishi Mela, Nov. 3-5, 2007 at Areraj, Bihar, p. 141-146.
3. Jha, V., Kargupta, A.N., Dutta, R.N., Jha, U.N., Mishra, R.K. and Sarwani, K.C. 1991. Utilization and conservation of *Euryale ferox* Salisb. In Mithila (North Bihar), India, *Aquatic Bot.*, **39:** 295-314.
4. Jha, V., Sarwani, K.C., Mishra, S. and Basak, P.K. 1994. Plants in Mithila rituals. Ethno-biology in human welfare: Abstract of the fourth International congress of ethno-biology, Lucknow, Uttar Pradesh, India, 17-21 Nov. 1994, p. 192.
5. Jha, S.N. and Prasad, S. 1994. Gorgon fruit or Makhana-its cultivation and processing. *Indian Horticulture*, 39(2): 18-20.
6. Kumar, L., Choudhary, A.K., Bhatt, B.P. and Singh, K.P. 2015. Genetic divergence in makhana (*Euryale ferox* Salisb.). *Indian J. Hort.*, **72(3):** 365-369.
7. Li, X.M; Ma, Y.L; and Liv, X.J., 2007. Effect of the *Lycium barbarum* polysaccharides on age related oxidative stress in aged mice. Journal of Ethanopharmacology, 111: 504-511.
8. Mandal, R.N. and Saha, G.S. 2012. Producing Makhana (*Euryale ferox* Salisb.) for its diverse benefits to depressed economic regions of India. *Water Garden Journal.*, **27:** 13-16.
9. Nath, B.K. and Chakraborty, A.K. 1985. Studies on amino acid composition of the seed of *Euryale ferox* Salisb. *Journal of Food Science and Technology*, India, **2(4):** 293.
10. Su, J.D. 1993. Studies of rapid determination of vitamin E and its application of natural food stuffs. *Journal of Food and Drug Analysis*, **1(1):** 61-70.
11. Zhao, H.R., Zhao, S.X., Sun, C.Q. and Guillaume. 1989. Glucosylsterols in extracts of *Euryale ferox* identified by high resolution NMR and mass spectrometry. *Journal of Lipid Research*, **30:** 1633-1637.

Genus *Neptunia*

Water Mimosa (*Neptunia oleracea* Lour.)

Introduction

Water mimosa or sensitive neptunia (*Neptunia oleracea* Lour.; 2n=56) is lesser known nutritive aquatic or semi-aquatic pantropical nitrogen-fixing perennial legumeleafy vegetable distributed in different wet lands and contribute a lot in the dietary nutrition of different tribals and other ethnic communities. It is also known as Puff, small leaf water sensitive plant, dwarf sensitive plant, neptunia, water dead and awake plant. It is called as Kharaishak by the rural peoples of Tripura, Ekaithibi to the Manipuri and Panilajuk to the Bengali communities. It is a seasonal vegetable producing young, soft shoots during May to August every year in NEH regions Young shoots of the species are used as a source of shoot vegetables. The shoots are commonly eaten raw, cooked or stir-fried in Thailand. In some area young leaves and young pods are also eaten. Water mimosa is rich in minerals and vitamin A contents. Per 100 g of edible parts contain moisture-89.4 g, protein-6.4 g, fat-0.4 g, carbohydrates-0.8 g, fiber-1.8 g, ash-1.2 g, calcium-387 mg, phosphorus-7 mg, iron-5.3 mg, vitamin A-5155IU, vitamin B_1-0.12 mg, vitamin $B_2$0.12 mg, niacin-3.2 mg, vitamin C-1.8 mg and energy-134kj (Paisooksantivatana, 1993). As medicinal properties it contain six chlorophyll related compounds in leaves which inhibit the activation of tumor promoter induced Epstein-Barr Virus (EBV). Photo-irradiation of phenophorbide a, the major active constituent, did not enhance this inhibitory activity, and thus ruled out any photosensitizing effect playing the major role (Nakamura *et al.*, 1996). Water mimosa contains 13.1 mg butylhydroxyanisol (BHA) equivalent/g dry weight, showing antioxidant activity. Fractionated extracts from water mimosa showed antioxidant activities (Thalang *et al.*, 2001). Besides leaves, pod of water mimosa possesses laxative property. Bhoomannavar *et al.* (2004) reported that alcoholic extract of water mimosa (800 mg/kg, p.o.) significantly increased the peristaltic movement, which was comparable to the standard senna glycosides (100 mg/kg, p.o.). This indicates the presence of anthraquinones, which may be responsible for the laxative activity. While evaluating the anti-inflammatory and anti-bacterial properties of stem of water mimosa, Lakshmayya *et al.* (2002) reported that alcoholic extracts of stem contained flavonoids, triterpenes and carbohydrates. Whereas acetone and ethyl acetate extracts showed the presence of flavonoids and the aqueous extracts demonstrated the presence of carbohydrates and saponins. All the extracts containing flavonoids were evaluated for antibacterial activity against *Staphylococcus aureus*, *Escherichia coli* and *Pseudomonas aeruginosa* and anti-inflammatory activity using formalin-induced rat hind paw oedema. None of the extracts showed any anti-inflammatory activity against the organisms tested. However, alcoholic extract at 200 mg/kg showed anti-inflammatory activity (7.3 per cent reduction in oedema volume at the 4th h), which is comparable to that of standard dicofenac sodium at 30 mg/kg (65 per cent reduction in oedema volume at the 4th h).

Water mimosa is an aquatic, nitrogen-fixing legume that releases nitrogen into water bodies. This can lead to increased algal blooms and encourage growth of other weeds such as water hyacinth, water lettuce and salvinia. Excess nitrogen also affects water quality and increases water-treatment costs. Floating aquatic plant stems often form thick foliage mats and is considered to be an invasive aquatic weed in some tropical waters where large mats may form that choke waterways, resulting in restricted water flow, reduced water quality, reduced fish activity and loss of some underwater and native wetland plants (Hannan and Csurhes, 2008).

Origin, History and Distribution

The exact origin of *N. oleracea* is unclear; however, it is generally accepted as being native to tropical Asia, Africa and South America (Windler, 1966). It grows wild and is cultivated as a vegetable throughout South-East Asia, particularly Thailand and Indo-China (Paisooksantivatana 1993). *N. plena* occur in the coastal regions of southern North America, Central America, northern South America, and tropical Asia (Windler 1966). It is not known when *N. oleracea* first arrived

Table 11.19: Biochemical Constituents of Fresh and Cooked Water Mimosa

Sample		*Biochemical Constituents (mg./g fresh weight); *Mean ± SD*			
Fresh		*Protein*	*Phenol*	*Sugar*	*Total Free Amino Acids*
	Twig	19.96 ± 0.17	0.12 ± 0.00	1.97 ± 0.05	66.16 ± 2.44
	Stem	20.27 ± 0.15	3.71 ± 0.08	1.96 ± 0.03	69.72 ± 1.53
	Leaf	10.78 ± 0.11	3.36 ± 0.02	2.10 ± 0.03	185.86 ± 1.93
Cooked	Twig	13.24 ± 1.47	0.79 ± 0.15	4.26 ± 0.09	36.36 ± 1.56
	Stem	13.50 ± 0.75	1.08 ± 0.16	5.21 ± 0.03	29.23 ± 1.09
	Leaf	9.63 ± 0.46	0.94 ± 0.13	4.15 ± 0.13	83.62 ± 1.66
	Curry juice	5.42 ± 0.27	0.72 ± 0.12	6.30 ± 0.10	10.78 ± 4.89

Source: Chakraborty and Sinha, 2011.

in Queensland, but the first record is possibly from a glasshouse in the Brisbane Botanic Gardens in 1979. The Queensland Herbarium has a record of a cultivated specimen of *N. plena* collected in 1964 from the Moreton district of Queensland. *N. oleraceae* is currently being used by some South-East Asian communities as a vegetable and is being sold in local farmers' markets in the Brisbane area.

Species Availability, Evolution and Diversity

Eleven species are in the genus with three aquatic. Commonly cultivated species are *Neptunia oleracea* lour. (*N. aquatic; N. prostrata* (Lam.) Baill.) (Africa, Asia, mexico, South America).

Synonym

(i) *Desmanthus stolonifera* D.C.

(ii) *Mimosa natans* L.F.

(iii) *Neptunia stolonifera* (D.C.) Guill. and Perr.

(iv) *Neptunia prostrata* (Lam.) Baill.)

(v) *Mimosa prostrata* Lam.

Botanical Description

Taxonomic Position

Kingdom	:	Plantae
Division	:	Magnoliphyta
Class	:	Magnolipsida
Order	:	Fabales
Family	.	Fabaceae
Sub-family	:	Mimosoideae
Tribe	:	Mimoseae
Genus	:	*Neptunia*
Species	:	*oleracea*
Bionomial name	:	*Neptunia oleracea* Lour.

Aerenchyma (white spongy air-conducting tissue that gives stems buoyancy forms on stems floating in water, but does not form on stems growing on land.

Table 11.20: Botanical Description

Growth Habit	Aquatic herb (5-6 feet tall), primarily found growing prostrate in wet soils near the water's edge or floating on the water in relatively still-water areas. A fast-growing plant, under favourable conditions young shoots may elongate at a rate of 5 - 7cm per day.
Root	Water mimosa takes root to the water courses. Tap root thick and becoming woody.
Stem	Stems 1.5- 2 m tall will spread in the water to 3-5' long. Creeping stems usually floating with the means of spongy aerenchyma. The sieve tube elements in the stem are around 190µ long and 13µ wide and have compound sieve plates at varying degrees of inclination. The developing sieve tube element has a single large spindle-shaped slime body, which presumably has an outer membrane. The slime body undergoes dispersal before or after full development of the sieve plate, but often nuclear degeneration occurs first. Distinct slime plugs are absent. Plastids and other granular bodies are attached to many of the strands, which are less than 0.5µ in diameter.
Leaf	Primary leaf segments have 8-40 small oblong leaflets arranged in opposite pairs. Leaves alternate, bipinnate, sensitive; stipules present but not apparent on floating stems; stipules obliquely ovate, 5-9 mm long, 3-5 mm wide, thin; petiole 2.5-9.0 cm long; rachis 1.1-4.2 (-6.5) cm long; pinnae 2-4 pairs; leaflets 7-22 pairs, oblong, 5-20 mm long, 1.5-4.0 mm wide, basal ones smaller, glabrous or with a few hairs on margins.
Floral biology	Tiny greenish-yellow flowers are densely crowded into feathery orbicular inflorescences. Inflorescence a spherical in heads 1.5-2.5 cm long; peduncles 6.5-23 (-30) cm long on an elongate peduncle. Flowers are actinomorphic, dimorphic: upper flowers bisexual lower male and often sterile; Calyx 1-3 mm long tubular, campanulate, lobes 5; petals 5, about 3-4 mm long, and yellow. Stamens 10; anthers eglandular at apex, even in bud and staminodes up to 17-21 mm long. Flowers bloom in summer.
Pod and pod set	Pods bent at an angle to the short basal stipe, shortly oblong, 1.3-2.7 (-3.8) cm long, and 1.0-1.2 cm wide.
Seeds	Seeds 5.0-5.5 mm long and 3.0-3.5 mm wide. New plant development through dispersal of seeds (Brenan, 1959).

Propagation occurred simply by dethatching pieces of stem that have already rooted into the water.

Genetic Resource Conservation

No improved cultivars are available and only local selections are grown. Despite of its rich genetic wealth in most of the South-East Asian countries, no concerted efforts have been made for collection, characterization and maintenance of germplasm. Efforts may be zeroed in to screen the genotypes having high yield potential along with phyto-chemical of nutritional and medicinal importance.

REFERENCES

1. Bhoomannavar, V.S., Hatapakki, B.C., Kumar, V.H. Setty, S.R. and Suresh, H.M. 2004. Laxative activity of pods of *Neptunia oleracea* in mice. *Indian Journal of Natural Products*, **20(1):** 43-45.

2. Chakraborty, S. and Sinha, R. 2011. *Neptunia prostrata* (Lamk.) Baillon: A socio-economic potential health diet for rural communities of Tripura. *VEGETOS*, **24(1):** 65-70.

3. Hannan-Jones, M. and Csurhes, S. 2008. Pest plant risk assessment: Water mimosa (*Neptunia oleracea*), Dead and Awake. *Neptunia plena*. Department of Primary Industries and Fisheries, Greensland p.14.

4. Lakshmayya, Joshi, N.H., Rao, N., Kumar, P., Agarwal, N. and Setty, S.R. 2002. Preliminary phytochemical, anti-inflammatory and anti-bacterial evaluation f stem of *Neptunia oleracea*. *Indian Journal of Natural Products*, **18(1):** 18-20.

5. Nakamura, Y., murakami, A., Koshimizu, K. and Ohigashi, H. 1996. Identification of pheophorbide and its related compounds as possible anti-tumor promoters in the leaves of *Neptunia oleracea*. *Bioscience, Biotechnology and Biochemistry*, **60(6):** 1028-1030.

6. Paisooksantivatana, Y. 1993. *Neptunia oleracea* Loureiro. In: Plant Resources of South East Asia-8. Vegetables. Siemonsma, J.S. and Piluek, K. (eds.), Pudoc Scientific Publishers, Wageningen.

7. Thalang, V.N., Trakoontivakorn, G., Nakahara, K. and Gassinee, T. 2001. Determination of antioxidant activity of some commonly consumed leafy vegetables in Thailand. *JIRCAS Journal*, **9:** 39-46.

7. Windler, D.R. (1966), 'A revision of the Genus *Neptunia* (Leguminosae)', *Australian Journal of Botany*, **14:** 379–420.

Genus *Hydrolea*

Hydrolea [*Hydrolea zeylanica* (L.) Vahl.]

Introduction

Hydrolea [*Hydrolea zeylanica* (L.) Vahl., 2n=18, 24] is nutritionally very important aquatic plant. Its young shoots are eaten as a vegetable in some parts of Tamil Nadu. It is rich in protein, carbohydrate and calcium. Per 100 g of edible portion contains 80 g moisture, 65 g protein, 0.8 g fat, 3.9 g minerals, 8.8 g carbohydrates, 68 kcal energy, 778 mg calcium, 100 mg phosphorus and 31.6 mg iron (Gopalan *et al.*, 2004). Plant possesses medicinal properties and leaves are used as a poultice for ulcers. In Cambodia as a traditional medicine, the leaves are prescribed for intestinal disorders. Its leaves possess soothing and healing properties.

Origin, History and Distribution

Plants are distributed in Indian continent, South-East Asia, Malaysia to tropical Australia.

Species Availability, Evolution and Diversity

The family includes only 11-20 species of the genus: *Hydrolea* L.:

1. *Hydrolea brevistyla* Verdc.
2. *Hydrolea corymbosa* JF Macbr. Ex. Ell.
3. *Hydrolea elatior* Schott
4. *Hydrolea elegans* AW Benn.
5. *Hydrolea floribunda* Kotschy and Peyre.
6. *Hydrolea glabra* Schumacher. and Thonn.
7. *Hydrolea macrosepala* AW Benn.
8. *Hydrolea megapotamica* Booby.
9. *Hydrolea multiflora* Mart.
10. *Hydrolea prostrata* Excell
11. *Hydrolea ovata* Nutt.
12. *Hydrolea paludosa* AW Benn.
13. *Hydrolea sansibarica* Gilge
14. *Hydrolea spinosa* L.
15. *Hydrolea uniflorae* Raf.
16. *Hydrolea urens*
17. *Hydrolea zeylanica* (L.) Vahl

Botanical Description

Other synonyms: *Nama zeylanica* L., Beloanthera opposittifolia Hasskarl; *Hydrolea arayatensis* Blanco; *H. inermis* Loureiro; *H. javanica* Blume; *H. zeylanica* (L.) Vahl var. *ciliata* Choisy; *Steris aquatica* N. L. Burman; *S. javanica* L. Taxonomic position is as under:

Division	:	Magnoliophyta
Class	:	Rosopsida
Sub-category	:	Asteridae
Rules	:	Solanales
Family	:	Hydroleaceae
Genus	:	*Hydrolea* BERCHT. and J. PRESL

Table 11.21: Botanical Description

Growth Habit	Herbs, erect to prostrate,
Root	Sometimes rooting from lower nodes, 10-60 (-100) cm tall
Stem	Often branched, glabrous or pubescent, and unarmed.
Leaf	Leaf blade lanceolate to ovate, 2-10 x 0.5-2.5 crn, glabrous or pubescent, base acute, and margin entire, apex acute.
Floral biology	Flowers in terminal panicles or clusters at branch apices, rarely solitary. Pedicel 1-3 mm, elongating after anthesis. Calyx lobes lanceolate, 4.5-8.0 mm, pubescent or glabrous. Corolla purple blue or deep purple green, 3-5 mm; tube white. Anthers scarlet. Capsule ovoid, skeptical or irregularly dehiscent, included in persistent calyx.
Seeds	Seeds oblong ovoid, 0.3-0.4 mm.

REFERENCES

1. Gopalan, C., Rama Sastri, B.V. and Balasubramanian, S.C. 2004. Nutritive value of Indian foods. National Institute of Nutrition, ICMR, Hyderabad, India.

Genus *Limnocharis*

Velvet Leaf [*Limnocharis flava* (L.) Buchenau]

Introduction

Velvet leaf [*Limnocharis flava* (L.) Buchenau; *L. laforestii* 2n=20 but *H. nymphoides*, 2n=16] is known with the several vernacular names like English-limnocharis, sawah flower rush, sawah lettuce, velvetleaf, yellow bur head, Spanish-cebolla de chucho. Basically it is a beautiful ornamental plant but in some South-East Asian countries velvetleaf is used as a source of food. In Java, the plant is a much esteemed vegetable, which is cultivated in rice fields and sold in markets for its young leaves, stems and flower clusters (Abhilash, 2004). In Vietnam, velvetleaf occurs naturally in many water bodies and is used as a vegetable in dishes. It is particularly popular in the Mekong Delta where it is known as *keo neo* or *cu neo*. It is frequently gathered by ethnic Vietnamese women and sold in local markets and by boat vegetable vendors. As a preparation, the young leaves and tops of the plant are boiled or cooked in mixed soups. The leaves of velvetleaf are also collected for household consumption in Bangladesh by women. The vegetable contains high amount of ca, Fe and β-carotene. Per 100 g of edible portion contains 1 g protein, 0.3 g fat, 0.5 g carbohydrates, and 5000 IU vitamin A.

Origin, History and Distribution

Velvetleaf is a native of tropical America. The earliest record of its introduction in the east dates back to 1866, when it was recorded to have been cultivated in the Botanic Garden at Bogor. In 1870, it was reported from banks of river Tjiliwung near Djakarta, which flows through the Botanic Garden at Bogor. It was introduced in Thailand in 1909, from where it reached Southern Myanmar. Much before, it got established in Eastern Asia. In 1989, it was introduced in Sri Lanka as an ornamental in the Botanic Gardens, 'Pera Deniya'. The native range of velvetleaf extends from Mexico to Bolivia, Paraguay and northern Argentina. As introduced species for ornamental use has been documented in the USA, Singapore, Sri Lanka and Indonesia (Abhilash, 2004). From its native range, *Limnocharis* has moved into parts of South East Asia, Africa and the south Pacific and recently it has entered in northern Australia. In India, the occurrence of this species was first reported from Kerala (Kamathy and Subramaniyam, 1967; Bahadur and Raizada, 1968).

Species Availability, Evolution and Diversity

The family Limnocharitaceae comprises of 3 genera of herbaceous aquatic plants:

i. *Limnocharis* and *Hydrocleys*- Naturally found in Neotropical regions

ii. Butomopsis- Naturally found in Paleotropical regions

Botanical Description

Taxonomic position of the *Limnocharis flava* (L.) Buchenau is as:

Domain	:	Eukaryota
Kingdom	:	Plantae
Sub- Kingdom	:	Viridaeplantae
Phylum	:	Tracheophyta
Sub- Phylum	:	Euphyllophytina
Infraphylum	:	Radiatopses
Class	:	Liliopsida
Sub- Class	:	Alismatidae
Sperorder	:	Alismatanae
Order	:	Alismatales
Family	:	Limnocharitaceae
Genus	:	*Limnocharis*
Specific epithet	:	*flava*
Variety	:	*minor*

Table 11.22: Botanical Description

Growth Habit	It is a glabrous, scapigerous and lactiferous plant. It has a short stout rhizome about 3 cm long and 3 cm in diameter
Root	It has numerous fibrous roots.
Leaf	The leaves are erect, sheathing and curvinerved, and rise in clusters above the water. The petioles are thick, triangular, 90 cm long and contain numerous lacunae (air spaces). The leaf sheaths narrow towards the top and the leaf blade is papery, light green, variable in shape rounded, ovate or broad elliptic and 6-20 cm long and almost as equal wide. The leaf apex is enerally apiculate with a small hydathode at the tip; the leaf base is cuneate to repandcordate; the leaf margin is undulated. Four to six of main nerves and a marginal one sub parallel and converging towards apices, numerous secondary nerves parallel to each other and nearly perpendicular to the midrib and dense fine reticulates

Floral biology	*Limnocharis flava* (L.) Buch. One to four peduncles which are axillary, erect, triangular, flattened at the base and 120 cm long. Peduncles bear, at the apex, a cluster of bisexual flowers 2-13, a ramet or both and normally produced flowers in a cincinnus (seorpioid cyme). Inflorescences are umbellate (consisting of flowers springing from a common centre and arranged in a cluster), with between two to 15 flowers supported by an involucre of bracts; sometimes one or two leaves are present between the flowers. The bracts are roundish to broad-elliptic with (usually) emarginated tops. The flowers are borne in an indeterminate umbel and each consists of 3 sepals, 3 yellow petals, and about 18 carpels surrounding by numerous stamens and staminodia. The androecium is centrifugally developed, and the last-formed members are staminodial; it is supplied by branching vascular systems. Carpels arise almost simultaneously, and a prominent residual flora apex remains. The carpels are partially conduplicately closed and are also primitive in possessing laminar placentation and in lacking differentiation of a style. The gynoecium is essentialy apocarpous, but there are slight fusions of adjacent carpels near their ventral margins where they are attached to the receptacle. It is suggested that the *Limnocharis* flower is the most primitive in the family (Kaul, 1967).
Anthesis and pollination	The anthesis of each flower lasted 5-13 hours. In natural populations, two species of beetles have been observed visiting the flowers in bloom. Several whorls of staminodes out-side of the stamens limits the pollen to disperse outward, and made the pollen deposit only on the stigma of the same flower. After the perianth closed, the inner perianth changed into liquid which created a suitable environment for pollen germination. The pollen-ovule ratio, which ranges from 650 to 1120 (930±180) reflects an economical self-pollination system. As the fruits matures, the inflorescence drop to ground or mud and the vegetative buds (*i.e.* vegetative components), developed at the apex of the inflorescence, grow into new vegetative shoots. One mother plant usally could produce 6-20 new vegetative shoots in a life cycle.
Seeds	One fruit of *L. flava* contains 470-640 seeds. Average seed set rate of about 80 percent. The seeds of *Limnocharis flava* could disperse a long distance and occupy new suitable habitats to establish new populations.

In order to analyze the karyotypic characteristics of the neotropical genera, root tips of *L. flava*, *L. laforestii* and *H. nymphoides* were cytologically prepared. The average karyotypes were constructed based on 10 metaphasic cells for each species. In all species, the chromosome length varied gradually and the highest quantity of chromatic material was found in *H. nymphoides* (total chromatin length, TCL=58.6 μm), as compared with *L. flava*, *L. laforestii* (TCL=51.0 and 41.9 μm, respectively). About 75 per cent of the chromosomes were acrocentric and all three species presented asymmetric karyotypes, with TF per cent rates equal to 17.8 per cent (*L. flava*), 15.3 per cent (*L. lafo*) and 19.7 per cent (*H. nymphoides*). Taking karyotypic asymmetry as a derivative character, it is concluded that, in the family Limnocharitaceae, the derivation may have been accompanied by an increase in reproductive elements (mainly carpels) and chromosomes number. Chromosome fragmentation would have led to both higher chromosome numbers and karyotypic asymmetry (Martins and Calligaris, 2002).

Augmentation of Genetic Resources and Crop Improvement

Some artificial treatments, such as emasculating and netting, emasculating and bagging, hand-pollination, netting and bagging, indicate that *L. flava* is selfing in general but also cross-compatible (Ping *et al.*, 2000).

Genetic Resource Conservation

Limnocharis flava is considered as one of the obnoxious weeds in several parts of the world and its use as a vegetable is restricted. Therefore no efforts have been made to collect, evaluate and characterize the available ecotypes. Efforts are needed to validate its nutritional, chemical and medicinal properties.

REFERENCES

1. Abhilash, 2004. Australian Weed Management. CRC.
2. Bahadur, K.N. and Raizada, M.B. 1968. *Limnocharis flava* (L) Bucharnauce=a new record for India Indane for 94, p. 641-644.
3. Kamathy, R.V. and Subramaniyam, K.J. 1967. *Bombay Nat. Hist. Soc.*, **64:** 389-390.
4. Kaul, R.B. 1967. Ontogeny and anatomy of the flower of *Limnocharis flava* (Butomaceae). *American Journal of Botany*, **54(10):** 1223-1230.
5. Martins, E.R.F. and Calligaris, K.P. 2002. Chromosomal studies on Neotropical Limnocharitaceae (Alismatales). *Aquatic Botany*, **74(1):** 33-41.
6. Ping, S.Z., Hao, G.Y. and Quan, H.S. 2000. Studies on the breeding system of *Limnocharis flava* (Butomaceae). *Acta Phytotaxonomica SinAlgaeica*, **38(1):** 53-59.

12

Genetic Resources of Rhizome Vegetables and Crop Improvement

Introduction

Zingiberaceae is major family of spices like small cardamom (*Elettaria cardamomum*), large cardamom (*Amomum subulatum*), Zinger (*Zingiber offinale*), Turmeric (*Curcuma domestica*), mango ginger (*Curcuma amada*), Alpinia (*Alpinia calcarata*), Greater Galangal, 'Lesser Galangal', Grain of Paradise, *etc.*, are of considerable economic importance. These are basically used in pharmaceutical and cosmetic industries. Alpinia oil has uses in perfumery, pharmaceutical, food and flavouring industries. In tribal areas plant parts of some Zingiberaceous species are used as carminative and an antidote to snakebites. Pharmaceutical studies show that plants parts of this family like ginger rhizomes are effective for intestinal disorder (Krishnamurty and Sreenivasamurthy, 1956) and salivary secretion for stimulating the vasometer and respiratory centres (Alley, 1961) for relaxing the tracheae and ileal smooth muscles (Reiter and Brandt, 1985) and for lowering serum and hepatic cholesterol levels (Gujral *et al.*, 1978). Recent investigations also indicate that pungent principals of ginger rhizomes, such as gingerols and shagaols, have pharmacological actions (Shoji *et al.*, 1982; Suekawa *et al.*, 1986). In addition, antioxygenic effect of the rhizomes on food has been found (Du Bios and Tressler, 1943; Sethi and Aggarwal, 1952; Chipault *et al;* 1955, 1956; Kihara and Ingue, 1962; Fujio, 1969). These spices are mainly propagated through rhizomes, thereby preserving the diversity on the one hand and restricting to some extent the evolution of new types. However, a large scale encroachments on forest lands and spreads of new varieties, are posing some threat to the diversity of these groups of crops. However, efforts are going on at various organizations in the country to conserve the biodiversity of these spices and also to utilize them.

Table 12.1: Some Important Rhizome Bearing Crops

Common Name	*Botanical Name*
Turmeric	*Curcuma longa*
Ginger	*Curcuma amada*
Ginger/Galangal (Lesser Galangal; Greater Galangal)	*Zingiber officinale*
Ginseng	*Panax ginseng*
Arrowroot	*Maranta arundinacea*
Canna	*Canna spp.*
Ti	*Cordyline fruticosa*
Cattail or bulrush	*Typha spp.*
Alpinia	*Alpinia calcarata*
Cardamom	*Elettaria cardamomum*
Lotus root	*Nelumbo nucifera*
Rengarenga, vanilla lily, and others	*Arthropodium spp.*

REFERENCES

1. Alley, M.M. 1961. *Proc. Pan Indian Ocean Sc. Congr.* Karachi, Pakistan 1960, 4th Section GPP11.

2. Chipault, J.R. Mizuna, G.R; Hawkins, J.M. and Lundberg, W.O. The antioxidant properties of spices in oil-in-water emulsions 1955. *Foof Res.*, **20:** 443-448.

3. Chipault J.R. Mizuna, G.R; Hawkins, J.M. and Lundberg, W.O. The antioxidant properties of spices in foods 1956. *Foof Res.*, **10:** 209-211.

4. Du Bios, C.W. and Tressler, D.K. 1943. *Proc. Inst. Food Technol.*, 202-207.

5. Fujio, H. 1969. *Nippon Shokulin Kogyo Gakki-Shi*, **16:** 241-246.

6. Gujral, S., Bhumra, H. and Swaroop, M. 1978. *Nutr. Rep. Int.*, **17:** 183-189.

7. Kihara, Y. and Ingue, T. 1962. *Nippon Shokulin Kogyo Gakki-Shi*, **9:** 290-293.

8. Krishnamurty, K. and Sreenivasamurthy, V. 1956. Bull. Centrl. Food Technol. Res. Inst. Mysore, **5:** 264-469.

9. Reiter, M. and Brandt, W. 1985. *Arzneium Ferch.*, **35:** 408-414.

10. Sethi, S.C. and Aggarwal, J.S. 1952. *J. Sci. Ind. Res.*, **B11:** 468-470.

11. Shoji, N.A., Iwasa, Takemoto, T., Ishida, Y., and Ohizumi, Y.1982. *J. Phakra Sci.*, **71:** 1174-1175.

12. Suekawa *et al.*, 1986. A *Nippon Yakurigaku Zaeshi*, **38:** 339-347.

Ginger (*Zingiber officinale* Rosc.)

Introduction

Ginger is used as fresh and after dry as warming spice. In Indian *Ayurvedic* medicine, ginger is used as an anti-inflammatory herb. Ginger has been under cultivation in India since times immemorial though the plant does not occur in a truly wild state and rarely flowers, though the cultivated plants on the West Coast of India flower quire frequently in October. Ginger has a long and well-documented history of both culinary and medicinal value and today ginger is grown throughout the tropics as a valuable commodity.

Origin, History and Distribution

Ginger (*Zingiber officinale* Rosc.) is native to India and China, originated in Southeast Asia but it has a long history of being cultivated in other countries, too (Purseglove, 1973). Incidentally, the centre of diversity of ginger is also in this region. It takes its name from Sanskrit word "singabera", which means "with a body like a horn" as in antlers. Ginger is mentioned in the writings of Confucius. It is also named in the Koran, the sacred book of the Muslims, indicating it was known in Arab countries as far back as 650 A.D. The history of Ginger goes back over 5,000 years when the Indians and ancient Chinese considered it a tonic root for all ailments. It is used used throughout the world, especially in Chinese (400 B.C.), Indian and Japanese medicinal care both in dried and fresh form. The Hindu epic 'Mahabharta' describes a meal where meat was stewed with ginger and other spices. It was one of the earliest spices known in Western Europe and used since the ninth century. It became so popular in Europe that it was included in every table setting, like salt and pepper. At an early date it was exported to Ancient Rome from India. It was used extensively by the Romans, but almost disappeared from the pantry when the Roman Empire fell. After the end of the Roman Empire, the Arabs took control of the spice trade from the east. Ginger became quite costly like many other spices. In medieval times it was commonly imported in a preserved form and used to make sweets. In the 15th century, ginger plants were carried on ships which are probably how they were introduced to the Caribbean as well as Africa. It was introduced into Jamaica and other islands of the West Indies by the Spaniards and ginger was exported from the West Indies to Spain in considerable quantity even during the year 1547 A.D. Ginger is well distributed and widely cultivated in many countries such as India, China, Nigeria, Fiji, Mauritius, Malaysia, Bangladesh, Taiwan, Thailand, Trinidad, Tobago Jamaica, Brazil, costa Rica, Ghana, Nigeria, and Sierra Leone, Japan, Indonesia, Australia, the Philippines, Sri Lanka, Soloman Islands, Thailand and West Indies islands. Presently in China, ginger is grown in about 13450 ha. Ginger is cultivated in about 14500 ha in Nigeria. Ginger is grown in about 9900 ha in Indonesia. The early movement of settlers across the length and breadth of Kerala and the story of shifting cultivation in north eastern India are well documented sociological events (Sasikumar *et al.*, 1999).

Ginger is grown throughout India but major states producing it are Kerala, Odisha, Karnataka, Arunachal Pradesh, West Bengal, Sikkim, Meghalaya, Manipur, Arunachal Pradesh, Tripura (North East States), Himachal Pradesh, Andhra Pradesh, Madhya Pradesh *etc.* are the leading Indian states in ginger production totaling to 70000 ha (Nair, 1974) but increase upto 10,0000 ha in 2018. Kerala alone, accounting for more than 33 per cent of the total ginger production in the country. However according to Rehman *et al.* (2009) ginger is main cash crop of north east region accounting for 49 per cent of India's ginger area and 72 per cent India's ginger production with highest productivity of 840.0 q/ha in Mizoram (Parthasarathy *et al.*, 2007).

Species Availability, Evolution and Diversity

Ginger is the one of 1,400 species of the Zingiberaceae family (natural order Scitamineae); there are more than 50 varieties of ginger in India alone. Baker (1894) described a total of 24 species from the Indo-Malayan region. Gamble (1925) recorded 7 species from South India as *Z. officinale* Rosc., *Z. roseum* Rosc., *Z. nimmonii* Dalz., *Z. wightianum* Thwaites, *Z. zerumbet* (Lin.) Smith,

Z. neesanum (Grah.) Ramamoorthy Syn. *Z. pupurem* Rosc. Kumar (1999) reported that *Z. clarkei* is endemic to Sikkim Himalayas. The other economically important (medicinal) *Zingiber* species found in Sikkim are *Z. capitatum* var. *elatum*, *Z. chrysanthum etc.* Kumar (1999). A putative wild *Zingiber officinale* is collected from Western Ghats of India; its occurrence in wild habitat is yet to be confirmed. Though *Zingiber officinale* Rosc., is the most economically important member of the genus *Zingiber* and there are other species of medicinal and ornamental value (Sabu and Dave, 2005).

Out of fourteen species of *Zingiber* reported to occur in India, only the certain species are of economic importance are showing in Table 12.3.

The rhizomes of *Z. elatum* Roxb. and *Z. chrysanthum* Rosc. wild herbs growing in the Tarai tract of Uttarakhand ; are used sometimes for the extraction of volatile oil. A type of ginger having rhizomes with bluish black tinge inside called black ginger is reported to have medicinal properties and is grown by the inhabitants of Mizoram just for their own use. It is also said to be sold at very high price probably due to high medicinal value. The tropical rai forest of Andaman and Nicobar Islans harboring 4 species of medicinal and aromatic wild gingers namely *Z. odoriferum* Bl., *Z. spectabile* Grioff, *Z. squarrosum* Roxb., and *Z. zerumbet* Rosc. ex Sm., whichare generally found growing in shady humid places. Most of these wild gingers possess an eye-catching colourful bract in ovoid conical or cylindrical spikes and are exhilarating the horticulturists to grow them as ornamental and decorative herbs in the gardens. These herbs are also being used for therapeutical purpose like that of the common, well-known ginger *Z. officinale* Rosc., a medicinal and a flavouring agent.

Table 12.2: Diversity in Economically Important Zingiber Species

Sl.No.	*Species*	*Use*	*Remark*
1	*Zingiber officinale* Rosc.	Spice and condiment, and medicine	Cultivation in India, Africa, Caribbean's and South-East Asian countries
2	*Z. mioga* Rosc.	Vegetable. Shoot and flower are edible	Japanese or 'Myoga ginger'
3	*Z. montanum* (Keonig) Link ex Diet (Syn. *Z. cassumunar Roxb.*)	Used in traditional medicine	Found in India, Sri Lanka and South-East Asian countries
4	*Z. zerumbet* (L.) Smith	Medicinal/ornamental use	'Shampoo ginger', common in tropical Asia and the USA
5	*Z. americanus* Bl.	Medicinal use	Grown widely in USA. Occur in the teak forests of and South-East Asia
6	*Z. aromaticum* Val.	Medicinal/ornamental/flavouring use	Tropical Asia
7	*Z. argenteum* J. Mood and I. Theilade	Ornamental	Endemic to Malaysia
8	*Z. bradleyanum* Craib	Ornamental	Cultivated in USA
9	*Z. chrysanthum* Rosc.	Ornamental	-
10	*Z. citriodorum* J. Mood and I. Theilade	Ornamental	Common in Thailand and the USA
11	*Z. clarkia* King ex Benth	Ornamental	Native of Sikkim, India, lateral spikes
12	*Z. collinsii* J. Mood and I. Theilade	Ornamental	Common in the USA and Europe
13	*Z. corallinum* Hance	Chinese medicine/ornamental	-
14	*Z. eborium* J. Mood and I. Theilade	Ornamental	'White ginger' or 'Ivory ginger' of Indonesia, now common in the West
15	*Z. griffithii* Baker	Ornamental	Malaysian species
16	*Z. gramineum* Noronha	Ornamental	'Palm the ginger' Cultivated in USA
17	*Z. junceum* Gagnepain	Ornamental	'Yelloe the Delight' Common in the USA
18	*Z. lambi* J. Mood and I. Theilade	Ornamental	Malaysian species
19	*Z. longipedunculatum* Ridley	Ornamental	Cultivated in Australia
20	*Z. malaysianum* C.K. Lin	Ornamental	'Midnight Beauty', Common in the USA
21	*Z. neglectum* Valet	Ornamental	Common in the USA
22	*Z. niveum* J. Mood and I. Theilade	Ornamental	Common in the USA
23	*Z. ottensii* Valet	Medicine/ornamental	Common in South-East Asia
24	*Z. pachysiphon* B.I. Burtt and R.M. Sem	Ornamental	Cultivated in Australia
25	*Z. rubens* Roxb.	Medicine/ornamental	Indo-Malaysian species introduced to USA
26	*Z. spectabile* Griff.	Medicine/ornamental	Malaysian species Common in the USA
27	*Z. vinosum* J. Mood and I. Theilade	Ornamental	Malaysian species introduced to USA

Source: Parthasarathy *et al.*, 2007.

Table 12.3: *Zingiber* Species of Economic Importance

Species	*Common Name*	*Description*
Zingiber officinale Rosc.	Ginger	This is the main source of ginger for commerce. Several commercial types are recognized and generally named after the localities where they are grown. Rhizomes are aromatic, thick lobed, pale yellowish differing in shape and size among the different cultivated types. Certain types of ginger like Japanese ginger and Martinique ginger are reported obtained from the species *Z. mioga* Rosc. and *Z. zerumbet* Rosc., respectively (Claus *et al.*, 1970)
Zingiber cassumunar Roxb.	Banada	A slender herb arising from a stout rootstock found throughout India. The herb is widely cultivated in tropical Asia for the rhizomes, which are used in medicine. The rhizomes are fleshy with scaly epidermis golden yellow inside and possess a camphoraceous odour with a warm spicy taste. They are used for flavouring food preparation and are credited with stimulant and carminative properties. The rhizomes are considered useful in diarrohea and colic and are sometimes used medicinally as a substitute for those of Z. officinale (Casey *et al.*, 1971, Chakravarty and Chakraborty, 1964).
Zingiber zerumbet Rosc.	Nar kachur	An erect herb upto 1.0 m high ariosing from a large tuberous aromatic rootstock found throughout India from the Himalayas Southwards. This plant is also widely cultivated in the Asian tropics for its rhizome, which are employed medicinally for cough, stomachache, asthama, worms and in leprosy and other skin diseases. The rhizomes possess strong aromatic and ginger like flavor with some bitterness. Its distribution can be seen in Dhanikhari; BSI experimental garden Haddo, BSI premises and in Mini zoo.
Zingiber odoriferum Bl.	-	It is a terrestrial, aromatic herb with 2-3 leaves in clumps, 3 m and above in length. Leaflets sub-opposite, 25-30 pairs, spikes conical, crimson-red turning greenish-red up to 50 cm long. The dark red pouch like bracts are basally white, spoon shaped, acuminate, embedding with solitary cream-yellow flower on the bracteoles with mucilage. Corolla white or pale yellow. The seeds are black, like beads, ellipsoid 0.6 x 0.3 cm with carunculate base, can be collected after 2 months of its flowering during August-October. Spike turns to dark brown when fully ripen. The stem juice is being used as a tranquiliser for honeybee. Its distribution can be seen in South Anadaman, Mini zoo, Haddo, Way to Aberdeen Bazar from BSI office.
Zingiber spectabile Griff.	-	It is a robust, aromatic herb, leaves in clumps of more than 2 m long and leaflets are 20-25 in pairs. The spike is erect, cylindrical up to 40 cm long. Bracts scarlet in honeycomb form, turning to red as ripen. The solitary pale yellow flowers bloom for a day, lies in between the primary yellow bracts and the secondary bracteoles. Fruit is cylindrical, covered with bracteoles in the pouch. Black beady eyes like ellipsoid seeds can be collected after 2 months of its flowering. Fruit wall remains persistent when dry. It could be grown as an ornamental. Its distribution can be seen in Chidiatappu to Kodaighat; BSI premise and Haddo.
Zingiber squarrosum Roxb.	-	It is a robust, aromatic herb, 2.0-2.5 m tall; rhizome is covered with mucronate red sheaths. Leaves broadly lancelate, acuminate, wavy margined, 35-50 cm long. Spikes obovate up to 8 cm long, bractsred pubscent, calyx tublar, pink, pubescent, corolla white, unequal, pubscent, pale red; flowering during August-October. The petioles are chewed when thirsty by the tribal 'Onges'. Its distribution can be seen in South Anadaman, Dhanikhari; Little Anadaman; Way to forest nursery.

Botanical Description

The family name of the ginger plants Zingiberaceae. A Swedish botanist named Linnaeus 1st gave ginger its scientific name, *Zingiber officinale*. The flesh of the ginger rhizome can be yellow, white or red in colour, depending upon the variety. It is covered with a brownish skin that may either be thick or thin, depending upon whether the plant was harvested when it was mature or young. The ginger rhizome has a firm yet striated texture and a taste that is aromatic, pungent and hot. The rhizomes are aromatic, thick-lobed, pale yellowish, differing in shape and size depending on the variety or cultivated type.

Table 12.4: Botanical Description

Growth Habit	Ginger is a herbaceous perennial plant which grows from underground tuberous aromatic stem (rhizomes).
Stem	The ginger plant grows from a creeping rhizome that sends up shoots of new growth. It grows to a height of approximately 60.00-120.00 cm. Pseudo-stems.
Leaf	Long slender leaves. The aerial leafy shoots are about 50.0 cm and the leaf blades are long and narrow.
Floral biology	The aerial parts of the ginger plant are flowers arising directly from the rhizome. Small yellowish green fragrant flowers with dark purple markings are bisexual. Pillai *et al.* (1978) reported that flowering start in October and continued till early December. The peak period is November.
Anthesis and pollination	The flower start opening at about 3.0 PM amnd anthesis is occurring simultaneously. It takes about 20-25 days from bud initiation to full bloom and 23-28 days to complete flowering.
Fruit and fruit set	The fruit, which is very rarely produced, in the form of a thin walled capsule.
Seeds	Capsule containing a number of small and black angular seeds.

Baker (1894) has recognized 4 sections in Zingiberaceae and *Zingiber officinale* is included in section II *Lampuzium* Haron. Holttum (1950) included *Zingiber* in the tribe Hedycheae along with other genera such as *Cucuma, Hedychium, Kaempferia*. Burtt and Smith (1983), however, disagreed with Holttum (1950) proposal and proposed *Zingiber* as a separate tribe.

In West Africa two main types of ginger are grown (Tindall, 1968). Graham (1936) reported that 5 kinds of ginger are recognized in Jamaica *viz.* St. Mary, Red Eye, Blue turmeric, Bull Blue and China Blue. Ridley (1912) reported 3 forms of ginger from Malaysia *viz.* 'Halyia Betle' (true ginger), 'Halyia Udang' (red ginger) and 'Halyia Bara' (yellow ginger). A red variety of ginger *Z. officinale* var. *rubra* (pink ginger) and a 'Withered Skin variety' have also been reported from Malaysia. Rosales (1938) reported two cultivars from the Philippines *viz.* 'Native' and 'Hawaiin'. 'Taffingiwa' is the common ginger in Nigeria, the next important being 'Yasum Bari' (black ginger) Ravindran *et al.* (2005).

Table 12.5: Horticultural Diversity of Ginger in China

Sl.No.	Category	Variety/Cultivar
1.	Sparse seedling/ big fleshy rhizome	Gandzhou, Shandong Laiwu
2.	Dense seedling/ fleshy type	Guangzhow, Zhejiang
3.	Edible/medicinal	Fujian bud, Hunan yellow heart, chicken claw ginger, Xingguo ginger
4.	Edible/processed	Guangzhou (fleshy type), Fuzhou giner (Purple ginger), Tongling (White ginger), Fujian bamboo ginger, Zunyi (big white ginger), Leifeng ginger
5.	Ornamental ginger	Laishe ginger
		Flower ginger
		Tea ginger
		Strong ginger
		Hengchum ginger
6.	Other types	Zaoyang (Hubei province)
		Zunji big (Giuzhou)
		Chenggu Yellow (Shaxi)
		Yulin round fleshy (Guangxi)
		Bamboo root ginger (Sichuan)
		Mainyang (Sichuan)
		Xuanchang (Ahuii)
		Yuxi yellow (Yunnan)
		Laiwu Slice ginger (Shandong)
		Yellow claw (Zhejiang)
		Taiwan fleshy (Taiwan)

Source: Ravindran *et al.*, 2005.

In Japan, the ginger cultivars are grouped into three types based on their stature (Ravindran *et al.*, 2005):

(i) Dwarf statured, early maturing, profuse tillering types with small rhizomes *e.g.* 'Kintoki'

(ii) Medium statured plants with moderate tillers and medium size rhizome *e.g.* 'Sanshu'

(iii) Tall plants with big rhizome and less tillers *e.g.* 'Oshoga'

Sanshu (4 x Sanshu) and Buderim Gold are tetraploid ginger varieties grown in Japan (Adaniya and Shirai, 2001) and Queensland, respectively.

Augmentation of Genetic Resources and Crop Improvement

In India good variability for yield and quality of ginger is encountered in Kerala and North East India. Though no sexual reproduction is reported in ginger, geographical spread accompanied by genetic differentiation into locally adapted population augmented by mutation could be the main factor responsible for diversity in this clonally propagated crop. Table ginger cultivars identified for various uses

Table 12.6: Ginger Cultivars and their Unique Traits

Sl.No.	End Use/Traits	Name of Cultivars
1	High yield (fresh)	Rio-de-Janerio, Suprabha, Varada, Mahima, Rejatha, Juggigan, Himachal, Himgiri, Maran, Wynad Local, Karakkal, Thingpui, Nadia, Suruchi
2	Bold rhizome	Mahima, Rejatha, Varada, Gorubathani, Bhaise, Suravi
3	High dry recovery	Varada, Mahima, Zahirabad, Jorhat, Kuruppampady, Ernad-Chernad, Maran, Mowshom, Rejaha,
4	Low fibre	Varada, China, Nadia, Poona, Zahirabad, Mahima
5	High oleoresin	Wynad Kunnmangalam, Ernad-Chernad, Nadan (Pulpally), Himachal, China, Rio-de-Janerio, Maran
6	High gingerol and shogaol	Wynad, Baharica, Rio-de-Janerio, Kunduli, Kunnmangalam, Ambalavayalan, Ernad-Chernad, Swathing Pui, Maran, Jorhat Local
7	High essential oil	Pulpally, Sabarimala, Valluvanad, Nadan (Pulpally), Rio-de-Janerio, Nadan (Pulpally), Thodupuzha, Himgiri, Ellakallan, Mananthody
8	High zingiberene and (6) gingerol	Baharica, Amaravathi
9	Salted ginger	Rejatha and Varada
10	Ginger candy	Varada

Source: Sasikumar, 2000.

Nepal is another country rich in cultivar diversity of ginger. Ginger is grown under different agro-ecological situations of Nepal which cover about 6082 ha. Though about 50 ginger cultivars are known in Nepal, some of them may not be genetically distinct as these cultivars are mainly recognized after their place of cultivation. Most of the Nepal ginger cultivars are bold types, known after their locality of cultivation such as 'Tehratum', 'Salyan', 'Naval Parasy', 'Palpa', Bhaktapur", 'Bhojpur', 'Illam' *etc.* (Sasikumar, 2000). Good variability in cultivated ginger is encountered in India and China (Parthasarathy *et al.*, 2007). In both these countries, most of the land races are known after their place of domestication and hence all of them may not be distinct genetically (Table 12.7).

In India following varietal diversity has been recorded in ginger.

Table 12.7: Cultivated/Varietal Diversity in Ginger

Sl.No.	Cultivar/Variety	Sl.No.	Cultivar/Variety
1	Anamika	26	Rajagarh Local
2	Assam	27	Rejatha*
3	Arippa	28	Rio-de-Janerio+
4	Bajpai	29	Sargi Guda
5	Burdwan	30	Saw Thing Laidum
6	China+	31	Singhihara
7	Ernad-Chernad	32	Seirra Leone-+
8	Ernad Manjeri	33	Suprabha*
9	Ellakallan	34	Suruchi*
10	Edappalayam	35	Suravi*
11	Himachal	36	Saying Makhim (Pink ginger)
12	Himgiri	37	Taiwan+
13	Jorhat	38	Taffingiva+
14	Juggigan+	39	Thang-chang
15	Jamaica+	40	Thingpui
16	Karakkal	41	Thing Laidum
17	Kuruppampady	42	Thodupuzha
18	Kunduli Local	43	Tura
19	Maran	44	Uttar Pradesh
20	Mahim	45	Valluvanad
21	Mahima*	46	Varada*
22	Mananthody	47	Vengara
23	Nadia	48	Wynad Local
24	Narasapattam	49	Wynad Kunnmangalam
25	Poona	50	Zahirabad

+= Exotic, *= Improved variety.

Source: Parthasarathy *et al.*, 2007

Indian ginger is very starchy and is almost as pungent as Jamaican Ginger, but is less agreeable in odor. Indian Ginger has a faint lemon-like odor due to the presence of a small quantity of citral.

There are three main type of Indian Ginger.

(i) **Cochin Ginger-**This type of ginger comes from Central Kerala, is the peeled type, light brown to yellowish grey externally

(ii) **Calicut Ginger-** This type of ginger comes from Malabar and bear orange or reddish brown, resembling African Ginger, but the periderm is usually removed; it is inferior to Cochin Ginger in quality.

(iii) **Calcutta Ginger-** This type of ginger is possibly the same as Cochin Ginger, is greyish brown to greyish blue externally.

Genetic Resource Conservation

Central Agricultural Research Institute, Port Blair

Table 12.8: Some High Yielding Cultivars/Races of Ginger

States	Area of Evaluation	High Yielding Cultivars/Races	Yield (t/ha)
Meghalaya	Nayabunglow	Thingpui	29.2
		Thinglaidon	25.1
		Poona	25.1
Meghalaya	Barapani	Poona	24.0
		Tura	24.0
		Maran	24.0
		Basar Local	24.0
Manipur	Longol hill	Poona	22.19
		Suprabha	19.14
Mizoram	Kolasib	Thinglaidon	15.18
		Nadia	15.82
		Jugijan	18.86
		Suprabha	9.6
Nagaland	Jharnapani	Nagaland	6.25
		Poona	6.15
		Nadia	6.31
Tripura	Lembucherra	Tripura Local	8.3
		Bhainse	20.08
		SG	14.48
Arunachal Pradesh	Basar	China	21.48
		Nadia	15.78
		Poona	13.80

Source: Annual Report of ICAR-Research Complex from 1977 to 1993.

(Andaman and Nicobar Islands) is maintaining 33 ginger germplasm possessing various useful horticultural traits. As a part of *ex situ* conservation, the germplasm collection of the wild ginger have been introduced/ maintained in the experimental garden-cum arboretum of the Botanical Survey of India (BSI) at Nayashar near Dhanikhari, Port Blair. In NEH Region ICAR-Research Complex, Barapani (Meghalaya) is also invlolved in collection evaluation and maintenance of gringer.

REFERENCES

1. Adaniya, S. and Shirai, D. 2001. *In vitro* induction of tetraploid ginger (*Z. officinale*) and pollen fertility and germinability. *SciHort., 83:* 277-287.

2. Baker, J.G. 1894. Scitamineae. In: Hooker, J.D. *Flora of British India*, **6:** 198-264.

3. Burt, B.L. and Smith, R.M. 1983. Zingiberaceae. In: Dasanayake, M.D. (Ed.) A revised Handbook to the flora of Ceylon. Vol IV Amerind Publication, New Delhi, p. 488-532.

4. Gamble, J.S. 1925. Flora of the Presidency of Madras. Vol.II. Botanical Survey of India, Kolkata.

5. Graham, J.A. 1936. Methods of ginger cultivation in Jamaica. *J. Jamaican Agric. Soc.*, **40:** 231-232.

6. Holttum, R.E. 1950. Zingiberaceae of the Malayan Peninsula. *Card. Bull. Singapore*, **13:** 1-249.

7. Kumar, S. 1999. A note on conservation of economically important Zingiberaceae of Sikkim, Himalaya. In: Biodiversity, Conservation and Utilization of Medicinal and Aromatic Plants. (Eds. Sasikumar, B. *et al.*, Indian Institute of Spices Research, Calicut (Kerala), India, p. 201-207.).

8. Nair, B.M. 1974. Export of spices from India. Ibid. pp. 68-70.

9. Parthasarathy, V.A., Sasikumar, B., Saji, K.V. and Prasath, D. 2007. Biodiversity of Zingiberaceous spices and their conservation. In: Agrobiodiversity, Vol.I. Crop Genetic Resources and Conservation (Eds. Kannaiyan and Gopalam, A.), Associated Publishing Co., New Delhi, p. 250-266.

10. Purseglove, J.W. 1973. Tropical Crops Monocotledons, Longman Group Ltd. London.

11. Tindall, H.D. 1968. Commercially grown vegetable. In: Commercial vegetable growing. Oxford Univ. Press, London

12. Ravindran, P.N., Nirmalbabu, K. and Shiva, K.N. 2005. Botany and Crop Improvement of Ginger. In: Ravindran, P.N. and Nirmal Babu, K. (Eds.) Ginger-the genus ginger, CRC Press, Boca Raton, p. 15-86.

13. Ridley, H.N. 1912. Spices. Mc Millan and Co. Ltd. London.

14. Rehman, H., Kuruppaiyan, R., Kishore, K. and Denzongpa, R. 2009. Traditional practices of ginger cultivation in North East India. *Indian J. Traditional Knowledge*, **8(1):** 23-28.

15. Sabu, M. and Dave, S. 2005. Other economically important Zingiber species. In: Ravindnran, P.N. and Babu, N.K. (Eds.) Ginger-the genus Ginger, CRC Press, Boca Raton, p. 533-545.

16. Sasikumar, B. 2000. Ginger in Nepal. *Indian Spices*, **36(2 and 3):** 7

17. Sasikumar, B., Krishnamoorthy, B., Saji, K.V., George, J.K., Peter, K.V. and Ravindran, P.N. 1999. Spices diversity and conservation of plants that yield major spices in India. *Plant Genetic Resources News Letter*, **118:** 19-26.

Black Ginger or Shampoo Ginger (*Zingiber zerumbet* Smith)

Introduction

The name Zingiber, actually derived from a Sanskrit word that refers to bull's horn. Black Ginger (*Zingiber zerumbet* Smith Syn. *Amomm zerumbet* L.) is known throughout the world as the 'Shampoo ginger' due to the presence of milky substance in the flower cones which are being used as a shampoo by some cultures. In the southern United States, it is known as the 'pinecone ginger' or'pinecone lily'. In India it is known as 'Karakashur' or 'Bitter ginger'. It reported that the herb possesses medicinal values therefore it is cultivated as a medicinal plant since old by various tribal groups of North Eastern India like the Naga group, the Meiteis, the Kulki-Chin-Mizo group, the Bodo group and so on. The rhizomes are much similar to ginger rhizome except its shape and size with deep blue in colour. The structure of leaves is just like that of turmeric. However, its leaves are small in size with light purple colour in midrib portion. The rhizome is the principle product like its cousin ginger but rhizome is very bitter with disagreeable taste. However, after certain traditional treatment it is used as spice by certain tribes in North Eastern India. The extract of rhizome is conventionally used in cough, asthma and stomach complaints (Jaganath and Ng, 2000). Some of the tribal people in Churachandpur district of Manipur used the black ginger rhizomes in the treatment of colic pain. It is also used as an anti-pyretic and analgesic and in the treatment of leprosy and other skin diseases (Somchit *et al.*, 2005). The rhizomes produce an essential oil which is used in the preparation of soap and other toilet items. The rhizome yields xeruminoe, a monocyclic sequiteopena ketone, which acts against a number of bacteria and viruses. Rhizome also yields an essential oil. Zerumbone can be isolated from rhizome and tetrahydrozerumbol and tetrahydric zerumbone can be prepared. The rhizomes of *Z. zerumbet*, particularly, have been regularly used as food flavouring and appetizer in various Malays' cuisines while the rhizomes extracts have been used in Malay traditional medicine to treat various types of ailments (*e.g.*, inflammatory- and pain-mediated diseases, worm infestation and diarrhea). In Hawaii, the spicy-smelling fresh rhizome were pounded and used as medicine for indigestion and other ailments. Th rhizomes can be stored in a cool, dark place to keep for use when needed. In traditional use, the rhizome was ground in a stone mortar with a stone pestle, was mixed with a ripe 'Noni fruit' and then used to treat severe sprains. The pulp was placed in a cloth and loosely bound around the injured area. For a toothache or a cavity, the cooked and softened rhizome was pressed into the hollow and left for as long as was needed. To ease stomach ache, the ground and strained rhizome material is mixed with water and drunk. Similarly, 'Awapuhi Pake' or ginger root is widely cultivated and eaten or made into a tea for indigestion as well as increased circulation of the blood and an increased sense of well-being.

Origin, History and Distribution

Black Ginger (*Zingiber zerumbet* Smith Syn. *Amomm zerumbet* L.) is originating in India, was distributed eastward through Polynesia and introduced to the

Hawaiian Islands in the canoes of early Polynesian settlers. It is a wild herb available in the warm and moist areas of the interior hilly regions of India. In India, it is cultivated by tribal people of north east states. It is locally known to the Malay as "Lempoyang," in Malaysia.

Species Availability, Evolution and Diversity

Considered as a moderately large genus of herbs belonging to the family Zingiberaceae, genus Zingiber is represented by approximately 141 species that are distributed mainly in Asia. In *Zingiber cassumunar* highest oil content was recorded in rhizome (1.05 per cent dry) followed by root (0.94 per cent) but no oil was found in leaf.

Botanical Description

Table 12.9: Botanical Description

Growth Habit	Plant is perennial in nature. The foliage grows to about 120.0-180.0 cm tall.
Stem	Leafy stems.
Leaf	Leaves are small in size (15.0-20.0 cm) with light purple colour in midrib portion. The 10-12 blade-shaped and grow in an alternate.
Floral biology	The flower spikes arise from the ground. They are long-lasting and make excellent flower arrangements. Amomng the leafy stems the conical or club-shaoped flower heads burst forth on separate and shorter stalks. The flower heads are reddish-green 3.0-10.0 cm long with overlapping scales, enclosing small yellowish white flowers that poke out a few at a time. As the flower heads mature, they gradually fill with an aromatic, slimy liquid and turn a brighter red colour. The flower stalks usually remain hidden beneath the leaf stalks.

REFERENCES

1. Jaganath, I.B. and Ng, L.T. 2000. Herbs: the green pharmacy of Malaysia. Vinpress Sdn. Bhd. and Malaysia Agricultural Research and Developmental Institute. 95-99.
2. Somchit, M.N., Shukriyah, M.H.N., Bustamam, A.A. and Zuraini, A. 2005. Anti-pyretic and analgestic activity of *Zingiber zerumbet*. *International Journal of Pharmacology*, 1**(3):** 277-280.

Genus *Curcuma*

Turmeric (*Curcuma* species)

Introduction

Turmeric (*Curcuma longa* X=21-basic chromosome number) and 2n=63-triploid) also known as Indian saffron finds extensive cultivation in the tropics. This is an important condiment such as turmeric and several medicinally useful as well as edible wild and cultivated species. It is gaining global importance as a potential source of new drug(s) to combat a variety of ailments as the genus contains molecules credited with many therapeutical, insects repellant and aromatic properties. The underground swollen stem known as rhizome is used as colourful spice (combining properties of a dye, cosmetic and drug). It has been a traditional food preservative and colour material (*curcumin*) in China, India and South-East Asia. Numerous domestic, therapeutic and medicinal properties of turmeric have been described recently (Chattopadhyay *et al.*, 2004; Tewari and Agarwal, 2004). Turmeric is grown in about 150000 ha in India. Several species like *Curcuma aromatic, C. harita, C. malabarica, C. zedoaria etc.* are used for extraction of starch having medicinal applications. Rhizomes and leaves yield essential oil. Oil content of leaves varies from 0.8 to 3.5 per cent while rhizomes yield 5 to 6 per cent on dry weight basis. It has β Phellandrene (1.0 per cent), Sabineno (0.6 per cent), Cineol (1.0 per cent), Borneol (1.0 per cent), Zingiberene (25 per cent) and Sesquiterpines (53 per cent). Curcumin is the colouring principal consists of curcumin I (93.7 per cent) and curcumin II (6 per cent) and curcumin III (0.3 per cent). Alleppey turmeric from Kerela, renowned for the high *curcumin* content is in short supply in trade circles.

Origin, History and Distribution

Turmeric is mainly of South Asian origin particularly north-east India and the centre of diversity of *Curcuma* also fall within this region. The plant grows in the wild in many places and is found in moist and cool habitats up to an altitude of 1,500 feet. Its species are mainly distributed in tropical and sub-tropical Asia extending to Australia. Turmeric is largely grown in India, Pakistan, Malaysia, Myanmar, Vietnam, Thailand, the Philippines, Japan, China, Korea, Sri Lanka, Nepal, South Pacific Islands, East and West Africa, Malagasy, Caribbean Islands and Central America but India is the major producer and exporter of this spice. Andhra Pradesh, Odisha, Tamil Nadu, Meghalaya, West Bengal, Maharashtra, *etc.*, are the major turmeric producing states in India. *C. zedoaria* Rosc. (Zedoary) is widely cultivated in many parts of India, Sri Lanka and China. In India, this species is growing wild in the Eastern Himalayas and in moist deciduous forests of coastal tracts of Karnataka. Sporadic cultivation of mango ginger (*C. amada*) and yellow zedoar (*C. zanthorrhiza*) is also noticed.

Species Availability, Evolution and Diversity

Curcuma is a genus of about 120 accepted species (Skornickova and Sabu, 2002) but some researchers in opinion of 80 species only found globally in the plant family Zingiberaceae that contains such species as turmeric and Siam Tulip. Apart from this, a large number of wild species numbering 29 also occur (Karthikeyan *et al.*, 1989). Since assembly of the genus *Curcuma* by Linnaeus in 1753 about 130 species have been described so far. Some of the species descriptions are without Latin diagnosis or type specimen, therefore the legitimate status of many species is suspicious and remains unclear. Out of which 40 species are distributed in India (Velayudhan *et al.*, 1999), including at least 10 species endemic to India. The name comes from Arabic *kurkum* meaning "turmeric". Genus *Curcuma* has 3 species *i.e. C. longa* (triploid), *C. domestica* (triploid) and *C. aromatica*. *C. aromatic* and *C. amada* are diploid in nature. It is now recognized that some of the species may be synonyms such as *C. zedoaria* syn. *C. xanthorrhiza* (India), *C. albicoma* syn. *C. sichunanensis*; *C. chuanyujin* syn. *C. kwangsiensis*, *C. wenyujin* syn. *C. aromatic* (China) *etc.* (Sasikumar, 2005). *Curcuma* species widely differ in their habitat. The habitat ranges from sea level (sandy costal) to high altitude such as above 200 m in the Western Ghats and the Himalayas. In India, *Curcuma* species diversity is concentrated in North-Eastern India, South India and Andaman and Nicobar Islands. The other countries rich in *Curcuma* species diversity are China, Thailand, Malaysia, the Philippines, Vietnam, *etc.* (Parthasarathy *et al.*, 2007).

Table 12.10: Taxon Name, Associated Authority and Common Name

Sl.No.	Name of taxon	Name of Authority	Common Name
1	*Curcuma aeruginosa*	Roxb.	
2	*Curcuma albicoma*	S.Q.Tong	
3	*Curcuma albiflora*	Thwaites	
4	*Curcuma alismatifolia*	Gagnep.	Siam tulip
5	*Curcuma amada*	Roxb.	Mango-ginger
6	*Curcuma amarissima*	Roscoe	
7	*Curcuma angustifolia*	Roxb.	
8	*Curcuma aromatica*	Salisb.	Wild turmeric
9	*Curcuma attenuata*	Wall. ex Baker	
10	*Curcuma aurantiaca*	van Zijp	
11	*Curcuma australasica*	Hook.f.	
12	*Curcuma bakeriana*	Hemsl. nomen illegit.	
13	*Curcuma bicolor*	Mood and K.Larsen	
14	*Curcuma bhatii*	(R.M.Smith) Skornickova and M.Sabu	
15	*Curcuma burttii*	K.Larsen and R.M.Smith	
16	*Curcuma caesia*	Roxb.	Black turmeric
17	*Curcuma ceratotheca*	K.Schum.	
18	*Curcuma chuanezhu*	Z.Y.Zhu nomen illegit.	
19	*Curcuma chuanhuangjiang*	Z.Y.Zhu nomen illegit.	
20	*Curcuma chuanyujin*	C.K.Hsieh and H.Zhang	
21	*Curcuma cochinchinensis*	Gagnep.	
22	*Curcuma codonantha*	Skornickova	
23	*Curcuma coerulea*	K.Schum.	
24	*Curcuma colorata*	Valeton	
25	*Curcuma comosa*	Roxb.	
26	*Curcuma coriacea*	Mangaly and M.Sabu	
27	*Curcuma decipiens*	Dalzell	
28	*Curcuma ecalcarata*	Sivar. and Balach.	
29	*Curcuma euchroma*	Valeton	
30	*Curcuma ecomata*	Craib	
31	*Curcuma elata*	Roxb.	
32	*Curcuma exigua*	N.Liu	
33	*Curcuma ferruginea*	Roxb.	
34	*Curcuma flaviflora*	S.Q.Tong	
35	*Curcuma glans*	K.Larsen and Mood	
36	*Curcuma gracillima*	Gagnep.	

Sl.No.	Name of taxon	Name of Authority	Common Name
37	*Curcuma grandiflora*	Wall. ex Baker	
38	*Curcuma haritha*	Mangaly and M.Sabu	
39	*Curcuma harmandii*	Gagnep.	
40	*Curcuma heyneana*	Valeton and van Zijp	
41	*Curcuma inodora*	Blatt.	
42	*Curcuma karnatakensis*	Amalraj	
43	*Curcuma kudagensis*	Velayudhan, V.S.Pillai and Amalraj	
44	*Curcuma kwangsiensis*	S.G.Lee and C.L.Liang	
45	*Curcuma lanceolata*	Ridley	
46	*Curcuma larsenii*	C.Maknoi and T.Jenjittikul	
47	*Curcuma latiflora*	Valeton	
48	*Curcuma latifolia*	Roscoe	
49	*Curcuma leucorrhiza*	Roxb.	
50	*Curcuma loerzingii*	Valeton	
51	*Curcuma longa*	L.	Turmeric
52	*Curcuma longispica*	Valeton	
53	*Curcuma malabarica*	Velayudhan, Amalraj and Muralidharan	
54	*Curcuma meraukensis*	Valeton	
55	*Curcuma mutabilis*	Skornickova, M.Sabu and Prasanthkumar	
56	*Curcuma neilgherrensis*	Wight	
57	*Curcuma nilamburensis*	K.C.Velayudhan *et al.*	
58	*Curcuma oligantha*	Trimen	
59	*Curcuma ornata*	Wall. ex Baker	
60	*Curcuma parviflora*	Wall.	
61	*Curcuma parvula*	Gage	
62	*Curcuma peethapushpa*	Sasidh. and Sivar.	
63	*Curcuma petiolata*	Roxb.	
64	*Curcuma phaeocaulis*	Valeton	
65	*Curcuma pierreana*	Gagnep.	
66	*Curcuma plicata*	Wall. ex Baker	
67	*Curcuma porphyrotaenia*	Zipp. ex Span.	
68	*Curcuma prakasha*	S.Tripathi	
69	*Curcuma pseudomontana*	J.Graham	
70	*Curcuma purpurascens*	Blume	
71	*Curcuma purpurea*	Blatt.	
72	*Curcuma raktakanta*	Mangaly and M.Sabu	
73	*Curcuma reclinata*	Roxb.	
74	*Curcuma rhabdota*	Sirirugsa and M.F.Newman	
75	*Curcuma rhomba*	Mood and K.Larsen	
76	*Curcuma roscoeana*	Wall.	
77	*Curcuma rubescens*	Roxb.	
78	*Curcuma rubrobracteata*	Skornickova, M.Sabu and Prasanthkumar	
79	*Curcuma sattayasaii*	A.Chaveerach and R.Sudmoon	
80	*Curcuma sichuanensis*	X.X.Chen	
81	*Curcuma singularis*	Gagnep.	
82	*Curcuma sparganiifolia*	Gagnep.	
83	*Curcuma stenochila*	Gagnep.	
84	*Curcuma strobilifera*	Wall. ex Baker	
85	*Curcuma sulcata*	Haines	
86	*Curcuma sumatrana*	Miq.	
87	*Curcuma sylvatica*	Valeton	

Sl.No.	Name of taxon	Name of Authority	Common Name
88	*Curcuma thalakaveriensis*	Velayudhan *et al.*	
89	*Curcuma thorelii*	Gagnep.	
90	*Curcuma trichosantha*	Gagnep.	
91	*Curcuma vamana*	M.Sabu and Mangaly	
92	*Curcuma vellanikkarensis*	K.C.Velayudhan *et al.*	
93	*Curcuma wenyujin*	Y.H.Chen and C.Ling	
94	*Curcuma wenchowensis*	Y.H.Chen and C.Ling	
95	*Curcuma xanthorrhiza*	Roxb.	Temu lawak
96	*Curcuma yunnanensis*	N.Liu and C.Senjen	
97	*Curcuma zedoaria*	(Christm.) Roscoe	Zedoary
98	*Curcuma zedoaroides*	A.Chaveerach and T.Tanee	

Velayudhan *et al.* (1999) recorded a total of 117 species (Table 12.11), obviously including the synonymous entities as separate species.

Table 12.11: *Curcuma* Species and their Origin as Proposed by Velayudhan *et al.* (1999)

Sl.No.	Species	Origin
1	*Curcuma aeruginosa* Roxb.	Myanmar
2	*C. albiflora* Thaw.	Sri Lanka
3	*C. amada* Roxb.	India
4	*C. amarissima* Rosc.	India
5	*C. Americana lam.*	-
6	*C. angustifolia* Dalz. and Gibs.	India
7	*C. angustifolia* Roxb.	India
8	*C. aromatica* Salisb.	India
9	*C. attenuata* Wall.	Myanmar
10	*C. australasica* hook. F.	Australia
11	*C. caesia* Roxb.	India
12	*C. cereyana* Wall.	India
13	*C. caulina* F. grah	India
14	*C. comosa* Roxb.	Myanmar
15	*C. cordata* Wall.	India
16	*C. cordifolia* Roxb.	-
17	*C. decipiens* Dalz.	India
18	*C. alata* Roxb.	Myanmar
19	*C. erubescens* Wall.	-
20	*C. ferruginea* Roxb.	India
21	*C. glaucophylla* Wall	-
22	*C. grandiflora* Wall.	-
23	*C. latifolia* Rosc.	India
24	*C. longa* Linn.	-
25	*C. longiflora* Salisb.	India
26	*C. montana* Roxb.	India
27	*C. musacea* Wall.	India
28	*C. neilgherrensis* Wight.	India
29	*C. officinalis* Salisb.	-
30	*C. oligantha* Trim	Sri Lanka
31	*C. ornata* Wall.	Myanmar
32	*C. pallida* Lour.	China
33	*C. parviflora* Wall.	Myanmar
34	*C. petiolaris* Hort.	-
35	*C. petiolata* Roxb.	Myanmar
36	*C. spicata* Wall	Myanmar
37	*C. porphyrotannica* Zijp.	-
38	*C. pseudomontans* F. Grah.	-
39	*C. purpurascens* Blume.	Java
40	*C. reclinata* Roxb.	India
41	*C. roscoeana* Wall.	Myanmar
42	*C. rotundata Linn.*	-
43	*C. rubescens* Roxb.	India
44	*C. rubricaulis* link.	-
45	*C. strobilifera* Will.	Myanmar
46	*C. strobinia* Wall.	India
47	*C. sumatrana* Miq.	Sumatra
48	*C. viridiflora* Roxb.	Sumatra
49	*C. xanthorrhiza* D. Dietr.	Malaya
50	*C. xanthorrhiza* Roxb.	-
51	*C. zedoaria* Rosc.	India
52	*C. zedoaria* Roxb.	-
53	*C. zerumpet* Roxb.	
54	*C. bakertiana* Hunsl.	New Guinea
55	*C. coccinia* Wall.	-
56	*C. sylvestris* Redl.	Malayasia
57	*C. urceolata* Rich.	-
58	*C. ceratotheca* Schum.	Celebes
59	*C. sessilis* Gag.	Burma
60	*C. sparganifolia* Gagnep.	Cambodia
61	*C. stenochyla* Ganep.	Cochin China
62	*C. cochinchinensis*	Indo-China
63	*C. harmnadi* Gagnep.	Cochin China
64	*C. lanceolata* Ridl.	
65	*C. leopoldii* Hot.	
66	*C. pierreana* Gagnep.	Indi-China
67	*C. singularis* Gagnep.	Indi-China
68	*C. thorelii*	Indi-China

Sl.No.	Species	Origin
69	*C. trichosantha* Gagnep.	Indi-China
70	*C. heyneana* Val. and Zijp.	Java
71	*C. loerzingii* Val.	Sumatra
72	*C. longispica* Val.	N. Guinea
73	*C. manga* Val. and Zijp.	Java
74	*C. ochrorhiza* Val. and Van Zijp.	Java
75	*C. phaeocaulis*	Java
76	*C. soloensis* Val.	Java
77	*C. sylvatica* Val.	-
78	*C. aurantiaca* Van Zijp.	Java
79	*C. elomata* Craib.	Siam
80	*C. latiflora* Val.	N. Guinea
81	*C. sulcata* Haines	India
82	*C. elata* Roxb.	Myanmar
83	*C. inodora* Blatter	India
84	*C. petiolata* Roxb.	Myanmar
85	*C. purpurea* Blatter	India
86	*C. zanthorrhiza* Roxb.	Amboinea
87	*C. zedoaria* (Bergius) Roscoe	-
88	*C. burtii* K. Lasru and R.M. Smith	Thailand
89	*C. kwangsiensis* S.G. Lee and C.F. Liang	China
90	*C. curcumorpha* Rao and Verma	India
91	*C. longiflora* (wall.) Rao and Verma	India
92	*C. ecalcarata* Sivar and Indu	India
93	*C. sichuanensis* Gagnep.	China
94	*C. alismatifolia* Gagnep.	Thailand
95	*C. gracilima* Gagnep.	-
96	*C. kunstleri*	-
97	*C. domestica* Val.	-
98	*C. speciosa* Link	-
99	*C. viridiflora* Roxb.	Java
100	*C. colorata* Val.	Java
101	*C. euchroma* Val.	Java
102	*C. ochrorrhiza* Val.	Java
103	*C. brog* Val.	Java
104	*C. agrostis* Sare	-
105	*C. ranadei* Prain	-
106	*C. haritha* Mangaly and Sabu	India
107	*C. raktakanta* Mangaly and Sabu	India
108	*C. kudagensis* Vel *et al.*	India
109	*C. thalakaveriensis* Vel *et al.*	India
110	*C. malabarica* Vel *et al.*	India
111	*C. karnatakensis* Amal *et al.*	India
112	*C. niodona*	-
113	*C. nilambuernsis* Amal *et al.*	India
114	*C. cannanorensis* Ansari and Nair	India
115	*C. vamana* Mangaly and Sabu	India
116	*C. lutea* Ansari and Nair	India
117	*C. peethapushpa* Sandh and Sivar	India

Source: Velayudhan *et al.* (1999).

Economically important *Curcuma* species are given in Table 12.12.

Table 12.12: Economically Important *Curcuma* Species

Species	Use
C. longa L. syn. *C. domestica* Val.	Spice, medicine, dye, religious local delicacies, insect repellant, aroma therapy and perfume.
C. amala Roxb., *C. mangga* Val. and Zijp.	Spice, medicine, pickles and salads
C. zedoaria Roxb.	Folk medicine, arrow root industry
C. ochrorrhiza Val. and Van Zijp.	Malayan traditional medicine
C. pierrema Gagnep.	Vietnamese traditional medicine
C. aromatic Salsb.	Medicine, toiletry articles, insect repellent
C. kwangsiensis S.G. Lec. and C.F. Liang syn. *C. chuanyujin*, *C. phaeocaulis* Val.	Chinese traditional medicine
C. caesia Roxb.	Spice and medicine
C. comosa Roxb.	Traditional medicine of Thailand
C. angustifolia Roxb. *C. zedoaria* Roxb., *C. caulina* F. Grah, C. pseudomontana F. Grah., *C. Montana* Roxb., *C. rubescens* Roxb., *C. leucorrhiza*, *C. xanthorrhiza* Roxb., *C. decipiens* Dalz. *C. malabarica* Vel. *et al.*, *C. raktaka* Mangaly and Sabu, *C. haritha*	Arrow root industry
C. aeruginosa Roxb.	-
C. alismatifolia Gagnep., *C. thorelii*, *C. roscoeana* Wall.	Ornamental (cut flower)

The genus *Curcuma* contains about 70 species of rhizomatous herbs distributed in India, Cambodia, Malayan Archipelago and North Australia. About 30 species occur in India, of which two species are useful for the production of starch *viz. C. angustifolia* (East Indian Arrowroot or Tikhur) and *C. zedoaria* (Kundu, 1967). In India, the species *C. angustifolia* (East Indian Arrowroot) occurs in the hilly tracts of Bengal, Maharashtra, Tamil Nadu and some of the lower Himalayan ranges.

Table 12.13: Distribution of Starchy *Curcuma* Species in India

Crops	Related Species	Distribution
Starchy *Curcuma* spp.	*Curcuma zedoaria* Rosc.	All over plains as wild and sometimes cultivated up to 3000 feet.
	C. malabarica	In coastal areas of Kerala
	C. raktakanta	In coastal areas of Kerala
	C. harita	In coastal areas and midlands of Kerala
	C. aerugenosa Roxb.	West Bengal
	C. caesia Roxb.	West Bengal
	C. ferrugennia	Foot hills of Himalayas in West Bengal
	C. montana Roxb.	In central India and Eastern Ghats of Andhra Pradesh

About 30 species occur in India, of which the following are of economic importance (Chaurasia *et al.*, 1974)

Table 12.14: Species and their Description

Species	*Common Name*	*Description*
Curcuma amada Roxb.	Mango Ginger	A species of which the rhizomes having characteristics odour of raw mangoes, occurring wild in parts of west Bengal, Konkan and Tamil Nadu. The rhizomes are used mostly in pickle preparation. It is considered as stomachic and carminative and applied over contusions and sprains.
Curcuma angustifolia Roxb.	East India Arrow Root	This occurs in the hilly tracts of Bengal, Bombay, Madaras and some of the lower Himalayan regions. The plants grow wild in many places particularly in most and cool situations at an altitude of about 1500 ft. it is cultivated to a small extent. The rhizome contains starch, which is used as a substitute for the true arrowroot powder.
Curcuma caesia	Black Zedoary	A native of West Bengal, where it is cultivated to some extent. Fresh tubers are pale yellow and aromatic and used as a cosmetic, sprains and bruises.
Curcuma zedoaria Rosc.	Zedoary	A species growing wild in Eastern Himalayas and in the moist deciduous forests of the coastal tracts of Kanara. It is native to northeast India and widely cultivated in many parts of India, Sri Lanka and China. It is closely resembles *C. longa* in appearance and rich in starch. The shoti starch of commerce is a product of extract from the tubers and used as a substitute for arrowroot. Shoti starch (flour) is produced in West Bengal on a cottage industry basis. The rhizomes possess stimulant and carminative properties.
Curcuma aromatica Salisb.	Willd Turmeric, Yellow Zedaoria	A species occurring wild throughout India and cultivated mainly in West Bengal and Andhra Pradesh. The rhizomes are light yellow (internally orange red) and possess campohoraceous odour. They are sometimes used as a substitute for C. longa, but not as a condiment. They are externally applied in combination as astringent to bruises and sprains.
Curcuma longa L. (*Curcuma domestica* Valenton)	Turmeric	Native of Southern Asia probably India and is cultivated extensively throughout the warmer parts of world. It is grown on a large scale in India, China, Mayanmar and East Indies. It is cultivated commercially in almost all the states in India particularly in Tamil Nadu, Kerala, Andhra Pradesh, West Bengal, Maharashtra and North Eastern Region. The rhizomes are short and thick constitute the turmeric of commerce. The turmeric is used both as a colouring matter and as a condiment. The characteristics yellow matter distributed throughout the plant especially concentrated in the rhizome.

Besides the above mentioned species, the starchy rhizomes of *Curcuma pseudomontana* Grah. (Konkan-Suderwans, Sindarbar) occurring in Konkan and Anamalai hills, C. leucorhiza Roxb. (Ben.-Tikar) and *C. rubescens* Roxb., found in bihar and West Bengal are utilized as source of starch in the region of their occurrence. Certain non-edible genotypes of turmeric have also been reported to exist in NEH region in feral form (Upadhyay *et al.*, 1982).

A New Species '*Curcuma kshonapatra*'

Velayudhan (2016) identified as one of the very distinct species after long duration studies *in situ* as well as *ex situ* conditions. It can be grouped along with the white staminode bearing species of the genus such as *C. albiflora* Thwaites, *C. oligantha* Trimen, *C. karnatakensis* Amalraj, Velay. and Mural. and *C. mutabilis* Skornick. M. Sabu and Prasanthk; which were reported from Western Ghats. It is found to be associated with a large population of a dominant species which is suspected as *C. albiflora* Thwaites. It is being described now as a new species due to some characteristic features suech as erect and thin tall plant types with few suckers that bear elliptic/lanceolate leaf lamina without wavy margin and slender purplish petioles as against more robust and profuse sucking companion species. Leaf lamina has purple midrib and closely arramnged prominent veins which radiates dark purple wash or purple violet tint below the juvenile leaves, which fade at maturity. The key character of the species *i.e.* purple midrib, is perhaps first of its kind in species (sect. *Nontuberosa*) having no sessile fingers from India. Usually, this character purple midrib is associated with the presence of sessile finger on the root-stock in the genus and hence it appears to have an evolutionary significance and is associated with a group of sessile tuber bearing species under the genus *Cucurcuma* such as *C. zedoria* (Christ) Roscoe, *C. zanthorrhiza* Roxb.; *C. ferruginea* Roxb.; *C. comosa* Roxb.; *C. latifolia* Roscoe; *C. amarissima* Roscoe, *etc.*, failing under the sect. Tuberosa Velay., Amalraj, Velay. and Mural. (Velayudhan, *et al.*, 1996). As the present species obviously falls under the sect. *Nontuberosa*. The purple colouration in the midrib region of the leasf and the purple tint of the tender leaves and petiole, purplish flower bracts, highly exerted and beaked flower buds are taken as the key character to delimit the present species from all other closely related species with white staminodes to propose a specific epithet *'kshonapatra'* which in Sanskrit means reddish purple tint on leaf similar to that of 'kshona' meaning blood.

C. albiflora Thwaites, *C. oliganthae* Trimen, *C. karnatakensis* Amalraj, Velay. and Mural. and *C. mutabilis* Skornick. M. Sabu and Prasanthk. *C. inodora* Blatt., characteribus vegetativorum aeriorum; absentia dactylorum sessilium; floribus maxime exertis et staminodiis albis arte affinis, sed costa folii purpurea; basi folii juvenis purpureo-tincta; spicis magnis, compactis, globosis ad ovalibus clare differt.

Root-stock conical to cylindrical, 7.2 x 1.3 cm branched with one or two secondary (lateral) mother rhizomes bearing suckers which resemble mother rhizome in shape but smaller, rhizome cortex flesh whitish cream, core pale yellow, flesh odourless, taste slightly astringent, less aromatic in smell; roots terminating into stipitate tubers, 10-15 in number, fusiform, rarely beaded, 3.0-3.5 cm x 2.0-2.2 cm in size, cortex dirty or transparent white, core white, without any smell and taste; plants erect, reaching up to 70 cm tall, bearing 2-4 suckers; petiole erect to semi-erect, 25-55 cm, sheath 15-30 cm long, margin purple extending to the leaf base, glabrous, ligule purple, pseudo-stem with 5-6 scale leaves at the base, 2.0 to 15.0 cm long; leaves 5-6, semi-erect to prostrate and drooping above middle at maturity, lamina 25-30 x 9-11 cm, elliptic-lanceolate, margin less wavy in tender ones and highly wavy in mature ones, veins arise from the midrib at 45° running closely parallel up to the margin, highly pronounced, young petiole with reddish tint, midrib is purple throughout along the channel spreading towards the leaf visible on the dorsal side towards the base especially in younger ones, base gradually tapering or acute, rarely oblique, acute-acuminate at the tip, obtuse or rounded at the base, green with reddish purple tint above middle, glabrous. Spike either lateral or central, the former arising from the side buds developing from the main root-stock or branches towards the top, each bud bifurcating into two, one being the inner one to form a spike and the other outer one which is usually in bud stage at the time of blooming to form a leafy shoot during the main season. Those associated with the shoot form scale leaves at the base and the latter borne central spike arising from the tuft of leaves during the main growing season. Spike 9.0-12.5 cm long without coma, oblong and compact with greenish purple or purple flower bracts bearing highly exerted and showy flowers with white staminodes. It is showy, oval to oblong, varies in size with age of the plant; peduncle 8-12 cm long, covered by whitish scaly bracts that are oblong to ovate towards the tip, sterile bracts vary in length (2.5 cm in the case of lower ones and 6 cm in the upper ones and width from 1.5 to 1.6 cm), whitish at the base tinged with red, purple or pink towards the top, glabrous, no coma bracts seen (usually flower buds arise from axils of all bracts)-, flower bracts 4-5 in number, shape vary from ovate to slightly oblong, narrow linear towards top with acute tip, purple violet towards tip, bracts hairy on both sides, length varying 2.2 to 6.5 and width from 1.5-1.6 cm, flowers 2 in a bract, highly exerted up to 7 cm long, buds whitish with light rose tinge and prominently beaked, calyx up to 2 cm long, hairy, three toothed, one side deeply cleft, middle tooth ovate and acute and shorter than the other lobes which are longer and acuminate, with purple tint, corolla tube up to 3.5 cm long, main lobe beaked, cucullate, shorter than the side lobes, whitish with pinkish tinge on the tip, 1.6 x 1.5 cm; side lobes oblong, 1.1-1.3 cm x 0.8-0.9 cm in size, rounded tip, staminodes white, side lobes unusually very large and exerted and obovate, rounded, 3.0-3.5 cm long and 0.8-1.0 cm wide; lip whitish, three lobed, mid lobe smaller than the side lobes in length and deeply emarginated, 1.7 x 1.5 cm in size, white with orange yellow median, stamen 1, anther basifixed, oblique, two celled, whitish, glabrous, 0.3 cm long, spurs 0.2 cm long, diverging, white, glabrous; ovary hairy, purple tinged up to 4 mm long; stigma not exerted; fruit a capsule with persistent calyx, 1 cm long, globose, containing dark seeds with white translucent arils Velayudhan (2016).

Botanical Description

The genus *Curcuma* belonging to the Hedycheae. Taxonomic position of turmeric is as under:

Class	:	Liliopsida
Subclass	:	Commelinids
Order	:	Zingiberales
Family	:	Zingiberaceae
Genus	:	*Curcuma*
Species	:	*longa/domestica/aromatica*

Turmeric is monocot triploid with 2n=63 with basic chromosome number (x) =21. Cytological studies show chromosome variation of 63, 74, 77, and 84. *C. aromatic* and *C. amada* are diploid with chromosome number 2n=42. The underground swollen stem is known as rhizome.

Table 12.15: Botanical Description

Growth Habit	It is herbaceous perennial, monocot and erects growth habit but grown ac annual Thc plant growo up to 1.0 meter.
Root	Roots are fibrous, sparse attached to rhizomes. Their function is nutritional supply to the developing rhizome.
Stem	Rhizome is thick, underground swollen short stem and consists of mother rhizome having primary and secondary branches known as blunt fingers. Rhizomes has rough dull surface with scales and root bits.
Leaf	Generally 6-10 leaves with long sheaths form pseudo-stem giving plant a height of 80-150 cm. The leaves are borne in a tuft and are about 60.0 cm tall. It bears green leaves with brownish-purple veins on either side of the midrib. They are thin, rather flaccid and light green in colour, lanceolate, acuminate, with a long leaf stalk.
Floral biology	Flowers are zygomorphic, bisexual, bracteates and epigynous are borne in cone shaped spikes in the tuft of leaves. The spikes are shorter than the leaves and supported by stout peduncles. They consist of a great number of thin greenish-white, ovate bracts. In each bract there are two flowers which are dense spikes. Flowers bear in axil of each bract and open at a time. Perianth in two whorls, androecium consists of 6 stamens, which are epipetalous. Ovary inferior, tricarpellary, trilocular, globose hairy at the top and placentation axile.

Anthesis and pollination	Flowering varies 1.2-95.8 per cent based on cultivar and climate. Flowering occurs in July-October. Inflorescence takes 8-12 days to bloom after the emergence and duration of opening lasts for 8-10 days. Anthesis timing 6.0-7.0 am and maximum at first hours. Pollination by insect only.Pollen fertility is 71-85 percent. Higher fertility and viability are higher in lower portion and low at middle and upper portion.
Rhizome	The rhizomes are large, fleshy, branched, and the inner part of which is pale yellowish brown in colour.
Fruit and fruit set	Fruit is thick walled trilocular capsule.
Seeds	Seeds are artilate. Seeds have two seed coats, outer one thick and inner one being thin. Process of seed development and ripening is completed in 22-22 days. Seeds are brownish black and filled with massive endosperm. Seed viability is low and immediate sowing is required to have effective germination.

The taxonomic classification and systematics of the genus may not be fool proof as there are many problems like lack of type specimens and illustration of old species and absence of important floral parts in herbariums, fleshy and perishable aerial portions *etc.* recent molecular taxonomic studies of 16 *Curcuma* species also pin point this fact (Shyamkumar and Sasikumar, 2005).

Botanical Classification

Velayudhan *et al.* (1996) proposed a new conspectus of the genus in India. These authors found that the Indian species can be accommodated in the two subgenera of Valeton (1918) *viz. Eucurcuma* and *Paracurcuma*. *Eucurcuma* contains 3 main sections, namely, *tuberosa*, *nontuberosa* and *stolonifera* based on the presence or absence of tubers and stolons, each section agasin encompassing one or more subsection based on floral, vegetative and rhizome characters. Under the subgenus *Paracurcuma*, only the anther spurs lacking species *viz. C. aurantiaca* and *C. ecalcarata* are included. Velayudhan *et al.* (1999) also proposed the existence of 21 distinct morphotypes in *C.longa*. These 21 morphotypes are grouped into six taxonomic varieties, namely, *C. longa* var. *typical*, *C. longa* var. *atypical*, *C. longa* var. *camphora*, *C. longa* var. *spiralifolia*, *C. longa* var. *musacifolia* and *C. longa* var. *platifolia*. Most of the Indian morphotypes belong to *C. longa* var. *atypia*.

Classification on the Basis of Maturity

Ginger varieties are classified on the basis of maturity:

(i) **Short duration varieties**- The short duration varieties are characterized by thick, long rhizome, low curcumin content and early maturity (180-200 days). For example G.L. Puram, Kasturi, Jabedi, Dughi, Dindigram, Udayagiri, Amalapuram *etc.* Some recently developed early duration varieties are Sudarshan, Suguna, Sugandham, Suvarna and Pant Peetabh. These varieties yield about 8.0-20.0 t (fresh) rhizome per hectare.

(ii) **Medium duration varieties**- Medium duration varieties, mature in about 230-240 days, and possess medium long, with close internodes, thick rhizome. *e.g.* Kesari, Kothapeta, Amruthapani, Rajapuri, Panamalur, Gorakhpur *etc.* Some recently developed medium duration varieties are Krishna, Suroma, Rashmi and Sonali. These varieties yield about 14.0-35.0 t (fresh) rhizome per hectare.

(iii) **Long duration varieties**- The long duration varieties are characterized by long, stout, smooth and hard rhizomes. These cultivars take about 9 months for maturity. For example Armoor, Duggirala, mydukur Tekurpet, Sugandham, Avanigadda, Ethamkula *etc.* Some recently developed late duration varieties are BSR-1, Ranga and Kanti. These varieties yield about 15.0-37.0 t (fresh rhizome) per hectare.

Augmentation of Genetic Resources and Crop Improvement

Turmeric varieties/cultivars such as Alleppey, Lekadong, Prabha, Prathibha, Kedaram, Alleppey Supreme, Wayanad Local, Mananthody, Aeing, Edappalayam, Roma, Sugandham *etc.* are rich in curcumin (5 per cent). Turmeric varieties also differ in their curcuminoid levels too. Variety x Location interaction is very high for curcumin content. Cultivar diversity of cultivated turmeric is the highest in India, South India, Odisha and north Eastern states are rich in cultivar diversity. A good number of varieties also exist in such countries as Bangladesh, Pakistan, Vietnam, Nepal, Myanmar, South Pacific Islands, Malagasi *etc.* Cultivar diversity of turmeric in India is given in table (12.16).

Table 12.16: Cultivar/Varietal Diversity in Turmeric

Sl.No.	*Cultivar/variety*	*Sl.No.*	*Cultivar/variety*
1	Armur	35	Lokhande
2	Alleppey	36	Lekadong
3	Alleppey Supreme*	37	Mundage
4	Avanigadda	38	Megha Turmeric*
5	Amruthapani	39	Mydukkur
6	Amalapuram	40	Nandyal
7	Balaga	41	Pattani
8	Bilaspur	42	Panamalur
9	BSR-1*	43	Pakistan
10	BSR-2*	44	Perumnadan
11	Bullapura	45	Prabha*
12	C-A-72 Udayagiri	46	Prathibha*
13	C-A-12	47	Rajapuri
14	CLL-324	48	Ranga*
15	CLL-328	49	Rasmi
16	Chinnanadan	50	Rajendra Sonia *
17	Chaya Paspu	51	Roma*

Sl.No.	Cultivar/variety	Sl.No.	Cultivar/variety
18	Co-1*	52	Sobha*
19	Deshi	53	Sona*
20	Duggirala	54	Shimla
21	Dunrigam	55	Sugantham*
22	Dughi	56	Suguna*
23	Erode Local	57	Suvarna*
24	Ethamukkala	58	Sudarsana*
25	Gorakhpur	59	Suranjana*
26	Guntur	60	Suroma*
27	GL Puram	61	Thekkurpetta
28	Jabedi (G-67)	62	TSundar
29	Kanti*	63	Thalachira
30	Kasturi Tanaka	64	Varna*
31	Kasturi	65	Vonimitta
32	Katpadi Local	66	Waynadan
33	Kedaram*	67	Yelachage
34	Kothapetta		

*Improved variety.

Source: Parthasarathy *et al.*, 2007.

A preliminary evaluation of 17 turmeric genotypes obtained from ICAR-NBPGR Regional Station, Vellanikkara Thrissur (Kerala) during 1991-1992.

Disease Resistance

A total of 237 accessions of various species of the genus *Curcuma* such as *Curcuma brog* Val. (2), *C. amarissima* Rosc. (5), *C. aromatic* Salisb. (34), *C. caesia* Roxb. (5), *C.comosa* Roxb. (2), *C. ferruginea* Roxb. (2), *C.haritha* Mangaly and Sabu (15), *C. latifolia* Rose (2), *C.leuicorrhiza* Roxb. (6), *C. aeruginosa* Roxb. (38), *C. Montana* Roxb. (1), *C. amada* Roxb. (49), *C. soloensis* Val. (1), *C. zanthorrhiza* Roxb. (45), *C. longa* var. *vanaharidra* Vela (1), *C. amada* var. *glabra* Vela. (7), and *C. raktakanta* Mangaly and Sabu (8) and two unidentified entities (14 accessions) collected from different eco-geographical regions of India and maintained at ICAR-NBPGR, Regional Station Vellanikkara (Kerala) were subjected to observation on incidence of Taphrina leaf blotch and *Colletotrichum* leaf spot diseases under natural epiphytotic conditions and results indicated a complete absence of Taphrina leaf blotch symptom in ten entities

Table 12.17: Yield and its Attributes in Turmeric

Inst. Code	Accession	Plant Height (cm)	No. of Tillers/ Plant	No. of Leaves/ Plant	Leaf Length (cm)	Leaf Width (cm)	Raw Yield (t/ha)	Curing (per cent)	Cured Yield (t/ha)	Curcumin Content (per cent)
VK88	IC 70130	126.33	1.88	23.00	56.33	14.33	32.88	20.76	6.83	5.02
VK	IC 70133	119.67	1.63	18.67	45.67	12.00	30.40	22.98	7.00	7.85
VK	IC 88752	141.00	1.47	15.00	53.67	15.00	29.97	17.71	5.33	6.98
VK	IC 70011	135.00	1.28	13.00	57.67	15.00	32.10	19.62	6.29	6.99
VK	IC 136894	138.00	1.28	12.33	47.00	11.67	29.40	22.22	6.51	5.13
VK	IC 88643	89.33	1.67	13.67	45.67	14.67	24.53	23.76	5.83	6.82
VK	IC 136887	156.00	1.73	17.33	63.67	17.00	22.20	20.35	4.51	7.23
VK	IC 88658	132.33	1.62	15.33	50.33	15.00	31.53	18.37	5.79	6.29
VK	IC 88662	98.00	2.00	24.00	50.33	14.83	38.83	21.68	8.43	6.88
VK	IC 136868	102.00	1.61	14.00	53.67	17.33	43.02	16.32	7.00	5.40
	Sugandham (c)	117.67.	1.14	12.67	54.67	15.00	25.42	17.43	4.44	5.25
NS= non-significant Significant at 5 per cent level		**23.71**	**NS**	**NS**	**NS**	**NS**	**3.52**	**0.64**	**0.78**	**0.22**

Source: Kurian and Nair, 1996.

Table 12.18: Promising Cultivars/Land Races of Turmeric

States	Location of Evaluation	Year of Evaluation	Promising Cultivars/Races	Yield (t/ha)
Meghalaya	Nayabunglow	1977-79	Daghi	28.4
			G.L. Puram	27.6
	Barapani	1982-1983	Manipur local	18.6
			Nagaland Local	18.8
	Barapani	1985-1986	Daghi	19.7
			G.L. Puram	18.3
			Lakadong	19.7
			Manipur Local	19.8

States	Location of Evaluation	Year of Evaluation	Promising Cultivars/Races	Yield (t/ha)
	Barapani	1988-1991	PCT-12	22.9
			PCT-13	23.7
			PCT-8	22.5
			G.L. Puram	22.8
	Barapani	1992-1993	PTS-10	33.4
			PTS-11	33.3
			VK-77	30.6
			VK-145	31.0
			RCT-1	30.0
Manipur	Mahadev hills	1981-1984	Duggirala	28.0
			G.L. Puram	23.6
			Chayapasupa	18.4
	Langol hills	1991-1994	Duggirala	29.5
			G.L. Puram	25.9
			PCT-11	21.6
			PCT-13	22.1
Mizoram	Kolasib	1983	Sel. No. 24	15.5
			Sugandhum	14.7
	Kolasib	1985	Sel. No. 24	20.5
			Avanigadda	19.5
			Dehradun Local	23.0
	Kolasib	1991-1993	Avanigadda	22.3
			Sel. No. 24	20.6
Nagaland	Jharnapani	1979	Nagaland Local	17.8
			Dehradun Local	19.0
			Sugandhum	15.6
	Jharnapani	1983-1985	G.L. Puram	27.6
			Kasturi Tanaka	27.4
			Lakadong	30.7
Sikkim	Tadong (Gangtok)		G.L. Puram	35.1
			Avanigadda	30.7
			Sugandhum	22.9
Arunachal Pradesh	Basar	1979	Sugandhum	18.6
			Dehradun Local	19.0
			Nagaland Local	18.4
	Basar		G.L. Puram	24.2
			Sugandhum	25.8

Source: Annual Report of ICAR-Research Complex from 1977 to 1993.

containing 97 accessions. Four entities with 14 accessions showed complete absence of symptom of both the diseases over the years.with respect to *Colletotrichum* leaf spot disease variation among the species and within the species over the years was noticed (Velayudhan, 2011).

Genetic Resource Conservation

ICAR-Central Agricultural Research Institute, Port Blair (Andaman and Nicobar Islands) is maintaining 5 turmeric germplasm possessing various useful horticultural traits. A total of 237 accessions of various species of the genus *Curcuma* such as *Curcuma brog* Val. (2), *C. amarissima* Rosc. (5), *C. aromatic* Salisb. (34), *C. caesia* Roxb. (5), *C.comosa* Roxb. (2), *C. ferruginea* Roxb. (2), *C.haritha* Mangaly and Sabu (15), *C. latifolia* Rose (2), *C.leuicorrhiza* Roxb. (6), *C. aeruginosa* Roxb. (38), *C. Montana* Roxb. (1), *C. amada* Roxb. (49), *C. soloensis* Val. (1), *C. zanthorrhiza* Roxb. (45), *C. longa* var. *vanaharidra* Vela (1), *C. amada* var. *glabra* Vela. (7), and *C. raktakanta* Mangaly and Sabu (8) and two unidentified entities (14 accessions) collected from different ecogeographical regions of India and maintained at ICAR-NBPGR, Regional Station Vellanikkara (Kerala).

Table 12.19: Species Showing Absence of Symptoms against *Taphrina* Leaf Blotch and *Colletotrichum* Leaf Spot

Species	Number of Collctions	Collection Identity
Species and collections showing absence of disease symptoms		
Curcuma species 3	5	IC266596, IC349879B, IC210466, 349850A, IC349879
C. caesia	5	IC313064, IC 313085, IC 266511, IC266608, IC 331735
C. comosa	2	IC313097, IC313092
C. ferrugenia	2	IC266513, IC313098
C. latifolia	2	IC266595, IC313093
C. montana	1	IC406441
Species and collections showing absence of *Colletotrichum* leaf spot only		
C. longa var. vanaharidra	1	IC360212
Species and collections showing absence of *Taphrina* leaf blotch only		
C. aeruginosa	33	IC88593, IC266532, IC88842, IC88843, IC88844, IC88845, IC88846, IC700009, IC70013, IC70015, IC70061, IC70072, IC88848, IC70110b, IC88849, IC88888, IC88917, IC88914, IC88912, IC313083, IC266538, IC88941, IC266546, IC266550, IC266551, IC266553, IC266561, IC266564, IC266566, IC266585, IC266588, IC266591, IC266524, IC406438, IC406439, IC349989, IC266531, IC29968, IC266511-B, IC210290, IC137087
C. amarissima	5	IC266531, IC29968, IC266511-B, IC210290, IC137087
C. aromatica	32	86850, IC88852, IC88929, IC88853, IC88856, IC88857, IC88858, IC88859, IC88860, IC88861, IC313087, IC266537, IC88938, IC88957, IC266543, IC266544, IC313090, IC406440, IC266569, IC266572, IC266577, IC266579, IC266580, IC266581, IC266582, IC266583, IC266590, IC360204, IC360205, IC210361A, IC266605, IC262716, IC34974, IC33215, IC349878
C. broh	2	IC29881, IC313086
C. latifolia	2	IC266595, IC313093
C. leucorrhiza	5	IC210263, IC210267, IC3210284, IC210325, IC210328

Source: Velayudhan, 2011.

REFERENCES

1. Chattopadhyay, I., Biswas, K., Bandhopadhyay, U. and Banerjee, R.K. 2004.Turmeric and curcumin: Biological actions and medicinal applications. *Current Science*, **87:** 44-53.
2. Karthikeyan, S., Jain, S.K., Nayar, M.P. and Sanjappa, M. 1989. *Florae Indicae Enmeratio Monocotyledonae.* Botanical Survey of India, Calcutta, India.
3. Kundu, B.C. 1967. Some edible rhizomatous and tuberous crops in India. *In. proc. Intl. Symp. Trop. Root Crops*, 124-130.
4. Kurian, A. and Nair, G.S. 1996. Evaluation of turmeric germplasm for yield and quality. *Indian J. Plant Genet. Resour.*, **9(2):** 327-329.
5. Parthasarathy, V.A., Sasikumar, B., Saji, K.V. and Prasath, D. 2007. Biodiversity of Zingiberaceous spices and their conservation. In: Agrobiodiversity, Vol.I. Crop Genetic Resources and Conservation (Eds. Kannaiyan and Gopalam, A.), Associated Publishing Co., New Delhi, p. 250-266.
6. Skornickova, J. and Sabu, M. 2002. The genus *Cucuma* L in India: Resume and future prospects. In: Das, A.P. (ed.). *Prospectives of Biology*. Bishen Singh Mahindrapal Singh, Dehra Dun, India, p. 45-51.
7. Sasikumar, B. 2005. Genetic resources of *Curcuma* diversity, characterization and utilization. *Plant Genetic Resources-Conservation and Utilization*, **3:** 230-251.
8. Shyamkumar, S. and Sasikumar, B. 2005. Molecular marker based genetic diversity analysis of *Curcuma* species from India. *Plant Genet. Res.*, 112-127.
9. Tewari, R.S. and Agarwal, A. 2004. Production technology of spices. International Book Distributing Co. Lucknow. p. 173-194.
10. Valeton, T.H. 1918. New notes on Zingiberaceae of Java and Malaya. *Bull. Jard. Butenzer Ser.*II, **27:** 1-81.
11. Velayudhan, K.C. 2011. Preliminary screening of Indigenous *Curcuma* germplasm against Taphrina leaf blotch and *Colletotrichum* leaf spot. *Indian J. Plant Genet. Resour.*, **24(3):** 277-282.
12. Velayudhan, K.C. 2016. The report of Curcuma kshonapatra (Zingiberaceae) from Karnataka, India. *Indian J. Plant Genet. Resour.*, **29(2):** 184-486.
13. Velayudhan, K.C., Amalraj, V.J. and Muralidharan, V.K. 1996. The conspectus of the genus *Curcuma* in India. *J. Econ. Taxa. Bot.*, 20: 345-382.
14. Velayudhan, K.C., Muralidharan, V.K., Amalraj, V.J., Gautam, P.L., Mandal, S. and Kumar, D. 1999. *Curcuma* genetic resources. Scientific Monograph No. 4, National Bureau of Plant Genetic Resources, New Delhi, p. 149.

Mango Ginger (*Curcuma amada* Roxb.)

Introduction

Mango-ginger (*Curcuma amada* Roxb.), commonly known as Am haldi (Sanskrit), Amada (Hindi), Mangai inchi (Bengali), Manga inch (Tamil), Amiya haldi (Malyalam) *etc.* as the name indicate that rhizome produce sweet smell of unripe mango, turmeric and ginger blend when crushed (Sharangi, 2009). Probably due to high usage in the preparations of pickles, sauce and various other culinary formulations, *C. amada* is the second most cultivated *Curcuma* species after *C. longa* (Syakumar and Sasikumar, 2007). It imparts mango flavor mainly due to the presence of car-3-ene and cis-ocimene (Achut and Bandyopadhyaya, 1984). The rhizomes of Mango-ginger is used as a food spice due to its mango flavor especially for the preparation of pickles, chutney, candy, preservatives, sauce, salads and in other culinary preparations. The rhizome has got excellent medicinal properties. Part of the rhizome is applied for the skin disease and itching. Tuber pastes made with the spirit and egg white is used as applicant for chronic rheumatism and bruises. Traces of essential oil and curcumin (0.1 per cent) are found in the rhizome. It is applied on the sprains, reducing muscular swellings *etc.* it has got a reputed position as a traditional medicinal herb. Ayurveda consider it is a sweet bitter, coolling, appetizer, antipyretic, aphrodisiac, anti-inflamatory, anti-mycobacterial, anti-hypercholesterolemic and anti-oxidant properties (Singh *et al.*, 2010). It is used as carminative and stomachic (Hussain *et al.*, 1992). It is also used as one of the ingredients of 'rasayanams' and number of carminatives and digestive 'churnams'. The plant is getting importance due to its antifungal properties found in volatile oil. It reported that the extract of rhizomes showed anti-inflammatory when it is tasted in albino rats (Majumdar *et al.*, 2000). On stem distillation of the dried rhizomes yield 1.0 per cent volatile oil, which constitutes d-U-pinene, ocimene, linalool, linalyl acetate and safrole (Rao *et al.*, 1989). The crop comes up well in open conditions, but can tolerate low levels of shade and therefore partially shaded situations can also be utilized for its cultivation. It can be well accommodated as an intercrop in coconut gardens and in rotation with other short duration crops like vegetables and also as a crop component in homesteads. The crop is easy to cultivate and free from major pests and diseases, which makes it attractive to the farmers. It is grown as a back yard crop in many households in Kerala.

Origin, History and Distribution

Mango-ginger is native of Malaya Archipelago, Siam and India. In India, it occurs as a wild in part of northeast, West Bengal, Odisha, Konkan region of Maharashtra, Madhya Pradesh, Chhattisgarh, and Tami Nadu. It can be also spotted in Kerala. Long back its use in traditional medicine is well reflected in ethno-botanical studies conducted mainly in Indian Subcontinent, Myanmar and Thialand.

Botanical Description

Mango-ginger is botanically related to neither mango nor ginger, but to turmeric (*Curcuma longa*), but has shorter crop duration of six months. Morphologically Mango-ginger plant is similar to turmeric, but has shorter crop duration of six months (Anonymous, 2006).

Table 12.20: Botanical Description

Growth Habit	It is rhizomatous herb with palmate and sessile tubers united to the sides of an ovate conic bud.which give rise to the
Leaf	Conic bud gives rise to the leaves. Leaves long, petiole in tufts, the blades are 30.0-35.0 cm long and 7.5-12.5 cm broad.
Floral biology	Conic bud gives rise to the spike. Flowers are autumnal spike. Flowering bracts 2.5 cm., greenish white, bract oblong, the lower half of the margins united to the backs of the two above, forming a pouch for a group of 4-5 flowers, coma longer and narrower, tinged with purple or light violet colour. Calyx nearly 13.0 mm long, unequally 3 toothed; corolla white or pale yellow, corolla tube funnel shaped. The plant closely resembles *Curcuma longa* (turmeric) but can be distinguished by cream coloured turmeric like rhizomes with the taste of fresh mango coupled with ginger flavor.
Rhizome	The rhizomes are pale yellow inside (central core) with lighter colour outside.Fiber content is less

Augmentation of Genetic Resources and Crop Improvement

'Amba' is a released variety from High Altitude Research Station, Pottangi, Odisha.

REFERENCES

1. Achut, S.G. and Bandyopadhyaya, C. 1984. Characterization of mango like aroma in *Curcuma amada* Roxb. *Journal of Agriculture and Food Chemistry*, **32:** 57-59.
2. Anonymous. 2006. http://www.kissankerala.net/kissan/kissancontents/mangoginger.htm
3. Hussain, A., Virmani, O.P., Popli, S.P., Misra, L.N. and Gupta, M.M. 1992. Dictionary of Indian Medicinal plants. Director, Central Institute of Medicinal and Aromatic Plants, Lucknow. 161-162.
4. Majumdar, A.M., Naik, D.G., Dandge, C.N. and Puntambekar, H.M. 2000. Anti-inflammatory activity of *Curcuma amada* Roxb. In albino rats. *Indian Journal of Pharmacology*, **32:** 375-377.
5. Rao, S.A., Bandaru, R. and Ramchandran, R. 1989. Volatile aroma components of *Curcuma amada* Roxb. *Journal of Agriculture and Food Chemistry*, **37:** 740-743.

6. Sharangi, A.B. 2009. Mangoginger: a two-in-one spice. *Indian Journal of Arecanut, Spices and Medicinal plants*, **11(2)**: 77-79.

7. Singh, K., Saikia, D., J.K., Shanker, K., Thakur, J.P. and Negi, A.S. 2010. A bioactive labmade diterpenoid from *Curcuma amada* Roxb. and its semisynthetic analogues as anti-tubercular agents. *European Journal of Medicinal Chemistry*, **45:** 4379-4382.

8. Syakumar, S. and Sasikumar, B. 2007. Molecular marker based genetic diversity analysis of *Curcuma* species from India. *Scientia Horticulturae*, **112:** 235-241.

Genetic Resources of Other Genus of Family Zingiberaceae

In family Zingiberaceae, few other species are economically important which are being given below:

(1) Genus Amomum

Species and its distribution- *Amomum* Roxb., is the second largest genus after *Alpinia Roxb.* in Zingiberaceae with about 180 species widely distributed in South-East-Asia (Tong, 1999). Many species of *Amomum* like *A. sbulatum*, *A. aromaticum*, *A. kepulaga*, *A. krervanth* and *A. tsao-ko* are used in traditional Chinese medicine and also considered as useful species in Indian medicine system. Genus *Amomum* occurs from the Himalayas through South-East-Asia to northern Australia and extends in to Central Pacific (Smith, 1985). *A. sbulatum* Roxb. (Large cardamom or Nepal cardamom) is distributed in tropical Asia, the Himalayas (Nepal, Bhutan) and India (Sikkim). It is cultivated in the sub-Himalayan region of north east India mainly in Sikkim and the adjoining areas. *A. aromaticum* Roxb. (Bengal cardamom) is a perennial herb distributed in North East India, Bangladesh, and Nepal. It is cultivated in Bengal, Assam and the north eastern region of the foot hills of the Himalaya. *A. kepulaga* (Round cardamom or Siamese cardamom) is endemic to western Java. It is cultivated in Java, Sumatra, Moluccas and South China. *A. krervanth* (Cambodian cardamom) is distributed in Thailand. *A. tsao-ko* (Tsao-ko cardamom) is endemic to Vietnam and South China. It is cultivated in South China and Vietnam.

Large cardamom is used as a spice in pickles, pulao, meat and vegetable dishes, used in industrial sector for flavouring tooth pastes, sweets, soft drinks, *etc.* it is also used as an ingredient to give a flavor to pan masala and curry powders (Varadasan and Biswas, 2002). It is said that in gulf countries large cardamom is used as a cheaper substitute for spicing tea in place of cardamom. Large cardamom is used as an ingredient in the indigenous and Unani medicine systems. Seeds and capsules of *A. aromaticum* are used as spice as well as for herbal medicinal preparation. The fruits of *A. kepulaga* are used as a spice. It is also used in many Chinese medicinal preparations. Seeds of *A. krervanth* is used as spice for flavouring curry and cakes and also used in traditional medicine. In Europe, it is used for flavouring sauces and liquor. The fruits and seeds of *A. tsao-ko* are used as a spice in South China and Vietnam as a substitute for true cardamom. It is also used as a local remedy for many ailments such as dyspepsia, diarrhea, indigestion, *etc.*

Botanical Description

Plants are generally evergreen herbs distributed in wet forests in light gaps as well as at forest margins. It is characterized by basal, compact usually cone like inflorescence, flowers become singly in the axils of prominent bracts, tubular bracteoles, a broad concave labellum slightly longer than the corolla lobes and a well-developed anther appendage bilobed, trilobed or entire. The fruits of *A. sbulatum* are dark brown, obovoid and three valved with numerous large seeds in each compartment. The seeds are held together by a thick sugary pulp. The properties of seeds are barely similar to that of true cardamom. Genetic resources- Ramsey, Sawney, Golsey, Varlangey and Bebo are the important cultivars of large cardamom (Gyatso *et al.*, 1980)

(2) Genus *Aframomum*

Species and its Distribution

Aframomum angustifolium, A. hanburyi, A. koranima and A. melegueta are the spice identified from this genus. *A. angustifolium* (Sonn) K. Schum (Madagascar, cardamom) is found distributed in E. Africa, Madagascar, Zanzibar, Mauritius, Pemba and the Seychelles. It is cultivated in the Pemba Islands. *A. hanburyi* K. Schum (Cameron cardamom) is native to West Africa. It is cultivated in Cameroon. *A. koranima* Perieira Ex Engl. (Koranima cardamom) is native of Ethiopia and Somalia. *A. melegueta* Schum (Grains of paradise, Guinea grain) of West Africa is also cultivated in Guinea, Sierra, Ghana, Benin, Nigeria, Cameroon and Angola.

The seeds of *A. angustifolium* are used as a spice like black pepper. The seeds of *A. hanburyi* are used as a spice as well as condiment. The fruits and seeds of *A. koranima* Perieira are used as a spice generally for flavouring coffee, tea, bread, butter, *etc.*, in the areas where it generally grown. The seeds of *A. melegueta* Schum are used as a spice to flavor all kinds of food in West African countries.

Botanical Description

Genus *Aframomum* is perennially large with leafy shoots and is fully developed at flowering time. Leaf sheaths always form a distinct pseudostem.

Inflorescence forms the base or forms rhizome. Flowers are large, zygomorphic. Calyx spathaceous, lateral petals are narrow, posterior is larger and hood like. Staminodes are larger and petalloid, fruits are usually red in colour. The plant of *A. melegueta* Schum., grows natueally at 1700-2000m altitude in Ethiopia and flowers from January to September.

(3) Genus *Alpinia*

Species and its Distribution

Genus *Alpinia* is the largest and most wide spread in the family Zingiberaceae with about 230 species occurring from Sri Lanka and the Western Ghats of India to China, Japan, South-East-Asia, the Pacific as far as Fiji, Samoa, and the Caroline Islands and Australia (Kress *et al.*, 2005). In this genus two species *i.e. A. galangal* and *A. officinarum* are grown as spices.

(i) *A. galangal* (L.) Willd (Greater galangal) is native of Indonesia, but has naturalized I many parts of South and South-East-Asian countries. It is widely cultivated in Bangladesh, South India and South-East-Asia. *A. galangal* is manly used in several medicinal preparations for skin disease, indigestion, dysentery enlarged spleen, *etc.* the plant is used to treat loss of appetite, upper abdominal pain, *etc.* It is also used as cattle medicine. The young rhizome is used as spice to flavor various dishes in Malaysia, Thailand, Indonesia and China.

(ii) *A. officinarum* (Lesser galangal) is generally distributed in South China and Japan. The rhizomes are used as spice for flavouring food. There is no clear cut distinction between the rhizomes of greater galangal and lesser galangal.

(3) Genus *Kaempferia*

Genus *Kaempferia galangal* L., a native of tropical Asia is distributed in the tropics and subtropics of Asia and Africa. It is cultivated in India, Sri Lanka, Malaysia, Moluccas, Philippines Islands and South-East-Asia.

Botanical Description

Kaempferia galangal L. is a short plant attaining about a height of about 30.0 cm and has fleshing, cylindrical aromatic root lubes. It is generally cultivated for the aromatic rhizome and is used as a spice. It is extensively used for flavouring rice for indigenous medical systems.

REFERENCES

1. Gyatso, K., Tshering, P. and Basnet, B.S. 1980. Large cardamom of Sikkim. *Publication of Dept. of Agric., Govt. of Sikkim*, **2(4):** 91-95.
2. Kress, W., Liu, A. Z., Newman, M. and Li, Q.J. 2005. The molecular phylogeny of Gingers. *American J. Bot.*, **92:** 167-178.
3. Smith, R.M. 1985. A review of Bornean Zingiberaceae: 1 (Alpineae). *Notes from the Royal Botanic Garden, Edunburgh*, **42:** 261-314.
4. Tong, S.Q. 1999. Zingiberaceae. In flora Yunnanica, Tomus 8, (ed). Zheng-Yi Wu, Beijing: Science Press, p. 528-651.
5. Varadasan, S. and Biswas, A.K. 2002. Large cardamom (*Amomum subulatum* Roxb.). In: Cardamom the genus Elettaria (Eds). Ravindran, P.N. and Madhusoodanan, K.J., Taylor and Francis Publishers, p. 299-329.

13

Genetic Resources of Salad/Garnish Vegetables and Crop Improvement

Genus *Apium*

Celery (*Apium* spp.)

Introduction

Celery (*Apium graveolens* L. var. *dulce* and *Apium graveolens* L. var. *rapa*, 2n=22) is also known as jawain ka Patta (Hindi); Ran Dhuni Sag (Bengali); Ajaman Pan (Gujarati) and Jauni Patra (Odisha). It is a very important leafy vegetable (rank second after lettuce) which leaves and stems (long fleshy leaf stalk somewhat look like leafy onion) are used as salads (Saini, 1996). Edible celery leaves contains 88.0 g moisture, 6.3 g protein, 0.6 g fat, 2.1 g minerals, 1.4 g fiber, 1.6 g carbohydrate, 37K. calorie energy, 230.0 mg calcium, 140.0 mg phosphorus, 6.3 mg iron, 3990 microgram carotene, 0.11 mg riboflavin, 1.2 mg niacin and 62.0 mg vit. C. Similarly, stems contain 93.5g moisture, 0.8 g protein, 0.1 g fat, 0.9 g minerals, 1.2 g fiber, 3.5 g carbohydrate, 18 K. calorie energy, 30.0 mg calcium, 38.0 mg phosphorus, 4.8 mg iron, 520 microgram carotene, 0.12 mg thiamine, 0.05 mg riboflavin, 0.3 mg niacin and 6.3 mg vit. C (Gopalan *et al.*, 1999). Apart from this, 100 g edible portion of leaves contain 52.0 mg maganesium, 35.3 mg sodium, 210.0 mg potassium, 0.30 mg copper, 102.0 mg sulpher and 19.0 mg chlorin. Leaves also contains high amount of lysine, tryptophan amd methionine among essential amino acids. As per quantification, leaves contain 25.0 mg arginine, 9.0 mg histidne, 15.0 mg lysine, 80.0 mg tryptophan, 280.0 mg phenyl alenine, 14.0 mg methionine, 210.0 mg threonine, 430.0 mg leucine, 240.0 mg isoleucine and 300.0 mg vallin. In celery apiin glucoside is very commonly available and principally, volatile oil contains 'terpene' but specific aroma is due to 'lactone sedanolid' (Susiatri and Siemonsma, 1993). Its wild plants were used for medicinal purpose 100 years ago. Celery is considered as tonic, stimulant, diuretic, carminative *etc.* Its use helps in digestion and also cures flatulence and other stomach troubles. The use celery seed in pills for relieving pain was described by Aulus Cormelius Celsius around AD 30. Celery is a healthy medicinal food, often used to reduce the need for salt in cooking. Celery seeds contain a compound, 3-n-butylphthalide, that has been demonstrated to lower blood pressure in rats. The seeds contain aliening and isoquerticin compounds that help blood vessels to expand, lowering blood pressure. Celery can reduce excess water buildup and decrease the uric acid that causes pain and inflammation in gout and arthritis. It is used in weight-loss diets, where it provides low-calorie dietary fiber bulk. Celery is often incorrectly thought to be a "negative-calorie food", the digestion of which burns more calories than the body can obtain. In fact, eating celery provides positive net calories, with digestion only consuming a small proportion of the calories taken in.

Some therapeutic uses are:

- Celery seeds and rock salts should be taken with warm water in stomachache and other stomach related troubles.
- Its use in morning and evening with water is useful in chronic fever, cold, *etc.*
- Celery has antispasmodic properties and is useful in treatment of asthama, bronchitis, pleurisy and tuberculosis.
- It is useful in diseases related to urine bladder, stomach and respiration pain.

The seeds are exported to European countries as condiment.

Origin, History and Distribution

Celery is native to marsh lands from Sweden to northern Africa and from southern Asia to Caucasus. The centre of origin of the genus *Apium* is considered to be the Mediterranean basin (Simonds, 1976). Celery seed is known for over 3000 years. It is likely that *A. graveolens* was first cultivated as a medicinal crop as early as 400 BC in Egypt and in the Roman Empire. As per the historical records, Jems (1953) describe in details about this crop in his Odyssey book which is considered 850 BC. Celery was domesticated as a true vegetable in Italy in the 16th century (de Vilmorin, 1950). Some people believed that celery plants were developed in lower part of Italy and from there it reaches to France and England. First time this crop was grown in France during 1623. With other information, it is also believed that celery originated from moisture prone area of Sweden and from there it reaches to Algeria, Ethiopia, Baluchistan and mountains of India for commercial cultivation. Still today, celery plant in their natural habitat can be spotted in California and New Zealand. It is now widely grown in USA, France and other European countries commercially and is available in the market throughout the year (Carlos F. Quiros, 1993). Celery introduced in India during 1930 at Amritsar (Punjab) with the mission to produce seed and export. In the high elevation of southern part of India, it is grown on limited scale for salad purpose (Yawalkar, 1969). In some pockets of northern India like Punjab and Uttar Pradesh celery is grown mainly as seed crop. It is cultivated in France, Great Britain, Hungary, Japan, India, the Netherlands and the United States. In India, it is mainly cultivated in Punjab, Haryana and Uttar Pradesh.

Species Availability, Evolution and Diversity

Under genus *Apium*, 20 species are reported which are either annual or biennial. Among them two species *i.e. Apium graveolens* var. *dulce* and *Apium graveolens* var. *rapa* are popular as vegetable crop. As per constituents and properties, *Apiumgraveolens* var. *dulce* has strong aroma and *Apiumgraveolens* var. *rapa* has sweet taste. Most of the species of *Apium* are diploid (2n=22) (Amseveler, 1929). Its triploids were developed by crossing *Apium prostratum* x *Parsle* (Whitaker, 1941). Celery is derived from *A. graveolens* L., which occurs wild as a marsh plant throughout temperate Europe and Asia (Purseglove, 1965).

Botanical Description

Celery belongs to the family Apiaceae.

Table 13.1: Botanical Description

Growth Habit	Celery is biennial and sometimes annual. Generally plants are erect.
Root	Fleshy roots.
Stem	Stem angled, grooved, and long striped.
Leaf	Leaves are compound and form leaflets. Leaf petiole is long. Leaf stalks are 15.0-38.0 cm long bearing three pairs and a terminal leaflet, coarsely serrated and tenately lobed and divided.
Floral biology	Flowers are 60.0-70.0 cm high, branched and leafy (Bailey, 1950). Inflorence are small white or green in colour which are borne in compound umbels. Flower has 5 petals and 5 stamens
Anthesis and pollination	Celery flowers are protandrous (degree of protoandry varies) with pollen being released 3-6 days before the stigma is receptive. At that time the anthers will drop and the two stigmas will unfold in an upright position, becoming receptive (Honma, 1959). The degree of protoandry varies which make reliable hybridization difficult to perform because of the possibility of accidental selfing. Celery has very high degree of out crossing, approximately 70 percent, due to to wind and insect pollination (Orton, 1984). The pollinators observed at Davis visiting celery flowers included a large number of species, covering ten families of wasps and bees and eight familier of flies (Quiros *et al*, 1986). In order to induce seed stalk development, plant requires a period of vernalization. Six to ten weeks at 5-8°C when they are atleast 4 weeks old (Honma, 1959 and for 6-8 months at -10°C (D' Antonio and Quiros, 1987). Due to a wide range of responses to the cold treatment, it is often difficult to synchronize the crosses since plants will flower at different time. However, pollen can be stored for 6-8 months at 10°C with a viability decline of approximately 20-40 percent, thus providing flexibility to perform crosses most of the time.
Fruit and fruit set	Fruits are bicarpelled. Fruit is rigiod, 1.0-1.5 mm long and 1.0 mm in diameter, and contains a small seed.
Seeds	Seeds are very small brown, slow in germination (Knott, 1960), 10.0 g seeds count about 26450 and seeds are somewhat bitter in taste. A 10 g seed counts about 26450.

Horticultural Classification

Classification on the Basis Plant Part Used

According to Carlos F. Quiros (1993) *Apium graveolens* L. comprises three distinct cultivated forms or taxonomic varieties, which are grown for different economic purpose:

- Celery (*Apium graveolens* var. *dulce*), common name celery: The plant of this type develop succulent, solid petioles which are consumed mostly raw in salads or as celery 'sticks' and are often cooked in soups and stews. Celery is widely gron in Europe and in the American province.

- ✰ Celeriac or Root celery (*Apium graveolens* var. *rapaceum*), common name celeric or root celery: It is characterized by the development of enlarged hypocotyl and root tissue resulting globe like structure, which has a strong flavor, is consumed mostly cooked in stews and soups or grated on salads. This crop is grown mostly in northern and Eastern Europe.
- ✰ Smallage or Leaf celery (*Apium graveolens* var. *secalinum*), common name smallage or leaf celery : The plants of this group have slender, leafy and often hollow petioles and are used mostly for their leaves as a condiments garnish and for medicinal purpose. Smallage is consumed mostly in Asia and in the Mediterranean countries. Annual forms of this variety are also grown for seed, which is used as conditments and for beverages.

Classification on the basis varieties behavior

Celery crops are two types, *viz.*

(i) Self blanched type

(ii) Green cultivars type

Self-Blanched Type

Self blanched type or golden self-blanch is tends to blanch easily and has less demand due to low vitamin content. They are Michigan Improved Golden, Cornell 19 (Thick petioles susceptible to brown spot), Cornell 619 (Thick petioles resistant to *Fusarium* yellow), Supreme Golden, Golden Plume or Wonderful, Golden Self-Blanching, Golden No. 15 and Florida Golden.

Green Cultivars Type

Green cultivars are gaining importance now days. They are Utah (dark green), Utah 52-70 (resistant to brown spot), Utah 10-B (have boron and magnesium deficiency), Giant Pascal, Emerson Pascal, Summer Pascal (resistant to brown spot) and Ford Hook. All green cultivars are resistant to *Fusarium* yellow. Emerson Pascal is resistant to early and late blights also, but has a tendency to bolt.

Augmentation of Genetic Resources and Crop Improvement

Cultivar 'Utah 52-70' is best for winter and Golden self-blanch for summer sowing.

Genetic control of mineral like Mg-inefficiency resistance in celery is controlled by single recessive gene and it may be possible reduced uptake. Similarly, B- inefficiency resistance is also is controlled by single recessive gene.

Table 13.2: Some of the Celery Varieties and their Horticultural Traits

Sl.No.	Varieties	Duration (Days)	Plant Height (cm)	Colour	Leaf Size (cm) L-B	Yield 3 m Line (kg)
1.	Utah-52-70	100	57	Dark green	17.4-9.2	2.738
2.	Golden Self Blanch	82	31	Medium yellow	14.6-14.6	1.533
3.	White Gem	100	35	Green	13.2-5.3	2.325
4.	Grant Red	70	67	Light pink	12.0-6.0	2.000
5.	Belder	67	18	Stem Dark green	12.0-5.3	1.700
6.	White Solid	68	32	Light green	14.7-5.8	1.300

Source: Peshin, 1998.

Genetic Resource Conservation

Celery germplasm are being maintained at Breeding Research Institute for Vegetables, Omou Check Republic; ICAR-National Bureau of plant Genetic Resources, New Delhi, Vavilov Institute for Plant Industry, Petersburg, Russia; Institute of Horticultural Research, Welsban, U.K. and North East Regional Plant Introduction Station, Geneva, USA (Susiatri and Siemonsma, 1993).

REFERENCES

1. Baily, L.H. 1950. The Standard Cylopedia of Horticulture. The Mcmillan Co; New York.
2. D' Antonio, V. and Quiros, C.F. 1987. Viability of celery pollen after collection and storage, *Hort. Sci.*, **22:** 479.
3. D' Antonio, V. and Quiros, 1987Gopalan, G., Rama Sastri, B.V., Balasubramanian, S.C., Rao, B.S.N., Deosthale, Y.G. and Pant, K.C. 1999. Nutritive value of Indian Foods. National Institute of Nutrition, ICMR, Hyderabad, India.
4. De Vilmorin, R.L. 1950. Pascal Celery and its Origin. J.N.Y. Bot. Gard; 51:39-41.
5. De Vilmorin, R.L. 1950. Pascal celery and its origin. *J.N.Y. Bot. Gard.*, **51:** 39.
6. Honma, S. 1959. A method for celery hybridization. *Proc. Am. Soc., Hortic. Sci.*, 73: 345.
7. James, W. 1953. The Odyssey.
8. Orton, T.J. and Arus, P. 1984. Out crossing in celery, *Euphytica*, **33:** 741.
9. Peshin, S.N. 1998. Promotion of Rare Exotic Vegetables. In: Production and Post-Harvest

Handling of Vegetables. Centre of Advanced Studies, Horticulture-Vegetable (ICAR), Dept. of Vegetable Crops, Dr. Y.S. P,U,H.F., Nauni-Solan, Himachal Pradesh.

10. Purseglove, J.W. 1965. Tropical Crops: Dicotledons, E.L.B.S. Longman London.
11. Quiros, C.F., Hulbert, S.B. and Durgan, M.E. 1986. Cytological and genetical studies of male sterile celery, *Euphytica*, **35:** 867.
12. Quiro, C.F. 1993. Celery, *Apium graveolens* L. In: Kalloo, G. and Bergh, B.O- (eds.) Genetic improvement of vegetable crops, pergamon, Oxford; p. 523-534.
13. Saini, G.S. 1996. Leafy salads vegetables. AVI Publishing Co., Westport, Connecticut.
14. Simonds, N.W. 1976. Evolution of crop plants. Longman, London and New York.
15. Susiatri, S. and Siemonsma, J.S., 1993. *Apiumgraveolens* L. In: PROSEO. Plant Resources of South East Asia and Vegetables. Pudoc Scirntific Publishers. Wageningen.
16. Whitaker, T.W. 1941. The occurrence of spontaneous triploid celery. *Proc. Amer. Soc. Hort. Sci.* **39:** 346-348.
17. Yawalkar, K.S. 1969. Vegetable Crops in India. Agri-Hort. Publishing House, Nagpur (Maharashtra).

Genus *Petroselinum*

Parsley (*Petroselinum crispum* Mill.)

Introduction

Parsley (*Petroselinum crispum* Mill.; 2n=22) is the most popular garden herb grown during the cool season. Due to nutritional and medicinal properties it is one of the most familiar and widely cultivated salad green vegetable worldwide. Parsley is very rich in vitamin A, C, iron, minerals, protein and sugars. The plant yields a volatile oil of which the active principle is 'apiol' (Franz and Glasl, 1976). Fresh leaves are rich in vitamin C, iron calcium and vitamin A. It contains (per 100g of edible portion : moisture 74.6 g, protein 2.2 g, fat 0.3 g, carbohydrate 1.5 g, vitamin C 90 mg, vitamin A5200 IU, iron 17.9 mg, thiamin 0.04 mg, riboflavin 0.18 mg, calcium 125 mg, niacin 0.5 mg and phosphorus 90 mg (Aykroyd, 1963). The leaves are used as a garnish and for flavouring sauces, soups, omlets, and stuffing. It is minced and stew over potatoes, carrot or other boiled vegetables (Singh and Singh, 2007). Crinkled or rosette (moss curled) leaves is also used as ornamental plant in pots and in small beds in the gardens. The essential oil in the leaves is superior and used as condiments and seasoning. Essential oil is also used for preparation of gripe water. Parsley tea is used as remedy to rheumatism. Field parsley is easy to prepare an unusual accompaniment to meat. The parsley smell is supposed to absorb the intoxicating fumes of wine and thus prevent any usual effects of over indulgence (Yamaguchi, 1983). The swollen root of turnip-rooted parsley is eaten as a cooked vegetable. The characteristic odour parsley is due to presence of 'apiin' present in all parts of the plant.

Origin, History and Distribution

Parsley is a native to southern Europe (Mediterranean region) and has been in cultivation for over 2000 years. It is found in wild form and cultivated all over the world. Parsley was first used for flavouring of dishes by the ancient Greeks and Romans. Its occurrence can be spotted throughout the tropics but tends to achieve out rather quickly near the equator. It is now well naturalized in most temperate countries of the world (Purseglove, 1984). UK, Italy and Asia are major growing area of parsley. This prized vegetable crop is not common in India (Swaider *et al.*, 1994).

Species Availability, Evolution and Diversity

Leaf Parsley

The two main groups of parsley used as herbs are:

i. **Curly leaf-** (*P. crispum* crispum group; syn. *P. crispum* var. *crispum*) and
ii. **Italian, or flat leaf-** (*P. crispum* neapolitanum group; syn. *P. crispum* var. *neapolitanum*);

Of these, the neapolitanum group more closely resembles the natural wild species. Flat-leaved parsley is preferred by some gardeners as it is easier to cultivate, being more tolerant of both rain and sunshine (Stobart, 1980), and has a stronger flavor (Huxley, 1992), though this is disputed while curly leaf parsley is preferred by others because of its more decorative appearance in garnishing. A third type, sometimes grown in southern Italy, has thick leaf stems resembling celery.

Root Parsley

Another type of parsley is grown as a root vegetable, the Hamburg root parsley (*P. crispum; radicosum* group, syn. *P. crispum* var. *tuberosum*). This type of parsley produces much thicker roots than types cultivated for their leaves. Although seldom used in Britain and the United States, root parsley is common in central and eastern European cuisine, where it is used in soups and stews, or simply eaten raw, as a snack (similar to carrots). Although root parsley looks similar to the parsnip, which is among its closest relatives in the family Apiaceae, and its taste are quite different.

Botanical Description

Kingdom	:	Plantae
Unranked	:	Angiosperms
Unranked	:	Eudicots
Unranked	:	Asterids
Order	:	Apiales
Family	:	Apiaceae
Genus	:	*Petroselinum*

The botanical description are given below:

Table 13.3: Botanical Description

Growth Habit	It is basically biennial or short lived, small bushy, or glabrous perennial and adopt well in cool season/ temperate (Choudhury, 1987) or an annual herb in subtropical and tropical areas. It attains the height of 30-70 cm.
Root	It has a swollen stout taproot used as a food store over the winter.
Leaf	Crinkled or rosette (moss curled) leaves in a dense tuft, dark green, shiny, 10–25 cm long, 3-pinnate, numerous leaflets 1-3 cm long, cuneate, lobed, often much crisped.
Floral biology	Flowering stem to 75 cm tall with sparser leaves. Flowers are umbel which is 2-10 cm in diameter and flat topped. Generally flowers are 2 mm diameter yellow to yellowish-green flowers.
Anthesis and pollination	It is a cross pollinated vegetable crop (Thompson and Kelly, 1972).
Seeds	Numerous small seeds developed. The seeds are ovoid, 2–3 mm long, with prominent styleremnants at the apex.

Horticultural Classification

There are three common varieties of parsley:

1. Italian or flat leaf parsley
2. Curly leaf parsley
3. Hamburg parsley

Augmentation of Genetic Resource and Crop Improvement

There are several varieties which are introduced and being cultivated are, Mass Curled, and Hamburg Parsley.

In 'Nepal Green Carpet' and 'New Curl Summer' developed by Takii Japan are being grown commercially.

REFERENCES

1. Aykroyd, W.R. 1963. ICMR Special Rept. Series No. 42.
2. Choudhury, B. 1987. Vegetables, National Book Trust, India, New Delhi, pp. 182-183.
3. Franz, C. and Glasl, H. 1976. The essential oils of parsley, II. Comparative investigations of fruit, leaf and root oils of some parsley varieties. *Qualitas Plantarum*, **25:** 231.
4. Huxley, A., 1992. New RHS (ed.) *Dictionary of Gardening* 3: 532. MacMillan, ISBN 0-333-47494-5
5. Purseglove, J.W. 1984. Tropical Crops-Dicotyledons. Vol. I and II (edn. 1 reprinted, 1984).
6. Singh, D.N. and Singh, S.P. 2007. Production Management of Underutilized Vegetables. Agrihortica Publication Junagadh, Gujarat, India, p. 191-198.
7. Stobart, T. (1980). *The Cook's Encyclopaedia*. MacMillan ISBN 0-333-33036-6.
8. Swaider, J.M., Ware, G.W. and Mc Collum, J.P. 1994. Producing Vegetable Crops. International Book Distributing Co. Charbagh, Lucknow (U.P.), India.
9. Thompson, H.C. and Kelly, W.C. 1972. Vegetable Crops. Tata Mc Graw-Hill Publishing Co. ltd., New Delhi, pp. 272-273.
10. Yamaguchi, M. 1983. World Vegetables Flis Horwood Limited.

Genus *Lactuca*

Lettuce (*Lactuca sativa* L.)

Introduction

Lettuce (*Lactuca sativa* L., 2n=2x=18) is unique salad crop among major vegetables (world's most important and widely-consumed leafy salad vegetable) for its nearly exclusive use as a fresh or raw products. It is an important gradient in a salad or a sandwich. The current word *lettuce*, originally from Middle English, came from the Old French *letues* or *laitues*, which derived from the Roman name (Chantrell, 2002). The word 'lactuca' is derived from the Latin word 'lac', meaning milk. Lettuce is derived from French word 'laiturte', meaning milky. Upon cutting, all lettuce types ooze milky substances and hence the name was derived. It is cultivated for its tender crisp leaves/heads. Lettuce is good source of vitamins particularly vitamin A and minerals like calcium and iron (Sharma *et al.*, 2001). High content of magnesium in its juice is exceptionally good for muscular tissues, nerves and the brain. In general per 100 g lettuce leaves contain 95.1 per cent moisture, 1.2 g protein, 0.2 g fat, 2.5 g carbohydrate, 900 IU vitamin A, 0.06 mg thiamin, 0.06 mg riboflavin, 0.4 mg niacin, 8mg ascorbic acid, 35 mg calcium, 26 mg phosphorus and 2.0 mg iron. Despite its cultivation under varied conditions it remains largely an underutilized crop. it thrives well during the cooler parts of growing season, and to be of the highest quality, their growth must be fairly rapid and continuous (Thompson, and Kelly, 1972).

Origin, History and Distribution

Lettuce (all four species *i.e. Lactuca sativa, L. serriola, L. saligna* and *L. virosa*) has originated in the Mediterranean region (basin) (Vavilov, 1935). It is believed that lettuce has been originated in Asia Minor, the Trans-Caucasus, Iran and Turkistan (Mehta, 1959). They occur most abundantly in the temperate regions of the world. *Lactuca sativa* originated in the eastern Mediterranean region about 7000 years ago, most likely in Egypt when the climate of that part of north eastern Africa was not as dry as it is today. Unfortunately, the available illustrated record found in ancient monuments is not complemented by dry or carbonized remains of leaves or seed of this crop in Egypt. Evidence from Egyptian tomb painting (as depicted by rosettes of a tall, large vegetable with subulate leaves painted on the walls of tombs at Thebes, 2500 B.C.) indicated its cultivation before 4500 BC (common oilseed crop in Upper Egypt). Keimer (1924) informed that early agriculture in Upper Egypt included lettuce as a seed crop for its edible oil and was being used in that country as such as recently as 1964. Subsequently, other refined edible oils more economically extracted from ground nuts, sunflower, olive, and maize has taken over throughout the Mediterranean region, particularly in the ancient Greeks and Roman civilizations. Lettuce was used by Persian royalty around 550 BC and was popular among Romans about the beginning of the Christian era. Salad lettuce was brought to the New World following Columbus's voyage in 1492 and by 1565 it was being grown abundantly in Haiti and the Caribbean. Around 6500 years ago lettuce was transported to Greece and subsequently spread throughout the Roman Empire and was known to Hippocrates, Aristotle and Pliny. It then spread to France, England and rest of Europe and then to the New World, where it was mentioned as early as 1494 (Hedrick, 1972; Ryder and Whitaker, 1976). In China, lettuce is known since 5th century but cultivation started in 7th century A.D., where a special cultivar (cv.) of the asparagus type (*L. serriola* f. *integrifolia*) was developed and reported by Fuchs in 1543 from Sinkiang. The local name of this cv. was 'Usum' or 'On-Sen'. Domesticated *L. indica* is also grown in china and Japan and latter introduced to Java and parts of Malaysia. Its leaves are usually boiled and eaten with chopsticks as a vegetable in Chinese cooking. After Columbus discovered the New World in 1942, lettuce cultivation quickly spread there. It was reported to be abundant in Bahamas by 1494, in Haiti by 1565 and Brazil by 1647. Lettuce was introduced into India either by the Portuguese or the British in the 16th country. Lettuce is a major salad crop in North America, Australia and most countries of Europe and South America. USA is the largest producer and consumer of lettuce. Apart from USA, other important lettuce growing countries include Australia, Japan, China, Israel, Northern Mexico, Chile, Argentina, Brazil and Peru (Pink and Eavan, 1993). Its popularity is increasing in Africa, the Middle East and Japan. In India it is grown in the kitchen gardens to meet the demand of continental hotels. But as a commercial it is becoming popular in northern part of the country and grown as winter season vegetables (Singh and Dhaliwal, 2002).

Species Availability, Evolution and Diversity

It represents 100 species of annual and perennial plants and is allied with other genera, such as chicory (*Cichorium intybus*) and dandelions. The common cultivated salad lettuce (*Lactuca sativa*), is one of the four species with 8, 9 and 17 pairs of chromosomes but common is (2n=18), the other three being *L. serriola, L. saligna* and *L. virosa*. The latter three species have now become obsolete and are rarely cultivated. Lettuce (all four species *i.e. Lactuca sativa, L. serriola, L. saligna* and *L. virosa*) are commonly found. Three main theories have been advanced concerning the origin of cultivated lettuce (Damania, 1975). The presence of latex in primitive lettuce also marked the plant as a fertility symbol (Damania, 1975)

i. It was domesticated from wild forms of *L. sativa*,
ii. That it arose as a result of a natural cross between *L. sativa* and *L. serriola*, and
iii. From hybridization among several species.

The latter theory is based on cytological and genetical evidence put forward by Lindquist (1960). However, the most commonly accepted theory of domestication of *Lactuca sativa* is that it evolved as a natural selection from its closest wild progenitor, *L. serriola*, which is a prickly biennial of the temperate regions (Purseglove, 1968). The cultivated vegetable lettuce exhibits a close relationship to a group of other *Lactuca* species. Its closest wild relatives *L. serriola* is found all over the Mediterranean region and the Near East, has colonized large areas as a weed much outside these territories. For instance, following the highly successful post Columbus colonization of the New world, *L. serriola* has become a proliferate weed in North America.

- ***Lactuca serriola***: Seems probable progenitor of cultivated lettuce. There are several wild species of lettuce, which are mostly used as source of resistance to insect pests and diseases. These wild species are:
- ***L. sativa***: It is most diverse species having several different forms that are nevertheless completely inter fertile. It is quite likely that this species has been derived through selection from hybrids between *L. serriola* and a third species *L. saligna* (Lindquist, 1960).
- ***L. serriola***:This has usually small seeds, reflexed involucre, spines, shoot formation at the stem base, relatively narrow leaves that may have

entire or toothed margins and a resetting tendency.

- ✰ **L. saligna:** It is similar to *L. serriola* in several respects but differ in having very narrow, toothed leaves and bearing sessile flowers on a spike compared to pedicelled flowers on panicle as in *L. serriola*.
- ✰ **L. virosa:** It is morphologically distinct. It has large flattened seeds; strongly resetting, blue green leaves, and includes biennial well as annuals forms.

Lactuca virosa is a preferred wild species from Europe—nutrient dense, good winter salad, bright purple-black seeds, much less spiny than the new world weedy types. Mature plants contain high concentrations of bitter, rubbery sap, which when milked from the plant and dried is known as 'lactucarium'. So concentrated, the sap is mildly sedative, antitussive, and analgesic. Early human selection for non-shattering seed heads, late flowering, non-prickly leavers, decrease in latex content and hearting character (the tendency of leaves to congregate in layers in a heart shaped head) is said to have led to its domestication as a leafy salad vegetable (Ryder and Whitaker, 1976).

Botanical Description

It belongs to the family Asteraceae (Compositae) having chromosome number, x=8, 9, 17).

Table 13.4: Botanical Description

Growth Habit	It is annual to biennial and attains the height of 120 cm. Plant give milky juice after cutting (Wilfred, 1924).
Root	It produces slender tap root, which grows on the surface of the soil and smaller secondary roots. Some varieties, especially those found in the United States and Western Europe, have long, narrow taproots and a small set of secondary roots. Longer taproots and more extensive secondary systems are found in varieties from Asia (Ryder and Williams, 1993).
Stem	A solitary erect stem, 30-1000 cm height.
Leaf	Large number of simple alternate leaves are produced of which outer leave are usually green to dark green and inner leave are light green to whitish which are tender and crisp. Inner tender leaves have better quality than outer leaves. There is one leaf per node along the stem. Basal leaves in rosette, undivided or runcinate-pinnatifid, shortly petiole, cauline leaves simple, ovate to orbicular in outline, entire cordate-amplexicaul, sessile and not held vertically. Apex blunt.
Floral biology	Inflorescence is a dense, corymbose, flat-topped panicle. Head or capitulum consisting of a large number of flowers (7-35 florets), ligules longer than involucre (Ferakova, 1977). Numerous small flowers remain closely together on a receptacle. Floret with a single lingulate blue to purple, white and yellow florets/petal, 5 stamens and 2 carpels.
Anthesis and pollination	There is very high degree of self-pollination. Pollen and ovule are usually highly fertile. Anthesis starts at sunshine and flower fully opens by late morning. Pollination takes place from 9.0-9.15 AM (Pal, 1975). The pistil grows up through the closely grouped stamens on which the pollen has already been shed. Some crossing (1.3-6.2 per cent) has been demonstrated.
Fruit and fruit set	Fruit is a typical achene and seed ex-albuminous. Achenes 6-8 mm, obovate, narrowed and often finely muricate at the apex, and greyish. Beak white, filiform, as long as body. Pappus white. The ovaries form compressed, obovate (teardrop-shaped) dry fruits that do not open at maturity, measuring 3 to 4 mm long. The fruits have 5–7 ribs on each side and are tipped by two rows of small white hairs. The pappus remains at the top of each fruit as a dispersal structure.
Seeds	Each fruit contains one seed, which can be white, yellow, gray or brown depending on the variety of lettuce.

Horticultural Classification

Different types of lettuce are recognized as sub-species or botanical varieties and are known under the following names:

- **i.** Head, variety *capitata*. Head type again has two types *viz.*, crisp head and butter head. Crisp head is most common of all the types. In UK and the USA, crisp head is referred to as 'iceberg' lettuce.
- **ii.** Cutting or leaf, variety *crispa*
- **iii.** Cos or romaine, variety *longifolia*
- **iv.** Asparagus or stem lettuce, variety *asparagina*

Among the cultivated types in USA, exhibit extensive variation of forms but four main types have been recognized (Purseglove, 1968), *viz.*,

- i. Asparagus type lettuce
- ii. Cabbage type lettuce
- iii. Cos type lettuce and
- iv. Curled type lettuce

In China a special cultivar (cv.) of the asparagus type (*L. serriola* f. *integrifolia*) was developed and reported by Fucha in 1543 from Sinkiang. The local name of this cv. was 'Usum' or 'On-Sen'. According to Vavilov (1992) this form of lettuce is practically unknown in South West Asia.

Classification on the Basis of Cultivars

Different types of lettuce (*Lactuca* species) are recognized as sub-species or botanical varieties and are known under the following names:

- ✰ Head, variety *capitata*,

  Head type again has been divided into two type's *viz.*:

- **(i) Crisp head type:** Crisp head is most common of all the types. In UK and the USA, crisp head is referred to as 'Iceberg' lettuce. The important cultivars are Great Lakes, Imperial
- **(ii) Butter head:** Bib, White Boston

✰ Cutting or leaf, variety *crispa*

✰ Cos or romaine, variety *longifolia* and

✰ Asparagus or stem lettuce, variety *asparagina*.

As early as 1904, Tracy classified the American types and recognized over 100 distinct varieties many of which still find mentioned in the seed catalogues. The classes and subclasses proposed by him are:

Class I: Butter head varieties

Subclass (i) Cabbage-heading varieties

Subclass (ii) Bunching varieties

Class II: Crisp head varieties

Subclass (i) Cabbage-heading varieties

Subclass (ii) Bunching varieties

Class III: Cosvarieties

Subclass (i) Spatulate-leaves varieties

Subclass (ii) Lanceolate leaves varieties

Subclass (iii) Lobed leaved varieties

Thomson (1951) and Rodenburg (1960) recognized six morphological type of lettuce:

- **(i) Crisp head type-**The cultivars are heading types with wrinkled non wrapper leaves, round large head (upto 1.0 kg), brittle-textured and tightly folded with green outer and white or yellowish inner leaves, having excellent shipping and handling abilities, var. New York, 515, Imperial 44 (Cabbage head), Great Lake and Alamo-1.
- **(ii) Butter head type-**It forms relatively small, loose heads, with broad, oily, crimpled, soft-textured leaves vary in colour, size and appearance, var. Big Boston, White Boston, May King (Cos), Salamandar and Waya Head.
- **(iii) Cos or Romaine type-**The cultivars have elongated leaves forming a leaf are finer and lighter in colour; coarse in appearance but eating quality is quite good, var. Eiffel Tower, Paris Island, Paris White (Cos) and Dark Green Valmaine.
- **(iv) Leaf or bunching type-**The cultivars are non-heading or leaf type; colour, heat sensitivity and quality vary but all produce a rosette or leaves, var. Slowbolt (leaf lettuce), Black Seeded Simpson, Prize Head, Australian, Chinese Yellow (Loose leaf), and Salad Bowl for outdoor growing and Grand Rapids for greenhouse production.
- **(v) Stem type-**It is also called 'celery lettuce' and is grown for its thick stem which is pealed and eaten either or cooked as vegetable. The leaves are edible but inferior in quality to the leaves of the other types, var. Celtuce, Saltues
- **(vi) Latin type-**It resembles butterhead in forming loose head but the leaves are elongated and similar to Cos lettuce. It is mainly grown in the Mediterranean countries and in South America var. Gallega, Coriolla, Verde, Criolla Blenca and Madrilene (Ryder, 1979).

Augmentation of Genetic Resources and Crop Improvement

A seeds catalog published in 1806 lists 16 cultivars already being offered in America (Hedrick, 1919). The moors in Spain developed several varieties, especially in Andalucía, when that region was under Arab domination. P.F. Knowles, a plant scientist working on oilseed crops in the Department of Agronomy at the University of California, Davis (US Davis), USA, travelled to Upper Egypt in 1958 to collect safflower (*Carthamus tinctorius*) and other oilseed crop genetic resources for his research. He was made aware of the oilseed producing lettuces of Egypt when he happened to visit the Plant Breeding Station at Giza (near Cairo) and met Mustapha Serry who was in-charge of the Oil Crops Section at that time. During the discussions Sery mentioned that the Station had a sample of an obsolete form of *Lactuca sativa*, was being grown sporadically in Upper Egypt for its seeds; from which edible oil was extracted. Sensing an opportunity for collection of germplasm potentially useful for crop improvement in the U.S., Knowles requested and received a subsample (collector's No. K1020, PI250020) seeds of these plants from the Station nursery. The seeds of this particular sample contained 30 per cent edible oil and the plant did not show any disease symptoms. Samples of the obsolete lettuce form which, turned out to be *L. serriola*, were cultivated exclusively for their relatively large seeds, and since the leaves are prickly and full of latex and bitter to taste this form is unpalatable as salad. The seeds of a prickly lettuce (*L. serriola*) having collector's No. K1058 reportedly grown in Upper Egypt for its oil in the same manner as safflower, were obtained from the warehouse of El Haq Menoufi el Said at Kenya. In Pakistan 'Cultivar No. 6420 (R), Telda and Kelvin (P) are commonly grown.

Interspecific Hybridization

All four species of lettuce *i.e. Lactuca sativa, L. serriola, L. saligna* and *L. virosa* are interfertile with each other thus facilitating exchange of desirable genes among them for breeding purpose. Due to its complete compatibility with cultivated lettuce, *L. serriola* accessions have been used to make crosses resulting in cvs such as, 'Calmar' and 'Valverde'. Prickly lettuce is reportedly also been utilized for improvement of iceberg type lettuce in California which produced 3 million tons seeds of this crop valued at approximately a billion dollars in 1995. All the species crosses readily with wild lettuce, *Lactuca serriola*. There are strong incompatibility barriers between the cultivated *Lactuca* species and the 17 chromosome species (Ryder and Whitaker, 1976). *L. serriola* are most closely related and cross readily with each other. Both cross with difficulty with other two species; *Lactuca sativa* should be used as the male in a cross with *L. saligna* (Lindquist, 1960). According to Thomson and Ryder (1961), the *sativa-virosa* cross yields sterile hybrid.

Resistance Breeding

There are several wild species of *Lactuca* which are mostly used as source of resistance to insect pests and diseases. The entire collection from Egypt was subsequently evaluated for disease resistance and it was discovered that the prickly lettuce sample collected at Kena, which was given the number PI251245, possessed resistance to lettuce mosaic. The resistance was located on a recessive allele named *mo*. Consequently, crosses with lettuce cvs. Were made and the desirable gene was transferred to four lettuce cultivars grown in the western U.S., *viz.*, Vanguard-75, Winter set Autumn Gold, and Salinas 88 (Ryder, 1979) resulting in huge financial saving to the farmers. This valuable sample (PI251245) and others have been multiplied and are now conserved in the USDA. A subsample (collector's no. K1020, PI250020) of *L. serriola* had been known to possess resistance to several diseases (especially downy mildew and mosaic virus) which affect the normal salad lettuce crop in Europe, north America, and Australia.

Resistance Source for Abiotic

There are certain known heat resistant varieties available which need to be introduced and evaluated in India. Such varieties are 'Summerling Buttercrunch' and 'Bibb (Butter head) and 'Gree Ice' (leaf). Some small varieties like 'Tom Thumb' (Butter head), 'Sweet Midget' (Cos) and 'Ruby' (leaf) should also be tried for growing in containers (Nath *et al.*, 2008).

Genetic Resource Conservation

This lettuce sample together with samples of safflower and other crops collected in Egypt was brought to the U.S. on 13 August 1958 where it entered the New Crops Research branch of the Crops Research Division of US Department of Agriculture (USDA) at Beltsville Maryland. *Lactuca* germplasm collection (LGCS), Salinas, California, which is the largest and most diverse collection of defined and characterized accessions of this genus in the world with over 10, 000 samples. The accessions were also duplicated for safety at the Western Regional Plant Introduction Station, Pullman, Washington (McGuire *et al.*, 1993). It is assumed that the samples Knowles collected in 1958 are no longer cultivated and possibly extinct. Hence his collection is very valuable for future breeding efforts.

Table 13.5: *Lactuca* Accessions at the USDA Regional Plant Introduction Station, Pullman, Washington

Sl.No.	*Species*	*Number*
1.	*L. altaica*	1
2.	*L. dregeana*	1
3.	*L. indica*	1
4.	*L. livida*	1
5.	*L. perennis*	5
6.	*L. quercina*	1
7.	*L. saligna*	16
8.	*L. sativa*	1,037
9.	*L. sativa* var. *capitata*	5
10.	*L. serriola*	159
11.	*L. serriola f.*	-
12.	*L. integrifolia*	1
13.	*L. undulate*	1
14.	*L. viminea*	1
15.	*L. virosa*	17
16.	*Lactuca* spp. (unidentified)	7
	Total	**1,254**

Source: Damania, 1977.

In USA, in addition to the collections at Salinas, California, and Pullman, Washington, UC Davis also maintains a sizable *Lactuca* collection. This collection was established in 1948 and includes mostly research and breeding material. Other major collections of this vegetable crop are located at: Centre for Genetic Resources (CGN), Wageningen, the Netherlands; Horticultural Research International (HRI), Welles Bourne, England; N.I. Vavilov Institute of Plant Industry (VIR), St. Petersburg, Russia; Research and Plant breeding Institute for vegetables, Olomouc, Czech Republic; and the Horticultural Sciences Department, Cornell University, New York, USA. It must be noted that, although *Lactuca* originated in the eastern Mediterranean, all the major genetic resources collections are located in the western hemisphere's major consumer countries. Survey of the International Plant Names Index revealed a total number of 538 *Lactuca* species, of which 357 referred to synonyms and basionyms whereas for another 51 the taxomic status and their belonging to the genus *Lactuca* was questionable. Of the remaining 130 species, 20 are generally considered to be part of the lettuce gene pool.

Accession data were collected from a total of 30 collections, including contributions from Armenia, Iran and the United States. Altogether, the data of 11,643 accessions were included in the ILDB. The main part of the ILDB is constituted by landraces, cultivars, research lines and breeding materials of cultivated lettuce (*Lactuca sativa* L.) represented with 10,198 accessions (88 per cent). A highly unbalanced distribution was observed for species belonging to the crop related wild gene pool as 96 per cent of the accessions are covered by only three species, namely *L. serriola* L., *L. saligna* L. and *L. virosa* L. Eight species, including *L. azerbaijanica* Rech. f. and *L. scarioloides* Boiss. that are part of the primary lettuce gene pool, even lack a single accession in the ILDB. Concerning the primary gene pool, *L. serriola* L. with 987 accessions is the only wild species that is well represented. The category "unknown and other taxa" included 167 accessions, comprising specimens of *L. biennis* (Moench) Fernald, *L. Canadensis* L., *L. dentata* Makino, *L. dissecta* D. Don, *L. floridana* (L.) Gaertn., *L. homblei* De Wild., *L. indica* L., *L. livida* Boiss. et Reut., *L. perennis* L., *L. raddeana* Maxim. and *L. tenerrima* Pourret, a number of hybrids and undetermined *Lactuca* species, and a few samples from the genera Chondrilla, Cicerbita, Mycelis and Steptorhamphus that are related to *Lactuca* (results not shown). Accessions of cultivated lettuce have been collected from all regions of the world, the majority originating from Europe and Asia and relatively few from Africa. All the main horticultural types' *i.e.* 'butterhead', 'crisphead', 'cutting', 'cos', 'stalk', and 'Latin lettuce', are well represented in the ILDB with a wide variety in geographic origin (results not shown). The more primitive horticultural type 'oilseed lettuce', which has a restricted cultivation in the Middle East, is represented by six *L. sativa* L. accessions, all originating from Egypt. No data on geographic origin was provided for 3,063 accessions of *L. sativa* L. (30 per cent). The majority of wild species of the primary gene pool have their distribution area in Asia, with the exception of *L. dregeana* DC. that is restricted to South Africa and *L. serriola* L. that can be found on all continents. For *L. serriola* L., being the only wild species of the primary gene pool that is well represented in the ILDB, relatively few samples were collected from Africa (14) and America (16), and none from Oceania. Like *L. serriola* L., also *L. saligna* L. and *L. virosa* L., constituting the secondary gene pool, show a cosmopolitan distribution. The predominance of accessions of these species from Europe and Asia in the ILDB was even more clear-cut than observed for *L. serriola* L. as no accessions from Africa and Oceania are included and America is only represented by a single sample of *L. virosa* L. The ten species of the tertiary gene pool are mainly distributed in Europe and Asia, and are represented by low numbers of accessions, or not at all, in the ILDB (van Treuren *et al.*, 2012.)

REFERENCES

1. Chantrell, Glynnis, (2002). (ed.) *The Oxford Dictionary of Word Histories*. Oxford University Press. p. 300. ISBN 0-19-863121-9.
2. Damania, A.B. 1975. Studies on high temperature dormancy in lettuce (*Lactuca sativa* L.) varieties. M.Sc. thesis, University of Birmingham, U.K.
3. Damania, A.B. 1975. Lettuce (*Lactuca* spp.): An underutilized crop. *Indian J. Plant Genet. Resour.*, **10(2):** 263-267.
4. Ferakova, V. 1977. The genus *Lactuca* L. In: Europe P.68. Univerzita Komenskeho, Bratislave.
5. Hedrick, U.P. 1919. Sturtevant's edible plants. New York, Department of Agriculture Annual Report, 27 (2/2), New York.
6. Hedrick, U.P. 1972. Sturtevant's notes on edible plants of the worlds. Dover Press, New York,
7. Keimer, L. 1924. Die Gartenpflanze im alten Agypten. Hamburg and Berlin, Germany.
8. Lindquist, K. 1960. On the origin of cultivated lettuce. *Hereditas*, **46:** 319-350.
9. Mehta, Y.R. 1959. Vegetable growing in Uttar Pradesh. Bureau of Agriculture Information, Uttar Pradesh (India).
10. Nath, P., Srivastava, V.K., Dutta, O.P. and Swamy, K.R.M. 2008. Vegetable Crops: Improvement and Production. PNASF, Bengaluru (Karnataka), pp. 173-175.
11. Pal, A.B. 1975. *Progressive Hort.*, **7:** 62-68.
12. Pink, D.A.C. and Eavan, M.K. 1993. Lettuce *Lactuca sativa* L. In: Genetic improvement of vegetable Crops (eds. G. Kalloo and B.O. Bergh). Pergamon Press, New York.
13. Purseglove, J.W. 1984. Tropical crops-Dicotyledons. Vol. I and II (edn. 1 reprinted, 1984).
14. Rodenburg, C. M. 1960. Varieties of lettuce. An International Monograph. Zwolle, The Netherlands: W.F.J. Tjeenk. Willink.
15. Ryder, E. 1979. Leaf salad Vegetables. AVI Publishing Co., Inc. Westport, Connecticut.
16. Ryder, E.J. and Whitaker, T.W. 1976. Lettuce. In: Simmonds, N.W. (ed.). Evolution of Crop Plants, pp. 39-41, Longman, New York, USA.
17. Ryder, J. and Williams, W. (1993). "New Directions in Salad Crops: New Forms, New Tools, and Old Philosophy". In Janick, J and Simon J.E. *New Crops*. Wiley. pp. 528–532.
18. Sharma, D.K., Chaudhary, D.R. and Pandey, D.P. 2001. Growth and yield of lettuce cv. Alamo-1 as influenced by dates of planting and plant density. *Veg. Sci.*, **28(1):** 38-39.
19. Singh, S. and Dhaliwal, M.S. 2002. Leafy and Salad Vegetables. In: Vegetables, Tuber crops and Spices (eds. S. Thamburaj and N. Singh), DIPA-ICAR, New Delhi.
20. Thompson, H.C. and Kelly, W.C. 1972. Vegetable Crops. Tata Mc Graw-Hill Publishing Co. ltd., New Delhi.

21. Thompson, R.D. and Ryder, E.J. 1961. Description of pedigrees of nine varieties of lettuce. USDA, *Tech. Bull.*, 1244.

22. Vavilov, N.I. 1935. The origin, variation, immunity and breeding of cultivated plants. *Chron. Bot.*, **13(1/6):** 1-366.

23. Van Treuren, R., Coquin, P. and Lohwasser, U. 2012. Genetic resources collections of leafy vegetables (lettuce, spinach, chicory, artichoke, asparagus, lamb's lettuce, rhubarb and rocket salad): composition and gaps. *Genet Resour and Crop Evol.*, 59:981-997; DOI 10.1007/s10722-011-9738-x.

24. Wilfred, M.R. 1924. The Botany of Crop Plants.

Genus *Coriandrum*

Coriander (*Coriandrum sativum* L.)

Introduction

In India coriander (*Coriandrum sativum* L., 2n=22, 24) is known with the several vernacular name like Kastumbari, Dhaniaka (Sanskrit); Dhania (Assam); Lanka Marich, Dhania (Bengali); Lanka Dhania (Odisha); Mirpakaya, Dhanialu (Telgu); Kothamali (Tamil); Kothamali (Malayalam); Kotambari (Kannad); Kothinbir Dhane (Marathi); Dhani (Gujarati) and Dhania (Punjabi). Its name is derived from the Greek words 'Koris', meaning bed bug because of the unpleasant, fetid, bug-like odour of the green unripened fruits. Both leaves and stems, may have the whole plants a strong pleasant aroma for which they are loved. It is grown for 'grain or seed' spice (important ingredient of curry powder, usually contributing the greatest quantity of all the spices followed by turmeric) and green leaf for flavouring soups and garnishing other dishes. The use of coriander can be traced back to 5,000 BC, making it one of the world's oldest spices. It was used as a spice in both Greek and Roman cultures, the latter using it to preserve meats and flavor breads. Coriander herb is widely available fresh in local market. It is also available as ground, frozen or preserved fresh in jars in super markets. Coriander leaves (cilantro) are featured in the culinary traditions of Latin American, Indian and Chinese cuisines. The early physicians including Hippocrates, used coriander for its medicinal properties, including as an aromatic stimulant. It is the only crop included in national crop forecasting system among seed spices. All parts of plant have pleasant aroma/ odor. Pruthi (2001) described that coriander leaves are rich source of vitamin C 310.0 mg (125-250 mg/100 g) and vitamin A (5200 IU/100 g). In general per 100g of edible leaves contains 2.13-3.30g protein,2.8g fibre, 92.21g moisture, 3.7-6.5 g carbohydrate, 0.52 g fats, 6748-10640 IU vitamin A, 0.149 mg vitamin B6, 0.8-1.1 mg niacin, 0.162 mg riboflavin, 0.067 mg thiamin, 27.0-135.0 mg vitamin C, 2.5mg vitamin E, vitamin K, 67.0 mg calcium, 0.225 mg copper, 1.77 mg iron, 0.426 mg manganese, 26.0 mg magnesium, 48.0 mg phosphorus, 521.0 mg potassium, 0.9 mg selenium, 46.0 mg sodium, 0.5 mg zinc and 23 Kcal energy (iloveindia.com, Pruthi, 2001). Seeds are also used to flavor tobacco and the essential oil is used in perfumes, cosmetics and soaps. The edible seeds contain 6.3 per cent moisture, 1.3 per cent protein, 0.3 per cent volatile oil, 22.0 per cent non-volatile ether extract, 19.6 per cent totalether extract, 31.5 per cent crude fibre, 24.0 per cent carbohydrates, 5.3 per cent mineral matter (total ash), 0.2 per cent acid soluble as, 0.8 per cent calcium, 0.006 per cent iron, 0.44 per cent phosphorus, 0.02 per cent sodium, and 1.2 per cent potassium. It also contains 0.26 mg vitamin B_1, 0.23 mg vitamin B_2, 3.2 mg niacin, 12.0 mg vitamin C and 175 IU vitamin A per 100 g. Coriander seeds contains 40 compounds and among them 'decyl' and nonyl aldehydes' occupied 80 per cent and alcohol 16 per cent (Pater and Fergusan, 1990). Seeds are used as carminative, diuretic, tonic, stomachic antibilious and refrigerant. In fusion of seeds with cardamom and caraway seeds is useful against flatulence, indigestion, vomiting and intestinal disorder. Roasted seeds are useful in dyspepsia. The green leaves and crushed seeds provide a spicy flavor in soups, curries, breads cakes and other dishes such as fish and poultry. Entire plant is used as appetizer in preparing fresh chutneys and sauces. Leaves are used as condiments in preparation of curry powder, pickling spices, seasoning and flavoring of pastries, coolies, buns, cakes and tobacco products. Leaves are used for flavoring curries and soups and also used as flavoring liquor, pork, meat, fish, salads, soda and syrup, gelatin dessert, pudding and candy preserve *etc.* The juice of fresh plant is applied to erythematic. The leaves are carminative, antibilious, diuretics, tonic, and stomachic. The early physicians, including Hippocrates, used coriander for its medicinal properties (Mathias, 1994), including as an aromatic stimulant. It is helpful as an aphrodisiac and for various disabilities of the female generative apparatus. Its oleoresin is used for flavouring beverages, pickels, sweets and sausages (Lal *et al.*, 2010). Coriander seeds have health supporting reputation (Saxena *et al.*, 2014) and decoction is used in different medicines to cure diseases related to idigestion, sore throat, common cold, dysentery, vomiting, urine and bilious complaints. Decoction may also be used in eye wash and in chronic conjunctivitis (Lal *et al.*, 2012). Seeds are chewed to remove foul breath and used as a purgative and roasted seeds are useful in dyspepsia. The paste of seeds is applied to relieve pain in cephalgia. Coriander essential oil is carminative, antiseptic, bactericidal, fungicidal and muscle relaxant. Its volatile oil is used to mask the bad smell of different allopathic medicines.

Some therapeutic uses are:

- ☆ To get relief from sunstroke, wet and squeeze coriander herb in water and add sugar candy and give it to the affected person.
- ☆ Coriander has an eroticizing effect due to its estrogen content and is also useful in rheumatism, neurologic conditions.
- ☆ Prepare decoction of coriander, dry ginger, sugar candy, *nagarmotha* in water and administered to pregnant woman or other patients to stop vomiting immediately.
- ☆ Cold decoction of coriander with sugar candy and honey is administered to provide relief from burning sensation of fever and thirst.
- ☆ Coriander leaves mixed with juice of apple, milk and *multani mitti* tones during summer.

Origin, History and Distribution

Coriander is native to the Mediterranean (Europe and Asia Minor) and Middle Eastern Regions. It has been known in Asian countries for thousands of years. It can be traced back to 5,000 BC, making it one of the world's oldest spices. It was cultivated in ancient Egypt (960 B.C.) and given mention in the Old Testament. Further, Persia grew coriander 3000 years ago and it added fragnance to the hanging gardens of Babylon. Hebrews knew it also, since in the Old Testament in the 'Book of Exodus', 'Manna' is described as being like white coriander seed. In the third century BC, Eromans also found coriander seed as an excellent seasoning for popular foods. The Russian Federation, India, Bulgaria, Morocco, Burma, Pakistan, Turkey, Palestine, Sri Lanka, Romania, Argentina, Iran, Africa and Holland (Diederichsen, 1996) are among the countries that commercially produce coriander seeds/grains. Coriander leaves (cilantro) are featured in the culinary traditions of Latin American, Indian and Chinese cuisines which indicate the earlier distributions. Coriander was used as a spice in both Greek and Roman cultures, the latter using it to preserve meats and flavor breads. Its use has been well mentioned in Sanskrit literature. It is extensively grown in Rajasthan, Madhya Pradesh, Uttar Pradesh, Gujarat, Andhra Pradesh, Chhattisgarh, Tamil Nadu and Karnataka. India is the largest producer and consumer of coriander in the world. India exports coriander seeds to Bahrain, Canada, Guyana, Indonesia, Jordan, Kuwait, Malaysia, Mauritius, Mozambique, Oman, Saudi Arabia, Singapore, South Africa, Srilanka, UAE, USA *etc.* The coriander powder is exported from India to Oman, Saudi Arabia, Singapore, South Africa, Srilanka, UAE, UK, USA *etc.* (Singhal, 1996).

Species Availability, Evolution and Diversity

Under genus *Coriandrum* two species are common:

(i) *Coriandrum sativum*

(ii) *Coriandrum tordylium* Fenzl. Bornm.

Botanical Description

Kingdom	:	Plantae–Plants
Subkingdom	:	Tracheobionta–Vascular plants
Superdivision Spermatophyta	:	Seed plants
Division Magnoliophyta	:	Flowering plants
Class Magnoliopsida	:	Dicotyledons
Subclass	:	Rosidae
Order	:	Apiales
Family	:	Apiaceae–Carrot family
Genus	:	*Coriandrum*
Species	:	*sativum*

Coriander is an annual herbaceous plant of the family Apiaceae or Umbelliferae, genus *Coriandrum* and species *sativum* with 2n=22.

Table 13.6: Botanical Description

Growth Habit	Annual bushy and smooth herb which attain 40.0-100.0 cm height. Plant possessing more number of primary and secondary branches.
Root	Deeper and branched roots.
Stem	The thin, slender, hollow and smooth stem is corymbosely branched in the upper part.
Leaf	Leaves are alternate, compound and petiole has a pair of stipules sheathing the stem base. The lower leaves are broad, pinnate and having long petioles with crenately lobed margins (lower leaves are 2.5-10.0 cm long and upper leaves divided in sub sects). The upper leaves are finely cut with linear lobes and are bi-pinnate or tri-pinnate.
Floral biology	The inflorescence is small compound umbel both terminal and leaf opposed and usually comprises about 5 smaller umbels. Flowers are pink or white flowers. The plant bear andromonoecious flower *i.e.* both staminate and perfect flowers (hermaphrodite) are produced on the same plant and even in the same umbellet (Pillai and Nambair, 1982). The corolla is made up of five petals. The flower possesses five stamens and the ovary is inferior and bilocular.
Anthesis and pollination	Generally, hermaphrodite flowers open early than male flowers. The flower opening (at 6.0 a,m. and reaches a maximum by 12.0 to 2.0 p.m., Hill, 1992) and anthesis proceed from the periphery inwards both within the compound umbel and within the umbel-lets. The time taken for complete anthesis for all flowers of an umbel is 3-4 days. Crop is highly cross pollinated and an out crossing rate is 19 per cent (Gladis *et al.*, 1996). Flowers are protandrous, with the anthers maturing before the stigma becomes receptive.

	Both autogamy and allogamy (pollination mainly through bees) prevail but the crop is highly croo-pollinated. Hymnopterans are most effective and efficient at carrying pollens. The visual impact of the compound umbel, presence of exposed nectar and availability of abundant pollens are the major attractants for insects visiting the umbels (Kaul *et al.*, 1989.). Winds seem to play no role in pollination of coriander. Petals are 5 in number and white or purple in colour. Stamens 5, and filaments are spread. Ovary 2 celled and inferior. The stigma remains receptive for five days and pollen remains viable for a maximum of 24 hours but higher fruit set recorded during the first three days.
Fruit and fruit set	Fruits are schizocarpi which form 2 mericarp. The fruits are globular, round, ribbed and small (3-4 mm diameter). Surface has longitudinal fine grooves and ribs alternately. When pressed, fruit easily breaks into 2 locules (meri-carp), convex outside and concave inside in each of which there is one seed. The fruits are mild, delicately fragnant due presence of volatile oil.
Seeds	Coriander seeds are very light weight, pale white to light brown in colour and have a mild flavor.

Botanical Classification

Based on size, two botanical groups are made:

(i) var. *vulgare*-Plant having diameter of 3.0-5.0 mm. It is large fruited types and mostly grown in tropical and subtropical countries (like Moracco and India) and contain low volatile oil are used extensively for grinding and blending purposes.

(ii) var. *microcarpum*-Plant having diameter of 1.5-3.0 mm. It is small fruited types and generally cultivated in temperate regions and usually has volatile oil content greater than 0.4 per cent. It is a small type which is generally grown in Russia and some other countries of eastern and central Europe contain volatile oil between 0.8 to 1.8 per cent and are highly valued as a raw material for preparation of essential oil.

Classification on the Basis of Seed Size

(i) Small seeded type-Seeds are small with less fibre and are most suited for growing in temperate regions. Volatile oil content is higher than bold seeded ones. The seed colour is light green.

(ii) Bold seeded type- Seeds are bold having more fibre and is most suitable for growing in tropical and subtropical regions. Volatile oil (Coriandrol) is lesser than small seeded ones and colour is light brown.

Augmentation of Genetic Resources and Crop Improvement

Introduction and Selection

Exotic type introduced by the Central Indian Medicinal Plant Organization (CIMPO) from Bulgaria have adapted very well to Indian conditions. Most of the released varieties have been developed as a result of simple selections. Selections with multipinnate leaves possessing heavy yield potentialities have been identified in CIMAP. Selections showing erect compact habit with purple stem having essential oil content between 1.6-1.7 per cent have been also isolated. CIMPO S-33 was released for its tall erect and basal branching habit. It is capable of yielding around 2.1 t/ha in duration of 143 days. This selection is small seeded with a higher essential oil content of 1.3 per cent.

The bold seeded varieties UD-1, UD-20 and UD-21 are selection made in Rajasthan. The variety UD-20 with a yield capacity of 1.2 t/ha was released again as RCR-20 from Rajasthan Agricultural University, while Gujarat Coriander-1 (GAU-1), and Gujarat Coriander-2 (GAU-2) are selection from Gujarat Agricultural University, Gujarat. A small seeded variety 'Karan' (UD-41) was also released from Rajasthan. In Madhya Pradesh, 'Gwalior-5385' and 'Katwari' type coriander selection were made. There are several genotypes like- Co. 1, Co. 2, Co. 3 (released by TNAU, T. Nadu), CS-287, Coriander-2, RCR-41, UD-20, UD-446, RCR-435, RCR-436, Sadhana, Swathi, Lam Sel.-2 (Thangaraj and Vijaykumar, 2001). Sindhu, Pant Haritima, Hisar Sugandh, Khushboo *etc.* are being cultivated by vegetable growers. Bihar has identified a variety 'Rajendra Swati'. Singh *et al.* (2016) evaluated 120 genotypes along with two checks (ND Cor.-2 and Hissar Anand) and observed significant differences for all the traits under study except days to 50 per cent flowering and but variance due to checks was non-significant for fruits/umbel while, highly significant for all characters. The genotypes ND Cor.-39 (14.39g) followed by ND Cor.-4 (13.33g), ND Cor.-92 (13.01g), ND Cor.-64 (12.85g) and ND Cor.-74 (12.77g) produced highest seed yield. The most desirable genotypes for characters other than seed yield were ND Cor.-5 and ND Cor.-7 for days to 50 per cent flowering, ND Cor.-80, ND Cor.-81 for plant height, ND Cor.-10, ND Cor.-9 for branches/plant, ND Cor.-25, ND Cor.-19 for fruiting nodes/plant, ND Cor.-40, ND Cor.-70 for umbels/plant, ND Cor.-20, ND Cor.-17 umbellates/umbel, ND Cor.-21, ND Cor.-16 for fruits/umbellate, ND Cor.-92, ND Cor.-75 for fruits/umbel, ND Cor.-69, ND Cor.-58 for umbel diameter and ND Cor.-34, ND Cor.-108 for test weigh the estimates of heritability in broad sense were high for all the characters except days to 50 per cent flowering. The estimates of heritability in broad sense ranged from -5.35 (days to 50 per cent flowering) to 99.73 (umbel diameter). Seed yield/plant exhibited significant and positive correlation with umbels/plant followed by non-significant and positive correlation with fruits/umbel, branches/plant and non-significant and negative correlation with umbel diameter and plant height. Singh *et al.* (2016) evaluated 15 genotypes and reported that genotype 9904 was superior in terms of seed yield/plant. The PCV was higher than GCV for all the traits. Heritability in broad sense was noticed high for all the traits except test weight (1000 seeds). Higher genetic advance was observed for plant height. High genetic advance as per cent of mean was found in number of seeds/umbel. The traits like plant height, primary branches per plant, number of umbels/plant, number of umbellates/umbel, umbel diameter, number of seeds/umbel and test weight showed positive significant correlation with seed yield/plant at phenotypic level. Number of seeds/umbel exhibited the highest positive direct effect at genotypic leve.

Hybridization

Coriander is basically a cross pollinated crop. Coriander reveals heterosis for plant height and yield. The characters of either of the parents are dominant in F_1 progeny. The crop does not exhibit inbreeding depression. The flowers are minute and manual emasculation and crossing is cumbersome.

Interspecific Hybridization

Through interspecific hybridization several tetraploid has been developed but the crosses with *Biofera* spp. was not successful (Weiss, 2002).

Mutation

The seeds of different country origin respond differently to gamma rays. Types from India and China respond very poorly to gamma rays. A dose up to 5 kr has a stimulating effect on various traits (Thangaraj and Vijaykumar, 2001).

Insect Resistance

Cultivar 'Sadhana' is tolerant to aphid, mites and fruit fly.

Disease Resistance

Cultivars'Co. 3', 'Gujarat Coriander-1', and 'Gujarat Coriander-2'are tolerant to wilt, powdery mildew and grain mould. Rcr-41 is resistant to stem gall.

Abiotic Stress

Coriander is capable of tolerating drought and heat. Due to its versatile adaptability, it is grown throughout the year for leaf purpose. Cultivar Co.2 is suitable for waterlogged, drought, saline and alkaline conditions. CS-287 is very much suitable foe rainfed condition. Most crop improvement programmes mainly aim at developing long duration varieties suitable for irrigated light soils (Bhandari and Gupta, 1993a; 1993b), sodic wastelands (Singh *et al.*, 2005) and special uses (Lopez *et al.*, 2007). Large-scale production and the development of better varieties suitable for rainfed vertisols are restricted by the lack of information about their genetic diversity, intra-specific variability and genetic relationships among this species. Due to paucity of genetic diversity, crop improvement of coriander for rainfed vertisols has become a challenge in India.

REFERENCES

1. Bhandari, M.M. and Gupta, A. 1993a. Cluster analysis for applications. Academic Press, New York
2. Bhandari, M.M. and Gupta, A. 1993b. Association analysis in coriander. *Indian J. Genet. Plant Breed.*, **53:** 66-70.
3. Diederichsen, A. 1996. Promoting the conservation and use of underutilized and neglected crops 3. Coriander (*Corianderum staivum* L.) *Int. Plant Genet. Res. Inst.* (IPGRI), Rome, Italy: 66.
4. Gladis, T., Diagne, O., Spahillari, M. and Hammer, K. 1996. Reproduction of medicinal and aromatic plants in the Gatersleben gene bank. *Proc. Int. Sym. Breed. Res. On Med. and Aromatic Plants.* Quadinburry, Germany, 30th June to 4th July.
5. Hill, A.F. 1992. Economic Botany. Tata McGraw-Hill Publishing Company Ltd., Bombay.
6. Kaul, A.K., Hamal, L.A. and Gupta, S.K. 1989. Pollination mechanism in *Corinderum sativum* Linn. (Apiaceae). *Proc. Indian Acad. of Plant Sci.*, **99(5):** 509-515.
7. Lal, G., Meena, S.S., Anwer, M.M., Mehta, R.S. and Maheria, S.P. 2010. Advance Production Technology of Coriander. A Technical bulletin published by National Research Centre on Seed Spices, Ajmer, India.
8. Lal, G., Anwer, M.M., Mehta, R.S., Meena, S.S. and Maheria, S.P. 2012. Major Seed Spices. Published by Agrotech Publishing Academy, Udaipur (Rajasthan), India.
9. Lopez, P.A., Widrlechner, M.P., Simon, P.W., Rai, S., Bailey, T.B. and Gardner, C.A. 2007. Screening coriander genepool for special uses.In: J. Janick and A. Whipkey (eds) Issues in New crops and New Uses. 2006 AAIC Annual Meeting and Sixth New Crops Symposium, October 14-18, 2006. San Diego, CA.
10. Mathias, M.E. 1994. Magic myth and medicine. Econ. Bot. 48:3-7.
11. Pillai, F.K.T. and Nambair, M.C. 1982. Cultivation and utilization of aromatic plants. *Bull., R.R.L.*, Jammu, p. 167-189.
12. Saxena, S.N., Verma, M., Kakani, R.K., Rathore, S.S., Saxena, R. and Sharma, L.K. 2014. Analysis of medicinally important compounds and antioxidant properties of Anise (*P. anisum*) seed extract and shoot callus. *Int. J. Seed Spices,* **4(1):** *55-62.*
13. Saxena, R., Saxena, S.N., and Soni, A.2014. Cryogenic grinding enhances analgesic and antipyretic activities of coriander (*C. sativum*). *Int. J. Seed Spices,* **4(1):** 14-18.

14. Singh, D., Chandrakar, U. and Kumar, R. 2016. Studies on genetic variability, heritability and character association in coriander (*Coriandrum sativum* L.) genotypes. 7th Indian Horticulture Congress (Abs: 58; Session II-Breeding of Horticultural Crops), held at ICAR-IARI, New Delhi, 15-18 November, p. 130.

15. Singh, D., Pandey, V.P., Maurya, P.K. and Singh, N.P. 2016. Studies on genetic variability in coriander (*Coriandrum sativum* L.). 7th Indian Horticulture Congress (Abs: 02; Session II-Breeding of Horticultural Crops), held at ICAR-IARI, New Delhi, 15-18 November, p. 87-88.

16. Singh, S.P., Katiyar, R.S., Rai, S.K., Tripathi, S.M. and Srivastava, J.P. 2000. Genetic divergence and its implication in breeding of desired plant type in coriander (*Coriandrum sativum* L.). *GENETIKA*, **37:** 155-163.

17. Singhal, V. 1996. Coriander. In: Indian Agriculture. Indian Economic Data Centre. New Delhi, p. 528-529.

18. Thangaraj, T. and Vijaykumar, M. 2001. Coriander (*Coriandrum sativum* L.). In: vegetables, Tuber Crops and Spices. (Eds. Thamburaj and Singh, 2001). Indian Council of Agricultural Research, New Delhi, p. 371-377.

19. Weiss, E.A. 2002. Spice crops: CAB International Wallingford, Oxon, U.K.

Genus *Eryngium*

Long Coriander or Spiny Coriander (*Eryngium foetidum* L.)

Introduction

Long coriander (*Eryngium foetidum* L.) is also known as Puerto Rican Coriander, Black Benny, Saw leaf herb, Mexican Coriander, Spiny Coriander and fireweed (English) and Bhandhania, Bhandhanaya (Hindi). It is a tropical perennial and annual herb which survived well under moist, shaded conditions near cultivated areas. Its scientific Latin name literally translates as "foul-smelling thistle". Common names include culantro, Mexican coriander and long coriander. In the United States, where it is not well known outside Latino and Caribbean communities, the name culantro sometimes causes confusion with *Coriandrum sativum* (also in Apiaceae), the leaves of which are known as cilantro, and of which culantro is said to taste like a stronger version.

The herb (leaf) is rapidly becoming an important import item into the US mainly due to the increasing ethnic immigrant populations who utilize it in their many varied dishes from around the world. It is closely related botanically to coriander but has distinct different appearance and much more potent volatile leaf oil.It is rich in iron, carotene, riboflavin and calcium. This plant has a strong aroma; similar to fresh coriander leaves. Generally, edible portion of long coriander contains 86-88 per cent moisture, 3.3 per cent protein, 0.6 per cent fat, 6.5 per cent carbohydrates, 1.7 per cent ash, 0.06 per cent phosphorus, and 0.02 per cent iron it also contain 10,460 IU vitamin A, 0.8 mg B_1, 60.0 mg B_2 and 150.0-200.0 mg vitamin C, per 100 g of edible portion (Bautista *et al.*, 1988). The essential oil from the leaves is rich in aliphatic aldehydes, most of which are β, β unsaturated compounds. The impact compound is E-2-dodecenal (60 per cent), furthermore 2, 3, 6-trimethylbenzaldehyde (10 per cent), dodecanal (7 per cent) and E-2-tridecenal (50 per cent) have been identified.Aliphatic aldehydes appear also in other spices with coriander-like scent (*e.g.*, Vietnamese coriander). Another essential oil can be obtained from the root, unsaturated alicyclic or aromatic aldehydes dominate (2, 3, 6-trimethylbenzaldehyde 40 per cent, 2-formyl-1,1,5-trimethyl cyclohexa- 2,5-dien-4-ol 10 per cent, 2-formyl-1,1,5-trimethyl cyclohexa-2, 4-dien-6-ol 20 per cent, 2, 3, 4-trimethylbenzaldehyde). In the essential oil from the seeds, sesquiterpenoids (carotol 20 per cent, β-farnesene 10 per cent), phenylpropanoids (anethole) and monoterpenes (β-pinene) were found, but no aldehydes.Long coriander is widely used in seasoning and marinating in the Caribbean. It is also used extensively in Thailand, India,Vietnam, and other parts of Asia as a culinary herb. Long coriander dries well, retaining good colour and flavor, making it valuable in the dried herb industry. It is sometimes used as a substitute for coriander, but has a much stronger taste. Like coriander, many people find the leaves distasteful, due to genetic traits. As the species name means 'fetid' in Latin, it is possible that Linnaeus also had noted this trait. '*Salsa*' is a spicy sauce prepared from tomatoes, garlic, onion, lemon juice, with liberal amounts of chillies. These constituents are fried and simmered together, mixed to a smooth paste and spiced with fresh herbs including long coriander.'*Salsa*' is usually consumed with 'tortilla chips' as an appetizer. Equally popular is '*sofrito*' or '*recato*', the name given to the mixture of seasonings containing long coriander and widely used in rice, stews, and soups (Wilson, 1991). There are several reports about recipes and cookings in Puerto Rico but basically '*sofrito*' consists of garlic, onion, green pepper, small mild peppers, and both cilantro and long coriander leaves. Ingredients are blended and can then be refrigerated for months. '*Sofrito*' is itself the major ingredient in a host of other recipes including eggplant pasta sauce, long coriander garlic butter, long coriander pesto, pineapple salsa, and gazpacho with herb yogurt.

Medicinally, the leaves and roots are used in tea to stimulate appetite, improve digestion, combat colic, soothe stomach pains, eliminate gases and act as an aphrodisiac. The plant is used in traditional medicines for fevers and chills, vomiting, diarrhea and in Jamaica for colds and convulsions in children (Honeychurch, 1980). The leaves and roots are boiled and the water drunk for pneumonia, flu, diabetes, constipation and malaria fever. The root can be eaten raw for scorpion stings and in India the root is reportedly used to alleviate stomach pains. The leaves themselves can be eaten in the form of chutney as an appetite stimulant (Mahabir, 1991). *Eryngium foetidum* is also known as *E. antihysterium*. The specific name *antihystericum* reflects the fact that this plant has traditionally been used for epilepsy.

Origin, History and Distribution

Long coriander is a leafy spice herb of tropical regions of world (America, South Asia, Pacific Islands, South Europe and Africa. It is a native to Mexico and South America, but is cultivated worldwide (Adams, 1970). Adams (1971) stated that this is Indigenous to tropical America and Caribbean islands, from southern Mexico to Panama through Brazil and from Cuba to Trinidad. It is mostly grown in tropical Africa, South Asia, warmer southern parts of Europe and Pacific islands (Wong *et al.*, 1994; Wagner *et al.*, 1999); and widely cultivated in Costa Rica and Puerto Rico for local consumption and export to the USA. Spiny coriander has been introduced into Florida and the Old-World tropics and become naturalized in many places. In the late 1800s and the beginning of 1900s, eryngo was introduced by Chinese into South-East Asia (Malaysia, Indonesia, Thailand, Vietnam, Singapore, Myanmar, Sri Lanka, Bangladesh and India) as a substitute of the coriander (*Coriandum sativum* L.) because of its similar pungent aroma (Ramcharan 1999; Boonsong, 2005; Seaforth and Tikasingh 2005; Chowdhury *et al.*, 2007). Today, it has been introduced to large parts of South East Asia (Indochina, Malaysia, Indonesia). In India, it is grown in whole northeast states. It is widely cultivated in Thailand and Vietnam. The global Range of this crop is 0-2000 m elevation.

Species Availability, Evolution and Diversity

Eryngium comprises over 200 tropical and temperate species (Willis, 1960) but *E. foetidum* is domesticated, cultivated and used extensively (Lawrence, 1967). Several other *Eryngium* species such as *E. maritimum* L., *E. creticum* Lam., *E. aquaticum* L., *E. campestre* L., *E. thyrsoideum* Boiss., *E. billardieri* F. Delaroche, *E. bungei* Boiss. and *E. caucasicum* Trautv., are known as cultivated plants globally (Pistrick 2001; Khoshbakht *et al.*, 2007). Out of which 113 cultivated species, excluding ornamentals, primarily used as medicinal plants (42.5 per cent); vegetables, salad plants and tuberous starch crops (21.2 per cent); spice plants (19.5 per cent); fodder plants (11.5 per cent); essential oil plants (4.4 per cent) and about 0.9 per cent hedge plants (Pistrick 2002). As a synonym it is *Eryngium antihystericum* Rottb.; *Eryngium foetidum* f. *comosum* Urb. and *Eryngium foetidum* f. *nudum* H.Wolff.

A significant variation in the plant morphology, composition of essential oil (>60 constituents reported) and secondary metabolites resulted from genetic variability and geographic location. Pharmacological investigations have demonstrated anthelmintic, anti-inflammatory, analgesic, anti-convulsant, anti-clastogenic, anti-carcinogenic, anti-diabetic and anti-bacterial activity. Northeastern regions of India, including Mizoram, bestowed with considerable genetic variability of vegetables and spices being adopted by local people for sustenance and intrinsically linked to their cultural and traditional systems (Singh *et al.*, 2013a). In Mizoram, it grows naturally in shaded-moist forest soils, Jhum lands and along moist-shaded pathways where heavy soils predominate.

Botanical Description

Long coriander belongs to family Apiaceae.

Table 13.7: Botanical Description

Growth Habit	*Eryngium foetidum*is a tropical perennial and annual herb. Spiny ornamental. Plants 8–40 cm high from a basal rosette.
Root	*Eryngium foetidum*commonly develop thick taproots which are evenly branched. Generally, taproot fusiform with fibrous roots.
Stem	Stem green
Leaf	Fleshy waxy leaves. The lanceolate or oblanceolate entir leaves 5-25 x 1.2-4 cm broad, arranged spirally around the short thick stem, form a basal leaves numerous, and rosette. Petiole short or obsolete, sheath up to 3 cm. The leaf margin is serrated, each tooth of the margin containing a small yellow spine. Venation pinnately reticulate, base cuneate to decurrent, apex obtuse, callous-margined, crenate to finely spinulose-serrate. Upper leaves sessile, opposite deeply spinulose-serrate to parted.
Floral biology	Whitish umbel inflorescence on a long stalk (25–50 cm) arising from the center of the leaf rosette. It bears a well branched cluster of blue flowers in cymose heads, the characteristics umbel inflorescence on a long stalk arising from the center of the leaf rosette (Morton, 1981). Inflorescence divaricately trifurcate; lateral branches often continuing to form a monochasium, heads numerous, short-pedunculate. Flower heads cylindrical, 5–12 x 3–5 mm; bracts 4–7, foliaceous, lanceolate, 1.5–3.5 x 0.4–1 cm, spreading to reflexed, margin 1–3-spinulose-serrate; bracteoles lanceolate, 1.5–1.8 x ca. 0.6 mm, brightly scarious-margined. The calyx is green while the corolla is creamy white in colour. Calyx teeth ovate-lanceolate, 0.5–1 mm, acute, equaling petals. Petals white or pale yellow. Styles erect, ca. 1.1 mm, exceeding calyx teeth. Fruit ovoid-globose, 1.1–1.3 mm, covered with tubercles.
Anthesis and pollination	Flowering start in December and continue for longer period.
Fruit and fruit set	Fruiting continue from April to December

REFERENCES

1. Adams, C.D. 1970. Flowering plants of Jamaica. Univ. West Indies, Mona, Jamaica, W.I.
2. Adams, C.D. 1971. Flowering plants of Jamaica. Univ. West Indies, Mona, Jamaica, W.I
3. Bautista, O.K., Kosiyachinda, S., Rahman, A.S.A. and Soenoeadji. 1988. Traditional vegetables of Asia. *Asian food J.*, **4:** 47-58.
4. Boonsong, E. 2005. Seed development and umbel order contribution in eryngio (*Eryngium foetidum* L.) seed yield and quality. Ph. D. Thesis, The Graduate School Kasetsart University, Chatuchank, Bangkok-10903, Thailand.
5. Chowdhury, J.U., Nandi, N.C. and Yusuf, M. 2007. Chemical constituents of essential oil of the leaves of *Erygium foetidum* from Bangladesh. *Bangladesh J. Sci. Ind. Res.*, **42:** 47-352.
6. Honeychurch, P.N. 1980. Caribbean wild plants and their uses. Letch worth Press, Barbados, W.I.
7. Khoshbakht, K., Hammer, K., Pistrick, K. 2007. *Eryngium caucasicum* Trautv. Cultivated as a vegetable in the Elburz Mountains (Northern Iran). *Genet Resour Crop Evol.*, **54:**445–448.
8. Lawrence, G.H.M. 1967. Taxonomy of vascular plants. Oxford and IBH Publication Company, Kolkata, p. 646
9. Mahabir, K. 1991. Medicinal and edible plants used by East Indians of Trinidad and Tobago. Chackra Publ. House, El Dorado, Trinidad, W.I.
10. Morton, J.F. 1981. *Umbelliferae*. In: Atlas of medicinal plants of Middle America. Springfield Illinois: Thomas, C.C.
11. Pistrick, K. 2001 Umbelliferae. In: Hanelt, P. (ed.) Institute of Plant Genetics and Crop Plant Research. Mansfield's encyclopedia of Agricultural and Horticultural Crops. Springer, Berlin, p. 1259–1328.
12. Pistrick, K. 2002. Current taxonomical overview of cultivated plants in the families' Umbelliferae and Labiatae. *Genet. Resour. Crop. Evol.*, **449:** 211-225.
13. Ramcharan, C. 1999. A much utilized, little understood herb. In: JaniK (Ed.). Perspectives on New Crops and New Uses. ASHS Press, Alexandria, VA.
14. Seaforth, C. and Tikasingh, T. 2005. Report: Study for the development of a 'Handbook of Selected Herb for Industry' CHBA/IICA, Trinidad, p. 44-47.
15. Singh, B.K., Pathak, K.A., Ramakrishna, Y. 2013a. Underutilized vegetable crops and spices of mizoram: needs exploration and utilization. In: Prakash, N., Roy, S.S., Sharma, P.K., Ngachan, S.V. (eds). Developing the potential of underutilized horticultural crops of hill regions. Today and Tomorrow's Printers and Publishers, New Delhi, p. 217–232
16. Wilson, P.L. 1991. Cilantro-heart of the ubiquitous sofrito. Food Section. San Juan Star, June, 1991. San Juan, Puerto Rico. p. 12-13.
17. Wagner, W.L., Herbst, D.R. and Sohmer, S.H. 1999. Manual of the flowering plant of Hawaii. University of Hawaii Press Honululu.
18. Wong, K.C; Feng, M.C; Sam, T.W. and Tan, G.L. 1994. Composition of the leaf and root oil of *Eryngium foetidum* L.J. Essent Oil. Res. 6: 369-374.

Genus *Mentha*

Mint or Pudina or Peppermint (*Mentha* Species)

Introduction

Mentha is also known as mint which is derived from Greek *míntha* (Linear B *mi-ta*) and falls in the genus of plants in the family Lamiaceae (mint family) (Harley *et al.*, 2004). It is very aromatic and possesses very good taste and due this reason mint is used in preparation soft drink adding raw mango or lemon. The leaf, fresh or dried, is the culinary source of mint. Fresh mint is usually preferred over dried mint when storage of the mint is not a problem. The leaves have a warm, fresh, aromatic, sweet flavor with a cool aftertaste, and are used in teas, beverages, jellies, syrups, candies, and ice creams. In Middle Eastern cuisine, mint is used on lamb dishes, while in British cuisine and American cuisine, mint sauce and mint jelly are used, respectively. Mint is a necessary ingredient in Touareg tea, a popular tea in northern African and Arab countries. Alcoholic drinks sometimes feature mint for flavor or garnish, such as the mint julep and the mojito. *Crème de menthe* is a mint-flavored liqueur used in drinks. Mint essential oil and menthol are extensively used as flavorings in breath fresheners, drinks, antiseptic mouth rinses, toothpaste, chewing gum, desserts, and candies, such as mint (candy) and mint chocolate. The substances that give the mints their characteristic aromas and flavors are menthol (the main aroma of peppermint and Japanese peppermint) and pulegone (in pennyroyal and Corsican mint). The compound primarily responsible for the aroma and flavor of spearmint is *L*-carvone. Mints are used as food plants by the larvae of some Lepidoptera species, including buff ermine moths. Mint was originally used as a medicinal herb to treat stomach ache and chest pains.[18] There are several uses in traditional medicine (Jamila and Mostafa, 2014) and preliminary research for possible use in treating irritable bowel syndrome (US Report). Menthol from mint essential oil (40–90 per cent) is an ingredient of many cosmetics and some perfumes. Menthol and mint essential oil are also used inaromatherapy which may have clinical use to alleviate post-surgery nausea (Hunt *et al.*, 2013; US Report). Although it is used in many consumer products, mint may cause allergic reactions in some people, inducing symptoms such as abdominal cramps, diarrhea, headaches, heartburn, tingling or numbing around the mouth, anaphylaxis or contact dermatitis (Bayat and Borici-Mazi, 2014; US Report). Mint oil is also used as an environmentally friendly insecticide for its ability to kill some common pests such as wasps, hornets, ants, and cockroaches (Bounds, 2010).

Origin, History and Distribution

The genus has a sub-cosmopolitan distribution across Europe, Africa, Asia, Australia, and North America (Brickell and Zuk, 1997). While the species that make up the *Mentha* genus are widely distributed and can be found in many environments, most grow best in wet environments and moist soils. It is extensively cultivated in USA, South East Europe and Latin American countries. In India, it is grown all over the country. Its commercial cultivation is restricted to erstwhile Uttar Pradesh and parts of Punjab because of high yield and superior quality of oil in terms of its chemical composition.

Species Availability, Evolution and Diversity

The species are not clearly distinct, and estimates of their number vary from 13 to 18 (Bunsawat *et al.*, 2004). The tribe contains about 65 genera, and relationships within it remain obscure (Harley *et al.*, 2004). The genus *Mentha* comprises of 40 species. Among the different species of menthe, only 3 species namely *Mentha arvensis* (Japanese mint), *Mentha spicata* (Spear Mint or Pahar Pudina) and *Mentha piperita* (Pepper Mint) are important as spices and condiments. The list below includes all of the taxa recognized as species in recent works on *Mentha*. Authors have disagreed on the circumscription of *Mentha*. Some authors have excluded *M. cervina* from the genus. *M. cunninghamii* has also been excluded by some authors, even in some recent treatments of the genus (Tucker and Naczi, 2007). In 2004, a molecular phylogenetic study indicated both of these species should be included in *Mentha* (Bunsawat *et al.*, 2004).

Common names are also given for species that have them. Synonyms, along with cultivars and varieties, are given in articles on the species.

Table 13.8: Common and Botanical Name

Sl.No.	Common name	Botanical name
1.	Water mint, marsh mint	*Mentha aquatica*
2.	Corn mint, wild mint, Japanese peppermint, field mint, banana mint	*Mentha arvensis*
3.	Asian mint	*Mentha asiatica*
4.	Australian mint	*Mentha australis*
5.	American wild mint	*Mentha canadensis*
6.	Hart's pennyroyal	*Mentha cervina*
7.	Bergamot mint, orange mint	*Mentha citrata*
8.	Wrinkled-leaf mint	*Mentha crispata*
9.	Dahurian thyme	*Mentha dahurica*
10.	Slender mint	*Mentha diemenica*
11.	Forest mint	*Mentha laxiflora*
12.	Horse mint	*Mentha longifolia* (syn. *Mentha sylvestris*)
13.	Peppermint	*Mentha piperita*
14.	Pennyroyal	*Mentha pulegium*
15.	Corsican mint	*Mentha requienii*
16.	Garden mint	*Mentha sachalinensis*
17.	Native pennyroyal	*Mentha satureioides*
18.	Spearmint, curly mint (a cultivar of spearmint); USDA Report	*Mentha spicata* (syn. *M. viridis, M. cordifolia*)
19.	Apple mint, pineapple mint (a variegated cultivar of apple mint)	*Mentha suaveolens*
20.	Gray mint	Mentha vagans

Source: Wikipedia; Retrieve on 05/05/2016.

The *mentha* x *piperita* hybrid is known as "chocolate mint". The mint genus has a large grouping of recognized hybrids. Synonyms, along with cultivars and varieties where available, are included within the specific species.

Table 13.9: Species Status

Sl.No.	Details
1.	*Mentha* x *dalmatica* (*M. arvensis* x *M. longifolia*)
2.	*Mentha* x *dumetorum* (*M. aquatica* x *M. longifolia*)
3.	*Mentha* x *gracilis* (*M. arvensis* x *M. spicata*)–Ginger mint, Scotch spearmint
4.	*Mentha* x *maximilianea* (*M. aquatica* x *M. suaveolens*)
5.	*Mentha* x *piperita* (*M. aquatica* x *M. spicata*) – Peppermint, chocolate mint
6.	*Mentha* x *rotundifolia* (*M. longifolia* x *M. suaveolens*)–False apple mint
7.	*Mentha* x *smithiana* (*M. aquatica* x *M. arvensis* x *M. spicata*)–Red raripila mint
8.	*Mentha* x *verticillata* (*M. aquatica* x *M. arvensis*)
9.	*Mentha* x *villosa* (*M. spicata* x *M. suaveolens* also called *M. nemorosa*)–Large apple mint, foxtail mint, hairy mint, woolly mint, Cuban mint, mojito mint, and *yerba buena* in Cuba
10.	*Mentha* x *villosonervata* (*M. longifolia* x *M. spicata*)–Sharp-toothed mint
11.	Mentha x gracilis and M. rotundifolia

Source: Wikipedia; Retrieve on 05/05/2016

The most common and popular mints for commercial cultivation are peppermint (*Mentha* × *piperita*), native spearmint (*Mentha spicata*), Scotch spearmint (*Mentha x gracilis*), and cornmint (*Mentha arvensis*) (Mint Report); also (more recently) apple mint (*Mentha suaveolens*).

Botanical Description

Mints will grow 10–120 cm tall and can spread over an indeterminate area. Due to their tendency to spread unchecked, some mints are considered invasive (Brickell and Cole, 2002).

Kingdom	:	Plantae
Unranked	:	Angiosperms
Unranked	:	Eudicots
Unranked	:	Asterids
Order	:	Lamiales
Family	:	Lamiaceae
Tribe	:	Mentheae
Genus	:	*Mentha* L.

Mentha is a member of the tribe Mentheae in the subfamily Nepetoideae. The other details are being given below:

Table 13.10: Botanical Description

Growth Habit	Exclusively perennial and rarely annual herbs. Mint grows all year round. They are fast-growing, extending their reach along surfaces through a network of runners.
Root	The roots of mints are very aggressive growers.
Stem	They have wide-spreading underground and over ground stolons (Aflatuni *et al.*, 2005) and erect, square (Rose, 1981) branched stems.
Leaf	The leaves are arranged in opposite pairs, from oblong to lanceolate, often downy, and with a serrated margin. Leaf colors range from dark green and gray-green to purple, blue, and sometimes pale yellow (Brickell and Zuk, 1997).
Floral biology	The flowers are white to purple and produced in false whorls called verticillasters. The corolla is two-lipped with four subequal lobes, and the upper lobe usually the largest.
Fruit and fruit set	The fruit is a nutlet which contains seeds
Seeds	Each nutlet has one to four seeds

Horticultural Classification

Mentha can be classified into:

1. *Mentha arvensis* (Japanese mint)-It is heavy feeder of major nutrients.
2. *Mentha spicata* (Spear mint or Pahar Pudina)
3. *Mentha piperita* (Pepper mint)

Augmentation of Vegetable Genetic Resource and Crop Improvement

Hybridization

Hybridization between some of the species occurs naturally. Many other hybrids, as well as numerous cultivars, are known.

Resistance to Abiotic Stress

All mints thrive near pools of water, lakes, rivers, and cool moist spots in partial shade. In general, mints tolerate a wide range of conditions, and can also be grown in full sun.

REFERENCES

1. Aflatuni, A.J.U., Ek, S. and Hohtola, A. 2005. "Variation in the amount of yield and in the extract composition between conventionally produced and micropropagated Peppermint and Spearmint". *Journal of Essential Oil Research*, **17 (1):** 66-70.
2. Bayat, R.; and Borici-Mazi, R. 2014. "A case of anaphylaxis to peppermint". *Allergy, Asthma and Clinical Immunology*, **10**: 6.
3. Bounds, G. 2010. "Death by Mint Oil: Natural Pesticides", *The Wall Street Journal*, July 30, 2009, accessed December 6, 2010.
4. Bradley, F. 1992. Rodale's All-new Encyclopedia of Organic Gardening. Emmaus, Pennsylvania, USA: Rodale Press. p. 390.
5. Brickell, C. and Cole, T. 2002. The American Horticultural Society: Encyclopedia of Plants and Flowers. New York, NY, USA: DK Publishing. p. 605.
6. Brickell, C. and Zuk, J.D. 1997. The American Horticultural Society: A-Z Encyclopedia of Garden Plants. New York, NY, USA: DK Publishing. p. 668.
7. Bunsawat, J., Elliott, N.E., Hertweck, K. L., Sproles, E., and Alice, L.A. 2004. "Phylogenetics of Mentha (Lamiaceae): Evidence from Chloroplast DNA Sequences". *Systematic Botany*, **29(4):** 959-64.
8. CFR Part 182 Substances Generally Recognized as Safe
9. Davidson, A. 1999. The Oxford Companion to Food. Oxford: Oxford University Press. p. 508.
10. GRAS FDA
11. Harley, R. M., Atkins, S., Budantsev, A. L. and Cantino, P.D. 2004. "Labiatae". In Kubitzki, Klaus; Kadereit, Joachim W. The Families and Genera of Vascular Plants VII. Berlin; Heidelberg, Germany: *Springer-Verlag.*, p. 167-275.
12. http://www.harighotra.co.uk/my-blog/minted
13. Hunt, R., Dienemann, J., Norton, H.J., Hartley, W., Hudgens, A., Stern, T. and Divine, G. 2013. "Aromatherapy as Treatment for Postoperative Nausea". *Anesthesia and Analgesia*, **117** (3): 597.
14. Jamila, F. and Mostafa, E. 2014. "Ethnobotanical survey of medicinal plants used by people in Oriental Morocco to manage various ailments".

Journal of Ethnopharmacology, **154**: 76.

15. Kew World Checklist of Selected Plant Families
16. Liddell, Henry George; Scott, Robert; *A Greek–English Lexicon* at the Perseus Project
17. "Mint, Economic Importance". The Canadian Encyclopedia. Retrieved 31 December, 2015.
18. "Mint". Herbsociety-stu.org. Retrieved 2013-07-14.
19. Ortiz, E. 1992. The Encyclopedia of Herbs, Spices and Flavorings. London: Dorling Kindersley. p. 36–7.
20. Palaeolexicon, Word study tool of ancient languages
21. "Peppermint oil". National Center for Complementary and Alternative Medicine, US National Institutes of Health. 2014. Retrieved 11 October 2014.
22. Quattrocchi, U. 1947. CRC World dictionary of plant names: Common names, Scientific Names, Eponyms, Sonyonyms, and Etymology. III M-Q. CRC Press. p. 1658.
23. Rose, F. 1981. The Wild Flower Key. Frederick Warne and Co. p. 310.
24. Sharon J.H. 2013. "A not-so-boring history of flooring". CSMonitor.com. Retrieved, 2013-07-14.
25. Tucker, A.O. and Naczi, R.F.C. 2007. "Mentha: An Overview of its Classification and Relationships". In Lawrence, Brian M. Mint: The Genus Mentha. Boca Raton, FL, USA: CRC Press, Taylor and Francis Group.
26. USDA GRIN: Mentha cordifolia Opiz ex Fresen.

14

Genetic Resources of Tree Vegetables and its Improvement

Genus *Moringa*

Drumstick (*Moringa oleifera*)

Introduction

Drumstick (*Moringa oleifera* Lam. Syn. *M. pterogosperma* Gaertn., 2n=28) is a fast growing multipurpose tree. The name drumstick derives from the shape of pod resembling the slender and curved stick used for beating the drum. Drumstick is synonym as 'Horse Radish Tree', 'West Indian Ben' and 'Never Die'. Probably the name 'radish tree' derives from the pendulous slender and thin shape of immature fruits of the tree resembling very much with the siliqua of radish. It is also commonly known as ben oil tree. In India this plant is known as Sobhjana (Sanskrit); Mungana, Saijan, Sahajan (Hindi); Sajeena (bengali); Achhaghadha, Shewagi (Marathi); Midhosaragewo,Saragewo (Gujarati); Maluga, Tela Munga (Telagu); Murgai (Tamil); Nuge (Kannad), Murina, Sigaru, Moringa (Malayalam); Saijana, Sahajan (Assam); Sajina (Odisha); Sainjana, Sonjana (Punjabi) and Munga Arak (Santhal). Presently, this vegetable can be spotted in each and every home yard. This is a backyard tree for daily use in more than two million homesteads of South, India. Presently, India is the largest producer of drumstick with an annual production of 2.0 -3.3 million tonnes of tender fruits from an area of 68,000 ha. Among the states Andhra Pradesh leads in both area and production (15,665 ha) followed by Karnataka (10,280 ha) and Tamil Nadu (7,408ha). In other states, it occupies an area of 8,625 ha.

Ancient Indian literature makes mention of drumstick as an interesting plant due to its widespread use in agriculture, medicine and industry. Basically drumstick is cultivated as multipurpose tree species like tender pods and its leaves and also flowers for vegetable in tropical and subtropical India (Kumar *et al.*, 2000). As pod, flower and tender leaves of drumstick is rich in vitamin and mineral matter, are used in vegetable preparation. Fresh tender leaflets are striped and used as a leafy vegetable and a pot herb that is cooked either alone or with lentils. Drumstick leaves contains 22-38 per cent dry matter, 5-8 per cent TSS and 0.7-1.5 mg/100 g chlorophyll. Tender leaves are also used to flavor ghee and to enhance the shelf-life of ghee. The pods of drumstick are very popular vegetables in South Indian cuisine and valued for their distinctly inviting flavor. The young pods, also known as "drumsticks" or "Moringa" by Indian communities, are cooked as vegetable (like cowpea pods). The pods are sold fresh and/or canned in many Asian grocery stores and markets. Drumstick is drought tolerant and easily adapted to varied ecosystems and farming systems, it occupies a unique and consistent position in the Indian vegetable industry. Because of its many different uses, its free flowering nature and the ease with which it can be cultivated, the area required for it is on the increase, as is demand for its products. The drumstick leaves and pods are used as vegetables. The leaves, flowers and pods are used as significant sources of vitamins A, B and C, riboflavin, nicotinic acid, folic acid, pyridoxine, calcium, iron, and alpha-tocopherol (Dahot, 1988). Leaves contain 4.0 per cent moisture, 38.4 per cent crude protein, 34.17 per cent fatty oil, 3.5 per cent fibre and 3.2 per cent mineral matter. Protein available in leaves has biological value and digestibility coefficient 41 per cent and 77 per cent, respectively. Tender leaves also contain 16 per cent non-protein nitrogen. After analysis of leaves Gopalan *et al.* (1999) reported that 100 g leaves contain 42.0 mg Maganesium, 259 mg potassium, 0.07 mg copper, 0.37 mg Manganese, 0.16 mg zinc, 0.010 mg chromium, 137.0 mg sulphr and 423.0 mg chlorine. The 100 g of edible pods contains 86.9 g moisture, 2.5 g protein, 0.1 g fat, 3.7 g carbohydrate, 4.8 g fiber, 2.0 g mineral matter, 30.0 mg calcium, 110.0 mg phosphorus 5.3 mg iron, 3.1 µg/g copper, 18.0µg/g iodine, 0.01 g oxalic acid, 184 IU carotene, 0.2 mg nicotinic acid, and 120.0 mg ascorbic acid. Alone leaves contains 75.0 g moisture, 6.7 g protein, 1.7 g fat, 13.4 g carbohydrate, 0.9 g fiber, 2.3 g mineral matter, 440.0 mg calcium, 70.0 mg phosphorus 7.0 mg iron, 1.1 µg/g copper, 51.0 µg/g iodine, 11,300 IU carotene, 0.8 mg nicotinic acid, and 220.0 mg Ascorbic acid. Apart from these, leaves contain 380.0 mg arginine, 140.0 mg histidine, 320.0 mg lysine, 100.0 mg tryptophan, 290.0 mg phenyl alanine, 110.0 mg methionine, 130.0 mg cystine, 250.0 mg threonine, 460.0 mg leucine, 280.0 mg isoleucine and 350.0 mg valine per gram nitrogen.

Almost all parts of the plant (roots, leaves, flowers, seeds etc.) have been used in the treatment of ailments in indigenous system of medicine (Ramachandran *et al.*, 1980). In Indian Ayurvedic traditional medicine and other folk remedies, it is very popular. Aqueous leaf extracts are reported to regulate thyroid hormone and can be used to treat hyperthyroidism (Tahiliani and Kar, 2000). Leaf extracts are also used to treat ulcer (Pal *et al.*, 1995). Reportedly, drumstick leaves and pods have a positive effect in reducing blood cholesterol (Ghasi *et al.*, 2000), and anti-tumor promoting activity (Guevara *et al.*, 1999). Furthermore, it is an important source of the chemo-preventives (Daxenbichler *et al.*, 1991) that can inhibit carcinogenesis. Drumstick is also being studied for its anti-inflammatory, antimicrobial, diuretic (Caceres *et al.*, 1991; Caceres *et al.*, 1992; Udupa *et al.*, 1994), antibiotic (Eilert *et al.*, 1981), hypotensive (Faizi *et al.*, 1998), and antimicrobial (Spiliotis *et al.*, 1998) properties. A compound found in the flowers and roots of the moringa tree, 'pterygospermin' has powerful antbiotic and fungicidal effects (Das *et al.*, 1957). The root bark contains two alkaloids: 'moringine' and 'moringinine'. Moringinine acts as cardiac stimulant, produces rise of blood-pressure, acts on sympathetic nerve-ending as well as smooth muscles all over the body, and depresses the sympathetic motor fibers of vessels in large doses only. Roots are used as spice like horse radish.

The seeds of *Moringa* are considered to be antipyretic, acrid, bitter (Oliveira *et al.*, 1999) and reported to show antimicrobial activity). Traditionally the drumsticks seeds have been used to purify domestic household water in rural areas in Sudan (Jahn, 1981, Jahn *et al.*, 1986). Women collecting water would tie the ground seeds and suspend in the turbid water overnight. The oil extracted from the seeds (called ben oil) was used as lubricant in watches, for edible purpose a practice that has now been discontinued. The kernel of the seed is rich in crude protein, fatty oil and fibre. Seeds after roasting can be used as groundnut. Seeds possess fatty acids in form of triglicerides and act as a strong antibiotic due to 4γ L rhmnosyl-oxybenzyl-isothiocyanate. The industrial uses of the drumstick tree include the use of its wood in paper (wood pulp suitable for newsprint, wrapping, printing and writing papers) and textile industries (rayon grade pulp for textiles-Guha and Negi, 1965), bark in the tanning industry, and the seeds in water purification (Bhattacharya *et al.*, 1982; Mayer and Stelz, 1993; Palada, 1996; Tauscher, 1994). More recently the ben oil has also been shown to be particularly effective in the manufacturing of soap producing a stable lather with high washing efficiency suitable for some African countries. The pressed cake obtained after oil extraction may be used as a fertilizer.

Origin, History and Distribution

Drumstick is well naturalized in the tropical and sub-tropical regions of the world. It is native to the Old World Tropics particularly to North West India (Indian subcontinents). Its wild races are occurring in the foothills of sub-Himalayan western regions from river 'Chinab' eastwards to 'Sarda' (North West Frontier Province now Khyber Pakhtunkhwa) and in the tarai tract of Northern India, and now grown worldwide in the tropics and sub-tropics. In Pakistan more occurrences can be spotted in the west (Anwar and Rashid, 2007). In Uttarakhand up to 700 m above mean sea level (msl), and also northern parts of Uttar Pradesh; its availability can be seen (Kanjilal, 1933). Drumstick was originally grown by the Dravidians and later by Aryans. Jahn (1996) has given the details of drumstick's introduction in China. He put forth the hypothesis that Indian and Malayan people first time brings drumstick to Makao from there it reach to Canton and later on its utility increases for supporting black pepper. The tree is widely distributed in Indian sub-continent, South East Asia, Egypt, the Phillippines, Ceylon, Thailand, Malaysia, Burma, Pakistan, Singapore, West Indies, Cuba, Jamaica and Nigeria. The cultivation of drumstick in India occurs mainly in the Southern states of Tamil Nadu, Karnataka, Kerala, and Andhra Pradesh. Tamil Nadu is the pioneering state as it has varied genotypes from Sri Lanka. Drumstick can be grown well in the humid tropics or hot dry lands and survives in the destitute (Morton, 1991). Palada and Changl (2003) reported that plant can tolerate wide range of rainfall (250-3000 mm) and soil pH (50-9.0).

Species Availability, Evolution and Diversity

The genus *Moringa* has more than 13 species (Pandey *et al.*, 2011) but Philips (1951) reported 4 species, while Puri (1942) reported 10 species, only. There are two common species *M. oleifera* Lam. and *M. concanensis*, the former being the vegetable species. A few of the species of *Moringa* related to *M. oleifera* are *M. pterygosperma* Gaertn. and *M. longifolia* and both are closest relative of *M. concanensis* Nimmo (occurs in India). The *M. oleifera* (the drumstick tree, horse radish tree, West Indian Bean) species is being distinguished by leaves usually tripinnate, leaflets 12-18 mm long; petioles yellow or white without red streakes and the tree being medium-sized. It is a fast growing and drought resistant species which is found widely in the sub Himalayan tracts of the northern India (Singh *et al.*, 2000). *M. concanensis* is distinguished by leaves bipinnate, leaflets 15-30 cm long, petioles with red streaks or reddish at base and the tree being large-sized (CSIR, 1962). Plants are small in nature. In West Indies and Indonesia, there are few forms which rarely flower and are principally cultivated for their foliage and other which can fruit abundantly and are cultivated for fruits.

The genus has centre of diversity in the Horn of Africa (National Research Council, 2006).

Botanical Description

Genus *Moringa* belongs to the family Moringaceae, a monogeneric family: *Moringa, M. oleifera* Lam. This family is distinguished by partial placentation, 3 valved fruit, elongated, non-dehiscent berry and winged seeds.

Table 14.1: Botanical Description

Growth Habit	Plant is medium sized trees (5.0-15.0 m in height) with soft white or brown wood like bark having thick fiber (Morton, 1991). Barks exudates white gums. Emerging new shoots are light green or purple in colour. Tree is fast growing.
Root	Tuberous roots
Stem	The tree itself is rather slender with drooping branches. Stem is soft woody and a thin crown.
Leaf	Leaves are spirally arranged, crowded at the ends of the branches, long petioled, incompletely tripinnate, on the rachises with glands between the pinnate and the leaflets and 25.0-45.0 cm long. Leaves deciduous alternate 2-3 pinnate; the pinnae and leaflets imparipinnate, opposite, glandular at base. Leaf is 50.0-60.0 cm and petiole is 4.0-15.0 cm long and thick. Leaflets are stalked, ovate or obovate; base acute, obtuse or rounded, often oblique; apex obtuse, rounded or emarginated; entire, dull green on both sides, lighter coloured somewhat angular, copular-cyathiform, oblique, green, densely shortly pubescent on both sides, unequal, 0.7-1.4 cm long, 0.25-0.5 cm wide.
Floral biology	White, creamy, yellow, streaked with red colour numerous flowers developed on 8.0-30.0 cm long branch. The flowers are bisexual, oblique, stalked, axillary and heteromorphic. Green in colour calyx tubular has 5 lobes. Corolla 5 unequal, yellowish, white with a greenish base, with thin veins, 1.0-2.0 cm long, stamens 5, stigma tubular and become narrow towards apex. The two hindmost and the two lateral ones reflexed, ovate or obovate, obtuse, with a canaliculated base, on the inner side hairy at the basque, otherwise glabrous, 1.0- 1.17 cm long, 0.5-0.6 cm wide; the foremost petal erect, obovate, obtuse, glabrous in the inside, on the outside with 2 longitudinal rows of hairs, 1.4-1.6 cm long, 0.6-0.8 cm wide. Perfect stamens are, alternating with 5 subulate staminodes densely pilose at the base, the hindmost stamen the longest, 0.8-0.9 cm long, and the other ones much shorter. Ovary is stalked, densely clothed with rather long, appressed hairs. Ovules, style thin, curved, white, shortly pubescent, hollowed at the apex. Capsules are pendulousk, linear, acuminate, obtusely trigonous, ribbed, 3-valved and 20.0-45.0 cm long. Valves are spongy thick and impressions of the seeds half-globose.
Anthesis and pollination	In north Indian plains drumstick flowers in the month of January and February. The plant is highly cross pollinated due to heteromorphism Plant is entomophilous, and bees or fly (*Xylocarpa latipes* and *X. pubescens*) being the pollinators which is very active during day and pollinate 6.0 AM morning to 3.0 PM evening. Pollen grain having size of 31.32-41.65 µm, germinate effectively and the size between 22.9-33.22 µm are generally sterile in nature. Fresh pollen grains germinate quickly. With the addition of 6 per cent sucrose and 15 ppm boric acid, pollens can be stored suitable up to 24 hours. Maximum fruit set (31.26 per cent) can be achieved using fresh pollens through hand pollination.
Fruit and fruit set	Pods are 10.0-50.0 cm long with triangle shape and green colour. In internal fruit development endogenous auxin and cytokinin play very important role (Iyer *et al.*, 1981; Nagar *et al.*, 1982). Vertical development of fruit is very fast. In 4 auxin components, two major compound *i.e.* indol acetic acid and indol aceto nitrile has been found in 30.0-40.0 cm long fruit. Auxin activity is very high when fruits are of 25.0-30.0 cm and reduced when pods attain the height of 50.0 cm. Palanisamy *et al.* (1985) reported that fruits continue to attain the growth up to 80-85 days after anthesis.
Seeds	Seed has short wings and elongate. Seeds are numerous, globular, about 1.0 cm diameter 3-winged; wings produced at the base and the apex, 2.0-2.5 cm long, 0.4-0.7 cm wide and scarious. Mature seed of 40 days usually germinate but less than 35 days mature fail to germinate. Seeds should be harvested from annual drumstick (seed propagated) after 100 days of anthesis and later on germination rate decrease. If mature seeds are properly treated with captan and kept in airtight plastic pouches germinate 40 per cent after one year while un-treated has less germination (4-14 per cent) as observed by Palanisamy *et al.* (1995).

Horticultural Classification

Based on morphological characters of the plant and the fruit characters other than the length of fruit, it was found that the several types could be broadly classified based on the branching habit of the tree and the pigmentations of the floral parts and the fruits (Irulappan, 1991). With regard to the branching habit, tree types could be broadly classified into 2 groups:

(i) Types having profuse branching habit in which the branches are straggly and drooping. It is preferred by growers on account of their larger yields.

(ii) Types having only a few erect branches.

Based on the pigmentation of the plant, floral parts and the fruit, 3 broad groups could be distinguished:

Table 14.2: Grouping of Drumstick and their Salient Feature

Group 1	The tender twigs are dirty brown, leaf stalk and peduncle of panicles tinged purple; stamens and ovary completely purple; fruits in tender stages reddish, turning green with reddish streaks running from base to tip at full maturity.
Group 2	Twigs are dirty brown, leaf stalk, nodes, peduncle, pedicel and their point of attachment deep purple; basal portion of sepals and petals tinged purple; stamens and ovary completely purple; fruits at tender stages deep reddish turning to black purplish on full maturity.
Group 3	The plant parts are green except the basal portions of the leaf stalk which are sometimes tinged purple; sepals whitish; petals either white or light yellow; stamens and pistil light yellow; fruits at tender stages light green with or without purple tinge turning to darker green on maturity.

The types which are under cultivation generally fall into the first or the third group. The second group is not found under large-scale cultivation presently because of the comparatively lower yields and the poor quality fruits.

In West Bengal, India two distinct genotypes are popular:

(i) Sajna- Flowers only once in a year (January-February)

(ii) Najna- Flowers 2 or 3 times in a year.

Genetic Resource Augmentation and Crop Improvement

Despite the rich nutritional content in fruit, leaf and its potential in industry, drumstick has been denied attention for improvement and commercial utilization particularly in northern India. Because of high heterozygosity, genotypes existing in varied geographical areas provide scope for crop improvement programmes (Sundarajan *et al.*, 1970) through broadening the genetic base. Moreover, germplasm with specific traits such as cluster-bearing habit, low tree canopy, tolerance to biotic and abiotic stress (Rajangam *et al.*, 2001) and high seed oil are the desirable traits in moringa improvement programmes.

Introduction and Selection

It is probable that drumstick germplasm collected from centre of origin can show greater genetic diversity and also superipr in having high functional property (Asian Vegetable Research and Development Centre, 2002; Bosch, 2004). In the last two decades improved annual cultivars (seed-propagated ones) have replaced around 60 per cent of perennial drumstick (stem-propagated) cultivating areas in south India (Rajangam *et al.*, 2001). In drumstick a few named varieties are grown. A type named 'Jaffna' grown in certain parts of South India produces fruits 60-90 cm long with a soft flesh 'Jaffna' of good taste. The other types are 'Chavakacherri Murunga', also a 'Jaffna type', bears fruits as long as 90-120 cm long and 'Chemmurunga' (red tipped fruits which is said to flower throughout the year and yielded heavy crops). Trees growing wild ('Kadu Murunga') usually bear small and inferior fruits. In Tirunelveli district of Tamil Nadu (India) a variety known as 'Palmurungai' with thick pulp and better taste and another called 'Puna Murungai' are grown. In Trichy district of Tamil Nadu (India), a variety known as 'Kodikal Murungai' producing very short fruits of 15-23 cm. Two varieties 'KM 1' and 'PKM 1' have been released as seed *Moringa* so far.

Table 14.3: Morphological Comparision of Wild and Cultivated Types of Drumsticks

Particulars	*Wild†*	*Cultivated**
Leaf	Petiole strictly green in colour	Petiole also violet-purple
Flower	White or pink in colour	White or yellowish white, rarely pink
Fruit	size 15-50 (-60) x 1.0-1.5 cm, thin, Slender, less fleshy, bitter; 15-20 seeds/fruit	(15-)20-120 (-150) x 1.3-2.5 (-3.3) cm, thick walled, fleshy, bitter-free; up to 26 seeds/fruit
Seed	Size (wings) 15-30 (32) x ()6-8-12 mm. seed coat white, mealy, fragile; wings soft, membranous, caducous	25-32 (-40) x 10-26 (-30) mm; seed coat black to brown at least in the ¾ area from centre, hard; wings smaller, persistent
Phenology	Deciduous (during winter); one time flowering (February-March) followed by fruiting (March-May0; leaf emergence (February-May.)	Evergreen (occasionally semi-deciduous during winter in north India); flowering almost throughout the year with less space in winter.

Source: Resmi *et al.* (2005); Parrotta (2009) and Radovich (2009).

Table 14.4: Fruit and Seed Characters of *Moringa oleifera* Germplasm

Sl. No.*	Fruit Length (cm)	Fruit Girth (cm)	Seeds per Fruit	100-Seed Weight (g)	Per cent Kernel	Kernel Kength (mm)	Kernel Dia-meter (mm)	Fruit Colour (Dry)	Kernel Shape	Wing Promi-nence	Seed-Coat Detach-ability	Remarks
1	33.34	3.44	15.40	14.70	69.68	10.32	6.41	LB	Ovate	++	†	-
2	40.27	2.85	18.70	12.08	76.45	10.19	7.26	RB	Ovate	++	†	Gregarious bearing
3	44.38	3.06	15.30	11.24	72.95	9.18	6.11	DB	Ovate	++	†	Gregarious bearing
4	28.31	2.53	16.60	10.22	79.35	8.91	5.88	LB	Ovate	+++	†	Gregarious bearing
5	35.48	2.92	14.10	8.64	67.81	7.41	3.36	RB	Ovate	+	††	Big-sized leaflet
6	22.67	2.34	13.50	10.32	82.21	6.76	5.05	RB	Ovate to round	++	†	Big-sized leaflet
7	37.40	2.83	14.20	9.38	76.57	8.47	5.07	LB	Ovate	++	††	-
8	44.55	2.79	14.30	9.04	76.96	9.23	6.72	RB	Ovate	+++	†	Gregarious bearing
9	45.19	2.97	16.10	10.04	73.71	8.04	4.18	RB	Ovate	+++	†	Gregarious bearing
10	35.34	2.47	15.30	9.98	81.17	10.28	6.85	LB	Ovate	+++	†	Less bitter fruits; Big-sized leaflet
11	47.18	2.96	14.20	13.78	76.92	10.39	8.52	LB	Ovate	++	††	-
12	25.34	2.20	13.20	12.29	68.93	10.95	5.88	LB	Ovate	+	††	-
13	50.64	2.66	13.60	7.42	72.78	6.33	4.65	LB	Ovate	++	†	-
14	40.32	2.46	13.70	11.68	80.90	10.12	6.52	LB	Ovate	++	†	-
15	45.21	2.75	14.90	12.22	77.33	9.55	6.84	DB	Ovate	+++	†	Fruit bitter free
16	47.89	2.57	13.90	7.89	79.05	8.78	5.54	LB	Ovate	+++	††	-
17	30.77	2.75	12.30	6.40	72.88	6.27	4.02	LB	Ovate to round	+++	†	-
18	37.38	2.81	12.10	10.24	70.31	8.86	5.13	LB	Ovate	+++	†	Big-sized leaflet
19	57.34	2.86	17.80	8.37	71.88	9.31	5.76	RB	Ovate	+++	††	-
20	60.44	3.13	16.40	8.44	72.04	10.11	6.07	LB	Spindle	++	†	-
21	42.53	2.68	13.50	11.34	67.80	9.50	5.33	LB	Spindle	++	†	-
22	30.56	5.62	15.30	16.78	76.98	10.27	16.46	DB	Round	+++	†††	Fruits fleshy
Mean ± S.E.	**40.57 ± 2.11**	**2.76 ± 0.06**	**14.72 ± 0.37**	**10.2 7 ±0.46**	**74.65 ± 0.98**	**8.97 ± 0.30**	**5.79 ± 0.26**	-	-			
Range	**22.67-60.44**	**2.20-3.44**	**12.10-18.70**	**6.40-14.70**	**67.80-82.21**	**6.27-10.95**	**3.36-8.52**	-	-			
CV (per cent)	**23.88**	**10.23**	**11.45**	**20.43**	**5.99**	**15.20**	**20.62**	-	-			

*Serial No. 1-22 indicates collector Nos.KP/PKS.RC-10-1 to KP/PKS.RC-10-22, respectively, last one is the cultivated type which is not considered for mean, range and CV; LB-Light brown, RB-Reddish brown, DB-Dark brown; †-Easy, ††-Slightly difficult, †††-Difficult

Source: Pradheep *et al.*, 2011.

- ☆ **'PKM 1'**-In 'KM 1'pods are short (32.0-37.0 cm long and 5.5-6.0 cm girth), pod weight (65.0-82.5 g); pods/tree/year (226-328); pod yield (14.69-26.90 kg/tree/year) and seeds/pod (10-13).
- ☆ **'PKM 2'**- In 'PKM 1'pod length (61.0-78.0 cm), pod girth (6.3-6.6 cm), pod weight (140.0-170.0 g), seeds/pod (19-20), pods/tree (186-215) and pod yield/tree (29.68-34.68 kg).

Genetic Resource Conservation

A total of 23 accessions including one cultivated types were collected and conserved at ICAR-National Bureau of Plant Genetic Resources, New Delhi. Fairly rich variability was observed with respect to several economic traits including cluster bearing, bitter content in fruits, prolific yield, *etc.* in wild/semi-wild *Moringa* germplasm. Decline in population of young plants and seedlings in their native range were observed (Pradheep *et al.*, 2011).

REFERENCES

1. Anwar, F. and Rashid, U. 2007. Physico-chemical characteristics of *Moringa oleifera* and seed oil from a wild provenance of Pkaistan. *Pak. J. Bot.*, **39:** 1443-1453.

2. Asian Vegetable Research and Development Centre. 2002. Progress Report. Asian Vegetable Research and Development Centre, Taipei, Taiwan, p. 106.

3. Bhattacharya, S.B., Das, A.K. and Banerji, N. 1982. Chemical investigation on the gum exudates from Sonja (*Moringa oleifera* Lam.). *Carbohydrate Res.*, **102:** 253-262.

4. Bosch, C.H. 2004. *Moringa oleifera* Lam. In: GJH Grubben and O.A. Denton (eds.). PROTA 2: Vegetables/Vegetables. [CD-ROM]. PROTA, Wageningen, Netherlands.

5. CSIR Report, New Delhi. 1962. Drumstick. *Wealth of India,* **6:** 425-428. Publication and Information Directorate, Council of Scientific and Industrial Research, New Delhi.

6. Dahot, M.U. 1988. Vitamin contents of the flowers and seeds of *Moringa oleifera*. *Pak. J. Biochem.*, **21(1-2):** 21-24.

7. Gopalan, G, Rama Sastri, B.V., Balasubramanian, S.C., Rao, B.S.N., Deosthale, Y.G. and Pant, K.C. 1999. Nutritive value of Indian Foods. National Institute of Nutrition, ICMR, Hyderabad.

8. Irulappan, I. 1991. Drumstick (*Moringa oleifera* Lam.). (In) *Recent Advances in Tropical Vegetable Production- Compendium of lectures*, p. 273-307. Peter, K.V., Joseph, Salikutty and Indira, P. (Eds.), Kerala Agricultural University, Vellanikkara (Kerala).

9. Iyer, R.J., Nayar, P.K. and Sirkar, P.K. 1981. Auxin in *Moringa pterygosperma* fruits. *Indian J. Experi. Biol.*, **19(5):** 487-489.

10. Jahn, S.A.A.1981. Traditional water purification in tropical developing countries: existing methods and potential applications. Manual No. 117, pub. GTZ., Eschborn, Germany.

11. Jahn, S.A.A., Musnad, H.A. and Burgstaller, H. 1986. The tree that purifies water: Cultivating multipurposeMoringaceae in the Sudan. *Unasylva.*, **38:** 23-28.

12. Kanjilal, P.C. 1933. Forest Flora for Pilibhit, Oudh, Gorakhpur and Bundelkhand (repr. Edn. 1982). Narendra Publishing House, New Delhi.

13. Kumar, S., Singh, R.C. and Kumar, S. 2000. *Ann. Agril. Res.*, **21:** 148-151.

14. Maclin, N.l. (2007). Seed viability studies in Drumstick (*Moringa oleifera* Lam.). M.Sc. Thesis; University of Agriculture Sciences, Dharwad (Karnataka).

15. Mayer, F. and Stelz, E. 1993. *Moringa stenopetala* provides food and low-cost water purification. *Agroforestry Today*, **5:** 16-18.

16. Morton, J.F. 1991. The horse radish tree, *Moringa pterigosperma* (Moringaceae). A boon to arid lands. *Econ. Bot.*, **45:** 318-333.

17. Nagar, P.K., Iyer, R.J. and Sircar, P.K. 1982. Cytokinins in developing fruits of *Moringa pterygosperma* Gartn. *Physiol. Pl.*, **55(1):** 45-50.

18. National Research Council. 2006. Lost Crops of Africa, Vegetables. Vol.II, The national Academic Press, Washington DC.

19. Palada, M.C. 1996. Moringa (*Moringa oleifera* Lam.): A versatile tree crop with horticultural potential in the subtropical United States. *HortSci.*, **31:** 794-797.

20. Palada, M.C. and Changl, L.C. 2003. Suggested cultural practices for *Moringa*. International Cooperators Guide AVRDC. AVRDC pub#03-545; www.avrdc.org.

21. Palanisamy, V., Kumarsen, K., Jayabharathi, M. and Karivartharaju, T.V. 1985. Studies on seed development and maturation in annual *Moringa*. *Veg. Sci.*, **12(2):** 74-78.

22. Palanisamy, V., Balakrishanan, K., Karivaratharaju, T.V. and Arumugam, R. 1995. Influences of seed treatments and containers on the viability of annual *Moringa* seeds. *South Indian Hort.*, **43(1/2):** 42-43.

23. Pandey, A., Prudheep, K., Gupta, R., Nayar, E.R. and Bhandari, D.C. (2011). Drumstick tree (*Moringa oleifera* Lam.): A multipurpose potential species in India. Genet. Resour. *Crop Evol.*, **58:** 453-460.

24. Parrotta, J.A. 2009. *Moringa oleifera* Lam. In: Roloff, A., Weisgerber, H., Lang, U. and Stimm, B. (eds.) *Enzyklopadie der Holzgewachse, Handbuch und Atlas der Dendrologie*, 40 Erg Lfg 6/05. Wiley-Vch Verlag Gmbh and Co. KGaA, Weinheim, p. 1-8.

25. Philips, E.P. 1951. *The Genera of South African Flowering Plants*, Edn. II, p. 358, Government Printer, Pretoria.

26. Pradheep, K., Singh, P.K., Pandey, A. and Bhandari, D.C. 2011. Collecting genetic resources of wild *Moringa oleifera* Lam. from Western Himalaya. *Indian J. Plant Genet. Resour.*, **24(3):** 292-298.

27. Puri, V. 1942. Studies in floral biology. II Floral anatomy of the Moringaceae with special reference to Gynaecium constitution. *Proceed. National Institute of Science,* **3:** 71-78.

28. Radovich, T. 2009. Farm and forestry production and marketing profile for drumstick (*Moringa oleifera* Lam.). In: Elevitch, C.R. (ed.). *Speciality Crops for Pacific Island Agroforestry*. Permanent Agricultural Resources, Holualoa, Hawaii.http://agroforestry.net/scps.version history Nov. 13. 2009 Accessed 10 Sept., 2010.

29. Rajangam, J., Azhakia-Manavalan, R.S., Thangaraj, T., Vijayakumar, A. and Muthukrishnan, N. 2001. Status of production and utilization of *Moringa* in southern India. Proceedings of a workshop *'Development Potential of Moringa Products'*. Dar es

Salaam, Tanzania, 29 th October to 2nd November, 2001.

30. Ramachandran, C., Peter, K.V. and Gopalakrishnan, P.K. 1980. Drumstick: A multipurpose Indian vegetable. *Eco. Bot.*, **34(3):** 276-284.

31. Resmi, D.S., Celine, V.A. and Rajamony, L. 2005. Variability among drumstick (*Moringa oleifera* Lam.) accessions from central and southern Kerala. *J. Trop. Agric.*, **43:** 83-85.

32. Sundarajan, J.S., Muthuswamy, S., Shanmugavelu, K.G. and Balakrishnan, R. 1970. *A guide on Horticulture* (2nd edn.). Velan Pathippagam, Coimbatore (Tamil Nadu), p. 261-262.

33. Tauscher, B 1994. Water treatment by flocculent compounds of higher plants. *Plant Res. Development,* **40:** 56-70.

Genus *Murraya*

Curry Leaf [(*Murraya koenigii* (L.) Sprengel]

Introduction

Curry leaf (*Murraya koenigii* (Linn.) Sprengal, 2n= 18) is also known as Narsinghe (Assamese), Barsanga (Bengali), and Currypatia (Punjabi). Curry leaf is aromatic and used as a flavourant (seasoning and flavouring of dishes) leafy vegetable by the early inhabitants (Dravidians). The leaves are used mainly for seasoning and flavouring of dishes. Curry leaf is rich in many of the minerals and vitamins (CSIR, 1962). Quantitatively it contain more vitamin A (12,600 IU)/100 g of leaf) than carrot root (3,150 IU)/100 g), more fat (1.0g/100 g of leaf) than the most of the common vegetables like cauliflower (0.40 g/100 g), cabbage (0.1 g/100 g), radish (0.3 g/100 g), and tomato (0.2 g/100 g). Per 100 g of fresh weight contain 63.8 per cent moisture, 6.10 per cent protein, 1.0 per cent fat, 16.0 per cent carbohydrate, 6.4 per cent fiber, 4.2 per cent minerals, 12,600 IU vitamin A, 2.0-3.0 mg nicotinic acid and 4.0 mg ascorbic acid. Apart from this leaves are wonderful source of vitamin B, C and E and vital minerals like iron, high enough to treat anemia, calcium and phosphorus (Josekutty, 2008). Leaves are slightly pungent and retain their flavor even after drying. Curry leaf loses flavor when cooked. To retain fresh flavor the leaves should not be removed from the branches until ready for use. The tree has many medicinal uses in the Ayurveda form of medicine, so popular in India. A volatile oil (sulphur containing essential oil), a crystalline glucocide 'Koenigin' from the leaves and a glucocide 'Murrayin' from the flowers are a few industrial products from the tree (Kirthikar and Basu, 1935).Ground curry leaf with matured coconut kernel and adequate spices form an excellent preserve and exported. The leaves, root and bark have stomach and carminative properties. The leaves and roots are slightly bitter, acrid, cooling and analgesic (Salikutty and Peter, 2001). The leaves and roots are useful in piles, inflammation and itching. External application of the pulped bark and root are reported to cure eruptions and bites by poisonous animals (Kumar *et al.,* 1999). The root is slightly purgative and the juice is used to relieve pain associated with kidney disorders. The green leaves are eaten raw to cure dysentery and diarrhea. An infusion of roasted leaves is used to control vomiting (Omanakutty, 2007)

Origin, History and Distribution

Curry leaf (*Murraya koenigii* (L.) Swreng. (syn. *Bergera koenigii* Koen; *Chaleas koenigii* Curz ex. Swingle) is the crop originated in the tropical tract of tarai region foot hills of Himalaya from Kumaon to Sikkim in Bengal, Assam, Deccan plateau and in Western Ghats (India). It is introduced and very common in Port Blair (Andaman and Nicobar Islands). In the East, its range extends into Myanmar (Burma), Malaysia, South Africa and Reunion Islands. The shrubby plant grown in long stretches (Salikutty and Peter, 1984) dry humid condition of India. The crop has been introduced to South India when the Dravidians moved to South with the arrival of Aryans from Persia, the present Middle East. The curry leaf is commonly found in forests often as a gregarious under growth throughout India and the Andaman Islands up to an altitude of 1,500 m. It occurs in wild form from Garhwall Hills in Uttar Pradesh to Sikkim, West Bengal, Assam, Central India, Western Ghats, Tamil Nadu and Kerala. The tree is also seen in Burma, Ceylon, China, Australia and Pacific Islands. The curry leaf tree adorns every household of South India. It is commercially cultivated in West Bengal, Assam, Deccan plateau, Western Ghats, Tamil Nadu, Andhra Pradesh, Karnataka and Kerala. It does survive throughout India (up to an altitude of 1500 m amsl) and a common 'tree spice' found in forests as gregarious undergrowth along the foot of the Himalayas (Thomas and Vradarsan, 2006). A highly fertile soil, vast reserve of underground water and symphony of temperate, subtropical and tropical climate favour large scale cultivation in south India (Salikutty and Peter, 1984) and West Bengal.

Species Availability, Evolution and Diversity

The genus *Murraya* has been given after Professor John Adam Murray of Gottingen. In two species *M. paniculata* (L.) Jack. Indigenous to Burma and *M. exotica* L. (syn. *Chaleas paniculta* Mant., *Camminum sinense* Rumph, *Marsana buxifolia* Sonnerat) an ornamental shrub indigenous to India. Ceylon and China are related to *M. koenigii*. The wood of *Murraya paniculata* sometimes referred to as Chinese boxwood or Andaman satinwood is popular in the wood trade. It is abundantly available in Port Blair (Andaman and Nicobar Islands). *Murraya exotica*, is a popular hedge plant and is well adapted for topiary work (Irulappan, 1991).

Botanical Description

The curry leaf tree belongs to the family Rutaceae. It is self-pollinated crop. The seeded nature, viability of seed and fertility of seedlings indicate the true diploid nature of the plant. The plant is essentially propagated through seeds. The sprouts from roots sometimes called root suckers are also used as propagating units with a limited success.

Table 14.5: Botanical Description

Growth Habit	Plant is an unarmed deciduous/semi-deciduous aromatic pubescent shrub or small to medium sized tropical/subtropical tree 3.0-5.0 m high (sometimes 4.0-6.0 m) with a slender trunk up to 40.0 cm diameter.
Root	The roots are wooly and widely spread. They produce many suckers.
Stem	Strong woody stem and branches covered with dark bark and closely crowded shady crown.
Leaf	The leaves are pinnate with alternate11-21 leaflets (each leaflet 2.0-4.0 cm long and 1.0-2.0cm broad), exstipulate, imparipinate 25.0-30.0 cm long usually glabrous, sometimes slightly pubescent when young. The leaves are dotted and very strongly aromatic.
Floral biology	Flowers usually hermaphrodite. The inflorescence is a terminal pedunculate, and many flowered compact corymbiform cymose panicle. The flowers are small, white, fragrant and about 1.0 cm long. Petals imbricate. Filaments subulate, stamens 10, anthers small. Ovules 1 or 2 in each cell, and style articulate at top of ovary,
Anthesis and pollination	Curry leaf is a self-pollinated crop.
Fruit and fruit set	Fruit indehiscent, shiny berries are edible but are poisonous. Fruits occur in close clusters. They are small, ovoid or sub-globose, apiculate, glandular minutely tuberculate, 1.0-2.0 cm in diameter with a thin pericarp.
Seeds	Seeds exalbuminous. Mucilaginous pulp enclosing 1 or 2 seeds. Cotyledons plain-convex

Botanical Classification

The curry leaf trees come in three different varieties, such as Dwarf, Regular, and Gamathi Dwarf varieties does not grow as tall but the branches spread. Light Green Leaves are the same size as the Regular variety but it has an aroma of its own. Gamathi variety is the most fragrant very thick leaf structure but slowest of all the curry trees.

Genetic Resource Augmentation and Crop Improvement

Farmers prefer local varieties, which have pink midrib. University of Agricultural Sciences, Dharwad (Karnataka) has recently released two cultivars *viz.* DWD-1 and DWD-2 which have an oil content of 5.22 per cent and 4.09 per cent, respectively. Both the varieties are having strong aroma. Yield of I (250-400kg/ha), II (1800 kg/ha), III (5400 kg/ha), IV (2500 kg/ha) and V (5000 kg/ha) can be obtained regularly in increasing years.

Genetic Resource Conservation

Curry leaf is abundantly available southern parts of the country with varying ecotypes. They varied with respect to growth behavior and aroma.

REFERENCES

1. CSIR, New Delhi. 1962. Curry leaf. Wealth of India. Publication and Information Directorate, Council of Scientific and Industrial Research, New Delhi. Vol. 6, p. 446-448.
2. Irulappan, I. 1991. Curry leaf [*Murraya koenigii* (L.) Spreng]-Rutaceae. (In) *Recent Advances in Tropical Vegetable Production, Compendium of Lectures*. Peter, K.V., Salikutty, Joseph and Indira, P. (Eds) Kerala Agricultural University, Vellanikkara (Kerala) p. 273-307.
3. Josekutty, K.A. 2008. Know the spice-Curry Leaf. *Spice India*, **21(5):** 28-31.
4. Kirthikar, K.R. and Basu, B.D. 1935. Indian Medicinal Plants, M/s Bishen Singh, Mahendra Pal Singh, Dehra Dun (Uttrakhand) edn 2, pp. 474-475.
5. Kumar, V.S., Sharma, A., Tiwari, R. and Kumar, S. 1999. *Murraya koenigii* (Curry leaf): a review. *Journal of Medicine and Aromatic Plant Sciences*, **21:** 1109-1144.
6. Omanakutty, M. 2007. Medicinal properties of curry leaf. *Spice India*, **20(5):** 4-5.
7. Salikutty, J. and Peter, K.V. 1984. Curry leaf. *Economic Botany*, **39:** 68-73.
8. Thomas, J. and Vradarsan, S. 2006. Curry leaf: to flavor or to Poisson. *Spice India*, **19(5):** 6-7.

Genus *Bambusa*

Bamboos (*Bambusa* Species)

Introduction

Common bamboo (*Bambusa vulgaris* and *Phyllostachys edulis*) is a very important evergreen grass plant for human, animals, birds *etc.* Common name in Hindi is Baans. Names in other Indian languages are Moongil Kuruthu (Tamil), Mulankoombu (Malayalam), Veduru Chiguru (Telugu), Bansberankur (Bengali), Kalkipan (Marathi), Baunsagaja (Oriya) and Baans

(Urdu and Punjabi). They are the young and new canes of the bamboo plant, which is extremely versatile. Since ancient times, the woody tree trunks of bamboo have been used to build entire houses, furniture, kitchenware and utensils. Bamboo has also laid the foundation for many bridges and musical instruments, apart from the shoots being widely used as food. Bamboo shoot or bamboo sprouts, are young stems that are harvested before they turn two weeks old and reach a height of 30.0 cm. Crispy and crunchy, bamboo shoots are valued for their delightfully supple texture and rich aroma. It is very nutritive and common dish among tribal. Snacks, curries, stir-fries, soups, salads and fried rice are just a few of the delicacies that make use of bamboo shoots.

The young shoots that emerge from the ground are tender and full of nutritional value. The ash content in the different edible bamboo species varied from 2.10 to 3.70 per cent with lowest and highest values in *C. hookeriana* and *G. rostrata*, respectively. The protein content determined for edible shoots of various bamboo species varied from 2.35 to 3.76 per cent. *D. hamiltonii* contained maximum protein content, which was at par with *C. hookeriana*. Out of 12 species, eight had a protein content of more than 3.0 per cent. Carbohydrate content ranged from 4.30 per cent (*S. dullooa*) to 5.2 per cent (*B. balcooa*). Acidity in young edible shoots ranged between 3.30 per cent and 5.23 per cent. *B. tulda* possessed maximum acidity among all the species; it was lowest in *G. rostrata*. Fat content was very low in all the edible bamboo species. However, it was found to be highest in tender shoots of *M. baccifera*. Lowest fat, however, was recorded in tender shoots of *D. sikkimensis*. Crude fiber content was recorded to be highest in edible shoots of *M. baccifera* (35.50 per cent), followed by *D. membranceous* (34.7 per cent) and *B.tulda* (28.50 per cent). Among various species, *C. hookeriana* exhibited the lowest crude fibre content (23.10 per cent). The food energy was highest in *D. sikkimensis* (16.90 MJ/kg) and lowest in *G. rostrata* (14.64 MJ/kg). Food energy of the rest of the species ranged in between these two extremes.

Table 14.6: Nutritional Value (per 100 g) of Raw Bamboo Shoots

Principle	*Nutritional Value*	*Percentage*
Energy	27 Kcal	1.4
Carbohydrates	5.2 g	4.0
Protein	2.60 g	5.0
Total Fat	0.3 g	1.0
Cholesterol	0.0 Mg	0.0
Dietary Fiber	2.2 g	6.0
Vitamins		
Folates	7.0 µg	2.0
Niacin	0.600 mg	4.0
Pantothenic acid	0.161 mg	3.0
Pyridoxine	0.240 mg	1.8
Riboflavin	0.070 mg	5.0
Thiamin	0.150 mg	12.0
Vitamin C	4.0 mg	7.0
Vitamin A	20.0 IU	<1.0
Vitamin E	1.0 mg	7.0
Vitamin K	0.0 µg	0.0
Electrolytes		
Sodium	4.0 mg	<1.0
Potassium	533.0 mg	11.0
Minerals		
Calcium	13.0 mg	1.5
Copper	0.190 mg	21.0
Iron	0.50 mg	6.0
Magnesium	3.0 mg	<1.0
Manganese	0.262 mg	11.0
Phosphorus	59.0 mg	8.0
Selenium	0.8 mg	1.5
Zinc	1.10 mg	1.0
Phyto-nutrients		
Carotene-ß	12.0 µg	-
Carotene-?	0.0 µg	-
Lutein-zeaxanthin	0.0 µg	-

Source: USDA Nutritional Nutrient database.

Table 14.7: Nutritional Quality of Fresh Shoots of Edible Bamboo Species in NEH Region

Bamboo species	*Ash (per cent)*	*Protein (per cent)*	*Carbo-hydrate (per cent)*	*Fat (per cent)*	*Crude Fibre (per cent)*	*Food Energy (MJ/kg)*	*Acidity**	*HCN (per cent)*
Bambusa balcooa	3.10 ± 0.02	3.31 ± 0.82	5.21 ± 0.10	1.01 ± 0.01	26.41 ± 1.17	15.52 ± 0.07	4.06 ± 0.03	0.020 ± 0.001
Bambusa polymorpha	2.70 ± 0.01	3.30 + 0.54	4.90 ± 0.12	1.00 ± 0.03	28.52 ± 1.40	15.44 ± 0.08	4.46 ± 0.05	0.021 ± 0.001
Bambusa tulda	3.00 ± 0.02	3.40 ± 0.25	4.72 ± 0.07	0.94 ± 0.02	24.64 ± 0.45	15.93 ± 0.10	5.23 ± 0.03	0.013 ± 0.001
Chimonobambusa hookeriana	2.10 ± 0.04	3.72 ± 0.22	4.83 ± 0.04	0.76 ± 0.02	23.13 ± 2.66	16.34 ± 0.07	4.20 ± 0.04	0.011 ± 0.001
Dendrocalamus giganteus	2.20 ± 0.06	2.84 ± 0.11	4.94 ± 0.03	0.84 ± 0.04	27.64 ± 2.87	16.90 ± 0.12	4.80 ± 0.06	0.010 ± 0.001
Dendrocalamus hamiltonii	2.80 ± 0.06	3.76 ± 0.25	4.81 ± 0.06	0.95 ± 0.03	25.45 ± 2.26	16.42 ± 0.06	4.52 ± 0.06	0.011 ± 0.003
Dendrocalamus longispathus	3.40 ± 0.03	2.66 ± 0.36	4.72 ± 0.05	0.77 ± 0.01	26.74 ± 4.81	16.63 ± 0.10	3.90 ± 0.05	0.020 ± 0.002
Dendrocalamus membranaceous	3.20 ± 0.05	3.42 ± 1.70	4.50 ± 0.13	1.02 ± 0.02	34.72 ± 2.32	16.84 ± 0.13	3.85 ± 0.03	0.020 ± 0.002

Bamboo species	Ash (per cent)	Protein (per cent)	Carbo-hydrate (per cent)	Fat (per cent)	Crude Fibre (per cent)	Food Energy (MJ/kg)	Acidity*	HCN (per cent)
Dendrocalamus sikkimensis	2.60 ± 0.02	3.11 ± 0.18	4.71 ± 0.06	0.68 ± 0.03	23.56 ± 3.91	16.90 ± 0.07	4.41 ± 0.01	0.013 ± 0.001
Gigantochloa rostrata	3.70 ± 0.04	3.62 ± 1.97	4.93 ± 0.11	0.91 ± 0.03	23.73 ± 2.12	14.64 ± 0.08	3.34 ± 0.04	0.020 ± 0.001
Melocanna baccifera	2.30 ± 0.04	2.46 ± 0.80	4.51 ± 0.09	1.03 ± 0.01	35.54 ± 3.12	15.82 ± 0.08	4.16 ± 0.04	0.012 ± 0.001
Schizostachyum dullooa	2.44 ± 0.04	2.35 ± 0.68	4.30 ± 0.08	0.98 ± 0.01	33.52 ± 3.12	16.10 ± 0.08	4.24 ± 0.04	0.010 ± 0.001
C.D. (5 per cent)	**0.12**	**0.28**	**0.13**	**0.10**	**2.70**	**0.46**	**0.46**	**0.003**

Mean ± S.D. *Estimated on fresh weight basis.

Table 14.8: Annual Supplement to Nutrition (kg/yr*) Provided by Major Edible Bamboo Species in NEH Region, irrespective of States

Species	Contribution to Edible Shoots (kg/yr)*	Ash	Crude Fiber	Protein	Carbohydrate	Fat
Bambusa balcooa	37,328.39	1,157.20	9,851.00	1,228.1	1,937.30	522.60
B. polymorpha	315.00	9.54	81.90	10.40	16.10	5.00
B. tulda	3,061.00	90.90	734.60	105.00	144.50	21.43
Dendrocalamus giganteus	87,520.00	1,899.21	24,173.0	2,433.1	4,297.20	630.10
D. hamiltoni	1,18,957.00	3,271.32	30,262.7	4,377.6	5,654.40	786.00
D. hookerii	21,173.10	669.07	7,353.42	722.00	948.60	444.60
D. longispathus	416.74	14.13	111.30	10.80	19.60	1.75
D. membranaceus	606.30	15.76	200.10	20.60	27.30	12.73
D. sikkimensis	82,019.20	2,134.84	19,258.1	2,575.4	3,830.30	524.90
Gigantochloa rostrata	2,316.60	72.28	532.80	72.70	108.20	14.80
Melocanna baccifera	45,301.20	1,041.93	16,852.0	1,105.4	2,043.10	434.90
Phyllostachys bambusoides	32,309.28	668.80	7,447.30	1,182.5	1,560.50	226.20
Schizostachyum dullooa	261.00	8.14	60.16	9.29	12.61	1.83
Teinostachyum weightii	4,388.50	163.25	1,037.90	159.50	213.90	49.15
Grand Total	4,35,073.31	11,216.3	1,17,956	14,012	20,813.61	**3,675.9**

* Estimated on fresh weight basis of young shoots.

Table 14.9: Annual Supplement (Kg/yr) to Minerals and Vitamins by Major Edible Bamboo Species in NEH Region

Bamboo Species	Calcium (kg/yr)	Potassium (kg/yr)	Phosphorus (kg/yr)	Magnesium (kg/yr)	Ascorbic Acid* (kg/yr)	Tryptophan	Methionine
Bambusa balcooa	522.60	10.080	347.15	15.30	1.410	410.61	190.37
B. polymorpha	4.41	0.091	2.93	0.13	0.012	3.15	1.58
B. tulda	40.09	0.670	19.89	1.29	0.160	42.55	10.41
Dendrocalamus giganteus	1,050.24	21.880	621.39	40.2	5.950	1,487.85	402.59
D. hamiltoni	1,724.87	24.981	678.05	47.58	15.320	1,451.27	666.16
D. hookerii	347.2	4.660	218.08	8.68	2.085	260.42	129.16
D. longispathus	5.21	0.101	2.33	0.18	0.022	2.83	1.08
D. membranaceus	9.09	0.150	3.03	0.25	0.031	4.12	1.58
D. sikkimensis	1,525.56	18.040	582.34	35.27	2.485	1,140.06	508.52
Gigantochloa rostrata	43.09	0.510	16.45	0.99	0.123	32.20	14.36
Melocanna baccifera	634.22	10.420	348.82	19.48	3.434	620.63	126.84
Phyllostachys bambusoides	429.71	7.110	216.47	13.25	1.958	365.09	248.78
Schizostachyum dullooa	3.47	0.060	1.75	0.10	0.015	2.94	2.01
Teinostachyum weightii	54.67	1.011	23.43	1.87	0.290	16.92	22.56
Grand Total	6,394.47	99.764	3,082.11	184.63	33.295	5,840.64	**2326.00**

* Estimated on fresh weight basis of young shoots.

HCN is one of the important anti-nutritional factors available in young bamboo shoots. HCN content was highest (0.02 per cent) in edible shoots of *Bambusa polymorpha*, followed by *B. balcooa* and *G. rostrata*. Among the various species, lowest HCN (0.01 per cent) was recorded in tender shoots of *D. giganteus* and *S. dullooa*.

Among the various micronutrients, calcium content was observed to be highest in tender shoots of *Dendrocalamus longispathus* (1.92g/100g), followed by *D. hamiltonii* (1.64g/100g). However, lowest value was noticed in edible shoots of *C. hookeriana* (1.26g/100g). Potassium content in edible bamboo species ranged from 0.021-0.033 g/100g, with the highest value in *B. balcooa* and lowest in *G. rostrata*, respectively. The phosphorus content determined in various species varied from 0.57 to 1.01 g/100 g with highest in *D. hamiltonii* and lowest in *M. baccifera*. The magnesium content in young shoots of various bamboo species ranged from 0.043 to 0.092 g/100 g. Ascorbic acid content was found to be highest in tender shoots of *D. hamiltonii* (9.452 mg/100 g), followed by *D. membranaceous* (8.345 mg/100 g), whereas it was lowest in tender shoots of *S. dullooa* (3.04 mg/100 g). Tryptophan content was highest in edible shoots of *C. hookeriana* (1.71 g/16 g N), followed by *D. hamiltonii* (1.46 g/16 g N). Among the various species, the lowest value of tryptophan was observed in *M. baccifera* (0.89 g/16 g N). Methionine content was also estimated in young edible shoots of various bamboo species and it was

Table 14.10: Determination of Minerals and Vitamins in Commercial Edible Bamboo Species of NEH Region

Bamboo Species	Ca (g/100 g)	K (g/100 g)	P (g/100 g)	mg (g/100 g)	Ascorbic Acid* (mg/100 g)	Tryptophan* (g/16 g N)	Methionine* (g/16 g N)
Bambusa balcooa	1.41 ± 0.08	0.033 ± 0.002	0.940 ± 0.011	0.043 ± 0.006	3.842 ± 0.321	1.13 ± 0.05	0.54 ± 0.02
Bambusa polymorpha	152 ± 0.07	0.032 ± 0.004	0.932 ± 0.021	0.041 ± 0.003	4.832 ± 0.410	0.94 ± 0.08	0.58 ± 0.03
Bambusatulaa	1.34 ± 0.04	0.024 ± 0.004	0.763 ± 0.031	0.046 ± 0.001	4.224 ± 0.340	1.05 ± 0.08	0.54 ± 0.02
Chimonobambusa hookeriana	1.26 ± 0.06	0.033 ± 0.005	0.751 ± 0.032	0.072 ± 0.007	6.860 ± 0.640	1.71 ± 0.10	0.61 ± 0.04
Dendrocalamus giganteus	1.52 ± 0.07	0.023 ± 0.007	0.874 ± 0.090	0.081 ± 0.004	7.743 ± 0.640	1.56 ± 0.10	0.62 ± 0.04
Dendrocalamus hamiltonii	1.64 ± 0.08	0.025 ± 0.007	1.012 ± 0.002	0.092 ± 0.007	9.452 ± 0.780	1.46 ± 0.06	0.64 ± 0.04
Dendrocalamus longispathus	1.92 ± 0.04	0.026 ± 0.002	0.865 ± 0.030	0.066 ± 0.006	7.451 ± 0.340	1.43 ± 0.07	0.66 ± 0.10
Dendrocalamus membranaceous	1.33 ± 0.10	0.024 ± 0.003	0.894 ± 0.100	0.072 ± 0.006	8.345 ± 0.560	0.98 ± 0.04	0.68 ± 0.03
Dendrocalamus sikkimensis	1.43 ± 0.06	0.023 ± 0.003	0.873 ± 0.030	0.065 ± 0.003	6.674 ± 0.342	1.40 ± 0.12	0.61 ± 0.05
Gigantochloa rostrata	1.35 ± 0.04	0.021 ± 0.009	0.814 ± 0.100	0.089 ± 0.002	7.645 ± 0.470	1.12 ± 0.03	0.81 ± 0.07
Melocanna baccifera	1.34 ± 0.04	0.023 ± 0.007	0.574 ± 0.004	0.046 ± 0.010	5.324 ± 0.451	0.89 ± 0.04	0.57 ± 0.09
Schizostachyum dullooa	1.32 ± 0.03	0.022 ± 0.004	0.752 ± 0.060	0.044 ± 0.003	3.044 ± 0.540	1.10 ± 0.08	0.67 ± 0.06
C.D. (5 per cent)	**0.057**	**0.0012**	**0.004**	**0.0001**	**0.804**	0.118	**0.046**

Mean ± S.D. *Estimated on fresh weight basis.

Table 14.11: Micronutrients in the Edible Portion of 14 Days Old Bamboo Shoot on Dry Matter Basis in the Bamboo Plant Species of North Eastern Region of India

Bamboo Species	Micronutrients (ppm)					
	Mn	Zn	Fe	Cu	Co	Mg
Bambusa arundinacea	7.1	0.9	14.1	0.1	ND	42.0
Bambusa balcoa	8.1	1.3	9.8	0.1	ND	42.4
Bambusa cacharensis	14.2	1.2	8.1	0.1	0.1	53.1
Bambusa nana	9.0	0.9	12.7	0.1	0.3	46.3
Bambusa nutans	10.7	1.2	10.4	0.1	0.4	54.5
Bambusa pallid Type I	12.5	1.2	14.6	0.1	0.2	54.5
Bambusa pallid Type II	15.6	1.3	11.9	0.2	ND	47.6
Bambusa polymorpha	8.9	1.4	5.6	0.1	0.1	61.3
Bambusa tulda	6.1	1.4	13.6	0.1	0.1	61.0
Bambusa variegata	6.9	0.6	22.7	ND	0.5	54.4
Dendrocalamus giganteus	4.1	0.9	12.6	0.1	0.3	37.6
Dendrocalamus hamiltonii	4.0	1.3	19.4	0.2	0.4	46.8
Dendrocalamus hookerii	7.2	1.3	8.4	0.1	ND	49.1
Dendrocalamus sahnii	8.7	0.9	8.0	0.1	0.2	46.5
Dendrocalamus sikkimensis	11.6	1.7	12.0	0.1	0.2	51.4

Bamboo Species	*Micronutrients (ppm)*					
	Mn	*Zn*	*Fe*	*Cu*	*Co*	*Mg*
Schizostachyum manii (*B. khasiana*)	9.0	2.1	6.2	ND	ND	46.8
Schizostachyum pergracile	17.6	3.5	23.1	0.2	0.5	50.1
Schizostachyum polymorphum	12.5	1.6	10.3	0.1	ND	57.3
Teinostachyum helferii	6.2	1.2	8.7	0.1	0.2	46.7
Hard jati	10.3	1.4	24.5	0.1	ND	50.9
Khupri	7.3	1.0	17.5	0.1	ND	52.4
Coefficient of variation (per cent)	**38.1**	**43.80**	**42.29**	**47.39**	**104.50**	**13.26**

highest in tender shoots of *G. rostrata* (0.81 g/16 g N). Lowest methionine content was in *B. tulda* (Table 14.8).

The determination of important aspects is essential which has been done in some edible bamboo species of NEH region, are as shown in Table 14.10.

The micronutrients analysis of the edible portion of the bamboo shoots on dry basis, showed a high variability (Table 14.11).

Detail descriptions are being given in Table 14.12.

Table 14.12: Nutritional Constituents

Low caloric contents	A 100 gram serving of bamboo shoots contains only 20 calories. Also, the carbohydrates found in bamboo shoots do not amount to more than 5.2 g/100 g serving.
Low sugar content	The amount of sugar found in bamboo shoots is about 2.5 g/100 g serving. This is less than amount of sugar found in many fruits and vegetables.
Negligible amount of fat	A serving of 100 g of bamboo shoots contains less than 0.3 g of fat. This fat consists of both saturated and unsaturated fats. Unsaturated fats are needed by the body and they can control the spread of bad Low Density Lipoprotein (LDL) cholesterol throughout the body.
Source of protein	A 100 g serving of bamboo shoots would have about 2.0-2.6 g of protein. The proteins found in bamboo consist of seventeen essential amino acids and two semi-essential amino acids.
Vitamins and minerals	Bamboo shoots contain vitamins such as vitamin A, vitamin B6, vitamin E, thiamin, riboflavin, niacin, folate and pantothenic acid. Minerals found in bamboo shoots include calcium, magnesium, phosphorous, potassium, sodium, zinc, copper, manganese, selenium and iron.
Dietary fiber	The amounts of dietary fiber contained in bamboo shoots make up 2.2 g out of 100 g serving.
Appetizing effects	Bamboo shoots contain high cellulosic materials which stimulate the appetite. The taste and texture of the bamboo shoots also make them a good appetizer.

Health Benefits of Bamboo Shoots

Bamboo shoots are an exotic food that is consumed in many countries in Asia and now it is slowly growing in demand from western countries. Bamboo shoots are known for their various health benefits, including those explained in greater detail given in Table 14.13.

Table 14.13: Deails of Health Benefits

Weight loss	Bamboo shoots are weight loss friendly. When we look at the amount of calories, carbohydrates and sugars contained in bamboo shoots, it is shocking to find that their presence is almost negligible. This makes it an ideal food for people who want to lose weight, but also who want their stomachs to be full.
Healthy skin and hair	Silica is an important compound that plays a vital role in maintaining your skin and hair healthy. Bamboo extract is the rich source of silica and hence its daily use is helpful in improving the condition of your skin and hair. Silica improves skin health by promoting better absorption of mineral like calcium, magnesium and potassium by your body. It has antioxidant properties that help in reducing wrinkles, fine lines and signs of aging.
Heart health	According to research studies, phyto-sterols and phyto-nutrients found in bamboo shoots are ideal for dissolving harmful Low Density Lipoprotein (LDL) in the body. This eases cholesterol out of arteries for the smooth supply and movement of blood throughout the body.
Control cholesterol	Consumption of bamboo shoots is also helpful in decreasing Low Density Lipoprotein (LDL) levels of cholesterol, with stable glucose levels. This is due to the fact that bamboo shoots contain negligible amounts of fat and very low calories. Research conducted by Park and Jhon at Washington State University showed that the consumption of bamboo shoots had favorable effects on cholesterol, lipids and bowel function.
Anti-cancer properties	Research studies conducted on bamboo shoots have indicated that leaves of bamboo shoots consists of phyto-sterols such as flavones, amylase and chlorophyll. Out of these, chlorophyll showed properties of controlling mutations and cancer.
Immune system	The vitamins and minerals in bamboo shoots are ideal for improving the body's immune system. The vitamins, minerals, and anti-oxidants present in bamboo shoots are essential for strengthening the body from the inside out.
High supply of dietary fiber	The amount of dietary fiber in bamboo shoots is high. Consuming sufficient amounts of dietary fiber is essential for easy digestion and healthy bowel movements. Including bamboo shoots in your meals can be a very good idea for losing weight as well. With lack of any physical activity during the night time, taking food with less calories and high fiber help lose weight easily, even while you sleep.

Anti-inflammatory properties	According to research conducted by Muniappan and Sundaraj, bamboo shoots possess anti- inflammatory and analgesic (pain-killing) properties. It helps in the healing of ulcers as well. Juice from bamboo shoots can also be used as a medicine for external wounds and ulcers.
Bone health	Bamboo shoot contains minerals like calcium, potassium and magnesium that are associated with the growth and development of healthy bones. The silica in bamboo is helpful in increasing bone mass and preventing conditions like osteoporosis.
Respiratory disorders	Bamboo shoots have been known to be effective against respiratory disorders. A decoction of the shoots can be prepared by boiling the shoots twice. The first boil should be for 5 minutes followed by a second boil for about 10 minutes. The decoction can be taken along with honey for the best effect.
Possible cure for poisoning	In Ayurvedic medicine, the ancient Indians science of medicine and lifestyle, it is believed that bamboo extracts contain anti-venomous properties. They are useful in case of both snake and scorpion bites.
Uterotonic properties	Traditional Chinese medicine believes that bamboo shoots can cause uterine contractions. It is used as a medicinal supplement during the last month of pregnancy when the delivery date is still pending. According to the research paper submitted by Gruber and O'Brien at the University of Vienna, bamboo is one of the many plants which have been listed among uterotonic plants.
Stomach disorders	Bamboo shoots are useful in treating stomach disorders. Apart from bamboo shoots, bamboo leaves are also suggested as a remedy for intestinal worms and stomach disorders as well.
Wound cleaning	Bamboo shoots are also used for cleaning wounds and sores.
Lowers blood pressure	Bamboo shoots contain high amounts of potassium. Potassium is highly beneficial as an electrolyte and is also very good for lowering and maintaining blood pressure. With all of these benefits from bamboo shoots, they are highly recommended for a healthy lifestyle.
Removing intestinal worms	Drinking the juice of bamboo shoot can help in killing and eliminating intestinal worms.

India is one of the leading countries of the world to China in bamboo production (Solanky *et al.*, 2004).Tribal communities of the region use this potential resource for food, shelter, furniture, handicraft and various ethno-religious purposes. Owing to short gestation period and recurring return, bamboo species has been considered more valuable for agro-forestry in Eastern Himalaya. Similarly, the edible delicacy of bamboo species is witnessed only in Eastern Himalaya (Bhatt and Bujarbaruah, 2004). The survey based information will be helpful in order to understand the rate of harvesting of young tender shoots for food and their role in the livelihood security of the tribal folk of the region so as to frame a comprehensive policy for mass a forestation of potential edible bamboo species in Eastern Himalaya, India. The sales of fresh bamboo shoots of *D. hamiltonii* was highest in Meghalaya and Nagaland state, while largest sales of *D. sikkimensis*was recorded in Manipur state. Tender shoots of *Melocanna baccifera* were sold in largest scale compared to the other bamboo species in Mizoram and Tripura state. The availability period of tender shoots in the market places ranged between 42-84 dyas/year. Fresh shoots of *D. hamiltonii* was consumed in largest quantity (1654.0 t/year) followed by *D. giganteus* (925.0 t/year) and *M. baccifera* (645.0 t/year), whereas fresh shoots of *Schizostachyum dullooa* was sold in smallest quantity (1.70 t/year). The annual net return of ca 27.14 million rupees from the bamboo shoot products sold as food items in the six states of North East Hill region of India, revealed bamboo as one of the potential sector for income generation. *D. hamiltonii,Melocanna baccifera, Bambusa balcooa, D. giganteus, C. hookeriana, D. sikkimensis, D. longispathus* and *G. rostrata* have been found most suitable bamboo species based on overall rank sum index, which were assessed on the basis of availability, growth and biomass performance, nutritive value, availability of minerals and HCN content.

Table 14.14: Physical Characteristics of 14 Days Old Edible Shoots of Important Bamboo Species

Species	*Length (cm)*	*Circum-ference (cm)*	*Fresh Weight/ Shoot (g)*	*Edible Portion/ Shoot (g)*
Dendrocalamus hookerii	53	47	3300	2054
Dendrocalamus giganteus	65	42	2900	1950
Bambusa balcoa	45	40	2350	1820
Dendrocalamus sikkimensis	75	29	2100	1440
Bambusa arundinacea	60	35	2250	1250
Bambusa tulda	61	31	1600	1020
Bambusa nutans	45	28	1300	950
Dendrocalamus hamiltonii	50	24	1000	600

Origin, History and Distribution

Basically bamboo is native to Asia (Indo-China and to the province of Yunnan in southern China) but it has been widely cultivated in many other places and has become naturalized in several parts of the world (Dieter, 1999). *B. vulgaris* is a species of the large genus Bambusa of the clumping bamboo tribe Bambuseae which are found largely in tropical andsubtropical areas of Asia, especially in the wet tropics (David, 1984). Due to its diverse adjustability to a wide range of climates, cultivation of bamboo shoots spread across different continents, from the tropical jungles of Chile to the mountain slopes of the Himalayas. But, owing to the weather conditions, bamboo shoots could not be cultivated in Canada, Europe, Antarctica and Western Asia, although successful efforts have been made in Northern Italy, which has mild winters and hot, humid summers. Bamboo shoots form an essential ingredient in the kitchen of China, India, Nepal, Indonesia, Vietnam, Philippines, Japan and Uganda. Today, bamboo shoots are commercially cultivated in Eastern and South Eastern Asia on extensive levels.

Bamboo species cover an area of around 10.03 m hectare, which contribute 12.8 per cent of the total forest cover (Anonymous, 2001). Many bamboo species also form pure patches in tropical to subtropical zones as they act as efficient early colonizers in secondary succession, resulting from shifting cultivation or other anthropogenic disturbances (Rao and Ramakrishnan, 1989). The diversity is distributed all across from coastal ares in th soputh to around 10, 000 feet above sea level in the Himalayan highlands (Kochhar and Rana, 1993). A detailed account of 113 species of bamboos found in India alongwith their distribution has been given by Varmah and Bahadur (1980).Tender shoots of *Dendrocalamus hamiltonii* is consumed in all the states of NEH region, whereas *Melocanna baccifera* is consumed in five states (Manipur, Meghalaya, Mizoram, Sikkim and Tripura).Tender shoots of other bamboo species like *Bambusa polymorpha, B. tulda* and *Schizostachyum dullooa* were commercially important in Tripura, whereas tender shoots of *D. membranaceous* and *Gigantochloa rostrata* were commercial in Nagaland. In Sikkim state *Chimonobambusa hookeriana* had high commercial value for its tender shoots while shoots of *D. sikkimensis* and *D. longispathus* was highly important in Manipur (Table 14.15).

The region presently discussed includes 12 States with politically defined territories (Table 14.16). Phytogeographically and biogeographically the region has distinct entities as northwest Himalaya, central Himalaya, eastern India and northeast India, characterized with 10 microhabitats for species and

Table 14.15: Commercial Edible Bamboo Species of NEH Region of India

Bamboo Species	*States and Altitude (m asl)*					
	Manipur	*Meghalaya*	*Mizoram*	*Nagaland*	*Sikkim*	*Tripura*
Bambusa balcooa	+	+	-	-	-	+
Bambusa polymorpha	-	-	-	-	-	+
B. tulda	-	-	-	-	+	+
Chimonobambusa hookeriana	-	-	-	-	+	-
Dendrocalamus giganteus	+	-	-	+	+	-
D. hamiltonii	+	+	+	+	+	+
D. longispathus	+	-	+	-	-	-
D. membranaceous	-	-	-	+	-	-
D. sikkimensis	+		-	-	-	-
Gigantochloa rostrata	+	-	-	+	-	-
Melocanna baccifera	+	+	+	-	+	+
Schizostachyum dullooa	-	-	-	-	-	+

Note: +,- represents occurrence and absence of the particular bamboo species in the respective states of NEH region.

Table 14.16: Key Areas of Conservation in Himalayan, Northeastern and Eastern Indian States

Sl.No.	*State*	*No. of KAC*	*Total Area (km²)*	*No. of Bamboo Species*
1.	Arunachal Pradesh	12 (NP2, WS10)	9 582.98	41
2.	Assam	10 (NP2, WS8)	1 920.58	41
3.	Himachal Pradesh	33 (NP2, WS31)	6 232.87	6
4.	Jammu and Kashmir	19 (NP4, WS15)	14 822.2	5
5.	Manipur	3 (NP2, WS1)	266.65	45
6.	Meghalaya	5 (NP2, WS3)	303.65	45
7.	Mizoram	5 (NP2, WS3)	810.00	30
8.	Nagaland	4 (NP1, WS3)	226.43	36
9.	Sikkim	5 (NP1, WS4)	1 011.10	30
10.	Tripura	4 (WS4)	603.62	19
11.	Uttar Pradesh (Himalaya)	11 (NP6, WS 5)	7 408.10	11
12.	West Bengal (incl. Himalaya)	20 (NP5, WS15)	2 779.30	30
	Total	**131 (NP30, WS101)**	**45 967.50**	**339 (20 genera)**

Note: KAC: Key areas of conservation (Protected areas); NP: National Parks; WS: Wildlife Sanctuary.

Source: Biswas 1988; 1995; 1997 and Wildlife Institute of India 1998.

genetic diversity represented broadly in i) Ladakh ii) Kedarnath iii) Nandadevi iv) Siwaliks v) Terai vi) Sikkim Himalaya vii) Bomdila - Subansiri viii) Namdapha ix) Tura-Khasi Range, and x) Lushai Hills. Agro-climatically the region has three zones, *viz.* western, eastern Himalaya and Lower Gangetic Plains. Total geographical area constitutes nearly 680 000 km^2 covering ca. 226 000 km^2 of forest cover with 46 000 km^2 of key areas of conservation (Table 14.16). The region has 131 key areas of conservation distributed over 45967.50 km^2 of geographical area with 20 genera represented by 117 species of bamboo. 30 National Parks and 101 Wild life sanctuaries spread over 12 states (Table 14.16).

Diversity of species in Himalayan region is known for the following species:

Himalayan region has 19 National Parks and 69 Wildlife sanctuaries. Status of occurrence, survey, species diversity, research needs have been indicated in the Table 14.17.

Table 14.17: Distribution of Species

Areas/Region	*Species*
Northwest Himalaya	*Dendrocalamus patellaris, D. strictus, Bambusa arundinacea, B. nutans, Drepanostachyum falcatum, Thamnocalamus spathiflorus, Himalayacalamus falconeri, Chimonobambusa (Sinarundinaria) jaunsarensis etc.*, Besides several other species are grown.
Eastern Himalaya	*Arundinaria gracilis, A. hirsuta (Sinarundinaria hirsuta), A. maling, A. racemosa, Bambusa balcooa, B. nutans, Drepanostachyum (Sinarundinaria) hookerianum, D. intermedium, D. jainianum, D. khasianum, D. polystachyum, D. suberectum, Himalayacalamus (Thamnocalamus) falconeri, Pleioblastus simonii, Schizostachyum arunachalensis, S. capitatum, S. dullooa, S. fuchsianum, S. helferi, S. latifolium, S. pergracile, S. polymorphum, Semiarundinaria (Sinarundinaria) pantlingii, Thamnocalamus aristatus etc.*
Northeastern and eastern India	*Arundinaria clarkei (Racemobambos clarkii), A. debilis, A. kurzii, A. racemosa, A. rolloana (Sinarundinaria rolloana), Bambusa arundinacea, B. auriculata, B. balcooa, B. burmanica, B. cacharensis, B. jaintiana, B. khasiana, B. kingiana, B. longispiculata, B. mastersii, B. nutans, B. oliveriana, B. pallida, B. polymorpha, B. pseudopallida, B. teres, B. tulda, Chimonobambusa callosa, C. griffithiana, Dendrocalamus calostachys, D. hamiltonii, D. hookeri, D. longispathus, D, patellaris, D. sikkimensis, Dinochloa compactiflora, D. gracilis, D. indica, D. maclellandii, Drepanostachyum (Sinarundinaria), D. kurzii, D. polystachyum, D. suberectum, Gigantochloa apus, G. hasskarliana, Melocanna baccifera, Neomicrocalamus (Racemobambos) clarkei, N. mannii, N. prainii, Oxytenanthera parvifolia, Phyllostachys mannii, Schizostachyum mannii, S. pallidum, S. wightii* and other species as mentioned under eastern Himalaya, *Sinarundinaria longispiculata, S. nagalandiana, Sinobambusa elegans etc.*

Species Availability, Evolution and Diversity

There are about 155 genera and 1300 species of bamboo worldwide. The status of genera and species are mentioned beolow:

Table 14.18: Status of Genera and Species

Regions	*Genera*	*Species*	*Source*
North-east region	16	58	Bahadur and Jain (1981)
India	23	125	Varmah and Bahdur (1980)
World	75	1250	Upreti and Sundriyal (2001)

Scattered research reports are available on bamboo resources in India (Haridashan *et al.*, 1987, Singh, 2002, Sarkar and Sundriyal, 2002, Solanki *et al.*, 2004). Bamboo is a term used for more than 60 genera and 1500 species of the family Poiaceae (Soberstrom and Calderon, 1979). India is rich in bamboo resourceshaving 19 genera. Out of 126 bamboo species available in India out of which 25 exotic species are introduced from Myanmar, China, Japan, Malaysia and Thialand (Arya and Arya, 1996), and NEH region possesses highest bamboo diversity, *i.e.* 78 species (Hore, 1998).

The different species can be spotted in different locations of the country

Table 14.19: Availability and Adoption of different Species of Bamboos

Sl.No.	*Species*	*M asl*	*zone*
1.	*Bambusa polymorpha, B. tulda* and *Schizostachyum dullooa*	10-250	Tropical zone
2.	*B. balcooa, Dendrocalamus hamiltonii* and *Melocanna baccifera*	10-1650	Tropical and subtropical zones
3.	*Chimonobambusa hookeriana, Dendrocalamus giganteus, D. longispathus, D. membranaceous, D. sikkimensis* and *Gigantochloa rostrata*	500-2000	Subtropical zone
4.	*B. balcooa, D. giganteus D. longispathus* and *D. sikkimensis*	-	Home garden
5.	Remaining species	-	Home garden and natural/ plantation forest

This diversity is distributed all across from coastal areas in the south to around 10,000 feet above sea level in the Himalayan highlands (Kochhar and Rana, 1993). A detailed account of 113 species of bamboos found in India along with their distribution has been given by Varmah and Bahadur (1980). As per survey, edible bamboo species of Eastern Himalaya include 12 which are economically most important on account of their edible delicacy and productivity.

Table 14.20: Bamboo Diversity at different Regions

Regions	*Genera*	*Species*	*Source*
North-east region	16	58	Bahadur and Jain (1981)
India	23	125	Varmah and Bahdur (1980)
World	75	1250	Upreti and Sundriyal (2001)

Among the seven sisters of NEH, Mizoram occupies largest forest area under bamboo (30.8 per cent), followed by Tripura (27.13 per cent) and Meghalaya (26.0 per cent). The total bamboo cover in NEH has been reported to be 38,197 sq. km (Trivedi and Tripathi, 1984; Anonymous, 1999). Within India, North Eastern Hill (NEH) region possesses largest species diversity. Out of 126 plant species available in India, nearly half of the variability is available in NEH region (Hore, 1998). This resource has also been considered valuable for agro forestry owing to its short gestation period and recurring return (Bhatt *et al.,* 2001).

Table 14.21: Major Edible Bamboo Species of Northeast India

Sl.No.	*State*	*Bamboo Species*
1.	Arunachal Pradesh	*Dendrocalamus asper, D. giganteus, D. hookerii, Phyllostachys bambusoides, Gigantochloa albocilliata*
2.	Assam	*Bambusa arundinacea, B. balcoa, B. cacharensis, Melocanna baccifera*
3.	Manipur	*Bambusa multiplex, D. hamiltonii, D. hookerii, D. giganteus, D. longispathus, Melocanna baccifera*
4.	Meghalaya	*Bambusa balcoa, D. hamiltonii, Melconna baccifera*
5.	Mizoram	*Bambusa balcoa, Melocanna baccifera, D. hamiltonii, D. hookerii*
6.	Nagaland	*D. gigantus, D. hamiltonii, D. hookerii, D. longispathus*
7.	Sikkim	*Chimonobambusa hookeriana, D. hamiltonii*
8.	Tripura	*Bambusa balcoa, D. hamiltonii, Melocanna baccifera*

The important bamboo species of Arunachal Pradesh are given below:

Table 14.22: Important Bamboo Species of Arunachal Pradesh

Botanical Name	*Local Name (Adi)*	*District of Occurrence*
Bambusa tulda Roxb.	Ejo	West Siang
Bambusa pallida Munro	Eso	West Siang
Bambusa nutans Roxb.	Dibang	Subansiri
Bambusa khasiana Munro	Tabun	West Siang
Bambusa arundinacea (Retz.) Willd	-	Subansiri
Bambusa balcoa Roxb.	Boluka	Subansiri
Dendrocalamus hamiltonii Nees and ex Munro.	Eni	West Siang, Subansiri
Dendrocalamus hookerii Munro	-	Tawang West, Kameng Subansiri
Dendrocalamus gigantus Munro	Surung/Aphi Bu	West Siang
Cephalastachyum pergracile Munro	Madang	East Siang

Table 14.23: Regionwise Bamboo Diversity in Key Areas of Conservation

KAC	*Status*	*Total Area*	*Known No. of Species*	*Remarks*
Himalayan Region				
States				
Jammu and Kashmir				
City Forest	NP	9.07	2	INA
Dachigan	NP	141.00	-	INA
Hemis	NP	4100.00	-	INA
Kistwar	NP	400.00	3	NFS
Baltal-Thajwas	WS	203.00		
Changthang	WS	4000.00	-	
Gulmarg	WS	180.00	-	-
Hirapora	WS	110.00	-	
Hokersar	WS	9.00	-	
Jasrota	WS	10.04	2	
Korakoram	WS	5000.00		
Lachipora	WS	80.00		
Limber	WS	26.00		
Nandini	WS	13.50	-	
Overa	WS	32.00	-	
Overa-Aru	WS	425.00	-	
Ramnagar Rakha	WS	12.75	4	
Surinsar Mansur	WS	39.13	-	
Trikuta	WS	31.73		
Himachal Pradesh				
Great Himalaya	NP	765.00	3	NFS
Pin Valley	WS	41.32	1	-
Chail	WS	108.54	1	NFS
Churdhar	WS	56.15	-	-
Daranghat	WS	42.00	-	-
Darlaghat	WS	140.00	-	-
Gamgul Siebehi	WS	108.85	-	-
Gobindsagar	WS	100.34	-	-
Kalatop-khajjair	WS	69.00	-	-
Kanawar	WS	54.00	-	-
Khokhan	WS	14.05	-	-
Kias	WS	14.19	-	-
Kibber	WS	1400.50	-	-
Kugti	WS	378.86	-	-
Lippa Asrang	WS	30.89	-	-
Majathal	WS	92.00	2	-

KAC	Status	Total Area	Known No. of Species	Remarks
Manali	WS	31.80	-	-
Naina Devi	WS	156.00	-	-
Nargu	WS	278.37	-	-
Pong Dam Lake	WS	307.29	-	-
Renuka	WS	4.02	3	-
Rupi Bahba	WS	125.00	-	-
Sangla (Raksham Chikul)	WS	650.00	-	-
Sainj	WS	90.00	-	-
Sechu Tuan Nala	WS	102.95	-	-
Shikari Devi	WS	214.00	-	-
Shilli	WS	2.13	1	US
Shimla Catchment	WS	10.25	4	-
Simbalbara	WS	19.03	2	NFS
Talra	WS	26.00	3	-
Tirhan	WS	61.12	-	-
Tundah	WS	64.22	-	-
Uttar Pradesh				
Corbett	NP	520.82	4	TS
Gangotri	NP	2390.00	-	-
Govind Pashu Vihar	NP	472.00	2	-
Nands Devi	NP	650.00	3	-
Valley of Flowers	NP	87.50	2	-
Askot Musk Deer	WS	600.00	-	INA
Govind Pashu Vihar	WS	481.00	4	-
Katernighat	WS	400.00	2	NFS
Kedarnath	WS	975.24	-	INA
Mussoorie	WS	11.00	5	-
Rajaji	NP	820.42	2	-
Sikkim				
Khangchendzonga	NP	850.00	-	INA
Fambong Lho	WS	51.76	15	NFS
Kyongnosla Alpine	WS	31.00	15	NFS
Maenam	WS	35.34	-	INA
Shingba	WS	43.00	5	NFS
Arunachal Pradesh				
Mouling	NP	483.00	5	NFS
Namdapha	NP	1985.23	2	NFS
D'ering Memorial	WS	190.00	5	-
Debang	WS	4149.00	15	NFS
Eagle Nest	WS	217.00	-	-
Itanagar	WS	140.00	25	-
Kamlang	WS	783.00	-	-
Kane	WS	55.00	-	-
Mehao	WS	281.00	10	NFS
Pakni	WS	861.00	13	-
Sessa Orchid	WS	100.00	-	-
Tale Valley	WS	337.00	10	NFS

KAC	Status	Total Area	Known No. of Species	Remarks
West (including Bengal Gangetic plains)				
Buxa	NP	117.10	10	NFS
Gorumara	NP	79.45	6	US
Neora Valley	NP	88.00	ca 10	NFS
Singalila	NP	78.60	7	NFS
Sunderban	NP	1330.10	5	NFS
Ballavpur	WS	2.00	-	NS
Bethuadhari	WS	0.67	-	NS
Bibhutibhushan	WS	0.64	-	NS
Buxa	WS	251.89	10	NFS
Chapramari	WS	9.49	ca 5	US
Haliday Island	WS	5.95	-	NS
Jaldapara	WS	216.51	5	US
Jonepokhri	WS	0.04	5	NFS
Lothian Island	WS	38.00	-	NS
Mahananda	WS	158.04	-	NS
Narendrapur	WS	0.10	-	NS
Raiganj	WS	1.30	5	US
Ramnabagan	WS	0.14	-	NS
Sajnakhali	WS	362.40	5	NS
Senchal	WS	38.88	6	NFS
NE Indian Region				
Assam				
Kaziranga	NP	430.00	19	-
Manas	NP	500.00	10	UHT
Barnadi	WS	26.00	10	-
Deepan Beel	WS	4.14	-	NS
Dibru-Saikhowa	WS	640.00	ca 20	NFS
Garampani	WS	6.00	10	US
Laokhowa	WS	70.00	-	INA, NS
Nameri	WS	130.00	10	NFS
Pobitra	WS	38.84	-	INA
Rajiv Gandhi (Orang)	WS	75.60	15	NFS
Barail (Proposed	WS	-	0	TS
Berak (Proposed)	WS	-	30	TS
Manipur				
Keibul-Lamjao	NP	40.00	ca 20	US
Sirohi	NP	41.80	ca 20	NFS
Yangoopokpi	WS	184.85	ca 10	NFS
Meghalaya				
Balphakram	NP	220.00	20	NFS
Nokrek	NP	49.44	10	US
Baghmara	WS	0.03	ca 5	INA
Nongkhyllem	WS	29.00	15	NFS
Siju	WS	5.18	ca 5	NFS
Mizoram				
Blue Mountain	NP	50.00	3	NFS
Murlen	NP	200.00	15	US
Dampa	WS	340.00	20	NFS

KAC	Status	Total Area	Known No. of Species	Remarks
Knwanglung	WS	50.00	-	INA
Ngenpui	WS	170.00	15	NFS
Nagaland				
Imtanki	NP	202.02	25	NFS
Fakim	WS	64.1	15	NFS
Puliebadze	WS	9.00	15	US
Rangapahar	WS	9.00	5	Hb/Lit
Tripura				
Gumti	WS	389.54	ca 10	Hb/lit
Rawa	WS	0.85	5	US/INA
Sepahijala	WS	18.53	-	INA
Trishna	WS	194.70	10	NFS

Hb (Based on herbarium record), INA (Information not available), KAC (Key areas of conservation), Lit. (Based on literature), NFS (Need for further survey), NS (Not surveyed), NP (National Park), TS (Thoroughly surveyed), US (Under surveyed), UHT (Under heavy threat), WS (Wildlife sanctuary).

Botanical Description

Kingdom	:	Plantae
(unranked)	:	Angiosperms
(unranked)	:	Monocots
(unranked)	:	Commelinids
Order	:	Poales
Family	:	Poaceae
Subfamily	:	Bambusoideae
Supertribe	:	Bambusodae
Tribe	:	Bambuseae
Subtribe	:	Bambusinae
Genus	:	*Bambusa*

The Bambuseae are a group of perennial evergreens in subfamily Bambusoideae, characterized by having three stigmata and tree-like behavior (Judd *et al.*, 2008), that are in turn of the true grass family Poaceae. The bambusoid taxa have long been considered the most "primitive" grasses, mostly because of the presence of bracteate, indeterminate inflorescences, pseudospikelets (units of the inflorescence in woody bamboos, consisting of one to many flowers and associated glumes, that rebranch to produce successive orders of spikelets and flowers with three lodicules (minute scales of the florets of grasses, found between the lemma and the sexual organs of the flower), six stamens, and three stigmas (Clark *et al.*, 1995). Bamboos are some of the fastest growing plants in the world (David). The pachymorph (sympodial or superposed in such a way as to imitate a simple axis) rhizome system of clumping bamboos expands horizontally by only a short distance each year. The shoots emerge in a tight or open habit (group), depending on the species; common bamboo has open groups. Regardless of the degree of openness of each species' clumping habit, none of the clumpers are considered invasive. New culms can only form at the very tip of the rhizome.

Horticultural Classification

The edible bamboo shoots are of two varieties:

- **Winter type shoots-** Winter shoots are small and soft.
- **Spring type shoots-** Spring shoots are larger and tougher compared to the winter types.

In the market both are available in fresh and canned forms.

Genetic Resources Augmentation and Crop Improvement

An attempt has been made to survey the NEH region for identification of bamboo rich localities, young bamboo shoot production and their market potential and nutritive values including HCN content.

Genetic Resource Conservation

On a conservative estimate the forest area with bamboos in India is about 9.57 m ha which is nearly 12.8 per cent of the total forest area of 75 m ha. Bamboo areas in the States of Arunachal Pradesh have 19 790 km^2, Assam 10 000 km^2, Manipur 2500 km^2, Tripura 750 km^2 and West Bengal 164 km^2 (Chakraborty 1988). Forests and rural areas of Northeastern and Eastern India comprising the political territories of the States of West Bengal, Sikkim, Arunachal Pradesh, Assam, Manipur, Meghalaya, Nagaland, Mizoram, Tripura have over 50 per cent of the bamboo species of the Indian floristic region.

Table 14.24: Key Areas for Conservation of Bamboo Diversity (km^2)

Sl. No.	State	Geographical Area	Forest Area	KAC
1.	Arunachal Pradesh	83,621	68,621	9,592.98
2.	Assam	78,438	24,061	1,920.58
3.	Himachal Pradesh	55,673	12,501	6,232.87
4.	Jammu & Kashmir	222,235	20,433	14,822.22
5.	Manipur	22,327	17,558	226.65
6.	Meghalaya	22,429	15,714	303.65
7.	Mizoram	21,081	18,576	810.00
8.	Nagaland	16,579	14,291	226.43
9.	Sikkim	7,069	3,127	1,011.10
10.	Tripura	10,486	5,538	603.62
11.	Uttar Pradesh Himalaya	51,125	16,999	7,408.10
12.	West Bengal (incl. Himalaya)	88,752	8,276	2,779.30
	Total	**679,815**	**225,695**	**45,967.50**

Note: KAC (Key Areas of Conservation).

Source: Forest Survey of India Report 1995; Wildlife Institute of India 1998.

Table 14.25: Field Plantation of *D. asper*

Sl.No.	Number of Plants	Place	State	Address
1.	2,000	T.F.R.I.	M.P.	T.F.R.I., P.O. R.F.R.C, Jabalpur
2.	700	Chhindwara	M.P.	Human Resource Development Research Institute, Chhindwara
3.	200	Bastar	M.P.	Distributed among farmers
4.	400	Korba	M.P.	District Forest Officer, Krba Division
5.	1200	Kanker	M.P.	Distributed among farmers
6.	450	S.F.R.I	M.P.	State Forest Research Institute, Jabalpur
7.	100	Mandla	M.P.	District Forest Officer
8.	40	Seonui	M.P.	District Forest Officer
9.	20	Kanour	U.P.	Forest Research Institute, Kanpur
10.	500	Allahabad	U.P.	I.C.F.R.E., Institute
11.	10	Dehradun	Uttarakhand	Forest Research Institute, Dehradun
12.	100	Jorhat	Assam	I.C.F.R.E., Institute
13.	700	Nagpur	Maharashtra	Divisional Manager, F.D.C.M., Nagpur
14.	715	Bengaluru	Karnataka	a-Forest Research Institute, Bengaluru
		Bengaluru	Karnataka	b-Institute of Wood Science Technology, Bengaluru
	300	New Delhi	Delhi	President House, N. Delhi

Source: Arya, and Arya, 1996.

Himalayan region consisting political boundaries of the States of Jammu and Kashmir, Himachal Pradesh, Uttar Pradesh (Himalayan part), Sikkim, North Bengal and Arunachal Pradesh is also rich with high altitude elements occurring under different ecophysiological conditions.

Table 14.26: Bamboos Growing in Botanical Gardens and Bamboosetum of Himalayan NE and Eastern India

Sl.No.	Species	IBG	FRI	VVk
1.	*Bambusa affinis*	+	-	-
2.	*B. arundinacea* (Syn. *B. Bambos*)	+	+ *	+
	var. *gigantea*	-	+	-
3.	*B. auriculata*	+ ***	-	-
4.	*B. balcooa*	+	+	+
5.	*B. burmanica*	+	+	+
6.	*B. copelandi*	-	+ ***	-
7.	*B. lineata*	+	-	+
8.	*B. longispiculata*	-	+	-
9.	*B. multiplex*	+	+	+
	var. (FRI)	+	-	-
10.	*B. nutans*	+	+ **	+
11.	*B. oliveriana*	+	-	-
12.	*B. pallida*	+	+	+
13.	*B. polymorpha*	+	+	+
14.	*B. rumphiana* (Syn. *B. atra*)	+	-	-
15.	*B. spinosa*	-	+	-
16.	*B. tulda*	+	+ *	+
	var. large	-	-	+
17.	*B. ventricosa*	+@	+@	-
18.	*B. vulgaris*	+	+	+
	var. *aureo-variegata*	+@	-	-
	var. *striata*	+	+	+
	green form	-	+	+
19.	*B. wamin*	+	+	+
20.	*B.* sp. (Hijo)	-	-	+@
21.	*B.* sp. (Seijusa)	-	-	+@
22.	*B.* sp. (Routa)	-	-	+@
23.	*B.* sp. (Maithang)	-	-	+@
24.	*B.* sp. (Tapii)	-	-	+@
25.	*B.* sp. (Chessa)	-	-	+@
26.	*B.* sp. (Nal Madhupur)	-	-	+@
27.	*B.* sp. (Mangal Madhupur)	-	-	+@
28.	*B.* sp. (Nal Chessa)	-	-	+@
29.	*Dendrocalamus asper*	-	+	+@
30.	*D. brandisii*	+	+	+
31.	*D. calostachyus*	-	+@	-
32.	*D. giganteus*	+	+	+
33.	*D. hamiltonii*	-	+ **	+
34.	*D. longispathus*	+	+	-
35.	*D. membranaceus*	-	+ *	+
36.	*D. sikkimensis*	+	+	+
37.	*D. somdevai*	-	+ **	-
38.	*D. strictus*	+	+	+
	var. (Fld.)	-	+@	-
39.	*D.* sp. (Selari)	-	-	+@
40.	*Drepanostachyum falcatum* (Syn. *Sinarundinaria falcata*)	-	+	-
41.	*Dinochloa compactiflora*	-	+ *	+
42.	*D. maclellandii*	-	+	-

Sl.No.	Species	IBG	FRI	VVk
43.	*Gigantochloa albociliata*	-	+	-
44.	*G. atroviolacea*	-	+	-
45.	*G. atter*	+	+	-
46.	*G. hasskarliana*	+@	-	+@
47.	*G. pseudoarundinacea*	-	+	-
48.	*G. rostrata*	+	+	-
49.	*G. verticillata* (Syn. *G. pseudoarundinacea*)	+@	-	-
50.	*Guadua angustifolia*	+	-	-
51.	*Melocanna arundina*	+	-	-
52.	*M. baccifera*	+ *	+ *	-
53.	*Ochlandra travancorica*	+	+	+ *
54.	*Oxytenanthera abyssinica*	+	+	+
55.	*O.* sp. (Medang)	-	-	+@
56.	*Phyllostachys aurea*	+	+@	-
57.	*P. marliacea*	+@	-	-
58.	*P. reticulata* (Syn. *P. bambusoides*)	+@	-	-
59.	*Pseudoxytenanthera thwaitesii*	+	-	-
60.	*Pseudosasa japonica*	-	+	+
61.	*Schizostachyum brachycladum*	+@	+	-
62.	*S. dullooa*	+	+	+
63.	*S. fuchsianum*	-	-	+
64.	*S. hasskarlianum*	+@	-	-
65.	*S. hespentinum*	+	-	-
66.	*S. pergracile*	+ *	+	+
67.	*Thyrsostachys oliveri*	+	+	+
68.	*T. regia*	+ ^^	+	-

Note: Based on Bahadur (1979); Biswas (1988, 1995, 1997); Sen and Naskar (1965); Bose *et al.* (1987); Haridasan *et al.* (1987); Beniwal and Haridasan (1988).

* Flowered and regenerated through good seeding. ** Flered but no seeding. *** Occowurrence doubtful for having not observed the species in the locality in recent past. Probably flowered, not seeded and died. @ Taxonomically dubious and needs further check up.

Abbreviation: IBG (Indian Botanic Garden, Howrah), FRI (Forest Research Institute, Dehra Dun) and VVK (Van Vigyan Kendra, Chessa, Arunachal Pradesh).

REFERENCES

1. Anonymous. 1984. Official Methods of Analysis 14th edn. Association of Official Analytical Chemist, Washington, DC.
2. Anonymous, 1999. *State of Forest Report*. Published by Forest Survey of India. Ministry of Environment and Forests, Govt. of India, p. 113.
3. Arya, I.D. and Arya, S. 1996. Introduction, mass multiplication and establishment of edible bamboo *Dendrocalamus asper* in India. *Indian J. Pl. Genet. Resour.*, **9(1):** 115-121.
4. Bahadur, K.N. 1979. Taxonomy of bamboos. *Indian J. For.*, **2(3):** 222-241.
5. Bhatt, B.P., Singh, R., Misra, L.K., Tomar, J.M.S., Singh, M., Chauhan, D.S., Dhyani, S.K., Singh, K.A., Dhiman, K.R. and Datta, M. 2001. Agroforestry research and practices: An overview. In: Steps towards Modernization of Agriculture in NEH Region (eds. Verma, N.D. and Bhatt, B.P.), ICAR Research Complex for NEH Region, Umiam, Meghalaya, India p. 365-392.
6. Bhatt, B.P. and Bujarbaruah, K.M. 2004. Potential Edible Bamboo Species of North Eastern Hill Region: Production Potential, Cost: Benefit Analysis, Nutritional Security and Management Strategies. ICAR Research Complex for NEH Region, Umiam, Meghalaya, India, pp. 100.
7. Beniwal, B.S. and K. Haridasan. 1988. Study on bamboos through establishment of Bamboosetum in Arunachal pradesh. *Ind. For.*, **114(10):** 650-655.
8. Biswas, Sas. 1988. Studies on bamboo distribution in northeastern region of India. *Ind. For.*, **114(9):** 514-531.
9. Biswas, S. 1995. Diversity and genetic resource of Indian bamboos and strategies for their conservation. pp. 29-34 *in* Proc. 1st INBAR Biodiversity, Genetic Resources and Conservation Working Group, 7-9 November 1994. (V. Ramanatha Rao and A.N. Rao, eds.). IPGRI-APO, Singapore.
10. Biswas, S. 1997. Diversity evaluation of Indian bamboos, *Dendroclamus strictus* (Roxb.) Nees with particular reference to isozyme pattern. Pp. 19-25 *in* Proc. Working Group Meeting on "Biodiversity, Genetic Resources and Conservation of Bamboos and Rattans". (V. Ramanatha Rao and A.N. Rao, eds.). IPGRI-APO, Serdang, INBAR, Beijing.
11. Bose, R.B., H.S. Pandey and A.K. Banerjee. 1987. Bamboos of the Indian Botanic Garden Bul. *Bot. Surv. India*, **29(1-4):** 24-42
12. Chakraborty, D. 1988. Utilization of bamboo as raw material in the handicraft industries in Tripura. *Ind. For.*, **114(10):** 635-636.
13. Clark, L.G., Zhang, W. and Wendel, J.F. 1995. A Phylogeny of the Grass Family (Poaceae) Based on ndhF Sequence Data. *Systematic Botany*, **20(4):** 436-460.
14. *David, F. (1984). The Book of Bamboo. Sierra Club Books.* ISBN 0-87156-825-X.
15. Dieter, Ohrnberger. (1999). The bamboos of the world. Elsevier, 1999, ISBN 978-0-444-50020-5. p. 279-280.
16. Haridashan, K., Benniwal, B.S., Deori, M.L. 1987. Bamboos in Arunachal Pradesh, distribution and utilization- A preliminary appraisal. *Arunachal Forest News*, **5(1):** 23-28.

17. Hore, D.K.1998. Genetic resources among bamboos of North East India. J. *Econ. Taxon. Bot.* **22(1)**:173-181.

18. Judd, W.S., Campbell, C.S., Kellogg, E.A., Stevens, P.F. and Donoghue, M.J. [eds.]. 2008. Plant Systematics: A Phylogenetic Approach, 296-301. Sinauer Associates, Inc., Sunderland, Massachusetts USA.

19. Kochhar, S. and Rana, R.S. 1993. Bamboo germplasm resources in India-documentation and conservation. *Indian J. Pl. Genet. Resour.*, **6(2):** 93-107.

20. Rao, K.S. and Ramakrishnan, P.S. 1989. Role of bamboos in nutrient conservation during secondary succession following slash and burn agriculture (*Jhum*) in North-East India. *J. Appl. Ecol.*, **26:** 625-633.

21. Sarkar, J. and Sundriyal, R.C. 2002. Indigenous uses, management and conservation of bamboo resource in Arunachal Pradesh, North East India. *Bamboo J.*, **19:** 24-39.

22. Sen, J. and J.N. Naskar. 1965. Non-herbaceous phanerogams of the Indian Botanic Garden, Calcutta. *Bull. Bot. Surv. India*, **7(1-4):** 31-61.

23. Singh, K.A. 2002. Boon of bamboo resources in North East India. In: Resource Management Perspective of Arunachal Agriculture. Singh, K.A. (ed). ICAR Research Complex for NEH region, A.P. Centre, Basar, Arunachal Pradesh, India, p. 69-112.

24. Soberstrom, T.R. and Calderon, C.E. 1979. A commentary on the bamboos (Poaceae: Bambusoideae). *Biotropica*, **11(3):** 161-172.

25. Solanki, K.R., Bujarbaruah, K.M., Bhatt, B.P. 2004. Bamboo: A potential Resource for Agroforestry and Social Forestry with special reference to NEH Region. ICAR Research Complex for NEH Region, Umiam, Meghalaya, India, p. 52.

26. Trivedi, S. and Tripathy, R.S. 1984. Bamboo as an important renewable resource of Northeast India. In: Resource potentials of North East India, Vol. II. Living Resources (edited by Tripathy, R.S.), p. 9-15, *Meghalaya Science Society*, Shillong, India.

27. Varmah, J.C. and Bahadur, K.N. 1980. Country report and status of bamboo research in India. *Ind. Forest Res.*, **6(1):** 1-28.

Genus *Parkia*

Tree Bean (*Parkia roxburghii* G. Don. or *Parkia timoriana* (DC.) Merr.)

Introduction

Tree bean (*Parkia roxburghii* G. Don.) is also known as Supota, Kharial (Hindi), Yongchak (Manipur), Urohi/ Yontak/Khorial/Manipur-urohi (Assam), Zongto (Mizoram), Unkam-pinching (Nagaland), Aoelgap (Garo hills); Barri-phang (Cachar); Zawngah (Mizo); Themuk-arang (Mikir); Unkamn-pinching (Naga); Shivalingada mara (Kannada) and Unkampinching (Marathi). This is lesser known nutritious, multipurpose and economically important tree species grown widely in NEH regions particularly in Manipur and Mizoram. The life span of this tree may be 80-90 years or even more. It is commonly grows in backyard of houses, *jhums* and forests (Sharma *et al.*, 1993) in Manpiur and adjoining area. At the age of 6 years the plant starts its production, however full bearing stage is after 10 years only. Flowers, tender pods and seeds of this plant are edible and are a good source of proteins, fats, carbohydrates, vitamins and minerals compared to other legumes (Seal, 2011; Singh *et al.*, 2009). The flowers are used as salad and pod as a vegetable in South East Asia, especially north Eastern India. Edible pods are available from December until March and consumption starts from the tender pods, when it is about 30 cm long until maturity. Long tender pods of this plant are most delicious vegetables of North East region of India and other Asiatic countries. In Manipur it is considered as the most costly vegetable. It is used in different forms of preparations, either fresh or cooked. The tree bean pods are consumed at different stages of maturity as per the consumer preference. Biochemical and nutritional values in the tender pods were superior to other stages (Salam *et al.*, 1998). The strong smelling tender pods are cooked as vegetable. Sometimes pods are cleaned and sundried for future use during off seasons. The seeds from mature pods are also eaten as 'dal' or 'chutney' after roasting. The immature seeds, the young leaves and the fleshy part of the flower stalk may be eaten as vegetable and half ripe pods are occasionally pickled. It is also consumed raw along with condiments or cooked with other vegetables or meat. The chemical analysis of immature pods showed 72.0 per cent water, 1.02 per cent fat, 7.7 per cent fiber, 12.68 per cent protein, 3.5 per cent carbohydrates and 3.0 per cent minerals (ash) content. The vitamin C is 7.38 mg per 100 g fresh weight basis. The nutritional values of immature pods indicated that these are a good source of protein and vitamin C. Protein content of the tree bean kernel was 28.8 per cent and that of pods 13.0-19.0 per cent. Similarly, fat content was higher (34 per cent) in the kernel and ranged from 1-16 per cent in the pods. Unsaturated acids Oleic, linoleic and linolenic acids made up 63-68 per cent of the total fat in the pods as well as the kernels. The total essential amino acids amounted to 33 per cent, 36 per cent and 39 per cent in the tender, immature, mature pods, respectively, and 42 per cent in the kernels. The great pungent smell of yongchak tells the presence of Thiazolidine-4-Carboxylic acid (TCA, Thioproline), a cyclic sulphur containing amino acids (Suvachittanont *et al.*, 1996). The essential amino acid pattern of the kernel was comparable to the FAO/WHO/UNO (1985) amino acid requirement for pre-schoolers. However, the amino acid scores of the tender, immature and mature pods were 64, 84 and 92, respectively, with sulphur amino acids as the limiting amino acids in all the pod samples (Longvah and Deosthale, 1998). The seeds as well as tender pods are known to cure stomach disorders and regulate liver function. Pods pounded in water are used in cleaning the face and head (Burkill, 1935; Quisumbing, 1951). The soil extracts of the plant possesses insecticidal properties (Salam *et al.*, 1995). The timber is also used as firewood. As medicinal properties, the seeds as well as tender pods are known to cure stomach disorders and regulate liver function. In Malaya, both seeds and pods are valued in medicine. Pods pounded in water are used in washing the head and the face (Burkill, 1935; Quisumbing, 1951). Bark and leaves are used in making lotions for skin diseases and ulcers. The bark is reported to be suitable for tanning; it is used for dyeing nets in Philippines. Fermentation of decoction of leaves to the rheumatic affected parts is beneficial. In favourable season a full grown plant bears 10,000-15,000 pods. On an average, a tree bean can fetch an income of Rs. 10,000-20,000/-per annum but Bhuyan (1996) reported that a single plant can yield approximately Rs. 8,000-10,000 per annum. Both the figure indicates its economic value. Tree bean has the capacity to fix nitrogen as well. It can produce 10.0-25.0 cubic meter of biomass in 10-15 years. It had no adverse effects on grain yield of soybean and linseed when grown in combination with them. It is also useful as a shade tree in plantations (Dhyani and Chauhan, 1990).

Origin, History and Distribution

Tree bean is commonly grows in backyard of houses, jhums and forests in India (Mizoram, Nagaland, Manipur, Arunachal Pradesh, Meghalaya as well as in forests of Cachar, North Cachar hills and Sibsagar of Assam and Tripura), (Kanjilal *et al.*, 1938), Bangladesh, Myanmar, Java, Thailand, Egypt and the Malaysian region. It is usually grows wild in plains and hills up to 500 m above sea level. *Parkia timoriana* is a widely distributed species and recorded from a variety of forest types on various soils up to 1300 m (Hopkins, 1994).

Species Availability, Evolution and Diversity

Tree bean is known to have 40 species widely distributed in tropical parts of the world, only 3 species occur in India (Willis, 1982). *Parkia roxburghii* hitherto denoted by different scientific names such as, *P. timoriana, P. javanica* and *P. biglobosa* (Helen, 1994). Edible wild form is found in Nagaland (Deb *et al.*, 2013). In Manipur, nine types of tree bean differing in seed size, seeds per pod, dry matter (per cent), specific gravity and taste, on the basis of organoleptic test were identified

Botanical Description

Tree bean belongs to the family Fabaceae (Leguminosae) sub-family Mimosoideae. The tree is non-deciduous but is a without a leaf for a brief period before flowering. Crop is propagated through seeds and cuttings. The chromosome number is 2n = 24, 26.

Table 14.27: Botanical Description

Growth Habit	Fast growing in nature. The tree bean grows 20.0-30.0 m tall with spreading branches, having white to brown or light-gray bark, spotted white. The lifespan of this tree may be 80–90 years or more.
Leaf	Bipinnate leaves are 20.0-40.0 cm long, rachis bearing 12-13 pairs of pinnae with 30-80 pairs of leaflets of size 0.6-1.2 cm x 0.1-0.2 cm.
Floral biology	The inflorescence head or capitulum arises terminally with clusters. The pear shaped flower heads are borne on 20.0-30.0 cm long peduncle and hanging at the top of long talks from the branches. Tiny flowers yellowish white, trimorphous-neutral at the base of club, male in the middle and female at the apex.
Anthesis and pollination	The yellowing drum-head like inflorescence is a combination of miniature of flowers and is found hanging in clusters from branches during the month of September-October. The flowers are pollinated by Old World fruit bats (Fleming, 2009), especially *Eonycteris spelaea*, which feeds on the nectar. Bats are more efficient pollinators of the flowers than are the insects that also visit (Bumrungsri, 2008.
Pod (Fruit) and pod set	Produces bright green in colour long, soft, tender beans called 'pod' (during November to December) and turned black when fully matured. Pods are formed in clusters of 10-15, remain suspended on 10-12 cm long stalks and are of 20.0-40.0 x 2.5-4.2 cm in size (Kumar *et al.*, 2002). They turn blackish when fully mature in March–April. At the age of 6 years the plant starts its production; however, full bearing stage is only after 10 years. During favourable season a full-grown plant bears 10,000-15,000 pods.
Seeds	Seeds germinate regularly and dormancy do not happens frequently, the Government of Manipur has already produced a new hybrid species of Parkia roxburghii which will fully developed at five years and bears fruit as well. The seeds bear transplanting well. Pod containing 12-20 seeds. Seeds are ovoid, 1.9-2.1 cm x 0.9-1.2 cm, enclosed in hard testa.

Genetic Resources Augmentation and Crop Improvement

There are no known varieties or cultivar of tree bean available in India. In Manipur, 9 types (Nungsiacharu, Kangchup, Thangmani, Kharum, Kangpokpi, Taothong, Narankonjin, Jiri and Tamenglong) of tree bean differing in seed size, seeds per pod, dry matter (per cent), specific gravity and taste, on the basis of organoleptic test were identified (Meitei and Singh, 1990). There is a vast decline of yongchak population in the valley of Manipur and no reason has left for the cause of becoming a vulnerable species.

Genetic Resource Conservation

The loss of Parkia roxburghii is due to the Verticillium wilt disease spreads all over the valley of Manipur. *Verticillium* species have been reported to cause wilt disease in various tree species (Bilgrami *et al.*, 1991).

REFERENCES

1. Bhuyan, T.C. 1996. *Parkia roxburghii*-the tree bean. *Rain Forest News.*, **1(1):** 1-2.
2. Burkill, I.H. 1935. A dictionary of the economic products of the Malay Peninsula. Govt. Printing Press, Calcutta.
3. Bumrungsri, S. 2008. The pollination ecology of two species of Parkia (Mimosaceae) in southern Thailand. *Journal of Tropical Ecology*, **24(5):** 467-475.
4. Deb C.R., Jamir, N.S. and Ozukum, S. 2013. A study on the survey and documentation of underutilized crops of three district of Nagaland, India. *Journal of Global Biosciences*, **2(3):** 67-70.
5. Dhyani, S.K. and Chauhan, D.S. 1990. Nitrogen fixing trees for agro-forestry in Meghalaya. *Indian J. Hill Fmg.*, **3(2):** 65-68.
6. Fleming, T.H. 2009. The evolution of bat pollination: a phylogenetic perspective. *Annals of Botany*, **104(6):** 1017–1043.
7. Helen, C.F.H. 1994. The Indo-Pacific Species of Parkia (Leguminosae, Mimosoidae). *Kew Bulletin*, **42(2):** 181.
8. Hopkins, H.C.F. 1986. Parkia (Leguminosae: Mimosoideae). *Fl. Neotrop*, **43:** 1-124.
9. Kanjilal, U.N., Kanjilal, P.C. and Das, A. 1938. Flora of Assam, Vol. II. Govt. of Assam Publication.
10. Meitei, W.I. and Singh, A.J. 1990. Organoleptic test of tree bean (*Parkia roxburghii* G. Don.). *Indian J. Hill. Fmg.*, **3(1):** 47-49.
11. Kumar, S.K., Suresh, V.R., Nagachen, S.V. and Singh, T.R. 2002. Tree bean: a potential multipurpose tree. Indian Horticulture, Oct.-Dec., p. 10-11.
12. Longvah, T. and Deosthale, Y.G. 1998. Nutrient composition and food potential of *Parkia roxburghii*, a less known tree legume from northeast India. *Food Chemistry*, **62(4):** 477-481.
13. Meitei, W.I. and Singh, A.I. 1990. Organoleptic test of tree bean (*Parkia roxburghii* G. Don.). *Indian J. Hill. Fmg.*, **3(1):** 47-49.
14. Quisumbing, E. 1951. Medicinal plants of the Philippines. Technical Bull. No. 16, Deptt. of Agriculture and Natural Resources, Manila (Philippines).
15. Seal, T. 2011. Nutritional omposition of wild edible fruits in Meghalaya state of India and their Ethnobotanical importance. *Res. J. Bot.*, **6:** 58–67.
16. Sharma, B.D., Hore, D.K. and Salam, J.S. 1993. *Parkia roxburghii-* A useful tree of North-Eastern India. *Indian J. Plant Genet. Resour.*, **6(2):** 171-173.
17. Singh, S. J., Singh, P. K., Dutta, B. K. and Sahoo, U. K. 2009. Chemical composition and nutritive indices in *Parkia roxburghii*, G. don, a leguminous plant of India. *Indian J. Agric. Biochem.*, **22:** 87-93.
18. Suvachittanont, W., Kurashima, Y., Esumi, H. and M. Tsuda. 1996. Formation of thiazolidine-4-carboxyiic acid (thioproline), an effective nitrite-trapping agent in human body, in *Parkia speciosa* seeds and other edible leguminous seeds in Thailand. *Food Chemistry*, **55 (4):** 359-363.
19. Willis, J.C. 1982. A dictionary of the flowering plants and ferns (8[th]ed.) Int. Book Distributers.

Tree Tomato (*Cyphomandra betacea*)

Introduction

Tree tomato (*Cyphomandra betacea* Cav., syn. *Solanum betaceum* Cav.) is also known as 'Tamarillo' or 'tamato de árbol. It is grown as a commercial crop for international export in New Zealand and Portugal. Due to high citric acid content, fruits are sour but fully ripe fruits are slightly sweet in taste due to accumulation of sugar (Rai and Bhujel, 1997). It also contains high amount of phosphorus, calcium, potassium, nitrogen, iron, β-carotene (pro-vitamin A), vitamin B-6, vitamin C (ascorbic acid), vitamin E and carbohydrates. Per 100 g of edible portion of fruit contains 82.7-87.8 per cent moisture, 1.5 g protein, 10.3 g carbohydrates, 0.06-1.28 g fat (ether extract), 1.4-4.2 g fiber, 0.223-0.445 g nitrogen, 0.61-0.84g ash, 3.9-11.3 mg calcium, 52.5-65.5 mg phosphorus, 0.66-0.94 mg iron, 0.371-0.653 mg carotene (calculated as Vit. A-540 I.U.), 0.038-0.137 mg thiamine, 0.035-0.048 mg riboflavin, 1.10-1.38 mg niacin (with seeds), 1.011 mg (without seeds) and 23.3-33.9 mg ascorbic acid. It has a unique flavor and is consumed as delicious chutney when raw or after roasting and peeling off the skin. The peeled fruits may be pickled whole, or may be substituted for tomatoes in a hot chilli sauce (Prohens *et al.*, 1998). Being high in pectin, the fruit is easily made into jelly but the fruit oxidizes and discolors without special treatment during processing. Peeled, sliced and seeded tree tomatoes, with lemon rind, lemon juice and sugar, can also be cooked to a jam. The peeled fruit can then be sliced and the slices added to stew or soups, or served with a sprinkling of sugar and perhaps with a scoop of vanilla ice cream. Chopped slices are blended with cream cheese and used as sandwich spread. The peeled, stewed fruits are combined with gelatin, milk, sugar and lemon juice, to make a dessert, which is then garnished with fresh tree tomato slices. The peeled fruits can be pureed in a blender or by cooking, strained to remove the seeds and then packed in plastic containers and frozen. Lemon juice may be added to the puree to enhance flavour (Sanwal, 2008).

The fruit is eaten by scooping the flesh from a halved, but in New Zealand children palpate the ripe fruit it is soft then bite off the stem end and squeeze the flesh directly into their mouths. When lightly sugared and cooled, the flesh makes a refreshing breakfast dish. In addition, they give a unique flavor when added to stews (such as Boeuf Bourguignon), hollandaise, chutneys and curries. They are also tasty and decorative in, for example, radicchio salads. Appetizing desserts using this fruit include bavarois and combined with apples in a strudel. In Colombia and Ecuador, fresh tamarillos are frequently blended together with water and sugar to make a juice.It is also available as a commercially pasteurized puree. As a natural remedy the tree tomato is considered a fruit that relieves stress and helps to cure migraines and severe headaches. It is used to create obesity because it is low in calories-as well as high blood pressure. It is used for colds, sore throats, liver diseases and diabetes and it is said to improve the immune and circulatory system (Morton, 1987).

Origin, History and Distribution

Tree tomato is native to the Andes of Peru, Chile, Ecuador and Bolivia (Bohs, 1989). It is cultivated in Argentina, Australia, Brazil, Colombia, Kenya, the United States, Venezuela, Portugal and Indonesia where it is known as "terong Belanda" or "Dutch eggplant". It has been introduced in most tropical highlands where it often occurs naturalized in the subtropics and also in the mild temperate areas. In India, it is popular in Sikkim, Darjeeling hills of West Bengal (Rai and Bhujel, 1997), Nilgiris (Tamil Nadu) as well as in Khasi Hills (Meghalaya) (Thakur *et al.*, 1988).

Botanical Description

Table 14.28: Botanical Description

Growth Habit	Small tree or shrub, half-woody, attractive, fast growing, and brittle tree attend the height of 2.0-6.0 m, rarely as much as 7.5 m.
Root	Shallow rooted
Leaf	The leaves are muskily odorous, evergreen, alternate, more or less heart-shaped at the base, ovate, pointed at the apex 10.0-35.0 cm long and 4.0-12.0 cm broad, thin, softly hairy, with conspicuous coarse veins (Rai and Bhujel, 1997).
Floral biology	Flowers born in small, loose clusters near the branch tips. The fragrant flowers are 1.25-2.0 cm long, have 5 pale-pink or lavender, pointed lobes, 5 prominent yellow stamens, and green-purple calyx.
Anthesis and pollination	The treetomato flowers are normally self-pollinated (Morton, 1987).
Fruit and fruit set	The fruit resembles a tomato; hence its name is given. The long-stalked, pendent fruit, borne singly, orin clusters of 3-12. It is egg-shaped but pointed at both ends and capped with the persistent conical calyx.In size it ranges from 5.0-10.0 cm long and 4.0-5.0 cm width. It has thin smooth skin and coloured surface (colour may be solid deep-purple, blood red, orange or yellow or red and yellow and may have faint dark, longitudinal stripes). Soft flesh (when ripe) colour varies from orange-red or orange to yellow or cream yellow, firm, succulent and the pulp surrounding the seeds in the two length wise compartments is soft, juicy, sub acid to sweet, it is black in dark pulp and red fruits, yellow in yellow and orange fruits.
	The fruit has a slightly resinous aroma and the flavor suggests a mild or under ripe tomato with a faintly resinous aftertaste.about 21-24 weeks after anthesis, fruits are considered sufficiently mature to harvest.
Seeds	Dark coloured seeds occupying about one third of the interior of fruit. The seeds are thin, nearly flat, circular, larger and harder than those of true tomato and distinctly bitter.

Genetic Resources Augmentation and Crop Improvement

In India, yet no variety has been released but outside genotypes were introduced time to time. Some popular cultivars existing in other countries are Ecuadorian Orange, Goldmine, Inca Gold, Oratia Red, Rothamer, Ruby Red, Solid Gold, Yellow, *etc.*

REFERENCES

1. Bohs, L.A. 1989. Ethnobotany of the genus *Cyphomandra* (Solanaceae). *Econ. Bot.*, **43(2):** 143-163.
2. Caceres, A., Cabrera, O., Morales, O., Mollinedo, P. and Mendia, P. 1991. Pharmacological properties of *Moringa oleifera*. 1: Preliminary screening for antimicrobial activity. *J. Ethnopharmacol.*, **33(3):** 213-216.
3. Caceres, A., Saravia, A., Rizzo, S., Zabala, L., De Leon, E. and Nave, F. 1992. Pharmacological properties of*Moringa oleifera*. 2: Screening for antispasmodic, antiflammatory and diuretic activity. *J. Ethnopharmacol.*, **363:** 233-237.
4. Das, D.R., Kurup, P.A., Rao, P.L. and Narasimha. 1957. Antibiotic principle from *Moringa pterygosperma*. VII. Antibacterial activity and chemical structure of compounds related to pterygospermin. *Indain J. Med. Res.*, **45:** 191-196.
5. Daxenbichler, M.E., Spencer, F.G., Carlson, D.G., Rose, G.B., Brinker, A.M. and Powell, R.G. 1991. 1991. Glucosinolate composition of seeds from 297 species of wild plants. *Phytochemistr.*, **30:** 2623-2638.
6. Eilert, U., Wolters, B. and Nashrstedt, A. 1981. The antibiotic principle of seeds of *Moringa oleifera* and *Moringa stenopetala*. *J. Medicinal Plant Res.*, **42:** 55-61.
7. Faizi, S., Siddiqui, B.S., Saleem, R., Aftab, K., Shaheen, F. and Gilani, A.H. 1998. Hypotensive constituents from the pods of *Moringa oleifera*. *Plant Med.*, **64(3):** 225-228.
8. Ghasi, S., Nwobodo, E. and Ofili, J.O. 2000. Hypocholesterolemic effects of crude extract of leaf of *Moringa oleifera* Lam. In high-fat diet fed wistar rats. *J. Ethnopharmaco.*, **69:** 21-25.
9. Guevara, A.P., Vargas, C., Sakurai, H., Fujiwara, Y., Hashimoto, K., Maoka, T., Kozuka, M., Ito, Y., Tokuda, H. and Nishino, H. 1999. An antitumor

promoter from *Moringa oleifera* Lam. *Mutation Res.*, **440:** 181-188.

10. Morton, J. 1987.Tree tomato In: Fruits of warm climates. (Eds: Julia F. Morton, Miami, F.L.), p. 437-440.
11. Pal, S.K. Mukherjee, P.K. and Saha, B.P. 1995. Studies on the antiulcer activity of *Moringa oleifera* leaf extract on gastric ulcer models in rats. *Phytotheropy Res.*, **9:** 463-465.
12. Prohens, J., Ruitz, J.J. and Nuez, F. 1998. In: A textbook on Pomology, Vol III (Ed. Chattopadhyay, T.K.), Kalyani Publishers, New Delhi, p. 205-215.
13. Rai, S. and bhujel, R.B. 1997. Tree tomato-future vegetable crop. *Indian Hort.*, p. 42.
14. Sanwal, S.K. 2008. Tree tomato In: Underutilized vegetable and spice crops. Published by AGROBIOS (India), p. 117-125.
15. Spiliotis, V., Lalas, S., Gergis, V. and Dourtoglou, V. 1998. Composition of antimicrobial activity of seeds of different *Moringa oleifera* varieties. *Pharmacol Lett.*, **8(1):** 39-40.
16. Tahiliani, P. and Kar, A. 2000. Role of *Moringa oleifera* leaf extract in the regulation of thyroid hormone status in adult male and female rats.*Pharmmacol Res.*, **41(3)**: 319-323.
17. Thakur, N.S.A., Sharma, Y.P. and Barwal, R.N. 1988. Tree tomato cultivation in Meghalaya. *Indian Farming*, 37: 3
18. Udupa, S.L., Udupa, A.L. and Kulkarni, D.R. 1994. Studies on the anti-inflammatory and wound healing properties of *Moringa oleifera* and Aegle marmelos. *Fitoterapia*, **65(2):** 119-123.

Genus *Sesbania*

Swamp Pea or Agathi (*Sesbania grandiflora* (L.) Agast]

Introduction

Swamp pea (*Sesbania grandiflora*) is also known as Agati, Agasti, Anari (Sanskrit); Buk, Agast, Basana, Hatian (Hindi); Agati, Agusta, bak, Bagphal (Bengali); Madga, Agasta, Shewar (Marathi); Agathio, Athathiop (Gujarati); Avasinana, Avesi (Telgu); Agathi, Pragathi (Tamil); Agse, Agache (Kannad); Akathi, Athi (Malayalam) and Buko, Agosti (Odishi). Agathi is a tropical, very fast growing plant and grow basically to provide support to black pepper and beetle vine. It is also useful to provide shade to developing coconut plant. Agathi [*Sesbania grandiflora* L. (Agast)] is a small, loosely branching, legume plant of tropical Asia (India, Indonesia, Malaysia, Myanmar and Philippines) which is very closely related to *S. formosa* (an Australian species).Tender leaves, flowers, immature pods and seeds are edible but flowers (white flowers are first choice to the red because mostly white flowers are less bitter and astringent than red) are more preferred. In the Philippines, white flower buds are a common vegetable, steamed or cooked in soups or stews after the removal of stamen and calyx where concentration of bitterness is more. The raw flowers are eaten as salad in Thailand. Young leaves (taste bitter, sour and mildly tart) are also eaten, usually chopped fine and steamed, cooked or fried. Leaves are rich source of vitamin C, calcium and protein. Available protein has 64 per cent biological value and digestibility coefficient at 5 per cent protein accumulation is 85 per cent. Leaves are demulcent in nature and have non-poisonous saponin. Per 100 g edible agathi leaves contains 73.1 g moisture, 8.4 g protein, 1.4 g fat, 3.1 g minerals, 2.2 g fiber, 11.8 g carbohydrates, 93 K calorie energy, 1130 mg calcium, 80 mg phosphorus, 3.9 mg iron, 5400 microgram carotene, 0.21 mg thiamine, 0.09 mg riboflavin, 1.2 mg niacine and 165 mg vitamin. C. With respect to essential amino acids, 360 mg arginine, 120 mg histidine, 250 mg lysine, 100 mg tryptophan, 380 mg phenyl alanine, 90 mg methionine, 90 mg cysteine, 300 mg threonine, 560 mg leucine, 390 isoleucine and 430 mg valine per 100 g edible portion (Gopalan *et al.,* 1999). It has 9000IU vitamin A. As medicinal properties, leaves are used as poultice on the wounds. Chewing of leaves, give relief to mouth problems. Bark contains tannin, fiber and gums. It also has ornamental and fodder values. jenerally 100 g flower contains 91.58 g water, 27kcal, 11kJ, 1.28 g protein, 0.04 g fat, 0.38 g ash, 6.73 g carbohydrates, 18 mg calcium, 0.84 mg iron, 12 mg magnesium, 30 mg phosphorus, 184 mg potassium, 15mgsodium, 0.8 µg selenium, 73 mg vitamin, 0.083 mg thiamin, 0.081 mg riboflavin, 0.43 mg niacin and 102 µg folate (USDA nutrient Database).

Origin, History and Distribution

Many researchers believed that Asia is the centre of origin specially India, Malaysia, and the Philippines. Its native range extends to Mayanmar and Australia. It is also true that till date no precise information is avbailable for its origin (World Agro Forestry Centre) but a closely related species *i.e. S. formosa* is available in Australia. On this basis, it is also concluded that it may be native to Indonesia. Agasthi (*Sesbania grandiflora*) is named after the Vedic sage, Agastya. Considered the father of Tamil literature, Agastya specialized in medicine, spirituality, and language, probably lived between 6th and 7th century BC. Agathi grows throughout Mexico and the south west USA, the warmer areas of South America, and parts of Africa. In West Africa this crop is being cultivated for the past 150 years. It's also an exotic plant in Cuba, the Dominican Republic, and Nepal. It is well distributed and cultivated in Benin, Burkina Faso, Cameroon, Chad, Cote d' Ivoire, Cuba, Djibouti, Dominican Republic, Eritrea, Ethiopia, Gambia, Ghana, Guadeloupe, Guinea, Guinea-Bissau, Haiti, Kenya, Liberia, Mali, Martinique, Mauritania, Mauritius, Mexico, Nepal, Niger, Nigeria, Puerto Rico, Senegal, Sierra Leone, Somalia, South Africa, Tanzania, Togo, Uganda, and United States of America.

In India agathi can be spotted largely in Tamil Nadu, Kerala and Maharashtra. Apart from this sporadically this crop can be seen in Punjab, Delhi, Gujarat, West Bengal, Andhra Pradesh, Assam, and Andaman Nicobar.

Species Availability, Evolution and Diversity

Sesbania is a tropical genus of about 50 species. The genus *Sesbania* is divided into at least four subgenera, of which two, *Sesbania* and *Agati*, contain species of potential agricultural value. The important species under the genus *Sesbania* includes:

1. *S. benthamiana*
2. *S. dalzielii*
3. *S.drummondii*
4. *S.exasperata*
5. *S.formosa*
6. *S.goetzei*
7. *S.grandiflora*
8. *S.hirtistyla*
9. *S.hobdyi*
10. *S.javanica*

Botanical Description

Swam pea belongs to the family Fabaceae. Chromosome numbers in 29 species and some details of karyotypes of a few species of the genus have already been reported. Karyotypes of five taxa of *Sesbania* species. *viz:*

Table 14.29: Species and Chromosome Number

Species	*Chromosome Number*	
S. bispinosa	n=6	2n=12
S. grandiflora	n=12	2n=24
S. procumbens	n=6	2n=12
S. sesban var. *bicolor*	n=6	2n=12
S. sesban var. *sesban*	n=6	2n=12

The occurrence of n=6 or 12 and 2n=12 or 2n=24 in the above taxa advocates that the genus is based on x=6. The trend of karyotype evolution in *Sesbania* is inferred to be from symmetry to asymmetry as a result of paricentric inversions or unequal translocations.

Table 14.30: Botanical Description

Growth Habit	It is soft wood tree plants which grow very fast. Plants attain the height of 8.0-15.0 m and 25-30 cm in diameter.
Root	Roots are normally heavily nodulated with large nodules and the tree can also develop floating roots.
Stem	Stem is tomentose, and unarmed
Leaf	Leaves are 15.0-30.0 cm long, alternate, compound, and pinnate (bear 12-20 pairs of oblong, round leaflets). Generally 3-4 cm long and about 1 cm wide and leaves borne only on terminal ends of branches. The leaves turn bright yellow before shedding.
Floral biology	Flower cluster hanging at leaf base have 2-5 large or giant pea like flowers (6.0-10.0 cm long and about 3 cm wide). The flowers are curved attractive juicy/ crisp and pink, red or white or creamy in colour.
Anthesis and pollination	The large-flowered hermaphroditic species appears to be pollinated by birds.
Pod and pod set	The pods are long (30-50 cm x 8 mm), compressed, linear, erect and clustered, double ridged on the dorsal side single ridge below, length and beaked.
Seeds	Individual fruit has 15-50 pale coloured seeds. The seed is bean like, elliptical, red brown, 6-8 in a pod, 3.5 mm, and each weighing 1g. One kilogram of seed weight contain about 17,000-30,000 seeds.

Botanical Classification

Botanically swam pea can be grouped on the basis of flower colour which is utilized as vegetables:

(i) Red colour flower- It is very useful in kitchen gardens (Chadha, 2000).

(ii) White colour flower-It is very useful for commercial cultivation

Genetic Resources Augmentation and Crop Improvement

In agathi, no varieties have been identified till now through systematic breeding programme. But based on the flower colour, agathi can be classified in to four distinct groups, which are as follows:

(i) Sita-This group produce white colour flower.

(ii) Peeta-It is a yellow flowered strain.

(iii) Neela-This group of plant produce blue colour flower.

(iv) Lohita-It is a red flowered strain

Twenty two accessions of seven *Sesbania* sp. *viz., S. emerus, S. rostrata, S. tetraptera, S. exasperata, S. grandiflora, S. sesban* and *S. virgata,* which are used for ruminant fodder, firewood, wood products, soil improvement, and human food were investigated with the aim of characterizing both inter and intra specific genetic variability, estimating genetic parameters for the characters evaluated and appraising the forage potential of the accessions. These were planted in a randomized complete block design with 22 treatments and four replications. Seventeen morphological and 17 agronomic characters were evaluated. Genetic parameters, coefficient of intraspecific genetic diversity (b_i) and coefficient of intraspecific genetic variation (CV_{gi}) were obtained for the species represented by more than one accessions. Highly significant differences were observed among as well as within species for most of the characters, showing considerable genetic variability. *S. exasperata* showed intraspecific genetic variability for the largest number of morphological characters. The same was observed for *S. sesban* for the agronomic characters. Most of the characters showed high genetic diversity, indicating the possibility of selecting superior genotypes. On the other hand, the values for the intra specific genetic variation, which indicates the magnitude of the existing genetic variability relative to the character mean, varied according to the species and character evaluated. Differences between annual and perennial species were observed for biomass production. The annual species had higher seed yield.

Interspecific Hybridization in *Sesbania* spp.

Interspecific crosses in all possible combinations were carried out using *Sesbania sesban* var. *nubica, S. goetzei* and *S. kiniensis* and resulting in pod and viable seed formation for the crosses *S. sesban* x *S. goetzei, S. sesban* x *S. kiniensis, S. goetzei* x *S. sesban* and *S. goetzei* x *S. kiniensis*. The two crosses with *S. kiniensis* as a female parent were unsuccessful. The hybrid plants established normally and produced viable seeds.

Genetic Resource Conservation

Sesbania are nitrogen-fixing legumes and are used in sub-Saharan Africa for soil improvement purpose, stakes, food fodder and fuel wood. The International Centre for Research in Agroforestry (ICRAE) has undertaken germplasm exploration and collection of *S. sesban* as one of the priority species for research and improvement in Western Kenya and Southern Africa. These were the additional collection and complementary to those made earlier by the International Livestock Centre for Africa (ILCA). These latter collections were conducted mainly in Tanzania which is considered a centre of diversity for the genus *Sesbania* in Africa. The seed storage of *Sesbania* is orthodox in nature. Due to the lack of seed coat, variability of the seeds can only be maintained when the species is stored in sealed containers at a temperature of 4 °C or less and moisture content of less than 10 per cent.

REFERENCES

1. Chadha, K.L. 2000. Spinach. In: Handbook of Horticulture. ICAR, New Delhi, p, 461-462.
2. Gopalan, G., Rama Sastri, B.V., Balasubramanian, S.C., Rao, B.S.N., Deosthale, Y.G. and Pant, K.C. 1999. Nutritive value of Indian Foods. National Institute of Nutrition, ICMR, Hyderabad, India.

Genus *Prosopis*

Khejri (*Prosopis cineraria*)

Introduction

Vegetation of Indian Thar region is ecologically important, though fragile. In fragile ecosystems like 'Thar Desert', pure cropping alone cannot be a dependable enterprise; hence desert dwellers with their traditional wisdom are integrating trees into farming since ages, in order to confer stability and generate assured income (Rathore, 2009). Most of the trees are drought-resistant, and are used for fuel, fodder and other products, when arable crops fail due to drought, a common feature of the hot arid regions. Historically Indian arid zone has been an agrarian pastoralist system, which was 'ecologically self-contained', 'economically safe producing' and 'socially complete' (Damodaran, 1992). It is now widely believed that the underlying heterogeneity that sustained it ecologically and economically has been systematically changed. This region has 682 species belonging to 352 genera and 87 families of flowering plants. *Prosopis cineraria* (L.) and *Ziziphus nummularia* (Bordi) are the two most important multipurpose woody components planted in traditional agro-forestry system of the region. *Prosopis cineraria*, locally called as Khejri or Jandi or 'King of the Desert' or 'Kalptaru of Desert' is an indigenous tree, which effectively stabilizes sand dunes and can withstand periodic burial (Gates and Brown, 1988). It is believed to be the best suitable agro-forestry species, due to its deep taproot system, positive allelopathy effect, soil fertility improvement and yield augmentation of understory crops. Khejri offers nutritious supplementary food (pod flour, fruits, seed, *etc.*), top feed (leaves and pods) and protection-cum-shelter for the benefit of humans and livestock during all the seasons, effectively during harsh periods. It offers lifeline to human beings during famine, as its powdered bark can be mixed with flour and made into cakes for consumption (Bhandari, 1974). Due to these reason this tree is worshiped by a large number of people in Rajasthan, especially by 'Bishnoi' community. Khejri wood is useful for house construction and for making cart and agricultural implements. The protein rich new leaf formation start during end of summer (June-August) assures fodder availability to livestock in the harsh summer periods. The leaves can be used for mulching and the pruned branches are used as fuel wood. Fruits are called pods (locally called as 'sangar' or 'sangri') have economic value due to the vegetable (major ingredient in Panchcuta, a traditional diet).

From a mature tree (20 years age), 25-30 kg leaf fodder (loong) and 10-15 kg tender pods ('sangar' or 'sangri') can be harvested annually. Tender pods (sangri) are sold at high cost both fresh and dehydrated (Rs 50-80 and 300-500/kg, respectively) and is a eye-catching vegetable.

Origin, History and Distribution

In AD 1730, this tree came into limelight when a brave lady (Amrita Devi) from a 'Bishnoi' clan of local Marwar region sacrificed her life along with her 3 daughters and 363 people for the sake of *Prosopis cineraria* trees. Soldiers of the then local ruler, Maharaja Abhay Singh, wanted to cut Khejri trees for construction of palace. But the 'Bishnoi' clan under the leadership of Amrita Devi hugged the trees and faced the axes of the soldiers. After the incident, the king realized his mistake and called back his soldiers. Even now, an annual fair in the memory of the 363 people killed, is held every year in the village Khejarli, where the massacre took place. Hence it became an integral part of the traditional agriculture and the life-line of the desert inhabitants. Farmers of these regions have been consciously maintaining and promoting a few randomly growing trees, most commonly the *Prosopis cineraria* (L.), in their fields with the aim of obtaining fodder from the foliage, vegetable from the pods and fuel from pruned branches, particularly during famine years (Ahuja, 1990). Khejri trees exist in the western part of India covering northeastern Gujarat (Bhuj), Rajasthan (Sanchore and Jodhpur), Haryana and Punjab, with density gradient from low to high in the same order.

Species Availability, Evolution and Diversity

Significant variability with respect to flowering and fruiting has been observed in this tree.

Score represents the quantity gradients, *viz.* very less (1) to plenty/full (10) flowering and fruiting. Wide ranges of vriations are recorded in immature pod traits (Table 14.32).

Score represents the quantity gradients, *viz.* very less (1) to plenty/full (10) flowering and fruiting (Table 14.33).

Table 14.32: Khejri Tree Density, Off-season (November-December, 2011) Flowering and Fruiting Variability over Various Locations

Location	*Khejri Tree Density (no./ha)*	*Off- season flowering*		*Off- season Fruiting*		*Site Condition*
		Per cent	*Score (1-10)*	*Per cent*	*Score (1-10)*	
Bhuj (Gujarat)	0-4	5	1	3	1	Saline soils, irrigated area; major crops: castor; agro-forestry system: Mixed MPTs (noom, Khejri, Babool, *etc.*) based.
Sanchore (Rajasthan)	8-12	8	1	6	2	Undulating sandy soil/terrain, irrigated area; major crops: castor, wheat, mustard: agro-forestry system: neem based is prevailing.
Jodhpur (Rajasthan)	16-38	22	2	20	4	Sandy loam soil, rainfed area; major crops: bajra, wheat, guar agro-forestry system: Khejri, Salvadora based.

Source: Krishnan and Jindal, 2015.

Table 14.33: Regular, Off-season Flowering and Fruiting Variability over Age, Height and Diameter at Breast Height (DBH) of Khejri Tree

Variability due to	*Group/class*	*Regular season (April June 2012)*				*off-flowering (November-December 2011)*			
		Flowering		*Fruiting*		*Flowering*		*Fruiting*	
		per cent	*Score range*	*per cent*	*Score range*	*per cent*	*Score range*	*per cent*	*Score range*
Age (years)	15	100	4-10	100	2-9	62	1-2	40	1
	20	100	4-10	100	4-10	33	1-4	24	1-3
	25	100	5-10	100	4-10	18	1-3	15	1-2
Height (m)	<8	95	4-10	95	2-8	15	1	15	1
	8-10	100	4-10	100	4-10	39	1-4	36	1-3
	>10	100	5-10	100	5-10	60	1-2	50	1
DBH (cm)	<20	98	4-10	98	2-9	10	1-3	10	1
	20-30	100	4-10	100	4-10	47	1-4	44	1-3
	>30	100	5-10	100	4-10	70	1-2	55	1-2

Source: Krishnan and Jindal, 2015.

Botanical Description

Khejri is a cross-pollinated leguminous tree species.

Table 14.34: Botanical Description

Growth Habit	Khejri is a small to medium-sized tree that can withstand extremes of temperature up to 48°C and less than 100 mm rainfall. Luxuriantly growth under extremely adverse agroclimate in hot arid regions.
Root	Deep taproot system
Leaf	The leaf formation starts during end of summer (June-August).
Pod and pod set	Fruiting generally start during summer (April-June). Fruits are called pods (locally called as 'sangar' or 'sangri').

Horticultural Clasification

Classification on the Basis of Pod Taste

1. Bitter pod
2. Acrid pod
3. Sweet pod

Classification on the Basis of Pod Tenderness

1. Hard pod
2. Semi-hard pod
3. Less tender pod
4. Tender pos

Classification on the Basis of Pod Shape

1. Round pod
2. Flat pod or straight
3. Wrinkled pod

Classification on the Basis of Fiber Content

1. Fibrous pod
2. Less or fibreless

Classification on the Basis of Pod Colour

1. Green pod
2. Light green pod
3. Dark green pod

Classification on the Basis of Pod Length

1. 6-10 cm length
2. 10-20 cm length
3. 20-29 cm length

Genetic Resource Augmentation and Crop Improvement

At ICAR-Central Institute of Arid Horticulture (CIAH) Bikaner, Rajasthan, from wide genetic variability, some genotypes that produce high quality pods were collected for *ex situ* evaluation and conservation. From them 'Thar Shobha' is recommended for commercial cultivation in 2007 and to develop plantations for uniform, better quality and high sangri production on large scale. Its plants are vigorous in growth and thornless in nature. Plant can tolerate extremes of high (>48°C) and low temperature (-4°C) under arid conditions. It is extremely drought tolerant; therefore, it has high potential of production under rainfed conditions. Its development ensures good harvesting of both fodder and pods from khejri plantations even when there is complete failure of monsoon rains. A five-year old bud grafted plant yields 4.25 kg sangria and 6.25 kg long in a year. The tender pods at marketable stages are 13.1-20.2 cm in length, 0.17-0.42 cm in width, 0.97-1.75 g in weight and seeds not-separable in the stages of vegetable use. Tender pods are light green, straight, roundish-flat, soft and sweet. The ripen dried pods are 13.5-22.5 cm in length, 0.49-0.65 cm in width, 1.65-2.15g in weight and number of seeds ranged from 11.15-16.85 per pod. The weight of 100 seeds is 5.02g (Samadia, 2016).

Genetic Resource Conservation

The popular/socio religious movements like Chipko or those of 'Bishnois' have been meeting the conservation efforts (Kaul, 1993) which call for the integration of trees in farm lands as an assured livelihood source. At ICAR-Central Institute of Arid Horticulture (CIAH) Bikaner, Rajasthan (India) 16 genotypes are being maintained in *ex situ*.

REFERENCES

1. Ahuja, L.D. 1990. In : Khejri in the India Desert-Its role in Agroforestry (eds Mann, H.S. and Saxena, S.K.), ICAR-CAZRI, Jodhpur, Monogr. No.11, pp 28-30.
2. Bhandari, M.M. 1974. Famine foods in the Rajasthan desert *Econ. Bot.*, **28(1):** 73.
3. Damodaran, A.1992. *Econ. Polit. Wkly.*, 419-424.
4. Gates, P. J. and Brown, K. 1998. *Acacia tortilis* and *Prosopis cineraria*: Leguminous trees for and areas *Outlook on Agric.*, **17:** 61-64.
5. Krishnan, P.R. and Jindal, S.K. 2015. Khejri, the king of Indian Thar Desert is under phenophase change. *Current Science*, **108(11):** 1987-1990.
6. Kaul, R.N. 1993. In: Biodiversity Conservation: Forests, Wetlands and Deserts (eds Frame, B., Victor, J. and Joshi, Y.), Tata Energy Research Institute, New Delhi, pp. 91-103.
7. Rathore, J.S. 2009. Drought and household coping strategies: A case of Rajasthan, *Indian J. Agric. Econ.*, **59(4):** 689-708.
8. Samadia, S.K. 2016. Thar Shobha: New khejri variety. *Indian Horticulture*, **61(4):** 1 and 12-13.

Genus *Cordia*

Gum Cherry (*Cordia myxa*)

Introduction

Gum cherry or Indian cherry of desert or lasoda (*Cordia myxa*, 2n=48) is a underutilized fruit tree and known by a wide variety of names including clammy cherries, glue berries, sebestem, snotty gobbles, lehsua, goonda/goondi, lasura, assyrian plum, naruvilli or Indian cherry (Hanelt, 2001; Samadia, 2007). In India, the tender bud stage fruits of local species are largely consumed as a vegetable, raw, and cooked. It is called as Assyrian Plum, Pidar, Panugeri, Naruvilli, Geduri, Spistan, Burgund dulu wanan and Ntege or pickled, and are known by many names, including *lasora* in Hindi. One such species is Fragrant Manjack (*C. dichotoma*), which is called *gunda* or *tenti dela* in Hindi and *lasura* in Nepali. Almost all parts of 'lasoda' are used for various purposes (Yadav and Goel, 2006). The fruit of the Fragrant Manjack is called *phoà-pò·-chí* (POJ: chhiû-chí) in Taiwan where they are eaten as pickled. The white, gooey inner pulp of the fruits is traditionally used to make glue. The half ripe lasura fruit makes a tasty broth which is hot in effect as per Ayurveda practitioners. The unripe fruit makes an excellent pickle too which is not so hot. In fact the preserve is quite affective against indigestion. The ripe fruit are full of vitamins and regular use is supposed to be helpful in good growth of hair. Lasura preparations are, thus, good for people whose constitution might have tendency to go baldy. In addition to fruit, lasura bark and roots are also very effective as a local remedy against cough, cold and various other ailments connected with indigestion and throat problems. "Nutritional Value of some Nonconventional Plant Foods of India," green fruit of lasura contains the following composition per 100g: 2 per cent Protein, 92 per cent Carbohydrate, 2 per cent Fat, 2 per cent Fiber, 394kcal, 6mg Iron, 275mg Phosphorous and 55mg Calcium.

Cordia species have been used in medicinal remedies for centuries. According to the book, "Indian Herbal Remedies," the man regarded as the father of surgery in India, Sushutra, prescribed *Cordia* fruits as a coolant and astringent, to be used for biliousness, cough and internal haemorrhage. The bark paste was used for treatment on spider bites and eruptive boils. In Unani medicine, the fruit supplemented purgatives as a way of counteracting their bad side effects. Other properties of the fruit include acting as a demulcent, expectorant, diuretic and antihelmintic. Lasura has a number of well-documented health benefits:

- ✰ A 2001 study published in "Nutrition" found that *Cordia myxa* fruit acted as an anti-inflammatory, remedying the effects of an inflamed colon, liver and plasma.
- ✰ According to a study published in "Phytotherapy Research," *Cordia myxa* decreased blood pressure, thus acting as a potent hypotensive agent.
- ✰ A 2005 study published in the "Indian Journal of Pharmaceutical Education and Research" affirmed the fruit's traditional use as a diuretic.
- ✰ According to a 2007 study published in the "Nigerian Journal of Natural Products and Medicine," the liver-protecting abilities displayed by the leaves of *Cordia dichotoma* were comparable to sylmarin, a well-known drug prescribed for liver damage.
- ✰ Study published in 2009 "The International Journal of Pharmacology" found that *Cordia dichomtoa*fruits possess significant anti-ulcer properties,.
- ✰ A 2009 study published in "Pharmacology Online" found that the leaves contain potent analgesic, cytotoxic and anti-bacterial activity.
- ✰ A 2011 study published in the "International Journal of Pharmaceutical Sciences and Research" affirmed lasora fruit's traditional use as an antidiabetic.

Origin, History and Distribution

It is an introduced tree in tropical Asia, Africa, Australia and the Americas (Warfa, 1998) and distributed in the western tropical Africa to Italy, South-East Asia, Australia, Europe and tropical Americas (Hamlet, 2001). Lasura originates along the Himalayan tract with its natural warmer habitat extending through the forests of India, Nepal, Sri Lanka and Myanmar. Today, *Cordia* trees are found in Taiwan, Thailand, Malaysia, China, Polynesia, and Australia. The species is indigenous to China and is widely cultivated in lower plains and tropical regions. It is found growing primarily in Asia, as well as, across the globe especially in tropical regions having the right type of geophysical environment. In the Indian subcontinent, it is seen coming up naturally and growing abundantly from Myanmar in the East to Afghanistan in the West. Its habitat starts at about 200 m above mean sea level in the plains and ascends right up to a height of about 1500 m in the hills. Lasura species (*C. myxa* and *C. gharaf*) grows in moist and dry forests/ arid lands (Dev *et al.*, 2016) of the country, spreading throughout the northern states of Himachal Pradesh, Rajasthan, Punjab, Maharashtra, and the Western Ghats. It is not as popular in the south of the country, but a single tree rests in Adyar's Theosophical Gardens for those wishing to find one. If we see the distribution status of species, *C. gharaf* and *C. myxa* occupied 29.09 per cent and 21.81 per cent, respectively. However, 28 sites of *Cordia* germplasm collection were made mostly distributed in Bhuj (53.57 per cent) followed by Mandvi (28.57 per cent), Rapar (10.71 per cent) and Bhachau (7.42 per cent).

Species Availability, Evolution and Diversity

The *Cordia* genus has over 300 species. *Cordia myxa* is very common in nature. Another related species of lasoda or goonda, having small fruits is *Cordia gharaf* Forsk. (gondi) distributed in arid semi-arid zones of the country. Other economically important species of family Boraginaceae are *C. rothii*, *C. macleodii*, *C. vestita* and *C. wallichii*. Many *Cordia* species have fragrant, showy flowers and are popular in gardens, although they are not especially hardy (Bennett, 2003). Not all of these species are edible—India's varieties described in the above entry are considered some of the tastiest. He also revealed that the wood of several *Cordia* species is commercially harvested. Ecuador laurel (*C. alliodora*), *ziricote* (*C. dodecandra*), Spanish elm (*C. gerascanthus*), and *C. goeldiana* are used to make furniture and doors in Central and South America.

Ziricote (http://www.lmii.com/CartTwo/thirdproducts.asp?CategoryName=±Backs±and±Sides and NameProdHeader=Ziricot), Luthiers Mercantile Ziricote and *bocote* (Ziricote) are sometimes used as wood for making acoustic guitar backs and sides. Well-known guitarist Richard Thompson currently plays a Lowden F-35C RT Richard Thompson Signature Model with *ziricote* back and sides (Prasad, 2009). Similarly, drums are made from *C. abyssinica*, *C. millenii*, and *C. platythrysa* due to the resonance of the wood (Tudge, 2007).

List of *Cordia* species available in nature are:

- *Cordia africana* Lam. – White Manjack
- *Cordia alliodora* (Ruiz and Pav.) Oken – Spanish Elm, Ecuador Laurel, Salmwood, Bocote (Neotropics)
- *Cordia boissieri* A.DC. – Anacahuita, Texas Olive (southern Texas, Northern Mexico)
- *Cordia curassavica* (Jacq.) Roem. and Schult. – Black Sage, Wild Sage
- *Cordia dentata* Poir. – White Manjack
- *Cordia dichotoma* G.Forst – Fragrant Manjack, Bird Lime Tree (Tropical Asia and Australasia)[6]
- *Cordia leucophlyctis* Hook f. – (endemic to the Galápagos Islands)
- *Cordia lutea* Lam. – Yellow Cordia (western South America, including the Galápagos Islands)
- *Cordia monoica*
- *Cordia myxa* L. – Assyrian Plum (South Asia)

Cordia domestica is treated as a separate species by some sources and as *C. myxa* var. *domestica* by others.[7] *Cordia obliqua* Willd. (The Clammy Cherry) has been placed in the "*Cordia myxa* complex", or treated as a synonym for *Cordia dichotoma*.

- *Cordia platythyrsa* Baker
- *Cordia rupicola* Urban – Puerto Rico Manjack (Puerto Rico)
- *Cordia sebestena* L. – Geiger Tree, Large-leaf Geigertree (Southern Florida, Greater Antilles, Central America)
- *Cordia sinensis* Lam. (=*C. gharaf*) – Grey-leafed Saucerberry
- *Cordia subcordata* Lam. – *Kou*, *Tou*, *Marer* (Africa, South Asia,Southeast Asia, northern Australia, Pacific Islands)
- *Cordia sulcata* DC. – Mucilage Manjack, Laylay, White Manjack, Wild Clammy Cherry

Its synonyms are as under:

- *Cerdana* Ruiz and Pav.
- *Cordiada* Vell.
- *Cordiopsis* Desv.
- *Lithocardium* Kuntze
- *Rhabdocalyx* Lindl.
- *Sebesten* Adans.
- *Sebestena* Boehm.

A large variability exists in plantation of lasoda in relation to morphological characters particularly plant height, spread, leaf size, fruit size; fruiting behaiour; quality parameters like fruit colour, pulp content, sweetness, pickling quality, seed and pulp ratio *etc.*

Botanical Description

The taxonomy of *Cordia* is complex and controversial. Gottschling *et al.* (2005) say this is partly due to 'extraordinarily high intraspecific variability" in some groups of species, making identification difficult, and partly due to new taxa having been "airily described on the basis of poorly preserved herbarium specimens' (Gottschling *et al.*, 2005). *Cordia myxa* is a species of flowering plant in the borage family, Boraginaceae.

Kingdom	:	Plantae
Unranked	:	Angiosperms
Unranked	:	Eudicots
Unranked	:	Asterids
Order	:	unplaced
Family	:	Boraginaceae
Subfamily	:	Cordioideae
Genus	:	*Cordia*
Species	:	*myxa*

Table 14.35: Botanical Description

Growth Habit	It is a medium-sized, deciduous (during unfavorable environmental conditions) and fast growing tree. Lasura is a fairly fast growing species. It matures in about 50 to 60 years by when its girth at the breast height is about 1 to 1.5 m. Its bole (main trunk) is generally straight and cylindrical, attaining a height of nearly 3 to 4 m or sometimes 15 m with dense foliage and crooked trunk.
Leaf	The leaves of lasura are simple, broad, ovate, sub-orbicular rounded-cordate, alternate, 3-5 nerved, 7-15 c long and stalked with the spread being 7 to 15 cm x 5 to 10 cm. In matter of external appearance these are glabrous above and pubescent below. The young leaves tend to be hairy. The fresh foliage is quite useful as fodder for cattle-more so during grass famines. These are also used for wrapping biddies and cheroots.
Floral biology	Lasura tree flowers during March-April. The inflorescence, mostly terminal, is, white in color. Flowers are short-stalked, bisexual, actinomorphic, hypogymous, pentamerous and white in colour, appear in loose corymbose cymes. Individual florets are nearly 5 mm in diameter. At places these are somewhat hairy and white. Being a deciduous plant, the species bears male and female flowers on the same tree. The calyx part of an independent flower is about 8 mm long and glabrous, but not pubescent. It splits irregularly at the opening of its bud into flower. The filaments are hairy. Bunches of light yellow coloured hermaphrodite fragrant flowers are borne axillary on current season's growth.
Anthesis and pollination	The stem bark is greyish brown, smooth or longitudinally wrinkled. Flowers are short-stalked, bisexual
Fruit and fruit set	The fruit, popularly called Sebestan (drupe), 1-3 cm long, yellowish brown and pink at ripening. The cherry-sized fruit of lasura start appearing during July-August. It is a kind of a drupe, light pale to brown or even pink in color. The appearance tends to darken when ripening sets in. Being full of viscid glue like mucilage, the pulp is somewhat translucent. When fully ripe the pulp becomes quite sweet in taste, viscid, and transparent surrounding the stone and is fully enjoyed by children during the season (Yadav and Goel, 2006; Barraso *et al.*, 2009). The pulp in a half ripe fruit can even be used as an alternative to paper glue in office work.
Seeds	The hard stone is 1-4 seeded.

Horticultural Classification

In general, two types of plants *viz.*,

i. **Large fruited type-** Large fruited cultivars have an average fruit weight of 8.55 g. fruits have comparatively more pulp thickness and therefore are suitable for consumption

ii. **Small fruited type-** Small fruited cultivars have fruit weight of 3 g and very prolific bearing.

Augmentation of Genetic Resources and Crop Improvement

Polyploidy (x=7, 8) is reported in lasoda which also widens the genetic base. Some elite types have been identified which have large fruits, small stone and less mucilage content. Large fruited types are recognized as 'Paras Gonda' in Gujarat and 'Naruvili' in Tamil Nadu. Similarly in Rajasthan 'Puskar Local' is big size with good fruit shape. Small fruited types of lasoda locally called as Gundi of which fruits are of very small size (about 1 cm), orange to light pink in colour at maturity but very much liked by the rural people of western Rajasthan and Gujarat. At ICAR-CHES, Bhubaneswar (Odisha) CHL-1 a promising lasoda genotype recorded the high content of TSS (17°Brix), pulp content and phenols. At CCSHAU, Hisar (Haryana), different genotypes were tried by Saini *et al.* (2002) and they have reported variability in plant height and spread. At ICAR-CIAH, Bikaner (Rajasthan) 'Selection-1' is performing well under irrigated hot arid ecosystem with respect to fruit size, pulp content and productivity.

Genetic Resource Conservation

This species is under severe neglect from research and extension despite their potential to grow over large areas. ICAR-CHES, Bhubaneswar (Odisha); ICAR-NBPGR, Jodhpur (Rajasthan) and ICAR-CIAH, Bikaner (Rajasthan) are involved in genetic resource conservation. At ICAR-CIAH, Bikaner (Rajasthan) under ex-situ conservation, more than 60 types of lasoda were collected and planted under field conditions to identify a few ideal cultivars. Kaushik and Dwivedi (2004) reported wide range of biodiversity in morphological and quality characters from 45 accessions of lasoda from Haryana. In lasoda there is no improved variety so far ripe fruits are collected from the trees and rubbed to remove the flesh. The healthy stones are dried in the shade and kept in tin containers. The stones can be stored for 1 year in airtight containers kept in a dry place to avoid insect attack. There are 4 000-7 000 stones/kg.

REFERENCES

1. Bennett, M. 2003. Pulmonarias and the Borage Family. Timber Press. p. 196–198. ISBN 978-0-88192-589-0.
2. Dev, R., Kumar, M.S., Kumar, S. and Dayal, D. 2016. Survey, collection and utilization of Grewia and Cordia species for human and animal nutrition in Arid Kachchh, Gujarat. 1st International Agro-biodiversity Congress (Abs. Session: Crop Wild Relatives: Back to the Wild to Save the Future), held at New Delhi; 6-9 November 2016, p. 323.
3. Gottschling, M., Miller, J.S., Weigend, M. and Hilger, H.H. 2005. "Congruence of a Phylogeny of Cordiaceae (Boraginales). Inferred from ITS1 Sequence Data with Morphology, Ecology, and Biogeography". *Annals of the Missouri Botanical Garden,* **92** (3): 425–437. JSTOR 40035480.
4. *Germplasm Resources Information Network. United States Department of Agriculture.* 2007-10-05.

5. Hamelt, P. 2001. Institute of Plant Genetics and Crop Plant Research (IPK) (Eds.). Mansfeld's encyclopedia of agricultural and horticultural crops. Springer, Berlin, p. 1874.

6. Kaushik, R.A. and Dwivedi, N.K 2004. Genetic diversity in lasoda. *Indian Horticulture*, **1:** 27.

7. Presad, A. (2009). "Richard Thompson" (PDF). Guitar Player: 50

8. Samadia, D.K. 2007. Variability and scope of improvement in Lasoda (*Cordia Myxa*). *Indian Journal of Agroforestry*, **9:** 111-115.

9. Saini, R.S; Kaushik, R.A. and Singh, S. 2002. Research note on the evaluation of lasora (*Cordia myxa Roxb.*) germplasm for vegetative growth charater under semi-arid conditions. *Haryana J.* Hort. Sci, 31 (1-2): 62-63.

10. Tudge, C.2007. *The Tree*. Random House. p. 237.

11. Warfa, A.M. 1988. *Cordia* (Boraginaceae) in NE Tropical Africa and Tropical Arabia. Acta Universitatis Upsaliensis-Comprehensive Summaries of Uppsala from the faculty of Science, 174: 1-80.

12. Yadav, P.K. and Goel, M. 2006. Lasoda (*Cordia Myxa*). In: Saroj, P.L., Awasthi, O.P. (Eds.). Advances in Arid Horticulture, Vol.II-Production Technology of arid and semiarid fruits. International book Distributing Co., Lucknow, p. 305-318.

Genus *Artocarpus*

Jackfruit (*Artocarpus heterophyllus* Lam.)

Introduction

Jackfruit (*Artocarpus heterophyllus* Lam.) is one of the most important minor fruit trees in tropical home gardens and perhaps the most widespread and useful tree for its fruit (as immature - vegetable and mature-fruit). Tender jackfruit leaves and young male flower clusters may be cooked and served as vegetables. The cuisines of India, Nepal, Bangladesh, Sri Lanka, Indonesia, Cambodia, Thailand and Vietnam use cooked young jackfruit (encyclopedia). In Indonesia, young jackfruit is cooked with coconut milk as gudeg. In many cultures, jackfruit is boiled and used in curries as a staple food. In northern Thailand, the boiled young jackfruit is used in the Thai salad called tam kanun. In West Bengal, the unripe green jackfruit called aechor or ichor is used as a vegetable to make various spicy curries and side dishes, and as fillings for cutlets and chops. It is especially sought after by vegetarians who substitute this for meat, hence is nicknamed as gacch-patha (tree-mutton). In Odisha jackfruit is called Panasa Katha and used to make Panasa Tarkari (Raw Jackfruit curry), it is also used to make achar (Pickel). In the Philippines, it is cooked with coconut milk (ginataang langka). In Réunion Island, it is cooked either alone or with meat, such as shrimp or smoked pork. In southern India, unripe jackfruit slices are deep fried to make chips. In Udipi cuisine, jackfruit is used make appa and addae. Because unripe jackfruit has a meat-like taste, it is used in curry dishes with spices, in Bihar, Jharkhand, Sri Lankan, Andhran, eastern Indian (Bengali) and (Odisha) and Keralan cuisines. The skin of unripe jackfruit must be peeled first, then the remaining whole jackfruit can be chopped into edible portions and cooked before serving. Young jackfruit has a mild flavor and distinctive meat-like texture and is compared to poultry. Meatless sandwiches have been suggested and are popular with both vegetarian and non-vegetarian populations. 100-g portion of edible raw jackfruit provides about 95 calories and is a good source of the antioxidant vitamin C, providing about 13.7 mg. Jackfruit seeds are rich in protein. The fruit is also rich in vitamin B6, potassium, calcium, and iron (Goldenburg, 2014).

Origin, History and Distribution

The actual place of origin of jackfruit is not known but it is believed as indigenous to the rainforests of Western Ghats and now has spread throughout the tropics. Today, it is under cultivation in many Asian, African and Southern American countries. It is commonly grown in India, Myanmar, Sri Lanka, Malaysia, China, Indonesia, New Guinea, Australia, Brazil, Chile, Uganda, Kenya, Mauritius and Philippines. Though planted in Hawaii prior to 1888, it is still rare there and in other Pactfic islands, as it is in most of tropical America and the West Indies. It was introduced into northern Brazil in the mid-19th Century and is more popular there and in Surinam than elsewhere in the New World. In 1782, plants from a captured French ship destined for Martinique were taken to Jamaica where the tree is now common, and about 100 years later, the jackfruit made its appearance in Florida, presumably imported by the Reasoner's Nursery from Ceylon. The United States Department of Agriculture's Report on the Conditions of Tropical and Semi-tropical Fruits in the United States in 1887 states: "There are but few specimens in the State. Mr. Bidwell, at Orlando, has a healthy young tree, which was killed back to the ground, however, by the freeze of 1886. "There are today less than a dozen bearing jackfruit trees in South Florida and these are valued mainly as curiosities. Many seeds have been planted over the years but few seedlings have survived, though the jackfruit is hardier than its close relative, the breadfruit (q.v.). There are over 11,000 acres (4,452 ha) planted to jack fruit in Ceylon, mainly for timber, with the fruit a much-appreciated by-product. The tree is commonly cultivated throughout Thailand for its fruit.

The largest area under jackfruit in India is in Assam, and is also widely grown in southern states *viz.*, Kerala, Tamil Nadu, Karnataka and Andhra Pradesh. It is cultivated in other states like Bihar, Odisha, Maharashtra, West Bengal, Tripura, Manipur and Arunachal Pradesh. In India, the total area under jackfruit is approximately 1, 02,000 ha. In Arunachal Pradesh, it is also grown in the homestead garden and in the forest area as underutilized fruit crop. In South India, the jackfruit is a popular food ranking next to the mango and banana in total annual production. There are more than 100,000 trees in backyards and grown for shade in betelnut, coffee, pepper and cardamom plantations. The total area planted to jackfruit in all India is calculated at 14,826 acres (26,000 ha). Government horticulturists promote the planting of jackfruit trees along highways, waterways and railroads to add to the country's food supply.

Species Availability, Evolution and Diversity

The genus *Artocarpus* contains about 50 species; most of which are native to Asia and of which about 15 produce edible stachy fruits. The most important species are the tropical:

Table 14.36: Common and Botanical Name of *Artocarpus*

Sl.No.	Common Name	Botanical Name
1.	Breadfruit	*Artocarpus altilis* (syn. *A. communis*; *A. incisus*)
2.	Jackfruit	*Artocarpus heterophyllus* (syn. *A. integer*, *A. intcgrifolius*)
3.	Chempedak	*Artocarpus integrifiloia* (syn. *A. polyphema*; *A. chempeden*)

The other *Artocarpus* species which produce edible fruits include *A. anisophyllus*, *A. elasticus*, *A. odoratissimus*, *A. rigidus*, *A. anisophyllus*, *A. fulvicortex*, *A. polyphema*, *etc.*

The Champedak, *A. integer* Merr. (syns. *A. champeden* Spreng., *A. polyphena* Pers.), is also known as *chempedak, cempedak, sempedak, temedak* in Malaya; *cham-pa-da in* Thailand, *tjampedak* in Indonesia;*lemasa* in the Philippines. The wild form in Malaya is called *bangkong* or *baroh*. The fruit is borne by a deciduous tree, reaching about 60 ft (18 m) in cultivation, up to 100 or 150 ft (30-45.5 m) in the wild. It is easy to distinguish from the jackfruit by the long, stiff, brown hairs on young branchlets, leaves, buds and peduncles. The leaves are often 3-lobed when young, are obovate oblong or elliptical when mature and 6 to 11 in (15-28 cm) long. The male flower spikes are only 2 in (5 cm) long and the fruit cylindrical or irregular, no more than 14 in (35.5 cm) long and 6 in (15 cm) thick, mustard-yellow to golden-brown, reticulated, warty, and highly odoriferous when ripe. In fact, it is described as having the "strongest and richest smell of any fruit in creation." The rind is thinner than that of the jackfruit and the seeds and surrounding pulp can be extracted by cutting open the base and pulling on the fruit stalk. The pulp is deep-yellow, tender, slimy, juicy and sweet. That of the wild form is thin, subacid and odorless. The tree is native and common in the wild in Malaya up to an altitude of 4,200 ft (1,300 m) and is cultivated throughout Malaysia and by many preferred to jackfruit. It is grown from seed or budded onto self-seedlings or jackfruit or other *Artocarpus* species. Seedlings bear in 5 years. The pulp is eaten with rice and the seeds are roasted and eaten. The wood is strong and durable and yields yellow dye, and the bark is rich in tannin.

The Lakoocha, *A. lakoocha* Roxb., is also known as monkey jack or *lakuchi* in India; *tampang* and other similar native names in Malaya; as *lokhat* in Thailand. The tree is 20 to 30 ft (6-9 m) tall with deciduous, large, leathery leaves, downy on the underside. Male and female flowers are borne on the same tree, the former orange-yellow, the latter reddish. The fruits are nearly round or irregular, 2 to 5 in (5-12.5 cm) wide, velvety, dull-yellow tinged with pink, with sweet sour pulp which is occasionally eaten raw but mostly made into curries or chutney. The male flower spike, acid and astringent, is pickled.

A native of the humid sub-Himalayan region of India, up to 4,000 ft (1,200 m), also Malaya and Ceylon, it is sometimes grown for shade or for its fruit. A specimen was planted at the Federal Experiment Station, Mayaguez, Puerto Rico, in 1921. There was a large tree in Bermuda in 1918. The wood, sold as *lakuch*, is heavier than that of the jackfruit, similar to teak, durable outdoors and under water, but does not polish well. It is used for piles, and in construction; for boats, furniture and cabinetwork. The bark contains 8.5 per cent tannin and is chewed like betelnut. It yields a fiber for cordage. The wood and roots yield a dye of richer color than that obtained from the jackfruit. Both seeds and milky latex are purgative. The bark is applied on skin ailments. The fruit is believed to act as a tonic for the liver.

The Kwai Muk, possibly *A. lingnanensis* Merr., was introduced into Florida as *A. hypargyraea* Hance, or *A. hypargyraeus* Hance ex Benth. The tree is a slow-growing, slender, erect ornamental 20 to 50 ft (6-15 m) tall, with much milky latex and evergreen leaves 2 to 5 in (5-12.5 cm) long. Tiny male and female flowers are yellowish and borne on the same tree, the female in globular heads to 3/8 in (1 cm) long.

The fruits are more or less oblate and irregular, 1 to 2 in (2.5-5.0 cm) wide, with velvety, brownish, thin, tender skin and replete with latex when unripe. When ripe, the pulp is orange-red or red, soft, of agreeable subacid to acid flavor and may be seedless or contain 1 to 7 small, pale seeds. The pulp is edible raw; can be preserved in sirup or dried. Generally, ripens from August to October in Florida. The tree is native from Kwangtung, China, to Hong Kong, and has been introduced sparingly abroad.

It was planted experimentally in Florida in 1927 and was thriving in Puerto Rico in 1929. It grows at an altitude of 500 ft (152 m) in China. Young trees are injured by brief drops in temperature to 28° to 30°F (-2.22°-1.11°C). Mature trees have endured 25° to 26°F (-3.89°-3.33°C) in Homestead, Florida; have been killed by 20°F (-6.67°C) in central Florida.

Jackfruit is a heterozygous, cross pollinated/ wind pollination resulted in wide range of variation with respect to yield and fruit quality. Collection, characterization and evaluation is still the most sought after method for releasing of new jackfruit varieties, owing to this vast amount of variability. Patil *et al.* (2016) evaluated 69 conserved accessions in the field genebank for superior fruit quality traits. Variability observed for fruit shape, average fruit weight (3 kg to 13.9 kg), length, rind thickness, skin and pulp colour, number of flakes per fruit (40 to 220), bulb weight, length and width, rachis length, total soluble solids (17.7 to 34.2 °B), seed weight, length, width, shape and colour. Further, these 69 genotypes were grouped into five clusters based on the genetic diversity analysis. The dendrogram obtained using phenotypic characters separated the accessions into five major clusters based on Euclidean distance. The Euclidean distance of tree demonstrated a genetic distance between the clusters I (G-70 and G-78) and cluster V (G-34, G-75, G-38 and G-51) were genetically apart based on the fruit traits.

Botanical Description

Jackfruit (family: Moraceae) is a heterozygous and mostly seed proapagted.

Table 14.37: Botanical Description

Growth Habit	It is a medium to big sized evergreen tree typically reaching 8 m to 25 m in height and is recognized by its fruit, the largest among cultivated plants. The canopy shape is usually conical or pyramidal in young trees and becomes spreading and domed in older trees. The canopy diameter of 5 years lod tree ranges from 3.5m to 6.7m and reach 10m or more in older trees. All parts of the tree exude sticky latex when injured.
Stem	Heavy side branching usually begins near the ground.
Anthesis and pollination	It is highly heterozygous and cross pollinated tree, normal wind-pollination.
Fruit and fruit set	The fruit is borne primarily on the trunk and interior part of main branches; usually weigh 4.5 kg to 30 kg, although a weight of 50 kg has been reported.
Seeds	Fruit contain 1 to 7 small and pale seeds

Botanical Classification

The jackfruit has innumerable types considering the fruit characteristics. The types differ widely among themselves in bearing, size, shape, quality and period of maturity. In India, mainly two types viz:

1. Soft flesh
2. Firm flesh

Some types with local names like 'Gulabi' (rose-scented),'Champa' (flavor like that of *Michelia* species), 'Hazari' (bearing large number of fruits), 'Rudrakshki' (roundish fruit) are also available. In South India, jackfruit are classified as of two general types:

1. **Koocha Chakka-**The fruit of which have small, fibrous, soft, mushy, but very sweet carpels (edible part)
2. **Koozha Pazham-**More important commercially, with crisp carpels of high quality known as 'Varika'.

These types are apparently known in different areas by other names such as:

1. 'Barka' or 'Berka' (soft, sweet and broken with the hands) and
2. 'Kapa' or 'Kapiya' (crisp and cut open with a knife).

In Arunachal Pradesh, generally jackfruits are classified as of two types based on the fruit surface:

1. Jackfruit having less prominent spine in the fruit
2. Jackfruit having prominent spine in the fruit

The equivalent types are known as Kha-nun nang (firm; best) and Kha-nun lamoud (soft) in Thailand; and as Vela (soft) and Varaka, or Waraka (firm) in Ceylon. The Peniwaraka, or honey jak, has sweet pulp, and some have claimed it the best of all. The Kuruwaraka has small, rounded fruits. Dr. David Fairchild, writing of the honey jak in Ceylon, describes the rind as dark-green in contrast to the golden yellow pulp when cut open for eating, but the fruits of his own tree in Coconut Grove and those of the Matheson tree which he maintained were honey jaks are definitely yellow when ripe. The Vela type predominates in the West Indies. Firminger described two types: the Khuja (green, hard and smooth, with juicy pulp and small seeds); the Ghila (rough, soft, with thin pulp, not very juicy, and large seeds). Dutta says Khujja, or Karcha, has pale-brown or occcasionally pale-green rind, and pulp as hard as an apple; Ghila, or Ghula, is usually light-green, occasionally brownish, and has soft pulp, sweet or acidulously sweet. He describes 8 varieties, only one with a name. This is Hazari; similar to Rudrakshi; which has a relatively smooth rind and flesh of inferior quality. The 'Singapore', or 'Ceylon', jack, a remarkably early bearer producing fruit in 18 months to 2 1/2 years from transplanting, was introduced into India from Ceylon and planted extensively in 1949. The fruit is of medium size with small, fibrous carpers which are very sweet. In addition to the summer crop (June and July), there is a second crop from October to December. In 1961, the Horticultural Research Institute at Saharanpur, India, reported the acquisition of air-layered plants of the excellent varieties, 'Safeda', 'Khaja', 'Bhusila', 'Bhadaiyan' and 'Handia' and others. The Fruit Experimental Station at Burliar, established a collection of 54 jackfruit clones from all producing countries, and ultimately selected 'T Nagar Jack' as the best in quality and yield. The Fruit Experimental Station at Kallar, began breeding work in 1952 with a view to developing short, compact, many-branched trees, precocious and productive, bearing large, yellow, high quality fruits, 1/2 in the main season, 1/2 late. 'Singapore Jack' was chosen as the female parent because of its early and late crops; and, as the male parent, 'Velipala', a local selection from the forest having large fruits with large carpers of superior quality, and borne regularly in the main summer season. After 25 years of testing, one hybrid was rated as outstanding for precocity, fruit size, off-season as well as main season production, and yield excelling its parents. It had not been named when reported on by Chellappan and Roche in 1982. In Assam, nurserymen have given names such as 'Mammoth', 'Everbearer', and 'Rose-scented' to preferred types.

Augmentation of Genetic Resources and Crop Improvement

A joint exploration and collection trip was conducted in parts of Western Ghats lying in states of Kerala, Karnataka and Tamil Nadu and total of 95 accessions comprising 'Koozha' (18), 'Navarikka' (3), 'Varikka' (59), 'Uudraksha Varikka' (2), 'Rudraksha koozha (2), wild, (2) others (7) were amassed by Muralidharan, *et al.* (1997).

In order to study the variability in jack fruit, field surveys were conducted in Biocholim, Canacona, pernem, Ponda and Bardez talukas in Goa state during May-July 2013, 2014 and 2015. Based on descriptor, fruit characters of the accessions were recorded. There was wide variation in horticultural traits like, stalk length (5.6 to 11.5 cm), fruit weight (3.62 to 19.36 kg), fruit length (31.2 to 68.2 cm), fruit girth (52.2 to 94.2 cm) rachis length (11.9 to 42.3 cm), rachis diameter (5.1 to 11.0 cm), number of bulbs per fruit (48 to 271), bulb weight without seeds per fruit (0.51 to 5.57 kg), weight of seeds per fruit (0.2 to 1.38 kg), weight of inedible waste per fruit (2.05 to 8.2 kg) *etc.* the bearing ranges from early to late season; fruit rind colour yellow to blackish green; fruit shape oblong/ellipsoidal/irregular; spine shape flat to pointed; spine density sparse to dense; fruit quality poor to excellent, bulb colour creamy white to deep orange; seeds with shapes ranging from elongate to reniform; seed surface plain or with striations (Devi *et al.*, 2016).

Table 14.38: Characterization of

Characters		*Wild*	*Koozha*	*Navarikka*	*Varikka*	*Rudraksha Koozha*	*Rudraksha Varikka*	*ICAR Project*
No. of collection		2	18	3	59	2	2	175
Fruit length	Range	32.0-37.0	28.0-68.0	32.0-58.0	23.0-72.0	28.0-30.0	21.0-28.0	22.0-84.0
	Mean	34.5	45.0	46.0	46.0	29.0	24.5	38.0
Fruit weight	Range	2.2-3.5	2.0-25.4	4.0-9.0	1.5-39.0	2.0-4.1	2.5-3.9	1.6-25.1
	Mean	2.9	8.4	7.5	10.4	3.6	3.2	9.7
Fruit circumference	Range	40.0-50.0	30.0-120.0	60.0-75.0	62.0-96.0	54.0-66.0	35.0-44.0	13.0-102.0
	Mean	45.0	52.0	68.0	73.0	60.0	40.0	41.5
Rind thickness	Range	2.4-1.8	1.0-3.0	1.3-2.0	1.5-3.0	0.8-1.0	0.8-1.0	0.4-3.1
	Mean	2.1	2.0	1.8	1.9	0.9	0.9	1.5
Core length	Range	23.0-24.0	13.0-15.0	16.0-38.0	12.0-56.0	13.0-14.0	0	11.0-61.0
	Mean	23.5	27.0	29.0	29.0	13.5	20.0	31.0
Core width	Range	4.5-5.0	4.5-12.0	6.0-9.0	3.5-12.0	5.0-6.5	-	3.2-14.0
	Mean	4.8	6.4	7.3	6.8	5.7	4.5	6.8
Flake length	Range	3.7-5.0	4.2-8.5	5.0-5.7	3.5-10.5	5.0-7.0	4.5-5.0	3.2-12.5
	Mean	4.4	6.1	5.4	6.7	6.0	4.7	6.2
Flake circumference	Range	6.0-6.1	6.0-13.0	8.0-9.2	4.0-13.0	9.0-13.0	8.0-10.0	-
	Mean	6.1	9.0	8.7	9.2	11.0	9.0	-
Weight of flake	Range	4.6-7.0	7.0-34.0	7.6-14.0	6.4-41.6	1.4-1.9	-	10.5-55.0
	Mean	5.8	17.0	10.8	19.0	17.0	23.0	27.8
Weight of seed	Range	4.6-5.0	5.0-15.4	7.0-11.0	3.4-13.4	6.2-7.0	3.0-8.4	1.2-14.2
	Mean	4.8	7.6	8.6	7.6	6.6	7.3	6.8
Seed length	Range	-	2.7-4.5	2.3-4.0	2.0-5.0	2.4-3.5	3.5-4.3	2.4-11.0
	Mean	3.0	3.5	3.8	3.2	2.9	3.9	3.6
Seed width	Range	-	1.2-3.0	2.0-2.3	1.2-3.0	1.4-2.0	2.0-2.2	1.3-9.0
	Mean	2.4	2.1	2.2	1.9	1.7	2.1	5.4

ICAR-IIHR, Bengaluru (Karnataka) evaluated 69 accessions of jackfruit and identified two superior accessions, one red flaked and another less gum type which were found suitable for commercial cultivation.

Genetic Resource Conservation

Red jackfruit is rare in Kerala and elsewhere, and has been spotted in Karnataka's two regions *i.e.* Toobugere hobli and Gubbi taluk in Tumkur district which are blessed with considerable percentage of red fleshed jackfruits. At Doddaballapur town, Bangalore (Karnataka) city and in Dabuspete and K.B. Cross of Tumkur, vendors selling red jackfruits are common. They very well know that red jackfruit attracts the consumers more that it's yellow cousin. Being a National Active Germplasm Site (NAGS) for jackfruit, ICAR-Indian Institute of Horticultural Research, Bengaluru (Karnataka) has collected 69 germplasm accessions from states where in Western Ghats lays (Patil *et al.*, 2016).

REFERENCES

1. Chellappan, K. and Roche J.I. 1982. Studies on clonal propagation of nutmeg (*Myristica fragrans* Houtt.) South Indian Hort: 30: 156-158.
2. Devi, S.P., Thangam, M., Porob, S.A. and Singh, N.P. 2016. Genetic diversity of jack fruit in Goa state. 1st International Agrobiodiversity Congress (Abstract), held at New Delhi; 6-9 Nov., 2016; p. 18.
3. Goldenberg, S. 2014. "Jackfruit heralded as 'miracle, food crop". The Guardian.
4. Muralidharan, V.K., Ganapathy, M.M., Velayudhan, K.C. and Amalraj, V.A. 1997. Collecting jackfruit germplasm in Western Ghats. *Indian J. Plant Genet. Resour.*, **10(2):** 227-231.
5. Patil, P., Sane, A. and Natduthodi, N. 2016. Evaluation of Jackfruit Germplasm accessions and genetic diversity analysis. 1st International Agrobiodiversity congress (Abs. Session Trait Discovery and Enhanced Use of PGR), 6-9 November, held at New Delhi, p. 186.
6. The encyclopedia of fruit and nuts, By Jules Janick, Robert E. Paull, p. 155.

15

Genetic Resources of Other Vegetable Crops and its Improvement

Genus *Passiflora*

Giant Granadilla (*Passiflora quadrangularis* L.)

Introduction

Giant Granadilla (*Passiflora quadrangularis* L. Syn. *P. macrocarpa* M.T. Mast.) is a fruit crop but it immature fruits are used as vegetable and considered as a potential future crop (Kumar *et al.,* 2005). It bears largest fruit in its genus therefore name is given as *'giant grandilla'* and sometimes often called merely *grandilla,* or parcha, Spanish names loosely applied to various related species; or it may be distinguished as grandilla real, *grandillagrande, parchagrandilla* or *parcha de Guinea*(Killip, 1938). In El Salvador, it is known as *grandilla de fresco* or *grandillapararefresco;* in parts of Colombia, it is *badea* or *corvejo;* in the State of Tachira, Venezuela, *badea;* in Bolivia, *grandilla* real or *sand1a de PasiAn.*

Giant Granadilla is a rich source of calcium, phosphorus and ascorbic acid. The unripe fruit is used as vegetable while the ripe one eaten as dessert.

Edible portions are mesocarp and endocarp of fruits which can be eaten as melons and as fruit salad. Flesh of the ripe fruits, with inner skin removed, is cut up and added to papaya, pineapple and banana slices in fruit salads, seasoned with lemon or lime juice. Jelly can be made from unpeeled flesh boiled for 2 hours and pulp simmered separately. Juice strained from both is combined and, with added sugar and lemon juice, is boiled until it jells. Its use as a flavouring agent in ice creams and sherbets is noteworthy. In Jamaica, fleshy tuberous roots are eaten as a substitute for yams. Leaves, skin and immature seeds contain a cyanogenic glycoside.

Table 15.1: Nutritive Value per 100 g of Edible Portion

Items	*Thick flesh*	*Arils and Seeds*	*Items*	*Thick flesh*	*Arils and Seeds*
Moisture	94.5 g	78.4 g	Iron	0.80 mg	2.93 mg
Protein	0.112 g	0.299 g	Carotene	0.004 mg	0.019 mg
Fat	0.15 g	1.29 g	Thiamine	-------	0.003 mg
Crude fiber	0.7 g	3. 6 g	Riboflavin	0.033 mg	0.120 mg
Ash	0.41 g	0.80 mg	Niacin	0.378 mg	--------
Calcium	13.8 mg	9.2 g	Ascorbic acid	14.3 mg	15.3 mg
Phosphorus	17.1 mg	39.3 mg			

Pulp contains passiflorine and if indulged in excessively, causes lethargy and somnolence. Raw root is emetic, narcotic and poisonous.

Origin, History and Distribution

The granadilla is generally agreed to be a native of tropical America, through the actual place of origin is unknown. It was growing in Barbados in 1750 and is present in several other Caribbean Islands and in Bermuda (Morton, 19870. It is commonly cultivated, and sometimes an escape from cultivation or truly wild, from Mexico to Brazil and Peru. It is also cultivated in the lowlands of India, Ceylon and the Philippines; in tropical Africa, and throughout Queensland. In tropical North Queensland it has run wild, growing lushly in forest areas. It flourishes and fruits heavily especially in the Cairns district. In Northern hill of India, the species was introduced quite a few centuries back. But the cultivation of species is absolutely located and the number of population is also very less obsolete use. They can hardly be seen in North Cachar Hill of Assam and foothill areas of Khasi Hill, adjacent to Bangladesh.

Botanical Description

Table 15.2: Botanical Description

Growth Habit	The vine is fast growing, large, coarse, herbaceous but woody at the base, arising from a fleshy root that becomes enlarged with age, and climbing trees to a height of 10.0-15.0 m.
Stem	It has thick 4 angled stems prominently winged on the angles and axillary tendrils to 30 cm long, flanked by leaf like, ovate or ovate-lanceolate stipules 2.0-3.5 cm long, sometimes faintly toothed.
Leaf	The alternater leaves are broad-ovate or oblong-ovate, 8.25-15.0 cm wide, 10.0-20.0 cm long; rounded or cordate at the base, abruptly pointed at the apex, sometimes toothed near the base; thin with conspicuous veins sunken on the upper surface, prominent beneath.
Floral biology	The solitary, fragrant flowers, up to 12.0-12.5 cm wide, have a bell-shaped calyx, the 5 sepals greenish or reddish green on the outside, white, pink or purple inside; the 5 petals, to 4.5 cm long, white and pink; the corona filaments 2-ranked, to 6.0 cm long, purple and white below, blue in the middle and pinkish blue above, around the typical complex of pistil, style and stigmas.
Fruit and fruit set	The pleasantly aromatic, melon like fruit is oblong-ovoid, 12.0-15.0 cm wide, and 10.0-30.0 cm long; may be faintly ribbed or longitudinally 3-lobed; has a thin, edible skin, greenish white to pale or deep yellow, often bluish with pink. Beneath it is a layer of firm, mealy, white or pink flesh, 2.5-5.4 cm thick of very mild flavor, and coated with a parchment like material on the inner surface. The central cavity contains some juice and masses of whitish, yellowish, partly yellow or purple-pink, sweet acid arils (commonly referred to as the pulp).
Seeds	Flattened oval, and purplish-brown seeds to 1.25 cm long.

Augmentation of Genetic Resources and Crop Improvement

There are various strains producing fruits of different sizes and quality. Wester stated that some are insipid, while one of superior flavor had originated at Cotabato. One strain with especially large fruits and good flavor was considered a separate species (*P. macrocarpa*), but it hybridize readily with smaller strains and there are intermediate types. An ornamental form 'Variegata' has leaves splashed with yellow.

REFERENCES

1. Killip, E.P. 1938. The American species of Passifloraceae. Publication of the foeld museum of National History. *Botanical Series*, **19:** 1-613.
2. Morton, J. 1987. Giant Granadilla. In: Fruits of warm climates. Julia F. Morton, Miami, Fl. p. 328-330.
3. Kumar, P.T., Peter, K.V. and Sajithbabu, D. 2005. Giant Granadilla-a crop for the future. *Indian Horticulture*, **50:** 35-36.

Genus

Asparagus (*Asparagus officinalis* Linn. var. *altilis*)

Introduction

Asparagus (*Asparagus officinalis* Linn. Var. *altilis*, 2n=20) is very important exotic hardy perennial vegetable and known worldwide for its nutritional and medicinal properties. It is known as Halgun, Halyun, Hillua and Shatawar. Asparagus is one of the most delicate, wholesome and appetizing products of the home garden (Thompson and Kelly, 1957). The fleshy green tender spears (shoots) of asparagus are both succulent and tender and have been considered a delicacy since ancient times. Asparagus has an agreeable flavor after it has benn boiled. The edible shoots should be consumed within 24 hours, otherwise they get spoil and lose their quality quickly. Wild asparagus (*Asparagus racemosus*) is a species of asparagus with a long history of use in India and other parts of Asia as a botanical medicine. Many medicinal qualities of wild asparagus have been associated with phytonutrients present in its roots, and especially one type of phytonutrients called saponins. Recent research has shown that the species of asparagus most commonly consumed in the U.S. (*Asparagus officinalis*) also contains saponins, not only in its root portion put also in its shoots. Saponins found in common, everyday asparagus include asparanin A, sarsasapogenin, and protodioscin. Asparagus even contains small amounts of the diosgenin. Profitable crops can be obtained for 15-20 years. Sawada *et al.* (1961) studied the botanical characteristics of asparagus and reported that the longer cutting season the greater are the number of spears and stalks produced. The number of crown buds and roots formed in the current year decreased with increase in the length of the cutting season. The food reserves in the fleshy roots were highest in treatment of no spears removed (Pandita and Bhan, 1984; Pandita *et al.*, 1987). For Greek and Romans it is an important crop, especially for the medicinal properties.

Origin, History and Distribution

Wild asparagus is native to coastal regions (eastern Mediterranean lands) of Europe and Asia minor. Some researches revealed that it is indigenous to temperate to Europe and Asia, where it has been in cultivation over 2000 years. Thus, asparagus is a vegetable much esteemed by Europeans (Shanmugavelu, 1989). Since the time of the ancient Greeks, who first gave the present name, 'asparagus' has been used as vegetable. It was brought to North America by European settlers in the 1700's. It widely distributed to tropical and subtropical region up to an altitude of about 1500 m (Velvan *et al.*, 2007). In Pakistan 14 different types of *Asparagus* has been found (Ali and Khan, 2009). Asaparagus is widely distributed in America, Greece, France, Holland, Germany, Spain, Hungry and England. The USA, Germany, Spain and France have large area under this crop. It is an important commercial crop in America (Kabir, 1993). It is newly emerged crop in India and its demand is gradually increasing. Recently, efforts have been made to grow this crop in temperate places of the country to make it commercialized.

Species Availability, Evolution and Diversity

Asparagus has more than 150 different species (Choudhury, 1967) but some other workers revealed that asparagus (*Asparagus officinalis* Linn.) has at least 1,150 species.

Table 15.3: Generic Species and Status Report/Reference

Generic Species	*Status Report/Reference*
Asparagus Tourn. ex L.,	Sp. Pl.: 313 (1753)-Accepted name
Asparagus abyssinicus Hochst. ex A. Rich.	Tent. Fl. Abyss. 2: 319 (1850)
Asparagus acerosus Thunb. ex Schult. and Schult.f.	In J.J.Roemer and J.A.Schultes Syst. Veg., ed. 15 bis 7: 337 (1829)
Asparagus acerosus Roxb.	Fl. Ind. ed. 1832, 2: 150 (1832), nom. illeg.
Asparagus acicularis	F.T.Wang and S.C.Chen, Acta Phytotax. Sin. 16(1): 93 (1978) Accepted
Asparagus acocksii Jessop	Bothalia 9: 74 (1966) Accepted name
Asparagus aculeatus Voss, Vilm.	Blumengärtn. ed. 3, 1: 1047 (1895)
Asparagus acutifolius L.	Sp. Pl.: 314 (1753) Accepted name
Asparagus acutifolius var. *achhalii* Valdés	Lagascalia 18: 327 (1996)
Asparagus acutifolius var. *ambiguus* (De Not.)	Nyman, Consp. Fl. Eur.: 716 (1882)
Asparagus acutifolius var. *brachyclados* Bornm.	Verh. K. K. Zool.-Bot. Ges. Wien 48: 641 (1898)
Asparagus acutifolius var. *brevifolius* (Tornab.)	Nyman, Consp. Fl. Eur.: 716 (1882)
Asparagus acutifolius var. *commutatus* (Ten.)	Nyman, Consp. Fl. Eur.: 716 (1882)
Asparagus acutifolius var. *eremicus* O.Bolòs and Vigo	Fl. Man. Paisos Catalans 4: 131 (2001)
Asparagus acutifolius var. *gracilis* Baker, J. Linn.	Soc., Bot. 14: 602 (1875)
Asparagus acutifolius var. *hookeri* Baker	J. Linn. Soc., Bot. 14: 602 (1875)
Asparagus acutifolius var. *mitis* P.Palau, Treb.	Inst. Catalana Hist. Nat. 7: 32 (1976 publ. 1977)
Asparagus acutifolius var. *orientalis* Baker	J. Linn. Soc., Bot. 14: 602 (1875)
Asparagus adscendens Roxb.	Fl. Ind. ed. 1832, 2: 153 (1832) Accepted name
Asparagus aethiopicus L.	Mant. Pl.: 1 (1770) Accepted name
Asparagus aethiopicus var. *angusticladus* Jessop	Bothalia 9: 69 (1966)
Asparagus aethiopicus var. *natalensis*	Baker in W. H. Harvey and auct. suc. (eds.), Fl. Cap. 6: 272 (1896)
Asparagus aethiopicus var. *ternifolius* Baker	Refug. Bot. 4: t. 261 (1871)
Asparagus aetnensis Tornab.	Atti Acc. Gioenia Sci. Nat. Catania, ser. 2, 12: 345 (1856)
Asparagus africanus Lam.	Encycl. 1: 295 (1783) Accepted name
Asparagus africanus var. *abyssinicus* (Hochst. ex A.Rich.) Fiori	Boschi Piante Legnose dell'Eritrea: 106 (1910)
Asparagus africanus var. africanus	Accepted name
Asparagus africanus var. biarticulatus De Wild. and T. Durand	Bull. Soc. Roy. Bot. Belgique 40: 31 (1901)
Asparagus africanus var. *concinnus* Baker	J. Linn. Soc., Bot. 14: 619 (1875)

Generic Species	*Status Report/Reference*
Asparagus africanus var. *microcarpus* Balf. f.	Proc. Roy. Soc. Edinburgh 12: 411 (1884)
Asparagus africanus var. *microphyllus*	Baker in W.H.Harvey and auct. suc. (eds.), Fl. Cap. 6: 266 (1896)
Asparagus africanus var. *puberulus* (Baker) Sebsebe	Fl. Trop. E. Afr., Asparagac.: 8 (2006) Accepted name
Asparagus africanus var. *pubescens* Baker	J. Linn. Soc., Bot. 14: 619 (1875)
Asparagus africanus var. *wrightii* Baker	J. Linn. Soc., Bot. 14: 619 (1875)
Asparagus aggregatus (Oberm.) Fellingham and N.L.Mey.	Bothalia 25: 206 (1995) Accepted name
Asparagus agul Forssk. ex Schult. and Schult.f.	In J.J.Roemer and J.A.Schultes, Syst. Veg., ed. 15 bis 7: 337 (1829)
Asparagus albanensis André	Rev. Hort. (Paris) 67: 249 (1895)
Asparagus albus L.	Sp. Pl.: 313 (1753) Accepted name
Asparagus alopecurus (Oberm.) Malcomber and Sebsebe	Kew Bull. 48: 73 (1993) Accepted name
Asparagus altilis (L.) Asch.	Fl. Brandenburg 1: 730 (1864)
Asparagus altilis subsp. *oxycarpus* (Steven) K.Richt.	Pl. Eur. 1: 230 (1890)
Asparagus altilis subsp. *polyphyllus* (Steven ex Ledeb.) K.Richt.	Pl. Eur. 1: 230 (1890)
Asparagus altilis subsp. *prostratus* (Dumort.) K.Richt.	Pl. Eur. 1: 230 (1890)
Asparagus altiscandens Engl. and Gilg	O.Warburg (ed.), Kunene-Sambesi Exped.: 196 (1903) Accepted
Asparagus altissim	Bull. Soc. Bot. France 2: 287 (1855) Accepted name
Asparagus altissimus var. *asperulus* Maire us Munby	Bull. Soc. Hist. Nat. Afrique N. 29: 453 (1938)
Asparagus altissimus var. *foeniculaceus (Lowe) Maire*	Bull. Soc. Hist. Nat. Afrique N. (1931)
Asparagus amarus DC.	Cat. Pl. Horti Monsp.: 81 (1813)
Asparagus ambiguus De Not.	Repert. Fl. Ligust.: 401 (1844)
Asparagus angolensis Baker, Trans. Linn.	Soc. London, Bot. 1: 254 (1878)
Asparagus angulofractus Iljin	In V.L.Komarov (ed.), Fl. URSS 4: 432 (1935) Accepted name
Asparagus angulofractus var. *scabridus Kitag.*	J. Jap. Bot. 13: 434 (1937)
Asparagus angusticladus (Jessop) J.-P.Lebrun and Stork,	Pl. Fleurs Afr. Trop. 3: 60 (1995) Accepted name
Asparagus aphyllus L.	Sp. Pl.: 314 (1753) Accepted name
Asparagus aphyllus subsp. *aphyllus.*	Accepted name
Asparagus aphyllus var. *bovei* Baker	J. Linn. Soc., Bot. 14: 600 (1875)
Asparagus aphyllus subsp. *orientalis* (Baker)	P.H.Davis, Fl. Turkey 8: 77 (1984) Accepted name
Asparagus aphyllus var. *stipularis* (Forssk.) Baker	J. Linn. Soc., Bot. 14: 600 (1875)
Asparagus aphyllus subsp. *stipularis* (Forssk.) K.Richt.	Pl. Eur. 1: 230 (1890)
Asparagus arborescens Willd. ex Schult. and Schult.f. in J.J.Roemer and J.A.Schultes	Syst. Veg., ed. 15 bis 7: 337 (1829) Accepted name
Asparagus aridicola Sebsebe, Fl. Trop. E. Afr.	Asparagac.: 17 (2006) Accepted name
Asparagus asiaticus Wight	Icon. Pl. Ind. Orient. 6: t. 2055 (1853), nom. illeg.
Asparagus asiaticus L.	Sp. Pl.: 313 (1753) Accepted name
Asparagus asiaticus var. *amharicus* Pic.Serm.	Miss. Stud. Lago Tana 7(1): 194 (1951)
Asparagus asiaticus var. *ellenbeckianus* Engl.	Sitzungsber. Königl. Preuss. Akad. Wiss. Berlin 40: 737 (1906), nom. nud.
Asparagus asiaticus var. *mitis* (A.Rich.) Chiov.	Nuovo Giorn. Bot. Ital. 26: 166 (1919)
Asparagus asiaticus var. *pauciflorus* Scott-Elliot	J. Linn. Soc., Bot. 29: 60 (1890)
Asparagus asiaticus var. *scaberulus* (A.Rich.) Engl.	Hochgebirgsfl. Afrika: 169 (1892)
Asparagus asparagoides (L.) Druce	Rep. Bot. Exch. Club Soc. Brit. Isles 3: 414 (1914) Accepted name
Asparagus aspergillus Jessop	Bothalia 9: 71 (1966) Accepted name
Asparagus baumii Engl. and Gilg in O.Warburg (ed.)	A. Kunene-Sambesi Exped.: 196 (1903) Accepted name
Asparagus bayeri (Oberm.) Fellingham and N.L. Mey.	Bothalia 25: 206 (1995) Accepted name
Asparagus bechuanicus Baker in D.Oliver and auct. suc. (eds.)	Fl. Trop. Afr. 7: 429 (1898) Accepted name
Asparagus benguellensis Baker	Trans. Linn. Soc. London, Bot. 1: 253 (1878) Accepted
Asparagus bequaertii De Wild.	Repert. Spec. Nov. Regni Veg. 12: 292 (1913) Accepted
Asparagus biflorus (Oberm.) Fellingham and N.L.Mey.	Bothalia 25: 206 (1995) Accepted name
Asparagus borealis S.C.Chen	Acta Phytotax. Sin. 19: 502 (1981)
Asparagus botschantzevii Vlassova	Bot. Zhurn. (Moscow and Leningrad) 65: 577 (1980) Accepted name
Asparagus botswanicus Sebsebe	Kew Bull. 63: 269 (2008) Accepted name

Generic Species	*Status Report/Reference*
Asparagus brachiatus Thulin	Nordic J. Bot. 22: 431 (2002 publ. 2003) Accepted name
Asparagus brachyphyllus Turcz.	Bull. Soc. Imp. Naturalistes Moscou 13: 78 (1840) Accepted name
Asparagus breslerianus Schult. and Schult.f.	In J.J.Roemer and J.A.Schultes, Syst. Veg., ed. 15 bis 7: 323 (1829) Accepted name
Asparagus brevifolius Tornab.	Atti Acc. Gioenia Sci. Nat. Catania, ser. 2, 12: 359 (1856)
Asparagus brevifolius Boiss. ex Baker	J. Linn. Soc., Bot. 14: 602 (1875), nom. illeg
Asparagus broussonetii J.Jacq.	Ecl. Pl. Rar. 1: 3 (1811)
Asparagus buchananii Baker	Bull. Misc. Inform. Kew 1893: 211 (1893) Accepted name
Asparagus bucharicus Iljin in V.L.Komarov (ed.)	Fl. URSS 4: 746 (1935) Accepted name
Asparagus burchellii Baker	J. Linn. Soc., Bot. 14: 618 (1875) Accepted name
Asparagus burjaticus Peschkova	Novosti Sist. Vyssh. Rast. 11: 86 (1974) Accepted name
Asparagus burkei Baker	J. Linn. Soc., Bot. 14: 607 (1875)
Asparagus buruensis Engl.	Bot. Jahrb. Syst. 45: 155 (1910)
Asparagus buxbaumius Gueldenst. ex Ledeb.	Fl. Ross. 4: 199 (1852)
Asparagus calcicola H.Perrier	Notul. Syst. (Paris) 5: 32 (1935) Accepted name
Asparagus capensis L.	Sp. Pl.: 314 (1753). Accepted name
Asparagus capensis var. *capensis.*	Accepted name
Asparagus capensis var. *litoralis* Suess. and Karling	Mitt. Bot. Staatssamml. München 2: 50 (1950) Accepted name
Asparagus capitatus Baker	J. Linn. Soc., Bot. 14: 607 (1875) Accepted name
Asparagus capitatus subsp. *capitatus.*	Accepted name
Asparagus capitatus subsp. *gracilis* Browicz	Fl. Iranica 165: 175 (1990) Accepted name
Asparagus caspius Schult. and Schult.f.	In J.J.Roemer and J.A.Schultes, Syst. Veg., ed. 15 bis 7: 322 (1829)
Asparagus caspius Hohen.	Enum. Pl. Talysch: 24 (1838), nom. illeg.
Asparagus chesneyi Baker	J. Linn. Soc., Bot. 14: 603 (1875)
Asparagus chimanimanensis Sebsebe	Kew Bull. 63: 271 (2008) Accepted name
Asparagus clareae (Oberm.) Fellingham and N.L.Mey.	Bothalia 25: 206 (1995) Accepted name
Asparagus cochinchinensis (Lour.) Merr.	Philipp. J. Sci. 15: 230 (1919) Accepted name
Asparagus cochinchinensis var. dolichoclados F.T.Wang and Tang	Bull. Fan Mem. Inst. Biol. 7: 291 (1937)
Asparagus cochinchinensis var. longifolius F.T.Wang and Tang	Bull. Fan Mem. Inst. Biol. 7: 291 (1937)
Asparagus cochinchinensis var. pygmaeus (Makino) Ohwi	Fl. Jap., ed. rev.: 1437 (1965)
Asparagus coddii (Oberm.) Fellingham and N.L.Mey.	Bothalia 25: 206 (1995) Accepted name
Asparagus collinus Jan ex Parl.	Fl. Ital. 3: 14 (1858)
Asparagus collinus Schur	Oesterr. Bot. Z. 10: 355 (1860), nom. illeg.
Asparagus commutatus Ten.	Fl. Napol. 3: 374 (1827).
Asparagus comorensis auct.	Wiener Ill. Gart.-Zeitung 1888: 158 (1888)
Asparagus compactus T.M.Salter	J. S. African Bot. 6: 165 (1940)
Asparagus compactus var. *nanus* Pynaert	Nursery Cat. (Éd. Pynaert - Van Geert) 1893-1894: 6 (1893)
Asparagus concinnus (Baker) Kies	Bothalia 6: 178 (1951) Accepted name
Asparagus confertus K.Krause	Bot. Jahrb. Syst. 51: 449 (1914) Accepted name
Asparagus confusus Scott-Elliot ex H.Perrier	Notul. Syst. (Paris) 5: 25 (1935)
Asparagus conglomeratus Baker in D.Oliver and auct. suc. (eds.)	Fl. Trop. Afr. 7: 428 (1898)
Asparagus consanguineus (Kunth) Baker	J. Linn. Soc., Bot. 14: 615 (1875)
Asparagus coodei P.H.Davis	Notes Roy. Bot. Gard. Edinburgh 41: 48 (1983) Accepted name
Asparagus cooperi Baker	Gard. Chron., n.s., 1: 818 (1874)
Asparagus corruda Scop.	Fl. Carniol., ed. 2, 1: 248 (1771)
Asparagus crassicladus Jessop	Bothalia 9: 75 (1966) Accepted name
Asparagus crispus Bojer	Hortus Maurit.: 350 (1837), sensu auct.
Asparagus crispus Lam.	Encycl. 1: 295 (1783)
Asparagus curillus Buch.-Ham. ex Roxb.	Fl. Ind. ed. 1832, 2: 152 (1832) Accepted name.
Asparagus cuscutoides Burch. ex Baker	J. Linn. Soc., Bot. 14: 606 (1875)
Asparagus dauricus Fisch. ex Link	Enum. Hort. Berol. Alt. 1: 348 (1821) Accepted name

Generic Species	Status Report/Reference
Asparagus dauricus var. elongatus Pamp.	Nuovo Giorn. Bot. Ital., n.s., 22(2): 264 (1915)
Asparagus declinatus Schousb.	Iagttag. Vextrig. Marokko: 173 (1800), nom. illeg.
Asparagus declinatus L.	Sp. Pl.: 313 (1753) Accepted name
Asparagus declinatus Sieber ex Baker	J. Linn. Soc., Bot. 14: 621 (1875), nom. inval.
Asparagus decumbens Jacq.	Pl. Hort. Schoenbr. 1: 51 (1797)
Asparagus deflexus Baker	Trans. Linn. Soc. London, Bot. 1: 254 (1878) Accepted name
Asparagus deltae Blatt.	J. Indian Bot. Soc. 6: 70 (1927)
Asparagus densiflorus (Kunth) Jessop	Bothalia 9: 65 (1966)
Asparagus densus Sol. ex Baker	J. Linn. Soc., Bot. 14: 616 (1875)
Asparagus denudatus (Kunth) Baker	J. Linn. Soc., Bot. 14: 606 (1875)
Asparagus denudatus subsp. *denudatus*	-
Asparagus denudatus var. *fragilis* Baker	J. Linn. Soc., Bot. 14: 606 (1875)
Asparagus denudatus subsp. *nudicaulis* (Baker) Sebsebe	Fl. Trop. E. Afr., Asparagac.: 10 (2006)
Asparagus dependens Thunb.	Prodr. Pl. Cap.: 66 (1794)
Asparagus devenishii (Oberm.) Fellingham and N.L.Mey.	Bothalia 25: 206 (1995)
Asparagus dichotomus Brouss. ex Webb and Berthel.	Hist. Nat. Iles Canaries 2(3): 327 (1846)
Asparagus dinteri Engl. and Krause	Bot. Jahrb. Syst. 45: 155 (1910)
Asparagus divaricatus Royle	Ill. Bot. Himal. Mts.: 393 (1840), nom. nud.
Asparagus divaricatus (Oberm.) Fellingham and N.L.Mey.	J. Linn. Soc., Bot. 14: 618 (1875)
Asparagus dolichorhizomatus J.M.Ni and R.N.Zhao	Acta Phytotax. Sin. 31: 378 (1993)
Asparagus draco L.	Sp. Pl. ed. 2: 451 (1762)
Asparagus dregeanus auct.	Gard. Chron., n.s., 8: 598 (1877)
Asparagus drepanophyllus Welw. ex Baker	Trans. Linn. Soc. London, Bot. 1: 254 (1878)
Asparagus drepanophyllus var. warneckei Engl.	Bot. Jahrb. Syst. 32: 97 (1902)
Asparagus dubius Decne.	Nouv. Ann. Mus. Hist. Nat. 3: 363 (1834)
Asparagus duchesnei L. Linden	Semaine Hort. 4: 471 (1900)
Asparagus dumosus Baker	J. Linn. Soc., Bot. 14: 608 (1875)
Asparagus ecklonii Baker	J. Linn. Soc., Bot. 14: 615 (1875)
Asparagus edulis (Oberm.) J.-P.Lebrun and Stork	Énum. Pl. Fleurs Afr. Trop. 3: 61 (1995)
Asparagus elephantinus S.M.Burrows	Bothalia 38: 23 (2008)
Asparagus engleri K.Krause	Bot. Jahrb. Syst. 51: 450 (1914)
Asparagus equisetifolius Webb and Berthel.	Hist. Nat. Iles Canaries 2(3): 328 (1846)
Asparagus equisetoides Welw. ex Baker	Trans. Linn. Soc. London, Bot. 1: 253 (1878)
Asparagus erectus (L.f.) Thunb.	Prodr. Pl. Cap.: 65 (1794)
Asparagus erinaceus Borzí and Mattei	Boll. Reale Orto Bot. Palermo 11: 238 (1912)
Asparagus esculentus Salisb.	Prodr. Stirp. Chap. Allerton: 252 (1796)
Asparagus euphraticus Kunth	Enum. Pl. 5: 73 (1850)
Asparagus exaltatus C.Sm. and Buch ex Webb and Berthel.	Hist. Nat. Iles Canaries 2(3): 326 (1846)
Asparagus exsertus (Oberm.) Fellingham and N.L.Mey.	Bothalia 25: 206 (1995)
Asparagus exuvialis Burch.	Trav. S. Africa 1: 432 (1822)
Asparagus exuvialis f. ecklonii (Baker) Fellingham and N.L.Mey.	Bothalia 25: 207 (1995)
Asparagus falcatus Thunb.	Fl. Jap.: 139 (1784), sensu auct.
Asparagus falcatus L.	Sp. Pl.: 313 (1753)
Asparagus falcatus var. *pubescens* Chiov.	Webbia 8: 9 (1951)
Asparagus falcatus var. *ternifolius* (Baker) Jessop	Bothalia 9: 70 (1966)
Asparagus fallax Svent.	Addit. Fl. Canar. 1: 4 (1960)
Asparagus fasciculatus R.Br.	Prodr. Fl. Nov. Holl.: 281 (1810), nom. illeg.
Asparagus fasciculatus Thunb.	Fl. Cap., ed. 1a, 2: 329 (1820)
Asparagus faulkneri Sebsebe	Fl. Trop. E. Afr., Asparagac.: 17 (2006)
Asparagus ferganensis Vved.	Fl. Uzbekistan. 1: 544 (1941)
Asparagus filicinus Buch.-Ham. ex D.Don	Prodr. Fl. Nepal.: 49 (1825)

Generic Species	Status Report/Reference
Asparagus filicinus var. *brevifolius* Baker	J. Linn. Soc., Bot. 14: 605 (1875)
Asparagus filicinus var. *brevipes* Baker	J. Linn. Soc., Bot. 14: 605 (1875)
Asparagus filicinus var. *giraldii* C.H.Wright	Gard. Chron., ser. 3, 44: 122 (1908)
Asparagus filicinus var. *lycopodineus* Baker	J. Linn. Soc., Bot. 14: 605 (1875)
Asparagus filicinus var. *megaphyllus* F.T.Wang and Tang	Bull. Fan Mem. Inst. Biol. 7: 290 (1937)
Asparagus filicinus var. *microclada* Hook.f.	Fl. Brit. India 6: 315 (1892)
Asparagus filicladus (Oberm.) Fellingham and N.L.Mey.	Bothalia 25: 207 (1995)
Asparagus filifolius Bertol.	Novi Comment. Acad. Sci. Inst. Bononiensis 5: 430 (1842)
Asparagus fiori Sennen	Mem. Real Acad. Ci. Barcelona, ser. 3, 20(14): 9 (1928)
Asparagus flagellaris (Kunth) Baker	J. Linn. Soc., Bot. 14: 614 (1875)
Asparagus flavicaulis (Oberm.) Fellingham and N.L.Mey.	Bothalia 25: 207 (1995)
Asparagus flavicaulis subsp. flavicaulis	-
Asparagus flavicaulis subsp. setulosus (Oberm.) Fellingham and N.L.Mey.	Bothalia 25: 207 (1995)
Asparagus fleckii Schinz	Bull. Herb. Boissier 4(App. 3): 43 (1896)
Asparagus flexuosus Thunb.	Prodr. Pl. Cap.: 66 (1794)
Asparagus flexuosus (Ledeb.) Krylov	Sist. Zametki Mater. Gerb. Krylova Tomsk. Gosud. Univ. Kuybysheva 9: 1 (1928), nom. illeg.
Asparagus floribundus Royle	Ill. Bot. Himal. Mts.: 393 (1840), nom. inval.
Asparagus foeniculaceus Lowe	J. Proc. Linn. Soc., Bot. 5: 44 (1864)
Asparagus fouriei (Oberm.) Fellingham and N.L.*Mey.*	Bothalia 25: 207 (1995)
Asparagus fractiflexus (Oberm.) Fellingham and N.L.Mey.	Bothalia 25: 207 (1995)
Asparagus fragilis Burch. ex Baker	J. Linn. Soc., Bot. 14: 606 (1875), pro syn.
Asparagus francisci K.Krause	Bot. Jahrb. Syst. 51: 449 (1914)
Asparagus fysonii J.F.Macbr.	Contr. Gray Herb. 56: 17 (1918)
Asparagus gaudichaudianus Kunth	Enum. Pl. 5: 71 (1850)
Asparagus gharoensis Blatt.	J. Indian Bot. Soc. 6: 70 (1927)
Asparagus gibbus Bunge	Mém. Acad. Imp. Sci. St.-Pétersbourg Divers Savans 2: 139 (1833)
Asparagus gillettii Chiov. ex J.Gillet	Bull. Misc. Inform. Kew 1941: 183 (1941)
Asparagus glaucus Kies	Bothalia 6: 229 (1951)
Asparagus glycycarpus Kunth	Enum. Pl. 5: 61 (1850)
Asparagus gobicus N.A.Ivanova ex Grubov	Bot. Mater. Gerb. Bot. Inst. Komarova Akad. Nauk S.S.S.R. 17: 9 (1955)
Asparagus gonoclados Baker	J. Linn. Soc., Bot. 14: 627 (1875)
Asparagus gourmacus A.Chev. ex Hutch. and Dalziel	Fl. W. Trop. Afr. 2: 352 (1936), nom. inval.
Asparagus gracilis Salisb.	Prodr. Stirp. Chap. Allerton: 252 (1796)
Asparagus gracilis Royle ex Baker	J. Linn. Soc., Bot. 14: 607 (1875), nom. illeg.
Asparagus graminifolius L.	Sp. Pl. ed. 2: 450 (1762)
Asparagus grandiflorus Willd. ex Webb and Berthel.	Hist. Nat. Iles Canaries 2(3): 327 (1846)
Asparagus graniticus (Oberm.) Fellingham and N.L.Mey.	Bothalia 25: 207 (1995)
Asparagus greveanus H.Perrier	Notul. Syst. (Paris) 5: 22 (1935)
Asparagus griffithii Baker	J. Linn. Soc., Bot. 14: 604 (1875)
Asparagus gypsaceus Vved.	Bot. Mater. Gerb. Bot. Inst. Komarova Akad. Nauk S.S.S.R. 9: 246 (1946)
Asparagus hajrae Kamble	J. Econ. Taxon. Bot. 20: 688 (1996)
Asparagus hedecarpus Andrews ex Baker	J. Linn. Soc., Bot. 14: 599 (1875)
Asparagus hereroensis Schinz	Bull. Herb. Boissier 4(App. 3): 43 (1896)
Asparagus hirsutus S.M.Burrows	Bothalia 38: 25 (2008)
Asparagus homblei De Wild.	Repert. Spec. Nov. Regni Veg. 12: 292 (1913)
Asparagus horridus L.f.	Suppl. Pl.: 203 (1782), nom. illeg.
Asparagus horridus L.	Syst. Veg. ed. 13: 274 (1774)
Asparagus hortensis Mill. ex Baker	J. Linn. Soc., Bot. 14: 598 (1875)
Asparagus humilis Engl.	Bot. Jahrb. Syst. 45: 154 (1910)

Generic Species	*Status Report/Reference*
Asparagus inarimensis Tornab.	Atti Acc. Gioenia Sci. Nat. Catania, ser. 2, 12: 368 (1856)
Asparagus inderiensis Blume ex Ledeb.	Fl. Altaic. 2: 44 (1830)
Asparagus insularis Hance	Ann. Sci. Nat., Bot., sér. 5, 5: 245 (1866)
Asparagus intangibilis Dinter	Repert. Spec. Nov. Regni Veg. 29: 269 (1931)
Asparagus intricatus (Oberm.) Fellingham and N.L.Mey.	Bothalia 25: 207 (1995)
Asparagus irregularis Baker	J. Linn. Soc., Bot. 14: 620 (1875)
Asparagus jacquemontii Baker	J. Linn. Soc., Bot. 14: 615 (1875)
Asparagus judtii Schinz	Bull. Herb. Boissier 4(App. 3): 44 (1896)
Asparagus juniperoides Engl.	Bot. Jahrb. Syst. 10: 3 (1888)
Asparagus kaessneri De Wild.	Repert. Spec. Nov. Regni Veg. 12: 293 (1913)
Asparagus kansuensis F.T.Wang and Tang ex S.C.Chen	Acta Phytotax. Sin. 16(1): 94 (1978)
Asparagus karthikeyanii (Kamble) M.R.Almeida	Fl. Maharashtra 5A: 166 (2009)
Asparagus kasakstanicus Iljin in V.L.Komarov (ed.)	Fl. URSS 4: 747 (1935)
Asparagus katangensis De Wild. and T.Durand	Bull. Soc. Roy. Bot. Belgique 40: 32 (1901)
Asparagus khorasanensis Hamdi and Assadi	Feddes Repert. 120: 421 (2009)
Asparagus kiusianus Makino	Bot. Mag. (Tokyo) 21: 161 (1970)
Asparagus klinghardtianus Dinter	Repert. Spec. Nov. Regni Veg. 29: 270 (1931)
Asparagus komarovianus Vved.	Bot. Mater. Gerb. Bot. Inst. Komarova Akad. Nauk S.S.S.R. 9: 246 (1946)
Asparagus krausii Baker	J. Linn. Soc., Bot. 14: 628 (1875), nom. superfl.
Asparagus kraussianus (Kunth) J.F.Macbr.	Contr. Gray Herb. 56: 17 (1918)
Asparagus krebsianus (Kunth) Jessop	Bothalia 9: 73 (1966)
Asparagus kuisibensis Dinter	Repert. Spec. Nov. Regni Veg. 29: 270 (1931)
Asparagus laetus Salisb.	Prodr. Stirp. Chap. Allerton: 253 (1796)
Asparagus laevissimus Steud. ex Baker	J. Linn. Soc., Bot. 14: 623 (1875)
Asparagus lanceus Thunb.	Prodr. Pl. Cap.: 66 (1794)
Asparagus laricinus Burch.	Trav. S. Africa 1: 537 (1822)
Asparagus laricinus var. katangensis De Wild. and T.Durand	Bull. Soc. Roy. Bot. Belgique 40: 33 (1901)
Asparagus lecardii De Wild.	Ann. Mus. Congo Belge, Bot., sér. 5, 1: 17 (1904)
Asparagus ledebourii Miscz.	Vestn. Tiflissk. Bot. Sada 12: 30 (1916)
Asparagus leptocladodius Chiov.	Atti Reale Accad. Italia, Mem. Cl. Sci. Fis. 11: 58 (1940)
Asparagus leptophyllus Schischk.	Izv. Tomsk. Gosud. Univ. 81: 434 (1928)
Asparagus levinae Klokov	Trudy Sil's'ko-Gosp. Bot. 1(3): 171 (1927)
Asparagus lignosus Burm.f.	Fl. Indica, Prodr. Fl. Cap.: 10 (1768)
Asparagus littoralis Steven	Bull. Soc. Imp. Naturalistes Moscou 30(2): 92 (1857)
Asparagus longicladus N.E.Br.	Bull. Misc. Inform. Kew 1921: 298 (1921)
Asparagus longiflorus Franch.	Nouv. Arch. Mus. Hist. Nat., sér. 2, 7: 110 (1884)
Asparagus longipes Baker	Bull. Misc. Inform. Kew 1901: 134 (1901)
Asparagus lowei Kunth	Enum. Pl. 5: 73 (1850)
Asparagus lownei Baker	J. Linn. Soc., Bot. 14: 601 (1875)
Asparagus lucidus Lindl.	Edwards's Bot. Reg. 30(Misc.): 29 (1844)
Asparagus lucidus var. dolichocladus Merr. and Rolfe	Philipp. J. Sci., C 3: 96 (1908)
Asparagus lucidus var. pygmaeus Makino	Bot. Mag. (Tokyo) 11: 281 (1897)
Asparagus lugardii Baker in D.Oliver and auct. suc. (eds.)	Fl. Trop. Afr. 7: 431 (1898)
Asparagus lujae De Wild.	Pl. Nov. Horti Then.: t. 34 (1905)
Asparagus lycaonicus P.H.Davis	Notes Roy. Bot. Gard. Edinburgh 41: 48 (1983)
Asparagus lycicus P.H.Davis	Notes Roy. Bot. Gard. Edinburgh 41: 49 (1983)
Asparagus lycopodineus (Baker) F.T.Wang and Tang	Bull. Fan Mem. Inst. Biol. 7: 291 (1937)
Asparagus lycopodineus var. sessilis F.T.Wang and Tang	Bull. Fan Mem. Inst. Biol. 7: 291 (1937)
Asparagus lynetteae (Oberm.) Fellingham and N.L.Mey.	Bothalia 25: 207 (1995)
Asparagus macowanii Baker	J. Linn. Soc., Bot. 14: 609 (1875)
Asparagus macowanii var. *zuluensis* (N.E.Br.) Jessop	Bothalia 9: 58 (1966)

Generic Species	Status Report/Reference
Asparagus macrorrhizus Pedrol, J.J.Regalado and López Encina	Fl. Iber. 20: 589 (2013)
Asparagus madagascariensis Baker	J. Linn. Soc., Bot. 16: 206 (1878)
Asparagus madecassus H.Perrier	Notul. Syst. (Paris) 5: 24 (1935)
Asparagus madecassus var. *madecassus.*	-
Asparagus madecassus var. *montanus* H.Perrier	Notul. Syst. (Paris) 5: 25 (1935)
Asparagus mahafalensis H.Perrier	Notul. Syst. (Paris) 5: 29 (1936)
Asparagus mairei H.Lév.	Repert. Spec. Nov. Regni Veg. 7: 339 (1909)
Asparagus mairei H.Lév.	Repert. Spec. Nov. Regni Veg. 11: 493 (1913), nom. illeg.
Asparagus mariae (Oberm.) Fellingham and N.L.Mey.	Bothalia 25: 207 (1995)
Asparagus marinus Rchb.	Icon. Fl. Germ. Helv. 10: t. 520 (1848)
Asparagus maritimus Pall.	Reise Russ. Reich. 2: 329 (1773), nom. illeg.
Asparagus maritimus (L.) Mill.	Gard. Dict. ed. 8: n.° 2 (1768)
Asparagus maritimus var. *breslerianus* (Schult. and Schult.f.) Boiss.	Fl. Orient. 5: 337 (1881)
Asparagus maritimus var. *laevis* Regel	Bull. Soc. Imp. Naturalistes Moscou 41(2): 270 (1869)
Asparagus mauritianus Lam. ex Desf.	Tabl. École Bot.: 20 (1804)
Asparagus maximus Voss	Vilm. Blumengärtn. ed. 3, 1: 1047 (1895)
Asparagus medeoloides (L.f.) Thunb.	Prodr. Pl. Cap.: 66 (1794)
Asparagus medeoloides var. *angustifolius* (Mill.) Baker in W.H.Harvey and auct. suc. (eds.)	Fl. Cap. 6: 273 (1896)
Asparagus medeoloides var. *falciformis* (Kunth) Baker in W.H.Harvey and auct. suc. (eds.)	Fl. Cap. 6: 273 (1896)
Asparagus meioclados H.Lév.	Repert. Spec. Nov. Regni Veg. 8: 59 (1910).
Asparagus meioclados var. trichoclados F.T.Wang and Tang	Bull. Fan Mem. Inst. Biol. 7: 290 (1937)
Asparagus merkeri K.Krause	Bot. Jahrb. Syst. 57: 238 (1921)
Asparagus micranthus Lindl.	Sketch Veg. Swan R.: 58 (1839)
Asparagus micranthus Siebold and Zucc. ex Baker	J. Linn. Soc., Bot. 14: 604 (1875)
Asparagus microphyllus (Baker) Kies	Bothalia 6: 178 (1951)
Asparagus microraphis (Kunth) Baker	J. Linn. Soc., Bot. 14: 612 (1875)
Asparagus migeodii Sebsebe	Fl. Trop. E. Afr., Asparagac.: 9 (2006)
Asparagus minutiflorus (Kunth) Baker	J. Linn. Soc., Bot. 14: 616 (1875)
Asparagus misczenkoi Iljin in V.L.Komarov (ed.)	Fl. URSS 4: 746 (1935)
Asparagus mitis A.Rich.	Tent. Fl. Abyss. 2: 319 (1850)
Asparagus mollis (Oberm.) Fellingham and N.L.Mey.	Bothalia 25: 207 (1995)
Asparagus monoclados Vved.	Bot. Mater. Gerb. Bot. Inst. Komarova Akad. Nauk S.S.S.R. 9: 247 (1946)
Asparagus monophyllus Baker	J. Linn. Soc., Bot. 14: 604 (1875)
Asparagus mozambicus Kunth	Enum. Pl. 5: 72 (1850)
Asparagus mucronatus Jessop	Bothalia 9: 56 (1966)
Asparagus multiflorus Baker	J. Linn. Soc., Bot. 14: 610 (1875)
Asparagus multituberosus R.A.Dyer	Bothalia 6: 442 (1954)
Asparagus munitus F.T.Wang and S.C.Chen	Acta Phytotax. Sin. 16(1): 91 (1978)
Asparagus myriacanthus F.T.Wang and S.C.Chen	Acta Phytotax. Sin. 16(1): 92 (1978)
Asparagus myriocladus Baker	J. Bot. 27: 43 (1889)
Asparagus namaensis Schinz	Bull. Herb. Boissier 4(App. 3): 44 (1896)
Asparagus natalensis (Baker) J.-P.Lebrun and Stork	Énum. Pl. Fleurs Afr. Trop. 3: 61 (1995)
Asparagus neglectus Kar. and Kir.	Bull. Soc. Imp. Naturalistes Moscou 14: 750 (1841)
Asparagus neglectus var. *dentatus* Vlassova	Bot. Zhurn. (Moscow and Leningrad) 65: 578 (1980)
Asparagus nelsii Schinz	Bull. Herb. Boissier 4(App. 3): 44 (1896)
Asparagus nelsonii Baker	J. Linn. Soc., Bot. 14: 617 (1875)
Asparagus nepalensis Baker	J. Linn. Soc., Bot. 14: 622 (1875)
Asparagus nesiotes Svent.	Index Seminum (Agron. Investig. Nat. Hispan. Inst.) 1968: 56 (1969)
Asparagus nesiotes subsp. *nesiotes*	-

Generic Species	*Status Report/Reference*
Asparagus nesiotes subsp. purpureiensis Marrero Rodr. and A.Ramos	Bot. Macaronés., IV, Ci. 18: 20 (1989)
Asparagus nitidus J.R.Forst. ex Baker in W.H.Harvey and auct. suc. (eds.)	Fl. Cap. 6: 263 (1896)
Asparagus nivenianus Schult. and Schult.f. in J.J.Roemer and J.A.Schultes	Syst. Veg., ed. 15 bis 7: 331 (1829)
Asparagus nodosus Sol. ex Baker	J. Linn. Soc., Bot. 14: 608 (1875)
Asparagus nodulosus (Oberm.) J.-P.Lebrun and Stork	Énum. Pl. Fleurs Afr. Trop. 3: 61 (1995)
Asparagus nudicaulis Baker in D.Oliver and auct. suc. (eds.)	Fl. Trop. Afr. 7: 428 (1898)
Asparagus obermeyerae Jessop	Bothalia 9: 73 (1966)
Asparagus officinalis L.	Sp. Pl.: 313 (1753)
Asparagus officinalis var. *altilis* L.	Sp. Pl.: 313 (1753)
Asparagus officinalis var. *campestris* Gren. and Godr.	Consp. Fl. Eur.: 716 (1882)
Asparagus officinalis var. *collinus* Nyman	Fl. France 3: 231 (1855)
Asparagus officinalis var. *decorticatus* Ledeb.	Fl. Altaic. 2: 43 (1841)
Asparagus officinalis var. *flexuosus* Ledeb.	Fl. Altaic. 2: 44 (1841)
Asparagus officinalis var. *gracilis* Ledeb.	Fl. Altaic. 2: 43 (1841)
Asparagus officinalis var. *maritimus* L.	Sp. Pl.: 313 (1753)
Asparagus officinalis var. *oxycarpus* (Steven) Nyman, Consp. Fl. Eur.	Suppl. 2: 299 (1890)
Asparagus officinalis var. *pallidus* Alef.	Landw. Fl.: 295 (1866)
Asparagus officinalis subsp. *polyphyllus* (Steven ex Ledeb.) Nyman	Consp. Fl. Eur.: 716 (1882)
Asparagus officinalis var. *prostratus* (Dumort.) Nyman	Consp. Fl. Eur.: 716 (1882)
Asparagus officinalis subsp. *prostratus* (Dumort.) Corb.	Nouv. Fl. Normandie: 568 (1894)
Asparagus officinalis var. *ruficeps* Alef.	Landw. Fl.: 295 (1866)
Asparagus officinalis var. *silvestris* Alef.	Landw. Fl.: 295 (1866)
Asparagus officinalis var. *strictus* Boiss.	Fl. Orient. 5: 335 (1884)
Asparagus officinalis var. *viridis* Alef.	Landw. Fl.: 295 (1866)
Asparagus oligoclonos Maxim.	Mém. Acad. Imp. Sci. St.-Pétersbourg Divers Savans 9: 286 (1859)
Asparagus oligoclonos var. purpurascens X.J.Xue and H.Yao	Bull. Bot. Res., Harbin 14: 242 (1994)
Asparagus oligophyllus Baker	J. Linn. Soc., Bot. 14: 604 (1875)
Asparagus oliveri (Oberm.) Fellingham and N.L.Mey.	Bothalia 25: 208 (1995)
Asparagus omahekensis K.Krause	Bot. Jahrb. Syst. 51: 447 (1914)
Asparagus orientalis Besser	Cat. Hort. Cremeneci: s.p. (1814)
Asparagus ovatus T.M.Salter	J. S. African Bot. 6: 166 (1940)
Asparagus oxyacanthus Baker	J. Linn. Soc., Bot. 14: 625 (1875)
Asparagus oxycarpus Steven	Bull. Soc. Imp. Naturalistes Moscou 30(2): 92 (1857)
Asparagus pachyrrhizus N.A.Ivanova ex N.V.Vlassova	Bot. Zhurn. (Moscow and Leningrad) 65: 577 (1980)
Asparagus palaestinus Baker, J. Linn. Soc.	Bot. 14: 602 (1875)
Asparagus pallasii Miscz.	Vestn. Tiflissk. Bot. Sada 12: 27 (1916)
Asparagus pallidus Salisb.	Prodr. Stirp. Chap. Allerton: 252 (1796)
Asparagus paragus Gueldenst. ex Ledeb.	Fl. Ross. 4: 197 (1852)
Asparagus parviflorus Turcz.	Bull. Soc. Imp. Naturalistes Moscou 27(2): 130 (1854)
Asparagus pastorianus Webb and Berthel.	Hist. Nat. Iles Canaries 2(3): 329 (1846)
Asparagus pastorianus f. longifolius G.Kunkel	Vieraea 8: 339 (1978 publ. 1980)
Asparagus patens K.Krause	Bot. Jahrb. Syst. 51: 448 (1914)
Asparagus pauli-guilelmi Solms in G.Schweinfurth	Be itr. Fl. Aethiop.: 203 (1867)
Asparagus pauli-guilelmi var. *daltonii* Baker	J. Linn. Soc., Bot. 14: 613 (1875)
Asparagus pauli-guilelmi var. katangensis De Wild. and T.Durand	Bull. Soc. Roy. Bot. Belgique 40: 33 (1901)
Asparagus pearsonii Kies	Bothalia 6: 179 (1951)
Asparagus pectinatus Redouté	Liliac. 7: t. 407 (1813)
Asparagus penduliflorus Zipp. ex Span.	Linnaea 15: 476 (1841)

Generic Species	Status Report/Reference
Asparagus pendulus (Oberm.) J.-P.Lebrun and Stork	Énum. Pl. Fleurs Afr. Trop. 3: 61 (1995)
Asparagus penicillatus H.Hara	J. Jap. Bot. 49: 134 (1974)
Asparagus persicus Baker, J. Linn. Soc.	Bot. 14: 603 (1875)
Asparagus petersianus Kunth	Enum. Pl. 5: 72 (1850)
Asparagus petitianus A.Rich.	Tent. Fl. Abyss. 2: 320 (1850)
Asparagus phyllacanthus Lam.	Encycl. 1: 296 (1783)
Asparagus pilosus Baker	J. Linn. Soc., Bot. 14: 610 (1875)
Asparagus planiuscula Burm.f.	Fl. Indica, Prodr. Fl. Cap.: 10 (1768)
Asparagus plocamoides Webb ex Steud.	Nomencl. Bot., ed. 2, 1: 150 (1840), nom. nud.
Asparagus plocamoides Webb ex Svent.	Index Seminum (Agron. Investig. Nat. Hispan. Inst.) 1968: 43 (1969)
Asparagus plumosus Baker	J. Linn. Soc., Bot. 14: 613 (1875)
Asparagus plumosus nanus Van Geert	Nursery Cat. (Auguste Van Geert) 83: 4 (1882)
Asparagus plumosus var. *tenuissimus* (Van Geert) L.H.Bailey	Cycl. Amer. Hort.: 106 (1906)
Asparagus poissonii H.Perrier	Notul. Syst. (Paris) 5: 21 (1935)
Asparagus polyphyllus Steven ex Ledeb.	Fl. Ross. 4: 198 (1852)
Asparagus popovii Iljin in V.L.Komarov (ed.)	Fl. URSS 4: 745 (1935)
Asparagus prostratus Dumort.	Fl. Belg.: 138 (1827)
Asparagus przewalskyi N.A.Ivanova ex Grubov and T.V.Egorova in V.I.Grubov and T.V.Egorova (eds.)	Rast. Tsentral. Azii 7: 81 (1977)
Asparagus pseudoscaber Grecescu	Consp. Fl. Roman.: 556 (1898)
Asparagus psilurus Welw. ex Baker	Trans. Linn. Soc. London, Bot. 1: 253 (1878)
Asparagus puberulus Baker	J. Linn. Soc., Bot. 14: 618 (1875)
Asparagus pubescens Baker	Trans. Linn. Soc. London, Bot. 1: 254 (1878)
Asparagus punjabensis J.L.Stewart	Punjab Pl.: 233 (1869)
Asparagus purpurascens M.Bieb. ex Baker	J. Linn. Soc., Bot. 14: 602 (1875)
Asparagus pygmaeus Makino	Bot. Mag. (Tokyo) 24: 73 (1910).
Asparagus qinghaiensis Y.Wan	Guihaia 11: 289 (1991)
Asparagus racemosus Willd.	Sp. Pl. 2: 152 (1799)
Asparagus racemosus var. *javanicus* (Kunth) Baker	J. Linn. Soc., Bot. 14: 624 (1875)
Asparagus racemosus var. *longicladodius* Chiov.	Malpighia 34: 530 (1937)
Asparagus racemosus var. *ruspolii* Engl.	Annuario Reale Ist. Bot. Roma 9: 245 (1902)
Asparagus racemosus var. *subacerosus* Baker	J. Linn. Soc., Bot. 14: 624 (1875)
Asparagus racemosus var. *tetragonus* (Bresler) Baker	J. Linn. Soc., Bot. 14: 624 (1875)
Asparagus racemosus var. *zeylanicus* Baker	J. Linn. Soc., Bot. 14: 623 (1875)
Asparagus radiatus Sebsebe	Kew Bull. 63: 275 (2008)
Asparagus ramosissimus Baker	Gard. Chron., n.s., 2: 6 (1874)
Asparagus recurvispinus (Oberm.) Fellingham and N.L.Mey.	Bothalia 25: 208 (1995)
Asparagus retrofractus C.Sm. ex Webb and Berthel.	Hist. Nat. Iles Canaries 2(3): 328 (1846), nom. illeg.
Asparagus retrofractus Forssk.	Fl. Aegypt.-Arab.: 73 (1775), sensu auct.
Asparagus retrofractus L.	Sp. Pl.: 313 (1753)
Asparagus richardsiae Sebsebe	Kew Bull. 63: 271 (2008)
Asparagus rigidulus Nakai	Bot. Mag. (Tokyo) 27: 213 (1913)
Asparagus rigidus Jessop	Bothalia 9: 64 (1966)
Asparagus ritschardii De Wild.	Contr. Fl. Katanga, Suppl. 3: 98 (1930)
Asparagus rivalis Burch. ex Schult. and Schult.f. in J.J.Roemer and J.A.Schultes	Syst. Veg., ed. 15 bis 7: 337 (1829)
Asparagus rivalis Burch. ex Kies	Bothalia 6: 179 (1951), nom. illeg.
Asparagus rogersii R.E.Fr.	Wiss. Erg. Schwed. Rhodesia-Kongo-Exp. 1911-1912, 1: 230 (1916)
Asparagus rottleri Baker	J. Linn. Soc., Bot. 14: 611 (1875)
Asparagus royleanus P.Daniel	Indian J. Forest. 4: 334 (1981 publ. 1982)
Asparagus ruber Burm.f.	Fl. Indica, Prodr. Fl. Cap.: 10 (1768)

Generic Species	Status Report/Reference
Asparagus rubicundus P.J.Bergius	Descr. Pl. Cap.: 88 (1767)
Asparagus rubricaulis (Kunth) Baker	J. Linn. Soc., Bot. 14: 616 (1875)
Asparagus rusciformis Brongn. ex H.Perrier	Notul. Syst. (Paris) 5: 30 (1935)
Asparagus sapinii De Wild	Compagnie du Kasaï: 367 (1910)
Asparagus sarmaticus Grint.	Bul. Grad. Bot. Univ. Cluj 15: 145 (1935 publ. 1936)
Asparagus sarmentosus L.	Sp. Pl.: 314 (1753)
Asparagus sarmentosus var. *comatus* Baker	J. Linn. Soc., Bot. 14: 625 (1875)
Asparagus sarmentosus var. *densiflorus* (Kunth) Baker	J. Linn. Soc., Bot. 14: 625 (1875)
Asparagus sarmentosus var. *kunthii* Baker	J. Linn. Soc., Bot. 14: 625 (1875)
Asparagus satawur James A.Murray	Drugs Pl. Sind: 24 (1881)
Asparagus sativus Mill.	Gard. Dict. Abr. ed. 4: s.p. (1754)
Asparagus saundersiae Baker	J. Bot. 27: 42 (1889)
Asparagus scaber Brign.	Fasc. Rar. Pl.: 22 (1810)
Asparagus scaber Lowe	Trans. Cambridge Philos. Soc. 4(1): 11 (1833), nom. illeg.
Asparagus scaber var. *littoralis* (Steven) Nyman	Consp. Fl. Eur.: 716 (1882)
Asparagus scaberulus A.Rich.	Tent. Fl. Abyss. 2: 320 (1850)
Asparagus scandens Thunb.	Prodr. Pl. Cap.: 66 (1794)
Asparagus scandens Gueldenst. ex Ledeb.	Fl. Ross. 4: 199 (1852), nom. illeg.
Asparagus scandens var. *deflexus* Baker	J. Linn. Soc., Bot. 14: 622 (1875)
Asparagus schoberioides Kunth	Enum. Pl. 5: 70 (1850)
Asparagus schoberioides var. *subsetaceus* Franch.	Nouv. Arch. Mus. Hist. Nat., sér. 2, 7: 112 (1884)
Asparagus schroederi Engl.	Bot. Jahrb. Syst. 32: 97 (1902)
Asparagus schumanianus Schltr. ex H.Perrier	Notul. Syst. (Paris) 5: 23 (1935)
Asparagus schweinfurthii Baker	J. Linn. Soc., Bot. 14: 616 (1875)
Asparagus scoparius Lowe	Trans. Cambridge Philos. Soc. 4(1): 11 (1833)
Asparagus scoparius var. *plocamoides* Bolle ex Pit. and Proust	Iles Canaries: 356 (1908), nom. nud.
Asparagus sekukuniensis (Oberm.) Fellingham and N.L.Mey.	Bothalia 25: 208 (1995)
Asparagus sennii Chiov.	Fl. Somala 2: 418 (1932)
Asparagus sessiliflorus Oett.	Trudy Bot. Sada Imp. Yur'evsk. Univ. 6: 80 (1905)
Asparagus setaceus (Kunth) Jessop	Bothalia 9: 51 (1966)
Asparagus setiformis Krylov	Sist. Zametki Mater. Gerb. Krylova Tomsk. Gosud. Univ. Kuybysheva 9: 2 (1928)
Asparagus shirensis Baker in D.Oliver and auct. suc. (eds.)	Fl. Trop. Afr. 7: 427 (1898)
Asparagus sichuanicus S.C.Chen and D.Q.Liu	Acta Phytotax. Sin. 22: 418 (1984)
Asparagus sidamensis Cufod.	Senckenberg. Biol. 50: 245 (1969)
Asparagus sieboldii Maxim.	Mém. Acad. Imp. Sci. St.-Pétersbourg Divers Savans 9: 287 (1859)
Asparagus simulans Baker, J. Linn. Soc.	Bot. 14: 609 (1875)
Asparagus sinensis Nois. ex Baker	J. Linn. Soc., Bot. 14: 607 (1875)
Asparagus sinicus (Miq.) C.H.Wright	J. Linn. Soc., Bot. 36: 103 (1903)
Asparagus somalensis Chiov.	Result. Sci. Miss. Stefan.-Paoli Somal. Ital. 1: 170 (1916)
Asparagus soongaricus Iljin in V.L.Komarov (ed.)	Fl. URSS 4: 432 (1935)
Asparagus spinescens Steud. ex Schult. and Schult.f. in J.J.Roemer and J.A.Schultes	Syst. Veg., ed. 15 bis 7: 333 (1829)
Asparagus spinescens Steud. ex Schult. and Schult.f. in J.J.Roemer and J.A.Schultes	Syst. Veg., ed. 15 bis 7: 334 (1829), nom. illeg.
Asparagus spinosissimus Kuntze	Revis. Gen. Pl. 3(2): 315 (1898)
Asparagus spinosissimus Dinter	Deut. Südw. Afrik.: 47 (1909), nom. illeg.
Asparagus sprengeri Regel	Trudy Imp. S.-Peterburgsk. Bot. Sada 11: 302 (1890)
Asparagus squarrosus J.A.Schmidt	Beitr. Fl. Cap Verd. Ins.: 165 (1852)
Asparagus stachyoides Spreng. ex Baker	J. Linn. Soc., Bot. 14: 624 (1875)
Asparagus stachyphyllus H.Lév. and Vaniot	Repert. Spec. Nov. Regni Veg. 5: 282 (1908)
Asparagus stellatus Baker	J. Linn. Soc., Bot. 14: 612 (1875)

Generic Species	Status Report/Reference
Asparagus steudneri Schweinf. ex Engl.	Abh. Königl. Akad. Wiss. Berlin 1891: 169 (1892)
Asparagus stipulaceus Lam.	Encycl. 1: 296 (1783)
Asparagus stipulaceus var. *spinescens* (Kunth) Baker	J. Linn. Soc., Bot. 14: 618 (1875)
Asparagus stipularis Forssk.	Fl. Aegypt.-Arab.: 72 (1775)
Asparagus stipularis var. *brachyclados* Boiss.	Fl. Orient. 5: 338 (1884)
Asparagus stipularis var. *horridus* Maire and Weiller	Fl. Afrique N. 5: 220 (1958)
Asparagus stipularis var. *tenuispinus* Holmboe	Stud. Veg. Cyprus: 52 (1914)
Asparagus straussii Hausskn. ex Bornm.	Beih. Bot. Centralbl. 24(2): 111 (1908)
Asparagus striatus (L.f.) Thunb.	Prodr. Pl. Cap.: 65 (1794)
Asparagus striatus De Wild.	Repert. Spec. Nov. Regni Veg. 12: 293 (1913), nom. illeg.
Asparagus striatus var. *dregei* Baker	J. Linn. Soc., Bot. 14: 622 (1875)
Asparagus striatus var. *linearifolius* Baker	J. Linn. Soc., Bot. 14: 622 (1875)
Asparagus suaveolens Burch.	Trav. S. Africa 1: 226 (1822)
Asparagus subfalcatus De Wild.	Pl. Bequaert. 1: 42 (1921)
Asparagus subscandens F.T.Wang and S.C.Chen	Acta Phytotax. Sin. 16(1): 92 (1978)
Asparagus subulatus Steud. ex Baker	J. Linn. Soc., Bot. 14: 614 (1875), nom. illeg.
Asparagus subulatus Thunb.	Prodr. Pl. Cap.: 66 (1794)
Asparagus sylvaticus Burch. ex Baker	Refug. Bot. 3: t. 214 (1870), nom. illeg.
Asparagus sylvaticus Waldst. and Kit.	Descr. Icon. Pl. Hung. 3: 223 (1812)
Asparagus sylvicola S.M.Burrows	Bothalia 38: 26 (2008)
Asparagus taliensis F.T.Wang and Tang ex S.C.Chen	Acta Phytotax. Sin. 16(1): 91 (1978)
Asparagus tamaboki Yatabe	Bot. Mag. (Tokyo) 7: 61 (1893)
Asparagus tamariscinus N.A.Ivanova ex Grubov	Bot. Mater. Gerb. Bot. Inst. Komarova Akad. Nauk S.S.S.R. 17: 10 (1955)
Asparagus tamariscinus var. *issykkulensis* N.V.Vlassova	Bot. Zhurn. (Moscow and Leningrad) 65: 579 (1980)
Asparagus tauscheri Fisch. ex Ledeb.	Fl. Altaic. 2: 44 (1830)
Asparagus tenuifolius Gilib.	Exerc. Phyt. 2: 459 (1792), opus utique oppr.
Asparagus tenuifolius Lam.	Encycl. 1: 294 (1783)
Asparagus tenuissimus Van Geert	Nursery Cat. (Auguste Van Geert) 83: 4 (1882)
Asparagus terminalis L.	Sp. Pl. ed. 2: 450 (1762), nom. superfl.
Asparagus ternatus Brongn. ex H.Perrier	Notul. Syst. (Paris) 5: 30 (1935)
Asparagus ternifolius Hook.f.	Bot. Mag. 126: t. 7728 (1900), sensu auct.
Asparagus tetragonus Bresler	Gen. Aspar.: 15 (1826)
Asparagus thunbergianus Schult. and Schult.f. in J.J.Roemer and J.A.Schultes	Syst. Veg., ed. 15 bis 7: 333 (1829)
Asparagus tibeticus F.T.Wang and S.C.Chen	Acta Phytotax. Sin. 16(1): 93 (1978)
Asparagus touranensis Hamdi and Assadi	Iranian J. Bot. 19: 44 (2013)
Asparagus transvaalensis (Oberm.) Fellingham and N.L.Mey.	Bothalia 25: 208 (1995)
Asparagus triacanthus Burm.f.	Fl. Indica, Prodr. Fl. Cap.: 10 (1768), nom. illeg.
Asparagus triacanthus Willd. ex Schult. and Schult.f. in J.J.Roemer and J.A.Schultes	Syst. Veg., ed. 15 bis 7: 334 (1829), nom. illeg.
Asparagus tricarinatus K.Koch	Linnaea 22: 266 (1849)
Asparagus trichoclados (F.T.Wang and Tang) F.T.Wang and S.C.Chen	Fl. Reipubl. Popul. Sin. 15: 111 (1978)
Asparagus trichophyllus Bunge	Enum. Pl. China Bor.: 65 (1833)
Asparagus trichophyllus var. *decorticatus* (Ledeb.) Regel	Bull. Soc. Imp. Naturalistes Moscou 41(2): 269 (1869)
Asparagus trichophyllus var. *flexuosus* (Ledeb.) Regel	Bull. Soc. Imp. Naturalistes Moscou 41(2): 270 (1869)
Asparagus trichophyllus var. *medius* Bong. and C.A.Mey.	Verz. Saisang-nor. Pfl.: 74 (1841)
Asparagus trichophyllus var. *trachyphyllus* Bong. *and C.A.Mey.*	Verz. Saisang-nor. Pfl.: 74 (1841)
Asparagus tuberculatus Bunge ex Iljin in V.L.Komarov (ed.)	Fl. URSS 4: 747 (1935)
Asparagus turkestanicus Popov in B.A.Fedtschenko and al.	Fl. Turkm. 1: 312 (1932)
Asparagus uhligii K.Krause	Bot. Jahrb. Syst. 57: 238 (1921)
Asparagus umbellatus Link in C.L.von Buch	Phys. Beschr. Canar. Ins.: 140 (1828)
Asparagus umbellatus var. *flavescens* Svent.	Addit. Fl. Canar.: 3 (1960)

Generic Species	*Status Report/Reference*
Asparagus umbellatus var. *gomerae* G.Kunkel	Cuad. Bot. Canaria 25: 21 (1975)
Asparagus umbellatus subsp. *lowei* (Kunth) Valdés	Lagascalia 9: 86 (1979)
Asparagus umbellatus var. *scaber* Baker	J. Linn. Soc., Bot. 14: 611 (1875)
Asparagus umbellatus subsp. *umbellatus*	-
Asparagus umbellulatus Bresler	Gen. Aspar.: 24 (1826)
Asparagus umbricola Kit. ex Kanitz	Reliq. Kitaib.: 121 (1863)
Asparagus undulatus (L.f.) Thunb.	Prodr. Pl. Cap.: 66 (1794)
Asparagus undulatus var. *rugosus* Baker	J. Linn. Soc., Bot. 14: 629 (1875)
Asparagus usambarensis Sebsebe	Fl. Trop. E. Afr., Asparagac.: 19 (2006)
Asparagus vaginellatus Bojer ex Baker	J. Linn. Soc., Bot. 14: 612 (1875)
Asparagus vaniotii H.Lév.	Cat. Pl. Yun-Nan: 164 (1916)
Asparagus verrucosus Nakai	Bot. Mag. (Tokyo) 37: 69 (1923)
Asparagus verticillaris Lam.	Encycl. 1: 297 (1783)
Asparagus verticillatus L.	Sp. Pl. ed. 2: 450 (1762)
Asparagus verticillatus subsp. *floribundus* (Baker) K.Richt.	Pl. Eur. 1: 230 (1890)
Asparagus verticillatus var. *floribundus* Baker	J. Linn. Soc., Bot. 14: 601 (1875)
Asparagus verticillatus var. *triphyllus* Bresler	Gen. Aspar.: 27 (1826)
Asparagus vicentinus Welw. ex Cout.	Herb. Gorg. Cat. 1: 277 (1914)
Asparagus virgatus Baker	Refug. Bot. 3: t. 214 (1870)
Asparagus virgatus var. *capillaris* Baker	J. Linn. Soc., Bot. 14: 606 (1875)
Asparagus volubilis (L.f.) Thunb.	Prodr. Pl. Cap.: 66 (1794)
Asparagus vulgaris Gueldenst. ex Ledeb.	Fl. Ross. 4: 197 (1852)
Asparagus vulgaris var. *maritimus* Bong. and C.A.Mey.	Verz. Saisang-nor. Pfl.: 74 (1841)
Asparagus vvedenskyi Botsch.	Novosti Sist. Vyssh. Rast. 1965: 268 (1965)
Asparagus warneckei (Engl.) Hutch. in J.Hutchinson and J.M.Dalziel	Fl. W. Trop. Afr. 2: 351 (1936)
Asparagus wildemanii Weim.	Bot. Not. 1937: 446 (1937)
Asparagus wrightii A.Gray	Mem. Amer. Acad. Arts, n.s., 6: 413 (1859)
Asparagus yanbianensis S.C.Chen	Acta Phytotax. Sin. 26: 139 (1988)
Asparagus yanyuanensis S.C.Chen	Acta Phytotax. Sin. 19. 501 (1981)
Asparagus yunnanensis H.Lév.	Cat. Pl. Yun-Nan: 164 (1916)
Asparagus zanzibaricus Baker	J. Linn. Soc., Bot. 14: 614 (1875)
Asparagus zeylanicus (Baker) Hook.f.	Fl. Brit. India 6: 317 (1892)
Asparagus zuluensis N.E.Br.	Bull. Misc. Inform. Kew 1912: 283 (1912)

The September 2014, the World Checklist of Selected Plant Families accepts 212 species of *Asparagus*, including:

- ☆ *Asparagus acutifolius*
- ☆ *Asparagus aethiopicus* (= *Asparagus sprengeri*, *Protasparagus aethiopicus*)–ground asparagus, asparagus fern, (*S. Afr.*) emerald fern, basket asparagus
- ☆ *Asparagus aethiopicus* 'Sprengeri' – Sprenger's asparagus
- ☆ *Asparagus africanus* (= *Protasparagus africanus*)–African asparagus
- ☆ *Asparagus asparagoides* (= *Myrsiphyllum asparagoides*)–smilax, African asparagus fern, (*Austr.*) bridal creeper
- ☆ *Asparagus cochinchinensis* – *Tian men dong* in Chinese, of which the tubers are used as a kind of herb in traditional Chinese medicine.
- ☆ *Asparagus declinatus*–foxtail asparagus fern, (*Austr.*) bridal veil (Biosecurity, 2014)
- ☆ *Asparagus capensis*
- ☆ *Asparagus curillus*–Himalayas
- ☆ *Asparagus densiflorus* (=*Protasparagus densiflorus*)
- ☆ *Asparagus falcatus*–large forest sicklethorn of southern Africa
- ☆ *Asparagus fallax*–esparraguera de monteverde. Macaronesian native flora.
- ☆ *Asparagus flagellaris*
- ☆ *Asparagus horridus*–Mediterranean, Arabian Peninsula, Macaronesia (World Checklist, 2014).

- ☆ *Asparagus laricinus*–katbos
- ☆ *Asparagus lignosus*
- ☆ *Asparagus macowanii* (= *Protasparagus macowanii*)
- ☆ *Asparagus maritimus*
- ☆ *Asparagus nesiotes* – esparraguera majorera. Macaronesian native flora.
- ☆ *Asparagus officinalis* – asparagus
- ☆ *Asparagus officinalis* subsp. *officinalis* – garden asparagus
- ☆ *Asparagus officinalis* subsp. *prostratus*
- ☆ *Asparagus pastorianus* – esparraguera espinablanca. Macaronesian native flora.
- ☆ *Asparagus persicus*
- ☆ *Asparagus racemosus*
- ☆ *Asparagus rubicundus*
- ☆ *Asparagus scandens*–krulkransie of the Western Cape, South Africa
- ☆ *Asparagus schoberioides*
- ☆ *Asparagus scoparius–esparragón raboburro*. Macaronesian native flora.
- ☆ *Asparagus setaceus* (=*Protasparagus setaceus, A. plumosus*)–lace fern, asparagus fern, florist's fern, (*Austr.*) climbing asparagus
- ☆ *Asparagus tenuifolius*
- ☆ *Asparagus umbellatus*–esparraguera común. Macaronesian native flora.
- ☆ *Asparagus vaginellatus* Bojer ex Baker
- ☆ *Asparagus virgatus*

Botanical Description

Asparagus (*Asparagus officinalis* Linn. var. *altilis*) belongs to the family Asparagaceae.

Table 15.4: Botanical Description

Growth Habit	It is an herbaceous dioecious herb growing 1.0-3.0 m heights. Some species are herbaceous and some woody, both erect and climbing forms are common. Wild asparagus produces a thick mat of fleshy roots that can grow 3 feet deep.
Root	Wild asparagus produces a thick mat of fleshy roots that can grow 3 feet deep.
Stem	Stems are smooth, green and 120 to 180 cm tall, with many green primary branches. Slender, green, needle-like secondary branches are produced in bundles off the main branches, appearing more like leaves than branches.
Leaf	Inconspicuous, reduced, scale-like leaves (bracts) occur at the tips of young shoots and at the bases of primary and secondary branches along mature stems. (Note: The bundles of green, needle-like structures along the main branches are actually finer branches).
Floral biology	Wild asparagus is dioecious, producing male and female flowers on separate plants. The male flowers are yellowish green and conspicuous, while the female flowers are found (on separate plants) and are less conspicuous. Flowers are bell-shaped, and situated on slender flower stalks in groups of 1 to 3 arising from the leaf bases.
Anthesis and pollination	The female flowers are pollinated by bees, and when fertilized give rise to berries, which are a conspicuous feature on the plants from July until autumn.
Fruit and fruit set	The fruit is three-celled berry produced on female plants, which becomes red as it matures.
Seeds	The seeds are 3-6, angular, large (3.0 mm or less in diameter) rounded at the back but more or less flattened on one side and black in colour (Kabir, 1993). The seeds can be extracted by macerating the berries in water.

It gives economic yield after three years and continues to grow upto several years.

Botanical Classification

The asparagus varieties are broadly divided into two groups:

(i) Spears overcast with purple and

(ii) Spear tips almost free from purple overcost

Augmentation of Genetic Resource and Crop Improvement

The seed of 39 introductions of Asparagus (11 from USA, 22 from New Zealand and 6 from South Africa) were evaluated at Regional Research Laboratory Farm Sanat Nagar, Srinagar (J&K) and among them four distinct complexes were recognized on the basis of number of spears/plant and weight of spear/plant. Majority of the introductions were in the complex IV. Marked differences were observed between complex I and complex II. The introductions SL-17, SL-21 and SL-29 were among top ranking for weight of spear/plant and spear yield/ha (Pandita and Bhan, 1996).

Interspecific Hybridization

Asparagus maritimus Mill. was found to be sexually compatible with the cultivated species *A. officinalis* L., and therefore can be considered to belong to the primary gene pool. Further information about inter-specific hybridization is presented by Gonza´lez Castan˜o´n and Falavigna (2008), but due to different results of the success of the crosses the gene pool structure of asparagus remained largely unclear.

Table 15.5: Evaluation of Asparagus Cultivars during 1993-1994

Cultivar	*Year*	*Spear Yield/Plant/ kg (g) (Male plant)*	*Spear Yield/Plant/ kg (g) (Female plant)*	*Average Spear Yield/Plant (g)*	*Number of Spear/ Plant*	*Spear Weight (g)*	*Spear Length (cm)*	*Spear Diameter (cm)*	*Plant Height (cm)*
Perfection	1993	403.55	280.45	342.00	14.2	24.09	23.56	1.82	123.33
	1994	416.39	297.18	356.78	16.5	26.27	23.30	2.20	131.52
Brpck Imperial-84	1993	611.80	374.40	492.60	12.4	29.69	23.66	2.26	100.00
	1994	627.26	395.50	511.20	13.6	41.63	23.20	2.40	105.90
Local	1993	207.15	169.50	188.30	9.6	19.56	23.33	1.04	86.66
	1994	229.63	189.76	209.70	10.2	20.43	23.10	1.60	95.80
Conovers Colossal	1993	245.10	184.90	215.00	10.7	19.97	23.30	1.95	96.66
	1994	264.98	197.87	231.42	11.3	22.47	23.10	1.90	100.40
UC-72	1993	239.80	173.60	206.70	10.5	19.67	23.13	1.93	61.66
	1994	258.84	192.42	225.63	11.4	21.77	23.00	1.80	96.15

Source: Sharma and Chaudhary, 1996.

REFERENCES

1. Ali, S.I. and Khan, S.W. 2009. Asparagaceae. In: Flora of Pakistan (eds.) Ali, S.I. and Quaisar, M., Inst. Plant Conser., Univ. Karachi, Karachi and Missouri Bot. Press, Missouri Bot. Garden, St. Louis, Missouri, USA. 217: 1-24.
2. Asparagus: World Checklist of Selected Plant Families (Royal Botanic Gardens, Kew), retrieved 2014-09-01.
3. "Bridal creeper, *Asparagus asparagoides*". CSIRO Division of Entomology. Archived from the original on 9 May, 2006.
4. Biosecurity. 2014. SA: Declared plants in South Australia, 2014 Accessed 1 September 2016.
5. Chase, M.W.; Reveal, J.L. and Fay, M.F. (2009), "A subfamilial classification for the expanded asparagalean families Amaryllidaceae, Asparagaceae and Xanthorrhoeaceae", *Botanical Journal of the Linnean Society*, **161 (2):** 132–136.
6. Choudhury, B. 1967. Vegetables. National Book Trust, India, New Delhi, (Edn. 7th reprint).
7. Gonza´lez Castan˜o´n ML, Falavigna A (2008) Asparagus germplasm and interspecific hybridization. *Acta Hort.*, **776:** 319–326.
8. Kabir, J. 1993. Asparagus. In: Vegetable Crops in India. (Eds. T.K. Bose, M.G. Som and J. Kabir). Naya Prokash Kolkata.
9. Kew World Checklist of Selected Plant Families
10. Malcomber, S. T. and Demissew, S. 1993."The Status of *Protasparagus* and *Myrsiphyllum* in the Asparagaceae", *Kew Bulletin*, **48 (1):** 63-78
11. Marloth, Rudolf. "The Flora of South Africa" 1932 Pub. Cape Town: Darter Bros. London: Wheldon and Wesley.
12. Pandita, P.N., Bhan, M.K. and Kaul, A. 1987. Asparagus a cash crop of Kashmir. *Indian Fmg*. **36(1):** 3-4.
13. Pandita, P.N. and Bhan, M.K. 1996. Genetic resources of Asparagus (*Asparagus officinalis* L.). *Indian J. Plant Genet. Resour.*, **9(1):** 97-103.
14. Pandita, P.N. and Bhan, M.K. 194. Prospects of Asparagus cultivation in Kashmir. *Asparagus Research News Letter*, **2(1):** 25-29.
15. Shanmugavelu, K.G. 1989. Production Technology of Vegetable Crops. Oxford and IBH Publishing Co. Pvt. Ltd., New Delhi.
16. Sharma, D.K. and Chaudhary, D.R. 1996. Cultivating Asparagus in Himachal Preadesh. *Indian Horticulture*, **41(2):** 8-9.
17. Velvan, S. Nagulendran, K.R., Mahesh, R. and Hazeenabegum, V.V. 2007. The chemistry, pharmacology and therauptical applications of *Asparagus racemosus*: A review. *Pharm. Rev.*, **1(2):** 350-360.
18. "World Checklist of Selected Plant Families". 2014.
19. "Bridal creeper". 2005. Weed of the month. CRC weed management. Archived from the original on. Retrieved 2006-04-30.

Genus *Caralluma*

Pimpa (*Caralluma edulis*)

Introduction

Pimpa (*Caralluma edulis* (Edgew.) Benth. and Hook.f.), is a very nutritious desert plant which tender shoots are used as vegetable; very commonly in Jaisalmer area of Rajasthan but it is now considered as a very rare herb of the extremely arid regions of the 'Thar Desert'. It is synonymically known as *Caralluma longidens* N.E.Br. (1892), *Caralluma vittata* N. E. Br. (1903), and *Caralluma mouretii* A. Chev. (1934), *Cryptolluma edulis* (Edgew.) Plowes (1995). *Caralluma edulis* is used raw as a vegetable in Mauritania. In India fresh stems are sold as a vegetable in local shops ('situn'). As a succulent herb it has ornamental value too. The stems are slightly acid. In a test in Pakistan to verify a possible hypoglycaemic effect of powdered *Caralluma edulis* roots, doses of 2–4 g/kg body-weight did not show any effect in normal or diabetic rabbits. There is no information on the phytochemistry of *Caralluma edulis*. Highly toxic glycosides have been found in several other *Caralluma* species (*e.g.* in *Caralluma acutangula* (Decne.) N.E.Br., *Caralluma gracilipes* K. Schum. and *Caralluma penicillata* (Defl.) N.E.Br.). *Caralluma edulis* is able to grow in dry, sandy locations it deserves more research attention concerning its phytochemistry, nutritive properties and cultivation possibilities.

Origin, History and Distribution

Caralluma edulis has been reported from Mauritania, Sudan, Eritrea, Somalia, Saudi Arabia, United Arab Emirates, Pakistan, India and possibly Iran and Afghanistan. In its distribution area it flowers and fruits year-round.

Species Availability, Evolution and Diversity

Pimpa (*Caralluma edulis* (Edgew.) Benth. and Hook.f.) belongs to the family Asclepiadaceae. The taxonomy of the genus and species is still in dispute. According to Plowes the correct name for this species is *Cryptolluma edulis* (Edgew.) Plowes and the first 3 synonyms mentioned here are considered to represent different, separate species. These are *Caralluma longidens* N.E.Br. (1892), *Caralluma vittata* N. E. Br. (1903), and *Caralluma mouretii* A. Chev. (1934), *Cryptolluma edulis* (Edgew.) Plowes (1995).

Table 15.6: Status of Taxon

Name	*Status*	*Source*
Boucerosia edulis Edgew.	Synonym	W C S P (in review)
Boucerosia stocksiana Boiss.	Synonym	W C S P (in review)
Caralluma edulis (Edgew.) Gravely and Mayur.	Synonym	TRO
Caralluma edulis (Edgew.) Benth. and Hook. f.	Synonym	TRO
Caralluma longidens N.E.Br.	Synonym	W C S P (in review)
Caralluma vittata N.E.Br.	Synonym	W C S P (in review)
Caudanthera edulis (Edgew.) Meve and Liede	Synonym	W C S P (in review)
Cryptolluma edulis (Edgew.) Plowes	Unresolved	W C S P (in review)
Desmidorchis stocksiana Kuntze	Unresolved	W C S P (in review)
Spiralluma longidens (N.E. Br.) Plowes	Synonym	TRO

Botanical Description

Kingdom	:	Plantae
unranked	:	Angiosperms
unranked	:	Eudicots
unranked	:	Asterids
Order	:	Gentianales
Family	:	Apocynaceae
Genus	:	*Caralluma*
Species	:	*edulis*
Botanical name	:	*Caralluma edulis*

Source: *Wikipedia*, 2016.

The detil descriptions are as under:

Table 15.7: Botanical Description

Growth Habit	Slightly succulent herb with slender, much branched. *Caralluma edulis* occurs usually in dry sandy locations, from sea-level up to 1300 m altitude.
Stem	Subterete stem 20–30 cm tall, at apex tapering to a pointed tip.
Leaf	Leaves opposite, simple, small, caducous, sometimes prominent, semi-persistent linear, acute and up to 1 cm long.
Floral biology	Flowers 1-2 together, extra-axillary near the stem apex, bisexual, 5-merous, about 8 mm in diameter, on slender curved pedicels 4-6 mm long; calyx-lobes ovate-lanceolate, 3-4 mm long, glabrous; corolla tube campanulate-urceolate, 7-8 mm long and broad, subglabrous, yellow-green, longitudinally ridged inside, corolla limb dark purple with a narrow disk and erect, oblong-triangular lobes 5 mm × 2.5 mm; outer corona lobes bipartite, forming 10 filiform horns 5 mm long, inner corona lobes longer than the outer ones, strap-shaped, incumbent on the anthers; anthers short, truncated, without appendages. Pollinium 1 in each anther cell. Follicles 8.5-11 cm long.

Natural propagation of this plant is extremely poor under the xeric conditions but the 4 to 6 multilple shoots were initiated from nodal explant on MS medium containg 0.25 mg l^{-1} IAA (Indol-3 acetic acid) and 2.5 mg l^{-1} BAP (Benzyle amino purine). The shoots elongated and multiplied on a medium supplemented with 0.25 mg l^{-1} IAA + 2.5 mg l^{-1} BAP and 2.5 mg l^{-1} GA_3 (Gibberellic acid). *In vitro* prodeuced shoots rooted on half strength MS medium containing 1.0 mg l^{-1} IBA (Indolebutyric acid) within 3 wweks. Plantlets were transplanted to pots and grown to maturity (Kaur *et al.*, 1992).

Genetic Resource Augmentation and Crop Improvement

Data were recorded on its growth, reproductive behaviour and germination. Initial and final germination of seeds of *C. edulis* (collected from plants raised by rooted-stems of Chandan area) recorded in nursery as 32.1 and 87.5 per cent on 4th and 6th days after sowing, respectively. The statistical analysis of characters under study revealed the presence of high co-efficient of variation (CV in per cent) in number of branches per plant (51.71) followed by number of pod (in pair) per plant (49.41) and least in collar diameter (10.05). The range and mean values of some important characters are as follows: plant height (38.0–62.0 and 50.7 cm), number of branches per plant (1.0–10.0 and 3.9), number of flowers per plant (2.0–7.0 and 4.8), number of seeds per pod (15.0–30.0 and 20.8) and seed length (4.9–8.7 and 7.0 mm), respectively. As it flowered and set seeds at nursery in our study, it has opened avenues for large scale multiplication and future cultivation of the species (Singh *et al.*, 2016).

Genetic Resource Conservation

Caralluma edulis is nowhere a common plant and it should be protected wherever it occurs naturally. Field surveys conducted in Jaisalmer in arid part of the Thar Desert in India during August to October for two consecutive years (2012 and 2013) revealed its occurrence in patches in villages such as Khuyiyala, Bahramsar dham, Deva-II and Chandan. Germplasm of *C. edulis* by means of rooted-stems collected from above sites was planted in nursery at Central Arid Zone Research Institute, Regional Research Station, Jaisalmer (Singh *et al.*, 2016). There is an urgent need to conserve the germplasm of this rare species in the desert are.

REFERENCES

1. Jansen, P.C.M., 2004. *Caralluma edulis* (Edgew.) Benth. ex Hook.f. [Internet] Record from PROTA4U. Grubben, G.J.H. and Denton, O.A. (Editors). PROTA (Plant Resources of Tropical Africa/Ressources végétales de l'Afrique tropicale), Wageningen, Netherlands. <http://www.prota4u.org/search.asp>.
2. Kaur, G., Rathore, T.S., Rao, S.R. and Shekhawat, N.S. 1992. *In vitro* micropropagation of *Caralluma edulis* (Edgew.) Benth. and Hook.f.-A rare edible plant species of Indian desert.
3. Singh, J.P., Kumar, S., Venkatesan, K. and Kulloli, R.N. 2016. Conservation status and utilization of *Caralluma edulis*: an important threatened medicinal plant species of the Thar Desert, India. *Genetic Resources and Crop Evolution*, **63(4):** 721-732.

Marsha

Botanical name: *Spilanthes acmella* Mure.

Family: Compositae, Asteraceae

It is found near jhum land and kitchen garden. It is mostly used as a pickle and sometimes eaten after cooking. It is an herb; flower is yellowish in colour while fruit is black in colour. In nature it is available throughout the year. An annual prostrate herb grows on damp soil, drains *etc.* whole plants is used for eco-friendly fishery.

Genus *Opuntia*

Cactus Pear or Prickly Pear (*Opuntia ficus-indica* (L.) Miller)

Introduction

Cactus pear or prickly pear (*Opuntia ficus-indica* (L.) Miller)

Origin, History and Distribution

Cactus pear or prickly pear is native to Mexico.

Species Availability, Evolution and Diversity

Family Cactaceae includes 400 species.

16

Varietal Wealth of Vegetable Crops

Vegetables are the most indispensable components of human diets which provide essential vitamins, minerals, dietary fibers, antioxidants, and other micronutrients. Apart from nutrition, they also contain sufficient amount of phytochemicals like anti carcinogenic principles and anti-oxidants (*e.g.* flavanoids, glucosinolates and isothiocynates). In India a number of vegetables are being grown due to diverse edaphic and climatic condition, well distinct cropping season, enriched with enormous genetic wealth *etc.* but its availability to the general mass for fulfilling the requirement of essential phyto-nutrients, is quite less (225 g/day/caput) than the recommended amount (310 g/day/caput) for better human health. Presently the production of vegetables is 187.50 mt, (2017-2018), about 3.5 per cent more than that in 2016-2017. The increase in vegetable production by increasing area under vegetable cultivation is limited due to continued decrease of land holdings. Without improved varieties/hybrids, increasing production and productivity through better agronomic package and practices is possible only up to certain limit. Although at state and national level, a large number of varieties and hybrids have been identified for commercial cultivation, many of them have not become popular among the farmers because of non-availability of sufficient quantity and quality of seeds and unorganized seed marketing system. On one hand resource poor farmers tend to grow old varieties/local land races due to several compulsion (unawareness, high cost, resource poor ness *etc.*), on the other, progressive vegetable growers replace their varieties more rapidly with the availability of better improved varieties/hybrids. Presently, a number of hybrids/improved varieties and locally adopted cultivars are popular among the farmers of different part of country.

Agro-Climatic Zones of India for AICRP (Vegetable Crops)

The need for growing more food, on a sustainable basis, to support the ever-increasing population demands a systematic appraisal of our natural resources including climate, soils, flora, and fauna. Since agriculture is highly location-specific, grouping the available land area in the country into different agro-ecological regions based on certain identifiable characteristics becomes all the more important. This may help the country to engage in more rational planning and optimizing resource use for the present and in preserving them for the future. An agro-climatic zone is a land unit in terms of major climates, suitable for a certain range of crops and cultivars. An ecological region is an area of the earth's surface characterized by distinct ecological responses to macro-climate, as expressed by soils, vegetation, fauna, and aquatic systems (FAO, 1983). The country has been divided into eight climatic zones for All India Coordinated Research Project (Vegetable Crops) is given in Table 1.

Table 1: Different Zones identified by ICAR for AICRP (Vegetable Crops) in the Country

Zone No.	Agro-climatic Zone	State
I.	Humid Western Himalayan Region	Jammu and Kashmir (J&K), Himachal Pradesh and Uttaranchal
II	Humid Bengal - Assam Basin	West Bengal and Assam
III.	Humid Eastern Himalayan and Bay of Islands	Sikkim, Meghalaya, Manipur, Nagaland, Mizoram, Tripura, Arunachal Pradesh and Andaman and Nicobar Islands
IV.	Sub-Humid Sutlej Ganga Alluvial Plain	Punjab, U.P., Bihar and Jharkhand
V	Sub- Humid to Humid Eastern and South Eastern Uplands	Chattisgarh., Odisha and Andhra Pradesh.
VI	Arid Western Plain	Rajasthan, Gujarat, Haryana and Delhi
VII	Semi-Arid Lava Plateau and Central High Lands	Madhya Pradesh and Maharashtra
VIII.	Humid to Semi-Arid Western Ghats and Karnataka Plateau	Karnataka, Tamil Nadu and Kerala

Varieties, Hybrids and Resistance Varieties Identified for Different Agro-climatic Zones of India

Majority of the vegetable growers are either unaware about the newly developed improved varieties/hybrids/resistance of vegetable crops or seeds of such varieties/hybrids/resistance, especially those bred at public sector are not readily available. A large number of high yielding and biotic stress resistant varieties (mainly disease resistant) have been identified through AICRP (Vegetable Crops) Central Varieties Release Commitee (CVRC), National and State Level Release Committee (SVRC) for release and commercial cultivation. The list of all varieties/hybrids/resistance identified through AICRP (Vegetable Crops) and description of most promising cultivars are being given briefly in the Table 2(a) and 2(b).

Table 2a. List of Varietal Wealth (Open Pollinated, Hybrids and Resistant) Identified through AICRP (VC)

Crop	Variety	Remark	Developing centre	Recommended zone	Year of identification	V/ H/R
Brinjal	H-4	Long	Hisar		1975	V
Brinjal	Pusa Purple Cluster	Long	IARI, New Delhi	IV,V,VI,VII	1975	V
Brinjal	Pusa Purple Long	Long	IARI, New Delhi	IV,VI,VII,VIII	1975	V
Brinjal	S-16	Long	Ludhiana	IV,VI	1975	V
Brinjal	S-4	Long	Ludhiana		1975	V
Brinjal	T-3	Round	Kalyanpur	-	1975	V
Cauliflower	327-14-8-3	Sep. maturity	IARI, New Delhi	-	1975	V
Cauliflower	351-4-1	Oct. maturity	IARI, New Delhi	-	1975	V
Cauliflower	Early Kunwari	Early group	Ludhiana	-	1975	V
Cauliflower	EC 12012	Jan. maturity	IARI, Katrain	-	1975	V
Cauliflower	Improved Japanese	Nov. maturity	IARI, New Delhi	-	1975	V
Cauliflower	Pusa Snowball	Jan. maturity	IARI, Katrain	-	1975	V
Cauliflower	Synthetic-1	Dec. maturity	IARI, New Delhi	-	1975	H
Chilli	G-4		Lam	-	1975	V
Chilli	G-5		Lam	-	1975	V
Muskmelon	Arka Jeet		IIHR	VIII	1975	V
Muskmelon	Arka Rajhans		IIHR	VIII	1975	V
Muskmelon	Durgpura Madhu		Durgapura	IV,VI,VII,VIII	1975	V
Muskmelon	Hara Madhu		Ludhiana	IV,VII	1975	V
Muskmelon	Pusa Sarbati		IARI, New Delhi	IV	1975	V
Muskmelon	SI-445 (Pusa Madhuras)		IARI, New Delhi	IV,VI,VIII	1975	V
Onion	Line- 106	White	IARI, New Delhi		1975	V
Onion	Line- 131	White	IARI, New Delhi		1975	V

Onion	Punjab Selection	Keeping quality	Ludhiana	-	1975	V
Onion	Pusa Ratnar	Red	IARI, New Delhi	-	1975	V
Onion	Pusa Red	Keeping quality	IARI, New Delhi	-	1975	V
Onion	S-48	White	Ludhiana		1975	V
Pea	Arkel	Early Season	IARI, New Delhi	-	1975	V
Pea	Bonneville	Mid Season	IARI, New Delhi	-	1975	V
Pea	GC-141	Mid Season	Gwalior	-	1975	V
Pea	GL-195	Mid Season	Gwalior	-	1975	V
Tomato	HS-101		Hisar	-	1975	V
Tomato	Pusa Ruby	Small fruited	IARI, New Delhi	-	1975	V
Tomato	S-12	Small fruited	Ludhiana	-	1975	V
Tomato	SL-120	Large fruited	IARI, New Delhi	-	1975	V
Tomato	Sweet-72	Large fruited	Gwalior	-	1975	V
Tomato	T-1	Large fruited	Kalyanpur	-	1975	V
Watermelon	Durgapura Meetha		Durgapura	IV,VI,VII,VIII	1975	V
Watermelon	Sugar Baby		IARI, New Delhi	V,VII,VIII	1975	V
Pea	Early December	Early Season	Gwalior	-	1977	V
Tomato	Pusa Early Dwarf	Small fruited	NBPGR	-	1977	V
Tomato	Sioux	Medium & large	IARI, New Delhi	-	1977	V
Cauliflower	K-1	Jan. maturity	IARI, Katrain	-	1979	V
Tomato	Punjab Chhuhara	Small fruited	Ludhiana	-	1979	V
Tomato	Sel-152	Small fruited	Ludhiana	-	1979	V
Brinjal	Arka Kusumkar	Green, Long	IIHR	VIII	1981	H
Brinjal	Arka Navneet	Round	IIHR	-	1981	H
Brinjal	PBR -129-5	Long	Pantnagar	IV	1981	V
Carrot	Sel-5		IARI, New Delhi	VII	1981	V
Cauliflower	114-S-1	-	Pantnagar	-	1981	V
Watermelon	Arka Jyoti		IIHR		1981	H
Brinjal	Azad Kranti	Long	Kalyanpur	I,VI	1983	V
Cowpea	L-1552		IARI, New Delhi	-	1983	V
Brinjal	ARU-1	Long	Pithoragarh	I	1985	V
Brinjal	PBR -91-2	Round	Pantnagar	-	1985	V
Cabbage	Sel-8		IARI, Katrain		1985	V
Cauliflower	Line 6-1-2-1	Dec. maturity	IARI, New Delhi	-	1985	V
Chilli	K-2		Coimbatore	-	1985	V
French Bean	VL-Boni-1		VPKAS	North Indian Hills	1985	V
Muskmelon	Sel -2		NBPGR		1985	V
Onion	N-2-4-1	Red	Rahuri	-	1985	V
Onion	N-257-9-1	White	Rahuri	-	1985	V
Pea	IP-3	Early Season	Pantnagar	-	1985	V
Pea	P-88	Early Season	Ludhiana	-	1985	V
Tomato	AC-238	Indet.	Pantnagar	-	1985	V
Tomato	KS-2	Det.	Kalyanpur	U.P.	1985	V
Brinjal	ARU-2C	Long	VPKAS	I,IV,VI,VIII	1987	V
Brinjal	K-202-9	Round	Anand	VI	1987	V
Brinjal	Kat-4	Long	IARI, Katrain	VIII	1987	H
Chilli	J-218		Jabalpur	I, IV, V, VI, VII	1987	V
Chilli	Muslawadi		Rahuri	V	1987	V
Chilli	X-235 (LCA-235)		Lam	IV, V, VI, VIII	1987	V
French Bean	Arka Komal		IIHR	I,VII,VIII	1987	V
French Bean	UPF-191		Pantnagar	IV,VIII	1987	V

Onion	Agri Found Dark Red	Kharif season	NHRDF	IV	1987	V
Onion	Arka Kalyan	Rabi Season	IIHR	IV,VI,VII,VIII	1987	V
Onion	Arka Niketan	Kharif season	IIHR	VII	1987	V
Onion	Line-102	Rabi Season	IARI, New Delhi	I, IV, VI, VII	1987	V
Pea	Lincon	Early Season	IARI, Katrain	I	1987	V
Pea	PM-2	Early Season	Pantnagar	I	1987	V
Pea	VL-3	Early Season	VPKAS	I,IV,VI	1987	V
Pumpkin	Arka Chandan		IIHR	VIII	1987	V
Pumpkin	Arka Suryamukhi		IIHR	VIII	1987	V
Pumpkin	CM-14		Vellanikara	IV,V,VII	1987	V
Pumpkin	Pusa Vishwas		IARI, New Delhi	I,VIII,IV,V	1987	V
Tomato	Arka Saurabh	Indet.	IIHR	I,IV,V,VI,VII,VIII	1987	V
Tomato	Arka Vikas	Indet.	IIHR	I,IV,V,VI,VII,VIII	1987	V
Tomato	CO-3	Det.	Coimbatore	I,IV,V,VI,VII	1987	V
Tomato	La Bonita	Det.	NBPGR	I,IV,V,VII	1987	V
Tomato	Pant T-3	Indet.	Pantnagar	I,II,IV,V,VI,VII,VII	1987	V
Tomato	Pb. Kesari	Det.	Ludhiana	I,IV,V,VI	1987	V
Watermelon	Arka Manik		IIHR	IV,VII,VIII	1987	V
Brinjal	Aruna	Small Round	Akola	VII	1988	V
Garlic	G-41		NHRDF	IV,VII	1988	V
Brinjal	BB-7		Bhubaneshwar	II, V	1990	R
Brinjal	BWR-12		IIHR	VIII	1990	R
Brinjal	H-7	Long	Hisar	IV,VI	1990	V
Brinjal	H-8	Round	Hisar	II, IV, V, VI	1990	V
Brinjal	NDB-25	Long	Faizabad	II, IV, VII	1990	V
Brinjal	Pusa Hybrid-6	Round	IARI, New Delhi	IV	1990	H
Capsicum	KT-1		IARI, Katrain	I	1990	H
Cauliflower	235-S	Nov. maturity	Pantnagar	II, VIII	1990	V
Cauliflower	Early Synthetic	Early group	IARI, New Delhi	IV, VIII	1990	H
Chilli	LCA 206-B		Lam	V,VI,VII,VIII	1990	V
Chilli	Sel-1		IIHR	V, VII, VIII	1990	V
Dolichos bean	Deepaliwal		Akola	V,VII	1990	V
Garlic	G-1		NHRDF	IV,VI,VII	1990	V
Muskmelon	Hybrid M-3		IARI, New Delhi	IV	1990	H
Okra	P-7		Ludhiana	All regions	1990	R
Okra	PB-57		Parbhani	All regions	1990	R
Okra	Sel-10 (A. Anamika)		IIHR	All regions	1990	R
Onion	VL-3	Rabi season	VPKAS	U.P. Hills	1990	V
Pea	JP-4	PM	Jabalpur	IV,VIII	1990	R
Pea	PRS-4	PM	Kalyanpur	IV,VI,VII	1990	R
Tomato	Sel-7	Det.	Hisar	I,IV,V,VII,VIII	1990	V
Bitter gourd	Priya		Vellanikara	II,VII,VIII	1992	V
Bottle gourd	Pusa Naveen		IARI, New Delhi	VII	1992	V
Brinjal	Pusa Hybrid-5	Long	IARI, New Delhi	IV, VII, VIII	1992	H
Brinjal	SM-6-7		Bhubaneshwar	VIII	1992	R
Cabbage	Pusa Synthetic		IARI, Katrain	IV, I, II	1992	H
Cabbage	Shri Ganesh Gol		Mahyco	V	1992	H
Carrot	Hybrid-1		Mahyco	I and VII	1992	H
Carrot	Pusa Meghali		IARI, New Delhi	VII	1992	V
Cauliflower	Pusa Hybrid-2	-	IARI, New Delhi	II and IV	1992	H

Cowpea	Sel-263		Ludhiana	IV	1992	V
Cowpea	Sel-61-B		IIHR	VII,VIII	1992	V
Okra	Sel-4 (A. Abhay)		IIHR	II	1992	R
Pea	JP- 83	PM	Jabalpur	VII	1992	R
Pea	VL-7	Early Season	VPKAS	IV	1992	V
Tomato	ARTH-3	Det.	Ankur	II,VII,VIII	1992	H
Tomato	ARTH-4	Indet.	Ankur	IV,VIII	1992	H
Tomato	BWR-5 (Arka Alok)	Bacterial wilt	IIHR	II	1992	R
Tomato	MTH-6	Indet.	Mahyco	VII,VIII	1992	H
Brinjal	ABH-1	Small Round	Anand	IV, VI, VII	1993	H
Brinjal	ARBH-201	Long	Ankur	IV, V, Vi, VII	1993	H
Brinjal	BB-26	Long	Bhubaneshwar	V	1993	V
Brinjal	MHB-10	Small Round	Mahyco	IV, VI, VII	1993	H
Brinjal	MHB-39	Small Round	Mahyco	IV, VI, VII	1993	H
Brinjal	NDBH-1	Round	Faizabad	IV, VI, VII	1993	H
Brinjal	Punjab Barsati	Long	Ludhiana	IV	1993	V
Cabbage	Nath-401		Nath Seeds	I, IV, V, VI, VII	1993	H
Cowpea	Sel-2-1		Faizabad	IV	1993	V
Garlic	G-50		NHRDF	IV	1993	V
Onion	Agri Found Light Red	Rabi season	NHRDF	VI, VIII	1993	V
Onion	Punjab Red Round	Rabi season	Ludhiana	IV	1993	V
Pea	Ageta-6	Early Season	Ludhiana	I,IV,VI	1993	V
Pea	FC-1 Ajeet)	PM (mid-season)	IIHR	VII	1993	R
Pea	PH-1	Mid-Season	Hisar	VII	1993	V
Pea	VL-6	Mid-Season	VPKAS	IV	1993	V
Tomato	FMH-2 (A.Vardhan)	Indet.	IIHR	I, VII	1993	H
Tomato	LE-79-5		Vellanikara	VII	1993	R
Tomato	Pusa Hybrid-2	Det.	IARI, New Delhi	I,IV,VI,VII	1993	H
Brinjal	ABH-2	Small Round	Anand	IV,VI	1995	H
Brinjal	DBR-8	Round	IARI, New Delhi	VI	1995	V
Brinjal	DBSR-31	Long	IARI, New Delhi	VI	1995	V
Brinjal	DBSR-44	Small Round	IARI, New Delhi	VI	1995	V
Brinjal	KS-224	Round	Kalyanpur	I,II,IV	1995	V
Brinjal	NDBH-6	Long	Faizabad	IV	1995	H
Brinjal	Sel-4	Long	Hyderabad	V	1995	V
Cabbage	BSS-32		Beejo Sheetal	VII	1995	H
Okra	HRB-55		Hisar	VI	1995	R
Onion	Agri Found Light Red		NHRDF	VIII	1995	V
Pea	NDVP-4	PM	Faizabad	IV	1995	R
Pea	PH-1	Mid-Season	Hisar	V	1995	V
Sponge gourd	Sel-99		IARI, New Delhi	IV,VI	1995	V
Tomato	BT-10	Bacterial wilt	Bhubaneshwar	V,VI	1995	R
Tomato	DTH-4	Det.	IARI, New Delhi	VII	1995	H
Tomato	A. Vardhan (FMH-1)	Indet.	IIHR	IV	1995	H
Tomato	KT-4	Indet.	IARI, Katrain	IV	1995	H
Tomato	NA-501	Det.	Nath Seeds	IV,VII	1995	H

Tomato	NA-601	Indet.	Nath seeds	VI,VII	1995	H
Tomato	Sel-1-6-4	Det.	Ludhiana	I	1995	V
Bitter gourd	MC-84		Vellanikara	VII	1996	V
Brinjal	AB-1	Round	Anand	III,VI,VII	1996	V
Brinjal	ABH-2	Small Round	Anand	VII	1996	H
Brinjal	BB-26	Long	Bhubaneshwar	V,VII	1996	V
Brinjal	BB-44		Bhubaneshwar	V,VII	1996	R
Brinjal	PLR-1	Small Round	Coimbatore	IV,VI,VII	1996	V
Brinjal	SM-6-6		Vellanikara	I,VII,VIII	1996	R
Chilli	Phule-C-5		Rahuri	VII	1996	V
Okra	HRB-9-2		Hisar	IV,VI	1996	R
Onion	Punjab Red	Rabi season	Ludhiana	VIII	1996	V
Onion	Punjab Round	Rabi season	Ludhiana	VII	1996	V
Pumpkin	Pusa Hybrid-1		IARI, New Delhi	IV, VI, VIII	1996	H
Tomato	BSS-20	Indet.	Beejo Sheetal	IV, VI, VIII	1996	H
Tomato	BT-12	Indet.	Bhubaneshwar	I,IV	1996	V
Tomato	DT-10	Indet.	IARI, New Delhi	IV,VI	1996	V
Tomato	Sel- 32	Det.	Hisar	II,VI,VII	1996	V
Brinjal	BB-13	Long	Bhubaneshwar	VIII	1997	V
Brinjal	Phule Hybrid-2	Small Round	Rahuri	VII	1997	H
Brinjal	Pusa Hybrid-9	Round	IARI, New Delhi	VI	1997	H
Cabbage	Nath-501		Nath Seeds	VII	1997	H
Chilli	ARCH-236		Ankur	IV	1997	H
Chilli	HOE-888		Sandoz	IV,VIII	1997	H
French bean	IIHR-909		IIHR	I	1997	V
Onion	PBR-5		Ludhiana	VI	1997	V
Pea	NDVP-8	Mid-Season	Faizabad	IV	1997	V
Tomato	H-24	TLCV (Moderate)	IIVR	V	1997	R
Bitter gourd	KBG-16		Kalyanpur	IV	1998	V
Bitter gourd	RHRBG-4-1		Rahuri	IV,VII	1998	V
Bottle gourd	PBOG-2		Pantnagar	VII	1998	H
Brinjal	CHBR-1	Round	HARP	IV	1998	V
Brinjal	DBSR-91	Small Round	IARI, New Delhi	VII	1998	V
Brinjal	Green Long	Green	Sabour	IV	1998	V
Brinjal	JB-15	Long	Jabalpur	I	1998	V
Brinjal	JB-64-1-2	Small Round	Jabalpur	VII	1998	V
Brinjal	JB-64-1-2	Small Round	Jabalpur	VII	1998	V
Brinjal	KS-331	Long	Kalyanpur	IV,V	1998	V
Cabbage	Quisto		Novartis	IV	1998	H
Cowpea	IIHR-6		IIHR	IV,VII	1998	V
Garlic	G-282		NHRDF	IV,VI,VII	1998	V
Okra	DVR-1		IIVR	IV, VII	1998	H
Okra	DVR-2		IIVR	VI	1998	H
Pea	NDVP-10	Mid-Season	Faizabad	IV	1998	V
Pea	VL-8	Mid-Season	VPKAS	I	1998	V
Tomato	Avinash-2	Det.	Novartis	VI	1998	H
Tomato	BRH-2	Bacterial wilt	IIHR	VII	1998	R
Tomato	HOE-303	Det.	Novartis	IV	1998	H
Tomato	KS-17	Indet.	Kalyanpur	IV	1998	V
Bottle gourd	PBOG-1		Pantnagar	I, VII	1999	H
Tomato	Sun-496	Indet.	Sungro	IV, II, VII	1999	H

Watermelon	MHW-6		Mahyco	-	1999	H
Bitter gourd	PBIG-1		Pantnagar	IV	2001	V
Bitter gourd	RHRBGH-1		Rahuri	All	2001	H
Bottle gourd	BG-L-C-2_1		Ludhiana	IV	2001	V
Bottle gourd	NDBH-4		Faizabad	All	2001	H
Bottle gourd	POBOG-61		Pantnagar	IV,VI	2001	V
Brinjal	ARBH-541	Long	Ankur	All	2001	H
Brinjal	D-2-88-6	Round	Sabour	VI & VII	2001	V
Brinjal	JBH-1	Round	Junagadh	All	2001	H
Brinjal	NDB-28-2	Long	Faizabad	IV	2001	V
Brinjal	PBH-6	Long	Pandey Beej	All	2001	H
Brinjal	Punjab Sadabahar	Long	Ludhiana	IV , VI	2001	V
Cauliflower	KT-25	Snowball group	IARI, Katrain	I	2001	V
Chilli	AKC-86-39		Akola	VII	2001	V
Chilli	BC-14-2		Bhubaneshwar	V, VI	2001	V
Chilli	RHRC-Cluster Erect		Rahuri	VII	2001	V
Cucumber	CHC-2		HARP	IV	2001	V
Cucumber	PCUCH-1		Pantnagar	All	2001	H
French Bean	CH-812	Bush	HARP	VII, III	2001	V
French Bean	CH-819	Pole	HARP	I	2001	V
Garlic	VLG-7		VPKAS	I	2001	V
Muskmelon	DMDR-1	CGMV	IARI, New Delhi		2001	R
Muskmelon	DMDR-2	Downy mildew + CGMV	IARI, New Delhi		2001	R
Okra	DVR-3		IIVR	All	2001	H
Okra	VRO-3		IIVR		2001	R
Pea	DPP-68	PM	Palampur	-	2001	R
Pea	KS-245	PM	Kalyanpur	-	2001	R
Pea	NDVP-12	Early Season	Faizabad	IV	2001	V
Pea	Organ Sugar Podded	Edible Podded	Ludhiana	VI	2001	V
Pea	VRP-2	Early Season	IIVR	VI	2001	V
Pea	VRP-3	Mid Season	IIVR	I	2001	V
Pumpkin	CM-350		Vellanikara	VII, VIII	2001	V
Pumpkin	NDPK-24		Faizabad	IV,VI	2001	V
Ridge gourd	CHRG-1		HARP	IV	2001	V
Ridge gourd	IIHR-7		IIHR	VIII	2001	V
Ridge gourd	PRG-7		Pantnagar	VII	2001	V
Tomato	BSS-20	Indet.	Beejo Sheetal	All	2001	H
Tomato	BT-116-3-2	Det.	Bhubaneshwar	V, VI	2001	V
Tomato	BT-20-2-1	Indet.	Bhubaneshwar	IV, VII	2001	V
Tomato	DVRT-2	Det.	IIVR	VI	2001	V
Tomato	KS-118	Det.	Kalyanpur	IV	2001	V
Tomato	NDT-3	Det.	Faizabad	IV, VII	2001	V
Tomato	NDT-9	Indet.	Faizabad	IV	2001	V
Bottle gourd	NDBG-104		Faizabad	IV	2002	V
Brinjal	BH-1	Round	Ludhiana	IV	2002	H
Brinjal	CHES-309	BW	HARP	I, VII	2002	R
Capsicum	Lario		Syngenta	I	2002	H
Chilli	PMR-57/88-K		IIHR	VII	2002	V
Chilli	Sungro-86-235		Sungro	IV, VIII	2002	H

Cucumber	CH-20		HARP	IV	2002	V
Cucumber	PCUC-28		Pantnagar	I, VII, VIII	2002	V
Muskmelon	DMDR-2	Downy mildew + CGMB	IARI, New Delhi	IV, V	2002	R
Muskmelon	MHY-5		Durgapura	VII	2002	H
Okra	DVR-4		IIVR	IV, V, VII	2002	H
Okra	VRO-4		IIVR	IV, V	2002	R
Pea	NDVP-250	PM	Faizabad	IV	2002	R
Tomato	ARTH-128	Indet.	Ankur	VII	2002	H
Tomato	CHTH-1	Det.	IARI, New Delhi	IV	2002	H
Tomato	DTH-8	Det.	IARI, New Delhi	IV	2002	H
Bitter gourd	Pusa Hybrid-2		IARI, New Delhi	IV, V, VI	2003	H
Brinjal	BH-2	Round	Ludhiana	IV, V	2003	H
Brinjal	VRBHR-1	Round	IIVR	IV, VI	2003	H
Capsicum	DARL-202		Pithoragarh	I, IV	2003	H
Cauliflower	DCH-541	Early group	IARI, New Delhi	II, IV	2003	H
Chilli	ARCH-228		Ankur	IV, V, VI	2003	H
Chilli	LCA334		Lam	III, IV, V, VII	2003	V
Cowpea	NDCP-13		Faizabad	II, III, IV, VII	2003	V
Garlic	DARL-52		Pithoragarh	I	2003	V
Garlic	G-323		NHRDF	VI	2003	V
Muskmelon	NDM-15		Faizabad	IV	2003	V
Okra	VRO-5		IIVR	VI	2003	R
Okra	VRO-6		IIVR	IV, V	2003	R
Pea	DPP-9411	PM	Palampur	I	2003	R
Tomato	KTH-2	Indet.	Kalyanpur	IV, V	2003	H
Bitter gourd	NBGH-167		Nirmal Seeds	IV	2004	H
Bottle gourd	NDBG-132		Faizabad	VI	2004	V
Brinjal	ABSR-2	Small Long	Anand	VII	2004	V
Brinjal	ARBH-786	Long fruited	Ankur	IV	2004	H
Brinjal	BB-64		Bhubaneshwar	IV, V, VII, VIII	2004	H
Brinjal	DBL-21	Long	IARI, New Delhi	IV	2004	V
Brinjal	IVBHL-54	Long fruited	IIVR	IV	2004	H
Brinjal	KS-235	Round	Kalyanpur	IV, V & VII	2004	V
Cauliflower	Summer King	Early group	Sungro	I, IV	2004	H
Cauliflower	SYCFH-202	Early group	Syngenta	IV, VII	2004	H
Chilli	ASC-2000-02		Anand	VII	2004	V
Cowpea	IVRCP-1		IIVR	IV	2004	V
Cucumber	Hybrid No. - 1		Century Seeds	I, IV, VII	2004	H
Dolichos bean	CHDB-1		HARP	IV	2004	V
Tomato	JKTH-3055	Determinate	J.K. Seeds	I, IV	2004	H
Tomato	KTH-1		Kalyanpur	IV	2004	H
Tomato	LE-415	Bacterial wilt	Vellanikara	I, V, VIII	2004	R
Tomato	Mani Laima	Det	ICAR Res. Com.for NEH Barapani	III	2004	V
Tomato	NDTS-2001-3	Det	Faizabad	IV	2004	V
Tomato	Nun-7730	Indeterminate	NunHems	I, IV	2004	H
Brinjal	BCTH-17	(NATP Prog.)	BCKV	II,V	2005	H
Brinjal	BWBH-3	BW resistance (NATP Prog.)	IIHR	All	2005	H
Brinjal	COBH-3	BW resistance (NATP Prog.)	TNAU, Coimbatore	VIII	2005	H

Brinjal	HABH-3	BW +Phomopsis resistance (NATP Prog.)	HARP	IV	2005	H
Brinjal	HABR-4	Round	HARP	IV	2005	V
Brinjal	IVBR-1	Round	IIVR	IV	2005	V
Brinjal	VNR-51	Small round	VNR	IV, VI	2005	H
Cabbage	KGMR-1		IARI, Katrain	I, IV	2005	H
Capsicum	KTCPH-3		IARI, Katrain	I,VI,VII	2005	H
Cauliflower	IVREC-2	Early	IIVR	IV	2005	V
Cauliflower	SYCFH-203	Early group	Syngenta	IV,V,VII	2005	H
Chilli	CCH-2		IIVR	II, IV,V,VI	2005	H
Chilli	CCH-3	Dual purpose (NATP Prog.)	IIVR	IV, V, VIII	2005	H
Chilli	KA-2		IIVR	IV	2005	V
Chilli	KCH-3	Pickle type fruit (NATP Prog.)	Kanpur	IV	2005	H
Chilli	MSH-149	CMS based (NATP Prog.)	IIHR	All	2005	H
Chilli	MSH-172	CMS based (NATP Prog.)	IIHR	All	2005	H
Cowpea	CHCP-2		HARP	VIII	2005	V
Cucumber	PCUCH-3		Pantnagar	I, IV	2005	H
French bean	IVFB-1		IIVR	I, VII	2005	V
Okra	HBH-142		Hisar	IV, V, VII, VII	2005	H
Okra	HRB-107-4		Hisar	VI,VIII	2005	R
Okra	IIVR-11		IIVR	V, VII	2005	R
Okra	NDO-10		Faizabad	IV	2005	R
Okra	SOH-152		Syngenta	IV, VII, VII	2005	H
Pea	CHP-2	Mid season	HARP	IV,VI	2005	V
Pea	KTP-8	PM	IARI, Katrain	I, IV, V	2005	R
Pea	VRP-5	Early	IIVR	I, IV, VIII	2005	V
Sponge gourd	CHSG-1		HARP	IV	2005	V
Sponge gourd	JSGL		Junagadh	VII	2005	V
Tomato	BCTH-4	TLCV resitance (NATP Prog.)	BCKV	II, V	2005	H
Tomato	BT-136	Indet	Bhubaneswar	II, IV	2005	V
Tomato	H-86	TLCV	IIVR	I,IV,V,VIII	2005	R
Tomato	HATH-3	BW& EB resistance (NATP Prog.)	HARP	IV	2005	H
Tomato	IIVR Sel-1	Det.	IIVR	V, VII	2005	V
Tomato	IIVR Sel-2	Indet	IIVR	IV	2005	V
Tomato	NTH-6	(NATP Prog.)	IIVR	IV, VI	2005	H
Tomato	TH01462	Determinate	Syngenta	I, II, IV, VI,VII	2005	H
Tomato	TLBRH-9	BW&TLCV resitance(NATP Prog.)	IIHR	All	2005	H
Tomato	VLT-32	Indet	VPKAS	IV	2005	V
Ash gourd	IVAG-90		IIVR	IV &VIII	2006	V
Ash gourd	PAG-72		Pantnagar	VIII	2006	V
Brinjal	HABL-1	Long	HARP	I	2006	V
Carrot	SKAUC-50		SKUA&T, Srinagar	I	2006	V
Muskmelon	IVMM-3		IIVR	IV	2006	V
Onion	HOS-1	Rabi season	Hisar	VI	2006	V
Onion	L-28	Rabi season	NHRDF	IV & VII	2006	V
Sponge gourd	KSG-14		Kalyanpur	IV	2006	V

Ash gourd	Pusa Ujwal		IARI, New Delhi	VIII	2007	V
Brinjal	HABH-17	Round	HARP	IV	2007	H
Brinjal	Navina	Long fruited	VNR	IV	2007	H
Brinjal	PB-66	Long	Pantnagar	VII, IV	2007	V
Cabbage	Green Emperor		Tokita Seeds	I	2007	H
Chilli	BC-25		Bhubaneshwar	VI, VII, V	2007	V
Chilli	LCA-353		Lam	V, VIII, IV	2007	V
Cowpea	IIVR CP-4		IIVR	IV, V, VII	2007	V
Okra	JNDOL-03-1		Junagadh	VII and VIII	2007	R
Okra	NBH-180		Nuzi Vedu Seeds	VII	2007	H
Okra	SOH-1016		Syngenta	IV,VII	2007	H
Onion	B-780-5-2-2	Rabi season	NRC (O&G)	VI	2007	V
Pea	PC-531	Mid season	Ludhiana	VI, VII, I	2007	V
Pea	VP-101	Early	VPKAS	IV, I	2007	V
Sponge gourd	PSG-40		Pantnagar	VII, I	2007	V
Bitter gourd	Vivek		Sungro	VIII	2008	H
Bottle gourd	SEL P-6		IARI, New Delhi	IV,VII	2008	V
Cauliflower	DC-76	Mid-late	IARI, New Delhi	I & VI	2008	V
Cauliflower	IVRMC-12	Mid-late	IIVR	IV	2008	V
Cowpea	Swarna Harita		HARP	II, IV, V, VIII	2008	V
Cowpea	VR-5		IIVR	IV, V, VII	2008	V
Muskmelon	GMM-3		Anand	IV & VII	2008	V
Okra	JNDOH 2-2		Junagadh	II, V, VI, VII, VIII	2008	H
Tomato	ARTH-734	Indeterminate	Ankur	VIII	2008	H
Tomato	HATH-5	Determinate	HARP	I	2008	H
Ash gourd	IVAG-3		IIVR	IV	2009	V
Bottle gourd	NDBG-619		Faizabad	VII	2009	V
Bottle gourd	Santosh-20		Krishidhan Seeds	IV	2009	H
Brinjal	HABH-8	Round	HARP	VIII	2009	H
Brinjal	PB-67	Long	Pantnagar	IV	2009	V
Brinjal	Rasika	Long fruited	Beejo Sheetal	IV	2009	H
Brinjal	Shamli	Long fruited	Seminis Seeds	IV	2009	H
Brinjal	VNR-51C	Small Round	VNR	IV	2009	H
Chilli	BSS-453		Beejo Sheetal	II	2009	H
Chilli	HS-HP-154		SKUA&T, Srinagar	VIII	2009	V
Chilli	IVPBC-535 (Paprika)		IIVR	VIII	2009	V
Chilli	PC-7		Pantnagar	V	2009	V
Dolichos bean	HADB-3	(Pole)	HARP	V	2009	V
French Bean	Arka Anoop	Bush	IIHR	I, V, VIII	2009	V
French Bean	HAFB-3	Bush	HARP	VI	2009	V
French bean	HAFB-4	Pole	HARP	VIII	2009	V
Garlic	AC-200		NRC(O&G)	VI	2009	V
Garlic	JNDG-213		Junagadh	IV	2009	V
Garlic	RAUG-5		Sabour	VII	2009	V
Onion	B-780-5-3-1		NRC(O&G)	VII	2009	V
Onion	PKV White		Akola	VI	2009	V
Onion	RHRO-S1		Rahuri	VI, VIII	2009	V
Pea	VP-233	PM(mid seaon)	VPKAS	I	2009	R
Pea	VRPMR-11	PM (mid-season)	IIVR	IV	2009	R
Pumpkin	HAPK-10		HARP	I	2009	V
Ridge gourd	HYRGH-5HB		Hyderabad	VII	2009	H

Ridge gourd	Pallavi		Sungro	V	2009	H
Tomato	BSS-488	Indeterminate	Beejo Sheetal	VI	2009	H
Tomato	HImsohna	Indeterminate	Syngenta	II	2009	H
Tomato	Tai-01458	Determinate	Syngenta	IV	2009	H
Bottle gourd	PBOG-89		Pantnagar	I, IV,VII	2010	V
Brinjal	DBL-02	Purple Long	IARI, New Delhi	I, IV, VI	2010	V
Brinjal	PB-70	Green Long	Pantnagar	IV, VI,VII, VIII	2010	V
Chilli	NCH-587		Nirmal Seeds	IV,VII	2010	H
Chilli	PC-56		Pantnagar	I, IV, V, VII, VIII	2010	V
Chilli	VNR-332		VNR	IV,VIII	2010	H
Dolichos bean (Pole)	IIVR Sem-8		IIVR	III, IV, V, VII	2010	V
Okra	JOH-05-9		Junagadh	V, VI, VII	2010	H
Sponge gourd	VR-1		IIVR	IV	2010	V
Tomato	BCTH-4	Determinate	Kalyani	I, IV	2010	H
Tomato	Pant T-10	Det.	Pantnagar	I, IV	2010	V
Ash gourd	KAG-1		Kalyanpur	VIII	2011	V
Brinjal	DBHL-20	Long	IARI, New Delhi	IV	2011	H
Brinjal	EPH-178	Round	Syngenta	IV	2011	H
Chilli	ACS-06-2		Anand	VIII	2011	V
Chilli	HH-41786		Syngenta	VII	2011	H
Chilli	VR-338		IIVR	II	2011	V
Cowpea (Bush)	Ankur Gomti		Ankur	V	2011	V
Cowpea (Bush)	VRCP-6		IIVR	IV	2011	V
Okra	JOL-2K-19		Junagadh	V	2011	R
Okra	VRO-22		IIVR	IV	2011	R
Tomato	ARTH-2104	Indet.	Ankur	IV	2011	H
Tomato	ATL-01-19	Det.	Anand	VI	2011	V
Tomato	PAU-2372	Indet.	Ludhiana	I	2011	V
Brinjal	VNR-218	Bacterial wilt	VNR Seed Company	II	2012	R
Brinjal)	PBHSR-31	Small Round	Ludhiana	IV & VI	2012	H
Brinjal)	PHBL-51	Long	Ludhiana	IV	2012	H
Chilli	BSS-378		Bejo Sheetal	VII	2012	H
Okra	AROH-631		Ankur	V & VII	2012	H
Okra	OH-597		Syngneta	VII	2012	H
Pea (Mid-season)	VP-434	powdery mildew	VPKAS, Almora	I & IV	2012	R
Tomato	VRTH-101	Det.	IIVR, Varanasi	I & IV	2012	H
Ash gourd	DAGH-14		IARI, New Delhi	IV	2013	H
Ash gourd	DAGH-16		IARI, New Delhi	VI&VIII	2013	H
Bitter gourd	Sel.-I		IARI, New Delhi	VI	2013	V
Bottle gourd	Anurag		Nuziveedu	IV	2013	H
Bottle gourd	NDBG 10		NUDA&T, Faizabad	IV	2013	V
Brinjal	HABR-21	Round	ICAR-RCER, Ranchi	IV	2013	V
Cabbage	KTCBH-81		Katrain	I	2013	H
Capsicum	DARL-70		DIBER, Pithorgarh	I	2013	V
Capsicum	PRCH-101		Ranichauri	I	2013	H
Chilli	VRN Vidya		VNR Seeds	IV	2013	H
Cucumber	Rajni-Cu-05		Syngenta, Seeds, Pvt. Ltd.	I	2013	H
Cucumber	Rajni-Cu-05		Syngenta, Seeds, Pvt. Ltd.	I	2013	H
Pea	GP 17	Early	IARI, New Delhi	I	2013	V

Pea	VRPE 25	Early	IIVR, Varanasi	IV	2013	V
Tomato	Bhagya		Nuziveedu	VI	2013	H
Tomato	VRT-0801	Det.	IIVR, Varanasi	IV	2013	V
Ash Gourd	DAG-12		IARI, New Delhi	VIII	2014	V
Brinjal	PBHL-52	Long	PAU, Ludhiana	IV	2014	H
Cauliflower	DC-31	Early	IARI, New Delhi	IV	2014	V
Cauliflower	SECF-102	Early	BAU, Sabour	VII	2014	V
Chilli	LCA-620		APHU HRS Lam	V	2014	V
Kale	KTK-64		IARI (RS), Katrain	I	2014	V
Okra	JOH-08-19		JAU, Junagadh	VI,VII	2014	H
Okra	VRO-25		IIVR	IV	2014	R
Tomato	DARL-68	Indeterminate	DIBER, Pithorgadh	III, IV	2014	V
Tomato	Punjab Ratta	Determinate	PAU, Ludhiana	IV	2014	V
Tomato Det.	Improved Bhagya	Det.	Nuziveedu Seeds	IV,VII	2014	H
Tomato Det.	KVTH-143	Det.	Krishidhan Seeds	I	2014	H
Brinjal	Nishant	Round	Advanta Seeds	IV	2015	H
Long Melon	VRSLM-16		IIVR-SPC, Sargatia	IV	2015	V
Tomato	IIHR-H-240		IIHR, Bangalore	VIII	2015	R
Pea (Mid)	IIHR-1	Mid	IIHR	I, IV, VIII	2016	R
Ridge gourd	VRRG-27		IIVR, Varanasi	IV	2016	V
Tomato	ATL-8-21	Determinate	Anand	VI	2016	V
Tomato	Kaveri-304 (KTH-304)	Indeterminate	Kaveri Seeds	IV	2016	H
Tomato	VRT-1201	Indeterminate	IIVR, Varanasi	VII	2016	V
Tomato	VRT-1202	Indeterminate	IIVR, Varanasi	VIII	2016	V

V=Variety; H=Hybrid; R=Resistant

The committee reviewed the data for the year, 2014-15, 2015-16 and 2016-17 thoroughly and following entries were identified for release and notification:

Sl.No.	*Crop*	*Trial*	*Code*	*Source*	*Zone*
Variety					
1.	Brinjal Long	Variety evaluation trial	2014/BRLVAR-3	IARI	VI, VII
2.	Brinjal Round	Variety evaluation trial	2014/BRRVAR-3	CHES, Bhubaneswar	V
3.	Cowpea (Bush Type)	Variety evaluation trial	2014/COPVAR-4	IARI	VI
4.	Dolichos Bean (Pole Type)	Variety evaluation trial	2014/DOLPVAR-1	IARI	VII
5.	Sponge gourd	Variety evaluation trial	2014/SPGVAR-1	IIVR	IV
6.	Cucumber	Variety evaluation trial	2014/CUCUVAR-1	IARI	IV
7.	Mustard Green	Variety evaluation trial	2014/MGVAR-1	Ranichauri	III
Hybrid					
1.	Chilli	Hybrid Evaluation Trial	2014/CHIHYB-3	-	IV
2.	Bitter gourd	Hybrid Evaluation Trial	2014/BIGHYB-2	-	IV
3.	Sponge gourd	Hybrid Evaluation Trial	2014/SPGHYB-2	IIVR	IV
Resistance varietal trial					
1.	Tomato (ToLCV)	Resistant Evaluation Trial	2014/TOLCVRES-5	-	VIII
2.	Okra (YVMV)*	Resistant Evaluation Trial	2014/OKYVRES-5	-	IV

Table 2b: List of Vegetable Varieties Developed by different Research Organizations and their Characteristics Features

1. Tomato (Open pollinated varieties)				
Variety	*Developing Centre*	*Recom-mened Zone*	*Year of identi-fication*	*Remarks*
S-12	PAU, Ludhiana (Punjab)	-	1975	It has been developed through irraditation. Plants are dwarf bushy and vigorous. Fruits are medium sized, compressed, round, uniform red at maturity, juicy and acidic having high yielding capacity. The yield ranges from 40.0-50.0 t/ha and seed yield is 500 g/q of fruits.
Pusa Ruby	ICAR-IARI, N. Delhi	All Zones	1975	This variety is derived through selection from the cross 'Sioux' x 'Improved Meeruthi'. Plants are indeterminate (80-85 cm), spreading type with less branches and hardy. Fruits are flattish small-medium, round, uniform red, slightly lobbed (4-5 locules) and possessing more acidity which is considered good for processing. Average yield is around 33.0 t/ha.This variety has exhibited yield potential of 63.2 t/ha under Ladakh conditions (Singh and Narayan, 2004). It is suitable for cultivation in autumn, winter and spring season. It is hardy and least affected by frost injury (Bhardwaj and Sharma, 1993). It is also suitable for rainfed conditions. It is best for processing (juice and ketchup preparation) and fresh market
HS-101	CCS HAU, Hisar, (Haryana)	-	1975	It is derivative of cross between Sel.2-3 and exotic cultivar and release by Central Sub-Committee on Release of Varieties in 1977 for Haryana, Uttar Pradesh, Gujarat and Tamil Nadu. It is suitable for winter season. Plants are tall (50 cm), multibranched, sturdy and determinate in growth habit. Fruits are borned in clusters (2-3 fruits/truss), round, small to medium in size, 3-4 loculed and uniform red on ripening. It bears 40-50 fruits/plant. Average yield is around 25-30 t/ha (Kalloo and Singh, 1976).
Pusa 120 (SL-120)	ICAR-IARI, N. Delhi	-	1975	Plants are determinate, spreading, late maturing and fruits are flattish round, medium-large, uniform red, and less acidic and less seeded; resistant to nematode. Average yield is around 30.5 t/ha.
Sweet-72	Regional Agricultural Research Institute Gwalior (M.P.)	-	1975	It is derived from a cross of Pusa Red Plum x Sioux (Wakankar, 1977). The plants are semi-erect in height and their height is 57 cm. the first picking starts 75-80 days after transplanting. The fruits are round oval and somewhat flattened in shape. They are attractive and scarlet red in colour. The fruit is firm, fleshy, possesses fine flavour and sweet in taste. It possesses fewer seeds. The numbers of locules are 4.6. The average weight of fruit is 75 g. Vitamin C content of this variety is very high (20.91 mg per 100 g pulp). It is very sweet in taste with sugar percentage of 1.62. The acidity content is 0.43 mg per 100 g pulp. An average yield of this variety is 20 t/ha.
Kalyanpur T-1	CSAUA&T, Kanpur, (Uttar Paradesh)	-	1975	It is a selection from local germplasm. Plants are 100-120 cm tall, erect and indeterminate in growth habit. Foliage is green in colour. Fruit is round, slightly pointed at the stigmatic end, 4-5 loculed, 4-5 fruits/truss, non-cracking, light green when unripe and red on ripening. It is susceptible to tomato mosaic virus. Seed yield is 80-100 kg/ha.
Pusa Early Dwarf	ICAR-IARI, N. Delhi	-	1977	This variety has been developed at ICAR-IARI, New delhi through selection from the cross Improved Meeruthi x Red Cloud. Plants are dwarf to indeterminate (80-85 cm), spreading and early maturing in nature. Fruits are flattish oblate (small round) medium large, red obscure, furrow with 5-6 locules. It is suitable for rainy season cultivation in plains and March sowing in hills. Average yield is around 30.0 t/ha.
Sioux (EC1151)	ICAR-IARI, Regional Station, Katrain (H.P.)	-	1977	This early variety has been developed through selection from an American introduction. Plants are indeterminate, early maturing, dwarf with compact fruiting. Fruits are whitish green colour when unripe and become uniform red when mature. Fruits are round, smooth, medium to large juicy, more acidic and less seeded. Average yield is around 29.5 t/ha. it is suitable for summer and spring season cultivation.
Sel-152	PAU, Ludhiana	-	1977	Small fruited cultivar which is very popular for early crop.
Punjab Chhuhara	PAU, Ludhiana, (Punjab)	-	1977	This variety has been developed through hybridization using parents Punjab Tropic and EC55055. Plants are determinate, dwarf (60.0 cm tall), bushy with dense luxuriant foliage. Fruits are small to medium round pear shaped having thick pericarp, fleshy, bilocular, less sour and less seeded. It is suitable for long distance transportation. It can retain marketable quality for a week after harvesting under ordinary conditions during summer. Average yield is around 23.0 t/ha. It is highly susceptible to Tomato Leaf Curl Virus (TLCV).
Azad T-2 (KS-2)	CSAUA&T, Kanpur (Uttar paradesh)	IV	1985	This variety has been developed through pedigree method from the cross HB-5 x Kuber. Plants are determinate, compact with leaves of curved (inwards) margin. Fruits are small, round with slightly furrowed, red and having 3-4 locules. Fruit possess 5.66 per cent of T.S.S., 3.68 mg/g of lycopene, and 15 mg/100 g vitamin C. First picking starts in 65-70 days after transplanting. Average yield is around 30.0-35.0 t/ha. It is moderately resistant to leaf curl virus.

Pant Bahar (AC-238)	GBPUA&T, Pantnagar (Uttarakhand)	-	1985	This variety has been developed through selection from the germplasm line (AC 238).Plants are indeterminate, bushy (90.0 cm height), profusely branched with light green foliage. Fruits are flattish-round, medium with 5-6 locules, slightly ridged, and have good storage and processing quality.First picking starts in 75-80 days after transplanting. Average yield is around 25.0 t/ha. It is suitable for cultivation in northern hills and plains. This variety has resistance to Verticillium and Fusarium wilt under field conditions. It has good storage and processing qualities. The variety could not become as popular as expected on account of relatively smaller fruit size and lack of consumer's interest.
Marutham (CO-3)	TNAU, Coimbatore (Tamil Nadu)	I, IV, V, VI, VII	1987	This variety has been developed through induced mutation by treating the seeds of CO-1 variety with EMS. Plants are bushy, determinate; 90 cm tall, erect, thick and main stem has strong primary branches with dark green foliage. Per plant 30-40 fruits can be obtained. Fruits are oval round-globular/flattish round, medium size bright smooth, four light grooves, parrot green (unripe), dark red (ripe), 5-6 locules slightly ridged, trilocular with thick pericarp, and born in cluster. Fruits possess high vitamin C and organic acid. First picking starts in 80-85 days after transplanting. It is capable of yielding as high as 45.0 t/ha under close spacing, in a duration of 110-115 days.
Punjab Kesari	PAU, Ludhiana (Punjab)	I, IV, V, VI	1987	This is an early maturing variety derived through selection from the cross Punjab Tropics x EC55055. Plants are determinate, erect, dwarf (60.0 cm) with profusely branched thick stem and dense foliage, which protect fruits from sunburn.Average weight of individual weight is 55 g.Average yield is around 56.25 t/ha.
La Bonita	ICAR-NBPGR, N. Delhi	I, IV, V, VII	1987	It is an introduction from USA. Plants are dwarf/semi determinate, heavy bearen and fruits are oblong (plum shaped) with thick pericarp. Average yield is around 30.0 t/ha. It is suitable for spring summer seasn. It is resistant to Fusarium wilt (Thomas *et al.*, 1976).
Pant T-3	GBPUA&T, Pantnagar (Uttarakhand)	I, II, IV, V, VI, VII, VII	1987	This variety has been developed through pure line selection and suitable for winter season cultivation in the Northern hills and plains. Plants are semi-determinate with thick round and hairy stem and dark green foliage.The leaves are medium in size and dark green in colour. Fruits are round, smooth and develop uniform red after ripening. First picking starts in 75-80 days after transplanting. The fruit weight is about 70 g. Average yield is around 30.0-35.0 t/ha. It is suitable for processing.
Arka Vikas (Sel-22)	ICAR-IIHR, Bengaluru (Karnataka)	I, IV, V, VI, VII, VIII	1987	This variety has been developed through selection from a variable population of American introduction (Tip Top) adapted to both rainfed and irrigated conditions. Plants are semi-determinate having narrow dark green foliage and good canopy. Fruits are medium large (80-90 g), firm, uniform and deep red, oblate with light green shoulder, adapted to both rainfed and irrigated conditions. Average yield is around 35.0-40.0 t/ha in 140 days of crop duration. It has yielded 53.25 t/ha in Ladakh region (Singh and Narayan, 2004). It is tolerant to mosaic. It is suitable for fresh market. It can be grown under rainfed conditions and is also suitable for both plains and hills.
Arka Saurabh (Sel-4)	ICAR-IIHR, Bengaluru (Karnataka)	I, IV, V, VI, VII, VIII	1987	This variety has been developed through selection from variable breeding line (V-605) introduced from Canada. Plants are semi-determinate (indeterminate) with broad light green foliage and good canopy. Fruits are thick, fleshy, firm, round, medium large (70-75 g) deep red and with nipple tip. It has an excellent blend of sugar and acidity and is very suitable for both fresh and processing purpose particularly for ketchup and juice making. Average yield is around 35.0 t/ha in 140 days of crop duration. It is resistant to fruit cracking, good transport quality and shelf life.
Hisar Arun (Sel-7)	CCS HAU, Hisar (Haryana)	I, IV, V, VII, VIII	1990	It is an early maturing variety developed through modified pedigree method from the cross Pusa Early Dwarf x K-1. Plants are determinate, dwarf, erect, cut leaves and synchronized clustered flowers. First picking starts in 60-65 days after transplanting. Fruits are medium round and per plant number of fruits are 15-20; having individual fruit weight of 65-70g.Fruits are deep red flesh and develop 4-6 locules. Average yield is around 15.4 t/ha.
Sel-1-6-1-4	PAU, Ludhiana (Punjab)	I	1995	This variety has been developed through selection from a cross between root knot nematode resistant line E 24-C8 and susceptible variety Punjab Chhuhara. Plants are determinate with dark green foliage, which protect fruits from sunburn. Fruits are pear shaped and medium size (40-50 g). Average yield is around 56.25 t/ha.
Hisar Lalit (Sel- 32)	CCS HAU, Hisar (Haryana)	II, VI, VII	1996	Plants are determinate. Fruits are flattish round, large, attractive and less seeded. Average yield is around 32.0 t/ha. It is suitable for spring and summer season cultivation and recommended for root knot nematode infested areas (Jain *et al.*, 1990.).
DT-10	ICAR-IARI, N. Delhi	IV, VI	1996	Plants are indeterminate, erect with thick stem and dark green broad foliage. Fruits are medium round, deep red with thick pericarp; suitable for processing and gives an average yield of 37.0 t/ha.

Utkal Urbasi (BT-12)	OUA&T, Bhubaneswar, (Odisha)	I, IV	1996	This variety has been derived from the cross Punjab Chhuhara x LE-79. Plants are indeterminate (70-85 cm) with good foliage, bear 5-8 fruits in cluster and fruits are pear shaped, red with green shoulder having 2-3 locules. Fruit has low seed but thick skin. It is resistant to bacterial wilt and tolerant to early blight. First picking starts in 100-115 days after sowing. Average yield is around 30.0-40.0 t/ha. Self-life of fruit is 10-12 days.
KS-17	CSAUA&T, Kanpur (Uttar Pradesh)	IV	1998	Determinate in nature and yield 28.0 t/ha in 105-110 days.
BT-116-3-2	OUA&T, Bhubaneswar (Odisha)	V, VI	2001	Determinate in nature and yield 24.8 t/ha in 100 days.
NDT-3	NDUA&T, Faizabad (Uttar Pradesh)	IV, VII	2001	Determinate in nature and yield 32.0 t/ha in 115 days.
KS-118	CSAUA&T, Kanpur (Uttar Pradesh)	IV	2001	Determinate in nature and yield 31.5 t/ha in 110 day.
Kashi Anupam (DVRT-2)	ICAR-IIVR, Varanasi (Uttarpradesh)	VI	2001	This variety has been developed through backcross pedigree method from a cross (Sel-7 (*Solanum lycopersicum*) x B-6013 (*S. hirsutum* f. *glabratum*). Plants are determinate and fruits are large spherical, attractive red with 5-6 locules. It has medium maturity and has been recommended for cultivation in Rajasthan, Gujarat, Haryana and Delhi. Average yield is around 55.0-60.0 t/ha.
BT-20-2-1	OUA&T, Bhubaneswar (Odisha)	IV, VII	2001	Indeterminate in nature and yield 25.0 t/ha in 100 days.
NDT-9	NDUA&T, Faizabad (Uttar Pradesh)	IV	2001	Indeterminate in nature and yield 22.5 t/ha in 125 days.
NDTS-2001-3	NDUA&T, Faizabad (Uttar Pradesh)	IV	2004	Determinate in nature and yield 27.0 t/ha in 118 days.
Mani Leima	ICAR-NEH, Barapani (Shillong) (Meghalaya)	III	2004	It is an early maturing, determinate and multibranched variety. Fruits are medium sized, oval, nipple shaped and bilocular. It is suitable for processing and long distance transportation. It is resistyant to fruit cracking and tolerance to leaf curl virus. It is also tolerant to moisture stress and suitable for spring summer season. Yield potential is around 63 t/ha. It is also released from State Variety Release Committee, Manipur (Yadav *et al.,* 2000).
IIVR Sel-1	ICAR-IIVR, Varanasi (Uttar Pradesh)	V, VII	2005	Plant determinate, fruits colour red, shape round weight 80 to 85 g. Yield 40.0-42.0 t/ha.
IIVR Sel-2	ICAR-IIVR, Varanasi (Uttar Pradesh)	IV	2005	Plants are indeterminate and fruits are slightly ovalfirm, thick pericarp, longer shelf life and 90 to 95 g fruit weight. Yield : 40.0-50.0 t/ha
BT-136	OUA&T, Bhubaneswar (Odisha)	II, IV	2005	Determinate in nature and yield 23.0 t/ha.
VLT-32	ICAR-VPKAS, Almora (Uttarakhand)	IV	2005	Determinate in nature and yield 25.0 t/ha.
Hisar Gaurav	CCS HAU, Hisar (Haryana)	IV	1996	Plants are straight, less branched and required staking. Fruits are large, flat orange and fleshy. It is recommended for cultivation in leaf curl virus infested area. Average yield is 32.0 t/ha.
Hisar Lalima (SEI-18)	CCS HAU, Hisar (Haryana)	IV	1992	It is developed through hybridization of Pusa Early Dwarf x HS-101. It is an early maturing and large fruited variety. Fruits are round, flat attractive red, fleshy, with 4-6 locules, thick pericarp, less prone to cracking and suitable for both the tomato seasons (autumn and spring) in the north India. Harvest index is very high because foliage is less and fruits are more in number. A average fruit weight is 75.0 g.

Tomato (Hybrids)				
Hybrids	*Developing Centre*	*Recom-mened Zone*	*Year of identi-fication*	*Remarks*
ARTH-4	Ankur Seeds Co., Nagpur (Maharashtra)	IV, VIII	1992	Plants are indeterminate/semi spreading with dark green foliage. Fruits are deep red and flat round. Flesh is red with 4.0-4.5 per cent TSS. It is moderately tolerant to drought. Harvesting starts after 80-85 days after sowing and continues up to 170-190 days. Yield potential is around 10.0-11.0 t/ha.
Gulmohar (MTH-6)	Mahyco Seed Co. Jalana (Maharashtra)	VII, VIII	1992	Plants are indeterminate. First picking starts in 60-67 days after transplanting. Fruits are red, square shape with average weight of 90.0 g. TSS is 4.17 per cent. Yield potential is around 9.0-10.0 t/ha.
ARTH-3	Ankur Seeds Co. Nagpur (Maharashtra)	II, VII, VIII	1992	Plants are determinate with medium dark green foliage. Fruits are squarish round and deep red. Flesh is red with 80-85 days after sowing and continues up to 160-180 days. Yield potential is around 8.0-9.5 t/ha.
Pusa Hybrid-2	ICAR-IARI, N. Delhi	I, IV, VI, VII	1993	This determinate hybrid has been developed and notified by the Central Varietal Release Committee (CVRC). Plants are compact, determinate with good foliage cover having prolific bearing. Fruits are medium, round to flattish round, firm, smooth and attractive with uniform red maturity. Fruit skin is orange red flesh. Plants are resistant to nematodes. It has been recommended for cultivation in humid western Himalayan regions, sub-humid Sutlej-Ganga alluvial plains and western plains, semi-arid plateau and has yield potential of 50.0-60.0 t/ha (Tewari *et al.,* 1996). It is highly tolerant to nematodes.
Arka Vardan (FMH-2)	ICAR-IIHR, Bengaluru (Karnataka)	I, VII	1993	This indeterminate hybrid has its parents IHR-550-3 and IHR-932. It has an average fruit weight of 140 g with oblate fruits, thick fleshed, green shouldered, and deep red with ashelf life of 13 days at room temperature yielding 75.0 t/ha in crop duration of 160 days. It is resitant to nematodes and is suited for fresh market.
NA-601	Nath Seeds Co. Aurangabad (Maharshtra)	VI, VII	1995	Determinate in nature and yield potential is 38.0 t/ha.
DTH-4	ICAR-IARI, N. Delhi	VII	1995	Determinate in nature and yield potential is 35.5 t/ha.
KT-4	Regional Station Katrain (H.P.)	IV	1995	This indeterminate variety has been developed from Res. Station Katrain (H.P.), IARI, N. Delhi.
NA-601	Nath Seed Co. (Maharashtra)	VI, VII	1995	Indeterminate in nature and yield potential is 31.0 t/ha.
Avinash-2	Novartis Seed Co. Pune (Maharshtra)	VI	1998	Plants are determinate and fruits are medium, thick walled, round, glossy, dark red with average weight of 70-90 g. Average yield is around 75t/ha.
HOE-303	Novartis Seed Co. Pune (Maharshtra)	IV	1998	Determinate in nature and yield potential 29.6 t/ha.
Sun-496	Sungro Seed Co. Shalimar bagh (Delhi)	II, IV, VII	1999	Indeterminate in nature and yield potential is 38.8 t/ha.
Meenakshi (BSS-20)	Beejo Sheetal Seed Co. Jalna (Maharashtra)	All	2001	Plants are indeterminate fruits are dark red, round global in shape with an individual fruit weight of 80 g. It is resistant to Fusarium wilt. It gives an average yield of 60-100 t/ha.
CHTH-1	ICAR-IARI, N. Delhi	IV	2001	Determinate in nature and yield potential is 34.0 t/ha.
ARTH-128	Ankur Seed Co. (Maharshtra)	VII	2002	Indeterminate in nature and yield potential is 37.0 t/ha.
DTH-8	ICAR-IARI, N. Delhi	IV	2002	Determinate in nature and yield potential 39.6 t/ha.
KTH-2	CSAUA&T, Kanpur (Uttara Pradesh)	IV, V	2002	Indeterminate in nature and yield potential is 32.0 t/ha.
CHTH-1	HARP, Ranchi (Jharkhand)	IV	2002	Determinate in nature and yield potential is 33.0 t/ha.
KTH-2	CSAUA&T, Kanpur (Uttar Pradesh)	IV, V	2003	Indeterminate in nature and yield potential is 30.0 t/ha.
JKTH-3055	J.K. Seeds Co. Hyderabad (Telangana)	I, IV	2004	Determinate in nature and yield potential is 41.0 t/ha.

KTH-1	CSAUA&T, Kanpur (Uttar Pradesh)	IV	2004	Indeterminate in nature and yield potential is 36.0 t/ha.
Nun-7730	Nunhems Seed Co. Hyderabad (Telangana)	I, IV	2004	Indeterminate in nature and yield potential is 40.0 t/ha.
TH-01462	Syngenta Seeds Co., Pune (Maharashtra)	I, II, IV, VI, VII	2005	Determinate in nature and yield potential is 41.5 t/ha.
TLBRH-9	ICAR-IIHR, Bengaluru (Karnataka)	IV, VI, VII	2005	Determinate in nature and yield potential is 35.5 t/ha.
HATH-3	ICAR-HARP, Ranchi (Jharkhand)	IV	2005	Determinate in nature and yield potential is 33.5 t/ha.
ARTH-734	Ankur Seeds Co. Nagpur (Maharshtra)	VIII	2008	Indeterminate in nature and yield potential is 42.0 t/ha.
HATH-5	HARP, Ranchi (Jharkhand)	I	2008	Determinate in nature and yield potential is 34.8 t/ha.
BSS-488	Beejo Sheetal Jalna (Maharshtra)	VI	2009	Indeterminate in nature and yield potential is 44.0 t/ha.
Himsona	Syngenta Seeds Co., Pune (Maharashtra)	II	2009	Indeterminate in nature and yield potential is 42.8 t/ha.
Tai-01458	Syngenta Seeds Co., Pune (Maharashtra)	IV	2009	Determinate in nature and yield potential is 43.5 t/ha.
BCTH-4	BCKV, Kalayani (West Bengal)	II, V	2010	Determinate in growth habit with uniparous cyme; less serrated leaf blade; roundish-flat, bright red colour, weighing 65 g with 3 locules and medium ribbing on fruit surface. Tolerant to tomato leaf curl virus under autumn winter condition. Average yield is 55 t/ha under autumn-winter season in West Bengal.
ARTH-104	Ankur Seeds Co. Nagpur (Maharashtra)	I, IV	2011	Indeterminate in nature and yield potential is 38.8-41.0 t/ha.
VRTH-101	ICAR-IIVR, Varanasi (Uttar Pradesh)	I, IV	2012	Determinate in nature and yield potential is 40.0-42.6 t/ha.
Bhagya	Nuziveedu Seed co. Hydrabad (Telangana)	VI	2013	Determinate in nature and yield potential is 41.0-45.8 t/ha.
Improved Bhagya	Nuziveedu Seed co. Hydrabad (Telangana)	IV, VII	2014	Determinate in nature and yield potential is 44.0 t/ha.

Tomato (Resistant varieties)

Variety	*Developing Centre*	*Recom-mened Zone*	*Year of identi-fication*	*Remarks*
Arka Alok (BWR-5)	ICAR-IIHR, Bengaluru (Karnataka)	II	1992	This variety is derived from selection of an introduced line (CL-114-5-1-0) from AVRDC (World Vegetable Centre), Taiwan. This is determinate and bacterial wilt resistant variety. Fruits are oblong with light green shoulder. Individual fruit weight is 100 g. Average yield is around 30-35 t/ha. It is resistant to bacterial wilt.
Arka Vishal (FMH-1)	ICAR-IIHR, Bengaluru (Karnataka)	V, VII	1993	It is a indeterminate hybrid developed between the cross of IIHR-837 x IIHR-932. The fruits are oblate, deep red on ripening, firm and green shouldered with good shelf life (13 days on room temperature). Average fruit weight is 140.0 g and yield is 75.0 t/ha. It is resistant to bacterial wilt and root knot nematodes. It is tolerant to fruit cracking.
Utkal Kumari (BT-10)	OUA&T, Bhubaneswar (Odisha)	V, VI	1995	This is indeterminate and bacterial wilt resistant variety. It is derived from the cross Sel.-22 x BT-2. Fruits are round red medium-large with 5-6 days shelf life. First picking starts in 90-95 days after sowing. Average yield is around 40.4 t/ha.

Hisar Anmol (H-24)	ICAR-Varanasi (Uttar Pradesh)	V	1997	Developed through back cross pedigree selection from the interspecific cross Hisar Arun (*Solanum lycopersium*) and *S. hirsutum* f. *glabratum*. Plants are determinate. Fruits are flattish-round, fleshy, medium and red. It gives an average yield of 40.0 t/ha. This variety is resistant to TLCV (Moderate) and being utilized worldwide for developing resistant variety. It is recommended for growing in the areas where this problem is serious.
BRH-2	ICAR-IIHR, Bengaluru (Karnataka)	VII	1998	Resistant to bacterial wilt.
LE-415	KAU, Vellanikkara Thrissur, (Kerala)	I, V, VIII	2004	LE-415 was developed through pure line selection at College of Horticulture, Vellanikkara Thrissur (Kerala). Plants are determinate with pubscent stem and normal leaf. Fruits are round, white in immature stage but turn red at the time of ripening and of medium size with average weight of 58 g; first picking starts in 75 days after sowing; field resistant to Bacterial wilt cracking; gives an average yield of 43.5 t/ha in 95-100 days of crop duration.
Kashi Vishesh (H-86)	CCS HAU, Hisar (Haryana) and ICAR-IIVR, Varanasi (Uttar Pradesh)	I, IV, V, VIII	2005	This is a TYLCV tolerant variety developed through backcross pedigree selection using x *S. hirsutum* f. *glabratum* B-6013, as a donor parent. It has determinate growth habit, dark green foliage and red spherical medium to large fruit weighing ~80 g. First harvest can be taken at 70-75 days after transplanting and it gives a total yield potential of abouit 50.0 t/ha.
(2) Brinjal (Open pollinated varieties)				
Hisar Pragati	CCSHAU, Hisar (Haryana)	0	1993	Suitable for summer season, fruits long bright dark purple colored
Pusa Purple Long	IARI, N. Delhi	IV, VI, VII, VIII	1975	This variety has been developed by selection from the Indigenous material collected from Punjab. It is an early maturing variety and is suitable for all growing seasons. Plants are dwarf (40-50 cm), semi-erect with light green foliage. First picking is possible 45 days after planting. Fruits are long (20-25 cm) shining purple, glossy and tender. Average yield is 28.0-35.0 t/ha.
Pusa Purple Cluster	ICAR-IARI, N. Delhi	IV, V, VI, VII	1975	It is an early maturing variety and developed by selection from the Indigenous material. Plants are tall, erect and compact with purple pigmentation on stem and leaves. First picking is possible 65 in days after transplanting. Fruits are deep purple long (10-12 cm) and born in cluster (4-9). Moderately resistant to bacterial wilt. It is recommended for hilly regions and mild climates of the country. Average yield is 30.0 t/ha.
S-16	PAU, Ludhiana (Punjab)	IV, VI	1975	Fruits are long, plump and shining purple. Average yield is about 28.7 t/ha.
Pusa Kranti	ICAR- IARI, N. Delhi	IV	1977	It has been developed by selection from a three-way cross (Pusa Purple Long x Hyderpur) x Wynad Giant. Plants are medium tall, upright and with purple pigmentation on leaves. Fruits are thick oblong (15-20 cm long), and dark purple with shining green calyx. Average yield is about 35.0 t/ha.
Pant Samrat (PB -129-5)	GBPUA&T, Pantnagar (Uttarakhand)	IV	1981	Fruits are medium long, dark purple and born in cluster. First picking can be done in 70 days after transplanting. It is tolerant to *Phomopsis* blight, bacterial wilt, shoot and fruit borer and jassids
Pant Samrat	GBPUA&T, Pantnagar (Uttarakhand)	IV, V, VI	1981	It is an early maturing variety developed by selection from the Indigenous material available around Haldwani (Foothills of the Himalayas). Plants are tall (80-120 cm), vigorous, with purple green broad leaves. Fruits are medium long, dark purple, attractive and fruits emerge in clusters. It takes about 70 days for first picking after transplanting. Average yield is 38.6 t/ha. Resistant to *Phomopsis* blight and bacterial wilt under field conditions; less infestation of shoot and fruit borer and jassids.
Arka Sheel	ICAR- IIHR, Bengaluru (Karnataka)	VIII	1981	It is a pure line selection from a 'Coorg type'. Plants are tall and fruits are medium long, and deep shining purple skin with green calyx. Fruits content less seed. Average yield is 38.0 t/ha in 120 days.
Azad Kranti	CSAUA&T, Kanpur (Uttar Pradesh)	I, VI	1983	Plants are medium tall and fruits are uniform oblong (15.20 cm) dark purple with green calyx and white flesh. Average yield is 30.0-42.5 t/ha.
PBR -91-2	GBPUA&T, Pantnagar (Uttarakhand)	-	1985	Plants are semi erect and fruits are round, purple, slightly tapering towards the bottom. Average yield is 30.0 t/ha.
ARU-1	DARL, Pithoragarh (Uttarakhand)	I	1985	Plants are upright and bear long purple fruits.
T-3	CSAUA&T, Kanpur (Uttar Pradesh)	-	1975	Fruits are round, light purple and whitish green colour at stigmatic end. Resistant to little leaf and bacterial blight. Average yield is 29.5-34.0 t/ha.
H-4	HAU, Hisar (Haryana)	-	1987	Plants are bushy with pigmented stem. Fruits are long to medium, thin, and dark purple in colour with light flesh.

ARU-2C	DARL, Pithoragarh (Uttarakhand)	I, IV, VI, VIII	1987	Fruits are cylindrical, long, bright, violet, and born in cluster of 4-6. This is a bacterial wilt tolerant variety. Average yield is 31.0 t/ha.
K-202-9	CSAUA&T, Kanpur (Uttar Pradesh)	VI	1987	Fruits bright purple, round and mature in 70-75 days after transplanting. Tolerant to fruit rot, wilt and blight
Aruna	PKV,Akola (Maharashtra)	VII	1988	Plants are prolific bearing and fruits are small round-oval with purple skin. Average yield is 41.2 t/ha and 22.0 t/ha during rainy and summer season, respectively.
H-7	HAU, Hisar (Haryana)	IV, VI	1990	Plants are compact, dwarf, bushy, indeterminate and erect. Fruits are long, and dark bright purple with creamy white flesh. Average yield is 33.36 t/ha.
NDB-25	NDUA&T, Faizabad (Uttar Pradesh)	II, IV, VII	1990	Long fruit, dark bright purple and white flesh. yield is about 30.50 t/ha.
Hisar Shyamal (H-8)	HAU, Hisar (Haryang)	II, IV, V, VI	1990	It is developed by crossing of Aushye x BR-112. Plants are erect with purple green foliage from flowering onwards.Fruits are round, dark bright purple with creamy white flesh and green purple pedicel. Fruits are round in shape. Average yield is 32.5 t/ ha. This variety is resistant to bacterial wilt and tolerant to little leaf. It is less prone to shoot and fruit borer.
Punjab Barsati	PAU, Ludhiana (Punjab)	IV	1993	It has been developed by selection from the cross 'Pusa Purple Cluster' x 'PH-4'. It is suitable for planting in rainy season and takes 65 days for first picking. Plants are dwarf, erect and thornless with green foliage. Fruits are medium long (18-20 cm), dark purple and produced in cluster. It is tolerant to fruit borer. Average yield is about 35.0 t/ha (Chadha and Sidhu, 1987).
Sel-4	APAU, Hyderabad (Telangana)	V	1995	Plants are erect, green stem and fruits are medium long, light purple, bear in cluster (3-5). Average yield is about 35.50 t/ha.
DBSR-31	ICAR-IARI, N. Delhi	VI	1995	Fruits are oval-round glossy purple, medium size, and 4-5 fruits/kg weight. It is of medium maturity duration (85 days for first harvest). Average yield is about 43.0 t/ha.
KS-224	CSAUA&T, Kanpur (Uttar Pradesh)	I, II, IV	1995	Plants are medium tall, compact with erect stem and fruits are round purple with green shining. Individual fruit weight is 300 g. Average yield is about 30-40 t/ha.
DBR-8	ICAR-IARI, N. Delhi	VI	1995	Plants are semi-upright and fruits are round (9.0 cm diameter), dark purple with 295 g of average weight. Average yield is about 41- 65 t/ha.
DBSR-44	ICAR-IARI, N. Delhi	VI	1995	Plants are 60.0-70.0 cm tall with green leaves. Fruits are small oval round, violet purple, born in cluster of 4-5 and having individual fruit weight of 125 g. Average yield is 25.0-30.0 t/ha.
AB-1	**GAU, Gujarat**	**III, VI, VII**	1996	**Round**
PLR-1	TNAU, Coimbatore (Tamil Nadu)	IV, VI, VII	1996	This variety is cluster bearing in nature and fruits are small round, egg shaped, drak purple and glassy in appearance. Average yield is about 25.1 t/ha.
BB-26	OUA&T, Bhubaneswar (Odisha)	V, VII	1996	Plants are tall erect and fruits are long, medium large, deep purple and tolerant to bacterial wilt and *Fusarium* wilt. Average yield is about 32.0 t/ha.
BB-13	OUA&T, Bhubaneswar (Odisha)	VIII	1997	Plants are medium tall, non-spiny and bear profuse fruits in cluster. Fruits are long, small medium purple with white flesh. Average yield is about 34.8 t/ha.
KS-331	CSAUA&T, Kanpur (Uttar Pradesh)	IV, V	1998	Plants are upright and fruits are long (15-19 cm). Average yield is about 40.0 t/ha.
JB-15	JNKVV, Jabalpur (Madhya Pradesh)	I	1998	Plants are upright and fruits are long (14.8 cm). Average yield is about 14.8 t/ha.
CHBR-1	HARP, Ranchi (Jharkhand)	IV	1998	Plants are prostrate (spreading type) and fruits are round and dark violet in colour. Average yield is about 75.6 t/ha.
DBSR-91	ICAR-IARI, N. Delhi	VII	1998	Plants are semi-upright and fruits are small round and purple. Average yield is about 56.0 t/ha.
JB-64-1-2	JNKVV, Jabalpur (Madhya Pradesh)	VII	1998	Plants are upright and fruits are small round long (6.0 cm). Average yield is about 31.5/ha.
Green Long	BAU, Sabour (Bihar)	IV	1998	Plants are upright and fruits are green long (15-19 cm). Average yield is about 40.0/ha.

Punjab Sada Bahar	PAU, Ludhiana (Punjab)	IV, VI	2001	It has been developed by selection from the cross Japanese Long x R-34. Plants are upright dwarf, erect and thornless with green foliage. It takes 75 days from transplanting to first picking. Fruits are long (20-25 cm), diameter 3-4 cm, thin and deep purplish to blackish in colour. Average yield is 60 t/ha. It is suitable for planting in all season.
NDB-28-2	NDUA&T, Faizabad (Uttar Pradesh)	IV	2001	Long-fruited variety. Average yield is about - 34.33 t/ha.
DBL-21	ICAR-IARI, N. Delhi	IV	2004	Long-fruited variety. Average yield is about 39.0 t/ha.
KS-235	CSAUA&T, Kanpur (Uttar Pradesh)	IV, V and VII	2004	Round-fruited variety. Average yield is about 31.83 t/ha.
ABSR-2	GAU, Anand, Gujarat	VII	2004	Small long fruits. Average yield is 29.88 t/ha.
HABR-4	HARP, Ranchi (Jharkhand)	IV	2005	Fruit are attractive, shiny purple in colour, and round in shape. Moderately resistant to bacterial wilt and average yield is 60-65 t/ha.
Kashi Prakash (IVBR-1)	ICAR-IIVR, Varanasi (Uttar Pradesh)	IV	2005	An open pollinated variety. Plants are semi upright, with green stems and leaves. Fruits oblong, attractive with light green spots, calyx spiny and average individual fruit weight 190 g. The picking starts 80-82 days after transplanting.Yield is about 65-70 t/ha.
HABL-1	HARP, Ranchi (Jharkhand)	I	2006	Developed through pure line selection. Fruit are medium long size (250 g), and shiny purple colour. Resistant to bacterial wilt. First harvest 75-80 days after planting.Yield is about 450-550 q/ha.
PB-66	GBPUA&T, Pantnagar (Uttarakhand)	VII, IV	2007	Long fruited cultivar. Average yield is about 38.9 t/ha.
BR-112	CCS HAU, Hisar (Haryana)	All Zones	1976	Early, plants are bushy, fruits round with bright purple colour and fleshy. Average yield is 25.0 t/ha.
PH-4	PAU, Ludhiana/ CCS HAU, Hisar (Haryana)	All Zones	1976	It is a cross of Hyderpur x Pusa Purple Long. Plant bushy, fruit medium long to long and thin, deep purple in color, flesh light green. Average yield is 25-30 t/ha.
Hisar Jamuni	CCS HAU, Hisar (Haryana)			It is product of Aushey round brinjal. Fruits oblong, bulging at the blossom end, 10-15 cm long, deep bright, purple blackish and dual season variety. It is very much suitable as ratoon crop in early spring. Average yield is 26-28 t/ha.

Brinjal (Hybrids)				
Hybrids	*Developing Centre*	*Recom-mened Zone*	*Year of identi-fication*	*Remarks*
Arka Kusumkar	ICAR-IIHR, Bengaluru (Karnataka)	VIII	1981	Plants are spreading and fruits are green, long, small and born in cluster of 5-7 and having very good appealing quality and taste. It matures in 110 days after transplanting. Average yield is 49 t/ha.
Arka Navneet	IIHR, Bengaluru (Karnataka)	-	1981	It is an early maturing F_1 hybrid developed between the cross of IIHR-22-1 x Supreme. This is a very high yielding round fruited hybrid with shining deep purple colour and green calyx. The fruits are shining dark purple, calys green, large, and oval. It is free from bitterness, skin attractive, and flesh soft with few seeds. Average fruit weight is 450 g and yield is 65.0-70.0 t/ha. It is recommended in south India under mild climate.
KT-4	ICAR- IARI Regional Station, Katrain (H.P.)	VIII	1987	Fruits are long, born in cluster of 3-5 fruits, cylindrical, tender and firm with purple colour. It has very good appealing quality and taste. It is resistant to wilt. Average yield is 37.1 t/ha.
Pusa Hybrid-6	ICAR-IARI, N. Delhi	IV	1990	Plants are non-spiny, semi erect plants having partial pigmentation on younger leaves and growing tips. Fruits are round, glossy, attractive and purple with partially pigmented peduncle. Single fruit weights 250.0-300.0 g. It takes 85-90 days from sowing to first picking. Average yield is 42-45 t/ha.
Pusa Hybrid-5	ICAR- IARI, N. Delhi	IV, VII, VIII	1992	It is an early bearing hybrid. Plants are vigorous and non-spiny with semi-erect branches. Leaves are green with bright pigmentation towards growing tips and younger leaves. Fruits are long, glossy, attractive and dark purple with partially pigmented peduncle. Individual fruit weight is 100 g. Average yield is 52-55 t/ha.
ARBH-201	Ankur Seed Co. (Maharashtra)	IV, V, VI, VII	1993	Long-fruited hybrid suitable for both the season. Average yield is 36.00 t/ha.
Narendra Hybrid Brinjal-1 (NDBH-1)	NDUA&T, Faizabad (Uttar Pradesh)	IV, VI, VII	1993	Plants are semi erect, non-spiny, and hardy green stems with light purple upper portion. Fruits are round attractive purple, soft and fleshy. This is moderately resistant to shoot and fruit borer, *Phomopsis* blight and little leaf. Average yield is 60-70 t/ha.

ABH-1	GAU Anand (Gujarat)	IV, VI, VII	1993	Small Round hybrid. Fruits are attractive. Average yield is about 42.85 t/ha.
Kalptaru (MHB-10)	Mahyco Seed Co. (Maharashtra)	IV, VI, VII	1993	Plants are erect, bushy, compact and spiny. Fruits are small round- oval. Average yield is 40-50 t/ha.
MHB-39 (F_1)	Mahyco Seed Co. (Maharshtra)	IV, VI, VII	1993	It is a cross of MBA-149 x MBR-1285. Fruits are small round, oval, shining, and reddish purple with non-spiny calyx born in cluster (4-6). It is an early bearing hybrid and takes about 60-65 days for first picking. Average yield is 35.0 t/ha in a crop duration of 130 days.
Narendra Hybrid Brinjal-6 (NDBH-6)	NDUA&T, Faizabad (Uttar Pradesh)	IV	1995	It is a cross of T-3 x Pant Samrat. Plants are non-spiny; semi erect, green stems with light purple colour. Fruits are long-oval, attractive, purple, soft and fleshy born in bunch. Individual fruit weight is around 150-200 fg. This is moderately resistant to shoot and fruit borer, phomopsis blight and little leaf. Average yield is 60.0 t/ha.
ABH-2	GAU, Anand (Gujarat)	IV, VI	1995	Small round fruits. Suitable for summer and rainy season. Average yield is about 39.45 t/ha. Resistant to wilt.
ABH-2	GAU, Anand (Gujarat)	VII	1996	Small round fruits. Fruits are attractive. Average yield 41.25 t/ha. It is resistant to wilt and leaf blight.
Phule Hybrid-2	MPKV, Rahuri (Maharashtra)	VII	1997	Small round fruits are very attractive. Can be grown in rainy and summer season. Average yield is 41.00 t/ha.
Pusa Hybrid-9	ICAR-IARI. N. Delhi	VI	1997	Fruits are oval round, glossy, dark purple, individual fruit weight 300 g and it takes 90 days from sowing to first picking. Average yield is 67 t/ha.
ARBH-541	Ankur Seed Co.	All	2001	Long fruited hybrid. Average yield is 45.00 t/ha.
PBH-6	Pandey Beej Co., Faizabad (Uttar Pradesh)	All	2001	Long fruited hybrid. Average yield is 41.00 t/ha.
JBH-1	Junagadh Agril. Univ. (Gujarat)	All	2001	Round fruited hybrid. Average yield is 40.50 t/ha.
BH-1	PAU, Ludhiana (Punjab)	IV	2001	This hybrid has been developed from the cross Punjab Jamuni Gola x Punjab Barsati at PAU, Ludhiana (Punjab). It is suitable for spring and autumn planting. Plants are medium tall, compact, thornless and have green foliage. First picking is possible in 52 days after planting. Fruits are oblong, purple and shining with green calyx. Average fruit weight is 300 g and yield is 62.5.0 t/ha.
BH-2	PAU, Ludhiana (Punjab)	IV, V	2002	This hybrid has been developed from the cross Punjab Neelam x Punjab Barsati at PAU, Ludhiana (Punjab). Plants are medium tall, spreading, thornless and have purplish green foliage. Fruits are oblong and deep purple with purple pigmentation on calyx. It is suitable for '*Bhartha*' making. Average fruit weight is 300 g and yield is 59.0 t/ha.
Kashi Sandesh (VRBHR-1)	ICAR-IIVR, Varanasi (Uttar Pradesh)	IV, VI	2002	Hybrid having semi-upright (height 71.0 cm) plant habit. Fruits are purple, size medium, shape round, fruit length. 12.4 cm, diameter 10.2 cm and weight 225.3 g. Picking starts in 76 days after transplanting. Yield is 75.0-80.0 t/ha.
Kashi Komal (IVBHL-54)	ICAR-IIVR, Varanasi (Uttar Pradesh)	IV	2004	It is popular hybrid with semi upright plant growth habit. Plant height is 90-100 cm, with light green stem and leaves. First flowering can be observed in 35-40 days after transplanting. Fruits are light purple, long, soft texture, average length 13 cm and diameter 3 cm. Picking starts in 65-70 days after transplanting. Yield is 75.0-80.0 t/ha (Kalloo *et al.,* 2006).
ARBH-786	Ankur Seed Co. (Maharashtra)	IV	2004	This hybrid has been developed by Ankur Seeds Co. Fruitys are long and slender with purple colour. Fruits are born in cluster. Individual fruit weight is around 60-70 g. first picking can be done in 100-110 days after transplanting. Yield is 75.0-80.0 t/ha.
VNR-51	VNR, Seed Co Raipur (Chhattisgarh)	IV, VI	2005	Small round fruits. Suitable for summer and rainy season. Average yield is 46.0 t/ha
BWBH-3	ICAR-IIHR, Bengaluru (Karnataka)	IV, VI	2005	It is Bacterial wilt resistant. Average yield 34.85 t/ha.
COBH-3	TNAU, Coimbatore (Tamil Nadu)	VIII	2005	This hybrid is suitable for costal area. Average yield is 45.0 t/ha.
HABH	ICAR-HARP, Ranchi (Jharkhand)	IV	2005	This hybrid is suitable for plains. Average yield is 40.0 t/ha
BCTH-17	BCKV, Kalyani (West Bangal)	II, V	2005	Very attractive hybrid. Suitable for humid area. Average yield is 42.0 t/ha

Navina	VNR Seeds Co., (Chhattisgarh)	IV	2007	Long fruited hybrid. Average yield is 46.0 t/ha.
HABH-17	ICAR-HARP, Ranchi (Jharkhand)	IV	2007	Round fruited hybrid. Average yield is 39.48 t/ha.

Brinjal (Resistant varieties)				
Variety	*Developing Centre*	*Recom-mened Zone*	*Year of identi-fication*	*Remarks*
BB-77	OUA&T, Bhubaneswar (Odisha)	II, V	1990	Plants are medium stature and fruits are oblong, medium and deep purple. It is resistant to Bacterial wilt. Average yield is 34.0 t/ha.
BWR-21	IIHR, Bengaluru (Karnataka)	VIII	1990	Plants are tall, compact and fruits are medium long with green purple calyx, tender and free from bitter principles with seed maturity. Crop is ready for picking in 150 days. Average yield is 30.0-45.0 t/ha
SM-6-7	OUA&T, Bhubaneswar (Odisha)	VIII	1992	Plants are non-prickly and spreading; fruits are oval and purple. Average yield is 20.0 t/ha
SM-6-6	KAU, Thrissur, Kerala	I, VII, VIII	1996	Plants are bushy and fruits are medium-long and light green. Average yield is 30.0 t/ha.
BB-44	OUA&T, Bhubaneswar (Odisha)	V, VII	1996	Plants are medium stature and fruits are green, long with white flesh. It is resistant to Bacterial wilt and *Fusarium* wilt. Average yield is 31.6 t/ha.
CHES-309	HARP, Ranchi (Jharkhand)	I, VII	2001	Very attaractive purple colour fruits. Average yield is 45.0 t/ha.
BB-64	OUA&T, Bhubaneswar (Odisha)	IV, V, VII, VIII	2004	This is wilt resistant variety. Average yield is 38.50 t/ha. Suitable for the year round cultivation.
(3) Chilli (Open pollinated varieties)				
Bhagya-lakshmi (G-4)	APHU, RARS Guntur, Lam (Andhra Pradesh)	-	1975	Plants are tall and straight with dense foliage.Fruits are olive green in colour and turn bright red on ripening. Fruits are about 8.8 cm long (medium long), firm attachment, having medium pungency and mild tolerance to leaf curl virus. Good capsacin content. Average green yield is 10.0-13.0 t/ha and 1.0-1.3 t/ha dry. It is very good as green chillies and dry powder use.
Andhra Jyoti (G-5)	APHU, RARS, Guntur, Lam (Andhra Pradesh)	-	1975	It is developed from a cross between (G-5) x (G-2 x 1331. Plants are tall with short, green smooth and stout 'gundu' type. Fruits are 5.1 cm long and 8.3 cm in girth. Fruits are conical shape and thick pericarp. It became bright red on ripening and possess good amount of capsaicin (0.65 mg/g). Average green yield is 8.3 t/ha. It is tolerant to thrips. It is very good as green chillies and dry powder use.
K-2	TNAU (Kovilpatti), T. Nadu	-	1985	Plants are tall and semi- spreading in nature. Fruits are medium long (7.3 x 5.5 cm) and green with blunt tip. Fruits became bright red on maturity. Crop duration is of 210 days. Average green yield is 23.5.0 t/ha. It is tolerant to thrips and fruit rot.
Pant-C-1	GBPUA&T, Pantnagar (Uttarakhand)	-	1977	It is a natural selection in a population of Local Kandhari which got naturally crossed by another variety NP-46A growing in adjoining area. Pant-C-1takes about 100 days to mature. Fruits are medium long (6-7 cm) which is erect, green, smooth, blunt apex, and highly pungent. First picking starts 60 days after transplanting. The green pod yield is 7.50 t/ha. The dry pod yield is about 1.5 t/ha. It is upright fruiting pods variety. The pods are highly pungent, relatively small in size, broader at the base and narrow towards the tip. It is moderately resistant to mosaic and leaf curl virus.
Jawahar Mirc-218 (J-218)	JNKVV, Jabalpur (Madhya Pradesh)	I, IV, V, VI, VII	1987	It is an early maturing variety (Chandra *et al.*, 1987). Plants are dwarf, spreading type and fruits are erect, long, thin and bright yellow with purplish tinge. Fruits became bright red and glossy on ripe. It is highly pungent. Good transport quality. Tolerant to leaf curl and fruit rot. Average green yield is 8.5-9.7 t/ha.
X-235 (LCA-235)	APHU, RARS, Guntur, Lam (Andhra Pradesh)	I, IV, V, VI, VIII	1987	It is an early maturing variety. Plants are compact, early and fruits are thin, medium long, green and highly pungent. Average green yield is 15.0 t/ha. It is tolerant to thrips. It is very much suitable for pickling and export. Average yield potential of green chillies is 18.5 t/ha.
Musalwadi	MPKV, Rahuri (Maharashtra)	V	1987	Plants are tall, dark green and spreading type. Fruits are smooth, medium long (6.86 x 0.95 cm), and dark green with black patches. Fruits became blood red on ripening. Average green yield is 17.5 t/ha. It is resistant to leaf curl.

Arka Lohit (Sel-1)	ICAR-IHHR, Bengaluru (Karnataka)	V, VII, VIII	1990	This variety has been developed by selection from an indigenous material.duration of crop is 180-210 days. Fruits are medium long sized, glossy, dark green, smooth, straight and highly pungent. Fruits became deep red on maturity. Average capsaicin content is 0.71 per cent. Average green yield is 28 t/ha. It is tolerant to powdery mildew and moisture stress.
LCA 206-B	APHU, RARS, Guntur Lam (Andra Pradesh)	V, VI, VII, VIII	1990	Plants are tall and fruits are green. Average green chilli yield is about 10.85 t/ha. Resistant to powdery mildew and leaf curl.
Jayanti (AKC-86-39)	PDMKV, Akola (Maharashtra)	VII	2001	Fruits are long, green, bigger size, good ascorbic acid and capsaicin content. Average green yield is 18.6 t/ha. Crop duration is of 215 days.
BC-14-2	OUA&T, Bhubaneswar (Odisha)	V, VI	2001	Fruits are medium size and deep at maturity. Average green yield is 11.8 t/ha.
RHRC-Cluster Erect	MPKV, Rahuri (Maharashtra)	VII	2001	Plants are erect and medium tall. Fruits develop in cluster. Fruits are attractive. Average yield is 19.00 t/ha.
PMR-57/88-K	ICAR-IIHR, Bengaluru (Karnataka)	VII	2002	Plants are tall. Fruits are dark green and attractive. Average yield is 13.00 t/ha. Resistant to powdery mildew and leaf curl.
LCA-334	APHU, RARS, Guntur, Lam (Maharashtra)	III, IV, V, VII	2002	This is an early maturing variety. Fruits are medium long and stout. It is ideal for green and red chilli. Average green yield is 18.5 t/ha.
ASC-2000-02	GAU, Anand (Gujarat)	VII	2004	Plants are medium tall. Suitable for round the year cultivation. Average yield is 12.0 t/ha.
Kashi Anmol (KA-2)	ICAR-IIVR, Varanasi (Uttar Pradesh)	IV	2005	Plants are determinate, bushy (60-75 cm), umbrella type and bear dark green, attractive and pendent fruits. It is highly pungent. First picking can be done 50-55 days after transplanting (early maturing). Crop duration is 150-160 days. Average green yield is 20 t/ha. It is very much suitable for green chilli and long distance transportation (Kalloo *et al.,* 2006)
LCA-353	APHU, RARS, Guntur, Lam (Andhra Pradesh)	IV, V, VII,	2007	This cultivar is tall and bear long fruits. Average green yield is 14.8 t/ha. Resistant to leaf cure virus.
BC-25	OUA&T, Bhubaneswar (Odisha)	V, VI, VII,	2007	Plants are tall. Average yield is about-22. t/ha.

Chilli (Hybrids)				
Hybrids	*Developing Centre*	*Recom-mened Zone*	*Year of identi-fication*	*Remarks*
HOE-888	Sandoz Seed Co. (Maharashtra)	IV, VIII	1997	Plants are tall. Fruits light green, Average yield of green chilli is 12.85 t/ha
ARCH-236	Ankur Seed Co., Nagpur (Maharashtra)	IV	1997	Fruits are elongated, undulated, pointed and green. Fruits became red on maturity. It is very good for fresh market and also possesses very good shelf life. It has high capsaicin content. Crop duration is 110-140 days.
Sungro-86-235	Sungro Seed Co. Salimar bagh (Delhi)	IV, VIII	2002	Plants are compact and umbrella type. Fruits are long (10-12 cm), thick skin, dark green and pendent. Fruits are medium pungent.
ARCH-228	Ankur Seeds Nagpur (Maharashtra)	IV, V, VI	2003	Fruits are medium long and attractive dark green in colour. Fruits became deep red in colour on the maturity. It has high capsaicin content with good colour value. Average green fruit yield is 41.5 t/ha.
CCH-2	ICAR-IIVR, Varanasi (Uttar Pradesh)	II, IV, V, VI	2005	Plants semi determinate (1-1.2 m), erect, nodal pigmentation on stem. Fruits light green, straight, length 11-12 cm. Recommended for green as well as red fruit production. Green fruit yield 24.0 t/ha
CCH-3	ICAR-IIVR, Varanasi (Uttar Pradesh)	IV, V, VIII	2005	Plants are medium tall. Average yield of green chilli is 14.5 t/ha.
MSH-149	ICAR-IIHR, Bengaluru (Karnataka)	All	2005	Plants are tall. Average yield is 11.8 t/ha.

MSH-172	ICAR-IIHR, Bengaluru (Karnataka)	All	2005	Plants are medium tall. Average yield is 12.46 t/ha.
KCH-3	CSAU and T, Kanpur (Uttar Pradesh)	IV	2005	Plants are tall. Average yield is 9.8 t/ha. as green chilli.
BSS-453	Beejo Sheetal Seed Co. Jalna (Maharashtra)	II	2009	Plants are tall. Average yield of green chilli is 76.8 t/ha.
NCH-587	Nirmal Seeds Co. Jalgana (Maharashtra)	IV, VII	2010	Plants are medium tall. Average yield of green chilli is 16.0 t/ha.
VNR-332	VNR Seed Co. Jalana	IV, VIII	2010	Plants are tall. Average yield of green chilli is 12.0 t/ha.
HH-41786	Syngenta Seed Co. Pune (Maharashtra)	VII	2011	Plants are tall & fruits are long with light green colour. Average yield of green chilli is 15.99 t/ha.
BSS-378	Beejo Sheetal Seed Co. Jalana (Maharashtra)	VII	2012	Plants are erect, medium tall and bear light green fruits. Average yield of green chilli is 15.5 t/ha.
VRN Vidya	VNR Seed Co. Raipur (Chhattishgarh)	IV	2013	Plants are erect and born dark green fruit. Average yield of green chilli is 13.5 t/ha.
Capsicum/Sweet pepper				
KT-1	ICAR-IARI Regional Station Katrain (H.P.)	I	1990	This hybrid is very early and high yielder with yellowish green fruits pointed at the tip end. Average green fruit yield is 32 t/ha. Suitable for growing in humid western Himalayan regions.
Pusa Deepti	ICAR-IARI Regional Station, Katrain (H.P.)	I	1991	This early to medium maturinh (6`0-70 days after transplanting) hybrid is very good bearer of quality fruits. Plants are vigorous, erect, medium and bushy. Fruits are smooth, erect, conical, light green with thick flesh, long (10 x 4 cm). Average green fruit yield is 20-23 t/ha. It is tolerant to bacterial leaf spot and anthracnose under field condition.
SEL-II	SKUA&T, Srinagar (J&K)	I	1997	Plants are interminate, medium and fruits are dark green and turn yellow at physiological maturity. Average green yield is 20-25 t/ha.
Lario	Syngenta/ Novartis, India Seed Co. (Maharashtra)	I	2002	This hybrid is profuse bearer. First picking starts 55 days after planting. Fruits are medium sized, dark green, thick walled, fleshy, blocky type and 3-4 lobes. Individual fruit weight is around 150 g. Fruits had long shelf life. Ideal for long distance transportation. It is tolerant to tomato mosaic virus. Average green fruit yield is 80 t/ha.
DARL-202	DARL, Pithoragarh (Uttarakhand)	I, IV	2003	Plants are medium tall. Very attractive fruit colour. Average yield is 25 t/ha.
KTCPH-3	ICAR-IARI Res. Station Katrain (H.P.),	I, VI, VII	2005	It is developed through crossing between Yolo Wonder x HC-201 by IARI Regional Station, Katrain (Himachal Pradesh). Fruits are blocky, smooth, and green with very thick flesh. Its yield potential is 27.5 t/ha.
PRCH-101	Ranichauri (Uttarakhand)	I	2013	Plants are tall and has green foliage. Average yield is 22 t/ha.

(4) Garden peas

Maturity Group	*Variety*	*Developing Centre*	*Recom-mened Zone*	*Year of identi-fication*	*Remarks*
Mid-Season	Bonneville	ICAR-IARI, N. Delhi	-	1975	An introduction from USA. This is a mid-season variety. Plants are medium tall (60-70 cm), flowers 55-60 days after sowing, two pods per peduncle and pods are light green, straight and well filled having 6-7 seeds per pod. Average green pod yield is 15.0 t/ha. Bold seeded, sweet and wrinkled at maturity. Its yield potential is 8.5 t/ha with 45 per cent shelling. It is highly susceptible to powdery mildew.
Mid-Season	Jawahar Matar-1 (GC-141)	Gwalior (M.P.)	-	1975	This is a mid-season variety derived from the cross T-19 x Greater Progress. Plants are dwarf (60.0 cm at fruiting and 82 cm at maturity), leaves are normal shaped, pods are big, attractive, green and 8-9 cm long. Average green pod yield is 12.0 t/ha.

Mid-Season	Jawahar Matar-4 (GC-195)	Gwalior (M.P.)	-	1975	This is a mid-season variety derived from the cross T-19 x Little Marvel. Plants are medium tall (55-60 cm) and first picking starts 75 days after seeding (Wakankar *et al.,* 1974). Pods are medium (7 cm long), and green. Average green pod yield is 9.8 t/ha.
Early Season	Arkel	ICAR-IARI, New Delhi	-	1975	An introduction from England. This is an early maturing variety. Plants are dwarf (35.0-45.0 cm), bear double pods at flower nodes and solitary at upper nodes. Pods are attractive, dark green, 8.8 cm long, curved towards suture at distal end (sickle shaped) and well filled with 7-8 ovules/pod. Shelling is 40percent.Average green pod yield is 7.5 t/ha in 50-55 days of crop duration.Wrinkled seeded, suitable for fresh maket and dehydration. It is tolerant to high temperature but susceptible to collar rot.
Early Season	Early December or Jawahar Matar-3	Gwalior (M.P.)	-	1977	Developed through T-19 x Early badger. This is very early maturing variety. Plants are dwarf (50 cm), produce 7.0 cm long light green, roundish-oval pods. Average green pod yield is 4.0 t/ha.
Early Season	Pant Uphar (IP-3)	GBPUA&T, Pantnagar (Uttarakhand)	-	1985	This variety of vegetable pea was developed through selection in germplasm lines maintained at this university. Plants are medium tall, thin stem and light green foliage with smaller leaf-lets. It bears white flowers. This variety is of intermediate duration between early cultivar Arkel and late maturing cultivar Bonneville. It was released to provide a vegetable variety which might fill the gap in time duration between the two dominant cultivars, Arkel and Bonneville. Pods are round having yellowish wrinkled seeds. The first green pod picking in this variety is available in 75-80 days after sowing. Average green pod yield is 10.0 t/ha. The shelling percentage is approximately 52 per cent. This has susceptibility to powdery mildewlike most of the check cultivars but has resistance to pea stem fly.
Early Season	P-88 -2-C	PAU, Ludhiana (Punjab)	-	1978	This is an early maturing variety selected from cross 'Pusa-2' x 'Morasis-55' (sister line of Punjab-87). Plants are medium tall, vigorous, erect with dark green foliage. First picking is possible in 100 days after sowing. Pods are dark green, long (8-10 cm) and slightly curved at the centre. The grains are bold and sweet. Average green pod yield is 15.0 t/ha and shelling out turn is 47 percent. It is resistance powdery mildew.
Early Season	Pant Maatar-2 (PM-2)	GBPUA&T, Pantnagar (Uttarakhand)	I	1987	This is an early maturing variety developed from a cross of 'Early Badger' x 'IP-3' (Pant Uphar) through pedigree method and suitable for cultivation in the hills and northern plains. Pods are small. The first green pod picking is possible after 60 days of sowing. This variety is equivalent to Arkel in maturity duration and other features and gives about 10 percent higher yield than Arkel. Average green pod yield is 10.0 t/ha.
Early Season	Lincoln	ICAR-IARI Regional station Katrain (H.P.)	I	1987	Plants are medium tall and double podded. Pods are green curved and 8-9 cm long. Average green pod yield is 5.50 t/ha.
Early Season	VL-3	VPKAS, Almora (Uttarakhand)	I,IV,VI	1987	This is medium maturing variety developed from a cross of' Old Sugar' x 'Early Wrinkled Dwarf 2-2-a'. Plants are determinate with light green foliage. Pods are light green, straight and 6.8 cm long. Average green pod yield is 10.0 t/ha. Susceptible to powdery mildew.
Early Season	VL-7	VPKAS, Almora (Uttarakhand)	IV	1992	This is an early maturing variety. Plants are determinate, dwarf and pods are light green, slightly curved towards suture at distal end. Average green pod yield is 23-25 t/ha.
Early Season	Matar Ageta-6 (E-6)	PAU, Ludhiana (Punjab)	I,IV,VI	1989	It was developed through pedigree selection from the cross 'Massey Gem' x 'Harabona'. Its plants are dwarf (40.0 cm), erect, vigorous and green. This is an extra early maturing variety and takes 45-50 days from sowing to marketable green pod stage. About 50 per cent of total yield is harvested in the very first picking. Each plant bears 12-15 pods and each pod contain up to six grains. Dry seeds are light green, smooth with slight dimples and contain high amount of protein. Average green pod yield is 6.0 t/ha and shelling out turn is 44.67 percent. It is tolerant to high temperature.
Mid-Season	Vivek-6 (VL-6)	ICAR-VPKAS, Almora (Uttarakhand)	IV	1993	This is medium maturing variety. Plants are dwarf and pods are straight, medium and light green. Average green pod yield is 11.0 t/ha (Sridhar *et al.,* 1999).
Mid-Season	Hisar Harit (PH-1)	HAU, Hisar (Haryana)	VII	1993	This is medium maturing variety (first picking after 70 days of seeding). Pods are green, large sickle shape and well filled. As per nature it bears single to double pod with a yield potential of 8.0-8.5 t/ha.
Mid-Season	NDVP-8	NDUA&T, Faizabad (Uttar Pradesh)	IV	1997	This is medium maturing variety. Average green pod yield is 10.0 t/ha.

Mid-Season	NDVP-10	NDUA&T, Faizabad (Uttar Pradesh)	IV	1998	This is medium maturing variety. Average green pod yield is 11.7 t/ha.
Mid-Season	VL-8	ICAR-VPKAS, Almora (Uttarakhand)	I	1998	This is medium maturing variety. Average green pod yield is 10.0 t/ha.
Early Season	Kashi Kanak (VRP-2)	ICAR-IIVR, Varanasi (Uttar Pradesh)	VI	2001	Early maturing variety developed through selection. Plant height 50-55 cm, foliage dark green, pod straight, light green, length 7-8 cm and filled with bold seeds. First picking 55-58 days after sowing. Yield is about 6-8 t/ha.
Early Season	NDVP-12	NDUA&T, Faizabad (Uttar Pradesh)	IV	2001	This is early maturing variety, Plant 55-59 cm. dark green foliage, straight pod and bold grain. Average yield is 8.5 t/ha.
Mid-Season	Kashi Arati (VRP-3)	ICAR-IIVR, Varanasi (Uttar Pradesh)	I	2001	Mid-season variety, starts flowering 45-50 days after sowing. Pods green, bold grain. Yield is about 8-12 t/ha.
Edible Podded	Organ Sugar Podded	PAU, Ludhiana (Punjab)	VI	2001	This is unique genotype in which whole puds can be consumed. Average yield 6.8 t/ha.
Early	Kashi Nandini (VRP-5)	IIVR, Varanasi (Uttar Pradesh)	I, IV, VIII	2005	This is an early maturing variety, derived through pedigree selection from the cross P-1542 x VT-2-1. Plants are erect, dwarf (45.0 cm), dark green and 7-8 pods/plant. Pods are long, attractive, well filled with high shelling percentage. Average green pod yield is 6.5 t/ha within 70 days. It is tolerant to leaf minor and pod borer.
Mid-season	CHP-2	ICAR-HARP, Ranchi (Jharkhand)	IV,VI	2005	Developed through pedigree selection. Plants are erect with dark green foliage. Average yield is 6-7 t/ha.
Early	VP-101	VPKAS, Almora (Uttarakhand)	I, IV	2007	Developed through selection. Suitable for cultivation in hilly area. Average green pod yield 8-10 t/ha.
Mid-season	PC-531	PAU, Ludhiana (Punjab)	I, VI, VII	2007	Plants erect and dark green foliage. Average yield is 6-7 t/ha.

Garden pea (Resistant varieties)

Variety	*Developing Centre*	*Recom-mened Zone*	*Year of identi-fication*	*Remark*
Azad P-2 (PRS-4)	CSAUA&T, Kanpur (Uttar Pradesh)	IV, VI, VII	1990	Developed through the cross between 'Bonneville' x '6587'. Plants are tall (130-150 cm), and erect with light green foliage. Pods are medium sized, light green, straight and born in clusters of two with 6-7 grains which are wrinkled and taste sweet. Average green pod yield is 12.0 t/ha in a crop duration of 90-95 days. It attains physiological maturity in 125-130 days. Seeds become wrinkled and brown after maturity. This is a powdery mildew resistant variety and suitable for cultivation under late sown conditions.
Jawahar Pea-4 (JP-4)	JNKVV, Jabalpur (Madhya Pradesh)	IV, VIII	1990	This is medium maturing and powdery mildew resistant variety. Pods are big, curved with 8 green sweet ovoules. Average green pod yield is 12-13 t/ha.
Jawahar Pea-83 (JP-83)	JNKVV, Jabalpur (Madhya Pradesh)	VII	1992	This is powdery mildew resistant variety and developed by selection from the double cross (JMI x JP-829) x (46 C x JP-501). Its plants are dwarf (50 cm); pods are large and curved, each containing about 8 sweet tasting grains. Average green pod yield is 12.0-13.0 t/ha.
NDVP-4	NDUA&T, Faizabad (Uttar Pradesh)	IV	1995	Plants are medium long and erect. Average yield is about 15-18 t/ha.
Palam Priya (DPP-68)	CSK HPKV, Palampur (Himachal)		2001	Plants are tall upto 94.0 cm. Pods are medium thick and well filled. Average green pod yield is 12-13 t/ha. It is tolerant to Fusarium wilt and resistant to powdery mildew.
KS-245	CSAUA&T, Kanpur (Uttar Pradesh)		2001	Plants are medium tall. Pods are well yield. Average yield is about 14.0 t/ha. Resistant to powdery mildew.

NDVP-250	NDUA&T, Faizabad (Uttar Pradesh)	V	2001	Plants are tall. Pods are straight and sweet in taste. Average yield is about 12.5 t/ha.
DPP-9411	CSKHPKV, Palampur (Himachal Pradesh)	I	2002	Plants are tall. Pods are long, well filled, sweet and tasty. Average yield is about 13.0 t/ha.
KTP-8	Regional Station, Katrain (H.P.) ICAR-IARI, N. Delhi	I, IV, V	2005	Plants are medium tall. Pods are straight, and well filled. Average yield is about 14.5 t/ha.
				(5) Cowpea
Pusa Komal (Sel-1552)	ICAR-IARI, New Delhi	IV	1983	It was developed by selection from the cross P-85-2 x P-426. Its plants are compact, bushy (35.0-40.0 cm tall) and flowers in 45 days after sowing. It is suitable for cultivation in spring summer and rainy seasons. The pods are light green, cylindrical, tender, and 25.0-30.0 cm long. First picking is possible in 60-65 days after sowing. It is resistant to bacterial blight and gives an average pod yield of 10.0-12.0 t/ha.
Arka Garima (Sel-61-B)	ICAR-IHHR, Bengaluru (Karnataka)	VII, VIII	1992	This is very popular variety in South India developed through back cross and pure line selection from cross Tuv-762 x *Vigna unguiculata* spp. *sesquipedalis.* Plants are tall, vigorous, bushy, spreading with small vines and light green leaves. Pods are light green, long (22.0-24.0 cm) thick, round fleshy and stringless. It is tolerant to heat and drought and gives an average pod yield of 18.0 t/ha.
Sel-263	PAU, Ludhiana (Punjab)	IV	1992	This is an early maturing variety and pedigree selected from a local genotypes collected from Bengaluru (Karnataka). Plants are photo-insensitive, dwarf and can be grown in both spring and rainy seasons. Its pods are green, thick, fleshy, tender and medium long (20.0 cm). It is resistant to cowpea golden mosaic viruses and gives an average pod yield of 10.0 t/ha.
Sel-2-1	NDUA&T, Faizabad (Uttar Pradesh)	IV	1993	It is cross between Cross Virginia x Iron Grey. Plants are bushy, 70-75 cm tall and pods are green, 25-30 cm length, black seeded; susceptible to *Pseudocercospors* and viruses.
Arka Samrudhi (IIHR-16)	ICAR-IIHR, Bengaluru (Karnataka)	IV, VII	1998	Pedigree selection from cross Arka Garima x Pusa Komal. This is an early maturing variety. Plants are bushy, 70-75 cm tall, and photo-insensitive. Pods are green, medium thick, medium long (15-18 cm), tender, fleshy without parchment, good cooking quality and gives an average pod yield of 19.00 t/ha.
Narendra Lobia-2 (NDCP-13)	NDUA&T, Faizabad (Uttar Pradesh)	II, III, IV, VII	2002	Pedigree selection from cross Sel-2 x Red Seeded. Plants are bushy, pods dark green and 28 cm long and get ready for harvest in 50 days. Average pod yield is 7.5-10.0 t/ha.
Kashi Shyamal (IVRCP-1)	ICAR-IIVR, Varanasi (Uttar Pradesh)	IV	2004	Plants dwarf, bushy (height 70-75 cm) and 3-4 branches per plant. Early flowering (40 days after sowing). First harvesting after 48 days after sowing. Produces 35-40 pods per plant. Pod yield is about 7.0-8.0 t/ha. Tolerant to golden mosaic virus.
Swarna Harita (CHCP-2)	ICAR-HARP, Ranchi (Jharkhand)	VIII	2005	Plants are dwarf type, pods are green, 35-40 cm long get ready for harvest in 55 days. Pod yield is about 10.0-12.5 t/ha.
Kashi Kanchan (VRCP-4)	ICAR-IIVR, Varanasi (Uttar Pradesh)	0	2007	This is bush type (50.0-60.0 cm height) photo-insensitive and early variety. It is suitable for sowing in both spring summer and rainy seasons. Flowering starts in 40-45 days and pods gets ready for harvest in 50-55 days after sopwing. It produces 40-45 pods per plant of 30.0-35.0 cm long. The pods are dark green, tender, pulp with less fibrous and free from parchment layer. The cultivar is resistant to golden mosaic virus, *Pseudocercospora cruenta* diseases. It gives green pod yield of about 15.0-20.0 t/ha.
Kashi Nidhi	ICAR-IIVR, Varanasi (Uttar Pradesh)	0	2013	Developed through pedigree selection (Kashi Unnati x Cowpea-263). Plants are dwarf and bushy. Pods get ready for harvest in 45 days; pods are 30-35 cm long, pulpy, dark green and parchment free. Average green pod yield is 12.5-15.0 t/ha. It is resistant to cowpea golden mosaic virus and *Pseudocercospora cruenta* diseases.
				(6) French bean (Bush type)
VL-Bauni-1	ICAR-VPKAS, Almora (Uttarakhand)	North Indian Hills	1985	Plants are dwarf (30-40 cm), flowers are white with purple ting and pods are pale green colour, medium long (12 cm), round, light green, attractive, string less fleshy and slightly curved. Pods are ready for harvesting in 50-60 days. It may be sown in the hills from march to July. Average green pod yield is 10.0-11 t/ha.
Arka Komal (Sel-9)	ICAR-IIHR, Bengaluru (Karnataka)	I, VII, VIII	1987	It has been developed through selection from an exotic material introduced from Australia for northern hills, semi-arid Lava plateau, central highlands and humid to semi-arid western Ghats and Karnataka plateau. Plants are photo-insensitive, bushy and erect with bright green stem and foliage. It takes 65-70 days to first green pod picking. Pods are straight, flat, tender, green, oblong and have good shelf life. Average green pod yield is 9.0 t/ha. Possesses resistance against anthracnose disease.

Pant Anupma (UPF-191)	GBPUA&T, Pantnagar (Uttarakhand)	IV, VIII	1987	It has been developed through selection from the local material. Plants are bushy (45.0 cm tall), dwarf, upright with green foliage and photosensitive in nature. Pods are fully covered by leaf canopy and thus get protected against sunlight. The pods are round, 12.0-15.0 cm long, smooth, tender, stringless and green. First picking is possible in 55-65 days after sowing. Average green pod yield is 9.0 t/ha. It is moderately resistant to bean common mosaic and angular leaf spot. It is suitable for hills and plains both.
Arka Suvidha (IIHR-909)	ICAR-IIHR, Bengaluru (Karnataka)	I	1997	It is a pedigree selection from a cross between Blue Crop x Contender. Plants are photo-insensitive, and bushy. Pods are straight, oval, light green, fleshy string less and crisp. Average green pod yield is 19.0 t/ha in about 70 days of crop duration. It is resistant to rust.
CH-812	ICAR-CHES, Ranchi (Jharkhand)	VII, III	2001	Developd through selection. Average yield is about 11.00 t/ha.
CH-819	CHES, Ranchi (Jharkhand)	I	2001	Developed through pedigree selection. Average yield is 10.5 t/ha.
Kashi Param (IVFB-1)	ICAR-IIVR, Varanasi (Uttar Pradesh)	I, VII	2005	Its plants are determinate in height, grow 70 cm high and are bush type, with dark green leaves. Its pods are attractive, fleshy, average length 14.7 cm, round and dark green. Its green pods yield is 12-14 t/ha and suitable for cultivation in Jammu and Kashmir, Himachal Pradesh, Uttarakhand, Madhya Preadesh and Maharashtra (Kalloo *et al.,* 2006)
				(7) Lablab bean/Dolichos bean
Deepaliwal	PKV, Akola (Maharshtra)	V, VII	1990	This variety was developed at Punjabrao Krishi Vidyapeeth, (Maharashtra). This variety is a high yielding, and pole type. Duration of maturity is 90 to 95 days.
CHDB-1	ICAR-HARP, Ranchi (Jharkhand)	IV	2004	Developed through pure line selection. Pod Straight, flat, green and fleshy pods (10-12 cm) having very good cooking quality. Mature seeds are light brown in colour. Tolerant to anthracnose, mosaic viruses and pod borer under field condition. First harvest 110-120 days after sowing. Average fresh pod yield is 350-400 q/ha.
Kashi Haritima	ICAR-IIVR, Varanasi (Uttar Pradesh)	III, IV, VII, VIII	2012	This is pole type cultivar and very high yielder. The average yield is 15-16 t/ha.
				(8) Okra (Hybrids)
Hybrid	*Developing Centre*	*Recom-mened Zone*	*Year of identi-fication*	*Remarks*
Shitla Uphar (DVR-1)	ICAR-IIVR, Varanasi (Uttar Pradesh)	IV, VII	1998	Plants medium tall, height 110-130 cm, and flowering starts at 38-40 days after sowing at 4-5 nodes. Fruits are green, 11-13 cm long at marketable stage. Yield is about 15-17 t/ha. Resistant to yellow vein mosaic virus and OLCV.
Shitla Jyoti (DVR-2)	ICAR-IIVR, Varanasi (Uttar Pradesh)	VI, VIII	1998	Plants are medium tall, and attain height 110-150 cm. Flowering starts on 30-40 days after sowing at 4-5 nodes. Fruit are green, and 12-14 cm long. Yield is about 18-20 t/ha. Resistant to YVMV and OLCV.
Kashi Bhairav (DVR-3)	ICAR-IIVR, Varanasi (Uttar Pradesh)	All	2001	Plants are medium tall with 2-3 branches. Fruits dark green with 10-12 cm length. Yield is about 20-22 t/ha. Resistant to YVMV and OLCV.
Kashi Mahima (DVR-4)	ICAR-IIVR, Varanasi (Uttar Pradesh)	IV, V, VII	2001	Plants tall, flowering starts at 36-40 days after sowing. Fruits green with 12-14 cm of length. Yield is about 200-220 q/ha. Resistance against YVMV and OLCV.
HBH-142	CCS HAU, Hisar (Harayana)	IV, V, VII, VIII	2006	Suitable for rainy season, high yielding hybrid, fruiting starts on third/fourth node, resistant to YVM disease. Yield is about 13.5-14.0 t/ha
SOH-152	Syngenta Seed Co. Pune (Maharashtra)	IV, VII, VIII	2005	High quality produce. Suitable for summer and rainy season. Free from YVMV & OLCEV. Average yield is 15.0 t/ha.
SOH-1016	Syngenta Seed Co. Pune (Maharashtra)	IV, VII	2007	Suitable for summer and rainy season. Dark green fruits. Average yield 16-18 t/ha. Free from YVMV.
NBH-180	Nuzi Veedu Seed Co. Hyderabad (Telangana)	VII	2007	Plants are medium tall and fruits are green and long. Free from YVMV & OLECV. Average yield is 14.5 t/ha.
JNDOH-02-2	GAU, Junagadh (Gujarat)	II, V, VI, VII, VIII	2008	Plants are tall. Fruits are green and thin. Average yield is 13.0 t/ha. Free from YVMV.
JOH-05-09	Junagadh (Gujarat)	V, VI, VII	----	Plants are medium tall and green foliage. Average yield is 14.0 t/ha.

Okra (Open pollinated resistant varieties)				
Resistant/ Tolerant	*Developing Centre*	*Recom-mened Zone*	*Year of identi-fication*	*Remark*
Sel-2	ICAR-NBPGR, New Delhi	All regions	1985	It is developed through selection. Fruits are long and tender. Average yield is 12.8 t/ha.
P-7	PAU, Ludhiana (Punjab)	All regions	1990	It is developed by selection from an inter-specific cross *A. manihot* ssp. *manihot* x *A. esculentus* cv. Pusa Sawani. Plants are medium tall (85-105 cm) with short internodes; leaves, stem and petiole are sparsely hairy. Fruits are medium long (15-20 cm), dark green, tender and 9-ridged. Fruit tip is slightly furrowed and blunt. Average green fruit yield of 11.0 t/ha in rainy season and 50.0 t/ha in spring season. It is resistant to jassid and cotton ball worm.
PB-57	MAU, Parbhani (Maharashtra)	All regions	1990	Plants are tall with mild purple tinge on the stem and petiole, leaves are dark green, deeply lobed and hairy; fruits are quick growing, dark green, hairy, 5-ridged and remain tender for a longer period. It attains marketable maturity in 55-60 days. It shows field tolerance to yellow vein mosaic virus.
Arka Anamika (Sel-10)	ICAR-IIHR, Bengaluru (Karnataka)	All regions	1990	It is developed by selection from an inter-specific cross *A. esculentus* (IIHR 20-31) x *A. tetraphyllus*. This is YVMV resistant variety and is suitable for cultivation both for spring and rainy seasons. Plants are medium tall (100.0 cm), upright open with short internodes and slightly pigmented at stem, petiole and lower leaves. Fruits are medium, dark green rough with five ridges. First picking is possible in 55 days aftyer sowing. Average green fruit yield of 12.5 t/ha. It is resistant to YVMV.
Arka Abhay (Sel-4)	ICAR-IIHR, Bengaluru (Karnataka)	II	1992	It is sister line of Arka Anamika and has been developed from an inter-specific cross *A. esculentus* (IIHR 20-31) x *A. tetraphyllus*. Plants are tall (170 cm) with branched stem, green stem with purple shade. Fruits are medium long, dark green having 5-6 ridges. Average green fruit yield of 10.5 t/ha. This is tolerant to fruit borer and resistant to yellow vein mosaic virus.
Hisar Unnat (HRB-55)	CCS HAU, Hisar (Haryana)	VI	1996	A variety developed in 1996 through a cross between Sel.-2 and Parbhani Kranti bears profusely branched-plants with green foliage and shiny dark green fruits. The plant and fruit both have soft spines. The variety at the time of release was reported resistant to yellow vein mosaic virus but now it has become susceptible.
Varsha Uphar (HRB-9-2)	CCS HAU, Hisar (Haryana)	IV, VI	1992	An early and high yielding variety developed in 1995 through a cross between Lam Selection and Parbhani Kranti following pedigree selection method bears dark green fast growing smooth surfaced fruits with five ridges well filled with seeds. Fruiting starts at 6–7th node and fruits become ready for first picking 47 days after sowing. Fresh pods yield realization is 12.5 t/ha. The variety at the time of release was reported resistant to yellow vein mosaic virus but now it has become susceptible.
Kashi Mohini (VRO-3)	ICAR-IIVR, Varanasi (Uttar Pradesh)	All the regions	2001	Plants are tall (110.0-140.0 cm) and flowers at 4-5th node during summer and 5-7th nodes during rainy season after 39-41 days of sowing. Fruits are green with five ridges, 11.3-12.6 cm long at marketable stage, suitable for summer and rainy season cultivation. Average green fruit yield of 13.0-15.0 t/ha. It is resistant to YVMV and ELCV under field conditions. It tolerates high temperature during summer season.
Kashi Mangali (VRO-4)	IIVR, Varanasi (Uttar Pradesh)	IV, V	2001	This variety has been developed through pure line selection. Plants are tall, height 120.0-125.0 cm, and flowers at 4-5th nodes after 40-42 days after sowing. Fruits are green with five ridges. Average green fruit yield of 13.0-15.0 t/ha. It is resistant to YVMV and ELCV under field conditions. It tolerates high temperature during summer season.
Kashi Vibhuti (VRO-5)	ICAR-IIVR, Varanasi (Uttar Pradesh)	VI, VIII	2002	Plants are dwarf, height 60.0-70.0 cm during rainy and 45.0-50.0 cm during summer seasons. Plants have 2-3 branches with short inter-nodal length. Flowering starts on 4th to 5th nodes after 38-40 days after sowing. A plant bears 18-22 fruits with 8-10 cm length at marketable stages. Yield is about 17-18 t/ha. This is resistant to YVMV and OLCV. It hass been released and notified during the XII meeting of Central Sub Committee on Crop Standard Notification and Release of Varieties for Horticultural Crops for the Cultivation in Rajasthan, Gujarat, Haryana and Delhi.
Kashi Pragati (VRO-6)	ICAR-IIVR, Varanasi (Uttar Pradesh)	IV, VII	2002	Plants are tall, height 130.0-175.0 cm, with 1-2 effective branches. First flower appears after 36-38 days after sowing on 4th nodes during rainy season and 3rd node during summer season. Fruits 8-10 cm in length at marketable stage and 23-25 per plants. Yield is about 13.0-14.0 t/ha during rainy and 10-12 t/ha during summer season. Resistant to YVMV and OLCV.
Kashi Lila (IIVR-11)	ICAR-IIVR, Varanasi (Uttar Pradesh)	VI, VII	2005	Plants are of medium height (110.0-130.0 cm), flowering startson 30-34 days after sowing. This is suitable for cultivation during rainy and summer season as early crop due to low temperature tolerance. Fruits are of five ridges, green and 13.0-15.0 cm long. This is resistant to YVMV and gives yield of 15.0-17.0 t/ha. This has been released and notified during the XXII meeting f Central Sub Committee on Crop Standard Notification and Release of Varieties for Horticultural Crops for the Cultivation in Chhattisgarh, Odisha, Andhra Pradesh, Rajasthan, Gujarat, Haryana and Delhi.

NDO-10	NDUA&T, Faizabad (Uttar Pradesh)	IV	2005	Plants are medium tall. Suitbale for rainy and summer season. Average yield is 15-16 t/ha. Free from YVMV.
HRB-107-4	CCS HAU, Hisar (Harayana)	VI, VIII	2005	Plants are tall and bears more fruits. Average yield is 14 t/ha.
JNDOL-03-1	GAI Junagarh (Gujarat)	VII, VIII	2007	Plants are medium tall. Average yield is 13 t/ha. Free from YVMV.
Kasi Kranti (VRO-22)	ICAR-IIVR Varanasi (Uttar Pradesh)	IV	2011	It is an early, medium tall 100-115 cm variety with short internodes along with single or double branch attached in narrow angle with main branch and tolerant to YVMV and OELCV under field condition. The fruits are available from 45-95 DAS and total yield is 14-15 t/ha.
JOL-2K-19	GAI Junagarh (Gujarat)	V	2011	Plants are tall. Fruits are attrative. Average yield is 15 t/ha.
(9) Onion (Open pollinated varieties)				
Variety	*Developing Centre*	*Recom-mened Zone*	*Year of identi-fication*	*Remark*
Punjab Selection	PAU, Ludhiana (Punjab)	IV, VII, VIII	1975	Plants are 55-65 cm tall with 6-9 leaves/plant. Bulbs are globular red, 5-6 cm in diameter, weight 50-70 g, firm with good keeping quality. Average green fruit yield of 30.0 t/ha.
Punjab-48(S-48)	PAU, Ludhiana (Punjab)	IV	1975	Bulbs are white, flatish-round, average weight of 80.0 g, T.S.S. 12-13 per cent, and good keeping quality. Average yield is 30.0-32.0 t/ha.
Pusa Red	ICAR-IARI, N. Delhi	-	1975	It has been developed by mass selection from the local material. Plants are 55-65 cm tall with 6-9 leaves/plant. Bulbs are purplish-red, medium sized, globular, flattish round, weight 70-90 g, less pungent and quite firm with good keeping quality. It takes 125-140 days from transplanting to harvesting. Average yield is 25.0-30.0 t/ha. Keeping quality of the variety is good.
Pusa Ratnar	ICAR-IARI, N. Delhi	-	1975	It has been developed by selection from segregating population of an F_1 hybrid 'Granex', introduced from the USA. Plants are 30.0 cm tall having dark green leaves with waxy bloom. Bulbs are bronze deep red, more exposed to the ground, obviate to flat-globular, less pungent and drooping necks with good storage quality. At maturity, bulbs are more exposed above ground. It takes 125 days from transplanting to harvesting. Average yield is 32.5-40.0 t/ha.
S-131	ICAR-IARI, N. Delhi	-	1977	White bulb bearing cultivar. Average yield is 28-30 t/ha.
N-257-9-1	MPKV, Rahuri (Maharashtra)	VII	1985	Bulbs are globular and white, suitable for rabi season cultivation. Average yield is 25.0 t/ha.
N-2-4-1	MPKV, Rahuri (Maharashtra)	IV, VII	1985	Plants are 40-45 cm in height, bulbs are globular (6.1 cm diameter), brick red, 12-13 per cent T.S.S., pungent and firm with good keeping quality. Suitable for rabi season. Average yield is 20.0 t/ha.
Pusa Madhvi (Line-102)	ICAR-IARI, N. Delhi	I, IV, VI, VII	1987	Rabi season cultivar and very popular due to better keeping quality. Bulbs are medium to large in size, light red in colour and flatish round in shape. Maturity 130-135 days after transplanting. The average yield potential is 30t/ha.
Arka Kalyan (Sel 14)	ICAR-IIHR, Bengaluru (Karnataka)	IV, VI, VII, VIII	1987	It has been developed by mass selection from the material collected from Maharashtra. This is a popular rabi season variety. Bulbs are globular, pink, flesh succulent, bulb weight of 130.0-190.0 g, 13 per cent T.S.S., moderately resistant to purple blotch. Average yield is 47.0 t/ha.
Arka Niketan	ICAR-IIHR, Bengaluru (Karnataka)	VII	1987	This is a popular kharif season variety. Plants are 60-70 cm height and takes 145 days from transplanting to reach marketable maturity. Bulbs are globular, pink, bulb weight of 100-180 g, highly pungent, 12-14 per cent T.S.S. thin neck, and good storage quality (can be stored up to 5 months at ambient temperature). Average yield is 45.0 t/ha. It can be grown in both *rabi* and *kharif* seasons.
Agri Found Dark Red	NHRDF, Nasik (Maharashtra)	IV	1987	It has been developed by mass selection from the local genetic stock of Nasik.This is a Kharif season variety and has the best storage quality. It takes 95-110 days from transplanting to harvesting. Bulbs are globular, dark red, medium large with tight skin, 4-6 cm in size, moderately pungent with 12-13 per cent TSS. Average yield is 30.0-40.0 t/ha.
VL-3	ICAR-VPKAS, Almora (Uttrakhand)	U.P. Hills	1990	This is a rabi season variety. Plants are 40-45 cm in height and bulbs are red, medium, globular and pungent. Average yield is 25.0 t/ha.
Punjab Red Round	PAU, Ludhiana (Punjab)	IV	1993	It has been developed by mass selection from the local material. This is an early maturing rabi season variety. Bulbs are shining red, globular, medium with tight neck. It is free from premature bolting and has good keeping quality. It takes 140-150 days from transplanting to harvesting. Average TSS content is 12.7 per cent. Average yield is 28.0-30.0 t/ha.

Agri Found Light Red	NHRDF, Nasik (Maharashtra)	VI, VIII	1995	This is a rabi season variety. Bulbs are light red, less pungent, firm tight skin, and 4-6 cm in size with 13 per cent of TSS. Average yield is 25.0-30.0 t/ha.
Punjab Naroya (PBR-5)	PAU, Ludhiana (Punjab)	VI	1997	It has been developed by mass selection from the local material. Plants are medium tall and green and bulbs are red, medium to large, round with thin tight neck. It takes 125 days from transplanting to harvesting. Average yield is 37.5 t/ha. It is tolerant to purple blotch and thrips.
L-28	NHRDF, Nasik (Maharashtra)	IV and VII	2006	Rabi season cultivar. Average yield is 30 t/ha.
Hisar Onion-2	CCS HAU, Hisar (Haryana)	VI	2006	It is rabi season onion variety. Bulbs bronze-red in color, flattish globular in shape, more pungent, and possess better shelf life. Average yield is 20-25 t/ha.
Hisar Onion-3	CCS HAU, Hisar (Haryana)	VI	2009	Bulbs globular shaped, tight skinned with thin neck, high yield, less bolting, good storage life. Average yield is 32-35 t/ha.
(10) Garlic				
Agrifound White (G-41)	NHRDF, Nasik (Maharashtra)	IV, VII	1988	It has been developed by clonal selection from the material collected from Bihar and release for commercial cultivation by NHRDF, Nasik (Maharashtra). Bulbs are compact, 20-25 cloves, silvery white skin, creamy flesh, TSS 41.0 per cent, dry matter 42.78 per cent, 3.5 to 4.5 cm diameter, 12-15 cm^2 size index with 20-25 cloves; good storage quality but susceptible to purple blotch and *Stemphylium* blight, gives an average yield of 13.00 t/ha.
Yamuna Safed (G-1)	NHRDF, Nasik (Maharashtra)	IV, VI, VII	1990	It has been developed by clonal selection from the material collected fron Delhi and release for commercial cultivation by NHRDF, Nasik (Maharashtra). Bulbs are compact with 25-30 cloves, 4.0-4.5 cm diameter, silvery white skin, creamy flesh, TSS 39.5 per cent and cloves are sickle shaped. Average yield is 15.0-17.5 t/ha. It is tolerant to purple blotch, *Stemphylium* blight and thrips.
Yamuna Safed-2 (G-50)	NHRDF, Nasik (Maharashtra)	IV	1993	Bulbs are compact, white creamy flesh, 10 bulbs weight 160.0-240.0 g, dry matter 40-41 per cent and TSS 38-40 per cent. Average yield is 15.0-17.5.0 t/ha. The variety is recommended for northern parts of India
Yamuna Safed-3 (G-282)	NHRDF, Nasik (Maharasthra)	IV, VI, VII	1998	Plant has larger leaves. The bulbs are creamy white and of big size (5.0-6.0 cm diameter) having 15-16 cloves per bulb. TSS is 38.0-42.0 per cent, dry matter 39.0-43.0 per cent with medium storability. Average yield is 18.0-20.0 t/ha.
VLG-7	ICAR-VPKAS, Almora (Uttarakhand)	I	2001	Bulbs are medium in size. Good keeping quality. Averaeg yield is 18.5 t/ha.
DARL-52	DARL, Pithoragarh (Uttarakhand)	I	2002	Bulbs are large and compact. Better storage quality. Cloves are white. Average yield is 20-28 t/ha.
G-323	NHRDF, Nasik (Maharashtra)	VI	2002	Bulbs are big (5.0-6.5 cm), creamy white with pinkish linage. One bulb contains 10-16 cloves. Average yield is 17.5-22.5 t/ha.
(11) Cauliflower (Open pollinated varieties)				
Early Kunwari	PAU, Ludhiana (Punjab)	IV	1975	This is an early maturity group variety. Curds are semi-spherical with even surface, and curd availability period is mid-September to mid-October. Average curd yield is about 10.0 t/ha.
Hisar-1	CCSHAU, Hisar (Haryana)	0	1976	Suitable for mid-late season, curds medium-large size and white in color. Average yield is 25 t/ha.
327-14-8-3	ICAR-IARI, New Delhi	-	1975	September maturity. Average yield is 26-28 t/ha.
351-4-1	ICAR-IARI, New Delhi	-	1975	October maturity. Average yield is 28-30 t/ha.
Improved Japanese	ICAR-IARI, New Delhi	-	1975	It is an introduction fro Israel. The plants are erect with narrow, bluish green leaves. It is November maturity crop. The curds are medium sized, compact and white. Average curd yield is about 20.0 t/ha.
EC 12012	ICAR-IARI, N. Delhi, Regional Station Katrain (H.P.)	-	1975	January maturity crop. Curds are white medium and compact. Average yield is about 22-25 t/ha.
Pusa Snowball K-1	ICAR-IARI, Regional Station Katrain (H.P.)	IV	1975	This is a January maturity group variety. It is late in maturity by about a week than Pusa Snowball-1. The leaves are puckered, serrated and light green in colour. Among Snowball group, its curd quality is the best; curds are compact, medium, snow white, even at delayed harvesting, and tolerant to black rot. Average yield is 20.0- 25.0 t/ ha. It is tolerant to black rot disease.

Pusa Snowball-1	ICAR-IARI, Regional Station Katrain (H.P.)		1979	It is developed from a cross between EC 12012 and EC 12013. The leaves are straight, upright. Inner laves cover the curd. Curds are available in the market during January-February in the north Indian Plains and during March-April in the hills. The curds are medium sized, compact and pure white. Average curd yield is about 22.5 t/ha.
Pant Subhra (114-S-1)	GBPUA&T, Pantnagar (Uttarakhand)	-	1984	This is a mid season variety (PI272775) which was evolved through simple recurrent selection from a local material. Curds are compact, slightly conical, creamish white in colour, non-ricey and non-leafy. It is suitable for rainy season. Curds do not fade colour under delayed harvesting and storage. This variety produces marketable curds by December-January. Average curd yield is about 27.5 t/ha in 120 days. The variety posses field resistant to black rot and curd blight disease. It is suitable for Bengal, Assam basin and Satlaj-Ganga Alluvial condition and transplanting is to be done by middle of July.
Line 6-1-2-1	ICAR-IARI, N. Delhi	-	1985	December maturity group. Curds are white and compact. Average yield is 20-22 t/ha.
Pant Gobhi-4 (235-S)	GBPUA&T, Pantnagar (Uttarakhand)	II, VII	1990	This is a November maturity group variety. Plants have medium long stem and sparse semi erect leaves. Curds are semi-spherical, creamish white, medium, compact and non-ricy. Average curd yield is about 14.3 t/ha.
Pusa Snowball (KT-25)	ICAR-IARI, Regional Station Katrain (H.P.)	I	2001	This is a late maturing variety of snowball group. Leaves are waxy, upright, slight bending towards inner side with puckered margins. Curds are very solid, medium sized, white good keeping quality. It is tolerant to *Sclerotinia* rot and black rot diseases. Average yield is 20.0- 30.0 t/ha.
Kashi Kunwari (IVREC-2)	ICAR-IIVR, Varanasi (Uttar Pradesh)	IV	2005	Early maturing group variety. Suitable time of sowing is end of June to July. Curds are semi-dome type, white compact; and fine texture. Average weight is about 300-450 g. Yield is about 30.0-35.0 t/ha.
Cauliflower (Synthetic varieties)				
Pusa Synthetic-1	ICAR-IARI, N. Delhi	-	1975	It is a synthetic variety comprising seven inbred lines. It is a December maturity variety. The plants are erect, 70-90 cm tall with narrow medium frame and bluish green leaves. The curds are creamy-white compact and partially self blanched. Average curd yield is about 22.0 t/ha.
Pusa Early Synthetic	ICAR-IARI, N. Delhi	IV, VIII	1990	This is an early maturing group variety. Plants are erect with bluish green leaves. Curds are small-medium, flat, creamy white and compact. This is resistant to *ricy-ness* and an average yield is 11.7 t/ha.
Pusa Synthetic	ICAR-IARI, N. Delhi	I, II, IV	1992	This is a synthetic and mid-season variety. Plants are erect with narrow-medium frame, having 24-28 varying colour leaves. Curds are creamish white to white, compact and curd availability period is mid-December to mid-January.
Cauliflower (Hybrid)				
Hybrid	*Developing Centre*	*Recom-mened Zone*	*Year of identi-fication*	*Remark*
Pusa Snowball Cauliflower Hybrid-1	ICAR-IARI, New Delhi	II, IV	1992	First indigenously developed cytoplasmic male sterility (CMS) based public sector hybrid. Curds are very compact, snow-white in colour with self-blanching habit. It is field tolerant to black rot disease. Avegrage curd weight is about 1.50 kg. Yield potential is around 50-55 t/ha.
DCH-541	ICAR-IARI, New Delhi	II, IV	2002	Plants are erect and foliage is waxy green. Curds are compact and yield is about 25-28 t/ha.
YCFH-202	Syngenta Seed Co. Pune (Maharashtra)	IV, VII	2004	Early group cultivar. Curds are compact and white. Average yield is 24 t/ha.
Summer King	Sungro Seed Co. Pune (Maharashtra)	I, IV	2004	It produces 400-500 g on an average curd weight. Whitish round shaped curd and ready for harvesting with 65 days after transplanting
SYCFH-203	Syngenta Seed Co. Pune (Maharashtra)	IV, V, VII	2005	Early group cultivar. Compact curds, very tasteful & attractive. Mature very early. Average curd yield is 22 t/ha.
(12) Cabbage (Open pollinated varieties)				
Variety	*Developing Centre*	*Recom-mened Zone*	*Year of identi-fication*	*Remark*
Pusa Mukta (Sel-8)	ICAR-IARI, Regional Station Katrain (H.P.)	II, IV	1985	It is developed from the cross EC24855 x EC10109. Plants are short stalk, medium frame, light green leaves which is slightly puckered with waxy margins, and curds are compact, flattish-round with loose wrapper leaf at the top and individual head weight of 1.5-2.0 kg. Average head yield is about 25.0-30.0 t/ha. It is resistant to black rot.

Cabbage (Hybrids)				
Hybrid	*Developing Centre*	*Recom-mened Zone*	*Year of identi-fication*	*Remark*
Shri Ganesh Gol	Mahyco Seed Co. (Maharashtra)	V	1992	Plants are vigorous and heads are round, compact, attractive, bluish green and become ready for harvest in 90-95 days after nursery sowing. Better siaying power in the field. It has good transportability. It gives an average yield of 30.0-35.0 t/ha. It is resistant to yellowing.
Nath Laxmi-401	Nath Seed Co. Aurangabad (Maharashtra)	I, IV, V, VI, VII	1993	Uniform compact heads with better shelf-life. It gives an average yield of 50.0-70.0 t/ha.
Bajrang (BSS-50)	Beejo Sheetal Seed Co. Jalna (Maharashtra)	VII	1995	It is an early maturing (takes 65 days to reach maturity) hybrid and release for commercial cultivation by Beejo Sheetal Seed Co. Its heads are dark green, small and compact weighing about 1.0 kg. Outer leaves are upright. It is tolerant to high temperature and resistant to *Fusarium* wilt.
Nath-501	Nath Seeds CO. Aurangabad (Maharashtra)	VII	1997	It is an early maturing (110 days to reach lavory) hybrid. Heads are uniform compact, round, better staying power and individual head weight is 3.0 kg. It gives an average yield of 70.0-80.0 t/ha.
Quisto (F_1)	Syngenta/ Novartis Seed Co. Pune (Maharashtra)	IV	1998	Plant strong, self supporting, sturdy, erect compact and well built. Bluish green, fleshy, broad leaves with serrated collars, 10-15 auxiliary leaves. Heads are compact, dark green, heavy, round and uniform. Ready for harvest in 75-80 days of transplanting. It can withstand in the field condition up to 70 days without split. Heads are round and dark green colour. Head has short, tender core with fine internal structure. Head weigh 1.0-3.0 kg. Very tasty and good for table, cooking or for process-industry. Tolerant to *Fusarium* wilt. Yield is 60-100 t/ha under optimal conditions.
S.Y.C.F.F.-2002	Syngenta Seed Co. Pune (Maharashtra)	IV	2004	Self supporting canopy of plant. Heads are compact, light greenn and uniform in maturity. Average yield is 38.0 t/ha.
KGMR-1	Res. Station Katrain (H.P.), ICAR-IARI, N. Delhi	I, IV	2005	Plant vigorous, strong, erect and leathery leaves. Heads are compact and possess very good cooling quality. Average yield is 40-42 t/ha. Free from Fusarum wilt.
Summer King	Sungro Seed Co. Shalimar bagh (N.Delhi)	I, IV	2005	Self-supporting plant canpoy. Heads are ready in 78 day after transplanting. Heads are compact. Yield is 40-42 t/ha.
Green Emperor	Tokita Seed Co.	I	2007	Vigorous plants suture. Heads are are compact and tasty. Better cooking quality. Average yield is 38 t/ha.
Cabbage (Synthetic Varieties)				
Synthetic	*Developing Centre*	*Recom-mened Zone*	*Year of identi-fication*	*Remark*
Pusa Synthetic	ICAR-IARI, Regional Station Katrain (H.P)	IV, I, II	1992	Plants with short stalk, cup shaped leaves, and early maturity in 60-70 days after transplanting. Heads are very compact and good cooking quality. Average weight is about 1.0 kg. This variety has yield potential of 35-46 t/ha.
(13) Carrot (Open pollinated varieties)				
Variety	*Developing Centre*	*Recom-mened Zone*	*Year of identi-fication*	*Remark*
Pusa Yamdagni (Sel-5)	ICAR-IARI, Regional Station, Katrain (H.P.)	VII	1975	This temperate type variety has been developed from the cross 'EC9981' x 'Nantes Half Long'. Plant top is medium, heavy and quick growing. Roots are 15.0-16.0 cm long, orange with self coloured core and slightly tapering stumpy to semi-stumpy ending. This variety is 7-10 days earlier than Nantes. Depending upon the region of cultivation, roots are ready for harvesting in 90-130 days after sowing. Average yield is 25.0-30.0 t/ha.
Pusa Meghali	ICAR-IARI, N. Delhi	VII	1992	This tropical type early variety has been developed and selected from the cross 'Pusa Kesar' x 'Nantes'. It has a short top, medium size roots (17.5 cm) smooth roots, orange flesh, self coloured core and stumpy to slightly tapering roots. It takes 110-120 days from sowing to harvesting. It is preferred for early sowing and can be marketed in 110-120 days. It produces 25.0-28.0 t/ha. It is capable of setting seed in plain areas.
SKAUC-50	SKUA&T, Srinagar (J&K)	I	2006	Plants are dark green and semi erect foliage. Roots are long, cylindrical semi-blunt to cracking and forcing with orange, sweet, fine textured flesh, gives yield of 25-27 t/ha.

Carrot (Hybrids)				
Hybrid	*Developing Centre*	*Recom-mened Zone*	*Year of identi-fication*	*Remark*
Hybrid-1	Mahyco Seed Co. Jalna (Maharashtra)	I, VII	1992	Roots are deep orange, uniform, length 7-9 inches, matures early, semi table for processing and suitable for fresh marketing.

(14) Muskmelon (Open pollinated varieties)				
Variety	*Developing Centre*	*Recom-mened Zone*	*Year of identi-fication*	*Remark*
Pusa Sarbati	ICAR-IARI, N. Delhi	IV	1975	It has been developed from a cross between Kutana and PMR-6, an introduction from the U.S.A. It is an early maturing variety. Fruits are round with netted skin; flesh is salmon orange, firm, thick, and 11-12 per cent T.S.S. with small seed cavity. It has been recommended for cultivation in the states of Punjab, Uttar Pradesh, Bihar, Madhya Pradesh and Maharashtra and has yield potential of 13.0-15.0 t/ha.
Hara Madhu	PAU, Ludhiana (Punjab)	IV, VII	1975	It is an inbred selection from the local material collected from Uttar Pradesh. Fruits are round, tapering towards the stalk end with open prominent green sutures and without netting. Average weight of fruit is 1.0 kg and do not slip from the stalk. Fruit has thin rind, smooth and pale yellow at maturity. Flesh is green, thick, very juicy with 12-15 per cent T.S.S. It has been recommended for cultivation in the states of Punjab, Uttar Pradesh, Bihar, Madhya Pradesh and Maharashtra and has yield potential of 12.5 t/ha.
Pusa Madhuras (Sel-45)	ICAR-IARI, N. Delhi	IV, VI, VIII	1975	This variety has been derived through selection from a germplasm. Fruits are roundish-flat, average weight of 1.0 kg, sparsely netted with pale green stripes; flesh is salmon-orange, juicy with 12-14 per cent T.S.S. but keeping quality is poor. First picking starts 90-95 days after sowing. Average yield is 15.0-17.0 t/ha. It has been recommended for cultivation in the states of Punjab, Uttar Pradesh, Bihar, Rajasthan, Gujarat, Haryana, Delhi, Madhya Pradesh, Karnataka, Tamil Nadu and Kerala
Arka Rajhans	ICAR-IIHR, Bengaluru (Karnataka)	VIII	1975	It is a single plant selection from the material collected (IIHR-107) from Rajasthan. Plants have densely hirsute stem, light green foliage with shallow lobes and bear 2-3 fruits. Fruits are round to slightly oval, medium-large, creamy white skin, weight varies from 1.25 to 2.0 kg, netted with shallow vein tracks at full slip stage; flesh is thick firm and white with 11-14 per cent T.S.S. Average yield is 28.5 t/ha in 85-90 days. It is resistant to powdery mildew. It has been recommended for cultivation in the states of Karnataka, Tamil Nadu, and Kerala.
Arka Jeet	ICAR-IIHR, Bongaluru (Karnataka)	VIII	1975	It is a selection from the local material collected from Uttar Pradesh. Fruits are small round with smooth orange flat ends, 300-500 g of weight, dark green skin, and thin golden orange at full slip stage. Excellent flavor and having T.S.S of 13-16 per cent and has yield potential of 13.0-15.0 t/ha in 90 days of crop duration. Average yield is 15.0-17.5 t/ha. It has been recommended for cultivation in the states of Karnataka, Tamil Nadu, and Kerala.
Durgpura Madhu	RRS, (RAU) Durgapura (Rajasthan)	IV, VI, VII, VIII	1975	This variety has been derived through selection from an early carly cultivar of Jaipur at ARS, Durgapura. It has been recommended for cultivation in the states of Punjab, Uttar Pradesh, Bihar, Rajasthan, Gujarat, Haryana, Madhya Pradesh, Karnataka, Tamil Nadu, Kerals and Maharashtra. Fruits are oblong with big seed cavity, average weight 500-600 g; skin pale green, flesh light green, and dry texture with 14.0 per cent T.S.S. Average yield is 15.0-20.0 t/ha.
Narendra Musk-melon-2 (NDM-15)	NDUA&T, Faizabad (U.P.)	IV	2002	It is developed through a selection from the cross between Mau-5 and Hara Madhu is identified for Punjab, Uttar Pradesh, Bihar and Jharkhand. Oval round shaped fruits with average weight 800-1000 g are smooth surfaced and light yellow in colour with green strips at maturity. Flesh is thick, dark orange and semi juicy with excellent musky lavor and TSS 10-13 per cent. It takes 70-75 days from sowing to first fruit harvesting and gives an average yield of 15.0-20.0 tonnes per hectare.
Kashi Madhu (IVMM-3	ICAR-IIVR, Varanasi (U.P.)	IV	2006	Kashi Madhu, a variety tolerant to powdery and downy mildew developed through a selection from local collection is medium in maturity. Its vines are medium and sparsely lobed and dark green leaves. Fruits are half-slip in nature. Fruits are round, rind thin, smooth, pale yellow at maturity and having prominent green sutures with an average weight of 700-750 g. Flesh salmon orange (Mango colour), thick, very juicy and having high TSS (13-14 per cent). Seeds are loosely packed in the seed cavity. Post harvest life is long with good transportability. Its average yield is 20.0-27.0 tonnes per hectare.

Muskmelon (Hybrids)				
Hybrid	*Developing Centre*	*Recom-mened Zone*	*Year of identi-fication*	*Remark*
Hybrid M-3	ICAR-IARI, N. Delhi	IV	1990	Vines are vigorous and medim spreading type. Average yield is 22-25 t/ha.
MHY-5	ARS, (RAU) Durgapura (Rajasthan)	VII	2001	This variety is developed through a cross between Durgapur Madhu x Hara Madhu. Its vines are vigorous and spreading about moderately spreading (about 2.0-2.5 m.) Fruits are round-flat, tapering at stem end, and colour green yellow with smooth skin. The flesh colour is light green with medium seed cavity, texture soft, and TSS 13-16 per cent. Average fruit weight is 700-800 g with average yield of 18.0-20.0 t/ha.
Muskmelon (Resistant varieties)				
Resistant	*Developing Centre*	*Recom-mened Zone*	*Year of identi-fication*	*Remark*
DMDR-1	ICAR-IARI, N. Delhi	CGMV	2001	Plants are vigorous and spreading (3.0 m). Average yield is 25 t/ha.
DMDR-2	ICAR-IARI, N. Delhi	Downy mildew + CGMV	2001	Plant spreads upto 3.5 m. Fruits are attractive and sweet with minium seed cavity. The average yield is about 20-22 t/ha.
(15) Watermelon (Open pollinated varieties)				
Variety	*Developing Centre*	*Recom-mened Zone*	*Year of identi-fication*	*Remark*
Durgapura Meetha	RAU, ARS, Durgapura (Rajasthan)	IV, VI, VII, VIII	1975	This variety has been developed through selection from the local material. Fruits are round with light green and thick rind, good keeping quality, weight varies from 7 to 8 kg; flesh is dark with 11 per cent T.S.S seeds with black tip and margin. It is a late maturing variety and takes about 125 days to reach marketable maturity. Average yield is 48.0-50.0 t/ha. It has been recommended for the cultivation in states of Punjab, Uttar Pradesh, Bihar, Rajasthan, Gujarat, Haryana, Delhi, Madhya Pradesh, Karnataka, Tamil Nadu and Kerala.
Sugar Baby	ICAR-IARI, N. Delhi	V, VII, VIII	1975	It is an introduction from U.S.A. It has medium long vines. It is an early maturing variety takes 70-75 days to reach maturity. Fruits are small to medium sized round with black rind, deep pink flesh with 11-13 per cent T.S.S. Fruit weight varies from 2.0-3.0 kg and gives an average yield of 17.5 t/ha. It has been recommended for cultivation in states of Chhattisgarh, Odisha, Andhra Pradesh, Madhya Pradesh, Karnataka, Tamil Nadu and Kerala.
Arka Manik	ICAR-IIHR, Bengaluru (Karnataka)	IV, VII, VIII	1987	It is developed from the cross between 'IIHR-21' and 'Crimson Sweet', an introduction from the U.S.A. It takes 100-110 days to reach fruit maturity. Fruits are round to oval with light green rind and dark green stripes; average weight of 6.0 kg, flesh is deep crimson, granular texture, pleasant aroma, high T.S.S. (12-15 per cent) with few small seeds. Average fruit yield is 60.0 t/ha. It is suitable for long distance transportation. It has resistance to powdery mildew, downy mildew and tolerance to anthracnose.
MHW-6	Mahyco Seed Co. Jalna (Maharshtra)	0	1999	Very popular hybrid. Plants spreds up 4-5 m. Fruits are attractivs. Average yield potential is 68 t/ha.
Watermelon (Hybrids)				
Hybrid	*Developing Centre*	*Recom-mened Zone*	*Year of identi-fication*	*Remark*
Arka Jyoti	ICAR-IIHR, Bengaluru (Karnataka)	0	1981	It is an F_1 hybrid developed between 'Crimson sweet'x'IIHR-20'. Fruits are round to oval, medium sized, dark green with blue angular stripes, average weight of 5.0-6.0 kg; flesh is bright crimson, granular texture with 12-14 per cent T.S.S. and excellent flavor. First picking is possible in 90 days after sowing. Average fruit yield is 70.0-72.0 t/ha.
MHW-6	Mahyco Seed Co. Jalna (Maharashtra)	IV, V, VIII	1999	Fruits are very attractive. Average yield potential is about 75 t/ha.

(16) Bitter gourd (Open pollinated varieties)				
Variety	*Developing Centre*	*Recom-mened Zone*	*Year of identi-fication*	*Remark*
Arka Harit	ICAR-IIHR, Bengaluru (Karnataka)	VIII	1986	It is a pure line selection from the local collection (IIHR-4) from Rajasthan. Vine is thin with light green deeply lobed foliage. Fruits are spindle shaped attractive, glossy green with smooth regular ribs and thick flesh and skin without tubercles.Average yield is 12.5-13.5 t/ha. Duration of crop is 100-110 days.
Kashi Mayuri (VRBTG-5)	ICAR-IIVR Varansi (U.P)	U.P	2018	It is cross derived cultivar. Fruits are medium long and dark green in colour. Dt is suit able for cultivator in plains and rverbed conditions. Average yield is 33 t/ha. It is resistant to mosaic virus under field contion. Suitable for pdyhouse cultivation.
Priya (VK 1)	KAU, Vellanikkara (Kerala)	II, VII, VIII	1992	It is selection from the local material. Bears long green spiny fruits tinged white at tip. The fruits are (40 cm) and 17 cm in girth with flesh thickness of 5-9 mm, average fruit weight 225 g; and each fruit contains 24 seeds weighing 6 g (100 seeds weight is 24 g). Average yield is 30.0-32.5 t/ha. It can be cultivated during January-August and September-December. It has been recommended for cultivation in the states of West Bengal, Assam, Madhya Pradesh, Maharashtra, Karnataka, Tamil Nadu and Kerala.
RHRBG-4-1	MPKV, Rahuri (Maharashtra)	IV, VII	1998	It is selection from local germplasm. Yield is about 22 t/ha.
KBG-16	CSAUA&T, Kanpur (Uttar Pradesh)	IV	1998	It is a pedgree selection. Plants are vigorous and attain the length of 3.8 m. Dark green fruit yield is 18-20 t/ha.
Pant Karela-1 (PBIG-1)	GBPU and T, Pantnagar (Uttarakhand)	:V	2001	This is pure line selection from inbreds of indigenous germplasm. It has been released by the Uttar Pradesh State Variety Release Committee in 1999 and identified in 2001 through All India Coordinated Research Project (Vegetable Crops). Plants are 2.0 m long and fruits are 15.0 cm long with tapering ends. It takes 55 days to first picking and the yield potential is 15.0 t/ha. It is suitable for cultivation in hills.
Bitter gourd (Hybrids)				
Hybrid	*Developing Centre*	*Recom-mened Zone*	*Year of identi-fication*	*Remark*
Pusa Hybrid-2	ICAR-IARI, N. Delhi	IV, V, VI	2002	This hybrid has been developed by heterosis breeding utilizing inbred Pusa vishesh and Pusa Do Mausami. Fruits are medium thick and long, glossy green, suitable for picking and dehydration. First pickling starts at 55-60 days after seed sowing. It is suitable for growing in spring summer season in the states of Punjab, Uttar Pradesh, Bihar, Jharkhand, Chhattishgarh, Odisha, Andhra Pradesh, Rajasthan, Gujarat, Haryana and Delhi and has yield potential of 20.00 t/ha. It gives 42 per cent and 40 per cent highor yiold than Pusa vishesh and Pusa Do Mausami, respectively.
NBGH-167	Nirmal Seed Co.	IV	2004	Very vigorous plants and has green foliage. Average yield is about 14 t/ha.
(17) Bottle gourd (Open Pollinated varieties)				
Variety	*Developing Centre*	*Recom-mened Zone*	*Year of identi-fication*	*Remark*
Kashi Ganga	ICAR-IIVR Varanasi (U.P.)	(U.P.)	2005	It is open pollinated variety. Vines are spreading (3.5m). Fruits are straight and 28 cm long. Yield is 48 t/ha.
Kashi Kiran	ICAR-IIVR Varanasi (U.P.)	(U.P.)	2018	It is round fruited open pollinated variety. Average fruit weight is 600-800 g. Yield potential is 52-55 t/ha.
Pusa Naveen	ICAR-IARI, N. Delhi	VII	1992	It is developed through selection from the local material. Suitable for *Zaid* and *Kharif* season in northern plains. Fruits are cylindrical, 30-35 cm long. Medium thick (6.0-7.0 cm diameter), straight and having an average fruit weight of 850 g. Average yield is30.0-35.0 t/ha.
Pant Lauki-3(PBOG-61)	GBPU and T, Pantnagar (Uttarakhand)	IV, VI	2001	This hybrid has been developed through pure line selection from a germplasm PBOG-61.It bears 40.0 cm long, cylindrical and light green fruits. First fruit harvest starts in 60 days. It is recommened for cultivation in states of Punjab, Uttar Pradesh, Bihar, Jharkhand, Rajasthan, Gujarat, Haryana and Delhi and has yield potential of 35.0-37.0 t/ha.
Narendra Jyoti (NDBG-104)	NDUA&T, Faizabad (U.P.)	IV	2002	It is developed from NDUA&T, Faizabad (Uttar Pradesh) and recommended for commercial cultivation in Uttar Pradesh, Punjab, Bihar and Jharkhand. Fruits are cylindrical, long, bottle shaped and early maturing. Having shallow neck and white flash. Average fruit weight is 1.0 kg. Average yield is about 30.0 t/ha.
NDBG-132	NDUA&T, Faizabad (U.P.)	VI	2004	This variety has been developed through local germplam. It is long fruited cultivar with slender attractive fruits. It produces first fruit after 60 days of sowing. It has been recommended for cultivation throughout the country and has yield potential of 43.0-45.0 t/ha.

				Bottle gourd (Hybrids)
Pant Sankar Lauki-1 (PBOGH-2)	GBPUA&T, Pantnagar (Uttarakhand)	VII	1998	This hybrid has been developed through heterosis breeding (PBOG-22 x PBOG-40). Fruits are medium size long (35.0 cm), green and somewhat cylindrical. Vine length is 5.5 m. First picking starts in about 60 days after seed sowing. This hybrid is suitable for hill and plain both of northern India. Average yield is 40.0-42.0 t/ha.
Pant Sankar Lauki-2 (PBOGH-1)	GBPUA&T, Pantnagar (U.P.)	IV, VI	1999	Fruits are 40.0 cm long and light green in colour, medium sized and somewhat cylindrical. This hybrid is suitable for rainy and summer season cultivation. First picking starts in about 65 days after seed sowing. Average yield is 40.0-42.5 t/ha and matures in 130 days. It has been recommended for cultivation in the states of Madhya Pradesh and Maharashtra.
Narendra Sankar Lauki-4 (NDBH-4)	NDUA&T, Faizabad (U.P.)	All	2001	It is in early flowering, prolific bearer with near cylindrical attractive fruits. It is suitable for cultivation throughout the year. Fruits remain soft even if they have grown out of normal size of picking. It has been recommended for cultivation throughout the country. It gives an average yield of about 40.0-60.0 t/ha.
Pusa Hybrid-2	IARI, New Delhi	IV, V, VI	2003	Fruits are 40.0 cm long, green, medium sized and cylindrical. First picking starts in about 65 days after seed sowing. It gives an average yield of 40.0-41.0 t.ha.
Kashi Bahar	ICAR-IIVR Varansi (U.P.)	U.P.	2005	It is a cross of VR-104 x VR-4. Plants are vigorous with green foliage. Fruits are 30. 0 cm long and cylindrical. First picking starts in 58 days. Yield potential is 52.0 t/ha.
				(18) Pumpkin (Open Pollinated Varieties)
Ambili (CM-14)	KAU, Vellanikkara (Kerala)	IV, V, VII	1987	This variety has been developed through selection from a local cultivar. Plants are spreading vigorously and fruits are flat round, medium (6.0 kg weight), green with shallow furrows on surface. It has yield potential of 34.0-36.0 t/ha in 95-110 days of crop duration. It has been recommended for cultivation in states of Kerala, Tamil Nadu and Karnataka.
Pusa Biswas	ICAR-IARI, N. Delhi	I, VIII, IV, V	1987	This variety has been developed through selection from a local collection (Line SM-107). Plants have vigorous vegetative growth, and dark green leaves with white spots including veins. Fruits are brown, spherical, average weight of 5.0 kg with thick and golden yellow flesh. Average yield is 40.0-42.5 t/ha in 120 days of crop duration. It has been recommended for cultivation in states of J&K, Himachal Pradesh and Uttarakhand, Punjab, Uttar Pradesh, Bihar, Chhattisgarh, Odisha, Andhra Pradesh, Madhya Pradesh and Maharshtra
Arka Chandan (IIHR-105-1-1)	ICAR-IIHR, Bengaluru (Karnataka)	VIII	1987	This variety has been developed through selection from a local collection (IIHR-105). Plants have vigorous veins, hairy stem, and green broad leaves without any pattern. Fruits are round with flat blossom end, medium fruit with 2-3 kg of weight, green with white patches on rinds (turn light brown at maturity). Flesh is thick, firm, bright orange with pleasant aroma, carotene rich and having high TSS (8-10 per cent). Average yield is 33.0 t/ha.
Arka Suryamukhi	ICAR-IIHR, Bengaluru (Karnataka)	VIII	1987	Plants have broad leaves and palmate veins and without pattern on leaf lamina. Fruits are small, round with flat end, deep orange with creamy white streaks on the rind, 1.0-1.5 kg of weight with orange flesh. Average yield is 36.0-38.5 t/ha in a crop duration of 120 days. It has been recommended for cultivation in the states of Karnataka, Tamil Nadu and Kerala.
Sooraj (CM-350)	KAU, Thrissur Kerala	VII, VIII	2001	This variety has been developed through selection from a local germplasm at KAU, Vellanikkara Thrissur (Kerala). It produces medium size globular fruits with orange coloured thick flesh. Average fruit weight is 3.0 kg. It is recommended for cultivation in the states of West Bengal, Assam, Karnataka, Tamil Nadu and Kerala and has yield potential of 35.0-40.0 t/ha.
Narendra Agrim (NDPK-24)	NDUA&T, Faizabad (U.P.)	IV, VI	2001	It is a short vined, small fruited and early maturing variety. Fruit shape is round and colour is dark green. Flesh colour is yellowish-white with flesh thickness of about 5 cm. It gives first picking in 55 days after seed sowing. Marketable green fruit weight is 1.5-2.0 kg, while ripened fruit weight is 3.0-4.0 kg. It is suitable for cultivation during both seasons. Average fruit weight is 7-8 kg when matured. Average yield is 30.0-35.0 t/ha. Storage life of the fruit is more than 6 months at ambient temperature.
				(19) Cucumber (Open pollinated varieties)
CHC-2	ICAR-HARP, Ranchi (Jharkhand)	IV	2001	Developed at Central Horticulture Experiment Station, Ranchi, and identified by XIX Group Meeting of Vegetable workers recommended for Zone-IV. Yield is 22-25 t/ha.
CH-20	ICAR-HARP, Ranchi (Jharkhand)	IV	2001	Plants are veigorous. Green foliage, medium long fruit of 15-18 cm. Average fruit yield is 13.0-15.0 t/ha.
Pant Khira-1 (PCUC-28)	GBPUA&T, Pantnagar (Uttarakhand)	I, VII, VIII	2001	This is selection from inbreds of indigenous germplasm. It is released by the Uttar Pradesh State Variety Release Committee also in 2001. The fruits are long (20.0 cm), cylindrical with light stripes. It takes 55-60 days after seed sowing to first picking and the yield potential is 15.0 t/ha. It is suitable for cultivation from March to July in plains and April to May in hills.

Cucumber (Hybrids)				
Hybrid	*Developing Centre*	*Recom-mened Zone*	*Year of identi-fication*	*Remark*
Pant Sankar Khira-1 (PCUCH-1)	GBPUA&T, Pantnagar (Uttarakhand)	All	2001	This is hybrid between PCUC-28 x PCUC-8. It is released by the Uttar Pradesh State Variety Release Committee also in 2001. Vines are about 120 cm long. Fruits are about 20 cm long, cylindrical and green with light stripes. First picking starts at about 50 days after seed sowing. It gives an average yield of 20.0 t/ha. It is suitable for cultivation in plains as well as hills.
Hybrid No. – 1	Century Seed Co. Mangolpuri (N. Delhi)	I, IV, VII	2004	Plants are vigorous with green foliage. Fruits are 18-20 cm long. It gives an average yield of 258-28 t/ha.
NBGH-167	Nirmal Seeds Co.	IV	2004	Plants are vigorous and spreading up to 3.5 m. Yield is 20-25 t/ha.
PCUCH-3	GBPUA&T, Pantnagar (Uttarakhand)	I, IV	2005	This hybrid has been developed through heterosis breeding (PCUC-15 x PCUC-25). Fruits are cylindrical, green and medium long (about 20 cm) with white spots. It has long fruiting period and first fruit can be harvestd in 50 days. It is moderately resistance to downy mildew and powdery mildew. It is recommended for cultivation in the states of J&K, Himachal Pradesh, Uttarakhand, Punjab, Uttar Pradesh and Jharkhand. This hybrid has the yield potential of 25.0-27.0 t/ha.
Rajni-Cu-05	Syngenta Seeds Pvt. Ltd. Pune (Maharshtra)	-	-	Very popular hybrid. Plants are vigorous and spread upto 3.5-4.5 m. Average yield potential is 28 t/ha. Free from cucumber mosaic virus. It is suitable for summer and rainy season cultivation.
(20) Ridge gourd (Open pollinated varieties)				
Variety	*Developing Centre*	*Recom-mened Zone*	*Year of identi-fication*	*Remark*
CHRG-1	HARP, Ranchi (Jharkhand)	IV	2001	Plants are very vigorous. Fruits are straight and slender. Average yield is 15-18 t/ha. Free from virus.
Pant Torai-1 (PRG-7)	GBPUA&T Pantnagar (Uttarakhand)	VII	2001	This variety is a pure line selection from an indigenous germplasm at GBUA and T, Pantnagar (Uttarakhand). The main shoot is 5.0 m long. Fruits are 15.0-20.0 cm long club-shaped. It takes about 65 days to first harvest. The variety is suitable for rainy season. It is recommended for cultivation in the states of Madhya Pradesh and Maharashtra and has yield potential of 10.0-12.0 t/ha.
IIHR-7	ICAR-IIHR, Bengaluru (Karnataka)	VIII	2001	Plants are vigorous. Fruits are very long, straight and tender. Average yield is 22 t/ha. Free from virus.
(21) Sponge gourd (Open Pollinated varieties)				
Kashi Divya (VRSG-1)	ICAR-IIVR, Varanasi (U.P.)	IV	2012	Sponge gourd variety 'Kashi Divya' was developed through hybridization and pedigree selection at IIVR, Varanasi. Kashi Divya was identified during 2010 and recommended for Zone IV (Punjab, Uttar Pradesh, Bihar and Jharkhand). Its yield potential is 20.0-25.0 t/ha which is 25 per cent more than the national check 'Pusa Chikani'. Due to medium size (20.0-25 cm) and straight, light green fruits, it is suitable for packing and distance marketing. Fruiting starts within 48 days at 8-9 nodes and bears 14-16 fruits/plant. It is tolerant to powdery mildew, downy mildew and anthracnose in field condition.
Sel-99	ICAR-IARI, N. Delhi	IV, VI	1995	Plants are vigorous. Average yield potential is 18-20 t/ha.
Swarna Prabha (CHSG-1)	ICAR-HARP, Ranchi (Jharkhand)	IV	2005	This variety has been developed through pure line selection. Fruits are medium in length (20-25 cm), each weighing 150-200 g, light greenish colour, pulp soft and contain less fibre. It is tolerant to powdery and downy mildew and resistant to leaf minor under field conditions. First harvest 70-75 days after sowing. It is recommended for July-September and February-April sowing in the states of Uttar Pradesh, Jharkhand and Bihar. Yield is about 20.0-25.0 t/ha.
JSGL	GAU, Junagarh (Gujarat)	VII	2005	Developed through pedigree selection. Suitable for summer and rainy season. Average yield is 16-20 t/ha.
Azad Torai-2 (KSG-14)	CSAUA&T, Kanpur (U.P.)	IV	2006	This short duration high yielding variety has been developed through selection. Its fruits are compact, green and rich in nutrients. It is recommended for commercial cultivation in the states of Punjab, Uttar Pradesh, Bihar and Jharkhand and has yield potential of 15.0-20.0 t/ha.
PSG-40	GBPUA&T, Pantnagar (Uttarakhand)	I, VII	2007	Develop through selection from the local germplasm. Plants are vigorous and green foliage. Average yields 18 t/ha.
(22) Ash gourd (Open pollinated varieties)				

Kashi Ujawal (IVAG-90)	ICAR-IIVR, Varanasi (U.P.)	I and VIII	2006	This variety is developed through selection. Fruits are globular, average fruit weight is 9.0-10.0 kg and less seeded. Crop duration is 130-140 days. This variety has been recommended for cultivation in the states of Punjab, Uttar Pradesh, Bihar, Jharkhand, Karnataka, Tamil Nadu and Kerala and has yield potential of 60.0-61.0 t/ha.
PAG-72	GBPUA&T, Pannagar (Uttarakhand)	VIII	2006	This variety is developed through pedigree selection. Fruits are of 8-10 kg having less seed. The aveage yield is about 65-68 t/ha.
Pusa Ujjwal (DAG-1)	ICAR-IARI, N. Delhi	VIII	2007	This variety is developed through selection.Vines are large, hairy and spreading Large, silvery green, hairy and lobed leaves with long petioles. Monoeciuos, male and female flowers are yellow and solitary, male flower with long pedicle but female flowers are almost sessile. The fruits are green in immature stage and white in full mature stage. They are oblong, ellipsoid, cylindrical, rind colour green, immature fruits are pubescent, cylindrical (average length-23.50 cm and girth-55.0 cm). Mature fruits have a dense pelt of minute hairs and covered with a white waxy bloom. The average fruit weight is 7.0 kg and average number of fruits per plant is 3.10. The fruit flesh is white and seeds are creamy white in colour. Average fruit yield is 48.05t/ ha (kharif season) and 41.0-42.5t/ha (summerseason). Its fruits are ideal for long distance transportation. It is recommended for cultivation in the states of Karnataka, Tamil Nadu and Kerala

Varietal Wealth of Vegetable Crops (Not Listed in Table 2)

Moreover, besides the most common and utilized vegetable varieties/hybrids/resistance, a very high number of locally known land races/varieties/hybrids release from "State Variety Release Committee or not release by any other agencies or organizations are also popular in the respective zones and being grown by vegetable growers. Locally known potential vegetable varieties in each crop are popular among the growers because these are more adopted in the area than released varieties/hybrids. The detail of these varietal wealth are given in Table 3.

Table 3: Land Races/Varieties/Hybrids Release from Central Varietal Release Committee, "State Variety Release Committee"or Not Release by any other Agencies

Variety	*Description*
	Crop Tomato
Kashi Amrit (DVRT-1)	This variety has been developed through backcross pedigree method from a cross (Sel-7 x *Solanum. hirsutum* f. *glabratum* B-6013) at ICAR- IIVR, Varanasi (U.P.). Plants are determinate in nature bearing medium to large fruits weighing ~100 g. It has potential of 62.0-63.5 t/ha in 115-120 days of crop duration.
Kashi hybrid-1 (VRTH-1)	This hybrid has been developed at ICAR- Indian Institute of Vegetable Research, Varanasi (U.P.). This is an early maturing hybrid having determinate growth habit. The average yield potential is 11 t/ha. As per the comparative analysis, this hybrid shows 87 per cent heterosis over standard check (ARTH-3). It posses field tolerance to tomato leaf curl virus and early blight.
Kashi hybrid-2 (VRTH-2)	This hybrid has been developed at ICAR- Indian Institute of Vegetable Research, Varanasi (U.P.). It is a short duration in nature and having maximum fruit production of 90 t/ha. As per the comparative analysis, this hybrid shows 54 per cent heterosis over standard check (ARTH-3).
Punjab Upama	Developed from PAU, Ludhiana (Punjab). Selection from segregating population. Its foliage cover is dense and leaflets are broad. The fruits are oval, medium in size, firm and deep red in colour. It is suitable for both table and processing purpose. Average fruit weight is 70.0 g and yield is 55.00 t/ha
Castle Rock	Developed from PAU, Ludhiana (Punjab). An introduction from USA. The plants have dark green foliage; fruits are oval, large, firm and deep red in colour. It is suitable for both table and processing purpose. Average fruit weight is 125.0 g and yield is 54.00 t/ha
Marglobe (EC 6870)	It is a mid season variety with indeterminate growth habit. Plant is 70-80 cm tall, vigorous, and erect. Leaves are large and broad with dense foliage. It has typically 4 or 5 flowers on truss, which bears 3 or 4 fruits. Fruits are large, nearly round, green shoulder, deep scarlet colour on ripening, 3-5 locules, juicy, mildly acidic, thick walled, smooth, solid and less seeded. Diameter of ripe fruit is about 7.5 cm and weighing 100-150 g.
Roma	This is an introduction from America. Plants are small with determinate growth habit and luxuriant foliage cover. Fruits are born in 4-5 clusters it produces yellowish green, pear shaped medium large fruit which becomes uniform red at ripening. It has ability to set the fruits in spring and summer seasons. Fruits are pear shaped, deep red coloured, free from green shoulder, small fruits small seed cavity and suitable for long distance transportation. Fruit has less seed. Average yield is 25.0-31.0 t/ ha. This variety possesses good keeping quality and good for processing.
Sioux	It is an exotic variety introduced from USA. It is indeterminate type and produces yellowish green, medium large fruit which becomes uniform red at ripening. It has ability to set the fruits in spring and summer seasons.
Ox Heart	Plant is 70-80 cm tall. Fruti is extremely solid with thick firm walls and small seed cavities with few seeds. Crop duration is 115-130 days.
Pusa Red Plum	It is derivative of a cross between the dwarf variety 'Red Cloud' and 'Merruti' and developed by ICAR-IARI, New Delhi. Fruits ripen a week earlier than cv. 'Pusa Ruby'. The variety does well in summer and in winter growing seasons.

Variety	*Description*
Pusa Sadabahar	It is developed from ICAR-IARI, New Delhi. Plants are determinate and dwarf. It is prolific bearer with smooth oval to round attractive fruits. This variety can set fruits even up to 30°C night temperature. It is most suited for cultivation in north Indian plains except rainy season. Its fruits are ideal for distant marketing. Average yield is 40.0 t/ha during main season. It is highly suitable for growing under wide range of night temperature (8-30 °C). This variety can be accommodated more in number per unit area. This variety can be grown through out the year in north Indian plains except rainy season *i.e.* July to September (Tiwari *et al.,* 2001).
Pusa Gaurav	Developed and selected from the cross 'Glamour' x 'Watch' at ICAR-IARI, New Delhi as segregating generation. The plants are determinate, dwarf/bushy with moderate foliage cover. Plant is very prolific bearer. The fruits are smooth, oval, elliptical (egg shaped), attractive and born in clusters. Fruits are firm, thick fleshed (0.6 cm) with two well-filled locules and long shelf life which facilitate easy transportation over long distance. It has better keeping quality at room temperature (Choudhury and Tiwari, 1993). Average yield is 40.0 t/ha. Whole fruit is suitable for processing and canning because it is without neck constriction and has higher TSS (6.0 percent).
Pusa Sheetal	It has been developed and selected from the cross Balkan (Bulgaria) x Temnorrsnij (USSR) at ICAR-IARI, New Delhi. Plants of this cultivar are compact with moderate foliage. Its fruits are flattish round, smooth, attractive and developed uniformly red colour on maturity. It is tolerant to low temperature (set fruit at very low temperature) and is recommended for early planting in north Indian plains (Tiwari *et al.,* 1990). Average yield is 36.3 t/ha.
Pusa Uphar	This indeterminate variety has been developed by ICAR-IARI, New Delhi. Its plants are erect, thick stemmed with dark green foliage. Plants are prolific bearers. It has higher fruit setting with desirable quality even in slightly adverse temperature. It bears in cluster of 4-5 fruits. Fruits are aatractive, medium sized (60 g), round, deep red with thick pericarp, uniform ripening and suitable for processing. It takes 100 days from transplanting to fruit ripening. Average yield is 35.0-40.0 t/ha. It is suitable for cultivation in Ganga alluvial and western plains (Tiwar *et al.,* 1999). It is slightly tolerant to fruit borer under field conditions.
Pusa Divya (F_1)	This hybrid has been developed by utilizing closed anther mutant with potato leaf marker gene (functional male sterility) as female parent. This hybrid has been released by the Central Variety Released Committee, especially for cultivation in Uttar Pradesh, Punjab and Bihar (Verma *et al.,* 2001). Plants are indeterminate with profuse branching. Fruits are firm thick, skinned, round to oval shaped, green with dark green stem end and turn dark red on ripening. Due to thick skin, fruits have longer shelf life with suitability for long distance transportation. Maturity period is 80-90 days from transplanting and 135-150 days from seed sowing in hills, however, this may be reduced in plain. An average yield is 42-45 t/ha. Less infestation to late blight and buckey rot.
Pusa Hybrid-1	This hybrid has been developed by ICAR-IARI, New Delhi. The plants are compact, determinate with good foliage cover having prolific bearing. Fruits are round, smooth and attractive. It is ready for market from late May to mid July. Pusa Hybrid-1 has an added advantage of providing tomatoes from June to mid July, the lean period for tomato in north India. It can set fruit up to a ptemperature of 28 °C during night. The average fruit weight is 63 g and yield is 32 t/ha (Tiwari, 1996).
Pusa Hybrid-4	This hybrid has been developed at ICAR-IARI, New Delhi and notified by the Central Varietal Release Committee (CVRC) during 1997. It has been developed by utilizing closed anther mutant with potato leaf marker gene (functional male sterility) as female parent. Plants are indeterminate with profuse branching. Fruits are firm, thick pericarp, round to oval green with dark green stem end and turn dark red on maturity. It has been recommended for cultivation in Uttar Pradesh, Punjab and Bihar and has yield potential of 45.0 t/ha.
Pusa Hybrid-8	This hybrid has been developed at ICAR-IARI, New Delhi. Plants are determinate with dark green foloiage. Fruits are round with thick skin. Suitable for long distance transportation. Generally, nursery raising and transplanting is done during October and November, respectively. The average fruit yield is 55-57 t/ha.
Best of All	It is an indeterminate and mid season variety. Plants are erect and height ranged from 100-200 cm. Fruits are medium in size, scarlet red in colour, round, juicy, small scar, thin skin, three locuied with green shoulder and born in cluster 4-5 fruits.
Kalyanpur Angoorlata	It is a selection from local germplasm at CSAU and T, Kalyanpur, Kanpur (Uttar Pradesh) and released for commercial cultivation in 1982. Plants are creeping in nature, 2-3 m tall and indeterminate growth habit. Leaves are medium, light green with spaced leaflets. Fruits are small, oblong, 8-10 fruits/trusses on an average, skin firm and thick and less juicy. Immature fruits are yellowish green and red on ripening. It is moderately resistant to leaf curl mosaic and resistant to root knot nematode. Fruit yield is 45-50 t/ha. Seed yield is 0.6-0.7 t/ha.
KS-1	It is deveopled from CSAU and T, Kalyanpur, Kanpur (Uttar Pradesh). It is a medium early variety having determinate growth habit. Fruits are medium in size, round, flattened and smooth. It is suitable for table purpose.
KS-3	It is deveopled from CSAU and T, Kalyanpur, Kanpur (Uttar Pradesh). It is mid early maturing variety having determinate growth habit. Fruits are medium in size, green shoulder, flattened, smooth and semi fleshy.
NDT-108	This variety has been developed from NDUA&T, Kumarganj, Fiazabad (Uttar Pradesh). Plants are dwarf. Fruits are like Punjab Chhuhara having good shelf-life. An average yield potential is 26-30 t/ha.
NDT-120	This variety has been developed from NDUA&T, Kumarganj, Fiazabad (Uttar Pradesh). Plants are dwarf, harvesting starts 75-80 days after transplanting. Fruits are attractive, big in size and red in colour. An average yield of this variety is 30-35 t/ha.
Pant T-1	This variety has been developed from GBPUA&T, Pantnagar (Uttarakhand). Plants are dwarf, straight and 100-120 cm tall. Fruits are round with nipple having 4-6 locules. Fruiting occure in cluster of 4-5 fruits in each cluster. This variety is tolerant to leaf curl virus. Average yield potential is about 40 t/ha.
TH-2312	Developed from PAU, Ludhiana (Punajb). It is a hybrid and developed between the cross of VFN x Punjab Chhuhara. Its plants are spreading with dense foliage; fruits are medium sized, oval-round, medium firm and deep red in colour. Average TSS content is 4.7 per cent. Average fruit weight is 70.0 g and yield is 65.00 t/ha. It is resistant to root knot nematodes.
TH-802	Developed from PAU, Ludhiana (Punjab). It is a hybrid and developed between the cross of Healani x Castle Rock. The plants are spreading and vigorous. Fruits are firm, smooth, round and uniformly deep red. Average TSS content is 4.7 per cent. It is suitable for table and processing purpose. Average fruit weight is 70.0 g and yield is 64.0 t/ha. It is resistant to root knot nematodes.

Variety	Description
TH-1	Developed from PAU, Ludhiana (Punjab). It is a dual purpose F_1 hybrid between W321 x I181, suitable for both fresh market and processing. The fruits are round, deep red and firm having TSS content of 5 per cent. Average fruit weight is 85.0 g and yield is 54.50 t/ha. It is moderately resistant to early blight.
Arka Ananya	This hybrid was developed by ICAR-IIHR Bengaluru (Karnataka) and released for commercial cultivation in 2005. Fruits are medium (65-70 g), having combined resistance to *ToLCV* and bacterial wilt. Deep red, firm fruits suitable for fresh market. Yield is 65-70 t/ha.
Arka Samrat	This high yielding hybrid was developed by ICAR-IIHR Bengaluru (Karnataka) and released for commercial cultivation in 2010. It is triple disease resistance to *ToLCV*, Bacterial wilt and early blight. Fruits oblate to high round, large (90-110 g.). Deep red, firm fruits, suitable for fresh market. Yield is 80-85 t/ha.
Arka Rakshak	This high yielding semi-determinate hybrid was developed by ICAR-IIHR Bengaluru (Karnataka) and released for commercial cultivation in 2010. It is triple disease resistance to *ToLCV*, Bacterial wilt and early blight. Fruits are square round, large (90-100 g). Firm fruits, light green shoulders, deep red, suitable for fresh market and processing. Yield 75-80 t/ha.
Arka Shresta (BRH-2)	This high yielding hybrid was developed by ICAR-IIHR, Bengaluru (Karnataka) by 15 SBSB x IIHR-1614 parents. Plants are semi-determinate with light green foliage. Fruits are firm, medium large, round with light green shoulder and are suitable for distant transportation. Average TSS content is 4.7 per cent. It is suitable for both fresh market and processing. Average fruit weight is 70 g and yield is 76-83 t/ha in crop duration of 140 days. It is resistant to bacterial wilt. Fruits can be stored for 17 days (from breaker to softening stage) at room temperature.
Arka Abhijit (BRH-1)	It is a determinate F_1 hybrid developed at ICAR-IIHR, Bengaluru. Plants are semi-determinate 75 cm tall and take 55-60 days to first fruit maturity. Fruits are round with light green shoulders, and firm fruit (individual fruit weight is 70.0 g) with 4.3 per cent TSS. Crop duration is 140 days. The fruit has better keeping quality *i.e.* 17 days at room temperature (from breaker to softening).This hybrid has high degree of resistance to bacterial wilt, suitable for both fresh market and processing. It has yield potential of 63-65 t/ha in crop duration of 140 days. It is resistant to bacterial wilt and suitable for fresh market.
Arka Vishal	It is an F_1 hybrid between IHR-837 x IHR 932 developed by ICAR-IIHR, Bengaluru (Karnataka). Its fruits are round, firm, deep red with green shoulder and have good shelf life. Average fruit weight is 140 g and yield is 75.0t/ha. It is tolerant to fruit cracking.
Arka Ahuti	It is a pure line selection of variety Ottawa-60 and released from ICAR-IIHR, Bengaluru (Karnataka). Plants are indeterminate in growth habit. Fruits are oval in shape, pulpy, tough skin and uniform ripening. Total soluble solid is 5.2 per cent. The yield potential is 45 t/ha in 135 days of crop duration.
Arka Abha	It is developed from ICAR- IIHR, Bengaluru (Karnataka). Plants are indeterminate in nature. Fruits are flat, round, green shoulder and average fruit weight is 70-80 g. an average yield is 35 t/ha. This variety is resiatant to bacterial wilt.
Arka Meghali	It is a crosss of Arka Vikas x IIHR-554, developed through pedigree method of selection and released from ICAR-IIHR Bengaluru (Karnataka). Plants are semi-indeterminate. Fruits are superior to Pusa Ruby variety in size (63 g), flesh thickness and fruit firmness.This variety is suitable for rainfed condition. The yield potential is 17.2-18.2 t/ha (Sadashiva *et al.,* 1996)
NDTR-1	It is a derivative from the multiple crosses between *Solanum lycopersicum x Solanum peruvianum* and released in 1978 from Karnataka. Plant is 60 cm tall, spreading type and indeterminate growth habit. Leaves are thick and dark green. Fruits are medium to large (70-80 g), round, 4-6 loculed, 3-4 mm flesh thickness and bright red on ripening. Seeds are light brown. Fruit yield is 17.5-25.0 t/ha. Seed yield is 0.12-0.14 t/ha/in 100-120 days of crop duiration.
CO-1	It is developed and released from TNAU, Coimbatore (Tamil Nadu). Plant is determinate and suitable for growing in southern India. Fruits are round with yellow stem end and uniform ripening. Average yield potential is 38 t/ha.
CO-2	It is developed through pure line (isolated from USSR type) and released from TNAU, Coimbatore (Tamil Nadu) which is well adopted in Tamil Nadu state. Plant is 70 cm tall, semi-dwarf, erect in growth habit with moderate branching. Plants are well spread and needs no staking. Stem is light green and foliage is dark green in colour. Flowering occurs 65-70 days and first picking starts 100 days after transplanting. Fruits are ovate, medium in size, smooth, green when unripe and bright red on ripening. Fruits are devoid of cracking. Plant contains 4-5 fruits/truss and 20-25 fruits/plant. It gives 41 t/ha yields in 6-8 harvesting.
COTH-1	It is a derivative of cross between IIHR-709 x LE-812 and developed at TNAU, Coimbatore (Tamil Nadu). Plants are determinate and 75 cm tall. Fruits are round to slightly oblong, medium sized (average weight 60 g) and bright red at maturity. First picking starts after 88-94 days of transplanting. It gives an average yield of 85-95 t/ha in 140 days of crop duration.
PKM-1	It is an induced mutant from Annanji tomato and released in 1978 from TNAU, Coimbatore (Tamil Nadu). Plants are dwarf (45 cm) and less spreading with dark green foliage. Flowering starts 55 days and first harvesting starts 95 days after transplanting. Fruits are flattish with 4-5 light grooves, 4-5 locules, 4-6 fruits/cluster, 30 fruits/plant, dark green when unripe and red at ripening with green at pedicel end. The mean fruit weight is 55.2 g, fruit length 3 cm and fruit width 4.6 cm. it is moderately resistant to pest and disease. It gives an average yield of 32 t/ha in 135 days of crop duration. The seed yield is 0.23- 0.25 t/ha.
Jawahar Tamatar-99	It is developed by Jawahar Lal Nehru Krishi Vishwavidyalaya (JNKVV), Jabalpur (M.P.) through selection from a germplasm collected from USA. Plants are determinate in growth habit. It is an early variety. First picking strats from 100-110 days after seed sowing. Fruits are round, hard and medium size with attractive yellow colour after ripening. This variety is suited for processing due to high TSS (5.3 per cent). It gives an average yield of 47.5-50.0 t/ha
RHRTH-57	This variety has been developed from a cross of 32-2 and 85-1 by MPKV, Rahuri (M.S.). Fruits are oval, red, medium size and medium stout with 4.1 per cent TSS and uniform ripening. It gives an average yield of 73 t/ha.
RHRTH-58	This variety has been developed from a cross of Dhanashree and RHRT-1 by MPKV, Rahuri (M.S.). Fruits areround, orange-red, medium sized, smooth surface, very sstout with uniform ripening and 3-9 per cent TSS. It gives an average yield of 66.95 t/ha.
Dhana Shree	This variety has been developed from MPKV, Rahuri (M.S.). It gives an average yield of 60 t/ha.
Bhagya Shree	This variety has been developed from MPKV, Rahuri (M.S.). It gives an average yield of 63 t/ha.

Variety	*Description*
Raja Shree	This hybrid has been developed from MPKV, Rahuri (M.S.) and released for commercial cultivation in 1994. Fruits are medium (90.0-100.0 g) firm with high pulp and content ascorbic acid (42-50 mg/100 g). Fruits are attractive orange coloured and oval round in shape. It is suitable for long distance transportation and comparatively less susceptible to the viral diseases (Wagh *et al.,* 1995).
Phule Raja	This Hybrid has been developed from MPKV, Rahuri (M.S.). It gives an average yield of 66 t/ha.
ATH-1	This hybrid has been developed from MPKV, Rahuri (M.S.) and notified by the Central Varietal Release Committee (CVRC) during 1989. Fruits are bright red and each weighing 90-100 g. This has yield potential of 45.0-47.0t/ha.
Sonali	This variety has been developed from an exotic collection at BSKKV, Daploi (M.S.). It is high yielding bacterial wilt resistant variety; therefore it is recommended for the bacterial wilt prone southern district of Konkan regions. It matures in 120-140 days and gives 50 t/ha fruit yield (Patil *et al.,* 1990)
Devgiri	This variety has been developed from, MAU, Parbhani (M.S.)
Vasundhara	This hybrid has been developed from, MAU, Parbhani (M.S.) and notified by the Central Varietal Release Committee (CVRC) during 2001. Plants are determinate with jasper red fruits, each weighing 70-75 g with good storability. It is resistant to virus and blight and has yield potential of 50.0-52.5 t/ha.
Parbhani Yash Shree	This variety has been developed from, MAU, Parbhani (M.S.) It gives an average yield of 66.8 t/ha.
Junagadh Ruby	It is a derivative of the cross between Inbred Big Silari x Pusa Ruby, release by Gujarat state in the year 1981. Plants are semi spreading in growth habit with purple attractive growth and more foliage. Leaves are medium in size and dark green. Fruits are born in cluster (4-6 fruits/truss), round without grooves, four loculed and orange red on ripening with pinkish red flesh. Days to first picking are 85-90 days. On an average, it gives 20 picking at short intervals. It does possess complete resistance against any disease and insect/pests/
Gujarat Tomato -1	This variety has been developed in the year 2002. It gives an average yield of 52 t/ha.
Gujarat Tomato -2	This variety has been developed in the year 2004. It gives an average yield of 58.5 t/ha.
Anand Tomato-3	This variety has been developed in the year 2008. It gives an average yield of 63 t/ha.
Punjab Tropic	It is a selection from the introduction of USA and released from Punjab Agricultural University, Ludhiana (Punjab).Plant is 100 cm tall, indeterminate growth habit with luxuriant growth. Stem and foliage are dark green. Flowering starts 95 days after sowing. Fruits are large (100 g), round, red on maturity, born in clusters (4-6 fruits/truss) and remain totally covered with foliage. It has late fruit setting with slow development of fruits. Yield potenyial is 22.5 t/ha in crop duration of 141 days.
TH-802	It is a derivative of cross between 'Healani' and 'Castle Rock' and released from Punjab Agricultural University, Ludhiana (Punjab). Plants are determinate, spreading type with broad leaflet and dark green foliage. Fruits are round, medium size (68 g of average fruit weight) with thick pericarp (6.4 mm) and 4.75 per cent TSS. Yield potenyial is 63.8 t/ha in crop duration of 140 days.
TH-2312	It is a hybrid derived from a cross of 'VFN-8' and 'Punjab Chhuhara' and developed from Punjab Agricultural University, Ludhiana (Punjab). Plants are dwarf and spreading with dense foliage. Fruits are oval-round, medium size (67.5 g average weight), medium firm with thick pericarp (5.4 mm) and 4.75 per cent TSS. It gives an average yield of 63.18 t/ha (6.46 per cent increase over 'Rupali' hybrid) in 150 days of crop maturity duration.
H.S.-102	It is a derivative of cross between 'S-12' x 'Pusa Early Dwarf' and developed and released from CCS HAU, Hisar (Haryana). This variety is characterized by extremely early, concentrated flowering and fruiting. Fruits are medium to small, round, juicy and thin pericarp. It is recommended for cultivation in Haryana state (Kalloo and Singh, 1976).
H.S.-110	It is a selection from an exotic genotype and released for commercial cultivation by CCS HAU, Hisar (Haryana) (Kalloo and Singh, 1982). Plants are determinate having potato leaf type. Fruits are large, smooth, fleshy and jointless. It is suitable for table purpose.
HTH-2-2	It is a hybrid developed by CCS HAU, Hisar (Haryana). The average yield of this hybrid (32.5-36.5 t/ha) is 22, 57, 67 and 43 per cent higher than hybrids Rashmi, DTH-8, ARTH-3 and variety Hisar Arun (Sel-7), respectively.
Swarna Deepti	This Hybrid has been developed by ICAR Research Complex for Eastern Region, Research Centre, Plandu, Ranchi (Jharkhand).
Swarn Lalima	This variety has been developed by ICAR Research Complex for Eastern Region, Research Centre, Plandu, Ranchi (Jharkhand). Develop through pure line selection from EC339074. Plant height is 70-80 cm, determinate growth habit. First harvest can be obtained 55-60 days after transplanting. Fruits are round, weight 120-125 g, deep red, with TSS 4-5°B and born in cluster of 3-4. Yield is about 60.0-70.0 t/ha. It is resistant to bacterial wilt.
Swarn Naveen	This variety has been developed by ICAR Research Complex for Eastern Region, Research Centre, Plandu, Ranchi (Jharkhand). Develop through pure line selection from EC386019. Plant height is 130-150 cm, indeterminate growth habit. First harvest can be obtained 60-65 days after transplanting. Fruits are round, weight 60-70 g, deep red, with TSS 4-5°B and born in cluster of 8-10. Yield is about 60.0-65.0 t/ha. It is resistant to bacterial wilt.
Swarn Baibhav (F_1)	This hybrid has been developed by ICAR Research Complex for Eastern Region, Research Centre, Horticulture and Agro-Forestry Research Programme (HARP), Plandu, Ranchi (Jharkhand) by the cross of CH-3 x CHRT-4). Plant height is 70-80 cm, and determinate growth habit. First harvest can be obtained 55-60 days after transplanting. Fruits are round, weight 140-150 g, deep red, firm, with TSS 4-5°B and born in cluster of 3-4. Yield is about 90.0-100.0 t/ha in 110-120 days of crop duration. It is suitable for long distance transportation and processing.

Variety	Description
Swarn Samridhi	This hybrid has been developed by ICAR Research Complex for Eastern Region, Research Centre, Plandu, Ranchi (Jharkhand). Hybrid developed between the cross of IC400215 x BL-333. Plant height is 100-120 cm, indeterminate growth habit. First harvest can be obtained 55-60 days after transplanting. Fruits are round, weight 70-80 g, red, firm, with TSS 5-6°B and born in cluster of 8-10. Yield is about 90.0-100.0 t/ha. It is resistant to bacterial wilt and early blight.
Swarn Sampada)	This hybrid has been developed by ICAR Research Complex for Eastern Region, Research Centre, Plandu, Ranchi (Jharkhand). Hybrid developed between the cross of IC400216 x HAT-193. Plant height is 70-80 cm, determinate growth habit. First harvest can be obtained 55-60 days after transplanting. Fruits are round, red, weight 70-80 g, red, firm, with TSS 5-6°B and born in cluster of 8-10. Yield is about 120.0-130.0 t/ha. It is resistant to bacterial wilt and early blight.
Swarn Vijay	This hybrid has been developed by ICAR Research Complex for Eastern Region, Research Centre, Plandu, Ranchi (Jharkhand). Yield is about 60.0-68.0 t/ha. It is resistant to viral disease.
Keckruth	It is a derivative of the cross of Keck Methi x Ruthger. Plant is 100 cm tall, and stem and foliage are green colour. Flowering takes place in 93 days after transplanting. Fruits are round, medium to large (70-80 g), deep red on maturity, 3-4 locules and pericarp 0.3 cm thick. Average yield potential is about 41 t/ha.
Keckruth Ageti	It is a derivative of the cross of Keckruth x x Ruthger. Plant is 100 cm tall, and stem and foliage are deep green coloure. Flowering takes place in 85 days after transplanting. Fruits are round, medium in size (40-50 g) with green shoulder, which disappears on ripening and 3-4 locules. Average yield is about 40 t/ha.
Shakati (LE-79)	It was developed through pure line selection at college of Horticulture, Vellanikkara, Kerala. Plants are determinate with pubscent stem and normal leaf. Fruits are round, white and of medium size with average weight of 58 g. First picking starts in 75 days after sowing. It is resistant to bacterial wilt and fruit cracking. It gives an average yield of 43.5 t/ha in a crop duration of 95-100 days. It is recommended for cultivation in the Jammu and Kashmir, Himachal Pradesh, Uttarakhand, Chhattisgarh, Odisha, Andhra Pradesh, Madhya Pradesh, Maharashtra, Karnataka, Tamil Nadu and Kerala.
Solan Sagun	This variety has been developed from YS Parmar University of Forestry and Horticulture, Solan (Himachal Pradesh).This is an early hybrid and the flowering starts 24 days after transplanting. It yields about 58.0 t/ha. It is quite suitable for low and mid hills of Himachal Pradesh (Sharma and Rattan, 1995).
Solan Gola	This variety has been developed from YS Parmar University of Forestry and Horticulture, Solan (Himachal Pradesh). Plants are indeterminate in growth habit. Fruits are round, attractive, medium size, red colour and thick pericarp. This variety is suited for transportation to distant markets. An average yield of this variety is 37.5 t/ha.
Yaswant-2	This variety has been developed from YS Parmar University of Forestry and Horticulture, Solan (Himachal Pradesh). Plants are indeterminate in growth habit and fruiting starts 75 days after transplanting. Fruits are uniform red in colour with thick pericarp. An average yield of this variety is 50 t/ha. This variety is resistant to buckeye rot.
Balkan	It is an introduction of Bulgaria. Plants are erect and compact with dark green leaves resembling as potato leaf. Stems are thick and hardy. Plant remains upright throughout the season. Stem internodes are much shorter, resulting in clustering of fruits. It is a prolific bearer (40-50 fruits/plant), dwarf (35-40 cm) with good foliage cover. The unripe fruits show green stems ends, whereas, the ripe fruits are slightly oval and possess two locules. Fruits are dark red with no greening or spot and medium to small in size. Plants can withstand in considerable cold. Due to its dwarf, compact and erect habit, more plants/unit area can be accommodated. Through, the fruits are medium small but attractive and firm, hence suitable for transportation. It is good for cultivation in early winter (*Rabi*). It can be used in hybridization for developing cold and frost tolerant varities (Saxena and Thomas, 1988).
VC-48-1	This variety was brought from 'World Vegetable Centre (old name Asian Vegetable Research and Development Centre), Taiwan and tested at Assam Agricultural University Jorhat (Assam). Plants are determinate and 80-90 cm tall with heavy foliage. Fruits are medium round and bright red in colour having 5.5 per cent TSS. This variety has been recommended for bacterial wilt infested area.
Krishna Nagar (S-20)	This variety has been released by Department of Agriculture (West Bengal). It is medium to late variety. Fruits for first picking are available in 70-80 days after transplanting. Fruits are medium-sized, round in shape and red in colour. It yields nearly 30-32 t/ha.
Floradade	This variety was brought from Florida. Plants are determinate and tall (45-50 cm). Fruits are big in size weighing 150-175 g/fruit. This variety is tolerant to early blight. It has a yield potential of 40-50 t/ha. This variety has very good keeping quality.
Manikhamnu	This variety has been released by ICAR-Research Comlex for Eastern Region, Umium, Barapani (Meghalaya). It is dwarf (34.5 cm), early maturing, determinate and multi branched. Fruits are medium size, round, smooth with 2-3 locules, uniform red at maturity and ready for harvesting in 70-75 cm days when transplanted in September and 85-90 days in case of spring summer planting. Flesh is thick and juicy with 5.4 per cent TSS. It is suitable for processing as well as fresh consumption purpose. It is resistant to fruit cracking and tolerant to leaf curl virus. The yield potential is 53.3 t/ha. This variety is released from State Variety Release Committee, Manipur (Yadav *et al.*, 2000).
Manithoibi	The plants are dwarf (35 cmm), determinate and multibranched. It is an early maturing variety and having better capacity to withstand in moisture stress condition. Fruits are oval, medium, bilocular, good shelf life and uniform ripening. This variety matures in 70-75 days when planted in September and 80-90 days in case of spring summer planting. The yield potential is 63.7 t/ha. This variety is released from State Variety Release Committee, Manipur (Yadav *et al.*, 2000).
Madhuri (BSS-39)	This Hybrid has been released by Beejo Sheetal Seed Pvt. Ltd. Jalana (Maharashtra). Plants are determinate and early maturity. Fruits are dark red, flat round with thick skin, average weight of 100 g, crack resistant and can withstand long transportation. It gives an average yield of 50-70 t/ha.
Megha (BSS-98)	This Hybrid has been released by Beejo Sheetal Seed Pvt. Ltd. Jalana (Maharashtra). Plants are determinate. Fruits are red with thick skin, good shelf life and transport quality. It gives an average yield of 50-70 t/ha.
Manisha (BSS-99)	This Hybrid has been released by Beejo Sheetal Seed Pvt. Ltd. Jalana (Maharashtra). It has good shelf life and transport quality. It gives an average yield of 80-100 t/ha.

Variety	Description
Mangla	The plant of this hybrid is determinate type. Fruits are medium-sized, and good in quality. First picking can be done in 70 days after transplanting. It gives an average yield of 75 t/ha.
Sheetal	The plant of this hybrid is determinate type. Fruits are medium-sized (120 g). Fruits are free from cracking. Plants start yielding about 70 days after transplanting. The harvesting duration is about 30 days. Average yield is about 75 t/ha.
Rashmi	It is determinate F_1 hybrid developed by Indo-American Hybrid Seeds Company, Bengaluru (Karnataka). The fruits are slightly oblong, deep red and each weighing about 55.0-60.0 g. Yield is about 90.0-95.0 t/ha.
Vaishali	It is determinate F_1 hybrid developed by Indo-American Hybrid Seeds Company, Bengaluru (Karnataka). Fruits are most suitable for processing. Fruits are medium sized (about 100 g) of high quality. Plant starts yielding in about 70 days after transplanting. The harvesting duration is 30 days. Fruits are round and dark red in colour, firm, capable of yielding 80.0-88.0 t/ha (Yawalkar, 1985).
Rupali	It is determinate hybrid developed by Indo-American Hybrid Seeds Company, Bengaluru (Karnataka). The fruits are oblong in shape with prominent nipples (styler end) and deep red in colour. Yield is about 60.0-70.0 t/ha.
Naveen	It is indeterminate F_1 hybrid developed by Indo-American Hybrid Seeds Company, Bengaluru (Karnataka). The fruits are oblong in shape with narrow base slightly broadened tip and deep red in colour. Yield is about 60.0-70.0 t/ha.
IAHS-88-2	It is indeterminate F_1 hybrid developed by Indo-American Hybrid Seeds Company, Bengaluru (Karnataka). This variety produces large fruits. Plants are tolerant to Fusarium and Verticilium wilt and root knot nematodes. This is suitable to supply tomatoes fresh in local markets.
Sonali	A determinate F_1 hybrid with oblong fruits. The average fruit weight is 85 g and fruits have a TSS of 3.0-3.5 brix.
BCTH-62	This is a highly pigmented big fruited tomato hybrid with slow ripening character. Semi-determinate in growth habit with partly biparous cyme. Fruits are oblong-heart shaped, deep red, 106 g in weight with 3-4 locules, 0.59 cm pericarp thickness, TSS 4.24 °Brix and acidity 0.6 percent. Average yield is 55 t/ha under autumn-winter season in West Bengal.
Brinjal	
Kashl Taru	It is developed at ICAR-Indian Institute of Vegetable Research, Varanasi (U.P.). Plants are tall and erect. Fruits are long, dark purple, glossy and tender. Individual fruit weight is 75-85 g. First picking starts 75-80 days after transplanting. It is suitable for summer and autumn seasons (Kalloo *et al.,* 2006).
Pusa Purple Round	It is developed at ICAR-IARI, New Delhi. The plants are very tall with a thick stem of greenish purple colour. Leaves are highly serrated and deep green in colour. Fruits are round with purple colour. Each fruit weighs on an average 130.0-140.0 g. It has a yield potential of 30.0-40.0t/ha.
Pusa Kranti	Plants are sturdy. Fruits are 15.0-20.0 cm long, dark purple with shining green calyx, and less seeded. It yields 37-40 t/ha.
Pusa Upkar	This is an early maturing variety has been developed by ICAR-IARI, New Delhi through selection from the cross GR x PB-91-1. Plants are thornless, moderately vigorous and branched. Young leaves and shoots show purple pigmentation. Fruits are medium sized (200 g), shining and dark purple. Average yield is 40.0 t/ha.
Pusa Ankur	This variety has been developed by ICAR-IARI, New Delhi through selection from the indigenous material. Plants are thornless, semi-erect, with purple pigmentation on younger leaves and branches. First picking is possible 45 days after transplanting. Fruits are smooth, oval-round, dark purple and glossy. Its fruits do not lose colour and tenderness even on delayed pickings. Average fruit weight is 60-80 g and yield is 34.0-38.5 t/ha.
Pusa Anmol	It is an F_1 hybrid developed at ICAR-IARI, New Delhi by crossing Pusa Purple Long x Hyderpur. It gives 80 per cent more yield than Pusa Purple Long. Yield potential is 38.5 t/ha
Pusa Bhairav	This variety has been developed by ICAR-IARI, New Delhi. Fruits are deep purple, glossy long, early and resistant to Phomopsis fruit rot (Kalda and Choudhury, 1991).
Pusa Anupam	This variety has been developed by ICAR-IARI, New Delhi. Fruits are cylindrical, medium long, purple born in cluster. It is resistant to bacterial wilt. Yield potential is 35 t/ha in a crop duration of 160 days.
Pusa Upkar	This variety has been developed by ICAR-IARI, New Delhi. Fruits are pendent, solitary round, medium size, glossy and dark purple. Individual fruit weight is 200 g. Yield potential is 40 t/ha.
Pusa Uttam	This variety has been developed by ICAR-IARI, New Delhi. Fruits are pendent, oval, medium large size, glossy and dark purple. Individual fruit weight is 250 g. Yield potential is 41 t/ha.
Pusa Bindu	This variety has been developed by ICAR-IARI, New Delhi. Fruits are small, oval, round glossy and violet purple. Solitary to partial cluster. About 14 fruits/kg weight. Yield potential is 30 t/ha.
Pusa hybrid-20	This variety has been developed by ICAR-IARI, New Delhi. Fruits are non-spiny, semi-erect light purple pigmentation partially have light purple partial pigmented younger leaves and new branches. Fruits are long, dark purple, glossy, weighing 90-100g. First picking starts 55 days after transplanting. Average yield is 52.5 t/ha.
Swarna Pratibha	This variety has been developed by ICAR Research Complex for Eastern Region, Research Centre, Plandu, Ranchi (Jharkhand). Develop through pure line selection from CH-309. Plant height is 60-70 cm erect and intermediate plant spread. First harvest can be obtained 55-60 days after transplanting. Fruit are long medium size (150-200 g) and shiny purple. Yield is about 60.0-65.0 t/ha. It is resistant to bacterial wilt.
Swarna Shree	This early variety has been developed by ICAR Research Complex for Eastern Region, Research Centre, Plandu, Ranchi (Jharkhand). Fruits are white round, less seeded and flesh spongy. It is moderately tolerant to bacterial wilt and shoot and fruit borer.

Variety	*Description*
Swarna Shakti (Hybrid)	This variety has been developed by ICAR Research Complex for Eastern Region, Research Centre, Plandu, Ranchi (Jharkhand). Hybrid developed between the cross of CH-309 x CH-157-16-4-1. Plant height is70-80 cm, erect intermediate growth habit and intermediate plant spread. First harvest can be obtained 55-60 days after transplanting. Fruits are oblong, weight 250-300 g, and attractive shiny light purple. Yield is about 70.0-75.0 t/ha. It is resistant to *Phomopsis* blight.
Swarna Shobha	Develop through hybridization (HAB-82 x HAB-27) followed by pedigree selection from HARP, Ranchi (Jharkhand). Plant height is 50-60 cm, prostrate growth habit and broad plant spread. First harvest can be obtained 55-60 days after transplanting. Fruits are round medium size, weight 300-350 g, and attractive milky white with purple tinge. Yield is about 50.0-60.0 t/ha. It is tolerant to bacterial wilt.
Swarna Shyamli	This variety has been developed by ICAR Research Complex for Eastern Region, Research Centre, Plandu, Ranchi (Jharkhand). The average yield is 50 t/ha. It is resistant to bacterial wilt.
Swarna Mani	This variety has been developed by ICAR Research Complex for Eastern Region, Research Centre, Plandu, Ranchi (Jharkhand).
Swarna Abhilamb	This variety has been developed by ICAR Research Complex for Eastern Region, Research Centre, Plandu, Ranchi (Jharkhand).
Swarna Aijay (Hybrid)	This variety has been developed by ICAR Research Complex for Eastern Region, Research Centre, Plandu, Ranchi (Jharkhand). Hybrid developed between the cross of CH-249 x HAB-792 from HARP, Ranchi (Jharkhand). Plant height is 50-60 cm, intermediate growth habit and broad plant spread. First harvest can be obtained 50-55 days after transplanting. Spines are visible on leap and calyx. Fruits are oblong, weight 100-120 g, and attractive light purple. Yield is about 70.0-75.0 t/ha. It is resistant to bacterial wilt and *Phomopsis* blight.
Swarna Neelima (Hybrid)	This variety has been developed by ICAR Research Complex for Eastern Region, Research Centre, Plandu, Ranchi (Jharkhand). Average yield is 55 t/ha. Resistant to bacterial wilt.
Azad B-3	It was developed at Chandrashekhar Azad University of Agriculture and Technology Kalyanpur, Kanpur (Uttar Pradesh). Fruits are medium thick, long, shiningpurple, glossy and attractive. It is very good for fresh market. The average yield is 35 t/ha.
Azad Hybrid	It was developed at Chandrashekhar Azad University of Agriculture and Technology Kalyanpur, Kanpur (Uttar Pradesh) by a cross Azad B_1 x Kaliyanpur Type 3 in which the prolific bearing character has been contributed by the Kaliyanpur Type 3 while colour, shape and fruit size by Azad B_1. The plants are semi erect with 60-70 cm in spread, branched, leaf lamina green and entire, early fruiting; fruits bold, bright purple, round, and green calyx. The fruits retain their purple colour for a longer period and possess better shelf life. The average yield is 45.0 t/ha. It is less prone to shoot and fruit borer infestation.
Krishna (hybrid)	This variety has been developed from MPKV, Rahuri (M.S.)
Phule Arjun (hybrid)	This variety has been developed from MPKV, Rahuri (M.S.)
Phule Harit	This variety has been developed from MPKV, Rahuri (M.S.)
Pragati	This variety has been developed from MPKV, Rahuri (M.S.)
Vaishali	This variety has been developed from MPKV, Rahuri (M.S.)
Aruna	This variety has been developed from PDKV, Akola (M.S.) through selection from the local material and suitable for cultivation in summer and rainy seasons. First picking is possible in 55-60 days after transplanting. Fruits are bright red, egg-shaped, born in cluster of 2-3 and each weighing 75-80 g. Average yield is 25.0-30.0 t/ha.
Anuradha	This variety has been developed from MAU, Parbhani (M.S.)
Junagadh Long	This early variety has been developed in the year 1984 by Gujarat Agricultural University, Gujarat. It is a cross between Dholi-5 x Panchmahal. Fruits are medium size, pinkish coloured long shaped fruits with tapering ends. First picking can be done in 50 days after transplanting. Average yield is 27 t/ha.
Junagadh Oblong	This variety has been developed in the year 1984 by Gujarat Agricultural University, Gujarat.
Gujarat Junagadh Brinjal-2	This variety has been developed in the year 2010 by Gujarat Agricultural University, Gujarat.
Junagadh Brinjal Green Round-1	This variety has been developed in the year 2006.
Anand Brinjal Hybrid-2	This variety has been developed in the year 1995.
Gujarat Brinjal Long-1	This variety has been developed in the year 2001.
Gujarat Brinjal Hybrid-2	This variety has been developed in the year 2001.
Gujarat Brinjal Long-1	This variety has been developed in the year 2002.
Gujarat Brinjal Oblong-1	This variety has been developed in the year 2004.

Variety	Description
HLB-12	This variety has been developed by CCS HAU, Hisar (Haryana).
Manjari Gota	The variety has a dwarf and spreading growth habit. The fruits are medium-large, round and purple coloured with white stripes. Upon maturity, the fruits attain a golden yellow colour. Average yield is 15.0-20.0 t/ha.
Punjab Chamkila	This variety has been developed by PAU, Ludhiana by selection from the local material. Average yield is 50 t/ha.
Punjab-8	This variety has been developed by Punjab Agril. Department, Punjab. Fruits are medium sized, light purple and early bearer. First harvesting of fruits can be done in 65 days after transplanting. Average yield is 30 t/ha.
Punjab Jamuni Gola	This early variety has been developed by PAU, Ludhiana by selection from the local material. Plants are medium tall, spreading and thornless. Fruits are round, purple and plumpy. First harvesting of fruits can be done in 65 days after transplanting. It is recommended for cultivation in autumn season and first picking is possible in 70 days after planting. Average yield is 27.5-31.5 t/ha.
Punjab Neelam	This variety has been developed by PAU, Ludhiana by selection from the cross S-16 x PBR-91-2. Plants are medium in height, erect, thornless and have green foliage with purple tinge on leaf blades and fruit calyx. Fruits are medium sized and dark purple. It is suitable for cultivation in autumn and spring season. First picking is possible in 65 days after planting. Average yield is 35.0 t/ha.
Punjab Nagina	This variety has been developed by PAU, Ludhiana by selection from the cross Aruna x Patiala Selection. Plants are dwarf, semi-erect with dark green and spineless leaves. Fruits are small weighing about 40-50 g, round, shining purple-black and born in clusters. First picking is possible in 55 days after transplanting. Average yield is 36.0 t/ha.
Punjab Barsati	This was developed at PAU, Ludhiana (Punjab). The plants grow to a height of 65-70 cm. Fruits oblong in shape bluish purple in colour. It has a potential yield of 25.0-30.0 t/ha. It is suitable for rainy and autumn season's cultivation.
HLB-12	This variety has been developed by CCS HAU, Hisar (Haryana) in 2001 and released by State Variety Release committee, Haryana for commercial cultivation in Haryana. It is suitable for summer season. Fruits long bright dark purple colored. It has a potential yield of 20 t/ha.
HLB-25	This variety has been developed by CCS HAU, Hisar (Haryana) in 2001 and released by State Variety Release committee, Haryana for commercial cultivation in Haryana. It is suitable for summer season. Fruits long bright dark purple colored. It has a potential yield of 35-37 t/ha. It is tolerant to high temperature and shoot and fruit borer
Pant Rituraj	This variety has been developed by pedigree selection from the cross T-3 x Pusa Purple Cluster at GBPUA&T, Pantnagar (Uttarakhand). It was released by the UPSVRC in 1984 and by the CVRC in 1985. Plants are medium tall (60-70 cm) and semi-erect. Leaves are dark green with occasional light purple pigmentation on young leaves. First picking is possible in 60 days after transplanting. Fruits are oval-round with slight tapering towards the bottom and violet purple. Average yield is 32.5 t/ha. It is suitable for planting both in winter and summer season and is a prolific bearer unlike Type 3 which is the parental cultivar and is shy bearer.
Arka Anand	A high yielding F_1 hybrid developed by ICAR-IIHR, Bengaluru in 2005 and released for commercial cultivation in 2005. It is resistant to bacterial wilt. Fruits are green, medium long with excellent cooking quality. Yields 55-60 t/ha in 145-150 days.
Arka Shirish	This variety has been developed by selection from the local material 'Irangeri' at ICAR-IIHR, Bengaluru (Karnataka). Fruits are extra-long, tender, light green and are born in cluster (6-8 fruits). It is recommended for cultivation in south India under mild climate. Average yield is 45.0-47.0 t/ha in a crop duration of 115 days.
Arka Sheel	This variety has been developed by ICAR-IIHR, Bengaluru (Karnata). Fruits are medium long, tenderand deep purple. Maturity 120 days after transplanting. The average yield is 48.5 t/ha.
Arka Kusumakar	This variety has been developed through pure line selection from the local material collected from Karnataka by IIHR, Bengaluru (Karnataka). Plants are dwarf and spreading with green stem and foliage. Fruits are medioum long, green, finger shaped and born in clusters of 5-7. It is recommended for cultivation in south India under mild climate. Average yield is 45.0 t/ha in 120 days.
Arka Neelkanth	Plants are tall and compact. Fruits are short with violet blue glossy skin, green purple calyx, tender flesh having slow maturing seeds and free from bitter flesh. Each fruit weighs 40 g on an average and are born in cluster in two flushes. Young leaves are dark green with purple leaf base and veins. Stem is purple green. This variety is resistant to bacterial wilt and has very good cooking and keeping qualities. It yields 40.0 t/ha in crop duration of 150 days.
Arka Nidhi	This variety has been developed by selection from the cross Dingrass Multiple Purpose x Arka Sheel at ICAR-IIHR, Bengaluru (Karnataka). Plants are tall, well branched and compact. Leaves are purplish when young. Fruits are blue-black, glossy and medium long with purplish calyx and are born in clusters. Fruits are free from bitter principle. The fruit weighs on average 43.0 g. Cooking and keeping qualities are good. Average yield is 48.5 t/ha in crop duration of 150 days. It is resistant to bacterial wilt.
Arka Keshav	This variety has been developed by selection from the cross Dingrass Multiple Purpose x Arka Sheel at IIHR, Bengaluru (Karnataka). Plants are tall and compact. Leaves are green; fruits are purple-red, glossy and long with green calyx and are born in cluster. Average yield is 45.0 t/ha. It is resistant to bacterial wilt.
Arka Navneet	It is high yielding hybrid. Fruits are large oval to oblong with deep purple shining skin with each fruit 450 g in weight. It has purple flowers with solitary beraring habit. Free from bitter principles with very good cooking qualities. Crop is ready for picking in 150-160 days. Average yield is 65.0-70.0 t/ha.
Arka Navneeth	It is an F_1 hybrid of IHR-221 x Supreme and the fruits are round to slightly oval big in size, black purple in colour, and each fruits weoghs 400-500 g. Average yield is 65.0-75.0 t/ha.
Shyamal	It is an important hybrid developed by Ankur Seeds Pvt. Ltd. Nagpur (Maharashtra). Plants are semi-spreading to spreading with vigorous growth. First picking is possible 100-110 days after planting. Fruits are oblong, purple with milky white flesh. Average yield is 70.0-75.0 t/ha.

Variety	*Description*
Dholi-5	This early variety has been developed by Rajendra Agricultural University, Samastipur (Bihar). Fruits are oval, stocky and dark purple. Individual fruit weight is 150 g. This variety is suitable for cultivation in kahrif and spring season. Average yield is 30 t/ha in a crop duration of 120 days.
Rajendra	This early variety has been developed by Rajendra Agricultural University, Samastipur (Bihar). Fruits are long, slight curve and heavy bearer. This variety is suitable for cultivation in kahrif and spring season. Average yield is 32.5 t/ha. It is tolerant to wilt and phomopsis blight.
Ramnagar Giant	This is a local ecotype and commonly found in the area of Ramnagar, Varanasi (Uttasr Pradesh). Fruits are round, very soft, light green, very big size and less seeded. It is very much suitable in preparation of local dish 'Bharta'. Average yield is 30 t/ha.
Surti Gota	This variety has been developed by Maharashtra, Agricultural Department. Fruits are medium to big, round, purple coulour, soft and watery in taste. Average yield is 20 t/ha.
Manjari Gota	This variety has been developed by Maharasthra, Agricultural Department. Fruits are medium round, rosy colour with white strips. Fruitys are very tasty. It is resistant to bacterial wilt. Average yield is 25 t/ha.
MHB-56	This hybrid has been developed by Mahyco Seed Co (Maharashtra). Fruits are oval, and greenish to yellow. Fruits mature in 130 days. Average yield is 30 t/ha.
MHB-63	This hybrid has been developed by Mahyco Seed Co (Maharashtra). Fruits are round with violet white streaks. Average yield is 55 t/ha.
MHB-70	This hybrid has been developed by Mahyco Seed Co (Maharashtra). Medium maturity duration. Fruits are very long and medium in fruit girth. Average yield is 57.5 t/ha.
MHB-80	This hybrid has been developed by Mahyco Seed Co (Maharashtra). Fruits are round with redish black colour. Fruit is known for rough skin but possessing good keeping quality. Average yield is 58.5 t/ha.
MHB-82	This hybrid has been developed by Mahyco Seed Co (Maharashtra). Fruits are big and heavy. Individual fruit weight is 150-200 g. Crops mature in 60-65 days. Average yield is 58.5 t/ha.
MHB-111(3)	This hybrid has been developed by Mahyco Seed Co. Fruits are long, blackish violet and less seeded. Average yield is 56.5 t/ha.
PLR-1	This hybrid is developed from Assam. Fruits are small, round and purple. Fruits are born in cluster of 2 or more. Average yield is 30 t/ha.
PPL-47 (F_1)	This hybrid is developed by Golden Seeds Co. Fruits are long, cylindrical and deep purple. Individual fruit weight is 90-100 g. Average yield potential is 36 t/ha.
PPL-321	This hybrid is developed by Golden Seeds Co. Fruits are long and light purple. Individual fruit weight is 80-100 g. First picking can be done in 55- to 65 days after transplanting. Average yield potential is 36 t/ha.
Nishant	This hybrid has been developed by Golden Seeds Co. Fruits are elongated oval, dark violet and glossy. Fruits are born in cluster of 3-4. Individual fruit weight is 100-120 g. Average yield potential is 45 t/ha.
Green Long	This hybrid has been developed by Golden Seeds Co. Fruits are long and slender. Fruits are born in cluster of 3-4. Individual fruit weight is 100-120 g. Average yield potential is 42 t/ha.
Preeti	This hybrid has been developed by Golden Seeds Co. Fruits are small round to oval and lustrous purple. Fruits are born in cluster of 3-5. Individual fruit weight is 40-60 g. First harvesting can be done in 40 days after transplanting. Average yield potential is 40 t/ha.
Krishna	This hybrid has been developed by Golden Seeds Co. Fruits are oval round and dark purple (blackish). Fruits are born in cluster of 2-3. Individual fruit weight is 80-100 g. First harvesting can be done in 40-50 days after transplanting. Average yield potential is 45 t/ha. It has potential to set fruit in adverse climatic condition.
New Kiran	This hybrid has been developed by Golden Seeds Co. Fruits are round and bright purple with solitary position. Individual fruit weight is 180-200 g. First harvesting can be done in 62 days after transplanting. Average yield potential is 40 t/ha.
310	This hybrid has been developed by Golden Seeds Co. Fruits are large sized, attractive round, and deep black with solitary position. Individual fruit weight is 200-250 g. First harvesting can be done in 60-65 days after transplanting. Average yield potential is 42 t/ha.
Green Moti	This hybrid has been developed by Golden Seeds Co. Fruits are small sized, attractive oval-round, and green with white streaks. Fruits are born in cluster of 3-4. Individual fruit weight is 30-40 g. First harvesting can be done in 40-42 days after transplanting. Average yield potential is 45 t/ha.
Rangeela	This hybrid has been developed by Golden Seeds Co. Fruits are attractive oval-round, and light purple with narrow white streaks. Fruits are born in cluster of more than 3. Individual fruit weight is 100-125 g. First harvesting can be done in 55 days after transplanting. Average yield potential is 45 t/ha.
Rangoli	This hybrid has been developed by Golden Seeds Co. Fruits are attractive oval-round, and lustrous purple with white streaks. Fruits are born in cluster of more than 3. Individual fruit weight is 100-125 g. Average yield potential is 35 t/ha.
Vijay	This hybrid has been developed by Golden Seeds Co. Fruits are attractive oval-round, and light purple with white streaks. Individual fruit weight is 80-100 g. First harvesting can be done in 45-50 days after transplanting. Average yield potential is 40 t/ha.
BE-701	This hybrid has been developed by Nunhem Seeds Co. Fruits are round, and attractive purple black shining. Individual fruit weight is 300 g. It possesses excellent flavoured pulp quality. It is partial resistant to mites. Average yield potential is 32 t/ha.
Sandhya-257	This early maturing (45-50 days after tranlsplanting) hybrid has been developed by Nunhem Seeds Co. Fruits are round, less seeded and attractive purple black shining. Individual fruit weight is 250-300 g. It possesses excellent shelf life. It is partial resistant to mites. Average yield potential is 33 t/ha.

Variety	Description
Black Boy	This hybrid has been developed by Nunhem Seeds Co. Fruits are oval-round and attractive purple black shining. Individual fruit weight is 200-250 g. It possesses excellent shelf life. It is partial resistant to mites. Average yield potential is 31 t/ha. It has partial resistant to sucking pests.
Jaya-1001	This hybrid has been developed by Nunhem Seeds Co. Fruits are round and attractive purple shining. Individual fruit weight is 240-50 g. Fruits are born in clusters. Fruits are very good for stuffing, frying and picles preparation. Average yield potential is 31.5 t/ha.
Kareena-1002 (F_1)	This early maturing (55 days after tranlsplanting) hybrid has been developed by Nunhem Seeds Co. Fruits are glossy, oval-round and attractive blackish purple shining. Individual fruit weight is 80-100 g. Average yield potential is 32 t/ha. It posses partial resistant to sucking pest and high temperature.
Shilpa-743	This hybrid has been developed by Nunhem Seeds Co. Fruits are glossy, long, tender and attractive blackish shining. Individual fruit weight is 100-125 g. Average yield potential is 34.5 t/ha. It posses partial resistant to sucking pest and high temperature. It posses partial resistant to sucking pest and bacterial wilt.
Yamini-177	This early maturing (45-50 days after tranlsplanting) hybrid has been developed by Nunhem Seeds Co. Fruits are oval and attractive blackish shining. Individual fruit weight is 100-150 g. It possesses excellent shelf life. It is partial resistant to mites. Average yield potential is 34.5 t/ha. It has partial resistant to sucking pests also.
BE-708	This hybrid has been developed by Nunhem Seeds Co. Fruits are elongated oval and purplish black. Individual fruit weight is 200-250g. Average yield potential is 35.5 t/ha.
Abhishek-256	This hybrid has been developed by Nunhem Seeds Co. Fruits are round and deep purplish black. Individual fruit weight is 300 g. Average yield potential is 34.5 t/ha.
BE-925	This hybrid has been developed by Nunhem Seeds Co. Fruits are glossy green, long cylindrical and straight. Individual fruit weight is 80-100 g. Average yield potential is 33 t/ha. It has partial resistant to sucking pests.
Chaya-3715	This hybrid has been developed by Nunhem Seeds Co. Fruits are half long and purplish black. Fruits are born in cluster of more than 3. Individual fruit weight is 80-100 g. Average yield potential is 32.5 t/ha. It is resistant to bacterial wilt and partially resistant to sucking pests.
BE-706	This early maturing (50-55 days after tranlsplanting) hybrid has been developed by Nunhem Seeds Co. Fruits are long (15-20 cm) and glossy purple. Average yield potential is 33.5 t/ha. It has wide adaptability.
Kanha- 091	This hybrid has been developed by Nunhem Seeds Co. Fruits are small (50 g), oval round and deep purple in colour. Fruits are born in cluster. Average yield potential is 31.5 t/ha. It has partial resistant to sucking pests.
Mohini	This hybrid has been developed by Nunhem Seeds Co. Fruits are round to oval and violet in colour with white stripes. Fruits are born in cluster and less seded. Average yield potential is 32.8 t/ha.
Super Mohini	This hybrid has been developed by Nunhem Seeds Co. Fruits are round, glossy, firm, attractive and pink with white stripes. Average yield potential is 37 t/ha. Individual fruit weight is 75-80 g. It has resistant to sucking pests.
Kamini	This hybrid has been developed by Nunhem Seeds Co. Fruits are round and pink with white stripes. Fruits are born in cluster and spines on calyx. Average yield potential is 36.5 t/ha. It has resistant to sucking pests.
Harshita-1346	This medium maturing (50-55 days after tranlsplanting) hybrid has been developed by Nunhem Seeds Co. Fruits are oval-round with white stripes, glossy and green in colour. Fruits are born in cluster of 2-3. Average yield potential is 35.5 t/ha. It has resistant to bacterial wilt.
Pinki1364	This medium maturing (50-55 days after tranlsplanting) hybrid has been developed by Nunhem Seeds Co. Fruits are medium large (15-16 cm) and attractive pink colour. Fruits are born in cluster and spines on calyx. Average yield potential is 36 t/ha. It has resistant to bacterial wilt
Mahakranti (PK-123)	This hybrid has been developed by Sungro Seeds Co. First picking can be done in 60 days after transplanting. Fruits are long club shaped and bright purple in colour. Individual fruit weight is 75 g. Average yield potential is 34.5 t/ha. It has tolerant to fruit borer.
Pragati (Chu-Chu)	This hybrid has been developed by Sungro Seeds Co. Fruits are oval-round and purple in colour. Individual fruit weight is 30 g. Fruits are born in cluster. First picking can be done in 45 days after transplanting. Average yield potential is 67.5 t/ha. It has tolerant to fruit borer.
Unnati (Chu-Chu)	This hybrid has been developed by Sungro Seeds Co. Fruits are oval-round and black in colour. Individual fruit weight is 40 g. Fruits are born in cluster. First picking can be done in 45 days after transplanting. Average yield potential is 61 t/ha. It has tolerant to fruit borer.
Green Long	This hybrid has been developed by Sungro Seeds Co. Fruits are long green. Individual fruit weight is 100 g. Fruits are born in cluster. First picking can be done in 45 days after transplanting. Average yield potential is 52 t/ha.
Kanhaiya	This hybrid has been developed by Sungro Seeds Co. Fruits are oval-round and dark black in colour. Individual fruit weight is 120 g. First picking can be done in 45 days after transplanting. Average yield potential is 60 t/ha. It has tolerance to jassids.
Navkiran	This hybrid has been developed by Sungro Seeds Co. Fruits are round and bright purple in colour. Individual fruit weight is 60 g. First picking can be done in 65 days after transplanting. Average yield potential is 46.5 t/ha. It us tolerant to fruit borer and jassids. It is recommended for rainy season.
Mukta	This hybrid has been developed by Sungro Seeds Co. Fruits are oval and dark purple in colour. Individual fruit weight is 100 g. First picking can be done in 65 days after transplanting. Average yield potential is 47.5 t/ha. It is recommended for rainy season and remains in fruiting for longer period.
PPL-74	This hybrid has been developed by Sungro Seeds Co. Fruits are cylindrical and dark purple in colour. Individual fruit weight is 90 g and less seeded. First picking can be done in 55 days after transplanting. Average yield potential is 47.5 t/ha.

Variety	Description
Ankur- 888	This hybrid has been developed by Ankur Seeds Co. Fruits are round and blackish purple and solitary bearing. Individual fruit weight is around 150-200 g. First picking can be done in 105- 115 days after seeding. Average yield potential is 44 t/ha.
Hirak	This hybrid has been developed by Ankur Seeds Co. Fruits are born in cluster.Fruits are oval round and green with light green strips. Individual fruit weight is around 150-200 g. First picking can be done in 105-115 days after seeding. Average yield potential is 45 t/ha.
Abhimannyu	This hybrid has been developed by Ankur Seeds Co. Fruits are oval round and purple with white strips. Fruits are born in cluster. Individual fruit weight is around 50-70 g. First picking can be done in 110-115 days after seeding. Average yield potential is 47.5 t/ha.
Ajay	This hybrid has been developed by Ankur Seeds Co. Fruits are born in cluster. Fruits are oval medium large size and purple with white green streaks. Average yield potential is 41.5 t/ha. It possesses good shelf life.
Kuldeep	This hybrid has been developed by Ankur Seeds Co. Fruits are oval medium large size and green purple. Average yield potential is 43.5 t/ha. It possesses good quality and shelf life.
Kaustubh	This hybrid has been developed by Ankur Seeds Co. Fruits generally emerge in single to pair. Fruits are oval and purple with white strips. Average yield potential is 43.5 t/ha. It possesses good quality and shelf life.
Utkarsh	This hybrid has been developed by Ankur Seeds Co. Fruits are born in cluster. Fruits are oval medium large size and purple. Average yield potential is 46.5 t/ha. It possesses good quality and shelf life.
Kushal	This hybrid has been developed by Ankur Seeds Co. Fruits generally emerge in clusters. Fruits are long, medium sized and light green in colour. Average yield potential is 37.5 t/ha.
Kirti	This hybrid has been developed by Ankur Seeds Co. Fruits are oval, medium to small sized and with light green strips. Average yield potential is 45.5 t/ha. It has good cooking quality
Shyamal	This hybrid has been developed by Ankur Seeds Co. Fruits are born in cluster. Fruits are oblong. Average yield potential is 56.5 t/ha. It possesses good cooking quality.
Abhishek	This hybrid has been developed by Ankur Seeds Co. Fruits are born in cluster. Fruits are oval round and purple with white strips. Individual fruit weight is around 55-65 g. Average yield potential is 43.8 t/ha.
Ajinky	This hybrid has been developed by Ankur Seeds Co. First picking can be done in 100-110 days after seeding. Fruits are oval round and green with purple strips. Individual fruit weight is around 50-70 g. Average yield potential is 43.8 t/ha.
Nakul	This hybrid has been developed by Ankur Seeds Co. Fruits are oval, and green with light green strips. Individual fruit weight is around 45-60 g. First picking can be done in 90-105 days after seeding. Average yield potential is 38.5 t/ha. It has good cooking quality.
NS-509	This hybrid has been developed by Namdhari Seeds Co. Fruits are glossy, attractive medium sized, dark purple with white stripes and solitary bearing. Individual fruit weight is around 70 g. Average yield potential is 42.5 t/ha. It has good cooking quality.
Apsara	This hybrid has been developed by Namdhari Seeds Co. Fruits are born in cluster. Fruits are round to oval, glossy, attractive medium sized, and bright purple with white stripes. Individual fruit weight is around 80 g. Average yield potential is 45 t/ha. It is suitable for growing throughout the year.
NS-1720	This early hybrid has been developed by Namdhari Seeds Co. Fruits are born in cluster. Fruits are oval, attractive medium sized and uniform purple. Individual fruit weight is around 60-70 g. Average yield potential is 46.5 t/ha.
Nisha	This early hybrid has been developed by Namdhari Seeds Co. Fruits are born solitary. Fruits are bell shaped, glossy, deep purple and uniform. Individual fruit weight is around 350 g. Average yield potential is 52 t/ha.
Sunder	This hybrid has been developed by Namdhari Seeds Co. Fruits are born solitary. Fruits are bell shaped, glossy, deep purple and uniform. Individual fruit weight is around 350-400 g. Average yield potential is 52 t/ha. It has very good adaptability in varying climatic conditions.
BSS-471	This hybrid has been developed by Bejo Sheetal Seeds Co. Fruits are oval to slightly high oval and purple with green as well as white stripes. Average yield potential is 41 t/ha.
BSS-472	This hybrid has been developed by Bejo Sheetal Seeds Co. Fruits are round to oval and purple with green as well as white stripes. Average yield potential is 37 t/ha. It is very much suitable for round the year cultivation.
Kalyan	This hybrid has been developed by Bejo Sheetal Seeds Co. Fruits are round to oblong and reddish purple with white stripes. First harvesting can be done in 65-70 days after transplanting. Average yield potential is 42.5 t/ha. It is very much suitable for round the year cultivation.
Arpan	This hybrid has been developed by Bejo Sheetal Seeds Co. Fruits are round to oblong and dark purple with white stripes. First harvesting can be done in 55-60 days after transplanting. Average yield potential is 41.5 t/ha. It is very much suitable for round the year cultivation.
BSS-331	This early maturing hybrid has been developed by Bejo Sheetal Seeds Co. Fruits are oval to oblong and purple. First harvesting can be done in 50 days after transplanting. Average yield potential is 42.1 t/ha.
BSS-332	This hybrid has been developed by Bejo Sheetal Seeds Co. Fruits are round to oval, thornless and dark purple. First harvesting can be done in 55-65 days after transplanting. Average yield potential is 41t/ha. It is very much suitable for round the year cultivation.
BSS-351	This hybrid has been developed by Bejo Sheetal Seeds Co. Fruits are oblong, thornless and black in colour. First harvesting can be done in 55-60 days after transplanting. Average yield potential is 39.5 t/ha. It is very much suitable for round the year cultivation.
BSS-426	This hybrid has been developed by Bejo Sheetal Seeds Co. Fruits are oval, small sized, chu-chu type and dark purple in colour. Average yield potential is 40 t/ha. It is very much suitable for round the year cultivation.

Variety	Description
BSS-427	This hybrid has been developed by Bejo Sheetal Seeds Co. First harvesting can be done in 55-60 days after transplanting. Fruits are oval, round and dark purple in colour. Average yield potential is 38 t/ha. It is very much suitable for round the year cultivation.
BSS-388	This profuse bearing hybrid has been developed by Bejo Sheetal Seeds Co. Fruits are born in cluster. Fruits are medium long, thick and dark purple in colour. Average yield potential is 39 t/ha. It is very much suitable for round the year cultivation. It is very much suitable for *Kharif* and autmn spring season.
BSS-431	This bearing hybrid has been developed by Bejo Sheetal Seeds Co. Fruits are medium long, thornless and purple in colour. First harvesting can be done in 55-60 days after transplanting. Average yield potential is 41.5 t/ha. It is very much suitable for round the year cultivation.
Kokila (BSS-387)	This bearing hybrid has been developed by Bejo Sheetal Seeds Co. Fruits are oblong chu-chu type and black in colour. First harvesting can be done in 55-60 days after transplanting. Average yield potential is 42.5 t/ha. It is very much suitable for round the year cultivation.
Sumo (BSS-425)	This profuse bearing hybrid has been developed by Bejo Sheetal Seeds Co. Fruits are oblong jumbo sized, less seeded and black in colour. Fruit is as big as 1 kg. First harvesting can be done in 70-75 days after transplanting. Average yield potential is 41t/ha. It is very much suitable for round the year cultivation.
Kajol	This profuse bearing hybrid has been developed by Bejo Sheetal Seeds Co. Fruits are round, medium sized, and black in colour. First harvesting can be done in 55-60 days after transplanting. Average yield potential is 41.8 t/ha.
Kajri (BSS-382)	This hybrid has been developed by Bejo Sheetal Seeds Co. Fruits are oval, medium to big sized, slightly oblong less seeded and black colour. First harvesting can be done in 70-75 days after transplanting. Average yield potential is 39 t/ha. It is very much suitable for round the year cultivation.
Nisha (BSS-383)	This hybrid has been developed by Bejo Sheetal Seeds Co. Fruits are medium long and dark purple in colour. First harvesting can be done in 55-60 days after transplanting. Average yield potential is 41 t/ha. It is very much suitable for round the year cultivation.
BSS-330	This hybrid has been developed by Bejo Sheetal Seeds Co. which is dual season in nature. Fruits are medium long, slightly thick and purple in colour. First harvesting can be done in 70 days after transplanting. Average yield potential is 46.5 t/ha.
BSS-334	This hybrid has been developed by Bejo Sheetal Seeds Co. Fruits are medium long to oblong and reddish black to black in colour. First harvesting can be done in 70-75 days after transplanting. Average yield potential is 46 t/ha. It is very much suitable for round the year cultivation.
BSS-465	This profuse bearing hybrid has been developed by Bejo Sheetal Seeds Co. which is dual season in nature. Fruits are long, shining and dark black in colour. First harvesting can be done in 60 days after transplanting. Average yield potential is 47 t/ha.
Preeti	This profuse bearing hybrid has been developed by Bejo Sheetal Seeds Co. which is dual season in nature. Fruits are round to oval, green with white stripe and slightly purple in colour. First harvesting can be done in 60-65 days after transplanting. Average yield potential is 41t/ha.
BSS-428	This hybrid has been developed by Bejo Sheetal Seeds Co. which is dual season in nature. Fruits are oval, green with white stripes, and small sized. First harvesting can be done in 60-65 days after transplanting. Average yield potential is 40.5 t/ha.
Raveena	This hybrid has been developed by Bejo Sheetal Seeds Co. which is dual season in nature. Fruits are long, green and thornless. First harvesting can be done in 65-70 days after transplanting. Average yield potential is 43.5 t/ha.
Green Long	This early fruiting hybrid has been developed by Century Seeds Co. In this hybrid prolong fruiting has been reported. Fruits are long, green, firm glossy, large sized and born in bunches of 2-3. Its flesh flavor is excellent. Average yield potential is 44.5 t/ha.
Kokila	This hybrid has been developed by Century Seeds Co. Fruits are long, less seeded and deep-purple black in colour. Average yield potential is 42.5 t/ha.
Vardhan	This hybrid has been developed by Century Seeds Co. Fruits are born in cluster. Fruits are attractive and deep-purple in colour. In dividual fruit weight is 40-50 g. Average yield potential is 38.5 t/ha.
Mahadeva	This hybrid has been developed by Century Seeds Co. Fruits are attractive, globular, less seeded and purplish black in colour. In dividual fruit weight is 300 g. It is good cropper and average yield potential is 41.5 t/ha.
Shiva	This hybrid has been developed by Century Seeds Co. Fruits are attractive, round, less seeded, medium in size and purplish black in colour. In dividual fruit weight is 220-260 g. Average yield potential is 39.5 t/ha.
Hybrid-1	This hybrid has been developed by Century Seeds Co. Fruits are cylindrically uniform, attractive, shiny and purplish black in colour. It remains continuous in fruiting for longer period. Flesh of the fruit is very good. Average yield potential is 42.5 t/ha.
Nisha	This is a very early maturing hybrid has been developed by Century Seeds Co. Fruits are born in cluster. Fruits are attractive, medium sized and bluish purple in colour. Average yield potential is 40.5 t/ha.
Choo-Choo	This is hybrid has been developed by Century Seeds Co. It is profuse bearing and continuous over a long period. Fruits are born in cluster of 3-5. Fruits are elongated small, glossy and purple black in colour. Average yield potential is 37 t/ha.
Nisha Improved	This hybrid has been developed by Century Seeds Co. Fruits are attractive, and medium sized. Fruits slow maturing. Average yield potential is 39.5 t/ha. It is well adapted to monsoon cultivation.
Kamdhenu	This hybrid has been developed by Century Seeds Co. Fruits are small and purple with white stripes. Average yield potential is 41.5 t/ha. It is a prolific cropper.
Sanyog	This hybrid has been developed by Century Seeds Co. Fruits are born in cluster. Fruits are dark blackish in colour and less seeded. Individual fruit weight is 42.8 g. Average yield potential is 41.5 t/ha. It is a prolific cropper.
Angora	This hybrid has been developed by Sandoz Seeds Co. Fruits are oval round, firm, fleshy and deep purple black. Individual fruit weight is 200-400 g. Average yield potential is 89.5 t/ha. It has good shelf life, ideal for long distance transport and export.

Variety	*Description*
Harit	This early maturing hybrid has been developed by Sandoz Seeds Co. First harvesting can be done 45-50 days after transplanting. Fruits are oblong-ovate and brilliant green with creamy streaks. Individual fruit weight is 40-50 g. Average yield potential is 55.5 t/ha.
Green Bunch	This early maturing hybrid has been developed by Sandoz Seeds Co. Fruits are born in cluster. Fruits are oval-round and brilliant green with creamy streaks. Individual fruit weight is 75-90 g. Average yield potential is 55.5 t/ha.
AUB-132	This hybrid has been developed by Sandoz Seeds Co. Fruits are oval-round, less seeded, thin skinned, solid flesh and brilliant black. Average yield potential is 100 t/ha.
Kanha	This hybrid has been developed by Sandoz Seeds Co. Plants are very vigorous. Fruits are long glossy and violet skin. Individual fruit weight is 90-110 g. Average yield potential is 110 t/ha.
Manju	This hybrid has been developed by Sandoz Seeds Co. Fruits are oval-round and light purple with white stripes. Individual fruit weight is 80-100 g. Average yield potential is 105 t/ha.
Mew-Mew	This medium maturing hybrid has been developed by Sandoz Seeds Co. Fruits are born in cluster. Fruits are small-round and lustrous violet in colour. Individual fruit weight is 60-80 g. Average yield potential is 80 t/ha.
Sanju	This medium maturing hybrid has been developed by Novartis Seeds Co. Fruits are oval round and 8-10 cm long. Fruits are born in cluster. Individual fruit weight is 90-100 g. Average yield potential is 75 t/ha.
Green Star	This hybrid has been developed by Sreemaa Seeds Co. Fruits are born in cluster. Fruits are long and cylindrical. Individual fruit weight is 150-200 g. Average yield potential is 35 t/ha. It is tolerant to bacterial wilt and root knot nematode.
Green Star Long	This hybrid has been developed by Sreemaa Seeds Co. Fruits are oval-long and green with white striped. Individual fruit weight is 150-200 g. Average yield potential is 41 t/ha. It is resistant to bacterial wilt.
Cross-99	This hybrid has been developed by Sreemaa Seeds Co. Fruits are oval and whitish green. Fruits are born in cluster of 4-5. Individual fruit weight is 200-250 g. Average yield potential is 38 t/ha. It is resistant to bacterial wilt and nematode. It is very much suitable for long distance transportation.
Uphar	This hybrid has been developed by Sreemaa Seeds Co. Fruits are long, cylindrical and dark purple. Individual fruit weight is 15-20 g. Average yield potential is 37.5 t/ha. It is resistant to bacterial wilt and root knot nematode. It is very much suitable for long distance transportation.
Sudha	This profuse bearerhybrid has been developed by Sreemaa Seeds Co. Fruits are born in cluster. Fruits are oblong and dark purple. Individual fruit weight is 100 g. Average yield potential is 70 t/ha. It is resistant to bacterial wilt and root knot nematode. It is very much suitable for long distance transportation.
Saptarathi	This hybrid has been developed by Sreemaa Seeds Co. Fruits are born in cluster of 8-10. Fruits are long, small and light purple. Individual fruit weight is 15-20 g. Average yield potential is 50 t/ha. It is resistant to bacterial wilt and root knot nematode. It is very much suitable for summer cultivation.
Vijaya	This hybrid has been developed by Sreemaa Seeds Co. Fruits are born in cluster. Fruits are oblong, and green white striped. Individual fruit weight is 100-150 g. Average yield potential is 38.5 t/ha. It is resistant to bacterial wilt and tolerant to shoot and fruit borer.
Blue Star	This heavy cropper hybrid has been developed by Sreemaa Seeds Co. Fruits are oval glossy and purple in colour. Individual fruit weight is 200-250 g. Average yield potential is 39.5 t/ha. It has wider adaptability. It is resistant to bacterial wilt and can be cultivated throughout the year.
Chandrika	This hybrid has been developed by Sreemaa Seeds Co. Fruits are round and dark green with white stripes. Individual fruit weight is 200 g. Average yield potential is 39 t/ha. It is resistant to bacterial wilt and root knot nematode.
Panchasakha	This hybrid has been developed by Sreemaa Seeds Co. Fruits are born in cluster of 8-10. Fruits are oval blue. Individual fruit weight is 200 g. Average yield potential is 37.5 t/ha. It is very much suitable for summer cultivation.
Mukta	This hybrid has been developed by Sreemaa Seeds Co. Fruits are oval, glossy and purple in coloure. Individual fruit weight is 100 g. Average yield potential is 34.5 t/ha. It is resistant to bacterial wilt and root knot nematode.
Purnima	This hybrid has been developed by Sreemaa Seeds Co. Fruits are born in cluster of 6-7. Fruits are cylindrical and whitish green. Individual fruit weight is 100-200 g. Average yield potential is 35 t/ha. It is very much suitable for summer cultivation. It is resistant to bacterial wilt and root knot nematode. It is very much suitable for long distance transportation.
Komal	This hybrid has been developed by Unicorn Seeds Co. Fruits are, solitary long, medium sized and dark purple in coloure. Individual fruit weight is 80-150 g. Average yield potential is 31.8 t/ha. It is resistant to bacterial wilt and root knot nematode.
Kajal	This hybrid has been developed by Unicorn Seeds Co. First picking start 60-65 days from transplanting. Fruits are oval-round and dark purple in coloure. Individual fruit weight is 300-400 g. Average yield potential is 32.7 t/ha.
Vijaya	This hybrid has been developed by Unicorn Seeds Co. Fruits are uniform, oval, firm glossy and attractive black in coloure. Individual fruit weight is 300-450 g. Average yield potential is 32.8 t/ha.
Ghanshyam	This early bearer hybrid has been developed by Unicorn Seeds Co. Fruits are solitary, oval, medium to big and dark purple in colour. Individual fruit weight is 300 g. Average yield potential is 32.5 t/ha. It is tolerant to fruit borer.
Shyam	This early bearing hybrid has been developed by Unicorn Seeds Co. First picking starts 45 days after transplanting.Fruits are born in cluster. Fruits are uniform, small, round oval and purple in coloure. Individual fruit weight is 30-45 g. Average yield potential is 32 t/ha. It is very much suitablr for stuffing.
Prabha	This early and prolific bearing hybrid has been developed by Unicorn Seeds Co. Fruits are round oval and violet in coloure with white streaks. Individual fruit weight is 80-100 g. Average yield potential is 34.2 t/ha. It is very much suitablr for stuffing.
Gowri	This hybrid has been developed by Unicorn Seeds Co. Fruits are born in cluster of 6. Fruits are medium size, long and shiny green. Individual fruit weight is 40-50 g. Average yield potential is 33.5 t/ha. It is very much suitablr for stuffing.

Variety	Description
Pratibha	This hybrid has been developed by Unicorn Seeds Co. Fruits are round and violet in coloure with white streaks. Individual fruit weight is 100-125 g. Average yield potential is 34 t/ha.
Divya	This hybrid has been developed by Unicorn Seeds Co. First picking starts 65 days after transplanting. Fruits are round and shiny purple. Individual fruit weight is 250-300 g. Average yield potential is 30.2 t/ha.
Kiran	This hybrid has been developed by Unicorn Seeds Co. Fruits are born in cluster. Fruits are round and shiny green with white spots. Individual fruit weight is 40-50 g. Average yield potential is 32.5 t/ha.
BJT-103	This hybrid has been developed by Unicorn Seeds Co. Fruits are long and black purple. Individual fruit weight is 100-130 g. Average yield potential is 36 t/ha.
BJT-146	This hybrid has been developed by Unicorn Seeds Co. Fruits are round and dark purple. Individual fruit weight is 500-600 g. Average yield potential is 34.5 t/ha.
Long-13	This early maturing hybrid has been developed by Indo-American Hybrid Seeds Co. Fruits are long and dark purple. Average yield potential is 45 t/ha.
Round-14	This hybrid has been developed by Indo-American Hybrid Seeds Co. Fruits are medium size, round and light purple. Average yield potential is 50 t/ha.
BCB-11	This variety has been developedby BCKV, Kalyani (West Bengal) from the local germplasm assemblage. Plants with indeterminate growth habit; oblong fruit shape with green colour with purple long tinge and having green calyx; fruiting pattern-solitary; fruit length 18.00 cm; fruit girth: 6.50 cm and average fruit weight is 320 g. Average fruit yield is 35 t/ha and moderately tolerant to bacterial wilt disease. This line has already been treated as National check variety.
BCB-30	This variety has been developed by BCKV, Kalyani (West Bengal) from the local germplasm assemblage. Plants having somewhat erect growth habit; round fruit shape with blackish purple colour. Fruiting pattern-solitary; fruit length 16.0 cm; fruit girth 10.30 cm and average fruit weight is 240 g. Average fruit yield is 30 t/ha.
Chilli	
Kashi Vishwanath	This hybrid has been developed from ICAR-Indian Institute of Vegetable Research, Varanai (U.P.). Plants are tall (120 cm), dull green stems and bear pendent and long (10-12 cm x 1.0-1.2 cm) fruit. Fruits highly attractive, yellowish green and turn bright red at maturity. Fruit surface is wrinkled. First picking of ripe red fruits at 90 days after transplanting. Crop duration is 220 days. It is suitable for dry fruit production (Kalloo *et al.,* 2006).
Kashi Early	This hybrid has been developed from ICAR-Indian Institute of Vegetable Research, Varanai (U.P.). Plants tall (110 cm), dull green stems and bear pendent and long (8-9 cm x 1.0-1.2 cm) fruit. Fruits highly attractive, dark green and turn bright red at maturity and pungent. Fruit surface is smooth. First picking of fruits at 90 days after transplanting. Crop duration is 220 days. Yield potential of green chilli is 30 t/ha. It is suitable for dry fruit production and green chillies (Kalloo *et al.,* 2006).
Hisar Shakti	This variety has been developed by CCS HAU, Hisar (Haryana) in 1995 and release by State Variety Release Comittee for commercial cultivation in Haryana. Fruits upright in clusters, pointed, thin, medium long, highly pungent, tolerant to leaf curl mosaic virus, fruit rot and powdery mildew. Yield potential of green chilli is 17-19 t/ha.
Hisar Vijay	This variety has been developed by CCS HAU, Hisar (Haryana) in 1995 and release by State Variety Release Comittee for commercial cultivation in Haryana. Early, fruit upright in clusters, thin, round at margin but with blackish tinge on fruit surface under low temperature, tolerance to leaf curl mosaic virus, fruit rot and powdery mildew. Yield potential of green chilli is 17-19 t/ha.
Azad Mirch-1	This is a very popular chilliy variety and developed from Chandra Sekhar Azad University of Agriculture and Technology, Kanpur (U.P.). Plants dwarf/bushy (35-45 cm) with dull green stem and 8-10 branches. Generally 12-14 clusters per plant. Fruits are smooth surface, erect, clustered green fruits (8-12 fruits in a cluster). Fruits turn bright red at matuirity. First picking can be done in 50 days after transplanting. Crop duration is 135-140 days. Yield potential of green chilli is 70 t/ha (Singh and Srivastava, 2006).
Arka Harita	This is a very popular hybrid developed by ICAR-IIHR, Bengaluru (Karnataka) in 2006. Tall plants with spreading growth habit. Fruits are 10.0 cm long and 1.0 cm in width having initial light green colour but in advanced stage it become dark green and turned to red on maturity. Resistant to powdery mildew and viruses. Yield 38.2 t/ha green chilli or 5.86 t/ha dry chilli.
Arka Meghana	This is a very popular hybrid developed by ICAR-IIHR, Bengaluru (Karnataka) in 2005. Fruits are light green 10.6 cm long and 1.2 cm width having initial light green colour but in advanced stage it become dark green and turned to red on maturity. Resistant to powdery mildew and viruses. Yield 33.5 t/ha green chilli or 5.00 t/ha dry chilli.
Arka Sweta	This is a very popular cultivar developed by ICAR-IIHR, Bengaluru (Karnataka) in 2005. Suitable for *Kharif and rabi* season under irrigated conditions. Fruit length 11.0-12.0 cm long and having width of 1.2-1.5 cm; smooth and medium pungent. Fruits light green and turn red on maturity. Tolerant to viruses. Duration of crop 160 days. Yield is around 35.0 t/ha (fresh) and 5.0 t/ha dry.
Arka Khyati	This is a very popular and high yielding hybrid developed by ICAR-IIHR, Bengaluru (Karnataka) in 2011 using CMS line for fresh market. Fruits are 12 x 1 cm in size, light green and turn deep red on maturity, medium pungent, fruits smooth and turn wrinkled after drying. It is tolerant to viruses. It yields around 40-45 t/ha (fresh) and 5.0-5.5 t/ha (dry) in 180 days.
Arka Abir	This is a very popular and high yielding cultivar developed by ICAR- IIHR, Bengaluru (Karnataka). Basically it is paprika variety suitable for oleoresis industry. Fruits are long, thin and green in colour. Fruits become red on maturity. It yields around 22 t/ha (fresh) and 2 t/ha (dry).
Jawahar Mirch-283	This variety has been developed by JNKVV, Jabalpur (Madhya Pradesh). Plants are dwarf (40-50 cm) with bushy habit. It produces 7-9 primary branches with dense foliage cover. The fruits are straight, firm, medium in length, thin (2.5-3.5 cm girth in the middle) and dark green in colour. Since its fruits are very attractive with high consumer preference, they fetch a premium price in the market. It is very much suitable for round the year cultivation in Madhya Pradesh (Chandra and Pandey, 2001).
JCA-154	This variety has been developed by JNKVV, Jabalpur (Madhya Pradesh). Plants are bushy. Fruits are long, dark green and turn to bright red when ripe. Crop duration is 115-120 days. It yields around 19.5 t/ha. It is most suitable for pickling making.

Variety	Description
Bharni	This variety has been developed by JNKVV, Jabalpur (Madhya Pradesh). Fruits are long (10 x 7 cm), deep green, straight and turn bright red on maturity. Fruitys are suitable for all purpose. It yields around 27 t/ha green chilli.
Japani Laungi	This variety has been developed by JNKVV, Jabalpur (Madhya Pradesh). Flowering starts in 60-65 days after transplanting. Fruits are long, green, clusterous (10-20) and upright in nature 'Jhumka' type. It yields around 21 t/ha green chilli. It is tolerant to pest and diseases.
Surya Rekha	This variety has been developed by JNKVV, Jabalpur (Madhya Pradesh). Flowering starts in 45-50 days after transplanting. Fruits are long; light green, single, erect and upward in nature. Fruits become maroon on ripe and red on drying. It is highly pungent. It yields around 15 t/ha. Vary much suitable for dual season.
Indira Mirch-1	This variety has been released by State Sub Seed Committee of Chhattisgarh State for cultivation in Chhattisgarh plains during 2010-11. The variety has attractive upright green fruits. Fruits are pungent and suitable for green purpose. Incidence of whitefly and mites were low and Aphid was moderate, whereas it is least affected by Leaf Curl and Anthracnose was found absent. It has yield potential of 8.90 t/ha. Recommended for rainfed and irrigated conditions, for Kharif and Spring-summer seasons.
Phule Jyoti	This variety has been developed from MPKV, Rahuri (M.S.). Fruits are medium long, green, clusterous and pendent. Fruits become red on maturity. It is tolerant to leaf curl and powdery mildew. It is also resistant to lodging. Average fresh green yield is 28.5 t/ha. It is dual type and suitable for kharif and summer planting.
Phule Mukta	This variety has been developed from MPKV, Rahuri (M.S.). Average yield potential is about 27.5 t/ha.
Phule Suryamukhi	This variety has been developed from MPKV, Rahuri (M.S.). Fruits are medium long, green, clusterous and erect. Fruits become red on maturity. It is tolerant to leaf curl and powdery mildew. Average fresh green yield is 28 t/ha. It is dual type and suitable for kharif and summer planting.
Phule Sai	This variety has been developed from MPKV, Rahuri (M.S.). Average yield is about 25.0 t/ha.
Parbhani Tejas	This variety has been developed from MAU, Parbhani (M.S.). Fruits are long green and of good quality. Fruits become red on maturity. It is suitable for green and dry purpose. Average yield potential is 26 t/ha (Sontakke *et al.,* 1995).
Surakta	This variety has been developed from PDKV, Akola (M.S.) Average is 24.0 t/ha.
Jayanti	This variety has been developed from PDKV, Akola (M.S.). Average yield is 25.5 t/ha.
Gujarat Vegetable Chilli-111	This variety has been developed in the year 2003. Average yield is 23.5 t/ha. Free from whitefly and powdery mildew.
Gujarat Vegetable Chilli-101	This variety has been developed in the year 2004. Average yield is 22.0 t/ha. Suitable for both the season, i.e., Kharif and Rabi. Resistant to mosaic virus.
Gujarat Vegetable Chilli-121	This variety has been developed in the year 2004. Average yield is 28.0 t/ha. Suitable for round the year cultivation. Free from powdery mildew and mosaic viruses.
Hisar Shakti (H-44)	This variety has been developed by CCS HAU, Hisar (Haryana). Plants are medium tall, perennial and widely branched. Fruits are medium long, clustered and highly pungent. It is resistant to mosaic and leaf curl virus. Average yield of green chilli is 23 t/ha.
Hisar Vijay	This variety has been developed by CCS HAU, Hisar (Haryana). Average yield is 21.6 t/ha.
Punjab Surkh	This variety has been developed at PAU, Ludhiana (Punjab) by selection from the material collected from Uttar Pradesh. The plants are medium tall (60.0 cm) and spreading with dark green foliage. First picking is possible in 60 days after transplanting. Fruits are medium (7.0 cm long) and dark green. Fruits become dark red with high capsaicin content of 0.8 per cent on dry weight basis. It is also rich in oleoresin and dry matter. Average yield is 24.2 t/ha. It is resistant to mosaic, wilt and die back.
Punjab Guchhedar	This variety has been developed at PAU, Ludhiana (Punjab) by selection from an exotic material collected from Indonesia. The plants are medium tall (60.0-65.0 cm), compact with dark green foliage. First picking is possible in 80 days after planting. Its fruits are 5.0 cm long with pointed apex, deep red erect, born in clusters of 8-12 and have destalking habit. Average capsaicin content of fruits is 0.98 per cent. Average yield is 15.0 t/ha. it is tolerant to fruit rot, leaf curl virus and resistant to mosaic.
Punjab Lal	This variety has been developed at PAU, Ludhiana (Punjab) by selection from the cross Perenniai x Long Red. Plants are medium tall (60-65 cm) and compact with dark green foliage. Fruits are medium long (4.25 x 0.79 cm), erect, dark red, highly pungent and are suitable for powder making. Fruits become dark red on maturity. Average capsaicin content of fruits is 1.0 per cent and fresh green fruit yield is 12.0 t/ha and 0.9-1.0 t/ha dry. It is resistant to CMV, TMV, leaf curl and moderately resistant to fruit rot and dieback.
CH-1	This hybrid has been developed at PAU, Ludhiana (Punjab) by crossing between MS-12 (male sterile) x Ludhiana Local Selection (LLS). Plants are tall (95.0-100.0 cm) and vigorous with light green foliage. Fruits are 6.62 cm long and red. Average capsaicin content of fruits is 0.8 per cent. Average yield is 25.0 t/ha. It is tolerant to viral and fungal diseases.
CH-3	This hybrid has been developed at PAU, Ludhiana (Punjab) by crossing between MS-12 (male sterile) x S-2530. Plants are medium tall (70.0 cm) and vigorous with dark green foliage. Fruits are 8.2 cm long, deep red and moderately pungent. Average capsaicin content of fruits is 0.51 per cent. Average yield is 27.50 t/ha. It is suitable for paste making.
Pusa Jwala	This variety has been developed at ICAR-IARI, New Delhi by selection from the cross NP-46A x Puri Red. Plants are dwarf (40 cm) and bushy with pendent fruit bearing habit. Fruits are long (12.0-15.0 cm), thin, curved at the tip, light green and very pungent. Its capsaicin content varies from 0.40-0.90 per cent and is suitable for oleoresin extraction. Average yield is 25.0 t/ha.
Pusa Sadabahar	Plants erect with dark green large foliage, has 6-8 cm long fruits and highly pungent in taste. It bears fruits in clusters (6-14 fruits/cluster). It has been reported to be field resistant to leaf curl virus complex (Pruthi, 1993; Tewari, 1991). It is suitable for both the seasons and get mature in 150-170 days. Average fresh green yield is 28 t/ha and 1.5-2.0 t/ha dry fruit.

Variety	*Description*
NP-46-A	It is developed through selection. Plants are dwarf and spreading type. Fruits are long (10.5 cm), thin, pendent, and wrinkled, samba type and green with less seeds. Fruits become red on ripening. Initially it was tolerant to mosaic but later on become susceptible. Capsaicin 0.53 mg/g of fruits. Yield is up to 12.5 t/ha. It is very much suitable for green and dry use.
Pant C-2	It is a cultivar derived from NP-46-A x Local strain and developed by GBPUA&T, Pantnagar (Uttarakhand). Plants are short bearing fruits which are highly pungent, solitary, and erect, 5.5 cm long with blunt tip. It is very good for fresh market. It is tolerant to leaf curl virus and mosaic. Average fresh green yield is 20.5 t/ha and 0.9-1.0 t/ha dry fruit.
Sindhur (CA-960)	This variety has been developed from Maharashtra. Fruits are long (8.2-9.0 cm), erect, and pericarp light green in colour which turn red at maturity, fruit surface smooth and light pungent. Capsaicin 0.81 mg/g of fruits. Crop duration is 180 days (Murthy and Bavaji, 1980). It is very much suitable for green purpose. Average fresh green yield is 20 t/ha.
Utkal Rashmi	This variety has been developed from Odisha University of Agriculture and Teachnology (OUAT), Bhubaneswar (Odisha). Fruits are long, medium girth, straight, green, pendent and having good pungency. Fruit become deep red on maturity. It is suitable for dry chilli. Crop duration is 130-135 days. Average fresh green yield is 29.5 t/ha. It is resistant to bacterial wilt and fruit borer.
Utkal Ava	This variety has been developed from Odisha University of Agriculture and Teachnology (OUAT), Bhubaneswar (Odisha). Fruits are medium size, straight, green and highly pungent. Fruit become deep red on maturity. It is suitable for dry chilli and as green for fresh market. Crop duration is 125-130 days. Average fresh green yield is 29.5 t/ha. It is resistant to bacterial wilt.
Utkal Ragini	This variety has been developed from Odisha University of Agriculture and Teachnology (OUAT), Bhubaneswar (Odisha). Fruit bearing is upright. Fruits are medium size, small, clusterous, green and highly pungent. Fruit become deep red on maturity. It is suitable as green for fresh market. Crop duration is 140-145 days. Average fresh green yield is 23 t/ha.
BC-24	This variety has been developed from Odisha University of Agriculture and Teachnology (OUAT), Bhubaneswar (Odisha). Fruit long bold, green, attractive, and pendent. It is suitable for green for dry chilli. Average fresh green yield is 25 t/ha. This variety possesses good transport quality.
CO-1	It is an improved variety has been developed by TNAU, Coimbatore (Tamil Nadu) and still popular among vegetable growers of Tamil Nadu. Fruits are medium long (7.3 x 1.5 cm), green and samba type. Fruits become bright red on maturity. It contain high amount of capsaicin. Crop duration is 210 days. Average fresh green yield is 22.5 t/ha.
CO-2	It is an improved variety has been developed by TNAU, Coimbatore (Tamil Nadu) and still popular among vegetable growers of Tamil Nadu. Fruits are oblong and green. Fruits become bright red on maturity. It contain high amount of capsaicin (0.56 mg/g). Average fresh green yield is 28 t/ha. It is dual purpose crop *i.e.* green and dry use and suitable for all season cultivation.
K-1	It is an improved variety has been developed by TNAU, Coimbatore (Tamil Nadu) through induced mutartion (K-1 by gamma irradiation). Fruits are medium long (6.6 x 3.8 cm) and green. Fruits become shiny red on maturity. It contain high amount of capsaicin. Crop duration is 215 days. Average fresh green yield is 29 t/ha. This cultivar is very much suitable for rainfed area (Pruthi, 1993).
MDU-1	It is an improved variety and has been developed by TNAU, Coimbatore (Tamil Nadu). Fruits appear in cluster (4-9). Fruits are long (8 cm) and green. Fruits become shiny red on maturity. It contain high amount of capsaicin (0.70 mg/g). Crop duration is 215 days. Average fresh green yield is 29 t/ha. This cultivar is very much suitable for rainfed area (Pruthi, 1993).
Ujwala	It is an improved variety and has been developed by Kerala Agricultural University, Vellanikkara (Kerala). Plant height medium (56 cm) and vigorous branching (11). This variety bears maximum fruits (86 fruits/plant). Good plant canopy development is key feature of the variety (Mini and Joseph, 1998.). Average fresh green yield is 22 t/ha.
LCA-304	It is an improved variety has been developed by Guntur, Lam (Andhra Pradesh). Fruits are medium long (7.8 cm) and green. Fruits become shiny red on maturity. Average fresh green yield is 18 t/ha. It is tolerant to virus and suitable for rainfed conditions.
LCA-324	It is an improved variety and has been developed by Guntur, Lam (Andhra Pradesh). fruits are long, thin and green. Fruits become shiny red on maturity. Average fresh green yield is 17.5 t/ha. It is tolerant to virus and suitable for rainfed conditions.
LCA-334	It is an improved and early maturing variety has been developed by Guntur, Lam (Andhra Pradesh). fruits are medium long, stout and green. Fruits become shiny red on maturity. Average fresh green yield is 18.5 t/ha. It is ideal for fresh green and red chillies.
HC-28	It is an improved variety and has been developed by Assam. Fruits are smooth, clustered, upright, straight, medium long (8-9 cm), thin and dark green. Slight blackish tinge on fruit surface under low temperature, a marker character. First picking can be taken in 63 days after transplanting for green and 82 days after transplanting for ripe fruits. Fruits become bright red on ripening. It is highly pungent and suitable for dry powder. Highly resistant to TMV, CMV, PVY and leaf curl. Apart from these it is also free from fruit rot, powdery mildew and dieback disease. This cultivar has been adjudged good for rainy and spring season. Average fresh green yield is 20.9 t/ha.
HC-44	Fruits are erect, clustered, medium long (8-9 cm) and light green in colour. Fruits become glossy red on maturity. It is highly pungent and rich in capsicin, oleoresin, dry matter and pigment content. It is suitable for dry powder. Its cultivation can be done in rainy and spring season. Average fresh green yield is 32.7 t/ha.
Puri Red	It is a local ecotype and still popular among vegetable growers of Odisha. Average yield is 20 t/ha.
Resham Patty	This variety has been developed from Madhya Pradesh. Fruits are long, thick, glossy and green. Fruits become deep red on ripening. Fruits are less seeded and mild pungent. It suitable for dry purpose. Average fresh green yield is 12 t/ha.
Kiran	This variety has been developed from Lam (Andhra Pradesh). Plants are compact. Fruits are long green and thin. Fruits become red on maturity. It is tolerant to thrips, mite and aphids. Average fresh green yield is 33.5 t/ha.
Aparna	This variety has been developed from Lam (Andhra Pradesh). Plants are compact. Fruits are medium long and green with blunt tip. Fruits turn to yellow on ripening. It is very good for capsaicin extraction. It is tolerant to salinity (Singh and Singh, 1996). Average fresh green yield is 33 t/ha.
Agni	This hybrid has been developed by Sandoz Seed Co. Pune (Maharshtra). Fruits are long (10.0 x 1.5 cm), attractive glossy and green. It is very good for long distance transportation. It is tolerant to TMV and PVY. Average fresh green yield is 30.5 t/ha.

Variety	Description
Angare	This hybrid has been developed by Sandoz Seed Co. Pune (Maharshtra). Fruits are long (12.0 x 1.5 cm), attractive cylindrical and pale green. It is moderately pungent. Average fresh green yield is 31 t/ha.
Bijlee	This early bearing hybrid has been developed by Sandoz Seed Co. Pune (Maharshtra). Fruits are long (13.0 x 2.5 cm), conical, thick walled attractive and green. Fruits are medium pungent. Fruits become red on maturity. It is very good for long distance transportation and also possesses longer shelf life. It is tolerant to TMV. Average fresh green yield is 31.2 t/ha.
Roshni	This hybrid has been developed by Novartis Seed Co. Pune (Maharshtra). Fruits are long (14.0 x 2.0 cm), thin skin, attractive cylindrical, glossy and dark green. Fruits become dark red on maturity. It is moderately pungent. Average fresh green yield is 31.4 t/ha. It is suitable for both fresh and dehydration.
Sholay	This hybrid has been developed by Novartis Seed Co. Pune (Maharshtra). Fruits are long (13.0 x 2.0 cm), thin walled, attractive cylindrical, glossy, uniform and dark green. It is moderately pungent. Average fresh green yield is 31.7 t/ha. It is suitable for both fresh and dehydration.
Agni Rekha	This hybrid has been developed by Novartis Seed Co. Pune (Maharshtra). Fruit bearing starts at 55 days after transplanting. Fruits are medium long (11.0 x 1.5 cm), solitary thic walled, lustrous smooth, uniform and faint green. Fruits become dark red on maturity. It is moderately pungent. It is very good for long distance transportation. Average fresh green yield is 31.5 t/ha. It is suitable for fresh market.
Flame Hot	This hybrid has been developed by Novartis Seed Co. Pune (Maharshtra). Fruit bearing starts at 55 days after transplanting. Fruits are small (8.0 x 1.5 cm), lustrous smooth, uniform and dark green. Fruits become deep red on maturity. It is strong pungent. It is very good for long distance transportation. Average fresh green yield is 31.8 t/ha. It is suitable for dry use.
Hot Line	This hybrid has been developed by Novartis Seed Co. Pune (Maharshtra). Fruits are long (14.0 x 1.5 cm), medium thick walled, smooth and lustrous faint green. Fruits become deep red on maturity. It is strong pungent. Average fresh green yield is 31.4 t/ha. It is suitable for dry use and powder.
Picador	This hybrid has been developed by Novartis Seed Co. Pune (Maharshtra). Fruit bearing starts at 45-50 days after transplanting. Fruits are long (20 x 8 cm), thick walled, tasty and green. Flavor of capsicum is prominent. Good keeping quality. It is tolerant to TMV. Average fresh green yield is 31.2 t/ha. It is suitable for pickling, salad and fry.
Akash	This hybrid has been developed by Indo-American Hybrid Seed Co. Bengaluru (Karnataka). Fruits are long, bright green, glossy and having good size. It is tolerant to foliar diseases. Average fresh green yield is 36.9 t/ha.
Utpan	This hybrid has been developed by Indo-American Hybrid Seed Co. Bengaluru (Karnataka). Fruits are long, bright green, glossy and having good size. It is tolerant to foliar diseases. Average fresh green yield is 32.7 t/ha.
Ruchi	This hybrid has been developed by Indo-American Hybrid Seed Co. Bengaluru (Karnataka). Fruits are long, bright green, glossy and having good size. It is tolerant to foliar diseases. Average fresh green yield is 32.5 t/ha.
No.6	This hybrid has been developed by Indo-American Hybrid Seed Co. Bengaluru (Karnataka). Fruits are medium long, green glossy and having good shape and size. It is tolerant to leaf curl and die back diseases. Average fresh green yield is 31.8 t/ha.
No.8	This hybrid has been developed by Indo-American Hybrid Seed Co. Bengaluru (Karnataka). Fruits are medium long, green glossy and having good shape and size. It is tolerant to leaf curl and die back diseases. Average fresh green yield is 32 t/ha.
No.9	This hybrid has been developed by Indo-American Hybrid Seed Co. Bengaluru (Karnataka). Fruits are medium long, green glossy and having good shape and size. It is tolerant to leaf curl and die back diseases. Average fresh green yield is 33.5 t/ha.
Kandhari	It is a local ecotype and still popular among vegetable growers. Average yield is 20.68 t/ha.
Pungent Pride	It is impropved cultivar and having better market acceptance. Resistant to bacterial wilt. Average yield is 18 t/ha.
Chery Red	It is impropved cultivar and having better market acceptance. Resistant to bacterial wilt. Average yield is 22 t/ha.
Divya Jyoti	This hybrid has been developed by Unicorn Seeds Co. Hyderabad (Telangana) Fruits are medium, green, long, solitary downward, slender and highly pungent. Average fresh green yield is 37.5 t/ha.
Shama	This hybrid has been developed by Unicorn Seeds Co. Hyderabad (Telangana) Fruits are long (20 x 2 cm), waxy light green and thick skin. It is very much suitable for fresh market, pickle and Bhaji preparation. Average fresh green yield is 38 t/ha.
Asha Jyoti	This hybrid has been developed by Unicorn Seeds Co. Hyderabad (Telangana) long (14.0 x 1.3 cm), medium thick skin and slender. Average fresh green yield is 38.7 t/ha.
Usha Jyoti	This hybrid has been developed by Unicorn Seeds Co. Hyderabad (Telangana) Fruits are long (16.3 x 1.3 cm), slender medium, green and pungent. Fruits become red at maturity. Fruits are very smooth and possess very good transport quality. It is suitable for both fresh markert and delay in maturing duration. Average fresh green yield is 37 t/ha.
Nava Jyoti	This hybrid has been developed by Unicorn Seeds Co. Hyderabad (Telangana) Fruits are long (15.0 x 1.2 cm), slim dark green and very pungent. Fruits become bright red at maturity. It is suitable for both fresh markert and drying. Average fresh green yield is 36.7 t/ha.
Red Flame	This hybrid has been developed by Unicorn Seeds Co. Hyderabad (Telangana) Fruits are long (16.0 x 1.8 cm), medium thick, light green and medium pungent. This hybrid has excellent fruit setting quality. Average fresh green yield is 38.9 t/ha.
Maha Jwala	This hybrid has been developed by Unicorn Seeds Co. Hyderabad (Telangana) Fruits are long (15.0 x 0.8 cm), slender, light green and highly pungent. Average fresh green yield is 37.2 t/ha.
Wonder Jyoti	This hybrid has been developed by Unicorn Seeds Co. Hyderabad (Telangana) Fruits are long (15.0 x 0.8 cm), slender, medium thick skin and shining. It is an excellent hybrid for both fresh and dry market. Average fresh green yield is 37.5 t/ha.
CHT-367	This hybrid has been developed by Unicorn Seeds Co. Hyderabad (Telangana) Fruits are medium long, green, upright, clustered and highly pungent. Average fresh green yield is 37.2 t/ha.

Variety	Description
CHT-298	This hybrid has been developed by Unicorn Seeds Co. Hyderabad (Telangana) Fruits are long (13.0 x 1.3 cm), light green to green and highly pungent. It is very much suitable for fresh market. Average fresh green yield is 37.6 t/ha.
CHT-347	This hybrid has been developed by Unicorn Seeds Co. Hyderabad (Telangana) Fruits are long (14.0 x 1.3 cm), light green to green and highly pungent. It is very much suitable for processing. Average fresh green yield is 37.9 t/ha.
CHT-380	This hybrid has been developed by Unicorn Seeds Co. Hyderabad (Telangana) Fruits are long (13.5 x 0.75 cm), green and highly pungent. It is very much suitable for fresh market. Average fresh green yield is 37.9 t/ha.
CHT-365	This hybrid has been developed by Unicorn Seeds Co. Hyderabad (Telangana) Fruits are long (14.5 x 1.5 cm), green and highly pungent. It is very much suitable for fresh market. Average fresh green yield is 36.5 t/ha.
CHT-399	This hybrid has been developed by Unicorn Seeds Co. Hyderabad (Telangana) Fruits are long (15 x 2 cm), light green and medium pungent. It is very much suitable for fresh market. Average fresh green yield is 37.2 t/ha.
CHT-248	This hybrid has been developed by Unicorn Seeds Co. Hyderabad (Telangana) Fruits are long (10 x 1 cm), green and highly pungent. It is very much suitable for fresh market and processing. Average fresh green yield is 37 t/ha.
MHP-58	This hybrid has been developed by MAHYCO. Jalna (Maharashtra) Fruits are long, deep green and pungent. First picking starts 75 days after transplanting. Average fresh green yield is 31 t/ha. This hybrid can be raised during kharif and spring season.
MHP-59	This hybrid has been developed by MAHYCO. Jalna (Maharashtra) Fruits are long, deep green and highly pungent. First picking starts 75 days after transplanting. Average fresh green yield is 30.5 t/ha. This hybrid is duel type.
Agnishikha	This hybrid has been developed by MAHYCO. Jalna (Maharashtra) Fruits are medium long (10 x 2 cm), thick skin, solitary and green. First picking starts 55 days after transplanting. Average fresh green yield is 31.5 t/ha. This hybrid is suitable for green and dry purpose.
MHP-62	This hybrid has been developed by MAHYCO. Jalna (Maharashtra) Fruits are long (12.5 cm), light green, wrinkles, pendent and moderately pungent. First picking starts 72 days after transplanting. Fruits turn red at maturity. It is very much suitable for fresh market. Average fresh green yield is 33.2 t/ha. This hybrid can be raised during summer season.
MHP-63	This hybrid has been developed by MAHYCO. Jalna (Maharashtra) Fruits are long (13 cm), light green, shiny, wrinkles, pendent and moderately pungent. First picking starts 70 days after transplanting. It is very much suitable for fresh market. Average fresh green yield is 33 t/ha. This hybrid can be raised during summer season.
MHCP-155	This hybrid has been developed by MAHYCO. Jalna (Maharashtra) Fruits are long (13 cm), shiny dark green, smooth, pendent and highly pungent. First picking starts 72 days after transplanting. Fruits turn red at maturity. It is very much suitable for fresh market as green and red purpose. Average fresh green yield is 31.5 t/ha. This hybrid can be raised during summer season.
MHCP-157	This hybrid has been developed by MAHYCO. Jalna (Maharashtra) Fruits are medium long (10.5 cm), shiny dark green, smooth, pendent and low pungent. First picking starts 70 days after transplanting. Fruits turn red at maturity. Average fresh green yield is 32 t/ha.
MHCP-157	This hybrid has been developed by MAHYCO. Jalna (Maharashtra) Fruits are medium long (10-11 cm), light green and medium pungent. First picking starts 65 days after transplanting. Fruits turn light red at maturity. Average fresh green yield is 33 t/ha.
MHCP-158	This hybrid has been developed by MAHYCO. Jalna (Maharashtra) Fruits are medium long (10-11 cm), light green and medium pungent. First picking starts 65 days after transplanting. Fruits turn light red at maturity. Average fresh green yield is 33 t/ha.
Tejaswini	This hybrid has been developed by MAHYCO. Jalna (Maharashtra) Fruits are long and green to deep green and highly pungent. First picking starts 75 days after transplanting. Average fresh green yield is 37.5 t/ha.
Sujata	This hybrid has been developed by MAHYCO. Jalna (Maharashtra) Fruits are light green and medium pungent. First picking starts 70 days after transplanting. Average fresh green yield is 36.4 t/ha. It is duel type.
Srrya	This hybrid has been developed by MAHYCO. Jalna (Maharashtra) Fruits are long, green, upright and highly pungent. First picking starts 75 days after transplanting. Average fresh green yield is 38.5 t/ha.
MHCP-159	This hybrid has been developed by MAHYCO. Jalna (Maharashtra) Fruits are medium long (10-11 cm), green, pendent and highly pungent. First picking starts 78 days after transplanting. Fruits turn red at maturity. Average fresh green yield is 33.7 t/ha.
CCH-1	This hybrid has been developed by Century Seed Co. Mangolpuri (Delhi) Fruits are medium long (6.6 cm), green and pungent. Fruits turn deep red at maturity. It has good VHC and capsaicin. It is moderately resistant to mosaic, leaf curl, wet rot and resistant to die back. Average fresh green yield is 34 t/ha.
CCH-2	This early medium type hybrid has been developed by Century Seed Co. Mangolpuri (Delhi) Fruits are long (12 x 1 cm), deep green, upright clusters and extremely pungent. Fruits turn scarlet red at maturity. Average fresh green yield is 33.5 t/ha.
Agnivaan	This early maturing (65-70 days after transplanting) hybrid has been developed by Century Seed Co. Mangolpuri (Delhi) Fruits are long (12 x 1 cm) and medium green. Fruits turn crimson at maturity. Average fresh green yield is 34.5 t/ha.
Agnidev	This hybrid has been developed by Century Seed Co. Mangolpuri (Delhi) Fruits are medium long (7 x 1 cm), solitary, pendent, excellent deep green colour and highly pungent. Maturity in 68-72 days after transplanting. Average fresh green yield is 34.2 t/ha. It is sutiable for dehydration.
Agnipath	This early medium type hybrid has been developed by Century Seed Co. Mangolpuri (Delhi) Fruits are long (10.0 x 16 cm), green, smooth, tight skin, medium thick wall and highly pungent. It is medium maturing variety (68-70 days after transplanting). Fruits turn red at maturity. It is sutiable for fresh market as green and dehydration. Average fresh green yield is 33.8 t/ha.
CCH-791	This hybrid has been developed by Century Seed Co. Mangolpuri (Delhi) Fruits are long (8 cm), thin and light green. It is medium maturing variety (70-75 days after transplanting). Fruits turn red at maturity. It is sutiable for fresh market as green and dehydration.Average fresh green yield is 35.8 t/ha.

Variety	*Description*
CCH-792	This hybrid has been developed by Century Seed Co. Mangolpuri (Delhi) Fruits are long (8-9 cm), medium thickness, and light green colour. It is medium maturing variety (75-80 days after transplanting). Fruits turn deep red at maturity. It is sutiable for fresh market as green and dehydration.Average fresh green yield is 34.8 t/ha.
Anaheim TMR-23	This hybrid has been developed by IAES. It is medium maturing variety (74-76 days after transplanting). Fruits are long (19 x 5 cm), bright green, smooth pods and medium pungent. Fruits turn red at maturity. It is resistant to Tobacco Mosaic Virus. Average fresh green yield is 32.5 t/ha.
Cayenne Large Red Thick	This hybrid has been developed by IAES. It is medium maturing variety (73-76 days after transplanting). Fruits are long (15 x 2 cm), two celled, medium thin wall, pendent and medium pungent. Fruits turn dark red at maturity. It is excellent for processing and fresh market. Average fresh green yield is 33.5 t/ha.
Hungarian Yellow Wax Hot	This early type hybrid has been developed by IAES. Fruits are long (15 x 2 cm); light yellow to orange, smooth conical, upright medium thick wall, and pendent. It is excellent for processing, fresh market and home garden. Average fresh green yield is 32.1 t/ha.
Cayenne Long Slim	This early-medium maturing type (65-70 days after transplanting) hybrid has been developed by IAES. Fruits are long (13 x 1 cm); dark green to red, thin walled two celled, slim, wrinkled and pendent. It is excellent for home garden. Average fresh green yield is 33 t/ha.
BSS-269	This hybrid has been developed by Bejo Sheetal Seeds Co. Jalna (Maharashtra) It is Hungarian wax type. Fruits are long (16 x 2 cm); medium thick skin, less wrinkled, light green and pungent. First picking starts 75 days after transplanting. Fruits become dark red at ripening. It is excellent for fresh market. Average fresh green yield is 35.9 t/ha.
BSS-356	This hybrid has been developed by Bejo Sheetal Seeds Co. Jalna (Maharashtra) It is Hungarian wax type. Fruits are long (15 x 3 cm); thick skin, rought green and pungent. First picking starts 80 days after transplanting. Fruits become red at ripening. It is excellent for fresh market. Average fresh green yield is 34.8 t/ha.
BSS-377	This hybrid has been developed by Bejo Sheetal Seeds Co. Jalna (Maharashtra) Fruits are long (14.0 x 1.4 cm); green smooth and pungent. First picking starts 85-90 days after transplanting. Fruits become dark red at ripening. It is excellent for fresh market and dry use. Average fresh green yield is 38.5 t/ha. It is tolerant to virus complex.
BSS-345	This hybrid has been developed by Bejo Sheetal Seeds Co. Jalna (Maharashtra) Fruits are long (13 x 1.1 cm); green smooth and medium pungent. First picking starts 65 days after transplanting. Fruits become dark red at ripening. It is excellent for dry use. Average fresh green yield is 41.5 t/ha.
BSS-219	This hybrid has been developed by Bejo Sheetal Seeds Co. Jalna (Maharashtra) Fruits are long (14.0 x 1.5 cm); green to light green, thick, smooth and medium pungent. First picking starts 70-75 days after transplanting. Fruits become dark red at ripening. It is excellent for fresh market as green and dry use. Average fresh green yield is 37.9 t/ha.
BSS-308	This hybrid has been developed by Bejo Sheetal Seeds Co. Jalna (Maharashtra) Fruits are long (14 x 1.8 cm); smooth, green to light green, thick skin, and pungent. First picking starts 75-80 days after transplanting. It is excellent for fresh market as green. Average fresh green yield is 43.5 t/ha.
Laxmi	This hybrid has been developed by Bejo Sheetal Seeds Co. Jalna (Maharashtra) Fruits are long (14.0 x 1.5 cm); smooth, blackish green, thick skin, and mild pungent. First picking starts 75 days after transplanting. It is excellent for fresh market as green and processing use. Average fresh green yield is 39.5 t/ha. It is tolerant to virus.
BSS-411	This hybrid has been developed by Bejo Sheetal Seeds Co. Jalna (Maharashtra) Fruits are long (15.0 x 1.5 cm); smooth, green, and medium pungent. First picking starts 70 days after transplanting. Fruits become dark red at ripening. It is excellent for dry chilli purpose. Average fresh green yield is 37.9 t/ha.
BSS-413	This hybrid has been developed by Bejo Sheetal Seeds Co. Jalna (Maharashtra) Fruits are long (18.0 x 1.4 cm); smooth, dark green, and medium pungent. First picking starts 75-80 days after transplanting. Fruits become dark red at ripening. It is excellent for dry chilli purpose. Average fresh green yield is 35.9 t/ha.
BSS-346	This hybrid has been developed by Bejo Sheetal Seeds Co. Fruits are long (14.0 x 1.2 cm); smooth, dark green, and medium pungent. First picking starts 75 days after transplanting. Fruits become dark red at ripening. It is excellent for dry chilli purpose. Average fresh green yield is 43 t/ha.
Arjun	This hybrid has been developed by Bejo Sheetal Seeds Co. Jalna (Maharashtra) Fruits are long (16.0 x 1.6 cm); less wrinkled, thick skin, green, and medium pungent. First picking starts 60 days after transplanting. Fruits become red at ripening. It is excellent for fresh use. Average fresh green yield is 43.8 t/ha.
BSS-213	This hybrid has been developed by Bejo Sheetal Seeds Co. Jalna (Maharashtra) It is Hungarian wax type. Fruits are long (16 x 3 cm); rough, green, thick skin, and medium pungent. First picking starts 75 days after transplanting. Fruits become dark red at ripening. It is excellent for fresh use. Average fresh green yield is 38.7 t/ha. It is suitable for all season.
BSS-414	This hybrid has been developed by Bejo Sheetal Seeds Co. Jalna (Maharashtra) Fruits are long (10.0 x 1.2 cm); light green to green, medium smooth and highly pungent. First picking starts 75-80 days after transplanting. Average fresh green yield is 36.5 t/ha. It is suitable for all season.
BSS-140	This early maturing hybrid has been developed by Bejo Sheetal Seeds Co. Jalna (Maharashtra) Fruits are long (10.0 x 1.2 cm); medium smooth surface, green to light green and highly pungent. First picking starts 60-70 days after transplanting. Fruits become red at ripening. It is excellent for fresh market. Average fresh green yield is 36.8 t/ha. It is very much suitable for rainy season.
BSS-141	This hybrid has been developed by Bejo Sheetal Seeds Co. Jalna (Maharashtra) Fruits are long (13.2 x 1.2 cm); dark green, less wrinkled, thin skined and highly pungent. First picking starts 70-75 days after transplanting. Fruits become dark red at ripening. It is excellent for fresh market and dry purpose. Average fresh green yield is 36.7 t/ha. It is very much suitable for all season. It is tolerant to virus complex and other diseases.

Variety	Description
BSS-202	This medium group maturing hybrid has been developed by Bejo Sheetal Seeds Co. Jalna (Maharashtra) Fruits are long (10.0 x 1.5 cm); green to dark green and highly pungent. First picking starts 70-75 days after transplanting. Fruits become dark red at ripening. It is excellent for fresh market. Average fresh green yield is 35.5 t/ha. It is very much suitable for rainy and summer season.
BSS-344	This hybrid has been developed by Bejo Sheetal Seeds Co. Jalna (Maharashtra) Fruits are long (10 x 1 cm); yellowish green, wrinkled, thin skined and highly pungent. First picking starts 70 days after transplanting. Fruits become red at ripening. It is excellent for fresh market and dry purpose. Average fresh green yield is 36 t/ha.
BSS-375	This hybrid has been developed by Bejo Sheetal Seeds Co. Jalna (Maharashtra) Fruits are long (13.0 x 1.2 cm); yellowish green, smooth, and highly pungent. First picking starts 75-80 days after transplanting. Fruits become dark red at ripening. It is excellent for fresh market and dry purpose. Average fresh green yield is 36.7 t/ha. It is very much suitable for all season. It is tolerant to virus and other diseases.
BSS-376	This hybrid has been developed by Bejo Sheetal Seeds Co. Jalna (Maharashtra) Fruits are long (12-14 cm); green, smooth, and medium pungent. First picking starts 75-80 days after transplanting. Fruits become dark red at ripening. It is excellent for dry purpose. Average fresh green yield is 35.9 t/ha. It is tolerant to virus and other diseases.
Garima	This hybrid has been developed by Bejo Sheetal Seeds Co. Jalna (Maharashtra) Fruits are long (15.0 x 1.2 cm); yellowish green, smooth and highly pungent. First picking starts 70-75 days after transplanting. Fruits become dark red at ripening. It is excellent for fresh market and dry purpose. Average fresh green yield is 37.8 t/ha.
Anmol	This hybrid has been developed by Bejo Sheetal Seeds Co. Jalna (Maharashtra) Fruits are long (16.0 x 1.3 cm); dark green, smooth, thin skined and pungent. First picking starts 75 days after transplanting. Fruits become dark red at ripening. It is excellent for dry purpose. Average fresh green yield is 36.2 t/ha.
BSS-305	This hybrid has been developed by Bejo Sheetal Seeds Co. Jalna (Maharashtra) Fruits are long (14.0 x 1.4 cm); dark green, medium wrinkled, thin skined and medium pungent. First picking starts 80-85 days after transplanting. It is excellent for dry purpose. It is tolerant to virus. Average fresh green yield is 36.4 t/ha.
BSS-306	This hybrid has been developed by Bejo Sheetal Seeds Co. Jalna (Maharashtra) Fruits are long (18.0 x 1.2 cm); attractive dark green, tough surface, and medium pungent. First picking starts 80-85 days after transplanting. It is excellent for dry purpose. It is tolerant to virus complex. Average fresh green yield is 36 t/ha.
BSS-355	This hybrid has been developed by Bejo Sheetal Seeds Co. Jalna (Maharashtra) Fruits are long (16.0 x 1.2 cm); attractive dark green, tough surface, and pungent. First picking starts 70 days after transplanting. Fruits become dark red at ripening. It is excellent for dry purpose. Average fresh green yield is 42.2 t/ha.
Suraj	This hybrid has been developed by Bejo Sheetal Seeds Co. Jalna (Maharashtra)Fruits are long (14.0 x 1.2 cm); attractive light green, wrinkled surface, thin skinned and pungent. First picking starts 75 days after transplanting. Fruits become bright red at ripening. It is excellent for fresh market and dry purpose. Average fresh green yield is 39.5 t/ha.
Savitri	This hybrid has been developed by Bejo Sheetal Seeds Co. Jalna (Maharashtra) Fruits are long (10 x 1 cm); wrinkled, yellowish green, thin skinned and highly pungent. First picking starts 70 days after transplanting. Fruits become red at ripening. It is excellent for fresh market and dry purpose. Average fresh green yield is 38 t/ha.
BSS-267	This hybrid has been developed by Bejo Sheetal Seeds Co. Jalna (Maharashtra) Fruits are long (10 x 1 cm); smooth to slightly wrinkled, light green to green, thin skinned and highly pungent. First picking starts 75-80 days after transplanting. It is excellent for dry purpose. Average fresh green yield is 37.5 t/ha.
BSS-268	This hybrid has been developed by Bejo Sheetal Seeds Co. Jalna (Maharashtra) Fruits are long (10 x 1 cm); smooth green and highly pungent. First picking starts 75 days after transplanting. It is excellent for fresh market and dry purpose. Average fresh green yield is 36.4 t/ha.
Gagan	This hybrid has been developed by Golden Seeds Co. Azadpur (Delhi) Fruits are small, green upright and highly pungent. First picking starts 65-70 days after transplanting. It is excellent for fresh market and dry purpose. Average fresh green yield is 28 t/ha. It is tolerant to mosaic and leaf curl virus.
Alankar	This prolific bearer hybrid has been developed by Golden Seeds Co. Azadpur (Delhi) Fruits are light green, pendent and highly pungent. First picking starts 70-75 days after transplanting. It is excellent for fresh market and dry purpose. Average fresh green yield is 30 t/ha.
Byadagi	This medium early maturing hybrid has been developed by Golden Seeds Co. Azadpur (Delhi) Fruits are long, light green, pendent, prominently wrinkled and medium pungent. It is excellent for fresh market and dry purpose. Average fresh green yield is 29 t/ha. It is tolerant to leaf curl virus.
A.K.-47	This medium early maturing hybrid has been developed by Golden Seeds Co. Azadpur (Delhi). First picking starts in 70 days after transplanting. Fruits are long, green, pendent, and medium to high pungency. It is excellent for fresh market and dry purpose. Average fresh green yield is 32 t/ha. It is tolerant leaf curl diseases and mosaic. Well adaptation to dry and wet conditions. It has been observed that this hybrid do well in hot regions and in long rainy period.
Farmers No.-1	This hybrid has been developed by Golden Seeds Co. Azadpur (Delhi) First picking starts 68 days after transplanting. Fruits are long (12-15 x 1.0-1.2 cm in size), dark green, pendent, and highly pungent. It is excellent for fresh market and dry purpose. Average fresh green yield is 32 t/ha. It is tolerant to virus and anthracnose disease.
Farmers Choice	This mid early maturity hybrid has been developed by Golden Seeds Co. Azadpur (Delhi) Fruits are long (12-14 x 1.3-1.5 cm in size), green shiny, pendent, and medium pungent. It is excellent for fresh market and dry purpose. Average fresh green yield is 30 t/ha. It is tolerant to virus diseases.
GC-39	This mid early prolific bearer hybrid has been developed by Golden Seeds Co. Azadpur (Delhi) Fruits are long (12-14 x 1.8-2.0 cm in size), light green, pendent, and light to medium pungent. It is very much suitable for dressing, stuffing and pickles. Average fresh green yield is 29 t/ha. It is tolerant to phytophthora root rot and bacterial diseases.

Variety	Description
S-16	This hybrid has been developed by Sungro Seeds Co. Shalimar Bagh (Delhi) Fruits are long (8-9 cm) pendent, thick skinned and highly pungent. Average fresh green yield is 27 t/ha. It is tolerant to thrips and mites.
New Vardan	This early maturing hybrid has been developed by Nunhems Seeds Co. Is is basically Cayenne type. Fruits are long (12.5 x 1.1 cm in size) pendent, light green, thin skinned and highly pungent. It is excellent for fresh market and dry purpose. Average fresh green yield is 24.5 t/ha. It is partial resistant to PVY, mites and stress.
Sona	This hybrid has been developed by Nunhems Seeds Co. It is basically Cayenne type. Fruits are long (8.0 x 0.8 cm in size) pendent, green, thin skinned and highly pungent. It is excellent for fresh market and dry purpose. Average fresh green yield is 32.5 t/ha. It is resistant to PVY.
Mashal	This early maturing hybrid has been developed by Nunhems Seeds Co. Is is basically Cayenne type. Fruits are long (15.0 x 1.2 cm in size) pendent, light green, strong, stall and mild pungent. It is excellent for fresh market. Average fresh green yield is 34.5 t/ha. It is resistant to PVY.
Red Star	This early maturing hybrid has been developed by Nunhems Seeds Co. It is basically Cayenne type. Fruits are long (14 x 1 cm in size) pendent, dark green and mild pungent. It is excellent for fresh market and dry purpose. Average fresh green yield is 43.5 t/ha. It is resistant to CVMV.
Sankranti	This early maturing hybrid has been developed by Nunhems Seeds Co. Is is basically Cayenne type and has strong framework. Fruits are long (13 x 1.2 cm in size) pendent, dark green and mild pungent. It is excellent for fresh market and dry purpose. Average fresh green yield is 43 t/ha. It is resistant to CVMV.
Dushhera	This medium early maturing hybrid has been developed by Nunhems Seeds Co. It is basically Cayenne type. Fruits are long (11.5 x 1.2 cm in size) pendent, green, thin skinned and highly pungent. It is excellent for fresh market, processing and dry purpose. It possesess very good storage quality. Average fresh green yield is 41.5 t/ha. It is resistant to PVY.
Devanur Deluxe	This hybrid has been developed by Nunhems Seeds Co. It is basically Cayenne type. Fruits are long (12.0 x 1.3 cm in size) pendent, green, medium skin thickness and highly pungent. It is excellent for fresh market and dry purpose. Average fresh green yield is 42.7 t/ha. It is resistant to CVMV.
Soldier	This hybrid has been developed by Nunhems Seeds Co. It is basically Cayenne type. Fruits are long (10 x 1cm in size) pendent, green, thin skinned and highly pungent. It is excellent for fresh market and dry purpose. Average fresh green yield is 43.5 t/ ha. It is resistant to CVMV.
Deepali	This medium early maturing hybrid has been developed by Nunhems Seeds Co. It is basically Cayenne type. Fruits are long (8 x 0.9 cm in size) green, thin skinned and highly pungent. It is excellent for fresh market and dry purpose. Average fresh green yield is 41.5 t/ha. It is resistant to CVMV.
Green Wonder	This medium maturing hybrid has been developed by Nunhems Seeds Co. It is basically anaheim type. Fruits are long (12.0 x 1.4 cm in size), pendent, light green, thick skinned and mild pungent. It is excellent for fresh market and dry purpose. Average fresh green yield is 41.5 t/ha. It is partial resistant to high temperature.
Rudra	This early maturing hybrid has been developed by Nunhems Seeds Co. It is basically anaheim type. Fruits are long (15.0 x 1.7 cm in size), pendent, ligh green, thick skinned and mild pungent. It is excellent for fresh market and dry purpose. Average fresh green yield is 42.2 t/ha.
Kranti	This early maturing hybrid has been developed by Nunhems Seeds Co. It is basically anaheim type. Fruits are long (13.5 x 1.4 cm in size), pendent, green, thick skinned and mild pungent. It is excellent for fresh market and dry purpose. Average fresh green yield is 44.5 t/ha.
Veeru	This early maturing hybrid has been developed by Nunhems Seeds Co. It is basically anaheim type. Fruits are long (12 x 2 cm in size), pendent, green, thick skinned and mild pungent. It is excellent for fresh market and dry purpose. Average fresh green yield is 42.2 t/ha.
Sikha	This early maturing hybrid has been developed by Nunhems Seeds Co. It is basically cayenne type. Fruits are long (8.0 x 0.8 cm in size), cluster bearing upright, pendent, green, thin skinned and very pungent. It is excellent for fresh market and dry purpose. Average fresh green yield is 43.5 t/ha.
Anjula	This early maturing hybrid has been developed by Nunhems Seeds Co. It is basically Jalopeno type. Fruits are long (8.0 x 0.8 cm in size), pendent, dark green, medium skinned and very pungent. It is excellent for processing. Average fresh green yield is 43.7 t/ha.
ARCH-226	This hybrid has been developed by Ankur Seeds Co. Nagpur (Maharashtra) Fruits are long, medium greem and shining. Fruits turn deep red at maturity and red colour remain stable. It is very much suitable for summer season cultivation. Average fresh green yield is 41.5 t/ha.
ARCH-001	This hybrid has been developed by Ankur Seeds Co. Nagpur (Maharashtra) Fruits are long, dark green, smooth and shining. Fruits turn deep red at maturity. It is very much suitable for summer season cultivation. It is excellent for fresh market and dry purpose. Average fresh green yield is 42.2 t/ha. It contain very high amount of capsaicin.
ARCH-112	This hybrid has been developed by Ankur Seeds Co. Nagpur (Maharashtra) Fruits are attractive, lustrous and light green shining. Fruits turn bright red at maturity and highly pungent. Average fresh green yield is 41.6 t/ha. It contain very high amount of capsaicin.
ARCH-162	This hybrid has been developed by Ankur Seeds Co. Nagpur (Maharashtra) Fruits are medium long, attractive, light green and wrinkled. Fruits turn dark red at maturity and highly pungent. Average fresh green yield is 41t/ha. It contain very high amount of capsaicin. It is excellent for fresh market and dry purpose.
ARCH-9	This hybrid has been developed by Ankur Seeds Co. Nagpur (Maharashtra) Plant express excellent flowering and fruiting. Fruits are medium size, attractive, green, compact and wrinkled. Fruits turn deep red at maturity and highly pungent. Average fresh green yield is 40.5 t/ha. This hybrid possesses very good transportation quality.

Variety	*Description*
ARCH-55	This hybrid has been developed by Ankur Seeds Co. Nagpur (Maharashtra) Fruits are medium long, attractive and lustrous light green wrinkled. Fruits turn bright red at maturity and highly pungent. Average fresh green yield is 43.5 t/ha. It contain medium amount of capsaicin but high amount of oleoresin. It is excellent for fresh market and dry purpose.
ARCH-55	This hybrid has been developed by Ankur Seeds Co. Nagpur (Maharashtra) Fruits are bold triangular shaped, blunt end and attractive green and medium pungent. Fruits turn dark red at maturity. Average fresh green yield is 42.5 t/ha. It contain medium amount of capsaicin but high amount of oleoresin. It is excellent for fresh market and dry purpose.
ARCH-83	This hybrid has been developed by Ankur Seeds Co. Nagpur (Maharashtra) Fruits are long, lustrous and dark green in colour. Fruits turn deep red at maturity. Average fresh green yield is 40.5 t/ha. It is excellent for red chilli.
Chilli-82	This hybrid has been developed by Ankur Seeds Co. Nagpur (Maharashtra) Fruits are medium long, smooth, lustrous dark green in colour and medium pungency. Average fresh green yield is 42.2 t/ha. It is excellent for fresh market.
NS-101	This hybrid has been developed by Namdhari Seeds Co. Bengaluru (Karnataka) Nagpur (Maharashtra) Fruits are long sized (12.0 x 1.5 cm), dark green, thick skined and moderately pungent. First picking is possible in 70 days after transplanting. Average yield is 40.1 t/ha. It is excellent for fresh market and dry purpose.
NS-1101	This hybrid has been developed by Namdhari Seeds Co. Bengaluru (Karnataka) Fruits are medium long sized (8 x 1 cm), light green, thin skined and highly pungent. First picking is possible in 70 days after transplanting. Fruits turn dark red at maturity. It is excellent for fresh market and dry purpose. Average yield is 41.5 t/ha. It is tolerant to virus and thrips.
NS-170	This hybrid has been developed by Namdhari Seeds Co. Bengaluru (Karnataka) Fruits are medium long sized (8.0 x 0.8 cm), light green, thin skined and highly pungent. First picking is possible in 75 days after transplanting. Fruits turn dark red at maturity. Average yield is 30.0 t/ha. It is excellent for fresh market and dry purpose. Average yield is 40.7 t/ha. It is tolerant to virus and thrips. This hybridis very much suitable for all season cultivation.
NS-555	This early bearer hybrid has been developed by Namdhari Seeds Co. Bengaluru (Karnataka) Fruits are long sized (13.0 x 2.25 cm), light green, medium thick skined and medium pungent. Fruits turn dark red at maturity. Average yield is 37.5 t/ha. It is excellent for fresh market and dry purpose.
NS-999	This early bearer hybrid has been developed by Namdhari Seeds Co. Bengaluru (Karnataka) Fruits are long sized (14 x 2 cm), light green, medium thick skined and medium pungent. Fruits turn dark red at maturity. Average yield is 39 t/ha.
Balwant	This early bearer hybrid has been developed by Namdhari Seeds Co. Bengaluru (Karnataka) Fruits are long sized (13.0 x 1.1 cm), light green, thick skined and medium pungent. Fruits turn dark red at maturity. Average yield is 38.5 t/ha. It is tolerant to viruses.
Angara	This early bearer hybrid has been developed by Namdhari Seeds Co. Bengaluru (Karnataka) Fruits are long sized (14.0 x 1.4 cm), green, thin skined and highly pungent. Fruits turn dark red at maturity. Average yield is 36.5 t/ha.
NS-3074	This hybrid has been developed by Namdhari Seeds Co. Bengaluru (Karnataka) First picking is possible in 70 days after transplanting. Fruits are long sized (13.0 x 1.2 cm), dark green, thin skinned, wrinkled and highly pungent. Fruits turn deep red at maturity. Average yield is 37 t/ha.
Goli	This hybrid has been developed by Namdhari Seeds Co. Bengaluru (Karnataka) First picking is possible in 70 days after transplanting. Fruits are long sized (11.0 x 1.2 cm), straight dark green, thin skinned, glossy and pungent. Fruits turn dark red at maturity. Average yield is 36.8 t/ha.
Akash	This solitary bearer hybrid has been developed by Namdhari Seeds Co. Bengaluru (Karnataka) First picking is possible in 95 days after transplanting. Fruits are long sized (6.0 x 0.8 cm), light green, thin skinned and highly pungent. Fruits turn red at maturity. Average yield is 39.5 t/ha.
BCC-1	This variety has been developed by BCKC, Kalyani (West Bengal) through selection from local germplasm assemblage. Plants having intermediate growth habit, immature fruit colour-green; ripe fruit colour-orange red; fruit shape-conical; fruit position-erect; fruit length 4.25cm; fruit width 1.20 cm; average fruit weight 3 g; and highly pungent. Average green fruit yield is 5 t/ha.
BCCB-Sel-4	This variety has been developed by BCKC, Kalyani (West Bengal) through selection from local germplasm assemblage. Plants having intermediate growth habit, immature fruit colour-green; ripe fruit colour red; fruit shape-long; fruit position-pendent; fruit length 6.20 cm; fruit width 1.20 cm; average fruit weight 1.65 g; and highly pungent. Average green fruit yield is 6.5 t/ha.
Capsicum/Sweet pepper/Bell pepper/Non Pungent Chilli	
Hybrid-1	This variety has been developed in the year 2011 in Gujarat. Average yield is 31 t/ha.
Chilli-112	This variety has been developed in the year 2011 in Gujarat. It is a non pungent chilli. Average yield is 28 t/ha.
Chilli-131	This variety has been developed in the year 2007 in Gujarat. Non Pungent Chilli. Average yield is 25.5 t/ha.
California Wonder	This is an introduction from USA and commercialized by Regional Station, ICAR-IARI, N. Delhi. It is most popular variety of India. Plants are medium tall, vigorous and upright. First picking is possible in 70-80 days after transplanting and it takes 170-195 days for maturity from seed to seed stage (Thakur and Gill, 1985). Generally 3-4 picking can be done. Fruits are blocky, large with 3-4 distinct lobes, dark green and medium thick fleshed. The colours turn crimson at ripe fruit stage. Individual fruit weight is 90-150 g. Average fruit yield is 28.5 t/ha.
California Supreme	This is an introduction. Fruits are dark green, blocky, thick and juicy flesh and 3-4 lobes. Fruits become dark red at maturity. It possess long shelf life and very good for long distance shipping. It is tolerant to tobacco mosaic virus. Average fruit yield is 45 t/ha.
Yolo Wonder	This unique genotypes identified and commercialized by Regional Station, ICAR-IARI, New Delhi. It is a prolific bearer, late maturing variety. Plants are dwarf. Fruits are blocky, drooping, medium flesh thickness, 3-4 lobes, sweet, fine and dark green in colour. Average green yield is 10.0-12.5 t/ha. Fruitys turn bright crimson maturity (Thakur and Gill, 1985).

Variety	Description
Arka Mohini	This variety has been developed at ICAR-IIHR, Bengaluru (Karnataka) by selection from exotic material introduced from the USA. Plants are dwarf with dark green foliage. Fruits are thick fleshed, 3-4 lobed, dark green, blocky and turn red on ripening. Average fruit weight is 80-100 g and yield is 20.0 t/ha. Suitable for both season of cultivation.
Arka Gaurav	This variety has been developed at ICAR-IIHR, Bengaluru (Karnataka) by selection from exotic material 'Golden Calwonder' from USA. Plants are tall with green foliage and erect fruit bearing habit. Fruits are dark green, thick fleshed, 3-4 lobed, green, blocky, and erect and turn orange-yellow on ripening. Average fruit weight is 70-80 g and yield is 23 t/ha. Suitable for both season of cultivation. It is tolerant to bacterial wilt diseases.
Arka Basant	This variety has been developed at ICAR-IIHR, Bengaluru (Karnataka) by selection from exotic material introduced from the Hungarian variety 'Soroksari'. Plants are tall with yellowish green foliage and have erect, indeterminate fruit bearing habit. It is prolific and continuous bearing in nature. Fruits are thick fleshed, blocky, erect 2-3 lobed, conical and creamy white at maturity stage and generally turn orange-red on maturity. Average fruit weight is 50-80 g and yield is 21 t/ha. Excellent keeping and cooking quality. Possess good aroma, crisp texture and good for stuffing. Suitable for both seasons growing. Fruits are export quality (Nath *et al.,* 1987).
Bharat	This hybrid has been developed by Indo-American Hybrid Seeds Co., Bengaluru (Karnataka). First picking is possible in 70 days after transplanting. Fruits are square shaped each side measuring 10-12 cm, dark green, 4-lobed tasty and thick fleshed. Average fruit weight is 150 g and yield is 35 t/ha. It is resistant to tobacco mosaic virus.
Bull Nose	The plants of this variety are tall erect and sturdy. Fruits are 8-10 cm long, dark green, thick walled, 3 lobes and flesh is thick. The yield potential of this variety is 15.0-20.0 t/ha.
Chinese Giant	It is an introduction from USA. Its plants are vigorous and prolific bearer; fruits are smooth, 3-4 lobes, medium flesh thickness with dark green skin colour (Thomas *et al.,* 1983). Average green yield is 12.0-15.0 t/ha in 125 to 150 days after transplanting.
Nishat-1	This variety is released from SKAUST (K) for commercial cultivation. Its plants are indeterminate, medium with dark green foliage. Fruits are dark green, 3-4 lobed and bell shaped. Average green yield is 20.0-25.0 t/ha.
World Beater	It is an introduction. Plant upright, fruit 4 lobed with medium thick flesh, mild and sweet. Average green yield is 12.5 t/ha.
IAHS-P-2	This is an early compact hybrid has been developed by Indo-American hybrid Seed Co., Bengaluru (Karnataka). Fruits are thick walled, tri-lobed, dark green to light green in colour, and slightly elongated. First harvesting is possible after 65 days of transplanting. Average fruit weight is nearly 140 g. Average green fruit yield is 20 t/ha. It is relatively heat tolerant.
Indira	This hybrid has been developed by Indo-American Hybrid Seeds Co., Bengaluru (Karnataka). Fruits are medium tall and vigorous with dark green foliage. Fruits are 3-4 lobed and glossy dark green with thick pericarp. First picking is possible 60-65 days after planting. Average fruit weight is 170 g and yield is 33.0 t/ha.
NS-436	This is an early bearer hybrid has been developed by Namdhari Seeds Co. Bengaluru (Karnataka). Fruits are dark green, very firm, attractive, 4-lobed, blocky and having good shape and size. Average green fruit yield is 30 t/ha.
Sindhoori	This hybrid has been developed by Namdhari Seeds Co. Bengaluru (Karnataka). Fruits are lustrous green, blocky, firm, smooth and having 8 x 8 cm size. Fruits turn yellow on maturity. Average green fruit yield is 32 t/ha.
Ratan	This hybrid has been developed by Namdhari Seeds Co. Bengaluru (Karnataka). Fruits are attractive, shiny dark green, blocky, lustrous green, firm, smooth and having 8 x 8 cm size. Fruits turn red on maturity. Average green fruit yield is 31 t/ha.
Arun	This is a prolific bearer hybrid has been developed by Namdhari Seeds Co. Bengaluru (Karnataka). Fruits are attractive, dark green, blocky and having 8 x 8 cm size. Fruits turn red on maturity. Average green fruit yield is 31.5 t/ha.
Gagan	This hybrid has been developed by Namdhari Seeds Co. Bengaluru (Karnataka). Fruits are beautiful light green, medium firm, thin walled blocky and having 8 x 8 cm siz. Fruits turn red on maturity. Average green fruit yield is 37 t/ha.
Vikas	This hybrid has been developed by Golden Seeds Co. Azadpur (Delhi) Fruits are large smooth, uniform, blocky, medium green thick wall, firm and four lobed. Individual fruit weight is 150-200 g. It is very much suitable for outdoor cultivation. Average green fruit yield is 32.7 t/ha.
Anupam	This is an early (55 days) hybrid has been developed by Nunhem Seeds Co. Fruits are bright green, smooth, thick wall and 3-4 lobed. Individual fruit weight is 75-80 g. It is very much suitable for outdoor cultivation. Average green fruit yield is 25.5 t/ha.
Alankar	This hybrid has been developed by Nunhem Seeds Co. Fruits are bell shaped, blocky and possessing excellent firmness. Average green fruit yield is 27 t/ha. It has been found to PVYs and Tomato Mosaic Virus.
Harit	This hybrid has been developed by Nunhem Seeds Co. Fruits are firm, green, blocky, smooth and thick wall. It is very much suitable for processing. Average green fruit yield is 26.5 t/ha
Nutan	This hybrid has been developed by Nunhem Seeds Co. Fruits are lobed, firm, green, and smooth surface. It is very much suitable for green house cultivation. Fruits turn yellow on maturity. Average green fruit yield is 26.5 t/ha
Tanvi	This hybrid has been developed by Unicorn Seeds Co. Fruits are dark green, elongated, blocky, long pendency and thick walled. Individual fruit weight is 200-250 g. Average green fruit yield is 21 t/ha
Tanvi Plus	This hybrid has been developed by Unicorn Seeds Co. Fruits are dark green, elongated, blocky, long pendency and thick walled. Individual fruit weight is 200-250 g. Average green fruit yield is 21.5 t/ha
BSS-89	This hybrid has been developed by Bejo Sheetal Seeds Co. First picking can be done in 65 days after transplanting. Fruits are dark green, blocky and 4-lobed. Individual fruit weight is 100 g. Average green fruit yield is 21.5 t/ha. It is suitable for all season growing.
BSS-399	This early type hybrid has been developed by Bejo Sheetal Seeds Co. Jalna (Maharashtra) First picking can be done in 55 days after transplanting. Fruits are green, blocky, thick skin and 4-lobed. Individual fruit weight is 100 g. Average green fruit yield is 24.5 t/ha. It is suitable for all season growing. It possess long distance transportation quality.

Variety	Description
BSS-401	This hybrid has been developed by Bejo Sheetal Seeds Co. Jalna (Maharashtra) Plants are very vigorous. Fruits are green and 4-lobed. Individual fruit weight is 120 g. Average green fruit yield is 24 t/ha.
Green Gold	This hybrid has been developed by MAHYCO Seeds Co. (Maharashtra). First picking can be done in 60-65 days after transplanting. Fruits are dark green, blocky, medium in size, thick skin, pendent and 3-4 lobed. Average green fruit yield is 25 t/ha.
Sun Gold	This hybrid has been developed by MAHYCO Seeds Co. (Maharashtra). First picking can be done in 65-70 days after transplanting. Fruits are long (10-12 cm), shiny dark green, blocky, medium in size, thick skin and 3-4 lobed. Average green fruit yield is 24 t/ha.
Bombay	This is an early bearer hybrid has been developed by Novartis India Seeds Co. Fruits are dark green, blocky, thick skin, glossy and 3-4 lobed. Individual fruit weight is 140 g. Average green fruit yield is 70 t/ha. It possess longer shelf life and ideal for distance transportation. It is tolerant to tobacco mosaic virus (TMV) and PVY.
Portos	This hybrid has been developed by Novartis India Seeds Co. Fruits are blocky and 4-lobed. It possess excellent long distance transportability. It is tolerant to tobacco mosaic virus (TMV) and PMVY. Average green fruit yield is 60 t/ha.
Tara	This hybrid has been developed by Novartis India Seeds Co. Fruits are medium large, dark green, thick wall and 4-lobed. Fruits turn to golden yellow on maturity. Individual fruit weight is 250 g. Average green fruit yield is 38.5 t/ha.
Zenith	This hybrid has been developed by Tropica (Green Co.). Fruits are dark green, 3-4 lobes, thick flesh and turn dark red at full maturity. Fruit size is 10 x 8 cm. It is very much suitable for export. Average green fruit yield is 37.5 t/ha.
King chilli	
Dorset Naga	Dorset Naga cultivar was developed as a hybrid of 'Naga Morich' (another name of Naga Jolokia) by a couple in Dorset, England. The cultivar Dorset Naga pepper (cultivar status in process) is grown in west Belington, Dorset, England (Anonymous, 2006b, Anonymous, 2006c).
Paprika	
Arka Abheer	This is a very good paprika variety. Plants are tall and spreading types producing light green wrinkled fruits which are on an average 10.0 cm long. It is resistant to powdery mildew and viruses. Yield potential is 38.0 t/ha (green chillies) and 5.0 t/ha (dry chilli).
Garden pea	
Kashi Shakti (VRP-7)	This variety is a member of medium maturity group, developed by ICAR-IIVR, Varanasi (Uttar Pradesh) derived through pedigree selection from the cross Harabona x NDVP-8. Plants are 80.0 cm long with dark green foliage, good attractive and well filled pods. Average yield is 7.5 t/ha.
Kashi Udai	An early maturing variety, and its plants grow 60 cm high. Its sowing can be done from October end to 25 November in plains. Plants with dark green foliage have short internodes and 8-10 pods/plant. Pods are 9-10 cm long, attractive with 8-9 bold ovules/pod with shelling quality of 48 per cent. The pods are picked up 60-65 days after sowing. Total green pod yield being 10-11t/ha.
Kashi Mukti	An early maturing variety and resistant to powdery mildew. Last week of October to November is considered ideal sowing time in plain regions of Uttar Pradesh. Plant height is 50-53 cm. The plants are slender with shorter leaves. Pods are 8.5-9.0 cm long, attractive filled with 8-9 bold and soft textured seeds with 48-49 per cent shelling and yield of 11-12 t/ha.
P-89 (PC-531)	It is an introduction from New Zealand and release for commercial cultivation by PAU, Ludhiana (Punjab) in 2007. It is main season variety (taking 150-160 days for maturity but first picking can be done in 85-90 days) and plants are medium tall, vigorous with 28-30 pods/plant. Pods are dark green, long, very attractive having 9-10 grains/pod. Pods are born in doubles. Shelling out turn is more than 55 percent. The green pod yield is 15.0 t/ha.
Punjab-87 (87-1)	It has been developed from the cross between Pusa-2 x Morrasis-55 and release for commercial cultivation by PAU, Ludhiana (Punjab) in 1978. Its plants are medium tall and dark green. First picking is possible in 100 days after sowing. Pods are dark green, 9-10 cm long and slightly curved at the end. The grains are bold, sweet and suitable for dehydration. Average yield is 13.8 t/ha and shelling out turn is 50 per cent.
Matar Ageta-7	It has been developed by PAU, Ludhiana (Punjab) and release for commercial cultivation in 2015. Its plants are dwarf and gives first picking within 60 days of sowing. Its pods are of medium in length, slightly curved from tip and are born singly or in doubles. First picking of pods is possible after 60 days of sowing. Each pod contains 7-8 seeds and shelling out turn is 48 percent. Dry seed colour is green with slight wrinkled. It gives more than 50 per cent yield in first picking and total yield is 9.0 t/ha.
Ageta	Plants are small, erect and green in colour. It is suitable for early sowing. First flower appears in about 25 days after seed sowing and it takes about 6 weeks for first picking. Two pickings are done. Average green pod yield is 5.0-5.5 t/ha.
T-19	It is double podded cultivar. Plants are medium tall in height and flowers in 55-60 days. Pods are yellowish green, slightly curved, 8.5 cm long and 5-7 seeded when fully developed.
NP-29	Developed at ICAR-IARI, New Delhi. It is double podded cultivar. Plants are medium long and 6-7 seeded when pod fully developed. Fully developed seeds are suitable for canning purpose.
Alaska	It is an early cultivar; flowers in 38 days, pods are born singly, light green in colour, 7 x 1.25 cm, and contain 5-6 small bluish green seeds. Shelling percentage is 42 (Mac Gillivray, 1961). Fully developed seeds are suitable for canning purpose.
Asauji	Selection from material collected from Amritsar. This smooth seeded is suitable for early sowing. Plants are dwarf and flower in 30-35 days. The pods are produced singly about 8.0 cm long, curved, dark green, narrow and appear around fully developed pods are 7 seeded. Average yield is 18 t/ha.

Variety	Description
Azad P-1	Developed from the cross between 6416 x 6405 by Vegetable Research Station, CSAUA&T, Kanpur (Uttar Pradesh). Plants are early, erect medium tall, dark green in colour. This is tall variety and plant height goes upto 106.0 cm. First flower appears in about 40 days after seed sowing. First picking can be obtained in 70 days after sowing, and 3-4 picking are done. Pods are large thick, well filled and attractive. Seeds become wrinkled when mature. Average yield is 22.0 t/ha. This variety escape to powdery mildew and pea rust.
Azad P-3	Developed by Vegetable Research Station, CSAUA&T, Kanpur (Uttar Pradesh). Pods are medium in size, thick, straight with green bold seeds, and gives an average yield of 8.0 t/ha.
Harbhajan (EC33866)	It is an introduction and suitable for all dry areas of north India. Susceptible to powdery mildew. Plants are dwarf. Pods are green and long. Seeds are small (13 g/100 seeds), round, wrinkled and yellow in colour. Flowering starts from 6th nodes. It is susceptible to powdery mildew.
Hara Bona	Pure line selection from the local field pea variety and release in 1980 from Punjab. Plant height is 40 cm and having dark green leaves. Flowering starts 21 days after sowing. First picking is 45-50 days after seeding. Pods are 7.8 cm long, dark green and 5-6 seeds per pod with 40 per cent of shelling. Green seeds are less sweet. Mature seeds are bold, round, dimpled from both sides and green in colour. Seed yield is 2 t/ha. This variety escape powdery mildew and pea rust.
Jawahar Matar-1	Developed from the cross between T-19 x Greater Progress. Wrinkled seeded. Its pod yield potential is 15 t/ha.
Jawahar Matar-2 (JM-2 or GC-477)	This variety has been developed from JNKVV, Jabalpur (Madhya Pradesh) through the cross between Russian-2 x Greater Progress. The plants are 60 cm tall. It has bigger pods having 9 green bigger sweet ovules. The green pod yield is around 10 t/ha. Shells are comparatively thicher therefore having better keeping quality and suitable for transportation. The matured seeds are green, and wrinkled with 100 seed weight of 17 g.
Jawahar Peas-4 (JP-4 and now JM-6)	Developed from the cross between Local Yellow Batri x (6588 x 46C) at JNKVV, Jabalpur (Madhya Pradesh). Suitable to hilly areas and can be planted in August (Tiwari and Narsinghani, 1986). The variety possesses medium size pods with 5-6 bigger size green ovules. It yields around 9 t/ha. The seeds are yellow and medium sizes (100 seed weight is 15 g).
Jawahar Matar-5 (JM-5)	It is released in 1980 from JNKVV, Jabalpur, Madhya Pradesh. Plants are 1.5-2.0 m in length with large internodes, greenish yellow foliage, two pods/axil and 5-6 seeds per pod. Fruiting begins from 13th node. Pods are comparatively thin walled. Mature seeds are yellow and wrinkled. Seed index varies from 10-20 g/100 seeds. Seed yield is 11 t/ha. This variety is immune to powdery mildew.
Jawahar Pea-15	It has been developed by selection from a three way cross (JMI x R-98 B) x JP-501A/C. Its plants are dwarf (50.0 cm), pods are large each containing about 8 grains. Average yield is 13.0t/ha. It is resistant to Fusarium wilt and powdery mildew.
Jawahar Peas-71 (JP-71)	Developed from the double cross between (Arkel x JP-829) x (JP-501 x JM-1) at JNKVV, Jabalpur, Madhya Pradesh. Powdery mildew resistance and belongs to mid season groups. The plants are 50 cm tall and pods are medium sized (7 cm) with 6-7 ovules. The seeds are green, wrinkled and bigger (100 seed weight 17 g). The yield of green pod is 12 t/ha.
Little Marvel	An introduction from England (cross of Chelsea Gem x Suttons Alaska. Plants are dwarf with dark green foliage. It flowers in 40 days and first blossom appears at 9-10th node. Pods are about 8 cm long, born singly thick-skinned dark green, straight and broad containing 5-6 seeds. Seeds are very sweet and fine in quality (Singh and Joshi, 1970). It gives shelling percentage of 40.
Madhu	Developed from the cross between 6126 x Sylvia. Susceptible to powdery mildew. Average green pod yield is 18 t/ha.
Meteor	An introduction from England. It is an early maturing variety. It is a round smooth seeded early variety and introduced from England. Plants are 35-40 cm tall, dark green and flowers born generally singly. Pods are dark green, 8.7 cm long, well filled with 7 seeds, and having percentage 45 shelling. Pods mature in 58-60 days and suitable for early October sowing (Choudhury and Ramphal, 1975).
Lucknow Boniya	It is early smooth-seeded cultivar which is very much suitable for the plains (Singh and Joshi, 1970). It is dwarf white seeded cultivar flowers in 40 days and first blossom appears at 8-9th node. The pod born singly, small, narrow, green and 4-5 seeded.
Madhu	It is of the cross 6126 x Sylvia, and released in 1973 in Uttar Pradesh. Plant height is 130-150 cm, foliage yellowish green. Days to flowering 50-55 days after seeding. Pods are 12-15 cm long, light green, smooth, membrane less, straight, 2-2.5 cm broad, 20-25 pods per plant and 5-6 seeds per pod. Mature seed smooth, round and of grey colour. This variety is susceptible to powdery mildew. Seed yield is 0.8-1.0 t/ha.
Vivek-6	Developed from the cross between Pant Upahar x VL Matar-3 and released by the Central Variety Release Committee in 1996. Tolerant to cold and moisture stress conditions. It is a medium maturing variety takes 135-140 days and 80-85 days to maturity in mid hills and plains, respectively. The plant are dwarf and vigorous with dark green foliage. The pods are smooth, straight, medium sized (6-7 cm) and completely filled. The yield is up to 12.38 t/ha green pod depending stress conditions. It is recommended for cultivation in Uttarakhand, Himachal Pradesh, Jammu and Kasmir, Haryana, Delhi and Rajastahn (Sri Dhar *et al.*, 1999).
Vivek Matar-10	Plants are quite distinct in both growth and yield attributes. Its plant height is 50-60 cm, branched 3-4, first flower appear at 7-8 node, green foliage, leaf normal, two white flowers/peduncle, dark green slightly curved pod, mostly double pods/peduncle, 13-15 pods/plant, 143 pods/kg, 8-9 cm pod length, 45 per cent shelling and greenish white wrinkled seeds. It shows 50 per cent flowering in 80-85 days in hills and 60-65 days in plains. It is recommended for sowing from first week of November to last week of November in midhills as well as in plains. The average green pod yield is 9-11 t/ha which mature in 120-130 days. It is resistant to white rot, wilt, leaf blight and moderately resistant to powdery mildew under field conditions and tolerant to pod-borer.
Arka Priya (IIHR-1)	It has been developed ICAR-IIHR Bengaluru (Karnataka) and released for commercial cultivation in 2012. It is mid-season variety. Pods mature in 65 days. Pods medium long (8 cm). Seeds are round, medium bold dark green and very sweet. Pod yield 12.0 t/ha in 90 days. Resistant to powdery mildew and rust.

Variety	Description
Arka Pramod (IIHR-18)	It has been developed ICAR-IIHR Bengaluru (Karnataka) and released for commercial cultivation in 2012. It is mid-season variety. Pods mature in 65 days. Pods are medium long (8.0 cm) and slightly flatish round. Seeds are bold, round, dark green very sweet. Pod yield is 12.0 t/ha in 90 days. Resistant to powdery mildew and rust.
Arka Ajit	It has been developed ICAR-IIHR Bengaluru (Karnataka) from the multiple cross involving Bonneville, IIHR-209 and Freezer 656. Pods are 8.0-9.0 cm long. Seeds are bold, sweet and shelling out turn is 55 percent. It takes about 90 days from sowing to first picking. Average yield is 10.0 t/ha. It is resistant to rust and powdery mildew.
Arka Karthik	It has been developed ICAR-IIHR Bengaluru (Karnataka) by selection from the cross Arka Ajit x IIHR-554. Plants are bushy and erect. Pods are 11.0-12.0 cm long with 8.0-10.0 sweet grains. It takes about 90 days from sowing to first picking. Average yield is 8.0t/ha. It is resistant to rust and powdery mildew.
Arka Sampoorna (IIHR53-6)	It has been developed ICAR-IIHR Bengaluru (Karnataka) by selection (in F_9 generation) from the multiple cross involving Bonneville, IIHR-209, Freeza-656 and Manao Sugar. Plants are bushy and erect. Pods are flat, sweet and crisp with the parchment layer. It attains marketable maturity in 80 days after sowing. Average yield is 8.0 t/ha.
Phule Priya	This variety has been developed from MPKV, Rahuri (M.S.) for commercial production in 2003 and suitable for *rabi* season sowing. Pods are straight and green in colour. It is a mid season variety (100-110 days duration). It is moderately resistant to powdery mildew under field condition. The average yield is 9.0-10.0 t/ha.
Sel.82	This variety has been developed from MPKV, Rahuri (M.S.) for commercial production in 1987and suitable for *rabi* season sowing. It is a mid season variety (105 days duration). Pods are straight, green in colour, high TSS and grains are bold. Average yield is 6.0-7.0 t/ha.
Hisar Harit	This variety has been developed by CCS HAU, Hisar (Haryana) through selection from the cross Bonneville x P-23. It takes relatively longer period for marketable maturity than typical early types. The plants are semi-dwarf and first picking is possible in 60 days after sowing. The pods are well-filled, sickle shaped, large and green. The seeds are green at maturity and show dimples on surface. Average yield is 9.0 t/ha.
Swarna Mukti	This variety has been developed by ICAR Research Complex for Eastern Region, Research Centre, Plandu, Ranchi (Jharkhand).
Swarna Amar	This variety has been developed by ICAR Research Complex for Eastern Region, Research Centre, Plandu, Ranchi (Jharkhand).
VL-7 (Ageti Matar-7)	This is an early season variety, developed by ICAR-VPKAS, Almora (Uttarakhand). Plants are determinate, dwarf with dark green foliage and white flowers. Pods are light green, slightly curved towards suture at distal end, medium, completely filled two pods in a bunch, and first picking in 120-125 days after sowing. Pods contain 6-7 seeds, are light green and unwrinkled, less incidence of *Alternaria* blight than Arkel. Physiological maturity completed in 140-150 days. It gives 23.0-25.0 t/ha pod yield with 42 per cent shelling.
Pant Sabji Matar-3	This is an early season variety developed at GBPUA&T, Pantnagar (Uttrakhand) derived from the cross 'Arkel' x 'GC-141' following pedigree selection. Plants are dwarf with dark green foliage. Pods are long, curvd, well filled and each pod contains 8-9 ovules. First picking starts 60-65 days after sowing. It gives 9.0 t/ha yield.
Ooty-1	This variety has been developed by Tamil Nadu Agricultural University, Tamil Nadu. The plants of this variety grow to a mean height of 50.1 cm. Average green pod yield potential can be ealized 12.1 t/ha in 90 days, which is 47 per cent higher than the ruling variety Bonnevile in the Nilgiris. The pods are dark green, long and smooth with a mean number of nine seeds, which possess good cooking quality. It is resistant to white fly attack and moderately resistant to pod borer and aphids. It is moderately resistant to powdery mildew disease. It is suitable for all the three seasons namely main (April-May), autmn (July-August) and irrigated (February-March).
NDVP-8	This mid season variety has been developed at NDUA&T, Faizabad (U.P.). It gives an average yield of 10 t/ha. It is recommended for release and cultivation in 1997 for Punjab, Uttar Pradesh and Bihar.
NDVP-10	This mid season variety has been developed at NDUA&T, Faizabad (U.P.). It gives an average yield of 11.7 t/ha. It is recommended for release and cultivation in 1998 for Punjab, Uttar Pradesh and Bihar.
Knwari	It is a smooth-seeded main season variety. This is a tall-growing double-podded cultivar. Flowering takes place 65-70 days and first blossoms appear at 15-17th node. Pods are about 8.5 cm long, yellowish green and 5-6 seeded with 40 per cent shelling.
Khapar Kheda	It is a tall growing double podded cultivar. Flowering takes place in 65-70 days and first blossom appears at 15-16th node. Pods are 5.5-6.0 cm long and 4-5 seeded with 50 per cent shelling. It is very popular in Madhya Pradesh.
Perfection New Line	It is a mid season and high yielding variety. The plant is medium-tall. Pod is about 8 cm, long dark green, sweet and well filled with wrinkled seeds. It is ready for first picking in 80 to 85 days (Nath, 1976).
Early Badger	It is a dwarf early wrinkled-seeded cultivar evolved at Wisconsin University (U.S.A.). It is suitable for growing in early October. The plant flower in 40-45 days and first blossom appears at 10-11th node and ready for harvesting in 60-65 days. The yellowish green pods are born singly, 7.5 cm long, well filled, 5-6 seeded sweet. It is a good canning cultivar having a shelling percentage 36. It is resistant to Fusarium wilt and tolerant to heat and drought (Choudhury, 1967).
Early Giant	It is a late maturing variety suited to hills of Himachal Pradesh. Plants are tall and therefore needs staking. Pods are 9-10 cm long, dark green with 9-10 bold grains. Seeds are wrinkled. Average yield is 14-15 t/ha.
Early Supperb	It is an introduction which is dwarf in nature but foliage is yellowish green. Plants are branched from base (Singh and Joshi, 1970). It is an early smooth seeded variety. It flowers in about 45 days and first blossom appears at 8-10th node. The pods are born singly, dark green and curved with 6-7 seeds. The shelling percentage is 40. Average yield is 13 t/ha.
Swarna Rekha	It is an early maturing selection from local material of Bihar. It matures in about 120 days. Plants are semi-spreading type with white flowers. Seeds are round, smooth and creamy in colour with black hilum. Average yield is 12-14 t/ha.

Variety	Description
Palam Priya	It is a medium maturity (110 days), medium tall (60-67 cm), and powdery mildew resistant garden pea. Its plants are vigorous with dark green foliage and branching habit. The pods are born in double. These are smooth, straight, light green, well filled with average length of 8-9 cm. The average seeds/pod are 7-8 with about 60 per cent shelling. Its green pods contain 15-16 per cent TSS. The seeds become wrinkled upon maturity with light bluish green colour. On an average it gives 4-5 green pod pickings in hills with and average pod yield of 12.5-15.0 t/ha.
Alderman	This is an excellent cultivar for home garden, shipment and freezing. Plants are dark green and 150 cm high. Pods are born singly, 11.25 cm long with 8-10 seeds and excellent quality. This cultivar is suitable for hills (Yawalkar, 1969).
Kelvedon Wonder	It is having dwarf plant. Flowering takes place 40 days after sowing and first blossom appears at 8-9th node. Pods are curved, born singly, green about 9 cm long and 6 seeded. The shelling percentage is about 40.
Thomas Laxton	It is a general-purpose cultivar with deep green foliage. Pods are of excellent quality, broad, and blunt, 8.25 cm long having cream and green coloured seed and suitable for freezing (Mac Gillivray, 1961).
Snow pea	
Swarna Tripti	This variety has been developed by ICAR Research Complex for Eastern Region, Research Centre, Plandu, Ranchi (Jharkhand) in 2008. The variety is found suitable for winter cultivation in plains and also as an off-season crop in Netarhat hills (>1,000 above mean sea level). This variety becomes ready for harvest in 80-85 days after sowing, and has high export potential. Its pods are sweet, flat and are green at harvest stage. Average yield is 14-16 t/ha.
Oregon Sugar Pod	This is an introduction from USA and very popular for commercial cultivation. The plants are tall (140.0 cm); pods are light green, broad, flat and 12.0-13.0 cm long. First picking is possible in 90 days after sowing. It has pod yield potential of 18.5 t/ha.
Oregon Sugar Pod-II	This is an introduction from USA. The plants are bushy (about 75-90 cm tall). The pods are 10.0-12.0 cm long, tender, crisp and suitable for-frying. It takes about 60 days from sowing to first picking.
Mithi Phali	It is an edible podded variety of main season developed by PAU, Ludhiana (Punjab) and was released for commercial cultivation in 1994. Its plants are tall, vigorous and take 90 days for first picking. Pods are light green and 12-13 cm in length. Shelling is not required and whole pod is edible. Its consumable yield is twice that of standard shelled pea varieties. It is rich in protein and total sugar. It has pod yield potential of 11.7 t/ha.
Arka Sampoorna	It was released from ICAR-IIHR, Bangaluru (Karnataka) in the year 2001. The plants are byshy and erect, pods are flat, parchment less, crisp and sweet. Yield potential of edible pod is around 8t/ha with crop duration of 80 days. The variety is resistanr to rust and powdery mildew.
Arka Apoorva	It was released from IIHR, Bangaluru in (Karnataka) the year 2013. It is pedigree selection from a cross between whole pod edible germplasm accession 'Oregon Sugar Pod' with 'IIHR-18' (dark green, bold and sweet seeds and resistant to powdery mildew and rust). The pods and seeds are light green and medium sweet. It gives about 20 per cent higher pod yield than Arka Sampoorna with pod yield potential of 10 to 11 t/ha in 90 days. The pods are of dual purpose. It is moderately resistant to rust and powdery mildew.
UN-53	It was released from ICAR-IIHR, Bangaluru (Karnataka) which gives 8-9 t/ha pod yield in crop duration of 90 days.
Mithi Phali	This variety is released by PAU, Ludhiana (Punjab). The plants are tall (140 cm) in height. The first picking can be taken in 90 days with subsequent picking at 15 days interval. Pods are light green with 12-13 cm pod length. It is rich in protein and total sugar. The average yield is 11.7 t/ha.
Other varities	Other popular varities of snow pea in European countries are 'Dwarf White Sugar', 'Mammoth Melting Sugar', 'Nofilla', 'Oregon Giant', 'Oregon Sugar Pod', 'Short' n Sweet', 'Snowflake', Snowman', and 'Sugar Daddy'.
Snap pea	
Sugar Snap	This is an introduction from USA. The plants are bushy (about 120.0 cm tall). The pods are 7.0-8.0 cm long, round, crisp and delicious. It takes about 70-75 days from sowing to first picking.
Sylavia	It is a variety from Sweden introduced by IARI, New Delhi. It is a tall growing cultivar, flowering in 60 days and first blossom appears at 14.0-16.0 cm node. Pods are born singly, yellowish, 12.0 cm long, sickle shaped without parchment and sweet. Pods have general appearance of a medium sized French bean pod. Staking is desirable and is suitable for late sowing.
French bean (Bush type)	
Contender	It is an introduction from U.S.A and release for commercial cultivation by ICAR-IARI, New Delhi. Plants are bushy in nature and bear pink flowers. Its pods are green, round, stringless, 12.0-18.0 cm long, meaty and slightly curved at tips. First picking is possible in 50-55 days after sowing. Average green pod yield is 6.0-9.5 t/ha. It is tolerant to powdery mildew and mosaic but susceptible to BCMV. Seeds are light brown.
Kashi Sampann (VRFBB-1)	It is developed at ICAR-IIVR, Varanasi (U.P.). The pods are light green in colour, stringless, fleshy and round in cross section. A plant bears 25-30 pods, and tolerant to GYMV. It has yield potential of 12-14 t/ha. Average green pod yield is 12.0-14 t/ha.
Kashi Rajhans (VRFBB-2)	It is developed at ICAR-IIVR, Varanasi (U.P.). The pods are green in colour, thin, stringless, fleshy and round in cross section. A plant bears 30-35 pods and tolerant to high temperature (25-30 °C). The seeds are small and white in colour.
VRFBB-91	It is developed at ICAR-IIVR, Varanasi (U.P.). The pods are light green in colour, thin, straight, stringless, fleshy and round in cross section. It is early genotype that bears 20-25 pods/plant during last week of November to mid-December. The seeds are medium in size and black in colour. It has green pod which yield potential is 10-11 t/ha.
Premier	This variety is suitable in the plains of Tamil Nadu. Pods light green in colour, stringless and straight. The seeds are black in colour and oval in shape. Average yield 7.0-10.0 t/ha. This is less susceptible to wilt and mosaic.

Variety	Description
Pusa Parvati	It is an early maturing variety developed by X-ray irradiation of 'Wax Pod' (an introduction from the U.S.A.) developed at ICAR-IARI Regional Station, Katrain (H.P.). Plants are early, (taking about 50 days to first green pod picking) and bushy in nature. The pods are green attractive, stringless, curved, flattish round, and 18.0-20.0 cm long. First picking is possible in 45 days after sowing. Pod yield potential is 9t/ha. It is susceptible to viral diseases.
Arka Anoop	It has been developed by selection from the cross 'Arka Bold' x 'Arka Komal'. This is a very good variety developed by ICAR-IIHR Bengaluru (Karnataka) and released for commercial cultivation in 2006. Plant is bushy and photo-insensitive in nature. Pods are long (17-18 cm), flat, smooth, crispy and straight with less parchment. Green pod yield is around 20.0 t/ha in about 70 days of crop duration. It is resistance to both rust and bacterial blight.
Arka Sarath	This is a very good variety developed by ICAR-IIHR, Bengaluru (Karnataka) and released for commercial cultivation in 2011. Plants are bushy and photo insensitive and it is suitable for both *Kharif* and *rabi*. Pods are perfectly round cross-section, stringless, smooth, crisp, fleshy with no parchment. It has green pod yield potential of 18.5 t/ha in 70 days.
Arka Bold	This is a very good variety developed by ICAR-IIHR, Bengaluru (Karnataka). It is photo-insensitive and resistant to rust. Pods are flat, stringless, fleshy, crisp and medium long. It has green pod yield potential of 15 t/ha in 70 days.
Phule Suyash	This variety has been developed from MPKV, Rahuri (M.S.) in the year 1997 and release for commercial cultivation in 2003. It is suitable for cultivation in *kharif*, spring and summer seasons. The pods are attractive, tender, and straight to slightly curved and smooth green. It has green pod yield potential of 16.8 t/ha.
Swarna Priya	This bush variety has been developed by ICAR Research Complex for Eastern Region, Research Centre, Plandu, Ranchi (Jharkhand). Average yield is about 15.5 t/ha.
Ooty-1 (FB-1)	This variety has been developed from Horticultural Research Station, Ooty, Tamil Nadu Agricultural University, Coimbatore (Tamil Nadu) which is suitable for high, middle and lower elevations of the Nilgiris and other similar hilly tracts. It produces light green attractive pods, which are highly fleshy and totally stringless. The pods have excellent cooking quality and it has been found to have good market preference. The plants have a vigorous growth habit and grow to a height of 60 cm. It has a yield potential of 14.3 t/ha in a duration of 90 days. It has recorded a 34.02 per cent higher yield than Arka Komal, a popular variety ruling in the area, and 50 per cent more than the local types.
Pant Anupama (UPF-191)	This variety developed at GBPUAT, Pantnagar (Uttarakhand). The plants are dwarf and dark green foliage with concentrated fruiting. It is prolific bearer, and pods are smooth, tender, non-stringy, medium long, round in cross section, straight, translucent and green in colour. The pods bear below the canopy, thus remain protected from sunlight. Pods are so tender and fibreless that these could be eaten as salad too. The seeds are dark brown. First picking is done in 55-65 days. The yield potential is 9 t/ha. It is moderately resistant to BCMV and resistant to rust.
Pant Bean-2 (UPF-626)	This variety developed at GBPUAT, Pantnagar (Uttarakhand) through pedigree method of a cross between Turkish Brown x Contender. The pods are flattish round, straight and non-stringy at edible stage. The seeds are also suitable as dry seed type (Rajmash). First green pod picking is done in about 60 days after sowing. Green pod yield potential is 9 t/ha. It is moderately resistant to BCMV and resistant to rust.
Bountiful	This variety is high yielding, suitable to grow from September to February in south India. On main stem, pods form in cluster. Pods are tender green, available for harvesting within a week of flowering. The pods remain tender for long period even when left unharvested (Shanmugavelu, 1989).
Jampa	This is Mexican variety, suitable to grow in Maharashtra. This is very early and pods are available for first picking 50 days after sowing. The pods are flat, smooth, non-stringy and pale green in colour. The seeds are black, smooth, small in size and flat in shape. It is resistant to wilt and can be successfully in warmer climate.
Giant Stringless	This variety was developed at ICAR-IARI, New Delhi. Variety is early; pods are green in colour, medium size, slightly curved and stringless. The seeds are glossy and yellowish brown. The yield potential is very high 13 t/ha.
Sel.5	This variety is developed at ICAR-IIHR Bengaluru (Karnataka). The pods are round stringless. It is suitable to grow in kharif season. Average yield is about 12.0 t/ha.
VL-Boni-1	This variety has been developed at ICAR-VPKAS, Almora (Uttarakhand). Plants are dwarf (30-40 cm) having white flower with purple tinged. Pods are found, light green, attractive, stringless and fleshy. First picking starts 45-60 days after seeding. It is moderately susceptible to foliar diseases. The yield potential of green pod is 10-12 t/ha.Suitable for multiple cropping and can be planted in the hills from March to July.
YED-1	It can give a green pod yield of 9.75 t/ha. Seeds are bold, attractive and dark-purple in colour.
Other variety	Swarna Priya, Black Prince, VL Bean-2, VL Bean-3, Aparna, IVFB-1, IVFB-2, IVRFB-1, VL-5, Sel-2 and Sel-4.
French bean (Pole type)	
Kentuky Wonder	It is an introduction from U.S.A and release for commercial cultivation by ICAR-IARI, New Delhi. Plants are pole type. Green pods are ready for harvesting in 60-65 days after seed sowing. Its pods are long (22.5 cm), flattish, meaty, curved, stringless and born in clusters. Seeds are glossy and yellowish-brown in colour. Average green pod yield is 10-12 t/ha. The variety is tolerant to wilt.
Canadian Wonder	This is a white seeded cultivar. Pods are straight, round in cross section, non-stringy, fleshy, green and 10-12 cm long. The variety is high yielding (13 t/ha) and widely adapted.
Pusa Himlata	This variety has been developed through selection from an exotic material and release for commercial cultivation by ICAR-IARI, New Delhi. Plants are medium tall (2.5 m). The pods are light green, medium sized (14.0 cm), round and stringless. First picking is possible in 60 days after sowing. It is also suitable for processing as green-shelled beans. Average green pod yield is 26.0 t/ha.

Variety	*Description*
Swarna Lata	This pole type variety has been developed by HARP, ICAR Research Complex for Eastern Region, Plandu, Ranchi (Jharkhand) through pure line selection. The pods are round, fleshy and stringless. First picking is possible in 55-60 days after sowing. Average green pod yield is 12.0-14.0 t/ha.
SVM-1	This variety has been developed from an interspecific cross involving *P. vulgaris* (Contender) x *P. multiflorus* (PBL-257). The pods are green, round, stringless and 13.0-14.0 cm long. First picking is possible in 65-75 days after sowing. Average green pod yield is 10.5-12.5 t/ha.
Laxmi	Pods attractive green, round, stringless with three pods (13-14 cm long) per bunch. Pods are ready for picking in 65-70 days. Average green pod yield is 16.0 t/ha. Seeds are white in colour with light yellow scar. It is tolerant to angular leaf spot disease.
TKD-1	It is a cross derivative selection between two pole types, *viz.*, Sel.1 and PV118. This variety has been developed from Horticultural Research Station, Thadiyankudisai, Tamil Nadu Agricultural University, Coimbatore (Tamil Nadu). Suitable for growing in the hill ranges of Tamil Nadu, *viz.*, Shevroys, Palani hills and Nilgiris. The crop duration is 90 days. Pods are less fibrous. The green tender pod yield ranged from 5-6 t/ha. Dry seed yield is 2.78 t/ha. Seeds turn white in the maturity used for cooking.
Birsa Priya	This variety has been developed from Birsa Agricultural University, Ranch (Jharkhand). Yield is 11 t/ha.
IC595238	A unique line with purple colour pod has been identified at ICAR-RC-NEH Region, Kolasib (Mizoram). The pods are ready for harvesting 60-65 days after sowing, round in cross section, fleshy, stringless, long (16.5-17.5 cm) and having about 8 seeds. It yields 13.5-15.0 t/ha in a total of 6-7 pickings with a crop duration of 95-105 days. The mature seeds are kidney-shaped, light brown and 100 seeds weight approximately 25.0 g.
KKL-1	It is a pole type best suited for elevation of 800-2,400 m. it has a yield potential of 70t/ha of green pods or 3 t/ha grains/ha.
French bean (Semi-Indeterminate type)	
Phule Surekha	This variety has been developed from MPKV, Rahuri (M.S.) in the year 1997 through pure line selection from an exotic material 'Jampa'. It is photo-and thermo-insensitive and is suitable for cultivation in *kharif*, spring and summer seasons. Plants are semi-determinate in nature. Its pods are straight, flat, long (9.2 cm) and green in colour. It is photo-and thermo-insensitive and is suitable for cultivation in *kharif*, spring and summer seasons. It has green pod yield potential of 15.0 t/ha. The grains are black when fully mature and also used for preparation of vegetable curries. The crop duration is 75 to 90 days. It is resistant to anthracnose, leaf crinkle, bean mosaic virus and wilt.
YCD-1	This variety has been developed from Horticultural Research Station, Yercaud, Tamil Nadu Agricultural University, Coimbatore (Tamil Nadu) through selection from indigenous material collected from Shevroy hiils. The variety is a dual purpose one, usefull as green pod and grain. It is adaptable to the rainfed situations of Shevroys region. The plants are semidwarf and grow to a height of 35.5 cm. It flowers in 35-50 days after sowing. The total crop duration is 105 days. It suitable for cultivation in Kharif (June-July) season. The pods are slightly flat, long (15.1 cm), broad (1.13 cm) and robust (7.76g). It possesses a higher green pod yield potential of 9.75 t/ha. It is field tolerant to root rot, rust, yellow mosaic and anthracnose.
Butter bean (*Phaseolus lunatus*)	
KKL-1	This variety has been developed from Horticultural Research Station, Kodaikanal, Tamil Nadu Agricultural University, Coimbatore (Tamil Nadu) through selection from a local type (collected from Villpatti). Butter bean possess high amount of protein coupled with fat content. It is pole type growing upto 2.42 m and bears pods in clusters. The pods are 11.6 cm long; beans seeds are 5 to 6 in number per pod. The pods are green when immature, turning creamy yellow with brownish purple streaks on the surface at maturity. The seeds are bold, globular, snow white in colour and with excellent cooking quality. The variety is suited for hilly regions of Tamil Nadu from an altitude of 1200 m to 2200 m above MSL. The crop will be ready for first harvest from 100 days after sowing. The harvest will continue upto 140 days. It yields 1.2-3.47 t/ha of ripe pods in 3 to 4 pickings.
Lab Lab Bean/Dolichos bean	
Pusa Early Prolific (PEP)	It is selection from the local material. 'Pusa Early Prolific' is a pole type early maturing variety developed form ICAR-IARI, New Delhi which bears medium long (9.3 cm), flat thin stringless pods in bunches. It is suitable for sowing in early autumn and early spring in the northern plains (Choudhury, 1972). The average yield is 10.0-12.0 t/ha.
Pusa Sem-2	It is a selection from the local material which is and release for commercial cultivation by ICAR-IARI, New Delhi. Plants are pole type and bears dark green foliage. Flowers appear on separate spikes above the plant canopy. Pods dark green, born in cluster, very tender, stringless, semi round, 16.0-17.0 cm long with 5-6 seeds per pod, 11-12 pods/spike and having excellent organoleptic quality. First picking in 120-125 days. Yields 15.0- 18.0 t/ha.Highly tolerant to viruses, anthracnose disease, and insects like aphids, jassid, pod borer and frost. It is recommended for sowing in June-July in north India (Sharma *et al.,* 1997).
Pusa Sem-3	It is a selection from the local material which is released for commercial cultivation by IARI, New Delhi. Plants are pole type and bears dark green foliage. Pods green, meaty, very tender, stringless, flat, 15.0 cm long with 5-6 seeds/pod, 9-10 pods/cluster; excellent organoleptic quality. First picking can be done in 115-120 days. Tolerant to viruses, anthracnose disease and insects like aphid, jassid and pod borer. Yield 17.0-18.0 t/ha (Sharma *et al.,* 1997).
Hima	Pole type variety developed at KAU, Velannikara (Kerala) by pure line selection from a local collection. It bears white flowers and light green broad straight pods and coffee brown coloured seeds. It is high yielding (14.82 t/ha), photosensitive variety suitable for July-August sowing.
Grace	Pole type variety developed at KAU, Velannikara (Kerala) by pure line selection from a local collection "Grace". It is photosensitive variety suitable for July-August sowing. It has purple stem purple leaf veins and lilac flowers producing medium broad, greenish purple, slightly curved pods. Yield 15.1 t/ha. Seeds are black.

Variety	Description
CO.1	Pole type variety 'Co.1' is released in the year 1980 by TNAU, Coimbatore, (Tamil Nadu) from the selection of local materials. Plant height of the cultivar is 60.0-70.0 cm, pods flat, green when tender and tan coloured at maturity. Crop duration is 140 days. It is known for high protein and fat content. The average yield is 28 t/ha. Free from virus.
CO.2	The cultivar has been developed by Tamil Nadu Agricultural university Coimbatore (Tamil Nadu) from cross Co.8 x Co.1 and released in Tamil Nadu in 1984. Plant height is 60.0 cm, photo insensitive, erect, and bushy with 5-6 branches. Pods flat, green and tetra-seeded. Average yield is 13.5 t/ha.
Swarna Utkrisht	Swarna Utkrisht is a pole type variety developed by ICAR-RCER Horticultur and Agroforestry Research Programme, Plandu, Ranchi (Jharkhand), through pure line selection from CHDB-1. Pods are straight flat, green fleshy (10.0-12.0 cm) having very good cooking quality. First picking can be done after 120 days and fresh pod yield is 35.0-40.0 t/ha. Mature seeds are light brown. It is tolerant to anthracnose, bean mosaic virus and pod borer.
Swarna Rituvar	This pole type variety has been developed by ICAR Research Complex for Eastern Region, Research Centre, Plandu, Ranchi (Jharkhand). Pods are attractive and tender. Average yield is 34.8 t/ha.
Arka Jay (IIHR Sel. 1)	It has been developed from ICAR-IIHR Bengaluru (Karnataka) by selection from the cross Hebbal Avare x IIHR 93. It is a photo-insensitive bush type variety, which is tolerant to both heat and drought. Pods are medium long, thin, light green and slightly curved having excellent cooking quality. It takes 75 days to reach marketable maturity and gives an average yield of 12.0 t/ha. It is tolerant to low moisture stress.
Arka Vijay (IIHR Sel. 2)	It has been developed at ICAR-IIHR, Bengaluru (Karnataka) by selection from the cross Hebbal Avare x IIHR 93. It is a dual-purpose bush type variety, which is tolerant to heat and drought. The pods are thin, short and dark green with characteristic aroma. It takes about 75 days to reach marketable maturity and gives an average yield of 12.0 t/ha. It is tolerant to low moisture stress.
Arka Soumya	It has been developed at ICAR-IIHR, Bengaluru (Karnataka) and released for commercial cultivation in 2010. Plants are medium tall, 50 per cent flowering in 45 days of seeding and pods are ready for harvest in 55 days. Pods are flat, light green, medium long (13-15 cm), medium width (1.5 cm). Yield is around 19.0-20.0 t/ha.
Arka Sambhram	It has been developed at ICAR-IIHR Bengaluru (karnataka) and released for commercial cultivation in 2010. Plants are medium tall, 50 per cent flowering in 40 days and pods are ready for harvest in 55 days. Pods are slender (1.0 cm width), and medium long (13-15 cm). Yield is around 19.0 t/ha.
Arka Amogh	It has been developed at ICAR-IIHR, Bengaluru (Karnataka) and released for commercial cultivation in 2010. Plants are medium tall, 50 per cent flowering in 40 days and pods are ready for harvest in 55 days. Pods are similar to Arka Jay and Konkan Bhushan. Yield is around 19.0-20.0 t/ha.
Arka Swagat	It has been developed at ICAR-IIHR Bengaluru (Karnataka) and released for commercial cultivation in 2011. It is suitable for growing during off season (June to November) and also during regular season (Rabi). Pods mature in 65 days. Pods are flat, green and medium long vegetable type with 4-5 seeds. Pod yield is R 26.7 t/ha in 120-130 days.
Konkan Bhusan	It has been developed by selection from the cross 'Hebbal Avare' x 'Wal- 1' at BSKKV, Dapoli (M.S.). It is bushy (60-75 cm tall) variety that bears about 10.0 cm long, tender, green and tubular pods. First picking is possible just in 60 days after sowing. The average yield is 8.0 t/ha.
Wal Konkan-1	This bushy variety is derivative of the cross 'Wal-2-K_2' x 'Wal 125-36' and released in Maharashtra for Konkan region by BSKKV Dapoli (M.S.) The plant is having long tendrils. The leaves are smooth, dark green in colour and plants flower 50-60 days after seeding. Duration of the crop is 14.68 per cent. It is resistant to yellow mosaic virus. Yield is about 9-10 t/ha.
Hebbal Avare-3	It has been developed by GKVK, Bengaluru (Karnataka) through pedigree selection of cross between Hebbal Avare-1 x US-67-13 and released in Karnataka in 1978. Plants are erect and attains the height of 65-75 cm. Determinate growth habit, photo-insensitive, flowers of white colour, pods green, 2-3 seeded, brown round seeds and small. Crop duration is 90-110 days. It is field bean. Yield is about 8 t/ha.
Hebbal Avare-4	It has been developed By GKVK, Bengaluru (Karnataka) selection from the cross 'Hebbal Avare' x 'Co.80'. Its plants are bushy compact and give first picking around 60 days after sowing. The average yield is 6.0t/ha in five pickings.
Wal-125-36	This variety is selection from local ermplasm of Gujarat and released for commercial cultivation in Gujarat state. Plant height 60-75 cm, and foliage green. Pods are born in cluster. Seeds are milky white. The crop duration is 110-120 days. Yield is 11 t/ha.
Phule Gauri	This variety has been developed from MPKV, Rahuri (M.S.) in the year 2001. It is good for *kharif* and *rabi* season. The growth habit is pole type and the crop duration is 180-200 days. The pods are whitish green in colour, tender, flat and slightly curved. It is tolerant to leaf miner and pod borer. The average yield is 28.0 t/ha.
Phule Ashwini	This variety has been developed from MPKV, Rahuri (M.S.) in the year 2001. The growth habit is pole type. Crop duration is 180-200 days. The pods are round, dark green, straight and slightly culved. Good for *rabi* season, less incidence of pod borer, moderately resistant to leaf spot disease, and good demand for local market. The average yield is 26.0 t/ha.
Phule Suruchi	This bush variety has been developed from MPKV, Rahuri (M.S.) and released for commercial cultivation in 2009. It is good for *kharif* and *rabi* season and the crop duration is 180-200 days. Pods are straight, slightly curved, and green with purple tinge at both the edges. It is tolerant to leaf miner and pod borer and leaf spot disease. The average yield is 8.87 t/ha.
Dasara	This variety has been developed from PDKV, Akola (M.S.). The average yield is about 19 t/ha.
Deepali	This variety has been developed from PDKV, Akola (M.S.). The average yield is about 15.5 t/ha.
Gujarat Indian Bean-1	This variety has been developed in the year 2003. The average yielld is 21 t/ha. This cultivar is free from mosaic virus and leaf spot.
Hisar Kirti (HD-18)	This variety has been developed from CCS HAU, Hisar (Haryana). It is an early maturing variety, which takes about 90 days from sowing to first pod harvesting. Its pods are flattened and dark green with good keeping quality. Its average pod yield is about 20-22 t/ha.

Variety	*Description*
BCDB-1	This variety has been developed from BCKV, Kalyani (West Bengal) through single plant selection, having dark green fleshy pod with purple suture, very high yielding (>40 t/ha) and pod protein content is 3.5 per cent. It is highly tolerant to bean yellow mosaic and *Cercospora* leaf spot.
BCDB-2	This variety has been developed from BCKV, Kalyani (West Bengal) through combination breeding involving bush and pole type parents, having light green pod with purple suture, high yielding ((>42 t/ha) and pod protein content is 3.5 per cent. It is tolerant to bean yellow mosaic and *Cercospora* leaf spot.
Utkal Ishani (BDB-2)	This pole type and high yielding cultivar has been developed from Odisha University of Agriculture and Technology, Bhubaneswar (Odisha) under AICRP (VC) programme adopting pure line selection. The duration of flowering and first harvest depend on the time of sowing of seeds. Flowers are white in colour. Pods are green, long, round type with 5-6 seeds per pod. Pods are born in clusters of 5-7 in numbers. The average yield 33 t/ha.
JDL-45	This pole type and high yielding cultivar has been developed from JNKVV, Jabalpur (Madhya Pradesh). Yield is about 39 t/ha.
JDL-37	This pole type and high yielding cultivar has been developed from JNKVV, Jabalpur (Madhya Pradesh). Yield is about 32 t/ha.
Thar Maghi	This variety has been developed by ICAR-CIAH, Bikaner (Rajasthan) provides early and high quality yields under the arid agro-climatic conditions of Rajasthan. Flowers are purple in colour. Plants are medium in growth habit and about 3.0 m tall. Pods are medium sized, soft and shining green in colour. Pod yield potential is 11-12 t/ha.
Thar Kartiki	This variety has been developed by ICAR-CIAH, Bikaner (Rajasthan) for early harvest in mid-October and higher yield under the arid agro-climatic conditions of Rajasthan. Plants are medium in growth habit and 2.75-3.25 m in height. Flowers are white in colour. Pods are medium sized, whitish shining green, sickle shape, smooth and soft. Pod yield potential is 10-11t/ha.
Kalianpur T-1	This is a pole type variety developed by Chandra Shekhar Azad Univiersity of Agriculture and Technology, Kanpur (U.P.). It bears green fruits and very much popular in eastern U.P. Average yield potential is 22 t/ha.
CO1 Mochai	It is high yielding, early culture with edible nature of green and dry bean. It is selection from DL-2539 through single plant selection (Annappan *et al.,* 1989). Average yield is 13 t/ha.
CO10 Avarai	This variety has been evolved from a gamma irradiated seeds (24 kr) of a bushy variety 'CO6' (derivative of a cross between 'DL244' (garden bean) and 'DL3196' field bean) and its progenies screened for green pod yield and quality. This variety is very high yielder than CO9 and possesses better values for sensory tests and biochemical analysis (Sivasubramanian *et al.,* 1989).
CO11 Avarai	It is derived from cross between CO9 (bush type) and White Yanaikathu (Pandal type), followed by selection during F6 generation as Colt-22. After different yield trialfinally named as CO11 Avarai (Ramasamy *et al.,* 1990).
Rewa	It has a protein content of 25.11 per cent. Average yield is 18 t/ha. Resistant to bean mosaic virus and leaf spot.
Other variety	CO3 (DL269), CO4 (DL453) and CO5 (DL692).
Cowpea	
Kashi Gauri (VRCP-2)	This is bush type, photo-insensirtive and early variety developed by ICAR-IIVR, Varanasi (Uttar Pradesh). It is suitable for sowing in both spring summer and rainy seasons. Flowering starts in 35-38 days and pods gets ready for harvest in 45-48 days after sowing. Pods are 25.0-30.0 cm long, light green, soft, fleshy and free from parchment layer. The cultivar is resistant to golden mosaic virus, *Pseudocercospora cruenta*, and gives green pod yield of about 10.0-12.0 t/ha. This has been identified by UP State Horticultural State Sub Committee and notified during XII meeting of Central Sub Committee on Crop Standard Notification and Release of Varieties for Horticultural Crops for the cultivation in U.P., Bihar and Jharkhand.
Kashi Unnati (VRCP-3)	This is bush type (40.0-45.0 cm height) photo-insensirtive and early variety developed pedigree selection (Cowpea-263 x Arka Garima) at by ICAR-IIVR, Varanasi (Uttar Pradesh). It is suitable for sowing in both spring summer and rainy seasons. Flowering starts in 30-35 days and pods gets ready for harvest in 40-45 days after sopwing. It produces 40-45 pods per plant with average pod length of 30.0-35.0 cm. It gives green pod yield of about 12.5-15.0 t/ha. The cultivar is resistant to golden mosaic virus, and *Pseudocercospora cruenta* diseses. This has identified by UP State Horticultural State Sub Committee and notified during XII meeting of Central Sub Committee on Crop Standard Notification and rRelease of Varieties for Horticultural Crops for the cultivation in U.P., Bihar and Jharkhand.
Sel-2-1	This is dwarf and bushy type cultivar developed through selection at NDUA&T, Kumarganj, Faizabad (U.P.). Pods are green and 25-30 cm long. It is black seeded type. It is susceptible to *Pseudocercospora cruenta* diseses and virus. Yield is 100 t/ha.
Bidhan Barbati-1	Advance generation selection of modified back cross EC243954 of culti group, *unguiculata* and EC305827 of culti group *sesquipedalis* at BCKV, Kalyani (W.B.). Plants are bushy compact, determinate, and photoinsensitive. The pods are green, tender, thick, fleshy and 25.0 cm long. Fleshy pod containing 3.87 per cent protein. First picking is possible in 55-60 days after sowing. It is resistant to cowpea mosaic and cowpea golden mosaic virus. It has yield potential of 13.5 t/ha (Hazra *et al.,* 2001).
Bidhan Barbati-2	Advance generation collection of cross V-70 of culti group, *biflora* and Sel Tm-3 of culti group *sesquipedalis* at BCKV, Kalyani (W.B.). Plants are semideterminate and photoinsensitive bearing 26.0 cm long, light green tender, solid and fleshy pods containing 4.0 per cent protein. Low incidence of cowpea mosaic and cowpea golden mosaic virus. Average yield is 16.0 t/ha (Hazra *et al.,* 2001).
BCCP-3	Developed from combination breeding. Plants are determinate in growth habit, and very early in flowering with green pod. Pod length 26.0 cm, pod weight 12.0 g with average pod yield is 15 t/ha. This variety is sutable for growing in spring-summer and rainy season and highly tolerant to *Cercospora* leaf spot and cowpea mosaic virus.
Birsa Sweta	This variety has been developed from Birsa Agricultural University, Ranchi (Jharkhand). It is selection from local cultivar. Suitable for rainy season. Average yield potential is 10.5 t/ha.
Pusa Phalguni	This cultivar has been developed through selection from a Canadian cultivar Dolique Du Tonkin by ICAR-IARI, New Delhi. It is bushy/dwarf type and takes 60 days for first picking and bears 10-12 cm long dark green pods. Average green pod yield is 5-6 t/ha.

Variety	Description
Pusa Barsati	Selection made from exotic introduction of Philippines and release for commercial cultivation by ICAR-IARI, New Delhi. It is a trailing type (semi-determinate) cultivar and suitable for rainy season in north Indian conditions under support system. Its pods are light green, pendent and medium long (26.0-28.0 cm). Crop is ready for first picking in about 55-60 days. Average yield is 9.0-9.5 t/ha.
Pusa Do-Phasali	It is a selection from a cross combination of 'Pusa Phalguni' with an exotic introduction 'Philippines Bush Cultivar' and developed by at ICAR-IARI, New Delhi. It is a photo-insensitive bushy cultivar suitable for in both spring-summer and rainy season. It takes 45 days for first picking. The pods are light green, semi pendent and medium long (18.0cm). Average yield is 7.5-8.0 t/ha.
Pusa Rituraj	It is a selection from ICAR-IARI, New Delhi. It is bushy in nature. It is highly photo-thermo-insensitive in nature, grown in summer as well as in rainy seasons. First picking can be done in 45-50 days after sowing. Pods are green 22-24 cm long and pod yield is 8.0-8.5 t/ha.
S-203	Cross of Sel-2 x Virginia. Recommended for double cropping system. Averge yield is 8-10 t/ha.
Indira Barbati Lal	This variety has been developed by IGKV, Raipur (Chhattisgarh) and released by State Sub Seed Committee of Chhattisgarh State for cultivation in Chhattisgarh Plains during 2010-11. The variety has attractive tender maroon (red) coloured pod. Moderately resistant to Mosaic and pod borers. It has yield potential of 6.15 t/ha. Recommended for rainfed and irrigated conditions, for Kharif and Spring-summer season.
Arka Mangla	This is a very good pole type variety having long pod (yard long bean) has been developed by ICAR-IIHR Bengaluru (Karnataka) and released for commercial cultivation in 2011. Pods round, stringless, crisp and long (65-75 cm). Pods are ready for 1st harvest in 60 days. It has the yield potential of 22.6 t/ha in 100 days.
Arka Suman	This bushy variety has been developed by pedigree selection from the cross Pusa Komal x Arka Garima and release for commercial cultivation from ICAR-IIHR, Bengaluru (Karnataka). The plants are erect, and bushy. The pods are green, medium thick, medium long (15.0-18.0 cm) tender and parchment free. It has the yield potential of 7.5-10.0 t/ha.
Konkan Wali	This bushy variety has been developed by Dr. BSKVV, Dapoli (Maharashtra) through selection from the yard long bean (*Vigna sesquipedalis*) germplasm and very much suitable for Konkan region. It produces light green tender pods suitable for vegetable purpose. It yields about 9.0-10.0 t/ha.
Anand Vegetable Cowpea-1	This variety has been developed in the year 2007. It has the yield potential of 8 t/ha. This cultivar is free from bean mosaic virus and leaf spot disease.
Swarna Sweta	This polwe type variety has been developed by ICAR Research Complex for Eastern Region, Research Centre, Plandu, Ranchi (Jharkhand). It has the yield potential 8-10 t/ha.
Swarna Mukut	This bush variety has been developed by ICAR Research Complex for Eastern Region, Research Centre, Plandu, Ranchi (Jharkhand). It has the yield potential of 6.5-8.8 t/ha.
Swarna Suphala	This pole type variety has been developed by ICAR Research Complex for Eastern Region, Research Centre, Plandu, Ranchi (Jharkhand). Plants are pole type. Pods are light green and 30-35 cm long. Pods get ready for harvest in 50-55 days and the green pod yield is 7.5-10.0 t/ha.
Bhagya-lakshmi	This bushy and short duration variety (flowering in 48 days) has been developed by KAU, Vellanikkara Thrissur (Kerala). It bears white flowers. The pods are light green, about 27.0 cm long and born in clusters of 2-4. Seeds are brown mottled. It has the yield potential of 6.48 t/ha. It is resistant to anthracnose disease.
Kanakamany	This dual purpose bushy variety has been developed by KAU, Vellanikkara Thrissur (Kerala) through pure line from PTB-1, is a semi trailing variety. The pods are light green, medium long (17.8 cm), dark green pods and having bold brown seeds. It gives an average yield of 7.0 t/ha.
Malika	This is a high yielding yard long bean variety has been developed by KAU, Vellanikkara (Kerala) through single plant selection from Thiruvananthapuram Local. Pods are long (43.5 cm), white podded without any purple or brown tip. Seeds are brown with characteristic white speck on one edge. It has the yield potential of 9.8 t/ha.
Lola	This variety (yard long bean) has been developed by selection from the local material'Kankari Local' and release for commercial cultivation from KAU, Vellanikkara Thrissur (Kerala). The pods are attractive white with green tinge, extra long (53.38 cm) and takes about 65 days from sowing to first picking. A total of 15-20 pickings are possible. Average green pod yield is 20.0 t/ha.
Sharika	This is a high yielding yard long bean variety has been developed by KAU, Vellanikkara Thrissur (Kerala) through single plant selection from Valiyavila Local. Pods are long (49.0 cm), white podded with purple tip. Average green pod yield is 10.62 t/ha. Seeds are black.
Vyjayanthi	This pole type variety has been developed by KAU, Vellanikkara Thrissur (Kerala) through pure line selection from Perumpadavam Local, is a long (50.62 cm) red podded variety with brown seeds. It gives an average yield of 12.62 t/ha.
KMV-1	This is a yard long bean variety suited to acid alluvial soils of Kuttanadu, developed by pure line selection from Manjeri Red Plain at KAU, Vellanikkara Thrissur (Kerala). It has trailing growth habit and pods are long (37.93 cm), light green with a brown tip giving an average yield of 13.5 t/ha. Seeds are reddish brown. The variety is suited for intercropping in coconut gardens.
Anaswara	It is high yielding, semi-trailing, dual purpose cowpea variety belonging to *Thandapayar* type at KAU, Vellanikkara Thrissur (Kerala). The pods are light green, bold seeded measuring a length of 28-32 cm, giving an average yield of 12.75 t/ha. Seeds are cream coloured. The variety can be grown in basins of 50-60 cm diameter, taken at a distance of 1.5-2.0 m without trailing on "Pandals".
Varun	It is a semi-trailing type developed by purline selection from a local collection from Manjapra at KAU, Vellanikkara Thrissur (Kerala). It has medium long pink pods and tolerance to mosaic and pod borer. It gives an average yield of 8.4 t/ha.

Variety	*Description*
Kairali	This is a semi-trailing type mosaic resistant variety developed from IC-45415 by pur line selection at KAU, Vellanikkara Thrissur (Kerala). Leaves are green with pink streak at base of petiole. It produces pink coloured medium long pods (22.78 cm) giving an average yield of 7.13 t/ha.
Vellayani Jyothica	It is selection from a local collection from Sreekaryam Local at KAU, Vellanikkara (Kerala). It is high yielding, long poded, light green variety giving an average yield of 19.33 t/ha. It is medium duration variety tolerant to Fusarium wilt.
Githika	It is a high yielding and mosaic resistant cultivar developed at KAU, Vellanikkara Thrissur (Kerala). It is a light green poded yard long bean variety having long thick fleshy pods (53.4 cm). It gives an average yield of 27.6 t/ha. Seeds are reddish brown.
PKM-1	This bushy and high yielding cultivar has been developed from Tamil Nadu Agricultural University, coimbatore (Tamil Nadu). It is cluster bearing type. The pods are light yellowish green, and long (40-45cm). Average green pod yield is 25.0 t/ha. Green pods can be harvested in 90-100 days.
Utkal Manika	This bushy and high yielding cultivar has been developed fromOdisha University of Agriculture and Technology, Bhubaneswar (Odisha) under AICRP (VC) programme and released by the SVRC during 2005. The pods are smooth, moderately long and fleshy. The tip of the pod is brown and seeds are light brown in colour. Harvesting of green pods starts 50-55 days after sowing. It shows moderate tolerance to aphids and YMV disease. It has become very popular among the farmers of Odisha. Average yield is 12.8 t/ha.
Durga Kranti (RCV-7)	This bush variety has been developed by RARI, Durgapura (Rajasthan), released in 1997 and notified in 2001. Plant determinate type, plant height 50-60 cm, light green pods of 17-20 cm length, white seed colour and 10-15 t/ha green pod yield.
Cluster bean	
Pusa Sadabahar	This variety has been developed by selection from the local material collected from Rajasthan and released for commercial cultivation by ICAR-IARI, New Delhi. The plants have single stem and is suitable for growing in both summer and rainy seasons. The pods are green, tender, fiberless and 12-13 cm long. First picking is possible in 45 days after sowing in summer season and in 55 days in rainy season. The average green pod yield is about 5.0 and 7.5 t/ha in summer and rainy season, respectively.
Pusa Mausami	This variety has been developed by selection from the local material and released for commercial cultivation by ICAR-IARI, New Delhi. The plants have branched stem and is suitable for sowing in rainy season. The pods are bright green, smooth and 10.0-13.0 cm long. It is late maturing variety and first picking is possible in 65-80 days after sowing depending upon the weather conditions and the region.
Pusa Navbahar	This variety has been developed by selection from the cross Pusa Sadabahar x Pusa Mausami. Its plants are single stemmed, about 1.0 m tall. The variety is suitable for sowing in both summer and rainy seasons. The pods are attractive green and 10.0-12.0 cm long. First picking is possible in 40 days after sowing.
Sharad Bahar	This variety has been developed by selection from the local material collected from Maharashtra and release by ICAR-NBPGR, New Delhi. The plants have branched stem and the variety is suitable for sowing in both summer and rainy seasons. The pods are attractive green, succulent and long. The average yield is 8 t/ha.
MDU-1	This variety has been developed by Agricultural College and Research Institute, Madurai, (Tamil Nadu Agricultural University, Coimbatore tamil Nadu). It has a cluster bearing nature (7-9 fruits/node) with long attractive green coloured fruits (150-175 fruits/plant) measuring 13-14 cm. Individual plant yield (pods) of 250-300 g. fruits are rich in fiber content (4.5/100g of the fruit) and the plants can tolerate powdery mildew. Average yield is 10 11 t/ha.
Kanchan Bahar (M-83)	This variety has been developed by RARI, Durgapura (Rajasthan), released in 1999 and notified in 2001. The variety is white flower, single stem plant 5-6 pods/cluster, 18-20 cluster/plant, and 7.5-10 t/ha green pod yield.
Thar Bhadavi	This variety has been developed by ICAR-CIAH, Bikaner (Rajasthan) through selection from the indigenous germplasm for cultivation in desert agro-climatic region of Rajasthan. The light green thin and medium long tender pods starts from 55 days after seed sowing as a rainfed crop. The average pod yield potential ranged from 6.5-12.5 t/ha under varying situations of crop production.
Vegetable Soybean	
Swarna Vasundhara	This vegetable type variety has been developed by ICAR Research Complex for Eastern Region, Research Centre, Plandu, Ranchi (Jharkhand). Average yield is 16-18 t/ha. Free from leaf spot & mosaic.
Broad bean	
Pusa Sumeet	It has been developed through selection from ICAR-IARI, New Delhi. Attractive dark green pods in cluster, first picking in 65-7 days. Average yield is 18.0 t/ha.
Swarna Safal (HAVFB-41; IC617240)	This vegetable type variety has been developed by ICAR Research Complex for Eastern Region, Research Centre, Plandu, Ranchi (Jharkhand) in 2015 for commercial cultivation in Jharkhand state. This cultivar proved its worthiness in respect of green pod yield (24.25 t/ha) with 24.04 per cent yield increase over the check variety 'Pusa Sumeet' (19.55 t/ha). The variety also exhibited its superior performance regarding green pod yield and quality at farmer's fields in three agro-climatic zones of Jharkhand. The line comes to 50 per cent flowering stage in 35-40 days after sowing. The green pods become ready for harvest in 89-94 days after sowing. The green pods are 3.0-3.3 g in weight and 5.4-6.0 cm in length. Swarna Safal was resistant to Chocolate spot (*Botrytis fabae*) and was least infested by pod borer under field conditions. Per 100 g edible portion, the green pods contained 2.95 g protein, 5.81 g carbohydrate, 25 mg calcium, 115 mg phosphorus, 194 mg potassium, 32 mg magnesium, 33 mg sulphur, 0.65 mg copper, 0.39 mg manganese and 0.11 mg iron (Pan *et al.,* 2016).
Swarna Suraksha	This vegetable type variety has been developed by ICAR Research Complex for Eastern Region, Research Centre, Plandu, Ranchi (Jharkhand) particulary for rainfed conditions though its performance was also better under irrigated ecology. It contains very less amount of anti-nutritional factor like tannin and phytate. The variety attains height of 70-90 cm, and plant type is bushy. It is medium duration cultivar matures in 110-115 days. It starts bearing pod at the height of 5.0 cm from the base. The (sweetness) mature seeds is containing 10.90 TSS.

Variety	Description
Lima bean (Bush type)	
Burpees Improved Bush	It is an introduction from U.S.A. The pods are flat, large (about 15.0 cm) born in cluster of 5 or 6, slightly curved are easy to shell. The average yield is about 20-25 t/ha.
Early Thorogreen	It is an introduction from U.S.A. It matures in 60 days (early genotype) and produce small green seeds. This cultivar is highly productive. The average yield 18-20 t/ha.
Lima bean (Pole type)	
KKI-1 Butter Bean	It has been developed by selection from the local material collected from Villpatti and release for vcommercial cultivation by HRS, Kodaikanal, TNAU, Coimbatore (Tamil Nadu). The pods are born in clusters are 11-12 cm long containing 5-6 seeds. First picking is possible in 100 days after sowing and a total of 3-4 pickings are possible. Average yield is 25-28 t/ha.
Carolina Sieva	Average yield is 26 t/ha. Free from leaf spot and virus. It is an introduction from U.S.A. It matures in 80 days and produce small greenish white seeds. It is tolerant to high temperature.
Swarna Poshan	This vegetable type variety has been developed by ICAR Research Complex for Eastern Region, Research Centre, Plandu, Ranchi (Jharkhand). Average yield is 25 t/ha.
Winged bean	
IIHR Selection-21	It has been developed by ICAR-IIHR, Bengaluru (Karnataka). Average yield is 18-20 t/ha.
IIHR Selection-60	It has been developed by IIHR, Bengaluru (Karnataka). Average yeild is 22-25 t/ha.
IIHR Selection-71	It has been developed by IIHR, Bengaluru (Karnataka). Average yield is 24 t/ha.
Revathi	It has been developed by selection from PT52 at KAU, Vellannikara (Kerala). It is a day neutral plant with trailing habit, violet flowers and medium sized, four ribbed green pods. Seeds are dark brown and large. Average yield is 20-22 t/ha.
WBC-2	It has been developed by ICAR Research Complex for NEH, Barapani, Umium (Megaalaya). Average yield is 18-20 t/ha.
Faba bean/Broad bean	
Pusa Sumeet	It has been developed by IARI, New Delhi. The plants are dwarf (75.0 cm tall). The pods are about 6.0 cm long and 1.3 cm thick, dark green and born in cluster. Average pod yield is 18.0 t/ha.
Selection BR-1	It is a black seeded variety. Commercially grown in Bihar. Yield potential of this varedy is 20 t/ha.
Selection BR-2	It is a yellow seeded variety. Commercially grown in Bihar. The average yield potential is 16 t/ha.
Imperial Long Pod	It is a long pod type. The average yield potential is 14-16 t/ha.
Masterpiece Green Long Pod	It is a long pod type. The average yield potential is 22 t/ha.
Imperial Green Long Pod	It is a long pod type. The average yield potential is 18-20 t/ha.
Red Epicure	It is a long pod type. The average yield potential is 22-24 t/ha.
Imperial White	It is a Windsor type. The average yield potential is 26 t/ha.
Giant Four Seeded Green Windsor	It is a Windsor type. The average yield potential is 24-56 t/ha.
Imperial Green Windsor	It is a Windsor type. The average yield potential is 24 t/ha.
Maris Beagle	It is an English synthetic variety composed of 5 inbred lines and released in 1974. The yield potential is 19 t/ha.
Throw's MS	It was the forst synthetic faba bean to be grown commercially in the U.K. The yield potential is 20 t/ha.
Soravi	It is synthetic varieties release n France. The average yield potential is 22 t/ha.
Avrissot	It is synthetic varieties release n France. The average yield potential is 24 t/ha.
Pigeon pea (Vegetable type)	
Anand Vegetable Pigeon pea-1	This protein rich variety has been developed in the year 2007 by Anand Agricultural University, Anand (Gujarat). Plant has determinate growth habit, bushy plant type with dark green foliage. The pods are green in colour with deep constriction between two seeds. Green pod yield is 8.3 t/ha.
Bitter gourd	
Kashi Urvashi	Develop from ICAR-IIVR, Varanasi and suitable for summer and rainy season cultivation. The fruits are dark green and have good keeping quality. Average yield is 22.0-24.0 t/ha.

Variety	*Description*
Pant Karela-2 (PBIG-1)	This variety has been developed through selection from a ermplasm PBIG-1 at GBPUA&T, Pantnagar (Uttarakhand). Its fruits are thin, 25 cm long, and dark green with tapering ends. First marketable fruit picking starts at 50 days after seed sowing. It has been recommended for cultivation in the states of Uttar Pradesh, Uttarakhand, Bihar and Chhattishgarh and has yield potential of 15.0-16.0 t/ha.
Coimbatore Long	This variety is released from TNAU, Coimbatore (Tamil Nadu). Fruit is long, tender, white, and suitable for rainy season cultivation. Average yield is 20.0 t/ha at proper cultural practices.
Coimbatore Green	It is selection from the local material and developed by TNAU, Coimbatore (Tamil Nadu). The fruits are extra-long (up to 60 cm) dark green, and each weighing 300-400 g. Yield is 15.0-20.0 t/ha.
Kashi Mayuri (VRBTG-5)	This variety has been developed by ICAR-IIVR, Varansi (U.P.) in 2018 through crossing (VRBTG-47 x VRBTG-89) and pedigree selection. Pants are very vigorous and produce more branches. It flowers in 45-20 dyas after seeding, and first harvesting can be done with in 50-60 days. Fruits are attractive, glossy green, medium long (18-20 cm) with discontinuous tubercles. The average yield is 20 t/ha in a crop duration of 100-110 days. It is tolerant to anthracnose and downy mildew. This variety is suitable for net house/polyhouse cultivation.
CO-1	This variety has been developed through selection from a local material collected from Thudiyaur and released from TNAU, Coimbatore (Tamil Nadu). Vines are medium long (130.0 cm) with 5-7 branches. First flowering starts in 45-50 days and marketable fruit is harvested in 55-60 days after sowing. A total of 6-8 pickings can be done. Fruits are dark green, and 25.0-30.0 cm long with prominent protuberance. Duration of the crop is 115 days with average yield of 14.0-15.0 t/ha.
CO-2	This variety is developed from TNAU, Coimbatore (Tamil Nadu) through induced mutation. It is early in flowering. Fruits are 35.0-40.0 cm long and thick. Each fruit weighs 40.0 g. It gives an average yield of 30.0 t/ha.
Pusa Do Mausmi	It is selection from the local material and released from ICAR-IARI, New Delhi. Suitable for both spring-summer and rainy season sowing. First picking starts after 55-60 days of sowing. Fruits are long medium, dark green, thick, and club shaped with 7-8 continuous ridges. Fruit length at edible stage is about 15.0-18.0 cm, and average fruit weight is10 g. The average fruit yield is 13.0 t/ha.
Pusa Vishes	It is selection from the local material of Uttar Pradesh and released from ICAR-IARI, New Delhi. Fruits are glossy green, thick, medium long in size. Vines short hence more number of plant can be accommodated per unit area. Average fruit weight is 120 g. Average yield is 15.0 t/ha. This variety is released from ICAR-IARI, New Delhi. Suitable for spring summer season; and first picking in 55-60 days after sowing.
Pusa Hybrid-1	This high yielding hybrid has been developed by ICAR-IARI, New Delhi. The fruits are attractive green medium long (13.5 cm), medium thick (5.0 cm diameter) with regular, smooth tubercles on the surface. First picking is possible in 50-55 days after sowing. Average fruit weight is 100 g and average yield is 22.0 t/ha. It is suitable for pickling and dehydration.
Punjab-14	It is selection from the local material and developed by PAU, Ludhiana (Punjab). Suitable for both rainy and summer season cultivation. Bushy plant habit and average fruit weight is about 35 g. Fruit colour is light green. Average yield is 12.5-13.0 t/ha.
Kalyanpur Sona	It is released by CSAUA&T, Kalyanpur, Kanpur (U.P.). Its vines are medium in growth. Fruits medium sized, green, and average yield is 12.0-13.0 t/ha in 160 days of crop duration. This variety has been recommended for cultivation in Uttar Pradesh.
Kalyanpur Baramasi	This has been developed at CSAUA&T Vegetable Research Station, Kalyanpur, Kanpur (U.P.). It is vigorous creeper. Fruits are long (30-50 cm) light green, thin and tapering. Yield potential is about 20.0 t/ha in about 120 days. It is reported to be tolerant to fruit fly and mosaic virus. Suitable for *kharif* season cultivation in the state of Uttar Pradesh.
Phule Green Gold (RHRBG-5)	This variety has been developed from the cross between Green Long and Delhi Local followed by pedigree selection at MPKV, Rahuri (M.S.). The fruits are dark green, 25-30 cm long and dented (raised tubercles). Average yield is 23.0 t/ha. It is tolerant to downy mildew.
Phule Ujwala	This variety has been developed from MPKV, Rahuri (M.S.). Average yield is 22-24 t/ha. Free from virus.
Phule Priyanka (RHRBGH-1)	This hybrid derived from a cross of RHRBG-5 and RHRBG-4, developed at MPKV, Rahuri (Maharashtra). Fruits are dark green, highly prickled surface and 20 cm long. This hybrid is suitable for cultivation under rainy and summer seasons. It is tolerant to downy mildew. It gives an average yield of 28.27 t/ha in 180 days of crop duration. It has been recommended for cultivation in Maharashtra.
Hirkani	This variety has been developed from MPKV, Rahuri (M.S.) through selection from the local germplasm. Fruits are dark green and 15.0-20.0 cm long. Each vine produces 20-22 fruits, each weighing 100-120 g. Duration of the crop is 160 days with an average yield of 20.0 t/ha.
Konkan Tara (DPLBG-2)	This variety has been developed from BSKKV, Dapoli (M.S.). Its fruits are green, medium long (15-16 cm) and spindle shaped with raised tubercles and good keeping quality with a shelf-life of 7-8 days at ambient temperature. It has been recommended for cultivation in Konkan region of Maharashtra and has yield potential of 23.0-25.0 t/ha.
NS 1024	This early maturing hybrid has been developed from Namdhari Seeds Co. Bengaluru (Karnataka). First picking is possible in 45-50 days after sowing. The plants are vigorous with prolific bearing habit. The fruits are 25-30 cm long, 2.5-3.0 cm thick with dark green shining skin and sharp dentation. It is a heavy yielder with good transportation qualities. Average fruit weight is 150-200 g.
MDU-1	This variety has been developed through mutation breeding (gamma irradiation of local cultivar MC-103) at TNAU, Coimbatore. Flowering starts at 60 days after sowing with a sex ratio of 1:20 of female and male flowers. Fruits are long with an average length of 40.34 cm and a girth of 1.7-5.4 cm having average fruit weight of 410 g. Each vine bears on an average of 16-18 fruits. It has been recommended for cultivation in Tamil Nadu and has yield potential of 30.0-32.5 t/ha.
Preethi (MC-84)	This variety has been developed at KAU, Vellanikara Thrissur (Kerala). It has light green vine of 7.0 m length. Fruits are white, spiny, and medium in size with 310 g. It has been recommended for cultivation in states of Madhya Pradesh, Maharashtra, Karnataka, Tamil Nadu and Kerala and has yield potential of 14.0-15.0 t/ha in 150 days of crop duration.

Variety	*Description*
	Bottle gourd
Kashi Ganga (DVBG-1)	This is an early variety derived from the cross IC-92465 x DVBG-151 at ICAR-IIVR, Varanasi (U.P.). It bears light green fruits of 30.0 cm length having 7.0 cm girth, fruit weighs about 800-900 g. It gives an average yield of about 48.0-55.0 t/ha. This is tolerant to anthracnose and suitable for cultivation during both kharif and summer seasons. It has been identified by Uttar Pradesh State Variety Release committee and notified by Central Variety Release Committee, New Delhi for commercial cultivation.
Kashi Bahar (VRH-1)	This poptential hybrid has been developed from ICAR-IIVR, Varanasi (U.P.). Plants are vigorous green. Single plant bears about 12 light green straight fruits of 30.0-32.0 cm length and 780-850 g weight. It is suitable for cultivation during both rainy and summer season as well as in riverbed areas. Fruit yield can be obtained about 50.0-55.0 t/ha. It is tolerant to anthracnose, downy mildew and *Cercospora* leaf spot under field condition. It has been identified by Uttar Pradesh State Variety Release committee and notified by Central Variety Release Committee, New Delhi for commercial cultivation.
Kshi Kiran (VRBG-4)	This variety has been developed by ICAR-IIVR, Varansai (U.P.) through hybridization (VRBG-11 x VRBG-67) and pedigree selection in 2018. Plants are vigorous and green foliage. Fruits are round, light green, smooth and very attractive. The average yield potential is 53.13 t/ha. It is tolerant to downy mildew, powdery mildew, anthrocnose and leaf mosaic. Suitable for riverbed cultivation & bower system. Suitable for both spring summer and rainy season.
Pusa Sandesh	It is developed by ICAR-IARI, New Delhi and suitable for summer and kharif season in northern plains. Vines are medium long with short internodes; fruits medium sized, green, round and weighing 600 g each. First picking starts after 55-60 days of seed sowing in kharif and 60-65 days of seed sowing in summer and gives an average yield of 32.0 t/ha.
Pusa Summer Prolific Round	It is developed through selection from the local material at ICAR-IARI, New Delhi. Fruit is round, green, 15-18 cm in girth, prolific bearer. Suitable for both summer and kharif seasons. First picking in 60-65 days after sowing with average yield of 25.0 t/ha.
Pusa Summer Prolific Long	It is developed through selection from the local ermplasm at ICAR-IARI, New Delhi and release for commercial cultivation. Suitable for both summer and Kharif season. Fruits are long (40-45 cm), green, and prolific bearer. First picking can be done in 60 days and yield potential is 25.0-27.5 q/ha.
Pusa Meghdoot	This hybrid has been developed at ICAR-IARI, New Delhi. It is early maturing and gives about 50 per cent higher yield than Pusa Summer Prolific Long (PSPL). Fruits are green, long, thin neck, slender, and suitable for cultivation in both summer and rainy season.
Pusa Manjari	This hybrid has been developed at ICAR-IARI, New Delhi. It is early maturing and gives about 50 per cent higher yield than Pusa Summer Prolific Round (PSPR). Fruits are round, green tender, and suitable for growing in summer and rainy seasons.
Pusa Hybrid-3	This hybrid has been developed from ICAR-IARI, New Delhi. It is suitable for cultivation in both spring-summer and *kharif* season. The fruits are cylindrical, attractive green, long and straight. First picking is possible in 55-60 days after sowing. Average fruit weight is 1.0 kg and yield is 42.5 t/ha in spring summer season and 43-47 t/ha in *kharif* season.
Pusa Sandesh	It is developed from ICAR-IARI, New Delhi. Vines are medium long, short internodes, fruits medium sized, green, round, weighing 600 g each and leaves deep oblate. Suitable for both summer and kharif season. Average yield is 32.0 t/ha. First picking 55-60 days in *kharif* and 60-65 days in summer.
Arka Bahar	It is developed through selection from the local material at ICAR-IIHR, Bengaluru (Karnataka). Fruits are cylindrical, straight, without crookneck, medium sized, average fruit weight is 1.00 kg, light green, very good cooking and keeping quality. Yield is 40.0-45.0 t/ha. Crop duration is 120 days. Tolerant to blossom end rot.
Punjab Round	It is developed from PAU, Ludhiana (Punjab) and recommended for cultivation under Punjab condition. Plants are vigorous and prolific bearer, fruit spherical, tender and shining. First picking starts 65 days after seed sowing. Recommended for cultivation under Punjab condition. It gives an average yield of 35.0-37.5 t/ha.
Punjab Komal	It is developed through a cross between LC-11 x LC-5 at PAU, Ludhiana (Punjab). Plants are vigorous and prolific bearer. Fruits are spherical/pear shaped, medium sized, tender, shining, each weighing about 625 g, light green and pubescent. First harvesting is possible in 70-80 days after sowing or 10-12 days after fruit setting. Average yield is 50.0 t/ha. Recommended for cultivation in Punjab. It is tolerant to cucumber mosaic virus (CMV).
Punjab Long (LC-2-1)	It is developed from PAU, Ludhiana (Punjab). It is an early variety, bears oblong, medium sized, light green, pubescent, and 10-12 fruits on a vine. It takes 70 days from sowing to first marketable fruit. It is tolerant to powdery and downy mildews. Its shape facilitates easy packing and transportation to distant markets. Its average yield is 45.0 t/ha.
Narendra Rashmi (NDBG-4)	It is developed from NDUA&T, Faizabad (Uttar Pradesh) and recommended for commercial cultivation in Punjab, Uttar Pradesh and Bihar. Plants are moderately tolerant to red pumpkin beetle, powdery mildew and downy mildew. Fruits are bottle shaped, having shallow neck and white flesh. Average fruit weight is 1.0 kg with 400-500 seeds. Average yield is 27.5-31.0 t/ha. This variety is suitable for cultivation during both summer and rainy season. It can also be sown in October-November in Eastern Uttar Pradesh.
Narendra Dharidar (NDBGH-208-1)	It is developed from NDUA&T, Faizabad (Uttar Pradesh) through pedigree selection method after hybridization and recommended forcommercial cultivation in Uttar Pradesh. Fruits are bottle shaped, striped and green in colour. The average yield potential is 35 t/ha.
Narendra Shishir	It is developed from NDUA&T, Faizabad (Uttar Pradesh) by hybridization (NDBG-108 x NDBG-1) followed by pedigree election and recommended for commercial cultivation in Uttar Pradesh. It is round fruited, winter type and multifid leaves. The variety bears fruits during winter when sown in July-August. It remains in fruiting for about 6 months. It is tolerant to anthracnose and downy mildew. Fruits are bottle shaped, striped and green in colour.
Narendra Sankar Lauki-4 (NDBGH-4)	This hybrid has been developed at NDUA&T, Faizabad (Uttar Pradesh). It is early in flowering, prolific bearer with near cylindrical attractive fruits. It is suitable for cultivation throughout the year. Fruits remain soft even if they have grown out of normal size of picking. It has been recommended for cultivation throughout the country and has yield potential of 40-45 t/ha.

Variety	Description
NDBGH-7	This variety has been developed at NDUA&T, Faizabad (Uttar Pradesh). Average yield is 38.7 t/ha
Samrat	This variety has been developed from MPKV, Rahuri (M.S.), through selection from the local material. The fruits are green, pubescent, 30-40 cm long and cylindrical. Average yield is 40 t/ha.
Anand Bottle gourd-1	This variety has been developed from GAU, Anand (Gujarat) in the year 2006. Average yield is 42 t/ha.
GH-22	This variety has been developed from CCS HAU, Hisar (Haryana). Average yield is 39 t/ha.
HBGH-35 (hybrid)	This hybrid has been developed from CCS HAU, Hisar (Haryana). Average yield is 41 t/ha.
Kalyanpur Long Green	This variety has been developed through selection from the local material at CSAUA&T, Kalyanpur, Kanpur (Uttar Pradesh). Vines are spreading type. The fruits are long with tapering blossom end semi round, smooth, green, tender and uniform fruit size. Average yield is 30.0 t/ha. It is suitable for rainy season.
Azad Nutan	This variety has been developed through selection from the local material at CSAUA&T, Kalyanpur, Kanpur (Uttar Pradesh). The fruits are attractive, shining light green and medium long. It is recommended for cultivation in both spring summer and rainy seasons but it performs better during rainy season. First picking is possible in 55-60 days after sowing. Average fruit weight is 1.0-1.5 kg. Average yield potential is 38 t/ha.
Azad Sankar Lauki-1	This hybrid has been developed through heterosis breeding at CSAU and T, Kanpur (Uttar Pradesh). Its vines are 3.0-3.5 m fruits are long (30.0 cm), cylindrical and green. It has yield potential of 58.0-60.0 t/ha.
CO-1	This variety has been developed by TNAU, Coimbatore (T. Nadu). Its vine and leaves are yellowish green. Fruits are round at the base with prominent bottleneck at the top, medium sized, and light green with average fruit weight of 2.0-2.5 kg. Average yield is 36.0 t/ha in 135 days crop duration. It has been recommended for cultivation in the states of Tamil Nadu.
Rajendra Chamatkar	This variety has been developed at RAU, Pusa (Bihar). First picking starts after 68-70 days of seed sowing. Plants are prolific bearing with long and cylindrical fruits weighing 1.35 kg. It has been recommended for cultivation in the states of Bihar and has yield potential of 17.5-20.0 t/ha.
PBOG-1	This open pollinated cultivar has been developed at GBPUA&T, Pantnagar (Uttarakhand) and released for commercial cultivation in U.P. Average yield is 38 t/ha.
PBOG-2	This open pollinated cultivar has been developed at GBPUA&T, Pantnagar (Uttarakhand) and released for commercial cultivation in U.P. Fruits are long (40 cm), and flute shaped with light green colour. First pocking starts in 65 days after transplanting. Average yield is 40 t/ha.
Pant Sankar Lauki-1 (PBOGH-2)	This hybrid has been developed through heterosis breeding (PBOG-22 x PBOG-40) at GBPUA&T, Pantnagar (Uttarakhand). Its fruits are medium size, long and somewhat cylindrical (35.0 cm long) and green. Vine length is 5.5 m. The first picking is possible in 60 days of seed sowing. It is suitable for planting in the plains as well as in hills of northern India and has yield potential of 40.0-42.5 t/ha.
Winter squash (*C. maxima*)	
Arka Suryamukhi	It is a variety of winter squash and evolved at ICAR-IIHR, Bangalore (Karnataka). Fruits are small (1-1.5 kg) round with flat ends and deep orange with prominent streaks on the rind. Flesh is firm and orango yellow in colour. Keeping and transport qualities are good. It is also resistant to fruitfly. Its yield potential is 36 t/ha in 100 days duration.
Arka Chandan	It is developed at ICAR-IIHR, Bengaluru (Karnataka). Fruits are round with pressed blossom end, sweet and rich in carotene content. Rind colour is green with white pathches when immature, which turn to light brown coloured with self coloured patches. Fruit size is medium, fruit weight is 2.0-3.0 kg, and crop duration is 115-120 days with average yield of 33.0 t/ha.
Pumpkin (*Cucurbita moschata*)	
Kashi Harit (IVPK-226)	This variety is derived from the cross between NDPK-24 x PKM through pedigree selection. It is identified by Uttar Pradesh State Variety Release Committee and notified through Central Seed Sub Committee for commercial cultivation. The plants of this variety have short vine (1.52-2.0 m), dark green leaves and white spots on veins and leaves. Fruits are green, spherical and 2.5-3.0 kg at green as well as mature stage. The equatorial and polar circumference of fruit is 52.0-55.0 cm. Single plant bears 3-4 fruits of light green colour with white splash. Five picking can be taken in the 70-75 days of crop duration and produce 30.0-35.0 t/ha. Fruit contains 3.14 mg β-carotene (at mature stage) in 100 g of edible portion. This is most popular variety in Uttar Pradesh, Madhya Pradesh, Bihar and Adjoining areas.
Anand Pumpkin-1	This variety has been developed in the year 2009. Plants are vigorous and possess dark green foliage. Fruits are tender fleshy and rich in β-caroten. Average yield is 38-42 t/ha.
Swarna Amrit	This promising cultivar has been developed from ICAR-HARP, Ranchi (Jharkhand). Average yield is 40 t/ha.
Pusa Vishwas	It is developed at ICAR-IARI, New Delhi by selection from a line SM-107. Plants have dark green yellow leaves with white spots including veins. Fruits are brown, spherical with thick golden flesh, average fruit weight is 5.0 kg. It is suitable for cultivation in lower, middle Gangetic plains and eastern plateau region. Average yield is about 40.0 t/ha.
Pusa Vikas	It is a small fruited, semi-dwarf to dwarf type variety evolved at ICAR-IARI, New Delhi. The fruits are small (2.0 kg), round and flat with yellow flesh. It is highly suitable for cultivation in spring-summer season in northern India. Its yield potential is about 30 t/ha.
Pusa Hybrid-1	A hybrid with medium vine length developed at ICAR-IARI, New Delhi is suitable for cultivation in Punjab, Delhi, Kerala and other parts of the northern India. Its fruits, which become mature for harvesting in about 100 days, are rounded flat and medium in size. The flesh is deep golden-yellow in colour. Its yield potential is about 52.0 t/ha.

Variety	Description
CO-1	It is developed at TNAU, Coimbatore (Tamil Nadu). It is a late maturing and has large sized globular fruits (7-8 kg) with a characteristic broad proximal end. The distal end is slightly tapering to form a tip. Immature fruits are dark-green in colour and turns to brownish-orange at maturity. Fruit number is 7-9/plant and flesh thickness is 4-5 cm. It yields 30.0 t/ha in crop duration of 160-180 days.
CO-2	This variety has been developed through selection from a local collection at TNAU, Coimbatore (Tamil Nadu). Its vines are moderately vigorous, less spreading and suitable for high density planting 2.2 x 1.9 m. It is an early maturing variety with small (1.5-2.0 kg) flat green fruits, slightly ridged with bright orange flesh, 10.8 per cent TSS, 3.4 per cent total sugar, 1.9 per cent reducing sugar, 9.2 per cent starch, 10.0 mg/100 g ascorbic acid and 0.14 per cent acidity. Average yield varies from 45.0-47.5 t/ha in crop duration of 135 days. Each plant produces 10-12 fruits.
Narendra Amrit (NDPK-130)	This variety has been developed at NDUA&T, Faizabad (Uttar Pradesh). It bears pistillate flower earlier than the staminate flower. Fruit shape is near round, colour pale green, mottled and turned almond colour at maturity. Flesh colour is yellowish-white with flesh thickness of about 6.0-7.0 cm with small seed cavity. It is suitable for cultivation during both seasons. Marketable green fruit weight is 3.0 kg, while ripened fruit weight is 6.0 kg and attends average fruit weight of 7-8 kg. The average yield varies from 35.0-40.0 t/ha. Storage life of the mature fruit is more than 6 months at ambient temperature.
Narendra Shankar -1 (NDPKH-1)	This variety has been developed at NDUA&T, Faizabad (Uttar Pradesh). Fruits shape is round-oblong with big sized fruits. This variety is recently identified through State Variety Release Committee. The average yield potential is 45-48 t/ha. This variety is free from mosaic.
Narendra Abhooshan	Narendra Pumpkin Hybrid-1 has been developed at NDUA&T, Faizabad (Uttar Pradesh). It is an early prolific bearing near round fruits with dark green stripes, turning deep orange at maturity. The variety is suitable for mid January to mid March sowing. It has been recommended for cultivation in Uttar Pradesh and has yield potential of 55.0-65.0 t/ha.
Narendra Agrim (NDPK-24)	This variety has been developed at NDUA&T, Faizabad (Uttar Pradesh). It has short vines, very early maturing, small fruited and highly homogeneous. Fruits are round and dark green. It bears female flowers four days earlier than male flowers and first picking could be taken within 55 days. Green fruits weigh 1.5-2.0 kg and mature fruits are of 3.0-4.0 kg weight. It is recommended for cultivation in the states of Punjab, Uttar Pradesh, Bihar, Jharkhand, Rajasthan, Gujarat, Haryana and Delhi and has yield potential of 30.0-35.0 t/ha.
Azad Pumpkin-1	This variety has been developed at CSAUA&T, Kanpur (Uttar Pradesh). It is spreading type with green hairy leaves, early fruiting, close fruit bearing, green fruit with yellowish stripes, thick flesh, less seeds and spherical flat. It has been recommended fo cultivation in Uttar Pradesh and has yield potential of 42.0-45.0 t/ha.
PP-113	It is evolved at PAU, Ludhian (Punjab) through selection from the local material. The fruits are spherical in shape, small sized and average weight is 2.5 kg. Fruit skin is mottled green when immature and orange-brown at maturity. The flesh colour is golden yellow and rich in β-carotene. Average yield is 41.0 t/ha.
Saras	This variety has been developed at KAU, Vellanikkara (Kerala). It has elongated, orange fleshed and medium size fruits each fruit weighing 2.7 kg. It has yield potential of 39.0-41.0 t/ha.
Summer squash (*Cucurbita pepo*)	
Punjab Chappan Kadoo No.1	This variety has been developed from PAU, Ludhiana (Punjab) through selection from the local material. The plants are bushy and erect with thick foliage. Leaves are non-lobed. Female flowers appear at a very early stage of plant growth and sometimes even before the male flowers. The fruits are attractive, light green, disc shaped, mildly ribbed with flat stem-end. First picking is possible in 60 days after sowing. Average fruit yield is 24.0 t/ha. It is tolerance to CMV, powdery and downy mildew and red pumpkin beetle under field condition. It is recommended for cultivation in the states of Punjab, Himachal Pradesh and Uttarakhand and has yield potential of 30.0-32.0 t/ha.
Patty Pan	It is an introduced variety from United States (USA), released by ICAR-IIHR, Bengaluru (Karnataka). Plants are erect with short internodes. Fruits are white, disc shaped with notches along the margins. Average fruit weight is 350.0-400.0 g. The average yield is 55.0 t/ha in 90-95 days crop duration.
Pusa Alankar	This early maturing bush type high yielding hybrid has been developed between the cross of EC207050 x Selection- 1 P1-81 from ICAR-IARI, New Delhi. The fruits are dark green with light coloured stripes and slightly tapered towards the stem-end. Fruits are 25-30 cm long. First picking is possible in 45-50 days after seed sowing. Average yield is 43-45 t/ha an increase of 50 per cent over commercial cultivar Australian Green.
Australian Green	It is an extra early Australian introduction purified by ICAR-IARI, Regional Station, Katrain and recommended for cultivation. It is a bushy variety. The fruits are dark green, 20.0-30.0 cm long with longitudinal greenish white stripes all over the fruit. Fruits are tender and delicious. It is suitable for late January to February sowing in northern plains, and bears 15-20 fruits per plant. Average yield is 20.0 t/ha.
Early Yellow Prolific	This variety has been developed at ICAR-IARI, Regional Station, Katrain (Himachal Pradesh). It is a bush type and early in flowering. Fruits are of medium size, warted and tapering towards stem end with light yellow skin turning orange yellow on maturity. Flesh is tender when it is consumed as vegetable. It is recommended for cultivation in the states of Punjab, Himachal Pradesh and Uttarakhand and has yield potential of 25.0-27.5 t/ha.
Cucumber	
Swarna Ageti	This variety has been developed through mutation breeding and selected in M-5 generation and release for commercial cultivation from ICAR-RCER-HARP, Ranchi (Jharkhand). First harvest can be obtained 45-50 days after transplanting. Fruits are cylindrical long, medum, weight 150-200 g, green and without prominent placental hollowness. It is resistant to powdery mildew. Yield is about 30.0-32.2 t/ha.Crop duration is 100-120 days.
Swarna Poorna	This variety has been developed at ICAR-RCER-HARP, Ranchi (Jharkhand). Its fruits are cylindrical, long, medium-sized (300 g), light green without placental hollowness. It is tolerant to powdery mildew. First picking starts at 55-60 days after sowing. It has yield potential of 31.0-33.5 t/ha in 100-120 days of crop duration.

Variety	*Description*
Swarna Sheetal	Develop through cross between Hot Season x long Green followed by pedigree and recurrent selection and release for commercial cultivation from ICAR-RCER-HARP, Ranchi (Jharkhand). First harvest can be obtained 60-65 days after transplanting. Fruits are cylindrical long, weight 200-250 g, greenish white and without prominent placental hollowness. Yield is about 20.0-30.0 t/ha. It is tolerant to powdery mildew.
Kalyanpur Green	This variety has been developed at CSAUA&T, Kalyanpur, Kanpur (U.P.). It is spreading type and early in fruiting. Fruits are thick, green but brown at maturity. It is suitable for cultivation in rainy season and has yield potential of 10.0-12.5 t/ha.
Swarna Poorna	It is developed from ICAR-RCER-HARP, Ranchi (Jharkhand). Develop through pure line selection from CH-20. Fruits are cylindrical long, weight 300 g, light green and without prominent placental hollowness. Yield is about 30.0-35.0 t/ha. It is tolerant to powdery mildew.
Straight-8	This variety has been developed through introduction from U.S.A by ICAR-IARI, Regional Station, Katrain, Kullu Valley (H.P.). It is an early cultivar (50-55 days) suited for hills, leaves dark green, fruit thick, 20-25cm long, straight with round end, medium green, flesh light green, crisp and heavy yielder. The average yield potential is 32 t/ha.
Japanese Long Green	This variety was developed by ICAR-IARI, Regional Station, Katrain (H.P.) through selection of introduced germplasm of Japan. It is a temperate variety suited to hills and lower hills, extra early with 45 days maturity, fruit 20-25 cm long, yellowish green, flesh light green, crisp, dark green surface with white spines and heavy yielder. Plants essentially require staking for obtaining straight fruits, and gives yield of 20.0-25.0 t/ha. This is popular among the farmers of J&K, H.P. and U.P. Now seeds of this variety are not being multiplied.
Sheetal	This improved early maturing variety has been developed from BSKKV, Dapoli (Maharashtra) by selection of local material and release for commercial cultivation in Konkan region in 1982. The vine grows up to 2.7 m with 10 branches on each vine and bears 24 fruits per plant in crop duration of 95 days. Fruits are green with medium length (25.0 cm) and weight of 300.0-350.0 g; gives yield of 30.0-40.0 t/ha during rainy season and 20.0-23.0 t/ha during summer season. It can be raised from June to September, while the summer crop may be sown in January-February.
Poinsette	It is an introduction from U.S.A. This variety was originally developed at Charleston (South Carolina) USA. So, in India it is an American introduction made by ICAR-IARI, Regional Station, Katrain (H.P.) and being multiplied by National Seed Corporation with the name of 'Poinsette'. Fruit are dark green, 20.0-25.0 cm long, resistance to downy mildew, powdery mildew, anthracnose and angular leaf spot. Its yield potential is 20.0-25.0 t/ha. This is popular among the farmers of J&K, H.P., Uttarakhand, Chhattishgarh, Orissa and A.P.
Pusa Uday	This popular variety has been developed by ICAR-IARI, New Delhi. It is released by State Variety Release Committee and recommended for cultivation in northern plains. The fruits are light green, medium sized (15 cm long) striped and soft-skinned. It takes 50-55 days from sowing to first picking in spring-summer season and 45-50 days in rainy season. Average fruit weight is 280-300 g and fruit yield 15.6 t/ha in spring summer season and 9.5 t/ha in rainy season. Crop duration is 100-110 days.
Pusa Sanyog	This hybrid (cross between Japanese gynoecious line and 'Green Long Naples') was released by ICAR-IARI Regional Station, Katrain, Kullu Valley (Gill *et al.,* 1973). Fruits are 28.0-30.0 cm long, cylindrical and dark green with yellow strips, and crisp flesh. First harvesting can be done after 50 days of sowing. This is early and high yielding (20.0 t/ha).
Himangi	It is developed from the cross between Poinsette and Kalyanpur Ageti at MPKV, Rahuri (Maharashtra) in the year 1992. Its fruits are greenish-white with very tender and juicy flesh. It is free from bronzing and takes 100-110 days to reach marketable maturity. Average yield is 19.0 t/ha.
Phule Priyanka	It is derived from a cross of RHRBG-5 and RHRBG-4, developed at MPKV, Rahuri (Maharashtra). Fruits are dark green, highly prickled surface and 20.0 cm long. This hybrid is suitable for cultivation under rainy and summer seasons. It is tolerant to downy mildew and resistant to bronzing. It gives an average yield of 19.27 t/ha in a crop duration of 100-110 days.
Phule Shubhangi (Sel-75-1-10)	This variety has been developed through pedigree selection at MPKV, Rahuri (Maharashtra). Its fruits are smooth and attractive green with white strips at apical end. It is tolerant to pod borer, anthracnose, leaf spot and leaf miner under field conditions and has yield potential of 30.0-32.5 t/ha.
Phule Prachi (F_1)	This hybrid has been developed at MPKV, Rahuri (Maharashtra) having gynoecious sex form suited to the tropical conditions. It is very much suitable for open field as well as polyhouse conditions. Colour of fruit is yellowish white. It has high yield potential and suitable for Kharif season. It gives an average yield of 36 t/ha.
PH-18	It is a cross between GYC-4 and Poona Khira. Fruits are yellow, with average weight of 146.0g. It gives an average yield of 58.60 t/ha.
Poona Khira	Fruits are small-sized and pale green. It is popular in Maharashtra and adjoining area of Karnataka and Andhra Pradesh. It gives an average yield of 18.27 t/ha.
CO-1	This popular variety has been developed by Tamil Nadu Agricultural University, Coimbatore (Tamil Nadu) through selection from a local genetic stock collected from Kanyakumari district. Fruits are long (60.0-65.0 cm), slightly curved tapering towards stalk end. The variety has an yield potential of 15.0-18.0 t/ha in a crop duration of 100 days.
KTCH-8	This early maturing hybrid has been developed at ICAR-IARI New Delhi by crossing between Poinsett and LC-3. Fruits are dark green, 25.0-30.0 cm long, cylindrical and skin bear white spines. Flesh is crisp. It gives an average yield of 18.2 t/ha.
KTCH-11	This hybrid has been developed by ICAR-IARI New Delhi using gynoecious line. Fruits are yellowish light green, 25.0-28.0 cm long, cylindrical, and skin bear brown spines. Flesh is crisp. It gives an average yield of 18.6 t/ha.
Pant Sanker Khira-1	This hybrid is developed by GBPUA&T, Pantnagar (Uttarakhand). Vines are about 120 cm long. Fruits are about 20 cm long, cylindrical and green with light stripes; first picking starts at about 50 days after sowing; gives average an yield of 20 t/ha; and suitable for planting in plains as well as hills. This is popular in Uttarakhand and U.P.
Priya	It is a hybrid cucumber of Indo-American Hybrid Seeds Co. Bengaluru (Karnataka) which is commercially available in India. The fruit is about 25 cm long, dark green, bitter free and slicing types. It is tolerant to mildew, mosaic, anthracnose and scab diseases.

Variety	Description
Gujarat Cucumber-1	This variety has been developed in the year 2004. It is tolerant to mosaic and anthracnose. The average yield potential is about 21 t/ha.
Balam Khira	'Balam Khira' is a primitive type/ecotype of genotype. Its fruits are non-bitter, small oval shaped, tender and pale green with brown spines. It is popular in Saharanpur region of Uttar Pradesh. It is akin to pickling types grown in Europe and U.S.A. Yield is 16-18 t/ha.
Solan Hybrid	This hybrid is released at YSPUH and F, Solan (H.P.) and suitable for growing in temperate regions of the country. Average yield potential is 22-24 t/ha.
Solan Green	This cultivar has been developed at YSPUH and F, Solan. Its fruits are light green and 11-15 cm long. It has yield potential of 18.0-20.0 t/ha. It is tolerant to mosaic, anthranose, powdery & downy mildew.
AAUC-1	This cultivar developed by Assam Agricultural University, Jorhat (Assam) and released for commercial cultivation.
AAUC-2	This cultivar developed by Assam Agricultural University, Jorhat (Assam) and released for commercial cultivation. Yield is 16 t/ha.
Gujarat Cucumber-1	This variety has been developed from a local collection at GAU, Anand (Gujarat). Its fruits are longer, tender and attractive with green strips, soft and non-hairy with longer shelf life. Gujarat Cucumber-1 has been recommended for cultivation in Gujarat and has yield potential of 24.0-26.0 t/ha.
Gherkin	
Ajax (hybrid)	This hybrid has been developed by Nunhems Seeds Pvt. Ltd. This is an early maturing F_1 hybrid that produces predominantly female flowers. The vines are vigorous growing. The fruits are straight, slightly tapered very firm, medium to dark green in colour, spiny and are born in clusters. It is tolerant to hot and humid weather conditions. In general, 200-250 numbers of fruits per kilogram has been recorded. The yield potential is about 28 t/ha.
Calypso	It is an gynoecious base F_1 hybrid introduced from U.S.A. It is pickling type, early maturing, and takes 50-55 days to reach harvestable maturity. The fruits are cylindrical, 2.0-2.5 cm thick and 8.0-10.0 cm long. The rind is medium green and bears white spines. The average yield potential is about 32 t/ha.
Stimora	This hybrid has been developed by Nunhems Seeds Pvt. Ltd. This is vigorously growing F_1 hybrid. The fruits are straight, cylindrical, very firm, dark green in colour and smooth. It is resistant to powdery mildew, scab and partially resistant to CMV. In general, 250-280 numbers of fruits per kilogram has been recorded.
Berendey	It is mid-ripening parthenocarpic hybrid of female type of flowering for winter spring and extended cycles. Plants are of high shade-resistance, medium length with medium branching. Each node contains 2-4 fruit sets. Fruits are 14-16 cm long, 3.0-4.0 cm in diameter, weighing 120-140 g, white spines, green, and intended for salads. The average yield is 28-32 kg/m^2. It is tolerant to powdery mildew and root moulds.
Gembit	It bears female type of flowers (gynoecious) and is a high level of parthenocarpy hybrid. Plants are long with medium branching. Rootage is vigorous, good foliated and leaves are intense green. Fruits are 14-16 cm long, green with light strips and cylindrical. Knobs are medium sized, frequent and each node has up to 4-6 ovaries. Fruits are good for storage and transportation. The hybrid is resistant to root rots, powdery mildew and olive spot and tolerant to downy mildew. It is intended for growing in low volume culture.
Kadet	It is a high shade-enduring, female type of flowering, with a high level of parthenocarpy hybrid. Plants are long; branching is medium and upper medium. Rootage is vigorous. Good foliated, leaves are intense green. Fruits are 19.0-21.0 cm long, dark green with short light strips reaching the fruit length and cylindrical. Knobs are medium sized and frequent. Each node has 1-2 (up to 4) ovaries and is intended for growing in low volume culture.
Kurazh	It is an early parthenocarpic hybrid of female type of flowering for planting in spring-summer and summer-autumn cycles in plastic houses and greenhouses. Transplanting of seedlings in greenhouses is possible from mid-February. Plants are of medium size with medium branching. Each node has 2-5 fruit sets and the plant can have up to 20-30 fruits simultaneously. Fruits are 13.0-15.0 cm long, 3.5-4.0 cm in diameter, of 120.0-140.0 g weight, dark green with light tuberous stripes, white spines on skin and good for multiple uses. The yield is in plastic tunnels is up to 25.0 kg/m^2. This hybrid is more or less resistant to powdery mildew and downy mildew.
Voyage	It is an early parthenocarpic hybrid of female type of flowering characterized by the unique precocity. Fruiting begins on the 38-40th day after emergence of first shoots. Hybrid is recommended for spring-summer and summer-autumn cycles in plastic houses and glasshouses. Plants are of medium size and with medium branching. This hybrid is characterized by good fruit-setting and quick fruit formation. Each node contains 1-2 fruits, which are 10-12 cm long, 3.0-3.5 cm in diameter, 90.0-110.0 g in weight, tuberous white spines on skin and for universal use. The yield of hybrid in plastic houses is 20-23 kg/m^2 and it is resistant to powdery mildew, tolerant to downy mildew and root moulds.
Pyzhik	It is an early parthenocarpic hybrid of gherkin for spring-summer and summer-autumn growing cycles in plastic houses and greenhouses. Each node contains 2-5 fruit sets (on side shoots up to 8). Plants are of medium size with medium branching and female type of flowering. Fruits are 10.0-12.0 cm long, 3.5-4.0 cm in diameter, 90.0-110.0 g in weight, dark green colour, frequently tuberous with light stripes and white spines and are suitable for universal use. The yield is up to 20.0 kg/m^2 and it is resistant to powdery and false powdery mildew.
Break	It is an early parthenocarpic hybrid of female type of flowering. It is recommended for spring-summer and summer-autumn growing cycles in plastic and glass houses. Plants are of medium size and the level of branching is medium. Each node contains 2-4 and more fruit sets. Fruits are of gherkin type, 9.0-11.0 cm long, and 3.3-3.8 cm in diameter, weighing 80.0-100.0 g, bright green in colour, tuberous with white spines and for universal use. The yield in plastic non-heated houses is 16-18 kg/m^2. The hybrid is more or less resistant to powdery mildew and downy mildew.

Variety	Description
Murashka	It is an early parthenocarpic hybrid of female type of flowering. Recommended for spring-summer and summer-autumn growing cycles in glass houses and plastic houses. Plants are medium size with medium level of branching. Each node contains 2-4 fruit sets. Fruits are 10.0-12.0 cm long, 3.5-4.0 cm in diameter of 90.0-110.0 g in weight, dark green with light green stripes, tuberous and have black spines. It has high picking qualities with yield up to 18 kg/m^2. The hybrid is resistant to powdery mildew andmore or less resistant to downy mildew.
Pechora	It is an early parthenocarpic hybrid of female type of flowering. Hybrid is recommended for spring-summer and summer-autumn growing cycles in plastic and glasshouses. Plants are of medium size and the level of branching is medium. Each node contains 2-4 ovaries. Fruits are 11.0-13.0 cm long, 3.5-4.0 cm in diameter, weighing 100.0-120.0 g, dark green, tuberous with brown spines and good for universal use. The yield is up to 18 kg/m^2 and it is resistant to powdery mildew and downy mildew.
Picnic	It is an early parthenocarpic hybrid spring-summer and summer-autumn growing cycles. Hybrid is notable for its simultaneous yield and is recommended for short cycle. Transplanting of seedling in green houses is possible from the end of February. Plants are of medium size with low branching level and female type of flowering. Each node contains 2-5 (up to 8-10) fruit sets with the main stem especially rich in fruit sets. Fruits are 10.0-12.0 cm long, 3.3-3.8 cm in diameter, of 90.0-100.0 g in weight, dark green, tuberous having white spines and can be used universally. Fruits are good at different maturity stages (immature big fruits, gherkins, pickle) and yields 18-20 kg/m^2 in plastic houses and are more or less resistant to powdery mildew and downy mildew.
Raisa	It is a high shade-tolerant variety with female type of flowering and high level of parthenocarpy. Plants are long with good branching, rooting is vigorous, good foliated and leaves are intense green. Fruits are 18.0-20.0 cm long, green with short light strips and cylindrical. Knobs are medium sized, frequent and each node has 1-2 (up to 3) ovaries. It is intended for growing in low volume culture.
Trembita	It is high level of parthenocarpy hybrid with female type of flowering. Plants are long with good branching. Rooting is vigorous, good foliated and leaves are intense green. Fruits are 15.0-17.0 cm long, green with short light strips and cylindrical. Knobs are medium sized, frequent and each node has up to 4-6 ovaries. Fruits are good for storage and transportation. The hybrid is resistant to root rots, powdery mildew and olive spot and tolerant to downy mildew. It is intended for growing in low volume culture.
Richmond green apple- #117	This cultivar developed at Morden Exp. Farm in Manitoba, Canada by acquiring a unique heirloom from Australia. Plants are short and bear hardy vines. The fruit are the size of a lemon, but are of a beautiful light green color. Fruits are uniform, very mild, sweet, juicy and excellent in eating. A good variety for salads or for dill pickles, and large yields. This variety is very early (45-50 days) and consider excellent for short season areas. Good yields of medium size fruits that is suitable for pickles or slicing.
Boston pickling- #114	This is an early variety, which start fruiting within 50 days. Vigorous vines give large yields of smooth green, crisp and good quality fruit which is excellent for pickles.
West Indian burr gherkins- #116	This is an early variety, which start fruiting within 65 days. *Cucumis anguria* not a true cucumber, but used much like it. This variety will not cross with *Cucumis sativus*. It bears long vines and hundreds of small beautiful tasty fruit, yields more than any other cucumber variety. These are becoming rare as they do great in hot humid weather. This cultivar was introduced to the USA in 1793 from Jamaica, and used pickled or boiled by the colonies in Jamaica.
Long melon	
Lucknow Early	This variety is very popular in Lucknow and its adjoining areas. The fruits are long, tender and fleshy.
Arka Sheetal	Arka Sheetal variety was developed by ICAR-Indian Institute of Horticultural Research, Bengaluru (Karnataka) through local selection from Lucknow Early. Fruits of this variety are medium long (22 cm), light green in colour and free from bitter principle. Fruits are tender and soft. This variety is of 90-100 days duration. It is an aromatic and crispy variety bearing stripes and individual fruit weighs around 80.0-90.0 g. The average yield is 35.0 t/ha.
Punjab Long Melon-1	This variety is released from PAU, Ludhiana. It is an early variety. The fruits are cylindrical long and light green in colour. The average yield is 22-24 t/ha. Free from downy mildew.
Chhoti Kakri	The vines are short in length and bear small size fruits. At initial stage the fruits have hairs on them. Yield is 18 t/ha.
Pant Kakri-1	This variety is developed through inbreeding and selection from indigenous collection at GBPU and T, Pantnagar (Uttarakhand) and released by Uttar Pradesh Variety Release Committee. The fruits are long, light green and straight. The harvesting starts in 50 days after seed sowing. It is suitable for spring-summer cultivation in plains. Seed to seed stage is about 90 days. The yield potential is about 24-26 t/ha.
Watermelon	
Arka Akash	This high yielding hybrid developed by ICAR-IIHR, Bengaluru (Karnataka) and released for commercial cultivation in 2009. Dark green with light green broken specks slightly deep foliage. Fruits are oblong/oval round, red flesh, juicy and very good taste. It has high TSS (12-13 Brix). Average fruit weight is around 6.5 kg with 1 fruit per vine. Yield 65-70 t/ha in 90-95 days of crop duration.
Arka Aiswarya	This high yielding hybrid developed by ICAR-IIHR, Bengaluru (Karnataka) and released for commercial cultivation in 2009. It has high T.S.S. (10-12°Brix). Fruits are oblong. Yield 80-90 t/ha.
Arka Muthu	This high yielding hybrid developed by ICAR-IIHR, Bengaluru and released for commercial cultivation in 2010. It has unique character of dwarf vine (vine length nearly 1.2 m), shorter internodal length and early maturing type (75-80 days). It has round to oval fruits with dark green stripes and deep red flesh. Average fruit weight is 2.5-3.0 kg with T.S.S. ranging from 12-14°Brix. Fruit yield 85-90 t/ha.
Arka Madhura	This high yielding cultivar has been developed at ICAR-IIHR, Bengaluru (Karnataka). Triploid seedless water melon (fully seedless) with light green rind colour and dark green stripes having longer shelf life and transport quality. Fruits are round with pink flesh colour and high T.S.S. (13-14° Brix). Suitable for year round production under protected condition. Round shaped. Fruit yield 50-60 t/ha in 120 days.

Variety	Description
Arka Manik	It is a multiple disease resistant variety developed through hybridization (IIHR-21 x Crimson Sweet) followed by pedigree selection at IIHR, Bengaluru. It has angular stem and light green deeply-lobed foliage. Its fruits are round-oval with light hreen rind and dark green stripes, eachweighing 6.0 kg, flesh is deep crimson, granular texture, pleasant aroma, TSS 12-15 per cent with few small seeds arranged in rows, seeds are small with dark brown specks on the testa. It has good storage and transport quality. It resistant to powdery and downy mildew and anthracnose. It has been recommended for cultivation in the states of Punjab, U.P, Bihar, M.P, Maharashtra, Karnataka, Tamiol Nadu and Kerala and has yield potential of 60.0-61.0 t/ha in110-115 days of crop duration.
HW-1	This variety has been developed by CCS HAU, Hisar (Haryana) in 1990. Fruits are small, round, green stripe fruits, deep red flesh, suitable for kitchen garden, and high TSS. Average yield potential is 27.5-30.0 t/ha.
Fuken	This variety has been developed by CCS HAU, Hisar (Haryana) in 1991. Fruits round, thick skinned red fleshed, 3-5kg fruits, good for transport, and high TSS. Average yield potential is 32.5-35.0 t/ha.
Improved Shipper	It is an introduction from U.S.A. and release by PAU, Ludhiana (Punjab). Its fruits are round with a quality characteristics. It is resistant to anthracnose. tripartite blossom end and takes 85-90 days to reach marketable maturity. Fruit weight varies from 6-10 kg. Its rind is thick and dark green. Its flesh is red and average TSS content is 8-9 per cent. Its seeds are uniform and light brown. It yields about 32.5 t/ha.
NS 295	This high yielding hybrid developed by Namdhari Seeds Co., Bengaluru (Karnataka). Fruits are blocky oblong with light green rind and dark green stripes; flesh is deep crimson with average TSS content of 12-13 per cent. It has good keeping quality and transportation qualities. Average fruit weight is 8 kg and yield is 80.0 t/ha.
Mayur	This high yielding hybrid developed by Namdhari Seeds Co., Bengaluru (Karnataka). The average yield is 75-85 t/ha.
Amrut	This high yielding hybrid developed by Mahyco Seed Co. The average yield potential is 85 t/ha.
Santrupthi	This high yielding hybrid developed by Mahyco Seed Co. The average yield potential is 78-80 t/ha.
Pusa Bedana	It is a triploid seedless watermelon developed at ICAR-IARI, New Delhi from the cross Tetra-2 (tetraploid) x Pusa Rasal (diploid.). Its fruits are medium dark green in colour with faint strips, shape triangular, rind tough, flesh red with white remnants of false seed. Each fruit weighing 3-4 kg and sometimes reach to 5.0-6.0 kg. This is the first known example of triploid in India. Its yield potential is 20.0-25.0 t/ha in 115-120 days. The seeds of this variety are not being multiplied presently.
Asahi Yamato	This is an introduction from Japan maintained by ICAR-IARI, New Delhi. Fruit shape is round, rind colours is pale green, smooth and without strips, flesh deep red, TSS 9-11 per cent. Average fruit weightis 6.0-7.0 kg. It can yield 20.0 t/ha in 90-100 days. The seed of this variety are not being multiplied presently.
New Hampshire Midget	This is an introduction from USA, maintained by ICAR-IARI, New Delhi. Fruit size is small, rind colour is light green with black strips, and flesh is red and sweet. Average fruit weight is 1.5-2.0 kg. It is suitable forhome gardening. The seed of this variety is not being multiplied presently.
Durgapura Kesar	It is also a late maturing varietyreleased by Rajasthan Agricultural Research Institute (ARS), Durgapura (Rajasthan) from cross between Durgapura Basanti x Durgapura Meetha and released by State Seed Varietal Evaluation Committee in 1977. Fruit shape is slightly oval; color light green with strips, and thick and brittle, flesh colour saffron, juicy and TSS 10-11 per cent. The fruit weighs about 4-5 kg with green rind having stripes. Flesh is yellow in colour, moderately sweet and seeds are large in size.
Durgapura Lal (RW-177-3)	The variety is developed at Rajasthan Agricultural Research Instituue, RAU, Durgapura (Rajasthan) from cross between Sugar Baby x KS 3566 and released by state seed varietal Evaluation committee in 2003. Fruit shape is round, colour dark green with dark thin linigh, ring hard flesh colour dark red, crisp, juicy, TSS 10-11 per cent, and average fruit weight is 4-5 kg. First picking starts 105-110 days after sowing. It gives an average yield of 35.0-45.0 t/ha. The variety is moderately resistant to blight and bud necrosis under field condition.
PKM-1	It is late maturing cultivar released by TNAU, Coimbatore (Tamil Nadu). The fruit is dark green, flesh pinkish red gets ready in 120-135 days yielding about 36.0-38.0 t/ha.
Madhu	It is a hybrid cultivar released by Indo-American Hybrid Seed Co., Bengaluru (Karnataka). The fruits are oval, light green coloured with dark green strips red fleshed. It is resistant to *Fusarium* wilt. The yield potential is 42 t/ha.
Milan	It is a hybrid cultivar released by Indo-American Hybrid Seed Co., Bengaluru (Karnataka). The fruits are oval with light grey green skin, tough rind and smooth surface. The flesh is red and sweet having good shipping quality. The yield is 44 t/ha.
Mohini	It is a hybrid cultivar released by Indo-American Hybrid Seed Co., Bengaluru (Karnataka) having rind and excellent shipping quality. It is resistant to Fusarium wilt. The yield potential is 42-46 t/ha.
Farrukhabadi	Very popular genotype in Farrukhabad (Uttar Pradesh). Fruits with dark green colour or pale green with black stripes, and oblong to round shape weighing 8 -10 kg with thick rind. The flesh colour would be variable from pale pink to pink with big, flat, black to brown coloured seeds. The yield potential is 44-49 t/ha.
Moradabadi	Very popular genotype in Moradabad (Uttar Pradesh). Fruits with dark green colour or pale green with black stripes, and oblong to round shape weighing 8 -10 kg with thick rind. The flesh colour would be variable from pale pink to pink with big, flat, and black to brown coloured seeds. The yield potential is 44-49 t/ha.
Faizabadi	Very popular genotype in Faizabad (Uttar Pradesh). Fruits with dark green colour or pale green with black stripes, oblong to round shape weighing 8 -10 kg with thick rind. The flesh colour would be variable from pale pink to pink with big, flat, and black to brown coloured seeds. The yield potential is 50 t/ha.
Matteera	In Rajasthan, Mateera cultivar is grown in rainy season around Bikaner. The yield potential is 52-55 t/ha.
Kotagolan	It is a local cultivar of Jamuna river bed. Its flesh is not sweet but it keeps well for over 2-3 months at ambient temperature during months of July-September. The flesh becomes dry and fibrous instead of getting fermented and rotten. Yield is 48-52 t/ha.
Honey Cream	Very popular watermelon cultivars of U.S.A. Average yield is 60 t/ha.

Variety	Description
Dixie Cream	Very popular watermelon cultivars of U.S.A. Average yield is 65 t/ha.
New Hampshire	Very popular watermelon cultivars of U.S.A. Average yield is 58-62 t/ha.
Midget	Very popular watermelon cultivars of U.S.A. Average yield is 70-75 t/ha.
Rhode Island Red	Very popular watermelon cultivars of U.S.A. Average yield is 70-72 t/ha.
Klondike (R7)	Very popular watermelon cultivars of U.S.A. It is resistant to*Fusarium* wilt. Average yield is 70 t/ha.
Congo	Very popular watermelon cultivars of U.S.A. It is resistant to*anthracnose* wilt. Average yield is 69-72 t/ha.
Cole's and Early	Very popular watermelon cultivars of U.S.A. Average yield is 70-75 t/ha.
Dixie Queen	Very popular watermelon cultivars of U.S.A. Average yield is 68 t/ha.
Fair Fax	Very popular watermelon cultivars of U.S.A. Average yield is 72 t/ha.
Black Lee	Very popular watermelon cultivars of U.S.A. It is resistant to*Fusarium* wilt. Average yield is 70 t/ha.
Early Kansas	Very popular watermelon cultivars of U.S.A. Average yield is 76 t/ha.
Early Blizona	Very popular watermelon cultivars of U.S.A. Average yield is 80 t/ha. Resistant to Fusarium welt.
Shipper	This American variety is an introduction, recommended by PAU, Ludhiana (Punjab). Its vines are long and vigorous; fruits are medium to large weighing 6.0-10.0 kg, round with characteristic tripartite blossom end. Flesh is red with moderate sweetness, TSS 8-9 per cent. Rind is thick and dark green. Seeds are uniform light brown. It has yield potential of 30.0-32.0 t/ha.
Muskmelon	
Gujarat Muskmelon-1	It is selection from a local germplasm collection (from Sabarkantha district in Gujarat) and released for commercial cultivation in 1975 by GAU, Anand (Gujarat). Fruits are small, tasty, lemon yellow, and smooth without suture and 5-6 fruits/plant. Flesh is light green with 2.1 cm thickness. The average weight of fruit is 982 g with a diameter of 10.0 cm. It has yield potential of 14.0-17.0 t/ha.
Gujarat Muskmelon-2	It is a selection from the local germplasm collected from Nagpur area of Maharashtra and released for commercial cultivation in 1975. In this variety, flowering starts within 32 days after sowing. Fruits are medium in size diameter (11.5 cm), tasty, 4-5 fruits/plant and average fruit weight is 1300 g. skin of the fruit is orange green with slight netting and no suture. Flesh is orange with 2-6 cm thickness. It has yield potential of 12.0-15.0 t/ha.
Gujarat Muskmelon-3	This variety has been developed in the year 2004 from a local collection at Anand Agricultural University, Anand. Its fruits are medium size, round to slightly oval, and yellowish brown spots on mature fruits with green strips. Flesh is green, juicy, very sweet, 12 per cent TSS, average fruit weight 800 g. It has been recommended for summer cultivation in irrigated areas of middle Gujarat and has yield potential of 11.0-12.5 t/ha.
Hisar Madhur	This variety has been developed by CCS HAU, Hisar (Haryana) in 1991 through pedigree selection between a cross of Pusa Sharbati x 75-34 (Kalloo *et al.*, 1993). Plant is indeterminate with spreading habit, foliage green, andromonoecious, fruits round with about 10 ribs, yellow brick red skin with green strips on ripening, orange flesh, small cavity, musky lavor, juicy, TSS 8.5 per cent and high yielding. It takes 75-80 days from sowing to first picking. Average yield potential is 9-10 t/ha.
Hisar Saras	This variety has been developed by CCS HAU, Hisar (Haryana) in 1991.It is early, fruits round, green flesh, netted, tolerant to downy mildew. Average yield potential is 8-10 t/ha.
Punjab Sunehri	It has been developed from the cross between Hara Madhu and Edisto, an introduction from the U.S.A. at PAU, Ludhiana (Punjab). The fruits are globular round having thick leathery and intensely netted brown rind. Sutures are absent. Flesh is thick, orange, medium juicy and flavoursome. Average TSS is 11 per cent and maximum goes up to 15 per cent. The maturity is 12 days earlier than that of Hara Madhu. The fruits develop full slip stage at maturity, have good keeping and transport qualities. Average fruit weight is 800 g and yield is 16.0 t/ha. It is tolerant to powdery and downy mildew.
Punjab Rasila (MR-12)	A variety moderately resistant to downy mildew and highly resistant to powdery mildew developed at PAU, Ludhiana (Punjab) through selection from a cross between WMR-29 x Hara Madhu and release for commercial cultivation in 1973 for Punjab state. Its vines are long and fruit setting is close to the base of the vine. Fruits are round in shape and light yellow in colour and develop full slip stage at maturity. Its average weight is about 600 g. The flesh is green, thick, juicy and flavoursome. The TSS is 10-11 per cent. It takes 73 days from transplanting to first picking. It gives an average yield of 15.0 -16.0 t/ha (Nandpuri *et al.*, 1995). It is suitable for riverbed cultivation.
Punjab Hybrid-1	It is an F_1 hybrid developed from the cross MS1 x Hara Madhu at PAU, Ludhiana (Punjab). The vines are vigorous and green. The fruits are nearly round, rind is light yellow, sutured and moderately netted. The fruits are invariably sweet. Flesh is thick, orange and juicy. Average TSS is 12 per cent. Fruits develop full slip stage. Average fruit weight is 1000 g and yield is 16.0-18.0 t/ha. It is moderately resistant to powdery mildew and fruit fly. It has been recommmened for cultivation in the states of Punjab, Bihar, Uttar Pradesh, Tamil Nadu and Kerala.
Pusa Hybrid-1	This hybrid has been developed from ICAR-IARI, New Delhi by using monoecious lines M-3 and Durgapura Madhu. Plants are vigorous and length of vine is 2.0-2.5 m. Fruits are round shaped with netted rind on the skin. Flesh is orange colour. It contains TSS 13.0-14 per cent. Average fruit weight is 800-1000g. An average yield of this variety is 17.0 tonnes per hectare in 75-80 of crop duration.
Pusa Rasraj	This hybrid has been developed from ICAR-IARI, New Delhi by using monoecious lines MS-1 x Hara Madhu. Fruits are slightly oval, smooth, and non-striped with green flesh. TSS is around 11-12 per cent. An average yield of this variety is 22.5 t/ha.

Variety	Description
MH-10	This hybrid has been developed at PAU, Ludhiana (Punjab) by cross combinations WI 998 x Punjab Sunehri. Fruits are juicy and having pleasant aroma. This hybrid is suitable for distant marketing. Average fruit weight is 400 g with TSS 9-10 per cent. Average fruit yield is 20.0 t/ha.
RM-43	It is developed at Rajasthan Agricultural Research Institute, Durgapura (Rajasthan) through a selection from locally collected germplasm and released at state level bears oblong pale green fruits with dark green 10-12 sutures, average weight 400-600 g, hard rind, small seed cavity, green flesh and high TSS (10-12 per cent). Fruits have good keeping quality and transportability. It gives an average yield of 15.0-20.0 t/ha.
RM-50	It is developed at Rajasthan Agricultural Research Institute, Durgapura (Rajasthan) through a selection from the cross Durgapura Madhu x Sel-1and released by State Varietal Release Committee in 2003. Its vigorous vines spreads up to 2.50 m. Ovate shaped fruits are pale green in colour having ten green sutures and contain high TSS (14-16 per cent). Average fruit weight is 500 g and yield potential is 15.0-20.0 t/ha. It has been recommended for cultivation in Rajasthan.
MHY-3	It is developed at Rajasthan Agricultural Research Institute, Durgapura (Rajasthan) through a selection from the cross Durgapura Madhu x Pusa Madhuras and was notified by the State Seed Varietals Evaluation Committee in 1999. Its vigorous vines spreads up to 3.00 m and fruits are round flat and smooth surfaced with yellow rind and light green stripes. Flesh colour is light green, texture soft and and contain high TSS (13-16 per cent). Average fruit weight is 700-800 g and yield 15.0-20.0 t/ ha. It has been recommended for cultivation in Rajasthan.
MHY-4	It is developed at Rajasthan Agricultural Research Institute, Durgapura (Rajasthan) through a selection from the cross Durgapura Madhu x Hara Madhu is recommended for zone VII (Madhya Pradesh, And Maharashtra). Its vigorous vines spreads up to2.5 m. fruits with smooth rind and medium seed cavity are round flat, tapering at stem end and green yellow in colour. The flesh colour is light green, texture soft and having high TSS (13-16 per cent). Average fruit weight is 700-800 g and yield is 15.0-20.0 t/ha.
NS 7455	It is an F_1 hybrid developed Namdhari Seeds Co. Bengaluru (Karnataka). The fruits are round and intensely netted with small seed cavity. The fruits have good shelf life and are suitable for shipping. The flesh is orange and juicy. Average TSS is 13-14 per cent. The average yield is 16-18 t/ha.
Gayatri	Unicorn seeds, Hyderabad (Telangana) The average fruit yield is 14-16 t/ha.
Deepti	Unicorn seeds, Hyderabad (Telangana). The yield potential is 18 t/ha.
Drake	Nunhems PGS seeds New Delhi. The yield potential is 21 t/ha.
MHC-2	Mahyco, Jalana (Maharashtra). The yield potential is 18-20 t/ha.
MHC-5	Mahyco, Jalana (Maharashtra). The average yield is 21-22 t/ha. Free from viruses.
MHC-6	Mahyco, Jalana (Maharashtra). The average yield is 18-20 t/ha.
Swarna	IAHS, Bengaluru (Karnataka). The average yield is 14-16 t/ha.
Sona	IAHS, Bengaluru (karnataka). The average yield is 15 t/ha.
Madhubala	Century Seeds, New Delhi. The average yield is 14 t/ha.
Lucknow Safeda	This variety has been developed at PAU, Ludhiana (Punjab). It is an early variety with medium vine growth. Fruits are small, roundish, flat, yellow, smooth and each weighing 300 g. Flesh is creamy white, less juicy and sweet with good keeping quality. It has yield potential of 11.0-13.0 t/ha.
Narendra Muskmelon-1 (NDM-2)	This variety has been developed at NDUA&T, Faizabad (Uttar Pradesh). Its fruits are round, large-sized with orange colour skin, and green striped at maturity. Flesh is dark orange, and thick with 9-11 per cent TSS. Single fruit weight is about 700-900 g. It is recommended for cultivation in Uttar Pradesh and has yield potential of 12.5-15. t/ha.
Pointed gourd	
Kashi Alnakar (VRPG-1)	A high yielding variety developed by ICAR-IIVR, Varanasi (Uttar Pradesh) release through State Variety Release Committee and notified through Central Variety Release Committee N. Delhi which produces medium sized light green spindle shape and attractive fruits with some strips at distal end. Individual plants bear 100-130 fruits and gives an average yield of 30.0-38.0 t/ha.
Kashi Suphal (VRPG-2)	A high yielding variety developed and release by ICAR-IIVR, Varanasi (Uttar Pradesh). It is an early maturing variety. Fruiting starts from 4th to 6th node onwards. Fruits are light green with soft flesh. It is high yielding (28.0-30.0 t/ha).
Kashi Amulya (VRP-89)	A high yielding and less seeded variety has been developed from ICAR-Indian Institute of Vegetable Research, Varanasi (Uttar Pradesh) in the year 2018, release by State Varietal Release Cormmettee & CVRC, N. Delhi. Very good quality fruit. Its average yield potential is 20-22 t/ha.
Swarna Alaukik	It is developed from ICAR-RCER Horticulture and Agro-Forestry Research Programme, Plandu, Ranchi (Jharkhand). A high yielding variety produces light green fruits with blunt ends. The fruits are 5-8 cm long, solid, thin skinned and good for vegetable as well as preparation of sweets. Its average yield is 23.0-28.0 t/ha on vertical staking. It is recommended for uplands and plateau regions of Bihar, Diara lands of Gangetic belts of Bihar and Uttar Pradesh and plains of Odisa and West Bengal. It has also been introduced in Telangana region of Andhra Pradesh.
Swarna Rekha	This variety has been developed through clonal selection at from ICAR-RCER Horticulture and Agro-Forestry Research Programme (HARP), Plandu, Ranchi (Jharkhand). A vigorously growing high yielding variety. Fruits are elongated, greenish-white, striped 8-10 cm long and tapering on both sides with soft seeds. Average yield is 20.0-23.0 t/ha on vertical bower system but on ground it yields 17.5-20.0 t/ha. Recommended for commercial cultivation in plains of Jharkhand, Odisa, plateau region of Bihar, diara lands and plains of Bihar, West Bengal, Eastern Uttar Pradesh, and Telangana region.
Swarna Suruchi	This variety has been developed ICAR-RCER Horticulture and Agro-Forestry Research Programme, Plandu, Ranchi (Jharkhand). Yield is 16 t/ha.

Variety	Description
Rajendra Parwal- 1	This variety has been developed by clonal selection and released by RAU, Bihar. It produces very attractive long green fruits with white stripes and tapering at both ends. Average fruit weight is 40 g with good long distance transportation quality and tolerant to fruit fly. The average yield is 15.0-17.0 t/ha. It is recommended for commercial cultivation in Uttar Pradesh and adjoining parts of Bihar (including both *diara* land and upland areas).
Rajendra Parwal- 2	This variety has been developed by clonal selection and ıeleased by RAU, Bihar. Fruits are drum shaped, big, whitish green with very light self stripes and soft. Average fruit weight is 30 g. It is suitable for cultivation in Bihar and Uttar Pradesh, average yield being 15.0-17.0 t/ha. This variety is highly suitable for '*diara*' cultivation. It is tolerant to vine and fruit rot as well as fruit fly.
Faizabad Parwal-3	This variety has been developed by clonal selection and released by NDUA&T, Faizabad (Uttar Pradesh). Its fruits are spindle shaped, green and less striped. They are excellent for culinary purpose. With an average yield of 12.5-15.0 t/ha, it is suitable for cultivation in eastern and western Uttar Pradesh.
Faizabad Parwal-4	This variety has been developed by clonal selection and released by NDUA&T, Faizabad (Uttar Pradesh). A high yielding variety, it is recommended for reclaimed sodic soils. The fruits are light green, spindle-shaped with tapering ends. It is recommended for bower system of cultivation.
Narendra Parwal-260	This variety is developed through clonal selection at NDUA&T, Faizabad. Fruits are light green, striped, 13.0-15.0 cm in length with thick flesh. It produces an average fruit yield of 22.5 t/ha, when trained on bamboo stakes. It is tolerant to wilt disease complex.
Narendra Parwal-307	This variety has been developed by clonal selection and released by NDUA&T, Faizabad (Uttar Pradesh). Fruit size is small round shaped and colour is dark green with strips. It has very good self life and gives an average fruit yield of 22.5 t/ha.
Narendra Parwal-604	This variety has been developed by clonal selection and released by NDUA&T, Faizabad (Uttar Pradesh). Fruits are medium sized, light green and without strips. It gives an average fruit yield of 22.5 t/ha.
CHES Hybrid-1	It is developed from ICAR-RCER Horticulture and Agro-Forestry Research Programme, Plandu, Ranchi (Jharkhand). It is the first pointed gourd hybrid developed in the country. Its solid, green striped fruits weight 30-35 g each, and the total yield being 28.0-32.0 t/ha. It is resistant to fruit fly.
CHES Hybrid-2	It is developed from ICAR-RCER Horticulture and Agro-Forestry Research Programme, Plandu, Ranchi (Jharkhand). It is high yielding hybrid, producing dark green striped fruits with an average fruit weight of 25-30 g, and average yield being 30.0-40.0 t/ha.
Chotta Hilli	Fruits medium, long (5.71 cm x 3.37 cm), oval to spindle-shaped swollen in the middle, greenish with prominent white stripes, blunt end and bulged at the stalk.
Dandali	Fruits are medium-sized (6.84 cm x 3.89 cm), egg-shaped, light green, stock end dispersed, striped and slightly grooved towards distal end. Average yield is 18-20 t/ha.
Hilli	Its fruits are oblong (9.65 cm x 3.08 cm), greenish white, white striped and tapering towards distal end with dispersed neck. Yield potential is 16-18 t/ha.
Shankolia	Fruits are medium long (7.56 cm x 2.94 cm), resembling to those of "shankh or shell". They taper towards ends, are greenish with white stripes, slightly beaked towards distal end and bulged towards stalk.
BCPG-3	This early variety has been developed by BCKV, Kalyani (West Bengal) through selection from the local germplasm. Plants having good vigour and medium viny in nature. Stem shape angular, tendril branched and coiled. Leaves serrated mund and medium in size (6.5-7.0 cm), pubescent, intermediate and sparse. Petiole length (3.5-5.0 cm), fruit spindle shaped, medium curved, fruit skin primary colour light green with white alternate stripe. Average fruit length 7.30 cm, girth 3.5 cm, and fruit weight 34 g. Average fruit yield is 62 t/ha.
BCPG-4	This early variety has been developed by BCKV, Kalyani (West Bengal) through selection from the local germplas. Plants having good vigour and medium viny in nature. Stem shape angular and pubescent, tendril branched and coiled. Leaves serrated oblong and medium in size (7.5-8.0 cm), Pubescent, intermediate and sparse. Petiole length (3.5-5.0 cm), fruit spindle shaped, medium curved, and fruit skin primary colour dark green with white alternate stripe. Average fruit length 8.78 cm, girth 3.52 cm, and fruit weight 44 g. Average fruit yield is 60 t/ha.
BCPG-5	This early variety has been developed by BCKV, Kalyani (West bengal) through selection from the local germplas. Plants having good vigour and medium viny in nature. Stem shape angular and pubescent, tendril branched and coiled. Leaves serrated oblong and medium in size (8-9 cm), Pubescent, intermediate and sparse. Petiole length (4-5 cm), fruit long shaped, and fruit skin primary colour dark green with prominent white alternate stripe having smooth skin. Average fruit length 9.5 cm, girth 3.75 cm, and fruit weight 46 g. Average fruit yield is 58 t/ha.
Ivy gourd	
Indira Kundru-5	This variety has been developed IGKVV, Raipur Chhattisgarh released by State Sub-Seed Committee of Chhattisgarh State for cultivation in Chhattisgarh Plains. Vines are perennial in nature. First picking can be taken after 75-85 days of vine cutting. Fruiting is obtained round the year (9 months). It has average yield of 101.9 t/ha at spacing of 1.5 x 1.5 m. Fruits are oblong in shape, green in colour with strips. Suitable for table purpose. Variety is tolerant to stem borer, powdery mildew, frost and drought.
Indira Kundru-35	This variety has been developed by IGKVV, Raipur (Chhattisgarh) and released by State Sub-Seed Committee of Chhattisgarh State for cultivation in Chhattisgarh Plains. Vines are perennial in nature. First picking can be taken after 75-85 days of vine cutting. Fruiting is obtained round the year (9 months). It has average yield of 93.88 t/ha at spacing of 1.5 x 1.5 m. Fruits are long in shape, light green in colour with strips. Suitable for table purpose. Variety is tolerant to stem borer, powdery mildew, frost and drought.
Sulabha (CG-23)	This variety has been developed by clonal selection from a germplasm (CG-23) at KAU, Vellanikkara (Kerala). Fruits are long (9.25 cm), pale green, with average fruit weight is 18.5 g. It takes 37 days from planting to first flowering. The first harvest can be done in 45-50 days after planting. Average fruit length is 9.25 cm. fruits are cylindrical, slightly green with continuous striation. Leaves are typical trilobed. It produces 1050 fruits/year and has yield potential of 40.0-42.5 t/ha.

Variety	Description
VRK 20	It is clonal isolation at ICAR-IIVR, Varanasi (Uttar Pradesh). A very early strain bears light green striped fruits of 6.0 cm length and 2.7 cm diameter. The individual fruit weight is about 20 g.
VRK 31	It is clonal isolation at ICAR-IIVR, Varanasi (Uttar Pradesh). Fruit is somewhat swollen in the middle but pointed at the upper portion. The individual fruit weight is 25-30 g.
VRK 35	It is clonal isolation at ICAR-IIVR, Varanasi (Uttar Pradesh). A high yielding strain bears highly striped fruits of medium size with fruit weight 15-18 g.
VRK 37	It is clonal isolation at ICAR-IIVR, Varanasi (Uttar Pradesh). The oval shaped fruits of this strain are round 1.90 cm in diameter with individual fruit weight 16 g.
VRK 49	It is clonal isolation at ICAR-IIVR, Varanasi (Uttar Pradesh). Fruits are oval in shape. It is high yielding variety with fruit weight about 15 g.
Ac.Mo.51	This variety was isolated at IGKV, Raipur (Sarnaik *et al.,* 1999). Fruits are light green and oval in shape. It is very high yielding variety (21.20 kg/plant).
Arka Neelanchal Sabuja	It is clonal isolation at ICAR-Central Horticultural Experiment Station, IIHR, Bengaluru, Aiginia, Bhubaneswar (Odisha). Fruits are dark green and nearly conical shaped with fragmented creamy stripes giving typical appearance of ivy gourd. Its yield ranges from 30-35 kg/plant. Fruits have excellent cooking quality and tolerate bruising during handling and transport. It can be grown in hot and humid agro-climatic condition of the Eastern and Southern States of India.
Arka Neelanchal Kunki	It is clonal isolation at ICAR-Central Horticultural Experiment Station,(IIHR, Bengaluru, Aiginia, Bhubaneswar (Odisha). This is an early fruiting variety which starts bearing in just after 40 days of transplanting. The variety produces light green, long slender, cylindrical fruits of 7.5-8.5 cm length with soft texture and has attractive stripes in skin. This variety has sequential fruiting habit and fruit develops by means of vegetative parthenocarpy hence, there is no need of male plants for pollination and more number of female plants can be accommodated per unit area. It has an excellent organoleptic quality as fried vegetable as well as vegetable curries. A single plant produces up to 20 kg fruits in one growing season. The fruits fetch good market price due to high consumer preference as they have good physical appearance and soft texture.
Ridge gourd	
Pusa Nasdar	It is medium to early maturing cultivar developed through selection from local collected material of Madhya Pradesh, at ICAR-Indian Agricultural Research Institute New Delhi and released for commercial cultivation. It starts flowering and fruiting after 60 days of seed sowing and bear 25-20 fruits per vine. Fruits are club shaped, medium in size, light green and having prominent ridged. Flesh yellowish white to green in colour, firm texture with a typical aroma. This variety is most suitable for rainy and summer season cultivation. Average yield is 16.5 t/ha.
CO-1	This important variety has been developed by Tamil Nadu Agriculture University, Coimbatore (Tamil Nadu) for commercial cultivation through and selection from local biotype collected from Aduthurai (Tamil Nadu). It is vigorous in vegetative growth (vine spread up to 5 meter), 2-3 branches at the base, stem light yellowish green, and foliage light green in colour. In general, it bears 10-12 fruits per vine (3.0-4.0 kg), having 60.0-75.0 cm long, 30.0 cm thick, and prominently ribbed with attractive green colour. Fruit weight 300.0-350.0 g, medium seeded (200-300 seeds/fruit), days to flowering 35-40 days, first harvesting begins after 55 to 60 days, and 6-8 picking can be taken. Duration of crop is 125 days. The yielding potential of this variety is 14.0-15.0t/ha.
CO-2	This is a pure line selection from LA-11 made at TNAU, Coimbatore (Tamil Nadu) and released in Tamil Nadu for commercial cultivation. It is suitable for sowing in June-July and December-January. Plants are moderately vigorous with less spreading vines, adaptable to all seasons. Fruits are very long (125.0 cm), slender, light green in colour with shallow 11 ridges (grooves). Individual fruit weight is around 350.0-380.0 g. Fruits bear less seeds. Crop mature in 110 days and first picking starts at 70 days after seed sowing. It yields 25.0-27.8 t/ha.
Arka Sumeet	It is a cross between IIHR-54 x IIHR-24 and developed by ICAR-IIHR, Bengaluru (Karnataka). As a parent IIHR-54 is an early fruiting lines and IIHR-24 is a long fruited. The advance generation selection was made during 1992 and released as Arka Sumeet for commercial cultivation. Vines are medium vigorous, grow up to 4.0-5.0 m length. Each plant produces an average 15 fruits as compared to Pusa Nasdar (10.0). Fruits are 55.0 cm long with 2.5 cm girth with prominent ridges and delicate aroma. It takes 52 days for first picking. Average fruit weight is 380.0 g, and crop duration is 120 days. Fruits are lush green, cylindrical with slow seed maturity. It is suitable for cultivation round the year in Karnataka. It yield around 53.0 t/ha in 120 days compared with Pusa Nasdar (28.3 tonnes/ha), the yield increase being 57.27 per cent over Pusa Nasdar. Anjanappa *et al.* (1999) reported that Arka Sumeet was considered to be most promising cultivar for the southern dry region of Karnataka. It is suitable for round the year cultivation in Karnataka.
Arka Sujat	This is developed from ICAR-IIHR Bengaluru (Karnataka) Arka Sujat is a selection from a cross between an early maturing medium fruited IIHR-54 and a long fruited IIHR-18. It has inherited desirable characters of both parents. Arka Sujat bears first female flower from 13-15th node in a medium vigorous vine 5.5 m long and start flowering after 40 days of seed sowing. Fruits are lush green cylindrical and tender, medium long (35.0-45.0 cm) with an average fruit weight of 350.0 g. Its yield potential is 63.0 t/ha. It has good transport and keeping quality. Duration of crop is 100 days. It is moderately resistant to downy mildew. It is also suitable for growing round the year in Karnataka.
IIHR-8	This is developed by ICAR-IIHR, Bangalore (Karnataka). The fruits are round and green in colour. Each vine produces 40-45 fruits. First harvest starts in 55 days after seed sowing. Average fruit weight is 100.0 g. It gives an average yield of 40.0 t/ha. Besides high yield, this variety is moderately resistant to downy mildew. The fruits are attractive, light green with good cooking quality especially for stuffing purposes.
Punjab Sadabahar	Punjab Sadabahar is a selection from 'Tori Jhinga Dharidar' line of Raja Seed Co. This variety has medium vine length (5.0-10.0 m), dark green foliage, fruits long (20.0 cm long and 3.0-5.0 cm broad), straight, soft angular and dark green in colour. Best sowing time of this variety is May to July for rainy season crop. Average yield per hectare is 10.0-12.0 t/ha. This variety was released by PAU, Ludhiana (Punjab) for commercial cultivation in Punjab and nearby area.

Variety	Description
PKM-1	This variety is developed by mutation breeding (genotype H-160) at Horticulture Research Centre, Periyakulam, TNAU, Coimbatore (T. Nadu). It is suitable for both rainy and spring summer sowing. Fruits of this variety are dark green, average fruit weight is 300.0 g. It yields 28.0 30.0 t/ha.
Swarna Manjari	This variety is developed by ICAR-RCER-HARP, Ranchi (Jharkhand). The fruits of this variety are medium sized (15-20 cm long), green and ridged. Individual fruit weight is about 150-200 g. Fruit harvesting starts 65-70 days after seed sowing. It yield 18.0-20.0 t/ha in 140-150 days of crop duration. This variety is moderately tolerant to powdery mildew.
Swarna Uphar	This variety is developed by ICAR-RCER-HARP, Ranchi (Jharkhand). The fruits of this variety are medium sized (15-20 cm long) green, ridged and having individual fruit weight of 150.0-200.0 g. Fruit harvesting starts 65-70 days after seed sowing. This variety is moderately tolerant to powdery mildew. It yield 20.0-30.0 t/ha in 140-150 days of crop duration.
Kalyanpur Dharidar	This variety is developed at Vegetable Research Centre (CSAU and T), Kalyanpur Kanpur (U.P.). The fruits of this variety bear prominent ridges. Fruits are light green in colour and fleshy. It is very early maturing variety. Vines are short and bear more number of pistillate flowers. It yields 20.0-25.0 t/ha.
Konkan Harita	This variety is developed at BSKKV, Dapoli (Maharashtra) through selection. It gives first harvest in 45 days after planting. Fruits are dark green in colour, 30.0-45.0 cm long and tapering at both ends. Fruits have good market value. A vine produces 11-12 fruits. The average yield potential is 24 t/ha.
Phule Sucheta	This variety has been developed from MPKV, Rahuri (M.S.). The average yield potential is 26 t/ha.
Gujarat Anand Ridge gourd-1	This variety has been developed in the year 2010. Vines are long-spreading and fruits are tender and attractive. The average yield potential is 22-24 t/ha.
Hisar Kalitori	This variety has been developed by CCS HAU, Hisar (Haryana) and release by State Variety Release Committee in 1995. Suitable for rain fed areas, early long thin and straight fruited variety, tolerant to powdery mildew. It has yield potential of 8-10 t/ha.
Deepthi	This variety has been developed through single plant selection from a local material, at KAU, Vellanikkara (Kerala). Fruits are green with intermediate lusture and tapered stem end, medium sized (23 cm long and 15 cm in girth) with finely wrinkled surface. First picking is done in 53 days after sowing. Average fruit weight is 165 g. It possesses field resistance to mosaic virus and downy mildew diseases. It has yield potential of 30.0-32.5 t/ha.
Haritham	This variety is developed at college of Horticulture, KAU Vellanikkara (Kerala). Fruits are cylindrical, bright green with typical ridges and tapering towards the base. Average length, girth and weight of fruit are 46.5 cm, 20.0 cm and 650 g, respectively. It gives an average yield of 13.0 t/ha.
Pant Torai-1	This variety is developed at GBPU and T, Pantnagar (Uttarakhand); derived through pure line selection from an indigenous germplasm. It bears 25.0 cm long and club shaped fruits. First picking starts about 65 days after sowing. It gives an average yield of 10.0 t/ha. It is very much suitable for rainy season cultivation.
Sponge gourd	
Pusa Chikni	This early variety has been released from ICAR-IARI New Delhi through selection from local germplasm collected from Patana (Bihar). Flowering starts in 60 days after seed sowing. It is suitable for summer and rainy season cultivation. Fruits are smooth, tender, dark green in colour and cylindrical in shape. Single plant bears 15-20 fruits. It has yield potential of 20.0-22.5 t/ha.
Pusa Supriya	This variety has been released from ICAR-IARI New Delhi through selection from old ruling cultivar 'Pusa Chikani'. Its vines are 2.5-2.8 m with 4-5 branches. Leaves are dark green, smaller in size with six lobed petiol tip pointed and simple. The first are pale green, medium long (15.0-20.0 cm), and straight but slightly curved at stem end and pointed at the blossom end. First picking is possible in 50-55 days after seed sowing. Average fruit weight is 90.0-100.0 g and fruit yield is about 8-9 t/ha in summer and 10.0-11.0 t/ha in rainy season. It is suitable for cultivation in both summer and rainy season. It is recommended for cultivation in the states of Punjab, Uttar Pradesh, Bihar, Jharkhand, Rajasthan, Gujarat, Haryana and Delhi.
Pusa Sneha	This variety has been released from ICAR-IARI New Delhi through selection from 'Pusa Supriya'. The fruits are dark green, 20.0-25.0 cm long and straight. First picking is possible in 50-55 days after seed sowing in spring summer season and 40-45 days in *kharif* season. Average fruit weight is 120.0 g and yield is 11.5-12.0 t/ha. It is suitable for cultivation in both summer and *kharif* season.
DSG-6	The genotype DSG-6 has been developed at ICAR-Indian Agricultural Research Institute, New Delhi, by selection from local material collected from village Bulandi, Hoogly district of West Bengal. The leaves are dark green, reniform, six-lobed, nearly glabrous, non-hairy, and smooth with long petioles. This variety yield 14.38 t/ha fresh fruit. It was screened under natural epidemic condition and found to be completely resistant to Tomato Leaf Curl New Delhi Virus (ToLCNDV) Begomovirus family Geminiviridae; in rainy season, which is conducive for white fly multiplication and the spread of disease (Munshi *et al.,* 2015).
PSG-9	This variety has been released from Punjab Agricultural University, Ludhiana (Punjab) through selection from local materials. Its fruits are smooth, cylindrical, long (15.0-18.0 cm) and dark green in colour. It takes about 60 days from sowing to reach edible maturity. It is suitable for planting in February-March in north Indian plains. Average yield is 16.0 t/ha.
Kalyanpur Hari Chikani	A selection from local material, 135-140 days duration, vine length 250.0-300.0 cm, stem and foliage dark green, fruit 20.0-25.0 cm long, 4.5 to 5.0 cm thick at edible stage, smooth and dark green, fruits distal end tapering; seeds black, ovate and smooth. This variety is partially resistant to fruit fly. Average yield is 18-20 t/ha.
Azad Torai Chikni -1	This variety has been developed through selection from a germplasm and released from CSAUA&T, Kanpur (Uttar Pradesh). It is an early maturing variety having smooth green fruits. This variety is suitable for rainy season and has yield potential of Uttar Pradesh 18.0-20.0 t/ha.
Azad Torai-2 (KSG-14)	This short duration high yielding variety has been developed at CSAUA&T, Kanpur (Uttar Pradesh). Its fruits are compact, green and rich in nutrients. It is recommended for cultivation in the states of Punjab, Uttar Pradesh, Bihar and Jharkhand and has yield potential of 15.0-20.0 t/ha.

Variety	Description
MSGH-1	This variety has been released from MAHYCO, Jalana (Maharashtra). The fruits are slender, cylindrical, 20.0-25.0 cm long with light green, smooth and shining skin. First picking is possible in 40-45 days after seed sowing. Its yield potential is 25.0 q/ha.
Phule Prajataka (GFE-SMG-108)	This variety has been developed through selection from a germplasm and released from MPKV, Rahuri (Maharashtra) for commercial cultivation. Fruits are medium green; sutures are dark green, straight, slender, tapering at ends and flesh is white. Its yield potential is 15.0 t/ha. It is highly tolerant to downy mildew, fruit fly and red pumpkin beetle and moderately resistant to wilt and bean mosaic virus.
Gujarat Sponge gourd-1	This variety has been developed from a germplasm at GAU, Anand (Gujarat) in the year 2005. Its fruits are medium-sized and light green. It is moderately tolerant to downy mildew and leaf miner. It has been recommended for cultivation in states of Madhya Pradesh and Maharashtra and has yield potential of 12.0-14.0 t/ha.
Swarna Prabha (CHSG-1)	This variety has been developed through pure line selection by HARP (ICAR-Research Complex for Eastern Region, Research Centre), Plandu, Ranchi (Jharkhand). Its fruits are medium in length (20.0-25.0 cm), each weighing 150-200 g, light greenish, pulp soft and contain less fibre. It is tolerant to powdery and downy mildew and resistant to leaf miner under field conditions. First harvest can be taken at 70-75 days after planting. It is recommended for July-September and February-April sowing in the states of Uttar Pradesh, Jharkhand and Bihar and has yield potential of 20.0-25.0 t/ha.
Rajendra Nenua-1	This variety is developed at BAU, Sabour (Bihar). Fruits are long, greenish white, smooth and thick. It is suitable for summer and rainy season cultivation. Its yield potential is 25.0 t/ha.
Snake gourd	
APAU Swetha	Variety has been developed by selection from APAU, Hyderabad (Telangana). Fruits are long with whitish background and green stripes. Yields 28.0-30.0 t/ha in 125-130 days.
CO-1	This early maturing variety (bear flowers within 70 days) has been developed through pure line selection from a germplasm collected from Alangulam (Tirunelveli) at TNAU, Coimbatore (Tamil Nadu). It is a high yielding, long (160-180 cm), dark green with white strips, smooth, whitish green flesh and straight fruited variety. Each vine bear 10-12 fruits on an average 4-5 kg/plant fruit may be obtained. It gives a yield of 18 t/ha with a crop duration of 135 days.
CO-2	This variety has been developed through pure line selection from a germplasm at TNAU, Coimbatore (Tamil Nadu). The vines are less spreading and hence seeds can be sown at a close spacing of 1.5 x 1.5 m. It bears short and stout fruit (30 cm length) and appears light greenish white. Individual fruit weight is 500 g. Its yielding capacity is about 35.0-37.0 t/ha in a crop duration of 105 to 120 days.
CO-4	It is developed by selection from TNAU, Coimbatore (Tamil Nadu). It is early maturing variety. First picking is possible in 70 days after sowing. The fruits are long (160-190 cm) and dark green with white stripes and light green flesh. Average fruit weight is 400 g.
MDU-1	This hybrid has been developed by crossing a long fruited 'Selection-1' and a short fruited 'Panripudal' at TNAU, Coimbatore (Tamil Nadu). The nature of this hybrid is less spreading. Plant produces 13 fruits/vine, which weigh about 7.15 kg. The fruits are medium long (70 cm) and green with white stripes. Fruits are generally long. It yields 30.0-35.0 t/ha with a crop duration of 143 days.
PKM-1	This variety has been developed by mutation breeding of local line (Puzhuthi Kathiri). The individual fruit weight is 500-800 g. It attained the fruit length of 155-170 cm. Its crop duration is 145 days. It may yield of 34.0 t/ha in crop duration of 150-155 days.
Baby	This variety has been developed through selection from the local material at KAU, Vellanikara. First picking is possible in 55 days after sowing. The fruits are small (30.0-40.0 cm long and 22.0 cm girth), elongated and attractive white. Average fruit weight is 475 g. This variety possesses good keeping quality and suitable for distant marketing. It has been recommended for cultivation in Kerala and has yield potential of 39.0-42.0 t/ha.
TA-19	It is developed by selection from KAU, Vellanikkara (Kerala). It takes 65-70 days to reach marketable maturity. The fruits are about 60 cm long, and light green with white stripes at styler end when immature. Average fruit weight is 600 g. Average yield is 35 t/ha.
Kaumudi	It is developed by selection from KAU, Vellanikkara (Kerala). Fruits are large (100.0 cm long and 29.9 cm girth), uniform in size with white acute tip, and average weight of fruit is 1.34 kg. It gives an average yield of 50.0 t/ha.
Manushree	This variety has been developed by selection from the local material at KAU, ARS, Mannuthi (Kerala). It produces male flowers in the sixth and female flowers in the seventh week after seed sowing. First picking is possible in 55 days after sowing. Its fruits are attractive, medium long (65-70 cm), and white with green stripes at the pedicel end. Average fruit weight is 750-800 g and yield potential is 60. t/ha.
Aakash Ganga	This hybrid cultivar is tall with excellent vigour. Milky white fruits with matching white strip born in abundance with an average diameter of 3.0-3.5 cm. fruits are nearly 125.0 cm long. Average yield potential is 42 t/ha.
Aakash Deep	This hybrid bears emerald green fruits with contrasting white strip. Fruits are only 50.0-60.0 cm long with 4.0 cm diameter. As vines naturally produce abundance of fruits, harvesting at early young stage would enable fruit set well into the mild winter months. Yield is 41-44 t/ha.
TA-19	Developed by Kerala Agricultural University, Velanikkara, Thrissur (Kerala). Fruits are tasty. Yield is 38-40 t/ha.
Konkan Sweta	This early maturing variety has been developed by selection from the local material at BSKKV, Dapoli (Maharashtra). Its fruits are medium long (90-100 cm) and whitish green. Flesh recovery is more if harvested timely, otherwise become hollow. Average yield is 15.0-20.0 t/ha in a crop duration of about 120-130 days. This variety is very much suitable for Konkan region.
Phule Vaibhav	This variety has been developed from MPKV, Rahuri (M.S.). The average yield is 32 t/ha. Free from virus.

Variety	Description
	Ash gourd
Pusa Shreyali	This promising hybrid has been developed and released by ICAR-IARI, New Delhi. Fruits are oval shape with white rind and white flesh. Average fruit weight is 9.0 kg and number of fruits per plant is 4-5. It is very much suitable for making sweet candy (Petha). Flesh thickness 5.0-5.5 cm and cavity diameter 9.5-10.0 cm. Average yield is around 45-48 t/ha.
Pusa Urmi	This promising hybrid has been developed and released by ICAR-IARI, New Delhi. Fruits are ellipsoid, with greenish white rind and white flesh. Average fruit weight is 10 kg and number of fruits per plant is 4-5. It is very much suitable for making sweet candy (Petha). Flesh thickness is 5.0-6.0 cm and cavity diameter is 9.5 cm. Average yield is around 40-42 t/ha.
PAG-3	It is a selection from the local material and released by PAU, Ludhiana (Punjab) for commercial cultivation in north Indian plains in February-March and May-June. The fruits are attractive, globular and medium sized. The immature fruits are hairy but smooth and waxy at maturity. It takes 145 days from seed sowing to reach maturity. Average fruit weight is 10-12 kg and yield is 70.0-75.0 t/ha. This variety is tolerant against disease and insect pest reaction.
CO-1	It is a selection from the local material and released by TNAU, Coimbatore (Tamil Nadu) for commercial cultivation. The vine is moderately vigorous, trailing (upto 4.0 m) with dark green leaves. Fruits are globular, green, large, oblong, 35.0 cm long, 22.0 cm girth with an average weight of about 6.8 kg. First harvesting is done at about 100 days after sowing. It gives an average yield of 20.0-25.0 t/ha in 140-150 days of crop duration.
CO-2	It is a selection from the local material and released by TNAU, Coimbatore (Tamil Nadu) for commercial cultivation. It is an early maturing variety and takes about 120 days to reach marketable maturity. The fruits are small, oblong, compressed on both ends, less seedy and with light green flesh. Average fruit weight is 3.0 kg and yield is 39.0 t/ha.
APAU Shakthi	It is a selection from the local material and released by APAU, Hyderabad (Telangana) for commercial cultivation. The fruits are cylindrical. It takes 140-150 days from sowing to marketable maturity. Average yield is 30.0-35.0 t/ha.
Indu	It is a selection from the indigenous collection and released by College of Horticulture, KAU, Velannikkara Thrissur (Kerala) for commercial cultivation. Fruits are round, 24.3cm long, 23.78 cm breadth with average weight of 4.82 kg; tolerant to mosaic disease and gives average yield of 24.5 t/ha.
Mudliar	It is a selection from the local material and released by TNAU, Coimbatore (Tamil Nadu) for commercial cultivation. Fruits are big and pale-green in colour. The yield potential is 36 t/ha. Free from virus and anthracnose.
KAU Local	It is a selection from the indigenous collection and released by College of Horticulture, KAU, Velannikkara (Kerala) for commercial cultivation and released by State Seed Sub-Committee. Fruits are medium sized, oval to oblong, green at tender stage. The fruits are 45.0-55.0 cm long, length: breadth ratio is 2.05 with average weight of 6.0-8.0 kg. Flesh thickness is 5.1-6.2 cm. it gives an average yield of 28.0 t/ha. The crop duration is 105-120 days for mature fruit production. Harvesting at tender stage is advisable for better yield and market demand for vegetable purpose.
Kashi Dhawal (IVAG-502)	It is a selection from the indigenous collection identified through UP State Variety Releasse Committee and notified by Central Seed Sub-Committee in 2005; released by IIVR, Varanasi (U.P.). the vine length is 4.5-5.0 m. fruits are oblong with average weight of 12.0-13.0 kg. Fruit flesh is white and seed arrangement is linear. It gives an average yield of 58.0-60.0 t/ha.
Kashi Ujwal (IVAG-90)	It is developed from ICAR-IIVR, Varanasi (U.P.). Individual fruit weight is 10-12 kg. Average yield potential is 30 t/ha. Very good for petha making.
Kashi Surbhi (IVAG-03)	This variety has been developed from ICAR-IIVR, Varanasi (U.P.) through the selection. Fruits are more fleshy and less seeded with linear arrangement. This variety yieldes 46% more in comparison to national check. The average yield potential is 60-70 t/ha. It is recommended for commercial cultivation for Zone-IV (Punjab, Uttar Pradesh Bihar and Jharkhand).
	Longmelon
Arka Sheetal	This variety has been developed through selection at ICAR-IIHR, Bengaluru (Karnataka). Fruits are medium long, very tender, dark green, crisp, fleshy, each weighing 90-100 g and free from bitter principle. It has yield potential of 21.0-25.0 t/ha in 100-110 days of crop duration.
Punjab Long melon-1 (H-10)	This variety has been developed through selection at Punjab Agricultural University, Ludhiana (Punjab). Fruits are long, thin and green. It has yield potential of 20.0-22.0 t/ha.
Pant Kakri-1	This variety is developed at GBPUA&T, Pantnagar (Uttarakhand) and released by Uttar Pradesh State Variety Release Committee. This is developed through inbreeding and selection from indigenous collection. The vines are vigorous with long, light green and straight fruits. The harvesting starts 50 days after sowing. Seed to seed stage is about 90 days. It is suitable for sowing in the plains from February to April.
Kamal Selection	This variety has been developed at CCS HAU, Hisar (Haryana). Fruits are tender and straight. It is also early in nature. The yield potential is 18-20 t/ha. Free from virus & powdery mildew.
	Roundmelon or Tinda
Hisar Tinda (HT-10)	This variety has been developed at CCS HAU, Hisar (Haryana) and release by State Variety Release Committee in 1991 for commercial cultivation in Haryana. It is an early maturing and high yielding variety. Fruits are round, medium in size, soft and flesh is greenish white. The yield potential is about 7.5-10.0 t/ha.
Arka Tinda	This variety has been developed by hybridization (T-3-4 x T-8-2) followed by pedigree selection at ICAR-IIHR, Bengaluru (Karnataka). It is an early, summer season variety. Fruits are round, light green; and skin is covered with soft hairs when tender. Average fruit weight is 60 g. It is tolerant to fruit fly and gives an average yield of 9.5-10.0 t/ha in 95-100 days.
Punjab Tinda (S-48)	This variety has been developed through selection from a local cultivar at PAU, Ludhiana (Punjab). Its plants are vigorous, bearing 8-10 fruits per vine. It bears 8-10 medium sized fruits per vine. Fruits are medium in size and surface of fruit is shining and pubescent. Its flesh is white, less seeded, tender and good in cooking quality. It takes about 60 days from sowing to marketable maturity. It has yield potential of 45.0-50.0 t/ha.

Variety	*Description*
	Snapmelon
Pusa Shandar	This variety has been developed at IARI, New Delhi. Fruits are oblong, cylindrical, each weighing 700 g., and skin creamy white with light flesh. It has yield potential of 40.0-42.05 t/ha (in summer) and 37.0-38.0 t/ha (in rainy season)
	Kachari (*Cucumis callosus*)
AHK-119	It is an early, high yielding selection from the land races of Kachari. The fruits are suitable for processing (dehydration). Fruits are small, egg shaped having 50-60 g weight, 5.0-6.0 cm length anf 4.0-4.5 cm diameter. Average fresh thickness is 0.4 cm and the seed cavity is 3.04 cm in width having too many seeds. The seeds are small and soft. It flowers very early; male flowers appear at 27-28 days and female flowers 38-40 days after sowing. Fruits are ready for harvesting 68-70 days after sowing and the vines continue to fruit up to 110-120 days. The vines are medium long (1.7-2.5 m) with 5-7 primary branches. On an average, 22 fruits are born per vine giving an yield of 9.5-10.0 t/ha. It is drought hardy and can be grown under high temperature conditions. It can be sown in February-March as an irrigated summer crop and in July as a rainfed crop under the arid condition (Samadia and Pareek, 2000).
AHK-200	It is very early maturing variety possessing drought tolerance characteristics. It has been selected from the land races growing in the arid region. The fruits are suitable for garnishing vegetables or as salad. Big-sized fruits are oblong in shape and develop tough, light yellow skin at maturity. Fruits are 100-120 g in weight, 7.5 cm in length and 5.1 cm in diameter. Flesh thickness is 1.0 cm and the seed cavity being 2.5 cm. seeds are few, small, soft and can be eaten as such. The flesh is sweet and tasty. The vines are medium in length (1.75-2.0 m) with 5-6 primary branches. Anthesis of the first male flowers starts after 26-27 days and female flower 32-33 days after sowing. The fruits are ready for harvesting 65-67 days after sowing and the vines continue to fruit for 90-100 days. On an average, 20 fruits can be harvested from single vine giving an yield of 11.5-12.0 t/ha. The variety is suitable for sowing in summer as well as in rainy season under arid environment (Samadia and Pareek, 2000).
	Tree tomato
Ecuadorian Orange	Fruit is medium orange in colour, and the size of fruit is large hen's egg. Pulp light orange, creamy in texture, less acid than the Ruby Red. Excellent for eating out of hand and also suited for culinary purposes.
Goldmine	A superior cultivar originating in new Zealand and recently introduced. Very large golden yellow fruit with golden, highly flavored flesh, less bland than 'Solid Gold', but not acidic. Have superb bearing qualities.
Inca Gold	A yellow fruited cultivar said to be less acid than the red types. When cooked the fruit is said to resemble the apricot in flavor.
Oratia Red	A large fruited red cultivar, and oval to rounded in shape, with a sharp acid flavor. Good quality for eating out of hand and excellent for jams and preserves.
Rothamer	Unusual large fruit and skin bright red. Flesh golden-yellow, flavor sweet and exotic. Seeds are dark red. Fruits start ripening from December to April. Delicious eaten out of hand. Vigorous and heavy bearing plant.
Ruby Red	Large, brilliant red fruit. Pulp dark red, tart and flavoured. Fair for eating out of hand, but very good for culinary use. If allowed to ripen for one to three weeks after picking, they will become less acid. The standard cultivar grown for export in New Zealand.
Solid Gold	Large, oval shaped fruit. Skin golden-orange in colour. Pulp soft, less acidic in flavor than Oratia Red. Very good for eating out of hand, with acceptable culinary qualities.
Yellow	The size and shape of fruit is like a large plum. Skin yellowish orange. Flesh yellow, with a milder flavor than the red types. The yellow form is the oldest in cultivation in New Zealand.
	Drumstick
PKM-1	Evolved through pure line selection from Tamil Nadu Agricultural University, Coimbatore (Tamil Nadu). Seed propagated and medium dwarf stature. In 'PKM 1' pods are short (32.0-37.0 cm long and 5.5-6.0 cm girth), pod weight (65.0-82.5 g); pods/tree/year (226-328); pod yield (14.69-26.90 kg/tree/year) and seeds/pod (10-13). Average yield is 50-54 t/ha. Suitable for ratoon crop, low incidence of insect pest and disease. Suitable for varied soil types (freely drained) in tropical plains (Sathashathi, 1997).
PKM-2	This is a hybrid derivative developed Tamil Nadu Agricultural University, Coimbatore (Tamil Nadu) by a cross between MP 31 x MP28. Medium tall stature, easily be propagated by seeds to cover extensive area within a limited time. 'Pod length (61.0-78.0 cm), pod girth (6.3-6.6 cm), pod weight (140.0-170.0 g), seeds/pod (19-20), pods/tree (186-215) and pod yield/tree (29.68-34.68 kg). Less seeded and more fleshy. Suitable for growing in tropical plains of India. It can be grown as an intermediate crop in coconut and tropical fruit orchards at pre-fruit-bearing stage (Sathashathi, 1997).
Thar Harsha	This is a selection from the PKM-1 polpulation and developed from ICAR-Central Horticultural Experiment Station (CIAH), Vejalpur (Gujarat). It is very much suitable for rainfed semi-arid conditions. The selection resembles more or less similar characteristics in respect of growth and quality with PKM-2 except pod colour, length of pod and colour of seeds. The plant growth was observed in term of height 405.6 cm and spreads 330.2 cm (E-W) and 390.4 cm (N-S). Its growth is medium having larger sized dark green leaves and produced 314 fruits/plant. Average length of pod was recorded 100.0 cm having average weight 155.30g/pod. It is highly drought to lerant and yield marketable pods during summer (April-May). This variety has been found ideal for table/export purpose due to attractive green colour of pod having more flesh with uniformity spaced seeds. It gives 85.90 per cent higher yields and 31.91 per cent number of fruits over PKM-1 under arid and semi-arid conditions. Physic-chemical attributes in terms of protein (9.3 g), Vitamin A (9783 IU) and Vitamin C content (246 mg)/100g of edible part was recorded under dry land conditions (Yadav *et al.*, 2016).
Jaffna Type	'Jaffna' is being grown in various parts of Southern India, produces fruits of 60.0-90.0 cm in length with soft, tasty flesh. It is suited foe coastal tracts of south India. This variety can yield 400 pods per tree per year from second year which increases to 600 pods per tree per year form third year (Kadar Mohideen and Shanmugavelu, 1982). Pods are 60.0-90.0 cm long.
Moolanoor Murangai	'Moolanoor Murangai' is commercially cultivated in some parts of south India. It produces medium size pod and yields 500-600 pods per tree per year.

Variety	Description
Chavakacheri Murungai	'Chavakacheri Murungai' is an ecotype of Jaffna moringa and produces about 400-600 pods per tree per year. Pods are 90.0-120.0 cm long.
Chem Murungai	'Chem Murungai' is an ecotype of Jaffna moringa and produces about 400-600 pods per tree per year. It flowers and fruits throughout the year (Sundarajan *et al.,* 1970).
Pal Murungai	'Pal Murungai' is an ecotype of Jaffna moringa and produces about 400-600 pods per tree per year.
Cha Muringai	'Cha Muringai' is popular cultivar of Sri Lanka. Plants are medium tall. It flowers and fruits throughout the year. Pods are quite long and tip is red.
Kattu Muringai	'Kattu Muringai' has been developed from State Department of Agriculture, Tamil Nadu. Plants are bushy which is helpful in pod harvesting. Seeds are sown for raising the crop. Generally after six months, plant starts fruiting. Pods are 25-30.0 cm long
Kodigal Muringai	'Kodigal Muringai' is very popular variety of Tamil Nadu. Plants are dwarf. Both leaves and pods are tasty. Pods are thick, fleshy and 20.0-25.0 cm long.
Poona Muringai	'Poona Muringai' is popular as back yard plant in Thiruvenelli Katavomen and Kanyakumari regions.
Yazphanam Muringa	'Yazphanam Muringa' is popular as back yard plant in Salem and Madurai regions.
Konkan Ruchira	Very much suitable for Konkan region.
Bitter brinjal	
'Turkish Orange' 'Turkish Italian'	The small, spherical fruit is about 6.0 cm in diameter. The skin is bright orange-red, and the flesh is seedy with an excellent sweet taste. Usually eaten when still green, it is delicious in Thai curries and caponata. A tall, hardy plant, it produces abundantly and is resistant to flea beetles. It can produce a crop within 75 days of sowing of seed.
'Sweet Red'	The small, very attractive fruit is about 2.5 cm in diameter, the skin is green with dark-green stripes, turning red with dark-red stripes at maturity, the flavor is strong, but non-bitter. It is usually eaten when small and green. A thorn less plant, growing up to 1.0 m tall, and it produces a crop within 125 days from sowing the seed.
'Small ruffled Red' 'Red Ruffles' 'Hmong Red'	The small, deeply creased fruit is about 5.0 cm in diameter; the skin is orange-red, and the flesh seed and bitter. It is prized in South East Asian cuisines for its bitter flavor. A tall plant, growing up to 1.0m in height, it bears fruit in clusters and is resistant to nematodes and light frosts. It can produce a crop of fruit within 100 days from sowing the seed.
Yam bean (Misrikand)	
Rajendra Misrikand-1 (RM-1)	This variety was released from Rajendra Agricultural University, Pusa (Bihar), is found most suitable under Bihar and West Bengal situations. It produces an average yield of 30.0-35.0 t/ha of marketable tuber which was double the yield of indigenous cultivars. It had maturity period of 120-135 days with shallow rooted fleshy and juicy tubers having swollen upper part being spherical and sharply tapering lower part.
L-19	Mexican lines introduced in India and performing well with respect to yield and other horticultural attributes.
L-20	Mexican lines introduced in India and performing well with respect to yield and other horticultural attributes.
Elephant Foot Yam	
Gajendra	It is developed through selection from the local materials at ICAR-CTCRI, Regional Center, Bhubaneswar (Odisha) and considered as ruling variety in India. The tubers are non-acridic, and well-shaped with light orange flesh. Its yield potential varies from 50.0-60.0 t/ha.
Sree Padma	A selection from indigenous collection of Wyanad district of Kerala was used in crossing and developed as hybrid 'Sree Padma' at ICAR-Regional Center CTCRI, Bhubaneswar (Odisha). Plants are medium tall; tuber skin colour is brown while flesh colour is cream and non-acrid. Very good keeping quality and field tolerant to collar rot. Average yield is 42.0 t/ha within 8-9 months of crop duration.
Sree Athira	First genetically improved elephant foot yam hybrid. Hybrid selection from the cross Am-15 x Am-45. Tuber skin colour is brown and tuber flesh colour is pink. Excellent keeping quality. Average yield is 40.5 t/ha within 9 months of crop duration.
Santragachi	Plants are vigorous in growth and produce corms having slight acridity. The corms are rough with light cream colour and bear several daughter corms which take about 3-4 growing seasons to attain the marketable size. The cultivar is suitable for the eastern part of India and gives yield of about 50.0-70.0 t/ha, depending on the size of the planting material.
Kovvur	It is a high yielding cultivar with vigorous growth habit. Corms are smooth, without acridity and have better quality of flesh. The corm growth is uniform but unlike cv. 'Santragachi' it does not produce daughter corms. The cultivar is grown commercially in the Southern part of India and corm yield exceed 100.0 t/ha.
Bidhan Kusum	It is mostly a single corm bearing, non-irritant, smooth corm type variety. It can be grown successfully twice a year. It is xcellent in cooking quality and in-built ability to tolerate major insect pests and diseases. Corm yield is 65.0 t/ha (Sen *et al.,* 2005).
NDA-9	Very good selection.
Cassava	
Shree Rekha	This cultivar has been developed by ICAR-CTCRI, Thiruvananthapuram (Kerala). It is a top cross hybrid produces long, conical tubers with light brown skin and cream coloured flesh. It matures in 10 months. Good cooking quality, suitable for both up land and low land conditions. It has recorded an average yield of 45.0-48.0 t/ha in 10 months.

Variety	Description
Shree Prabha	This cultivar has been developed by ICAR-CTCRI, Thiruvananthapuram (Kerala). It is a top cross hybrid produces short, conical tubers with brown skin and yellow flesh. It is suitable for both up land and low land conditions. The tubers with good cooking qualities have won the admiration of both the growers and the consumers. It has recorded an average yield of 40.0-45.0 t/ha in 10 months.
Shree Vijaya	This cultivar has been developed by ICAR-CTCRI, Thiruvananthapuram (Kerala). It is a short duration cultivar and having good culinary quality. Suitable for low land cultivation as a rotation crop in a paddy based cropping system.
Shree Jaya	This cultivar has been developed by CTCRI, Thiruvanathapuram (Kerala). Early maturity variety with good culinary quality. Suitable for low land cultivation as a rotation crop in a paddy based cropping system.
Shree Visakham	This cultivar has been developed by ICAR-CTCRI, Thiruvananthapuram (Kerala). It is a cross between indigenous collection and exotic collection. It is medium to tall and pre-dominantly non-branching type. Number of tubers 8 to 9 per plant. Tubers are conical shaped and compactly arranged. Skin of tuber is brown and flesh is light yellow. Starch is 26 per cent and cooking quality is fair, and rich in carotene. Resistant to *Cercospora* and mites. Crop duration is 270-300 days. Suitable for Assam, Andhra Pradesh, Karnataka, Kerala, Maharashtra, North Eastern Region and Tamil Nadu.
Sree Sahya	This hybrid has been developed by ICAR-CTCRI, Thiruvananthapuram (Kerala) by multiplying crossing of two indigenous and three exotic collections to have a plant of tall and compact structure with high yield and better quality. Emerging leaves have light sepia colour with purple coloured petiole. Tubers are long and cylindrical. Tuber skin is light brown, has whitish flesh and post-harvest life of 3-4 days. Tolerant to csaava mosaic diseases (CMD), *Cercospora* and mites. Crop duration is 300-320 days. Suitable for Assam, Andhra Pradesh, Karnataka, Kerala, North Eastern Region and Tamil Nadu.
Sree Prakash	This cultivar has been developed by ICAR-CTCRI, Thiruvananthapuram (Kerala) from a selection of indigenous line. The plant is short, non-branching with dark leaves. Tubers are medium, arranged in radial form and shallow bulking type. The variety is highly tolerant to *Cercospora* leaf spot. Starch content is 29-31 per cent with fair cooking quality. Crop duration is 210-240 days.
H-97	This hybrid has been developed between the cross of Manjavella and Accession No. 300 by CTCRI, Thiruvananthapuram (Kerala). The plants are medium tall (1.5-2.0 m) and characterized by the light sepia colour of the emerging leaf. The tubers are conical, medium sized (25.0-35.0 cm) with light brown skin and white flesh with 27-29 per cent starch content. The yield ranges from 25.0-30.0 t/ha.
H-119	This cultivar has been developed by ICAR-CTCRI, Thiruvananthapuram (Kerala). Performs well under Coimbatore condition with shorter crop condition (7.5 months), higher yield, non-branching and better consumer preference.
H-165	This hybrid has been developed between the cross of Chadayamangalam Vella and Kalikalan by ICAR-CTCRI, Thiruvananthapuram (Kerala). The plants are medium tall (1.5-2.5 m), generally non-branching and bear thick conical tubers. Tubers have golden brown skin and cream flesh. Starch content of the tubers is 23-25 per cent. It yields 33.0-38.0 t/ha.
H-226	This hybrid has been developed between the cross of Malyan-4 and Ethakkakaruppan by ICAR-CTCRI, Thiruvananthapuram (Kerala). The plants are tall (2.0-2.5 m), occasionally branching and the leaves have a characteristic green colour. The tubers are medium sized (25.0-35.0 cm) with creamy skin, light purple rind and white flesh and 28-30 per cent starch. It yields 30.0-35.0 t/ha.
Taro or Arvi (*Colocasia esculenta* (L.) Schott.	
Sree Rashmi	This cultivar has been developed by ICAR-CTCRI, Thiruvananthapuram (Kerala) from a selection of local material collected from Kerala. The plants are tall (100.0-150.0 cm), erect, with medium broad leaves and purplish green petiole and sheath. All parts of the plant namely leaf, petiole, corm and cormels (side tubers) are non-acrid. The corms are big and cylindrical and cormels are medium sized. The cormels have good cooking quality and taste; starch content 15 per cent and protein 2.5 per cent. It yields 15.0-20.0 t/ha within 7-8 months.
Shree Pallavi	This cultivar has been developed by ICAR-CTCRI, Thiruvananthapuram (Kerala) from a selection of the material collected from Meghalaya. The plants are tall (100.0-150.0 cm) and produces comparatively bigger sized corms and several small cormels. The cormels have good cooking quality and taste; starch content of 16-17 per cent and protein 2.0-3.0 per cent. It yields 15.0-18.0 t/ha within 6-7 months.
Shree Kiran	This first taro hybrid has been developed by ICAR-CTCRI, Thiruvananthapuram (Kerala). The cormels have good cooking quality. It yields 16.0-18.0 t/ha within 6-7 months.
Mukta Keshi	A tall variety with large number of small sized tubers. Protein 2-3 per cent. Suitable for Odisha. It yields 15.0-18.0 t/ha within 6-7 months.
Pani Saru-1	Clonal selection from variety 'Kantilo Local' suitable for water logged conditions of Odisha.
Vallabh Nikki	It is high yielding variety developed by selection from germplasm maintained at SVPB University of Agriculture and Technology, Meerut (U.P.). It is resistant to leaf blight. The number of leaves is 4-6 while corms 24-26. Yield is about 28.10 t/ha (Singh and Singh, 2008).
Bilaspuri Arum	This cultivar has been developed by JNKVV, Jabalpur Centre Bilaspur (M.P.) and very much suitable for commercial cultivation in Bilaspur region. The corms are small, round; and each weighing 50-100 g. The flesh is white with low acridity. Average yield is 30.0 t/ha.
ML-1	This genotype has been selected from local germplasm of Meghalaya by ICAR Research Complex for NEH Region, Umiam (Meghalaya). Leaf colour is light green and petiole colour is purple green. Plant height is around 75.0 cm. leaf length varies from 20.0-25.0 cm and leaf breadth varies from 15.0-20.0 cm. The germplasm having medium sized corm length 10.0-12.0 cm and diameter 5.0-6.0 cm and more numbers of small to medium sized cormel (length 6.0-7.0 cm and diameter 2.5-3.0 cm). On an average this line gives yield of 25.0-27.5 t/ha. This genotype is comparatively tolerant to leaf blight of taro.

Variety	Description
ML-1	This genotype has been selected from local germplasm of Meghalaya by ICAR Research Complex for NEH Region, Umiam (Meghalaya). Leaf colour is light green and petiole colour is purple green. Plant height is around 60.0 cm. leaf length varies from 25.0-30.0 cm and leaf breadth varies from 20.0-25.0 cm. The germplasm having medium sized corm length 10.0-12.0 cm and diameter 5.0-6.0 cm and 7-8 small to medium sized cormel. On an average this line gives yield of 20.0-22.5 t/ha. This genotype is comparatively tolerant to leaf blight of taro.
ML-9	This genotype has been selected from local germplasm of Meghalaya by ICAR Research Complex for NEH Region, Umiam (Meghalaya). Both the leaves and petioles are having light green colours. Plant height is around 60.0 cm. Leaf length varies from 15.0-20.0 cm and leaf breadth varies from 12.0-15.0 cm. The germplasm having medium sized corm (length 10.0-12.0 cm and diameter 5.0-6.0 cm) and small to medium sized cormel (length 6.5-6.5 cm and diameter 3.0-3.5 cm). On an average this line gives yield of 20.0-22.0 t/ha. This genotype is comparatively tolerant to leaf blight of taro.
Satamukhi	Plants attained the height of 100.0-150.0 cm. main corm is medium size and outer cormlets are many which are quite firm. It is very good in taste and cooked easily. Average yield is 15.0-20.0 t/ha. It recommended for commercial cultivation in Andhra Pradesh, Bihar and West Bengal.
Jhankri	Clonal selection and high yielding variety.
Sonajuli	Clonal selection and high yielding variety.
CO-1	Clonal selection and high yielding variety.
Bhavpuri	Clonal selection and high yielding variety.
Panisaru-2	Clonal selection from local material.
Indira Arvi-1	Clonal selection and high yielding variety.
Narendra Arvi-1	Clonal selection and high yielding variety.
Narendra Arvi-2	Clonal selection and high yielding variety.
Rajendra Arvi-1	Clonal selection and high yielding variety.
BCC-1	Clonal selection and high yielding variety.
BCC-17	Clonal selection and high yielding variety.
Konkan Haritparni	This variety has been developed from BSKKV, Dapoli (M.S.). Very much suitable for Konkan region.
Chinese potato or Coleus	
Shree Dhara (CP-58)	'Shree Dhara' has been developed by ICAR-CTCRI, Thiruvananthapuram (Kerala). This cultivar has an average yield potential of 25.0t/ha within 5 months of maturity. Total dry matter is about 28.5 per cent and starch content is 16.5-20.0 per cent.
Nidhi	The Kerala Agricultural University Thrissur (Kerala) has released a variety of coleus with the name 'Nidhi' for general cultivation in the state. It has an average tuber yield of 27.0 t/ha. Its tubers are oblong in shape and they mature in 135 days after planting (Jyothi *et al.,* 2005).
White Yam or Guinea Yam (*Dioscorea rotundata* Poir.)	
Shree Subhra	'Sree Subhra' has been developed by clonal selection from 'Iwo', an introduction from Africa at ICAR-CTCRI, Thiruvananthapuram (Kerala). The tubers are cylindrical with brown skin and white flesh having starch content of 21.0-23.0 per cent. Tuber yield varies from 35.0-40.0t/ha.
Shree Priya	'Sree Priya' has been developed by clonal selection from 'Umidika', an introduction from Africa at ICAR-CTCRI, Thiruvananthapuram (Kerala). The plant produces 2-3 long tubers with good cooking quality and having starch content of 20-21 per cent. Tuber yield varies from 35.0-40.0 t/ha.
Shree Dhanya	'Shree Dhanya' has been developed by clonal selection at ICAR-CTCRI, Thiruvananthapuram (Kerala). Its plants are dwarf (50.0 cm) and bushy and do not need staking. The tuber yield varies from 20.0-25.0t/ha.
Greater Yam	
Shree Keerthi	'Sree Keerthi' has been developed by clonal selection at ICAR-CTCRI, Thiruvananthapuram (Kerala). Its tubers are conical with brown skin and white flesh. It possesses good cooking quality. The starch content varies from 20-22 per cent. Yield potential is 30.0t/ha.
Shree Roopa	'Shree Roopa' has been developed by clonal selection at ICAR-CTCRI, Thiruvananthapuram (Kerala). The tubers are digitate with black skin and white flesh. Starch content varies from 17-19 per cent. Tubers yield varies from 25.0-30.0 t/ha.
Shree Shilpa	It has been developed from ICAR-CTCRI, Thiruvanathapurm (Kerala). The yield is 36 t/ha.
Odisha elite	It has been developed from ICAR-CTCRI- Regional station Bhupaneshwar (Odisha). The yield is 32 t/ha.
Shree Swathy	This variety has been developed from ICAR-CTCRI. Thiruvanathapuram (Kerala). Yield is 38 t/ha.
CO-1	This is developed from TNAU, Coimbatore (T. Nadu), Average yield is 35 t/ha.
Konkan Ghorkand	This variety has been developed from BSKKV, Dapoli (M.S.). Very much suitable for Konkan region. The average yield potential is 28.8 t/ha.

Variety	Description
Lesser Yam (*Dioscorea esculenta* (Lour) Burkill)	
Shree Latha	'Shree Latha' has been developed by clonal selection from Puerto Rico introduction at ICAR-CTCRI, Thiruvananthapuram (Kerala). Its tubers are smooth with brown skin and white flesh. Tuber content 18.4 per cent starch. It possesses excellent cooking quality. Average tuber yield is 20.0t/ha.
Shree Kala	'Sree Kala' has been developed by clonal selection at ICAR-CTCRI, Thiruvananthapuram (Kerala) from the material introduced from Puerto Rico. Its tubers are smooth with brown skin and white flesh. It possess excellent cooking quality. Average yield is 20.0 t/ha.
Konkan Kanchan	This variety has been developed from BSKKV, Dapoli (M.S.). Very much suitable for Konkan region. The average yield 38.8 t/ha.
Giant Taro or Giant Elephant Ear	
Variegata	'Variegata' is selection with leaves decorated with creamy white or grayish splotches.
Violacea	'Violacea' has pale violet leaves.
Curry leaf	
DWD-1(Suwasini)	Developed from UAS, Hebbal, Bengaluru (Karnataka). Leaves are shining dark green aromatic with oil content of 5-22 per cent. The clone is sensitive to low temperature.
DWD-2	It is an open pollinated seedling progeny developed by UAS, Dharwad (Karnataka). Leaves are pale green having lesser aroma (5.09 per cent). It is not sensitive to low temperature and is superior in bud burst internodal length, shoot length and weight of new shoots. Due to winter insensitive nature, and farmers get extra income.
Senkaampu	It is local cultivar with purplish petiole and highly aromatic leaves.
Amaranths	
CO-1 (*A. dubius* Mart ex Thell.)	'CO.1' has been developed through selection from local material (Tirunelveli) at Tamil Nadu Agricultural University, Coimbatore (T. Nadu) in the year 1968. It lacks vigour but make rapid growth after about one month of sowing. Its leaves are dark green, broad with ridged surface; stem is round, glossy green and succulent. Leaf stem ratio = 2. It yields 10.0 t/ha in 30-35 days after seed sowing. It is especially suitable for late harvesting. Seeds are small and black. It is resiatant to leaf spot caused by *Rhizoctonia* species.
CO-2 (*A. tricolor*)	'CO.2' has been developed through selection from local material at Tamil Nadu Agricultural University, Coimbatore (Tamil Nadu). This cultivar is particularly suited for early harvest. Its leaves are glossy green, lanceolate, slightly elongated and stem is green and succulent. Average yield is 10.0 t/ha.
CO-3 (*A. tristis*)	'CO.3' has been developed through selection from local material at Tamil Nadu Agricultural University, Coimbatore (T. Nadu) in the year 1981. This is also locally named as 'Araikeerai' in Tamil. Since it is suitable for periodical cuttings, it can be retained for more than three months. It becomes ready for picking 20 days after sowing and there after at weekly intervals. It yields as high as 30.0 t/ha.
CO-4 (*A. hypochon-driacus*)	CO-4 has been developed through selection from local material at Tamil Nadu Agricultural University, Coimbatore (T. Nadu). It is grown as dual purpose (leaf and grain). Leaves are green. Average yield is 80.0 t/ha.
CO-5 (*A. tricolor*)	'Co.5' has been developed through selection from an introduced material at Tamil Nadu Agricultural University, Coimbatore (Tamil Nadu). The leaves are double shaded (green and pink) and have rosette appearance in early stages. The stem is also tender and consumable. First harvesting is possible 25 days after seed sowing. Average yield is 40.0 t/ha.
Pusa Chhoti Chauli (*A. blitum*)	'Pusa Chhoti Chauli' has been developed through selection from local material at ICAR-IARI, New Delhi. The plants are erect, thin stems and dwarf. Leaves are small size, green in colour and well to cutting. It is best suited for early summer and can be grown in rainy season also.
Pusa Lal Chauli (*A. tricolor*)	Developed at ICAR-IARI, New Delhi. 'Pusa Lal Chauli' is suitable for kitchen gardening as well as for commercial cultivation in northern plains. Since the colour of the stem and leaves are bright red (magenta). It is suitable for ornamental purpose also. The red dye extracted from its plants could be used as natural food additive. The dye can also be used in textile or woolen industry for dyeing. Suitable for growing in both summer and rainy season. On an average, it yields 45.0-50.0 t/ha.
Pusa Badi Chauli (*A. tricolor*)	'Pusa Badi Chauli' has been developed through selection from local material at ICAR-IARI, New Delhi. It stem is thick and green; leaves are large and green, responds to cutting and has longer growing period. First cutting is possible 35 days after sowing. It is best suited for summer season. Average yield is 30.0 t/ha.
Pusa Kiran (*A. tricolor* x *A. tristis*)	'Pusa Kiran' has been developed from the naturally occurring inter-specific cross between *A. tricolor* x *A. tristis* at ICAR-IARI, New Delhi. It has predominantly the characteristics of *A. tricolor*. The leaves and stems are glossy green. First harvesting is possible 20-25 days after sowing. It is a heavy yielder and is possible for cultivation in rainy season. Average yield is 35.0 t/ha.
Pusa Kirti (*A. tricolor*)	'Pusa Kirti' has been developed through selection from local material at ICAR-IARI, New Delhi. It is recommended for commercial cultivation in summer season. The leaves are green with broad ovate 6.0-8.0 cm long and 4.0-5.0 cm wide. Petiole is 3.0-4.0 cm long. It is ready for first harvest in 30-35 days after sowing. Its subsequent cuttings may be taken at 10-12 days intervals. Average yield is 50.0-55.0 t/ha.
Arka Arunima (*A. tricolor*)	'Arka Arunima' has been developed by pure line selection from 'IIHR 49' at ICAR-IIHR Bengaluru (Karnataka). Its leaves are rich in calcium and iron and low in anti-nutrient factors like oxalate and nitrates. This variety is resistant to white rust. It can be grown throughout the year successfully. However, it comes well during *kharif* and *rabi*-summer seasons. It yields 27.4 t/ha of greens (Varalakshmi *et al.*, 2001).

Variety	Description
Arka Suguna (*A. tricolor*)	'Arka Suguna' has been developed by pure line selection from an exotic material introduced from Taiwan at ICAR-IIHR Bengaluru (Karnataka). Leaves are light green, succulent and broad. First harvesting is possible 20-30 days after seed sowing and total of 5-6 cuttings are possible. Average yield is 25.0-30.0 t/ha. It is moderately resistant to white rust.
Arka Samraksha	Arka Samrakshais a high yielding cultivar and has been developed by ICAR-IIHR Bengaluru (Karnataka) which was released for commercial cultivation in 2010. High yielding amaranth variety with high antioxidant activity of 499 mg (AEAC units) and minimum nitrate content is 27.3 mg and 1.34 g of oxalates per 100 g fresh weight of leaves. It is a pulling type amaranth variety with green leaves and stem. It yields 10.9 t/ha in 30-35 days duration.
Arka Varna	Arka Varna is a high yielding cultivar and has been developed by ICAR-IIHR Bengaluru (Karnataka) which was released for commercial cultivation in 2010. High yielding amaranth variety with high antioxidant activity of 417 mg (AEAC units), nitrate content of 37.6 mg and 1.42 g of oxalates per 100 g fresh weight of leaves. It is a pulling type amaranth variety with green leaves and pink stem, yields 10.6 t/ha in 30-35 days duration.
Renu Shree (*A. tricolor*)	This cultivar has been developed by Kerala Agricultural University, Thrissur (Kerala). Green leaves and red stem, multicut, leaf: stem ratio=1.5 and yield is around 15.5 t/ha.
Krisna Shree (*A. tricolor*)	This cultivar has been developed by Kerala Agricultural University, Thrissur (Kerala). Reddish green leaves, multicut, leaf stem ratio=1.10 and yield is 14.8t/ha.
Arun (*A. tricolor*)	This cultivar has been developed by Kerala Agricultural University, Thrissur (Kerala). This is a multicut variety. Marron red leaves, leaf length 25 cm, leaf width 11 cm. Average yield is 20.1 t/ha.
Mohini	This cultivar has been developed by Kerala Agricultural University, Thrissur (Kerala). 'Mohini' is a medium branching variety with erect growth habit. It attains a height of 82.0-89.0 cm at flowering stage. Its stem is thick, green with red basal parts. The leaves are dark green, elliptical with acute tip.
Bathua	
Pusa Bathua-1	'Pusa Bathua-1' has been developed from ICAR-IARI, New Delhi. Plants grow up to a height of 2.25 m; leaves and stems purplish green, leaves tender; first picking in 40-45 days after seeding. Leaves are 10.5 cm long and 3.0 cm broad. Tender stem girth is 0.45 cm and internodal length is 2.5 cm. It remains in vegetative phase up to 150 days. Plants start flowering after 175-180 days of seeding. Yield 30.0 t/ha.
Magenta	The leaves of this cultivar are magenta coloured, and it is considered by some people to be one of the best tasting of all potherbs.
Ootty -1	Ootty -1 has been developed from TNAU, Coimbatore (Tamil Nadu). It grows to a height of 38-40 cm and yields about 28.9t/ha. In 55 days in hill and 17.0 t/ha in 50 days in plans. Leaves are dark green and unfolded leaves are pinkish in colour. It shows resistance to *Cercopspora* leaf spot, *Collectotrichum* leaf spot, and *Macrophomina* foot rot.
Fenugreek	
Hisar Sonali	This variety has been developed by CCS HAU, Hisar (Haryana). Plants are bushy, semi erect with bold yellow attractive seed, moderate tolerant to leaf spot and rot disease. Maturity period is 140-150 days with an average seed yield of 1.9 t/ha.
Hisaer Mukta (HM34)	This variety has been developed by CCS HAU, Hisar (Haryana) through a pureline mutant from an indigenous collection from U.P. This is a quick growing seed type suitable to grow in Haryana, Rajasthan and Gujarat under both irrigated and rainfed conditions.
Hisar Suvarna (HM103)	It has been developed by pur line selection from the local material and released for commericial cultivation from CCS HAU, Hisar (Haryana). It is a quick growing dual purpose variety suitable for cultivation throughout the country. The plants are tall and erect with more number of branches. It takes 130-140 days maturity period and yield 1.6 to 1.9 t/ha bold seeds with typical yellow colour. This variety is tolerant to *Cercospora* leaf spot and powdery mildew diseases. Geen leaf yield is 20 t/ha.
HM-219	This variety has been developed by CCS HAU, Hisar (Haryana). The average yield of green leaf is 18-20 t/ha.
Hisar Manohar	This variety has been developed by CCS HAU, Hisar (Haryana). The average yield of green leaf is 22 t/ha.
Hisar Madhavi (HM350)	This variety has been developed by CCS HAU, Hisar (Haryana) through single plant selection from the line PME46-1. It is quick growing dual purpose, plant erect, tall with more branches. Pods are long, dark green leaves, typical yellow colour. It yield 1.9 to 2.0 t/ha seeds. It is suitable to grow in Hryana, Rajasthan and Gujarat. It is tolerant to powdery mildew and downy mildews.
Pusa Early Bunching	It has been developed by selection from the local material and released for commericial cultivation from ICAR-IARI, New Delhi. It is quick growing variety (first cutting can be made after 55-60 days of sowing) and has been developed by selection from the local material. Plants are upright and suitable for bunching. Leaves are green and flowers are white. Average yield is 7.0-8.0 t/ha. It takes 125 days for seed maturity.
Kasuri Methi	It has been developed by selection from the local material and released for commericial cultivation from Department of Agriculture, Punjab. Plants are dwarf and give rosette appearance. Flowers are yellow and produce small and sickle shaped pods.
Pusa Kasuri Selection	It has been developed by selection from the local material and released for commericial cultivation from ICAR-IARI, New Delhi. It is slow growing variety. Leaves are small (remain in rosette condition during most of its vegetative growth period); flowers are orange yellow and pods are small sickle shaped. Leaves of this variety have special fragrance. It becomes ready for first harvest after 60-65 days of sowing.
Kasuri Supreme	It has been developed by selection from the local material and released for commericial cultivation by SKUAST, Jammu. The variety is characterized by broad dark green leaves. It is a multi-cut variety and can be sown from September to January. The green yield potential is 400 kg/ha. With average leaf yield of 25.0 t/ha. While seed yield potential is 8-10 q/ha.
Prabha (NLM)	It is the most suitable variety for Rajasthan and Gujarat developed by College of Agriculture, Jobner, Rajasthan. It is a pure line selection from a local collection of Nagpur district. Semi erect growth, leaves soft, sweet with dark green colour, can tolerate many pests and diseases (Sharma and Bhati, 1984). Potential leaf yield is 25.0-28.0 t/ha.

Variety	Description
Lam Selection-1	It is selection from germplasm collection and release by Regional Agricultural Research Station at Lam, Guntur of N.G. Ranga Agricultural University Hyderabad (Telangana). Plant is bushy in nature with medium height. Leaves are better in quality. It is a high yielding variety (Grain yield is 7.0-8.0 q/ha and leaf yield is upto 12.0 t/ha) under irrigated condition. It contains 5.3 per cent of protein in seed (Rao and Babu, 1981).
CO-1	It is dual purpose variety released by TNAU, Coimbatore (Tamil Nadu). The crop matures in 80-90 days. Its grain yield is 0.68 t/ha and green leaf yield is 9 t/ha. Seed protein is 20-22 per cent and leaf contains 15.9 per cent protein (Ram Das *et al.*, 1982). Due to short duration nature it suits well for intercropping and relay cropping.
CO-2	Plants are medium sized, high yielding, less susceptible to root-rot, and suitable to grow in Coimbatore, Dharampuri, Guddalore, Ramanathapura, Karur, Tiruchirapalli, Perambalur and Tirunelveli districts of Tamil Nadu, Maturity period is 95-100 days with an average yielding of 14 t/ha.
Rajendra Kranti (RMT-16)	This variety is a selection and released by Dholi station of Rajendra Agricultural University, Bihar. The plants are medium tall and bushy. The crop is ready for harvest in 120 days. The leaf yield potential is 1200-1400 kg/plant. It is very much suitable for raising as a pure crop or as an intercrop.
RMT-1	It is a high yielding selection from SKM College of Jobner, Rajasthan. It yields 1.5 t/ha seed and ready for harvest in 150 days. The seed contain 21 per cent of protein. It is tolerant to root rot and moderately resistant to powdery mildew.
RMT-143	Plants are medium sized and seeds are typical yellow coulour. Tolerant to powdery mildew. Maturity period is 140-150 days with an average yield of 1.6 t/ha seed, suitable to grow in heavier soils of Chittor, Bhilwara and Jodhpur area (Rajasthan). It was released in 1997.
Ajmer Fenugreek Mutant-1 (AFg-M-2)	This cultivar developed from ICAR-National Research Centre for Seed Spices Ajmer (Rajasthan). It is a rosette leaf type mutant in which the leaves come up in small bunches on the top, making easy for its nipping, in comparision to other varieties in which there is need to detach the leaves from stem for vegetable use. This mutant carry leave bunches that can be directly chopped and used as vegetable.
Vegetable pod type fenugreek (AM-327-3)	This cultivar developed from ICAR-National Research Centre for Seed Spices Ajmer (Rajasthan). It is vegetable pod type bears fibrous broad pods bearing very few undersized seeds making it best for use as vegetable at green stage just like cowpea and other beans whose green pods are consumed traditionally.
Ajmer Fenugreek-5 (Afg-5)	This cultivar developed from ICAR-National Research Centre for Seed Spices Ajmer (Rajasthan). The market availability of fenugreek leaves is limited to winter months in India beaing a *rabi* crop. Recent research has shown remarkable results of cultivating fenugreek in harsh summer months. In the harsh summer months, Afg-5 entry has shown late flowering, the bolting period has been recorded to be around 36 days after sowing and days to 50 per cent flowering was after 46 days under green shade net covered walk in tunnels equipped with drip, in these trials.
Ajmer Fenugreek-3 (Afg-3)	This cultivar developed from ICAR-National Research Centre for Seed Spices Ajmer (Rajasthan). A recently national leleased variety of fenugreek possesses high medicinal values due to presence of high 1, 4 hydroxy isoleucine, which is an anti-diabetic compound. Leaves of this variety can be marketed with a trade value of product bearing high medicinal value.
Green seeded fenugreek	This cultivar developed from ICAR-National Research Centre for Seed Spices Ajmer (Rajasthan). This genotype has green coloured seeds rather than golden yellow seeds. These genotypes are prominent for harvesting micro green seeds at various stages, because in genotypes bearing yellow or golden coloured seed the need of micro greens can be met only at the grain filling stage but in these green coloured seed genotypes a late harvesting period is available as it sustains the green colour after harvesting. These micro greens also pffer a vegetyable for table need as salads.
Methi No. 47	It is an improved variety of Maharashtra, which is high yielding and containing more vitamin C in the leaves.
Methi No. 14	It is developed by Maharashtra Agriculture Department. It is high yielding and contains more vitamin C in the leaves.
HM-10	Plants are bushy, bold, yellow attractive seds moderate tolerant to leaf spot disease, maturity period is about 140-150 days with an average yield of 2.01 t/ha seed.
HM-57	It is suitable variety for Haryana and it is capable of yielding 1.0 tonne of grain/ha. and 15-18 t/ha. green leaf.
UM-50	It is superior cultivar for the greens and seed yield.
UM-112	The plants of this variety are very long and cultivated for both seed and leaf purpose. It contains more amount of diosgenin, which is used for preparation of medicines related to embryo distoetion.
IC74	It is a local cultivar having erect greens, more branching type plants, and good as greens as well as seed.
Barbara	It is having erect plants, yellow seeds and high trigonellin content in seeds.
Water spinach	
Hong Kong	Hong Kong is less affected to low temperature. In India it has has given poor yield.
Chingquet-1	Chingquet-1is less affected to low temperature. In India it has has given poor yield.
Pak Quet	'Pak Quet' is a white flowered cultivar is very much suitable for summer cultivation because its flourish well under dry conditions. It is very good yielder and produce leaves and stems of good quality. It is a multicut cultivar and develops several side branches.
Leek	
Miner	An autumn or early winter variety derived from crosses and family selection in the variety 'St. George', has long stem and gives high yield.
Splendid	Cultivated throughout the season. It mature in 105 days, it is a rapid and vigorous grower, making a good size before fall (important since it is very winter tender). The 18.0-21.0 cm medium green, and non-bulbous stalks are easily blended.
Renova	Derived by selection from the variety 'Regions' has darker leaves, longer stalk, and fewer tendencies to form bulb and is more uniform in size and shape.

Variety	Description
Special 050	Suitable for forcing
Regius 930/57	It is a cross between 'St. George' x 'Special'. F_1 hybrid is good cropper especially in late autumn and early winter.
Varna	It is early season short duration and tall "bunching" type developed for thick direct seeding to produce clumps of slender plants.
King Richard	It is a summer leek with long, slender stems that stay sweet tender and white all season long. It can be sown densely to grow mini-leeks for use as a garnish or for soups or salads. It has some frost resistance but will not overwinter.
American Flag	It matures in 130 days. It has a mild, sweet, and onion-like flavor. It is very hardy, and will overwinter in milder climates. The 18.0-23.0 cm long stems blanch snowy white and topped with blue green leaves resemble giant scallions.
Beet leaf or Palak or Spinach beet (*Beta vulgaris* var. *bengalensis* Hort.)	
All Green	It has been develop by selection from the local material and released for commercial cultivation by ICAR-IARI, New Delhi. It produces uniform green tender leaves and multicut (6-7) at 15-20 days interval. High yielding variety (12.5 t/ha). During winter season it produces seed stalks in 75 days.
Pusa Jyoti	It has been develop by polyploidy breeding using 2 per cent colchine solution (24 hours) for doubling chromosome number of the cultivar 'All Green' and released for commercial cultivation by ICAR-IARI, New Delhi after selection (subsequent generations upto 10 years). Plants are vigorous, quick growing, regenerate quickly, tender, multicut (6-8) and yield is 29.0 t/ha. Its leaves are large sized, thick, tender, dark green, succulent and crisp. Plants are very vigorous, quick growing and regenerate quickly after each cutting. It gives 6-8 cuttings and yields about 49.0 t/ha.
Pusa Bharathi	It has been develop by selection from the variety S44-1 and released for commercial cultivation by ICAR-IARI, New Delhi. It is suitable for summer and kharif. Leaves are cordate 25.0 cm long and 15.0 cm breadth, green, smooth and tender. First harvesting is possible in 30-40 days after sowing. It is rich in vitamin C and b-carotene content. Average yield is 50.0 t/ha.
Pusa Harit	It has been develop from an interspecific cross involving 'Sugar Beet" and 'Local Palak' and released for commercial cultivation by ICAR-IARI, New Delhi. Suitable for round the year cultivation. Leaves are upright, green slightly crinkled, thick, giant size leaves, and late bolting. First picking is possible in 40-50 days after sowing. It is a heavy yielder (15.0-20.0 t/ha). Tolerant to alkaline soil condition.
Pusa Palak	It has been developed from the segregating population of 'Swiss Chard' (*Beta vulgaris* var. *cicla*) x 'Local Palak' and released by ICAR-IARI, New Delhi after selection. It has uniform green colour leaves (without any purple pigmentation) with late bolting habit.
Arka Anupama	It has been developed by selection from the cross 'IIHR-10' x 'IIHR-8' and released for commercial cultivation from ICAR-IIHR, Bengaluru (Karnataka). Leaves are medium large, dark green, succulent, crinkled and attractive. It has vigorous initial growth, regenerates faster and is a slow bolter. First cutting is possible in 35 days after seed sowing and subsequent at 10-12 days intervals. It is a heavy yielder (41.0 t/ha). It is resistant to *Cercospora* leaf spot under field condition.
Jobner Green	It has been developed by selection from a spontaneous mutation identified from a local material, selection No. 5 and released for commercial cultivation from Department of Horticulture, University of Agricultural Sciences Udaipur, Jobner Campus (Rajasthan). Uniform green, large thick succulent tender leaves and yield is around 25.0-30.06 t/ha.
Ooty-1	It has been developed by selection from the material introduced from Himachal Pradesh and released for commercial cultivation from TNAU, Ooty (Tamil Nadu). Leaves are green, 50.0-60.0 cm long, and 8-10 cm broad. First harvesting is possible in 45 days after seed sowing and subsequent 4 cuttings at 15 days intervals. It is a high yielding (14.0-15.0 t/ha) cultivar and contain very high amount of carotene. Suitable for round the year cultivation. Can withstand frost.
Banerjee's Giant	It has been developed through crossing (local palak and beet root) and mass selection by S.P. Banerjee in West Bengal. It has green large thick succulent leaves and stem. Very popular variety of West Bengal. It produces large thick leaf and succulent stem and fleshy root.
HS-23	It has been developed by mass selection from the local material and released for commercial cultivation from f CCS HAU, Hissar (Haryana). Large, dark green and thick leaves are tender and juicy. It is quick growing variety and 1st cutting can be done within 30 days after seed sowing. A total of 6-8 cutttings at two-week intervals can be obtained.
Punjab Green	It has been develop by selection from the local material and released for commercial cultivation by PAU, Ludhiana (Punjab). Plants are semi-erect; leaves are succulent, shining dark green, thick, long and broad and free from sourness. There is mild purple pigment on the stem. First cutting is possible in 30 days after sowing. It is slow bolter and has low oxalic acid. Average yield is 30.0 t/ha.
Palak 51-16	It is late bolting cultivar developed by Department of Agriculture, Maharashtra. It yields about 16.0 t/ha of green leaves. It is late bolting cultivar.
C-13	It is composite variety developed at SKUAST, Jammu (J&K) by mixing of 13 high yielding germplasm lines of beet leaf. The crop is ready for first cut after 35 days of sowing. It is multi-cut variety and can be harvested 4-5 times during the rabi season under subtropical conditions of Jammu Province. Its yield potential is 40.0-50.0 t/ha and superiority over the variety 'All Green' is 15-20 percent.
Silver	The improved variety 'Silver' supplied 10-15 t/ha of leaves in two picking in july and September-October. Two additional picking may be taken if the root of the existing crops is sown in poly house.
Pant Composite	A *Cercospora* leaf spot resistant variety developed at GBPUA&T, Pantnaghar (Uttarakhand). It produces tender, juicy uniform green colour leaves.
Red Kanak	It has red blood colour, grow very fast, soft and succulent leaves, generally cultivated in winter season. It is ready for harvest 25 days after seed sowing. One time harvest is common with this variety.
White Kosala	It can be grown during rainy and winter season. It is locally grown genotype and considered as a common leafy vegetable in states like Odisha, Bengal, Bihar, Jharkhand and parts of Chhattisgarh.

Variety	Description
Light Red Kosala	It can also be grown during rainy and winter season. It is locally grown genotype and considered as a common leafy vegetable in states like Odisha, Bengal, Bihar, Jharkhand and parts of Chhattisgarh.
F_1 Upright	It has shiny, very dark green leaves and 18-20 leaves per plant, very tasty, easy for harvesting. Resistant to downy mildew and rich in minerals.
Sungro All Green	It has uniform green, smooth, tender leaves. Starts first cutting 25-30 days after sowing. Give 5-6 cuttings at 15-20 days interval. A late bolter has very good market preference.
Spinach (*Spinacea oleracea* L.)	
Virginia Savoy	It is an introduction from U.S.A. It prickly seeded cultivar having blistered dark green large leaves with a round tip. The plant is upright and vigorous. It is ready for marketing in 30-35 days. The yield potential of variety is 8.0-10.0 t/ha.
Early Smooth Leaf	It is an introduction from U.S.A. It is smooth seeded exotic cultivar producing small light green leaves. Leaves have a pointed apex. Days to marketing are 25-30 and yield potential is 7.0-8.0 t/ha.
Marathon	It is a Savoy type. Plants are large and produce dark green leaves. Yield potential is 8.0 t/ha.
Lettuce (*Lactuca sativa*)	
Punjab Lettuce No.1	It has been develop by selection from the exotic material and released for commercial cultivation by PAU, Ludhiana. The leaves are light green, shining and crisp. It is non-heading type and bears loose leaves. It takes 45 days from sowing to first harvest of fully developed leaves. Harvesting of leaves is done from November to March. Average yield is 9.0 t/ha. It gives high yield under suitable package of practices and in favourable climatic condition.
Great Lakes	It is an introduction from U.S.A. It is a crisp-head type and recommended for cultivation by the ICAR-IARI, New Delhi. The head is large, firm with green, crisp and tender leaves. Outer leaves are blistered. Harvesting of leaves is started in 88 days after sowing. It is fairly resistant to tip burn but susceptible to powdery mildew (Choudhury, 1967).
Alamo-1	It is an introduction from U.S.A. It is a heading type variety. Heads are compact and average head weight is around 500 g. Leaves are dark green, curly and crisp. It attains harvesting maturity in 90-100 days after sowing. Average yield is 23.0 t/ha.
Slobolt	It is a leafy type.
Chinese Yellow	It is a leafy type. Early leaf variety, light green crisp and tender leaves (Anonymous, 1998).
Imperial 859	It is a crisp-head type.
White Boston	It is a butter head type.
Dark Green	It is a cos type.
Revolution	It has sensational red colour, very slow bolting, resistant to Bremia NL-1-16 and good vigour are important features of this Lollo Rosa type variety. Good for outdoor and indoor cultivation.
Bergamo	This Lollo bionda type variety is bigger and slow in bolting. Fine serated leaves are very uniform in shape and have a high quality appearance. Resistant to Bremia NL-1-4, Bremia NL-1-6, Bremia NL-1-10, Bremia NL-1-13, Bremia NL-1-15 and suitable for glass house and outdoor cultivation.
Fast Fall	This black-seeded loose-leaf lettuce tolerant to disease. This variety performs large and vigorous plant, produce attractive highly filled crisp, light green leaves, popular all around Asia for fresh leaves consumption and can be harvested after 40-45 days.
Dublin	This is a new medium sized variety with a fresh medium green colour. The head is filled quickly and it generates a firm fare sized ball, which is well closed at the base. Resistant to Bremia NL-1-6.
Red Sun	A very attractive and fine tasting leaf lettuce. Loose leaf type, with maturity 45-50 days after sowing. An all-year round growing type, with distinct glossy red colour leaves. Good resistance to bolting. Displays excellent development at high temperatures.
Panama	Panama is a loose leaf type, very resistant to diseases, and light green coloured leaves. The variety is slow to bolt.
Minetto	Crisp head, Batavia type, medium-small size. Leaves are crispy, deeply notched and medium-green. Black seeded variety. It takes 25 days in nursery and 34 days after transplanting harvesting can be done. Minetto is resistant to heat and humidity (it can form heads even in tropical low lands) and show very high tolerance to tip burn.
X-01	It is a butter head type. The head is medium size; leaves are slightly wavy and medium green colour. Black seeded. Maturity 55 to 60 60 days after sowing. It is tolerant to heat and is slow bolting.
Mansang	It is a new romaine type cultivar developed from a cross between Late Texa green and Ithaca by the lettuce breeding team of National Institute of Highland Agriculture (NIHA), RDA in 2002. This new cultivar has brown seed colour and grey-green leaves. Bolting is a little later than 'Chaser Green'. Average market yield is 33.9 t/ha. It shows a good taste, resistant to lettuce mosaic virus and tolerant to soft rot caused by *Erwinia carotovora* sub. sp. *carotovora.*
Adam	It is a crisp head lettuce cultivar of NIHA, Korea Republic. It is cross of 'Montello' and 'Salad Crisp'. It has good growth, compactness, head weight and high yield of 45.2 t/ha.
Cauliflower (Open pollinated variety and hybrids)	
A. Extra Early group-Temperature requirement for curd initian and development 20 °C -27 °C.	
Pusa Meghana	It is an extra early variety developed at ICAR-IARI, New Delhi from heterozygous material collected from Hazipur (Bihar) using recurrent selection method. It produces white, compact, small to medium sized curds weighing about 350-400 g. curds are ready for harvesting in early September and takes only 95 days to come to its maturity. Average yield is 12.5 t/ha. Suitable for commercial cultivation under hot and humid weather conditions (temperature range for curd development is 22-27°C in northern plains).

Variety	*Description*
Early Selection	It is an early cauliflower. Seeds are sown in nursery during May-June then after 25 days old seedling are transplanted in main field it mature within 70-75 days after transplanting.
Parijat	An extra early variety, maturing in 50-60 days after transplanting snow white curd. Suitable for early spring in western and southern region. Summer sowing in northern region.
Himdhara	It is extra early type and cultivated in rainy season. It matures within 65-75 days after planting. Curd size bigger *i.e.* 700-800 g, curd uniform and compact, and required narrow spacing for higher return.
Himangini	It is extra early type and cultivated in summer and rainy season. It matures within 55-60 days after planting. Curds are more compact and can be procured for more days in the field
Himlata	It is tolerant to high temperature and grown in summer. It matures within 50-55 days after planting. It has highly compact curd and has permanent white colour. It is popular and well known to throughout country.
Himdev	It is extra early type grown in summer and rainy seasons. It matures 35-40 days after planting. It requires low spacing for high return. Curd size is medium and size varies from 25 g to 300 g.
Basant (F_1) NS245	A good hybrid with open type plants. The hybrid has good white curd colour with semi dome shape and good firmness. Each curd weighs 1.0-1.5 kg with a maturity of 55-60 days. It is suitable for late summers and can perform well in the southern states (latitude range of 13-26 degree).
NS60N (F_1)	A typical summer cauliflower with early maturity. The hybrid has open canopy with medium vigour. Each curd weighs 0.75-1.0 kg with good white colour and firmness. The curds are dome shaped with maturity of 55-60 days. It has wide adaptability.
B. Early group-Temperature requirement for curd initian and development 20°C-25°C.	
Pusa Katiki	One of the early released variety from the ICAR- IARI, New Delhi, having medium sized plants and bluish green leaves and medium sized curds.
Pusa Kartik Sankar	This is a hybrid variety developed by ICAR-IARI, New Delhi. Curds of this variety are compact, retentive white, medium sized weighing about 457 g. The mean temperature required for curd formation is 22-25°C. It is resistant to downy mildew. It gives an average yield of 12-15 t/ha.
Pusa Deepali	It is early group variety developed through selection from the local material from Punjab and released by ICAR-IARI, New Delhi for commercial cultivation in north Indian plains. The plants are medium tall, and erect with narrow frame. Leaves are waxy green with rounded apex. Inner leaves partially cover the curd. The curds are round, creamy-white and medium sized, weighing 500-700 g. Average curd yield is 15.0 t/ha.
Arka Kranti	It is early group variety developed through selection from the local material collected from Bihar at ICAR-IIHR, Bengaluru (Karnataka). Curds are white and compact and attain harvesting maturity sixty days after planting. Average curd yield is 22.0-25.0 t/ha.
Arka Spoorthi	Early group cauliflower variety 'Arka Spoorthi' is developed by ICAR-IIHR, Bengaluru (Karnataka) and released for commercial cultivation in 2011. It has very good curd quality and seed set under tropical conditions. It takes 54-56 days from sowing to 50 per cent curd initiation. Average curd weight 60-62 days from sowing to 50 per cent curd initiation. Average curd weight is 332 g and curd yield is 16.0 t/ha in 65-75 days duration. Moderately resistant to *Alternaria* leaf spot, downy mildew and susceptible to black rot.
Arka Vimal	Early group cauliflower variety 'Arka Vimal' is developed by ICAR-IIHR, Bengaluru (Karnataka) and released for commercial cultivation in 2011. It has very good curd quality and seed set under tropical conditions. It takes 60-62 days from sowing to 50 per cent curd initiation. It takes 60-62 days from sowing to 50 per cent curd initiation. Average curd weight is 374 g and curd yield is 18.7 t/ha in 75-80 days duration. Moderately resistant to *Alternaria* leaf spot, downy mildew and susceptible to black rot.
Pant Gobhi-2	Early group composite cultivar developed by GBPUA&T, Pantnagar (Uttarakhand) through selection from the local material and is suitable for cultivation in north Indian plains. The marketable curds of this variety are available in October in the plains. For this purpose nursery sowing is to be done by the middle of June. The curds are medium, compact and yellowish in colour. Average curd yield is 10.0 t/ha.
Pant Gobhi-3	It is early synthetic variety developed by intercrossing of early maturing eight inbred lines by GBPUA&T, Pantnagar (Uttarakhand) in 1989. This is classified as September maturity variety. The plants have long stem with sparse and semi-erect leaves. Its marketable curds can be available in September in the northern plains and for this nursery sowing is to be done around middle of May. The curds are semi-spherical, non-ricey, creamish-white and medium compact. Average curd yield is 15.0 t/ha.
Pusa Deepali	This variety is developed by ICAR-IARI, New Delhi through inbreeding from the local collection. It is hot weather cauliflower and temperature requirement is 20-25 °C during curd initiation and development. The variety is recommended for cultivation in entire north Indianm plains. Plants are medium tall, erect, bluish green and waxy leaves, curds white to creamy, very compact, medium sized, and weighing between 500-700 g. Retentive white and medium in size. Curds get ready in 100-120 days. Due to its erect habit and narrower frame (40-50 cm) it is possible to accommodate 35-40 per cent more plant per unit area. It is highly tolerant to riceyness, and fuzziness when grown in proper climate.
Summer Special-10	Early maturing hybrid suitable for sowing after mid may to July in northern India and during February to April in southern region. Curds mature in 45-50 days after transplaning. The plants are erect with medium frame. Curds are white, dome shaped, compact and weigh 800-900 g and have good self blanching habit.
Rimzim	Early maturing hybrid. Curds can be harvested approximately in about 55-60 days after transplantation. Curds are well protected by inner folding leaves, white in colour, semi-dome shaped and weigh 700-1000 g; excellent adaptability to different agroclimatic regions. Suitable for rainy season cultivation. It is recommended to plant at wider spacing to obtain uniform curds of larger size.
Sharavani	Early maturing hybrid, mature in about 60 days after transplantation. Curds are pure white and of excellent quality and dome shaped, weigh 800-1000 g. and upright leaves of the plant provide good protection to the curds. Tolerant to soft rot (*Erwinia* sp.) and suitable for rainy season cultivation.

Variety	Description
Hybrid-212	Plant green, maturity 45 days after transplanting, curd white, flattish dome, maximum curd weight 900 g, and suitable for early sowing (tropical condition).
Pawas	It is most suitable early hybrid, which can be grown from middle May to August. It produces average curd weight 700-800 g and ready within 60 days after transplanting.
Ati Shighra (F_1) NS-1351	Early summer cauliflower with open plant habit. Curds are medium white in colour and semi dome in shape with good firmness. Each curd weighs 0.5 kg and matures in 50 days. Comes up well in North and North East India of prevailing temperature 22-23 °C.
C. Medium group/Early mid group	
Pusa Himjyoti	It is a mid-late group maturing variety developed through pure line selection from local material 'MG-2-4 (Maghi Sri Ganga Nagar) Selection at ICAR- IARI, New Delhi and release for commercial cultivation in hills from April to July. In north Indian plains it is suitable to grow under December maturity group. The leaves are erect and bluish green with waxy coating. It takes 62-75 days from transplanting to curd formation. The curds are medium sized, white, compact and self-blanched, weighing 500-600 g. Average curd yield is 20.0 t/ha.
Pusa Hybrid-2	This is the first F_1 hybrid (a self-incompatible line-female parent and Sel.1-3-18-19-male parent) released by public sector organization and developed at ICAR-IARI, New Delhi. Plants are semi erect with bluish green upright leaves. It matures from mid November to mid December. Curds are creamy white, very compact and weighing on an average 907 g. It gives curd yield of 23.0 t/ha. The curd plant ratio is 10 per cent higher because the curds are very compact and with lesser number of leaves compared with 'Improved Japanese'. It is resistant to powdery mildew. Temperature between 16-20 °C is ideal for curd formation. Optimum sowing time is late July.
Improved Japanese	It is an introduction from Israel with long, erect, stem long, narrow leaves and medium to large sized curds which is compact and white. A mid season variety, crop in about 90-95 days. The average curd yield is 14.5 t/ha.
Pusa Sharad	This Early-Mid group cauliflower variety developed through selection from local material at ICAR-IARI, New Delhi. The plants are semi erect and open type plants with small stalk. Leaves are glabrous, bluish-green, long petiole oblong with narrow apex and waxy margins. Prominent midrib and ear like lobes at the base of the lamina. Its marketable curd becomes ready 85 days after transplanting. It is about two weeks early in maturity over the Improved Japanese. The curds are retentive white, knobby, very semidome shaped, compact and each weighing 750-1000 g. Average curd yield is 25.0 t/ha. The optimum temperature for its curd initiation and development is 16-20 °C. Seed should be sown by the end of July to early August depending upon the locality and climatic condition.
Pusa Shubhra	It is a mid-late group maturing variety developed through hybrid population of triple cross involving MG-2-3, 15-1-1 and D-96 at ICAR-IARI, New Delhi. The plants are tall with long stalk; the leaves are light bluish-green and twisting backwards from the middle. The curds are medium sized weighing 700-800 g, flat compact and white. Average curd yield is 25.0 t/ha. It has field tolerance to black rot and curd blight.
Pant Shubhra	This cultivar is developed by GBPUA&T, Pantnagar (Uttarakhand). Outer leaves semi-erect, inner leaves partially cover the curd, compact, slightly conical, non-ricey, and slightly creamish white curds. The average curd yield varies from 18.5-22.5 t/ha.
Pant Gobhi-4	This cultivar is developed by GBPUA&T, Pantnagar (Uttarakhand) through selection from local material. It has medium long stem, sparse, semi erect leaves, semi spherical creamish-white, medium, compact, non-ricey curds. This is November maturity group, and takes 115 days for marketable curds. Average curd yield is 14.0 t/ha.
Swarna	It can be grown from September to December in plain and round the year in the hills. Curds are white, compact and ready within 80-85 days after transplanting. It yielded 1.0-2.5 kg on an average curd weight. The yield potential is 16 t/ha.
Shewta	A medium maturity hybrid (70-75 days after transplanting) suitable for tropical and subtropical regions. Plants have thick leaves which provide protection to the curd. Solid and tender curds are white in colour, dome shaped and weighing approximately 500-600 g.
Chandramukhi (OP)	Plant vigorous and upright, first picking 75-80 days after transplanting. Curds set at temperature 15-17 °C, dome shaped, snow white, average weight 1.5 kg, typical self blanching type and keeps the curd white for longer period.
Basanti (OP)	Plant erect and vigorous, maturity 85-90 days after transplanting, curd set at temperature 29 °C, shape dome, snow white, average weight 2.0 kg, and most suitable for medium planting in North India.
Somersat	It has got a better heat tolerance, and matures in 65-70 days with curd weight of 2.0 kg each. Curds are very well protected with leaves. Recommended for August to December sowing. It posses high level tolerant to *Fusarium* wilt.
Amazing	It is uniform mid early cauliflower which gives non hairy white curd, mature in 70-80 dys after transplanting. Compact with self-blanching leaves, which gives snow white curd with weight of 1.0-1.5 kg. Recommended for August to November sowing in the plains and tolerant to *Fusarium* wilt.
Hybrid-218	Plants deep green, maturity 60 days after transplanting, curd white, flattish dome, average weight 650 g, and suitable for areas having milder climate.
NS-60 (F_1)	It is a very popular hybrid with wide adaptability. It is a mid season tropical cauliflower with semi blanches. The hybrid has very good attractive curds with compact dome shape and good firmness. Curds are milky white in colour and weigh 1.25-1.5 kg each and maturing in 60-65 days after transplanting.
NS-50 (F_1)	A mid season maturity hybrid. Heads are compact, uniform and deep dome shaped. Holds shape very well even when harvesting is delayed. Uniform heads with tolerant to downy mildew and black rot. Sutable for processing, fresh market and home garden.
Puspanjali	An ealy-mid variety, leaves are medium sized upright, greenish blue colour, curds are solid dome shaped and fine texture, and weigh 0.8 kg., and mature in 70 days transplanting.

Variety	*Description*
D. Mid Late Group-Temperature requirement for curd initiation and development is 12-16 °C.	
Punjab Giant-26	It is a mid-late group maturing variety developed through selection from local material at PAU, Ludhiana (Punjab) and release for commercial cultivation in Punjab and Haryana. The leaves are medium to large with waxy margins. The curds are medium sized, compact, yellowish-white and large. Average curd yield is 27.5 t/ha.
Punjab Giant-35	This variety is developed at PAU, Ludhiana (Punjab). Plants are vigorous with a height of 65.0 cm and spread is 75.0 cm. it has medium sized white compact curds. This variety is mainly recommended for Punjab and Haryana.
Pusa Shubra	A selection from a triple cross (MGS-2-3 x 15-1-1) x (D-96) developed ICAR-IARI, New Delhi. plants erect with somewhat long stalk and light bluish-green leaves. Light bluish-green twisting backwards from the middle. Curds are medium size flat, compact, white, each weighing 700-800 g. Once ready, they should be harvested without delay. They have a tendency to loose, yet least prone to ricyness. The best temperature for curd formation is 12-16 °C. In plains nursery sowing can be done in late August and transplanting by late September are consider ideal. The curd yield is about 25 t/ha.
Marghashira	It has a pure white and big size curd. Seeds are sown in nursery during September. Generally 25 days old seedlings are transplanted in the main field. It matures within 80-90 days after transplanting.
Hisar-1	This variety has been developed by CCS HAU, Hisar (Haryana). It is suitable for mid late season under Haryana conditions. Medium to large sized white curds. It yields 25 t/ha
Amarabati	It is highly profitable variety and mature within 65-70 days. Curds are white and compact. Seeds sowing should be done in September. It is medium group cauliflower. Yield potential is 40.5 t/ha.
Aparajita	It has big white and compact curd. Seeds are sown in nursery during September and curds are harvested in December. It is medium type cauliflower. Yield potential is 40 t/ha.
Goodman	Mature in 65 days after transplanting, head weight 2.0 kg and compact, snow white curd. Seeds sown in September-October and tolerant to *Fusarium*, and self blanching leaves structure. The leaves are spreading and cover the head very nicely.
Fastman	It matures in 60-65 days after transplanting. Very compact and snow white curds with curd weight of 1.0-1.5 kg. Suitable for cultivation during winter season in plains. Resistant to *Fusarium.* It is an early maturing cauliflower which gives nice white and non-hairy curds.
Lateman	It matures in 80-90 days after transplanting. Very compact self blanching white curd, and with upright growing leaves. It gives good quality curds having attractive white colour with average weight of 2.0 kg. Recommended for September to November sowing.
NS-66 (F_1)	Mid late season cauliflower with excellent uniform and quality curd. Plants are very vigorous with medium blanch. Curds are compact and smooth. They are dome shaped and have excellent firmness and density. Each curd weighs 1.5-1.75 kg with excellent attractive pure white colour and matures in 70-75 days.
NS-75 (F_1)	This hybrid cauliflower is very much suitable for early winter season. It has very vigorous plants with dark coloured foliage and an excellent self blanch covering the curds. Each curd weighs 1.5-2.0 kg, which are dome shaped and admirable snow white colour. Curds are compact with good firmness and density. Curd matures in 80-85 days.
NS-76 (F_1)	This cauliflower has very vigorous plants with good blanch. Curds are compact and dome shaped with excellent milky colour, firmness and density. The curds weigh 1.5-2.0 kg each. It is mid season cauliflower maturing in 75-80 days after transplanting.
Romanesco	It is a fractal in structure, with conical curd forms of small peaks of buds. It has slightly waxy, large, grey green leaves, bigger in curd size (0.6-0.75 kg each) complex, symmetrical curd features whorls of pointy, vivid green florets, and self blanching type. A good source of vitamin C for anti-ageing wound healing, and decreasing blood cholesterol. It takes 120-125 days for curd formation after transplanting. Yield potential is 35.1 t/ha.
Cassius	This is vigorous growing hybrid, which gives a strong and healthy plant. It is having very nice self-blanched leaves which give snow white curds. Mature in 65-75 days, good compact curd with average weight of 2.0 kg. Recommended for August to December sowing. Tolerant to Fusarium and other diseases. Yield potential is 38-39 t/ha.
Anarkali	Upright plant. Sown in September-October, and average weight 1.5 kg. Curds are being harvested within 65-70 days. Curds are white and dome shaped. Yield potential is 40 t/ha.
Puspagiri (F_1)	Suitable for western and southern region of India and maturing in 55-60 days after transplanting. Curds are very compact, medium sized and of snow white colour. Yield potential is 40-45 t/ha.
Himsagar (F_1)	Suitable for western and southern region of India and maturing in 65-70 days after transplanting. Curds are very compact, medium sized and of snow white colour. Yield potential is 48-50 t/ha.
Ooty-1	This variety is suitable for growing in hilly regions of Tamil Nadu above 1800 msl. It has a yield potential of 46 t/ha in 110-120 days.
E. Late group- Temperature requirement for curd initiation and development is 10-16°C.	
Snowball-16	It is a late group maturing variety developed through inter-varietal cross of EC12012 x EC12013 (from the exotic materal) at ICAR-IARI, New Delhi. The plants are straight; upright leaves, and tightly covering the curd. The curds are medium sized, compact and pure snow-white in colour. Suitable time for sowing in northern India is the first fortnight of October. Average curd yield is 15.0-20.0 t/ha.
Pusa Snowball -2	It has upright leaves covering curd tightly and slightly puckered at the margins. The curds are solid, white with very good staying power and slightly raised in the centre. It takes 110-135 days from sowing to curd harvesting. It yielded 15-12 t/ha in hills and 40-50 t/ha in plains.
Pusa Snowball K-1	It is field resistant to black-rot. Selection made at Regional Station, Katrain (H.P.) of ICAR-IARI, New Delhi from an exotic material. It has upright, waxy, light green leaves, self-blanched, snow white, and hi-dome curds. Becoming increasingly popular with the growers because of its curd quality and late maturing nature. Period of fresh availability is extended by at least 15 days.

Variety	Description
Pusa Snowball K-25	It has been developed by mass pedigree method of selection from cross of heading broccoli EC103576 x Pusa Snowball-1. Its curds are very solid, compact, medium sized, and retentive shining, white with very good staying power. Inner leaves cover the curd very tightly. Leaves are waxy, slightly bending towards inner sides with puckered margins. It takes 175-180 days from transplanting to first harvesting at Katrain and 105 to 130 days in different parts of the country. It is a very late variety and average yield being 27.5-30.0 t/ha.
Dania	This variety is developed at Kalimpong, and successful in the eastern region, more particularly the eastern hills. It has very sturdy plants with waxy leaves and medium deep curds. It is sensitive to fluctuating environment. Yield potential is 28-34 t/ha.
Dania Kalimpong	More commonly grown in Eastern parts of India. Plants sturdy, waxy leaves; curds medium deep and it is less sensitive to fluctuations of environment. Yield potential is 35 t/ha.
Punjab Giant-26	This variety is developed at Punjab Agricultural university, Ludhiana (Punjab). Main season variety, foliage green, plant height 57 cm with a spread of 75 cm. curds solid, white, and medium in size. Yield is 38 t/ha.
Punjab Giant-35	This variety is developed at Punjab Agricultural university, Ludhiana (Punjab). The leaves are erect and taller than Punjab Giant-26. The curds are snow white in colour, compact and large sized. The yield potential is 32.5 t/ha.
Himyug	It is a late type and seed is sown during late November and December in nursery and 25 days old seedlings are transplanted in main field. Leaves are dark green and curds are more white and compact. It matures within 70-80 days after trans planting. Curds are bigger in size *i.e.* 1.1 kg.
Garima	Early maturing vigorous hybrid, land curds can be harvested approximately 45-50 days after transplanting. Leaves are bluish green coloured. Curds are dome shaped, white and weigh 700-800 g, and suitable for trans planting from middle of February to end of May. Tolerant to black rot and heat. Yield potential is 38.5 t/ha.
Sharad Safedhi-70	Top quality hybrid in early snow ball group. Curds are snow white, dome shaped and relatively large weighing 1.2-1.5 kg and upright leaves provide excellent protection to the developing curds. The hybrid has flexible adaptability to different climatic conditions.
Mahima-80	Late maturing hybrid of mid snow ball group. Curds ready for harvest in about 80-85 days from transplantation. The plants vigorous, upright and have excellent self blanching habit. Curds are dome-shaped, pure white and weigh 2-3 kg and fairly tolerant to black rot disease. The bluish gree coloured waxy leaves provide some protection from insect pests also. This hybrid is recommended for winter season cultivation.
Sharmilee-85	Vigorous growing hybrid of late snowball group. Curds can be harvested in about 85-90 days after transplanting. Curds are snow white, high dome shaped and weigh 1.75-2.0 kg. This hybrid is recommended for winter season cultivation in the plains and for summer season in the hills.
Shivangi (F_1)	Snowball type cauliflower with very vigorous plant habit. It is suited very well for late season sowing. This hybrid has very good blanch giving very good protection to the curds. Curds are very compact and dome shaped with excellent firmness. Each curd weighs 1.5-2.0 kg and is of milky white colour. Matures in 90-95 days.
NS-84 (F_1)	This hybrid is of winter type late cauliflower with vigorous plant habit and an excellent bloom. Plants are uniform with compact curds and superior blanch. Each curd weighs 2.0 kg and is of a superb pure white colour and deep dome shape. It matures in 85-90 days after planting.
NS-90 (F_1)	It is late season cauliflower with large heads bred for winter season and suitable for sowing in areas with long winters. It has very vigorous plants with excellent blanch giving good tolerance to cold. Curds are compact, and deep dome shaped with excellent pure white colour. Each curd weighs 2.0-2.5 kg and matures in 90 days.
Other hybrids	Flanca, Sebal, Gija and Nedcha which matures in 3-5 months.
Cabbage (Open pollinated varieties and hybrids)	
A. Early maturity group- Heads can be harvested in 45-75 days.	
Golden Acre	It is an introduction 'Copenhagen Market' from U.S.A. and selection made on the basis of head quality and ealy maturity (taskes 60-75 days to reach maturity after transplanting). The plants are small, short stemmed and compact with few small and cup shaped outer leaves. The heads are small, round and compact, weighing 1.0-1.5 kg. The harvesting should not be delayed beyond 10 days of head-formation; otherwise the heads have a tendency to burst open. Average yield is 30.0 t/ha.
Pride of India	The plant type of this variety is similar to Golden Acre but it is about a week later in maturity. It is an early maturing (takes 70 days to rech edible maturity) variety developed by selection from Copenhagen Market and release for commercial cultivation by ICAR-IARI, Regional Research Station, Katrain (H.P.). The heads are small sized (1.0-1.5 kg), round and compact. Average yield is 20.0 t/ha.
Sel-8	A selection from an intervarietal cross of EC10109 x EC24855 made at Regional Station, Katrain (H.P.) of ICAR-IARI, New Delhi. The variety is characterized by a short-stalk, medium sized leaves, light green with waxy margins and slightly puckered leaf-blade. Heads slightly flattish round weighing 1.5-2.0 kg. It is also available a week later than the 'Golden Acre'. It ia resistant to black rot.
Copenhagen Market	It has bigger head than 'Goldn Acre' and takes 75-80 days for maturation. It is popular particularly in West Bengal and some nearby areas. Late varieties are grown on limited area for bulk supplies. Now in most the area it has been replaced by 'Golden Acre' but it is still grown on a limited scale in some pockets. Head size is 2.0-2.5 kg with heavy outer frame.
Pusa Ageti	It is an early maturing tropical variety developed by selection from an F_1 hybrid TKCBH-28 which was introduced from Taiwan and release for commercial cultivation by ICAR-IARI, New Delhi under high temperature conditions. It takes 70-90 days to reach maturity. The leaves are rayish-green and upright with waxy margins. The heads are flatish-round and compact weighing 600-1200 g. The yield varies from 11.5-35.0 t/ha depending upon the growing season.
Pusa Sambandha	An early maturity variety developed by ICAR-IARI, New Delhi which has wide adaptability. Heads are round and compact. It is very much suitable for close planting.

Variety	*Description*
Green Express	Marketed by the National Seed Corporation (NSC) New Delhi. Even before the NSC started marketing the variety, it had become popular in West Bengal, where its seeds had been smuggled from Bangladesh. Medium sized plants with few outer leaves. The head size depends upon the time of planting. Those planting earlier can attain head size up to 3 kg while in the subsequent plantings the head size decreases.
Kalimpong Eclipse Ball	This variety is also developed at the HRS, Krishnanagar and the VRS, Kalimpong. This is an early variety. Heads are ready for harvesting 60-80 days. Heads are oval in shape, big sized and compact.
Green Boy	It is marketed by National Seed Corporation New Delhi. It is an introduction from Japan. Early in habit, medium compact heads weighing 2.0-3.0 kg and is very popular throughout India.
CH-21	Extra early, dark green and ball headed hybrid with solid texture. Heads weigh 750 -1000 g with very short core. Inner leaves are very dense and outer excellent wrappers. Excellent field holding ability enables harvest at will to catch best market. Tolerant to cabbage yellow and Tip burn.
Kranti	It is developed by Mahyco. It is ready for harvesting within 60 days after transplanting. It is good for close space planting. Heads are 0.8-1.2 kg weight, round and compact.
Mitra	It is developed by Sungro Seed Co, Shalimar Bagh, N. Delhi. An early hybrid with excellent head to plant ratio. Heads are fresh green uniform, and compact, ball shaped with weight around 0.8-1.2 kg each. Good field retention.
Early Drumhead	It is an introduction from Denmark. It is early maturing variety. Foliage is light green. The heads are large, flat and pale green with inward enclosing the head loosely. Average yield is 25.0t/ha.
Sudha (BBS-115)	It has been developed by Bejo Sheetal Seed Co. Ltd. Jalna (M.S.). Heads are highly, compact, good interior structure, matures in 75 days and can be grown round the year. Average weight of head is 2-3 kg. It is resistant to Fusarium wilt.
	Female parent (FRICO-11): Medium late maturity; leaves are bluish and plant structure is good. It has flattish round and medium heads.
	Male parent (LAWIC): Early in maturity; bluish leaves, compact, round and medium heads.
BBS-44	It has been developed by Bejo Sheetal Seed Co. Ltd. Jalna (M.S.). Heads are roundish, average weight 2.5 kg; matures in 75 days but can stand for 20 days in field after maturity; grows well both in cold and hot weather; resistance to Fusarium wilt.
	Female parent (BOL-20): Medium early in maturity, bluish leaves; compact, round and big heads.
	Male parent (BIX-12): Early in maturity; bluish leaves, compact, round and medium heads.
Hara Gola	This early maturing hybrid produces globe shaped head with extra fine quality crispy and sweet flesh. Individual head weight is about 1.0-1.25 kg
Vaibhav	Early maturing hybrid with excellent field holding ability makes it best choice. Globe shaped compact heads are green and weight is around 1.0-1.5 kg. It response well to organic inputs.
Evergreen	Very promising and early maturing hybrid with relatively very good field standing ability. Head is round, firm, compact and weighs about 1.0-1.2 kg.
Super Cross	Very promising and early maturing hybrid with relatively very good field standing ability. Head is semi flat, very attractive and weighs about 1.3-1.7 kg. It response well to organic inputs and can tolerate both hot and wet condition.
Harini	Early maturing hybrid with excellent adaptability for upland/high altitude areas. Head is semi-flat, compact and weighs around 1.2 kg. This variety has good field holding ability and is suitable for long distance shipping.
Red Green	This newly introduced hybrid cabbage produces red colour round shaped head which are firm and compact. Slightly upstanding in habit. Matures in 70 days from transplanting. Highly tolerant to wet conditions in the field. It is suitable for fresh market and processing.
Primero	It is red cabbage. Mature in 65-70 days after transplanting. Attractive round head and very compact. Its standing ability is 30-40 days after maturity. Head weight 1.0-1.2 kg and used for salad decoration. It gives attractive red colour heads.
Benson	This is an early hybrid and good for intercropping in sugarcane. Crop matures in 45-50 days after transplanting. Head round with less outer leaves, and medium compact. It can stand one week in the field after maturity. Average head weight is 800 g and suitable for August to December sowing.
Fresco	It produce uniform round, compact head, and mature in 70-75 days after transplanting. It can stand 25 days in the field after maturity. Head weight varies from 2.0-3.0 kg. It can be planted in all three season and gives good heads even in the summer season. It gives good taste and an attractive head.
Gonzales	It produces very attractive, round, and compact head weighing 0.8-1.0 kg each, which have less outer leaves. It matures in 50 days after transplanting. It has standing ability up to 20 days after maturity. Suitable for August to December sowing in plains.
Morris	It is a hybrid and matures in 65-70 days. Heads are round shape and compact. Head weight is 65-70 days maturing, round shape and compat hybrid. Head weight is1.0 to 1.5 kg and is recommended specially for sowing in summer season.
Fieldman	It is very durable hybrid and can give quality head in all seasons. Head is round with less leaves and very compact with green leaves. Head weight is 1.2-1.2 kg and has standing ability 30 days after maturity. Crop matures in 60-65 days after transplanting. Suitable for Kharif, rabi and summer crop and good for long-distance transport.
General	This is a 60 days cabbage with 1.5 kg head weight. It is medium compact with 15 days standing ability and tolerant to *Fusarium* wilt.
Invento	It produces attractive round, and compact with short core head, which mature in 60-65 days after transplanting. Average head weight is 1.0 kg and has standing ability of 30 days after maturity. It is suitable for August to December sowing in the plains. It is good for transport and resistant to *Fusarium* wilt.

Variety	Description
Bronco	This is very uniform, and high yielding hybrid which can be grown round the year and specially recommended for sowing in summer and rainy season. It produces, very compact, round and firm head weighing 1.5-2.5 kg each. It mature in 70-75 days after transplanting and can stand 30 days in the field without bursting. It is resistant to *Fusarium* wilt and highly tolerant to *Xanthomonas*.
Governor	This is 75 days maturity, and round shape hybrids. Average head weight is 1.5 kg and can stand in the field for 30 days after maturity. Having multiple disese tolerance and can be sown round the year.
Rotonda	It is round, and very compact hybrid, which matures in 75 days after transplanting. Average head weight 1.5 kg with good field standing ability. It has good tolerance to *Fusarium* wilt and can be sown round the year.
Pragati	An early maturing hybrid with compact round heads and green outer leaves. Late bursting and excellent performance under spring and fall season
Green Ball	An early maturing hybrid, compact, round heads with green leaves, and wide possess adaptability. Tolerant to black rot and brusting. Average head weight is 600-700 g.
Green Gold	It is green type and head mature within 55 to 65 days after transplanting. Head weight varies from 1.5-2.0 kg.
Green Diamond	It is round type and early variety. It matures in 60 days after transplanting. Heads are compact and uniform in maturity and size. It is highly beneficial for farmers and greenish type head with weight from 1.0-1.5 kg.
Green Diamond (Improved)	An extra early variety maturing in 50-55 days after transplanting. Heads are white coloured which is suitable for early spring, sowing in western and southern region of India. It is also suitable for summer sowing in northern region of India.
Green-Samrat (F_1)	A mid season hybrid weighing 2.5-3.0 kg of medium large, round and compact head. Ready for harvesting in 70-75 days after transplanting. An excellent genetic material for strong and distance transportation.
Gold Star (F_1)	An excellent hybrid of cabbage. Heads are very solid and don't burst even after several days of maturity. Dark green coloured leaves have a protective layer. Maturity 60-65 days after transplanting. Average head weight is 1.5 kg.
Supper Ball-50	Strong and vigoruous early maturing hybrid with excellent head and plant ratio. Heads are compact, round with gray green outer waxy leaves and weigh 1.0-1.5 kg. Outer waxy leaves provide good protection against the pest like diamond back moth. Excellent field holding ability and wide adaptability makes this hybrid quite popular. Tolerant to black rot, black leg, tip burn and diamond back moth.
Gray Ball-60	Early maturing vigorous hybrid ready for harvest in 55-58 days after transplanting. Heads are very compact, round, gray green and weigh 1.0-1.2 kg. Good field holding capacity and tolerant to black rot and tip burn diseases.
Goody Ball	An excellent hybrid with bluish green leaves and compact round heads, weighing 0.8-1.2 kg. Field holding capacity is excellent and heads do not burst for a long time after maturity. Suitable for planting from July to January in northern India and throughout the year in southern parts.
Big Ball-75	A relatively late maturing hybrid, ready for harvest in 70-75 days after transplanting. The heads are very compact, dark green, predominantly globe-shaped and weigh 1.5-2.0 kg. The hybrid has excellent field holding capacity and very good field tolerance to black rot and fungal disease.
Quick Ball-45	Extremely early maturing hybrid, ready for harvest after 45 days of establishment of transplanted seedlings. The plants are medium sized and heads are round, compact, dark green and weigh 0.8-1.0 kg, suitable for plantig from the second week of July and onwards and up to December in the northern parts of the country and throughout the year in southern parts. Suitable for high density planting.
September	It is an introduction from Germany and is popular in Nilgiri hills. It has bluish green foliage, large frame, very compact heads, and each weighing 4-6 kg; better staying power and matures in 110 days. It is resistant to *Fusarium* wilt and highly tolerant to *Xanthomonas*.
Golden Ball	Early maturing hybrid, ready for harvest after 50-55 days of transplanting. The heads are uniformly round, compact, green weigh 1.0-1.2 kg with dark green wrapper leaves. Field holding capacity is very good. The hybrid has very good field tolerance to major fungal diseases and black rot.
Kopen Hegen Market Imported	It is cultivated in more acerage in Odisha. Its heads are medium in size and vary from 1.0-3.0 kg., oval green and compact. Seeds are sown in nursery during August and September. It matures within 75 days after transplanting.
Pashan	It is early type and head mature within 55-65 days. It is resistant to root rot and can tolerate rain, so it can be cultivated in rainy season.
Priya (F_1)	A tropical type hybrid, medium vigorous plant with very uniform growth and maturity, round and very compact head. Heads are of 1.0-1.5 kg in size. Ready to mature in 60-70 days after transplanting. It yields 70-105 t/ha.
Gold Star (F_1)	It matures within 60 days after transplanting of seedlings. It is dark green and compact head type. Heads are very solid and do not burst after several days of maturity. Average head weight is 1.5 kg. It yields about 20-25 t/ha.
Red Baby	A medium maturity hybrid, compact heads can be harvested in 65-70 days after transplanting of seedlings. Heads are round, and uniformly dark red with strong waxy outer leaves and weigh 1.0-1.2 kg. Inside colour uniformly dark red and layers finely arranged with a short core. Possess good storage ability. Tolerant to black rot disease and head bursting.
Mahima (F_1)	Categorized under round early group of cabbage. Good head firmness. Round shaped, light green coloured and compact heads approximately weighing 0.8-1.0 kg. Head maturity in 60-65 days after transplanting with a field holding capacity of 8-10 days.
Simran (F_1)	Categorized under round early group of cabbage. Good standing ability. Round shaped green coloured and compact heads approximately weighing 1-2 kg. Head maturity in 60-65 days after transplanting with a field holding capacity of 12-15 days.

Variety	*Description*
Sumithra	Early maturing hybrid and heads can be harvested in about 58-60 days after transplanting. Compact heads are round, dark green and weigh 0.9-1.2 kg. It is tolerant to black rot disease.
Bobby	A medium early maturing vigorous hybrid. Heads can be harvested in about 70-75 days after transplanting of seedlings. Heads are round, compact with grey green leaves and weigh 2.0-3.0 kg. Its versatile nature makes it ideal for long distance transportation. Tolerant to black rot, tip burn and diamond back moth.
NS-22 (F_1)	It is developed by Namdhari Seed Co. Bengaluru (Karnataka). It is tropical type hybrid, vigorous plants, uniform growth and maturity. Heads are round with very good firmness and density. The heads weigh 1.5-2.0 kg each with good field holding capacity and are suitable for spring and summer sowing and matures in 70 days after transplanting.
NS-25 (F_1)	It is developed by Namdhari Seed Co. Bengaluru (Karnataka). This early hybrid has upright habit, uniform compact heads and good wrapper leaves. The heads are round and weigh 1.0-1.25 kg each. It matures in 55 days. This hybrid is suitable for close planting because of small compact heads. It has very good field holding capacity.
NS-26 (F_1)	It is developed by Namdhari Seed Co. Bengaluru (Karnataka). Very good hybrid with vigorous plants and green foliage with good wrapper leaves. The round heads are very compact with good density and firmness. They weigh 1.5-2.0 kg each with good field holding capacity. This hybrid is suitable for spring and summer sowing with a maturity period of 65-70 days.
Mansi	An early maturing hybrid has compact, and round heads with green leaves. Wide adaptability. Intermediate level of partial resistance to black rot, bursting and high temperature. Average head weight is 900-1250g.
Bharat (45)	This promising hybrid is round shape and early maturing (55-60 days). Medium sized heads are bluish green in colour, partial tolerant to diamond back moth, high temperature and weighing about 1.25-1.50 kg.
Ria (105)	Medium maturing hybrid (65-70 days). The round heads are very compact dark green in colour with good taste. Good field holding capacity. Average head weight is 1.5 to 2.0 k.
Summer Queen (F_1)	Plants are strong, compact with 10-12 auxiliary leaves. Heads are attractive green, semi-flat, weighing about 2.0 - 2.5 kg. Heads will be ready in 55-60 days after transplanting. It is suitable for growing round the year, especially for summer. This is good hybrid for hot humid climate. It is tolerant to soft rot and cracking.
Vishwas (F_1)	It is very early maturing hybrid. Leaves dark green, fleshy broad and open. Heads are medium size, moderately compact, round and uniform green coloured, weighing 0.8-1.0 kg each. Heads will be ready in 45-50 days after transplanting. Uniform maturity which make single harvest possible. Matured head does not split up to week. It is suitable for close planting (70, 000-80,000 plants/ha), could be plant densely. Yield 50-60 t/ha under optimal conditions. Tolerant to *Fusarium* wilt.
Combiner (F_1)	It is developed by Novartis (Syngenta) Seed Co; Pune (Maharashtra). Compact, vigorous and strong plant with 8-10 auxiliary leaves. Heads round, green, compact, weighing approximately 1.0-1.25 kg each. Ready for harvest in 60-65 days sfter transplanting. Most suitable for cool and dry season. Tolerant to head bursting. Excellent standing ability of 45-60 days after the crop is ready.
Equatoria (F_1)	Strong plant with vigorous growth and 8-10 auxiliary leaves. Head flattened attractive green coloured, glob-shaped head with height 14-16 cm, diameter 20-22 cm and weight 2.5-3.0 kg, and head is ready for harvest within 70-75 days sfter transplanting. Tolerant to head bursting and low temperature.
Green Globe (F_1)	Plants vigorous, and compact with 8-10 auxiliary leaves. Head is round, attractive green and weight is 1.0-1.25 kg. Extremely early maturing. Head is ready within 50-55 days after transplanting. Suitable for growing round the year. Tolerant to cracking, makes the harvesting period more adjustable.
KS-Cross (F_1)	Plants strong, and sturdy vigorous with 10-12 auxiliary leaves. Head is green, semi flat, 13-15 cm in height and 19-21 cm in diameter, weighing about 1.0-1.5 kg with attractive appearance. Very early maturing and ready within 56-60 days after transplanting. Tolerant to bacterial soft rot, black rot and suitable for growing round the year and especially for hot and humid climate.
S-107	It comes to maturity in 70-75 days after transplanting. Heads are flattened globe, green, very good firm, average weight 2.0kg, and excellent field holding capacity.
Hybrid-123	It matures in 45 days sfter transplanting. Heads are ball type, dark green, solid excellent firm, average weight 1.2 kg and excellent field holding capacity.
Hybrid-124	Plants come to maturity in 50-60 days sfter transplanting. Heads are flattish round, light green, good solid, average weight 1.2 kg and moderate field holding capacity.
Hybrid-125	Plants come to maturity in 65 days sfter transplanting. Heads are round, light green, good solid, average weight 1.4 kg and good field holding capacity.
Hybrid-126	Plants come to maturity in 65 days sfter transplanting. Heads are round, light green, good solid, average weight 1.6 kg and good field holding capacity.
Hybrid-133	Plants come to maturity in 65 days sfter transplanting. Heads are flattish round, bluish green, good solid, average weight 1.5 kg and excellent field holding capacity.
Super Hero	It matures in 65-75 days sfter transplanting. Heads are semi-flat green, firmness excellent, average weight 2 kg and tolerant to pests and diseases.
S-101	It is extra early maturing hybrid. It matures in 55 days sfter transplanting. Heads are ball shaped, glossy green, firmness excellent, average weight 1.2 kg and tolerant to pests and diseases.
Improved Savitri (OP)	It is developed by Sungro India Ltd. New Delhi. It matures in 65 days after transplanting. Heads are round, green, firmness excellent, average weight 1.2 kg and suitable for tropical condition.

Variety	*Description*
	Medium maturity group
Pusa Drumhead	Selection made at Regional Station, Katrain (Himachal Pradesh) of ICAR-IARI, New Delhi from an introduced line 'EC6774' of Japan. It is the earliest in this group. The heads are flat, slightly loose weighing 3-5 kg. The outer leaves are light green with prominent midrib. It has field resistance to black-leg disease (*Phomalingum*). Heads are ready in 80-90 days after transplanting.
Bahar	It is developed by Pro-Agro Seeds Co. New Delhi medium early maturing hybrid. Medium sized heads, round and compact with short core and bluish outer leaves. Excellent performances under low land summer fall cropping. Tolerant to black rot, soft rot and least damage from diamond back moth.
Unnati	A late maturing hybrid (75-85 days), deep green leaves with large plant frame, flattish heads are compact, good for storage and long distance transportation. Excellent field holding ability. Average head weight 3.0-4.0 kg.
Geetanjali	It matures within 80-90 days after transplanting. Head weight varies from 2.5-3.0 kg. It is highly suitable for cultivation in Odisha and West Bengal.
Konark	Extremely good material for North East region, West Bengal and Odisha. A mid season hybrid with flattish-round solid head weighing 2.5-3.0 kg with large frame and strong outer leaves. Excellent field holding ability. More adopted for sowing during rainy season in these regions. Head is crisp, tasty and excellent for fresh market.
Raj-2	A dfark bluish green ball headed hybrid suitable for fresh market and long trans-shipment. Maturity 80-85 days after transplanting. Heads are medium large grows upto 4.0 kg which depends on field conditions. Leaves are thick and core short. Tolerant to black rot. This hybrid has been bred for a wide adaptability.
Meenakshi	It is deep green and head size is upto 5.0 kg and can be kept in field for more days without any damage. Generally head size varies from 2.5-3.0 kg. It matures in 70-80 days after transplanting.
Hari Rani Gol	It is being marketed by Mahyco Seed Co. (Maharashtra) and is very popular in the southern states especially in the Nilgiri hills. It has a medium-sized frame with light green foliage. The heads are flattish round and individual head weighing around 2.5 to 3.0 kg.
Varun	A medium maturing hybrid. Heads are medium compact globe shaped, medium green in colour and individual head weighing around 1.0-1.5 kg. It has excellent field holding ability. It is also adaptable for high density planting and long distance shipping.
Hema II	An outstanding hybrid produces flattened, globe shaped heads of weight 1.5-2.5 kg, goot tolerance to bursting and gives an extended harvesting time.
Bhima	A versatile late maturing hybrid ready for harvest in about 75-80 days after transplanting. Heads are very compact and solid, short cored, flattened globe in shape, weigh 2.5-4.0 kg and covered with dark green wrapper leaves. Tolerant to black rot disease. Excellent field holding capacity allows flexible harvest. Ideally suited for all season cultivation in southern India, especially Karenataka.
Vijay	Medium maturing variety (80-85 days after transplanting). Semi round, heads are compact, and dark green in colour with good taste. Excellent field holding capacity allows flexible harvest. Ideal for processing and fresh use. Tolerant to yellow and black rot.
Dublin	This is a new medium sized variety with a fresh medium green colour leaves. The head is filled quickly and it generates a firm fair sized ball, which is well closed at the base.
Hero	It matures in 80-85 days after transplanting. Heads are flat, bluish green, firmness excellent, and average weight is 2.0 kg. Excellent field holding capacity allows flexible harvest. Very strong resistance against pests and diseases.
Quisto (F_1)	Plant strong, self supporting, sturdy, erect compact and well built. Bluish gree, fleshy, and broad leaves with serrated collars, 10-15 auxiliary leaves. Head compact, dark green, heavy, round and uniform. Ready for harvest in 75-80 days after transplanting. Short and tender core with fine internal structure. Head weight is around 1.0-3.0 kg. Matured head does not split for 90 days. Very tasty and good for table, cooking or for process-industry. Tolerant to *Fusarium* wilt. Yield is 60-100 t/ha under optimal conditions
Gideon	It gives very compact bluish green, round head with pointed upward leaves. Maturity 75-80 days after transplanting and standing ability is 35 days after maturity. Average head weight is 2.0 kg. Good for round the year production and suitable for long storage. It is resistant to *Fusarium* and *Xanthomonas*.
Ramada	It is suitable for long storage and transport. It has good vigour and suitable for round the year planting. It produces flattish round and very compact head with 2-3 kg individual fruit weight. It matures in 80-85 days after transplanting and can stand 25-30 days in the field after maturity. It is resistant to *Fusarium* wilt.
Shristi	Late maturing hybrid, extremely suitable for planting in southern plains. Heads are relatively large, compact, semi-globe to flat shape with attractive outer dark green leaves and individual fruit weight is around 2.5-3.5 kg. Heads mature in 80-85 days after transplanting and have excellent field holding capacity. It can be grown throughout the year in the plains of southern parts of the country. Tolerant to black rot and black leg disease.
Melisa	It matures in 75-80 days after transplanting. It has wrinkled leaves with dark green outer leaves and a dense interior structure and has good taste. It is healthy with very good disease tolerance. It is suitable for August-November sowing.
Drumhead Savoy	It has dark green foliage which is much blistered or wrinkled. The quality is superior but it is not commercially accepted in our country.
Red Cabbage	It is developed by Bejo Sheetal Seed Co. Jalna (Maharashtra). It has distinct coat of wax and produces a head of 1-2 kg, taking about 90 days after transplanting to head formations. All the red cabbage cultivars are tolerant to diamond back moth insect which is serious pest of cabbage.
Field Rocket	It gives flattish round, compact head, with well covered leaves, and mature in 85-90 days after transplanting. It can stand 25 days in the field after maturity. Head weight is around 2.5-3.0 kg. It can be sown round the year except very hot summer. It is good for transport. Resistant to *Fusarium* wilt.

Variety	*Description*
Jersey Wakefield	A pointed head cultivar is preferred by some growers but it has very limited market. It takes about 55-60 days after transplanting to form head. The outer leaves are few and slightly crinkled. Average individual fruit weight is 1.0-1.5 kg. It has got better staying power. This has also preferred for raw eating due to better taste.
Green Samrat (F_1)	A mid season hybrid has the individual head weight of 2.5-3.0 kg with bigger size. It has dark green and compact medium large round head. Heads mature within 80-90 days after transplanting. An excellent hybrid for sowing and distance transportation. It yields 70-100 t/ha. Tolerant to black rot and *Fusarium* wilt.
MCV-1	It is developed by ICAR-IARI Regional Station Katrain (Himachal Pradesh) Heads become ready for harvesting after 80 days from transplanting. It has round shape, bigger head (1.5-2.5 kg) and compact. The foliage is dark green with waxy margins.
NS-21 (F_1)	It is developed by Namdhari Seed Co. Bengaluru (Karnataka). Very vigorous hybrid with an excellent plant stand. Foliage is dark bluish green in colour with strong uniform head formation. Heads are flat shape with individual fruit weight of 1.5-2.0 kg. It is tolerant to several insect pests. Produces, very compact dense heads of very good firmness and short core. It matures in 75-80 days from transplanting.
S-103	It matures in 80-85 days from transplanting. Heads are ball shaped bluish green, excellent firm and individual average head weight 2.5 kg. It is suitable for long distance transportation. It also possess outstanding field holding ability.
601	A medium-early maturing hybrid is suitable for fresh market and long transportation. Heads are ball shaped and having individual head weight of 1.0-1.5 kg. Heads are firmed, shored cored with attractive bluish green colour and excellent field holding ability.
Bejo 1857	It is an 85-90 days maturing hybrid. The head is flattish round and individual head weight is around 1.5-2.0 kg with attractive green colour. It is very compact and can stand for 30 days in the field without bursting. It has long storage ability. It can be stored at room temperature after harvesting for 25 days. It is recommended fort round the year sowing, as it has good tolerance to *Fusarium* wilt.
Late maturing group- Heads are formed in 90-120 days after transplanting	
Late Large Drumhead	It has large flat head somewhat loose, wider frame and takes 115-120 days for maturity after transplanting. It is being marketed by different seed companies.
Kalimpong Eclipse Drumhead	This variety is developed at the Horticultural Research Station, Krishna Nagar and at the Vegetable Research Station, Kalimpong, Darjeeling. This variety is suitable for West Bangal. It is comparatively late variety and heads become ready for harvesting in 100 to 110 days. The heads are large, flat and compact.
September	An introduction from the German Democratic Republic and a most popular open pollinated variety in the Nilgiri hills. It is characterized by a large frame, bluish-green leaves with serrated margins. It has a long stalk and the plants fall on one side after the head-formation has completed. The heads are flattish oblong, very tight and individual head weight id around 4-6 kg. Highly susceptible to black rot. It takes 96-100 days from transplanting to head formation.
Hari Rani Gol	It is a good hybrid for medium late maturity. It produces medium sized, ball shaped dark green, and solid heads with good wrapper leaves. Each head weight 1.5-2.0 kg. It can be harvested after 95-100 days of transplanting.
Ratna	An outstanding hybrid with medium late maturity. Big size heads are dark green in colour and very compact. Heads weigh approximately 1.5-2.5 kg.
Suvarna (BSS-32)	It has been developed by Bejo Sheetal Seeds Pvt. Ltd. Jalna (Maharashtra), Jalna. Heads are round, average head weight 3 kg; strong smooth outer leaves; and matures in 110 days.
	Female Parent (ET-2-2)-Medium late in maturity, blue leaves, vigorous growth; round and medium heads.
	Male Parent (BOL-13)-Late in maturity, blue leaves, flattish round and big heads.
Cabbage (Synthetic)	
Pusa Sambandh	It is a synthetic variety, developed by ICAR-IARI Regional Station, Katrain (Himachal Pradesh). It has wide adaptability with round, compact heads, early narrowframe, and suitable for close planting. Average yield is 80-90 t/ha.
Pusa Synthetic	It is developed by ICAR-IARI Regional Station Katrain (Himachal Pradesh). Average yield is 95-100 t/ha. Resistant to balck rot.
Savoy group-This group is also not popular in India. All the cultivars have blistered leaves. The group has all the shapes like pointed, round and flat. The commonly catalogued cultivar is Chieftain. It has fairly light green foliage and large rather smooth, flatround head.	
Other cabbage (Open pollinated varieties and hybrids)	
Regalia	It is developed by Sakata.
Stone	It is developed by Sakata.
Sel.-507-4	It is a tropical type cabbage and developed by ICAR-IARI, Regional Station, Katrain (Himachal Pradesh) which do not require chilling for seed production. Therefore seed production (good seed yield) of cabbage is possible under Delhi conditions. Plants bear acceptable head quality.
Sel.-528	It is a tropical type cabbage and developed by ICAR-IARI, Regional Station, Katrain (Himachal Pradesh) which do not require chilling for seed production. Therefore seed production (good seed yield) of cabbage is possible under Delhi conditions. Plants bear acceptable head quality.
Sel.-507-22	It is a tropical type cabbage and developed by ICAR-IARI, Regional Station, Katrain (Himachal Pradesh) which do not require chilling for seed production. Therefore seed production (good seed yield) of cabbage is possible under Delhi conditions. Plants bear acceptable head quality.
AC 238	Resistance to cabbage yellows (*Fusariumb oxysporum* f. sp. *conglutinans*).
Spitz Kool	Resistance to cabbage yellows (*Fusariumb oxysporum* f. sp. *conglutinans*). It is also resistance to black rot (*Xanthomonas campestris*).

Variety	Description
EC 93559	Resistance to cabbage yellows (*Fusariumb oxysporum* f. sp. *Conglutinans*).
EC 10109	It is resistance to black rot (*Xanthomonas campestris*).
Sel.-8	It is resistance to black rot (*Xanthomonas campestris*).
Green Land Hammer	It is resistance to black rot (*Xanthomonas campestris*).
MR-1	It is an introduction from the U.S.A., carries multiple resistance to sclerotinia rot, wire stem (*Rhizoctonia solani*), downy mildew, black rot and soft rot (*Erwinia carotovora*).
Red Pickling	It is red type and resistant (preferential) to cabbage caterpillar (*Pieris brassicae*) and susceptible to aphids (*Brevicoryne brassicae*)
Red Rock Mammoth	It is red type and resistant (preferential) to cabbage caterpillar (*Pieris brassicae*) and susceptible to aphids (*Brevicoryne brassicae*).
All Season	It is green type and susceptible to to cabbage caterpillar (*Pieris brassicae*) and resistance to aphids (*Brevicoryne brassicae*).
KK Cross	It is developed by Takii. It is green type and susceptible to to cabbage caterpillar (*Pieris brassicae*) and resistance to aphids (*Brevicoryne brassicae*).
OS Cross	It is developed by Takii.
Green Cornet	It is developed by Takii.
BC-26	It is developed by Novartis (Syngenta).
BC-34	It is developed by Novartis (Syngenta).
BC-35	It is developed by Novartis (Syngenta).
Murketer	It is developed by Novartis (Syngenta).
GS-403	It is developed by Golden Seeds Co.
Goldy Ball	It is developed by Golden Seeds Co.
No. 23	It is developed by Mahyco.
No.47	It is developed by Mahyco.
No. 120	It is developed by Mahyco.
No. 139	It is developed by Mahyco.
C.B. 296	It is developed by Indo-American
Krishna	It is developed by Indo-American
IAHS-5	It is developed by Indo-American
I.A. Express	It is developed by Indo-American
IRCH-5	It is tolerant to both aphids (*Brevicoryne brassicae*) and caterpillar (*Pieris brassicae*).
Nath Indu	It is developed by Nath Seed Co.
NL-138	It is developed by Nath Seed Co.
BBS-116	It is developed by Bejo Sheetal Seed Co.
BBS-207	It is developed by Bejo Sheetal Seed Co.
Hybrid-123	It is developed by Sungro Seed Co.
Chinese Cabbage	
ASVEG-1'	Locally known as Kiew Maysa (Hybrid-62) or 'April Jade'or some time Xia Feng, Nai Re-1, and Xia Bai-62, made this genotype at an affordable price during off season debeloped by World Vegetable Centre (AVRDC) Taiwan and very much popular in Vietnam. This variety has been tropicalized, improved and tested for Thailand conditions by ARC and KU researchers. It can grow and form head during hot rainy season assuring farmers of the needed income. This variety gives an average yield of around 29 t/ha during the hot rainy season, around 39 t/ha on the rainy season and as high as 51 t/ha during the cool season.
77M(2/3)-41-0-2-0-0	Locally known as Nai Re-4 developed by World Vegetable Centre (earlier AVRDC) Taiwan and very much popular in Vietnam.
Hybrid-58	Locally known as FU Bao, Nai Re-1 and Xia Bai-59 debeloped by World Vegetable Centre (earlier AVRDC) Taiwan and very much popular in Vietnam.
Hybrid-59	Locally known as FU Bao, Nai Re-3 and Xia Bai-58 debeloped by World Vegetable Centre (AVRDC) Taiwan and very much popular in Vietnam.
Chinese Cabbage/ Sarson No.1	It is developed by PAU, Ludhiana in the year 1987. It is a non-heading type with light green, broad leaves and semi-erect plant habit. Its midrib is white, succulent and tender. It takes about 30 days for the first cutting after transplanting. It is quite rejuvenating, giving 6-8 cuttings. Its average yield is 40.0 t/ha. It is field resistant to *Alternaria* leaf spot.
Tianjin	It is grown in Tianjin area of China since very long time. It has good quality, low cellulose content, disease resistance, high yield better storage and transport quality.
Naebyung 60 days	It is an early cultivar, has significantly more vitamin C (36.8 mg/100 g fresh weight) than other cultivars.

Variety	*Description*
Raseosamgye	This cultivar is having highest percentage (85 per cent) of marketable heads at harvest in comparision to others.
Olympic	This cultivar is having highest percentage (85 per cent) of marketable heads at harvest in comparision to others.
Nepos	This cultivar perform well and do not produce bolter when planting is done in cold glass houses from second week of March and harvesting is done in third week of May.
Nestor	This cultivar perform well and do not produce bolter when planting is done in cold glass houses from second week of March and harvesting is done in third week of May.
Nerva	This cultivar perform well and do not produce bolter when planting is done in cold glass houses from second week of March and harvesting is done in third week of May.
Spectrum	This cultivar perform well and do not produce bolter when planting is done in cold glass houses from second week of March and harvesting is done in third week of May.
Chiifu	This cultivar is very much popular in China having good morphological character and performs well in their growing region.
Kaga	This cultivar is very much popular in China having good morphological character and performs well in their growing region.
Hotoren	This cultivar is very much popular in China having good morphological character and performs well in their growing region.
Aichi	This cultivar is very much popular in China having good morphological character and performs well in their growing region.
Kenshin	This cultivar is very much popular in China having good morphological character and performs well in their growing region.
Peking	It is fairly late maturing Japanese variety, released in late 1979. It has a tight head weighing about 3.0 kg and few leaves.
Chihili	It produces a tall narrow and cylindrical head.
Wong Bok	It is short and very chunky in shape.
Hybrid 59	It is an early maturing, higher yielding and more heat tolerant than others.
Yuan Pao-2	It has been developed for commercial cultivation.
65 Days	It produces more than 80 tonnes per hectare.
Michihli	It is heading type give yield of 54.2 t/ha. Cylindrical head and tightly folded dark green leaves, medium maturity and uniform head. White midribs very crispy and tender.
Manha (F_1)	Cheongbang x Yungching 30 Days and very good performer with respect to yield and other horticultural attributes. It has moderate resistance to *Erwinia carotovora* and turnip mosaic virus.
Sambok (F_1)	It is cros of Yungching 60 Days x Ta Feng Feng Lu nd very good performer with respect to yield and other horticultural attributes. This hybrid shows moderate resistance to *Erwinia carotovora* and turnip mosaic virus.
Taoyung Yashu-2	This F_1 hybrid is responsive to nitrogen fertilizer and high temperature and it is high yielder.
Sumo	It has dark green foliage, leaves hairless and shiny. Mature in 55 to 60 days after sowing. Head oval shape and weighing 1.0-2.0 kg. The crisp head has a pleasant flavor and is ideal for salad and good substitute of head lettuce. Resistant to virus, soft rot and downy mildew.
Tropic Early	It produces marketable heads within 40 days after transplanting in tropical low land areas. The head is very firm and the leaves are hairless, thick and very juicy. It is suitable for fresh consumption, frying and processing. It is tolerant to bacterial soft and mosaic virus.
Rex (CC-24)	Rex is adapted and cultivated in the tropical highlands. It matures in 65 days after sowing. The head is oval shape and light green in colour with an internal orange yellow colour and very attractive. It is resistant to club root disease and shows good tolerance to soft rot and downy mildew.
Broccoli (*Botrytis* group)	
Pusa Broccoli (KTS-1)	It is developed by selection from the exotic material and release for commercial cultivation by ICAR-IARI, Regional Research Station, Katrain (H.P.). It is medium maturing variety. Its plants are medium tall (65-70 cm long), foliage waxy and dark green with slightly waxy margins. It takes 125-140 days to reach marketable maturity after sowing and 90-105 days after transplanting in hills and reduced by 5-10 days in plains. The heads are compact/solid, light green, small buds, slightly raised at the centre and each weighing 350-450 g. The main head size and weight are about 6 cm and 15.4 cm (height x diameter). It is resistant to mild frost and snowfall (Verma and Sharma, 1998).
Palam Samridhi	It is developed by selection from the exotic material and release for commercial cultivation by CSK, HPAU, Palampur (H.P.). It has medium plant type with medium sized, dark green leaves. It is a green sprouting type with heads born on long stalk. Its heads are medium sized, compact, each weighing 300-400 g. The subsequent tender sprouts arising from leaf axils. Heads are ready for harvesting in 85-90 days after transplanting. Average yield is 15.0-20 t/ha (Kalia, 2002).
Palam Kanchan	It is developed by selection from the exotic material and release for commercial cultivation by CSK, HPAU, Palampur (H.P.). It has long, broad, bluish green upright entire margin leaves with prominent white midrib and veins. Its heads are large, compact, attractive and yellowish-green in colour. Harvesting is possible in 140-145 days after transplanting. Heads remain free from yellow eys and bracts. Average yield is 25.0-27.5 t/ha (Kalia and Singh, 2004).
Palam Vichitra	It is developed by selection and release for commercial cultivation by CSK, HPAU, Palampur (H.P.). It has medium open, dark green leaves with redish tinge. Purple streaks are prominent on stems at seedling as well as adult plant stage. The head is medium, compact, and purple coloured and attractive. It attains marketable maturity in 115-125 days after transplanting with an average yield potential is 22.5-25.0 t/ha (Kalia and Jamwal, 2006).

Variety	Description
Punjab Broccoli No.1	It is developed by selection from the exotic material and release for commercial cultivation by PAU, Ludhiana (Punjab). The leaves are small, waxy and dark green with smooth leaf surface. Plant branched, bears large number of sprouts in the axil of leaves. It is early maturing and takes 65 days to reach marketable maturity. The heads are compact and dark green with bluish tinge. Average yield is 17.5 t/ha.
Lucky	It matures in approximately in 60-65 days after transplanting. Head weight is 600-800 g and gives compact curd with fine buds. It can be planted in winter season in the plains.
Fiesta	It matures in 65-70 days after transplanting. Head weight is 800-1000 g. It is very compact and has fine buds and very high yield and can be planted in winter season in the plains.
Early Green	Other cultivars: Premium crop, Green Duke, and Green Comet
Brussels Sprouts	
Hilds Ideal	It is a popular cultivar in Indian conditions. Its plant grows up to 60 cm and produces 45-55 sprouts, each weighing 7-8 g and the sprouts are compact and possess good flavor. A negligible percentage of loose sprouts at the base of the plants is also marketable. It takes about 115 days for first picking. The yield per plant is 25-400 g in four pickings at an interval of 10 days. Delay in picking will lower the sprout quality (Sharma and Gill, 1986).
Jade Cross (F_1)	A high yielding late hybrid is very much popular and takes 90 days after transplanting. Sprouts are firm, dark green, closely packed on long stems, and can be grown under wide range of growing conditions (Gill, 1993).
Diablo	It matures in 120-150 days after transplanting, during winter season. In order to get uniform sprouts, cut the top of the plant as soon as sprouts at the bottom start to develop. It can be boiled eaten with cheese spread. Plant height is 90 cm, winter variety, gives round solid green sprouts, which are well spaced on the stem. It was developed by Bejo Sheetal Seeds Co., Jalana (Maharashtra).
Cacptain Marvel	It is an early season cultivar.
Prince Marvel	It is a mid-season cultivar.
Kales (*Brassica oleracea* var. *acephala* L.)	
Khanyari	It is a summer Kale. Stem tall, erect, glabrous unbranched; leaves with two sub-terminal, alternate sub-opposite small lobes and green in colour; frost tolerant yield potential is 35.0-45.0 t/ha.
Kawdari	It is a summer Kale. Stem tall, erect, glabrous unbranched; leaves petiolate, shorter than lamina, petiole broad base and sheathed, green in colour yield and frost intolerant. Yield potential is 40.0-50.0t/ha.
Kashur Haak (Hanz Haak)	It is a biennial Kale.Stem green, thin reduced and erect, glabrous with racemose type of branching leaves; simple, petiolate, exstipulate green; hardy with a yield potential of 50.0-52.05 t/ha.
G.M. Dari Haak or Wand Haak	It is a winter Kale. Stem green, condensed, glabrous; leaves verticillate, petiolate, exstipulate, green tolerant to frost with a yield potential of 50.0-700 t/ha.
Anchar Haak	It is a summer Kale/winter Kale. Stem short, turns cyclindrical with maturity glabrous purplish green or green in colour; leaves petiolate, exstipulate, purplish green to green in colour and susceptible to frost with a yield potential of 30.0-45.0 t/ha.
Dwarf Green Scotch	It is a dwarf variety. Plants are 40.0 cm tall and having good tasty green leaves.
Dwarf Blue Scotch	It is a dwarf variety. Plants are 40.0 cm tall and having good tasty bluish leaves.
Hamberg Market	It is a medium tall variety. Plants attain the height of 40.0-80.0 cm.
Scotish Type	'Scotish Types' are tall in nature and attains the the height of more than 80.0 cm. Leaves fall quickly and give poor yield. Very much suitable for Scotish region.
Siberian Type	'Siberian Type' are tall in nature and attains the the height of more than 80.0 cm. Leaves fall quickly and give poor yield. Very much Siberian for Scotish region.
Collard (*Brassica oleracea* var. *Viridis* L.)	
Georgia	It is non-bunching type and cultivated commercially at large scale. Plants attain the height of 70.0-90.0 cm. Leaves are bluish green and leaf overlap to each other. First harvesting can be done after 80 days of seeding.
Vates	Plants are resistance to low temperature and process of bolting is very slow. Leaves retain their green colour for longer period. Plants are erect, dense and attain the height of 40.0-80.0 cm. It is known for high yield. First harvesting can be done after 75 days of seeding.
Morris Heading	Plants are resistance to low temperature and process of bolting is very slow. Leaves retain their green colour for longer period. Plants are erect, dense and attain the height of 75.0-85.0 cm. Leaves are savoy type which form cabbage type head, large and medium green in colour.
Lousiana Sweet	'Lousiana Sweet' has been developed by Testing Centre, Lousiana State University. Leaves are Georgia type, thick and soft (tender). First harvesting can be done after 75 days of seeding.

Variety	*Description*
	Rhubarb
Rhubarb Chard	Leaves are broad and veins are red in colour.
Lucullus	This cultivar is yellowish green in colour.
	Dill or Sowa
Mehsana	'Mehsana' is popular in Gujarat.
Rubi Local	'Rubi Local' is popular in Gujarat.
Pratapgarh Local	'Pratapgarh Local' is popular in Rajasthan.
	Coriander
Pant Haritima	It is developed from GBPUA&T, Pantnagar (Uttarakhand) through selection from the germplasm and released by UPSVRC in 1993. The plants are relatively taller (1.5 m) and bear 8 to 9 branches. Dual purpose, highly fragment leaves. Resistant to stem gall forming fungus (*Protomyces macrosporous*) which causes small tumor like swellings at all the herbaceous parts of the plant. Its yield potential is 12.5-14.0 t.ha of green leaves and 15.0-18.0t/ha of dry seeds.
CO-1	It is developed from TNAU, Coimbatore (Tamil Nadu). Small plants and small grains suited for rainfed condition.
CO-2	It is developed from TNAU, Coimbatore (Tamil Nadu). Dual purpose, suitable for waterlogged, drought, saline and alkaline conditions.
CO-3	It is developed from TNAU, Coimbatore (Tamil Nadu).
CS-287	It is developed from TNAU, Coimbatore (Tamil Nadu). Early, suitable for for irrigated and rainfed condition, tolerant to wilt, powdery mildew and grain mould.
Gujarat Coriander-1	It is developed from Junagadh (Gujarat). Semi-spreading, dark green leaves, no lodging and shattering and suitable for early sowing. This cultivar is tolerant to wilt and powdery mildew.
Gujarat Coriander-2	It is developed from Junagadh (Gujarat). Semi-spreading, dark green leaves, no lodging and shattering and suitable for early sowing. This cultivar is tolerant to wilt and powdery mildew.
RCr-41	It is developed from RAU Jobner, Rajasthan. Plants are erect with thick stem. It is suitable for rainfed condition. Resistant to stem gall.
RCr-435	It is developed from RAU Jobner, Rajasthan. Plants are erect, bushy and fall under medium maturity group.
RCr-436	It is developed from RAU Jobner, Rajasthan. Plants are semi dwarf and fall under early maturity group. It is suitable for heavy soil of Rajasthan and other parts of country.
UD-20	It is developed from RAU Jobner, Rajasthan. Plants are erect with thick stem. It is suitable for rainfed condition. Resistant to stem gall..
UD 446	It is developed from Jobner, Rajasthan. This cultivar is dual purpose and suitable for rainfed condition. It is moderate resistant to wilt, and stem gall
Sadhana	It is developed from Guntur, Andhra Pradesh. Plants are semi-erect, dual purpose and suitable for rainfed condition. This cultivar is tolerant to wilt, aphid, mites and fruit fly.
Swathi	It is developed from Guntur, Andhra Pradesh. Plants are semi-erect and suitable for rainfed condition. It escapes from powdery mildew. This cultivar is tolerant to wilt, powdery mildew, grain mold, and white fly.
Sindhu	It is developed from Guntur, Andhra Pradesh. It is a fine seeded cultivar and rich in essential oil. It is suitable for rainfed condition and intercropping.
Hisar Sugandh	It is developed from CCS HAU, Hisar (Haryana). It is a multicut, dual purpose, and highly aromatic cultivar. It can be grown during spring, winter and autumn season.
	Mustard leaf
Selection-1	This cultivar has 12.9 per cent dry matter and 28.3 per cent crude protein.
Selection-902	This cultivar has 15 per cent dry matter and 70 per cent digestibility value.
	Indian Sorrel or Gogu
Victor	Edible and very good taste.
Rico	Edible and very good taste.
Archer	Edible and very good taste.
Temprono	Edible and very good taste.
	Beetroot
Detroit Dark Red	This is an introduction from U.S.A. The roots are round with smooth, uniform and deep red skin. Flesh is dark blood red, tender and fine textured with light red zoning. The tops are small; and leaves are dark green with maroon tinge. It takes 80-100 days to reach maturity.
Crimson Globe	This is an introduction from U.S.A. Bright green leaves having maroon shade. The roots are globular to flat, medium red with small shoulders; and flesh is dark crimson-red with indistinct zoning. Tops are medium to tall, large, and bright green with maroon shades. It is heavy yielder (Anonymous, 1998).

Variety	*Description*
Crossby Egyptian	This is an introduction from U.S.A. The roots are flat to globular with smooth skin. The flesh colour is dark purple-red with some zoning. The top is medium tall, and green with red veins. It 55-60 days to reach marketable maturity.
Early Wonder	This is an introduction from U.S.A. The roots are flattened globe with rounded shoulders and smooth, with dark red skin. The flesh is dark red with light red zoning. The leaves are large with red veins. The roots attain marketable maturity after 55-60 days of sowing.
Ooty-1	This is a selection from half-sib progeny of local type (DC-3) and developed from Udhagamandalam, TNAU, Coimbatore (Tamil Nadu) in 1997. The roots are long and slightly tapering with attractive deep orange colour and suitable for growing in hilly areas at altitude above 1800 m. Its skin is thin and flesh colour is blood-red. It takes 120-130 days (medium maturing) to reach marketable maturity and suitable for growing in through out the year. The central core is thin, fleshy and palatable. The roots are rich in β carotene. Average root yield is 45.0-49.1 t/ha and 700-1000 kg/ha seeds. The crop is free from pre-mature bolting and resistant to powdery mildew, leaf spot and drought.
Edmond	Moderate size, handsome, rounded, smooth, deep, red, good grain and flavor.
Eclipse	Uniform globular, bright red, fine grained and sweet. One of the best quick growing early beets.
Egyptian Blood	Top quite small, roots fair size, rich deep red, and considered a standard early variety.
Mahyco Lal	Top is medium with erect leaf crown, and maturity 65-70 days. Root globe with close tail, deep red, and crisp flesh. Uniform with smooth texture, and retain original quality without ringing. Root 125-130 g each. Yield potential is 40-45 t/ha and resistant to premature bolting.
Action	This is an early maturing hybrid, which become ready in 50-55 days from sowing. The average beet weights 100-150 g. It can stand in the field for 30 days after maturity, without cracking or losing colour and taste. The size keeps increasing upto 400 g.
Pablo	It matures in 55-60 days from sowing and weight 120-150 g each. On maturity, if not harvested, keeps growing up to 45 g, without cracking or losing taste and colour and does not develop fibre. It can be grown very well in all three seasons.
Lalima	Vigorous growing early maturing hybrid. Uniform roots can be harvested in about 60 days after planting. Tops are erect with bright green and glossy foliage. Roots are top to globe shaped, smooth, dark red, very sweet and free from zones. The hybrid has good tolerance to *Cercospora* leaf spot, *Rhizoctonia* root rot, downy and powdery mildew and drought. Yield potential is 45-50 t/ha, suitable for both fresh market and processing industry, especially for pigment extraction.
Ruby Queen	Is is a vigorous growing and early maturing variety. Uniform smooth shouldered beet with small tap root which holds its shape well even when crowded. The interior colour is deep red. Roots can be harvested in about 50-55 days after planting. Roots are crisp, sweet, round and dark red. Yield potential is 40-45 t/ha.
Bolivar	Very strong plant type produces round uniform beet, with smooth skin and deep red internal colour. Upright leaves facilitate early harvest. Very slow bolting. Good for fresh market and processing. Roots dark red, 200-250 g each, diameter 12-14 cm with round shape. First harvest within 45 days with good keeping quality.
Ruby Queen Improved	Plant vigorous, and light green leaves with red vines petiole. Roots crisp, sweet, round, dark red and ready for harvest after 55 days of sowing. Slices are very dark red and free from white rings.
Corumba	It has very uniform, and has medium long cylindrical roots. Excellent internal and external colour with good shelf life.
Other varieties	Golden Tankard, Golden Yellow, Mammoth, Mammoth Long Red, Seneca Detroit, King Red, Detroit Red, Detroit Perfect, Rubia, Formanova, Little Ball, Kogel and Red Ball, which are common in temperate countries.
Carrot (Tropical or Asiatic type)	
PC-34	This tropical variety has been developed through selection from the local material at PAU, Ludhiana (Punjab) and released for commercial cultivation. It takes 90 days from sowing to harvesting maturity. The roots are deep red and tapering with self colour core. The roots are attractive, long having 3.15 cm diameter. Average TSS content is 8.8 per cent and yield is 51.0 t/ha.
Hisar Gairic (HC-1)	This tropical variety has been developed through mass selection from the local material at CCS HAU, Hisar (Haryana) by Kalloo *et al.* (1993). Plant top is medium heavy, with light green leaves. The roots are attractive, long (18.1 cm and 3.9 cm across), tapering, attractive, light brick red with thin and self-coloured core. Its pecualarity is less fiber and forking. It is good for early sowing. It contains 96.2 mg carotene in 100 g of fresh weight. It takes 120 days from sowing to harvesting. Average root yield is 28.5 t/ha. This variety is free from insect and pest.
HCB-1	This black variety has been developed by CCS HAU, Hisar (Haryana). Black, long, and smooth rooted variety, which is good for the preparation of juice, kanji, candy and coloured vine. Its root to shoot ratio is 2.1 Average root yield is 22.3 t/ha.
HCP-2	This variety has been developed by CCS HAU, Hisar (Haryana). Long, smooth, purple and self-core colour root variety gives higher yield. Its root to shoot ratio is 2.25. The variety will fulfill the demand of deep colour roots. Average root yield is 27.7 t/ha.
Selection-233	This tropical type variety has been developed through selectuion from the cross of Nantes x No. 29 by PAU, Ludhiana (Punjab). Plant top is medium heavy; leaves are erect and dark green. Roots are orange colour, 20.0 cm long, cylindrical with round shoulders, semi-blunt end and self coloured core. Average root yield is 44.0 t/ha.
Selection-21	It is a tropical type variety developed through selection from local material. Its roots are medium long, and red with self coloured core. Average root yield is 17.5.0 t/ha.
Pusa Kesar	This tropical type variety has been developed through selectuion from the cross of 'Local Red' x 'Nantes Half Long' by ICAR-IARI, New Delhi and released for commercial cultivation. Plant top is small. Roots are medium sized (18.0 cm), scarlet red with thin and self-colour core. It is suitable for early sowing from August to early November and attains marketable maturity in 90-110 days after sowing. Average root yield is 30 t/ha. It is rich source of carotene (38.0 mg/100 g of edible portion).

Variety	Description
Pusa Vasuda	This first tropical type hybrid has been developed using CMS by ICAR-IARI, New Delhi and released for commercial cultivation. Roots are self-coloured, red, sweety, juicy and rich in total carotenoids, lycopene, TSS and minerals. It attained edible maturity in 80-90 days after seeding. Average root yield is 40 t/ha.
No.29	It is an Asiatic type developed through selection from Local material and released from PAU, Ludhiana. It is Desi variety with long tapering light red and thin roots.
Sel. 21	Roots are medium long, red with self coloured core and less barked. The yield potential of this variety is 17.5 t/ha.
Sel. 233	It is derivative of a cross of Nantes x No. 29 and released by PAU, Ludhiana (Punjab) in 1978. It is a Desi type with all good qualities of Nantes. Tops are small. Leaves are dark green and 30.0-33.0 cm in size. Roots are 15.1 cm long, smooth, semi cylindrical orange with light orange core colour, less cracking and forking. Harvesting can be delayed without bolting and impairing its edible qualities. It takes about 90-100 days for root formation.
Desi Red	It has medium foliage; mature in 80-90 days after sowing. Roots tapering, 15-20 cm long, appealing red, uniform red, internal colour, core small and self coloured. High yielder 34 t/ha.
Corumba	It has very uniform, medium long and cylindrical roots. Excellent internal and external colour with good shelf life. It is good for cold storage.
NS820	It is open pollinated (OP) cultivar. It is high yielding; early vigorous plants very uniform long, straight and cylindrical roots. Very small self coloured core. Crisp flesh with good sweetness and excellent quality. It has very good storage and transport qualities.
Namdhari Kuroda	It is open pollinated (OP) cultivar. It is an early vigorous, and high yielder with uniform long cylindrical roots. Roots are smooth, deep orange with very small self coloured core. Flesh is very sweet, crisp and endowed with good storage and transport qualities.
Samurai Red	It is medium sized vigorous plant with green leaves. Roots are 30-35 cm long, dark red with self coloured core, ready for harvest after 90 days of sowing. Roots are very uniform, smooth, sweet and juicy.
Kuroda Improved	A very uniform foliage and root developing variety. Deep orange colour, self coloured core, 18-20 cm long, round shoulder with stump roots, ready for harvest after 80-90 days of sowing. Ideal for summer sowing to harvest in fall and winter.
Samsoon-196	It is Nantes type, very high yielding, and having soft edible core. It gives long carrots with vigorous upright growing leaves, having good taste. Maturity mid early and tolerant to powdery mildew.
Bangor	It matures approximately in 100 days. It is Borlikum type carrot, tolerant to powdery mildew and very high yielding.
Oranza	It is having edible core and good for slicing; having good yield and dry matter content is 9 to 12 per cent. Maturity mid early and Nantes types.
Carson	It matures in approximately in 110 days from sowing. Shape is Chantenay type, having healthy foliage and gives attractive coloured carrots.
Premia (F_1)	It is a very good and early hybrid which is generally grouped under Nantes types. It is suitable for fresh market. Short foliage and produces long (15-18 cm), slender cylindrical carrots with rounded shoulders. Its external and internal colour isan attractive orange red and its core is fine. Resistance to cracking is very good. Normal yield ranges between 45-49 t/ha under optimum growing conditions.
Carrot (Temperate type or European type)	
Chantenay	This is temperate type (European type) variety is an introduction from France and release for commercial cultivation by ICAR-IARI Regional Station, Katrain (Himachal Pradesh). It is preferred for cultivation during winter monthes. It is suitable for canning and storage. Roots are 12.0-15.0 cm long and 3.0-5.0 cm in diameter, conical, attractive orange scarlet in colour with self-coloured core and smooth with stump end. Core is self-colour; flesh is tender, sweet and fine textured. Roots are ready for harvesting in 110-120 days after sowing and yield 28.0 t/ha.
Pusa Nayanjyoti	This is temperate type (European type) hybrid has been developed by ICAR-IARI, New Delhi. It is very much suitable for early sowing (April-August) conditions in the hills and late sown conditions (November-December) in the north Indian plains as well as other regions. Roots are orange, smooth, uniform, cylindrical and stumpy with small indistinct self-coloured core. Early in maturity takes 80-85 days to reach marketable stage. It possesses high β-carotene content (7.552 mg/100 g fresh weight). Average yield is around 38-40 t/ha.
Royal Chantenay	This is widely adopted cultivar and is primarily grown for processing, but is well suited for home garden.
Red Cored Chantenay	It similar to Chantenay variety, quite popular and extensively grown.
Early Nantes	Roots are almost cylindrical terminating abruptly in small thin tail, 12.0-15.0 cm long, and orange flesh with self-coloured core. It takes 90-100 days for root formation and yield 27.0 t/ha.
Nantes	It is European or temperate type cultivar, which can be grown in the plans of India for root production but not for seed production. Its seeds are produced in the hills of India. The roots are half-long, slim well shaped, and cylindrical with stumed end forming a small thin tail, deep orange red cortex and core. It is suitable for cultivation during winter months. It takes 105 days for root formation and yield 29.0 t/ha. It does not keep well because of its thin skin and fine texture. This is not suitable for seed production in plains.
Nantes Half Long	This temperate type variety has been developed through selectuion from ICAR-IARI, Regional Research Station, Katrain (H.P.). Roots are cylindrical, stumpy, well shaped with abrupt tail. Roots are orange scarlet with self-colour core, flesh sweet and fine textured with good flavor. It takes 110-120 days to reach marketable maturity.
Ooty No.-1	This temperate type medium maturing variety has been developed through selectuion from half-sib progeny of the local type 'DC-3' and released from Horticultural Research Station, Udhagamandalam, TNAU (Tamil Nadu) in 1977. Roots are long (25 cm), deep orange, slightly tapering and with thin self-colour central core. Harvesting is possible 100-110 days after sowing. Average root yield is 49.0 t/ha. It is resistant to powdery mildew and leaf spot.

Variety	*Description*
Danvers	It is European cultivar whose leaves are very big size (12.5-15.0 cm length and 2.3-4.5 cm diameter). The roots are yellow in colour. This variety is suitable for both fresh as well as for processing purposes.
Zeno	This is temperate type popular variety of Nilgiri hills which was intropduced from German Democratic Republic and recommended by the State Department of Horticulture, Ooty (Tamil Nadu) for cultivation in Tamil Nadu. Its roots are 15.0-17.0 cm long with conical shape in lower portion. The exposed portion of the root becomes pinkish green, medium top and self coloured. It takes about 110-120 days after seed sowing.
Imperator	It is a cross between 'Nantes' and 'Chanteny' and extensively grown for fresh market. It is a mid-season to late maturing cultivar with large and strong foliage. The foliage is large. Roots are 15.0-17.5 cm long and 2.5-4.5 cm in diameter with short tapered end, deep orange cotex and slightly less pigmented core.
Chaman	This variety is developed from SKU and T, Srinagar. Roots are long, cylindrical and semi blunt; tolerant to cracking and yields 25-27 t/ha.
Radish (Tropical or Asiatic type)	
Hisar Sel-1	This variety has been developed by CCS HAU, Hisar (Haryana) and is more popular in Haryana. Its roots are straight, white and ready to harvest after 50-55 days of sowing. It is sown from September to October. Yield potential to the tune of 30.0-32.5 t/ha.
Pusa Desi	It is selection from a local material. It is a tropical and subtropical cultivar, suitable for sowing from middle of August to early October in the northern plains. Roots are 30.0-35.0 cm long, tapering, white with green stem end and mildly pungent in taste. Leaves are medium sized, cut and green in colour. Suitable for sowing from August to October. It matures in 50-55 days. Yield potential to the tune of 30.0 t/ha. It is tolerant to slightly higher temperature.
Pusa Chetki	This variety (tropical type) has been developed by ICAR-IARI, New Delhi through selection from an exotic material introduced from Denmark (Choudhury and Sirohi, 1975). It is tolerant to high temperature and therefore, it is, recommended for growing in summer months *i.e.* from April to August. Plant top is light green with a few leaves. Leaves are medium in size, green, soft, erect with entire lamina. Roots are small (15.0-20.0 cm long), medium thick (3.5 cm), pure white and less pungent or sweet in taste with blunt end. Roots are ready for harvesting 40-45 days after sowing. Average root yield is 26.0 t/ha. It is suitable for sowing from mid March-August. The cultivar sets seeds profusely in the plains because it bolts very early in the month of late October or early November. It might be grown as a summer and monsoon crop from April to September. During summer sowing, though the yield will be comparatively low, but it is compensated by the high price, it fetches at that time.
Pusa Reshmi	It is Asiatic type and developed by ICAR-IARI, New Delhi. Roots are 30.0-35.0 cm long, tapering, white with green stem end and mildly pungent in taste. Leaves are small to medium sized and green in colour. Suitable for sowing from October to December. It matures in 55-60 days. Yield potential to the tune of 35.0-40.0 t/ha.
Pusa Himani	It is developed through hybridization between a Temperate Type (black) and a popular Asiatic Type and released from ICAR-IARI, Regional Station, Katrain (Singh *et al.,* 1971). Roots are 30.0-35.0 cm long, tapering or semi stump, white with green stem end and mildly pungent in taste. Leaves are small in size and dark green in colour, suitable for sowing from December-February in plains and round the year in the hills. It matures in 60-65 days. Yield potential to the tune of 30.0-32.5 t/ha.
Punjab Ageti	This is an early season variety (tropical type) derived from the cross Punjab Safed x Local Red and realeased by PAU, Ludhiana (Punjab). Plant top is medium heavy and leaves are spreading with leaf lamina. Roots are medium long (25 cm), thick, tapering and mildly pungent. The upper half is red and the lower half is white. Roots are ready for harvesting 50 days after sowing. Average root yield is 31.5 t/ha.
Punjab Pasand	This variety (tropical type) has been developed by PAU, Ludhiana (Punjab) through selection from Pusa Chetki. It is suitable for main and spring seasons. It is quick growing and attains marketable maturity in 45 days after sowing. Plant top is light with a few leaves. Leaves are erect with complete leaf lamina. Roots are medium long (25.0-30.0 cm), 3.0-4.0 cm thick, pure white, smooth and semi-stumped. Average root yield is 54.0 t/ha.
Punjab Safed	This variety (tropical type) has been developed by PAU, Ludhiana (Punjab) through cross of White-5 x Japanese White. Plant top is medioum heavy; leaves are spreading with cut leaf lamina. Roots are pure white, tapering, smooth, 30.0-40.0 cm long, and 3.0-5.0 cm thick with broad shoulders and mildly pungent. Roots are ready for harvesting in 60 days after sowing. Average root yield is 40.0 t/ha.
Japanese White	This tropical type variety is an introduction from Japan 'Shiroaguri-Kyo' and released by ICAR-IARI, New Delhi. It is recommended for late sowing in the main season (November-December) in North Indian plains and July to September in hills. Plant top is medium and leaves have deep cuts, medium size and grayish green in colour. Roots are 25.0-30.0 cm long, 5.0 cm thick, cylindrical, slightly compressed from middle, smooth, pure white skin with blunt end and mildly pumgent. Flesh is snow white, crisp and solid. Roots are ready for harvesting in 60-65 days after sowing. Average root yield is 35.0-40.0 t/ha. Suitable for sowing from October to December in plains and July to September in hills.
Kashi Sweta (IIVR-1)	This variety (tropical type) has been developed by ICAR-IIVR, Varanasi (U.P.) through selection from the Pusa Chetki genotype. It is suitable for sowing in summer, spring (autumn) and winter season. It is quick growing variety and reaches marketable maturity in 30-35 days after sowing. Roots are 25.0-30.0 cm long, 3.5-4.0 cm thick, straight, and tapering with blunt end. Average yield is 50.0 t/ha in main season.
Kashi Hans (IIVR-2)	This variety (tropical type) has been developed by ICAR-IIVR, Varanasi (U.P.) through selection. It is suitable for planting from September to February and harvesting within 40-50 days after sowing. Due to late pithiness, it can stand the field upto 10-15 days after ready for harvesting. Its foliage is soft and smooth. Roots are straight and tapering types, 35.0-40.0 cm long and 3.5-4.2 cm in diameter. Average yield is 45.0 t/ha in main season.
Mino Early	This tropical type variety is an introduction from Japan. It is a quick growing variety and roots attain harvesting maturity in 40 days after sowing. Roots are pure white, 30.0-40.0 cm long, slightly tapering and mildly pungent. Average yield is 29.0 t/ha in main season.

Variety	Description
Early 40 Days	This tropical type variety is an introduction from Japan. It is a quick growing variety and roots attain harvesting maturity in 40 days after sowing. Roots are about 30.0 cm long, white, cylindrical with semi-blunt end and mildly pungent. It is suitable for sowing in late spring for early summer harvest.
Palam Hriday	This tropical type variety is a selection from the exotic material. Plant top is small, leaves are dark green and hairy with entire lamina. Roots are oblong with stump end. Top half of the root is green and lower haif is creamy. Tail and flesh is pinkish. The roots attain marketable maturity 50-60 day's aftyer sowing. Average yield is 30.0 t/ha in main season.
Rapid Red White tipped	Roots of this variety are small, and round or globular. It had typical roots of red upper portion and lower white portion. It is suitable for kitchen garden or container garden from October to January. It matures in 25 day's after sowing. Average yield is 23.0 t/ha in main season.
Arka Nishant	It is Asiatic variety and developed by an improvement over a collection (IIHR-72- Singapore collection) at ICAR-IIHR, Bengaluru (Karnataka) adopting 10 cycles of mass selection. It takes approximately 45-55 days to mature. Roots are medium sized (25.0 cm x 3.4 cm); root has marble white colour, crisp texture and pleasant aroma. It is free from bolting, pithiness, splitting and forking. Each root weighs approximately 30.0-40.0 g. Root shoot ratio is 1:1. It has mild pungency and yield potential is 20.0-30.0 t/ha.
IIHR-1-1	Roots are thin long (30.0 cm), each root weighing about 300.0 g and free from pre-mature bolting, pithiness, splitting and forking and mildly pungent. Root surface is smooth and shining white. It is highly resistant to white rust and takes about 45 days to reach harvest maturity.
CR-45	It is a high yielding progeny from cross of Japanese White Long x Local White developed by SKUAST, Jammu (J&K). Roots are mildly pungent, white, and long with short tail. The crop is ready for marketing in 45 days when sown in the month of October. Root length is 30±5 cm. If kept in field for longer period then size of the root may go up to 60.0 cm attaining weight up to 1.5 kg. The yield potential following integrated nutrient management with vermicompost is above 80.0 t/ha.
CO-1	It is selection from germplasm type (RS-44-2-1) and released from TNAU, Coimbatore (Tamil Nadu) in 1981. The roots are milky white, long (22.0 cm), thick (12.5 cm girth) and each weight 220.0 g. The roots contain 10.9 mg/100 g of vitamin C per 100 g and 1.20 per cent crude fiber. It bolts and sets seeds in the plains and hence very easy for the farmers to produce seeds for themselves. Roots have better consumer's preference. Top to root ratio is 0.90. It is susceptible to rust and aphids. This variety is suitable for growing round the year under different cropping systems. The crop duration ranges from 40 to 45 days. It yields 20.0-25.0 t/ha.
Kalianpur No.-1	It is a selection from local material released in 1982. Stem and foliage is green, smooth, top heavy, leaves long, broad and less lobed. Roots are 22.0-23.0 cm long, smooth, crispy and white with green shoulder, thick and tapering. It is free from mustard sawfly, aphids and white rust.
Nandauni	This variety is a selection from a local material and very popular in Himachal Pradesh. The roots are long, tapering, smooth and light pink in colour. The tops are dark green with cut leaves.
Scarlet Globe	Foliage is 10-15 cm tall. Root is round small, 2 cm in diameter and bright red. Flesh is crisp and white. It takes 25-30 cm from sowing to root formation and delayed harvesting will result in pithiness.
Scarlet Long	It is an early variety, having light green foliage with 15-20 cm length. Roots are long and tapering to a point. Flesh is white and crisp.
Long White (F_1)	It is developed by Tropica Green Co. which is a multinational seed company. It is popular hybrid and can be grown both in high land and lowlands. Excellent root quality, uniform appearance with white skin with upright and dark green foliage. Root 600-8000 g each, highly uniform and 22-26 cm long. It is tolerant to heat and drought. Resistant to downy mildew, Fusarium yellow and other diseases. Mature in 45-50 days.
Longo (F_1)	It is developed by Tropica Green Co. multinational which is a seed company. Slow bolting hybrid with excellent root enlargement at low temperatures. Highly uniform roots of 35-40 cm long, 7-8 cm in diameter and 1.0-1.2 kg in weight. It is resistant to Fusarium wilt. Suitable for fresh market.
Shogunmino (F_1)	This hybrid (F_1) developed by Unicorn Seeds Company. Roots are uniform, long (30-40 cm), very slow in pithiness development and can grow well under high temperatures.
Summer Queen (F_1)	This hybrid (F_1) developed by Sungrow Seeds Company New Delhi. Roots are straight and tapering, white coloured 30 cm long, late in pithiness. Matured in 55 days and performed best in summer and rainy season.
Hill Queen (F_1)	This hybrid (F_1) developed by Sungrow Seeds Company New Delhi. It matures in 55-65 days after sowing, roots straight, slightly tapering, lustrous white, 40-45 cm long very late in pithiness. Suitable for planting in Oct.-Nov. Roots have very little fibre, fetches premium price in the market.
Komal (F_1)	This hybrid (F_1) developed by Sungrow Seeds Company New Delhi specially for areas where leaves are eaten, roots are less payment, leaves soft and palatable. It is suitable for planting from February to May. Roots cylindrical, and stump, white, upper part some times has greenish tinge, lemgth 20-25 cm, and very late in pithiness.
Other varieties	Jaunpuri Giant (Jaunpur, U.P.); Newari, Bombay Red, Nadauni, Baramasi (Himachal Pradesh).
Radish (Temperate type or European type)	
White Icicle	This temperate type variety is an exotic introduction and suitable for late sowing in the plains. Plant top is small, and leaves are dark green. Roots are small (12.0-15.0 cm long) and 2.0-3.0 cm in diameter, icicles (pillar) shaped, straight and tapering. It is quick growing variety and attains marketable maturity after 4 weeks (30 days) of sowing. Roots become pithy if not harvested within a couple of days after reaching maturity. Average yield is 50.0 t/ha in main season. It is suitable for kitchen garden or container garden from October to January. It is suitable for October and November sowing in the plains.

Variety	*Description*
Pusa Himani	This temperate type variety has been developed by selection from the cross Black Radish (temperate type) x Japanese White (Asiatic type) at ICAR-IARI, Regional Station, Katrain, H.P. (Singh *et al.,* 1971). It is suitable for commercial cultivation in the north Indian plains in the second fortnight of January. Plant top is light, leaves are small sized, spreading, dark green and hairy. Roots are 30.0-35.0 cm long, 10.0-12.0 cm in girth, white, medium thick, semi-stumped and are characterized by green shoulder. The skin is pure white. Roots are ready for harvesting after 55 days of sowing. The flesh is crisp and sweet-flavoured with mild pungency. Average yield is 40 t/ha.
Scarlet Globe	This temperate type variety is an exotic introduction. Roots are small (8.0-10.0 cm long), round (2.0 cm diameter), and bright red in colour. Flesh is crisp and white. Roots are ready for harvesting in 25-30 days from sowing and develop pithiness if harvesting is delayed even for a few days.
Scarlet Long	It is an early variety, having light green foliage with 15.0-20.0 cm length. Roots are long and tapering to a point. Flesh is white and crisp.
Chinese Pink	It is an introduction identified by Dr. Y.S. Parmar University of Horticulture and Forestry, Nauni (Solan) Himachal Pradesh. Roots are 12.0-15.0 cm long, semi stumpy to stumpy roots, pink with white colour towards the tip and takes about 50-55 days from sowing to root formation. It is a good cultivar for the hills but does well in plains with mild climate. It requires low temperature for seed production.
Nadauni	This is developed from a local material and very popular in Himachal Pradesh. The roots are long, tapering, smooth and light pink in colour. The tops are dark green with cut leaves.
Rapid Red White Tipped	It is an extra early European variety, which matures in 25-30 days. It is bunching and table type. Roots are small, globular, and bright red with white tip. Flesh is pure white, crisp and snappy. Tops are short, mildly pungent in flavor and suitable for November-December sowing in the plains.
Turnip (Tropical or Asiatic type)	
L-1	This variety has been developed by PAU, Ludhiana (Punjab) from the cross PTWG x 4-White. Plant top is medium, leaves are complete with serrated margins and dark green in colour. Roots are round, pure white, smooth, rat-tailed and crisp with mild flavor. It takes 45-60 days to reach marketable maturity. Average yield is 26 t/ha.
4-White	It is an early maturing deshi type. Its roots are white, round and medium and developed in about 60 days. The average yield is about 20 t/ha.
4-Red	This variety has been developed by Department of Agriculture, Punjab through the selection of local material. Plant top is small and leaves are erected with complete lamina. Roots are flatish-round, rat tailed, red in colour and medium in size. Roots are strongly flavoured and reach marketable maturity in 45-50 days after sowing. Average yield is 20.0 t/ha.
Punjab Safed-4	This is an early Asiatic variety and commonly grown in Punjab and Haryana. The roots are pure white, round and medium sized having mild taste.
Early Milan Red Top	This tropical type variety is an introduction from Italy. Plant top is small with 4-6 sessile leaves. Roots are flat with purplish red top and white underneath. Flesh is pure white, crisp and mildly pungent. It is an extra early variety and roots are ready for harvesting 45 days after sowing.
Pusa Kanchan	This tropical type (Asiatic type) variety has been developed through selection from the cross Local Red Round (Asiatic type) x Golden Ball (temperate type) at ICAR-IARI, New Delhi (Singh, 1963) and release for commercial cultivation. Plant top is small. Roots are round with red skin and are ready for harvesting 50-55 days after sowing. Flesh is creamy-yellow with excellent flavor and taste. Root yield varies from 25.0-40.0 t/ha.
Pusa Sweti	It is an Asiatic cultivar suitable for sowing from August to October. The roots are attractive white and mature in 45-50 days from sowing. It was released from ICAR-IARI, New Delhi (Choudhury, *et al.,* 1976).
Turnip (Temperate type)	
Golden Ball	This temperate type variety is an introduction from England and recommended by ICAR-IARI Regional Station, Katrain (Kullu Valley) himachal Pradesh. Plant top is small; leaves are erect, deeply serrated and light green in colour. Roots are round, medium sized with yellow skin and yellow flesh. It reaches marketable maturity in 70-75 days after sowing. Average yield is 20.0-25.0 t/h.
Purple Top White Globe	This temperate type variety is an introduction from U.S.A. and recommended by ICAR-IARI Regional Station, Katrain (Kullu Valley) Himachal Pradesh. It is suitable for cooler months. Plant top is medium heavy; leaves are erect, deeply serrated and dark green. Roots are large, round, smooth with bright purple top and white base. Roots attain marketable size in 60 daysv after sowing. Flesh is white, firm, crisp and mildly sweet flavoured. Average yield is 20.0 t/ha.
Pusa Chandrima	This temperate type variety has been developed by ICAR-IARI, Regional Research Station, Katrain (H.P.) through crossing of Asiatic x European type and release for commercial cultivation. It is suitable for sowing from October to December in plains. It has been developed by selection from the cross Snow Ball x Japanese White. Plant top is medium heavy and leaves have moderately deep cuts. Roots are medium to large, globular, smooth and pure white. Flesh is sweet and tender. Roots are ready for harvesting in 50-60 days after sowing. Average yield is 17.5 t/h. It is biennial cultivar and does not produce seeds in the plain of India (Gill *et al.,* 1971).
Pusa Swarnima	This temperate type early variety has been developed through selection from the cross of Japanese White (Asiatic type) x Golden Ball (European type) by ICAR-IARI, Regional Research Station, Katrain (H.P.) and release for commercial cultivation. Roots are flatish round, 6-7 cm long and 7-8 cm in diameter with creamy-yellow skin, amber flesh and mild taste. Tops are medium heavy with moderately cut leaf lamina. It is superior to 'Golden Ball' in seed yield and less susceptible to turnip malformation, a mycoplasma disease. It is earlier than Golden Ball by about a fortnight gives 40 per cent higher yield (Gill, 1980). Harvesting is possible in 65-75 days after sowing. Average yield is 28.0 t/h.

Variety	*Description*
Early Milan Top	A temperate variety recommended by vIARI Regional Station, Katrain (Kullu Valley) Himachal Pradesh. Leaves are small and light-green. It takes about 65-70 days from sowing to root formation. Roots are round and medium sized, fleshy and light yellow. It is very shy seeder and highly susceptible to turnip malformation.
Snow Ball	It is ready in 55 days. Leaves are small and light green. Roots are round and medium sized with light yellow flesh.
Knol khol	
White Vienna	It is an introduction from Austria. It is an early maturing variety and takes 55-60 days to reach marketable maturity. Plants are dwarf with light gree foliage. Knobs are globular, light green and flesh is creamy white. Average yield is 20.0-25.0 t/ha.
Early White Vienna	It is an introduction from Austria. It is developed by selection from White Vienna and takes 45-50 days from transplanting to marketable maturity. The knobs are globular-round and green with fewer leaves. The flesh is creamy-white, tender and crisp. Average yield is 20.0 t/ha.
Purple Vienna	It is an introduction from Austria. Plants are medium sized with purplish-blue leaves and stem. It takes 55-60 days from transplanting to edible maturity. Knobs are globular-round, large with purple skin and light green flesh. Average yield is 20.0-25.0 t/ha.
Early Purple Vienna	It is an introduction from Austria. It is developed by selection from Purple Vienna. It is also an early maturing variety and takes 45-50 days to reach marketable maturity. Foliage is purple; Knobs are globular-round, large with purple skin and light green flesh. Average yield is 20.0-25.0 t/ha.
Large Green	It is developed by CSK HPKV, Palampur (Himachal Pradesh) and release foe commercial cultivation in Himachal Pradesh and nearby states. It takes about 60-70 days to reach marketable maturity. Plants are vigorous with dark green foliage. Its knobs are flatish-round and dark green with white flesh. Average yield is 22.5-25.0 t/ha.
Palam Tender Knob	It is an improved variety of knol khol bred through mass selection at Department of Vegetable Science, HPAU, Palampur (H.P.). It has small light green foliage. It forms turnip-shaped knobs, which remain split free. These are round, flat, stringless, light green, sweet and fleshy. These knobs attain marketable size of 5-7 cm of diameter weighing 250-300 g in about 80 days. Its average yield potential is 25-30 t/ha.
King of Market	It is an early variety having dark green foliage and round flat knobs. Leaf sheath largeand well spread over the knob.
King of North	It is an early variety, and dark green foliage with broad leaf blade. Knobs are flattish round and dark green in colour. Leaf sheath is large and well spread over knob. It takes 65-70 days for knob formation.
G-40	It is an early selection of White Vienna which is ready for marketing in 35-40 days under subtropical conditions of Jammu. Knobs are green, tastier and leass fibrous. It is suitable for pickling. The yield potential of the variety is 40.0-50.0 t/ha.
P-40	Knobs are purple and the variety is ready for marketing in 40 days. It is preferred for consumption in hilly areas of J&K. Average yield is 30.0-40.0 t/ha.
Szentesi Faber	It is a summer cultivar, has a thin, pale green skin and crisp flesh, early bearing, produce a higher marketable yield and show greater resistance to *Alternaria* leaf spot.
Roggli's	Outdoor white and is medium type, resistant to bolting and tuber do not become fibrous soon.
Goliath	It is a late cultivar maturing between 70-100 days after planting.
Korist	It matures in 55-60 days after sowing. The maximum weight is 150 g and has good standing ability up to 30 days after maturity. It keeps growing even after 65 days and can be grown upto 300 g in size without cracking and without fiber development.
Kolibri	It is attractive red coloured knol khol with more sweet taste. It matures in 55-60 days after sowing. Approximately weight is 150 g and has good standing ability upto 30 days after maturity. It is tolerant to crcking and there is no fibre.
Kossak	It is a big sized knol khol, which mature in 65-70 days after sowing. The approximate weight of tuber is 1.0 kg.
Champion (F_1)	This hybrid variety grown in dry and cool climates. It gives high yield and very uniform, green bulb of globe shape and weight about 600-700 g at maturity. Very good bulb quality and maturity 60-65 days after seeding. It possesses erect growing habit and very good shelf life.
Dakini (F_1)	Plants are erect medium and tuber mature early in about 57 days after sowing. Heavy yielder, pale green bulb of semi-globe shape. Weight 500-600 g at maturity.
EWV Improved	It is developed by Tokita Seed Co., India. Pale skin, white flesh and crisp knobs are flat round to roundish. Ready for harvest after 55 days of transplanting. Average knob weight about 200-250 g. knobs are with less foliage and smooth surface.
Fennel	
Hisar Swarup	This variety has been developed by CCS HAU, Hisar (Haryana).
Onion	
Arka Swadista	This variety has been developed by ICAR-IIHR, Bengaluru (Karnataka) and release for commercial cultivation in the year 2010. It is white onion variety for fermented preservation has uniform white colour, bulbs are oval globe in shape, TSS 18-20 per cent, dry matter content 15-18 per cent, edible bulb 98 per cent, weight 35-40g, and small in size (3.0-3.5 cm) suitable for bottle preservation. It gives bulb yield of 6-18 t/ha in 105 days.
Arka Ujjwal	This variety has been developed by ICAR-IIHR, Bengaluru (Karnataka) and release for commercial cultivation in the year 2010. It is multiplier onion variety with uniform bright dark red bulb colour, compound bulb with flat shape, bulb size 4-5 cm, number of bulblets/bulb 3-5, bulb weight 40-45 g, TSS 16-18 per cent, dry matter content 14-16 per cent and bulb yield 20-25 t/ha in 85 days.

Variety	*Description*
Arka Vishwas	This variety has been developed by ICAR-IIHR, Bengaluru (Karnataka) and release for commercial cultivation in the year 2010. Arka Vishwas has the bulb colour dark red, flat globe shape, small size of bulb diameter 3-4 cm, bulb weight 40.0 g, TSS 16° brix, bulb yield 30.0 t/ha, duration 115 days, suitable for growing in *rabi* season and suitable for export.
Arka Sona	This variety has been developed by ICAR-IIHR, Bengaluru (Karnataka) and release for commercial cultivation in the year 2010. Arka Sona has the yellow bulb colour, globe shape, big size of bulb diameter 6.5-7.0 cm, bulb weight 140.0 g, low TSS 10° Brix, bulb yield 45 t/ha, duration 120 days, suitable for growing in *rabi* season and suitable for export.
Arka Akshay	This variety has been developed by ICAR-IIHR, Bengaluru (Karnataka) and release for commercial cultivation in the year 2011. It is a tri-parental synthetic variety with dark red globe shaped bulbs. Average bulb weight is 115 g. yield 45.0 t/ha in 130 days.
Arka Bheem	This variety has been developed by ICAR-IIHR, Bengaluru (Karnataka) and release for commercial cultivation in the year 2011. It is a tri-parental synthetic variety with red to pinkish red elongated globe shaped bulbs. Average bulb weight is 120 g. Yield 47.0 t/ha in 130 days.
Arka Bindu	This variety has been developed by mass selection from the local material and realeased from ICAR-IIHR, Bengaluru (Karnataka) for commercial cultivation. It takes 100 days from transplanting to reach marketable maturity. Bulbs are flattish-round, small, deep red and highly pungent with TSS content of 14-16 per cent. Average yield is 25.0 t/ha.
Arka Pragathi	This variety has been developed by mass selection from the local material and realeased by ICAR-Indian Institute of Horticultural Research, Hessaraghatta Lake, Bengaluru (Karnataka) for commercial cultivation. It takes 130 days from transplanting to reach marketable maturity. Bulbs are medium sized, globular, highly pungent, pink with tight neck and good keeping quality. Average yield is 40.0 t/ha.
Arka Kirthiman	It is an F_1 hybrid between MS65 x Selection13-1-1 and developed by ICAR-Indian Institute of Horticultural Research, Hessaraghatta Lake, Bengaluru (Karnataka). It takes 120-130 days from transplanting to reach marketable maturity. Bulbs are medium sized, red and globular with thin neck and long shelf life. It is suitable for planting in both *rabi* and *kharif* seasons. Average yield is 47.0 t/ha. It is tolerant to to purple blotch, basal rot and thrips.
Arka Lalima	It is an F_1 hybrid between MSO-48 x Sel-14-1-1 and developed by ICAR-Indian Institute of Horticultural Research, Hessaraghatta Lake, Bengaluru (Karnataka). Bulbs are deep red, globe, weight firm texture, red flesh with 12 per cent TSS; shelf life of 5 months at ambient condition; good transport quality; resistant to biotic stresses under field condition; free from split and bolter; crop duration 140 days, bulb weight 110-130 g, TSS 10-12 per cent, and yield 45-50 t/ha. It can be grown in Kharif and Rabi season. Recommended for Andhra Pradesh, Madhya Pradesh and Odisha regions.
Arka Pitamber	This variety has been developed by selection from the cross UD-102 x IHR-396 and realeased by ICAR-Indian Institute of Horticultural Research, Hessaraghatta Lake, Bengaluru (Karnataka) for commercial cultivation. Bulbs are medium sized, globular, mildly pungent and attractive yellow with thin neck and good keeping quality. Average bulb weight is 80 g and TSS content is 10 per cent and yield is 35.0 t/ha. It is moderately resistant to purple blotch.
Bangalore Rose	It is developed from ICAR-Indian Institute of Horticultural Research, Hessaraghatta Lake, Bengaluru (Karnataka). Small onion, bulb uniform in colour, shape, size and neck thickness is thin.
N-2-4-1	This variety has been developed by MPKV, Rahuri (M.S.). It is a deep red coloured variety suitable for kharif season cultivation. Average potential yield is 15.0-20.0 t/ha.
N-53 (Niphad-53)	This variety has been developed by mass selection from the local material and realeased by MPKV, Rahuri (M.S.) for commercial cultivation. Bulbs are bell shaped, medium sized and attractive. Bulbs are purplish at harvesting, turn deep red at maturity and are moderately pungent. It takes 150-160 days to reach marketable maturity. Average yield is 15.0-20.0 t/ha.
Baswant-780	This variety has been developed by MPKV, Rahuri (M.S.). Bulbs are attractive red, globular, TSS 11-12 per cent, keeping quality good, and average bulb yield 20-25t/ha. Crop duration is 100-120 days. It is very much suitable for *Kharif* season.
N-257-9-1	This variety has been developed by MPKV, Rahuri (M.S.). Globe shaped bulbs with attractive white colour high yield potential with good keeping quality. Suitable for cultivation in rabi season.
BL Piaz-3	It is developed by MPKV, Rahuri (M.S.) which has wide adaptability. Medium sized closed neck, round shape and tight skinned, light red bulbs weighing 80-150 g and measuring 5 cm across. Maturity 160-170 days after planting. Resistant to purple blotch. Low incidence of thrips. It is a cross from Line BYG-2207A and local collection.
Phule Safed	This variety has been developed by MPKV, Rahuri (M.S.). The average yield is 36 t/ha.
Phule Suvarna	This variety has been developed by MPKV, Rahuri (M.S.). The average yield is 34 t/ha.
Phule Samarth	This variety has been developed by MPKV, Rahuri (M.S.). The average yield is 38.5 t/ha.
Akola Safed	This variety has been developed by PDKV, Akola (M.S.). The average yield is 38.5 t/ha
Bhima Dark Red	This variety has been developed by DOGR, Pune (M.S.). This onion variety from DOGR has been identified for release in Chhattisgarh, Delhi, Gujarat, Haryana, Karnataka, Madhya Pradesh, Maharashtra, Odisha, Punjab, Rajasthan and Tamil Nadu for *kharif* season. Average marketable yield is 20-22 t/ha. It is recommended particularly due to its attractive dark red flat globe bulbs. It attains maturity within 95-100 DAT.
Bhima Red	This variety has been developed by DOGR, Pune (M.S.). This variety already recommended for *rabi* season in Maharashtra and Madhya Pradesh, is also recommended for release for*kharif* season in Delhi, Gujarat, Haryana, Karnataka, Maharashtra, Punjab, Rajasthan and Tamil Nadu. It can also be grown in late *kharif*. Maturity is 105-110 DAT during *kharif* and 110-120 DAT during late *kharif* and *rabi* seasons. The average marketable yield in *kharif* season is 19-21 t/ha, in late *kharif* season is 48-52 t/ha and it is 30-32 t/ha in *rabi* season. It can be stored up to 3 months in *rabi*.

Variety	Description
Bhima Super	This variety has been developed by DOGR, Pune (M.S.). A red onion variety from DOGR has been identified for release for *kharif* season in Chhattisgarh, Delhi, Gujarat, Haryana, Karnataka, Madhya Pradesh, Maharashtra, Odisha, Punjab, Rajasthan and Tamil Nadu. It can also be grown in late *kharif.* It is reported to have an average yield of 20 - 22 t/ha in *kharif* and 40 - 45 t/ha in late *kharif.* Bulbs attain maturity within 100-105 days after transplanting (DAT) in *kharif* and 110 -120 DAT in late *kharif.* It produces mostly single centered bulbs.
Bhima Raj	This variety has been developed by DOGR, Pune (M.S.). The average yield is 38.8 t/ha.
Bhima Shakti	This variety has been developed by DOGR, Pune (M.S.). It is red onion variety for late kharif and rabi season. Yield is 36.5 t/ha.
Bhima Kiran	This variety has been developed by DOGR, Pune (M.S.). This white red variety for rabi season. Yield is 34 t/ha.
Bhima Sweta	This variety has been developed by DOGR, Pune (M.S.). This white onion variety already recommended for *rabi,* is also recommended for release for *kharif* in Chhattisgarh, Gujarat, Karnataka, Madhya Pradesh, Maharashtra, Odisha, Rajasthan and Tamil Nadu. TSS is around 11-12 oB and it matures within 110-120 DAT. It has medium keeping quality and can be stored up to 3 months. Average marketable yield during *kharif* season is 18 - 20 t/ha and in *rabi* is 26-30 t/ha.
Bhima Shubra	This variety has been developed by DOGR, Pune (M.S.). This white onion variety has been recommended for Chhattisgarh, Gujarat, Karnataka, Madhya Pradesh, Maharashtra, Odisha, Rajasthan and Tamil Nadu for *kharif* season. It is also recommended for late *kharif* in Maharashtra. It matures in 110-115 DAT during *kharif* and 120-130 DAT in late *kharif.* TSS is 10-12 OB. It is a medium storer with a capacity to tolerate environmental fluctuation. Average marketable yield during *kharif* is 18 - 20 t/ha and during late *kharif* 36-42 t/ha.
Gujarat White Onion-1	This variety has been developed by GAU, Gujarat in the year 2000. Average marketable yield is 32-35 t/ha.
Hisar Onion-4	This variety has been developed by CCS HAU, Hisar (Haryana) in 2009. Bulbs globular and rose red in colour, less bolting (5.4 per cent), high TSS (14.2 per cent), and dark green leaves.
PRO-6	It has been developed by mass selection from the exotic material 'EC383494' and release for commercioal cultivation by PAU, Ludhiana (Punjab). Bulbs are medioum to large, deep red and round with thin and tight neck. It takes 125 days from transplanting to harvesting. It has good keeping quality. Average TSS content is 12 per cent and yield is 40.0 t/ha.
Punjab White	It has been developed by mass selection from exotic material'English White Spanish' and release for commercioal cultivation by PAU, Ludhiana (Punjab). The bulbs are white, medium to large, round with tight neck and less pungent. It has TSS content of 15 per cent and is suitable for dehydration. Average yield is 33.7 t/ha.
S-48	It has been developed by PAU, Ludhiana (Punjab). Plant height is 40-60 cm. Bulb flattish round and white in colour, good keeping and storage quality, texture and flavor. Suitable for dehydration. TSS is 12-13 per cent and bulb yield is 25-30 t/ha in 140 days.
Pusa Madhvi	It has been developed from the local material collected from Uttar Pradesh by selfing and massing and release for commercioal cultivation by ICAR-IARI, New Delhi. Bulbs are light red, medium to large and flattish-round with good keeping quality and TSS content of 11-12 per cent. It takes 130-135 days from transplanting to harvesting. Average yield is 31.0 t/ha.
Pusa White Round	It has been developed from the local material and release for commercioal cultivation by ICAR-IARI, New Delhi. The bulbs are white, roundish- flat with average TSS of 12-13 per cent. It is suitable for green onion production and dehydration. Average yield is 30.0 t/ha. The variety has good storability and is suitable for dehydration.
Pusa White Flat	It has been developed and release for commercioal cultivation by ICAR-IARI, New Delhi. It has good potential as vegetable green onion. The bulb is white in colour with flattish round shape. It has TSS value 11-12 per cent with drying ratio of 9:1, so it is suitable for dehydration purpose. It has good storability. The bulb yield ranges between 30.0-35.0 t/ha.
Agrifound light red (ALR)	It has been developed from the local material and release for commercioal cultivation by NHRDF, Nasik (Maharashtra). Bulbs are light red and globular in shape with TSS content of 13 per cent. It has less incidence of pre-mature bolting and has good keeping quality. It takes 140 days from transplanting to harvesting. Average yield is 30.0-32.5 t/ha.
Kalyanpur Red Round	It has been developed from the local material and release for commercioal cultivation by CSAUA&T, Kalyanpur Kanpur (Uttar Pradesh). It takes 120 days from transplanting to harvesting. Bulbs are bronze-red, globular, and moderately pungent with TSS content of 13-15 per cent. Average yield is 25.0-30.0 t/ha.
Early Grano	It is an introduction fro U.S.A. It takes 95-110 days from transplanting to harvesting. Bulbs are yellow, large and globular with TSS content of 6-7 per cent. It has less pungent and suitable for salad making and export but has poor shelf life. Average bulb weight is 150 g and yield is 50.0-60.0 t/ha.
Brown Spanish	Bulb solid, reddish brown, oblong, mild pungent with high TSS and good keeping quality. A long day type suitable for cultivation at an altitude of 1000 m and above.
Udaipur-101	This variety has been developed from Rajasthan Agricultural University, Udaipur (Rajasthan). Bulbs are deep red, flattish-globular and sweet with less pungency. Good for salad purpose. Yield 20-30 t/ha. Matures in 150-160 days after sowing and TSS is 12-14 per cent.
Udaipur-102	This variety has been developed from Rajasthan Agricultural University, Udaipur (Rajasthan). Bulb white, round to flat, and size medium to large. TSS is 12 per cent and average bulb yield 30-35 t/ha in crop duration of 120 days.
Udaipur-103	This variety has been developed from Rajasthan Agricultural University, Udaipur (Rajasthan). Bulbs are less red than Udaipur-101, oblate-globular in shape, sweet but slightly more pungent. Yield is 25-30 t/ha. Maturity is in 150-165 days and TSS is 10.5-13.0 per cent.
Gujarat White	Flattish round, medium in size and good keeping quality. The average yield is 32-34 t/ha.
VL Hybrid Piaz- 67	This variety has been developed from VPKAS, Almora (Uttarakhand). Uniform colour and medium size high yielder hybrid. Recommended for cultivation in Uttar Pradesh (hills), Uttarakhand plains and other places.

Table 4: Impropved Cultivars of Ginger

Variety	*Fresh Yield (t/ha.)*	*Maturity (Days)*	*Oleoresin*	*Essential Oil (Per cent)*	*Crude Fiber (Per cent)*	*Dry Recovery (Per cent)*
Rejatha	23.2	300	-	1.7	3.3	23.0
Majhma	22.4	200	-	2.4	4.0	19.0
IISR-Varada	22.6	200	6.7	1.8	3.3-4.5	20.7
Suprabha	16.6	229	8.9	1.9	4.4	20.5
Suruchi	11.6	218	10.0	2.0	3.8	23.5
Suravi	17.5	225	10.2	2.1	4.0	23.5
Himgiri	13.5	230	4.3	1.6	6.4	20.6
Rio-de-Janeiro	17.6	190	10.5	2.3	5.6	20.0

Source: ICAR-Indian Institute of Spice Research, Kerala.

Table 5: Potato Varieties Developed by ICAR-Central Potato Research Institute (CPRI), Shimla (H.P.)

Variety	*Parantage*	*Year*	*Tuber Characters*	*Area of Adaptation*
Kufri Kisan	Up-to-date x Sd.16	1958	Large, round, white, deep eyes with prominent eye brows flesh white. Tubers tend to crack under high doses of nitrogen	Released for cultivation in North Indian Plains
Kufri Kuber	(*S. curtilobum* x *S. tuberosum*) x *S. andigenum*	1958	Medium, oval, tapering towards crownend, white, medium-deep eyes, flesh white. Poor keeping quality	An early variety released for cultivation in North Indian Plains and Plateau Region (Punjab, Bihar, and Maharashtra)
Kufri Kumar	Lumbri x Katahdin	1958	Medium, oval, tapering towards heel end, white, fleet eyes, flesh white	Released for cultivation in North Indian hills
Kufri Kundan	Ekishirazu x Katahdin	1958	Medium, round-oval, flattened, white, medium, deep eyes, flesh yellow	Released for cultivation in Himachal Pradesh and hills of Uttrakhand
Kufri Neela	Katahdin x Shamrock	1963	Medium, round, white, medium-deep eyes, flesh white	Released for cultivation in South Indian Hills
Kufri Sindhuri	Kufri Red x Kufri Kundan	1967	It is medium late variety and takes 110-120 days to mature. The tubers are medium, round, red skin, medium deep eyes, and dull white flesh. The tubers become hollow under high fertility condition. It is moderately resistant to early blight. It possesses very good keeping quality. Average yield is 30.0 t/ha.	Released for cultivation in North Indian Plains and plateau region
Kufri Jyoti		1968	It is suitable for planting in the spring season. Plants are tall, erect, compact and vigorous with green foliage and white flowers. It matures in about 100 days. Tubers are large, oval, and white with fleet eyes and white flesh. It is resistant to wart disease. Average yield is 30.0 t/ha.	This variety was released for northern and southern hills
Kufri Alankar	Kennebec x ON 2090	1968	Large, oblong, white, fleet eyes, flesh dull white, tubers turn purple on exposure to light	Early variety released for cultivation in North Indian hills and plateau region of peninsular India
Kufri Chamatkar	Ekishirazu x Phulwa	1968	Medium, round, white, medium-deep eyes, flesh yellow	Released for cultivation in North Indian Plains
Kufri Jeevan	M 109-3 x Seedling 698-D	1968	Medium, oval, white, fleet eyes, flesh pale yellow	Released for cultivation in North Indian Hills
Kufri Chandra-mukhi	Seedling 4485 x Kufri Kuber	1968	This is suitable for cultivation in North India and Plateau Region of peninsular India. It matures in 80-90 days. Its plants are medium tall and spreading with gray green foliage. It has good keeping quality and attractive white, large and smooth tubers with fleet eyes. Average yield is 25.0t/ha.	Released for cultivation in North Indian Plains and plateau region
Kufri Khasigaro	Taborky x Seedling 698-D	1968	Medium, round-oval, white, deep eyes, flesh pale yellow	Released for cultivation in North Eastern Hills
Kufri Naveen	3070-D (4) x Seddling 692-D	1968	Medium, oval, white, fleet eyes, flesh pale yellow	Released for cultivation in North Indian Hills
Kufri Neelamani	Kufri Kundan x 134-D	1968	Medium, oval, flattened, white, fleet eyes, tubers turn purple on exposure to light, flesh yellow	Released for cultivation in South Indian Hills

Variety	Parantage	Year	Tuber Characters	Area of Adaptation
Kufri Sheetman	Craigs Defiance x Phulwa	1968	Medium, oval, white, fleet eyes, flesh dull white	Released for cultivation in North Western Plains
Kufri Muthu	3046 (1) x M 109-3	1971	Large, round-oval, white, medium deep eyes, flesh white	Released for cultivation in South Indian Hills
Kufri Deva	Craigs Defiance x Phulwa	1973	Medium, round, white, with purple splashes, deep and picked purple eyes, flesh white	Released for cultivation in North Indian Plains
Kufri Badashah	Kufri Jyoti x Kufri Alankar	1979	It matures in about 90-100 days. Plants are medium compact, vigorous with grey green foliage and white flowers. Tubers are large oval, white, fleet eyes; tubers turn purple on exposure to light, and having flesh dull white. Average yield is 31.5 t/ha.	Released for cultivation in North Indian Plains and plateau region
Kufri Bahar	Kufri Red x Gineke	1980	Large, round-oval, white, medium deep eyes, flesh white	Released for cultivation in North Indian plains
Kufri Lalima	Kufri Red x AG 14 (Wis. X 37)	1982	It matures in about 90-100 days. Plants are tall, erect, medium compact, vigorous with green foliage with purplish flowers. Large to medium, round red, medium deep eyes with white flesh. Average yield is 22.0 t/ha. It has good keeping quality and floury texture.	Released for cultivation in North Indian plains
Kufri Sherpa	Ultimus x Adina	1983	Tubers are medium, round, flattened, white, medium deep eyes, and having flesh pale yellow	Released for cultivation in North Bengal Hills
Kufri Swarna	Kufri Jyoti x $(VTn)^2$ 62.33.3	1985	The plants are tall, erect, medium compact and vigorous with white flowers. It matures in about 120 days and produces large to medium, oval, white, flesh white with fleet eyes tubers. The variety is highly resistant to late blight and cyst nematodes. Average yield is 27.0-28.0 t/ha.	Released for cultivation in South Indian Hills
Kufri Megha	SLB/K-31 x SLB/Z-73	1989	Medium, round, oval, white, fleet eyes, flesh white	Released for cultivation in North Eastern Hills
Kufri Giriraj		1992	The plants are medium tall, semi erect, medium compact and vigorous with light purple flowers. It matures in 100 days and produces tubers that are medium to large, white and oval with fleet eyes. The variety is resistant to late blight. Average yield is 20.0-25.0 t/ha.	Released for cultivation in hilly regions of Himachal Pradesh.
Kufri Jawahar	Kufri Neelamani x Kufri Jyoti	1996	It takes 80 days to reach maturity. Its plants are short, erect, compact and vigorous with light green foliage and white flowers. Its tubers are oval, medium sized with fleet eyes and good keeping quality. This is moderately resistant to late blight.it is suitable for inter-cropping and mechanized cultivation. Average yield is 35.0 t/ha.	Released for cultivation in North Indian plains (Punjab, Haryana) and plateau region
Kufri Ashoka	EM/C-1020 x Allerfuheste Gelbe	1996	It takes 75-80 days to reach maturity. Its plants are medium tall and spreading with glossy-green foliage. The tubers are oval to oblong, large sized with fleet eyes. Its yield is 27.5 t/ha.	Released for cultivation in Indo-Gangetic plains of central and eastern Uttar Pradesh Bihar and west Bengal.
Kufri Sutlej	Kufri Bahar x Kufri Alankar	1996	Tubers are medium-large, smooth, round-oval, creamy white, fleet eyes and flesh pale yellow. It matures in about 90-95 days. It is moderately resistant to late blight. Average yield is about 35.0 t/ha	Released for cultivation in North Indian plains
Kufri Chipsona-1	Mex.750825 x MS/78-79	1998	It takes 100 days to reach maturity. Plants are medium tall, semi-erect and vigorous. Tubers are medium to large, oval, white, fleet eyes, with dull white flesh. It is low in reducing sugars, low phenols and high dry matter. It is resistant to late blight. Average yield is 25.0-30.0 t/ha.	Released for cultivation in North Indian plains
Kufri Chipsona-2	F-6 x QB/B-92-4	1998	It takes 100 days to reach maturity. Tubers are medium, round, white, fleet eyes, with light yellow flesh. The tubers are low in reducing sugars and phenols but rich in dry matter. It is tolerant to late blight and frost. Average yield is 25.0-30.0 t/ha.	Released for cultivation in North Indian plains
Kufri Giriraj	SLB/J-132 x EX/A 680-16	1998	Medium to large, oval, white, fleet eyes, flesh white	Released for cultivation in North Indian Hills

Variety	Parantage	Year	Tuber Characters	Area of Adaptation
Kufri Pukhraj	Craigs Defiance x JEX/B-687	1998	It takes 75 days to reach maturity. Its plants are tall, erect, medium, compact and vigorous with grey green foliage and white flowers. Its tubers are medium sized, oval, slightly tapered with fleet eyes. The variety is resistance to early blight, moderately resistant to late blight and has an average yield of 32.5 t/ha.	Released for cultivation in North Indian plains (Punjab, Uttar Pradesh, Madhya Pradesh, Karnataka, Gujarat and Maharashtra) and plateau region
Kufri Anand	Kufri Ashoka x PH/F-1430	1999	Large to medium, oval-long flattened white, fleet eyes, flesh white	North Indian Plains
Kufri Kanchan	SLB/Z-405 (a) x Pimpernel	1999	Medium, oval, red, eyes medium deep, flesh white	North Bengal Hills
Kufri Arun	Kufri Lalima x MS/82-797	2005	Medium to large, oval, red, shallow eyes, flesh light yellow	Released for cultivation in North Indian Plains
Kufri Pushkar	QB/A 9-120 x Spatz	2005	Medium, oval, white, shallow eyes, flesh light yellow	Released for cultivation in North Indian Plains
Kufri Shalija	Kufri jyoti x EX/A 680-16	2005	Medium, oval, white, eyes fleet, flesh dull white	Released for cultivation in North Indian Plains
Kufri Himalini	I-1062 x Bulk pollen	2006	Medium, oval, white, shallow eyes, flesh pale yellow	Released for cultivation in North Indian Hills
Kufri Chipsona-3	MP/91-86 x Kufri Chipsona-2	2006	Medium, oval, white, fleet eyes, flesh cream	Released for cultivation in North Indian Plains
Kufri Surya	Kufri Lauvkar x LT-1	2006	Medium, oval, light yellow shallow, flesh dull white	Released for cultivation in North Indian plains and plateau region
Kufri Sadabahar	MS/81-145 x PH/F-1545	2008	Medium, oval, white, cream, eyes shallow, flesh white	Released for cultivation in North Central Plains
Kufri Himsona	MP/92-35 x Kufri Chipsona-2	2008	Medium, round, white cream, shallow eyes, flesh cream	Released for cultivation in North Indian Hills
Kufri Girdhari	Kufri Megha x Bulk Pollen (10 genotypes)	2008	Medium, oval, white cream, shallow eyes, flesh white	Released for cultivation in Indian Hills
Kufri Khyati	MS/82-638 x Kufri Pukhraj	2008	Medium, oval, white cream, eyes medium, flesh cream	Released for cultivation in North Indian plains and plateau region
Kufri Frysona	MP/92-30 x MP/90-94	2009	Medium to large, long-oblong, eyes shallow, flesh white	Released for cultivation in North Indian Plains

Source: Naik *et al.*, 2012.

Table 6: Sweet Potato Varieties Released from ICAR-CTCRI, Thiruvananthapuram (Kerala)

Variety	*Maturity (Days)*	*Starch (Per cent)*	*Yield (t/ha)*	*Special features*
H-41	120-130	19.0	20.4-37.2	It is released in 1971 by Central Sub-committee on Release of Varieties for cultivation in the States of Tamil Nadu, Kerala and Karnataka. Tubers are sweet with excellent cooking quality. It is responsive to fertilizers but suteptifle to sweet potato weevil.
H-42	120	16.0-18.0	22.0-25.0	This is a hybrid between indigenous cultivar 'Vella Damph' and the American 'Triumph' and developed at ICAR-CTCRI, Thiruvanathapuram (Kerala). The plants are semi-spreading with unifoliate leaves and serrated leaf margins. Tubers are fusiform, medium sized with pink skin and white flesh. The crop matures in 120 days after planting. The tuber cooks well, taste sweet and is almost free from fibre. It is tolerant to potato weevil.
Varsha (H-268)	120	-	20.0-25.0	This is a hybrid from a double cross involving two exotic *viz.*, ACC No. 39 x ACC No.3 and two indigenous cultivars ACC No. 1871 x ACC No. 1130. The plants are semi-spreading with 3-5 lobed dark green leaves. The tubers are big (14-20 cm long and 6-8 cm in diameter), fusiform, with pink skin and light yellow flesh. Tubers cook well and are sweet in taste. This can withstand drought to a certain extent.
H-368	115-130	24.0-26.8	20.0-25.0	This is a double cross hybridand developed at ICAR-CTCRI, Thiruvanathapuram (Kerala). The plants are semi-spreading with 3-5 lobed dark green leaves. The tubers are big, fusiform with pink skin and light yellow flesh.
Sree Nandini	100-105	25.0-26.0	20.0-25.0	It is an early maturing selection from an open pollinated progenies 'S-32', an introduction from Sudan. The plants are semi spreading and leaves are unifoliate with entire margins and shy flowering. The tubers are medium sized with white skin and white flesh. The tuber neck is medium long. It possesses very good culinary quality.

Variety	*Maturity (Days)*	*Starch (Per cent)*	*Yield (t/ha)*	*Special features*
Sree Vardhini	100-105	13.0-15.0	20.0-25.0	It is a selection from an open pollinated progeny of 'S-13' and developed at ICAR-CTCRI, Thiruvanathapuram (Kerala). It is an early maturing variety and takes 100-105 days to reach marketable maturity. The plants are semi-spreading and leaves are simple with 3-5 shallow lobes. The plants are shy flowering. Tubers are medium sized with pink skin and light yellow flesh having carotene content of 1200 IU.
Sree Rethna (X-108-2)	90-105	22.0-23.0	20.0-26.0	This hybrid is derivative of a cross of S-187 x Sree Vardhini, released in 1998 for Kerala state. Plants are spreading type having light greenish brown vine colour. Emerging leaf colour brown with light green shoulder having light greenish brown petiole colour. Tuber is spherical in shape with purple skin, orange flesh and excellent cooking quality. It is susceptible to sweet potato weevil. This variety possess excellent cooking quality.
Sree Bhadra (S-1010)	90	18.0-20.	20.0-27.0	It is a seedling selection from the introduced seed population from Nigeria, released in 1996. Plants are semi- spreading with greenish brown vine colour. Leaves are broad cordate with dark from colour of emerging leaf. Tuber is spherical in shape with light pink skin, cream flesh and excellent cooking quality. This variety is susceptible to sweet potato weevil. It is an early maturing with excellent cooking quality.
Gouri (85-16)	110-120	16.0-17.0	19.0	It is derivative of cross H-219 x H-42, released in 1998. Plants are semi-spreading with slightly purple vine colour. Leaves are similar with 5-6 lobes having slightly purple emerging leaf colour and green petiole colour. Purple red skin and orange flesh variety with high β-carotene content. Can tolerate mid-season moisture stress. Suitable for Kharif and Rabi season.
Sankar	120	17.0-18.0	14.0	It is progeny of H-219 x S-73 released in 1998. Plants are spreading type with green vine colour. Emerging leaves are green with purple in colour. Tubers are Red skin, cream flesh, medium duration variety and excellent cooking quality. Suitable for irrigated and rainfed conditions.
Sree Arun	90	19.0-21.0	20.0-28.0	Spreading, pink skin, cream flesh and good cooking quality
Sree Varun	90	21.0-22.0	20.0-28.0	Spreading, cream skin, cream flesh and good cooking quality.
Sree Kanaka	75-85	14.0-15.0	10.0-15.0	Spreading, cream skin, and orange flesh.
Kalinga	105-110	27.0-29.0	26.0-28.0	Purple red skinned tuber with creamy flesh. Suitable for rainfed as well as irrigated uplands. Dual purpose variety can be used for food, animal feed and for starch extraction.
Goutam	105-110	24.0-26.0	18.32-32.0	Ovate tubers, white skin and cream flesh. Tolerant to mid-season stress. Suitable for rainfed as well as irrigated, and medium to uplands and hilly areas.
Gouri (85-16)	110-120	24.3	19.0	It is derivative of cross H-219 x H-42, released in 1998. Plants are semi-erect with slightly purple vine colour. Leaves are similar with 5-6 lobes having slightly purple emerging leaf colour green petiole colour. This is a medium duration variety with high carotene content, can tolerate mid season stress. This variety is suitable for *kharif* and *rabi* seasons. This variety is widely adopted for Odisha region.
Sourin	105-110	24.0-26.0	16.0-32.0	Round elliptic tubers with red skin and creamy flesh. Suitable for rainfed as well as irrigated, medium to uplands in both *kharif* and *rabi* season.
Kishan	110-120	29.0-30.0	16.0-26.0	Long elliptic tubers with purple skin and white flesh. Suitable for rainfed as well as irrigated, medium to uplands and hilly areas.
Kiran	110-120	24.5	26.0-29.0	This variety has been recommended for cultivation in Andhra Pradesh and Telangana
CO-1	135	24.2	28.3	It is clonal selection of IB-3 collected from Tiruchinapalli district and released from TNAU, Coimbatore (Tamil Nadu). The leaves are moderate and less spreading with medium lobed dark green. Each vine produces 3.2 tubers per vine weighing 1.12 kg and the tubers have pink skin and white flesh. Raw tuber taste is saltish sweet while it is sweet on cooking with firm consistency. Tubers contain 10.6 per cent TSS, 1.2 per cent acidity, 1.6 per cent reducing sugars, 9.7 per cent total sugar and 20 mg per 100 g of ascorbic acid. The clone is tolerant to root weevil in which infestation has been registered only 11-15 percent.
	110	29.5 per cent	32.0	It is clonal selection of 'IB-3' from seedling progeny obtained from open pollinated seeds of a type 'IB-37' in germplasm bank. Vines are dwarf and erect, tubers are medium sized with light pink skin and white flesh contain carotenoids of 3.2 mg per 100 g. The variety has better consumer's appeal and grower preference. The root weevil incidence is only 10-15 percent.

Variety	*Maturity (Days)*	*Starch (Per cent)*	*Yield (t/ha)*	*Special features*
CO-3	105-110	30.72	43.6	It is a seedling clone (IB 2837) obtained from the seeds of random mating population of IB 758 and released from TNAU, Coimbatore (Tamil Nadu). The tubers are medium sized with red skin and deep orange flesh. Tubers are markedly rich in carotenoids (13.28 mg/100 g), ascorbic acid (21.1 mg/100 g) and amylopectin (81.31 percent). The clone has better consumer's preferance than the other clones. The root weevil (*Cylas formicarius*) incidence ranges from 0 to 3 percent.
Samrat	100-105	29.3	18.0-24.0	This is a selection from the open pollinated seedlling progeny. The tubers are fusiform with light pink skin and white flesh. It is an early maturing variety. The variety is resistant to sweet potato weevil.
Punjab Sweet potato-21	140-145	-	19.0-20.0	It is selection from the local material and developed by PAU, Ludhiana. The vines are medium long and lobed. Tubers are 20.0 cm long, 4.0 cm thick having deep red skin and white flesh. Average tuber weight is 75.0 g.
Kanjangad	110-120	-	15.0-20.0	Semi-spreading, reddish purple skin, light yellow flesh and good cooking quality.
Konkan Ashwani	-	-	-	Spreading habit and average tuber weight is 70-75 g.
Jawahar Shakarkand-114	125	27.8	22.5-30.0	This variety has been developed from clonal selection method of breeding from a material obtained from U.S. Vegetable Breding Lab, South Carolina, America and released from JNKVV, Jabalpur for Madhya Pradesh state.
Jawahar-145	122	26	23.5-24.0	This variety has been developed from clonal selection method of breeding from a material obtained from U.S. Vegetable Breding Lab, South Carolina, America and released from JNKVV, Jabalpur for Madhya Pradesh state.
Kalmegh	110	24	26.5-32.5	This is an extra early variety developed from Rajendra Agricultural University, Dholi (Bihar). Tubers are round with light brown colour.
Pusa Safed	115	26	28.2-30.1	It is a selection of Taiwan variety FA-17. The tubers are medium size with good cooking quality.
Pusa Lal	129	23.7	29.6-31.0	It is selection of Japani variety Norin. Tubers are medium size with red colour skin and white flesh. This variety has good keeping quality.
Pusa Sundheri	124	26	30.0	The variety has high carotene content, importd from USA and selected at ICAR-IARI, New Delhi. Flesh colour of tuber is orange.
Rajendra Sakarkand-5 (X-5)	105-120	26	25.0-30.0	It is a derivative of cross between Cross-4 x H-5 and released by Central Sub-Committee on Crop Standard, Notification and Release of Varieties in 1985 for Bihar state. It is suitable for *rabi* season. Plants are spreading with fewer canopies. Vines are 1.8 cm thick in girth and hairy. Leaves are green and thick with 4-6 lobes. Days to tuber initiation are 35-40 days. Tubers are cylindrical with thin rind, tubers skin and flesh both white; 4-6 per plant tubers with average tuber weight of 595 g. It is tolerant to frost, water logging for a week period and sweet potato weevil.
Rajendra Sakarkand-35	105-120	27	25.0-30.0	This variety is recommended for cultivation in Assam, Karnataka and West Bengal. The tubers are dark brown colour with white flesh.
Rajendra Sakarkand-43	110-120	26	20.0-25.0	This variety is recommended for cultivation in Assam, Karnataka, Maharashtra and Kerala states.
V.L. Sakarkand-6	135-140	27	23.0-25.0	It is an introduction from U.S.A. under the name B-129, released in 1974 primarily Uttar Pradesh for hilly areas. Vines are spreading type with light green foliage. Tubers are elongated, slight bent, and purple skin with light yellow flesh. Length of the tuber is 15.0-18.0 cm and girth of tuber is 12.0-14.0 cm.
Indira Madhur	110-120	26	26.0	This variety was developed at Indira Gandhi Krishi Vishwa Vidyalaya, Raipur (Chhattishgarh) through clonal selection from the material received from CIP. It has vine length of 100-120 cm and having semi-spreading growth habit. It is rich in carotene (about 5524 IU/100 g edible portion). It is soft and easy to cook having excellent taste. Recommended for commercial cultivation in the state of Chhattishgarh.
Indira Naveen	90-100	27	33.8	This variety was developed at Indira Gandhi Krishi Vishwavidyalaya, Raipur (Chhattishgarh) through clonal selection from 'Sree Vardhani'. This variety has semi-spreading growth habit with vine length of 80-100 cm. Recommended for commercial cultivation in the state of Chhattishgarh. It is soft and having excellent cooking quality.

Table 7: Casava Varieties Released from ICAR-CTCRI, Kerala

Variety	*Maturity (Months)*	*Starch (Per cent)*	*Yield (t/ha)*	*Special features*
'H-97'	10	27.0-31.0	25.0-35.0	It is hybrid between Manjavella and Accession No.300 which is developed at ICAR-CTCRI, Thiruvanathapuram (Kerala). The plants are medium tall (1.5-2.0 m) and develop semi branching habit. It is characterized by the light sepia colour of the emerging leaf. The tubers are conical, medium sized (25.0-35.0 cm) with light brown skin and white flesh.
'H-165'	8-9	23.0-25.0	33.0-38.0	It is hybrid between 'Chadayamangalam Vella' and 'Kalikalan'which is developed at ICAR-CTCRI, Thiruvanathapuram (Kerala). The plants are medium tall (1.5-2.5 cm), generally non-branching, bear thick conical tubers and early maturity type. Tubers have golden brown skin and cream flesh.
'H-226'	10	28.0-30.0	30.0-35.0	It is hybrid between 'Malyan-4' and 'Ethakkakruppan' which is developed at ICAR-CTCRI, Thiruvanathapuram (Kerala). The plants are tall (2.0-2.5 m) occasionaly branching and the leaves have a characteristic green colour. The tubers are medium sized (25.0-35.0 cm) with creamy skin, light purple rind and white flesh.
'Sree Visakham' (H-1687)	10	25.0-27.0	35.0-38.0	It is hybrid between Accession No. 1501 and S 2312 which is developed at ICAR-CTCRI, Thiruvanathapuram (Kerala). Its plants are predominantly non-branching and Tall (2.0-2.5 m). Tubers are medium sized, conical with brown skin, cream rind and light yellow flesh. Yellow flesh content 466 IU carotene. It is tolerant to cassava mosaic virus.
'Sree Sahya' (H-2304)	10-11	39.0-31.0	35.0-40.0	It is developed at ICAR-CTCRI, Thiruvanathapuram (Kerala). Plants are tall (2.0-2.5 m), predominantly non-branching with dark brown petioles and a prominent spiny stipular marks. Tubers are medium long with light brown skin, cream rind and white flesh
'Sree Prakash' (S-856)	7	29.0-31.0	30.0-35.0	It is an early maturing variety and developed at ICAR-CTCRI, Thiruvanathapuram (Kerala). The plants are erect and relatively short (1.0-1.5 m), generally non-branching with highly retentive of leaves. Suitable for low land as a rotation crop. It is tolerant to cassava leaf spot.
'Sree Harsha' (2-14)	10	38.0-41.0	35.0-40.0	Erect branching, good cooking quality and susceptible for Cassava Mosaic Disease (CMD)
'Sree Jaya' (CI649)	6-7	24.0-27.0	26.0-30.0	Erect branching, early maturing and having excellent cooking quality. Suitable for low land as a rotation crop. It is susceptible for Cassava Mosaic Disease (CMD).
'Sree Vijaya' (CI-731)	6-7	27.0-30.0	25.0-28.0	Erect branching and early maturing. It is susceptible to spider mites and scale insects. Suitable for low land as a rotation crop.
'Sree Rekha' (TCH-1)	8-10	28.0-30.0	45.0-48.0	It is a top cross hybrid which is developed at ICAR CTCRI, Thiruvanathapuram (Kerala).Erect branching and having long, conical tubers with light brown skin, creamy flesh and good cooking quality. Suitable for both upland and low land conditions.
'Sree Prabha' (TCH-2)	8-10	25.0-29.0	40.0-45.0	It is a top cross hybrid which is developed at ICAR-CTCRI, Thiruvanathapuram (Kerala).Semi-branching and having short, conical tubers with brown skin, yellow flesh and good cooking quality. Suitable for both upland and low land conditions.
M4	10		18-23	Non-branching and good table variety It is susceptible to spider mites.
Vellayani Hraswa	6	24.0	25-30	Branching and having good cooking quality. Suitable for both upland and low land conditions.
Adira-1	7-8	23.8	25.0-30.0	It is a sweet cultivar with only 50 ppm HCN in the tuber. It is recommended for small-scale farms, where cassava is intercropped with other food crops.
Adira-2	8-9	28.0	29.0-31.0	It is high yielding cultivar. Tuber is medium in starch content and high in HCN. It is tolerant to cassava bacterial blight (CBB).
Adira-4	9-10	27.0	35.0-45.0	It is bitter cultivar with approximately 90 ppm HCN, but capable of producing high yield and widely adopted. It is suitable for large planting of industrial use.
Colombia (PR-C62)	10-12	29.0	46.0	It is an edible purpose cultivar.
Golen Yellow (PR-C24)	8-10	28.0	43.0	This cultivar gives yield both under open and partially shaded conditions prevalent in coconut garden. Tubers are low in HCN and suitable for table purpose
Kadabao (PRC-13)	10-12	27.0	42.0	This cultivar also performs well under open and partial shade. This cultivar has high starch content and fairly low HCN.
Faroka	9-10	26.0	19.0	It is suitable for dry areas. The tubers are low in HCN with good table qualities.
Kalikalan	7-8	25.0	32.0	It is early maturing cultivar with good table quality of tubers. It is highly susceptible to CMD. It is well adapted to low fertility conditions.
Karuthakaliyan	9-10	27.0	32.0	This cultivar is spreading type. It produces non-bitter tuber. It is susceptible to CMD, bacterial leaf spot (BLS) and mites.

Variety	Maturity (Months)	Starch (Per cent)	Yield (t/ha)	Special features
Rayong-1	8-9	25.0	25.0	It is widely adopted bitter cultivar. Since tuber has high dry matter and starch content, it is well accepted as an industrial cultivar.
Rayong-2	9-10	27.0	28.0	It was selected from hybrid seeds of CM305 introduced from CIAT, Colombia. Its fresh root yield dry matter content and starch content are similar to Rayong-1. This cultivar is table type, good for making fried chips and other convenient foods. It has yellow flesh content, high carotene and very low HCN.
Rayong-3	9-10	24.0	29.0	It is selection from CM407 which was introduced from CIAT, Colombia. In tuber dry matter and starch content are superior to Rayong-1. These qualities fetch higher price for Rayong-3.
TMS-30001	8-9	25.0	22.0	It is early maturing and sweet cultivar. Tubers contain low HCN and well accepted by consumers. It is resistant to CMD but susceptible to root rot and insect pest.
TMS-30555	9-10	28.0	40.0	It is high yielding and widely adopted cultivar. It is resistant to CMD and CBB.
TMS-30572	8-9	25.0	40.0	It is widely adopted cultivar, because of high yield and good quality of tubers. It is resistant to CMD and CBB. It is susceptible to cassava mealy bug and cassava green mites.
TMS-50395	8-9	26.0	33.0	It is resistant to CMD, CBB and tolerant to cassava green mite. It is widely adopted though the tubers are high in HCN.

Table 8: Varieties Released for Commercial Cultiavtion

Variety	Special Features
Arvi (*Colocasia*)	
Sree Rashmi (C-149)	This accession is a natural triploid and released for cultivation from Kerala State by Varietal Release Committee. This is a tall, erect, purplish with green petiole accession, collected from Kerala. The cormels are medium size, conical and have bluish and mucilageous pulp. Leaves are free from acridity, thus rendering all vegetative parts edible. Cormel contains 15.25 per cent starch and 2.5 per cent protein. It is a late maturing (29-30 weeks) with recorded yield of 15.0-19.0 t/ha.
Sree Pallavi (C-266)	It is also triploid. This accession is from Meghalaya and released for cultivation in Kerala by the State Varietal Release Committee. Plants are erect, tall with green petiole. This is also late maturing (29-30 weeks) with yield potential of 14.0-18.0 t/ha. Cormel is small (8.85 g mean weight), club shaped and edible. Cormels alone are found edible.
Thamarannah (C-9)	It is popular variety cultivated locally in Kerala for the mainly big cormels (20-40 g).
Thamgri	It is suitable for rainfed crop in Odisha for local cultivation.
Apra Kochai	It is grown in Madhya Pradesh in Bilaspur region. The yield potential of this cultivar is 15.0-30.0 t/ha.
Bangau Banda	It is very common in north India, having dark purple leaves and petioles.
Ban Kachu	This wild variety commonly known in Bengal as Bay Kachu. It is more important for leaves and stalks than tubers.
Char Kachu	This wild variety commonly important for their leaves and stalk than their tubers.
Kaka Kachu	It is promising variety, which produces side corms and is fleshy with better taste. The yield varies from 22.0-40.0 t/ha.
Kalla Kachu	It is wild variety more valued for their leaves and stalks than for their tubers.
Koni Kachu	This variety belongs in Eddoe group. This variety produces many daughter tubers, very similar to poultry eggs. It is very good for currying and has a good cooking quality. It gives a yield of 12.-25.0 t/ha in a soil rich in humus amnd potash.
Sar Kachu	It is a very good variety for aquatic situation and recommended for Bengal.
Punch Mukhi	This variety belongs in Eddoe group. The word Punch Mukhi means a type of colocasia with five-month or five tubers attached together. It can be kept for more than a year after harvest. It is grown successfully in the hills above an elevation of 1200 m and has very good cooking quality. The yield varies from 25.0-40.0 t/ha.
Sat Mukhi	Its tuber shape is rectangular, medium in size, brown skin, pulp (flesh) white in colour and matures in 175-190 days. The yield potential of this variety is 10.0-15.0t/ha. It is widely grown in Andhra Pradesh, Kerala and Karnataka.
Munshikuntla Kachu	It is just like wooden pegs in shape and size. It is very soft and highly palatable. The yield potential of this variety is 20.0-35.0 t/ha.
Kovur	It is reported to be high yielding.
Ladhra	It is very common in north India
Dasheen	This kind of Colocasia produces only one main corms are elongated and fleshy having nodes and internodes and only one growing but at the top.
Deshi Banda	This is very common in north India
Indira Arvi-1	It has been developed and released from IGAU, Raipur (Chhattishgarh) in 2003. It is good table variety having good cooking quality and taste. It is suitable for cultivation in both irrigated and rainfed areas. The yield potential of this variety is 33.0 t/ha. In 210-240 days of crop duration. Keeping in view its wider adaptability, high yielding capacity and disease and pest tolerance, this variety is suitable for cultivation in the whole of Chhattishgarh.

Variety	*Special Features*
Khasi Bagga	It contains more acridity as compared to other varieties. The yield varies from 14.0-16.0 t/ha.
Naga Kaju	This is famous variety of Nagaland having bigger szed mother tubers and small daughter tubers. The yield varies from 12.0-28.0 t/ha and takes more time for boiling.
Elephant Foot Yam (*Amarphophallus paeoniifolius*)	
Gajendra	A local collectionfrom Kovuur region of Andhra Pradesh developed by Reginal Centre of ICAR-CTCRI, Bhubaneswar (Odisha) and it is recommended for all India cultivation. The yield potential of this variety is 50-60 t/ha and the tubers are smooth, well shaped and non-acrid. The flesh is light orange coloured.
Aerial Yam (*Dioscorea bulbifera* L.)	
Kpnkan Kalika (KKVDb-1)	Developed by Regional Agricultural Research Station, Karjat, Balasaheb Sawant Konkan Krishi Vidyapeeth, Dapoli, Dist. Ratnagiri (Maharashtra). It yield marketable bulb about 4.85 t/ha but total aerial bulb is 5.27 t/ha. It is resistant to leaf blight diseses (Mhaskar *et al.,* 2016).
Asaparagus	
Perfection	It is early, spears large, green, succulent, heavy cropper and uniform in maturity (Anonymous, 1998).
Conovers Colosal	Light green or whitish spears. They are high yielder and commonly grown for marketing purpose.
Mammoth White	Light green or whitish spears. They are high yielder and commonly grown for marketing purpose.
Palmettes	This is most important and very high yielder. The tips of spears before exposer to the light are purplish and the spears become dark green in sunlight (Krarup, 1989).
Argentenil	This is most important and very high yielder. The tips of spears before exposer to the light are purplish and the spears become dark green in sunlight (Krarup, 1989).
Mary Washington	This is most important and very high yielder. The tips of spears before exposer to the light are purplish and the spears become dark green in sunlight (Krarup, 1989).
Violet Dutch	It is the oldest variety which may be regarded as the form from which all the modern types have been derived.
White Germa	It is selected in Germany and widely grown by Germans. It is rounded tip and good quality spears.
Palmetto	It has a little difference in appearance than other varieties and considerably resistant to rust and extensively grown throughout America.
Scheweizinger	It is a commercial variety and has a greater vigour when planted by clones.
Lara	It is an early hybrid which evolved through breeding. It gives a very high yield and quality spears.
Mira	It is an early hybrid which evolved through breeding. It gives a very high yield and quality spears.
Aneto	It is an inter-clone hybrid, high yielding and less heterogenous.
Desta	It is an inter-clone hybrid, high yielding and less heterogenous.
Faribo	It is high yielder and produce good quality spears.
UC-711	It is superior hybrid than other and gives maximum yield and high return.
Gijnlim	This asparagus hybrid is very early in production with medium thick, and straight spears of excellent quality. A total yield by means of 100 per cent male plants gives 20 per cent higher production compared to female plants and gives the advantages of no production of seed berries and seedlings under normal conditions.
Paradise	It is a selection of Mary Washington developed in California are said to be superior to the standard strain in some regions.
Jersey Queen	A clone cultivar selected from Mary Washington. It produces high quality spear and maximum yield (Ellison *et al.,* 1989).
Mc Grady	It produces spears of 8-20 mm diameter in comparision to cv. Jersy Giant followed by UC-157.
Tilt	It produces spears of 8-20 mm diameter in comparision to cv. Jersy Giant followed by UC-157.
Lucullus 1883	It is white type cultivar.
Limbrass-126	It is green type cultivar generally grown in Poland (Knaflewaski, 1988).
Leek	
Borde Pearl	It is high yielding and resistant line to bolting.
Bluevetia Early	It is very much suitable for summer and early autumn culture (Geriya, 1973).
Copenhagen Market	It is short-necked type.
Elephant	It is short-necked type.
Gennivillies	It is long-necked type and resistant to frost.
St. George	It is long-necked type and resistant to frost.
Miner	An autumn or early winter variety derived from crosses and family selection in the variety 'St. George' has long stem and gives high yield.
Splendid	Cultivated throughout the season
Renova	Derived by selection from the variety 'Regions' has darker leaves, longer stalk and less tendency to form bulb and is more uniform in size and shape.
Special 050	It is sutable for forcing.
London Flag	American Seed Men varieties.

Variety	*Special Features*
American Flag	American Seed Men varieties.
Elephant or Monstrous Carentan	American Seed Men varieties.
Giant Musselburgh	American Seed Men varieties.
Lyon	American Seed Men varieties.
White Alberta	It is Belgium variety and suitable for winter season
Alberta	It is a very high yielding variety.
Blizzard	It is Belgium variety and suitable for winter season
Carina	It is Belgium variety and suitable for winter season. It produces spears in high number
Longa	It is Canada a variety
Odin	It is Canada a variety
Kilima	It is Canada a variety
Goliath	It is Canada a variety
Siberia	It is Canada a variety
Artico	It is Canada a variety
France	It is a variety developed by Tropica seeds (Green Co.)
Large Long Summer	It is a variety developed by Tropica seeds (Green Co.)
Lincoln	It is a variety developed by Bejo Sheetal Seeds Co. Jalna (Maharashtra)
Carlton	It is a variety developed by Nunhems Seeds Co.
Roxton	It is a variety developed by Nunhems Seeds Co.
Parton	It is a variety developed by Nunhems Seeds Co.
Regins 930/57 (St. George x Special)	F_1 hybrid is good cropper especially in late autumn and early winter
London Flag	Tall with a thick stem, and leaves broad. It is most generally cultivated.
Musselburgh	It is a very long and Hardy; leaves broad and tall, stems long and thick.
Large Roven	Its leaves are very dark green, broad and thick. Stem rather short, but remarkably thick. It is said to grow as thick as man's arm. It is found to be the best kind of forcing as it acquires a sufficient thickness of stem sooner than any other varieties.
Lincoln	It matures in 80 days after transplanting. It is early Bulgarian Giant type leek, which gives uniform long and thick stems. It is suitable for rabi season planting.
Carlton	This hybrid variety is very fast in growth, leaves are erect and green in colour. Long and short length spear have excellent uniformity.
Roxton	Long and short length spear have excellent uniformity. The growth of this hybrid is very fast with green to dark green leaf colour.
Parton	This hybrid produces medium shoft length of excellent uniformity. Attains medium growth. Leaves are erect and dark green leaf colour.
France	It shows medium maturity. The leaves are dark-blue green. Shanks are 18-20 cm long and 2.8 to 3.3 cm in diameter.
Large Long Summer	It is very early type and matures in four months. Foliage, medium green in colour. Plant long, medium sized shanks (20-25 cm long and 2.3 to 3.0 cm diameter), widely adapted and very fast growing.
Palam Paushtik	It has long, broad, flat garlic like bluish green leaves and long white pseudostem thickening cylindrically to attain 3-4 cm diameter at harvesting. It takes about 140-150 days to reach marketable stage. It has mild digestive flavor due to organosulphur compounds. It can be consumed like green onions after cooking with potatoes or in any other form of Indian dishes. Besides, it can be enjoyed raw as salad and in soup. It has healthful digester nutritive substances, especially protein, fibre, carbohydrates, vitamins and minerals. Its average yield is 30-35 tonnes/ha.
Parsnip	
Popular cultivars	Cobham Improved Marrow, Tender Harris Model, Andover, Lancer, Alba, Unicorn, White Gem, All American, Fullback and Hollow Crown
Parsley	
Hamburg	It is a variety of turnip-rooted parsley. It has tough stem. Deeply cut plain leaves and moss curled. It is dwarf with finley cut serrated and deeply curled leaves and grown for its edible root. It is grown in the vicinity of the large cities, where it is utilized by the foreign population.
Parsley Petra	It matures in 75 days after transplanting. It can be sown from August to January. It is vary uniform with very attractive curled leaves. High dry matter content and tolerant to bolting.
Other Hybrids	Festible 68, Hybrid. No. 41 and Hybrid No.-42 (Madzharova, 1971).
Celery	
Standard Bearer	It is acclimatized strains at Regional Station, Katrain (H.P.) ICAR-IARI, New Delhi. It is an early, and medium tall plant, produce medium pink stems with longitudinal streaks, stalk solid and of good size and flavor.

Variety	Special Features
Wrights Grove Giant	This cultivar is medium late, tall growing, high yielding produces large white stalks of fine quality
Fordhook Emperor	It is late cultivar, plant dwarf and stocky, produces solid, white thick, broad and tender stalks.
Triumen	It matures in 85 to 90 days, has dark green coloured attractive foliage and medium long petioles. It has good standing ability.
Celery Paris	It is very resistant to heat and rain. The foliage is tall and dark colours. Stalks are thick smooth and fleshy. Seeds are very small (more than 2500 seeds per gram). Plants can be harvested 2-3 months after transplanting.
Celery France Leaf	It has vigorous dark green and sturdy foliage with long, thick and smooth stalks.
Other cultivars	Utah-52-70, Golden Self Blanch, White Gem, Grant Red, Belder and White Solid

Table 9: Improved Varieties of Coriander Developed by Various Institutes

Variety	Special Features	Developing Source	Leaf Yield (kg/ha)
CO-1	Tall plant suitable for both leaf and seed production. Small grains suited for rainfed condition. Since water requirement is not as high and cultivated as irrigated as well as rainfed. Crop duration is 110 days. Seed yield is 0.5 t/ha.	TNAU, Coimbatore (Tamil Nadu)	400.0-500.0
CO-2	This is dual purpose, suitable for waterlogged, drought, saline and alkaline conditions. The yield potential for seed production in 90-100 days is 0.6-0.8 t/ha and the green leaf production is 1.0 t/ha in 40 days after sowing. The oil content of seed is 0.3 per cent.	TNAU, Coimbatore	520.0-790.0
CO-3	This is dual purpose, suitable for irrigated condition of Rabi and rainfed *Kharif* season. It is tolerant to Fusarium wilt, powdery mildew and grain mould. The yield potential for seed production is 0.8-1.0 t/ha. Seed colour is attractive brownish yellow.	TNAU, Coimbatore (Tamil Nadu)	650.0-1250.0
CS-287	Early (78 days duration), small bold grain and very much suitable for irrigated and rainfed condition. Tolerant to wilt, powdery mildew and grain mould. Seed yield is 0.5 t/ha.	TNAU, Coimbatore (Tamil Nadu)	600.0-750.0
Gujarat Coriander-1	Semi-spreading (plant growth medioum), dark green leaves, no lodging and shattering. It is suitable for early sowing and unirrigated conditions. The variety is bold seeded having higher yield potential and seed yield is about 1.1 t/ha in 110 days. Tolerant to wilt, and powdery mildew. Doing well in Gujarat.	Junagadh, Gujarat	1100.0-1200.0
Gujarat Coriander-2	Semi-spreading (plant growth medioum), dark green leaves, no lodging and shattering. It is suitable for early sowing and unirrigated conditions. It is bold seeded having higher yield potential of about 1.4 t/ha seed yield in 120 days duration. Tolerant to wilt, and powdery mildew	Junagadh, Gujarat	1450.0-2000.0
RCR-41	Plants are tall erect with thick stem. Seeds are round and small. It is very much suitable for irrigated tracts. Oil content is 0.25 per cent. It is tolerant to wilt and resistant to stem gall (*Protomyces mecroporus*). Seed yield potential is about 1.2 t/ha. It is late maturing (130-140 days) variety.	RAU, Jobner, Rajasthan	1200.0-1900.0
RCR-47	Plants grow in medium height, resistant to stem gall (*Protomyces mecroporus*) disease. Seed yield is about 1.2-1.4 t/ha in 120-140 days. This variety is suitable for irrigated areas.	RAU, Jobner, Rajasthan	1200.0-1900.0
RCR-435	It is developed through half-sib recurrent selection from a local collection of Jalore districts and identified for release in 1995 for irrigated condition. Erect, bushy, attains quick early growth and fall under medium maturity group. It bears medium-sized grains (14.0 g/1000). It matures in 110-130 days and produces average yield of 1.05t/ha.	RAU, Jobner, Rajasthan	1000.0
RCR-436	It is developed through half-sib recurrent selection from a local collection of Kota and identified for release in 1995 for limited moisture condition. Plants are semi dwarf, attains quick early growth, early maturity, and suitable for heavy soil of Rajasthan. It posses very good bearing habit. It bears bold-sized grains (16.0 g/1000). It matures in 90-100 days and produces average seed yield of 1.19 t/ha under limited moisture conditions.	RAU, Jobner, Rajasthan	1200.0
UD-446	It is developed through half-sib recurrent selection from a local collection of Jaipur district and identified for release in 1997 for irrigated condition. The plants are leafy and erect with higher number of grains per umbel. Dual purpose, for rainfed conditions. Seeds of this cultivar are medium in size (13.5 g/1000). It mature in 110-130 days and produces an average seed yield of 1.2 t/ha. Moderate resistant to wilt, and stem gall (*Protomyces mecroporus*).	Udaipur, Rajasthan	1200.0-1900.0
Kiran (UD-14)	This variety is suitable for irrigated areas, resistant to stem gall (*Protomyces mecroporus*). Seed yield is about 1.2 t/ha in 130-140 days. Oil content is 0.25 percent.	Udaipur, Rajasthan	

Variety	*Special Features*	*Developing Source*	*Leaf Yield (kg/ha)*
UD-20	Plant growth is medium, bold grain, suitable for growing in unirrigated areas. Seed yield is about 1.2 t/ha in 120 days.	Udaipur, Rajasthan	
Sadhana (CS-4)	Semi dwarf/erect, dual purpose (seed and leaf) and suitable for rainfed condition of Rajasthan, Telangana and Andhra Pradesh. Tolerant to wilt and resistat to aphid, mites and fruit fly. Seed yield, 1.0-1.2 t/ha in 115 days. Seed contain 0.28 per cent oil.	Guntur, A.P.	1025.0-1300.0
Swathi (Lam CS-6)	It is eary to medium duration maturing crop which helps in escaping powdery mildew incidence. Fine seeded and rich in oil (0.3 per cent). It is suitable for rainfed and intercropping.	Guntur, A.P.	885.0-1080.0
Lam CS-2	It is a medium tall (80 cm) variety with more branching and bushy in nature. Mericarp is straw coloured. Crop duration is about 105 days. The average seed yield is 1.3-1.4 t/ha. It is reasonably tolerant to pest and diseases as compared to local varieties. It is suitable for cultivation in Andhra Pradesh, Rajasthan and Gujarat.	Guntur, A.P.	1025.0-1300.0
Lam CS-7	It is a selection from north India, acclimatized to condition of Andhra Pradesh. It is a dwarf cultivar with more leaves of good flavor. Average seed yield is 0.9 t/ha. The essential oil content of seed is also high (0.3-0.48 per cent).	Guntur, A.P.	1000.0-1200.0
Sindhu	Plants grow to medium height. Dual purpose, fine seeded and rich in oil. It is suitable for rainfed condition of Andhra Pradesh and intercropping. High yielding (1 t/ha seed yield). It is tolerant to disease and pests.	Guntur, A.P.	1050.0-1200.0
Pant Haritima	Plant growth is heavy to medium and late maturing in nature. Basically it is dual purpose (seed and leaf) and bears highly fragmented leaves. Seeds are very fine. Seed yield is 1.5-2.0 t/ha. Resistant to stem gall (*Protomyces mecroporus*) disease.	GBPUA&T, Pantnagar (Uttarakhand)	1500.0-1800.0
Khushboo	Multicut and dual purpose variety. It is highly aromatic and suitable for spring, winter and autumn type.	SKUAST, Jammu	1000.0-1200.0
Arka Isha	This variety was released for commercial cultivation in 2011. It is high yielding multicut type of coriander (3 cuttings can be taken). Plants are bushy, leaves broad and leaf lobes short late flowering (50 days after sowing). First cutting at 40 days after sowing and the subsequent cuttings are at 15 days interval. Plants are harvested by pulling and start in 40 days after sowing. Leaf essential oil yield is 0.083 per cent with good aroma.	ICAR-IIHR, Bengaluru (Karnataka)	3700.0
Hisar Anand	It is medium season cultivar.	CCS HAU, Hisar (Haryana)	
Hisar Surabhi	It is and early season cultivar.	CCS HAU, Hisar (Haryana)	
Hisar Sugandh	It is an early variety and resistant to stem gall (*Protomyces mecroporus*).	CCSHAU, Hisar (Haryana)	1400.0-1500.0
DH-206	It is high yielding and possess very good aroma.	CCS HAU, Hisar (Haryana)	
Hisar Bhoomit	It is dual purpose cultivar.	CCS HAU, Hisar (Haryana)	
DH-5	It is developed through selection from a local collection and identification in 1993 for irrigated condition. The plants are medium tall and bushy. It matures in 120-130 days and produces an average yield 1.8-2.0t/ha under good management. Seeds are round, attractive and medium in size (13.8 g/1000).	CCS HAU, Hisar (Haryana)	
Naraul Selection	Plant growth medium, water requirement is less, bold seeded and high yielding (1.5-2.0 t/ha). Mostly grown in Haryana.	CCS HAU, Hisar (Haryana)	
NP (D) 92	It is selection from local material grown around Delhi area. Seeds are small, and light green in colour having less fibre.	-	
Caribe	This hybrid has got broad leaves, nice aroma and is tolerant to bolting. This hybrid has got broad leaves, nice aroma and is tolerant to bolting. Mature in 50 days after transplanting. It can be planted in *Kharif*, rabi and in mild summer. It gives multiple cuttings, escaping powdery mildew incidence. The seed yield is about 0.9-1.0 t/ha in 95-105 days. Essential oil about 0.3 per cent. Tolerant to powdery mildew.	Bejo Sheetal Seeds, Jalna, Maharashtra	
Katwan type	It is recommended for Madhya Pradesh. It is very much popular among the farmers due to maximum green leaf cutting, late maturity and taller growth habit. Crop duration is 150-170 days. The average seed yield is 1.0-1.2 t/ha.	-	
Gwalior No. 5365	This variety is resistant to Fusarium wilt, medium seed size. Produces 0.8 to 1.0 t/ha seed yield in 100-130 days. Generally grown in Madhya Pradesh.	-	

Variety	*Special Features*	*Developing Source*	*Leaf Yield (kg/ha)*
Chimpa	Plant grows tall, late maturing, and high seed yield, 1.8-2.0 t/ha in 180-210 days. It is introduction from Bulgaria. Suitable for cultivation in Madhya Pradesh.	-	
Merrocan	Seed yield is 1.0-1.2 t/ha in 140-160 days. Oil content is high (0.34 per cent). Suitable for growing in Madhya Pradesh.	-	
Rajendra Swati	Plant grows medium in height, and resistant to stem gall (*Protomyces mecroporus*), disease and fruitfly. Variety isan early maturing. Seed yield is about 1.2-1.4 t/ha in 100 days. Seeds are rich in oil content. It is suitable for intercropping.	RAU, Pusa, Dholi Campus, Bihar	

Plant Genetic Resource and Crop Improvement Base Glossary

Access: Acquisition (right to use) biological resources and/or knowledge associated thereof, and their derivatives or, as applicable, intangible components for purpose of research, conservation, biological prospecting and industrial application or commercial use, among others.

Accession: A distinct, uniquely identifiable sample of seeds/propagules representing a landrace, cultivar, breeding line or a population, which is maintained in a genebank for conservation and use.

Accession number: A unique identifier assigned to an accession, when it is held in a genebank. This number is never assigned again to another accession even after loss of the accession.

Active collection: A germplasm accession that is used for regeneration, multiplication, distribution, characterization and evaluation. Active collections are maintained in short-to-medium-term storage and usually duplicated in a base collection maintained in medium to long term storage.

Adaptation: The process by which individuals (or parts of individuals), populations, or species change form or function in such away to better survive under given environmental conditions.

Additional declaration: A statement that is required by an importing country to be entered on a phytosanitary certificate and which provides specific additional information on a consignment in relation to regulate pests.

Adventitious regeneration: Development of shoots or roots from regions (such as callus, leaf segments, intermodal segments) other than usual points of origin (buds, shoot bases) or embryos from sources other than zygotes.

Agricultural biodiversity: Also referred as 'agrobiodiversity', encompasses the variety and variability of animals, plants and micro-organisms necessary to sustain key functions of an agro-ecosystem, its structure and processes for, and in support of, food production and food security. It covers, inter alia, crop varieties, including forage and fodder plants and trees, animal breeds, including fish, molluscs, bird species and insects, as well as fungi, yeasts and micro-organisms such as algae and diverse bacteria.

Agro-ecosystem: The organisms and environment of an agricultural area considered as an ecosystem.

Allele or Allelomorph: One of several alternative forms of a gene occupying the same locus on a particular homologous chromosome. When the alternates exceed two, the alleles form a multiple allelic series. Multiple alleles arise by repeated mutations of a gene, each with different effects.

Allelic frequency: A measure of commonness of an allele in a population. It is used to describe the frequencies of polymorphic genes statistically correlated with temporal variation of environmental factors.

Asynapsis: Failure of pairing of homologous chromosome during meiosis.

Autogamy: Self-fertilization.

Avirulent: Inability of a pathogen to produce a disease on its host.

Axenic culture: A culture without foreign or undesired life forms.

Backcross: A cross of a hybrid to either of its parent. In genetics, a cross of a hweterozygote to a homozygous recessive.

Backcross Breeding: A system of breeding whereby recurrent backcrosses are made to one of the parents of a hybrid, accompanied by selection for a specific character or characters.

Balance: The condition in which genetic components are adjusted in proportions that give satisfactory development. Balance applies to individuals and populatios.

Barcode: A computerized coding system that uses a printed pattern or bars on labels to identify ermplasm accessions. Barcodes are read by optically scanning the printed pattern and using a computer program to decode the pattern.

Base collection: The widest and most complete collection of ermplasm accessions. It is stored for long-term, more for conservation *per se*, and used only for filling gaps in the active collection.

Basic number: The number of chromosomes in ancestral diploid ancestors of polyploids, represented by x.

Benefit sharing: Means sharing of benefits arising from use of biological resources and associated knowledge based on prior informed consent and mutually agreed terms, with contracting party providing such resources.

Biased sampling: Deliberate collecting of desired phenotypes from the observable diversity in a population at a given time.

Biochemical markers: Genetic markers that can be classified on the basis of protein and/or isozymes.

Biological diversity: The variability among living organisms from all sources including, *inter-alia*, terrestrial, marine and other aquatic ecosystems and the ecological complex of which they are part. It includes diversity within species, between species and of ecosystems.

Biological material: Material containing genetic information and capable of reproducing itself or being reproduced in a biological system.

Biological resources: Include genetic resources, organisms or its parts, populations, or any other biotic component of ecosystems with actual or potential use or value for humanity.

Biotechnology: Any technological application that uses biological systems, living organisms, or derivatives thereof, to make or modify products or processes for specific use. It includes a number of individual techniques, *inter-alia*, recombinant DNA-molecule manipulation, protein engineering, cell fusion, nucleotide synthesis, monoclonal antibody use and production, product recovery, and unique fermentation techniques.

Biotype: A group of individual with the same genotype. Biotypes may be homozygous or heterozygous.

Bivalent: A pair of homologous chromosomes united in the first meiotic division.

Breeder seed: Seed produced by the agency sponsoring a variety and used to produce foundation seed.

Breeding: The art and science of changing plants or animals genetically.

Bulk Breeding: The growing of genetically diverse populations of self-pollinated crops in a bulk plot with or without mass selection, followed by single-plant selection.

Bulking: Combining several samples in a collection and managing them as a single accession on the basis of similarity in phenotypic characters, habitat, geographic locations, *etc.*

Callus: A least differentiated, unorganized, proliferative mass of cells usually formed as a wound response.

Carrier: A heterozygous individual not expressing a recessive trait but capable of passing it on to its offspring.

Cell suspension: Cells or aggregate of cells which multiply while suspended in liquid medium.

Centre of diversity: A geographical area where a plant species first developed its distinctive properties (in farmer's field or in the wild). A primary centre of diversity is the region of true origin (often referred to as the centre of origin), and secondary centres of diversity are regions of subsequent spread of a crop.

Certified seed: Seed used for commercial crop production produced from foundation, registered, or certified seed under regulation of a legally constituted agency.

Character: An attribute of an organism resulting from the interaction of a gene or genes with the environment.

Characterization: Recording of highly heritable qualitative traits which are easily observed and expressed in all environments; helpful for distinguishing accessions from each other. Standard descriptors and descriptor states are normally used for description of traits.

Chromosomes: Structural units of the nucleus which carry the genes in linear order. Chromosomes undergo a typical cycle in which their morphology changes drastically in various phases of the life cycle of the organisms.

Clonal propagule: The vegetative plant parts such as cuttings, grafts, or tissue culture that can be collected in plants for raising a new plant.

Clone: A group of genetically identical individuals that result from asexual, vegetative multiplication, and are, thereof, a genetic duplicate (true-to-type) of its parent or A group of organisms descended by mitosis from a common ancestor.

Coarse grid survey: The preliminary survey and collecting wherein sampling is made at wide intervals (25-50 km, depending upon the terrain), with a few samples taken from each site (*vs.* fine grid survey)

Collecting: Includes identifying, locating, acquiring, organizing, cataloguing and post-collection handling the Germplasm of importance as PGR.

Collection: A group of ermplasm accessions maintained under defined conditions for a specific purpose.

Collector: A specialist who explore, survey and collects ermplasm in the form of plans, seeds, propagules, *etc.* and records related information on diversity distribution, use, environmental features, *etc.*

Combining ability: General, average performance of a strain in a series of crosses. Specific deviation from from performance predicted on the basis of the general Combining ability.

Conservation: The practice that permits the perpetuation of a resource. Seed conservation refers to the practice of protecting the abundance of biological diversity of plant genes and seeds from loss and extinction, and ensuring the preservation of those genetic resources and seeds for the use of future generations.

Contaminant: Soil, fungal spores, fruiting bodies, plant debris, live/dead/dormant insects/stages thereof.

Contamination: Presence in a commodity, storage place, conveyance or container, of pests or other regulated articles, not constituting an infestation.

Convention on Biological Diversity: An international convention adopted in June, 1992 during the United Nations Conference on Environment and Development held in Rio de Janeiro, Brazil. According to Article 1 of the CBD, the Convention aims at "*the conservation of biological diversity the sustainable use of its components and the fair and equitable sharing of the benefits arising out of the utilization of genetic resources, including by appropriate access to genetic resources and by appropriate transfer of relevant technologies, taking into account all right over those resources and to technologies, and by appropriate funding*".

Core set: A subset of accessions (~10 per cent) from the entire collection that captures most of the available diversity in the species which would represent with minimum of repetitiveness, the genetic diversity of a crop and its wild relatives.

Country of origin (of a consignment of plant): Country from where the plants were received.

Critical difference (CD): The difference between effects of different treatments and used to explain the variation due to treatment effect.

Crop-specific survey: Survey to collect maximum observable diversity in a single crop occurring in the region to be explored.

Cross-pollination: The process of pollen transfer from the anthers of one flower to the stigma of another flower by different means.

Cryopreservation or cryostorage: Storage of cells, tissues, organs or organisms, in a viable condition, at ultra-low temperatures usually below -100 °C. in the genebank context, it is used for storage of buds, shoots tips, other meristematic and embryogenic tissue, non-orthodox seeds or their explants, orthodox seeds (in special cases), pollen and somatic embryos, in liquid nitrogen (-196 °C) or its vapour phase.

Cultivar: A variety defined by a set of common features (morphological, phonological, *etc.*) of a plant that has been created or selected intentionally and maintained through cultivation.

Database: An organized set of interrelated data assembled for a specific purpose and held in one or more storage media.

Descriptor: An easily identifiable and measurable trait or characteristic of a plant observed in an accession, which is used to facilitate data classification, storage, retrieval and use.

Descriptor list: A collection of all individual descriptors of a particular crop or species.

Description: The process whereby a substance is released from or through the surface of another substance.

Diallel cross: The crossing in all possible combinations of a series of genotypes.

Dihybrid: Heterozygous with respect to two genes.

Dihybrid cross: A genetic cross that takes into account the effect of alleles at two separate loci (two different genes).

Dioecious: Plants in which staminate and pistillate flowers occur on different individuals.

Disease: A departure from normal metabolism and a reduction of its normal potential for growth and reproduction.

Distribution: The process of supplying samples of germplasm accessions to breeders, researchers, curators and other bonafide users.

Documentation: Presentation of record data on Germplasm accessions in a standard format that describes structure, purpose, operation, maintenance and other requirements.

Dominance: Intra-allelic interaction such that one allele manifests itself more or less, when heterozygous, than its alternative allele.

Dominant allele: An allele that is always expressed when present, regardless of whether the organism is homozygous or heterozygous for that gene.

Donor: An institution or individual responsible for donating germplasm.

Donar parent: The parent from which one or a few genes are transferred to the recurrent parent I backcross breeding.

Domestication: The process by which plants, animals or microbes selected from the wild adapt to a special habit created for them by humans, bringing a wild species under human management. In a genetic context, the process in which changes in gene frequencies and performance arise from a new set of selection pressures exerted on a population.

Dormancy: The state in which certain live seeds do not germinate, even under optimal suitable conditions.

Double cross: A cross between two F_1 hybrids.

Ecosystem: A dynamic complex of plant, animal and micro-organism communities and their abiotic environment interacting as a function unit.

Ecotype: A population of a species that survives as a distinct group through environmental selection and isolation, maintains its genetic identity and usually is linked to one or more environmental factors in which it is favoured (mostly with reference to wild species).

Emasculation: Removal of the anthers from a flower.

Endangered: When used in the context of the IUCN Red List, a taxon is classified as 'Endangered' when there is very high risk of extinction in the wild, in the immediate future.

Endemic: Native to, and restricted to, a particular geographical region. Highly endemic species, those with very restricted natural ranges, or especially vulnerable to extinction if their natural habitat is eliminated or significantly disturbed.

Epistasis: Dominance of one gene over a non-allelic gene. The gene suppressed is said to be hypostatic. More generally, the term epistasis is used to describe all types of interallelic interaction whereby manifestation at any locus is affected by genetic phase at any or all loci.

Epiphytotic: An unarrested spread of a plant disease.

Equilibrium moisture content: The moisture content at which seed is in equilibrium with the relative humidity of the surrounding air.

Eradication: Application of phytosanitary measures to eliminate a pest from an area.

Ethnobotany: All studies (containing plants) which describe local people's interaction with the natural environment.

Evaluation: The recording of quantitative traits whose expression is often influenced by environmental factors; it provides an assessment of the potential of Germplasm for use in breeding/research.

Expressivity: The degree of manifestation of a genetic character.

***Ex situ* conservation:** The conservation of biological diversity outside its natural habitat. In case of PGR, it may compare facilities such as botanical gardens, seed genebanks, *in vitro* genebanks, cryo genebanks, field genebanks *etc.*

Exotic accession: Accession introduced or imported from a foreign country.

Exotic species: A species which is not native to the region in which it occurs.

Explant: An excised plant part implanted on a medium for growth or maintenance.

Extinct: The taxa (*e.g.* population, subspecies and species) not found even after repeated searches of known and likely areas.

Fecundity: Actual reproductive rate of an organism or population.

Field: A plot of land with defined boundaries within a place of production on which plants are grown.

Filial generation: The generation that receives gametes from the parental generation.

F_1: The first generation of a cross.

F_2: The second filial generation obtained by self-fertilization or crossing F_1 individuals.

F_3: Progeny obtained by self-fertilization of F_2 individuals.

Facultative: Parasites which can grow and live in environment other than living host tissue.

Family: A group of individuals directly related by descent from a common ancestor.

Fertility: Ability to produce viable offspring.

Fertilization: Fusion of the nuclei of male and female gametes.

Fine grid survey: An intensive sampling at sites (10-20 km) for survey and collecting (where earlier coarse grid survey has been done).

Foundation seed: Seed stock produced from breeder seed under the direct control of an Agricultural experiment station. Foundation seed is the source of certified seed, either directly or through registered seed.

Fumigation: Treatment with a chemical agent that reaches the plant or area wholly or primarily in a gaseous state.

Gene: The basic unit of heredity transmitted from generation to generation during sexual or asexual reproduction. Genetically, the term is used in relation to the transmission and inheritance of particular identifiable trait.

Gene flow: The movement of genes by pollen (dispersal of gametes), seeds (through zygote) and plants from one population to another. The exchange of genes (in one or both directions) at a low rate between two populations, due to the dispersal of gametes or individuals from one population to another, also called migration.

Genebank: Is a facility where genetic resources (genetic material) are conserved under suitable conditions to prolong their lives.

Gene frequency: The proportion in which alternative alleles of a gene occur in a population.

Genepool: The sum total of all the genes and combinations of the genes that occur in a population of organisms of the same species. The term genepool in reference to wild relatives and related taxa is classified into primary, secondary, tertiary depending upon crossability.

Genetic conservation: The collection, maintenance, storage and sustainable management of genetic resources aimed at ensuring their continued existence, evolution and availability for future generations. Also referred as 'gene conservation' and 'genepool conservation'.

Genetic diversity: The genetic variability (variety of genetic traits) within a population or a species, arising due to number and relative abundance of alleles. It can be assessed at three levels:(a): Diversity within breeding populations(b): Diversity between breeding populations(c): Diversity within the speciesGenetic diversity occurs at gene (molecular), individual, population, species and ecosystem levels.

Genetic drift: Random change in allele frequencies in a population from one generation to the next because of small population size, and not due to selection, migration or mutation. Smaller the population, greater the genetic drift, with the result that some alleles are lost, and genetic diversity is reduced. Minimization genetic drift is an important consideration for conservation of genetic resources.

Gene interaction: Modification of gene action by a non-allelic gene or genes.

Genetic material: Any material of plant, animal, microbial or other origin containing functional units of heredity.

Genetic resource: Genetic material of actual or potential economic, scientific or social value contained within and among species. In a domesticated species, it is the sum of all the genetic combinations produced in the process of evolution. The term includes modern cultivars and breeds, traditional cultivars and breeds, special genetic stocks (breeding lines, mutants, *etc.*), wild relatives of domesticated species and genetic variants of wild species.

Genetic resource for food and agriculture: A genetic material of actual or potential value for food and agriculture.

Genetic shift: Change in the performance of an accession/genotype if grown over a long period in areas outside its adaption.

Genetic stocks: Plants or populations generated and/or selected for genetic studies and represent a unique class of extremely valuable Germplasm.

Genetic variation: The occurrence of differences among individuals of the same species, arising due to variation in alleles, genes or genotypes. Genetic variation is brought about by a change in genes, as distinct from difference due to environmental factors.

Genome: All the genetic material (DNA sequences) in a single (haploid) set of chromosomes of an organism. The genetic material inherited from either parent.

Genomics: Study of genomes of organisms, including intensive efforts to determine the entire DNA sequence and genetic mapping.

Genotype: The genetic constitution of an individual plant or organism as distinguished from its appearance or phenotype. The genotype interacts with the environment to produce the phenotype.

Genotyping: The process of identifying the genetic make-up of an organism by using molecular methods (DNA sequence level).

Germination: The biological process that leads to the emergence and development of a seedling from a seed, to a stage where the aspect of its essential structure indicates whether it is able to develop further into a satisfactory plant under favourable conditions. According to ISTA rules, only seedlings showing normal morphology are considered to have germinated.

Germplasm: The genetic material which forms the physical basis of heredity and which is transmitted from one generation to the next by means of germ cells. Often synonymous with genetic material.

When applied to PGR, it includes plants in whole or in parts and their propagules including seeds, vegetative parts, tissue cultures, cell cultures, genes and DNA-based sequences that are held in a repository or collected from wild, and are utilized in genetic studies or breeding programmes for crop improvement.

Genome: A set of chromosomes corresponding to the haploid set of a species.

Germplasm exchange: Mutual give and take of germplasm or plant genetic resources from all available.

Glasshouse: A glass building used for growing plants under controlled environmental conditions.

Grain: A commodity class for seeds intended for processing or consumption and not for planting.

G x E interaction: Part of phenotypic variation which is the result of interaction between genotype and environment.

Habitat: Part of an ecosystem with conditions in which an organism naturally occurs or can establish.

Herbarium: Herbarium is a collection of dried plant specimens identified and arranged using a standard system of classification. All parts of the plants (vegetative and reproductive) are represented in a herbarium specimen; information on locality habitat used *etc.* is recorded on the herbarium label.

Heritability: The proportion of observed variability which is due to heredity, the remainder being due to environmental causes. More strictly, the proportion of observed variability due to the additive effects of genes.

Heterosis: Hybrid vigour such that an F_1 hybrid falls outside the range of parents with respect to some character or characters. Usually applied to size, rate of growth, or general thriftiness.

Heterozygosity: Dissimilar allelic combination that provides genetic reserves and potential plasticity and permits a large proportion of the individuals to exhibit combination of phenotypic properties near the optimum.

Heterozygous: Having unlike alleles one or more corresponding loci (opposite of homozygous).

Homologous chromosomes: Chromosomes those are similar in morphology (shape and form) and genetic constitution.

Homology chromosomes: Applied to whole chromosomes or parts of chromosomes which synapse or pair in meiotic prophase.

Homozygous: Possessing a pair of identical alleles for a particular locus (gene).

Host resistance: The result of genetic manipulation of the host which renders it less susceptible to pathogen that would or do attack the host.

Hygroscopic: Tendency to absorb water in the form of water vapour.

Hybrid: The product of a cross between genetically unlike parents.

Import permit: An official document authorizing importation of a commodity/resource in accordance with specified phytosanitary requirements, into the country from outside.

Inbred line: A line produced by continued inbreeding. In plant breeding, a nearly homozygous line usually originating by continued self-fertilization, accompanied by selection.

Inbreeding: Inbreeding is defined as the production of offspring through mating between related individuals.

Inbreeding depression: Loss in vigour due to mating among closely related individuals.

Independence: The relationship between variables when the variation of each is uninfluenced by that of others, that is, correlation is zero.

Indigenous knowledge: The local knowledge, unique to a given culture or society.

Indigenous species: A species that is assumed to be intrinsically part of the ecosystem of ones country, owing to having developed or arrived in the area long before record of such matters was kept, having arrived by natural means (unaided by human action) *etc.*

Infestation: Presence of a living pest of the plant or plant product concerned. Infestation includes infection.

***In situ* conservation:** The conservation of ecosystems and natural habitats and the maintenance and recovery of viable populations of species in their natural surroundings and, in the case of domesticated or cultivated species, in the surroundings where they have developed their distinctive properties. Until recently, it was narrowly used to describe conservation of genetic resources in their natural surroundings, normally protected from human interference. However, it is increasingly used to designate conservation on the farm, where genetic resources are developed, bred and maintained.

Isogenic lines: Two or more lines differing from each other genetically at one locus only. Distinguished from clones, homozygous lines, identical twins, *etc.* which are identical at all loci.

Isolation: The separation of one group from another so that the mating between or among groups is prevented.

Intellectual property rights (IPR): Legal rights that are conferred to the owner of an intellectual creation. The IPR are granted by means of protection through appropriate legislation based on the type of creation that generally include patents, copyright, trademark, industrial design, geographical indications, trade secrets, protection of layout design of integrated circuits and protection of new plant varieties. The IPR protection entitles the owner or his assignee the exclusive right to fully utilize the invention/ creation for commercial gain generally for a fixed period of time.

Interception (of a pest): The detection of a pest during inspection or testing of an imported consignment.

Intermediate seeds: A category of non-orthodox seeds that are more tolerant of desiccation than recalcitrant seeds, though that tolerance is much more limited as compared to orthodox seeds, and they generally lose viability more rapidly at low temperature. Seed viability is influenced by initial seed quality, stage of maturation, and initial exposure after collection to conditions favorable for germination. Seeds can be stored in liquid nitrogen for long-term, with specialized handling and processing.

International Plant Protection Convention (IPPC): A multilateral treaty overseen by FAO to secure coordinated action to prevent and control the introduction and spread of pest of plants and plant product.

International Standard for Phytosanitary Measures (ISPM): Adopted by the members of IPPC, comprise standards, guidelines and recommendations for applying phytosanitary measures by WTO members under the Agreement on application of Sanitary and Phytosanitary Measures.

International Treaty on plant genetic Resources for Food and Agriculture: Popularly known as The Treaty, is a comprehensive international agreement in harmony with Convention on Biological Diversity which aims at guaranteeing food security through the conservation exchange and sustainable use of the world's PGRFA as well as the fair and equitable benefit sharing arising from its use. It has a Multilateral System of access and benefit sharing among those countries that ratify the treaty for food and forage crops listed in Annex 1 to The Treaty.

***In vitro* active genebank:** Represents *in vitro* germplasm collections under slow growth (by imposing growth restrictive conditions) to enhance subculture duration. The conservation is thus cost-effective and involves cyclic flow of material following subculture and recovery and facilitating distribution of cultures to users.

***In vitro* base genebank:** Represents *in vitro* germplasm collections under suspended growth (under cryopreservation) for long-term and the germplasm is normally not distributed to users.

***In vitro* collecting:** Plant where seeds or cuttings are not available to a collector or transport is not practical tissue collected using standard *in vitro* methods for obtaining and transporting ermplasm. This methodology is generally practiced for collecting wild endangered species for propagation, tissue banking particularly for species with recalcitrant seeds or for species making a few or no seeds.

***In vitro* conservation:** Maintenance of germplasm in a relatively stable form under more or less defined nutrient conditions in an artificial environment.

***In vitro* culture:** The cultivation of plant organs or entire plants on artificial nutrient medium in glass or plastic containers. *In vitro* cultures are used for micropropagation, virus elimination via meristem culture and slow-growth storage. *In vitro* culture is also used as a preparatory phase to cryopreservation as well as for recovery phases after cryopreservation.

***In vitro* genebank:** A genebank involving conservation of vegetatively propagated genetic resources under tissue culture conditions involving normal growth, slow growth or suspended growth (under cryopreservation).

Laminated aluminum foil packets: Packets constructed of a laminate consisting of an inner layer of polyethylene, a middle layer of aluminum foil and an inner layer of polyester.

Landrace: A crop cultivar or animal breed that evolved with and has been genetically improved by traditional agriculturalists, but has not been influenced by modern breeding practices. It comprises population of individuals that have become adapted to specific local environmental conditions in which it evolved and is usually genetically heterogeneous.

Line breeding: A system of breeding in which a number of genotypes, which have been progeny tested in retrospect to some character or group of characters, are composited to form a variety.

Linkage: Association of characters in inheritance due to location of gemnes in proximate on the same chromosome.

Linkage map: Map of position of genes in chromosomes determined by recombination relationships.

Linkage value: Recombination fraction expressing the proportion of crossovers versus parental types in a progeny. The recombination fraction can vary from zero to one half.

Liquid nitrogen: It is the liquid produced industrially by fractional distillation of liquid air. It boils at -196°C and freezes at -210 °C. Being a cryogenic fluid it causes rapid frost-bites and when insulated from ambient heat it can be stored and transported in a vacuum flask.

Locus: The position occupied by a gene in a chromosome.

Long-term storage/conservation: The storage of Germplasm for a long period, such as in case collections and duplicate collections. The period of storage before seeds need to be regenerated varies, but is at least several decades and possibly a century or more. Long-term conservation takes place at sub-zero temperatures.

M_1, M_2, M_3—: Symbols used to designate first, second, third, *etc.* generations after treatment with a mutagenic agent.

Male sterility: Absence or non-function of pollen in plants.

Mass pedigree method: A system of breeding in which a population is propagated in mass generation is propagated from the aggregate of their seeds.

Material Transfer Agreement (MTA): A contractual arrangement (can legally binding) in commercial and academic research partnerships involving the transfer of tangible biological materials (such as germplasm, microorganisms and cell cultures) from a provider to a recipient. The MTA sets conditions for access, use, commercialization and sharing of benefits derived from the use of the materials provided.

Mating system: Any number of schemes by which individuals are associated in pairs leading to sexual reproduction. Random; assortment of pairs is by chance. Genetic assertive mating; mating together of individuals more closely related than individuals mating at random. Genetic disassortative mating; mating together of individuals less closely related than individuals mating at random. Phenotypic assortative mating; mating individuals more like in appearance than the average. Phenotypic disassortive mating; mating of individuals less like in appearance than individuals mating at random.

Medium-term storage/conservation: The storage of Germplasm in the medium-term such as in active and working collection, it is generally assumed that little loss of viability will occur for approximately ten years. Medium-term storage/conservation takes place at temperatures between 0°C and 10°C.

Mini core set: A subset of germplasm accessions comprising 10 per cent of the core set representing the total variability in the core set.

Modifying genes: Genes that affect the expression of a non-allelic gene or genes.

Moisture content (wet-weight basis): The weight of free moisture divided by the weight of water plus dry matter, expressed as a percentage.

Molecular markers: Specific fragments of DNA that can be identified within the whole genome and are used for measuring genetic variation, trait identification, fingerprinting *etc.*

Monitoring: The periodic checking of accessions for viability and quantity, to ascertain whether to continue storing the accessions or to produce a fresh stock of seeds by regenerating or replacement from other source.

Monitoring interval: The period of storage between monitoring tests.

Monoecious: Staminate and pistillate flowers born separately on the same plant.

Monohybrid cross: A genetic cross that takes into account the effect of alleles at a sngle locus (single gene).

Most original sample (MOS): A sample of seeds that have undergone the lowest number of regeneration since the material was acquired by a genebank, as recommended for storage as a base collection. it may be a sub-sample of the original seed lot or a seed sample from the first regeneration cycle if the original seed lot required regeneration before storage.

Multilateral system: Established under the ITPGRFA for efficient, effective and transparent access to PGRFA and to share in a fair and equitable way the benefits arising from utilization of these resources on a complementary and mutually reinforcing basis.

Multiplication: The representative sample of an accession grown to multiply the quantitative of conserved material for distribution.

Mutation: A sudden heritable variation in a gene or in a chromosome structure.

Mutually agreed terms: Terms used for access to genetic resources; involves two or more parties at least one party will be supplying the genetic resource and other acquiring them. The terms under which the recipient acquires genetic resources are likely to affect its freedom to supply the same resources or their derivatives to a third party.

National identity: The unique national identity number generated in the centralized database of ICAR-NBPGR, prefixed letter IC or EC denotes indigenous collection and exotic collection, respectively. It serves as the identification number in accession table of the centralized database.

Native species: A species which is a part of the original flora of the area, occurring naturally in the given area or region.

Novel and unique trait: Novel unique distinct and stable trait(s) which has been the basis for registration of germplasm.

Non-orthodox seeds: Seeds which are not desiccation-tolerant, do not undergo maturation drying and are shed at water content in the range of 0.3-4.0 g g^{-1}. Any loss of water rapidly results in decreased vigour and viability, and seeds die at relatively high water contents. Non-orthodox seeds comprise two categories-recalcitrant and intermediate. Seeds do not survive drying to any large degree, and are thus not amenable to long-term storage (usually not more than few months), although the critical moisture level for survival varies among species.

Obligate: Parasite that cannot multiply in nature without a host.

Obligate resistance: Resistance determined by one or few genes whose effects are readily detectable.

Orthodox seeds: Seeds that can be dried to low moisture content and stored at low temperatures (without damage) to increase seed longevity.

Outcrossing: Controlled or natural mating among unrelated individuals of sexually propagated species.

Parasite: Lives in or on another organism and obtains nutrients from it.

Parent generation (P): The generation that supplies gametes to the filial generation.

Parthenogenesis: Development of an organism from a sex cell in respect to some characteristic.

Parameter: A numerical quantity which specifies a population in respect to some characteristic.

Passport data: Basic information about the origin of an accession, such as details recorded at the collecting site, pedigree or other relevant information that assists in the identification of an accession.

Pathogen: A living microorganism such as a virus, bacterium or fungus that causes disease in another organism.

Pathway: Any means that allows the entry or spread of a pest.

Pedigree: The record of the ancestry of a genetic line or variety.

Pedigree breeding: A system of breeding in which individual plants are selected in the segregating generations from a cross on the basis of their desirability judged individually and on the basis of a pedigree record.

Penetrance: The frequency with which a gene produces a recognizable effect in the individuals which carry it.

Pest: Any species, strain or biotype of plant, animal or pathogen regarded as injurious or harmful (*e.g.* weeds, insects, nematodes, fungi, bacteria, viruses).

Pest risk analysis: The process of evaluating biological or other scientific and economic evidence to determine whether a pest should be regulated and the strength of any phytosanitary measures to be taken against it.

Phenophases: An observable stage or phase in the life cycle of a plant that can be defined by start and end point. Phenophases generally have duration of a few days or weeks.

Phenotype: The external appearance of an organism, that results from the interaction of its genetic composition (genotype) with the environment.

Phenotyping: The method of determining differences in the appearance of an individual at phenotypic level.

Phytosanitary certificate: An official paper document or its official electronic equivalent, issued by an authorized officer at the country of origin of consignment or re-export, consistent with the model certificates of the IPPC, attesting that a consignment meets phytosanitary import requirements.

Phytosanitary measures: Any legislation, regulation or official procedure having the purpose to prevent the introduction and/or spread of quarantine pests, or to limit the economic impact of regulated non-quarantine pests.

Phytotron: An enclosed research greenhouse used for studying interaction between plants and the atmosphere.

Phytoxins: Substances produced or formed by host plants in response to injury, physiological stimuli, infectious agents, or their products that accumulate to levels which inhibit the growth of microorganisms. Some include toxic substances produced to repel insects and nematodes.

Plant debris: Dried plant parts or pieces thereof other than seed.

Plant exploration: Plant exploration is the survey and search for plants with specific aim and intention.

Plant genetic resources: Genetic material of plants of value for present and future generations of people. In crops, it is the sum of all the genetic combinations produced in the process of evolution and breeding. The term includes modern cultivars, traditional cultivars, landraces, special genetic stocks (breeding lines, mutants, *etc.*), wild relatives of domesticated plants and genetic variants of wild species.

Plantlets: Rooted shoots derived from plant parts other than seed.

Plant for planting: Plants intended to remain planted to be planted or replanted.

Planting (including replanting): Any operation for the placing of plants in a growing medium, or by grafting or similar operations, to ensure their subsequent growth, reproduction or propagation.

Plant quarantine: All activities designed to prevent the introduction and/or spread of quarantine pests, or to ensure their official control.

Pollination: The process in which pollen is transferred from an anther to a respective stigma by pollinating agents such as wind, insects, birds, bats, or the opening of the flower itself.

Poly cross: Open pollination of a group of genotypes (generally selected), in isolation from other compatible genotypes, in such a way as to promote random mating.

Polygenic: Determined by several genes whose effects are readily detectable.

Population: In general-a group of individual plants or animals that share a geographic area or region and have common traits. In genetics-a community of individuals which share a common gene pool. In statistics- a hypothetical and infinitely large series of potential observations among which observations may actually constitute a sample.

Post-entry quarantine: Quarantine applied to a consignment after entry.

Primitive cultivar: An early cultivated form of a crop species evolved from a wild population.

Prior informed consent: Recognizes indigenous people's inherent and prior rights to their lands and resources and respects their legitimate authority to require that third parties enter into an equal and respectful relationship with them based on the principle of informed consent.

Progeny test: A test of the value of a genotype based on the performance of its offspring produced in some definite system of mating.

Propagule: Any structure with the capacity to give rise to a new plant, whether through sexual or asexual (vegetative) reproduction. This includes seeds, spores, and any part of the vegetative body capable of independent growth if detached from the parent.

Protected areas: A clearly defined geographical space recognized, dedicated and managed through legal or other effective means to achieve the long-term conservation of nature with associated ecosystem services and cultural values.

Protoandry: Maturation of anthers before pistils.

Protogyny: Maturation of pistils before anthers.

Providers and recipients: Many include the government sector [*e.g.* government ministers, government agencies (National, regional or local), including those responsible for administration of national parks and government land]; commerce or industry (*e.g.* pharmaceutical, food and agriculture, horticulture, and cosmetics enterprise); research institutions (*e.g.* universities, genebanks, botanic gardens, microbial collections); custodians of genetic resources and traditional knowledge holders (*e.g.* Associates of healers, indigenous peoples or local communities, peoples' organizations, traditional farming communities); and other [*e.g.* private land owners(s), conservation group(s) *etc.*]

Pure line: A strain homozygous at all loci, ordinarily obtained by successive self-fertilization in plant breeding.

Qualitative character: A character in which variation is discontinuous.

Quantitative character: A character in which variation is continuous so that classification into discrete categories is not possible.

Quarantine: Official confinement of regulated articles (introduced germplasm) for observation and research or for further inspection, testing or treatment to ensure that it does not carry diseases or pests injurious to the importing country.

Quarantine pest: A pest of potential economic importance to the area endangered thereby and not yet present there, or present but not widely distributed and being officially controlled.

Quarantine inspection: Official examination of plants, plant products or other regulated articles to determine if pests are present or to determine compliance with phytosanitary regulations.

Quarantine interception (of a consignment): The refusal or controlled entry of an imported consignment due to failure to comply with phytosanitary regulations.

Random: Arrived at by chance without discrimination.

Randomization: Process of making assignments at random.

Rare: Taxa with few plants/small populations that are not at present endangered but are at risk of loss.

Recalcitrant seeds: Seeds that are not desiccation-tolerant; which do dry during physiological maturity and are shed at water contents in the range of 0.3-4.0 g g^{-1}. The loss of water rapidly results in decreased vigour and viability, and seed death at relatively high water contents, although the critical moisture level for survival varies among species. Hence, they are usually not amenable to long-term storage.

Recessive: The member of an allelic pair which is not expressed when the other (dominant) member occupies the homologous chromosome.

Recessive allele: An allele that is only expressed when the organism is homozygous for that allele and not expressed when heterozygous (when paired with a dominant allele).

Reciprocal crosses: Cross in which the sources of the male and female gametes are reversed.

Recombination: Formation of new combinations of genes as a result of segregation in crosses between genetically different parents. Also, the rearrangement of linked genes due to crossing over.

Recurrent parent: The parent to which successive backcrosses are made in backcross breeding.

Recurrent selection: A method of breeding designed to concentrate favourable genes scattered among a number of individuals by selecting, each generation, among the progeny produced by mating of the selected individuals (or their selfed progeny) of the previous generation.

Reference set: An ideal set of best promising accessions for a given trait used for allele mining, association genetics mapping and cloning gene(s).

Regeneration: Grow-out test of a seed accession to obtain a fresh sample with high viability and numerous seeds. It is renewal of accession such as that it possess the same characteristics as the original sample and is a very critical operation in genebank management.

Regeneration standard: The percentage seed viability at or below which the accession must be regenerated to produce fresh seeds.

Registered seed: The progeny of foundation seed normally grown to produce certified seed.

Relative humidity: The ratio of the amount of water vapor in the air at a specific temperature to the maximum amount that the air could hold at the temperature expressed as a per cent. It differs from absolute humidity, which is the amount of water vapour present in a unit volume of air, usually expressed in kilograms per cubic meter.

Repatriation of germplasm: Restoration of germplasm to the source countries.

Rescue collecting: Germplasm collecting in an area where genetic diversity available is under threats or loss due to natural (floods, earthquakes, tsunami *etc.*) or manmade (construction of dams, barriers *etc.*) factors.

Rogue: A variation from the standard type of a variety or strain. Roguing-removal of undesirable individuals to purify a stock.

Resistance: The restriction of development of a pathogenic agent or parasite. Can vary in degree from immunity (no development) to only slight retardation relative to a so called susceptible reaction.

S_1, S_2m, S_3—: Symbols for designating first, second, third, *etc.* selfed generations from an ancestral plant (SO)

Safety duplicates: A duplicate of a base collection stored under similar conditions for long-term conservation, but at a different location to insure against accidental loss of material from the base collection.

Sample: A part of a population used to estimates the characteristics of the whole.

Sampling: Drawing a part of the whole lot *i.e.* a few individuals out of a population mainly with an assumption that a sample represents the diversity available in the source population.

Seed dormancy: Block to the completion of germination of an intact viable seed under favourable conditions.

Seed-borne: The pest present on, in or along with a seed.

Seed-transmitted: The pest present in or with a seed and transmitted to the next generation of growing seedlings.

Seed viability: The capacity of seeds to germinate under favourable conditions.

Segregation: Separation of parental from maternal chromosomes at meiosis and consequent separation of genes relating to the possibility of recombination in the offspring.

Selection: Any process, natural or artificial, which permits a change in the genetic structure of population in succeeding generations. Natural selection is the differential reproduction and survival of individuals, such that poorly adapted offspring do not continue in the population. Artificial selection occurs when human manipulation yield specific traits deemed desirable.

Self-pollination: The process of pollen transfer from anthers to stigma of the same flower.

Self-fertilization: Fusion of male and female gametes from the same individual.

Self-incompatibility: Genetically controlled physiological hindrance to Self-fertilization.

Shoot meristem: An undifferentiated mass of cells as a shiny dome like structure within the shoot tip, usually measuring less than 0.1 mm in diameter.

Shoot tip: Comprises shoot apical meristem along with one to several primordial leaves generally measuring 0.1 to 1.0 mm in diameter.

Single cross: A cross between two genotypes, usually two inbred lines, in plant breeding.

Slow growth: Refers to reducing growth rate of plant tissues without detrimental effects on them.

Somatic embryogenesis: Origin of plantlets from somatic cells by a developmental pathway similar to that of zygotic embryogenesis.

Somatic embryo: Embryos developed from somatic cells by a developmental pathway similar to that of zygotic embryos.

Species: A population of organisms. The unit of taxonomic classification into which genera are subdivided. A group of similar individuals different from other similar arrays of individuals. In sexually reproducing orghanisms, the maximum interbred group isolated from other species by barrieers of sterility or reproducing incapacity.

Species diversities: Species represented as a measure of the diversity of the area.

Standard Material Transfer: The FAO has developed and adopted in 2006 a Standard Material Transfer Agreement (SMTA) as required for the implementation of the ITPGRFA. It specifies the conditions for exchange of germplasm between signatories (parties) with members of the ITPGRFA.

Storage life: The numbers of years that a seed/ propagule can be stored before death occurs.

Strain: A group of similar individuals within a variety

Subculture: A process which involves cutting the tissue or explant into smaller segments (propagules) and transferring them to fresh culture medium.

Synthetic variety: A variety produced by crossing a number of genotypes selected for good combining ability in all possible hybrid combinations, with subsequent maintenance of the variety by open pollination.

Taxon (plural of taxa): A formal taxonomic unit or category at any level in a classification (family, genus, species *etc.*)

Taxonomy: Classification of plants in hierarchical groups called taxa (singular taxon).

Threatened species: Any species which is likely to become endangered within the near future, all throughout or a significant portion of its range.

Tissue culture medium: A solid or liquid combination of nutrients and water which meet the requirements of cells tissue or organs grown *in vitro.*

Top cross: A cross between a selection, line, clone, *etc.*, and a common pollen parent which may be a variety, inbred line, single cross, *etc.* The common pollen parent is called the top cross or tester parent. In corn, a top cross is commonly an inbred-variety cross.

Trait: A recognizable quality or attributes resulting from interaction of a gene or a group of genes with the environment.

Variation: The occurrence of differences among individuals due to differences in their genetic composition and/or the environment in which they were raised.

Variety: A recognized division of a species, next in rank below sub-species; it is distinguishable from another variety by a set of heritable characters (such as flower colour, leaf colour etc.). The term is considered synonymous with cultivar.

Viability: A test on a sample of seeds from an accession that is designed to estimate the viability of the entire accession.

Virulence: Capacity of a pathogen to incite a disease.

Voucher: Plant material such as herbarium specimen, seed sample, and economic product preserved for future reference.

Vulnerable: When used in the context of the IUCN Red List a taxon is classified as 'Vulnerable' when facing a high risk of extinction in the wild in the immediate future.

Weed: A plant that is growing where it is not 'wanted' generally used to describe plants which colonize readily and compete for resources with a crop.

Wild relatives: Plant species that are taxonomically related to crop species and serve as potential sources for genes in breeding of new variety of those crops.

Wild species: Species occurring in relatively undisturbed or less disturbed habitat and areas that have not been subject to breeding to alter them from their native (wild) state.

X1, X2, X3—-: Symbols denoting first, second, third, etc. generations from irradiated ancestral plants (XO).

Source: ICAR-NBPGR, 2016 (eds.); and Bardwaj, 2012 (ed.)

REFERENCES

1. ICAR-NBPGR. 2016. Guidelines for Management of Plant Genetic Resources "Glossary". ICAR-National Bureau of Plant Genetic Resources, New Delhi, India. 142+xxivp.
2. Bharadwaj, D.N. 2012. Breeding of Field Crops "Glossary". Published by AGROBIOS (India).

References

1. Annappan, R.S., Sivasubramanian, P., Rangaswamy, P., Sundrapandian, G., Ramanathan, T. and Iyemperumal, S. 1989. CO 1 Mochai-A high yielding field lablab. *Madras Agricultural Journal*, **76:** 640-642.
2. Anonymous, 1998. Varieties and hybrids. Division of Vegetable Crops, ICAR-IARI, New Delhi.
3. Anonymous, 2006b. Dorset claims world's hottest chilli. Daily Telegraph, 2006-04-01.
4. Anonymous, 20062c. The chilli so hot you need gloves. The Times, 2006-04-01.
5. Bhardwaj, M.L. and Sharma, P.P. 1993. Evaluation of tomato genotypes for cold tolerance. Golden Jubilee Sym. On Hort. Res. Changing Scenario, held on May 24-28 at ICAR-IIHR, Bengaluru (Karnataka), India.
6. Chadha, M.L. and Sidhu, A.S. 1987. Punjab Barsati, Variety of brinjal. *Indian J. Hort.*, **40(1 and 2):** 60-64.
7. Chandra, A. and Pandey, R.P. 2001. Jawahar Mirch-283: A summer-set chilli. *Indian Hort.*, **46(1):** 20-21.
8. Chandra, A., Singh, K., Nair, P.K.R. and Bhalle, U. 1987. Jawahar Mirh-218: A new bright red pungent chilli. *Indian Hort.*, **31(4):** 7-8.
9. Choudhury, B. 1967. Vegetable. National Book Trust of India, New Delhi.
10. Choudhury, B. and Ramphal. 1975. *Indian Hort.*, **19:** 25-27.
11. Choudhury, B., Sirohi, P.S. and Sharma, J.C. 1976. 'Pusa Sweti'- An early maturing tropical turnip. *Indian Hort*. **21:** 7.
12. Choudhury, B. 1972. Vegetables, National Book Trust, India, New Delhi, p. 14.
13. Choudhury, B. and Tiwari, R.N. 1993. Tomato 'Pusa Gaurav': an all purpose variety. *Indian Hort*. **37(4): 14-15**.
14. Ellison, J.H., Kinelski, J.J. and Garrison, S.A. 1989. *HortSci*. **37:** 868.
15. Geriya, R.F. 1973. *Plant Breed. Abstr*. **44(4):** 2812.
16. Gill, H.S. 1980. 'Pusa Swarnima'- A turnip variety with great promise. *Indian Hort.*, **24:** 4.
17. Gill, H.S. 1993. Improvement of Cole Crops. In: Advances in Horticulture Vol. 5-Vegetable Crops. Part-I (Eds. Chadha, K.L. and Kalloo, G.). Malhotra Publishing House, New Delhi, India.
18. Gill, H.S., Singh, J.P. and Bhullar, B.S. 1971. 'Pusa Chandrima'-A high yielding turnip. *Indian Hort.*, **16:** 19-20.
19. Hazra, P., Chattopadhyay, A., Dasgupta, T. and Som, M.G. 2001. Bidhan Barbati-1 and Bidhan Barbati-2: New high yielding and non-viny vegetable cowpea. *Indian Hort.*, **46(1):** 13-15.
20. Hedau, N.K., Agarwal, P.K., Bhatt, J.C., Tuti, M.D., Chadrashekra, C. and Subbanna, A.R.N.S. 2015. VL Lahsun-2: New long day garlic. *Indian Horticulture*, May-June. **60(3):** 12.
21. Jain R., Bhutani, R.D., Kalloo, G., Bhatti, D.S. and Singh, K. 1990. *Indian Hort.*, **36:** 24.
22. Jyothi, M.L., Narayanan Kutty, M.C. and Balachandran, P.V. 2005. Popularizing cultivation of Chinese potato in Kerala. *Indian Hort.*, **50(2):** 12-13.

23. Kader Mohideen, M. and Shanmugavelu, K.G.1982. Studies on the performance of annual drumstick at Coimbatore, *SHF*. **30(1):** 95-98.

24. Kalda, T.K. and Choudhury, B. 1991. Note on evaluation of 'Pusa Bhairav' a resistant variety to Phomopsis fruit rot disease in egg plant. Scientific Hort., 1: 145-150.

25. Kalia, P. 2002. Palam Samridhi: A new variety of sprouting broccoli, *Indian Hort.*, **47(1):** 24-25.

26. Kalia, P. and Jamwal, R.S. 2006. Palam Vichitra: A new broccoli. *Indian Hort.*, **51(3):** 16-17.

27. Kalia, P. and Singh, Y. 2004. Palam Kanchan: A new nutritious broccoli. *Indian Hort.*, **49(2):** 36.

28. Kalloo, G., Rai, M., Kumar, R., Prasanna, H.C., Singh, M., Kumar, S., Singh, B., Ram, D., Pandey, S., Lal, H., Pandey, P. K. Rai, S., Pandey, K.K. and Satpathy, S. 2006. New Vegetable varieties from IIVR, Varanasi, *Indian Hort.*, **51(3):** 22.

29. Kalloo, G. and Singh, K. 1976. *Haryana Journal of Horticultural Science*, **11:** 261-263.

30. Kalloo, G. and Singh, K. 1982. *Journal of Horticultural Science*, **5:** 92-95

31. Knaflewaski, M. 1988. *Gemuse*, **24:** 238-239.

32. Krarup, A. 1989. *Asparagus Research Newsletter*, **7:** 12-13.

33. Mac Gillivray, J.H. 1961. Vegetable production, McGraw hill book Co., New York.

34. Madzharova, D. 1971. The new parsley varieties 'Festival 68 B'. *Lgarski Plodve. Zelenchustii I Konservi*, **4:** 13.

35. Mhaskar, N.V., Chavan, S.A., Haldankar, P.M. and mahadkar, U.V. 2016. "Konkan Kalika": Promising unique cultivar of aerial yam and wealth for farmers of Konkan Region of Maharashtra. 7th Indian Horticulture Congress (Abs. 96), held at ICAR-IARI, New Delhi, 15-18 November, p. 74.

36. Mini, C. and Joseph, A. 1998. Effect of seed spacing in nursery on seedling vigour and field performance in chilli (*Capsicum annuum* L.). *Veg. Sci.*, **25(2):** 124-126.

37. Munshi, A.D., Islam, S., Kumar, R., Mandal, B., Behera, T.K. and Sureja, A.K. 2015. Sponge gourd DSG-6 to combat ToLCNDV. *Indian Hort.*, **60(3):** 7-8.

38. Murthy, N.S.C. and Bavaji, J.N. 1980. Sindoor: A new dual purpose chilli strain. *Indian Cocoa, Arecanut and spices J.* **4(2):** 36-38.

39. Naik, P.S., Luthra, S.K. and Kumar, V. 2012. Potato (*Solanum tuberosum* L.) In: *Breeding of Field Crops* (Ed. Bharadwaj, D.N.). Published by Agrobios (India), p. 660-707.

40. Nath, P. 1976. Vegetables for tropical region.ICAR Publicatio Series No.2.

41. Nath, P., Velayudhan, S. and Singh, D.P. 1987. Vegetables for tropical region. Indian Council of Agricultural Reasearch, New Delhi.

42. Pan, R.S., Singh, A.K., Das, B., Naik, S.K. and Maurya, S. 2016. Swarna Safal: An Improved vegetable Faba bean variety for Jharkhand. 1st International Agrobiodiversity Congress (Abstract), held at New Delhi; p. 17.

43. Patil, B.P., Rajput, J.C., Palve, S.B., Hambhole, A.S. and Diwakar, M.P. 1990. *Indian Hort.*, 35: 6-7.

44. Pruthi, J.S. 1993. Major spices of India. Crop Management and Post harvest Terchnology. Indian Council of Agricultural Research, New Delhi.

45. Ram Das, S., Armugam, R., Sayed, S., Shah, H. and Muthuswami, S. 1982. 'CO-1' fenugreek-A high yielding grain cum-green fenugreek. *South Indian Hort.*, **30:** 242-243.

46. Ramasamy, P., Balasubramanian, M., Gnanam, R. and Rangasami, P. 1990. CO11 Avarai, (*Lablab purpureus* var. *typicus* (L.) Sweet.). *Madras Agricultural Journal*, **77:** 121-124.

47. Rao, T.S. and Babu, M.K. 1981. Lam Sel.1. A high yielding fenugreek type for Andhra Pradesh. *Indian Cocoa, Arecanut and Spices J.*, **4:** 83-86.

48. Sadashiva, A.T., Deshpandey, A.A. and Reddy, M.K. 1996. 'Arka Meghali' : a new rainfed tomato. *Indian Hort.*, **412:** 6-7.

49. Sathashathi, A. 1997. Variability studies in annual moringa, Ph.D. Thesis, Tamil Nadu Agricultural University, Coimbatore, Tamil Nadu (India).

50. Saxena, R.K. and Thomas, T.A. 1988. *Indian Hort.*, **33:** 9-10.

51. Sen, H., Chattopadhyay, A. Tarafdar, J. and Das, S. 2005. Bidhan Kusum: a new elephant foot yam. *Indian Hort.* **50(3):** 24-25.

52. Sharma, H.R. and Rattan, R.S. 1995. Evaluation of some F_1 hybrids of tomato under mid hill conditions of U.P. *Natl. Sym. on Recent Development in Vegetable Improvement*, held at Indira Gandhi Krishi Vishwa Vidyalaya, Raipur (M.P.) on 2-5 Feb., 1995.

53. Sharma, R.K. and Bhati, D.S. 1984. Prabha. A promising variety of fenugreek forRajasthan and Gujarat. *Indian Cocoa, Arecanut and Spices J.* **8:** 4-15.

54. Sharma, R.K., Tewari, R.N. and Pachauri, D.C. 1997. Pusa Sem-2 and Pusa Sem-3: Promising lalblab beans. *Indian Hort.* **41(4):** 46.

55. Sharma, S.R. and Gill, H.S. 1986. 'Hilds Ideal': A promising cultivar of Brussels sprouts. *Indian Horticulture*, **31:** 17-18.

56. Singh, A.K. and Narayan, R. 2004. Variability studies in tomato under cold arid condition of Ladakh. *The Horticulture Journal*, **17(1):** 67-72.

57. Singh, J.P., Gill, H.S., Swarup, V. and Arora, R.S. 1971.'Pusa himani' radish. *Indian Hort.*, **16(1):** 13-14.

58. Singh, H.B. 1963. Carrot and turnip varieties 'Pusa Kesar' and 'Pusa Kanchan'. *Indian Hort.* **7:** 24.

59. Singh, H.B. and Joshi, B.S. 1970. Pulse crop of India. Indian Council of Agricultural Research, New Delhi, p. 183-200.

60. Singh, S.K. and Srivastava, J.P. 2006. A new clustered chilli: 'Azad Mirch-1'. *Indian Hort.*, **51(5):** 8.

61. Sivasubramanian, P., Annappan, R.S., Iyemperumal, S., Sreerangaswamy, S.R. and Pandiyan, M. 1989. CO10-A dual purpose lablab. *Madras Agricultural Journal*, **48:** 144-154.

62. Sontakke, M.B., Gunjakar, S.N., Purandare, N.D., Anserwadakar, K.W. and Shinde, N.N. 1995. A new rainfed chilli variety'Parbhani Tejas' for Marathwada region. *Natl. Symp. on 'Recent Development in Vegetable Improvement'* held at Indira Gandhi Krishi Vishwa Vidyalaya, Raipur (M.P.) on February 2-5, 1995.

63. Sri Dhar, Mani, V.P., Ved Prakash and Lal, S. New pea variety VL-6 1999. *Indian Hort.* **44:** 19.

64. Sundarajan, J.S., Mutghusamy, S., Shanmugavelu, K.G and Balakrishnan, R. 1970. A Guide to Horticulture, Velan pathippagam, Coimbatore. pp. 261-262.

65. Tiwari, A. and Narsinghani, V.G. 1986. Promising varieties of kharif peas with powdery mildew resistance. *Veg. Sci.*, **13:** 137-139.

66. Tewari, R.S. 1975. Grow more vegetable this winter. *Indian Farmers Digest*, **8(1):** 83-57.

67. Thakur, P.C. and Gill, H.S. 1985. Produce quality seed of sweet pepper. *Seeds and Farms*, **11(12):** 25-27.

68. Tiwari, R.N., Choudhury, B. and Pachauri, D.C. 1990. *Indian Hort.*, **35:** 4-5.

69. Tiwari, R.N., Pachauri, D.C. and Sharma, R.K. Tomatoes from ICAR-IARI-Tomato' Pusa Hybrid-1' and Pusa Hybrid-2 1996. *Indian Hort.*, **40:** 26-27.

70. Tiwari, R.N., Pachauri, D.C. and Sharma, R.K. 2001. "Pusa Sadabahar": A new tomato variety. *Indian Hort.*, **48(3):** 11 and 24.

71. Tiwari, R.N., Pachauri, D.C. and Sharma, R.K. 1999. New tomato: Pusa Uphar. *Indian Hort.*, **44:** 9-10.

72. Tewari, V.P. 1991. A multipurpose perennial chilli: "Pusa Sadabahar". *Indian Hort.*, **35(4):** 29-31.

73. Thomas, T.A., Saxena, R.K. and Hardas, M.W. 1976. 'L-Bonita': A promising introduction in tomato. *Indian Horticulture*, **23(4):** 9-10.

74. Thomas, T.A., Singh, R. and Prasad, R. 1983. Genetic resources for improvement of vegetable crops. *Indian Hort.*, **31:** 59-73.

75. Varalakshmi, B., Pratap Reddy, V.V., Naik, G., Celia Chalam, V. and Pal, D.K. 2001. 'Arka Arunima': a new, multicut amaranth. *Indian Horticulture*. **46(1):** 29-30.

76. Verma, T.S., Joshi, S., Sharma, R.K. and Pachauri, D.C. 2001. *Indian Hort.*, **6:** 1-3.

77. Verma, T.S. and Sharma, S.C. 1998. Pusa Broccoli 'KTS-1': New broccoli. *Indian Hort.*, **43(3):** 16-29.

78. Wagh, R.S., Patil, R.S., Kale, P.N., Pujari, C.V. and Kesar, B.G. 1995. 'Raja Shree': A new high yielding tomato F_1 hybrid. *Natl. Sym. on Recent Development in Vegetable Improvement*, held at Indira Gandhi Krishi Vishwa Vidyalaya, Raipur (M.P.) on 2-5 Feb., 1995.

79. Wakankar, S.M. 1977. Tomato 'Sweet-72'. *Indian Hort.*, **22(3):** 11.

80. Wakankar, S.M., Yadav, L.N. and Kelkar, G.M. 1974. Path coefficient analysis for some characters in soybean. *JNKVV, Research Journal* **8:** 192-195.

81. Yadav, D.S., Rai, N. and Yadav, R.K. 2000. *Indian Horti.*

82. Yadav, L.P., Raja, S., Apparao, V.V., Singh, S. and Sharma, B.D. 2016. Thar Harsha: A new potential variety of drumstick for dryland. (Abs. 70; Session: Breeding of Horticultural Crops II). 7th Indian Horticulture Congress-2016, held at ICAR-IARI, New Delhi, p. 139.

83. Yawalkar, K.S. 1969. Vegetabl Crops of India, Agri-Horticultural Publishing House, Nagpur (Maharashtra)

84. Yawalkar, K.S. 1985. Vegetabl Crops of India, Agri-Horticultural Publishing House, Nagpur (Maharashtra).

Index

A

B

C

D

H

I

J

K

L

Q

R

S

T

U

V

W

X

Y

Z